2025 IEEE 16th International Conference on ASIC (ASICON 2025)

Kunming, China
21-24 October 2025

Pages 1–490

IEEE Catalog Number: CFP25442-POD
ISBN: 979-8-3315-3918-4

Copyright © 2025, IEEE

All Rights Reserved

Copyright and Reprint Permissions:

Abstracting is permitted with credit to the source. Libraries are permitted to photocopy beyond the limit of U.S. copyright law for private use of patrons those articles in this volume that carry a code at the bottom of the first page, provided the per-copy fee indicated in the code is paid through Copyright Clearance Center, 222 Rosewood Drive, Danvers, MA 01923.

For other copying, reprint or republication permission, write to IEEE Copyrights Manager, IEEE Service Center, 445 Hoes Lane, Piscataway, NJ 08854. All rights reserved.

*** This is a print representation of what appears in the IEEE Digital Library. Some format issues inherent in the e-media version may also appear in this print version.

IEEE Catalog Number: CFP25442-POD

ISBN (Print-On-Demand):   979-8-3315-3918-4

ISBN (Online):   979-8-3315-3917-7

ISSN: 2162-7541

Additional Copies of This Publication Are Available From:

Curran Associates, Inc
57 Morehouse Lane
Red Hook, NY 12571 USA

Phone: (845) 758-0400
Fax: (845) 758-2633
E-mail: curran@proceedings.com
Web: www.proceedings.com

TABLE OF CONTENTS

Generation, Modulation and Application of Spintronic Markov Chain Signal 1
Xihui Yuan, Jiajia Jian, Zheng Chai, Xue Zhou, Weidong Zhang, Jian Fu Zhang, Tai Min

Opportunities for Advanced Logic Technology with Dual-sided Integrations: From Lateral to Vertical
Transistors 5
Yanbang Chu, Yu Liu, Runsheng Wang, Ming Li, Heng Wu

Si Hybrid Tunnel FET-CMOS Foundry Platform for Ultra-low-Power Circuit Applications 9
Qianqian Huang, Kaifeng Wang, Ru Huang

Si-MoS$_2$ Heterogeneous CFET for Ultra-low Power Logic Technology Scaling 13
Zehua Wang, Wenzhong Bao, Peng Zhou, Jing Wan

Impact of Off-state Stress on the Reliability of 14nm nFinFETs 16
Wendi Wei, Kun Chen, Chen Wang, Yaolin Wang, Zhao Yang, Zhiteng Zhang, Zhuming Wang
Qingqing Sun, David Wei Zhang

Performance Comparison Between Bulk-Si and FDSOI Nanosheet GAAFETs 20
RS.He, BX.Gan, S.Cristoloveanu, Y.Xu, J.Wan

Improving EUV Patterning Fidelity and Aberration Control through Source-Mask Co-Optimization 24
Qi Wang, Qiang Wu, Ying Li, Xianhe Liu, Yanli Li

Enhancement of HfO$_2$-Based Ferroelectric Thin Film Performance via Interface and Defect Engineering 28
Xiao Yu, Peiyuan Du, Huan Liu, Dongya Li, Fei Yu, Bing Chen, Ran Cheng, Mengnan Ke, Yan Liu
Genquan Han

Threshold Voltage Swing Caused by Intense Phonon-Electron Interaction in High-k Dielectrics 32
Jinchen Wei, Mansun Chan

A Ti/ITO Bilayer Gate Electrode Strategy for Improving Subthreshold Swing of Oxide Transistors 36
Chuanlin Sun, Tingchen Yi, Han Gao, Jiakang Zhang, Junchen Dong, Kai Zhao, Dedong Han, Xing Zhang

Fabrication of High-Performance β-Ga$_2$O$_3$ MOSFETs via Ohmic Contact Optimization 40
Hui Li, Qihao Zhang, Haodong Fu, Jianguo Li, Dongyuan Zhai, Yi Zhao, Jiwu Lu

Experimental Study on 1.2kV/40mΩ SiC MOSFET with Integrated JBS Diode 44
Moufu Kong, Qizhi Feng, Hongfei Deng, Yufeng Dong, Wei Han, Xuequan Yin, Jiakai You

A Review of Active Gate Drivers for SiC Power MOSFETs 48
Yuchu Ge, Wei Jia Zhang

A Quadruple RESURF LDMOS with Enhanced Hot-Carrier-Induced Degradation Immunity 52
Wenliang Liu, Ming Qiao, Penglong Xu, Chunxia Ma, Feng Lin, Bo Zhang

Investigation of Dual-Mode R$_{on}$ Degradation Mechanisms in LOCOS-Based LDMOS 56
Wenliang Liu, Ming Qiao, Penglong Xu, Chunxia Ma, Feng Lin, Bo Zhang

Advanced Gate Driver Solutions for Fast-Switching SiC Power Device Applications 60
Yu Qing, Zhihao Yan, Zekun Zhou, Jiaxing Mao, Zijun Zhou, Yun Dai, Rongxing Lai, Yue Shi, Bo Zhang

Design Considerations for Smart Gate Drivers 64
Wai Tung Ng, Jingyuan Liang, Wentao Cui, Chun Yin Au Yueng, Namjee Kim, Wei Jia Zhang

Investigation of Threshold Voltage Instability in GaN HEMTs Using Rapid Ramp Sweeping Technique 68
Diangang Hu, Yutian Gan, Shufu Yu, Sai Liu, Lirong Zhang, Weijing Wu, Hongyu Yu

Design and Kirk Effect Improvement of 30V NLDMOS Base on 0.18μm BCD Platform 72
*Qi Ding, Ning Ning, Renxiong Li, Jun Huang, Yutuo Guo, Yu Wang, Kunqin He, Yaxin Liu, Ziyi Zeng
Ming Qiao, Lulu Peng, Bo Zhang*

Device Modeling Based on Residual Neural Network with Ensemble-Based Active Learning 76
*Hongfei Su, Yutong Wu, Jithish Jayarajan, Bharatha Kumar Thangarasu, Nagarajan Mahalingam
Fanyi Meng, Kaixue Ma, Kiat Seng Yeo*

Performance Benchmark of Gate-All-Around Nanosheets Transistors Based on DTCO Simulation 80
*Chunlei Wu, Jian Ma, Hanzhi Gu, Yueyuan Yu, Yiming Xia, Jiayi Wu, Boqian Shen, Qingqing Sun
David Wei Zhang*

The impact of Back-Gate biasing and layout on temperature sensitivity of transistors in FD-SOI CMOS
technology 84
Yann Deval, Maxime Guillot, Hervé Lapuyade, François Rivet

Considerations for Low Temperature High-Performance Computing: Design Technology
Co-optimization, Device Variation, and Hot Carrier Injection 88
*Yuanxi Liao, Keyang Zhu, Ran Tao, Yongyu Wu, Qinglan Jiang, Yaxiong Liu, Kai Xu, Chen Shen, Dong Liu
Dawei Gao, Ran Cheng*

Feature clustering-driven data augmentation in multi-level hotspot detection for integrated circuits
based on GAN 92
*Pengyu Ren, Bojie Ma, Yajuan Su, Xiaojing Su, Xin Hong, Yuqin Wang, Yujie Jiang, Zhanzi Chen
Tianao Chen, Yayi Wei*

A New TCAD Simulation Framework for Strain-Aware Quantum Tunneling Current Modeling 96
Jian Ma, Chunlei Wu, Hanzhi Gu, Yueyuan Yu, Yiming Xia, Jiayi Wu, Qingqing Sun, David Wei Zhang

DTCO-based Hybrid Rail 8T Complementary FET SRAM Design towards advanced node 100
Yutian Zhang, Khawar Sarfraz, Mansun Chan

Design of VCM Motor Coil Based on Five Factor Integration 104
Shengxian Quan, Huihong Zhang, Yuejun Zhang, Qiang Wang, Guanglong Xu, Jinsheng Yang

Full-spiking Bio-inspired Target Detection Vision Algorithm based on Gating Attention Prediction for
DVS and SPAD Sensors 108
Lengjun Yang, Xingyu Xiang, Yiyao Wen, Jian Liu, Nanjian Wu, Liyuan Liu, Shuangming Yu

The Quest for Reliable AI Accelerators: Cross-Layer Evaluation and Design Optimization 112
Meng Li, Tong Xie, Zuodong Zhang, Runsheng Wang

A Lightweight Hardware Defense Against DSE-Based Trojans in NN Accelerators 116
Yujing Wu, Chao Guo, Youhua Shi

SpykSim: A Cycle-Level Full-System Simulator for Systolic SCNN Accelerators 120
Wanwan Zhao, Yichu Yao, Qiang Niu, Qian Li, Chen Zhao

A Data-Efficient Deep Reinforcement Learning Algorithm and FPGA Accelerator for Real-Time Robot Motion Control Applications 124
Wenhao Huang, Rao Fu, Aiwu Ruan, Huiyun Li, Chongyang Zhu

GraphFlow-PIM: Annotated Execution Graphs of DNN Workloads across Diverse PIM Configurations 128
Syeda Munazza Marium, Song Chen

Optimizing LLM inference for FPGAs 132
Jorge R De Freitas, Jose G. F. Coutinho, Ce Guo, Suleyman Demirsoy, Wayne Luk, Zhiqiang Que

Fine-Grained Layer Scheduling and Mapping for Chiplet-Based LLM Inference 136
Hongyang Gu, Lei Xu, Haochen Zhao, Naifeng Jing

A 16×16 High-Utilization Systolic Array Hardware Accelerator for Long-Sequence Flash-Attention Computation in Transformer 140
Zhenkun Li, Liji Wu, Yi Yang, Tianling Ren, Le Wu, Xiangmin Zhang

Sparse Approximation of Softmax: Hardware-Efficient Acceleration for Long Sequence Inference 144
Lanqi Ma, Zifeng Zhao, Xiaoxing Wu, Gengsheng Chen, Wenbo Yin

A Hybrid Processing-in-Memory and Computing-in-Memory Architecture for Large Language Model Inference in Edge Devices 148
Yujia Sun, Ruicong Zhang, Yuanfeng Chen, Qiang Zhou, Xiaoyong Xue, Xiaoyang Zeng

MCDC: A Memory-efficient and Computation-efficient Architecture for Deformable Convolutions 152
Zhiyi Shu, Xinhua Shi, Jun Han

Hardware-Efficient Lightweight Feature Map Compression for Convolutional Neural Networks 156
Bing Wu, Shan Cao, Zhiyuan Jiang

A Study on Dwell Time Impacts in Charge-trapping 3D NAND Flash Memory 160
Yining Zhou, Ruidong Li, Xuepeng Zhan, Guangkuo Yang, Yujiao Ding, Xinghao Wang, Pengpeng Sang Peng Guo, Jixuan Wu, Jiezhi Chen

Access Mode Impacts on 3D Charge-trapping (CT) QLC (4bit/cell) Raw NAND Chip 164
Guangkuo Yang, Ruidong Li, Yining Zhou, Yujiao Ding, Xinghao Wang, Pengpeng Sang, Peng Guo Xuepeng Zhan, Jixuan Wu, Jiezhi Chen

The Influence of Radiation on Reliability of Cold Data in 3D CT NAND Flash Memory 168
Haitao Dong, Xinghao Wang, Haotian Li, Xuesong Zheng, Pengpeng Sang, Xuepeng Zhan, Jixuan Wu Jiezhi Chen

Process Co-Optimization of Void Suppression in ULK Dielectric Layers for 28 nm RRAM Arrays Towards High-density Integration 172
Zhenchao Sui, Yanqing Wu, Xing Zhang

A Study on Performance Enhancement of TiO_2/HfO_2 Memristors through Rapid Thermal Annealing 175
Yifan Wu, Yuzhe Hu, Yuewei Qu, Pengpeng Sang, Jixuan Wu, Xuepeng Zhan, Jiezhi Chen

Investigation of Self-Heating Effects in InGaZnO Vertical Channel Transistors for DRAM Application 179
Zhuoran Kong, Yizhan Liu, Jinfeng Kang, Xiaoyan Liu

Broadband Characterization of Ferroelectric Domain Switching Hysteresis Loops in $TiN/Hf_{0.5}Zr_{0.5}O_2/TiN$ Thin-film Capacitors 183
Wen Di Zhang, Xian Yu Hu, An Quan Jiang

On the Reliability of Sub-10nm Ultra-thin Ferroelectric HZO Thin Film 187
Xiaopeng Li, Yang Feng, Pengpeng Sang, Xuepeng Zhan, Jixuan Wu, Jiezhi Chen

High Dielectric Permittivity in Size-scaled $Hf_{0.5}Zr_{0.5}O_2$ Thin-film Capacitors 191
Wendi Zhang, Anquan Jiang

1.2 V Operation of $Hf_{0.5}Zr_{0.5}O_2$ Ferroelectric Thinfilm Capacitors for Low-Power ASICs after
Interfacial-layer Engineering 195
Xianyu Hu, Di Hu, Wendi Zhang, Bowen Shen, Anquan Jiang

Reliability Enhancement of $Hf_{0.5}Zr_{0.5}O_2$-Based Ferroelectric Capacitors via Argon Plasma Treatment 199
Hongrui Zhang, Xiu Yang, Rongzong Shen, Yian Ding, Haoji Qian, Jiajia Chen, Chengji Jin, Ran Cheng
Bing Chen, Xiao Yu, Yan Liu, Genquan Han

Cryogenic Ferroelectricity (10-298 K) of Superlattice and Solid-solution HZO Films 203
Yiming Xia, Chunlei Wu, Jian Ma, Hanzhi Gu, Yueyuan Yu, Jiayi Wu, Qingqing Sun, David Wei Zhang

Systematic Review of Write Reliability in Spin-Transfer Torque Magnetic Random-Access Memory 207
Yuhao Chen, Yiming Qu, Ziyuan Chen, Choonghyun Lee, Yi Zhao

Emerging Magnetoresistive Memories 211
Viktor Sverdlov, Nils Petter Jørstad, Bernhard Pruckner, Mario Bendra, Wolfgang Goes
Siegfried Selberherr

Comprehensive Characterizations of Polarization Switching Dynamics in HfO_2-based FRAM across a
Broad Temperature Spectrum 215
Yilin Hou, Jixuan Wu, Xiaopeng Li, Xiaoyu Dou, Yaoyu He, Pengpeng Sang, Xuepeng Zhan, Yuqi Gao
Linhui Hu, Feng Wang, Yushi Hu, Qian Tao, Jiezhi Chen

WO_x interlayer employed to improve the imprint effect on $HfZrO_2$ ferroelectric capacitors 219
Zibo Dong, Zeping Weng, Jianguo Li, Lijian Chen, Ziyuan Chen, Yi Zhao, Daolin Cai

Understanding the Physical Mechanism of Endurance Cycling in Antiferroelectric Memories 223
Y. Qu, Y. Huo, Y. Ding, L. Chen, Z. Weng, Y. Zhao

Accelerated polarization switching speed and durable endurance enabled by confined domain size and
solid defect migration barrier in FE/AFE multilayer stacked $Hf_xZr_{1-x}O_2$ ferroelectric capacitor 227
Wenhao Wu, Xinyu Xie, Zeping Weng, Yi Zhao, Jiabin Qi

A High-Speed Dual-Entropy Sources True Random Number Generator Implemented on FPGA 231
Yizhi Liu, Jierui Liao, Hao Xing, Pengpeng Sang, Jixuan Wu, Jiezhi Chen, Xuepeng Zhan, Xiangye Wei

A Lightweight Arbiter PUF Design Based on Threshold Loss in Transmission Gates 235
Haoxuan Yan, Yitian Su, Qiwen Wu, Yong Ding, Hui Li, Yuejun Zhang

Destruction-Free Soft PUF Architecture: Merging Security and Efficiency in 4T2M TCAM Without
Data Migration 239
Shimao Ren, Pengjun Wang, Bo Chen, Zhenhong Chen

An RRAM-Based Soft PUF Achieving Near-Zero BER through Skewed Voltage Masking 243
Xinrong Yang, Pengjun Wang, Cailong Jin, Yixin Lu

A Performance Enhancement Strategy for Strong PUF Circuits to Improve IoT Authentication Security 247
Dong Lu, Pengjun Wang, Gang Li, Hao Ye

A Sequential Obfuscation PUF Resistant to Machine Learning Attacks Based on AES Key Expansion 251
Xuejiao Ma, Yimeng Jin, Shuyang Ren, Ziyu Zhou

Hardware-Efficient Doppler Estimation and Compensation in PDSCH for 5G Non-Terrestrial Networks 255
Chih-Chen Chen, Yi-Shan Huang, Chung-Lun Tu, Shyh-Jye Jou

FPGA Bitstream Modification Attacks on CRYSTALS Kyber 259
Lei Chen, Jiahao Lu, Tianze Huang, Aobo Li, Shengfei Gu, Ang Hu, Dongsheng Liu

BIND: A Batch Cache-Invalidation Framework Based on Doorbell Mechanism 263
Jialin Liu, Zhiyuan Zhang, Chao Fu, Jun Han

High-Throughput Multiplier-Free FPGA Implementation for Pure-Number Discrete Fractional Complex
Hadamard Transform 267
Chengqi Zhao, Zi-Chen Fan, Shan Cao, Susanto Rahardja

Design and Implementation of a Bilateral Filtering Accelerator Based on RISC-V 271
Zhengyao Shi, Yushan Dai, Angyang Li, Jian Mei, Lei Deng, Rui Yin

Skip-Zero Strategy: A Latency and Power Optimization for SRT Divider 275
Ke Xu, Ping Yin, Jun Han

A High-precision Stochastic Computing Multiplier with Co-optimization of Area and Latency 279
Qiang He, Yudi Zhao, Zhihuai Zhang, Xiaofei Nie, Shisheng Xiong, Kai Zhao

A Low-Overhead Fault-Tolerant Design for Quantized CNN Accelerators 283
Shanqiang Yang, Chenxu Wang, Lexiang Shen, Xinlei Su, Min Luo, Tianliang Xu, Ruoshi Li, Siyuan Wang

A Real-Time and Reconfigurable Pre-Driver Design for ABS Solenoid Valve Applications 287
Zhinan Li, Yitian Su, Shaochen Han, Huihong Zhang, Yuejun Zhang, Cang Liu

Real-Time Highly Flexible Wheel Speed Sensing Interface IP Design 291
Yitian Su, Zhinan Li, Haoxuan Yan, Zhenkai Zhou, Yuejun Zhang, Cang Liu

Design of Secure Storage Circuit Based on Reversible Logic XOR-Toffoli Gate 295
Yiting Guo, Yuejun Zhang, Shutong Zhang, Mengfan Xu, Zhenkai Zhou, Hui Li

A Hierarchical Approximate Floating Point MAC Unit with Precision-Adaptive Self-Configuration 299
Xianghui Fu, Yike Wang, Chaojie Wei, Yu Gong

High-Performance Radiation-Hardened Flip-flop for Reliable Systems 303
Jie Li, Xiaoming Teng, Yufeng Zhang

Analysis and Design of Regulating Rectifier with Multiple Outputs for Wirelessly Powered Biomedical
Devices 307
Quanrong Zhuang, Junyi Sun, Bo Li, Jie Lu, Yi Shi, Hao Qiu

Noise Notch Frequency Design for EMI Mitigation in DC-DC Converters Using Digital-to-Time
Converter 310
Yasunori Kobori, Yifei Sun, Guiyi Dong, Nobukazu Tsukiji, Ramin Khatami, Takuya Arafune
Shogo Katayama, Anna Kuwana, Jianglin Wei, Haruo Kobayashi

A 240nA-1μA Quiescent SIMO Converter Featuring 3mV Undershoot under 30mA/μs Transients 311
Yuhua Chen, Qianhui Liu, Yixing Wang, Yimeng Zhang, Yuming Zhang

A 280-nA, 85.8% Efficiency Boost Converter with Optimal Inductor Current in Burst Mode for Brain Stimulation 315

Dejian Li, Xin Jin, LianXi Liu, Gang Dong, Xufeng Liao, Shihao Xiao, Xincai Liu

A High-Efficiency Low-Ripple Buck Converter with Adaptive Load Frequency Control 319

Tao Ren, Xufeng Liao, Gefu Wang, Jiatong Wu, Lianxi Liu

A 400V High-speed Level-Shifting Gate Driver with Adaptive Signal-Path Disconnection for 278V/ns dv/dt Immunity in Soft-Switching Converters 323

Yile Xie, Hanyu Shi, Ting Yi, Zhiliang Hong

GaN-based complementary logic sawtooth generator for smart power ICs 327

Yutao Geng, Ji Shu, Tao Chen, Yan Cheng, Yat Hon Ng, Kevin J. Chen

Battery Charger Designs for Low-Voltage Energy Harvesting Based on the Return-on-Investment Concept (Invited) 331

W. Saito, A. Higuchi, T. Yamano, T. Tanzawa

High-Efficiency Energy Extraction Interface for Piezoelectric Energy Harvesting 335

Chenghao Zhang, Junkai Chen, Jingjie Huang, Yue Shi, Zekun Zhou, Bo Zhang

PWM Scheme Selection Strategy for Fast Ramp-Up DC-DC Boost Converters in SSD Applications 339

Yuji Kanayama, Toru Tanzawa

A 98.5% Efficiency Single-Mode Buck-Boost Converter with All-1.8-V-Switch and Non-Stopping Output Current Delivery 343

Qianhui Liu, Yuhua Chen, Yixing Wang, Yuming Zhang, Yimeng Zhang

Design of a Fully Integrated Low Dropout Linear Regulator with Bandgap Reference 347

Fan He, Jiao Liu, Yiyun Mao, Haoyuan Gao, Xianhui Wang, Yubing Zhang, Hao Xu, Na Yan

A 455mV-Hysteresis, 120 nA, Bandgap less Power-on-Reset Circuit for IoT in 40nm CMOS 351

Mingzong Lin, Chaoran Chen, Jian Xu, Yue Lin, Wei Li, Hongtao Xu

An Anti-Single Particle Effect Over Temperature Protection Circuit Based on Dual Detectors 355

Ping Luo, Hong Zhao, Hao Wang, Fulin Yao, Kai Luo

A Fast Start-Up and Low-Power 32-kHz Crystal Oscillator for Real-Time Clock and Frequency Calibration 359

Jie Zheng, Qiang Li, Hao Min

A 0.067 mm² PNP-Based Temperature Sensor with ±0.6°C (3σ) Inaccuracy from −20°C to 80°C 363

Letian Li, Peilin Xiao, Xuyang Lu

A Cryogenic Voltage Reference with Diode-Based Sensing and Substrate Resistor Compensation Compensation in 180-nm CMOS Process 367

Yixin Zhang, Hanze Liu, Jing Li, Zhong Zhang, Ning Ning, Qi Yu

77K Modeling and Implementation of a Cryogenic OTA for Infrared Sensors 371

Zhuokai Wang, Lei Deng, Rui Yin, Jian Mei, Jiaming Zhang, Zhicheng Shi

Design and Analysis of PI Controller for Resonant Drive Circuits with AGC-PI Architecture 374

Yichen Lu, Tao Yin, Ying Liu, Jian Liu, Nanjian Wu, Liyuan Liu

Design of 11MHz Isolated Current Sense Amplifier Based on FDDA and Current Feedback Frequency Modulation Loop 378

Xinghong Chen, Jiahui Liu, Shaowei Zhen, Hongwei Shen, Wei Yang, Yongwang Ma, Bo Zhang

A Charge Pump Powered Current Sense Amplifier with -20 V to 40 V Input Common-Mode Range 382

Dejian Li, Hongwei Shen, Jinzhao Li, Jiahui Liu, Lixing Wang, Shaowei Zhen, Bo Zhang

The Sequency Domain: a new Approach for Radio Frequency Front End 386

François Rivet, Pierre Ferrer, Maxandre Fellmann, Nathalie Deltimple, Hervé Lapuyade, Eric Kerhervé Yann Deval

300-GHz Phased-Array Transceiver in 40-nm CMOS with Interpolated Feeding and OTA Metrics 390

Minoru Fujishima

A Polar-Modulation OFDM Backscatter System for Passive IoT Communication 394

Qijing Xiao, Xin Hu, Weixiao Wang, Yuxuan Luo, Bo Zhao

A Compact Q/V Band Bidirectional Phase Shifter with 0.32° Phase Error 398

Congrui Li, Yan Wang, Lei Zhang

A 300GHz Coherent Radiator Array with Multi-functional Antenna in 65nm CMOS 401

Houyi Yan, Kaizhe Guo

Design of a 300GHz Wideband On-Chip Antenna in 28nm CMOS 405

Jinghao Zhang, Chen Jiang

ATSim: A Fast and Accurate Simulation Framework for 2.5D/3D Chiplet Thermal Design Optimization 409

Qipan Wang, Tianxiang Zhu, Jiajia Cui, Yicheng Wei, Linxiao Shen, Zhe Cheng, Runsheng Wang Ru Huang, Yibo Lin

Radio Frequency Integrated Circuits Generated by AI-based Design Automation 413

Ruoyu Wang, Meijun Hou, Jun Wu, Hongtao Xu, Ye Lu

Advancing Sparse Matrix Solvers via Exploring More Parallelism and Random Sketching 417

Wenjian Yu, Jiawen Cheng, Baiyu Chen

Snow Ablation Optimizer Accelerator Based on High Level Synthesis 421

Maoshuo He, Renjing Hou, Zirui Li, Kang Zhao

An MLIR-Based Framework for Efficient Dynamic Circuits Generation 425

Yuxuan Guan, Jiangnan Li, Lingli Wang

HybridEPP: Hybrid Numerical and Symbolic Error Probability Propagation in Logic Network 429

Gaopeng Shen, Chang Wu

Success-Rate Improvement of Analog Circuit Topology Generation by Large Reasoning Model 433

Koutaro Hachiya, Kentaro Yoshikawa, Atsushi Kurokawa

Fast Thermal-driven 3D Fixed-outline Floorplanning By Learning-based Thermal Analysis 437

Yikai Liu, Jindong Zhou, Jiayi Li, Pingqiang Zhou

Systematic design for coupled heterogeneous accelerators 441

Tim Todman, Wayne Luk

TorchLitho 2.0: Differentiable Lithography Simulation Engine for Large-Scale Layouts 445

Shuo Yin, Su Zheng, Ziyang Yu, Bei Yu

Hierarchical Residual Fitting for Enhanced S-Parameter Accuracy in Devices Exhibiting Complex Delay 449

Jiaxin Wei, Haonan Wang, Ting-Jung Lin, Lei He

Hybrid Model-Based Hardware Acceleration for Diesel Engine NOx Emission Prediction 453

Xinlei Su, Shanqiang Yang, Tianliang Xu, Xiaozhen Yan, Jianfeng Li, Tian Rong, Chenxu Wang Yuhang Wang, Zhiwei Han

Extending Straight-Through Estimation for Robust Neural Networks on Analog CIM Hardware 457

Yuannuo Feng, Wenyong Zhou, Yuexi Lyu, Yixiang Zhang, Zhengwu Liu, Ngai Wong, Wang Kang

A Parallel Level-Set Based Approach for Etching Topography Simulation in Process Emulation 461

Yin Cheang Ng, Xin Wen, Boyuan Yu, Wenjian Yu

Some Signal Processing Techniques for Testing Wireless Communication LSIs 465

Koji Asami

Voltage-Domain vs. Time-Domain: Trade-offs in High-Speed Applications 466

Haoyu Li, Sai-Weng Sin, Rui P. Martins, Mingqiang Guo

A Pitch-Matched Transceiver ASIC with Element ADC and Continuous-Time Gain Compensation for 3D Ultrasound Probes 471

Jing Li, Tianci Zhang, Li Dai, Yingchen Liu, Jinlai Fu, Zhongshan Wang, Penghao Jiang, Yihu Yu Zhong Zhang, Kejun Wu, Ning Ning, Qi Yu

A Low-Power-Consumption Capacitance to Digital Converter with Novel Calibration Technology 475

Xiwen Zhu, Yufeng Zhang, Xiaoming Teng, Yihan Wang

A 10-bit 4 GS/s 67.79-dBc SFDR Switched-Capacitor DAC with Reservoir Capacitor-based Reference Generation 479

Yitao Wang, Meng Xu, Qiang Pan, Jize Liu, Yuekang Guo, Jing Jin

A 16-channel Neural Signal Acquisition Analog Front-End with Foreground Calibration for High-Precision Backend SAR ADC 483

Chun Feng, Junfeng Tang, Longhao Chen, Songping Mai, Xian Tang

An 18-bit 1MS/s SAR ADC with Weight-Fitting Digital Calibration and High-Linearity Capacitor Array Design 487

Baoyi Zheng, Guoao Wang, Zongmin Wang, Jin Qian, Bosen Liu, Zhaohang Bing, Tieliang Zhang

A digital front-end self-calibration algorithm for SAR ADC 491

Fuming Liu, Jie Ding, Jiangfeng Wu, Yongzhen Chen

Bitwise Bayesian Optimization for SAR ADC Calibration 495

Yu Shi, Shen Ye, Yihang Luan, Jiahao Wang, Ting Yi

An Area-Efficient C2C SAR ADC with Hybrid Switching Mode for Ultrasound Miniature Probes 499

Tianci Zhang, Jinlai Fu, Li Dai, Dongxu Li, Yingchen Liu, Jing Li, Zhong Zhang, Ning Ning, Qi Yu

A Reconfigurable 9-to-14b 15MS/s 4th-Order NS-SAR ADC with Self-Calibrated Open-loop FIA 503

Chaoran Chen, Mingzong Lin, Jian Xu, Yue Lin, Wei Li, Hongtao Xu

A 16-bit 4-MS/s Deadlock-free Asynchronous SAR ADC Using High-level First Transmission Gate 507

Xiaokun Zhou, Baijie Zhang, Xu Cheng

A Differential SAR-SS ADC with Gain-Scaled Ramp Quantization for High-Speed CMOS Image Sensors 511

Nanbo Chen, Jingyang Chen, Gang Wang, Peng Feng, Jian Liu, Nanjian Wu, Liyuan Liu

A 79.2dB-SNDR 12.5MHz-BW Pipelined SAR ADC with Analog-Domain Gain Error Shaping 515

Qiaoyu Hu, Guolong Fu, Yanbo Zhang, Zhangming Zhu

A Deep Reservoir Computing System based on IGZO Electrical-Double-Layer Transistors 519

M. Han, Y. Chen, H. Cui, Y. Wang, C. Wan

High-density and High-reliability (H²DR) RRAM for Energy-efficient AI Computing 521

Yimao Cai, Yiyun Chen, Lin Bao, Ling Liang, Zheng Zhou, Zongwei Wang

The Digital Coupled Ring Oscillator Ising Machine 525

Yue Han, Ranjith R Unnithan, Robin Evans, Efstratios Skafidas

Nanocrystal-Si Flash Memory-based Engergy-efficient Multi-bit Compute-in-Memory Design for Edge Neural Networks 530

Xianping Liu, Jian Huang, Zihan Zheng, Xinrui Zhang, Ruibin Zhou, Zhiyi Yu, Zhongyuan Ma, Kunji Chen Yuhan Wang, Jian Cheng, Peng Zhang

A Multi-level RRAM-based Ising Machine for Solving Combinatorial Optimization Problems 534

Zhenchao Sui, Xiaoxin Xu, Chengshuo Yu, Jingxin Deng, Xu Zheng, Chengyue Li, Hailan Yi, Jianguo Yang Xing Zhang

DRAM-Centric Near-Data Processing: A Survey of Architectures, Technologies, and Trends 538

Taoran Shen, Yujia Sun, Tingyi Xu, Li Xiong, Xiaoyong Xue, Xiaoyang Zeng

Challenges and Trends of SRAM based Floating Point Computing-in-Memory Circuits 542

Yuchen Tang, Yanqi Zhang, Zhichao Liu, Xing Wang, Defa Wu, Huaiwen Zhang, Yeqi Sun, Xin Si

Mapping of Graph Convolution Network on Sparse-Aware Computing-In-Memory Macros 546

Guoxiang Li, Tianhang Zhou, Xinyu Qu, Zecheng Zhou, Yufei Ma

CDCC: A High-Efficiency SRAM-Based Charge-Domain Compute-in-Memory Macro with Complement Compensation Design for AI Applications 550

Wanting Zhou, Zihao Xuan, Song Chen, Yi Kang

HPD: Hybrid Projection Decomposition for Robust State Space Models on Analog CIM Hardware 554

Yuannuo Feng, Wenyong Zhou, Yuexi Lyu, Hanjie Liu, Zhengwu Liu, Ngai Wong, Wang Kang

ADC-Free RRAM-Based XNOR-Bitcount Architecture for Hand Gesture Recognition 558

Lixun Wang, Yuejun Zhang, Qikang Li, Liang Wen

ESD Reliability Roadmap Considerations for 3D Heterogeneous Integration Microsystems (Invited) 562

Zijin Pan, Xunyu Li, Weiquan Hao, Runyu Miao, Zijian Yue, Albert Wang

Tiny Chiplets Enabled by Packaging Scaling: Opportunities in ESD Protection and Signal Integrity 566

Emad Haque, Pragnya Sudershan Nalla, Jeff Zhang, Sachin S. Sapatnekar, Chaitali Chakrabarti, Yu Cao

Time-dependent Dielectric Breakdown in Advanced MOSFET: From Theoretical Models to Experimental Findings 570

Chu Yan, GuoQiXin Huang, Yiming Qu, Yi Zhao

Mechanical Stress Induced by Temperature Cycling: Impact of MOSFET Placement on Bandgap Reference Voltage Offset 574

Fengbo Zhang, Yancong He, Zhinong Liu, Shuang Jiao, Yang Li, Zhigang Ji

Experimental and Theoretical Study of Single Event Latchup in a 3D TLC NAND Flash Memory Under Heavy Ion Irradiation 578

Xinghao Wang, Haitao Dong, Yujiao Ding, Yining Zhou, Haotian Li, Xuesong Zheng, Yuhang Wang
Pengpeng Sang, Jixuan Wu, Xuepeng Zhan, Chaoming Liu, Jiezhi Chen

A Data Hierarchy-Based Adaptive Testing Method for Integrated Circuit Parameter Sets 582

Kaiming Hao, Yan Li, Xu Cheng, Qiong Wu, Wenfa Zhan, Yujie Huang

Microstructural Evolution and Reliability Analysis of RDL Copper Interconnects under High-Temperature Conditions 586

Peng Xu, Lan Li, Jialu Huang, Yu Yao, Hengchang Bi, Jiang Xia, Zongyi Li, Zuoyuan Dong, Xing Wu

Reliability Screening for Yield Improvement in IC Design Industry: Progress, Challenges and Prospects 590

Yixian Wang, Xiaoxiao Qiu, Zhigang Ji

Impact of Thermal Shock on the Threshold Voltage and Transconductance of FinFET I/O Devices 594

Yaolin Wang, Kun Chen, Wendi Wei, Zhao Yang, Zhiteng Zhang, Zhuming Wang, Chen Wang
David Wei Zhang

Effects of Total Ionizing Dose on ESD Performance in High-Voltage SCR with Double Snapback Characteristics 598

Yujie Liu, Xiangliang Jin

A New Surge Protection Circuit with Low Dynamic Leakage Current 602

Zhiqiang Hu, Ran Ye, Qiao Kang, Ke Cui, Hao Luo, Weipeng Ye, Siyang Liu, Weifeng Sun

Reliability Enhancement in HfO$_2$-Based FeRAM: Circuit-Level Solutions for Insufficient Polarization and Memory Window Degradation 606

Changnan Shi, Taoran Shen, Li Xiong, Shuyang Lv, Yuanfeng Chen, Xiaoyong Xue, Xiaoyang Zeng

A PVT-Tolerant Quick Startup CMOS Crystal Oscillator With Chirp-Assisted Fixed Injection 610

Hao Luo, Yue Lin, Jian Xu, Hongtao Xu

A 20 Gb/s/Wire Short-Reach Simultaneous Bi-Directional Transceiver with DuoBinary Coding for Die-to-Die Interface in 28 nm CMOS 614

Bohui Bai, Fangxu Lv, Zhengbin Pang, Geng Zhang, Ruixiao Kuai, Liangyong Yuan, Ruotian Yin
Jiliang Liu

A 56Gb/s PAM4 Transceiver Based on BSS-LMS Algorithm With 3-Taps Adaptive TX FFE 618

Xianchao Zeng, Fangxu Lv, Liquan Xiao, Jiaqing Xu, Zhouhao Yang, Liangyong Yuan, Cewen Liu
Xiaoyue Hu, Yingjie Zhang

A Low-Power Gm-Boosted VCO with Multi-Transformer in 40nm CMOS 622

Zilong Wu, Bowen Chen, Yue Lin, Hongtao Xu

A 64 Gbps 10 mW 0.0081 mm² Inverter-Based CTLE Employing Power-Efficient Split Biasing Topology in 40 nm CMOS 626

Fang Ding, Huzhi Tang, Ke Wu, Yuekang Guo, Jing Jin, Jianjun Zhou

A Low-Power Area-Efficient Serializer for CMOS Image Sensors 630

Jingyang Chen, Nanbo Chen, Gang Wang, Peng Feng, Jian Liu, Nanjian Wu, Liyuan Liu

A 6b 14GHz Phase Interpolator with 2-Stage Injection-Locked Ring Oscillators in 28nm CMOS 633
Danqi Ding, Bingyi Ye, Weixin Gai

Boosting Self-Powered Properties of 2D Material-Based Photodetectors via Asymmetry Engineering 637
Ran Huo, Han Zhang, Yihong Sun, Shijun Ou, Changming Pi, Mansun Chan, Changjian Zhou

Interfacial adhesion enhancement enabled mechanically durable flexible organic optoelectronics 641
Ziqi Wang, Xiangzhe Li, Huimin Wu, Kai Wang, Sixing Xiong, Jin Qian

Image Flare Removal via Stable Diffusion Framework 645
Jiazheng Lian, Ruoxi Zhu, Jiaming Liu, Ming'e Jing, Xiaoyang Zeng, Yibo Fan

AI-Assisted Droplet Splitting on a Parallel-Plate Optoelectrowetting Chip 649
Junyan Tian, Shang Gao, Tengpu Zhu, Enqing Liu, Gaifang Chen, Jia Zhou

A Sub-1mV Voltage-Variation Pixel Power Supply Architecture with Radiation-Hardened Built-In
LDO for Pixel Readout ASIC 653
Lei Li, Jinxiang Wang, Yini Hong, Yuxiao Zhao, Yongsheng Wang

A CMOS Pixel with Gradient-Doped PPD and LOFIC for 1.7 ns Charge Transfer Time and 92 dB
Dynamic Range 657
Tianjing Qiu, Jinglei Du, Junli Zhang, Peng Feng, Jian Liu, Nanjian Wu, Liyuan Liu

Stacked 2D materials Nanopore Sensors 660
Candong Zhao, Qinjie Pan, Guangyi Yang, Peng Cheng, Fuwei Zhuge, Yuhui He

On-chip Contact Angle Sensor Using Coplanar Capacitors for Digital Microfluidic Systems 664
Akira Tsuchiya, Hayato Fukui, Tsubasa Furuta, Toshiyuki Inoue, Keiji Kishine

Selective manipulations of droplets on photo-driven microfluidic chip with virtual electrowetting
channels 667
Gaifang Chen, Enqing Liu, Junyan Tian, Shang Gao, Jia Zhou

A MEMS Rectenna for RF energy harvesting around 2.4GHz 671
Liu Xiaoqiang, Wang Tiancong

Design and Implementation of Shared Storage Communication Architecture for MCCSIP-RAA 674
Longmei Nan, Yu Jin, Yiran Du, Tao Chen, Lin Chen, Yanjiang Liu, Wei Li, Weiquan Sang

A Fully Quantized LeNet-5 accelerator for Edge Computing with Quantization-Aware Training 678
Yushan Dai, Angyang Li, Jian Mei, Rui Yin

A High-Voltage And High-Precision Operational Amplifier 682
Juan Wei, ZongLin Li, HongRui Che, FuMei Liang, DaGang Li

A 12-bit 1MS/s SAR ADC design for high-temperature MEMS accelerometers 686
Yanlin Mo, Min Qi

A Low-Noise Ultrasound Analog Front-End with Low Gain Error Time-Gain Compensation 690
Xiangchen Wan, Fan Ye

Adaptive Frequency Modulation Buck Converter Based on Valley Current Mode ACOT Control 694
Bowen Jiang, Hong Ren, Ningning Wang

A High-Voltage Level Shifter for BMS Chip in EV with 0-80V Input Range 698
Kunning Mao, Liji Wu, Jing Hu, Zhiwei Li, Haifeng Chen, Xiangmin Zhang

A New Circuit for Generating Half of VDD 702
    Li Zeng, Ming Wang, Peng Bo, Zhangwen Tang

A Boost DC-DC Converter with Low Power and High Efficiency for Portable Device Applications 706
    Jing Cao, Bingjie Chen, Hongfei Ye, Jianhua Feng

A Hybrid Complex-Filtering Scheme with High Image Rejection and Efficient Channel Selection for
Low-IF Receivers 711
    Yue Yin, Guanlin Zhang, Haobo Qi, Haodong Lu, Xinbing Zhang, Ziting Feng, Ye Zhang

Design of a Nonlinear Temperature Compensated Bandgap Reference in 55nm Process 715
    Hezhuang Nie, Ningning Li, Jian Mei, Rui Yin

A Novel Light-load Control Method For Switching Converters in Portable Devices 718
    Jie He, Shuyu Zhang, Langyuan Wang, Suyi Yao, Kejia Zhu

Adaptive on-time Control Buck Converter Based on Phase-Locked-loop and Dynamic Calibration of
DC Offset 722
    Xinyu Zhang, Sujuan Liu, Kun Liu, Bingxue Zhang, Yahua Shi

Design of a Fast Transient Response LDO Circuit Based on Transient Enhancement Structure 726
    Xudong Sun, Sujuan Liu, Kun Liu, Junchao Zhao

A 32-MHz FLL-Based RC Oscillator with PVT Compensation Using Frequency Tripler 730
    Ikhwan Kim, Yajie Qin

A Constant On-Time Buck Converter with VCO-based DC Offset Calibration Technique 733
    Bingxue Zhang, Sujuan Liu, Kun Liu, Xinyu Zhang, Yahua Shi

A Low-Power High-Precision Impedance Measurement Circuit Using DC Servo Loop for Closed-Loop
DBS Systems 737
    Ziqi Tan, Yijun Ye, Yutao Mao, Hui Wu, Xiaofei Kuang, Jie Yang, Mohamad Sawan

A Low-Power BJT-Based Thermal Shutdown Circuit with Hysteresis for BMS chip in EV 741
    Zonghuan Wu, Xiangmin Zhang, Liji Wu

Design of CRFF-B Loop Filter Architecture for Wideband Continuous Time Sigma-Delta Modulators
in CMOS 28 nm 745
    Zhihao Hou, Yuqi Fan, Yifei Gao, Chuan Liu, Chuan Qin, Maliang Liu, Yintang Yang

An Open-Loop Residue Amplifier with SSF Structure Achieving 69dBc SFDR for High-Speed and
High-Precision PSAR ADCs 749
    Chengjun Liu, Deng Luo, Hanbing Liu, Chengchao Mou, Bin Liang, Yaqing Chi, Jianjun Chen, Kai Tang
    Jing Xiao, Ming Tao

A 14-bit R-2R DAC with All-Digital Foreground Calibration based on Redundant LSB 753
    Hanbing Liu, Deng Luo, Chengjun Liu, Chengchao Mou, Bin Liang, Yaqing Chi, Jianjun Chen, Kai Tang
    Jing Xiao, Ming Tao

A Multi-Channel Reconfiguration and Combination Technique for Timing Mismatch Calibration in
Time-Interleaved ADCs 757
    Jize Liu, Jinwei Wu, Jiayi Chen, Xinqi Liu, Yuekang Guo, Jing Jin

A 12-bit 620 MS/s Pipelined-SAR ADC with Feedforward Compensation Closed-loop Residual Amplifier in 28 nm CMOS 761
Shuai Liu, Yi Hu, Guoyu Li, Congyang Sun, Yidong Yuan, Hao Xu, Na Yan

A novel RA architecture and digital calibration method for SAR-assisted pipeline ADCs 765
Jieqiong Zeng, Hao Min

An 8-bit 0.4-mW 740-μm^2 DS Digital-to-Analog Converter in 28nm CMOS with 60.89-dBc SFDR 769
Xiongfeng Bi, Bingyi Ye, Weixin Gai

A K-Band CMOS Switched-Type Attenuator with Temperature Compensation Technique 773
Xiaodong Zhao, Kai Zhang

An Ultra-Wideband 1.5–18.5 GHz MMIC Phase Shifter in 0.25-μm GaAs Technology 777
Bo Fu, Xuan Ding, Xuesong Han, Xiao Ding

A 0.2–7.3-GHz Compact LNA with Super Linearity for 5G NR in 22-nm CMOS Technology 781
Kaiyun Deng, Zan Zhou, Yingqi Liu, Haoyu Dong, Haigang Feng

An Area-Efficient Bi-directional Cascode PA-LNA For 5G NR in 28-nm CMOS 785
Yue Wu, Wei Li, Shijiao Dong, Hongtao Xu

A 223M-235MHz Fully-Integrated Differential Class-E Power Amplifier with 45.5% PAE and 22.8dBm 789
Chaoyang Zheng, Yanxiang Chen, Jianhua Lu, Yan Ma, Zhiliang Hong, Yumei Huang

A 16~46-GHz, >77-dB IRR, Low-Amplitude and Phase-Error IQ Generator with Self-Adaptive I/Q Calibration in 28-nm CMOS 793
Lijiang Zhang, Wei Li, Bowen Yu, Chengzhang Cai, Yue Wu, Bowen Chen, Yue Lin, Hongtao Xu

Impact of Process Parameter Variations on the Random Values of SRAM-Based PUFs 797
JinJin Shao, Ruiqiang Song, Chunmei Hu, Biwei Liu, Bin Liang, Yaqing Chi, Yaohua Wang

MIVO: Operator-Level On-Chip Memory System with Dynamic Bank Scheduling for Many-Core Neural Processing Unit 801
Xinghao Zhu, Zifeng Zhao, Xiaoxing Wu, Gengsheng Chen, Xiaofang Zhou

DyQRA: A Deadlock-free Routing Algorithm for Large-Scale Mesh NoCs 805
Haoxiang Sun, Aoyun Feng, Hongfei Ye, Jianhua Feng

An Enterprise Solid-State Drive Controller Supporting Spin-transfer Torque Magnetoresistive Random Access Memory 810
Chao Song, Qihao Liu, Yunzhe Wang, Rufa Su

An IO Die with Collective-Aware Routing and In-Situ Processing for Data Synchronization in Multi-Chiplet Systems 814
Qi Luo, Chen Mu, Chixiao Chen, Xin Chen, Shiwei Liu

An STT-MRAM Last Level Cache Management Method Based on Write Intensity Prediction for GPUs 818
Yujie Pu, Qiaoran Zhang, Shitong He, Fanchen Wu, Chen Zhao

Towards Scalable and High-Throughput NTT Acceleration On Hybrid-Bonding Architecture 822
Wenxuan Zhang, Yi Sun, Xinglong Yu, Yifan Zhao, Jun Han

A Low-Cost Multiplier-Free Accelerator for Binary Neural Network 826
    Z. W. You, J. H. Wu, R. C. Ma, G. C. Qiao

Design of a MobileNetV2 FPGA Accelerator for Low-Power Real-Time Identification of Plant
Nematodes 830
    Ying Zhu, Pengjun Wang, Qikang Li, Huihong Zhang

A 65nm Analog-Computing Chip With Reconfigurable Charge-Pump-Based Adders for
5.26nJ/Decision Retrainless Keyword-Spotting 834
    Lichen Feng, Rundong Cai, Lin Wu, Zhangming Zhu

HAMP: Head-Aware Mixed-Precision Token Pruning and Quantization for Efficient ASR 838
    Xiaoxing Wu, Xinghao Zhu, Lanqi Ma, Gengsheng Chen, Wenbo Yin

A Compressed Sensing Spiking Neural Network System for Radar-Based HGR 842
    Liyu Qian, Zikai Zhu, Yuhan He, Jie Lu, Yaojie Sun, Lirong Zheng, Zhuo Zou

FlexiCore-DNN: A Configurable and Templated Architecture for End-to-End FPGA Acceleration of
Deep Neural Networks 846
    Rao Fu, Wenhao Huang, Aiwu Ruan, Huiyun Li, Yongqing Wang

A 7-bit 6.25-GHz Low Power High Linearity DPC for CDR Applications 850
    Jingsong Cui, Kai Li, Chengyu Yang, Jiahao Lu, Hao Li, Ang Hu, Dongsheng Liu

Design of Low-Voltage Differential Signaling Driver for Image Sensor 854
    Zhongwei Lin, Ningning Li, Angyang Li, Jian Mei, Rui Yin, Jiaming Zhang, Zhicheng Shi

Exploring The Further Fracturability of Intel ALM 858
    Chenyu Jiang, Xianfeng Cao, Lingli Wang

A General and Modular FPGA Hardware Architecture for Enhanced Scalability and Flexibility 862
    ZiRui Qin, ZhiNan Li, YaBo Xiao, Hui Zhang, Cang Liu

Pipelined Parallel Design of SIFT Algorithm on FPGA 866
    *Yuanhao Zhang, Tianliang Xu, Jianfeng Li, Zhenbin Lv, Shanqiang Yang, Chenxu Wang, Yuhang Wang
    Bo Chu, Zhiwei Han*

Design and Implementation of an FPGA-based MIPI DSI Interface for Micro-LED Displays 870
    Runfeng Yao, Xinyi Liu, Kaisong Zhu, Jinbo Liang, Zhaojun Liu

A Sub-100µs-Latency Visual-Cortex-Mimicking Heterogeneous Multi-Core Edge Neuromorphic
Processor Enabling On-Chip High-Accuracy Learning 874
    Junxian He, Ying Jiang, Zhengqing Zhong, Mingju Chen, Liyuan Liu, Cong Shi

An Energy-Optimized FPGA Implementation for Convolutional Neural Networks Accelerator 878
    Yujie Zhu, Jianxuan Yin, Jingjing Liu, Jianhua Zhang

A Lightweight Low-Latency Hardware Architecture for Dual Attention Super-Resolution Network 882
    Haocan Jiang, Aiying Guo, Jianhua Zhang, Jingjing Liu

A Scalable Channel-Parallel Accelerator for Spiking Neural Network 886
    Yuchun Wu, Lingling Miu, Jingjing Liu, Jianhua Zhang

A precise current-controlled resistor and its applications in zero-pole tracking frequency compensation for LDO 890

Guanting Liu, Guijuan Zhao, Feng Shi, Xiaohuan You, Shuhai Chen

ASSVD: A Self-Supervised Surgical Video Desmoking Network with Sparse Attention 894

Yinna Zhu, Wanyi Zhou, Zijing Zhang, Gengsheng Chen, Wei Xu

A 71 TOPS/W 24.2 TOPS/mm² 14nm SRAM CIM Macro with a Capacitor-less ADC for Edge AI 898

Zexing Chen, Siyao Jia, Chixiao Chen

Data-Centric Automatic Design Migration of Low Voltage CMOS Bandgap Reference Circuit 902

Shun-Qi Dai, Yuan Lei, Bei-Ping Yan

Innovative Detection Capacitor Utilization in ESD Power Clamp Circuits for HBM Residual Voltage Suppression 906

Zelong Huang, Guangyi Lu, Haoyu Xia, Qi Wu, Haiming Wang

High Efficient Efuse Full Process Burning Solution Based on ATE 909

Qian Zhai, Yichen Xiao, Xin Song, Haobin Wang, Yuyuan Wang, Xuxin Chen

Study of Reliability Screening Method to Improve the DPPM of IC Products 913

Yancong He, Zhiyong Yang, Zhinong Liu, Shuang Jiao, Chuyuan He, Yixian Wang, Zhigang Ji

Weight Bit Sensitivity Analysis and FPRH-Based Hardening Strategy for CNN Accelerators 917

Jinghao Chen, Shanqiang Yang, Tianliang Xu, Congan Xu, Yuehong Gong, Chenxu Wang

An effective method for low-contrast high-noise lithography SEM image contour extraction 921

Ruirui Zhang, Gongyan Ye, Xianhe Liu

Design of A Dual-Mode Analog Front-End Circuit Applied in the Voice Activity Detection System 925

Zirui Dong, Xuhaohan Wang, Fan Ye

Research on Radiation-Hardened High-Voltage Gate Driver Circuit Based on 0.8μm 1200V Bulk Silicon BCD Process 930

Xiaohui Li, Yi Zhang, Qiang Wang, Qiankun Xiong, Bo Zhang, Ming Qiao

Parameter identification of single-phase inverter digital twin system 934

Ao Shen, Hui Li, Jie Kang, Jia Hao Lv

Optimization of Three-dimensional High-k Superjunction under Non-Punch-Through Mode: Theoretical Modeling and Comparison 938

Zhentao Xiao, Chenxing Wang, Zonghao Zhang, Haimeng Huang

Smart Adaptive Perception for High-Precision Lightweight Infrared UAV Detection and Tracking 942

Shiyu Mei, Lei Deng, Rui Yin

Design and validation of fluorescence lifetime solving algorithm for fiber-optic temperature sensor 946

Yuxuan Yang, Xiangliang Jin

A Multi-Cycle Pulse Transfer Timing Scheme for Enhancing Charge Efficiency in CMOS Image Sensors 949

Zhenhao Zhang, Chiang Zhu, Haiyang Liu, Peng Peng, Sikai Wang, Junjie Hao, Xiaona Zhu

Design of RF Microsystem Based on Silicon-based Stereoscopic Integration Technology 953

Xiaoqing Zhang, Lei Shi, Mengmeng Yin, Cui Jing, Dexi Liu

A Novel Pretreatment Approach to High-quality SiO_2 Surface Applied for C2W Cu/SiO_2 Hybrid Bonding 957

Han Jiang, Xianlong Wang, Ziyu Liu, Yabin Sun

Approximately Timed Scalable DSP Model Based on SystemC 961

Yongwang Qin, Sheng Liu, Yang Zhang, Xing Hu, Chen Shangqian

Microscopic Mechanisms of Bias Temperature Instability Induced by Defects in Si/SiO_2/HfO_2 Gate Stacks: A DFT and NEGF Study 965

Yantao Huang, Yunzhi Lin, Yixin Zhang, Junlong Li, Xiaoxu Kang, Fengying Yao, Shaojian Hu, Qing Shi Tao Wu

Mechanism of Leakage Current Enhancement Induced by La Doping in HfO_2 Gate Stacks: A DFT Investigation 969

Yunzhi Lin, Yantao Huang, Yixin Zhang, Qing Shi, Fengying Yao, Junlong Li, Shaojian Hu, Xiaoxu Kang Tao Wu

Layout-Aware Performance Analysis of the CFET based NAND2 constructed Ring Oscillator 973

Junjie Hao, Chiang Zhu, Huawei Tang, Xiaona Zhu, Shaofeng Yu

2025 IEEE 16th International Conference on ASIC (ASICON)

www.asicon.org

ASICON 2025

PROGRAM

Oct. 21st - Oct. 24th , 2025

Crowne Plaza Kunming City Centre, Kunming, China

◆IEEE Beijing Section

2025 IEEE 16th International Conference on ASIC (ASICON)

ASICON 2025

Oct. 21st - Oct. 24th , 2025
Crowne Plaza Kunming City Centre,
Kunming, China

Sponsored by
IEEE Beijing Section
Fudan University
Yunnan University
National IC Innovation Center

Supported by
IEEE CASS
IEEE EDS Shanghai Chapter
IEEE SSCS Shanghai Chapter
IET Shanghai Network

Organized by
Fudan University

Welcome to ASICON 2025

On behalf of the ASICON 2025 organizing committee and the General Co-Chairs, it is our distinct honor and pleasure to warmly welcome all attendees to the 16th Advanced Semiconductor Integrated Circuits—ASICON 2025. This year, we gather in the vibrant city of Kunming, Yunnan Province, from October 21 (for Tutorial) to October 24, with our Opening Ceremony at 8:30 AM on October 22 in the Crowne Plaza Kunming City Center Hotel.

ASICON holds a special place in the international semiconductor community, and 2025 is a significant year for our conference: we are proudly celebrating the 30th anniversary of ASICON, which was established in 1994. For three decades, ASICON has grown and thrived as a venue for innovators, researchers, and industry leaders to exchange ideas, build collaborations, and drive the progress of integrated circuit technology.

This year's program reflects ASICON's enduring commitment to excellence and innovation. We will feature in-depth tutorials on the first day, along with an inspiring series of keynote speeches delivered by world-renowned academic and industry leaders during the plenary sessions. ASICON continues to bring together VLSI circuit designers, system integrators, IC manufacturers, device engineers, ASIC users, CAD/CAE tool developers, and academics from across the globe, providing a unique environment for sharing pioneering research, technical advancements, and emerging trends in Advanced Semiconductor Integrated Circuits.

In addition to marking our significant milestone, 2025 is also the centennial anniversary of the invention of the Field-Effect Transistor (FET)—a breakthrough that has fundamentally shaped the world of electronics and paved the way for the vast landscape of modern integrated circuits. We are excited to honor the 100-year legacy of the FET and recognize its profound impact on our field and society.

Over the last thirty years, ASICON has played a pivotal role in advancing both academic research and industrial applications, building bridges between scholars and practitioners, and fostering lifelong professional networks. As we embark on this celebratory gathering, we look forward to nurturing the spirit of innovation, collaboration, and inspiration that defines ASICON.

We sincerely appreciate your participation and invaluable contributions, which make this conference a continuing success. Let us unite, share, and innovate as we celebrate ASICON's 30th anniversary and the centenary of FET—two landmarks for our community and our future together.

We look forward to meeting you in Kunming for ASICON 2025!

General Co-Chairs of ASICON 2025

Jan Van der Spiegel

Hao Min

Bin Zhao

Yong Lian

Kun Yue

Oct. 21st, 2025

Conference Committee

Gerneral Co-Chairs

Name	Affiliation	Country/Area
Jan Van der Spiegel	University of Pennsylvania	USA
Hao Min	Fudan University	China
Bin Zhao	IEEE EDS	China
Yong Lian	Shanghai Jiao Tong University	China
Kun Yue	Yunnan University	China

Steering Committee Co-Chairs

Jan Van der Spiegel	University of Pennsylvania	USA
Mengqi Zhou	IEEE Beijing Section	China
Ting-Ao Tang	Fudan University	China

Advisory Committee Co-Chairs

Chenming Hu	UC Berkeley	USA
Hiroshi Iwai	Yang Ming Chiao Tung University	Taiwan, China
Cor Claeys	IMEC & KU Leuven	Belgium
Qianling Zhang	Fudan University	China

Program Committee Co-Chairs

Fan Ye	Fudan University	China
Na Yan	Fudan University	China
Francois Rivet	University of Bordeaux	France

Haruo Kobayashi	Gunma University	Japan
Jyi-Tsong Lin	Sun Yat-sen University	Taiwan, China
Yimao Cai	Peking University	China
Huaqiang Wu	Tsinghua University	China
Yi Zhao	Huada Semiconductor Co., Ltd.	China
Hao Xu	Fudan University	China

Organizing Committee Co-Chairs

Huihua Yu	Fudan University	China
Lixing Yu	Yunnan University	China

Publicity Co-Chairs

Wei Xu	Fudan University	China
Rui Yin	National IC Innovation Center	China
Jiting Sheng	Fudan University	China

Publication Chair

Mengqi Zhou	IEEE Beijing Section	China

Secretary-General

Xiaona Zhu	Fudan University	China

Technical Program Committee Members

Analog and RF Circuits Subcommittee

Wenning Jiang	Fudan University	China
Hao Xu	Fudan University	China
Wei-Zen Chen	Yang Ming Chiao Tung University	Taiwan, China

Tai-Cheng Lee	Taiwan University	Taiwan, China
Feng Zhan	Institute of Microelectronics, CAS	China
Haruo Kobayashi	Gunma University	Japan
Ang Simon	University of Arkansas	USA
Mo Huang	University of Macau	Macao, China
Fei Song	Ubilinx technology, Inc	USA
Nanjian Wu	Institute of Semiconductor, CAS	China
Wenjun Zhang	Intel	USA
Liang Qi	Shanghai Jiaotong University	China
Shuang Song	Zhejiang University	China
Chao Chen	Delft University of Technology	Netherlands
Zhiming Xiao	Nankai University	China
Hao Gao	Eindhoven University of Technology	Netherlands

Digital Circuits and SOC Subcommittee

Jun Han	Fudan University	China
Xiaoyong Xue	Fudan University	China
Yibo Fan	Fudan University	China
PengCheng Xiao	Fudan University	China
Deepu John	University College Dublin	United Kingdom
Pengjun Wang	Wenzhou University	China
Dongsheng Liu	Huazhong University of Science and Technology	China
Shaoyun Wang	NextInput，Inc.	USA
Chua-Chin Wang	Sun Yat-Sen University	Taiwan, China
Gerald Sobelman	University of Minnesota	USA
Na Gong	University of South Alabama	USA
Shyh-Jye Jou	Yang Ming Chiao Tung University	Taiwan, China
Tzu-Hsien Sang	Yang Ming Chiao Tung University	Taiwan, China
Liang Liu	Lund University	Sweden
Kyeong-Sik Min	Kookmin University	Korea
Makoto Ikeda	University of Tokyo	Japan
Zhiyi Yu	Sun Yat-sen University	China

Chuan Zhang	Southeast University	China

CAD Techniques Subcommittee

Fan Yang	Fudan University	China
Jun Tao	Fudan University	China
Sheldon Tan	University of California, Riverside	USA
Gang Qu	University of Maryland	USA
Bei Yu	Chinese University of Hong Kong	Hongkong, China
Ahmed Jerraya	CEA Tech	France
Mansun Chan	Hong Kong University of Science and Technology	Hongkong, China
Xingang Wang	Skyworks Solutions, Inc.	USA
Xiaoqing Wen	Kyushu Institute of Technology	Japan

Process and Devices Subcommittee

Yuan Dong	Shanghai University	China
Chen Wang	Fudan University	China
Xiaoxi Li	Xidian University	China
Zhigang Ji	Shanghai Jiao Tong University	China
Chunlei Wu	Fudan University	China
Yanli Li	Fudan University	China
Xiaona Zhu	Fudan University	China
Kuei-Shu Chang-Liao	Tsinghua University	Taiwan, China
Chao-Sung Lai	Chang Gung University	Taiwan, China
Masaharu Kobayash	The University of Tokyo	Japan
Kuan-Neng Chen	Yang Ming Chiao Tung University	Taiwan, China
Ching-Ting Lee	Cheng Kung University/ Yuan Ze University	Taiwan, China
Pei-Wen Li	Chiao Tung University	Taiwan, China
Weisheng Zhao	Beihang University	China
Simon Ang	University of Arkansas	USA
Wai Tung Ng	University of Toronto	Canada
Simoen Eddy	IMEC	Belgium

Kazuhiko Endo	Advanced Industrial Science and Technology (AIST)	Japan
Jianfu Zhang	Liverpool John Moores University	United Kingdom
Weidong Zhang	Liverpool John Moores University	United Kingdom
Ya-Hong Xie	UCLA	USA
Moufu Kong	University of Electronic Science & Technology of China	China

Generation, Modulation and Application of Spintronic Markov Chain Signal

Xihui Yuan [1,2,3], Jiajia Jian [1], Zheng Chai [1]*, Xue Zhou [1], Weidong Zhang [4], Jian Fu Zhang [4], Tai Min [1]

[1] State Key Laboratory for Mechanical Behavior of Materials, and School of Materials Science and Engineering, Xi'an Jiaotong University, Xi'an, China, [2] School of Microelectronics, Xidian University, Xi'an, China, [3] HangZhou Institute of Technology, Xidian University, HangZhou, China, [4] School of Engineering, Liverpool John Moores University, UK

* Email: zheng.chai@xjtu.edu.cn

Abstract—Stochastic processes are widely used in many real-world fields. As a basic type of stochastic process, Markov chain (MC) represents a stochastic transition model in which the probability of each event depends only on the state attained in the previous state. The hardware generation and arbitrary modulation of MC signals are urgently needed because of the large energy and circuit consumption of existing software methods. However, the hardware MC remains challenging due to the difficulty of modulating randomness in devices. In this work, a novel MC hardware generation method based on a single magnetic tunnel junction (MTJ) device is proposed and further verified via electrical experiments and mathematical derivation. The crucial MC model parameters such as transition matrix and average dwell time (ADT) can be flexibly modulated by changing the probabilistic switching voltage in the proposed waveform. Furthermore, based on the MC generation method and inspired by the divide-and-conquer strategy, a new hardware architecture to estimate eigenvector of an n×n stochastic matrix where n is the power of 2 is proposed using only $\log_2 n$ MTJ devices. This method provides a new hardware solution for the generation, modulation and application of stochastic signals in semiconductor IC chips.

Keywords—*Markov chain, magnetic tunnel junction, stochastic switching, stochastic matrix, eigenvector*

I. INTRODUCTION

Stochastic processes universally exist and widely utilized in various fields such as mathematics, computer science and artificial intelligence (AI). As a fundamental stochastic process, Markov chain (MC) [1] represents a sequence of possible events, in which the probability of the next event depends only on the state attained in the present event, and is independent of the past events. The MC is different from the generic random bit stream generation where the random variables are totally independent. The transition relationship among the states of MC can be revealed by a Markov transition matrix A which is a square matrix with elements representing the transition probability from a state to another state. The dimension of transition matrix is equal to the amount of states in MC. And the stationary distribution π of MC indicates the long-term probability of MC model in each state according to (1).

$$\pi \cdot A = \pi \qquad (1)$$

From a mathematical perspective, the transition matrix of an MC is a stochastic matrix (SM) characterized by non-negative elements with each row summing to one. And the stationary distribution is also known as the eigenvector of SM. The well-known application of MC and SM in computer science is Google's PageRank algorithm which measures the importance of webpages by calculating the eigenvector [2].

Moreover, the MC signal is formed by segments of alternating consecutive appearing states. The lengths of the segments, often called dwell times (DT), are randomly but exponentially distributed. The average dwell time (ADT) represents DTs' distribution and it is desirable that ADTs could be adjusted to any target sets, for practical applications such as the stabilization of stochastic systems in robust control theory and fuzzy systems [3].

Most existing MCs are generated by software algorithms. However, for the Internet-of-Things (IoT) applications, it is desirable that MCs could be directly generated by hardware devices, for the performance, area, and power considerations. Although randomness of scaling/emerging devices such as random telegraph noise [4] [5] and variability in a resistive-switching memory (RRAM) [6] has been utilized for MC generation, their transition matrix and ADTs are non-modulable. Spintronic devices including spin torque transfer magnetic memory (STT-MRAM), spin orbit transfer (SOT)-MRAM, and domain wall Hall cross-bars, have also been employed for stochasticity-related tasks such as Boltzmann machine [7], spiking neural network [8] and true random number generator [9] [10]. Previous studies have demonstrated the generation of telegraphic switching signals in MRAM devices, leveraging magnetic fields and electrical pulses [11]. However, the implementation of such techniques in integrated circuit (IC) chips are challenging.

In this work, we proposed a MC generation and modulation method based on the probabilistic switching of single magnetic tunnel junction (MTJ) under a specially designed waveform, verified by mathematical derivation and experimental demonstration. Furthermore, the proposed spintronic MC signal is used in hardware architecture for estimate the eigenvector of SM with the required devices' number reducing to $\log_2 n$, significantly minimizing hardware requirements. This method offers a promising solution for the development of compact and efficient circuits, addressing the needs of emerging applications.

II. DEVICES AND EXPERIMENTS

The device is a bottom pinned perpendicular magnetization anisotropy (PMA) MTJ with a diameter of 78 nm. **Fig. 1a** demonstrates its magneto resistive switching between the low-resistance parallel (P) state, and the high-resistance anti-parallel (AP) state with the cross section scanning electron microscopy (SEM) image (**inset of Fig. 1a**). **Fig. 1b** shows the sigmoid-like dependence of switching probability on pulse amplitudes with fixed pulse width of 500 ns, which can be explained by the Néel-Brown equation [12]:

$$P(t) = 1 - exp(-t/\tau) \qquad (2)$$

979-8-3315-3918-4/25 $31.00 © 2025 IEEE

where $P(t)$ is the thermal switching probability, and τ is the relaxation time:

$$\tau = f_0^{-1} exp(E_b/k_B T) \qquad (3)$$

where f_0 is the attempt frequency, E_b is the energy barrier, and T is the temperature. All electrical measurements were conducted using the pulse measurement units (PMUs) in a Keithley 4200 semiconductor characterization system.

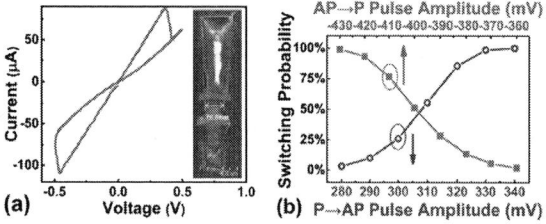

Fig.1. (a) The I-V curve of magneto-resistive switching. Inset: cross-section SEM image of the MTJ. (b) The sigmoid-like dependence of switching probability on pulse amplitudes with pulse width of 500 ns.

III. RESULTS AND DISCUSSIONS

A. Generation of Spinntronic Markov Chain Signal

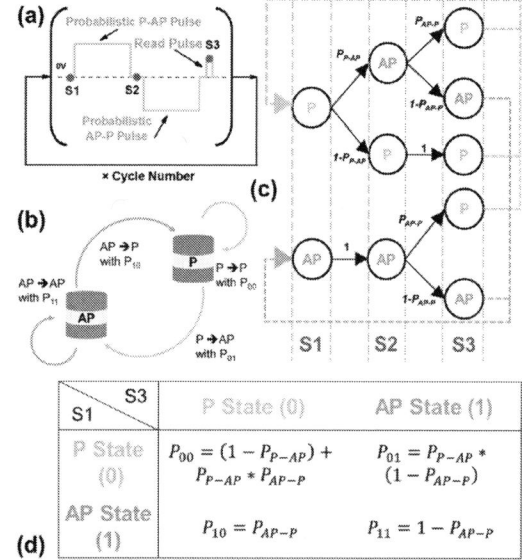

Fig.2. (a) Schematic of the three-pulse waveform. (b) Schematic of state transition. (c) When S1 is at P/AP state, the resulted S2 and S3 with corresponding probabilities. (d) State transition matrix of S1 and S3 in one cycle. P is defined as 0 and AP as 1; e.g., P_{00} means the probability of S3 = P (or 0) when S1 = P (or 0).

Combined with the non-volatile property and bipolar probabilistic switching phenomenon of MTJ device, a three-stage waveform (**Fig. 2a**) comprising a pulse for probabilistic P-to-AP transition (P_{P-AP}), a pulse for probabilistic AP-to-P transition (P_{AP-P}), and a read pulse is proposed to generate MC signal in single MTJ device. Specifically, the states before the first and second pulse are named as S1 and S2, respectively. Current measurement is done by the read pulse to get S3. S3 obtained from all cycles form a sequence, which is the generation result.

Regarding the relation between S3 in one cycle and the next cycle, there are 4 cases: (1) AP → AP; (2) P → P; (3) AP → P; (4) P → AP, as summarized in **Fig. 2b**. The

corresponding transition probabilities are P_{11}, P_{00}, P_{10} and P_{01} in which the 0 and 1 correspond to the state P and AP, respectively. **Fig. 2c** schematizes the resulted S2 and S3 with the corresponding probabilities. Since each S3 in the MC is influenced by, and only influenced by the S1 in the same cycle (that is, S3 in the last cycle), the generated bit stream conforms to the definition of a standard MC. This is further evidenced by **Fig. 2d**, which is the Markov transition matrix of the generated result. Apparently, the different rows in the transition matrix support that the generated bits are not independent random variables [13]. For preliminary demonstration, 10,000 cycles i.e. three-pulse waveforms are repeatedly applied onto the MTJ to generate 10,000 bits.

To validate the effectiveness and generalizability of the proposed MC generation approach, the quality of the MC signal is evaluated by comparing the counted stationary distribution with the calculated stationary distribution. In the two-level MC signal based on MTJ, the stationary distribution consists of AP state and P state stationary distribution, named as π_{AP} and π_P, respectively. On the one hand, the calculated π_{AP} and π_P are the eigenvector of the transition matrix when the eigenvalue is 1. And the transition probabilities included in the matrix is obtained using (4):

$$P_{ij}^k = \frac{n_{ij}^k}{n_{ii}^k + n_{ij}^k} \quad (i,j = 0/1) \qquad (4)$$

where P_{ij}^k is the state transition probabilities (the probability transfer from state i to state j in the k cycles), k is the cycle number, and n_{ij}^k refers to the amount transfer from state i to state j in the k cycles. On the other hand, the counted π_{AP} and π_P are counted directly from the cycling results. Note that the lower the relative error between counted and calculated stationary distributions, the higher the quality of the generated MC signal. For example, for a MC signal generated at V_{P-AP} =310 mV and V_{AP-P} = -390mV (**Fig. 3a**), along with cycling, all state transition probabilities P_{ij}^k approach their steady values, P_{ij} (**Fig. 3b**), as evidenced by the decreasing mismatch rates between P_{ij}^k and P_{ij} for all transition probabilities (**Fig. 3c**).

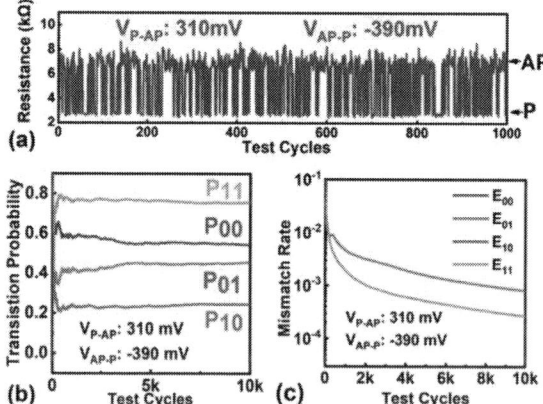

Fig.3. (a) MC signal generated at V_{P-AP} =310 mV and V_{AP-P} = -390mV. (b) The state transition probabilities approach to the steady values with the cycle number. (c) The mismatch rate gradually decreases after more cycle. The E_{11} and E_{01} overlap with the E_{10} and E_{00}, respectively.

Fig. 4a and **Fig. 4b** summarize the π_{AP} and π_P with different combinations of pulse amplitude: the AP/P states' stationary distribution of the generated MC increases/decreases with a

979-8-3315-3918-4/25 $31.00 © 2025 IEEE

higher V_{P-AP}/a lower V_{AP-P}. Compared with the counted values, the calculated stationary distributions are quite close, with an averaged relative error of only 0.017% for AP state (**Fig. 4c**) and 0.013% for P state (**Fig. 4d**), respectively.

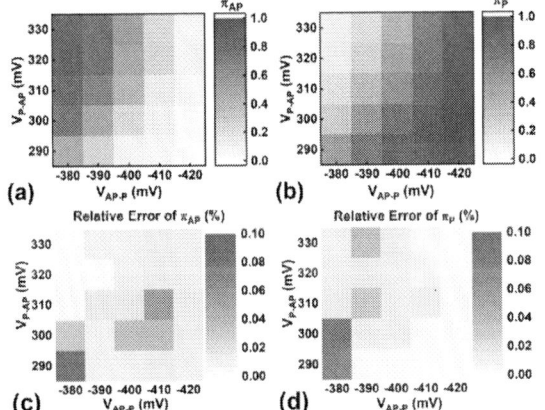

Fig.4. Summary of the counted (a) AP and (b) P stationary distribution π_{AP} and π_P, with (c-d) an averaged relative error of only 0.017% for AP state and 0.013% for P state compared to the calculated AP/P stationary distribution.

B. Modulation of Spintronic Markov Chain Signal

In a MC signal, t_i denotes the dwell time, i.e. the time that the MC remains in one state i (in this work, i represents the state P or AP) until the next transition, while τ_i denotes the averaged value of t_i. **Fig. 5** partially demonstrates the generated MC, together with the t_i and τ_i. Mathematically, the dependence of τ_P and τ_{AP} on the transition probabilities can be derived in the following:

$$\tau_{AP} = \frac{1}{P_{AP-P}} \quad (5)$$

$$\tau_P = \frac{1}{P_{P-AP} \cdot (1 - P_{AP-P})} \quad (6)$$

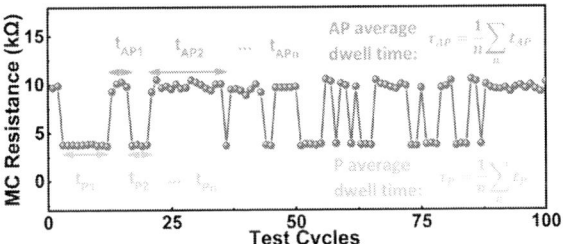

Fig.5. Partial demonstration of the generated MC, with the dwell time (t_P and t_{AP}) and their average values (τ_P and τ_{AP}).

According to the formula derivation above, the τ_{AP} is only determined by P_{AP-P}, while τ_P is determined by both P_{AP-P} and P_{P-AP}. Such dependency has been experimentally verified in **Fig. 6**: with a fixed V_{AP-P} of -420mV, when V_{P-AP} increases, τ_P gradually decreases, but τ_{AP} remains constant (**Fig. 6a**); if V_{P-AP} is fixed at 300mV and V_{AP-P} increases, τ_P increases but τ_{AP} decreases (**Fig. 6b**), both in agreement with the trend of theoretical values after calibration. Such dependency is further supported by the distribution of τ_P and τ_{AP} with V_{AP-P} and V_{P-AP} covering a wider range, as visualized in the 3D plots in **Fig. 6c** and **6d** based on experimental data collected from a range of V_{P-AP} and V_{AP-P}. ADTs cover a wide range as the switching probability varies from 0 to 1.

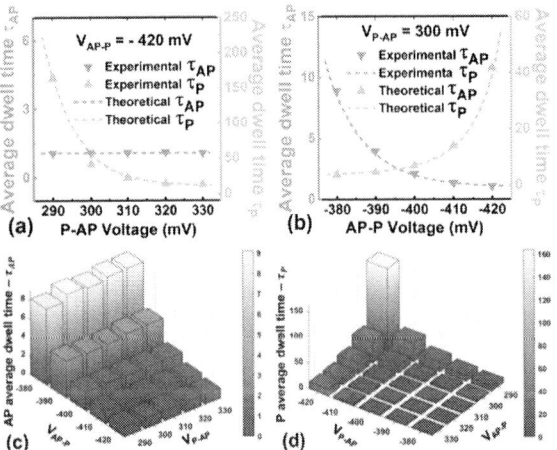

Fig.6. The dependence of τ_{AP} and τ_P on (a) V_{P-AP} and (b) V_{AP-P} is similar to the theoretical counterpart, with fixed V_{AP-P} and V_{P-AP}, respectively. The (c) τ_{AP} and (d) τ_P across a range of V_{P-AP} and V_{AP-P}.

The different dependences of τ_P and τ_{AP} on the pulse conditions provides the inspiration that a target set of τ_{AP} and τ_P combination can be reached following a two-step procedure (**Fig. 7**): (i) Since τ_{AP} is not affected by P_{P-AP}, for a target τ_{AP}, a corresponding P_{AP-P} can be obtained, which is given by a suitable V_{AP-P} from the sigmoid-like dependence. (ii) Next, for the target τ_P, since P_{AP-P} is already determined, a suitable P_{P-AP} can be calculated given by the V_{P-AP}. In this way, the two-step procedure facilitates the electrical modulation of any arbitrary ADT pairs of MC generated from a single MTJ entirely under electrical operation [14].

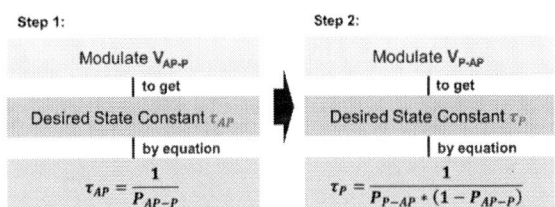

Fig.7. The two-step procedure to modulate ADT in a MC: (i) the target τ_{AP} is achieved by setting V_{AP-P}, regardless of V_{P-AP}. (ii) with fixed V_{AP-P} and determined τ_{AP}, the target τ_P is achieved by setting V_{P-AP}.

C. An application of Spintronic Markov Chain Signal

Beyond the PageRank algorithm, the eigenvector of SMs play a crucial role in data analysis, probability theory and statistics. Traditionally, eigenvectors are computed using software-based algorithms and field-programmable gate arrays (FPGAs), which needs complex circuit and larger energy consumption. Despite the memristive crossbars are also employed to compute the eigenvectors, this approach still results in a polynomial increase (proportional to n^2) in crossbar size as the dimension n of the SM grows. Consequently, there is an urgent need for hardware approaches with less device count to compute eigenvectors.

Drawing inspiration from the spintronic MC property, we proposed a MTJ based eigenvector estimation hardware architecture, reducing the amount of consumed device from n^2 to $\log_2 n$. The practical operations are outlined as follows, corresponding to the steps in **Fig. 8a**:

979-8-3315-3918-4/25 $31.00 © 2025 IEEE

(1) Convert the n×n SM into $\log_2 n$ 2×2 sub-SMs by using the divide-and-conquer (D&C) strategy. For example, **Fig. 8b** and **Fig. 8c** demonstrates how a 4×4 SM and 8×8 SM is converted into two and three 2×2 sub-SMs, respectively.

(2) Setup the waveforms for generating each 2-state MC that corresponds to each 2×2 sub-SM.

(3) Apply each waveform to an MTJ. The generated 2-state MC signals will merge into an n-state MC because the bottom electrodes of the MTJs are connected.

(4) Count the percentage of each state in the n-state MC and take it as the eigenvector of the n×n SM [15].

More generally, an n×n SM, where n is a power of 2, can be converted into $\log_2 n$ 2×2 sub-SMs. To evaluate the accuracy of this method, calculated eigenvector π_{MTJ} and experimental eigenvector π_o are compared using the cosine similarity, a widely used mathematical metric in statistics and machine for evaluating the similarity between two vectors. Cosine similarity is defined as follows:

$$cosim = \frac{\pi_{MTJ}\pi_O^T}{\|\pi_{MTJ}\|\|\pi_O\|} \quad (7)$$

where $\|\cdot\|$ is the Euclidean norm. A value closer to one indicates greater similarity between the two vectors.

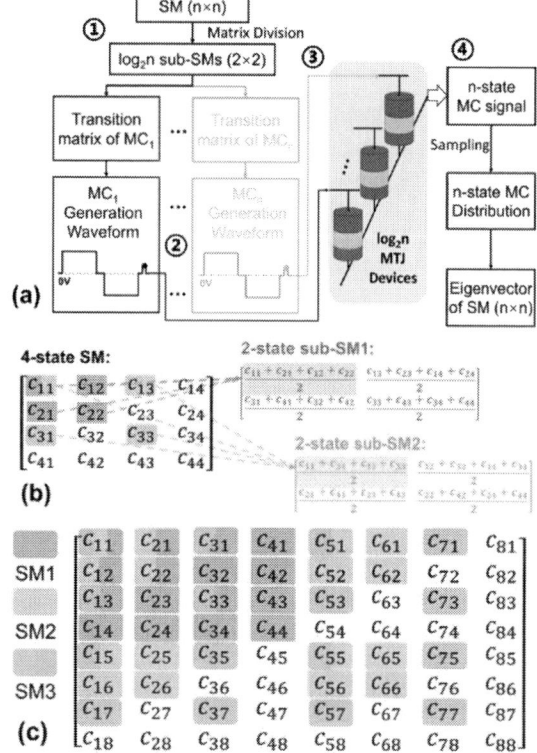

Fig.8. (a) Flow chart of the proposed eigenvector estimation method. (b-c) Demonstration of how the 4×4 (8×8) SM is divided into two (three) 2×2 sub-SMs.

The accuracy of this method for SMs of different dimensions is evaluated, by randomly generating an n×n SM, getting the eigenvector with this method, and comparing the resultant π_{MTJ} with the reference π_o. For statistical purpose,

this process is repeated 1,000 times for each dimension: 4×4, 8×8, 16×16 or 32×32: The box plot in **Fig. 9a** and the probability distribution function (PDF) in **Fig. 9b** illustrate the distribution of cosine similarity between π_{MTJ} and π_o. Clearly, as the dimension increases, the distribution becomes narrower, with values clustering closer to 100%, which might be attributed to the information loss during matrix conversion and variability of MTJs.

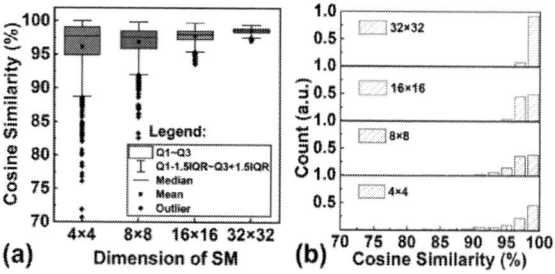

Fig.9. (a) Box plot and (b) probability distribution function (PDF) of the cosine similarities between π_o and π_{MTJ} with n = 4, 8, 16, and 32, which gradually increase with larger matrix dimension, approaching ~99% for 32×32 SMs. The Q and IQR represent the quartile and interquartile range, respectively.

IV. CONCLUSIONS

In this paper, a hardware generation, modulation and application method of spintronic MC signal based on MTJ device is proposed and demonstrated by exploiting its probabilistic switching behavior. The high-quality MC signals with near-zero relative error between counted and calculated stationary distribution are generated by changing the switching voltages in the special waveform. Verified by the formula derivation and experimental demonstration, a two-step procedure to control the ADT of MC is developed. And in the proposed hardware approach, only $\log_2 n$ MTJ devices are used to estimate the eigenvector of n×n SM. This approach provides a hardware solution for generating stochastic signals and addressing complex mathematical problems in the IoT era.

ACKNOWLEDGMENT

This work is supported by the National Natural Science Foundation of China (No. 62104188 and 12327806), the National Key R&D Program of China (Grant No. 2022YFB4400200).

REFERENCES

[1] P. A Gagniuc, Amsterdam: John Wiley & Sons, 2017.

[2] J. Wang, et al., IEEE Transaction on Fuzzy Systems, 2021.

[3] L. Page, et al., The Web Conference, 1999.

[4] F. M. Puglisi, et al., IEEE Transactions on Electron Devices, 2018.

[5] Z. Chai, et al., IEEE Symposium on VLSI Technology, 2016.

[6] H. Tian, et al., Nature Communications, 2018.

[7] W. A. Borders, et al., Nature, 2019.

[8] B. R. Zink, et al., Journal of Applied Physics, 2018.

[9] J. Jian, et al., IEEE Electron Device Letter, 2025.

[10] X. Yuan, et al., IEEE Electron Device Letter, 2025.

[11] B. R. Zink, et al., IEEE Transactions on Electron Devices, 2019.

[12] W. F. Brown, Phys. Rev., 1963.

[13] X. Yuan, et al., IEEE Electron Device Letter, 2023.

[14] X. Yuan, et al., IEEE Electron Device Letter, 2024.

[15] X. Yuan, et al., IEEE Electron Device Letter, 2025.

Opportunities for Advanced Logic Technology with Dual-sided Integrations: From Lateral to Vertical Transistors

Yanbang Chu[+], Yu Liu[+], Runsheng Wang, Ming Li, Heng Wu*

School of Integrated Circuits, Peking University, Beijing 100871, China, †email: hengwu@pku.edu.cn

[+]These authors contribute equally

Abstract— With the great advancement of semiconductor processes, the CMOS integration on both sides of wafer provides brand new scope for advanced logic technologies by fully utilizing the wafer backside, emerging as one of the key knobs to further extend the Moore's law. Benefited from it, lateral transport transistors can be stacked in a back-to-back fashion to form the Flip FET (FFET), doubling the transistor density. Furthermore, similar concepts can be applied to vertical transistors to realize the dual-sided vertical FET (DSVFET), with better device symmetry and area efficiency. In this work, we systematically reviewed the applications of dual-sided integration for advanced logic devices, from the lateral transistors to the vertical ones. Specifically, to meet the urgent requirement of high energy efficiency computing (EEC) for future AI applications, the DSVFET was thoroughly investigated in terms of process development, electrical characteristics, circuit design and chip-level evaluation. Thanks to the adjustable vertical structure, DSVFET holds great features of both reduced parasitic and area. Area scalability of DSVFET is also confirmed at standard cell (~30%) and block (~40%) level. The block level evaluation validates the energy efficiency benefit of DSVFET at nominal Vdd range of EEC by ~15% energy delay product gain. This work validates the great potential of dual-sided integration in the next-generation logic technology.

Keywords—Dual-sided Integration, Flip FET (FFET), Vertical FET (VFET), Design Technology Co-Optimization (DTCO), Energy Efficient Computing (EEC)

I. INTRODUCTION

The rapid rise of AI ecosystem has evolved into several branches for diverse applications, leading the semiconductor industry to provide different technology options accordingly for design-technology co-optimization. Apart from the high performance computing (HPC) and general purpose computing (GPC) [1], energy efficiency [2] is of great importance, calling for the innovative path for energy efficient computing.

Recently, dual-sided integration emerges as a critical technology enabler for advanced logic. Backside power delivery network (BSPDN) helps to relax the routing congestion and optimize the power integrity in IC [3], firstly stepping onto the backside of wafer. Beyond the BSPDN, the dual-sided active device integration will push the semiconductor technology to a new stage by higher integration density. FFET features the back-to-back integrated lateral devices, providing an innovative dual-sided device architecture for HPC and GPC [4]. The effective area scaling of FFET holds less BEOL parasitic, benefitting the chip energy efficiency. While the stacked lateral device nature leaves the device parasitic un-optimized, like lateral GAA (LGAA), complementary FET (CFET), in contrast to the requirement of EEC. The vertical FET (VFET), benefited from

the adjustable vertical structure, holds optimized device parasitic, which is the key to EEC. As a result, the technology paths to HPC and EEC diverge, as in Fig. 1. The lateral transport transistor will follow the path from FinFET to Flip FET, providing higher device integration density and computing power for HPC applications, while the vertical transport transistor will provide higher energy efficiency required by the EEC.

This paper will firstly introduce the latest progress of dual-sided integrated lateral transport transistor in the view of process and scalability for HPC, then a new vertical transistor architecture optimal for EEC, namely DSVFET, is reviewed from process to block level PPA analysis.

II. FFET: A NOVEL DUAL-SIDED INTEGRATION PLATFORM FOR LATERAL TRANSPORT TRANSISTOR

FFET is an innovative dual-sided integration platform for lateral transport transistors, featuring the back-to-back integration of lateral transport transistors with self-aligned active [4] and dual-sided interconnects. The basic process flow of FinFET-based FFET is shown in Fig. 2 for illustration purpose. After the frontside FinFET is formed, the wafer is bonded onto a carrier wafer, followed by wafer flipping and substrate thinning. Then the backside active is revealed selectively to form the backside FinFET. In the end, the back-to-back integrated FinFETs with dual-sided BEOL interconnects are achieved.

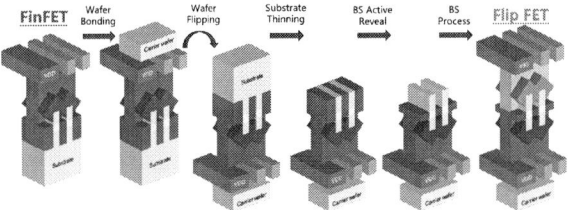

Fig. 2. Basic process flow of FinFET-based Flip FET.

Several advantages can be realized from this architecture. Firstly, it relaxes the stringent requirement for high aspect ratio (AR) processes, which is the key challenge for other transistor stacking architectures like CFET. Secondly, the frontside and backside transistors' process flows remain almost the same as the standard non-stacked transistor except the via formation through the frontside and backside. Besides the greatly relaxed process challenges, FFET also features much more straightforward multi-Vt processes, following the standard process flows. Furthermore, in the viewpoint of the circuit design, the FFET also holds the natural split gate structure, providing much more flexible circuit design (such as transmission gates).

Meanwhile, it's also worth noting that FFET is a novel 3D transistor stacking methodology rather than just a transistor structure. As discussed previously, FFET enables the stacking of lateral transport transistors on both sides of wafer, indicating that the structural knobs on conventional lateral transistors are also applicable for FFET to further push the scaling boundaries.

Fig. 3 shows the scaling roadmap of FFET via structural innovations. FinFET, GAAFET, forksheet FET and CFET can all act as the basic transistor structures to be stacked in FFET, providing the progressive potentials to do dual-sided integration of lateral transport transistor. Besides, the self-aligned gate can be introduced along with the self-aligned active, significantly relaxing the tight overlay requirement from the footprint shrinking. Taking all above together, FFET, with doubled transistor density, manufacturing friendly low AR processes and great circuit design flexibility, emerges a competitive technology pathway for future HPC.

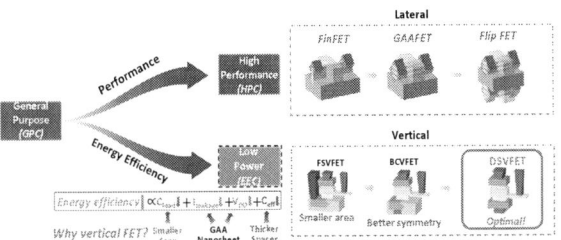

Fig. 1. Different computing paradigms of advanced logic. Lateral devices like FinFET, GAAFET and FFET are suitable for HPC, while VFET exceeds lateral devices in EEC for its adjustable device geometry in the vertical direction. To push VFET in EEC application, DSVFET is proposed with the best device geometry, smallest footprint and dual-sided design.

979-8-3315-3918-4/25 $31.00 © 2025 IEEE

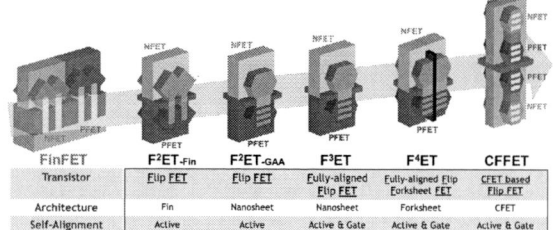

Fig. 3. The scaling roadmap of FFET from the stacking of fin to nanosheet, forksheet and CFET, providing great potential to stack the lateral transistors.

III. FROM VERTICAL TO LATERAL: DUAL-SIDED VFET

The bottleneck to parasitic optimization of lateral devices is the highly restricted device footprint, which is critical to EEC. VFET, stacking the S/D/G vertically unlike lateral devices, stands as the only structure capable of improving energy efficiency at the device level in comparison to LGAA, CFET and even FFET, making it a great candidate for future EEC (Fig. 1) [6]. The energy efficiency optimization of VFET is resulted from the flexibility in adjusting top and bottom spacer of VFET in the vertical direction. This results in smaller parasitic RC without area waste [7], unlike the larger overlap area between G and S/D in FinFET or LGAA. VFET also features the gate-all-around (GAA) channel, providing less leakage and lower V_{DD}. More importantly, VFET has smaller footprint thanks to its zero diffusion break (ZDB) [8], reducing the BEOL load capacitance and resistance.

However, the asymmetric S/D and extra area waste with high aspect ratio (AR) contact stop VFET from practical usage [9]. For here, following the line of optimizing the device asymmetry in Fig. 1, DSVFET was proposed and comprehensively benchmarked with Frontside VFET (FSVFET), Backside Contact VFET (BCVFET) and FinFET in process and device performance.

What's more, the vertical placed S/D/G makes VFET naturally suitable for dual-sided routing, therefore DSVFET integrated with dual-sided designs was carried on circuit and block level with DTCO methodology to unlock its substantial potential for EEC.

A. Device Fabrication for DSVFET

Fig. 4 summarizes the process flow of DSVFET. After the vertical nanosheet (NS) etching, gate processes are skipped considering thermal problems and replaced by depositing STI and front spacer. The front epitaxy (epi) and contact are formed followed by wafer bonding and flipping. The active wafer's substrate is thinned down by CMP stopping on the STI, which is then removed to reveal the vertical NS. In following, the gate processes and back spacer formation are done. Finally, the back epi is grown directly on the tip of vertical NS in low temperature to form the symmetric S/Ds of DSVFET, followed by contact formation.

Experimentally, the integrated processes were also demonstrated step by step following the process flow discussed above, as in Fig.

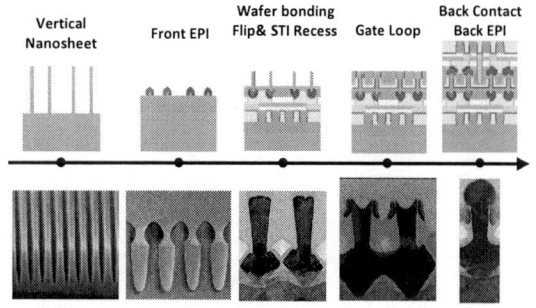

Fig. 4. The DSVFET process flow, taking 2NS inverter as an example. Experimental results are shown following the process flow: vertical nanosheet formation, frontside S/D epi growth, wafer bonding/flipping and vertical channel reveal, gate formation and backside S/D formation.

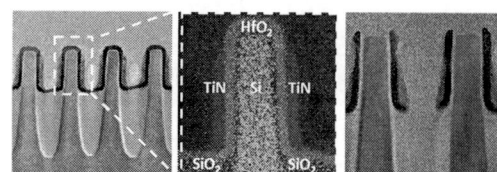

Fig. 5. Integration process of gate structure of DSVFET. From right to left: gate structure after recess; ESD image for the recessed gate structure; top S/D recess.

4. After finishing the front epi and contact, the wafer was bonded, flipped and thinned down, then STI recess was done to expose the back NS, followed by ALD HKMG. Gate recess should be precisely controlled to reveal the tip of the back NS and regulate the gate length. Details of gate stack is shown in Fig. 5. In the end, the back epi and contact were formed, delivering the world's first structure validation of DSVFET with aggressive CGP of 50 nm and L_g of 19 nm. It clearly demonstrates the symmetric S/D epi and contact with well-formed gate structure, all the loops are compatible with the standard CMOS processes, proving the DSVFET process feasibility.

B. Asymmetry and Performance Optimization

A Physical TCAD model of FSVFET was first established and calibrated to ref. [8], with matched structure and characteristics, showing well-fitting results (Fig. 6).

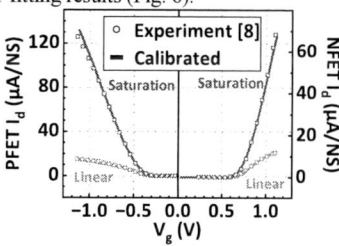

Fig. 6. I_d-V_g curves of experiments [8] and calibration. Linear: $V_{DD} = 0.05$ V. Saturation: $V_{DD} = 0.65$ V.

The calibrated model was then extended to BCVFET and DSVFET with design rules (shown later in Tables I-II) in advanced nodes. Fig. 7 compares three types of VFETs' structures in forward (FWD, front epi as drain) mode, among them DSVFET has the straight current flow direction, so does the best device symmetry and smallest parasitic especially at the front S/D region.

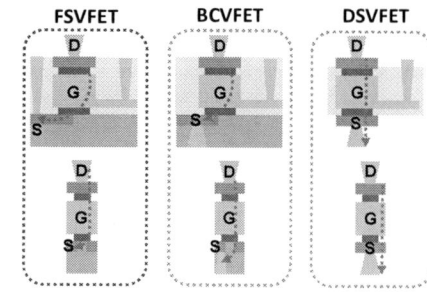

Fig. 7. Device structure cross sections with current flow for the three VFETs.

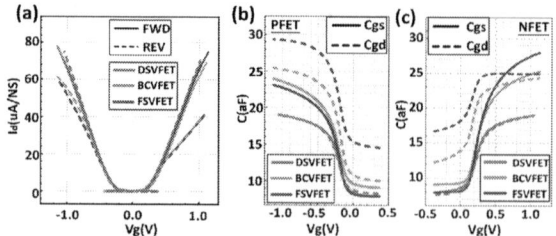

Fig. 8. Simulated I_d-V_g and C-V curves of the three VFETs. The DSVFET shows the best FWD and REV symmetry thanks to the symmetric S/D structure.

979-8-3315-3918-4/25 $31.00 © 2025 IEEE

It is consistent that I_d-V_g (Fig. 8(a)) and C_{gd}/C_{gs}-V_g curves (Fig. 8(b-c)) of DSVFET in forward and reverse (REV, back epi as drain) modes almost coincide, showing the best symmetry. What's more, it also has the largest I_d and smallest parasitic capacitance due to the dual-sided structure, overcoming the main disadvantages of traditional FSVFET. Note that, the C_{gd} at Vg in Fig. 8(b-c) of the three types of VFET are different due to the different fabrication processes sequence, meaning the three types of VFETs see different thermal processes, resulting in changed capacitance. In addition, this is verified by the similar C_{gd} at Vg = 0 V of the three types of VFETs as in Fig. 8(b-c). because only structure capacitance matters in this situation. This means the different C_{gd} at Vg comes from different channel capacitance, not directly related to the structure parasitic.

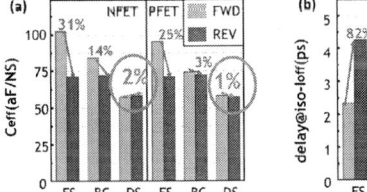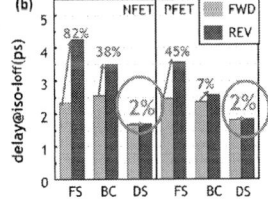

Fig. 9. C_{eff} (a) and delay (b) of forward and reverse modes for three types of VFET. FS: FSVFET. BC: BCVFET. DS: DSVFET.

Some important parameters were also extracted from the electric curves, showing minor C_{eff} differences of 2%/1% for N/P (Fig. 9(a)) and 2%/2% delay at iso-I_{off} difference for N/P (Fig. 9(b)) in forward and reverse modes of DSVFET. While for FSVFET and BCVFET, there exists more than 10% difference due to the unsymmetric structure, posing inconvenience for circuit design.

C. Novel Dual-sided Design and PPA Analysis

The S/D placed on the dual sides of wafer makes VFET naturally compatible with dual-sided design. Here, considering the symmetric characteristics of DSVFET, backside for local routing and frontside for inter cell routing was introduced for better design flexibility without concerning the current flow direction in FWD or REV mode, and this design can satisfy all basic standard cells' routing requests (17 types of standard cell). Note that dual-sided power is required in DSVFET. Though Super via (SV) could be used to introduce frontside Vss to backside, this will bring huge RC penalties. Therefore dual-sided power was introduced, with frontside power rails connected to the backside by power tap cells [10].

Moreover, VFET also has some special points for the layout design.

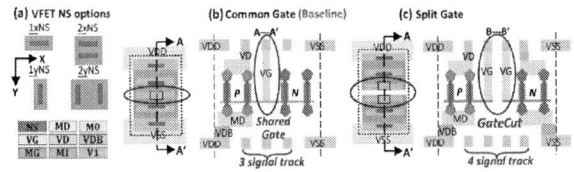

Fig. 10. (a) DSVFET's NS design options. (b) Layouts and cross sections of CG design. CG design is the baseline due to area benefits. (c) Layouts and cross sections of SG design. SG has one more track than CG.

TABLE I   2NS DSVFET and FinFET design rules

Parameter (nm)		2xNS	2yNS	FinFET
Fin Pitch		36	36	26
CH	SG	182	143	130
	CG	156	117	
M0 CD		14	12	12
M0 Pitch		28	22	22
M1 Pitch		28	24	24
L_g		20	20	16
Fspacer Thk		10	10	-
Bspacer Thk		10	10	-
CGP		36	36	48
NS Length		26	26	-
NS/Fin CD		6	6	6
W_{eff}		128(2NS)	128(2NS)	116(1Fin)

The NS number (1NS，2NS) and NS placement orientation (xNS, yNS) of DSVFET were studied, as in Fig. 10(a). Besides, for key sequential logics with transmission gate (TG), common gate (CG, Fig. 10(b)) and split gate (SG, Fig. 10(c)) should also be considered [9], adding more diversity in DSVFET layouts.

Note that SG design has one more track than CG, meaning larger layout area in the standard cell design. But in sequential logics like TG, SG may have smaller area because CG design would sacrifice areas to introduce N/P gate signal to outside separately. Interestingly this would not happen in DSVFET due to the special design in CG.

For fair comparison, 2NS DSVFET and 1Fin FinFET were used due to their similar W_{eff}, with design rules listed in Table I with the same M2 pitch, considering the minimum space for isolation and the gear ratio between metal layers.

A 15-stage FO3 RO with distributed BEOL load extracted from STA and P&R critical path statistics of a RISC-V core was used for the evaluation [10] at V_{DD} = 0.35 V, which is a commonly operating voltage for EEC according to [2]. From RO level evaluation in Fig. 11, power at iso-frequency of 2xNS-CG is 16.2%/39% less than FinFET with/without BEOL load, respectively. It's further reduced by 38% on 2yNS-CG with BEOL load due to its smaller cell area in the y direction, resulting in smaller wire parasitic. DSVFET performs better than FinFET especially at lower V_{DD} due to its vertical structure and smaller footprint, fully showing its suitability for EEC.

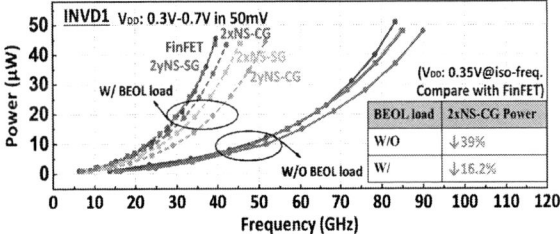

Fig. 11. Power vs. frequency of different 2NS DSVFETs and FinFET. DSVFET performs better considering the BEOL load.

D. The Block-level Scalability

DSVFET can be further scaled with 1NS. Based on PPA results of 2NS above, CG design performs better than SG, therefore 1xNS and 1yNS with CG design and PowerVia were studied as in Fig. 12.

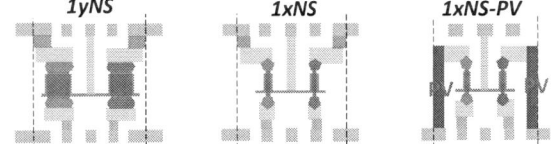

Fig. 12. Cross sections of three 1NS DSVFETs, new design of PV is introduced.

For the PV, it was only introduced in 1xNS due to area penalty in 1yNS. In the form of a trench filled with metal, PV can replace power tap cells in dual-sided power design [11-13], saving extra area at the block level. It also helps eliminate the frontside power rail, relaxing the M0 pitch (Table 2) because it can introduce backside power to the frontside inside standard cells. Standard cells of the three types of

TABLE II   1NS-CG DSVFET design rules

Parameter (nm)	1yNS	1xNS	1xNS-PV
Fin Pitch	36	36	36
CH	117	104	104
M0 CD	12	12	F12/B14
M0 pitch	22	22	F22/B28
M1 Pitch	24	28	28
L_g	20	20	20
Fspacer Thk	10	10	10
Bspacer Thk	10	10	10
NS Length	26	26	26
W_{eff}	64 (1NS)	64 (1NS)	64 (1NS)

1NS DSVFET were designed and compared with 2NS designs and FinFET. 1NS cell areas are all smaller than FinFET (Fig. 13), proving its scaling potential.

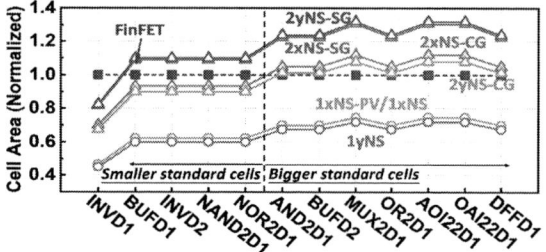

Fig. 13. Standard cell areas comparison between DSVFET and FinFET.

PPA analysis was carried out on circuit and block level to evaluate DSVFET comprehensively. The circuit level PPA (Fig. 14(a)) further indicates 41% lower power at iso-frequency than FinFET for 1xNS without BEOL load and 60% for 1xNS-PV with BEOL load. The block level benchmark of DSVFET and FinFET was conducted on a 32bit RISC-V core, following the standard flow of synthesis, place & route and timing/power signoff. The results show 1NS design all performs better than FinFET as illustrated in Fig. 14(b).

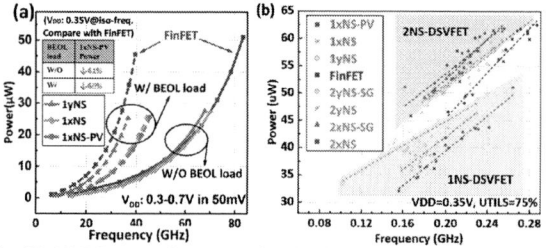

Fig. 14. (a) RO power-frequency of 1NS DSVFET and FinFET. (b) Block level benchmarks based on 32-bit RISC-V core for DSVFET and FinFET.

In terms of the block area, 1xNS-PV is 39.6% smaller than FinFET at the same utilization and 32.5% smaller for minimum area achieved (Fig. 15(a)), as further validated by the layout comparison in Fig. 15(b), 1NS designs all achieve more than 30% area benefits.

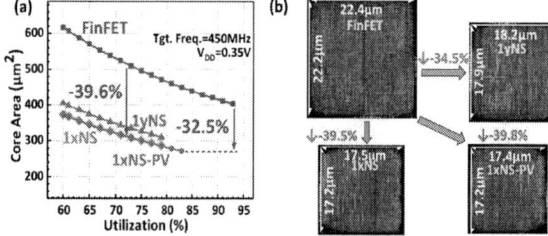

Fig. 15. Block area readouts on 1NS DSVFET and FinFET: (a) Area benefits by 1NS DSVFET, as further validated by the core layout (b).

Key figures of merit for EEC were also extracted and shown in Fig. 16. 1xNS-PV clearly outperforms FinFET at V_{DD} from 0.3 to 0.5 V in the typical EEC operation voltage range, with 15% EDP

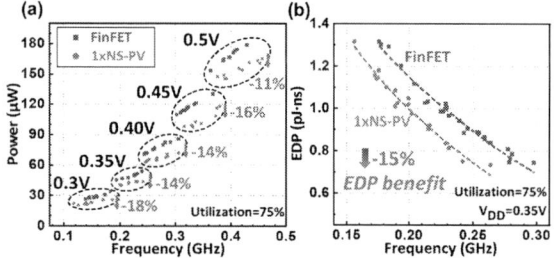

Fig. 16. (a) 1xNS-PV DSVFET exceeds FinFET at various V_{DD}. (b) EDP vs. frequency.

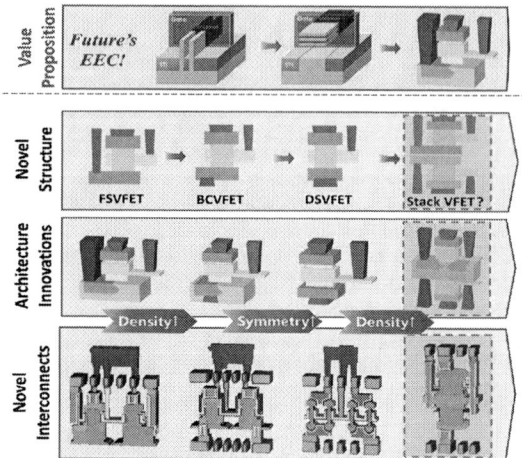

Fig. 17. Roadmap of VFET from aspects of the structure and interconnect.

benefit at iso-frequency at the chip level, fully revealing its suitability in EEC.

Fig. 17 shows the roadmap of VFET [14]. It features clear benefits in EEC from the vertical transport nature compared to FinFET and LGAA. From FSVFET, to BCVFET, then to DSVFET, dual-sided process and integration are introduced with device symmetry improvement and much less parasitics. As for the vertical transistor architecture beyond the DSVFET, stacked VFET is one of the candidates, while further investigation is needed.

IV. CONCLUSION

In this work, we reviewed the applications of dual-sided integration for advanced logic devices, from the lateral transistors to the vertical ones. For lateral transistors, FFET, as the integration platform with great extendibility and relaxed process requirement, is examed for HPC and GPC. We also proposed an innovative DSVFET device architecture based on VFET as the candidate for future EEC. The DSVFET holds the symmetric S/D structure, reduced device parasitic, smaller area usage and natural capability of dual-sided routing, standing out for the EEC application. Key process modules of DSVFET were demonstrated experimentally. In terms of DTCO, standard cell layout design methodology was developed, confirming the PPA benefit at circuit level. Further, chip level PPA analysis shows that DSVFET with 1xNS-PV holds the best energy efficiency and area scalability, convincing the potential of DSVFET in EEC applications. The roadmap of VFET scaling was also introduced. Scaled DSVFET, stacked VFET and dual-sided inter-cell signal routing act as future enablers, waiting for further investigation.

ACKNOWLEDGMENT

This work was supported in part by the National Key R&D Program of China under Grant 2023YFB4402200; in part by the 1+1 Project under Grant QYJS-2023-2301-B; in part by the NSFC under Grant 92464206.

REFERENCES

[1] B. Choi et al., 2021 IEEE VLSI, 2021, pp. 1-3.
[2] D. -H. Kim et al., IEEE EDL, vol. 45, no. 9, pp. 1673-1676, Sept. 2024.
[3] A. Jourdain et al., 2022 IEEE ECTC 2022, pp. 1531-1538
[4] H. Wu et al., 2025 IEEE VLSI , 2025, pp. 1-3
[5] W. Peng et al., IEEE VLSI. 2025, , pp. 1-3.
[6] A. Veloso et al., 2019 IEEE IEDM, 2019, pp. 11.1.1-11.1.4.
[7] G. Tsutsui et al., 2022 IEEE IEDM, 2022, pp. 34.4.1-34.4.4.
[8] H. Jagannathan et al., 2021 IEDM, 2021, pp. 26.1.1-26.1.4,
[9] J. Jeong, S. Lee et al., IEEE TED, vol. 72, no. 1, pp. 75-82, Jan. 2025.
[10] B. -S. Kim et al., 2024 IEEE VLSI, 2024, pp. 1 -2.
[11] L. Liebmann et al., 2021 IEEE IEDM, 2021, pp. 3.1.1-3.1.4.
[12] M. Radosavljević et al., 2023 IEEE EDM, 2023, pp. 1-4.
[13] S. Yang et al., 2023 IEEE VLSI, 2023, pp. 1-2.
[14] Y. Liu et al., 2025 IEEE VLSI, 2025, pp. 1-3.

Si Hybrid Tunnel FET-CMOS Foundry Platform for Ultra-low-Power Circuit Applications

Qianqian Huang*, Kaifeng Wang, Ru Huang

School of Integrated Circuits, Peking University, Beijing 100871, China
* Email: hqq@pku.edu.cn

Abstract—This work demonstrates the recent progress on 55 nm Tunnel FET(TFET)-CMOS hybrid integration platform and its ultra-low-power circuit applications. By integrating the bulk-Si-based novel dopant-segregated TFET (DS-TFET) with large I_{ON} and record high I_{ON}/I_{OFF} ratio, as well as the novel laminated isolation technology into the CMOS baseline technology, energy-efficient TFET-CMOS hybrid circuits are experimentally realized. 1Kbit TFET-Gated-Ground SRAM is implemented and demonstrated in MCU always-on domain, showing sub-100nA ultra-low leakage. Moreover, a novel DS-TFET-like device (AsyFET) with ultralow off-state leakage current and bidirectional conductivity is further demonstrated as the write transistor of 2T0C eDRAM, leading to the long retention of 3.9 s in 55nm technology node. The TFET-CMOS hybrid platform of this work demonstrates the great potential for cutting-edge power-dieting applications.

Keywords—*Tunnel FET, Low power, MCU, eDRAM*

I. INTRODUCTION

The fast-expanding market of Artificial Intelligence of Things (AIoT), which is being considered as one of the most promising, disruptive and cutting-edge driving force for AI applications, has stimulated a lot of innovations from device to circuit level [1-2]. Despite of diverse requirements from different AIoT chips, low power consumption is always the first-priority [3-4]. Tunnel FET (TFET) has already demonstrated its superiority in terms of power consumption among emerging device technologies, while suffering from the fundamental low on-state current (I_{ON}) issue [5]. So far, many industry-level manufacturers have demonstrated lots of TFET devices with novel structures or new materials to improve the device I_{ON} [6], while the device I_{ON}/I_{OFF} ratio and CMOS compatibility are still poor for high-volume production. For Si TFETs, the I_{ON} is usually less than 1μA/μm, which cannot satisfy the real time requirement of AIoT applications. Besides, the conventional TFETs also face challenges like poor device complementarity [7] and large device variation [8]. Moreover, for circuit, the experimental demonstration of TFET-based circuits is extremely scarce, except a few inverter studies based on nanowire TFET structures [9]. According to our previous work [8], for bulk Si TFETs on CMOS foundry platform, heavily-doped sources or drains from adjacent TFETs may induce parasitic leakage paths through substrate, leading to abnormal functionality and increased static power consumption in TFET-based circuits. Therefore, no experimental demonstration of TFET-CMOS hybrid circuits based on foundry platform has been reported.

To solve the above issue, we proposed a novel dopant-segregated TFET (DS-TFET) structure co-integrated with baseline CMOS technology in 12-inch foundry platform, and experimentally demonstrated the record I_{ON}/I_{OFF} ratio without CMOS device performance penalty. For large-scale circuit application, we propose a laminated well technology to solve the electrical isolation requirement between neighboring devices without area penalty. Therefore, the first bulk TFET-based circuits and TFET-CMOS hybrid circuits on the

standard CMOS foundry platform are experimentally demonstrated [10]. Based on the developed TFET-CMOS hybrid integration technology based on the standard 55 nm foundry platform, the first TFET-MCU with a 1Kbit TFET-Gated-Ground SRAM in always-on domain is experimentally demonstrated [11], as well as the first TFET-based 2T0C eDRAM with long retention [12-13], showing its great potential for ultralow-power AIoT applications.

II. COMPLEMENTARY DOPANT-SEGREGATED TFET

In our previous work, we have successfully manufactured conventional Si TFET device structure using 130nm CMOS baseline platform [8]. Although it has demonstrated superior CMOS compatibility, the device I_{ON}/I_{OFF} ratio can still not satisfy the AIoT requirements due to the ion implantation limitation in improving junction abruptness and the relatively large gate oxide thickness. To further optimize TFET performance, based on the standard 55 nm CMOS foundry platform which is more commonly utilized towards low-cost and low-power AIoT applications, a novel device structure, called DS-TFET, with asymmetric sidewalls and self-aligned dopant segregation junction is proposed as illustrated in Fig. 1. Fig. 2 shows the process flow of DS-TFET [11]. Compared with standard CMOS technology, only a few masks and processes are added or modified without introducing new materials. Both n- and p-type DS-TFETs show the similar drive current capability with both large I_{ON} of almost 10 μA/μm, which is even larger than recently reported III-V TFETs [6] (Fig. 3a).

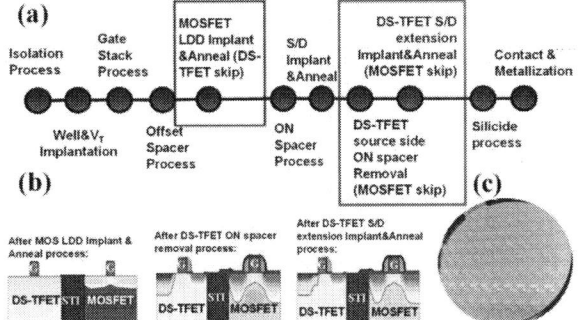

Fig. 1. (a) Schematic structures of C-DS-TFETs on standard 55 nm CMOS baseline platform. (b) Schematic surface energy band of C-DS-TFETs when BTBT is turned on.

Fig. 2 (a) Process flow for co-integrating C-DS-TFETs and CMOS. (b) Simulated structures by TCAD tools. (c) Photo of 300 mm wafer with C-DS-TFETs.

This work was supported by NSFC (62374009) and 111 Project (B18001).

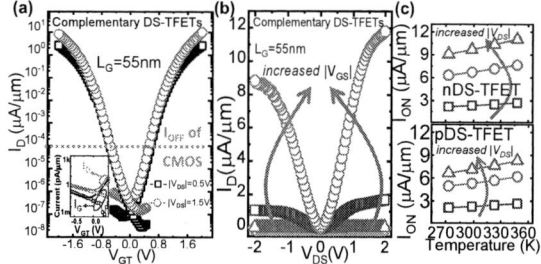

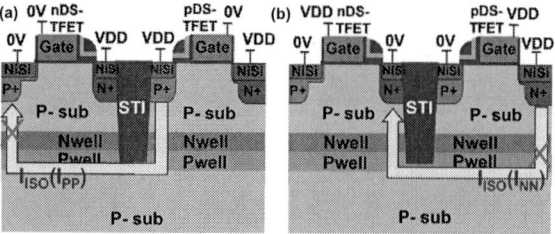

Fig. 6 Schematic of C-DS-TFET with laminated isolation well which can suppress the (a) I_{PP} and (b) I_{NN} leakage.

Fig. 3 Measured (a) transfer curves and (b) output curves of fabricated C-DS-TFETs on the same wafer based on standard CMOS platform. (c) The measured positive I_{ON} dependence on temperature confirms the band-to-band tunneling current.

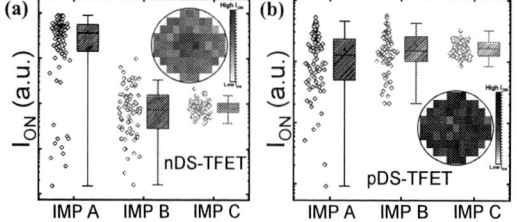

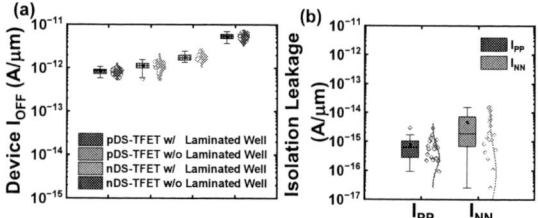

Fig. 7 Measured (a) device I_{OFF} with and without laminated isolation well, (b) isolation leakage current between adjacent devices with laminated wells.

Fig. 4 Measured I_{ON} distributions in (a) nDS-TFETs and (b) pDS-TFETs. Insets are 2D mapping of measured I_{ON} in optimized DS-TFETs with proposed implantation (IMP C).

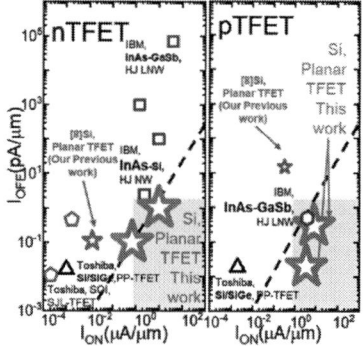

Fig. 5 Performance comparison of fabricated TFETs by industry-manufacturers. The dashed line indicates I_{ON}/I_{OFF} equals to 10^7.

Moreover, the I_{OFF} of DS-TFET is limited at gate leakage current level and can be 2 decades lower than the standard CMOS with the same gate length. The output curves with weak super-linear onset behavior (Fig. 3b) and the positive I_{ON} dependence on temperature (Fig. 3c) further confirm the tunneling current rather than thermal emission current. Furthermore, there is no performance penalty of CMOS by utilizing the proposed fabrication process. As for the device variability, according to our previous work [8], in standard TFETs, the source doping gradient is the dominant variation source due to its direct effect on BTBT junction area and may induce a trade-off between device performance and variability. For DS-TFET, the introduced salicide process for source tunnel junction becomes the new dominant variation source [14]. Therefore, steep source gradient will result in large I_{ON} distribution on 300 mm-wafer. As shown in Fig. 4, by further reasonable implantation optimization [15], the device variation can be significantly reduced while still maintaining the large I_{ON} compared with conventional TFETs. As a result, the fabricated complementary DS-TFET (C-DS-TFET) can be monolithically integrated on CMOS baseline platform and demonstrate the record I_{ON}/I_{OFF} ratio among TFETs by industry-manufacturers (Fig. 5).

III. LAMINATED WELL ISOLATION TECHNOLOGY

For large-scale low-power circuit application, the device isolation technology should be carefully concerned for bulk Si TFET. Since the channel surface of TFET must be lightly doped for ultra-low leakage current, TFETs cannot be fabricated with the standard twin-well technology. However, the lightly doped substrate will result in a large parasitic leakage current (I_{PP} or I_{NN}) between adjacent P+ or N+ regions, and this leakage will increase the static power consumption of TFET circuits and even cause abnormal functionalities [8]. Moreover, due to the intrinsically asymmetrical doped source and drain in TFET, an isolation well with single dopant type cannot simultaneously isolate both I_{NN} and I_{PP} leakage. To solve this problem, a laminated isolation well technology is proposed for complementary TFETs in bulk substrate as demonstrated in Fig. 6 [10]. To determine the specific process conditions of ion implantation for laminated isolation wells, a DTCO workflow is also demonstrated to avoid go through all process conditions [10]. Experimental results in Fig. 7 demonstrate that this isolation well can successfully suppress both I_{NN} and I_{PP} which can be lower than the I_{OFF} of TFET device for almost 3 decades. Moreover, the laminated isolation well can also enable the significant reduced layout distance between TFETs for area reduction.

Benefiting from the laminated isolation well design which can isolate both I_{NN} and I_{PP} leakage simultaneously, the experimental C-DS-TFET-based circuits are demonstrated with correct functionality without area penalty [10]. Furthermore, the power supply voltage of the inverter can also be reduced (Fig. 8). Besides the inverter, the other basic circuits such as NAND, AND, NOR XOR and XNOR are also experimentally fabricated by using C-DS-TFET devices with laminated isolation wells, and their logical functions are experimentally verified [10]. To monolithically integrate the laminated isolation well for DS-TFET with standard CMOS, the additional TFET isolation well lithography and ion implantation process should be performed after the CMOS well implantation process. The thermal budget from the CMOS FEOL process is utilized to activate the impurity of the isolation well, eliminating the need for additional annealing process (Fig. 9), enabling the experimental implementation of hybrid TFET-CMOS circuits. Based on our proposed energy-

979-8-3315-3918-4/25 $31.00 © 2025 IEEE

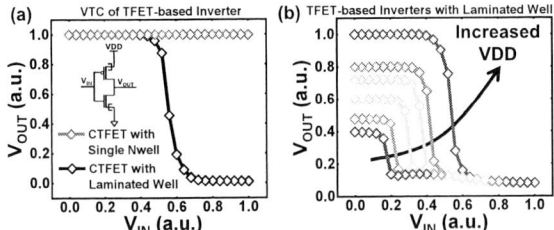

Fig. 8 Measured (a) VTC of TFET-based inverters with laminated wells and single Nwell, (b) VTC of TFET-based inverter under different V_{DD}.

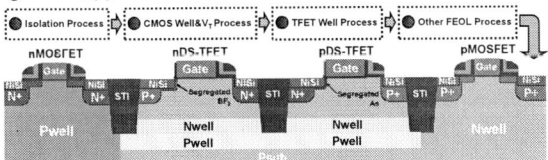

Fig. 9 Key process flow and schematic structure for co-integrating C-DS-TFETs with laminated wells and standard CMOS.

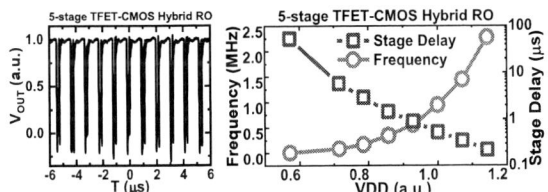

Fig. 10 Measured output voltage waveform, output frequency and stage delay of a typical 5-stage hybrid TFET-CMOS RO.

efficient TFET-CMOS circuit topology design with CMOS-comparable operation speed [2], the TFET-CMOS hybrid logic gates and 5-stage ring oscillator (RO) are experimentally demonstrated (Fig. 10), indicating its great potential of TFET-CMOS hybrid system for power-constraint AIoT applications.

IV. TFET-MCU BASED ON FOUNDRY PLATFORM

The monolithic integration of TFET and CMOS on the same platform makes it possible to take advantage of low-power TFET and high-performance CMOS simultaneously. It can be expected that by using TFET-CMOS hybrid system, such as co-integrating a TFET-based intelligent wake-up chip with a CMOS-based high-performance system (Fig. 11), both ultra-low power, high accuracy and real-time can be achieved for AIoT nodes in random-sparse-event (RSE) scenarios. For a standby-power-sensitive microcontroller unit (MCU), it always requires long-term data retention in sleep mode, and the SRAM standby power accounts for most of the power consumption. Therefore, using ultra-low-leakage SRAM, such as TFET-SRAM, is effective to reduce the overall power consumption of MCU. To evaluate the leakage current of MCU system embedding DS-TFET, two MCU full chips (TFET MCU and CMOS MCU) are fabricated based on 55 nm CMOS foundry platform [11], integrating ARM Cortex-M3 as processor. CMOS MCU (Fig. 12a) adopts conventional 6T-SRAM, while the TFET MCU (Fig. 12b) utilizes the novel TFET-Gated-Ground SRAM [11]. The fabricated 1Kbit TFET-SRAM achieves 6nA leakage current with a DS-TFET tail transistor. The TFET-MCU with TFET-SRAM in deep sleep mode achieves 75 nA leakage current, which is lower than that in CMOS MCU by 88% (Fig. 12c).

V. TFET-BASED 2T0C EDRAM

Data-intensive computing increasingly demands on-chip memory with larger capacity. Compared with 6T-SRAM, capacitor-less 2T0C eDRAM shows the higher density with

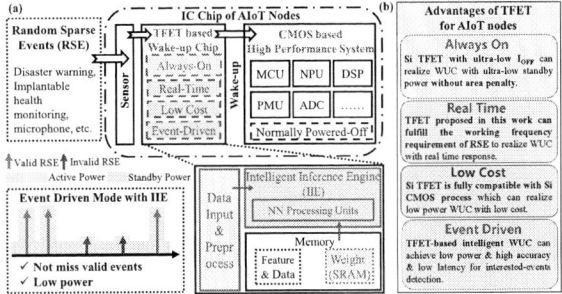

Fig. 11 (a) A TFET-based intelligent wake-up chip (WUC) co-integrated with a CMOS-based high performance system (HPS). (b) The advantages of TFET for AIoT nodes with always on, real time, low cost and event driven requirements.

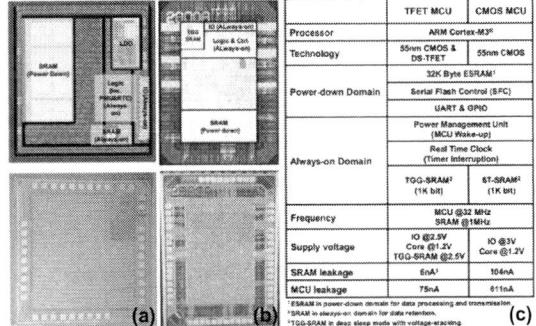

Fig. 12 (a) Layout and die photo of CMOS-MCU chip with traditional 6T-SRAM in always-on domain. (b) Layout and die photo of TFET-MCU chip with TGG-SRAM in always-on domain. (c) Comparison summary table of TFET-MCU and CMOS-MCU.

smaller memory cell footprint, while the smaller storage capacitor also puts forward more stringent requirements on the transistor I_{OFF} and logic platform compatibility. However, despite possessing ultra-low leakage current, traditional TFET with gated p-i-n structure is a unidirectional conducting device which cannot be utilized as the write transistor of DRAM. By modulating the implantation condition of DS-TFET, a gate-controlled source Schottky junction is introduced for bidirectional conductivity. This modulated TFET-like device with both ultralow I_{OFF} and bidirectional conductivity, called Asymmetrical FET (AsyFET), can be used as the write transistor of DRAM. As shown in Fig. 13, the proposed 2T0C AsyFET-DRAM cell utilizes a P-type AsyFET as the write transistor for long retention and a standard N-type MOSFET as the read transistor for fast read [12-13].

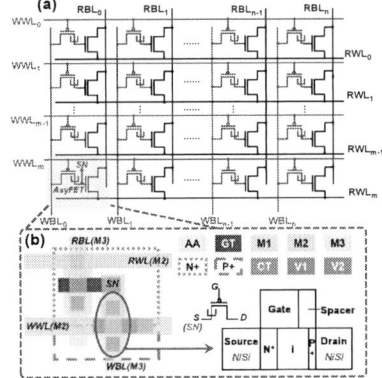

Fig. 13 Proposed AsyFET-based DRAM. (a) Schematic of AsyFET-DRAM array. (b) Layout of AsyFET-DRAM cell and schematic of AsyFET device.

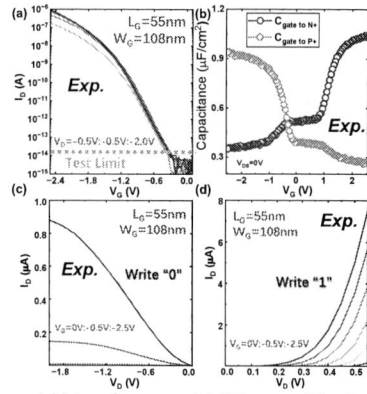

Fig. 14 Measured (a) transfer curves, (b) CV curves, (c) output curves when writing "0" and (d) "1" of the fabricated AsyFET.

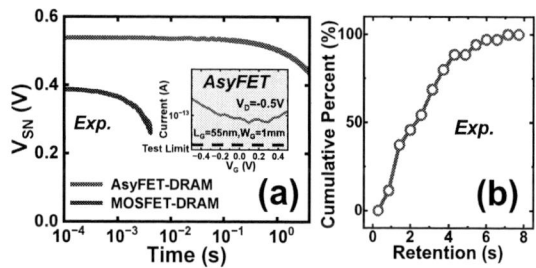

Fig. 15 (a) **Measured** retention of AsyFET-DRAM with AsyFET of $W_G=1\mu m$ and MOSFET-DRAM with MOSFET of $W_G=189nm$. (b) **Measured** retention distribution of AsyFET 2T0C cell across wafer.

The designed AsyFET is fabricated on standard 55 nm Si CMOS foundry platform with the process flow similar with DS-TFET [13]. Benefiting from the laminated isolation well technology, the leakage current between AsyFETs and MOSFETs can also be successfully suppressed without area penalty. The fabricated AsyFET device with the gate length of 55 nm shows ultralow I_{OFF} below test limit, as well as the asymmetrical CV and bidirectional conduction characteristics (Fig. 14). As shown in Fig. 15, the measured retention time of AsyFET 2T0C DRAM cell can be as long as about 3.9s. The retention of AsyFET-eDRAM across the 12-inch wafer is also measured. The retention time is in 1s~7s range across the whole wafer. As a comparison, the retention time of all-MOSFET-based 2T0C DRAM cell fabricated on the same wafer is about 3.3 ms, which is 1000× shorter than that of fabricated AsyFET-DRAM cell even with smaller W_G of the write transistor (Fig. 15). To obtain the device I_{OFF}, AsyFET with very large W_G (about 1mm) is also fabricated on the same wafer and the measured I_{OFF} is about $10^{-17}A/\mu m \sim 10^{-16}A/\mu m$, which is significantly lower than standard MOSFET. The experimental results of AsyFET-eDRAM demonstrates the longest retention among silicon-based gain cell eDRAM, and even comparable with 2T0C eDRAM based on oxide semiconductor, indicating its great potential for high-speed, low-power and high-density on-chip memory applications.

VI. SUMMARY

The foundry platform of monolithically integrating the proposed complementary DS-TFET and CMOS on 55nm technology node are presented, as well as the bulk TFET-based circuits and hybrid TFET-CMOS circuits. Without CMOS performance penalty, the fabricated n- and p-type Si DS-TFETs show the significant performance enhancement with high I_{ON} and record I_{ON}/I_{OFF} ratio among TFETs by

industry-manufacturers. By utilizing a novel laminated isolation well technology, the parasitic leakage between adjacent TFET devices can be significantly suppressed without area penalty, promoting the realization of large-scale circuits based on hybrid TFET-CMOS foundry platform. Based on the developed platform, a TFET-MCU is presented, achieving 75nA leakage current in deep sleep mode. Moreover, the DS-TFET is modified into a novel AsyFET device for eDRAM, and the fabricated AsyFET-DRAM shows 1000× longer retention than conventional MOSFET-DRAM on the same wafer. The results demonstrate the great potential of the hybrid TFET-CMOS foundry platform for cutting-edge power-dieting circuit applications.

ACKNOWLEDGMENT

The authors would gratefully acknowledge SMNC and SMIC for the assistance in the device and circuit fabrication.

REFERENCES

[1] L. Ye et al., "The Challenges and Emerging Technologies for Low-Power Artificial Intelligence IoT Systems," in IEEE Transactions on Circuits and Systems I: Regular Papers, vol. 68, no. 12, pp. 4821-4834, Dec. 2021.

[2] Z. Wang et al., "Ultra-Low-Power and Performance-Improved Logic Circuit Using Hybrid TFET-MOSFET Standard Cells Topologies and Optimized Digital Front-End Process," in IEEE Transactions on Circuits and Systems I: Regular Papers, vol. 68, no. 3, pp. 1160-1170, March 2021.

[3] Z. Wang et al., "12.1 A 148nW General-Purpose Event-Driven Intelligent Wake-Up Chip for AIoT Devices Using Asynchronous Spike-Based Feature Extractor and Convolutional Neural Network," IEEE International Solid-State Circuits Conference (ISSCC), 2021, pp. 436-438.

[4] M. Cho et al., "17.2 A 142nW Voice and Acoustic Activity Detection Chip for mm-Scale Sensor Nodes Using Time-Interleaved Mixer-Based Frequency Scanning," IEEE International Solid-State Circuits Conference - (ISSCC), 2019, pp. 278-280.

[5] Q. Huang et al., "A novel Si tunnel FET with 36mV/dec subthreshold slope based on junction depleted-modulation through striped gate configuration," IEEE International Electron Devices Meeting (IEDM), 2012, pp. 8.5.1-8.5.4.

[6] C. Convertino et al., "Sub-Thermionic Scalable III-V Tunnel Field-Effect Transistors Integrated on Si (100)," IEEE International Electron Devices Meeting (IEDM), 2019, pp. 37.1.1-37.1.4.

[7] C. Wu, R. Huang, Q. Huang, J. Wang and Y. Wang, "Design Guideline for Complementary Heterostructure Tunnel FETs With Steep Slope and Improved Output Behavior," in IEEE Electron Device Letters, vol. 37, no. 1, pp. 20-23, Jan. 2016.

[8] Q. Huang et al., "First foundry platform of complementary tunnel-FETs in CMOS baseline technology for ultralow-power IoT applications: Manufacturability, variability and technology roadmap," IEEE International Electron Devices Meeting (IEDM), 2015, pp. 22.2.1-22.2.4.

[9] G. V. Luong et al., "Complementary Strained Si GAA Nanowire TFET Inverter With Suppressed Ambipolarity," in IEEE Electron Device Letters, vol. 37, no. 8, pp. 950-953, Aug. 2016.

[10] K. Wang et al., "First Foundry Platform Demonstration of Hybrid Tunnel FET and MOSFET Circuits Based on a Novel Laminated Well Isolation Technology," IEEE 53rd European Solid State Device Research Conference (ESSDERC), 2023, pp. 13-16.

[11] Y. Hou#, K. Wang# et al., "A Sub-100nA Ultra-low Leakage MCU Embedding Always-on Domain Hybrid Tunnel FET-CMOS on 300mm Foundry Platform," IEEE International Electron Devices Meeting (IEDM), 2023, pp. 1-4.

[12] K. Wang et al., "A Novel CMOS FEOL-Embedded Asymmetrical FET for Capacitor-less and Long-Retention DRAM Based on 300mm Logic Foundry Platform", IEEE European Solid-State Electronics Research Conference (ESSERC), 2024, pp. 13-16.

[13] K. Wang et al., "Logic-Compatible Asymmetrical FET for Gain Cell eDRAM With Long Retention and Fast Access Speed", in IEEE Journal of the Electron Devices Society, vol. 13, pp. 237-244, 2025.

[14] K. Wang et al., "Analysis and Characterization Approach of Variation Behavior for Dopant-Segregated Tunnel FETs With Self-Aligned Drain Underlap", IEEE Trans. Electron Devices, vol. 72, no. 4, pp. 2051-2058, 2025.

[15] K. Wang et al., "Variation-aware optimization of salicide-enhanced tunnel FET technology based on 300 mm foundry platform", Science China Information Sciences, 2025.

Si-MoS$_2$ Heterogeneous CFET for Ultra-low Power Logic Technology Scaling

Zehua Wang [1], Wenzhong Bao [1], Peng Zhou [1], and Jing Wan*[1]

[1] State Key Laboratory of Integrated Chips and Systems, College of Integrated Circuit & Micro-Nano Electronics, Fudan University

* Email: 24110720066@m.fudan.edu.cn, jingwan@fudan.edu.cn

Abstract—In this work, a heterogeneous complementary field-effect transistor(CFET) is designed and studied using Sentaurus TCAD. The mobility mismatch issue is address by combining FDSOI pFET with multilayered MoS$_2$ nFET. The results demonstrate good voltage transfer characteristics and potential for ultra-low power integration. Moreover, the structure and electrical characteristics of heterogeneous CFET and Si-CFET are compared systematically. Our device shows lower parasitic effect and leakage current. Our CFET inverter exhibits 18%/10% lower pull-up/down delay reduction compared with Si-CFET and 0.015fJ power consumption at V$_{DD}$=0.6V.

Keywords—CFET, 2D materials, gate-all-around, heterogeneous integration, ultra-low power

I. INTRODUCTION

The Complementary Field-Effect Transistor (CFET) consists of vertically stacked p-type and n-type metal-oxide-semiconductor transistors reduces device footprint by approximately 34%-42% [1] compared to conventional layouts. Thereby it is used to increase integration density and performance for sub-3nm CMOS nodes. However, as device size shrink further, quantum effects become more pronounced. For instance, when the Si channel thickness is scaled down to 3 nm or below, surface roughness scattering causes severe mobility degradation [2], compromising device performance. Transition metal dichalcogenides (TMDs) have atomically smooth surfaces and maintain high carrier mobility at scaled size [4-5], making it a promising pathway to extend Moore's Law.

A heterogenous CFET structure comprising a bottom silicon-on-insulator (SOI) p-type transistor and a top molybdenum disulfide (MoS$_2$) n-type transistor has been previously reported [6]. In this work, the scaling potential of heterogenous CFET is studied at the 2 nm node through Sentaurus TCAD simulations, and the performance is compared with Gate-All-Around (GAA) CFET under various operating voltages systematically.

II. DEVICE STRUCTURE AND SIMULATION SETUP

A. Device Structure

The schematic structure of the heterogeneous CFET is shown in Fig 1(a), consisting of a bottom FDSOI pFET and a top MoS$_2$ nFET. Table 1 lists key structural parameters for the simulated 2 nm node CFET, which is based on the IRDS specifications for high-performance logic devices [7]. The top channel comprises a 1.4 nm-thick bilayer MoS$_2$, while the bottom device is a conventional p-type FDSOI MOSFET with adjusted thickness of 2nm SOI to suppress the short channel effect and match the On-current of MoS$_2$ nFET. The mobility and carrier lifetime of bilayer MoS$_2$ material is

adjusted for the calibration of transfer characteristics. The physical gate lengths of both pFET and nFET are set to 12 nm. A 2 nm HfO$_2$ layer serves as the gate dielectric. Moreover, metals with work functions of 4.53 eV and 4.83 eV are used as gate electrodes for pFET and nFET, respectively. A 13 nm-thick oxide spacer is inserted between pFET and nFET to ensure electrical isolation.

Table I. Structural parameter reference from 2nm technology node in IDRS

Structural Parameter	Quantities(nm)	
	Heterogeneous CFET	Si-CFET
Physical gate length(L$_G$)	12	
Spacer width(L$_{sp}$)	3	
Via width(L$_{via}$)	10	
Gate oxide thickness(T$_{ox}$)	2	
Device physical width(W)	30	
Channel thickness(T$_{si}$)	2	5
Channel width(W$_{ch}$)	30	10
Effective channel width(W$_{eff}$)	30	85 for nET 75 for pFET
p/nFET separation(L$_d$)	13	30

B. Simulation Setup

For device physics, models of holes gradient quantum correction and Lombardi surface mobility are used to accurately simulate carrier transportation in the thin SOI layer. Models for bandgap narrowing, electric field-dependent mobility, doping-dependent mobility, and doping-dependent Shockley-Read-Hall (SRH) recombination are activated.

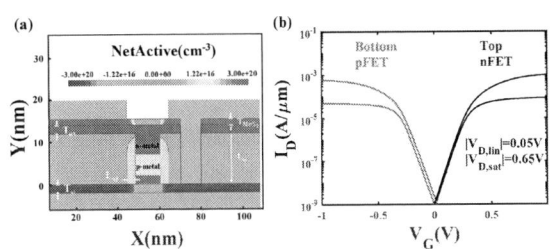

Fig. 1. (a) Simulated structure of the heterogeneous CFET. (b) The transfer curves of the MoS$_2$ nFET and FDSOI pFET

To eliminate crosstalk, all terminals of the top nFET are floated when measuring the bottom pFET C-V curve, and vice versa. For voltage transfer characteristics(VTC) and transient response simulations, the drain terminals of pFET and nFET are connected through a metal via as the output node, while the pFET source is biased at V$_{DD}$ and the nFET source is grounded.

979-8-3315-3918-4/25 $31.00 © 2025 IEEE

III. RESULT AND DISSCUSSION

A. Electrical Characteristics of Heterogeneous CFET

The simulated transfer curves are demonstrated in Fig. 1.(b). Owing to MoS_2's superior threshold voltage tunability and high carrier mobility [3] at scaled dimensions, the threshold voltages and On- current of pFET and nFET exhibit excellent matching.

Fig. 2.(a) shows VTC of the heterogeneous CFET at different V_{DD}. A maximum gain of ~33.2 V/V is achieved at $V_{DD} = 0.6$ V, $V_{in} = V_{DD}/2$, indicating excellent On-current I_{on} matching between two FETs and ultra-low-voltage operation capability. Furthermore, Fig. 2.(b) plot the V_{DD} current versus input voltage sweep. Benefiting from the low leakage of FDSOI and high on/off ratio of MoS_2, the device exhibits low leakage current and static power consumption.

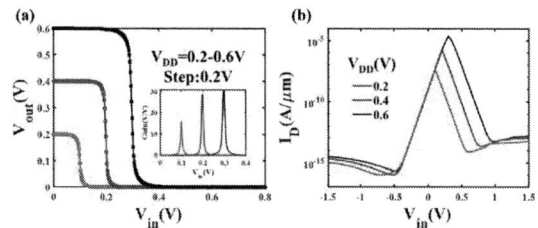

Fig. 2. (a)The voltage transfer characteristics and gains of heterogeneous CFET inverter under V_{DD}=0.2-0.6V. (b)The change of supply current I_D with the sweep of V_{in} under various V_{DD}.

B. Structures and Parasitic Parameter Comparison

A systematic comparison between Si-GAA CFET and heterogeneous CFET elucidates the advantages of the latter for large-scale integration. Fig. 3(a) shows a 3D bird's-eye view of Si-CFET. Fig. 3.(b) and (c) show its cross-section with structural parameters. The key structural parameters of Si-CFET are aligned with heterogeneous CFET for fair comparison as listed in Table 1. The Si-CFET substrate is a standard p-type SOI, with alternating epitaxial Si (5 nm)/SiGe (10 nm) superlattice structures. SiGe is selectively etched to form nanosheets. A 30 nm oxide layer is deposited to isolate the top nFET from the bottom pFET.

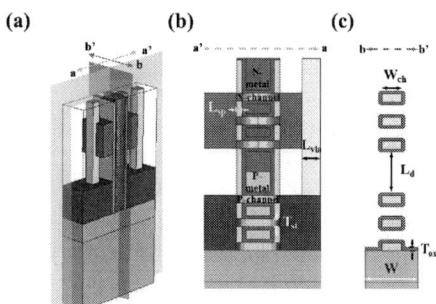

Fig. 3. (a) 3-D schematics of the Si-CFET. (b)Cross section view of Si-CFET. (c)Cross section view of the nanosheet.

The C-V curves of pFET are obtained when the source and drain of nFET is floating (same for nFET). For VTC and transient analysis, metal via are added to link the drain of bottom pFET and top nFET as the outputs. Notably, current density and parasite capacitance of both CFETs are

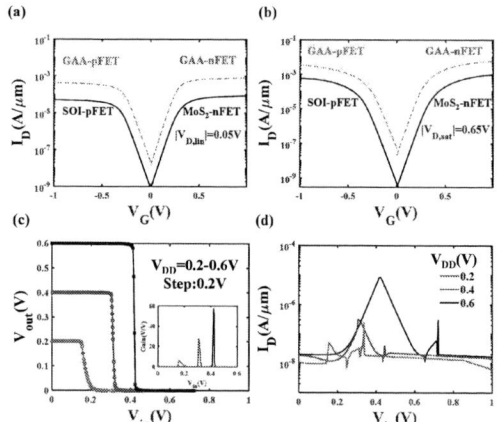

Fig. 4. The comparison of transfer curves under different bias condition: (a) $|V_D|$=0.05V, (b) $|V_D|$=0.65V. (c) The voltage transfer characteristics and gains of Si-CFET under V_{DD}=0.2-0.6V. (c) The change of supply current I_D of Si-CFET with the sweep of V_{IN}.

normalized to the device physical width W for equal footprint.

Fig. 4.(a) and (b) compare the transfer characteristics of two CFET configurations. The On-current of the MoS_2 nFET reaches ~0.530mA/μm, which is in consistent with reported values employing semimetal contacts for contact resistance optimization [8]. The On-current of FDSOI nFET is ~0.368mA/μm, demonstrating good match with MoS_2 nFET. The On-current I_{on} of GAA pFET/nFET are 5.02×/6.47× higher than heterogeneous CFET at $|V_D|$=0.65V, which reaches ~1.84 mA/μm and ~3.43 mA/μm for pFET and nFET respectively owing to the extra effective gate width from the gate-all-around structure and 3-level-stacked nanosheets.

Although the GAA pFET/nFET exhibit higher On-current compared with the heterogeneous CFET due to the tri-layer nanosheets, it exhibits ~10× higher leakage. Fig. 4.(c) shows the VTC of Si-CFET. The VTC and gains of Si-CFET are distorted at low supply voltage of V_{DD}=0.2V, while heterogeneous CFET maintains functionality. Heterogeneous CFET static leakage (~10^{-13} A) is five orders of magnitude lower than Si-CFET (~10^{-8} A) under positive V_{in} bias as shown in Fig. 5.(d). Owing to the higher On/Off ratio of MoS_2 nFET, the leakage current of heterogeneous CFET demonstrates ultralow magnitude on the order of 10^{-15} A/μm, which is expected to drastically reduces the static power consumption in large-scale ICs.

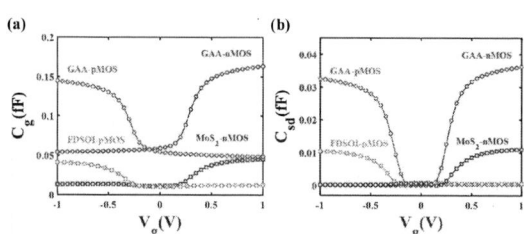

Fig. 5. (a) Comparison of capacitance of two device configurations. (a)Gate capacitance(C_g). (b)Source-Drain parasitic capacitance(C_{sd})

979-8-3315-3918-4/25 $31.00 © 2025 IEEE

Fig. 5.(a) and (b) compare the gate and parasitic capacitances under equal device footprint. With a total height of 123 nm (5.6× taller than heterogeneous CFET's 22 nm), GAA pFET gate capacitance (C_g) is 0.14fF at V_G=1V, which is ~3.52× higher than FDSOI pFET (0.041fF at V_G=1V), while the C_g of GAA nFET is 0.16fF at V_G= -1V, which is ~3.63× higher than MoS_2 nFET (0.045fF at V_G=-1V). The source-drain parasitic capacitance (C_{sd}) of GAA pFET is 0.033fF, which is ~3.25× higher than FDSOI pFET's 0.01fF, while the C_{sd} of GAA nFET is 0.036fF, which is ~3.27× higher than MoS_2 nFET's 0.011fF.

The taller structure of Si-CFET leads to increased parasitic capacitance and more complex fabrication process. Moreover, the performance of Si-CFET requires more delicate co-optimization through design levels and cell levels [10,11], which presents new challenges for device fabrication. Heterogeneous CFET addresses this issue through lower fabrication complexity and simpler device structure.

C. Transient Characteristics Comparison

To evaluate digital circuit performance, a 1 GHz pulse is applied to both devices' gates. The transient output of heterogeneous CFET is shown in Fig. 6.(a). The propagation delay is calculated as the time difference between output and input at the stage of 0.5V_{DD} [9]. Energy consumption per clock cycle is computed as: $\int_0^T V_{DD} I_D dt$, where I_D is the pFET source current and T is the period of the clock signal.

Heterogeneous CFET exhibits lower power consumptions as illustrated in Fig. 6.(d). At V_{DD}=1.2V, the energy consumption of the heterogeneous CFET per cycle is about 0.084 fJ. At V_{DD}=0.6V, the energy consumption is about 0.015fJ, which is much lower than Si-CFET due to minimal leakage and lower propagation delay.

IV. CONCLUSION

The scaling ability of heterogeneous CFET is analyzed and compared with Si-CFET systematically. Heterogeneous CFET offers lower fabrication costs through structural simplicity. TCAD simulations reveal ~3.24-3.63× lower parasitic capacitance and ideal leakage characteristics versus Si-CFET. Transient analysis demonstrates superior propagation delay, supporting high-frequency digital applications. Additionally, heterogeneous CFET achieves much lower power consumption than Si-CFET.

ACKNOWLEDGMENT

This work was supported by the National Natural Science Foundation of China (62474052), National Key R&D Program of China (2021YFA1200500), Key Technology R&D Plan of Shanghai(25CL2900100) and Natural Science Foundation of Shanghai (23ZR1405900).

REFERENCES

[1] H. Kükner et al., "Double-Row CFET: Design Technology Co-Optimization for Area Efficient A7 Technology Node," in 2024 IEEE International Electron Devices Meeting (IEDM), Dec. 2024, pp. 1–4. doi: 10.1109/IEDM50854.2024.10873524.

[2] A. Agrawal et al., "Silicon RibbonFET CMOS at 6nm Gate Length," in 2024 IEEE International Electron Devices Meeting (IEDM), Dec. 2024, pp. 1–4. doi: 10.1109/IEDM50854.2024.10873367.

[3] K. Uchida, H. Watanabe, A. Kinoshita, J. Koga, T. Numata, and S. Takagi, "Experimental study on carrier transport mechanism in ultrathin-body SOI nand p-MOSFETs with SOI thickness less than 5 nm," in Digest. International Electron Devices Meeting, Dec. 2002, pp. 47–50. doi: 10.1109/IEDM.2002.1175776.

[4] B. Radisavljevic, A. Radenovic, J. Brivio, V. Giacometti, and A. Kis, "Single-layer MoS2 transistors," Nature Nanotechnology, vol. 6, no. 3, pp. 147–150, Mar. 2011, doi: 10.1038/nnano.2010.279.

[5] D. Akinwande et al., "Graphene and two-dimensional materials for silicon technology," Nature, vol. 573, no. 7775, pp. 507–518, Sep. 2019, doi: 10.1038/s41586-019-1573-9.

[6] L. Tong et al., "Heterogeneous complementary field-effect transistors based on silicon and molybdenum disulfide," Nature Electronics, vol. 6, no. 1, pp. 37–44, Jan. 2023, doi: 10.1038/s41928-022-00881-0.

[7] International Roadmap for Devices and Systems (IRDS). IRDS.[Online]. Available: https://irds.ieee.org/editions/2020

[8] P.-C. Shen et al., "Ultralow contact resistance between semimetal and monolayer semiconductors," Nature, vol. 593, no. 7858, pp. 211–217, May 2021, doi: 10.1038/s41586-021-03472-9.

[9] S. -G. Jung, D. Jang, S. -J. Min, E. Park, and H. -Y. Yu, "Performance Analysis on Complementary FET (CFET) Relative to Standard CMOS With Nanosheet FET," IEEE Journal of the Electron Devices Society, vol. 10, pp. 78–82, 2022, doi: 10.1109/JEDS.2021.3136605.

[10] S. Liao et al., "First Demonstration of Monolithic CFET Inverter at 48nm Gate Pitch Toward Future Logic Technology Scaling," in 2024 IEEE International Electron Devices Meeting (IEDM), Dec. 2024, pp. 1–4. doi: 10.1109/IEDM50854.2024.10873334.

[11] C. -Y. Huang et al., "3-D Self-aligned Stacked NMOS-on-PMOS Nanoribbon Transistors for Continued Moore's Law Scaling," in 2020 IEEE International Electron Devices Meeting (IEDM), Dec. 2020, p. 20.6.1-20.6.4. doi: 10.1109/IEDM13553.2020.9372066.

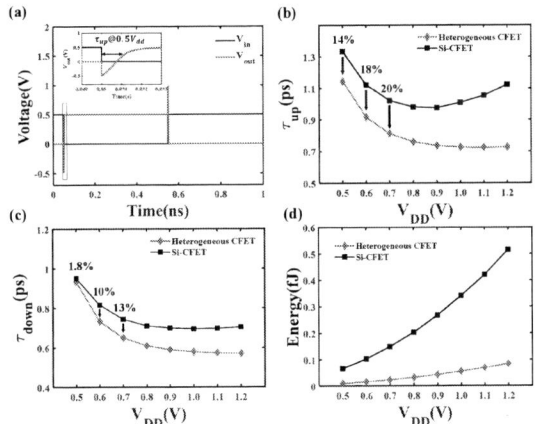

Fig. 6. (a) Method for extracting propagation delay and transient response of heterogeneous CFET under 1GHz input signal. (b) Pull-up delay, (c) Pull-down delay caparison of two CFET configurations. (d)The power consumption caparison of two CFET configurations.

Fig. 6.(b) and (c) compare the propagation delay of both CFET structures. Heterogeneous CFET exhibit significantly lower pull-up delay than Si-CFET, which is 14% lower at V_{DD}=0.5V and 18% at V_{DD}=0.6V attributed to substantially reduced parasitic effects. Furthermore, despite the On-current disparity between GAA nFET and MoS_2 nFET, the heterogeneous CFET still demonstrates 1.8% and 10% greater pull-down delay in the aforementioned operating conditions.

Impact of Off-state Stress on the Reliability of 14nm nFinFETs

Wendi Wei[1], Kun Chen[*1,2], Chen Wang[*1,2], Yaolin Wang[1], Zhao Yang[1], Zhiteng Zhang[1], Zhuming Wang[1], Qingqing Sun [1,2], and David Wei Zhang [1,2]

[1] School of Microelectronics, Fudan University, Shanghai 200433, China
[2] National Integrated Circuit Innovation Center, Shanghai 201203, China

* Email: 23212020159@m.fudan.edu.cn, chen_w@fudan.edu.cn

Abstract— Off-state stress increasingly threatens the reliability of advanced-node CMOS devices and has become a critical concern for circuit performance. Experimental data demonstrate that the threshold voltage (V_{th}) of short-channel FinFETs decreases initially and subsequently increases during off-state stress. Decreases initially and subsequently increases during off-state stress, revealing that two competing mechanisms contribute to this non-monotonic behavior. In the early phase, hole injection dominates, while with prolonged stress, impact ionization becomes the primary mechanism of degradation.

Keywords— Hot Carrier, Impact Ionization, Off-state stress, FinFET

I. INTRODUCTION

The transition to advanced 3D FinFET transistors and high-κ metal-gate (HKMG) technologies has prompted extensive reliability research. However, most prior research has primarily focused on the degradation characteristics under on-state stress conditions, where both the gate and the drain of the device are in a high-voltage state. Conventional theories attribute that on-state degradation, to high-field hot carriers that generate interface defects and accumulate trapped charge, leading to phenomena such as hot carrier degradation (HCD). [1-4]. Off-state stress describes the bias condition in which the gate voltage is below the threshold (≈ 0 V) while the drain is held near V_{dd}, under this bias only leakage current flows through the channel. Off-state stress has traditionally been considered harmless for these devices because the off-state carrier population is limited, and noticeable degradation appeared only under extreme drain bias. However, with continued device scaling, the transverse electric field in the gate–drain overlap builds a localized high-field region even at nominal off-state bias. Under a high drain voltage (V_d), a small number of carriers at the channel edge or in the junction region are accelerated by a high electric field to the point where electron-hole pairs are generated through impact ionization. These energetic carriers, as known as hot carriers, not only cause increase in the drain current (I_d), but also create defects at the interface or inside the gate oxide. The resulting defects lead to a series of performance degradations, including V_{th} shift, I_{dast} degradation, g_m degradation, and deterioration of SS [5, 6]. At the same time, some hot carriers may penetrate the gate medium or interface defects via hole injection, thereby triggering more complex physical mechanisms.

From a circuit design perspective, such degradation can have significant implications for basic logic cells. For instance, in CMOS inverters, shifts in threshold voltage (V_{th}) can lead to reduced switching speeds, resulting in increased delays. In SRAM memory cells, degradation resulting from off-state stress has been observed to increase leakage current, impacting the device's static power consumption.

In this paper, the performance degradation characteristics of FinFET devices under off-state stress are systematically investigated, focusing on the evolution of key electrical parameters. A pronounced non-monotonic V_{th} of shift is observed in short-channel devices under the off-state stress. In the early stage, the hole injection mechanism dominates the degradation process, leading to a decrease in the V_{th}. As stress time increases, the hot carrier impact ionization mechanism gradually becomes more significant, driving the V_{th} upward. In the meantime, the g_m and SS degradation is more pronounced in short-channel FinFETs than in long-channel counterparts because the stronger transverse field enhances interface-defect generation. The mechanisms has been further clarify by Charge-pumping measurements and TCAD simulations, confirming negligible hole impact ionization at the source and significant electron impact ionization near the drain during off-state stress. These insights deepen the understanding of off-state degradation and provide guidance for future advanced IC design.

II. EXPERIMENT DETAIL

In this work, commercial 300mm 14nm bulk FinFET devices with ultra-low threshold voltage were used for HCD characterization. A typical fin height is approximately 40nm, a fin width is around 10nm, the smallest effective channel length is 16nm. The device consists of 6 fins, and the effective gate area is approximately 8.64×10^{-3} μm^2. The gate stack features an interface layer of SiO_2 measuring about 0.5 nm and a high-k dielectric layer of HfO_2 that is approximately 2 nm thick.

During actual operation, the n-FinFET inside a CMOS inverter cell alternates between positive-bias temperature instability (PBTI) and off-state stress, as illustrated in Fig. 1(a). Therefore, a V_{DD} gate bias should also be applied after the off-state stress interval to more accurately emulate its performance under actual circuit conditions. The quasi-static MSM (Measure-Stress-Measure) stress testing procedure is divided into two stages, as shown in the waveform in Fig. 1(b). In stage I, the gate stress was set to 0 ($V_g = 0$) and the drain stress was set to the corresponding stress voltage ($V_d = V_{dstr}$) for 1000 seconds to simulate the degradation process induced by off-state stress in the inverter. In stage II, the gate stress bias was adjusted to V_{DD} ($V_g = V_{DD}$) and the drain stress bias to 0 ($V_d = 0$ V) for a further 1000 seconds to mimic PBTI-assisted recovery under circuit operation. It is worth noting that the positive gate bias during stage II has a negligible effect on device degradation. Therefore, the analysis focused on residual degradation behavior due to off-state stress during the recovery process. This stress-recovery testing method effectively separates off-state degradation from recovery dynamics, enabling accurate assessment of its impact on device performance.

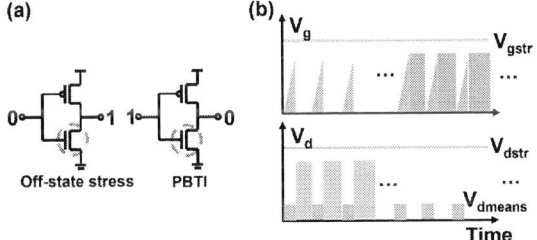

Fig. 1. (a) nFinFET stress bias condition during CMOS inverter operation. (b) Dual stage Off-state stress degradation test stress waveform, consist of off-state stress stage and PBTI-assisted recovery stage

III. RESULT AND DISCUSSION

A. V_{th} and I_{dsat} Degradation of Short- and Long-channel FinFET

In order to produce notable degradation during the given test duration, an overdrive operating voltage stresses of $V_{DD}=V_{dstr}=2.4$V, 2.5V, 2.6V is select to short-channel FinFET devices and $V_{DD}=V_{dstr}=3.1$V, 3.2V, 3.3V to the long-channel counterparts, then the resulting electrical degradation was monitored. Fig. 2. (a) shows that the V_{th} of the short-channel FinFETs first shifts negatively at the beginning of the off-state stress and later begins to shift positively as the stress time grows, exhibiting pronounced non-monotonicity behavior. By contrast, V_{th} of long-channel FinFET shifts monotonically positive, as illustrated in Fig. 2(b). These channel-length-dependent trends imply a competition between distinct degradation mechanisms that dominate at different stress durations. As for the recovery stage, when the V_g is biased at the operating voltage V_{DD}, both long-channel and short-channel devices exhibit similar behavior, the V_{th} recover rapidly once the stress is removed, but continues to degrade even after complete recovery.

Fig. 2(c)(d) plot the I_{dast} versus stress time for short- and long-channel devices, respectively. Both device types show a significant I_{dast} degradation during the off-state stress stage. In particular, short-channel devices do not exhibit the complex non-monotonic behavior observed in V_{th} response. It can also be seen that during the recovery stage, I_{dsat} did not really recovers but continues to decline slightly, while V_{th} shows substantial recovery. These observations further indicate that V_{th} "recovery" does reverse the degradation; instead, the partial rebound of V_{th} represents only one aspect of a broader, more complex degradation process. The continued I_{dsat} degradation during the recovery phase indicates that the off-state stress degradation process is likely dominated by permanent oxide traps rather than recoverable interface traps.

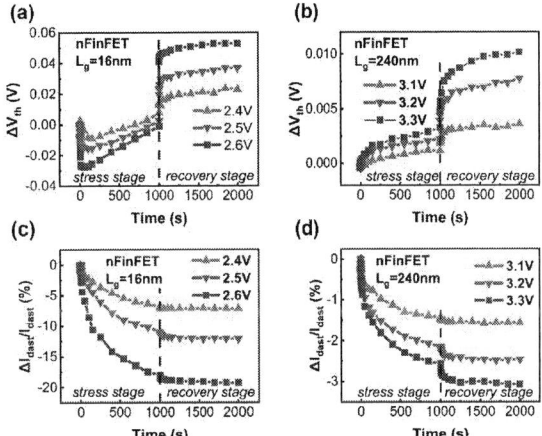

Fig. 2. V_{th} degradation of short-channel FinFET (a) and long-channel FinFET (b) under off-state stress. I_{dsat} degradation of short-channel FinFET (c) and long-channel FinFET (d) under off-state stress.

B. g_m and SS Degradation of Short- and Long-channel FinFET

The maximum transconductance (g_{mmax}) can reflect the change in channel carrier mobility under stress in the device[7,8]. Fig. 3(a) presents g_{mmax} during off-state stress and subsequent recovery for short-channel devices. The g_{mmax} degradation of the short-channel FinFET is very significant compared to the degradation of the long-channel FinFET shown in Fig. 3(b). It can be concluded that carrier mobility degradation of short-channel FinFET under off-state stress is very severe, and it's due to impact ionization to produce more oxide traps and interface defects.

To further determine whether short-channel FinFET produce more interface defects, the degradation of the SS of the device under off-state stress has been extracted. SS is highly sensitive to interface defect density; consequently, its variation predominantly reflects the creation of interface defects. As shown in fig. 3(c), as the stress time increases, the ΔSS of short-channel FinFET gradually increases until saturation. The amount of SS degradation with stress time for long-channel FinFET is very marginal, with no observable saturation, compared to that of short-channel FinFET, as shown in Fig. 3(d). This phenomenon indicates that the generation of interface defects in long-channel FinFET is significantly reduced because the equivalent transverse electric field of long-channel FinFET is smaller than that of short-channel FinFET, and it is difficult for carriers to obtain enough energy for impact ionization to occur.

C. Degradation Mechanisms

There are three physical mechanisms could involve in the degradation in FinFETs under off-state stress: (1) hole trapping in gate-oxide traps within the source–drain overlap region; (2) impact ionization of high-energy electrons near the drain that breaks Si–H bonds and forms interface/oxide defects; and (3) impact ionization of high-energy holes near the source, which can also generate interface defect. Among three of these mechanisms, the hole impact ionization should be negligible according to previous studies, the amount [5,6]. To validate this claim, a charge-pumping (CP) measurement is taken before and after a negative-bias temperature instability (NBTI) stress. If nFinFET generate enough interface defects to affect the electrical parameters of the device when holes are collisionally ionized at the source under the high electric field of the gate source, it will result much larger CP current. As shown in Fig. 4(a), The absence of any CP current shift confirms that hole-induced interface defect generation at the source is quite small, which also means that the physical mechanisms of hole impact ionization contributed to off-states degradation is limited.

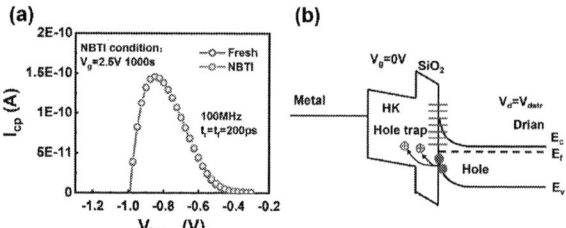

Fig. 4. (a) The charge pumping test curve of nFinFET device at a frequency of 100MHz, where the applied NBTI stress condition is V_g = 2.5V, stress time = 1000s, the stress curve before and after basically overlaps, indicating that no interface defect is generated. (b) Energy band diagram at the overlap of gate and drain under off-state stress.

To validate the physical picture, the ionization collision process of the short-channel FinFET was simulated using TCAD. To ensure the accuracy of the simulation results, the simulation model was carefully calibrated and optimized based on the electrical characteristics obtained from the measured DC I-V data of the FinFET device, achieving excellent agreement. Several advanced physical models were also employed to accurately simulate the complex mechanisms during off-state stress: the Avalanche (CarrierTempDrive) model was used to simulate the electron impact ionization rate and the impact ionization process, while the Band to Band model is used to simulate BTBT [6].

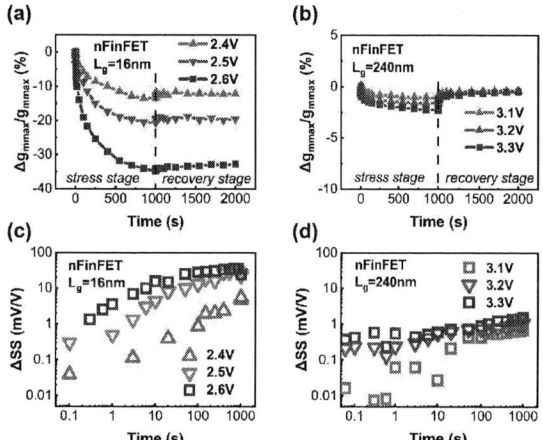

Fig. 3. g_m degradation of short-channel FinFET (a) and long-channel FinFET (b) under off-state stress. SS degradation of short-channel FinFET (c) and long-channel (d) under off-state stress.

The observed shift of the V_{th} from negative to positive with increasing off-state stress time, as shown in Fig. 2(a), can be explained by two competing physical mechanisms: hole injection and electron impact ionization. At the initial stage of off-state stress application, when the V_g is set to 0 and the V_d is high, the energy band immediately bending at the drain-gate overlap in the nFinFET device, as illustrated in Fig. 4(b), allowing for band-to-band tunneling that injects holes. These holes then get trapped in the gate dielectric, leading to a decrease in V_{th}. However, when high stress is applied to the drain, a high transverse electric field is also formed between the source and drain of the device. This effect is especially pronounced in short-channel FinFET; as the channel size decreases, the impact of the electric field becomes more significant. As the stress time increased, the strong transverse field starts to produce hot electrons that trigger impact ionization and generate additional electron–hole pairs, resulting in significant leakage currents. These secondary carriers become 'hot' when accelerated by the electric field, and they can collide with other carriers, further generating more electron-hole pairs and leading to hot carrier degradation (HCD)[9-12]. These extra electron-hole pairs generate defects at the interface between the gate dielectric and the channel. Over time, these defects gradually accumulate, leading to an increase in the absolute value of V_{th}.

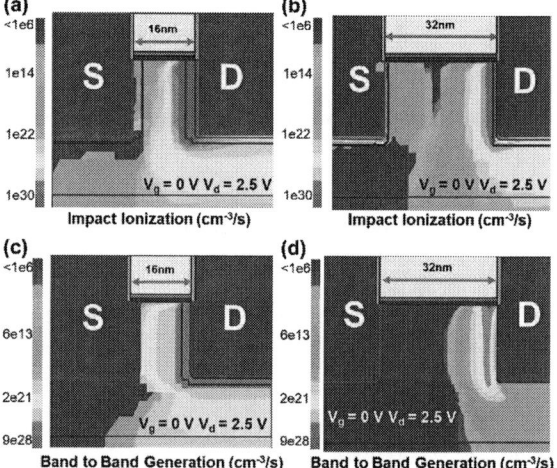

Fig. 5: TCAD simulation result of distribution of impact ionization caused by off-state stress in FinFET channel length L_g = 16nm (a) and inFET with channel length L_g = 32nm (b); Distribution diagram of inter band tunneling caused by off-state stress in FinFET with channel length L_g = 16nm (c) and FinFET with channel length L_g = 32nm (d)

The effect of a high electric field in the gate-drain overlap region results in the formation of a significant depletion region on the drain side. Since when the gate-drain voltage (V_{ds}) exceeds 1.2 V, band-to-band tunnelling (BTBT) may occur [11]. With the help of TCAD, it is possible to create detailed maps of physical processes. Figures 5(a) and 5(b) illustrate the impact ionization rate, revealing a pronounced activity that extends from the drain into the mid-channel region for the 16 nm device. In contrast, the 32 nm device exhibits this activity to a lesser extent. Figures 5(c) and 5(d) show the BTBT generation rate for both

979-8-3315-3918-4/25 $31.00 © 2025 IEEE

channel lengths. For the short-channel device, BTBT tunneling primarily occurs in the channel-drain overlap region and extends well into the the mid-channel. However, this phenomenon is much less pronounced in the longer channel device. TCAD simulation results confirm that secondary carriers are generated in FinFETs even when they are in the off state. Due to the reduced channel lengths, these carriers are found not only near the drain but also in the mid-channel region.

IV. CONCLUSION

This work investigated off-state-induced degradation in advanced 14nm nFinFET devices. Key electrical parameters - V_{th}, I_{dast}, g_m, and SS —were systematically characterized under off-state stress. A pronounced non-monotonic V_{th} evolution was observed on short-channel devices. Detailed analyses, including charge-pumping measurements and TCAD simulation, demonstrate that this behavior arises from the time-dependent competition between hole injection and electron impact ionization. First, the large gate–drain voltage promotes hole injection into the gate oxide and is trapped in the gate dielectric, lowering V_{th}. On the other hand, BTBT and punch-through current intensify under the strong drain field, yielding non-negligible leakage current and elevated carrier density in the off state. The accelerated carriers become hot carriers that collide to create additional electron–hole pairs, thereby triggering. As stress time increases, defects gradually accumulate, leading to an increase in the absolute value of V_{th}. These findings could enhance the predictive accuracy of circuit-level aging models and help mitigate off-state reliability concerns in designs utilizing advanced FinFET technologies.

REFERENCES

[1] YU Z, ZHANG J, WANG R, et al. New insights into the hot carrier degradation (HCD) in FinFET: New observations, unified compact model, and impacts on circuit reliability; proceedings of the 2017 IEEE International Electron Devices Meeting (IEDM), 2017 [C].IEEE

[2] WANG R, SUN Z, LIU Y Y, et al. Understanding Hot Carrier Reliability in FinFET Technology from Trap-based Approach; proceedings of the 2021 IEEE International Electron Devices Meeting (IEDM), 2021 [C].IEEE

[3] MAHAPATRA S, SHARMA U. A Review of Hot Carrier Degradation in n-Channel MOSFETs—Part II: Technology Scaling [J]. IEEE Transactions on Electron Devices, 2020, 67(7): 2672-81.

[4] SUN Z, YU Z, ZHANG Z, et al. Investigation on the Lateral Trap Distributions in Nanoscale MOSFETs During Hot Carrier Stress [J]. IEEE Electron Device Letters, 2019, 40(4): 490-3.

[5] SUN Z, WANG Z, WANG R, et al. Investigation of the Off-State Degradation in Advanced FinFET Technology—Part II: Compact Aging Model and Impact on Circuits [J]. IEEE Transactions on Electron Devices, 2023, 70(3): 921-7.

[6] SUN Z, WANG Z, WANG R, et al. Investigation of the Off-State Degradation in Advanced FinFET Technology—Part I: Experiments and Analysis [J]. IEEE Transactions on Electron Devices, 2023, 70(3): 914-20.

[7] CHO M, ROUSSEL P, KACZER B, et al. Channel Hot Carrier Degradation Mechanism in Long/Short Channel n-FinFETs [J]. IEEE Transactions on Electron Devices, 2013, 60(12): 4002-7.

[8] GUPTA A, GUPTA C, VEGA R A, et al. Reliability Modeling and Analysis of Hot-Carrier Degradation in Multiple-Fin SOI n-Channel FinFETs With Self-Heating [J]. IEEE Transactions on Electron Devices, 2019, 66(5): 2075-80.

[9] CECCARELLI E, MANNING K, MAXWELL S, et al. GIDL Increase Due to HCI Stress: Correlation Study of MOSFET Degradation Parameters and Modelling for Reliability Simulation; proceedings of the 2019 IEEE International Reliability Physics Symposium (IRPS), 2019 [C].IEEE

[10] LEE N H, BAEK D, KANG B. Effect of off-State Stress and Drain Relaxation Voltage on Degradation of a Nanoscale nMOSFET at High Temperature [J]. IEEE Electron Device Letters, 2011, 32(7): 856-8.

[11] HAUSER M J, SRINIVASAN P, VALLETT A, et al. Parasitic Drain Series Resistance Effects on Non-conducting Hot Carrier Reliability; proceedings of the 2022 IEEE International Reliability Physics Symposium (IRPS), 2022 [C].IEEE

[12] KIM K, CHUNG I, SUN D, et al. Study on off-state hot carrier degradation and recovery of NMOSFET in SWD circuits of DRAM; proceedings of the 2016 IEEE International Integrated Reliability Workshop (IIRW), 2016 [C].IEEE

Performance Comparison Between Bulk-Si and FDSOI Nanosheet GAAFETs

RS.He[1], BX.Gan[1], S.Cristoloveanu[2], Y.Xu*[2], J.Wan*[1]

[1] State Key Laboratory of Integrated Chips and Systems, College of Integrated Circuits and Micro-Nano Electronics, Fudan University, Shanghai 200433, China
[2] Guangdong Greater Bay Area Institute of Integrated Circuit and System, Guangzhou 510300, China
* Email: 24110720131@m.fudan.edu.cn, jingwan@fudan.edu.cn, xuyong@giics.com.cn

Abstract—**This work presents a TCAD-based comparative study of Bulk-Si and FDSOI nanosheet GAAFET at the 2nm technology node. We developed 3-D device models and performed systematic simulations. The results demonstrate that FDSOI structures maintain similar DC characteristics while achieving 5-10% lower gate capacitance compared to Bulk-Si counterparts. Circuit verification using BSIM-CMG compact models confirms that 11-stage ring oscillators on FDSOI substrates exhibit higher oscillation frequency, with 16-18% lower delay at identical energy and 35-56% energy reduction at equivalent delays. This validates the superior performance and energy efficiency of FDSOI GAAFET for sub-3-nm IC designs.**

Keywords—*GAAFET; FDSOI; Bulk-Si; parasitic capacitance; ring oscillator*

I. INTRODUCTION

As CMOS technology approaches its physical scaling limits, FinFETs face significant challenges including short-channel effects (SCEs) beyond the 5nm node [1-4]. Gate-all-around FETs (GAAFETs) have emerged as the most promising successor due to their superior gate controllability. This work demonstrates that substrate selection critically impacts device performance: while Bulk-Si GAAFETs suffer from parasitic capacitance issues [5], FDSOI-based devices leverage the inherent buried oxide layer (BOX layer) isolation to achieve enhanced immunity to interference and radiation, lower power consumption, and improved electrostatic integrity [6-8].

Existing studies on FDSOI GAAFETs have primarily focused on individual parameter characterization [9,10], with limited comprehensive performance evaluation. In this work, 3-D TCAD models are systematically developed for both Bulk-Si and FDSOI GAAFET architectures to comparatively analyze their electrical characteristics, followed by rigorous circuit-level validation through 11-stage ring oscillator (RO) simulations using BSIM-CMG models. The results demonstrate FDSOI GAAFETs' advantages in parasitic capacitance reduction and propagation delay improvement.

II. DEVICE STRUCTURE AND SIMULATION SETUP

In this study, Synopsys Sentaurus TCAD tools were utilized to conduct three-dimensional device simulations [11]. Through a comprehensive process-device co-simulation flow, we systematically compare and analyze the impacts of FDSOI and bulk-Si substrates on device performance.

A. Device Architecture Design

Based on the International Roadmap for Devices and Systems (IRDS 2023) size projections for the 2-nm node [12], we simulated nanosheet GAAFETs in 3-D. The doping concentrations were set as:

- The Bulk-Si substrate is doped at 2×10^{18} cm^{-3} to prevent punch-through effects.

- The FDSOI substrate is lightly doped at 1×10^{15} cm^{-3} to maintain full depletion characteristics.

- The channel region is doped at 1×10^{15} cm^{-3}.

- The source/drain regions are doped at 3×10^{20} cm^{-3}.

The bulk-Si and FDSOI GAAFET 3-D models were separately established in Sentaurus Process (Fig. 1), with their 2-D channel cross-sections illustrated in Fig. 2. Only the substrate differed between two structures, as shown in Table I (the Bulk-Si model excludes embedded strain components).

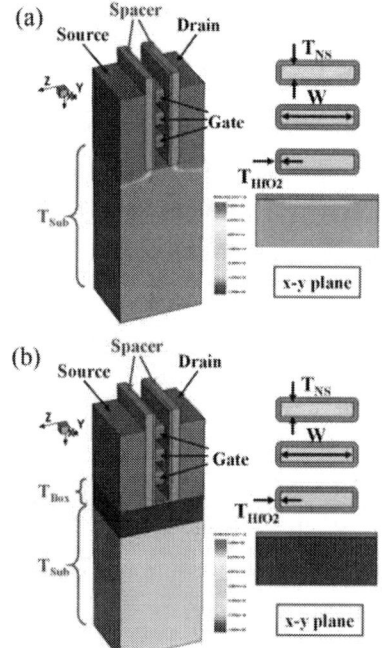

Fig.1. 3-D views and x-y plane channel cross-section of (a) Bulk-Si GAAFET and (b) FDSOI GAAFET (not to scale).

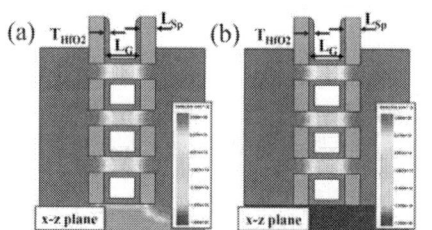

Fig.2. 2-D cross-sections along x-z plane of (a) Bulk-Si GAAFET and (b) FDSOI GAAFET (not to scale).

TABLE I.   STRUCTURAL PARAMETERS OF GAA TRANSISTOR

	Bulk-Si	FDSOI
T_{Sub} (nm)	120	100
T_{Box} (nm)	—	20
L_G (nm)	14	14
L_{Sp}(nm)	6	6
W_{NS} (nm)	30	30
T_{NS} (nm)	6	6
T_{Hf02} (nm)	2	2

B. Simulation Model Configuration

We performed numerical simulations using Synopsys Sentaurus Device, establishing a comprehensive physical model system to accurately characterize the electrical properties of GAAFETs. The key physical models including density gradient theory, Canali's and Philips mobility models, SRH and Auger recombination, non-local tunneling, and self-heating coupling were employed in the simulations. The work functions were calibrated to meet IRDS HP specifications (I_{OFF} = 10 nA/μm):

- Bulk-Si GAAFET: $\Phi_{m,P}$ = 4.892 eV, $\Phi_{m,N}$ = 4.397 eV.

- FDSOI GAAFET: $\Phi_{m,P}$ = 4.889 eV, $\Phi_{m,N}$ = 4.399 eV.

We adopted contact resistivity values of 1×10^{-9} Ω·cm² for n-FETs and 7×10^{-10} Ω·cm² for p-FETs, consistent with advanced node implementations [13].

III. RESULTS AND ANALYSIS

In this section, we systematically compare the DC/AC characteristics of Bulk-Si and FDSOI GAAFETs, and quantitatively analyze the advantages of the FDSOI structure in terms of propagation delays based on 11-stage RO circuit simulations.

A. Comparison of Output and Transfer Characteristics

Through TCAD simulation, we obtained the I-V characteristics of the Bulk-Si GAAFET and the FDSOI GAAFET. Fig.3 shows the I_D-V_G curves in linear scale. At V_D = 0.05 V and 0.65 V, the transfer curves of both substrates almost overlap. With V_G = 0.65 V, the P-type FDSOI GAAFET demonstrates ~2.5% higher I_{ON} than Bulk-Si, while the N-type shows ~1.4% improvement, and current characteristics remain nearly identical. The logarithmic plot (Fig. 4) reveals similar subthreshold slopes (SS) and nearly identical drain-to-barrier lowering (DIBL) for both devices. Key performance parameters are summarized in Table II.

The I_D-V_D of both devices are compared in Figure 4. FDSOI shows slightly higher current than Bulk-Si, and all curves overlap generally, consistent with the transfer characteristics. The universality of these conclusions was verified through three sets of simulations at V_D = 0.45V, 0.55V, and 0.65V to exclude any potential bias from a single operating point.

The results show that the FDSOI GAAFET exhibits essentially identical DC characteristics to its Bulk-Si counterpart, establishing an ideal foundation for subsequent comparative AC performance analysis.

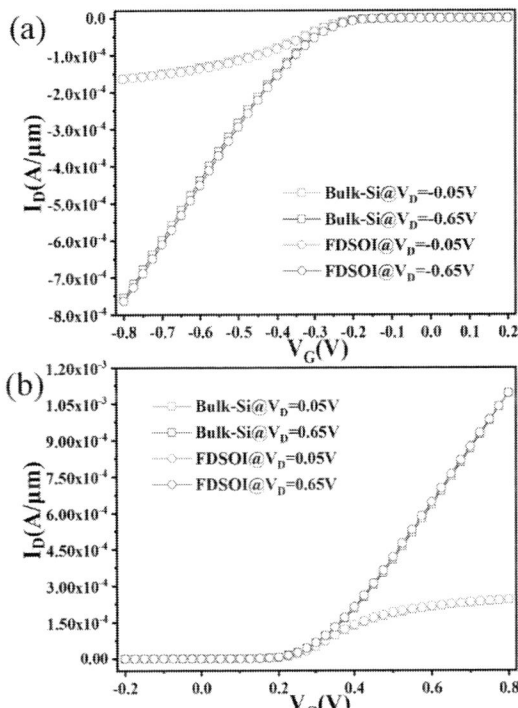

Fig.3. Comparison of I_D-V_G characteristics between Bulk-Si and FDSOI of (a) P-type and (b)N-type (linear scale).

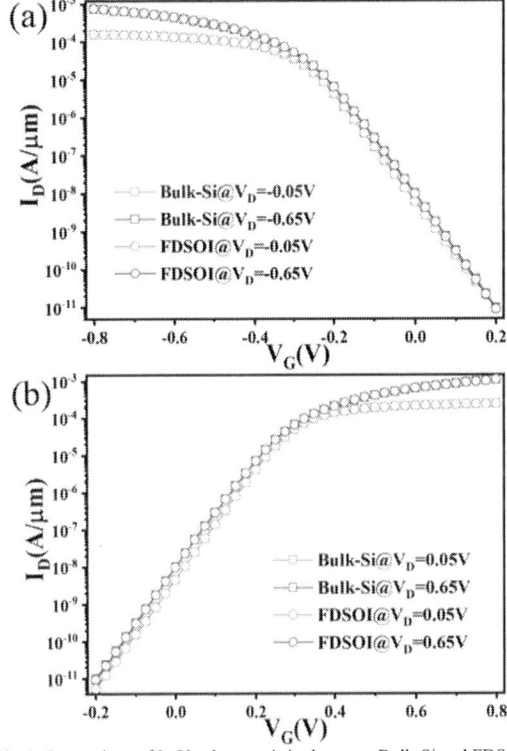

Fig.4. Comparison of I_D-V_G characteristics between Bulk-Si and FDSOI of (a) P-type and (b)N-type (logarithmic scale).

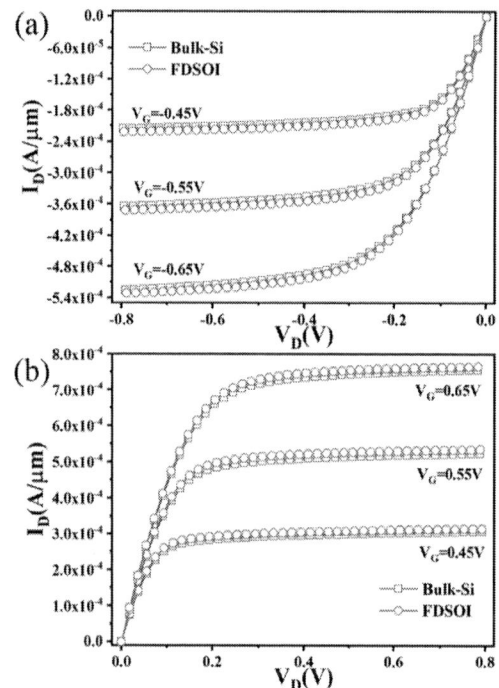

Fig.5. Comparison of I_D-V_D characteristics between Bulk-Si and FDSOI of (a) P-type and (b)N-type.

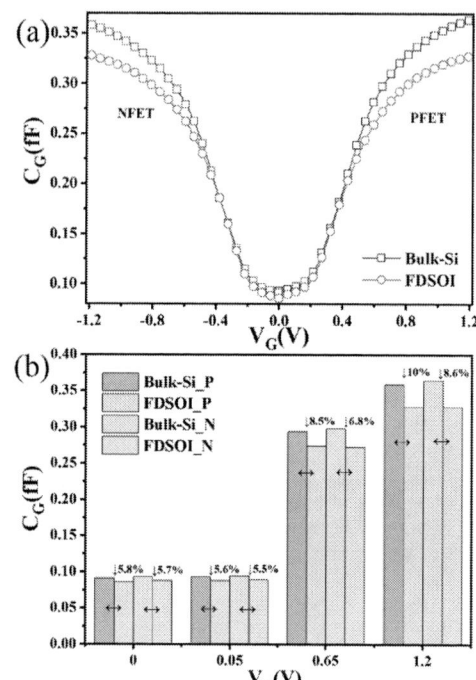

Fig.6. Comparison of C-V characteristics between Bulk-Si and FDSOI GAAFETs at 1 MHz.

TABLE II. COMPARISON OF KEY PERFORMANCE PARAMETERS OF BULK-SI AND FDSOI GAAFET

	Bulk-Si GAA FET		FDSOI GAA FET	
	P-type	N-type	P-type	N-type
V_{th} (V)	-0.115	0.114	-0.115	0.114
I_{on}@0.65V (µA/µm)	518.3	750.6	531.2	761.1
I_{off} (nA/µm)	10	10	10	10
DIBL (mV/V)	26.7	33.3	26.7	33.3
SS (mV/decade)	65.5	65.9	61.0	65.3

B. Comparison of capacitance characteristics

Fig. 6(a) compares the C-V characteristics of Bulk-Si and FDSOI GAAFETs at 1 MHz. We extracted capacitance values at four characteristic voltages: V_G=0V (reference state), 0.05V (linear region), 0.65V (saturation region), and 1.2V (high-voltage region) for clearer comparison, as shown in Figure 6(b). The FDSOI GAAFET exhibits a 5-10% capacitance reduction, attributed to the complete device isolation enabled by the BOX layer that effectively minimizes parasitic capacitance components.

C. Ring Oscillator (RO) Circuit Verification

Based on TCAD simulation data, we extracted BSIM-CMG model parameters using the Model Builder Program (MBP) and established an HSPICE simulation environment for circuit-level performance verification of an 11-stage RO.

Figure 7(a) displays the output voltage time-domain waveform of the 11-stage RO at a supply voltage of 0.65V. The FDSOI GAAFET demonstrates an oscillation period of 47.35ps, representing a 16% reduction compared to Bulk-Si's 39.80ps, corresponding to a ~19% improvement in oscillation frequency. Figure 7b compares single-stage energy-delay

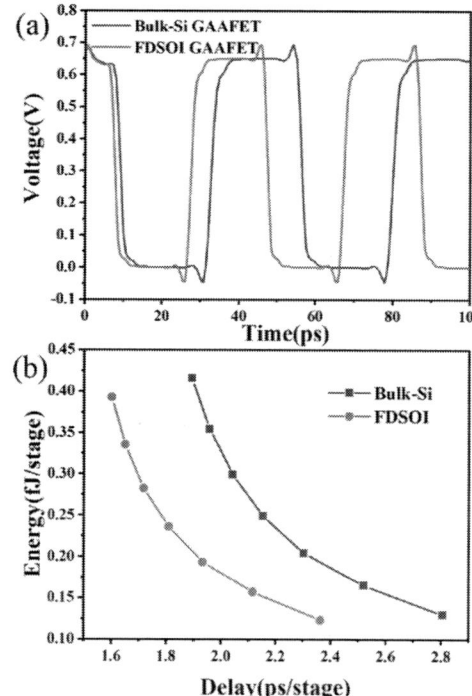

Fig.7. Performance comparison of 11-stage ring oscillators (a) Output voltage waveforms of FDSOI GAAFET versus Bulk-Si (at VDD = 0.65 V) and (b) Single-stage energy-delay characteristics of both devices.

characteristics in the ring oscillator. The FDSOI GAAFET shows 16-18% lower delay at identical energy conditions and 35-56% reduction at equivalent delays versus bulk-Si GAAFET, demonstrating superior performance across the operating voltage range. This enhancement stems from parasitic capacitance suppression by the BOX layer, confirming FDSOI's advantages for advanced nodes.

IV. CONCLUSION

Through systematic simulation analysis, this work reveals the performance characteristics of FDSOI GAAFET compared to Bulk-Si GAAFET at the 2nm node. While maintaining comparable gate control capability, the FDSOI GAAFET achieves slight improvement in drive current and effective reduction in parasitic capacitance through its unique BOX layer design. Circuit-level simulations further confirm that this structure can significantly enhance the performance of RO. However, this study assumes unstrained channels for both technologies. In practice, Bulk-Si typically employs embedded strain techniques for carrier mobility enhancement, whereas such methods are fundamentally incompatible with FDSOI's ultra-thin body structure. For FDSOI PFETs, SiGe channel integration provides an effective solution by significantly enhancing hole mobility. Future work should evaluate strained Bulk-Si and SiGe-channel FDSOI PFETs for comprehensive mobility benchmarking. The simulation methodology and conclusions established in this work provide theoretical basis for device selection in advanced nodes, promoting practical applications of nanosheet transistors.

ACKNOWLEDGMENT

This work was supported by the Key Technology R&D Plan of Shanghai(25CL2900100), Natural Science Foundation of Shanghai (23ZR1405900), National Natural Science Foundation of China (62474052), Guangdong Province Research and Development in Key Fields from Guangdong Greater Bay Area Institute of Integrated Circuit and System (No.2021B0101280002) and Guangzhou City Research and Development Program in Key Field (No.20210302001)), and in part by Guangdong Key Laboratory of Integrated Circuit Technology and Products Based on Fully Depleted Silicon On Insulator. (Corresponding authors: Yong Xu, and Jing Wan).

REFERENCES

[1] C. Pan, B. Smith, J. Doe, and A. Lee, "Technology/system codesign and benchmarking for lateral and vertical GAA nanowire FETs at 5-nm technology node," IEEE Transactions on Electron Devices, vol. 62, no. 10, pp. 3125-3132, Oct. 2015.

[2] S. Kim, M. Park, K. Lee, and T. Chen, "Investigation of device performance for fin angle optimization in FinFET and gate-all-around FETs for 3 nm-node and beyond," IEEE Transactions on Electron Devices, vol. 69, no. 4, pp. 2088-2093, Apr. 2022.

[3] G. Bae, H. Bae, M. Son, and J. Park, "3nm GAA technology featuring multi-bridge-channel FET for low power and high performance applications," in 2018 IEEE International Electron Devices Meeting (IEDM), San Francisco, CA, Dec. 2018, pp. 28.7.1-28.7.4.

[4] S. Kim, D. Kim, J. Lee, and S. Park, "Reliability assessment of 3nm GAA logic technology featuring multi-bridge-channel FETs," in 2023 IEEE International Reliability Physics Symposium (IRPS), Mar. 2023, pp. 1-8.

[5] S.-H. Chen, C.-W. Chang, T.-Y. Huang, and M.-J. Tsai, "ESD diodes in a bulk Si gate-all-around vertically stacked horizontal nanowire technology," in 2016 IEEE International Electron Devices Meeting (IEDM), Dec. 2016, pp. 35.4.1-35.4.4.

[6] L. Cao, W. Zhang, Y. Wang, and X. Liu, "Investigation of fabricated CMOS FishboneFETs and TreeFETs with strained SiGe nano-fins on Bulk-Si substrate," IEEE Electron Device Letters, vol. 44, no. 9, pp. 1396-1399, Sep. 2023.

[7] "A comparative study on electrical characteristics of bulk, SOI, and DG MOSFET," in Lecture Notes in Electrical Engineering. Singapore: Springer, 2023, pp. 51-59.

[8] D. Dhiman and V. Kumar, "Performance analysis and optimization of 10nm SOI-FinFET using high-k dielectric materials," in 2023 International Conference on Advances in Power, Signal, and Information Technology (APSIT), Jun. 2023, pp. 1-4.

[9] H. H. Radamson, J. Zhu, L. Thylen, and M. Ostling, "CMOS scaling for the 5 nm node and beyond: Device, process and technology," Nanomaterials, vol. 14, no. 10, p. 837, May 2024.

[10] Z. Liu, Q. Wang, R. Chen, and P. Li, "Analysis of drain current variability components in extremely narrow GAA silicon nanowire MOSFETs of 4nm width," IEEE Journal of the Electron Devices Society, in press.

[11] Sentaurus Device User Guide, Synopsys, Inc., Mountain View, CA, USA, 2023.

[12] 2023 IRDS Update, IEEE, 2023. [Online]. Available: https://irds.ieee.org

[13] H. Wu, L. Zhang, Y. Chen, and W. Liu, "Parasitic resistance reduction strategies for advanced CMOS FinFETs beyond 7nm," in 2018 IEEE International Electron Devices Meeting (IEDM), Dec. 2018, pp. 35.4.1-35.4.4.

Improving EUV Patterning Fidelity and Aberration Control through Source-Mask Co-Optimization

Qi Wang* [1,2], Qiang Wu [1,2], Ying Li [1,2], Xianhe Liu [1,2], Yanli li* [1,2]

[1] School of Micro-Electronics, Fudan University, No. 825 Zhangheng Road, Shanghai 200433, PR China
[2] National Integrated Circuit Innovation Center, Shanghai 201203, China

* Email: wangqi_fd@fudan.edu.cn, li_yanli@fudan.edu.cn

Abstract—**Extreme Ultraviolet (EUV) lithography faces significant challenges in maintaining patterning fidelity and mitigating aberrations at advanced technology nodes. This paper presents a source-mask co-optimization (SMO) framework designed to improve resolution, contrast, and process robustness in EUV lithography. By simultaneously optimizing the illumination source and mask geometries, our approach maximizes pattern transfer accuracy while mitigating the impact of aberrations. We demonstrate through rigorous simulations that SMO significantly improves critical dimension control and expands the process window, including enhanced exposure latitude (EL) and reduced mask error factor (MEF). Furthermore, we analyze the interplay between source shaping and mask corrections under practical aberration conditions, providing key insights for achieving higher yield in sub-7nm node manufacturing.**

Keywords—Extreme Ultraviolet (EUV) lithography, Pattern Fidelity, Aberration Control, Source-Mask Co-Optimization (SMO) ,Lithography Imaging Model

I. INTRODUCTION

Extreme ultraviolet lithography (EUVL) has become the cornerstone of semiconductor manufacturing at advanced technology nodes, enabling the continued scaling of integrated circuits beyond the 7 nm regime [1,2]. However, as feature sizes approach fundamental physical limits, the patterning process faces escalating challenges in maintaining fidelity, particularly in critical layers such as the metal and cut layer at the Back-End-Of-the-Line (BEOL), where geometric complexity and sensitivity to aberrations converge [3,4]. These challenges manifest as degraded critical dimension uniformity (CDU), increased line edge roughness (LER), and constrained process windows - all of which directly impact device performance and yield [5].

Source-mask co-optimization (SMO) has emerged as a powerful computational lithography technique to address these limitations by simultaneously optimizing illumination sources and mask patterns [6,7]. While SMO has demonstrated success in improving exposure latitude (EL) and reducing mask error factor (MEF) for conventional 193 nm deep ultraviolet (DUV) layers [8], its application to the EUV layer presents unique challenges due to the layer's intricate pattern geometry and heightened sensitivity to wavefront aberrations [9]. Recent studies have highlighted the need for advanced optimization frameworks that can account for these nonlinear interactions between source, mask, and aberrations [10].

In this work, we present a comprehensive SMO methodology specifically designed to enhance EUVL patterning fidelity for logic device applications. Our approach systematically analyzes the complex interplay between source illumination characteristics, mask topology, and optical aberrations to develop robust solutions that improve CDU while maintaining manufacturability. Through rigorous computational modeling, we demonstrate significant improvements in both imaging performance and process window compared to conventional optimization techniques. These results provide critical insights into aberration-aware SMO strategies, offering a viable pathway for next-generation semiconductor manufacturing.

II. SIMULATIONS AND SETUPS

Hardware foundations of Source-Mask Co-Optimization (SMO) are two paired facet mirrors subsystems. Pupil facet mirrors provide real-time control of source intensity distribution, allowing SMO algorithms to explore non-intuitive illumination shapes like freeform pupils that improve image contrast by more than 10%. Field facet mirrors complement this by maintaining dose uniformity during scanning, essential for preserving SMO-predicted edge placement accuracy across full-chip exposures[11]. Recent implementations of these mirrors to adaptive SMO could dynamically adjust for both dense and isolated features within a single exposure source[12]. The illumination source was parameterized as a pixelated map (64×64 grid), with each pixel's intensity independently adjustable during optimization. The optical system was configured with a 0.33 numerical aperture (NA) EUV scanner, operating at a wavelength of 13.5 nm, to replicate current industry-standard lithography tools[13].

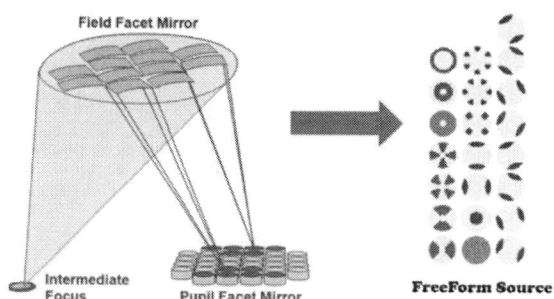

Fig.1. Schematic of field and pupil facet mirror systems enabling source-mask optimization (SMO).

Software basic of our source-mask co-optimization (SMO) framework includes a comprehensive simulation approach combining rigorous electromagnetic modeling with lithographic process emulation. We developed a high-fidelity EUV imaging simulator that incorporates all critical physical phenomena, including wavefront aberrations (represented by 49-term Zernike polynomial expansions),

979-8-3315-3918-4/25 $31.00 © 2025 IEEE

flare effects (5% DC flare baseline), and mask 3D topography. We employed a evaluation algorithm to explore the solution space efficiently, with cost functions weighted for CD, MEF, and EL. EUV-layer patterns optimization from the mask are incorporated both main features and sub-resolution assist features (SRAFs). The resist model was carefully calibrated to account for chemical amplification effects, with parameters including 50 nm thickness, complex refractive index (n = 1.0 + 0.025i), and 4 nm photo-acid diffusion length. Other input conditions include non-polarized illumination and 55 mJ/cm² exposure dose.

Then We present a detailed simulation methodology for evaluating Source-Mask Optimization (SMO) in EUV lithography, including the complete SMO workflow, critical optical and mask parameters, and quantitative analytical approaches for assessing pattern fidelity enhancement.

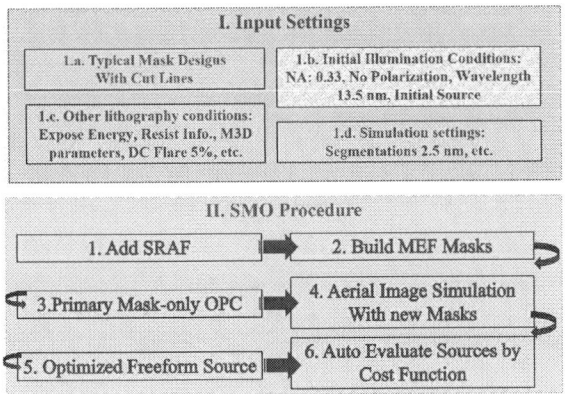

Fig.2. Step-by-step Flow of Source-Mask Co-Optimization.

This computational framework above combines several critical components for accurate Source-Mask Optimization[14]. First, the EUV source is modeled using a hybrid approach that integrates discrete point sources with Köhler illumination principles, capturing both spatial and angular radiation characteristics essential for EUV patterning. The mask representation employs rigorous 3D modeling to account for multilayer reflectivity, feature topography effects, and assist structures, validated across eight distinct pattern types. Optical propagation is simulated through combined rigorous coupled-wave analysis (RCWA) and finite-difference time-domain (FDTD) methods, accurately modeling diffraction effects while incorporating system aberrations up to Z49. Finally, the resist model incorporates stochastic effects including photoacid diffusion and development threshold variations, calibrated against experimental data from 18 nm-thick chemically amplified resists.

The effectiveness of the model is demonstrated through its ability to co-optimize source and mask parameters while maintaining computational cost. The source model achieves no intensity error compared to measured pupil images, while the 3D mask representation accounts for shadowing effects at 6° incidence angles. Optical simulations show agreement with aerial image measurements, and the resist model reproduces experimental critical dimensions within high accuracy. This integrated approach enables simultaneous optimization of illumination conditions (σ=0.2-0.9) and mask geometries while predicting pattern fidelity metrics (EL, MEF and CD variations) crucial for advanced node development.

The optimization process began with Quasar illumination modes as initial conditions, then iteratively refined the source configuration to maximize the process window. We selected representative EUV geometries (including both dense and isolated features) to capture real-world patterning challenges while maintaining computational efficiency. These test patterns provide a comprehensive yet tractable evaluation set, avoiding the prohibitive computational costs of full-chip simulation while preserving all essential lithographic characteristics for thorough parameter space exploration.

III. RESULTS AND ANALYSIS

The mask is represented with high fidelity, taking into account the detailed geometry of the patterns, including any assist features designed to enhance pattern fidelity. The simulation framework employs a typical representation of the mask design in EUV metal layer, accounting for material properties and mask three-dimensional effect. This study includes eight kinds of patterns at 2.1 nm technology nodes in Fig.3.

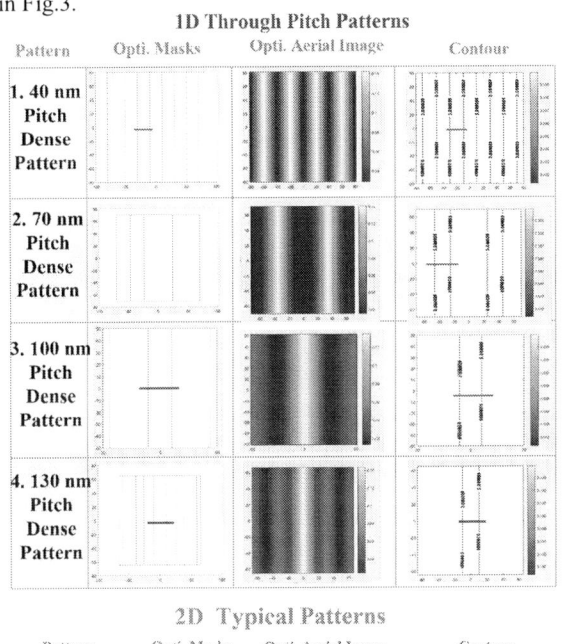

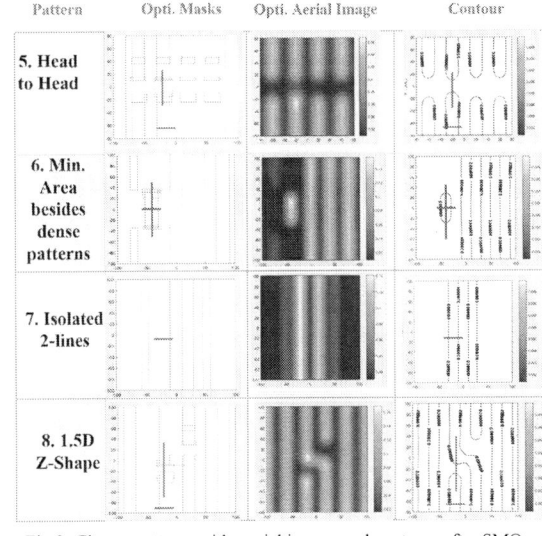

Fig.3. Chosen patterns with aerial images and contours after SMO.

Three key lithographic metrics were employed to evaluate the SMO-optimized process performance: critical dimension errors, exposure latitude (EL, defined as the dose range maintaining $\pm$ 10% CD variation), and mask error factor (MEF). These metrics were integrated into a multi-objective cost function that guided the co-optimization process, with weighting factors adjusted to prioritize pattern-specific requirements at the 2.1 nm node.

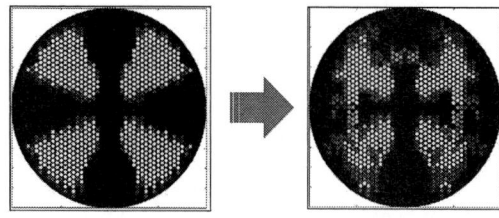

Fig.4. Source evolution with 3 times iterations..

The results demonstrate significant performance improvements achieved through joint source-mask optimization. As shown in Fig. 4, the expanded process window quantified through three key metrics validates the enhanced robustness of the optimized lithographic process. These metrics collectively verify the improved pattern fidelity and process stability attained through systematic co-optimization.

Critical Error (CD, nm)				
Mask	Initial		Optimized	
1.TP40	20.00		20.00	
2.TP70	20.08		20.01	
3.TP100	20.12		19.93	
4.TP130	20.84		20.01	
5.HtH	18.35	21.27	19.99	20.51
6.Min Area	20.94	57.40	20.39	55.71
7.isolated 2lines	19.35		19.68	
8.1.5D Zshape	18.82	19.05	20.00	19.75
Exposure Lattitude(EL)				
Mask	Initial		Optimized	
1.TP40	20.03%		20.33%	
2.TP70	17.81%		17.94%	
3.TP100	16.89%		18.41%	
4.TP130	19.04%		19.22%	
5.HtH	19.22%	13.63%	20.44%	15.11%
6.Min Area	21.05%	40.42%	21.25%	42.68%
7.isolated 2lines	18.92%		19.04%	
8.1.5D Zshape	17.99%	13.30%	19.76%	14.46%
Mask Error Factor(MEF)				
Mask	Initial		Optimized	
1.TP40	1.70		1.69	
2.TP70	1.76		1.73	
3.TP100	2.03		2.04	
4.TP130	2.29		2.24	
5.HtH	1.79	-3.90	1.76	-3.84
6.Min Area	1.55	2.85	1.65	2.72
7.isolated 2lines	1.71		1.78	
8.1.5D Zshape	1.79	-3.56	1.72	-3.51

Specification:
Critical Dimension (CD) Variation Tolerance : $\leq \pm 5\%$
Exposure Latitude (EL) Tolerance : 1D $\geq$18% , 2D $\geq$13%
Mask Error Effects (MEF) : 1D $\leq$ 2 , |2D| $\leq$ 4

Out of Spec.Degree		Optimization capability	
a little bit worse	worse	slightly better	significantly better

Fig.5. process window analysis with initial and final source.

The analysis demonstrates consistent improvements in critical dimension error, exposure latitude (EL), and mask error factor (MEF) across all eight test patterns when comparing initial and optimized illumination conditions. As shown in Fig. 5, the SMO-optimized source achieves overall enhancement in key metrics, confirming that source-mask co-optimization is essential for meeting stringent process window requirements in EUV patterning at advanced nodes.

To further evaluate aberration robustness, we incorporated Zernike coefficients into the optical model and iteratively refined the SMO process to minimize pattern sensitivity. This study specifically examines coma aberration (Z8) at 0.25 nm, 0.5 nm, and 1 nm RMS levels, representing typical wafer-level variations in 0.33 NA EUV systems.

Fig. 6 demonstrates that EUV SMO effectively compensates for coma aberrations (Z8) at lower magnitudes (<0.5 nm RMS), maintaining CD errors within production specifications ($\leq \pm 5\%$). However, for Z8 $\geq$ 0.5 nm RMS, the residual placement errors exceed the tolerance threshold (highlighted in red, Fig. 6), revealing a fundamental limitation in SMO's ability to correct larger wavefront distortions. These findings indicates the need for next-generation, aberration-aware SMO algorithms to address the challenges of sub-2 nm node EUV patterning.

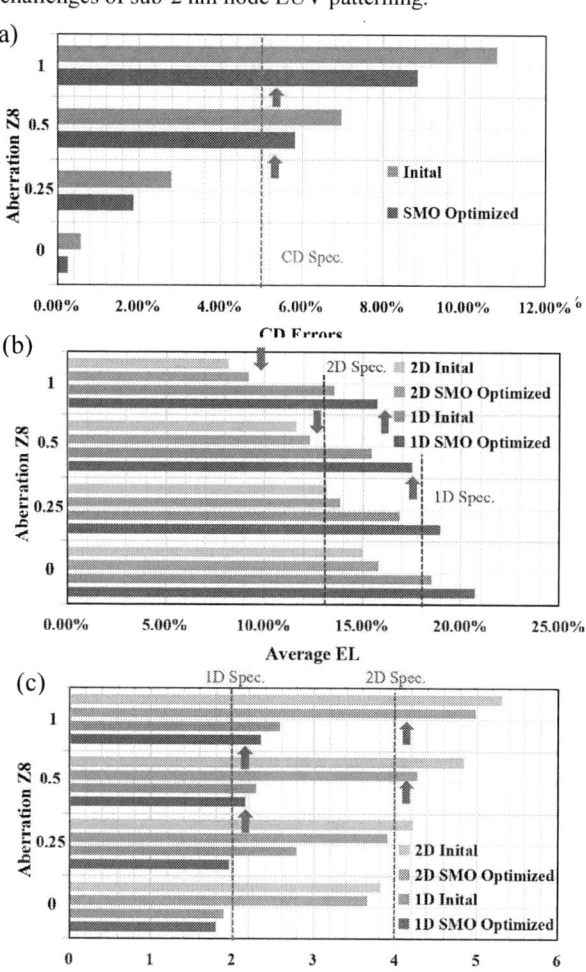

Fig.6. process window analysis with initial and final source. (a) CD errors.(b) exposure latitude. (c) Mask Error Factor.

Based on a series of simulations above, it reveals that Source-Mask Optimization (SMO) provides a powerful computational approach to mitigate lithographic performance degradation caused by optical aberrations in advanced nodes. Our systematic analysis demonstrates SMO's effectiveness in compensating for coma (Z8) aberrations while expanding the usable process window. Production-ready correction for SMO sufficiently addresses typical scanner-level aberrations(≤ 0.25 nm Z8 in EUV tools), while hardware limits requires complementary strategies (e.g., in-situ aberration monitoring) for out-of-spec conditions. Recent breakthroughs in computational lithography (e.g., NVIDIA cuLitho's GPU-accelerated inverse algorithms) suggest that next-gen DICO platforms will achieve faster convergence through machine learning-guided SMO, aberration-aware OPC with better correction resolution and 3σ CDU prediction accuracy via full-stack resist/etch modeling.

IV. CONCLUSION

This study demonstrates that Source-Mask Optimization (SMO) significantly enhances Extreme Ultraviolet Lithography (EUVL) performance by simultaneously improving critical metrics — reducing critical dimension errors, increasing exposure latitude, and lowering the mask error factor (MEF). Effectively compensating for coma aberrations (Z8) is up to 0.25 nm RMS in EUV layer patterning. Our comprehensive simulation framework reveals fundamental limitations in current SMO at higher aberration levels (≥ 0.5 nm RMS), establishing clear boundaries for current computational correction capabilities. Future work will expand SMO verification to full-chip intellectual property (IP) blocks under production conditions, where the emerging Design and Equipment, Material, and Process Co-Optimization (DICO) paradigm shows particular promise. The ongoing advancement of Design-Technology DICO approaches, combined with rigorous physical process modeling, will prove indispensable for addressing the escalating requirements of next-generation semiconductor fabrication.

ACKNOWLEDGMENT

We thank Dr. Jiang Yan from National Integrated Circuit Innovation Center for his support and discussion.

REFERENCES

[1] Zhang, Z., Li, S., Wang, X., & Cheng, W. Fast heuristic-based source mask optimization for EUV lithography using dual edge evolution and partial sampling. Optics Express, 29(14), 22778-22795. (2021).

[2] Hsu, S., Li, Z., Chen, L., Gronlund, K., Liu, H. Y., & Socha, R.(2009,December). Source-mask co-optimization: optimize design for imaging and impact of source complexity on lithography performance. Proc. SPIE In Lithography Asia (Vol. 7520, pp. 101-111). .(2009)

[3] S.Z. Zhang, S. Li, X. Wang, W. Cheng, Y. Qi, " Source mask optimization for extreme-ultraviolet lithography based on thick mask model and social learning particle swarm optimization algorithm," Optics Express Vol. 29, Issue 4, pp. 5448-5465 (2021)

[4] Ma, X., Wang, Z., Chen, X., Li, Y., & Arce, G. R. Gradient-based source mask optimization for extreme ultraviolet lithography. IEEE Transactions on Computational Imaging, 5(1), 120-135.(2018).

[5] Zhang, Z., Li, S., Wang, X., Cheng, W., & Qi, Y. Source mask optimization for extreme-ultraviolet lithography based on thick mask model and social learning particle swarm optimization algorithm. Optics Express, 29(4), 5448-5465.(2021).

[6] Armeanu, A., Philipsen, V., Jiang, F., Fenger, G., Lafferty, N., Gillijns, W. & Sturtevant, J.Enabling enhanced EUV lithographic performance using advanced SMO, OPC, and RET. Proc. SPIE In International Conference on Extreme Ultraviolet LithographyVol. 10809, pp. 85-93. .(2019).

[7] Chae, Y. J., Kim, M. W., Yu, D. K., Son, S. W., Yeung, M., & Oh, H. K. (2024, November). Improving process window and resolution through source polarization in High-NA EUV. Proc. SPIE. In Photomask Technology Vol. 13216 (2024)

[8] X. Liu, R. Howell, S. Hsu, K. Yang, K. Gronlund, F. Driessen, H.Y. Liu, S. Hansen, "EUV source-mask optimization for 7 nm node and beyond," Proceedings Volume 9048, Extreme Ultraviolet (EUV) Lithography V; 90480Q (2014)

[9] Zhang, S., Zhang, L., Gai, T., Gao, P., & Wei, Y. Aberration analysis and compensate method of fully connected neural network in deep ultraviolet lithography. Optical Engineering, 60(12), 123105-123105.(2021).

[10] Li, G., Zhang, C., & Zheng, Z. Computing method for extreme ultraviolet lithography imaging with aberration consideration. Proc. SPIE. In Optoelectronic Imaging and Multimedia Technology XI Vol. 13239, pp. 259-265 (2024).

[10] Li ZQ, Dong LS, Ma X et al. "Fast source mask co-optimization method for high-NA EUV lithography," Opto-Electron Adv 7, 230235 (2024).

[11] Y. Pan, X. Ma, S. Zhang, J. Garcia-Frias, G.R. Arce, " Efficient informatics-based source and mask optimization for optical lithography," Applied Optics Vol. 60, Issue 27, pp. 8307-8315 (2021)

[12] W. Gao, C.K. Chen, J. Zimmermann, "Computational evaluation of critical logic metal layers of pitch 20-24nm and aberration sensitivity in high NA EUV single patterning," Proceedings Volume 12495, DTCO and Computational Patterning II; 1249509 (2023)

[13] W Gao, B Zhu, TB Chiou, SE Tseng, W Lin, CK Chen, J Zimmermann, A Yen, "Computational lithographic study of 0.55 NA EUV single patterning for metal layers of the 2nm logic node and beyond," Proceedings Volume 12052, DTCO and Computational Patterning; 120520G (2022)

[14] D.S. Nam, J.H. Ser, N. Seong, X. Li, S. Hsu, A. Yen, "Mask and illumination optimization for low-k1 EUV lithography," Proceedings Volume 12325, Photomask Japan 2022: XXVIII Symposium on Photomask and Next-Generation Lithography Mask Technology; 1232502 (2022)

Enhancement of HfO₂-Based Ferroelectric Thin Film Performance via Interface and Defect Engineering

Xiao Yu [1,2]*, Peiyuan Du[1,2], Huan Liu[1,2], Dongya Li[1], Fei Yu[1,2], Bing Chen[1,2], Ran Cheng[3], Mengnan Ke[4], Yan Liu[2], and Genquan Han[1,2]

[1] Hangzhou Institute of Technology, Xidian University, Hangzhou 311231, China [2] School of Microelectronics, Xidian University, Xi'an 710071, China, [3]College of Integrated Circuits, Zhejiang University, Hangzhou 311200, China.
[4]Institute for Multidisciplinary Sciences, Yokohama National University, Yokohama 240-8501, Japan
* Email: yuxiao@xidian.edu.cn

Abstract—We have successfully demonstrated improved performance and reliability in HfO₂-based ferroelectric thin films through HfO₂/ZrO₂ superlattice structures, TiO₂ seed layers and helium ion (He^{2+}) implantation. The HfO₂/ZrO₂ superlattice ferroelectric thin film incorporating a 1-nm TiO₂ seed layer, exhibits remarkable remanent polarization (P_r) and a low coercive field (E_c), along with stable endurance exceeding 10^9 cycles. He^{2+}-implanted HZO capacitors also demonstrate reduced Ec, faster switching speeds, and enhanced endurance, featuring wake-up and fatigue-free characteristics. Furthermore, the He^{2+} implantation stabilizes oxygen vacancies (V_O), ensuring a uniform distribution of V_O and traps, and preventing further V_O generation during cycling. This study contributes valuable insights into the optimization of HfO₂-based ferroelectric films for non-volatile memory applications.

Keywords—Ferroelectric, Superlattice, TiO₂ seed layer, Reliability, He ion, Ion bombardment

I. INTRODUCTION

The discovery of HfO₂-based ferroelectric materials has garnered significant attention due to their high operating speed, low power consumption, excellent scalability, and CMOS process compatibility [1]. However, achieving optimal ferroelectric performance typically requires high-temperature annealing, which can degrade the quality of HfZrOₓ (HZO) films. Reliability remains a critical challenge for HfO₂-based devices, particularly in terms of limited endurance and undesirable cycling-induced effects. Defects, especially oxygen vacancies(V_O) play a crucial role in both ferroelectric behavior and device reliability [2]. To address these issues, various strategies have been proposed, including interface engineering, ferroelectric layer structure optimization, and doping engineering [3], [4], [5].

In this study, recent progress in performance and reliability enhancement through interface and defect engineering are discussed. It has been demonstrated that an HfO₂/ZrO₂ superlattice combined with TiO₂ seed layers can lower the crystallization thermal budget to below 400 °C, enabling robust endurance performance at Back-end-of-line (BEOL)-compatible temperatures. In addition, a uniform, CMOS-compatible helium ion (He^{2+}) implantation technique has been developed, effectively enhancing ferroelectricity by reducing the coercive field (E_c) and enabling faster switching speeds. The underlying mechanism for these improvements has been attributed to defect redistribution and modulation of the interfacial layer thickness.

II. INTERFACE ENGINEERING OF HfO₂/ZRO₂ SUPERLATTICE WITH TiO₂ SEED LAYER

Ferroelectric devices processed at temperatures below 400 °C are highly desirable for their BEOL compatibility in CMOS integration. However, the crystallization of solid-solution HZO typically requires annealing temperatures of ~500 °C. To overcome this limitation, we have demonstrated metal-ferroelectric-metal (MFM) capacitors incorporating HfO₂/ZrO₂ superlattice (SL_HZO) structures deposited by atomic layer deposition (ALD), with and without the insertion of a 1 nm TiO₂ seed layer between the bottom electrode (BE) and the ferroelectric film. Here, the 10-nm-thick superlattice consisted of eight periodic stacks of ~6.25Å HfO₂ and 6.25 Å ZrO₂, deposited at 250 ºC. Then the samples were treated by rapid thermal annealing (RTA) at 300 to 600 ºC for 30 s in a nitrogen atmosphere, respectively. As shown in Fig. 1, the key fabrication steps and cross-sectional high-resolution transmission electron microscopy (HRTEM) images of the TiO₂/SL_HZO sample clearly reveal the periodic superlattice configuration within the 10-nm stack, as well as the formation of a polycrystalline structure in the SL_HZO films [6].

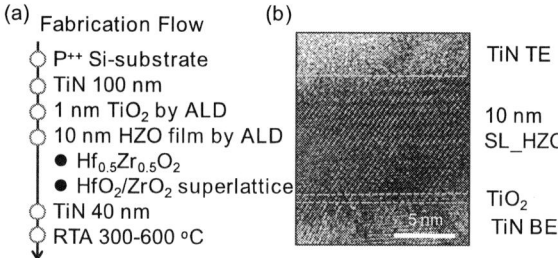

Fig. 1. (a) The device key fabrication process flow. (b) The HRTEM image of TiO₂/SL_HZO sample [6].

A. Electrical and Physical Properties of the Superlattice Ferroelectric Capacitor with TiO₂ Seed Layer

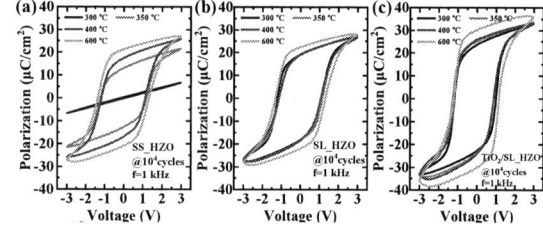

Fig. 2. The evolution of the *P-V* loops for (a) TiN/SS_HZO/TiN, (b) TiN/SL_HZO/TiN, and (c) TiN/TiO₂/SL_HZO/TiN with different annealing temperatures (300-600 ºC)[6].

Figs. 2(a)-(c) present the polarization-voltage (*P-V*) characteristics of three samples annealed at temperatures from 300 ºC to 600 ºC. For the HfO₂/ZrO₂ superlattice samples, an annealing temperature as low as 300 ºC is sufficient to activate ferroelectricity, whereas the device with the SS_HZO exhibits no ferroelectricity at this temperature. These results indicate that superlattice ferroelectric structures can achieve ferroelectric phase formation at 300 °C, making them highly suitable for the BEOL integration [7]. For all the samples, the remanent polarization (P_r) increases with annealing temperature, likely due to the higher fraction of ferroelectric phase formed at elevated temperatures, as previously reported [8], [9]. Across the entire RTA range, TiO$_2$/SL_HZO devices exhibit the highest P_r, indicating that the superlattice structure combined with a TiO$_2$ seed layer can deliver robust ferroelectricity even at low annealing temperatures. Moreover, the TiO$_2$-seeded samples consistently show a markedly lower E_c throughout the RTA range. This reduction in E_c is advantageous for low-voltage operation, which is particularly desirable in applications such as FeFETs and FeRAM.

To clarify the origin of the enhancement in ferroelectricity, Grazing incident X-ray diffractometry (GIXRD) was performed to analyze the crystal structure. Calculations based on the experimental data reveal that the TiO$_2$-seeded structure exhibits larger orthorhombic grains, which are favorable for polarization enhancement. Moreover, since the E_c is inversely proportional to the grain size [10], [11], the larger grains in TiO$_2$-seeded samples account for their consistently lower E_c. Furthermore, X-ray photoelectron spectroscopy (XPS) was performed (Fig. 3), as oxygen vacancies are a critical factor influencing the ferroelectricity of HZO films. The full width at half maximum (FWHM) values in Figs. 3(a) and (b) are identical. Deconvolution of the O 1s peaks reveals that the non-lattice oxygen content in TiO$_2$-seeded samples decreases from 12.50% to 8.65%, indicating a reduced oxygen vacancy concentration. In addition, XPS spectra of Hf 4f and Zr 3d (Figs. 3(c) and (d)) show more pronounced sub-oxidation states in the SL_HZO sample compared to the TiO$_2$/SL_HZO sample, consistent with its higher oxygen vacancy concentration.

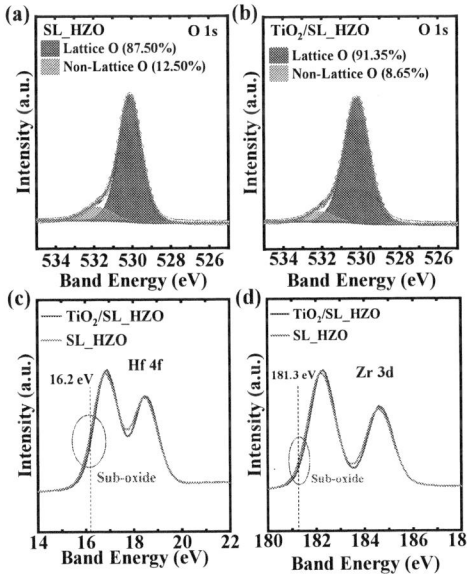

Fig. 3. O 1s of XPS spectra for (a) SL_HZO sample and (b) TiO$_2$/SL_HZO sample. (c)-(d) The Hf4f and Zr3d XPS spectra of the SL_HZO and TiO$_2$/SL_HZO samples[6].

B. Reliability Characteristics and Benchmarking

To evaluate the practical reliability of the fabricated HfO$_2$/ZrO$_2$ superlattice ferroelectric capacitors, endurance and retention tests were conducted under different electric fields. As shown in Fig. 4(a), endurance testing of SL_HZO

and TiO$_2$/SL_HZO samples annealed at 350 °C was performed using 1 MHz bipolar switching pulses. Both devices exhibited outstanding endurance up to 10^9 cycles, maintaining a high $2P_r$ of 28.3 μC/cm^2 with only 26.4% degradation, confirming robust reliability. The retention characteristics, shown in Fig. 4(b), demonstrate stable polarization up to 10^4 s, with over 10-year retention projected by linear extrapolation. Benefited by the reduced oxygen vacancies, TiO$_2$/SL_HZO device shows better retention properties.

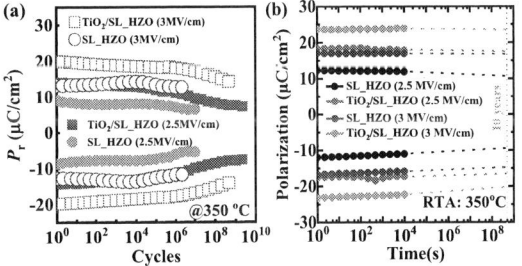

Fig. 4. (a) Endurance and (b) retention characteristic of SL_HZO and TiO$_2$/SL_HZO devices at 350 °C annealing temperature [6].

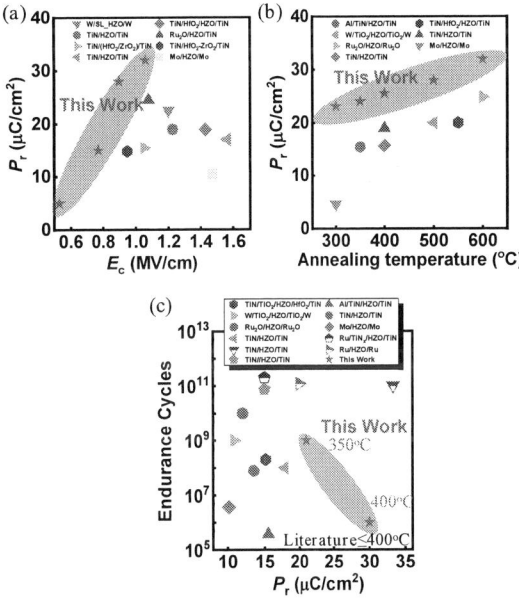

Fig. 5. (a) Benchmark of P_r versus E_c shows that TiO$_2$/SL_HZO exhibits a low E_c and a larger P_r. (b) a summary of the optimal P_r exhibited at different annealing temperatures. (c) compared to the best endurance verse P_r results in previous work, the TiO$_2$/SL_HZO device demonstrates superior endurance characteristics [6].

To benchmark the ferroelectric performance of the fabricated HZO-based thin-film devices against prior studies, Pr verse Ec are employed as key figures of merit, as depicted in Fig. 5(a). The results indicate that the proposed devices exhibit competitive ferroelectric properties, achieving a higher Pr while maintaining a lower Ec. Notably, the proposed device annealed at a BEOL-compatible temperature below 400 °C demonstrates a significantly higher Pr compared with previously reported low-temperature counterparts. Fig. 5(c) further summarizes the comparison of Pr and endurance cycles with recent reports for devices processed at ≤400 °C. The incorporation of TiO2 seed layers proves to be an effective and versatile strategy to enhance polarization switching, endurance, and overall device performance, especially in ultra-thin ferroelectric films. This approach suppresses oxygen vacancy formation and promotes improved crystallization, making it broadly applicable to various oxide-based ferroelectric systems and device architectures[12], [13].

III. DEFECT ENGINEERING OF HZO FERROELECTRIC FILMS BY HE ION IMPLANTATION

Helium ion implantation can introduce intrinsic point defects that suppress leakage current, delay polarization saturation under low electric fields, enhance polarizability under high electric fields, and improve breakdown strength [4]. Here, a uniform CMOS-compatible He^{2+} implantation process was employed. Following HZO deposition by ALD, He^{2+} implantation was performed using a commercial ion implantation system at doses of 0 (w/o He^{2+}), 10^{11} cm^{-2} and 10^{12} cm^{-2}, respectively [14]. The HRTEM images of the He^{2+}-implanted MFM structures (Fig. 6) reveal that, with increasing implantation dose, the crystalline phase distribution of the polycrystalline HZO becomes more uniform, accompanied by grain size enlargement. Post-implantation, most He ions are retained within the bottom electrode and silicon substrate [5].

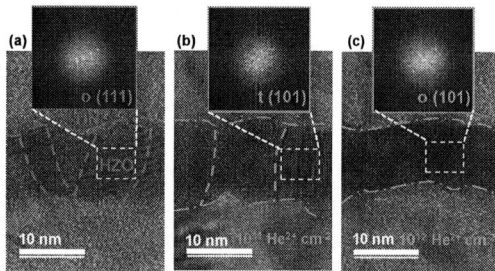

Fig. 6. The HRTEM results of (a) w/o He^{2+} implanted, (b) 10^{11} He^{2+} cm^{-2} and (c) 10^{12} He^{2+} cm^{-2} device. The crystalline phase distribution of the polycrystalline HZO becomes more uniform after implanted [14].

A. Electrical Properties and Switching Dynamics of He^{2+} implanted Ferroelectric Capacitors

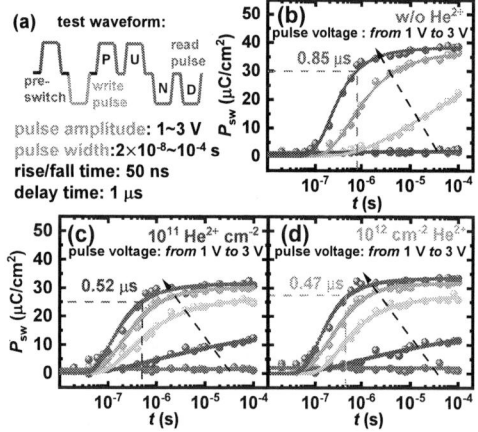

Fig. 7. (a)The switching polarization test waveform and the test results of (b) w/o He^{2+} implanted, (c) 10^{11} He^{2+} cm^{-2} and (d) 10^{12} He^{2+} cm^{-2} device. He^{2+} dose enhance the switching speed of the devices [14].

To evaluate the electrical properties and switching speed of the HZO devices, P-V characteristics and the relationship between switching polarization (P_{sw}) and pulse width were measured (Fig. 7). He^{2+} implantation does not degrade the

breakdown field of the MFM capacitors, and an optimal implantation dose exists that effectively reduces E_c while minimizing P_r degradation. The extracted switching times were 0.85 μs, 0.52 μs, and 0.47 μs for the He^{2+} doses of 0, 10^{11} cm^{-2}, and 10^{12} cm^{-2}, respectively, indicates that increasing the He^{2+} dose enhance the device switching speed. This improvement can be attributed to the higher orthorhombic-phase fraction and larger grain size induced by implantation. Grain size variations within the thin film significantly influence the local electric field distribution, as differences in grain size lead to fluctuations in local field direction, thereby affecting the polarization switching time.[15], [16].

B. Reliability improvement and the mechanism of He^{2+} implanted Ferroelectric Capacitors

The endurance characteristics of the devices were assessed under identical program/erase (P/E) cycling conditions, employing an electric field of 3 MV/cm and a cycling frequency of 1 MHz, as illustrated in Fig. 8(a). As indicated by the results in Fig. 7, the devices are capable of operating at 1 MHz, achieving approximately 80% polarization switching within a pulse width of 0.5 μs. The He^{2+}-implanted devices exhibited markedly enhanced endurance, sustaining up to 10^{10} cycles without any observable wake-up effect and showing negligible fatigue prior to 10^8 cycles. In the absence of environmental perturbations, the post-implantation endurance is projected to exceed 10^{11} cycles. Regarding retention performance, the He^{2+}-implanted devices preserved P_r without measurable degradation for durations over 10^4 s. These results underscore the efficacy of He^{2+} implantation in substantially enhancing the reliability of ferroelectric HZO films.

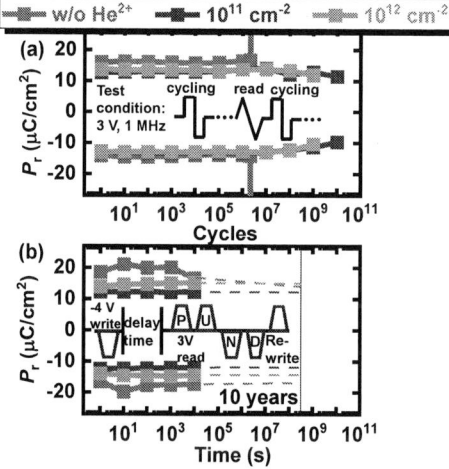

Fig. 8. (a) Endurance and (b) retention characteristics of all devices. The endurance has been improved to 10^{10} cycles after He^{2+} implanted without fatigue effect before 10^8 cycles [14].

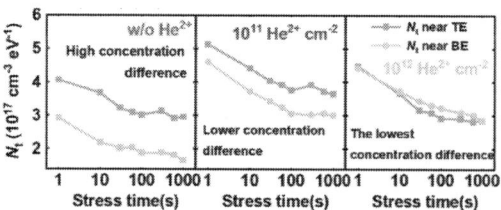

Fig. 9. The extract N_t results from C-f measurement of w/o He^{2+} implanted, 10^{11} He^{2+} cm^{-2} and 10^{12} He^{2+} cm^{-2} device [14].

In addition, an in-depth analysis of interfacial layer defects was performed using a multi-V C-f technique to evaluate the bulk trap density, as depicted in Fig. 9. The extracted trap density (N_t) values were associated with stress-induced trap generation. As the He^{2+} implantation dose increased, the defect distribution became more uniform, verifying that He^{2+} implantation leads to a homogeneous defect distribution. These results highlight the efficacy of He^{2+} implantation in mitigating defect-related reliability issues. Furthermore, we propose a mechanism explaining how He^{2+} implantation improves the ferroelectric properties of the HZO film, as depicted in Fig. 10. The He^{2+} implantation process impacts two pivotal factors: the thickness of the interfacial layer and the redistribution of internal defects, both of which significantly improve the film's endurance. Despite the fact that the sample thickness remains consistent due to the uniform application of the ALD process, we posit that the reduction in E_c is primarily due to the stress associated with the TiO_2 interfacial layer and its superior interfacial characteristics. This reduction in E_c is attributed to the increased thickness of the TiO_2 layer, which promotes the formation of the o-phase, diminishes oxygen vacancies, alleviates stress and interface defects, and optimizes charge distribution. [17], [18].

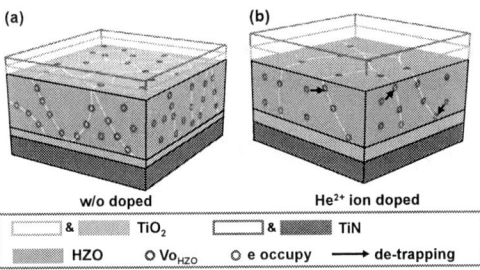

Fig. 10. The proposed model of He^{2+} implantation that involves FE polarization and interfacial trap[14].

IV. CONCLUSION

In conclusion, we have demonstrated performance enhancement in HfO_2-based ferroelectric devices through interface and defect engineering. The incorporation of TiO_2 facilitates lower crystalline annealing temperatures, higher polarization, reduced E_c, and faster switching speeds in superlattice HZO structures, which are attributed to larger grain sizes and diminished oxygen vacancies. Additionally, He ion-doped HZO capacitors exhibit marked improvements, including decreased E_c, accelerated switching speeds, and enhanced endurance, effectively suppressing wake-up and fatigue effects. The enhancement mechanism involves the introduction of uniform point defects via He ion implantation, which promotes the formation of large orthorhombic phase grains and traps charged defects. These findings underscore the promising potential for optimizing HfO_2-based devices through targeted interface and defect engineering strategies.

ACKNOWLEDGMENT

The authors acknowledge support from the National Key Research and Development Project (Grant No. 2023YFB4402303), the Fundamental Research Funds for the Central Universities (Grant Nos. YJSJ25013 and ZYTS25038) the National Natural Science Foundation of China (Grant No. 62374151, 62204226, 62025402, 62090033, 92464205, 62174146), Major Program of Zhejiang Natural Science Foundation (Grant No. DT23F0402). Zhejiang Province Key R&D programs (No. 2025C01193).

REFERENCES

[1] J. Muller *et al.*, "Ferroelectric hafnium oxide: A CMOS-compatible and highly scalable approach to future ferroelectric memories," in *2013 IEEE International Electron Devices Meeting*, Washington, DC, USA: IEEE, Dec. 2013, p. 10.8.1-10.8.4. doi: 10.1109/IEDM.2013.6724605.

[2] J. Zhou, Y. Guan, M. Meng, P. Hong, S. Ning, and F. Luo, "Improving the endurance for ferroelectric $Hf_{0.5}Zr_{0.5}O_2$ thin films by interface and defect engineering," *Applied Physics Letters*, vol. 124, no. 9, p. 092904, Feb. 2024, doi: 10.1063/5.0194207.

[3] S. Saremi *et al.*, "Electronic Transport and Ferroelectric Switching in Ion-Bombarded, Defect-Engineered $BiFeO_3$ Thin Films," *Adv Materials Inter*, vol. 5, no. 3, p. 1700991, Feb. 2018, doi: 10.1002/admi.201700991.

[4] J. Kim *et al.*, "Ultrahigh capacitive energy density in ion-bombarded relaxor ferroelectric films," *Science*, vol. 369, no. 6499, pp. 81–84, Jul. 2020, doi: 10.1126/science.abb0631.

[5] S. Kang *et al.*, "Highly enhanced ferroelectricity in HfO_2-based ferroelectric thin film by light ion bombardment," *Science*, vol. 376, no. 6594, pp. 731–738, May 2022, doi: 10.1126/science.abk3195.

[6] H. Liu *et al.*, "Back-End-of-Line Compatible HfO_2/ZrO_2 Superlattice Ferroelectric Capacitor With TiO2 Seed Layer for Enhanced Ferroelectricity," *IEEE Transactions on Electron Devices*, vol. 72, no. 2, pp. 665–670, Feb. 2025, doi: 10.1109/TED.2024.3524953.

[7] J. Hur, Y.-C. Luo, N. Tasneem, A. I. Khan, and S. Yu, "Ferroelectric Hafnium Zirconium Oxide Compatible With Back-End-of-Line Process," *IEEE Trans. Electron Devices*, vol. 68, no. 7, pp. 3176–3180, Jul. 2021, doi: 10.1109/TED.2021.3072610.

[8] V. Gaddam, D. Das, T. Jung, and S. Jeon, "Ferroelectricity Enhancement in $Hf_{0.5}Zr_{0.5}O_2$ Based Tri-Layer Capacitors at Low-Temperature (350 °C) Annealing Process," *IEEE Electron Device Letters*, vol. 42, no. 6, pp. 812–815, Jun. 2021, doi: 10.1109/LED.2021.3075082.

[9] D. Das, V. Gaddam, and S. Jeon, "Ferroelectricity in $Al_2O_3/Hf_{0.5}Zr_{0.5}O_2$ Bilayer Stack: Role of Dielectric Layer Thickness and Annealing Temperature," *Journal of Semiconductor Technology and Science*, vol. 21, no. 1, pp. 62–67, Feb. 2021, doi: 10.5573/JSTS.2021.21.1.062.

[10] T. Mimura, T. Shimizu, and H. Funakubo, "Ferroelectricity in $YO_{1.5}$-HfO_2 films around 1 μm in thickness," *Appl. Phys. Lett.*, vol. 115, no. 3, Jul. 2019, Art. no. 032901, doi: 10.1063/1.5097880.

[11] M. Materano, P. Lomenzo, M. Mulaosmanovic, M. Hoffmann, A. Toriumi, T. Mikolajick, and U. Schoroeder, "Polarization switching in thin doped HfO_2 ferroelectric layers," *Appl. Phys. Lett.*, vol. 117, no. 26, Dec. 2020, Art. no. 262904, doi: 10.1063/5.0035100.

[12] Y. Lin *et al.*, "Improving Edge Dead Domain and Endurance in Scaled $HfZrO_x$ FeRAM," in *IEDM Technical Digest*, Dec. 2021, p. 6.4.1-6.4.4. doi: 10.1109/IEDM19574.2021.9720692.

[13] S. Guo *et al.*, "Low Operation Voltage, High-Temperature Reliable, and High-Yield BEOL Integrated $Hf_{0.5}Zr_{0.5}O_2$ Ferroelectric Memory Arrays," *IEEE Transactions on Electron Devices*, vol. 71, no. 6, pp. 3645–3650, Jun. 2024, doi: 10.1109/TED.2024.3394460.

[14] P. Du *et al.*, "A Comprehensive Study on the Polarization, Reliability, and Switching Dynamics of Helium Ion-Bombardment $HfZrO_x$ Ferroelectric Films," *IEEE Transactions on Electron Devices*, vol. 72, no.7, pp.3521-3527, Jul. 2025, doi: 10.1109/TED.2025.3566357.

[15] K. Li *et al.*, "A Comparative Study on the Polarization, Reliability, and Switching Dynamics of HfO_2-ZrO_2-HfO_2 and ZrO_2-HfO_2-ZrO_2 Superlattice Ferroelectric Films," *IEEE Trans. Electron Devices*, vol. 70, no. 4, pp. 1802–1807, Apr. 2023, doi: 10.1109/TED.2023.3248538.

[16] Y. Li *et al.*, "Switching dynamics of ferroelectric HfO_2-ZrO_2 with various ZrO2 contents," *Applied Physics Letters*, vol. 114, no. 14, p. 142902, Apr. 2019, doi: 10.1063/1.5093793.

[17] H. Joh, T. Jung, and S. Jeon, "Stress Engineering as a Strategy to Achieve High Ferroelectricity in Thick Hafnia Using Interlayer," *IEEE Trans. Electron Devices*, vol. 68, no. 5, pp. 2538–2542, May 2021, doi: 10.1109/TED.2021.3068246.

[18] X. Wang *et al.*, "Oxygen Vacancy Modulation With TiO2 Stack Interface Engineering for Ferroelectric $Hf_{0.5}Zr_{0.5}O_2$ Thin Films," *IEEE Electron Device Lett.*, vol. 45, no. 1, pp. 100–103, Jan. 2024, doi: 10.1109/LED.2023.3330784.

Threshold Voltage Swing Caused by Intense Phonon-Electron Interaction in High-k Dielectrics

Jinchen Wei[*1], Mansun Chan[1]

[1]Department of Electronic and Computer Engineering, The Hong Kong University of Science and Technology, Clear water bay, Kowloon, Hong Kong

* Email: jweiba@connect.ust.hk

Abstract—The dynamic charge trapping/releasing behavior of oxygen vacancies (V_O) in the high-k dielectric layer of advanced multi-gate devices is investigated through a combination of first-principles calculations, non-radiative multi-phonon (NMP) model and TCAD simulations. Under the effect of phonon-electron interactions, the extremely large electron captures cross-sections (ECCS) for V_O defects illustrates easy trapping/releasing of electrons, resulting in uncontrollable charge quantity in the dielectric layer and significant leakage current. Consequently, significant V_{th} swing of the operated device can be expected due to the electron capture and emission, thus degrading the reliability. These findings elucidate the fundamental mechanisms behind reliability degradation in scaled devices and establish a defect optimization foundation to optimize fabrication processes in future technology nodes.

Keywords— high-k dielectric layer, V_O defects, phonon-electron interactions, reliability, V_{th} swing

I. INTRODUCTION

With the continuous scaling of integrated circuits, the reliability of transistor in post-Moore period is significantly impacted by the stability of gate controllability through the dielectric [1-3]. One of the most competitive gate dielectric materials is the amorphous HfO_2 with high dielectric constant (18~22) and remarkable Si-compatibility [4-7]. However, the further scaling of HfO_2 gate dielectric also brings serious challenges to the reliability of transistors. It has been reported that the formation of intrinsic point defects in especially for FinFET and gate-all-around field effect transistor (GAAFET) with multiple channel surfaces wrapped by HfO_2 [8-11]. Previous studies have demonstrated that oxygen vacancies (V_O) are the dominant defect in monoclinic and amorphous HfO_2 (a-HfO_2) under O-poor growth conditions [12, 13]. High density of V_O will act as carrier traps to capture and localize electrons and holes from the conductive channel or electrode [14, 15], thus significantly affecting the reliability of devices. Nevertheless, it remains unclear about how the dominant V_O defects electrically affect the devices from a dynamic perspective. For instance, the process of electron and hole capture by the defects in a-HfO_2 can significantly impact the performance on the devices, while the non-radiative multiphonon process of the carrier capture/emission is still not as clear as the cases in Si bulk and Si/SiO_2 interface [16, 17]. Therefore, it is highly demanded to unveil the electrical properties of V_O defect together with its carrier-trapping process in the high-k dielectric film to overcome the reliability challenges of scaled Hf-based devices.

Herein, we investigate the underlying physical mechanisms on reliability influenced by V_O defects in a-HfO_2 gate dielectrics with an n-type GAAFET as a typical example. The calculated defect formation energies and transition energy levels demonstrate the possibility of high density positively-charged V_O defects in a-HfO_2 dielectric layer, which can cause significant leftward shift of threshold voltage (V_{th}) of GAA according to 3D TCAD simulations. The non-radiative multi-phonon (NMP) model shows that V_O defects have large electron capture cross-sections (ECCS) and thus can capture electrons quickly, then the V_O defects become neutral and V_{th} shifts rightward. Therefore, V_O centers keep capturing and releasing electrons, which cause the swinging of V_{th} and also increase the leakage current, harming the reliability of GAAFET. Our results provide a brand-new understanding of trapping mechanisms of electrons from the conduction band of the dielectric layer, which should be considered in the future studies for reliability improvement.

II. COMPUTATIONAL DETAILS

The influence of V_O defect is investigated in the 1-nm thick a-HfO_2 layer in the model of GAAFET, as shown in Fig 1(a). The amorphous HfO_2 structures are generated from a 96-atom supercell with monoclinic structure through melt-and-quench process using the LAMMPS code [18] with MBKS potentials[19], as shown in Fig. 1(b). The formation energies and transition energy level of all the V_O defects were calculated with the formula [20, 21]:

$$\Delta H_f(q, E_F) = E(q) - E_h + E_O + \mu_O + q(E_F + \Delta V) \quad (1)$$

where q is the charge state of defect, $E(q)$ and E_h illustrate the total energy of the defected and defect-free structures, respectively. E_O and μ_O illustrate the elemental phase and chemical potential of O atom. E_F is the Fermi level referenced to the valence band maximum (VBM) and ΔV includes the potential alignment and finite supercell size correction.

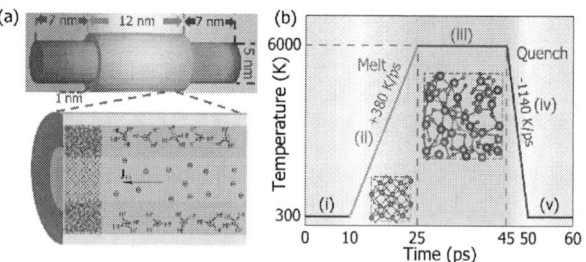

Fig. 1. The model and structure used in the simulation. (a) The model of 3D GAAFET with its schematic section diagram. (b) The melt-and-quench generation process of amorphous HfO_2 using the optimized MBKS potential.

The transition rate was calculated within the NMP model via the Huang's formula [22]:

$$k_{j \to i} = \frac{1}{\hbar} \left(-\frac{\pi k T'}{S \hbar w_s} \right)^{\frac{1}{2}} \left(\sum_l \langle i | u_l | j \rangle^2 \left(\frac{1}{w_l^2} \right) \right) e^{-\frac{\left(W_{ji} - S \hbar w_s \right)^2}{4 k T' S \hbar w_s}} \quad (2)$$

where $\langle i|u_l|j\rangle^2$ is the electron-phonon coupling matrix element and w_l is the frequency of the phonon l. W_{ji} is static energy difference between initial state j and final state i, and $S\bar{h}w_s$ is the relaxation energy. T' is the working temperature. The cross section σ can be derived from the transition rate $k_{j\rightarrow i}$, following

$$\sigma = \frac{S(T')\times V\times k_{j\rightarrow i}}{v_t} \quad (3)$$

where $S(T')$ is the Sommerfeld factor, V is the volume of a-HfO$_2$ supercell and v_t is the thermal velocity of carriers.

The first-principles calculations were performed based on the density functional theory as implemented in the VASP package with 520 eV energy cutoff of plane-wave basis [23]. The HSE06 hybrid functional [24] was adopted for its accuracy of calculating band gap and defect levels. The effects of the electron capture/emission process by V$_O$ defects on the reliability of Hf-based GAAFET were evaluated by the 3D TCAD simulation. The parameters of the device are shown in Fig. 1(a) and both source and drain are n-type doped silicon with the concentration of 10^{19} cm^{-3}.

III. RESULTS AND DISCUSSION

In a-HfO$_2$, all O sites are non-equivalent, so O vacancies on these sites are different. The coordination number (CN) and radial distribution function (RDF) of the simulated a-HfO$_2$ are shown in Fig. 2(a) and Fig. 2(b), respectively. Due to the non-equivalence of these V$_O$ defects, we calculated the formation energies of all the 64 V$_O$ configurations in the 96-atom a-HfO$_2$ model, and listed the formation energies of the neutral states under O-poor condition (low O chemical potential) in Fig. 2(c). The values range from 0.50 to 1.88 eV. Since lower formation energy indicates higher density, we select V$_{O3-c}$ (three coordinate O vacancy) and V$_{O4-c}$ with the lowest formation energy as prototypes and plots their formation energy changed with Fermi level under O-poor growth condition in Fig. 2(d).

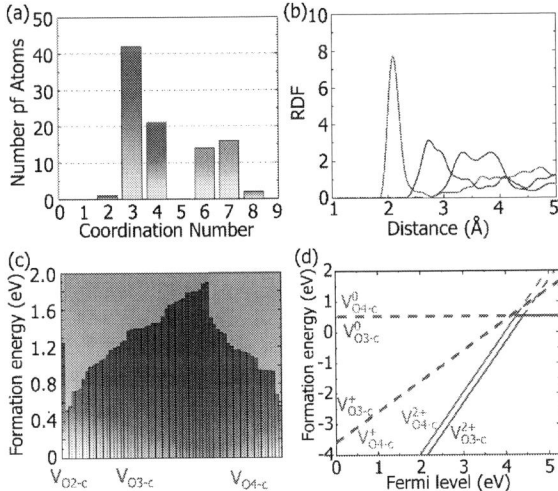

Fig. 2 The non-equivalent O site and O vacancies in a-HfO$_2$. (a)The coordination number distribution in a-HfO$_2$. The O and Hf are marked in red and blue, respectively. (b) the RDF of amorphous HfO$_2$, (c) formation energy distribution of 64 neutral V$_O$ defects in a-HfO$_2$ under O-poor condition, (d) the calculated formation energy changed with Fermi level of V$_{O3-c}$ and V$_{O4-c}$ in a-HfO$_2$ under O-poor condition.

The transition energy levels are derived from Fig. 2(d) and plotted in Fig. 3(a). The neutral V$_O$ in a-HfO$_2$ can have very low formation energy under O-poor condition (0.50 eV), much lower than that in monoclinic (m-) HfO$_2$ (1.07 eV) [13].

Accordingly, the density of neutral V$_O$ in a-HfO$_2$ (~10^{19} cm^{-3} for 800 K growth temperature) is much higher than that in m-HfO$_2$ (~10^{17} cm^{-3}, estimated from the formation energy in Ref. [13]) under O-poor condition.

As shown in Fig. 3(a), the (2+/0) transition energy levels of both types of V$_O$ defects are much higher than the Fermi level which generally locates near the middle of the bandgap of a-HfO$_2$. Therefore, the V$_O$ centers usually exist in the form of the positively-charged states and the density of ionized V$_O{}^{2+}$ defects can be higher than the neutral one due to the lower formation energy. The massively formed V$_O{}^{2+}$ defects in the dielectric layer can contribute external electric field and thus significantly influence the capability of gate voltage in the control of channel, thus reducing the reliability of devices. For example, the V_{th} of n-type MOSFET will be left-skewed compared to the defect-free case because of the V$_O{}^{2+}$ defects.

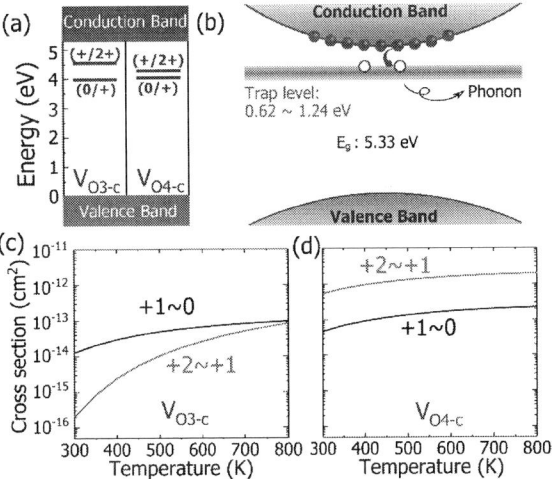

Fig. 3 The non-radiative multi-phonon model and calculated electron capture cross sections for a-HfO$_2$. (a) The transition energy level of V$_{O3-c}$ and V$_{O4-c}$ in a-HfO$_2$, (b) the non-radiative multi-phonon model with V$_O$ defects for a-HfO$_2$, and the state-to-state electron capture cross sections of (c) V$_{O3-c}$ and (d) V$_{O4-c}$ in a-HfO$_2$ at temperature from 300 K to 800 K.

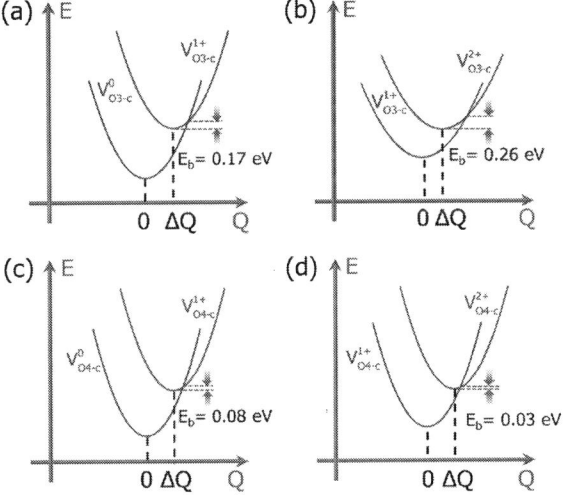

Fig. 4 The configuration coordinate diagrams of +2 to +1 and +1 to 0 non-radiative state transition for V$_{O3-c}$ and V$_{O4-c}$ under the harmonic approximation. The low energy barrier which indicates the large cross section for electrons are marked according to the crossing point.

Since the transition energy levels of V_O defects are quite close to the conduction band minimum (CBM), it is highly possible that the electrons can be quickly captured by the positively-charged V_O centers and thus neutralize the positive charged state if they are tunneled onto the conduction band of a-HfO$_2$ during the operation of the device. Therefore, we simulate the electron trapping process via the NMP model (seen in Fig. 3(b)) and calculate the electron-carrier capture cross section in the 300-800 K temperature range, which is shown in Fig. 3(c) and 3(d). The magnitude of ECCS for V_O defects ranges from 10^{-16} cm^2 to 10^{-11} cm^2. Considering that the hole capture cross section in V_O defect Si/SiO$_2$ interface (lower than 10^{-17} cm^2 at 300 K [16]) is much lower than the ECCS of V_O defects in a-HfO$_2$, it is quite easy for positively-charged V_O defects to capture electrons, especially for V_{O4-c} defects with ECCS larger than 10^{-14} cm^2. The large ECCS of V_O defects are attributed to very small calculated transition barriers, which determines the rate of NMP transition process [25]. For V_{O3-c} defects, the transition barriers from +2 to +1 state and from +1 to 0 state are 0.17 eV and 0.26 eV, respectively, while that from +2 to +1 state and from +1 to 0 state are 0.08 eV and 0.03 eV for V_{O4-c} defects. The transition barriers of V_{O4-c} defects is significantly lower than that of V_{O3-c} defects, since V_{O4-c} defects have a much larger electron-phonon coupling effect.

With high density of V_O^{2+}, the a-HfO$_2$ dielectric layer can be viewed as a large electron tank and the quantity of captured electrons can electrically influence the performance and reliability of devices. Assuming the density of V_O^{2+} in a-HfO$_2$ is 10^{20} cm^{-3}, the equivalent positive fixed charge in the layer can vary from 0 to 2×10^{20} cm^{-3}. Considering all the V_O are ionized and thus positively-charged, the threshold voltage V_{th} drops almost by half (from 0.31 V to 0.17 V) compared with the case when each defects captures two electrons. As a result, there is a significant uncertainty window for V_{th} when the device is under operation. Meanwhile, the leakage current I_{off} increases by more than two orders of magnitude with increasing density of V_O defects which contribute to the electron tunneling.

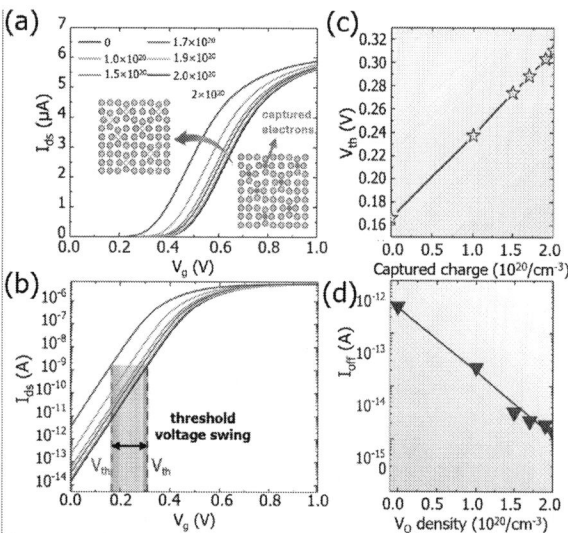

Fig. 5 I_{ds}-V_{gs} characteristic curve of GAAFET under (a) linear coordinate and (b) logarithmic coordinates with different captured electron density distinguished by different colors. The area of V_{th} with varied captured electron density is displayed in (c), while (d) shows the trends of I_{off} influenced by the implemented V_O.

Synthesizing the results, the physical mechanism of the swinging V_{th} of the devices can be established. Based on the fact that (i) most V_O defects are in +2 state because of the high defect level, (ii) V_O^{2+} defects have lower formation energy and thus higher density than other state and (iii) ECCS of V_{O4-c} from +2 to +1 state is much larger, we assume that all the V_O defects in a-HfO$_2$ are completely ionized and then schematically plot the circling processes in Fig 6, describing the capturing/releasing mechanisms by V_O^{2+} centers in a-HfO$_2$ dielectric layer.

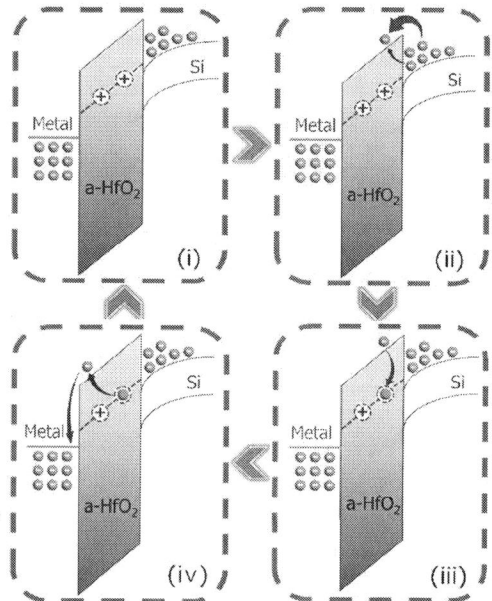

Fig. 6 The mechanism of the swinging threshold voltage caused by V_O defects in a-HfO$_2$.

Before capturing electrons, all the V_O defects are positively-charged since the Fermi level is much lower than the defect level, as shown in Fig. 6(i), and the V_{th} states leftwards because of the charged defects. When GAAFET operates under large gate voltages, the channel electrons may leak into the a-HfO$_2$ gate dielectric layer under strong electrical field, as displayed in Fig. 6(ii). Since V_O^{2+} defects have large ECCS, they can capture the leaked electrons and thus become V_O^+ defects, as shown in Fig. 6(iii). As a result, the leftward shifted I_{ds}-V_{gs} curve will rebound to right, i. e., V_{th} increases. However, since the competitive relationship of ECCS of +1 state to 0 state and ECCS of +2 state to +1 state can be inverse for different type of V_O defects (Fig. 3c-d), the electron emission rate and second electron capture rate are comparable for V_O^+ defects. Therefore, it is expected the V_O^+ center will either capture and become neutral, leading to increased V_{th}; or release the electron and become +2 charged (Fig. 6(iv)), causing further drop of V_{th} (i.e., shift leftward). On the other hand, due to the electric field from the gate, electrons are continuously captured and released by V_O defects while hopping towards and finally injecting into the metal layer.

In this manner, the V_{th} of device can swing back and forth because of the electron capturing and releasing from V_O center in the dielectric layer accompanying with higher leakage current and thus degrading the reliability of GAAFET. Therefore, effective suppression of V_O formation is paramount for achieving reliable operation in scaled multi-gate device technologies.

IV. CONCLUSION

Two types of V_O^{2+} defects can massively form in a-HfO$_2$, inducing the leftward shift of V_{th}. The large ECCS illustrates that they can capture electrons quickly and become +1 or +2 charged, causing the rightward shift of V_{th}. Then, the emission of the electrons from the high defect level to the CBM of a-HfO$_2$ will shift V_{th} leftward. As a result, the V_O defects consistently captured and released electron in a-HfO$_2$ and degrade the reliability of GAAFET through causing the V_{th} swinging and leakage current. In order to overcome this reliability issue, the quality of dielectric layer, especially in modern scaled multi-gated devices should be strictly ensured with lower V_O density during the process of device fabrication.

ACKNOWLEDGMENT

This work was supported by the General Research Fund (GRF) 16201223 from the Research Grants Council (RGC) of Hong Kong.

REFERENCES

[1] A. S. Oates, "Reliability issues for high-k gate dielectrics," in *IEEE International Electron Devices Meeting 2003*, 8-10 Dec. 2003 2003, pp. 38.2.1-38.2.4, doi: 10.1109/IEDM.2003.1269429.

[2] K. Tselios *et al.*, "On the Distribution of Single Defect Threshold Voltage Shifts in SiON Transistors," *IEEE Transactions on Device and Materials Reliability*, vol. 21, no. 2, pp. 199-206, 2021, doi: 10.1109/TDMR.2021.3080983.

[3] K. Tselios *et al.*, "Impact of Single Defects on NBTI and PBTI Recovery in SiO$_2$ Transistors," in *2022 IEEE International Integrated Reliability Workshop (IIRW)*, 9-14 Oct. 2022 2022, pp. 1-5, doi: 10.1109/IIRW56459.2022.10032748.

[4] Y. Liu, Y. Peng, and G. Han, "Non-Volatile FETs with Amorphous (Al$_2$O$_3$, HfO$_2$, ZrO$_2$, etc.) Gate Insulators," in *2021 5th IEEE Electron Devices Technology & Manufacturing Conference (EDTM)*, 8-11 April 2021 2021, pp. 1-3, doi: 10.1109/EDTM50988.2021.9420964.

[5] M. K. Mahadevaiah *et al.*, "Reliability of CMOS Integrated Memristive HfO$_2$ Arrays with Respect to Neuromorphic Computing," in *2019 IEEE International Reliability Physics Symposium (IRPS)*, 31 March-4 April 2019 2019, pp. 1-4, doi: 10.1109/IRPS.2019.8720552.

[6] H. Park *et al.*, "The Effect of Nanoscale Nonuniformity of Oxygen Vacancy on Electrical and Reliability Characteristics of HfO$_2$ MOSFET Devices," *IEEE Electron Device Letters*, vol. 29, no. 1, pp. 54-56, 2008, doi: 10.1109/led.2007.911992.

[7] E. P. Gusev *et al.*, "Ultrathin high-K gate stacks for advanced CMOS devices," in *International Electron Devices Meeting. Technical Digest (Cat. No.01CH37224)*, 2-5 Dec. 2001 2001, pp. 20.1.1-20.1.4, doi: 10.1109/IEDM.2001.979537.

[8] Z. Chen, "Gate-all-around nanosheet transistors go 2D," *Nature Electronics*, vol. 5, no. 12, pp. 830-831, 2022/12/01 2022, doi: 10.1038/s41928-022-00899-4.

[9] A. R. Trivedi *et al.*, "A Simulation Study of Oxygen Vacancy-Induced Variability in HfO$_2$ Metal Gated SOI FinFET," *IEEE Transactions on Electron Devices*, vol. 61, no. 5, pp. 1262-1269, 2014, doi: 10.1109/ted.2014.2313086.

[10] H. Zhu, B. Ye, C. Tang, X. Li, Q. Sun, and D. W. Zhang, "Improving Low-Frequency Noise in 14-nm FinFET by Optimized High-k/Metal Gate Thermal Processing," *IEEE Electron Device Letters*, vol. 42, no. 8, pp. 1112-1115, 2021, doi: 10.1109/LED.2021.3091488.

[11] K. M. A. H. Fahim, M. J. H. Khalid, M. Z. A. Mia, and M. Rasheduzzaman, "Study of 3-nm Cylindrical GAAFETs with Variations in High-k Dielectric Gate-oxide Materials," in *2022 IEEE Symposium on Industrial Electronics & Applications (ISIEA)*, 16-17 July 2022 2022, pp. 1-5, doi: 10.1109/ISIEA54517.2022.9873651.

[12] C. Tang and R. Ramprasad, "Point defect chemistry in amorphous HfO$_2$: Density functional theory calculations," *Physical Review B*, vol. 81, no. 16, 2010, doi: 10.1103/PhysRevB.81.161201.

[13] J. Wei, L. Jiang, M. Huang, Y. Wu, and S. Chen, "Intrinsic defect limit to the growth of orthorhombic HfO$_2$ and (Hf,Zr)O$_2$ with strong ferroelectricity: First-principles insights," *Advanced Functional Materials*, vol. 31, no. 42, p. 2104913, 2021, doi: 10.1002/adfm.202104913.

[14] J. H. Stathis and S. Zafar, "The negative bias temperature instability in MOS devices: A review," *Microelectronics Reliability*, vol. 46, no. 2-4,

pp. 270-286, 2006, doi: 10.1016/j.microrel.2005.08.001.

[15] T. Grasser *et al.*, "The Paradigm Shift in Understanding the Bias Temperature Instability: From Reaction–Diffusion to Switching Oxide Traps," *IEEE Transactions on Electron Devices*, vol. 58, no. 11, pp. 3652-3666, 2011, doi: 10.1109/ted.2011.2164543.

[16] P. Junsung and H. Sung-Min, "First Principles Approach to Analyze Defect-induced Multiphonon Transition at the Si-SiO$_2$ Interface," (in Korean), *Journal of Semiconductor Technology and Science*, vol. 19, no. 2, pp. 145-152, 2019, doi: 10.5573/JSTS.2019.19.2.145.

[17] Y.-Y. Liu, F. Zheng, X. Jiang, J.-W. Luo, S.-S. Li, and L.-W. Wang, "Ab Initio Investigation of Charge Trapping Across the Crystalline Si/Amorphous-SiO$_2$ Interface," *Physical Review Applied*, vol. 11, no. 4, p. 044058, 04/18/ 2019, doi: 10.1103/PhysRevApplied.11.044058.

[18] A. P. Thompson *et al.*, "LAMMPS - a flexible simulation tool for particle-based materials modeling at the atomic, meso, and continuum scales," *Computer Physics Communications*, vol. 271, p. 108171, 2022/02/01/ 2022, doi: 10.1016/j.cpc.2021.108171.

[19] J. P. Trinastic, R. Hamdan, Y. Wu, L. Zhang, and H. P. Cheng, "Unified interatomic potential and energy barrier distributions for amorphous oxides," *Journal of Chemical Physics*, vol. 139, no. 15, p. 154506, Oct 21 2013, doi: 10.1063/1.4825197.

[20] S. H. Wei and S. B. Zhang, "Chemical trends of defect formation and doping limit in II-VI semiconductors: The case of CdTe," *Physical Review B*, vol. 66, no. 15, p. 155211, Oct 15 2002, Art no. 155211, doi: 10.1103/PhysRevB.66.155211.

[21] J. Ma, S.-H. Wei, T. A. Gessert, and K. K. Chin, "Carrier density and compensation in semiconductors with multiple dopants and multiple transition energy levels: Case of Cu impurities in CdTe," *Physical Review B*, vol. 83, no. 24, 2011, doi: 10.1103/PhysRevB.83.245207.

[22] H. K. a. R. Avril, "Theory of light absorption and non-radiative transitions in F-centres," *Proceedings of the Royal Society, Series A*, vol. 204, pp. 406-423, 1950, doi: 10.1098/rspa.1950.0184.

[23] G. Kresse and J. Furthmuller, "Efficiency of ab-initio total energy calculations for metals and semiconductors using a plane-wave basis set," *Computational Materials Science*, vol. 6, no. 1, pp. 15-50, Jul 1996, doi: 10.1016/0927-0256(96)00008-0.

[24] J. Heyd and G. E. Scuseria, "Efficient hybrid density functional calculations in solids: Assessment of the Heyd-Scuseria-Ernzerhof screened Coulomb hybrid functional," *Journal of Chemical Physics*, vol. 121, no. 3, pp. 1187-1192, Jul 2004, doi: 10.1063/1.1760074.

[25] W. Goes *et al.*, "Identification of oxide defects in semiconductor devices: A systematic approach linking DFT to rate equations and experimental evidence," *Microelectronics Reliability*, vol. 87, pp. 286-320, 2018/08/01/ 2018, doi: https://doi.org/10.1016/j.microrel.2017.12.021.

A Ti/ITO Bilayer Gate Electrode Strategy for Improving Subthreshold Swing of Oxide Transistors

Chuanlin Sun[1], Tingchen Yi[1], Han Gao[2], Jiakang Zhang[3], Junchen Dong[3], Kai Zhao[3], Dedong Han[1], Xing Zhang[1]*

[1]School of Integrated Circuits, Beijing Advanced Innovation Center for Integrated Circuits,
Peking University, Beijing, China
[2]School of Microelectronics, Southern University of Science and Technology, Shenzhen, China
[3]School of Information & Communication Engineering, Beijing Information Science and Technology University,
Beijing, China

*Email: zhx@pku.edu.cn

Abstract—**Subthreshold swing (SS) plays a critical role in determining the electrical performance of oxide transistors and circuits. Here, we propose a Ti/ITO bilayer gate electrode strategy to improve SS of InSnO/ZnO (ITO/ZnO) transistors. We found that the ITO layer can effectively suppress interfacial oxidation between the gate electrode and the Al_2O_3 dielectric, therefore increasing the effective gate capacitance ($C_{ox\text{-}eff}$) and reducing the interface trap density (N_{Trap}). Compared with conventional Ti-gated devices, ITO-gated and Ti/ITO-gated devices exhibit a 35% reduction in SS and improved device stability. Furthermore, the inverters based on Ti/ITO-gated devices were fabricated, achieving a maximum voltage gain ($Gain_{max}$) of 215.3 and a static power consumption ($P_{s,\,max}$) of 0.9 μW at $V_{DD} = 3$ V. Compared with ITO-gated inverters, $Gain_{max}$ increased by 18% and $P_{s,\,max}$ decreased by 65%. This work provides a practical electrode engineering strategy for achieving low-power, high-performance oxide transistors and circuits.**

Keywords—oxide transistors, gate electrode, subthreshold swing, interfacial oxidation, inverters

I. INTRODUCTION

Oxide transistors are considered promising candidates for next-generation electronic devices due to their excellent uniformity on large-area substrates, low-temperature fabrication processes, and compatibility with CMOS technology[1-2]. Various integrated circuit (IC) applications based on oxide transistors have been developed, including flexible microprocessors[3], monolithic 3D-integrated static random-access memory (SRAM)[4], and neuromorphic computing circuits[5]. Subthreshold swing (SS) is a key electrical metric for oxide transistors, as it determines the switching speed of devices and directly affects the static and dynamic power consumption of circuits. Therefore, reducing SS is essential for achieving high-density, low-power ICs based on oxide transistors.

In oxide transistors, SS is related to the speed at which carriers fill defects and the gate control capability[6]. By reducing bulk defects in the active layer and gate dielectric, as well as the interface defects between the active layer/gate dielectric, the carrier filling speed can be effectively improved, thereby reducing SS[7]. In addition to the bulk defects and interface defects in oxide semiconductors, the gate electrode material also affects device SS. Due to the different conductivity, adhesion, and oxidation resistance among different gate materials[8], the effective gate capacitance ($C_{ox\text{-}eff}$) and interface trap density (N_{Trap}) of the device are affected to varying degrees, thereby influencing SS and stability.

Currently, most of the research on oxide transistors focuses on improving the source/drain electrodes and channel materials, while less attention is paid to gate electrodes and their interface oxidation behavior. This has become a limiting factor for further improving the performance of oxide transistors.

In this work, we proposed a Ti/InSnO (Ti/ITO) bilayer gate electrode strategy to improve the SS of ITO/ZnO transistors. The mechanism by which the ITO layer in the gate electrode suppresses oxidation reaction at the gate electrode/gate dielectric interface was analyzed. The effects of Ti, ITO, and Ti/ITO gate electrodes on the performance of devices, especially SS, were also systematically compared. Furthermore, we fabricated the inverters based on the Ti/ITO-gated devices and investigated the variation trends of the inverter voltage gain and static power consumption.

II. DEVICE FABRICATION PROCESS

ITO/ZnO transistors were fabricated on Si/SiO$_2$ substrates, and the process flow is shown in Fig. 1. Single-layer Ti, ITO, and bilayer Ti/ITO gate electrodes were deposited by sputtering. A 10-nm Al_2O_3 film deposited by atomic layer deposition (ALD) was used as the gate dielectric. The active layer consisted of an ITO/ZnO film, with a 6-nm bottom ITO layer deposited by sputtering and a 20-nm top ZnO layer deposited by ALD. A sputtering-deposited Al film served as the source/drain (S/D) electrode. Notably, the total thickness of all the electrodes is fixed at 100 nm, and the thickness of each layer of the double-layer electrode is 50 nm.

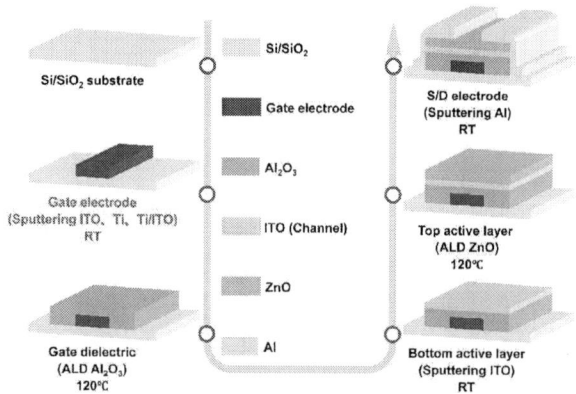

Fig. 1. Fabrication process flow of ITO/ZnO transistors.

According to the gate electrode materials, the devices were designated as Ti-gated devices, ITO-gated devices, and Ti/ITO-gated devices. The channel width/length (W/L) ratio was 10 μm/10 μm. Capacitance-voltage and current-voltage curves were measured by a semiconductor characterization system (Keithley 4200-SCS) and a semiconductor device analyzer (Agilent B15000A), respectively.

III. RESULTS AND DISCUSSION

Fig. 2(a) shows the transfer curves of the ITO/ZnO transistors. The ITO-gated and Ti/ITO-gated devices exhibit a similar turn-on voltage (V_{ON}) of -1.2 V, while the Ti-gated device shows a V_{ON} of -1.7 V. This difference is attributed to the lower work function of the Ti film (4.33 eV) compared to that of the ITO film (4.50 eV)[9], resulting in a more negative V_{ON} for the Ti-gated device[10]. Fig. 2(b) presents the output curves of the ITO/ZnO transistors. No current crowding is observed in the linear region, and the saturation current of the ITO-gated device is slightly higher than that of the Ti-gated and Ti/ITO-gated devices. The inset of Fig. 2(b) presents the S/D contact resistance (R_c) extracted using the transmission line method (TLM). Since all devices share the same structure and Al S/D electrode, they exhibit comparable R_c values.

Fig. 2(c) demonstrates the variation trend of SS of ITO/ZnO transistors as a function of drain current (I_{DS}). The ITO-gated and Ti/ITO-gated devices exhibit similar SS values, both of which are significantly lower than that of the Ti-gated device. Fig. 2(d) presents the average SS of 20 ITO/ZnO transistors with different gate electrodes calculated over I_{DS} range of 10^{-12} A to 10^{-8} A. Notably, the average SS of ITO-gated and Ti/ITO-gated devices is approximately 35% lower than that of Ti-gated devices. Overall, ITO/ZnO transistors exhibit superior performance, particularly in terms of SS, when ITO or Ti/ITO films are used as gate electrodes.

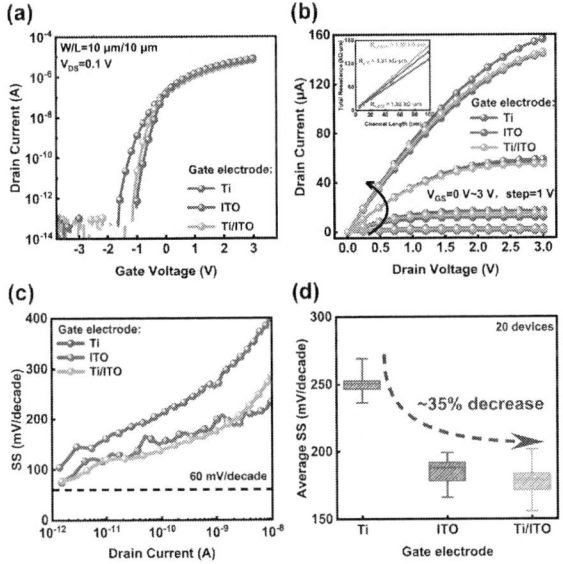

Fig. 2. Electrical performance of ITO/ZnO transistors, (a) transfer curves, (b) output curves, (c) SS, and (d) average SS.

To investigate the mechanism by which the gate electrode influences the electrical properties of ITO/ZnO transistors, X-ray energy dispersive spectroscopy (EDS) was conducted on the gate electrode/gate dielectric/active layer region. Fig. 3(a) presents the EDS results of the Ti-gated device, where both O

and Al elements are found to diffuse into the Ti gate electrode. The O diffusion region is notably larger than that of Al, suggesting an interfacial oxidation reaction occurring at the interface between the Ti gate electrode and the Al_2O_3 gate dielectric layer. Fig. 3(b) shows the EDS results of the ITO-gated device. A distinct interface is observed between the ITO gate electrode and the Al_2O_3 gate dielectric, with no detectable elemental diffusion. This indicates that no interfacial oxidation reaction occurs between the ITO gate electrode and the Al_2O_3 dielectric layer.

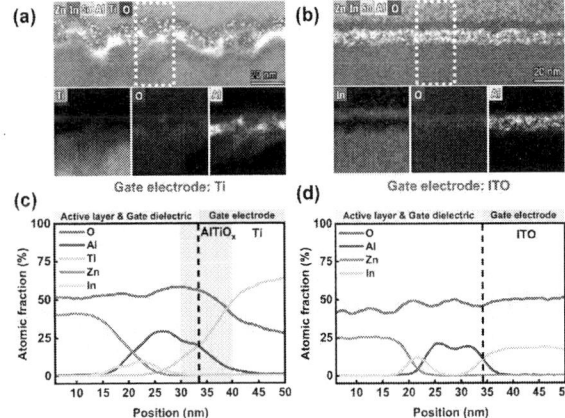

Fig. 3. EDS images of the gate electrode/gate dielectric/active layer region of (a) Ti-gated and (b) ITO-gated devices. Spectra of main elements content variation with depth of (c) Ti-gated and (d) ITO-gated devices.

We also analyzed the elemental composition of the Ti-gated and ITO-gated devices. Within the etching depth range of 30 ~ 40 nm, the Al diffusion region in the Ti-gated device (Fig. 3(c)) is larger than that in the ITO-gated device, and is accompanied by a high O content. This suggests that Al undergoes an oxidation reaction within the Ti gate electrode, resulting in the formation of an $AlTiO_x$ compound at the Ti/Al_2O_3 interface. In the Ti-gated devices, this $AlTiO_x$ compound forms a series connection with the Al_2O_3 gate dielectric, thereby reducing the $C_{ox\text{-}eff}$ and leading to a degradation in SS. In contrast, for the ITO-gated device (Fig. 3(d)), no oxidation region is observed at the ITO/Al_2O_3 interface.

To evaluate the impact of interfacial oxidation at the gate electrode/gate dielectric interface on the SS of ITO/ZnO transistors, we examined the gate capacitance characteristics of the devices. The gate capacitance mechanism is classified into two types based on whether oxidation occurred at the gate interface: oxidized (Fig. 4(a)) and unoxidized (Fig. 4(b)) gate electrodes. In the capacitance model, C_{Bulk} is the depletion capacitance of the active layer, C_{Int} is the capacitance associated with interface states between the active layer and the gate dielectric, C_{GO} is the capacitance of the oxidation product (i.e., $AlTiO_x$) formed at the interface, and C_{ox} is the capacitance of the Al_2O_3 gate dielectric. Among these, C_{GO} is in series with C_{ox}.

To obtain $C_{ox\text{-}eff}$ ($1/C_{ox\text{-}eff} = 1/C_{ox} + 1/C_{GO}$), metal-insulator-metal (MIM) capacitors were fabricated and used the unoxidized Mo as the reference material[11]. Fig. 4(c) presents the structures of MIM capacitors and the statistical results of their measured capacitances. The capacitances of the $Mo/Al_2O_3/Mo$ and $Mo/Al_2O_3/ITO/Mo$ were measured to be C_{MAM} = 648.3 nF/cm² and C_{MAIM} = 607.7 nF/cm², respectively.

ITO, as an n-type conductive oxide film[12], exhibits self-capacitance ($C_{ITO-self}$) under applied bias. Based on (1), the $C_{ITO-self}$ was calculated to be 9679.8 nF/cm².

$$\frac{1}{C_{MAIM}} = \frac{1}{C_{MAM}} + \frac{1}{C_{ITO-self}}$$

$$\rightarrow C_{ITO-self} = \frac{C_{MAM} \cdot C_{MAIM}}{C_{MAM} - C_{MAIM}} \qquad (1)$$

$$\frac{1}{C_{ITO}} = \frac{1}{C_{ITO-self}} + \frac{1}{C_{ox-eff-ITO}}$$

$$\rightarrow C_{ox-eff-ITO} = \frac{C_{ITO-self} \cdot C_{ITO}}{C_{ITO-self} - C_{ITO}} \qquad (2)$$

The capacitance of the ITO/Al$_2$O$_3$/Mo capacitors (C_{ITO}) was measured to be 610.0 nF/cm². Therefore, based on the obtained $C_{ITO-self}$ and (2), $C_{ox-eff-ITO}$, representing the C_{ox-eff} of the ITO/Al$_2$O$_3$/Mo capacitor, was calculated to be 651.0 nF/cm².

It is worth noting that $C_{ox-eff-ITO}$ closely matches the C_{MAM}, indicating that no oxidation reaction occurs at the ITO/Al$_2$O$_3$ interface. In addition, we measured the C_{ox-eff} of the Mo/Al$_2$O$_3$/Ti capacitor ($C_{ox-eff-Ti}$) and found it to be 593.8 nF/cm². This result confirms that the formation of AlTiO$_x$ at the Ti/Al$_2$O$_3$ interface leads to a reduction in the C_{ox-eff} of devices.

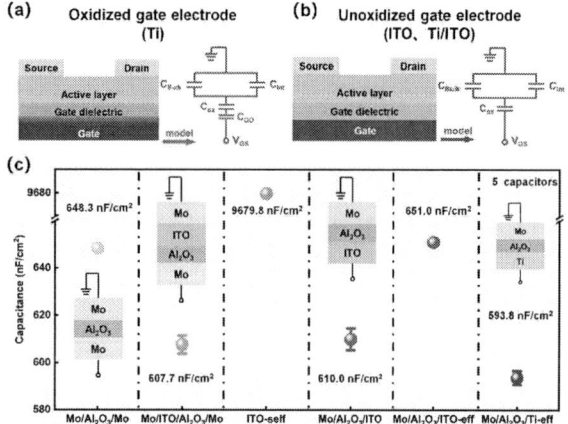

Fig. 4. Capacitance models of ITO/ZnO transistors with (a) oxidized and (b) unoxidized gate electrode. (c) Structures and statistical capacitances of MIM capacitors.

Based on the average SS and C_{ox-eff}, we calculated the trap state density (N_{Trap}) of the ITO/ZnO transistors using the following equation[13]:

$$N_{Trap} = \left(\frac{SS}{\frac{k_B T}{q} \ln 10} - 1 \right) \cdot \frac{C_{ox-eff}}{q} \qquad (3)$$

Where k_B, T, and q represent the Boltzmann constant, temperature in Kelvin, and elementary charge, respectively. The ITO-gated and Ti/ITO-gated devices show similar N_{Trap} values of 7.3×10^{12} and 8.4×10^{12} eV^{-1}cm^{-2}, while the Ti-gated devices show a higher N_{Trap} of 1.2×10^{13} eV^{-1}cm^{-2}. This implies that using the ITO or Ti/ITO electrodes effectively suppresses the formation of interfacial trap states in ITO/ZnO transistors. Stability of the ITO/ZnO transistors was also assessed. Negative bias stress (NBS) of -1 V and positive bias stress (PBS) of +1 V were applied to the gate electrode for 1000 s. The ITO-gated and Ti/ITO-gated devices demonstrate better stability than the Ti-gated devices (Fig. 5(a) and 5(b)), as evidenced by smaller V_{ON} shifts ($|V_{ON}|$). This trend is consistent with the differences in N_{Trap}.

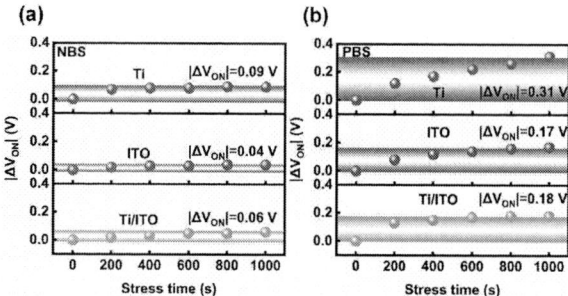

Fig. 5. $|\Delta V_{ON}|$ of ITO/ZnO transistors with Ti, ITO, and Ti/ITO gate electrodes under (a) NBS and (b) PBS.

Finally, we investigated the performance of inverters based on ITO/ZnO transistors with different electrodes. Fig. 6(a) depicts the circuit diagram of the inverter, which consists of two ITO/ZnO transistors configured as the load and the driver transistor. Their respective W/L ratios are 100 μm/10 μm and 10 μm /10 μm, as shown in Fig. 6(b). For the inverters with Ti/ITO gate electrodes, voltage transfer curve (VTC), voltage gain (Gain) curve, and static power consumption (P_s) curve are presented in Fig. 6(c) under an applied supply voltage (V_{DD}) of 3 V.

Fig. 6(d) shows the maximum voltage gain (Gain$_{max}$) of the inverters. The inverters with Ti/ITO gate electrodes exhibit a Gain$_{max}$ of 215.3 at a V_{DD} of 3 V, which is 18% and 46% higher than those of the ITO-gated and Ti-gated inverters, respectively. This enhanced gain is mainly attributed to the improved SS, which boosted the switching speeds of the ITO/ZnO transistors. Meanwhile, Fig. 6(e) presents the maximum static power consumption ($P_{s, max}$) of the inverters. Ti/ITO-gated inverters show a $P_{s, max}$ of 0.9 μW at V_{DD} of 3 V, which is 65% lower and 81% higher than those with ITO or Ti gate electrodes, respectively. This result is primarily attributed to the lower resistivity of Ti films compared to ITO films. In general, these findings demonstrate that the Ti/ITO bilayer gate electrode strategy is very promising for high-performance oxide transistors and circuits.

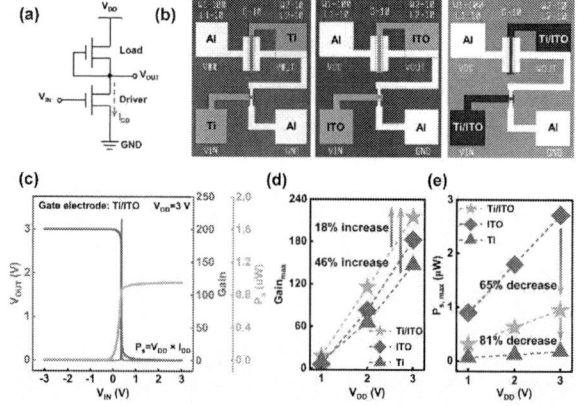

Fig. 6. (a) Circuit diagram and (b) optical images of the inverters. (c) VTC, voltage gain curve, and static power consumption curve of Ti/ITO-gated inverters at V_{DD} = 3 V. Comparison of (d) Gain$_{max}$ and (e) $P_{s, max}$.

IV. CONCLUSIONS

In summary, we propose a Ti/ITO bilayer gate electrode strategy to improve SS of the ITO/ZnO transistors. By characterizing the material components and analyzing MIM capacitors, we demonstrate that the ITO layer plays a critical role in suppressing oxidation reaction between the gate electrode (ITO or Ti/ITO) and the Al_2O_3 gate dielectric. Compared with Ti-gated devices, ITO-gated and Ti/ITO-gated devices exhibit a 35% reduction in SS as well as improved stability. We also investigated the inverters based on the ITO/ZnO transistors. Notably, the inverters with Ti/ITO gate electrodes exhibit a $Gain_{max}$ of 215.3 and a $P_{s, max}$ of 0.9 μW at a V_{DD} of 3 V, which are 18% higher and 65% lower, respectively, than those of the inverters with ITO gate electrodes. These results highlight the potential of the Ti/ITO bilayer gate electrode strategy for achieving low-power-consumption oxide transistors and circuits.

V. ACKNOWLEDGMENT

This work was supported by the Beijing Nova Program (20230484256, 20240484536), the R&D Program of Beijing Municipal Education Commission (KM20231123201), and Shenzhen Science and Technology Innovation Committee (KQTD 2020082011310-5004).

VI. REFERENCES

[1] A. Z. Yan, C. L. Wang, J. L. Yan, Z. Z. Wang, E. Y. Zhang, and T. L. Ren, "Thin-Film Transistors for Integrated Circuits: Fundamentals and Recent Progress," Advanced Functional Materials, vol. 34, pp. 2304409, 2024.

[2] K. Myny, "The development of flexible integrated circuits based on thin-film transistors," Nature Electronics, vol. 1, pp. 30-39, 2018.

[3] J. Biggs, J. Myers, J. Kufel, E. Ozer, S. Craske, and S. White, "A natively flexible 32-bit Arm microprocessor," Nature, vol. 595, pp. 532–536, 2021.

[4] J. A. Zhang, W. T. Wang, J. H. Zhu, C. X. Wang, T. Y. Zhu, and M. Zhang, "Ultraflexible Monolithic Three-Dimensional Static Random Access Memory," ACS Nano, vol. 18, pp. 3362-3368, 2024.

[5] Y. Jang, J. Park, J. M. Kang, and S. Y. Lee, "Amorphous InGaZnO (a-IGZO) Synaptic Transistor for Neuromorphic Computing," ACS Applied Electronic Materials, vol. 4, pp. 1427-1448, 2022.

[6] A. Sharma, N. K. Chourasia, N. Pal, S. Biring, and B. N. Pal, "Role of Electron Donation of TiO Gate Interface for Developing Solution-Processed High-Performance One-Volt Metal-Oxide Thin Film Transistor Using Ion-Conducting Gate Dielectric," Journal of Physical Chemistry C, vol. 123, pp. 20278-20286, 2019.

[7] Y. C. Zhang, G. He, L. N. Wang, W. H. Wang, X. F. Xu, and W. J. Liu, "Ultraviolet-Assisted Low-Thermal-Budget-Driven α-InGaZnO Thin Films for High-Performance Transistors and Logic Circuits," ACS Nano, vol. 16, pp. 4961-4971, 2022.

[8] R. D. Fan, W. Sun, C. M. Li, Y. H. Chen, H. P. Xie, and H. P. Zhou, "Physically and Chemically Stable Molybdenum-Based Composite Electrodes for p-i-n Perovskite Solar Cells," Advanced Materials, vol. 36, 2024.

[9] S. D. Nehate and K. B. Sundaram, "Work Function Extraction of Indium Tin Oxide Films from MOSFET Devices," ECS Journal of Solid State Science and Technology, vol. 77, pp. 1905-1910, 2017.

[10] M. S. Ozório, D. H. Vieira, G. L. Nogueira, C. S. Martin, N. Alves, and C. J. L. Constantino, "Effect of the gate electrodes/water interface on the performance of ZnO-based water gate field-effect transistors," Materials Science in Semiconductor Processing, vol. 151, 2022.

[11] I. H. Kang, S. H. Hwang, Y. J. Baek, S. G. Kim, Y. L. Han, and B. S. Bae, "Interfacial Oxidized Gate Insulators for Low-Power Oxide Thin-Film Transistors," Acs Omega, vol. 6, pp. 2717-2726, 2021.

[12] S. J. Won, M. S. Huh, S. Park, S. Suh, T. J. Park, and H. J. Kim, "Capacitance and Interface Analysis of Transparent Analog Capacitor Using Indium Tin Oxide Electrodes and High-Dielectrics," Journal of the Electrochemical Society, vol. 157, pp. G170-G175, 2010.

[13] L. Y. Su, H. Y. Lin, H. K. Lin, S. L. Wang, L. H. Peng, and J. J. Huang, "Characterizations of Amorphous IGZO Thin-Film Transistors With Low Subthreshold Swing," IEEE Electron Device Letters, vol. 32, pp. 1245-1247, 2011.

Fabrication of High-Performance β-Ga₂O₃ MOSFETs via Ohmic Contact Optimization

Hui Li [1#], Qihao Zhang [1#], Haodong Fu [1], Jianguo Li [3], Dongyuan Zhai [1], Yi Zhao [2], Jiwu Lu*[1,2]

[1] College of Electrical and Information Engineering, Hunan University, Changsha, Hunan 410082, China
[2] Research Center of Integrated Circuits, Huada Semiconductor, Shanghai 201203, China
[3] College of Information Science and Electronic Engineering, Zhejiang University, Hangzhou 31007, China

* Email: lujiw@hdsc.com.cn

Abstract—**This work presents an improved ohmic contact formation for β-Ga₂O₃ MOSFETs. A dielectric-capped ion implantation process is proposed to replace the conventional multi-step implantation and etching sequence, which features simplicity and cost-effectiveness without compromising performance. The process yields a low specific contact resistivity (≈ 6.5×10⁻⁶ Ω·cm²). MOSFETs fabricated using this method exhibit low on-resistance, and the incorporation of a gate field-plate structure further improves the voltage withstand capability. The results verify the effectiveness of the ohmic contact optimization approach and provide a practical reference for scalable fabrication of high performance β-Ga₂O₃ power devices.**

Keywords—β-Ga₂O₃, ohmic contact, ion implantation, MOSFET, gate field plate

I. INTRODUCTION

β-Ga₂O₃ has attracted significant attention in recent years as an ultra-wide bandgap semiconductor, owing to its wide bandgap (~ 4.8 eV), high breakdown electric field, and the potential for low-cost, wafer-scale single-crystal growth [1]-[3]. These advantages make it a strong candidate for high-voltage power devices [4]. However, the low intrinsic carrier concentration [5] and relatively high Schottky barrier height [6] of β-Ga₂O₃ severely limit its application in high-frequency and high-voltage environments. Thus, it is crucial to develop ohmic contacts with low resistance and good thermal stability. In 2013, Sasaki et al. [7] demonstrated that heavy n-type doping significantly reduces contact resistance. Using quintuple Si ion implantation, they achieved a specific contact resistivity (ρ_c) of 4.6×10⁻⁶ Ω·cm². Later, Yao et al. [8] reported

that among various metals, Ti forms the most favorable contact with β-Ga₂O₃, and high-quality ohmic contacts can be realized by combining heavy doping with defect-repair annealing.

Ion implantation, with its precise controllability over dopant dose and depth, is widely adopted in the fabrication of β-Ga₂O₃ ohmic contacts [7], [9]. However, the resulting dopant profile typically follows a Gaussian distribution [10], where the surface dopant concentration is lower than the peak concentration (located at ~13 nm depth) [7]. To address this, Higashiwaki et al. [11] proposed removing the ~13 nm low-concentration surface layer after multiple ion implantations. This approach yielded a ρ_c of 8.1×10⁻⁶ Ω·cm². Since then, surface etching has become a conventional step in ohmic contact formation of β-Ga₂O₃ by ion implantation. However, multiple high-dose, high energy implantations inevitably damage the lattice, while the subsequent etching process can create extra defects which requires additional repair [12]. Yang et al. [13] revealed that annealing at 450 °C effectively heals etching-induced damage. Smith et al. [14] further found that annealing can alleviate contamination caused by the lift-off process. Therefore, optimizing the ion implantation process to achieve low dose, low damage, high activation efficiency and good ohmic contact properties has become a critical challenge.

In this work, we propose an ALD-deposited Al₂O₃ dielectric capping layer which prevents etching-induced damage before Si ion implantation. Furthermore, only two low-dose implantation steps are required to achieve both sufficient doping concentration and the desired doping depth.

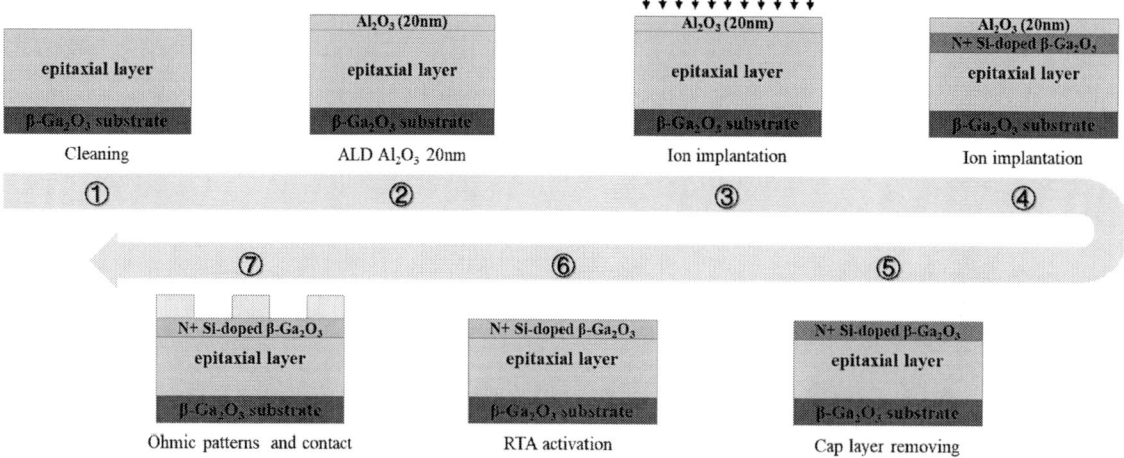

Fig. 1. Process schematic of the optimized ion implantation technique for fabricating β-Ga₂O₃ ohmic contacts.

#These authors contribute equally, and they are listed in alphabetical order.

After high-temperature annealing activation, high-quality ohmic contacts can be obtained without the additional ~13 nm damaged layer etching step. Based on this optimized ohmic contact formation scheme, β-Ga_2O_3 metal-oxide-semiconductor field-effect transistors (MOSFETs) were realized, incorporating a gate field-plate structure to achieve both a low on-resistance and a high breakdown voltage. It thus provides a lower cost approach for large-scale manufacturing of high-performance β-Ga_2O_3 power devices.

II. OHMIC CONTACT PROCESS OPTIMIZATION

In this work, we developed an optimized Si ion implantation technique for fabricating low-resistance ohmic contacts on β-Ga_2O_3. The overall fabrication flow is shown in Fig. 1. Each step is designed to address specific challenges in conventional ohmic contact formation. The β-Ga_2O_3 surface was first cleaned to remove contamination. A 20 nm Al_2O_3 layer was then deposited by ALD to serve as an implantation cap layer, protecting the surface from direct ion damage. After Si ion implantation, the Al_2O_3 layer was removed using TMAH solution to ensure a clean surface. High-temperature annealing at 950 °C and 1000 °C for 2 minutes in pure nitrogen was used to activate dopants and repair implantation damage. Ti/Au ohmic contacts were formed by e-beam evaporation and lift-off. Finally, a 470 °C anneal for 1 minute was performed to study the effect of alloying on contact resistivity ρ_c.

Traditional processes for β-Ga_2O_3 ohmic contacts often rely on multi-energy, high-dose multiple implantations. These are usually followed by etching to expose the peak doping region. However, the etching step can cause severe surface damage and increase defect density, degrading contact quality. To overcome these issues, we carried out Monte Carlo simulations to optimize the implantation parameters under an Al_2O_3 cap layer. The results are shown in Fig. 2. With only two implantation steps, a heavily doped region exceeding 1×10^{20} cm^{-3} is formed within ~40 nm from the surface. This process eliminates the need for post-implantation etching. It reduces process complexity and minimizes defect generation, providing a robust and efficient route for β-Ga_2O_3 ohmic contact fabrication.

III. RESULTS AND DISCUSSION

A. Ohmic Contact Resistivity

Contact resistivity ρ_c was extracted using a circular transmission line model (c-TLM), which employs concentric

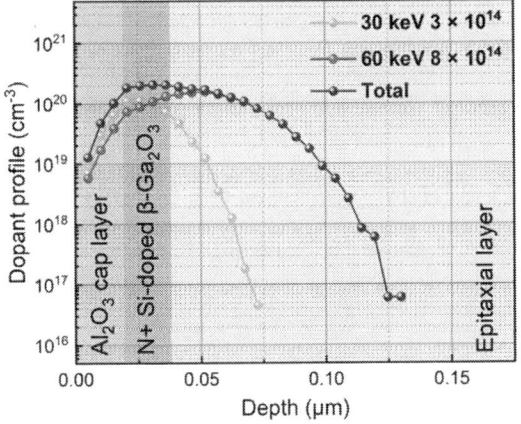

Fig. 2. Monte Carlo simulation of Si impurity distribution for ion implantation with an Al_2O_3 cap layer.

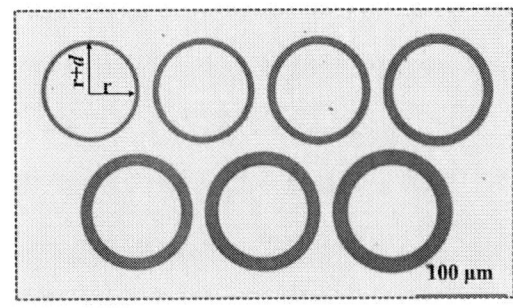

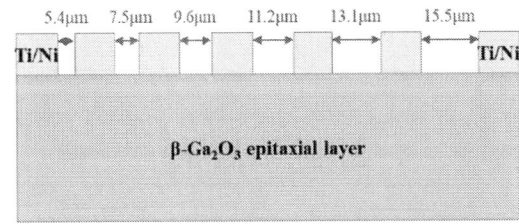

Fig. 3. Top-view micrograph and cross-sectional schematic of the circular transmission line model (c-TLM) test structure.

circular electrodes and requires no additional lateral isolation, thus simplifying device fabrication. Multiple embedded circular electrodes with varying spacing form independent test units. The linear I-V characteristics indicates ohmic behavior, enabling the accurate extraction of ρ_c. As shown in Fig. 3, the actual electrode spacing of the c-TLM structure is presented.

Fig. 4 presents the measured I-V characteristics of β-Ga_2O_3 under different activation temperatures and post-annealing conditions. Across all spacings, the linear I-V characteristics confirm the formation of ohmic contacts, and the total resistance increases with spacing as expected. The insets reveal a proportional relationship between R_{total} and $\ln[(r+d)/r]$, where R_{total} at $d = 0$ corresponds to the metal/β-Ga_2O_3 contact resistivity ρ_c.

At 950 °C activation [Fig. 4(a) and 4(c)], the extracted ρ_c values for the as-fabricated and post-annealed samples are 7.1×10^{-6} $\Omega\cdot$cm^2 and 6.5×10^{-6} $\Omega\cdot$cm^2, respectively. These are in line with reported benchmarks [7], [11], indicating that our optimized implantation scheme yields low ρ_c even without surface etching. The slight improvement after annealing suggests that, in the absence of etching damage, direct metal deposition on the activated surface is already sufficient to achieve low contact resistance. For 1000 °C activation [Fig. 4(b) and 4(d)], ρ_c decreases from 1.0×10^{-5} $\Omega\cdot$cm^2 to 8.2×10^{-6} $\Omega\cdot$cm^2 after annealing. This reduction points to effective healing of implantation-induced defects and mitigation of localized high-resistance regions. In addition, Ti diffusion toward the semiconductor interface during annealing likely enhances current transport [15].

Overall, high temperature thermal annealing around 900-1000 °C is needed to activate Si dopants in β-Ga_2O_3. However, excessive thermal budgets at this high temperature may promote undesirable dopant redistribution, which can counteract the benefits. Based on these results, 950 °C emerges as the more favorable activation temperature, offering both low ρ_c and good thermal stability for Si-implanted β-Ga_2O_3 ohmic contacts.

#These authors contribute equally, and they are listed in alphabetical order.

979-8-3315-3918-4/25 $31.00 © 2025 IEEE

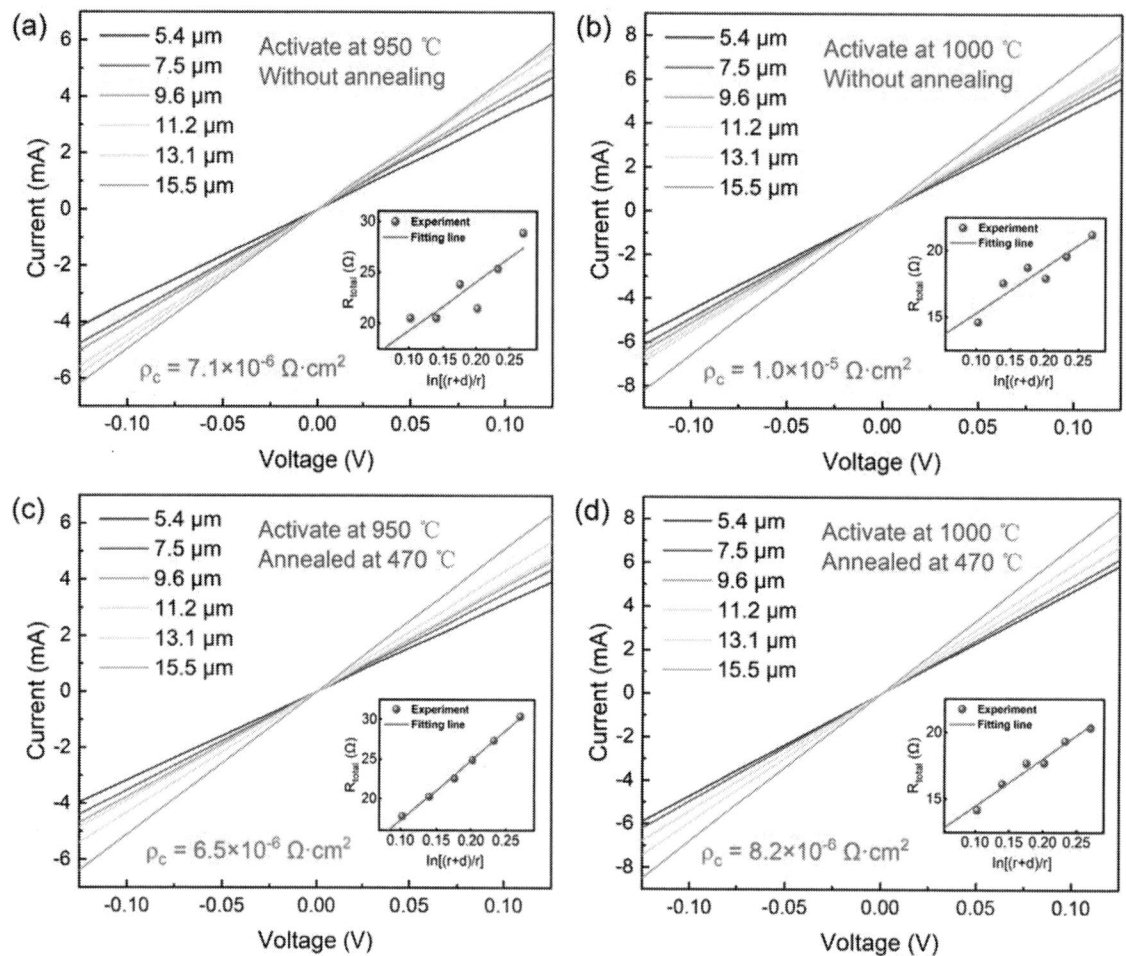

Fig. 4. I-V characteristics of c-TLM contacts at 950 °C and 1000 °C, before (a, b) and after (c, d) a 470 °C alloying anneal.
Insets: R_{total} vs. $\ln[(r+d)/r]$.

B. Power Device

Based on the optimized ohmic contact, β-Ga_2O_3 MOSFET was fabricated to evaluate the practicability of this process. The device fabrication incorporated the refined ion implantation and ohmic contact techniques to ensure low contact resistance and reliable performance.

The β-Ga_2O_3 substrates were cleaned thoroughly to remove surface contaminants. Device isolation was formed by Cl-based ICP etching to prevent leakage. The source and drain regions were implanted using the optimized parameters from the improved process. Post-implantation treatments included plasma ash and chemical cleaning to repair damage. Ti/Ni bilayer ohmic contacts were deposited without rapid thermal annealing, leveraging the optimized contact process. A 20 nm ALD-grown Al_2O_3 layer was used as the gate dielectric. The gate metal Ni was patterned to expose ohmic contacts by selectively etching the dielectric. Finally, a PECVD SiO_2 gate field plate was formed to reduce electric field concentration at the gate edge. The device structure is shown in Fig. 5(a).

The epitaxial substrate used for device fabrication had a doping concentration of approximately $1.3 \sim 3.0 \times 10^{17}$ cm^{-3} in the epitaxial layer, with an unintentionally doped (UID) layer thickness of 300 nm. To evaluate the electrical performance of the fabricated β-Ga_2O_3 MOSFETs, output characteristics were measured using an Agilent B1500A parameter analyzer. Fig. 5(b) shows the output curves of a device with a gate-to-drain spacing (L_{gd}) of 20 μm. The gate voltage (V_G) was swept linearly from -26 V to +10 V in 2 V steps. The device exhibits typical MOSFET behavior with clear linear and saturation regions. From the linear region, the specific on-resistance ($R_{on,sp}$) was extracted to be 87.2 m$\Omega \cdot$cm^2. Transfer characteristics (I_D-V_G) were measured at $V_{DS} = 10$ V, showing an on/off current ratio of 2.8×10^8. Using linear extrapolation, the threshold voltage (V_{TH}) was found to be -16.9 V. The maximum transconductance (g_{max}) was measured to be 2.9 mS/mm. Fig. 5(c) presents the breakdown characteristics. The device transitions from a low-leakage off state to a high-leakage breakdown state, with a breakdown voltage (V_{BR}) up to 2221 V. During breakdown testing, the gate voltage was held at -40 V to ensure the device remained in the off state.

IV. CONCLUSION

This work addresses the challenges in ohmic contact fabrication for β-Ga_2O_3 power devices by proposing a dielectric-capped ion implantation optimization that simplifies conventional multi-energy implantation and eliminates etching-related damage. The method achieves uniform high doping concentration and low contact resistivity, enhancing ohmic contact performance and stability. MOSFETs fabricated using this optimized ohmic contact

---

#These authors contribute equally, and they are listed in alphabetical order.

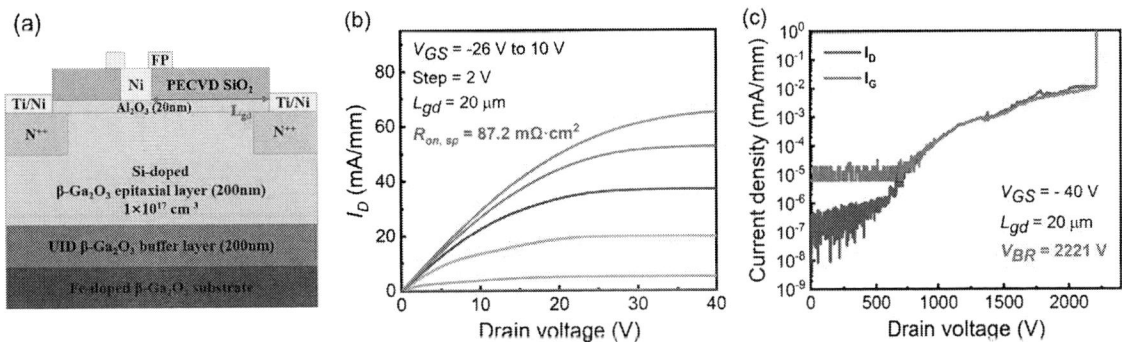

Fig. 5. (a) Device schematic, (b) Output characteristics, (c) Breakdown characteristics.

process exhibit excellent electrical characteristics and increased breakdown voltage, demonstrating the positive impact of process optimization on device performance. This study provides strong technical support for realizing high-quality ohmic contacts in β-Ga₂O₃ power devices.

ACKNOWLEDGMENT

This work was supported in part by the National Natural Science Foundation of China under Grant 52177179, and in part by the Jie Bang Headed Project of Changsha City, Hunan Province, China under Grant No. kq2501006.

REFERENCES

[1] M. Higashiwaki, "β-Ga₂O₃ material properties, growth technologies, and devices: a review," AAPPS Bull, vol. 32 (3), 2022.

[2] K. Ghosh and U. Singisetti, "Ab initio velocity-field curves in monoclinic β-Ga₂O₃," J. Appl. Phys., vol. 122 (3), pp. 035702, 2017.

[3] M. Slomski, N. Blumenschein, P. P. Paskov, J. F. Muth, and T. Paskova, "Anisotropic thermal conductivity of β-Ga₂O₃ at elevated temperatures: Effect of Sn and Fe dopants," J. Appl. Phys., vol. 121 (23), pp. 235104, 2017.

[4] M. Baldini, Z. Galazka, and G. Wagner, "Recent progress in the growth of β-Ga₂O₃ for power electronics applications," Materials Science in Semiconductor Processing, vol. 78, pp. 132-146, 2018.

[5] N. Ma, N. Tanen, A. Verma, Z. Guo, T. Luo, H (Grace) Xing, et al., "Intrinsic electron mobility limits in β-Ga₂O₃," Appl. Phys. Lett., vol. 109 (21), pp. 212101, 2016.

[6] M. Mohamed, K. Irmscher, C. Janowitz, Z. Galazka, R. Manzke, R. Fornari, "Schottky barrier height of Au on the transparent semiconducting oxide β-Ga₂O₃," Appl. Phys. Lett., vol. 101 (13), pp. 132106, 2012.

[7] K. Sasaki, M. Higashiwaki, A. Kuramata, T. Masui and S. Yamakoshi, "Si-Ion Implantation Doping in β-Ga₂O₃ and Its Application to Fabrication of Low-Resistance Ohmic Contacts," Applied Physics Express, vol. 6(8), pp. 086502, 2013.

[8] Y. Yao, R. F. Davis, L. M. Porter, "Investigation of Different Metals as Ohmic Contacts to β-Ga₂O₃: Comparison and Analysis of Electrical Behavior, Morphology, and Other Physical Properties," Journal of Electronic Materials, vol. 46(4), pp. 2053-2060, 2016.

[9] A. Nikolskaya, E. Okulich, D. Korolev, A. Stepanov, D. Nikolichev, A. Mikhaylov, et al., "Ion implantation in β-Ga₂O₃: Physics and technology," J. Vac. Sci. Technol. A, vol. 39 (3), pp. 030802, 2021.

[10] M. H. Wong, C.-H. Lin, A. Kuramata, S. Yamakoshi, H. Murakami, Y. Kumagai, et al., "Acceptor doping of β-Ga₂O₃ by Mg and N ion implantations," Appl. Phys. Lett., vol. 113 (10), pp. 102103, 2018.

[11] M. Higashiwaki, K. Sasaki, T. Kamimura, M. H. Wong; D. Krishnamurthy, A. Kuramata, et al., "Depletion-mode Ga₂O₃ metal-oxide-semiconductor field-effect transistors on β-Ga₂O₃ (010) substrates and temperature dependence of their device characteristics," Applied Physics Letters, vol. 103(12), pp. 123511, 2013.

[12] H.-C. Huang, Z. Ren, A. U. Bhuiyan, Z. Feng, Z. Yang, X. Luo, et al., "Wet etch, dry etch, and MacEtch of β-Ga₂O₃: A review of characteristics and mechanism," Journal of Materials Research, vol. 36 (21), pp. 4360-4377, 2021.

[13] J. Yang, F. Ren, R. Khanna, K. Bevlin, D. Geerpuram, L.-C. Tung, et al., "Annealing of dry etch damage in metallized and bare (-201) Ga₂O₃. J. Vac. Sci. Technol. B, vol. 35 (5), pp. 051201, 2017.

[14] K. T. Smith, C. A. Gorsak, A. Kalra, B. J. Cromer, K. Azizie, D. M. Dryden, et al., "Non-alloyed ohmic contacts to (010) β-Ga₂O₃ with low contact resistance," Appl. Phys. Lett., vol. 123 (24), pp. 242101, 2023.

[15] P.-W. Hsieh, C.-C. Chi, C.-M. Wu, K.-Y. Hsiao, M.-Y. Lu, "Unveiling interface engineering dynamics between Ti and Ga₂O₃ nanowire," Applied Surface Science, vol. 670, pp. 160612, 2024.

#These authors contribute equally, and they are listed in alphabetical order.

Experimental Study on 1.2kV/40mΩ SiC MOSFET with Integrated JBS Diode

Moufu Kong[1,*], Qizhi Feng [1], Hongfei Deng [1], Yufeng Dong[1], Wei Han[2,*], Xuequan Yin[2], Jiakai You[2]

[1] State Key Laboratory of Electronic Thin Films and Integrated Devices of China, University of Electronic Science and Technology of China, Chengdu, 611731, Sichuan, China

[2] The 54th Research Institute of China Electronics Technology Group Corporation, Shijiazhuang, 050081, Hebei, China

* Corresponding authors: Moufu Kong and Wei Han, kmf@uestc.edu.cn & whan@semi.ac.cn

Abstract—**In this work, a novel SiC junction barrier Schottky (JBS) diode integrated MOSFET structure (SiC J-MOSFET) is proposed and manufactured to improve the reverse conduction function and enhance the reverse recovery performance of the device while saving chip area and application cost. The experimental test results shows that the proposed device achieves a breakdown voltage of 1320V with a very low specific on-resistance ($R_{on,sp}$) of 3.29mΩ·cm². The integrated JBS diode of the SiC J-MOSFET exhibits a significantly lower on-state voltage, measuring only 1.7 V at a 10A current capacity. This represents a 54% reduction compared to that of the conventional planar SiC MOSFET. Additionally, the proposed SiC J-MOSFET demonstrates a reverse recovery time (t_{rr}) of 54 ns, with a reverse recovery charge (Q_{rr}) of 159 nC. In contrast, the conventional SiC MOSFET shows a longer trr of 89 ns and a higher Q_{rr} of 287 nC, marking reductions of 39% and 44.6%, respectively. These findings highlight promising commercial applications for the SiC J-MOSFET.**

Keywords—SiC MOSFET; junction barrier Schottky diode; reverse recovery; bipolar degradation

I. INTRODUCTION

Silicon carbide (SiC) power MOSFETs have been maintaining a very considerable market demand in recent years for many applications [1], such as photovoltaic inverters, household appliances, new energy vehicles, and airplanes [2][3]. In practical circuit applications, when using the PN body diode of a SiC MOSFET as a reverse current diode, the turn-on voltage of the body diode is much higher than that of the silicon-based diode due to the wide energy bandgap of the SiC material, resulting in a larger diode turn-on loss of the SiC MOSFET in reverse current. Not only that, SiC MOSFET body PN diode in reverse recovery will produce a relatively large loss and longer reverse recovery time. At the same time, due to defects in the SiC material wafers used in industry, bipolar degradation effects occur in any type of SiC device under bipolar (PN junction body diode of the MOSFET in conduction) conditions [4][5]. This effect is mainly triggered by the presence of base plane dislocations (BPD) on the SiC crystal [6][7][8]. During bipolar operation, the energy released from the recombination of electrons and holes causes the spread of stacking layer dislocations at the BPD. This results in a reduced effective device area and a narrower current path, leading to degraded device performance and potential device failure [9]. Although we can use SiC Schottky diodes (SBDs) in anti-parallel with the external SiC MOSFETs to be used as reverse recovery diodes

to circumvent these problems, each SiC MOSFET needs to be connected in anti-parallel with a SiC SBD, which greatly increases the number of components required for the system, which increases the cost and size of the whole system, and the reliability of the whole system will be reduced. The external connections between MOSFETs and SiC SBDs are increased, which does not satisfy the requirements of miniaturization, high power density, and high reliability of power electronic devices demanded today [10]. Although a number of researchers have proposed some schemes for integrating SBDs in SiC MOSFET devices, all of them additionally increase the chip area or fabrication process steps [11][12][13][14]. Another approach involves integrating the SiC MOSFET and SBD on a single chip, but this also increases packaging complexity and costs.

In response to the above problems, a structure with an integrated junction barrier Schottky (JBS) diode in the three-dimensional direction of the device cell is proposed, whose unique design allows the integration of the JBS diode to be realized without increasing additional chip area, which enables the device to further enhance the performance of the body diode on the basis of maintaining the original performance of the MOSFET device. The integrated JBS diode can achieve a lower on-state voltage drop, effectively suppress bipolar degradation effect and reduce the reverse recovery current in the body diode of the proposed SiC J-MOSFET. This helps reduce both the reverse recovery time (t_{rr}) and reverse recovery charge (Q_{rr}), thereby lowering the overall reverse recovery losses and improving the reliability of the device.

II. DEVICE STRUCTURE AND BASIC MECHANISM

As illustrated in Fig. 1(a), a new monolithic SiC J-MOSFET device structure is demonstrated. And Fig. 1(b), (c) and (d) show the cross-section views of the device in the three-dimensional (3D) direction along aa', bb' and cc', respectively. As can be seen from Fig.1(b), the aa' cross-section reveals a view consisting only of the N+ region and ohmic contact metal without the P+ injection layer. The bb' cross-section demonstrates a structure containing the P+ injection layer in combination with the ohmic contact metal Ni [15][16][17] (Fig.1(c)). And the cc' cross-section presents a structure with the JFET region in combination with the Schottky contact metal Ti [18][19][20] (Fig.1(d)). By this periodic arrangement of the 3D structure cell, the integration of a JBS diode into the device without increasing additional chip area is realized. The innovation of this design is that it cleverly utilizes structural changes within the device to achieve functional integration of different types of diodes. In

This work was supported in part by the Central Guiding Local Science and Technology Development Special Project of Sichuan (2024ZYD0310) and the Key R & D project of science and technology plan of Sichuan province (Grant 2023YFG0005).

979-8-3315-3918-4/25 $31.00 © 2025 IEEE

conventional device design, the SiC Schottky barrier diodes usually require a separate fabrication process and layout area, whereas in the proposed design, the JBS diode is combined with a conventional MOSFET structure by designing the layout of the J-MOSFET, which not only saves the chip area, but also improves the integration and compactness of the device.

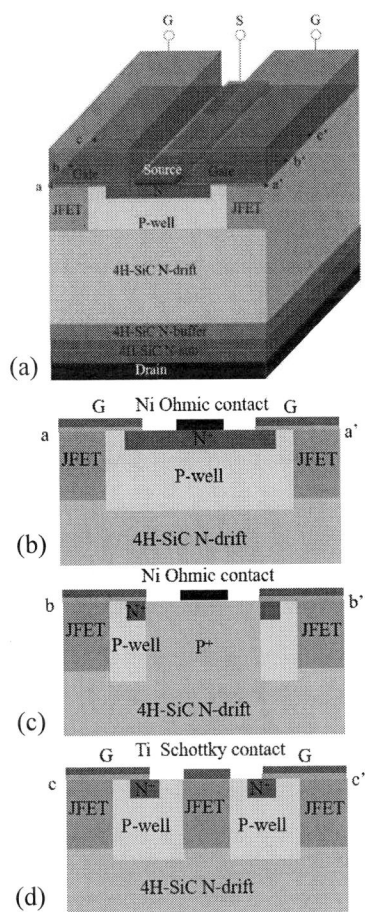

(a)

(b)

(c)

(d)

Fig. 1. (a) The 3D cell structure of the proposed SiC-JMOSFET; Schematic diagram of the device cell structure along (b) aa' cross-section, (c) bb' cross-section and (d) cc' cross-section.

The layout of the proposed SiC J-MOSFET utilizes a strip cell design with an integrated periodically aligned P+ region connected to the Schottky structure, as depicted in Fig. 2(a). The conventional planar gate SiC MOSFET also utilize a strip shaped cell while periodically aligning the P+ region (Fig. 2(b)). The cell size, active region area, and total device area are consistent for both devices. This layout achieves the goal of integrating the JBS diode with essentially no change in MOSFET on-state performance.

Fig. 2(a) presents a schematic illustration of the architectural layout for a silicon carbide (SiC) power integrated circuit. This design is derived from the conventional planar vertical structure of the metal-oxide-semiconductor field-effect transistor (MOSFET) process, maintaining consistency with the necessary lithographic layers required for fabrication. Fig. 2(b) offers a micrographic view of the proposed SiC power integrated circuit. Within this micrograph, a silver-white windowing represents the

exposure of individual electrodes. These electrodes are sequentially identified as the source, gate, driver circuit input (paired with the power electrode), and drain, moving from left to right in the image.

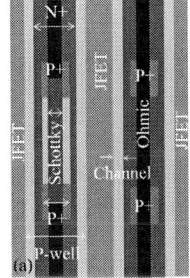 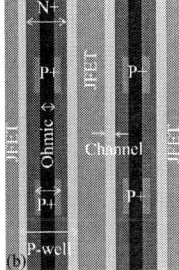

Fig. 2. Strip cell layout schematic diagram of (a) proposed SiC J-MOSFET and (b) the conventional SiC MOSFET.

III. SIMULATION AND MEASURED RESULTS AND DISCUSSION

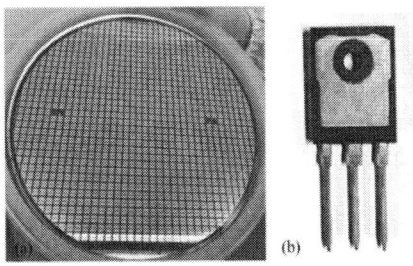

Fig. 3. (a) manufactured SiC J-MOSFET wafer; (b) packaged SiC J-MOSFET device.

Based on the aforementioned 3D cell structure and layout design, the proposed SiC J-MOSFETs are manufactured based on the foundry's existing mature process flow. And Fig. 3(a) shows the one of the fabricated wafers. The wafers are scribed and the J-MOSFET chips are packaged using TO-247-3L package, as shown in Fig 3(b).

The electrical performance testing for the both SiC J-MOSFET and the conventional MOSFET devices are conducted. Fig. 4 demonstrates the test results of transfer characteristic curves of the two devices. In these curves, the test conditions are set such that $V_{GS} = V_{DS}$, and the test results show that the SiC J-MOSFET device has a threshold voltage $V_{GS(th)} = 3.1$ V at $I_{DS} = 10$mA , which is almost the same as that of the conventional SiC MOSFET. This result shows that the SiC J-MOSFET device does not change the $V_{GS(th)}$ due to the integration of the JBS diode.

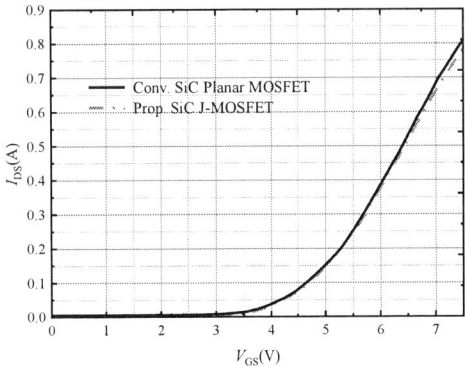

Fig. 4. Comparison of transfer characteristic curves of two devices

979-8-3315-3918-4/25 $31.00 © 2025 IEEE

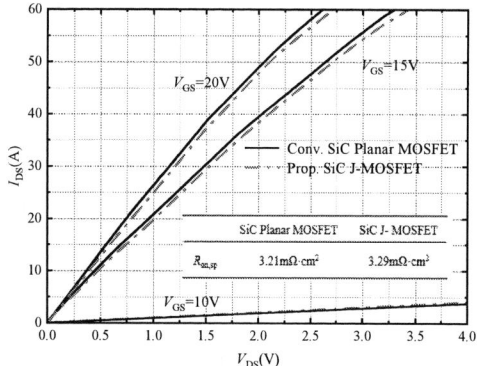

Fig. 5. Comparison of the output curves of the two devices

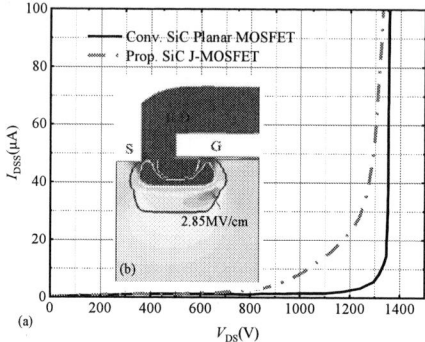

Fig. 6. (a) Comparison of breakdown characteristic curves of two devices; (b) Electric field distribution of Proposed SiC J-MOSFET at breakdown.

As depicted in Fig. 5, the output characteristics were tested at different V_{GS} values, specifically at V_{GS} = 10V, 15V, and 20V.These test results show that the current capability at V_{DS}=2V of the SiC J-MOSFET is only slightly lower (<5%) compared to the SiC planar gate MOSFET, which is mainly due to that, during the design process of the SiC J-MOSFET, The proposed SiC J-MOSFET maintains the same chip active area as the conventional SiC MOSFET, which resulted in a corresponding reduction of the source ohmic contact area, and by calculating the specific on-resistance ($R_{on,sp}$) at V_{DS} = 0.5V of the two devices, the performance difference between the two devices can be evaluated more precisely. Under our test conditions, the Ron,sp of the SiC J-MOSFET is 3.29 mΩ·cm^2, which is only a 2.5% increase compared to the conventional SiC MOSFET flowed on the same process platform. It shows that the proposed SiC J-MOSFET does not significantly change the $R_{on,sp}$ of the device due to the integration of JBS diode.

Fig. 6 illustrates the breakdown characteristic curves of the proposed SiC J-MOSFET and the conventional SiC MOSFET. It can be observed from the figure that the leakage current of the SiC J-MOSFET is only slightly higher than that of the conventional planar gate SiC MOSFET. This is due to the fact that the narrowing of the Schottky barriers at the interface of the Schottky contacts under reverse bias conditions leads to a certain degree of tunneling leakage current. However, the SiC J-MOSFET design takes the protective measure of deep P+ injection on both sides of the Schottky contacts, which effectively suppresses the growth of leakage current at high blocking voltage. As a result, although the leakage current of the proposed SiC J-MOSFET is slightly higher, according to the breakdown voltage judgment standard of commercial devices with drain current up to 100uA, its breakdown voltage still reaches 1320V, which is comparable to that of the SiC planar gate MOSFET's 1360V, and both satisfy the product specification's requirement for a minimum withstand voltage ($V_{(BR)DSS}$) of more than 1200V. This design optimization ensures that the performance of both devices meets the needs of high-voltage application scenarios.

Fig. 7(a) displays the test curves depicting the reverse conduction characteristics of both the proposed SiC J-MOSFET and the conventional SiC MOSFET. At I_{SD} =10A, the on-state voltages (V_F) for the SiC J-MOSFET and the conventional SiC MOSFET are 3.7 V and 1.7 V (when V_{GS}= -5 V), respectively, representing a substantial improvement of 54%.

Fig. 7(b) shows the electric current path when the SiC J-MOSFET body diode turns on, and it can be found that the electric current flows only from the Schottky contact interface, and this improvement effectively suppresses the turn-on behavior of the PN junction diode in the SiC MOSFET body, and effectively avoids the bipolar degradation effect brought about by the conduction of the PN junction diode in the conventional SiC MOSFETs.

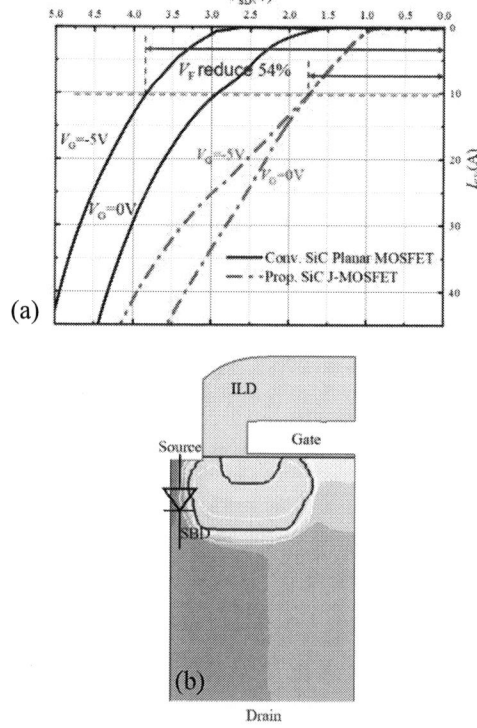

Fig. 7. (a) Comparison of body diode conduction characteristics of two devices; (b) Proposed SiC J-MOSFET body diode current path during conduction

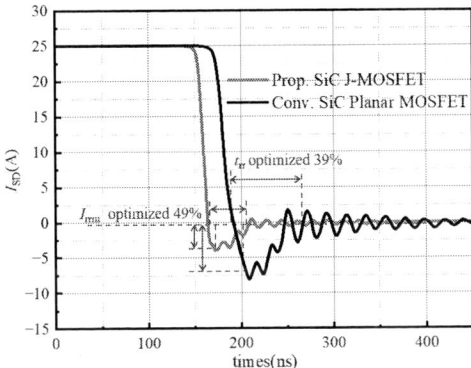

Fig. 8. Comparison of reverse recovery characteristics of two devices

Fig. 8 depicts the the reverse recovery test results of the two SiC MOSFET. And from the results, the significant improvements in reverse recovery time (t_{rr}) and reverse recovery charge (Q_{rr}) of the proposed SiC J-MOSFET are evident when compared to those of the SiC planar MOSFET. The trr is reduced by 39%, decreasing from 89 ns to 54 ns compared to the conventional SiC MOSFET. Similarly, the Q_{rr} of the proposed SiC J-MOSFET shows a reduction of 44.6%, decreasing from 287 nC to 159 nC. The design of the integrated JBS diode optimizes the charge storage and recovery process, thereby accelerating the transition from the on-state to the off-state of the body diode. This improvement is important for increasing power conversion efficiency and reducing energy loss during reverse recovery.

IV. CONCLUSION

In this paper, the on-chip integration of SiC JBS diode with SiC MOSFET is realized by changing the layout in response to the issues in the application of SiC MOSFET. Through the test and comparison between SiC J-MOSFET and traditional SiC MOSFET, the forward characteristics of MOSFETs are basically unchanged and the body diode characteristics are greatly optimized, which indicates that the proposed SiC J-MOSFET improves the performance of the body diode, avoids the bipolar degradation effect, and reduces the cost of the application. Finally, a series of reliability tests were conducted on the proposed SiC J-MOSFET to investigate the degradation of various performance metrics. The results validated the excellent reliability of the proposed SiC J-MOSFET.

REFERENCES

[1] J. Millán, P. Godignon, X. Perpiñà, A. Pérez-Tomás and J. Rebollo, "A Survey of Wide Bandgap Power Semiconductor Devices," in IEEE Transactions on Power Electronics, vol. 29, no. 5, pp. 2155-2163, May 2014.

[2] C. Ota, J. Nishio, K. Takao, and T. Shinohe, "V_F Degradation of 4H-SiC PiN Diodes Using Low-BPD Wafers," Materials Science Forum, vol. 778–780, pp. 851–854, Feb. 2014.

[3] M. Östling, R. Ghandi and C. -M. Zetterling, "SiC power devices — Present status, applications and future perspective," 2011 IEEE 23rd International Symposium on Power Semiconductor Devices and ICs, San Diego, CA, USA, 2011, pp. 10-15.

[4] M. Skowronski and S. Ha, "Degradation of hexagonal silicon-carbide-based bipolar devices," Journal of Applied Physics, vol. 99, no. 1, Jan. 2006.

[5] P. Pirouz and A. Galeckas, "(Invited) Degradation of SiC Bipolar Devices: A Review of Likely Causes and Recent Advances in its Understanding," ECS Transactions, vol. 41, no. 8, pp. 225–236, Oct. 2011.

[6] C. Ota, J. Nishio, K. Takao, and T. Shinohe, "V_F Degradation of 4H-SiC PiN Diodes Using Low-BPD Wafers," Materials Science Forum, vol. 778–780, pp. 851–854, Feb. 2014.

[7] R. E. Stahlbush, N. A. Mahadik, A. J. Lelis and R. Green, "Effects of Basal Plane Dislocations on SiC Power Device Reliability," 2018 IEEE International Electron Devices Meeting (IEDM), San Francisco, CA, USA, 2018, pp. 19.4.1-19.4.4.

[8] N. Kawabata et al., "Effects of basal plane dislocation density in 4H-SIC substrate on degradation of Body-Diode forward voltage," Materials Science Forum, vol. 858, pp. 384–388, May 2016.

[9] T. Kimoto et al., "Understanding and reduction of degradation phenomena in SiC power devices," 2017 IEEE International Reliability Physics Symposium (IRPS), Monterey, CA, USA, 2017, pp. 2A-1.1-2A-1.7.

[10] D. Martin, P. Killeen, W. A. Curbow, B. Sparkman, L. E. Kegley and T. McNutt, "Comparing the switching performance of SiC MOSFET intrinsic body diode to additional SiC schottky diodes in SiC power modules," 2016 IEEE 4th Workshop on Wide Bandgap Power Devices and Applications (WiPDA), Fayetteville, AR, USA, 2016, pp. 242-246.

[11] W. Sung and B. J. Baliga, "Monolithically Integrated 4H-SiC MOSFET and JBS Diode (JBSFET) Using a Single Ohmic/Schottky Process Scheme," in IEEE Electron Device Letters, vol. 37, no. 12, pp. 1605-1608, Dec. 2016.

[12] W. Sung and B. J. Baliga, "On Developing One-Chip Integration of 1.2 kV SiC MOSFET and JBS Diode (JBSFET)," in IEEE Transactions on Industrial Electronics, vol. 64, no. 10, pp. 8206-8212, Oct. 2017.

[13] X. Li et al., "SiC Trench MOSFET With Integrated Self-Assembled Three-Level Protection Schottky Barrier Diode," in IEEE Transactions on Electron Devices, vol. 65, no. 1, pp. 347-351, Jan. 2018.

[14] M. Kong, Y. Duan, B. Zhang, R. Yan, B. Yi, and H. Yang, "A novel 4H‑SiC accumulation mode MOSFET with ultra‑low specific on‑resistance and improved reverse recovery capability," IET Power Electronics, vol. 16, no. 14, pp. 2369-2377, Sep. 2023.

[15] S. Y. Han et al., "Ohmic contact formation mechanism of Ni on n-type 4H–SiC," Applied Physics Letters, vol. 79, no. 12, pp. 1816–1818, Sep. 2001.

[16] A. V. Kuchuk et al., "Ni-Based Ohmic contacts TON-Type 4H-SIC: the formation mechanism and thermal stability," Advances in Condensed Matter Physics, vol. 2016, pp. 1–26, Jan. 2016.

[17] S. J. Yang, C. K. Kim, I. H. Noh, S. W. Jang, K. H. Jung, and N. I. Cho, "Study of Co- and Ni-based ohmic contacts to n-type 4H-SiC," Diamond and Related Materials, vol. 13, no. 4–8, pp. 1149–1153, Apr. 2004.

[18] D. Perrone, M. Naretto, S. Ferrero, L. Scaltrito, and C. F. Pirri, "4H-SIC Schottky barrier diodes using MO-, TI- and NI-Based contacts," Materials Science Forum, vol. 615–617, pp. 647–650, Mar. 2009.

[19] K. Shili, M. B. Karoui, R. Gharbi, M. Abdelkrim, M. Fathallah, and S. Ferrero, "Series resistance study of Schottky diodes developed on 4H-SiC wafers using a contact of titanium or molybdenum," Microelectronic Engineering, vol. 106, pp. 43–47, Feb. 2013.

[20] M. Vivona, F. Giannazzo, and F. Roccaforte, "Materials and processes for Schottky contacts on silicon carbide," Materials, vol. 15, no. 1, p. 298, Dec. 2021.

A Review of Active Gate Drivers for SiC Power MOSFETs

Yuchu Ge, Wei Jia Zhang*

Department of Electronic and Computer Engineering
The Hong Kong University of Science and Technology

*Email: yc.ge@connect.ust.hk, eewjzhang@ust.hk

Abstract—**Active gate drivers (AGDs) have emerged as a promising field to enhance the efficiency and performance of power MOSFETs by dynamically regulating the hard-switching slew rate, enabling calibrated trade-off among efficiency and EMI suppression, and providing device protection. This paper reviews recent AGD topologies for silicon carbide power MOSFETs, covering the segmented gate-drive strategy, the slew rate adjustment techniques, and the protection schemes for reliability. This paper primarily focuses on topics including short-circuit protection, current balancing, and thermal management in parallel-connected power systems. The benefits, limitations, and applications of each approach are compared and analyzed.**

Keywords—active gate driver, SiC power MOSFET gate driver, segmented control, circuit reliability, EMI Suppression.

I. INTRODUCTION

With the rapid surge in demand for AI computing, electric vehicles (EVs), and renewable energy, there is an increasing need for fast-switching, high-temperature-compliant, and high-power-density modules. Silicon carbide (SiC) has gained increasing traction from both research and industry due to its superior material properties, including high channel mobility, high current handling capability, and high breakdown voltage [1]. These characteristics define SiC as an ideal candidate for next-generation power switches.

In high-power applications, driving SiC power MOSFETs presents significant challenges due to demanding operating conditions. To facilitate high-frequency and high-voltage switching, the advanced gate driving techniques for SiC power MOSFETs have become increasingly essential. As shown in Fig. 1, recent developments in gate drivers have primarily focused on optimizing the switching performance through intelligent slew rate (dV_{DS}/dt and dI_{DS}/dt) control schemes and enhancing the power device reliability through advanced protection circuitries and health monitoring features.

A. Optimizing the Switch Performance

The fast turn-on and turn-off processes of the SiC power MOSETs inherently lead to high voltage and current slew rates. These rapid transitions result in significant electromagnetic interference (EMI), current ringing, voltage over/undershoot, and crosstalk between the high-side and the low-side devices, thereby degrading the system performance. When involving low on-resistances devices, their switching loss constitutes a substantial portion of the total power loss in power converters [1]. The challenges in gate driver design lie in finding ways to suppress EMI, ringing, and crosstalk while maintaining low switching loss. The conventional fixed driving methods can slow down switching to limit the dV_{DS}/dt and dI_{DS}/dt but often lead to an increase in switching loss.

To achieve optimal trade-offs among switching speed, EMI suppression, and energy efficiency, various active gate drivers (AGDs) designs have emerged as promising solutions. Compared with conventional gate drivers, AGDs offer dynamic gate driving strength during the turn-on and turn-off transients, allowing flexible slew rate control to minimize EMI and ringing with less sacrifice of switching efficiency.

B. Enhancing the Device Reliability

The extreme operating conditions of SiC devices introduces additional challenges to their reliability. High-frequency and high-power operation increases the risk of overcurrent events [2], while the small die size and low thermal mass of the SiC power MOSFETs necessitate tighter control of their junction temperatures. Furthermore, under heavy-load conditions with parallel-connected devices, internal and external non-idealities lead to current imbalance and junction temperature variations [3]. These factors can compromise the SiC power MOSFETs' reliability. To mitigate these risks, AGDs and the protection circuitries, are essential for safe and reliable operation.

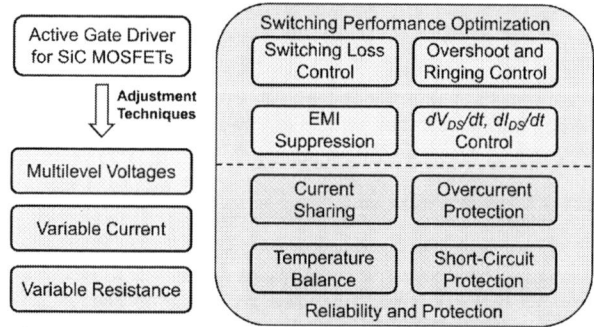

Fig. 1. Functional block diagram of active gate driver.

This paper reviews and summarizes recent researches on active gate driver for SiC power MOSFETs. Section II discusses the segmented control strategy employed during switching processes. Section III examines the mainstream tuning techniques used in AGDs. Section IV highlights recent advancements in protection and monitoring features.

II. SEGMENTED CONTROL STRATEGY

The SiC power MOSFETs' turn-on (turn-off) behavior , similar to those of their silicon counterparts, can be divided into four stages, as illustrated in Fig.2. During the interval t_1–t_3, the device experiences high drain-source voltage and current (V_{DS} and I_{DS}). Exacerbated by the high switching frequency, significant power loss will be incurred. Steep dV_{DS}/dt and dI_{DS}/dt further introduce voltage and current over/undershoot, ringing, and EMI issues. The turn-off process essentially mirrors the turn-on process in the reverse sequence. The trade-offs among switching speed, over-/undershoot, EMI, and switching loss pose significant challenges for the conventional gate driving schemes. Therefore, segmenting the switching process and applying AGD with dynamic driving strength is highly beneficial for optimizing the device performance.

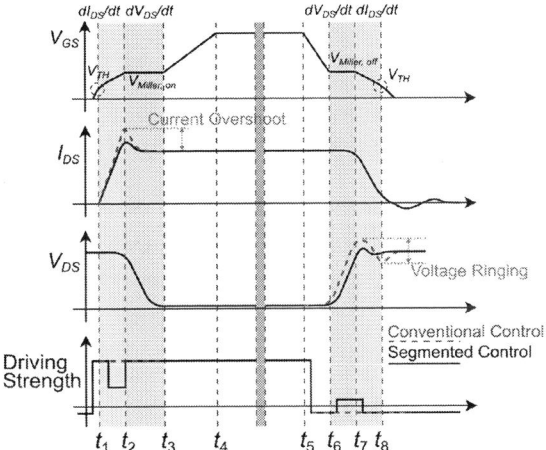

Fig.2. Segmented control strategy illustrated with SiC power MOSFET switching waveforms.

The waveform in Fig.2 illustrates the principle of segmented driving. Optimal performance relies on the precise timing and the careful adjustment of driving strength. Open-loop AGDs predetermine the driving strength and sequence through simulation or empirical tuning [4-6]. While this approach lowers the cost and complexity by eliminating a high-speed feedback control, it lacks adaptability to varying operating conditions . In contrast, closed-loop AGDs actively sense the switching voltage or current signals and apply real-time corrections using high-bandwidth analog or digital controls. As shown in the block diagram on the right side of Fig.3, various feedback techniques have been reported. For example, AGDs in [7, 8] detect the slope of V_{GS} to identify the interval between the start and the end of the Miller plateau, reducing gate driving strength during this interval to mitigate EMI, and increasing driving strength after this interval to reduce switching loss.

Additionally, the parasitic inductive voltage between the Kelvin source and the power source is monitored to estimate dI_{DS}/dt, aiding the identification of current rise phase in t_1–t_2 [9-11]. Other approaches involve directly sampling the V_{DS} or I_{DS} to determine phases associated with dV_{DS}/dt and dI_{DS}/dt [12, 13]. These detected signals are used in the timing control loop to adjust the driving strength on-the-fly.

One way to enhance segmented control AGD is by developing mathematical models. Researchers have proposed cost functions based on key performance metrics to determine the optimal time-domain segmentation strategy for each switching phase under various operating conditions. For example, the cost function in [14] considers both switching loss and dV_{DS}/dt, enabling trade-off by controlling the segment timing and driving strength. This approach assigns equal importance to both metrics. Meanwhile, [15] utilizes a cost function which includes energy loss, dV_{DS}/dt and dI_{DS}/dt, with adjustable weights to suit different operating conditions. The ADG in [16] also divides the switching process into multiple time segments and introduces three types of cost functions, which incorporate correction factors to penalize deviations from target values. To address the challenge of weight assignment, the method in [12] uses an exhaustive search to identify the maximum overshoot voltage across all operating conditions, then creates a look-up table to select the best weights for each scenario. Ultimately, the trade-offs between various metrics are critical to gate driver design, and achieving the optimal driving strength allocation remains a significant challenge.

III. ACTIVE GATE DRIVER FOR SiC POWER MOSFETs

Current AGDs can be broadly classified into three types based on the technique to adjust their driving strength. Each type provides a distinct control mechanism and slew rate resolution during the switching transients. In this section, various state-of-the-art AGDs are analyzed.

A. Multilevel Voltage Control

Multilevel voltage driving is an area of interest in advanced gate drivers. A conventional voltage-source driver (VSD) utilizes two power rails to set the gate voltages (V_{GS}) during the turn-on and turn-off, respectively. In contrast, multilevel voltage driving employs more than two discrete voltage levels, allowing segmented adjustment throughout the switching process. Lower drive voltages are applied during damping phases to suppress EMI, while higher voltages are used for fast gate charging/discharging to minimize loss.

Fig.3 (1.a) illustrates a typical variable voltage drive scheme with an intermediate power rail , providing an additional V_{GS} option during both the turn-on and turn-off processes separately [14]. Similar approaches with additional power rails are also used in [17-18]. During switching, the use of an adjustable LDO could allow a variable intermediate voltage (V_{INT}), enabling finer control. The ADG in [15] implements a digitally controlled 128-level adjustable V_{INT}, allowing higher adjustment resolution between EMI and efficiency across varying operating conditions. An alternative technique for generating multilevel voltages utilizes

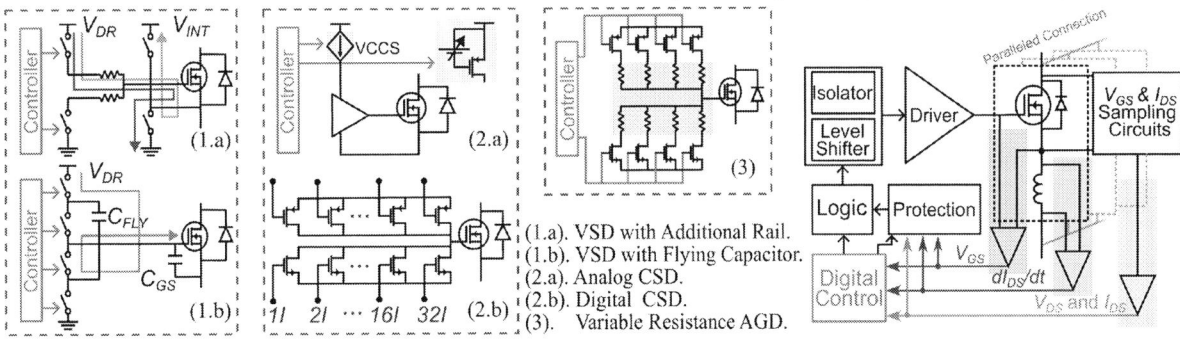

(1.a). VSD with Additional Rail.
(1.b). VSD with Flying Capacitor.
(2.a). Analog CSD.
(2.b). Digital CSD.
(3). Variable Resistance AGD.

Fig.3. Block diagram of the gate driver for SiC power MOSFETs and recent active gate driving control approaches.

979-8-3315-3918-4/25 $31.00 © 2025 IEEE

capacitive voltage division [19], as shown in Fig.3(1.b), a flying capacitor connects in series with the gate capacitance, producing multiple V_{GS} levels for enhanced switching control.

Multilevel gate driver provides an effective operating principle and high adjustment resolution. However, when implementing these drivers into integrated circuits, it is essential to ensure that the power rails are designed with adequate sourcing and sinking capability.

B. Variable Current Control

Current-source based drivers (CSD) provide direct regulation of the gate current which allows precise and responsive control of switching transients. Their simplified integration, compared to the voltage-driven approaches, has led to wide adoption in recent research[8, 10, 20, 21].

During switching, the gate drive current is closely correlated with the dV_{DS}/dt. For example, [5] proposes an analog approach that adjusts the gate current by modulating the V_{GS}, as shown in fig.3 (2.a). Alternatively, current sources can be constructed using bipolar junction transistors, with different current levels set by adjusting the values of external resistors [6]. These solutions offer simple adjustment for different devices and changing operating conditions, based on external pre-setting trigger references. As open-loop controls, they may yield suboptimal trade-offs outside their calibrated ranges [22]. To address this limitation, ADGs reported in [12] and [23] incorporate a high-bandwidth peak voltage sampling module as feedback. In each switching cycle, the peak V_{DS} is captured, and the resulting error signal is fed back to dynamically adjust the gate drive current in real time.

Fig.3 (2.b) illustrates digital current-source-based AGDs [8, 20, 21] which employ segmented gate current control across different switching phases. The gate current modulation is realized using parallel MOSFETs with aspect ratios scaled in binary. By digitally selecting various MOSFETs combinations within pull-up and pull-down arrays, the gate current achieves 6-bit resolution. This approach enables precise and flexible gate current shaping during critical switching intervals.

C. Variable Resistance Control

The variable resistance methods modulate the gate resistance during the switching transients to provide variable driving strength. This approach is compact, low cost, and integration-friendly, but typically lacks fine adjustment and can introduce additional power loss in the gate current path.

As shown in Fig.3 (3), a common implementation of a variable resistance involves adding resistors and switches in series with the gate node. By selectively enabling the switches, the gate resistance can be adjusted [24]. Recent works [7][25] have further simplified the design by utilizing the on-resistance of MOSFETs in place of discrete resistors, reducing circuit complexity and enhancing the potential for on-chip integration. Additionally, recent studies such as [13] have combined variable resistance control with multi-voltage rails, enabling more versatile and adaptable driving scheme.

IV. PROTECTION AND MONITORING SCHEME

A. Short-Circuit Detection and Protection

SiC power MOSFETs inherently have a lower overcurrent withstand time (less than 3 µs), compared to IGBTs (less than 10 µs) [2]. Therefore, fast short-circuit detection circuits are crucial to the longevity of the SiC power MOSFETs. Short-circuit (SC) in SiC power MOSFETs are generally classified into Hard Switching Faults (HSF) and Faults Under Load (FUL). HSF denotes events where the device is turned on into an already short-circuited load, whereas FUL refers to short-circuit events that arise during device conduction.

Common detection methods include: (1) dI_{DS}/dt detection, which senses the voltage across parasitic inductance generated by steep change in drain current; this method is sensitive to high bandwidth SC but exhibits reduced sensitivity to overcurrent events with low dI_{DS}/dt; (2) Desaturation detection (DESAT), which identifies SC or overcurrent (OC) conditions by monitoring rapid changes in V_{DS}; while simple to implement, it requires a blanking time to prevent false triggering, limiting its effectiveness in fast SiC applications; (3) SenseFET detection, Which uses a current mirror-like structure to replicate the drain current in a separate FET for measurement, enabling accurate and fast current sensing, but necessitating custom devices matched to the main power MOSFET; and (4) Rogowski coils and Current Transformer, which are non-contact, galvanically isolated methods suitable for large current measurement, though their size and cost reduce their practicality in compact applications.

The limitations of traditional detection techniques have driven recent research toward hybrid and active detection strategies. For example, combining DESAT and dI_{DS}/dt sensing has been proposed to achieve effective detection of both SC and OC events [26]. In [27] dI_{DS}/dt sensing is paired with V_{GS} detection to address both HSF and FUL fault types. In [28], DESAT with adaptive blanking times adjusted based on the previous cycles' V_{DS} and V_{GS} significantly improves response speed. Another DESAT method adjusts the blanking time according to temperature changes with a temperature-dependent reference [29]. Beyond electrical signal, temperature-based SC protection has been explored in [30], where authors argue that the root cause of SC failure in SiC power MOSFETs is primarily due to thermal runaway. By implementing instantaneous temperature monitoring and power estimation, the SC events can be identified with greater reliability and broader fault coverage.

B. Reliability in Parallel-Device Applications

Junction temperature imbalance and poor current-sharing performance are common challenges when operating devices in parallel, both can significantly degrade device lifetime and the system reliability. Junction temperature imbalance is typically a consequence from parasitic elements, PCB asymmetry, and uneven ambient temperature across devices. Dynamic balancing of junction temperature can be achieved by regulating the current through each parallel device. A multilevel topology has been proposed to flexibly adjust V_{GS} values, thereby altering on-state resistance of SiC power MOSFETs to redistribute current and compensate for temperature differences [19]. Alternatively, junction temperature control can also be implemented by adjusting the turn-on delay of individual devices using a digital buffer; turning on devices at higher junction temperature later decreases the switching loss, thereby providing thermal compensation [3].

Poor current-sharing can result from factors such as parasitic elements, device mismatches, and inconsistent gate drive propagation delays. To address this, [11] utilizes closed-loop adjustment of the gate drive current, achieving real-time current balancing among parallel SiC power MOSFETs.

979-8-3315-3918-4/25 $31.00 © 2025 IEEE

Moreover, a local digital control of the isolated gate driver, as proposed in [9], can generate individual delay signals to calibrate start and end time points for rising current. Additionally, segmented-control AGD approaches presented in [21], dynamically adjusts both the driving strength and the on-state V_{GS}, enhancing the current-sharing performance across both switching and conduction phases.

V. CONCLUSION

AGD technology plays a crucial role in optimizing switching performance and enhancing the operational reliability of SiC-based systems. The surveyed methodologies highlight how slew rate control can achieve an optimal balance between efficiency and EMI mitigation, while also providing a foundation for advanced reliability features. However, despite significant advancements, generalizability and reliability of AGDs under different operation conditions (temperature, load current…) remain key challenges, often restricting their commercial implementations. Future development should focus on integrated, cost-effective AGDs designs with precise sensing and rapid feedback control to fully unlock the potential of SiC power MOSFETs in high-power applications.

REFERENCES

[1] X. Yuan, I. Laird, and S. Walder, "Opportunities, Challenges, and Potential Solutions in the Application of Fast-Switching SiC Power Devices and Converters," IEEE Transactions on Power Electronics, vol. 36, no. 4, pp. 3925-3945, 2021.

[2] S. Mocevic et al., "Comparison and Discussion on Shortcircuit Protections for Silicon-Carbide MOSFET Modules: Desaturation Versus Rogowski Switch-Current Sensor," IEEE Transactions on Industry Applications, vol. 56, no. 3, pp. 2880-2893, 2020.

[3] J. Brandelero, J. Ewanchuk, and S. Mollov, "Selective Gate Driving in Intelligent Power Modules," IEEE Transactions on Power Electronics, vol. 36, no. 1, pp. 898-910, 2021.

[4] S. Acharya et al., "Active Gate Driver for SiC-MOSFET-Based PV Inverter With Enhanced Operating Range," IEEE Transactions on Industry Applications, vol. 55, no. 2, pp. 1677-1689, 2019.

[5] Y. Ding et al., "A Cost-Efficient Active Gate Driver for Seamless Slew Rate Control of SiC MOSFETs," IEEE Transactions on Power Electronics, vol. 39, no. 10, pp. 12558-12569, 2024.

[6] X. Wang, H. Wu, and V. Pickert, "A cost-efficient Current-Source Gate Driver for SiC MOSFET Module and its Comparison with Voltage-Source Gate Driver," in 2020 IEEE 9th International Power Electronics and Motion Control Conference (IPEMC2020-ECCE Asia), 2020, pp. 979-984.

[7] W. T. Cui et al., "A Dynamic Gate Driver IC with Automated Pattern Optimization for SiC Power MOSFETs," in 2022 IEEE 34th International Symposium on Power Semiconductor Devices and ICs (ISPSD), 2022, pp. 33-36.

[8] T. W. Wang et al., "Active Gate Driver IC Integrating Gate Voltage Sensing Technique for SiC MOSFETs," IEEE Transactions on Power Electronics, vol. 39, no. 7, pp. 8562-8571, 2024.

[9] Y. Wen, Y. Yang, and Y. Gao, "Active Gate Driver for Improving Current Sharing Performance of Paralleled High-Power SiC MOSFET Modules," IEEE Transactions on Power Electronics, vol. 36, no. 2, pp. 1491-1505, 2021.

[10] C. W. Kuo et al., "Closed-Loop Gate-Sensing Active Driver IC with Adaptive Delay Compensation Technique for Silicon Carbide Power MOSFETs," in 2024 36th International Symposium on Power Semiconductor Devices and ICs (ISPSD), 2024, pp. 462-465.

[11] X. Wang et al., "An Active Gate Driver for Dynamic Current Sharing of Paralleled SiC MOSFETs," in 2021 IEEE Energy Conversion Congress and Exposition (ECCE), 2021, pp. 5407-5411.

[12] X. Chen et al., "A Novel Control Strategy for Optimal Tradeoff between Overshoot and Switching Loss Based on Double Closed-Loop Self-Regulating Active Gate Driver," IEEE Transactions on Power Electronics, vol. 39, no. 10, pp. 13033-13043, 2024.

[13] Q. Li et al., "Active Gate Driver With the Independent Suppression of Overshoot and Oscillation for SiC MOSFET Modules," IEEE Transactions on Industrial Electronics, vol. 72, no. 3, pp. 2325-2335, 2025.

[14] Y. Yang, Y. Wen, and Y. Gao, "A Novel Active Gate Driver for Improving Switching Performance of High-Power SiC MOSFET Modules," IEEE Transactions on Power Electronics, vol. 34, no. 8, pp. 7775-7787, 2019.

[15] S. Zhao, et al., "An Intelligent Versatile Model-Based Trajectory-Optimized Active Gate Driver for Silicon Carbide Devices," IEEE Journal of Emerging and Selected Topics in Power Electronics, vol. 8, no. 1, pp. 429-441, 2020.

[16] J. Wiesemann and A. Mertens, "An Isolated Variable-Resistance Active Gate Driver for Use in SiC-Driven Inverters," in IECON 2021 – 47th Annual Conference of the IEEE Industrial Electronics Society, 2021, pp. 1-6.

[17] S. Zhao et al., "Adaptive Multi-Level Active Gate Drivers for SiC Power Devices," IEEE Transactions on Power Electronics, vol. 35, no. 2, pp. 1882-1898, 2020.

[18] H. B. Ekren et al., "Four Level Voltage Active Gate Driver for Loss and Slope Control in SiC MOSFETs," in 2022 IEEE 13th International Symposium on Power Electronics for Distributed Generation Systems (PEDG), 2022, pp. 1-6.

[19] J. Liang, L. Sun, W. T. Cui, W. T. Ng, M. Iwamoto, and H. Nishio, "A Multi-Level Active Gate Driver for Achieving Thermal Balance in Parallel Connected Power MOSFETs," in 2025 IEEE Applied Power Electronics Conference and Exposition (APEC), 2025, pp. 1108-1113.

[20] Y. Liang, K. Hata, and M. Takamiya, "Fully Integrated Closed-Loop Active Gate Driver IC With Real-Time Control of Gate Current Change Timing by Gate Current Sensing," in IEEE Applied Power Electronics Conference and Exposition (APEC), 2025, pp. 1084-1089.

[21] K. Horii et al., "Single-Input Dual-Output Digital Gate Driver IC Automatically Equalizing Drain Current Variations of Two Parallel-Connected SiC MOSFETs," IEEE Transactions on Power Electronics, vol. 40, no. 1, pp. 467-485, 2025.

[22] M. M. Alam, S. Khalid, and N. Ho, "Operation Point based Optimization of Switching Losses with Current-Source Gate Driver for SiC-based Power Modules," in 2024 IEEE Applied Power Electronics Conference and Exposition (APEC), 2024, pp. 2502-2509.

[23] Z. Gao, J. Zhang, Y. Huang, R. Guan, and Y. Zhou, "A Closed-Loop Active Gate Driver of SiC MOSFET for Voltage Spike Suppression," IEEE Open Journal of Power Electronics, vol. 3, pp. 723-730, 2022.

[24] G. Engelmann, T. Senoner, and R. W. D. Doncker, "Experimental investigation on the transient switching behavior of SiC MOSFETs using a stage-wise gate driver," CPSS Transactions on Power Electronics and Applications, vol. 3, no. 1, pp. 77-87, 2018.

[25] M. Wang et al., "A Smart Gate Driver for SiC Power MOSFETs with Aging Compensation and Ringing Suppression," in 2021 33rd International Symposium on Power Semiconductor Devices and ICs (ISPSD), 2021, pp. 67-70.

[26] T. Li et al., "Short-circuit and overcurrent protection scheme of SiC MOSFET based on combined protection method," in PCIM Asia 2022; International Exhibition and Conference for Power Electronics, Intelligent Motion, Renewable Energy and Energy Management, 2022, pp. 1-8.

[27] Z. Li, B. Ji, K. Tan, and W. Cao, "A Review of Short Circuit Detection Sensor for SiC MOSFET," in 2022 9th International Forum on Electrical Engineering and Automation (IFEEA), 2022, pp. 1-5.

[28] T. Liu, X. Cheng, and W. T. Ng, "A Self-Adjustable Blanking Time Short Circuit Protection Circuit for SiC Power MOSFETs," in 2023 20th China International Forum on Solid State Lighting & 2023 9th International Forum on Wide Bandgap Semiconductors (SSLCHINA: IFWS), 2023, pp. 153-156.

[29] Y. Quan et al., "Double-Voltage Short-Circuit Protection Circuit for SiC MOSFETs with Temperature Compensation," in 2024 IEEE Workshop on Wide Bandgap Power Devices and Applications in Europe (WiPDA Europe), 2024, pp. 1-5.

[30] Y. Wen, Y. Yang, and Y. Li, "A Novel Short-Circuit Protection Scheme for Silicon Carbide (SiC) MOSFET Module Considering Operation Temperature," IEEE Transactions on Power Electronics, vol. 40, no. 8, pp. 10661-10671, 2025.

A Quadruple RESURF LDMOS with Enhanced Hot-Carrier-Induced Degradation Immunity

Wenliang Liu
State Key Laboratory of Electronic Thin Films and Integrated Devices
University of Electronic Science and Technology of China
Chengdu, China
liuwenliang.97@foxmail.com

*Ming Qiao
State Key Laboratory of Electronic Thin Films and Integrated Devices
University of Electronic Science and Technology of China
Chengdu, China
qiaoming@uestc.edu.cn

Penglong Xu
Process Intergration Technology Development Center
CSMC Technologies Co. Ltd.
Wuxi, China
XUPENGLONG@csmc.crmicro.com

Chunxia Ma
Process Intergration Technology Development Center
CSMC Technologies Co. Ltd.
Wuxi, China
MACX@csmc.crmicro.com

Feng Lin
Process Intergration Technology Development Center
CSMC Technologies Co. Ltd.
Wuxi, China
linfeng@csmc.crmicro.com

Bo Zhang
State Key Laboratory of Electronic Thin Films and Integrated Devices
University of Electronic Science and Technology of China
Chengdu, China
zhangbo@uestc.edu.cn

Abstract—In this work, a 200 V shallow trench isolation (STI) based quadruple REduced SURface Field (RESURF) lateral double-diffused MOSFET (LDMOS) is proposed. Compared to the conventional LDMOS, this novel structure is characterized by additional P-N-P-N layers in the N-type drift region. The implantation doses and energies of these layers are thoroughly optimized using the technology computer-aided design (TCAD) tools. Notably, at the on-state stress condition, perpendicular electrical field and impact ionization (I.I.) generation rates of the proposed device at silicon/oxide (Si/SiO₂) interface are considerably relieved, thus limiting the hot carries injection (HCI) effects. Furthermore, a comprehensive R_{on} degradation model is proposed and validated by comparison with the experiments.

Keywords—reduced surface field, PNPN LDMOS, shallow trench isolation, hot carrier injection, R_{on} degradation

I. INTRODUCTION

Lateral double-diffused MOSFET (LDMOS) transistors are widely used as power switches, amplifiers, drivers in various integrated circuit applications such as ac-dc LED driver, flat panel display, switched-mode power supply, and so on [1], [2]. Two significant parameters in the design consideration are specific on-state resistance ($R_{on,sp}$) and off-state breakdown voltage (V_B) and the REduced SURface Field (RESURF) technique is frequently used to optimize the tradeoff between $R_{on,sp}$ and V_B [3], [4]. However, during the switching operation applications, high drain voltages can result in high electric field and high impact ionization (I.I.) rates within the device, leading to severe hot carrier injection (HCI) reliability issues [5], [6]. For conventional LDMOS, the highly doped N-type drift region (N-Drift) tends to direct the majority carriers to flow along the silicon/oxide (Si/SiO₂) interface [7], [8]. As a result, carriers gaining enough kinetic energy can be injected into the above oxide layer, then become fixed charges and potentially break the Si-H bonds [9], [10], causing the electrical parameters to degrade and posing significant challenges to the long-term reliability of LDMOS devices.

This work was supported by the National Natural Science Foundation of China under Grant 62174024.

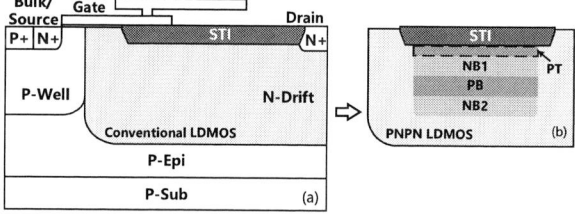

Fig. 1. Schematic cross-sectional views of the (a) conventional LDMOS, and (b) proposed PNPN LDMOS with embedded P-N-P-N layers in the N-type drift region.

To address this limitation, a novel quadruple RESURF LDMOS (PNPN LDMOS) is proposed. Compared to the conventional device, this design introduces additional P-N-P-N layers in the N-Drift, which only requires one additional mask. After optimizing the implantation doses and energies of P-N-P-N layers, the proposed PNPN LDMOS experimentally exhibits enhanced HCI immunity with considerably decreased on-state resistance (R_{on}) degradation.

II. DEVICE STRUCTURE AND PROCESS

The schematic cross-sectional views of conventional and proposed PNPN LDMOS devices are illustrated in Fig. 1 (a) and (b). The devices under investigation feature a gate width of 40 µm, a gate oxide thickness of 12 nm. For both devices, the operational drain-source voltage (V_{ds}) is 200 V and gate-source voltage (V_{gs}) is 5 V. As shown in Fig. 1(b), these embedded layers in PNPN LDMOS are represented by PT (P-top), NB1 (N-bury1), PB (P-bury) and NB2 (N-bury2). The introduction of PT layer can effectively relieve the perpendicular electrical field and I.I. generation rates at the Si/SiO₂ interface. Besides, layers NB1 and NB2 can provide low R_{on} paths for majority carriers. According to the RESURF principle, a full depletion is required to achieve high V_B. While the introduction of P-N-P-N layers complicates the charge distribution among these layers, N-Drift and the underlying epitaxy layer. Failure to achieve charge balance within the device leads to premature breakdown, compromising the device's performance. Therefore, careful optimization of the implantation doses and energies of P-N-P-N layers is essential to ensure charge balance of this novel structure.

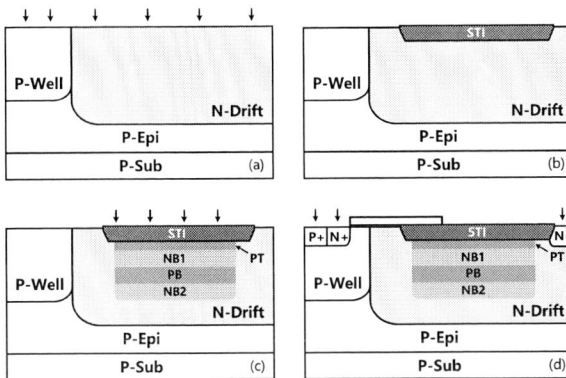

Fig. 2. The key manufacturing process flows of the proposed PNPN LDMOS. (a) P-Well and N-Drift implantation and diffusion. (b) The STI formation. (c) The implantation of NB2, PB, NB1 and PT layers in sequence, using one same mask. (d) The formation of gate oxidation, poly deposition and N+/P+ implantation.

The key manufacturing process flows of PNPN LDMOS are illustrated in Fig. 2. The process begins with a p-type substrate (P-Sub) as the base wafer. Next, a p-type epitaxy layer (P-Epi) with a resistivity of 45 $\Omega \cdot$cm is grown on the P-Sub. In the second stage, phosphorus and boron ions are sequentially implanted into the P-Epi layer. Then followed by a thermal diffusion process to form the N-Drift and p-type well (P-Well) regions, as shown in Fig. 2(a). Following this, the shallow trench isolation (STI) structure is formed through etching, deposition and thermal processes, as shown in Fig. 2(b). The STI structure is essential for device isolation, which can reduce the parasitic capacitance and leakage currents while improving the overall electrical performance. As shown in Fig. 2(c), the subsequent critical step introduces the P-N-P-N layers into N-Drift by multiple ion implantation processes, which follows a deep-to-shallow sequence to avoid the potential blocking effects that earlier-implanted ions may impose on the subsequent ions. Then the process continues with gate oxidation, polysilicon deposition and N+/P+ formation. Compared to conventional LDMOS fabrication processes, the key difference of PNPN LDMOS processes is an additional implantation step to introduce the P-N-P-N layers in N-Drift. Notably, this modification requires only one additional mask, making the proposed process flow both efficient and cost-effective.

III. RESULTS AND DISCUSSION

Extensive simulation works are conducted using Sentaurus technology computer-aided design (TCAD) tools to validate and optimize the key parameters of the P-N-P-N layers. Specifically, the implantation energies (E_{PT}, E_{NB1}, E_{PB} and E_{NB2}) and doses (D_{PT}, D_{NB1}, D_{PB} and D_{NB2}) for each layer are systematically adjusted to ensure an optimal device performance. The ion concentration distributions for each layer are illustrated in Fig. 3 (a) – (d), respectively. The peak concentration depth is determined by the implantation energy. In other words, as the implantation energy increases, the peak concentration shifts deeper, signifying the necessity for precise calibration to achieve a desired dopant profile.

As shown in Fig. 3(a), when a lower implantation energy (100 KeV) is applied for PT layer, the peak concentration of boron (N_{PT}) is driven into the above STI region, preventing a proper formation of the PT layer. Nevertheless, applying a higher implantation energy (200 KeV) for PT layer will sacrifice the low R_{on} conduction path provided by NB1 layer.

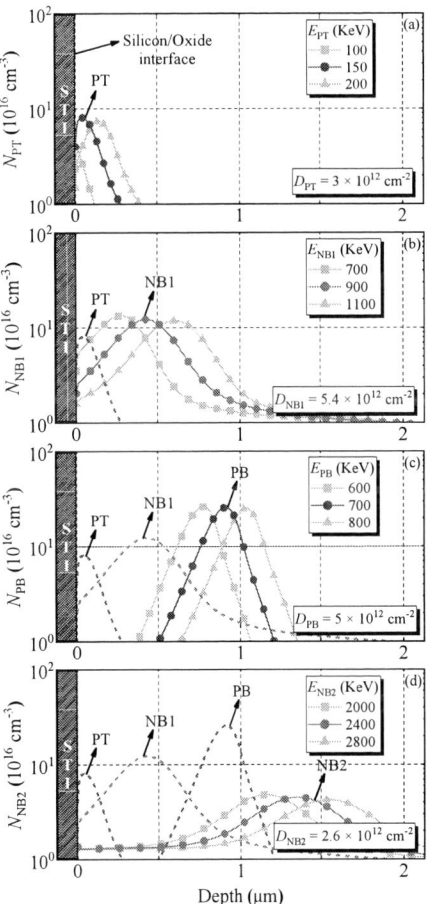

Fig. 3. The corresponding ion concentration distributions after implantation at different energies of (a) PT, (b) NB1, (c) PB and (d) NB2 layers, represented by E_{PT}, E_{NB1}, E_{PB} and E_{NB2}, respectively.

Therefore, this trade-off highlights the critical need to ensure a proper dopant distribution and the implantation energy of PT layer is optimized to 150 KeV. Similarly, as shown in Fig. 3(b), the implantation energy for NB1 layer presents the same challenge. A low NB1 implantation energy (700 KeV) can lead to the inversion of PT layer, preventing the formation of PT layer. On the other hand, an excessively high energy (1100 KeV) prevents a full depletion between the PT and NB1 layers, risking potential premature breakdown. As a result, the implantation energy of NB1 layer is optimized to 900 KeV. Moreover, the same principle is applied to the implantation

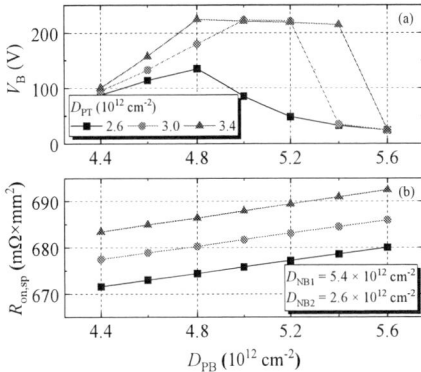

Fig. 4. Simulated (a) V_B and (b) $R_{on,sp}$ versus D_{PB} with different D_{PT}, at D_{NB1} = 5.4 × 10^12 cm^-2 and D_{NB2} = 2.6 × 10^12 cm^-2.

energies for PB and NB2 layers, as shown in Fig. 3(c) and Fig. 3(d), respectively. Consequently, the implantation energies of PB and NB2 layers are optimized to 700 KeV and 2400 KeV to ensure a proper peak dopant profile.

Furthermore, the implantation doses of P-N-P-N layers are also optimized considering the trade-off between $R_{on,sp}$ and V_B. As illustrated in Fig. 4(a), for any given dose of PT layer, there is an optimized dose of PB that maximizes V_B. The trend shows that as the dose of PB layer increases, V_B initially rises, reaching a peak then declines. This phenomenon is explained by the mechanism of charge imbalance induced premature breakdown: both excessively low and high dose of PB will result in charge imbalance, leading to premature breakdown and thus a V_B reduction. The high V_B corresponds to the condition that a proper charge balance is obtained. On the other hand, as shown in Fig. 4(b), $R_{on,sp}$ increases progressively with the dose rise of both PT and PB layers. This phenomenon is explained by the enhanced compensation effect by PT and PB layers, where increasing D_{PT} and D_{PB} sacrifices the low R_{on} conduction paths provided by NB1 and NB2 layers, and thereby an increased $R_{on,sp}$.

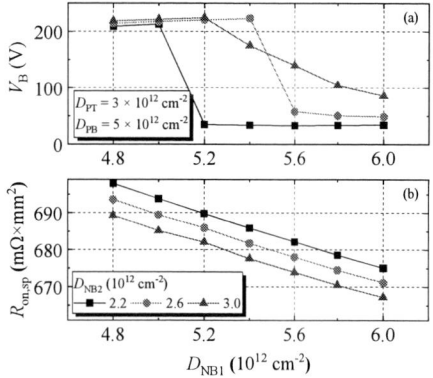

Fig. 5. Simulated (a) V_B and (b) $R_{on,sp}$ versus D_{NB1} with different D_{NB2}, at $D_{PT} = 3 \times 10^{12}$ cm^{-2} and $D_{PB} = 5 \times 10^{12}$ cm^{-2}.

Fig. 5(a) further demonstrates that for any given dose of NB2 layer, an optimized dose of NB1 can be determined that achieves the maximum V_B. As the dose of NB1 increases, V_B follows a similar trend of rising initially then decreasing once charge imbalance begins to dominate. In this case, the physical mechanism is the same to that observed in Fig. 4(a), where charge balance is critical to obtain an optimized V_B. Moreover, Fig. 5(d) shows that $R_{on,sp}$ decreases as the dose increases of both NB1 and NB2 layers, which is attributed to the introduction of more majority carriers within the conduction paths, thereby effectively reducing $R_{on,sp}$. This indicates that by appropriately increasing the dose of both NB1 and NB2 layers, the compensation effect can be optimized to reduce $R_{on,sp}$, while maintaining a high V_B.

TABLE I.   EXPERIMENTAL DEVICE ELECTRICAL PARAMETERS

Parameter	Conventional	PNPN LDMOS
$R_{on,sp}$ (mΩ×mm^2)	737.2	680.1
V_B (V)	258.3	257.8
V_{th} (V)	0.98	0.98

The experimental device parameters are summarized in Table. I. Obviously, due to the introduction P-N-P-N layers into N-Drift, the $R_{on,sp}$ of PNPN LDMOS can be decreased around 7.75 %, while maintain the same V_B. Besides, the threshold voltage (V_{th}) of the proposed PNPN LDMOS remains the same as the conventional device.

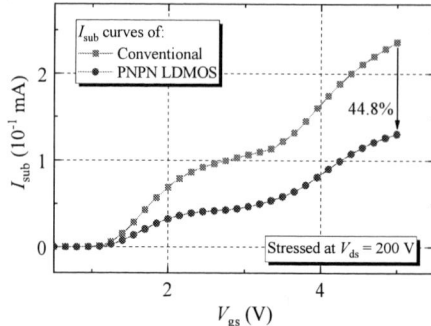

Fig. 6. The experimental bulk current (I_{sub}) curves as a function of V_{gs} for the conventional and proposed PNPN LDMOS devices.

The experimental bulk current (I_{sub}) curves, stressed at V_{ds} = 200 V, for both conventional and PNPN LDMOS devices are illustrated in Fig. 6. It's well known that I_{sub} consists of holes generated by impact ionization [11]. As a result, the measurement of I_{sub} curves provides an excellent indicator for the generation rate of electron-hole pairs within the device [12], [13]. Obviously, for both devices, the stress condition of V_{gs} = 5 V exhibits the maximum I_{sub}. Therefore, the following degradation mechanisms are investigated under the stress condition of V_{gs} = 5 V and V_{ds} = 200 V. Notably, compared to conventional LDMOS, the I_{sub} curve of PNPN LDMOS exhibits a reduction of around 44.8 % at V_{gs} = 5 V, indicating a considerable reduction of impact ionization rates within PNPN LDMOS.

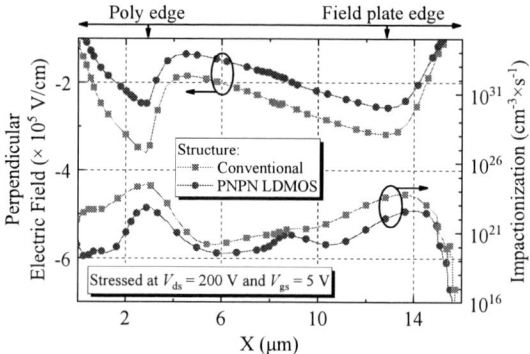

Fig. 7. The perpendicular electrical fields and I.I. generation rates distributions along the Si/SiO$_2$ interface for both devices, at the stress condition of V_{ds} = 200 and V_{gs} = 5 V.

The perpendicular electrical field and I.I. generation rates distributions along the Si/SiO$_2$ interface, under V_{gs} = 5 V and V_{ds} = 200 V, are illustrated in Fig. 7. Accordingly, for both devices, the main peak of perpendicular electrical fields are located at the poly-edge and field plate edge, and their values are negative (the direction is pointing to the device surface), which are in favor of the injection of hot holes [14]. These injected holes can attract mirror negative charges and then increase the majority carrier density in N-Drift, leading to the decrease of R_{on}. Meanwhile, the I.I. generation rates at these two locations are also high, suggesting large numbers of interface states generation here. Opposite to hot holes injection, the interface states generation degrades the mobility of majority carrier, leading to the increase of R_{on}. As shown in Fig. 7, compared to conventional LDMOS, both the perpendicular electrical field and I.I. generation rates of PNPN LDMOS are considerably relieved. As a result, the injection of hot holes and generation of interface states along the Si/SiO$_2$ interface of PNPN LDMOS are relieved.

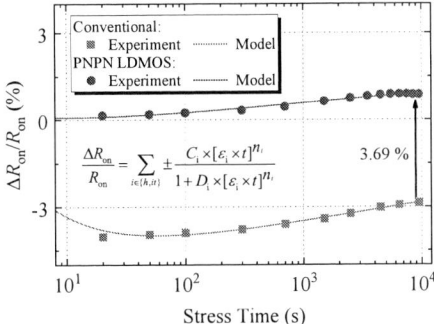

Fig. 8. The experimental R_{on} degradation data (dots) versus stress time for both devices, and the degradation data have been fitted with the R_{on} degradation model (lines).

TABLE II. MODEL PARAMETERS

Parameter	Conventional	PNPN LDMOS
C_h	0.172	0.622
D_h	$C_h/0.27$	$C_h/0.78$
C_{it}	0.273	0.692
D_{it}	$C_{it}/1.035$	$C_{it}/0.792$
n_h	0.875	0.29
n_{it}	0.016	0.278

The experimental R_{on} degradation data for both devices, under $V_{gs} = 5$ V and $V_{ds} = 200$ V, are illustrated in Fig. 8. For conventional LDMOS, the R_{on} decreases first due to the injection of hot holes (driven by high perpendicular electrical field), and rapidly saturates. After that, the generation of interface states dominates the R_{on} degradation, exhibiting a gradual R_{on} increase. On the other hand, for PNPN LDMOS, due to the reduction of hot holes injection (relieved perpendicular electrical field), the R_{on} degradation is primarily dominated by interface states generation and exhibits a gradual R_{on} increase. Notably, the R_{on} degradation slope of PNPN LDMOS is also slighter than that of the conventional LDMOS, which is because of the mitigated generation of interface states.

Overall, the R_{on} degradation results from the combined effects of hot hole injection and interface states generation. Additionally, the R_{on} degradation can be modeled by the following equation [9], [15]:

$$\frac{\Delta R_{on}}{R_{on}} = \frac{C_{it} \times [\varepsilon_{it} \times t]^{n_{it}}}{1 + D_{it} \times [\varepsilon_{it} \times t]^{n_{it}}} - \frac{C_h \times [\varepsilon_h \times t]^{n_h}}{1 + D_h \times [\varepsilon_h \times t]^{n_h}} \quad (1)$$

where the first and second term represent the interface states generation and hot holes injection caused R_{on} degradation, respectively. The parameters C and D competitively determine the magnitude and saturation effects of R_{on} degradation, while n describes the time dependency, and ε is the acceleration factor of R_{on} degradation. For interface states generation, the sign in front is positive, indicating the positive R_{on} degradation. In contrast, for hot holes injection, the sign in front is negative, corresponding to the negative R_{on} degradation. Based on the experimental degradation data in Fig. 8, the R_{on} degradation model is calculated for both devices, and Table. II exhibits all the model parameters used in fitting. Obviously, the experimental data and degradation model in Fig. 8 exhibits a good agreement, proving that the degradation mechanisms discussed above are responsible for the R_{on} degradation.

IV. CONCLUSION

In this work, a novel quadruple RESURF LDMOS with embedded P-N-P-N layers is proposed. Compared to conventional device, the perpendicular electrical field and impact ionization generation rates of proposed PNPN LDMOS at the Si/SiO$_2$ interface are considerably relieved. Therefore, for PNPN LDMOS, an enhanced HCI immunity is obtained, exhibiting a R_{on} degradation less than 0.87 % after stressing 10^4 seconds at $V_{ds} = 200$ V and $V_{gs} = 5$ V, and showing a reduction of $R_{on,sp}$ around 7.75 %, due to the introduction of P-N-P-N layers in the N-Drift.

REFERENCES

[1] F. Udrea, G. Deboy, and T. Fujihira, "Superjunction Power Devices, History, Development, and Future Prospects," *IEEE Transactions on Electron Devices*, vol. 64, no. 3, pp. 713–727, Mar. 2017, doi: 10.1109/TED.2017.2658344.

[2] A. Yoo, J. C. W. Ng, J. K. O. Sin, and W. T. Ng, "High performance CMOS-compatible super-junction FINFETs for Sub-100V applications," in *2010 International Electron Devices Meeting*, Dec. 2010, p. 20.7.1-20.7.4. doi: 10.1109/IEDM.2010.5703402.

[3] M. Imam *et al.*, "Design and optimization of double-resurf high-voltage lateral devices for a manufacturable process," *IEEE Trans. Electron Devices*, vol. 50, no. 7, pp. 1697–1701, Jul. 2003, doi: 10.1109/TED.2003.814981.

[4] B. Duan, Z. Cao, S. Yuan, and Y. Yang, "Complete 3D-Reduced Surface Field Superjunction Lateral Double-Diffused MOSFET Breaking Silicon Limit," *IEEE Electron Device Lett.*, vol. 36, no. 12, pp. 1348–1350, Dec. 2015, doi: 10.1109/LED.2015.2493080.

[5] S. Liu *et al.*, "A review on hot-carrier-induced degradation of lateral DMOS transistor," *IEEE Trans. Device Mater. Rel.*, vol. 18, no. 2, pp. 298–312, Jun. 2018, doi: 10.1109/TDMR.2018.2833490.

[6] D. Varghese, P. Moens, and M. A. Alam, "on-State Hot Carrier Degradation in Drain-Extended NMOS Transistors," *IEEE Transactions on Electron Devices*, vol. 57, no. 10, pp. 2704–2710, Oct. 2010, doi: 10.1109/TED.2010.2059632.

[7] S. Reggiani *et al.*, "TCAD simulation of hot-carrier and thermal degradation in STI-LDMOS transistors," *IEEE Trans. Electron Devices*, vol. 60, no. 2, pp. 691–698, Feb. 2013, doi: 10.1109/TED.2012.2227321.

[8] J. F. Chen, J. R. Lee, K.-M. Wu, T.-Y. Huang, and C. M. Liu, "Effect of drift-region concentration on hot-carrier-induced $r_{\rm on}$ degradation in nLDMOS transistors," *IEEE Electron Device Lett.*, vol. 29, no. 7, pp. 771–774, Jul. 2008, doi: 10.1109/LED.2008.2000610.

[9] P. Moens and G. Van Den Bosch, "Characterization of total safe operating area of lateral DMOS transistors," *IEEE Trans. Device Mater. Rel.*, vol. 6, no. 3, pp. 349–357, Sep. 2006, doi: 10.1109/TDMR.2006.882212.

[10] S. E. Tyaginov *et al.*, "Interface traps density-of-states as a vital component for hot-carrier degradation modeling," *Microelectron. Reliab.*, vol. 50, no. 9–11, pp. 1267–1272, Sep. 2010, doi: 10.1016/j.microrel.2010.07.030.

[11] E. Riedlberger, C. Jungemann, A. Spitzer, M. Stecher, and W. Gustin, "Comprehensive analysis of the degradation of a lateral DMOS due to hot carrier stress," in *2009 IEEE International Integrated Reliability Workshop Final Report*, South Lake Tahoe, CA, USA: IEEE, Oct. 2009, pp. 77–81. doi: 10.1109/IRWS.2009.5383027.

[12] L. Lu *et al.*, "Improved hot-carrier reliability of an ultralow RON,sp SOI-LDMOS by linearly doped technology for automotive application," *IEEE Trans. Electron Devices*, vol. 71, no. 1, pp. 935–939, 2024, doi: 10.1109/TED.2023.3338171.

[13] A. N. Tallarico *et al.*, "Investigation of the hot carrier degradation in power LDMOS transistors with customized thick oxide," *Microelectron. Reliab.*, vol. 76–77, pp. 475–479, Sep. 2017, doi: 10.1016/j.microrel.2017.07.043.

[14] Weifeng Sun *et al.*, "Hot-Carrier-Induced On-Resistance Degradation of n-Type Lateral DMOS Transistor With Shallow Trench Isolation for High-Side Application," *IEEE Trans. Device Mater. Relib.*, vol. 15, no. 3, pp. 458–460, Sep. 2015, doi: 10.1109/TDMR.2015.2429739.

[15] P. Moens, J. Mertens, F. Bauwens, P. Joris, W. De Ceuninck, and M. Tack, "A Comprehensive Model for Hot Carrier Degradation in LDMOS Transistors," in *2007 IEEE International Reliability Physics Symposium Proceedings. 45th Annual*, Phoenix, AZ, USA: IEEE, Apr. 2007, pp. 492–497. doi: 10.1109/RELPHY.2007.369940.

Investigation of Dual-Mode R_{on} Degradation Mechanisms in LOCOS-Based LDMOS

Wenliang Liu
State Key Laboratory of Electronic Thin Films and Integrated Devices
University of Electronic Science and Technology of China
Chengdu, China
liuwenliang.97@foxmail.com

*Ming Qiao
State Key Laboratory of Electronic Thin Films and Integrated Devices
University of Electronic Science and Technology of China
Chengdu, China
qiaoming@uestc.edu.cn

Penglong Xu
Process Intergration Technology Development Center
CSMC Technologies Co. Ltd.
Wuxi, China
XUPENGLONG@csmc.crmicro.com

Chunxia Ma
Process Intergration Technology Development Center
CSMC Technologies Co. Ltd.
Wuxi, China
MACX@csmc.crmicro.com

Feng Lin
Process Intergration Technology Development Center
CSMC Technologies Co. Ltd.
Wuxi, China
linfeng@csmc.crmicro.com

Bo Zhang
State Key Laboratory of Electronic Thin Films and Integrated Devices
University of Electronic Science and Technology of China
Chengdu, China
zhangbo@uestc.edu.cn

Abstract—In this work, the hot carrier injection (HCI) degradation mechanisms in a 100 V local oxidation of silicon (LOCOS)-based lateral double-diffused MOSFET (LDMOS) are thoroughly investigated. Experiments show that the on-state resistance (R_{on}) degradation of the device undergoes distinct mechanisms under the two worst gate stress conditions. To investigate these degradation mechanisms, the charge pump (CP) experiments and comprehensive TCAD simulations are applied, revealing that the degradation is dominated by the interplay between two HCI effects: the interface states (N_{it}) generation and hot holes injection.

Keywords—Local Oxidation of Silicon, hot carrier injection, R_{on} degradation, Charge pump technique

I. INTRODUCTION

Lateral diffused MOS (LDMOS) transistors are widely used in automotive systems, display drivers and power management ICs due to their compatibility with standard CMOS fabrication processes [1], [2]. However, under high-bias operation conditions, these devices are particularly vulnerable to hot carrier injection (HCI) issues. Energetic electrons and holes, accelerated by high electric fields can impact the silicon/oxide (Si/SiO₂) interface, leading to carrier trapping and Si-H bond breaking [3], [4]. As a result, these trapped charges and interface states (N_{it}) degrade the critical device parameters, critically compromising the long-term reliability of power LDMOS transistors.

Researches reveal that the HCI degradation mechanism depends on device architecture. According to Tallarico et al. [5], in LOCOS-based LDMOS, degradation is primarily driven by high-energy electrons generating N_{it} near the source-side bird's beak, correlating strongly with impact ionization (I.I.). In Mao's study [6], for ultra-high voltage (700V) triple RESURF devices, optimizing the p-type buried layer dose reduces degradation by suppressing the critical peak electric fields and I.I. near the source-side bird's beak, without compromising breakdown voltage. Furthermore, an improved degradation model, accounting for saturated hot hole injection, proposed by Zhang et al. [7], offers precise lifetime prediction

This work was supported by the National Natural Science Foundation of China under Grant 62174024.

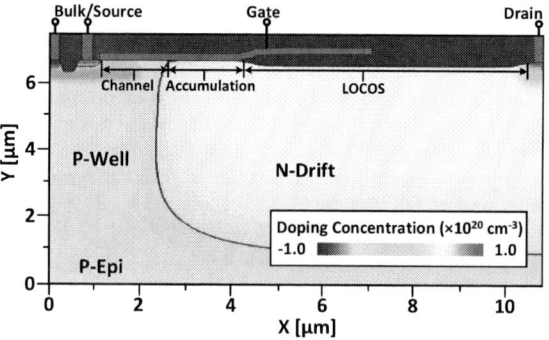

Fig. 1. Schematic cross-sectional view and 2D doping concentration distribution of the investigated LOCOS-based LDMOS device.

and demonstrates that hot hole injection degrades the LDMOS device.

In this work, the degradations of on-state resistance (R_{on}) and threshold voltage (V_{th}) in 100 V LOCOS-based LDMOS device under various gate stress conditions are experimentally investigated. Notably, the stress condition corresponding to the gate-source voltage (V_{gs}) that induces the maximum substrate current (I_{sub}) results in the most pronounced positive R_{on} degradation, while the stress condition of maximum V_{gs} achieves the most negative R_{on} degradation. Obviously, the contrasting degradation trends indicate the presence of distinct mechanisms between these two stress conditions. Despite extensive prior studies, the underlying mechanisms driving negative R_{on} degradation under the maximum V_{gs} stress condition in LOCOS-based LDMOS device remain not fully investigated. Therefore, both the charge pump (CP) technique and technology computer-aided-design (TCAD) simulations are employed to analyse the degradation mechanisms under these two stress conditions.

II. DEVICE STRUCTURE AND PROCESS

The schematic cross-sectional view and 2D doping concentration distribution of the device under test (DUT) are illustrated in Fig. 1. The p-type epitaxial layer (P-Epi) and gate oxide thicknesses are 9 μm and 12 nm, respectively. The

979-8-3315-3918-4/25 $31.00 © 2025 IEEE

operational voltages of DUT are 100 V for drain-source voltage (V_{ds}) and 5 V for V_{gs}. The source and body electrodes are grounded throughout the degradation experiments, and the degradation of V_{th} and R_{on} are monitored by periodically interrupting the degradation procedure. To extract V_{th} and R_{on}, the on-state drain current is measured at $V_{ds} = 0.1$ V, with V_{gs} ranging from 0 to 5 V and R_{on} is measured at $V_{ds} = 0.1$ V and $V_{gs} = 5.0$ V.

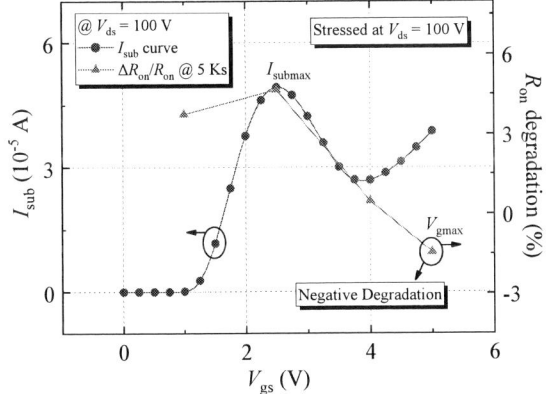

Fig. 2. Experimental I_{sub} as a function of V_{gs} and R_{on} degradation values at various V_{gs} stress conditions for the LOCOS-based LDMOS device stressed at $V_{ds} = 100$ V.

The experimental I_{sub} and R_{on} degradation values at various V_{gs} stress conditions for DUT, stressed at $V_{ds} = 100$ V, are shown in Fig. 2. The I_{sub} is measured from bulk electrode with source grounded and bias the drain at 100 V then sweep the gate voltage from 0 to 5 V. During the degradation tests, drain stress voltage is fixed at 100 V and the gate stress voltages are fixed at 1.0, 2.5 (I_{submax} stress condition), 4.0 and 5.0 (V_{gmax} stress condition) V, respectively. Under the stress condition of I_{submax}, the device exhibits the most positive R_{on} degradation, approximately 4.7 %. In contrast, under the V_{gmax} stress condition, the device shows the most negative R_{on} degradation, around -1.5 %. Notably, the underlying dominant mechanisms

responsible for R_{on} degradation differ under these two stress conditions. Therefore, the primary purpose of this paper is to investigate the degradation mechanisms under the two distinct stress conditions that result in the maximum positive and negative R_{on} degradation, respectively.

The R_{on} degradation curves under the I_{submax} and V_{gmax} stress conditions are shown in Fig. 3(a). Under the I_{submax} stress, the R_{on} degradation curve exhibits a gradually increasing trend. Initially (within 10 seconds), the increase rate is relatively slow, then accelerates (approximately between 10 ~ 1 K seconds), and slows down again after around 1 K seconds. In contrast, under the V_{gmax} stress condition, the R_{on} degradation curve first decreases rapidly (within 10 seconds), followed by a continuous increase. As for the V_{th} degradation illustrated in Fig. 3(b), the degradations remain minimal under both V_{gs} stress conditions (less than 0.1%). This indicates that the channel region exhibits negligible hot carrier degradation in both V_{gs} stress conditions.

III. RESULTS AND DISCUSSION

In this section, both the CP technique and comprehensive TCAD simulations are conducted to verify the degradation mechanisms under the two distinct stress conditions. The CP experiments with a constant pulse amplitude and variable base voltage (V_{base}) are conducted, proving that HCI effects in the channel and accumulation regions are negligible. Then the dominant HCI effects and their specific locations in field oxide region are thoroughly investigated and clarified using the extensive TCAD simulations, which incorporates the following several key physical models: the Enormal, DopingDep and HighFieldSaturation models for Mobility; the SRH (DopingDep TempDep) and Avalanche (Unibo) for Recombination; and the OldSlotboom and Thermodynamic models are also applied.

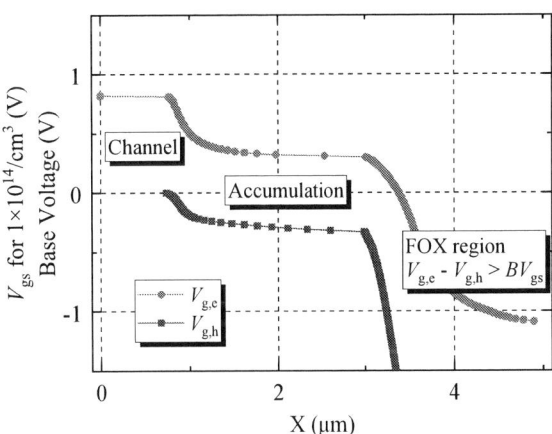

Fig. 4. Simulated $V_{g,e}$ and $V_{g,h}$ curves, assigning the channel, accumulation and field oxide regions.

As shown Fig. 4, the $V_{g,e}$ and $V_{g,h}$ curves as a function of location along the Si/SiO$_2$ interface are simulated to assign CP current (I_{cp}) contributions from different regions. The values of $V_{g,e}$ and $V_{g,h}$ correspond to the gate voltages to induce 1×10^{14} cm^{-2} electrons and holes at the Si/SiO$_2$ interface [8], [9], [10], respectively. The device contains three regions: the field oxide region, the accumulation region and the channel region. It is worth noting that $V_{g,h}$ is below -18 V in the field oxide region (not shown in Fig. 4) due to the thick LOCOS oxide (> 360 nm). As a result, a gate voltage exceeding 17 V ($V_{g,e} - V_{g,h}$ in the field oxide region) is required to pump out the interface

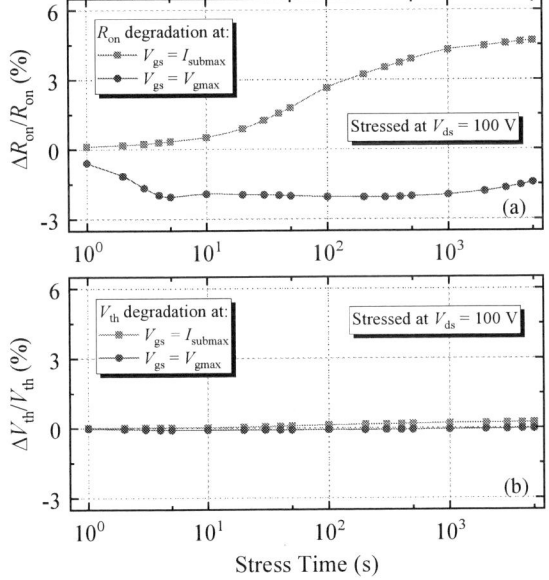

Fig. 3. Experimental (a) R_{on} and (b) V_{th} degradation curves for the LOCOS-based LDMOS device stressed at $V_d = 100$ V and $V_{gs} = I_{submax}$ and V_{gmax} conditions.

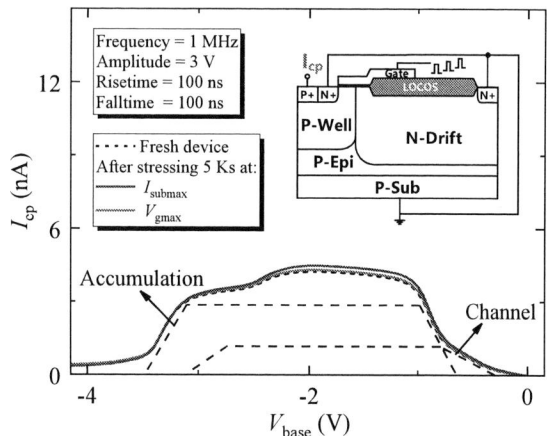

Fig. 5. Measured CP current curves of the LOCOS-based LDMOS device before and after stressing 5 K seconds under $V_{ds} = 100$ V and $V_{gs} = I_{submax}$ and V_{gmax} stress conditions, respectively.

damage in field oxide region. However, the maximum applied gate voltage is limited by the gate oxide breakdown voltage (BV_{gs}), which is less than 12 V. Consequently, the interface damage in field oxide region cannot be measured by the CP experiments.

The CP experiments in base mode have been performed, with the drain and source grounded, while monitoring the I_{cp} from bulk electrode. As shown in the inset figure of Fig. 5, the CP pulses are applied at gate electrode to periodically switch the device surface between inversion and accumulation states. The amplitude and frequency of CP pulse are set to 3 V and 1 MHz, respectively. The rise time and fall time of CP pulse are both set to 100 ns. In this way, the I_{cp} curves, before and after stressing 5 K seconds at $V_{ds} = 100$ V and under $V_{gs} = I_{submax}$ and V_{gmax} stress conditions, have been measured and shown in Fig. 5. Generally, a leftward or rightward shift of the I_{cp} curve indicates the injection of positive or negative charges in the oxide layer, respectively, and an amplitude increase of the I_{cp} curve reflects the generation of N_{it} [11]. Nevertheless, as shown in Fig. 5, the I_{cp} curves barely shift after stressing 5 K seconds at both $V_{gs} = I_{submax}$ and V_{gmax} stress conditions, indicating that the channel and accumulation regions are negligibly impacted by the HCI effects.

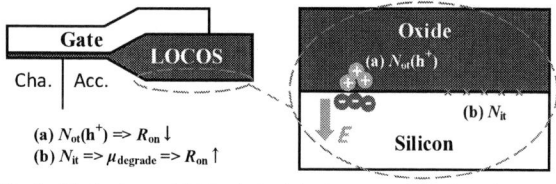

Fig. 6. Mechanisms of (a) hot holes injection, attracting mirror charges at the surface of N-Drift. The direction of electric field is represented by green arrow. (b) The generation of N_{it} caused by hot carriers.

In this LOCOS-based LDMOS device, two primary mechanisms are dominating the R_{on} degradation. As illustrated in Fig. 6, the first mechanism is hot holes injection, driven by the perpendicular electric field, and the second mechanism is generation of N_{it}. Along the Si/SiO₂ interface, hot carriers generated by high I.I. create N_{it}, which can degrade the mobility of majority carriers and result in an increased R_{on}. Simultaneously, driven by the strong perpendicular electric field, hot holes are injected into SiO₂ layer. These injected holes attract electrons near the drift region surface, increasing

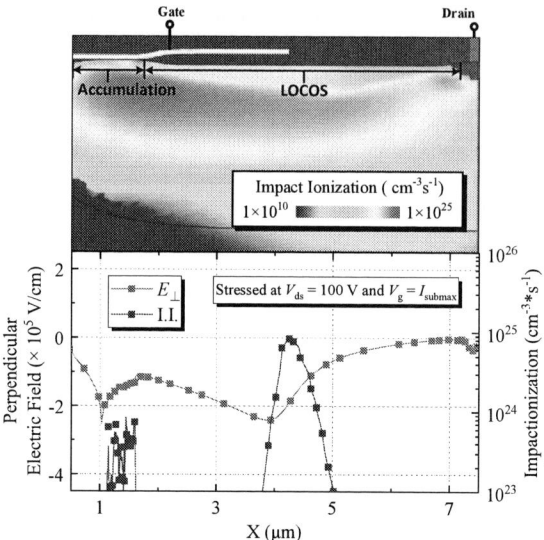

Fig. 7. Simulated 2D distribution of I.I., surface perpendicular electric field and I.I. under the I_{submax} stress condition.

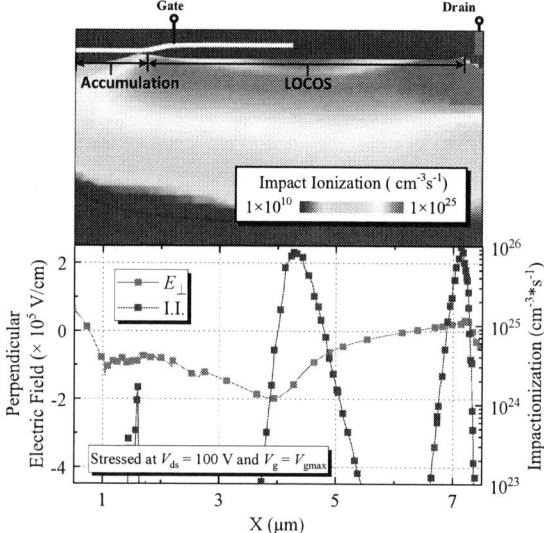

Fig. 8. Simulated 2D distribution of I.I., surface perpendicular electric field and I.I. under the V_{gmax} stress condition.

the effective doping concentration of n-type drift region and leading to a reduction in R_{on}. Consequently, the interplay of these two competing mechanisms dominates the overall degradation behavior of the LOCOS-based LDMOS device.

The simulated 2D distribution of I.I., surface perpendicular electric field and I.I., under the I_{submax} stress condition, are shown in Fig. 7. The 2D I.I. distribution indicates that I.I. mainly located near the source side bird's beak of the device. Notably, according to the surface I.I. distribution, a significant number of electron-hole pairs are generated under the poly-gate. These generated hot carriers contribute to the formation of a high density of N_{it} at the Si/SiO₂ interface, which in turn degrades the carrier mobility and leads to an increase of R_{on}. Moreover, due to the presence of a negative perpendicular electric field along the field oxide region, the hot holes can be injected into the above SiO₂ layer [12], [13]. These injected holes attract mirror electrons at the surface of n-type drift region, increasing its effective doping concentration and reducing the R_{on}.

979-8-3315-3918-4/25 $31.00 © 2025 IEEE

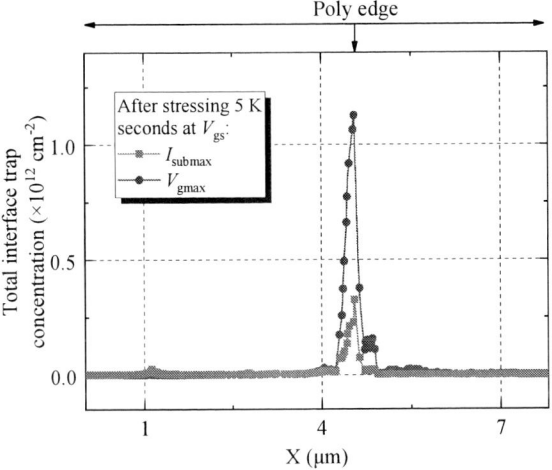

Fig.9 Simulated N_{it} generation along the Si/SiO2 interface under V_{ds} = 100 V and V_{gs} = I_{submax} and V_{gmax} stress conditions.

Accordingly, the N_{it} generation follows a time-dependent behavior described by the equation: $N_{it} = PN_0(1-e^{-kt})$ [14], [15], where P is the probability of defect generation, N_0 is the maximum number of interface bonds, and k is the reaction rate constant. Furthermore, the trap degradation model supported by Sentaurus Device is applied to track the N_{it} generation along the SiO$_2$ interface, capturing the depassivation process where hot carriers interact with the Si-H bonds and introduce the formation of Si- dangling bonds. The simulated N_{it} generation curves as a function of location along the Si/SiO$_2$ interface, stressed at V_{ds} = 200 V and V_{gs} = I_{submax} and V_{gmax}, are shown in Fig. 9. Obviously, significant numbers of N_{it} are generated under the poly edge at both the I_{submax} and V_{gmax} stress conditions. Besides, due to a higher I.I. distribution under the poly edge, the number of N_{it} generated under V_{gmax} stress condition is around 3 times more than the I_{submax} stress condition. Therefore, the R_{on} degradation mechanisms under I_{submax} stress condition can be summarized as follows: (1) In the first stage (within 10 seconds), N_{it} is slowly accumulated and its impact on R_{on} degradation is suppressed by the hot holes injection, causing a mild rise in R_{on}. (2) After that (between 10 ~ 1 Ks), as stress time increases, the injection of hot holes gets saturated and N_{it} grows more rapidly, accelerating the degradation of R_{on}. (3) Finally (after 1 Ks), the generation of N_{it} reaches saturation, and the increase of R_{on} slows down again. Clearly, the above explanations align well with the R_{on} degradation trends shown in Fig. 3(a).

Moreover, the simulated 2D distribution of I.I., surface perpendicular electric field and I.I., under the V_{gmax} stress condition, are shown in Fig. 8. Caused by kirk effect, the peak of I.I. is transferred to drain side of the device at V_{gmax} stress condition. Notably, as illustrated in the surface I.I. distribution, the I.I. beneath poly-gate is one order of magnitude higher than that under I_{submax} stress condition, generating a significant number of electron-hole pairs. Besides, due to the presence of perpendicular electric field along the field oxide region, a large number of hot holes are injected into the above oxide layer. Therefore, the R_{on} degradation mechanisms under V_{gmax} stress condition can be summarized as follow: (1) Initially (within 10 seconds), driven by perpendicular electric field, a large number of hot holes are injected into the above oxide layer, causing a rapid decrease of R_{on}. (2) After that (10 ~ 5 Ks), the injection of hot holes gets saturated and N_{it} tends to dominate, leading to the gradual increase of R_{on}.

IV. CONCLUSION

In this work, the HCI degradation mechanisms of a 100 V LOCOS-based LDMOS device have been comprehensively investigated under the two worst gate stress conditions (I_{submax} and V_{gmax}). The CP experiments confirm minimal HCI damage in both the channel and accumulation regions, and TCAD simulations further approve the degradation location in the field oxide region. The findings reveal that R_{on} degradation is dominated by the interplay between two HCI effects: N_{it} generation and hot holes injection. Overall, understanding and managing the competitional interplay between these two mechanisms are essential in improving the HCI immunity of this LOCOS-based LDMOS device.

REFERENCES

[1] S. Liu et al., "A review on hot-carrier-induced degradation of lateral DMOS transistor," IEEE Trans. Device Mater. Rel., vol. 18, no. 2, pp. 298–312, Jun. 2018, doi: 10.1109/TDMR.2018.2833490.

[2] X. Yang et al., "Enhancing Reliability of Bipolar-CMOS-DMOS Technology Through Delicately STI Optimization," in 2025 Conference of Science and Technology of Integrated Circuits (CSTIC), Mar. 2025, pp. 1–3. doi: 10.1109/CSTIC64481.2025.11018036.

[3] S. Liu, W. Sun, C. Zhang, T. Huang, and Q. Qian, "Model of hot-carrier degradation for lateral IGBT device on SOI substrate," Electronics Letters, vol. 49, no. 7, pp. 497–499, 2013, doi: 10.1049/el.2012.4036.

[4] J. F. Chen, Kuen-Shiuan Tian, Shiang-Yu Chen, Kuo-Ming Wu, J. R. Shih, and K. Wu, "An Investigation on Anomalous Hot-Carrier-Induced On-Resistance Reduction in n-Type LDMOS Transistors," IEEE Trans. Device Mater. Rel., vol. 9, no. 3, pp. 459–464, Sep. 2009, doi: 10.1109/TDMR.2009.2025770.

[5] A. N. Tallarico et al., "Hot-carrier degradation in power LDMOS: selective LOCOS- versus STI-based architecture," IEEE J. Electron Devices Soc., vol. 6, pp. 219–226, 2018, doi: 10.1109/JEDS.2018.2792539.

[6] K. Mao, H. Nie, and Y. Yao, "Effect of p-type buried layer dose on hot carrier degradation of R_{ON} in 700 V triple RESURF nLDMOS," IEEE Electron Device Lett., vol. 37, no. 3, pp. 242–244, Mar. 2016, doi: 10.1109/LED.2016.2518303.

[7] C. Zhang, Y. Li, Z. Li, X. Fu, and Z. Chen, "An improved hot-carrier lifetime evaluation method for the n-type LDMOS with hot-hole injection," IEEE Trans. Electron Devices, vol. 65, no. 8, pp. 3567–3571, 2018, doi: 10.1109/TED.2018.2842117.

[8] L. Lu et al., "Improved hot-carrier reliability of an ultralow RON,sp SOI-LDMOS by linearly doped technology for automotive application," IEEE Trans. Electron Devices, vol. 71, no. 1, pp. 935–939, 2024, doi: 10.1109/TED.2023.3338171.

[9] B. Djezzar and H. Tahi, "Using Oxide-Trap Charge-Pumping Method in Radiation-Reliability Analysis of Short Lightly Doped Drain Transistor," IEEE Transactions on Device and Materials Reliability, vol. 10, no. 1, pp. 18–25, Mar. 2010, doi: 10.1109/TDMR.2009.2030414.

[10] P. Moens, G. Van den bosch, and G. Groeseneken, "Hot-carrier degradation phenomena in lateral and vertical DMOS transistors," IEEE Transactions on Electron Devices, vol. 51, no. 4, pp. 623–628, Apr. 2004, doi: 10.1109/TED.2004.824688.

[11] S. Liu et al., "Hot-Carrier-Induced Degradations and Optimizations for Lateral DMOS Transistor With Multiple Floating Poly-Gate Field Plates," IEEE Trans. Electron Devices, vol. 64, no. 8, pp. 3275–3281, Aug. 2017, doi: 10.1109/TED.2017.2711276.

[12] S. Liu et al., "Lateral DMOS with partial-resist-implanted drift region for alleviating hot-carrier effect," IEEE Trans. Device Mater. Rel., vol. 17, no. 4, pp. 780–784, Dec. 2017, doi: 10.1109/TDMR.2017.2765687.

[13] Weifeng Sun et al., "Hot-Carrier-Induced On-Resistance Degradation of n-Type Lateral DMOS Transistor With Shallow Trench Isolation for High-Side Application," IEEE Trans. Device Mater. Relib., vol. 15, no. 3, pp. 458–460, Sep. 2015, doi: 10.1109/TDMR.2015.2429739.

[14] S. Reggiani et al., "TCAD simulation of hot-carrier and thermal degradation in STI-LDMOS transistors," IEEE Trans. Electron Devices, vol. 60, no. 2, pp. 691–698, Feb. 2013, doi: 10.1109/TED.2012.2227321.

[15] D. Varghese et al., "off-state degradation in drain-extended NMOS transistors: interface damage and correlation to dielectric breakdown," IEEE Trans. Electron Devices, vol. 54, no. 10, pp. 2669–2678, Oct. 2007, doi: 10.1109/TED.2007.904587.

Advanced Gate Driver Solutions for Fast-Switching SiC Power Device Applications

Yu Qing, Zhihao Yan, Zekun Zhou, Jiaxing Mao, Zijun Zhou, Yun Dai, Rongxing Lai, Yue Shi, Bo Zhang*

State Key Laboratory of Electronic Thin Films and Integrated Devices,
University of Electronic Science and Technology of China, Chengdu, China
Email: zhangbo@uestc.edu.cn

Abstract—**Silicon carbide (SiC) MOSFETs, as core components in high-power, high-frequency, and high-temperature power electronic systems, are highly reliant on gate drive technologies to realize their performance potential. This paper systematically reviews recent advancements in SiC MOSFET gate drive technologies, with a focus on three key domains: galvanic isolation, slew rate control, and reliability enhancement. Ultimately, the paper outlines the future development directions of advanced SiC MOSFET gate drive technologies, providing a reference for further performance optimization.**

Index Terms—**SiC MOSFETs, advanced gate driver, galvanic isolation, slew rate control, reliability enhancement**

I. INTRODUCTION

In recent years, silicon carbide (SiC), as a typical representative of wide-bandgap semiconductor materials, has been widely recognized as the core development direction of next-generation power semiconductors. However, traditional Si MOSFET and IGBT drivers cannot fully exploit the performance of SiC MOSFETs. Therefore, research on advanced gate driver for SiC MOSFETs is of great significance. In this paper, the research progress of advanced gate driver technologies for SiC MOSFETs in recent years will be demonstrated, with a state-of-the-art technical guide furnished. Section II mainly introduces galvanic isolation technology and related research. Section III focuses on slew rate control technology. Section IV covers reliability enhancement measures. Section V presents the conclusion. Section VI is the acknowledgment.

II. GALVANIC ISOLATION

A. Galvanic Isolator Structure

In modern wide-bandgap semiconductor drive systems, magnetic isolation is prevalent for its high isolation voltage and data rate. However, traditional coil structures limit these parameters, prompting recent research on coil optimization [1–3]. As shown in Fig. 1a, early CMOS-fabricated transformers use inter-layer dielectrics but have low breakdown voltage (BV < 2.5 kV) [2, 3]. As shown in Fig. 1b, [3] proposes a vertically stacked structure with polyimide, achieving 7 kVrms BV but requiring special processes and suffering limited CMTI. As shown in Fig. 1c, [1] introduces a lateral resonant coupler with 650 kV/µs CMTI and 3.34 kVrms BV[1].

In contrast, due to good compatibility with the CMOS process, the development of capacitive isolation is highly synchronized with that of the CMOS process. Almost all relevant literature are based on standard CMOS process [4, 5], using the inter-layer dielectric (usually silicon dioxide, with a typical breakdown field strength of 500 V/µm) as the insulation material.

B. Modulation Schemes

In addition to the structure of galvanic isolator, signal modulation schemes are also significant factors limiting data rates and CMTI. Common modulation schemes include carrier modulation and pulse edge modulation. Carrier modulation greatly enhances immunity to noise and interference by introducing high-frequency carriers. As shown in Fig. 2, the most common carrier modulation scheme is On-Off Keying (OOK) modulation [1, 4]. To improve CMTI, [1, 4] propose a fully differential cross-coupled pre-amplifier which enhances DM gain while effectively suppressing CM interference. To reduce power consumption, a novel ring oscillator that adjusts carrier frequency according to pulse width is proposed in [6], as shown by AFC OOK in Fig. 2.

In contrast, although pulse edge modulation offers the advantages of simple structure and high speed, it suffers from poor robustness against noise and interference[2, 5, 7]. As shown in Fig. 2, several common edge modulation methods are presented. To address this vulnerability, improved modulation schemes are presented in [7] as shown by pulse count* in Fig. 2.

III. SLEW RATE MANAGEMENT

SiC MOSFETs exhibit fast switching transients characterized by high dv/dt and di/dt, which can induce serious electromagnetic interference (EMI), voltage overshoot, and ringing. To mitigate these issues, a segmented drive strategy is employed, in which different driving capabilities are applied during each stage of the switching process. This allows independent control of the slew rates (dv/dt and di/dt), thereby optimizing the trade-off between device performance and system reliability under specific applications.

In segmented gate driving, segmented timing control is crucial, defining trigger nodes to coordinate gate-driving sequences. As shown in Fig. 3, segmented timing control signals can be generated by real-time monitoring of drain–source voltage, gate–source voltage, gate voltage and drain current and used for open-loop control of switching timing, as demonstrated in [8–11]. However, complex logic controls circuit

---

979-8-3315-3918-4/25 $31.00 © 2025 IEEE

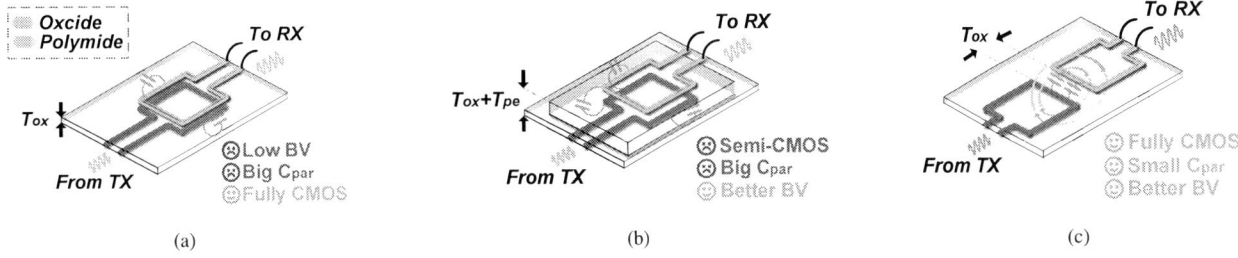

Fig. 1. Magnetically coil structures (a) Fully-integrated CMOS (b) Stacked with DAF (c) Lateral magnetic coupling[1].

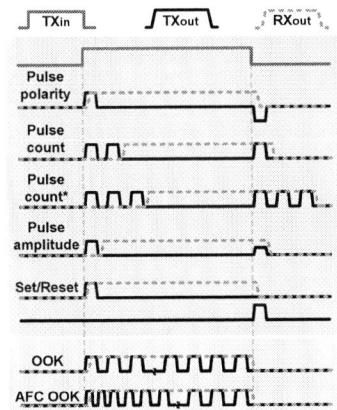

Fig. 2. Various modulation schemes (Pulse edge modulation at the top; Carrier modulation at the bottom).

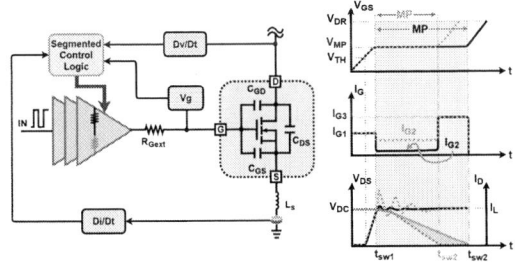

Fig. 3. Basic segmented driver architecture.

introduce uncontrollable delay, impairing timing accuracy. In high-frequency scenarios, such delay cause slew rates to exceed target ranges, compromising device reliability and system efficiency. To address this, a closed-loop strategy using phase-locked loops (PLLs) is introduced, with dynamic feedback enabling real-time phase correction. An analog PLL-based solution [12] performs dynamic delay compensation but has a narrow modulation range. Digital PLLs offer wider modulation but suffer limited resolution and higher power consumption [11]. To overcome these, [8] presents an innovative mixed-signal auto-timing (MSAT) technique combining analog PLLs with digital delay lines, achieving high accuracy and wide

modulation range.

Driving strength modulation, complementary to segmented timing control, dynamically adjusts driving voltage/current for precise dv/dt and di/dt control. Existing methods (fixed or programmed adjustable, [9, 10]) lack adaptability to load or temperature changes. [8] proposes a novel CLEC technique with common-mode noise sampling, enabling closed-loop driving adjustment, high adaptability, and precise slew rate control.

IV. RELIABILITY ENHANCEMENT

The intrinsic fragility of SiC MOSFETs presents significant reliability challenges in practical applications due to their prolonged operation under severe voltage and current stresses. Conventional gate drivers, failing to adapt to both device characteristics and application requirements, may cause critical failures—including shoot-through faults from spurious turn-on and device degradation . Thus, it is imperative to develop reliability enhancement strategies tailored to the unique characteristics of SiC MOSFETs to ensure their long-term safe and stable operation.

A. Spurious Turn-on Suppression

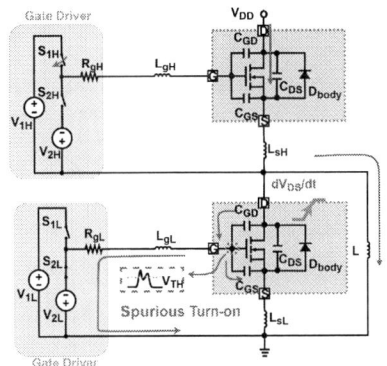

Fig. 4. Mechanism causing spurious turn-on.

In the half-bridge topology, the high-speed switching of SiC MOSFETs poses a significant risk of unintended conduction, leading to shoot-through events. The mechanism of spurious turn-on is illustrated in Fig. 4. To address this issue, negative turn-off voltage and active Miller clamping are commonly employed as effective suppression techniques.

Negative-voltage turn-off operation serves as a primary defense against unintended turn-on in SiC MOSFETs, enhancing noise immunity by effectively extending the threshold voltage margin, which typically requires an additional negative voltage supply [13].

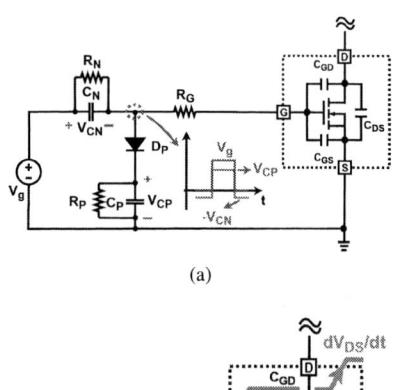

(a)

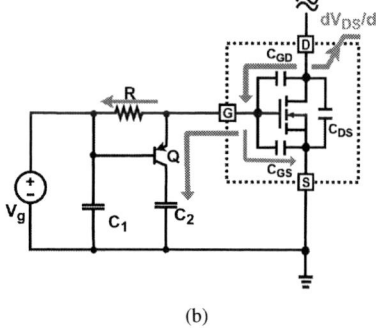

(b)

Fig. 5. Spurious turn-on suppression circuit. (a) RCD level shifter; (b) Active Miller clamp spurious turn-on suppression circuit.

To further simplify the power architecture, the approach presented in [14, 15] utilizes RCD-based level shifters to generate the required negative gate voltage, enabling operation from a single power rail. The RCD level shifter is illustrated in Fig. 5a. Another commonly used method for suppressing spurious turn-on is active Miller clamping. Its core principle is to activate a low-resistance path when spurious turn-on occurs, thereby reducing the magnitude of the crosstalk voltage. As shown in Fig. 5b, [16] uses a transistor in series with a large auxiliary capacitor.[17] modifies the structure in [16] by adding a diode in the clamping path, mitigating the impact on switching speed, but the auxiliary capacitor still suffers from the lack of a discharge path. To resolve this, [14] improves circuit reliability by paralleling a resistor with the auxiliary capacitor to provide a discharge path.

B. Rapid Short-Circuit Protection

SiC MOSFETs have relatively poor short-circuit withstand capability, with most exhibiting a short-circuit withstand time of only 3–5 µs. Therefore, it is necessary to develop Rapid Short-Circuit Protection to address their low short-circuit withstand capability.

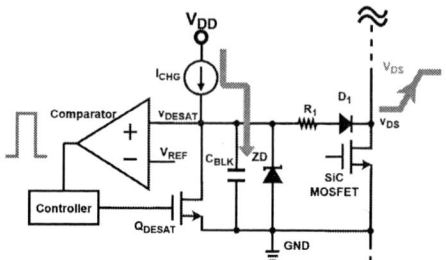

Fig. 6. Current source desaturation detection circuit.

The most widely used method for short-circuit protection is desaturation detection. This method utilizes the characteristic that the drain-source voltage of SiC MOSFETs rises rapidly during a short circuit to detect this voltage. When the drain-source voltage exceeds the preset threshold, the circuit charges the blanking capacitor, thereby generating a fault signal. Desaturation detection can be classified into current-source and voltage-source topologies based on the charging method of the blanking capacitor. Fig. 6 depicts the current-source topology.

[18] proposes an FPGA-based short-circuit protection circuit based on voltage-source desaturation detection, with a response time of 1.5 µs. [19] adopts voltage-source desaturation detection. To reduce the detection delay caused by the blanking capacitor, it employs a 600-ns digital delay module, thereby further reducing the detection delay. [20] utilizes a current-source desaturation circuit. Building on the traditional current-source circuit, it incorporates two resistors and one diode, enabling adjustable detection delays by tuning resistor parameters. However, it is prone to interference from high drain voltage slew rates, leading to reduced reliability.

V. CONCLUSION

This paper systematically reviews recent advances in SiC MOSFET gate driver technologies from three perspectives: galvanic isolation, slew rate control, and reliability enhancement. Despite these developments, the technology still faces multiple challenges in modern power electronics under extreme operating conditions, including voltage withstand capability, switching speed, common-mode noise, and reliability. This review aims to provide relevant designers with a reference for understanding the technical context and facilitate innovative breakthroughs in the field.

VI. ACKNOWLEDGMENT

This work was supported in part by the National Natural Science Foundation of China under Grant 62074028, in part by the Sichuan Natural Science Foundation under Grant 23NSFSC0359, in part by the Chunhui Cooperative Research Program of the Ministry of Education of China under Grant HZKY20220583.

REFERENCES

[1] M. Javid, K. Ptacek, R. Burton, and J. Kitchen, "A 650 kV/µs common-mode resilient cmos galvanically isolated

communication system," *IEEE Transactions on Circuits and Systems I: Regular Papers*, vol. 69, no. 2, pp. 587–598, 2022.

[2] S. Kaeriyama, S. Uchida, M. Furumiya, M. Okada, and M. Mizuno, "A 2.5kV isolation 35kV/μs CMR 250Mbps 0.13mA/Mbps digital isolator in standard CMOS with an on-chip small transformer," in *2010 Symposium on VLSI Circuits*, 2010, pp. 197–198.

[3] S. Uchida, S. Kaeriyama, H. Nagase, K. Takeda, Y. Nakashiba, T. Maeda, and K. Ishihara, "A face-to-face chip stacking 7kV RMS digital isolator for automotive and industrial motor drive applications," in *2014 IEEE 26th International Symposium on Power Semiconductor Devices & IC's (ISPSD)*, 2014, pp. 442–445.

[4] S.-Y. Li, W.-C. Hung, T.-W. Wang, Y.-T. Hsu, K.-H. Chen, K.-L. Zheng, Y.-H. Lin, S.-R. Lin, and T.-Y. Tsai, "20.1 a high common-mode transient immunity GaN-on-SOI gate driver for high dv/dt SiC power switch," in *2023 IEEE International Solid-State Circuits Conference (ISSCC)*, 2023, pp. 302–304.

[5] I. Altoobaji, A. Hassan, M. Ali, Y. Audet, and A. Lakhssassi, "Capacitively isolated 400 Mbps data transfer system with 2 ns propagation delay and 5 kV /μs common mode transient immunity," in *2024 22nd IEEE Interregional NEWCAS Conference (NEWCAS)*, 2024, pp. 283–287.

[6] D. Pan, Z. Xiong, Q. Lu, F. Miao, L. Wu, and L. Cheng, "A 250-Mb/s on-chip capacitive digital isolator with adaptive frequency control," *IEEE Solid-State Circuits Letters*, vol. 7, pp. 231–234, 2024.

[7] S. Mukherjee, A. N. Bhat, K. A. Shrivastava, M. Bonu, B. Sutton, V. Gopinathan, G. Thiagarajan, A. Patki, J. Malakar, and N. Krishnapura, "25.4 A 500Mb/s 200pJ/b die-to-die bidirectional link with 24kV surge isolation and 50kV/μs CMR using resonant inductive coupling in 0.18μm CMOS," in *2017 IEEE International Solid-State Circuits Conference (ISSCC)*, 2017, pp. 434–435.

[8] R. Lai, Y. Yang, Y. Dai, J. Wu, Y. Shi, Z. Zhou, B. Zhang, H. Li, and X. Peng, "A quad-slope smart gate driver with mixed-signal auto-timing technique for power devices segment control," in *2024 36th International Symposium on Power Semiconductor Devices and ICs (ISPSD)*, 2024, pp. 347–350.

[9] S. Zhang, C. Liu, X. Li, D. Zhang, Y. Yang, R. Min, Q. Tong, and H. Peng, "Minimal switching loss three-stage active gate driving strategy based on dynamic dv/dt switching model for GaN HEMT," *IEEE Transactions on Power Electronics*, vol. 40, no. 3, pp. 4142–4155, 2025.

[10] W. Cui, W. Zhang, J. Liang, M. Wang, H. Nishio, H. Sumida, H. Nakajima, and W. Ng, "Slope sensing for optimum dynamic gate driving of SiC power MOS-FETs," in *2021 33rd International Symposium on Power Semiconductor Devices and ICs (ISPSD)*, 2021, pp. 199–202.

[11] D. Zhang, K. Horii, K. Hata, and M. Takamiya, "Digital gate driver IC with fully integrated automatic timing control function in stop-and-go gate drive for IGBTs," in *2023 IEEE Applied Power Electronics Conference and Exposition (APEC)*, 2023, pp. 1225–1231.

[12] Y. Chen and D. B. Ma, "A 10-mhz closed-loop emi-regulated gan switching power converter using emulated miller plateau tracking and adaptive strength gate driving," *IEEE Journal of Solid-State Circuits*, vol. 56, no. 2, pp. 531–540, 2021.

[13] S. Zhao, A. Dearien, Y. Wu, C. Farnell, A. U. Rashid, F. Luo, and H. A. Mantooth, "Adaptive multi-level active gate drivers for SiC power devices," *IEEE Transactions on Power Electronics*, vol. 35, no. 2, pp. 1882–1898, 2020.

[14] J. Ye, M. Wang, S. Cui, C. Zhang, and L. Li, "An improved crosstalk suppression driver topology for SiC MOSFET with fast switching transient in the phase-leg configuration," *IEEE Transactions on Power Electronics*, vol. 40, no. 6, pp. 8448–8467, 2025.

[15] F. Gao, Q. Zhou, P. Wang, and C. Zhang, "A gate driver of SiC MOSFET for suppressing the negative voltage spikes in a bridge circuit," *IEEE Transactions on Power Electronics*, vol. 33, no. 3, pp. 2339–2353, 2018.

[16] Y. Zushi, S. Sato, K. Matsui, Y. Murakami, and S. Tanimoto, "A novel gate assist circuit for quick and stable driving of SiC-JFETs in a 3-phase inverter," in *2012 Twenty-Seventh Annual IEEE Applied Power Electronics Conference and Exposition (APEC)*, 2012, pp. 1734–1739.

[17] H. Li, Y. Zhong, R. Yu, R. Yao, H. Long, X. Wang, and Z. Huang, "Assist gate driver circuit on crosstalk suppression for SiC MOSFET bridge configuration," *IEEE Journal of Emerging and Selected Topics in Power Electronics*, vol. 8, no. 2, pp. 1611–1621, 2020.

[18] S. Ji, M. Laitinen, X. Huang, J. Sun, W. Giewont, F. Wang, and L. M. Tolbert, "Short-circuit character-ization and protection of 10-kV SiC mosfet," *IEEE Transactions on Power Electronics*, vol. 34, no. 2, pp. 1755–1764, 2019.

[19] X. Huang, S. Ji, J. Palmer, L. Zhang, D. Li, F. Wang, L. M. Tolbert, and W. Giewont, "A robust 10 kV SiC MOSFET gate driver with fast overcurrent protection demonstrated in a MMC submodule," in *2020 IEEE Applied Power Electronics Conference and Exposition (APEC)*, 2020, pp. 1813–1820.

[20] Z. Liu, J. He, D. Jiang, X. Sun, G. Yang, H. Liu, P. Li, and R. Qu, "A hybrid desaturation fast detection circuit for bridge leg short-circuit faults," *IEEE Transactions on Power Electronics*, vol. 39, no. 8, pp. 9221–9229, 2024.

Design Considerations for Smart Gate Drivers

Wai Tung Ng*[1], Jingyuan Liang[1], Wentao Cui[1], Chun Yin Au Yueng[1], Namjee Kim[1], Wei Jia Zhang[2]

[1] The Edward S. Rogers Sr. Dept. of Electrical & Computer Engineering, University of Toronto, 10 King's College Road,
Toronto ON, Canada M5S 3G4
[2] Department of Electrical and Computer Engineering, The Hong Kong University of Science and Technology,
Clearwater Bay, Hong Kong, China

* Email: ngwt@ece.utoronto.ca

Abstract—**Many useful functions can be incorporated into smart gate driver ICs. Dynamic gate driving and dead-time correction techniques can be used to suppress unwanted ringing oscillation and eliminate body diode or reverse conduction, respectively. Advanced packaging techniques such as silicon interposer can offer highly compact assemblies, effectively minimize parasitic inductances. However, when taking thermal performance into consideration the most compact die placement in power semiconductor modules may not offer the best overall performance. Smart gate driver and thermal considerations remain the most practical co-design approach.**

Keywords—GaN and SiC Power Transistors, Smart Gate Drive Circuits, Thermal Management

I. INTRODUCTION

Modern wideband gap (WBG) power transistors exhibit unprecedented current handling, breakdown voltage and high-speed switching capabilities. In particular, gallium nitride (GaN) and silicon carbide (SiC) power transistors occupy unique power electronic application spaces according to their current/voltage ratings and mission requirements [1]. While power ratings, power conversion efficiency, reliability and cost are some of the important considerations, the suppression of electro-magnetic interference (EMI) and effective thermal management are also critical for the implementation of compact systems. To fully exploit the performance of GaN and SiC power transistors, dedicated gate driving circuit is required to switch these devices on and off quickly without unwanted ringing oscillation [2]. This paper reviews some of the critical co-design considerations between smart gate driving techniques and advanced liquid cooled packaging.

II. DYNAMIC GATE DRIVE

Power transistors are typically designed to have large current handling capabilities, from several to 10's or even 100's A. This would lead to large W/L ratio and hence a large gate capacitance (C_{GS}) often in the 100's pF to several nF range. In a typical PCB layout, the gate driver with an output resistance of R_G, PCB traces with a lumped parasitic inductance of L_G and the C_{GS} of the power MOSFET can be modeled as an equivalent *RLC* circuit as shown in Fig. 1(a). The damping factor, denoted by ζ, is a crucial parameter in the design and analysis of oscillatory systems. It quantifies the degree of damping in a system and directly affects how quickly and smoothly the system returns to equilibrium after being disturbed. In electrical circuits, such as those involving power MOSFETs, the damping factor is influenced by the resistance, inductance, and capacitance within the circuit [1]. If a voltage step change is applied to the gate driver, the gate voltage could experience either under or over damp conditions, leading to either unwanted ringing oscillations or excessively slow switching speed as shown in Fig. 1(b). Gate ringing

could cause severe electromagnetic interference (EMI) while slow gate transition will increase switching loss.

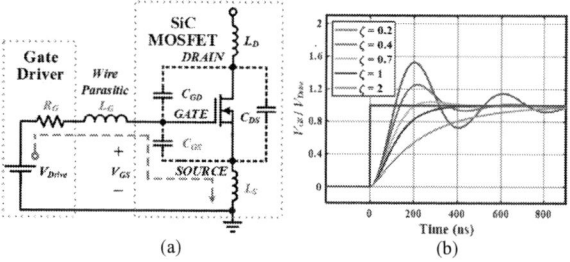

Fig. 1. (a) An equivalent RLC circuit for the gate loop of a power transistor, (b) effect of different damping factor (ζ), leading to unwanted gate ringing.

As the power transistor switches between on/off states, they also transition between cut-off, saturation and triode operations as shown in Fig. 2. This also leads to significant changes to the actual values in C_{GS}. This is evident by the Miller plateau as illustrated during time intervals t_2-t_3 and t_6-t_7 by the green V_{Gate} curves. During this time, the transistor operates in saturation mode as the gate voltage is above the threshold voltage and with an appreciable amount of drain voltage. The apparent gate capacitance is increased as a result of Miller effect, hence the name.

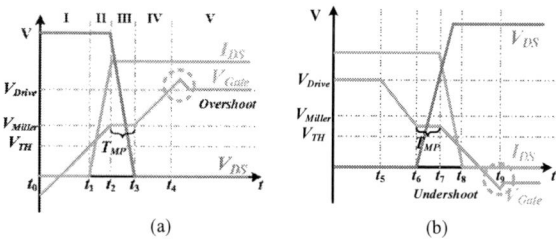

Fig. 2. The different intervals during the (a) turn-on and (b) turn-off transient for a power MOSFET.

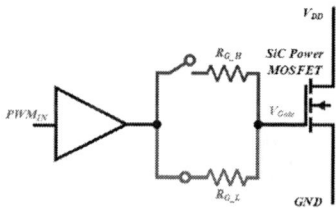

Fig. 3. A gate driver circuit for a SiC power MOSFET with selectable pull up (R_{G_H}) and pull down (R_{G_L}) resistors.

Due to the rapid changes in C_{GS}, it is difficult to select a single gate resistance value that would provide the proper damping factor. As a quick solution, most commercially gate

979-8-3315-3918-4/25 $31.00 © 2025 IEEE

driver ICs provide a fixed low output resistance. The user is required to add external pull up (R_{G_H}) and pull down (R_{G_L}) resistors (see Fig. 3) to control the turn-on/off speeds. However, this technique cannot accommodate the changes in C_{GS} during switching. As a result, excessive EMI and/or switching loss would still occur.

Dynamic or adaptive gate driving was developed to provide the ability to change the gate driver's output resistance on-the-fly. At the beginning of the gate transient, a low R_G is used to speed up the transition. At some point during the Miller plateau, a high R_G is selected to increase damping as shown in Fig. 4. Other methods using more that two resistance levels have also been shown to be effective but with higher control complexity.

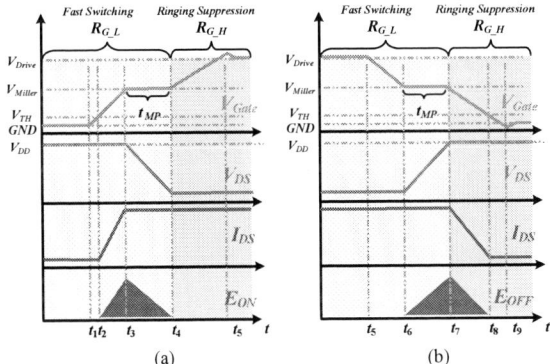

Fig. 4. A two-step dynamic gate driving technique. A low equivalent gate resistance is first used to provide fast (a) turn-on and (b) turn-off, followed by a high gate resistance to suppress ringing.

The implementation of dynamic gate driver consists of a combination of integrated parallel output stages that can be activated or disabled with precise timing, thereby providing the optimum R_G at all times Fig. 5. The die photo of this dynamic gate driver IC is shown in Fig. 6(a). The double pulse test topology in Fig. 6(b) is often used to examine the turn-on/off characteristics at high current.

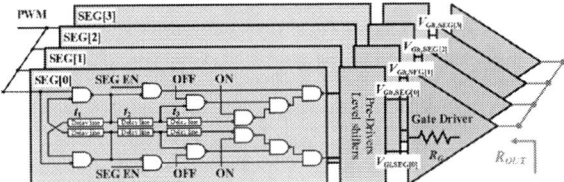

Fig. 5. A segmented gate driver circuit with multiple output stages to provide different combinations of output resistance [3].

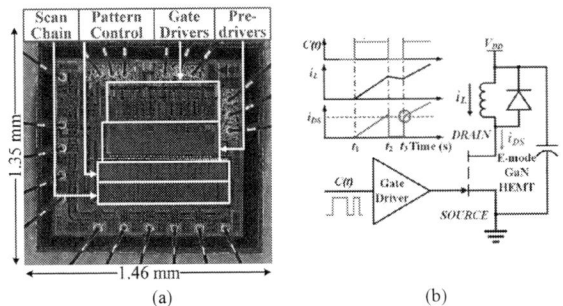

Fig. 6. (a) Micrograph of the dynamic gate driver implemented in 0.18 μm BCD technology. (b) A typical double pulse testing topology [3].

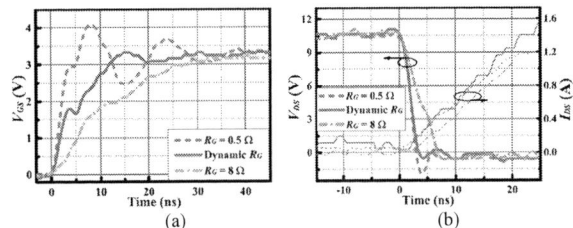

Fig. 7. Gate voltage waveforms for dynamic R_G and fixed values of 0.5 and 8 Ω, showing the trade-offs between gate ringing and switching speed [3].

The benefits of dynamic gate drive can be clearly observed through the comparison waveforms in Fig. 7. However, the effectiveness of dynamic gate drive remains difficult to realize as the timing of when to switch to different effective gate resistances must be manually fine tuned with sub-nano second precision for each operating condition. Recently, a dynamic gate driver that can detect the end of the Miller plateau and using it as a timing reference was demonstrated. Since the Miller plateau is a function of various operating conditions, including the supply voltage and load current, it can be used to automatically adjust the timing for selecting the appropriate gate resistance. The block level diagram of this gate driver is as shown in Fig. 8. A tunable switched capacitor (SC) filter that can detect the change in slope of the gate voltage waveform is used to identify the Miller plateau region.

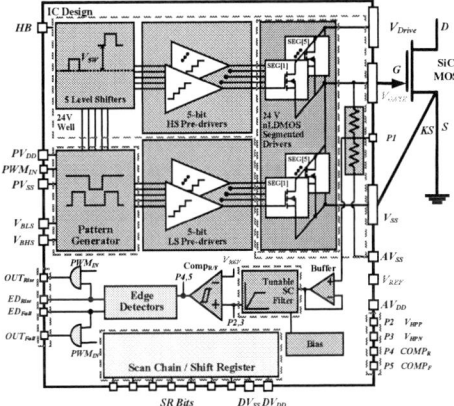

Fig. 8. Block level diagram of a dynamic gate driver with auto-timing [4].

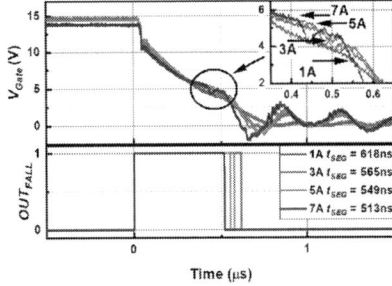

Fig. 9. Dynamic gate drive waveform with auto-timing for different drain currents. The pulse width for the bottom waveform is a direct indication of the optimum time to switch to different R_G [4].

Using this approach, the timing for the dynamic gate driving can be automatically determined even with varying operating conditions. This feature is an important step in promoting the wide adoption of dynamic gate driving technique.

III. DEADTIME CORRECTION

In order to prevent shoot-through current between the high-side (*HS*) and low-side (*LS*) transistors in the output stage, it is important to make sure that one is completely turn-off before the other is activated. Traditionally, an excessively long fixed dead-time (e.g., a few ns) is inserted between the *HS* and *LS* gate signals. Unfortunately, this long dead-time would result in body diode conduction in the case of Si or SiC power MOSFETs and reverse conduction in GaN HEMTs. In both cases, the power loss will be either the load current multiplied by one diode drop (0.7 V) or the reverse gate-drain voltage (higher that the threshold voltage). Furthermore, the switching speeds of the *LS* and *HS* transistors are dependent on the load current. As a result, a constant fixed dead-time is not appropriate in applications with constantly varying loading conditions. Numerous dead-time correction techniques have been reported. An effective approach to ensure optimum condition is to directly measure the duration of the body diode or reverse conduction and subtract it from the current dead-time [5]. The presence of this unwanted conduction can be observed as a negative voltage at the output switching (*SW*) node. However, this node has a very large voltage swing, making a simple voltage divider ineffective. A clamping circuit using a pass-transistor with a constant gate bias can be used to shield the sensing circuit from the high voltage swing while allowing the low negative voltage to be detected. The block level diagram of a complete gate driver IC with dynamic gate driver and dead-time correction is shown in Fig. 10(a).

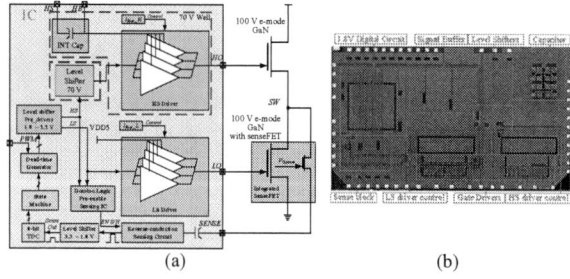

(a)   (b)

Fig. 10. (a) Block level diagram of a gate driver IC with dynamic gate driver and dead-time correction. (b) Micrograph of the corresponding gate driver IC implemented using a 0.18 μm BCD technology [5].

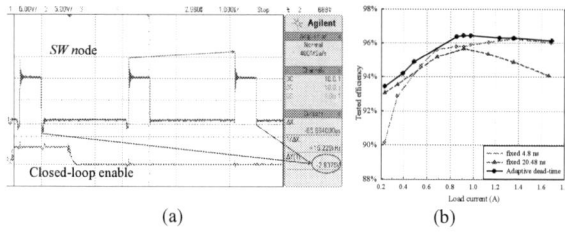

(a)   (b)

Fig. 11. (a) Switching (*SW*) node waveform of a DC-DC converter showing the elimination of reverse conduction after the closed-loop dead-time correction is enabled. (b) Power conversion efficiency, showing the benefit of continuous dead-time correction over a wide range of load current [5].

The operation of the dead-time correction can be observed on the *SW* node waveform of a DC-DC converter in Fig. 11(a). Before the closed-loop function is enabled, reverse conduction can be witnessed as negative pulses. After dead-time correction is enabled, reverse conductions at turn-on and turn-off are eliminated. With continuous dead-time correction, switching loss can be reduced, the power conversion

efficiency in Fig. 11(b) also shows significant improvement across a wide range of load current.

In the next section, we will discuss how thermal design considerations can affect the overall electrical performance of power semiconductor modules.

IV. PACKAGING CONSIDERATIONS

The TO-220 package is one of the most common ways to house power transistors. This traditional assembly relies on long thick bond wires to provide low resistance connections as illustrated in Fig. 12(a). Unfortunately, this topology also introduces parasitic inductances, especially when using multiple packages to form the half-bridge output stage as shown in Fig. 12(b). The switching waveforms in Fig. 13 show significant ringing oscillations at the *SW* node, *HS* and *LS* gates with unwanted EMI.

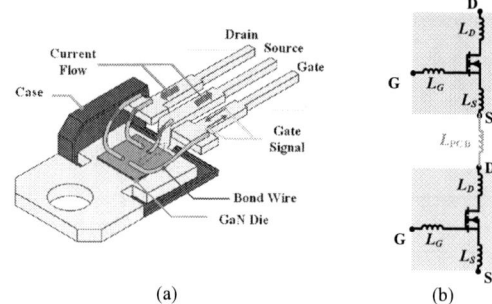

(a)   (b)

Fig. 12. The internal structure of the TO-220 package [6]. (b) The parasitics that need to be considered in a half bridge configuration [6].

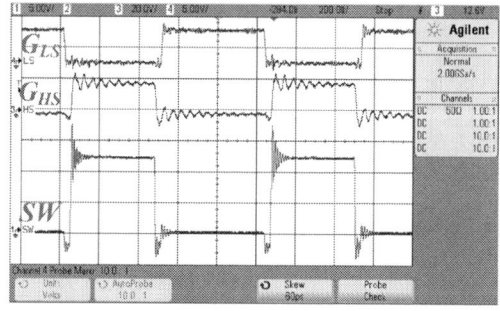

Fig. 13. Voltage waveforms at the *SW* node, *HS* and *LS* gate terminals of a half-bridge using two GaN power HEMTs [6].

(a)   (b)

Fig. 14. (a) A finished silicon wafer with multiple interposers. (b) A diced silicon power interposer intended for a DC-DC converter.

To mitigate the ringing oscillations and the unwanted reverse conductions, the dynamic gate drivers could be good solutions. However, another approach is to use advanced packaging techniques such as the power silicon interposer or intelligent power modules. Power interposer is essentially a miniaturized PCB built on silicon wafers as shown in Fig. 14. The idea is to take advantage of the excellent thermal conductivity of silicon and the lithography with much better resolution when compared to traditional FR4 PCBs. This would allow the components to be placed much closer together, reducing the lengths of the connection traces. Hence the parasitic inductance can be minimized. This would result in very clean waveform at the *SW* node. An example is as shown in Fig. 15. As a result, the need for smart gate drivers could be eliminated. However, this may not be the case in higher power applications.

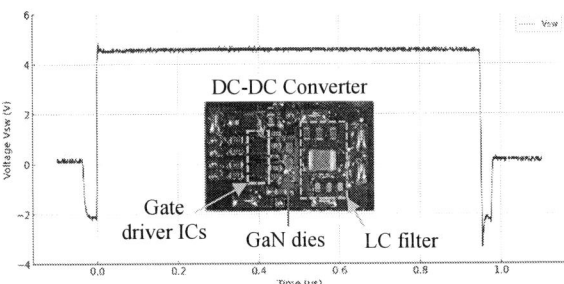

Fig. 15. A DC-DC converter implemented on a silicon interposer platform without any significant ringing oscillation.

With thermal power dissipation taken into account, the ultra compact form factor may not necessary be the optimum configuration. GaN and SiC power transistors have very low specific on-resistance (on-resistance normalized to unit area). For a given current or on-resistance rating, these WBG power transistors would have die sizes that are much smaller than comparable silicon power MOSEFTs. The simulated temperature profiles in a GaN power module with different active metal brazed (AMB) ceramic substrate sizes and die placements indicate that the most compact layout does not necessarily translate to the most efficient thermal paths.

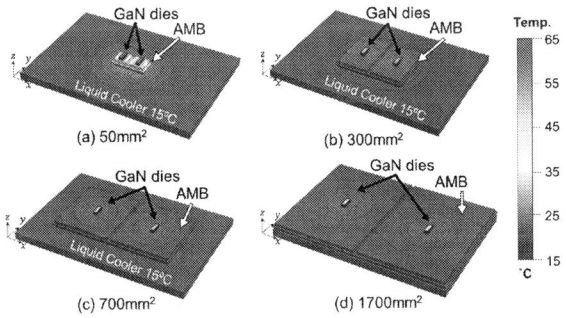

Fig. 16. Simulated temperature profiles for a GaN power module with different AMB substrate sizes and die placements [7].

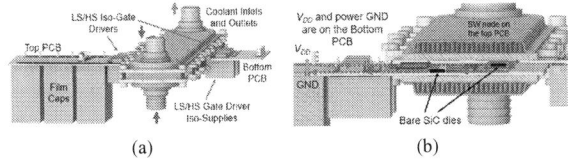

Fig. 17. (a) A 3-phase liquid-cooled SiC power module (showing only one phase). (b) The cross-sectional view showing lengthy power loops.

A 3-phase liquid-cooled SiC power module is shown in Fig. 17(a). Due to power dissipation considerations, the SiC power transistor dies are placed relative far apart according to

the design requirements as specified in Fig. 16. As a result, the power and gate loops can be quite long, accumulating parasitic inductances in the several nH range. This would lead to unacceptable amount of overshoot and ringing at the *SW* node as shown in Fig. 18. To accommodate both thermal and electrical requirements, it is not only essential to minimize the length of the power loops, but also necessary to make use of dynamic gate driving to suppress the unwanted ringing.

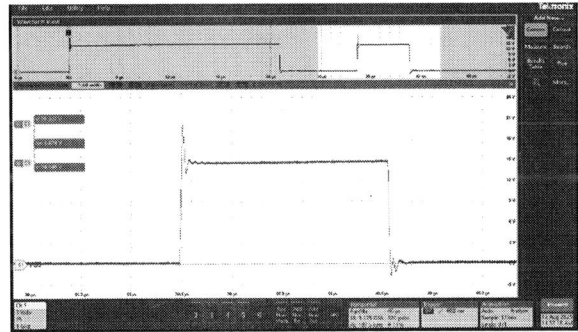

Fig. 18. A double pulse testing waveform showing significant overshoot and ringing during turn-on transient.

V. CONCLUSIONS

This paper introduced several important functionalities in smart gate driver ICs to address issues such as EMI and reverse/body diode conduction. The origin of ringing is explained with the aid of the equivalent *RLC* circuit in the gate loop. The timing requirements to implement effective dynamic gate drive technique is also outlined. Additionally, a smart gate driver capable of performing one-step dead-time correction is introduced. By optimizing the dead-time over a wide range of output current, the best power conversion efficiency can be obtained. Finally, advanced packaging techniques such as silicon interposer can be used to minimize parasitic inductance in low power applications. However, when taking thermal performance into account, an ultra compact form factor may not necessarily offer the best performance. In most practical applications, the finite power loops will still require the assistance of smart gate drivers.

REFERENCES

[1] M. Buffolo *et al.*, "Review and Outlook on GaN and SiC Power Devices: Industrial State-of-the-Art, Applications, and Perspectives," *IEEE Trans. Electron Devices*, vol. 71, no. 3, pp. 1344–1356, Mar. 2024.

[2] L. F. Gomez-Rivera, et al., "Analysis of a Gate-Driving Technique for Enhancing the GaN Power Transistors Performance," *2023 IEEE 2nd Industrial Electronics Society Annual On-Line Conference (ONCON)*, SC, USA, 2023, pp. 1-7, doi: 10.1109/ONCON60463.2023.10430548.

[3] W.J. Zhang, et al., "A Smart Gate Driver IC for GaN Power HEMTs With Dynamic Ringing Suppression," *IEEE Trans. Power Electronics*, vol. 36, no. 12, pp. 14119 – 14132, Dec. 2021.

[4] W.T. Cui, et al., "A Dynamic Gate Driver with Auto-Patterning to Reduce Ringing and Switching Loss," *ISPSD 2025*, Kumamoto, Japan, June 1-5, 2025.

[5] W.J. Zhang, et al., "An Integrated Gate Driver for E-Mode GaN HEMTs with Active Clamping for Reverse Conduction Detection," ISPSD 2019, pp. 83-86, Shanghai, May 19-23, 2019.

[6] Q. Deng, et al., "Thermal Management for Buck Converters Using Integrated Packaged GaN HEMTs," PCIM Asia, Nov. 16-18, 2020, Shanghai China.

[7] N.J. Kim, et al, "A Direct Bond Fabrication Process for Compact GaN Power Modules on Liquid Coolers for EV Applications," *ISPSD 2021*, Nagoya, Japan, May 30 - Jun 3, 2021.

Investigation of Threshold Voltage Instability in GaN HEMTs Using Rapid Ramp Sweeping Technique

Diangang Hu [1], Yutian Gan [2], Shufu Yu [1], Sai Liu [1], Lirong Zhang [1], Weijing Wu*[1], Hongyu Yu*[2,3]

[1] State Key Laboratory of Luminescent Materials and Devices, Guangdong Basic Research Center of Excellence for Energy and Information Polymer Materials, South China University of Technology, Guangzhou, China
[2] School of Microelectronics, Southern University of Science and Technology, Shenzhen 518055, China
[3] School of integrated Circuit, Shenzhen Polytechnic University, Shenzhen, 518055, China.

* Email: wuwj@scut.edu.cn, yuhy@sustech.edu.cn

Abstract—In this work, a comprehensive V_{th} instability test system was developed which integrated three simultaneous operational modes: conventional DC characterization, common fast sweeping measurement and innovative rapid ramp sweeping technique. This system was driven by the gate driver IC output under elevated V_{DS} conditions, thereby intensifying the electrical stress parameters. The testing results validated the rapid ramp sweeping method, showing remarkable congruence with established DC measurement techniques. Moreover, in the rapid ramp sweeping measurement，a synchronous ramp mode was implemented to ensure $V_{DS} = V_{GS}$ condition, which effectively enhanced the test accuracy for V_{th} instability assessment. The results further demonstrate that ramp rate (dv/dt) variation leads to threshold voltage overestimation, establishing it as a crucial parameter in dynamic V_{th} characterization. These findings highlight the considerable potential of rapid ramp sweeping measurements for assessing V_{th} instability in GaN HEMTs that balance measurement accuracy and dynamic stress conditions.

Keywords—Threshold voltage instability, Rapid ramp sweeping, GaN HEMTs

I. INTRODUCTION

Gallium nitride high electron mobility transistors (GaN HEMTs) employing p-type GaN gate have become indispensable components in power electronics systems. These devices offer the normally-off operation essential for safe power switching application, and the exceptional electron transport properties inherent to the AlGaN/GaN heterostructure channel [1], [2]. However, threshold voltage (V_{th}) instability remains a fundamental reliability limitation in p-GaN gate HEMTs, hindering their broader commercialization [3], [4].

Charge trapping in p-GaN gated devices induces substantial and often undesirable V_{th} shifts during operation, leading to uncontrolled drain current variations that compromised system performance and safety. Positive bias temperature instability exhibited particularly complex behavior, including bidirectional V_{th} shifts under stress conditions [5]. Quasi-static DC measurement remains the standard V_{th} characterization method per JEDEC standard JEP183A. This approach provides high measurement accuracy but requires extensive testing time and fails to capture V_{th} variations under dynamic switching conditions [6], [7]. To address these challenges, fast sweeping technique was promoted to rapidly reduce the measurement sweep time to

the microsecond range, significantly improving dynamic characterization capabilities [8], [9].

In previous studies, the gate of GaN HEMTs were driven by a single signal generator during both stress and measurement phases. Although such signal generator enables flexible waveforms generation, its limited current output capability fails to meet the high-drive requirements of GaN HEMTs under practical operating conditions during stress phase [10]. In conventional characterization methods, DC and fast sweeping methods operate independently, with V_{DS} typically held constant during fast sweeping measurements. Accurately characterizing V_{th} shift under realistic high-frequency switching conditions remains a significant challenging.

In this study, a integrated stress-measurement system utilizing a gate driver IC to apply stress conditions was developed, enabling comprehensive characterization of V_{th} instability through DC, common fast sweeping and rapid ramp sweeping measurements. Unlike common fast sweeping measurement, a synchronous ramp mode with $V_{DS}=V_{GS}$ is incorporated, facilitating enhanced analysis of the underlying physical mechanisms. This work established a precise and reliable methodology for evaluating V_{th} instability in GaN HEMTs that balance measurement accuracy and dynamic stress conditions.

II. EXPERIMENT SETUP

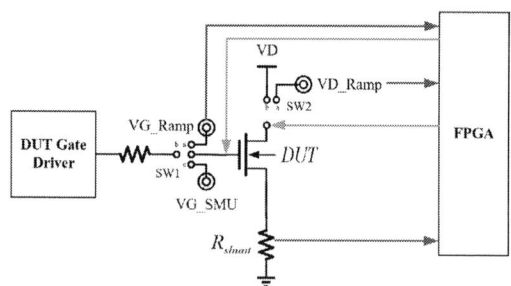

Fig. 1. schematic diagram of the V_{th} instability test system

Fig.1 shows the schematic diagram of the V_{th} instability test system. The gate driver UCC27517 outputs a square wave signal with amplitude ranging from 4.5V to 18V, a peak drive current of 4A and a frequency capability exceeding 1 MHz. The VG_Ramp provides the rapid ramp sweeping signal by

979-8-3315-3918-4/25 $31.00 © 2025 IEEE

function generator, while the VG_SMU (for comparison) supplies the DC sweeping signal with Source Measure Unit (SMU). During sweeping, the FPGA's ADC acquires gate voltage (V_{GS}) and drain current (I_{DS}) in real-time (using R_{shunt} for I_{DS} sensing) to plot I_{DS}-V_{GS} curve to extract V_{th}. The outputs of the gate driver, VG_Ramp, and VG_SMU are selected via SW1 – a single-pole triple-throw (SP3T) switch with an extremely low on-resistance (< 0.1Ω). The high voltage power supply VD provides drain voltage, and VD_Ramp by function generator can synchronously outputs a rapid ramp signal such that $V_{DS} = V_{GS}$. The outputs of VD and VD_Ramp are selected by SW2 – a single-pole double-throw (SPDT) switch. SW1 and SW2 switching is controlled by digital I/O signals from the FPGA circuit. Strict isolation is maintained between the gate drive circuit and the power loop. The device under test (DUT) is a commercial 650V Schottky-type p-GaN gate HEMT.

Figure 2 depicts the stress-measure procedure. The DUT gate driver continuously applies a 7V, 100 kHz square wave stress signal to the device, with V_{th} measured every 10 minutes. Three measurement methods are implemented: 1) DC method (VG_SMU output with VD fixed at 1 V, Fig. 2a); 2) Common fast sweeping method (VG_Ramp output with VD fixed at 1 V, Fig. 2b); 3) Rapid ramp sweeping method (VG_Ramp output with VD_Ramp supplies the drain voltage, while VD_Ramp = VG_Ramp, Fig. 2c).

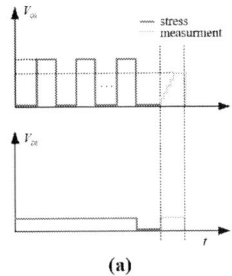

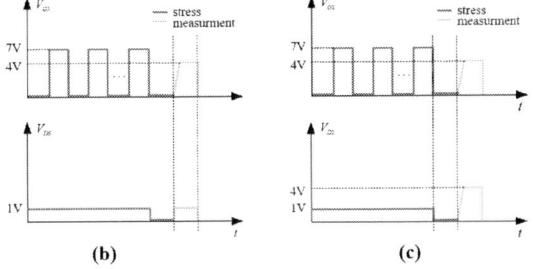

Fig. 2. V_{th} *stress-measure procedure. a) DC method, b) common fast sweeping method, c) rapid ramp sweeping method*

III. RESULT AND DISCUSSION

Throughout the measurement sweeping, synchronous acquisition of V_{GS} and I_{DS} waveforms enables threshold voltage extraction. V_{th} is determined as the V_{GS} value measured at the precise occurrence of $I_{DS} = 11$ mA (according to the GaN HEMT device datasheet) as show in Fig. 3. All reported V_{th} data are obtained under controlled 30°C conditions to eliminate thermal influence on threshold voltage characteristics.

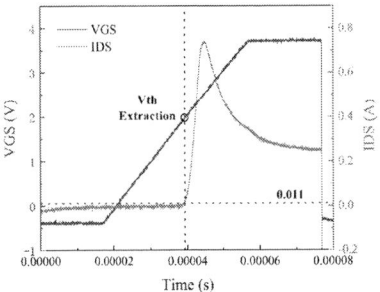

Fig. 3. *threshold voltage extraction*

According to the stress-measure process depicted in Fig. 2, V_{th} measurements obtained via DC, common fast sweeping, rapid ramp sweeping methods were compared across four devices. The observed trends and relationships showed strong consistency (Fig. 4), confirming reproducibility beyond device-specific variations. Compared to DC method, both common fast sweeping and rapid ramp sweeping methods reduce measurement-induced stress. The ramp rate is 0.1 V/μs with V_{GS} from 0V to 4V. The rapid ramp sweeping method yields results more closely aligned with DC measurements than common fast sweeping, while exhibiting certain discernible differences.

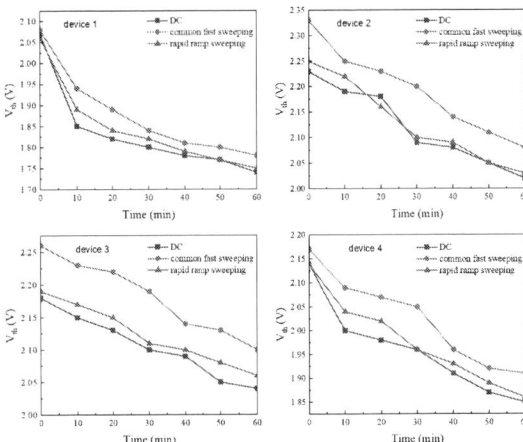

Fig. 4. V_{th} *measurements obtained via DC, common fast sweeping, rapid ramp sweeping methods*

With rapid ramp sweeping method, the two-dimensional electron gas (2DEG) electric potential follows the drain ramp voltage. When V_{GS} exceeds V_{th}, the channel forms and the 2DEG conducts, initiating I_{DS} flow. The channel on resistance R_{on} is extremely low (130 mΩ), thus V_{DS} is low as determined by $V_{DS} = I_{DS} \cdot R_{on}$. This measurement methodology ensures V_{DS} dynamically follows V_{GS} throughout the sweeping process. Upon reaching the threshold voltage point, V_{DS} rapidly decrease as show in Fig. 5 due to device turning-on, maintaining the device under low-field conditions during the entire V_{th} extraction process.

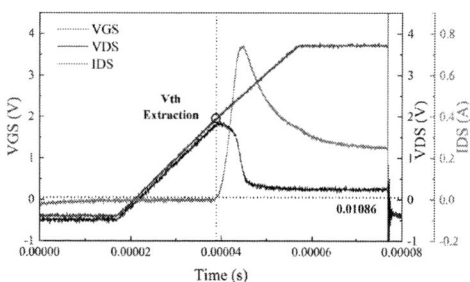

Fig. 5. V_{DS} rapidly decrease as device turn-on

The fixed V_{DS} in common fast sweeping method sustains high-field conditions. a) V_{DS}-Induced 2DEG Channel Non-Uniformity: Elevated drain electric fields cause partial depletion of 2DEG electrons beneath the AlGaN barrier, increasing channel resistance [11]. Higher V_{GS} is needed to achieve the target current, manifesting as elevated V_{th}. b) Drain-Induced Barrier Lowering (DIBL): High V_{DS} reduces the drain potential barrier, enhancing electron injection from source to drain. The elevated drain potential attracts 2DEG electrons, creating preferential conduction pathways near the drain terminal. Consequently, increased gate voltage is required to establish full channel control, resulting in higher V_{th}[12], [13].

In rapid ramp sweeping method, we further investigated the impact of varying sweep time on threshold voltage measurements with ramp rate dv/dt of 0.2 V/μs, 0.1 V/μs, and 0.08 V/μs respectively. As shown in Fig. 6, the 0.1 V/μs slope yielded results most closely approaching DC measurements.

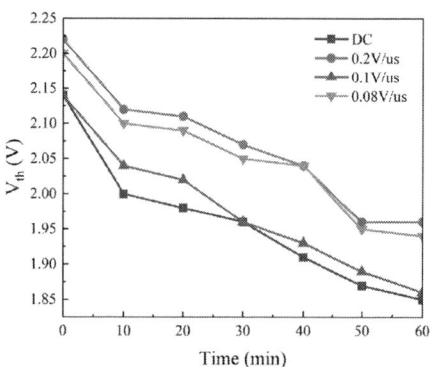

Fig. 6. impact of varying sweep time on threshold voltage measurements

Excessive ramp rate (0.2 V/μs) induces V_{th} overestimation due to: a) Carrier response delay: High ramp rates cause the gate voltage to change too rapidly for the 2DEG to respond instantaneously due to carrier mobility limitations [14]. The resultant delay in channel formation creates a lag in I_{DS} rise relative to the V_{GS} increase. Consequently, at the threshold current criterion (I_{DS} = 11 mA), the recorded V_{GS} exceeds the intrinsic V_{th} value. b) Incomplete gate capacitor charging: The gate-channel capacitance requires finite time to charge [15]. At excessive ramp rates, the capacitor voltage fails to settle to the applied V_{GS}, resulting in reduced effective gate overdrive voltage ($V\_eff < V\_GS$). Since I_{DS} depends directly on V_eff, elevated applied V_{GS} becomes necessary to achieve the target current (11 mA), thereby overestimating the measured V_{th}.

However, slow ramp rates (0.08 V/μs) also induce V_{th} overestimation which should ascribed to: a) Interface trap charge accumulation: At low ramp rates, gate oxide interface traps have sufficient time to capture/release carriers, creating additional charge screening effects. This effectively degrades gate control over the channel, necessitating higher V_{GS} to establish adequate channel current [16], [17]. Consequently, measured V_{th} becomes artificially elevated. b) Self-Heating thermal effects: Slow ramps permit lattice heating (>30°C rise in 10μs simulations), reducing mobility through phonon scattering [18]. This slows the I_{DS} rising, necessitating elevated applied V_{GS} to achieve the target current, thereby introducing positive error in V_{th} measurement.

IV. CONCLUSION

This work presents a fast dynamic stress methodology for precise characterization of V_{th} instability in GaN HEMTs. The proposed stress-measurement system employs a gate driver output to achieve enhanced stress conditions. Three measurement methods are implemented, including conventional DC method, common fast sweeping and advanced rapid ramp sweeping method. Comparative analysis reveals that both common fast sweeping and rapid ramp sweeping method significantly reduce measurement-induced stress compared to DC measurement, owing to their accelerated ramp rates. Notably, the rapid ramp sweeping method demonstrates superior agreement with DC measurements results while maintaining distinct advantages in dynamic characterization. This improved correlation stems from the maintenance of low-field conditions throughout the V_{th} extraction process in rapid ramp sweeping. Furthermore, the experimental results identify ramp rate (dv/dt) as a critical parameter in dynamic characterization, with higher rates leading to measurable V_{th} overestimation. These findings establish rapid ramp sweeping as an effective approach for balancing measurement accuracy and dynamic stress conditions in GaN HEMT characterization.

ACKNOWLEDGMENT

This work was supported by National Natural Science Foundation of China (Grant No: 62274082), Research on mechanism of Source/Drain ohmic contact and the related GaN p-FET (Grant No: 2023A1515030034), Study on the reliability of GaN power devices (Grant No: JCYJ20220818100605012), Research on novelty low-resistance Source/Drain ohmic contact for GaN p-FET (Grant No: JCYJ20220530115411025), and Research on GaN-Based Devices for Industrial Applications (Grant No. 6025312001K).

REFERENCES

[1] M. Ishida, T. Ueda, T. Tanaka, and D. Ueda, "GaN on Si technologies for power switching devices," IEEE Trans. Electron Devices, vol. 60, no. 10, pp. 3053–3059, Oct. 2013. DOI: 10.1109/TED.2013.2268577

[2] J. Wei, Z. Zheng, G. Tang, H. Xu, G. Lyu, L. Zhang, J. Chen, M. Hua, S. Feng, T. Chen, and K. J. Chen, "GaN power integration technology and its future prospects," IEEE Trans. Electron Devices, vol. 71, no. 3, pp. 1365–1382, Mar. 2024, doi: 10.1109/TED.2023.33 41053.

[3] L. Sayadi, G. Iannaccone, S. Sicre, O. Häberlen, and G. Curatola, "Threshold voltage instability in p-GaN gate AlGaN/GaN HFETs," IEEE Trans. Electron Devices, vol. 65, no. 6, pp. 2454–2460, Jun. 2018, doi: 10.1109/TED.2018.2828702

[4] Y. Shi, Q. Zhou, Q. Cheng, P. Wei, L. Zhu, D. Wei, A. Zhang, W. Chen, and B. Zhang, "Bidirectional threshold voltage shift and gate leakage in 650 V p-GaN AlGaN/GaN HEMTs: The role of electron-trapping

and hole-injection," in Proc. IEEE 30th Int. Symp. Power Semiconductor Devices ICs (ISPSD), Chicago, IL, USA, May 2018, pp. 96–99, doi: 10.1109/ISPSD.2018.8393611

[5] P. Lagger, M. Reiner, D. Pogany, and C. Ostermaier, "Comprehensive study of the complex dynamics of forward bias-induced threshold voltage drifts in GaN based MIS-HEMTs by stress/recovery experiments," IEEE Trans. Electron Devices, vol. 61, no. 4, pp. 1022–1030, Apr. 2014. DOI: 10.1109/TED.2014.2303853

[6] G. P. Lansbergen et al., "Threshold voltage drift (PBTI) in GaN D-mode MISHEMTs: Characterization of fast trapping components," in Proc. IEEE Int. Rel. Phys. Symp., Jun. 2014, p. 6, doi: 10.1109/IRPS.2014.6861111.

[7] D. Heh, C. D. Young, and G. Bersuker, "Experimental evidence of the fast and slow charge trapping/detrapping processes in high-κ dielectrics subjected to PBTI stress," IEEE Electron Device Lett., vol. 29, no. 2, pp. 180–182, Feb. 2008, doi: 10.1109/LED.2007.914088

[8] X. Li et al., "Observation of dynamic V_{TH} of p-GaN gate HEMTs by fast sweeping characterization," IEEE Electron Device Lett., vol. 41, no. 4, pp. 577–580, Apr. 2020, doi: 10.1109/LED.2020.2972971.

[9] E. Canato, F. Masin, M. Borga, E. Zanoni, M. Meneghini, G. Meneghesso, A. Stockman, A. Banerjee, and P. Moens, "μs-range evaluation of threshold voltage instabilities of GaN-on-Si HEMTs with p-GaN gate," in Proc. IEEE Int. Rel. Phys. Symp. (IRPS), Monterey, CA, USA, Mar. 2019, pp. 1–6, doi: 10.1109/IRPS.2019.8720549

[10] U. K. Mishra, P. Parikh, and Y.-F. Wu, "AlGaN/GaN HEMTs-an overview of device operation and applications," Proc. IEEE, vol. 90, no. 6, pp. 1022–1031, Jun. 2002, doi: 10.1109/JPROC.2002.1021567

[11] S. Pan, S. Feng, X. Zheng, X. He, X. Li, and K. Bai, "Effects of temperature and bias voltage on electron transport properties in GaN high electron-mobility transistors," IEEE Trans. Device Mater. Rel., vol. 21, no. 4, pp. 494–499, Dec. 2021, doi: 10.1109/TDMR.2021.3109088.

[12] M. Nuo, J. Wei, M. Wang, J. Yang, Y. Wu, Y. Hao, and B. Shen, "Gate/drain coupled barrier lowering effect and negative threshold voltage shift in Schottky-type p-GaN gate HEMT," IEEE Trans. Electron Devices, vol. 69, no. 7, pp. 3630–3635, Jul. 2022

[13] HULT B, BERGSTEN J, CASTILLO R F D D, Darakchieva V. Malmros Anna, Hjelmgren Hans, "Characterization of Drain-Induced Barrier Lowering in GaN HEMTs Using a Drain Current Injection Technique," IEEE Transactions on Electron Devices, 2024, 71(12): 7383-7389.

[14] O. Katz, A. Horn, G. Bahir, and J. Salzman, "electron mobility in an AlGaN/GaN two-dimensional electron gas. I. Carrier concentration dependent mobility," IEEE Trans. Electron Devices, vol. 50, no. 10, pp. 2002–2008, Oct. 2003

[15] CHO T, PARK J, JUNG S, KANG M, "Analysis and Modeling of Intrinsic Capacitance in Enhancement Mode GaN HEMT," IEEE Journal of the Electron Devices Society, 2025, 13: 638-641.

[16] SHI L, LIU T, ZHU S, QIAN J, JIN M, MADDI H L R, "Effects of Oxide Electric Field Stress on the Gate Oxide Reliability of Commercial SiC Power MOSFETs" proceedings of the 2022 IEEE 9th Workshop on Wide Bandgap Power Devices & Applications (WiPDA), 2022: 45-48.

[17] ZHU J, HOU B, CHEN L, ZHU Q, YANG L, ZHOU X, "Threshold voltage shift and interface/border trapping mechanism in Al2O3/AlGaN/GaN MOS-HEMTs," proceedings of the 2018 IEEE International Reliability Physics Symposium (IRPS), 2018: P-WB.1-1-P-WB.1-4.

[18] ZHANG C, WANG M, XIE B, WEN C P, WANG J, HAO Y, "Temperature Dependence of the Surface- and Buffer-Induced Current Collapse in GaN High-Electron Mobility Transistors on Si Substrate," IEEE Transactions on Electron Devices, 2015, 62(8): 2475-2480.

Design and Kirk Effect Improvement of 30V NLDMOS Base on 0.18μm BCD Platform

Qi Ding[1,2], Ning Ning[1,2], Renxiong Li[2], Jun Huang[2], Yutuo Guo[2], Yu Wang[2], Kunqin He[2], Yaxin Liu[2], Ziyi Zeng[2], Ming Qiao*[1], Lulu Peng*[2], Bo Zhang[1]

[1]State Key Laboratory of Electronic Thin Films and Integrated Devices, University of Electronic Science and Technology of China, Chengdu, P.R. China
[2]United Microelectronics Center Co., Ltd, Chongqing, P.R. China

Email: dingxiaoqi_0804@163.com, qiaoming@uestc.edu.cn, lulu.peng@cumec.cn

Abstract—This paper introduces a 30V full isolated N type laterally diffused metal oxide semiconductor (LDMOS) by employing techniques of reduced surface field (RESURF), field plates, and field isolation to achieve the required breakdown voltage (BV) and specific on-resistance ($R_{on,sp}$). To widen the safe operation area (SOA) and alleviate the Kirk effect that causes significant drain current increasing without saturation, two methods are adopted: extending the LOCOS length (Ls) and adding N-type well (NW) as a buffer layer at the drain side. Compared to simply extending Ls for suppressing the drain current increasing, thereby mitigating the Kirk effect and flattening the drain current-drain voltage (Id-Vd) curve, adding NW at drain side further balances the electric field and reduces $R_{on,sp}$.

Keywords—LDMOS, breakdown voltage, $R_{on,sp}$, Kirk effect

I. INTRODUCTION

The rapid advancement of industrial, automatic electric systems, and smart applications has significantly increased the importance of power integrated circuits (PICs) in fields such as power converters, power controllers, and power management ICs. Single-chip integration is a crucial development direction for PICs [1]. Due to its ease of integration with traditional CMOS processes, laterally diffused metal oxide semiconductor (LDMOS) used in bipolar, CMOS, and LDMOS (BCD) process has become widely adopted [2]. Breakdown voltage (BV) and specific on-resistance ($R_{on,sp}$) are significant for the performance of LDMOS devices. Additionally, safe operation area (SOA) and high voltage-high current reliability which need suppression of drain current expansion pose a bottleneck for device qualification [3,4].

Kirk effect is a detrimental phenomenon in PICs that manifests when the gate and drain voltages are high, leading to a dramatic increase and unsaturation in drain current. This effect degrades both SOA and hot carrier injection (HCI) performance [5,6]. To mitigate this issue, heavier doping concentration at the drain side and longer drift region can be usually employed [6,7].

In this paper, a fully isolated 30V N type LDMOS (NLDMOS) is designed. The design incorporates techniques of reduced surface field (RESURF), field plates, and field isolation to achieve the desired BV and $R_{on,sp}$. By adjusting the length of LOCOS (Ls) in drift region and adding N-type well (NW) at the drain side in the design, the suppression of Kirk effect is maximized while minimizing its negative impact on $R_{on,sp}$. Throughout this work, TCAD simulations are

This work was supported in part by the National Natural Science Foundation of China under Grant 62174024

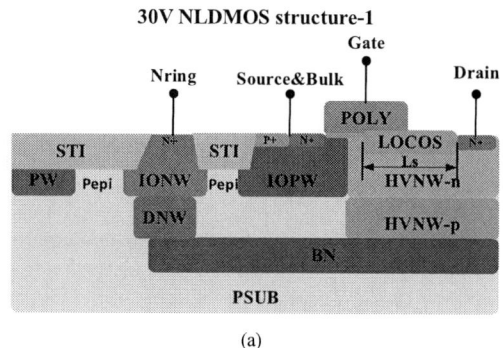

(a)

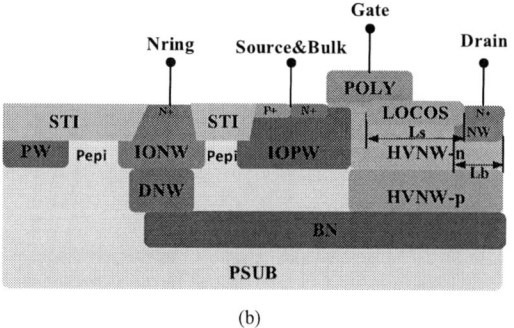

(b)

Fig.1. The two structures of the full isolated 30V NLDMOS design: (a) the structure without NW buffer layer, (b) the structure with NW buffer layer.

conducted to demonstrate device performances and help to analyze the underlying phenomena and mechanisms.

II. DEVICE STRUCTURE AND MECHANISM

Figure 1 illustrates the two structures of a 30V NLDMOS. The device has four terminals: gate, bulk/source, drain, and N-type isolation ring (Nring). LOCOS is employed under the gate and over the drift region instead of STI, which optimizes $R_{on,sp}$ due to its shallow depth oxidation. N type high voltage well (HVNW-n) and P type high voltage well (HVNW-p) are two types of implantation with one mask; deeper P-type implantation contrasts with shallower N-type implantation in the drift region, forming a RESURF structure that optimizes BV and $R_{on,sp}$. P type epitaxy (Pepi) is added between Nring and bulk, as well as between Nring and substrate to improve BV between these terminals, ensuring good isolation performance when Nring is on high potential. Structure-1 is

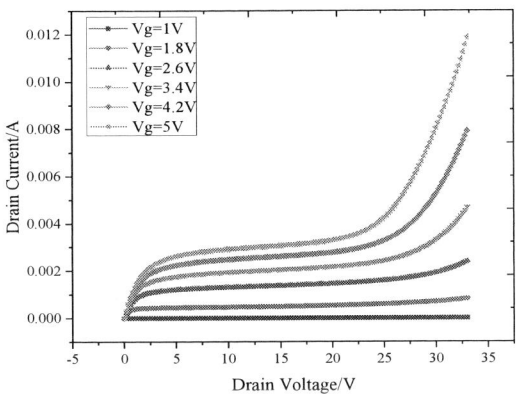

Fig. 2. Id-Vd curves of structure-1 at the baseline condition (Ls=0.8μm).

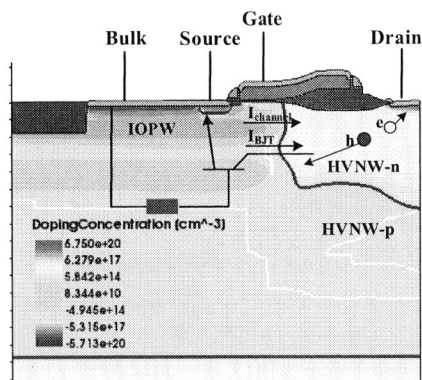

Fig. 3. The NLDMOS simulation with doping concentration and the schematic of Kirk effect mechanism.

the NLDMOS without NW buffer, while structure-2 is with NW. Comparing both structures, structure-2 exhibits lower $R_{on,sp}$ and weaker Kirk effect.

The N-type isolation ring offers flexible electrical connection methods. Typically, it can be either zero voltage or equal to the drain (connected to drain). Under these two connection methods for structure-1, when Ls=0.8μm, the BV is about 43V to 45V, meeting the requirements of a 30V NLDMOS application.

Figure 2 shows the Id-Vd curves at baseline condition (Ls=0.8μm) for structure-1 under different gate voltages. The Kirk effect is obvious when the gate voltage exceeds 4V and the drain voltage surpasses 25V.

Figure 3 shows simulation of structure-1 with doping concentration and illustrates the schematic of the Kirk effect. As shown in Figure 3, the Kirk effect arises due to the parasitic BJT turning on (source, bulk, and drift region acts as emitter, base, and collector respectively). Ideally, drain current should originate solely from channel current coming from source to

drain, however, the total drain current comprises two sources: channel current ($I_{channel}$) and parasitic BJT current (I_{BJT}). When the drain voltage is high, excessive electron-hole pairs are generated under intense electric field and impact ionization. Holes flow to the device bulk, causing a voltage drop between source and bulk interior. If the hole current or the bulk resistance becomes large enough, this voltage drop satisfies the PN junction turning on condition of source-bulk, leading to parasitic BJT turning on. The parasitic BJT current flows are from emitter to collector (source to drift region and drain), significantly increasing the drain current. With the gate voltage increasing, the drain side has more saturation current, therefore, more electron-hole pairs will be generated and the Kirk effect behaves more seriously.

Figure 4 illustrates the current density conditions of structure-1 under Vd=20V (no Kirk effect) and Vd=33V (with Kirk effect) when Vg=5V. Comparing figure 4(a) and 4(d), high drain voltage results in increased currents not only in the channel and drift region, but also in the bulk. According to figures 4(b) and 4(c) or 4(e) and 4(f), the primary current in the bulk originates from holes. Analyzing Figure 4(b) and 4(e),

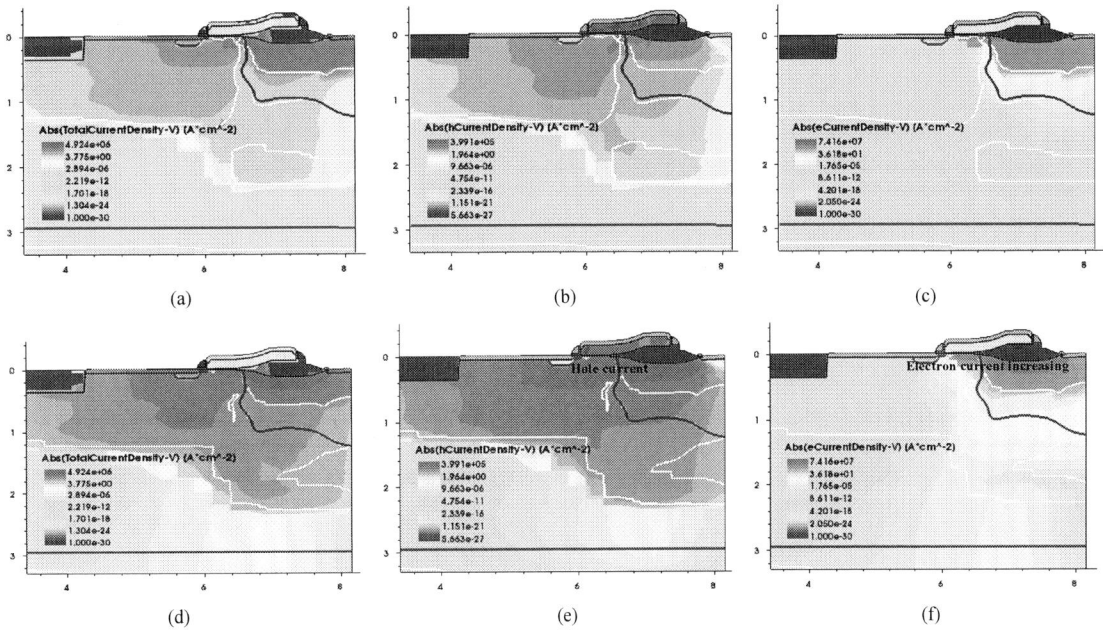

Fig. 4. The current density condition when Vg=5V: (a) total current density at Vd=20V, (b) hole current density at Vd=20V, (c) electron current density at Vd=20V, (d) total current density at Vd=33V, (e) hole current density at Vd=33V, (f) electron current density at Vd=33V.

TABLE I   THE SPECIFIC EXPERIMENT CONDITIONS

Experiment	Ls/μm	Experiment	Ls/μm	NW width (Lb)/μm
1	0.6	*7*	1	0.2
2	0.8	*8*	1	0.4
3	1	*9*	1.2	0.2
4	1.2	*10*	1.2	0.4
5	1.4	*11*	1.4	0.2
6	1.6	*12*	1.4	0.2

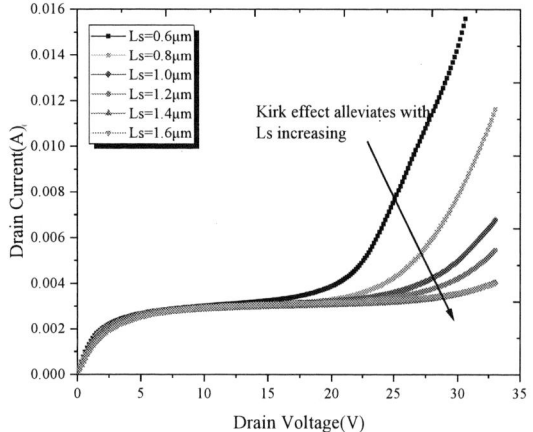

Fig.5. NLDMOS Id-Vd curve performance of structure-1 under different Ls.

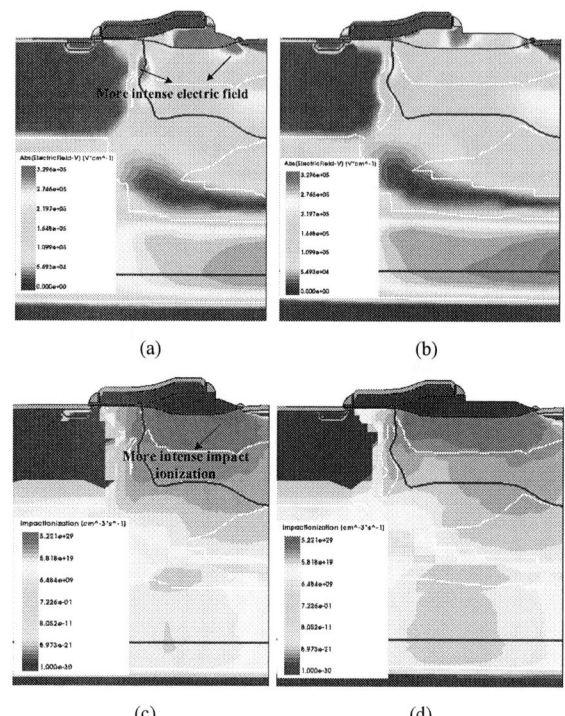

(a) (b)

(c) (d)

Fig. 6.Electric field and impact ionization of structure-1: (a) electric field when Ls=0.8μm, (b) electric field when Ls=1.4μm, (c) impact ionization when Ls=0.8μm, (d) impact ionization when Ls=1.4μm.

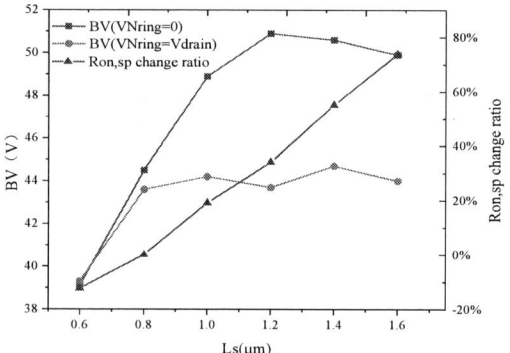

Figure 7 BV and $R_{on,sp}$ change ratio of structure-1 with Ls variation.

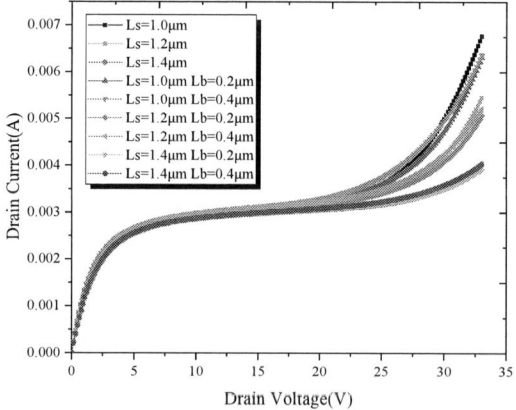

Fig. 8. The Id-Vd curves comparison between structure-1 and structure-2 under different Ls and Lb.

when Kirk effect occurs, a mass of hole current flows from the drift region to the bulk, leading to the parasitic BJT turning on. Consequently, the turned on BJT exhibits an increased electron density current from emitter to collector (source to drift region and drain), resulting in higher drain current.

III.   RESULTS AND DISCUSSION

In order to optimize the Id-Vd curve and flatten it at the saturation region, the methods of Ls variation and adding NW as the buffer layer at drain side are conducted by simulation. The specific conditions are outlined in Table I. These experiments can be divided into two groups: Group 1 (Experiments 1-6) involves varying Ls without NW (structure-1), while Group 2 (Experiments 7-12) varies both Ls and NW width (structure-2).

Figure 5 shows the Id-Vd curves for structure-1 under different Ls when the gate voltage is 5V. As Ls increasing, the drain current surge point moves towards higher drain voltage, while the maximum drain current decreases. The Kirk effect alleviates because that the increased length of the drift region allows more area to share the total drain voltage, and reduces the electric field peak and impact ionization at the drain side. Under the condition of Ls=1.4μm and Ls=1.6μm, the device has similar drain current performance, therefore, Ls=1.4μm is enough for this device Id-Vd curve optimization.

Figure 6 illustrates electric field distribution and impact ionization of structure-1 under Ls=0.8μm and Ls=1.4μm. Under the condition of Ls=0.8μm, electric fields and impact ionizations are more intense at the LOCOS end of the drain side and the PN junction of the device bulk and drift region. On one hand, the drain side intense electric field and impact ionization can stimulate more electron-hole pairs to provide

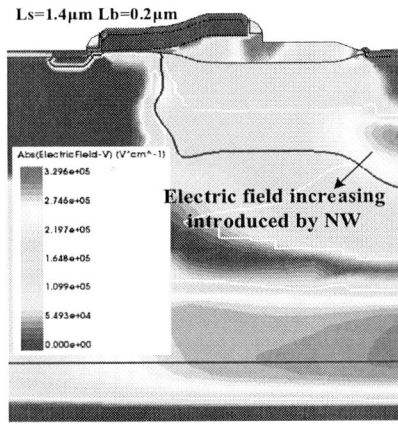

Ls=1.4μm Lb=0.2μm

Electric field increasing
introduced by NW

Fig.9. Electric field distribution under structure-2 within Ls=1.4μm and Lb=0.2μm.

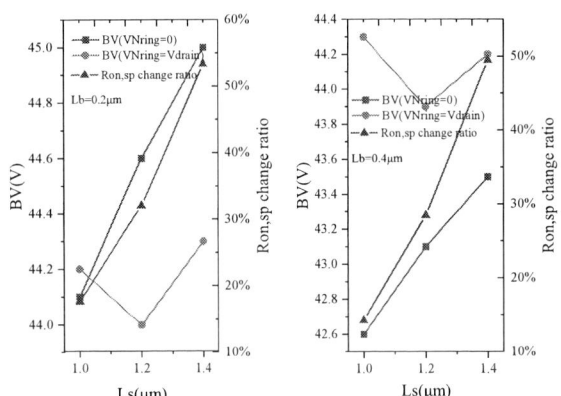

Fig. 10. BV and $R_{on,sp}$ change ratio with Ls variation under different Lb.

holes for parasitic BJT turning on; on the other hand, the PN junction of bulk and drift region is corresponding to the parasitic BJT base and collector, the more intense electric field will generate high collector current, also contributes to the drain current increasing.

Figure 7 shows BV and $R_{on,sp}$ change ratio of structure-1 with Ls variation. When Nring voltage is 0V, because the breakdown path is from the horizontal direction and helps the depletion region extend, BV increases with Ls increasing; when Nring voltage equals to drain voltage, because the breakdown path is from the vertical direction with the fixed device depth, BV is stable when Ls is larger than 0.8μm. However, $R_{on,sp}$ increases within Ls extending under the two Nring connection methods, negatively impacting device performance.

To further mitigate Kirk effect and $R_{on,sp}$ increasing, structure-2 introduces NW as a buffer layer at the drain side that could balance electric field and provide more current carriers, as shown in figure 1(b). Figure 8 is the Id-Vd curves of structure-1 and structure-2 under different Ls and NW width (Lb) which are corresponding to the Group 2 in table I experiment 7~12. When adding NW at drain side, the Kirk effect will also alleviate slightly.

Figure 9 is the electric field distribution under structure-2 within Ls=1.4μm and Lb=0.2μm. Comparing figure 9 and figure 6(b), NW introducing has the following effects: 1) making drain concentration distribution variation more uniform, so that the drain side electric field intensity has further reduction; 2) making drain vertical concentration and electric field increasing at the drift region bottom, helping to reducing the electric field intensity at drain surface.

Figure 10 is the BV and $R_{on,sp}$ change ratio (comparing with baseline condition: Ls=0.8μm without NW buffer) with Ls variation under different Lb. Adding NW buffer reduces BV when Nring voltage is zero, but makes minimal effect when Nring voltage equals to drain voltage. Regardless of the connection method, BV can still meet the application requirements for 30V NLDMOS. Adding NW buffer layer at drain side could further reduce $R_{on,sp}$, and longer Lb achieves lower $R_{on,sp}$. Thus, adding NW as a buffer layer at the drain side in structure-2 could effectively alleviate the performance of $R_{on,sp}$ degradation which is introduced by flattening the Id-Vd curve.

IV. CONCLUSION

In this paper, a 30V full isolated LDMOS is designed and improved to flatten the Id-Vd curve; the mechanisms of drain current unsaturation and Kirk effect optimization are discussed. Techniques of RESURF, field plates, and field isolation are employed to achieve higher BV and lower $R_{on,sp}$. Because of the serious Kirk effect when gate voltage is higher than 4V, the methods of extending Ls and adding NW at drain side are adopted. Because of longer Ls sharing the drain voltage and reducing the electric field, and also NW introducing balancing the drain side electric field, the Kirk effect is alleviated effectively. Furthermore, introducing NW at drain side, can also reduce the $R_{on,sp}$ and relieve the performance degeneration coming from extending Ls. Consequently, the performances of BV, $R_{on,sp}$, and SOA are balanced preferably.

REFERENCES

[1] B. Zhang, W. T. Zhang, L. Zhu, J. Zu, M. Qiao, and Z. J. Li, "Review of technologies for high-voltage integrated circuits," Tsinghua Science and Technology, vol. 27, no. 3, pp. 495-511, June 2022.

[2] J. Huang, N. Ning, R. X. Li, Q. Ding, Y. T. Guo, Y. Wang, K. Q. He, Y. X. Liu, and L. L. Peng, "An Ultra-Low Specific On-Resistance LDMOS With Segmented LOCOS In 0.18 μm BCD Process Platform," 2023 IEEE 15th International Conference on ASIC (ASICON), Nanjing, China, 2023, pp. 1-3.

[3] A. Mishra, S. Gupta, A. Gupta, S. K. Boeila, M. Monishmurali, M. Shrivastava, Avinash K. Singh, and Amit K. Singh, "Inverted SOA and Transient Non-Linearity of LDMOS Devices With RESURF-Implant," 2022 IEEE International Conference on Emerging Electronics (ICEE), Bangalore, India, 2022, pp. 1-4.

[4] A. Kuwana, J. I. Matsuda, and H. Kobayashi, "Analysis of Switching Characteristics of Wide SOA and High Reliability 100 V N-LDMOS Transistor with Dual RESURF and Grounded Field Plate Structure," 2021 IEEE 14th International Conference on ASIC (ASICON), Kunming, China, 2021, pp. 1-4.

[5] H. L. Chou, P. C. Su, J. C. W. Ng, P. L. Wang, H. T. Lu, C. J. Lee, W. J. Syue, S. Y. Yang, Y. C. Tseng, C. C. Cheng, C. W. Yao, R. S. Liou, Y. C. Jong, J. L. Tsai, J. Cai, H. C. Tuan, C. F. Huang, and J. Gong, "0.18 μm BCD Technology Platform with Best-in-Class 6 V to 70 V Power MOSFETs," 2012 24th International Symposium on Power Semiconductor Devices and ICs (ISPSD), Bruges, Belgium, 2012, pp. 402-404.

[6] A.W. Ludikhuize, "Kirk effect limitations in High Voltage IC's," 1994 6th International Symposium on Power Semiconductor Devices and ICs (ISPSD), Davos, Switzerland, 1994, pp. 249-252.

[7] Z. Z. Xu, T.Tian, M. X Fang, W. Song, Y. T Zhang, Z. Q Fang, D. H. Liu, H. L. Chen, and W. S. Qian, "Substrate Current Improvement And Investigation In Low Voltage Power LDMOS With A Novel Design," 2024 Conference of Science and Technology for Integrated Circuits (CSTIC), 2024, pp. 1-4.

Device Modeling Based on Residual Neural Network with Ensemble-Based Active Learning

Hongfei Su[1], Yutong Wu[1], Jithish Jayarajan[2], Bharatha Kumar Thangarasu[1], Nagarajan Mahalingam[1],
Fanyi Meng[1], Kaixue Ma[1], Kiat Seng Yeo*[1,2]

[1] Tianjin University, School of Microelectronics, No. 92, Weijin Road, Tianjin, China 300072
[2] Singapore University of Technology and Design, 8 Somapah Road, Singapore 487372

* Email: suhongfei@tju.edu.cn, yeokiatseng@tju.edu.cn

Abstract—Accurate semiconductor device modeling requires extensive I-V characterization data, creating an expensive bottleneck in design cycles. We propose a neural framework integrating residual networks (ResNet) with ensemble-based active learning to minimize data requirements and improve the prediction accuracy of the model. ResNet captures the global trend and local details of transistor I-V behavior more accurately through skip connections and residual calculations. Experiments show that the prediction accuracy of ResNet is 6% higher than that of multilayer perceptron. Our approach uses a five-model ensemble to estimate model uncertainty—the prediction divergence of the five models caused by the limited training data. These estimates guide the selection of the most informative samples in an iterative active learning process. Our framework reduces the required training data volume by 25% (from 3720 to 2782 samples) compared to multilayer perceptron, while simultaneously improving prediction accuracy. The mean absolute error (MAE) drops from 5.3660e-07 to 4.9107e-07 (an 8% reduction), and the mean absolute relative error (MARE) decreases from 0.36% to 0.28% (a 26% improvement). This approach offers significant cost and time savings for semiconductor companies and research laboratories, where device characterization remains a major bottleneck.

Keywords—Machine learning, device modeling, Residual neural network, Ensemble-based Active Learning.

I. INTRODUCTION

Circuit simulation serves as the foundation for efficient analysis and design of integrated circuits, where accurate and efficient device models are critical prerequisites for predicting circuit performance and reliability [1]. For decades, physics-based compact models (CMs)-such as the BSIM series, PSP, and HiSIM - have dominated the field of device modeling due to their clear physical interpretability, computational efficiency, and seamless integration into circuit simulators (e.g., SPICE). However, as transistor dimensions continue to scale down to the nanometer regime, emerging devices exhibit secondary effects such as short-channel effects and quantum mechanical phenomena, which are challenging to capture precisely using traditional analytical equations derived from simplified physical assumptions [2]. Moreover, to accommodate novel device architectures and complex effects, model complexity has increased drastically, making model development and parameter extraction (PE) extremely cumbersome. Consequently, compact models face significant limitations in advanced technology nodes.

In recent years, machine learning (ML) techniques, particularly artificial neural networks (ANNs), have demonstrated remarkable potential in semiconductor device modeling. Unlike traditional physics-based approaches, data-driven ML methods learn nonlinear mappings between input parameters (e.g., geometric dimensions, bias voltages) and output characteristics (e.g., current, capacitance) directly from large-scale experimental or simulated datasets [3]-[7]. Research has shown that these methods not only enable end-to-end device behavioral modeling but also facilitate parameter extraction [8] and design space optimization [9]. Existing literature confirms that ANNs not only overcome the limitations of conventional modeling approaches but also establish a new research paradigm in device modeling.

The pioneering application of ANNs in transistor modeling dates back to Litovski's work in the early 1990s [10]. While traditional multilayer perceptrons (MLPs) perform well in predicting static I-V characteristics, they exhibit significant errors in higher-order derivatives (e.g., transconductance gm and output conductance gds). To address this limitation, [1] and [7] introduced first and second order derivative terms into the loss function, improving transconductance prediction accuracy. Additionally, due to the wide dynamic range of transistor currents, data preprocessing, particularly logarithmic transformation of outputs, significantly enhances accuracy in the subthreshold region. Ref. [11] employs neural networks to learn the residual between physical model predictions and actual values, thereby improving the model's accuracy through residual correction. Ref. [12] proposes a densely connected deep neural network, where features from any preceding layer are directly propagated to all subsequent layers. This densely interconnected architecture enhances information flow and gradient propagation, leading to improved prediction accuracy and model stability.

However, the current(*Ids*) of MOSFET devices span several orders of magnitude and exhibit strong nonlinearity. Traditional MLP model may suffer from insufficient capture of local features and gradient instability, leading to reduced accuracy. Additionally, conventional MLP models require extensive training data obtained from TCAD simulations or experimental measurements, which inevitably prolongs the development cycles and increases costs for semiconductor companies and research institutions. Therefore, deep learning-based device modeling approaches warrant further exploration and advancement.

In this work, we employ residual neural network (ResNet) with an ensemble-based active learning framework. ResNet facilitates the approximation of complex mappings through task decomposition while maintaining exceptional training stability. In practical applications, ResNet demonstrates particular suitability for addressing problems characterized by intricate nonlinear relationships and wide dynamic ranges, owing to its capacity for phased, progressive output refinement. In contrast, MLPs often struggle to simultaneously accommodate multi-scale feature variations. These advantages render ResNet significantly superior in domains requiring high - precision fitting, such as physical

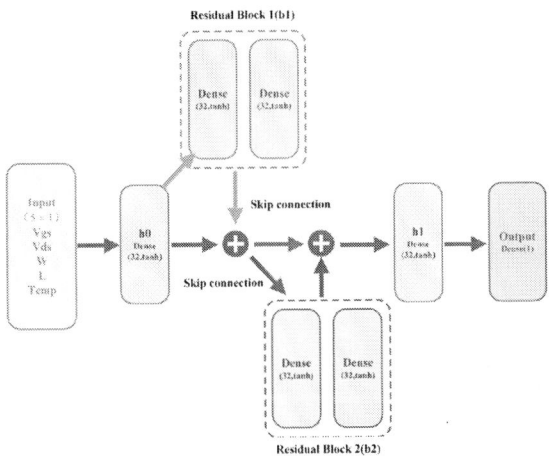

Fig. 1. The schematic view showing the structure of ResNet model.

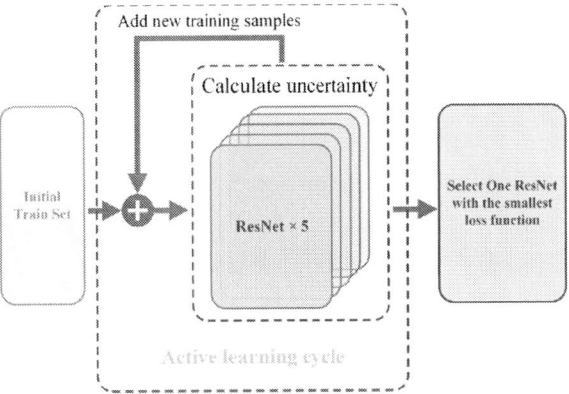

Fig. 2. The schematic view showing the structure of ensemble-based active learning.

modeling. Meanwhile, to enhance modeling efficiency and fully exploit the device characteristics contained in the training set, we adopt an ensemble-based active learning strategy to achieve high-information data selection. The remainder of this paper is organized as follows. Section II presents the architectural details of the proposed model. Section III discusses the performance of our approach and provides comprehensive comparisons with conventional modeling methods. Section IV concludes the paper.

II. DEEP-LEARNING IV MODEL

A. ResNet Architecture

ResNet decomposes the target function into two components:

$$y = F(x) + x \tag{1}$$

The conventional MLP architecture enforces the learning of the complete mapping y. When the span between y and x is very large, traditional networks struggle to simultaneously capture both large-scale and small-scale nonlinear relationships. In contrast, ResNet transmits the large-span trend directly through x, while $F(x)$ focuses on learning local nonlinear fluctuations. This approach is particularly suitable for nonlinear modeling of device current characteristics. Due to its unique capability of decoupling multi-scale features and the nonlinear residual learning mechanism, ResNet can accurately capture the complex cross-scale current transport behaviors in device physics.

Fig. 1 illustrates the architecture of the ResNet model. The ResNet model comprises two hidden layers (h0, h1) and two residual blocks (b1, b2). Both hidden layers h0 and h1 contain 32 neurons, while each residual block consists of two hidden layers (each with 32 neurons). Through skip connections, the residual blocks directly incorporate the original input into the output prediction, with their two layers learning residual features.

The input of the neural network comprises voltage bias (*Vgs* and *Vds*) and device parameters (channel width *W*, channel length *L*, and Temperature *T*)，which are available for circuit designers. Since the current (*Ids*) of MOSFET varies by several orders of magnitude with complex non-

linear characteristics during device operation (particularly in the subthreshold region), a transformation function is required to ensure the accuracy of the ResNet model. The transformation function is defined as follows:

$$Ids = I_0 \cdot Vds \cdot 10^{\,y} \cdot W/L \tag{2}$$

I_0 is the scaling factor, and y is the output of the ResNet. The term *Vds* ensures that *Ids* is 0 when *Vds* = 0V, and *W/L* serves as a physical constraint to enhance the model's convergence capability and prediction accuracy.

In order to improve the accuracy of the ResNet model, it is very important to design an appropriate loss function. The loss function is shown as follows:

$$NMSE = \frac{\sum_{i=1}^{N}(y_i - \hat{y}_i)^2}{\sum_{i=1}^{N} y_i^2} \tag{3}$$

$$LogMSE = \frac{\sum_{i=1}^{N}(log(y_i) - log(\hat{y}_i))^2}{N} \tag{4}$$

$$Loss = \alpha \cdot [10 \cdot LogMSE(\,Ids\,) + NMSE(\,Ids\,)] +$$
$$\beta \cdot NMSE(\,gm\,) + \gamma \cdot NMSE(\,gds\,) \tag{5}$$

The loss function comprises three terms, which impose constraints on the current, *gm*, and *gds*, respectively. Here, *α*, *β*, and *γ* denote the weighting coefficients for the corresponding terms. The current constraint term employs *LogMSE* for subthreshold current fitting and *NMSE* for strong inversion region current fitting. For both the gm and gds constraint terms, *NMSE* is adopted for fitting.

B. Ensemble-Based Active Learning Architecture

In the development of semiconductor devices, the acquisition of TCAD simulation data and test data often consumes a significant amount of time, leading to prolonged research and development cycles and reduced efficiency. Active learning can reduce the annotation overhead by selecting the unlabeled samples that are "the most valuable" for model training and adding them to the training set for training through manual annotation. Ensemble-based active learning enables robust uncertainty quantification, thereby facilitating the selection of maximally informative samples for model training.

Fig. 2 presents the ensemble-based active learning framework. The training begins with a small labeled dataset

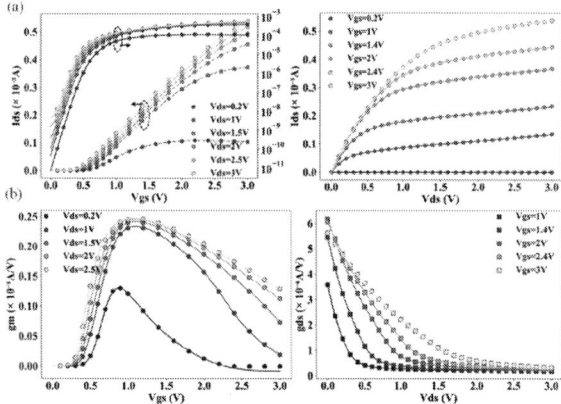

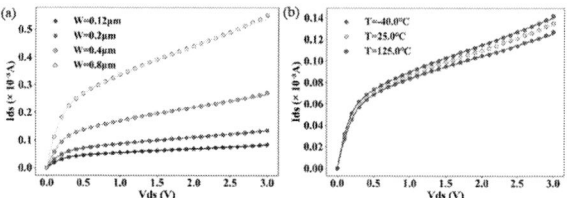

Fig. 5. Comparison of ResNet-predicted (lines) and Actual (dots) I-V Characteristics under different channel widths and temperatures (a). under different channel widths, (b). under different temperatures.

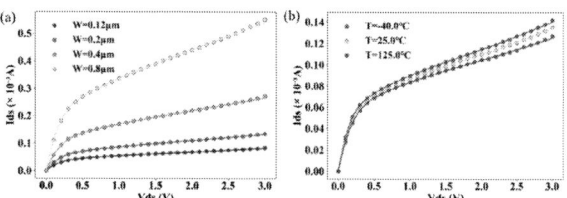

Fig. 6. Comparison of ResNetEns-predicted (lines) and Actual (dots) I-V Characteristics under different channel widths and temperatures (a). under different channel widths, (b). under different temperatures.

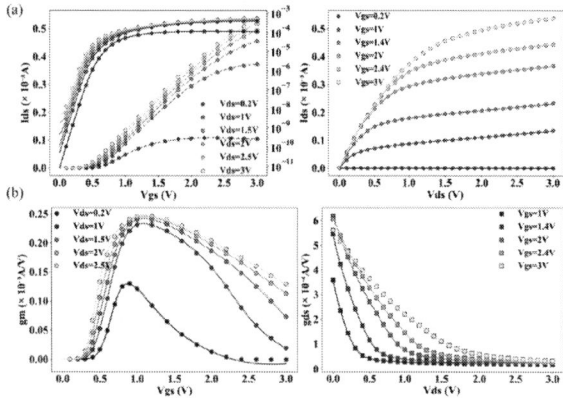

Fig. 3. Comparison of ResNet-predicted (lines) and actual (dots) I-V characteristics under different bias voltages with W=200 nm, L=40 nm, T=25 ℃ on the test set (a). Ids versus Vgs and Vds, (b). gm and gds versus Vgs or Vds.

Fig. 4. Comparison of ResNetEns-predicted (lines) and actual (dots) I-V characteristics under different bias voltages with W=200 nm, L=40 nm, T=25 ℃ on the test set (a). Ids versus Vgs and Vds, (b). gm and gds versus Vgs or Vds.

to simultaneously train five ResNet with identical architectures but different initializations. The uncertainty of sample points is measured by the divergence of the predicted outputs of the five models, so as to select samples more reliably.

Uncertainty quantification is performed through two metrics:

1. Variance, which is used to detect uncertain points in high-current regions.

2. Coefficient of variation (CV, Equation (6)), which is applied to identify uncertain points in subthreshold regions.

$$CV = \frac{Var(y_i)}{mean(y_i)} \quad (6)$$

Each active learning cycle expands the training set by 100 points (50 points selected by variance and 50 points selected by CV). After completing all active learning cycles, we select the ResNet model with the minimum loss function for deployment and evaluation. Therefore, although our active learning solution uses five models in the training stage, our deployment only requires one ResNet, which does not increase the time and resource consumption of model simulation.

TABLE I. COMPARISON AMONG MLP, RESNET, AND RESNETEns

Model	Training Samples	MAE	MARE
MLP	3720	5.3660e-07	0.36%
ResNet	3720	5.3101e-07	0.33%
ResNetEns	2782	4.9107e-07	0.28%

III. RESULT AND DISCUSSIONS

To demonstrate the effectiveness of proposed model, we conducted validation using 40 nm MOSFET device. The training dataset was derived from simulation using TSMC's 40 nm process technology library. The input parameters include bias voltages (Vgs, Vds) and device parameters (W, L, and T). The scanning range of Vds is 0 V to 3 V with a step of 0.1 V, the scanning range of Vgs is 0.1 V to 1.0 V with a step of 0.1 V and 1.2 V to 3.0 V with a step of 0.2 V, the W/L combinations are 120/40, 200/40, 200/80, 400/40, and 800/40 (unit: nm), the temperature(T) scanning points are –40°C, 25°C, and 125°C.

To enhance model accuracy, we employed L-BFGS optimizer for model training. L-BFGS, a quasi-Newton optimization algorithm, is particularly suitable for large-scale nonlinear optimization problems and offers better precision compared to the Adam optimizer. The learning rate was set to 0.1, with a maximum iteration limit of 30 per epoch, and the total training epochs were set to 1500. For the proposed ResNet with ensemble based active learning (ResNetEns), the initial training phase was set to 500 epochs. Subsequently, 10 rounds of active learning were performed, with each round comprising 100 epochs, and a total of 1500 epochs.

Fig. 3 and Fig. 4 comparatively present the predicted versus target I-V characteristics of the ResNet and ResNetEns models, respectively, under varying Vgs and Vds bias conditions. Fig.5 and Fig. 6 present comparative analyses between the predicted and target current for the ResNet and ResNetEns models under varying temperature and channel width conditions, respectively. The precise characterization of these parameters is crucial for accurate device modeling. As evidenced by the curve comparisons, the model predictions exhibit excellent agreement with the target values, confirming the robustness of the proposed approach.

To further validate the performance of the proposed method, we conducted comparative experiments with traditional MLP approaches. In our experiments, we configured three models: an MLP model, a ResNet model, and proposed ResNet with ensemble-based active learning (ResNet$_{Ens}$). The MLP, ResNet, and ResNet$_{Ens}$ models maintained identical width and depth. For MLP and ResNet models, the training and testing sets contained 3720 and 5580 samples respectively. Regarding ResNet$_{Ens}$, the initial training set comprised 1782 samples, which was subsequently expanded to 2782 samples during active learning cycles, with a fixed test set of 6518 samples.

The results are presented in Table I. The ResNet achieve comparable MAE in I-V prediction to conventional MLP, but demonstrated a 6% improvement with MARE of 0.33%. The proposed ResNet$_{Ens}$ exhibit an MAE of 4.9107e-7, representing an 8% reduction in error compared to standard MLP, along with a prediction MARE of 0.28% — a 22% enhancement over standard MLP methods. Our proposed approach achieve superior I-V characterization while using 25% fewer training samples than traditional methods, which can significantly reduce the time cost of TCAD simulations or experimental testing, thereby accelerating model development.

IV. CONCLUSION

In this paper, a neural network integrating ResNet with ensemble-based active learning approach is proposed for device I-V characterization. ResNet uses skip connections to preserve overall trends, while residual blocks handle complex nonlinear corrections. Compared to standard MLP models, ResNet is demonstrated to be better suited for device I-V characterization, with experimental results showing a 6% reduction in MARE. To reduce TCAD simulation or testing time and accelerate model development and research, our ensemble-based active learning scheme quantifies uncertainty and selects high-value samples. Using 25% fewer samples than standard MLP requires, this approach achieves an MAE of 4.9107e-7 (8% lower than standard MLP) and a MARE of 0.28% (26% improvement over standard MLP). This method not only offers higher modeling accuracy but also provides significant time and cost savings. In future, we will further reduce the parameters of the model to achieve a more lightweight model and promote the simulation speed. Meanwhile, we will train the model using measurement data with noise and conduct circuit-level simulation to verify the reliability and robustness of the model.

REFERENCES

[1] C.-T. Tung and C. Hu, "Neural network-based BSIM transistor model framework: Currents, charges, variability, and circuit simulation," IEEE Trans. Electron Devices, vol. 70, no. 4, pp. 2157-2160, Apr. 2023.

[2] J. Choi H. Jeong and S.Woo, "Enhancement and Expansion of the Neural Network-Based Compact Model Using a Binning Method," in IEEE J. Electron Devices Soc., vol. 12, pp. 65-73, 2024.

[3] Y. -S. Yang, Y. Li and S. R. R. Kola, "A Physical-Based Artificial Neural Networks Compact Modeling Framework for Emerging FETs," in IEEE Trans. Electron Devices, vol. 71, no. 1, pp. 223-230, Jan. 2024.

[4] M. Reuter J.Wilm and A. Kramer, "Machine Learning-Based Compact Model Design for Reconfigurable FETs," in IEEE Journal of the Electron Devices Society, vol. 12, pp. 310-317, 2024.

[5] C. Park S. Lee, and J. Park,, "Large-Scale Training in Neural Compact Models for Accurate and Adaptable MOSFET Simulation," in IEEE Journal of the Electron Devices Society, vol. 12, pp. 745-751, 2024.

[6] C. Akbar, Y. Li, and W. L. Sung, "Machine learning aided device simulation of work function fluctuation for multichannel gate-all-around silicon nanosheet MOSFETs," IEEE Trans. Electron Devices, vol. 68, no. 11, pp. 5490-5497, Nov. 2021.

[7] M. -Y. Kao, H. Kam, and C. Hu, "Deep-learning-assisted physics-driven MOSFET current-voltage modeling," IEEE Electron Device Lett., vol. 43, no. 6, pp. 974-977, Jun. 2022.

[8] Z. Yang, A. D. Gaidhane, K. Anderson, G. Workman and Y. Cao, "Graph-Based Compact Model (GCM) for Efficient Transistor Parameter Extraction: A Machine Learning Approach on 12 nm FinFETs," in IEEE Trans. Electron Devices, vol. 71, no. 1, pp. 254-262, Jan. 2024.

[9] S. Guglani, J. Patel, and A. Dasgupta, "Artificial neural network surrogate models for efficient design space exploration of 14-nm FinFETs," in Proc. Int. Conf. Device Res. Conf. (DRC), Columbus, OH, USA, 2022, pp. 1-2.

[10] H. Huang and C. X. Wu, "Approximation capabilities of multilayer fuzzy neural networks on the set of fuzzy-valued functions," Inf. Sci., vol. 179, no. 16, pp. 2762-2773, Jul. 2009.

[11] X. Tang, Z. Li, L. Zeng, H. Zhou, X. Cheng and Z. Yao, "Device Modeling Based on Cost-Sensitive Densely Connected Deep Neural Networks," in IEEE J. Electron Devices Soc., vol. 12, pp. 619-626, 2024.

[12] M.-Y. Kao, H. Kam, and C. Hu, "Deep-learning-assisted physics-driven MOSFET current-voltage modeling," IEEE Electron Device Lett., vol. 43, no. 6, pp. 974-977, Jun. 2022.

Performance Benchmark of Gate-All-Around Nanosheets Transistors Based on DTCO Simulation

Chunlei Wu[1,2,3*], Jian Ma[1], Hanzhi Gu[1], Yueyuan Yu[1], Yiming Xia[1], Jiayi Wu[1], Boqian Shen[1], Qingqing Sun[1,2,3], David Wei Zhang[1,2,3]

[1] School of Microelectronics, Fudan University, Shanghai 200433, China
[2] Shanghai Integrated Manufacturing Innovation Center Co., Ltd, Shanghai 201203, China
[3] Jiashan Fudan Institute, Jiaxing 314100, China

* Email: wuchunlei@fudan.edu.cn

Abstract—In this paper, stacked nanosheet GAAFETs based on advanced FinFET technology platform has been studied utilizing design-technology co-optimization (DTCO) simulation. Through benchmarking the device power performance of GAAFETs against baseline FinFETs of the same layout footprint, the power, performance and area (PPA) advantages of GAAFET structure in advanced FinFET technology nodes has been evaluated. The results reveal that under the same layout footprint, the GAAFET with nanosheet width W_{NS} of 39nm can achieve ~12% performance improvement and ~22% power reduction compared to baseline FinFET. While under the same power-performance merits, GAAFET with W_{NS} of 25nm can achieve ~13% area layout reduction and ~43% standby power reduction. The results would provide practical guidance for the development of advanced semiconductor device manufacturing technology.

Keywords—Gate-All-Around (GAA), Nanosheet, FinFET, Design-technology co-optimization (DTCO), Footprint efficiency.

I. INTRODUCTION

Stacked nanosheet Gate-All-Around (GAA) field-effect transistor (GAAFET) has been recognized as the alternative logic device for 3 nm node and beyond applications, owing to its superior electrostatics and enlarged effective channel width (W_{eff}) per active footprint compared with that of FinFET [1-2]. The planar-like sheet width design of GAAFET enables the non-digital n/p-FETs width optimization and therefore improved layout efficiency. Extensive research has been devoted to GAAFET adoption in terms of integration technology development as well as power-performance improvement [3-6]. Despite above advantages, the main process flows of FinFET can also be adopted for GAAFET, rendering a non-disruptive extension of device scaling.

However, as device dimensional scaling and fabrication has become increasingly complex in advanced nodes, adoption of GAAFET in sub-3 nm nodes poses great difficulties and roadblocks. For sub-7 nm nodes, dimensional scaling is losing steam due to limitations in patterning [7], the continued area scaling has been achieved mainly by track height reduction driven by design-technology co-optimization (DTCO) [8-9]. In this case, comprehensive study on the power-performance evaluation of GAAFET based on advanced FinFET technology platform is still of great practical significance.

In this paper, AC and DC characteristics of GAAFETs based on advanced FinFET technology platform has been studied through DTCO simulation, with the goal to investigate and benchmark the power-performance of GAAFET against FinFET of the same layout footprint, so as to assess the power, performance and area advantages of GAAFET structure. The results would provide practical guidance for the development of advanced semiconductor device manufacturing.

II. SIMULATION METHOD

The DTCO flow used is schematically shown in Fig.1. The technology specifications (tech. spec.) are desired electrical qualification of a given technology, while the design specifications (design spec.) are criterions of circuit figure-of-merits (FOMs). DTCO requires the technology computer-aided-design (TCAD) simulation, SPICE model extraction, and circuit SPICE simulation teams in a concerted effort to co-optimize power, performance, area, and cost (PPAC) of the given technology. The TCAD simulation part performs device process simulations based on given process data including both process flow design and physical dimension parameters like oxide thickness, gate length, gate pitch, etc., providing device electrical data such as I-V, C-V curves, and parasitic RC data. The SPICE extraction part is to extract accurate device model and parasitic RC netlists based on above TCAD-based electrical characterization. Then circuit simulation can be performed for PPAC assessments until the design specifications are met.

Device and process TCAD simulations of both FinFETs and GAAFETs are carried out using Sentaurus tools based on the technology specifications. The self-consistent calculation is achieved though Drift-Diffusion transport equation combined with Poisson equation and density-gradient quantum correction. The Auto-orientation Inversion and Accumulation Layer mobility (IALMob) as well as high field saturation velocity models were included. Dynamic Nonlocal Band-to-Band model, Shockley-Read-Hall (SRH) recombination models were adopted. For the strain models, a multivalley electron and hole mobility model was used considering the mass anisotropy and valley/band energy change with orientation and stress. The transistors' SPICE

979-8-3315-3918-4/25 $31.00 © 2025 IEEE

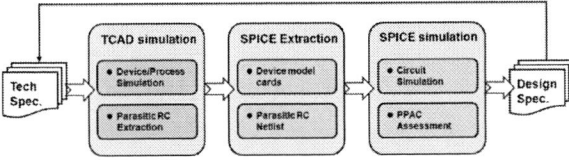

Fig. 1 (a) Used DTCO flow combing TCAD with SPICE simulations.

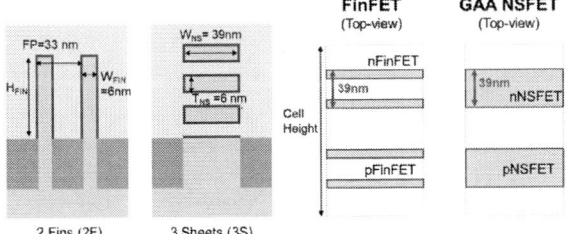

Fig. 2 Cross-view and top-view of the studied double-fin FinFETs and three-layer stacked nanosheets GAAFET structures based on the same technology specifications.

TABLE I.   PARAMETERS USED IN SIMULATION

Parameters	Values [unit]
Gate length, L_G	20 (nm)
Spacer length, L_{SP}	8 (nm)
Interfacial oxide thickness, T_{IL}	0.5 (nm)
High-k dielectric thickness, T_{HK}	1.5 (nm)
Fin Height, H_{FIN}	50 (nm)
Fin Width, W_{FIN}	6 (nm)
Fin Pitch, FP	33 (nm)
Contact Poly Pitch, CPP	57 (nm)
Nanosheet width, W_{NS}	39 (nm)
Nanosheet thickness, T_{NS}	6 (nm)
Nanosheet space, T_{SP}	11 (nm)
S/D recess depth, T_{RE}	5 (nm)
Sub-gate height, H_{SUB}	5 (nm)
R_{BEOL}	300 (Ohm.μm)
C_{BEOL}	2.74×10^{-16} (F/μm)

netlists are then obtained using BSIMProplus tools by extraction of BSIM-CMG model cards. Finally, the simulations of the FinFET and GAAFET based circuits are conducted with SPICE.

III. RESULTS AND DISCUSSION

A. n/p-FET device benchmark

Device structures of the studied three-layer stacked nanosheets GAAFET and double-fin FinFET are shown schematically in Fig. 2. Identical structure dimensions including contact poly pitch (CPP) of 57 nm, gate length (L_G) of 20 nm, gate spacer length (L_{SP}) of 8 nm, interfacial oxide thickness (T_{IL}) of 0.5 nm, high-k dielectric thickness (T_{HK}) of 1.5 nm is used, following 7 nm node ground rules [10], as also listed in Table I. The nanosheet width (W_{NS}) in GAAFETs is set to be 39 nm (=Fin pitch+Fin width) if not otherwise stated,

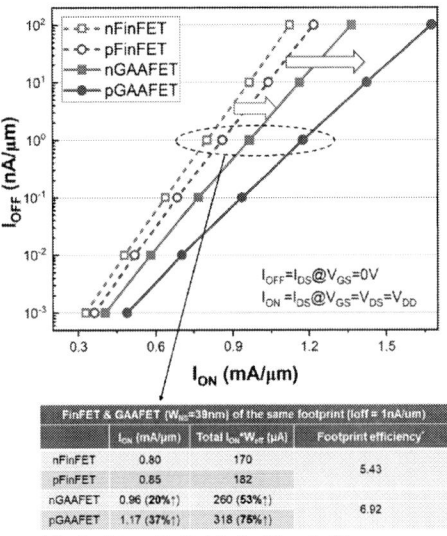

FinFET & GAAFET (W_{NS}=39nm) of the same footprint (Ioff = 1nA/μm)			
	I_{ON} (mA/μm)	Total I_{ON}*W_{eff} (μA)	Footprint efficiency*
nFinFET	0.80	170	5.43
pFinFET	0.85	182	
nGAAFET	0.96 (20%↑)	260 (53%↑)	6.92
pGAAFET	1.17 (37%↑)	318 (75%↑)	

*: Footprint efficiency = W_{eff} / (Fin pitch + Fin width) or = W_{eff} / W_{NS}

Fig. 3 Device performance comparison results of three-layer stacked nanosheets GAAFETs (W_{NS}= 39 nm) and double-fin FinFETs [11-12] of the same footprint. The supply voltage V_{DD} is 0.7 V.

so as to keep the same layout footprint as in baseline double-fin FinFETs. The epitaxial $Si_{0.5}Ge_{0.5}$ (pFET) and Si (nFET) S/D stressors are considered, here defect-free S/D epitaxy is assumed in both FinFET and GAAFET.

Fig. 3 shows the performance of GAAFETs and FinFETs of the same footprint. As can be seen in the figure, compared to FinFETs of identical footprint, GAAFETs can achieve not only reduced off leakage current I_{OFF} but also boosted on current I_{ON}. For constant off leakage (I_{OFF}=1 nA/μm), the on current I_{ON} of n/p-GAAFETs can be increased by 20% and 37% respectively, compared to that of n/p-FinFETs. In lateral nanosheets GAAFET, the main transport surface orientation is changed from (110) to (100). The boosted I_{ON} in n-GAAFET can be attributed to the higher electron mobility in (100), while the boosted I_{ON} in p-GAAFET can be attributed to high stress sensitivity of (100) hole mobility [3]. Moreover, since GAAFETs get larger effective channel width W_{eff} owing to improved footprint efficiency, the total driving capability I_{ON} * W_{eff} of n/p-GAAFET can be increased by 53% and 75%, as shown in the inset table.

Fig. 4 (a) and (b) show the total effective current I_{eff} and total effective capacitance C_{eff} of GAAFETs in comparison with that of FinFETs. As shown in the figure, while increasing the nanosheet width W_{NS} enhances the I_{eff} of GAAFETs, it also leads to higher C_{eff} as a penalty. The trade-off between I_{eff} and C_{eff} should always be take into consideration in device performance optimization. However, the sensitivity of I_{eff} and C_{eff} to nanosheet width can be different. In general, the performance advantage of GAAFETs tends to be realized at devices with wider nanosheets and smaller sheet-to-sheet space, so as to achieve higher I_{eff} and smaller C_{eff}.

The device delay characteristics of GAAFETs and FinFET has been estimated based on total I_{eff} and C_{eff}, as shown in Fig. 5. Reduced RC delay has been obtained in GAAFETs with wider nanosheet width ($W_{NS} \geq 25$ nm) compared to that of baseline FinFET.

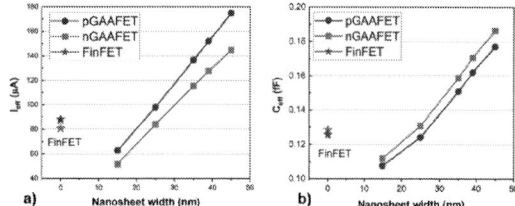

Fig. 4 (a) The total effective current I_{eff}, and (b) total effective capacitance C_{eff} of n/p-GAAFETs in a single stage inverter as a function of nanosheet width W_{NS}.

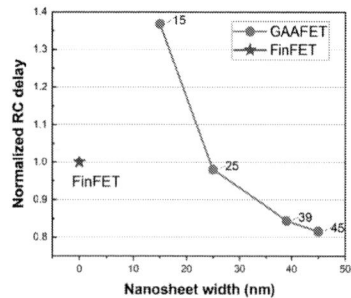

Fig. 5 The RC delay characteristics $((C_{eff} \times V_{DD})/I_{eff})$ comparison between FinFET and GAAFETs as a function of nanosheet width W_{NS}. GAAFETs with wider nanosheet width ($W_{NS} \geq 25$ nm) show better performance.

B. 11-stage RO benchmark

Based on above TCAD simulations and obtained SPICE netlists, GAAFET and FinFET based ring oscillator (RO) circuit simulation has been conducted for power performance evaluation. Since the studied GAAFETs and FinFETs are based on the same technology platform, identical back-end-of-line (BEOL) parasitic resistance R_{BEOL} and load capacitance C_{load} has been assumed. The total BEOL parasitic resistance of baseline double-fin FinFETs is $R_{BEOL} = 300$ $\Omega \cdot \mu m / 0.2 \mu m = 1.5$ kΩ [10]. While the load capacitance C_{load} of light load (50 CPP) and heavy load (300 CPP) is 50×0.057 $\mu m \times 2.74e{-}16$ F/$\mu m = 0.78$ fF, and $300 \times 0.057 \mu m \times 2.74e{-}16$/$\mu m = 4.68$ fF respectively, thus C_{load} of 1.0 ~5.0 fF has been considered.

Fig. 6 (a) and (b) present the frequency and active power of 11-stage RO circuit as a function of load capacitance C_{load} respectively. As can be seen in the figure, the frequency of RO chain decreases significantly with increasing C_{load}. While the active power consumption of different RO chains varies very little with increasing C_{load}. This is because that the frequency of RO circuit decreases with increasing C_{load}, thus resulting in nearly unchanged active power $P_{active} = Freq. *C*V^2$. Fig. 7 compares the power-performance characteristics of FinFET and GAAFETs with different nanosheet width W_{NS}. It can be observed that GAAFET with $W_{NS} = 39$ nm shows higher working frequency compared to FinFET of the same footprint, about 12% performance improvement is achieved under the same power consumption condition (30 μW), and about 22% power reduction is achieved under the same working frequency condition (4 GHz). Besides, it can also be noticed that GAAFET with $W_{NS} = 25$ nm can achieve well-matched power-performance characteristics with the baseline FinFET,

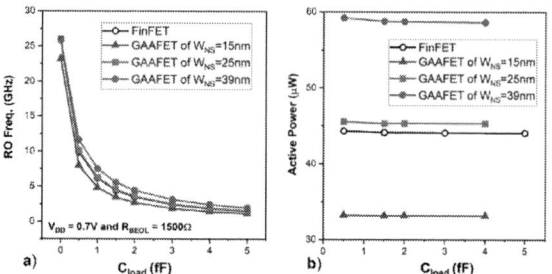

Fig. 6 (a) Frequency performance, and (b) Active power consumption comparison between FinFET and GAAFETs based 11-stage RO circuit. The supply voltage V_{DD}=0.7 V.

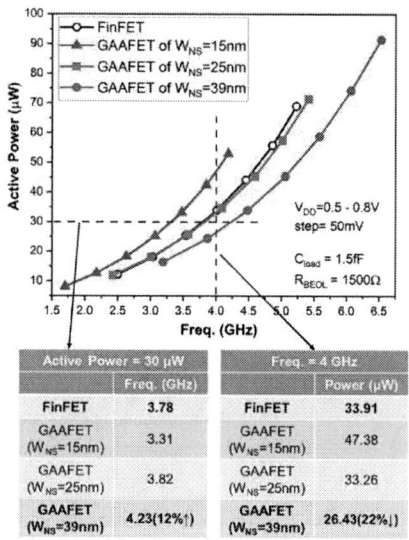

Fig. 7 The power- performance comparison between FinFET and GAAFETs based 11-stage RO circuit.

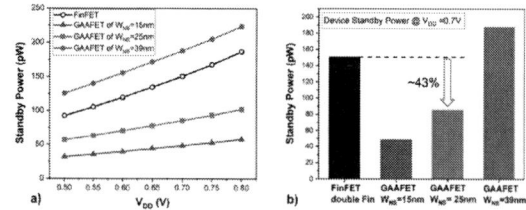

Fig. 8 The standby power consumption comparison between FinFET and GAAFETs based 11-stage RO circuit, under (a) varied V_{DD} from 0.5 V to 0.8 V, (b) V_{DD}=0.7 V.

which suggests area reduction advantages of GAAFET (about 13%) without the penalty of power-performance degradation.

Fig.8 compares the standby power consumption of FinFET and GAAFETs based RO chains. It can be seen that GAAFET with $W_{NS} = 25$ nm shows significantly lower standby power consumption (reduced by 43%), which can be attributed to superior gate control of GAA structure and accordingly reduced leakage current. The results would provide practical guidance for the development of advanced semiconductor device manufacturing technology.

IV. CONCLUSIONS

In this paper, stacked nanosheets GAAFETs based on advanced FinFET technology platform has been studied by DTCO simulation. The results reveal that under the same layout footprint, the GAAFET with nanosheet width W_{NS} of 39nm can achieve ~12% performance improvement and ~22% power reduction compared to baseline FinFET. While under the same power-performance merits, GAAFET with W_{NS} of 25nm can achieve approximately 13% area layout reduction and about 43% standby power reduction.

ACKNOWLEDGMENT

This work was supported by NSFC under Grant 62304048.

REFERENCES

[1] N. Loubet, T. Hook, P. Montanini, et al., "Stacked nanosheet gate-all-around transistor to enable scaling beyond FinFET," in Proc. Symp. VLSI Technol., Jun. 2017, pp. T230–T231, doi: 10.23919/VLSIT.2017.7998183.

[2] G. Bae, D.-I. Bae, M. Kang, et al., "3nm GAA Technology featuring Multi-Bridge-Channel FET for Low Power and High-Performance Applications," in 2018 IEEE International Electron Devices Meeting, San Francisco, CA, USA, 2018, pp. 28.7.1-28.7.4, doi: 10.1109/IEDM.2018.8614629.

[3] T. Liu, D. Wang, Z. Pan, et al., "Novel Postgate Single Diffusion Break Integration in Gate-All-Around Nanosheet Transistors to Achieve Remarkable Channel Stress for N/P Current Matching," IEEE Trans. Electron Devices, vol. 69, no. 3, pp. 1497-1502, Mar. 2022, doi: 10.1109/TED.2021.3139579.

[4] S. Mochizuki, M. Bhuiyan, H. Zhou, et al., "Stacked Gate-All-Around Nanosheet pFET with Highly Compressive Strained $Si_{1-x}Ge_x$ Channel," in 2020 IEEE International Electron Devices Meeting, San Francisco, CA, USA, 2020, pp. 2.3.1-2.3.4, doi: 10.1109/IEDM13553.2020.9372041.

[5] C. Tsai, Y. Chen, C. Tu, et al., "First Demonstration of Multi-VT Stacked $Ge_{0.87}Sn_{0.13}$ Nanosheets by Dipole-Controlled ALD WN_xC_y Work Function Metal with Low Resistivity and Thermal Budget $\leq$ 400 °C," in Proc. Symp. VLSI Technol., Jun. 2021, pp. T1-2.

[6] S. Kim, K. Lee, S. Kim, et al., "Investigation of Device Performance for Fin Angle Optimization in FinFET and Gate-All-Around FETs for 3 nm-Node and Beyond," IEEE Trans. Electron Devices, vol. 69, no. 4, pp. 2088-2093, Apr. 2022, doi: 10.1109/TED.2022.3154683.

[7] M. Garcia Bardon, Y. Sherazi, P. Schuddinck, et al., "Extreme scaling enabled by 5 tracks cells: Holistic design-device co-optimization for FinFETs and lateral nanowires," in 2016 IEEE International Electron Devices Meeting (IEDM), San Francisco, CA, USA, 2016, pp. 28.2.1-28.2.4, doi: 10.1109/IEDM.2016.7838497.

[8] G. Yeap, S.S. Lin, Y.M. Chen, et al., "5nm CMOS Production Technology Platform featuring full-fledged EUV, and High Mobility Channel FinFETs with densest $0.021\mu m^2$ SRAM cells for Mobile SoC and High-Performance Computing Applications," in 2019 IEEE International Electron Devices Meeting (IEDM), San Francisco, CA, USA, 2019, pp. 36.7.1-36.7.4, doi: 10.1109/IEDM19573.2019.8993577.

[9] S. -Y. Wu, "Key Technology Enablers of Innovations in the AI and 5G Era," in 2019 IEEE International Electron Devices Meeting (IEDM), San Francisco, CA, USA, 2019, pp. 36.3.1-36.3.4, doi: 10.1109/IEDM19573.2019.8993613.

[10] International Roadmap for Devices and Systems (IRDS), Mar. 2019, [online] Available: https://irds.ieee.org/editions/2018.

[11] S. -Y. Wu, C. Y. Lin, M. C. Chiang, et al., "A 7 nm CMOS platform technology featuring 4th generation FinFET transistors with a $0.027um^2$ high density 6-T SRAM cell for mobile SoC applications," 2016 IEEE International Electron Devices Meeting (IEDM), San Francisco, CA, USA, 2016, pp. 2.6.1-2.6.4, doi: 10.1109/IEDM.2016.7838333.

[12] S. -Y. Wu, C. Y. Lin, M. C. Chiang, et al., "Demonstration of a sub-$0.03 um^2$ high density 6-T SRAM with scaled bulk FinFETs for mobile SOC applications beyond 10nm node," 2016 IEEE Symposium on VLSI Technology, Honolulu, HI, USA, 2016, pp. 1-2, doi: 10.1109/VLSIT.2016.7573390.

The impact of Back-Gate biasing and layout on temperature sensitivity of transistors in FD-SOI CMOS technology

Yann Deval, Maxime Guillot, Hervé Lapuyade, François Rivet

University of Bordeaux, CNRS, Bordeaux INP, IMS, UMR5218, F-33400 Talence, France

Abstract—FD-SOI technology offers degrees of freedom in analog and digital circuit design through back-gate biasing. This work investigates the influence of back-gate voltage and layout on the temperature sensitivity of transistors this, through both simulation and measurement. A temperature-compensated voltage reference, from a previous work of ours, is employed as a test vehicle. We compare three layout strategies to evaluate spatial effects on temperature drift. The measurement campaign, conducted using a dedicated datalogger and temperature-controlled setup, confirms simulation about temperature-induced variations. The results underline the importance of careful back-gate management in FD-SOI for robust design.

Index Terms—FD-SOI, Back-Gate biasing, temperature dependency, voltage reference, temperature compensation, thermal sensitivity.

Introduction

Temperature sensitivity is a critical parameter in the design of electronic circuits. In CMOS technology, the threshold voltage of transistors exhibit significant temperature dependency, which can lead to performance degradation or failure in circuits. Therefore, understanding and mitigating the effects of temperature on these components is essential for reliable circuit operation. This work builds upon previous work of ours that developed a temperature-compensated voltage reference circuit, which serves as a test vehicle for this paper. The goal is to investigate how back-gate voltage and back-gate layout strategies influence the temperature sensitivity of transistors.

The paper is structured as follows: Section I provides an overview of transistors in 28nm FD-SOI CMOS technology, including the concept of back-gate biasing and its implications for temperature sensitivity, equations on the temperature dependency of threshold voltages, and the theoretical background on how these parameters affect circuit performance will be given. Section II introduces the voltage reference circuit used as a test vehicle, detailing its topology and operation. Section III presents the results of simulations and measurements, comparing different layout strategies and their effects on temperature drift. Finally, the paper concludes with a discussion of the findings and their implications for future circuit design in FD-SOI technology.

I. Transistors in 28nm FD-SOI technology

In FD-SOI CMOS technology, transistors are built on a thin fully depleted silicon layer separated from the bulk substrate by an insulating layer. A cut view of a Low Threshold Voltage

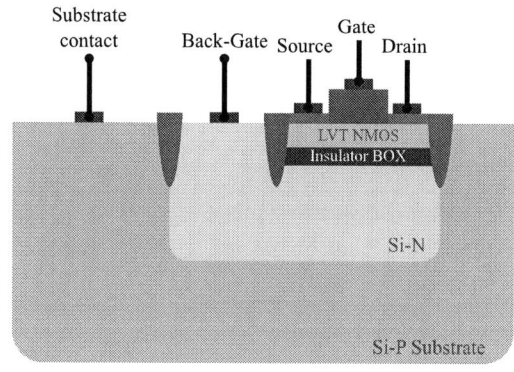

Fig. 1. Cross-sectional illustration of a LVT-NMOS transistor in 28nm FD-SOI technology [2].

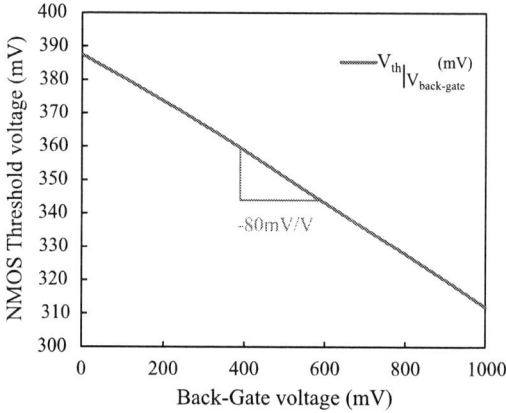

Fig. 2. Dependency of the threshold voltage of an NMOS transistor on back-gate biasing.

NMOS (LVT-NMOS) transistor in 28nm FD-SOI technology is shown in Figure 1. On this figure, a fourth terminal is visible, this terminal is connected to the N-Well surrounding the transistor and is know as the back-gate [1].

The back-gate allows for additional control over the transistor's characteristics and can be used to control it's threshold voltage. For each 1V variation in back-gate voltage, the threshold voltage shifts by about 80mV. This tunability allows precise adjustment of the transistor's performances. The Figure 2 illustrates the dependency of the threshold voltage of an NMOS transistor on back-gate biasing.

A. Temperature dependency of electronic components

The threshold voltage of transistors dependent on temperature. Figure 3 illustrates the variation of the threshold voltages of two RVT-NMOS with temperature. On this figure, the threshold voltage of two NMOS transistors is shown, one with a back-gate bias of 0V and the other with a back-gate bias of 1V. The comparison between the two threshold voltages shows that the back-gate biasing has a neglectable impact on the temperature sensitivity of the transistor.

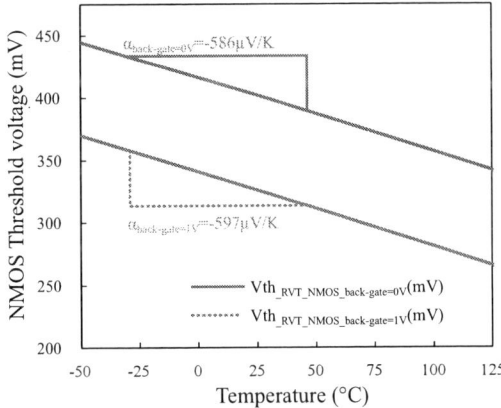

Fig. 3. Dependency of threshold voltages on temperature (Cadence simulations).

Assuming a linear relationship, the threshold voltage variation with temperature can be expressed as shown in Equation 1.

$$V_{th}(T) = \alpha(T - T_0) + V_{th0} \tag{1}$$

With:

- T and T_0 : operating and reference temperature.
- α: thermal sensitivity of the threshold voltage in V/°C. For transistors, α is negative.
- V_{th}: threshold voltage at temperature T.
- V_{th0}: threshold voltage at temperature T_0.

Equation 2 expresses α as a function of temperature T, derived from Equation (1).

$$\alpha = \frac{V_{Th} - V_{th0}}{T - T_0} \tag{2}$$

In the technology used in this paper, $\alpha_{Transistor} \approx -600\mu V/°C$ [3]. In this paper, the assumption is made that resistors' temperature dependency is negligible compared to the one of the transistors and diodes. Therefore $\alpha_{Resistor} = 0ppm/°C$.

B. Theorical temperature dependency of the threshold voltage

The threshold voltage V_{Th} of the transistor is given by Equation 3, taken from [1]:

$$V_{Th} = \Delta\varphi_m + \Psi_{S,th} + BF(\Psi_{S,th} - V_{BS} + \varphi_{fb}) + \frac{h^2}{8qm_{conf}t_{Si}^2} \tag{3}$$

With:

- $\Delta\varphi_m$ — the metal work function
- $\Psi_{S,th} = \frac{kT}{q} \ln\left(\frac{C_{OX}kT/Q}{qn_it_{Si}}\right)$ — the silicon body side surface potential
- φ_{fb} — the flat band voltage
- h and t_{Si} — dimensions of the transistor
- m_{conf} — effective mass of confined carrier
- $BF = \left(\frac{1}{C_{OX}} \Big/ \left(\frac{1}{C_{Si}} + \frac{1}{C_{BOX}}\right)\right)$ — capacitor ratio [4]

The temperature sensitivity of the transistor's threshold voltage is called α and can be expressed with Equations 4 to 5.

$$\alpha = \frac{dV_T}{dT} = \frac{dV_T}{d\Psi_{S,th}} \cdot \frac{d\Psi_{S,th}}{dT} \tag{4}$$

$$\alpha = \frac{k}{q} \left(\ln\left(\frac{C_{OX}kT/Q}{qn_it_{Si}}\right) + 1\right)(1 + BF) \tag{5}$$

With $n_i^2 \propto T^3 e^{-qV_g/kT}$ [5], for this reason, α is always negative.

II. TEST VEHICLE VOLTAGE REFERENCE CIRCUIT

A. Voltage reference topology

As a test vehicle, a temperature-compensated voltage reference circuit is used. This circuit was developed in a previous work [6] and is based on the compensation of the Complementary To Absolute Temperature (CTAT) NMOS's threshold voltage with a Proportionnal To Absolute Temperature (PTAT) current flowing through a resistor. The topology of the voltage reference is illustrated in Figure 4.

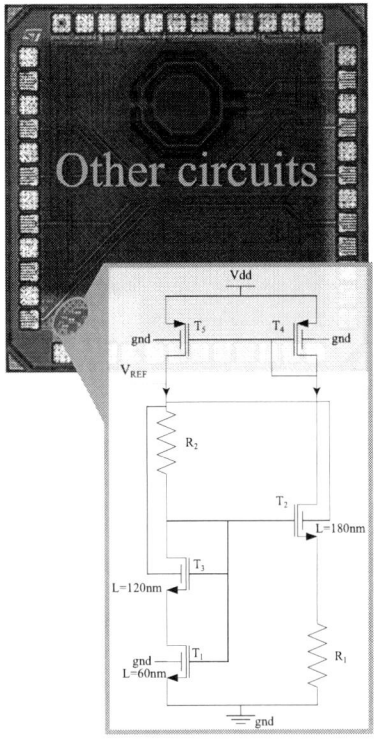

Fig. 4. Voltage reference circuit, schematic view and chip photograph.

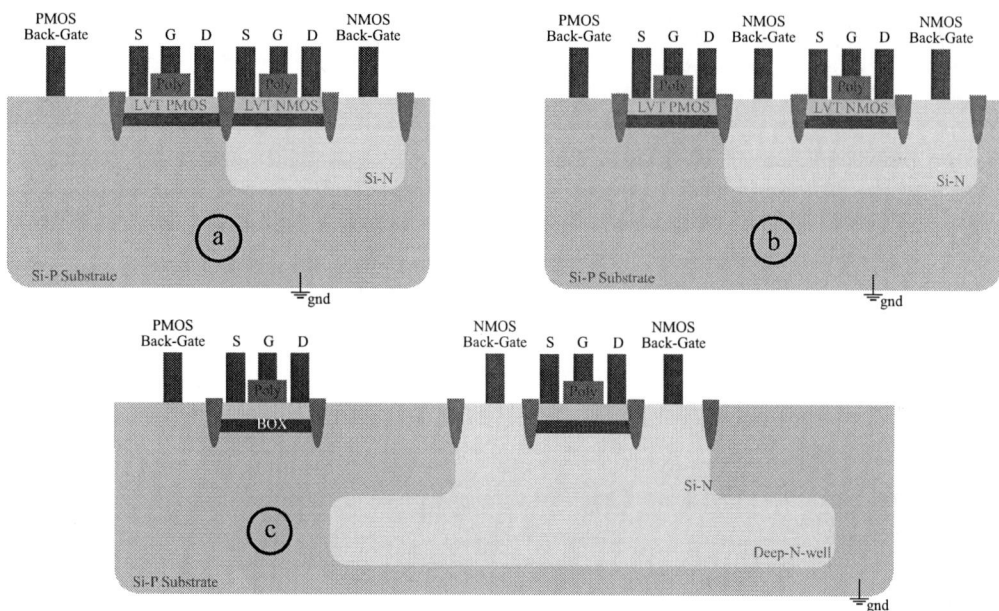

Fig. 5. Three layout strategies of the voltage reference circuit.

Ideally, the voltage reference is designed to provide a constant output voltage V_{REF} that is independent of temperature. The output voltage is given by Equation 6, where V_{th1} is the threshold voltage of transistor T_1, and I_{PTAT} is the PTAT current flowing through resistor R_2. The goal is to ensure that the temperature coefficient of the output voltage is zero, as expressed in Equation 7.

$$V_{REF} = V_{th1} + I_{PTAT} \times R_2 \qquad (6)$$

$$\frac{dV_{REF}}{dT} = \alpha_1 + \frac{dI_{PTAT}}{dT} \times R_2 = 0 \qquad (7)$$

Compared to the circuit presented in [6], the resistor values R_1 and R_2 have been adjusted to better observe the dominance of either the PTAT or CTAT behavior. This adjustment results in a parabolic-shaped output voltage as a function of temperature, providing clearer insight into the compensation mechanism.

B. Layout strategies of the voltage reference

The final aim of the voltage reference circuit is to operate into nuclear power plants, where both temperature and radiation dose rate can vary significantly. For the purpose of radiation hardness, this circuit has been layouted in three different ways. The differences between these three layout strategies are illustrated in Figure 5.

In this figure, the three layout strategies are illustrated: the first strategy (a) connects the back-gate through a single contact to the N-Well, kwown as a "bottom" back-gate in Cadence. In the second strategy (b), the back-gate connection surrounds each transistor using a guard-ring that ties the N-Well. The third strategy (c) is similar to the second, but includes a Deep-N-Well beneath the back-gate.

As a result, the parametric extraction gives differents value of voltage reference while components are identical. The simulation results of the three layout strategies are shown in Figure 6.

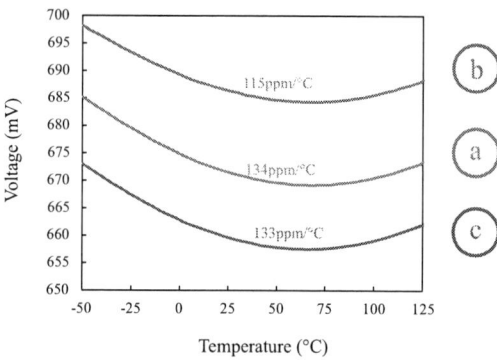

Fig. 6. Simulation results of the three layout strategies.

III. MEASUREMENT SETUP & CAMPAIGN

The measurement campaign was conducted using a dedicated datalogger and temperature-controlled setup. The datalogger is *ADC-24 Precision Data Logger* from *Pico Technology*, which is capable of measuring voltages with a resolution of 24 bits onto 16 channels. A climatique chamber was used to control the temperature of the circuit, which was placed inside the chamber. The measurement campaign was conducted over a range of temperatures from -50°C to 125°C, with a step of 25°C and with 15 minutes step duration. One custom PCB has been designed to host the voltage reference circuit and another one for the signals and power management, this second PCB is not intended to go into the climatic chamber or any other extreme environment. The measurement setup is illustrated in Figure 7.

Fig. 7. Measurement setup, the main PCB and main chip are into the climatic chamber while the signals and power management PCB is outside, connected to a laptop.

The need to devide the PCB into two parts is due to the fact that the PCB with the integrated circuit will have to be placed into a climatic chamber, which is not suitable for numerous COST components such as the datalogger, the power supply, and the laptop, this is why the signals and power management PCB is connected to the main PCB through a passthrough in the climatic chamber wall, it is supply by a 5V usb, allowing it to be powered by a laptop outside the climatic chamber. The laptop is also used to treat data from the datalogger (not on pictures). On the other side, the main PCB, inside the climatic chamber do not have any COST components except for connectors and a few capacitor used for decoupling. The main PCB is on the picture on the right side of Figure 7, it is in the climatic chamber, the D sub 25 connector can be seen on the right side of the PCB, this connector is used to connect the main PCB to the signals and power management PCB. Also, some white coaxial connector can be seen, those connectors are used for othe circuits embedded in the chip and are not presented in this paper.

The results of the measurement campaign are shown in Figure 8. The results confirm the simulation results, showing that the back-gate biasing and layout strategies have a significant impact on the temperature sensitivity of the voltage reference circuit.

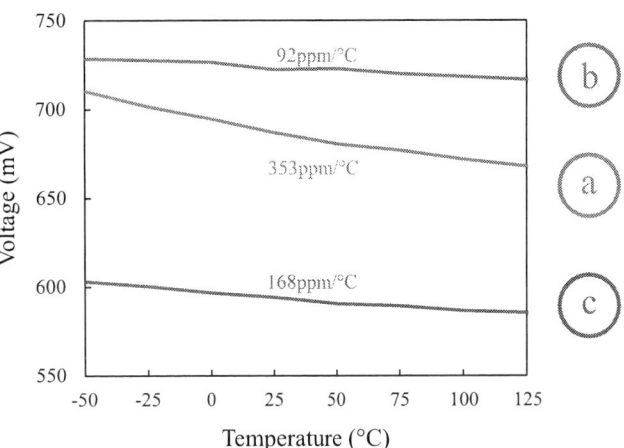

Fig. 8. Measurement results of the three layout strategies.

Again, the three layout strategies are compared, although simulation results indicate only minor differences between them, the measurement results demonstrate a pronounced influence of back-gate biasing and layout on the temperature sensitivity of the voltage reference circuit. This highlights the importance of experimental validation, as layout-dependent effects may not be fully captured in simulation.

IV. CONCLUSION

This work demonstrate the significant influence of back-gate biasing and layout strategies on the temperature sensitivity of transistors in 28nm FD-SOI CMOS technology. Through both simulation and measurement, it was shown that careful management of the back-gate connection and layout can substantially affect the thermal behavior of a voltage reference circuit. The results highlight the necessity of considering back-gate effects early in the design process to ensure robust and reliable circuit performance.

REFERENCES

[1] S. Clerc, T. Gilio, and A. Cathelin, *The Fourth Terminal Benefits of Body-Biasing Techniques for FDSOI Circuits and Systems*, Jan. 2020.

[2] A. Cathelin, "Fully Depleted Silicon on Insulator Devices CMOS: The 28-nm Node Is the Perfect Technology for Analog, RF, mmW, and Mixed-Signal System-on-Chip Integration," *IEEE Solid-State Circuits Magazine*, vol. 9, no. 4, pp. 18–26, 2017. [Online]. Available: http://ieeexplore.ieee.org/document/8110867/

[3] M. Guillot, Y. Deval, H. Lapuyade, and F. Rivet, "Temperature Compensation in FD-SOI Transistors: A Novel Approach with 27% Performance Improvement via Parasitic Diode Biasing," in *2024 31st IEEE International Conference on Electronics, Circuits and Systems (ICECS)*, 2024, pp. 1–4.

[4] O. Rozeau, M.-A. Jaud, T. Poiroux, and M. Benosman, "Surface potential based model of ultra-thin fully depleted SOI MOSFET for IC simulations," in *IEEE 2011 International SOI Conference*, 2011, pp. 1–22.

[5] K. K. N. S.M. Sze, *Physics and Properties of Semiconductors A Review*, 2006, pp. 5–75.

[6] M. Guillot, Y. Deval, H. Lapuyade, and F. Rivet, "A 860mV and 73.5Ppm/°C Voltage Reference that Relies on Back-Gate Biasing Techniques in 28nm FD-SOI Technology," in *2024 31st IEEE International Conference on Electronics, Circuits and Systems (ICECS)*, 2024, pp. 1–4.

979-8-3315-3918-4/25 $31.00 © 2025 IEEE

Considerations for Low Temperature High-Performance Computing: Design Technology Co-optimization, Device Variation, and Hot Carrier Injection

Yuanxi Liao[1,#], Keyang Zhu[1,#], Ran Tao[1,2,3], Yongyu Wu[1,2], Qinglan Jiang[4], Yaxiong Liu[4], Kai Xu[1,2,3], Chen Shen[4], Dong Liu[1], Dawei Gao[1,2], Ran Cheng[*,1]

[1]Zhejiang University, China. [2]Zhejiang ICsprout Semiconductor, China. [3]ZJU-Hangzhou Global Scientific and Technological Innovation Center, China. [4]Suzhou Peifeng Tunan Semiconductor Co., Ltd., China. [#]Equal Contributors [*]Email: chengran@zju.edu.cn

Abstract—This work analyzes the major issues in realizing a low temperature (LT) high performance computing system, namely, the increased threshold voltage V_T, exacerbated device-to-device variation (DDV), and the aggravated hot carrier degradation. A comprehensive LT design-technology co-optimization (DTCO) methodology is demonstrated for 55 nm CMOS, utilizing a calibrated TCAD simulation framework and a channel doping engineering strategy to modulate V_T. Substantial improvement in on-current and circuit speed for logic gates can be achieved. The impact of DDV on SRAM static noise margin is studied, revealing improved nominal performance but increased variability at LT. In addition, the LT HCD exhibits a non-monotonic temperature and stress time dependence, governed by multiple trap mechanisms. System-level evaluation of a RISC-V core confirms that co-optimization of channel doping and supply voltage enables significant speed improvement, highlighting the potential of LT CMOS for next-generation high-performance computing.

Keywords— DTCO, HCI, High Performance Computing, Low Temperature CMOS, Variation

I. INTRODUCTION

Low-temperature (LT) CMOS has attracted great attention for its improved performance and promising potential for high-performance computing of data centers and quantum integrations [1]–[5]. Thanks to higher mobility, lower subthreshold swing (SS) and off-leakage I_{off} at temperatures below 100 K [Fig. 1(a)] [3], low-temperature (LT) CMOS technology has attracted great attention for its potential application in high-performance computing [4],[5]. Despite of the several improved device parameters that are related to the circuit performance, as listed in Fig. 1(b), the threshold voltage V_T or V_{th}, which is a primary parameter for the digital circuit design, increases as the working temperature T decreases. At a fixed power supply V_{DD}, the on-current I_{on}, which is proportional to the gate overdrive (V_{DD} - V_T), reduces at lower T, leading to a degradation of the propagation delay. On the other hand, if the gate overdrive keeps unchanged, the power consumption will be sacrificed due to the raised V_{DD}. Therefore, process optimization is essential to modulate V_T for CMOS operating at LT before it can be used to improve the speed and power performance of the processing circuits.

Furthermore, current research on LT CMOS suggests that the device-to-device variation (DDV) and the hot carrier injection (HCI) may aggravate at LT due to the reduced scattering and increased carrier energy [4],[6], respectively. The former leads to greater challenges in the building of PDK for LT circuit design, while the latter results in more severe reliability issues for LT circuits.

In this work, the methodology for LT design technology co-optimization (DTCO) featuring a TCAD-SPICE tool at 77 K and a demonstration of using this tool for the DTCO of a 55 nm RISC-V core has been presented. With careful design of V_{DD} and channel doping engineering (CDE), lower power consumption and higher energy efficiency can be achieved, based on the DTCO scheme. The impact of DDV on the SRAM signal noise margin (SNM), which is a major concern for the computing unit, is primarily studied. Lastly, the HCI induced degradation (HCD) at LT and its involved

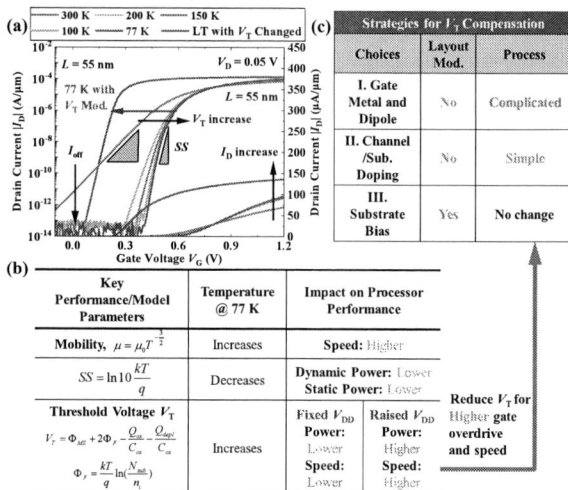

Fig. 1. (a) The I_D-V_G characteristics for a 55 nm n-FET measured from 300 K to 77 K. I_D, SS and I_{off} can be greatly improved while V_T increases as T reduces. It is necessary to optimize the LT process to reduce V_T at 77 K. (b) Change of key device parameters at 77 K and their impace on the processor performance. (c) Comparison of several V_T modulation strategies in terms of design and process development complexity.

mechanisms for advanced node transistors are also reported in this paper.

II. DEVICE INFORMATION AND CHARACTERIZATION

For DTCO and DDV analysis: The CMOS devices at 55 nm technology node are fabricated by ICsprout Semiconductor Ltd with a standard 12 inch foundry level production line. Polysilicon gate with a 1.6 nm-thick SiO_2 dielectric layer was adopted as the gate stack structure. The fabricated CMOS devices with L_G from 55 nm to 1 μm are used for the calibration of the TCAD process model.

For HCD analysis: The devices-under-test (DUTs) are 14 nm bulk Si n-channel FinFETs fabricated by the foundry following the standard Si process flow. HfO_2-based HKMG stack and raised source/drain processes were adopted in the FinFET technology. The effective oxide thickness (EOT) for the gate stack is ~1 nm. The fin height is 40 nm, and the fin width is 10 nm.

Tektronix Keithley 4200A-SCS semiconductor analyzer and Lakeshore cryogenic probe station were used for the electrical data acquisition at various temperatures ranging from 15 K to 300 K. Fig. 1(a) shows the drain current-gate voltage (I_D–V_G) characteristics for a 55 nm n-FET at various temperatures. For HCI analysis: The threshold voltage V_{th} degradation was characterized at various temperatures from 15 K to 300 K. The threshold voltage V_{th} shift ΔV_{th} at linear region (V_D = 50 mV) was extracted by constant current method. The FinFETs were stressed at channel-hot-electron (CHE) conditions where $V_G = V_D$ = 2 V. So far, there are several strategies for the V_T modulation [5],[8],[9]. Fig. 1(c) summarizes the change in layout and process complexity for the three commonly used choices for the V_T modulation. As

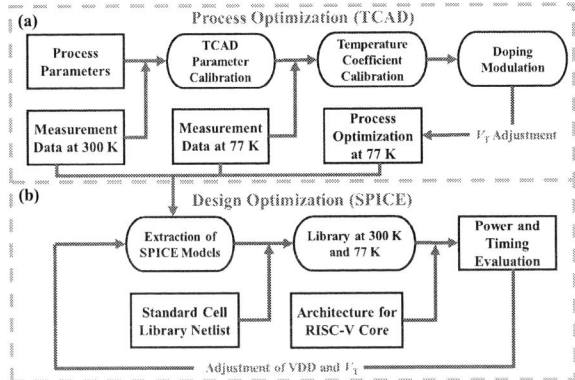

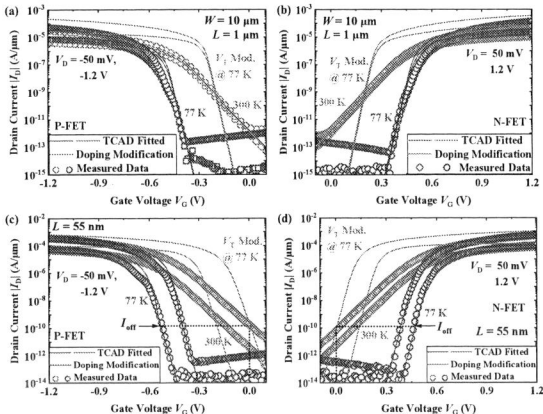

Fig. 2. The TCAD-to-SPICE flow for the evaluation of power and speed performance of the low-temperature high-performance processor optimized with the CDE V_T modulation technique.

Fig. 5. (a)-(d) Measured (dots) and TCAD (lines) simulated I_D-V_G characteristics of devices at 300 K and 77 K. The green lines are the simulated transfer characteristics with CDE for V_T modulation: (a) P-FET, (b) n-FET with W/L = 10 μm/1 μm; (c) p-FET, (d) n-FET with W/L = 500 nm/55 nm.

III. RESULTS AND DISCUSSION

A. Methodology and Tools for Cryogenic DTCO

Choice I increases the process complexity and Choice III needs a more complicated layout design for the additional substrate bias, CDE (Choice II) is the most reasonable choice for its good balance between the design and process complexity. To pave the way for the DTCO of LT high performance processing chips, device-to-circuit modeling and simulation tools at LT should be developed to accelerate the technology development for LT CMOS applications. The TCAD-to-SPICE simulation has been used for the DTCO engineering process [10], for its function in precise optimization of the power, performance, and area (PPA) through physics-based device modeling and circuit co-simulation.

As shown in Fig. 2, an LT TCAD-SPICE simulation tool has been developed based on Mozz TCAD-SPICE tool suite at 300 K [11], to study the effectiveness of the process optimization, namely, the V_T modulation by channel doping engineering, on the performance improvement of an LT processing core. The TCAD model at 300 K and 77 K are calibrated using the physical and

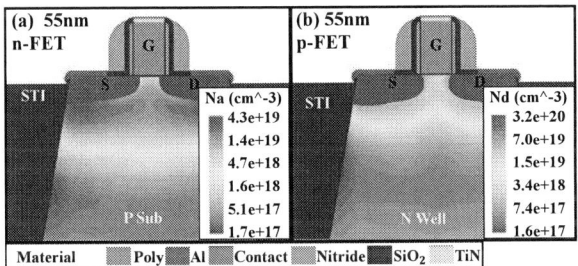

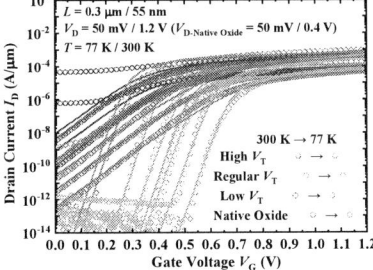

Fig. 3. TCAD device structures and channel doping profiles of (a) n-FET and (b) p-FET. Gate length L = 55 nm

Fig. 4. I_D-V_G characteristics of devices with different CDE dose at 300 K and 77 K. The simulated data (lines) fits well with the measured data (dots) after the model calibration.

measured parameters from the foundry fabricated 55 nm CMOS. The effect of CDE on the device electrical characteristics has been simulated and experimentally calibrated.

The work flow of the device-to-circuit simulation and calibration is shown in Fig. 2. Fig. 3 shows the device structure for the TCAD simulation and the channel doping profile of CMOS after the calibration with the 55 nm fabricated devices [(a) n-FET, (b) p-FET]. The simulated and measured I_D-V_G characteristics of the device are compared in Fig. 4. In the simulation, the temperature coefficients of phonon scattering were adjusted to fit the change of drain current as T decreases. The reduced phonon scattering at low temperatures is the key reason for the increase of I_{on}, as it is proportional to the effective mobility.

B. Cryogenic Device Characteristics

Fig. 5 shows the simulated and measured I_D-V_G characteristics for the p- and n-FET at L_G = 55 nm and 1 μm. Due to the reduction of phonon scattering at LT, I_{on} of the long channel device increases by 2.73X for NMOS, and 2.72X for PMOS from 300 K to 77 K, and for the short channel device, I_{on} increases by 1.95X for NMOS, and 2.00X for PMOS, at the same gate overdrive of 0.6 V. As compared with the long channel devices, in the short channel devices, saturation velocity and velocity overshoot are the main factors to affect I_{on}, which is less sensitive to temperature [13]. Therefore, I_{on} improvement due to T reduction is smaller for the short channel devices. For the long and short channel devices, SS is reduce by 40.9% and 22.7% from 300 K to 77 K, respectively, which can help to reduce the dynamic power consumption P_D. From 300 K to 77 K, V_T increases greatly due to the bandgap widening and surface potential increase. For long channel devices, $|V_T|$ increases by 154 mV for n-FET and 188 mV for p-FET, while for short channel devices, $|V_T|$ increases by 181 mV for n-FET and 218 mV for p-FET.

As V_T increases at lower T, I_{on} and the circuit speed will be reduced if V_{DD} does not increases accordingly. To optimize I_{on} and reduce the cell delay, the channel doping of both the n- and p-FETs was modulated to reduce $|V_T|$ at 77 K. Here, the primary goal is to tune the V_T to a point that the I_{off} at $|V_G|$ = 0 V and $|V_D|$ = 1.2 V for the CDE device at 77 K is the same as that of the control device at 300 K, as illustrated in Fig. 5(c) and (d). Based on this goal, the V_T modulation for n-FET is 384

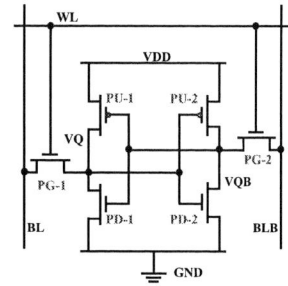

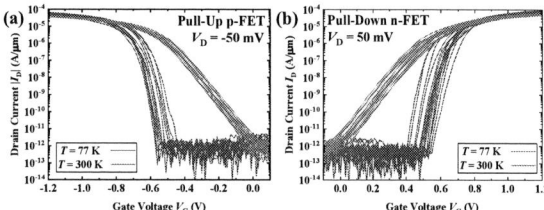

Fig. 6. Schematic of the 6T SRAM Bitcell.

Fig. 7. I_D-V_G variation test for (a) pull-up p-FETs and (b) pull-down n-FETs. For the devices, $W/L = 10\ \mu m/55\ nm$. Over 20 devices were tested for each type of FETs.

mV and that for p-FET is 399 mV. With the channel doping engineering by TCAD process simulation, I_{on} exhibits an improvement of 2.58X for 55 nm n-FET and 2.37X for 55 nm p-FET at an unchanged V_{DD} of 1.2 V

C. The Impact of Variation on Device and SRAM

As the fastest on-chip memory register for read and write operations, SRAM is of great importance in the digital computing units. To study the impact of DDV on the performance of SRAM, massive device characterization and SPICE simulation were carried on based on the 55 nm SRAM bitcells. Fig. 6 illustrates the circuit schematic of a 6T SRAM test structure fabricated using the 55 nm node standard process. Over 20 sets of pull-up (PU) and pull-down (PD) transistors at room temperature and 77K were characterized, as show in Fig. 7. V_T of the PU and PD transistors is a critical parameter that directly determines the static noise margin (SNM) and, consequently, the data stability of the cross-coupled inverter pair in a 6T SRAM cell. It can be observed that reducing the temperature not only leads to an increasing V_T but also aggravate the DDV of V_T.

To further study the impact of DDV on SRAM, SPICE model was developed for each pair of characterized PU and PD devices. These models were then employed to simulate the SNM of the 6T SRAM bitcells. The statistical results of the SNM are presented in Fig. 8. Compared to the room temperature case (0.2305 V²), the cryogenic operation exhibits a higher average SNM (0.2832 V²). This enhancement is primarily attributed to the smaller SS and the higher V_T at low temperatures, which improves the switching speed of the transistors, thereby accelerating the state transition of the cross-coupled inverters within the bitcell. It should be noted that, although cryogenic operation improves the nominal SNM, it also introduces greater variability in the SNM distribution, primarily resulting from the wider distribution of V_T at LT.

D. The Worse Hot Carrier Degradation for LT CMOS

For LT HCI, most of the low-temperature HCI studies only focus on two temperature points: 77 K and 4 K. It is reported that the hot carrier degradation of FinFET shows a non-universal temperature dependence at high temperatures (300 K ~383 K). However, comprehensive temperature-dependent

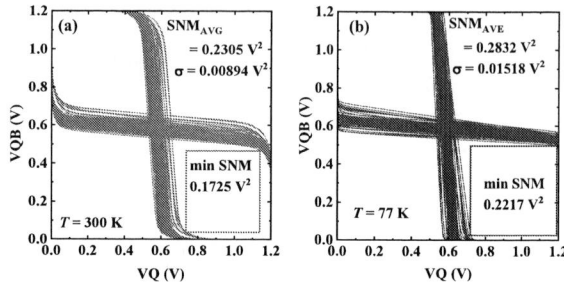

Fig. 8. SNM variation of 6T SRAM bitcells with the PU and PD devices were measured at (a) 300 K and (b) 77 K.

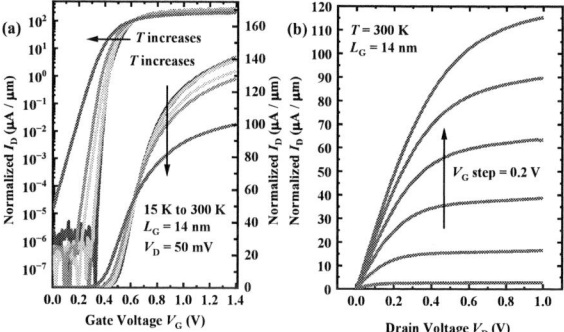

Fig. 9. (a) The I_D-V_G characteristics of a 14 nm FinFET device from 15 to 300 K. As temperature increases the threshold voltage Vth increases, and the drain current I_D decreases. (b) The I_D-V_D characteristics of a 14 nm FinFET device at 300 K.

studies on the HCI behavior of advanced node devices at LT are rarely reported. Fig. 9(a) shows the change of I_D-V_G characteristics as the temperature decreases from 300 K to 15 K for 14 nm n-FinFETs. As shown in Fig. 10(a)-(c), the HCI degradation exhibits a non-monotonical change at both the stress time and T changes, suggesting multiple trap mechanism controls the degradation at different temperature domains. An HCI degradation model including the effects from multiple traps (SP, MP, and oxide traps) is established [6]. By model fitting, the HCI degradation due to different trap components is separated and compared. As shown in Fig. 10(d), SP and oxide traps increase with decreasing temperature, while MP shows a non-monotonical change as T decreases. As the stress time increases, the dominant traps change from MP to SP and finally switch to the oxide traps. As temperature decreases, SP and oxide traps take more parts to the HCI-induced performance degradation.

E. The Core-level Evaluation for LT CMOS

The measured CMOS device and capacitor data were used to calibrate the LT TCAD-SPICE model and evaluate the circuit performance of a RISC-V core. BSIM4-based SPICE model was extracted from the I-V and C-V data at 77 K and 300 K. Standard cell libraries at 77 K and 300 K were consequently built for the circuit simulation. The average delay of various logic cells at 300 K, 77 K and 77 K with doping engineering are shown in Fig. 11. For the devices without CDE, V_T increases as T reduces, leading to a reduction of gate overdrive (V_{DD} -V_T) and a degraded I_{on}. Therefore, the average cell delay is longer for all types of cells at 77 K. With the channel doping adjustment, V_T is effectively shifted to be even lower than the value at 300 K. Therefore, the gate overdrive can be increases for an unchanged V_{DD}. For the standard cells at 77 K with channel doping engineering, the cell delay can be greatly shortened, as shown in Fig. 11.

979-8-3315-3918-4/25 $31.00 © 2025 IEEE

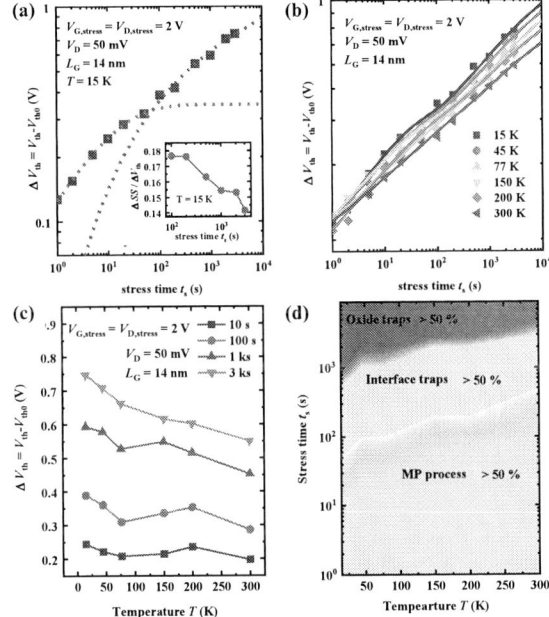

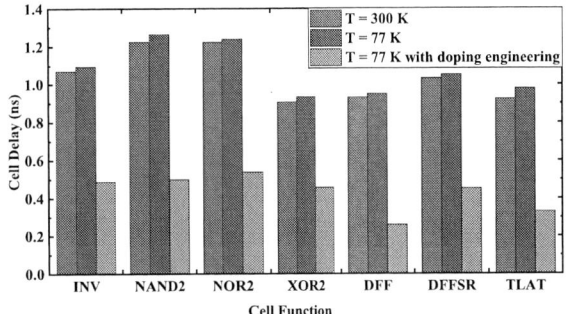

Fig. 11. Propagation delay for different cells in the digital standard cell library at 300 K, 77 K and 77 K with CDE. The core power supplies V_{DD} is fixed at 1.2 V.

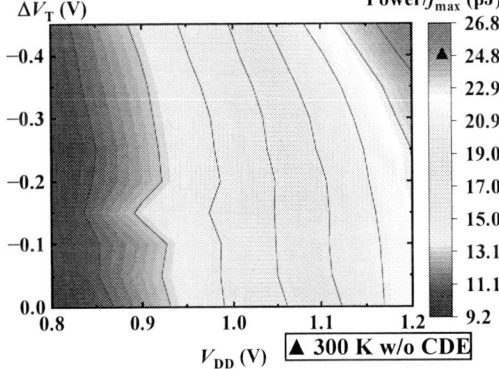

Fig. 10. (a) ΔV_{th} at 15 K shows a two-stage trend as t_s increases. The subplot shows $\Delta SS/\Delta V_{th}$ at 15 K varies with t_s. (b) ΔV_{th} varies with stress time at different temperature. The dots are experimental data and the lines are fitted curves. (c) ΔV_{th} as a function of T under different stress times t_s. The temperature dependence of ΔV_{th} changes with T and t_s. (d) Dominant defect mechanisms (contribute more than 50%) in different conditions.

To compare the circuit performance for devices with and without CDE process optimization, the power and timing simulation were performed for CMOS technologies with and without CDE. Open-source RISC-V core PicoRV32 [12] was used to evaluate the speed and power consumption of the processing unit. As the channel doping engineering is a process-level optimization, no change was made on the core architecture and the layout of digital IP. With the CDE process optimization, I_{on} increases significantly, leading to a f_{max} increase by ~1.4X. The core at 77 K with CDE shows a slightly worse energy efficiency, partially due to the greatly improved I_{on} at a reduced V_T. V_{DD} optimization needs to be performed to achieve a balance between the speed and power consumption. Fig. 12 plots the contour profile of power efficiency (Power/f_{max}) as V_{DD} and CDE changes. Large CDE/higher $|\Delta V_T|$ leads to worse power efficiency. Therefore, V_{DD} should be lowered to suppress the power efficiency degradation at LT. In addition, with slight reduction of V_{DD}, the power efficiency can be always kept low. For V_{DD} above 1.1 V, moderate CDE can result in a high energy efficiency and excellent outperformance of f_{max} over that at 300 K. Therefore, aggressive channel doping adjustment can be explored to achieve better frequency.

IV. CONCLUSION

In this work, the considerations on constructing a high performance and durable LT computing system are discussed, in terms of DTCO, variation and HCI reliability. A practical DTCO flow is demonstrated using a calibrated TCAD-SPICE framework to optimize a 55 nm technology at 77 K. Channel doping engineering can effectively counter the detrimental rise in V_T, significantly improving on-current and circuit speed. The analysis of DDV and SRAM revealed that non-DTCO cryogenic operation improves the nominal SNM but also exacerbates its variability. Furthermore, HCD at LT for advance node transistors are characterized. HCD aggravates at

Fig. 12. The contour profile of power efficiency (Power/fmax) as VDD and ΔV_T changes. Large CDE leads to worse power efficiency. Therefore, VDD should be lowered to suppress the power efficiency degradation at LT.

LT but exhibits a non-monotonic dependence on temperature and stress time. Finally, the system-level evaluation confirms that with co-optimized CDE and V_{DD}, a ~1.4X improvement in maximum frequency can be achieved, underscoring the great potential of LT CMOS for high-performance and energy-efficient computing systems.

ACKNOWLEDGMENT

The authors acknowledge supports from the Zhejiang Provincial Natural Science Foundation (No. LDQ24F040001), National Natural Science Foundation of China (No. 92464205, 62474164, T2388102), and Zhejiang Province Jianbing Key R&D programs (No. 2025C01193).

REFERENCES

[1] F. Jazaeri et al., *Proc. 26th Int. Conf. Mixed Design Integr. Circuits Syst. (MIXDES)*, Jun. 2019, pp. 15.

[2] B. Patra et al., in *IEEE Int. Solid-State Circuits Conf. (ISSCC)*, 2020, pp. 304.

[3] Y. Sun et al., *Trans. Elec. Dev.*, vol.71, no. 1, pp. 107, 2024.

[4] D. Prasad et al., *IEDM*, pp. 23.5.1, 2022.

[5] H. -L. Chiang et al., *IEDM*, pp. 13.2.1, 2021.

[6] J. Qu et al., *Int'l. Relia. Phy. Symp.*, pp. P70.TX, 2024.

[7] Z. Yu et al., in *IEEE Elec. Dev. Techno. and Manu. Conf.*, pp. 34, 2018.

[8] H. Bohuslavskyi et al., *Trans. Elec. Dev.*, vol. 65, no. 9, pp. 3682, 2018.

[9] E. Schriek et al., *IEEE Sol. Sta. Cir. Lett.*, vol. 3, pp. 310, 2020.

[10] S. M. Amoroso et al., *SISPAD*, pp. 35, 2020.

[11] Peifeng Tunan Semiconductor Co., Ltd, https://www.pftn-semi.com/en/posts/solutions/dtco-logic.html.

[12] YosysHQ, "PicoRV32 - A Size-Optimized RISC-V CPU." https://github.com/YosysHQ/picorv32.

979-8-3315-3918-4/25 $31.00 © 2025 IEEE

Feature clustering-driven data augmentation in multi-level hotspot detection for integrated circuits based on GAN

Pengyu Ren[1,2,3], Bojie Ma[1,2,3], Yajuan Su[1,2,3*], Xiaojing Su[1,2,3], Xin Hong[1,2,3], Yuqin Wang[1,2,3], Yujie Jiang[1,2,3], Zhanzi Chen[1,2,3], Tianao Chen[1,2,3], and Yayi Wei[1,2,3*]

[1]EDA Center, Institute of Microelectronics of the Chinese Academy of Sciences, Beijing, China
[2]School of Integrated Circuits, University of Chinese Academy of Sciences, Beijing, China
[3]State Key Laboratory of Fabrication Technologies for Integrated Circuits, Institute of Microelectronics, Chinese Academy of Sciences, Beijing, China
[*]Email: suyajuan@ime.ac.cn, weiyayi@ime.ac.cn

Abstract—With the development of technology node, hotspot detection of layout has become a critical step in integrated circuit (IC) physical design flow. The machine learning-based method has become a competitive candidate due to high efficiency. However, the inherent imbalance in datasets of layout hotspots hinder the performance of the detection model. This study proposes a feature clustering-driven data augmentation framework based on generative adversarial network (GAN) to mitigate the imbalance within datasets, thus enhancing the multi-level hotspot detection. By pattern feature extraction, hotspots are clustered into different categories, and four augmentation methods are conducted on them to present performance differences. Experimental results demonstrate that our framework significantly enhances detection accuracy by comparison with conventional data augmentation techniques, with the highest average enhancement ratio of 8.4%.

Keywords—hotspot detection, generative adversarial network, feature clustering-driven data augmentation

I. INTRODUCTION

Hotspots are layout patterns with narrow lithography process windows (PW). These patterns are susceptible to distortion during lithography and subsequent processes, and result in disfunctions of chips. Consequently, hotspot detection has emerged as a significant technique to ensure the manufacturability and reliability of integrated circuits.

Conventional hotspot detection methods heavily depend on lithography simulation, which consumes lots of time [1]. In contrast, machine learning-based hotspot detection is considered a promising alternative due to its advantage of speed and accuracy [2]. Several studies focus on two-level hotspot detection which only distinguishes hotspots and non-hotspots, and have attempted to enhance the classifier performance by improving data utilization efficiency, such as upsampling and random-mirror flipping [3]–[5]. With the increasing complexity and diversity of hotspots, more precise hotspots information is required, and the multi-level hotspot detection is often used to divide hotspots into various levels, thus helping physical designers to fix hotspots according to their severity levels [6]. However, the insufficient quantity and imbalanced distribution in datasets severely affects the performance of classifiers.

In this paper, we propose a feature clustering-driven data augmentation framework based on GAN to enhance the performance of multi-classifier, and the effectiveness of the framework is validated on ICCAD2012 dataset [7]. The remainder of the paper is organized as follows: Section II describes the feature clustering-driven data augmentation framework. Section III presents the experimental results and analysis. Section IV provides the conclusion and outlook of this paper.

II. FEATURE CLUSTERING-DRIVEN DATA AUGMENTATION FRAMEWORK

The framework of feature clustering-driven data augmentation is shown in Fig. 1(a). The main stages of our framework include data preparation, feature clustering, GAN training, augmented dataset, and multi-classifier training.

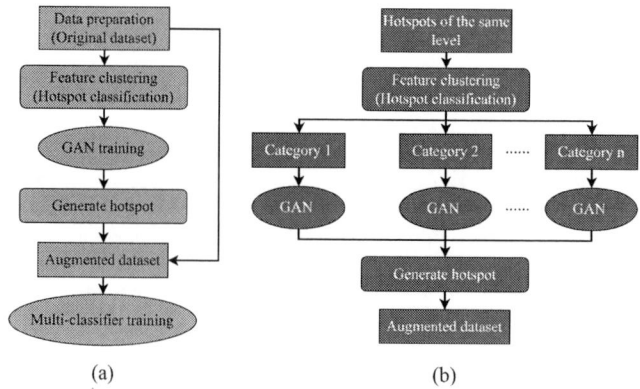

Fig. 1. (a)Framework of feature clustering-driven data augmentation, and (b)more details about how to augment the dataset

A. Data Preparation

In this stage, hotspot clips of OASIS format are converted into image format through a density-based feature extraction to be compatible with the GAN and detection model [6], and the hotspots in original datasets are classified into different levels based on their severity.

979-8-3315-3918-4/25 $31.00 © 2025 IEEE

B. Feature Clustering

For hotspot datasets, even within the same hotspot level, there are different categories of hotspots with distinct features. Therefore, depending on the K-medoids cluster algorithm [8], we cluster hotspots into several categories before training GAN to make data augmentation more efficiently, as shown in Fig. 1(b).

C. GAN Training

Also shown in Fig. 1(b), each hotspot category is fed into GAN [9], and the trained model will generate corresponding hotspots to establish augmented datasets. It should be noted that we reuse the GAN to make it learn the feature of only one category of hotspots at a time.

D. Augmented Dataset

The trained GAN model generates hotspots of the corresponding category to augment hotspots data, which are combined with the original dataset to get the augmented dataset. Our framework gives four augmentation methods which include category-proportional, category-balanced, level-balanced and high augmentation ratio data augmentation.

Category-proportional means that the hotspots of each category within the same level in the original dataset are augmented at the same proportion, while category-balanced means that the aforementioned hotspots are augmented to the same amount. Similarly, level-balanced ensures that hotspots of each level are augmented to the same amount while maintaining balanced amount across all categories. High augmentation ratio refers to augmenting hotspots at a higher augmentation ratio based on category-balanced. The above descriptions can be quantified in the experiments in Section III.

E. Multi-classifier Training

A multi-classifier based on convolutional neural network (CNN) from Gai et al [6] is rebuilt as the detection model with negligible modification. Original and augmented datasets are conducted on the classifier for training, respectively, and the effectiveness of our framework is validated by testing the multi-classifier on the testing sets.

III. EXPERIMENTS

A. Experimental Setup and Preparation

Experiments are implemented in Python with TensorFlow 2.1 on a Linux server with an 8-core 3.4GHz CPU, GTX1080 GPU, and 32GB memory. The schematic diagram of the experimental setup and preparation is shown in Fig. 2.

Two original datasets derived from ICCAD2012 are used for experiments, as listed in Table I. The severity of hotspots become worse with the increase of level order. ICCAD2012 Benchmark1 hotspot dataset is used as the experiment benchmark or control group, and LMC hotspot dataset is used as the experimental group. For LMC hotspot dataset, it is obtained with the lithography manufacturability check (LMC) tool on the ICCAD2012, which is why we named it LMC. In GAN training stage, we choose CycleGAN due to its advantages

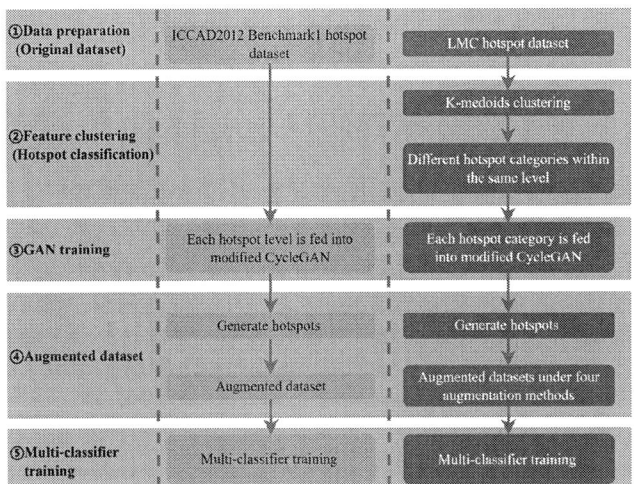

Fig. 2. The schematic diagram of the experimental setup and preparation

in unpairing image-to-image transformation and representing high-resolution complex layout patterns [10], [11], and we modify the model to make it suitable for generative work.

TABLE I
ICCAD2012 BENCHMARK1 AND LMC HOTSPOT DATASET

Hotspot Level	ICCAD2012 Benchmark1 Hotspot Dataset		LMC Hotspot Dataset	
	Training Set	Testing Set	Training Set	Testing Set
L0	340	319	500	1000
L1	60	133	500	800
L2	39	91	500	600

As the experiment benchmark or control group, we skip feature clustering stage on ICCAD2012 Benchmark1 hotspot dataset and hotspots of each level are directly fed into GAN for training and generating. For LMC hotspot dataset, as our framework demonstrates, hotspots of each level are first clustered into four categories, and each hotspot category is fed into GAN. Two original datasets and their corresponding augmented datasets are shown in Tables II and III. Notably, for LMC hotspot dataset, we utilize four augmentation methods to present the classifier performance difference, and the details about these methods are described in Section II.

TABLE II
ICCAD2012 BENCHMARK1 HOTSPOT DATASET AUGMENTATION

Hotspot Level	Original Dataset	Augmented Dataset	Augmentation Ratio
L0	340	500	47%
L1	60	200	233%
L2	39	100	156%

B. Results and Discussions

The multi-classifier is trained on the original and augmented datasets, respectively, and tested on the testing set. As shown in Figures 3 and 4, classification results are illustrated as confusion matrices, and the results are the average over ten

TABLE III
LMC HOTSPOT DATASET AUGMENTATION

Hotspot Level	Original Dataset					Augmented Dataset					
						Category-proportional/Category-balanced/Level-balanced/High augmentation ratio					
	C1	C2	C3	C4	Total	C1	C2	C3	C4	Total	Augmentation Ratio
L0	122	139	65	174	500	488/500/500/750	556/500/500/750	260/500/500/750	696/500/500/750	2000/2000/2000/3000	300%/300%/300%/500%
L1	142	63	133	162	500	426/375/500/500	189/375/500/500	399/375/500/500	486/375/500/500	1500/1500/2000/2000	200%/200%/300%/300%
L2	163	141	135	61	500	326/250/500/250	282/250/500/250	270/250/500/250	122/250/500/250	1000/1000/2000/1000	100%/100%/300%/100%

independent runs. The left side of these figures shows the ground truth level of hotspots in the testing set, and the top side shows the predict level of the multi-classifier for these hotspots. Taking Fig. 3(a) as an example, for the L0 hotspots in the testing set, the multi-classifier predicts or classifies 291.2 of them as L0 hotspots, 19.3 as L1 hotspots, and 8.4 as L2 hotspots. Understandably, a larger value along the diagonal of the confusion matrices indicates greater prediction or classification accuracy for the multi-classifier, demonstrating its improved classification performance. Combined with Fig. 3 and Fig. 4, it is clear that the multi-classifier trained on augmented datasets has more correct predictions at all hotspot levels, which indicates that the modified GAN can generate effective hotspots for data augmentation.

In order to visually compare the multi-classifier's performance across different training sets, particularly those generated by the four augmentation methods listed in Table III, we present the classification accuracy and enhancement ratio in Fig. 5. The average enhancement ratio is also calculated in Fig. 6. As shown in Fig. 5, the multi-classifier trained on augmented datasets has higher classification accuracy at all hotspot levels. Combined with Fig. 6, further analysis presents that based on feature clustering, all four augmentation methods almost obtain higher enhancement ratios for all hotspot levels, thus resulting in average enhancement ratios more than benchmark's 3.1 %.

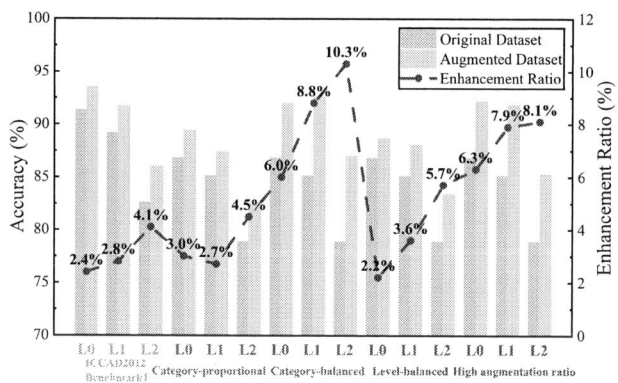

Fig. 5. Classification accuracy and enhancement ratio of different original and augmented datasets

Fig. 3. Confusion matrices of ICCAD2012 Benchmark1 hotspot dataset augmentation: (a)Original dataset, and (b)Augmented dataset

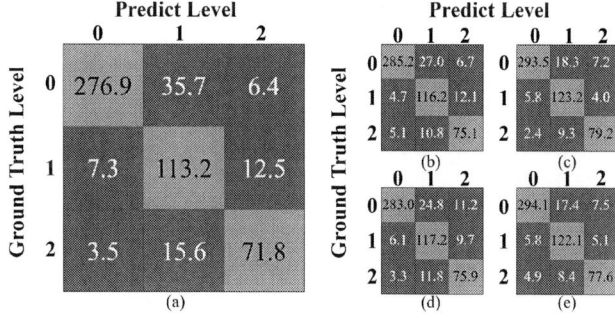

Fig. 4. Confusion matrices of LMC hotspot dataset augmentation: (a)Original dataset, (b)Category-proportional, (c)Category-balanced, (d)Level-balanced, and (e)High augmentation ratio

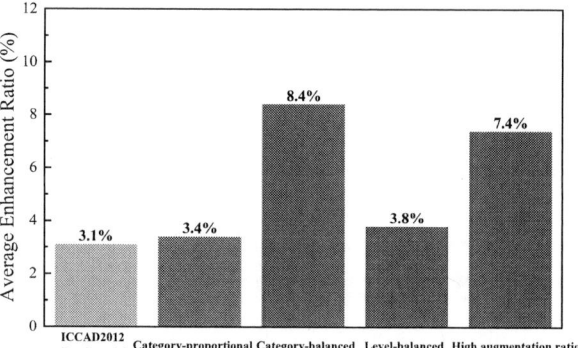

Fig. 6. Average enhancement ratio of different augmented datasets

For the category-proportional augmentation, there are negligible performance improvements, with an average enhancement ratio of 3.4%. The details of this augmentation method in Section II reveal that hotspots of each category in augmented dataset are proportional to those in original dataset, and the imbalance between categories still exists, hindering classifier performance. Category-balanced augmentation addresses this problem by balancing the amount of hotspots in different categories, and the classifier performance is significantly enhanced, with the highest average enhancement ratio of 8.4%. Level-balanced augmentation is conducted to explore the influence of data distribution of hotspot levels on classifier performance. Results show that L1 and L2 have slightly higher enhancement ratios than the experiment benchmark, yet L0 exhibits a lower ratio, which is likely attributed to the mismatch distribution of hotspot levels between the training and testing sets. With an average enhancement ratio of 7.4%, the high augmentation ratio method achieves performance enhancement second only to the category-balanced, and the degradation may derive from altered feature distributions which caused by increased hotspots.

Above observations demonstrate the imbalance within datasets, especially between hotspot categories indeed exists and deteriorates classifier performance severely, and our framework mitigates the imbalance problem in datasets with an appropriate augmentation ratio, thus enhancing the multi-level hotspot detection.

IV. CONCLUSION

This paper proposes a feature clustering-driven data augmentation framework based on GAN to enhance the multi-level hotspot detection. Based on feature clustering, four augmentation methods are conducted on datasets. Experimental results show that for any augmentation method, our framework significantly enhances classifier performance compared to conventional augmentation methods, with an average enhancement ratio more than the benchmark's 3.1%. Further analysis demonstrates that with an average enhancement ratio of 8.4%, the category-balanced data augmentation obtains excellent performance enhancement under an appropriate augmentation ratio, which gives a guidance for augmentation ways. This study reveals that our feature clustering-driven data augmentation framework can mitigate the imbalance within datasets, especially between different hotspot categories, and provides a superior approach to enhance the hotspot detection compared to conventional ones.

ACKNOWLEDGMENT

This work was financially supported by the Strategic Priority Research Program of Chinese Academy of Sciences (Grant No. XDA0330303), National Natural Science Foundation of China (No. 62204257 and 62274181). We appreciate the support from University of Chinese Academy of-Sciences(118900M032), fundamental Research Funds for the Central Universities(E2ET3801). We acknowledge the support from Youth Innovation Promotion Association Chinese Academy of Sciences (No. 2021115)and frontier technology collaboration project (No. QYJS-2023-2900-B) as well.

REFERENCES

[1] G. Yeric, B. Cline, S. Sinha, D. Pietromonaco, and R. Aitken, "The past present and future of design-technology co-optimization," in *IEEE Custom Integrated Circuits Conference*, 2013.

[2] D. Z. Pan, "Machine learning for ic design and technology co-optimization in extreme scaling," in *International Symposium on VLSI Design, Automation and Test*, 2018.

[3] H. Yang, J. Su, Y. Zou, B. Yu, and E. F. Y. Young, "Layout hotspot detection with feature tensor generation and deep biased learning," in *the 54th Annual Design Automation Conference 2017*, 2017.

[4] Y. Chen, Y. Lin, T. Gai, Y. Su, Y. Wei, and D. Z. O. Pan, "Semisupervised hotspot detection with self-paced multitask learning," *IEEE Transactions on Computer-Aided Design of Integrated Circuits and Systems*, 2020.

[5] H. Yang, L. Luo, J. Su, C. Lin, and B. Yu, "Imbalance aware lithography hotspot detection: A deep learning approach," 2017.

[6] T. Gai, T. Qu, X. Su, S. Wang, L. Dong, L. Zhang, R. Chen, S. Yajuan, W. Yayi, and T. Ye, "Multi-level layout hotspot detection based on multi-classification with deep learning," 2021.

[7] J. A. Torres, "Iccad-2012 cad contest in fuzzy pattern matching for physical verification and benchmark suite," *IEEE*, 2012.

[8] L. Kaufmann and P. J. Rousseeuw, "Clustering by means of medoids," *North-Holland*, 1987.

[9] I. J. Goodfellow, J. Pouget-Abadie, M. Mirza, B. Xu, D. Warde-Farley, S. Ozair, A. Courville, and Y. Bengio, "Generative adversarial nets," in *Proceedings of the 28th International Conference on Neural Information Processing Systems - Volume 2*, NIPS'14, (Cambridge, MA, USA), p. 2672–2680, MIT Press, 2014.

[10] J. Y. Zhu, T. Park, P. Isola, and A. A. Efros, "Unpaired image-to-image translation using cycle-consistent adversarial networks," *IEEE*, 2017.

[11] W. Sim, K. Lee, D. Yang, J. Jeong, J. S. Hong, S. Lee, and H. Lee, "Automatic correction of lithography hotspots with a deep generative model," *Proceedings of SPIE*, vol. 10961, no. 000, 2019.

A New TCAD Simulation Framework for Strain-Aware Quantum Tunneling Current Modeling

Jian Ma[1], Chunlei Wu[1,2,3*], Hanzhi Gu[1], Yueyuan Yu[1], Yiming Xia[1], Jiayi Wu[1], Qingqing Sun[1,2,3], David Wei Zhang[1,2,3]

[1] School of Microelectronics, Fudan University, Shanghai 200433, China
[2] Shanghai Integrated Manufacturing Innovation Center Co., Ltd, Shanghai 201203, China
[3] Jiashan Fudan Institute, Jiaxing 314100, China

* Email: wuchunlei@fudan.edu.cn

Abstract—A novel strain-aware band-to-band tunneling (BTBT) simulation framework has been proposed, so as to enable high efficiency simulation analysis of stress effects in quantum tunneling current. In order to achieve a tight coupling of band structure deformation simulation and tunneling current simulation, a transformation algorithm for strain-aware BTBT model correction has been developed, which can include the stress effects on material parameters and accordingly tunneling probability automatically. Based on the proposed strain-aware tunneling simulation framework, the impacts of channel stress on electrical characteristics of GAA TFETs has been discussed. The results show that, unlike the case in MOSFETs, n-type and p-type TFETs shows no obvious preference for the tensile or compressive stress type regarding to current enhancement. In general, the biaxial <110> compressive stress is preferred for n-type TFETs, while biaxial <110> tensile stress is preferred for p-type TFETs. The conclusions are helpful for evaluation of tunneling current in strained devices.

Keywords—Band-to-band tunneling (BTBT), Strained silicon, Stress effects, Nanoscale devices, TCAD simulation

I. INTRODUCTION

The relentless scaling of CMOS technology according to the Moore's law has left almost no space for further reduction of the physical channel length, without incurring penalties of excessive off leakage current [1]. In nanoscale devices, apart from the most frequently discussed drain-induced barrier lowering and punch through leakage, the quantum band-to-band tunneling (BTBT) current can no longer be neglected [2]. The direct source-to-drain tunneling current, as well as the gate induced drain leakage (GIDL) tunneling current has grown to be the primary leakage components in extremely scaled MOSFETs [3]. Moreover, the emerging steep slope tunnel FET (TFET), which promises sub-60mV/dec subthreshold swing at room temperature, utilizes quantum tunneling mechanism for device switching [4]. Therefore, modeling of band-to-band tunneling (BTBT) current in technology computer-aided design (TCAD) device simulators needs urgently enhancement.

Since the band-to-band tunneling probability exhibits exponentially dependence on bandgap (E_g) and effective mass (m^*) of electrons and holes [5]. In order to achieve higher tunneling current in TFETs, device materials with low band gap as well as reduced effective mass are required. Apart from adoption of narrow-bandgap materials such as III-V

semiconductors [6], strained silicon technology has emerged as another effective approach to enhance tunneling probability without introducing process compatibility problems [7]. The strained silicon can lift the sub-band degeneracy of the conduction and valence bands, leading to carrier repopulation in the sub-bands and consequently reduction of bandgap and effective carrier mass [8]. However, most commercial TCAD simulation tools use semi-classical models for tunneling current calculation [9], which requires the user to manually specify a set of model parameters aligning with the specific experimental data. These simulation algorithms exhibit no direct reflection of material parameters such as E_g or m^*, making TCAD simulations of tunneling current to be "stress-independent" and fail to capture the strain characteristics of quantum tunneling. Existing simulations of strained BTBT predominantly rely on complex first-principles calculations or the non-equilibrium Green's function (NEGF) methods [10]-[11].

In this work, a novel TCAD simulation framework with automatically tight coupling of strained sub-band simulation with tunneling current simulation has been proposed, so as to enable strain-aware characterization of quantum tunneling current.

II. SIMULATION METHODOLOGY

A. Strain-independent Tunneling Simulation Problem

The nonlocal BTBT model is used in Sentaurus TCAD simulation and the BTBT generation rate G is calculated by Kane's model at the room temperature [12]

$$G = A\left(\frac{F}{F_0}\right)^P \exp\left(-\frac{B}{F}\right) \qquad (1)$$

where F is the electric field and F_0 is 1V/m, P is the degeneracy factor of 2 or 2.5 corresponding to direct and indirect tunneling respectively. The parameters A and B are coefficients related to material parameters

$$A = \frac{g(m_v m_c)^{\frac{3}{2}}(1+2N_{op})D_{op}^2(qF_0)^{5/2}}{2^{21/4}h^{5/4}\rho\varepsilon_{op}[E_g(300K)+\Delta_C+\Delta_V]^{7/4}} \qquad (2)$$

$$B = \frac{2^{7/2}\pi m_r^{1/2}[E_g(300K)+\Delta_C+\Delta_V]^{3/2}}{3qh} \qquad (3)$$

Where m_c and m_v are the effective masses of the conduction and valence bands, $m_r = (m_c m_v / m_c + m_v)$ and g are the reduced

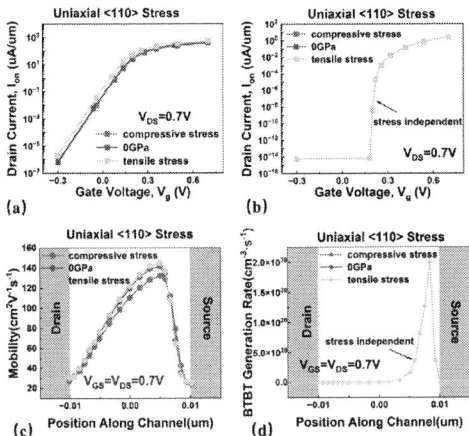

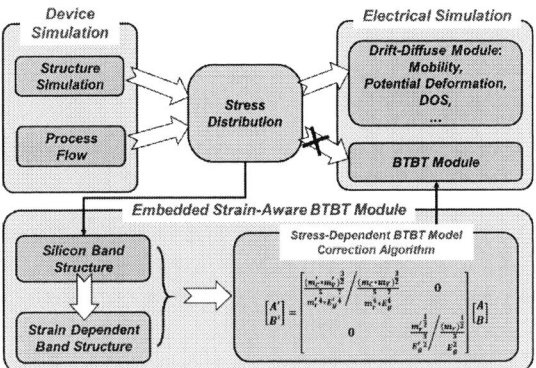

Fig. 2. The proposed strain-aware quantum tunneling simulation framework

Fig. 1. Transfer characteristics of n-type (a) MOSFET and (b) TFET under uniaxial strain condition, (c) Mobility and (d) BTBT generation rate under uniaxial strain condition.

tunneling mass and degeneracy factor, p is the mass density, D_{op}, ε_{op}, and N_{op} are the deformation potential, energy, and number of optical phonons, respectively.

In practical simulations, however, A and B adopted in BTBT models are fixed values obtained through empirical fitting to experimental data. Thus, the BTBT generation rate G would stay the same for certain materials regardless of applied strain condition. This modeling method with predetermined A and B coefficients overlooks the strain-induced material band structure deformations such as the tunneling effective mass and bandgap, leading to severe underestimated tunneling current in strained devices.

Fig. 1 presents the transfer characteristics of n-type MOSFET and TFET with uniaxial strain along the <110> direction. It can be observed that, the drift-diffusion current in MOSFET exhibits evident variations under different stress condition, indicating strong stress sensitivity of the mobility models. While the tunneling current in TFET remains unchanged regardless of stress conditions, suggesting the lack of due consideration of stress effects in the BTBT models.

B. Strain-Aware Tunneling Simulation Framework

To address the above illustrated problem, a strain-aware BTBT simulation framework based on Sentaurus TCAD simulation tools is proposed, supporting a seamless simulation flow from band structure deformation simulation to device tunneling current simulation, as depicted in Fig. 2.

Firstly, the device structure and process simulations are performed to get the stress distribution within the device. Then, import the localized stress data into the Sentaurus Band Structure simulator for band structure deformation calculation using empirical pseudopotential methodology (EPM). The obtained stress-dependent band structure parameters are then extracted and fed into a transformation matrix to generate strain-dependent BTBT model parameters automatically. Finally, the obtained corrected model parameters are embedded into the BTBT models for tunneling current simulation. The proposed strain-aware BTBT simulation algorithm can be well implemented in Sentaurus TCAD simulation platform, supporting analysis of impacts of

different strain conditions (e.g., uniaxial, biaxial) on the tunneling current characteristics in scaled devices.

III. APPLICATION RESULTS AND DISCUSSION

A. Strain-Dependent Band Structure Simulation

Applying strain will change the six-fold degeneracy of the silicon conduction band valleys in k-space, splitting them into two-fold and four-fold degenerate valleys. The longitudinal and transverse effective masses of conduction band electrons under uniaxial and biaxial strains along the <110> direction are presented in Figs. 3(a) and (b), respectively.

It can be seen in the figure, for uniaxial stress in the <110> plane, the transverse electron effective mass tends to decrease with tensile stress whereas increase with compressive stress. While the longitudinal electron effective mass remains nearly unaffected by uniaxial <110> stress. For biaxial stress, both longitudinal and transverse electron effective masses tends to decrease with tensile stress and increase with compressive stress. The longitudinal electron effective mass exhibits greater sensitivity to biaxial <110> stress than the transverse counterpart. Fig. 3(c) and (d) depict the effective masses of valence band holes under uniaxial and biaxial stress along the <110> direction respectively. For uniaxial stress in the <110> plane, both tensile and compressive stress will cause the light hole (LH) effective mass increasement. For biaxial <110> stress, however, both the effective masses of heavy and light holes exhibit low sensitivity to stress.

Fig. 4 depicts the impacts of uniaxial and biaxial <110> stress on the bandgap of silicon. A sharp reduction of bandgap can be observed under <110> compressive stress condition, irrespective of whether uniaxial or biaxial stress. While reduced bandgap under tensile stress can also be observed only with weaker dependence of stress.

B. Strain-Aware Band-to-Band Tunneling Modeling

On the ground of the above band structure deformation simulations, a stress-dependent BTBT model correction algorithm has been developed, which can include the stress effects on material and accordingly on tunneling probability automatically. The transformation matrix can be described as

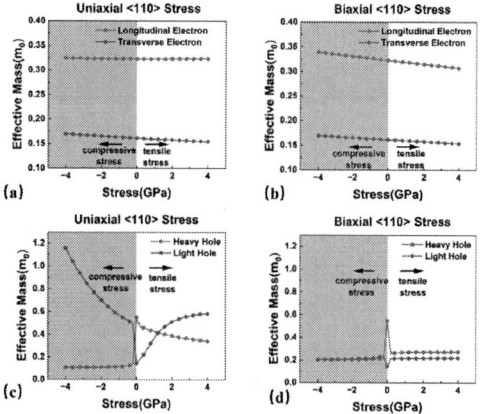

Fig. 3. Impact of stress on (a)-(b) Electron effective masses, (c)-(d) Hole effective masses

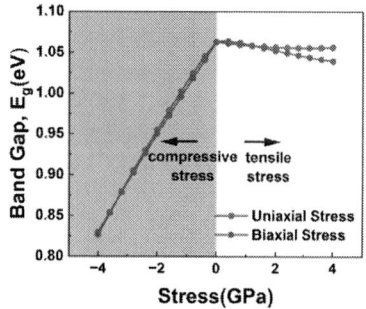

Fig. 4. Impact of stress on silicon bandgap.

$$
\begin{bmatrix} A' \\ B' \end{bmatrix} = \begin{bmatrix} \dfrac{(m'_C * m'_V)^{\frac{3}{2}}}{m'^{\frac{5}{4}}_r * E'^{\frac{7}{4}}_g} \Big/ \dfrac{(m_C * m_V)^{\frac{3}{2}}}{m^{\frac{5}{4}}_r * E^{\frac{7}{4}}_g} & 0 \\ 0 & \dfrac{m'^{\frac{1}{2}}_r}{E'^{\frac{3}{2}}_g} \Big/ \dfrac{(m_r)^{\frac{1}{2}}}{E^{\frac{3}{2}}_g} \end{bmatrix} \begin{bmatrix} A \\ B \end{bmatrix} \qquad (4)
$$

Fig. 5 presents the stress dependence of BTBT generation coefficients A, B and total generation rate G derived from the correction algorithm. As can be seen in Fig. 5(c), strong stress dependence of BTBT generation rate G has been obtained, enhanced tunneling probability can be observed under both tensile and compressive stress (uniaxial or biaxial <110>). It should to be noted that biaxial <110> tensile stress and uniaxial <110> compressive stress is preferred for more effective tunneling probability enhancement.

C. Strain-Aware Tunneling Current Calculation

Fig. 6 shows the transfer characteristics of n-type and p-type tunnel field-effect transistors (TFETs) applying uniaxial <110> and biaxial <110> stress. As can be seen in the figures , both compressive and tensile stress can effectively reduce the turn-on voltage and enhance the on-state current (I_{ON}) in tunnel devices. What's more, unlike the case in MOSFETs, n-type and p-type TFETs shows no obvious preference for the tensile or compressive stress type. The n-type TFETs can achieve enhanced on current under both tensile and compressive stress, the same goes for p-type TFETs. This can be attributed to the monotonically increase of BTBT generation rate with increasing stress as derived in Fig. 5 (c).

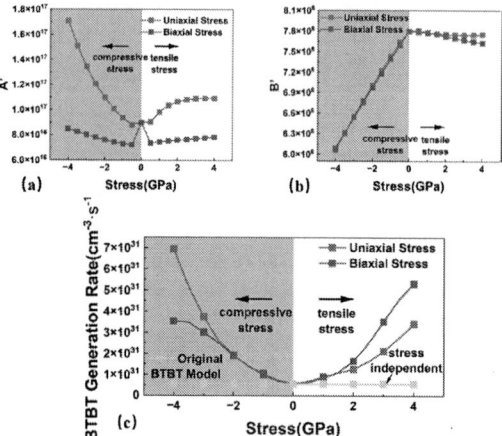

Fig. 5. Impact of stress on (a)-(b) Electron effective masses, (c)-(d) Hole effective masses

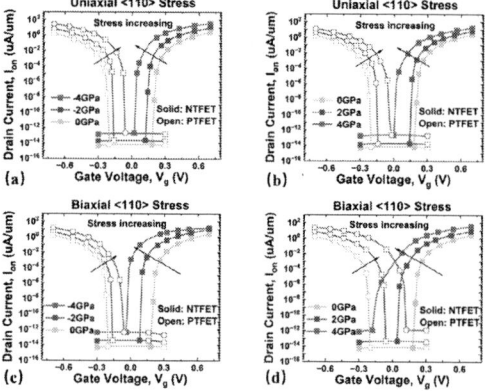

Fig. 6. Transfer characteristics of n- and p-type TFETs under (a)-(b) uniaxial Stress, and (c)-(d) biaxial Stress along <110> direction.

Fig. 7 summarizes the impacts of stress on the electrical characteristics of n- and p-type TFETs. As can be seen in Fig.7 (a), the threshold voltage (V_{th}) of both n- and p-type TFETs decreases monotonically with increasing stress values. The n-type TFETs seem to be more sensitive to the biaxial compressive stress, while p-type TFETs seem to be more sensitive to the biaxial tensile stress. Regarding to the subthreshold swing (SS) (Fig. 7 (b)), a distinct SS degradation can be observed for n-type devices subjected to biaxial <110> tensile stress. The on-state current (I_{ON}) is a critical performance metric for TFETs. As shown in Fig.7 (c), for n-type TFETs, the most significant I_{ON} enhancement is achieved under biaxial tensile stress and uniaxial compressive stress, but the biaxial tensile stress leads to substantial increase of off-state current (I_{OFF}) (Fig .7 (d)). For p-type TFETs, the uniaxial compressive stress and biaxial tensile stress deliver the optimal I_{ON} improvement, but the biaxial tensile stress introduces higher I_{OFF}. In general, the biaxial compressive stress is preferred for n-type TFETs, while biaxial tensile stress is preferred for p-type TFETs.

IV. CONCLUSION

In this work, a novel strain-aware band-to-band tunneling (BTBT) simulation framework has been proposed, so as to enable high efficiency simulation analysis of stress effects in quantum tunneling current. A transformation algorithm for

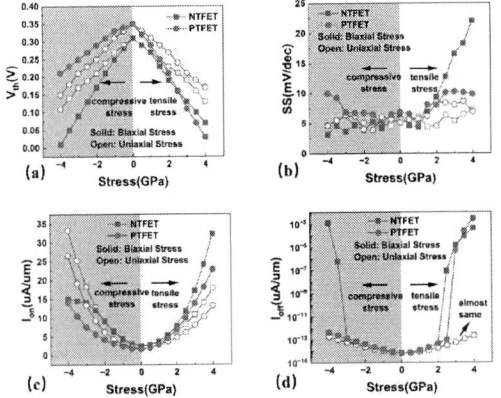

Fig. 7. Impact of stress on the electrical characteristic of n- and p-type TFETs. (a) Threshold voltage (V_{th} =V_{GS} @ I_{DS}=10 μA/μm), (b) Subthreshold Swing (SS) (average SS for I_{DS} = 10^{-3}~10^{-9} μA/μm), (c) On state current (I_{ON}=I_{DS} @V_{GS}=V_{DS}=0.7 V), (d) Off state current (I_{OFF}=I_{DS} @V_{GS}=0V).

strain-aware BTBT model correction has been developed, which can include the stress effects on material parameters and accordingly tunneling probability automatically. Based on the proposed strain-aware tunneling simulation framework, the impacts of channel stress on electrical characteristics of GAA TFETs has been discussed. The results show that, unlike the case in MOSFETs, n-type and p-type TFETs shows no obvious preference for the tensile or compressive stress type regarding to current enhancement. In general, the biaxial <110> compressive stress is preferred for n-type TFETs, while biaxial <110> tensile stress is preferred for p-type TFETs. The proposed strain-aware simulation framework provides an efficient approach for evaluation of quantum tunneling current of strained devices in advanced nodes.

ACKNOWLEDGMENT

This work was supported in part by NSFC under Grant 62304048, in part by the Science and Technology Commission of Shanghai Municipality under Grant 21TS1400800.

REFERENCES

[1] K. J. Kuhn, "Considerations for Ultimate CMOS Scaling," in IEEE Transactions on Electron Devices, vol. 59, no. 7, pp. 1813-1828, July 2012, doi: 10.1109/TED.2012.2193129.

[2] R. Kim, U. E. Avci and I. A. Young, "CMOS performance benchmarking of Si, InAs, GaAs, and Ge nanowire n- and pMOSFETs with Lg=13 nm based on atomistic quantum transport simulation including strain effects," 2015 IEEE International Electron Devices Meeting (IEDM), Washington, DC, USA, 2015, pp. 34.1.1-34.1.4, doi: 10.1109/IEDM.2015.7409824.

[3] Z. Jiang et al., "Comprehensive Simulation Study of Direct Source-to-Drain Tunneling in Ultra-Scaled Si, Ge, and III-V DG-FETs," in IEEE Transactions on Electron Devices, vol. 64, no. 3, pp. 945-952, March 2017, doi: 10.1109/TED.2017.2656921.

[4] R. Pandey, S. Mookerjea and S. Datta, "Opportunities and Challenges of Tunnel FETs," in IEEE Transactions on Circuits and Systems I: Regular Papers, vol. 63, no. 12, pp. 2128-2138, Dec. 2016, doi: 10.1109/TCSI.2016.2614698.

[5] Ionescu, Adrian M., and Heike Riel. "Tunnel field-effect transistors as energy-efficient electronic switches." nature 479.7373 (2011): 329-337.

[6] R. Kotlyar, U. E. Avci, S. Cea, R. Rios, T. D. Linton, K. J. Kuhn, I. A. Young; Bandgap engineering of group IV materials for complementary n and p tunneling field effect transistors. Appl. Phys. Lett. 18 March 2013; 102 (11): 113106.

[7] T. Krishnamohan, D. Kim, S. Raghunathan and K. Saraswat, "Double-Gate Strained-Ge Heterostructure Tunneling FET (TFET) With record high drive currents and ≪60mV/dec subthreshold slope," 2008 IEEE International Electron Devices Meeting, San Francisco, CA, USA, 2008, pp. 1-3, doi: 10.1109/IEDM.2008.4796839.

[8] R. People, "Physics and applications of GexSi1-x/Si strained-layer heterostructures," in IEEE Journal of Quantum Electronics, vol. 22, no. 9, pp. 1696-1710, September 1986, doi: 10.1109/JQE.1986.1073152.

[9] K. Ahmed, Mirza Mohammad Monzure Elahi and M. Shofiqul Islam, "A compact analytical model of band-to-band tunneling in a nanoscale p-i-n diode," 2012 International Conference on Informatics, Electronics & Vision (ICIEV), Dhaka, Bangladesh, 2012, pp. 521-524, doi: 10.1109/ICIEV.2012.6317361.

[10] A. Afzalian, G. Doornbos, T. -M. Shen, M. Passlack and J. Wu, "A High-Performance InAs/GaSb Core-Shell Nanowire Line-Tunneling TFET: An Atomistic Mode-Space NEGF Study," in IEEE Journal of the Electron Devices Society, vol. 7, pp. 88-99, 2019, doi: 10.1109/JEDS.2018.2881335.

[11] H. Lee, Y. Cho, S. Jeon and M. Shin, "First-Principles-Based Quantum Transport Simulations of Interfacial Point Defect Effects on InAs Nanowire Tunnel FETs," in IEEE Transactions on Electron Devices, vol. 68, no. 11, pp. 5901-5907, Nov. 2021, doi: 10.1109/TED.2021.3112395.

[12] Kane, Evan O. "Theory of tunneling." Journal of applied Physics 32.1 (1961): 83-91.

DTCO-based Hybrid Rail 8T Complementary FET SRAM Design towards advanced node

Yutian Zhang*, Khawar Sarfraz, Mansun Chan

Department of Electronic and Computer Engineering, The Hong Kong University of Science and Technology,
Clear water bay, Kowloon, Hong Kong

* Email: yzhangkq@connect.ust.hk

Abstract—As technology scaling advances toward the sub-1nm era, Complementary Field-Effect Transistor (CFET) architectures offer a promising path to sustain SRAM density and performance. The inherent 3D integration capability of CFET technology proves particularly advantageous for implementing 8T SRAM configurations at advanced nodes, overcoming the structural limitations of conventional 6T designs. However, the footprint of the current 8T CFET SRAM is still much larger than 6T SRAM. This work proposes a CFET-compatible hybrid-rail 8T SRAM design that addresses interconnect and layout challenges, achieving a 27% reduction in cell area compared to conventional dual-port 8T CFET SRAM. Furthermore, it demonstrates reduced parasitic effects on the bitline (BL) and wordline (WL). Through a Design Technology Co-Optimization approach (DTCO), SPICE simulations reveal that the hybrid-rail 8T SRAM achieves significant improvements in terms of power and delay. The results show that hybrid-rail 8T SRAM offers superior area efficiency, energy savings, and performance, establishing it as a compelling solution for high-density, low-power memory in advanced-node applications.

Keywords—CFET, DTCO, SRAM, Backside Interconnect

I. INTRODUCTION

As Moore's Law approaches physical limitations, novel transistor architectures and fabrication processes have emerged to enable continued device scaling. The transition from planar to three-dimensional transistor topologies has facilitated enhanced device density, improved power efficiency, and superior performance characteristics. Among these developments, CFET has emerged as a particularly promising candidate for post-2nm technology nodes [1]. This vertically integrated architecture, implementing stacked nFET and pFET, enables higher functional density within equivalent footprints. As CMOS technology scales toward fundamental physical limits, SRAM remains indispensable for modern computing systems due to its unmatched access speed and operational reliability in high-performance applications. While conventional 6-transistor SRAM remains widely adopted for its area efficiency and fast access characteristics, significant research efforts have focused on CFET-based 6T SRAM implementations, demonstrating substantial reductions in both cell area and parasitic parameters [3].

However, CFET-based 6T SRAM implementations confront fundamental limitations arising from both the vertical integration constraints of CFET technology and inherent 6T SRAM topology characteristics. The asymmetric composition of 6 transistor cells permits only partial CFET stacking, resulting in suboptimal area utilization equivalent to four discrete transistors. The inherent single-port architecture of 6T SRAM fundamentally constrains data throughput, exacerbating access latency during frequent read/write operations. Meanwhile, dual-port solutions remain essential for addressing concurrency demands such as high-performance computing and neural accelerators [4].

Comparative analysis reveals that 8T SRAM architectures exhibit significantly improved area efficiency when implemented using CFET technology [5]. While the 8T SRAM cell's balanced transistor configuration (4 nFETs and 4 pFETs) effectively exploits CFET vertical stacking advantages, its actual footprint (34.01% larger than 6T baseline) reveals unresolved scaling bottlenecks.

Although conventional SRAM optimization methods exist, application-specific optimization techniques for 8T CFET SRAM remain unexplored. This work develops a novel DTCO-based layout optimization scheme specifically targeting 8T CFET SRAM performance enhancement. Leveraging this framework, a significant reduction in cell area, coupled with a concurrent decrease in power dissipation is demonstrated.

II. LAYOUT CONSTRAINTS AND INTERCONNECT-AWARE SRAM DESIGN

In the default structure of CFET, PMOS devices are placed in the bottom tier to maintain high channel stress [3] as shown in Fig. 1. In the original 2nm nanosheet process, front-end signal/power rails result in 10-track cell height. In the 6T SRAM structure (Fig. 2(a)), though BSPDN improves routing, cell area scaling remains limited by pFET-nFET minimum

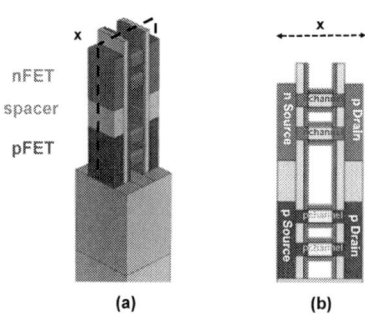

Fig. 1. 3D schematic and the cross section of CFET includes pFET at bottom layer and nFET at top layer.

979-8-3315-3918-4/25 $31.00 © 2025 IEEE

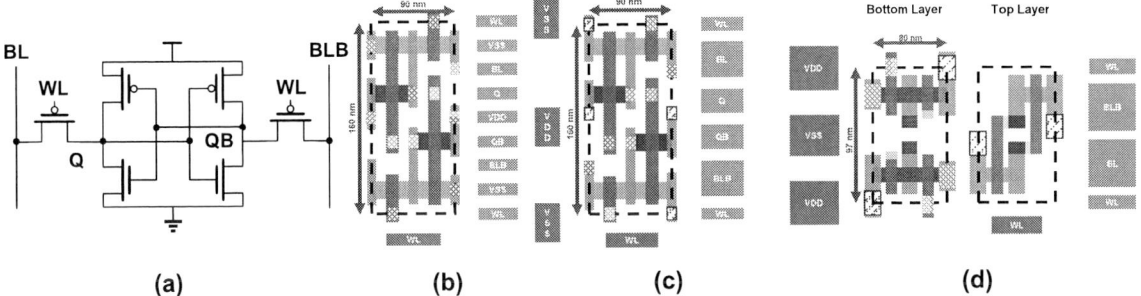

Fig. 2. (a) Schematic of 6T SRAM highlighting PMOS access transistor in CFET implementation. (b) Layout of 2nm node Nanosheet 6T SRAM without buried power rail. (c) Layout of 2nm node Nanosheet 6T SRAM with buried power rail. (d) 1nm sequential 6T CFET SRAM layout with buried power rail.

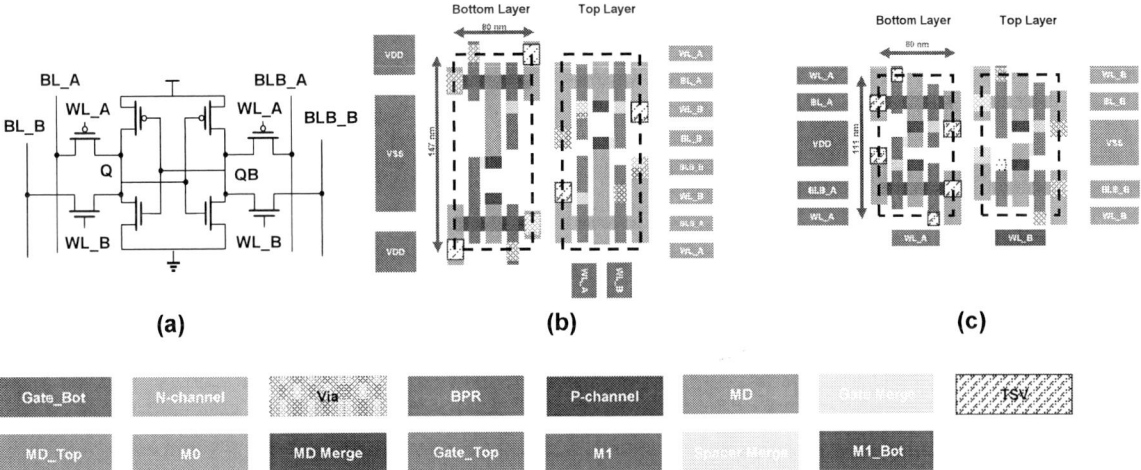

Fig. 3. (a) Dual-port 8T SRAM schematic enabling simultaneous read and write operations for improved throughput. (b) 1nm dual-port 8T CFET SRAM layout with buried power rail. (c) 1nm dual-port 8T CFET SRAM layout with hybrid rail.

spacing (Fig. 2(b) and Fig. 2(c)). Owing to the vertically stacked structure of CFETs, the cell height of 6T CFET SRAM can be reduced by up to 43%. Consequently, in the 6T SRAM schematic, the six transistors are distributed as 2 nFETS and 4 pFETs, which explains why the access transistors are PMOS in Fig. 2(a). To further optimize routing, the gate merge layer technique developed by IMEC is adopted [3], eliminating the need for an additional metal layer to connect source/drain terminals to the gate (Fig. 2(d)). Key process parameters are summarized in Table 1.

A conventional 8T SRAM incorporates two additional PMOS transistors compared to the 6T version. These transistors decouple the read and write paths, effectively mitigating read disturbance issues. However, implementing this structure using CFET technology poses challenges. To minimize area, CFET SRAM designs align transistors sharing source/drain terminals in the same row, requiring a sequential CFET structure and complex routing to connect bottom-tier sources to top-tier gates. Due to the asymmetry of conventional 8T SRAM, a super via is required to connect the Q/QB nodes from the bottom-tier PMOS to the source/drain of the top-tier NMOS. This introduces extra parasitic capacitance and resistance.

Another 8T structure is dual-port 8T SRAM shown in Fig. 3(a), which adds an access port atop the 6T cell to enable concurrent read and write operations, thereby increasing bandwidth and data throughput [2]. Here, port A is used for read operations and port B is used for write operations. Compared to a traditional 8T layout, the area is reduced by 18% owing to its symmetric structure (Fig. 3(b)). Furthermore, the symmetric layout eliminates the need for super vias to interconnect Q/QB and signal lines.

Despite these advantages, the height of both 8T SRAM layouts cannot be efficiently scaled due to the large number of signal ports. For instance, an 8T dual-port SRAM requires eight routing tracks solely for signal lines. To address this issue, a hybrid-rail scheme is introduced to improve layout density in Fig. 3(c). In this scheme, VSS is routed through the buried rail and VDD through the top rail, allowing vias and Through-Silicon Vias (TSVs) to be aligned at the same location for both nFET and pFET. All signals connected to nFET are routed through the top rail, while pFET-related signals are routed through the bottom rail.

As a result, the number of routing tracks in the dual-port 8T SRAM is reduced from 8 to 6, leading to a 27% reduction in cell area, only 12.61% larger than a conventional 6T CFET SRAM. This also results in a nearly 50% reduction in the

TABLE I.   1 NM DESIGN RULE

1 nm Design Rules	
Mx pitch (nm)	16
M1 pitch (nm)	20
M0 pitch (nm)	16
Gate Pitch (nm)	40

TABLE II.   NANOSHEET CFET PARAMETERS

CFET Parameters	Value
Gate Length (nm)	12
Nanosheet width (nm)	20
Nanosheet thickness (nm)	6
Extension length (nm)	5
nFET/pFET space (nm)	20
Channel doping conc.(cm^{-3})	1×10^{16}
Substrate doping conc.(cm^{-3})	1×10^{15}
Source/drain doping conc.(cm^{-3})	1×10^{20}
Effective oxide thickness (nm)	0.84

number of masks required as the layout of the buried rail mask is identical as the top layer. While this approach can also be applied to traditional 8T SRAM layouts, its impact is less significant due to the inherent asymmetry of the conventional design.

III. SRAM DESIGN FLOW AND SIMULATION RESULTS

The overall SRAM DTCO flow (Fig. 4) is illustrated as below. The process begins with the extraction of device parameters from 3D TCAD simulations, where current-voltage (I–V) characteristics are obtained (Fig. 5). These electrical characteristics are then translated into SPICE-compatible models, which are subsequently used for circuit-level simulations in HSPICE.

To accurately capture parasitic effects within SRAM, such as parasitic resistance and capacitance, data from process emulation is utilized. These values are extracted using a 3D field solver, which accounts for the detailed layout and material properties (Fig. 6). The key device parameters for the 1nm technology node used in this study are summarized in Table 2, followed by the IRDS Roadmap [7].

In terms of capacitance (Fig. 7), the hybrid-rail design exhibits a 24% reduction in BL capacitance and a 29.4% reduction in WL capacitance compared to the traditional dual-port 8T SRAM configuration. For resistance (Fig. 8), the BL resistance remains nearly identical between the two 8T SRAM variants, while the WL resistance in the hybrid-rail structure is 27% lower than that of the conventional 8T CFET SRAM. These improvements in parasitic parameters are primarily attributed to the reduction in the number of routing tracks, as well as the shorter interconnect lengths enabled by the optimized alignment of signal and power rails.

HSPICE is used to evaluate the performance of various SRAM structures by analyzing key metrics such as static noise

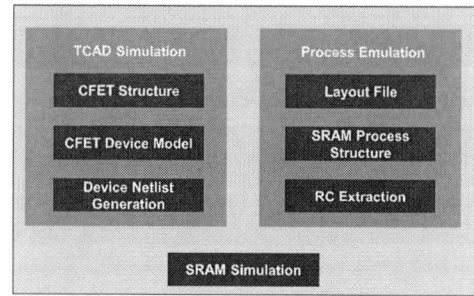

Fig. 4. DTCO-based SRAM design and simulation flow.

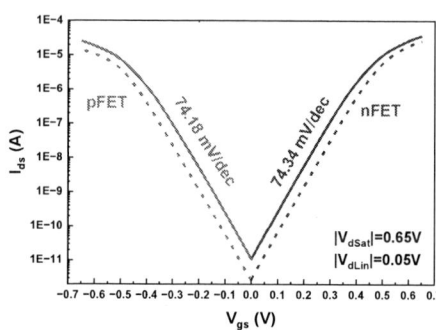

Fig. 5. Id–Vg characteristics of CFET devices at the 1nm technology node, derived from TCAD simulations.

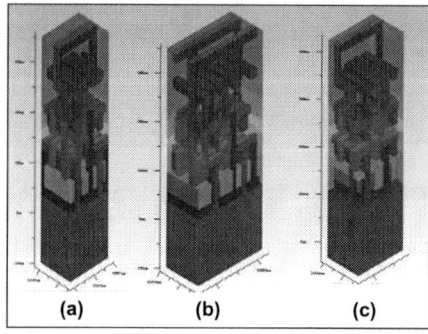

Fig. 6. Process emulation of 3 CFET layout structure: (a) sequential 6T CFET SRAM, (b) sequential BPR-based 8T CFET SRAM and (c) Hybrid 8T CFET SRAM.

margin, read margin, and write margin across different designs. Since the simulations are static, the influence of parasitic parameters on the results is minimal. Additionally, dual-port 8T SRAM utilizes the same read and write paths as the conventional 6T SRAM, resulting in comparable read and write margins between the two architectures.

The power consumption of each structure during different operations is also evaluated (Fig. 9). Static power consumption is nearly identical among the two 8T SRAMs and the baseline 6T SRAM because the influence of parasitic components is limited in static conditions, and the two additional access transistors in the 8T SRAM do not introduce significant leakage current. In terms of read power, the 6T SRAM consumes less power due to its smaller transistor count.

979-8-3315-3918-4/25 $31.00 © 2025 IEEE

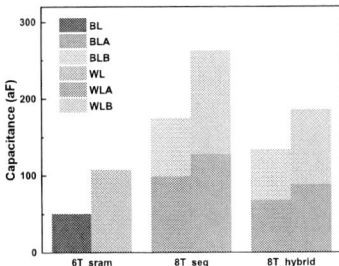

Fig. 7. Extracted capacitance of all different signal lines in three layout structures.

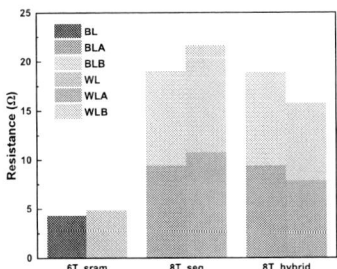

Fig. 8. Extracted resistance of all different signal lines in three layout structures.

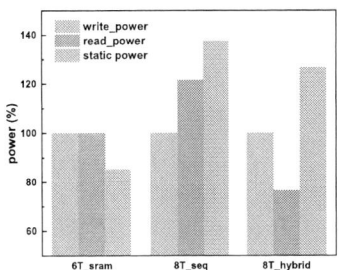

Fig. 9. Normalized power analysis of 3 different structures in 3 operation modes.

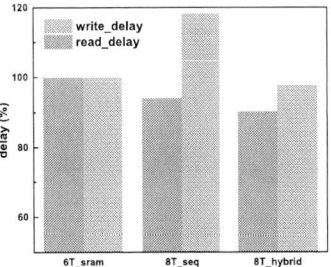

Fig. 10. Normalized delay analysis of 3 different structures in 3 operation modes.

The hybrid-rail 8T SRAM achieves 37% lower read power than the conventional dual-port 8T layout, making it the most energy-efficient design among all evaluated structures. This is attributed to its reduced WL capacitance and lower VDD rail resistance, which minimizes IR drop during the read operation and enhances energy efficiency. For write power, the hybrid-rail 8T SRAM also outperforms BPR-based 8T SRAM by 5.8%. Its lower parasitic resistance and capacitance contribute to reduced energy consumption during write operations.

Delay performance is also examined (Fig. 10). Write delay is defined as the time from the WL reaching 50% of its swing to the Q or QB node reaching 50%. Read delay is defined as the time from the WL reaching 50% to the voltage difference between the two BL reaching 100 mV. As can be seen, the Hybrid rail 8T SRAM achieves the lowest read delay due to its lower RC parasitic, with a 10.9% and 4.0% improvement over the 6T SRAM and the conventional 8T SRAM. Unlike the BPR-based design, it avoids the high resistance introduced by TSVs during read operations. For write delay, the hybrid-rail 8T SRAM shows a 17.41% improvement over the BPR-based counterpart due to its reduced parasitic effects. In summary, the hybrid-rail 8T SRAM proves to be superior to the dual-port 8T SRAM in terms of area efficiency, power consumption, and timing performance.

IV. CONCLUSION

A CFET-compatible hybrid-rail 8T SRAM design is proposed and evaluated through DTCO methodology. Compared with traditional dual-port 8T SRAM, it achieves reduced parasitic effects, improved power efficiency, and shorter read/write delays. The optimized layout reduces cell area and enhances routing density, making it a promising candidate for advanced-node, high-performance memory applications.

ACKNOWLEDGMENT

This work was supported by the General Research Fund (GRF) 16201223 from the Research Grants Council (RGC) of Hong Kong.

REFERENCES

[1] J. Ryckaert et al., "The complementary FET (CFET) for CMOS scaling beyond N3," in 2018 IEEE Symposium on VLSI Technology, Jun. 2018, pp. 141–142. doi: 10.1109/VLSIT.2018.8510618.

[2] H.-H. Liu et al., "CFET SRAM DTCO, interconnect guideline, and benchmark for CMOS scaling," IEEE Transactions on Electron Devices, vol. 70, no. 3, pp. 883–890, Mar. 2023, doi: 10.1109/TED.2023.3235701.

[3] H.-H. Liu et al., "CFET SRAM with double-sided interconnect design and DTCO benchmark," IEEE Trans. Electron Devices, vol. 70, no. 10, pp. 5099–5106, Oct. 2023, doi: 10.1109/TED.2023.3305322.

[4] K. Nii et al., "2RW dual-port SRAM design challenges in advanced technology nodes," 2015 IEEE International Electron Devices Meeting (IEDM), Washington, DC, USA, 2015, pp. 11.1.1-11.1.4, doi: 10.1109/IEDM.2015.7409673.

[5] D. Abdi et al., "Area-efficient CFET dual-port SRAM with backside interconnect," in 2024 IEEE European Solid-State Electronics Research Conference (ESSERC), Bruges, Belgium: IEEE, Sep. 2024, pp. 21–24. doi: 10.1109/ESSERC62670.2024.10719562.

[6] IRDS 2016–2022 Reports. Accessed: May. 24, 2025. [Online]. Available: https://irds.ieee.org/edition.

Design of VCM Motor Coil Based on Five Factor Integration

Shengxian Quan[1], Huihong Zhang*[1], Yuejun Zhang*[1], Qiang Wang*[2], Guanglong Xu[2], Jinsheng Yang[2],

[1] Faluty of Electrial Engineering and Computer Science, Ningbo University, Zhejiang, 315211, China.
[2] Ningbo Huayuan Electronic Technology Co., Ltd, Ningbo, Zhejiang,315476, China.

* Email: 15395714228@163.com, zhanghuihong@nbu.edu.cn, zhangyuejun@nbu.edu.cn, king.wang@wakan.cn

Abstract—As the core actuator of precision positioning systems, Voice Coil Motors (VCM) face significant challenges in the coordinated optimization of inference stability, heat dissipation performance, and space utilization. It is difficult to for the traditional VCM design method to solve the complex coupling relationships among the five factors of size, line width, line spacing, copper thickness, and resistance. This study proposes an innovative design framework based on multi physics field modeling and intelligent optimization: establish the VCM coupling model of five factors mentioned above，reveal the mechanism to balance these parameters; Develop a simulation system that integrates electromagnetic and thermal field analysis, quantifying the interdependent relationship among thrust fluctuations, thermal resistance, and coil density; NSGA-II multi-objective optimization algorithm is adopted to achieve collaborative optimization of four key performances. The results show that the optimal VCM design scheme achieves an 18.8% increase in inductance, a decrease of 22.6% in temperature rise, resistance value reduced by 10.7% and a reduction of 9.1% in force ripple.

Keywords—VCM, coupled system, Parametric modeling, optimization algorithm

I. INTRODUCTION

With the development of technology and processes, precision positioning systems such as semiconductor lithography, high-precision detection equipment, micro nano operation platforms have an urgent demand for high-precision, high stability, and high response speed of driving actuators. VCM (Voice Coil Motor) have become the mainstream actuators for camera components in many fields of precision instruments and equipment due to their advantages of small size and high precision, and their performance has a direct impact on the quality of camera data acquisition [1]. Electromagnetic thrust is the core output performance indicator of VCM. In precision positioning systems, it is not only required to have sufficient thrust, but also to have extremely high stability and reliability [2], especially the thrust fluctuation must be strictly limited to a very small range. Thermal resistance is a key parameter for measuring heat dissipation capability, which directly affects coil temperature rise, reliability, and long-term stability. Meanwhile, coil density is a direct reflection of space utilization, which directly affects the compactness and output per unit volume of the

This work was supported in the Key R&D Program of Ningbo Science and Technology Yongjiang 2035 under Grant 2024Z139, in part by the Ningbo University Student Research and Innovation Program (2025SRIP1315).

motor [3]. How to design VCM coils that can simultaneously provide high-precision, low fluctuation thrust and good heat dissipation performance while meeting strict spatial constraints is the core challenge in the current precision drive field [4]. Based on the above challenges, this project systematically models the relationships of key parameters (size D, line width W, line spacing S, copper thickness T, resistance value R) of VCM coils. Multi objective optimization is carried out based on simulation analysis results of different specifications of coil combinations to achieve the optimal design of VCM motor coils under high-precision thrust requirements of precision positioning systems. An integrated simulation analysis framework is proposed and constructed to accurately characterize the interaction among the five key parameters of size, line width, line spacing, copper thickness, and resistance value. Explore their comprehensive effects on electromagnetic thrust (including fluctuations), thermal resistance, and coil density [5]. Based on the above coupling model, a systematic simulation analysis platform has been developed to efficiently evaluate the performance under different parameter combinations (coil specifications). Establish a five factor coupling model of size line width line spacing copper thickness resistance.

A. Parameters Coupling Mechanism

The five parameters of VCM motor coils form a multidimensional coupling network through electromagnetic fields, thermodynamics, and geometric constraints. Including the basic coupling effect of size, the size parameter serves as the basic boundary of the design space and affects other parameters through geometric constraints, such as spatial capacity constraints, where the maximum number of turns is limited by the ratio of available area to the area occupied by a single turn. Large size provides a larger heat dissipation surface area, but increases the inertia of motion, while small size limits wiring space and exacerbates heat accumulation. The outer diameter Do of the coil determines the strength of the edge effect of the permanent magnetic field, and the distortion of the edge magnetic field is the main source of thrust fluctuations [6]. The five parameters of VCM motor coil form a closed-loop coupling system through physical laws, and its core principles are reflected in the electromagnetic geometric coupling principle and the multi parameter coupling of thermal resistance network. The relationship between coupling modeling and analysis methods is shown in Fig. 1.

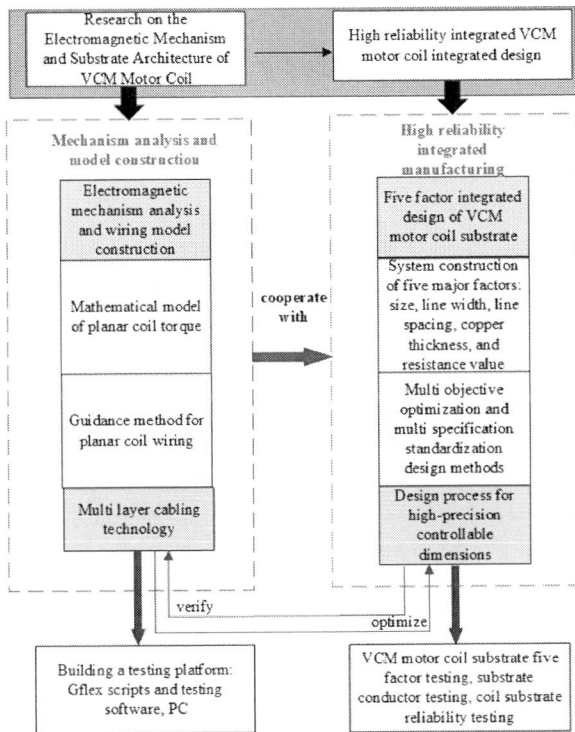

Fig. 1. VCM motor coil research scheme.

In the precise design of voice coil motors (VCM), the principle of electromagnetic geometric coupling reveals the essential relationship between the physical configuration of the coil and its electromagnetic performance. This principle states that the geometric parameters of the coil (size, line width, line spacing) directly determine the thrust output characteristics and motion stability by regulating the magnetic field spatial distribution and current path [7]. The number of distributable turns N of a coil is strictly limited by geometric space, and its mathematical model can be expressed as:

$$N_{\max} = \frac{A_{coil}}{(W+S)^2} = \frac{D_x \times D_y}{(W+S)^2} \tag{1}$$

The sizes D_x, D_y have a double-edged sword effect. Although increasing the coil size can improve the turn capacity ($N \propto D^2$), it leads to an increase in the inertia of the moving parts and a decrease in the system response speed. In high-speed precision positioning systems, a balance must be struck between thrust gain and dynamic performance. Increasing W can reduce resistance ($R \propto 1/W$), but it occupies wiring space ($N\downarrow$); Expanding S can improve heat dissipation and insulation strength, but also reduce the number of turns ($N\downarrow$). When the coil adopts a spiral structure, the outer diameter expression is:

$$D_o = D_i + 2N(W+S) \tag{2}$$

This length gradient will cause a radial non-uniform distribution of current density, thereby distorting the symmetry of the magnetic field. A wide wire causes current to gather at the edge of the conductor, and its edge magnetic flux density can be modeled as:

$$B_{edge} = \frac{u_0 I}{2\pi r} exp\left(-\frac{\tau}{W}\right) \tag{3}$$

In this formula, τ is the distance from the edge, r is the radius of the magnetic field, u_0 is the electromagnetic density coefficient, which is related to the material. This fluctuation will cause even harmonic enhancement, with a second harmonic amplitude of $B_2 \propto W^{0.8}$, and will also exacerbate thrust fluctuations. ΔF is positively correlated with the amplitude of $B2/B4$ harmonics. Increasing the line spacing can weaken the magnetic coupling between conductors and reduce the magnetic field gradient:

$$\nabla B \approx \frac{u_0 I}{4\pi}\left[\frac{1}{\left(d-\frac{S}{2}\right)^2} - \frac{1}{\left(d+\frac{S}{2}\right)^2}\right] \tag{4}$$

Where d is the distance from the observation point to the center of the conductor. Based on force analysis, establish a mathematical model between the current passing through a rectangular planar coil and important parameters such as coil thrust, magnetic field strength, and coil voltage. Establish a relationship curve between mechanical stroke and coil current, and use key constraints such as starting current and slope to preliminarily determine the number of turns, resistance value, and coil thickness of the planar coil. The analysis method is shown in Fig. 2.

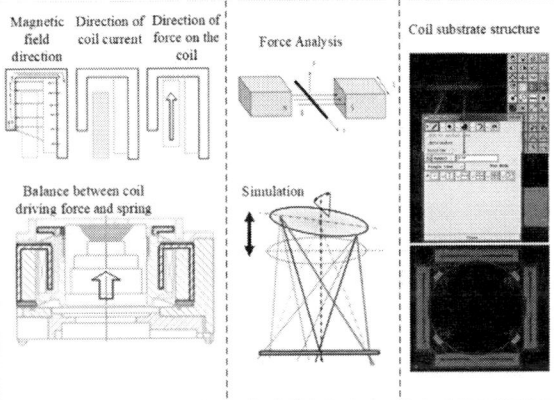

Fig. 2. Analysis of electromagnetic mechanism.

B. Multi-Parameter Coupling of Thermal Resistance Network

In the design of VCM motor coils, the electric thermal energy conversion layer is the core link that determines system reliability [7], which reveals the conversion mechanism of electrical energy to thermal energy and the reverse constraint of thermal management on electromagnetic performance, forming a closed-loop game relationship of "Joule heat generation-thermal resistance conduction-temperature rise degradation". Under the requirement of fixed thrust, reducing the line width/spacing can improve space utilization, but it leads to a sharp increase in current density $J = I/(W \cdot T)$, resulting in a square fold increase in Joule heating [8]. Its physical essence is described as follows:

When current passes through a coil conductor, free electrons collide with the lattice to convert kinetic energy into thermal energy, and their power density is determined by Joule Lenz's law:

$$P_j = I^2 R \tag{5}$$

The resistance value R is a function of geometric parameters shown in equation 6, and Joule heating is not an independent variable, but the result of the combined effect of size, line width, copper thickness, and line spacing, which can be expressed as:

$$R = \rho_{cu}\frac{L}{A_c} = \rho_{cu}\frac{\pi N(D_i + D_o)}{W \cdot T} \tag{6}$$

C. Technical Scheme of Two-stage Optimization Framework

Using a combination of stratified sampling and spatial filling, an improved Latin hypercube sampling method is employed to generate uniformly distributed sample points within a five dimensional hypercube. The sample distribution is optimized using the minimum maximum distance criterion, with 500 basic samples generated in the first round. Based on preliminary simulation results, 200 encrypted samples are added to the performance sensitive area. The Voronoi diagram is used to identify sparse regions in the parameter space, and the degree of influence of each parameter is quantified through global sensitivity analysis. A second-order polynomial response surface model is established to capture nonlinear relationships and construct a multi-objective optimization framework. The core framework of the algorithm is shown in Fig. 3.

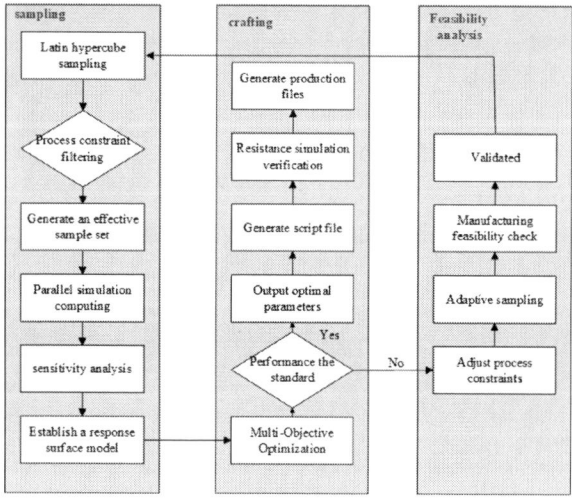

Fig. 3. Two-stage optimization core flow chart.

D. NSGA-II Algorithm Core Formula

Normalize the objective function, unify the dimensions of thrust fluctuation, thermal resistance, and coil density, and incorporate them into the objective function using penalty function method:

$$F_{total} = F_{original} + 10^6 \sum_{j=1}^{2} max\left(0, -g_j(x)\right) \tag{7}$$

Provide basic physical parameter constraints as shown in Table I, and subsequent simulations and experiments must be strictly based on this numerical range.

TABLE I.   PARAMETER DESIGN

Parameter	Symbol	Value	Unit
Copper resistivity	ρCu	1.68×10^{-8}	$\Omega \cdot m$
maximum current density	J_max	12×10^{6}	A/m^2
Working current	I	0.5	A
Thermal conductivity coefficient	κ	150	$W/m \cdot K$
Target coil density	-	180	circle /cm^2

Through the synergistic effect of the above formula, NSGA-II achieved thrust fluctuation suppression, thermal resistance control, and high-density achievement in five factor optimization.

II. EXPERIMENTAL DESIGN AND RESULTS

To verify the effectiveness of the five factor integration model, this section provides a detailed introduction to the experimental platform and implementation method, including database architecture design, storage size, design values for the five major factors including size, line width, line spacing, copper thickness, and resistance, recording simulation/measured data such as force, temperature rise, and efficiency, and environmental temperature and humidity, equipment manufacturing metadata:

A. Visualization and Results Analysis

From the Fig.4, it can be clearly seen that the performances have been comprehensively improved. The inductance optimization scheme has an average increase of 21.1%, the temperature rise has been saved by 23% under the 20 °C environment, the resistance has decreased by 10%, and the thrust fluctuation has decreased by 35% at 80 °C, inductance increased to 5.7 uH, resistance reduced by 10.7%, thrust fluctuation reduced to 8.6%. The results show that the optimal VCM design scheme achieves an 18.8% increase in inductance, a decrease of 22.6% in temperature rise, resistance value reduced by 10.7% and a reduction of 9.1% in force ripple.

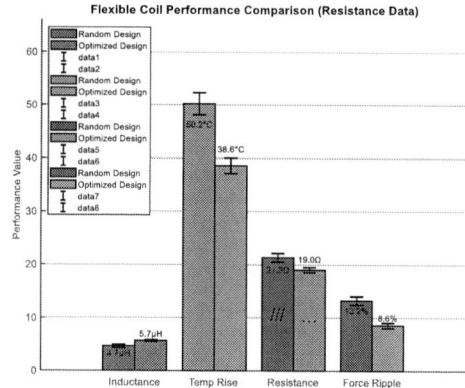

Fig. 4. Comparison of performance indicators of coils.

It can be seen from Fig. 5 that the temperature rise, resistance and thrust fluctuation have been improved to a certain extent at different temperatures.

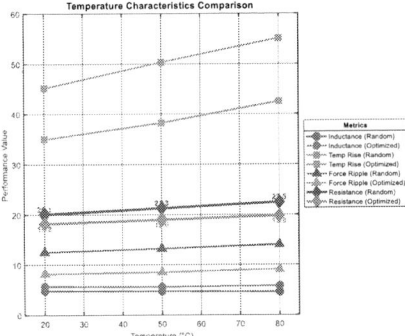

Fig. 5. Comparison of temperature characteristic curves.

B. Motor Coil Wiring Design Diagram

Based on the obtained data, the physical coil is designed as shown in Fig. 6. The wiring is completed through the designed line width and spacing, and the coil resistance is also controlled within a good range. This VCM design scheme successfully realizes the optimal balance between electromagnetic performance and thermal reliability. It can be widely used in the thermal management design of various high power density electromagnetic devices.

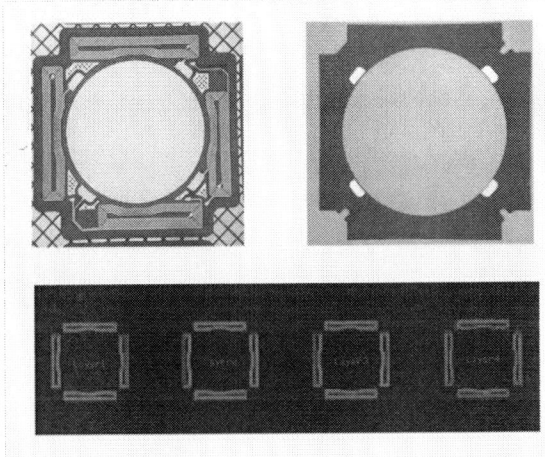

Fig. 6. Design of wiring for motor coil substrate.

III. CONCLUSION

Through the comparative analysis of random design and optimal design, the performance improvement of flexible coil is systematically evaluated. Key findings indicate consistent performance enhancements in the 10% -50% range of all key metrics. Visualization includes bar graph with explicit value label, temperature curve with error range and improvement percentage graph, which visually verifies these improvements. These results emphasize the effectiveness of the optimization method in balancing performance gain and practical constraints, and provide a reliable framework for coil design optimization.

REFERENCES

[1] J. Mu, H. Zhang, "Design and optimization of a large-air-gap voice coil motor with enhanced thermal management for magnetic levitation vibration isolation in a vacuum," Actuators, vol. 14, no. 6, pp. 301-301, June 2025.

[2] R. Lin, Y. Li, Z. Xu, C. Cheng, X. Gao, W. Sun, J. Qian, "Dynamic rate-dependent hysteresis modeling and trajectory prediction of voice coil motors based on TF-NARX neural network," Microsystem Technologies, vol. 29, no. 9, pp. 1319-1331, July 2023.

[3] Y. Guo, J. Zhou, Z. Gao, B. Feng, M. Xu, "Integrated design and experiment of a micro-vibration isolation and pointing platform for large space optical payloads based on voice coil motors," Sensors, vol. 25, no. 4, pp. 1179-1179, February 2025.

[4] J. Gong, J. Luo, "Rapid and precise zoom lens design based on voice coil motors with tunnel magnetoresistance Sensors," Applied Sciences, vol. 16, no. 14, pp. 6990-6990, August 2025.

[5] Z. Song, H. Joon Ahn, "Identification and compensation of the flexure-induced motion error with a voice coil motor," International Journal of Precision Engineering and Manufacturing, vol. 25, no. 8, pp. 1683-1688, June 2024.

[6] K. S. Hoon, J. Y. Gon, C. Y. Man, "Design of a finger-sized voice coil motor for high-speed scanners," International Journal of Precision Engineering and Manufacturing, vol. 24, no. 2, pp. 209-217, December 2022.

[7] H. Wu, F. Zhang, H. Gao, "A simulation study of static electromagnetic characteristics of voice coil motor injector," IFAC Papers OnLine, vol. 54, no. 10, pp. 494-499, January 2021.

Full-spiking Bio-inspired Target Detection Vision Algorithm based on Gating Attention Prediction for DVS and SPAD Sensors

Lengjun Yang [1,2], Xingyu Xiang [1,2], Yiyao Wen [1,2], Jian Liu [1,2], Nanjian Wu [1,2], Liyuan Liu [1,2], Shuangming Yu*[1,2]

[1] State Key Laboratory of Semiconductor Physics and Chip Technologies, Institute of Semiconductors,
Chinese Academy of Sciences
2 University of Chinese Academy of Sciences

* Email: yushuangming@semi.ac.cn

Abstract—This paper presents a visual algorithm for target detection based on dynamic and single-photon sensors, along with the system architecture used to deploy the algorithm on a hardware platform. The system integrates two subsystems that fuse the event stream signals from the dynamic vision sensor and the spiking frame signals from the single-photon sensor. Leveraging attention predictions generated by the former subsystem, it performs attention-based gating control and recognition enhancement for the latter. This approach improves recognition accuracy while reducing data processing volume. The system demonstrates strong performance in target detection task. Since both sensors output sparse spiking signals, a full-spiking system can be designed through thoughtful engineering. This approach features low-power consumption, low latency, low computation, and hardware-friendliness. Compared to conventional algorithms, our method achieves comparable performance while significantly reducing both parameter count and computational load. Especially, our method achieves an approximately 14% reduction in parameter count compared to YOLOv5m.

Keywords—dynamic vision sensor, single-photon avalanche diode, target detection, full-spiking system

I. INTRODUCTION

Target detection tasks in high-speed scenarios demand high-performance, high-speed cameras to avoid motion blur and detection failures caused by fast-moving objects, as well as highly accurate and efficient vision algorithms to rapidly process large volumes of image data. Traditional methods typically utilize high-frame-rate CMOS cameras and Convolutional Neural Networks (CNNs), such as YOLO networks. However, due to the massive data volume from high-frame-rate images and large size of CNN models, such systems suffer from high latency and significant computational energy consumption[1]. In edge device scenarios with constrained hardware resources, traditional methods struggle to be efficiently deployed on hardware while maintaining high accuracy and speed.

The human visual system excels at capturing high-speed moving targets with low power consumption and can be regarded as an exemplary model of an edge system. The human retina contains two types of ganglion cells responsible for dynamic and static information, respectively. Magnocellular cells handle motion perception and spatial localization, capable of sensing fast-moving objects, while Parvocellular cells handle object recognition and color analysis, capable of resolving fine details[2]. Inspired by this human visual system, we employ an event camera (or Dynamic Vision Sensor, DVS) with dynamic sensing capability and a high-performance Single-Photon Avalanche Diode (SPAD) with static sensing capability to mimic this biological perception process. Each pixel unit in a DVS only detects intensity changes. Consequently, this type of sensor images only moving objects in a scene and cannot provide detailed grayscale feature information about objects. However, the DVS performs well in moving object detection[3]. In contrast to traditional CMOS sensors, SPAD sensors offer higher frame rates, higher dynamic range, and lower noise in low-light conditions[4]. More importantly, both DVS and SPAD output spike-based imaging data, where dynamic scene information and grayscale information are represented by spike frequencies. It makes them suitable for a full-spiking bio-inspired design emphasizing low power consumption and low computation.

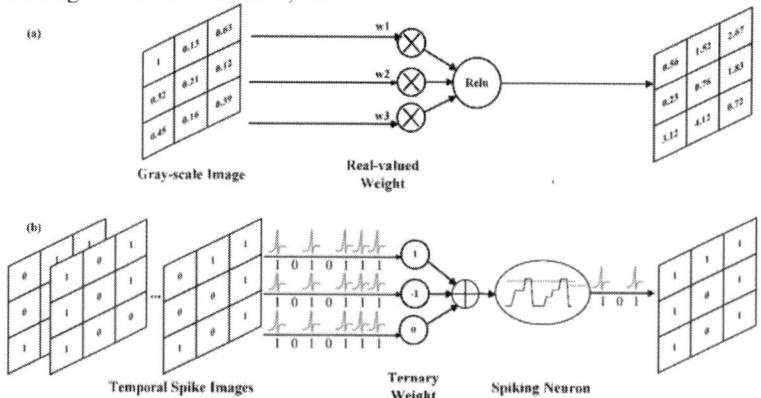

Fig. 1. Comparison between the real-valued network and the ternary spiking network. (a) The real-values network. (b) The ternary spiking network.

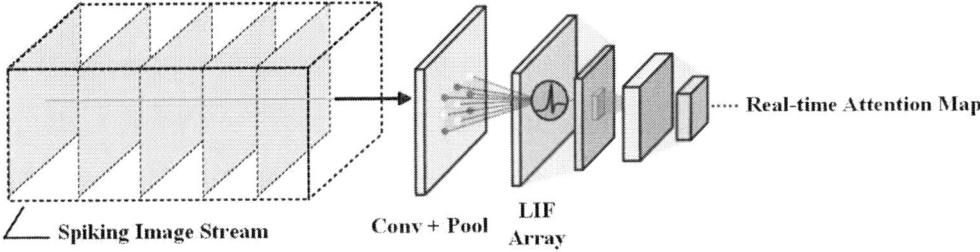

Fig. 2. The SNN structure used to get real-time attention map

To fully leverage the imaging characteristics of DVS and SPAD, this paper proposes utilizes distinct networks to process data from each sensor. One network processes the dynamic information from the DVS to obtain a real-time attention map of the scene and controls the gating of the SPAD network. Simultaneously, the other network using SPAD image data for target detection, incorporating this attention map. Furthermore, we propose a method employing a Spiking Neural Network (SNN) utilizing ternary low-bit quantization to process the sensor data[5]. As shown in Fig. 1, this network simplifies numerous real-valued Multiply-Accumulate (MAC) operations into coincidence operations and spike-counting operations, saving significant multiplier resources and memory accesses. This facilitates deployment on resource-constrained edge devices. In Section II, we introduce the proposed algorithmic model structure and workflow. Section III describes the deployment architecture of this system on hardware platforms such as FPGAs followed by Section IV presenting the simulation results on DND21[6] and DDD20[7] datasets.

II. ALGORITHM AND MODELING

A. Attention Prediction with DVS

Event-based cameras are built upon a 2D grid of specialized pixels that detect changes in photocurrent, specifically, variations in brightness, on their receiver circuits. A pixel will output an event spike under the following condition:

$$\Delta I(x, y, t) \geq pC \qquad (1)$$

Where $\Delta I(x, y, t)$ represents the change of photocurrent of pixel *(x, y)* at time *t*, *p* is the event polarity (either 1 or -1), and *C* is a fixed constant determined by the sensor.

Based on this mechanism, when objects move within a scene, the sensor outputs a stream of events in the form of (x, y, p, t). These events contain Coordinates of occurrence, Polarity (indicating brightness increase or decrease) and Timestamp of occurrence. By accumulating events within a specific time window, we obtain spike frames composed exclusively of values {1, 0, -1}. Since brightness changes primarily occur at the edges of moving objects in the scene, these spike frames capture only contour information of moving subjects. This establishes DVS as intrinsic edge extractors, which help us focus attention on dynamic regions of interest, precisely the elements demanding critical attention such as moving vehicles and pedestrians in autonomous driving scenarios.

The SNN shown in Fig. 2 efficiently processes sequential spike frames in the temporal domain. The network continuously ingests real-time spiking frames and outputs compact attention prediction maps that dynamically updates with the latest input. Leveraging the inherent memory properties of SNNs, this output depends not only on current inputs but also on preceding spiking frames.

Through training, the generated attention prediction maps identify regions of highest interest, typically moving objects critical to the current task. For instance, in autonomous driving scenarios, attention prioritizes nearby moving vehicles rather than pedestrians on sidewalks or even birds in flight.

The system activates the SPAD and its corresponding SNN recognition network only when significant attention regions exist (i.e., when values in attention prediction matrices exceed a predefined threshold). Otherwise, the recognition network remains in a dormant state to minimize overall power consumption. Additionally, multiplying the raw SPAD spiking image with an upsampled and binarized attention map increases sparsity in non-critical regions. Empirical evidence confirms this operation simultaneously boosts recognition accuracy and reduces power consumption in the SPAD processing network.

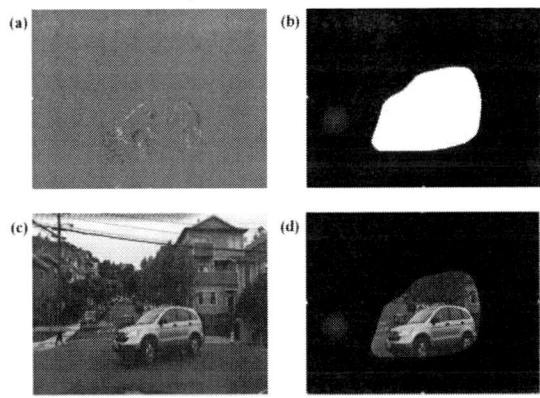

Fig. 3. Demonstration of the fusion of attention map and the SPAD imaging. (a) DVS image. (b) The upsampled and binarized attention map output of the DVS network. (c) SPAD image (For better demonstration, the image is recovered to a gray-scale image). (d) Fusion result by multiplying (b) and (c).

B. Quantization of Ternary SNN

Ternary quantization is a method that reduces neural network weights to three discrete values: {1, -1, 0}. This approach compresses 32-bit floating-point weight data into merely 2-bit storage, significantly reducing both storage requirements and computational overhead. The technique proves exceptionally efficient for full-spike scenarios, where multi-bit multiplication operations simplify into spike coincidence operations. Notable advancements in this field have been achieved by multiple research teams[5]. Below is our adaptation based on characteristics of spiking frame within SNNs after briefly introducing ternary quantization fundamentals

The process of ternary quantization can be presented as follows:

$$I * W \approx \alpha(I * \widetilde{W}) \qquad (2)$$

where I is the input of the network, α is the scaling factor, W is floating-point full-precise weights and $\widetilde{W}$ is quantized ternary weights which are obtained as follows:

$$\widetilde{W}_i = f(W_i, \Delta) = \begin{cases} +1 & if \quad W_i > \Delta \\ 0 & if \quad |W_i| < \Delta \\ -1 & if \quad W_i < -\Delta \end{cases} \qquad (3)$$

Here Δ is a positive threshold parameter and it can be computed as follows:

$$\Delta^* \approx 0.75E(|W|) \approx \frac{0.75}{n}\sum_{i=1}^{n}|W_i| \qquad (4)$$

Based on the result above, the optimal α can be computed as followed:

$$\alpha_\Delta^* = \frac{1}{|I_\Delta|}\sum_{i\in I_\Delta}|W_i| \qquad (5)$$

where $I_\Delta = \{i\,|\,|W_i| > \Delta\}$ and $|I_\Delta|$ represents the number of elements in this set. For detailed derivations of the above formulations, readers may refer to [5], where valuable contributions were made in the field of ternary weight quantized networks.

In SNNs, the output of convolutional layers is fed into activation neurons (Integrate-and-Fire model, IF or Leaky Integrate-and-Fire model, LIF), which convert continuous real-valued membrane potentials back into spikes. We adjust the neuronal firing threshold by dividing it by the scale factor α from (5), followed by fixed-point quantization, to establish a new threshold voltage. Crucially, during actual computation, the weights employed are restricted to {1, 0, -1} rather than using α and -α.

$$activation = I * \widetilde{W} \qquad (6)$$

$$U_{thr}^* = (U_{thr}/\alpha)_{quan} \qquad (7)$$

As previously discussed, accumulated images of DVS contain values {1, 0, -1}, while network weights also occupy the ternary space {1, 0, -1}. These still require 2-bit storage rather than 1-bit, and their multiplication necessitates 2-bit multipliers, leaving room for further optimization. Our approach builds upon three key observations:

1) For DVS outputs, we prioritize motion contours over event polarity.

2) In SNNs, negative activations contribute minimally to neuron (IF or LIF) firing.

3) When convolutional outputs exceed the neuron's threshold voltage, their firing impact approximates that of inputs precisely at threshold.

First, based on observation 1, we replace all -1 values in DVS spiking images with 1. This negligibly affects attention prediction performance while substantially reducing image storage and multiplication hardware requirements.

Additionally, leveraging observations 2 and 3, we truncate MAC results before neuronal integration. We discard negative values of MAC results enabling sign-bit elimination in

hardware and clamp outputs exceeding threshold voltage to the threshold value. This dual truncation allows finite-bit representation of results (4 bits for a 3×3 conv kernel), reducing hardware power consumption and computation cycles. It can be confirmed that these operations introduce negligible impact on network performance while significantly saving hardware resources for storing convolutional results and implementing neuronal activation through experiments.

III. HARDWARE IMPLEMENTATION AND TIMING

Based on Section II, we naturally derive a concise hardware deployment scheme. Fig. 4 illustrates the system's overall work flow:

1) DVS Event Preprocessing: Raw DVS event streams are transformed into spike frames containing only {0,1} values through accumulation and other preprocessing.

2) Attention Prediction: These frames are fed into a lightweight spiking neural network, generating real-time attention prediction maps.

3) SPAD Gating: Attention heatmap controls activation of SPAD sensors and associating network while multiplying the raw SPAD spiking image with the upsampled and binarized attention map.

4) Target Detection: Processed images enter a spiking convolutional spiking neural network for object detection.

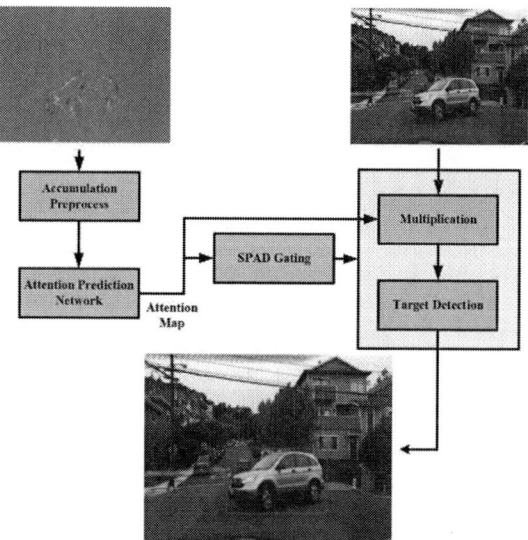

Fig. 4. Overall work flow diagram of the system.

The extensive MAC operations in convolution use abundant AND gates rather than multi-bit multipliers. Results are tallied by counters and compared against a fixed-point threshold voltage. Given the predominantly low-bit-width spike-based computations, pipeline implementation requires fewer registers, avoiding frequent DDR memory accesses for intermediate data. This design yields faster processing and lower power consumption.

The timing schematic for the system is shown in Fig. 6. The SPAD sensor and its corresponding processing network are controlled by an enable signal generated from the result of attention prediction map. Crucially, the SPAD subsystem enters an off state when no targets require detection, thereby conserving power.

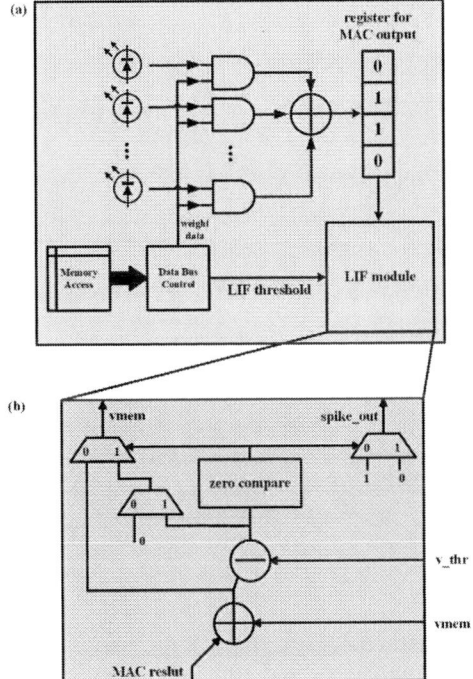

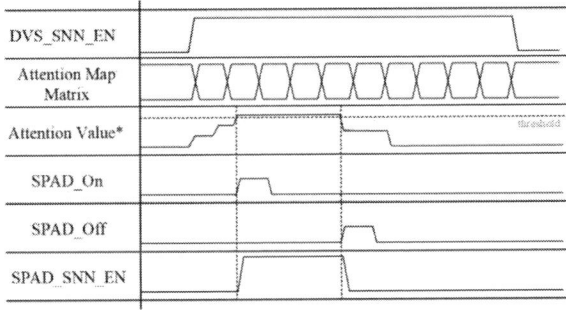

Fig. 5. Architecture of hardware deployment. (a) Architecture of simplified MAC units. (b) Detailed structure of LIF module.

DVS_SNN_EN

Attention Map Matrix

Attention Value*

SPAD_On

SPAD_Off

SPAD_SNN_EN

Fig. 6. Timing schematic of Attention Network's Gating Control over SPAD. (*The Attention Value is computed by multiplying sum of elements of the attention map matrix by specified scaling factor.)

TABLE I. PERFORMANCE COMPARISON ON DDD20[7]

Methods	Precision	mAP	Param storage
YOLOv4-Tiny (frame images)	85.9%	77.2%	26M(float32)
YOLOv5m (frame images)	95.2%	80.1%	21M(float32)
OURS (SPAD only)	84.1%	74.9%	15M(2bit)
OURS (SPAD+DVS)	92.4%	79.6%	19M(2bit)

IV. SIMULATED RESULTS

We selected two datasets to evaluate the effectiveness of our proposed attention prediction algorithm. DND21[6] is the dataset for DVS background denoising and used by us for our attention prediction network. DDD20[7] dataset includes event streams and frame images of driving scenarios and frame images undergo pulse-frequency modulation (PFM) processing to emulate SPAD imaging results. We use the measure of precision, mAP, and parameter storage to evaluate the performance of our method and YOLO and the results are shown in Table I. Empirical testing shows our method achieves comparable performance to YOLO network and significantly reduces the parameter count.

A FPGA-implemented version of our algorithm is being deployed, integrated with physical DVS and SPAD sensor chips to enable real-world system performance testing. This practical validation will further demonstrate our method's distinct advantages for resource-constrained edge hardware scenarios. These implementation details will be detailed in future publications.

V. SUMMARY

In this paper, we proposed a full-spiking bio-inspired target detection algorithm based on attention prediction for DVS and SPAD. By leveraging DVS data for attention prediction, we significantly reduce the computational load of neural networks in target detection, consequently decreasing the overall network parameter count. Furthermore, capitalizing on the spiking characteristics of the sensors, we achieve full-spiking implementation via ternary quantization, resulting in low power consumption and hardware-friendly deployment. Simulation on the DDD20 dataset empirically demonstrate our method's effectiveness in reducing model complexity while enabling deployment across diverse resource-constrained hardware platforms.

ACKNOWLEDGMENT

This work is supported by National Key Research and Development Program of China (2024YFE0201500), the National Natural Science Foundation of China under Grant 62274154 and U21A20504.

REFERENCES

[1] S. Bianco, R. Cadene, L. Celona and P. Napoletano, "Benchmark Analysis of Representative Deep Neural Network Architectures," in *IEEE Access*, vol. 6, pp. 64270-64277, 2018.

[2] C. Posch, T. Serrano-Gotarredona, B. Linares-Barranco and T. Delbruck, "Retinomorphic Event-Based Vision Sensors: Bioinspired Cameras with Spiking Output," in *Proceedings of the IEEE*, vol. 102, no. 10, pp. 1470-1484, Oct. 2014.

[3] G. Gallego, "Event-based vision: A survey," 2019, arXiv:1904.08405.

[4] D. Bronzi, F. Villa, S. Tisa, A. Tosi and F. Zappa, "SPAD Figures of Merit for Photon-Counting, Photon-Timing, and Imaging Applications: A Review," in *IEEE Sensors Journal*, vol. 16, no. 1, pp. 3-12, Jan.1, 2016.

[5] F. Li, B. Liu, X. Wang, B. Zhang, J. Yan, "Ternary Weight Networks," 2022, arXiv:1605.04711v3.

[6] S. Guo and T. Delbruck, "Low Cost and Latency Event Camera Background Activity Denoising," in *IEEE Transactions on Pattern Analysis and Machine Intelligence*, vol. 45, no. 1, pp. 785-795, 1 Jan. 2023.

[7] Y. Hu, J. Binas, D. Neil, S. -C. Liu and T. Delbruck, "DDD20 End-to-End Event Camera Driving Dataset: Fusing Frames and Events with Deep Learning for Improved Steering Prediction," *2020 IEEE 23rd International Conference on Intelligent Transportation Systems (ITSC)*, Rhodes, Greece, 2020, pp. 1-6.

The Quest for Reliable AI Accelerators: Cross-Layer Evaluation and Design Optimization

Meng Li[123*], Tong Xie[21], Zuodong Zhang[4], and Runsheng Wang[234*]

[1]Institute for Artificial Intelligence & [2]School of Integrated Circuits, Peking University, Beijing, China
[3]Beijing Advanced Innovation Center for Integrated Circuits, Beijing, China
[4]Institute of Electronic Design Automation, Peking University, Wuxi, China

Abstract—As the CMOS technology pushes to the nanoscale, aging effects and process variations have become increasingly pronounced, posing significant reliability challenges for AI accelerators. Traditional guardband-based design approaches, which rely on pessimistic timing margin, sacrifice significant performance and computational efficiency, rendering them inadequate for high-performance AI computing demands. Current reliability-aware AI accelerator design faces two core challenges: (1) the lack of systematic cross-layer analysis tools to capture coupling reliability effects across device, circuit, architecture, and application layers; and (2) the fundamental trade-off between conventional reliability optimization and computational efficiency. To address these challenges, this paper systematically presents a series of reliability-aware accelerator designs, encompassing (1) aging and variation-aware dynamic timing analyzer, (2) accelerator dataflow optimization using critical input pattern reduction, and (3) resilience characterization and novel architecture design for large language models (LLMs). By tightly integrating cross-layer reliability modeling and AI workload characteristics, these co-optimization approaches effectively achieve reliable and efficient AI acceleration.

I. INTRODUCTION

Recently, AI has emerged as a game-changing technology and has revolutionized many different applications, including reliability-critical tasks such as autonomous driving etc. With high computational requirements, AI workloads, particularly large language models (LLMs), have been widely deployed on customized accelerators such as TPU-like systolic arrays. However, as Moore's law pushes to the nanoscale, these accelerators are susceptible to hardware faults. These faults, including timing errors, arise from various reliability issues, such as process, voltage, temperature, and aging (PVTA) variations [1]–[4]. Therefore, it is crucial to enable both reliable and efficient AI acceleration.

However, ensuring reliability in AI accelerators presents significant challenges. First, reliability issues exhibit strong cross-layer coupling and lack accurate modeling. For example, at the device level, voltage stress induces threshold-voltage shifts (ΔV_{th}); at the circuit level, these shifts manifest as specific timing-error patterns and timing error rates (TERs); and at the application level, such errors degrade AI model accuracy (Acc). These relationships are highly complex and difficult

This work was supported in part by NSFC (62495102, 92464104, 62125401), National Key Research and Development Program (2024YFB4505004), Beijing Municipal Science and Technology Program (Z241100004224015), Beijing Outstanding Young Scientist Program (JWZQ20240101004), and the 111 Project (B18001).

*Corresponding author: {meng.li,r.wang}@pku.edu.cn

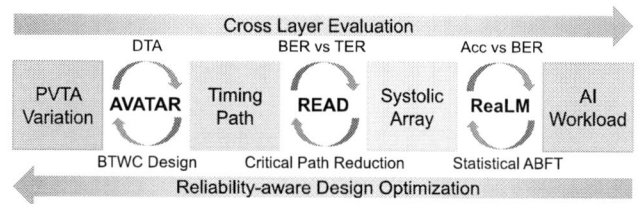

Fig. 1. Overview of our recent works.

to model precisely. Second, existing mitigation methods fail to balance reliability and efficiency. Worst-case voltage margins eliminate timing errors but incur heavy overhead, while reduced margins increase bit error rates (BERs) and degrade model performance. Classical techniques such as Razor flip-flops [5] and algorithm-based fault tolerance (ABFT) [6] are either unscalable to large accelerators or incur frequent yet unnecessary recovery, making them impractical for compute-intensive AI workloads. To address these challenges, this paper summarize our recent efforts for cross layer evaluation and design optimization techniques [7]–[9], as depicted in Fig. 1.

II. AGING- AND VARIATION-AWARE DTA

A. Motivation

As technology scaling amplifies timing guardbands, *better-than-worst-case* (BTWC) design seeks to improve efficiency by exploiting dynamic timing margins. Conventional dynamic timing analysis (DTA) tools add worst-case aging and variation guardbands, inflating margins and diminishing BTWC gains. AVATAR [7] addresses this by incorporating transistor aging and process variation directly into DTA, eliminating unnecessary guardbands and preserving performance and efficiency.

B. AVATAR Algorithm

The core idea of AVATAR is to integrate aging analysis into the DTA process, as illustrated in Fig. 2. The algorithm consists of three main steps: (1) gate-level aging and variation model characterization, (2) workload analysis to estimate ΔV_{th} for each transistor, and (3) event-based DTA.

In Step 1, the aging model takes as inputs the timing arc, input slew, output load, per-transistor ΔV_{th}, V_{DD}, and temperature, and outputs aged cell delay and transition time using a first-order Taylor expansion. The variation model is based on POCV analysis with LVF data. In Step 2, zero-delay gate-level simulation is performed to obtain the toggle rate of each internal net, after which the aging model is

979-8-3315-3918-4/25 $31.00 © 2025 IEEE

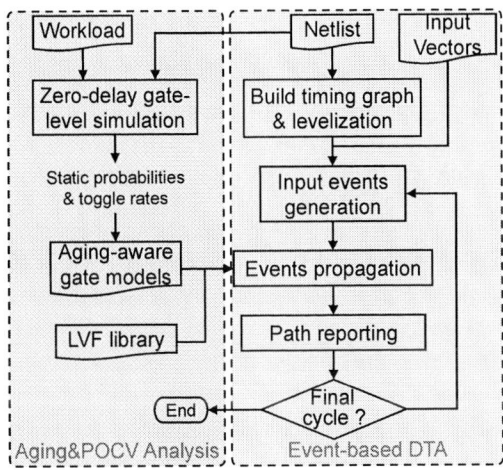

Fig. 2. Aging- and variation-aware DTA flow of AVATAR.

TABLE I
PERFORMANCE IMPROVEMENT FROM APPLICATION-BASED DVFS BASED
ON THE CORNER-BASED DTA AND AVATAR

Benchmark	Corner-based DTA [10], [11]		AVATAR (this work)	
	Max Freq. (MHz)	Impro. (vs STA)	Max Freq. (MHz)	Impro. (vs STA)
SHA	948	13.75%	1020	22.38%
AES_CBC	883	5.99%	951	14.10%
FIR	915	9.82%	986	18.35%
BubbleSort	1290	55.38%	1380	65.36%
Motion_Detection	958	15.00%	1030	23.97%
CNN	868	4.18%	936	12.30%
Convolution	868	4.19%	936	12.28%
2d_Filter	936	12.33%	1050	26.37%
MatrixMult	916	9.89%	989	18.63%
DCT	1170	40.77%	1270	52.15%

applied to compute the ΔV_{th} of each transistor. In Step 3, events are defined as digital switching activities on pins. Event-based DTA begins by constructing a timing graph for event propagation, followed by cycle-by-cycle analysis that includes input event generation, event propagation, and path reporting, with the arrival times at all timing endpoints recorded.

C. Experiments

To evaluate the benefits of AVATAR, we determine the application-specific $V_{\min}/f_{\max}$ using two methods. (1) Corner-based DTA with extra guardbands [10], [11]. Dynamic delay is computed as $delay \times (1 + total\_guardband)$, where the aging guardband is assumed to be 15% and the random variation guardband 5% at nominal V_{DD}. The trend of guardband versus V_{DD} is characterized using FO4 delay as a representative cell. (2) AVATAR-based analysis. AVATAR inherently accounts for both aging and variation, eliminating the need for extra guardbands. The final delay is calculated as $\mu(delay) + 3\sigma(delay)$. Table I shows the performance improvement by applying the maximum frequency at the nominal 0.8V.

III. CRITICAL INPUT PATTERN REDUCTION

A. Computing Sequence and Timing Error

READ [8] proposes a reliability-enhanced accelerator dataflow optimization technique that can effectively reduce

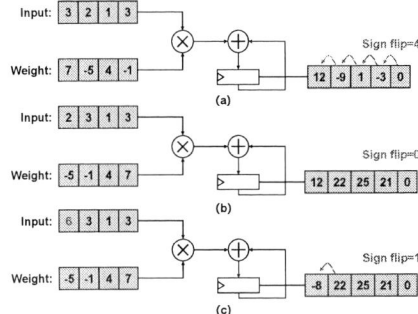

Fig. 3. A 1×4 convolution calculated in different orders. Reordering weights does not change the computing result, but avoids the critical input pattern.

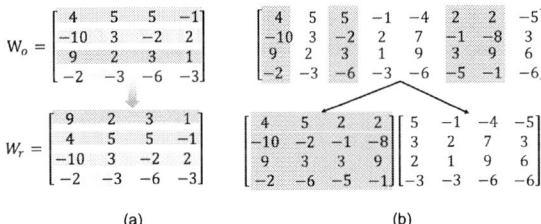

Fig. 4. (a) Input channel reordering. (b) Output channel clustering.

timing errors. For a given cycle, timing errors are primarily determined by two factors: (1) the input pattern, which dictates the activated paths, and (2) the operating conditions (e.g., temperature, voltage, aging) that influence the delay of these paths. While operating conditions are often uncontrollable, only a small subset of paths, i.e., critical paths, typically exceed the clock period. Thus, reducing the activation rate of these critical paths is key to lowering the timing error rate.

The basic processing element (PE) in our study is a multiplier-and-accumulator (MAC) unit. Using the 8-bit multiplier and 24-bit accumulator of a TPU MAC as an example, dynamic timing analysis of the synthesized unit reveals that the most frequent critical input patterns are those that flip the sign bit of the partial sum (PSUM). For instance, computing $3 \times (-2) + 2 = -4$ produces the 2's complement encoding 111111111111111111111100, where the sign bit flip triggers a long carry chain in the accumulator, activating critical paths. A similar effect occurs when PSUM transitions from negative to positive. Consequently, reducing the PSUM sign-flip rate is an effective strategy for minimizing timing error rates.

B. Input Channel Reordering and Output Channel Clustering

Given that the ReLU function predominantly results in non-negative outcomes, our heuristic prioritizes non-negative weight computations, as shown in Fig. 4 (a). This is achieved via a channel-wise reordering algorithm that sorts operations by the fraction of positive weights within the same channel.

To enhance effectiveness when the number of columns A_c is large, we adopt a *cluster-then-reorder* strategy: output channels are first clustered, and then input channels within each cluster are reordered. As illustrated in Fig. 4 (b), instead of directly segmenting W, we first cluster the output channels into groups and then segment W into two sub-matrices. Clus-

979-8-3315-3918-4/25 $31.00 © 2025 IEEE

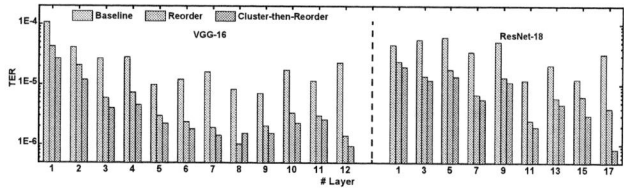

Fig. 5. Timing error rate comparison for different reliability-enhanced algorithms on ResNet-18 and VGG-16.

tering groups channels with similar sign patterns facilitates input-channel reordering within each sub-matrix.

Formally, we define the **sign difference (SD)** between two n-dimensional vectors $\mathbf{x}$ and $\mathbf{y}$ as the Manhattan distance of their sign vectors. Minimizing the SD within each sub-matrix simplifies input-channel reordering and reduces sign bit flips. The output-channel clustering problem is thus formulated as minimizing $SD(W_{T_i})$ across all clusters, and we adopt the balanced KNN on the weight sign matrix by the Manhattan metric to solve this problem.

C. Experiments

As shown in Fig. 5, the direct reordering and cluster-then-reorder algorithms achieve average TER reductions of $4.9\times$ and $7.8\times$, respectively. The cluster-then-reorder approach consistently delivers greater reductions across most layers and exhibits superior performance in later network layers, where the number of output channels is larger.

IV. STATISTICAL ALGORITHM-BASED FAULT TOLERANCE

A. LLM Resilience Characterization

ReaLM [9] characterizes the error resilience of LLMs and proposes statistical ABFT to adaptively corrects only those critical errors. In this section, we aim to answer six questions. **Q1.1**: How does resilience vary across different layers of LLMs? **Q1.2**: What is the bit-wise resilience of LLMs? **Q1.3**: How does resilience vary among different components within LLMs during the prefill stage? **Q1.4**: What is the correlation between fault magnitude and frequency in impacting LLM performance? **Q2.1**: How does the resilience of LLMs compare between the prefill and decode stages? **Q2.2**: How does resilience vary among different computational components within LLMs during the decode stage?

Fig. 6 illustrates the resilience characterization for these questions. For **Q1.1**, *while the layers exhibit comparable resilience behaviors, the earlier layers are more vulnerable to errors.* For **Q1.2**, we conclude that *while errors at lower bits have a negligible impact on model performance, errors at higher bits can reach a saturation point due to re-quantization.*

For **Q1.3**, we find that *network components followed by normalization operations exhibit much worse resilience.* On the other hand, components like QKV are much more resilient. So, we can divide these components into two categories: the sensitive components, such as O and Down projection, and the resilient components, such as QKV.

Q1.4 is about the error magnitude and frequency. As depicted in Fig. 6 (g)(h), we find that given a total sum of

errors, as the error frequency increases and error magnitude decreases, the resilient components exhibit a non-monotonic resilience behavior. *Resilient components can tolerate both sporadic large and frequent small errors. But for sensitive components, even a few large errors can severely degrade performance.* This finding aligns with our earlier observations.

For **Q2.1** and **Q2.2**, Fig. 6 (i) and (j) reveal that *the prefill stage exhibits greater sensitivity to errors than the decode stage.* And Fig. 6 (k) and (l) further confirm that *the resilient and sensitive components remain consistent across both stages.*

B. Statistical Algorithm-Based Fault Tolerance

Our statistically based error detection strategy aims to correct only those critical errors. Fig. 7 rephrases the tradeoff between error magnitude and frequency. Given a specific task, we first define an acceptable performance degradation threshold. This threshold defines a critical region where computational errors significantly impact LLM performance. The recomputation process is only triggered if the observed errors fall inside the critical region, thereby avoiding unnecessary error recovery process.

Fig. 8 illustrates the design of our statistical ABFT circuit. We first take the weight stationary dataflow as an example. A column of PEs is added on the right, storing weight checksums for computing $e^T W X$. Meanwhile, a row of adders is appended at the bottom, which accumulates the output checksum. In addition, we introduce a statistical unit to capture the error statistics and determine the need for recovery. Once the observed errors fall inside the critical region, the recomputation process will be triggered. We also support the output stationary dataflow. A column of adders is appended to the left side to compute the weight checksum, and a row of PEs is added at the bottom to calculate the output checksum.

C. Experiments

For circuit overhead comparison, compared to the unprotected systolic array, our design introduces only about 1.4% area and 1.8% power overhead, making it lightweight and practical for real-world deployment. We then evaluate the LLM performance and energy savings in Fig. 9. We take K in OPT-1.3B on WikiText-2 and V projection in LLaMA-3-8B on HellaSwag as examples. For the OPT-1.3B, our design achieves a sweet point at 0.72 V and saves 23% energy compared to prior-art methods. And for LLaMA-3-8B, it finds a sweet point at 0.70 V and saves 24% energy.

V. CONCLUSION

Reliability issues in AI accelerators arise from cross-layer coupling across the device, circuit, architecture, and application levels. Accurate reliability modeling can significantly reduce overly conservative design margins, while cross-layer co-optimization effectively mitigates reliability constraints. Many hardware faults may be masked, and AI applications are inherently fault-tolerant, eliminating the need to design for worst-case conditions. Allowing limited timing violations thus opens opportunities for substantial improvements in power, performance, and area.

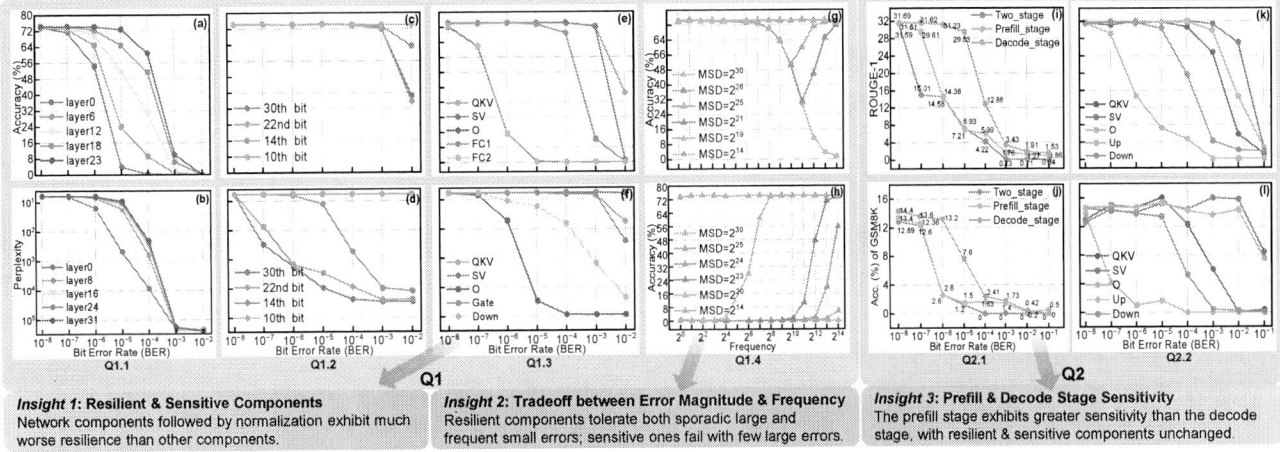

Insight 1: Resilient & Sensitive Components
Network components followed by normalization exhibit much worse resilience than other components.

Insight 2: Tradeoff between Error Magnitude & Frequency
Resilient components tolerate both sporadic large and frequent small errors; sensitive ones fail with few large errors.

Insight 3: Prefill & Decode Stage Sensitivity
The prefill stage exhibits greater sensitivity than the decode stage, with resilient & sensitive components unchanged.

Fig. 6. **Q1.1**: (a)(b) Layer-wise resilience of different LLMs on different tasks. **Q1.2**: Bit-wise error resilience. (c) Error injection on K. (d) Error injection on O. **Q1.3**: (e)(f) Sensitivity to errors in different LLM components. **Q1.4**: Relationship between error frequency and magnitude. (g) Resilient components like K. (h) Sensitive components like O Given MSD, the error magnitude decreases as the error frequency increases. **Q2.1**: (i)(j) Comparison between the prefill stage and decode stage. **Q2.2**: (k)(l) Impact of error injection across network components: O and Down remain highly sensitive. (a)(c)(e)(g)(h) are evaluated with OPT-1.3B on LAMDABA; (b)(d)(f) with LLaMA-2-7B on WikiText-2; (i)(k) with LLaMA-2-7B on X-Sum; (j)(l) with LLaMA-2-7B on GSM8K.

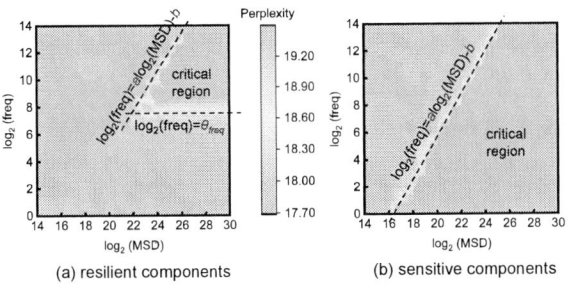

(a) resilient components   (b) sensitive components

Fig. 7. Our statistical ABFT strategy only corrects errors falling inside the critical region.

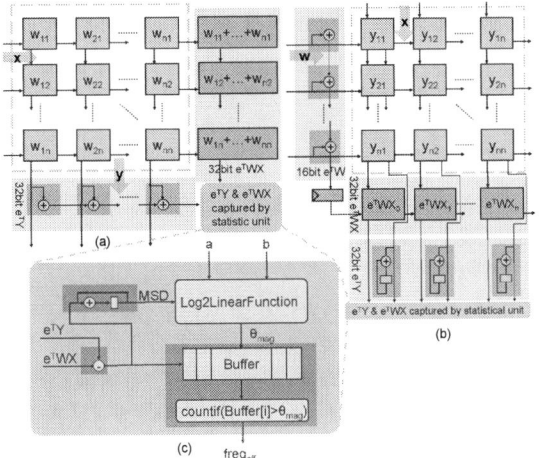

Fig. 8. Architecture design of statistical ABFT on systolic array: (a) ABFT implementation for weight stationary dataflow; (b) ABFT implementation for output stationary dataflow; and (c) customized statistical units.

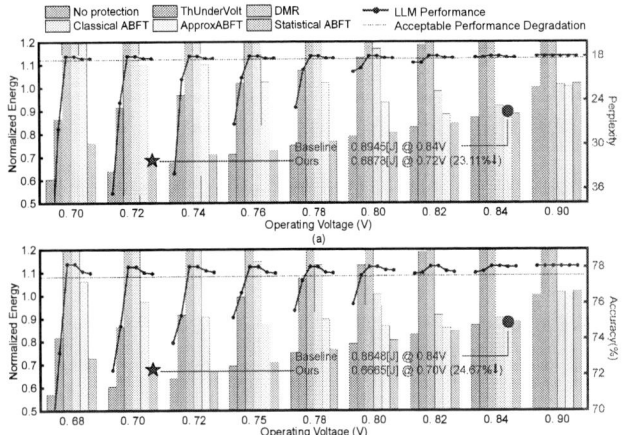

Fig. 9. LLM performance and total energy savings comparison. (a) OPT-1.3B on WikiText-2. (b) LLaMA-3-8B on HellaSwag. Our method achieves competitive performance with minimal protection overhead, significantly reducing costs versus existing methods, with maximal energy savings at 0.72V and 0.70V, respectively. The dash line marks the acceptable performance degradation.

REFERENCES

[1] R. Huang *et al.*, "Variability-and reliability-aware design for 16/14nm and beyond technology," in *Proc. IEDM*, pp. 12–4, IEEE, 2017.

[2] R. Wang *et al.*, "Can emerging computing paradigms help enhancing reliability towards the end of technology roadmap?," in *Proc. IRPS*, o. 1–7, IEEE, 2021.

[3] R. Wang *et al.*, "Cross-layer design for reliability in advanced technology nodes: An eda perspective," in *Proc. ICSICT*, pp. 1–4, IEEE, 2022.

[4] Z. Sun *et al.*, "Challenges of design for reliability in advanced cmos technology: From single-mode to mixed-mode mechanisms," in *Proc. IC-SICT*, pp. 1–1, IEEE, 2024.

[5] D. Ernst *et al.*, "Razor: A low-power pipeline based on circuit-level timing speculation," in *Proc. MICRO*, pp. 7–18, IEEE, 2003.

[6] K.-H. Huang *et al.*, "Algorithm-based fault tolerance for matrix operations," *IEEE Transactions on Computers*, pp. 518–528, 1984.

[7] Z. Zhang *et al.*, "Avatar: an aging-and variation-aware dynamic timing analyzer for application-based dvafs," in *Proc. DAC*, pp. 841–846, 2022.

[8] Z. Zhang *et al.*, "Read: Reliability-enhanced accelerator dataflow optimization using critical input pattern reduction," in *Proc. ICCAD*, pp. 1–9, IEEE, 2023.

[9] T. Xie *et al.*, "Realm: Reliable and efficient large language model inference with statistical algorithm-based fault tolerance," in *Proc. DAC*, pp. 703–709, 2025.

[10] H. Cherupalli *et al.*, "Exploiting dynamic timing slack for energy efficiency in ultra-low-power embedded systems," in *Proc. ISCA*, 2016.

[11] J. Constantin *et al.*, "Exploiting dynamic timing margins in microprocessors for frequency-over-scaling with instruction-based clock adjustment," in *Proc. DATE*, pp. 381–386, 2015.

979-8-3315-3918-4/25 $31.00 © 2025 IEEE

A Lightweight Hardware Defense Against DSE-Based Trojans in NN Accelerators

Yujing Wu, Chao Guo, Youhua Shi*

School of Fundamental Science and Engineering, Waseda University, Tokyo, Japan
*Email: yujing_wu@akane.waseda.jp, youhua.shi@islab.cs.waseda.ac.jp

Abstract—The emergence of hardware Trojans (HTs) auto-generated by compromised Design Space Exploration (DSE) frameworks poses a new class of stealthy, on-chip threats to neural network (NN) accelerators. To address this vulnerability, for which no practical defense exists, we propose Kernel Integrity Verification (KIV), the first run-time hardware countermeasure designed to neutralize such attacks. KIV employs a lightweight, side-channel resistant verification core to perform targeted, in-situ integrity checks on a few pre–identified vulnerable weights immediately before their use in computation. Experimental results on a state-of-the-art accelerator confirm that KIV effectively mitigates on-chip parameter tampering caused by hardware Trojans. For VGG-11 and ResNet-18, the classification accuracy is restored from 0.11% and 0.1% to 70.29% and 69.51%, respectively. The area overhead of KIV remains below 0.41% of the overall design.

Index Terms—Hardware Trojan (HT), Neural Network Accelerators, Design Space Exploration (DSE), Hardware Security.

I. INTRODUCTION

Automated AI-accelerator generation frameworks such as Vitis AI [1] and Gemmini [2] have emerged to bridge the gap between high-level neural-network models and efficient hardware designs, dramatically lowering the bar for deploying custom ASICs and FPGAs. Yet, despite their growing adoption in both industry and academia, the security of these end-to-end toolchains remains almost entirely unexplored.

Recent work showed that an attacker who inserts malicious code into the design space exploration (DSE) stage can stealthily tag vulnerable convolution parameters and embed a dormant hardware Trojan (HT) [3]. Once triggered, this HT forces the accelerator to misclassify almost all inputs into a single attacker-chosen label (an "N-to-1" attack). A follow-up study formulated a unified threat model demonstrating equivalent compromises across both pipeline-based and systolic-array architectures [4]. These findings reveal that—in the absence of dedicated safeguards—today's AI-hardware compilers can themselves become powerful vectors for hardware-level attacks.

Embedding hidden Trojans directly into the generated RTL represents a fundamentally new attack paradigm. Unlike earlier off-chip assaults (e.g., bit-flip schemes [5], [6]), this approach tightly couples malicious algorithms with on-chip logic, offering far greater flexibility and stealth. Consequently, standard defenses–whether based on cryptographic hashes or checksums [7]–[9] or on model-level anomaly detectors [10]–are powerless against such intrusions. To date, no mitiga-

This research is partially supported by the Telecommunications Advancement Foundation.

tion strategy has been proposed to secure these automated accelerator-generation flows against Trojan insertion.

This paper introduces Kernel Integrity Verification (KIV), the first hardware-level countermeasure specifically designed to neutralize DSE based attacks. KIV is a lightweight, run-time module that provides flexible, tiered defense postures—from a blind hash-based verification to a full-knowledge, side-channel resistant comparison—to surgically verify the integrity of a few highly sensitive kernel weights. Critically, KIV can perform on-the-fly data correction, enabling near-total accuracy recovery and ensuring operational resilience even while under attack. We summarize our key contributions as follows:

- We propose KIV, the first hardware-level defensive framework specifically engineered to counteract HT embedded by compromised DSE toolchains. Our architecture provides run-time, in-situ detection and features a configurable mechanism for on-the-fly data correction.
- We introduce a novel tiered security architecture that enables the defense to be tailored to different trust models. We define and implement two distinct postures: a full-knowledge defense employing dynamic masking obfuscation for maximum side-channel resistance and a blind defense based on securely provisioned hash verification for low-trust scenarios.
- We validate KIV on a state-of-the-art accelerator, demonstrating its ability to mitigate the impact of HT at runtime and restore model accuracy to near-original levels, with minimal hardware overhead and no impact on overall latency.

II. THREAT MODEL AND SECURITY PRELIMINARIES

This section first outlines the DSE-based HT attack, which is the threat our work addresses. Next, we define the threat model.

A. Threat Landscape: DSE-Based Hardware Trojans

Recent work has demonstrated that by injecting malicious code into the DSE stage of an automated AI-accelerator synthesis pipeline, attackers can stealthily identify the convolutional weights whose perturbation induces the largest increase in a chosen target-class confidence. When combined with a hardware Trojan, this DSE-based manipulation enables an N-to-1 misclassification attack, in which the majority of input images are misclassified into the attacker's chosen category at deployment time [3], [4]. Two prominent methods, Sensitive Filter Exploration (SFE) and Cross-layer Sensitive Filter

Exploration (C-SFE), formalize this search as

$$\widehat{W}^* = \arg\max_{\widehat{W}} \sum_{i=1}^{I} \big(f_{\text{target}}(x_i, \widehat{W}) - f_{\text{target}}(x_i, W)\big),$$
$$\text{s.t.} \quad \widehat{W}_l \in [\min(W_l), \max(W_l)], \; \forall l$$

$$(1)$$

where $f_{\text{target}}(x_i, W)$ is the model's confidence in the target class for input x_i under parameters W. After the attack, the parameter set $\widehat{W}$ differs from W in that N weights have been replaced with new values derived using heuristic algorithms such as Evolutionary Algorithms (EA). Here, l denotes the index of the model layer. Depending on the accelerator architecture, the Trojan may be confined to a single convolutional stage–minimizing hardware overhead in pipelined designs– or spread across layers in systolic-array implementations to exploit logic reuse.

B. Threat Model

Our threat model, illustrated in Figure 1, considers an attack vector where the hardware generation platform itself is compromised. A malicious actor can exploit the hardware build flow to inject a stealthy HT into the final VLSI design, using the user's own model as a template. At runtime, this HT activates within the deployed accelerator to intercept and manipulate critical kernel weights, causing targeted misclassifications.

To counter this, we introduce the KIV module. It performs real-time integrity checks on the weight data stream, operating under one of two user-defined security postures: a 'Blind' verification against pre-provisioned fingerprints or a 'Full knowledge' check using dynamic masking. Upon detecting a malicious alteration, KIV ensures output integrity by raising a security alert or, optionally, performing on-the-fly data correction.

III. DESIGN AND IMPLEMENTATION DETAILS

A. Architectural Integration of the KIV Module

Depending on the attacker's capabilities, hardware Trojans can be inserted into various hardware units within the accelerator. For example, as illustrated in Figure 2, the attack described in [4] targets the memory transfer unit by injecting modifications immediately before weights are written to on-chip memory. To block this attack, we integrate KIV directly into the systolic-array compute unit, co-locating the verification logic immediately before the weights are used for computation. This arrangement performs in-situ integrity checks at the last possible moment. This fine-grained placement provides two key advantages: 1) It ensures that the sensitive weights are validated immediately before computation, neutralizing any Trojan along the entire data path from off-chip memory. 2) It incurs minimal latency and does not increase the system's overall delay.

B. Tiered Defense Architecture and Configuration

To accommodate diverse security requirements, we designed the KIV framework with a tiered defense architecture, enabling a system designer to provision different security IPs based on the end-user's trust model and requirements.

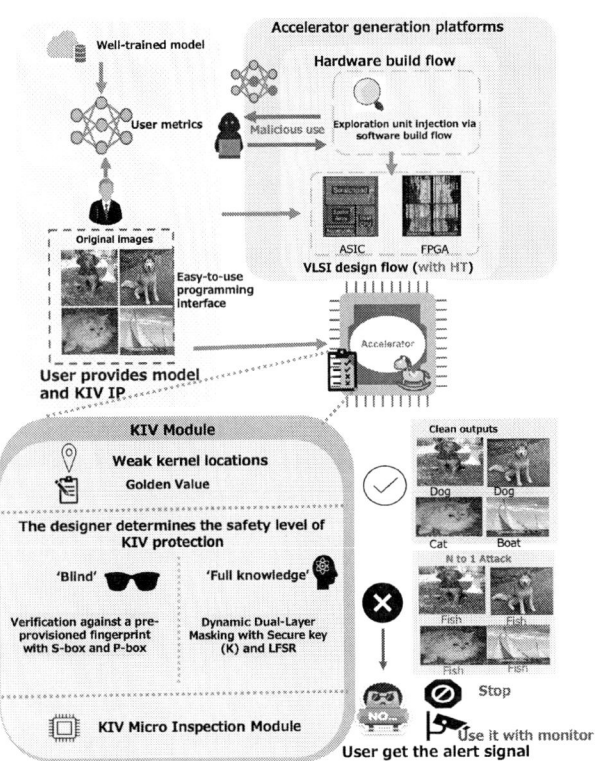

Fig. 1. The overall architecture of our proposed framework. It illustrates the threat model, where a malicious user can inject Trojans during the hardware design flow, and our defense mechanism, the KIV (Kernel Integrity Verification) module. The KIV module provides two configurable defense postures ('Blind' and 'Full knowledge') to detect attacks and ensure output integrity during runtime.

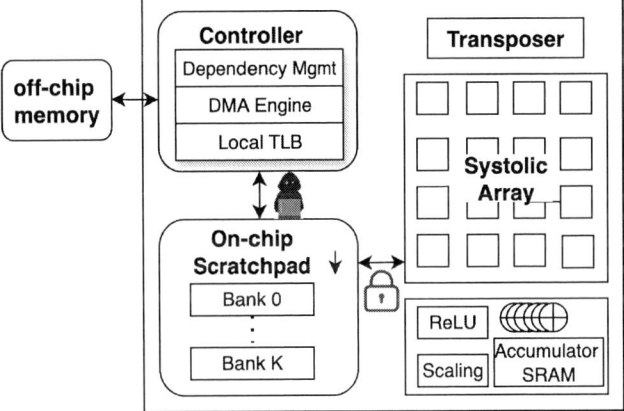

Fig. 2. Strategic placement of the KIV defense module within a systolic array architecture.

1) Defense Posture 1: Full-Knowledge (Location and Value Trust)

At the highest level of trust, the defender possesses complete knowledge of the final model, including the exact correct values of the vulnerable weights, denoted by G. In this scenario, the defense employs dynamic masking obfuscation comparison. The hardware architecture of this verification core is depicted in Figure 3. The KIV module stores an obfuscated

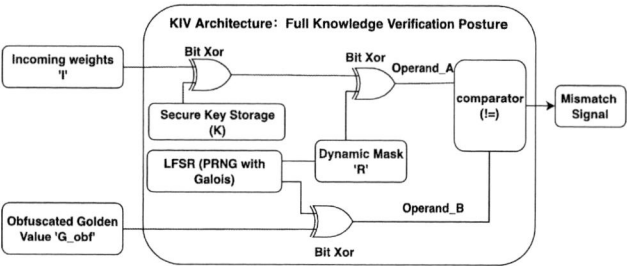

Fig. 3. Hardware architecture of the KIV verification core for the Full-Knowledge defense posture.

version of the correct values of the vulnerable weights. This can be written as $G_{obf} = G \oplus K$, where K is a pre-defined secret key. At runtime, it performs an integrity check shielded by a dynamic mask R generated by a Linear Feedback Shift Register (LFSR), ensuring the comparison is a randomized, data-dependent operation rather than a direct check on a static secret, this can be written as $(I \oplus K) \oplus R \stackrel{?}{=} G_{obf} \oplus R$, where I represents the value of the target weight provided as input at runtime. This posture offers maximal security, defending against reverse engineering by obfuscating the stored data and thwarting side-channel attacks by using the dynamic mask to decouple physical emissions from the secret values during the comparison process.

2) Defense Posture 2: Blind Verification for the End-User

This posture establishes a "blind" operational model for an end-user with low authority, who is prevented from knowing the true values of the model's weights. The core of this defense is a mechanism we call 'Securely Provisioned Hash Verification', which distinctly separates the roles of the Intellectual Property (IP) owner and the end-user.

This posture establishes a "blind" operational model for an end-user, separating the roles of the system-provisioning IP owner and the end-user. The entire workflow is formalized in Algorithm 1. The process begins with a one-time, offline provisioning phase, where the IP owner uses a secure interface to generate and store cryptographic "fingerprints" (hashes) of sensitive weights into a lockable, on-chip register file. During the subsequent real-time Detection phase, the KIV module operates as an autonomous black box for the end-user. It computes a hash from the incoming data stream and verifies it against the pre-provisioned fingerprint, allowing for integrity verification without exposing the secret values. The overall security of this model thus relies on the integrity of the owner's provisioning environment and the collision resistance of the hash function.

The blind verification posture requires a module to generate a unique, collision-resistant "fingerprint" for each kernel weight. For this, we designed a Lightweight Keyed Hash Engine implemented as a compact Substitution-Permutation Network (SPN), detailed in Figure 4. The entire engine is built from purely combinatorial logic, ensuring it introduces no additional clock cycle latency to the data path. Its security is derived from three stages. First, a key-mixing XOR layer $(I \oplus K_{hash})$ prevents offline pre-computation attacks. Second, a non-linear substitution layer, using resource-efficient parallel 4×4 S-Boxes [11], provides cryptographic confusion. Finally,

---

Algorithm 1: KIV Blind Defense Process

Data: IHRF (Integrity Hash Register File): An on-chip, lockable map // Stores trusted fingerprints

Data: K_{hash}: A secret key for the hash engine

Function *Lightweight Hash(data, key)*
    return Computed Hash

Procedure *Provision Fingerprints(Correct Weights Map)*
    // Phase 1: Secure Provisioning **(Offline)**
    foreach (location, G) in Correct WeightsMap **do**
        trusted Hash $\leftarrow$ Lightweight Hash(G, K_{hash})
        IHRF[location] $\leftarrow$ trusted Hash
    Lock(IHRF) // Lock IHRF to be read-only

Function *Verify Integrity(Incoming Weight, Location)*
    // Phase 2: Real-time Detection **(Online)**
    if Location $\in$ dom(IHRF) **then**
        live Hash $\leftarrow$
            Lightweight Hash(Incoming Weight, K_{hash})
        stored Hash $\leftarrow$ IHRF[Location]
        if live Hash $\neq$ stored Hash **then**
            return ATTACK_DETECTED
    return OK

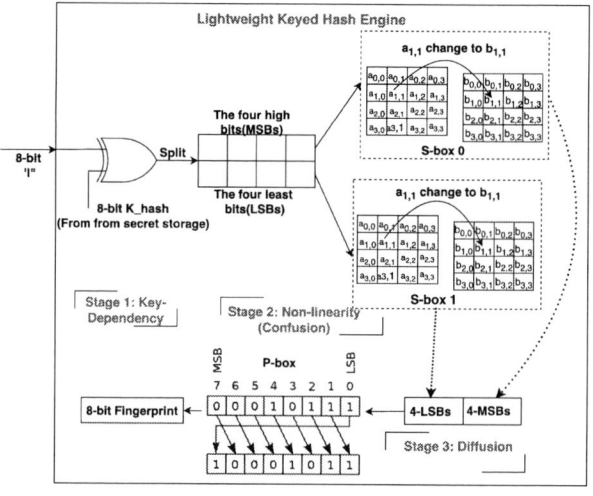

Fig. 4. Internal architecture of the Lightweight Keyed Hash Engine. The 8-bit input is processed through three stages: (1) Keyed XOR for key-dependency, (2) Parallel 4x4 S-Boxes for non-linear confusion, and (3) a P-Box for bit-wise diffusion to generate the final fingerprint.

a bit-permutation P-Box [11] provides diffusion at zero logical cost. This SPN [11] structure makes finding hash collisions computationally infeasible for an adversary without the secret key, thus producing a trustworthy credential for real-time integrity verification.

C. Alarm and Runtime Recovery Procedures

In our work, we adopt both the SFE [3] and C-SFE [4] method presented for sensitive weight identification. When KIV detects anomalies in the weights, we have defined two optional post-processing methods: 1) Alarm-based suspension: pause the accelerator by zeroing out all outputs of the model's final classification layer. 2) Runtime accuracy recovery: inspired by our finding that zeroing out the compromised

979-8-3315-3918-4/25 $31.00 © 2025 IEEE

TABLE I
ACCURACY RECOVERY ON IMAGENET
THE ATTACK MODIFIES 18/132.8M PARAMETERS FOR VGG-11
AND 27/11.7M FOR RESNET-18.

Model	Baseline Acc. (%)	Trojan Acc. (%)	Recovered Acc. (%)
VGG-11	70.38	0.11	70.29
ResNet-18	69.76	0.10	69.51

TABLE II
HARDWARE OVERHEAD COMPARISON OF A BASELINE DESIGN WITH
BOTH KIV DEFENSE POSTURES

Resource	Baseline Design	Posture 1 (Masking)	Posture 2 (Hash)
LUT	174,104	174,510 (+0.23%)	174,819 (+0.41%)
LUTRAM	11,537	11,565 (+0.24%)	11,581 (+0.38%)
FF	125,277	125,440 (+0.13%)	126,612 (+1.07%)
BRAM	601	601 (0%)	601 (0%)
DSP	222	222 (0%)	222 (0%)

weights identified by these methods restored the model's accuracy–which had effectively collapsed (e.g., to only 0.1% on ImageNet)–to nearly its original level (see Section IV-A for details), we simply set any KIV-flagged weights to zero at runtime.

IV. EXPERIMENTAL EVALUATION

A. Defensive Efficacy and Accuracy Recovery

To evaluate our KIV defense framework, we simulated a DSE-HT style attack on VGG-11 [12] and ResNet-18 [13] models pre-trained on ImageNet [14]. The attack is highly precise, modifying a minuscule fraction of weights (18 out of 132.8M for VGG-11 and 27 out of 11.7M for ResNet-18) to be effective. We use Gemmini [2] as our hardware test platform. It features a 16×16 compute array and is deployed on a Xilinx Alveo U50 accelerator card.

Upon detection of an integrity violation by the KIV module, its primary and default response is to raise a security alert and halt further computation. This fail-safe mechanism prevents the accelerator from producing any potentially compromised output. However, for mission-critical applications where operational continuity is paramount, KIV can be configured for a resilience-oriented response. This latter strategy, termed Vulnerable Kernel Ablation, involves setting the values of all pre-identified vulnerable kernels to zero upon a positive detection. The comprehensive results evaluating the efficacy of this specific recovery strategy are summarized in Table I. For example, with the VGG-11 model, when KIV is not deployed, the activation of the hardware Trojan reduces the model's accuracy from 70.38% to 0.11%. When KIV is deployed, it detects the anomalous kernels at runtime and zeros out the tampered ones, successfully restoring the accuracy to 70.29

B. Hardware Overhead Analysis

The hardware overhead for both proposed defense postures is summarized in Table II. The results confirm that both implementations are exceptionally lightweight. Our full-knowledge defense (Posture 1), which utilizes dynamic masking, has a near-zero footprint, with its primary cost being only 163 Flip-Flops (FFs). The more complex blind defense (Posture 2), which requires a hash engine, still incurs a minimal overhead, primarily consuming 715 LUTs and 1335 FFs. Neither posture consumes additional BRAM or DSP blocks, validating KIV as a low-cost solution adaptable to different security requirements.

V. CONCLUSION

This paper introduced Kernel Integrity Verification (KIV), the first hardware-level defense framework against the emerging threat of Trojans embedded by compromised Design Space

Exploration (DSE) toolchains. We presented a lightweight, run-time architecture that offers flexible, tiered security postures to perform in-situ, side-channel resistant integrity verification on critical kernel weights. Our experimental results validate that the KIV framework can effectively neutralize these stealthy attacks and enable near-total accuracy recovery with minimal hardware cost, providing a practical and essential safeguard for the increasingly automated AI hardware design pipeline.

REFERENCES

[1] Xilinx. "Optimal artificial intelligence inference from edge to cloud." https://www.xilinx.com/products/design-tools/vitis/vitis-ai.html (accessed Jan. 5, 2025).

[2] H. Genc et al., "Gemmini: Enabling systematic deep-learning architecture evaluation via full-stack integration," in Proc. 2021 58th ACM/IEEE Des. Automat. Conf. (DAC), San Francisco, CA, USA, Dec. 2021, pp. 769–774.

[3] C. Guo, M. Yanagisawa, and Y. Shi, "DSE-based hardware Trojan attack for neural network accelerators on FPGAs," IEEE Trans. Neural Netw. Learn. Syst., 2024, doi: 10.1109/TNNLS.2024.3482364.

[4] C. Guo and Y. Shi, "A Novel Security Threat Model for Automated AI Accelerator Generation Platforms," IEEE Access, vol. 13, pp. 61237–61249, Apr. 2025.

[5] A. S. Rakin, Z. He, and D. Fan, "Bit-flip attack: Crushing neural network with progressive bit search," in Proc. 2019 IEEE/CVF Int. Conf. Comput. Vis. (ICCV), Seoul, Korea, Oct. 2019, pp. 1211–1220.

[6] A. S. Rakin, Z. He, J. Li, F. Yao, C. Chakrabarti, and D. Fan, "T-BFA: Targeted bit-flip adversarial weight attack," IEEE Trans. Pattern Anal. Mach. Intell., vol. 44, no. 11, pp. 7928–7939, Nov. 2022.

[7] J. Li, A. S. Rakin, Z. He, D. Fan, and C. Chakrabarti, "RADAR: Run-time adversarial weight attack detection and accuracy recovery," in Proc. 2021 Des. Autom. Test Eur. Conf. Exhib., Grenoble, France, Feb. 2021, pp. 790–795.

[8] Q. Liu, W. Wen, and Y. Wang, "Concurrent weight encoding-based detection for bit-flip attack on neural network accelerators," in Proc. 2020 IEEE/ACM Int. Conf. Comput. Aided Des., San Diego, CA, USA, Nov. 2020, pp. 1–8.

[9] Y. Guo, L. Liu, Y. Cheng, Y. Zhang, and J. Yang, "ModelShield: A generic and portable framework extension for defending bit-flip based adversarial weight attacks," in Proc. 2021 IEEE 39th Int. Conf. Comput. Des., Storrs, CT, USA, Oct. 2021, pp. 559–562.

[10] Z. He, A. S. Rakin, J. Li, C. Chakrabarti, and D. Fan, "Defending and harnessing the bit-flip based adversarial weight attack," in Proc. 2020 IEEE/CVF Conf. Comput. Vis. Pattern Recognit., Seattle, WA, USA, Jun. 2020, pp. 14083–14091.

[11] National Institute of Standards and Technology (NIST), FIPS 197: Advanced Encryption Standard (AES), Federal Information Processing Standards Publication 197, Gaithersburg, MD, USA, Nov. 2001

[12] K. Simonyan and A. Zisserman, "Very deep convolutional networks for large-scale image recognition," 2014, arXiv:1409.1556.

[13] K. He, X. Zhang, S. Ren, and J. Sun, "Deep residual learning for image recognition," in Proc. 2016 IEEE Conf. Comput. Vis. Pattern Recognit., Jun. 2016, pp. 770–778.

[14] J. Deng, W. Dong, R. Socher, L. J. Li, K. Li, and F. F. Li, "Imagenet: A large-scale image database," in Proc. 2009 IEEE Conf. Comput. Vis. Pattern Recognit. (CVPR), Miami, FL, USA, Jun. 2009, pp. 248–255.

SpykSim: A Cycle-Level Full-System Simulator for Systolic SCNN Accelerators

Wanwan Zhao [1], Yichu Yao [1], Qiang Niu [1], Qian Li [1], Chen Zhao*[1]

[1] School of Computer Science and the Embedded Systems Integration Engineering Research Center, Northwestern Polytechnical University, Xi'an, Shaanxi 710072, China

* Email: zhaoww@mail.nwpu.edu.cn, chenzhao@mail.nwpu.edu.cn

Abstract—**We present SpykSim, the first cycle-level full-system simulator for spiking convolutional neural network (SCNN) accelerators based on systolic array. SpykSim integrates the DRAMsim3 memory simulator to accurately model off-chip DRAM behavior and data movement between memory and compute units. We evaluate three feature-map data layouts in DRAM: channel-first (CHW), channel-last (HWC), and channel-middle (HCW). Experimental results show that the HCW layout aligns most effectively with our output-stationary dataflow, significantly reducing the overall total number of DRAM accesses. We also propose a multi-bank input buffer with ping-pong scheduling to exploit data reuse in convolutional sliding windows. This design pipelines memory accesses and computation, greatly speeding up the im2col transformation and reducing off-chip DRAM reads. In a 16×16 systolic array, SpykSim achieves a 51% improvement in inference speed and a 63% reduction in DRAM read requests for the SCNN11 model compared to a baseline design. SpykSim provides a valuable tool for architectural exploration and optimization of SCNN designs.**

Keywords—Full-system simulator, spiking convolutional neural network, systolic array, im2col, accelerator architecture

I. INTRODUCTION

Spiking Convolutional Neural Networks (SCNNs) combine the low-power features of spiking neural networks (SNNs) with the powerful feature extraction of traditional CNNs. SCNNs use discrete spike signals to transmit information and accumulate temporal information in neuron membrane potentials, exhibiting strong biological plausibility. Recent work has shown SCNNs can achieve competitive accuracy on image classification tasks [1]. Efficient hardware acceleration is critical to fully realizing the potential of SCNNs. Systolic arrays are widely used in deep learning accelerators due to their high parallelism and regular dataflow [2]. They process data in a pipelined grid of processing elements (PEs) with local communication and enable high PE utilization and scalability. Using systolic arrays for SCNN accelerators holds promise for high-performance, low-power neuromorphic computing.

However, there are currently no dedicated simulators for SCNN accelerators. Existing CNN accelerator simulators (e.g., STONNE [3], SCALE-Sim [4]) do not support SCNNs or integrate accurate DRAM timing model. In practice, off-chip DRAM access often dominates latency and energy, especially when the model is large and on-chip buffer capacity is limited. If DRAM behavior is not accurately modeled, system performance may be misestimated, and design effectiveness and scalability could be compromised.

To address these gaps, we propose SpykSim, a full-system, cycle-accurate simulator for systolic SCNN accelerators.

SpykSim integrates DRAMsim3 [5] to model real DRAM timing and energy consumption. Our contributions can be summarized as follows:

- We develop SpykSim, a full-system, cycle-accurate simulator for systolic SCNN accelerators. It integrates DRAMsim3 to accurately model off-chip DRAM timing and energy consumption, enabling detailed analysis of data movement across the memory hierarchy.

- We evaluate the impact of feature-map data layout on DRAM access efficiency in SCNN accelerators. Specifically, we compare three layouts in DRAM: CHW, HWC, and HCW. Among these, HCW (Height-Channel-Width) provides the best balance under an output-stationary dataflow strategy.

- We design a multi-bank input buffer with ping-pong scheduling. By retaining the R−1 overlapping rows of input data between consecutive convolution windows (where R is the kernel height), the design minimizes redundant off-chip DRAM accesses. Moreover, multiple banks can be read in parallel to accelerate the im2col transformation, substantially improving overall inference speed.

The remainder of this paper is structured as follows. Section II describes the overall architecture of SpykSim, including its compute, memory, and control modules. Section III introduces the design of the multi-bank input buffer and discusses its impact on data reuse and preprocessing efficiency. Section IV analyzes various DRAM data layouts and their effects on off-chip memory access efficiency. Section V presents experimental results that demonstrate the performance benefits of our proposed techniques. Section VI concludes the paper.

II. SYSTEM ARCHITECTURE

The architecture of SpykSim is shown in Fig. 1, which can be divided into three main components: compute module, memory module, and control module.

A. Compute Module

The compute module performs the core operations of the SCNN and consists of three main submodules: the systolic array module, the membrane potential (MP) update module, and the optional pooling module. The heart of the compute module is an N×N systolic array that implements output-stationary convolution and fully-connected layers. This module draws inspiration from the TPU-like architecture implemented in the STONNE simulator [3]. In this dataflow, each PE is assigned to an output activation position. Input spikes propagate horizontally across rows, while weights

979-8-3315-3918-4/25 $31.00 © 2025 IEEE

flow vertically down columns. Each PE performs a multiplication between the incoming spike and its local weight, which is computationally efficient due to the binary nature of the spike, and accumulates the result into MP of the corresponding output neuron. After accumulation, a threshold comparator triggers a spike if MP exceeds a preset threshold then resets the MP. This simulates the behavior of Integrate-and-Fire (IF) spiking neuron [6]. The optional pooling module aggregates and arranges partial results across multiple passes, enabling on-chip max-pooling through bitwise OR operations. This approach eliminates the need for off-chip transfers of intermediate data.

B. Memory Module

The memory hierarchy adopts a three-level storage structure: off-chip DRAM modeled by DRAMsim3, on-chip buffers for intermediate data staging, and local register files for per-element computations. DRAMsim3 provides accurate modeling of DRAM timing and energy behavior. On-chip buffers include a multi-bank input buffer, a ping-pong weight buffer for hiding load latency, an output buffer, a neuron state buffer for MPs, a dedicated im2col buffer for holding flattened input windows used in convolution, and an accumulation SRAM that stores partial output results over multiple systolic array passes before pooling is performed. Registers within each PE and pooling unit offer low-latency local storage for partial sums. Notably, the multi-bank input buffer enables parallel row access and efficient incremental updates, as described in Section III.

C. Control Module

The control module coordinates the overall dataflow of the accelerator. A dedicated DMA engine manages burst transfers between off-chip DRAM and on-chip buffers ensuring efficient data movement. The convolution scheduler regulates the timing of input and weight delivery to the systolic array and handles the accumulation of partial outputs. When pooling is enabled, a pooling controller ensures and reduces outputs accordingly. These components guarantee correct timing and effective data reuse.

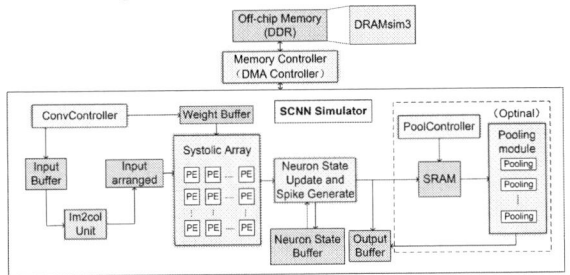

Fig. 1. The architecture of the SpykSim simulator.

SpykSim accepts a hardware configuration file specifying parameters such as array size, buffer capacities, DRAM data layout, and neuron thresholds, as well as an SCNN topology file defining the network layers and their dimensions. The simulator generates detailed statistics, including per-layer cycle counts, off-chip access counts for inputs, weights, outputs, and neuron states, along with overall energy and latency estimates. These outputs facilitate comprehensive performance evaluation for various architectural designs.

III. MULTI-BANK INPUT BUFFER

In the proposed architecture, the input feature map is loaded in row-wise blocks to accommodate sliding-window overlaps and limited on-chip buffer capacity. A multi-bank input buffer is employed to exploit this data reuse, thereby reducing off-chip DRAM traffic and accelerating the im2col transformation.

A. Reducing DRAM Accesses

For a convolutional kernel of height R, R consecutive rows of the input feature map are loaded from off-chip DRAM into the on-chip buffer as a block. Corresponding weights for the target output channels are also loaded, with the number of channels limited by the number of columns in the systolic array. The array then processes this block to produce one row of the output feature map. Since adjacent convolution windows overlap by R−1 rows, conventional designs reload nearly the same data with each slide. To eliminate this redundancy, the input buffer is organized into R banks, allowing selective updates.

When the window slides, the lower R−1 rows are retained, and only one new row is fetched. A ping-pong scheme alternates the updated bank. This incremental loading reduces the data read per window from R × C × Y to C × Y, where C is the number of input channels and Y is the width of the input feature map. Fig. 2 compares the traditional ping-pong buffer, which reloads all R rows, with our multi-bank design that fetches only one row. This significantly reduces off-chip DRAM reads.

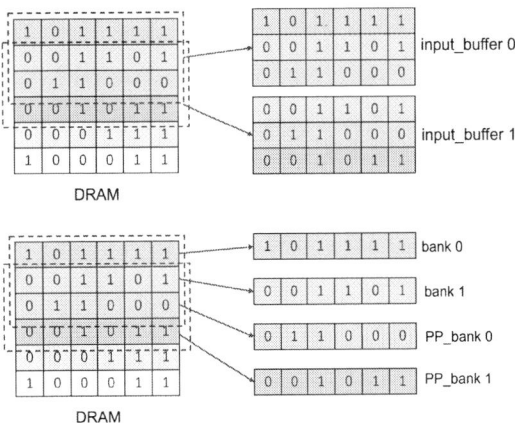

Fig. 2. Illustration of the multi-bank input buffer mechanism. Instead of reloading all R rows on each window slide, only the new row is fetched.

B. Accelerating the im2col Process

In the im2col unit proposed in [7], flattening the first convolution window requires sequentially reading all elements from the input buffer, taking R × S × C cycles, where S is the kernel width. For the second window, R × (S−1) × C elements overlap with the first window and can be reused through internal shifting, so only R × C new elements need to be read, requiring R × C cycles. This data movement is illustrated in Fig. 3.

In the proposed architecture, since multiple banks can be accessed in parallel each cycle, the first window can be constructed in S×C cycles and subsequent windows in only C cycles. This significantly accelerates the im2col transformation and improves overall inference performance.

979-8-3315-3918-4/25 $31.00 © 2025 IEEE

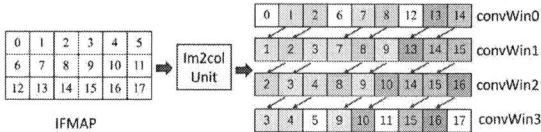

Fig. 3. Illustration of data movement in the im2col process.

IV. DRAM DATA ORGANIZATION STRATEGIES

The layout of feature maps in DRAM has a significant impact on memory access efficiency during SCNN inference. Fig. 4 illustrates the three typical DRAM data layouts evaluated in this work: CHW, HWC and HCW. While Fig. 5(a) shows the sequence of output block generation under the output-stationary dataflow.

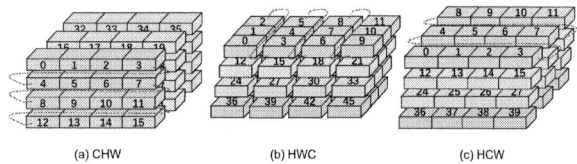

Fig. 4. Comparison of DRAM storage orders for 3D feature maps under three layout strategies. (a) CHW, (b) HWC and (c) HCW.

In the CHW layout, each channel's entire 2D feature map is stored contiguously in memory. This enables efficient burst writes when storing the accumulated outputs of multiple channels, as shown in Fig. 5(b). However, input tiles span multiple channels, resulting in fragmented memory access patterns, as illustrated in Fig. 4(a). In contrast, the HWC layout makes input spikes within a spatial block contiguous and highly suitable for burst reads, as shown in Fig. 4(b). Yet, after convolution, the output channels become interleaved in memory, leading to poor continuity during DRAM write-back and degraded write efficiency, as illustrated in Fig. 5(b).

The HCW layout combines the strengths of the previous two. It supports burst-efficient reads, comparable to HWC, as shown in Fig. 4(c). More importantly, when the systolic array generates multiple output channels for a row in one pass, the outputs are stored contiguously, as shown in Fig. 5(b). This results in more efficient burst writes compared to HWC.

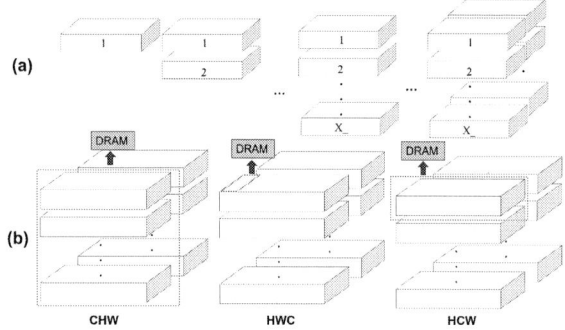

Fig. 5. (a) The sequence of generated computation result blocks and (b) the blue wireframe part is the data continuously stored in DRAM.

Overall, the HCW layout inherits HWC's high input-read performance and achieves near-ideal output-write efficiency close to that of CHW, thereby minimizing total DRAM transactions.

V. EXPERIMENTAL RESULTS

SpykSim was evaluated on several SCNN models (e.g., LeNet [8], VGG16 [9], MobileNet [10], and spiking variants SCNN6 [11], SCNN9 [12], SCNN11 [12]) across different systolic array sizes and feature-map data layouts.

A. Impact of Three Data Layout on Performance

Across all evaluated SCNN models and array sizes, the HCW consistently results in the fewest DRAM access requests. Fig. 6 compares DRAM total requests for CHW, HWC, HCW layouts.

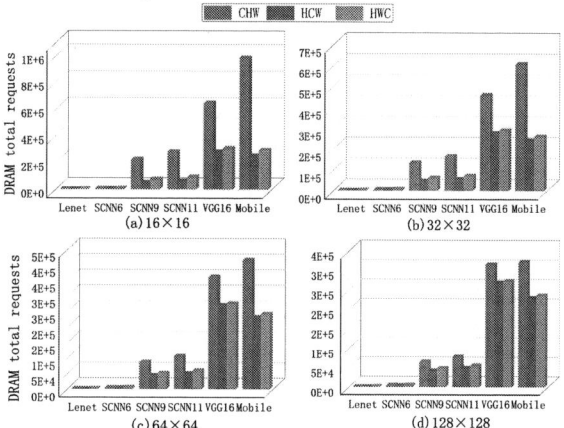

Fig. 6. Total DRAM accesses (read and write) for CHW, HWC, and HCW layouts across all evaluated models and different systolic array sizes.

Using the 16×16 array as an example, HCW offers read efficiency comparable to HWC and significantly better than CHW (Fig. 7(a)), while also achieving write efficiency superior to HWC and close to that of CHW (Fig. 7(b)).

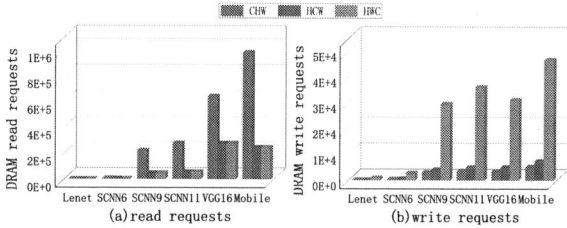

Fig. 7. DRAM read and write requests for CHW, HWC, HCW with a 16×16 array.

As the array size increases, HCW's write request count gradually approaches that of CHW. At a 128×128 array size, HCW achieves comparable write efficiency to CHW for the SCNN9 and SCNN11 models, as shown in Fig. 8.

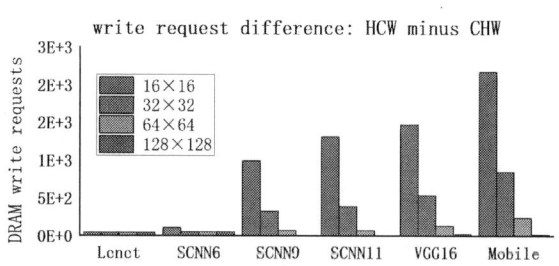

Fig. 8. Write requests difference between HCW and CHW.

B. Impact of Multi-bank Input Buffer on DRAM Access

Under the HCW layout, using the multi-bank input buffer significantly reduces DRAM read requests. As shown in Fig. 9, we compare the DRAM read request counts between the baseline ping-pong buffer and the multi-bank scheme across a range of SCNN models and systolic array sizes. The multi-bank design effectively exploits spatial reuse and reduces DRAM read requests by approximately 50% to 63% in most configurations. The benefit becomes even more pronounced in larger networks, which typically contain more convolutional layers and thus offer more reuse opportunities.

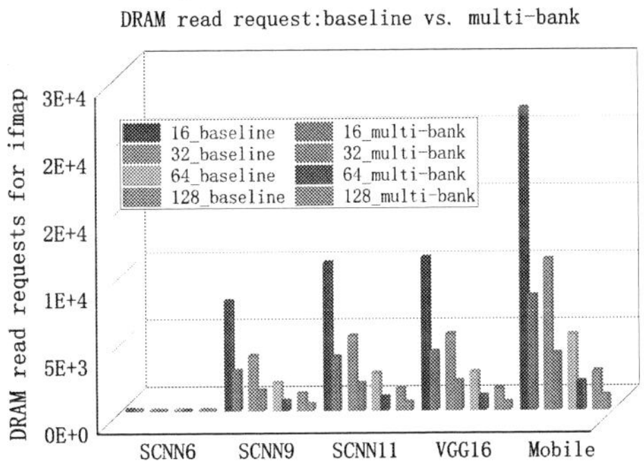

Fig. 9. DRAM read requests for various SCNN models, comparing the traditional ping-pong buffer vs. the multi-bank input buffer.

C. Impact of Multi-bank Input Buffer on Inference Speed

Using the same HCW layout, we evaluate the total inference cycles with both the traditional input buffer and the multi-bank design across various network models and systolic array configurations, as illustrated in Fig. 10. Since the multi-bank structure supports parallel data reads, it greatly accelerates the im2col data rearrangement process, thereby improving overall inference speed. For the SCNN11 model running on a 16×16 systolic array, the multi-bank input buffer achieves up to a 51% reduction in total inference cycles.

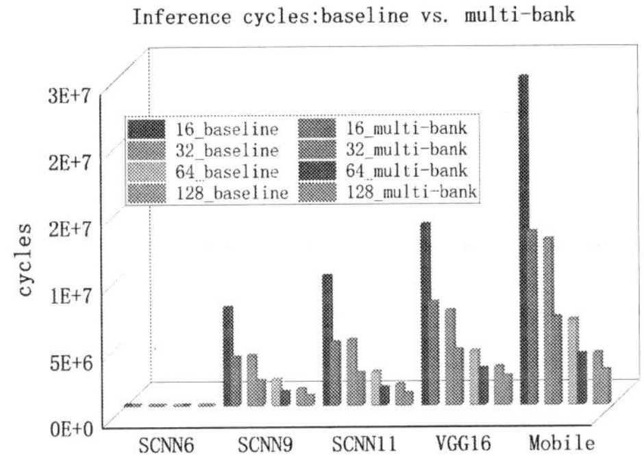

Fig. 10. Total inference cycles for different models with baseline vs. multi-bank buffer.

These experimental results validate the effectiveness of our architectural and data layout optimizations in SpykSim. The combination of the HCW layout and multi-bank input buffer substantially reduces DRAM access requests and significantly enhances inference performance.

VI. CONCLUSION

We introduced SpykSim, a cycle-level full-system simulator for SCNN accelerators based on systolic arrays, with integrated DRAM behavior modeling. SpykSim enables detailed evaluation of architectural choices and memory-system interactions. Our analysis shows that a channel-middle (HCW) DRAM layout significantly reduces off-chip transactions under output-stationary dataflow. In addition, we propose a multi-bank input buffer with ping-pong scheduling to leverage convolutional data reuse. By retaining overlapping rows in place and supporting parallel reads, this design reduces DRAM access and accelerates the im2col transformation. Experimental results demonstrate that these techniques reduce DRAM reads by over 60% and cut inference time by up to 51%. Overall, SpykSim provides an effective tool and design methodology for developing low-power, high-performance neuromorphic accelerators.

ACKNOWLEDGMENT

This work was supported in part by the STI2030-Major Project under Grant 2022ZD0208805; in part by the Fundamental Research Funds for the Central Universities under Grant 24GH0201339 and Grant WH00001161.

REFERENCES

[1] G. Shen, D. Zhao, and Y. Zeng, "Backpropagation with biologically plausible spatiotemporal adjustment for training deep spiking neural networks," Patterns, vol. 3, no. 6, Art. no. 100522, Jun. 2022.

[2] Y.-H. Chen, T. Krishna, J. S. Emer, and V. Sze, "Eyeriss: An energy-efficient reconfigurable accelerator for deep convolutional neural networks," IEEE J. Solid-State Circuits, vol. 52, no. 1, pp. 127–138, Jan. 2017.

[3] F. Muñoz-Martínez, J. L. Abellán, M. E. Acacio, and T. Krishna, "STONNE: Enabling cycle-level microarchitectural simulation for DNN inference accelerators," in Proc. IEEE Int. Symp. Workload Characterization (IISWC), Storrs, CT, USA, 2021, pp. 201–213.

[4] A. Samajdar, J. M. Joseph, Y. Zhu, P. Whatmough, M. Mattina, and T. Krishna, "A systematic methodology for characterizing scalability of DNN accelerators using SCALE-Sim," in Proc. IEEE Int. Symp. Performance Analysis of Systems and Software (ISPASS), Boston, MA, USA, 2020, pp. 58–68.

[5] S. Li, Z. Yang, D. Reddy, A. Srivastava, and B. Jacob, "DRAMsim3: A cycle-accurate, thermal-capable DRAM simulator," IEEE Comput. Archit. Lett., vol. 19, no. 2, pp. 106–109, Jul.–Dec. 2020.

[6] L. F. Abbott, "Lapicque's introduction of the integrate-and-fire model neuron (1907)," Brain Res. Bull., vol. 50, no. 5–6, pp. 303–304, Nov.–Dec. 1999.

[7] S. Q. Wang, L. Wang, Y. Deng, et al., "SIES: A novel implementation of spiking convolutional neural network inference engine on field-programmable gate array," J. Comput. Sci. Technol., vol. 35, no. 2, pp. 475–489, Mar. 2020.

[8] Y. LeCun, L. Bottou, Y. Bengio, and P. Haffner, "Gradient-based learning applied to document recognition," Proc. IEEE, vol. 86, no. 11, pp. 2278–2324, Nov. 1998.

[9] K. Simonyan and A. Zisserman, "Very deep convolutional networks for large-scale image recognition," in Proc. Int. Conf. Learn. Represent. (ICLR), 2015, arXiv:1409.1556.

[10] A. G. Howard, M. Zhu, B. Chen, D. Kalenichenko, W. Wang, T. Weyand, M. Andreetto, and H. Adam, "MobileNets: Efficient convolutional neural networks for mobile vision applications," 2017, arXiv:1704.04861.

[11] W. Ye, Y. Chen, and Y. Liu, "The implementation and optimization of neuromorphic hardware for supporting spiking neural networks with MLP and CNN topologies," IEEE Trans. Comput.-Aided Des. Integr. Circuits Syst., vol. 42, no. 2, pp. 448–461, Feb. 2023.

[12] J. Li, G. Shen, D. Zhao, Q. Zhang, and Y. Zeng, "FireFly v2: Advancing hardware support for high-performance spiking neural network with a spatiotemporal FPGA accelerator," IEEE Trans. Comput.-Aided Des. Integr. Circuits Syst., vol. 43, no. 9, pp. 2647–2660, Sept. 2024.

A Data-Efficient Deep Reinforcement Learning Algorithm and FPGA Accelerator for Real-Time Robot Motion Control Applications

Wenhao Huang[1], Rao Fu[1], Aiwu Ruan*[1], Huiyun Li*[2], Chongyang Zhu[1]

[1] State Key Laboratory of Electronic Thin Films and Integrated Devices, University of Electronic Science and Technology of China, Chengdu, China

[2] Faculty of Computility Microelectronics, Shenzhen University of Advanced Technology, Shenzhen, China

* Email: 202321310907@std.uestc.edu.cn, ruanaiwu@uestc.edu.cn, huiyun.li@suat-sz.edu.cn

Abstract—Deep reinforcement learning (DRL) has revolutionized end-to-end self-learning in most AI tasks, but existing algorithms like Proximal Policy Optimization (PPO) suffer from data inefficiency and slow convergence, posing challenges for real-time robotic control. This paper presents the Decay Proximal Policy Scaler (DPPS), a novel DRL algorithm that enhances data efficiency through dynamic clipping decay and smooth policy transitions. Additionally, a CPU-FPGA hybrid controller architecture is proposed, mapping neural network computation graphs to dedicated hardware units for parallel inference and training. Experiments in robotic control tasks show that DPPS achieves higher final scores than PPO in the same training steps. The FPGA accelerator reduces latency by 143.0x compared to Cortex-A9 and improves IPC by 38.24x compared to modern desktop CPU implementations for single-data calculation, while maintaining 21.23x energy efficiency, demonstrating its suitability for real-time motion control.

Keywords—*deep reinforcement learning, neural network, robotic controller, hardware acceleration*

I. INTRODUCTION

Deep Reinforcement Learning (DRL), which integrates deep learning's feature extraction with reinforcement learning's decision-making, has revolutionized end-to-end self-learning in robotic control tasks, gaining significant traction after breakthroughs in complex control scenarios. Among policy-based DRL methods, Proximal Policy Optimization[1] (PPO) has emerged as a cornerstone due to its stability, yet it faces a critical limitation: data inefficiency. PPO's fixed clipping mechanism often leads to slow convergence and suboptimal policies, requiring excessive environmental interactions—an acute challenge for real-time robotic control where low latency and rapid adaptation are paramount.

While prior work has proposed improvements to PPO's algorithmic efficiency[2] -[4], a complementary challenge lies in hardware acceleration for real-time deployment. Traditional neural network accelerators typically adopt a layer-by-layer computation model, where hardware units sequentially process each network layer. This approach, though simple, struggles to meet the strict real-time requirements, such as robotic motion control, where millisecond-level latency and parallel processing are non-negotiable.

This paper addresses both algorithmic and hardware bottlenecks:

Algorithmically, we introduce the Decay Proximal Policy Scaler (DPPS), which enhances PPO with dynamic clipping and smooth policy transitions to accelerate convergence and improve data efficiency.

Hardware-wise, we present a FPGA controller architecture designed for real-time DRL inference and training. Unlike conventional accelerators, our FPGA implementation fully maps the neural network's computation graph to dedicated hardware units, enabling parallel execution of forward and backward passes. This design leverages customized linear, activation, and data flow units (e.g., Fork/Concatenate) to optimize throughput and reduce latency, the key to real-time robotic control.

By integrating DPPS with the FPGA accelerator, we demonstrate superior performance in robotic control tasks compared to PPO, achieving faster policy convergence and higher control scores. The hardware architecture specifically addresses the real-time constraints of robotic AI, making it suitable for applications requiring low-latency decision-making in dynamic environments.

The DPPS algorithm demonstrates significant performance improvements over traditional PPO, achieving higher final scores and enhanced data efficiency through dynamic clipping decay and smooth policy transitions. In robotic control tasks, DPPS consistently showed faster policy convergence and superior control scores. Complementing this, the proposed FPGA accelerator dramatically boosts computational efficiency, reducing latency by 143.0x compared to Cortex-A9, improving Instructions Per Cycle (IPC) by 38.24x for single-data calculation over modern desktop CPUs, and maintaining an impressive 21.23x energy efficiency. This combined approach of DPPS and its dedicated FPGA accelerator makes it highly suitable for real-time robot motion control applications.

II. ALGORITHM MODEL

The proposed DPPS algorithm introduces an adaptive framework to enhance policy stability and accelerate convergence in DRL. At its core, DPPS reimagines PPO's static clipping mechanism through two pivotal innovations: a dynamic decay mechanism for exploration control and a scaled limiting function for policy update refinement.

A. Dynamic Clipping with Decay

PPO employs a fixed clipping range ε(typically 0.2) to constrain policy updates, which can prematurely limit exploration and excessive divergence[5]. DPPS addresses this rigidity by introducing a time-dependent clipping parameter($\varepsilon_t^{\text{DPPS}}$).

$$\varepsilon_t^{\text{DPPS}} = \begin{cases} \varepsilon_0, & t \leq T \cdot \text{DPPS\_r0} \\ \dfrac{\varepsilon_0}{1 - \text{DPPS\_r0}} \cdot \dfrac{T-t}{T}, & T \cdot \text{DPPS\_r0} < t \leq T \end{cases} \quad (1)$$

This mechanism dynamically adjusts the clipping range based on the time elapsed since the beginning of the training process. The parameter `DPPS_r0` controls the rate of decay, and ε_0 represents the initial clipping range. For early training, the clipping range remains a wide clipping window ε_0 to encourage exploration. As the training progresses, the clipping range decreases, allowing the algorithm to focus more on exploitation.

B. Smoothing Policy Transition

While PPO uses a hard clip function $r_t(\theta)$, DPPS replaces it with a differentiable, curvature-controllable function $\mathcal{R}^{\text{DPPS}}$:

$$\mathcal{R}^{\text{DPPS}}(r_t(\theta), \varepsilon_t, \mu) =$$
$$\begin{cases} 1 - \varepsilon_t - \mu \tanh(\varepsilon_t) - \mu \tanh(r_t(\theta) - 1) & r_t \leq 1 - \varepsilon_t \\ 1 + \varepsilon_t + \mu \tanh(\varepsilon_t) - \mu \tanh(r_t(\theta) - 1) & r_t \geq 1 + \varepsilon_t \\ r_t(\theta) & \text{otherwise} \end{cases} \quad (2)$$

This function introduces a smooth transition between the clipped and unclipped regions, enabling the policy to perform additional optimizations. The parameter μ controls the smoothness of the transition. This smooth, adaptive containment of policy updates reduces oscillations and avoids the "policy out-of-bounds" issue of PPO's rigid clipping.

Combining the smoothed $\mathcal{R}^{\text{DPPS}}$ and Generalized Advantage Estimator[5] $\hat{A}_t$, the policy objective in DPPS attains low-variance advantage calculation, and ensures policy updates remain within adaptive, stable bounds:

$$L^{\text{policy}} = \hat{\mathbb{E}}_t \big[\min(r_t(\theta) \hat{A}_t, \mathcal{R}^{\text{DPPS}} \hat{A}_t) \big] \quad (3)$$

C. Clipping Value Objective

The value objective of PPO $v_\psi(s_t)$ is also clipped in DPPS by ε_t to prevent overfitting to noisy rewards:

$$\mathcal{F}_\psi^{\text{CLIP}} = \begin{cases} v_{\psi_{\text{old}}} - \varepsilon_t & (v_\psi(s_t) - v_t) \leq -\varepsilon_t \\ v_{\psi_{\text{old}}} + \varepsilon_t & (v_\psi(s_t) - v_t) \geq \varepsilon_t \\ v_\psi(s_t) & \text{otherwise} \end{cases} \quad (4)$$

For the value objective, DPPS also adopts the clip mechanism of PPO:

$$L^{\text{value}} = \frac{1}{2} \hat{\mathbb{E}}_t \Big[\min \big((\mathcal{F}_\psi^{\text{CLIP}} - r_t)^2, (v_\psi(s_t) - r_t)^2 \big) \Big] \quad (5)$$

III. HARDWARE ACCELERATION

Considering that the motion control networks in robot AI applications are typically characterized by compact architectures, real-time performance and low latency emerge as paramount requirements for seamless interaction with dynamic environments. Traditional neural network accelerators, often adhering to a layer-by-layer computational paradigm, face inherent limitations in meeting such demands. These systems rely on sequential processing of network layers, with hardware units reusing resources for different operations under software-controlled data flow, which introduces latency and hampers parallel execution—critical bottlenecks for robotic motion control and other time-sensitive tasks.

To address these challenges, the proposed FPGA accelerator fully maps the computation graph of the neural

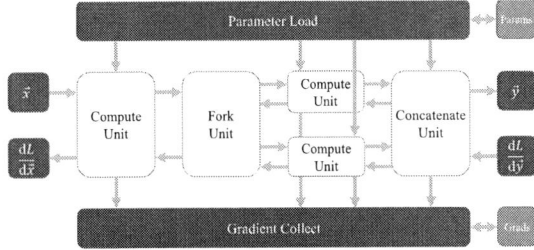

Fig. 1: Architecture of FPGA Accelerator

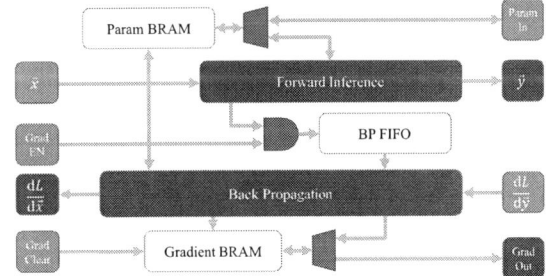

Fig. 2: Basic structure of all computation units

network to a dedicated hardware architecture. This approach ensures that each operation in the computation graph is handled by a specific hardware unit, enabling parallel execution and optimized data flow.

To achieve this goal, High-Level Synthesis (HLS) was utilized in this study. This approach bridges the gap between software algorithm design and hardware implementation, significantly reducing the time and effort required for hardware development.

Our accelerator takes advantages of customized hardware units. These units are specifically designed to handle the operations within the computation graph of the neural network, ensuring optimized execution of both forward and backward passes. The architecture of the FPGA accelerator is depicted in Fig. 1, consisting of parameter and gradient management units, different compute units, and flow control units.

All the compute units in the accelerator share the same basic architecture as depicted in Fig. 2. This unified architecture streamlines the development process while optimizing hardware resource utilization.

The core parts in the basic architecture are the forward inference and back propagation module, which takes charge of forward and backward passes in computation respectively. Both modules require access to parameter BRAMs, if any. During network training, additional input related data are required for back propagation, which will be stored in the BP FIFO. The calculated gradients are stored in gradient BRAMs.

To enhance resource utilization, the BRAM Multiplexer (BRAM Mux) is employed to time-division multiplex BRAM buses, enabling one bus access by multiple components (e.g., forward propagation and parameter loading). This design reduces hardware overhead in network loading and computation phase.

A. Linear Unit

The linear unit is designed to handle the General Matrix Multiplication, or GEMM operation:

$$\vec{y} = \mathbf{W}\vec{x} + \vec{b} \qquad (6)$$

By adopting loop unrolling optimization, this unit is able to leverage the parallel processing capabilities of FPGAs, utilizing dedicated DSP blocks and BRAMs to ensure high throughput. To balance the resource usage and performance, we decide to set loop unroll parameter to 16.

In the forward pass, this unit takes data from input port and params from BRAM, calculating the matrix multiplication and addition, while caching input into BP FIFO for weight gradient calculation. In the backward pass, the linear unit computes the gradients with respect to the weights, bias, and input, making use of weight in the param BRAM and inputs generated in the forward pass from the BP FIFO.

The combination of pipelining, local storage, and parallel processing makes the linear unit a powerful and efficient component of the FPGA accelerator, capable of handling the computationally intensive tasks of neural networks with high performance and low latency.

B. Activation Units

The activation unit is designed to handle the non-linear transformations that are essential for introducing complexity and enabling learning in neural networks. This unit, depending on the actual activation function such as tanh, and sigmoid, utilizes dedicated DSP blocks to ensuring high throughput and minimal resource occupation.

In the forward pass, this unit takes data from input port, calculating the activated data, sending it to both output port and BP FIFO. In the backward pass, the activation unit computes the gradients of the activation functions with respect to their inputs, using both gradient inputs and cached outputs.

C. Flow Control Units

The fork and concatenate units are designed to manage data distribution and merging within the neural network's computation graph.

The fork unit handles the duplication and distribution of data streams to multiple subsequent operations. In the forward pass, this unit controls the handshake signals in data bus to input data streams, ensuring that each downstream receives the necessary data without delays. In the backward pass, the fork unit aggregates gradients from multiple sources.

The concatenate unit handles the merging of multiple data streams into a single output. In the forward pass, it utilizes control signals and optional FIFOs to merge data streams, ensuring that the combined output is ready for subsequent operations without delays. By calculating the pipeline delay of each data path, the usage of FIFOs for synchronization is determined completely by the computation graph. In the backward pass, this unit splits the gradients into multiple output streams.

By efficiently managing the branching and merging of data streams, the flow control units contribute significantly to the seamless data flow and optimized execution of the neural network, making them indispensable for the high-performance operation of the FPGA accelerator.

D. Parameter and Gradient Management units

The parameter and gradient management units are designed to handle the storage, retrieval, and transfer of neural network parameters and gradients, forming the data management backbone of the FPGA accelerator.

The parameter load unit retrieves weights and biases from external DDR memory through AXI buses and caches them in param BRAMs of computation units, minimizing latency from repeated external memory accesses.

The gradient collect unit extracts computed gradients from gradient BRAMs and transfers them to the DDR via AXI interfaces for gradient decent.

In the training pipeline, these units dynamically manage parameter updates during backpropagation, ensuring that gradient computations from the linear and activation units are efficiently stored and applied to update weights. The integration of pipelined data transfer, BRAM caching, and bus multiplexing makes the Parameter and Gradient Management Units crucial for maintaining low-latency, high-throughput operation in the FPGA accelerator, supporting real-time training and inference for deep reinforcement learning tasks.

IV. EXPERIMENTS AND RESULTS

A. DPPS Algorithm

As mentioned in section II, the corner coefficient `DPPS_r0` affects the exploration rate and convergence of the algorithm. Therefore, it is necessary to determine an appropriate `DPPS_r0` prior to performing the experiment. HalfCheetah is used as the environment. Five seed experiments are carried out corresponding `DPPS_r0 = 0.2` to 0.8 stepped by 0.1, as is showed in Fig. 3.

It can be seen that in Fig. 3, when timestep = 0.5e7, i.e. 5 million episodes, the curve of `DPPS_r0 = 0.2` has some changes in score compared with other settings. This is due to

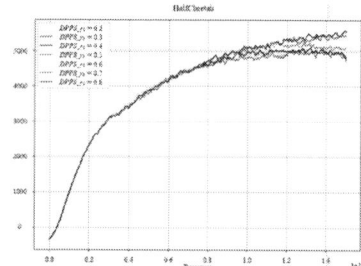

Fig. 3: Learning curve in different hyperparameter `DPPS_r0` setting

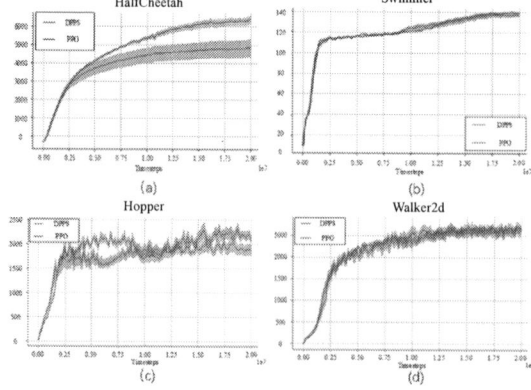

Fig. 4: Comparison of two algorithms on several MuJoCo environments, training for 20 million timesteps, 10 random seeds.

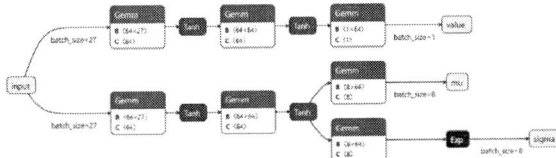

Fig. 5: Example computation graph for experiment

the lower `DPPS_r0`, which makes the algorithm tend to converge and hope to obtain the optimal value at an early stage, but higher `DPPS_r0` have higher scores in the end. We choose `DPPS_r0 = 0.2` to do the following tests.

Four cases from OpenAI's open-source database Gym are chosen as the experimental environments, which are robotics applications such as Half-Cheetah, Swimmer, Hopper, and Walker2d.

Fig. 8 illustrates that, under most circumstances, the score of DPPs is higher than that of PPO, and the policy shock of DPPs is less than that of PPO. This indicates DPPs is more stable, and can achieve better final scores compared to PPO.

B. FPGA Accelerator

The performance of the FPGA accelerator was evaluated on a ZYNQ-7000 series SoC platform (ZYNQ7020CLG400-2) with the PYNQ operating system. Comparative experiments were conducted against CPU and CPU-GPU he implementations of PyTorch on a Cortex-A9 dual-core processor, an AMD 5800H octa-core processor with a NVIDIA RTX 3060 GPU to assess time and energy efficiency.

Considering the hardware implementation of the accelerator will depend on the architecture of the neural network, we take the same network of the algorithm experiment as an example, depicted in Fig. 5. The accelerator's hardware resource consumption is summarized in Table I. Compared to [6], our architecture consumes more BRAMs, but achieved less consumption of LUTs, registers and DSPs.

Table II shows the time consumption tests of the accelerator, focusing on evaluating the time consumption of single point forward/backward propagation. To acquire these data, we deploy the same network as in Fig. 5, and run forward inference and back propagation for 100 times each, calculating the average time elapsed in each run.

From Table II we can see that, when running a single point propagation, FPGA has a 143.0x IPC speed up compared to Cortex-A9, 38.24x compared to AMD 5800H, and 72.60x compared to NVIDIA RTX 3060.

Table II. Resource utilization of the accelerator

Platform	LUTs	Registers	BRAMs	DSPs
This work	26086	27631	80.5	163
[6]	12864	54336	32	768

Table III. Results of single point time consumption test

Platform	Freq. (MHz)	Time consumption(us)		
		Forward Inference	Back Propagate	Total
This work	125	219.14	208.45	427.58
Cortex-A9 x2	533.33	4725.20	9605.76	14330.96
AMD 5800H x8	4200	205.20	281.46	486.67
NVIDIA RTX 3060	1780	687.50	1492.50	2180.00

Table I. Energy efficiency comparison of accelerator

Platform	Power (W)	Accelerator energy efficiency ratio		
		Forward Inference	Back Propagate	Total
Accelerator	2.40	1x	1x	1x
Cortex-A9 x2	1.52	13.66x	29.19x	21.23x
AMD 5800H x8	67.92	30.22x	38.21x	32.21x
NVIDIA RTX 3060	83.71	109.43x	249.73x	177.83x

To evaluate energy efficiency, while the algorithm was running on different devices, we collected ten sets of power consumption data respectively. After discarding the maximum and minimum values, we averaged the remaining eight sets to obtain the power consumption results presented in Table III.

The FPGA accelerator demonstrates exceptional energy efficiency, standing out as the optimal solution compared to other platforms. This superior efficiency is a direct result of its customized hardware architecture, which mitigates energy waste by utilizing dedicated DSP blocks for compute-heavy tasks and efficient BRAM for data caching. Consequently, the architecture's low power draw and high performance make it perfectly suited for embedded AI applications that demand both real-time responsiveness and energy-conscious design.

V. CONCLUSION

This work introduces DPPS, a data-efficient DRL algorithm, and a corresponding FPGA accelerator for real-time robot control. By replacing PPO's static clipping with a time-dependent decay mechanism and a differentiable policy constraint function, DPPS optimizes exploration-exploitation balance, leading to faster convergence and higher performance in robotic control tasks. The FPGA architecture, featuring computation graph mapping, addresses the real-time latency requirements of dynamic environments, while achieving efficient resource utilization and high energy efficiency. Future work will focus on extending the architecture to more complex neural networks and real-world robotic systems.

REFERENCES

[1] J. Schulman, F. Wolski, P. Dhariwal, A. Radford, and O. Klimov, "Proximal policy optimization algorithms," *arXiv preprint arXiv:1707.06347*, 2017.

[2] C. Li, W. Dong, L. He, M. Cai and D. Wang, "Intelligent decision for joint operations based on improved proximal policy optimization," *Scientific Reports*, vol. 15, no. 1, p. 9418, 2025.

[3] M. Farsang and L. Szegletes, "Decaying clipping range in proximal policy optimization," in *2021 IEEE 15th International Symposium on Applied Computational Intelligence and Informatics (SACI)*, 2021, pp. 000 521--000 526.

[4] J. S. Byun, B. Kim and H. Wang, "Proximal policy gradient: PPO with policy gradient," arXiv preprint arXiv:2010.09933, 2020.

[5] J. Schulman, P. Moritz, S. Levine, M. I. Jordan, and P. Abbeel, "High-Dimensional Continuous Control Using Generalized Advantage Estimation," in *4th International Conference on Learning Representations, ICLR 2016, San Juan, Puerto Rico, May 2-4, 2016, Conference Track Proceedings*, Y. Bengio and Y. LeCun, Eds. 2016. [Online]. Available: http://arxiv.org/abs/1506.02438

[6] H. Taha and A. Abdelhadi, "HEPPO: Hardware-Efficient Proximal Policy Optimization--A Universal Pipelined Architecture for Generalized Advantage Estimation," *arXiv preprint arXiv:2501.12703*, 2025.

GraphFlow-PIM: Annotated Execution Graphs of DNN Workloads across Diverse PIM Configurations

Syeda Munazza Marium, Song Chen

School of Microelectronics, University of Science and Technology of China, Hefei 230026, China
Email: munazza@mail.ustc.edu.cn, songch@ustc.edu.cn

Abstract—Mapping deep neural networks (DNNs) onto Processing-in-Memory (PIM) architectures remains a central challenge in energy-efficient AI due to complex couplings between computation, memory, and scheduling. We introduce GraphFlow-PIM, a large-scale graph-structured benchmark dataset derived from the NICE-PIM design space exploration framework. GraphFlow-PIM provides 5,594 annotated execution graphs covering 18 DNN models and 314 PIM hardware configurations, each encoding fine-grained computational steps, memory operations, energy- and cycle-accurate performance metrics. By representing end-to-end DNN executions as heterogeneous graphs, the dataset enables learning-based analyses of PIM workloads with fidelity and generality previously unattainable. We demonstrate that graph neural networks (GNNs) trained on GraphFlow-PIM achieve robust performance on both standard splits and cross-architecture generalization tests. Notably, a GraphSAGE model obtains approximately 70% node-classification accuracy on unseen hardware architectures, significantly outperforming baseline GCN, GAT, and MLP models. Moreover, a GNN-based predictor achieves 98% accuracy in selecting optimal data layout formats and replaces complex solvers in approximately 70% of partitioning scenarios, as validated by NICE-PIM's simulator outputs. These results establish GraphFlow-PIM as a reusable and foundational resource for the community, facilitating further research in DNN-to-PIM co-design, benchmarking of novel PIM mapping algorithms, and exploration of robust, transferable machine learning techniques for PIM systems.

Keywords GraphFlow-PIM, DNN, GNN, NICE-PIM

I. INTRODUCTION

The growing complexity of deep neural network (DNN) models has shifted AI's primary efficiency bottleneck from computation to data movement [1]. Processing-in-memory (PIM) architectures mitigate this by integrating compute logic within memory arrays, drastically reducing off-chip data transfers. However, mapping DNN workloads optimally onto PIM hardware remains daunting each; layer (convolutional, fully-connected, etc.) may require a custom partitioning and scheduling strategy, and even small suboptimal decisions can degrade performance and energy efficiency.

Existing automation tools like NICE-PIM [2], Timeloop/Accelergy [3], and tensor virtual machine (TVM) [4] have tackled DNN-to-PIM mapping, but each faces limitations. Exhaustive analytical design space exploration (DSE) as in timeloop becomes intractable for large design spaces; operator-level mappers ignore inter-layer dependencies; and integer linear programming (ILP)-based schedulers (e.g., in NICE-PIM's data scheduler)

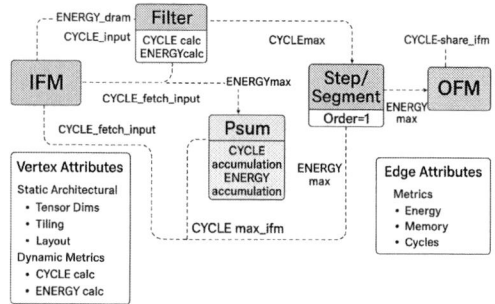

Fig. 1. Node edge attribute flow diagram.

struggle with complexity growth. Other optimization approaches [5], [6] also fall short, lacking robustness and interpretability essential for real-world deployment. These challenges have motivated learning-based approaches, such as graph neural networks (GNNs), which aim to learn holistic mapping strategies. However, progress is hindered by the lack of structured datasets that capture realistic PIM execution behavior.

To address this, we introduce GraphFlow-PIM: a large-scale benchmark that systematically converts energy- and cycle-accurate execution traces from NICE-PIM into richly attributed graphs representing DNN workloads on diverse PIM architectures. By leveraging NICE-PIM's exhaustive exploration of partitioning, scheduling, and architectural options, it provides comprehensive coverage of PIM-node arrays, PE dimensions, buffer sizes, and DNN types enabling robust benchmarking for both classical and learning-based co-design. Fig. 1, illustrates the node and edge attribute flow in GraphFlow-PIM's execution graphs, showing the heterogeneous node types and edge semantics.

This work, motivated by two central challenges *(i)* faithfully encoding PIM execution of DNN workloads as graphs and *(ii)* enabling learned predictors that transfer across previously unseen hardware architectures introduces GraphFlow-PIM, benchmark that unlocks systematic, learning-based DNN-to-PIM co-design.

II. GRAPHFLOW-PIM DATASET: STRUCTURE AND GENERATION

A. Dataset Overview

GraphFlow-PIM (v1.0) encompasses 5,594 graph samples.

TABLE I. SUMMARY OF GRAPHFLOW-PIM DATASET (V1.0).

Statistic	Value
Total graphs	5,594
Distinct DNN architectures	18
PIM hardware configurations	314
Nodes per graph (min – max)	19 – 419
Edges per graph (min – max)	68 – 19,400
Node types (heterogeneous)	8
Edge types (heterogeneous)	5

It comprises data drawn from 18 distinct DNN architectures running on 314 PIM hardware configurations, as shown in TABLE I.The DNNs include classic models (e.g., AlexNet, VGG, ResNet-50), modern architectures (Googlenet/Inception variants, YOLOv5), and multilayer perceptrons, providing a broad range of layer types and topologies. The PIM configurations span various memory sizes, number of PIM nodes (e.g., 4×4 vs. 16×16 node arrays), processing element (PE) array dimensions, and memory hierarchies. According to TABLE II, GraphFlow-PIM outscales canonical graph-learning benchmarks in node count, edge density, and structural heterogeneity [7-11].

B. PIM Hardware Configuration Space:

For each DNN workload in our dataset, we instantiate the hardware configuration from a baseline 3D-stacked DRAM-PIM architecture. This baseline uses a 28 nm process, 256 DRAM banks arranged in a 16×16 array, 8 MiB per bank, a 128-bit data width, and a 48 mm² logic die area. In every dataset entry, we independently vary the number of PIM nodes, the dimensions of the PE array, the sizes of on-chip buffers (for input, output, and weights), and the NoC flit address ratio, while all other architectural parameters are held constant.

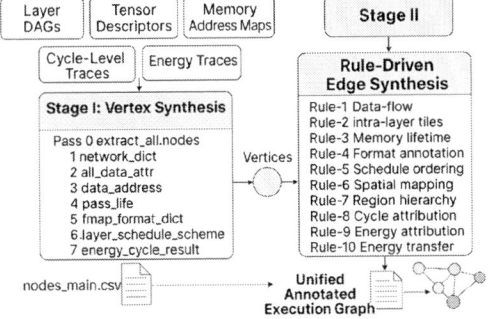

Fig. 2. Overview of the GraphFlow-PIM pipeline with two-stage node synthesis and edge construction.

C. Reusability for PIM Research:

GraphFlow-PIM can be reused and leveraged by other PIM and DSE-PIM researchers. The dataset is structured in an open format (graph `.pkl` files plus accompanying `.csv` hardware configuration information) designed for easy integration into new workflows. By providing a standardized, energy- and cycle-accurate set of PIM execution graphs, GraphFlow-PIM enables fair comparisons of novel mapping algorithms, GNN-based predictors, and other co-design tools on a common ground. It effectively serves as a community benchmark analogous to (OGB) [12] in graph learning or ImageNet in vision [13], lowering the barrier to entry and accelerating new research in PIM co-design.

III. METHODS: DATA GENERATION & GRAPH CONSTRUCTION

A. Derivation from NICE-PIM:

NICE-PIM is a comprehensive simulation and design space exploration (DSE) platform for mapping deep neural networks onto processing-in-memory hardware. Building on this foundation, GraphFlow-PIM systematically maps each DNN in the dataset to specific PIM hardware instances using NICE-PIM's automated DSE pipeline. This process exhaustively explores valid mapping options including layer partitioning, parallelism, and data layout and logs detailed records of execution cycles, energy, memory traffic, and scheduling. We extended NICE-PIM with enhanced logging modules to capture all relevant events, then transformed these multi-modal logs into graph-structured data. As a result, GraphFlow-PIM maintains cycle-accurate fidelity to actual execution, while formatting the data for graph-based learning. Fig. 2, illustrates the graph generation pipeline.

B. Graph Structure (Nodes & Edges Definition):

Each execution instance in GraphFlow-PIM is modeled as a heterogeneous, directed multigraph.We followed a two-stage process: Node synthesis and edge construction. Nodes represent both DNN layers and related data entities such as input/output feature maps, weights, partial sums, buffers, and composite compute blocks capturing all states and operations of the workload. Every node is annotated with static properties (tensor shape, data type, layer or operation type, index, tiling) and dynamic metrics (latency, energy, memory footprint, partitioning). Directed edges encode communication and scheduling dependencies, spanning data movement, accumulation, memory transfer, and synchronization. Edge types are distinguished by semantic role (data flow, control, inter-tile) and further annotated with transfer size,

TABLE II. COMPARATIVE GRAPH STATISTICS ACROSS POPULAR GNN BENCHMARKS AND GRAPHFLOW-PIM DATASET.

Dataset	#Graphs	Nodes $\mu \pm \sigma$	Edges $\mu \pm \sigma$	Density $\mu \pm \sigma$	Max Deg $\mu \pm \sigma$
ENZYMES [8]	600	32.63 ± 15.28	124.27 ± 51.00	0.3198 ± 0.2231	6.09 ± 1.00
PROTEINS [8]	1113	39.06 ± 45.76	145.63 ± 169.20	0.4244 ± 0.3960	5.79 ± 1.25
MUTAG [9]	188	17.93 ± 4.58	39.59 ± 11.37	0.2769 ± 0.0701	3.01 ± 0.07
NCI1 [8]	4110	29.87 ± 13.56	64.60 ± 29.87	0.1779 ± 0.0752	3.34 ± 0.48
IMDB-BINARY [10]	1000	19.77 ± 10.06	193.06 ± 211.20	1.0412 ± 0.4868	18.77 ± 10.06
COLLAB [10]	5000	74.49 ± 62.30	4914.43 ± 12877.83	1.0179 ± 0.6095	73.49 ± 62.30
Computers [11]	1	13752.00 ± 0.00	491722.00 ± 0.00	0.0052 ± 0.0000	2992.00 ± 0.00
Photo [11]	1	7650.00 ± 0.00	238162.00 ± 0.00	0.0081 ± 0.0000	1434.00 ± 0.00
Reddit [7]	1	232965.00 ± 0.00	114615892.00 ± 0.00	0.0042 ± 0.0000	21657.00 ± 0.00
GraphFlow-PIM [This work]	5594	176.20 ± 151.88	3916.08 ± 4923.69	0.2283 ± 0.0915	71.85 ± 60.16

979-8-3315-3918-4/25 $31.00 © 2025 IEEE

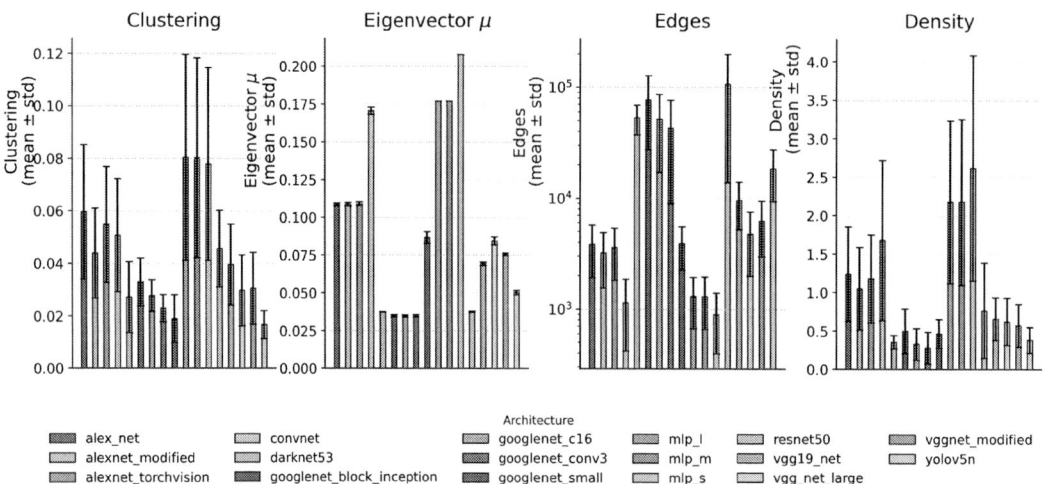

Graph Metrics by Architecture

Clustering · Eigenvector μ · Edges · Density

Architecture

▨ alex_net	▨ convnet	▨ googlenet_c16	▨ mlp_l	▨ resnet50	▨ vggnet_modified
▨ alexnet_modified	▨ darknet53	▨ googlenet_conv3	▨ mlp_m	▨ vgg19_net	▨ yolov5n
▨ alexnet_torchvision	▨ googlenet_block_inception	▨ googlenet_small	▨ mlp_s	▨ vgg_net_large	

Fig. 3. Demonstration of clustering, eigenvector centrality, edge count, and density metrics across all architectures on PIM.

TABLE III. CROSS-ARCHITECTURE TRANSFERABILITY BENCHMARK RESULTS.

Held-Out Architecture	# Train	# Test	GCN	GAT	GraphSAGE	MLP
alex_net	5287	307	0.346	0.341	0.697	0.554
alexnet_torchvision	5284	310	0.347	0.369	0.648	0.558
darknet53	5280	314	0.366	0.525	0.492	0.597
googlenet_c16	5285	309	0.400	0.423	0.377	0.605
googlenet_small	5281	313	0.415	0.379	0.467	0.600
mlp_m	5280	314	0.320	0.320	0.522	0.520
resnet50	5282	312	0.359	0.511	0.472	0.593
yolov5n	5280	314	0.388	0.414	0.452	0.618
alexnet_modified	5280	314	0.348	0.369	0.654	0.556
convnet	5298	296	0.340	0.378	0.622	0.567
googlenet_block_inception	5282	312	0.395	0.413	0.383	0.608
googlenet_conv3	5290	304	0.404	0.433	0.383	0.610
mlp_l	5280	314	0.320	0.320	0.520	0.520
mlp_s	5280	314	0.316	0.317	0.486	0.526
vgg_net_large	5282	312	0.369	0.397	0.616	0.577
vggnet_modified	5280	314	0.346	0.386	0.613	0.567
googlenet	5287	307	0.425	0.424	0.363	0.604
vgg19_net	5280	314	0.348	0.367	0.634	0.563

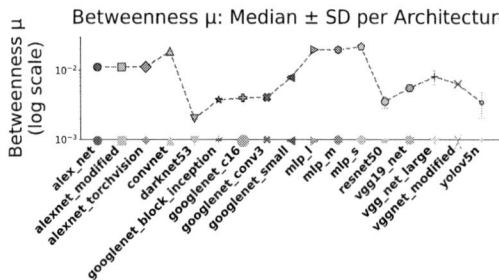

Fig. 4. Betweenness centrality distribution across architectures.

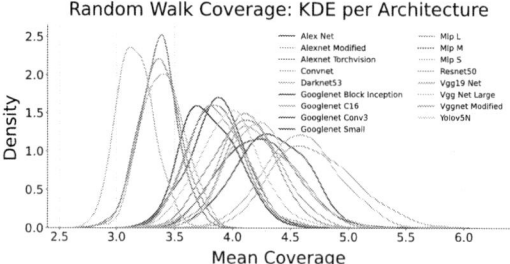

Fig. 5. Random walk coverage analysis across architectures.

energy cost, latency, and ordering. This richly attributed graph enables energy- and cycle-accurate analysis and robust benchmarking of DNN-to-PIM execution.

IV. EXPERIMENTAL METHODOLOGY AND KEY RESULTS

A. Experiments (Replicating NICE-PIM Outputs & Introducing New Analyses):

We performed extensive experiments on GraphFlow-PIM, explicitly distinguishing between tests that replicate NICE-

PIM schemes such as feature map format and partitioning scheme and new analyses uniquely enabled by our dataset. For validation, we reproduced key NICE-PIM results using our graph-based methodology, confirming that GraphFlow-PIM reliably recovers established outcomes. Beyond replication, we conducted new studies, most notably a cross-architecture generalization experiment where GraphSAGE achieved $\sim 70\%$ accuracy on previously unseen hardware, outperforming strong graph learning baselines GCN, GAT, and MLP chosen for their representative architectures in

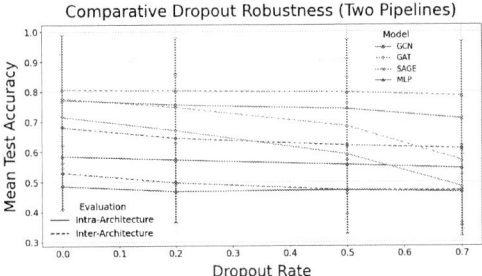

Fig. 6. Dropout robustness of all models under intra- and inter-architecture evaluation.

graph neural network research (GCN: spectral convolution, GAT: attention, GraphSAGE: neighborhood aggregation, MLP: non-graph baseline). Robustness was further evaluated under heavy dropout regularization to probe model stability. In both experiments, our models predict node-level mapping labels. Collectively, these analyses highlight practical challenges in learning-based mapping and demonstrate GraphFlow-PIM's versatility as a research resource.

- **Dataset Structural Diversity:** GraphFlow-PIM's internal diversity is quantified through a range of graph-theoretic metrics, including clustering coefficient, eigenvector centrality, edge count, and density (Figure 3), betweenness centrality Fig. 4, and random-walk coverage Fig. 5, MLP architectures consistently exhibit the highest clustering and density, while CNNs are generally more sparse and modular. Notably, architectural differences shape distinct communication patterns, as reflected by the spread in betweenness and Random-walk kernel density estimate (KDE) coverage across models.

- **Robustness and Transferability Tests:** We trained GCN, GAT, GraphSAGE, and MLP models to predict node-level mapping labels from NICE-PIM data. Training used subsets, with testing on held-out graphs. In leave-one-architecture-out tests TABLE III, Graph-SAGE led in transfer accuracy, averaging 55–60% and reaching 70% on AlexNet, outperforming others.

- **Cross-Architecture Generalization:** We investigated the effect of dropout regularization a widely used technique that randomly silences a portion of neurons during training to reduce overfitting on node classification accuracy, evaluating both random and cross-architecture splits. As illustrated in Fig. 6, while all GNNs performed robustly under moderate dropout in intra-architecture scenarios, GraphSAGE consistently maintained superior accuracy compared to baseline models under cross-architecture transfer and more aggressive dropout conditions.

- **Replacing Analytical Solvers with GNN Predictors:** A GraphSAGE model predicted optimal feature map formats and valid partitioning schemes for AlexNet. The GNN predictor achieved 98% format prediction accuracy and 70% feasibility on proposed mappings as validated by NICE-PIM's simulator, demonstrating the practical utility of learning-based methods.

V. CONCLUSION

We presented GraphFlow-PIM, a novel graph-based dataset for analyzing DNN executions on PIM hardware, systematically derived from the NICE-PIM DSE tool. The dataset offers energy- and cycle-accurate, richly annotated DNN-to-PIM mappings. We described its graph construction and demonstrated its scientific value through replication, cross-architecture generalization, and robustness experiments. GraphFlow-PIM is a robust, reusable resource for the PIM research community, and we believe it will help enable direct energy and cycle optimization with learning-based approaches. A future journal release will include an expanded dataset, more experiments, and enhanced methods.

ACKNOWLEDGMENT

This work was funded by the National Natural Science Foundation of China under Grant 92473114. We thank the Information Science Center of the University of Science and Technology of China for hardware/software services and the ANSO scholarship for young talent.

REFERENCES

[1] M. Horowitz, "1.1 computing's energy problem (and what we can do about it)," in 2014 IEEE international solid-state circuits conference digest of technical papers (ISSCC), 2014: IEEE, pp. 10-14.

[2] J. Wang, M. Ge, B. Ding, Q. Xu, S. Chen, and Y. Kang, "Nicepim: Design space exploration for processing-in-memory dnn accelerators with 3-d stacked-dram," IEEE Transactions on Computer-Aided Design of Integrated Circuits and Systems, vol. 43, no. 5, pp. 1456-1469, 2023.

[3] Parashar et al., "Timeloop: A systematic approach to dnn accelerator evaluation," in 2019 IEEE international symposium on performance analysis of systems and software (ISPASS), 2019: IEEE, pp. 304-315.

[4] T. Chen et al., "TVM: An automated End-to-End optimizing compiler for deep learning," in 13th USENIX Symposium on Operating Systems Design and Implementation (OSDI 18), 2018, pp. 578-594.

[5] M. J. Rasch et al., "Hardware-aware training for large-scale and diverse deep learning inference workloads using in-memory computing-based accelerators," Nature communications, vol. 14, no. 1, p. 5282, 2023.

[6] P. Das, P. R. Sutradhar, M. Indovina, S. M. P. Dinakarrao, and A. Ganguly, "Implementation and evaluation of deep neural networks in commercially available processing in memory hardware," in 2022 IEEE 35th International System-on-Chip Conference (SOCC), 2022: IEEE, pp. 1-6.

[7] W. Hamilton, Z. Ying, and J. Leskovec, "Inductive representation learning on large graphs," Advances in neural information processing systems, vol. 30, 2017.

[8] C. Morris, N. M. Kriege, F. Bause, K. Kersting, P. Mutzel, and M. Neumann, "Tudataset: A collection of benchmark datasets for learning with graphs," arXiv preprint arXiv:2007.08663, 2020.

[9] K. Debnath, R. L. Lopez de Compadre, G. Debnath, A. J. Shusterman, and C. Hansch, "Structure-activity relationship of mutagenic aromatic and heteroaromatic nitro compounds. correlation with molecular orbital energies and hydrophobicity," Journal of medicinal chemistry, vol. 34, no. 2, pp. 786-797, 1991.

[10] P. Yanardag and S. Vishwanathan, "Deep graph kernels," in Proceedings of the 21th ACM SIGKDD international conference on knowledge discovery and data mining, 2015, pp. 1365-1374.

[11] Shchur, M. Mumme, A. Bojchevski, and S. Günnemann, "Pitfalls of graph neural network evaluation," arXiv preprint arXiv:1811.05868, 2018.

[12] W. Hu et al., "Open graph benchmark: Datasets for machine learning on graphs," Advances in neural information processing systems, vol. 33, pp. 22118-22133, 2020.

[13] J. Deng, W. Dong, R. Socher, L.-J. Li, K. Li, and L. Fei-Fei, "Imagenet: A large-scale hierarchical image database," in 2009 IEEE conference on computer vision and pattern recognition, 2009: Ieee, pp. 248-255.

Optimizing LLM inference for FPGAs

Jorge R De Freitas*, Jose G. F. Coutinho*, Ce Guo*, Suleyman Demirsoy[†], Wayne Luk*, Zhiqiang Que*,

[†] Altera, UK, suleyman.demirsoy@altera.com

* Imperial College London, UK. {jorge.de-freitas22, jgfc, c.guo, w.luk, z.que}@imperial.ac.uk

Abstract—**Large Language Models (LLMs) deliver state-of-the-art performance but demand high computation and memory, making deployment in resource-limited settings challenging. Field-Programmable Gate Arrays (FPGAs) offer parallelism and efficiency, yet most prior FPGA accelerators rely on low-level, platform-specific flows that hinder portability. This work presents oneLLM, to our knowledge, the first FPGA-based LLM inference design using Intel's oneAPI, enabling a unified high-level programming model across CPUs, GPUs, and FPGAs. Our deeply pipelined, multi-kernel hardware architecture connects specialized kernels via oneAPI pipes for on-chip streaming, reducing host–device communication. Implemented on an Intel Agilex 7 FPGA, it achieves 3 times faster than a CPU implementation, and 8.8 times faster than a non-pipelined baseline while meeting resource constraints, demonstrating the potential of portable FPGA development for LLM acceleration. Code available at https://github.com/custom-computing-ic/llm-oneapi-fpga.**

I. INTRODUCTION

Large Language Models (LLMs) deliver state-of-the-art results in question answering, code generation, and dialogue, but their hundreds of millions to billions of parameters usually require large GPU clusters for training and deployment [1, 2]. Bringing LLM inference to resource- and latency-constrained settings such as edge and embedded devices is challenging due to high compute demand, memory bandwidth pressure, and energy cost [3]. Although pruning, quantization, and distillation can shrink models, transformer workloads still need highly parallel and memory-efficient execution to meet high throughput and low latency targets. GPUs provide strong throughput but often at high power and with limited support for deeply customized dataflows, whereas FPGAs offer fine-grained parallelism, reconfigurable datapaths, and tight coupling between computation and memory [4], making them attractive for edge LLM inference.

Despite this potential, deploying LLMs on FPGAs presents several technical challenges. First, complex transformer operations such as attention mechanisms, normalization layers, and position encodings must be mapped into hardware-friendly kernels without exceeding device resource limits. Second, designers must balance performance and resource utilization to ensure that multi-stage inference pipelines fit within the FPGA's logic, memory, and DSP budgets [5]. Third, communication overheads between processing stages and between the FPGA and host can significantly degrade throughput if not carefully managed. Finally, many prior FPGA-based accelerators rely on low-level, platform-specific design flows, which increase development complexity and limit portability across different hardware targets.

Recent works have explored FPGA-based LLM acceleration from various angles. Llamaf [6] targets embedded FPGAs with a lightweight Llama2 [7] architecture, while HLSTransform [8] applies AMD high-level synthesis (HLS) optimizations for energy-efficient Llama2 inference on Amazon FPGA Cloud. The TinyLLM co-design approach [9] integrates programmable logic with software execution to balance flexibility and performance, and recent surveys [10] have listed key trends and design strategies for small language models. While these approaches have demonstrated promising results, they are often based on custom design flows or low-level hardware development, which can limit portability across different computing platforms.

In this work, we present oneLLM which leverages Intel's oneAPI framework to enable portable, high-level FPGA development for LLM inference. To the best of our knowledge, this is the first study to explore LLM inference on FPGAs using the oneAPI framework, providing a unified programming model that can target CPUs, GPUs, and FPGAs while still delivering competitive performance.

We make the following contributions in this paper:

1) Design and optimization (Section III): We propose oneLLM, a deeply pipelined, multi-kernel hardware architecture for Llama2 inference on FPGAs that uses oneAPI pipes to alleviate communication overheads between kernels.

2) Implementation (Section IV): We implement the entire forward pass of Llama2 models in SYCL, decomposing it into specialized hardware kernels (e.g., MatMul, RMSNorm, RoPE) connected via a streaming dataflow targeting an Intel Agilex 7 FPGA, and develop a cycle-accurate performance model to guide architectural exploration. In addition, we open-source the code to benefit the wider research community.

3) Evaluation (Section V): Our design is 3 times faster than CPU and achieves 8.8 times higher throughput than a non-pipelined baseline while fitting the target FPGA resources.

II. RELATED WORK

Several recent studies have examined LLaMA2 inference on FPGAs. Llamaf [6] targets embedded devices with a lightweight Llama2 architecture, while HLSTransform [8] uses AMD HLS for energy-efficient inference on FPGA cloud platforms. TinyLLM [9] adopts a hardware–software co-design to balance flexibility and performance, and a recent survey [10] outlines trends in small language model deployment for resource-constrained systems. Other FPGA-based LLM accelerators address different challenges: FlightLLM [11] develops a complete FPGA mapping flow, EdgeLLM [12] builds

979-8-3315-3918-4/25 $31.00 © 2025 IEEE

a CPU–FPGA heterogeneous edge accelerator, An agile framework [13] enables rapid design exploration, Ref. [14] maximizes memory bandwidth utilization on embedded FPGAs, and MEADOW [15] improves memory efficiency for low-power edge LLM inference. Broader architectural approaches include GLITCHES [16], a GPU–FPGA collaborative system. In contrast, our work is the first to explore LLM inference on FPGAs using Intel's oneAPI framework, delivering a portable, high-level development flow and a deeply pipelined architecture that mitigates inter-kernel communication overheads while achieving competitive performance on Intel Agilex 7 devices.

III. DESIGN AND OPTIMIZATION

The core contribution of oneLLM is a device-agnostic accelerator architecture for transformer inference on FPGAs. The main challenges are, first, sustaining high tokens-per-second under tight logic, on-chip memory, and external bandwidth budgets, and second, avoiding host-to-device launch and transfer overhead while keeping a deeply pipelined dataflow stable across heterogeneous operators with mismatched latencies. Our design introduces two key features to address these issues. First, we build a persistent multi-kernel pipeline that keeps intermediate tensors on chip and connects attention, feed-forward, normalization, and residual paths through streaming FIFOs with explicit back-pressure. This removes per-kernel invocation costs and host intervention, limits external memory traffic to inputs and outputs, and enables latency balancing by tuning FIFO depths. Because the pipeline is expressed as a graph of communicating kernels rather than a monolithic block, it exposes pipeline parallelism and maps to any FPGA platform that offers synthesizable streaming FIFOs and persistent kernels; an RTL realization would use similar streaming interfaces with on-chip FIFOs. Second, we adopt a resource-aware decomposition and quantization strategy: operators are partitioned to minimize cross-kernel state and buffer duplication, and weights used in matrix multiplications are quantized with symmetric per-group INT8, while sensitive parameters such as RMSNorm scale and bias remain in `float32`. This combination delivers most of the bandwidth and storage benefits of low precision with minimal accuracy loss and without exceeding device capacity.

We implement these principles with the Intel oneAPI programming model. oneAPI's `pipe` primitive provides a high-level abstraction for point-to-point FIFO channels in the FPGA fabric, allowing us to describe the pipelined dataflow in C++ and synthesize it into an efficient hardware realization. The result is a productive path from architecture to implementation that preserves portability while achieving high throughput.

The primary objective was to maximize throughput, measured in tokens per second (tok/s), while ensuring the entire design could be successfully placed and routed on the target device. Our exploration systematically progressed through several design iterations, culminating in a deeply pipelined multi-kernel architecture. Our initial approach centred on a single, monolithic kernel designed to execute an entire layer of the model's forward pass. The hypothesis was that this would

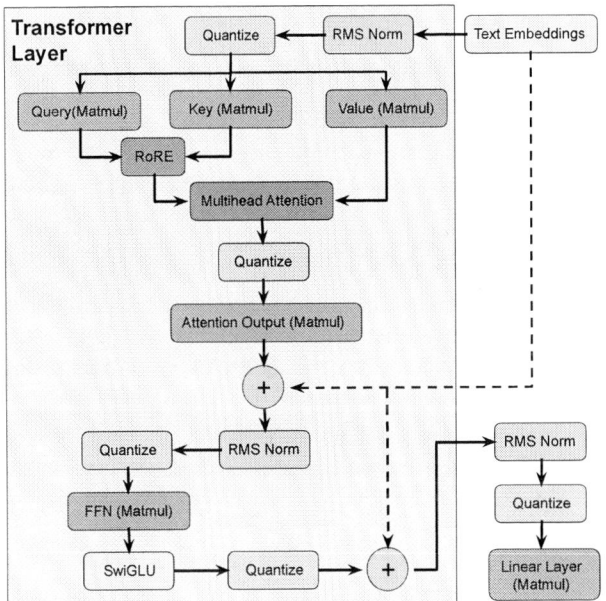

Fig. 1. Pipelined multi-kernel dataflow for transformer inference. The sequence starts from text embeddings and passes through 12 repeated Transformer layers (Llama 2 110M). Within each layer, RMSNorm precedes (i) a multi-head attention block, comprising Query/Key/Value matrix multiplications, RoRE, and the attention-output matmul, and (ii) a feed-forward network with SwiGLU. Dashed arrows indicate residual connections. Green "Quantize" boxes mark where weights used by the subsequent matrix multiplications are quantized (Q8_0); all other parameters, especially RMSNorm scale and bias, remain `float32`. The overall implementation divides attention, FFN, and the final linear projection into cooperating kernels that form a pipeline, allowing overlap of compute while preserving the standard Transformer data dependencies.

minimise data transfer overhead between the host system and the FPGA accelerator. However, this design proved to be impractical due to excessive resource utilization, exceeding the logic and memory capacity of a target FPGA. A subsequent refinement involved integrating tiled matrix multiplication and buffering the run state on-chip. While this improved raw computational speed, it further increased the design's footprint, rendering it unsuitable for deployment. These initial findings prompted a pivot to a multi-kernel strategy. In this approach, the forward pass was decomposed into a series of smaller, more manageable kernels. While this resolved the resource utilization issue, it introduced a new performance bottleneck. The frequent data transfers between the host and the device for each kernel, coupled with the invocation latency of launching each kernel, severely degraded overall performance. The final and most successful architecture, illustrated in Figure 1, retains the multi-kernel structure but eliminates the host-device communication bottleneck by connecting the kernels using oneAPI's pipe primitives. This creates a deeply pipelined hardware implementation directly on the FPGA. Data flows from one kernel to the next through dedicated on-chip channels, removing the need for intermediate data to travel back to host memory. This design not only mitigates the data transfer overhead but also substantially reduces the effective invocation latency, leading to a significant improvement in end-to-end throughput.

IV. IMPLEMENTATION

A. Performance Estimation Model

To predict the performance of the pipelined kernel groups, we developed a cycle-accurate performance model. In the proposed multi-kernel dataflow with coarse grained pipelining, the processing of the cascaded kernels can be overlapped. The second kernel does not need to wait for the whole results of first kernel to be ready. Since we are using same unrolling for all the computational intensive kernels, just one result vector from the first iteration of the former kernel is sufficient to start the calculation of the next kernel.

The complete cycle number for each individual kernel k is then calculated based on its II, its specific iteration count N_k, and its pipeline depth (latency to first output) D_k reported by oneAPI, and the start offset of the kernel Offset_k:

$$\text{Cycles}_k = \text{II}_k \times (N_k - 1) + D_k + \text{Offset}_k \quad (1)$$

And the Offset_k is:

$$\text{Offset}_k = \begin{cases} \text{Offset}_{k-1} + D_{k-1}, & \text{if } k \geq 1 \\ 0 & \text{if } k = 0 \end{cases} \quad (2)$$

Once each kernel's execution time is calculated and the amount of overlap/offset between consecutive kernels is identified, the total latency of the design can be estimated by adding up the execution times across all the kernels in Fig. 1 and removing the overlaps.

B. Implementation with oneAPI

The final design was implemented using Intel oneAPI 2024 (DPC++) with C++ High-Level Synthesis (HLS), targeting an Intel Agilex 7 AGFB014R24A2E2V FPGA. The implementation instantiates the architecture described in Section III, decomposing the Llama 2 forward pass into 37 distinct kernels connected by on-chip pipes. The current build supports three Tiny Llama 2 configurations of 15M, 42M, and 110M parameters by adjusting tensor shapes and tiling factors while reusing the same kernel graph and performance model. Focusing on smaller models is practically relevant: widely used encoder models such as BERT base at 110M and ALBERT (lite BERT) at 68M deliver near state of the art results across many NLP tasks and remain in broad deployment [17]. Each kernel implements a self-contained operation such as tiled matrix multiplication ('Matmul'), RMS normalization ('RMSNorm'), rotary position embedding ('RoPE'), SwGLU and residual add. Kernels communicate through typed, bounded pipes. Key hardware parameters for the target platform include a memory clock speed of 480 MHz and a transceiver bandwidth of 58 Gbps across 8 lanes. Although this implementation targets Agilex 7 with oneAPI, nothing in the implementation prevents mapping to other FPGAs. The key ideas are persistent kernels and synthesizable streams with back-pressure. oneAPI pipes would be replaced with the corresponding streaming primitives in another toolchain, and the performance model remains unchanged because it is expressed in terms of per-stage II and latency.

V. EVALUATION

This section evaluates the performance and resource costs of our implemented design. We first compare our final pipelined architecture against the earlier design iterations to quantify the impact of our internal architectural choices. We then benchmark our work against state-of-the-art CPU, GPU, and FPGA implementations from related literature to position our contribution within the wider field.

A. Performance and Resource Utilization Results

Table I compares resource utilization and throughput for four major design iterations, combining estimates from our performance model based on the oneAPI compiler report and achievable 480MHz clock frequency with 4 DDR banks. It details the hardware resource consumption for each design. All results assume a target FPGA board equipped with four DDR4 memory channels. As shown, the two single-kernel variants are too large to be fit on the target FPGA, whereas the multi-kernel designs fit within the device's constraints. Splitting the design into multiple kernels without pipes fits well but the performance is low, only 7.95tok/s. Adding pipes to form a pipelined multi-kernel design raises throughput to 70.31tok/s ($8.8\times$ over the no-pipes version) while remaining within the device limits, demonstrating a performance-resource trade-off. This performance improvement comes at the cost of increased resource utilization, with the pipelined design consuming 81% more ALUTs and 123% more FFs. Despite this increase, the design remains efficient, utilizing 83% of available ALUTs.

B. Comparative Analysis

We compare our final design with the work of HLSTransform [8], which also targets a 110M Llama-2 model. Table II presents a performance comparison across different hardware platforms. Our FPGA implementation achieves a throughput of 70.31 tok/s, which is more than triple the performance of the Intel Xeon CPU baseline. As expected, our design does not outperform the NVIDIA 3090 GPU, which benefits from a much higher clock rate and memory bandwidth. The RTX 3090 has 24GB VRAM running at 1219 MHz with a base core clock of 1395 MHz.

When comparing against the prior FPGA design (HLSTransform [8] on Xilinx VU9P) which achieves 57.11 tok/s using 1.18 M LUTs, our oneLLM on Intel Agilex 7 AGF014 achieves 70.31 tok/s while using 0.81 M LUTs. It has about 23% higher throughput than the prior FPGA. The novel aspect of our work is the end-to-end pipelining of the entire forward pass, which minimizes host-device interaction to a single invocation. This contrasts with other multi-kernel approaches that may require more frequent synchronization with the host [18], incurring latency penalties.

Our analysis shows the current design is limited by external memory bandwidth rather than compute. The streaming pipeline back-pressures on DDR4, and the Matmul stage stalls

TABLE I
RESOURCE UTILIZATION AND PERFORMANCE COMPARISON OF INTERNAL DESIGN APPROACHES TARGETING A LLAMA 2 110M MODEL. PERCENTAGES SHOW UTILIZATION RELATIVE TO THE DEVICE MAXIMUM. VALUES OVER 100% ARE IN BOLD.

Design	ALUTs	FFs	MLABs	RAMs	DSPs	Throu.(tok/s)	Fit?
Single Kernel (no buffers)	1,522,445 (**156%**)	1,980,829 (67%)	37,840 (**155%**)	260,744 (**3667%**)	1,262 (28%)	11.51	No
Single Kernel (with buffers)	2,711,593 (**278%**)	3,312,105 (**111%**)	73,746 (**303%**)	3,868 (54%)	1,606 (36%)	14.79	No
Multiple Kernels (no pipes)	535,424 (55%)	596,346 (20%)	8,244 (34%)	5,237 (74%)	1,040 (23%)	7.95	Yes
Multiple Kernels (pipelined)	811,962 (83%)	1,152,251 (39%)	21,314 (87%)	6,147 (86%)	2,714 (60%)	**70.31**	Yes
Device Resources Max	**974,400**	**2,971,920**	**24,360**	**7,110**	**4,510**	-	

TABLE II
PERFORMANCE COMPARISON WITH PREVIOUS WORK

Platform	Design	Throu. (tok/s)	LUT Count
CPU [8]	Intel Xeon E5-2686 v4	23.21	-
GPU [8]	NVIDIA RTX 3090	107.00	-
FPGA [8] HLSTransform	Xilinx VU9P	57.11	1.18M
FPGA (oneLLM) This Work	Intel Agilex 7 AGF014	**70.31**	0.81M

when input data cannot be supplied fast enough. Consequently, the final throughput is upper-bounded by the effective bandwidth delivered by the four DDR4 channels. Accordingly, adding compute or further unrolling will not increase throughput until memory traffic per token is reduced or the external memory bandwidth is increased, such as using High Bandwidth Memory (HBM). In our future work, we plan to raise bandwidth utilization, such as using wider AXI bursts, or alignment, or bank interleaving across the four channels, and deeper double-buffered prefetch. We also plan to cut bytes per token via more fine-grained tiling/on-chip reuse, operator fusion, lower-precision quantization, sparsity/structured pruning. Overall, the results show that our architecture is not only resource-efficient but also offers a higher performance ceiling than HLSTransform [8] on comparable hardware. This work establishes a strong baseline for high-performance LLM inference on Intel FPGAs using the oneAPI framework.

VI. CONCLUSIONS

This work presents the first study of LLaMA2 inference on FPGAs using Intel's oneAPI framework, showing that portable, high-level development flows can achieve competitive performance for LLMs on reconfigurable hardware. Our deeply pipelined multi-kernel design, connected via oneAPI pipe primitives, reduced inter-kernel communication overheads and delivered a 8.8 times throughput gain over a non-pipelined baseline while meeting Intel Agilex 7 resource constraints. Future work includes scaling pipelined LLM inference on FPGA clusters, such as ESSPER [19], improving memory bandwidth (e.g, HBM) and utilization, automating the approach using metaprogramming (e.g, MetaML [20]), and extending it to support larger models with compression techniques.

Acknowledgement. The support of the United Kingdom EPSRC (grant number UKRI256, EP/V028251/1, EP/N031768/1,

EP/S030069/1, and EP/X036006/1), Altera and Intel is gratefully acknowledged.

REFERENCES

[1] T. Brown et al., "Language models are few-shot learners," *Advances in neural information processing systems*, vol. 33, pp. 1877–1901, 2020.

[2] B. Workshop et al., "Bloom: A 176b-parameter open-access multilingual language model," *arXiv preprint arXiv:2211.05100*, 2022.

[3] Z. Wan et al., "Efficient large language models: A survey," *arXiv preprint arXiv:2312.03863*, 2023.

[4] Z. Que, H. Fan et al., "LL-GNN: Low latency graph neural networks on FPGAs for high energy physics," *ACM Transactions on Embedded Computing Systems*, 2024.

[5] Z. Que et al., "Recurrent neural networks with column-wise matrix–vector multiplication on FPGAs," *IEEE Transactions on Very Large Scale Integration (VLSI) Systems*, 2021.

[6] H. Xu, Y. Li, and S. Ji, "Llamaf: An efficient llama2 architecture accelerator on embedded FPGAs," in *IEEE 10th World Forum on Internet of Things (WF-IoT)*, 2024.

[7] H. Touvron et al., "Llama 2: Open foundation and fine-tuned chat models," *arXiv preprint arXiv:2307.09288*, 2023.

[8] A. He, D. Key, M. Bulling, A. Chang, S. Shapiro, and E. Lee, "HLSTransform: Energy-Efficient Llama 2 Inference on FPGAs Via High Level Synthesis," *arXiv preprint arXiv:2405.00738*, 2024.

[9] M. Muller et al., "Co-design of a TinyLLM using Programmable Logic and Software on an FPGA," in *IEEE 67th International Midwest Symposium on Circuits and Systems (MWSCAS)*, 2024.

[10] C. Van Nguyen et al., "A survey of small language models," *arXiv preprint arXiv:2410.20011*, 2024.

[11] S. Zeng et al., "FlightLLM: Efficient large language model inference with a complete mapping flow on FPGAs," in *Proceedings of the 2024 ACM/SIGDA International Symposium on Field Programmable Gate Arrays*.

[12] M. Huang et al., "EdgeLLM: A highly efficient CPU-FPGA heterogeneous edge accelerator for large language models," *IEEE Transactions on Circuits and Systems I: Regular Papers*, 2025.

[13] L. Chen et al., "An agile framework for efficient llm accelerator development and model inference," in *Proceedings of the 43rd IEEE/ACM International Conference on Computer-Aided Design*, 2024.

[14] J. Li et al., "Pushing up to the limit of memory bandwidth and capacity utilization for efficient LLM decoding on embedded FPGA," in *2025 Design, Automation & Test in Europe Conference (DATE)*. IEEE.

[15] A. Moitra, A. Ghosh, S. Agarwal, A. Amarnath, K. Swaminathan, and P. Panda, "Meadow: Memory-efficient dataflow and data packing for low power edge llms," *arXiv preprint arXiv:2503.11663*, 2025.

[16] F. Yang et al., "GLITCHES: GPU-FPGA LLM Inference Through a Collaborative Heterogeneous System," in *IEEE High Performance Extreme Computing Conference (HPEC)*. IEEE.

[17] A. Rogers, O. Kovaleva, and A. Rumshisky, "A primer in BERTology: What we know about how BERT works," *Transactions of the association for computational linguistics*, 2021.

[18] turingmotors, "Swan: A lightweight language model execution environment using fpga," https://github.com/turingmotors/swan, 2024.

[19] K. Sano et al., "ESSPER: Elastic and scalable FPGA-cluster system for high-performance reconfigurable computing with supercomputer Fugaku," in *Proceedings of the International Conference on High Performance Computing in Asia-Pacific Region*, 2023.

[20] Z. Que et al., "MetaML-Pro: Cross-Stage Design Flow Automation for Efficient Deep Learning Acceleration," *arXiv preprint arXiv:2502.05850*, 2025.

Fine-Grained Layer Scheduling and Mapping for Chiplet-Based LLM Inference

Hongyang Gu, Lei Xu, Haochen Zhao, Naifeng Jing*

Department of Micro-Nano Electronics, Shanghai Jiao Tong University

* Email: guhongyang@sjtu.edu.cn, sjtuj@sjtu.edu.cn

Abstract—The inference of long-context large language models (LLMs) has been increasingly deployed on chiplet-based accelerators due to their scalable computational capabilities. However, existing coarse-grained scheduling approaches fail to effectively leverage data reuse opportunities within the fine-grained topology of chiplet architectures, leading to substantial resource under-utilization. In particular, prior work employs layer pipeline (LP) and spatial partitioning for parallel execution, where long-path data transfers—such as inter-chiplet and off-chip communication—become performance and energy bottlenecks. In this work, we introduce a fine-grained scheduling strategy that fuses adjacent operators within tiled subdomains to improve on-chip data locality and reduce long-path communication overhead. Building on this strategy, we develop a mapping framework that systematically models the interplay among fusion strategies, tiling granularity, and loop ordering, enabling comprehensive design space exploration for enhanced data reuse and minimized die-to-die (D2D) communication. It enables more Pareto-optimal tradeoffs under the constraints of the chiplet package, such as on-chip buffer size and long-path transfer overhead. Comprehensive evaluation demonstrates that our method shows 11.4% average and up to 45.8% energy reduction.

Keywords—Large Language Models, Chiplets, Fine-grained Scheduling, Operator Fusion

I. INTRODUCTION

Large language models (LLMs), consisting of billions of parameters and supporting long token sequences, have driven significant advances in natural language processing, generative AI, and multi-modal reasoning. However, their intensive computational and memory demands pose significant challenges for resource-limited monolithic dies, which face inherent constraints in scalability, power efficiency, and on-chip memory capacity. To address these limitations, chiplet-based accelerators have emerged as a promising solution [4]. By enabling modular scaling of compute and memory resources connected via interconnects, chiplets provide abundant aggregated resources that better meet the requirements of large-scale LLM inference. This architecture not only enhances scalability but also offers improved flexibility and resource utilization compared to traditional monolithic designs.

However, the main challenges stem from the increased scale enabled by chiplet technology and costly die-to-die (D2D) links, which pose significant difficulties for efficiently scheduling LLM operations [6]. Layer pipeline (LP), as leveraged in Tangram [3], distributes workloads across chiplets to enable parallel execution, but introduces considerable inter-chiplet communication overhead. Spatial mapping (SPM), proposed by Gemini [2], determines the allocation of layer partitions to specific clusters to minimize inter-chiplet communication. However, the lack of on-chip

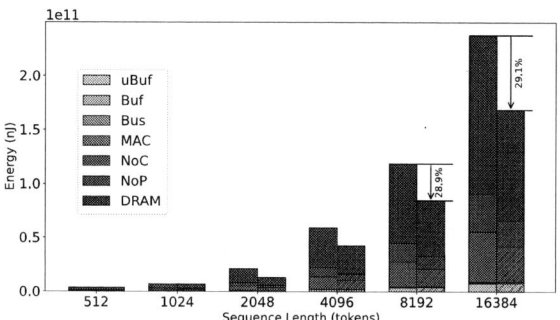

Fig. 1. Experimental results for mapping BERT. For each sequence length, the left bar shows Gemini, and the right bar shows our tiling-aware layer fusion scheduling.

data reuse leads to huge off-chip data transfer, dominating both energy consumption and execution latency.

Layer fusion, which schedules dependent layers in the layer-sequential (LS) manner to buffer inter-layer data for reuse, has been recognized as a promising optimization to reduce off-chip communication. However, redundant data retention across fused layers can lead to excessive pressure on limited local buffer capacity on the chiplet cluster, ultimately diminishing the benefits of fusion and potentially degrading performance [1].

We identify that the key drawback lies on the mismatch between **coarse-grained scheduling methods** and **fine-grained resource fabric in chiplet-based architectures**.

While existing methods schedule workloads at the full-layer or full-batch level, the chiplet fabric consists of numerous small, distributed compute and memory units interconnected with varying communication costs. This mismatch prevents effective exploitation of local data reuse within chiplets, leading to costly long-path data transfers across chiplets and off-chip memory, thereby increasing energy consumption and latency [5].

These limitations become more pronounced as the sequence length of LLMs increases. As illustrated in Fig. 1, the energy cost of NoP and DRAM grows significantly faster than that of intra-chiplet components, such as MACs and buffers.

To solve this problem, we propose a fine-grained scheduling method that enables fusion across arbitrary tensor dimensions. As shown in Fig. 1, our approach achieves over 29.1% reduction in energy consumption compared to Gemini as sequence length increases, which substantially reduces communication energy associated with Network-on-Chip (NoC) and Network-on-Package (NoP) hops as well as DRAM access. Building upon this, we develop a DNN

979-8-3315-3918-4/25 $31.00 © 2025 IEEE

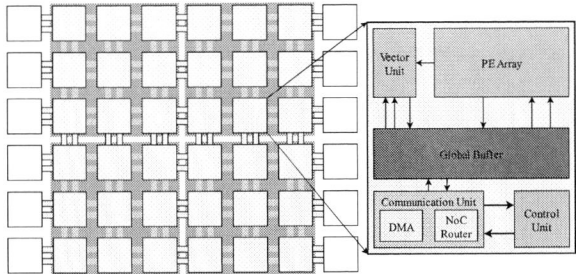

(a) 6x6 Mesh with Xcut = Ycut = 2   (b) Computing Core

Fig. 2. An example of 6×6 mesh with $X_{cut} = Y_{cut} = 2$. Gray-shaded regions represent chiplet packages. Thick black lines mark D2D links. DRAM routers at mesh edges handle off-chip memory.

mapping framework that systematically models the interactions among fusion strategy, tiling granularity, and loop order. By exploring different combinations of these factors, our framework facilitates more Pareto-optimal trade-offs under various chiplet resource constraints, such as on-chip buffer capacity and long-path communication overhead. Evaluation on Gemini evaluator demonstrates that our approach achieves 11.4% average and up to 45.8% energy reduction.

II. FINE-GRAINED LAYER SCHEDULING FRAMEWORK

A. Chiplet-Based Accelerator Architecture

The chiplet architecture consists of multiple computing cores interconnected via a mesh NoC within each chiplet. To enable scalable inter-chiplet communication, a set of D2D interfaces is placed along the periphery of each chiplet, with the number of interfaces matching the number of computing cores on each edge as show in Fig. 2(a). This design allows chiplets to form a larger mesh topology when integrated. As illustrated in Fig. 2(b), each computing core comprises a communication unit (including DMA and router), a control unit, a global buffer, a PE array, and a vector unit. The global buffer in each computing core is accessible by other cores for valid reads and writes, facilitating efficient data sharing. The PE array is responsible for executing matrix multiplication or convolution operations, while the vector unit handles vector/scalar computations and post-processing. This architectural organization supports high-throughput parallel computation and efficient on-chip data movement.

B. Framework Overview

To represent the design space of the fine-grained layer fusion scheduling for chiplet-based DNN accelerators, we use a recursive Resource Allocation Tree (RA Tree) structure [1] as shown in Fig. 3(a), which consists of three types of nodes:

- **T-Cut nodes**: indicating time-division allocation of hardware resources, where child nodes execute sequentially using the same set of clusters.

- **S-Cut nodes**: indicating space-division allocation, where child nodes are assigned different subsets of hardware clusters and execute in parallel.

- **Tile nodes**: representing specific loop tiling execution plan on clusters that correspond to the actual computing cores.

Given a DNN model, hardware architecture, and a specified optimization objective—typically the energy-delay product (EDP)—our framework searches for the optimal RA

Tree to schedule layers across chiplets. The overall process is outlined as follows:

*1) **Outer-level fusion segment Search**: Based on the estimated costs, we apply dynamic programming to identify beneficial layer fusion boundaries, which define the **fusion segments**. Each fusion segment corresponds to a T-Cut subtree within the RA tree.*

*2) **Intra-segment Optimization**: For each fusion segment, we search for the optimal **loop order** and **tiling factors** that minimize traffic and maximize reuse in on-chip buffers by a fast polyhedral-based cost model.*

*3) **Cluster Allocation and Placement**: For each subtree, we estimate the compute workload and proportionally assign computing cores using the stripe-based placement strategy and XY routing [3].*

*4) **Mapping via Gemini Backend**: The generated RA Tree is passed to Gemini's LP SPM mapper, which performs detailed spatial placement mapping to produce the final execution plan as illustrated in Fig. 3(b).*

C. Outer-level Fusion Segment Search

To reduce algorithmic complexity, we assume the execution order follows the topological sorting of the network layers, assigning an index to each layer accordingly.

To enable a fast and thorough exploration of the layer fusion design space, we formulate the layer fusion schedule problem as a dynamic programming optimization. Let $C[i]$ denote the minimal cumulative scheduling cost for layers from 0 to i under the fusion combination, and let $c(j, i)$ represent

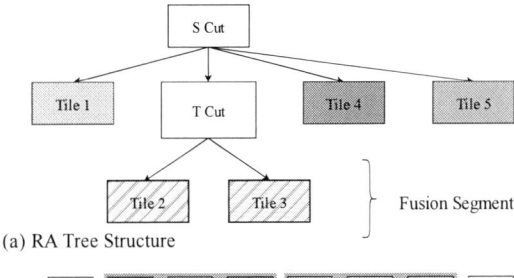

(a) RA Tree Structure

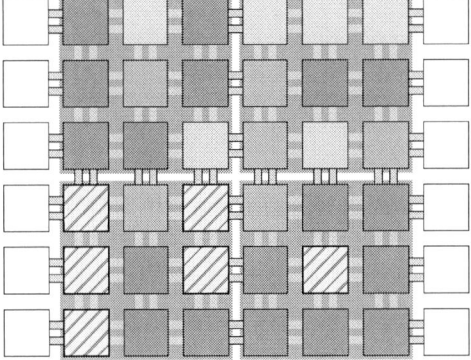

(b) Mapping on 6x6 Mesh with Xcut = Ycut = 2

Fig. 3. (a) The scheduling is represented as the RA tree structure that consists of T-Cut, S-Cut and Tile nodes. (b) An example mapping on a 6×6 mesh. Cluster colors show allocated tile nodes; diagonal hatching indicates multiple tile nodes from a fusion segment mapped together.

979-8-3315-3918-4/25 $31.00 © 2025 IEEE

the cost of fusing layers j through i under the optimal tiling factor combination $\mathcal{B}(j, i)$. The recurrence relation is defined as:

$$C[i] = \min_{0 \leq j \leq i}\{C[j-1] + c(j,i) \mid \mathcal{B}(j,i) \neq 0\} \quad (1)$$

with the boundary condition $C[-1] = 0$.

A fusion segment is defined as a contiguous sequence of layers [j, i] that are fused together for joint scheduling and optimized tiling. At each iteration i, the algorithm enumerates all candidate fusion segments [j, i] and selects the tiling factor that yields the minimal cumulative cost. A fusion segment is considered valid only if an admissible tiling factor combination exists, i.e., $\mathcal{B}(j, i) \neq 0$, ensuring feasibility of the scheduling.

This procedure iterates over all layers $i = 0, 1, \ldots, N-1$, concurrently recording the optimal tiling factor and fusion boundary configuration for each subproblem. To improve computational efficiency, results are memoized using a unique key derived from the fusion bitmask, thereby preventing redundant computations.

D. Intra-segment Optimization

Given a fusion segment, we optimize the loop order and tiling factors using a fast polyhedral-based cost model that efficiently estimates costs. We minimize the inter-chiplet traffic while respecting buffer capacity constraints, taking into account the tensors within the fusion segment. Let $\mathcal{X} = \{d_1, d_2, \ldots, d_n\}$ denote the set of all loop dimensions across the fused layers.

1) Loop Order of the Fusion Segment

In a fusion segment, all loop dimensions across layers are organized into a unified loop nest, denoted by an ordered set $\mathcal{D} = (d_1, d_2, \ldots, d_n)$, where $d_i \in \mathcal{X}, \forall i \in \{1, \ldots, n\}$, d_1 is the innermost loop and d_n the outermost. This ordering defines a total precedence relation, such that $d_i \prec d_j$ if d_i is nested deeper than d_j. The loop order $\mathcal{D}$ determines the execution and tiling schedule across the segment, directly impacting data reuse and communication efficiency [7].

2) Tile Size of the Tensor

For a tensor F with dimensions $\{d_{t1}, d_{t2}, \ldots, d_{tk}\}$ and tiling factors $\{t_{d_{tk}}\}$ where $t_d \in Z_{>0}, \forall d \in \mathcal{X}$, the base tile size is given by:

$$\text{TileSize}(F) = \prod_{d \in \text{Dims}(F)} \frac{|d|}{t_d} \quad (2)$$

3) Inter-chiplet Traffic Scaling

For certain non-batch dimensions that cannot be fully reused, the required data must be fetched repeatedly within the nested loops, leading to the scaling of the inter-chiplet traffic.

Formally, let the ordered loop dimensions be $\{d_1, d_2, \ldots, d_k\}$ from innermost (d_1) to outermost (d_k). Then the total on-chip traffic scaling factor for a fmap F can be expressed as:

$$\text{TrafficScale}(F) = \begin{cases} 0, & \text{if } F \in \mathcal{F}_{\text{intermediate}} \\ \prod_{\substack{d \in \mathcal{L}(F) \\ d_1 \prec d \prec d_m}} t_d, & \text{otherwise} \end{cases} \quad (3)$$

where $\mathcal{L}(F)$ is the set of loop dimensions whose loop order in the fusion segment lies between d_1 and d_m, and t_d is

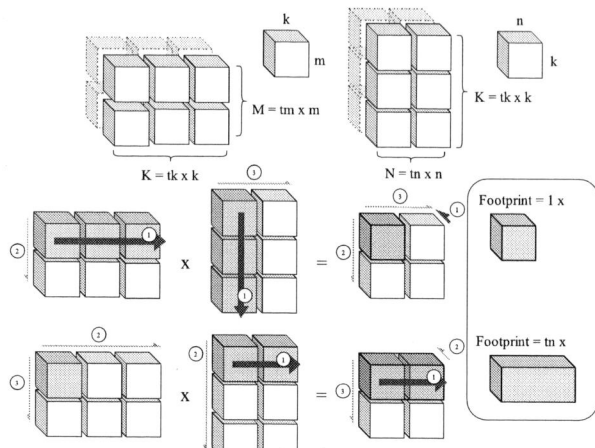

Fig. 4. Consider multiplying the matrix A (M×K) by B (K×N) with tiling, where K is the reduction dimension. With loop order MNK (output stationary), the output tile footprint is fixed at 1× tile. However, with loop order MKN (input stationary), since partial sums along N remain until completion, the output tile expands along the N dimension to enable immediate reuse in the fused layer. The loop order is indicated by the number in circles.

the tiling factor for dimension d. For those intermediate fmaps which is consumed and produced by layers within the same fusion segment, we exclude them from inter-chiplet traffic accounting.

4) Footprint for Intermediate Tensors

To guarantee that intermediate tensor tiles—specifically those passed from one layer to the next within a fusion segment—can be buffered, the footprint quantifies the required buffer size. As shown in Fig. 4, the footprint expands along outer loop dimensions that are tiled, which are in the ifmap's dimension set $\mathcal{D}_\ell$ but not in the ofmap's dimension set $\mathcal{D}_\mathcal{F}$.

Formally, upon encountering any $d \in \mathcal{D}_\ell \setminus \mathcal{D}_\mathcal{F}$, the footprint is:

$$\text{Footprint}(F) = \text{TileSize}(F) \times \prod_{d \in \mathcal{E}} t_d \quad (4)$$

where $\mathcal{E} \subseteq \mathcal{D}_\ell \setminus \mathcal{D}_\mathcal{F}$ are the relevant outer loop dimensions after expansion, and t_d are their tiling factors.

5) Fusion Segment to RA tree Transformation

To guide the Gemini mapper in performing cluster-level mapping based on the optimized loop order and tiling factors, we transform each fusion segment, which represents operator-level tiling, into a corresponding RA subtree that encodes the hardware scheduling strategy.

Each optimized fusion segment is transformed into a T-Cut node, with each fused layer tile mapped to a Tile node under it. These Tile nodes share a common placement—cluster allocation and partitioning—under the Gemini mapper. Within a layer segment, both unfused Tile nodes and fused T-Cut nodes are grouped under an S-Cut node as Fig. 3(a), which are assigned to clusters using a stripe-based strategy based on the estimated workload [3].

For row-wise operators such as softmax, we adopt the common row-wise decomposition approach, which enables their integration into tensor operators such as batched matrix multiplication [8].

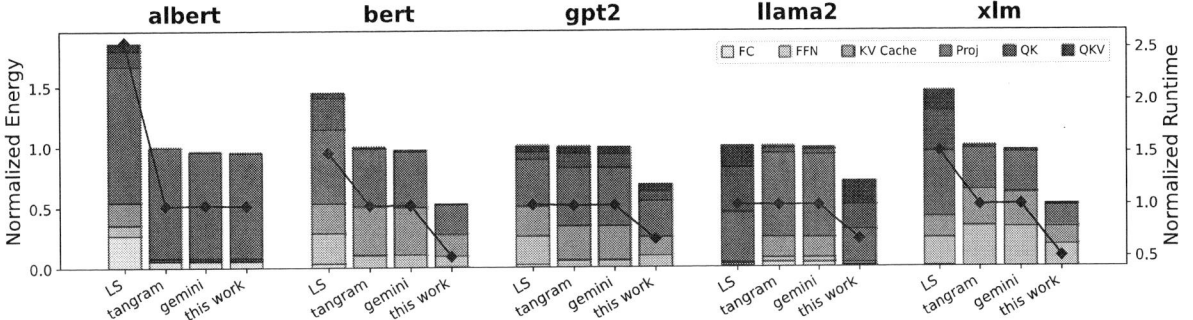

Fig. 5. Comparison with existing chiplet mapping strategies including LS, Tangram and Gemini. All designs are implemented using a 6 × 6 computing core mesh on a chiplet-based architecture with chiplet partitioning ($X_{cut} = Y_{cut} = 2$) as Fig. 2(a). Each computing core contains 16 × 16 MAC units with a 2 MB global

III. EXPERIMENTAL RESULTS

A. Experiment Setup

We conduct design space exploration with E^2D metric and evaluate our proposed mapping strategy on the Gemini evaluator, configured with a 6×6 mesh Simba architecture [4]. The baseline system targets a 7 nm technology node and operates at 1 GHz. Chiplets communicate via NoP with 4 GB/s bandwidth, while intra-chiplet communication is supported by 8 GB/s NoC. Each computing core integrates 256 MACs with 2 MB global buffer.

B. Evaluation across different LLM models

TABLE I. AVERAGE ENERGY REDUCTION (\%) OF OUR METHOD OVER DIFFERENT LLMS WITH SEQUENCE LENGTH FROM 128 TO 4096

Strategy	albert	bert	gpt2	llama2	xlm	**Average**
Gemini	0.19	26.97	8.58	5.59	15.39	**11.35**
Tangram	1.22	28.85	11.57	6.62	18.98	**13.45**
LS	12.65	52.14	43.93	15.72	57.05	**36.30**

For example, as shown in Fig. 5, our approach reduces energy consumption by 45.8% and latency by 49.0% on BERT, compared to Gemini. On GPT-2, which exhibits more non-uniform memory access patterns, our method achieves a 31.1% reduction in energy and a 31.0% reduction in latency. On average, our method achieves 11.4% and 13.5% energy reduction compared to Gemini and Tangram, respectively, and delivers the most significant gain of 36.3% over the LS strategy as shown in Table 1.

TABLE II. ENERGY CONSUMPTION OF BERT MODELS WITH VARYING BATCH SIZES, STRATEGIES, AND SEQUENCE LENGTHS (10^{10} PJ)

Model	Strategy	Sequence Length					
		128	256	512	1024	2048	4096
bert (batch=256)	LS	2.54	4.86	9.49	18.75	37.29	74.38
	Tangram	1.05	1.95	3.75	12.91	36.97	73.71
	Gemini	1.02	1.87	3.51	12.62	36.68	73.71
	This work	1.02	1.87	3.51	6.86	25.38	50.62
bert (batch=64)	LS	0.69	1.28	2.46	4.81	9.52	18.93
	Tangram	0.32	0.55	1.02	3.35	9.44	18.76
	Gemini	0.27	0.55	0.96	3.27	9.36	18.71
	This work	0.27	0.55	0.84	1.83	6.55	13.01
bert (batch=16)	LS	0.23	0.38	0.70	1.32	2.57	5.06
	Tangram	0.13	0.20	0.34	0.95	2.55	5.02
	Gemini	0.12	0.20	0.32	0.93	2.53	5.01
	This work	0.12	0.20	0.32	0.57	1.84	3.60

C. Scalability with Batch Size and Sequence Length

Table 2 demonstrates that our proposed method consistently outperforms the baseline Gemini strategy in energy consumption, with the advantage becoming more pronounced as batch size decreases and sequence length increases. This trend occurs because smaller batches reduce reuse opportunities, while longer sequences increase tensor sizes and data movement costs. Under these conditions, our fine-grained layer scheduling better exploits data locality, cuts redundant transfers, and improves on-chip buffer efficiency.

IV. CONCLUSION

In summary, we propose a fine-grained layer scheduling method to reduce long-path communication overhead in chiplet-based accelerators for LLMs. Our mapping framework explores the tradeoffs between fusion, tiling, and scheduling under hardware constraints. Experiments show our method reduces energy use by an average of 11.4%, up to 45.8%. This approach effectively boosts efficiency in scalable chiplet accelerators.

REFERENCES

[1] J. Cai, Y. Wei, Z. Wu, S. Peng, and K. Ma, "Inter-layer Scheduling Space Definition and Exploration for Tiled Accelerators," in *Proceedings of 50th Annual International Symposium on Computer Architecture (ISCA)*, Jun. 2023, pp. 1–17.

[2] J. Cai, Z. Wu, S. Peng, Y. Wei, Z. Tan, and G. Shi, "Gemini: Mapping and Architecture Co-exploration for Large-scale DNN Chiplet Accelerators," in *2024 IEEE International Symposium on High-Performance Computer Architecture (HPCA)*, Mar. 2024, pp. 156–171.

[3] M. Gao, X. Yang, J. Pu, M. Horowitz, and C. Kozyrakis, "TANGRAM: Optimized Coarse-Grained Dataflow for Scalable NN Accelerators," in *Proceedings of the 24th International Conference on Architectural Support for Programming Languages and Operating Systems (ASPLOS)*, Apr. 2019, pp. 807–820.

[4] Y. S. Shao *et al.*, "Simba: Scaling Deep-Learning Inference with Multi-Chip-Module-Based Architecture," in *Proceedings of the 52nd Annual IEEE/ACM International Symposium on Microarchitecture (MICRO)*, Oct. 2019, pp. 14–27.

[5] S. Zheng *et al.*, "TileFlow: A Framework for Modeling Fusion Dataflow via Tree-based Analysis," in *56th Annual IEEE/ACM International Symposium on Microarchitecture (MICRO)*, Oct. 2023, pp. 1271–1288.

[6] Z. Huang, S. Fan, C. Tang, X. Lin, S. Deng, and Y. Liu, "Hecaton: Training Large Language Models with Scalable Chiplet Systems," Nov. 27, 2024, *arXiv: arXiv:2407.05784*.

[7] S. Zheng *et al.*, "Chimera: An Analytical Optimizing Framework for Effective Compute-intensive Operators Fusion," in *2023 IEEE International Symposium on High-Performance Computer Architecture (HPCA)*, Feb. 2023, pp. 1113–1126.

[8] L. Xu, Z. Mo, Q. Wang, J. Jiang, and N. Jing, "Enabling Multiple Tensor-wise Operator Fusion for Transformer Models on Spatial Accelerators," in *Proceedings of the 61st ACM/IEEE Design Automation Conference (DAC)*, Jun. 2024, pp. 1-6.

A 16×16 High-Utilization Systolic Array Hardware Accelerator for Long-Sequence Flash-Attention Computation in Transformer

Zhenkun Li, Liji Wu*, Yi Yang*, Tianling Ren*, Le Wu, Xiangmin Zhang

[1] School of Integrated Circuits, Tsinghua University, Beijing, China
[2] Beijing National Research Center for Information Science and Technology, Beijing, China

* Email: lijiwu@tsinghua.edu.cn , yiyang@tsinghua.edu.cn, RenTL@tsinghua.edu.cn

Abstract—This paper presents a Flash-Attention accelerator design methodology based on a 16×16 high-utilization systolic array architecture for long-sequence Transformer applications. By reformulating the Flash-Attention algorithm into a blocked matrix computation pattern combined with an improved softmax architecture and on-chip memory optimization strategy, an accelerator system operating at 200MHz is implemented on a Xilinx Virtex-7 XC7VX690T-2FFG1761C FPGA platform. Experimental results demonstrate that the accelerator achieves an average speedup of 4.6× compared to conventional CPU implementations while maintaining a mean squared error (MSE) of 10^{-7} order magnitude and structural similarity (SSIM) above 0.98. Furthermore, we present: (1)a dynamic weight reloading mechanism for small systolic arrays, improving processing elements utilization to 79.01% in typical NLP application scenarios; (2)a hybrid-precision quantization-based matrix computation optimization scheme preserving model accuracy under 8/16-bit integer quantization. This research provides an effective hardware solution for lightweight Transformer deployment at the edge computing domain.

Keywords—Flash-Attention, Hardware Accelerator, Systolic Array, Transformer, Neural Network

I. INTRODUCTION

In recent years, many Large Language Models (LLMs) based on the Transformer architecture have emerged, such as GPT, DeepSeek, Qwen, etc. These growth of artificial intelligence technologies has intensified the demand for edge-side model deployment. Breakthroughs in knowledge distillation[1] and pruning techniques have established theoretical foundations for neural network compression. Against this backdrop, systolic arrays (SA) have emerged as a preferred architecture for deep learning accelerators due to their unique dataflow optimization characteristics.[2] The successful implementation of Google TPUs validates the architectural advantages in matrix computation: its regularized PE interconnection supports efficient synthesis and P&R, and its computing paradigm, comparing to SIMD, is instruction-free. That delivers 42 TOPS/W energy efficiency, which is 25~29 times superior to contemporary GPUs[3].

Despite the success of systolic arrays in CNN architectures[4], attention accelerators still face some critical challenges: (1) Limited sequence length—existing solutions suffer from intermediate matrix storage and computational efficiency issues when processing long sequences; (2) Low processing elements utilization—because of SAs need clock cycles to load weights, when typical Transformer accelerators are computing, more than 50% PEs are not really working. This results in severe hardware idling; (3) Exponential resource escalation—when QKV matrix dimensions improve, conventional accelerators experience exponential LUT and Flip-Flop consumption growth. To address these issues, this study makes the following contributions:

- Proposes a Flash-Attention accelerator architecture: We construct an execution engine combining "systolic array + nonlinear computing units" to make it possible for edge devices to efficiently compute long sequence attention (more than 1024 tokens).

- Develops a new weight loading strategy to implement a high utilization systolic array. Develops a scheduling algorithm that theoretically achieves more than 95% PE utilization in matrix multiplication, maintaining 79.01% measured utilization in real Transformer computing scenarios.

- Designs a time-multiplexed small-scale systolic array. Employs a 16×16 weight-fixed systolic array with pipelined dataflow, reducing computing resources consumption for deployment on edge devices while preserving a 16-bit precision.

II. BACKGROUND

A. Algorithm: Flash-Attention

The core mechanism of Flash-Attention employs matrix partitioning method (also known as the matrix tiling method) to decompose large scale softmax operations into iterative subtasks. By dynamically updating the output matrix in each iteration, it avoids storing intermediate results from long-sequence QK^T matrix multiplication, effectively reducing memory footprint and data loading times[5].

Although matrix tiling techniques are widely adopted in neural network computations, the softmax's characteristic in attention mechanisms introduces data dependencies that prevent direct decomposition of intermediate matrices. Flash-Attention solves this through an iterative formulation of softmax computation (1), transforming the regular softmax formula into an incremental updating process. This approach enables online calculation of the final results, overcoming traditional attention's dependency on the entire intermediate matrix storage.

$$m_i = \max\left(m_{i-1}, x_i\right)$$

$$l_i = \sum_{j=0}^{i} e^{x_j - m_i}$$

$$= \left(\sum_{j=0}^{i-1} e^{x_{j-1} - m_i}\right) + e^{x_i - m_i}$$

$$= \left(\sum_{j=0}^{i-1} e^{x_{j-1} - m_{i-1}}\right) e^{m_{i-1} - m_i} + e^{x_i - m_i} \quad (1)$$

979-8-3315-3918-4/25 $31.00 © 2025 IEEE

$$= l_{i-1}e^{m_{i-1}-m_i} + e^{x_i-m_i}$$

$$softmax_{0:i}(x_k) = \frac{e^{x_k}}{\sum_{j=0}^{i}e^{x_j}} = \frac{e^{x_k-m_i}}{\sum_{j=0}^{i}e^{x_j-m_i}} = \frac{e^{x_k-m_i}}{l_i}$$

$$= \frac{e^{x_k-m_{i-1}}}{l_{i-1}}(e^{m_{i-1}-m_i})\frac{l_{i-1}}{l_i}$$

$$= (e^{m_{i-1}-m_i})\frac{l_{i-1}}{l_i}softmax_{0:i-1}(x_k)$$

Building on this, the output matrix O computation also adopts a tiling strategy. By decomposing the partitioned product of softmax and V matrix into a row-wise iteration form (2), combined with NumPy indexing notation, V[j,:] denotes matrix row j, the output matrix is expressed in an iterative form. This partitioned computation method eliminates the need to store full QK^T intermediate matrix in the whole attention process, significantly lowering memory footprint and matrix loading requirements.

$$O_i = \left(\sum_{j=0}^{i}\frac{e^{x_j-m_i}}{l_i}\right)V[j,:]$$

$$= \left(\sum_{j=0}^{i-1}\frac{e^{x_j-m_i}}{l_i}\right)V[j,:] + \frac{e^{x_i-m_i}}{l_i}V[i,:]$$

$$= \left(\sum_{j=0}^{i-1}\frac{e^{x_j-m_{i-1}}}{l_{i-1}}\right)V[j,:](e^{m_{i-1}-m_i})\frac{l_{i-1}}{l_i}$$

$$+ \frac{e^{x_i-m_i}}{l_i}V[i,:]$$

$$= O_{i-1}\left((e^{m_{i-1}-m_i})\frac{l_{i-1}}{l_i}\right) + \frac{e^{x_i-m_i}}{l_i}V[i,:] \quad (2)$$

As shown in Fig. 1, the attention computation workflow based on tiling matrix multiplication takes the 4 matrices Q, K, V, and O into two groups: QO and KV. According to block matrix computation principles, the two matrices in each group maintain the same positions in the tile matrix during each iteration cycle.

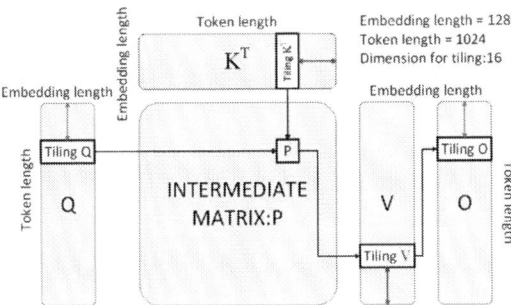

Fig. 1. Schematic of Flash-Attention Calculation Process

B. Regular Systolic Arrays

The dataflow of a standard weight-fixed systolic array is illustrated in Fig. 2(a), using the multiplication of two 3×3 matrices A and B to calculate C as an example: the weight matrix B (typically corresponding to model weights) is statically imported into PE units, while input matrix A moves horizontally. To satisfy data alignment requirements for matrix multiplication, input matrix A must be imported into the systolic array after 90° clockwise rotation.

The weight update operation for matrix B is only triggered after the last data element a_{33} exits the array (Fig. 2(b)). This update strategy, however, causes unacceptable issues for accelerators: pipeline bubbles theoretically reduce the array

utilization to 37.5%, but if we update it prematurely, it will cause computational errors. Notably, when the first output element c_{11} exits the array, the weight b_{11} has completed its final computational task, making this moment best for initiating weight updates to ensure continuous PE computation, but this would require a more complex hardware design on our work.

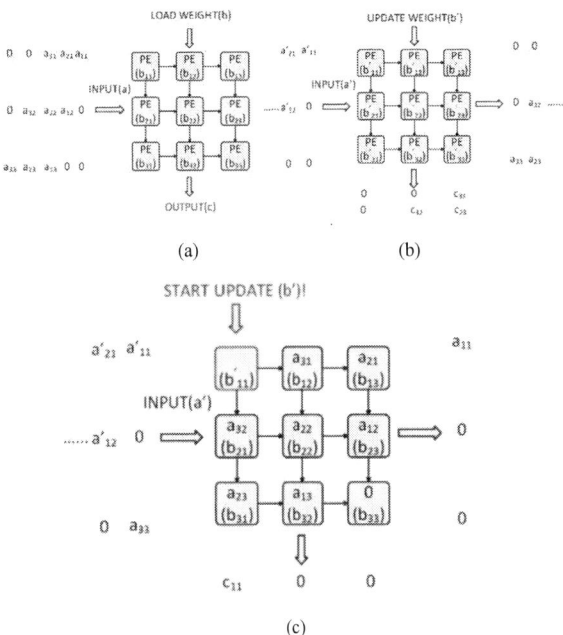

Fig. 2. (a) Dataflows in Conventional Systolic Array (b)Pipeline Bubbles in Conventional Systolic Array (c) Dynamic-Updating-Strategy for SA

Figure 2(c) illustrates the weight update strategy employed in this work. Adopting this dynamic weight update scheme represents an effective approach to enhance the utilization rate of PEs in the SA. Furthermore, it can be demonstrated that once the pipeline is fully enabled, this method achieves near-100% PE utilization when output, while minimizing the clock cycles are required in matrix multiplications.

III. HARDWARE DESIGN

According to the algorithm of Flash-Attention, to efficiently compute attention for long sequences on FPGA, we should update the attention results using the iterative tiling matrix approach in (2). To satisfy this computational paradigm, the system requires the following extra hardware resources:

- Extra mi/li memories: Implemented through on-chip BRAM for intermediate maximum and normalization factor storage.

- Optimized softmax computing unit: Compute and output mi/li parameters in each iteration cycle and apply them to the current computation flow, which incurs additional LUTs and Flip-Flops resource consumption.

- Coefficient computing unit: Calculates O-matrix update coefficients within each iteration cycle, mapped to additional DSP computing units.

Furthermore, to achieve hardware acceleration goal with such a small-scale systolic array, it should be highly utilized.

The systolic array we designed in this work employs a dynamic weight matrix update method, significantly improving the utilization rate of the array.

A. Hardware Architecture

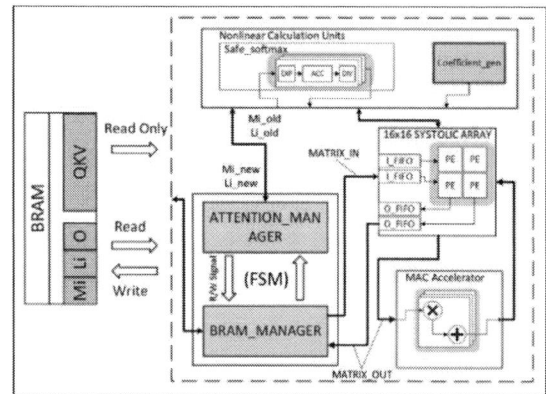

Fig. 3. Hardware Architecture of the Accelerator

The Flash-Attention-based hardware accelerator architecture is shown in Fig. 3. Compared to traditional architecture, there are three newly added modules marked in red boxes. The matrices to be processed are stored in on-chip BRAM, and the BRAM controller (BRAM_MANAGER) schedules dataflows to the systolic array during the Attention computation. According to the systolic array structure we implemented, and the block matrix calculation strategy we adopted, each read/write operation reads/writes a 16×16 data block from BRAM, and each iteration completing 16×128 matrix operations in all 64×64 iterative cycles.

In the Flash-Attention process, the scale operation involves vector-matrix multiplications. Considering the startup latency of the SA, these computational tasks are handled by the newly added independent Multiply-Accumulate units (MACs), in order to avoid idle cycles in SA and improve overall computational efficiency.

B. Systolic Array

As illustrated in Fig. 4(a), the dynamically updatable PE unit designed in this study incorporates dual weight registers and a control port (LOAD SIGNAL) in addition to the conventional MAC unit. The control signal propagates through PEs using a systolic-style broadcasting mechanism: upon receiving the LOAD_SIGNAL instruction, a PE weight register updates; before signal arrival, the PE continues computations with the original weight parameters.

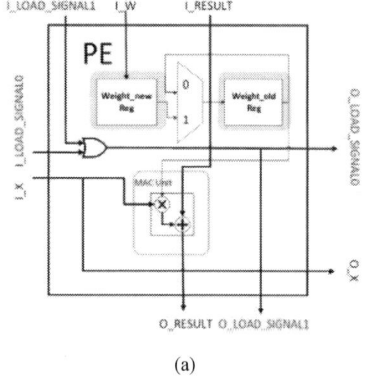

(a)

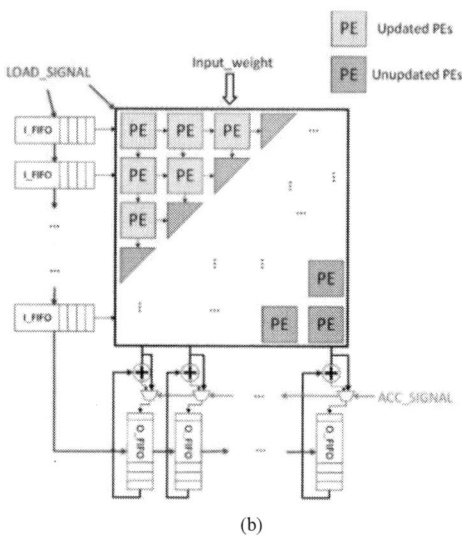

(b)

Fig. 4. (a)Architecture of PEs, Include 2 Weight Registers. (b)The Complete Architecture of Systolic Array and its Update Strategy.

Fig. 4(b) illustrates the complete systolic array architecture, including peripheral FIFOs and control signals. The important control signal LOAD_SIGNAL serves as a trigger for PE units, input FIFOs, and output FIFOs, the PEs and the I/O FIFOs propagating in a systolic style cascade. This systolic style enabling strategy eliminates bubbles in the pipeline dataflow, ensuring SA computing efficiencies.

To fit the Attention computation patterns, this design incorporates a 1-bit ACC_SIGNAL (Accumulate Signal) to control output FIFO write operations via a multiplexer (MUX). Because of the tiling matrix strategy, during QK^T matrix multiplication (16×128(Q) × 128×16(K^T)), an adder must be selected at the output FIFO's input to accumulate partitioned matrix computation results. In contrast, during P×V matrix multiplication (16×16(P) × 16×128(V)), the SA outputs are independent, so there is no need for accumulation. It is because there are two computation patterns in attention, so we design SA's peripheral circuits that work in dual mode. This dual-mode computing requirement is managed through the 1-bit ACC_SIGNAL control.

C. Non-linear Computation Units

The nonlinear computation unit consists of the softmax unit and the coefficient generator unit. Among them, the exponential function unit is the most important challenge part. Leveraging neural network's robustness to reduce computational errors, this design optimizes hardware cost instead of improving precision, increasing throughput while relaxing accuracy requirements. A mapping-LUT-shifting e^x approximation algorithm[6] is adopted as follows (3):

$$e^x = 2^{x \cdot log_2^e}$$
$$= 2^{x_1} \times 2^{x_2},$$
$$x \cdot log_2^e = x_1 + x_2 \quad (x_1 \in \mathbb{Z}, x_2 \in (-1,1))$$
$$e^x = LUT(2^{x_2}) \ll x_1 \tag{3}$$

First, e^x is converted to base 2 representation: $2^{x \cdot log_2 e}$. The exponent is split into integer and fractional parts, x_1 and x_2. The integer part x_1 is used in bit shifting, and the fractional part x_2 is used to retrieve 2^{x_2} approximations from a lookup table. This decomposition compresses the table range to (0,1),

979-8-3315-3918-4/25 $31.00 © 2025 IEEE

which is easy to increase clock frequency, because this method reduces combinational delay to satisfy 200MHz timing requirements.

To optimize critical paths, three to four stages DFF pipelining registers are inserted in long data paths of the softmax and coefficient computing modules. Measurements show that can reduce maximum combinational levels from 8 to 3 stages, achieving timing closure at 200.80MHz.

IV. FPGA VERIFICATION

This study selects representative Transformer architectures for evaluation. Statistical analysis of multi-head attention parameters in mainstream LLMs, e.g., DeepSeek, Qwen, reveals that average to single-head attention, the embedding lengths are predominantly 128. Based on this, the QKV matrix column dimension is set to 128. To achieve long-sequence computation while balancing storage efficiency, 1024 tokens is selected as the baseline test length, because longer sequences can be supported as long as data precision is sufficient.

For precision requirements in long-sequence computation, conventional INT8 quantization in neural network accelerators cannot satisfy error accumulation constraints during multiple iterations. Therefore, we adopted an INT16 and INT8 hybrid quantization to meet the requirements. Through 8/16-bit hybrid-precision design, a balance between resource consumption and computational accuracy is achieved. Validation on normalized random datasets demonstrates a mean squared error (MSE) of 10^{-7} magnitude with a structural similarity (SSIM) averaging of 98.3%.

Considering the hardware resource cost on FPGA, this design achieves 66.7% on-chip BRAM resource savings (because we make 1024×1024 intermediate matrix storage requirements free) through Flash-Attention algorithm, while eliminating data dependency in traditional large-dimension softmax layers via incremental iterate operations. This architectural improvement enables parallel processing of adjacent block softmax computation and reducing overall computation latency. The detailed resource utilization metrics are summarized in Table I.

TABLE I. UTILIZATION REPORT FOR HARDWARE ACCELERATOR

	LUT	FF	BRAM	DSP
Available	433200	866400	1470	3600
Using	91633	168507	236	544
Utilization	21.15%	19.45%	16.05%	15.11%

The implemented accelerator operates at 200MHz on a Xilinx Virtex-7 XC7VX690T-2FFG1761C FPGA board, and software verification runs on AMD Ryzen 5900HX platform, with a nominal frequency of 3.30GHz). As shown in Table II, the Flash-Attention-based FPGA accelerator completes a full attention computation in 9.83ms, compared to 70.54ms for the PyTorch-optimized CPU software implementation. And at the meanwhile, the power consumption is reduced to 14.32%, making it a good choice for low-power processing. With a mean squared error (MSE) maintained at 10^{-7} magnitude, our

design achieves a 4.6× speedup while preserving structural similarity (SSIM) of 98.66%, as shown in Fig 5.

TABLE II. COMPARISON OF ATTENTION COMPUTING RESULTS

Devices	FPGA	CPU
Frequency	200MHz	3.30GHz
Time	9.83ms	45.05ms
Speedup	4.6×	1×
Power	6.442w	45w
SSIM	Over 98%	/

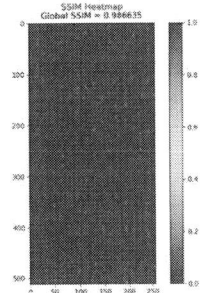

Fig. 5. The Heatmap of Output Matrix

V. CONCLUSIONS

This study presents a Flash-Attention accelerator that enables efficient edge-side deployment of long-sequence attention computation. Through hardware design innovations, this design effectively solve the three challenges in edge-side Transformer acceleration: (1)Our work overcomes regular attention accelerator's sequence length limitations, supporting 1024-token processing while maintains precisions; (2) Dynamic weight updating improves systolic array utilization to 79.01%, a 41.51% enhancement over traditional schemes, achieving 4.6× speedup with resource reduction; (3) The quantization method we take and the pipelined dataflow can reduce the on-chip resources consumption, establishing a technical framework for edge-side Transformer accelerators which support long-sequence input.

REFERENCES

[1] G. Hinton, O. Vinyals, and J. Dean, "Distilling the knowledge in a neural network," arXiv preprint arXiv: 1503.02531, 2015.

[2] Lu, Siyuan, Meiqi Wang, Shuang Liang, Jun Lin, and Zhongfeng Wang. "Hardware accelerator for multi-head attention and position-wise feed forward in the transformer." In 2020 IEEE 33rd International System-on Chip Conference (SOCC), 2020, pp. 84-89.

[3] N. P. Jouppi et al., "In-datacenter performance analysis of a tensor processing unit," in Proc. Int. Symp. Comput. Archit., 2017, pp. 1–12.

[4] L. Liu and S. Brown, "Leveraging Fine-grained Structured Sparsity for CNN Inference on Systolic Array Architectures," 2021 31st International Conference on Field-Programmable Logic and Applications (FPL), Dresden, Germany, 2021, pp. 301-305.

[5] T. Dao, D. Fu, S. Ermon, A. Rudra, and C. Ré, "FlashAttention: Fast and memory-efficient exact attention with IO-awareness," in Proc. Int. Conf. Neural Inf. Process. Syst., 2022, pp. 16344–16359.

[6] Wang, Meiqi, Siyuan Lu, Danyang Zhu, Jun Lin, and Zhongfeng Wang. "A high-speed and low-complexity architecture for softmax function in deep learning." In 2018 IEEE asia pacific conference on circuits and systems (APCCAS), 2018, pp. 223-226

979-8-3315-3918-4/25 $31.00 © 2025 IEEE

Sparse Approximation of Softmax: Hardware-Efficient Acceleration for Long Sequence Inference

Lanqi Ma [1], Zifeng Zhao [1], Xiaoxing Wu[1], Gengsheng Chen[1,2], Wenbo Yin*[1]

[1]College of Integrated Circuits and Micro-Nano Electronics, Fudan University, Shanghai 200433, China
[2]Jiashan Fudan Institute, Jiaxing, Zhejiang Province 314100, China

* Email: lqma24@m.fudan.edu.cn, wbyin@fudan.edu.cn

Abstract—The Softmax function, a fundamental nonlinear activation in machine learning, underpins the performance of modern top-performing large language models (LLMs). However, as sequence lengths scale, its computational demands become a significant performance bottleneck, stemming from complex exponentiation and division operations coupled with sequential data dependencies. Critically, existing sparse implementations exhibit inherent incompatibility with efficient two-stage workflows, resulting in excessive complexity. This paper introduces a hardware-friendly sparse approximate Softmax algorithm that employs shift-based approximation to replace arithmetic operations and a runtime maximum-based sparsity strategy to accelerate computation. Additionally, our hardware design utilizes selective storage and data reuse methods to reduce buffer requirements while improving performance. Evaluations on BERT-based model across the GLUE benchmark show that our approach achieves an average sparsity of 83% while incurring negligible accuracy loss. Following hardware deployment, the proposed accelerator yields 1.707× speedup over conventional implementations. Our design achieves 999.54 GOPS/mm² area efficiency and 1,221.96 GOPS/W power efficiency at 1.7 GHz in 28nm technology.

Keywords—Softmax, Large Language Model, Hardware Implementation

I. INTRODUCTION

Transformer-based Large Language Models (LLMs) have advanced rapidly demonstrating remarkable performance across diverse domains including Natural Language Processing (NLP), Computer Vision (CV), multimodal interaction, and programming assistance. The attention mechanism, serving as the core component in Transformer, fundamentally relies on the Softmax function to convert raw Query-Key scores into valid probability distributions.

However, this critical dependency on Softmax introduces scaling limitations. As Transformer-based LLMs continue to evolve toward larger scales, their computational and memory requirements grow substantially. While quantization serves as a common-used approach for reducing deployment costs, Softmax presents distinctive acceleration barriers. Using quantized data in exponential computations induces substantial precision degradation due to heightened arithmetic sensitivity. Moreover, Softmax exhibits significant computational complexity from exponentiation and division operations. These constraints make Softmax a performance bottleneck, particularly acute during long sequence processing, posing substantial challenges for efficient accelerator design.

Softmax computation involves subtractive normalization to prevent exponent overflow, typically comprises a multi-stage process: Stage 1 identifies the maximum value within the input vector, Stage 2 computes the exponential summation and Stage 3 completes probability normalization via division. Softermax [1] proposed online normalization, merging the first two stages through maximum detection and concurrent exponential accumulation, significantly reducing latency. Although numerous prior works have explored approximation or sparsity for Softmax, existing sparse implementations exhibit fundamentally incompatibility with efficient two-stage Softmax workflows, resulting in excessive complexity.

In this paper, we propose a sparse Softmax approximation method to achieve both computational efficiency and seamless integration into efficient two-stage workflows. We incorporate dynamic sparsification in Stage 1 by evaluating divergence between the immediate global maximum and the current processing block, thereby eliminating high-deviation data to reduce computational load. Additionally, we employ shifted approximation with effective error compensation to lower computational complexity. For hardware design, our optimized computational flow exclusively stores sparse-processed valid data during Stage 1 while retaining computed shift values for Stage 2 reuse. This strategy alleviates memory requirements while accelerating computation and enhancing power efficiency. In summary, our innovations are as follows:

- We introduce a sparse Softmax approximation optimized for two-stage computational workflows. This approach eliminates exponential and division operations while bypassing redundant computations.

- We implement a dedicated hardware design that eliminates multipliers, dividers, and LUTs. Selective storage and data reuse techniques are employed to reduce buffer requirements while enhancing overall hardware performance.

- We conduct comprehensive experimental evaluations. Results demonstrate 83% average sparsity with negligible accuracy degradation without requiring fine-tuning or retraining. Achieving 1.707× speedup over baseline designs, our hardware implementation delivers 999.54 GOPS/mm² area efficiency and 1,221.96 GOPS/W power efficiency.

II. BACKGROUND

In Transformers, the original Softmax function is defined by (1), where x_i denotes the i-th element of the input vector x:

$$Softmax(x_i) = \frac{e^{x_i}}{\sum_{j=1}^{n} e^{x_j}} \qquad (1)$$

To ensure numerical stability, each compute element is shifted by its maximum value prior to exponentiation:

$$Softmax(x_i) = \frac{e^{x_i - max(x)}}{\sum_{j=1}^{n} e^{x_j - max(x)}} \qquad (2)$$

The core innovation of online normalization lies in maintaining and propagating the exponential sum computed under prior maxima, which enables exact current summation via adjustments during maximum shifts as formalized below:

979-8-3315-3918-4/25 $31.00 © 2025 IEEE

$$e^{x_{i-1} - max_i} = e^{x_{i-1} - max_{i-1}} \cdot e^{max_{i-1} - max_i} \quad (3)$$

where max_{i-1} denotes the maximum value among the first i-1 elements. Therefore, the summation is concurrently computed during maximum detection, where Sum_i denotes the exponential summation of the first i elements:

$$Sum_i = Sum_{i-1} \cdot e^{max_{i-1} - max_i} + e^{x_i - max_i} \quad (4)$$

Numerous studies have been proposed to accelerate Softmax computation. Ref. [2] replaced runtime exponential computations with pre-stored lookup table (LUT) values. However, this method incurs significant memory overhead, and LUT storage scales exponentially with precision demands. Another strategy converts base-e exponentials to base-2 equivalents [3], enabling hardware-efficient implementation via bit-shift operations with notably fewer hardware resources. For division approximation, a prevalent technique involves converting the dividend to the nearest power-of-two value [4-6], replacing division with bitwise right-shifting operations. As direct substitution introduces significant approximation errors, compensation techniques mitigate precision loss with marginal overhead. For instance, Ref. [4] employed 1-bit error compensation, Ref. [5] extended it to n-bit precision, while Ref. [6] adopted a simpler 3-bit direct approximation.

Furthermore, several studies have explored sparse computation techniques to achieve computational speedup. Ref. [7] introduced mean-based thresholding during maximum detection—calculating the global mean to eliminate sub-mean elements. ULSeq-TA [8] leveraged group and global maxima for multiple sparsification rounds before summation and normalization, reducing computational burden. However, current sparse methods exhibit excessive complexity and inherently conflict with streamlined two-stage Softmax architectures. We resolve this by integrating a lightweight yet effective sparse processing module into stage 1, achieving enhanced computational performance. Additionally, our approach dramatically reduces buffer requirements by storing only sparse-processed data, effectively mitigating the storage overhead induced by growing sequence length.

III. PROPOSED SPARSE APPROXIMATE SOFTMAX

A. Sparse Method

From the functional form perspective, the exponential operator amplifies input divergences, consequently producing output distributions where most probabilities become numerically negligible.

Within the online normalization framework processing G elements in parallel per iteration, each group evaluation yields a local group maximum m_g. This local maximum, along with the propagated historical maximum m_{la}, establishes the updated global maximum through $m_{glb} = Max(m_{la}, m_g)$. After determining the updated global maximum m_{glb} and group maximum m_g, the algorithm executes a threshold-driven sparsification procedure as shown in Fig.1. If the difference between m_{glb} and m_g exceeds the predefined threshold τ, this implies that all elements within the current group fall below the significance threshold relative to the global maximum. Consequently, the entire group is discarded by setting all values to zero. Otherwise, we introduce a fine-grained processing: individual elements within the group are compared directly against m_{glb}, and any value deviating from it by more than τ is truncated to zero while preserving others.

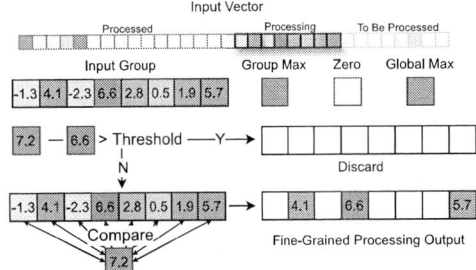

Fig. 1. Group Threshold Sparsification Flow

Following sparsification, the modified computational flow exclusively processes preserved non-zero elements rather than full vectors, dramatically reducing computational workload by eliminating redundant operations on zeroed values.

B. Sparse Approximate Softmax

To reduce computational overhead in exponentiation, a logarithmic base transformation is commonly employed, converting e^x into equivalent base-2 operations:

$$e^x = 2^{x \cdot log_2 e} \quad (log_2 e \approx 1.442695) \quad (5)$$

This conversion enables replacing exponentiation with efficient bit-shift operations, expressed as:

$$2^{x \cdot log_2 e} \approx 2^{1.4375x} \approx 1 \ll \lfloor 1.4375x \rfloor \quad (6)$$

The substitution of $x \cdot log_2 e$ with $1.4375x$ is strategically employed to eliminate hardware multipliers, leveraging efficient binary shift-add operations for implementation:

$$1.4375x = x + x \gg 1 - x \gg 4 \quad (7)$$

Building upon the exponentiation optimization paradigm, division operations are similarly streamlined by converting divisors into powers of two. This extends the shift-based computation approach to eliminate hardware overhead:

$$\frac{1}{d} = \frac{1}{2^k \cdot (1+s)} = \frac{1}{1+s} \gg k \quad (8)$$

Assuming $s = 0.abcde\ldots$ is represented with p bits, the interval $[0,1)$ is partitioned into 2^p sub-intervals. For any s in a sub-interval, $1/(1+s)$ is approximated by precomputed value at the left endpoint. For small p, this approximation is efficiently implemented via decoder (e.g., p=3 requires only an 8-entry decoder). However, this method introduces approximation errors. Under uniform data distribution, the compensated result after applying average-error correction is:

$$\frac{1}{1+s} = \frac{1}{1+t2^{-p}} - \frac{\int_{t2^{-p}}^{(t+1)2^{-p}} \left(\frac{1}{1+t2^{-p}} - \frac{1}{1+s} \right) ds}{(t+1)2^{-p} - t2^{-p}}$$
$$= 2^p \ln \frac{1 + (t+1)2^{-p}}{1 + t2^{-p}} \quad s \in [\frac{t}{2^{-p}}, \frac{t+1}{2^{-p}}) \quad (9)$$

As empirically demonstrated in Fig.2, the error distribution of compensated results significantly diminishes and concentrates closer to zero compared to exact values, confirming the efficacy of the compensation technique.

Given the exponential dividend form, the entire division computation simplifies to bit-shifting after approximation:

$$\frac{e^{x_i - max}}{d} \approx \frac{2^{1.4375(x_i - max)}}{2^k(1+s)} \approx \frac{1}{1+s} \gg (k + \lfloor 1.4375(max - x_i) \rfloor) \quad (10)$$

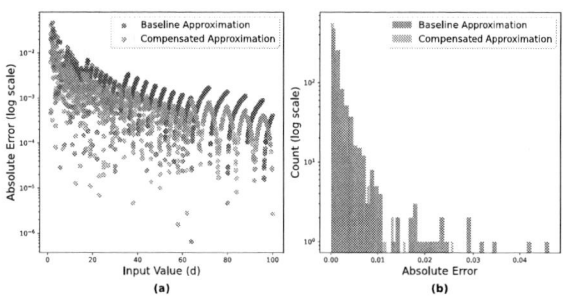

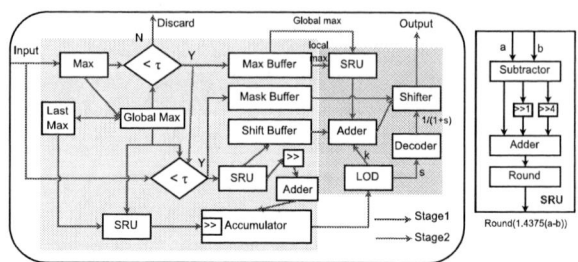

Fig. 3. Proposed hardware architecture

Fig. 2. Error comparison under approximation (p=3 bit). (a)Absolute error comparison. (b)Error distribution

The entire computational workflow for sparse approximation Softmax is formally specified in Algorithm 1. The input is an N-dimensional vector, while the output is a sparsified vector of length N_s. During Stage 1, group-wise sparsification eliminates elements based on their deviation from the global maximum, followed by shift-based approximation of exponential summation. Stage 2 then performs shift-implemented approximate division on the retained non-zero data to derive final results.

Algorithm 1: Proposed Sparse Approximation Softmax

Input:$x_1,x_2,x_3,\cdots,x_N$:Input Activation

Output: $y_1,y_2,y_3,\cdots,y_{N_s}$:Softmax Output

1   $m_{glb}\leftarrow-\infty, m_g\leftarrow-\infty, m_{la}\leftarrow-\infty$

2   $d\leftarrow0, N_g\leftarrow ceil(N/G), \tau\leftarrow threshold, S_g\leftarrow ceil(N_s/G)$

3   **for** $i\leftarrow1$ **to** N_g **do**   ◇Stage 1

4      $m_g\leftarrow Max(x_{(i-1)G+1},x_{(i-1)G+2},\cdots,x_{i\cdot G})$

5      $m_{glb}\leftarrow Max(m_{la},m_g)$

6   **if** $m_{glb}-m_g>\tau$ **then**

7         $d\leftarrow d\gg\lfloor1.4375m_{glb}-m_{la}\rfloor$

8   **else**

9         $mask_{ij}\leftarrow\left((m_{glb}-x_{(i-1)G+j})<\tau\right), j\leftarrow1$ to G

10        $d\leftarrow d\gg\lfloor1.4375(m_{glb}-m_{la})\rfloor+$
           $\sum_{j\leftarrow1\text{ to }G}(1\gg\lfloor1.4375(m_{glb}-x_{j+(i-1)G})\rfloor)\cdot mask_{ij}$

11      $m_{last}\leftarrow m_{global}$

12   **end for**

13   $2^k(1+s)\leftarrow d$ ◇Stage 2

14   $ds\leftarrow decoder(s)$

15   **for** $i\leftarrow1$ **to** S_g **do**

16      $y_{ij}\leftarrow ds\gg k+\lfloor1.4375(m_{glb}-\hat{x}_{ij})\rfloor$ $j\leftarrow1$ to G

17   **end for**

IV. HARDWARE ARCHITECTURE

We design a dedicated hardware architecture that implements the proposed Softmax with the detailed structure illustrated in Fig. 3. In the diagram, blue traces show Stage 1 paths and red traces indicate Stage 2 microarchitecture.

Processing begins with comparator trees identifying per-group maxima in Stage 1. Each group maximum sequentially updates the global maximum through comparison with the prior global value. Groups where the global-group maximum difference exceeds a preset threshold are discarded. For retained groups, the updated global maximum is stored in the Max Buffer. Concurrently, each input element is subjected to

a precise comparison with the global maximum. Elements satisfying the absolute difference threshold criterion trigger parallel operations: Their positional identifiers are recorded in the mask buffer, while both the element value and global maximum proceed through specialized Scale and Round Unit (SRU) processing modules. These SRUs implement deterministic computation sequences beginning with difference calculation, followed by 1.4375× scaling via sequential arithmetic right shifts (single-bit and four-bit) with integrated rounding. The resulting shift magnitude is stored in the shift buffer while simultaneously directing right-shift operations on the unit value. These shifted results are then feed into the adder tree for arithmetic summation, with the summed value subsequently updating the accumulator. Independently, a dedicated SRU processes the difference between current and previous global maxima to derive offset shifts. These offsets dynamically reconfigure the accumulator through arithmetic right-shifting before new group summation data integration. Upon completion of all group processing, the stabilized accumulator initiates Stage 2 computation.

Stage 2 initiates by processing the final accumulator value through a leading one detector (LOD) to extract the dominant exponent. Subsequent bits decode into an approximate reciprocal value. Output reconstruction leverages precomputed shift counts and locally buffered maxima from Stage 1, deriving compensation shifts from the difference between final and intermediate global maxima. The sum of original shifts, compensation shifts, and the dominant exponent yields the composite shift magnitude, which right-shifts the decoded approximation to generate final outputs.

Our design achieves substantial hardware efficiency by strategically optimizing buffer utilization—selectively storing only threshold-qualified data while replacing raw input buffering with compressed metadata representations. Specifically, group maxima and computed shift values supplant conventional input storage, dramatically reducing buffer requirements. Complementing this approach, shift operators eliminate traditional multipliers, dividers, and LUTs, yielding substantial area savings while maintaining computational precision.

V. EVALUATION

A. Experimental Setup

To validate our algorithm, we employed a pretrained BERT-based model from HuggingFace libraries on the GLUE multi-task benchmark. We replaced the original Softmax function with our custom solution while preserving all other components. The hardware architecture was implemented in SystemVerilog HDL, synthesized through Synopsys Design Compiler under TSMC 28nm technology, with power analysis performed using PrimeTime PX.

TABLE I.	ACCURACY RESULTS ON GLUE TASKS								
Approach	CoLA	SST-2	MRPC	STS-B	QQP	MNLI	QNLI	RTE	WNLI	Average
Baseline	58.56	91.86	89.08	89.09	87.59	83.98	90.63	64.62	53.52	78.77
Ours(1,∞)[a]	58.56	92.32	89.43	89.08	87.60	83.97	90.54	65.34	53.52	78.93
Ours(3,∞)[a]	58.57	92.09	89.93	89.07	87.61	83.97	90.66	66.79	53.52	79.13
Ours(3,6)[a]	58.81	91.97	89.77	89.11	87.61	84.00	90.68	64.62	53.52	78.90
Ours(3,4)[a]	59.05	92.20	89.06	89.04	87.57	83.80	90.46	65.34	53.52	78.98
Ours(3,2)[a]	59.06	91.74	89.35	88.35	86.69	83.05	89.68	63.18	53.52	78.29

a. (p, τ): Bit-width p of decoder input and threshold τ

B. Model Performance

Table I presents the model's accuracy results across various tasks, where the baseline denotes results without Softmax function replacement. We employ 8-element group processing with 16-bit fixed-point arithmetic. Under approximation-only configuration (no sparsification, $\tau=\infty$), decoder input bit-widths of 1 and 3 resulted in a maximum accuracy degradation not exceeding 0.09%. Analysis of the mean accuracy across the nine tasks indicates a marginal improvement using our approximation method, with performance superior at a bit-width of 3 compared to 1. With sparsification incorporated at fixed 3-bit width, progressive threshold tightening ($6\rightarrow4\rightarrow2$) demonstrates that optimal performance occurs at a threshold setting of 4, slightly below the non-sparse configuration of 79.13%. Notably, a maximum accuracy reduction of 1.44% relative to the baseline is observed at the threshold of 2. These results collectively validate that our approximation-sparsification co-design imposes negligible accuracy penalties and may help models focus more effectively on critical features.

Fig.4 illustrates sparse rate comparisons across threshold settings, where sparse rate denotes the percentage of zero elements after sparse processing. Blue bars represent experimental sparse rates and orange bars denote ideal sparse rates derived from final global maximum. The minimal gap between both rates at every threshold demonstrates that our runtime maximum-based sparsification in Stage 1 achieves excellent effects. While higher thresholds gradually reduce sparse rates, they consistently maintain high levels—reaching 83.0% average sparsity at the threshold setting of 4.

C. Hardware Implementation Results

Our hardware implementation employs 16-bit fixed-point arithmetic with parallelism degree of 8, decoder input bit-width $p=3$, and threshold setting of 4. Under this configuration, average sparsity reaches 83%, delivering 1.707× speedup versus non-sparsified two-stage Softmax and 1.217× acceleration against the approach in [8]. Our design handles vector lengths up to 32,768 with maximum clock frequency

TABLE II.	OUR HARDWARE DESIGN VS. PREVIOUS WORKS		
Design	[4]	[5]	Ours
Technology	65nm	40nm	28nm
Parallelism (n)	1	8	8
Frequency (GHz)	0.2	1.67	1.7
Max Input Length (N)	N/A	8192	32768
Average Latency (cycles)	2	0.25	0.147
Area (μm^2)	2492.4	20119	11570
Power (mW)	0.19	9.57	9.464
Throughput (GOPS/s)	0.1	6.68	11.56
Normalized Area Efficiency (GOPS/mm^2)	462.50	677.6	999.54
Normalized Power Efficiency (GOPS/W)	1221.80	997.16	1221.96
Remark		Need Retraining	

reaching 1.7 GHz. The implementation occupies 11,570 μm² area and consumes 9.464 mW power, achieving remarkable hardware efficiency with 999.54 GOPS/mm² area efficiency and 1,221.96 GOPS/W power efficiency. Table II compares our design against prior works, where average latency represents cycles per element. For equitable comparison, area efficiency and power efficiency metrics are normalized to 28nm technology, confirming our implementation achieves superior results in both normalized efficiency categories.

VI. CONCLUSION

In this paper, we propose a co-designed sparse Softmax accelerator implementing group-based threshold sparsification (83% average sparsity) and shift-based approximation units replacing exponential/division operations, achieving negligible accuracy degradation without requiring fine-tuning Our hardware architecture in 28nm technology operates at 1.7 GHz, delivering 1.707× speedup versus conventional designs with 999.54 G/mm² area efficiency and 1,221.96 G/W power efficiency while supporting sequences up to 32,768 elements for large language model acceleration.

REFERENCES

[1] J. R. Stevens, R. Venkatesan, S. Dai, B. Khailany and A. Raghunathan, "Softermax: Hardware/Software Co-Design of an Efficient Softmax for Transformers," 2021 58th ACM/IEEE Design Automation Conference (DAC) pp. 469-474.

[2] X. Dong, X. Zhu, and D. Ma, "Hardware Implementation of Softmax Function Based on Piecewise LUT," in IEEE International Workshop on Future Computing (IWOFC), 2019, pp. 1–3.

[3] Zhang, Yuan et al. "Base-2 Softmax Function: Suitability for Training and Efficient Hardware Implementation," in IEEE Transactions on Circuits and Systems I: Regular Papers, 69(9), pp. 3605–3618, 2022.

[4] W. Wang, S. Zhou, W. Sun, P. Sun and Y. Liu, "SOLE: Hardware-Software Co-design of Softmax and LayerNorm for Efficient Transformer Inference," 2023 IEEE/ACM International Conference on Computer Aided Design (ICCAD). IEEE, pp. 1-9.

[5] Li, W. et al. "Hardware-oriented algorithms for softmax and layer normalization of large language models." Sci. China Inf. Sci. 67, 200404 (2024).

[6] M. -H. Hsieh, X. -H. Li, Y. -H. Huang, P. -H. Kuo and J. -D. Huang, "A Hardware-Friendly Alternative to Softmax Function and Its Efficient VLSI Implementation for Deep Learning Applications," 2024 IEEE International Symposium on Circuits and Systems (ISCAS) pp. 1-5.

[7] N. A. Koca, A. T. Do and C. -H. Chang, "Hardware-efficient Softmax Approximation for Self-Attention Networks," 2023 IEEE International Symposium on Circuits and Systems (ISCAS), 2023, pp. 1-5.

[8] J. Wang, L. Zhang, X. Li, H. Yang and Y. Liu, "ULSeq-TA: Ultra-Long Sequence Attention Fusion Transformer Accelerator Supporting Grouped Sparse Softmax and Dual-Path Sparse LayerNorm," in IEEE Transactions on Computer-Aided Design of Integrated Circuits and Systems, 43(3), pp. 1-1.

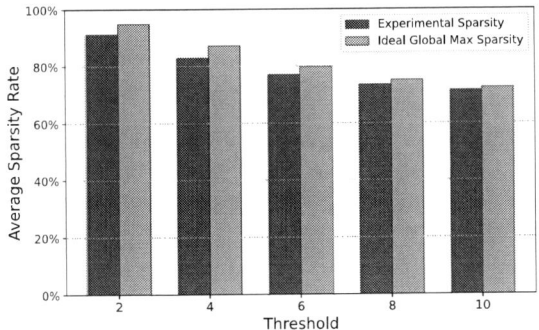

Fig. 4. Average Sparsity Rate Comparison at Different Thresholds

A Hybrid Processing-in-Memory and Computing-in-Memory Architecture for Large Language Model Inference in Edge Devices

Yujia Sun[1], Ruicong Zhang[1], Yuanfeng Chen[3], Qiang Zhou*[3], Xiaoyong Xue*[1,2,3], Xiaoyang Zeng[1]

[1] State Key Lab of Integrated Chips and Systems, College of Integrated Circuits and Micro-Nano Electronics, Fudan University, Shanghai, China
[2] School of Microelectronics, Fudan University, Shanghai, China
[3] TRANSCPUTING Technology LTD, Shanghai 201203, China

Email: xuexiaoyong@fudan.edu.cn, alex.zhou@transcputing.com

Abstract—**Deploying Large Language Models (LLMs) on resource-constrained edge devices is critically challenged by the "memory wall" bottleneck, where energy-intensive data movement between processors and memory dominates system costs. This paper introduces a hybrid Processing-in-Memory (PIM) and Computing-in-Memory (CIM) architecture that synergistically integrates DRAM-based PIM and SRAM-CIM to accelerate Transformer inference. Our key innovation lies in strategically partitioning the attention computation between the two technologies: SRAM-CIM efficiently handles the dynamic Value (V) cache operations prone to access conflicts in DRAM-PIM, while DRAM-PIM accelerates other linear layers. To maximize efficiency, we co-design a hierarchical channel-bank data mapping scheme and an asymmetry-aware workload assignment. Implemented and verified in 28nm technology, our architecture achieves a 1.51 times speedup and 1.24 times reduction in energy consumption compared to DRAM-PIM-only baselines in the focused Transformer layers for Llama2-7B inference, demonstrating potential for edge deployment.**

Keywords—*Processing-in-Memory, Computing-in-Memory, Transformer, edge inference*

I. INTRODUCTION

Large Language Models (LLMs) have emerged as a pivotal research focus in artificial intelligence, demonstrating unprecedented capabilities in natural language processing and generation [1]. These models excel not only in comprehension tasks like text classification [2], but also in complex generative applications such as content creation [3] and code synthesis [4]. Their transformative potential has spurred growing interest in edge deployment to enable personalized, privacy-preserving LLM agents while reducing reliance on cloud infrastructure [5].

Modern LLMs adopt a multi-layer Transformer decoder architecture with autoregressive computation, where each output token serves as input for subsequent generation. During this process, key (K) and value (V) vectors are computed through matrix multiplication and stored as KV cache to avoid redundant computation in multi-head attention (MHA) layers. This cache dynamically expands with each new token.

The advancement of deep learning relies heavily on large-scale datasets and computational resources. While scaling model parameters enhances performance, it introduces critical bottlenecks: a substantial memory footprint and intensive data movement. Conventional von Neumann architectures suffer

This work was supported by STI 2030-Major Projects (2022ZD02092 00), in part by the National Natural Science Foundation of China (6227403 8), the Science and Technology Commission of Shanghai Municipality (24 JD1400200), and ZTE Industry-University-Institute Cooperation Funds under Grant (IA20241120003).

particularly during autoregressive decoding, where the arithmetic intensity drops to ~2 under INT8 quantization, making data transfer energy dominate over actual computation. Edge devices further exacerbate this issue by employing single-batch inference to limit KV cache size, sacrificing batch parallelism that alleviates bandwidth pressure in cloud scenarios. The in-memory computing architecture addresses the memory bandwidth bottleneck by integrating processing units within memory, enabling the memory itself to perform certain computations. This approach mitigates the data movement challenges inherent in traditional compute-memory separation architectures.

Among various memory technologies, DRAM-based Processing-in-Memory (PIM) [6], [7] offers advantages for edge-side LLM inference: its gigabyte-scale capacity far exceeds that of SRAM, allowing it to store large model parameters and reduce data transfers directly; its read/write speed and bandwidth significantly outperform NAND Flash; and its compatibility with existing DRAM technology makes mass production more cost-effective than emerging memory solutions, making it suitable for consumer devices.

DRAM-based PIM architectures achieve computational acceleration by integrating processing units at the bank level, enabling parallel vector multiplication across banks to optimize generalized matrix-vector (GEMV) operations. This approach primarily benefits from the high parallelism of simple multiply-accumulate (MAC) operations, demonstrating particular efficacy for linear layers characterized by static weight matrices, deterministic dataflow patterns, and repetitive computation modes.

However, a fundamental limitation emerges when applying DRAM-PIM to attention computation in large language model (LLM) inference. The dynamic nature of KV matrices—which are incrementally updated with each new token—introduces two conflicting requirements:(1) Storage: V cache grows row-wise through sequential token appends. (2) Computation: Attention mechanisms demand column-wise access for efficient matrix operations. This creates a critical access pattern mismatch that disrupts DRAM-PIM's architectural invariants, thereby limiting acceleration efficiency.

To address this, a hybrid architecture can be adopted for acceleration: offloading the computation that impairs DRAM-PIM's parallel efficiency to SRAM-based Computing-in-Memory (CIM). Though SRAM-CIM has limited capacity, it eliminates refresh overheads, offers faster read/write speeds, and delivers higher energy efficiency and compute density [8], making it well-suited for small-scale attention computations.

In this work, we propose a hybrid architecture based on DRAM-PIM and SRAM-CIM for accelerating edge-side LLM inference. Our key contributions include:

- A hybrid in-memory acceleration architecture consisting of SRAM-CIM based on outer-product computation for attention value and DRAM-PIM for other linear layer computation in LLM inference.

- A hierarchical channel-bank data mapping scheme for maximized computational parallelism and minimized data movement overhead.

- A hardware-asymmetry-aware workload assignment for simultaneous compute feeding and data prefetching

II. MOTIVATION

A. Transformer

The core computational unit of the LLM is the Transformer Block, which consists of two main components: the Multi-Head Attention (MHA) layer and the Feed-Forward Network (FFN) layer.

Transformer inference layers exhibit distinct computational characteristics: In the MHA layer, the K and V matrices must be stored and dynamically updated (as KV cache) to support autoregressive generation. As new tokens are processed, the K matrix is generated and stored row-wise, then retrieved row-wise for QK^T computation, while the V matrix is generated and stored row-wise but retrieved column-wise for the subsequent attention value computation $S \cdot V$, where $S = \text{Softmax}(QK^T/\sqrt{d_k})$. In contrast, the FFN layer employs fixed weights and requires no dynamic updates—its computation simply retrieves and processes data column-wise. The heterogeneous computational patterns across Transformer inference layers motivate hybrid accelerator architectures, where specialized designs targeting distinct layer-wise characteristics offer a viable approach for efficient large model acceleration.

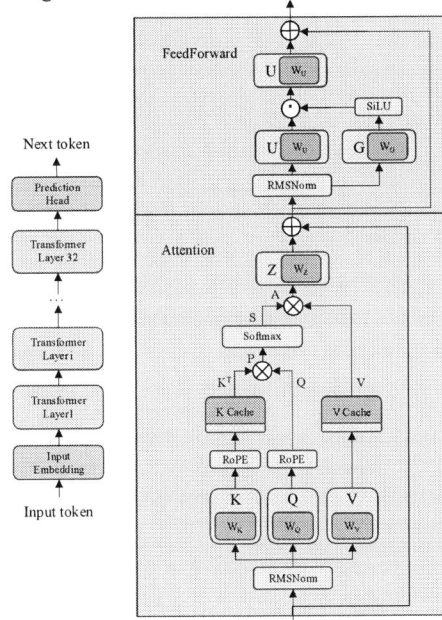

Fig. 1. Transformer architecture

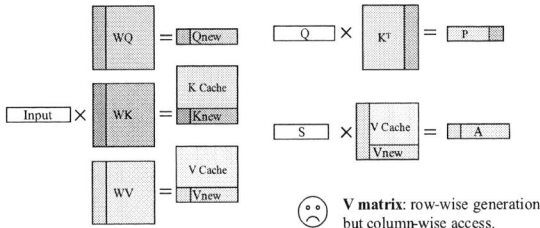

Fig. 2. Row-column access conflict in V cache during LLM inference

B. LLM Inference on DRAM-PIM-only Architecture

DRAM-PIM architectures typically integrate simple computing units within DRAM arrays to perform near-data processing, thereby reducing data movement overhead. However, deploying LLM inference on conventional DRAM-PIM faces a critical challenge. There exists an inherent conflict between the row-wise appending of KV cache data and column-wise access for computation. If new tokens are appended row-wise, column-wise computation requires cross-row access, triggering frequent row activations and precharges that severely degrade throughput. Conversely, if the V matrix is stored in column-major order, inserting a new token's row vector into the V cache requires a full reorganization of the cache: blocks are read to on-chip memory, transposed, and rewritten column-wise, incurring substantial latency overhead.

III. HARDWARE DESIGN

A. DRAM-PIM Design

For each DRAM channel, every bank is equipped with a Processing Unit (PU), with 8 banks sharing two Global Buffers (GBs). The primary computing module within each PU comprises 32 multipliers, a 5-layer adder tree, and one accumulator. A single MAC operation can complete the multiplication of two 32-dimensional vectors in each PU - one vector originates from the bank itself, while the other comes from the GB. Before computation, vectors are first copied to the GBs and broadcast to all 8 PUs across banks. These banks activate and compute simultaneously to enhance parallelism. Within one channel, a single MAC operation can execute a $1 \times 32 \times 8$ GEMV computation in parallel across 8 banks.

B. SRAM-CIM Design

As illustrated in Fig. 3, the SRAM-CIM architecture employs a highly parallel computational structure wherein each SRAM-CIM macro integrates a 512-byte activation buffer (AB), eight independent weight SRAM banks (each organized as 512 rows $\times$ 128 columns with 8-bit width per column), and 128 computing units (CUs) each featuring an 8-bit multiplier and accumulator.

The weight matrix is stored row-wise across these SRAM banks, eliminating the transpose requirement for V matrices with computation proceeding based on outer-product computation as follows: during each clock cycle, one selected bank delivers full 128-element row to the corresponding CUs, where these weight elements undergo parallel multiplication with activation values broadcast from the AB. This process propagates results through 512 successive accumulation cycles within the CUs' registers to generate complete $1 \times 512 \times 128$ GEMV outputs. The architecture scales efficiently through bank-interleaved execution—sequentially

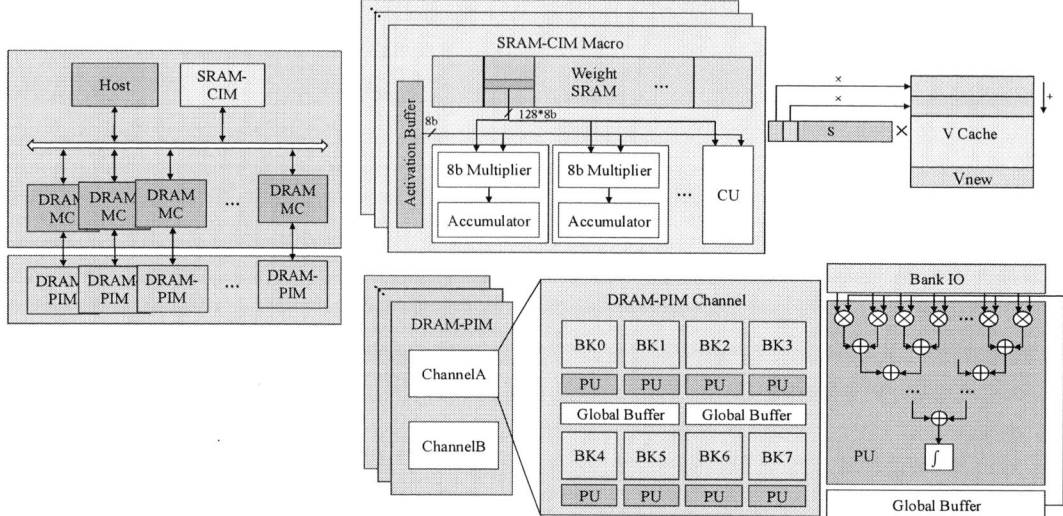

Fig. 3. Proposed hybrid acceleration architecture consisting of SRAM-CIM and DRAM-PIM

cycling through all eight banks while sustaining accumulator states—eliminating intermediate data flushing and enabling seamless computation of larger 1×4096×128 GEMV operations.

IV. HARDWARE MAPPING AND WORKLOAD ASSIGNMENT

A. Data Mapping Scheme

Fig. 4 illustrates the Channel-Bank level data mapping scheme designed to maximize computational parallelism for the fixed computation pattern and data sources in DRAM-PIM. Our optimized mapping strategy operates at two hierarchical levels.

At the channel level, we partition attention computation by mapping each head entirely to a single channel, thereby minimizing inter-channel data movement. Correspondingly, the V, $W_{Q/K/V}$, and $W_{G/U}$ matrices are column-wise partitioned across channels, while K^T, W_Z/W_D matrices adopt row-wise partitioning.

At the bank level, we implement specialized mapping schemes ("Single-Bank" or "All-Bank") according to data characteristics. All static weight matrices and K^T matrices employ All-Bank mapping, where columns are cyclically distributed across all 8 banks in a channel (Bank i stores columns i+8j, j=0,1,...column_dim/8). This enables parallel MAC operations across Banks for continuous GEMV generation. The dynamically updated K matrix maintains consistency with this scheme - as K grows row-wise during inference, its transpose K^T naturally follows the column-wise

All-Bank storage pattern without requiring reorganization. For activation vectors/matrices and the V matrix, we adopt Single-Bank mapping that sequentially fills individual Banks. This approach optimizes data locality, enables efficient bulk transfers to GBs or the outside of DRAM, and minimizes bank switching overhead during memory operations.

The scheme effectively balances computational parallelism with access efficiency for different data types in transformer inference.

B. Workload Assignment

Each attention head is assigned to one DRAM-PIM channel and one SRAM-CIM macro for computation.

However, the number of channels and macros may not match (e.g., 16 DRAM-PIM channels versus 8 SRAM-CIM macros). To address this resource asymmetry, 8 DRAM-PIM channels are dedicated to computing outputs for 8 distinct attention heads. These outputs are directly mapped 1:1 to the 8 CIM macros, ensuring perfect alignment and eliminating buffering overhead. Concurrently, the remaining 8 PIM channels are utilized to proactively load the V cache into the SRAM-CIM. Hardware utilization is maximized by keeping all 16 PIM channels and all 8 CIM macros active. This overlapping of data prefetch with ongoing head computation partially hides V cache access latency. The scheme also simplifies the critical dataflow path for delivering attention head results to the CIM engines.

V. EVALUATION

The hardware configuration parameters are listed in Table I. The DRAM-PIM's processing unit (PU) operates at 500 MHz, constrained by the DRAM's tCCD (2 ns), while the SRAM-CIM frequency is set to 500 MHz as well. The proposed architecture is implemented in Verilog and synthesized using Synopsys Design Compiler, with area and power measurements based on TSMC 28nm process node. Additionally, a decoder-only Llama2-7B model is utilized for evaluation. The model is quantized to 8-bit. Due to limitations in the evaluation methodology, certain variations in the results may exist.

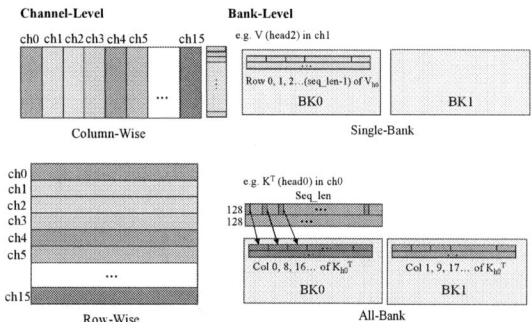

Fig. 4. Channel-Bank level data mapping scheme

Baseline1: Since MAC operations cannot retrieve operands from multi-row random addresses, DRAM-PIM baseline with on-chip matrix reorganization is utilized to evaluate computational-data access pattern alignment.

Baseline2: Another baseline, where the computation for each attention head is uniformly distributed across all banks of all DRAM-PIM channels, is employed for comparative evaluation against the proposed data mapping scheme.

TABLE I.   HARDWARE CONFIGURATION

Memory	Configuration		
DRAM	Organization	Die number	8
		Channel/die	2
		Capacity/die	8Gb
		Banks/die	16
		Row number/Bank	32768
		Bits/Row	2048×8
	Timing(ns)	tRAS=42, tRP=18, tRCD=18, tRRD=10, tCCD=2	
	Bandwidth	136GB/s	
SRAM	Organization	Macro number	8
		Bank/macro	8
		Capacity/Bank	64KB

A. Area Evaluation

In the DRAM-PIM configuration, each PU integrates 32 multipliers, 31 adders, 1 accumulator and supporting logic (including registers and control circuitry), occupying 0.062 mm² (scaled to 1z-nm DRAM process). In addition, each SRAM-CIM macro contains 128 computing units composed of multipliers and accumulators, which collectively occupy approximately 45% of the macro area.

B. Energy Consumption and Performance Comparison

Fig. 5 demonstrates that the proposed hybrid architecture achieves a 2.17× energy consumption improvement and 2.25× speedup in the S·V computation layer compared to the PIM-only baseline. The improvement derives from reduced DRAM data access and enhanced computational efficiency of the SRAM-CIM subsystem.

Fig. 6 demonstrates reduced per-layer energy/latency versus baseline 2 using our Channel-Bank mapping scheme. This approach confines computation within channels, minimizing inter-channel communication—particularly benefiting data-local QK^T layers and high-dimension FFN linear layers.

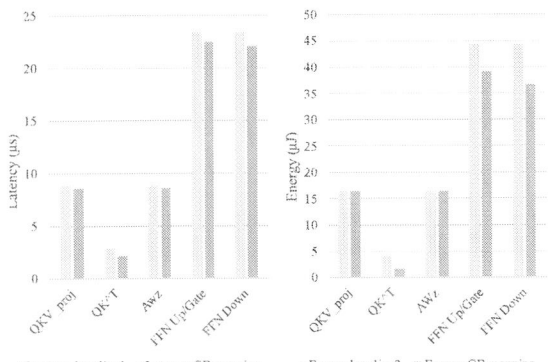

Fig. 6. Per-layer energy and performance comparison between the proposed architecture with and without Channel-Bank mapping scheme(baseline2)

Based on comparative evaluation results of the S·V layer and other Transformer layers, the proposed hybrid architecture achieves a 34% reduction in overall latency and 19% lower energy consumption compared with baselines.

VI. CONCLUSION

This work presents a hybrid DRAM-PIM/SRAM-CIM architecture for efficient edge LLM inference. By offloading dynamic V cache operations prone to access conflicts to SRAM-CIM and accelerating other linear layers with DRAM-PIM, coupled with a hierarchical data mapping and asymmetry-aware workload assignment, our design resolves critical memory bottlenecks. It achieves 1.51× speedup and 1.24× energy reduction in the focused Transformer layers for Llama2-7B inference compared to the DRAM-PIM-only baselines, demonstrating potential for edge deployment.

REFERENCES

[1] A. Vaswani, N. Shazeer, N. Parmar, J. Uszkoreit, L. Jones, A. N. Gomez, et al., "Attention is all you need," in Advances in neural information processing systems, vol. 30, pp. 5999–6009, 2017.

[2] S. Garg and G. Ramakrishnan, "BAE: BERT-based adversarial examples for text classification," in Proceedings of the 2020 Conference on Empirical Methods in Natural Language Processing, pp. 6174–6181, 2020.

[3] D. Leiker, S. Finnigan, A. R. Gyllen, and M. Cukurova, "Prototyping the use of large language models (LLMs) for adult learning content creation at scale," CEUR Workshop Proceedings, vol. 3487, pp. 3–7, 2023.

[4] H. Khlaaf, P. Mishkin, J. Achiam, G. Krueger, and M. Brundage, "A hazard analysis framework for code synthesis large language models," unpublished.

[5] Z. Yu, S. Liang, T. Ma, Y. Cai, Z. Nan, D. Huang, et al., "Cambricon-LLM: a chiplet-based hybrid architecture for on-device inference of 70B LLM," in 2024 57th IEEE/ACM International Symposium on Microarchitecture (MICRO), pp. 1474–1488, 2024.

[6] J. Park et al., "AttAcc! Unleashing the power of PIM for batched transformer-based generative model inference," in Proceedings of the 29th ACM International Conference on Architectural Support for Programming Languages and Operating Systems, vol. 2, pp. 103–119, 2024.

[7] M. Zhou, W. Xu, J. Kang, and T. Rosing, "TransPIM: a memory-based acceleration via software-hardware co-design for transformer," in 2022 IEEE International Symposium on High-Performance Computer Architecture (HPCA), pp. 1071–1085, 2022.

[8] C. Wolters, X. Yang, U. Schlichtmann, and T. Suzumura, "Memory is all you need: an overview of compute-in-memory architectures for accelerating large language model inference," unpublished.

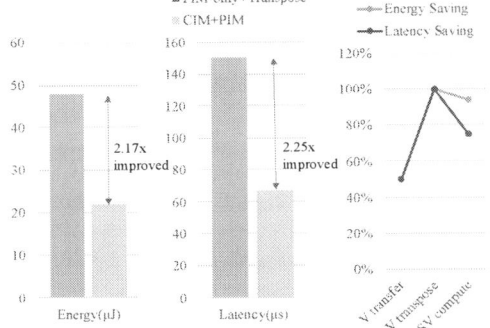

Fig. 5. Energy and performance comparison of the S·V layer between the proposed hybrid architecture and PIM-only baseline1

MCDC: A <u>M</u>emory-efficient and <u>C</u>omputation-efficient Architecture for <u>D</u>eformable <u>C</u>onvolutions

Zhiyi Shu*[1], Xinhua Shi [1], Jun Han*[1]

[1] State Key Laboratory of Integrated Chips and Systems, Fudan University

* Email: zyshu23@m.fudan.edu.cn, junhan@fudan.edu.cn

Abstract—**Deformable convolutional networks (DCNs) are widely used in video processing tasks, such as super-resolution and frame interpolation, due to their superior inter-frame content alignment capability. By dynamically adjusting kernel sampling positions via learnable offsets, DCNs better adapt to object shapes and sizes, achieving superior performance. However, this flexibility introduces significant memory overhead and latency due to input resampling, offset storage, and expanded receptive fields. Additionally, irregular memory access and cache missing from pixel resampling increase memory consumption and create access conflicts, posing key challenges for efficient hardware implementation. To address these issues, we propose OS-DCN, an offset-sharing deformable convolution operator, along with its corresponding hardware accelerator featuring dual optimization in both memory and computation. MCDC is designed and evaluated under TSMC 22nm library. It costs 6.95KB SRAM and a power of 31.259mW, while achieving an area efficiency of 4.96TOPs/mm² , which is higher than previous DCN processors.**

Keywords—Deformable Convolutional networks, Offset-sharing Strategy, Memory-efficient Architectures, Computation-efficient Architectures.

I. INTRODUCTION

Deformable convolutional networks (DCNs) play an important role in video processing tasks such as super-resolution (SR) and frame interpolation (FI), owing to their superior inter-frame content alignment capability. By dynamically generating learnable offsets, DCNs overcome the limitation of fixed sampling locations in conventional convolutions, enabling adaptive feature extraction from relevant spatial locations. This capability is particularly critical for handling moving objects in video sequences. Numerous algorithms [1] [2] [3] have achieved significant gains in video performance through the incorporation of deformable convolutions, which are capable of exploring contextual information in images. However, pixel resampling, along with the need for offset storage and expanded receptive fields, incurs significant memory overhead and increased latency. Although recent approaches [3] [4] [5] targeting DCN architectures have reduced computation and storage costs through bit-shift operations and optimized convolution workflows, their overall efficiency gains are limited by suboptimal joint optimization of memory and computation.

To alleviate these problems, we propose OS-DCN, an offset-sharing deformable convolution operator, and MCDC, a memory-efficient and computation-efficient hardware

architecture. The main contributions are summarized as follows:

1) We propose OS-DCN, a novel deformable convolution operator that employs an offset-sharing strategy to reduce offsets storage requirements by 93.75% and computational complexity by 41.53%, while limiting PSNR degradation to just 0.143dB.

2) We propose MCDC, a hardware-optimized architecture for OS-DCN that jointly enhances memory and computation efficiency. It employs a coefficient generator array that reuses bilinear interpolation coefficients across input channels, reducing computational complexity by 12.62%. Additionally, it implements the offset-sharing strategy in hardware, reducing on-chip offset SRAM usage by 91.41%.

3) We evaluate our design under TSMC 22nm CMOS library. MCDC costs 6.95KB SRAM and a power of 31.259mW, while achieving an area efficiency of 4.96TOPs/mm² , and can process more pixels at 500MHz, which is more efficient than previous DCN processors.

II. OFFSET-SHARING DEFORMABLE CONVOLUTION OPERATOR

A. Deformable Convolution

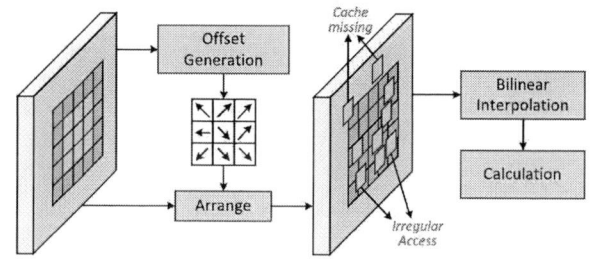

Fig. 1. Deformable convolution calculation flow

By dynamically adjusting kernel sampling positions through learned offsets, DCNs adapt to targeted object shapes and sizes. As shown in Fig.1, the DCN operation shifts the kernel's sampling positions on the input feature map (FM), expanding the receptive fields from 3×3 to 5×5 , which is determined by the range of offsets. To calculate as Eq.1, DCN first generates spatial offsets through convolutions to determine adaptive sampling positions. These offsets often produce fractional coordinates, potentially causing irregular memory access and cache missing. It then estimates feature

979-8-3315-3918-4/25 $31.00 © 2025 IEEE

values at these non-integer locations using bilinear interpolation, before finally computing the outputs by convolving the interpolated features with their corresponding weights.

$$y(p_0) = \sum_n W(p_n) * X(p_0 + p_n + \Delta p_n) \qquad (1)$$

B. Offset Aggregation Characteristic

A key character of DCN is that each output pixel requires a group of offsets (totaling $2 \times kernel\_size \times kernel\_size$). As shown in Eq.2, this results in a direct correlation between offset storage demands and feature map dimensions, leading to a dramatic increase in on-chip memory requirements for large feature maps. Consequently, a major hardware design challenge is to efficiently balance offset storage and operator performance.

$$Sum_{offset} = 2 \times Output_x \times Output_y \times K_x \times K_y \qquad (2)$$

Fortunately, we observe the DCN exhibits a remarkable offset aggregation characteristic. As shown in Fig.2, although each output pixel corresponds to a unique set of offsets, these offsets collectively form spatial point clouds. For example, when the kernel size is 3×3, it will form nine point clouds, corresponding to nine kernel positions. Therefore, it is practical and effective to implement a shared offset scheme across all output feature pixels within specific feature maps.

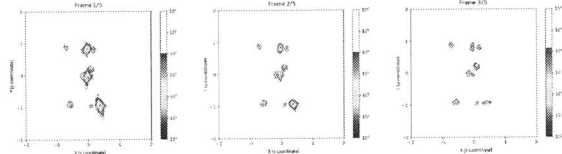

Fig. 2. Offset aggregation characteristic

C. OS-DCN

Inspired by the offset-sharing concept in grouped channels from [3], the critical offset spatial aggregation characteristic motivates our proposal of a new offset-sharing strategy. Our solution leverages the natural spatial clustering of offsets to unify and share them across local pixel blocks, which simplifies the hardware's offset access and storage logic, leading to superior efficiency gains.

TABLE I. OFFSET-SHARING STRATEGY EXPERIMENTS

Tile size		16×16			32×32			
Block size		4×4	8×8	16×16	4×4	8×8	16×16	32×32
	origin	27.508			27.1803			
Center	+ offset sharing	26.680	26.557	26.376	26.383	26.318	26.247	26.169
	+ frame sharing	26.465	26.387	26.247	26.152	26.125	26.108	26.044
Max_pool	+ offset sharing	25.757	25.307	24.849	25.408	24.949	24.500	24.036
	+ frame sharing	25.531	25.090	24.606	25.168	24.723	24.253	23.733
Avrg_pool	+ offset sharing	27.035	26.994	26.981	26.689	26.646	26.633	26.634
	+ frame sharing	26.887	26.934	26.946	26.535	26.580	26.599	26.603

* All PSNR values are reported in dB.
* Tile size refers to the partitioned feature map dimensions, while block size indicates the pixel group size for offset-sharing.
* Center uses the center pixel offsets for the group.
* Max_pool/Avrg_pool operation generates the shared pixel offsets for the group.

As shown in Table.I, we conduct multiple sets of ablation experiments to explore the effect of different configurations

on the offset-sharing strategy. Experimental results demonstrate that: (1) Larger block sizes degrade performance by increasing offset variance within shared regions; (2) Multi-frame implementations exhibit reduced sharing efficiency due to additional temporal variations; and (3) Among various pixel-selecting methods, average pooling works best.

Based on the above experiments, we further explore high-performance deformable convolutional operators employing offset-sharing and propose OS-DCN, which achieves 29.399dB on SPMC [6] datasets, while reducing computational complexity by 0.42× within a marginal 0.14 dB degradation compared to standard DCN.

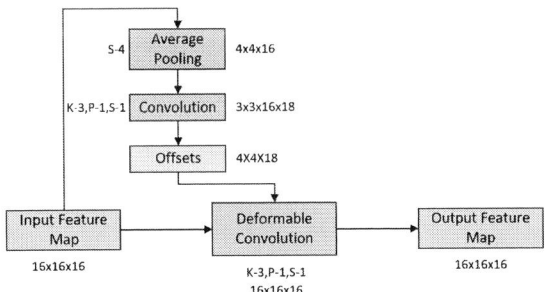

Fig. 3. The structure of OS-DCN

III. HARDWARE ARCHITECTURE

In this section, we will detail the hardware architecture of the proposed MCDC, which is both memory-efficient and computation-efficient, achieving high array utilization and low on-chip memory.

A. Architecture Overview

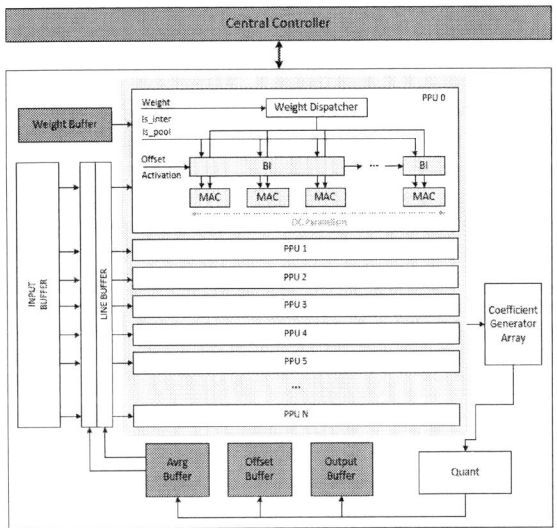

Fig. 4. The overall architecture of the proposed MCDC

The overall architecture of MCDC is shown in Fig.4. MCDC consists of multiple pixel-processing units (PPUs), a coefficient generator array (CGA), a quant module, a central controller, and several on-chip buffers to store weights, activations and offsets on-chip separately.

MCDC employs a layer-fusion approach, supporting average pooling, standard convolution, and deformable convolution. When receiving a host CPU request, the controller parallelizes pixel computation across multiple PPUs. Each PPU has a bilinear interpolation (BI) unit for pixel resampling and maps output channels (OCs) to dedicated BI and MAC units, enabling parallel processing of 8 OCs. Two control flags (is_inter and is_pool) determine whether interpolation or pooling is required. If interpolation is needed, pixels first undergo resampling through BIs before being processed by MACs for convolution or pooling. After all PPUs complete computation, output pixels are routed to the CGA or quant module for post-processing and buffering.

B. Coefficient Generator Array

Based on the offset-sharing strategy, we propose a coefficient generator array (CGA) to eliminate redundant computations. As shown in Fig.5, the CGA directly processes offsets from the preceding layer, generating the four bilinear interpolation coefficients: $(1-a) \times b$, $a \times (1-b)$, $(1-a) \times (1-b)$, and $a \times b$. This is done because the offsets are shared in the input channel direction, and pixel points at the same location in different input feature maps will have the same interpolation process to calculate these coefficients. By caching the computed CGA factors directly instead of offsets, the BI engine eliminates the need for repeated factor calculations, enhancing overall performance. For multi-channel convolutions, CGA reduces computational complexity significantly while preserving interpolation accuracy.

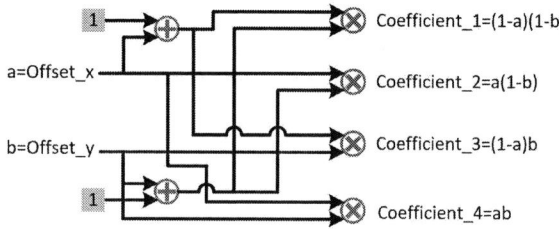

Fig. 5. The hardware architecture of the coefficient generator unit

C. Box-bounding Bilinear Interpolation Engine

As demonstrated in Fig.6, BI resamples input pixels using coefficients derived from fractional offsets. The BI unit fetches pixels from the line buffer via offset-controlled multiplexers, computes interpolation in a single cycle, and forwards results to the MAC unit.

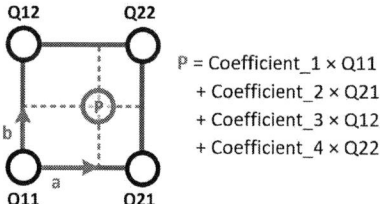

Fig. 6. The calculation of bilinear interpolation

Moreover, our hardware implementation incorporates a bounding box that constrains all deformation offsets to magnitudes below 1, based on the observation that over 95% of offsets naturally fall within the [-1,1] range. This choice provides three key benefits: it limits the expansion of the receptive fields to prevent excessive on-chip memory demands; enables efficient single-cycle pixel fetching through

optimized multiplexer selection in BIs while maintaining low power consumption; and preserves computational efficiency with negligible performance loss.

D. Dual-mode MAC Unit

Unlike conventional designs, our dual-mode MAC unit is designed to support both average pooling and convolution operations through configurable computation modes. Two dedicated multiplexers, which are shared among MACs within a PPU, dynamically adjust the receptive fields to accommodate different kernel sizes (4×4 for pooling and 3×3 for convolution). For pooling (triggered when is_pool is true), it computes the average through summation followed by right-shifting. This dual-mode design maintains computational efficiency while supporting both fundamental operations.

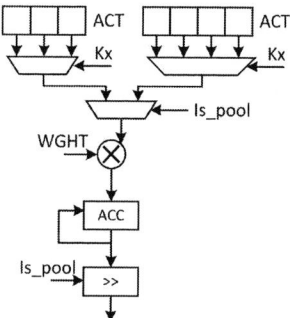

Fig. 7. The hardware architecture of the MAC unit

E. Quant Module

To optimize hardware efficiency, MCDC employs symmetric quantization by representing both weights and activations as 8-bit integers (INT8). This design choice significantly reduces memory requirements and bandwidth requirements compared to floating-point implementations.

For the offset coefficients, we conduct extensive quantization experiments to balance computational accuracy with hardware efficiency. Our analysis uses Peak Signal-to-Noise Ratio (PSNR) as the evaluation standard. As shown in Fig.8, the coefficient should be implemented as 5-bit integers within an acceptable 0.086dB quality degradation.

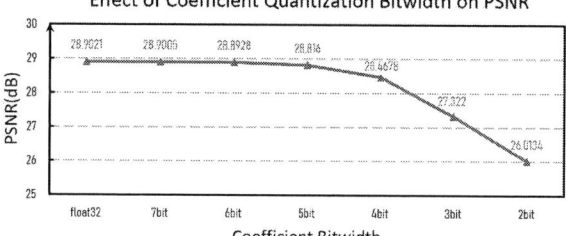

Fig. 8. The results of offset quantization experiments

IV. EXPERIMENT

In this section, we perform the training experiments on DCN to find the best structure, and then verify the efficiency of the corresponding hardware design on ASIC.

A. OS-DCN Training Strategy

We implement our DCN operator using the PyTorch framework and train it on SPMC [6] datasets. As Table.II demonstrated, compared to the standard convolution (SC), DCN achieves superior performance, albeit with a $1.53\times$ increase in computational complexity. In order to utilize the aggregation of offsets, we propose the offset-sharing strategy. Through the comparative analysis of s4conv and avrgpool approaches for block pixels reuse, we identify the optimal configuration for preserving performance.

Our final proposed structure, OS_DCN_avrgpool (as illustrated in Fig.3 and equivalent to OS-DCN), achieves a 0.103 dB PSNR improvement over SC. Compared to conventional DCNs, it achieves a 41.53% reduction in computational complexity while limiting PSNR degradation to just 0.143dB.

TABLE II.   ALGORITHM EXPERIMENTS

Algorithm	PSNR(dB)	$\triangle$ Computation
SC	29.296	+0
DCN	29.542	$+1.53\times$
OS_DCN_s4conv	29.329	$+0.54\times$
OS_DCN_avrgpool	29.399	$+0.48\times$

* s4conv: standard convolution with a stride of 4.

B. Computation Reduction

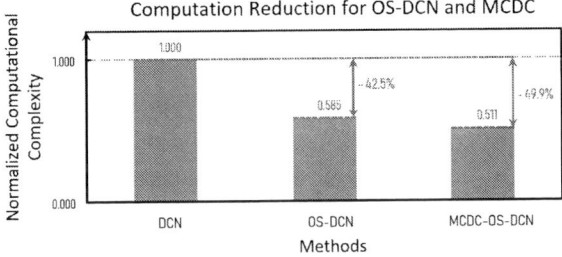

Fig. 9.   The results of computation reduction

TABLE III.   COMPARISON RESULTS WITH THE STATE-OF-THE-ART DESIGN

	[7]	[8]	Ours
Technology	14nm[1]	TSMC 40nm	TSMC 22nm
Frequency	200MHz	200MHz	500MHz
Power(mW)	9.60×10^3	438-700	31.26
SRAM(KB)	-	994	6.95 20.06[2]
Area Efficiency (GOPS/mm^2)	-	222.2 734.5[3]	4956.5[3]
Pixels/s	793.6M	1240M	1836M

[1] Implemented in Intel Stratix 10GX 2500.
[2] Feature map size is 48 × 48.
[3] The area efficiency ratio is normalized from 40nm to 22nm technology node.
[4] GOPS include interpolation.

By incorporating the offset-sharing strategy and our novel coefficient-generation scheme, MCDC achieves substantial improvements in computational efficiency for hardware implementation. As demonstrated in Fig.9, OS-DCN effectively aggregates offsets, achieving 42.5% computational efficiency. Furthermore, by optimizing hardware through

CGA, we eliminate redundant interpolation computation across input channels, reducing overall computational complexity by 49.9%.

C. Comparison Result

The proposed MCDC is implemented using Chisel, a hardware construction language. The number of PPU is set to 16, and the OC parallelism in each PPU is set to 8. We evaluate the architecture under TSMC 22nm library using Synopsys Design Compiler 2017.09-SP4 and compare the proposed MCDC with the state-of-the-art design in Table.III. MCDC demonstrates superior memory efficiency by minimizing on-chip SRAM requirements while processing input feature maps of the same size. Furthermore, it can process more pixel values per second, outperforming other designs in computational efficiency.

V. CONCLUSION

In this paper, we propose OS-DCN, a high-performance DCN operator, employing the offset-sharing strategy, and MCDC, a memory-efficient and computation-efficient hardware architecture for OS-DCN. MCDC is designed and evaluated under TSMC 22nm library. It costs 6.95KB SRAM and a power of 31.259mW, while achieving an area efficiency of 4.96TOPs/mm^2, which is more efficient than previous DCN processors.

ACKNOWLEDGMENT

This work was supported by the National Natural Science Foundation of China under Grant 61934002 and 62234008.

REFERENCES

[1] Y. Tian, Y. Zhang, Y. Fu, and C. Xu, "TDAN: Temporally-deformable alignment network for video super-resolution," in Proc. IEEE/CVF Conf. Comput. Vis. Pattern Recognit. (CVPR), Jun. 2020, pp. 3357–3366.

[2] Q. Huang, D. Wang, Z. Dong, and et al., "Codenet: Efficient deployment of input-adaptive object detection on embedded fpgas," in ACM/SIGDA International Symposium on Field-Programmable Gate Arrays, 2021, pp. 206–216.

[3] J. Guan et al., "Memory-Efficient Deformable Convolution Based Joint Denoising and Demosaicing for UHD Images," in IEEE Transactions on Circuits and Systems for Video Technology, vol. 32, no. 11, pp. 7346-7358, Nov. 2022, doi: 10.1109/TCSVT.2022.3182990.

[4] S. Zhang, W. Mao and Z. Wang, "An Efficient Accelerator of Deformable 3D Convolutional Network for Video Super-Resolution," 2022 IEEE Computer Society Annual Symposium on VLSI (ISVLSI), Nicosia, Cyprus, 2022, pp. 110-115, doi: 10.1109/ISVLSI54635.2022.00032.

[5] S. Li, S. Cao, L. Hui, Z. Jiang, Y. Sun and S. Xu, "A Computation-efficient Deformable Convolution Network Accelerator via Hardware and Algorithm Co-Optimization," 2022 IEEE Workshop on Signal Processing Systems (SiPS), Rennes, France, 2022, pp. 1-6, doi:10.1109/SiPS55645.2022.9919242.

[6] Tao, Xin, et al. "Detail-revealing deep video super-resolution." Proceedings of the IEEE international conference on computer vision. 2017.

[7] S. Zhang, W. Mao and Z. Wang, "An Efficient Accelerator Based on Lightweight Deformable 3D-CNN for Video Super-Resolution," in IEEE Transactions on Circuits and Systems I: Regular Papers, vol. 70, no. 6, pp. 2384-2397, June 2023, doi: 10.1109/TCSI.2023.3258446.

[8] K. -P. Lin, J. -H. Liu, J. -Y. Wu, H. -C. Liao and C. -T. Huang, "VISTA: A 704mW 4K-UHD CNN Processor for Video and Image Spatial/Temporal Interpolation Acceleration," 2023 IEEE International Solid-State Circuits Conference (ISSCC), San Francisco, CA, USA, 2023, pp. 48-50, doi: 10.1109/ISSCC42615.2023.10067857.K

Hardware-Efficient Lightweight Feature Map Compression for Convolutional Neural Networks

Bing Wu, Shan Cao, Zhiyuan Jiang

School of Communication and Information Engineering, Shanghai University, Shanghai 200444, China
Email: {Wubing-0112, cshan, jiangzhiyuan}@shu.edu.cn

Abstract—Convolutional neural networks (CNNs) are widely used in computer vision and other fields, while their model parameters and intermediate feature map sizes continue to grow significantly. When deployed on resource-constrained embedded platforms, the frequent data movement of feature maps causes substantial energy consumption and memory bandwidth pressure that demands urgent solutions. To address this challenge, this paper proposes a lightweight feature compression scheme that effectively compresses feature maps through dynamic uniform quantization combined with exponential-Golomb entropy coding. The corresponding hardware architecture is introduced for this scheme, which further improves its hardware efficiency through techniques such as partial comparison, division lookup tables, and multi-way parallelism. On 8/16-bit quantized models, our solution achieves a compression ratio of 2.76×∼4.64×, outperforming existing approaches.

Keywords—Neural networks, compression, quantization, FPGA implementation.

I. Introduction

Convolutional neural networks (CNNs) have achieved remarkable progress in fields such as computer vision, speech recognition, and natural language processing. However, when deploying high-performance CNN models to resource-constrained edge devices for inference, we still face significant challenges. The main issue lies in the fact that CNN inference generates a large number of intermediate feature maps and model parameters, which typically far exceed the on-chip memory capacity of edge devices. As a result, the system is frequently forced to transfer data between computing units and off-chip memory. Research [1] indicates that in the total energy consumption of CNN inference, off-chip memory access accounts for up to about 70%, with feature maps (FMs) transferred between layers being the main source of energy consumption.

To reduce off-chip memory access energy from feature map storage, various compression strategies have been proposed. Spatial methods like 2D-DCT [2], [3] exploit spatial correlations by retaining key low-frequency components, though their efficiency degrades in deeper networks with weaker correlations. Channel-wise approaches using PCA [4], [5] reduce dimensionality through principal components but suffer from high computational overhead and storage needs for projection matrices. Hybrid training-based methods [6] combine 1D-DCT with custom masks, but introduce training overhead and limit generalizability.

To address the limitations of the aforementioned schemes, we propose a FM compression method based on block-wise dynamic quantization combined with variable-length coding. This approach can be used during the CNN network inference process to reduce energy consumption and memory storage requirements. The main contributions of this paper include:

- A lightweight FM compression scheme is first presented, which boasts a high compression ratio and requires no fine-tuning. This scheme divides the original feature map into fixed-size blocks, applies uniform quantization based on the maximum value of each block, and accomplishes compression through Exponential-Golomb coding.

- The corresponding hardware architecture is then introduced for the proposed compression scheme. The proposed architecture uses partial bit comparison and division look-up table approximation to reduce hardware resources while improving the compression ratio. Additionally, it adopts a pipelined design and multi-path parallel quantization encoding to enhance throughput.

- The proposed hardware compression algorithm is implemented on FPGA. Experimental results demonstrate that the proposed compression achieves 2.76×∼4.64× compression ratios across various CNN networks while delivering 3.47 Gbps throughput with minimal hardware resource consumption.

II. Motivation

To efficiently deploy neural networks on resource-constrained hardware platforms, full-precision (32-bit floating-point) models typically need to be quantized to lower bit-widths. Existing research primarily focuses on the quantization of both network weights and activations [1][2][3], with methods generally categorized into quantization-aware training (QAT) and post-training quantization (PTQ). Weights can be quantized to extremely low bit-widths (2 bits) [7] while maintaining model accuracy in these quantization methods. However, due to the diverse sparsity patterns of activations in different applications, excessive quantization of activations (below 8 bits) [8] often leads to significant accuracy degradation. Therefore, activations are typically quantized to 8 bits [9] or higher in practice.

Nevertheless, quantized activations (FMs) still exhibit high sparsity. In hardware platforms such as neural processing units (NPUs), where on-chip memory resources are limited, intermediate FMs are stored in off-chip memory. FMs are frequently transferred between NPUs and off-chip memory, resulting in significant energy and latency overhead. Therefore,

---

979-8-3315-3918-4/25 $31.00 © 2025 IEEE

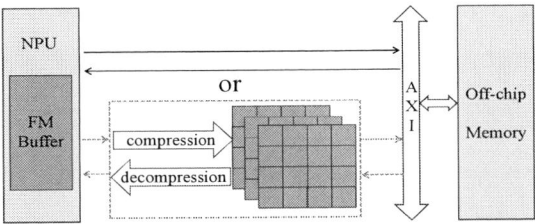

Fig. 1. Illustration of system position of the proposed FM compressor.

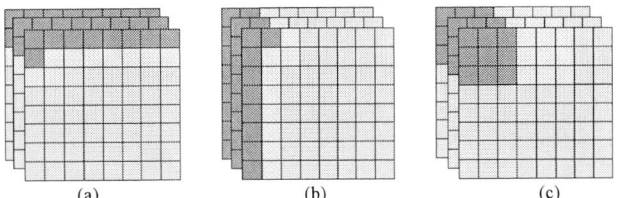

Fig. 2. Block-wise partitioning methods. For visualization convenience, a 3×3 block is used to represent an 8×8 block. Blocks are divided by (a) rows, (b) columns, and (c) space.

it is crucial to fully exploit the inherent sparsity of activations during run-time processing. As shown in Fig. 1, a hardware FM compressor unit is considered, positioned between the NPU and the system bus for the compression and decompression of FM data. For effective real-time FM compression, the compressor needs to achieve a high compression ratio while maintaining low latency and minimal hardware resource usage. This remains a challenge with current compression techniques, as discussed in the Introduction section. In response to this, a lightweight yet highly efficient compression scheme, as well as its hardware architecture, is introduced in following sections.

III. PROPOSED COMPRESSION SCHEME

In this section, a lightweight compression scheme is introduced. The scheme first employs block-wise dynamic quantization to reduce the data range and enhance sparsity, then utilizes a variable-length coding algorithm that excels in small-value encoding to effectively minimize data representation, thereby achieving efficient compression.

A. Block-wise Dynamic Uniform Quantization

To efficiently quantize FM data, the FMs are first divided into smaller blocks. As illustrated in Fig. 2, a three-dimensional FM structure typically has three partitioning strategies, where color-coded blocks represent the corresponding partitioned segments. Blocks can be divided into rows, columns, or spaces. Since the subsequent quantization flow is agnostic to the specific blocking method, our quantization scheme consistently and efficiently handles row-wise, column-wise, or spatially partitioned data, ensuring compatibility with various CNN accelerator input dataflows and demonstrating the flexibility of our solution.

Following partitioning, the quantization in each block is processed as

$$X_q = \text{round}\left(x_i \times \frac{\text{block}_{\max}}{N-1} \right) \quad (1)$$

where $\text{block}_{\max}$ denotes the block-specific maximum, N signifies the predefined quantization level, round represents the rounding operation, while x_i and X_q denote the original and quantized values, respectively. After fixed-level quantization, lossless entropy encoding processes the quantized data, storing both block maxima and encoded output. During decompression, the original feature maps are reconstructed through inverse quantization applied to the decoded data using the stored maximum values.

B. Unsigned Exponential-Golomb Entropy Coding

The prevalent sparse encoding algorithms [6] focus on efficiently compressing zero values, while neglecting the redundancy in small non-zero values. However, following the dynamic uniform quantization, the quantized FM data exhibit a high proportion of zero values and small non-zero values. Accordingly, we employ the unsigned Exponential-Golomb coding (UE) for lossless compression. As a method with variable-length coding capabilities, UE is demonstrated to have particular advantages for encoding small numerical values.

The UE encoding procedure is as follows. After block partitioning and dynamic quantization (Section A), the FMs are converted to quantized data, then the data are first incremented by 1 (for lossless decoding), converted to binary, and the minimum bit count n is determined. The binary sequence is prefixed with $n-1$ leading zeros, then packed into 8-bit byte segments with zero-padding if needed.

The decoding process is equally straightforward: Since all values undergo plus 1 operation during encoding, the actual values processed during decoding are no less than 1. By simply parsing the number of leading zeros, the bit width of the subsequent valid data can be determined, and subtracting 1 from this value restores the original data.

IV. HARDWARE ARCHITECTURE

The hardware architecture corresponding to the proposed compression scheme is presented, including both the compressor and the decompressor, as illustrated in Fig. 3. The blue arrows indicate the compression process. Firstly, the original FMs are read from NPU and sequentially transmitted to both the block max finder module and the original data FIFO. Upon determining the maximum value, the quantization process (requiring one multiplication-division operation) is performed in the Average Quant/De-Quant unit, the results of which are fed into the UE encoding unit. Since the decoding process requires the maximum value parameter, this parameter is stored in external memory along with the encoded data. The red arrows represent the decompression process: first obtaining the low-bit quantized value through UE decoding, followed by inverse quantization using the stored maximum value.

A. Approximation-based Maximum Finder Unit

As shown in Fig. 4, a pipelined structure for the block-wise maximum value finder unit is presented. This unit takes

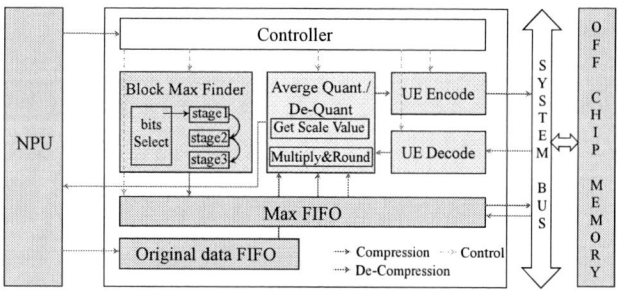

Fig. 3. Hardware architecture of the compressor and de-compressor.

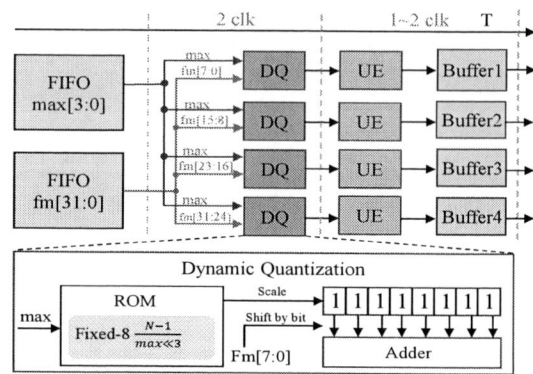

Fig. 5. The quad-path pipelined structure for the dynamic quantization and entropy coding units.

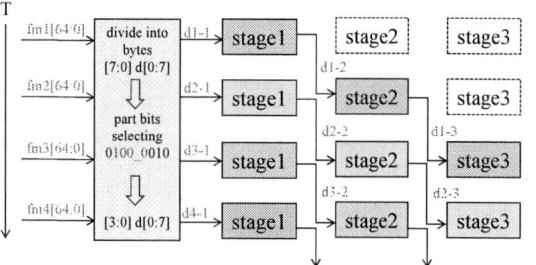

Fig. 4. The three-stage pipeline structure of partial bits compared.

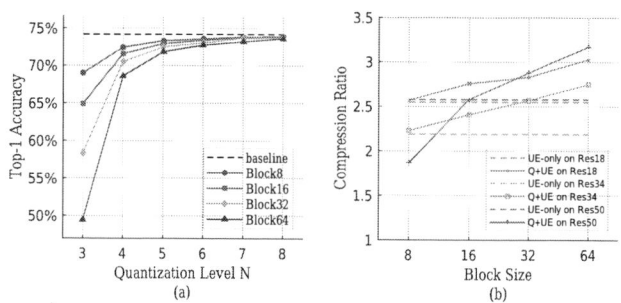

Fig. 6. (a) The effect of block size and quantization level on accuracy. (b) The effect of block size on compression ratio.

8-byte data as input, performs pairwise comparisons through a three-stage pipeline, and outputs the maximum value. To simplify the pipelining structure and reduce the latency of the critical path, approximate comparators are designed, which compare only the higher-order 4 bits between the input data. The accuracy loss is shown to be negligible compared to a full-precision scheme through experimental evaluation.

B. Pipelined Quantization and Coding Parallelism

Following the maximum finder, the quantization and coding are processed for each element. To match the rate of maximum finder and reduce latency, four-way dynamic quantization (DQ) and UE coding are computed in parallel, as shown in Fig. 5. A fixed-point reciprocal is first retrieved corresponding to the maximum value in the maximum FIFO, generating the scale parameter. Then, the FM data are fetched from the FM FIFO, which is then bit-shifted according to each binary weight of the scale, with shifted results accumulated to achieve multiplicative equivalence.

To simplify hardware implementation, the DQ unit replaces division operations with precomputed reciprocal lookup tables, while multiplications are implemented using bit-shifting and adder circuits. In this way, the computation latency is significantly reduced for division and multiplication operations. Furthermore, by matching the rate of the maximum finder unit with 3-cycle latency, the DQ unit with 2-cycle latency, and the UE unit with 1~2 cycle latency, the illustrated quad-path parallel architecture eliminates serial processing bottlenecks, thereby enhancing system throughput.

V. EXPERIMENTAL RESULTS

To evaluate the efficiency of the proposed compression scheme, the compressor is tested on 8-bit quantized ResNet-18/34/50 and 16-bit quantized VGG-16, using 50,000 validation images from ImageNet ILSVRC2012. The compression operation is applied after each ReLU activation layer in all networks.

A. Comparisons of compression algorithm

The proposed algorithm is compared with the baseline (the 8-bit quantized model by PTQ) in terms of top-1 accuracy and compression ratio. As shown in Fig. 6(a), the baseline achieves a 74.16% accuracy. As the quantization level N is increased from 3 to 8, the accuracy initially improves rapidly and then stabilizes. A similar pattern is observed with block size. At N8, all block sizes exceed 73.5% accuracy. As shown in Fig. 6(b), the quantization and UE coding (Q+UE) improves the compression ratio by 2.34× ~3.17×, compared with the UE-only method. The proposed scheme shows a consistent advantage in compression ratio, yet varying magnitudes across networks, reflecting its architectural adaptability.

The performance of the proposed compression for different models on ImageNet is compared in Table I. The compression ratio is evaluated for three block-wise methods, including the column block (CB), the row block (RB), and the space block (SB). VGG16 achieves the highest compression ratios (4.64×)

TABLE I. COMPRESSION RATIO OF DIFFERENT MODELS ON
IMAGENET

Model	Average Accuracy Drop (%)	Compression Ratio		
		CB	RB	SB
ResNet-18	0.41	3.04×	3.03×	3.00×
ResNet-34	1.19	2.76×	2.75×	2.74×
ResNet-50	0.57	3.18×	3.17×	3.15×
VGG16	0.13	4.64×	4.55×	4.52×

TABLE II. COMPRESSION RATIO COMPARISON OF RESNET50
FOR DIFFERENT ALGORITHMS

Compression Method	Accuracy Drop (%)	Compression Ratio
TNNLS'23 [2]	0.40	2.31×
ISCAS'21 [6]	0.39	2.90×
TCASI'23 [4]	0.30	2.74×
ASICON'23 [10]	–	2.35×
Ours	0.47	**3.15×**

with minimal accuracy drop (0.13%), while ResNet variants show relatively lower but consistent compression performance across all block methods. The marginal accuracy degradation (<1.2%) across all models confirms the effectiveness of these compression techniques for practical deployment.

Table II compares the compression performance of different algorithms on ResNet50, where our proposed method achieves the highest compression ratio of 3.15× with a marginal accuracy drop of 0.47%. While existing approaches demonstrate competitive compression ratios (2.31×-2.90×), our technique provides a 8.6-36.4% improvement in compression efficiency.

B. Comparisons of compression hardware implementation

The hardware implementation is conducted on the Xilinx Zynq-7000 FPGA to verify the effectiveness of the proposed hardware architecture based on the optimal parameter combination of block size = 64, quantization level $N = 8$. As shown in Table III, our work achieves superior hardware efficiency, using only 1.9K LUTs (47.5% of PCA) with 32.4μJ energy consumption (48.4% lower than PCA) and 3.47 Gbps throughput. While the resource consumption of TECO is undisclosed, our design has a 15.7% throughput improvement and 45.9% energy reduction compared to TECO. This optimized implementation maintains peak performance with minimal resource overhead.

TABLE III. COMPARISON OF RESOURCE UTILIZATION AND
PERFORMANCE WITH OTHER ALGORITHMS

Method	LUTs	BRAMs	Energy (μJ)	Throughput (Gbps)
PCA [11]	4K	-	62.8	0.66
TECO [2]	-	-	59.9	3.0
Ours	1.9K	2.5	32.4	3.47

VI. CONCLUSION

This paper proposes a lightweight FM compression algorithm based on block-wise dynamic uniform quantization combined with UE entropy coding, which effectively reduces energy consumption during CNN inference. The proposed algorithm achieves high compression ratios across various CNN models. We have completed the hardware architecture design and implementation, with future work focusing on deploying the hardware algorithm in practical real-time inference frameworks for further validation.

ACKNOWLEDGMENT

This work was supported in part by the National Natural Science Foundation of China (NSFC) under Grants 62271300 and 12141107, in part by the Shanghai Municipal Science and Technology Commission under grants 24DP1501100 and 24DP1500600.

REFERENCES

[1] Fengbin Tu, Weiwei Wu, Shouyi Yin, Leibo Liu, and Shaojun Wei. Rana: Towards efficient neural acceleration with refresh-optimized embedded dram. in 2018 acm/ieee 45th annual international symposium on computer architecture (isca'18). *IEEE, 340s352*, 2018.
[2] Yubo Shi, Meiqi Wang, Tianyu Cao, Jun Lin, and Zhongfeng Wang. Teco: A unified feature map compression framework based on transform and entropy. *IEEE Transactions on Neural Networks and Learning Systems*, 2023.
[3] Zhuang Shao, Xiaoliang Chen, Li Du, Lei Chen, Yuan Du, Wei Zhuang, Huadong Wei, Chenjia Xie, and Zhongfeng Wang. Memory-efficient cnn accelerator based on interlayer feature map compression. *IEEE Transactions on Circuits and Systems I: Regular Papers*, 69(2):668–681, 2021.
[4] Chenjia Xie, Zhuang Shao, Ning Zhao, Yuan Du, and Li Du. An efficient cnn inference accelerator based on intra-and inter-channel feature map compression. *IEEE Transactions on Circuits and Systems I: Regular Papers*, 70(9):3625–3638, 2023.
[5] Feng Xiong, Fengbin Tu, Man Shi, Yang Wang, Leibo Liu, Shaojun Wei, and Shouyi Yin. Stc: Significance-aware transform-based codec framework for external memory access reduction. In *2020 57th ACM/IEEE Design Automation Conference (DAC)*, pages 1–6. IEEE, 2020.
[6] Yubo Shi, Meiqi Wang, Siyi Chen, Jinghe Wei, and Zhongfeng Wang. Transform-based feature map compression for cnn inference. In *2021 IEEE International Symposium on Circuits and Systems (ISCAS)*, pages 1–5. IEEE, 2021.
[7] Jungwook Choi, Swagath Venkataramani, Vijayalakshmi Viji Srinivasan, Kailash Gopalakrishnan, Zhuo Wang, and Pierce Chuang. Accurate and efficient 2-bit quantized neural networks. *Proceedings of Machine Learning and Systems*, 1:348–359, 2019.
[8] Ron Banner, Yury Nahshan, and Daniel Soudry. Post training 4-bit quantization of convolutional networks for rapid-deployment. *Advances in Neural Information Processing Systems*, 32, 2019.
[9] Kang Zhao, Sida Huang, Pan Pan, Yinghan Li, Yingya Zhang, Zhenyu Gu, and Yinghui Xu. Distribution adaptive int8 quantization for training cnns. In *Proceedings of the AAAI Conference on Artificial Intelligence*, volume 35, pages 3483–3491, 2021.
[10] Hang Xu, Chenjia Xie, Xin Lu, Li Du, and Yuan Du. Memory-efficient compression based on least-squares fitting in convolutional neural network accelerators. In *2023 IEEE 15th International Conference on ASIC (ASICON)*, pages 1–4. IEEE, 2023.
[11] B Chmiel, C Baskin, R Banner, E Zheltonozhskii, Y Yermolin, A Karbachevsky, AM Bronstein, and A Mendelson. Feature map transform coding for energy-efficient cnn inference. arxiv 2019. *arXiv preprint arXiv:1905.10830*.

979-8-3315-3918-4/25 $31.00 © 2025 IEEE

A Study on Dwell Time Impacts in Charge-trapping 3D NAND Flash Memory

Yining Zhou[1], Ruidong Li[2], Xuepeng Zhan[1], Guangkuo Yang[1], Yujiao Ding[1], Xinghao Wang[1],
Pengpeng Sang[1], Peng Guo[3], Jixuan Wu[*,1], Jiezhi Chen[*,1]

[1] School of Information Science and Engineering, Shandong University, P. R. China; [2]Cloud Computing Equipment
Industry Innovation Co., Ltd., P. R. China; [3]Shandong Sinochip Semiconductors Co. Ltd, P. R. China.

* Email: jixuanwu@sdu.edu.cn, chen.jiezhi@sdu.edu.cn

Abstract—To evaluate the role of dwell time in enhancing the reliability of 3D charge-trapping (CT) NAND flash memory, this work investigates key usage scenarios systematically, including read disturb (RD), high-temperature data retention (HTDR), and cross-temperature operations. The analysis of raw bit error rate (RBER) reveals that the increased dwell time can effectively suppresses RD-induced errors in the initial stage and exhibits a controlled degradation trend thereafter, suggesting a potential recovery mechanism. Under HTDR conditions, extended dwell time leads to a marked reduction in RBER and improved stability, with benefits gradually saturating over longer periods. Additionally, dwell time can help mitigating cross-temperature degradation effects. These findings highlight the dwell time as an efficient parameter to optimize 3D NAND flash memory.

Keywords—3D NAND, dwell time, read disturb, data retention, cross-temperature

I. INTRODUCTION

With the rapid development of data-intensive applications like data centers, edge computing, and artificial intelligence, the global data scale has experienced explosive growth [1],[2]. To accommodate this trend, 3D NAND flash memory has become a mainstream non-volatile storage solution due to its high density and cost efficiency. As fabrication technologies advance, the number of stacked layers in 3D NAND continues to increase, with more than 400 layers already demonstrated in industry [3]. Meanwhile, storage technologies such as multi-level cell (MLC), triple-level cell (TLC), and quad-level cell (QLC) have evolved significantly, pushing flash memory toward higher density. Among them, TLC NAND strikes the balance between the capacity and the cost, and has been widely adopted in commercial applications. TLC features one erase state and seven program states (A, B, ..., G), corresponding to eight threshold voltage (V_{th}) distributions, as illustrated in Fig. 1. Due to the narrow voltage margins between adjacent states, TLC NAND is particularly vulnerable to read errors arising from the overlap of V_{th} distributions, including both up-shift and down-shift bit errors. Although multi-bit storage and 3D stacking improve storage capacity, they also introduce complex failure mechanisms, presenting new challenges to 3D NAND flash reliability.

In practical usage, dwell time has been identified as a key factor affecting the reliability of 3D NAND flash memory [4],[8]. Dwell time is defined as the time interval between two consecutive program/erase (P/E) cycles [5]. Previous studies have shown that longer dwell times promote charge de-trapping, which can partially alleviate the wear-out effects of P/E cycling and slow down device degradation [5]-[8]. While prior works have mostly focused on the impact of dwell time on endurance degradation [4],[9], some have also examined its effects on data retention in floating-gate [10] and charge-

trapping (CT) flash memories [11]. In addition, the influence of erase-to-program interval on cross-temperature behavior has been investigated[12], although it differs from the dwell time definition adopted in this study. In modern 3D CT NAND applications, data access patterns have become increasingly dynamic, including high-frequency access (hot data) [13], low-frequency access with long-term retention (cold data) [14], and operations across varying temperatures. However, the impact of dwell time on scenarios such as read disturb (RD), high-temperature data retention (HTDR), and cross-temperature effects remains underexplored. Therefore, a systematic investigation into the influence of dwell time on these reliability aspects is of great significance for improving the robustness and stability of 3D NAND flash memory.

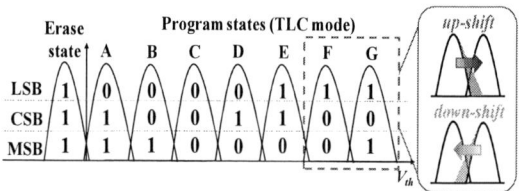

Fig. 1 Coding method and V_{th} distributions in 3D TLC NAND flash memory.

II. CHARACTERIZATION METHODS

In this work, a 96-layer stacked 3D CT TLC NAND flash chip is characterized. According to the standard of 3000 P/E cycles within three years at 35 °C, and accelerated equivalence at 85°C, the interval between two P/E cycles is approximately 58s, and other time points were selected for comparison on this basis. Previous studies have shown that the self-recovery effect is only associated with time intervals in the last 10% of P/E cycles [5],[15]. Therefore, all dwell time in this work is applied only during the last 300 P/E cycles. After P/E cycling, RD tests at room temperature, data retention tests at high-temperature (i.e., HTDR), and cross-temperature tests are performed, and the data are evaluated based on the raw bit error rate (RBER), as shown in Fig. 2. The high-temperature data retention test involves high-temperature baking only during the retention period, while programming and reading are both at room temperature. Cross-temperature tests include two operation modes: high-temperature program followed by low-temperature read (HPLR), and low-temperature program followed by high-temperature read (LPHR). For comparative analysis, two same-temperature tests are added: high-temperature program and read (HPHR), and low-temperature program and read (LPLR). In these tests, the high-temperature is set at 85°C and the low-temperature is set at 25°C.

979-8-3315-3918-4/25 $31.00 © 2025 IEEE

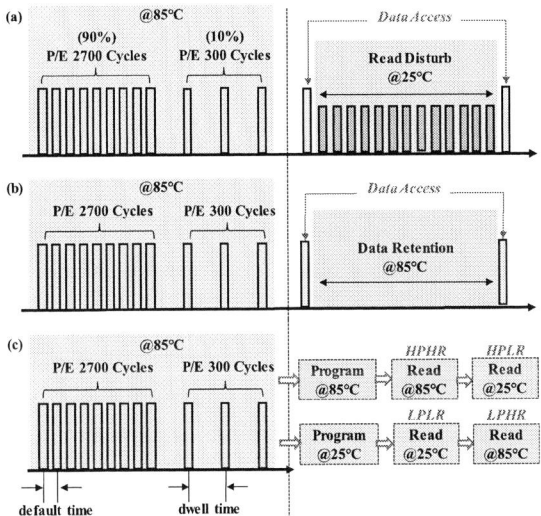

Fig. 2 Experimental setup to study the impacts of dwell time on (a) RD, (b) HTDR, and (c) cross-temperature.

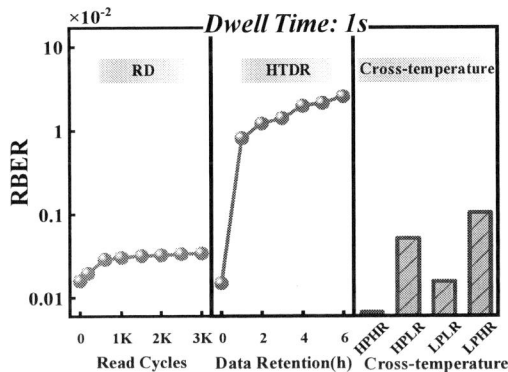

Fig. 3 RBER behaviors under a fixed dwell time of 1s for RD, HTDR, and cross-temperature.

III. RESULTS AND DISCUSSIONS

In this section, the dwell time impacts on 3D NAND reliability are systematically analyzed under representative scenarios: RD, HTDR, and cross-temperature effects. Fig. 3 presents a comparison of RBER at a fixed dwell time of 1s across these scenarios. The results show distinct reliability characteristics, with HTDR showing higher RBER due to the high-temperature impacts. The subsequent analysis examines how dwell time variations impact each scenario in detail.

A. Dwell Time Impacts on RD

Fig. 4 proves that the error behavior of RD is closely related to dwell time. As dwell time increases, the RBER of RD initially decreases and then increases again. In other words, while a shorter dwell time can help suppress the RBER of RD, an excessively long dwell time may lead to reliability degradation. This phenomenon indicates that the RD characteristics of 3D NAND flash memory depend on the dwell time during the P/E phase to a certain extent. By optimizing the interval between P/E operations, the RBER of RD can be effectively mitigated, thereby improving the reliability of 3D NAND flash memory.

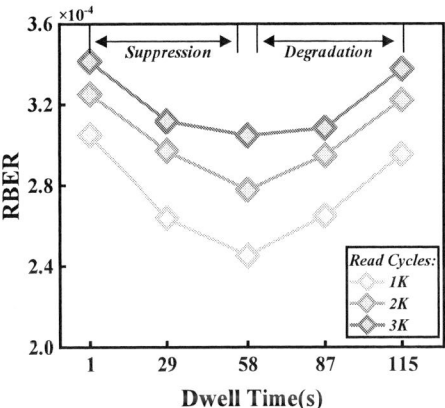

Fig. 4 The trend of RBER variations with the dwell time under different read cycles.

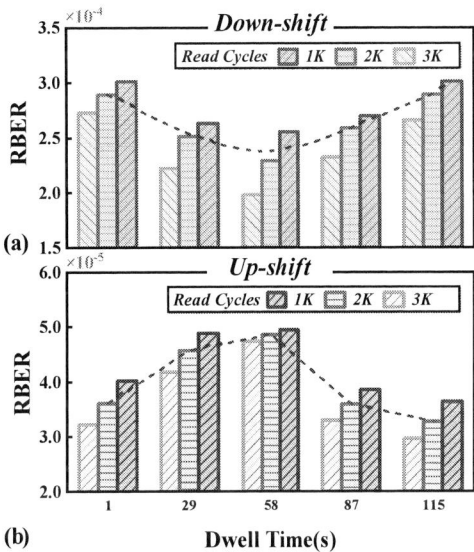

Fig. 5 The trend of RBER from (a) down-shift and (b) up-shift with the dwell time under different read cycles.

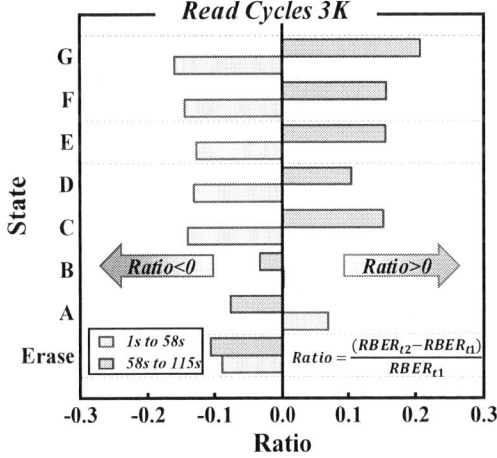

Fig. 6 Comparisons of the ratio of RBER changes in each state after 3K read cycles during the decreasing and increasing intervals.

Then, the RBER from V_{th} shifts is analyzed and the results are shown in Fig. 5. After RD, down-shift error bits are an order of magnitude higher than up-shift error bits, which is similar to the performance of RD after conventional P/E endurance. It is worth noting that as dwell time increases, the down-shift error bits exhibit a similar trend to the total RBER, decreasing initially and then increasing, while the up-shift error bits exhibit the opposite trend, with an initial increase followed by a decrease. Fig. 6 separates the different states after 3K read cycles and analyzes the RBER change ratio during the decreasing interval (from 1s to 58s) and increasing interval (from 58s to 115s). During the decreasing interval, the error bits of higher program states, which are prone to down-shift, decrease significantly, indicating the down-shift error bits are suppressed, while during the increasing interval, these error bits of higher states increase again. In comparison, lower program states exhibit limited error variation, thus having a minor impact on the total RBER.

B. Dwell Time Impacts on HTDR

The same scheme is used to explore the impacts of different dwell time on HTDR, and the results are shown in Fig. 7. As dwell time increases, the RBER of HTDR decreases rapidly initially, showing a significant suppression effect. The interesting thing is, as dwell time continues to increase, its improvement effect on HTDR gradually saturated, and the RBER tends to stabilize with negligible variations. This suggests that the benefit of longer dwell times could reach a saturation point, beyond which further extension provides limited reliability gain.

Furthermore, down-shift error bits and up-shift error bits are analyzed as shown in Fig. 8. Down-shift error bits are four orders of magnitude higher than up-shift error bits and dominate the total RBER trend. As dwell time increases, down-shift error bits initially decrease significantly and tend to stabilize under longer dwell time. In contrast, the up-shift error bits remain at a low level and do not fluctuate significantly. Similarly, the RBER change ratios of different states after 6 hours of high-temperature retention are analyzed during the decreasing interval (from 1s to 29s) and the stable interval (from 29s to 115s), as shown in Fig. 9. During the decreasing interval, the error bits of all states exhibit a noticeable reduction, while in the subsequent stable interval, no significant variation is observed in any of states.

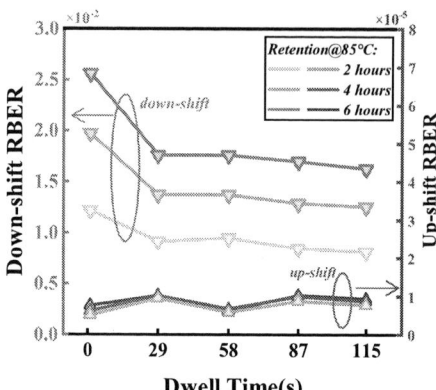

Fig.8 The trend of RBER from down-shift and up-shift with the dwell time under different retention time.

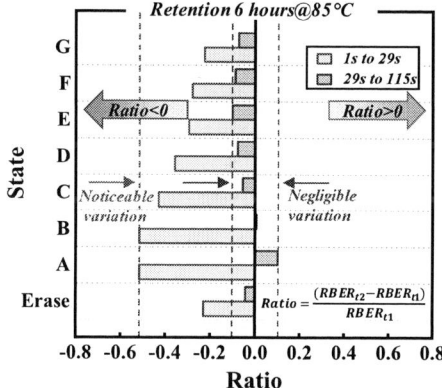

Fig. 9 Comparisons of the ratio of RBER changes in different states after retention 6 hours during the decreasing and increasing intervals.

C. Dwell Time Impacts on Cross-temperature

The impact of different dwell time on cross-temperature is shown in Fig. 10. Cross-temperature degrade data reliability compared to same-temperature program and read operations, as consistently observed in all characterized chips. It is worth noting that as dwell time increases, the negative impact of cross-temperature on data reliability gradually diminishes, indicating that appropriately extending the P/E interval helps mitigate the effects caused by cross-temperature.

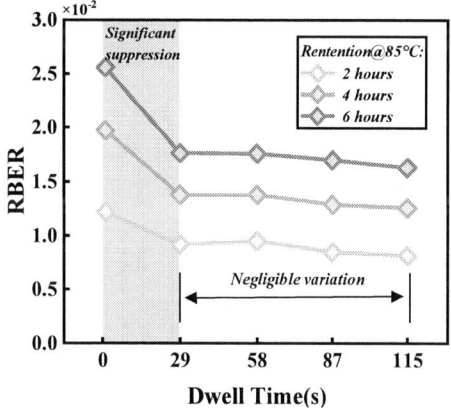

Fig. 7 The trend of RBER variations with the dwell time under different retention time.

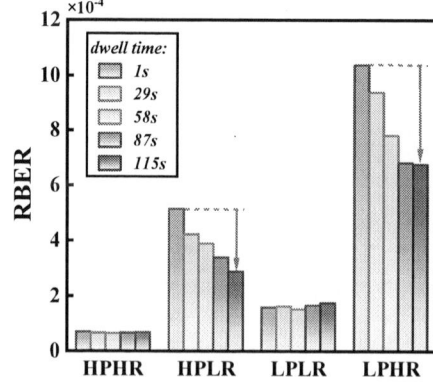

Fig. 10 Impacts of the dwell time on RBER under HPHR, HPLR, LPLR, and LPHR conditions.

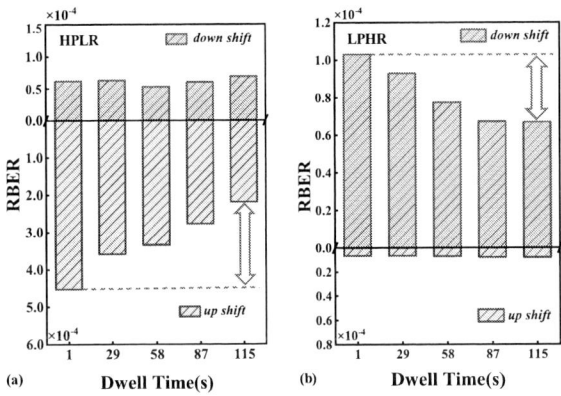

Fig. 11 Trend of RBER from down-shift and up-shift with the dwell time at (a) HPLR and (b) LPHR.

To gain deeper insights, further analyses of up-shift error bits and down-shift error bits are conducted. When the read temperature is lower than the program temperature, the total RBER is mainly driven by up-shift error bits. With a longer dwell time, these up-shift error bits gradually diminish, resulting in a lower total RBER, as shown in Fig. 11(a). Conversely, when the read temperature is higher than the program temperature, down-shift error bits become the dominant source of RBER. As dwell time increases, these down-shift error bits are gradually reduced, leading to a lower RBER, as shown in Fig. 11(b). These results confirm that the adjustment of dwell time consistently exhibits optimization effects during the cross-temperature program-read stage, which is important for improving the stability of 3D NAND in temperature fluctuation scenarios.

IV. Conclusion

In this work, we systematically investigated the impacts of dwell time on the reliability of 3D CT NAND flash memory under three typical usage scenarios: RD, HTDR, and cross-temperature operations. Experimental results reveal that with increasing dwell time, the RBER of RD initially decreases and then rises, indicating a temporary suppression effect. For HTDR, longer dwell times significantly reduce RBER in the early stages, but the improvement gradually saturates. In cross-temperature conditions, extended dwell time effectively alleviates the degradation caused by temperature differences between program and read operations. These findings suggest that appropriately managing dwell time can serve as a promising approach for enhancing the reliability of 3D NAND flash memory across diverse application environments.

Acknowledgment

This work was supported by China Key Research and Development Program under Grant (2023YFB4402500, 2023YFB4402400), National Natural Science Foundation of China (U23B2040, U2441248,62034006,92264201), Natural Science Foundation of Shandong Province (ZR2023LZH007, ZR2023QF054, TSQN202306059), and MIND project (MINDXZ202407).

References

[1] G. Yang *et al.*, "High-Precision Error Bit Prediction for 3D QLC NAND Flash Memory: Observations, Analysis, and Modeling," *IEEE Trans. Comput.*, vol. 74, no. 4, pp. 1392-1404, April 2025.

[2] Z. Shao, R. Zhao, S. Yuan, M. Ding, and Y. Wang, "Tracing the evolution of AI in the past decade and forecasting the emerging trends," *Expert Syst. Appl.*, vol. 209, 2022, Art. no. 118221.

[3] S. -S. Park *et al.*, "30.1 A 28Gb/mm^2 4XX-Layer 1Tb 3b/Cell WF-Bonding 3D-NAND Flash with 5.6Gb/s/Pin IOs," in *Proc. IEEE Int. Solid-State Circuits Conf.*, San Francisco, CA, USA, 2025, pp. 1-3

[4] X. Fang et al., "Impacts of operation intervals on program disturb in 3D charge-trapping triple-level-cell (TLC) NAND flash memory," in *Proc. IEEE Electron Devices Technol. Manuf. Conf.*, Chengdu, China, 2021, pp. 1–3.

[5] Y. Luo, S. Ghose, Y. Cai, E. F. Haratsch, and O. Mutlu, "HeatWatch: Improving 3D NAND flash memory device reliability by exploiting self-recovery and temperature awareness," in *Proc. Int. Symp. High Perform. Comput. Archit.*, Vienna, Austria, 2018, pp. 504–517.

[6] W. Qi, G. Dong, and T. Zhang, "Exploiting heat-accelerated flash memory wear-out recovery to enable self-healing SSDs," in *Proc. 3rd USENIX Conf. Hot Topics Storage File Syst.*, 2011, pp. 1–5.

[7] D. Resnati, G. Nicosia, G. M. Paolucci, A. Visconti, and C. Monzio Compagnoni, "Cycling-induced charge trapping/detrapping in Flash memories—part I: Experimental evidence," *IEEE Trans. Electron Devices*, vol. 63, pp. 4753–4760, Dec. 2016.

[8] T. Ren, Q. Li, Y. Lv, M. Ye, N. Guan and C. Jason Xue, "Near-Free Lifetime Extension for 3-D nand Flash via Opportunistic Self-Healing," *IEEE Trans. Comput. Aided Des. Integr. Circuits Syst.*, vol. 43, no. 11, pp. 4226-4237, Nov. 2024.

[9] R. Shirota *et al.*, "Improvement of oxide reliability in NAND Flash memories using tight endurance cycling with shorter idling period," in *Proc. Int. Rel. Phys. Symp.*, 2015, pp. MY.12.1–MY.12.5.

[10] D. Wei, X. Li, L. Niu, L. Qiao, and X. Peng, "The influence of NAND flash self-recovery effect on retention error," in *Proc. 8th Int. Conf. Instrum. Meas. Comput. Commun. Control.*, Harbin, China, 2018, pp. 1764–1769.

[11] M. D. Sciacca, T. Kosuru and N. Papandreou, "Cycling Condition Impacts on 3D QLC NAND Reliability," in *Proc. IEEE Electron Devices Technol. Manuf. Conf.*, Bangalore, India, 2024, pp. 1-3.

[12] D. Wu, H. You, X. Wang, S. Zhong and Q. Sun, "Experimental Investigation of Threshold Voltage Temperature Effect During Cross Temperature Write–Read Operations in 3-D NAND Flash," *IEEE J. Electron Devices Soc.*, vol. 9, pp. 22-26, 2021.

[13] Y. Cai, Y. Luo, S. Ghose, and O. Mutlu, "Read disturb errors in MLC NAND Flash memory: Characterization, mitigation, and recovery," in *Proc. 45th IEEE/IFIP Int. Conf. Dependable Syst. Netw.*, Jun. 2015, pp. 438–449..

[14] Y. Kong, M. Zhang, X. Zhan, R. Cao, and J. Chen, "Retention correlated read disturb errors in 3-D charge trap NAND flash memory: Observations, analysis, and solutions," *IEEE Trans. Comput-Aided Des. Integr. Circuits Syst.*, vol. 39, no. 11, pp. 4042–4051, Nov. 2020.

[15] N. Mielke, H. P. Belgal, A. Fazio, Q. Meng, and N. Righos, "Recovery effects in the distributed cycling of flash memories," in *Proc. IEEE 44th Annu. Int. Rel. Phys. Symp.*, 2006, pp. 29–35.

Access Mode Impacts on 3D Charge-trapping (CT) QLC (4bit/cell) Raw NAND Chip

Guangkuo Yang[1], Ruidong Li[2], Yining Zhou[1], Yujiao Ding[1], Xinghao Wang[1], Pengpeng Sang[1], Peng Guo[3], Xuepeng Zhan[1], Jixuan Wu[*,1], Jiezhi Chen[*,1]

[1] School of Information Science and Engineering, Shandong University, P. R. China; [2]Cloud Computing Equipment Industry Innovation Co., Ltd., P. R. China; [3]Shandong Sinochip Semiconductors Co. Ltd, P. R. China.

* Email: jixuanwu@sdu.edu.cn, chen.jiezhi@sdu.edu.cn

Abstract—**Unlike traditional sequential access patterns, artificial intelligence (AI) inference tasks often involve frequent non-sequential data access. Thereby, to address the reliability challenges posed by AI inference workloads in large-capacity storage systems, this study investigated the impacts of various data access modes by using 3D Charge-trap (CT) type QLC NAND flash memory. Experimental results demonstrate that non-sequential access can reduce the raw bit error rate (RBER) by ~26% at the fresh state, while the benefit decreases to ~17% after 1K Program/Erase (P/E) cycles. Our findings indicate that, aiming at AI-oriented storage architectures with robust reliabilities, read operations should be well optimized to develop reliability-aware strategies.**

Keywords—Reliability, Access mode, 3D QLC NAND flash, RBER.

I. INTRODUCTION

The rapid proliferation of artificial intelligence (AI) applications, particularly the widespread adoption of Large Language Models (LLMs), has imposed unprecedented demands on high-throughput, low-latency, and highly reliable storages [1]-[3]. As AI inference workloads increasingly depend on the retrieval of massive model parameters stored in non-volatile memory (NVM) [2], [3], NAND flash memory has become a cornerstone of modern data infrastructure. This shift is largely attributed to NAND flash's superior storage density, energy efficiency, and cost-effectiveness per bit [4], [5]. However, the irregular and high-frequency access patterns characteristic of AI workloads poses serious reliability challenges for NAND flash. In large-scale Transformer-based models, inference often involves frequent and randomized accesses to weight parameters—access behaviors that deviate significantly from the sequential access patterns for which NAND architectures are optimized. Despite this growing mismatch, the studies examining the reliability impact of such access modes remain relatively scarce. Furthermore, as the number of bits per cell (bit density) increases, the threshold voltage (V_{th}) margins between adjacent states continue to shrink (Fig. 1a), exacerbating the read disturb effect [6], [7]. This issue is especially pronounced in quad-level cell (QLC) devices, where narrower voltage windows make them more vulnerable to even minor disturbances, thereby amplifying the impacts of access patterns on cell reliability and underscoring the urgency of addressing this problem.

In NAND flash memory, errors can be broadly classified into two categories: (1) V_{th} up-shift errors, caused by a shift to higher V_{th} values (Fig. 1b), typically resulting from channel charge injection [8], charge de-trapping from the tunneling oxide layer (TNL) [9], and spatial charge redistribution [10]; and (2) V_{th} down-shift errors, arising from shifts to lower V_{th} values (Fig. 1c), primarily driven by charge leakage and lateral charge spreading due to multiple cells sharing a common

charge trapping layer (CTL) [11]-[13]. Current research on access-mode-related reliability primarily follows three approaches: (i) physical-mechanism-based modeling, which develops high-precision error prediction models by simulating sequential page-level accesses and capturing detailed error bit distributions [2]; (ii) systematic characterization of read disturb effects, particularly in multi-level cell (MLC) devices, revealing the influence of factors such as read voltage offsets [7]; and (iii) collaborative management techniques, such as multidimensional read disturb mitigation strategies that intelligently adjust read voltages to alleviate reliability degradation [14], [15].

In this work, we systematically investigate the impacts of various access modes on the reliability of 3D QLC NAND flash, with a particular focus on non-sequential access patterns. Four representative access modes are analyzed in detail, and the underlying mechanisms associated with non-sequential behavior are thoroughly characterized. The results provide a deep insight into the reliability challenges posed by AI-driven storage workloads and lay a solid foundation for the development of fault-tolerant scheduling strategies and access-aware management techniques.

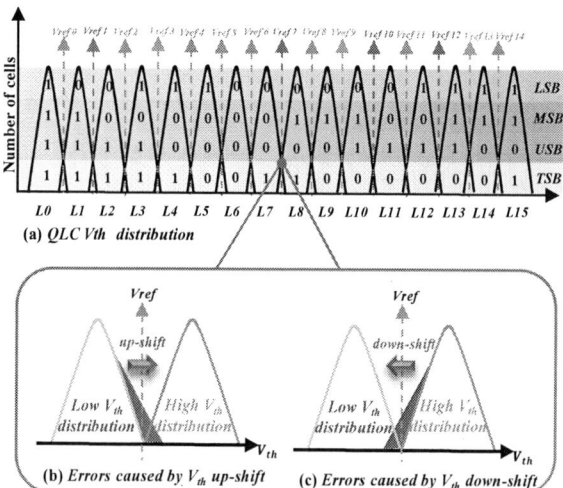

Fig. 1 (a) Threshold voltage distribution in QLC NAND flash. (b) Illustration of V_{th} up-shift induced errors. (c) Illustration of V_{th} down-shift induced errors.

II. CHARACTERIZATION METHODS

In this work, we systematically evaluate the impacts of access modes in 3D CT QLC NAND flash memory by testing the raw NAND chip. Algorithm 1 shows the process of the experiment in detail. All operations are performed at room temperature, and pseudo-random numbers are used for writing data. Raw bit error rate (RBER) and V_{th} offsets are studied by comparing write data with read data. Fig. 2 illustrates the four access modes: Mode 0 denotes page-by-page sequential read, Mode 1 denotes page-by-page reverse order read, Mode 2 denotes random page read, and Mode 3 denotes worldline-by-worldline (WL) reverse order read.

Algorithm 1: The process of experiment

Definitions:

PEn: the number of PE cycles;

RDm: the number of read cycles;

MaxPEn: the max number of program/erase cycles;

MaxRDm: the max number of read cycles;

AccessMode: different access modes;

RT: room temperature;

Process:

set temperature RT;

while AccessMode do

  for PEn=1 to MaxPEn do

    Execute erase/program operation;

    if PEn==1 or PEn==100 or PEn%500==0 then

      for RDm=1 to MaxRDm do

        Execute read operation;

        if RDm==1 or RDm%100==0 then

          Execute read dump operation;

        end if;

      end for;

    end if;

  end for;

end while;

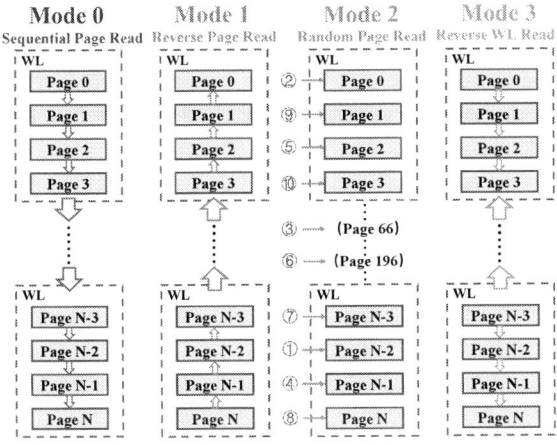

Fig. 2 Execution order of read operations in the four access modes.

III. EXPERIMENTAL OBSERVATIONS AND ANALYSIS

After executing various program/erase (PE) cycles, read operations are performed using four distinct access modes, with the measured RBER shown in Fig. 3. Firstly, it is found that all four modes exhibited a strong power-law relationship between RBER and cycles. This indicates that the power-law behavior is intrinsic with no dependence on the page access order. It is known that both sequential and random accesses could induce read disturb, where the dominant electrical stress arises from repeated stressing of the pass-through voltage (V_{pass}) in un-selected WLs [2], [7], [16]. Since the charge injection is related to stress-induced leakage current (SILC)

originated from P/E-related dielectric degradation [17], which is independent of the access sequence, thus the power-law trend will not be affected by the read access mode. Secondly, across all P/E cycle counts, the RBER of Modes 1/2/3 is significantly lower than that of Mode 0, and Mode 2 shows intermediate performance between Mode 0 and Mode 1. Thirdly, at a fixed number of P/E cycles, the rate of RBER degradation with more read cycles is significantly slower in Modes 1/2/3 in comparison to Mode 0. This trend is attributed to both the order of V_{read} application and the cumulative V_{pass} stress prior to V_{read}, in conjunction with the structural characteristics of 3D V-NAND. Specifically, the shape of the memory hole and defect distributions in different layers could be the intrinsic mechanisms [18]-[20]. Moreover, at the fixed read cycle, RBER suppression of Modes 1 and 3 diminishes as increasing P/E cycles. For instance, at 3K read cycles, Mode 1/3 shows a 26.49% RBER reduction over Mode 0 in the fresh state, but this advantage drops to 16.62% after 1K P/E cycles. This suggests that the cell degradation induced by P/E stressing turns to be a main contributor to RBER and the impacts of access mode is minimized.

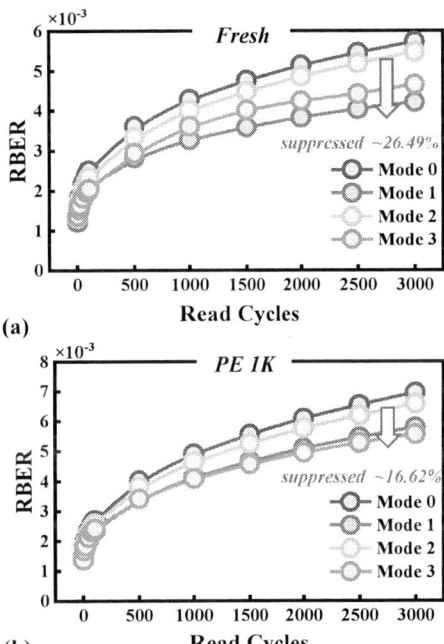

Fig. 3 The relationship between read cycles and RBER in different access modes. (a) Fresh state, (b) 1K PE cycles.

Further analysis from the perspective of V_{th} shift reveals that, regardless of whether the device is at the fresh state or after 1K P/E cycles, Mode 1 and Mode 3 consistently exhibit suppression effects on V_{th} down-shift compared to Mode 0 as increasing the number of read cycles. However, they also exhibit a more pronounced degradation trend in V_{th} up-shift. Taking the case of 3K read cycles as an example, the suppression in V_{th} down-shift for Mode 1/3 relative to Mode 0 reaches 31.84% in the fresh state, and drops to 19.40% after 1K P/E cycles, as shown in Fig. 4. In the V_{th} up-shift, the degradation is significantly reduced from 39.77% in the Fresh state to 13.41%, as shown in Fig. 5. For the V_{th} down-shift, error bits suppression is mainly attributed to the space charge compensation, and the same reason can explain V_{th} up-shift phenomena because the charge compensation also induced the unwanted charges and cause error bits.

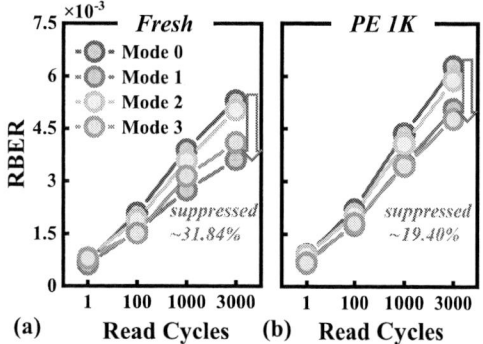

Fig. 4 The relationship between read cycle and RBER of V_{th} down-shift (a) Fresh state, (b) 1K PE cycles.

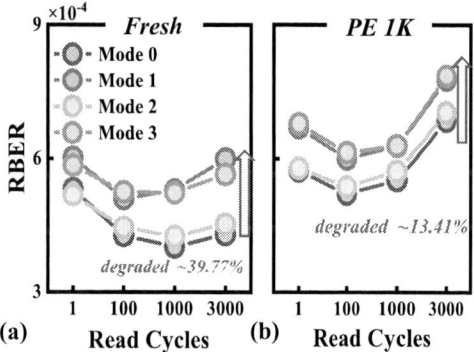

Fig. 5 The relationship between read cycle and RBER of V_{th} up-shift. (a) Fresh state, (b) 1K PE cycles.

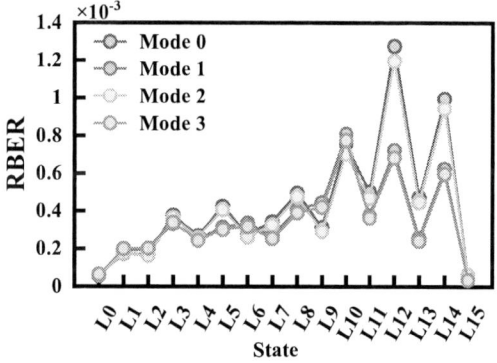

Fig. 6 The RBER of each state for different access modes in 3K read cycles after 1K PE cycles.

Further analysis is conducted from the perspective of state. Taking 3K read cycles after 1K P/E cycles as an example, as shown in Fig. 6, it can be observed that the RBER of Mode 2 is significantly higher than that of Mode 1/3 in the high state, indicating stronger suppression of the latter. In order to quantify the difference, the RBER of Mode 1/2/3 in the V_{th} up/down-shift is studied: for V_{th} down-shift, as shown in Fig. 7(a), Mode 1/3 is significantly lower than that of Mode 0 in the high states while Mode 2 is essentially consistent with Mode 0, indicating that the optimized read mode is quite effective in suppressing the V_{th} down-shift. While for V_{th} up-shift shown in Fig. 7(b), Mode 1/3 is higher than Mode 0 in the low states, while Mode 2 is basically the same as mode 0. This means that the optimized read order is more effective for V_{th} up-shift, especially at the low states from L1 to L3.

Finally, analyzing from the perspective of the layer, taking 1K P/E cycles as an example, as shown in Fig. 8, as read cycles increase, the RBER differences between different stacked layers are gradually significant. The comparison reveals that the interlayer RBER differences between Mode 1/3 are significantly smaller than those between Mode 0/2, and this is more obvious when read cycles increase, which means that the access mode can affect the interlayer interference.

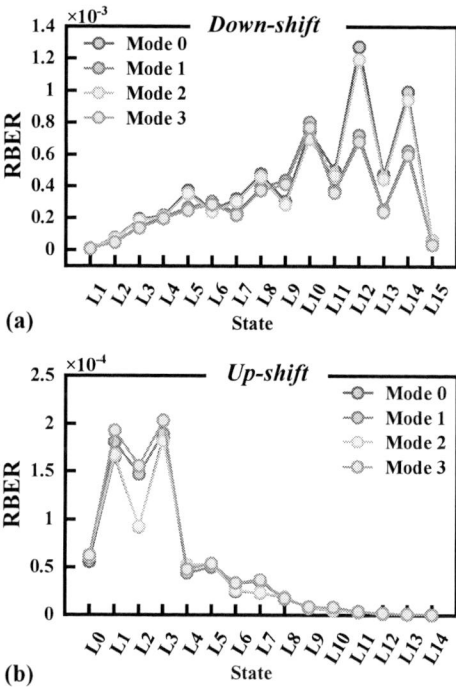

Fig. 7 The V_{th} up/down-shift for different access modes in 3K read cycles after 1K PE cycles. (a) down-shift, (b) up-shift.

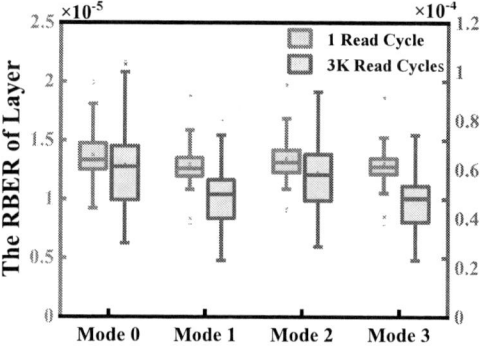

Fig. 8 Characterization of the RBER of layers for each access mode under different read cycles based on 1K PE cycles.

IV. CONCLUSION

In this work, we investigate the impact of access modes on the reliability of 3D QLC NAND chip. It is shown that non-sequential access modes can suppress ~26% RBER in early stages compared to the traditional sequential reads, while this effect gradually diminishes with P/E cycling. This indicates that device degradation progressively reduces the influence of access modes. Moreover, it is observed that specific access patterns can mitigate layer-to-layer interference in 3D NAND. Our findings provide new insights for the design of reliability-aware memory system.

979-8-3315-3918-4/25 $31.00 © 2025 IEEE

ACKNOWLEDGMENT

This work was supported by China Key Research and Development Program under Grant (2023YFB4402400, 2023YFB4402500), National Natural Science Foundation of China (62034006, U23B2040, U2441248, 92264201), Natural Science Foundation of Shandong Province (ZR2023LZH007, ZR2023QF054, tsqn202306059), and MIND project (MINDXZ202407).

REFERENCES

[1] Y. Jin, S. Kim, T. J. Ham, and J. W. Lee, "Architecting a flash-based storage system for low-cost inference of extreme-scale DNNs," *IEEE Trans. Comput.*, vol. 71, no. 12, pp. 3153–3164, Dec. 2022.

[2] G. Yang *et al.*, "High-Precision Error Bit Prediction for 3D QLC NAND Flash Memory: Observations, Analysis, and Modeling," *IEEE Trans. Comput.*, vol. 74, no. 4, pp. 1392-1404, Apr. 2025.

[3] J. Song, "AI Revolution Driven by Memory Technology Innovation," *in Proc. IEEE International Solid-State Circuits Conference*, San Francisco, CA, USA, vol.68, pp.26-36, 2025.

[4] A. Goda, "NAND Flash Innovations and Future Scaling," *in Proc. IEEE International Electron Devices Meeting*, San Francisco, CA, USA, pp. 1–4, 2024.

[5] S. I. Shim, "Trends and Future Challenges of 3D NAND Flash Memory," *in Proc. IEEE International Memory Workshop*, 2023.

[6] Y. Cai, S. Ghose, E. F. Haratsch, Y. Luo, and O. Mutlu, "Error characterization, mitigation, and recovery in flash-memory-based solid-state drives," *Proc. IEEE*, vol. 105, no. 9, pp. 1666–1704, Sep. 2017.

[7] Y. Cai, Y. Luo, S. Ghose, and O. Mutlu, "Read Disturb Errors in MLC NAND Flash Memory: Characterization, Mitigation, and Recovery," *in Proc. IEEE International Conference on Dependable Systems and Networks*, Rio de Janeiro, Brazil, pp. 438-449, 2015.

[8] Y. Cai, Y. Luo, E. F. Haratsch, K. Mai, and O. Mutlu, "Data retention in MLC NAND flash memory: Characterization, optimization, and recovery," *in Proc. Int. Symp. High Perform. Comput. Archit.*, Burlingame, CA, USA, pp. 551–563, 2015.

[9] Y. Luo, S. Ghose, Y. Cai, E. F. Haratsch, and O. Mutlu, "HeatWatch: Improving 3D NAND flash memory device reliability by exploiting self-recovery and temperature awareness," *in Proc. Int. Symp. High Perform. Comput. Archit.*, Vienna, Austria, pp. 504–517, 2018.

[10] X. Jia et al., "Impact of cycling induced intercell trapped charge on retention charge loss in 3-D NAND flash memory," *IEEE J. Electron Devices Soc.*, vol. 8, pp. 62–66, 2020.

[11] Y. Kong et al., "Comprehensive investigations on data pattern dependences in charge-trap (CT) 3D NAND flash memory," *in Proc. IEEE Int. Conf. Solid-State Integr. Circuit Technol.*, Kunming, China, pp. 1–3, 2020.

[12] Y. Kong, M. Zhang, X. Zhan, R. Cao, and J. Chen, "Retention correlated read disturb errors in 3-D charge trap NAND flash memory: Observations, analysis, and solutions," *IEEE Trans. Comput-Aided Des. Integr. Circuits Syst.*, vol. 39, no. 11, pp. 4042–4051, Nov. 2020.

[13] R. Shirota et al., "New method to analyze the shift of floating gate charge and generated tunnel oxide trapped charge profile in NAND flash memory by program/erase endurance," *IEEE Trans. Electron Devices*, vol. 62, no. 1, pp. 114–120, Jan. 2015.

[14] K. Ha, J. Jeong and J. Kim, "An Integrated Approach for Managing Read Disturbs in High-Density NAND Flash Memory," *IEEE Trans. Comput-Aided Des. Integr. Circuits Syst.*, vol. 35, no. 7, pp. 1079-1091, July.2016.

[15] K. Ha, J. Jeong, and J. Kim, "A read-disturb management technique for high-density NAND flash memory," *in Proc. ACM Asia-Pac. Workshop Syst.*, Singapore, 2013.

[16] C. Zambelli, P. Olivo, L. Crippa, A. Marelli, and R. Micheloni, "Uniform and concentrated read disturb effects in mid-1X TLC NAND flash memories for enterprise solid state drives," *in Proc. IEEE Int. Rel. Phys. Symp.*, pp. 5, 2017.

[17] JEDEC, "Failure mechanisms and models for semiconductor devices," JEDEC Solid State Technology Association, Arlington, VA, USA, Tech. Rep. JEDEC 122H, Oct. 2016.

[18] D. Kojima and K. Takeuchi, "Error suppression of last-programmed word-line for real usage of 3D-NAND flash memory," *in Proc, IEEE Trans. Circuits Syst.*, Daegu, Korea, pp. 1–4, 2021.

[19] C. Kim et al., "A 512-gb 3-b/cell 64-stacked wl 3-d-nand flash memory," *IEEE Journal of Solid-State Circuits*, vol. 53, no. 1, pp. 124–133, 2017.

[20] W.-J. Tsai et al, "Polycrystalline-silicon channel trap induced transient read instability in a 3D NAND flash cell string," *in IEDM Tech. Dig.*, pp. 11.3.1–11.3.4, *Dec.* 2016.

The Influence of Radiation on Reliability of Cold Data in 3D CT NAND Flash Memory

Haitao Dong[1], Xinghao Wang[1], Haotian Li[1], Xuesong Zheng[2,3], Pengpeng Sang[1], Xuepeng Zhan*[1],
Jixuan Wu *[1], and Jiezhi Chen[1]

[1] School of Information Science and Engineering (ISE), Shandong University, China
[2] School of Astronautics, Harbin Institute of Technology, Harbin, China
[3] China Aerospace Components Engineering Center, Beijing, China

*E-mail: zhanxuepeng@sdu.edu.cn, jixuanwu@sdu.edu.cn

Abstract— This study investigates the reliability degradation of cold data in 3D charge-trap (CT) flash memory under radiation exposure. When subjected to a wider range of radiation doses, the threshold voltage (V_{th}) degradation no longer strictly follows the conventional power-law evolution model, indicating the emergence of complex coupled mechanisms. Furthermore, temperature-dependent behavior reveals that V_{th} negative shifts are significantly more pronounced at low temperatures compared to high temperatures. This is primarily attributed to the elevated internal electric field that enhances hole generation. At the same time, at low temperatures, the mobility of holes is drastically reduced, severely hindering the annealing of trapped holes. The resulting accumulation of holes leads to a large V_{th} shift. These findings highlight the importance of both radiation dose and temperature effects in assessing the reliability of cold data storage in 3D CT flash memories.

Keywords—Flash memory, radiation degradation, cold data, temperature

I. INTRODUCTION

The high storage density of 3D NAND flash memory not only gained an important market in commercial applications but also are widely used in aerospace field [1]. Compared to earth surface environments, aerospace exhibits more complex radiation conditions. Protons and heavy ions induce single-event effects, while X-rays and gamma rays contribute to total ionizing dose (TID) effects. TID effects introduce positive charge trapping in SiO_2 regions, such as the shallow trench isolation (STI) and oxide blocking layer [2]. This leads to V_{th} shifts and the generation of interface defects at the channel surface, consequently degrading channel mobility. In addition to radiation exposure, space environments are characterized by extremely large temperature variations. For example, the Lunar surface experiences temperature swings from approximately 93K at night to 153K during the day [3], while Martian temperatures range from 130K to 343K [4]. These thermal extremes place stringent requirements on the operational stability and reliability of flash memory devices across a broad temperature range.

Cold data refers to information with a low access frequency, such as backups and archived images [5]. These memory cells are continuously exposed to radiation over extended periods, making them particularly susceptible to TID-induced degradation. Under radiation exposure, TID effects induce both a negative V_{th} shift and a broadening of threshold voltage distributions [6]. This distribution widening increases the overlap between adjacent programmed states, elevating the risk of read-out errors. Thereby, for the high-density storage formats like triple-level-cell (TLC) and quad-level-cell (QLC), they are more vulnerable to TID effects than single-level-cell (SLC) devices, requiring lower doses to reach equivalent bit error rates.

In this study, we investigate the degradation behavior of TLC charge trap flash memory cells across a wide temperature range (200K to 450K). Using TCAD simulations, we examine V_{th} degradation under both irradiated and non-irradiated conditions, analyze the commonalities and distinctions in their degradation pathways, and propose a mechanism to explain the exacerbated degradation observed at low temperatures.

II. DEVICE STRUCTURE AND SIMULATION SETTINGS

The structure of the CT flash memory device used in this study is illustrated in Fig. 1. All physical models used in the simulations incorporate temperature-dependent terms. A nonlocal tunneling model is employed at the tunneling interface to simulate carrier transport across the dielectric. For the Si_3N_4 charge trapping layer, the Poole–Frenkel emission model is activated to account for field-enhanced thermal emission of carriers. The electron trap energy level is assumed to be 1.31 eV below the conduction band edge, with a trap density of 3×10^{19} cm^{-3}. All additional simulation parameters are listed in Table I.

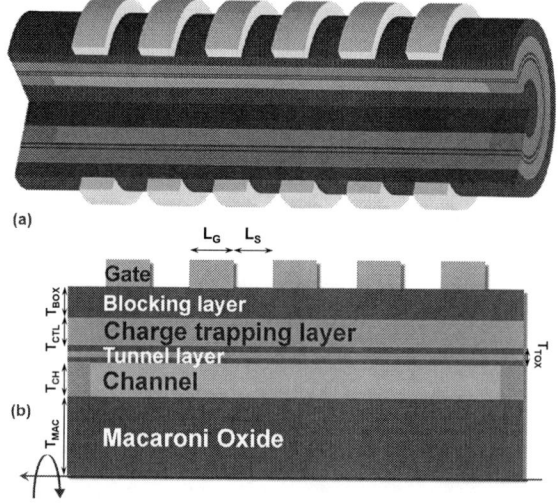

Fig. 1. (a) Schematic diagram of the 3D NAND charge trap flash memory structure. (b) Cross-sectional view illustrating the layer composition and device geometry.

In SiO_2 region, only donor-type defects are considered. The electron and hole mobilities at 300 K are set to $\mu_n = 20$

cm²/V·s and $\mu_p = 10^{-5}$ cm²/V·s, respectively [8]. The temperature dependence of electron mobility follows the default model parameters, while the variation of hole mobility with temperature is calibrated based on reported experimental data [9].

TABLE I. STRUCTURE AND CT LAYER MATERIAL PARAMETERS

Parameter	Value
Thickness of blocking (T_{BOX})	8nm
Thickness of charge trap layer (T_{CTL})	8nm
Thickness of Tunnel (ONO)(T_{OX})	1.5\1\1.5nm
Thickness of channel (T_{CH})	10nm
Thickness of oxide filler (T_{MAC})	40nm
Length of gate (L_G)	40nm
Length of spacer (L_S)	40nm

III. RESULTS AND DISCUSSION

A. V_{th} degradation with radiation

To investigate the V_{th} degradation with radiation, a total ionizing dose of 1×10^8 rad with a dose rate of 1000 rad/s is applied [10]. The V_{th} evolutions for the L0-L7 state (ranging from the lowest to the highest programmed states in a TLC cell) are shown in Fig. 2. Due to the nature of deep level defects, a high temperature of 450 K was used to accelerate the degradation process, and the comparison with the data at 300K is shown in Fig. 2 (a). In the non-radiation environment, the V_{th} degradation follows a power-law dependence, and this trend remains consistent across different temperatures. As illustrated in Fig. 2(d), the degradation process under radiation exposure can be categorized into three distinct phases: a linear phase, a sublinear phase, and an approximate saturation phase. In contrast, Fig. 2(b) does not exhibit clear inflection points as observed in Fig. 2(d), indicating that radiation not only accelerates the degradation process but also fundamentally alters its degradation behavior.

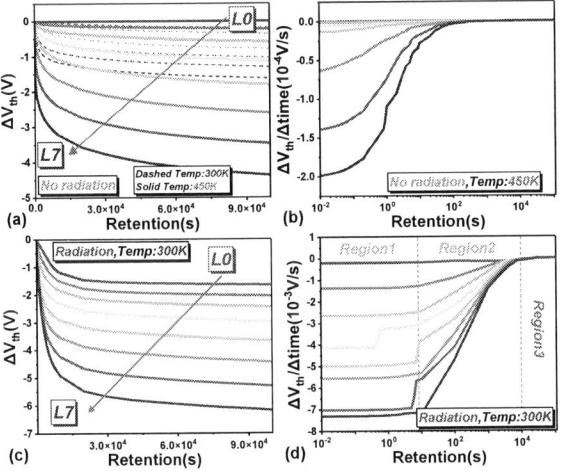

Fig. 2. (a) Evolution curve of ΔV_{th} under 300K radiation environment. (b) The first-order difference of Fig. 2(a). (c) Evolution curve of ΔV_{th} under 450K non-radiation environment. (d) The first-order difference of Fig. 2(c).

As shown in Fig. 3(a), the reason for the linear degradation interval in the radiation environment is that the accumulation of holes in the Macaroni layer and blocking layer increases linearly with time at low absorbed doses, and the temperature difference does not affect the linear pattern.

And from Fig. 3(b), it can be seen that the ΔV_{th} loss with radiation is much larger than that without radiation during retention process. Therefore, the degradation of ΔV_{th} in the linear region is basically caused by radiation introduced holes. Temperature and electronic defect energy levels can affect the degradation rate under non-radiation retention process, so changes in these factors will affect the size of the linear region in Fig. 2(b).

$$\Delta V_{th} = a_1 t^{b_1} + c \tag{1}$$

$$\Delta V_{th} = a_1 \exp\left(-\left(\frac{t}{b_1}\right)\right) + a_2 \exp\left(-\left(\frac{t}{b_2}\right)\right) + C_0 \tag{2}$$

$$\Delta V_{th} = \Delta V_{th,x}\left(1 - \exp\left(-\left(\frac{t}{\tau}\right)^{\beta}\right)\right) \tag{3}$$

$$\Delta V_{th} = \frac{abt^{1-c}}{1 + bt^{1-c}} \tag{4}$$

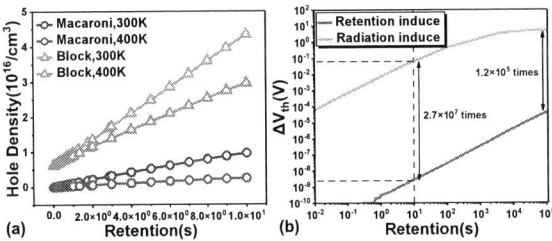

Fig. 3. (a) Hole concentration in the Macaroni layer and blocking layer within the linear degradation interval under radiation environment. (b) The V_{th} loss of L7 at 300K during retention with and without radiation.

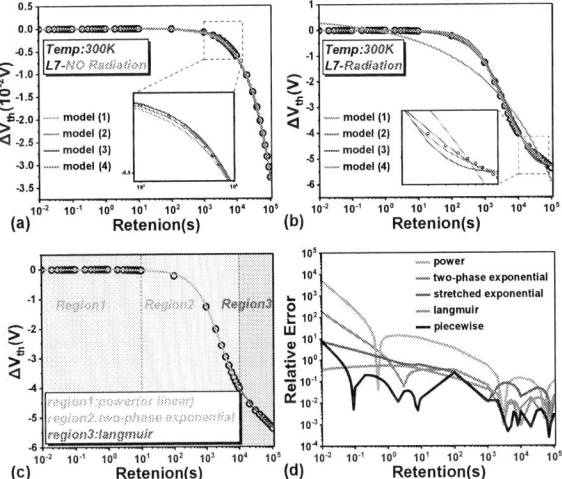

Fig. 4. Fitting of degradation process of L7 at 300K (a) No radiation (b) Radiation. (c) Piecewise function fitting curve. (d) The relative error of all fitting methods under radiation of L7 program state at 300K.

To fit the degradation data of the L7 state with and without radiation at 300 K, four models are employed: the power-law model, two-phase exponential model, empirical stretched exponential model, and the Langmuir model. From Fig. 4(a), it can be seen that all four models can fit the degradation process without radiation well. In Fig. 4(b), the power function is no longer suitable, while the other three models still fit well, but there is a significant deviation at higher absorbed doses. The three degradation intervals

979-8-3315-3918-4/25 $31.00 © 2025 IEEE

identified in Fig. 2(b) are individually fitted, and the results are presented in Fig. 4(c). The data in region 1 can be fitted using both linear or power-law functions. The relative fitting errors for the curves shown in Fig. 4(b) and (c) are summarized in Fig. 4(d). Models (3) and (4) demonstrate relatively better accuracy; however, applying a piecewise fitting approach can further reduce the error.

The above rules are also applicable to data from other programmed states. Based on the above analysis, it is speculated that the overall degradation process of V_{th} under radiation no longer strictly follows a power-law relationship due to the superposition of holes in the radiation environment. However, at low absorbed doses, the power-law relationship can still describe the degradation process.

B. V_{th} degradation analysis over a wide temperature range

We analyze the V_{th} degradation characteristics of CT flash memory across a wide temperature range (Fig. 5-6), and investigate the underlying mechanisms (Fig. 7-9). Examining the V_{th} degradation for the L7 state across temperatures, Fig. 5(a) shows the evolution of ΔV_{th} during retention under non-radiation environment, demonstrating a clear temperature dependence: higher temperatures result in more pronounced degradation, consistent with thermally activated charge loss processes. In contrast, Fig. 5(b) shows that under radiation exposure, V_{th} degradation is significantly accelerated at all temperatures. However, the magnitude of ΔV_{th} no longer follows a monotonic trend with temperature. In certain dose intervals, degradation is faster at elevated temperatures, while in others, low-temperature conditions lead to more severe degradation. Notably, at higher accumulated doses, the ΔV_{th} observed at 200K surpasses that at 450 K, indicating a different low-temperature degradation process radiation mechanism.

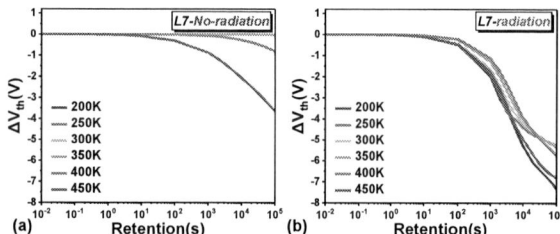

Fig. 5. ΔV_{th} degradation of L7 at different temperatures in a (a) non-radiation environment (b) radiation environment.

ΔV_{th} values extracted at four distinct doses (Fig. 6) show that higher programmed states produce larger degradation magnitudes under identical conditions. However, the observation of increased degradation severity at lower temperatures is strictly limited to higher accumulated doses, as evident in Fig. 6(d). For the low program states (L0, L1), this lower temperature aggravated degradation pattern is absent at lower doses, as shown in Fig. 6(a)-(c).

The aggravated V_{th} loss observed for higher programmed states in radiation environment arises from two principal mechanisms. On the one hand, lateral migration (LM) is significantly influenced by programmed state. Fig. 7(a) and 7(b) depict the absolute electric field distribution along cross-sections through the center of the CT layer and blocking layer, respectively. Higher programmed states correspond to steeper electron concentration gradients and larger lateral electric

fields within the CT layer. Fig. 7(c) illustrates the relationship between Poole-Frenkel emission probability and electric field at 300K [7], indicating that increased electric fields enhance trapped electron emission rates, thereby aggravating LM losses. On the other hand, it comes from the holes introduced by radiation. Fig. 7(d) shows the dependence of holes yield on electric field in SiO_2 under gamma irradiation [2]. Higher electric fields increase holes generation, leading to more hole accumulation for a given absorbed dose.

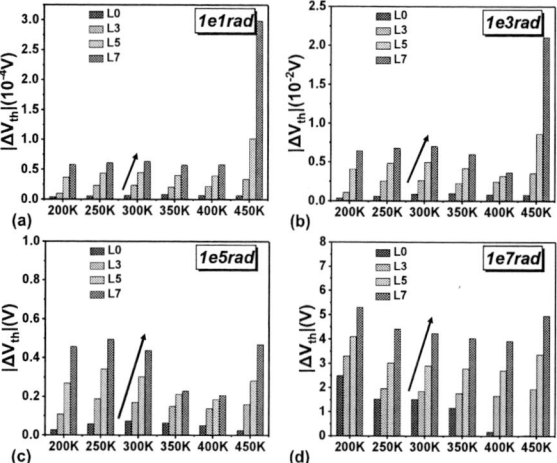

Fig. 6. ΔV_{th} degradation under different absorbed doses (a) 1e1rad (b) 1e3rad (c) 1e5rad (d) 1e7rad.

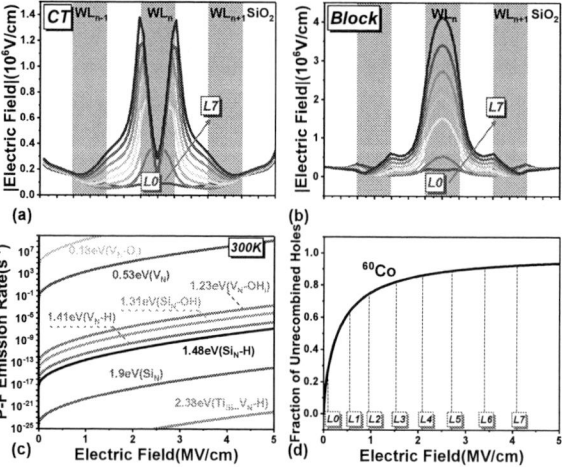

Fig. 7. The absolute electric field on the cross-section along the channel direction at the center of the (a) CT layer (b) blocking layer. (c) P-F emission rate of different defect energy levels with electric field at 300K (d) Holes yield of gamma rays in SiO_2 with electric field.

The aggravated degradation at low temperatures primarily stems from enhanced hole accumulation. Analyzing L0 as the object, Fig. 8(a) and 8(b) show the hole concentration distribution on the cross-section perpendicular to the channel direction. It can be observed that at 10^3 rad in Fig. 8(a), the blocking layer exhibits higher hole concentration at 200K than at 400K, while the Macaroni layer demonstrates the inverse relationship. From Fig. 8(c)-(d), it can be seen that due to the low temperature at 200K, the electric field inside the Macaroni layer is smaller than that at 400K temperature, while the electric field in the blocking layer is significantly larger than that at 400K. This explains why the hole concentration distribution at 200K in Fig. 8(a) occurs. The inconsistent relationship of internal electric field magnitudes at low doses leads to an unclear pattern between the $|\Delta V_{th}|$ of L0 and temperature in Fig. 6(a)-(c).

However, at an accumulated dose of 10^7 rad in Fig. 8(b), both the Macaroni layer and blocking layer exhibit significantly higher hole concentrations at 200K than at 400K. This effect stems from the electric field increase in the Macaroni layer at about 5×10^6 rad in Fig. 8(c). The field strength surpasses the corresponding values at 400K at subsequent doses, accelerating hole accumulation in the Macaroni layer at low temperatures.

The continuous growth of the electric field at low temperatures and high doses is caused by the accumulation of holes in the Macaroni layer near the channel interface, as shown in Fig. 9(a). As dose increases, a large number of holes will accumulate in the Macaroni layer near the channel at 200K, with similar behavior occurring in the blocking layer. But at 400K (Fig. 9(b)), accumulated hole concentrations in both layers remain substantially lower than at 200K.

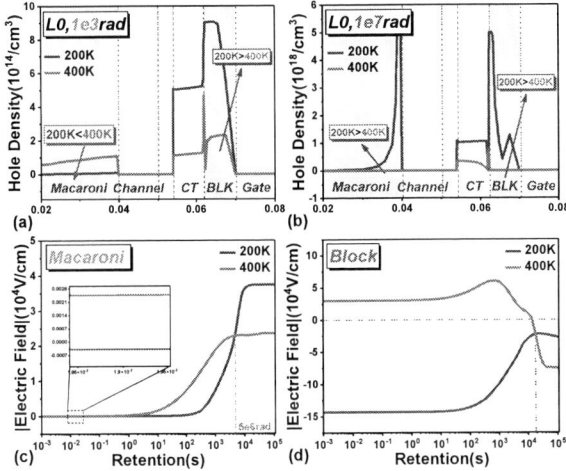

Fig. 8. The hole concentration distribution on the cross section perpendicular to the channel direction of the L0 erased state at (a) 1e3 rad (b) 1e7 rad. The average electric field of the L0 erased state changes during the entire irradiation process of the (c) Macaroni layer (d) blocking layer.

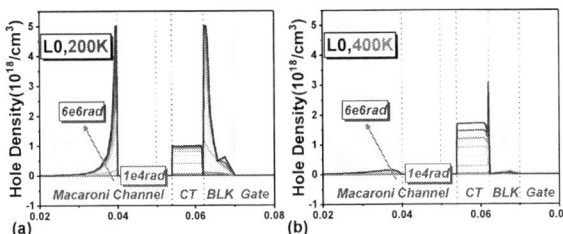

Fig. 9. The hole concentration distribution on the cross section perpendicular to the channel direction of the L0 erased state at (a) 200K (b) 400K.

Low temperature will greatly reduce the migration process of holes in SiO₂, causing the holes to be basically frozen in local place. Hole mobility at 200K is drastically lower than at 400K [9]. Additionally, most radiation-induced hole traps in SiO₂ correspond to shallow E_δ' defects [11]. Holes captured by these defects readily de-trap and loss

(through channel/gate loss) at higher temperatures, leading to reduced accumulation shown in Fig. 9(b). Here, holes buildup in the Macaroni layer at 200K gradually modify the internal electric field distribution. As radiation dose increases, this accumulation leads to a condition where the electric field within the Macaroni layer at 200 K surpasses that at 400 K. Therefore, L0 eventually evolves into the situation shown in Fig. 6(d) where degradation becomes more severe with decreasing temperature.

CONCLUSION

In this work, we investigated the combined effects of radiation and temperature on the V_{th} degradation behavior of 3D CT NAND flash memory. The V_{th} degradation induced by radiation exhibits a non-power-law evolution, indicating a fundamentally different mechanism from that in environments without radiation. Moreover, we identified that aggravated degradation at low temperatures is primarily caused by suppressed hole mobility, which limits the dissipation of trapped holes. This leads to substantial charge accumulation and a pronounced V_{th} shift. These findings provide critical insights into the reliability of CT flash memory under space-related conditions and offer guidance for future radiation-hardened memory design.

ACKNOWLEDGMENT

This work was supported by China Key Research and Development Program under Grant (2023YFB4402500, 2023YFB4402400), National Natural Science Foundation of China (U2441248, U23B2040, 62034006, 92264201), Natural Science Foundation of Shandong Province (ZR2023LZH007, ZR2023QF054, tsqn202306059), and MIND project (MINDXZ202407).

REFERENCES

[1] S. Gerardin et al., "A Heavy-Ion Beam Monitor Based on 3-D NAND Flash Memories," in IEEE Transactions on Nuclear Science, vol. 68, no. 5, pp. 884-889, May 2021.

[2] T. R. Oldham and F. B. McLean, "Total ionizing dose effects in MOS oxides and devices," IEEE Trans. Nucl. Sci., vol. 50, no. 3, pp. 483–499, Jun. 2003.

[3] J. D. Cressler, "Radiation effects in SiGe technology," IEEE Trans. Nucl. Sci., vol. 60, no. 3, pp. 1992–2014, Jun. 2013.

[4] T. S. Balint, J. A. Cutts, E. A. Kolawa, and C. E. Peterson, "Extreme environment technologies for space and terrestrial applications," Proc. SPIE, vol. 6960, pp. 36–47, Apr. 2008.

[5] I. Shin, "Hot/cold clustering for page mapping in NAND flash memory," IEEE Trans. Consumer Electron., vol. 57, no. 4, pp. 1728–1731, Nov. 2011.

[6] M. A. Kumar, M. Raquibuzzaman, M. Buddhanoy, T. Boykin, and B. Ray, "Origin of post-irradiation Vth-shift variability in 3D-NAND memory array," IEEE Trans. Nucl. Sci., pp. 1–1, 2024.

[7] J. Wu, D. Han, W. Yang, S. Chen, X. Jiang, and J. Chen, "Comprehensive investigations on charge diffusion physics in SiN-based 3D NAND flash memory through systematical Ab initio calculations," in 2017 IEEE International Electron Devices Meeting (IEDM), San Francisco, CA, USA: IEEE, Dec. 2017, p. 4.5.1-4.5.4.

[8] V. Vasudevan and J. Vasi, "A simulation of the multiple trapping model for continuous time random walk transport," Journal of Applied Physics, vol. 74, no. 5, pp. 3224–3230, Sep. 1993.

[9] R. C. Hughes, "Hole mobility and transport in thin SiO₂ films," Applied Physics Letters, vol. 26, no. 8, pp. 436–438, Apr. 1975.

[10] G. Borghello et al., "Dose-rate sensitivity of 65-nm MOSFETs exposed to ultrahigh doses," in IEEE Transactions on Nuclear Science, vol. 68, no. 8, pp. 1482-1487, May 2021.

[11] C. J. Nicklaw, Z.-Y. Lu, D. M. Fleetwood, R. D. Schrimpf, and S. T. Pantelides, "The structure, properties, and dynamics of oxygen vacancies in amorphous SiO₂," IEEE Trans. Nucl. Sci., vol. 49, no. 6, pp. 2667–2673, Dec. 2002

Process Co-Optimization of Void Suppression in ULK Dielectric Layers for 28 nm RRAM Arrays Towards High-density Integration

Zhenchao Sui[1,2*], Yanqing Wu[1], Xing Zhang[1*]

[1] School of Software and Microelectronics, Peking University, Beijing 102600, China
[2] Semiconductor Manufacturing Beijing Corporation, Beijing 101102, China

* Email: zhenchao.sui@stu.pku.edu.cn, zhx@pku.edu.cn

Abstract—**In this work, we demonstrate a process co-optimization approach to effectively suppress void defects in ultra-low-κ (ULK) dielectric layer during 1T1R resistive random-access memory (RRAM) array integration into 28 nm CMOS logic platforms. By employing a high-fill-capability dielectric thickening layer (CIP2, continuous improvement process), we enhance the sidewall coverage while maintaining stable resistive switching. Additionally, tailored adjustments of etching parameters address morphology issues from material substitution, ensuring compatibility with baseline electrical parameters. This strategy enables significant void reduction in the ULK layer and seamless CMOS-RRAM integration, facilitating scalable fabrication of high-reliability RRAM devices without sacrificing electrical performance.**

Keywords—*28 nm, RRAM, ULK dielectric, void defects, process integration*

I. INTRODUCTION

Resistive random-access memory (RRAM), as one of the most promising emerging non-volatile memory technologies, has attracted extensive research interest due to its simple structure, fast programming speed, and low power consumption. The compatibility with conventional CMOS processes enables high integration density, showing great potential for applications in neuromorphic computing, such as analog artificial neural networks and spiking neural networks [1,2,3]. Currently, mass-produced RRAM devices predominantly utilize the 1T1R (1 transistor and 1 resistor) architecture, driven by planar MOSFETs. In this configuration, the RRAM element is embedded in a sandwich-like structure between adjacent back-end-of-line (BEOL) metal layers.

In CMOS BEOL processes, copper is widely used as a material for via and interconnect in sub-130 nm technology nodes, with dielectric layers providing electrical insulation. To mitigate RC delay, 28 nm technology node employs ultra-low-κ (ULK) dielectric materials as inter-metal isolation layers, effectively reducing parasitic capacitance and enhancing chip performance. However, ULK materials exhibit inherent limitations such as porous microstructure, low mechanical strength, and poor step coverage [4]. When integrating RRAM structures into CMOS platforms, the high-aspect-ratio RRAM array trenches present significant challenges for ULK filling. organo-silicate precursors struggle to penetrate these structures, resulting in micro-void aggregation at trench bottoms that evolves into 3D void defects. These defects not only compromise structural integrity but also induce severe metal line deformation, eventually leading to copper bridge failures. Furthermore, conventional inspection methods fall short to effectively detecting such nanoscale voids, and reliability risks remain in the end-use stages, potentially causing intermittent errors or catastrophic failures. However, process adjustment to mitigate these effects without co-optimization can typically lead to the compromise of the BEOL baseline electric parameters such as contact resistance, sheet resistance, and capacitance.

This work systematically investigates void defects and optimization methods in ULK dielectric layers of 28 nm logic-based 1T1R RRAM arrays. By optimizing dielectric filling processes and adapting etching parameters to address morphological variations caused by material modifications, we achieve co-optimization of material compatibility and process stability. The approaches here can significantly reduce void dimensions without degrading electrical performance, providing a robust pathway toward high-density RRAM-CMOS integration.

II. EXPERIMENTS AND RESULTS

Figure 1a and 1b show the optical image and the layout of the RRAM chip, respectively. Figure 1c and 1d show the transmission electron microscope (TEM) images of individual RRAM cells and the structural voids. The RRAM array is integrated between metal layers M2 and M3 using standard 28 nm CMOS logic technology. Distinct structural defects in top electrode (TE) line and TE dot RRAM configurations showing continuous linear voids with Cu deformation in TE line versus periodic void clusters with Cu bridges in TE dot. These defects escalate risks of interlayer shorts and performance degradation via electromigration, causing long-term reliability issues. Figure 1e shows the schematic view of the voids in the RRAM structures.

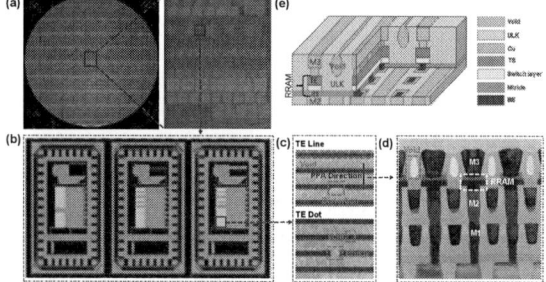

Fig. 1. (a) Optical image of the RRAM chip; (b) Layout of the RRAM chip; (c) Top view of both TE Line and TE Dot types of RRAM arrays; (d) TEM cross-section image of the RRAM arrays; (e) Schematic view of the void generation in stack layers of RRAM array.

979-8-3315-3918-4/25 $31.00 © 2025 IEEE

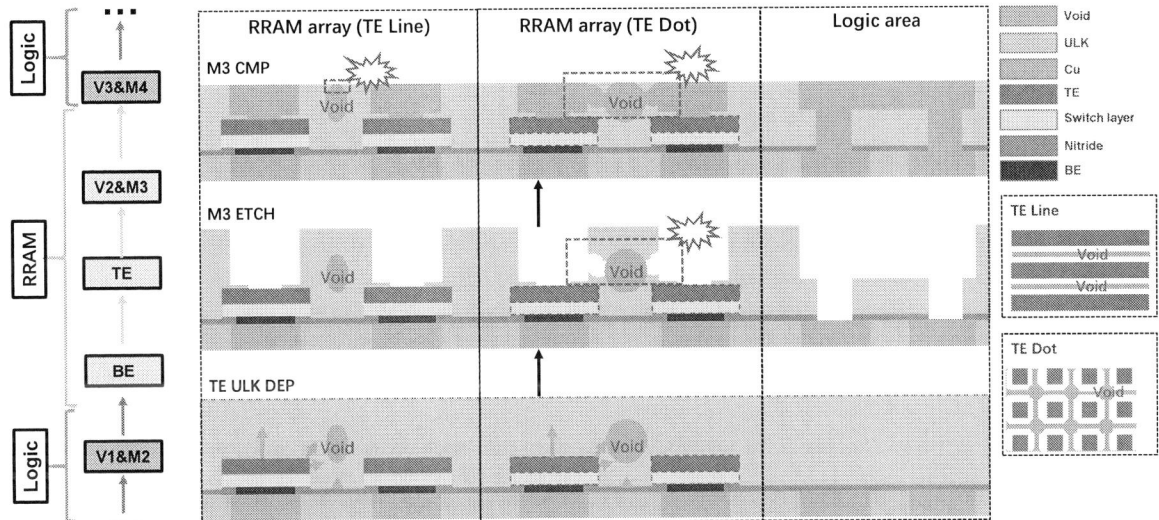

Fig. 2. Schematic diagram of flow and void formation processes in RRAM arrays.

Figure 2 shows a schematic diagram of void formation in RRAM arrays. The conventional back-end Damascene process used in traditional logic circuits involves depositing an ULK dielectric layer on a flat M2 metal layer interface, followed by etching and subsequent M3 metal layer filling. This process does not involve material filling in high-aspect-ratio structures. In contrast, the high-aspect-ratio trench structures in RRAM arrays impose specific requirements on ULK deposition processes. Due to the limited step coverage of ULK materials within three-dimensional trenches, unfilled regions are prone to aggregate, leading to structural void defects.

The mechanism of void formation during ULK material filling (TE ULK DEP) can be attributed to the multi-physics coupling effects of filling dynamics, material properties, and process parameters. Constrained by the spatial morphology of high-aspect-ratio trenches between RRAM arrays, organo-silicate precursors are transported to the trench bottom through Knudsen diffusion (gas-phase transport dominated) [5] and surface diffusion (adsorbed-state migration) in both TE line and TE dot architectures [6]. The concentration gradient exhibits a nonlinear distribution in the three-dimensional trench structures. This non-uniformity causes the faster surface reaction rate of precursors at the trench entrance than

those at the bottom and sidewalls, triggering a "bread-loafing" self-limiting growth phenomenon [7]. Specifically, precursors preferentially form an arched deposition layer at the trench opening, where premature closure traps unfilled regions at the bottom, forming void defects.

In the TE line configuration, anisotropic diffusion along the TE line direction leads to spindle-shaped linear voids. These structural defects aggravate Cu line deformation during subsequent etching and metal filling processes, increasing the risk of Cu bridge formation between adjacent interconnects. For the TE dot architecture, the grid-like trenches formed by periodic TE islands hinder precursor diffusion in the X/Y directions of the grid. This results in clustered void defects at grid intersection nodes. Such defects are susceptible to etch-through during metal line dielectric etching. During Cu interconnect electroplating, voids and etch-through channels may be filled by copper, ultimately leading to Cu bridge failure.

To address ULK deposition challenges, we propose two solutions:

Gradient deposition (CIP1, continuous improvement process): a staged deposition strategy increases initial layer (IL) thickness at reduced rates to ensure bottom coverage, followed by parameter-tuned gradient deposition to prevent top sealing.

High-fill-capability dielectric thickening (CIP2): a high-fill-capability dielectric layer significantly reduces the aspect ratio of TE trenches, effectively suppressing void formation during subsequent ULK filling.

As shown in Figure 3a and 3b, comparative TEM analyses of dielectric films under baseline (BL), CIP1-optimized, and CIP2-optimized conditions reveal that CIP2 delivers remarkable improvements in void suppression, reducing X/Y-direction void sizes and void-top-to-M2 (Void-TM) space by a factor of 65%/44% and 10% versus BL, while maintaining the unchanged void height in Figure 3c. This demonstrates CIP2's superior trench sidewall coverage capability over ULK, primarily attributed to its higher surface migration rate and enhanced step coverage during chemical vapor deposition

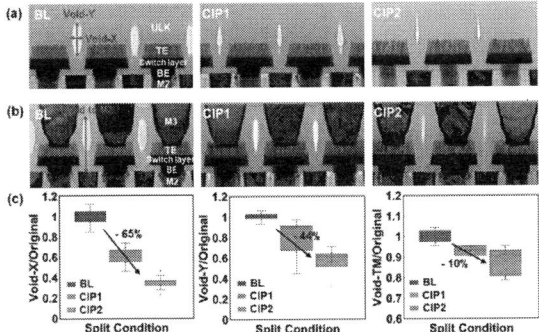

Fig. 3. Process split condition for void performance improvement. TEM image of BL, CIP1 and CIP2 condition after TE ULK CMP process (a) and after M3 CMP process (b); (c) Void size of BL, CIP1 and CIP2 condition.

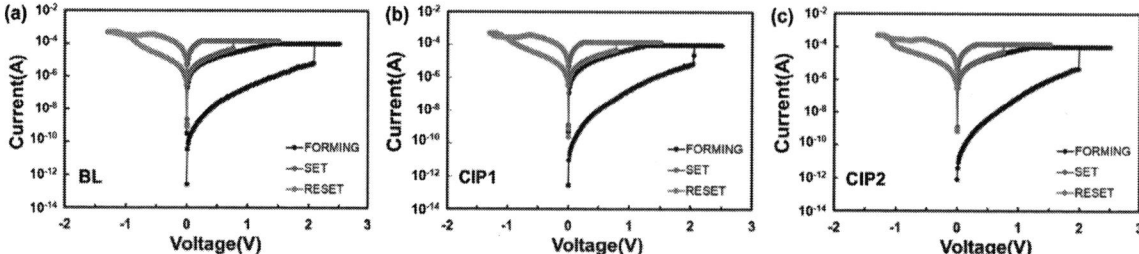

Fig.4. RRAM bit cell I-V curves of BL (a), CIP1 (b) and CIP2 (c) condition.

(CVD). These characteristics effectively suppress delayed sidewall deposition, providing a critical process solution for void-free filling in high-aspect-ratio structures.

To investigate the impact of dielectric film modifications on RRAM switching performance, we conducted electrical characterization of devices under BL, CIP1, and CIP2 conditions, with corresponding I-V curves shown in Figure 4. All devices underwent a forming operation by applying a DC voltage sweep from 0 to 2.5 V with a compliance current of 100 μA. Voltage application was terminated upon reaching the compliance current to prevent hard breakdown or prolonged high-voltage operation. Figure 4 shows that devices under all three conditions exhibit a sharp current increase to 100 μA at ~2.1 V, transitioning to a low-resistance state (LRS). Post-forming SET/RESET cycling demonstrates similar switching behavior across all conditions, enabling reversible transitions between high-resistance states (HRS) and LRS. This confirms that dielectric film adjustments have a negligible impact on the resistive switching characteristics of RRAM devices.

While dielectric film optimization addresses void-related issues caused by insufficient trench filling in RRAM arrays, material substitution introduces new process challenges. Variations in chemical bonding states and doping ratios between dielectric materials significantly influence plasma etching kinetics. For instance, in fluorine-based plasma environments, the reactivity difference between fluorine radicals and silicon atoms results in a 15%–20% slower etching rate for CIP2 compared to ULK. This etching selectivity leads to insufficient sidewall etching at trench bottoms, creating via necking defects that degrade electrical performance in logic circuits. To resolve this problem, we adjusted etching parameters based on material-specific characteristics to enhance sidewall etching reactions at trench bottoms, as shown in the TEM image in Figure 5a. Post-etch electrical characterization (WAT data) confirms that optimized CIP2 conditions yield critical electrical parameters, including contact resistance for V2 chain/Kelvin structures, sheet resistance of M3 metal lines, and M3 capacitance, all within the 3σ range of BL conditions as shown in Figure 5b. This achievement represents an effective co-optimization strategy in material compatibility and process stability.

III. CONCLUSION

This study resolves ULK void defects in 28nm 1T1R RRAM arrays through gradient deposition (CIP1) and high-fill-capability dielectric integration (CIP2). CIP2 demonstrates superior void suppression via enhanced sidewall growth, reducing lateral void dimensions while maintaining vertical profiles. Electrical tests confirm stable RRAM operation (Forming at ~2.1V, intact SET/RESET switching). Concurrent etch optimization eliminates material-induced via defects, achieving full compatibility with baseline electrical parameters (contact resistance, sheet resistance, capacitance). The proposed co-optimization strategy provides a scalable solution for RRAM-CMOS integration, overcoming ULK filling bottlenecks while ensuring device reliability.

ACKNOWLEDGMENT

The authors gratefully acknowledge the technical support of RRAM integration and module team of SMBC to this work.

REFERENCES

[1] J. T. Zhou, F. X. Cai, Q. W. Wang, B. Chen, S. Gaba, and W. D. Lu, "Very low-programming-current RRAM with self-rectifying characteristics," IEEE Electron Device Lett., vol. 37, pp. 404-407, April 2016.

[2] B. J. Choi, A. C. Torrezan, J. P. Strachan, P. G. Kotula, A. J. Lohn, M. J. Marinella, Z. Y. Li, R. S. Williams, and J. J. Yang, "High-Speed and Low-Energy Nitride Memristors," Adv. Funct. Mater., vol. 26, pp. 5290-5296, May 2016.

[3] S. P, C. Li, H. Jiang, W. W. Xia, H. L. Xin, J. J. Yang, and Q. F. Xia., "Memristor crossbar arrays with 6-nm half-pitch and 2-nm critical dimension," Nature Nanotech., vol. 14, pp. 35-39, August 2019.

[4] V. McGahay, "Porous dielectrics in microelectronic wiring applications," Materials, vol. 3, pp. 536-562, January 2010.

[5] T. F. Liang, and Q. Li, "Accurate modeling of Knudsen diffusion in nanopores using a physical-based boundary model," J. Appl. Phys., vol. 126, pp. 084304, August 2019.

[6] A. Ledesma-Durán, S. I. Hernández, and I. Santamaria-Holek, "Effect of surface diffusion on adsorption-desorption and catalytic kinetics in irregular pores. I. local kinetics," J. Phys. Chem. C, vol. 121, pp. 14544-14556, June 2017.

[7] T. J. Kunene, L. K. Tartibu, K. Ukoba, and T. C. Jen, "Review of atomic layer deposition process, application and modeling tools," Materials Today: Proceedings, vol. 62, pp. S95-S109, February 2022.

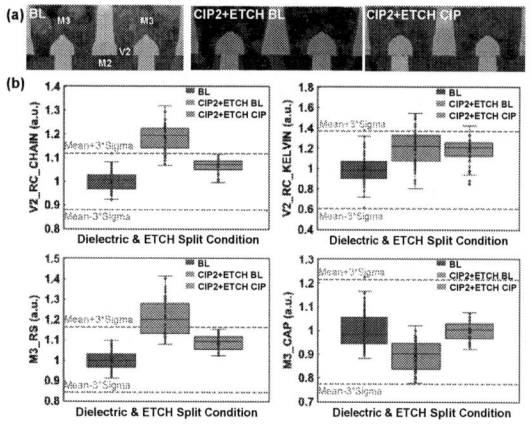

Fig. 5 TEM image (a) and WAT data (b) of BL, CIP2+ETCH BL and CIP2+ETCH CIP condition in logic area.

A Study on Performance Enhancement of TiO_2/HfO_2 Memristors through Rapid Thermal Annealing

Yifan Wu, Yuzhe Hu, Yuewei Qu, Pengpeng Sang, Jixuan Wu, Xuepeng Zhan* and Jiezhi Chen

School of Information Science and Engineering (ISE), Shandong University, Qingdao, China.

* Email: zhanxuepeng@sdu.edu.cn

Abstract—**Enhanced memristive performances are of great importance to construct energy-efficient neural networks (NN). In this work, with process optimizations of rapid thermal annealing (RTA), performances of TiO_2/HfO_2 memristor are improved and used for weight updating. By using Ag as the top electron (TE), it is demonstrated that photoelectric efficiency could be improved, which facilitates the function of optical bias mapping. With induced optical bias in deep NN architecture, higher recognition accuracy (99.1%) as well as lower energy consumption (44.21 pJ) are achieved successfully. Our results can provide great potential for developing highly energy-efficient and reliable memristor-based NN.**

Keywords—Memristor, Neural network, Optical bias, RTA

I. INTRODUCTION

The exponential growth of artificial intelligence, particularly in deep neural networks (DNNs), has led to an unprecedented demand for computational power. The traditional von Neumann architecture, which relies on separate processing and storage units, has been hindered by the well-known "memory wall" bottleneck [1]. Frequent data is transferred between CPU/GPU and off-chip memory consumes over 60% of system energy and severely constrains throughput, rendering them for energy-constrained edge devices and large-scale AI deployments. To overcome these limitations, emerging nonvolatile memories like resistive random-access memory (RRAM, or memristor) are regarded as transformative hardware paradigms. Memristor enables in-memory computing by leveraging analog crossbar arrays to perform matrix-vector multiplication (MVM) - the core operation of neural networks - directly within memory cells via Ohm's and Kirchhoff's laws [2]. With the non-von Neumann approach, the data transfer overhead is eliminated, which achieves orders-of-magnitude improvement in energy efficiency (>10 TOPS/W). This can support large-scale parallel analog computing [3], [4]. Consequently, memristor-based neuromorphic computing is a promising solution for maintaining the scalability of next-generation AI systems [5].

Normally, the memristor-based hardware neural networks confront the problem like lacking effective implementations of bias terms, where the bias term (b) is important in the forward propagation equation of the backpropagation (BP) algorithm [6], [7]:

$$a^l = f\left(W^l * a^{l-1}\right) + b^l \qquad (1)$$

Here, the W is the wight value, and a means the output value of the weight layer.

In most memristor-based hardware neural networks, the bias term is usually implemented by varying voltage on fixed resistors [8], [9], [10]. Prezioso M *et al* achieved RRAM-based bias term by exploiting the multi-level cell (MLC) characteristics of memristors [11]. The electrical implementation of bias terms is restrained by high energy consumption, significant inter-line crosstalk, and limited bandwidth [12]. Much attention has been devoted to developing optical implementation to address these issues. However, an optical bias strategy is rarely reported, especially on TiO_2/HfO_2 memristor, which are generally limited by poor memristive characteristics and device reliability.

This study enhances TiO_2/HfO_2-based memristors' performance for energy-efficient hardware neural networks. Two device structures are employed whose performances are enhanced by using rapid thermal annealing (RTA): For TiN(TE)/TiO_2/HfO_2/TiN memristor (synaptic weight updates), the RTA significantly improves reliability, conductance update linearity, and memory window. For Ag(TE)/TiO_2/HfO_2/TiN memristor, the RTA boosts photoelectric efficiency, enabling its use as an optical bias term. It shows significantly lower energy consumption (44.21 pJ) compared to electrically biased networks by integrating these RTA-treated devices into a deep neural network architecture. Moreover, superior recognition accuracy (99.1%) is achieved with only 2.81% network scale improvement compared to results of network without bias term.

II. MEMRISTOR-BASED HARDWARE NEURAL NETWORK

A. Theoretical Background

The BP algorithm is composed of three stages: forward propagation, error backpropagation, and weight update. During the forward process, the output formulas for each layer are as follows [7]:

$$a^l = f(W^l * a^{l-1}) + b^l \qquad (1)$$

Memristor-based hardware neural networks exploit the unique electrical properties - dynamic resistance modulation via charge flux and non-volatility - to directly map synaptic weights (W) onto memristor conductance (G). In crossbar arrays, input voltages (V_j) are applied along rows, generating output currents ($I_i = \sum G_{ij} * V_{ij}$) via Kirchhoff's law to perform MVM. In this

979-8-3315-3918-4/25 $31.00 © 2025 IEEE

way, the multiply-accumulate (MAC) operation is efficiently achieved.

The implementation of the optical bias term (*b*) can be achieved by connecting a photovoltaic device in parallel on the bit line. The photovoltaic device does not require a connection to the word line, whose control is accomplished through optical pathways.

In the process of weight updating, iterative updates of network weights can be achieved based on the long-term potentiation (LTP) and long-term depression (LTD) characteristics of memristors. LTD and LTP arise from the bipolar modulation mechanism of their resistance in response to electrical stimuli. Specifically, when a positive voltage pulse is applied, the conductance gradually increases (LTP), while the application of a negative voltage pulse results in a gradual decrease in conductance (LTD). This behavior is highly analogous to Hebbian plasticity observed in biological synapses [13]. When the weight gradient ΔW^l derived from backpropagation is positive, a positive voltage pulse is generated to trigger the LTP effect, resulting in an increase in memristor conductance to facilitate weight augmentation. Conversely, a negative voltage pulse is applied to induce the LTD effect, thereby achieving weight reduction. Weight gradients during backpropagation are calculated as follows [7]:

$$\delta^l = (W^{l+1})^T * \delta^{l+1} * f'(z^l) \qquad (2)$$

$$\Delta W^l = -\eta \frac{\partial L}{\partial W^{l+1}} = \eta \delta^l * z^{l-1} \qquad (3)$$

Here, δ is the error term, z is the output value of each layers，and L is the loss function, which is the total error between the model's predicted values and the true values.

B. Synaptic Memristors for Weight Mapping

The memristors with structure of TiN(TE)-TiO$_2$-HfO$_2$-TiN are utilized as weight layer units in neural networks for the mapping and updating of weights. To handle issues such as poor reliability and low updating linearity, RTA under different conditions is conducted. The performance variations are investigated to select the optimal device for constructing the weight array. The device structure and fabrication process are illustrated in Fig. 1, with labels: D1 (un-annealed), D2 (RTA: 60s-300°C), D3 (RTA: 60s-400°C), D4 (RTA: 60s-500°C).

Fig. 1. The structure and fabrication process of TiN /TiO$_2$/HfO$_2$/TiN cell and Ag/TiO$_2$/HfO$_2$/TiN cell.

Fig. 2 presents the forming voltage distribution of both TiN(TE) and Ag(TE) memristors. The forming voltage distribution of the unannealed device is highly dispersed, which is unsuitable for its application in arrays. After the RTA process,

the forming voltage distribution of the devices becomes more concentrated.

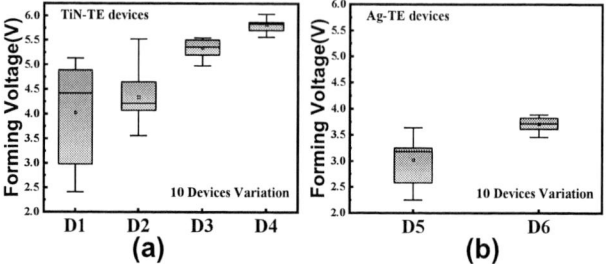

Fig. 2. Forming voltage distributions of (a) TiN (TE)/TiO$_2$/HfO$_2$/TiN devices and (b) Ag(TE)/TiO$_2$/HfO$_2$/TiN devices.

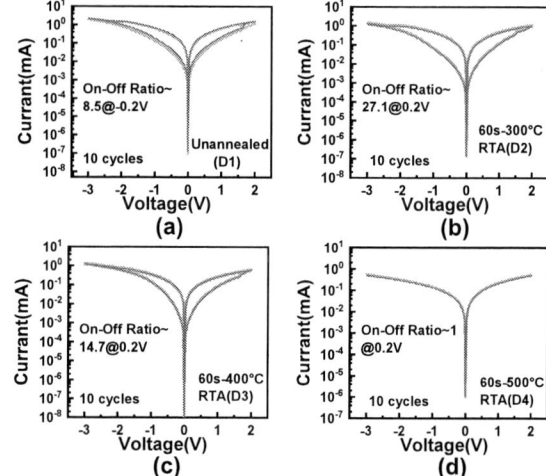

Fig. 3. IV curves of samples (a) D1, (b) D2, (c) D3 and (d) D4.

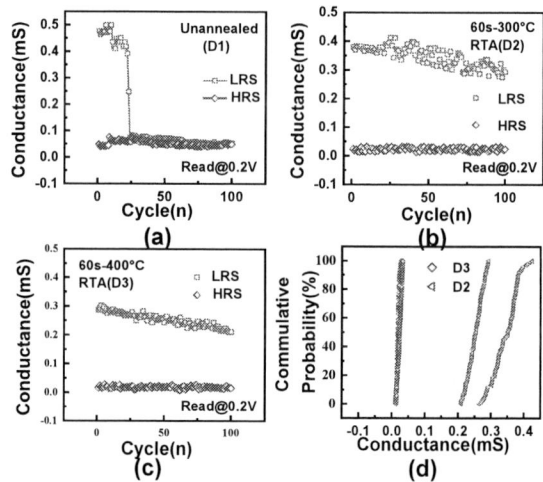

Fig. 4. Endurance characteristics of (a) D1, (b) D2 and (c) D3; (d) Cumulative probability distributions of different resistance states of D2 and D3.

I-V characteristics (Fig. 3) revealed that switching ratios initially increased, then decreased with annealing temperature. The highest ratio occurred at 300°C (D2). At 500°C (D4), the I-V window collapses, indicating loss of memristive behavior due to HfO₂/TiO₂ crystallization inhibiting oxygen vacancy migration.

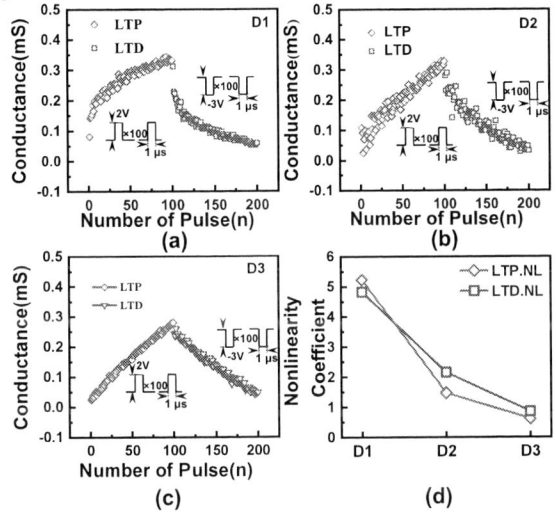

Fig. 5 LTP/LTD characteristics of (a) D1, (b) D2, and (c) D3; (d) nonlinearity coefficient of different resistance states of D1, D2, and D3.

Due to the loss of memristive properties in D4(500°C), endurance and LTD/LTP tests are performed on D1， D2， and D3 (Fig. 4&Fig. 5). D1 sustained <20 DC cycles, while annealed devices exceeded 100 cycles. D3 exhibited the most stable low-resistance state (LRS), evidenced by tighter distributions and slower degradation (Fig. 4. (d)). 100 voltage pulses (1µs, +2V/-3V, 50% duty cycle) were applied for LTP/LTD test. All devices showed conductance modulation with varying linearity. Nonlinearity coefficients of D1, D2, and D3 are shown in Fig. 5. (d). The nonlinearity coefficient is calculated using the following equations [14]:

$$G_{LTP} = B(1 - e^{(-\frac{P}{A})}) + G_{min} \qquad (4)$$

$$G_{LTD} = -B(1 - e^{(-\frac{P-P_{max}}{A})}) + G_{max} \qquad (5)$$

$$B = \frac{(G_{max} - G_{min})}{1 - e^{(-\frac{P_{max}}{A})}} \qquad (6)$$

Here, the G_{max}, G_{min}, and P_{max} are the maximum conductance, minimum conductance, and pulse numbers used to switch.

The D1, D2, and D3 memristors exhibit LTP/LTD nonlinearity coefficients of 5.32/4.74,1.24/2.11, and 0.67/0.93, respectively. In the process of updating the weight at the neural network, the linearity of the devices in the array directly determines the weight mapping accuracy [15]. The high linearity LTP/LTD characteristics of D3 make it more suitable for application in neural networks. Despite D2's higher switching ratio, D3 demonstrates superior reliability and linearity, making it optimal for weight arrays.

C. Photovoltaic-effect memristors for optical bias term

For emulating synaptic behaviors, photovoltaic-effect memristors (TiN-HfO₂-TiO₂-Ag(TE)) share the same materials and structures except the top electrode. Devices (D5: unannealed; D6:60s-350°C) were fabricated as shown in Fig. 1. Under 520nm laser illumination, both D5 and D6 exhibited a

photovoltaic effect, generating volatile photocurrent. Photovoltaic effect efficiency, defined as the value of the internal current generated under certain illumination conditions, is improved by RTA (seen in Fig. 6. (b)). This can obviously reduce optical input power consumption. As the optical power increases (seen in Fig. 6. (a)), the photogenerated current of the device also rises, exhibiting a high degree of linearity and a response time of less than 10 ns. This device effectively addresses the challenges associated with traditional photovoltaic devices based on perovskite materials, which often suffer from high latency and low light intensity-voltage linearity issues [16], [17]. In the neural networks, this device generates corresponding internal photogenerated currents by receiving light of varying power levels. Unlike traditional electrical biasing methods, this approach eliminates the need for connections to word lines, thereby avoiding the introduction of additional crosstalk pathways and effectively reducing the intensity of crosstalk.

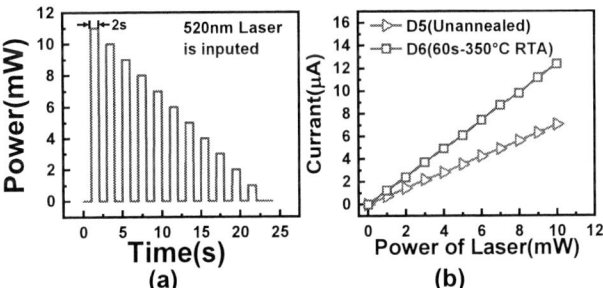

Fig. 6. (a) Power–time curve of input laser with wavelength of 520 nm; (b) the optical currents under different laser power.

III. NEURAL NETWORK PERFORMANCE

An annealed memristor-based DNN with optical biasing is constructed using D3 (weights) and D6 (bias), which is shown in Fig. 7. D3, connected to word lines and bit lines(WL&BL), perform weight mapping and updating. D6, connected only to bit lines, generate photocurrent under 520nm illumination as the bias term.

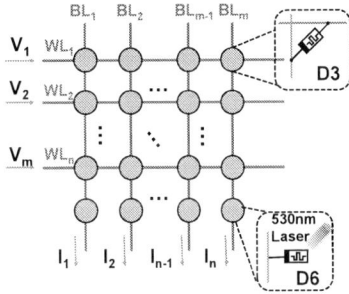

Fig. 7. Structure of weight array with optical bias terms.

Training on MNIST over 50 epochs (Fig. 8. (a-c)), it demonstrates a 5% higher recognition accuracy (99.1%) for the optically biased neural network compared to a bias-free network. The accuracy is comparable to the results of electrically biased networks.

Energy and network scale analyses (Table 1) reveal that the optical bias terms consume quite small additional energy and network scale, which shows significantly lower per-device consumption compared to the network with electrical bias term.

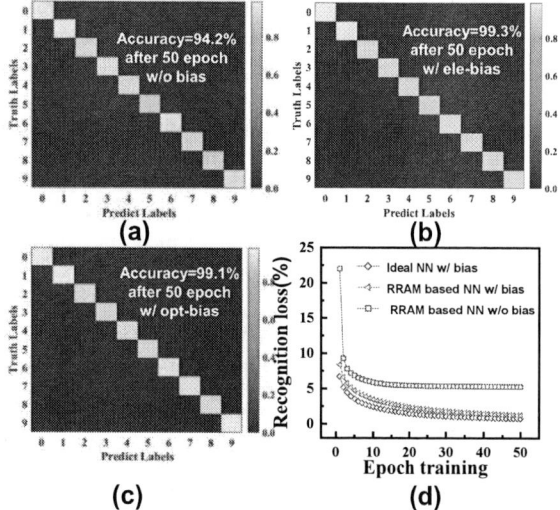

Fig. 8. Confusion matrices of the classification with probability of the predicted output digit for neural networks (a) without bias term, (b) with electrical bias term and (c) with optical bias term. (d) The recognition loss with epochs of the three neural networks.

TABLE I.   COMPARISONS OF THREE NETWORKS

		W/o bias	Ele-bias	Opt-bias
Network scale	Total	3800	3910	3910
	Bias term	0	110	110
Energy consumption of each device for bias term		0	20.18nJ	44.21pJ
Recognition Accuracy		94.2%	99.3%	99.1%

IV. Conclusion

TiO$_2$/HfO$_2$-based memristors are studied by adopting RTA process for performance enhancement. On the one side, in TiN(TE)/TiO$_2$/HfO$_2$/TiN memristor for synaptic weight updating, improved reliability and synaptic modulation characteristics are demonstrated; On the other side, in Ag(TE)/TiO$_2$/HfO$_2$/TiN structure, photoelectric efficiency is substantially improved, enabling its application in optical bias mapping. In comparison to networks using electrical bias, superior recognition accuracy (99.1%) and significantly reduced energy consumption (44.21pJ) can be achieved by implementing the optical bias. Our results shed light to the high-performance and low-power memristor-based neural networks.

ACKNOWLEDGMENT

This work was supported by National Key Research and Development Program of China (2023YFB4402400, 2023YFB4402500), Natural Science Foundation of China (U2441248, 92264201, U23B2040), Natural Science Foundation of Shandong Province (ZR2023LZH007), and MIND Project (MINDXZ202407).

REFERENCES

[1] Horowitz M. 1.1 computing's energy problem (and what we can do about it)[C]//2014 IEEE international solid-state circuits conference digest of technical papers (ISSCC). IEEE, 2014: 10-14.

[2] J. Xia Q, Yang J J. Memristive crossbar arrays for brain-inspired computing[J]. Nature materials, 2019, 18(4): 309-323.

[3] Chen W H, Li K X, Lin W Y, et al. A 65nm 1Mb nonvolatile computing-in-memory ReRAM macro with sub-16ns multiply-and-accumulate for binary DNN AI edge processors[C]//2018 IEEE International Solid-State Circuits Conference-(ISSCC). IEEE, 2018: 494-496.

[4] Shafiee A, Nag A, Muralimanohar N, et al. ISAAC: A convolutional neural network accelerator with in-situ analog arithmetic in crossbars[J]. ACM SIGARCH Computer Architecture News, 2016, 44(3): 14-26.

[5] Wong H S P, Salahuddin S. Memory leads the way to better computing[J]. Nature nanotechnology, 2015, 10(3): 191-194.

[6] Aziza H, Zambelli C, Hamdioui S, et al. On the Reliability of RRAM-Based Neural Networks[C]//2023 IFIP/IEEE 31st International Conference on Very Large Scale Integration (VLSI-SoC). IEEE, 2023: 1-8.

[7] Rumelhart D E, Hinton G E, Williams R J. Learning internal representations by error propagation[EB/OL].(1985-9)

[8] Wan W, Kubendran R, Schaefer C, et al. A compute-in-memory chip based on resistive random-access memory[J]. Nature, 2022, 608(7923): 504-512.

[9] Chen P Y, Peng X, Yu S. NeuroSim: A circuit-level macro model for benchmarking neuro-inspired architectures in online learning[J]. IEEE Transactions on Computer-Aided Design of Integrated Circuits and Systems, 2018, 37(12): 3067-3080.

[10] Ielmini D, Wong H S P. In-memory computing with resistive switching devices[J]. Nature electronics, 2018, 1(6): 333-343.

[11] Prezioso M, Merrikh-Bayat F, Hoskins B D, et al. Training and operation of an integrated neuromorphic network based on metal-oxide memristors[J]. Nature, 2015, 521(7550): 61-64.

[12] Liu S, Feng J, Tian Y, et al. Thermo-optic phase shifters based on silicon-on-insulator platform: state-of-the-art and a review[J]. Frontiers of Optoelectronics, 2022, 15(1): 9.

[13] Prezioso M, Merrikh-Bayat F, Hoskins B D, et al. Training and operation of an integrated neuromorphic network based on metal-oxide memristors[J]. Nature, 2015, 521(7550): 61-64.

[14] P.-Y. Chen et al., "Mitigating effects of non-ideal synaptic device characteristics for on-chip learning," in Proc. IEEE/ACM Int. Conf. Comput.-Aided Design (ICCAD), Nov. 2015, pp. 194–199.

[15] Rao M, Tang H, Wu J, et al. Thousands of conductance levels in memristors integrated on CMOS[J]. Nature, 2023, 615(7954): 823-829.

[16] Tang S, Yan J, Chen L, et al. Circuit modeling and analysis of hysteresis effect of perovskite photovoltaic cells[J]. Solar Energy Materials and Solar Cells, 2024, 278: 113182.

[17] Ou W, Liang J, Guo J, et al. High‐Efficiency Fabry‐Pérot‐Resonance‐Based Color‐Tunable Bifacial Perovskite Solar Cells for Building Integrated Photovoltaics[J]. Advanced Energy Materials, 2

Investigation of Self-Heating Effects in InGaZnO Vertical Channel Transistors for DRAM Application

Zhuoran Kong [1], Yizhan Liu [1], Jinfeng Kang[2], Xiaoyan Liu*[2]

[1] School of Software and Microelectronics, Peking University, Beijing, 100871, China
[2] School of Integrated Circuits, Peking University, Beijing, 100871, China

* Email: liuxiaoyan@pku.edu.cn

Abstract—Self-heating effect (SHE) in vertical transistors with indium-gallium-zinc-oxide (IGZO) channels is analyzed using finite element method (FEM) simulations. Due to the asymmetric structure of the transistor, the source/drain position affects the thermal characteristics. Self-heating effect is worse when the drain is placed at the bottom. Changes in the thermal environment also affect the temperature, and the presence of capacitor has an effect on the heat flux distribution. The presence of thermal crosstalk in transistor arrays under different conditions is investigated, and the temperature rise can be mitigated by increasing the pitch between transistors. Additionally, the transient thermal response of the devices are investigated to provide guidelines for layout design and thermal management for DRAM design.

Keywords—self-heating effect, vertical channel transistors, InGaZnO, FEM, thermal crosstalk

I. INTRODUCTION

With technology nodes scaling down, vertical devices have become a research hotspot due to their advantages in high-density integration and current drivability [1]. Particularly in dynamic random-access memory (DRAM), the $4F^2$ structure based on vertical transistors has been proposed to overcome the challenges for conventional $6F^2$ DRAM, such as floating-body effect and row hammering. Indium gallium zinc oxide (IGZO) has attracted attention as a channel material which has excellent electrical properties, including high electron mobility and low leakage current [2]. However, the scaling of devices and low thermal conductivity of IGZO enhanced the self-heating effects (SHEs), potentially affecting device reliability and performance.

In this work, we investigated the self-heating effect of IGZO vertical channel transistors (VCT) using finite element method (FEM) including the temperature rise and heat flux distribution. The input power is obtained from the TCAD simulation [3]. As the structure of transistor is asymmetric, the SHE behavior is different. And temperature has an effect on electrical characteristics. The self-heating effect also changes when changing the thermal environment and considering the presence of the capacitor. Moreover, for the memory cell array, the thermal crosstalk makes thermal management more complex, and the influence of cell pitch is evaluated. Finally, the dynamic thermal response characteristics of VCT are investigated through transient simulations, and the thermal effect for the structure operating under different frequencies is analyzed.

This work was supported by the Shenzhen Science and Technology Program KQTD20200820113105004.

II. DEVICE STRUCTURE AND SIMULATION METHOD

A. Device Structure

Fig.1 (a) shows the structure of a IGZO VCT. The channel is a frustum of a cone with a wide top and a narrow bottom, surrounded with gate oxide [2]. The IGZO VCT is the gate all around (GAA) structure. The electrodes are connected upwards to the metal wire by contact via. The bottom electrode is surrounded by shallow trench isolation (STI), and connected to the substrate via STI-all-around (SAA) [4]. The length parameters and relevant thermal conductivity used in simulation are listed in Table I and Table II respectively..

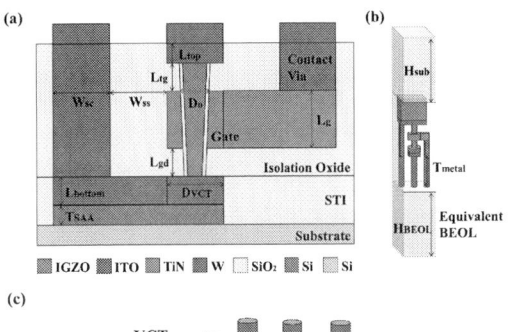

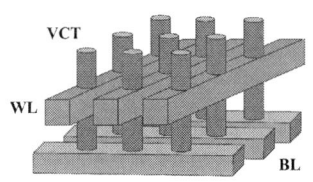

Fig. 1.   (a) The cross section of a single InGaZnO vertical channel transistor. (b) Entire thermal environment. (c) The simplified structure of a 3*3 VCT array.

TABLE I.   LENGTH PARAMETERS OF IGZO VCT

Symbol	Description	Value
L_g	Length of gate	46.7 nm
L_{top}	Length of top electrode	15.5 nm
L_{bottom}	Length of bottom electrode	22.6 nm
D_o	Diameter of gate oxide	26 nm
D_{VCT}	Diameter of VCT	51.1 nm
W_{ss}	Width between gate and contact	51.1 nm
W_{sc}	Width of contact at the side	51.1 nm
L_{tg}	Length between top electrode and gate	25.3 nm

979-8-3315-3918-4/25 $31.00 © 2025 IEEE

Symbol	Description	Value
L_{gd}	Length between gate and bottom electrode	27 nm
T_{SAA}	Thickness of SAA	200 nm
T_{metal}	Height of the first layer of metal wires	400 nm
H_{BEOL}	Height of equivalent BEOL layer	3.6 μm
H_{sub}	Height of substrate	30 μm

TABLE II.   RELEVANT THERMAL CONDUCTIVITY

Material	Sections	Value[W/(m·K)]
IGZO	Channel	1.2
TiN	Gate	19.2
ITO	Source/Drain	14
W	Contact via	170
SiO_2	STI/isolation oxide/Gate oxide	1.4
Si	STI-all-around (SAA)	18
Si[bulk]	Substrate	148

B. Simulation Methods

The electrical characteristic of the IGZO VCT is simulated using the TCAD tool [3]. The transfer characteristic curve is shown in Fig. 2. The curves obtained by switching the top and bottom electrodes match. The results are agreed with the experiment data [2]. From the figure, when the drain is at the bottom, the threshold voltage is -0.026 V, and the subthreshold swing is 81.8 mV/decade. The input power of this transistor is calculated to be 2.6 μW

To simplify the complex interconnects, the second layer of metal wires and above is equivalent to a layer with thermal conductivity of 10 W/(m·K), as shown in Fig. 1 (b). Two 300 K heat sinks are positioned at the top of equivalent back end of line (BEOL) and the bottom of the substrate [5]. Other external boundary conditions are set as thermal isolation.

III. SELF-HEATING EFFECT IN IGZO VCT

A. Thermal Response of a Transistor

Based on the discussion above, thermal simulation of the transistor is carried out and the results are shown in Fig. 3. Due to the asymmetrical structure of the transistor, the temperature rise is different when the source/drain positions are changed. When drain is at the bottom, the maximum temperature rise is 20 K. When drain is at the top, the maximum temperature rise is 18 K. The reason is the varying diameters of channel. The drain is the main heat generating area, and when the drain is at the bottom, the contact surface between channel and the drain is smaller, resulting in worse heat dissipation.

Fig. 4 illustrates the heat flux distribution in drain bottom and drain top cases. The substrate is the main heat dissipation path, accounting for more than 55% in both cases, followed by BEOL. When the drain is changed from the bottom to the top, it is farther away from the substrate and closer to BEOL, so the heat flux through the substrate has reduced while the heat flux of BEOL has increased. Since the diameters of channel are not constant, the heat flux at the gate is also slightly reduced.

Considering the temperature rise, the electrical characteristics of the transistor are simulated again using the

TCAD tool. When the drain is at the bottom, the average temperature of the transistor is 307.12 K. The obtained transfer characteristic curve has a threshold voltage of -0.043 V, and a subthreshold swing of 83.1 mV/decade. Compared to the previous curve, the threshold voltage decreases and the subthreshold swing increases. So higher temperature leads to poorer electrical characteristics.

B. Transient State Self-Heating Effect of a Transistor

1) Response of a Transistor with fixed Input Power: Single IGZO VCT is simulated with input power of 2.6 μW, and the drain is at the bottom. The transient temperature response is shown in Fig. 5. The maximum temperature is 319.64 K, which is the same as it under steady state conditions. The temperature rises through a process and reaches its maximum value at 6.31 μs.

2) Influence of Input Power Frequency: The transient temperature response obtained by applying input power with different frequencies is shown in Fig. 6. The duty cycles are 0.5. When the input power frequency is 1 MHz, the maximum

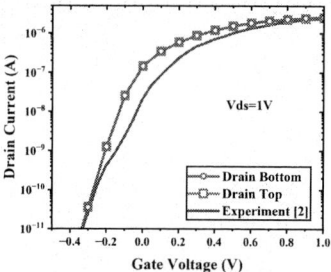

Fig. 2.   The transfer characteristic curve of IGZO VCT

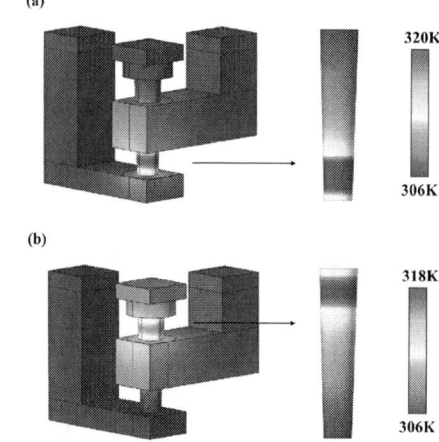

Fig. 3.   The temperature distribution of a transistor when the drain is (a) at the bottom and (b) at the top.

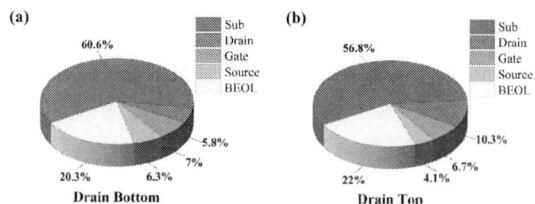

Fig. 4.   The heat flux distribution of a transistor when the drain is (a) at the bottom and (b) at the top.

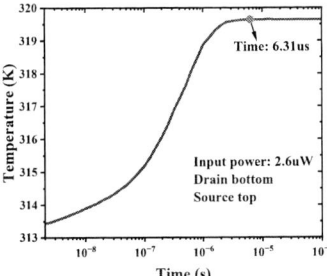

Fig. 5. The transient temperature response of a transistor with steady input power.

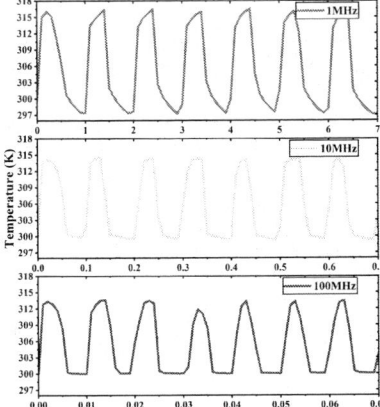

Fig. 6. The transient temperature response of a transistor with transient input power with frequency of 1 MHz, 10 MHz and 100 MHz.

temperature is 316.55 K. And when the input power frequency is 10 MHz and 100 MHz, the maximum temperature is 314.45 K and 313.59 K. The higher the frequency, the smaller the maximum temperature. Therefore, the device reliability a high input frequency needs to be taken into account.

C. The Effect of Thermal Environment

To investigate the effect of the thermal environment on the self-heating effect, the thicknesses of the substrate and BEOL are varied respectively. The change of the maximum temperature is observed as shown in Fig. 7.

As the thicknesses of both the substrate and BEOL increase, the maximum temperature increases. The temperature changes more significantly when the thickness of BEOL changes because of the lower equivalent thermal conductivity. Reducing the thicknesses of the substrate and BEOL is beneficial in reducing the self-heating effect of the transistor.

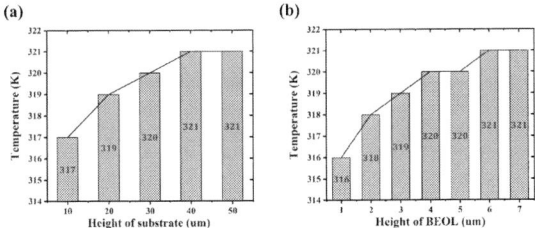

Fig. 7. The maximum temperature at (a) different substrate thickness and (b) different BEOL thickness.

D. The Effect of Capacitor

In DRAM, the transistor is connected to a capacitor generally, so the effect of the presence of the capacitor is further investigated. Referring to the structure of oxide-semiconductor channer transistor DRAM [2], a 1.35 um thick layer with thermal conductivity of 18.9 W/(m·K) is used to equate the capacitor using FEM simulation. When the drain is at the top, the equivalent is added to the bottom of the source. Thermal simulation is performed at the same input power as the previous steady-state experiment, and the heat flow percentage of the substrate is 56.38%. The percentage decreases due to some of the heat flowing through the capacitor to the bit line. When the drain is at the bottom, the equivalent layer is added to the top of the source and the heat flow of the substrate is 60.75% which increases. Because of the high thermal conductivity of metal wires, more heat flows to the substrate as the capacitor diverts some of the heat.

IV. SELF-HEATING EFFECT IN VCT ARRAY

The discussions and results above are based on a single transistor. In this section, a 3*3 array is designed to investigate the thermal crosstalk.

A. Thermal Crosstalk of VCT Array

In the structure of VCT array, the thermal environment of individual transistor is extended and word lines (WL) and bit lines (BL) are connected [6]. A simplified schematic is shown in Fig. 1 (c). The temperature rise of each transistor is investigated when one transistor and a line of transistors is on, as shown in Fig. 8.

a) *Case 1:* When the transistor at the center is on, the temperature distribution is shown in Fig. 9. The maximum temperature rise is 15 K. It can be obtained that the temperature of neighboring transistors rises by 1.64 - 1.94 K, and the temperature of diagonal transistor rises by 1.50 - 1.55

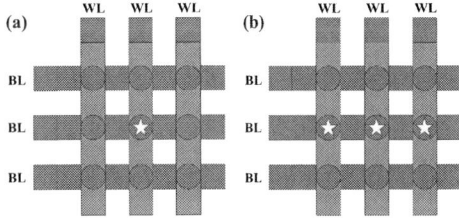

Fig. 8. Top view of the VCT array, where the star symbol indicate the transistor is on. (a) The transistor at the center is on. (b) Three transistors on the center BL are on.

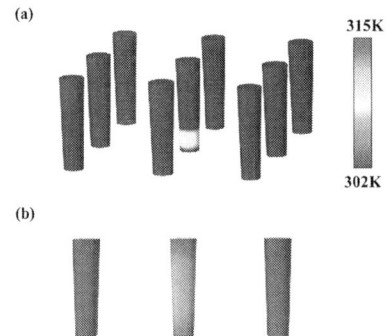

Fig. 9. The temperature distribution of a 3*3 VCT array when a transistor is on. (a) Three-dimension view of the entire array. (b) YZ view of a BL lines located on an edge.

979-8-3315-3918-4/25 $31.00 © 2025 IEEE

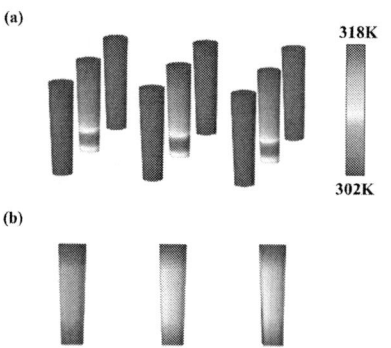

318K

302K

Fig. 10. The temperature distribution of a 3*3 VCT array when three transistors are on. (a) Three-dimension view of the entire array. (b) YZ view of a BL lines located on an edge.

K. Therefore, the thermal crosstalk is more severe when the distance is closer.

    a)   *Case 2:* When three transistors on the center BL are on, the temperature distribution is shown in Fig. 10. The maximum temperature rise is 18 K, higher than the previous case. The temperature of the remaining transistors rises by 4.80 – 5.01 K, and the temperature of the middle transistor in each BL is the highest, which shows that the thermal crosstalk is more severe.

B. The Influence of Pitch

The simulations of the array above are based on the structure that the pitch between neighboring transistors is twice the diameter of one transistor (D_{VCT}). When the pitch is changed, the maximum temperature (T_{max}) changes. The results are shown in Table III. When the pitch is increased, the maximum temperature decreases and the thermal crosstalk is reduced.

TABLE III.   THE MAXIMUM TEMPERATURE AT DIFFERENT PITCH

Pitch	1.5 D_{VCT}	2 D_{VCT}	2.5 D_{VCT}	3 D_{VCT}
Tmax (K)	19	18	17	17

V. CONCLUSION

The self-heating effect of vertical channel transistors with IGZO channel are simulated based on 3D FEM simulation. The SHE of the transistor and its thermal crosstalk behavior in multi-device arrays are systematically analyzed by steady state and transient thermal simulations. The results show that the SHE leads to significant temperature rise. Changes of the thermal environment affect heat dissipation, and the thermal crosstalk in the arrays is worse with decreasing pitch. The study provides guidelines for layout design and thermal management for DRAM design.

ACKNOWLEDGMENT

This work was supported by the Shenzhen Science and Technology Program KQTD20200820113105004.

REFERENCES

[1] S. Chen et al., "Self-heating effects of vertical gate-all-around transistors: analysis and modeling," in IEEE Transactions on Electron Devices, vol. 71, no. 11, pp. 6478-6485, Nov. 2024.

[2] S. Fujii et al., "Oxide-semiconductor channel transistor DRAM (OCTRAM) with 4F2 architecture," 2024 IEEE International Electron Devices Meeting (IEDM), San Francisco, CA, USA, 2024, pp. 1-4.

[3] T. Fung et al., "Two-dimensional numerical simulation of radio frequency sputter amorphous In–Ga–Zn–O thin-film transistors," J. Appl. Phys., V. 106, pp. 084511-1.

[4] I. Myeong, J. Jeon, M. Kang and H. Shin, "Analysis of self heating effect in vertical-channel field effect transistor," 2019 20th International Conference on Thermal, Mechanical and Multi-Physics Simulation and Experiments in Microelectronics and Microsystems (EuroSimE), Hannover, Germany, 2019, pp. 1-5.

[5] L. Cai, W. Chen, G. Du, X. Zhang and X. Liu, "Layout design correlated with self-heating effect in stacked nanosheet transistors," in IEEE Transactions on Electron Devices, vol. 65, no. 6, pp. 2647-2653, June 2018.

[6] D. Son, I. Myeong, H. Kim, M. Kang, J. Jeon and H. Shin, "Analysis of electrothermal characteristics of GAA vertical nanoplate-shaped FETs," in IEEE Transactions on Electron Devices, vol. 65, no. 7, pp. 3061-3064, July 2018.

Broadband Characterization of Ferroelectric Domain Switching Hysteresis Loops in TiN/Hf$_{0.5}$Zr$_{0.5}$O$_2$/TiN Thin-film Capacitors

Wen Di Zhang, Xian Yu Hu, An Quan Jiang*

College of Integrated Circuits and Micro-Nano Electronics Innovation, Fudan University, Shanghai 200433, China
* Email: 19212020017@fudan.edu.cn, aqjiang@fudan.edu.cn

Abstract—For the fabrication of ferroelectric memories, it is necessary to measure voltage-dependent domain switching times and frequency-dependent coercive voltages. Traditionally, it uses two separate characterization techniques of Positive-Up-Negative-Down and modified Sawyer-Tower circuits in limited timescales. Here we exhibit a powerful technique to characterize both domain switching time and coercive voltage over a wide frequency range of 50 MHz–100 Hz on 10 nm–thick TiN/Hf$_{0.5}$Zr$_{0.5}$O$_2$/TiN capacitors. The broadband domain switching hysteresis loops were transformed from domain switching current transients versus time under bipolar write voltage pulses in various rise/fall times. From the domains switching hysteresis loops, we derived time-dependent polarization reversal at various write voltages. The experimental data in Hf$_{0.5}$Zr$_{0.5}$O$_2$ show the frequency-dependent coercive field that increases quickly from 1.4 MV/cm to 3.5 MV/cm with the lowering of measuring periodicity from 5 ms to 20 ns. The fundamental physics of the huge coercive field could be related to the presence of interfacial-layers between TiN and Hf$_{0.5}$Zr$_{0.5}$O$_2$ that slow down the domain switching speed. Our characterization technique offers a practical tool to characterize write/read voltages and times of ferroelectric memories in fast operation speeds.

Keywords—*Hf$_{0.5}$Zr$_{0.5}$O$_2$, Ferroelectric Thin Film, Domain switching time, Domain switching hysteresis loop, Interfacial layer, Coercive field*

I. INTRODUCTION

A nonvolatile Ferroelectric random-access memory is promising for the high-speed operation and low energy consumption [1, 2]. While commercial lead zirconate titanate (PZT) memory has a low coercive field ($\sim$ 0.2 MV/cm) and high endurance ($> 10^{14}$ cycles), the relatively thick film thicknesses (80–200 nm) and high processing temperatures ($\geq$ 650°C) arouse several challenges in the high-density integration compatible to the silicon-based technologies [3]. Atomic-layer deposition (ALD) of HfO$_2$-based thin films can offer the advantages of low-temperature synthesis and conformal growth in three-dimensional structures on silicon surfaces, thereby circumventing the high temperature requirements for deposition of ferroelectric perovskite oxides. This has led to applications in volatile/nonvolatile high-storage-density memories and extremely size-scaled CMOS devices for use at low operating voltages [4–7]. However, the high intrinsic coercive field ($\sim$ 1.5 MV/cm) that varies with the operation time is generally estimated from the dielectric displacement–electric field (D–E) hysteresis loops using the

modified Sawyer-Tower circuits at frequencies below 1 MHz. For the nanosecond operating memory, this measuring frequency range is insufficient. Instead, a Positive-Up-Negative-Down (PUND) characterization technique is proposed to test the switched polarization as function of the applied voltage and pulse width upon many programming cycles from which one could estimate the memory operation voltage and time at the nanosecond scale [8-11]. This method is time-consuming. It is wondered if there is a simple method to measure broadband D-E hysteresis loops from which one can immediately obtain operation voltages of the memories. Unfortunately, this technique is still lacking in the literature.

For the energy-efficient memory, the operation voltage is generally required below 1.2 V. However, the coercive voltage for the Hf$_{0.5}$Zr$_{0.5}$O$_3$ (HZO) thin film capacitor could be over 3 V if the operation time is shorter than 10 ns [12].This could be due to the presence of interfacial layers between HZO and electrodes [13].The characterization of broadband D-E hysteresis loops would be helpful for understanding of the interfacial-layer formation.

II. EXPERIMENTAL

Continuous Hf$_{0.5}$Zr$_{0.5}$O$_2$ thin films were grown by ALD (TFS 200, Beneq) at 200°C on a Si substrate coated with 10-nm-thick TiN bottom electrodes, where hafnium tetrachloride (HfCl$_4$) and zirconium tetrachloride (ZrCl$_4$) precursors were used with water as the oxidizing reactant and argon acting as a purging gas. A 1:1 ratio between the alternately laminated HfO$_2$ and ZrO$_2$ atomic layers ($\sim$1 nm) was used. X-ray reflection and Scanning Transmission Electron Microscopy high-angle annular dark-field image indicated the film thickness of 10 nm [12]. TiN and W top electrode layers were deposited by sputtering (PVD-75, Kurt J. Lesker) at room temperature. After photoresist layer patterning of the top electrodes using ultraviolet photolithography (NQX4006, Neutronix-Quintel), the top electrodes were etched into squares in side lengths of 30 µm via ion milling using SF$_6$ and O$_2$ plasmas with gas flows of 15 and 5 sccm, respectively, at an output power of 50 W for 30 s in a reactive ion etching system (RIE-10NR, Samco, Japan). The amorphous HZO that was finally crystallized at 550°C for 30 s. The capacitor dimensions were verified using planar-view scanning electron microscope (Sigma HD, Zeiss) images. The film is a mixture of orthorhombic, tetragonal and monoclinic phases, as investigated using synchrotron in-plane grazing-incidence X-ray diffraction

979-8-3315-3918-4/25 $31.00 © 2025 IEEE

pattern (XRD) at a wavelength of 0.6887 Å located at the Shanghai Synchrotron Radiation Facility [12].

The high-frequency (D–E) hysteresis loops were transformed from the measurements of the domain switching current transients versus time. Under the application of bipolar switching voltages $V_{appl}(t)$ to the top electrode of the HZO capacitor with the bottom electrode remaining grounded, the domain switching/nonswitching currents (I_{sw}) versus time (t) through a resistor R in series with a ferroelectric capacitor (C_F) were measured using an oscilloscope (HDO6054, LeCroy, USA) with 12-bit voltage resolution and a bandwidth of 1 GHz, was sketched in Fig. 1(a). Square switching voltage pulses $V_{appl}(t)$ were supplied using a two-channel Agilent 81110A pulse generator with adjustable rise/fall times (2 ns–10 ms).

III. RESULTS AND DISCUSSION

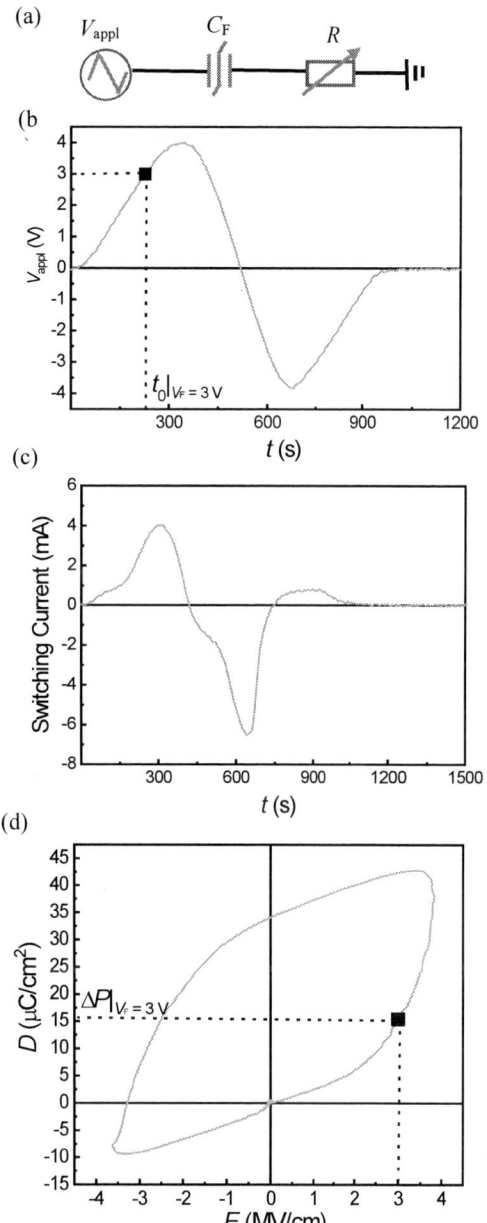

The circuit's resistor-capacitor (RC) time constant in Fig. 1(a) can be adjusted using an in-series resistance of $R = 100\ \Omega$–$1\ M\Omega$. Figs. 1(b) and 1(c) show the input voltage pulses $V_{appl}(t)$ and domain switching current transient $I_{sw}(t)$. Therefore, the voltage drop $V_F(t)$ across C_F is $V_{appl}(t) - I_{sw}(t)R$. Through time integration of the domain switching current transient, we derived time-dependent dielectric displacement $D(t)$. From $D(t)$ and $V_F(t)$ curves, we derived the D–E hysteresis loop, as shown in Fig. 1(d).

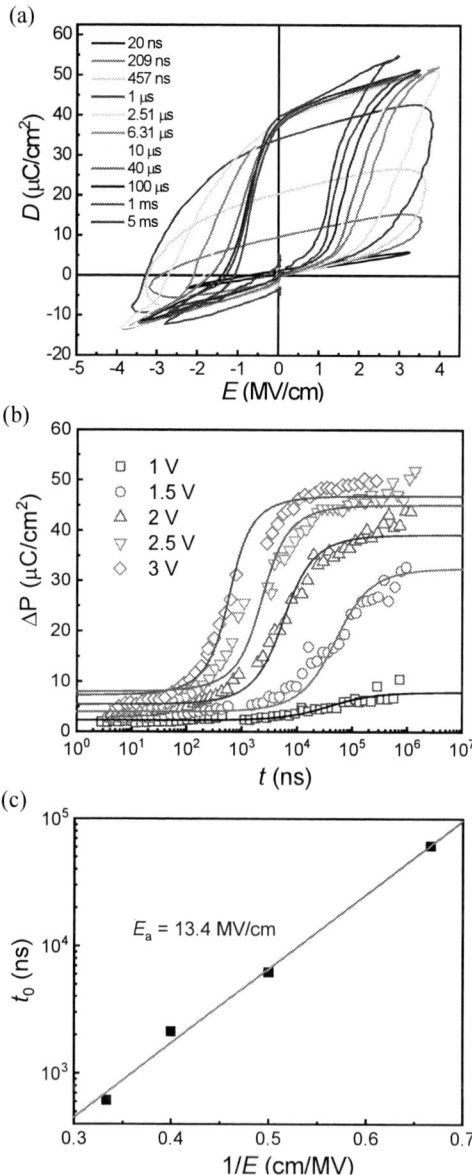

Fig. 1. (a) Experimental setup for the pulse characterization. (b) Input bipolar voltage pulses. (c) Domain switching current transient versus time. (d) The transformed D-E hysteresis loop.

Fig. 2. (a) The transformed D–E hysteresis loops at various periodicities. (b) Time dependence of the switched polarization under various applied voltages derived from frequency-dependent D–E hysteresis loops. The solid lines are the best fits of the data according to Eq. (1). (c) The inverse of electric field dependence of domain switching time fitted by the solid line according to Eq. (2).

Fig. 2(a) shows D–E hysteresis loops at periodicities ranging from 20 ns to 5 ms, as transferred from domain switching current transients using the method shown in Figs. 1(a)–(d). The characteristic frequency is much higher than the upmost limit of 1 MHz by most commercial ferroelectric testers using the modified Sawyer-Tower circuits. From $V_{appl}(t)$, $D(t)$ and $V_F(t)$ transients, we can further estimate the time dependence of the switched polarization $\Delta P|_{V_F = \text{const}}$, using the method shown by the square symbols in Fig. 1(b)–(d) when $V_F = 3$ V. This technique is critical important to estimate memory write/read times and voltages, and is more convenient than the traditional PUND testing through the step-by-step increase of the write pulse width under a constant write voltage [8-10].

Figs. 2(b) shows time dependence of the switched polarization over a large time interval at various write voltages. When V_F increases from 3 V to 1.5 V, domain switching time increases from 617 ns to 61.3 μs, besides broadening of the time distribution. When $V_F \leq 1$ V, most polarizations are nearly unswitched in demonstration of high voltage operations of a fast HZO memory.

According to domain nucleation-limited-switching model[11], the nucleation time (t_0) of the reversed domain in the independent region of the film decisive for the ultimate speed of domain switching is assumed to be far longer than that of domain growth. If the hatching times of the reversed nuclei that are preferentially located at around lattice imperfections of the film, such as defect centers, surfaces, and grain boundaries, obey the Breit-Wigner distribution [14], we have the following equation [8]:

$$\Delta P(t) = 2P_s \left[\frac{1}{\pi} arctan \left(\frac{\ln t - \ln t_0}{\Delta \ln t_0} \right) + \frac{1}{2} \right] \quad (1)$$

where $\Delta \ln t_0$ is the distribution width and P_s is the saturation polarization. The solid lines in Fig. 2(b) are the best fits of the data according to Eq. (1) from which we estimated t_0. Fig. 2(c) shows the inverse of electric field dependence of domain switching time. Generally, t_0 obeys the Merz law in the form of [15]

$$t_0 = \tau_0 \exp\left(E_a / E \right) \quad (2)$$

where E_a is the activation field. The solid line in Fig. 2(c) is the best fit of the data according to Eq. (2) from which we derived that $E_a = 13.4$ MV/cm. The E_a value is almost 3 times that of the PZT thin films [16], highlighting a higher driving force of domain motion in HZO than in PZT.

To understand the huge E_a in HZO, it is necessary to measure the periodicity (T) dependence of the mean coercive field (E_c). E_c can be estimated from symmetrized D–E hysteresis loops along the D axis in Fig. 3(a) that intercept with a dashed line fitted by the equation of $D = \varepsilon_0 \varepsilon E$, where ε_0 is vacuum permittivity and ε is dielectric constant. It seems that E_c in HZO is ten times that in PZT. This could be due to the partial oxidation of TiN top electrode in contact with the HZO film during thermal crystallization that increases the electrode resistivity (ρ_{on}) [13]. E_c is determined in the D-E hysteresis loop and can be described in the form of [12]

$$E_c = E_{c0} + \sqrt{8P_s E \rho_{on}/T} \quad (3)$$

where E_{c0} is the intrinsic coercive field. Fig. 3(b) shows the

E_c–$T^{-1/2}$ dependence for the capacitor. The plot is almost linear and converges into $E_{c0} = 1.4$ MV/cm as $T^{-1/2} \to 0$, as shown by the solid line fit of the data according to Eq. (3). From the slope of the linear fit, we estimated $\rho_{on} = 59$ Ω·m for the interfacial layer. Therefore, this technique is useful to characterize write/read voltages of broadband memories. It is believed that the reduction of the interfacial-layer resistance in TiN/HZO/TiN capacitors could decrease the operation voltage of a fast HZO memory.

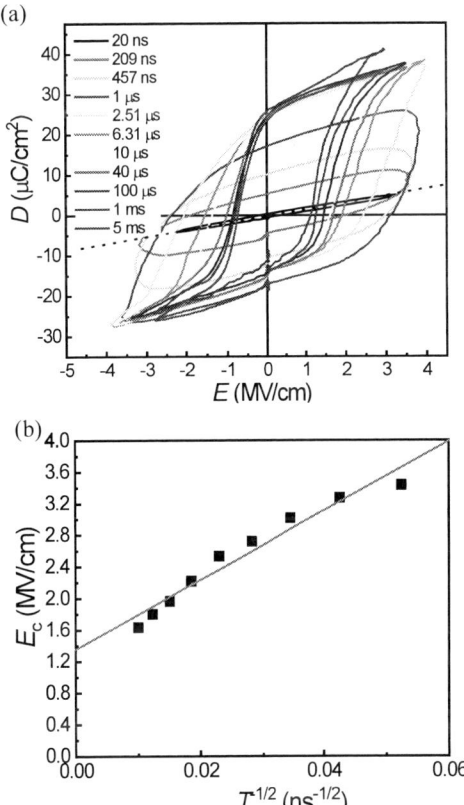

Fig. 3. (a) The symmetrized D–E hysteresis loops along the D-axis at various periodicities. (b) E_c-$T^{-1/2}$ dependence fitted by the solid line according to Eq. (2).

IV. CONCLUSION

We demonstrated the broadband measurements of D–E hysteresis loops in a wide frequency range to break through the upward limit of commercial ferroelectric testers that use Sawyer-Tower circuits. All these loops can be transformed from domain switching current transients directly using the PUND characterization technique. From $D(t)$ and $V_F(t)$ transients at various periodicities, we derived time dependences of the switched polarization at various write voltages over a broad time interval of six orders of magnitude. Using this powerful characterization technique, we found write voltage in HZO that increases quickly with shortening of the operation time. The fundament physics of so large a coercive voltage at a high frequency could be related to the presence of interfacial layer induced by the partial oxidation of TiN top electrode that increases the activation field of domain nucleation. To enable energy-efficient HZO fast memory in the future, more attention

979-8-3315-3918-4/25 $31.00 © 2025 IEEE

would be paid to the suppression of interfacial-layer resistance and minimize the domain nucleating energy barrier.

ACKNOWLEDGMENT

One of us (A. Q. J.) thanks the financial supports by the National Key Research and Development Program of China (grant number 2024YFA1409500), Shanghai Science and Technology Innovation Action Plan (grant number 24CL2900900), and the National Natural Science Foundation of China (grant number 62174034).

REFERENCES

[1] B. Zeng, L. Yin, R. Liu, C. Ju, Q. Zhang, et al., "Multiple polarization states in $Hf_{1-x}Zr_xO_2$ thin films by ferroelectric and antiferroelectric coupling," Adv. Mater., vol. 37, art no. 2411463, February 2025.

[2] Y. Sun, S. Zhang, Q. Liu, Y. Li, H. Lu, et al., "Back-End-of-Line Compatible 2T1C Memory Cell With InGaZnO Thin-Film Transistors and $Hf_{0.5}Zr_{0.5}O_2$-Based Ferroelectric Capacitors," IEEE T. Electron Dev., vol. 72, pp. 1-7, March 2025.

[3] W. Wang, K. Takai, T. Eshita, M. Nakabayashi, K. Nakamura, et al., "Development of a high-endurance ferroelectric capacitor for FeRAM in automotive and industrial applications." IEEE T. Electron Dev., vol. 72, pp. 629-634, February 2025.

[4] S.-C. Chang, N. Haratipour,; S. Shivaraman, T. L. Brown-Heft, J. Peck, et al., "Anti-ferroelectric $Hf_xZr_{1-x}O_2$ capacitors for high-density 3-D Embedded-DRAM," International Electron Devices Meeting (IEDM), vol. 20, pp. 605-608, December 2020.

[5] S. K. Kim and P. Mihaela, "Future of dynamic random-access memory as main memory," MRS Bull., vol. 43, pp. 334–339, May 2018.

[6] G. Ribes, J. Mitard, M. Denais, S. Bruyere, F. Monsieur, et al., " Review on high-k dielectrics reliability issues," IEEE T. Device Mat. Re., vol. 5, pp. 5–19, March 2005.

[7] Q. Luo, J. G. Yang, R. R. Cao, H. L. Ma, Y. Yang, et al., "A highly CMOS compatible hafnia-based ferroelectric diode," Nat. Commun., vol. 11, art no. 1391, March 2020.

[8] A. Q. Jiang, J. W. Fei, Y. Y. Lin, and T. A. Tang, "Formulization of long-time domain switching around the coercive field from imprint measurements on ferroelectric thin films," J. Appl. Phys., vol. 103, pp. 124112, June 2008.

[9] I. Stolichnov, A. Tagantsev, N. Setter, J. S. Cross, and M. Tsukada, "Crossover between nucleation-controlled kinetics and domain wall motion kinetics of polarization reversal in ferroelectric films," Appl. Phys. Lett., vol. 83, art no. 3362, October 2003.

[10] I. Stolichnov, A. K. Tagantsev, E. Colla, and N. Setter, "Physical model of retention and temperature-dependent polarization reversal in ferroelectric films," J. Appl. Phys., vol. 98, art no. 084106, October 2005.

[11] A. K. Tagantsev, I. Stolichnov, and N. Setter, "Non-Kolmogorov-Avrami switching kinetics in ferroelectric thin films," Phys. Rev. B, vol. 66, art no. 214109, December 2002.

[12] W. D. Zhang, Z. Z. Song, S. Q. Tang, J. C. Wei, Y. Chen, et al., "Ultrahigh dielectric permittivity in $Hf_{0.5}Zr_{0.5}O_2$ thin-film capacitors," Nat. Commun., vol. 16, art no. 2679, March 2025.

[13] W. Hamouda, A. Pancotti, C. Lubin, L. Tortech, C. Richter, et al., "Physical chemistry of the $TiN/Hf_{0.5}Zr_{0.5}O_2$ interface," J. Appl. Phys., vol. 127, art no. 064105, February 2020.

[14] E. K. H. Salje and M. A. Carpenter, "Local symmetry breaking in the bulk and in domain boundaries: Breit-Wigner damping of phonons and acoustic resonances," Ferroelectrics, vol. 433, pp. 111-122, January 2012.

[15] W. J. Merz, "Domain formation and domain wall motions in ferroelectric $BaTiO_3$ single crystals," Phys. Rev., vol. 95, pp. 690-698, January 1954.

[16] Y. H. Shin, I. Grinberg, I. W. Chen, and A. M. Rappe, "Nucleation and growth mechanism of ferroelectric domain-wall motion," Nature, vol. 449, pp. 881–884, October 2007.

979-8-3315-3918-4/25 $31.00 © 2025 IEEE

On the Reliability of Sub-10nm Ultra-thin Ferroelectric HZO Thin Film

Xiaopeng Li, Yang Feng, Pengpeng Sang, Xuepeng Zhan, Jixuan Wu*, Jiezhi Chen*

School of Information Science and Engineering, Shandong University, Qingdao, China;

* Email: jixuanwu@sdu.edu.cn; chen.jiezhi@sdu.edu.cn

Abstract—$Hf_xZr_{1-x}O_2$ (HZO) ferroelectric (FE) thin films are promising for next-generation non-volatile memory and computing-in-memory due to their CMOS compatibility, fast switching speed, low power, and excellent thickness scalability. However, as the film scales to sub-10nm, the applicability of conventional reliability mechanisms remains unclear. This work reviews recent progress on the reliability of sub-10nm HZO FE films across a wide temperature range (4-420K). On the basis of experimental observations and first-principles calculations, the temperature-dependent evolution of defect dynamics and phase transition behaviors are analyzed in-depth, and the dominant mechanisms under different thermal conditions are discussed. Specifically, high-temperature operations could modulate defect behaviors, facilitating wakeup and fatigue suppression, while low-temperature promotes tetragonal-to-orthorhombic (t-to-o) phase transformation by lowering the energy barrier, enhancing the initial polarization. With these insights, we propose a series of defect-phase co-optimization strategies tailored to specific thermal regimes. This work provides valuable insights into the reliable design and process optimization of ultra-thin HZO ferroelectric devices to achieve ultimate performance.

Keywords— Ferroelectric HZO, reliability, wakeup effect, phase transition, oxygen vacancy.

I. INTRODUCTION

The rapid development of artificial intelligence (AI), edge computing, and cloud data centers has significantly increased the demand for high-throughput processing and energy-efficient memory access. This has driven innovation in emerging non-volatile memory (NVM) technologies such as RRAM, PCRAM, and FERAM, which are designed to overcome the energy consumption and scalability limitations of conventional SRAM and Flash memory. Among these technologies, HfO_2-based ferroelectric (FE) devices have garnered significant attention due to their fast switching speed, low operating voltage, excellent thickness scalability, and outstanding compatibility with standard CMOS processes [1-2]. In particular, Zr-doped HfO_2 ($Hf_xZr_{1-x}O_2$, HZO) has emerged as a leading candidate among doped HfO_2 systems, owing to its relatively low crystallization temperature and broad compositional tunability. Over the past decade, substantial efforts have been devoted to understanding the fundamental physical mechanisms, including defect dynamics and phase transformation behavior, leading to remarkable improvements in device performance and integration. Notably, the demonstration of high-density 3D FE-NAND and 32 Gb FE-based NVDRAM devices has validated the commercial viability of HZO ferroelectrics, enabling their deployment in monolithic 3D architectures, neuromorphic computing, and embedded NVM systems [3-5].

In addition, as integrated circuits advance toward ultra-low-power and high-density architectures, the push for monolithic 3D integration demands further reductions in operating voltage and feature size. In this context, film thickness scaling has emerged as one of the most efficient and widely adopted strategies (**Fig. 1**). Beyond enabling lower operating voltage, thickness scaling also shortens ALD process cycles, improves manufacturing throughput, and enhances process controllability and uniformity [6-7]. Recently, there have been many reports on advancements in thickness scaling. For example, Huang et al. achieved sub-5nm films through stress-enhanced electrode engineering [8], while Lee et al. reported a 3nm HZO film capable of operating at just 0.5V [6], highlighting the feasibility of scaling.

However, despite these encouraging results, significant reliability challenges emerge as film thickness is reduced to sub-10nm. Ultra-thin HZO films often suffer from degraded FE properties, pronounced wakeup and fatigue effects, increased leakage currents, and severe retention loss, all of which critically impact device stability and practical application [6, 9-11]. These challenges suggest that conventional reliability mechanisms developed for thicker films may no longer be applicable at the sub-10nm scale, necessitating deeper exploration of the underlying physical mechanisms. Considering that both defect dynamics and phase stability are highly temperature-dependent, comprehensive characterization across wide temperature ranges could serve as a powerful approach for probing these reliability issues and elucidating the underlying defect- and phase-related mechanisms.

In this work, we conduct a systematic investigation on the reliability of sub-10nm HZO FE films over a wide temperature range (4-420K). By combining experimental observations with first-principles calculations, we explore a series of temperature-dependent reliability phenomena, including high-temperature-induced re-wakeup and re-initialization, as well as temperature-modulated wakeup behaviors in ultra-thin 3 nm FE films. These results offer mechanistic insights into the synergistic interplay between defect dynamics and phase transitions, providing theoretical guidance for the performance optimization of ultra-thin HZO-based FE devices.

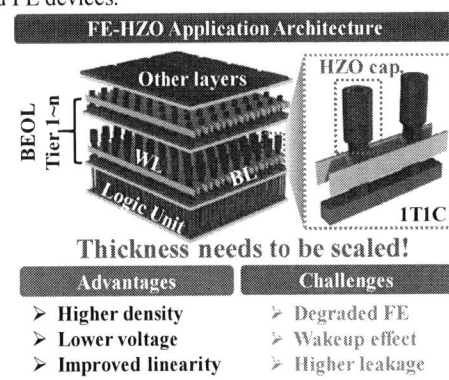

Fig. 1. Advantages and challenges in scaling HZO FE film thickness.

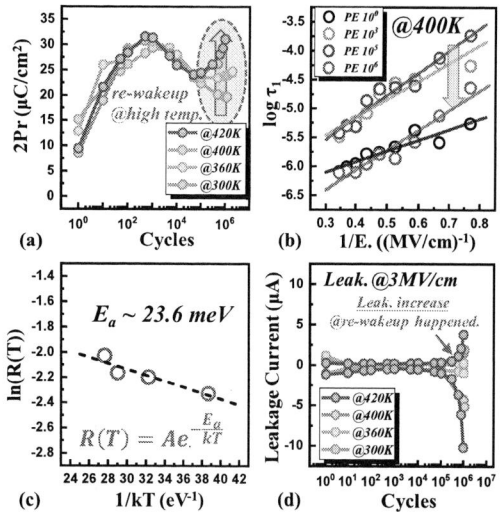

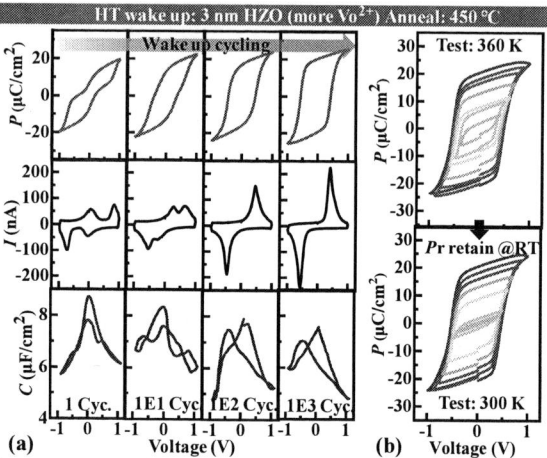

Fig. 4. (a) Evolution of PV, IV, and CV loops during HTEC for 3nm HZO with more V_O/V_O^{2+}, suggesting the potential t-to-o phase transition. (b) PV loops after HTEC, achieving $2Pr > 40\ \mu C/cm^2$ at 1V, which are well retained upon returning to room temperature [16].

Fig. 2. (a) Cycling behaviors of 2Pr at different temperatures. (b) log τ_1 versus 1/E at 400K. (c) The plot of the temperature-dependent Arrhenius function and Ea extraction. (d) Leakage current versus cycling [12].

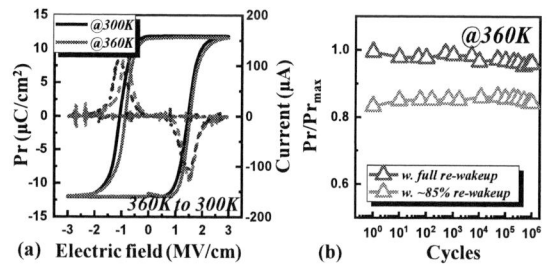

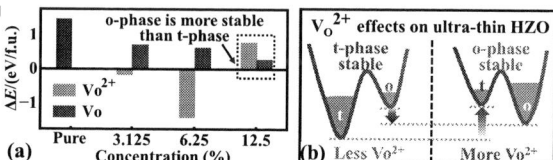

Fig. 5. (a) Simulations reveal the phase engineering effects of V_O and V_O^{2+}. (b) Unveiled effects of increased V_O^{2+} on stabilizing the o-phase over the t-phase in the ultra-thin FE film, facilitating t-to-o phase transition [16].

Fig. 3. (a) PUND read after re-wakeup at 360K and back to 300K. (b) Pr stability at different re-wakeup degrees [12].

II. DEFECT DYNAMICS AND PHASE TRANSITIONS AT HIGH-TEMPERATURE

High-temperature reliability testing not only facilitates accelerated evaluation of device performance but also serves as a powerful tool for uncovering defect- and phase-related mechanisms that are otherwise difficult to observe at room temperature (RT). During high-temperature experiments, a distinct re-wakeup phenomenon is observed during thermal cycling, as illustrated in **Fig. 2(a)** [12]. Unlike conventional fatigue behavior characterized by a gradual decline in remanent polarization (Pr), Pr in this case unexpectedly recovers, reaching levels comparable to the wakeup stage. Detailed defect-related analyses provide further insights into this behavior: the polarization switching speed increases during re-wakeup (**Fig. 2(b)**); the thermal activation energy during the fatigue stage approximates that associated with FE domain pinning (**Fig. 2(c)**); and the leakage current increases during re-wakeup (**Fig. 2(d)**). Given the critical role of defects in polarization switching dynamics, domain pinning, and trap-assisted tunneling (TAT) [13-14], these results reveal a complex interplay between FE domain pinning and de-pinning effects driven by leakage currents at grain boundary. Specifically, the field-cycling-dependent defect generation and FE domain pinning at grain boundaries

contribute to Pr degradation and reduced switching speed. However, faster defect generation at elevated temperatures increases leakage current via TAT and enhances electron injection, which promotes FE domain de-pinning, thereby enabling recovery of Pr and switching speed. This dynamic competition between defect trapping-induced pinning and injection-driven de-pinning provides a plausible mechanism for the observed re-wakeup behavior. As shown in **Fig. 3**, considering the potential of the re-wakeup phenomenon to extend device lifetime, we further design experiments to evaluate its stability and repeatability. The results indicate that the recovered Pr remains stable even after the device returns to RT. Moreover, the device exhibits significantly improved fatigue resistance in subsequent cycling.

Building on these findings, and considering that ultra-thin HZO films typically suffer from severe initial FE degradation and wakeup difficulty due to an increased tetragonal (t-) phase fraction [15], we propose a high-temperature electrical cycling (HTEC) scheme to enhance Pr in 3nm HZO films. As shown in **Fig. 4**, the evolution of FE characteristics (P-V, I-V, and C-V) under HTEC, along with the stable retention of the enhanced Pr after returning to RT, demonstrates effective wakeup and robust Pr stabilization in ultra-thin HZO [16]. Given that 3 nm HZO films exhibit pronounced anti-ferroelectric behavior at the initial stage, the observed wakeup cannot be solely attributed to defect dynamics. To gain further insight, we perform first-principles calculations to investigate the effect of oxygen vacancy (V_O/V_O^{2+}) on the t-to-orthorhombic (o-) phase transition. The results reveal that an increased concentration of V_O^{2+} significantly raises the free

979-8-3315-3918-4/25 $31.00 © 2025 IEEE

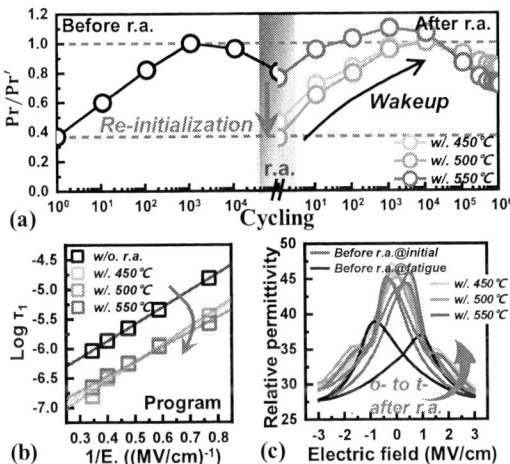

Fig. 6. Cycling behaviors of normalized Pr with re-annealing at different temperatures after cycling. (b) Effect of re-annealing on the polarization switching speed. (d) C-V loops before and after re-annealing [21].

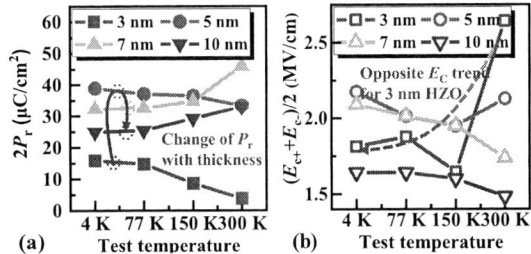

Fig. 7. Summary of (a) 2Pr vs. Temperature and (b) Ec vs. Temperature for various HZO thicknesses, where 3/5nm HZO shows a distinct trend over 7/10nm HZO [24].

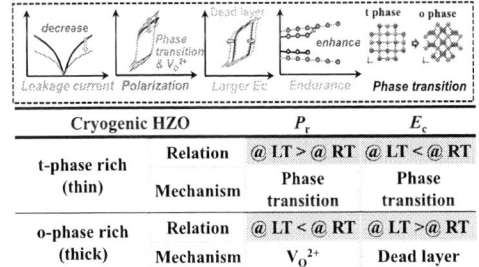

Cryogenic HZO		P_r	E_c
t-phase rich (thin)	Relation	@ LT > @ RT	@ LT < @ RT
	Mechanism	Phase transition	Phase transition
o-phase rich (thick)	Relation	@ LT < @ RT	@ LT > @ RT
	Mechanism	V_O^{2+}	Dead layer

Fig. 8. Summary of behaviors and their mechanisms of HZO at cryogenic temperature [24].

energy of the t-phase while reducing that of the o-phase, thus stabilizing the o-phase and facilitating the electric-field-induced t-to-o phase transition (**Fig. 5**). The HTEC scheme effectively increases the concentration of Vo^{2+} at elevated temperatures, thereby enabling this phase transformation.

Moreover, Vo generated by oxygen bond breakage during field cycling is known to be a major contributor to fatigue, breakdown, and other reliability concerns in HfO_2-based FE thin films [17]. To address such defect-related degradation, strategies such as high-temperature annealing, thermal cycling, and self-heating have been widely validated in other memory systems, particularly Flash memory [18-19]. However, in HfO_2-based ferroelectrics, high-temperature treatments have primarily been employed to assess device robustness or enhance ferroelectricity, with limited exploration of their restorative effects and underlying mechanisms [20]. To bridge this gap, we investigate the effect of high-temperature re-annealing at various cycling stages (initial, wakeup, and fatigue) on the reliability of sub-10nm HZO FE films [21]. It is found that FE properties (Pr, coercive electric field (Ec), polarization switching speed, wakeup/fatigue effect, and symmetry) can be improved after re-annealing, all of which are critical for FERAM applications (**Fig. 6**). Notably, we observe a re-initialization phenomenon, where post-cycling re-annealing restores device performance to the initial stage. Considering the polycrystalline nature of HZO film, thermodynamic stability of various phases (e.g., higher stability of the t-phase at elevated temperatures), Curie temperature, and effectiveness of defect repair, it is considered that temperature-dependent phase transition and non-switchable region repairing could be the dominant mechanisms [22-23]. These results indicate that re-annealing could effectively improve the performance of HZO FE devices and shed light on reliability optimizations.

III. DEFECT DYNAMICS AND PHASE TRANSITIONS AT LOW-TEMPERATURE

Additionally, the rapid advancement of emerging technologies such as data centers and quantum computing has introduced exciting new opportunities for FE applications under cryogenic conditions. This development necessitates a

more comprehensive understanding of FE behavior at low-temperature. To address this need, we systematically evaluate the low-temperature reliability of sub-10nm HZO capacitors with thicknesses of 3, 5, 7, and 10nm [24]. Notably, the 3 and 5nm films exhibit markedly different temperature-dependent trends in Pr and Ec compared to the 7 and 10nm counterparts, as shown in **Fig. 7**. Specifically, as temperature decreases, the 3/5nm HZO films demonstrate enhanced initial Pr and reduced Ec, whereas the 7/10nm films show the opposite behavior. To elucidate these observations, we analyze the temperature-dependent Gibbs free energy of the t-/o-phase and consider the influence of the dead layer. The analysis reveals distinct thickness-dependent dominant physical mechanisms in HZO films operating under cryogenic conditions. In t-phase-dominated films (3/5nm), the increase in initial Pr is attributed to a temperature-assisted t-to-o phase transition, while the reduction in Ec is linked to a decreased dielectric constant of the HZO film. Conversely, in o-phase-dominated films (7/10nm), the decline in Pr is primarily caused by Vo-induced FE domain pinning, and the increase in Ec results from a reduced probability of TAT, as summarized in **Fig. 8**.

IV. SUMMARY

This work reviews unique reliability behaviors of HZO FE films at the sub-10nm regime. By combining wide-temperature-range (4K-420K) reliability characterization with first-principles calculations, the temperature-dependent phase transition mechanisms and defect dynamics that govern device performance are investigated systematically. A series of defect-phase co-optimization strategies is proposed for different thermal conditions, effectively addressing issues such as wakeup and fatigue. These findings deepen the understanding of thickness-dependent reliability in HZO films and provide valuable guidance for process optimization and robust device design under extreme environmental conditions.

ACKNOWLEDGMENT

This work was supported by China Key Research and Development Program under Grant (2023YFB4402500, 2023YFB4402400), National Natural Science Foundation of China (Nos. 62034006, 92264201, U23B2040, 624B2090), Natural Science Foundation of Shandong Province (ZR2023LZH007, ZR2023QF054), China Postdoctoral Science Foundation (GZC20231435), TaiShan Scholars (TSQN202306059), and MIND project (MINDXZ202407).

REFERENCES

[1] T. Mikolajick, U. Schroeder, P. D. Lomenzo, S. Slesazeck, and S. Lancaster, "Ferroelectric Materials and Their Applications for Next-Generation Integrated Devices," in 2024 IEEE International Electron Devices Meeting (IEDM), 2024, IEEE, pp. 1-4.

[2] U. Schroeder, M. H. Park, T. Mikolajick, and C. S. Hwang, "The fundamentals and applications of ferroelectric HfO2," Nature Reviews Materials, vol. 7, no. 8, pp. 653-669, 2022.

[3] S. Yoon et al., "QLC programmable 3D ferroelectric NAND Flash memory by memory window expansion using cell stack engineering," in 2023 IEEE Symposium on VLSI Technology and Circuits (VLSI Technology and Circuits), 2023, IEEE, pp. 1-2.

[4] N. Ramaswamy et al., "NVDRAM: A 32Gb Dual Layer 3D Stacked Non-volatile Ferroelectric Memory with Near-DRAM Performance for Demanding AI Workloads," in 2023 International Electron Devices Meeting (IEDM), 2023, IEEE, pp. 1-4.

[5] A. Calderoni et al., " Voltage Reduction (1.4V) and Array Scaling (41nm) of Ferroelectric NVDRAM for Low-Power and High-Density Applications," in 2025 IEEE Symposium on VLSI Technology and Circuits (VLSI Technology and Circuits), 2025, IEEE, pp. 1-3.

[6] M. Lee et al., "BEOL Compatible Ultra-Low Operating Voltage (0.5 V) and Preconfigured Switching Polarization States in Effective 3 nm Ferroelectric HZO Capacitors," in 2024 IEEE Symposium on VLSI Technology and Circuits (VLSI Technology and Circuits), 2024, IEEE, pp. 1-2.

[7] K. Tahara et al., "Strategy toward HZO BEOL-FeRAM with low-voltage operation ($\leq$ 1.2 V), low process temperature, and high endurance by thickness scaling," in 2021 Symposium on VLSI Technology, 2021, IEEE, pp. 1-2.

[8] F. Huang et al., "Dimensional Scaling of Ferroelectric Properties of Hafnia-Zirconia Thin Films: Electrode Interface Effects," ACS Nano, vol. 18, no. 27, pp. 17600-17610, 2024.

[9] K. Ito et al., " Revealing wake-up mechanism in ultra-thin ferroelectric HZO: Domain de-pinning triggered by oxygen vacancy annihilation exhibiting optimal wake-up frequency," in 2025 IEEE Symposium on VLSI Technology and Circuits (VLSI Technology and Circuits), 2025, IEEE, pp. 1-3.

[10] J. Lyu, T. Song, I. Fina, and F. Sánchez, "High polarization, endurance and retention in sub-5 nm Hf 0.5 Zr 0.5 O 2 films," Nanoscale, vol. 12, no. 20, pp. 11280-11287, 2020.

[11] F. Huang et al., "First Observation of Ultra-high Polarization (~ 108 μC/cm²) in Nanometer Scaled High Performance Ferroelectric HZO Capacitors with Mo Electrodes," in 2023 IEEE Symposium on VLSI Technology and Circuits (VLSI Technology and Circuits), 2023, IEEE, pp. 1-2.

[12] X. Li et al., "Temperature-dependent Defect Behaviors in Ferroelectric Hf 0.5 Zr 0.5 O 2 Thin Film: Re-wakeup Phenomenon and Underlying Mechanisms," in 2022 International Electron Devices Meeting (IEDM), 2022, IEEE, pp. 32.3. 1-32.3. 4.

[13] X. Li et al., "Effect of ALD temperature on the polarization switching dynamics of BEOL-compatible Hf0. 5Zr0. 5O2 ferroelectric film," Appl. Phys. Lett., vol. 126, no. 15, 2025.

[14] M. H. Park, Y. H. Lee, T. Mikolajick, U. Schroeder, and C. S. Hwang, "Review and perspective on ferroelectric HfO2-based thin films for memory applications," MRS Commun., vol. 8, no. 3, pp. 795-808, 2018.

[15] C. Wang et al., "Evolution of pronounced ferroelectricity in Hf 0.5 Zr 0.5 O 2 thin films scaled down to 3 nm," J. Mater. Chem. C, vol. 9, no. 37, pp. 12759-12767, 2021.

[16] Y. Feng et al., "Record-high Pr (2Pr > 40 μC/cm²) in 3 nm (Physical) Ferroelectric HZO Annealed at 450 °C: High-T (85 °C) Electrical Cycling and Oxygen Vacancy Engineering," in 2025 IEEE Symposium on VLSI Technology and Circuits (VLSI Technology and Circuits), 2025, IEEE, pp. 1-3.

[17] Y. Jeong et al., "Oxygen Vacancy Control as a Strategy to Enhance Imprinting Effect in Hafnia Ferroelectric Devices," IEEE T. Electron Dev., vol. 70, no. 1, pp. 354-359, 2023. doi:10.1109/TED.2022.3223886.

[18] H. Lue et al., "Radically extending the cycling endurance of Flash memory (to> 100M Cycles) by using built-in thermal annealing to self-heal the stress-induced damage," in 2012 International Electron Devices Meeting, 2012, IEEE, pp. 9.1. 1-9.1. 4.

[19] J. Han, M. Kebaili, and M. Meyyappan, "System on microheater for on-chip annealing of defects generated by hot-carrier injection, bias temperature instability, and ionizing radiation," IEEE Electr. Device L., vol. 37, no. 12, pp. 1543-1546, 2016.

[20] N. Tasneem, Z. Wang, H. Chen, S. Yu, W. Chern, and A. Khan, "Immediate Read-After-Write Capability in p-Type Ferroelectric Field-Effect Transistors and Its Evolution With Fatigue Cycling," IEEE T. Device Mat. Re., vol. 23, no. 1, pp. 142-146, 2023.

[21] X. Li et al., "Re-Annealing-Induced Recovery in 7nm Hf 0.5 Zr 0.5 O 2 Ferroelectric Film: Phase Transition and Non-Switchable Region Repair," IEEE Electr. Device L., vol. 44, no. 8, pp. 1288-1291, 2023.

[22] T. Xin et al., "Atomic visualization of the emergence of orthorhombic phase in Hf 0.5 Zr 0.5 O 2 ferroelectric film with in-situ rapid thermal annealing," in 2022 IEEE Symposium on VLSI Technology and Circuits (VLSI Technology and Circuits), 2022, IEEE, pp. 343-344.

[23] Y. Zheng et al., "In-situ atomic visualization of structural transformation in Hf 0.5 Zr 0.5 O 2 ferroelectric thin film: from nonpolar tetragonal phase to polar orthorhombic phase," in 2021 Symposium on VLSI Technology, 2021, IEEE, pp. 1-2.

[24] D. Zhang et al., "Unveiling cryogenic performance (4 to 300 K) towards ultra-thin ferroelectric HZO: Novel kinetic barrier engineering and underlying mechanism," in 2024 IEEE Symposium on VLSI Technology and Circuits (VLSI Technology and Circuits), 2024, IEEE, pp. 1-2.

High Dielectric Permittivity in Size-scaled $Hf_{0.5}Zr_{0.5}O_2$ Thin-film Capacitors

Wendi Zhang, Anquan Jiang*

College of Integrated Circuits and Micro-Nano Electronics Innovation, Fudan University, Shanghai 200433, China
* Email: 19212020017@fudan.edu.cn, aqjiang@fudan.edu.cn

Abstract—**Hafnium oxide (HfO_2) materials have been extensively studied owing to the advantages of good scalability and compatibility to the CMOS fabrication processes. However, there are several challenges including a high coercive voltage and a low dielectric permittivity for applications in nonvolatile/volatile memories. In this work, we demonstrated the size-scaled $TiN/Hf_{0.5}Zr_{0.5}O_2/TiN$ capacitors in reduced coercive voltages and low leakage current densities. When the lateral dimension of the capacitor is reduced from 50 to 0.3 μm, the dielectric permittivity enhances from 32 to 74 in a low dielectric loss of 0.009 at 1 kHz without the occurrence of the ferroelectric-to-nonferroelectric transition. Synchrotron X-ray diffraction patterns show the dependence of the enhancing dielectric permittivity on the increasing concentration of an orthorhombic phase, showing exceptional prospects for DRAM/CMOS implementations.**

Keywords—*$Hf_{0.5}Zr_{0.5}O_2$, Ferroelectric thin films, High dielectric permittivity, Size scaling effect, Coercive field reduction*

I. INTRODUCTION

Ferroelectric materials are highly promising due to their large spontaneous polarizations and high domain switching speeds, enabling their use as both logic and memory devices in low power consumption [1, 2]. However, as the devices are scaled down to the nanoscale dimension, maintaining the ferroelectricity of traditional perovskite-structure materials becomes very challenging. Doped hafnium oxide (HfO_2) offers the potential for sub-10 nm scaling using CMOS-compatible processes like atomic layer deposition (ALD) and sputtering [3-5]. Notably, the discovery of ferroelectricity in this material has reignited interest in ferroelectric random-access memory (FeRAM) and high-storage-density dynamic random-access memory (DRAM). However, its high coercive field (> 1 MV/cm) and low dielectric permittivity (~32) pose challenges for achieving ultrafast domain switching speeds at low operating voltages in FeRAMs and large charge storage densities in DRAMs [6, 7]. To overcome these obstacles, several methods have been proposed, including the reduction of HfO_2 film thicknesses [8], the optimization of the ferroelectric (FE) layer deposition techniques [9], and the insertion of interfacial layers between HfO_2 and electrodes in high-k values [10].

There are various approaches to enhance the dielectric permittivity in hafnium-based materials: 1) doping with multiple elements (e.g., La, Zr, Al) [11], 2) utilizing field cycling effects [12], 3) designing nanolaminates and superlattices [13, 14], and 4) leveraging phase compositions near morphotropic phase boundary (MPB) [7]. Recent studies on Zr-doped hafnium-based micro-capacitors after near-edge ion implantation shows an ultrahigh dielectric

permittivity of 921 after the occurrence of a ferroelectric-to-nonferroelectric transition under bipolar high electric-field cycling [15, 16], attractive for the realization of high charge- and energy-storage devices and energy-efficient transistors. Consequent high-resolution Scanning Transmission Electron Microscopy (STEM) images showed ordered oxygen vacancies to occur within the distorted orthorhombic grains that increase the leakage current density on the order of 6.98×10^{-5} A/cm^2 at 1 MV/cm [17]. For most DRAMs, it is expected that the leakage density can be reduced to the level of 1.8×10^{-7} A/cm^2 of a ferroelectric capacitor [17].

Here we show the possibility of the dielectric enhancement to occur in the size-scaled ferroelectric capacitors without the requirement of a ferroelectric-to-nonferroelectric transition. The reduction of coercive voltages and enhancement of dielectric permittivity could occur in crystallized $Hf_{0.5}Zr_{0.5}O_2$ (HZO) thin-film capacitors upon downscaling of their lateral capacitor sizes from 30 to 0.3 μm when their remanent polarizations keep constant. The strengthened dielectric response is seemly correlated with the increased orthorhombic phase and can be described by a universal law.

II. DEVICE FABRICATION AND STRUCTURAL ANALYSIS

A. Device Fabrication

The 10-nm-thick HZO ferroelectric films were deposited on TiN-coated silicon substrates, where the HZO films were formed by ALD (TFS 200, Beneq) at 200 °C using hafnium tetrachloride ($HfCl_4$), zirconium tetrachloride ($ZrCl_4$), and water as precursors. Subsequently, a 5-nm-thick TiN layer and a 50-nm-thick tungsten (W) layer were sequentially deposited by physical vapor deposition (PVD-75, Kurt J. Lesker) to cover the ferroelectric layer. Then, the TiN-capped films underwent rapid thermal annealing at 550°C for 30 seconds in N_2 atmosphere to induce the ferroelectric phase. Finally, the top electrodes were patterned using electron-beam lithography (JEOL 6300FS) and dry etching system (RIE-10NR, Samco, Japan). The square capacitors with side lengths (l) of 0.3-50 μm were fabricated.

B. Structural Analysis

The capacitor side lengths were individually verified using planar-view scanning electron microscope (SEM; Sigma HD, Zeiss) images. Figure 1(a) shows the SEM image of the arrays of HZO capacitors in side lengths of 500 nm and periodicities of 1 μm. The cross-sectional STEM image in Fig. 1(b) revealed the HZO thin-film thickness of 10 nm where the atomic sharpness of distinct interfaces near electrodes was observed. The phase compositions of the

979-8-3315-3918-4/25 $31.00 © 2025 IEEE

size-scaled capacitors were analyzed through the X-ray diffraction patterns (XRD) that were acquired using synchrotron radiation with an incident wavelength of 1.2398 Å.

C. Electrical Characterization

For the measurements of domain switching transient currents, bipolar square voltage pulses were applied using an Agilent 81110A two-channel pulse generator. Following the application of negative/positive poling voltages to the top electrodes of HZO capacitors with bottom electrodes grounded, domain switching/non-switching currents were monitored across an in-series resistor (R) using a LeCroy HDO6054 oscilloscope (USA) featuring 12-bit voltage resolution and 1 GHz bandwidth. The current transients can be transformed into polarization-voltage (P-V) hysteresis loops [15]. Figs. 1(c) and (d) show the current transients versus time of a TiN/HZO/TiN capacitor under different applied voltages. The difference in ferroelectric domain switching and nonswitching currents using opposite presetting pulses confirms the appearance of the ferroelectricity in the HZO thin film.

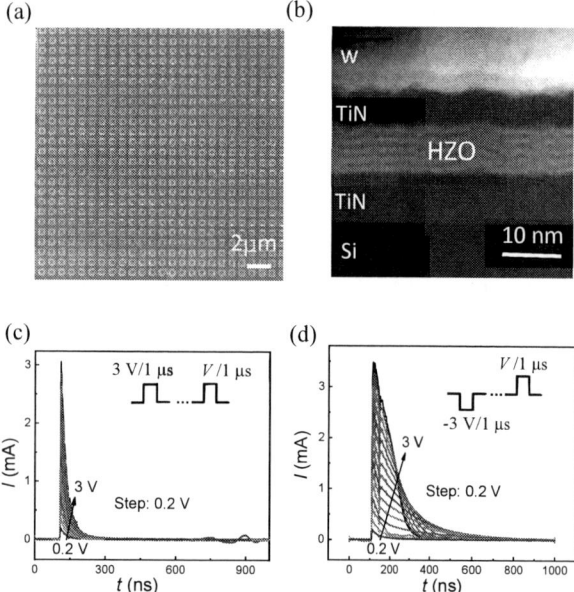

Fig. 1. (a) The planar SEM image of the arrayed HZO capacitors in side lengths of 500 nm and periodicities of 1 μm. (b) The cross-sectional TEM photograph of the TiN/HZO/TiN thin-film capacitor. (c) Nonswitching/Switching current transients versus time with increasing V from 0.2 to 3 V in steps of 0.2 V for a ferroelectric capacitor in the diameter of 30 μm using the pulses sketched in the insets.

III. RESULTS AND DISCUSSION

Figs. 2(a) and (b) show the frequency-dependent P-V hysteresis loops measured on etched and unetched HZO capacitors when $l = 50$ μm. The coercive voltage (V_c) in the former capacitor gradually increases with enhancing frequency, whereas for the unetched capacitors in Fig. 2(b) V_c is comparably lower at high frequencies. Figs. 2(c) and (d) show the square root of frequency dependences of coercive voltages for the etched/unetched capacitors when $l = 50$ μm and 30 μm, respectively. The quick increase of V_c in high frequencies could be due to the partial oxidation of

TiN top electrode in contact with HZO during thermal crystallization that increases the electrode resistance (R_i) [18]. R_i can increase V_c in the form of [15]:

$$V_c^{(a)} = V_c + \sqrt{8R_i P_r S V_m f} \ . \tag{1}$$

The solid lines in Figs. 2(c) and (d) are the best fits of the data according to Eq. (1). From the slopes of the solid lines, we found that R_i of the etched capacitor is two-to-three times that of the unetched capacitor.

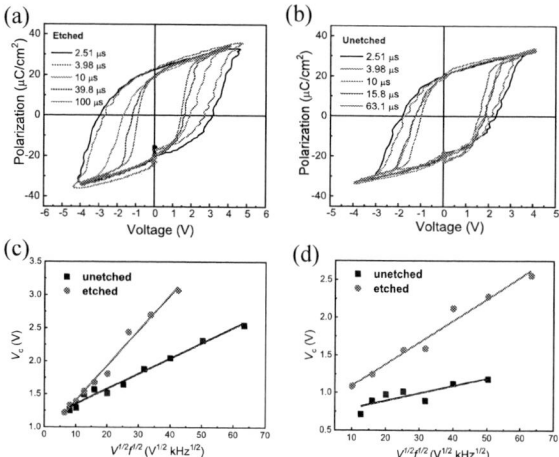

Fig. 2. (a, b) P-V hysteresis loops at various periodicities for etched/unetched HZO capacitors in side lengths of 50 μm. (c, d) Corresponding V_c-$f^{1/2}$ plots of etched/unetched capacitors when $l = 50$ μm and 30 μm, respectively. The solid lines are the best fits of the data according to Eq. (1).

Fig. 3(a) shows the P-V hysteresis loops of the unetched HZO capacitors with dimensions ranging from 50 to 0.3 μm. All loops are asymmetric along the voltage axis. This could be attributed to the non-uniform electric field distribution induced by injected charges within the thin films. However, after the wakeup treatments at bipolar voltages of +/−4 V for cycling numbers of 10^5, all loops become more symmetric after the erasure of internally trapped charges, as demonstrated in Fig. 3(b). Fig. 3(c) and (d) show the frequency dependences of their dielectric permittivities (ε) and loss tangents (tanδ). ε exhibits a significant increase from 32 to 74 at 1 MHz upon the lateral shrinkage of l from 50 to 0.3 μm with loss tangents lower than 0.01 at 1 kHz. In contrast, their remanent polarizations keep nearly constant between 24 and 26 μC/cm^2, as inferred from P-V loops in Fig. 3(b). The dielectric dispersion behavior can be described by a universal law expressed by the following equation [15]:

$$\varepsilon(\omega) \propto (i\omega)^{n-1} , \tag{2}$$

where $0 < n \leq 1$. Based on the equivalent circuit model incorporating the electrode resistance R_i in series with an ideal capacitor, we derived [17]:

$$\tan\delta = \omega\tau' + \text{ctan}\frac{n\pi}{2} , \tag{3}$$

where $\tau' = CR_i$ and C is the capacitance. The solid lines in Figs. 3(c) and (d) are the best fits of the data. From these fits,

979-8-3315-3918-4/25 $31.00 © 2025 IEEE

we derived $n = 0.90$ and $R_i = 0.76$–0.98 kΩ, in agreement with those in the previous reports [15, 17].

To further elucidate the microscopic mechanism of V_c and ε enhancements in the size-scaled capacitors, Fig. 4(a) shows the synchrotron XRD patterns of the etched HZO capacitors in various side lengths. The continuous HZO film without etching shows overlapping of orthorhombic O (111) and tetragonal T (011) reflections with the area ratio of O:T $= 0.77:0.23$, as determined from Gauss functional fittings of all peaks. As the capacitor size is progressively decreased, the peak intensity of the O phase increases in contrast to the weakening of the T phase. When $l = 0.3$ μm, the T (011) reflection disappears. It seems that the mixed T phase could reduce V_c in the continuous film in Fig. 2(b).

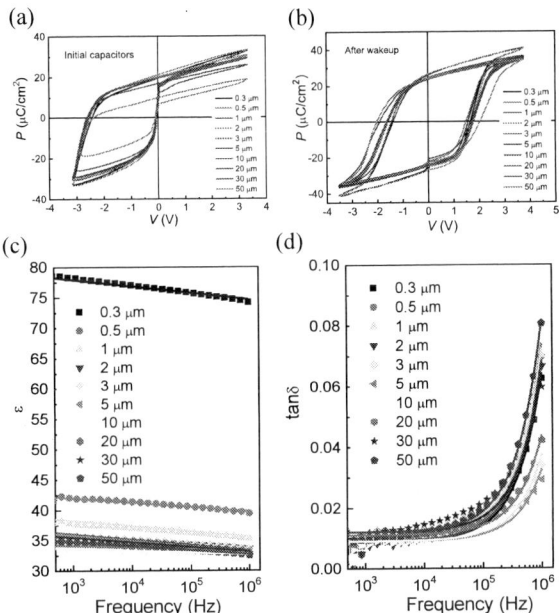

Fig. 3. (a) Frequency-dependent D-E hysteresis loops at 1 MHz for the initial HZO capacitors. (b) P-V hysteresis loops after wakeup treatments of the size-scaled capacitors in cycling numbers of 10^5 under applied bipolar voltages of $+/-4$ V in a repeat frequency of 1 MHz. (c, d) Frequency dependences of ε and tanδ for the size-scaled capacitors.

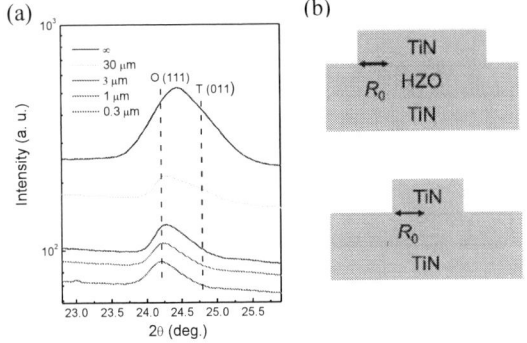

Fig. 4. (a) The XRD patterns of the size-scaled HZO capacitors. (b) Schematic diagram of high-ε phases near edging areas of top electrodes.

To understand the size scaling effect on ε within a ferroelectric capacitor, we proposed a large amount of oxygen vacancies to appear near the edging region of a

crystalline TiN/HZO/TiN thin-film capacitor due to the etching damage on HZO near the periphery of the top electrode. These oxygen vacancies ($V_O^{\cdot\cdot}$) can diffuse into the inner region in a penetration length of R_0 beneath the top electrode area that increases ε [15]. Upon the shrinkage of top electrode size, the area fraction of the peripheral damaging region increases, as schematically illustrated in Fig. 4(b), so that the overall dielectric permittivity becomes much higher for a smaller capacitor in Fig. 3(c). This dielectric enhancement doesn't require the ferroelectric-to-nonferroelectric transition that suppresses the leakage current [17].

IV. CONCLUSION

We fabricated etched and unetched $Hf_{0.5}Zr_{0.5}O_2$ thin-film capacitors. For the continuous HZO thin film without etching, its coercive voltage is much lower, especially at high frequencies. This may be attributed to the reduced resistance of the TiN electrode in contact to HZO. For the size-scaled capacitors after etching of top electrodes, a significant enhancement in dielectric permittivity was observed without change of the remanent polarization. The dielectric permittivity can highly reach 74 when $l = 0.3$ μm, and the dielectric response can be described by a universal law. Finally, the phase structures of the etched HZO capacitors were analyzed from the synchrotron XRD patterns. With the shrinkage of the capacitor size, it is found that of the O-phase concentration increases in contrast to the reduction of the T phase. The higher dielectric permittivity in a smaller capacitor is believed to originate from the increasing contribution of the peripheral region under the top electrodes. This dielectric enhancement without the requirement of the ferroelectric-to-nonferroelectric transition could suppress the large leakage current that invoked in ion-implanting HZO capacitors [15, 17], providing critical guidance for boosting low-energy efficient DRAM and CMOS devices[19].

ACKNOWLEDGMENT

One of us (A. Q. J.) thanks the financial supports by the National Key Research and Development Program of China (grant number 2024YFA1409500), Shanghai Science and Technology Innovation Action Plan (grant number 24CL2900900), and the National Natural Science Foundation of China (grant number 62174034).

REFERENCES

[1] B. H. Kim, S. H. Kuk, S. K. Kim, J. P. Kim, Y. J. Suh, et al., "Low Operating Voltage and Immediate Read-After-Write of HZO-Based Si Ferroelectric Field-Effect Transistors with High Endurance and Retention Characteristics," Adv. Electron. Mater., vol. 10, art no. 2300327, January 2024.

[2] M. H. Park, Y. H. Lee, H. J. Kim, Y. J. Kim, T. Moon, et al., "Ferroelectricity and antiferroelectricity of doped thin HfO_2-based films," Adv. Mater., vol. 27, pp. 1811-1831, March 2015.

[3] T. S. Böscke, J. Müller, D. Braeuhaus, U. Schröder, U. J. Böttger, "Ferroelectricity in hafnium oxide thin films," Appl. Phys. Lett., vol. 99, art no. 102903, September 2011.

[4] B. Y. Kim, B. S. Kim, S. D. Hyun, H. H. Kim, Y. B. Lee, et al., "Study of ferroelectric characteristics of $Hf_{0.5}Zr_{0.5}O_2$ thin films grown on sputtered or atomic-layer-deposited TiN bottom electrodes," Appl. Phys. Lett., vol. 117, art no. 022902, July 2020.

[5] P. Nukala, M. Ahmadi, Y. Wei, S. De Graaf, E. Stylianidis, et al., "Reversible oxygen migration and phase transitions in hafnia-based ferroelectric devices," Science, vol. 372, pp. 630-635, May 2021.

[6] C. Alessandri, P. Pandey, A. Abusleme, A. Seabaugh, "Switching dynamics of ferroelectric Zr-doped HfO$_2$," IEEE Electron Device Lett., vol. 39, pp. 1780-1783, September 2018.

[7] S. Oh, H. Jang, H. Hwang, "Accurate Evaluation of High-k HZO/ZrO$_2$ Films by Morphotropic Phase Boundary," IEEE Electron Device Lett., vol. 45, pp. 28-31, November 2023.

[8] K. Tahara, K. Toprasertpong, Y. Hikosaka, K. Nakamura, H. Saito, et al., "Strategy toward HZO BEOL-FeRAM with low-voltage operation ($\leq$ 1.2 V), low process temperature, and high endurance by thickness scaling," In2021 Symposium on VLSI Technology, vol. 13, pp. 1-2, June 2021.

[9] E. Yu, X. Lyu, M. Si, P. D. Ye, K. Roy, "Interfacial layer engineering in sub-5-nm HZO: Enabling low-temperature process, low-voltage operation, and high robustness," IEEE T. Electron Dev., vol. 70, pp. 2962-2969, May 2023.

[10] C. Y. Chan, K. Y. Chen, H. K. Peng, Y. H. Wu, "FeFET memory featuring large memory window and robust endurance of long-pulse cycling by interface engineering using high-k AlON," In2020 IEEE Symposium on VLSI Technology, vol. 16, pp. 1-2, June 2020.

[11] A. Kashir, M. G. Farahani, H, "Hwang. Towards an ideal high-κ HfO$_2$-ZrO$_2$-based dielectric," Nanoscale, vol. 13, pp. 13631-13640, June 2021.

[12] S. Kim, S. H. Lee, M. J. Kim, W. S. Hwang, H. Soo, et al., "Method to achieve the morphotropic phase boundary in Hf$_x$Zr$_{1-x}$O$_2$ by electric field cycling for DRAM cell capacitor applications," IEEE Electron Device Lett., vol. 42, pp. 517-520, February 2021.

[13] A. Kashir, M. Ghiasabadi Farahani, S. Kamba, M. Yadav, H. Hwang, "Hf$_{1-x}$Zr$_x$O$_2$/ZrO$_2$ Nanolaminate Thin Films as a High-κ Dielectric," ACS Appl. Electron. Mater., vol. 3, pp. 5632-5640, December 2021.

[14] S. S. Cheema, N. Shanker, L. C. Wang, C. H. Hsu, S. L. Hsu, et al., "Ultrathin ferroic HfO$_2$-ZrO$_2$ superlattice gate stack for advanced transistors," Nature, vol. 604, pp. 65-71, April 2022.

[15] W. D. Zhang, Z. Z. Song, S. Q. Tang, J. C. Wei, Y. Cheng, et al., "Ultrahigh dielectric permittivity in Hf$_{0.5}$Zr$_{0.5}$O$_2$ thin-film capacitors," Nat. Commun., vol. 16, art no. 2679, March 2025.

[16] A. K. Jonscher, "Dielectric Relaxation in Solids," Chelsea Dielectrics Press, pp. 62-101, 1983.

[17] W. D. Zhang, B. Li, W. W. Wang, X. Y. Wang, Y. Cheng, and A. Q. Jiang, "Ultrahigh dielectric permittivity of a micron-sized Hf$_{0.5}$Zr$_{0.5}$O$_2$ thin-film capacitor after missing of a mixed tetragonal phase," Nano-Micro Lett. vol. 18, art no.6, July 2025.

[18] W. Hamouda, A. Pancotti, C. Lubin, L. Tortech, C. Richter, et al., "Physical chemistry of the TiN/Hf$_{0.5}$Zr$_{0.5}$O$_2$ interface," J. Appl. Phys., vol. 127, art no. 064105, February 2020.

[19] C. H. Cheng, A. Chin, "Low-leakage-current DRAM-like memory using a one-transistor ferroelectric MOSFET with a Hf-based gate dielectric," IEEE Electron Device Lett., vol. 35, pp. 138-140, December 2013.

1.2 V Operation of $Hf_{0.5}Zr_{0.5}O_2$ Ferroelectric Thin-film Capacitors for Low-Power ASICs after Interfacial-layer Engineering

Xianyu Hu [1], Di Hu [1], Wendi Zhang [1], Bowen Shen*[2], Anquan Jiang*[1]

[1] College of Integrated Circuits and Micro-Nano Electronics Innovation, Fudan University, Shanghai 200433, China
[2] School of Integrated Circuits, Tsinghua University, 100084, Beijing, China
* Email: 23112020073@m.fudan.edu.cn, hud21@m.fudan.edu.cn, 19212020017@fudan.edu.cn,
shenbowen@tsinghua.edu.cn, aqjiang@fudan.edu.cn

Abstract—**This paper demonstrates a 1.2V operation of $Hf_{0.5}Zr_{0.5}O_2$ (HZO) ferroelectric thin-film capacitors for low-power ASICs by optimizing interfacial-layer resistivity. Thermal annealing at 800°C during HZO fabrication reduces the interfacial layer resistivity by a factor of 2.6 while maintaining the intrinsic coercive field at ~0.92 MV/cm. The Positive-Up-Negative-Down characterization showed stable polarization switching at 1.2 V, significantly lower than those of conventional HZO devices. The calculation from the frequency dependence of coercive field indicates the fundamental physics related to the lower interfacial-layer resistance that minimizes energy loss and enhances domain switching speed, making it promising for energy-efficient ferroelectric memory. Our work addresses the high-voltage operation limit in hafnia-based ferroelectrics, offering a practical solution for next-generation low-power integrated circuits.**

Keywords—$Hf_{0.5}Zr_{0.5}O_2$ (HZO), Ferroelectric capacitors, Low-power ASICs, Interfacial-layer resistivity, Thermal annealing, Coercive field reduction

I. INTRODUCTION

A nonvolatile ferroelectric random-access memory (FeRAM) is promising for the high-speed operation and low energy consumption[1, 2]. While commercial lead zirconate titanate (PZT) FeRAM has the low coercive field (~0.2 MV/cm) and high endurance (>10^{14} cycles), its thickness ranges from 80 to 200 nm and processing temperatures are over 650°C, which hinder compatibility with the BEOL on the basis of scaled silicon technologies[3]. In contrast, $Hf_{0.5}Zr_{0.5}O_2$ (HZO)-based ferroelectric thin films using atomic layer deposition offer sub-10 nm scalability, and low thermal budgets (~400°C), making them ideal for the application as an embedded memory[4]. However, their high intrinsic coercive fields (~1.5 MV/cm) that increase with increasing operation speeds require operating voltages ≥3 V for 10 nm-thick films, exceeding the sub-1.5 V requirement of advanced logic nodes. Furthermore, HZO's breakdown field ($E_{BD} \approx$ 3–4 MV/cm) is close to its operating field at nanosecond-range speeds, causing early dielectric failure[5]. Although thickness scaling reduces the operating voltage, the increased crystallization temperatures and pinched hysteresis loops in ultrathin HZO films (<6 nm) pose additional reliability challenge[6].

There exist interfacial layers between HZO films and electrodes. Recent studies highlight interfacial-layer engineering as a key strategy to mitigate these issues[7]. In this work, we demonstrated a HZO thin film capacitor operable at +/−1.2 V after optimizing interfacial-layer resistivity via a thermal annealing process at 800°C. This thermal treatment reduces the W/HZO interfacial-layer resistivity by 2.6×, effectively lowering the coercive field while retaining robust ferroelectricity. This approach unlocks FeRAM integration in energy-constrained ASICs, meeting the low-voltage demands of IoT edge devices and AI accelerators.

II. DEVICE FABRICATION AND STRUCTURAL ANALYSIS

A. Device Fabrication

The 8 nm-thick HZO ferroelectric thin films were fabricated on $Si/SiO_2/W$ substrates (Fig. 1a). A 200 nm-thick thermally grown SiO_2 layer served as the isolation dielectric, followed by a 30 nm-thick tungsten (W) bottom electrode deposited via ion-beam-assisted PVD to minimize interfacial roughness (<0.6 nm RMS). The HZO layer was grown at 250°C via ALD using TDMAHf and TDMAZr precursors with ozone oxidation, achieving a stoichiometric Hf:Zr ratio of 49.8:50.2 ± 0.3% (EDX mapping, Fig. 1b). A top W electrode was patterned via lift-off with sub-100 nm alignment accuracy. Post-deposition thermal treatment involved two-step annealing: (1) Crystallization at 500°C (30 s, N_2) to nucleate the ferroelectric orthorhombic (o-) phase, and (2) Interfacial-layer optimization at 800°C (1 s, N_2). This dual-annealing strategy reduces interfacial-layer resistivity by 2.6× compared to single-step crystallization, enabling full polarization switching at 1.2 V.

B. Structural Analysis

Cross-sectional TEM (Fig. 1b) reveals the sharp W/HZO interface. Complementary EDS mappings demonstrate uniform Hf/Zr atomic ratios (50.2%:49.8% ± 0.3%) across the 8 nm-thick film.

C. Wake-up effect

Figs. 1c–d show the evolution of *P-E* hysteresis loops and polarization enhancement of 800°C-annealed HZO capacitors under wake-up cycling. The initial pinched hysteresis loop (Fig. 1c) reveals the coexistence of metastable antiferroelectric (AFE) and ferroelectric (FE) phases. After wake-up, the antiferroelectric-to-ferroelectric transition increases remanent polarization ($2P_r$: 23 → 46

$\mu C/cm^2$) to the near-theoretical value of the polycrystalline HZO films[8].

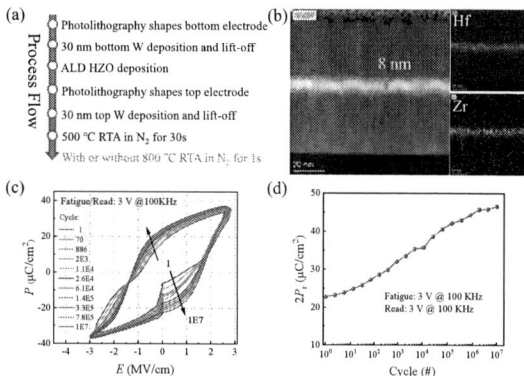

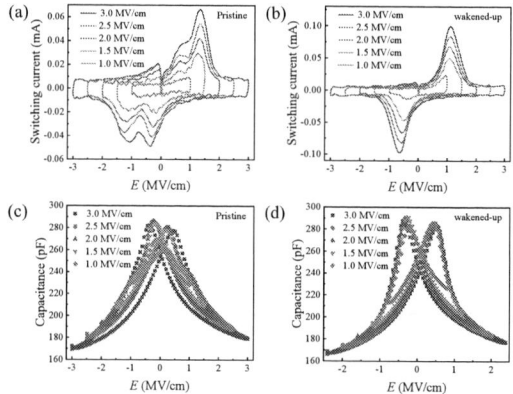

Fig. 1. Structural and Electrical Characterization of 8 nm-thick HZO Ferroelectric Thin-film Capacitor. (a) Schematic of the fabrication process flow. (b) Cross-sectional TEM image (left) and EDS elemental mapping (right). (c) P-E hysteresis loops after various wake-up cycles. (d) $2P_r$ as a function of cycling number.

III. RESULTS AND DISCUSSION

Figs. 2a and b compare the electrical field dependences of domain switching current at various amplitudes of electric fields of 800°C-annealed HZO capacitors before (pristine) and after wake-up cycling. In the pristine state, the current-electric field (I-E) curve (Fig. 2a) exhibits two distinct current peaks at 1.4 MV/cm and 0.8 MV/cm (−1.2 MV/cm and −0.4 MV/cm), indicative of coexisting antiferroelectric and ferroelectric phases. This metastable antiferroelectric phase is further corroborated by the appearance of asymmetric, pinched capacitance-electric field (C-E) loops in Fig. 2c, where the distorting butterfly shape at 1 MV/cm underscores the dominance of non-ferroelectric contribution. Such behavior renders the device unsuitable for low-voltage FeRAM operation due to reduced remanent polarization from the antiferroelectric.

After wake-up (3 V, 10^7 pulses, 100 kHz), there is only a single sharp domain switching current peak to emerge either at 1 MV/cm or at −0.8 MV/cm within each I-E curve in Fig. 2b. Meantime, the current magnitude increases by 1.43× (from 0.07 mA to 0.10 mA at 3 MV/cm), signifying enhanced domain alignment. Concurrently, the C-E loop (Fig. 2d) reveals a symmetric butterfly shape with a coercive field (E_c) of 0.7 MV/cm, confirming the AFE-to-FE transition. Remarkably, the device at 1 MV/cm maintains a $2P_r$ value of 10.2 $\mu C/cm^2$ (Fig. 3e)—comparable to PZT-based devices at equivalent voltages. This transition is attributed to high-field-driven defect redistribution at the W/HZO interface, which passivates oxygen vacancies and suppresses charge trapping[9, 10].

Figs. 3a–e show P-E hysteresis loops at various periodicities of 800°C-annealed HZO under varying operational electric fields (3–1 MV/cm). After wake-up cycling, the P-E hysteresis loops reveal various $2P_r$ values across scaled voltages. At 3 MV/cm, the $2P_r$ reaches 46.02 $\mu C/cm^2$, demonstrating near-saturation polarization. As the field decreases to 2.5 MV/cm and 2 MV/cm, the $2P_r$ values remain high at 44.2 $\mu C/cm^2$ and 42.0 $\mu C/cm^2$, respectively, with minimal degradation rates ($\Delta 2P_r < 5\%$ per MV/cm reduction), highlighting excellent field scalability. Remarkably, even at an ultralow field of 1 MV/cm, a $2P_r$

value of 10 $\mu C/cm^2$ is retained (Fig. 3f), surpassing the sensing threshold ($2P_r > 5$ $\mu C/cm^2$) for embedded DRAM applications[11].

Fig. 2. Domain Switching Characteristics of Pristine and Wakened-Up Devices Under Variable Electric Fields. (a–b) I-E hysteresis loops of pristine (a) and wakened-up (b) devices measured with electric field amplitude sweeping from 3 MV/cm to 1 MV/cm in steps of 0.5 MV/cm. (c–d) C-E loops of pristine (c) and wakened-up (d) devices in the same field range.

The E_c dependence on $T^{-1/2}$ (Figs. 4a–b) for two devices of various electrode areas with (w) and without (w/o) 800°C annealing obeys the following relationship[12]:

$$E_c = E_{c0} + \sqrt{\frac{8P_r E_a \rho_{on}}{T}} \quad (1)$$

where E_{c0} is the intrinsic coercive field, P_r is remanent polarization, E_a is the amplitude of the applied field, and ρ_{on} is interfacial resistivity. The solid lines in Figs. 4a–b are the best fits of the data according to Eq. (1) from which we observed an approximately equal E_{c0} between w/o and w samples as $T \rightarrow \infty$ (0.95 and 0.92 MV/cm).

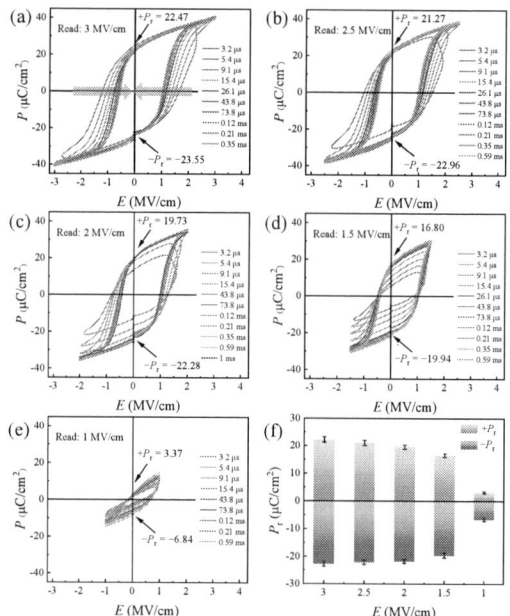

Fig. 3. Frequency-Dependent P-E Hysteresis Loops for Wakened-Up HZO Devices. (a–e) P-E hysteresis loops at 3 MV/cm (a), 2.5 MV/cm (b), 2 MV/cm (c), 1.5 MV/cm (d) and 1 MV/cm (e). (f) Dependences of positive and negative P_r on applied electric field.

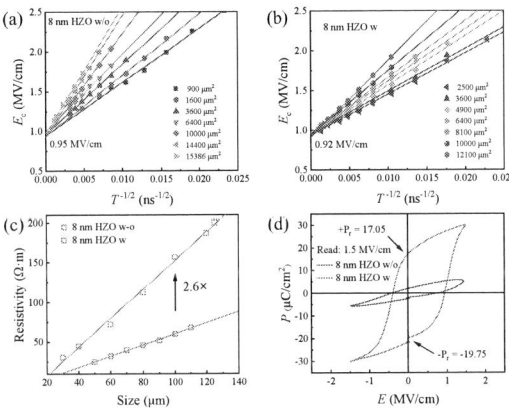

Fig. 4. The Comparison of Resistivity Between w/o and w HZO Capacitors Estimated from P-E Loops. (a–b) E_c-$T^{-1/2}$ dependences for the HZO w/o (a) and HZO w (b) capacitors in various electrode areas. (c) Capacity size dependence of resistivity for HZO w/o and w devices. The solid lines are linear fits of the data. (d) P-E hysteresis loops for HZO w and w/o capacitors under 1.5 MV/cm (1.2 V) low-field operation at 1KHz.

Fig. 4c shows size-dependent resistivity estimated from the linear fits in Figs. 4a–b. The two dependences are almost linear, and ρ_{on} in the latter for 800°C annealed devices shows a 2.6-fold reduction (from 156 $\Omega\cdot$m to 60 $\Omega\cdot$m under 100-μm capacitor size). This reduction could be correlated with optimized W/HZO interface contact.

The advantage of annealing is starkly evident in the Fig. 4d after comparison of P-E hysteresis loops of the two capacitors under an operation voltage (electric field) of 1.2 V (1.5 MV/cm). 800°C-annealed capacitors exhibit complete polarization switching with a $2P_r$ value of 36.8 μC/cm², while non-annealed devices show pinched loops when $2P_r < 5$ μC/cm². This disparity stems from annealing-induced interfacial resistivity reduction, which lowers the energy consumption for ferroelectric domain switching under equivalent applied voltage.

Table I summarizes E_c and operational voltage (V_{op}) among conventional HZO devices, our 800°C-annealed HZO films, and PZT-based FeRAMs. Conventional HZO devices exhibit $E_c \geqslant 1.5$ MV/cm and $V_{op} > 3$ V due to interfacial oxygen vacancy pinning and coexisting metastable antiferroelectric phase. In contrast, our 800°C-annealed HZO achieves that $E_c = 0.92$ MV/cm (1KHz) when $V_{op} = 1.2$ V, comparable to the voltages of industrial PZT FeRAMs (1.2–3 V). The 1.2 V operation aligns with sub-1.5 V logic nodes for advanced ASICs, bridging the voltage gap between conventional HZO films (> 3V) and legacy PZT-based FeRAMs.

TABLE I.   COMPARISON THE E_c OF PZT AND HZO

Device	E_c (MV/cm, 1KHz)	t_{FE} (nm)	V_{op} (V)
PZT[13]	0.2	80-200	1.2-3
HZO[14]	1.5	10	3
HZO (This work)	0.92	8	1.2

IV. CONCLUSION

We demonstrated a low-voltage operation of HZO ferroelectric thin film capacitors after interfacial-layer engineering. By 800°C thermal annealing of the capacitor during fabrication, the resistivity at the W/HZO interface is reduced by 2.6× while maintaining the equal E_{c0}. This optimization enables stable polarization switching at 1.2 V, achieving a $2P_r$ of 36.8 μC/cm²—comparable to legacy PZT-based FeRAMs. The annealed devices exhibit large remanent polarization ($\Delta 2P_r < 5\%$/MV/cm) at wide-range electric fields (3–1 MV/cm). This work paves the way for integrating ferroelectric memory into low-power IoT edge devices and AI accelerators, one step forward low-voltage requirement of next-generation logic devices.

ACKNOWLEDGMENT

This work was supported by the National Key Research and Development Program of China (grant number 2024YFA1409500), Shanghai Science and Technology Innovation Action Plan (grant number 24CL2900900), and the National Natural Science Foundation of China (grant number 62174034).

REFERENCES

(1) Zeng, B.; Yin, L.; Liu, R.; Ju, C.; Zhang, Q.; Yang, Z.; Zheng, S.; Peng, Q.; Yang, Q.; Zhou, Y.; Liao, M. Multiple Polarization States in $Hf_{1-x}Zr_xO_2$ Thin Films by Ferroelectric and Antiferroelectric Coupling. *Advanced Materials* **2024**, 2411463.

(2) Sun, Y.; Zhang, S.; Liu, Q.; Li, Y.; Lu, H.; Zhang, X.; Wei, Y.; Si, M.; Jiang, H.; Liu, Q. Back-End-of-Line Compatible 2T1C Memory Cell With InGaZnO Thin-Film Transistors and $Hf_{0.5}Zr_{0.5}O_2$-Based Ferroelectric Capacitors. *IEEE Transactions on Electron Devices* **2025**, 1-7.

(3) Wang, W.; Takai, K.; Eshita, T.; Nakabayashi, M.; Nakamura, K.; Oikawa, M.; Sato, N.; Suezawa, K.; Okita, Y.; Ozawa, S.; et al. Development of a High-Endurance Ferroelectric Capacitor for FeRAM in Automotive and Industrial Applications. *IEEE Transactions on Electron Devices* **2025**, 72 (2), 629-634.

(4) Yu, J.; Guo, S.; Zhang, J.; Jin, X.; Wu, C.; Zhao, M.; Li, H.; Guo, C.; Xu, K.; Tian, Y.; et al. 3D Trench $Hf_{0.5}Zr_{0.5}O_2$-Based 32 Kbit 1T1C FeRAM Chip with 2/5 ns Write/Read Speed, Low Power Consumption (0.605 pJ/bit) and Prominent High-Temperature Reliability (Baking @ 175°C). *International Electron Devices Meeting (IEDM)* **2024**, 1-4.

(5) Toprasertpong, K.; Tahara, K.; Hikosaka, Y.; Nakamura, K.; Saito, H.; Takenaka, M.; Takagi, S. Low Operating Voltage, Improved Breakdown Tolerance, and High Endurance in $Hf_{0.5}Zr_{0.5}O_2$ Ferroelectric Capacitors Achieved by Thickness Scaling Down to 4 nm for Embedded Ferroelectric Memory. *ACS Applied Materials & Interfaces* **2022**, 14 (45), 51137-51148.

(6) Park, H. S.; Choi, S.; Kim, K. D.; Yeom, M. K.; Lee, S. H.; Ryoo, S. K.; Hwang, C. S. Optimization of the 4 nm-Thick $Hf_{1-x}Zr_xO_2$ Film with Low Operating Voltage and High Endurance for Ferroelectric Random Access Memory. *ACS Applied Electronic Materials* **2024**, 6 (9), 6826-6836.

(7) Shijie Jia, J. L., * Qiong Yang, Renci Peng, Junhui Wang, Fei Yan, Shubin Wen, Zhipeng Wang, Jin Huang, Keyu Bao, Xuanling Liu, Min Liao, Jie Jiang,* and Yichun

Zhou*. Developing HZO‑Based Superlattices to Enhance Fatigue‑Resistance by Charge Injection. *Advanced Functional Materials* **2025**, *2501470*.

(8) Huang, F.; Saini, B.; Wan, L.; Lu, H.; He, X.; Qin, S.; Tsai, W.; Gruverman, A.; Meng, A. C.; Wong, H. S. P.; et al. Dimensional Scaling of Ferroelectric Properties of Hafnia-Zirconia Thin Films: Electrode Interface Effects. *ACS Nano* **2024**, *18* (27), 17600-17610.

(9) Peng, Y.; Xiao, W.; Liu, Y.; Jin, C.; Deng, X.; Zhang, Y.; Liu, F.; Zheng, Y.; Cheng, Y.; Chen, B.; et al. HfO_2-ZrO_2 Superlattice Ferroelectric Capacitor With Improved Endurance Performance and Higher Fatigue Recovery Capability. *IEEE Electron Device Letters* **2022**, *43* (2), 216-219. DOI: 10.1109/led.2021.3135961.

(10) Shen, B.; Yang, B.; Shi, M.; Sun, W.; Liu, H.; Ding, Y.; Gao, B.; Qian, H.; Wang, Y.; Tang, J.; Wu, H. Review on Ferroelectricity and Atomic Characterization of $Hf_{0.5}Zr_{0.5}O_2$ in FeRAM. *ACS Applied Electronic Materials* **2025**, *7* (11), 4675-4702.

(11) Chang, S.-C.; Haratipour, N.; Shivaraman, S.; Brown-Heft, T. L.; Peck, J.; Lin, C.-C.; Tung, I. C.; Merrill, D. R.; Liu, H.; Lin, C.-Y.; et al. Anti-ferroelectric $Hf_xZr_{1-x}O_2$ Capacitors for High-density 3-D Embedded-DRAM. In International Electron Devices Meeting (IEDM), 2020.

(12) Zhang, W. D.; Jiang, A. Q. Size-Scaling Effect on Domain Switching Time and Coercive Field of $TiN/Hf_{0.5}Zr_{0.5}O_2/TiN$ Thin-Film Capacitors. *IEEE Transactions on Electron Devices* **2023**, *70* (12), 6324-6328.

(13) Jiang, A. Q.; Zhang, D. W. Estimation of film-electrode contact resistance and domain switching time from ferroelectric polarization-voltage hysteresis loops. *Thin Solid Films* **2013**, *545*, 145-148.

(14) Zhang, W. D.; Jiang, A. Q. Improvement of Polarization Retention at Low and High Temperatures for $Hf_{0.5}Zr_{0.5}O_2$ Thin-Film Capacitors. *IEEE Electron Device Letters* **2024**, *45* (7), 1181-1184.

Reliability Enhancement of $Hf_{0.5}Zr_{0.5}O_2$-Based Ferroelectric Capacitors via Argon Plasma Treatment

Hongrui Zhang[1], Xiu Yang[1], Rongzong Shen[2], Yian Ding[1], Haoji Qian[1], Jiajia Chen[1*], Chengji Jin[1], Ran Cheng[3], Bing Chen[1], Xiao Yu[1*], Yan Liu[1], Genquan Han[1]

[1] HangZhou Institute of Technology, Xidian University, HangZhou, China,[2] Shenzhen Xinkailai Technology Co., Ltd,.
[3]School of Micro-Nano Electronics, Zhejiang University, 310000, Hangzhou, China.

* Email: chenjiajia@xidian.edu.cn, yuxiao@xidian.edu.cn

Abstract—We have experimentally demonstrated that Argon (Ar) plasma treatment at the top interface of ferroelectric capacitors with a titanium nitride (TiN)/ $Hf_{0.5}Zr_{0.5}O_2$(HZO)/TiN structure significantly enhances device reliability. Key improvements include increased endurance, elimination of the wake-up effect, and suppression of imprint characteristics. These enhancements are attributed to the formation of a smoother interfacial layer (IL) induced by Ar plasma treatment, which leads to a more uniform distribution of oxygen vacancies, thereby enabling the wake-up free characteristic. In addition, structural and chemical analyses using transmission electron microscopy (TEM), energy dispersive spectrometer (EDS), and X-ray photoelectron spectroscopy (XPS) confirm the improved IL quality with Ar plasma treatment, leading to the endurance enhancement and imprint suppression.

Keywords—FE capacitors, Ar plasma, HZO, oxygen vacancies.

I. INTRODUCTION

Ferroelectric (FE) HfO_2 holds tremendous promise for the next-generation high-density, non-volatile memory [1,2]. However, reliability issues such as wake-up, fatigue, and imprint effects can substantially affect the device performance [3,4]. Various studies have demonstrated that the redistribution and generation of oxygen vacancy (V_O) in FE dielectric leads to wake-up and fatigue [5,6]. Here, the interfacial layer (IL) between dielectric and electrode typically exhibits a high concentration of V_O for the pristine state, and tends to generate additional V_O during electrical cycling [7,8,9]. It is also reported that the plasma treatment, such as NH_3 during the fabrication process is one of the effective ways to improve the quality of the IL, while N_2 plasma treatment enhances the electrical properties of the film such as leakage current suppression [10,11]. However, the impact of the plasma treatment on IL-related reliability of the ferroelectric capacitors, and its underlying mechanisms have not been fully studied yet.

In this paper, we investigate the impact of argon (Ar) plasma treatment on the $Hf_{0.5}Zr_{0.5}O_2$(HZO) film of the metal-ferroelectric-metal (MFM) capacitor. Significant reliability enhancement of the device is exhibited including enhanced endurance, wake-up-free operation, and suppression of fatigue. The underlying mechanisms are examined through electrical properties and physical characterization. It is revealed that Ar plasma treatment effectively reduces the surface roughness, improves the uniformity of V_O distribution, and suppressed the V_O generation, resulting in device reliability improvement.

II. EXPERIMENTAL

Figs. 1(a) and (b) show the schematic structure of the MFM capacitors and their fabrication process flow, respectively. First, titanium nitride (TiN) film was deposited on N-type silicon wafers for 50 nm using direct current (DC) sputtering as the bottom electrodes. Then, 10 nm-thick HZO film was deposited via atomic layer deposition (ALD) at 250 °C. After ALD deposition, the samples were treated with Ar plasma for 1 minute using radio frequency (RF) power of 100 W under Ar flow of 2 sccm and 5 sccm, respectively. After depositing a 50 nm-thick TiN top electrode by DC sputtering, rapid thermal annealing (RTA) was carried out in N_2 ambient at 400 °C for 30 s to crystallize the ferroelectric films. Here, control samples without Ar plasma treatment were also fabricated. The electrical properties of the proposed devices are characterized with the measurement setup illustrated in Fig. 1(c).

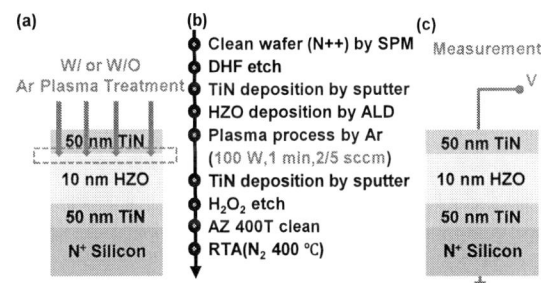

Fig. 1 (a) Cross-sectional schematic illustrations of the MFM capacitors with and without Ar plasma process. (b) Key fabrication process flow. (c) Measurement setup for electrical properties.

III. RESULTS AND DISCUSSION

A. Electrical Properties and Interface Characterization

Fig. 2(a) shows the comparison of positive-up, negative-down (PUND) tests for the capacitors with plasma under different flow (marked as W/ P2 and W/ P5), and without Ar plasma (W/O plasma). P5 capacitors and those without Ar plasma treatment exhibited comparable remnant polarization (P_r). Therefore, we selected devices with plasma treatment in 5-sccm Ar flow (W/ P5) in the following analysis to reveal the impact of Ar plasma. *C-V* characteristics plotted in Fig. 2(b) show that both capacitors with or without plasma treatment exhibit FE butterfly-like features [12]. The cross-sectional transmission electron microscopy (TEM) images shown in Fig. 3 indicate that devices with plasma treatment exhibit a smoother IL. Besides, the corresponding energy dispersive spectrometer (EDS) profiles shown in Fig. 4

exhibit a more abrupt oxygen element distribution at the IL with Ar plasma treatment. These physical characteristics suggest that the top TiN/HZO interface can be passivated with Ar plasma treatment, resulting in smoother surface and less oxygen scavenging of TiN electrode from HZO film.

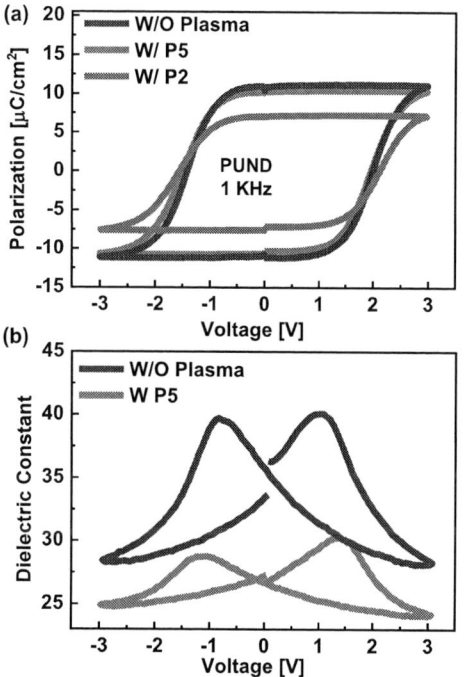

Fig. 2 (a) *P-V* curves measured by PUND test with frequency of 1 KHz and (b) *C-V* characteristics of the capacitors with and without Ar plasma process.

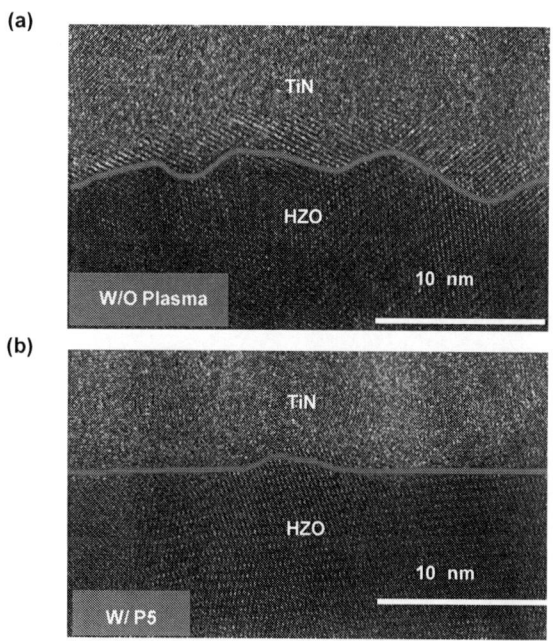

Fig. 3 Cross-sectional TEM images for the MFM capacitors (a) without and (b) with Ar plasma process with RF powers of 100 W for 1 min under Ar flow of 5 sccm (P5). A smoother TiN/HZO interface is achieved with Ar plasma process.

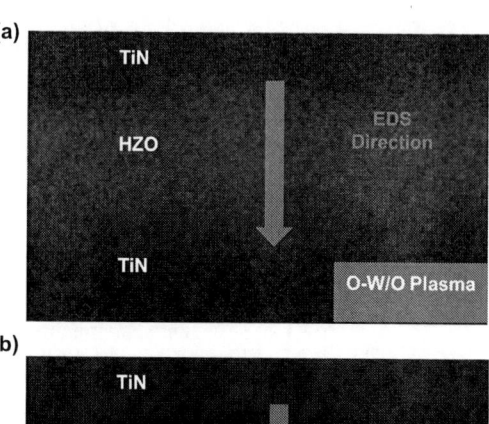

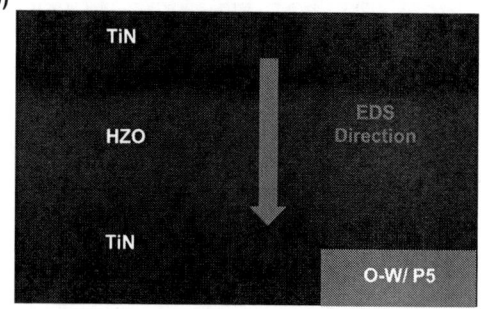

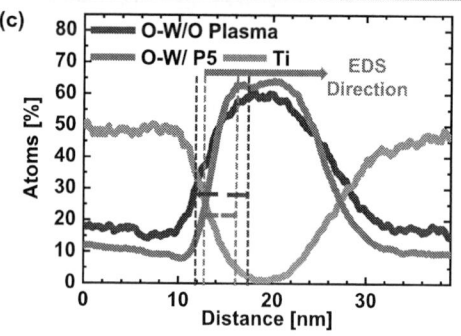

Fig. 4 Cross-sectional EDS mapping of oxygen atoms (a)with Plasma treatment and (b) without Plasma treatment (c)line cut of the EDS mapping along the vertical direction of devices with and without Ar plasma process, the dashed lines mark the interface layers between the top electrode and the HZO film. A more abrupt O distribution is observed.

B. Reliability Properties

The endurance of devices with/without Ar plasma treatment was characterized and shown in Fig. 5. Under triangular electrical cycling with 3 V amplitude and 1 kHz frequency, the capacitor with Ar plasma treatment exhibits wake-up-free behavior. The value of $2P_r$ showed only slight degradation after 10^7 cycles by 11.4 $\mu C/cm^2$, which is significantly less than the ones without treatment (17 $\mu C/cm^2$), indicating a suppressed fatigue effect. Wake-up-free behavior and enhanced endurance can be attributed to a more uniform V_O distribution at the pristine state, and the suppressed V_O generation during cycling, respectively [13]. Additionally, as shown in Fig. 5(b), the device with Ar plasma treatment maintains stable leakage currents and dielectric constant for devices with Ar plasma during cycling up to 10^6 cycles, indicating the robust resilience to V_O-induced degradation.

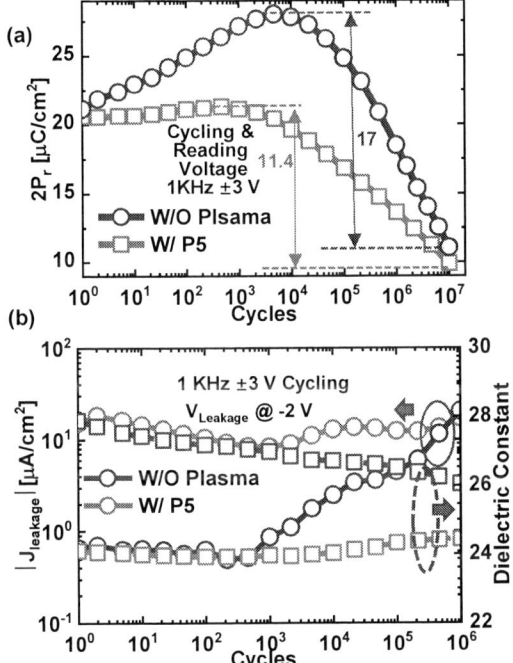

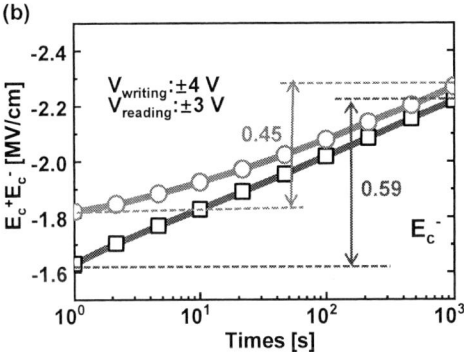

Fig. 6 Retention characteristic of devices. the amplitudes of writing and reading voltages are (a) 4 V and 3 V (b) -4 V and -3 V, respectively. the frequency of reading voltage is 1 kHz.

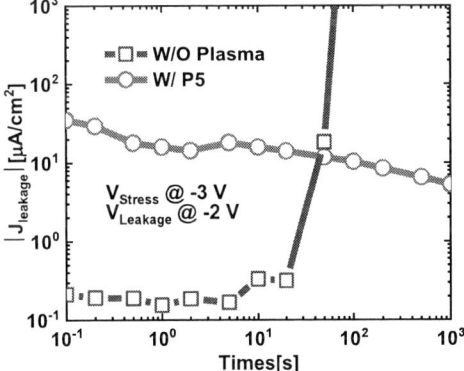

Fig. 7 Leakage currents for devices with and without Ar plasma process under -3 V electrical stress for different times (0-1000 s).

Fig. 5 Endurance characteristic of devices with and without Ar plasma process, (a) $2Pr$ values vs. voltage stress cycling, with the amplitude of 3 V and cycling and reading frequencies of 1 KHz. Better endurance without wake-up is obtained with Ar plasma. (b) Leakage current and the corresponding dielectric constant during cycling. Stable leakage and dielectric constant can be observed with Ar plasma process.

Besides, the retention characteristics of devices with/without Ar plasma treatment were evaluated, as shown in Fig. 6. After 1000 s, the devices with Ar plasma exhibit smaller shifts in the coercive field (E_c) with positive/negative E_c shifts($\Delta E_c^{+/-}$) of 0.18/0.45 MV/cm, respectively. In contrast, the ones without Ar plasma treatment showed larger shifts with $\Delta E_c^{+/-} = 0.38/0.59$ MV/cm. These reduced shifts can be attributed to a lower V_O concentration in the IL, indicating that the effective IL passivation is achieved through Ar plasma treatment. This conclusion is further supported by the stable leakage current observed during applied voltage stress, as shown in Fig. 7.

C. Underlying Mechanisms for Reliability Improvement

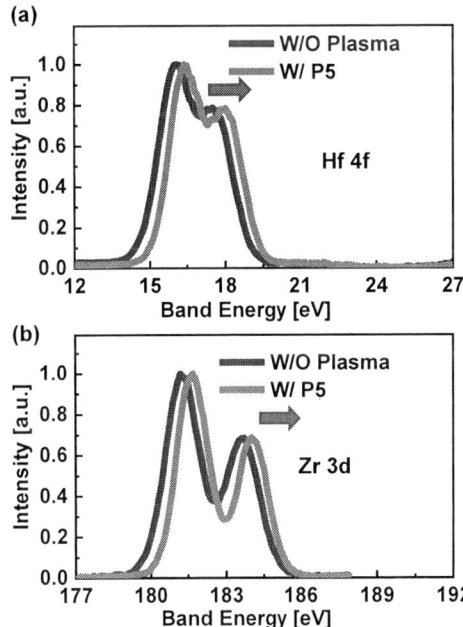

Fig. 8 (a) Hf 4f and (b) Zr 3d XPS spectra of HZO films for devices with and without Ar plasma process.

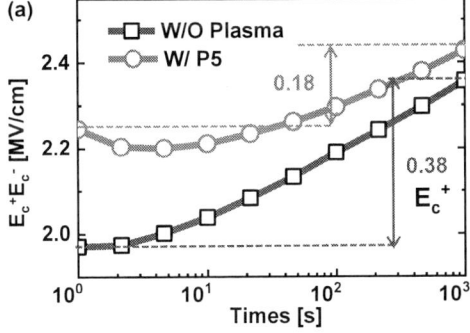

979-8-3315-3918-4/25 $31.00 © 2025 IEEE

To reveal the underlying mechanism for reliability improvement after Ar plasma treatment, X-ray Photoelectron Spectroscopy (XPS) of the HZO surfaces with and without plasma process are shown in Fig. 8. Note that the peaks for Hf 4f and Zr 3d shift towards a higher binding energy region in the XPS spectra, indicating the less lattice distortion and dangling bonds at IL [14]. These results proved that the improved interfacial quality is also the main reason of the suppressed imprint effect.

Based on the electrical properties and surface physical and chemical analysis, the mechanism of the V_O evolution for the devices with/without Ar plasma treatment is proposed and summarized in Fig. 9. Plasma treatment of HZO prevents the cycling-induced increase and redistribution of V_O observed in untreated devices, which degrades the IL and increases leakage. In contrast, plasma-treated devices maintain stable V_O distribution and leakage current during cycling. This inherent V_O stability eliminates the wake-up effect and significantly enhances IL quality, directly resulting in superior endurance.

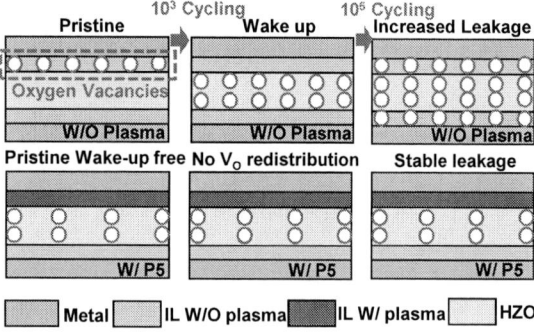

Fig. 9 Schematic diagrams of the mechanisms for the reliability improvement for devices with Ar plasma process. The plasma treatment will provide a uniform V_O distribution with less concentration at the pristine state, and suppress the V_O generation and redistribution during cycling.

CONCLUSIONS

The impact and the mechanisms of Ar plasma treatment on FEHZO film have been characterized and addressed. The devices with plasma exhibit wake-up free, enhanced endurance, and suppressed imprint characteristics. By electrical characterization and TEM, EDS, XPS analyzation, the uniform V_O distribution and suppressed V_O generation with a well-passivated interface formed by plasma treatment is attributable to the reliability improvement.

ACKNOWLEDGMENT

The authors acknowledge support from the National Key Research and Development Project (Grant No. 2023YFB4402303), the Fundamental Research Funds for the Central Universities (Grant No. ZYTS25038), the National Natural Science Foundation of China (Grant No. 62374151, 62204226, 62174146, 92464205, 62474164), Major Program of Zhejiang Natural Science Foundation (Grant No. DT23F0402, LDQ24F040001, LTGC24F040001). Zhejiang Province Key R&D programs (No. 2025C01193).

REFERENCES

[1] K. T. Chen et al., "Non-Volatile Ferroelectric FETs Using 5-nm $Hf_{0.5}Zr_{0.5}O_2$ With High Data Retention and Read Endurance for 1T Memory Applications", in IEEE Electron Device Letters, vol. 40, no. 3, 2019, pp. 399-402.

[2] B. Prasad, V. Thakare, A. Kalitsov, Z. Zhang, B. Terris, R. Ramesh, "Large Tunnel Electroresistance with Ultrathin $Hf_{0.5}Zr_{0.5}O_2$ Ferroelectric Tunnel Barriers", Adv. Electron. Mater. 2021, 7, 2001074.

[3] E. D. Grimley, T. Schenk, T. Mikolajick, U. Schroeder, & J. LeBeau, "Atomic structure of domain and interphase boundaries in ferroelectric HfO_2", Adv. Mater. Interfaces. 2018, 5, 1701258.

[4] P. Jiang, Q. Luo, X. Xu, T. Gong, P. Yuan, Y. Wang, Z. Gao, W. Wei, L. Tai, H. Lv, "Wake-Up Effect in HfO_2-Based Ferroelectric Films", Adv. Electron. Mater. 2021, 7, 2000728.

[5] J. Chen et al., "Impact of Oxygen Vacancy on Ferroelectric Characteristics and Its Implication for Wake-Up and Fatigue of HfO_2-Based Thin Films", in IEEE Transactions on Electron Devices, Sept. 2022, vol. 69, no. 9, pp. 5297-5301.

[6] A. Kashir, S. Oh, H Hwang, "Defect Engineering to Achieve Wake-up Free HfO_2-Based Ferroelectrics", Adv. Eng. Mater. 23: 2000791.

[7] J. Chen et al., "Controlling the Ferroelectricity of Doped-HfO_2 via Reversible Migration of Oxygen Vacancy", in IEEE Transactions on Electron Devices, April 2023, vol. 70, no. 4, pp. 1789-1794.

[8] C. Liu et al., "Role of Oxygen Vacancies in Electric Field Cycling Behaviors of Ferroelectric Hafnium Oxide", 2018 IEEE International Electron Devices Meeting (IEDM), San Francisco, CA, USA, 2018, pp. 16.4.1-16.4.4.

[9] L. Chen, Z. X. Liang, S. X Shao, Q. Q. Huang, K. C. Tang, R. Huang, "First direct observation of the built-in electric field and oxygen vacancy migration in ferroelectric $Hf_{0.5}Zr_{0.5}O_2$ film during electrical cycling", Nanoscale, 2023, 15, 7014-7022.

[10] K. Y. Chen, P. H. Chen, R. W. Kao, Y. X. Lin and Y. H. Wu, "Impact of Plasma Treatment on Reliability Performance for HfZrOx-Based Metal-Ferroelectric-Metal Capacitors", in IEEE Electron Device Letters, Jan. 2018, vol. 39, no. 1, pp. 87-90.

[11] N. J. Seong, S. G. Yoon, S. J. Yeom, H. K. Woo, D. S. Kil, J. S. Roh, H. C. Sohn, "Effect of nitrogen incorporation on improvement of leakage properties in high-k HfO_2 capacitors treated by N_2-plasma", Appl. Phys. Lett. 26 Sept. 2005; 87 (13): 132903.

[12] Y. J. Lin, C. Yu. Teng, S. J. Chang, Y. F. Liao, C. Hu, C. J. Su, Y. C. Tseng, "Role of electrode-induced oxygen vacancies in regulating polarization wake-up in ferroelectric capacitors", Applied Surface Science, 2020, vol 528, 147014, ISSN 0169-4332.

[13] Y. Zhou, D. J. Apo, S. Priya, "Dual-phase self-biased magnetoelectric energy harvester", Appl. Phys. Lett. 4 November 2013, 103 (19): 192909.

[14] D. Zhao, Z. Chen, X Liao, "Microstructural evolution and ferroelectricity in HfO_2 films", Microstructures, vol. 2, pp. 2022007, 2022,

Cryogenic Ferroelectricity (10-298 K) of Superlattice and Solid-solution HZO Films

Yiming Xia[1], Chunlei Wu[1,2,3*], Jian Ma[1], Hanzhi Gu[1], Yueyuan Yu[1], Jiayi Wu[1], Qingqing Sun[1,2,3], David Wei Zhang[1,2,3]

[1] School of Microelectronics, Fudan University, Shanghai 200433, China
[2] Shanghai Integrated Manufacturing Innovation Center Co., Ltd, Shanghai 201203, China
[3] Jiashan Fudan Institute, Jiaxing 314100, China

* Email: wuchunlei@fudan.edu.cn

Abstract—In this work, the cryogenic performance of hafnium oxide based ferroelectric (FE) capacitors has been systematically characterized from 298 K down to 10 K, targeting the temperature ranges relevant to cryogenic logic and memory applications. The ferroelectricity and reliability properties of metal-ferroelectric-metal (MFM) devices using superlattice HZO (SL-HZO) and solid-solution HZO (SS-HZO) films have been characterized and analyzed. The measurement results show that both SS-HZO and SL-HZO based capacitors show degraded remnant polarization but improved endurance at cryogenic temperature. Significantly improved endurance from 10^2 cycles at 298 K to more than 10^8 cycles at 10 K has been achieved under 4 MV/cm. Moreover, evident suppression of the wake-up and fatigue effects at cryogenic temperatures has also been demonstrated. The results provide key insights into the feasibility of ferroelectrics for cryogenic memory applications.

Keywords—*Cryogenic, Ferroelectric (FE), $Hf_xZr_{1-x}O$ (HZO), Superlattice, Non-volatile memory*

I. INTRODUCTION

The relentless scaling of semiconductor devices has pushed main stream flash memory technology to its physical limits and brought to light many emerging memory concepts. Among which, ferroelectric (FE) capacitors and transistors have attracted most attentions owing to the distinct bistable and switchable FE polarization properties. The introduction of the ferroelectric random-access memory has been motivated by the realization of one transistor and one capacitor (1T1C) memory based on FE capacitors using available FE materials at the time. Traditional perovskite-based ferroelectrics such as PZT ($Pb(ZrTi)O_3$) and SBT ($SrBi_2Ta_2O_9$) have been found to lose ferroelectricity at scaled dimensions, and face   great challenges of poor process compatibility with standard CMOS technology[1], [2]. Therefore, the discovery of ferroelectricity in hafnium oxide (HfO_2) films has triggered intensive researches on FE capacitors and transistors for non-volatile memory (NVM) applications [1], [3].

Moreover, emerging technologies such as cryo-electronics, quantum computing and high-performance computing are now driving exploration of cryogenic non-volatile memory solutions [4]-[6]. Thanks to the excellent scalability and CMOS process compatibility, as well as fast switching speed and low power consumption, FE capacitors

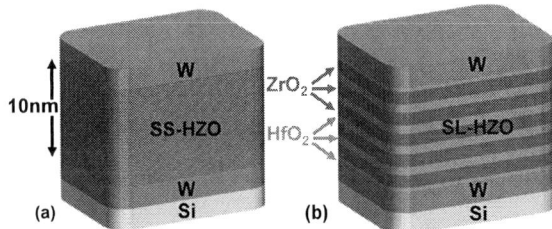

Fig. 1 The schematics of (a) solid-solution $Hf_{0.4}Zr_{0.6}O_2$ (SS-HZO), and (b) HfO_2-ZrO_2-HfO_2 superlattice (SL-HZO) based MFM capacitors.

using hafnium zirconium oxide (HZO) have been extensively investigated for emerging memory applications [1], [3]. However, most reported studies of HZO based FE capacitors still focus on enhancing ferroelectricity or improving FE reliability at room temperature (RT) conditions [7], [9] . The assessment of the ferroelectricity as well as reliability characteristics including wake-up and cycling fatigue effects at cryogenic temperatures is in ample necessity [14]-[17].

In this work, we have investigated the cryogenic properties of metal-ferroelectric-metal (MFM) capacitors based on both solid-solution and superlattice HZO films. The ferroelectricity as well as endurance performance of FE capacitors are characterized from 298 K down to 10 K. The results would provide key insights into the feasibility of ferroelectrics for cryogenic memory applications.

II. DEVICE FABRICATION

To investigate the cryogenic properties of HZO based ferroelectric capacitors, Metal-Ferroelectric-Metal (MFM) capacitors were fabricated. First, a 100-nm-thick tungsten (W) was deposited on Si substrates via physical vapor deposition (PVD) as the bottom electrode (BE). Then the ferroelectric films of the HfO_2-ZrO_2-HfO_2 superlattice with Hf:Zr = 2:3 periodicity and $Hf_{0.4}Zr_{0.6}O_2$ solid solution (all 10nm) were deposited via atomic layer deposition (ALD) at 250°C, using $Hf[N(C_2H_5) CH_3]_4$ and $Zr[N(CH_3)_2]_4$ as precursors. A 100 nm-thick W top electrode (TE) was deposited subsequently by PVD. All samples were exposed to post-deposition annealing (PDA) at 450°C for 30s in N_2 ambient, if not otherwise stated, for crystallization and stabilization of FE layer.

979-8-3315-3918-4/25 $31.00 © 2025 IEEE

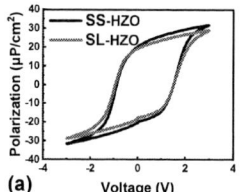 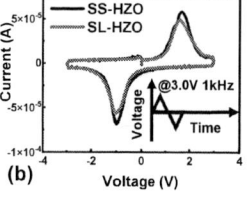

Fig. 2 The (a) P-V and(b) I-V characteristics of 10 nm SS-HZO and SL-HZO based MFM capacitors. The inset shows the measurement setup.

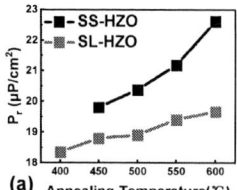 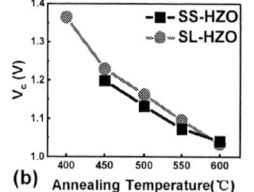

Fig. 3 (a) Remnant polarization P_r and (b) coercive voltage V_C of SS-HZO and SL-HZO films with different annealing temperatures.

Schematical illustrations of the solid-solution $Hf_{0.4}Zr_{0.6}O_2$ (SS-HZO) and HfO_2-ZrO_2-HfO superlattice (SL-HZO) are shown in Fig. 1 (a) and (b) respectively. The capacitor area is defined by the top electrode patterning as 100 μm × 100 μm. Electrical characterization was performed using a Keithley 4200-SCS semiconductor parameter analyzer in a Lakeshore CRX-VF cryogenic probe station.

III. CHARACTERIZATION RESULT AND DISCUSSION

A. Room Temperature Ferroelectricity

All MFM devices were firstly subjected to a wake-up operation at room temperature (RT). The wake-up operation consisted of a 1 kHz bipolar signal with one hundred triangular pulses and a peak amplitude of 3.0 V at RT. Fig. 2 (a) and (b) present the polarization-voltage (P-V) hysteresis loops and current-voltage (I-V) characteristics of the MFM capacitors based on SS-HZO and SL-HZO. After the wake-up operation at RT, the devices were exposed to a 1 kHz triangular bipolar signal with a peak amplitude of 3.0 V. As can be seen in the figure, the SL-HZO and SS-HZO based capacitors exhibited similar FE characteristics, while SS-HZO capacitor showed higher remnant polarization $2P_r$ of 40 μC/cm² compared to that of SL-HZO capacitor (38 μC/cm²).

Fig. 3 presents the ferroelectric properties of SL-HZO and SS-HZO capacitors as a function of annealing temperature. It can be seen in the figures that both SL-HZO and SS-HZO capacitors showed degraded ferroelectricity for low-temperature annealing, exhibiting both reduced polarization (P_r) and increased coercive voltage (V_c). For SS-HZO capacitors, lowest annealing temperature of 450 °C was required for ferroelectric phase crystallization. For SL-HZO capacitors, however, a lower FE crystallization temperature of 400 °C was achieved, which can be attributed to the stress introduced by the lattice and thermal mismatch at the multi-layers interfaces during annealing, thereby enabling lower annealing temperature for the stabilization of ferroelectric orthorhombic phase [10], [13].

Furthermore, acceleration measurements of endurance at RT were performed by applying programmed pulse amplitudes of ± 1.75 V, ± 2.25 V and 4 V, as depicted in

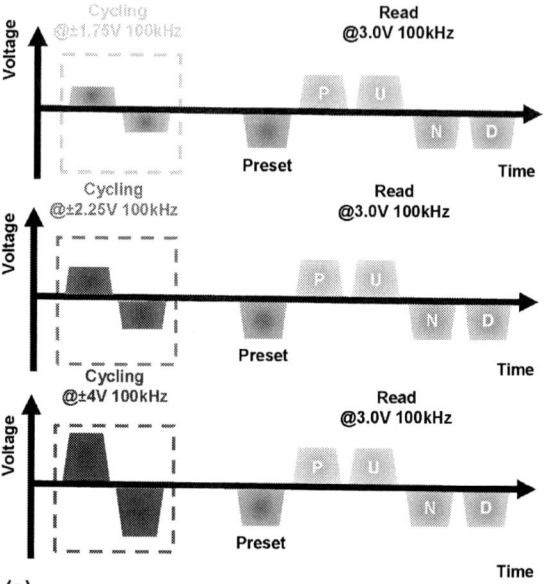

(a)

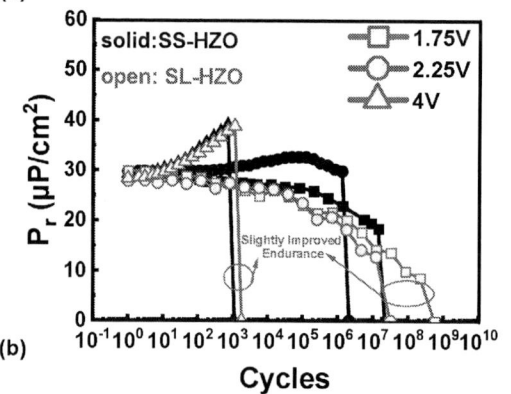

Fig. 4 (a) Schematic of endurance cycling and PUND read scheme. (b) Endurance characteristics of SS-HZO and SL-HZO MFM capacitors at RT.

Fig. 4 (a). The read scheme of ± 3 V positive-up-negative-down (PUND) at 100 kHz was used. Fig. 4 (b) illustrates the endurance properties of the SS-HZO and SL-HZO capacitors at RT under different applied biases. Breakdown can be observed in both MFM capacitors under higher cycling bias of 4 V (4 MV/cm) when cycled up to around 10^3 cycles. For lower cycling bias of 1.75 V (1.75 MV/cm), however, improved endurance of 2×10^7 and 6×10^8 was achieved in SS-HZO and SL-HZO capacitors respectively, though fatigue behavior can be observed. The improved endurance of SL-HZO MFM capacitors can be attributed to the multi-layer structure, which introduces dimensional confinement so as to suppress the generation and migration of defects and oxygen vacancies during cycling [10], [13].

B. Cryogenic Ferroelectricity

To further investigate the cryogenic ferroelectricity of SS-HZO and SL-HZO films, the capacitors were cooled to 10 K for certain electrical measurements. Fig. 5 presents the wake-up behavior of MFM capacitors at cryo-temperature in comparison with that at RT. It can be observed in the figure that the polarization is significantly reduced at 10 K

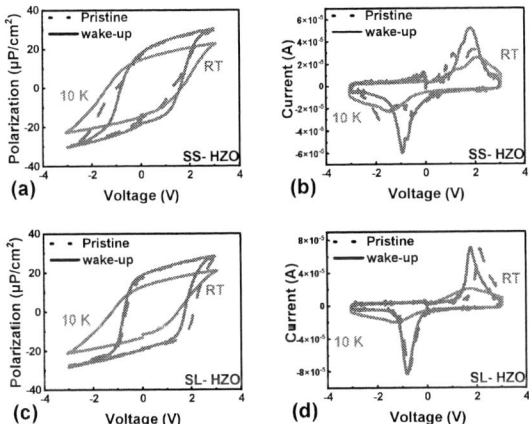

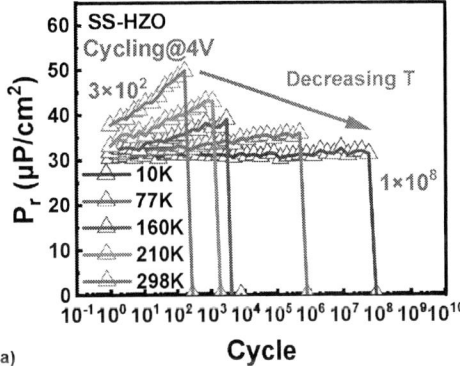

Fig. 5 Different wake-up characteristics of (a-b) SS-HZO and (c-d) SL-HZO based MFM capacitors at RT and 10 K.

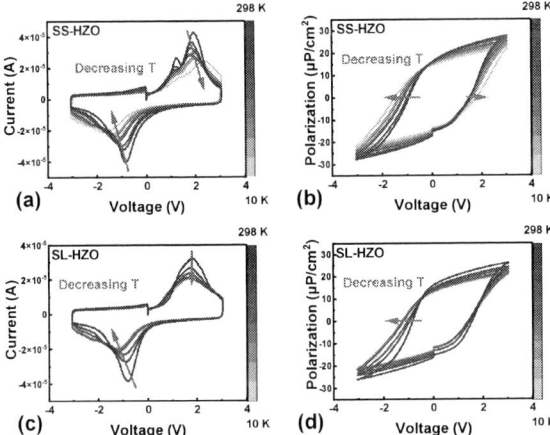

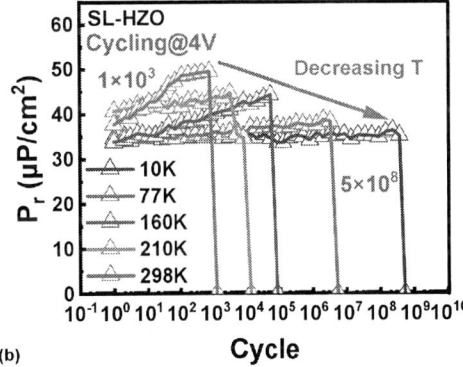

Fig. 6 Evolution of the I-V and P-V characteristics of (a-b) SS-HZO and (c-d) SL-HZO based MFM capacitors from 10 K to 298 K after wake-up cycling at 10 K.

Fig. 7 Endurance characteristics of (a) SS-HZO and (b) SL-HZO based MFM capacitors at varied temperatures from 10 K to 298 K.

compared to that at RT. Moreover, unlike the case at RT, the polarization and coercive voltage of both SS-HZO and SL-HZO capacitors showed no visible change after one hundred wake-up cycling at 10 K. The wake-up behavior in HZO capacitors can be attributed to the diffusion and redistribution of the oxygen vacancies by repetitive electric field cycling [14], [17]. The wake-up cycling at cryogenic temperature, however, becomes invalid owing to suppressed migration of oxygen vacancies. What's more, as can be seen in Fig. 5(b), the SS-HZO based capacitors exhibited double-peak current loops in the pristine state, which implies high defects or oxygen vacancies concentration in SS-HZO film. The SL-HZO based devices, however, were able to maintain the single-peak current loops at both RT and 10 K, suggesting suppressed generation and migration of defects in the superlattice structure.

To study the temperature dependence of ferroelectricity, the I-V and P-V characteristics of SS-HZO and SL-HZO capacitors has been evaluated across a wide temperature range of 10-298 K, as shown in Fig. 6. Evident reduction of FE polarization can be observed in all devices as the temperature is lowered, which may be caused by the

expansion of the dead layer at the interfaces at cryogenic temperature [18]. In SS-HZO devices, a gradual increase of the negative coercive voltage V_{c-} and a slightly increase of the positive coercive voltage V_{c+} can be observed with decreasing temperature. In SL-HZO devices, however, a gradual increase of only the negative coercive voltage V_{c-} was observed, while the positive coercive voltage V_{c+} remain nearly constant with decreasing temperature. The increased fields for polarization switching can be caused by enhanced domain wall pinning at low temperatures [18]. While the different temperature dependence of V_{c-} and V_{c+} may result from the asymmetric interfaces properties at the bottom and top interfaces.

Acceleration measurements of endurance at cryogenic temperatures have been performed at 1 kHz, as shown in Fig. 7. Both SS-HZO and SL-HZO capacitors exhibit significant endurance improvement from 10^2 cycles at 298 K to 10^8 cycles at 10 K, as well as greatly improved polarization fatigue behavior ($\Delta P_r/P_{r,\ pristine} < 6\ \%$) at 10K. This further indicates the suppression of generation and migration of defects and oxygen vacancies in HZO films at cryogenic temperature.

IV. CONCLUSION

In this work, the ferroelectricity as well as reliability properties of metal-ferroelectric-metal (MFM) capacitors using superlattice HZO (SL-HZO) and solid-solution HZO (SS-HZO) films have been analyzed from 298 K down to 10 K. The room temperature measurement results show that SL-HZO based MFM capacitors can achieve slightly improved endurance and enable lower annealing temperature (400℃), attributing to the dimensional confinement and the stress introduced by the superlattice structure. The cryogenic temperature measurement results show that there's a trade-off between the ferroelectricity and endurance performance improvement at cryogenic temperature, both SS-HZO and SL-HZO based capacitors show degraded remnant polarization P_r but improved endurance. Evident suppression of the wake-up and fatigue effects at cryogenic temperatures has also been observed, owing to the suppression of generation and migration of defects and oxygen vacancies. The results provide key insights into the feasibility of ferroelectrics for cryogenic memory applications.

REFERENCES

[1] M. H. Park, Y. H. Lee, T. Mikolajick, U. Schroeder, and C. S. Hwang, "Review and perspective on ferroelectric HfO₂-based thin films for memory applications," *MRS Communications.*, vol. 8, no. 3, pp. 795–808, Sep. 2018.

[2] T. Mikolajick, U. Schroeder, and S. Slesazeck, "The Past, the Present, and the Future of Ferroelectric Memories," *IEEE Transactions on Electron Devices.*, vol. 67, no. 4, pp. 1434–1443, Apr. 2020.

[3] T. S. Böscke, J. Müller, D. Bräuhaus, U. Schröder, and U. Böttger, "Ferroelectricity in hafnium oxide thin films," *Applied Physics Letters.*, vol. 99, no. 10, p. 102903, Sep. 2011.

[4] S. G. Kirtania, K. A. Aabrar, A. I. Khan, S. Yu, and S. Datta, "Cold-FeFET as Embedded Non-Volatile Memory with Unlimited Cycling Endurance," in *2023 IEEE Symposium on VLSI Technology and Circuits (VLSI Technology and Circuits).*, Jun. 2023, pp. 1–2.

[5] B. Patra et al., "Cryo-CMOS Circuits and Systems for Quantum Computing Applications," *IEEE Journal of Solid-State Circuits.*, vol. 53, no. 1, pp. 309–321, Jan. 2018.

[6] C.-Y. Liao et al., "Multibit Ferroelectric FET Based on Nonidentical Double HfZrO₂ for High-Density Nonvolatile Memory," *IEEE Electron Device Letters.*, vol. 42, no. 4, pp. 617–620, Apr. 2021.

[7] M. Pešić et al., "Physical Mechanisms behind the Field-Cycling Behavior of HfO₂-Based Ferroelectric Capacitors," *Advanced Functional Materials.*, vol. 26, no. 25, pp. 4601–4612, 2016.

[8] A. G. Chernikova et al., "Improved Ferroelectric Switching Endurance of La-Doped $Hf_{0.5}Zr_{0.5}O_2$ Thin Films," *ACS Appl. Mater& Interfaces.*, vol. 10, no. 3, pp. 2701–2708, Jan. 2018.

[9] M. H. Park, H. J. Kim, Y. J. Kim, W. Jeon, T. Moon, and C. S. Hwang, "Ferroelectric properties and switching endurance of $Hf_{0.5}Zr_{0.5}O_2$ films on TiN bottom and TiN or RuO_2 top electrodes," *physica status solidi (RRL) – Rapid Research Letters.*, vol. 8, no. 6, pp. 532–535, 2014.

[10] Y. Peng et al., "HfO_2-ZrO_2 Superlattice Ferroelectric Capacitor With Improved Endurance Performance and Higher Fatigue Recovery Capability," *IEEE Electron Device Letters.*, vol. 43, no. 2, pp. 216–219, Feb. 2022.

[11] K. Li et al., "A Comparative Study on the Polarization, Reliability, and Switching Dynamics of HfO_2-ZrO_2-HfO_2 and ZrO_2-HfO_2-ZrO_2 Superlattice Ferroelectric Films," *IEEE Transactions on Electron Devices.*, vol. 70, no. 4, pp. 1802–1807, Apr. 2023.

[12] B. Cui et al., "Back-End-of-Line Compatible HfO_2/ZrO_2 Superlattice Ferroelectric Capacitor With High Endurance and Remnant Polarization," *IEEE Electron Device Letters.*, vol. 44, no. 6, pp. 1011–1014, Jun. 2023.

[13] Y.-K. Liang et al., "ZrO_2-HfO_2 Superlattice Ferroelectric Capacitors With Optimized Annealing to Achieve Extremely High Polarization Stability," *IEEE Electron Device Letters.*, vol. 43, no. 9, pp. 1451–1454, Sep. 2022.

[14] Y. Xing et al., "Improved Ferroelectricity in Cryogenic Phase Transition of $Hf_{0.5}Zr_{0.5}O_2$," *IEEE J. Electron Devices Soc.*, vol. 10, pp. 996–1002, 2022.

[15] D. Zhang et al., "Unveiling Cryogenic Performance (4 to 300 K) Towards Ultra-Thin Ferroelectric HZO: Novel Kinetic Barrier Engineering and Underlying Mechanism," in *2024 IEEE Symposium on VLSI Technology and Circuits (VLSI Technology and Circuits).*, Jun. 2024, pp. 1–2.

[16] J. Hur, Y.-C. Luo, Z. Wang, S. Lombardo, A. I. Khan, and S. Yu, "Characterizing Ferroelectric Properties of $Hf_{0.5}Zr_{0.5}O_2$ From Deep-Cryogenic Temperature (4 K) to 400 K," *IEEE Journal on Exploratory Solid-State Computational Devices and Circuits.*, vol. 7, no. 2, pp. 168–174, Dec. 2021.

[17] M. Karthik Ram, H. Dahlberg, and L.-E. Wernersson, "Cryogenic Ferroelectricity of HZO Capacitors on a III-V Semiconductor," *IEEE Electron Device Letters.*, vol. 45, no. 10, pp. 1827–1830, Oct. 2024.

[18] Y. Goh, S. H. Cho, S.-H. K. Park, and S. Jeon, "Crystalline Phase-Controlled High-Quality Hafnia Ferroelectric With RuO_2 Electrode," *IEEE Transactions on Electron Devices.*, vol. 67, no. 8, pp. 3431–3434, Aug. 2020.

Systematic Review of Write Reliability in Spin-Transfer Torque Magnetic Random-Access Memory

Yuhao Chen [1,4], Yiming Qu [2,3], Ziyuan Chen [1], Choonghyun Lee [1], Yi Zhao [1,3,*]

[1] College of Information Science and Electronic Engineering, Zhejiang University, Hangzhou 310027, China
[2] School of Integrated Circuits, East China Normal University, Shanghai 200241, China
[3] Huada Semiconductor Co., Ltd. Shanghai 200241, China
[4] Zhejiang Li-ryder Technology Co., Ltd., Hangzhou 311300, China

* Email: yizhao@zju.edu.cn

Abstract—As a promising candidate for next-generation memory, the commercialization of Spin-Transfer Torque Magnetic Random-Access Memory (STT-MRAM) is still fundamentally challenged by its write reliability. The central metric to describe this reliability issue is the Write Error Rate (WER). This review examines the physical mechanisms governing WER and highlights its stochastic nature rooted in thermal fluctuations. It also analyzes the influence of device parameters and reliability degradation phenomena such as back-hopping. Key strategies for WER testing, characterization, and mitigation, such as Error-Correcting Codes (ECC), are also included. To achieve the ultra-low WER for mass production, an integrated approach is required combining accurate modeling with co-optimization of materials, device structure design, and write schemes.

Keywords—STT-MRAM, Magnetic Tunnel Junction (MTJ), Write reliability, WER, Stochastic switching, Back-hopping (BH)

I. INTRODUCTION

The von Neumann computing architecture, characterized by the physical separation of processing and memory units, faces significant bottlenecks known as the "memory wall" and "power wall," which are particularly acute in the era of data-intensive applications [1]. To address this challenge, memory-centric computing architectures have emerged with emerging non-volatile memory (NVM) technologies and have been playing a crucial role. Compared to traditional memories, NVMs such as RRAM, PCM, and MRAM generally offer advantages like high density, low power consumption, high speed, and non-volatility, making them ideal candidates for building future high-performance computing systems [2].

Spin-transfer torque magnetic random-access memory (STT-MRAM) stands out among emerging NVMs. It stores data by using a spin-polarized current to switch the magnetization state of a magnetic tunnel junction (MTJ) [3]. Its key advantages include high density and scalability, compatibility with advanced CMOS processes [1, 2], high speed with read/write latency in the nanosecond range, high endurance far exceeding traditional flash memory, and non-volatility enabling zero standby power. These features make STT-MRAM a strong candidate to replace SRAM and Flash, with the potential to become a universal memory [4].

The Write Error Rate (WER), defined as the probability that an MTJ fails to switch successfully under a given write pulse, is a core metric for evaluating the write reliability of STT-MRAM. It directly impacts data storage integrity and

memory yield. Commercial applications require a WER below 10^{-9}. This demand also makes characterization and verification very time-consuming and extremely challenging [5, 6].

The primary challenge of WER characterization in STT-MRAM is the inherent stochasticity of its writing process. The write operation is a probabilistic event influenced by thermal fluctuations, causing the switching time to exhibit a statistical distribution with a long tail [2, 7]. As a result, some write operations may fail under a given write pulse, leading to write errors. Furthermore, anomalous switching phenomena such as back-hopping (BH), also degrade write reliability [8, 9].

It is critical to address these write reliability issues for the large-scale adoption and wide acceptance of STT-MRAM. Therefore, a thorough understanding and accurate characterization of WER and its relationship with various physical parameters and process variations are fundamental for optimizing device design, improving write strategies, and ultimately ensuring the overall yield and reliability of STT-MRAM systems.

II. PHYSICAL BASIS OF THE STT-MRAM WRITING PROCESS

A. STT-MRAM Cell Structure and Operating Principle

Fig. 1 illustrates the typical structure of a STT-MRAM cell with one select transistor and the core storage element MTJ. Usually, one MTJ consists of two ferromagnetic layers separated by an ultrathin tunnel barrier, such as Magnesium Oxide (MgO) [2, 6]. One of the ferromagnetic layers has a fixed magnetization direction and is known as reference layer (RL) or the pinned layer (PL). The other ferromagnetic layer, whose magnetization can be switched by a spin-polarized current, is called the free layer (FL).

The operation of an MTJ is based on the tunneling magnetoresistance (TMR) and spin-transfer torque effects.

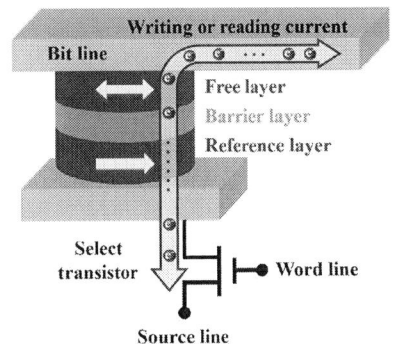

Fig. 1. Typical STT-MRAM cell structure.

This work was supported by the National Key Research and Development Program of China (2020AAA0109001) and the Chenguang Program of Shanghai Education Development Foundation and Shanghai Municipal Education Commission (23CGA35).

979-8-3315-3918-4/25 $31.00 © 2025 IEEE

When the magnetization of FL and RL are aligned, the MTJ is in a low-resistance, i.e., parallel (P) state. When their magnetizations are opposed, it is in a high-resistance, i.e., anti-parallel (AP) state [2]. This difference in resistance is used to represent binary "0" and "1". The read operation is performed by sensing the resistance value.

The write operation is achieved through the STT effect. When a sufficiently large spin-polarized current flows perpendicularly through the MTJ, the spin angular momentum of the electrons is transferred to the FL, exerting a torque on it. This spin-transfer torque can drive the magnetization of the FL to switch into either the P or AP state, depending on the direction of the current [1, 2]. For instance, electrons flowing from the RL to the FL switch the MTJ to the parallel state. In contrast, the current in the opposite direction switches the MTJ to the anti-parallel state.

B. Stochasticity of the Writing Process and the Origin of WER

The magnetization switching in STT-MRAM is an inherently stochastic process. The fundamental source of this randomness is thermal fluctuation. At room temperature, the thermal energy causes a random effective magnetic field, known as the thermal field. This field continuously exerts a random torque on the free layer's magnetic moment, interfering with the deterministic switching torque generated by the spin-polarized current [6, 7]. Consequently, the success of magnetization switching under a given write pulse becomes a probabilistic event. Thermal fluctuations can either assist or impede the switching, leading to a distribution of switching probabilities [10].

As for the evaluation of write reliability, the WER is typically plotted as a function of write voltage, covering AP2P and P2AP write scenarios, as exemplified in Fig. 3(a). When the write voltage is sufficiently low, the MTJ cannot be written, and the WER is 1. As the write voltage increases to near the critical value, the MTJ begins to switch, but write failures still occur. The WER decreases as the write voltage increases, until it drops to 10^{-6}. To achieve a very low WER and verify the robust write performance from characterizations, the write pulses are required to have a sufficiently large amplitude or long time, which in turn increases power consumption and write latency.

In addition to basic thermal fluctuations, there are more complex physical phenomena affecting the WER, especially at high write voltages. A key phenomenon is BH, as shown in Fig. 3(c). After successfully switching under the high write voltage, the magnetization of FL unexpectedly jumps back to its initial state and then exhibits a write error [12]. A deeper understanding of WER degradation will be elucidated in the next section.

III. KEY FACTORS INFLUENCING STT-MRAM WER

The switching of the free layer in a MTJ involves an energy barrier, as depicted in Fig. 2. The process of overcoming this energy barrier is influenced by operating conditions. The magnitude of the energy barrier is affected by device physical parameters. A comprehensive study of these factors is crucial for optimizing the overall write reliability in STT-MRAM.

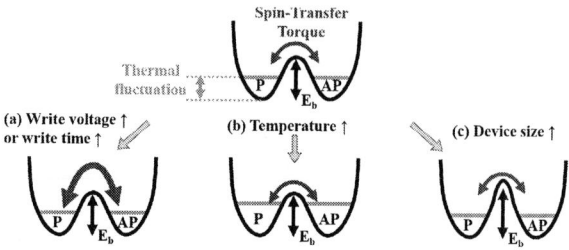

Fig. 2. (a) Higher write voltage or a longer pulse width reduces WER. (b) Higher temperature intensifies thermal fluctuation. (c) Larger device size raises the energy barrier.

A. Impact of Operating Conditions on WER

Write Pulses: The write pulse, including voltage and duration, is one of the most direct factors affecting WER characteristics. Fig. 3(a) and (d) give experimental examples illustrating the impact of write voltage and time on WER, respectively. Increasing the write voltage results in a lower WER [6, 13]. This is because a higher voltage enhances the flow of spin-polarized carriers, generating a larger spin-transfer torque, as shown in Fig. 2(a). However, it is inappropriate to blindly increase the write voltage for a better WER performance. Excessively high voltages will inevitably increase power consumption and the high electric field across the barrier layer risks the breakdown. As for the pulse width, increasing the duration of the write pulse helps to reduce the WER. A longer write time provides a longer duration of spin-transfer torque for the FL to complete its reversal, thus improving the write success rate, but simultaneously reduces the write speed [4]. Notably, a short write time with a higher write voltage can effectively reduce the WER variation caused by device parameter dispersion [11].

Operating Temperature: The temperature dependence of the WER is shown in Fig. 3(b), illustrating that the WER

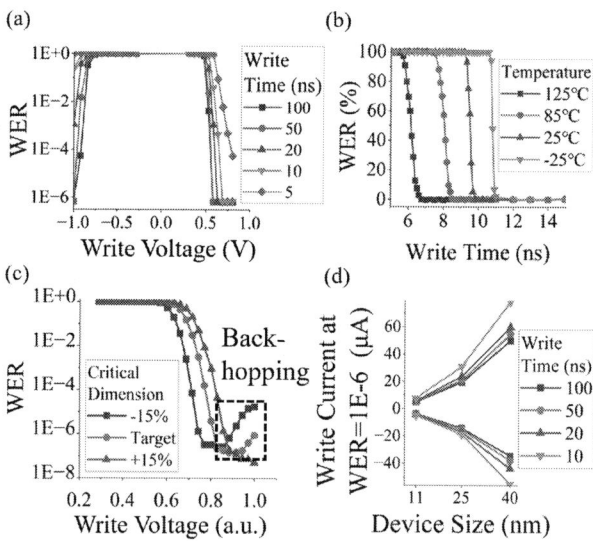

Fig. 3. (a) Comparison of WER under different write pulse widths. When no error is observed at the time that the pre-defined upper boundary for counting is reached, the WER is defined as the inverse of the total count [18]. (b) Comparison of WER under different temperatures and different write times [23]. (c) WER and the BH at different critical dimensions [24]. (d) Device size affects the write current [13].

979-8-3315-3918-4/25 $31.00 © 2025 IEEE

decreases as the temperature increases. As illustrated in Fig. 2(b), the elevated temperatures can assist magnetic switching by enhancing thermal fluctuation, thereby reducing the WER [23]. On the other hand, excessive thermal fluctuation reduces the thermal stability of the MTJ. This exacerbates the BH phenomenon, where the magnetization switches back to its initial state after a successful write [12, 14, 24]. This complex behavior means that while higher temperatures can lower the WER at moderate write voltages, they can also increase it at high write voltages due to the onset of the BH phenomenon.

External Magnetic Field: Previous studies have shown that an assisting magnetic field applied along the easy axis of magnetization can effectively lower the WER [5]. Furthermore, a recent study highlights the significance of field alignment. When the external field is not parallel to the easy axis, the WER deteriorates sharply, as shown in Fig. 4 [15]. This places stricter requirements on the magnetic shielding design for STT-MRAM.

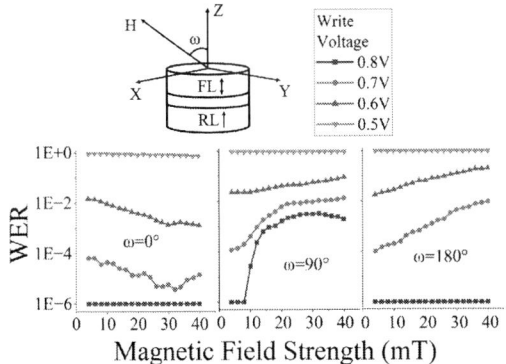

Fig. 4. WER as a function of the field magnitude at different external magnetic field angles for a 10 ns write pulse [15].

B. Impact of Intrinsic Device Parameters and Process Conditions on WER

Device Dimensions and Intrinsic Parameters: Reducing the MTJ diameter is an effective way to increase storage density, but it also lowers the energy barrier for FL reversal, as shown in Fig. 2(c). This means that a lower write current is required to achieve the same WER, as shown in Fig. 3(d) [13]. Additionally, MTJs with smaller device dimension are also more susceptible to BH, as shown in Fig. 3(c) [16]. Furthermore, intrinsic magnetic parameters such as magnetic anisotropy, saturation magnetization, damping constant, and spin polarization factor collectively determine the critical switching current and thermal stability, thus fundamentally affecting the WER [5, 6, 11]. One of the possible origins of the BH is found to be the destabilization of the RL which leads to perpetual switching of both layers [19]. Therefore, optimizing its material parameters to make RL more stable can effectively suppress this phenomenon [9, 14].

Process Variation: Minor fluctuations during manufacturing can cause variations in MTJ physical parameters, such as device size. This leads to a wide distribution of WER and severely impacts the manufacturing yield [2, 13]. Theoretical analysis indicates that a normal distribution of device parameters results in a log-normal distribution of WER [11]. Therefore, effective control of process variation is essential to achieve an ultra-low WER. Moreover, certain specific write error modes, such as low-probability bifurcated switching (LPBS) illustrated in Fig. 5,

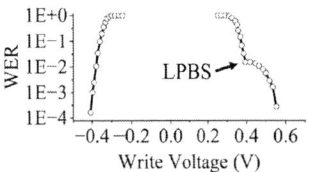

Fig. 5. A typical LPBS phenomenon, which is commonly observed in early-stage fabrication [5].

are also related to non-uniform magnetization and process defects within the device itself [5].

IV. WER CHARACTERIZATION STRATEGIES

The accurate experimental characterization of WER in STT-MRAM is essential to assess its reliability, guide device optimization, and ensure its viability in commercial applications. For instance, high-performance memory like cache requires extremely low error rates. However, directly testing such a rare event by simple counting is prohibitive in terms of time and equipment cost. Consequently, a range of sophisticated strategies has been developed in WER characterization.

Typically, the MTJ characterization is performed on a single MTJ device in the array. With specialized high-speed equipment and test schemes, a large number of write-read cycles are applied to the MTJ to accumulate sufficient statistics and detect a possible write error. As illustrated in Fig. 6, a basic cycle involves three steps. Firstly, set the MTJ to a known initial state (P or AP) with a reset pulse; secondly, apply a write pulse with a specific pulse width and voltage to switch the MTJ to the opposite state (AP or P); and finally, identify if the MTJ state has been successfully switched by verifying with a low-bias read operation [13]. To achieve a test limit of, for example, 10^{-6} WER, at least one million write cycles must be executed for each data point, which underscores the demanding nature of such tests.

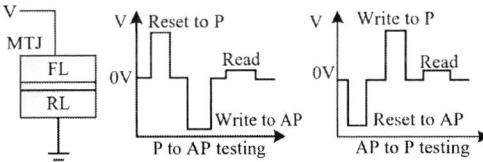

Fig. 6. WER test waveform that will be repeated millions of times in the characterization.

For instance, Nowak et al. systematically measured the WER of 655 devices with diameters ranging from 11 nm to 50 nm, successfully obtaining statistically significant data at an error level of 10^{-6} [13]. Similarly, Gao et al. developed a high-throughput parallel characterization system and successfully characterized WER at the 10^{-6} error level [18, 22]. To probe an even deeper error rate, Hu et al. successfully pushed the WER testing of a single device with a 5-nanosecond pulse down to the 10^{-10} level, which is crucial for verifying the ultimate performance of STT-MRAM in high-speed applications [19]. To precisely control and characterize device behavior in the sub-nanosecond regime, researchers have also developed specialized test platforms, such as Li-ryder Oe Probe system and M3000 MRAM analyzer, as shown in Fig. 7 [20]. These advanced characterization solutions can capture physical details that are invisible to conventional test methods.

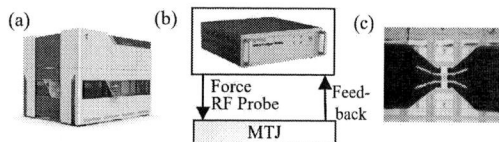

Fig. 7. (a) Li-ryder Oe Probe system. (b) Li-ryder M3000 MRAM analyzer. (c) RF probes and layout configuration.

V. RELIABILITY ENHANCEMENT AT THE CIRCUIT AND ARCHITECTURAL LEVELS

To address the reliability challenges posed by the inherently stochastic nature of write operations in magnetic memories, researchers have developed various techniques at the circuit and architectural levels to realize a lower WER. For instance, Zhao et al. designed a Pre-Charge Sense Amplifier for STT-RAM, which enables a more reliable read margin, thereby preventing inadvertent switching during read operations [7]. Noguchi et al. employed a write-verify-write technique to reduce the WER [17]. Furthermore, Error-Correcting Codes (ECC) serve as the final line of defense for ensuring system-level reliability. By introducing redundant bits, ECC can detect and correct a limited number of errors. This process ensures data integrity at the system level. Building upon this, Sayed et al. optimized ECC by leveraging the inherent properties of the MTJ. They designed a novel error-correcting code that utilizes the asymmetrical AP to P and P to AP switching characteristics of STT-MRAM and combined it with an "early write termination" strategy [21]. The resulting errors from this incomplete write are subsequently corrected by ECC. This trade-off significantly reduces average write latency and power consumption, achieving a co-optimization of performance and reliability.

VI. CONCLUSIONS

STT-MRAM is a key next-generation memory technology, offering high density, high speed, and low power consumption. The core challenge to its commercialization is write reliability, which stems from the inherent stochasticity of the writing process. The WER is affected by factors such as write pulse parameters, temperature, and fabrication process variations. To characterize the low WER required for commercial applications, academia and industry have developed advanced methods and equipment. Furthermore, researchers have developed various techniques at the circuit and architectural levels to realize a lower WER.

REFERENCES

[1] Y. Li, T. Bai, X. Xu, Y. Zhang, B. Wu, and H. Cai, et al., "A survey of MRAM-centric computing: from near memory to in memory," IEEE Trans. Emerg. Topics Comput., vol. 11, no. 2, pp. 318–330, 2023.

[2] P. Girard, Y. Cheng, A. Virazel, W. Zhao, R. Bishnoi, and M. B. Tahoori, "A survey of test and reliability solutions for magnetic random access memories," Proc. IEEE, vol. 109, no. 2, pp. 149–169, 2021.

[3] A. V. Khvalkovskiy, D. Apalkov, S. Watts, R. Chepulskii, R. S. Beach, and A. Ong, et al., "Basic principles of STT-MRAM cell operation in memory arrays," J. Phys. D: Appl. Phys., vol. 46, no. 7, p. 074001, 2013.

[4] D. Apalkov, B. Dieny and J. M. Slaughter, "Magnetoresistive random access memory," in Proceedings of the IEEE, vol. 104, no. 10, pp. 1796-1830, 2016.

[5] T. Min, Q. Chen, R. Beach, G. Jan, C. Horng, and W. Kula, et al., "A study of write margin of spin torque transfer magnetic random access memory technology," IEEE Trans. Magn., vol. 46, no. 6, pp. 2322–2327, 2010.

[6] D. Das and X. Fong, "A Fokker–Planck approach for modeling the stochastic phenomena in magnetic and resistive random access memory devices," IEEE Trans. Electron Devices, vol. 68, no. 12, pp. 6124–6131, 2021.

[7] W. S. Zhao, Y. Zhang, T. Devolder, J. O. Klein, D. Ravelosona, and C. Chappert, et al., "Failure and reliability analysis of STT-MRAM," Microelectron. Reliab., vol. 52, no. 9–10, pp. 1848–1852, 2012.

[8] J. Sun, M. C. Gaidis, G. Hu, E. J. O'Sullivan, S. L. Brown, and J. J. Nowak, et al., "High-bias backhopping in nanosecond time-domain spin-torque switches of MgO-based magnetic tunnel junctions," J. Appl. Phys., vol. 105, no. 5, p. 07D109, 2009.

[9] C. Abert, H. Sepehri-Amin, F. Bruckner, C. Vogler, M. Hayashi, and D. Suess, "Back hopping in spin-transfer-torque devices, possible origin and counter measures," Phys. Rev. Applied, vol. 9, no. 5, p. 054010, 2018.

[10] Z. Wang, Y. Zhou, J. Zhang, and Y. Huai, "Write error rate in spin transfer torque magnetic random access memory," Spin, vol. 2, no. 3, p. 1240001, 2012.

[11] H. Imamura, H. Arai, and R. Matsumoto, "Distribution of write error rate of spin-transfer-torque magnetoresistive random access memory caused by a distribution of junction parameters," J. Magn. Magn. Mater., vol. 563, p. 170012, 2022.

[12] T. Min, J. Z. Sun, R. Beach, D. Tang, P. Wang, "Back-hopping after spin torque transfer induced magnetization switching in magnetic tunneling junction cells", J. Appl. Phys., vol. 105, pp. 07D126, 2009.

[13] J. J. Nowak, R. P. Robertazzi, J. Z. Sun, G. Hu, J.-H. Park, and J. Lee, et al., "Dependence of voltage and size on write error rates in spin-transfer torque magnetic random-access memory," IEEE Magn. Lett., vol. 7, pp. 1-4, 2016.

[14] S. Yuan, M. Taouil, M. Fieback, H. Xun, E. J. Marinissen, and G. S. Kar, et al, "Device-aware test for back-hopping defects in STT-MRAMs," in 2023 Design, Automation & Test in Europe Conference & Exhibition (DATE), Antwerp, Belgium, pp. 1-6, 2023.

[15] N. V. Meeren, S. Van Beek, M. G. Monteiro, F. Garcia-Redondo, J. Chatterjee, and A. Kumar, et al., "Magnetic immunity of STT-MRAM: external magnetic field orientation impact on writing reliability," in Proc. 2024 IEEE Int. Electron Devices Meeting (IEDM), pp. 1-4, 2024.

[16] S. H. Han, J. H. Lee, K. S. Suh, K. T. Nam, D. E. Jeong, and S. C. Oh, et al., "Reliability of STT-MRAM for various embedded applications," 2021 IEEE International Reliability Physics Symposium (IRPS), Monterey, CA, USA, pp. 1-5, 2021.

[17] H. Noguchi, K. Ikegami, S. Takaya, E. Arima, K. Kushida, and A. Kawasumi, "7.2 4Mb STT-MRAM-based cache with memory-access-aware power optimization and write-verify-write / read-modify-write scheme," 2016 IEEE International Solid-State Circuits Conference (ISSCC), pp. 132-133, 2016.

[18] S. Gao, B. Chen, N. Xu, Y. Qu, and Y. Zhao, "Probing write error rate and random telegraph noise of MgO based magnetic tunnel juction using a high throughput characterization system," in Proc. 2019 IEEE Int. Rel. Phys. Symp. (IRPS), pp. 1–4, 2019.

[19] G. Hu, J. J. Nowak, M. G. Gottwald, J. Z. Sun, D. Houssameddine, and J. Bak. et al., "Reliable Five-Nanosecond Writing of Spin-Transfer Torque Magnetic Random-Access Memory," in IEEE Magnetics Letters, vol. 10, pp. 1-4, 2019.

[20] "Zhejiang Liryder Technologies Co., LTD", Jan., 2023. [Online]. Available: https://www.liryder.com/en/. [Accessed Jun. 20, 2025].

[21] N. Sayed, M. Ebrahimi, R. Bishnoi, and M. B. Tahoori, "Opportunistic write for fast and reliable STT-MRAM," in Design, Automation & Test in Europe Conference & Exhibition (DATE), Lausanne, Switzerland, pp. 554-559, 2017.

[22] Z. Zhang, S. Gao, Y. Zhao, X. Yang, J. Zhao, and S. He, "High-efficiency array-level MRAM parameters extraction with the device-in-series test structure," J. Appl. Phys., vol. 131, no. 4, pp. 554-559, 2022.

[23] B. Wu, Y. Cheng, J. Yang, A. Todri-Sanial and W. Zhao, "Temperature impact analysis and access reliability enhancement for 1T1MTJ STT-RAM," in IEEE Transactions on Reliability, vol. 65, no. 4, pp. 1755-1768, 2016.

[24] J. Tan, J.H. Lim, J.H. Kwon, V.B. Naik, N. Raghavan b, and K.L. Pey, "Role of temperature, MTJ size and pulse-width on STT-MRAM bit-error rate and backhopping," Solid-State Electronics, vol. 183, p. 108032, 2021.

Emerging Magnetoresistive Memories

Viktor Sverdlov
Christian Doppler Laboratory for Nonvolatile Magnetoresistive Memory and Logic
Institute for Microelectronics, TU Wien
Vienna, Austria
sverdlov@iue.tuwien.ac.at

Nils Petter Jørstad
Christian Doppler Laboratory for Nonvolatile Magnetoresistive Memory and Logic
Institute for Microelectronics, TU Wien
Vienna, Austria
jorstad@iue.tuwien.ac.at

Bernhard Pruckner
Christian Doppler Laboratory for Nonvolatile Magnetoresistive Memory and Logic
Institute for Microelectronics, TU Wien
Vienna, Austria
pruckner@iue.tuwien.ac.at

Mario Bendra
Institute for Microelectronics, TU Wien
Vienna, Austria
bendra@iue.tuwien.ac.at

Wolfgang Goes
Silvaco Europe Ltd.
Cambridge, United Kingdom
wolfgang.goes@silvaco.com

Siegfried Selberherr
Institute for Microelectronics, TU Wien
Vienna, Austria
selberherr@iue.tuwien.ac.at

Abstract—Nonvolatile CMOS-compatible spin-transfer torque (STT) and spin-orbit torque (SOT) magnetoresistive random access memories (MRAMs) exhibit high speed, endurance, and long retention when compared to competing technologies. These advanced devices are composed of multiple magnetic layers, which are separated by tunnel barriers and non-magnetic metalic spacers. To effectively model magnetization dynamics in such complex structures, we employ a coupled spin and charge transport approach accurately capturing spin accumulation and the torques acting on ferromagnetic layers. We apply appropriate boundary conditions at the interfaces to evaluate spin and charge transport in metallic spin valves and magnetic tunnel junctions. Our approach has demonstrated versatility in several areas, including the accurate evaluation of the operation of ultra-fast multilayer STT-MRAM, achieving magnetic field-free switching in SOT-MRAM with a heavy metal/ferromagnetic stack, and managing magnetization in the strained noncollinear antiferromagnet Mn_3Sn. By integrating an Mn_3Sn layer with a ferromagnetic layer, we enable electrical control over magnetization, thus creating opportunities for future field-free SOT-MRAM devices.

Keywords— Spin and charge drift-diffusion, spin torques, magnetic tunnel junctions, TCAD, STT-MRAM, composite free layer, SOT-MRAM, field-free switching, noncolinear antiferromagnets.

I. INTRODUCTION

Non-volatile memristive devices offer great potential in overcoming the limitations of traditional CMOS-based memories by enabling innovations for automotive systems, Internet of Things devices, and AI accelerators [1]. While resistive random access memory (RRAM) provides low costs, high density, and good scalability and integration [2], magnetoresistive random access memory (MRAM) delivers high reliability and superior power, performance, and area in extreme environments.

The key element of an MRAM cell is a magnetic tunnel junction (MTJ) [3], [4], [5]. It consists of two single-domain ferromagnetic layers separated by a tunnel barrier. MTJ resistance is high for antiparallel (AP) and low for parallel (P) ferromagnetic magnetization arrangement. Binary information is stored in an MTJ configuration by assigning, i.e., a logical zero to an MTJ in P and a logical one in AP configuration. The magnetization of one of the layers, the reference layer (RL), is fixed. The magnetization of the free layer (FL) can be changed by an external magnetic field or by other means. In advanced CoFeB|MgO|CoFeB MTJ with perpendicular magnetization (p-MTJs), the magnetoresistance ratio reaches a few hundred percent at room temperature [6].

The binary data can be reliably read by sensing the MTJ resistance. The resistance of an MTJ is comparable to that of a MOSFET, which makes MRAM electrically compatible with CMOS circuitry, eliminating the need to convert the spin (magnetization) orientations into electric (current, voltage) signals with the help of additional amplifiers.

MRAM development [7] has been given a significant boost in the last decade as the ability to write information purely electrically without an external magnetic field was discovered. Electric current running through an MTJ becomes spin polarized in the RL. It delivers a spin-transfer torque (STT) [8], [9] on the FL magnetization and can flip it from P to AP configuration. FL magnetization reversal is achieved by inverting the current. STT-MRAM is non-volatile and offers low power consumption, fast speed (1-10 ns), long retention (~10 years), and high endurance (10^{12}). MRAM is thus perfectly suited for stand-alone and embedded memory applications. STT-MRAM is scalable beyond the flash memory limits [10]. It also requires fewer additional masks for fabrication compared to flash memory. STT-MRAM for last-level caches implemented with 16 nm FinFET [11] and 16 nm fully depleted silicon-on-insulator technology [12] for embedded applications is commercially available. Major foundries have already demonstrated a perspective of 8 nm STT-MRAM for automotive applications [13] and plan to proceed beyond [14]. To reduce the MTJ footprint to a record small 2.3 nm and to simultaneously boost the out-of-plane magnetic anisotropy, the FL was assembled of several elongated CoFeB nanopillars with MgO layers in between [15].

Despite its remarkable performance demonstrated by STT-MRAM, including its applicability for last-level caches [11], STT-MRAM reliability remains a critical concern [16], especially for fast applications, when large electric currents run through the tunnel barrier. Optimizing the MTJ stack for a lower write current for high-density memory could be done at the cost of the retention time [17]. To further increase the STT-MRAM operating speed beyond one nanosecond, the second fixed ferromagnetic RL separated from the FL by a non-magnetic metal spacer (NMS) is included in the MRAM cell [18]. This double spin-torque magnetic tunnel junction (dsMTJ) structure [19] with two RLs reduces the switching current by a factor of two compared to the one of a traditional MTJ structure. However, the current is still running through the MTJ. In STT-MRAM, increasing the current is the only option to meet the requirements for memory operations in higher-level caches.

979-8-3315-3918-4/25 $31.00 © 2025 IEEE

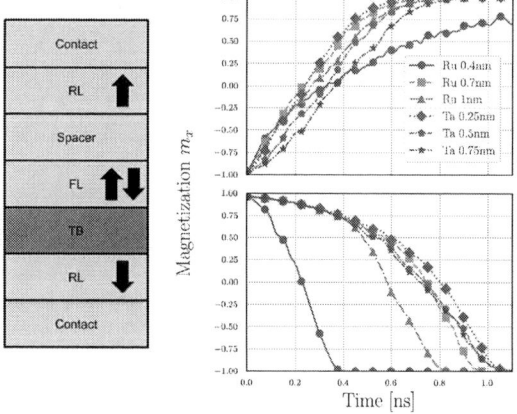

Fig.1. **Left**: dsMTJ multilayer design. The color coding represents RLs in red, the FL in cyan, the tunnel barrier and the NMS in orange, HL in purple, and non-magnetic contacts in gray. **Right**: Switching performance for dsMTJ multilayer devices utilizing a Ru/Ta-NMS under varying NMS thicknesses and IEC strength.

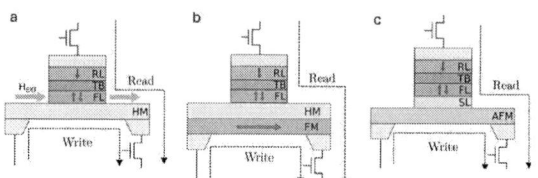

Fig.2 SOT-MRAM cells consisting of FL and RL, separated by a TB, sitting on top of: (**a**) the HM with SHE/REE; (**b**) an SOT-MRAM cell with an additional in-plane ferromagnetic layer under the HM; (**c**) an antiferromagnetic layer with the magnetic spin Hall effect. A spacer layer (SL) is introduced between the FL and the HM to mitigate the exchange coupling.

Spin-orbit torque (SOT) MRAM is another example of fast, electrically addressable, and nonvolatile magneto-resistive memory with high endurance [19]. SOT-MRAM is promising for replacing SRAM even in high-level caches [20]. In analogy to STT-MRAM, SOT-MRAM employs an MTJ to store the information. To read the data, a small read current is applied through the MTJ to sense its resistance.

In contrast to STT-MRAM, the FL of an SOT-MRAM cell is grown on a SOT stack. The spin currents delivering the SOT on the FL are generated through the spin Hall effect (SHE) in the SOT stack and through the Rashba-Edelstein effect (REE) at the heavy metal (HM)/FL interface [21]. SOT provides a fast and efficient way to manipulate the magnetization of the FL [19]. Importantly, the write current generating SOTs flows only within the SOT stack. Since the current does not flow through the MTJ, it can be quite high to alleviate fast FL switching. However, to achieve the deterministic magnetization reversal of a perpendicularly magnetized FL, an external in-plane magnetic field is required [22].

Optimizing STT- and SOT-MRAM for optimal performance is an important task, as it requires achieving a proper balance between the thermal stability (retention) and speed of operation (power) and reading. TCAD tools are essential to efficiently assist in developing and optimizing emerging devices. Modeling advanced MRAM cells requires an accurate evaluation of spin currents and torques in complex MTJs with composite FLs and several RLs. We employ a recently developed coupled spin and charge transport approach to accurately evaluate spin torques in multilayered structures [23] in the presence of interfacial exchange coupling [24]. We account for the torques generated by the SHE and REE [25] and integrate them in a finite element method based micromagnetic simulation environment ViennaSpinMag [26]. We confirm the versatility of our approach by showing reliable sub-ns switching in multilayer ultra-scaled STT-MRAM, deterministic field-free switching in SOT-MRAM with a HM/ferromagnet SOT stack and with the magnetic SHE, as well as magnetization control in strained noncollinear antiferromagnet (AFM) Mn$_3$Sn.

II. MODELING ULTRA-SCALED STT-MRAM

Achieving reliable sub-nanosecond MTJ switching is crucial for next-generation nonvolatile memory for computing, automotive, and storage applications. DsMTJs [18] (Fig.1,

left) enhance traditional MTJs by integrating an additional RL. We employ our fully three-dimensional finite element modeling approach [23], [24], [25], [26] which combines the drift-diffusion spin and charge transport with the Landau-Lifshitz-Gilbert equation:

$$\frac{\partial \mathbf{m}}{\partial t} = -\gamma\mu_0 \mathbf{m} \times \mathbf{H}_{\text{eff}} + \alpha \mathbf{m} \times \frac{\partial \mathbf{m}}{\partial t} + \frac{1}{M_S}\mathbf{T_S} \qquad (1)$$

Here $\mathbf{m} = \mathbf{M}/M_S$ is the position and time-dependent normalized magnetization, M_S is the saturation magnetization, α is the Gilbert damping, γ is the gyromagnetic ratio, and μ_0 is the vacuum permeability. The effective field $\mathbf{H}_{\text{eff}}$ includes the magnetic anisotropy field, the exchange, as well as the demagnetization and stray fields. The coupled spin and charge drift-diffusion transport model is employed to determine the spin accumulation $\mathbf{S}$ and the corresponding torques $\mathbf{T_S}$ acting on the magnetization. Continuous boundary conditions for the spin accumulation and the spin current densities [23] are employed at the interfaces between a ferromagnet layer and an NMS. In order to accurately describe the STT acting on the magnetization in a nanometer-sized multilayer MTJ for a given charge current density $\mathbf{J_C}$, the spin current density through the tunnel barrier is described by [23]:

$$\mathbf{J_{S,TB}} = -\frac{\mu_B}{e}\frac{\mathbf{J_C} \cdot \mathbf{n}}{1 + P_{RL}P_{FL}\mathbf{m_{RL}} \cdot \mathbf{m_{FL}}}\Big(P_{RL}\mathbf{m_{RL}} + P_{FL}\mathbf{m_{FL}}$$
$$+ \frac{1}{2}\big(P_{RL}P_{RL}^{\eta} - P_{FL}P_{FL}^{\eta}\big)\mathbf{m_{RL}} \times \mathbf{m_{FL}}\Big) \qquad (2)$$

Here $\mathbf{m_{RL}}$ and $\mathbf{m_{FL}}$ are the normalized magnetization vectors of the RL and FL at the TB interfaces, respectively, and the in-plane Slonczewski interface polarization parameters P_{RL}, P_{FL} and the out-of-plane factors $P_{RL}^{\eta}, P_{FL}^{\eta}$ are introduced.

Fig.1, right panel, shows the impact of a Ru or a Ta NMS of various thicknesses on the dsMTJ performance. Optimal sub-ns performance is achieved with a Ru NMS with 1 nm thickness, exhibiting antiferromagnetic interlayer interface exchange coupling of 0.65 mJm^{-2}. For a Ta NMS, the best performance is observed at 0.25 nm thickness with FM coupling of 0.4 mJm^{-2}. The fastest switching from the AP to the P state is achieved with a Ru NMS at 0.4 nm, corresponding to an AFM coupling of 2.1 mJm^{-2} (Fig.1, right panel). A Ru NMS is characterized by a longer spin-flip length of 4 nm and demonstrates better performance compared to a Ta NMS, with a shorter spin-flip length of 1.7 nm, since a slower spin accumulation decay in the NMS results in larger torques on the FL from the additional RL.

III. SPIN-ORBIT TORQUES FOR ADVANCED MRAM

Spin-orbit torque provides rapid and energy-efficient manipulation of magnetic states in emerging SOT-MRAM

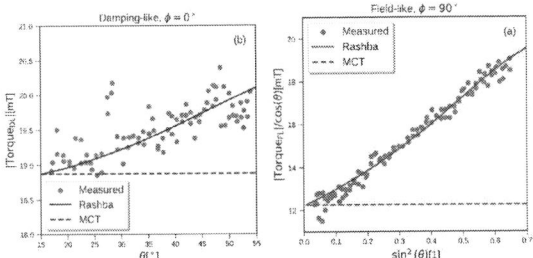

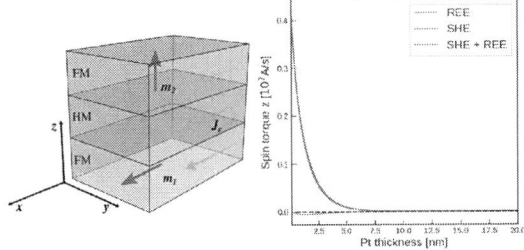

Fig.3. Angular dependence of SOTs in a Pt(3 nm)/Co(0.6 nm) bilayer with a 0.38×10^{12} Am^{-2} electrical current along the x-direction. The torques were computed using a spin drift-diffusion model with boundary conditions for interfacial SOC and fitted to experimental data [25]. The damping-like (**left**) and field-like (**right**) torques could be reproduced only, if the interfacial REE is considered.

Fig.4 Out-of-plane spin current originating from the lower FM/HM interface (Fig.2b) and its dependence on the HM thickness (right panel).

[19]. Several flavors of SOT-MRAM with different SOT tracks are shown in Fig. 2. Torques $\mathbf{T_S}$ acting on the FL magnetization $\mathbf{m}$ are caused by the nonequilibrium spin accumulation $\mathbf{S}$ [22]:

$$\mathbf{T_S} = -\frac{D_e}{\lambda_J^2}\mathbf{m} \times \mathbf{S} - \frac{D_e}{\lambda_\varphi^2}\mathbf{m} \times (\mathbf{m} \times \mathbf{S}) \quad (3)$$

Here, λ_J is the exchange length in the FL, and λ_φ is the spin dephasing length. The spin accumulation is obtained with the coupled spin and charge drift-diffusion approach [24], [25], [26], generalized to include the SHE via the spin current $\overline{\mathbf{J}}_\mathbf{s}$:

$$\overline{\mathbf{J}}_\mathbf{s} = -D_e\nabla\mathbf{S} - \theta_{SH}\boldsymbol{\varepsilon}\mathbf{J_c} \quad (4)$$

The last contribution is due to the SHE effect, where $\boldsymbol{\varepsilon}$ is the unit antisymmetric tensor and θ_{SH} is the spin Hall angle. The SHE generates the spin current orthogonal to the charge current, witch a spin polarization normal to both the spin and charge currents. The spin current impinging on the FL generates the SOT. However, purely SHE-generated torques are not sufficient to explain observed angular dependences [27]. The spin drift-diffusion approach must be complemented with the corresponding boundary conditions to include the interfacial REE. Effects of interfacial spin-orbit interaction on the three-dimensional spin transport at the FL/HM interface at $z=0$ are conveniently described with the spin-dependent scattering potential-barrier $V(\mathbf{r})$ [21], [25]:

$$V(\mathbf{r}) = \frac{\hbar^2 k_F}{m}\delta(z)\left[u_0 I + \boldsymbol{\sigma}\left(u_{ex}\mathbf{m} + u_R\, \mathbf{z} \times \frac{\mathbf{k}}{k_F}\right)\right] \quad (5)$$

Here u_0, u_{ex}, u_R are the dimensionless parameters describing the spin-independent potential, the exchange interaction, and the Rashba spin-orbit coupling at the interface, respectively. $\delta(z)$ is the Dirac delta function, k_F is the Fermi wave vector, $\mathbf{k}$ is the wave vector on the Fermi surface, $\boldsymbol{\sigma}$ is the vector of Pauli matrices, m is the effective mass, and $\mathbf{m}$ is the magnetization direction in the ferromagnet at a certain position at the interface. Fig. 3 demonstrates that with the interfacial SOC included, the experimentally measured angular dependences of SOTs at the Pt/Co interface are successfully reproduced [25].

Even with interfacial SOTs included, deterministic switching of a perpendicular FL requires an external magnetic field. A promising approach is to leverage unconventional SOTs in FL/NM/FM trilayers [28], [29] (Fig. 2,left). Spin currents with out-of-plane spin polarization are generated due to scattering from the magnetization-dependent po-

tential-barrier at the lower NM/FM interface [29]. This current reaches the upper FL (Fig. 4, right) and ensures the sub-ns deterministic switching.

Employing a noncollinear antiferromagnet (nc-AFMs) like Mn$_3$Sn as a spin-polarizing layer in SOT-MRAM (Fig. 2c) offers an alternative path to generate nonconventional torques due to out-of-plane z-polarized spin currents generated by the magnetic SHE (MSHE) [30]. Fig. 5 (left) displays the angular dependence of the SOTs acting on the magnetization in a Mn$_3$Sn/FL bilayer. The z-polarized spin current breaks the mirror symmetry by shifting the magnetization out of plane (bright spots). The magnetization relaxes up (down) depending on the current direction after the current is off, making the switching deterministic. Thus, the MSHE-assisted deterministic switching can be accelerated into the sub-ns regime (Fig. 5, right).

The Mn$_3$Sn layer itself could serve for information storage. Epitaxial tensile strain distorts the Mn$_3$Sn lattice (Fig. 6, left) grown on W epitaxially [31]. This lifts the sixfold degeneracy (Fig. 6, right) and induces a small net magnetic moment $\mathbf{m}_{net}$, resulting in two stable states, with $\mathbf{m}_{net}$ pointing either up or down (Fig. 6, left).

The direction of the net magnetization $\mathbf{m}_{net}$, or, to be precise, the octupole moment, can be manipulated by the SOT [31], [32], generated by charge currents in W. Spin currents polarized normal to the Kagome plane (Fig.6, left) push the net magnetization out of plane. The Mn spins shown in Fig. 6, while staying in-plane due to the exchange interaction, start precessing around the out-of-plane net magnetization. The precession persists if the SOTs are strong enough to bring the spin configuration over the potential barriers. The precession proceeds in either clockwise or anticlockwise directions depending on the direction of the charge current. The in-plane magnetic field lifts the degeneracy between the two potential barriers, as shown in Fig.7 . This can interrupt the precession due to the rise of one of the barriers, while allowing the net magnetization to flip over the decreased barrier (Fig. 7, right). Therefore, Mn$_3$Sn can also be used as a material for FL in a memory device [31], [32], whose state can be electrically manipulated and detected.

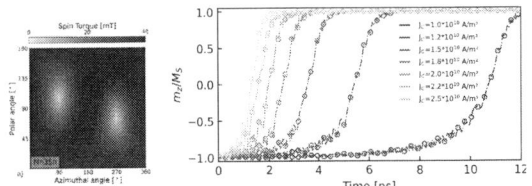

Fig.5 **Left**: MSHE breaks the mirror symmetry and ensures the deterministic switching in a Mn$_3$Sn(5 nm)/CoFeB(1 nm) bilayer. **Right**: MSHE-assisted switching of CoFeB layer.

979-8-3315-3918-4/25 $31.00 © 2025 IEEE

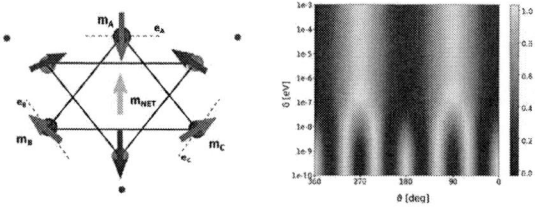

Fig.6 **Left**: Spins on the Kagome lattice of Mn₃Sn. Tensile strain δ breaks the 6-fold degeneracy (**right**) and develops a small net magnetization **m**net points up or down (**left**).

Manipulating $\mathbf{m}_{net}$ of a Mn₃Sn layer by spin currents opens an interesting perspective. Initially, a perpendicular CoFeB FL can be switched by the MSHE from Mn₃Sn. Then, the unconventional torques due to the spin-polarized currents from CoFeB [29] prompt $\mathbf{m}_{net}$ of Mn₃Sn to precess. The Oersted field generated by a current through the Mo NMS between CoFeB and Mn₃Sn layers [33] enables deterministic switching of the Mn₃Sn layer. Utilizing enhanced electric control over magnetizations in a Mn₃Sn/Mo/CoFeB trilayer structure holds significant promise for future SOT-MRAM devices.

IV. CONCLUSION

The effectiveness of the spin- and charge-coupled transport approach in evaluating the spin-transfer torques acting on ferromagnets within advanced multilayer MRAM cells is demonstrated through an analysis of ultra-fast STT-MRAM and advanced SOT-MRAM. The findings indicate that nc-AFM Mₙ3Sn is a promising candidate for emerging devices as a free switching layer. Additionally, the enhanced electric magnetization control in a Mn₃Sn/NMS/CoFeB trilayer structure presents exciting possibilities for the creation of innovative SOT-driven devices.

ACKNOWLEDGMENT

Financial support by the Austrian Federal Ministry of Labour and Economy, the National Foundation for Research, Technology and Development and the Christian Doppler Research Association is gratefully acknowledged.

REFERENCES

[1] M. Hellenbrand, I. Teck, J.L. MacManus-Driscoll, "Progress of Emerging Non-volatile Memory Technologies in Industry," MRS Communications 14, pp. 1099–1112 (2024).

[2] J. Xie, A.F. Yekta, F.A. Mamun, et al., On-chip Direct Synthesis of Boron Nitride Memristors. Nat. Nanotechnol. (2025) https://doi.org/10.1038/s41565-025-01988-z

[3] H. Aikawa, J. Song, T. Nagase, et al., "Reliable Memory Operation with Low Read Disturb Rate in the World Smallest 1Selector-1MTJ Cell for 64Gb Cross-Point MRAM," in Proc.IEDM Conf.,p.1-4 (2024).

[4] T.Y. Lee, M. Lee, M.K. Kim, et al., "World-most Energy-Efficient MRAM Technology for Non-Volatile RAM Applications," in Proc. IEDM Conf., pp.10.7.1-10.7.4 (2022).

[5] S. Ikegawa, K. Nagel, F.B. Mancoff, et al., "High-Speed (400MB/s) and Low-BER STT-MRAM Technology for Industrial Applications", in Proc. IEDM Conf., pp. 10.4.1-10.4.4 (2022).

[6] I.S. Ikeda, K. Miura, H. Yamamoto, et al., "A Perpendicular-anisotropy CoFeB–MgO Magnetic Tunnel Junction," Nature Mater. 9, 721 (2010).

[7] D. Apalkov, B. Dieny, J.M. Slaughter, "Magnetoresistive Random Access Memory," Proceedings of the IEEE 104, p. 1796 (2016).

[8] J. Slonczewski, "Current-Driven Excitation of Magnetic Multilayers," J. Magn. Magn. Mater. 159, L1-L7 (1996).

[9] L. Berger, "Emission of Spin Waves by a Magnetic Multilayer Traversed by a Current," Phys. Rev.B 54, pp.9353-9358 (1996).

[10] https://www.eenewseurope.com/en/tsmc-moves-to-mram-for-scratch-pad-memory/ Accessed on 15.8.2025.

[11] J. G. Alzate, U. Arslan, P. Bai, et al., "2Mb Array-Level Demonstration of STT-MRAM Process and Performance towards L4 Cache Applications," in Proc. IEDM Conf., pp.2.4.1–2.4.4 (2019).

[12] Y.-D. Chih, C.-C. Chou, Y.-C. Shih et al., "Design Challenges and Solutions of Emerging Nonvolatile Memory for Embedded Applications," in Proc. IEDM Conf., pp.2.4.1–2.4.4 (2021).

[13] S. Ko, J. Shim, J. H. Park, et al, "Key Technologies of Scaling Embedded MRAM to 8nm Logic and Beyond for Automotive Application", in Proc. IEDM Conf., 2024, pp. 1-4 (2024).

[14] https://www.eenewseurope.com/en/tsmc-looks-to-5nm-mram-plans-first-european-design-centre/ / Accessed on 15.8.2025.

[15] B. Jinnai, J. Igarashi, K. Watanabe, et al., "High-Performance Shape-Anisotropy Magnetic Tunnel Junctions down to 2.3 nm", in Proc. IEDM Conf., pp. 24.6.1–24.6.4 (2020).

[16] H. Jung, Y.J. Song, S. Ko, et al., "Overview of Reliability in Scaling Embedded STT-MRAM," in Proc. 2025 IEEE International Reliability Physics Symposium (IRPS), pp.1-6 (2025).

[17] W.-S. Khwa, Y.-L.Lu, S.Q. Zhang, et al., "MRAM Design-Technology-System Co-Optimization for Artificial Intelligence Edge Devices," in Proc. IEDM Conf., pp.1-4 (2024).

[18] G. Hu, C. Safranski, J. Sun, et al., "Double Spin-Torque Magnetic Tunnel Junction Devices for Last-Level Cache Applications," in Proc. IEDM Conf., pp. 10.2.1-10.2.4 (2022).

[19] S. Hu, X. Qiu, C. Pan, et al., "Frontiers in All Electrical Control of Magnetization by Spin-Orbit Torque," Journal of Physics: Condensed Matter, vol. 36, no. 25, p. 253001, 2024.

[20] V.D. Nguyen, G. Talmelli, M. Gama Monteiro, et al., "Achieving 1ppm Write-Error Rate in SOT-MRAM with Synthetic Antiferromagnetic Free Layer," in Proc. IEDM Conf., pp.1-4 (2024).

[21] V.P. Amin, P.M. Haney, and M.D. Stiles, "Interfacial Spin–orbit Torques", Journal of Applied Physics, vol. 128, no. 15, p. 151101 (2020).

[22] I.M. Miron, K. Garello, G. Gaudin, et al., "Perpendicular Switching of a Single Ferromagnetic Layer Induced by In-plane Current Injection," Nature, vol. 476, p. 189 (2011).

[23] S. Fiorentini, M. Bendra, J. Ender, et al., "Spin and Charge Drift-diffusion in Ultra-scaled MRAM Cells," Sci.Rep. 12, p. 20958 (2022).

[24] M. Bendra, R.L. de Orio, S. Selberherr, et al., "Advanced Modeling and Simulation of Multilayer Spin–Transfer Torque Magnetoresistive Random Access Memory with Interface Exchange Coupling," Micromachines 15, pp. 568-1–568-14 (2014).

[25] N.P. Jørstad, S. Fiorentini, J. Ender, et al., "Micromagnetic Modeling of SOT-MRAM Dynamics,"PhysicaB: Cond.Mat. 676, 415612 (2024).

[26] https://www.iue.tuwien.ac.at/viennaspinmag Accessed on 15.8.2025

[27] K. Garello, I. M. Miron, C. O. Avci, et al., "Symmetry and Magnitude of Spin-orbit Torques in Ferromagnetic Heterostructures," Nat.Nanotechnol. 8, pp. 587–593, 2013.

[28] J.Ryu, R. Thompson, J.Y. Park, et al., "Efficient Spin-orbit Torque in Magnetic Trilayers Using All Three Polarizations of a Spin Current," Nat.Electron. 5, pp. 217–223 (2022).

[29] V. P. Amin, G.G. Baez Flores, A.A. Kovalev, and K.D. Belashchenko, "Direct and Indirect Spin Current Generation and Spin-orbit Torques in Ferromagnet/Nonmagnet/Ferromagnet Trilayers," Phys.Rev.B 110, p. 214427 (2024).

[30] S. Hu, D.-F. Shao, H. Yan, et al., "Efficient Perpendicular Magnetization Switching by a Magnetic Spin Hall Effect in a Noncollinear Antiferromagnet," Nat.Commun. 13, p. 4447 (2022).

[31] T. Higo, K. Kondou, T. Nomoto, et al., "Perpendicular Full Switching of Chiral Antiferromagnetic Order by Current," Nature 607, pp. 474 - 479 (2022).

[32] J.-Y. Yoon, P. Zhang, C.-T. Chou, et al., "Handedness Anomaly in a Non-collinear Antiferromagnet Under Spin–orbit Torque," Nat.Materials 22, pp. 1106 – 1114 (2023).

[33] J.-Y. Yoon, Y. Takeuchi, R. Takechi, et al., "Electrical Mutual Switching in a Noncollinear-antiferromagnetic–ferromagnetic Heterostructure." Nat.Commun. 16, p. 1171 (2025).

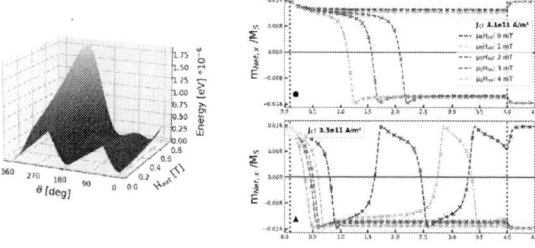

Fig.7 **Left**: Energy as a function of in-plane $\mathbf{m}_{net}$ orientation and magnetic field points up or down. Right: time dependence of m_{xnet}, for two values of current and for several magnetic fields.

Comprehensive Characterizations of Polarization Switching Dynamics in HfO₂-based FRAM across a Broad Temperature Spectrum

Yilin Hou[1], Jixuan Wu[1, *], Xiaopeng Li[1], Xiaoyu Dou[1], Yaoyu He[1], Pengpeng Sang[1], Xuepeng Zhan[1],
Yuqi Gao[2], Linhui Hu[2], Feng Wang[2], Yushi Hu[3], Qian Tao[3, *], Jiezhi Chen[1, *]

[1]School of Information Science and Engineering, Shandong University, 266237 Qingdao, China; [2]GTA Semiconductor
Co., Ltd., 200131 Shanghai, China; [3]Wuxi Smart Memories Technologies, Co., Ltd., 214000 Jiangsu, China

*E-mail: jixuanwu@sdu.edu.cn; qtao@smartmem.cn; chen.jiezhi@sdu.edu.cn

Abstract—This work presents a comprehensive evaluation of polarization behaviors in HZO-based ferroelectric capacitors across a broad temperature (−40 °C to 125 °C). By mapping remanent polarization across the whole wafer and selecting mass devices for in-depth reliability studies, it is demonstrated that electric field cycling could partially mitigate spatial inhomogeneity with defect redistribution. In addition, detailed characterizations reveal complex interplays among thermal activation, electric field strength, and polarization saturation. Importantly, time-dependent switching dynamics show that fatigue-induced defect accumulation will slow switching speed significantly and suppress temperature-dependent benefits. Our findings shed light on localized reliability mechanisms in HZO ferroelectric capacitors and highlight the need for multi-scale strategies to guide integration co-optimization.

Keywords—ferroelectric devices, Hf₀.₅Zr₀.₅O₂ (HZO), wafer-level reliability

I. INTRODUCTION

Ferroelectric materials have attracted significant attention in nonvolatile memory (FRAM, FeFET), reconfigurable logic, and neuromorphic computing, attributed to their intrinsic nonvolatility, ultra-fast read/write operations, and low power consumption [1–5]. Among these, HfO₂-based ferroelectric devices have emerged as leading candidates in advanced ferroelectric research, primarily due to their excellent CMOS compatibility, scalability, as well as integrability [6–7]. However, the ferroelectric properties of HfO₂-based materials are highly sensitive to fabrication conditions, including annealing temperature, electrode material, doping concentration, and film thickness, leading to significant spatial nonuniformity in electrical characteristics at wafer scale [8-10]. Such nonuniformity not only compromises device-to-device uniformity but may also trigger location-dependent reliability degradation—for instance, differences in the degradation of polarization switching and fatigue behaviors under various temperatures and voltages. These issues have become bottlenecks limiting large-scale integration, yield improvement, and stable performance. Although, many works have focused on single-device or array reliability [11-16], the exploration of the correlation between wafer-level property distributions and localized degradation mechanisms remains rare. Specifically, systematic investigations in high-performance regions under extreme conditions—wide temperature ranges, high electric field stress, and cycling endurance—remain insufficient, highlighting a critical research gap in this field.

In this work, we investigate a Hf₀.₅Zr₀.₅O₂ (HZO) ferroelectric capacitor (FeCap) wafer by analyzing the remanent polarization (Pr) distribution across multiple regions to assess spatial variation in electrical performance. Based on this mapping, a set of representative high-performance devices is selected for detailed reliability studies under various voltage stress and temperature conditions. Further analyses were carried out on the evolution of polarization switching behavior during electric field cycling, aiming to identify underlying dominating reliability mechanisms in this region. By linking the macroscopic spatial performance with localized dynamic responses, this work points out that low-Pr regions may become the bottleneck restricting overall reliability, and proposes a multi-scale evaluation approach that can provide valuable insights into process optimization, testing strategies, and predictive reliability analysis for ferroelectric devices.

II. EXPERIMENTAL RESULTS AND DISCUSSIONS

The schematic of the FRAM structure and Scanning Transmission Electron Microscopy (STEM) image of the critical cell are shown in Fig. 1, which depicts a transistor and a capacitor structure with TiN/HZO (8 nm)/TiN stacked layers. The transistors were fabricated with 180 nm node technology, while the HZO FeCaps are integrated with the back-end-of-line (BEOL) process. Measurements were performed using positive-up-negative-down (PUND) technique to effectively isolate the ferroelectric switching charge by subtracting contributions from the leakage current and the dielectric displacement currents at 2.4 V and 2.5 kHz, following an electric field cycling pretreatment at 2.4 V, 125 kHz square wave.

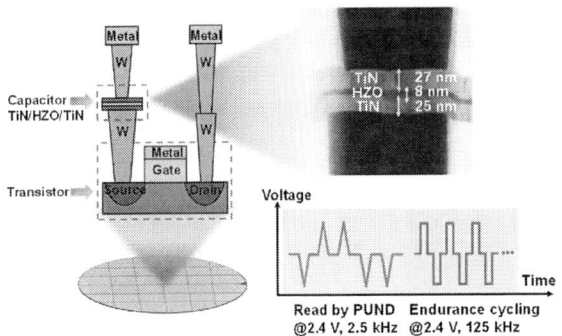

Fig. 1. Schematic of the FRAM structure, STEM image and the measurement method of FeCaps in this work.

A. Wafer-Level Polarization Dynamics

Fig. 2 presents normalized counts of polarization distributions across the wafer. In the initial stage (prior to cycling), the maximum polarization difference (ΔPr_{max}) is approximately 10.63 $\mu C/cm^2$, indicating a certain degree of spatial non-uniformity. After 10^4 cycles, this difference decreases to 8.82 $\mu C/cm^2$. This improvement is attributed to field-driven defect redistribution, such as oxygen vacancy migration, which mitigates domain pinning and promotes more uniform switching behavior. Notably, the extreme edge regions may exhibit relatively inferior performance, potentially due to stress concentration or local thickness variations caused during deposition or annealing. These results highlight that electric field cycling can effectively enhance wafer-level polarization uniformity, critical for high-yield ferroelectric device fabrication. Fig. 3 shows the 2Pr distributions across randomly selected regions (A–D) of the wafer, revealing pronounced spatial heterogeneities. Region B exhibits a relatively concentrated distribution profile, whereas Region C demonstrates a broader dispersion, indicative of distinct microstructural or interfacial characteristics. The imprint ($Vc_+ + Vc_-$) distribution at different positions during the wake-up stage (Fig. 4) shows that the maximum difference of imprint is less than 0.05 V, indicating minimal wafer-level variation in the imprint effect. This uniformity suggests the imprint effect is relatively less sensitive to process-induced variations compared to Pr.

Collectively, electric field cycling significantly improves wafer-level polarization uniformity and reduces imprint variation, confirming its role in stabilizing ferroelectric performance. Nevertheless, spatial non-uniformities—especially in edge regions—persist due to residual structural and processing-related heterogeneities.

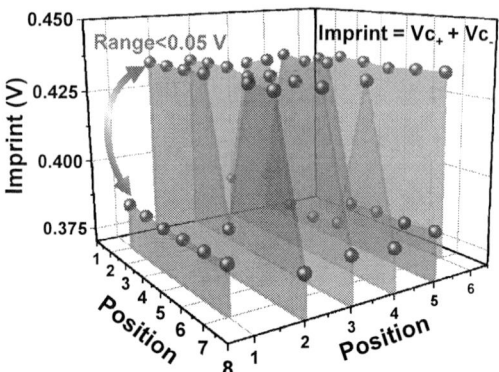

Fig. 4. Imprint effects measured at different positions across the wafer (after 10^4 cycles), with minimal overall variation.

B. Voltage and Temperature Dependent Characteristics

Fig. 5 illustrates a 3D bar chart of $2Pr_{max}$ under different voltages and temperatures, measured at a representative high-performance site near the wafer center. The X-axis denotes the operating voltage (1.6–2.4 V), the Y-axis represents the testing temperature (−40 to 125 °C), and the Z-axis corresponds to the maximum 2Pr value observed during the cycling process (up to 10^7 cycles). A clear trend is observed along the voltage axis: at all temperatures, $2Pr_{max}$ increases steadily with increasing voltage. For instance, at 25 °C, $2Pr_{max}$ rises from ~13.7 $\mu C/cm^2$ at 1.6 V to ~37.4 $\mu C/cm^2$ at 2.4 V. This enhancement stems from the increased electric field, which enhances domain switching kinetics, facilitates depinning, and activates deeper ferroelectric domains. Moreover, the increment of $2Pr_{max}$ in the high-voltage region is smaller than that in the low-voltage region, indicating polarization trends to approach saturation as voltage continues to increase. $2Pr_{max}$ exhibits distinct temperature dependences under different voltages. At lower voltages (1.6–1.8 V), a distinct inverse correlation exists between temperature and polarization: $2Pr_{max}$ decreases progressively with increasing temperature. For example, at 1.8 V, $2Pr_{max}$ drops from 21.7 $\mu C/cm^2$ at −40 °C to 18.1 $\mu C/cm^2$ at 125 °C. This is mainly because low-field-induced polarization is temperature-sensitive, with domain stability declining at elevated temperatures. In addition, elevated temperature can accelerate oxygen vacancies migration, releasing pinned domains; however, the insufficient electric field fails to reorient them

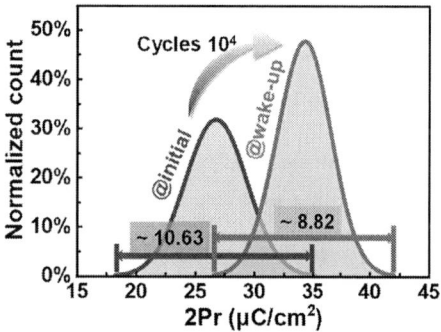

Fig. 2. Normalized counts of polarization distributions for FRAM at initial and wake-up stages.

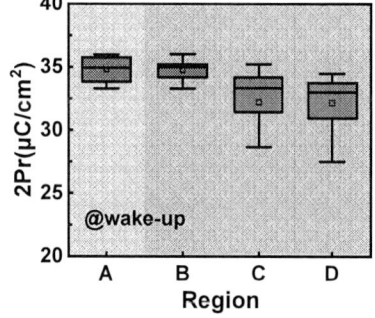

Fig. 3. 2Pr distributions across the wafer in different regions (A–D).

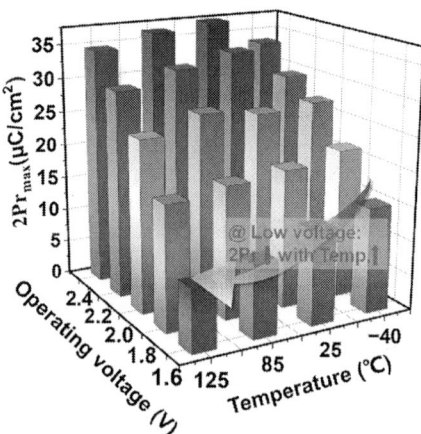

Fig. 5. The 3D bar chart visualizes the relationships among various parameters: operating voltage, temperature and polarization.

979-8-3315-3918-4/25 $31.00 © 2025 IEEE

into a stable polarized state. Another contributing factor is the enhanced depolarization field and increased leakage current caused by higher temperatures, which are not fully compensated by the limited driving field at low voltages. Notably, at higher voltages (2.2–2.4 V), the temperature response becomes more nuanced and nonmonotonic. $2Pr_{max}$ decreases with increasing temperature above room temperature (e.g., from 34 μC/cm² at 25 °C to 30 μC/cm² at 125 °C for 2.2 V), but the polarization at −40 °C is the lowest among all temperatures (e.g., 29.7 μC/cm² at −40 °C vs. 34.6 μC/cm² at 125 °C for 2.4 V). At cryogenic temperatures, suppressed oxygen vacancy mobility impedes the internal field redistribution that occurs during cycling, hindering the wake-up behavior and leading to incomplete domain switching even under strong fields. Additionally, the lack of thermal activation restricts defect de-trapping and charge compensation, further reducing Pr. As temperature rises to 25–85 °C, these constraints are gradually alleviated, resulting in enhanced polarization due to improved dipole kinetics and field-driven defect migration. However, at 125 °C, thermal degradation effects (e.g., increased leakage and enhanced depolarization) re-emerge as the dominant factor, causing a mild polarization drop.

These results indicate the existence of an optimal temperature window at high voltages, where thermal energy is sufficient for domain switching without inducing significant degradation. Overall, ferroelectric polarization performance depends on a complex interplay between electric field and thermal energy. At low fields, thermal effects degrade polarization due to insufficient driving energy; at high fields, thermal energy facilitates switching by enabling defect movement and dipole rotation. This highlights the importance of operating voltage selection in temperature-sensitive ferroelectric applications, where low-voltage operation may suffer from reliability degradation at elevated temperatures.

C. Polarization Switching Dynamics Under Different Stages and Temperatures

To further evaluate the impacts of stress-induced structural evolution on the polarization switching dynamics, we investigated the time-dependent ferroelectric switching behavior under three representative operational stages: the initial stage (pre-cycling), the wake-up stage (after 10^4 cycles), and the fatigue stage (after 10^6 cycles). Measurements were performed at two characteristic temperatures, 25 °C and 85 °C, using voltage pulses ranging from 0.9 V to 2.4 V with pulse widths spanning from 10 ns to 1 ms, as shown in Fig. 6. Across all conditions, the switching speed increases with voltage amplitude, consistent with the field-dependent domain wall motion predicted by the nucleation-limited switching (NLS) model [17–18]. At higher voltages ($\geqslant$2.1 V), nearly complete polarization reversal is achieved within 1 μs. In contrast, at low voltages (e.g., 0.9 V), substantial switching delays are observed, with incomplete reversal even at the 1 ms scale. As shown in Fig.7(a), in the fatigue stage, switching speeds at 25 °C and 85 °C converge, suggesting that temperature-induced kinetic acceleration is effectively suppressed in the presence of severe structural degradation. The fatigue stage exhibits a significant slowdown in switching, particularly under short pulse widths (Fig. 7b). For instance, under 1.8 V at 25 °C, the normalized polarization $\Delta P(t)/2P_s$ reaches only ~17.7% at 100 ns in the fatigue stage, compared to ~53.4% in the wake-up stage. This pronounced delay is attributed to fatigue-induced defect accumulation (e.g., oxygen vacancies

and interface traps), which introduces strong pinning centers that hinder domain nucleation and propagation.

To quantitatively assess these trends, Fig. 8 presents the extracted switching time to reach 80% reversal across different voltages, temperatures (25 and 85 °C), and device stages. At 25 °C and 1.8 V, the time in the fatigue stage is approximately 2.5× longer than that in the initial stage, and nearly 3× longer than that in the wake-up stage. At 85 °C, the time exhibits a nonmonotonic evolution: it decreases significantly from the initial to the wake-up stage, but then increases again in the fatigue stage, reaching a level comparable to the initial condition. This demonstrates that at

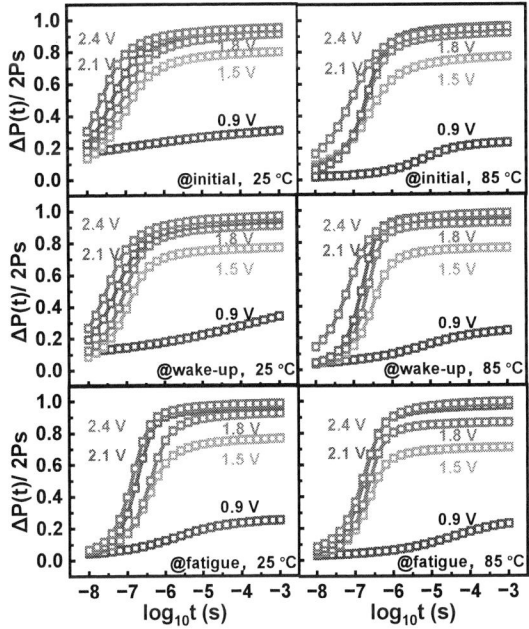

Fig. 6. Time-dependent switched polarization $\Delta P(t)/2P_S$ under various operation conditions.

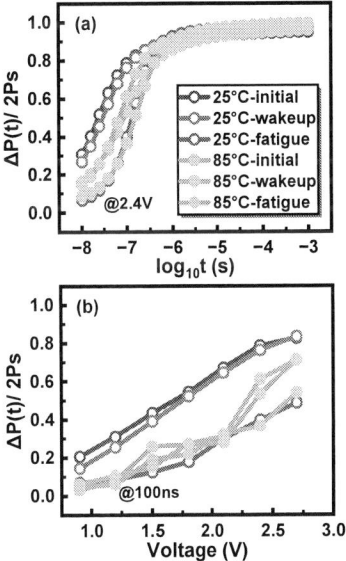

Fig. 7. (a) Time-dependent switched polarization $\Delta P(t)/2P_S$ for different stages and temperatures at 2.4 V. (b) Switched polarization $\Delta P(t)/2P_S$ as a function of the external voltage for different stages and temperatures at 100 ns.

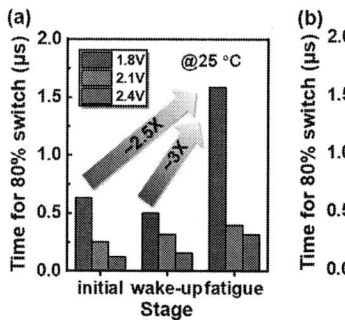 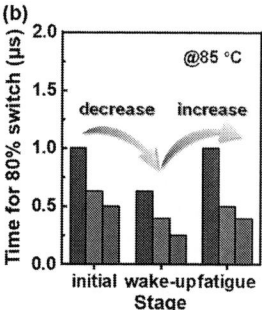

Fig. 8. The time for 80% switching under different voltages and stages at (a) 25 °C; (b) 85 °C.

elevated temperatures, fatigue can fully negate the kinetic advantages introduced by the wake-up effect. In summary, polarization switching dynamics are intricately modulated by electric field amplitude, operational temperature, and cycling number. While moderate cycling numbers improve switching behavior via defect redistribution (wake-up), fatigue-induced degradation becomes dominant at high temperatures, substantially compromising fast-switching capability. These results highlight the critical importance of accounting for thermal and reliability effects in the design of high-speed ferroelectric devices.

III. CONCLUSION

This study reveals the critical role of spatial inhomogeneity and electrical stress-induced degradation in determining HfO_2-based ferroelectric device reliability. At high-performance regions, voltage and temperature dependent studies uncover nonmonotonic polarization response, driven by the interplay between field-induced switching and thermal effects. Time-dependent analysis further highlights that, while moderate cycling enhances kinetics (wake-up), prolonged stress induces fatigue suppressing switching speed, especially at elevated temperatures. These findings underscore multi-scale evaluation necessity to inform process optimization and predictive reliability design in advanced FRAM technologies.

ACKNOWLEDGMENT

This work was supported by China Key Research and Development Program (2023YFB4402400), National Natural Science Foundation of China (92264201, U23B2040, 62034006), National Natural Science Foundation of Shandong Province (GZC20231435, TSQN202306059), and MIND project (MINDXZ202407).

REFERENCES

[1] L. Jung, J. Lee, S. Oh, and H. Hwang, "Superior retention (>1 year, 85 °C) and memory window (1.8 V) using ultra-thin HZO FTJ with OTS selector for X-point memory applications," in IEDM Tech. Dig., Dec. 2023, pp. 1–4.

[2] J. Okuno et al., "SoC compatible 1T1C FeRAM memory array based on ferroelectric $Hf_{0.5}Zr_{0.5}O_2$," in Proc. IEEE Symp. VLSI Technol., 2020, pp. 1–2.

[3] Y. Feng et al., "First demonstration of BEOL-compatible 3D vertical FeNOR," in Proc. IEEE Symp. VLSI Technol., 2024, pp. 1–2.

[4] W. Xu et al., "A novel small-signal ferroelectric capacitance-based content addressable memory for area-and energy-efficient lifelong learning," IEEE Electron Device Lett., vol. 45, no. 1, pp. 24–27, Jan. 2024.

[5] S. Takagi et al., "Physical reservoir computing using HZO-based FeFETs for edge-AI applications," in IEDM Tech. Dig., Dec. 2023, pp. 1–4.

[6] S. J. Kim, J. Mohan, S. R. Summerfelt, and J. Kim, "Ferroelectric $Hf_{0.5}Zr_{0.5}O_2$ thin films: A review of recent advances," JOM, vol. 71, no. 1, pp. 246–255, Jan. 2019.

[7] U. Schroeder, M. H. Park, T. Mikolajick, and C. S. Hwang, "The fundamentals and applications of ferroelectric HfO_2," Nature Rev. Mater., vol. 7, no. 8, pp. 653–669, Mar. 2022.

[8] A. M. Walke et al., "La doped HZO-based 3D-trench metal-ferroelectric-metal capacitors with high-endurance (>10^{12}) for FeRAM applications," IEEE Electron Device Lett., vol. 45, no. 4, pp. 578–581, Apr. 2024.

[9] P. Jiang et al., "A 256 Kbit $Hf_{0.5}Zr_{0.5}O_2$-based FeRAM Chip with Scaled Film Thickness (sub-8 nm), Low Thermal Budget (350 °C), 100% Initial Chip Yield, Low Power Consumption (0.7 pJ/bit at 2 V write voltage), and Prominent Endurance (>10^{12})," in IEDM Tech. Dig., Dec. 2023, pp. 1–4.

[10] Y. Lin et al., "Highly reliable, scalable, and high-yield HfZrOx FRAM by barrier layer engineering and post-metal annealing," in IEDM Tech. Dig., Dec. 2022, pp. 1–4.

[11] S. Jindal et al., "Temperature-dependent field cycling behavior of ferroelectric hafnium zirconium oxide (HZO) MFM capacitors," IEEE Trans. Electron Devices, vol. 69, no. 7, pp. 3990–3996, July 2022.

[12] X. Wang et al., "Impact of charges at ferroelectric/interlayer interface on depolarization field of ferroelectric FET with metal/ferroelectric/interlayer/Si gate-stack," IEEE Trans. Electron Devices, vol. 67, no. 10, pp. 4500-4506, Oct. 2020.

[13] J. Li et al., "High endurance (>10^{12}) via optimized polarization switching ratio for $Hf_{0.5}Zr_{0.5}O_2$-based FeRAM," Appl. Phys. Lett., vol. 122, no. 8, Feb. 2023, Art. no. 082901.

[14] J. Okuno et al., "Investigation of recovery phenomena in $Hf_{0.5}Zr_{0.5}O_2$-based 1T1C FeRAM," IEEE Journal of the Electron Devices Society, vol. 11, pp. 43-46, 2023.

[15] Z. Liu et al., "Role of charge injection/de-trapping in imprint behavior of ferroelectric $Hf_{0.5}Zr_{0.5}O_2$ thin film," Appl. Phys. Lett., vol. 125, no. 7, August 2024, Art. no.072904.

[16] B. Cui et al., "Unveiling the role of local stress in enhancing ferroelectric properties and endurance of HfO_2/ZrO_2 superlattice structures," IEEE Electron Device Lett., vol. 46, no. 1, pp. 107-110, Jan. 2025.

[17] J.Y. Jo et al., "Domain switching kinetics in disordered ferroelectric thin films," Phys. Rev. Lett., vol. 99, no. 26, pp. 267602-267605, Dec. 2007.

[18] W. Wei et al., "In-depth understanding of polarization switching kinetics in polycrystalline $Hf_{0.5}Zr_{0.5}O_2$ ferroelectric thin film: A transition from NLS to KAI," in IEDM Tech. Dig., Dec. 2021, pp. 19.1.1–19.1.4.

WO$_x$ interlayer employed to improve the imprint effect on HfZrO$_2$ ferroelectric capacitors

Zibo Dong [1], Zeping Weng [2], Jianguo Li [2, 4], Lijian Chen [3], Ziyuan Chen [3], Yi Zhao *[4] and Daolin Cai *[1, 4]

[1] School of Integrated Circuits, East China Normal University, Shanghai 200241, China
[2] State Key Laboratory of Silicon and Advanced Semiconductor Materials, Zhejiang University, Hangzhou 310027, China
[3] College of Information Science and Electronic Engineering, Zhejiang University, Hangzhou 310027, China
[4] Huada Semiconductor Co. LTD, Shanghai 201210, China

* Email: caidl@hdsc.com.cn

Abstract—HfO$_2$-based ferroelectric thin films have emerged as promising candidates for nonvolatile memory due to their excellent compatibility with CMOS. However, their reliability is severely limited by the imprint effect. In this work, a WO$_x$ interlayer is introduced into a W/Hf$_{0.5}$Zr$_{0.5}$O$_2$ (HZO)/W ferroelectric capacitor to mitigate the imprint effect by suppressing interfacial defects and providing oxygen ions. The thickness and composition of the WO$_x$ interlayer are precisely controlled through O$_2$ plasma oxidation during atomic layer deposition (ALD). The optimized WO$_x$ interlayer effectively alleviates oxygen scavenging by the W electrodes, reducing oxygen vacancy generation in the ferroelectric layer, thereby enhancing the remnant polarization (2P$_r$) from 33 μC/cm^2 to 45 μC/cm^2 and achieving an endurance of 10^8 cycles. High-temperature (85 °C) imprint tests demonstrate that the WO$_x$ interlayer acts as an oxygen reservoir, neutralizing oxygen vacancies generated at the ferroelectric interface during the electrical pulsing. This study provides an effective interface engineering strategy for HfO$_2$-based ferroelectric devices and highlights the critical role of WO$_x$ interlayers in improving device reliability.

Keywords—HfZrO$_2$ ferroelectric thin films, imprint effect, WO$_x$ interlayer, oxygen vacancies, interface engineering

I. INTRODUCTION

In 2011, the first report of ferroelectricity in Si-doped HfO$_2$ thin films overcame the compatibility barrier between traditional perovskite ferroelectric films and CMOS technology, significantly advancing the field of ferroelectric memory [1]. This discovery has made HfO$_2$-based ferroelectric thin films a key research focus on non-volatile memory (NVM) applications. However, HfO$_2$-based ferroelectric films still suffer from the imprint effect. As one of the critical reliability issues in ferroelectric memory devices, imprint can lead to read/write operation failures, increase of operating voltage, and degradation of retention characteristics [2].

The imprint effect is generally attributed to the asymmetric charge redistribution at the ferroelectric thin film interface and the associated built-in electric field. Several models have been proposed to explain imprint behavior, such as the domain-pinning model [3], charge-injection model [4, 5], and interfacial dead-layer model [6]. Among these, the charge-injection model has gained broad acceptance due to its ability to explain the time-dependent and polarization-dependent characteristics of imprint. The model demonstrates that an applied electric field polarizes the ferroelectric layer, thereby generating a strong electric field across the interfacial

dielectric layer between the ferroelectric film and electrode. This strong electric field drives charge injection. With the continuous accumulation of injected charges at the interface, an asymmetric charge distribution forms within the ferroelectric film, resulting in imprint. Furthermore, the trapping of electrons at oxygen vacancies near the interface is considered the dominant mechanism behind imprint in HfO$_2$-based devices [7].

This work introduces a WO$_x$ interlayer into the conventional W/HZO/W capacitor structure and optimizes its thickness and composition. The WO$_x$ layer suppresses defect generation in the ferroelectric layer, enhances dipole switching, and increases remnant polarization. Furthermore, the WO$_x$ interlayer serves as an oxygen reservoir that compensates for oxygen vacancies generated at the HZO interface during electrical cycling by supplying oxygen ions. This mechanism reduces electron trapping and effectively mitigates imprint effects in HZO ferroelectric capacitors. Notably, at elevated temperatures, oxygen ions in WO$_x$ exhibit higher mobility toward the HZO layer, facilitating recombination with oxygen vacancies and further improving imprint.

II. EXPERIMENTS

The fabrication process of the ferroelectric capacitor is illustrated in Fig. 1. First, a 100-nm-thick tungsten (W) bottom electrode (BE) was sputter-deposited on a heavily doped p-type Si substrate. The W BE was subsequently subjected to controlled oxidation using O$_2$ plasma in the atomic layer deposition (ALD) chamber to form a WO$_x$ interlayer. The oxidation times were 0 min (WO$_x$-0), 10 min (WO$_x$-10), 20 min (WO$_x$-20), and 30 min (WO$_x$-30), respectively. A 10-nm-thick HZO film was deposited ALD at 270 °C, with TEMAH and TEMAZ as precursors. The device structure was completed by depositing a 100-nm-thick W top electrode, followed by rapid thermal annealing (RTA) at 500 °C for 60 s in N$_2$ atmosphere. Finally, the devices were patterned, and aluminum (Al) was deposited as the back contact.

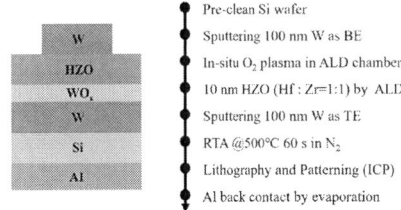

Fig. 1. Schematic diagram of the cross-sectional structure and preparation process flow of W/WO$_x$/HZO/W ferroelectric capacitor.

This work was supported in part by the National Science and Technology Major Project under Grant 2020AAA0109001, and in part by the ECNU/HDSC Integrated Circuit Engineering Technology Joint Laboratory.

III. RESULTS AND DISCUSSION

X-ray photoelectron spectroscopy (XPS) analysis was performed on the W BE subjected to varying O_2 plasma treatment durations (Fig. 2). Initially, the W metal undergoes native oxidation in ambient conditions, forming an ultrathin oxide layer (0.36 nm) with a high density of oxygen vacancies. The synergistic effect between oxygen vacancies and the inherent oxygen scavenging property of W metal induces oxygen vacancy defects in the ferroelectric layer, consequently degrading device performance [8]. With prolonged O_2 plasma treatment duration, the WO_x interlayer exhibits progressive thickness augmentation accompanied by a marked reduction in oxygen vacancy concentration. The relatively lower bond dissociation energy of W-O (720 kJ/mol) compared to Hf-O (801 kJ/mol) and Zr-O (766 kJ/mol) results in a lower oxygen vacancy formation energy in WO_3 than in HfO_2/ZrO_2 [9]. Consequently, oxygen vacancies preferentially form in the WO_x interlayer rather than in the HZO ferroelectric layer during electrical pulsing. This unique property allows an optimized WO_x interlayer to act as an effective oxygen reservoir, compensating for oxygen vacancies in the ferroelectric layer.

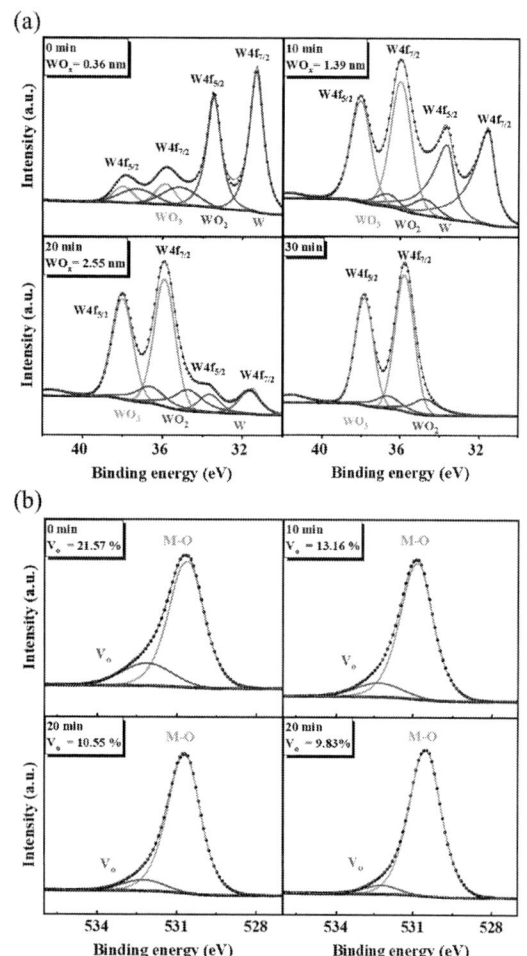

Fig. 2. XPS analysis of W 4f (a) and O 1s (b) fine spectra for W substrates with varying O_2 plasma treatment times.

The fabricated HZO ferroelectric capacitors were characterized using a Keysight B1500 semiconductor parameter analyzer for polarization-voltage (P-V) and current-voltage (I-V) measurements. The obtained P-V and I-V curves are shown in Fig. 3. The results demonstrate that introducing a WO_x interlayer between HZO and the BE effectively enhances the remnant polarization. The $2P_r$ value significantly increases from 33 $\mu C/cm^2$ (WO_x-0) to 45 $\mu C/cm^2$ (WO_x-20). This improvement is attributed to the WO_x interlayer mitigating the oxygen scavenging property of W metal and suppressing the generation of defects in the ferroelectric layer, thereby facilitating dipole flipping. However, with prolonged O_2 plasma treatment (30 min), the composition and thickness of WO_x change, leading to a reduced $2P_r$ of 36 $\mu C/cm^2$ (WO_x-30). This degradation may stem from excessive oxygen content in WO_x. Such excessive oxygen content promotes interfacial reactions with the HZO ferroelectric layer and forms an additional oxide dead layer, which consequently impedes efficient polarization switching [10]. The WO_x interlayers formed by 10-minute and 20-minute O_2 plasma treatments enhance dipole switching, as evidenced by the negative shift of current peaks in the I-V curves under positive bias. In contrast, the 30-minute O_2 plasma treatment restricts interfacial dipole switching, corresponding to a positive shift in the current peak. The phenomenon that the peak position in the I-V curve shifts with changes in O_2 plasma treatment time strongly demonstrates the dual effects of WO_x on dipole switching at the interface.

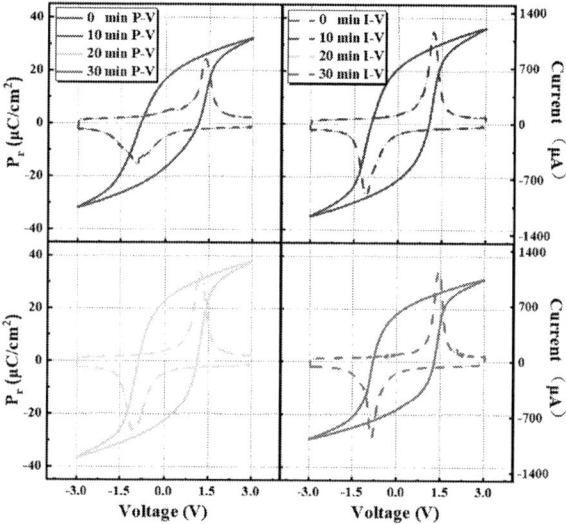

Fig. 3. The P-V and I-V characteristics of the fabricated ferroelectric devices.

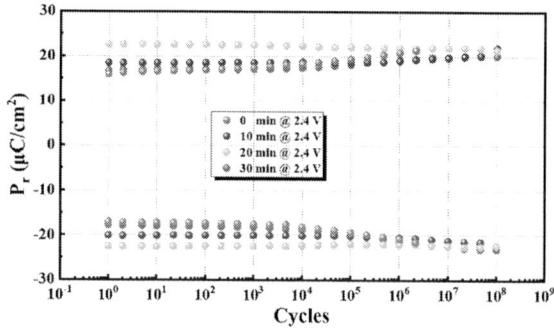

Fig. 4. Endurance test of the fabricated ferroelectric devices.

The endurance test of ferroelectric capacitors was conducted at a voltage amplitude of 2.4 V (Fig. 4). The oxygen scavenging property of the W electrode generates substantial interfacial defects, which readily form leakage paths. As a result, the untreated device (WO$_x$-0) breaks down after 10^6 cycles. In contrast, the optimal WO$_x$ interlayer effectively provided oxygen ions, neutralizing both intrinsic defects in the ferroelectric layer and pulse-generated oxygen vacancies, resulting in an endurance of up to 10^8 cycles. Although the WO$_x$-30 device also demonstrated an endurance of 10^8 cycles, it exhibits significant a wake-up effect due to the presence of an additional parasitic oxide dead layer between the WO$_x$ and HZO ferroelectric layers.

To investigate the effect of the incorporated WO$_x$ interlayer on the imprint characteristics of the devices, imprint tests were performed under both room temperature and elevated temperature (85 °C) conditions. A positive-polarity waveform with an electric field intensity of 3 MV/cm was applied to characterize the imprint behavior. The imprint electric field (E$_{imp}$) was calculated using Equation (1):

$$E_{imp} = \frac{(+E_c) + (-E_c)}{2} \quad (1)$$

where +E$_c$ and -E$_c$ denote the positive and negative coercive electric fields, respectively. Fig. 5(a) illustrates the imprint effect of the device at room temperature. Compared with the device without O$_2$ plasma treatment, the introduction of WO$_x$ interlayers shows a certain improvement in the imprint effect of the device. In particular, the WO$_x$-20 device exhibits significantly enhanced imprint mitigation under high-temperature conditions (Fig. 5(b)) compared to the WO$_x$-0 device. These results provide stronger evidence that an optimized WO$_x$ interlayer can serve as an oxygen reservoir, supplying oxygen ions to neutralize oxygen vacancies generated at the interface during pulsing. The WO$_x$ interlayer formed by 30-minute O$_2$ plasma treatment alleviates the oxygen scavenging property of the W metal electrode. However, its excessive oxygen content leads to an additional oxide dead layer between WO$_x$ and the HZO ferroelectric layer, resulting in severe imprint degradation. Figs. 5(c) and 5(d) show the imprint-induced P-V curve shifts in WO$_x$-0 and WO$_x$-20 devices at 85 °C.

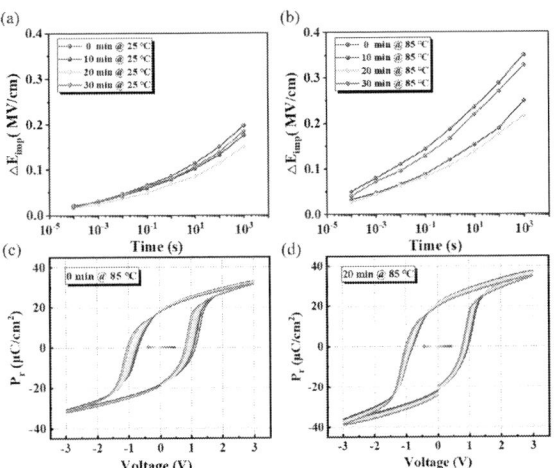

Fig. 5. Shift of coercive electric field of the device under different holding times at room temperature (a) and high temperature (b); the imprint-induced P-V curve shifts in WO$_x$-0 (c) and WO$_x$-20 (d) devices at 85 °C.

Fig. 6 illustrates the plausible mechanism derived from the aforementioned microstructural and electrical analyses. On the untreated substrate (WO$_x$-0), a high density of oxygen

vacancies exists. When HZO thin films are deposited, oxygen ions in the ferroelectric layer react with the W electrode due to the presence of oxygen vacancies and the oxygen scavenging property of the W metal. It introduces numerous defects into the HZO film and degrades device reliability. The optimized WO$_x$ interlayer effectively suppresses the oxygen scavenging property of the W metal. Moreover, the partial oxygen ions present in WO$_x$-20 can passivate the intrinsic oxygen vacancies in the ferroelectric layer, further enhancing reliability. Under positive bias stress, the combined effects of the electric field and the difference in oxygen vacancy formation energy facilitate the migration of oxygen ions from WO$_x$-20 toward the ferroelectric interface. It effectively neutralizes oxygen vacancies induced by pulses. This phenomenon becomes more pronounced at elevated temperatures. The reduction of oxygen vacancies at the ferroelectric interface suppresses electron injection, thereby mitigating the imprint effect. However, in the WO$_x$-30 device, excessive oxygen content leads to the formation of an oxide dead layer at the ferroelectric interface, limiting the improvement in imprint characteristics.

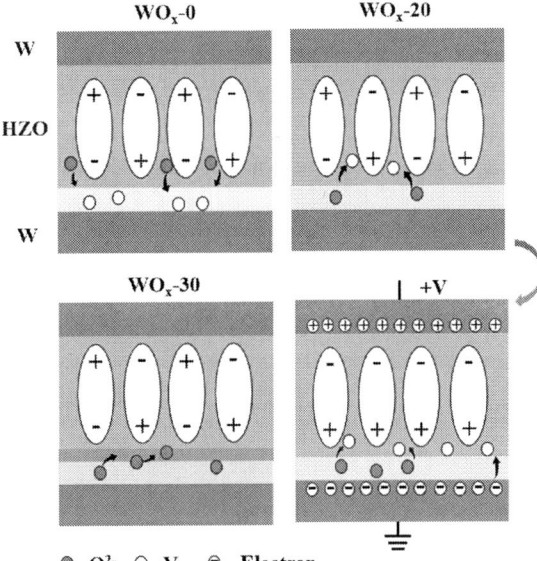

Fig. 6. The schematic illustrates the critical role of the WO$_x$ interlayer in enhancing device reliability.

IV. CONCLUSIONS

In this work, we systematically investigate the mechanism of imprint improvement in W/HZO/W ferroelectric capacitors by introducing a WO$_x$ interlayer. Through optimization of the thickness and composition of the WO$_x$ interlayer, it simultaneously serves as an interface defect suppressor and an oxygen reservoir. This enhances the 2P$_r$ value from 33 μC/cm^2 to 45 μC/cm^2 and achieves an endurance of 10^8 cycles. High-temperature (85 °C) measurements reveal that the WO$_x$ interlayer exhibits more pronounced imprint mitigation at elevated temperatures. This is because the accelerated migration of oxygen ions from WO$_x$ to the HZO layer effectively neutralizes the oxygen vacancies induced by pulses. This work provides an interfacial engineering strategy for imprint suppression in HfO$_2$-based ferroelectric devices and demonstrates the critical role of the WO$_x$ interlayer in enhancing reliability.

REFERENCES

[1] T. S. Böscke, J. Müller, D. Bräuhaus, U. Schröder, and U. Böttger, "Ferroelectricity in hafnium oxide thin film," Appl. Phys. Lett., vol. 99, no. 10, p. 102903, Sep. 2011.

[2] Z. Liu, K. Toprasertpong, Z. Cai, M. Takenaka, and S. Takagi, "Role of charge injection/de-trapping in imprint behavior of ferroelectric $Hf_{0.5}Zr_{0.5}O_2$ thin film," *Appl. Phys. Lett.*, vol. 125, no. 7, p. 072904, 2024.

[3] Y. Zhou, H. K. Chan, C. H. Lam, and F. G. Shin, "Mechanisms of imprint effect on ferroelectric thin films," *J. Appl. Phys.*, vol. 98, no. 2, p. 024111, Jul. 2005.

[4] A. K. Tagantsev, I. Stolichnov, N. Setter, and J. S. Cross, "Nature of nonlinear imprint in ferroelectric films and long-term prediction of polarization loss in ferroelectric memories," *J. Appl. Phys.*, vol. 96, no. 11, pp. 6616–6623, 2004.

[5] P. Yuan et al., "Microscopic mechanism of imprint in hafnium oxide-based ferroelectrics," *Nano Res.*, vol. 15, no. 4, pp. 3667–3674, 2022.

[6] P. D. Lomenzo, C. Richter, T. Mikolajick, and U. Schroeder, "Depolarization as driving force in antiferroelectric hafnia and ferroelectric wake-up," *ACS Appl. Electron. Mater.*, vol. 2, no. 6, pp. 1583–1595, 2020.

[7] S. Yu et al., "Inhibiting the imprint effect of the TiN/HZO/TiN ferroelectric capacitor by introducing a protective HfO_2 layer," AIP Adv., vol. 14, no. 8, p. 085012, 2024.

[8] G. Segantini et al., "Interplay between strain and defects at the interfaces of ultra-thin $Hf_{0.5}Zr_{0.5}O_2$-based ferroelectric capacitors," *Adv. Electron. Mater.*, vol. 9, no. 10, p. 2300171, 2023.

[9] Y. Luo, *Bond Dissociation Energies.* Boca Raton, FL, USA: CRC Press, 2012, pp. 65–98.

[10] X. Wang, S. Slesazeck, T. Mikolajick, and M. Grube, "Modulation of oxygen content and ferroelectricity in sputtered hafnia-zirconia by engineering of tungsten oxide bottom electrodes," *Adv. Electron. Mater.*, vol. 10, no. 6, p. 2300798, 2024.

Understanding the Physical Mechanism of Endurance Cycling in Antiferroelectric Memories

Y. Qu [1], Y. Huo [2], Y. Ding [2], L. Chen [1,3], Z. Weng [3], and Y. Zhao*[1]

[1] Research Center of Integrated Circuits, Huada Semiconductor, Shanghai, 201210, China
[2] School of Communication and Electronic Engineering, East China Normal University, Shanghai 200241, China
[3] College of Information Science and Electronic Engineering, Zhejiang University, Hangzhou 310027, China

* Email: yizhao@hdsc.com.cn

Abstract—Anti-ferroelectric HfZrO films have received extensive attention for potential applications to advanced memories, due to good scalability and superior endurance. However, the failure mechanisms are not thoroughly understood yet. This work investigated endurance using bipolar and unipolar stress. In addition, unipolar stress waveforms with various amplitudes and frequencies were employed in characterizations to make stress conditions closer to real eDRAM operation. Furthermore, the device-to-device variation during endurance was analyzed. The results indicate that charge injection at the HfZrO/electrode interface dominates the degradation during endurance cycling.

Keywords—anti-ferroelectric, polarization, endurance, charge injection, device-to-device variation, eDRAM

I. INTRODUCTION

Anti-ferroelectric (AFE) HfZrO is considered as one of the promising candidates as the capacitor material for next-generation high-speed and high-density embedded DRAM (eDRAM) [1-4]. While enjoying the same good scalability with Hf-based ferroelectrics, it has the advantages of a lower coercive field and improved endurance than ferroelectrics [5-6]. This makes it possible for low-power and high-reliability applications. However, the endurance of AFE HfZrO thin films has not been competitive with the commercial DRAM standard (10^{14}~10^{16} cycling). Up to now, only a few papers have published the endurance of 10^{14} cycles through real experiment characterizations [7]. Recently, some literature discussed the degradation behaviors under bipolar and unipolar stress modes [7-8]. The endurance degradation mechanisms during the cycling process still lack an in-depth understanding, which must be clarified before real applications in the near future.

II. DEVICE AND EXPERIMENT SETUP

A. AFE Device

The sample used in this work is the W/TiN/HZO/TiN structure, as schematically shown in Fig. 1(a). Metal-ferroelectric-metal (MFM) capacitor devices are fabricated on the heavily-doped p⁺ Si substrate. The TiN top electrode and different thicknesses of HZO film are deposited by ALD. By adjusting the cycle ratio of two precursors, the Hf:Zr ratio is carefully controlled to 1:4 and confirmed through XPS

This work was supported by the National Natural Science Foundation of China (62204086) and the Chenguang Program of Shanghai Education Development Foundation and Shanghai Municipal Education Commission (23CGA35), and the ZJU-HDSC Joint Research Center.

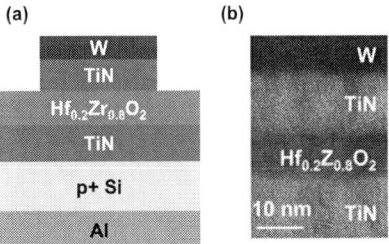

Fig. 1 (a) The schematic of fabricated anti-ferroelectric (AFE) devices. (b) The HRTEM cross-section 8 nm AFE capacitor.

characterization. Then, RTP at 450 °C in N_2 atmosphere is adopted as the annealing condition for the film crystallization. Following that, the patterning and Al back contact are processed to form AFE capacitors. More fabrication details can be found in our previous work [9]. Fig. 1 (b) shows the HRTEM cross section of an AFE capacitor with $Hf_{0.2}Zr_{0.8}O_2$ film of 8 nm thickness.

B. Device Characterization

Through bipolar triangle pulses with a rising/falling time of 10 μs, the polarization charge can be calculated by integrating the current trace. Fig. 2(a) shows the applied

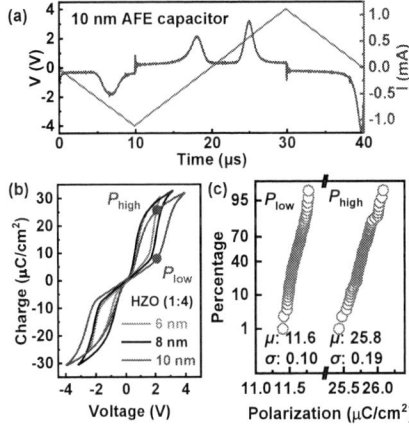

Fig. 2. (a) P-V loop of the AFE capacitors shows scalability at 10 nm thickness. (b) The example of applied triangle voltage V-t and traced current I–t curves to characterize polarization in AFE devices. (c) The device-to-device variation of P_{high} and P_{low} state for 60 AFE capacitors at the pristine state.

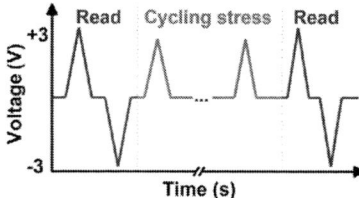

Fig. 3 The illustration of the unipolar endurance measurements adopted in this study.

voltage waveform and the traced current response of the AFE capacitor under test. The corresponding polarization-voltage (P-V) loop in Fig. 2(b) illustrates the extraction of P_{high} and P_{low} for "1" and "0" storage, respectively. To avoid the impact of variation from fabrication, the device-to-device variation of 60 AFE capacitors was also characterized at the pristine state. And the distribution of P_{high} and P_{low} is statistically shown in Fig. 2(c). At the P_{low} point, the AFE device has not been polarized yet. The stored charge is contributed by the dielectric constant, and the mean value μ of polarization is 11.6 μC/cm². At the P_{high} point, except for the stored charge, the polarization charge also contributes, and the μ of the polarization is 25.8 μC/cm². The standard deviation σ of P_{high} and P_{low} is 0.1 and 0.19 μC/cm², implying the variation induced by polarized charge is stronger than that of the dielectric itself. As for the endurance characterization, a ±3 V triangle pulse is used to measure P-V loops, and the bipolar or unipolar triangle train with a 50% duty cycle is applied as stress, as exampled in Fig. 3.

III. RESULT AND DISCUSSION

A. Cycling Degradation Under Bipolar and Unipolar stress

For an overall view of the degradation behavior during cycling, the degradation under bipolar (−3~+3 V) and unipolar (0~+3 V) stress was investigated first. In Fig. 4(a) of the bipolar case, it is clear that both positive and negative loops shift towards the zero electric field in the P-E evolution. However, only the positive part degrades under unipolar stress, as shown in Fig. 4(b). This shift of P-V loop will inevitably lead to the memory window P_r instability in AFE memories. The comparison in Fig. 4(c) shows that the P_r degrades earlier and faster during bipolar cycling than the unipolar.

Fig. 5 compares the leakage current degradation for the above two cases. The increase of leakage during ±3 V bipolar cycling may be related to new defects generation, which

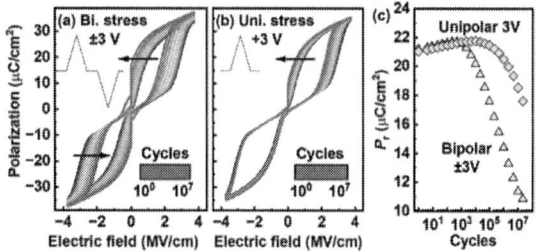

Fig. 4 Under (a) ±3 V bipolar stress and (b) +3 V unipolar stress, P-E evolution show different behavior.(c) Memory window key parameter P_r degrades earlier and faster under the bipolar stress than the unipolar cases.

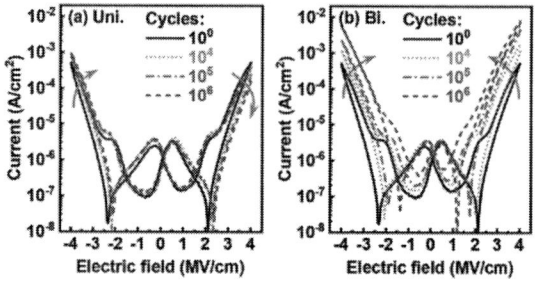

Fig. 5 The leakage current of 10 nm AFE capacitors during the 10^6 endurance cycling with (a) unipolar +3 V stress and (b) ±3 V bipolar stress.

induces the breakdown once accumulated. In contrast, for the +3 V unipolar cases, the leakage increases at negative high E but decreases at positive high E. This phenomenon may be due to the charge injection at the interface between the HZO interfacial layer and the electrode. Specifically, when applying a positive bias, the negative charges at the bottom IL/HZO interface decrease the voltage drop across the bottom IL. It makes the leakage current shift positively along the voltage axis, leading to a lower leakage current.

B. Endurance during eDRAM-like Cycling

The following mainly focuses on unipolar cases and positive loops, since it is closer to the real eDRAM application. Endurance results under unipolar stress were studied with various amplitude V_{stress}, covering below and over coercive voltage. By comparing the P-V evolution of different stress voltages, it can be directly observed that the whole positive loop shifts to the left under +3V stress, while only the switching part shifts in the other two cases with lower V_{stress}.

To further investigate the degradation of the switching part, the I-V curves during unipolar cycling are given in Fig. 7. For V_{stress} = 3 V, the whole positive I-V curve shifts. While V_{stress}<3 V, part of the P-V loops degrades, and the other peak appears. Moreover, this current peak at the smaller V_{stress} shifts to a lower E, and it remains at a larger V_{stress} [10]. This degradation behavior under unipolar stress is because part of AFE domains has a polarization voltage shift during their repeat switching.

Except for the various amplitudes of V_{stress}, the dependence of stress frequency was also explored. Since V_{high} decreases more than V_{low}, the bias V_{bias} has a degradation with cycling, as summarized by ΔV_{bias} in Fig. 8(a). Here, ΔV_{bias} is defined

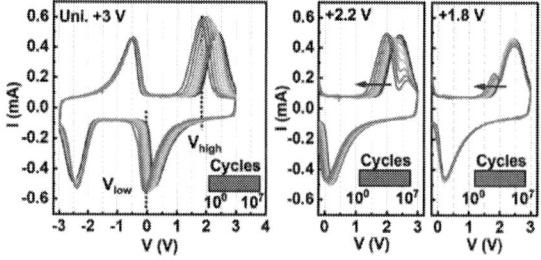

Fig. 6 Corresponding I-V curves during unipolar cycling with various amplitude V_{stress} [10]. At V_{stress} of 3 V, V_{high} and V_{low} shift toward zero bias. For V_{stress}<3 V, V_{high} appears to have the other peak, which is consistent with the part of a shift in P-V loops.

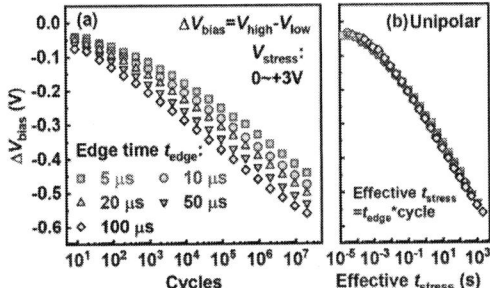

Fig. 7 (a) The bias voltage degradation ΔV_{bias} versus stress cycle number with different rising/falling edges, i.e. different frequencies. (b) The ΔV_{bias} data of different frequencies merges at the same effective stress time.

by $V_{high} - V_{low}$, where V_{high} and V_{low} are the corresponding voltages shown in Fig. 6. It can be seen from Fig. 7(a) that as the edge time t_{edge} decreases, that is, the frequency of the cycling stress waveform increases, the degradation of bias voltage ΔV_{bias} shift tends to decrease. This indicates that when applied to memory with a higher operating speed, AFE capacitors may have less degradation during the endurance cycling, which is very friendly to eDRAM applications. While from another perspective of effective stress time, the ΔV_{bias} trends become coincident in Fig. 7(b), illustrating that it is dominated by the effective stress time rather than the number of stress cycles alone.

To study the mechanism of endurance degradation, the device-to-device variation among 60 AFE devices during endurance cycling has also been experimentally analyzed. When a positive voltage unipolar stress is applied to, the $P_{high} - P_{low}$ during the cycling degradation is obtained at a constant voltage using the method shown in Fig. 8(a). Fig. 8(b) shows the extracted standard deviation value $\sigma(P_{high} - P_{low})$ as a function of endurance cycles. Under positive unipolar stress, the $\sigma(P_{high} - P_{low})$ of the positive hysteresis voltage loop increases with the cycling number, while the variation of the negative voltage hysteresis loop does not change significantly.

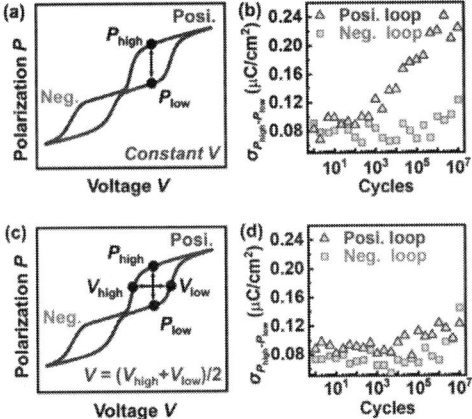

Fig. 8 (a) The P_{high} and P_{low} are extracted at constant voltage. (b) Corresponding standard deviation σ of $P_{high} - P_{low}$ increases during +3V cycling. (c) Considering endurance degradation, V is corrected by $(V_{high} + V_{low})/2$ for P_{high} and P_{low} extraction. (d) standard deviation of $(P_{high} - P_{low})$ shows few change, implying no new defects generation.

However, during the endurance cycling, the remanent polarization also changes, and the increase of variation may be related to the decrease of remanent polarization.

Therefore, considering this, the device-to-device variation has been re-examined. As schematically shown in Fig. 8(c), the coercive voltage V_{high} and V_{low} are first calculated, and the P_{high} and P_{low} are then extracted at the voltage of $(V_{high} + V_{low})/2$. During the endurance cycling, when the voltage of P_{high} and P_{low} are synchronously shifted due to degradation, almost no $\sigma(P_{high} - P_{low})$ change could be observed in Fig. 8(d), suggesting there are few defects generated during the unipolar endurance cycling.

C. Physical Mechanism of Endurance Cycling

During the positive unipolar stress, the endurance test results in the hysteresis loop of positive voltage shift left to the zero field. Fig. 9 schematically shows the mechanism of the AFE devices under a low V_{stress} during the endurance cycling. When V_{stress} is insufficient to make all domains polarized, only part of them is polarized. Under the positive voltage stress, the polarization direction of the polarized part is downward, generating a large electric field on the interface layer (IL). Driven by this field, the electrons accumulated on the bottom electrode will inject and then accumulate at the IL, thereby causing a voltage drift. Meanwhile, in the unpolarized regions, the macroscopic polarization is zero, so the electric field on the interface layer of this part of the region is not sufficient to cause charge injection. This mechanism well explains the experimental phenomenon that voltage drift occurs only in the switching regions during the stress cycling.

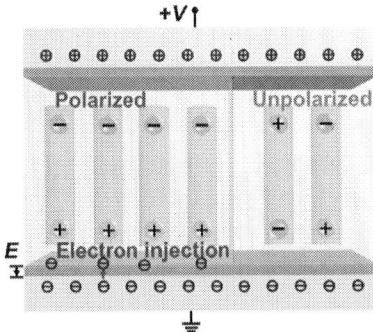

Fig. 9 Schematic of degradation mechanism under a low V_{stress} during the endurance cycling.

When electrons are injected and accumulate at the IL, it will also cause changes in the internal electric field distribution of the AFE device. The accumulation of electrons will cause the energy band at the corresponding physical position to rise. Fig. 10(a) shows the energy band diagrams before and after the endurance cycling with positive stress. After the stress, the electric field on IL decreases relative to the initial device. It is well known that the current density flowing through the device is related to the electric field on the dielectric. Hence, when the electric field at the IL decreases, it will lead to a reduction in leakage current. When the unipolar cycling stress is negative, the band diagram changes to Fig. 10(b). This negative stress could lead to the accumulation of electric charges, which in turn increases the

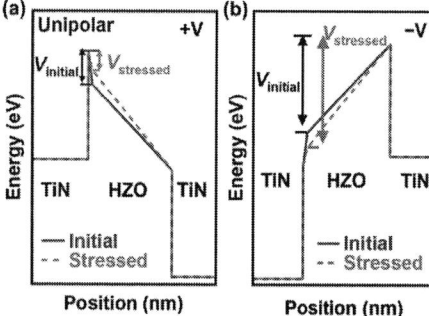

Fig. 10 The band diagram of AFE devices before and after the unipolar cycling stress. (a) With the accumulation of injected electrons at the IL, the electric field E on IL decreases, resulting a smaller leakage current at the positive stress (b) Oppositely, the band diagram change at the negative stress will result in a larger leakage current.

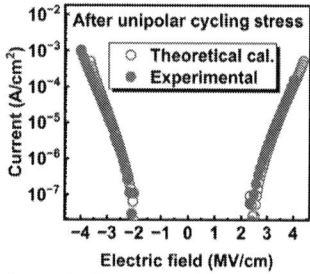

Fig. 11 After unipolar stress, the comparison results of J-E curves calculated according to the charge injection theory and the experimental data.

electric field in the dielectric on the other side, thereby increasing the leakage current of the AFE devices.

To verify the aforementioned mechanism of the leakage current change caused by the unipolar stress, we first obtained the voltage change caused by the injected charge from P-V curves. With this, one can calculate the theoretical value of the current after degradation caused by charge injection based on the I-V curves in the initial state. The theoretical calculation results show consistent with the experimental data, as shown in Fig. 11. This confirms the proposed theory that the change of J-E curves under unipolar stress is caused by the electron injection and accumulation resulting from the electric field. Through these experiments and theoretical analysis, this study further verified the influence of electron injection and accumulation on AFE device performance under unipolar stress.

IV. CONCLUSION

This paper reports the endurance result under different polarities, amplitudes, and frequencies in anti-ferroelectric HfZrO devices. It is confirmed that bipolar stress enhances the defect generation rate, leading to a higher breakdown probability. For the unipolar stress conditions, more characterizations have been conducted since it is closer to real eDRAM operation. Experimental data at lower stress amplitudes and different rise/fall edges reveals that its endurance degradation is mainly induced by charge injection

at the interface between the HZO interfacial layer and the electrode. The statistics data of over 60 devices also excludes the possibility of new defect generation during unipolar endurance cycling.

REFERENCES

[1] S.-C. Chang, N. Haratipour, S. Shivaraman, T. L. Brown-Heft, J. Peck, C.-C. Lin, I.-C. Tung, D. R. Merrill, H. Liu, C.-Y. Lin, F. Hamzaoglu, M. V. Metz, I. A. Young, J. Kavalieros, and U. E. Avci, "Anti-ferroelectric $Hf_xZr_{1-x}O_2$ capacitors for high-density 3-D embedded-DRAM," in *IEDM Tech. Dig.*, Dec. 2020, pp. 2811–2814.

[2] S.-C. Chang, N. Haratipour, S. Shivaraman, C. Neumann, S. Atanasov, J. Peck, N. Kabir, I.-C. Tung, H. Liu, B. Krist, A. Oni, S. Sung, B. Doyle, G. Allen, C. Engel, A. Roy, T. Hoff, H. Li, F. Hamzaoglu, R. Bristol, M. Radosavljevic, B. Turkot, M. Metz, I. Young, J. Kavalieros, and U. Avci, "FeRAM using anti-ferroelectric capacitors for high-speed and high-density embedded memory," in *IEDM Tech. Dig.*, Dec. 2021, pp. 3321–3324,

[3] M. Pesic, S. Knebel, M. Hoffmann, C. Richter, T. Mikolajick, and U. Schroeder, "How to make DRAM non-volatile? anti-ferroelectrics: a new paradigm for universal memories," in *IEEE International Electron Devices Meeting*, San Francisco, CA., December, 2016, pp. 11.6.1.1–11.6.4.

[4] A. M. Walke, M. I. Popovici, S. H. Sharifi, E. C. Demir, H. Puliyalil, J. Bizindavyi, F. Yasin, S. Clima, A. Fantini, A. Belmonte, G. S. Kar, J. V. Houdt, "La doped HZO-Based 3D-trench metal-ferroelectric-metal capacitors with high-endurance ($>10^{12}$) for FeRAM applications," *IEEE Electron Device Lett.*, vol. 45, no. 4, April 2024, pp. 578–581.

[5] M. Pesic, U. Schroeder, S. Slesazeck, and T. Mikolajick, "Comparative study of reliability of ferroelectric and anti-ferroelectric memories," *IEEE Trans. Device Mater. Rel.*, vol. 18, no. 2, June 2018, pp. 154–262.

[6] M. H. Park, Y. H. Lee, H. J. Kim, Y. J. Kim, T. Moon, K. D. Kim, H. Müller, A. Kersch, U. Schroeder, T. Mikolajick, and C. S. Hwang, *Adv Mater.* "Ferroelectricity and antiferroelectricity of doped thin HfO2-based films," vol. 27, no. 11, March 2015, pp. 1811–31.

[7] Z. Weng, Z. Lan, Y. Ding, Y. Qu, and Y. Zhao, "Orthorhombic-I phase and related phase transitions: mechanism of superior endurance ($> 10^{14}$) of HfZrO anti-ferroelectrics for DRAM applications," in *IEEE International Reliability Physics Symposium (IRPS)*, Grapevine, TX, April 2024, pp. P11.EM.1–P11.EM.4.

[8] K.-Y. Hsiang, J.-Y. Lee, Z.-F. Lou, F.-S. Chang, Y.-C. Chen, Z.-X. Li, M. H. Liao, C. W. Liu, T.-H. Hou, P. Su, and M. H. Lee, "Fatigue mechanism of antiferroelectric $Hf_{0.1}Zr_{0.9}O_2$ toward endurance immunity by opposite polarity cycling recovery (OPCR) for eDRAM," *IEEE Trans. Electron Devices*, vol. 70, 2023, no. 4, pp. 2142–2146.

[9] Z. Weng, L. Zhao, C. Lee, and Y. Zhao, "Phase transitions and anti-ferroelectric behaviors in $Hf_{1-x}Zr_xO_2$ films," *IEEE Electron Device Lett.*, vol. 44, 2023, no. 10, pp. 1780–1783.

[10] Y. Ding, Y. Huo, C. Yan, Z. Weng, J. Li, Z. Lan, X. Yu, C. Yan, Z. Weng, Y. Qu, and Y. Zhao, "Modeling of endurance degradation of anti-ferroelectric $Hf_{1-x}Zr_xO_2$ capacitors," in *IEEE Conference of Science and Technology for Integrated Circuits (CSTIC)*, Shanghai, China, March 2024, pp. 1–4.

Accelerated polarization switching speed and durable endurance enabled by confined domain size and solid defect migration barrier in FE/AFE multilayer stacked $Hf_xZr_{1-x}O_2$ ferroelectric capacitor

Wenhao Wu [1,2,#], Xinyu Xie [3,#], Zeping Weng [3], Yi Zhao [2], Jiabin Qi [2]*

[1] School of Communication and Electronic Engineering, East China Normal University, Shanghai, China
[2] Research Center of Integrated Circuits, Huada Semiconductor Co., Ltd., Shanghai, China
[3] State Key Laboratory of Silicon and Advanced Semiconductor Materials, Zhejiang University, Hangzhou, China

* Email: qijb@hdsc.com.cn

Abstract—$Hf_xZr_{1-x}O_2$ (HZO) film-based ferroelectric capacitor is a serious contender for the next-generation nonvolatile memories. However, inevitable fatigue and non-uniform polarization reversal hinder the industrial applications. In this work, the HZO capacitors are constructed by stacking the ferroelectric (FE) / antiferroelectric (AFE) layer. It is found that the FE/AFE multilayer ferroelectric capacitor with overlapped stack structure exhibits outstanding comprehensive performance: larger remnant polarization (Pr), reduced coercive field (Ec), enhanced polarization switching speed and improved cycle reliability (over 10^9 @ 2 MV/cm). The mechanism of the AFE insertion layer on domain size and pinning effect is revealed, which are correlated with switching speed and endurance, respectively. The stacked structure can induce the confined small-size domains, reducing the domain wall migration distance to accelerate the switching speed. Moreover, the solid defect migration barrier property of the AFE layer hinders defect agglomeration, thus inhibiting the domain pinning effect. This work provides a better understanding of the switching dynamics and reliability in ferroelectric memories from a theoretical perspective.

Keywords—HZO, ferroelectric, multilayer, stack structure

I. INTRODUCTION

Ferroelectric (FE) memory based on $Hf_xZr_{1-x}O_2$ (HZO) material has attracted widespread attention for various applications (*i.e.* machine learning and artificial intelligence era) due to its low power consumption, scalability, non-volatility, and CMOS process compatibility. The criteria for evaluating the performance of FE memory is the polarization switching speed and endurance, which directly determines the operation speed and the life of practical devices[1]. In detail, the FE domain states in HZO film play an important role in the regulation of switching speed and endurance, and always affected by grain boundaries, oxygen vacancies, and lattice distortions[2]. The large domain size controlled by the lattice leads to a longer migration distance of reverse domain walls, reducing switching speed at particular external electric field and pulse periods[3]. The fatigue effect induced by domain wall pinning after oxygen vacancy redistribution leads to poor cycling reliability[4]. Recently, antiferroelectric (AFE) HZO film with the advantages of low coercive field (Ec) and high endurance has substantially developed, which would help to find a way to enhance not only switching speed but also cycle reliability for the real-world utilization ferroelectric devices. Although several reports have proposed that the performance of FE capacitors could be improved by

bilayer AFE/FE structure [5, 6], research in this area has still left gaps. There is no clear understanding between stack structure and favorable properties (*i.e.* ferroelectricity, endurance, and polarization switching speed) of FE/AFE multilayer stacked devices.

In this work, the FE/AFE multi-layer stacked HZO capacitors are fabricated by varying a quantity of FE and AFE HZO depositions. By exploring the ferroelectric performance, polarization switching speed, endurance / retention properties of these devices, the benefits of AFE HZO layers for switching speed and endurance in multilayer stack structure is investigated. Based on this, the mechanisms of accelerated switching speed and durable endurance are revealed by domain size and domain wall pinning analysis. This work provides a new perspective and a viable solution for achieving high speed and reliable FE capacitors.

II. EXPERIMENTS

Metal-insulator-metal (MIM) capacitors were fabricated by stacking the FE layer (Hf:Zr=1:1) and AFE layer (Hf:Zr=1:4) in overlapped way, as shown in Fig. 1. W bottom electrode was deposited through DC sputtering on an n-type Si substrate. Plasma-enhanced atomic layer deposition (PEALD) with a growth rate of 0.1 nm/cycle was used to deposit HZO layer in three conditions: 1) 10 nm FE HZO, 2) 10 nm HZO stacked with two 3 nm FE layers and a 4 nm AFE layer, 3) 10 nm HZO continuously stacked with three 2 nm FE layers and two 2 nm AFE layers. The temperature of the Hf and Zr precursor was maintained at 150 °C, and the Hf/Zr ratio was controlled by alternating deposition of HfO_2 and ZrO_2 cycles. A 50 nm W layer was deposited as the top electrode for all three conditions. The crystallization was achieved by rapid thermal annealing at 500 °C in N_2 ambient for 60 s. The above samples were labeled as FE-HZO, 2FE-HZO/1AFE-HZO and 3FE-HZO/2AFE-HZO, respectively. The electrical measurement was performed by the Agilent B1500A semiconductor parameter analyzer.

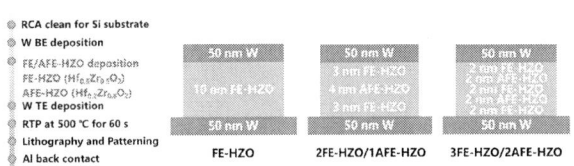

Fig. 1. Key process flow for the fabrication of MIM capacitors and the schematic cross-section of three types of MIM capacitors.

This work is supported by ECNU/HDSC Integrated Circuit Engineering Technology Joint Laboratory.

\# These authors contributed equally and are co-first authors.

III. RESULTS AND DISCUSSION

Fig. 2 presents the C-V curves of FE-HZO, 2FE-HZO/1AFE-HZO and 3FE-HZO/2AFE-HZO at 200 kHz, and the three samples all experience 10^4 wake-up cycles. The curves of FE-HZO and 3FE-HZO/2AFE-HZO exhibit typical ferroelectric characteristics with analogous "butterfly" peaks under double voltage sweep. By comparison, the curve of 2FE-HZO/1AFE-HZO is more likely to have four peaks and shows a larger capacitance value. It is well known that the capacitance value is affected by FE phase. The HZO films contain various phases, including t-, o- and m-phases, and the t-phase has the largest dielectric constant [6]. Because of the same thickness of three samples, the dielectric constant of the 2FE-HZO/1AFE-HZO is the largest, indicating a large t-phase ratio in the film. The o-phases in FE-HZO and 2FE-HZO/1AFE-HZO devices occupy a larger proportion due to the smaller capacitance value in C-V curves.

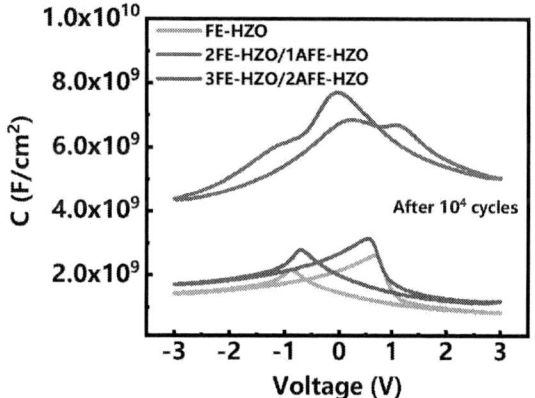

Fig. 2. The C-V curves of FE-HZO, 2FE-HZO/1AFE-HZO and 3FE-HZO/2AFE-HZO at 200 kHz after 10^4 wake-up cycles.

The P-E hysteresis loop is an important characterization method to evaluate the FE properties, especially showing the Pr and Ec of HZO thin films. However, the polarization switching speed is affected by the width of applied pulse on HZO films considering the non-uniform polarization reversal. Fig. 3 (a) and (c) show the P-E loops for the three samples, obtained by positive up negative down measurements with a rise/fall time of 1 μs and 10 μs. The 3FE-HZO/2AFE-HZO shows an excellent ferroelectricity with a rise/fall time of 1 μs, indicating its ability of rapid polarization switching. By comparison, the FE-HZO and 2FE-HZO/1AFE-HZO samples exhibit weak ferroelectricity, which means the domain polarization switching is not experienced completely at the pulse width of 1 μs. As shown in Fig. 3(b), The positive coercive electric (E_{c+}) value is 2.89, 2.13, 1.36 MV/cm, respectively, while the negative coercive electric (E_{c-}) is -2.89, -1.21, -1.76 MV/cm, respectively. The positive and negative remnant polarization (Pr^+ and Pr^-) value of the 3FE-HZO/2AFE-HZO sample are 17.9 and -17.3 μC/cm². To further analyze the polarization switching, we adjust the rise/fall time to 10 μs. Here, the FE-HZO shows outstanding ferroelectric performance, and the 3FE-HZO/2AFE-HZO sample exhibits an almost equal Pr and a smaller Ec, as shown in Fig. 3(c). Fig. 3(d) lists the Ec and Pr of all samples, the Ec values decrease with the AFE HZO layers, showing the Ec of 3FE-HZO/2AFE-HZO reaches 0.95 MV/cm. Therefore, we conclude that the multilayer stacked structure enhances the polarization switching speed and maintains the outstanding ferroelectric properties. Notably, the pulse amplitude also affects the polarization switching, and the switching variability at different voltages is verified by varying the triangular wave pulse amplitude. P-E loops and corresponding I-V results of 3FE-HZO/2AFE-HZO with different voltages are depicted in Fig. 3(d) and (e). The sample also presents the outstanding ferroelectricity under the 2 V testing condition, evident from the sharp current peaks during polarization switching.

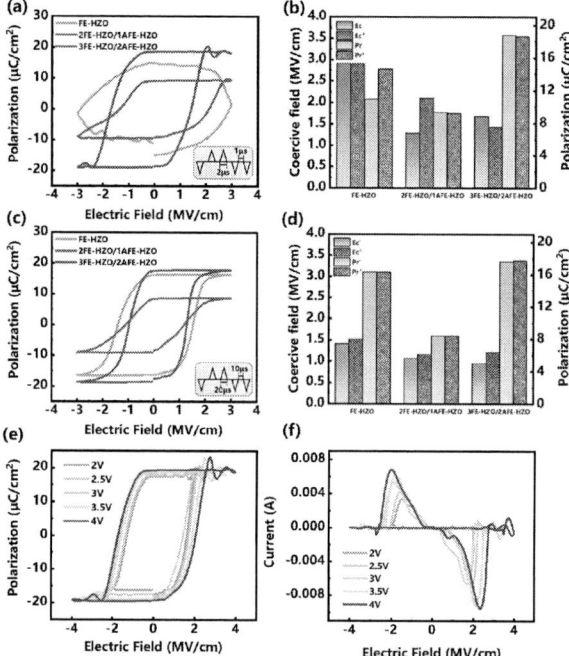

Fig. 3. (a, c) The P-E loops obtained under PUND pulses with a rise/fall time of 1 μs and 10 μs. (b, d) The Ec and Pr statistics of FE-HZO, 2FE-HZO/1AFE-HZO and 3FE-HZO/2AFE-HZO. (e, f) The P-E and I-V loops for 3FE-HZO/2AFE-HZO, employing a triangular waveform characterized by a rise and fall time of 1 μs.

As depicted above, the reduced Ec and sharper switching current peaks collectively indicate the accelerated switching dynamics observed in the multilayer stacked films. In order to clearly compare the variability in polarized switching speed of three samples, the switching dynamics at different pulse amplitudes and widths are characterized. Fig. 4 (a), (b) and (c) displays the evolution of the switched polarization (ΔP/Ps) with the duration time at different write amplitudes. The 2FE-HZO/1AFE-HZO and 3FE-HZO/2AFE-HZO exhibit an evidently faster switching behavior than the FE-HZO film. The 3FE-HZO/2AFE-HZO sample achieves more than 80% domain switching under 2 MV/cm stimulation, whereas the FE-HZO only reaches 60% domain switching. Here, the relation between ΔP/Ps and duration time aligns well with nucleation-limited switching (NLS) model, which considers that polarization switching is limited by the domain nucleation [7]. The ΔP/Ps at a given time (t) is expressed as [7]:

$$\frac{\Delta P}{2Ps}(t) = \int_{-\infty}^{\infty} [1 - \exp\{-(t/\tau)^n\}] F(log\tau) d(log\tau)$$

Where n and τ are the effective dimension characteristic switching time, respectively, and $F(log\tau)$ is the distribution function following the Lorentzian distribution function for the model as below:

$$F(log\tau) = \frac{A}{\pi}\left[\frac{\omega}{(log\tau - log\tau_1)^2 + \omega^2}\right]$$

Where A is a normalized constant, ω and $log\tau_1$ are the half width at half-maximum and center of the distribution. The Lorentzian distribution functions used to fit the switching kinetics of three samples are extracted in Fig. (d), (e) and (f). The 3FE-HZO/2AFE-HZO sample displays a smaller ω under 3 MV/cm stress, suggesting that the domains in this film tend to form a quicker and uniform domain reversal. Commonly, $log\tau_1$ of the Lorentzian distribution curves is also used to assess the domain switching speed [8]. Notably, the $log\tau_1$ of the 3FE-HZO/2AFE-HZO gradually diminishes with the increasing stacking layers, signifying that increased layers expedite domain switching speed. Although 2FE-HZO/1AFE-HZO has an insufficient Pr value, it also shows a narrower distribution peak than FE-HZO, confirming that the stacking structure contributes to domain switching.

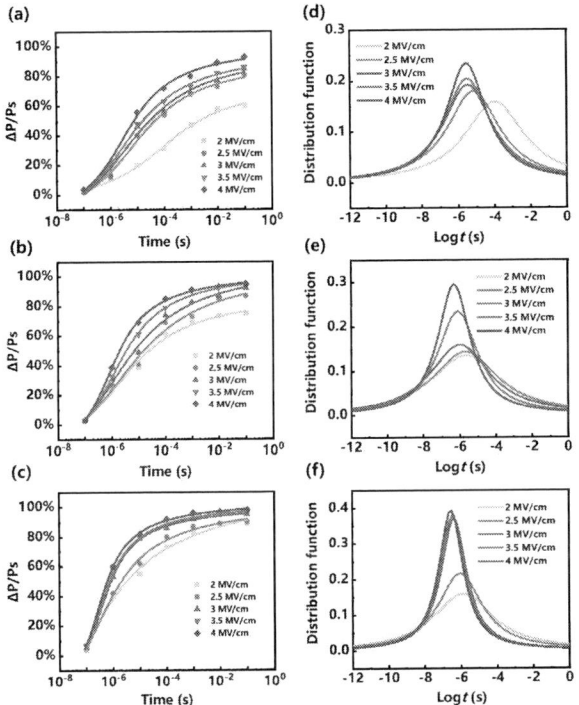

Fig. 4. Switching dynamic characterizations of FE-HZO (a), 2FE-HZO/1AFE-HZO (b) and 3FE-HZO/2AFE-HZO (c). Lorentzian distribution functions. (d) FE-HZO, (e) 2FE-HZO/1AFE-HZO, (f) 3FE-HZO/2AFE-HZO.

At present, some studies have explored the relationship between domain and polarized switching speed under external electric field, proving that the velocity depends on the domain wall migration [3]. The switching process includes the nucleation and growth of the reverse domains, and the sideways spreading out of the domain to fill the entire films is accompanied by the domain wall migration. When the domain size is too large, corresponding to a longer switching region, the domain wall should move a longer distance before achieving the reverse of the whole films. Then, the insufficient domain wall migration velocity directly weakens the polarization switching speed. Fig. 5 (a) shows the relationship of polarization switching speed and domain size.

The films of 2FE-HZO/1AFE-HZO and 3FE-HZO/2AFE-HZO generate small-size domains to accelerate the polarization switching speed, which benefits from the reduction of domain formation enthalpy by the stacking structure. Specifically, the AFE-HZO layer acting as an insert layer causes chemical composition fluctuations at the AFE-HZO and FE-HZO interfaces, thus reducing the domain formation enthalpy containing the electrostatic and elastic energy. As the AFE-HZO layer increases, smaller domains are generated within the film, which directly regulate the domain density. The summary of the effects of stacking structure on polarization switching speed is shown in Fig. 5 (b).

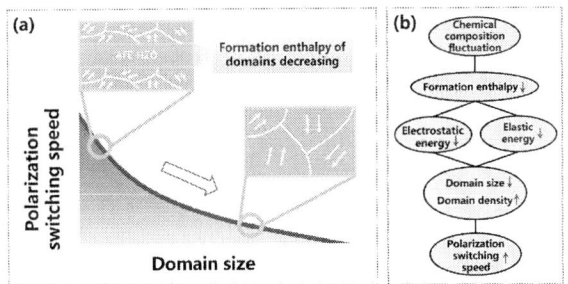

Fig. 5. (a) Schematic of polarization switching speed and domain size. (b) Summary of the effects of stacking structure on polarization switching speed.

Endurance is an important parameter for the application of FE thin films in the memory field. As shown in Fig. 6 (a), the testing was performed under a 2 MV/cm or 3 MV/cm cycling pulse with a pulse width of 1 μs, and the PUND testing was used to determine the 2Pr values. Fig. 6 (b) presents the endurance characteristics of the FE-HZO and 3FE-HZO/2AFE-HZO capacitors. Since FE-HZO sample achieves below 40% domain switching under 2 MV/cm condition, corresponding to the Pr value of less than 10 μC/cm², so we only show the endurance performance at 3MV/cm in the figure. The result demonstrates that the wake-up effect exists in the FE-HZO film and the breakdown occurs after 10^5 cycles. In contrast, the 3FE-HZO/2AFE-HZO sample suffers a breakdown after 10^7 cycles under 3MV/cm stress, confirming that the FE/AFE stacking structure elevates the endurance. The results of the switching kinetics show that the 3FE-HZO/2AFE-HZO sample also achieves fast switching under 2 MV/cm stimulation, and the endurance at 2MV/cm condition is also characterized, as shown in the orange curve. Notably, the 3FE-HZO/2AFE-HZO shows an excellent endurance behavior and experiences a breakdown after 6×10^9 cycles. The initial 2Pr value of the sample also reaches 38 μC/cm² and does not drop substantially during cycling. Therefore, it is believed that increasing AFE layers is conducive to promote the endurance behavior of FE capacitors.

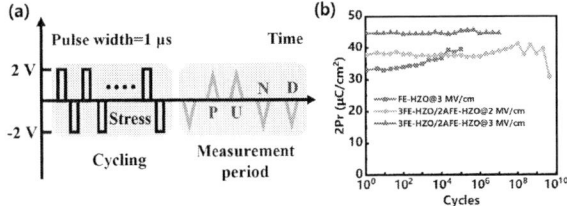

Fig. 6. (a) Schematic of endurance testing pulse. (b) Endurance characteristics of the FE-HZO and 3FE-HZO/2AFE-HZO capacitors.

The retention characteristics is another essential function for nonvolatile memory. Fig. 7 shows the retention characteristics of FE-HZO and 3FE-HZO/2AFE-HZO capacitors at 25 °C and 85 °C. A triangle pulse with an electric field of 3 MV/cm was applied for the retention measurements. The Pr values present 16 and 20 μC/cm^2 of FE-HZO and 3FE-HZO/2AFE-HZO capacitors up to 10^4 second and slightly degradation extended to 10 years. Even at 85 °C, the 3ZO/2AFE-HZO sample undergoes weak degradation, demonstrating that the FE/AFE stacking structure does not weaken the polarization retention. Therefore, the capacitor with stacking structure shows advantages in polarization switching speed, endurance and retention characteristics, as well as application potential in the memory field.

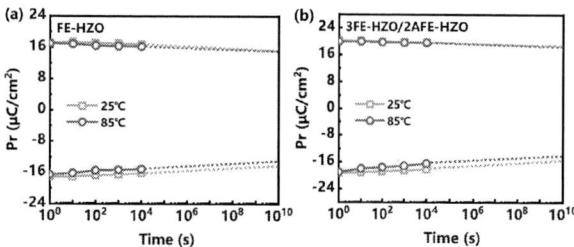

Fig. 7. Polarization retention characteristics at 25 °C and 85 °C. (a) FE-HZO capacitor. (b) 3FE-HZO/2AFE-HZO capacitor.

The endurance results show that the fatigue of the capacitor with FE/AFE stacking structure occurs at 4×10^9 cycles and the breakdown behavior appears at 6×10^9 cycles, which is significantly better than FE-HZO capacitor. According to the previous reports, the FE fatigue and degradation are dominated by the domain pinning. Fig. 8 (a) illustrates the endurance fatigue of FE-HZO caused by domain wall pinning and the redistribution of defects under the electric field cycling. Initially, the defects (especially oxygen vacancies) are concentrated at the interface between FE-HZO and metal electrodes, which play a negative role in FE fatigue. The defects generally aggregate to pin the domain wall motion as the cyclic electric field repetitively reverses FE polarization. When most of the domain walls are pinned, the FE-HZO film has severe fatigue effect until breakdown. Defect agglomeration is a complex process, and the migration barrier directly determines the evolution process of oxygen vacancies. The average of migration barrier of oxygen vacancies is ~0.355 eV in FE-HZO film and ~1.859 eV in AFE-HZO [9]. Thus, the endurance enhancement based on the stacked structure can be attributed to the high defects migration barrier in the AFE insertion layer, resulting in the defects dispersion in the film. Fig. 8 (b) presents the endurance fatigue of 3FE-HZO/2AFE-HZO and defects evolution under the electric field cycling, especially showing the defects migration barriers distribution in the film. Predictably, the defects at the interface cannot traverse the AFE layer with a high migration barrier, indicating that the scattered defects pin a few domain walls. Besides, due to the small-size domains induced by the AFE insertion layers, domain pinning slightly affects the switching of other independent domains, inhibiting the fatigue and degradation of the film.

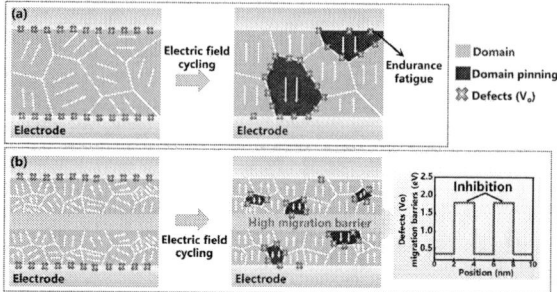

Fig. 8. Endurance fatigue caused by domain wall pinning and the redistribution of defects under the electric field cycling. (a) FE-HZO capacitor. (b) 3FE-HZO/2AFE-HZO capacitor.

IV. CONCLUSION

The polarization switching speed and endurance of the capacitors featuring HZO films with FE/AFE stacking structure are evaluated. It is found that the 10 nm HZO sample continuously stacked with three 2 nm FE layers and two 2 nm AFE layers exhibits the highest Pr of 20 μC/cm^2, lowest Ec of 0.97 MV/cm and best cycle reliability of 6×10^9 @ 2 MV/cm. Based on the switching dynamics analysis, it is revealed that the small-size domains induced by the AFE insertion layers lead to the reduction of domain wall migration distances, thereby benefiting the enhancement of the switching speed in multilayer stacked capacitors. The endurance improvement is attributed to the AFE layer with a high defects migration barrier, which suppresses defects agglomeration to hinder domain pinning. This work provides a guide for the construction of high-speed and reliable FE devices.

ACKNOWLEDGMENT

The authors would like to thank the ECNU/HDSC Integrated Circuit Engineering Technology Joint Laboratory for technical support.

REFERENCES

[1] R. Bian, R. He, E. Pan, "Developing fatigue-resistant ferroelectrics using interlayer sliding switching", Science, eado1744, 2024.

[2] K. Y. Hsiang, J. Y. Lee, Z. F. Lou, et al. "Fatigue Mechanism of Antiferroelectric $Hf_{0.1}Zr_{0.9}O_2$ Toward Endurance Immunity by Opposite Polarity Cycling Recovery (OPCR) for eDRAM", IEEE Transactions on Electron Devices, 70(4): 2142-2146, 2023.

[3] X. Lu, J. Zhu, X. Zhang, et al. "Effects of poling on the switching properties of $SrBi_2Ta_2O_9$ films", Applied physics letters, 80(16): 2961-2963, 2002.

[4] X. Wang, T. Xu, F. Xuan, et al. "Effect of the oxygen vacancy on the ferroelectricity of 90° domain wall structure in $PbTiO_3$: A density functional theory study", Journal of Applied Physics, 126(17), 2019.

[5] C. Shi, N. Mao, K. Zhang, et al. "Domain-dependent strain and stacking in two-dimensional van der Waals ferroelectrics", Nature communications, 14(1): 7168, 2023.

[6] G. Park, A. H. Nguyen, M. C. Nguyen, et al. "Tailoring of Ferroelectric Coercive Field and Polarization with Ferroelectric and Antiferroelectric $Hf_xZr_{1-x}O_2$ Bilayer Structure", IEEE Electron Device Letters, 2024.

[7] J. Y. Jo, H. S. Han, J. G. Yoon, et al. "Domain switching kinetics in disordered ferroelectric thin films", Physical review letters, 99(26): 267602, 2007.

[8] C. Zhou, L. Ma, Y. Feng, et al. "Enhanced polarization switching characteristics of HfO_2 ultrathin films via acceptor-donor co-doping", Nature Communications, 15(1): 2893, 2024.

[9] H. Qian, R. Shen, H. Zhang, et al. "Dynamic evolution of oxygen vacancies during cycling in antiferroelectric $Hf_xZr_{1-x}O_2$", Applied Physics Letters, 124(24), 2024.

A High-Speed Dual-Entropy Sources True Random Number Generator Implemented on FPGA

Yizhi Liu [1], Jierui Liao [1], Hao Xing [1], Pengpeng Sang [1], Jixuan Wu [1], Jiezhi Chen [1]
Xuepeng Zhan *[1], Xiangye Wei *[2]

[1] School of Information Science and Engineering (ISE), Shandong University, Qingdao, China.
[2] TAF Circuits Co., Ltd. Suzhou, China.

* Email: zhanxuepeng@sdu.edu.cn; weixiangye@taf-circuits.com

Abstract—True random number generators (TRNGs) play a crucial role in hardware security, which can provide unpredictable randomness and strengthen system security. In this paper, a TRNG with dual-entropy sources is proposed based on metastability of DFF & MUX and jitter of RO, which can complement each other. TAF-DPS clock generators are adopted to generate multichannel time-varying signals, which themselves contain randomness. The final output was verified by using the NIST randomness test suite. The circuit design is implemented on FPGA. Moreover, the proposed TRNG achieves a random sequence generation rate of 250 Mbps and shows strong robustness against temperature variation. Our findings may have the potential to improve the security of hardware devices and systems.

Keywords—TRNG, dual entropy sources, metastable state, RO jitter, TAF-DPS

I. INTRODUCTION

With the rapid expansion and intelligent evolution of the Internet of Things (IoT) ecosystem, the issue of cybersecurity is becoming increasingly critical and a widely recognized concern. Random numbers serve as a core element of the information security architecture [1]. Their unpredictability is essential for safeguarding critical processes like data encryption, key negotiation, and identity authentication [2]. In the field of random number generation technology, based on the differences in randomness sources and generation mechanisms, there are two major technical systems: Pseudo-Random Number Generators (PRNGs) and True Random Number Generators (TRNGs). Traditional PRNGs rely on deterministic iterative operations with initial seed states. This algorithm-based generation mechanism exposes them to the high risk of attacks [3]. The PRNG based on the improved memristive hopfield neural network (MHNN) has contributed to improving the quality of randomness, while it still faces the problem of requiring a large amount of Look-Up Table (LUT) resources [4].

In contrast, as shown in Fig. 1, hardware-based TRNGs, which harvest physical randomness as entropy sources, can provide inherent randomness. The generated random sequences are truly random and cannot be predicted, which are becoming the fundamental primitives in security and cryptography. So far, considerable research has been conducted on TRNG designs utilizing a single signal entropy source. Their practical applications face certain challenges, including security risks, area overhead, low generation rate, *etc*. For example, ring oscillator (RO)-based TRNGs are vulnerable to electromagnetic interference (EMI) and environmental fluctuations (temperature, *etc*.) [5,6]. Moreover, to achieve the required random number quality,

employing ROs as the sole entropy source typically incurs considerable hardware resource cost [7]. Structures based on metastable phenomena (such as flip-flop metastability) show the disadvantage of low bit rates despite excellent randomness [8]. Additionally, designs relying on analog signals often occupy a large chip area. Notably, existing TRNG solutions generally lack efficient dynamic configurability. This makes it difficult to meet the requirements of diverse application scenarios in IoT devices [9-11]. However, robust multi-entropy-source TRNGs capable of simultaneously achieving high generation rates and flexible configurability remain rarely reported.

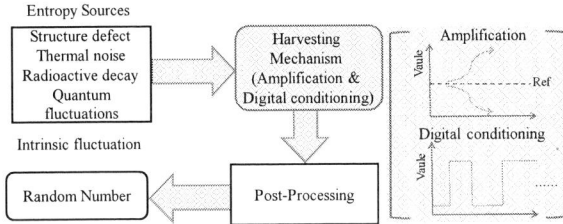

Fig. 1. TRNGs with randomness derived from noise, structural defects, and manufacturing imperfections, *etc.*

This paper proposes a TRNG based on dual-entropy sources for Field-Programmable Gate Array (FPGA) implementation. Metastability of DFF & MUX and jitter of RO are utilized as entropy sources. This design compensates for the low output rate of metastability-based entropy sources and eliminates the need for additional post-processing in RO-based entropy extraction. The randomness of the proposed TRNG output data was validated by using the National Institute of Standards and Technology (NIST) standard. Through experimental verification, the FPGA-based design has strong robustness, and the generation rate can reach up to 250Mbps. In addition, the all-digital design is convenient for integration. The frequency characteristics of Time-Average-Frequency Direct Period Synthesis (TAF-DPS) enhance the configurability of the circuit and may improve the anti-cracking capability. Our findings may be promising for constructing dual-entropy sources TRNG that can be applied under different scenarios.

II. BACKGROUND AND RELATED TECHNOLOGY

A. Background of TAF-DPS

The theory and practice of TAF-DPS clock technology have been significantly evolving for years [12]. Works in this area explored alternative timing architectures to overcome limitations in conventional phase-locked loops (PLLs) and direct digital synthesizers (DDSs). As an advanced technique for generating digital clocks and frequency synthesis, the TAF-DPS clock features arbitrary frequency generation (AFG), instantaneous frequency switching (IFS), and non-fixed cycle clock. These characteristics make it suitable to be used in the field of information security. Fig. 2(a) depicts the frequency synthesis process of the TAF-DPS circuit. Two types of cycles, T_A and T_B, which are expressed in equations (1) & (2), are created to synthesize the output waveform.

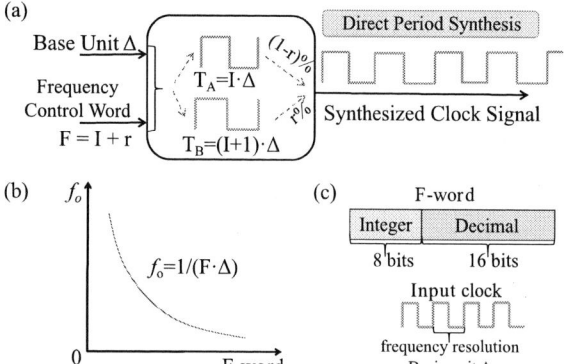

(a)

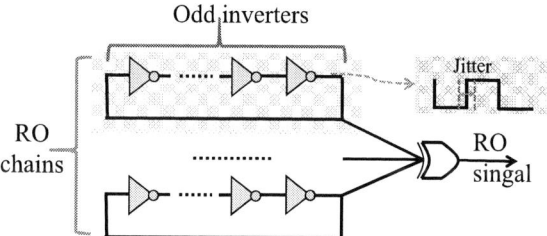

(b) (c)

Fig. 2. (a) Working principle of TAF-DPS combined T_A & T_B waveform synthesis. (b) The relationship between output frequency & F-word. (c) The configuration of F-word and Δ.

$$T_A = I \cdot \Delta \qquad (1)$$

$$T_B = (I + 1) \cdot \Delta \qquad (2)$$

The base unit Δ is the frequency resolution, and its value is equal to the period of the input high-frequency clock. The output frequency/period is given by equation (3), where F (comprising an integer 'I' and a decimal 'r') represents the frequency control word. The decimal 'r' controls the occurrence of T_A and T_B.

$$T_{TAF}=1/f_{TAF}=F \cdot \Delta=(1+r)\cdot \Delta=(1-r)\cdot T_A+r\cdot T_B \qquad (3)$$

Fig. 2(b) shows the relationship between TAF-DPS circuit output frequency and F-word. Fig. 2(c) shows the configuration of the F-word and Δ, where Δ represents the period of the input high-frequency clock in TAF-DPS.

B. Entropy sources implementation

The RO consisted of inverters that are commonly employed as an entropy source in TRNGs due to intrinsic manufacturing variations among individual components. These process-induced discrepancies lead to stochastic timing jitter. Fig. 3 shows the RO structure, composed of multiple parallel RO chains, each implementing an odd-numbered inverters. The RO signal is completed by subjecting it to the XOR logic of these RO chains.

In digital circuits, metastability arises when sampling asynchronous signals, which potentially leads to logic errors and indeterminate output states. To address this issue, common countermeasures include dual-stage synchronizers (*e.g.*, two-flop synchronizers) and asynchronous first-in-first-out (FIFO) buffers. Notably, metastability can also be exploited as a robust entropy source for random number

generation. Fig. 4 illustrates the principles behind the occurrence of the two forms of metastable states. Sequential logic metastability occurs when an asynchronous signal violates the setup time (T_{su}) or hold time (T_h) of a D flip-flop (DFF). This happens if the D input signal transitions too close to the clock edge, which causes the DFF to enter an unstable state. In this case, the output may settle to an "incorrect" logic level. This condition is referred to as a timing violation. Combinational logic metastability occurs when the selection signal (SEL) of the MUX transitions from input A to input B, while the data at B undergoes a simultaneous transition. The output may exhibit glitch or temporarily an indeterminate intermediate state characterized by a voltage between valid logic levels. Due to the high gain and positive feedback inherent in CMOS circuitry, the output will eventually converge stochastically to either a valid high or low logic level.

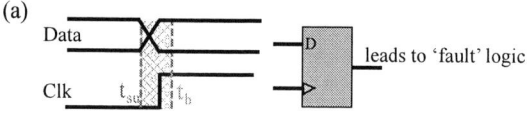

Fig. 3. The generation mechanism of the RO entropy source with odd inverters in one RO chain.

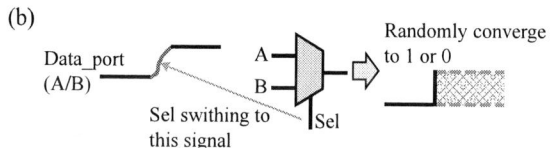

Fig. 4. Metastable entropy source of the proposed TRNG. Principle of (a) Sequential logic metastability. (b) Combinational logic metastability.

III. CIRCUIT PROPOSED

A. Introduction of proposed circuit

Fig.5 depicts the proposed TRNG circuit. A compact design of RO module is employed, which consists of three parallel RO chains. Each chain comprises the minimum required three-inverter configuration. These RO signals, which contain random jitter, can be seen as different clock domain signals. Metastability can occur when asynchronous signals are applied to a DFF because it may violate T_{su} or T_h requirements. Since the TAF-DPS F-word is parallel, the DFF outputs should be serial-to-parallel converted to regulate their frequency, which produces a time-varying output waveform. When the MUX SEL signal switches, simultaneous transitions of data from different paths may cause the output to randomly converge to either a high or low logic level. This behavior introduces combinational logic metastability into the TAF-DPS output, which also increases the time-varying frequency uncertainty of the signal.

979-8-3315-3918-4/25 $31.00 © 2025 IEEE

Multiple circuit structures, as described above, can be applied. Finally, the XOR logic gate mixes these signals, and the DFF is responsible for sampling and two-stage synchronization to generate the random sequence.

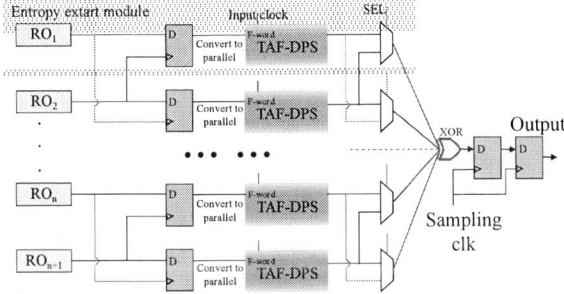

Fig. 5. The circuit schematic diagram of the proposed TRNG with two entropy sources and TAF-DPS as the frequency generator.

B. Parameter configuration and principle explanation

The proposed TRNG circuit is implemented on the Xilinx Kintex-7 FPGA system. To eliminate the flaw of a single chain of the inherent deficiencies or biases that RO may have, it is configured as three chains with three inverters in each chain structure. DFFs and MUXs are adopted to add sequential and combinational logic metastability, respectively. The SEL of MUX can be a fixed-frequency clock signal. By leveraging two forms of metastability mechanisms, this design effectively exploits and maximizes the inherent characteristics of metastability to enhance entropy generation, which ensures superior randomness. TAF-DPS maps the harvested entropy onto diverse frequency outputs to optimize the exploitation of the generated randomness. Meanwhile, the inverse correlation between the F-word and output frequency, combined with TAF-DPS inherently non-periodic waveform generation, enhances resistance against reverse engineering attacks. In addition, conventional random number generators regulate output rates by truncating raw entropy sources with DFFs, whereas the TAF-DPS-based frequency generation method allows raw data to be fed into the F-word (bit-width adjustable). This approach not only maximizes entropy utilization but also preserves the intrinsic characteristics of the entropy source, which enables higher output frequency. The RO module, DFF, TAF-DPS, and MUX form an entropy extraction module. In the proposed design, four modules are adopted, which is motivated by the need to balance random number generation rate with circuit resource utilization.

IV. RESULTS AND ANALYSIS

A. Randomness verification and resource statistics

The NIST SP 800-22 test suite was employed to evaluate the mathematical properties of the random numbers generated by the proposed circuit. This standard comprises 15 core statistical tests, where the results are assessed using p-values. A sequence is considered statistically random if the p-value is greater than or equal to 0.01. Table I presents the NIST SP 800-22 test results for 1 M-bit data generated by the proposed TRNG. The statistical analysis demonstrates that all computed P-values exceed the 0.01 significance threshold, which confirms the sequence compliance with randomness criteria.

Fig. 6 presents the bitmap of the proposed TRNG output (250 MHz) at various scaling factors. The absence of

discernible patterns in the bitmap demonstrates the system effective randomness generation. Quantitative analysis reveals an efficient generation time of approximately 1 ms for a 512×512 matrix. The implementation results demonstrate efficient resource usage, with only 410 LUTs (0.07%) and 356 FFs (0.09%).

TABLE I.   NIST RANDOMNESS TESTS

NO.	NAME	P-value	Result
1	Approximate Entropy	0.088	PASS
2	Block Frequency	0.534	PASS
3	Cumulative Sums	0.205	PASS
4	FFT	0.122	PASS
5	Frequency	0.122	PASS
6	Linear Complexity	0.740	PASS
7	Longest Run	0.350	PASS
8	Overlapping Template	0.067	PASS
9	Rank	0.740	PASS
10	Runs	0.534	PASS
11	Serial1/Serial2	0.631	PASS
12	Universal	0.402	PASS
13	NonOverlapping Template	0.350	PASS
14	Random Excursions	0.546	PASS
15	Random Excursions Variant	0.496	PASS

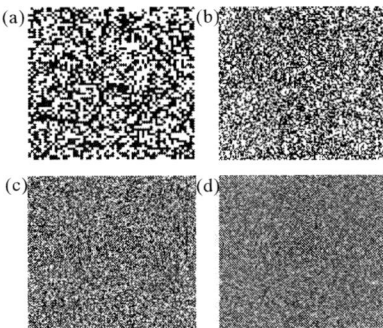

Fig. 6. TRNG output bitmaps (0: white, 1: black) in (a) 64×64, (b) 128×128, (c) 256×256, and (d) 512×512 arrays.

B. Robustness test

To further evaluate the robustness of the proposed TRNG architecture, experiments were conducted under different temperatures (20°C to 80°C with a step of 10°C) and different implementation strategies (corresponding to different circuit placement positions) on the FPGA. Fig. 7 demonstrates that the entropy value maintains a consistently high level (all >0.99996) across all test conditions and successfully passes NIST statistical tests. This confirms that the TRNG circuit structure has the merit of good robustness.

C. Comparisons with other work

A comparative analysis between the proposed TRNG circuit and existing TRNG architectures is presented in Table II. Experimental results, including NIST test, robustness assessment, and bitmap analysis, validate the high randomness and robustness of the TRNG output. Compared to conventional TRNGs based on single entropy sources (e.g., RRAM, RO, or metastability), the proposed dual-entropy sources TRNG demonstrates potential improvements in generation rate, configurability, and robustness.

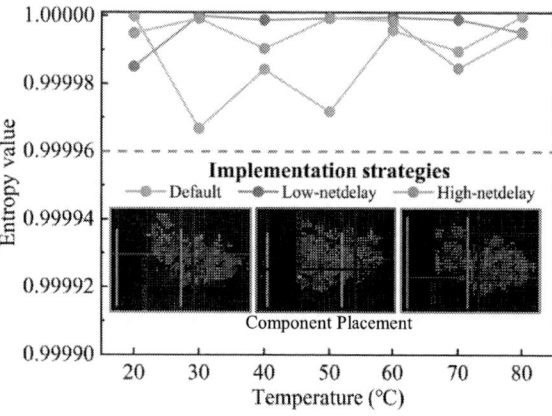

Fig. 7. Entropy value of the proposed TRNG output under temperature fluctuations and implementation strategies.

TABLE II.   THE BENCHMARK OF THIS WORK AND OTHER WORKS

Ref.	TED '24[13]	ICTA '23[14]	ESL '25[15]	ASICON '17[16]	IEEE access '24[17]	This work
Multi entropy	×	×	√	×	×	√
Post-process circuit	×	×	×	√	×	×
Generation rate	7Mbps	5Mbps	200Mbps	30Mbps	225Mbps	250Mbps
NIST test	Pass	Pass	Pass	Pass	Pass	Pass
Entropy source	SNGCT Selector	RRAM	RO jitter & RTN	Metastability	DD-cell	RO jitter & Metastability
Configurability	N/A	√	×	N/A	N/A	√
Robustness verified	√	×	√	×	√	√

V. CONCLUSION

In conclusion, a TRNG is proposed based on metastability of DFF & MUX and jitter of RO as entropy sources with TAF-DPS implemented on FPGA. A maximum generation rate of 250 Mbits/s is achieved, and the randomness was verified by NIST tests. Our design takes advantage of the metastable state and preserves the original characteristics of these two entropy sources. Robustness has been verified through experimental methods. The frequency characteristics of TAF-DPS enhance its resistance to cracking and provide high configurability. Our findings might facilitate advancements in information security.

ACKNOWLEDGMENT

This work was supported by National Key Research and Development Program of China (2023YFB4402500, 2023YFB4402400), Natural Science Foundation of China (U2441248), Natural Science Foundation of Shandong Province (ZR2023LZH007), and MIND Project (MINDXZ202407).

REFERENCES

[1] D. E. Kouicem, A. Bouabdallah, and H. Lakhlef, "Internet of things security: A top-down survey," *Comput. Netw.*, vol. 141, pp. 199–221, Aug. 2018, doi: 10.1016/j.comnet.2018.03.012.

[2] K. Seyhan and S. Akleylek, "Classification of random number generator applications in IoT: A comprehensive taxonomy," *J. Inf. Secur. Appl.*, vol. 71, p. 103365, Dec. 2022, doi: 10.1016/j.jisa.2022.103365.

[3] K. Bhattacharjee and S. Das, "A search for good pseudo-random number generators: Survey and empirical studies," *Comput. Sci. Rev.*, vol. 45, p. 100471, Aug. 2022, doi: 10.1016/j.cosrev.2022.100471.

[4] F. Yu, Z. Zhang, H. Shen, Y. Huang, S. Cai, and S. Du, "FPGA implementation and image encryption application of a new PRNG based on a memristive Hopfield neural network with a special activation gradient," *Chin. Phys. B*, vol. 31, no. 2, p. 020505, Jan. 2022, doi: 10.1088/1674-1056/ac3cb2.

[5] Z. Zhang and T. Su, "Behavioral Analysis and Immunity Design of the RO-Based TRNG under Electromagnetic Interference," *Electronics*, vol. 10, no. 11, p. 1347, Jun. 2021, doi: 10.3390/electronics10111347.

[6] Y. Ma, J. Lin, and J. Jing, "On the Entropy of Oscillator-Based True Random Number Generators," in *Topics in Cryptology – CT-RSA 2017*, vol. 10159, H. Handschuh, Ed., in Lecture Notes in Computer Science, vol. 10159. , Cham: Springer International Publishing, 2017, pp. 165–180. doi: 10.1007/978-3-319-52153-4_10.

[7] S. Choi, Y. Shin, and H. Yoo, "Analysis of Ring-Oscillator-based True Random Number Generator on FPGAs," in *2021 International Conference on Electronics, Information, and Communication (ICEIC)*, Jeju, Korea (South): IEEE, Jan. 2021, pp. 1–3. doi: 10.1109/ICEIC51217.2021.9369714.

[8] Y. Lu, C. Cao, Y. Liu, H. Liang, L. Yao, and L. Ma, "High throughput true random number generator based on dynamically superimposed hybrid entropy sources," *Integration*, vol. 102, p. 102380, May 2025, doi: 10.1016/j.vlsi.2025.102380.

[9] Z. Fu et al., "An Overview of Spintronic True Random Number Generator," *Front. Phys.*, vol. 9, p. 638207, Apr. 2021, doi: 10.3389/fphy.2021.638207.

[10] K. Chen, N. Li, Y. Luo, and Y. Yao, "High-performance hardware primitives based on sub-10 nm nanodiodes for cryptography applications," *J. Mater. Chem. C*, vol. 12, no. 44, pp. 17878–17889, 2024, doi: 10.1039/D4TC02206H.

[11] H. Rohail and R. Ramzan, "A High Throughput True Random Number Generator using Metastability and Chaos," in *2022 20th IEEE Interregional NEWCAS Conference (NEWCAS)*, Quebec City, QC, Canada: IEEE, Jun. 2022, pp. 40–44. doi: 10.1109/NEWCAS52662.2022.9842126.

[12] L. Xiu, "Clock Technology: The Next Frontier," *IEEE Circuits Syst. Mag.*, vol. 17, no. 2, pp. 27–46, 2017, doi: 10.1109/MCAS.2017.2689519.

[13] J. Guy, E. Ambrosi, C.-H. Wu, and X. Bao, "Ultrafast ~7 Mbps True Random Number Generator Based on SNGCT Selector," *IEEE Trans. Electron Devices*, vol. 71, no. 4, pp. 2794–2800, Apr. 2024, doi: 10.1109/TED.2024.3371424.

[14] Z. Cheng et al., "A High-Throughput and Configurable TRNG Based on Dual-Mode Memristor for Stochastic Computing," in *2023 IEEE International Conference on Integrated Circuits, Technologies and Applications (ICTA)*, Hefei, China: IEEE, Oct. 2023, pp. 108–109. doi: 10.1109/ICTA60488.2023.10364295.

[15] Y. Liu et al., "High-Speed True Random Number Generator With Multiple Entropy Sources: Ring Oscillator Jitter and Random Telegraph Noise," in *IEEE Embedded Systems Letters*, doi: 10.1109/LES.2025.3549110.

[16] C. Li, Q. Wang, J. Jiang, and N. Guan, "A metastability-based true random number generator on FPGA," in *2017 IEEE 12th International Conference on ASIC (ASICON)*, Guiyang: IEEE, Oct. 2017, pp. 738–741. doi: 10.1109/ASICON.2017.8252581.

[17] R. D. Sala and G. Scotti, "Exploiting the DD-Cell as an Ultra-Compact Entropy Source for an FPGA-Based Re-Configurable PUF-TRNG Architecture," *IEEE Access*, vol. 11, pp. 86178–86195, 2023, doi: 10.1109/ACCESS.2023.3304901.

979-8-3315-3918-4/25 $31.00 © 2025 IEEE

A Lightweight Arbiter PUF Design Based on Threshold Loss in Transmission Gates

Haoxuan Yan[1], Yitian Su[1], Qiwen Wu[1], Yong Ding[2], Hui Li[3], Yuejun Zhang*[1]

[1] Faculty of Electrical Engineering and Computer Science, Ningbo University, Ningbo 315211, China
[2] College of Integrated Circuits, Zhejiang University, Hangzhou 311200, China
[3] Dahua Technology Co., Ltd, Hangzhou 311200, China

* Email: 2411100030@nbu.edu.cn, zhangyuejun@nbu.edu.cn

Abstract—With the rapid expansion of Internet of Things (IoT) devices, the demand for lightweight and secure hardware authentication in area-limited chips is growing rapidly. As a widely used strong physical unclonable function (PUF), the conventional arbiter PUF (APUF) faces challenges related to hardware overhead, limited randomness, and vulnerability to attacks, which hinder its use in resource-constrained applications. To overcome these shortcomings, this work introduces a novel arbiter PUF architecture utilizing transmission gates that exploit threshold voltage loss, replacing traditional multiplexers. This approach effectively increases path delay differences, thereby improving response randomness. Each delay element in the proposed design comprises only 10 MOS transistors, offering a 58.3% reduction in device count compared to conventional APUFs. Simulations based on TSMC 28nm technology show that the proposed PUF passes 9 out of 15 NIST randomness tests without requiring any response post-processing. The fully custom layout occupies just 1.77 μm², reducing area usage by 51.2% compared to the baseline. These results validate the proposed design as a viable solution for low-cost and secure PUF implementations in IoT and other resource-constrained environments.

Keywords—Threshold Voltage Loss, Physical Unclonable Function, Arbiter PUF, Hardware Security, Low-Area Design

I. INTRODUCTION

With the rapid development of Internet of Things (IoT) technology, how to ensure information security in lightweight applications with limited hardware resources has become a growing concern. Traditional secure chips typically store secret keys in non-volatile memory (NVM), making them vulnerable to invasive and side-channel attacks. Moreover, their relatively high hardware cost makes them unsuitable for resource-constrained IoT devices.

To address these challenges, physical unclonable functions (PUF) have emerged as low-cost hardware primitives capable of resisting physical attacks. The responses of a PUF are derived from intrinsic manufacturing process variations in transistors and circuits, which are used to generate unique challenge-response pairs (CRPs), functioning like a digital fingerprint of the chip. Since these CRPs are not stored in NVM but are dynamically generated upon each activation, PUF-based systems significantly reduce the risk of

This work was supported in part by the Zhejiang Provincial Science and Technology Program "Pioneer and Leading Goose + X" Project under Grant 2025C01063, in part by the National Natural Science Foundation of China under Grant 62474100, in part by the Major Special Project of China Innovation Challenge (Ningbo) from Ningbo Science and Technology Program under Grant 2024T016, in part by the CCF-Huawei Populus Grove Fund, and in part by the Ningbo University Student Research and Innovation Program (2025SRIP1309).

key leakage via physical attacks. With their inherent unclonability, unpredictability, and hardware efficiency, PUFs have been widely applied in secure identification, authentication, and key generation for IoT systems [1][2][3].

PUFs can be broadly classified into strong and weak types based on the number of CRPs they are capable of producing. Strong PUFs, such as arbiter PUFs (APUFs) and XOR PUFs, support a large-often exponential-number of CRPs, making them well-suited for applications like authentication and device identification. In contrast, weak PUFs, including SRAM and DRAM PUFs, generate a limited number of CRPs and are typically used for key generation or low-overhead cryptographic operations. Over the last twenty years, researchers have proposed numerous PUF architectures. Among them, APUF stands out as one of the earliest and most extensively explored strong PUF types. Its advantages-including structural simplicity, a large CRP space, and ease of implementation on platforms such as FPGAs-have made it the basis for many derived designs and enhancements [3][4].

However, conventional APUFs still suffer from limitations in area efficiency, randomness, and resistance to modeling attacks, which hinder their applicability in large-scale IoT deployments. To overcome these issues, this paper proposes a compact APUF design based on threshold-loss in transmission gates. By introducing non-ideal voltage transfer behavior to amplify path delay variations, the proposed structure improves response randomness without significant hardware overhead. Simulation results demonstrate that, compared to conventional APUFs, the proposed design achieves substantial reductions in transistor count and layout area while maintaining superior randomness and suitability for secure, resource-constrained applications.

II. THEORETICAL BACKGROUND

A. Conventional Arbiter PUF

A basic APUF consists of two symmetric signal propagation paths and a final arbiter [5]. As shown in Fig. 1, each stage is composed of a pair of multiplexers (MUXes) controlled by a challenge bit C_i. When $C_i=1$, the two signals are swapped (crossed); when $C_i=0$, the signals propagate in parallel. Due to manufacturing process variations, the delays of the two paths differ slightly, causing the rising edges to arrive at the arbiter at different times. The response R is then generated by comparing the arrival times of the two signals after propagating through the delay paths.

Assuming the delays of the upper and lower paths are t_{up} and t_{down}, respectively, the arbiter outputs the final response R as:

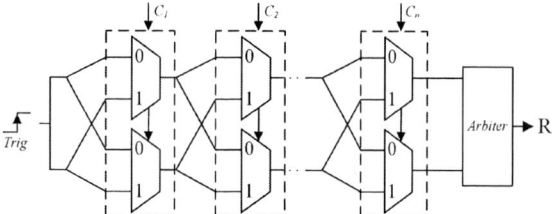

Fig. 1. Basic structure of a conventional APUF.

$$R=\begin{cases} 1, & t_{\text{up}} < t_{\text{down}} \\ 0, & t_{\text{up}} \geq t_{\text{down}} \end{cases} \quad (1)$$

Ideally, for uniformly distributed challenge inputs C, the responses R should also be evenly distributed between 0 and 1. In practice, however, the delay difference at each stage is affected by process variations, transistor aging, and capacitive mismatch, which may lead to systematic bias. The overall delay difference Δt of an APUF can be expressed as a weighted sum of the challenge vector:

$$\Delta t = \sum_{i=1}^{n} w_i \phi_i(C) \quad (2)$$

Here, $\phi_i(C) \in \{-1,1\}$ denotes the signal path selection at the i-th stage, and w_i is the actual delay difference at that stage. Theoretically, if the w_i values follow a zero-mean symmetric distribution and ϕ_i are independent and uniformly distributed, the output R should be unbiased. However, in real-world APUFs, mismatches in internal transistors of the MUXes may cause certain stages to consistently favor one path, resulting in a systematic offset in the total delay Δt.

Moreover, because the delay difference is linearly dependent on the challenge input, the APUF becomes vulnerable to machine learning attacks. Therefore, designing an APUF structure that offers better randomness and stronger resistance to modeling attacks-while maintaining low hardware overhead-is of great significance for secure authentication in resource-constrained IoT devices.

B. Threshold Loss in MOSFETs

The delay unit proposed in this paper is based on the threshold loss behavior of transmission gates. As shown in Fig. 2, when an NMOS transistor has a gate voltage $V_G=V_{DD}$, it turns on and can transmit a low-level signal (V_{SS}) with full voltage swing. However, when transmitting a high-level signal (V_{DD}), the output voltage V_{out} cannot exceed $V_{DD}-|V_{th,n}|$, because the gate-source voltage V_{GS} falls below the threshold $|V_{th,n}|$, turning the device off. As a result, the signal is clipped, and V_{out} stalls at $V_{DD}-|V_{th,n}|$, failing to reach full logic high.

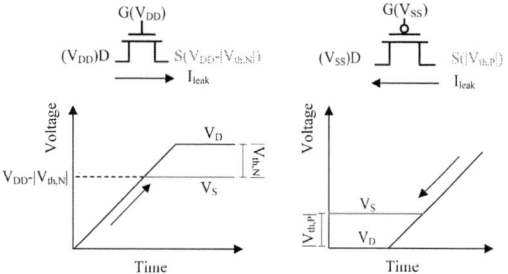

Fig. 2. Threshold loss in NMOS and PMOS.

Similarly, for a PMOS transistor transmitting a low-level signal (V_{SS}), threshold loss also occurs: once the output voltage falls below $|V_{th,p}|$, the transistor turns off due to insufficient gate-source voltage, and the output cannot reach full logic low. Therefore, both NMOS and PMOS devices suffer from threshold-induced signal degradation, which causes incomplete voltage swing during logic transmission.

III. PROPOSED PUF

A. Structure of the Proposed

This section introduces a lightweight APUF architecture based on threshold loss in transmission gates. The delay units of the proposed design are constructed using only a single type of MOSFET-either NMOS or PMOS. As shown in Fig. 2, the transistor-level design of the conventional and proposed APUFs is illustrated. By reordering the combination of NMOS and PMOS delay stages, the proposed design supports two variants: rising-edge-sensitive APUF (Fig. 3a) and falling-edge-sensitive APUF (Fig. 3b).

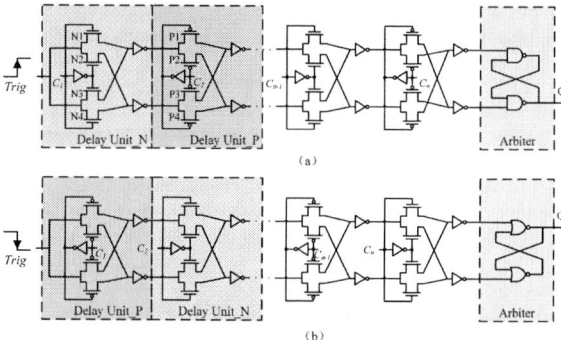

Fig. 3. Transistor-level structures of delay units: (a) Rising-edge-sensitive design; (b) Falling-edge-sensitive design.

Each proposed delay unit integrates three inverters-two for waveform recovery and one for path control-alongside four transmission transistors (either NMOS or PMOS), totaling 10 MOS devices. By contrast, conventional delay elements typically include eight inverters and eight pass transistors, summing up to 24 MOSFETs per unit. This results in a 58.3% reduction in transistor count per stage. Path configuration is governed by the challenge bit C_i, which controls the switching of specific transistors. In the NMOS-based design, when $C_i=1$, transistors N1 and N4 are activated to form a parallel connection, while $C_i=0$ enables N2 and N3 for cross transmission. In the PMOS-based design, $C_i=1$ activates P1 and P4 for parallel routing, while $C_i=0$ turns on P2 and P3 to maintain cross flow.

If the delay units are arranged in an NMOS-PMOS (NP) configuration, the resulting APUF becomes sensitive to rising edges. The two symmetric signal paths ultimately enter an SR latch composed of NAND gates. If the rising edge from the upper path arrives first, the output is logic 1; otherwise, if the lower path's edge arrives first, the output is logic 0, thus completing the arbitration. Conversely, if the units are arranged in a PMOS-NMOS (PN) configuration, the APUF becomes sensitive to falling edges. In this case, the outputs of the two paths enter an SR latch composed of NOR gates. The falling edge that arrives first determines the final output.

B. Signal Propagation in the Proposed Delay Unit

In the proposed delay unit, delay variation primarily originates from the threshold voltage loss in NMOS and PMOS transistors. As illustrated in Fig. 4, a simplified schematic and Monte Carlo waveform are presented using the rising-edge-sensitive configuration as an example. Initially, the NMOS is conducting, and at t=0, a low-level signal propagates through it with a full voltage swing. At t=20ns, the input signal Trig switches from low to high. Due to the NMOS threshold loss, the voltage at node A drops from a full '1' to a degraded level of $V_{DD}-|V_{th,n}|$, resulting in waveform distortion and added delay. This degraded high signal is then fed into an inverter to restore the voltage swing, while preserving the induced delay. The restored output (*OUT1*), now low, passes into the PMOS stage. Because of PMOS threshold loss, the voltage at node B no longer reaches ground but is clipped at $|V_{th,p}|$, causing additional degradation and delay. During the period t =0 to 20 ns, the input remains low and signal propagation proceeds normally. After t =20ns, as Trig transitions high, both NMOS and PMOS experience threshold clipping: node A is limited to $V_{DD}-|V_{th,n}|$, and node B to $|V_{th,p}|$. These degraded transitions cumulatively introduce delay at the final output node *OUT2*.

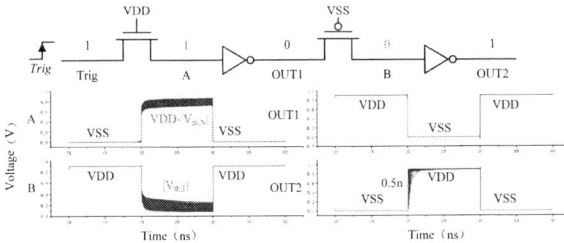

Fig. 4. Simplified circuit and simulation waveform of the proposed delay unit.

Simulation results show that NMOS and PMOS threshold loss significantly expands the delay distribution at *OUT2*, enhancing arbiter output randomness. Unlike conventional APUFs that rely on ideal full-swing signals and symmetric paths-leading to limited delay variation and systematic bias-the proposed design introduces nonlinear delay effects and greater sample diversity. The structure pushes inverters to operate near switching thresholds, making their speed highly sensitive to process variations. These variations are amplified into measurable delay differences, improving uniqueness and response unpredictability.

More importantly, the non-ideal and asymmetric signal propagation breaks the linear correlation between challenge and delay, which is the basis of many machine learning attacks. As a result, the proposed design achieves stronger modeling resistance while maintaining low hardware overhead.

IV. EXPERIMENTAL RESULTS AND ANALYSIS

Both the conventional APUF and the proposed PUF were implemented using a full-custom layout under the TSMC 28nm CMOS process in Cadence Virtuoso. Simulations were conducted to evaluate the performance of the designs. The results are summarized as follows.

A. Hardware Overhead

In the conventional APUF, each delay unit requires 24 MOS transistors. In contrast, the proposed design replaces

multiplexers with transmission gates, such that each signal path requires only a single transistor. As a result, each delay unit contains just 10 MOS transistors. As shown in Fig. 5, the layout area of the proposed PUF under TSMC 28nm technology is $1.77\,\mu m^2$, compared to $3.62\,\mu m^2$ for the conventional APUF, achieving a 51.2% reduction in layout area. This improvement in area efficiency makes the proposed PUF more suitable for resource-constrained applications such as IoT edge devices.

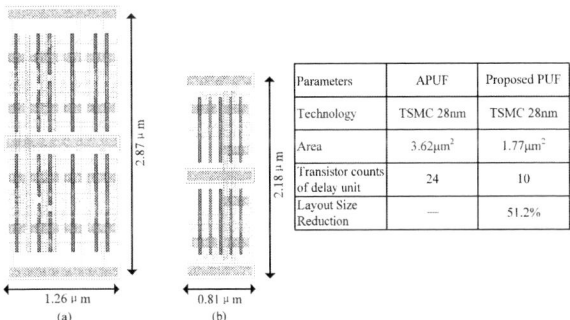

Parameters	APUF	Proposed PUF
Technology	TSMC 28nm	TSMC 28nm
Area	$3.62\mu m^2$	$1.77\mu m^2$
Transistor counts of delay unit	24	10
Layout Size Reduction	—	51.2%

Fig. 5. Full-custom layouts. (a) the conventional APUF; (b) the proposed PUF.

B. Randomness

Randomness is one of the most critical metrics for evaluating PUF performance. Ideally, a PUF should produce binary responses where the probability of '0' and '1' is equal, i.e., 50%.

To comprehensively assess the statistical randomness of the proposed PUF, this work employs the NIST Statistical Test Suite developed by the National Institute of Standards and Technology (NIST) [6]. This widely adopted benchmark suite includes 15 sub-tests designed to evaluate various aspects of bitstream randomness. In the NIST test, a sub-test is passed if the P-value exceeds 0.01, indicating no significant deviation from randomness at 99% confidence. Based on Monte Carlo simulations, 48K response bits were tested. As shown in Table I, the proposed PUF passes 9 out of 15 tests, while the conventional APUF passes only 2, demonstrating its superior randomness.

TABLE I. NIST RANDOMNESS TEST RESULTS FOR CONVENTIONAL AND PROPOSED PUFS

NIST test	APUF		Proposed PUF	
	P-value	PASS?	P-value	PASS?
Frequency	0.0000*	NO	0.5118	YES
Block Frequency	0.0000*	NO	0.5313	YES
Runs	0.0000*	NO	0.5004	YES
Longest Runs	0.0000*	NO	0.4842	YES
FFT	0.0298	YES	0.4614	YES
Non Overlapping Template	0.0000*	NO	0.7175	YES
Serial	0.0571	YES	0.5257	YES
Approximate Entropy	0.0000*	NO	1.0000	YES
Cumclative Sums	0.0000*	NO	0.5501	YES

*means P-value is very small, close to 0

C. Stability

Stability is another key metric reflecting a PUF's reliability under environmental variations. A stable PUF should yield consistent responses despite temperature and voltage fluctuations. In this study, stability is quantified using Bit Error Rate (BER), where a lower BER indicates higher reliability.

Monte Carlo simulations were performed on both the proposed and conventional PUFs under varying conditions: temperature from -20°C to 80°C and voltage from 0.7 V to 1.2 V. For each condition, 10K response bits were compared to those at standard conditions (27°C, 0.9 V). As shown in Fig. 7, the proposed PUF's worst-case voltage-induced BER is 9.8%, slightly worse than the conventional APUF's 9.0% (Fig. 7a). For temperature variation, both designs show a worst-case BER of 1.5% (Fig. 7b). These results suggest that the proposed design achieves comparable stability while offering better area efficiency and randomness.

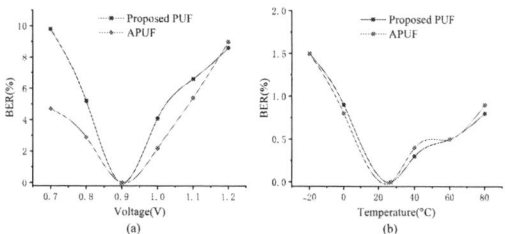

Fig. 6. Stability analysis of APUF and proposed PUF: (a) BER under supply voltage from 0.7 V to 1.2 V; (b) BER under temperature range from -20 °C to 80 °C.

D. Uniqueness

Uniqueness is an essential metric that evaluates a PUF's capability to differentiate between various chips or physical instances. It reflects the variation in responses generated by distinct devices when subjected to the same challenge input and is commonly assessed using the inter-chip Hamming distance (HD).

From a theoretical perspective, for a response sequence of length n, the expected average HD between any two chips should approach $n/2$, indicating that around 50% of the bits differ. This level of response diversity corresponds to ideal uniqueness, ensuring high distinctiveness and unpredictability across devices. The inter-chip Hamming distance is determined using the following equation:

$$HD_{\text{inter}} = \frac{2}{k(k\text{-}1)} \sum_{i=1}^{k\text{-}1} \sum_{j=i+1}^{k} \frac{HD(R_i, R_j)}{n} \times 100\% \qquad (3)$$

Where k is the number of PUF instances, and $HD(R_i, R_j)$ represents the Hamming distance between the responses R_i and R_j.

As shown in Fig. 7, the proposed PUF achieves an average inter-chip Hamming distance of 0.496, close to the ideal 50%, indicating strong uniqueness.

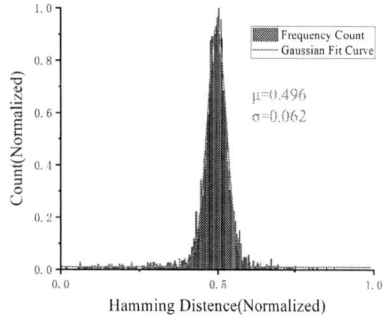

Fig. 7. Uniqueness of the proposed PUF.

E. Performance Comparison

As shown in Table II, the proposed PUF is compared with a conventional APUF across key metrics. Under the same process, it uses significantly fewer MOS transistors per delay unit and achieves a 51.2% reduction in layout area. It also passes 9 out of 15 NIST randomness tests 7 more than the conventional APUF. Furthermore, it outperforms the baseline in uniqueness, stability, and autocorrelation, making it a more secure and resource-efficient solution for low-power hardware security applications.

TABLE II. PERFORMANCE COMPARISON BETWEEN APUF AND PROPOSED PUF

Parameters	APUF	Proposed PUF
Technology (nm)	28nm	28nm
Transistor counts of delay unit	24	10
Core area (µm2)	3.62µm2	1.77µm2
Pass counts of NIST subtests	2	9
Worst BER (%)	9	9.8
Temperature range (°C)	-20-80	-20-80
Voltage range (V)	0.7-1.2	0.7-1.2

V. CONCLUSION

This work introduces a compact strong APUF architecture that leverages MOSFET threshold voltage loss. The proposed delay unit occupies just 1.77 µm², which represents a 51.2% reduction in layout area compared to standard APUF implementations. It successfully passes 9 out of 15 tests in the NIST randomness suite-7 more than the conventional design-and achieves an average inter-chip Hamming distance of 49.6%, indicating high uniqueness. In terms of stability, the worst-case bit error rates are 8.6% under voltage variation and 1.5% under temperature changes, which are comparable to traditional counterparts. Overall, the proposed design offers reduced hardware overhead, improved randomness, and enhanced security, making it well suited for secure authentication and key generation in IoT and other area-constrained systems.

REFERENCES

[1] J. Liu, Y. Zhao, Y. Zhu, C. -H. Chan and R. P. Martins, "A weak PUF-assisted strong PUF with inherent immunity to modeling attacks and ultra-low BER," IEEE Trans. Circuits Syst. I, Regul. Pap., vol. 69, no. 12, pp. 4898-4907, December 2022.

[2] X. Zhao, C. Xie, Q. Zhao, and X. Pan, "A dual-entropy-superposed PUF with in-cell entropy sign-based stabilization," IEEE Trans. Circuits Syst. I, Regul. Pap., vol. 69, no. 1, pp. 284–296, January 2022.

[3] Z. Zhou, P. Wang and G. Li, "Bagua protocol: a whole-process configurable protocol for IoT sensing devices security based on strong PUF," IEEE Internet Things, vol. 11, no. 1, pp. 805-819, January, 2024.

[4] E. Hunt Schroeder and T. Xia, "12-nm stable pre-amplifier physical unclonable function with self-destruct capability," IEEE Trans. Very Large Scale Integr. (VLSI) Syst., vol. 31, no. 6, pp. 840–850, January 2023.

[5] Y. Wang, G. Zhang, X. Mei, and C. Gu, "A high-reliability, non-CRP-discard arbiter PUF based on delay difference quantization," IEEE Trans. Circuits Syst. I, Regul. Pap., vol. 72, no. 2, pp. 573–585, February 2025.

[6] C. -Y. Lee, K. Bharathi, J. Lansford and S. P. Khatri, "NIST-Lite: randomness testing of RNGs on an energy-constrained platform," IEEE Int. Conf. Comput. Des. VLSI Comput. Process., Storrs, CT, USA, 2021, pp. 41-48

Destruction-Free Soft PUF Architecture: Merging Security and Efficiency in 4T2M TCAM Without Data Migration

Shimao Ren[1], Pengjun Wang*[1], Bo Chen [1], Zhenhong Chen [1]

[1] College of Electrical and Electronic Engineering, Wenzhou University, Wenzhou 325035, China

* Email: 23461249007@stu.wzu.edu.cn ,wangpengjun@wzu.edu.cn

Abstract—Magnetoresistive random-access memory (MRAM) based ternary content addressable memory (TCAM) has emerged as a promising candidate for internet of things (IoT) and embedded systems. However, it remains vulnerable to emerging security threats such as side-channel attacks and data tampering. This work proposes a dual-mode soft physical unclonable function (PUF) based on a 4T2M TCAM architecture, leveraging the intrinsic resistance mismatch within the TCAM array. The proposed design eliminates reliance on conventional data migration by employing a non-destructive readout scheme. To further improve efficiency, the system incorporates the dual-mode hamming weight clustering (DM-HWC) algorithm with Hash indexing for optimized challenge-response address generation. Simulation results in 65nm CMOS demonstrate that the PUF's quantization circuitry adds only 263 transistors (3.4% area overhead). The PUF achieves 49.9%/50.1% inter-chip uniqueness, passes NIST SP 800-22 randomness tests, and enables 6.86× acceleration in address clustering (3.23×10^{-5} s vs 2.21×10^{-4} s) for 32×32 arrays.

Keywords—PUF, Hardware security, TCAM, MRAM

I. INTRODUCTION

With the rapid growth of Internet of Things (IoT) and embedded systems, Magnetoresistive random-access memory (MRAM) based ternary content addressable memory (TCAM) has been widely deployed in mission-critical applications such as packet classification and routing table lookup, due to its non-volatile storage and parallel search capabilities [1], [2]. Moreover, MRAM TCAM plays a pivotal role in data center cache hierarchies and network intrusion detection systems, offering low-latency and high-throughput performance. However, as system complexity increases and security threats evolve, conventional MRAM TCAM implementations exhibit critical vulnerabilities, including susceptibility to side-channel attacks and data tampering [3]. Furthermore, stringent constraints on power consumption and chip area in sensor nodes and portable devices call for innovative solutions that simultaneously enhance security and maintain resource efficiency.

Physical unclonable function (PUF) have emerged as lightweight security primitives for device authentication and key generation [4]. Existing MRAM PUF implementations can be broadly categorized into two types. Hard PUF designs utilize dedicated circuitry to exploit process variations for response generation—such as comparing resistance across multiple MRAM paths [5] or selecting and comparing MRAM cell groups with challenge bits [6]. However, these approaches often incur significant area and power overhead. In contrast, soft PUFs leverage intrinsic physical characteristics of

This work was supported in part by National Natural Science Foundation of China (Grant 62174121 and Grant 62234008).

standard MRAM cells to generate responses without compromising memory functionality, offering minimal hardware overhead [7], [8]. Nevertheless, majority of soft PUF architectures require data migration to support PUF operation [9], which introduces additional buffer logic and poses a risk of transient data exposure.

To address these challenges, this work proposes a dual-mode soft PUF design based on a 4T2M TCAM architecture with three key innovations:

(1) *Stable Entropy Source:* The proposed PUF leverages intrinsic resistance mismatches in TCAM cells as a stable entropy source, achieving robust randomness with only 3.4% area overhead. A voltage comparator-based quantization circuit is integrated to precisely capture match-line (ML) voltage characteristics.

(2) *Non-Destructive Readout:* A novel readout mechanism is introduced to generate PUF responses without requiring data migration or memory overwriting, thereby eliminating destructive operations and mitigating security risks.

(3) *Fast Clustering Algorithm:* We propose the dual-mode hamming weight clustering (DM-HWC) algorithm with hash indexing, which is developed for 32×32 TCAM arrays, efficiently identifying candidate rows with uniform hamming weights to support reliable and addressable PUF extraction.

The remainder of this paper is organized as follows: Section II analyzes the data migration issues in traditional intrinsic PUFs; Section III details the architectural design and fast search algorithms of the 4T2M TCAM Soft PUF; Section IV presents experimental simulation results and performance analysis; Section V concludes the paper.

II. CONVENTIONAL DATA MIGRATION CHALLENGES

Traditional soft MRAM PUF implementations (illustrated in Fig. 1 using TCAM as an example) typically employ data migration protocols to avoid overwriting stored data during cryptographic key generation. This process involves transferring data to external buffers prior to PUF operation and restoring it afterward.

However, such approaches introduce considerable hardware overhead and security vulnerabilities. First, additional buffer memory for temporary data storage significantly increases chip area and power consumption, which is especially problematic for resource-constrained IoT devices. Second, security vulnerabilities emerge during data transfer: Fault injection and timing analysis attacks may exploit transmission pathways [10], while residual magnetization states or resistance traces during restore opera-

979-8-3315-3918-4/25 $31.00 © 2025 IEEE

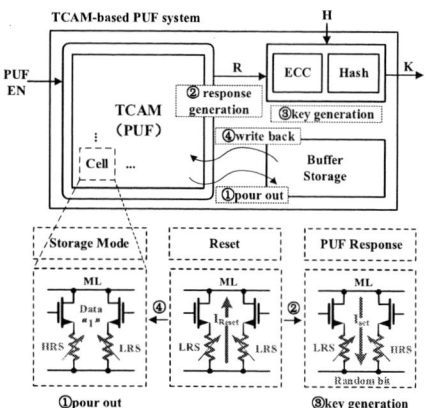

Fig. 1. Operational mechanism of traditional memory-based PUF systems (using TCAM as an example).

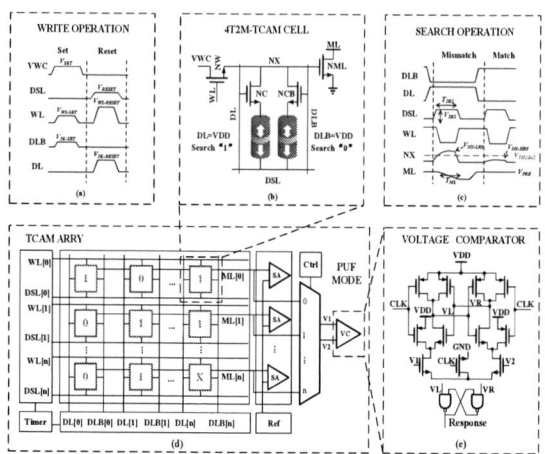

Fig. 2. The proposed 4T2M TCAM PUF architecture without data migration.

-tions can facilitate feature extraction attacks. In high- speed operation, synchronization challenges further degrade system stability, as bias voltage discrepancies and dynamic environmental interference can disrupt migration timing.

To overcome these limitations, we propose a dual-mode soft PUF design based on 4T2M TCAM architecture [11], which eliminates the need for data buffers and mitigates associated risks by leveraging a non-destructive readout mechanism that enables migration-free PUF response generation.

III. 4T2M TCAM SOFT PUF DESIGN

A. 4T2M TCAM Soft PUF Architecture

The 4T2M TCAM cell utilizes a symmetric topology comprising four NMOS transistors and dual magnetic tunnel junctions (MTJs). Its unit structure and operational mechanism are illustrated in Fig. 2(a)-(c): During the write phase, the word line (WL) is driven to the operating voltage, and differential programming signals are applied to the data line (DL) and its complement (DLB) to store information. In the content-matching phase, a standard-amplitude V_{DSL} pulse drives the cells, inducing charging dynamics at the critical node (NX) that vary across cells due to intrinsic process variations. By comparing the NX voltage to the conduction threshold ($V_{TH\text{-}NML}$) of transistor NML, the ML triggers match or mismatch states, thus enabling search functionality.

The overall 4T2M TCAM Soft PUF architecture, depicted in Fig. 2(d)-(e), operates as follows: A pre-generated PUF response activates target rows, initializes the DL/DLB in differential search mode, and applies a standard V_{DSL} pulse to quantify NX voltage profiles. Under identical storage configurations, process-induced resistance mismatches create subtle NX voltage deviations across cells, that directly modulate ML levels. A 32:2 multiplexer routes ML outputs to voltage comparators, enabling direct extraction of PUF responses without data migration or overwriting. The non-destructive read mechanism obviates the need for auxiliary buffer circuits, reduces area and power overhead, and prevents transient data exposure during access operations. This approach establishes a hardware-based root of trust by inherently enforcing security primitives at the architectural level.

In dynamic storage environments where TCAM contents are frequently updated, the proposed PUF adaptively gener-

Algorithm: DM-HWC algorithm

Require:
TCAM: M×N matrix with entries {'0', '1', 'X'}
Ensure:
       Hash maps index_A (X→0) and index_B (X→1) **for** instant 1's count queries
\# Phase 1: Preprocessing
1: Initialize count_A ← [0]^M, count_B ← [0]^M   ▷ Dual counters
2: Initialize index_A, index_B ← empty hash maps
3: **for** i ← 0 to M-1 **do**
4:   **for** j ← 0 to N-1 **do**
5:     **if** TCAM[i][j] = '1' **then**
6:       count_A[i] += 1 ; count_B[i] += 1 ▷ Case B: count '0'
7:     **else if** TCAM[i][j] = 'X' **then**
8:       count_B[i] += 1   ▷ Case B only: X→1
9:   ▷ Build index_A (X→0)
10: index_A[count_A[i]] ← index_A.get(count_A[i], []).append(i)
11: ▷ Build index_B (X→1)
12: index_B[count_B[i]] ← index_B.get(count_B[i], []).append(i)
13: **end for**

\# Phase 2: Query Function
Function Query(index, target):
1: return index.get(target, []).copy()   ▷ O(1) retrieval with null safety

---

-ates new responses from refreshed data, enhancing its challenge response pair (CRP) capacity through intrinsic hardware reconfigurability. By leveraging the intrinsic parallel search capability of TCAM, a fast clustering algorithm is employed to identify candidate rows with uniform hamming distance (HD) within a single clock cycle. These selected rows are then subjected to all-1 and all-0 search operations to enable non-destructive PUF response extraction.

B. DM-HWC Algorithm Design

The row address selection for PUF response generation in 4T2M TCAM arrays must be based on stored cell values, with systematic alignment to the array's physical characteristics and PUF security requirements. The DM-HWC algorithm employs a two-phase methodology: 1) Preprocessing phase for hash table construction, and 2) Query phase for instant candidate retrieval. The logical matrix is transformed by interpreting 'X' as programmable values ('0' or '1'), enabling row-wise HD calculation through vectorized summation. For N×N TCAM arrays, the algorithm partitions rows into 8-bit

 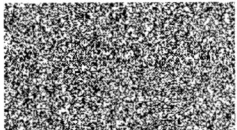

CASE1 Black[1] White[0]   **CASE2 Black[1] White[0]**

Fig. 3.   Grayscale mapping of PUF response generation.

TABLE I.   MTJ PARAMETERS AND VARIABLES

Parameters	Description	Value
TMR	Tunnel magnetoresistance ratio	2
tox	Thickness of the oxide barrier	0.85 nm
Dev-TMR	Variation percentage of TR when RV=2	3%
Dev-tsl	Variation percentage of tsl when RV=2	3%
Dev-tox	Variation percentage of tox when RV=2	3%
STO-dev	Variation percentage of switching duration when STO=2	3%
RV	Device parameters follow a Gaussian distribution	2
STO	Switching duration follows a Gaussian distribution	2
RA	Resistance area product	5Ω μm
a	Length of surface long axis	32 nm
b	Width of surface short axis	32 nm

TABLE II.   NIST800-22 RANDOMNESS TEST

Test item	Stream length	Proportion	P-value	Pass
Frequency	2560	9/10	0.350485	100
Block Frequency	2560	9/10	0.213309	100
Cumulative Sums	2560	9/10	0.004301	100
Longest Run	2560	10/10	0.739918	100
Rank	2560	8/10	0.002043	100
FFT	2560	8/10	0.350485	100
Runs	2560	10/10	0.534146	100
Non Overlapping Template	2560	10/10	0.122325	100
Serial	2560	10/10	0.213309	100
Linear Complexity	2560	9/10	0.066882	100

segments, calculates HD via precomputed LUTs for complete segments, and performs direct summation for residual bits.

While conventional bitwise traversal methods exhibit $O(N^2)$ complexity with scalar operations, our algorithm optimizes practical efficiency through three key techniques: 1) Precomputed weight tables, 2) Vectorized bit-block processing, and 3) Dual hash indexing. This architecture enables single-clock-cycle candidate row identification through $O(1)$ hash table lookups after the initial preprocessing.

IV. SIMULATION RESULTS AND PERFORMANCE COMPARISON

To validate the performance of the proposed data-migration-free 4T2M TCAM Soft PUF architecture, Monte Carlo simulations were conducted using 65nm CMOS technology and a compact STT-MRAM model on the Cadence Virtuoso platform. Key parameters of the MTJ are summarized in Table I.

Two operational modes (CASE1/CASE2) were configured by programming the TCAM contents to evaluate PUF response characteristics before and after reconfiguration. The randomness of PUF responses was rigorously verified

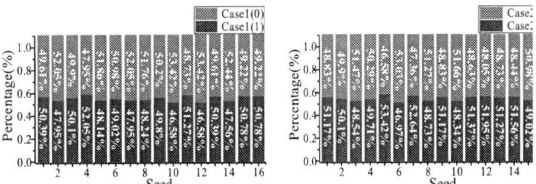

Fig. 4.   Uniformity of PUF responses.

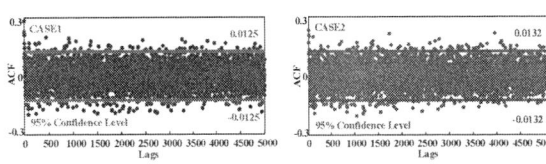

Fig. 5.   Autocorrelation of measured bit cells.

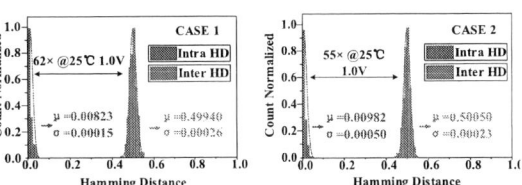

Fig. 6.   Measured normalized hamming distance.

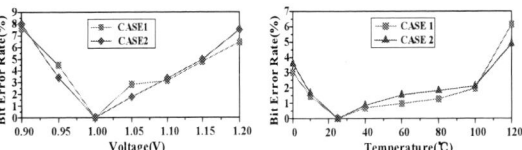

Fig. 7. Reliability of PUF responses under temperature/voltage variations.

through the NIST SP 800-22 test suite, with 25,600-bit response sequences successfully passing all 10 statistical tests. Visual confirmation of high unpredictability was achieved by mapping PUF responses to grayscale images, as shown in Fig. 3. Fig. 4 demonstrates a near-ideal 50% 0/1 bit distribution, confirming the uniformity. The autocorrelation coefficients of the 25,600 PUF bits remain below 0.132% (95% confidence interval, Fig. 5), validating response independence. Inter-chip HD reach 49.9%/50.1% pre- and post-reconfiguration (Fig. 6), with worst-case intra-chip HD limited to 0.82%/0.98%. The inter/intra-chip HD ratio of 62:1/55:1 confirms the uniqueness and native reliability.

Under the data-migration-free architecture, the reliability of the PUF was verified through a series of experiments, as shown in Fig. 7. Temperature adaptability: Within the temperature range of 0°C to 120°C, the worst-case bit error rate (BER) were measured at 6.2%/4.9% (at 120°C), with temperature sensitivities of 0.71/0.77 per 10°C. Voltage stability: For supply voltages ranging from 0.9 V to 1.2 V, the maximum BER reached 7.6%/8% (at 0.9 V), with voltage sensitivities of 4.96/5.17 per 0.1 V.

Fig. 8 compares the execution time of traditional character-wise scanning with that of the DM-HWC algorithm across 10,000 randomized TCAM instances. For 32×32 arrays, the proposed DM-HWC algorithm achieves 6.86× acceleration, completing in $3.23×10^{-5}$ seconds compared to the traditional method's $2.21×10^{-4}$ seconds. Acceleration

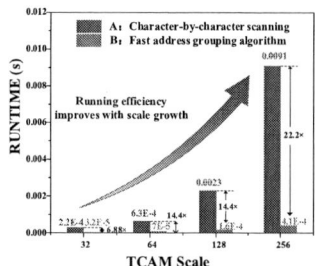

Fig. 8. Traditional bitwise traversal algorithm vs. DM-HWC algorithm.

TABLE III. PERFORMANCE COMPARISON

	TCE' 23[12]	SCI REP' 24[13]	VLSI' 19[8]	DATE' 15[14]	THIS WORK
Tech	45 nm	28 nm	40 nm	65 nm	65 nm
Temp (°C)	-40-100	-25-100	-45-100	0-100	0-120
VDD (V)	0.8-1.2	0.65-0.85	0.9-1.1	0.9-1.1	0.9-1.2
BER/10 °C	1.37%	0.34%	0.40%	0.84%	0.71/0.77
BER/0.1 V	0.2%	1.75%	3.36%	5.92%	4.96%/ 5.17%
Inter- HD	49.99%	49.96%	50.20%	49.62%	49.9%/ 50.1%
Storage mode	No	No	Yes	Yes	Yes
Buffer Storage	Yes	Yes	Yes	Yes	No

ratios exhibit superlinear growth, reaching 22.20× for 256×256 arrays. The stabilization of acceleration growth rates for N > 64 aligns with the asymptotic behavior of $O(N^2)$ complexity, enabling real-time key generation in large-scale PUF systems.

As shown in Table III, the 4T2M TCAM-PUF eliminates the need for buffer storage (Buffer Storage = No) while maintaining full storage functionality (Storage Mode = Yes). The 32×32 array core occupies 5,958 transistors, with only 203 additional transistors (a 3.4% overhead) allocated for quantization circuits through shared TCAM match lines. Dynamic power consumption is reduced by leveraging native TCAM read operations instead of cache-intensive workflows. Competitive inter-chip HD of 49.9% and 50.1%, comparable to those of dedicated PUF implementations, further validate the effectiveness of TCAM resistance mismatch as a source of entropy.

V. CONCLUSION

The proposed destruction-free 4T2M TCAM soft PUF exploits intrinsic resistance mismatch in TCAM cells to avoid hardware overhead and security risks associated with conventional data migration. By preserving native TCAM functionality and incorporating preprocessing-enhanced clustering with hash indexing, the design achieves 6.2%/4.9% worst-case BER across 0-120°C and 7.6%/8% BER under 0.9-1.2V supply voltages, while maintaining 49.9-50.1% interchip HD. NIST-validated randomness and accelerated address generation ($3.23×10^{-5}$ s for 32×32 arrays, scaling to 22.20× speedup for 256×256 arrays) confirm the scheme's efficacy. These results validate an optimal trade-off between area

efficiency, power consumption, and security robustness, presenting a viable authentication scheme for resource-constrained IoT and embedded systems.

ACKNOWLEDGMENT

This work was supported in part by National Natural Science Foundation of China under Grant 62174121 and Grant 62234008.

REFERENCES

[1] E. Garzón, M. Lanuzza, A. Teman, and L. Yavits, "AM4: MRAM crossbar based CAM/TCAM/ACAM/AP for in-memory computing," IEEE Journal on Emerging and Selected Topics in Circuits and Systems, vol. 13, no. 1, pp. 408-421, Mar. 2023, doi: 10.1109/JETCAS. 2023.3243222.

[2] Q. K. Trinh, Q. M. Duong, X. T. Do, V. P. Hoang, H. G. Vu, and V. N. Dinh, "A novel in-memory matching circuit based on non-volatile resistive memory," Proc. Int. Conf. IC Design Technol. (ICICDT), pp. 97–100, Hanoi, Vietnam, 2022, doi: 10.1109/ICICDT56182.2022. 9933127.

[3] Y. Chiu, W. Khwa, C. Yang, S. Teng, H. Huang, and F. Chang, "A CMOS-integrated spintronic compute-in-memory macro for secure AI edge devices," Nat. Electron., vol. 6, pp. 534–543, 2023, doi: 10.1038/ s41928-023-00994-0.

[4] L. Ni, P. Wang, Y. Zhang, G. Li, L. Ding, and J. Zhang, "PI PUF: A processor-intrinsic PUF for IoT," Comput. Electr. Eng., vol. 105, p. 108540, 2023, doi: 10.1016/j.compeleceng.2022.108540.

[5] Y. Hu, L. Wu, Z. Chen, Y. Huang, X. Xu, and K. Li, "STT-MRAM-based reliable weak PUF," IEEE Trans. Comput., vol. 71, no. 7, pp. 1564–1574, Jul. 2022, doi: 10.1109/TC.2021.3095657.

[6] S. Khaleghi, P. Vinella, S. Banerjee, and W. Rao, "An STT-MRAM-based strong PUF," Proc. IEEE/ACM Int. Symp. Nanoscale Archit. (NANOARCH), pp. 129–134, Beijing, China, 2016, doi: 10.1145/295 0067.2950080.

[7] A. Nejat, F. Ouattara, M. Mohammadinodoushan, B. Cambou, K. Mackay, and L. Torres, "Practical experiments to evaluate quality metrics of MRAM-based physical unclonable functions," IEEE Access, vol. 8, pp. 176042–176049, Sep. 2020, doi: 10.1109/ACCESS.2020. 3024598.

[8] S. Ben Dodo, R. Bishnoi, S. Mohanachandran Nair, and M. B. Tahoori, "A spintronics memory PUF for resilience against cloning counterfeit," IEEE Trans. Very Large Scale Integr. (VLSI) Syst., vol. 27, no. 11, pp. 2511–2522, Nov. 2019, doi: 10.1109/TVLSI.2019.2931481.

[9] L. Zhang, X. Fong, C.-H. Chang, Z. H. Kong, and K. Roy, "Optimizing emerging nonvolatile memories for dual-mode applications: data storage and key generator," IEEE Trans. Comput.-Aided Design Integr. Circuits Syst., vol. 34, no. 7, pp. 1176–1187, Jul. 2015, doi: 10.1109/ TCAD.2015.2427259.

[10] A. Dhaman and U. Suman, "A survey on security techniques for cloud data migration," Proc. Int. Conf. Adv. Comput. Res. Sci. Eng. Technol. (ACROSET), pp. 1–5, Indore, India, 2024, doi: 10.1109/ACROSET6 2108.2024.10743269.

[11] L. Y. Huang, M. F. Chang, C. H. Chuang, C. C. Kuo, C. F. Chen, and G. H. Yang, "ReRAM-based 4T2R nonvolatile TCAM with 7x NVM-stress reduction, and 4x improvement in speed-wordlength-capacity for normally-off instant-on filter-based search engines used in big-data processing," Symp. VLSI Circuits Dig. Tech. Papers, pp. 1–2, Honolulu, HI, USA, 2014, doi: 10.1109/VLSIC.2014.6858404.

[12] Z. Kahleifeh, H. Thapliyal, and S. M. Alam, "Adiabatic/MTJ-based physically unclonable function for consumer electronics security," IEEE Trans. Consum. Electron., vol. 69, no. 1, pp. 1–8, Feb. 2023, doi: 10.1109/TCE.2022.3201247.

[13] M. J. Adel, M. H. Rezayati, M. H. Moaiyeri, A. Amirany, and K. Jafari, "A robust deep learning attack immune MRAM-based physical unclonable function," Sci. Rep., vol. 14, art. no. 20649, Oct. 2024, doi: 10.1038/s41598-024-71730-7.

[14] E. I. Vatajelu, G. Di Natale, M. Indaco, and P. Prinetto, "STT MRAM-based PUFs," Proc. Des. Autom. Test Eur. Conf. Exhib. (DATE), pp. 872–875, Grenoble, France, 2015, doi: 10.7873/DATE.2015.0505.

An RRAM-Based Soft PUF Achieving Near-Zero BER through Skewed Voltage Masking

Xinrong Yang, Pengjun Wang*, Cailong Jin, Yixin Lu

College of Electrical and Electronic Engineering, Wenzhou University, Wenzhou 325035, China

* Email: 24461251013@stu.wzu.edu.cn , wangpengjun@wzu.edu.cn

Abstract—Internet of Things (IoT) devices face escalating security threats but are constrained by limited hardware resources, making it difficult to adopt conventional cryptographic solutions due to high area overhead. Physical Unclonable Functions (PUF) offer a lightweight alternative by leveraging inherent process variations to extract entropy, yet most existing designs rely on dedicated circuitry, resulting in extra area cost. This work presents an soft PUF architecture based on Resistive Random-Access Memory (RRAM) enabling near-zero-overhead integration. The design reuses the memory array's entropy with minimal control logic changes, employs a dual-region indexing algorithm for fast challenge address generation, and introduces a voltage-skew-based masking strategy to enhance reliability without additional area. Experimental results show that the proposed RRAM-PUF passes the NIST SP800-22 randomness tests. Compared to exhaustive search, the indexing algorithm achieves a 5.9× speedup in address generation. Moreover, the masking strategy ensures zero bit error rate across a wide temperature range, demonstrating the design's suitability for secure, resource-efficient deployment in IoT edge devices.

Keywords—*PUF, Hardware security, Reliability enhancement, RRAM*

I. Introduction

With the rapid development of Internet of Things (IoT) technologies, a vast number of terminal devices are being rapidly integrated into various networks. However, due to their frequent deployment in open and untrusted environments, these devices are highly susceptible to diverse security threats, including physical tampering, side-channel attacks, and firmware replacement. Although traditional cryptographic techniques offer strong security guarantees, their high overhead in terms of area, power consumption, and computational complexity makes them unsuitable for resource-constrained IoT devices.

The Physical Unclonable Function (PUF), as an emerging lightweight hardware security primitive, provides effective mechanisms for device authentication and key generation in constrained environments. Depending on whether they rely on dedicated circuit structures, PUFs can be divided into two types: hard PUFs and soft PUFs. Hard PUFs introduce deliberate design-induced randomness, such as the Arbiter PUF (APUF) and Ring Oscillator PUF (RO-PUF), but typically incur significant area overhead [1]. In contrast, soft PUFs leverage entropy sources inherent to the existing circuit design, offering natural advantages in area and cost, such as SRAM PUF [2]. However, as CMOS devices approach physical scaling limits, soft PUFs based on conventional memories face challenges such as high susceptibility to environmental variations and growing power bottlenecks. Distinct from these approaches, PUFs based on Resistive

Random-Access Memory (RRAM) are widely regarded as promising next-generation entropy sources due to their low power consumption, high integration density, and intrinsic write-time stochasticity. In recent years, the academic community has proposed a variety of soft PUF design schemes based on RRAM. For example, Li et al. introduced a concealable RRAM-PUF that exploits differential switching behaviors in parallel RRAM cells as the entropy source. They used Temporal Majority Voting (TMV) and masking techniques to reduce the raw Bit Error Rate (BER) to 0%. However, under high-temperature conditions (373 K), the BER degraded to 4% [3]. Pang et al. proposed an RRAM-PUF based on a differential readout method, which achieved a BER reduction to 1% by applying Reliability-Enhancement Design (RED) and oxide stack engineering. Nevertheless, this method is highly dependent on process variation, and the RED operation necessitates reprogramming, resulting in loss of regular memory functionality [4]. Although these designs offer improved reliability over traditional RRAM-PUF, they do not fully resolve the above limitations.

To address these challenges, this work proposes a highly reliable soft PUF design based on RRAM. The main contributions are as follows:

(1)***RRAM-PUF Architecture:*** By utilizing the existing RRAM memory array for entropy extraction, the proposed design integrates PUF functionality with minimal hardware modification. Only 42 additional logic gates are required in the peripheral control circuitry, resulting in an overall area overhead of just 1.3%, thereby achieving a nearly zero-overhead PUF integration.

(2)***Dual-Region Indexing Algorithm:*** The proposed design employs a dual-region indexing algorithm to enable fast generation of PUF response addresses. The algorithm achieves an average address generation time of 4.4×10^{-5} s, representing a 5.9× speed improvement compared to the conventional approach with an average time of 2.61×10^{-4} s.

(3)***Voltage-Skew-Based Masking Strategy:*** A zero-overhead masking strategy is introduced to enhance the reliability of the PUF. With the masking approach, the BER of the PUF responses is reduced to 0% under wide temperature (−40 °C to 120 °C) and voltage (1 V to 1.4 V) range.

II. Soft PUF Design Based on RRAM

A. RRAM-PUF Architecture

The proposed 32×32 (1 KB) RRAM array is divided into left, right, and current-averaged reference regions, as shown in Fig. 1. To extract PUF responses without affecting original memory functionality, two operational modes are designed and controlled via a multiplexer. In memory mode, the sense amplifier reads the stored data by comparing the leakage current between the selected bitline (BL) and the reference BL. In PUF mode, all rows are activated, and the response is generated by comparing leakage currents of selected BLs from

This work was supported in part by National Natural Science Foundation of China (Grant 62174121 and Grant 62234008).

979-8-3315-3918-4/25 $31.00 © 2025 IEEE

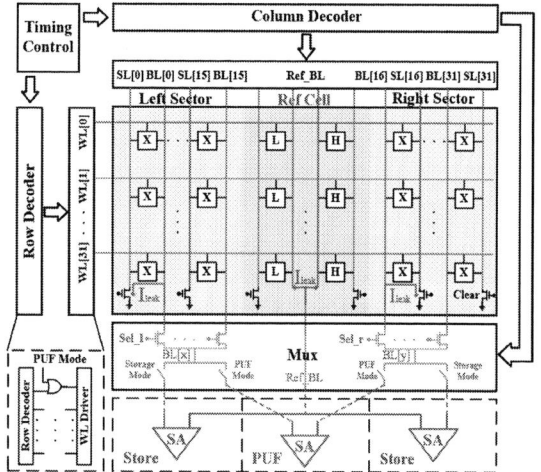

Fig. 1. RRAM-PUF architectures.

the left and right regions. To enable full-row activation in PUF mode, the proposed architecture adds 32 OR gates to the original row decoder, along with simple mode control logic, totaling approximately 42 logic gates. The additional area overhead is estimated as:

$$A_{ratio} \approx \frac{32 \times A_{or} + 10 \times A_{logic}}{A_{rram}} \quad (1)$$

In 65 nm CMOS technology, the chip area of a 1 KB RRAM array (A_{rram}) is approximately 1157 μm² [5]. According to the standard definition, the area of one logic gate equivalent (A_{logic}) is about 0.25 μm². A 2-input OR gate implemented using 65 nm static CMOS logic occupies around 0.375 μm². Therefore, the area ratio (A_{ratio}) is approximately 1.3%, enabling near-zero-overhead integration of PUF functionality.

In the proposed architecture, the core entropy source of the PUF stems from minute differences in leakage current between BLs in the left and right regions. These differences are primarily caused by random process variations in RRAM devices and their associated control NMOS transistors. To accurately model the resistance characteristics of RRAM cells, we adopt a resistance expression based on an asymmetric barrier tunneling model [6]:

$$R_r = \frac{V_{read} \cdot h}{2e} \times \left\{ eV + \frac{1}{\alpha} ln \left[\frac{1 + e^{\alpha(\Phi - \beta eV)}}{1 + e^{\alpha(\Phi + (1-\beta)eV)}} \right] \right\}^{-1} \quad (2)$$

Where the barrier height Φ, voltage V, and modulation parameters α and β are all affected by process fluctuations. Considering the series connection of a selection transistor R_{on} with each RRAM cell, as well as a control NMOS transistor R_N' in series with a bank of 32 parallel RRAM cells, the total resistance from the BL of the n-th column to ground can be expressed as:

$$R_{total} = \frac{1}{\sum_{i=1}^{n}(R_r^{-1} + R_{on}^{-1})} + R_N' \quad (3)$$

Consequently, the leakage current for the n-th column is:

$$I_{leak,n}(t) = \frac{V_{DD}}{R_{total,n}} e^{-\frac{t}{\tau_n}} \quad (4)$$

Under a constant supply voltage V_{DD}, the leakage current $I_{leak,n}(t)$ on the BL exhibits a non-linear exponential decay

---

Algorithm: Fast searching the number of '0's in a RRAM

Require:
    M*N RRAM storage arrays

Ensure:
    Construct zero-count index: {zero_count: [col_indices]}
    Support fast retrieval of columns sharing the same zero count.
    Enable efficient address pair matching across zones via GetExcitationPairs

  # Part 1: Build Index
1. Function BuildIndex(RRAM, start, end):
2.   Initialize index as empty dictionary
3.   For col = start to end do
4.   Count zeros in RRAM[:, col] → zero_count
5.   If zero_count not in index then index[zero_count] = [col]
6.   Else append col to index[zero_count]
7.   End for, Return index

  # Part 2: Generate Excitation Address Pairs
8. left_index = BuildIndex(RRAM, 1, N/2)
9. right_index = BuildIndex(RRAM, (N/2)+1, N)
10. Function GetExcitationPairs(left_index, right_index):
11.   Return all (l, r) where l in left_index[c], r in right_index[c], for common c

---

relationship with the total resistance R_{total}. Since total resistance is determined by a series-parallel network of multiple cells, even minor variations in individual cells can be significantly amplified through the mechanisms of conductance summation and exponential decay, thereby enhancing the distinguishability and reliability of the PUF entropy source.

B. Dual-Region Indexing Algorithm

To generate reliable challenge addresses, it is necessary to select column pairs based on the number of logical 0's in each column. Traditional approaches typically rely on full-table scanning of the entire RRAM array, resulting in significant query latency. To address this, a dual-region indexing algorithm is proposed in this work. By leveraging preprocessing and hash tables, the algorithm constructs *left_index* and *right_index* dictionaries that categorize column indices based on their zero count. During the query phase, column indices corresponding to a specified zero count (*zero_count*) are retrieved and combined via Cartesian product to form address pairs. The separation of left and right regions, along with parallel processing, significantly accelerates the generation of PUF challenge addresses in large-scale arrays.

C. Voltage-Skew-Based Masking Strategy

The proposed design employs a precharge-type clamped current-mode sense amplifier to sensing the resistance of two selected BLs, as shown in Fig. 2(b). During the precharge phase, both output nodes are charged to the same high voltage. In the evaluation phase, process variations cause a mismatch in the total pull-down resistance of the two BLs, and a cross-coupled feedback mechanism amplifies the resulting voltage difference between the output nodes, thereby generating the PUF response.

To enhance the response reliability of the proposed PUF system, a voltage-skew-based masking strategy is introduced to eliminate PUF cells with minimal process variation. This is achieved by applying a bias to the clamping voltage of the

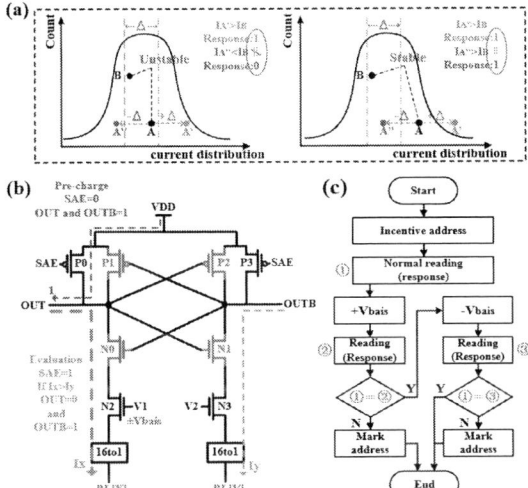

Fig. 2. (a) Masking strategy implementation principle. (b) Sense amplifier operation schematic. (c) Masking strategy flowchart.

sense amplifier to identify unstable cells, as shown in Figure 2(c) for the operating process. Figure 2(a) shows the implementation principle, where the applied voltage offset introduces a mismatch between the BL voltages, $\Delta V_{BL}=|V_{BLL}-V_{BLR}|$, resulting in a current difference $\Delta I_{BL}=|I_{BLL}-I_{BLR}|$. When the initial current difference ΔI is small, the applied bias may induce current reversal, leading to read errors. By repeatedly applying positive and negative voltage offsets, cells whose responses vary across multiple readouts are marked as unstable. Their addresses are recorded as helper data, and only stable cell responses are retained during key reconstruction to ensure PUF reliability.

III. PERFORMANCE EVALUATION

To evaluate the performance of the proposed RRAM-PUF, this work employs 65-nm CMOS technology and an RRAM compact model, conducting Monte Carlo simulations via Cadence Virtuoso to characterize key performance metrics including randomness, uniqueness, and reliability.

A. Randomness

PUF randomness refers to the statistically independent responses exhibited by different PUF instances fabricated under the same process, owing to microscopic structural variations. To evaluate the randomness of the proposed PUF, the NIST SP 800-22 statistical test suite, developed by the National Institute of Standards and Technology (NIST), was employed. As shown in Table I, the proposed PUF successfully passed 11 tests. Additionally, autocorrelation analysis was conducted on the response sequence generated by the PUF to quantify the degree of internal correlation. Fig. 3(a) shows the autocorrelation test result for a 25,600-bit response sequence, where the autocorrelation function (ACF) value is 0.0125 within a 95% confidence interval, which is close to the ideal value of zero. Fig. 3(b) presents the spatial distribution of the PUF response bits, with black pixels indicating logic '1'. The absence of discernible patterns confirms the strong randomness of the response.

B. Uniqueness

PUF uniqueness refers to the degree of difference in the responses generated by different PUF instances when subjected to the same set of challenges under identical

TABLE I. NIST800-22 TEST RESULT

NIST tests	Stream length	Proportion	P value	Pass
Frequency	5120	9/10	0.3505	Yes
Block Frequency	5120	10/10	0.5341	Yes
Cumulative Sums	5120	9/10	0.3505	Yes
Runs	5120	10/10	0.7399	Yes
Longest Run	5120	10/10	0.9114	Yes
Rank	5120	9/10	0.5341	Yes
FFT	5120	10/10	0.1223	Yes
Non-Overlapping Template	5120	10/10	0.7399	Yes
Overlapping Template	5120	10/10	0.1223	Yes
Serial	5120	10/10	0.2133	Yes
Linear Complexity	5120	8/10	0.5341	Yes

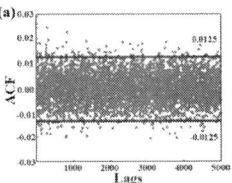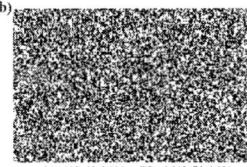

Fig. 3. (a) Autocorrelation test. (b) PUF response space distribution.

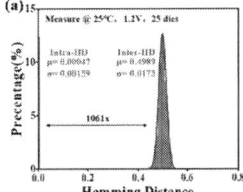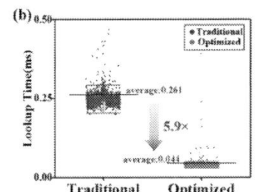

Fig. 4. (a) Uniqueness test. (b) Algorithm performance test.

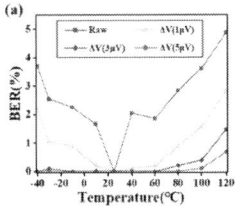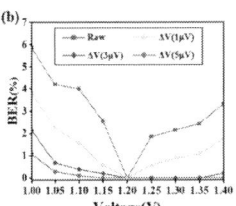

Fig. 5. (a) Wide-temperature BER. (b) Wide-voltage BER.

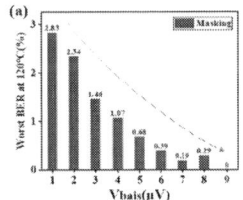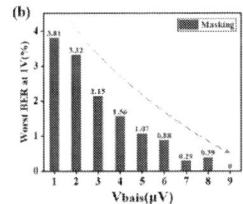

Fig. 6. (a) Variation of BER with Bias Voltaget at 120°C. (b) Variation of BER with Bias Voltaget at 1V.

fabrication and testing conditions. Fig. 4 presents the simulated Inter-chip Hamming distance (Inter-HD) and Intra-chip Hamming distance (Intra-HD) values for 25 different PUF instances under a temperature of 25 °C and a supply voltage of 1.2 V. The results show that the average Inter-HD and Intra-HD are 0.4989 and 0.0047, with standard deviations close to the ideal value, indicating excellent uniqueness.

TABLE II. PERFORMANCE COMPARISON

	TNANO' 25[7]	IEDM' 20[8]	TED' 19[9]	This Work
Technology	55nm	N/A	90nm	65nm
Inter-HD	0.5005	0.5004	0.51	0.4989
Intra-HD	7.2E-4	4.3E-4	N/A	4.7E-4
Temperature Range(°C)	0~125	25~125	25~85	-40~120
Worst case BER	1.2% /0.034%	8.4%	29% (average)	4.9%
Stabilization Method	NO	In-Cell Reprograming	temperature compensation	Masking
BER after Stabilization	N/A	0.014%	2%	0
Storage Mode	NO	YES	YES	YES

C. Reliability

PUF reliability is the ability of a PUF to consistently produce stable responses under repeated evaluations or varying conditions for the same challenge input. Fig. 5 depict the variation of BER across a wide range of environmental conditions, including temperatures from −40 °C to 120 °C and supply voltages from 1 V to 1.4 V. The improvement in BER is positively correlated with the magnitude of the applied bias voltage. After eliminating unstable cells identified under a 5 μV bias, the resulting BER—indicated by the red lines in both figures—approaches 0% across the entire tested temperature and voltage ranges.

To further evaluate the effectiveness of the masking strategy in enhancing response reliability, Fig. 6(a) illustrates that under extreme temperature conditions (120 °C), increasing the bias voltage to 9 μV reduces the BER of the proposed PUF to 0%. Similarly, as shown in Fig. 6(b), under the lowest supply voltage condition (1 V), the BER also drops to 0% when a 9 μV bias voltage is applied.

D. Algorithm Acceleration Performance

To validate the efficiency improvement of the proposed dual-region indexing algorithm in response address generation, this work compares the search performance of the traditional linear search method and the proposed hash-based indexing approach. Fig. 4(b) illustrates the distribution of search times over 1000 iterations for both algorithms on a 32×32 RRAM array. The proposed method achieves an average search time of $4.4×10^{-5}$ s, significantly lower than the $2.61×10^{-4}$ s required by the linear search method, resulting in a 5.9× performance gain. These results confirm the superior lookup efficiency of the proposed indexing technique.

E. Comparison with Related Works

The PUF proposed in this work has good uniqueness and original BER. The lowest worst-case BER was achieved over a wider temperature range, as shown in Table II. The implemented masking strategy further suppresses the BER to 0%. Unlike [9], which suffers from increased drift sensitivity due to fine-grained resistance quantization. [7] although it has higher uniqueness and reliability, it lacks storage PUF integration, while our design provides enhanced security through dynamic response generation. It simultaneously

supports secure functionality and memory reuse, delivering a balanced and scalable solution.

IV. CONCLUSION

This work presents a highly reliable RRAM-based PUF leveraging the inherent resistance randomness of RRAM cells as its entropy source. A dual-region preprocessing scheme combined with hash-based mapping enables a low-overhead, fast address search algorithm. To combat sensitivity to environmental fluctuations, a voltage-skew-based masking strategy effectively filters unstable cells via bias voltage application. Experimental results show seamless integration with minimal area overhead. The masking achieves 0% BER across wide environmental variations. The address generation improves search efficiency by 5.9× over exhaustive methods. Passing NIST SP 800-22 tests, this design offers a novel physical security solution for IoT edge devices, combining memory reuse, high security, and low hardware cost, expanding RRAM's secure application potential in resource-constrained settings.

ACKNOWLEDGMENT

This work was supported in part by National Natural Science Foundation of China under Grant 62174121 and Grant 62234008.

REFERENCES

[1] L. Ni, P. Wang, Y. Zhang, G. Li, L. Ding, and J. Zhang, "PI PUF: A processor-intrinsic PUF for IoT," Comput. Electr. Eng., vol. 105, p. 108540, 2023, doi: 10.1016/j.compeleceng.2022.108540.

[2] S. Baek, G. H. Yu, J. Kim, C. T. Ngo, J. K. Eshraghian, and J. P. Hong, "A Reconfigurable SRAM Based CMOS PUF With Challenge to Response Pairs," IEEE Access, vol. 9, pp. 79947-79960, 2021, doi: 10.1109/ACCESS.2021.3084621.

[3] J. Li, Y. Cui, C. Wang, W. Liu, and S. Kvatinsky, "A Concealable RRAM Physical Unclonable Function Compatible with In-Memory Computing," 2024 Design, Automation & Test in Europe Conference & Exhibition (DATE), Valencia, Spain, 2024, pp. 1-6, doi: 10.23919/DATE58400.2024.10546697.

[4] Y. Pang, H. Wu, B. Gao, N. Deng, D. Wu, and R. Liu, "Optimization of RRAM-Based Physical Unclonable Function With a Novel Differential Read-Out Method," IEEE Electron Device Letters, vol. 38, no. 2, pp. 168-171, Feb. 2017, doi: 10.1109/LED.2016.264723.

[5] M. F. Chang, C. W. Wu, C. C. Kuo, S. J. Shen, K. F. Lin, and S. M. Yang et al., "A 0.5V 4Mb logic-process compatible embedded resistive RAM (ReRAM) in 65nm CMOS using low-voltage current-mode sensing scheme with 45ns random read time," 2012 IEEE International Solid-State Circuits Conference, San Francisco, CA, USA, 2012, pp. 434-436, doi: 10.1109/ISSCC.2012.6177079.

[6] B. Lin, Y. Pang, B. Gao, J. Tang, D. Wu, and T. W. Chang, "A Highly Reliable RRAM Physically Unclonable Function Utilizing Post-Process Randomness Source," IEEE Journal of Solid-State Circuits, vol. 56, no. 5, pp. 1641-1650, May 2021, doi: 10.1109/JSSC.2021.3050295.

[7] Y. Cui, J. Li, C. Gu, C. Wang, and W. Liu, "A Multi-Mode Configurable Physical Unclonable Function Based on RRAM With Adjustable Programmable Voltage," IEEE Transactions on Nanotechnology, vol. 24, pp. 166-177, 2025, doi: 10.1109/TNANO.2025.3552433.

[8] J. Yang, D. Lei, D. Chen, J. Li, H. Jiang, and Q. Ding, "A Machine-Learning-Resistant 3D PUF with 8-layer Stacking Vertical RRAM and 0.014% Bit Error Rate Using In-Cell Stabilization Scheme for IoT Security Applications," 2020 IEEE International Electron Devices Meeting (IEDM), San Francisco, CA, USA, 2020, pp. 28.6.1-28.6.4, doi: 10.1109/IEDM13553.2020.9372107.

[9] G. S. Lee, G. H. Kim, K. Kwak, D. S. Jeong, and H. Ju, "Enhanced Reconfigurable Physical Unclonable Function Based on Stochastic Nature of Multilevel Cell RRAM," IEEE Transactions on Electron Devices, vol. 66, no. 4, pp. 1717-1721, April 2019, doi: 10.1109/TED.2019.2898455.

A Performance Enhancement Strategy for Strong PUF Circuits to Improve IoT Authentication Security

Dong Lu, Pengjun Wang*, Gang Li, Hao Ye

College of Electrical and Electronic Engineering, Wenzhou University, Wenzhou 325000, China.

* Email: 24451250014@stu.wzu.edu.cn, wangpengjun@wzu.edu.cn

Abstract—With the rapid development of Internet of Things (IoT) technology and the increasing number of security attacks, IoT security has become an important research area. Physical unclonable function (PUF) circuits provide an effective solution for device authentication. In this study, a performance enhancement strategy based on arbiter PUF (APUF) is proposed to improve the response stability and randomness of XOR-APUF and multiplexer PUF (MPUF). Firstly, machine learning (ML) based on the logistic regression (LR) algorithm is used to establish mathematical models of all underlying APUFs, The same noise is injected into these mathematical models to systematically screen out challenges that can produce stable responses. Subsequently, the intersection of challenges that generate stable responses for all underlying APUFs is obtained, followed by randomness screening. Finally, the performance-enhanced challenges are used for the IoT authentication protocol. Experimental results show that the proposed method greatly improves the stability and randomness of PUF responses. Screened challenges can effectively lower the authentication threshold, thus strengthening the security of IoT authentication.

Keywords—Physical unclonable function, machine learning, stability, randomness, authentication protocol.

I. INTRODUCTION

The development of the Internet of Things (IoT) has enabled a large number of smart devices to be interconnected. However, security vulnerabilities in device authentication have become a critical constraint. Traditional authentication mechanisms based on passwords or digital certificates are vulnerable to risks such as cracking and impersonation, making them difficult to address complex cyber threats. Physical unclonable function (PUF) technology using manufacturing process variations to generate unique digital fingerprints with inherent advantages such as non cloning and cost-effective. These properties greatly enhance the security of authentication [1]. PUFs are categorized into two types, weak PUF and strong PUF, based on the number of their challenge-response pairs (CRPs). Strong PUF has rich CRPs and is particularly suitable for authentication applications. Weak PUF is highly respected for their high reliability and is mainly used in key generation scenarios.

To address the strong PUF environment sensitivity problem, the current common solution is to dynamically adjust the authentication threshold to accommodate response fluctuations. However, this compromise strategy reduces the security level and is vulnerable to machine learning (ML) modeling attacks. Wu proposed a lightweight PUF based on response feedback. this PUF achieves near-optimal uniformity, uniqueness and reliability [2]. Prob-PUF uses the characteristics of nanoscale transistors to realize resistance

against ML attacks [3]. Zhou proposed an ML-based to enhance APUF stability method, which resulted in a significant improvement of APUF stability [4]. Ferens proposed a selective CRP defense technique that selectively generates challenges instead of randomly generating them in order to negatively influence the parameters of an attacker trained ML model [5].

Due to the structural properties of XOR-APUF and multiplexer PUF (MPUF), this paper proposes a hierarchical optimization method based on dynamic noise regulation. The underlying APUFs are individually modeled to screen out their stable challenge sets. Then, the intersection of the stable challenge sets of all underlying APUFs is computed, and the response sets corresponding to the screened stable challenge sets are screened for randomness. Experimental results show that by precisely controlling the noise threshold at each stage of stability screening and performing randomness screening on the screened stable responses, the response stability of 64-bit XOR-APUFs and MPUFs can be significantly improved. The randomness reaches the ideal value of 50%, thus overcoming the limitation of traditional stability enhancement techniques that typically compromise randomness. Using the screened CRPs for authentication, the threshold of authentication protocol can be effectively optimized to achieve exponential enhancement of anti-attack capability. The method can effectively enhance the stability and randomness of complex PUFs with only standard environment training and without additional hardware, which has important application value. It provides a reliable technical guarantee for

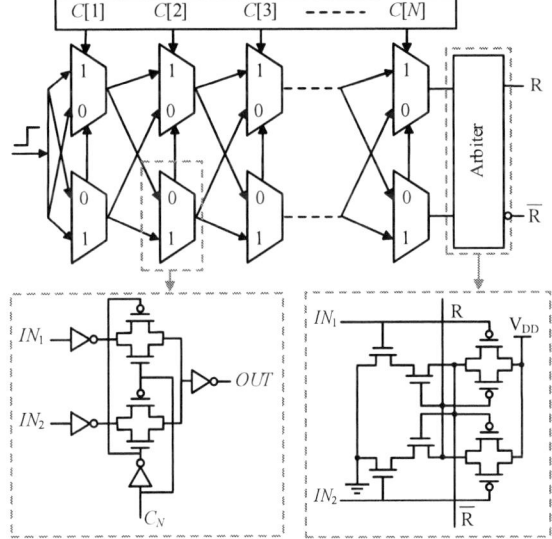

Fig. 1. APUF structure.

This work was supported in part by National Natural Science Foundation of China (Grant 62234008, Grant 62174121 and Grant 62374117).

the practical application of complex PUF structures.

II. THEORETICAL FOUNDATION

A. Strong PUF

The classic strong PUF architectures mainly include APUF, XOR-APUF, and MPUF [6]. APUF is realized based on circuit delay characteristics, which internally consists of symmetric delay path pairs and arbiters, with random differences in path delays due to manufacturing process deviations. The input challenge vector selects the combination of paths. After the signal propagates through different paths, the arbiter generates the response according to the arrival order. The physical randomness of the delay difference ensures that the response is unpredictable and replicable, which is the core mechanism for realizing hardware unclonability. Its structure is shown in Fig. 1 with the following mathematical expression:

$$R_{APUF} = sgn(\Delta) = sgn(\vec{\omega} \cdot \vec{\phi}) \tag{1}$$

$sgn(.)$ denotes the signum function, $\vec{\omega}$ represents the delay vector derived from the delay differences at each hierarchical level through mathematical operations, and $\vec{\phi}$ denotes the challenge vector computed from the input challenge. The linear characteristic of APUF renders it significantly vulnerable to ML modeling attacks. XOR-APUF achieves nonlinear expansion by combining k independent APUF units in parallel. The XOR-APUF operation effectively breaks the linear separability of APUF, thereby enhancing resistance against modeling attacks. Its mathematical formulation is as follows:

$$R_{XOR-APUF} = XOR_{i=1}^{x}\left[sgn(\vec{\omega}_i \cdot \vec{\phi})\right] \tag{2}$$

The MPUF adopts a dynamic selection mechanism based on multiplexer, by combining responses generated by multiple underlying APUFs in a non-linear way, breaking the constraints of the traditional linear relationship, and making the overall output with more complex and unpredictable characteristics. This greatly enhances the ability to resist modeling attacks, effectively defends against modeling attacks implemented by attackers through the collection of CRPs, and improves security. The mathematical representation is as follows:

$$R_{MPUF} = RD_{\sum_{i=1}^{s} RS_i \cdot 2^{i-1}} \tag{3}$$

RD_i signifies the response corresponding to the input side of the multiplexer, while RS_i represents the response associated with the selection side of the data selector.

B. Screening Strategy

The responses of XOR-APUF and MPUF are obtained by performing logical operations on the responses of underlying APUFs, and the mapping relationship between the challenge-response of APUFs makes them easy to be modeled by ML [7]. Based on these characteristics, multiple underlying APUFs are constructed on FPGA, and each APUF is modeled separately. The same random challenge set was repeatedly input 30 times to each APUF on the FPGA and a stable set of 10000 CRPs was screened out as the training set for logistic regression (LR)-based ML. The original random challenge set is input into each underlying APUF on the FPGA 30 times repeatedly to obtain the responses of the underlying APUFs. Through logical operations on the responses of the underlying APUFs, the original response sets of XOR-APUF and MPUF are obtained, and their original stability is calculated. After completing the modeling of the underlying APUFs, a noise factor α is added to these models. The original challenge set to be screened is input into the models to obtain the responses of the underlying APUF models at this noise level, and the process is repeated 30 times. In order to obtain the target PUF responses, a logical operation corresponding to the target PUF is performed on all underlying APUFs responses at the same noise level. Then, the stability of the resulting response is used as a criterion for screening the stability of the original response set. Through this screening process, a stable challenge set at the current noise level is obtained. The noise factor α is gradually increased until all unstable responses in the original

---

Algorithm 1 Screening Strategy

Input: LR-based ML: $PUFmodel_k$
PUF type: T
Challenge set: $C[n,b]$
Noise factor: α
Testing times: N
Output: Screened challenge set: C_{IDR}

1: **for** each $k \in [1,2,...,m]$ **do**
2:   **for** each $i \in [1,2,...,n]$ **do**
3:     **for** each $j \in [1,2,...,N]$ **do**
4:       $noise = \alpha$;
5:       $R_{k\text{-}ij} = PUFmodel_{k\text{-}\alpha}(C_i)$;
6:     **end for**
7:     **if** $R_{k\text{-}i}[1] \sim R_{k\text{-}i}[j]$ Congruent **then**
8:       $C_{k\text{-}S} \leftarrow$ append($C[i,:]$);
9:       $R_{k\text{-}S} \leftarrow$ append($R_{k\text{-}i}[1]$);
10:     **end if**
11:   **end for**
12: **end for**
13: $C_{IDS} = C_{1\text{-}S} \cap C_{2\text{-}S} \cap ... \cap C_{m\text{-}S}$;
14: $R_{sumIDS} = package[R_{1\text{-}S}, R_{2\text{-}S}, \cdots, R_{m\text{-}S}]$;
15: Select $C_T(.) \leftarrow T$
16: $R_{IDS} = C_T(R_{sumIDS})$;
17: $Diff = count(R_{IDS} = 0) - count(R_{IDS} = 1)$
18: $Target = Diff > 0 ? 0 : 1$
19: $k = |Diff|$
20: Targetindices $\leftarrow \{i \mid R_{IDS}[i] = Target\}$
21: Removeindices $\leftarrow$ UniformSelect(Targetindices, k)
22: $C_{IDR} = C_{IDS} - \{C_{IDS}[i] \mid i \in$ Removeindices$\}$

(a)

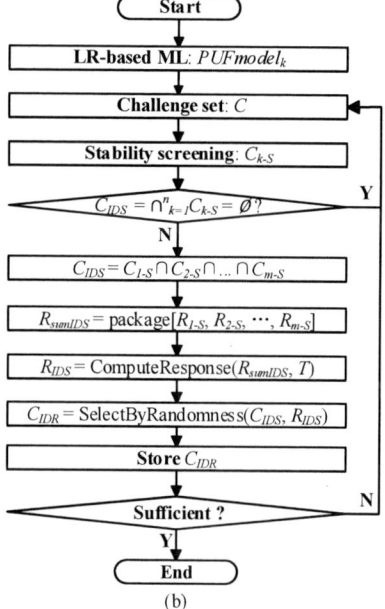

(b)

Fig. 2. The corresponding algorithm and workflow of the screening strategy (a) algorithm, (b) workflow.

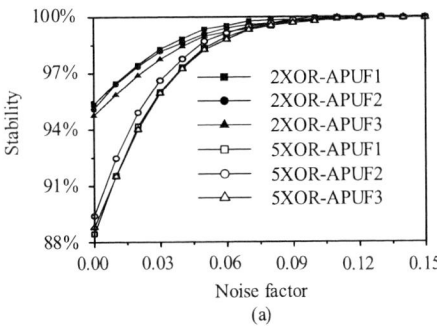
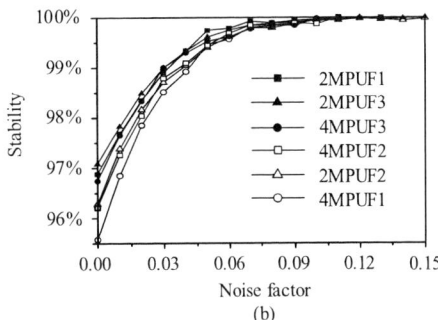

Fig. 3. Stability improvement of (a) XOR-APUF, (b) MPUF.

response set are screened. Following this, randomness screening is conducted to obtain a challenge set that combines high stability and high randomness. As shown in Fig. 2. Fig. 2(a) is the algorithm and Fig. 2(b) is the process.

III. EXPERIMENTAL RESULTS

In order to verify the effectiveness of the proposed screening strategy, we have implemented all the underlying APUF of XOR-APUF and MPUF on FPGA by modifying the layout. We have used MATLAB to generate the challenge set and input it into the APUF circuit on FPGA to get the response set and verified the proposed strategy by Python programming.

A. The improvement of stability

The experimental results are shown in Fig. 3. Fig. 3(a) demonstrates the stability enhancement of the XOR-APUF, and Fig. 3(b) shows the stability improvement of the MPUF, which verifies the applicability of the method to different structures. When $\alpha = 0$, the original stability of the target PUF is obtained, and the stability is significantly improved as α increases. The experiments show that the method can effectively enhance the stability of XOR-APUF and MPUF.

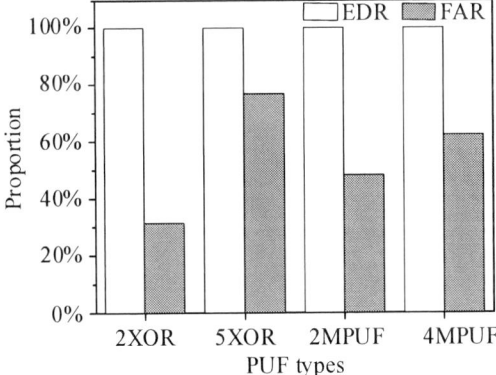

Fig. 4. EDR and FAR of different PUFs.

To validate the method overhead, the correct detection rate for unstable challenge (EDR) and false detection rate for stable challenge (FAR) of the screening process were counted. As the noise factor α increases, the EDR gradually rises and eventually reaches 100%. The FAR increases with the increase of α, which is consistent with the structural characteristics of the target PUF, but it ultimately remains within a reasonable range. The results are shown in Fig. 4.

B. Enhancement of Randomness

Since the deviation of randomness increases the success

rate of ML modeling attacks, and theoretically. The closer the 0 and 1 distributions of PUF responses are, the better. Therefore, we perform randomness screening on the screened stable CRPs. After random screening, the response randomness of the stable challenge set has reached the expected 50%. As shown in Fig 5, where Fig 5(a) depicts the randomness of the screened 2XOR-APUF, and Fig 5(b) illustrates the randomness of the screened 2MPUF. The screened challenge set of this method lowers the IoT authentication threshold and greatly improves the security of XOR-APUF and MPUF in applications. Meanwhile, the CRPs still maintains high usability after the randomness screening of the stable challenge set.

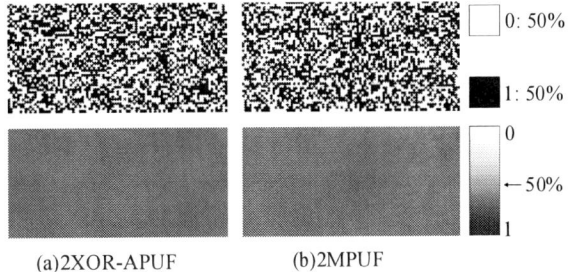

(a)2XOR-APUF (b)2MPUF

Fig. 5. Randomness improvement of (a) XOR-APUF, (b) MPUF.

C. Analysis of the Protocol's Resistance to Attacks

In the process of IoT device authentication, using unstable challenges will significantly increase the authentication threshold, thereby increasing the attacker's success rate. Additionally, there is a risk of data leakage during device authentication. If the CRPs are used for authentication without any concealment, there is a threat that attackers may use the obtained CRPs to conduct ML modeling attacks. This paper proposes a high-security authentication protocol based on screened high-performance CRPs, which effectively reduces the authentication threshold. Meanwhile, the protocol avoids the transmission of response information over public channels during the authentication process, thereby enhancing security. The specific process is shown in Fig 6.

In order to verify the security of the proposed authentication protocol at a formal level, we use ProVerif for formal verification. The key information of the proposed protocol is judged by the running results whether it is secure or not, and whether the authentication process meets the standard or not. The verification results are shown in Fig 7. All private information in the protocol has been effectively protected. In addition, all the processes of the protocol are verified for reliability.

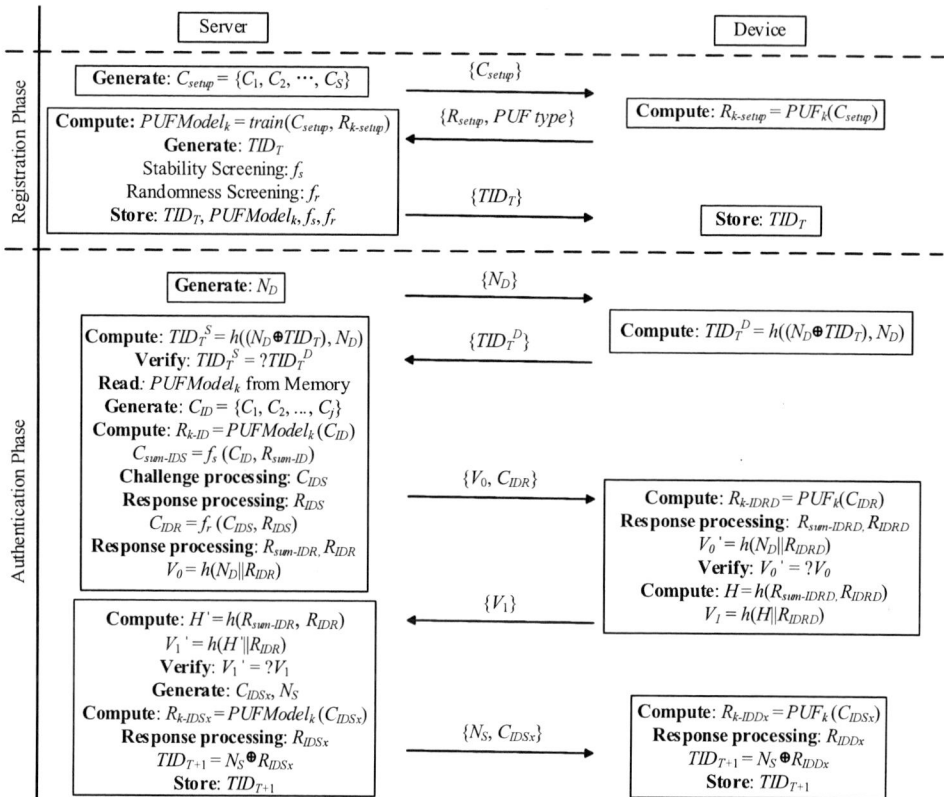

Fig. 6. Process of the proposed protocol.

```
Verification summary:
Query not attacker(TIDt[]) is true.
Query not attacker(Rsetup[]) is true.
Query not attacker(PUFtype[]) is true.
Query not attacker(RIDR[]) is true.
Query not attacker(RIDRD[]) is true.
Query not attacker(RIDSx[]) is true.
Query not attacker(RIDDx[]) is true.
Query not attacker(TIDSnew[]) is true.
Query not attacker(TIDDnew[]) is true.
Query not attacker(modelk(Cs[],modeltrain(Cs[],Rks[]))) is true.
Query inj-event(TIDVerified) ==> inj-event(ServerSentTID) is true.
Query inj-event(DeviceVerified) ==> inj-event(ServerSentV0s) is true.
Query inj-event(ServerVerified) ==> inj-event(DeviceSentTIDtD) is true.
```

Fig. 7. The ProVerif result of the proposed protocol.

IV. CONCLUSION

In this study, a screening strategy based on underlying APUFs is proposed to address the stability and randomness issues of XOR-APUF and MPUF. The approach employs a LR-based ML algorithm to establish an accurate mathematical model for underlying APUFs. By designing a controllable noise injection mechanism, the common stable challenge set and corresponding responses of all underlying APUFs are obtained. Subsequently, responses from underlying APUFs are calculated according to the target PUF type to derive the stable responses of the target PUF, and a random screening process is applied to the screened stable challenge set. The performance of the complex PUFs is significantly improved. Experimental results confirm that this method achieves 100% response stability and 50% response randomness under standard environmental conditions without additional hardware overhead, while maintaining the ideal effective availability of CRPs. This provides an efficient and reliable solution for security authentication of IoT devices, particularly suitable for lightweight security requirements in resource-constrained scenarios.

ACKNOWLEDGMENT

This work was supported in part by National Natural Science Foundation of China (Grant 62234008, Grant 62174121 and Grant 62374117).

REFERENCES

[1] A. Abdulaziz and A. Saif, "Physical unclonable functions (PUF) for IoT devices," *ACM Computing Surveys* 55.14s (2023): 1-31.

[2] L. Wu, Y. Hu, K. Zhang, W. Li, X. Xu, and W. Chang, "Flam-PUF: A response–feedback-based lightweight anti-machine-learning-attack PUF," *IEEE Transactions on Computer-Aided Design of Integrated Circuits and Systems* 41.11 (2022): 4433-4444.

[3] P. Ren, Y. Xue, L. Jing, L. Zhang, R. Wang, and Z Ji, "A strong physical unclonable function with machine learning immunity for Internet of Things application," *Science China Information Sciences* 67.1 (2024): 112404.

[4] Z. Zhou, G. Li, and P. Wang, "A challenge-screening strategy for enhancing the stability of strong PUF based on machine learning," *Microelectronics Journal*, vol. 131, pp. 105667, 2023.

[5] M. Ferens, E. Dushku, and S. Kosta, "When random is bad: selective CRPs for protecting PUFs against modeling attacks," *IEEE Transactions on Computer-Aided Design of Integrated Circuits and Systems*, vol. 44, no. 5, pp. 1648-1661, May 2025.

[6] Y. Wang, G. Zhang, X. Mei, and C. Gu, "A high-reliability, non-CRP-discard arbiter PUF based on delay difference quantization," *IEEE Transactions on Circuits and Systems I: Regular Papers*, vol. 72, no. 2, pp. 573-585, Feb. 2025.

[7] N. Sayadi, P. H. Nguyen, M. v. Dijk, and C. Jin, "Breaking xor arbiter PUFs with chosen challenge attack," *IEEE Transactions on Information Forensics and Security*, vol. 20, pp. 4971-4984, 2025.

A Sequential Obfuscation PUF Resistant to Machine Learning Attacks Based on AES Key Expansion

Xuejiao Ma*[1,2], Yimeng Jin[1], Shuyang Ren[1], Ziyu Zhou[2]

[1] School of Data Science and Artificial Intelligence, Wenzhou University of Technology, Wenzhou, 325024, China
[2] Faculty of Electrical Engineering and Computer Science, Ningbo University, Ningbo, 315211, China

*Email: maxuejiao@wzu.edu.cn, maxuejiao@wzu.edu.cn

Abstract—Physical unclonable functions (PUFs) are lightweight primitives for security device authentication. However, strong PUFs struggle with resilience against machine learning (ML) attacks, hardware efficiency, and reliability. To address these limitations, we propose a sequential obfuscation PUF (SO_PUF) that dynamically obfuscates challenges using the AES key expansion algorithm. This design incorporates temporal dependencies, making the response contingent on both the current challenge and all previous ones. As a result, the static challenge-response mappings are disrupted and the attack resistance is significantly enhanced. Evaluation on 1 million samples demonstrates that SO_PUF achieves exceptional attack resistance. It maintains prediction accuracy below 51.4% against four major ML attack models (LR, SVM, ANN, RNN). Moreover, compared to state-of-the-art defense PUFs, the SO-PUF achieves lower hardware overhead while exhibiting improved randomness, uniqueness, and reliability.

Keywords—*physical unclonable function (PUF), machine learning (ML), sequential obfuscation, AES key expansion.*

I. INTRODUCTION

Physical unclonable functions (PUFs) leverage inherent process variations in IC manufacturing to generate unique, random, and tamper-resistant keys stored as challenge-response pairs (CRPs) [1]. According to the number of generated CRPs, PUFs are classified into weak and strong PUFs [2]. Weak PUFs, whose entropy is tied to isolated structures provide a limited number of CRPs per unit area and are thus suitable for high-sensitivity applications such as key storage. [1]. Strong PUFs reconstruct entropy via multi-stage circuits, generating exponentially more CRPs and inherently resisting cloning attacks, making them ideal for large scale device authentication [3-4].

However, strong PUFs exhibit inherent correlations in their responses, rendering them susceptible to machine learning (ML)-based modeling attacks. For example, a 64-bit arbiter PUF (APUF) can be predicted with 99.9% accuracy using 18,050 CRPs [3]. Existing defense strategies include adding nonlinear circuits [5], XORing responses from multiple APUFs, inserting responses into challenges [6], multiplexing responses [7], external challenge or response obfuscation modules [8], self-obfuscation (where responses obfuscate challenges or vice versa) [9], [10]. These approaches often compromise reliability or increase hardware overhead.

To address these problems, this work proposes a sequential obfuscation PUF (SO_PUF) that dynamically obfuscates challenges by using AES key expansion algorithm. The core innovation lies in replacing the round constant (RC) with challenges. As subkeys depend iteratively on the initial key and prior rounds, each obfuscated challenge relates to all preceding challenges. To prevent key theft, SO_PUF

This work was supported in part by the National Natural Science Foundation of China under Grant 62404155, in part by the General Scientific Research Project of Zhejiang Provincial Department of Education Grant Y202352235.

integrates into device authentication protocols, synchronizing the initial key during registration. Without modifying PUF internals or adding complex circuits, SO_PUF ensures high reliability, low overhead, enhanced ML resistance, and strengthened protocol security. The principal contributions are as follows:

1) Sequential obfuscation mechanism: A sequential obfuscation mechanism establishes a cryptographic linkage between response R_t and temporally adjacent challenges (C_t, C_{t-1}, ⋯, C_1). The mechanism significantly increases complexity for ML-based PUF modeling attacks.

2) Resource-Efficient unified architecture: The PUF employs a single obfuscation module without requiring multiple PUFs, minimizing hardware overhead, and ensuring broad applicability.

3) High ML resistance: With 1 million CRPs, the prediction accuracy for LR, SVM, and ANN remains below 50.5%, while RNN achieves only 51.33% accuracy.

II. RELATED WORK

A. APUF and Its Variants

As shown in Fig. 1, the APUF consists of two symmetric signal paths, each with n multiplexers. Challenge C = (c_1, c_2, ⋯, c_n), where $c_i = 0$ or 1 ($i = 1, 2, ⋯, n$), controls whether the delay signals cross or pass in parallel, quantifying the delay difference between the two reconfigurable paths. The response R is determined by an arbiter. The delay difference is modeled as $\Delta = \omega^T \Phi$, where $\omega = (\omega_1, \omega_2, ⋯, \omega_n, \omega_{n+1})$ is the delay vector and $\Phi = (\Phi_1, \Phi_2, ⋯, \Phi_1, 1)$ is the feature vector with $\Phi_i = \prod_{i=j}^{n}(1 - 2c_i), j = 1, ..., n$. The response is $R_{APUF} = sgn(\Delta) = sgn(\omega^T \Phi)$. The variants of APUF include: (1) x-XOR PUF: XORed outputs of x APUFs; (2) (x, y)-IPUF: incorporates responses from x-XOR PUFs into the challenges of a y-XOR PUF configuration; (3) (n, k)-MPUF: uses APUF outputs as data /select inputs to a multiplexer.

B. ML Attack Algorithms

ML-based PUF modeling attacks follow this workflow: collect CRPs, partition them into training and test sets, and apply ML algorithms to extract features and define loss functions. The model learns challenge-response mappings through iterative parameter updates minimizing the loss function

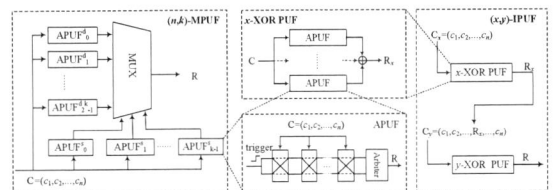

Fig. 1 APUF and its variants

on the training set, with final evaluation and potential hyper parameter tuning conducted on the test set.

To validate our attack resistance, we employ four ML algorithms for PUF modeling: (1) Logistic regression (LR) uses the sigmoid function σ to transform linear regression output into a probability in [0, 1], modeled as$(\hat{R} = 1|\Phi) = \sigma(\omega\Phi) = (1 + e^{-\omega\Phi})^{-1}$. It optimizes the log-likelihood function $L = -\frac{1}{N}\sum_{i=1}^{N}[R_i\log\sigma(\omega\Phi) + R_i\log(1 - \sigma(\omega\Phi))]$. LR demonstrates high effectiveness when dealing with linear PUFs; (2) Support vector machine (SVM) aims to find an optimal hyperplane to separate different classes by the $f(\Phi) = \omega\emptyset(\Phi) + b$, with kernel functions (linear or RBF) enabling high dimensional data mapping for linear or nonlinear classification; (3) Artificial neural network (ANN) learns CRP mappings through layered nonlinear transformations. Mathematically, given an input vector x, the output of a neural network layer y^l is calculated as $y^l = \text{ReLU}(W^l x^{l-1} + b^l)$, where W^l and b^l are the weight matrix and bias vector of the l-th layer respectively, and ReLU is the activation function. Updating weights via backpropagation to minimize loss functions and being capable of approximating complex nonlinear CRP relationships; (4) Recurrent neural network (RNN) is similar to ANN, but it introduces hidden layer in the network to capture the temporal dependencies of sequential data. At time step t, the hidden state h_t is computed as: $h_t = \sigma(W_{hh}h_{t-1} + W_{xh}x_t + b_h)$, where W_{hh} and W_{xh} are weight matrices, b_h is the bias, and σ denotes an activation function (e.g., tanh). The output y_t is generated by $y_t = W_{hy}h_t + b_y$. Parameters are updated via backpropagation through time to model sequential CRPs.

C. AES Key Expansion

The AES key expansion algorithm generates 10 round subkeys for AES-128 from a 4×4-byte initial key matrix W_0 as shown in Fig. 2(a). Each 32-bit word $w[i]$ (i=4, 5, ..., 44) is derived from previous key values according to Eq. (1):

$$w[i] = \begin{cases} w[i-4]\oplus w[i-1], & \text{if } i \bmod 4 \neq 0 \\ w[i-4]\oplus T(w[i-1]), & \text{otherwise} \end{cases} \quad (1)$$

T-function processes 4-byte inputs, denoted as (B_0, B_1, B_2, B_3) in Fig.2(b) through three nonlinear operations. First, a cyclic left-shift operation is applied, yielding the result (B_1, B_2, B_3, B_0). Second, byte-wise substitution via a s-box is performed, producing (B_1', B_2', B_3', B_0'). Third, integration with RC is executed by XORing with RC_j (a Galois field $GF(2^8)$), generating the final output $w[j]'$. As a result, AES key expansion iteratively diffuses the initial key across all subkeys and enhances cryptographic security .

III. PROPOSED SEQUENTIAL OBFUSCATION PUF

A. Principle of Sequential Obfuscation PUF

We replace the RC with challenge by utilizing the core mechanism of the AES key expansion algorithm. Then we extract specific words from generated subkeys as obfuscated challenges. As illustrated in Fig. 3, the obfuscation process for m sets of 32-bit challenges operates as follows: each challenge set sequentially substitutes the corresponding RC value. For clarity, subkeys are denoted as W_i, with component words $w_{i,j}$(where i=1~n, j=0~3, n denotes the total number of rounds). Fig.3 demonstrates the first two rounds of the simplified process. Specifically, the initial word of each subkey acts as the obfuscated challenge, with the first C'_1 derived from first word of the first subkey $w_{1,0}$ and the second C'_2 from that

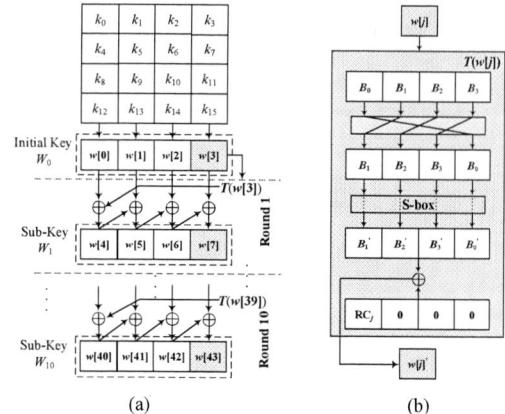

(a)   (b)

Fig. 2 AES-128 key expansion principles: (a) Subkey generation process; (b) T-function implementation

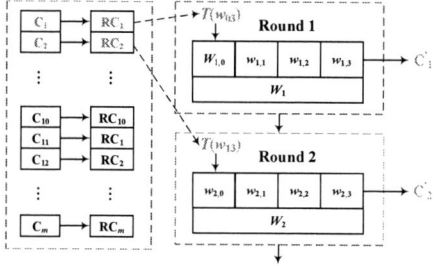

Fig. 3 Principle of challenge obfuscation (first 2 rounds)

of the second subkey $w_{2,0}$. Two critical dependencies emerge: the first, C'_1 depends on C_1 and the initial key owing to $w_{1,0} = T(w_{0,3}) \oplus w_{0,0}$; the second, C'_2 depends on C_2 and C_1 owing to $w_{2,0} = T(w_{1,3}) \oplus w_{1,0}$. For extended $32k$-bit challenges (k>1), partition the input into k sequential 32-bit blocks and apply block-specific RC values (RC_j) during processing.

This dependency propagates cumulatively (e.g., C'_m relies on C_m and all preceding challenges C_1~C_{m-1}). As formalized, each obfuscation round has two inputs (W_{i-1}, C_i) and two outputs (W_i, C'_i) for i=1, 2, ..., m. The obfuscated challenge is derived as:

$$C'_i = C'_{i-1}\oplus T(w_{i-1,3}) = C'_{i-1}\oplus f(W_{i-1}, C_i) \quad (2)$$

where C'_{i-1} is extracted from W_{i-1}, and f represents the T-function parameterized by W_{i-1} and C_i. The subkey updates as:

$$W_i = [w_{i,0}, w_{i,1}, w_{i,2}, w_{i,3}], w_{i,j} = w_{i-1,j}\oplus w_{i,j-1} \quad (3)$$

with each term dependent on prior-round values for iterative propagation.

B. Sequential Obfuscation PUF Design

To proactively prevent leakage of the initial key W_0, this design first synchronizes the encrypted W_0 during the registration phase of the PUF-based device authentication protocol. Subsequently, challenge obfuscation is performed in the device-server mutual authentication phase. The complete architecture is depicted in Fig. 4.

1) Input stage: The randomly generated initial key W_0 and m sets of challenges C = (C_1, C_2, ..., C_m) are serially fed into the sequential obfuscation (SO) structure.

2) Reusable module (RM): The SO structure internally employs m iterations of RM. To change the default, adjust the template as shown in Fig.4. RM features two input ports (A, B)

979-8-3315-3918-4/25 $31.00 © 2025 IEEE

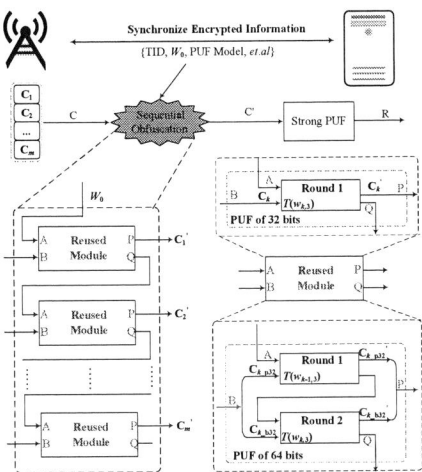

Fig. 4 Strong PUF design with sequential obfuscation

and two output ports (P, Q).

3) Bit-width adaptation: a) The 32-bit PUFs: Port A receives the previous subkey, Port B receives the current challenge. Port P outputs the obfuscated challenge, Port Q outputs the current subkey; *b)* g×32-bit PUFs (g>1): Each challenge is split into g segments, for 64-bit PUFs (g=2): challenge segmented into 2 groups. Processed over two RM; *c)* Outputs merged at Port P.

4) Iteration: The RM is reused m times to generate m obfuscated challenges C′.

5) Final output: C′ drives the strong PUF to produce the response R.

Compared to traditional schemes, the SO scheme derives the current obfuscated challenge from both the present and all prior challenges. Consequently, the PUF response generated with this challenge correlates with current and historical challenges. This sequential mechanism mimics the memory of sequential circuits, complicating response-challenge links and greatly improving ML-attack resilience.

C. The Resistance of SO_PUF to ML Attack

All anti-attack tests were executed on a single-threaded CPU without GPU, utilizing scikit-learn and TensorFlow1 libraries. Seven PUFs were implemented on the FPGA, with 1 million CRPs collected for each PUF. ML modeling used a fixed 200K CRP test set, with training sets spanning 1K~800K CRPs (8 log-spaced intervals).

As shown in Fig. 5, all four ML models show rising prediction accuracy with larger training sets, reducing attack resistance for conventional PUFs. However, SO_PUF is unaffected. At the largest training size, it maintains the lowest prediction rate ($\approx$50%). Experimental results demonstrate that the seven PUFs show attack resistance capabilities: 2-XORPUF < (2,1)-MPUF < 5-XORPUF < (3,3)-IPUF < 8-XORPUF. Fig.5 demonstrates the superior attack resistance of the SO_PUF through two findings: first, across all algorithms and training sizes, SO_PUF achieves the lowest attack prediction rate, with a maximum value of 51.33%. Second, its prediction accuracy remains stable regardless of training data volume, avoiding the degradation typically observed in other attack-resistant PUF designs. Even RNN which is specifically optimized for sequential data analysis, achieves merely 51% accuracy.

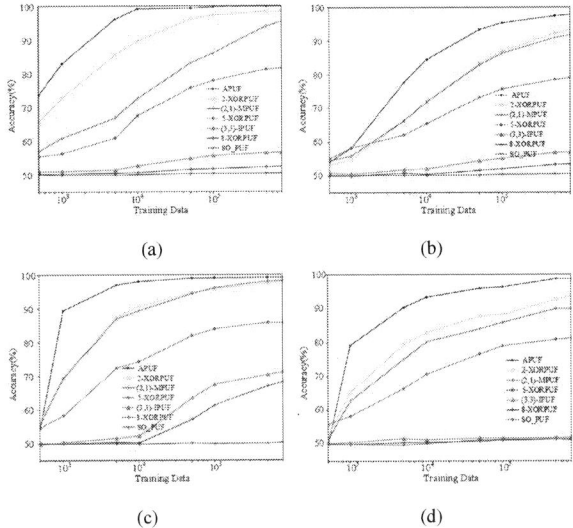

(a)   (b)

(c)   (d)

Fig. 5 Prediction accuracy of four ML attack on eight PUFs: (a) LR; (b) SVM; (c) ANN; (d) RNN

IV. PERFORMANCE ANALYSIS OF PROPOSED PUF

A. Experimental Platform

All evaluations were run on an Intel Core i7-1260P (2.10 GHz)/16GB RAM platform. CRPs were first collected via FPGA-based PUFs through UART-JTAG serial communication linking MATLAB to the FPGA development board in Fig. 6, followed by performance evaluation of the PUFs using the collected CRPs.

B. Randomness

Employing the NIST SP800-22 for randomness testing, we evaluated APUF and SO_PUF with 1 million randomly generated challenges and corresponding responses. Then response data was divided into 50 groups. The results demonstrate that SO_PUF passes 6 test items, while APUF passes 5 as shown in Table I, indicating the SO_PUF exhibits superior randomness.

C. Uniqueness and Reliability

Uniqueness and reliability were evaluated using the average inter hamming distance (HD) and intra HD, represented by the red histogram and fitted blue curve in Fig. 7, respectively. The normalized HD distribution follows a Gaussian distribution with mean 0.5002 and standard deviation 0.021. The SO_PUF achieves 50.02% uniqueness, 99.83% reliability and intra-chip HD of 0.00175. Moreover, the ratio of inter chip HD to intra chip HD of 286 demonstrates enhanced distinguishability.

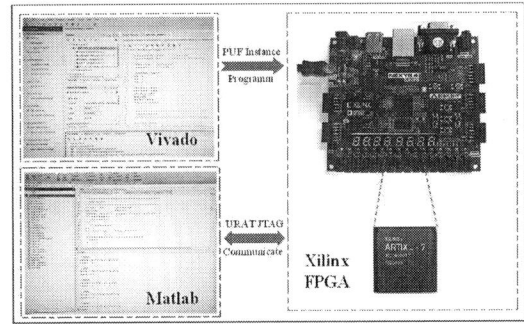

Fig. 6 FPGA-based CRP collection platform

TABLE I. NIST Test

Statistical Test	APUF			SO_PUF		
	P-value	Prop	Pass?	P-value	Prop	Pass?
Frequency	0.0000*	8/50	No	0.0000*	2/50	No
Block Frequency	0.0001	49/50	No	0.0000*	49/50	No
Cumulative Sums	0.0000*	9/50	No	0.0000*	4/50	No
Runs	0.0000*	34/50	No	0.0000*	26/50	No
LongestRun	0.0269	48/50	Yes	0.0457	49/50	Yes
Rank	0.0757	50/50	Yes	0.6579	50/50	Yes
FFT	0.2897	48/50	Yes	0.0966	50/50	Yes
Overlapping Template	0.3191	48/50	Yes	0.1538	50/50	Yes
Universal	0.0000*	50/50	No	0.0000*	50/50	No
Approximate Entropy	0.0000*	46/50	No	0.0000*	45/50	No
Serial1	0.3838	50/50	Yes	0.7792	50/50	Yes
Serial2	0.1223	50/50		0.8832	50/50	
Linear Complexity	0.0015	48/50	No	0.4190	50/50	Yes

*:P-value < 0.001 (statistically significant)

To further validate reliability, we applied p=10000 challenge sets and each was repeated q=50 times to the same PUF instance in Fig.8. Reliability is computed as in Eq. (4), the SO_PUF exhibits marginally lower reliability than APUF but surpassed other strong PUFs.

$$Reliability = \frac{1}{p}\sum_{i=1}^{p}(\prod_{j=2}^{q}(1-(R_{i,1}-R_{i,j})^2)) \quad (4)$$

D. Hardware overhead

We summarize the hardware overhead, including lookup tables (LUTs), D-flip-flops (DFFs), and RAM blocks as shown in Table II, comparing without SO scheme (WO) against with SO scheme (WI). Resource overhead for the PUF control circuitry is included.

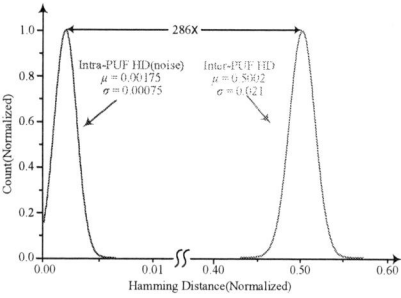

Fig. 7 Normalized inter HD and intra HD

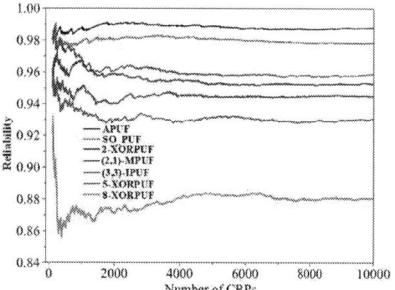

Fig. 8 Reliability test on various PUFs

TABLE II. Hardware overhead

Type of PUF	LUT		DFF		Num of RAM	
	WO	WI	WO	WI	WO	WI
APUF	287	385	345	474	1	3
2-XORPUF	357	455	346	475	1	3
5-XORPUF	561	659	349	478	1	3
8-XORPUF	767	865	352	481	1	3
(2,1)-MPUF	425	523	347	476	1	3
(3,3)-IPUF	632	730	352	481	1	3

The results show uniform increases of 98 LUTs, 129 DFFs, and 2 RAM blocks across all PUFs using the SO scheme. Significantly, SO_PUF adds minimal resources compared to lightweight APUF while maintaining consistent hardware costs independent of PUF complexity, confirming broad applicability.

V. CONCLUSION

This paper proposes a novel SO_PUF defense against modeling attacks. Its core uses the AES key expansion algorithm to create a sequential obfuscation circuit through iterative encryption, making each response depend on both the current challenge and prior ones. SO_PUF effectively resists ML attacks, keeping prediction accuracy below 51.4% even with 1 million CRPs, outperforming other designs. Moreover, SO_PUF demonstrates enhanced randomness, uniqueness, and stability while maintaining lower hardware overhead. The design paradigm offers an innovative approach and enhances PUF security and broadening its applicability.

REFERENCES

[1] Z. Zhou, P. Wang, G. Li, S. Hu and Y. Zhang, "Improving the Stability of APUF to 100% Without Extra Hardware Overhead for Enhancing the Performance of Security Authentication Protocols"[J], IEEE Internet of Things Journal, 2025, 12(12), pp. 19818-19832.

[2] X. Ma, P. Wang, G. Li, Z. Zhou, " Machine learning attacks resistant strong PUF design utilizing response obfuscates challenge with lower hardware overhead"[J], Microelectronics Journal, 2023(142), pp. 1-11.

[3] Z. Zhou, P. Wang and G. Li, "Bagua Protocol: A Whole-Process Configurable Protocol for IoT Sensing Devices Security Based on Strong PUF"[J], IEEE Internet of Things Journal, 2024, 11(1), pp. 805-819.

[4] N. Hassan and U. Chatterjee, "Machine Learning Attacks on Challenge-Response Obfuscations in Strong PUFs"[C], IEEE International Symposium on Hardware Oriented Security and Trust (HOST), Tysons Corner, VA, USA, 2024, pp. 361-372.

[5] R. Kumar and W. Burleson, "On design of a highly secure PUF based on non-linear current mirrors"[C], 2014 IEEE International Symposium on Hardware-Oriented Security and Trust (HOST), Arlington, VA, USA, 2014, pp. 38-43.

[6] P.H. Nguyen, D.P. Sahoo, C. Jin, et al,"The Interpose PUF: Secure PUF Design against state-of-the-art Machine Learning Attacks"[C], IACR Transactions on Cryptographic Hardware and Embedded Systems, 2019, pp.243-290.

[7] H. Wang, C. Wan and H. Jin, "Efficient Modeling Attack on Multiplexer PUFs via Kronecker Matrix Multiplication"[J], IEEE Transactions on Computer-Aided Design of Integrated Circuits and Systems, 2025, Early Access.

[8] G. Yan, M. Hua, F. Said. Al. Sarawi, D. Abbott, and C. Damith. Ranasinghe, "PUF-FSM: A Controlled Strong PUF"[J]. IEEE Transactions on Computer-Aided Design of Integrated Circuits and Systems,2018, 37(5), pp. 1104 - 1108.

[9] L. Xu, L. Zhang, P. I. Mak, R. P. Martins and M. K. Law, "Fully Symmetrical Obfuscated Interconnection and Weak-PUF-Assisted Challenge Obfuscation Strong PUFs Against Machine-Learning Modeling Attacks,"[J], IEEE Transactions on Information Forensics and Security, 2024, 19, pp. 3927-3942.

[10] J. Zhang and C. Shen, "Set-Based Obfuscation for Strong PUFs Against Machine Learning Attacks"[J], IEEE Transactions on Circuits and Systems I: Regular Papers, 2021, 68(1), pp. 288-300.

Hardware-Efficient Doppler Estimation and Compensation in PDSCH for 5G Non-Terrestrial Networks

Chih-Chen Chen , Yi-Shan Huang , Chung-Lun Tu and Shyh-Jye Jou

Institute of Electronics, National Yang Ming Chiao Tung University, HsinChu,Taiwan

Email: wl03140431@gmail.com, jerryjou@nycu.edu.tw

Abstract—This paper proposes a hardware-efficient Doppler estimation and compensation scheme tailored for the Physical Downlink Shared Channel (PDSCH) stage of 5G NR-based Non-Terrestrial Networks (NTN). Unlike conventional methods that focus on synchronization signal stages, the proposed approach utilizes the Demodulation Reference Signal (DM-RS) embedded in PDSCH to dynamically track and correct Doppler shift. A delay-and-correlate-based method is adopted to estimate the fractional Doppler, while an FFT-based cross-correlation structure is used for integer Doppler estimation. The entire design avoids operations like division and square root to favor fixed-point implementation. Simulation results under LEO satellite TDL channel models show that the proposed system ensures residual Doppler shift within ±0.1 ppm, while achieving low hardware complexity.

Keywords—5G NTN, Doppler shift, Carrier Frequency Offset, Frequency Compensation

I. INTRODUCTION

The 5G NR standard incorporates advanced synchronization mechanisms during the cell search stage to address initial frequency offsets and establish coarse synchronization. Primary Synchronization Signal (PSS) and Secondary Synchronization Signal (SSS) are employed to estimate timing and frequency offsets, ensuring successful initial access. Since precise synchronization is essential for decoding PDSCH, the cell search process must effectively compensate frequency offsets and Doppler variations, ensuring stable and robust communication even in dynamic satellite-based NTN environments. Several studies [1]–[7] have proposed various synchronization schemes for 5G NR systems, focusing on the utilization of PSS and SSS sequences during the cell search stage. These methods leverage the known properties of the synchronization signals to estimate timing offsets and coarse frequency offsets effectively. Nevertheless, these methods are limited to the cell search phase and do not adequately address the dynamic and residual Doppler effects encountered during data transmission phases, such as the Physical Downlink Shared Channel (PDSCH). Such residual impairments hinder reliable demodulation, thereby requiring advanced strategies for accurate frequency offset compensation.

To address these challenges, this work focuses on enhancing the performance of 5G NTN systems by proposing advanced Doppler frequency offset compensation methods tailored for the PDSCH stage. Inspired by the work in [8], which proposed a Doppler estimation framework leveraging the cyclic prefix for fractional offset and DM-RS for integer offset, we recognize its effectiveness in accurately estimating the Doppler shift under various NTN channel conditions. However, their fractional Doppler estimation poses challenges for real-time hardware implementation due to high complexity. Additionally, the integer Doppler offset is estimated through time-domain correlation with a reference waveform, which requires additional 1024 points FFT/IFFT—another barrier for hardware realization.

To improve hardware feasibility, this work proposes a hardware-efficient fractional Doppler estimation algorithm. Furthermore, we perform integer Doppler estimation in the frequency domain, using frequency-aligned DM-RS correlation to simplify. This design maintains estimation performance while significantly reducing complexity, making it more suitable for 5G NTN.

II. PROPOSED SYSTEM MODEL

We consider a downlink system with 3GPP Release 17 NTN specifications. The system assumes a LEO satellite at 600 km altitude, transmitting over a Ka-band carrier at 20 GHz. The Doppler shift due to satellite motion can reach up to ±480 kHz. In addition, oscillator imperfections can introduce Carrier Frequency Offset (CFO), leading to a combined frequency error that must be corrected at the receiver. The subcarrier spacing (SCS) is set to 60 kHz, and the total bandwidth is 50 MHz. This corresponds to 66 Physical Resource Blocks (PRBs) with totaling 792 subcarriers. A 1024-point FFT is adopted to provide sufficient guard bands. The effective sampling rate becomes 61.44 MHz. DM-RS is inserted in specific symbols, following the PDSCH mapping structure (DRMS configuration type = 1 or 2) defined in TS 38.211. The proposed system flow of 5G NTN is illustrate in Fig. 1.

To analytically characterize the impact of time-varying Doppler shifts in 5G NTN downlink, we adopt the Doppler signal model for an L-tap SISO channel proposed in [8]. After an analog-to-digital conversion with a sampling interval T_s with the assumption of negligible higher-order variations within a slot, the received discrete signal becomes:

$$y(n) = \sum_{m=0}^{L-1} h(m) \cdot x(n-m) \exp\left(j2\pi \sum_{p=0}^{n}(\alpha + \beta p)T_s\right) + w(n) \quad (1)$$

where α and β represent the Doppler shift and the Doppler rate, respectively, $h(m)$ is channel impulse response, and $w(n)$ is white noise. In practical systems, the observed frequency offset is a combination of both Doppler shift and Carrier Frequency Offset (CFO), as shown in (2), meaning that compensation mechanism must jointly estimate and correct both impairments.

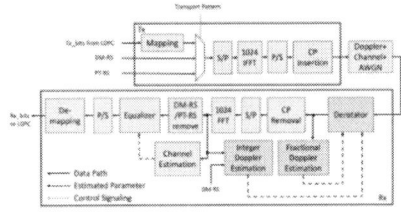

Fig. 1. Proposed system flow of 5G NTN

979-8-3315-3918-4/25 $31.00 © 2025 IEEE

$$y(n) = \sum_{m=0}^{L-1} h(m) \cdot x(n-m) \exp\left(j2\pi \sum_{p=0}^{n} (\alpha + \beta p)T_s\right) \exp(j\theta_{CFO}n) + w(n) \quad (2)$$

III. PROPOSED DOPPLER ESTIMATION

A. Fractional Doppler Estimation

A well-known approach for estimating fractional Doppler is the Delay-and-Correlate (DC) algorithm proposed in [10]. We modify the algorithm for hardware friendly implementation.

$$DC(m) = \left|\sum_{i=m}^{m+L-1} y(i)y^*(i+N)\right|^2 \quad (3)$$

where N denotes the FFT size, L represents the CP length, and $m \in \{1 \ldots N_{sample} - L + 1\}$.

A normalized form known as Normalized Autocorrelation (NAC) has been proposed in [8] to improve robustness and reduce false alarm rates. From a hardware implementation perspective, we simplify the *NAC* method with *NAC*2, as described in [9], by replacing the absolute value operation with a squared magnitude and transforming the division into a subtraction:

$$NAC2(m) = \left|\sum_{i=m}^{m+L-1} y(i)y^*(i+N)\right|^2 - \lambda \sum_{i=m}^{m+L-1}|y(i)|^2 \sum_{i=m}^{m+L-1}|y(i+N)|^2 \quad (4)$$

where λ is a threshold value. Due to lower implementation complexity, we adopt the modified DC method shown in (3). After the valid peak location (*validLoc*) is identified, the fractional phase offset ($\hat{\theta}_f$) can be derived as:

$$\hat{\theta}_f = \frac{\arg(DC(validLoc))}{N} \quad (5)$$

B. Integer Doppler Estimations

Integer frequency offset (IFO) introduces a circular shift of subcarrier positions in the frequency domain, which significantly distorts the demodulation process in OFDM systems. To address this, the estimation method proposed in [11] operates in the frequency domain by leveraging the cross-correlation between received and reference signals. But this method applies a magnitude operation prior to correlation, which eliminates the phase information inherent in the signal. This can lead to incorrect estimations when the signal exhibits phase distortions or low SNR. To overcome this issue, we retains phase characteristics by removing the magnitude operation and utilizing complex-valued correlation:

$$\tilde{\varepsilon} = \max_{\tilde{\varepsilon}}\left\{\left|\sum_{k=1}^{w} Y_k X_{k-\tilde{\varepsilon}}^*(n)\right|\right\} \quad (6)$$

where X is the reference signal, Y is the received signal, k is the subcarrier index, and w is the number of data subcarriers.

However, we observe that when operating under the TDL-C channel, the calculation based on (6) suffers from severe degradation in frequency estimation accuracy. The phase variation across the frequency band tends to be linear in the LOS channel. Due to the lack of complex phase variations caused by multiple delayed paths, the correlation calculations suffer from degraded resolution. To address the challenges posed by channel effects, [12] proposes a method that utilizes neighboring subcarrier multiplication to eliminate the

influence of the channel response. Based on this approach, we develop an algorithm adapted to our target environment:

$$\tilde{\varepsilon} = \max_{\tilde{\varepsilon}}\left\{\left|\sum_{k=1}^{W} A_k D_{k-\tilde{\varepsilon}}^*(n)\right|\right\} \quad (7)$$

$$D_k = X_{p,k}^* X_{p,k-2} \ or \ X_{p,k}^* X_{p,k-1} \quad (8)$$

$$
\begin{aligned}
A_k &= Y_{p,k}^* Y_{p,k-2} \ or \ Y_{p,k}^* Y_{p,k-1} \\
&= H_{k-\varepsilon}^* X_{p,k-\varepsilon}^* H_{k-\varepsilon-2} X_{p,k-\varepsilon-2} \ or \ H_{k-\varepsilon}^* X_{p,k-\varepsilon}^* H_{k-\varepsilon-1} X_{p,k-\varepsilon-1} \\
&\approx |H_{k-\varepsilon}|^2 X_{p,k-\varepsilon}^* X_{p,k-\varepsilon-2} \ or \ |H_{k-\varepsilon}|^2 X_{p,k-\varepsilon}^* X_{p,k-\varepsilon-1} \\
&= D_{k-\varepsilon}|H_{k-\varepsilon}|^2
\end{aligned} \quad (9)
$$

where D_k represents the cross-correlation of two adjacent locally generated reference signals (DRMS configuration type = 1 or 2), A_k represents the cross-correlation of two adjacent received subcarrier symbols (DMRS configuration type = 1 or 2). Since the channel responses for adjacent subcarriers are similar in both TDL-A and TDL-C channels, this modification effectively eliminates the impact of channel fading.

Specifically, the DMRS sequence consists of $\pm\frac{1}{\sqrt{2}}$ combinations, leading to four possible patterns: $\{+\frac{1}{\sqrt{2}}+\frac{1}{\sqrt{2}}j, +\frac{1}{\sqrt{2}}-\frac{1}{\sqrt{2}}j, -\frac{1}{\sqrt{2}}+\frac{1}{\sqrt{2}}j, -\frac{1}{\sqrt{2}}-\frac{1}{\sqrt{2}}j\}$. Therefore, each term in (8) for D_k can only take one of four possible values $\{+1, -1, +j, -j\}$. Based on the DMRS configuration type, the possible values of D_k can be pre-computed and stored, thus significantly reducing the real-time computational complexity during on-line operation. Accordingly, the term $A_k D_{k-\tilde{\varepsilon}}^*$ in (7) can be rewritten as:

$$A_k D_{k-\tilde{\varepsilon}}^* = \begin{cases} a + bj, & if \ D_k = +1 \\ -a - bj, & if \ D_k = -1 \\ b - aj, & if \ D_k = +j \\ -b + aj, & if \ D_k = -j \end{cases} \quad (10)$$

where we assume $A_k = Y_{p,k}^* Y_{p,k-1} = a + bj$.

This transformation allows the original complex multiplication to be replaced by a simple multiplexer. By selecting different IQ components of A_k according to the precomputed D_k value, the correlation result can be obtained without explicit complex multiplication. This substitution significantly reduces the hardware complexity associated with the estimation process. Finally, based on the estimated integer shift $\tilde{\varepsilon}$, the integer phase rotation $\hat{\theta}_i$ can be derived from (11), as expressed in (12).

$$e^{-j2\pi f_i n T_s} = e^{-j2\pi(\tilde{\varepsilon} \times \Delta f) n T_s} = e^{\frac{-j2\pi(\tilde{\varepsilon} \times \Delta f)n}{N \times \Delta f}} = e^{\frac{-j2\pi \tilde{\varepsilon} n}{N}} \quad (11)$$

$$\hat{\theta}_i = \frac{-2\pi \tilde{\varepsilon}}{N} \quad (12)$$

where f_i denotes the integer Doppler frequency, Δf represents the subcarrier spacing, T_s is the sampling time, n is the sample index, and N is the FFT size.

I. PROPOSED HARDWARE ARCHITECTURE

Fig. 2 illustrates the hardware architecture for Doppler estimation and compensation. Since the proposed algorithm is symbol-based (1024-point data), so SRAM is adopted to store the data. Moreover, to accommodate different DMRS types, five separate 1024x32 SRAM blocks are utilized to store data

979-8-3315-3918-4/25 $31.00 © 2025 IEEE

arriving at different time instances. The D register is controlled by control signals (clear, i_valid, f_valid) that decide whether the data in the register should be cleared or validated. A multiplexer (MUX), controlled by control signals, is used to choose between the estimated fractional phase $\hat{\theta}_f$ or integer phase $\hat{\theta}_i$. These selected values are then passed to the phase accumulator, which subsequently feeds into a CORDIC module to realize the phase compensation.

Fig. 3 illustrates the block diagram of the fractional Doppler estimation module, which is based on the DC method described in (3). The inputs are the current symbol(r1024) and the buffered symbol(r0) stored in SRAM, which are conjugated and multiplied to compute the correlation. A moving average is then applied to the correlation results. These correlation results are stored in SRAM and the maximum value is identified through comparison logic. The corresponding phase angle is then computed using the CORDIC module, and the final average phase offset $\hat{\theta}_f$ is obtained by right-shifting the CORDIC output by 10 bits(N=1024), as defined in (5).

Fig. 4 illustrates the block diagram of the integer Doppler estimation module, which is based on (7). In this design, the input Y[k] is taken from the FFT output. The structure is designed to be serial in, following the estimation framework presented in [12], which requires only a single multiplier, resulting in reduced hardware complexity. Additionally, the three replicated structures in the lower part are used to select one from three candidates. The shift takes values in $(-1, 0, 1)$. Based on (10), the result of $A_k D_{k-\hat{\varepsilon}}^*$ is easily obtained via a simple MUX (P_selector) , eliminating the need for multiplication. We know that D_k can take four possible values: $\{+1, -1, +j, -j\}$, or 0 (indicating non-DMRS subcarrier positions). The values 0, 1, 2, 3, and 4 are used to represent $D_k = +1, -1, +j, -j, and \ 0$, respectively. These values are then sent to the P_selector for selection. The lookup part can be pre-stored, the four possible 2-bit values $(+1, -1, +j, -j)$ are stored in SRAM in advance. When D_k=0, the MUX makes selection accordingly using the control signal. Since the three P_selector control signals D_k are sequentially related, a 3×3-bit shift buffer is used to store them. According to (7), the maximum value is selected, and the final phase offset $\hat{\theta}_i$ is calculated by shifting right the result by 9 bits.

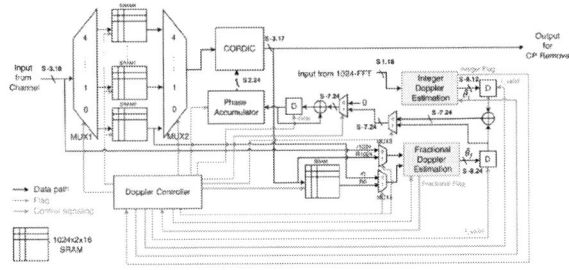

Fig. 2. Fixed point process flow of Doppler estimation and compensation.

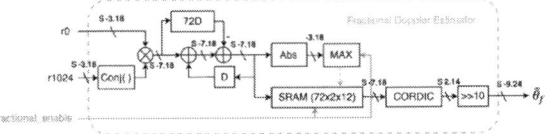

Fig. 3. Block diagram of fractional Doppler estimation.

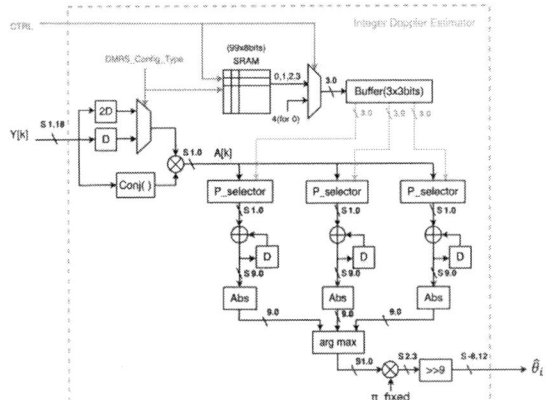

Fig. 4. Block diagram of integer Doppler estimation.

II. SIMULATION RESULTS AND ANALYSIS

The performance results of a 5G PDSCH link are presented for frequency range 2 (FR2) with carrier frequency of 20 GHz and subcarrier spacing of 60 kHz. The channel models used for link-level simulations are NTN- TDLC5.

Three different algorithms—DC, NAC, and NAC2—are implemented and compared as shown in Fig. 5. While all three methods demonstrate similar detection performance in simulation and can keep the residual frequency offset within 0.1ppm. The DC method in (3) exhibits lower implementation complexity. TABLE I. summarizes the hardware complexity comparison between DC and NAC2. The DC method demonstrates a 46.67% reduction in adders, 45.45% reduction in multipliers, and 6.24% reduction in registers compared to the NAC2 method. It also avoids the high computational complexity introduced by the division and square root operations required in the NAC method, thereby improving hardware efficiency by reducing the number of required resources. For this reason, we adopt the modified DC method for the fractional Doppler estimation, as it performs a better balance between performance and hardware simplicity.

Fig. 6 shows the performance comparison of integer frequency shift estimation using different methods under NTN-TDLC5 channel. Method 1 corresponds to (6), while Method 2 corresponds to (7). From the simulation results, it can be observed that Method 2 achieves 100% estimation accuracy for the integer Doppler shift in NTN-TDLC5 channel because it successfully mitigates the influence of channel effects.

To compare the efficiency of the proposed frequency-domain method with the time-domain approach presented in [8], we conduct an analysis of the hardware complexity, and the summary is shown in 0The time-domain method [8] requires converting the DM-RS from frequency to time domain via an IFFT. Additionally, an extra FFT is needed to convert the result back to the frequency domain and extract the corresponding bin value. The proposed method avoids the need for 1024 points IFFT/FFT, and the additional computations in (8) can be pre-computed and the results are pre-stored in SRAM.

TABLE I. HARDWARE COMPLEXITY REDUCTION

Algorithm	Adder Reduction (%)	Multiplier Reduction (%)	Register Reduction (%)
DC	46.67%	45.45%	6.24 %

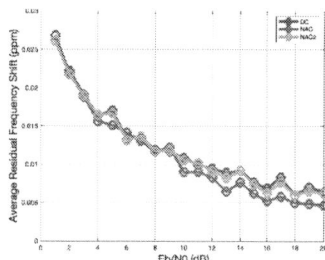

Fig. 5. Comparisons of average residual frequency offset across fractional Doppler estimation algorithms.

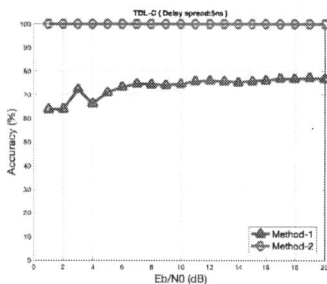

Fig. 6. Comparison of integer frequency offset estimation accuracy under NTN-TDLC5.

TABLE II.   HARDWARE COMPLEXITY COMPARISON

Operation	Proposed	[8]
Multiplier	10	4
Additional FFT/IFFT	x	1x1024 points FFT 1x1024 points IFFT
Memory Buffer	99x8bits SRAM (pre-computed)	x

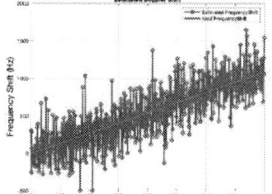 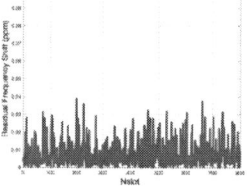

Fig. 7. Estimated Doppler shift versus ideal Doppler trend   Fig. 8. Residual frequency shift after compensation

Fig. 7 shows that the estimated Doppler frequency shift closely follows the ideal curve. Fig. 8 indicates the residual frequency offset remains within ±0.1 ppm, indicating that the proposed method effectively mitigates frequency shifts under NTN-TDLC5 multi-path fading and time-varying Doppler conditions. Unlike prior studies that focus on algorithm, this work presents a system-level receiver design that balances Doppler tracking performance and hardware feasibility. A hardware-efficient framework is proposed for residual Doppler compensation in PDSCH, addressing a not well explored aspect of 5G NTN systems.

III. CONCLUSIONS

In this paper, we propose a Doppler estimation and compensation framework tailored for 5G Non-Terrestrial Networks (NTN). The proposed solution effectively mitigates Doppler shift and frequency offset while being suitable for hardware implementation. We integrate both fractional and integer Doppler estimation techniques. The fractional Doppler reduces 46.67% in adders, 45.45% in multipliers, and 6.24% in registers, while avoiding costly operations such as division and square root. The integer Doppler estimation is performed using frequency-domain DMRS-based correlation, eliminating the need for 1024 points IFFT/FFT by time-domain estimation. Additionally, extra computations need to be calculated and the results are pre-stored in SRAM. The fixed-point simulation results confirm that the proposed method could achieve a residual frequency offset within ±0.1 ppm, ensuring reliable performance for practical deployments.

REFERENCES

[1] D. Wang, Z. Mei, H. Zhang and H. Li, "A Novel PSS Timing Synchronization Algorithm for Cell Search in 5G NR System," in *IEEE Access*, vol. 9, pp. 5870-5880, 2021.

[2] S. Yoneda, M. Sawahashi and S. Nagata, "Physical Cell ID Detection Probability in the Presence of CFO Including Doppler Shift for NR TN and NTN," *2023 VTS Asia Pacific Wireless Communications Symposium (APWCS)*, Tainan city, Taiwan, 2023, pp. 1-5.

[3] D. Inoue, K. Ota, M. Sawahashi and S. Nagata, "Physical Cell ID Detection Using Joint Estimation of Frequency Offset and SSS Sequence for NR Initial Access," *2021 IEEE 93rd Vehicular Technology Conference (VTC2021-Spring)*, Helsinki, Finland, 2021, pp. 1-6.

[4] S. Yoneda, M. Sawahashi and S. Nagata, "Comparisons of Physical Cell ID Detection Methods with Carrier Frequency Offset Compensation for Millimeter-Wave Bands," *2022 IEEE 96th Vehicular Technology Conference (VTC2022-Fall)*, London, United Kingdom, 2022, pp. 1-6.

[5] Q. Sultan, S. Ha, S. Choi and Y. S. Cho, "Downlink Synchronization in New Radio (NR) Non-Terrestrial Networks (NTN)," *2022 13th International Conference on Information and Communication Technology Convergence (ICTC)*, Jeju Island, Korea, Republic of, 2022, pp. 1351-1353.

[6] M. Assaf and O. G. Ponomarev, "Efficient and Low Complexity Frequency Synchronization in NR-5G Downlink," *2023 25th International Conference on Digital Signal Processing and its Applications (DSPA)*, Moscow, Russian Federation, 2023, pp. 1-6.

[7] Z. Zhang, B. Gong, J. Wang, C. Zhang and C. Pan, "A Joint Frequency Offset Estimation Algorithm for 5 G NTN," *2024 IEEE International Symposium on Broadband Multimedia Systems and Broadcasting (BMSB)*, Toronto, ON, Canada, 2024, pp. 1-6.

[8] S. Tadavarty and N. K. Chavali, "Estimation and Compensation of Doppler in 5G NR Based Non-Terrestrial Networks," *2022 IEEE International Conference on Advanced Networks and Telecommunications Systems (ANTS)*, Gandhinagar, Gujarat, India, 2022, pp. 135-140.

[9] K. L. Chiu, P. H. Shen, B. R. Lin, W. H. Hsiao, S. J. Jou and C. C. Huang, "Design of Downlink Synchronization for Millimeter Wave Cellular System Based on Multipath Division Multiple Access," in *IEEE Transactions on Circuits and Systems I: Regular Papers*, vol. 67, no. 9, pp. 3211-3223, Sept. 2020.

[10] T. Z. Wei, S. J. Jou and M. T. Shieu, "Memory reduction ICFO estimation architecture for DVB-T," *ISCAS 2006*, pp.3406 -3409.

[11] D. G. Urdaneta, C. F. Dias, F. K. Pereira, L. S. d. Moraes, A. Távora and E. R. d. Lima, "An Implementation of a Low Complexity Integer Carrier Frequency Offset estimator for OFDM," *2022 IEEE International Symposium on Circuits and Systems (ISCAS)*, Austin, TX, USA, 2022, pp. 1695-1698.

[12] D. C. Alves-Tamagno, E. de Lima and R. R. Lopes, "A Low Complexity ICFO Estimator and Compensator for IEEE 802.15.4g MR-OFDM PHY: Algorithm Proposal and Hardware Implementation," *2018 IEEE 29th Annual International Symposium on Personal, Indoor and Mobile Radio Communications (PIMRC)*, Bologna, Italy, 2018, pp. 1-7.

[13] 3GPP, *User Equipment (UE) Radio Transmission and Reception; Part 2 : Range 2 Standalone.* TS 38.101-2 v19.0.0, Mar. 2025.

[14] 3GPP, *User Equipment (UE) conformance specification; Radio transmission and reception; Part 5: Satellite access Radio Frequency (RF) and performance.* TS 38.521-5 v18.2.0, Jun. 2024.

FPGA Bitstream Modification Attacks on CRYSTALS Kyber

Lei Chen[1], Jiahao Lu[1], Tianze Huang[1], Aobo Li[1], Shengfei Gu[1], Ang Hu[1,2], Dongsheng Liu*[1,2]

[1]School of Integrated Circuits, Huazhong University of Science and Technology, Wuhan, China
[2]JinYinHu Laboratory, WuHan 430040, China

* Email: dsliu@mail.hust.edu.cn

Abstract—**Quantum computers pose a significant threat to current public-key cryptography, driving the urgent need for post-quantum cryptography (PQC). While PQC algorithms, such as the standardized key encapsulation mechanism (KEM) CRYSTALS-Kyber, are designed to resist quantum attacks, their hardware implementations, particularly on Field Programmable Gate Arrays (FPGA), remain susceptible to physical attacks. This paper explores bitstream modification attacks specifically on a Xilinx Artix-7 xc7a35 FPGA, targeting Kyber's Number Theoretic Transform (NTT) unit. By manipulating the bitstream to nullify NTT twiddle factors, we demonstrate significant weaknesses in both key generation and encapsulation phases, leading to predictable secret keys and enabling man-in-the-middle attacks. We detail the bitstream extraction and modification process, including twiddle factor identification and CRC updates for covertness. Finally, we discuss potential countermeasures.**

Keywords—Post-quantum cryptography (PQC), CRYSTALS-Kyber, FPGA, Number Theoretic Transform (NTT), Bitstream modification attacks

I. INTRODUCTION

Post-quantum cryptography represents the future of public key cryptographic systems, engineered to withstand the formidable computational power of quantum computers. Traditional algorithms, vulnerable to quantum attacks such as Shor's algorithm [1] and Grover's algorithm [2], have prompted the National Institute of Standards and Technology (NIST) to initiate a standardization process. This effort culminated on August 13, 2024, with the release of three federal information processing standards (FIPS), marking a significant milestone in securing cryptographic implementations against quantum threats [3].

Among the standardized solutions, CRYSTALS-Kyber (hereafter referred to as Kyber) [4] stands out as a key encapsulation mechanism based on the module learning with errors (MLWE) problem. Kyber leverages the inherent algorithmic security of lattice-based cryptography, providing a robust foundation that balances efficiency and resilience against quantum attacks. However, its physical implementations remain susceptible to side-channel and fault attacks that could expose sensitive key material [5]. Initial vulnerabilities were highlighted in 2018 when Ravi et al. demonstrated the first successful fault attack on Kyber [6].

Building on this foundation, recent research has advanced the understanding of bitstream attacks, particularly targeting FPGA configurations[7]. A bitstream in this context refers to the binary configuration file that defines the functionality of an FPGA, programming its logic gates and interconnections to implement specific hardware designs, such as cryptographic algorithms like Kyber. Bitstream attacks exploit vulnerabilities in this configuration data, often through unauthorized access, modification, or reverse-engineering, to compromise the security of the implemented system. Recent studies have further escalated concerns by demonstrating methods to bypass AES encryption processes typically used to protect FPGA bitstreams, exposing critical weaknesses in hardware security [8]. This article extends that work by exploring novel bitstream modification attacks, aiming to further assess and mitigate these emerging threats in Kyber implementations.

II. KYBER KEM PROTOCOL

The Kyber KEM protocol is a post-quantum cryptographic scheme designed to securely establish a shared secret key between two parties, Alice and Bob. It can be transformed into a chosen-ciphertext attack (CCA)-secure public-key encryption (PKE) system through the application of a Fujisaki-Okamoto transform.

1) Key generation phase

Initiating the process, Alice generates a public/private key pair. She selects a random matrix $\mathbf{A}$ and a secret key polynomial vector $\mathbf{s}$, complemented by a noise polynomial vector $\mathbf{e}$. The public key is computed as $pk = \text{Encode}_{12}((\hat{\mathbf{A}}^T \circ \hat{\mathbf{s}}+\hat{\mathbf{e}})||\rho)$, where $\hat{\mathbf{e}}$ denotes $\mathbf{e}$ in the NTT domain , and ρ is a seed used to generate $\mathbf{A}$. Alice then transmits pk to Bob.

2) Encapsulation phase

Receiving *pk*, Bob begins the encapsulation (Encaps) phase. He regenerates the matrix $\mathbf{A}^T$ (transpose of $\mathbf{A}$) and computes a polynomial $v = \mathbf{t}^T\mathbf{r} + e_2 + \text{Decompress}_q(m,1)$, where $\mathbf{t}$ derives from pk, $\mathbf{r}$ is random coins, e_2 adds noise, and m is a message. The ciphertext is formed as $c = (\mathbf{u}\,||v)$, with $\mathbf{u} = \mathbf{A}^T\mathbf{r} + \mathbf{e}_1$, where $\mathbf{e}_1$ is additional noise. Bob sends c to Alice.

3) Decapsulation phase

Upon receiving c, Alice enters the decapsulation (Decaps) phase. She uses her private key, computing $v - \mathbf{s}^T \cdot \mathbf{u}$ to recover the original message m. Under the CCA security model, Alice recomputes the ciphertext c' using pk and m, comparing it with c. A match confirms the shared secret key's integrity.

III. ATTACK SCHEME

This attack scheme targets the Number Theoretic Transform unit, a critical component of CRYSTALS-Kyber's polynomial multiplication, by manipulating the FPGA bitstream to alter the behavior of Block RAM (BRAM) units. In Kyber, the NTT relies on BRAM to store the 128 twiddle factors used in butterfly operations, which are essential for efficient polynomial multiplication. By strategically modifying the bitstream, we tamper with specific BRAM

979-8-3315-3918-4/25 $31.00 © 2025 IEEE

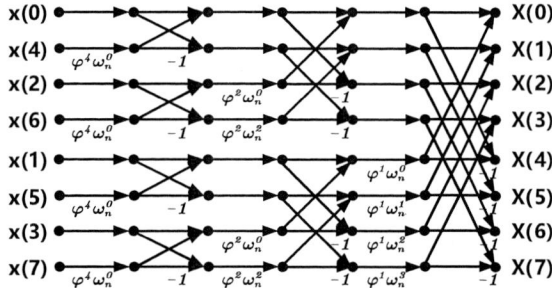

Fig. 1. Diagram of 8-Point NTT with Negative Wrapped Convolution [9]

blocks to set these twiddle factors to zero, effectively disrupting the butterfly computations. Specifically, we alter the bitstream to modify the contents of targeted BRAM units, ensuring that the twiddle factors retrieved during NTT execution are nullified. This tampering disrupts the butterfly operations, which Kyber organizes into odd-even groupings to enhance computational efficiency. The diagram of 8-Point NTT with Negative Wrapped Convolution is shown in Fig. 1. Consequently, the NTT output depends solely on the first and second polynomial coefficients, significantly reducing the transform's complexity. As a result, both the secret key generated during the key generation phase and the random coins r produced in the encapsulation phase become highly predictable, severely weakening Kyber's security and potentially enabling an attacker to deduce the shared key K with minimal computational effort.

A. Attacks on Key Generation phase

In Kyber's hardware design, the storage spaces for NTT twiddle factors, INTT (Inverse NTT) twiddle factors, and point-wise multiplication (PWM) intermediate data are typically independent, ensuring that this attack on the NTT does not affect other components like PWM or INTT during key generation. Consequently, the attack does not impact Bob's side, as Bob performs the encapsulation phase using the received public key pk*, which appears valid since the matrix $\widehat{\mathbf{A}}$ is generated independently and remains unaffected. An attacker, Eve, can exploit this by intercepting the ciphertext c $= (\mathbf{u} \,\|v)$ sent by Bob, where $\mathbf{u} = \mathbf{A}^{\mathrm{T}}\mathbf{r} + \mathbf{e_1}$ and $v = \mathbf{t}^{\mathrm{T}}\mathbf{r} + \mathbf{e_2} + \mathrm{Decompress_q}(m,1)$. Using the limited set of possible secret keys $\mathbf{s}$, Eve can compute $\widehat{\mathbf{t}} - \widehat{\mathbf{A}} \circ \widehat{\mathbf{s}} = \widehat{\mathbf{e}}$. The correct $\mathbf{s}$ is identified when the resulting $\mathbf{e}$ falls within the expected range of the centered binomial distribution (CBD), accounting for potential offsets due to encoding and decoding operations. With the correct $\mathbf{s}$ and ciphertext c, Eve can then derive the shared key K by replicating Alice's decapsulation process.

However, this bitstream modification attack on the key generation phase typically impacts the decapsulation phase as well, due to architectural resource sharing in most practical implementations. During decapsulation, Alice computes m = $\mathrm{Compress_q}\left(v - \mathrm{INTT}\left(\widehat{\mathbf{s}}^{\mathrm{T}} \circ \mathrm{NTT}(\mathbf{u})\right), 1\right)$, which relies on the NTT unit. If the NTT is tampered with, the computation of NTT($\mathbf{u}$) is disrupted, preventing Alice from recovering the message m correctly and causing the verification of the shared key K to fail. Therefore, this attack is only feasible when the NTT units for key generation and decapsulation are implemented independently. Only under this architectural condition can an attacker selectively compromise the key generation process while leaving the decapsulation path intact thus enabling Alice to accept the shared key as valid. From Alice's perspective, the limited set of possible secret keys $\mathbf{s}$

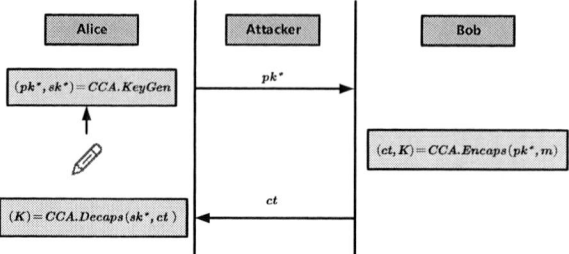

Fig. 2. Attack on CCA-Secure KEM with Weak Key Generation

makes it feasible to detect the attack by inspecting the internal $\mathbf{s}$ values, as their reduced entropy deviates significantly from the expected distribution, providing a clear signal of potential tampering or unauthorized access.

However, Bob cannot detect the attack, as the received pk* appears legitimate due to the unaffected generation of matrix $\widehat{\mathbf{A}}$. This asymmetry enables the attacker to operate covertly on Alice's side while exploiting the weakened key generation process to recover the shared key K by intercepting the ciphertext c sent by Bob. The entire process is illustrated in Figure 2.

B. Attacks on Encapsulation phase

The attack targets the Encapsulation phase of Kyber, focusing on the NTT unit to disrupt the computation of NTT(r), while leaving other operations, such as INTT and PWM, unaffected. Eve tampers with the NTT unit by setting the twiddle factors to zero, simplifying $\widehat{\mathbf{r}}$ and affecting $\mathbf{u}$. Similar to the attacks on key generation phase, Eve recovers $\mathbf{r}$ by solving: $\mathbf{u} - \mathrm{INTT}(\widehat{\mathbf{A}}^{\mathrm{T}} \cdot \widehat{\mathbf{r}}) = \mathbf{e_1}$. With $\widehat{\mathbf{r}}$, $\widehat{\mathbf{t}}$, and v, Eve computes the message m: m = $\mathrm{Decompress_q}\left(v - \mathrm{INTT}(\widehat{\mathbf{t}}^{\mathrm{T}} \circ \widehat{\mathbf{r}}) - \mathbf{e_2}, 1\right)$. The small range of $\mathbf{e_2} \sim \mathrm{CBD_{n_2}}$ ensures it does not impact m's decoding, a clever aspect of the attack that eliminates the need for $\mathbf{e_2}$'s exact value. However, recovering $\mathbf{r}$ from v, $\mathbf{t}$, and m alone, without $\mathbf{e_2}$, is an intractable Learning With Errors problem, highlighting the necessity of using $\mathbf{u}$ to obtain $\mathbf{r}$. Eve then derives Bob's shared key K* using m and pk via a key recovery process.

If Eve sends the tampered ct* directly to Alice, it fails verification during Alice's Decapsulation phase—not because her decapsulation process is affected, but because the shared key derived by Alice does not match Bob's. During decapsulation, Alice computes: m = $\mathrm{Compress_q}\left(v - \mathrm{INTT}\left(\widehat{\mathbf{s}}^{\mathrm{T}} \circ \mathrm{NTT}(\mathbf{u})\right), 1\right)$. Using the tampered ct* = $(\mathbf{u} \,\| v)$, Alice can recover m correctly due to the unaffected $\mathbf{u}$. However, Alice uses m to derive the shared key K, which differs from Bob's K* because Bob's ct* was generated with a tampered NTT($\mathbf{r}$), while Alice's decapsulation assumes an uncompromised process. This mismatch in shared keys causes the verification to fail. To maintain covertness, Eve employs a man-in-the-middle strategy. Eve intercepts Bob's ct*, extracts m, and generates a valid ciphertext ct_1 using the same m without NTT tampering. Eve sends ct_1 to Alice, whose uncompromised decapsulation succeeds, yielding a shared key K_1. This results in Eve possessing two shared keys: K* for communication with Bob and K_1 for communication with Alice. The attack process is illustrated in Figure 3, showing Eve's interception of ct*, recovery of m, and reconstruction of ct_1 to deceive Alice while communicating with Bob using the tampered ct*. Bob seems unaware of the tampering, as the generated ct* and K* appear random.

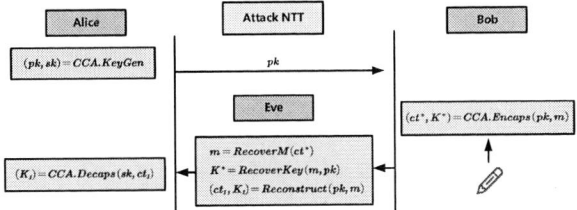

Fig. 3. Attack on CCA-Secure KEM with Weak Encapsulation

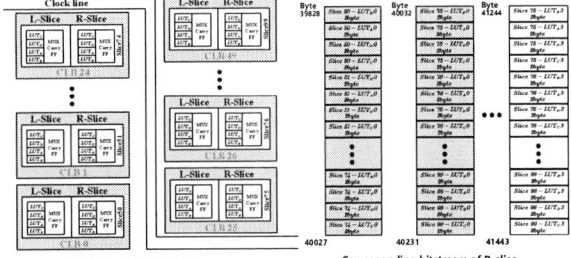

Fig. 4. Bitstream Configuration Structure

IV. BITSTREAM INFORMATION EXTRACTION AND MODIFICATION

The process of extracting and modifying bitstream information in an FPGA, specifically the Artix-7 xc7a35, enables targeted manipulation of cryptographic primitives such as the Number Theoretic Transform twiddle factors stored in BRAM. The Artix-7 xc7a35 FPGA is configured with 404 bytes per slice, accommodating 50 slices per block, as illustrated in the figure 4. For more details, please refer to Reference [10], [11]. The bitstream data exhibits a structured regularity, which can be leveraged to identify and alter specific memory regions, including those containing the NTT twiddle factors.

A. Automated Bitstream Generation for Mapping Analysis

To investigate the mapping relationship between a 12-bit wide and 128-deep twiddle factor storage space and the bitstream, we propose an automated bitstream generation approach. BRAM can be initialized using Coefficient (COE) files. We will create two sets of COE files to explore the horizontal and vertical mapping regularities independently.

Set A COE files will be designed such that only the memory address 0 contains a non-zero 12-bit value. Furthermore, within this 12-bit value, only a single bit will be set to '1'. This set will comprise 12 unique COE files, each with a different bit position set to '1', allowing us to analyze the horizontal mapping pattern. Set B COE files will be configured such that only the least significant bit of the 12-bit width may be set to '1'. Across the 128 memory depth locations, only one address will contain a non-zero value. This set will consist of 32 unique COE files, enabling the investigation of the vertical mapping pattern. Using these specifically crafted COE files, we can automate the process of generating different bitstream files within Vivado. This automation will be achieved using a Tool Command Language (TCL) script available at [12].

B. Structural Analysis and Twiddle Factor Identification

The NTT twiddle factors require a 12-bit width and a 128-depth storage space within the bitstream, as depicted in Figure 5 (A). This configuration is mapped across the Configurable Logic Blocks (CLBs), with each 404-byte block representing a slice configuration. To extract the mapping pattern, we analyze instances where the same address contains varying positions of '1' bits. This approach, as shown in Figure 5 (A), allows us to derive the 12-bit width mapping regularity by enumerating all possible combinations of '1' positions.

Additionally, the depth mapping regularity can be determined by observing changes in the '1' bit positions across the 128-depth dimension. These patterns, visible in the bitstream differences highlighted in Figure 5 (B), enable us to trace how depth variations correspond to sequential shifts in the bitstream. Thus, the twiddle factors such as 1, 17, 289, and 1584 can be mapped to their target sequences within the bitstream by applying these identified width and depth transformation rules.

C. Modification of Twiddle Factors

Once the target sequences are located, the modification process involves locking the position of each twiddle factor in the bitstream and setting the corresponding 12-bit values to zero. Experimental analysis reveals that every four twiddle factors are grouped together within the bitstream. Given the 128 twiddle factors required for the NTT, this necessitates repeating the zeroing operation 32 times. After modifying one group of four twiddle factors, the next group to be modified is found 404 bytes ahead in the bitstream, matching the FPGA's slice layout. During the mapping of the 12-bit wide data at the same address of BRAM, some sections might be skipped and not mapped, possibly due to configuration data or interference in the mapping process. To address this uncertainty, an approach is employed: all possible positions of these skipped sections are enumerated, and all the resulting candidate sequences are searched within the bitstream. This exhaustive method ensures accurate identification of the target locations. Following the alteration of the twiddle factors, the bitstream's Cyclic Redundancy Check (CRC) must be updated to maintain integrity and prevent detection. The CRC modification can be performed using the Python script available at [13].

V. COUNTERMEASURES

To bolster the security of Kyber KEM against bitstream manipulation targeting its Number Theoretic Transform, a multi-faceted approach is crucial. Beyond basic entropy checks, employing statistical analysis on the randomness of the secret key and the distribution of the random polynomial r can detect subtle anomalies. Furthermore, robust twiddle factor integrity verification through multi-level hashing and cryptographic commitments adds a strong layer of defense. Obfuscation techniques, such as dynamic randomization of twiddle factor placement and the insertion of data-dependent dummy factors, significantly hinder reverse engineering efforts. Integrating these advanced countermeasures with secure boot, memory protection, and continuous monitoring creates a more resilient system against tampering.

VI. CONCLUSION

This paper investigated bitstream modification attacks on the CRYSTALS-Kyber key encapsulation mechanism implemented on FPGAs. By demonstrating how manipulating the Number Theoretic Transform unit's twiddle factors can nullify its function, we exposed significant vulnerabilities in both key generation and encapsulation, potentially leading to predictable secret keys and man-in-the-middle attacks. We detailed the bitstream extraction and modification process on a Xilinx Artix-7 xc7a35 FPGA, including twiddle factor

979-8-3315-3918-4/25 $31.00 © 2025 IEEE

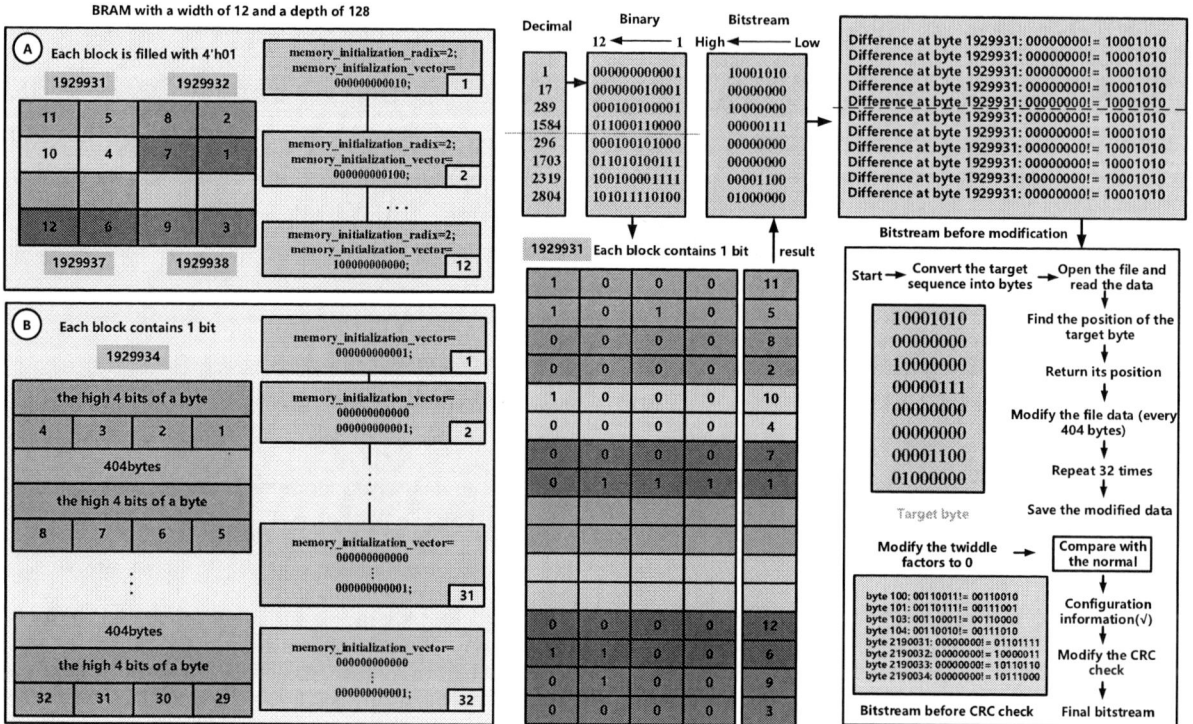

Fig. 5. Detailed Process of Bitstream Extraction and Modification for NTT Attack

identification and necessary CRC updates for covertness. Finally, we proposed potential countermeasures to enhance the resilience of Kyber implementations against these emerging hardware-level threats, emphasizing the need for robust security measures ensuring both algorithmic strength and hardware implementation integrity.

ACKNOWLEDGMENT

This work is supported by the National Natural Science Foundation of China (No. 62134002), the Postdoctoral Fellowship Program of CPSF under Grant Number GZC20240541, the Fund of National Key Laboratory of Multispectral Information Intelligent Processing Technology (Grant No. 202410487401), the Open Fund of the State Key Laboratory of Spintronics Devices and Technologies (Grants No. SPL-2406), the Innovation Project of JinYinHu Laboratory. (Grant No. 2024JYH011401).

REFERENCES

[1] P. W. Shor, "Polynomial-time algorithms for prime factorization and discrete logarithms on a quantum computer," SIAM Review, vol. 41, no. 2, pp. 303-332, 1999.J. Clerk Maxwell, A Treatise on Electricity and Magnetism, 3rd ed., vol. 2. Oxford: Clarendon, 1892, pp.68–73.

[2] L. K. Grover, "A fast quantum mechanical algorithm for database search," in 28th ACM Symposium on Theory of Computing, pp. 212-219, 1996.

[3] National Institute of Standards and Technology, "Post-Quantum Cryptography FIPS Approved," NIST Computer Security Resource Center (CSRC), Aug. 13, 2024.

[4] R. Avanzi, J. Bos, L. Ducas, E. Kiltz, T. Lepoint, V. Lyubashevsky, J. M. Schanck, P. Schwabe, G. Seiler, and D. Stehlé, "CRYSTALS-Kyber algorithm specifications and supporting documentation," NIST PQC Round 3, 2020.

[5] P. Ravi, A. Chattopadhyay, J. P. D'Anvers, and A. Baksi, "Side-channel and fault-injection attacks over lattice-based post-quantum schemes (kyber, dilithium): Survey and new results," ACM Transactions on Embedded Computing Systems, 2022.

[6] P. Ravi, D. B. Roy, S. Bhasin, A. Chattopadhyay, and D. Mukhopadhyay, "Number 'not used' once-practical fault attack on pqm4 implementations of NIST candidates," in Constructive Side-Channel Analysis and Secure Design: 10th International Workshop, COSADE 2019, Darmstadt, Germany, April 3-5, 2019, Proceedings 10, pp. 232-250, Springer, 2019.

[7] M. Moraitis, "FPGA Bitstream Modification: Attacks and Countermeasures," in IEEE Access, vol. 11, pp. 127931-127955, 2023.

[8] M. Einder, A. Moradi, and C. Paar, "The unpatchable silicon: A full break of the bitstream encryption of Xilinx 7-series FPGAs," in 29th USENIX Security Symposium (USENIX Security 20), pp. 1803-1819, 2020.

[9] W. D. Zhao, "Yingyong yu gemicheng de ke chonggou shulun bianhuan yingjian shixian yanjiu [Research on Reconfigurable Number Theoretic Transform Hardware Implementation for Lattice Cryptography]," Master's thesis, Huazhong University of Science and Technology, Hubei, China, 2021.

[10] M. Jeong, J. Lee, E. Jung, Y. H. Kim and K. Cho, "Extract LUT Logics from a Downloaded Bitstream Data in FPGA," 2018 IEEE International Symposium on Circuits and Systems (ISCAS), Florence, Italy, 2018, pp. 1-5.

[11] Xilinx, "7 Series FPGAs Configuration User Guide." Vol. 1.11. September 2016.

[12] L. Chen, "Bitstream automation TCL script," GitHub repository, 2025. [Online]. Available: https://github.com/chenleigood/bitstream_auto.tcl [Accessed: May 9, 2025].

[13] B. Gardner, "Python module for FPGA bitstream parsing and CRC computation," GitHub Gist, 2025. [Online]. Available: https://gist.github.com/bggardner [Accessed: May 10, 2025].

BIND: A Batch Cache-Invalidation Framework Based on Doorbell Mechanism

Jialin Liu[*1], Zhiyuan Zhang [1], Chao Fu[†2], Jun Han[†*1]

[1] State Key Laboratory of Integrated Chips and Systems, Fudan University
[2] Shao-Chip Laboratory   †Corresponding authors

* Email: jialinliu23@m.fudan.edu.cn, junhan@fudan.edu.cn

Abstract—With the growing demand for processing ability of network applications in server chips for Software-Defined Data Plane (SDP) applications, traditional cache architectures are facing challenges in efficient strategies of managing I/O data allocation. Previous optimizations for I/O invalidation in deep packet processing (DPP) introduce increased latency and limits the system bandwidth and CPU utilization efficiency. To address this issue, this paper proposes BIND, a batch cache-invalidation framework based on a doorbell mechanism, which enables the CPU to invalidate sequentially addressed cachelines with a single instruction, significantly reducing the latency of invalidation operations. By implementing BIND's key structures on an FPGA and modeling the entire framework in gem5, we evaluate the performance of two selected DPP tasks across varying packet sizes using gem5-FPGA co-simulation. Experimental results demonstrate that BIND delivers peak performance improvements of 119% in throughput and a maximum latency reduction of 61% compared to state of art.

Keywords—SDP, DPDK, Cache, CMO, FPGA

I. INTRODUCTION

Network Function Virtualization (NFV) has increasingly been adopted by network operators due to its flexibility, and many practices have fully demonstrated the economic efficiency and high performance of NFV solutions in large-scale data processing [1] [2]. Software Data Plane (SDP) plays a key role in NFV performance. High-performance SDP imposes new requirements on server chips for stable network I/O and high-throughput capability, which traditional Linux kernel network stacks can no longer meet.

Many prior works have optimized network I/O frameworks, such as software solutions like DPDK [3] that can bypass the OS kernel. Moreover, IDIO [6] and Hyperdata [7] have proposed optimizations for the cache system in deep packet processing (DPP) scenarios, employing key techniques such as data prefetching and I/O buffer invalidation. IDIO directly invalidates packet buffers that will no longer be accessed using Cache Maintenance Operation (CMO) instructions, thereby reducing redundant write-backs. However, IDIO iteratively applies CMO instructions to invalidate each cacheline individually when processing large packets, leading to increased latency and ultimately constraining system bandwidth and CPU efficiency.

To address this issue, this paper proposes BIND, a batch cache-invalidation framework based on doorbell mechanism, which allows the CPU to invalidate multiple sequentially addressed cache blocks with a single instruction, significantly reducing invalidation latency. Our main contribution contains:

- We propose a batch optimization based on doorbell mechanism for those DPP tasks needing cache invalidation, which significantly improve the average packet latency and overall throughput.

- We propose a set of APIs in DPDK and enhancement to gem5 modeling to support the doorbell mechanism and batch cache-invalidation operation.

- We conduct comprehensive experiments entailing performance and area consumption based on FPGA implementation and gem5-FPGA co-simulation.

II. BACKGROUND

A. Data Plane Development Kit

Data Plane Development Kit is an open-source software library specifically designed to enhance high-speed packet processing capabilities for data plane applications. Through key optimizations including the Environment Abstraction Layer, memory pools, ring queues, and Poll Mode Drivers, DPDK bypasses the Linux kernel during packet transmission and reception [3]. This architecture eliminates the overhead of interrupts and context switching, significantly improving packet processing performance and making DPDK widely adopted in SDP.

B. Fetch-Drop Packet Processing Model

Network forwarding tasks can be classified into shallow packet processing (SPP) and deep packet processing based on whether they require access to packet payloads. Unlike SPP tasks such as L2 forwarding that only process packet headers, DPP applications must examine the packet payload to implement their specialized functions. Typical DPP applications (e.g. traffic monitors in network security systems) analyze packet contents to verify compliance with security policies and identify potential attacks, subsequently discarding the packets. Another important category is key-value store (KVS) services that parse payloads to extract lookup keys, then allocate new packet buffers for response data while releasing the original request buffers. Their common behavior can be abstracted as a fetch-drop model, where "fetch" represents payload access operations and "drop" denotes buffer release after processing without forwarding the payload. In our experimental evaluation, we use Fetch-Drop and micro-KVS implemented with DPDK as benchmarks to validate the effectiveness of BIND.

C. IO Buffer Invalidation

Fig. 1 shows the typical I/O buffer movements among memory system in Fetch-Drop tasks. IO buffers are first allocated at last-level cache (LLC) DDIO ways [4] after DMA write ❶❷❸. Then the buffers are fetched by CPU and brought into the mid-level cache (MLC) ❹. Considering a typical DPDK NIC ring size of 1024 and the working set exceeding MLC capacity (typically 1MB) when processing 1514-byte packets (1514*1024≈1.5M), eviction of IO buffer-related cachelines to LLC becomes inevitable. Thus, the IO buffers either remain in the MLC or are evicted to

LLC/DRAM (**5**-a, **5**-b) due to cache capacity limitation, between processor access and subsequent NIC refills. These write-backs stress the coherent bus and exacerbate the contention for shared LLC resources. To address this issue, IDIO proposes invalidating IO buffer-related cachelines after packet processing to resolve resource contention. This invalidation is implemented using ARM's CMO instruction *DC IVAC* in IDIO, which invalidates specified cachelines without requiring dirty line write-back to lower-level caches or DRAM (**5**-c). This approach perfectly aligns with Fetch-Drop-type tasks since released packet data will never be accessed again.

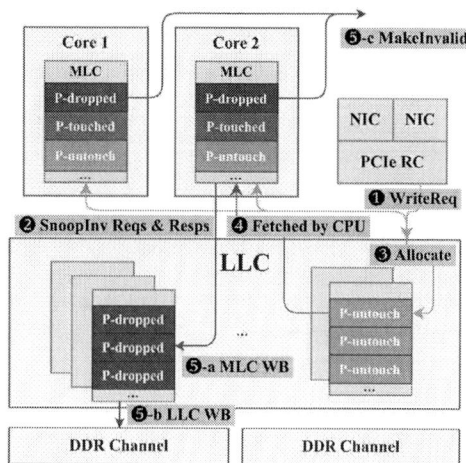

Fig. 1. Packet Buffer Movement Among Memory System

D. Limitations

However, IDIO's current implementation employs a loop to execute cache-invalidation instructions for each cache block to be dropped (Loop-Invalidation). For a 1514-byte ethernet packet, this can lead to 24 invalidation operations, severely constraining the overall system performance. Our experimental results show that even though Loop-Invalidation can avoid the throughput degradation caused by increased write-backs when the system bus pressure increases (from 512-byte to 1024-byte), it results in a bandwidth loss of up to 60% (under 1-core 1514-byte) when the bus pressure is low.

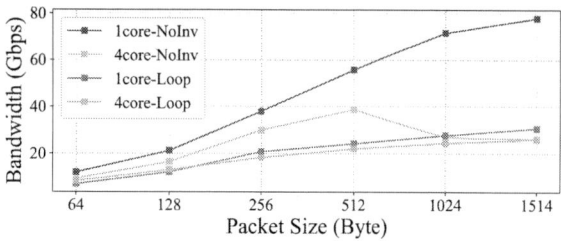

Fig. 2. Per-core bandwidth degradation of Loop-Invalidation

While existing instructions like ARM's *DC ISW* can invalidate cache at way or set granularity, they are unsuitable for high-speed networking scenarios where we only need to invalidate specific data structures. Designing a custom instruction may meet our requirements, however, it introduces unnecessary microarchitecture-wise complexity. Hence, there exists a critical need for in-cache high-efficiency IO buffer invalidation mechanism that can handle large network packets without the great performance penalties.

III. DESIGN

BIND is a hardware-software cooperative mechanism that enables batch invalidation of contiguous cache blocks through a single instruction, significantly improving system performance for Fetch-Drop-type tasks. Fig. 3 shows both the main architecture and typical workflow of BIND.

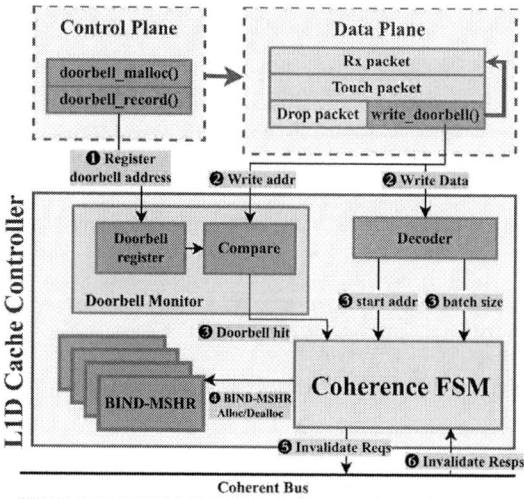

Fig. 3. Architecture of BIND framework

A. Software

During initialization, the Fetch-Drop application should allocate a dedicated doorbell for each worker thread. Each doorbell's address should then be registered in the L1 data (L1D) cache controller of the corresponding CPU core. To achieve this, we implement the following APIs:

- **doorbell_malloc:** Allocates a cacheline-aligned (e.g. 64B) memory region using DPDK's API to prevent false sharing. The returned virtual address serves as the doorbell address.

- **doorbell_record:** Retrieves the physical address of the doorbell via DPDK's API and programs this address into the in-cache register using architecture-specific methods **①** (e.g. ARM's mcr instruction with CP15 co-processor [5]).

At runtime, the DPDK program executes a loop that sequentially receives packets, touches them, and then discards them. The packet-discarding part is implemented as:

- **write_doorbell:** This API encodes the starting physical address and batch size into a 64-bit value, where bits [63:6] stands for starting physical address and bits [5:0] stands for batch size. This design enables BIND to invalidate up to 63 contiguous cache lines. Then it triggers the doorbell by writing the encoded value to the doorbell address, initiating a batch invalidation **②**.

B. Hardware

The L1D cache should monitor incoming write requests and determine whether the doorbell is triggered. Upon detecting a doorbell trigger, the cache needs to fetch the request address and batch size for batch invalidation **③**. Given that a maximum-sized ethernet packet (e.g. 1514-byte) spans at least 24 cache lines—far exceeding the typical MSHR capacity in L1D cache—we propose the following architectural enhancements to MSHRs for BIND.

- **Doorbell Register:** A 64-bit register storing the doorbell address, with one instance for each L1D cache.

- **Doorbell Monitor:** Monitors write requests from the CPU core and compares the request address with the doorbell register's value to determine if a doorbell has been triggered.

- **BIND-Decoder:** Extracts the starting address and batch size from BIND-encoded write data.

- **BIND-MSHR:** Efficiently tracks multiple contiguous cachelines by enhancing the traditional MSHR with two new registers: *a) Batch Mode register* (1-bit) is set when the batch mode is enabled, otherwise it is unset and *b) Pending Counter register* (6-bit) tracks the status of a batch invalidation transaction. The counter is initialized to the batch size when a BIND-MSHR is newly allocated ❹ and decrements by one upon receiving a invalidate response ❻. Transaction completes when the counter reached zero, indicating all responses have been received, and the BIND-MSHR is then deallocated.

IV. METHODOLOGY

In this section, we focus on the hardware feasibility and the overhead of BIND. We implement BIND-MSHR on a Virtex-7 HTG-710 FPGA and use gem5 simulator for full-system (FS) modeling. The Chimera co-simulation framework [8], which combines gem5 and FPGA, is used for validation and evaluation.

A. FPGA-Based Hardware Prototyping

The FPGA implementation, as shown in Fig. 4, integrates XDMA for PCIe communication, an AXI bridge for request routing, asynchronous input/output buffers implemented as FIFOs to handle clock domain crossing between the fixed-frequency XDMA interface and custom logic, and BIND-compliant MSHRs for cache coherence management. Incoming gem5 requests are received by XDMA and undergo AXI address filtering. Non-coherence operations such as PCIe initialization writes are filtered out, while valid requests are buffered in the asynchronous input FIFO. Upon receiving a request, the BIND-MSHR module first identifies the gem5 trigger type, then performs corresponding updates to MSHR entries. Any necessary responses are generated and queued in the output FIFO for gem5 to retrieve.

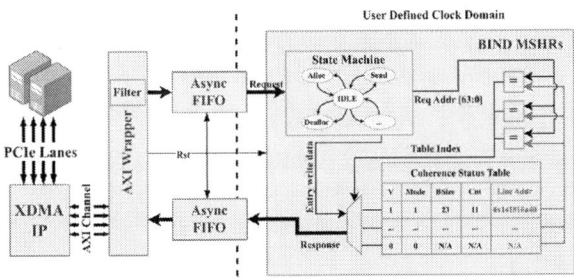

Fig. 4. FPGA prototype of BIND-MSHR

TABLE I. presents the post-implementation FPGA resource utilization comparison between the MSHR prototype and BIND (with 8 entries). While BIND demonstrates

marginally higher register consumption compared to the prototype—attributed to the integration of Batch Mode and Pending Counter registers—this overhead remains negligible in practical implementation scenarios.

TABLE I. FPGA UTILIZATION OF 8-ENTRY BIND-MSHR

Resource	Utilization	
	Prototype	*BIND*
LUT	140179	137895
F7 MUX	28054	27421
Registers	214970	215034

B. gem5 Simulation Platform

We modify the cache controller, MSHR and the coherent crossbar logic in gem5 to align with the architectural specifications outlined in Section III. To integrate with the Chimera co-simulation framework, we also implement FPGA triggers in the following critical MSHR operation:

- MSHR allocation and deallocation
- Coherent request generation based on MSHR states
- Other MSHR table read/write operations

After triggering the FPGA, gem5 pauses if responses are needed to verify correctness or continues simulation otherwise. Full-System mode is used to boot a Linux OS on gem5, enabling native execution of DPDK applications within the simulated network environment. The key configuration parameters of gem5 are shown in TABLE II.

TABLE II. KEY CONFIGURATION PARAMETERS OF GEM5

Parameters	Values
Core ISA, freq, Superscalar	aarch64, 2GHz, 8
ROB/IQ/LQ/SQ entries	384/128/128/128
I/D/L2/L3 (total size, assoc)	32KiB, 2/64KiB, 8/1MiB, 8/10MiB, 20
I/D/L2/L3 (latency, MSHRs)	2CC, 4/2CC, 4/5CC, 32/24CC, 128
DRAM size	4 Channels, 16GiB DDR3 1600
DDIO ways	2 ways in LLC
Operating system	Linux 6.1.57
Network HW	Intel Gigabit NIC e1000
DPDK configuration	Ver. 20.11.3, 1024-entry RX queue

We use a hardware load generator in gem5 [9] to send traffic to the system under test. In bandwidth testing, we measure whether the system experiences packet loss at different transmission rates to estimate the saturated bandwidth. For latency testing, we use load generator to measure the loop-back latency for each packet under different packet-transmission rate.

C. Benchmark

- **Fetch-Drop:** This benchmark simulates deep packet processing tasks that require packet dropping. Upon receiving packets, it iteratively reads 8-byte segments of the payload until reaching the end, then discards the packet buffer using either Loop-Invalidation or Batch Invalidation.

- **Micro-KVS:** A memory-intensive benchmark that simulates an in-memory key-value store service. Each core allocates a configurable-sized table. For each received packet, it performs two hash computations to obtain the table index and the number of rehash iterations, then read/write the indexed entry. The next index is derived by rehashing the current index.

V. EXPERIMENTAL RESULTS

A. Fetch-Drop Bandwidth

In the Fetch-Drop bandwidth test as shown in Fig. 5, the BIND framework achieves a 119% higher bandwidth than Loop-Invalidation under single-threaded 1514-byte traffic. Larger packet size brings better performance gain because BIND's batch-processing mechanism significantly reduces the overhead of CMOs by consolidating multiple high-latency invalidation instructions into a single.

However, when four cores simultaneously process 1514-byte packets, the memory subsystem becomes a bottleneck, leading to increased latency during packet payload fetching. This diminishes the relative speedup of BIND during drop-phase, resulting in a stagnant benefit ratio, yet maintains a 30% performance advantage over Loop-Invalidation.

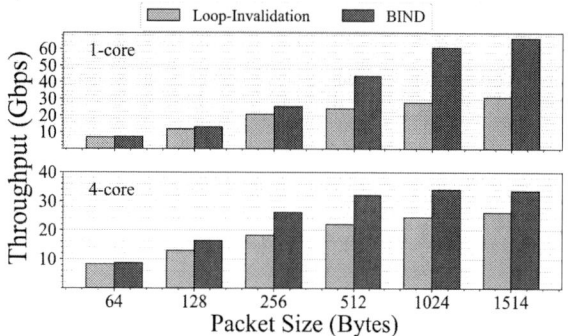

Fig. 5. Bandwidth of Fetch-Drop Benchmark

B. Micro-KVS Bandwidth & Latency

In the Micro-KVS benchmark as shown in Fig. 6, BIND achieves a throughput improvement ranging from 6% to 77%, where the enhancement becomes more pronounced as the packet size increases. BIND also reduces average packet processing latency as shown in Fig. 7 by 35% under 256-byte heavy-load traffic and 61% under 1514-byte light-load traffic. Under light-load traffic, the BIND framework demonstrates increasing performance gains as the packet size grows. However, under heavy-load traffic, the memory subsystem becomes the bottleneck and BIND cannot provide such performance gain as under light-load traffic.

It's noteworthy that the latency curve exhibits a slight decline at low packet-transmission rates due to DPDK's TX path batching, where TX queues are less frequent to reach the batch size under light load and introduce additional waiting latency for packet's transmission.

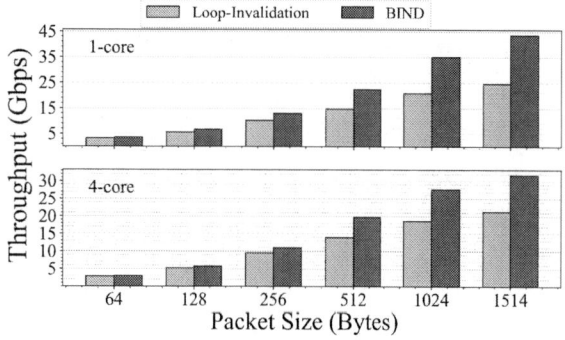

Fig. 6. Bandwidth of Micro-KVS Benchmark

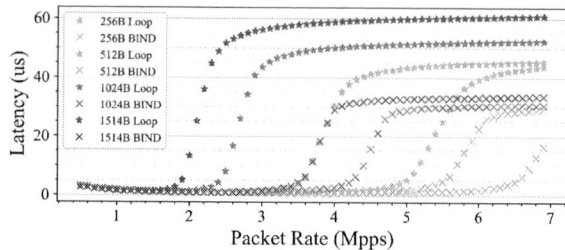

Fig. 7. Latency of Micro-KVS Benchmark under 256-byte, 512-byte, 1024-byte and 1514-byte

VI. CONCLUSION

The processing capabilities and stability requirements of SDP for DPP tasks are increasingly stringent. Prior work introduced Loop-Invalidation to eliminate redundant write-backs to the LLC, but this approach introduced performance bottlenecks. To overcome this limitation, we propose BIND, a hardware-software co-optimized batch cache-invalidation framework that mitigates the performance degradation caused by Loop-Invalidation. BIND not only addresses the inefficiencies of Loop-Invalidation but also introduces a novel approach to scalable, low-latency cache management in next-generation network data planes.

ACKNOWLEDGEMENT

This work was supported by the National Natural Science Foundation of China under Grant 61934002 and 62234008.

[1] J. de J. Gil Herrera and J. F. Botero Vega, "Network Functions Virtualization: A Survey," IEEE Latin America Transactions, vol. 14, no. 2, pp. 983–997, Feb. 2016.

[2] K. Kaur, V. Mangat, and K. Kumar, "A comprehensive survey of service function chain provisioning approaches in SDN and NFV architecture," Computer Science Review, vol. 38, p. 100298, Nov. 2020.

[3] Intel Corporation. Data Plane Development Kit.https://www.intel.com /content/www/us/en/developer/topictechnology/networking/dpdk.htm l.

[4] Intel Corporation. "Intel Data Direct I/O Technology (Intel DDIO): A Primer". https://www.intel.com/content/www/us/en/io/data-direct-i-o-technology-brief.html

[5] ARM Corporation. CP15. https://developer.arm.com/documentation/d en0013/d/ARM-Processor-Modes-and-Registers/Registers/Coprocess or-15

[6] M. Alian et al., "IDIO: network-driven, inbound network data orchestration on server processors," IEEE/ACM International Symposium on Microarchitecture (MICRO), Oct. 2022, pp. 480–493.

[7] H. Golestani and T. F. Wenisch, "HyperData: a data transfer accelerator for software data planes based on targeted prefetching," in 2021 IEEE 39th International Conference on Computer Design (ICCD), Storrs, CT, USA: IEEE, Oct. 2021, pp. 326–334.

[8] C. Fu, Z. Wang, and J. Han, "Chimera: A co-simulation framework combining with gem5 and FPGA platform for efficient verification," in 2024 34th International Conference on Field-Programmable Logic and Applications (FPL), Sep. 2024, pp. 133–139.

[9] J. Umeike, S. Agarwal, N. Lazarev, and M. Alian, "Userspace Networking in gem5," in 2024 IEEE International Symposium on Performance Analysis of Systems and Software (ISPASS), May 2024, pp. 179–191.

High-Throughput Multiplier-Free FPGA Implementation for Pure-Number Discrete Fractional Complex Hadamard Transform

Chengqi Zhao[1], Zi-chen Fan[2], Shan Cao[1*], Susanto Rahardja[3]

[1]School of Communication and Information Engineering, Shanghai University, Shanghai 200444, China
[2]School of Marine Science and Technology, Northwestern Polytechnical University, Xi'an 710072, China
[3]Infocomm Technology Cluster, Singapore Institute of Technology, Singapore 138683.
*Email: cshan@shu.edu.cn

Abstract—With the growing demand for efficient real-time signal processing, the Walsh-Hadamard Transform (WHT) is valued for its hardware efficiency, but lacks flexibility. The Fast Fractional Hadamard Transform (FHT) addresses this with tunable orders, yet introduces costly complex multipliers. This paper explores the Pure-Number Fractional Complex Hadamard Transform (PN-FCHT), a multiplier-free extension, and proposes a resource-efficient 1024-point FPGA architecture tailored for real-time signal processing. A fully multiplier-free butterfly unit is designed, replacing complex multiplications with sign-controlled additions. In parallel, a dual-data-group pipelined scheme is adopted to improve throughput. Implemented on Xilinx Virtex-7, the design reduces DSP usage by 86.4%, BRAM by 75%, and LUT by 4.7% compared to the baseline Fast-FHT. Operating at 285.38 MHz, it achieves a throughput of 135.63 M/s, 4.24 times higher than the reference, while preserving 16-bit fixed-point accuracy.

Keywords—*PN-FCHT, FPGA, multiplier-free, butterfly unit, fast transform.*

I. INTRODUCTION

Discrete transforms serve as the cornerstone of modern digital signal processing (DSP), underpinning key functionalities in image compression, wireless communications, and cryptographic systems. Among them, the Walsh-Hadamard Transform (WHT) stands out for its hardware efficiency due to a multiplier-free structure based solely on additions and subtractions [1] [2]. However, WHT's integer-order nature limits its spectral adaptability, making it unsuitable for applications requiring fractional domain analysis. The Discrete Fractional Hadamard Transform (FHT) was then proposed by introducing fractional-order control through eigenvalue modulation [3]. Unfortunately, FHT relies on eigenvalue decomposition, involving numerous complex multiplications that result in high computational complexity and irregular structure, severely limiting its practical deployment on hardware platforms. Subsequent improvements, such as the Pseudo-Fractional Hadamard transform (Pseudo-FHT) [4] and fast algorithms for FHT (Fast-FHT) [5], improved computational efficiency, but inherited the FHT reliance on complex multipliers [6] [7].

The recently proposed Pure-Number Fractional Complex Hadamard Transform (PN-FCHT) [8] offers a paradigm shift by expressing transform coefficients exclusively as ± 1 and $\pm j$ ($j = \sqrt{-1}$). This pure-number property theoretically eliminates complex multiplications, reducing hardware complexity to sign-controlled additions and real-imaginary swapping operations. Such structural simplicity makes PN-FCHT highly promising for low-power, high-throughput applications in embedded signal processing and real-time computing. However, current studies on PN-FCHT are predominantly theoretical, and there is a lack of reported hardware implementations to verify its efficiency when deployed in practical systems.

To bridge the gap from algorithm to hardware, we propose the first complete FPGA architecture for 1024-point PN-FCHT, optimized for high-throughput, low-resource applications. By fully exploiting the pure-number property of PN-FCHT, the design maps efficiently to fixed-point hardware with minimal overhead. Key features include:

- Multiplier-Free Butterfly Unit: All complex multipliers are replaced by sign-controlled additions and real-imaginary swaps using ± 1 and $\pm j$ coefficients, resulting in a fully combinational, fixed-point-friendly structure.
- Dual-Data-Group Pipelining: A parallel pipeline processes two interleaved data groups simultaneously, ensuring continuous data flow and improved stage utilization.

Implemented on a Xilinx Virtex-7 FPGA device, the design achieves 135.63 M/s throughput at 285.38 MHz while consuming only 6 DSPs, 6 BRAMs, and 7682 LUTs. Compared to Fast-FHT, it reduces DSP usage by 86.4% and increases throughput by 4.24 times, making it well suited for real-time, low-power applications such as radar and compressed sensing.

II. PRELIMINARIES OF PN-FCHT

A. Fractional Definition and Decomposition

The PN-FCHT extends the classical Walsh-Hadamard Transform (WHT) to the fractional domain while preserving a pure-number coefficient set, enabling low-complexity implementations in hardware. The 2^n-point PN-FCHT matrix $\mathbf{H}_n^\alpha$ [9] is defined as:

$$\mathbf{H}_n^\alpha = \mathbf{V}_n \widetilde{\mathbf{\Lambda}}_n^\alpha \mathbf{V}_n^H, \qquad (1)$$

979-8-3315-3918-4/25 $31.00 © 2025 IEEE

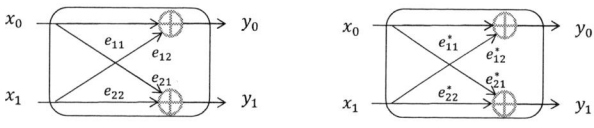

(a)   (b)

Fig. 1. Signal flow diagrams of (a) $\mathbf{V}_1$ and (b) $\mathbf{V}_1^H$.

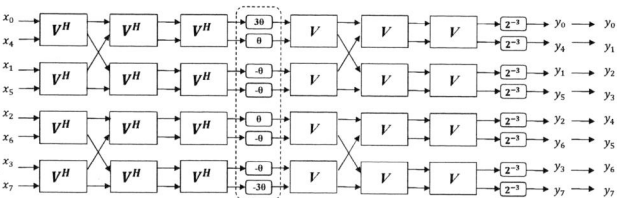

Fig. 2. Data flow diagram of the 8×8 PN-FCHT architecture.

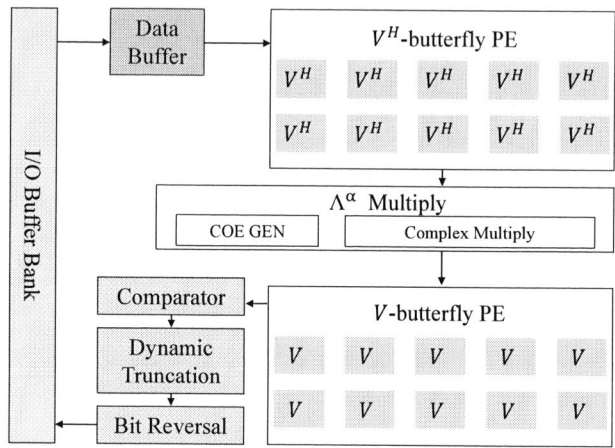

Fig. 3. Overall architecture of the 1024-point PN-FCHT implementation.

where $\widetilde{\mathbf{\Lambda}}_n^\alpha = \frac{1}{2^n}\mathbf{\Lambda}_n^\alpha$ is the normalized eigenvalue matrix, and

$$\mathbf{\Lambda}_n^\alpha = \mathrm{diag}(\lambda_0^\alpha, \lambda_1^\alpha, \ldots, \lambda_{2^n-1}^\alpha) \tag{2}$$

contains eigenvalues raised to the fractional order α. The eigenvector matrix $\mathbf{V}_n$ is recursively constructed using Kronecker products as:

$$\mathbf{V}_n = \prod_{i=0}^{n-1} \left(\mathbf{I}_i \otimes \mathbf{V}_1 \otimes \mathbf{I}_{n-i-1}\right), \tag{3}$$

where the base matrix $\mathbf{V}_1$ is given by:

$$\mathbf{V}_1 = \begin{bmatrix} \mu_1 & \mu_1 \\ -\mu_1\mu_2 j & \mu_1\mu_2 j \end{bmatrix}, \quad \mu_1, \mu_2, \mu_3 \in \{1, -1, j, -j\}. \tag{4}$$

This recursive formulation leads to a multi-stage decomposition framework composed of 2×2 butterfly operations [10]. Such a structure provides algorithmic regularity and enables scalable mapping to parallel or pipelined hardware architectures.

B. Pure-Number Butterfly and Scalable Architecture

The fundamental computation in PN-FCHT is the butterfly unit $\mathbf{V}_1$, with matrix elements e_{ij} ($i, j \in \{1, 2\}$) strictly from $\{\pm 1, \pm j\}$, as illustrated in Fig. 1. This enables multiplier-free implementation using only additions, sign inversion, and real-imaginary swapping [11]. The operation is defined as:

$$\begin{bmatrix} y_1 \\ y_2 \end{bmatrix} = \underbrace{\begin{bmatrix} a & b \\ c & d \end{bmatrix}}_{\mathbf{V}_1} \begin{bmatrix} x_1 \\ x_2 \end{bmatrix}, \quad a, b, c, d \in \{\pm 1, \pm j\}. \tag{5}$$

The structure of the full PN-FCHT arises from Kronecker decomposition:

$$\mathbf{D}_n = \prod_{i=0}^{n-2} \mathbf{I}_i \otimes \mathbf{V}_1^H \otimes \mathbf{I}_{n-2-i}, \tag{6}$$

with each stage comprising parallel $\mathbf{V}_1^H$ units. These stages are modular and regular, allowing unified control and efficient

pipelining. Rotation factors in $\mathbf{V}_1^H$ are encoded using 2-bit values:

$$00 \to +1, \quad 01 \to +j, \quad 10 \to -1, \quad 11 \to -j,$$

which minimizes routing and control overhead.

The butterfly and its Hermitian form follow simple signal paths. Fig. 2 further shows an 8×8 PN-FCHT pipeline, composed of forward $\mathbf{V}_1$ stages, diagonal modulation, and backward $\mathbf{V}_1^H$ stages. Identical butterfly units are reused across layers, supporting straightforward scalability and deep pipelining for real-time FPGA applications.

III. HARDWARE ARCHITECTURE OF PN-FCHT

A. Overview of the Overall Architecture

This paper implements the PN-FCHT transformation with a scale of 1024 points and supports uninterrupted processing of two consecutive 1024-point data groups. The overall architecture consists of a data buffer, a 10-stage $\mathbf{V}$-butterfly network, an eigenvalue multiplication unit, a 10-stage $\mathbf{V}^H$-butterfly network, followed by a dynamic most significant bit (MSB) truncation unit and a bit-reversal unit. The complete hardware design framework is shown in Fig. 3.

The system adopts a fully pipelined structure to enable continuous high-throughput processing. The input data are first buffered and then fed into the 10-stage $\mathbf{V}^H$-butterfly network for the forward transform. The intermediate results are modulated by a diagonal eigenvalue matrix, where only 11 unique eigenvalues are stored and cyclically accessed from buffer. The data is then processed by the 10-stage $\mathbf{V}$-butterfly network to complete the inverse operation. The dynamic MSB truncation unit adaptively retains 16 effective bits by detecting frame-wise peaks, whereas the bit-reversal unit restores the natural output order for downstream compatibility.

B. Dual-buffer Ping-Pong Scheme

To ensure continuous high-throughput execution of butterfly operations, a dual-port memory system is implemented using

979-8-3315-3918-4/25 $31.00 © 2025 IEEE   268

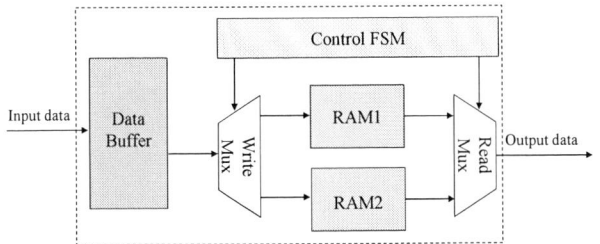

Fig. 4. Dual-buffer architecture with ping-pong buffering for butterfly data scheduling.

TABLE I. INSTRUCTION SCHEDULING ACROSS PIPELINE STAGES

	Stage 1	Stage 2	Stage 3	Stage 4	Stage 5	Stage 6	Stage 7
inst1	BUF	$\mathbf{V}^H$	MULT	$\mathbf{V}$	OUT		
inst2			BUF	$\mathbf{V}^H$	MULT	$\mathbf{V}$	OUT

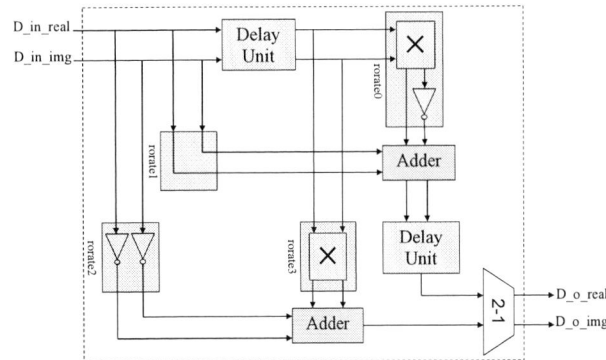

Fig. 5. Radix-2 butterfly processing element with sign-controlled rotation.

two buffer blocks (RAM1 and RAM2) in a ping-pong configuration. Fig. 4 illustrates the architecture, which enables concurrent data read and write operations across processing cycles.

The input data stream is temporarily buffered in a dedicated Data Buffer, with access and scheduling governed by a centralized Control FSM. The FSM manages write-enable signals, buffer fill levels, and memory selection logic for both writing and reading phases. A Write Multiplexer directs incoming data alternately into RAM1 and RAM2 based on the current processing phase. During computation, a Read Multiplexer extracts pre-stored data from the inactive buffer and feeds it to the butterfly pipeline. This ping-pong mechanism decouples read/write paths, eliminating data hazards and ensuring uninterrupted streaming.

C. Pipelined Instruction Scheduling

To further highlight the throughput advantage, Table I illustrates the execution timeline of two consecutive instructions across six pipeline stages. The pipeline consists of input buffering (BUF), $\mathbf{V}^H$-butterfly ($\mathbf{V}^H$), eigenvalue multiplication (MULT), $\mathbf{V}$-butterfly ($\mathbf{V}$), and dual-cycle output (OUT). With the dual-buffer ping-pong scheme, while one buffer is being written with new input data, the other buffer concurrently supplies data to the processing pipeline.

This decoupling of read/write access allows new instructions to enter the pipeline without stalling. As shown in the table, Instruction 2 starts as soon as Instruction 1 advances to the next stage. Once the pipeline is filled, the architecture delivers one output per clock cycle, significantly enhancing throughput compared to sequential or single-buffer implementations.

D. Butterfly Network

The $\mathbf{V}$-butterfly network consists of 10 cascaded stages performing element-wise complex additions and sign-controlled rotations. Each butterfly unit supports $0°$, $90°$, $180°$, and $270°$ operations using only sign and part-swapping logic, thereby eliminating complex multipliers. This simplified structure reduces the critical path delay by over 40% compared to conventional multiplier-based designs.

The internal architecture of a radix-2 butterfly processing element is illustrated in Fig. 5. Input data is routed through four controlled rotation paths (rotate0–rotate3), which apply fixed-angle rotations using sign inversion and component swapping, without multipliers. Delay units are placed before and after key operations to align signal timing across parallel branches. The processed signals are then combined by two complex adders, and the final output is selected via a 2-to-1 multiplexer. This modular structure supports efficient pipelining and reuse across all butterfly stages.

The $\mathbf{V}^H$-butterfly network mirrors the $\mathbf{V}$-butterfly structure but applies conjugate rotation coefficients. Delay elements and arithmetic logic are reused, reducing hardware complexity. The unified control logic synchronizes rotation selection with pipeline stages, enabling one output per clock cycle.

IV. IMPLEMENTATION RESULTS

The proposed 1024-point PN-FCHT architecture was implemented on a Xilinx Virtex-7 XC7VX485T-FFG1157-1 FPGA using fixed-point (Q1.15) arithmetic under a 200 MHz clock constraint. To assess its efficiency, we conducted three levels of evaluation: (1) benchmarking against existing Fast Walsh-Hadamard Transform (Fast-WHT), Pseudo-FHT, and Fast-FHT, (2) comparing a custom RTL design with an HLS-generated version as a self-baseline, and (3) performing an ablation study on the dual-data-group pipeline scheme designs synthesized under the same HLS flow. These experiments validate the architectural advantages of PN-FCHT in terms of resource usage, speed, and throughput.

A. Comprehensive Analysis and Performance Summary

To comprehensively evaluate the PN-FCHT architecture, we compare it with three representative designs: the floating-point Fast-WHT [12], Pseudo-FHT [4], and Fast-FHT [5]. Since no official hardware implementation of Pseudo-FHT and Fast-FHT has been reported, we synthesized its released software code using the same HLS flow to obtain fair and consistent results for 1024-point transforms. The key implementation metrics are summarized in Table II.

TABLE II. COMPARISON OF DIFFERENT HARDWARE IMPLEMENTATIONS

Metric	Fast-WHT [12]	Pseudo-FHT [4]	Fast-FHT [5]	PN-FCHT
Points	1024	1024	1024	1024
Accuracy	float32	Q1.15	Q1.15	Q1.15
LUTs	17536	8255	8060	7682
FFs	12940	5277	6090	10073
DSPs	40	81	44	6
BRAMs	42	24	24	6
Latency (μs)	11.12	49.29	32.03	11.25
Max Freq (MHz)	100	84.75	130.30	285.38
Throughput (M/s)	91.96	20.78	31.97	135.63

TABLE III. COMPARISON OF THE PROPOSED ARCHITECTURE WITH THE HLS DESIGN

Metric	HLS Design	Proposed Design
LUTs	8301	7682
FFs	10389	10073
DSPs	125	6
BRAMs	141	6
Latency (μs)	14.57	11.25
Max Freq (MHz)	175.4	285.38

The proposed PN-FCHT architecture achieves the best overall performance in terms of resource efficiency, latency, frequency, and throughput. In particular, it eliminates the need for DSPs and minimizes buffer usage while maintaining competitive latency and surpassing other methods in throughput. This reflects the effectiveness of both the algorithmic formulation and the structurally optimized RTL design.

B. Comparison to HLS Auto-Synthesized Implementation

As no prior hardware implementations of PN-FCHT exist in the literature, we evaluate our architecture against an HLS auto-synthesized baseline. For 1024-point transforms, Table III shows that the proposed design reduces DSP usage by 95.2% and BRAM consumption by 95.7%, while utilizing 7.5% fewer LUTs. The architecture simultaneously achieves a 62.7% higher maximum clock frequency and 22.8% lower latency. These improvements result from structural optimizations, including multiplier-free butterfly units and a modular dataflow architecture, confirming the advantages of the proposed architecture for large-scale PN-FCHT implementations.

C. Ablation Study on Data Path Optimization

To evaluate the impact of the proposed dual-data-group parallel architecture, an ablation experiment was conducted. As shown in Table IV, the design provides 51.8% higher throughput. This performance gain requires only modest resource additions: a 1.6% LUT increase and two additional BRAM blocks. The results validate that parallel data paths with overlapped memory-computation operations provide substantial throughput benefits for real-time PN-FCHT applications without significant resource expansion.

TABLE IV. COMPARISON OF SINGLE VS. DUAL-DATA PROCESSING PERFORMANCE

Metric	Single Data	Dual-Data Parallel
LUTs	7560	7682
FFs	10011	10073
DSPs	6	6
BRAMs	4	6
Latency (μs)	11.46	15.1
MAX Freq(MHz)	279.88	285.38
Throughput (M/s)	89.35	135.63

V. CONCLUSION AND FUTURE WORK

In this work, the PN-FCHT algorithm is efficiently implemented on the Xilinx Virtex-7 FPGA with a hardware architecture customized to its structural properties. Key optimizations include simplified butterfly units, dynamic bit-width control, and modular dataflow scheduling. Experimental results show that the proposed 1024-point design achieves superior area, power, frequency, and throughput performance compared to existing Fast-WHT, Fast-FHT, and pseudo-FHT architectures. These results validate the practical value of PN-FCHT in high-throughput applications such as compressed sensing, image processing, and radar, with future work focusing on scaling and energy-efficient deployment.

REFERENCES

[1] Monir T Hamood and Said Boussakta. Fast walsh–hadamard–fourier transform algorithm. *IEEE Transactions on Signal Processing*, 59(11):5627–5631, 2011.

[2] Lakshmi Priya GG and S Domnic. Walsh–hadamard transform kernel-based feature vector for shot boundary detection. *IEEE Transactions on Image Processing*, 23(12):5187–5197, 2014.

[3] Ran Tao, Jun Lang, and Yue Wang. The multiple-parameter discrete fractional hadamard transform. *Optics Communications*, 282(8):1531–1535, 2009.

[4] Dorota Majorkowska-Mech and Aleksandr Cariow. Discrete pseudo-fractional hadamard transform and its fast algorithm. *IEEE Signal Processing Letters*, 27:1195–1199, 2020.

[5] Zi-Chen Fan, Di Li, and Susanto Rahardja. Efficient computation for discrete fractional hadamard transform. *IEEE Transactions on Circuits and Systems I: Regular Papers*, 2024.

[6] Aleksandr Cariow, Dorota Majorkowska-Mech, Janusz P Papliński, and Galina Cariowa. A new fast algorithm for discrete fractional hadamard transform. *IEEE Transactions on Circuits and Systems I: Regular Papers*, 66(7):2584–2592, 2019.

[7] Aleksandr Cariow and Dorota Majorkowska-Mech. Fast algorithm for discrete fractional hadamard transform. *Numerical Algorithms*, 68:585–600, 2015.

[8] Zi-Chen Fan, Di Li, and Susanto Rahardja. Pure number discrete fractional complex hadamard transform. *IEEE Signal Processing Letters*, 30:1087–1091, 2023.

[9] Soo-Chang Pei and Min-Hung Yeh. Discrete fractional hadamard transform. In *1999 IEEE International Symposium on Circuits and Systems (ISCAS)*, volume 3, pages 179–182. IEEE, 1999.

[10] Shafiqul Hai and Tella Rajashekhar Reddy. Fpga implementation of an image classifier using pipelined fft architecture. *IEEE Embedded Systems Letters*, 2024.

[11] Zi-Chen Fan, Di Li, and Susanto Rahardja. Real-valued discrete fractional hadamard transform: Fast algorithms and implementations. *IEEE Transactions on Circuits and Systems I: Regular Papers*, 2025.

[12] A Manjarres Garcia, C Osorio Quero, J Rangel-Magdaleno, José Martínez-Carranza, and D Durini Romero. Parallel-pipeline fast walsh-hadamard transform implementation using hls. In *2021 International Conference on Field-Programmable Technology (ICFPT)*, pages 1–4. IEEE, 2021.

Design and Implementation of a Bilateral Filtering Accelerator Based on RISC-V

Zhengyao Shi [1], Yushan Dai [1], Angyang Li [1], Jian Mei*[1,2], Lei Deng [2], Rui Yin [1,3]

[1] College of Integrated Circuits and Micro-Nano Electronics, Fudan University, Shanghai, China
[2] National Integrated Circuit Innovation Center, Shanghai, China
[3] Jiashan Fudan Institute, Jiaxing, China

* Email: zyshi25@m.fudan.edu.cn, meijian@shnicic.com

Abstract—As a crucial image processing algorithm, bilateral filtering has been widely applied in image denoising, smoothing, and edge preservation. However, due to its high computational complexity, implementing bilateral filtering via Micro Control Unit (MCU) suffers from low computational efficiency. The Reduced Instruction Set Computer V (RISC-V) architecture, features open-source nature, modularity, and strong extensibility. It not only realizes the general functions of MCU but also enables the design of coprocessors for specific application and computation scenarios. This paper proposes a bilateral filtering accelerator based on RISC-V. The hardware acceleration of the bilateral filtering is implemented as the coprocessor of RISC-V, and a corresponding custom instruction set is designed. Experimental results demonstrate that the coprocessor can correctly execute bilateral filtering. Compared with software implementation, the coprocessor implementation achieves an 11-14× performance improvement, significantly enhancing computational efficiency.

Keywords—*RISC-V, coprocessor, bilateral filtering, FPGA*

I. INTRODUCTION

Bilateral filtering is a crucial image processing algorithm [1], finding extensive applications in image denoising, smoothing, and edge preservation [2]. By concurrently considering spatial proximity and pixel value similarity, it effectively reduces noise while retaining critical image features [3]. However, the bilateral filtering suffers from high computational complexity because of its complex nonlinear weighted averaging operations [4], and the traditional serial processing via Micro Control Unit (MCU) leads to low computational efficiency [5]. The Reduced Instruction Set Computer V (RISC-V) architecture boasts advantages of open-source nature, modularity, and strong extensibility [6]. It not only realizes the general functions of MCU but also enables the design of custom instruction sets for specific application scenarios, processing specific computations through coprocessors [7]. This is a key advantage over mainstream commercial architectures like ARM and x86. The Hummingbird E203 RISC-V core, a lightweight yet efficient processor core, is well-suited for embedded systems [8] where power consumption and resource utilization must be meticulously balanced. Within the E203 processor core, the Nuclei Instruction Co-unit Extension (NICE) mechanism enables coprocessor expansion [9].

This paper presents a bilateral filtering accelerator built upon the E203 RISC-V core. The accelerator is integrated as a coprocessor via the NICE interface, enabling hardware

acceleration of bilateral filtering through custom instructions. Principal contributions of this work are as follows:

1) Hardware design and implementation: A hardware implementation of bilateral filtering was developed, featuring a sixteen-stage pipeline and First-In-First-Out (FIFO) line buffers to enable streaming input and output of pixel.

2) NICE interface integration: The hardware design was adapted to NICE interface specifications, allowing it to function as a coprocessor for the E203 core and be controlled via custom instructions.

3) Custom instruction set design: A custom instruction set was designed for the coprocessor, adhering to the NICE instruction format, and a corresponding instruction decoder was implemented in hardware.

4) FPGA implementation and validation: The RISC-V core and proposed design were deployed on a Xilinx Zynq7020 Field Programmable Gate Array (FPGA).

Experimental results confirm the accuracy of the bilateral filtering implementation. The coprocessor implementation achieves an 11-14× speedup over the software. We also complete the layout design of the accelerator based on the HLMC 55nm process for the subsequent tape-out process.

II. BILATERAL FILTERING

As a nonlinear filtering algorithm with the dual characteristics of noise suppression and edge preservation, bilateral filtering constructs a joint weight function that integrates spatial domain weights and range domain weights. In the spatial domain, bilateral filtering employs a two-dimensional Gaussian function (1) to measure the spatial distance between pixels.

$$G_{\sigma_s}(x, y) = \frac{1}{2\pi\sigma_s^2} e^{-\frac{x^2+y^2}{2\sigma_s^2}} \qquad (1)$$

With the central pixel as the origin, the weight of this function decays exponentially as the distance increases. For instance, when σ_s is set to a relatively small value, only the neighboring pixels are assigned high weights, thus preserving the fine details of the image. Conversely, an increase in σ_s expands the scope of smoothing.

In the range domain, bilateral filtering constructs the Gaussian function (2) based on the grayscale differences. Taking the grayscale value $I(x, y)$ of the central pixel as the reference, pixels with significant grayscale differences receive lower weights. This mechanism enables bilateral filtering to perceive the boundaries of objects in images.

This work was supported by Shanghai Pudong New Area Science and Technology Development Fund Industry-Academia-Research Collaboration Special Funding PKX2024-D06.

979-8-3315-3918-4/25 $31.00 © 2025 IEEE

$$G_{\sigma_r}(I(x,y)) = \frac{1}{2\pi\sigma_r^2} e^{-\frac{|I(x,y)-I(i,j)|}{2\sigma_r^2}} \qquad (2)$$

The joint weight function (3) multiplies the weights of the spatial and range domains, so it can achieve both image smoothing and edge preservation.

$$w(x,y,i,j) = G_{\sigma_s}(x,y)G_{\sigma_r}(I(x,y)) \qquad (3)$$

The filtered output pixel value $I_{BF}(x,y)$ is calculated using the following formula.

$$I_{BF}(x,y) = \frac{\sum_{i,j\in\Omega} w(x,y,i,j)I(i,j)}{\sum_{i,j\in\Omega} w(x,y,i,j)} \qquad (4)$$

where Ω represents the filtering matrix centered at (x,y). Essentially, this calculation process is a weighted average of the pixel values within the filtering matrix.

III. PROPOSED WORK

A. Hardware Design of Bilateral Filtering

Based on the principle of the bilateral filtering algorithm, this paper presents a hardware design featuring a sixteen-stage pipeline architecture. Prior to the filtering process, a spatial domain lookup table and a range domain lookup table are pre-computed and solidified in read-only memory in FPGA. The spatial domain lookup table has a size of 3×3, while the range domain lookup table has a size of 256×1. Pre-computation reduces the exponential and division operations during the weight calculation process. The specific filtering process is shown in Fig. 1 and described as follows:

1) Spatial domain weight calculation: Within a 3×3-pixel matrix, the spatial domain weights are determined by the relative position relationships between each pixel and the central pixel. Since these relative positions remain constant throughout the filtering process and do not vary with pixel values, the spatial domain weights can be directly obtained from the pre-computed 3×3 spatial domain lookup table data.

2) Range domain weight calculation: The calculation of range domain weights relies on the grayscale value differences between each pixel and the central pixel, and it changes dynamically with pixel values. The specific steps are as follows. First, the absolute differences in grayscale values are calculated in parallel, which can be efficiently achieved through parallel data selector. Subsequently, based on the calculated differences, the corresponding weight values are

retrieved from the 256×1 range domain lookup table to form a 3×3 range domain weight matrix. The calculation process above serves as the first stage of the pipeline.

3) Joint weight calculation: The joint weights, serving as the weight matrix ultimately used for filtering, also have a size of 3×3. This matrix is obtained by element-wise multiplication of the spatial domain weight matrix and the range domain weight matrix, and this operation constitutes the second stage of the pipeline.

4) Weighted sum calculation: The third stage of the pipeline is responsible for calculating the sum of all elements in the joint weight matrix (weight sum). The calculated result is registered for one clock cycle. Meanwhile, in the third and fourth stages, the joint weight matrix is multiplied with the corresponding pixel matrix data and accumulated to obtain the accumulated value (pixel weighted sum).

5) Output pixel calculation: The fifth to sixteenth stages of the pipeline complete the calculation of the final output pixels. The division operation between pixel weighted sum and weight sum is implemented using a divider IP core. The resulting value is normalized to confine the pixel values within the range of $0-255$.

This paper implements the calculation of a 3×3 image matrix based on the FIFO line buffer. A data processing chain is constructed through two level FIFO buffer units. The first level FIFO receives the first line data of the image. Driven by the clock signal, the second line data enter the second level FIFO. This hierarchical buffering mechanism ensures that the pixel data of three consecutive lines are time aligned and can be synchronously output within the same clock cycle, thus forming a 3×3 image matrix for operations. The architecture constructing an efficient streaming data processing chain, where one pixel is input and output at each clock edge. The working principle of the FIFO line buffer are detailed in Fig. 2, providing a stable data processing foundation for matrix operations based on 3×3 matrix.

B. Design of Custom Instruction Set

Domain specific architecture serves as an advanced architectural paradigm for specialized computing tasks. The collaborative extension of the main processor and coprocessor accelerator enhances computational efficiency in specific domains. The NICE coprocessor, built upon the Hummingbird E203 RISC-V core, functions as an independent computing unit with an instruction-driven control model. By executing custom instructions, this coprocessor enables memory-cache interaction, specific numerical computations, data encryption, etc. In terms of interface design, the NICE coprocessor features four independent channels: request channel, memory request channel, response channel, and memory response channel. This architecture allows the coprocessor to operate independently from the main core process.

The application of the NICE coprocessor relies on custom RISC-V instruction sets. To achieve hardware acceleration for the bilateral filtering algorithm, this work designs three

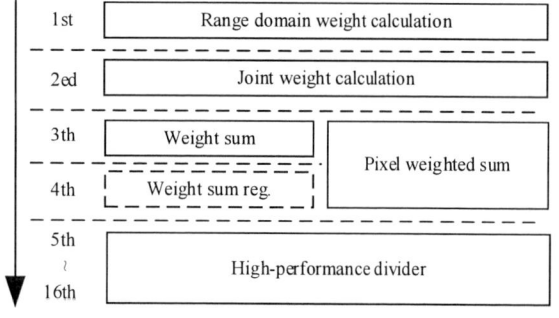

Fig. 1. Bilateral filtering pipeline architecture.

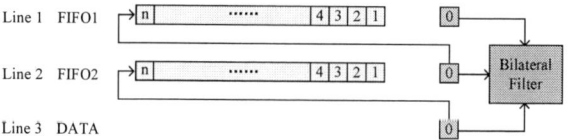

Fig. 2. FIFO line buffer.

TABLE I. INSTRUCTION SET DESIGN

Instruction Name	Load_buffer	Store_buffer	Bilateral_filter
ID	1	2	3
Opcode	0x7b	0x7b	0x7b
Funct3	010	010	011
rd	00000	00000	%0
rs1	%1	%1	%1
rs2	Unused	Unused	Unused
Funct7	000_0001	000_0010	000_0011

custom instructions, which is shown in Table I. Instruction 1 enables data reading from memory to registers, Instruction 2 enables data writing from registers to memory, and Instruction 3 executes the bilateral filtering accelerator to process original pixel data and output filtered pixels. In the instructions, rd denotes the destination register, while rs represent source registers. The funct3 field specifies available registers through encoding: rs1 stores the address of data to be processed in Instructions 1 and 3, and the write-back address in Instruction 2; rd in Instruction 3 stores the filtered pixel values. The funct7 and opcode fields define specific instruction functions. Register addresses use 5-bit encoding allocated by the compiler, requiring specification of data to be stored in registers when programming assembly instructions. Notably, Instruction 3 also leverages memory request and response channels to complete memory read-write operations, ensuring the algorithm's full execution.

C. Overall Architecture

The accelerator architecture proposed in this design is illustrated in Fig. 3, which exclusively depicts the modules

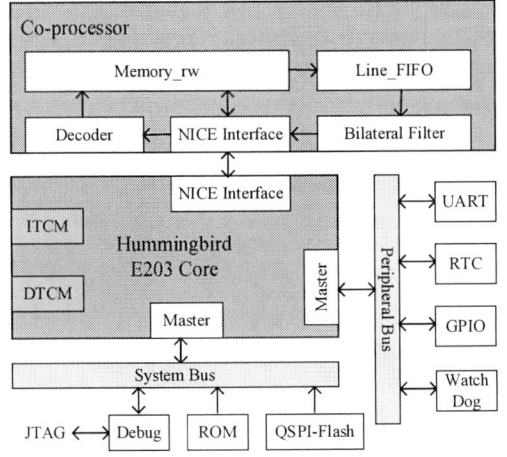

Fig. 3. Overall architecture.

employed in this implementation. The NICE coprocessor is integrated into the E203 core, enabling users to develop custom hardware coprocessor units. The workflow is structured as follows: First, the design undergoes synthesis and implementation processes to generate a bitstream, which is programmed into an FPGA to instantiate the RISC-V core. Subsequently, a C-language program is compiled and downloaded via the JTAG interface into the Quad SPI Flash. Upon system initialization, the RISC-V transfers all instructions from the Flash to the Instruction Tightly Coupled Memory (ITCM) and loads data into the Data Tightly Coupled Memory (DTCM) before commencing program execution. When encountering custom coprocessor instructions, the coprocessor's decoder identifies the instruction type and executes the corresponding operations through hardware. Data accesses to the DTCM are handled by the main processor upon request via the NICE interface, with results returned to the coprocessor through the same interface. Within the peripheral bus system, the UART interface is utilized to transmit filtering results and timing metrics to the host computer for analysis.

IV. EXPERIMENTAL RESULTS

A. Experimental Setup

The proposed design is implemented on a Zynq-7020 with the model number XC7Z020CLG400-2. The ITCM and DTCM are configured with a size of 64 K-Byte each, realized by the on-chip Block RAM resources of the FPGA. In terms of peripherals, a Quad SPI Flash (model: W25Q128) with a capacity of 16 M-Byte is integrated into the system. The Nuclei RISC-V Debugger is utilized as the JTAG downloader for programming and debugging. Vivado 2018.3 is employed for the hardware design and implementation. Meanwhile, Nuclei Studio 2019.9 serves as the integrated development environment, facilitating the software development lifecycle, including compilation, downloading, and debugging.

B. Filtering Results

To verify the accuracy and consistency of the implementation of the bilateral filtering, this study conducted experiments using a software implementation on the RISC-V core, a hardware acceleration on the RISC-V coprocessor, and a Python bilateral filtering function on the host computer. Images with size of 640×480 were processed under the same bilateral filtering parameter settings ($\sigma_s = 0.8$, $\sigma_r = 100$). The experiments show that the filtering results output by the three different implementation methods are same and correct.

C. Performance Evaluation

This paper conducts an experimental study on the performance of the accelerator. The experimental results are presented in Table II. The experiment uses the images with size of 640×480, a clock frequency set at 32 MHz, and is based on different compiler optimization levels (O2, O1, O0). The results show the coprocessor implementation achieves an

TABLE II. PERFORMANCE EVALUATION

Comparison Items	O2			O1			O0		
	Instruction Count	Clock Cycles	Elapsed Time(ms)	Instruction Count	Clock Cycles	Elapsed Time(ms)	Instruction Count	Clock Cycles	Elapsed Time(ms)
Software	88280909	195835875	6120.422	84816737	196700702	6147.461	297001072	465396638	14544.982
Coprocessor	10584279	12462747	389.495	13178904	16007772	500.305	25158479	31551723	986.084

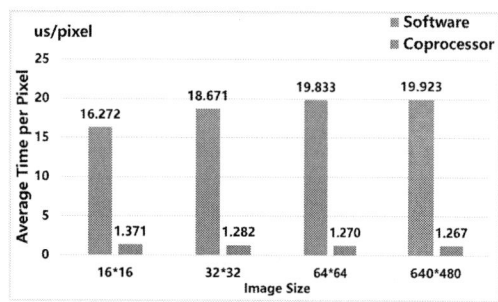

Fig. 4. Performance of different image sizes.

TABLE III.   RESOURCE UTILIZATION

Items	Software		Coprocessor	
	Utilization	*Utilization %*	*Utilization*	*Utilization %*
LUT	9665	18.2	11005	20.7
LUTRAM	16	0.09	16	0.09
FF	9090	8.54	10074	9.47
BRAM	32	22.9	33.5	23.9
DSP	0	0	9	4.09
IO	28	22.4	28	22.4
BUFG	5	15.6	5	15.6

11-14× performance boost, and the average clock cycles per instruction is also reduced.

This paper also analyzes the filtering processing speed for four different image sizes, and the relevant results are shown in Fig. 4. The experiment is carried out under the conditions of compiler optimization level O2 and clock frequency of 32 MHz. The results show that with the increase of image size, the coprocessor implementation has more advantages.

The timing results show that the maximum stable clock frequency of the E203 RISC-V core on the Zynq7020 is about 32 MHz. This paper is to verify that the integration of the bilateral filtering accelerator does not affect it. After optimizing the critical path by integrating a high-performance divider IP core, the updated timing report indicates that under the system clock frequency of 32 MHz, the worst negative slack is 0.389 ns. This result meets the timing closure criteria.

The resources utilization results show that the coprocessor implementation incurs higher resource consumption, with data detailed in Table III. This phenomenon can be attributed to two factors. First, the coprocessor needs to integrate instruction decoding and data access control logic. Second, large lookup tables in the bilateral filter and the multi-stage pipeline design increase the utilization of FPGA resources.

D. Layout Design Result

This paper completes the layout design of the accelerator based on the HLMC 55nm process, as shown in Fig. 5, laying the groundwork for the subsequent tape-out process. The total area of the layout is 1mm², and the post-synthesis layout area is 0.608mm². Among them, the Static RAM, which is generated using a memory compiler, consists of 32K ITCM

Fig. 5. Layout deisgn.

and 32K DTCM. The layout report results show that under the condition of a clock frequency of 100MHz, the timing meets the timing closure criteria. The post-simulation results based on the post-synthesis netlist are verified to be correct.

V. CONCLUSION

This paper presents a RISC-V based bilateral filtering accelerator, which achieves hardware acceleration of the algorithm through a custom coprocessor. Experimental results validate that the coprocessor correctly implements bilateral filtering with an 11–14 × performance improvement. By leveraging RISC-V's open-source and extensible architecture, the design effectively addresses the high computational complexity of bilateral filtering.

REFERENCES

[1] Chen B H, Tseng Y S, Yin J L, "Gaussian-adaptive bilateral filter," IEEE Signal processing letters, 2020, 27: 1670-1674.

[2] Akbar S A, Verma A, "Analyzing Noise Models and Advanced Filtering Algorithms for Image Enhancement," arxiv preprint arxiv: 2410.21946, 2024.

[3] R. G. Gavaskar and K. N. Chaudhury, "Fast Adaptive Bilateral Filtering," in IEEE Transactions on Image Processing, vol. 28, no. 2, pp. 779-790, Feb. 2019.

[4] F. Spagnolo, P. Corsonello, F. Frustaci and S. Perri, "Design of Approximate Bilateral Filters for Image Denoising on FPGAs," in IEEE Access, vol. 11, pp. 1990-2000, 2023.

[5] Ye Anlong, Ma Lingkun, Qu Zongyi, "An Acceleration Scheme for LMS Algorithm Based on RISC-V," Integrated Circuits and Embedded Systems, 2025, 25(05): 52-59.

[6] R. Höller, D. Haselberger, D. Ballek, P. Rössler, M. Krapfenbauer and M. Linauer, "Open-Source RISC-V Processor IP Cores for FPGAs — Overview and Evaluation," 2019 8th Mediterranean Conference on Embedded Computing (MECO), Budva, Montenegro, 2019, pp. 1-6.

[7] E. Gholizadehazari, T. Ayhan and B. Ors, "An FPGA Implementation of a RISC-V Based SoC System for Image Processing Applications," 2021 29th Signal Processing and Communications Applications Conference, Istanbul, Turkey, 2021, pp. 1-4.

[8] Zhao Chenxu, Qu Yingjie, Wang Haiting, "Design of License Plate Recognition Coprocessor Based on RISC-V Extended Instructions," Integrated Circuits and Embedded Systems, 2025, 25(04): 47-53.

[9] H. Zhenbo, "Step-by-Step Guide to CPU Design," Beijing: Posts & Telecom Press, 2018: 55-62.

Skip-Zero Strategy: A Latency and Power Optimization for SRT Divider

Ke Xu [1], Ping Yin [2] and Jun Han [1]

[1] State Key Laboratory of ASIC and System, School of Microelectronics, Fudan University, Shanghai, China
[2] China Mobile (SuZhou) Software Technology Co.,Ltd, Suzhou, Jiangsu, China

Email: 24212020033@m.fudan.edu.cn, junhan@fudan.edu.cn

Abstract—Division operations are widely used in various scientific computing tasks, such as in gradient descent computations within machine learning algorithms, serving as a core numerical operation for learning rate scaling and gradient normalization. However, the inherently high latency and low throughput of division operations pose significant challenges to the design of efficient hardware for machine learning training. Although numerous efforts have been made to optimize hardware dividers, the inherently long delay of division operations still often becomes a bottleneck in critical data paths.

In this work, we propose an optimization strategy named Skip-Zero Strategy, which effectively reduces the average number of iterations by skipping those in which the partial quotient is zero. Based on this strategy, we design a RISC-V SRT divider and implement it on the Vivado 2019.1 platform targeting the xc7a35tcpg236-1 FPGA. Our divider achieves an average operation frequency of 11.39 MOP/s and an average energy efficiency of 421.8MOP/J. Compared to the dividers in Xuantie-910 and Rocket-Chip under identical test conditions, our design achieves ×1.13 and ×1.59 in average operation frequency and ×2.93 and ×1.18 in average energy efficiency, respectively.

Keywords—hardware divider, SRT algorithm, RISC-V, Chisel

I. INTRODUCTION

In recent years, with the rapid advancement of machine learning, gradient descent has garnered increasing attention as a widely adopted optimization algorithm. Within gradient descent, division operations serve as core numerical components for learning rate scaling and gradient normalization. With a typical latency of tens of clock cycles [3], division often lies on the critical data path due to its high latency and strong data dependency. Consequently, reducing the latency and energy consumption of division instructions is of great importance for improving processor performance in such application scenarios.

Compared to the basic restoring algorithm, the Sweeney-Robertson-Tocher (SRT) [1][2] algorithm offers two major advantages. First, the logic for partial quotient selection in SRT has a relatively simple hardware structure and a short critical path, which facilitates the design of high-frequency circuits. Second, the SRT algorithm also enables both high-radix division and variable-cycle execution, making it suitable for high-performance applications. As a result, the SRT algorithm has been widely adopted in processor designs.

Numerous studies have been conducted to optimize SRT dividers, including techniques such as variable-cycle execution, high-radix algorithms, overlapping operations, quotient carry-save schemes, and partial quotient & partial remainder register reuse [3].

To further improve the performance of hardware division, we propose an optimization technique which exploits the low logical delay associated with zero partial quotients, referred to as the Skip-Zero Strategy. The main contributions of this work are summarized as follows:

1) We propose an optimization strategy named Skip-Zero Strategy with its theoretical analysis and hardware implementation. This strategy can significantly reduce the number of division cycles without noticeable increases in hardware cost or critical path delay.

2) We design a parameter-configurable, variable-cycle divider to evaluate the impact of different configuration parameters on performance. FPGA-based simulation results demonstrate that, under the optimal configuration, our divider achieves ×1.13 and ×1.59 in average operation frequency compared to the dividers in Xuantie-910 and Rocket-Chip, respectively. Moreover, average energy efficiency is advanced to ×2.93 and ×1.18, respectively.

II. ALGORITHM DESCRIPTION

This section briefly introduces the SRT division algorithm and then presents our proposed Skip-Zero Strategy, which aims to reduce the number of iterations in SRT division. For clarity, we focus on radix-2 SRT division in our explanation, as the principles are similar for higher-radix versions.

A. SRT Division Algorithm

As shown in the pseudo-code in Fig. 1, the SRT algorithm determines the partial quotient in each iteration by examining only the leading bits of the partial remainder and the divisor, whereas the restoring algorithm requires an actual subtraction to decide the partial quotient. Consequently, the SRT algorithm significantly simplifies the logic for partial quotient selection and makes high-radix division feasible.

As shown in Fig. 1, the SRT algorithm assumes that both the divisor and the dividend are unsigned numbers. For signed division, additional sign handling logic is needed at the beginning and end of the computation.

The core step in SRT iteration is selecting the partial quotient. For clarity, the partial remainder R referred to in this paper is consistent with the pseudo-code in Fig. 1, representing the result of left-shifting the conventional partial remainder by several bits.

Taking N-*bit* radix-2 SRT as an example, assume the divisor D is an M-*bit* unsigned number ($D[M-1]=1$), and the initial partial remainder R (the dividend) is an $(N+1)$-bit non-negative signed number (N-bit unsigned number). After the k-th iteration, the partial remainder R is expected to

979-8-3315-3918-4/25 $31.00 © 2025 IEEE

consistently be an $(N + 1)$-bit signed number. Thus, the goal of radix-2 SRT is to find a partial quotient digit $q_k \in \{-1, 0, 1\}$ such that the updated partial remainder R' satisfies:

$$-2^N \le R' = [(R - q_k \cdot D \cdot 2^{N-M}) << 1] < 2^N \quad (1)$$

$$-2^{N-1} \le R - q_k \cdot D \cdot 2^{N-M} < 2^{N-1} \quad (2)$$

In the other word, the goal of radix-2 SRT is to find a q_k to ensure that R fits within an $(N+1)$-*bit* signed representation.

Similarly, for radix-2^b SRT algorithm, the partial quotient digit $q_k \in (-2^n, 2^n)$ must satisfy:

$$-2^{N-b} \le R - q_k \cdot D \cdot 2^{N-M} < 2^{N-b} \quad (3)$$

These bounds allow the construction of a mathematical partial remainder table, where the appropriate q_k can be determined by consulting the leading bits of the partial remainder R and the divisor D. For radix-2/4/8 SRT, Fig. 2 presents the partial quotient selection logic. It shows that, in radix-2 SRT, only the two most significant bits of R are needed to determine q_k, which significantly simplifies hardware implementation.

It can be mathematically shown that, for a radix-2^b SRT algorithm, the partial quotient q_k can be determined by inspecting the significant $(b + 1)$ bits of the partial remainder and the significant b bits of the divisor D, denoted as $q_k = q_{R,D}$. It is straightforward to observe that there is $q_{R^-,D} = -q_{R,D}$, where R^- s the one's complement of R. Therefore, the partial quotient lookup table only needs to store entries for non-negative R values. Fig. 2 presents the partial remainder tables for radix-2, radix-4, and radix-8 SRT algorithms. It can be seen that, for radix-2 SRT, q_k can be obtained by examining only the two most significant bits of R, which is advantageous for hardware implementation.

As shown in Fig. 1, the iteration terminates After the partial remainder R is left-shifted $(N - M + 1)$ times. At this point, the accumulated quotient is computed as $Q = \sum(q_k \cdot 2^{N-M-k})$. Then, R is right-shifted by $(N - M + 1)$ bits, and it is guaranteed that $|R| < D$. If $R \ge 0$, the final quotient and remainder are $quo = Q$ and $rem = R$; otherwise, algorithm is done with $quo = Q - 1$ and $rem = R + D$. This final stage is named remainder correction stage.

B. Skip-Zero Strategy

As observed from Fig. 2 under radix-2 SRT, assuming the four possible cases are equally probable, the partial quotient $q_k = 0$ occurs in half of the cases. In such cases, the updated partial remainder becomes $R' = R << 1$, which does not require an adder to compute. Consequently, the operation delay in these cases is significantly shorter than in others that involve arithmetic computation.

Motivated by this observation, we propose an optimization strategy. When $R[N: N - 1]$ indicates $q_k = 0$, we perform a pseudo radix-2 SRT iteration in which we left-shift R by 1-bit and directly proceed to the next iteration in which checking $R[N - 1: N - 2]$ for the next partial quotient q_{k+1}. If $q_{k+1} = 0$ as well, this process is repeated, with the number of speculative skips bounded by hardware-defined limits. In the hardware implementation, the evaluations of $R[N: N - 1]$ and $R[N - 1: N - 2]$ can be performed in parallel to further reduce the overall latency. We refer to this approach as the Skip-Zero Strategy, and refer to the associated logic illustrated in Fig. 3 as the Skip-Zero Unit.

Restoring Algorithm	SRT Algorithm

`// N bit unsigned R & D`
`restore(R, D) {`
`    D = D << N;`
`    Q = 0;`
`    // iteration`
`    for (k = 0; k < N; k++) {`
`        R << 1;`
`        // base version`
`        R = R - D;`
`        if (R > 0) {`
`            Q[N-1-k] = 1;`
`        } else {`
`            Q[N-1-k] = 0;`
`            // restore remainder`
`            R = R + D;`
`        }`
`        // opt version (1 adder)`
`        // Q[N-k] = R - D > 0;`
`        // R = Q[N-k] ? R - D : R;`
`    }`
`    R = R >> N;`
`    return (Q, R);`
`}`
	`//N bit unsigned R & D`
	`srtRadix2(R, D) {`
	`    M = log2floor(D);`
	`    D = D << (N - M);`
	`    Q = 0;`
	`    // iteration start`
	`    for (k = 0; k < N-M+1; k++) {`
	`        Q << 1;`
	`        // lookup qk in Radix-2`
	`        if (R[N: N-1] == 2'b01)`
	`            qk = 1;`
	`        else if (R[N: N-1] == 2'b10)`
	`            qk = -1;`
	`        else`
	`            qk = 0;`
	`        R = R - D;`
	`        Q = (Q + qk);`
	`        R << 1;`
	`    }`
	`    // iteration done`
	`    R >> (N - M + 1);`
	`    // correction stage`
	`    if (R < 0) {`
	`        R = R + D;`
	`        Q = Q - 1;`
	`    }`
	`    return (Q, R);`
	`}`

Fig. 1. Pseudo-code of restoring algorithm and SRT algorithm.

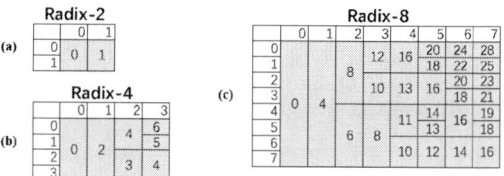

Fig. 2. The partial quotient of SRT algorithm in (a) radix-2; (b) radix-4; (c) radix-8. For radix-2^b, The horizontal axis represents $R[N-1: N-1-b]$, and the vertical axis represents $D[N-1: N-1-b]$. Near the end of the iteration process, high-radix iterations need to degrade into lower-radix iterations. This detail is omitted here due to space limitations.

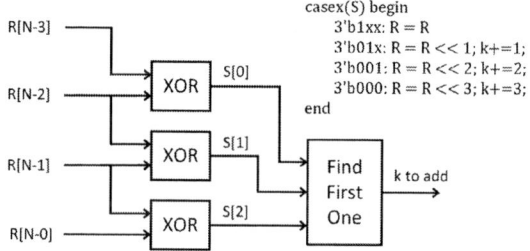

Fig. 3. The maximum number of consecutive zero-skipping steps is determined using xor gates and the logic detecting the leading one.

Although similar strategy has been explored in prior work [4], their strategy is limited to a fixed iteration radix of 2 and confined to restrictive conditions, which depend on examining the 4 most significant bits. In contrast, our method enables more fine-grained speculation by utilizing pseudo radix-2 SRT steps regardless of the base radix of iterations. Moreover, our strategy allows skipping multiple consecutive zero quotient digits, with each skip requiring only 2 MSBs to be examined, which enables greater flexibility in application and a more lightweight hardware implementation.

The performance gain of the Skip-Zero Strategy can be analytically estimated. Assuming each iteration in the division process has an equal probability of producing a zero partial quotient under radix-2 SRT, if the hardware supports a S-level Skip-Zero Unit, the expected number of bits that can be skipped in a single cycle is:

$$\sum_{n \in [1, S]} (1 / 2^n) = 1 - 1 / 2^S \tag{4}$$

Given a radix-2^b SRT iteration, the overall acceleration factor η is:

$$\eta = 1 + (1 - 1 / 2^S) / b \tag{5}$$

It is worth emphasizing that in radix-2 SRT, introducing only a 1-level Skip-Zero Unit can yield up to a 50% improvement in iteration throughput.

III. HARDWARE DESIGN

A. Parameterized Design

To evaluate the effectiveness of the proposed optimization strategy in practical hardware design, we design and implement a parameter-configurable, variable-cycle integer divider based on Chisel, a hardware construction language. The specific configurable parameters are listed in Table I. Functional verification of the divider is conducted using Chisel's built-in simulation framework.

Our proposed divider supports parameterized configuration, which not only facilitates the exploration of optimal configuration parameters, but also enables a systematic evaluation of the impact of individual parameters on divider performance, as demonstrated in Section IV. The parameterized design allows the divider to be flexibly adapted to various application scenarios, enabling trade-offs among hardware cost, operating speed, and power consumption.

B. Architecture Overview

Fig. 5 illustrates the hardware block diagram of the proposed divider. The design shown in the diagram includes three iteration units, which means $l = 3$. A Skip-Zero Unit is placed before each iteration unit. The overall operation flow of the divider is presented in Fig. 4.

C. Additional Feature

As shown in Fig. 4, our design incorporates a special-case detection stage after the sign processing phase to handle RISC-V defined corner cases such as division by zero and overflow. This ensures compatibility with RISC-V processor integration.

TABLE I.   CONFIGURABLE PARAMETERS TABLE

Parameter	Description
data width (w)	The width of dividend, divisor, quotient and remainder
iteration radix (b)	specifies the radix for each iteration is 2^b.
level of iteration units (l)	The number of cascaded iteration units (the number of iterations per cycle).
skip-zero strategy (s)	The level of Skip-Zero Unit.

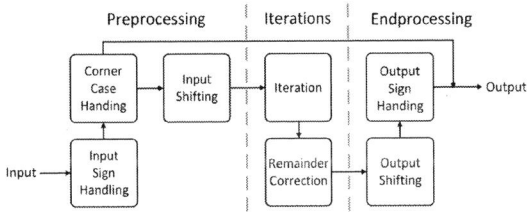

Fig. 4.   The overall operation flow of the divider.

Our design supports variable-cycle execution. By employing the logic detecting the leading one to identify the MSB of the dividend and the divisor, an additional pre-shifting stage is introduced before the iteration begins. This allows the divider to skip several initial iterations where the partial quotient would be deterministically zero. As shown in Fig. 4, this stage is integrated with the sign and exception handling stage, and these three stages correspond to the *Preprocessing* block in Fig. 5.

When multiple iterations are executed within a single clock cycle, the divisor D remains unchanged across iterations, while the partial remainder R varies in each iteration. To efficiently support this behavior, our design includes a single shared partial remainder table. As illustrated in Fig. 5, the row corresponding to D is selected from the table, and each iteration unit independently selects the partial quotient q_k from the row based on the current value of R. This approach avoids the excessive resource consumption associated with large partial remainder tables in high-radix designs and reduces the latency of the lookup logic.

As shown in Fig. 5, The final iteration unit differs from the preceding ones by incorporating additional logic to handle the case where the partial remainder becomes negative after the last iteration. When multiple iterations are executed within a single clock cycle, this design allows remainder correction stage will not occupy an entire clock cycle.

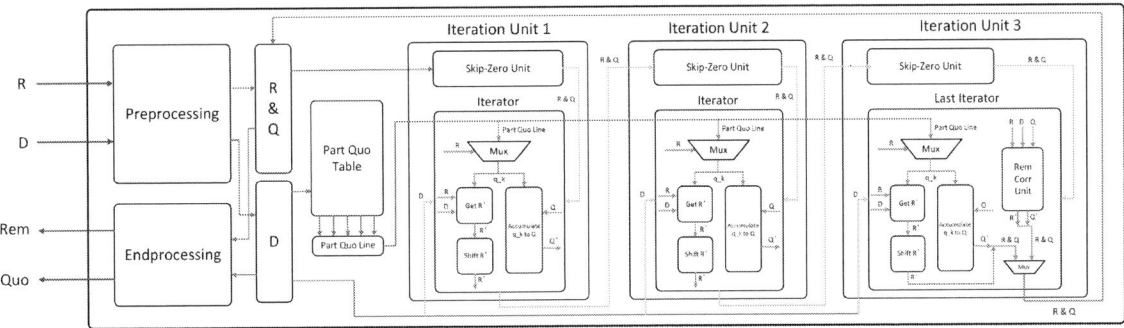

Fig. 5.   The hardware block diagram of the divider. the level of Skip-Zero Unit (s) and the radix of iteration unit (2^b) are not shown.

Moreover, given that the bit width of the partial remainder decreases while the accumulated quotient grows during the division process, we combine them into a single register to reduce register usage (the partial quotient & partial remainder register reuse). This optimization is not explicitly depicted in Fig. 5.

IV. EVALUATION

A. Experiment

We evaluate the performance of our divider using Vivado 2019.1 and software-based simulation. To ensure consistency with comparison targets, the divider's bit-width is fixed at 64 bits. The Vivado synthesis is performed on the xc7a35tcpg236-1 FPGA device using default settings to measure the divider's frequency and power.

To validate the effectiveness of the Skip-Zero Strategy in reducing the number of execution cycles, we estimate the average number of cycles through software-based modeling. The simulation follows these assumptions:

- Due to the variable-latency design, the first iteration always produces a non-zero partial quotient regardless of radix. In subsequent iterations, the probability of performing a pseudo radix-2 SRT iteration is 1/2.

- The effective bit width of the absolute values of the dividend and divisor are assumed to be uniformly distributed over 0 to 64. This assumption leads to half of the quotients being zero; to correct this bias, we assign such cases a weight of 0.25 rather than 1.

- Signed and unsigned divisions are assumed to occur with equal probability. For signed divisions, the signs of the dividend and divisor are independently and uniformly distributed. This assumption only affects the estimation of average cycle counts for the Rocket-Chip divider.

To distinguish different configurations of our design, we denote a divider as (B, L, S), where the radix is 2^B, the level of iteration units is L, and the level of Skip-Zero Unit is S. As the experiment result shown in Fig. 6, the divider achieves optimal synthesis performance under the $(1, 4, 1)$ configuration, with an average operation frequency of 11.39 MOP/s and an average energy efficiency of 421.8MOP/J.

Contrary to common intuition, radix-2 design performs better than radix-4 design. This is because the Skip-Zero Strategy with $S = 1$ introduces almost no degradation in frequency while significantly improving performance and as shown in (1), $S = 1$ configuration achieves a 50% speedup in radix-2 iterations, while only 25% in radix-4 iterations.

B. Comparative Results

To further evaluate the effectiveness of the Skip-Zero Strategy, we conducted synthesis and software-based simulations of the dividers in Xuantie-910 [5] and Rocket-Chip [6][7] under identical conditions. Fig. 6 compares the performance metrics of our proposed dividers against the dividers in Xuantie-910 and Rocket-Chip. Compared with the dividers in Xuantie-910 and Rocket-Chip, our proposed divider achieves ×1.13 and ×1.59 in average operation frequency and ×2.93 and ×1.18 in average energy efficiency, respectively.

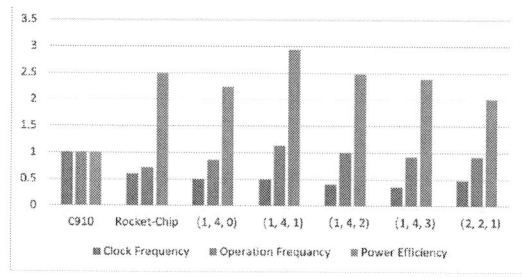

Fig. 6. Normalized divider clock frequency, average operation frequency, average power efficiency. Rocket-Chip divider is also parameter-configurable, we only show the optimal configuration.

V. CONCLUSION

We propose a Skip-Zero Strategy based on the SRT division algorithm. By employing pseudo radix-2 SRT iterations to skip low-latency operations where the partial quotient is zero. With small number of Skip-Zero Unit level, this approach effectively not only reduces the average number of iterations but also has negligible impact on the critical path delay.

An integer RISC-V divider implementing the Skip-Zero Strategy with parameterizable configuration is designed and synthesized. Compared with the dividers in Xuantie-910 and Rocket-Chip under identical test conditions, our design with the optimal configuration achieves ×1.13 and ×1.59 in average operation frequency and ×2.93 and ×1.18 in average energy efficiency, respectively.

ACKNOWLEDGMENT

This work was supported by the National Natural Science Foundation of China under Grant 61934002 and 62234008.

REFERENCES

[1] J. E. Robertson, "A new class of digital division methods," *IRE Transactions on Electronic Computers*, vol. EC-7, no. 3, pp. 218–222, Sep. 1958.

[2] K. D. Tocher, "Techniques of multiplication and division for automatic binary computers," *Quarterly Journal of Mechanics and Applied Mathematics*, vol. 11, no. 3, pp. 364–384, 1958.

[3] U. S. Patankar and A. Koel, "Review of basic classes of dividers based on division algorithm," *IEEE Access*, vol. 9, pp. 23035–23069, 2021.

[4] P. Montuschi and L. Ciminiera, "Reducing iteration time when result digit is zero for radix-2 SRT division and square root with redundant remainders," *IEEE Transactions on Computers*, vol. 42, no. 2, pp. 239–246, Feb. 2002.

[5] C. Chen, X. Xiang, C. Liu, Y. Shang, R. Guo, D. Liu, *et al.*, "Xuantie-910: A commercial multi-core 12-stage pipeline out-of-order 64-bit high performance RISC-V processor with vector extension: Industrial product," in *Proc. 2020 ACM/IEEE 47th Annual International Symposium on Computer Architecture (ISCA)*, Valencia, Spain, May 2020, pp. 52–64.

[6] J. Zhao, B. Korpan, A. Gonzalez, and K. Asanovic, "SonicBoom: The third generation Berkeley out-of-order machine," in *Proc. 4th Workshop on Computer Architecture Research with RISC-V (CARRV)*, Valencia, Spain, May 2020, pp. 1–7.

[7] K. Asanovic, R. Avizienis, J. Bachrach, S. Beamer, D. Biancolin, C. Celio, *et al.*, *The Rocket Chip Generator*, EECS Dept., Univ. of California, Berkeley, CA, USA, Tech. Rep. UCB/EECS-2016-17, Apr. 15, 2016. [Online]. Available: https://aspire.eecs.berkeley.edu/wp/wp-content/uploads/2016/04/Tech-Report-The-Rocket-Chip-Generator-Beamer.pdf

A High-precision Stochastic Computing Multiplier with Co-optimization of Area and Latency

Qiang He[1], Yudi Zhao*[1], Zhihuai Zhang[2,3], Xiaofei Nie[1], Shisheng Xiong[2,3], Kai Zhao[1]

[1]School of Information and Communication Engineering, Beijing Information Science & Technology University, Beijing 102206, China
[2]Zhangjiang Laboratory, Shanghai, 201210, China
[3] Innovation School of Future Information, Fudan University, Shanghai, 200433, China
*Email: zhaoyd@bistu.edu.cn

Abstract—Stochastic computing (SC) is a new computing paradigm, which has the advantages of small area and low power consumption. However, the conversion units require a significant amount of hardware, and the operation period is relatively long. This paper proposes a novel high-precision and fast SC multiplier that performs calculations by partitioning the multiplier and generating parallel bitstreams. The proposed multiplier requires only 16 computational cycles to perform an 8-bit multiplication, which represents a reduction of 93.75% compared to Classic SC methods. The design is validited via FPGA testing. The proposed multiplier has increased area-efficiency by 1.9 times over existing optimal SC multipliers. Energy consumption per multiplication operation is 52.38% lower than existing optimal designs. The area is reduced by 80.21% compared with the binary multiplier. The SC multiplier is designed as a Multiply-Accumulate (MAC) unit for image processing. The experimental results demonstrate that the proposed SC multiplier design performs well in several image processing algorithms.

Keywords—*Stochastic Computing, Approximate Computing, Fault-tolerant Applications, High-precision, Multiplier.*

I. INTRODUCTION

Stochastic computing (SC) uses stochastic bitstreams for approximate computations, with small area and low power consumption [1]. Initially proposed by Gaines et al., it's now utilized in image processing and neural networks [2]-[10]. Before computation, SC requires the use of a stochastic number generator (SNG) to convert binary streams into stochastic bitstreams. The computation results also need to be converted back into binary format using a counter. The value represented by a stochastic bitstream corresponds to the probability of a '1' in the stream, which is also the normalized value of the binary number. Hence, SC is also referred to as probabilistic computing. Fig. 1(a) illustrates the basic structure of a SNG, which generates stochastic bitstreams by continuously comparing random numbers with binary values. Fig. 1(b), (c), and (d) respectively depict the scaling adder, adder, and multiplier units of SC [11]-[12]. Fig. 1(d) shows the SC multiplier, implemented with an AND gate, needing a lengthy bitstream for accuracy [13]. In SC, the length of a stochastic bitstream corresponds to the number of computational cycles for a single operation. The number of computational cycles for a stochastic bitstream grows exponentially with the bit-width of the binary number it represents [14]. For a 4-bit binary number, one operation typically requires 2^4 clock cycles, whereas an 8-bit binary number requires 2^8 clock cycles. A corresponding bit-width SNG is also needed for conversion. SNGs consume significant area in SC systems, and the computational cycles imply considerable latency.

This paper presents an SC multiplier based on multiplier partitioning to address these issues. The proposed SC multiplier design completes calculations in fewer cycles. The SC multiplier uses low-bit-width SNG for bitstream conversion, reducing hardware and power consumption.

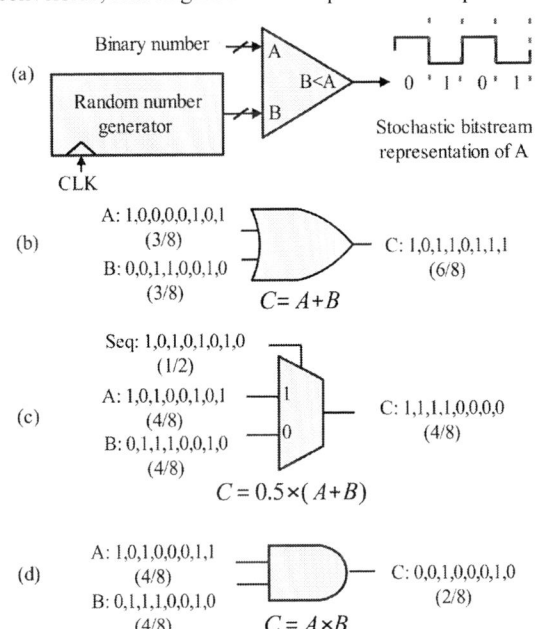

Figure 1. Basic stochastic computing units: (a) Stochastic number generator, (b) Adder, (c) Scaling adder, and (d) Multiplier.

The remainder of this paper is structured as follows: Section 2 introduces related advanced SC multiplier designs; Section 3 presents the proposed SC multiplier; Section 4 shows the experimental results of the multiplier; and finally, Section 5 concludes the paper.

II. RELATED WORKS

Fig. 2 introduces several design schemes of SC multipliers. Fig. 2(a) shows the classic SC multiplier [1], which utilizes two SNGs to convert each multiplier separately, and finally employs a counter for the reverse conversion. Fig. 2(b) uses a down counter to perform multiplication [9], requiring only one SNG. It is a form of Binary-Interfaced Stochastic Computing (BISC). A down counter is initialized with the other multiplier, completing the computation when it counts down to zero. This multiplier can also be configured as a Matrix-Vector Multiplier (MVM) for neural network acceleration. Fig. 2(c) shows the amplitude and frequency

979-8-3315-3918-4/25 $31.00 © 2025 IEEE

encoding (AFE) SC design [15], leveraging binary weights with an accumulator for conversion. This AFE method effectively reduces the length of random bit streams. However, when performing 8-bit multiplication, it still requires 29 clock cycles to meet the precision requirements of some applications.

III. THE PROPOSED SC MULTIPLIER

The proposed SC multiplier takes advantage of multiplier partitioning, enabling binary-to-bitstream conversion with low-bitwidth SNGs. It then produces several parallel stochastic bitstreams through Weighted AND and OR truncated operations. Fig. 3 shows the 8-bit implementation of the proposed SC multiplier. To implement an m-bit multiplication, only $2m/2$ clock cycles are needed, with a SNG of $m/2$ bits used. The above design is for even m; for odd m, a multiplier design of $m+1$ bits is required.

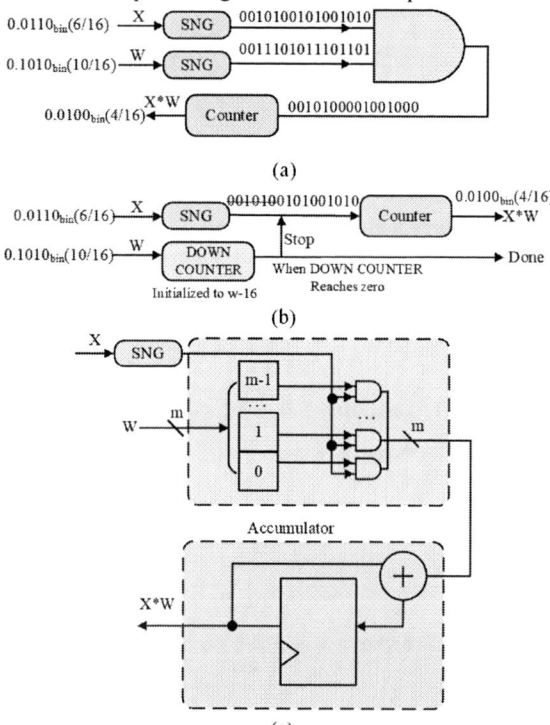

Figure 2. Several designs of SC multipliers: (a) Classic, (b) BISC and (c) AFE SC.

A. Multiplier Partitioning and Bitstream Conversion

Fig. 3(a) illustrates the details of multiplier partitioning and bitstream conversion. An 8-bit multiplier X is divided into two 4-bit multipliers. A 4-bit SNG is used to compare and generate stochastic bitstreams $f_1(i)$ and $f_0(i)$. Where $f_1(i)$ has a weight of 2^4, and $f_0(i)$ has a weight of 1. In the example shown in Fig. 3(a), $P(f_1(i)) = 7/16$ and $P(f_0(i)) = 12/16$ represent the normalized values of the two binary numbers obtained after partitioning X. The Fig. 3(a) shows that even using counter as SNG, the binary number is still accurately converted into stochastic bit streams.

B. Weighted AND

Fig. 3(b) shows the implementation circuit of the weighted AND gate multiplication. stochastic bitstreams $f_1(i)$ and $f_0(i)$ obtained from Fig. 3(a) are input separately into the Weighted

AND gates. The $f_n(i)$ undergoes weighted AND multiplication with each bit of the other multiplier Y. Each bit of Y and $f_n(i)$ has a weight. If the weight of $f_n(i)$ is 2^j, then after the AND operation with each bit of Y, the resulting weight is multiplied by 2^j. The bitstream $f_n(i)$ and the multiplier Y, when input into the Weighted AND gate, generate $S_n(i)$.

C. OR truncated

Fig. 3(c) shows the implementation circuit for the OR gate addition and truncation. Since the weight of $f_1(i)$ is 2^4 times that of $f_0(i)$, the weight of $S_1(i)$ is also 2^4 times that of $S_0(i)$. This means that the lower 4 bits of $S_1(i)$ have the same weight as the lower 4 bits of $S_0(i)$. Perform an OR operation on bits with equal weights and then truncate the last 4 bits of $S_0(i)$. When $S_1(i)$ and $S_0(i)$ are input into the OR truncated operation, an 8-bit $S'(i)$ is obtained. The $S'(i)$ can be considered as 8 parallel stochastic bit streams. These 8 parallel bits can be concatenated to form a binary number, which can be directly utilized by the accumulator. The SC multiplication result is obtained after $S'(i)$ passes through the accumulator.

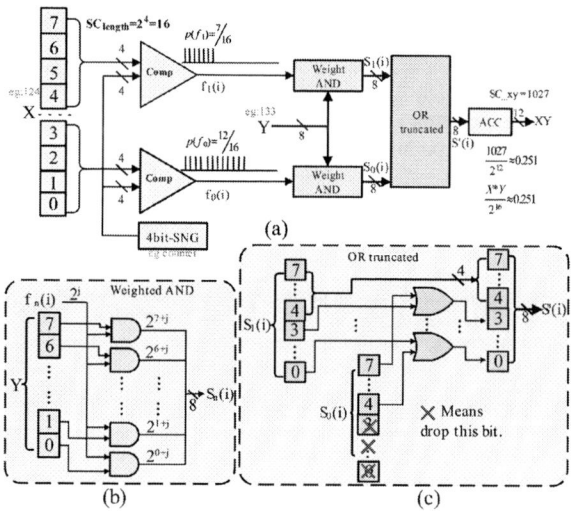

Figure 3. The 8-bit implementation of the proposed multiplier: (a) The multiplier partitioning and bitstream Conversion, (b) Weighted AND and (c) OR truncated.

TABLE I. MAE AND LATENCY OF DIFFRENT STOCHASTIC METHODS.

Method			Classic [2]	BISC [9]		AFE SC [15]		Proposed
SNG			LFSR	LFSR	Sobol	LFSR	Sobol	Sobol/ Counter
Bit width	4 bit	MAE	0.04	0.04	0.04	0.04	0.04	**0.02**
		SL	16	17	15	11	5	**4**
	8 bit	MAE	0.0083	0.008	0.008	0.008	0.008	**0.0053**
		SL	256	483	122	184	29	**16**
	10 bit	MAE	0.0042	0.004	0.004	0.004	0.004	**0.0026**
		SL	1024	2117	279	732	56	**32**
SNG has the same bit width as input data, SL means the Stream Length.								

IV. EXPERIMENTAL ANALYSIS

We implemented the proposed SC multiplier in circuitry and conducted a detailed comparison with other SC multipliers. The Mean Absolute Error (MAE) is used to represent the accuracy of the multiplier, and its calculation formula is as follows:

$$MAE(N) = \frac{\sum_{X=0}^{2^N-1}\sum_{Y=0}^{2^N-1}|SC(XY)-P(XY)|}{2^{2N}} \quad (1)$$

The proposed SC multiplier was implemented on a Field-Programmable Gate Array (FPGA), using the Vivado 2020.2 tool for synthesis and implementation. The power consumption reported is the dynamic power consumption from the power report.

A. Error Experiment

The accuracy of SC multipliers is typically influenced by the length of the stochastic bitstreams and the SNG. The Classic method employs a Linear Feedback Shift Register (LFSR) for its SNG, while the BISC and AFE SC methods use either an LFSR or a Sobol as SNG. The Sobol sequence, a type of low-discrepancy pseudo-random number sequence [16], can effectively reduce the length of bitstreams in SC. The proposed SC multiplier can use a counter or Sobol SNG for bitstream conversion. When using a counter, computations can be concluded early, similar to BISC. Table I. presents the MAE and required bitstream lengths for various SC multipliers across different bit-widths. For 8-bit multiplication, the bitstream length required by proposed SC multiplier decreases by 93.75%, 86.89%, and 44.83% compared to the Classic, BISC, and AFE SC methods, respectively. Moreover, the accuracy of our proposed SC multiplier is superior to other SC multipliers. This implies that other SC multipliers would require longer bitstreams to achieve the same level of accuracy as proposed SC multiplier.

TABLE II.   HARDWARE MEASURMENTS OF DIFFERENT METHODS WITH 10-BIT DATA.

Method	Area (LUT)	Latency (Clk)	Power (mw)	Energy (nJ)	Area-Efficiency
Classic[2]	24	1024	7	17.92	40.7
BISC[9]	36	279	7	4.88	99.6
AFE SC[15]	31	56	6	0.84	850.3
Proposed	19	32	5	0.40	1644.7
Binary	96	1	12	0.03	10416.7

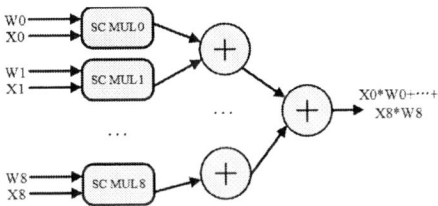

Figure 4. The MAC unit based on the SC multiplier.

B. Hardware Consumption

Table II. shows the hardware consumption of several SC multipliers at 10-bit. All SC multipliers are synthesized at 400 MHz. Using LUTs to represent area, Latency to indicate the number of clocks needed to complete one computation, Energy for the power consumed to complete one computation, and Area-Efficiency to denote the area efficiency. The calculation formula for Area-Efficiency is as follows:

$$\text{Area-Efficiency} = \frac{10^6}{LUT \times latency} \quad (2)$$

The proposed SC multiplier reduces the area by 80.21% compared to the binary multiplier. The proposed SC multiplier outperforms the Classic in terms of area, latency, and energy consumption. Compared with BISC and AFE SC, the proposed SC multiplier reduces the area by 65.38% and 9.52%, respectively, and decreases the energy consumption by 91.8% and 52.38%, respectively. Meanwhile, the area efficiency is enhanced by 16.5 times and 1.9 times, respectively.

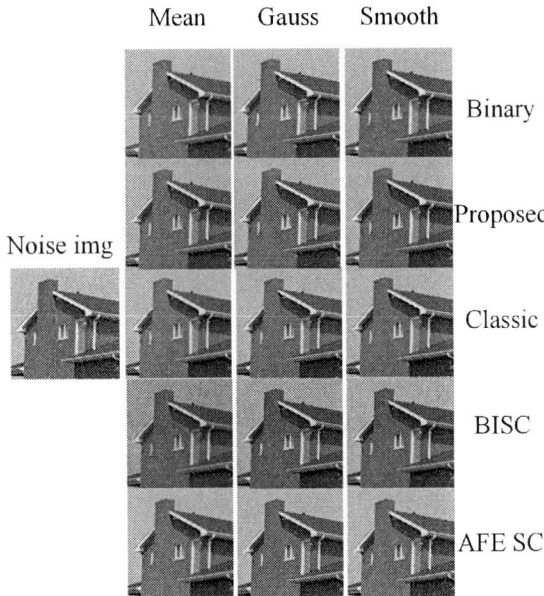

Figure 5. The practical effects of several filtering operators under different methods.

TABLE III.   THE AVERAGE PSNRs AND SSIMs OF MULTIPLIER FILTERS.

Method	Algorithm	house	orangutan	Cameraman
Classic[2]	Gaussian	25.09/0.7232	25.32/0.8362	25.94/0.7918
	Mean	23.01/0.8307	23.27/0.8916	23.95/0.8689
	Smoothing	28.53/0.8617	28.45/0.9260	29.09/0.8901
BISC[9]	Gaussian	26.47/0.9225	26.35/0.9625	31.81/0.9724
	Mean	23.46/0.9293	24.29/0.9508	23.81/0.9285
	Smoothing	31.91/0.9698	31.70/0.9867	26.31/0.9299
AFE SC[15]	Gaussian	39.91/0.9866	40.63/0.9915	38.71/0.9838
	Mean	38.73/0.9911	40.30/0.9940	35.93/0.9526
	Smoothing	37.08/0.9853	38.89/0.9914	39.52/0.9850
Proposed	Gaussian	53.68/0.9987	53.66/0.9994	53.71/0.9988
	Mean	37.16/0.9955	37.19/0.9975	37.39/0.9918
	Smoothing	55.55/0.9993	55.41/0.9997	55.51/0.9994

C. Image Applications

Fig. 4 illustrates the MAC architecture based on the SC multiplier, which has been implemented for image filtering applications. The 8-bit SC multiplier was employed in the MAC design. The image data are sourced from [17]. The images are processed using the Smooth, Gauss, and Means operators, with identical noise added to the images before processing. Based on the binary computation results, the Peak Signal-to-Noise Ratio (PSNR) and Structural Similarity (SSIM) metrics are calculated to evaluate the practical performance of the SC multipliers in image processing. Fig. 5 presents the actual effects of the house image after image processing. Table III. shows the SSIM and PSNR of several images processed using the SC multiplier. The proposed SC multiplier outperforms Classic, BISC, and AFE SC in implementing several image processing algorithms. The proposed SC multiplier achieves a PSNR greater than 50 dB

979-8-3315-3918-4/25 $31.00 © 2025 IEEE

when implementing the Gaussian and Smoothing algorithms. After being processed by the proposed SC multiplier, all images have an SSIM greater than 0.99 compared to those processed by the binary multiplier.

V. SUMMARY

This paper proposes a novel SC multiplier, which can reduce the bit-stream length while ensuring the required accuracy. The proposed SC multiplier reduces the bit-width of the SNG by partitioning the multiplier. The weighted AND operation ensures the accuracy of multiplication. The OR truncated reduces the resources required for the accumulator unit. When implementing 10-bit multiplication, the proposed SC multiplier reduces the area by 33.33% and the energy consumption by 97.77% compared with the classic one. The area is reduced by 80.21% compared with the binary multiplier. The proposed SC multiplier can be used for neural networks and image processing. It outperforms advanced SC multiplier designs in image processing algorithms such as smoothing, Gaussian, and Mean.

ACKNOWLEDGMENT

The research is supported by National Natural Science Foundation of China (92264105), the R&D Program of Beijing Municipal Education Commission (KM202211232007), the Beijing Municipal Natural Science Foundation (4232067), and the Young Elite Scientists Sponsorship Program by BAST (BYESS2023313).

REFERENCES

[1] A. Alaghi and J. P. Hayes, "Survey of Stochastic Computing," *ACM Trans. Embed. Comput. Syst.*, vol. 12, no. 2s, pp. 1–19, May 2013.

[2] B. R. Gaines, "Stochastic computing systems," in Advances in *Information Systems Science, J. T. Tou, Ed. New York, NY, USA: Springer*, 1969, pp. 37–172.

[3] X. Jia, H. Gu, and Y. Liu "An Energy-Efficient Bayesian Neural Network Implementation Using Stochastic Computing Method," *IEEE Trans. Neural Netw. Learning Syst.*, vol. 35, no. 9, pp. 12913–12923, Sep. 2024.

[4] J. Zhu, J. Wu, and Z. Yang, "An Area and Energy-Efficient Systolic Array Accelerator Architecture for Deep Neural Networks Using Stochastic Computing," in *IEEE Trans. VLSI Syst., vol.* 33, no. 6, pp. 1582-1595, June. 2025.

[5] A. Hu, W. Li, D. Lyu and G. He, "Efficient Parallel Stochastic Computing Multiply-Accumulate (MAC) Technique Using Pseudo-Sobol Bit-Streams," in *IEEE Trans. Nanotechnology*, vol. 23, pp. 170-179, Feb. 2024.

[6] M. Li Y. Hu, and T. Zhang, "Not your father's stochastic computing (SC)! Efficient yet Accurate End-to-End SC Accelerator Design," in *Proc. IEEE 15th Int. Conf. ASIC (ASICON), Nanjing, China*, Oct. 2023, pp. 1–4.

[7] H. Sim and J. Lee, "Cost-effective stochastic MAC circuits for deep neural networks," *Neural Netw.*, vol. 117, pp. 152–162, Sep. 2019.

[8] P. Schober, M. H. Najafi and N. TaheriNejad, "High-Accuracy Multiply-Accumulate (MAC) Technique for Unary Stochastic Computing," in *IEEE Trans. Comput.*, vol. 71, no. 6, pp. 1425-1439, June. 2022.

[9] H. Sim and J. Lee, "A New Stochastic Computing Multiplier with Application to Deep Convolutional Neural Networks," in *Proc. 54th ACM/EDAC/IEEE Design Autom. Conf. (DAC)*, Jun. 2017, pp. 1–6.

[10] W. Liu, S. Xiao, and Y. Liu, "SC-PLR: An approximate spiking neural network accelerator with on-chip predictive learning rule," *IEEE Trans. Biomed. Circuits Syst.*, vol. 18, no. 5, pp. 1156–1165, Oct. 2024.

[11] C.F.Frasser, P.Linares-Serrano, and I.D.d.l.Ríos, "Fully parallel stochastic computing hardware implemen tation of convolutional neural networks for edge computing applica tions," *IEEE Trans. Neural Netw. Learn. Syst.*, vol. 34, no. 12, pp. 10408–10418, Dec. 2023.

[12] S. Liu and J.Font-Rosselló, and S. Liu, "From Multipliers to Integrators: A Survey of Stochastic Computing Primitives," *IEEE Trans. Nanotechnology*, vol. 23, pp. 238–249, Mar. 2024.

[13] A. Alaghi and J. P. Hayes, "Exploiting correlation in stochastic circuit design," in *Proc. IEEE 32st Int. Conf. Computer Design (ICCD), Asheville, NC, USA*, Oct. 2013, pp. 39–46.

[14] A. Alaghi, W. Qian, and J. P. Hayes, "The promise and challenge of stochastic computing," *IEEE Trans. Comput.-Aided Design Integr. Circuits Syst.*, vol. 37, no. 8, pp. 1515–1531, Aug. 2018.

[15] Y. Chen and H. Li, "Stochastic Computing Using Amplitude and Frequency Encoding," *IEEE Trans. VLSI Syst.*, vol. 30, no. 5, pp. 656–660, May 2022.

[16] S. Liu and J. Han, "Energy efficient stochastic computing with sobol sequences," in *Proc. Design, Autom. Test Europe Conf. Exhib. (DATE)*, Mar. 2017, pp. 650–653.

[17] "The USC-SIPI image database." 1981. [Online]. Available: https: //sipi.usc.edu/database/.

A Low-Overhead Fault-Tolerant Design for Quantized CNN Accelerators

Shanqiang Yang [1], Chenxu Wang*[1,2], Lexiang Shen [1], Xinlei Su [1], Min Luo [1], Tianliang Xu [1], Ruoshi Li [1], Siyuan Wang [1]

[1] Harbin Institute of Technology, Weihai 264209, China
[2] Shandong Provincial Key Laboratory of Marine Electronic Information and Intelligent Unmanned Systems, Weihai 264209, Shandong, China

* Email: 23b921024@stu.hit.edu.cn, wangchenxu@hit.edu.cn

Abstract— Convolutional neural network (CNN) accelerators are susceptible to accuracy degradation due to weight perturbations caused by single-event upsets (SEUs), which threaten the reliability of mission-critical systems. Unlike previous high-cost strategies that focused on floating-point models or used triple modular redundancy (TMR) storage, this paper conducts a fine-grained fault analysis of INT8 quantization traffic sign recognition networks and finds that the directionality characteristics of SEUs have a significant impact on inference results. When the sign bit is 0, 0 to 1 flips are more likely to cause severe errors, while when the sign bit is 1, 1 to 0 flips are more likely to cause severe errors. Inspired by this, this paper proposes a fault-tolerant mechanism based on the voter and logical XNOR gate (VXNOR), which guides fault perturbations toward zero convergence through majority voting of the sign bit and selective correction of higher bits, thereby reducing their destructive impact on network behaviour. Large-scale injection experiments on the LeNet-5 and GTSRB datasets demonstrate that VXNOR achieves a protection efficiency of 94.1% while maintaining model accuracy and significantly reduces hardware overhead compared to existing methods, making it suitable for deployment in resource-constrained embedded systems.

Keywords—CNN Accelerator, SEU, Fault Injection, Fault-Tolerant Design, Reliability

I. INTRODUCTION

With the rapid development of artificial intelligence technology, convolutional neural networks (CNNs) have been widely applied in image processing applications due to their powerful data processing and learning capabilities, particularly in safety-critical applications such as aerospace, healthcare, and autonomous driving [1]. Thanks to the high flexibility and parallel processing capabilities of modern Field Programmable Gate Arrays (FPGAs), they are often used as hardware acceleration platforms for deploying CNNs. However, FPGAs based on SRAM are composed of a large number of memory cells that are highly sensitive to radiation effects. Single-event upsets (SEUs) caused by radiation effects can result in the model processing incorrect data, altering its output and classification results and potentially modifying the behavior of critical applications, which may lead to catastrophic consequences [2]. There have been tragic accidents involving vehicles equipped with intelligent driving systems colliding with trucks, with one possible cause being that a white truck was mistakenly identified as a

cloud [3]. There is an urgent need to enhance the reliability of FPGA-based CNNs in safety-critical applications.

Traditional fault tolerance methods for FPGAs mainly rely on hardware redundancy technology. Triple modular redundancy (TMR) is the gold standard for safety-critical systems, but it requires an additional 200% resource overhead, which is not feasible for resource-constrained embedded systems [4].

Previous literature has shown that the impact on CNN performance varies depending on the location of the error, and CNN models can tolerate minor errors in internal states or parameters [5]. In recent years, researchers have begun to attempt selective radiation-hardening designs targeting key SEUs that affect CNN performance in FPGAs.

Libano et al. [6] conducted neutron radiation experiments and fault injection simulation experiments on CNNs. Their research found that different layers have different levels of tolerance for errors. They, therefore, proposed a selective TMR targeting the most vulnerable layer to improve reliability.

Liu et al. [7] found that certain errors had no impact on the model results by examining the characteristics of the MLP neural network algorithm. They proposed a selective recalculation technique based on unreliable classification results, which saved more than 60% of the recalculation overhead compared to all redundant calculations.

Reference [8] proposes a minimal fault-tolerant method for low-bit floating-point neural networks, which mitigates the impact of 0-to-1 SEUs by performing DMR on highly sensitive bits in the exponent and mantissa and performing bitwise "AND" operations. However, this strategy relies on floating-point characteristics and is difficult to transfer directly to fixed-point networks, such as INT8.

This paper proposes a lightweight selective fault-tolerant technique based on an accurate analysis of INT8 weights on CNNs. This technique is referred to as selective hardening based on XNOR gates. The core idea of this technique is to move erroneous weights toward zero by combining DMR with XNOR gates for high-sensitivity bits. By preserving the sign bit TMR for low-sensitivity bits and protecting only sensitive weight layers, the hardware overhead is further reduced. Experimental results on the LeNet-5 model and GTSRB dataset demonstrate that this method enhances the model's fault tolerance capabilities with lower resource consumption and critical path overhead compared to other methods.

II. SEU-SENSITIVITY ANALYSIS

A. Use-case Neural Network architecture

This study employs the classic LeNet-5 convolutional neural network as the evaluation model, with the task objective being German Traffic Sign Recognition (GTSRB). The primary consideration for selecting the GTSRB dataset is its significant application value in the field of autonomous driving: accurate traffic sign recognition is critical for the safe operation of autonomous vehicles, and any recognition errors could lead to severe traffic accidents. The GTSRB dataset includes 43 categories of German traffic signs. The training set comprises 21,312 color images of 32×32 pixels, and the test set contains 5,328 images, encompassing various safety-critical types, including speed limit signs, warning signs, and prohibition signs.

The LeNet-5 model consists of two convolutional layers (CONV-1 and CONV-2) and three fully connected layers (FC-1, FC-2, and Output) with 64,811 quantized parameters. An FP32 model is first trained, achieving 98.89% accuracy. It is then quantized to the INT8 format with minimal accuracy loss (98.87%), demonstrating effective quantization.

Fig. 1 shows the distribution histogram of the weights of each layer after quantization. Although INT8 supports the full numerical range of [-128, 127], the actual weights exhibit a highly concentrated distribution with a peak near zero. This concentrated distribution is a result of the regularization mechanism during deep learning training and also provides important evidence for subsequent selective protection strategies.

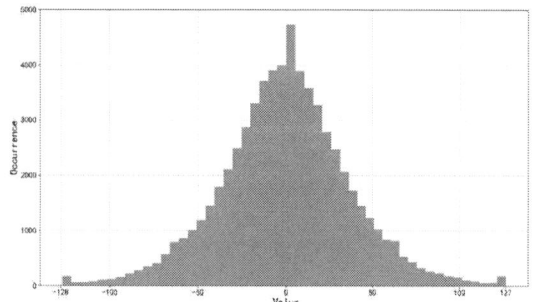

Fig. 1. Histograms for weights.

B. Fault Injection Framework

This paper considers SEUs that occur during inference and inject faults into the network weight data. The fault injection process is shown in Fig. 2. For weight parameters, fault injection is limited to the convolutional layer and the fully connected layer. Once the fault is injected, the model inference is performed using the new parameters after updating the faulty weight parameters. The fault injection adopts a systematic bit flip strategy to inject SEU errors into the INT8 format weights of the LeNet-5 model layer by layer and bit by bit.

SEU injection activities are exhaustive, exploring potential fault bits at each layer. The results obtained are compared with the accuracy inferred by the fault-free model of the CNN. If there is a significant decrease in accuracy after fault injection, that layer is marked as a sensitive layer, and the corresponding bit is marked as a sensitive bit.

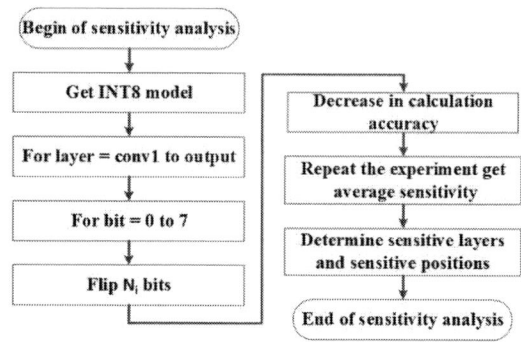

Fig. 2. Fault injection flow diagram.

C. Impact of the SEU direction

This section analyzes the impact of SEU direction on neural network accuracy, i.e., the case where a bit flips from 0 to 1 or from 1 to 0. To better understand this phenomenon, we first collected all weight value distributions during the fault-free model execution process, as shown in Fig. 1, and then performed fault injection analysis on these values.

Fig. 3 shows the effect of SEUs on the weight range of the INT8 format. As can be seen from the figure, for the INT8 format, faults in the higher bits have a more serious impact than those in the lower bits because the higher bits have greater numerical weight. When the higher bits are upset, the change in value is more significant than when the lower bits are upset.

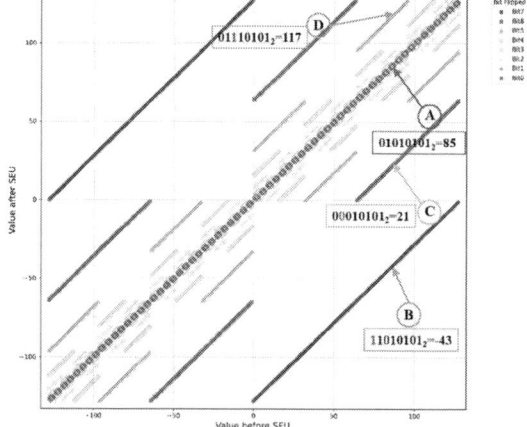

Fig. 3. SEUs impact on the int8 format.

To further analyze the impact of SEU, consider the following example: Take the value 85, represented as 01010101_2 in INT8 format, marked as point A in Fig. 3. When the sign bit (the leftmost bit) undergoes a bit flip, it changes from 01010101_2 to 11010101_2. Since negative numbers are represented using two's complement, this value represents -43, marked as point B. When the second-highest significant bit (the 6th bit) flips from 1 to 0, the weight value decreases toward 0, and the fault weight value becomes 21, marked as point C. When the 5th bit changes from 0 to 1, the value suddenly increases, and the faulty data value becomes 117, marked as point D. Finally, if a lower bit changes from 0 to 1, such as the 1st bit flipping from 01010101_2 to 01010111_2, the value becomes 87, with a relatively small

979-8-3315-3918-4/25 $31.00 © 2025 IEEE

change. This aligns with the typical operating range of a neural network and has minimal disruptive impact on the network.

The histogram indicates that most weights are concentrated near zero, with ~85% within the range of [-20, 20]. In INT8 two's complement, the direction of faults significantly affects value deviation. For positive weights (sign bit = 0), a 1 to 0 flip reduces the magnitude, pushing the value toward zero; for negative weights (sign bit = 1), a 0 to 1 flip has a similar converging effect. Conversely, 0 to 1 flips in positive numbers or 1 to 0 flips in negative ones drive weights away from zero, causing more severe network disruption. This asymmetric fault impact differs from floating-point formats, where 1 to 0 flips generally reduce magnitude regardless of sign.

Leveraging this directional behavior enables lightweight protection by guiding faulty weights toward zero. TABLE I confirm this: bit-7 flips drop accuracy to 53.54% in CONV-1 and show a 22.1% average loss; bits 6 and 5 reduce performance by 13.6% and 7.7%, respectively, while lower bits cause <3% degradation. Bits 1 and 0 have negligible effects, validating the "high-bit sensitivity" pattern.

TABLE I.   FAULT INJECTION IMPACT TABLE

Layers	Accuracy(%)				
	CONV-1	CONV-2	FC-1	FC-2	FC-3
bit7	53.54	62.44	83.55	93.16	91.24
bit6	67.09	71.09	94.41	97.13	97.02
bit5	76.68	86.62	96.86	97.96	97.76
bit4	90.91	94.31	97.46	97.83	97.71
bit3	96.93	97.14	97.64	98.10	97.80
bit2	97.48	97.55	98.09	98.30	97.84
bit1	98.01	98.31	98.75	98.69	98.80
bit0	98.70	98.74	98.85	98.86	98.79

III. PROPOSED METHODOLOGY

A. VXNOR Hardening Mechanism

This section introduces hardware protection technology using VXNOR (a Voter block based on the logical XNOR gate), which is driven by the error analysis presented in Chapter 2. The primary objective is to reduce the fault value toward zero. The VXNOR technology is shown in Fig. 4.

Based on the analysis results in Chapter II, the design of the VXNOR protection mechanism follows the following core principles:

- Sign bit TMR protection: Since the INT8 sign bit is the most sensitive and single-bit flips cause significant accuracy degradation, TMR protection is applied to it. To minimize storage overhead, two backup copies are innovatively stored in the originally less sensitive bit positions (bit1 and bit0), thereby avoiding the need for additional storage resources.

- High-sensitivity bit VXNOR protection: VXNOR protection targets bit6 and bit5 using DMR for error detection. When inconsistencies are detected, the faulty bit is assigned the current sign bit value to shift

weights toward zero. This strategy exploits the directional characteristics: both positive and negative weights converge toward zero when corrected with their respective sign bit values.

- Low-sensitivity bit processing: Bits 4-2 remain unprotected, utilizing the CNN's natural fault tolerance. Bits 1 and 0 directly use the voted sign bit values, enhancing zero-convergence effects.

Fig. 4 shows the detailed implementation of the 8-bit weighted VXNOR protection circuit. The entire protection mechanism requires only minimal additional hardware resources: three majority voters for the sign bit TMR, two XNOR gates for consistency detection of high-sensitivity bits, and corresponding multiplexers. The critical path introduces only a delay of one XNOR gate and one MUX, which has a negligible impact on system performance.

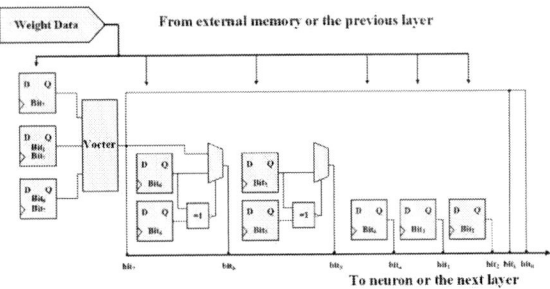

Fig. 4. VXNOR protection hardware implementation.

When a fault occurs, the decision logic of the VXNOR voter is shown in TABLE II. If the sign bit is 0 (positive number), the fault weight position is set to 0; if the sign bit is 1 (negative number), the fault weight position is set to 1. This design ensures that the weight value after a fault always approaches zero, minimizing the interference of SEUs on network functionality.

TABLE II.   VXNOR TRUTH TABLE

Sign	DMR bit		Output bit
	Sensitive bit	Duplicated bit	
0	0	0	0
0	0	1	0
0	1	0	0
0	1	1	1
1	0	0	0
1	0	1	1
1	1	0	1
1	1	1	1

B. Selective Layer Protection Strategy

Based on the large-scale SEU injection experiments in Section III-A, it can be observed that the vulnerability of each layer of the network to bit flips exhibits a strongly heterogeneous distribution: CONV-2 and the output layer are most susceptible to misflips being amplified to classification results, followed by CONV-1. In contrast, the two fully connected layers, relying on feature sparsity and subsequent nonlinear filtering, can almost absorb errors through the model's redundancy. If fault-tolerant logic is uniformly embedded in all layers, it not only fails to improve accuracy but also dilutes the limited hardening budget. Therefore, this paper follows the "sensitive layer priority" principle, localizing the sign bit TMR and bit 6/bit 5 VXNOR protection to the CONV-2 and output layer and extending it to the CONV-1 when resources permit; FC-1 and FC-2 remain in their original implementations. This strategy employs a joint sensitivity ranking of channels and bits, inserting lightweight voters before weight access paths that are most likely to cause decision collapse while avoiding additional processing for layers with strong error self-suppression capabilities. This maximizes overall accuracy recovery effects without disrupting the original data flow and timing.

IV. EXPERIMENTAL RESULTS

On the Zynq-7020, use Vivado HLS/Vivado 2021.1 to synthesize INT8-quantized LeNet-5 (GTSRB baseline accuracy of 98.2%). Perform fault injection and inference testing on the protected model.

A. Protection Performance

Under a 10% SEU injection, the accuracy of the unhardened model drops to 30.12%; however, by adopting selective VXNOR, it can be restored to 94.10%, representing a 64% improvement over the baseline and only 4.8 percentage points lower than the fully TMR (98.87%). Even with a fault injection rate increased to 20%, the method described in this paper still performs well. As shown in TABLE III, results demonstrate that VXNOR can maintain robustness comparable to TMR under extreme fault rates while significantly reducing area.

TABLE III.   PERFORMANCE COMPARISON OF REINFORCEMENT METHODS

SEU rate	Accuracy(%)		
	No	*TMR*	*VXNOR(Our)*
10%	30.12	98.87	94.10
20%	10.30	97.62	92.83

B. Hardware Overhead Analysis

Fig. 5 shows the area increment trends of the two mechanisms as the protection bit width is gradually expanded.

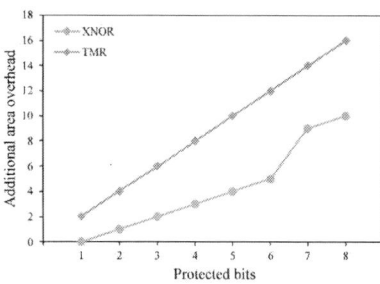

Fig. 5. VXNOR comparison against TMR hardening methods in additional area overhead.

The slope of TMR is almost twice that of VXNOR: when protecting up to bit 7, TMR requires twice the additional logic, while VXNOR only requires an additional 1.25 times. This gap stems from VXNOR's minimalist structure, which

requires the least additional hardware resources and adds only one XNOR gate and one 2:1 MUX on the critical path, resulting in the lowest consumption of flip-flops and LUTs; its arbitration logic is a purely combinational path, having almost no impact on timing. In summary, VXNOR maintains high fault tolerance while reducing area and clock overhead to less than half that of TMR, making it particularly suitable for resource-constrained edge CNN accelerator scenarios.

V. CONCLUSION

In this paper, we first conduct a comprehensive analysis of the SEU impact on the INT8 quantized LeNet-5 network for traffic sign recognition, which can be deployed on FPGA accelerators. We observe sensitivity differences between different layers of the CNN. The CONV-1 and CONV-2 layers are the most sensitive to SEU in the weight parameters. Through the study of SEU injection into weight features, we observe that CNNs are more tolerant to faults close to zero. Based on this observation, we propose VXNOR, which first performs TMR on the most critical bits with zero additional overhead, followed by DMR on bits 6 and 5. Finally, a Voter module based on XNOR gates is applied to control the direction of faults, causing faulty weights to approach zero. Experimental results show that for models using the INT8 data format, a protection efficiency of 95.46% is achieved with minimal additional hardware overhead.

ACKNOWLEDGMENT

This work was mainly supported by Major scientific and technological innovation projects of Shandong Province of China, with Grant No.2022ZLGX04. The research presented in this paper is also partially supported by the NSF project of China with granted No. U2106202 and Shandong Provincial Natural Science Foundation with Grant ZR2023MA074.

REFERENCES

[1] S. K. Venkataramanaiah, J. Meng, H.-S. Suh, I. Yeo, J. Saikia, and S. K. Cherupally, "A 28-nm 8-bit Floating-Point Tensor Core-Based Programmable CNN Training Processor With Dynamic Structured Sparsity," IEEE J. Solid-State Circuits, vol. 58, no. 7, pp. 1885–1897, Jul. 2023.

[2] W. Guillemé, A. Kritikakou, Y. Helen, C. Killian, and D. Chillet, "VANDOR: Mitigating SEUs into Quantized Neural Networks," in 2024 IEEE 30th International Symposium on On-Line Testing and Robust System Design (IOLTS), Rennes, France: IEEE, Jul. 2024, pp. 1–6.

[3] H. Wang, W. Shao, C. Sun, K. Yang, D. Cao, and J. Li, "A Survey on an Emerging Safety Challenge for Autonomous Vehicles: Safety of the Intended Functionality," Engineering, vol. 33, pp. 17–34, Feb. 2024.

[4] M. Nicolaidis, "Design techniques for soft-error mitigation," in 2010 IEEE International Conference on Integrated Circuit Design and Technology, Jun. 2010, pp. 208–214.

[5] Y. Li, Y. Liu, M. Li, Y. Tian, B. Luo, and Q. Xu, "D2NN: a fine-grained dual modular redundancy framework for deep neural networks," in Proceedings of the 35th Annual Computer Security Applications Conference, San Juan Puerto Rico USA: ACM, Dec. 2019, pp. 138–147.

[6] F. Libano, B. Wilson, J. Anderson, M. J. Wirthlin, C. Cazzaniga, and C. Frost, "Selective Hardening for Neural Networks in FPGAs," IEEE Trans. Nucl. Sci., vol. 66, no. 1, pp. 216–222, Jan. 2019.

[7] S. Liu, P. Reviriego, and F. Lombardi, "Selective Neuron Re-Computation (SNRC) for Error-Tolerant Neural Networks," IEEE Trans. Comput., vol. 71, no. 3, pp. 684–695, Mar. 2022.

[8] W. Guillemé, A. Kritikakou, Y. Helen, C. Killian, and D. Chillet, "HTAG-eNN: Hardening Technique with AND Gates for Embedded Neural Networks," in Proceedings of the 61st ACM/IEEE Design Automation Conference, San Francisco CA USA: ACM, Jun. 2024, pp. 1–6.

A Real-Time and Reconfigurable Pre-Driver Design for ABS Solenoid Valve Applications

Zhinan Li [1], Yitian Su [1], Shaochen Han [1], Huihong Zhang [*1], Yuejun Zhang [1], Cang Liu [*23]

[1] Faculty of Electrical Engineering and Computer Science, Ningbo University, Ningbo 315211, China
[2] Ningbo Yonghua Innovation Science and Technology Development Co., Ltd, Ningbo, Zhejiang, 315211, China
[3] Department of Electronic Engineering, Tsinghua University, Tianjin, 300467, China

* Email: 2411100222@nbu.edu.cn, zhanghuihong@nbu.edu.cn, liucang@nbyhcx.com

Abstract—The Anti-lock Braking System (ABS) solenoid valve pre-driver chip, as a key component in controlling the operation of the entire ABS system, has its driving capability directly determining the reliability of the ABS. However, the current ABS solenoid valve pre-driver chips are all used as electronic control units to control the solenoid valve. The command operation time interval is relatively long, which results in weaker real-time performance. To address the real-time and flexibility issues of the ABS solenoid valve pre-driver chip, this paper proposes a design scheme for a digital pre-driver circuit based on the collaboration of High-Side Driver (HSD) and Low-Side Driver (LSD). In the proposed scheme, a hardware-based fault transmission mechanism is introduced to reduce the instruction input operations, while adding parameter configuration registers, thus enabling the design of a high real-time, high-flexibility ABS pre-driver circuit. Experimental results show that the proposed design scheme requires only one clock cycle for the instruction transmission between HSD and LSD, significantly improving real-time performance and flexibility. Compared to mainstream chips, the proposed scheme operates at a frequency of 100 MHz, with configurable fault filtering time for each fault, and an output precision of 16 bits.

Keywords—ABS, Pre-driver circuit, Real-time optimization, HSD, LSD

I. INTRODUCTION

With the rapid development of the automotive industry and the increasing speed of vehicles, people are paying more attention to automotive safety performance. The Anti-lock Braking System (ABS), as a key technology for enhancing vehicle active safety, aims to prevent wheel lock-up during the braking process by automatically adjusting the braking force [1]. In the complex technological system of ABS, the solenoid valve, as the core component of the actuator, plays a critical role. The driving capability of the ABS system's solenoid valve directly determines the precision and response speed of dynamic brake force adjustment [2]. The core control circuit of the ABS solenoid valve, the ABS chip, monitors the wheel's motion status in real-time, analyzes complex road surface information, and dynamically adjusts the brake force distribution. Under extreme conditions, it effectively maintains the tire-road friction, thereby

This work was supported in part by the Ningbo University "Double First-Class" Cooperation Special Directional Entrusted Scientific and Technological Cooperation Project under Grant HX2024000574, HX2025000106, in part by the Cixi Science and Technology Program under Grant CZ2025006, in part by the Major Special Project of China Innovation Challenge (Ningbo) from Ningbo Science and Technology Program under Grant 2024T016, in part by Yongjiang Talent Project Youth Innovation Project 2024A-275-G, and in part by the Ningbo University Student Research and Innovation Program 2025SRIP1309.

effectively suppressing secondary risks such as loss of steering control and brake-side slipping [3].

The application of High-Side Driver (HSD) and Low-Side Driver (LSD) modules in ABS chips is widespread. The integration of these two modules enables intelligent control to a certain extent [4][5]. Common ABS control chips with HSD and LSD in mainstream solutions include the 33SB0400, SC900719, and L9388, among others. Since the solenoid valve needs to respond within a short time during operation, common driver chips communicate with the external system through an SPI bus for connectivity and communication. Additionally, when a fault occurs in the solenoid valve, the separation of control for HSD and LSD results in weaker real-time performance. Moreover, when faced with varying operating conditions, the internal configuration is limited, leading to lower flexibility.

To address the related issues with the solenoid valve driver chips, this paper proposes a digital pre-driver circuit for driving ABS solenoid valves. The driver circuit includes both the HSD and LSD sections. To address the real-time performance issue, this paper introduces a hardware-based fault transmission mechanism to reduce instruction operations, hardware-implementing software functions to enhance the real-time response of instructions during fault occurrences. Compared to the chips mentioned, additional configuration registers will be added to enhance the flexibility of pre-driver circuit.

The remainder of this paper is organized as follows: Section II introduces the working principles of the ABS solenoid valve. Section III elaborates on the design concepts and methods for each module in detail. Section IV presents the simulation results and compares them with related work to validate the effectiveness and superiority of the proposed scheme. Finally, the conclusion of this paper is drawn.

II. PRELIMINARIES

A. Working Principle of the ABS Solenoid Valve

The ABS system monitors the wheel's angular acceleration and slip ratio in real-time through wheel speed sensors. When a tendency for wheel lock-up is detected, the control circuit triggers the high-frequency switching solenoid valve. The magnetic flux generated by the solenoid coil drives the valve core to overcome the spring force, achieving millisecond-level opening and closing. By adjusting the duty cycle of the inlet/outlet valves, a pulsed hydraulic pressure is formed for brake fluid. This process involves three stages of dynamic control: when the slip ratio exceeds the limit, the inlet valve is first closed to maintain pressure. If the slip continues to increase, the outlet valve is opened to release pressure. Once the tire's traction is restored, the inlet valve is reopened to increase pressure, thereby maintaining optimal braking performance within the best slip ratio range.

Fig. 1 shows the schematic diagram of the ABS solenoid valve integration circuit. During the operation of the ABS solenoid valve, the HSD is primarily responsible for switching the high-side Metal-Oxide-Semiconductor Field-Effect Transistor (MOSFET), ensuring that the solenoid valve receives power during its operation. The LSD is primarily responsible for switching the low-side power MOSFET and controlling its current state. Specifically, the LSD outputs a controllable pulse wave to the Gate terminal of the low-side power MOSFET. By continuously adjusting the on/off state of the low-side MOSFET, it controls the current, thereby regulating the valve pressure. From the circuit perspective, the current control logic is primarily handled by the LSD. When a fault occurs in the low-side MOSFET, the low-side MOSFET output should be immediately turned off. However, there will still be current within the ABS solenoid valve, which needs to be dissipated through the high-side MOSFET to release this energy.

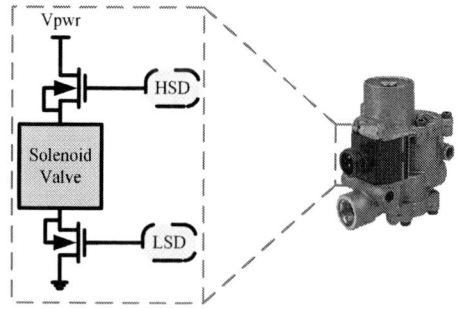

Fig. 1. Circuit schematic diagram of ABS solenoid valve.

III. PROPOSED PRE-DRIVE CIRCUIT

A. Pre-driver Circuit

The pre-driver circuit (Fig.2), as the core execution unit for controlling the ABS solenoid valve, performs its key function through the collaborative action of the HSD and multiple LSD output control signals. After conversion by the analog circuit, it achieves precise gate driving of the power MOSFETs. From the circuit topology perspective, the pre-driver module adopts a dual-channel isolated drive architecture. The HSD is responsible for controlling the switching state of the high-side MOSFET, while the LSD outputs pulse width modulation (PWM) signals to control the conduction and switching of the low-side MOSFET, as well as regulating the current through the solenoid valve.

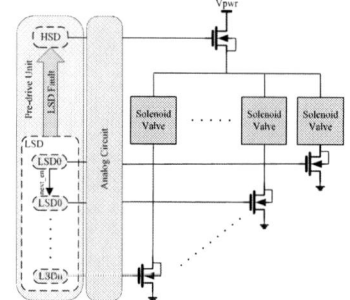

Fig. 2. Pre-driver circuit architecture.

As the main module for controlling the current through the solenoid valve, the LSD's operational stability directly impacts the safety of power devices and actuators. Compared to the HSD, the LSD is responsible for more complex dynamic detection tasks. Solenoid valve is an inductive load, and when the LSD experiences an abnormal sudden shutdown, the internal energy of the solenoid valve cannot be dissipated. Therefore, the fault needs to be reported to the HSD, which must respond accordingly to prevent damage to both the solenoid valve and the MOSFET.

B. HSD Architecture

The HSD adopts a layered design, integrating control logic, fault detection, and power driving functions. It stores parameters (such as enable signals, output status, and fault signal filtering time) in configuration registers through the AMBA bus. Additionally, it uses multiplexers to dynamically select signal paths. After processing the control signals through the analog circuitry, they drive the high-voltage power devices, directly controlling the switching of large power loads such as the solenoid valve. At the same time, the load current is monitored in real-time to ensure operation within safe thresholds.

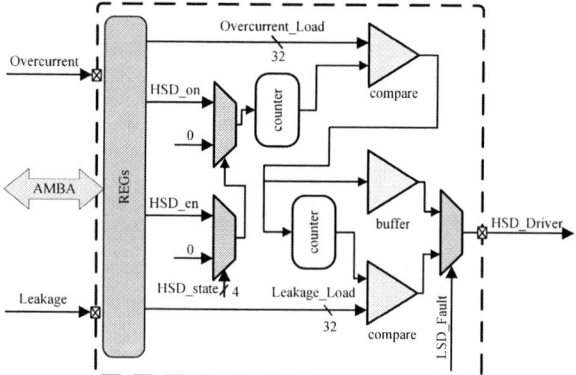

Fig. 3. HSD circuit architecture.

In terms of signal flow, the module begins with register configuration, including the loading and configuration of signals such as *HSD_en*, *HSD_on*, and *Overcurrent_Load* for the HSD digital control circuit. As shown in Fig. 3, the actual digital signal controlling the external power section is *HSD_on*. Before enabling it, the enable signal *HSD_en* must be activated, and only then can the *HSD_on* signal be controlled. After activating the *HSD_on* signal, load leakage detection is performed by comparing the duration of the signal from the Source terminal of the high-side MOSFET, input through the analog circuit to the digital control circuit, with the configured 32-bit load value. Based on this comparison, a decision is made whether to activate the external power driving source. In the 32-bit load value, the high 16 bits are used as the shielding time, while the low 16 bits are used as the filtering time. During the activation process, overcurrent detection is simultaneously enabled. The voltage difference between the Drain and Source terminals of the high-side MOSFET is level-shifted and used as an overcurrent signal. This signal is then passed through the analog circuit to the digital control circuit. The duration of the signal is compared with the configured load value to determine if an overcurrent fault has occurred. Similarly, in the 32-bit load value, the high 16 bits represent the shielding time, and the low 16 bits represent the filtering time. During

979-8-3315-3918-4/25 $31.00 © 2025 IEEE

the fault response stage, the fault signal is reported to the system, and manual reset or clearing instructions are required to clear the latch. If the fault persists, the system enters a latch state to prevent further damage, significantly enhancing the system's flexibility. Finally, the HSD outputs the processed signal through the analog circuit to the gate terminal of the high-side MOSFET, completing the power supply to the solenoid valve and forming a closed-loop control from instruction to execution.

It is worth noting that there is also a fault signal input from the LSD, which can directly control the *HSD_Driver*. When a fault occurs in the LSD, since the solenoid valve is an inductive load, interrupting the LSD path generates an inverse induced electromotive force. Therefore, the HSD needs to delay for a period before shutting down to safely dissipate the energy. Fig. 3 shows that when the LSD generates a fault signal, the output driving signal is passed through a delay timer before being output to the high-side MOSFET, preventing damage to the low-side MOSFET.

C. LSD Architecture

The LSD(Fig.4) also adopts a layered design, integrating control logic, fault detection, and current control functions to form a complete high-reliability system. The module configures key parameter thresholds via an AMBA bus, including fault signal filtering time, driving signal frequency, duty cycle, and other parameters. These parameters define the boundary conditions for the system's normal operation. The control signals generated by the *LSD_Driver* are adjusted through analog circuits to regulate the working state of the solenoid valve, ensuring real-time constraints.

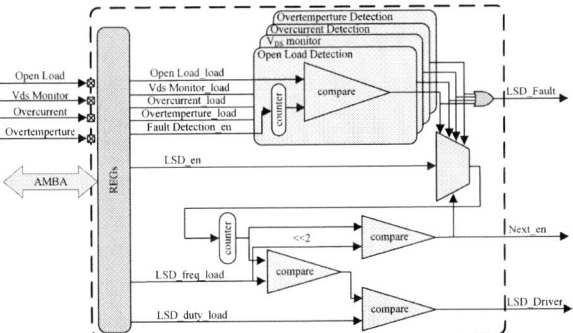

Fig. 4. LSD circuit architecture.

In terms of fault detection, the LSD reserves four external fault detection interfaces: open-circuit detection, source-drain voltage monitoring, overcurrent detection, and overtemperature detection. In the detection strategy section, the LSD digital control circuit uses a parallel comparison approach, synchronously monitoring different risk sources through separate channels. Overcurrent and overtemperature detection are similar to HSD fault detection. Both involve continuously monitoring external analog signals input to the digital control circuit during the driving process. When the duration exceeds the 16-bit filtering load value configured in the register, an overcurrent or overtemperature fault is generated. This triggers feedback to the *LSD_en* signal, setting it to 0 and shutting down the *LSD_Driver*. The open-circuit detection and source-drain voltage monitoring differ. Both fault detection sources involve monitoring the voltage between the Drain and Source terminals of the low-side MOSFET, which is input through the analog circuit into the

LSD. However, open-circuit detection occurs when the *LSD_Driver* is turned off, while source-drain voltage monitoring is performed when the *LSD_Driver* is turned on.

Another key feature of this module is the regulation of current through the LSD using PWM signals. During the register configuration process, configuring the 16-bit *LSD_freq_load* register and the 16-bit *LSD_duty_load* register determines the frequency and duty cycle of the pulse signal. Then, through the coordinated operation of the counter and comparator, the original clock signal is divided, generating the desired pulse signal. The pulse signal is output to the analog section for level conversion, and then applied to the Gate terminal of the LSD power-driving MOSFET. Through the continuous switching of the MOSFET via the PWM signal, fine control of the current output is achieved. It is worth noting that the entire ABS system requires multiple solenoid valve drivers. To reduce the issue of uneven braking force caused by sharp fluctuations in the hydraulic system pressure. The control signals of multiple solenoid valves are staggered by 1/4 of the cycle, ensuring that each solenoid valve operates at different time points, thereby maintaining the stability of the braking system.

IV. PERFORMANCE ANALYSES

A. Function Simulation

The real-time and configurable solenoid valve pre-driver circuit proposed in this paper is first behaviorally simulated using the joint simulation of Synopsys' VCS and Verdi tools. This simulation aims to verify the flexibility of internal register data and the real-time coordination between HSD and LSD during operation.

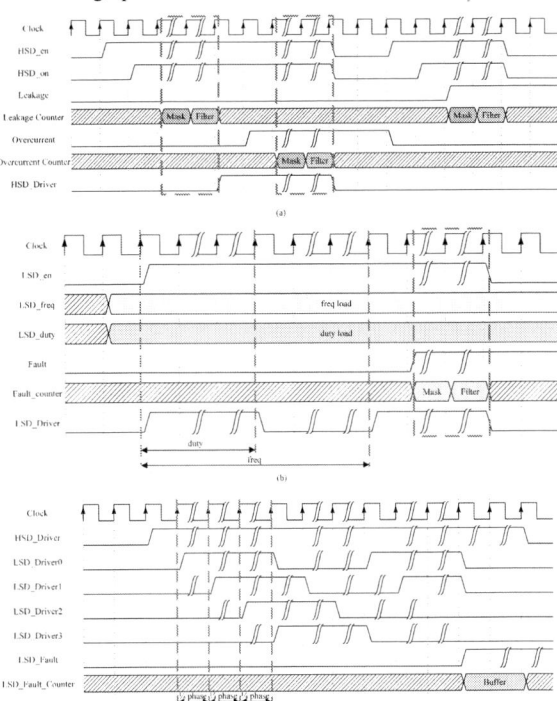

Fig. 5. Simulation waveform (a) HSD simulation (b) LSD simulation (c) Pre-driver circuit simulation.

Fig. 5(a) shows the functional simulation waveform of the HSD. Before activating the HSD, a leakage detection is performed. The system has set shielding time and filtering

979-8-3315-3918-4/25 $31.00 © 2025 IEEE

time. If no leakage is detected, the *HSD_Driver* is activated; otherwise, the *HSD_Driver* is not allowed to be turned on. During the activation process, overcurrent detection is simultaneously performed. When the overcurrent signal is input and its duration exceeds the shielding time and filtering time set in the register, a fault occurs, and the *HSD_Driver* is shut down. Fig. 5(b) shows the functional simulation waveform of the LSD. Before starting the LSD, the output frequency and duty cycle need to be configured in the register. Once the enable signal is activated, the output control is completed. For fault detection, an internal filtering time is set. When a fault signal is input, fault detection counting is triggered. If the count exceeds the filtering time load value, the *LSD_Driver* is turned off. Fig. 5(c) shows the functional simulation waveform of the entire pre-driver system. From the perspective of the solenoid valve, the startup process requires first activating the *HSD_Driver* as the power input, followed by the activation of the *LSD_Driver* to complete the entire current loop. During operation, it is observed that the output of each LSD is phase-shifted by 1/4 of the cycle. Additionally, when an LSD fault occurs, the *HSD_Driver* output is delayed before being shut down.

B. Performance Analysis

The logic synthesis part is completed in the TSMC 65nm environment using the Synopsys Design Compiler tool, while the layout is carried out using the Synopsys IC Compiler tool. The pre-driver circuit uses one HSD module and one LSD module. Its layout and related data are shown in Fig. 6, occupying an area of 0.034mm² with an average power consumption of 0.8454mW.

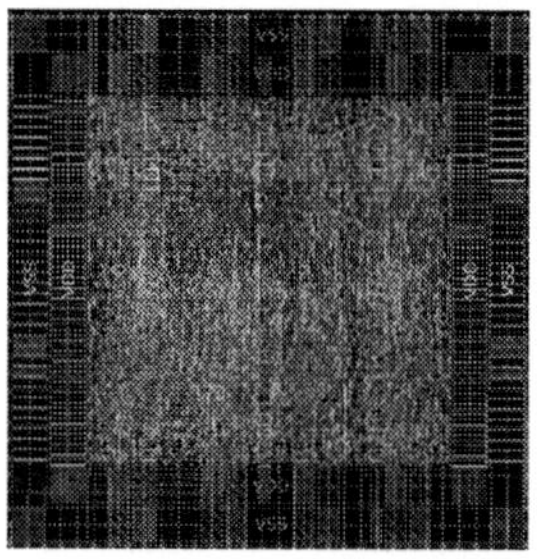

Pre-drive Circuit Feature	
Technology	TSMC 65nm
Area	0.034mm²
NAND	5.026Kgates
Voltage	1V
Frequency	100MHZ
Power	0.8454mW

Fig. 6. Pre-driver circuit layout design.

As shown in Table I, compared to other mainstream chips, when a low-side MOSFET fault occurs, it is necessary to control the HSD to delay shutdown. The circuit proposed in this paper requires one clock cycle to issue the corresponding instruction. Compared to mainstream chips, the frequency is higher, which leads to better real-time performance in the solution presented in this paper. Additionally, there are more filter time registers for external faults, and the output offers higher precision and greater flexibility.

TABLE I. PERFORMANCE COMPARISON

	33SB0400	L9388	SC900719	This work
Operating Frequency	10Mhz	16Mhz	10Mhz	100Mhz
Interface	SPI	SPI	SPI	AMBA
Every Fault Filter Time Configurable	NO	NO	NO	YES
Output Accuracy	8bit	8bit	8bit	16bit

V. CONCLUSION

This paper addresses the limitations of conventional ABS solenoid valve pre-driver chips regarding real-time performance and flexibility, proposing a digital pre-driver circuit design solution based on the cooperative control of HSD and LSD. Through a hardware-based fault response mechanism and an AMBA bus dynamic configuration architecture, the proposed solution effectively resolves the issue of delayed fault handling caused by SPI bus communication latency in existing schemes. Furthermore, it overcomes the limitation of traditional fixed filter time configurations that cannot adapt to complex operating conditions. In the future, the circuit proposed in this paper will be mounted onto a RISC-V core to further enhance real-time performance, and the analog circuit portion between the pre-driver and the MOSFETs will be completed.

REFERENCES

[1] H. Yao, Q. Li and J. Leng. "Physics-informed multi-step real-time conflict-based vehicle safety prediction," Accid. Anal. Prev., vol. 182, no. 2023, pp. 106965, March 2023.

[2] H. Koylu, E. Tural, "Experimental study on braking and stability performance during low speed braking with ABS under critical road conditions," Eng. Sci. Technol., vol. 24, no. 5, pp. 1224-1238, July 2021.

[3] M. Mokarram, A. Khoei and K. Hadidi. "A fuzzy anti-lock braking system (ABS) controller using CMOS circuits," Microprocess. Microsy., vol.70, no. 2019, pp. 47-52, July 2019.

[4] J. Ma, Y. Gu, X. Hou, et al, "First demonstration of all-SiC half-bridge gate driver with high-side floating substrate region," 2024 36th Int. Symp. on Power Semiconductor Devices and ICs (ISPSD), Bremen, Free Hanseatic City of Bremen, Germany, 2024, pp. 502-505.

[5] A. A. Antonov, M. S. Karpovich and V. Y. Vasilyev, "Dual 4-A high-speed low-side gate driver IC for GaN and Si MOSFETs and IGBTs," 2022 IEEE 23rd Int. Conf. of Young Professionals in Electron Devices and Materials (EDM), Gorno-Altaisk, Republic of Altai, Russian Federation, 2022, pp. 378-382.

Real-Time Highly Flexible Wheel Speed Sensing Interface IP Design

Yitian Su[1], Zhinan Li [1], Haoxuan Yan[1], Zhenkai Zhou[1], Yuejun Zhang*[1], Cang Liu*[2,3]

[1] Faculty of Electrical Engineering and Computer Science, Ningbo University, Ningbo 315211, China
[2] Ningbo Yonghua Innovation Science and Technology Development Co., Ltd, Ningbo, Zhejiang,315211, China
[3] Department of Electronic Engineering, Tsinghua University, Tianjin, 300467, China

* Email: 2411100232@nbu.edu.cn, zhangyuejun@nbu.edu.cn, liucang@nbyhcx.com

Abstract—The Wheel-Speed Sensor (WSS) is a critical component of an Anti-lock Braking System (ABS), directly affecting vehicle handling and safety. Conventional solutions use the Serial Peripheral Interface (SPI) to relay WSS pulses to the Electronic Control Unit (ECU), where software routines count and timestamp these pulses-incurring millisecond-scale latency and slow fault diagnosis. To address these limitations, this paper presents a real-time, highly flexible, high-speed WSS IP design. The IP hardware-parallelizes the capture of high and low-level pulse durations, applies programmable digital filtering to remove spurious edges, and employs a multi-stage comparator tree to generate precise wheel-speed readings and fault flags on the fly. All filtering thresholds are tunable at runtime via the AMBA APB bus. the scheme is tested and verified under the TSMC 65nm process. Experiments show that under a 100 MHz clock, each decode and decision cycle incurs only a 10 ns delay, effectively eliminating latency bottlenecks. The ability to configure on the fly significantly boosts system flexibility and improves the reliability of fault diagnosis, outperforming traditional methods.

Keywords—*ABS, WSS, APB, Real-Time, Flexible*

I. INTRODUCTION

With the rapid evolution of the automotive industry and ever-increasing vehicle speeds [1], the Anti-Lock Braking System (ABS) [2] has become indispensable for enhancing driving safety and stability. By preventing wheel lock-up during emergency braking, ABS lets drivers maintain steering control and shorten stopping distances [3], thereby significantly reducing accident risk.

The Wheel-Speed Sensor (WSS) acts as the ABS's primary "sensory organ," capturing wheel speed pulses and direction in real time to drive anti-lock control [4]. As a critical safety subsystem [5], WSS must provide robust, rapid fault diagnosis: upon detecting any anomaly, the system immediately alerts the driver and initiates fail-safe measures to maintain braking performance and ensure vehicle safety with minimal latency [6]. Yet, popular ABS controllers such as the 33SB0400, SC900719, and L9388 still send square wave outputs over Serial Peripheral Interface (SPI) to the Electronic Control Unit (ECU), where software timers or input capture modules count pulses and decode data-incurring millisecond-level latency, risking missed pulses under high

This work was supported in part by the Ningbo University "Double First-Class" Cooperation Special Directional Entrusted Scientific and Technological Cooperation Project under Grant HX2025000106, HX2024000574, in part by the Cixi Science and Technology Program under Grant CZ2025006, in part by the Major Special Project of China Innovation Challenge (Ningbo) from Ningbo Science and Technology Program under Grant 2024T016, in part by the Ningbo University Student Research and Innovation Program (2025SRIP1314), and in part by Yongjiang Talent Project Youth Innovation Project (2024A-275-G).

speed or multi-channel loads, and overtaxing the Central Processing Unit (CPU). Additionally, off-the-shelf sensors use fixed filtering parameters with no online adjustability, hindering optimization for changing conditions or sensor aging and limiting diagnostic speed and accuracy.

To overcome these limitations, this paper proposes a real-time, highly flexible WSS IP that is deeply integrated into the system-on-chip (SoC). Using the AMBA APB bus, the CPU directly accesses speed and fault data in the registers and dynamically adjusts the filter duration and pulse width threshold. All digital filtering and Pulse Width Modulation (PWM) decoding are executed in parallel in hardware, enabling real-time reporting of fault and wheel speed information, with updated speed and fault status available within a single clock cycle. By moving the entire process to hardware, CPU resources are completely freed up, enabling the system to provide continuous, reliable diagnostic functions even under extreme conditions—laying a solid foundation for the rapid response and reliability required by next-generation intelligent vehicle active safety systems.

The main contributions of this work are summarized as follows:

1) We have integrated the entire wheel information, digital filtering, fault detection, and PWM decoding into parallel hardware, achieving nanosecond-level parameter updates and microsecond-level speed accuracy, thereby eliminating traditional latency bottlenecks.

2) The core directly accesses registers via APB, enabling flexible adjustment of filtering time and PWM pulse width thresholds during operation to adapt to different sensor characteristics and driving conditions.

3) Dedicated hardware detectors monitor over-temperature, open-circuit, and overcurrent events, while a latch mechanism stores the last valid speed output during power transients or faults, ensuring stable wheel information extraction even in harsh environments.

The rest of the paper is organized as the follows. In Section II, we illustrate the working principle of the wheel speed sensor and the PWM protocol. In Section III, the designed hardware architecture for WSS Interface Hardware Circuit. In Section IV, we provide the implementation results and comparisons. Finally, Section V draws conclusions.

II. SYSTEM FRAMEWORK DESIGN

A. ABS System Framework

The ABS system integrates four wheel-speed sensing interfaces, as shown in Fig. 1, all managed by a Reduced Instruction Set Computing-V (RISC-V) core-leveraging its

979-8-3315-3918-4/25 $31.00 © 2025 IEEE

open Instruction Set Architecture (ISA) and compact instruction set to orchestrate real-time data flow and dynamic configuration. Communication between the core and IPs occurs over the AMBA APB bus, its low-latency, low-overhead register accesses and simplified topology enable seamless runtime parameter tuning, minimal bus congestion, and straightforward integration into on-chip systems.

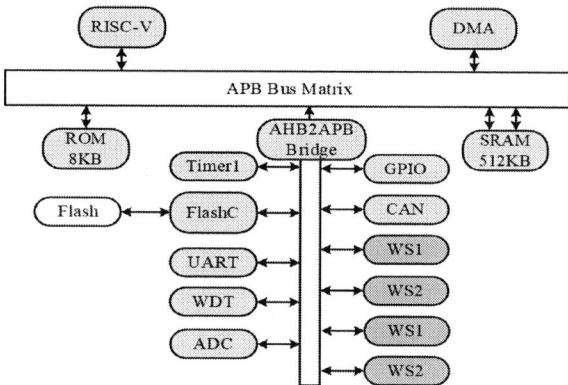

Fig. 1. ABS SoC architecture.

B. Wheel Speed Sensor Principle and PWM Protocol

Variable reluctance wheel speed sensors detect wheel rotation by mounting an iron pole wheel on the hub so that it spins within a magnetic circuit formed by a permanent magnet and a pick-up coil, as shown in Fig. 2 As each pole approaches the coil, the magnetic reluctance drops sharply and flux concentrates; as a gap passes, reluctance rises and flux disperses, inducing an Alternating Current (AC) voltage in the coil proportional to rotational speed. Traditional designs rectify this voltage into a fixed duty cycle square wave, containing only speed information. In contrast, a PWM protocol maintains an output period equal to half the magnetic cycle, issuing a pulse at each zero crossing and encoding speed via duty cycle modulation. This approach not only conveys wheel speed with high precision but can also append auxiliary data-such as rotation direction, mounting position, and air gap status-providing ABS systems with richer, more reliable wheel speed signals.

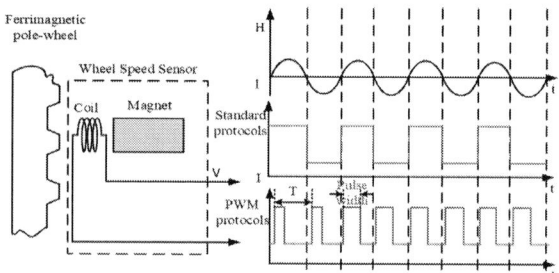

Fig. 2. Wheel-speed sensor structure and output current waveform diagram.

When measuring speed using PWM signals, the rotational speed can be calculated from the complete cycle duration T and the number of magnetic pole pairs N:

$$v = \frac{60}{N \times T \times 2} \qquad (1)$$

Here, v represents rotational speed (revolutions per minute), T is the duration of one complete PWM signal cycle

(seconds), and the factor of 2 arises because one PWM pulse is generated per two magnetic pole zero-crossings. This method retains the real-time performance of traditional frequency measurement while enabling simultaneous transmission of comprehensive status information.

III. WHEEL SPEED SENSING CIRCUIT DESIGN

A. Module Framework

Fig. 3 shows the hardware architecture diagram of the wheel information acquisition and fault detection system. The peripheral modules on the left side include the system's V_PWR power supply, current monitoring for detecting circuit current and identifying abnormal current faults, and wheel speed sensors that collect physical signals related to wheel speed as data input sources. The middle blue functional module includes a fault detection module that receives current monitoring and sensor signals to determine system faults, a wheel speed information processing module that performs preliminary processing on sensor raw signals to extract wheel speed features, a wheel speed counter that converts processed signals into specific wheel speed data, and registers (Regs), which temporarily stores wheel speed and fault data, acting as a data relay station between hardware and the processor, and interacts with the processor on the right via the AMBA bus; the green area on the right is the core RISC-V processor, which reads wheel information and fault data from the registers at high speed via the AMBA bus. The signal and data flow is Sensor → Wheel Speed Information Processing → Regs, Current Monitoring → Fault Detection → Regs. The registers and RISC-V processor communicate bidirectionally via the AMBA bus. The entire system is based on the RISC-V processor, utilizing sensor data acquisition and hardware module processing (wheel speed + fault detection), and employs the AMBA bus for data interaction, enabling real-time wheel speed monitoring and rapid fault diagnosis.

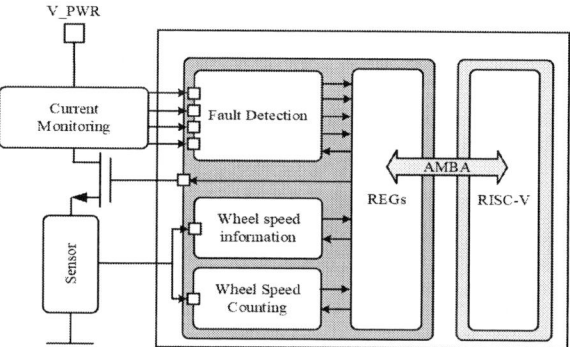

Fig. 3. Wheel speed sensor module architecture.

B. Wheel-Speed Decoding Module

The threshold-based decoding process is shown in Fig. 2 By modulating the PWM duty cycle, a single pulse stream encodes multiple wheel speed states—air gap conditions, rotation direction, wheel lock, stationary state, and signal validity. This method efficiently provides the ABS controller with rich multidimensional wheel speed information, enabling high-fidelity capture and real-time response to complex wheel dynamic behavior.

979-8-3315-3918-4/25 $31.00 © 2025 IEEE

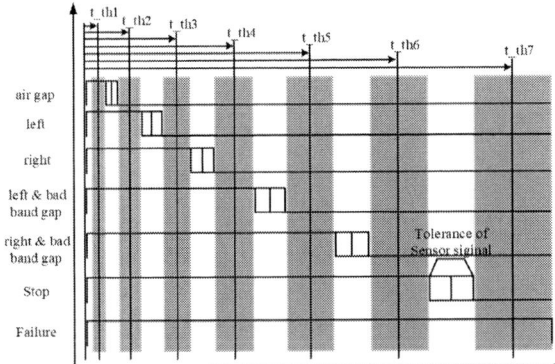

Fig. 4. Decoding threshold time plot.

As shown in Fig. 5, the sensor's PWM pulses first pass through a digital filter with a configurable time window, accurately removing spikes and transient noise. The filtered high-level periods are then accumulated in a parallel counter channel, with each falling-edge event latching the current count value. A first-stage comparator filters for valid pulse widths, which are then routed via a multiplexer to a second-stage comparator bank. Here, eight preset thresholds (i_ws_th1-i_ws_th8) stored in the register module are compared in parallel. Under a 100 MHz system clock, the complete wheel-state decision occurs within a single clock cycle. The decoded status is written directly into the REGs and transmitted to the CPU over the APB bus.

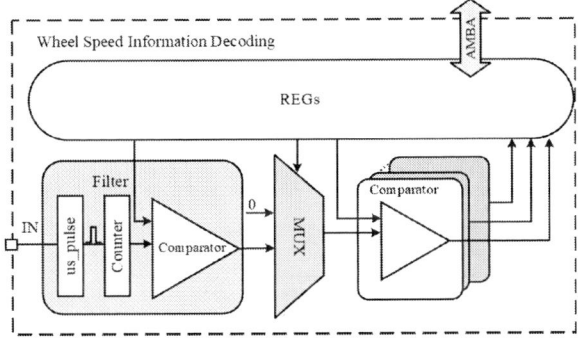

Fig. 5. Wheel speed decoder module.

C. Speed Counting Module

Fig. 6 shows the fully hardware-driven speed counting process: incoming wheel speed information first enters a configurable digital filter to remove transient glitches. Subsequently, dual-channel counters operate in parallel, with one accumulating timing during high-level periods and the other counting during low-level periods. Both are updated at each clock edge to ensure complete capture of the entire signal cycle. Upon completion of the pulse cycle, the module invokes the conversion formula stored in the register to convert the accumulated high/low-level count values into real-time vehicle speed. To ensure the accuracy of wheel speed information under extreme conditions, the system also integrates a fault latch mechanism: when an open-circuit or overcurrent fault caused by power supply transients is detected, the system immediately latches the last valid vehicle speed output and blocks subsequent pulse inputs until the fault is cleared and the latch is reset, preventing the generation of false speed measurement data.

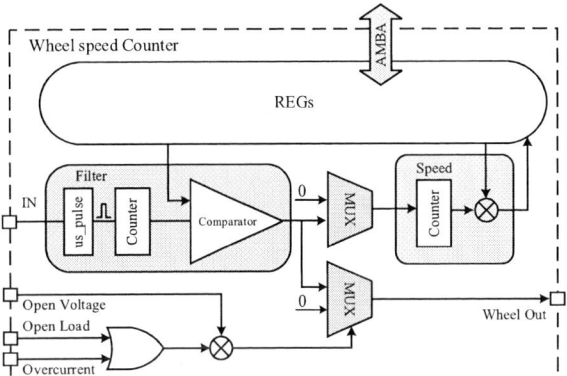

Fig. 6. Wheel speed counting module.

D. Fault Detection Module

As illustrated in Fig. 7, the fault-detection flow: signals from over-temperature, open-circuit, and over-current sensors first pass through a digital filter using register-configured thresholds to remove noise, then feed parallel high-precision counters that tally fault pulses almost instantly. These counts are compared against preset thresholds, and any sustained over-limit condition sets a fault flag-written to the APB slave status register within the same clock cycle. Simultaneously, the switch-enable logic drives a multiplexer to cut power, protecting the sensor. This hardware-only path delivers true "trigger-and-respond" fault handling with near-zero latency.

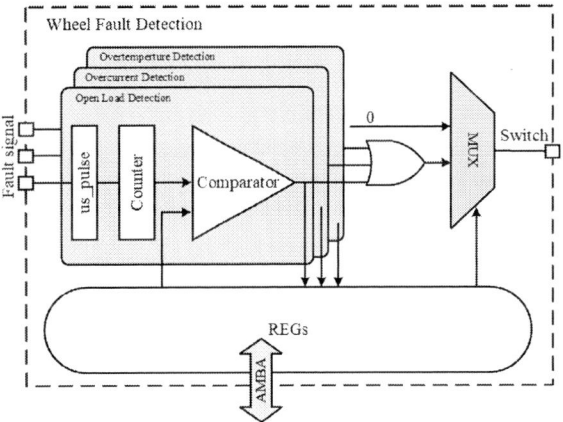

Fig. 7. Fault detection module.

IV. EXPERIMENTAL RESULTS AND ANALYSIS

We validated the proposed high-flexibility, real-time wheel-speed sensing IP in two stages. First, behavioral co-simulation of a flexible real-time wheel speed sensing IP in Synopsys VCS and Verdi, verifying its online dynamic configurability for filtering thresholds and fault parameters with nanosecond-level parsing. Using TSMC 65nm technology, through Design Compiler logic synthesis and IC Compiler placement and routing, the feasibility and reliability of the IP in terms of area, timing, and power consumption were comprehensively evaluated.

A. Functional Simulation Verification

The simulation results are presented in Fig. 8. In the fault handling test (a), the system first suppresses transient noise during startup, then counts high-level fault pulses within the original filter window. Once the count exceeds the preset threshold, protection is triggered immediately; the filter threshold can be updated in real time within a single AMBA bus cycle, forming a hardware-level loop of "fast detection and dynamic reconfiguration." In the signal purification test (b), raw wheel speed pulses first pass through the Filter Counter to eliminate spurious edges, then are sent to the Compensate Counter to adjust for timing drift and EMI. Through this two-stage filtering process, a high-precision "Clean Wheel Signal" is generated. In the PWM decoding test (c), the Width Counter measures the high-level duration of each pulse and classifies it as invalid, air gap, or gear edge (indicating rotation direction). Meanwhile, the Wheel Counter tracks complete "gear-air gap" cycles and converts rising edge timestamps into frequency, collectively producing comprehensive wheel motion data.

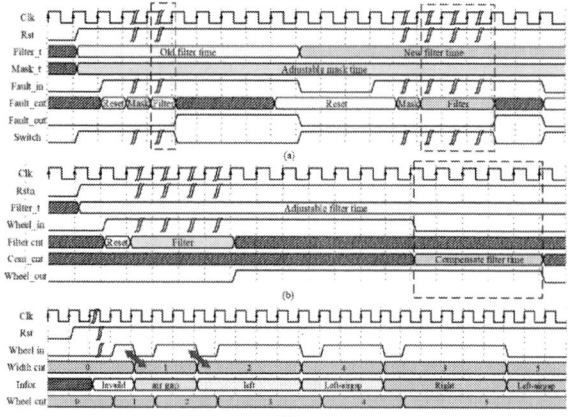

Fig. 8. Function waveform simulation: (a) Fault handling; (b) Signal purification; (c) PWM decoding.

The layout and feature of the Wheel speed sensor interface circuit core are illustrated in Fig. 9. Under TSMC 65nm process, the area of the rejection sampling circuit is 0.0173mm²; the equivalent gates (NAND) are 14.13Kgates; the voltage is 1V; the frequency is 100MHZ; and the power is 0.3096mW.

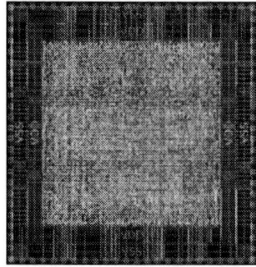

Wheel Speed Sensor Sampler feature	
Technology	TSMC 65nm
Area	0.0173mm²
NAND	14.13Kgates
Voltage	1V
Frequency	100MHZ
Power	1.4421mW

Fig. 9. Wheel speed layout design.

Table 1 compares the proposed wheel speed sensing interface IP with three traditional solutions (L9388 TR, SC900719, 33SB0400). Traditional products use SPI, with fixed filtering times and max 16 MHz operating frequency. The proposed design, based on AMBA bus, supports online dynamic adjustment of filtering times and thresholds, reaching 100 MHz, meeting ABS' real-time needs. It is compatible with automotive SoCs, simplifying integration and breaking traditional bottlenecks of fixed parameters and delayed response.

TABLE I.   COMPARISON RESULTS WITH OTHER CHIPS

	Interface	Filter time adjustable or not	Operating Frequency
This Work	AMBA	yes	100MHz
L9388-TR	SPI	no	16MHz
SC900719	SPI	no	10MHz
33SB0400	SPI	no	10MHz

V. CONCLUSION

This paper presents a real-time, highly flexible wheel-speed sensing IP that offloads pulse capture, digital filtering, multi-channel hardware fault detection, and PWM decoding entirely into hardware. During operation, the core can dynamically adjust parameters via the APB bus, enabling adaptive tuning to diverse sensor characteristics, environmental noise, and operating conditions. Compared to traditional approaches, this design frees CPU resources while ensuring high speed-sensing accuracy and fault-diagnosis reliability. In future work, the proposed IP will be integrated onto a RISC-V platform to further validate its full functionality.

REFERENCES

[1] F. Wang. "Research on hydraulic ABS control algorithm based on support vector machine model," 2023 2nd Int. Conf. on AI and Autonomous Robot Syst. (AIARS), Bristol, United Kingdom, 2023, pp. 566-570.

[2] F. Yao, X. Lin, Z. Wu, G. Li and F. Jiang. "Active disturbance rejection control for automotive Anti-lock Braking System," China J. Highw. Transp., Xian, Shanxi, China, vol. 34, no. 3, pp. 235-244, March 2021.

[3] G. Xu and B. Li. "Design of ABS functional safety architecture based on ISO" 2025 6th Int. Conf. on Electr. Electron Inf. and Commun. Eng. (EEICE), Shenzhen, Guangdong, China, 2025, pp. 1408-1412.

[4] L. Sun, J. Wang and Y. Huang. "Fault diagnosis of ABS Wheel Speed Sensor based on improved BP neural network," 2023 Global Rel and Prognostics and Health Manage Conf. (PHM-Hangzhou), Hangzhou, Zhejiang, China, 2023, pp. 1-6.

[5] A. Li, C. Niu, B. Yang and X. Huang, "Research on remote fault intelligent prediction system of ABS ring gear based on internet of vehicles," 2022 6th CAA Int. Conf. on Vehicular Control and Intell. (CVCI), Nanjing, Jiangsu, China, 2022, pp. 1-5.

[6] D. G, P. Manohar and S. M. Pasha, "Development of an algorithm for braking force distribution to avoid wheel locking in ABS," 2020 Int. Conf. on Recent Trends on Electr., Infor., Commun. & Technol. (RTEICT), Bangalore, India, 2020, pp. 199-203

Design of Secure Storage Circuit Based on Reversible Logic XOR-Toffoli Gate

Yiting Guo [1], Yuejun Zhang[*1], Shutong Zhang[1], Mengfan Xu[1], Zhenkai Zhou[1], Hui Li[*2]

1 Faculty of Electrical Engineering and Computer Science, Ningbo University, Zhejiang, 315211, China.
2 Dahua Technology Co., Ltd Hangzhou, China

* Email: 226000910@nbu.edu.cn, zhangyuejun@nbu.edu.cn, li_hui@dahuatech.com

Abstract—Solving the trouble of hardware data being stolen from storage device, this paper proposes a secure storage architecture underpinned by a reversible XOR–Toffoli gate. Firstly, the key is generated by the power-up characteristic of the static random-access memory physical unclonable function (SRAM PUF). Then, the reversibility inherent in the reversible logic input–output mapping is leveraged to decide bit-flips in private data: the PUF response is compared against the constant "1". This completes the encryption and decryption of the private data. Finally, the encrypted data is deposited into the storage unit to achieve data security storage. The circuit under the TSMC 28 nm process was simulated and verified. Experimental evaluation reveals a brute-force resistance of 3.67×10^{60} years, evidencing excellent capability against violent attacks. The SRAM PUF derived key demonstrates 99.48 % randomness and attains 51.02 % uniqueness, successfully satisfying NIST randomness test Across the operational temperature range of 20 °C to 80 °C and the supply voltage span of 0.7 V to 1.2 V, the worst bit error rate (BER) is constrained to 6.2%, affirming its viability using for hardware secure storage.

Keywords—SRAM PUF, Reversible Circuit, Secure Storage.

I INTRODUCTION

In IoT deployments, ubiquitous interconnection via the public Internet or local area networks renders electronic devices susceptible to a broad spectrum of active and passive security threats [1]. Although conventional encryption and authentication primitives furnish rudimentary safeguards, their security guarantees remain fundamentally bounded in the face of continuously advancing attack strategies. Most electronic devices store keys in Electrically Erasable Programmable Read-Only Memory (EEPROM), and the core of that encryption technology is encrypting the data stored in memory using algorithms [2]. Empirical studies, however, demonstrate that ensuring the confidentiality of non-volatile memory is intrinsically challenging, thereby its susceptibility to a broad spectrum of active and passive assaults markedly elevates the probability of cryptographic-key exfiltration. The physical unclonable function (PUF) technique generates unique and unrepeatable hardware fingerprints using process deviations inherent in the manufacturing of integrated circuits, and this technique obviates the need for persistent storage or on-wire transmission, enabling the application of hardware secure storage [3].

This work was supported in part by the National Natural Science Foundation of China under Grant 62474100, in part by the Major Special Project of China Innovation Challenge (Ningbo) from Ningbo Science and Technology Program under Grant 2024T016, in part by the Zhejiang Provincial Science and Technology Program "Pioneer and Leading Goose + X" Project under Grant 2025C01063, in part by the Ningbo University "Double First-Class" Cooperation Special Directional Entrusted Scientific and Technological Cooperation Project under Grant HX2024000574, HX2025000106.

In order to solve the problem of storage information leakage caused by the complexity of traditional encryption and decryption techniques, the design of a secure storage circuit with low complexity is necessary. Leveraging the unique power-up fingerprint of the SRAM PUF for key generation, and exploiting the characteristic of the reversible XOR–Toffoli gate, the private data is encrypted directly, consequently, only the ciphertext must be stored [4]. In view of the aforementioned factors, the design uses the SRAM PUF to generate the key, utilizes the Reversible Logic XOR-Toffoli Gate to encrypt the data directly, and subsequently transmits it to the storage unit, thereby enhancing the security of private data storage.

II DESIGN OF SECURE STORAGE ARRAYS BASED ON REVERSIBLE LOGIC XOR-TOFFOLI GATE

Initially, the PUF circuit is powered up to generate the cryptographic key. The key and the private data are then jointly fed into the reversible logic array, which performs direct encryption. The resulting ciphertext is committed to the storage array for secure retention. Upon subsequent data decryption, the PUF is powered up again to regenerate the key. The encrypted data are then read from the storage array and passed through the reversible logic array again to be decrypted.

A. Design of Reversible Logic XOR-Toffoli Gate

Conventional logic circuits are inherently irreversible, so each computational step annihilates both information and the energy bound to it. In the 1970s, Bennett verified that circuits can be designed using reversible logic, thus avoiding both the loss of bits of information and energy dissipation [5]. Today, reversible logic has been applied to a segment of hardware circuit design, including reversible logic full adders and reversible ternary multipliers [6][7]. As an emerging technology, reversible logic circuits have found expanding deployment in quantum computing, CMOS systems, digital signal processing, communications, and computer graphics.

In a conventional Toffoli gate, the target bit toggles if and only if both control bits are asserted as 1; otherwise, the target bit remains unchanged [8]. The architecture adopted here employs the XOR-Toffoli gate, whose reversible logic is defined as follows.

$$\overline{C} = C \oplus (A \oplus B) \qquad (1)$$

C flips when A and B differ and remains unchanged only when both are 1. A and B act as auxiliary bits to control the change of C.

979-8-3315-3918-4/25 $31.00 © 2025 IEEE

The modified reversible logic XOR-Toffoli gate is shown in Fig. 1 (a) below. And Fig. 1 (b) depicts the encryption flow within the unit, wherein PUF denotes the cryptographic key, Data the plaintext, and Code the resultant ciphertext. During encryption, the value of the PUF determines whether Data is flipped or kept unchanged to perform the encryption function. During decryption, the PUF is powered up again and the encrypted data is read from the storage array and passed into the reversible logic XOR-Toffoli gate. The initial Data can then be obtained.

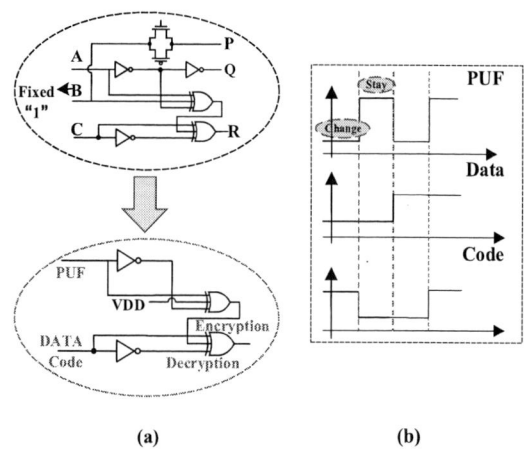

(a) **(b)**

Fig. 1. Reversible logic XOR-Toffoli gate improved design (a) Circuit structure; (b) Encryption process.

B. SRAM Memory Cell

Fig. 2 shows the SRAM memory cell. This design uses a basic SRAM memory cell including pre-charge, write control, read pipe and latch structure. During the write process, the DATA_code is input. The write and WL lines are pulled high by the timing control. The transmission gate then opens and the data is written from M1 and M2 to the latch structure. During the read process, PRE is first pulled low, bit lines are pre-charged to VDD and then WL is turned on, enabling data to be read out.

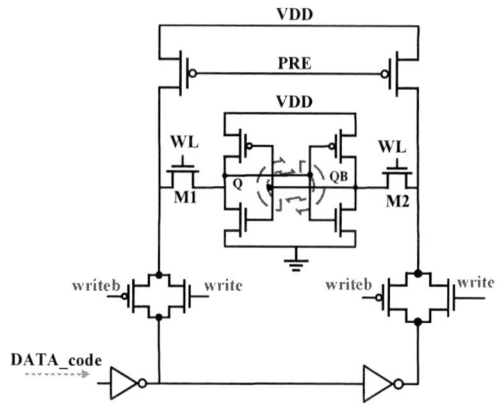

Fig. 2. Memory cell design.

C. Design of Secure Storage Circuit Array

The overall block diagram is shown in Fig. 3 below. The array architecture incorporates the following components: (i) a timing-control module; (ii) an SRAM PUF module; (iii)a reversible logic XOR–Toffoli module; (iv) a storage module. Timing controlled PUF activation yields the key, which

together with the data is synchronously delivered to the reversible logic XOR-Toffoli module for encryption; the encrypted data are then written to the storage module under read write control, enabling secure retention.

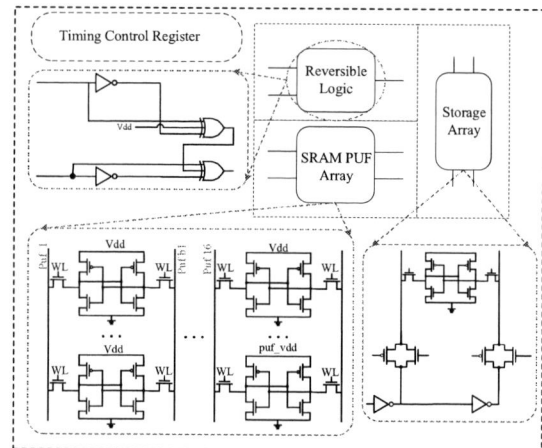

Fig.3. Block diagram of secure storage circuit array structure.

D. Secure Storage Timing Control

During the encryption process, the key is generated from the PUF cell, the private data Data and the key PUF are synchronously passed into the input of the reversible logic XOR-Toffoli to realize encryption, and then the write and WL signal lines are turned on to achieve the secure storage of the encrypted data into the SRAM memory unit.

During the decryption process, firstly the bit line is pre-charged, then the WL is turned on and the encrypted data Code is read out from the BLB, Data signal line is reused, encrypted data is passed in, puf is powered up again, PUF and Code are synchronously input into the reversible logic unit, decryption result is obtained, and finally output by DATAOUT.

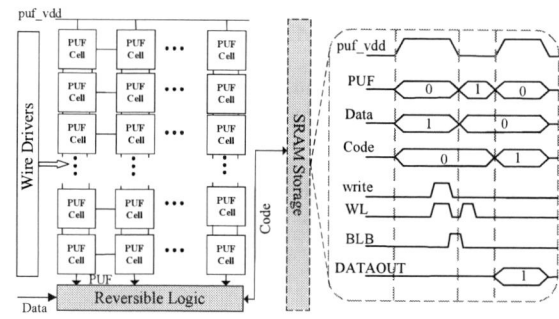

Fig. 4. Timing diagram for secure storage circuit.

III EXPERIMENTAL RESULTS AND ANALYSES

An SRAM PUF based encryption–decryption circuit employing a reversible logic XOR-Toffoli gate was designed and verified in Cadence Virtuoso with a 28nm CMOS process. The circuit was simulated on the Spectre platform, and the simulation results are as follows:

The work is based on the 28nm process, and the reversible logic cells and arrays of this design were drawn using a fully customized layout. In addition, the area of the reversible logic cell was normalized, with the results shown in Fig. 5 below.

979-8-3315-3918-4/25 $31.00 © 2025 IEEE

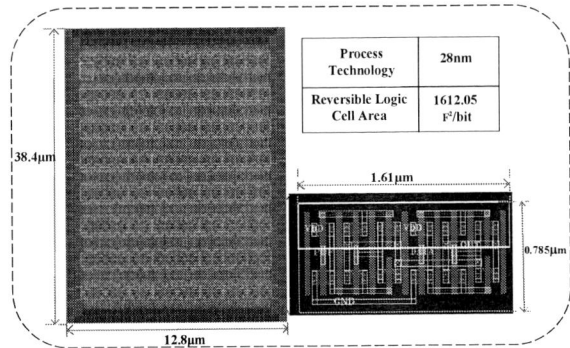

Fig. 5. 16×16 Array layout design & reversible logic cell layout design.

A. Simulation Results of Encryption and Decryption

The encryption and decryption results of the array are displayed in Fig. 6 below. The red box represents the encryption results, the blue box represents the decryption results. In the proposed timing design, the input data is represented by the binary number "1011". In the event of PUF being set to 1, the data remains unchanged; conversely, if PUF is set to 0, the data is inverted. The encrypted data is "0100", a result which adheres to the established encryption rules. At the final decryption, the PUF is powered up again, and the data obtained is "1011", which is consistent with the initial data.

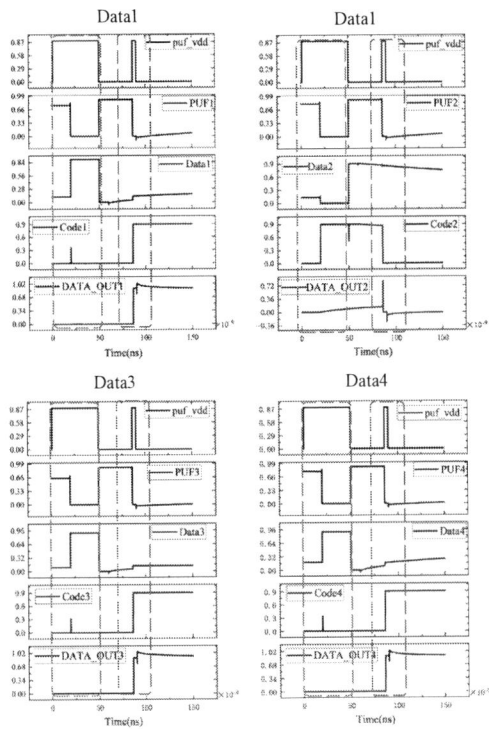

Fig. 6. Encryption and decryption results.

B. Randomness

Randomness indicates whether the ratio of "0" and "1" in the SRAM PUF output is balanced, in other words, whether the probability of "0"and "1" is close to 50%.

$$Randomness = ((1 - |2P(r=1) - 1|) \times 100\%) \qquad (2)$$

P in Equation (2) represents the probability that the PUF power-up result is 1. For the 8000 keys generated by the PUF circuit, they are statistically calculated and the test results are shown in Fig. 7 below, which shows that the probability of a response of 1 in the PUF is 49.74%, so the corresponding randomness is seen to be 99.48%.

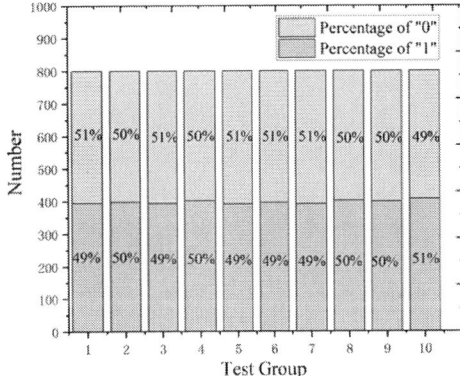

Fig. 7. Randomness test.

As can be seen in Table I below, this measurement divides the keys generated by the PUF into 10 groups of 800 data each, all of which pass the NIST test, which shows that the PUF has a high degree of randomness.

Table I. NIST test

Test Name	Stream Length	Average P_value	Pass?
Cumulative Sums	800	0.8898	Pass
FFT	800	0.0402	Pass
Frequency	800	0.6387	Pass
Linear Complexity	800	0.7103	Pass
Non-Overlapping Template	800	0.4064	Pass
Overlapping Template	800	0.3440	Pass
Rank	800	0.6883	Pass
Serial	800	0.5099	Pass

C. Uniqueness

Uniqueness is the degree of difference between PUFs. Under ideal conditions, different PUFs respond to the same excitation with 50% of both 0s and 1s in their results. It is generally quantified by the Hamming distance.

The average inter-slice Hamming distance $E(HD_{Inter})$ for k PUFs is shown in Equation (3):

$$HD_{Inter} = \frac{2}{k(k-1)} \sum_{i=1}^{k-1} \sum_{j=i+1}^{k} \frac{HD(R_x R_y)}{N} \times 100\% \qquad (3)$$

Rx and Ry represent the N bits response outputs generated by the xth and yth PUF circuits, respectively. The results of this average Hamming distance inter-distance distribution and Gaussian fitting of the PUF is shown in the Fig. 8 below, which demonstrates that the mean value $\mu = 0.5102$ and the standard deviation $\sigma = 0.1454$, and thus the uniqueness of the PUFs is 51.02%.

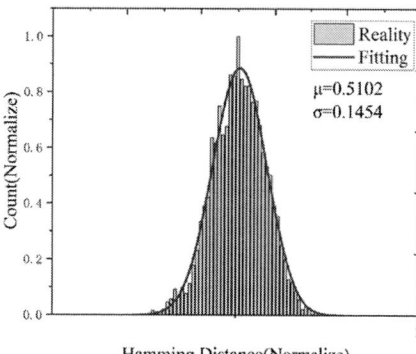

Fig. 8. Hamming Distance test.

D. Stability

Stability can be quantified by measuring how much the PUF changes for the same excitation at different temperatures and voltages. In this work, we compared the results with other results by performing Monte Carlo simulations of the designed PUF at different temperatures and different voltages, where the data at 0.9V/27°C was used as a benchmark. As shown in the Fig. 9 below, the BER of the PUF is 2.6% at -20°C and 1.10% at 80°C; at a voltage of 0.7V, the BER is 2.40%, and at 1.2V the BER is 6.20%.

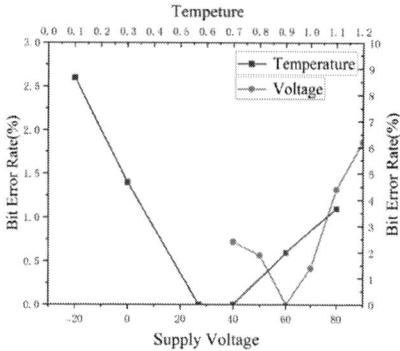

Fig. 9. Stability test.

E. Resistance to Brute-force Attacks

Brute-force attack is an iterative process whereby the system response is obtained by traversing the input excitation. The correct key is then obtained by analysing the multiple excitation responses. The majority of contemporary hardware attacks are predicated on brute force attacks, and when combined with the chip characteristics, they continue to present a significant threat to the chip. The anti-violence attack is analysed as follows in Equation (4).

$$Y_{Cracking} = 2^M \times T_1 \times d \qquad (4)$$

The term $Y_{Cracking}$ denotes the total cracking time, T_1 signifies the time required for a single attack, M represents the number of key bits, and d denotes the number of computers attacking the chip in parallel. Assuming the time of one attack T_1=1ns, d=1, and M as a variable, the relationship between the breaking time and the key length M is obtained as follows in Fig. 10.

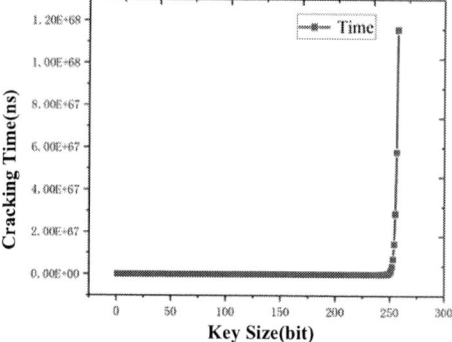

Fig. 10. Analysis of resistance to brute force attacks.

With a 256 bits key, brute force decryption is estimated at 3.67×10^{60} years, demonstrating formidable resistance to violent attacks.

V CONCLUSION

This paper introduces a secure storage circuit that employs a reversible logic XOR-Toffoli gate. A key derived from PUF drives the gate array to perform encryption and decryption, after which only the ciphertext is committed to the storage unit, enabling hardware tamper-evident data storage. Experimental results confirm that the proposed secure storage circuit simultaneously guarantees confidential retention and faithful reconstruction of private data, thereby substantiating its applicability to hardware-based secure storage applications.

REFERENCE

[1] B. M. S. Bahar Talukder, F. Ferdaus and M. T. Rahman, "Memory-based PUFs are vulnerable as well: A Non-Invasive attack against SRAM PUFs," IEEE Trans. Inf. Forensics Security, vol. 16, pp. 4035-4049, July. 2021.

[2] Y. Feng, Jixuan Wu, Xuepeng Zhang, Jing Liu, Zhaohui Sun and Junyu Zhang, "A novel encrypted computing-in-memory (eCIM) by implementing random telegraph noise (RTN) as keys based on 55 nm NOR flash technology," IEEE Electron Device Lett., vol. 43, no. 9, pp. 1455-1458, September. 2022.

[3] B. Karpinskyy, Y. Lee, Y. Choi, Y. Kim, M. Noh and S. Lee, "Physically unclonable function for secure key generation with a key error rate of 2E-38 in 45nm smart-card chips," IEEE Solid-State Circuits Conf. (ISSCC), San Francisco, CA, USA, 2016, pp. 158-160.

[4] S. Vyas, N. K. Dumpala, R. Tessier, and D. E. Holcomb, "Improving the efficiency of PUF-based key generation in FPGAs using variation-aware placement," IEEE Field Program Logic Conf. (FPL), Appl Lausanne, Canton de Vaud, Switzerland, 2016, pp. 1-4.

[5] U. Kumar, L. Sahu, and U. Sharma, "Performance evaluation of reversible logic gates," IEEE ICT Business Ind. Gov. (ICTBIG), Indore, Madhya Pradesh, India, 2016, pp. 1-4.

[6] S. S. Devi and V. Bhanumathi, "Design of reversible logic based full adder in current-mode logic circuits," Microprocessors and Microsystems, vol.76, pp.103110, July. 2020.

[7] M. M. Panahi, O. Hashemipour, and K. Navi, "A Novel Design of a Multiplier Using Reversible Ternary Gates," IETE Journal of Research, vol. 67, no.6, pp. 744-753, January, 2021.

[8] C. Ganesh, A. S. Kumar, P. Santhosh, A. Ramya, C. S. Kumar and P. Thivani, "A Novel Design of Area Efficient Full Adder Architecture Using Reversible Logic Gates," IEEE Devices, Circuits Syst Conf. (ICDCS), Coimbatore, Tamil Nadu, India, 2024, pp. 107-111

A Hierarchical Approximate Floating Point MAC Unit with Precision-Adaptive Self-Configuration

Xianghui Fu[1,2], Yike Wang[1,2], Chaojie Wei[1,2], Yu Gong[1,2] *

[1]School of Integrated Circuits, Nanjing University of Aeronautics and Astronautics, Nanjing 211106, China
[2]Key Laboratory of Aerospace Integrated Circuits and Microsystem, MIIT, Nanjing 211106, China
*Corresponding Author: gongyu@nuaa.edu.cn

Abstract—The floating point Multiply-Add-Computing (FP-MAC) is one of the most resource-consuming and frequently used operation in image processing. In this paper, a precision-adaptive FP-MAC unit is proposed with self-configurable feature, to achieve the balance of precision, performance, and power consumption. For the computing strategy optimization, a precision-adaptive algorithm is deployed for floating point MAC operation, which composed of precision requirements analysis for input, working mode and rounding strategy selections for MAC unit. For the circuit design implementation, a bit-width self-adaptive MAC unit with configurable working mode is proposed to make full use for the precision-adaptive algorithm. Compatible to the IEEE 754 standard, the proposed MAC unit achieves 47.70 dB PSNR and 3.15×10^5 MAC/s throughput @FP16, verified using Lena images and 3×3 Gaussian filters, and shows great improvement in area and latency.

Index Terms—Multiply-Add Computing, Floating-Point Arithmetic, Approximate Computing, Image Processing

I. INTRODUCTION

In the domain of image processing, the demand for efficient floating-point computation grows rapidly, especially for operations involving convolution-based image filtering. The fused multiply-add (FMA) operations, also namely multiply-add computing (MAC) operations, are fundamental to convolution computations, with requirements of high energy efficient acceleration [1], [2]. While the IEEE 754 half-precision (FP16) format offers notable benefits in reducing memory usage and bandwidth, traditional floating point MAC (FP-MAC) units often lack adaptability and demonstrate suboptimal energy efficiency in hardware implementations [3].

In image filtering tasks, input data frequently presents characteristics such as sparsity and low precision sensitivity, making it a suitable candidate for approximate computation [4]. Since convolution is intrinsically composed of repeated multiply-accumulate operations, MAC units can directly accelerate this process, making image filtering an ideal scenario for evaluating the efficiency and precision of FP-MAC architectures.

Recent research has investigated techniques such as approximate arithmetic, dynamic precision adjustment, and low-level hardware optimization to improve MAC performance [4], [5]. Most of the work either focus on algorithm-level software optimization or sacrifice numerical accuracy, limiting their practical deployment in hardware-limited environments. The deployment of FP-MAC or FP-FMA are also proposed on platforms of SIMDs [5] and FPGAs [6] . For neural networks, MAC units are also significant while improving energy efficiency, for both inference[7] and training[8]. To further improve the performance and flexibility for FP-MACs, different rounding approaches and reconfigure designs are proposed[10][11].

To address the challenges of FP-MAC while considering the real computing precision requirements, this paper presents a hierarchical approximate FP-MAC design, targeting image filtering applications. The proposed design incorporates a mantissa-aware approximation strategy and supports hardware-level adaptive computation to achieve Precision-Adaptive and Self-Configuration, aiming to improve both energy efficiency and computational accuracy. The rest of the paper is organized as follows. Section II introduces the fundamental algorithm principle for precision controlling, Section III illustrated the architecture of proposed MAC unit. The experiments are conducted in Section IV and the paper is concluded in Section V.

II. PRECISION-ADAPTIVE MAC ALGORITHM DESIGN

A. Overall Algorithm Framework

The precision-adaptive floating point MAC algorithm employs a self-configurable computational strategy that dynamically optimizes operational parameters through real-time analysis of input data precision requirements. The design objective is to achieve multi-dimensional optimization of precision, performance, and power consumption while maintaining IEEE 754 standard compatibility.

The overall framework comprises four core state:

1) **Precision Requirements Analysis**:analysis of input data characteristics and precision demands;
2) **Working Mode Selection**: Self-configurable mode selection based on precision requirements;
3) **Bit-width Adaptive MAC Module**: Dynamic bit-width adjustment for optimal resource utilization;
4) **Multi-strategy Rounding Control** : Rounding strategy selection based on precision and performance trade-offs.

The entire MAC computational process can be expressed as:

$$MAC_{result} = Round(Accumulate(Multiply_{adaptive}(a,b),c))$$

where $Multiply_{adaptive}$ represents the precision-adaptive multiplication with self-configurable bit-width, $Accumulate$ represents the adaptive accumulation process, and $Round$ represents multi-strategy rounding control.

B. Precision-Adaptive MAC Algorithms

Algorithm 1 implements the complete computational logic for precision-adaptive MAC operations, adopting a five-stage self-configurable pipeline design: precision analysis, working mode selection, bit-width adaptive computation, multi-strategy rounding, and configuration update.

C. Precision-Adaptive Strategies

1) Bit-width Self-Adaptive Multiplication Strategy: Based on [7], the self-adaptive multiplication strategy of bit width selects the optimal bit partitioning scheme based on the numerical characteristics of the 10-bit mantissa. In response to the shortcomings of [7], we have proposed four completely new partitioning strategies,this strategy optimizes computational efficiency by dynamically adjusting the operational bit width based on precision requirements.

979-8-3315-3918-4/25 $31.00 © 2025 IEEE

Four partitioning strategies are implemented for the 10-bit mantissa processing:

- **H2L8 Partitioning:** Optimized for high-bit sparse numerical patterns with minimal computational complexity.

$$P_{H2L8} = (A_h \cdot 2^8 + A_l) \cdot (B_h \cdot 2^8 + B_l)$$
$$= A_h B_h 2^{16} + (A_h B_l + A_l B_h) 2^8 + A_l B_l \quad (1)$$

- **H4L6 Partitioning:** Balanced scheme providing optimal precision-performance trade-off.

$$P_{H4L6} = (A_h \cdot 2^6 + A_l) \cdot (B_h \cdot 2^6 + B_l) \quad (2)$$

- **H6L4 Partitioning:** Adaptive configuration suitable for diverse numerical patterns.

$$P_{H6L4} = (A_h \cdot 2^4 + A_l) \cdot (B_h \cdot 2^4 + B_l) \quad (3)$$

- **H8L2 Partitioning:** High-precision mode for critical computational scenarios.

$$P_{H8L2} = (A_h \cdot 2^2 + A_l) \cdot (B_h \cdot 2^2 + B_l) \quad (4)$$

where A_h and A_l represent the high-bit and low-bit segments of operand A, respectively. Similarly, B_h and B_l represent the corresponding segments of operand B.

Algorithm 1 Precision-Adaptive MAC Compute Unit

1: **Input:** multiplicand, multiplier, accumulator, precision_requirement
2: **Output:** MAC computation result
3: // Stage 1: Precision Requirements Analysis
4: precisionLevel ← AnalyzePrecisionRequirement(multiplicand, multiplier, accumulator)
5: dataPattern ← ExtractDataCharacteristics(multiplicand, multiplier)
6: // Stage 2: Working Mode Selection
7: workingMode ← SelectWorkingMode(precisionLevel, dataPattern)
8: bitWidth ← DetermineBitWidth(workingMode, precisionLevel)
9: // Stage 3: Bit-width Adaptive Computation
10: **if** workingMode == "high_precision" **then**
11:    multiplicationResult ← FullPrecisionMultiply(multiplicand, multiplier, bitWidth)
12: **else if** workingMode == "balanced" **then**
13:    multiplicationResult ← PartitionedMultiply(multiplicand, multiplier, bitWidth)
14: **else**
15:    multiplicationResult ← ApproximateMultiply(multiplicand, multiplier, bitWidth)
16: **end if**
17: // Stage 4: Adaptive Accumulation
18: intermediateResult ← AdaptiveAccumulate(multiplicationResult, accumulator, bitWidth)
19: // Stage 5: Multi-strategy Rounding
20: roundingStrategy ← SelectRoundingStrategy(workingMode, precisionLevel)
21: result ← ExecuteRounding(roundingStrategy, intermediateResult)
22: // Configuration Update
23: UpdateSelfConfiguration(workingMode, precisionLevel)
24: **return** result

As shown in Fig. 1, experimental comparison of partitioning strategies demonstrates that the combination of H6L4 and H2L8 partitioning achieves optimal performance. The H6L4 partitioning provides balanced performance with good adaptability for general computational scenarios, while H2L8 partitioning delivers maximum efficiency for high-throughput applications, creating complementary effects in the self-configurable MAC unit. The H4L6 configuration offers intermediate performance suitable for moderate precision requirements, and H8L2 provides maximum precision for critical applications.

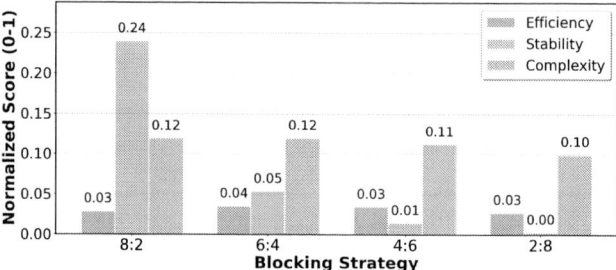

Fig. 1 Performance comparison of bit-width partitioning strategies

2) Working Mode Selection Strategy: The self-configurable MAC unit operates in three distinct working modes:

- **High-Precision Mode:** Full bit-width computation ensuring maximum accuracy
- **Medium-Precision Mode:** Adaptive bit-width selection balancing precision and performance
- **Speed-Optimized Mode:** Reduced bit-width computation optimizing for throughput

The working mode selection follows:

$$Mode = \begin{cases} \text{High-Precision} & \text{if } R_{prec} > 0.95 \\ \text{Medium-Precision} & \text{if } 0.85 \leq R_{prec} \leq 0.95 \\ \text{Speed-Optimized} & \text{if } R_{prec} < 0.85 \end{cases} \quad (5)$$

where R_{prec} represents the normalized precision requirement.

3) Bit-width Self-Adaptive Accumulator Management: As illustrated in Fig. 2, the bit-width self-adaptive accumulator dynamically adjusts its operational width based on accumulation count and precision requirements. The optimization analysis demonstrates consistent performance across multiple configurations, with Configuration C1 achieving optimal balance between precision guarantee and switching overhead.

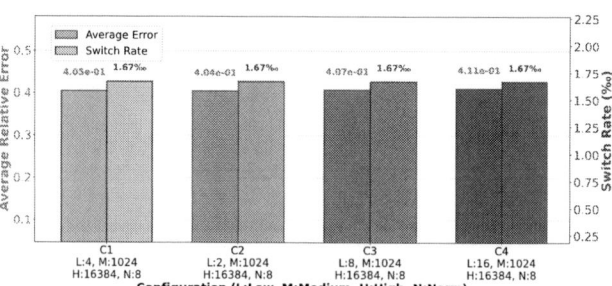

Fig. 2 Performance analysis of bit-width self-adaptive accumulator

The self-adaptive accumulator width W_{acc} is determined by:

$$W_{acc}(N_{acc}, M_{work}) = \begin{cases} 16 & \text{if } N_{acc} \leq 4, M_{work} = \text{HT} \\ 24 & \text{if } 4 < N_{acc} \leq 16, M_{work} = \text{B} \quad (6) \\ 32 & \text{if } N_{acc} > 16, M_{work} = \text{HP} \end{cases}$$

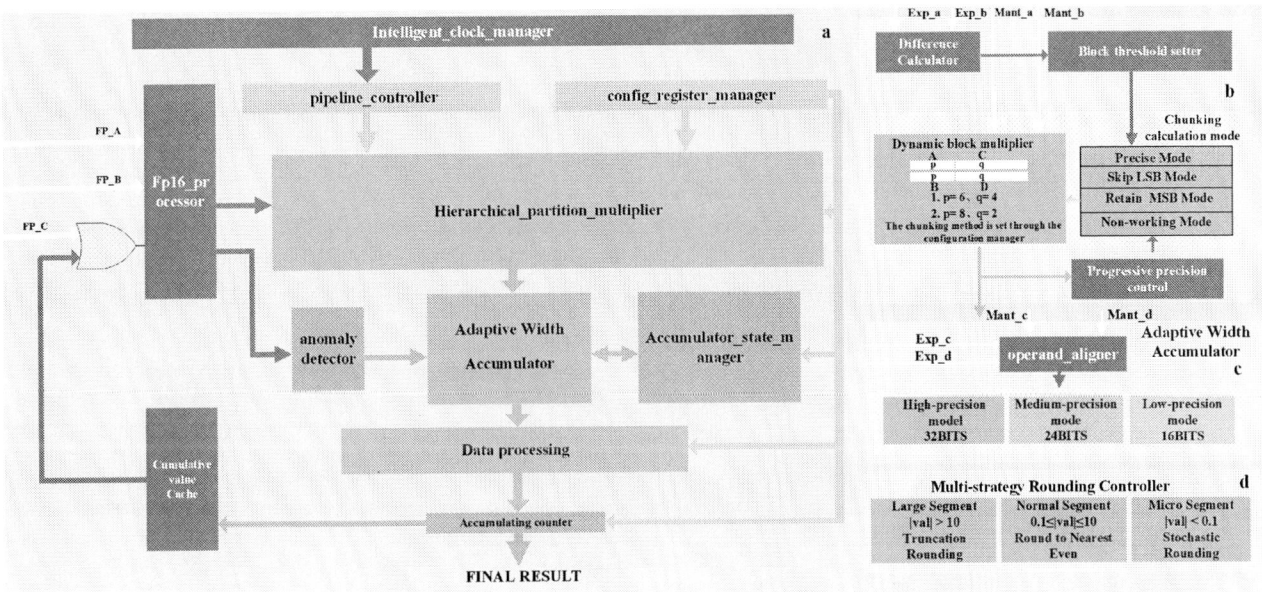

Fig. 3 Hierarchical Adaptive MAC Hardware Architecture

where N_{acc} is the accumulation count and M_{work} is the current working mode (HT: High-Throughput, B: Balanced, HP: High-Precision).

The bit-width adaptation factor follows:

$$A(N_{acc}) = 2^{\lfloor \log_2(N_{acc}/4) \rfloor + 1}, \quad A \in \{1, 2, 4, 8\} \quad (7)$$

4) Multi-strategy Rounding Control: The precision-adaptive MAC unit integrates four rounding strategies, self-configured based on working mode and precision requirements:

- **Round to Nearest Even:** IEEE 754 compliant for high-precision mode

$$Round_{even}(x) = \begin{cases} \lfloor x \rfloor & \text{if } frac(x) < 0.5 \\ \lceil x \rceil & \text{if } frac(x) > 0.5 \\ \lfloor x \rfloor & \text{if } frac(x) = 0.5, \lfloor x \rfloor \text{ even} \end{cases} \quad (8)$$

- **Truncation Rounding:** Optimized for high-throughput mode

$$Round_{trunc}(x) = \lfloor x \rfloor \quad (9)$$

- **Stochastic Rounding:** Adaptive precision preservation

$$Round_{stoch}(x) = \lfloor x \rfloor + Bernoulli(frac(x)) \quad (10)$$

- **Adaptive Rounding:** Self-configurable based on data characteristics

$$Round_{adapt}(x) = Round_{method}(x, \\ SelectMethod(x, M_{work})) \quad (11)$$

The rounding strategy selection is self-configured according to:

$$Strategy_{round} = f(M_{work}, R_{prec}, DataPattern) \quad (12)$$

where the function f implements the self-configuration logic based on current working mode, precision requirements, and input data characteristics.

III. DESIGN AND IMPLEMENTATION OF A BIT-WIDTH SELF-CONFIGURATION MAC UNIT HARDWARE ARCHITECTURE

Based on the theoretical foundation established in Chapter 2, this chapter focuses on the hardware design and implementation of the hierarchical approximate floating point MAC unit. The emphasis is on practical hardware circuit design, pipeline architecture, and key module implementations.

A. Overall Hardware Architecture Design

The overall architecture adopts a modular pipeline design to achieve efficient hardware implementation. As illustrated in Fig. 3.a, the system comprises eight core modules: Intelligent clock manager, FP16 processor, pipeline controller, Configuration register manager, Cumulative value cache, Anomaly detector, Accumulator state manager, and Accumulation counter.

The computational flow follows a systematic pipeline: input data pre-processing through Fp16 processor, dynamic configuration management via Configuration register manager, core MAC operations through the three key computational modules (detailed below), and result output with anomaly detection. The Intelligent clock manager provides adaptive clock gating for power optimization, while the pipeline controller coordinates the entire operation sequence.

B. Key Module Technical Implementation

1) Hierarchical Partition Multiplier Design: The Hierarchical partition multiplier implements the core partitioning strategies from Chapter 2. As shown in Fig. 3.b, it comprises five main components: Difference Calculator, Block threshold setter, Dynamic block multiplier, Chunking calculation mode, and Progressive precision control.

The Difference Calculator analyzes input exponent differences to determine optimal partitioning strategy. Four calculation modes are implemented: precise mode, skip LSB mode, Retain MSB mode, and Non-working mode. The Dynamic block multiplier supports two validated configurations: high 6-bit/low 4-bit and high 8-bit/low 2-bit partitioning, dynamically selected based on precision requirements. The Progressive precision control manages adaptive precision adjustment, ensuring optimal accuracy-performance trade-offs.

2) Adaptive Width Accumulator Design: The Adaptive width accumulator implements precision-adaptive accumulation from Chapter 2. As illustrated in Fig. 3.c, it includes an operand aligner and three precision modes: High-precision (32 bits), Medium-precision (24 bits), and Low-precision (16 bits).

Input data undergoes alignment through the operand aligner based on exponent differences. The system then dynamically selects among the three precision modes according to accumulation count and numerical range requirements. This adaptive selection mechanism optimizes hardware resource utilization while maintaining IEEE 754 compatibility and implementing the mathematical formulations from Chapter 2.

3) Multi-strategy Rounding Controller Design: The Multi-strategy rounding controller implements magnitude-based rounding control from Chapter 2. As shown in Fig. 3.d, it employs three rounding strategies based on numerical ranges:

- **Large Segment** ($|val| > 10$): Truncation Rounding for computational efficiency
- **Normal Segment** ($0.1 \leq |val| \leq 10$): Round to Nearest Even for IEEE 754 compliance
- **Micro Segment** ($|val| < 0.1$): Stochastic Rounding to prevent bias accumulation

This magnitude-based strategy selection ensures optimal precision across the full dynamic range while providing IEEE 754 compatibility and enhanced precision for specific application requirements.

IV. EXPERIMENTAL EVALUATION

A. Validation of proposed Precision-Adaptive FP-MAC Algorithms

This experiment used the classic Lena grayscale image with a resolution of 128×128 pixels. A 3×3 Gaussian filter with $\sigma = 0.8$ was applied for smoothing. Six computational strategies were evaluated, FP32 reference standard, FP16 standard implementation, MAC adaptive mode, MAC high precision mode, MAC progressive precision mode, MAC speed-optimized mode.

As shown in Fig. 4, visual comparisons reveal distinct performance characteristics across different computation modes. The FP16 standard implementation achieves excellent visual fidelity with a PSNR of 47.70 dB, appearing nearly identical to the FP32 reference. Key features such as facial contours and texture details are well preserved.

The high precision mode produces comparable quality with a PSNR of 47.38 dB, maintaining sharp edges and detailed structures. The progressive precision and speed-optimized modes deliver reasonable visual quality (PSNR around 44 dB), with slight degradation in high-frequency details.

In contrast, the adaptive mode exhibits the most significant quality drop, with a PSNR of 26.30 dB. Despite the loss of fine details, the overall structure of the image remains visually recognizable and the smoothing effect is retained.

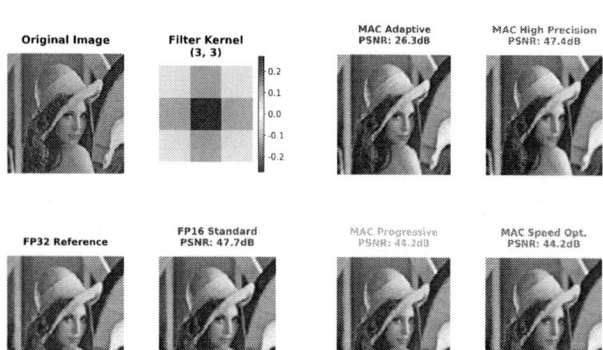

Fig. 4 Comparison of the proposed mac algorithm's Gaussian 3x3 filtering image

B. Implementation of a Bit-width Self-configuration MAC unit

This section employs Synopsys Design Compiler for hardware synthesis based on TSMC 28 nm HPC+ process library, targeting 500 MHz operating frequency at 0.9 V supply voltage and 85 °C temperature condition. The synthesis strategy utilizes **compile_ultra** mode with clock gating, datapath optimization, and power optimization enabled. The implementation results show the proposed design of FP-MAC is with small footprint of 926 μm^2, with low latency of 0.28 ns, working at the frequency of 40MHz.

V. CONCLUSION

This paper presents a hierarchical adaptive half-precision fused multiply-add unit design achieving multi-dimensional optimization of precision-performance-power through numerical feature-driven adaptive strategies. A hierarchical algorithm based on 10-bit mantissa analysis and 6-stage pipeline hardware architecture are designed, with software validation showing FP16 standard achieving a PSNR of 47.70 dB. Future work requires FPGA implementation for actual data, timing optimization, application extension, intelligent strategy selection, and unified multi-precision framework construction.

ACKNOWLEDGE

This work is supported by the National Natural Science Foundation of China (Grant No. 62304107), and the Fundamental Research Funds for Central Universities (Grant No. NS2024027).

REFERENCES

[1] J. de Fine Licht, C. Pattison, A. N. Ziogas, D. Simmons-Duffin, and T. Hoefler, "Fast Arbitrary Precision Floating Point on FPGA," Proceedings of the 30th IEEE International Symposium on Field-Programmable Custom Computing Machines (FCCM), 2022.

[2] E. Calore and S. Schifano, "Energy-Efficiency Evaluation of FPGAs for Floating-Point Intensive Workloads," in Parallel Computing: Technology Trends, IOS Press, 2020. DOI:10.3233/APC200085.

[3] Z. Ebrahimi, M. Zaid, M. Wijtvliet, and A. Kumar, "RAPID: Approximate Pipelined Multipliers and Dividers for High-Throughput and Energy-Efficiency," IEEE Transactions on Circuits and Systems II: Express Briefs, vol. 69, no. 2, pp. 676-680, 2022.

[4] Tan, Hongbing, et al. "Multiple-mode-supporting floating-point FMA unit for deep learning processors." IEEE Transactions on Very Large Scale Integration (VLSI) Systems 31.2 (2022): 253-266.

[5] Yagi, H. et al. "Acceleration of interactive multiple precision arithmetic toolbox MuPAT using FMA, SIMD, and OpenMP." MIMS EPrint, 2022.

[6] Rai, Himanshu, et al. "FPUGen: A FrameWork to Generate Custom Floating Point FMA Accelerators on FPGAs." 2025 38th International Conference on VLSI Design and 2024 23rd International Conference on Embedded Systems (VLSID). IEEE, 2025.

[7] Yao, Yuan, et al. "TangramFP: Energy-Efficient, Bit-Parallel, Multiply-Accumulate for Deep Neural Networks." 2024 IEEE 36th International Symposium on Computer Architecture and High Performance Computing (SBAC-PAD). IEEE, 2024.

[8] Ali, Sami Ben, Silviu-Ioan Filip, and Olivier Sentieys. "A Stochastic Rounding-Enabled Low-Precision Floating-Point MAC for DNN Training." 2024 Design, Automation and Test in Europe Conference and Exhibition (DATE). IEEE, 2024.

[9] Liu, Haotian, et al. "A 3-D Multi-Precision Scalable Systolic FMA Architecture." IEEE Transactions on Circuits and Systems I: Regular Papers (2024).

[10] Murillo, Raul, et al. "Energy-efficient MAC units for fused posit arithmetic." 2021 IEEE 39th International Conference on Computer Design (ICCD). IEEE, 2021.

[11] Dias, Guilherme, et al. "Dynamic Reconfigurable FPU for Next-Generation Transprecision Computing." 2025 IEEE 16th Latin America Symposium on Circuits and Systems (LASCAS). IEEE, 2025.

High-Performance Radiation-Hardened Flip-flop for Reliable Systems

Jie Li*[1], *Member*, *IEEE*, Xiaoming Teng[1], Yufeng Zhang[1], *Member*, *IEEE*

[1] School of Astronautics, Harbin Institute of Technology, Harbin 150001, China

* Email: lijie0221@hit.edu.cn

Abstract—To address the performance overhead and design complexity challenges of fault-tolerant D flip-flop (DFF) cells when hardening performance-critical circuits such as pipeline, this paper proposes a High-Performance Radiation-Hardened Flip-flop (HPRH-FF) cell. The design employs pulsed-clock temporal redundancy to remove the SET mitigation delay from critical timing paths, thereby minimizing performance degradation. Simultaneously, latch-based spatial redundancy is used to enable real-time correction of both SET and SEU faults while reducing area and power overhead. Simulation results confirm that the HPRH-FF cell features low delay overhead and achieves significantly reduced Area-Power-Delay Product (APDP) compared to conventional radiation-hardened flip-flops.

Keywords—Flip-flop, Radiation-Hardened, SET, SEU, High-Performance

I. Introduction

High-energy particles prevalent in space environments can induce radiation effects when striking electronic devices, leading to critical reliability issues [1]. As fundamental building blocks of sequential circuits, D flip-flops (DFFs) are widely used in integrated circuits. When exposed to space radiation, DFFs become vulnerable to soft errors originating from two mechanisms: Single-Event Upsets (SEUs) caused by direct particle strikes on the storage nodes in DFFs; Single-Event Transients (SETs) generated and propagated in combinational logic and captured by sequential elements.

In larger technology nodes, SEUs constitute the dominant soft-error threat, while SETs in combinational logic pose minimal risk due to the electrical, logical and latch-window masking ways [2]. As transistor feature sizes scale down, reduced node capacitance and critical charge increase both the generation probability and propagation length of SET pulses in combinational logic, while simultaneously raising their capture probability by DFFs. Consequently, SET-induced soft errors in combinational logic have become comparable to SEU threats in advanced technology nodes, emerging as a critical factor in radiation-hardening methodologies [3].

In DFFs hardening designs, Triple Modular Redundancy (TMR) effectively mitigates SEUs. When extended to temporal redundancy as Temporal TMR (T-TMR), it additionally addresses SETs in combinational logic. Alternatively, the BISER cell [4] employs Dual Modular Redundancy (DMR) architecture, reducing replication overhead versus TMR. It replaces majority voters with a C-element that maintains correct output during input disagreement, thereby blocking error propagation. BISER cell similarly incorporates temporal redundancy for SET mitigation. DAD-FF cell [5] enhances BISER's delay unit with an adjustable-delay structure, enabling dynamic tuning of detectable SET pulse widths through delay calibration without circuit layout modifications. TH-FF [6] integrates

delay and C- elements within its master-slave latches, concurrently blocking SETs from combinational logic and preventing internal SEUs. SEM cell [7] employs a data speculative mechanism to eliminate temporal-redundancy-induced delays from critical timing paths, significantly reducing DFF timing overhead and mitigating performance penalties associated with conventional hardening. However, its implementation requires additional clock cycles and error-recovery circuitry to resolve soft errors, increasing design complexity.

In summary, radiation hardening DFFs targets concurrent mitigation of both SEU and SET faults, with ongoing research focusing on minimizing the area, power, and performance overhead inherent to hardened designs. To address these challenges across diverse application scenarios, this paper presents a High-Performance Radiation-Hardened Flip-flop (HPRH-FF) cell. The main contributions of this paper are as follows.

- Implementation of spatial redundancy using level-sensitive latches to reduce area and power overhead.

- Application of pulsed-clock temporal redundancy that removes SET mitigation delays from critical timing paths, minimizing performance degradation.

- Enabling real-time SET/SEU mitigation without additional error-recovery circuitry, significantly reducing design complexity.

II. HPRH-FF Design

A. Architecture

The circuit architecture of the proposed HPRH-FF is shown in Fig. 1. It comprises: a conventional edge-triggered master-slave register, two level-sensitive redundancy latches (R1-latch and R2-latch) and a clock module. The majority voter is functionally integrated into the R2-latch to enable inherent fault tolerance.

Data enters at input D and propagates sequentially through the master-slave register, R1 latch, R2 latch (with embedded majority voter). Output Q is governed by the R2 latch's integrated voter, which arbitrates the final state based on values from all three storage elements.

HPRH-FF utilizes three clock inputs named CLK, CKR1 and CKR2 respectively. CLK denotes the conventional clock used in non-hardened D flip-flops, which drives the master-slave register. CKR1 and CKR2 are derived pulsed clocks that gate the R1 latch and R2 latch respectively. Fig. 2(a) illustrates the timing relationships between CLK, CKR1, and CKR2.

B. Normal Operation

As illustrated in Fig. 2(a), clocks CLK, CKR1, and CKR2 share identical period T with phase synchronization,

979-8-3315-3918-4/25 $31.00 © 2025 IEEE

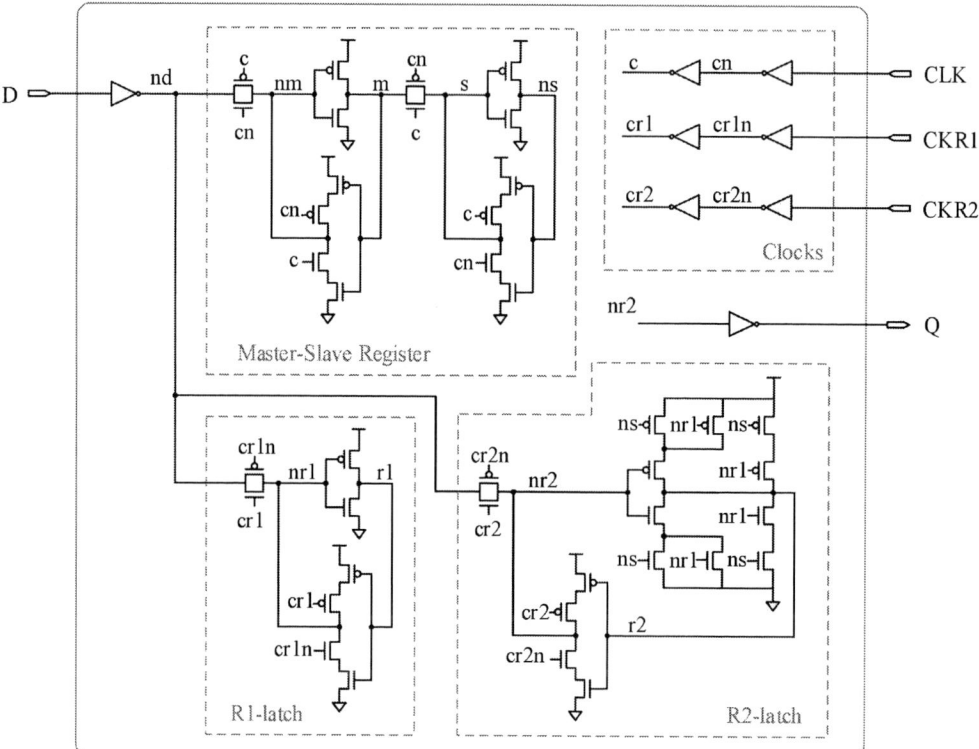

Fig. 1. Circuit-level schematic of the proposed HPRH-FF.

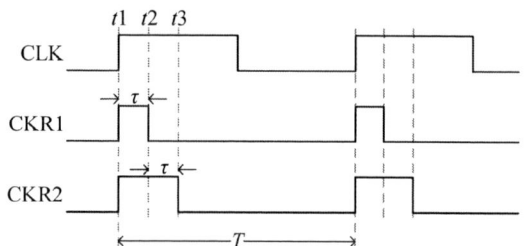

Fig. 2. Timing relationship between clocks.

exhibiting coincident rising edges at time $t1$. The CLK signal maintains a 50% duty cycle (high/low = $T/2$), while CKR1 and CKR2 operate as pulsed clocks with distinct pulse widths: CKR1 features pulse width τ (falling edge at $t2$), and CKR2 features pulse width 2τ (falling edge at $t3$). At $t1$, input data D is sampled by the master-slave register while CKR1 and CKR2 simultaneously transition high, forcing R1 and R2 latches into transparent mode. This enables D to propagate through both latches to Q. Subsequently, at $t2$, the falling edge of CKR1 triggers R1 latch to enter latching mode, capturing the data value. Finally, at $t3$, CKR2's falling edge transitions R2 latch to latching mode, completing the storage sequence.

C. Error Tolerance

The spatial redundancy implemented through the master-slave register, R1 latch, and R2 latch provides inherent SEU mitigation within the flip-flop structure. Concurrently, the temporally staggered latching operation of R1 and R2 latches filters SET pulses propagating from upstream combinational logic. For SET pulses narrower than the phase offset (τ) between latch closure events, the transient disturbance

affects at most one of the three storage elements, while the other two retain uncorrupted values. This configuration ensures the majority voter outputs correct data at Q blocking erroneous latching in the flip-flop. Conversely, when encountering SET pulses exceeding τ width — persisting continuously from $t1$ to $t2$ — both the master-slave register and R1 latch become simultaneously corrupted during their shared vulnerability window. With two of three storage elements compromised, the voter fails to recover the correct state, propagating erroneous values to Q. The maximum tolerable SET pulse width T_{pw} is therefore bounded by:

$$T_{\mathrm{pw}} \leq \tau \tag{1}$$

Equation (1) demonstrates that larger τ values expand the tolerable SET pulse width, thereby enhancing the circuit's SET fault tolerance. However, increasing τ adversely impacts hold-time constraints, establishing a critical design trade-off between radiation hardening and timing closure. Based on characterization data in [8], this work adopts $\tau = 300$ ps for subsequent simulations and comparison under a 65 nm technology library.

D. Simulations

Fig. 3 demonstrates the circuit implementation of HPRH-FF, where the highlighted regions are used for SPICE-based evaluation of both normal operation and SET/SEU fault tolerance. Critical and non-critical paths in combinational logic stage $j+1$ are emulated by varying the number of inverters in these chains, enabling controlled analysis of propagation delays under radiation-induced transients.

1) Normal Operation: Initial simulations validate the functional correctness of the proposed HPRH-FF, as demonstrated in Fig. 4 (1-5ns). Crucially, both clock-to-Q

979-8-3315-3918-4/25 $31.00 © 2025 IEEE

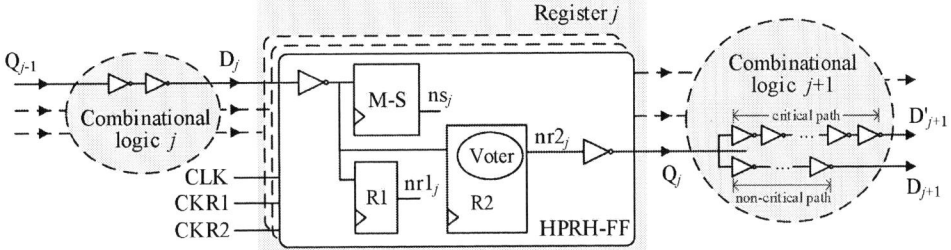

Fig. 3. The proposed HPRH-FF in circuits, where the solid line part is the circuit for SET and SEU simulations.

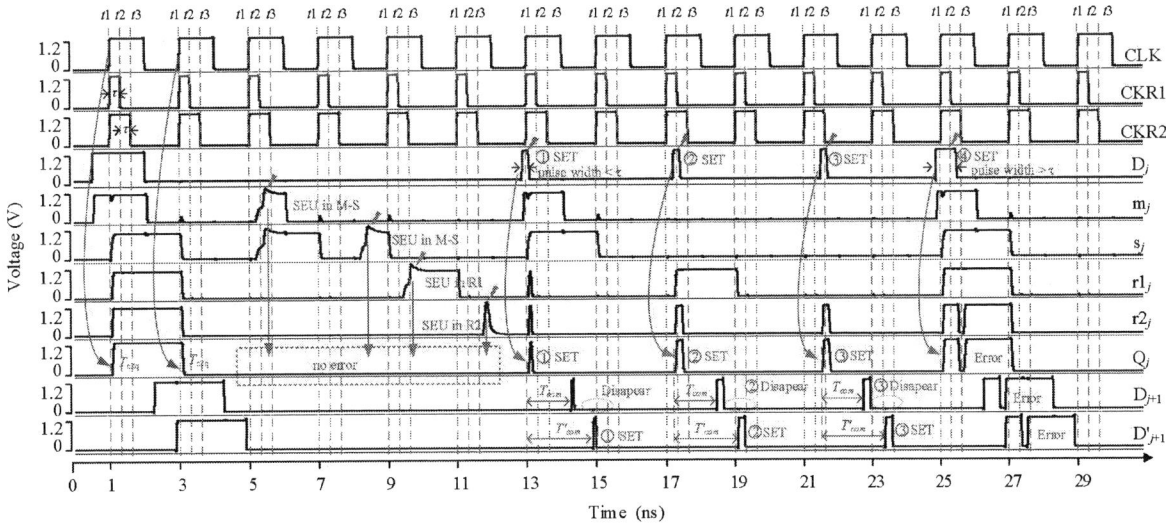

Fig. 4. SPICE simulation timing diagrams showing normal operation and SEU/SET fault tolerance.

delay remains unaffected by the latching interval τ, confirming that temporal redundancy is excluded from critical timing paths. This design consequently achieves minimal latency overhead, making it suitable for high-performance radiation-hardened systems.

2) SEU Tolerance: As shown in Fig. 4 (5-13ns), when SEUs were individually injected into the master-slave register, R1 latch, and R2 latch storage nodes, the design consistently maintained correct output states. HPRH-FF provides inherent SEU mitigation capability.

3) SET Tolerance: SET faults injected in Combinational Logic j (Fig. 3) generate transient pulses at input D_j of the HPRH-FF. As shown in Fig. 4, three interference scenarios occur when SET pulse width is below τ:

a) Scenario I (@13ns in Fig. 4) occurs when the SET pulse coincides with the CLK rising edge, causing the master-slave register to capture erroneous data; however, after τ elapses, D_j recovers its correct value, enabling R1 and R2 latches to store valid data at t2 and t2 respectively, with the majority voter consequently producing correct Q_j output.

b) Scenario II (@17ns in Fig. 4) features SET corruption exclusively in R1 latch during its latching transition, while the master-slave register and R2 latch retain uncorrupted values due to the pulse's absence during their sampling windows, resulting in error-masked Q_j output.

c) Scenario III (@21ns in Fig. 4) shows the embedded voter preventing error propagation despite SET occurrence

at R2's latching edge, maintaining Q_j integrity.

Transient pulses at Dj propagate to Qj without latching, subsequently manifesting as combinational SETs in downstream stages. When propagating through non-critical paths (delay Tcom), pulses disappear before the next clock edge at Dj+1, preventing erroneous latching. Critical path propagation (delay T'com) may deliver pulses coinciding with sampling windows at Dj+1, yet these remain processable as input SETs rather than storage errors. Notably, our aggressive simulation parameters (clock period almost equals to critical path delay) exceed practical design norms where timing margins reduce critical-path triggering probability, with consecutive critical-path occurrences being statistically improbable [9], thus confining most SETs to local resolution.

Consequently, when the SET pulse width is smaller than τ, SETs are effectively filtered by the HPRH-FF architecture, preventing soft errors from latching. When the pulse width of SET is larger than τ (exemplified @25ns in Fig. 4), SET pulse simultaneously corrupts the master-slave register and R1 latch, producing erroneous Qj. This failure mode necessitates τ-value optimization: increasing τ enhances SET tolerance but incurs hold time penalties, requiring application-specific design tradeoffs.

III. COMPARISON

The area, power, delay, Area-Power-Delay Product (ADPD) overheads different radiation hardened flip-flop cells are compared with our proposed flip-flop; the unprotected flip-flop with a standard structure (i.e., Std. DFF)

TABLE I. TABLE TYPE STYLES

Cells	Area (Transistors)	Power (vs. Std. DFF)	Delays (ps)		APDP (vs. Std. DFF)
			T_{c2q}	ΔT_s	
Std.DFF	24	1	84	0	1
T-TMR	90	3.12	129	600	101.54
BISER	56	1.88	200	300	26.11
DAD-FF	54	1.69	126	300	19.28
TH-FF	83	1.65	86.8	300	26.28
SEM	96	3.53	89.2	0	14.99
HPRH-FF	**56**	**2.06**	**79.2**	**0**	**4.53**

is also implemented as a baseline to show the relative overhead of different hardened cells. Performance metrics, including area, power, delay overhead, and error-tolerance capability, were characterized through SPICE simulations in a 65nm CMOS technology node. Transistor sizing adheres to the standard cell library specifications, with simulations conducted at 1.2V supply voltage. Stimulus signals were buffered through inverter chains to drive flip-flop inputs, while outputs loaded a fanout-of-4 (FO4) inverter. Dynamic power dissipation was measured at 20% switching activity.

As evidenced by the comparative data in Table 1, the proposed HPRH-FF exhibits the lowest APDP overhead among radiation-hardened flip-flop cells. Traditional radiation-hardened designs, including T-TMR, BISER, DAD-FF, and TH-FF, introduce SET mitigation delays into their setup time overhead, resulting in excessive propagation latency that incurs significant performance penalties when deployed in hardened circuits, rendering them unsuited for high-performance applications. Both SEM and the proposed HPRH-FF maintain minimal delay overhead, offering lower performance costs for radiation hardening. Crucially, HPRH-FF achieves lower area and reduced power consumption than SEM while eliminating additional error-recovery circuitry, thereby providing superior implementation advantages for high-performance radiation-hardened systems.

IV. CONCLUSION

This paper proposes a High-Performance Radiation-Hardened Flip-flop (HPRH-FF) cell. It eliminates SET-induced delays from critical timing paths, making it suitable for hardened designs with high-performance. Simulation results confirm that the HPRH-FF cell features low delay overhead and achieves significantly reduced APDP overhead compared to conventional radiation-hardened flip-flops. Additionally, lower application complexity provides superior implementation advantages for high-performance radiation-hardened systems than similar cells.

ACKNOWLEDGMENT

This work was supported in part by the Heilongjiang Postdoctoral Fund (Grant No. LBH-Z24144).

REFERENCES

[1] Y. Xu, Z. Liu, Y. Wang, N. Bai and Y. Liu, "A DICE Flip-Flop Design by Resetting Redundancy Hardening for Single Event Upset Tolerance," IEEE Transactions on Device and Materials Reliability, doi: 10.1109/TDMR.2025.3570099.

[2] L. A. Garcia-Astudillo, L. Entrena, A. Lindoso, H. Martín, P. Martin-Holgado and M. Garcia-Valderas, "Analyzing Reduced Precision Triple Modular Redundancy Under Proton Irradiation," IEEE Transactions on Nuclear Science, vol. 69, no. 3, pp. 470-477, March 2022

[3] Y. Chen, Y. Zhuang, "Analysis and Design of a Delay-Locked Loop with Multiple Radiation-hardened Techniques," Circuits Syst Signal Process, vol. 42, pp. 130–146, 2023.

[4] S. Mitra, M. Zhang, N. Seifert, T. Mak, and K. S. Kim, "Built-in soft error resilience for robust system design," in Proc. IEEE Int. Conf. Integr. Circuit Design Technol., May 2007, pp. 1–6.

[5] D. Y.-W. Lin and C. H.-P. Wen, "DAD-FF: Hardening designs by delay-adjustable D-flip-flop for soft-error-rate reduction," IEEE Trans. Very Large Scale Integr. (VLSI) Syst., vol. 28, no. 4, pp. 1030–1042, Apr. 2020.

[6] Y. Q. Li et al., "A 65 nm temporally hardened flip-flop circuit," IEEE Trans. Nucl. Sci., vol. 63, no. 6, pp. 2934–2940, Dec. 2016

[7] N. D. P. Avirneni and A. K. Somani, "Low overhead soft error mitigation techniques for high-performance and aggressive designs," IEEE Trans. Comput., vol. 61, no. 4, pp. 488–501, Apr. 2012.

[8] S. Jagannathan, Matthew J. G., Bharat L. B., et al., "Independent Measurement of SET Pulse Widths From N-Hits and P-Hits in 65-nm CMOS," IEEE Transactions on Nuclear Science, vol. 57, no. 6, pp. 3386-3391, Dec. 2010.

[9] V. Subramanian, M. Bezdek, N. D. Avirneni and A. Somani, "Superscalar Processor Performance Enhancement through Reliable Dynamic Clock Frequency Tuning," 37th Annual IEEE/IFIP International Conference on Dependable Systems and Networks (DSN'07), Edinburgh, UK, 2007, pp. 196-205.

Analysis and Design of Regulating Rectifier with Multiple Outputs for Wirelessly Powered Biomedical Devices

Quanrong Zhuang, Junyi Sun, Bo Li, Jie Lu, Yi Shi, Hao Qiu *

School of Electronic Science and Engineering, Nanjing University

* Email: haoqiu@nju.edu.cn

Abstract—Wireless power transfer (WPT) to implanted devices is an elegant solution that can obviate the need for batteries. As more functions are implemented in the implanted devices, more than one output voltages are typically requested for the optimal performance. In this talk, first, we summarized several representative designs of the regulating rectifier with multiple outputs. Being aware of the shortcomings of previous works, we proposed a design that can support the simultaneous charging of multiple outputs in a half cycle. Experimental results verified a peak power conversion efficiency of 92.2 % and output power of 131 mW. These performance indexes are the best compared to previous works.

Keywords—Charge distribution, dual outputs, implantable medical devices (IMD), single-stage regulating rectifier, wireless power transfer (WPT).

I. INTRODUCTION

Wireless power transfer to implanted devices is an elegant solution that can obviate the need for batteries [1-5]. As more functions are implemented in the implanted devices, more than one output voltages are typically requested for the optimal performance [6]. For example, a low output voltage (V_{OUT2}) is used for biopotential recording whereas a high output voltage (V_{OUT1}) is used in the back-end circuitry for neurostimulation. The requirements on V_{OUT1} and V_{OUT2} are different [7]. The front-end recording circuitry requires a stable and accurately regulated V_{OUT2} whereas the back-end stimulation circuitry necessitates a high bandwidth to support a high load current at V_{OUT1}.

An obvious design is to use two stages comprising a rectifier followed by a dual-output regulator [8-9]. On the other hand, the single-stage dual-output (SSDO) design [10-15] is advantageous with an improved power conversion efficiency (η_{REC}) and a reduced form factor. Fig. 1 shows several representative single-stage multi-output rectifier topologies. In [10], a 125kHz rectifier comprising four PMOS active diodes was proposed. Since the half-wave rectification is employed, it suffers from low output power (P_{OUT}) and high ripple voltages. Additionally, V_{OUT1} and V_{OUT2} must be close to each other, otherwise η_{REC} will degrade fast. The rectifier in [11] suffers from a low P_{OUT} due to its half-bridge topology. In [12], V_{OUT2} with the half-wave rectification still suffers from a high ripple voltage. Moreover, owing to a shared total charging time in a half cycle, that for V_{OUT1} is decreased and thus degrades its P_{OUT}. In [13], implemented by six PMOS active diodes, the rectifier employs the time-multiplexing charging for three outputs in a half cycle. As a result, V_{OUT1} suffers from a limited charging time and thus degrades its P_{OUT}. Furthermore, all above topologies are implemented

This work was financially supported by the National Natural Science Foundation of China (62374082, 62341408).

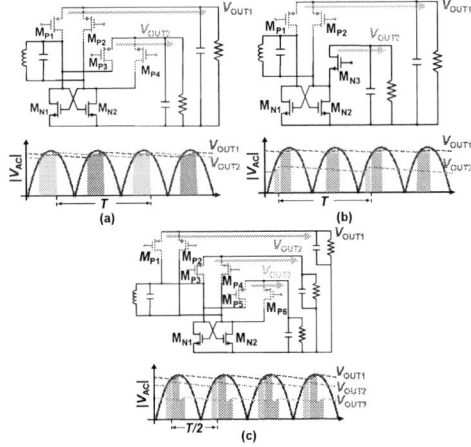

Fig. 1. Topologies and key waveforms of conventional multiple-output rectifiers charged in the conventional time-multiplexing manner. (a) V_{OUT1} and V_{OUT2}: half-wave rectification. (b) V_{OUT1}: full-wave rectification; V_{OUT2}: half-wave rectification. (c) V_{OUT1}, V_{OUT2}, and V_{OUT3}: full-wave rectification.

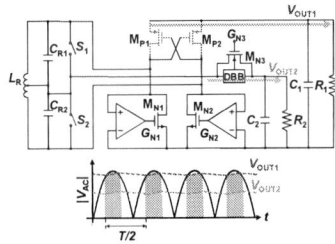

Fig. 2. Topology and key waveforms of the proposed SSDO rectifier.

using PMOS active diodes, which could induce larger power loss including switching loss and conduction loss [16].

Being aware of the shortcomings of these works, we proposed to implement a 6.78MHz SSDO rectifier using only three NMOS active diodes, which achieved a high peak η_{REC} of 92.2%. Since it supported the simultaneous charging of V_{OUT1} and V_{OUT2} in a half cycle rather than in a time-multiplexing manner, a high P_{OUT} of 131mW was obtained. Furthermore, owing to the proposed charge distribution (CD) operation mode, the rectifier eliminated a large drop voltage during a large load transient at V_{OUT2}. As a result, with two small off-chip output capacitors (150nF & 200nF), V_{OUT1} and V_{OUT2} were successfully regulated at 3.3V and 1.6V, respectively. The ripple voltages of V_{OUT1} ($V_{OUT1, ripple}$) and V_{OUT2} ($V_{OUT2, ripple}$) were as small as 50mV and 75mV, respectively.

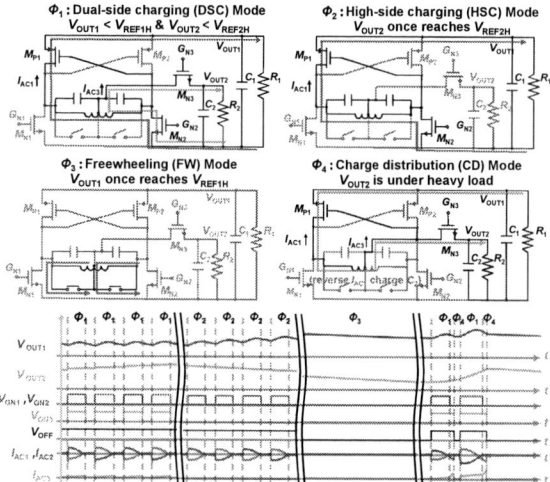

Fig. 3. Operational principle and corresponding waveforms in four operation modes of the proposed SSDO rectifier.

II. OPERATION PRINCIPLE OF PROPOSED SSDO RECTIFIER

As shown in Fig. 2, the proposed SSDO rectifier consists of a pair of cross-connected PMOS transistors (M_{P1} and M_{P2}), three NMOS active diodes (M_{N1}, M_{N2}, and M_{N3}), and two short switches (S_1 and S_2). M_{P1} and M_{P2} do not contribute to the switching loss as their input parasitic capacitances are considered as part of the resonant capacitor at the rectifier input. Their conduction loss is minimized through using a large transistor size. As is shown in Fig. 3, the rectifier can be reconfigured into four operation modes: dual-side charging (DSC) (Φ_1), high-side charging (HSC) (Φ_2), freewheeling (FW) (Φ_3), and CD (Φ_4) modes. The selectable modes depend on the condition of the load current at V_{OUT2} (I_2). When I_2 is under light condition, the rectifier operates in DSC, HSC, and FW modes. In the DSC mode, M_{N2}, M_{N3}, and M_{P1} are turned on. V_{OUT1} and V_{OUT2} can be simultaneously charged in a half cycle through current I_{AC1}/I_{AC2} and I_{AC3}, respectively, and thus both increases. In the HSC mode, M_{N2} and M_{P1} are turned on, and only V_{OUT1} is charged through I_{AC1}/I_{AC2}. In the FW mode, S_1 and S_2 are turned on and both V_{OUT1} and V_{OUT2} decrease. By detecting the voltage levels of V_{OUT1} and V_{OUT2}, the controller generates corresponding logic states to regulate V_{OUT1} and V_{OUT2} into hysteresis windows [V_{REF1L}, V_{REF1H}] and [V_{REF2L}, V_{REF2H}], respectively. When I_2 is under heavy condition, the rectifier operates in DSC, HSC, FW, and CD modes. In CD mode, at the end of the DSC mode, by turning off M_{N2}, a current path from V_{OUT1} to V_{OUT2} emerges and the proposed CD mode starts. Through this current path, the accumulated charge at V_{OUT1} during the DSC mode helps alleviate the problem of insufficient charging current for V_{OUT2}. V_{OUT2} could be regulated within the hysteretic window [V_{CD}, V_{REF2H}]

III. MEASUREMENT RESULTS

The proposed SSDO rectifier was fabricated in a 180nm CMOS process and occupied a chip area of 1.63mm². Fig. 4 shows the photographs of the proposed SSDO rectifier. The maximum load currents I_1 at V_{OUT1} and I_2 at V_{OUT2} were designed as 33mA and 15mA, respectively, for a maximum P_{OUT} of 131mW. Under different combinations of light/heavy load conditions for I_1 and I_2, V_{OUT1} and V_{OUT2} were successfully regulated at 3.3V and 1.6V, respectively, within corresponding hysteresis windows (Fig. 5). It is noted that,

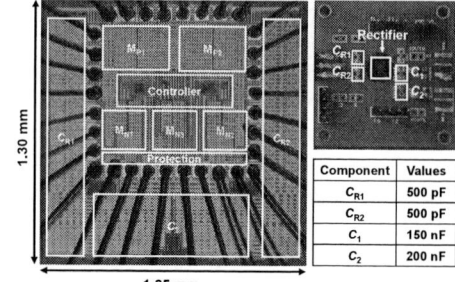

Fig. 4. Photographs of proposed SSDO rectifier and PCB prototype.

Component	Values
C_{R1}	500 pF
C_{R2}	500 pF
C_1	150 nF
C_2	200 nF

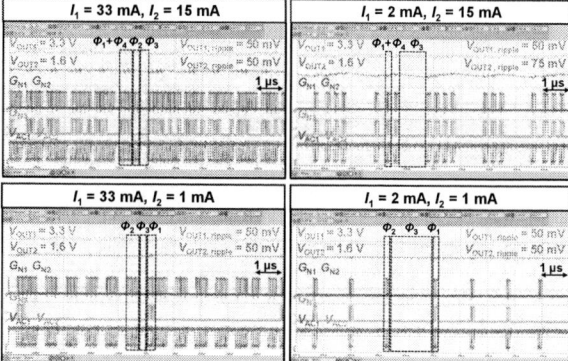

Fig. 5. Measured steady-state waveforms under different combinations of load conditions at V_{OUT1} and V_{OUT2}.

Fig. 6. Measured load transient waveforms.

$V_{OUT1, ripple}$ and $V_{OUT2, ripple}$ were as small as 50mV and 75mV with two small output capacitors C_1 of 150nF and C_2 of 200nF, respectively. In the steady state, G_{N1}, G_{N2}, and G_{N3} notified the rectifier's four operation modes. Measured load transient waveforms in Fig. 6 demonstrated negligible load transient response and unnoticeable cross regulation between two outputs. When I_2 changed from 1mA to 15mA at I_1 of 33mA, owing to the proposed CD mode, a negligible drop voltage was guaranteed. Fig. 7 shows the measured η_{REC} of the rectifier. By replacing PMOS to NMOS active diodes and reducing its number to three in the power stage, the proposed rectifier achieved a peak η_{REC} of 92.2% at P_{OUT} of 72.6mW. When I_2 is under the heavy condition and I_1 changes from heavy to light conditions, a high η_{REC} is maintained owing to the proposed CD mode. Fig. 8 and Table I show the performance comparisons. Compared to previous SSDO rectifiers, this work achieved the highest η_{REC} and P_{OUT}. The figure of merit (FoM) for $V_{OUT1, ripple}$ and $V_{OUT2, ripple}$, defined as $[I_1/(fC_1V_{OUT1, ripple}) + I_1/(fC_2V_{OUT2, ripple})] \times 100\%$, is used for comparison. The FoM in this work is the best.

IV. CONCLUSION

We presented a 6.78 MHz SSDO regulating rectifier for wireless charging of IMDs. Instead of being charged in the conventional time-multiplexing manner, the proposed SSDO

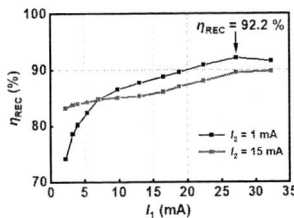

Fig. 7. Measured I_1 dependence η_{REC} of at different I_2.

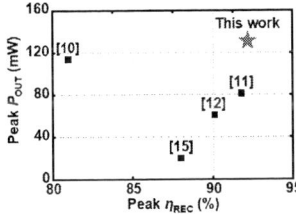

Fig. 8. P_{OUT} and η_{REC} performance comparisons with previous works.

TABLE I.   COMPARISONS WITH PREVIOUS WORKS ON MULTIPLE-OUTPUT RECTIFIERS

	ISSCC 2024 [13]	TCAS-I 2019 [10]	CICC 2023 [11]	ISSCC 2023 [12]	JSSC 2024 [15]	This work
Technology	0.25 μm CMOS	0.18 μm CMOS	0.18 μm BCD	65 nm CMOS	65 nm CMOS	0.18 μm CMOS
Number of power stages	Single stage	Single stage	Single stage	Single stage	Single stage	Single stage
Frequency (MHz)	2.0	0.125	13.56	40.68	13.56	6.78
LC tank connection	Parallel	Parallel	Parallel	Parallel	Parallel	Parallel
Simultaneous charging of multiple outputs	No	No	No	No	No	Yes
No. and type of active diodes	6 PMOS	4 PMOS	1 PMOS + 1 NMOS	2 PMOS + 1 NMOS	4 PMOS + 2 NMOS	3 NMOS
No of outputs	3	2	2	2	2	2
Output voltage	4.5 V 3.5 V – 1.0 V 3.5 V – 1.0 V	2.0 V – 1.6 V 2.0 V – 1.6 V	3.6 V 1.8 V	2.2 V 1.1 V	2.5 V 1.2 V	3.3 V – 2.2 V 1.6 V – 1.0 V
Peak P_{OUT}	135.6 mW	114 mW P_{OUT1} = 60 mW P_{OUT2} = 54 mW	91 mW P_{OUT1} = 64.8 mW P_{OUT2} = 16.2 mW	60.5 mW P_{OUT1} = 55 mW P_{OUT2} = 5.5 mW	20 mW	131 mW P_{OUT1} = 109.9 mW P_{OUT2} = 24.1 mW
Peak η_{REC}	90.82 % @ 84.6 mW V_{OUT1} = 4.5 V, V_{OUT2} = 3.3 V, V_{OUT3} = 2.5 V	81 % @ 42 mW V_{OUT1} = 2.0 V, V_{OUT2} = 1.6 V.	91.8 % @ 33.5 mW V_{OUT1} = 3.6 V, V_{OUT2} = 1.8 V.	90.1 % @ 56.1 mW V_{OUT1} = 2.2 V, V_{OUT2} = 1.1 V.	88 % @ 5 mW V_{OUT1} = 2.5 V, V_{OUT2} = 1.2 V.	92.2 % @ 72.6 mW V_{OUT1} = 3.3 V, V_{OUT2} = 1.6 V.
Output Capacitor	10 μF ×3	10 μF × 2	3.9 nF × 2	470 nF & 220 nF	1 μF × 2	150 nF & 200 nF
Ripple voltage	N / A	V_{OUT1}: 80 mV V_{OUT2}: 60 mV	V_{OUT1}: 340 mV V_{OUT2}: 300 mV	V_{OUT1}: 200 mV V_{OUT2}: 100 mV	V_{OUT1}: 100 mV V_{OUT2}: 100 mV	V_{OUT1}: 50 mV V_{OUT2}: 75 mV
Load transient recovery time	Negligible	540 μs	Negligible	Negligible	< 100 μs	Negligible
Cross regulation	Unnoticeable	Unnoticeable	Unnoticeable	Unnoticeable	Unnoticeable	Unnoticeable
FoM (%)	N / A	70	43	1.6	0.73	80

FoM is defined as $[f_1 \cdot f(fC_1 V_{OUT1, ripple}) + f_2 \cdot f(fC_2 V_{OUT2, ripple})] \times 100\%$
Noted that the SSDO rectifier design [17-18] for series LC tank connection are not listed here.

rectifier can support the simultaneous charging of V_{OUT1} and V_{OUT2} in a half cycle. Implemented using only three NMOS active diodes, the rectifier can save power loss that would be caused by using PMOS ones. Fabricated in a 180 nm CMOS technology, the proposed SSDO rectifier occupies 1.63 mm^2. Measurement results verified two regulated outputs at 3.3 V and 1.6 V. And the corresponding FoM is the best compared with all previous works. The rectifier also achieves competitive peak P_{OUT} of 131 mW and peak η_{REC} of 92.2 %.

ACKNOWLEDGMENT

This work was financially supported by the National Natural Science Foundation of China (62374082, 62341408, T2221003), and the National Natural Science Foundation of China for Excellent Young Scholars (Overseas), the Engineering Research Center of Opto-Electro Materials and Chip Techniques, Nanjing University, Nanjing 210023, and the Interdisciplinary Research Center for Future Intelligent Chips, Nanjing University, Suzhou 215163, China. Corresponding authors: Yi Shi and Hao Qiu (yshi@nju.edu.cn, haoqiu@nju.edu.cn).

REFERENCES

[1] G. L. Barbruni, P. M. Ros, D. Demarchi, S. Carrara and D. Ghezzi, "Miniaturised wireless power transfer systems for neurostimulation: a review," *IEEE Trans. Biomed. Circuits Syst.*, vol. 14, no. 6, pp. 1160–1178, Dec. 2020.

[2] Q. Duan, C. Chen, X. Han and L. Cheng, "A 40.68-MHz active rectifier using an inverter-based conduction-time generator for wirelessly powered implantable medical devices," *IEEE Trans. Circuits Syst. II, Exp. Briefs*, vol. 69, no. 11, pp. 4334–4338, Nov. 2022.

[3] S. -W. Hong, "A resonant current-mode wireless power and data receiver for loosely coupled implantable devices," *IEEE J. Solid-State Circuits*, vol. 55, no. 12, pp. 3200–3209, Dec. 2020.

[4] H. Qiu, T. Sakurai, M. Takamiya, "A 6.78-MHz multiple-transmitter wireless power transfer system with efficiency maximization by adaptive magnetic field adder IC," *IEEE J. Solid-State Circuits*, vol. 57, pp. 2390–2403, Jun. 2022.

[5] Q. Zhuang, J. Sun, X. Zhang, B. Li, Y. Shi and H. Qiu, "A 6.78 MHz wireless power and data transfer system achieving simultaneous 52.6% end-to-end efficiency and 4.0 Mb/s forward data delivery with interference-free rectifier," in *Proc. IEEE Symp. VLSI Circuits*, Jun. 2024, pp. 1–2.

[6] M. Kiani, "Wireless power transfer and management for medical applications: wireless power," *IEEE Solid-State Circuits Mag.*, vol. 14, no. 3, pp. 41–52, Summer 2022.

[7] R. Erfani, F. Marefat and P. Mohseni, "A dual-output single-stage regulating rectifier with PWM and dual-mode PFM control for wireless powering of biomedical implants," *IEEE Trans. Biomed. Circuits and Syst.*, vol. 14, no. 6, pp. 1195–1206, Dec. 2020.

[8] Y. Lu, M. Huang, L. Cheng, W. -H. Ki, S. -P. U and R. P. Martins, "A dual-output wireless power transfer system with active rectifier and three-level operation," *IEEE Trans. Power Electron.*, vol. 32, no. 2, pp. 927–930, Feb. 2017.

[9] C. -Y. Wu, X. -H. Qian, M. -S. Cheng, Y. -A. Liang and W. -M. Chen, "A 13.56 MHz 40 mW CMOS high-efficiency inductive link power supply utilizing on-chip delay-compensated voltage doubler rectifier and multiple LDOs for implantable medical devices," *IEEE J. Solid-State Circuits*, vol. 49, no. 11, pp. 2397–2407, Nov. 2014.

[10] Q. W. Low and L. Siek, "A single-stage dual-output tri-mode AC–DC regulator for inductively powered application," *IEEE Trans. Circuits Syst. I, Reg. Papers*, vol. 66, no. 9, pp. 3620–3630, Sep. 2019.

[11] T. Lu, Z.-Y. Chang, J. Jiang, K. Makinwa, and S. Du, "A 13.56 MHz fully integrated 91.8% efficiency single-stage dual-output regulating voltage doubler for biomedical wireless power transfer," in *Proc. IEEE Custom Integr. Circuits Conf. (CICC)*, Apr. 2023, pp. 1–2.

[12] Z. Luo, J. Liu, and H. Lee, "30.9 A 90%-efficiency 40.68 MHz single-stage dual-output regulating rectifier with ZVS and synchronous PFM control for wireless powering," in *Proc. IEEE Int. Solid-State Circuits Conf. (ISSCC)*, Feb. 2023, pp. 454–456.

[13] H.-S. Lee, K. Eom, and H.-M. Lee, "27.3 A 90.8%-efficiency SIMO resonant regulating rectifier generating 3 outputs in a half cycle with distributed multi-phase control for wirelessly-powered implantable devices," in *IEEE Int. Solid-State Circuits Conf. (ISSCC)*, Feb. 2024, pp. 448–450.

[14] D. -H. Yao, T. -N. Liu, M. Takamiya and P. -H. Chen, "A 6.78-MHz wireless power transfer system with dual-output resonant current-mode regulating rectifier and transmission power regulation," *IEEE Trans. Circuits Syst. I: Reg. Papers*, vol. 70, no. 12, pp. 4986–4998, Dec. 2023.

[15] Y. Liu, Y. Yao and W. -H. Ki, "A 13.56-MHz single-input dual-output wireless power and data transfer system for bio-implants," *IEEE J. Solid-State Circuits*, vol. 59, no. 8, pp. 2557–2567, Aug. 2024.

[16] Z. Luo, J. Liu and H. Lee, "A high-efficiency 40.68-MHz single-stage dual-output regulating rectifier with ZVS and synchronous PFM control for wireless powering," *IEEE J. Solid-State Circuits*, vol. 59, no. 8, pp. 2418–2429, Aug. 2024.

[17] Y. Chen, Y. Luo, J. Guo, X. Tang and D. Chen, "A 2-W, 90% efficiency single-stage dual-output wireless power receiver with 0.1 to 700-mA output current range through dynamic delay compensation and bootstrap adaptive body biasing circuit," in *Proc. IEEE Asian Solid-State Circuits Conf. (A-SSCC)*, Nov. 2023, pp. 1–3.

[18] J. Lin, Y. Lu, C. Zhan and R. P. Martins, "A single-stage dual-output regulating rectifier with hysteretic current-wave modulation," *IEEE J. Solid-State Circuits*, vol. 56, no. 9, pp. 2770–2780, Sep. 2021.

Noise Notch Frequency Design for EMI Mitigation in DC-DC Converters Using Digital-to-Time Converter

Yasunori Kobori *[1], Yifei Sun [2], Guiyi Dong [3], Nobukazu Tsukiji **[4], Ramin Khatami[3],
Takuya Arafune[3], Shogo Katayama[3], Anna Kuwana[3], Jianglin Wei[5], Haruo Kobayashi***[3]

[1] Maebashi Institute of Technology, Maebashi, Gunma, Japan, [2] Shenyang University of Chemical Technology, China
[3] Gunma University, Japan, [4] National Institute of Technology (KOSEN), Gunma College, Japan. [5] Yibin University, China

Email: *yasu.kobo1028@gmail.com, **tsukiji@gunma.kosen-ac.jp, ***koba@gunma-u.ac.jp

This paper introduces our innovative band-selective frequency technology designed to tackle Electromagnetic Interference (EMI) noise in DC-DC switching converters for communication devices [1-6]. EMI noise generated by DC-DC switching converters (Fig. 1(a)) often necessitates the use of noise spread spectrum technology to comply with regulations and minimize bulky filters and shielding requirements. However, traditional methods may inadvertently allow noise to infiltrate the signal band (Fig. 1(b)).

To resolve this issue, selective notch frequency technology has been developed. This approach generates spectrum bands with notch characteristics, ensuring minimal noise within the signal received frequency range (Fig. 1(c)). It utilizes pulse coding control technology with the configuration depicted in Fig. 2. By employing a 1-bit digital-to-time converter, which replaces the conventional 1-bit DAC, it functions similarly to a $\Delta\Sigma$ modulator. The technology is adaptable and can be implemented using pulse width coding (PWC) (Fig. 3), pulse phase coding (PPC), or a combination of both methods (PWPC) [6].

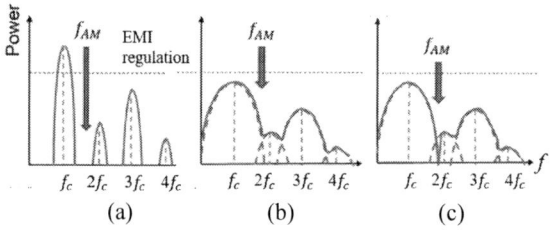

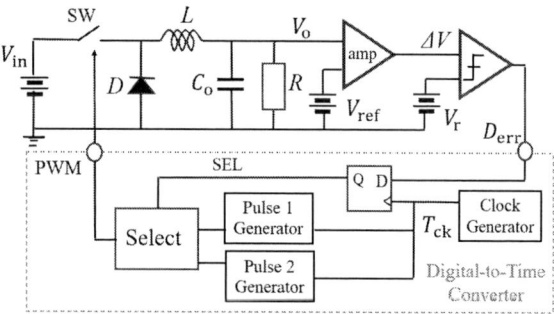

Fig. 1. EMI from DC–DC converter. (a) Original spectrum without clock modulation. (b) With conventional noise spread spectrum method. (c) With band selective noise spread spectrum method.

Fig. 2. Switching converter with pulse coding control.

Fig. 4 demonstrates the measured power spectrum from our prototype, showcasing notch frequencies at $N/(W_H - W_L)$ for N = 1, 2, … This formula is derived theoretically as well. Our latest efforts are focused on extending this technology for application in single-inductor dual-output converters [7].

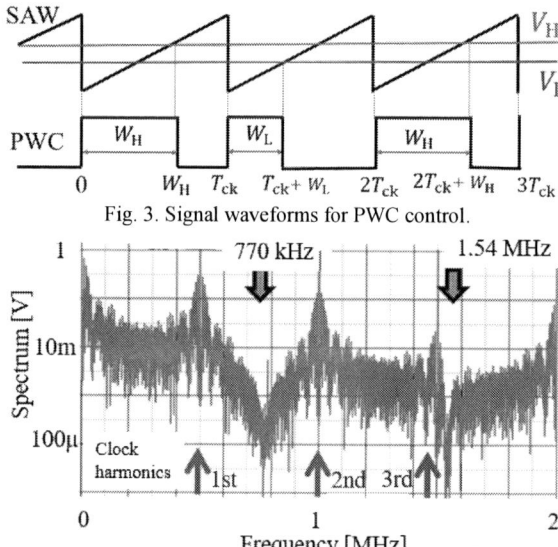

Fig. 3. Signal waveforms for PWC control.

Fig. 4. Spread spectrum of PWM signal with PWC control [6]

REFERENCES

[1] R. Khatami, H. Kobayashi, Y. Kobori, "Delta-Sigma Digital-to-Time Converter For Band-Select Spread Spectrum," Key Engineering Materials (2015).

[2] Y. Kobori, T. Arafune, N. Tsukiji, H. Kobayashi, "Selectable Notch Frequencies of EMI Spread Spectrum Using Pulse Modulation in Switching Converter," IEEE ASICON (Nov. 2015).

[3] Y. Sun, Y. Kobori, H. Kobayashi, "Full Automatic Notch Generation in Noise Spectrum of Pulse Coding Controlled Switching Converter," IEEE ICSICT (Nov. 2018).

[4] Y. Sun, Y. Kobori, A. Kuwana, H. Kobayashi, "Pulse Coding Controlled Switching Converter that Generates Notch Frequency to Suit Noise Spectrum," IEICE Trans. Communications (Nov. 2020).

[5] G. Dong, S. Katayama, Y. Sun, Y. Kobori, A. Kuwana, H. Kobayashi, "Notch Frequency Generation Methods in Noise Spread Spectrum for Pulse Coding Switching DC-DC Converter," IEEE LASCAS (Mar. 2022).

[6] Y. Kobori, Y. Sun, H. Kobayashi, "Selective Notch Frequency Technology for EMI Noise Reduction in DC-DC Converters: A Review," Sensors, Vol. 25, Issue 10, 3196 (May 2025).

[7] Y. Kobori, H. Kobayashi, "Single-Inductor Multi-Output DC-DC Switching Converters Using Exclusive Control Method," Digital Technologies Research and Applications, Vol. 4 Issue 1 (2025).

A 240nA-1μA Quiescent SIMO Converter Featuring 3mV Undershoot under 30mA/μs Transients

Yuhua Chen [1], Qianhui Liu [1], Yixing Wang [1], Yimeng Zhang [*1] Yuming Zhang [1]

[1] School of Microelectronics, Xidian University, Xian710000 Xi'an, China

* Email: zhangyimeng@xidian.edu.cn

Abstract—This paper proposes a reconfigurable multi-mode single-input multiple-output (SIMO) converter with an adaptive switching PCCM–MDSCM–DCM (ASPMD) control strategy for wireless sensor nodes powered by a battery and piezoelectric energy harvester. The proposed strategy enables dynamic mode transitions based on per-channel load conditions, achieving fast transient response and minimal cross-regulation in multi-output operation. A current-starved oscillator and quasi-dynamic comparator are employed to reduce the controller's quiescent current below 1 μA. Implemented in 0.18-μm BCD process, the converter exhibits a 2.94 mV undershoot and 0.9 μs recovery time under a 100 μA-to-30 mA load step. The system achieves peak end-to-end efficiencies of 88.9% and 78.5% under light and heavy loads, respectively.

Keywords—*SIMO converter, Low quiescent current, Wireless sensor nodes, Piezoelectric energy harvester (PEH).*

I. INTRODUCTION

Wireless sensor nodes (WSNs) play a key role in the Internet of Things (IoT). However, due to limited energy storage and constraints on size and weight, most battery-powered WSNs cannot achieve perpetual operation. To address this, energy harvesting (EH) technologies have emerged, aiming to prolong battery life and ultimately enable energy-autonomous systems. Among various EH methods, piezoelectric (PZT) energy harvesters (PEHs) have gained popularity due to their high power density, scalability, and relatively high output voltage [1]. PEHs extract energy from ambient mechanical vibrations. To convert the alternating output into regulated DC power and maximize harvested energy for supplying a variable load, a typical piezoelectric interface consists of a rectifier, maximum power point tracking (MPPT) circuitry, and a DC-DC converter. As shown in Fig. 1, conventional architectures often cascade the rectifier, MPPT controller, and battery, storing the harvested energy before delivering it to the load. However, this cascaded structure introduces multiple stages of loss, significantly degrading the end-to-end energy extraction efficiency.

Moreover, WSNs commonly require simultaneous power delivery to multiple loads such as sensors, microcontrollers units (MCUs), and transceivers (TRx). SIMO converters have thus become a preferred solution for efficient energy management in these systems. Due to the highly dynamic load profile of WSNs—operating at tens of microwatts during extended sleep periods and activated for only 1% of the total operating time with power consumption rising to tens of milliwatts—SIMO converters must support a wide dynamic load range while providing fast transient response to ensure system reliability and energy efficiency [2], [3]. Conventional

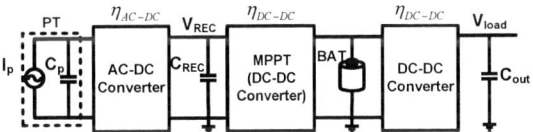

Fig. 1. Cascaded energy harvesting interface

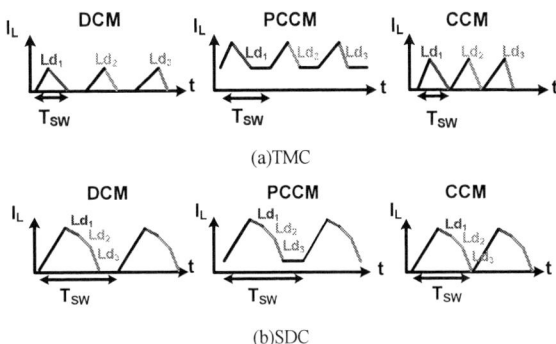

(a)TMC

(b)SDC

Fig. 2. Conventional energy distribution schemes

energy distribution schemes for single-inductor multiple-output converters mainly include time-multiplexed control (TMC) and sequential discharge control (SDC). As shown in Fig. 2(a), the TMC mode supplies power to only one output channel during each inductor charge-discharge cycle, requiring multiple cycles to supply all outputs. The inductor current must fall to zero at the end of each cycle, resulting in discontinuous energy delivery to each channel. This approach avoids cross-regulation issues among outputs but is suitable only for light-load conditions. In contrast, as depicted in Fig.2(b), the SDC mode charges the inductor once per cycle and then discharges it sequentially to multiple output channels. This mode supports a wide dynamic load range but suffers from significant cross-regulation due to continuous inductor current discharge to all channels [4].

II. ARCHITECTURAL OF PROPOSED SIMO CONVERTERS

Based on the design challenges above, This paper proposes a reconfigurable multi-mode SIMO converter. The converter operates in three modes—piezoelectric harvesting, energy recovery, and battery supply—based on load conditions, thereby improving overall energy utilization. The ASPMD control stategy is introduced to dynamically switch between PCCM, MDSCM, and DCM using load current feedforward. This enhances transient response and suppresses cross-regulation among multiple outputs. The system also

979-8-3315-3918-4/25 $31.00 © 2025 IEEE

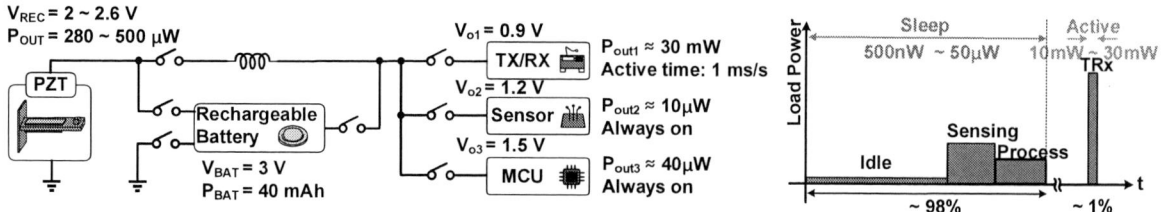

Fig. 3. Reconfigurable dc–dc converter in WSN

integrates P-SSHI bias-flip circuits and efficient MPPT techniques to ensure effective rectification and energy extraction from the piezoelectric source [5].

A. Reconfigurable multi-mode SIMO converter

As shown in Fig. 3, the wireless sensor node comprises a sensor and MCU that operate continuously with ultra-low power (~50 µW), while the TRx is duty-cycled and only activated during data transmission, leading to a peak load of 30 mW. Given that the rectified power from the piezoelectric energy harvester (PEH) is typically a few hundred microwatts, and the average system load in standby is only tens of µW, the rectifier directly powers the load under light-load conditions, with surplus energy stored in a battery. During heavy-load operation when harvested power is insufficient, the battery supplies the load to maintain functionality.

The proposed reconfigurable multi-mode SIMO converter utilizes a single shared inductor and a power stage consisting of eight switches, enabling adaptive operation across four distinct modes depending on load conditions: As shown in Fig. 4(a), when all loads are in light-load mode, the rectifier directly powers the outputs, and the converter operates in buck mode (PSBK). In Fig. 4(b), during ultra-low-power light-load operation where output power is only intermittently required, the harvested energy is stored in the battery, and the converter operates in boost mode (PSBT). Fig. 4(c) illustrates the transition to high-load mode, during which the rectifier alone cannot meet power demands, and the converter switches to battery-supplied buck mode (BSBK). Additionally, as shown in Fig. 4(d), the system supports an Inductor Energy Storage Channel (IESC) mode, which is detailed in Section B.

Compared with conventional cascaded piezoelectric harvesting architectures, the proposed SIMO converter significantly improves end-to-end energy extraction efficiency by enabling direct energy delivery from the rectifier to the loads during most of the time when the node operates under light-load conditions, thereby avoiding the losses associated with energy transfer between the rectifier and the battery.

B. ASPMD Control Strategy

Fig. 5 shows the architecture of the proposed single-inductor multiple-output (SIMO) converter for wireless sensor nodes. The system adaptively switches between the rectified voltage from a piezoelectric transducer (VREC) and a battery, depending on real-time load demand. To support efficient power delivery across a wide load range (µW–mW), an adaptive control scheme—ASPMD—is employed to dynamically transition among pseudo-continuous conduction mode (PCCM), multi-channel discontinuous single-channel continuous mode (MDSCM), and discontinuous conduction mode (DCM), as illustrated in Fig. 6(a–c).

The ASPMD integrates voltage hysteresis feedback and

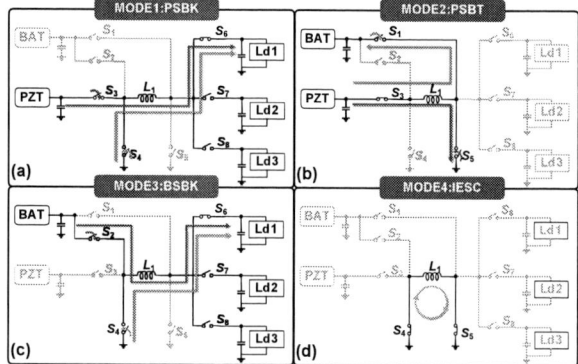

Fig. 4. Four operating modes of proposed SIMO converter

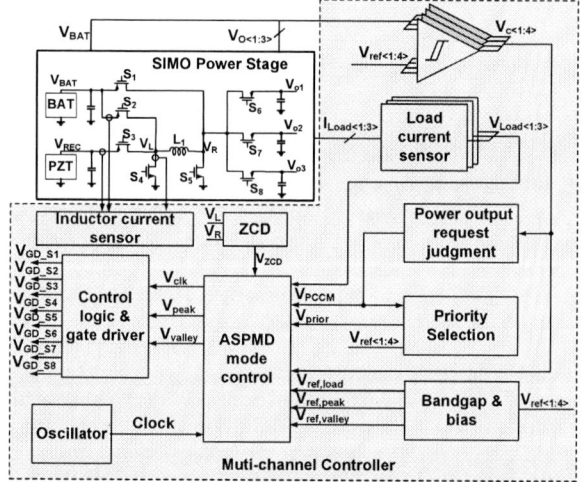

Fig. 5. System architecture of the proposed SIMO converter

current feedforward to identify load conditions and control the inductor current profile. Each output voltage is compared against its reference using a hysteresis comparator, while output current is sampled and compared to a 200 µA threshold to distinguish between light- and heavy-load conditions. Based on this, the controller adjusts the inductor peak current (I_{peak}) and valley current (I_{valley}) thresholds in real time. In PCCM [Fig. 6(a)], triggered by single-channel heavy load, the system disables the global clock and adopts event-driven control, storing residual inductor current across cycles to reduce ripple and improve transient response. Under multi-channel heavy load, the system enters MDSCM [Fig. 6(b)], ensuring continuous current within each channel while resetting the inductor to zero between channels to suppress cross-regulation. When all outputs are in light load, the system switches to DCM [Fig. 6(c)], reducing the clock to 50 kHz and enabling sequential delivery with full inductor discharge, minimizing control overhead.

979-8-3315-3918-4/25 $31.00 © 2025 IEEE

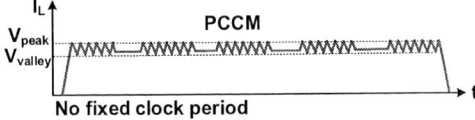

(a)Single-channel heavy-load condition

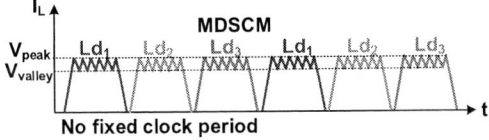

(b) Muti-channels heavy-load condition

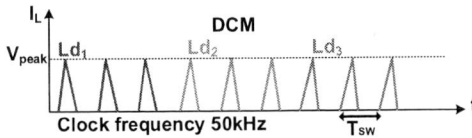

(c) All channels light-load condition

Fig. 6. Operating modes of the proposed ASPMD control

By adaptively selecting the optimal mode and adjusting inductor current thresholds, ASPMD achieves fast transient response, reduced cross-regulation, and improved energy efficiency for dynamic multi-output operation.

III. LOW-POWER DESIGN OF ASPMD CONTROLLER

In typical power converter systems, the control circuit's power consumption is primarily attributed to the oscillator and comparator blocks. To meet the stringent low-power requirements of wireless sensor nodes (WSNs), the proposed ASPMD controller integrates dedicated circuit-level optimizations for both components.

A. Current-starved ring oscillator

A current-starved ring oscillator is employed, as illustrated in Fig. 7. It operates with an ultra-low bias current of only 6 nA. The inverter stages are current-limited via current mirrors, which suppress large transient currents during logic transitions—an issue commonly found in conventional ring oscillators—thereby significantly reducing dynamic and static power consumption. To ensure robust startup and prevent the oscillator from entering a degenerate state, a startup circuit is designed. Upon power-up, transistor M_{16} momentarily pulls the V_{OUT} node to V_{DD} before shutting off, initiating stable oscillation. The oscillator operates at 50 kHz with a total power consumption of only 75 nW.

B. Quasi-dynamic comparator

Given that the ASPMD control strategy dynamically switches between high and low switching frequencies based on load conditions, conventional static hysteresis comparators are unsuitable due to their relatively high static current at high frequencies. Although dynamic comparators offer low static power, they require an external high-frequency, high-precision clock, which incurs additional overhead. To resolve this trade-off, a quasi-dynamic (QD) comparator [6] is adopted, as shown in Fig. 8. This architecture integrates a differential hysteresis comparator with a clockless regenerative latch, combining the benefits of both static and

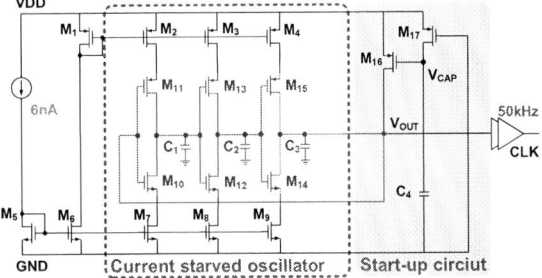

Fig. 7. Current-starved oscillator

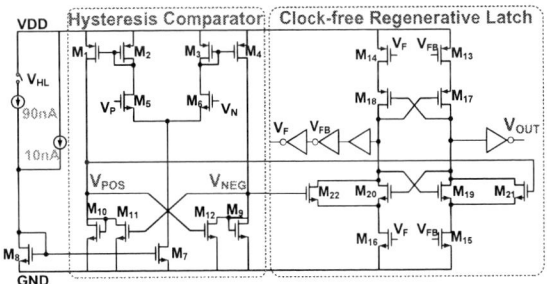

Fig. 8. Quasi-dynamic comparator

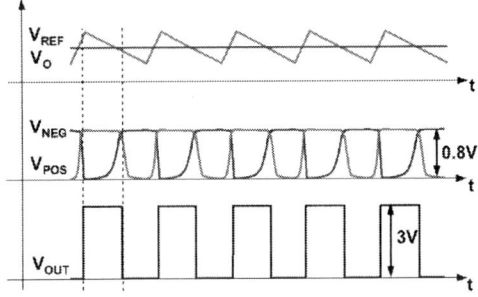

Fig. 9. Waveforms of QD comparator

dynamic designs—namely, low static power and high-speed operation. The latch is triggered by a self-generated internal signal derived from the comparator output, eliminating the need for an external clock and preventing any direct power-to-ground conduction path, thus avoiding static current consumption. Under light-load conditions (10 μA), the comparator consumes only 154 nW, while under full-load conditions (30 mA on all three outputs), total comparator power remains as low as 2.21 μW, validating its suitability for ultra-low-power wireless sensing applications.

IV. SIMULATION RESULTS

The proposed SIMO converter is implemented using the 0.18 μm BCD process. Fig. 10 (a-c) shows the waveforms of the SIMO converter operating in three modes under ASPMD control. Fig. 11. illustrates the output voltage waveform when the three loads of the SIMO converter simultaneously transition from 100 μA to 30 mA. It can be observed that the output voltage undershoot is 2.94 mV, the recovery time is 0.9 μs, and the output voltage ripple across all channels remains below 9 mV under heavy load conditions.

Fig. 12. presents efficiency of the proposed SIMO converter under different load conditions. Notably, the results

979-8-3315-3918-4/25 $31.00 © 2025 IEEE

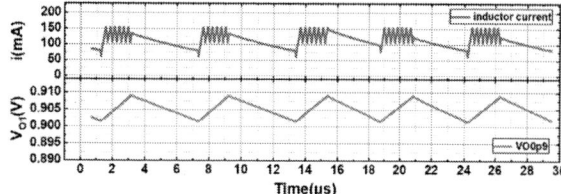

(a) Single-channel heavy-load condition

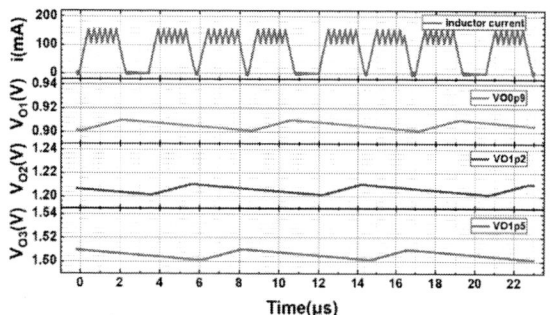

(b) Muti-channels heavy-load condition

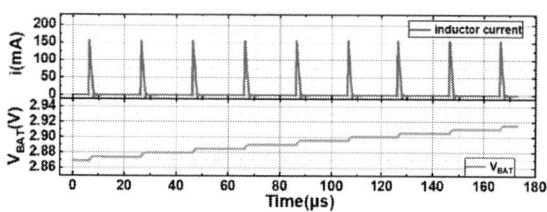

(c) All channels light-load condition

Fig. 10. Waveforms of the proposed ASPMD control

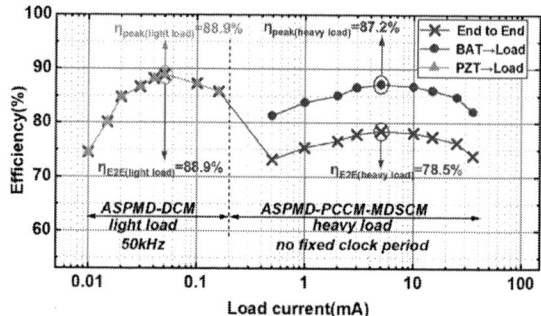

Fig. 12. Efficiency under different load conditions

V. CONCLUSION

This paper presents a multi-mode SIMO piezoelectric energy harvesting interface based on the ASPMD control strategy, which adaptively switches between PCCM, MDSCM, and DCM modes to achieve low cross-regulation and fast transient response under varying load conditions. Low-power design techniques, including a current-starved oscillator and a quasi-dynamic comparator, reduce total control circuit quiescent current to sub-1 μA, making the system well-suited for wireless sensor nodes. Implemented in 0.18 -μm BCD process, the chip demonstrates an undershoot of only 2.94 mV and a recovery time of 0.9 μs during 100 μA-30 mA load transients. Measured end-to-end efficiency reaches up to 88.9% under light loads and 78.5% under heavy loads, validating the proposed interface's high efficiency and low-power operation.

ACKNOWLEDGMENT

This work was supported by the National Natural Science Foundation of China (Grant NO.62234010).

REFERENCES

[1] X. Yue et al, "A Single-Stage Bias-Flip Regulating Rectifier With Fully Digital Duty-Cycle-Based MPPT for Piezoelectric Energy Harvesting," in IEEE Journal of Solid-State Circuits, vol. 60, no. 3, pp. 850-860, March 2025.

[2] S. S. Amin et al, "MISIMO: A multi-input single-inductor multi-output energy harvester employing event-driven MPPT control to achieve 89% peak efficiency and a 60,000x dynamic range in 28nm FDSOI," ISSCC , pp. 144-146, 2018.

[3] Y. -S. Noh, et al, "A Reconfigurable DC-DC Converter for Maximum Thermoelectric Energy Harvesting in a Battery-Powered Duty-Cycling Wireless Sensor Node," in IEEE Journal of Solid-State Circuits, vol. 57, no. 9, pp. 2719-2730, Sept. 2022.

[4] T. -H. Yang et al., "A 94.3% Peak Efficiency Adaptive Switchable CCM and DCM Single-Inductor Multiple-Output Converter With 0.03 mV/mA Low Crosstalk and 185 nA Ultralow Quiescent," in IEEE Journal of Solid-State Circuits, vol. 57, no. 9, pp. 2731-2740, Sept. 2022.

[5] X. Yue et al, "A Bias-Flip Rectifier With Duty-Cycle-Based MPPT for Piezoelectric Energy Harvesting," in IEEE Journal of Solid-State Circuits, vol. 59, no. 6, pp. 1771-1781, June 2024.

[6] S. -Y. Tay et al, "A 50-MHz Pulse-Width Modulator Embodying Low-Loss Quasi-Dynamic Comparators for Very High-Frequency DC–DC Converters," in IEEE Journal of Solid-State Circuits, doi: 10.1109/JSSC.2025.356200.

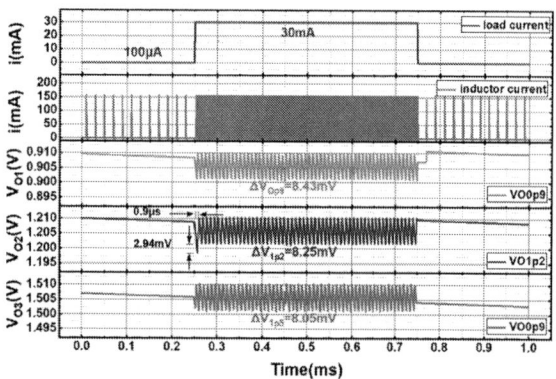

Fig. 11. Transient response from 100 μA to 30 mA

indicate that when the load operates in light-load mode, the E2E efficiency aligns with the efficiency from the piezoelectric rectifier to the load, achieving a peak efficiency of 88.9%. This result highlights the effectiveness of the proposed SIMO converter in achieving high E2E efficiency from the energy harvester to the load. Under heavy-load conditions, the peak efficiency from the battery to the load is 87.2%, while the E2E efficiency in this case corresponds to the cascaded efficiency of energy transfer from the piezoelectric harvester to the battery and from the battery to the load, reaching a peak efficiency of 78.5%.

979-8-3315-3918-4/25 $31.00 © 2025 IEEE

A 280-nA, 85.8% Efficiency Boost Converter with Optimal Inductor Current in Burst Mode for Brain Stimulation

Dejian Li[1], Xin Jin[1], LianXi Liu[2], Gang Dong[2], Xufeng Liao[2], Shihao Xiao[2] and Xincai Liu*[2]

[1] Beijing Smart-Chip Microelectronics Technology Co., Ltd., Beijing, 100192, China
[2] Key Laboratory of Analog Integrated Circuits, Xidian University, Xi'an 710071, China.

* Email: 23111110538@stu.xidian.edu.cn

Abstract—A highly efficient and accurate boost converter is presented in this letter for brain stimulation applications. By analyzing the dominant loss, this work employs a hysteretic architecture with asynchronous peak current control to enhance efficiency under light-load conditions. The peak current is further optimized to improve overall efficiency. Additionally, a novel hysteretic structure incorporating two comparators is proposed to enhance output voltage accuracy. The converter is implemented in a 0.18 μm BCD process, occupying a core area of 1.304 mm². It operates over an input voltage range of 2.6–4.2 V and delivers a regulated 10, 15, or 20 V output. The prototype achieves a peak conversion efficiency of 85.8% across a wide load current range from 10 μA to 10 mA, while consuming an ultra-low quiescent current of only 280 nA.

Keywords—asynchronous, boost converter, burst mode, high efficiency, low quiescent current

I. INTRODUCTION

Recently, brain pathologies have evolved into a critical biomedical challenge significantly compromising human health, where real-time electrophysiological monitoring and immediate therapeutic intervention are pivotal for optimizing clinical outcomes. Portable home-use transcranial stimulation systems are emerging as an effective neurotherapy solution, offering both treatment flexibility and improved patient adherence. However, these battery-operated implantable pulse generators face stringent power constraints: they must deliver high-voltage biphasic stimulation pulses (about 10 V) while predominantly operating in quiescent or microampere-level load regimes. This necessitates power management ICs capable of maintaining high power conversion efficiency across light-load ranges with high output voltage.

II. PROPOSED CONVERTER

A. Architecture Selection

The converter for the brain stimulation application works in light-load. Under light-load conditions, the dominant efficiency-limiting losses include gate driving loss, conduction loss, quiescent power dissipation, and switching overlap loss.

Compared to synchronous rectification architectures, the diode-based asynchronous structure offers several advantages. First, it eliminates the need for a gate driving circuit for the synchronous power switch, thereby effectively reducing switching and driving losses. Second, it avoids the requirement for a zero-current detection circuit, which further lowers quiescent power consumption. Finally, it prevents inductor current reversal caused by delayed zero-crossing detection, thereby minimizing unnecessary energy backflow

losses. These benefits collectively enhance the overall efficiency, especially under light-load conditions for the brain stimulation application.

In addition, the selection of the light-load control mode has a direct impact on system efficiency. Given that brain stimulation systems often operate under light-load or even near-no-load conditions, commonly adopted control schemes include pulse frequency modulation (PFM), pulse skipping modulation (PSM), and burst-mode operation. These techniques aim to improve efficiency while maintaining system stability. Among these, burst-mode operation effectively reduces switching losses and is therefore more suitable for light-load conditions. However, conventional burst mode still faces two key challenges in brain stimulation applications. Firstly, due to the fixed hysteresis window ($\triangle V_{HYS}$ of the comparator and the feedback voltage divider network, the output ripple ($\triangle V_{RIPPLE}$) increases significantly with higher output voltage, which may exceed the allowable range and degrade system accuracy. Secondly, when burst mode operates with continuous conduction mode (CCM), both the conduction losses during active periods and the quiescent power consumption during sleep intervals become more pronounced under light-load conditions, leading to a reduction in overall efficiency.

B. Proposed Architecture

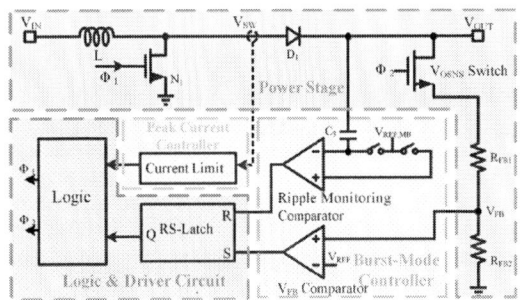

Fig. 1. The structure diagram of the overall circuit

Thus, this work proposed an asynchronous burst-mode for the brain stimulation application as shown in Fig.1. It includes the power stage, the burst-mode controller, the peak current controller, the bias circuit, and the logic&driver circuit. The hysteretic structure incorporating two comparators is proposed to enhance output voltage accuracy, which is realized by the burst-mode controller. The peak current controller realizes the peak current DCM operation, further improving the efficiency.

979-8-3315-3918-4/25 $31.00 © 2025 IEEE

III. DUAL-COMPARATOR BURST ARCHITECTURE

A. Principle of the Proposed Dual-Comparator Burst

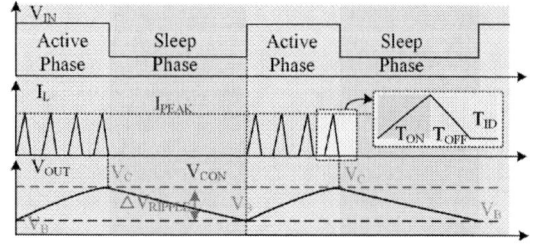

Fig. 2. Key waveforms of dual-comparator burst.

In the conventional burst-mode architecture, the hysteretic window ΔV_{HYS} is regulated by a single hysteretic comparator. And ΔV_{HYS} is related to the output ripple ΔV_{RIPPLE}.

$$\Delta V_{RIPPLE} = \Delta V_{HYS} \times \frac{R_{FB1} + R_{FB2}}{R_{FB2}} \quad (1)$$

where R_{FB1} and R_{FB2} are the feedback resistances in Fig. 1. For a 20 V output voltage with a 1 V reference, the feedback resistor ratio $(R_{FB1}+R_{FB2})/R_{FB2}$ must be set to 20. As a result, a 30 mV output ripple requires the comparator hysteresis window to be as small as 1.5 mV. Achieving such a precise hysteresis window is technically challenging. For instance, a comparator offset voltage of only 1 mV can increase the effective hysteresis window to 2.5 mV, leading to a significant ΔV_{RIPPLE} of 50 mV. This severely degrades the output accuracy and highlights the sensitivity of the system to comparator imperfections.

To address the aforementioned issues, this paper proposes a burst-mode architecture employing dual-comparator. The key signal waveforms are illustrated in Fig. 2. V_{FB} comparator compares the feedback-scaled output voltage V_{FB} with the reference voltage, thereby setting the DC operating point of the output. The ripple monitoring comparator directly samples the ripple component of the output voltage, decoupling the hysteresis window from the feedback resistors and thus preventing degradation in output accuracy.

With the proposed dual-comparator burst architecture, the converter operates in two distinct modes: the active phase and the sleep phase, as illustrated in Fig. 2. In the active phase, the converter functions in discontinuous conduction mode (DCM) with peak current control. During the on-time (T_{ON}), switch N_1 is turned on, and the inductor current increases. When the inductor current reaches the preset peak current (I_{LP}), the peak current controller turns off N_1, allowing the inductor current to decrease through diode D_1, while the output voltage rises accordingly. The T_{OFF} and T_{ID} durations during the asynchronous rectification phase are determined by the logic circuit, aiming to prevent the converter from operating in continuous conduction mode (CCM), which would otherwise lead to reduced efficiency. During this phase, the feedback voltage is continuously monitored by the V_{FB} comparator. Once the feedback voltage exceeds the reference voltage, indicating that the output voltage has surpassed the threshold voltage V_C, the converter transitions into the sleep phase. In the sleep phase, the output voltage gradually decreases due to load consumption. A ripple monitoring comparator is activated to monitor the output voltage, while the rest of the circuitry is powered down to minimize power consumption. When the output voltage drops to the lower threshold V_B, the ripple monitoring comparator triggers the converter to return to the active phase. One key advantage of the dual-comparator burst architecture is that the offset voltage ΔV_{HYS} is not amplified by the feedback network. This prevents degradation of regulation accuracy and ensures stable and precise output voltage control.

B. Ripple Monitoring Comparator

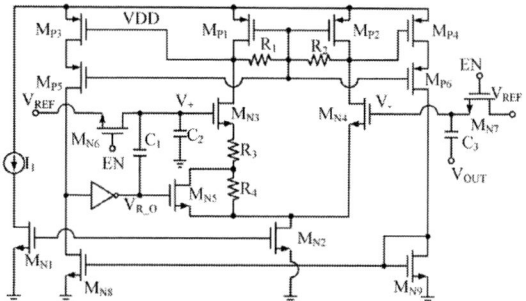

Fig. 3. Schematic of ripple monitoring comparator.

The ripple monitoring comparator designed in this work is shown in Fig. 3. The output voltage is directly coupled to the input of the comparator through capacitor C_3. M_{N3} and M_{N4}, R_1 and R_2, as well as M_{P1} and M_{P2}, are all symmetrically designed with identical parameters. Resistors R_3 and R_4 introduce a controlled offset in the comparator, which determines the value of ΔV_{RIPPLE}. When EN is high (EN = 1), the boost converter operates in the active phase, and transistor M_{N6} and M_{N7} are turned on, placing the comparator in a reset state. When EN is low (EN = 0), the boost converter enters the sleep phase, M_{N6} and M_{N7} are turned off, releasing the reset state, and the comparator begins to monitor the output voltage (V_{OUT}). In the active phase, M_{N6} and M_{N7} are turned on, and V_{REF} is applied to both the V_+ and V_- inputs of the comparator. Due to the presence of R_3 and R_4, the comparator output voltage V_{R_O} remains at a low logic level. In the sleep phase, M_{N6} and M_{N7} are turned off. Due to the charge storage effect of C_2, the V_+ terminal remains at V_{REF}. With the assistance of C_3, the V_- terminal retains V_{REF} as a baseline while superimposing the ripple component of V_{OUT}. When V_{OUT} drops by ΔV_{RIPPLE}, the ripple detection circuit generates a transition in the output R_O. When R_O switches from low to high, capacitor C_1 boosts the voltage at the V_+ terminal to accelerate the comparison process. Meanwhile, M_{N5} turns on to short-circuit R_4, thereby introducing hysteresis and preventing oscillation.

C. V_{FB} Comparator

The designed V_{FB} comparator is shown in Fig. 4. It compares V_{FB} with V_{REF} and generates the output signal V_{EC_OUT}, which is then used by the subsequent logic control circuit to regulate the conduction of the power switch. In the active phase, when CLK is high (CLK = 1), the comparator is enabled. Once V_{FB} exceeds V_{REF}, V_{EC_OUT} transitions from a low to a high logic level. Meanwhile, capacitor C_1 introduces an AC hysteresis effect, which helps prevent oscillation during the switching transition. In the sleep phase, when CLK is low (CLK = 0), the comparator is turned off, and its quiescent current is nearly zero. Based on the operating state of the proposed feedback voltage comparison circuit, the utilization of high-power modules during the sleep phase can be significantly reduced, thereby lowering the overall system power.

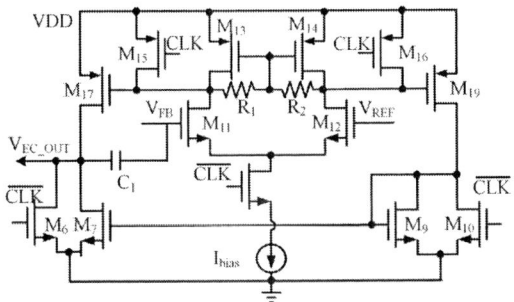

Fig. 4. Schematic of V_{FB} comparator

IV. OPTIMAL PEAK CURRENT

A. Optimal I_L Analysis

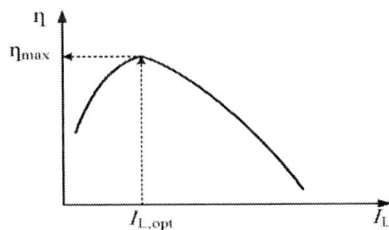

Fig. 5. η vs I_L

To achieve optimal η, it is essential to minimize power losses. This work analyzes the power loss components of the proposed architecture. According to [2], The losses can be categorized into two types: (1) losses during the active phase, including driving loss, conduction loss, switching loss, and static loss; (2) static loss during the sleep phase. Based on this classification, an analytical expression for η can be derived:

$$\eta = 1 - \frac{1}{V_{IN}}[R_{eq,a}I_L + \frac{V_{IN}I_{Q,a} + f_s(C_G V_{IN}^2 + C_A V_{OUT}^2)}{I_L} + V_{OUT}t_c f_s + R_{eq,s}I_{IN}] - I_{Q,s}/I_{IN} \quad (2)$$

where $R_{eq,a}$ is the equivalent resistance during the active phase, $I_{Q,a}$ is the quiescent current in active mode, f_s is the switching frequency, C_G is the gate parasitic capacitance of power transistor N_1, C_{SW} is the parasitic capacitance at the switching node, t_c is the average turn-on/off transition time, $R_{eq,s}$ is the equivalent resistance during the sleep phase, and $I_{Q,s}$ is the quiescent current in sleep mode. By taking the derivative of η with respect to the inductor current I_L, the relationship shown in Fig. 5 can be obtained. It can be observed that selecting an appropriate I_{LP} helps reduce power loss, thereby improving the overall efficiency. Based on the derivative analysis, The optimal peak current ($I_{LP,opt}$) corresponding to the maximum efficiency is:

$$I_{LP,opt} = \sqrt{\frac{V_{IN}I_{Q,a} + f_s(C_G V_{IN}^2 + C_{SW} V_{OUT}^2)}{R_{eq,a}}} \quad (3)$$

By substituting the relevant parameters, the $I_{LP,opt}$ is calculated to be 250 mA.

B. Current Limiting

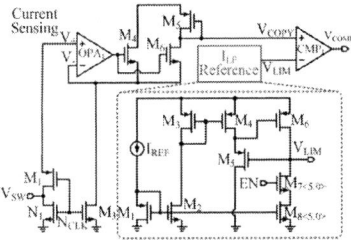

Fig. 6. Schematic of current limiting

Based on the above derivation, to achieve optimal efficiency, I_{LP} must be controlled. Therefore, the current limiting circuit shown in Fig. 6 is proposed. The circuit consists of a current sensing block and a I_{LP} reference block. In the current sensing circuit, the W/L of the power MOS N_1 is K_1 times that of the replica transistor M_3. When the gate voltage signal N_{CLK} rises, N_1 turns on and the inductor begins charging, while M_1 and M_3 turn on synchronously. At this point, the operational amplifier OPA_1 together with transistor M_4, M_5, and M_6, forms a feedback loop to ensure the voltage at the V_- equals that at the V_+ (i.e., $V_- = V_+ = V_{SW}$), while maintaining the current I_{M6} at a preset value. The current I_{M6} in branch M_6 can thus be expressed as:

$$I_{M6} = \frac{I_L}{K_1 K_2} \quad (4)$$

where K_2 represents the current ratio between M_3 and M_6. The current I_{M6} is converted to the voltage V_{COPY} through MOSFET M_5. When I_L reaches 250 mA, V_{COPY} drops to the reference voltage V_{LIM}, causing the output of comparator CMP_1 to toggle. The reference voltage V_{LIM} is determined by the peak current definition circuit and can be expressed as:

$$V_{LIM} = V_{IN} - V_{SG6} - V_{SG5} \quad (5)$$

In the peak current definition circuit, a fixed current is copied from I_{M1} to flow through I_{M5} and I_{M6} to generate the reference voltage V_{LIM}. To ensure optimal I_{LP}, a current adjustment circuit is implemented. By enabling or disabling MOS $M_{7<5:0>}$ and $M_{8<5:0>}$, the current I_{M6} is adjusted, thereby changing V_{SG6} and subsequently tuning V_{LIM}.

V. SIMULATION ANALYSIS

Fig. 7 shows the layout of the proposed circuit, with a core area of 1.64×0.88 mm². Fig. 8 presents the simulated waveforms of key signals under steady state operation. As observed, during the sleep phase, when V_{OUT} drops by 28.9 mV, the output of the ripple monitoring comparator toggles, triggering the converter to enter the active phase. In the active phase, the current limiting circuit constrains the peak inductor current I_L to 249.6 mA. Following a charge-discharge cycle, as V_{OUT} rises to its peak, the output of the V_{FB} comparator toggles, causing the converter to return to the sleep phase.

Fig. 9 shows the variation of η with load current under different input voltages. It can be observed that the system achieves a conversion efficiency of 85.8% at $V_{IN} = 4.2$ V, $V_{OUT} = 10$ V, and $I_{OUT} = 10$ mA. Fig. 10 illustrates the dependence of efficiency on I_{LP} under various V_{IN} conditions. The results indicate that when I_{LP} is set to around 250 mA, the

979-8-3315-3918-4/25 $31.00 © 2025 IEEE   317

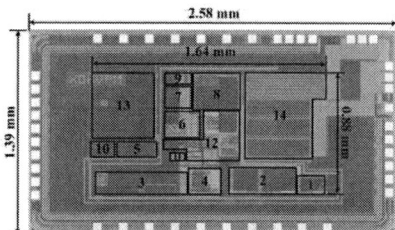

1.Curr Gen 2.Curr Bias 3.BGR 4.Mem Buffer 5.I_L Samp 6.I_L Cmp
7.Err Cmp 8.Ripp Det 9.VOSNS 10.OVP 11.OTP 12.Logic&UVLO
13.Gate Driver 14.Power MOS&SBD

Fig. 7. Layout

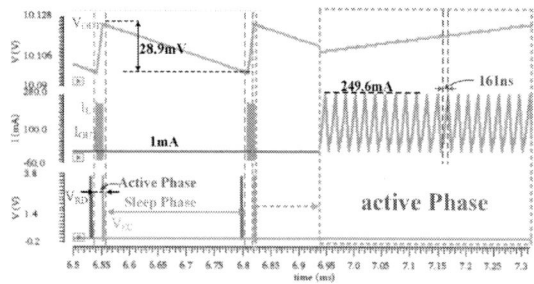

Fig. 8. Waveform of steady state

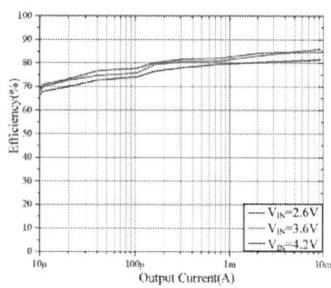

Fig. 9. Load efficiency with different inputs for providing an output voltage of 10 V

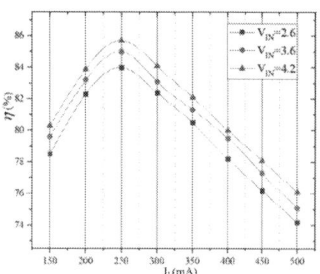

Fig. 10. Experimental efficiency versus I_{LP} for different V_{IN}

system maintains high efficiency across all input voltage levels.

TABLE I. THE PERFORMANCE COMPARISON TABLE

Parameter	[3]	[4]	[5]	This work
Technology	0.13 μm	0.18 μm	-	0.18 μm
V_{IN}	5.5 V	1.8-3.3 V	0.7-5.5 V	2.6-4.2 V
V_{OUT}	12 V	3.0-4.5 V	1.8-5.5 V	10-20 V
I_{LOAD}	70 μA - 0.6 A	0.1 mA - 0.7 A	10 μA - 0.3 A	10 μA - 0.01 A
Efficiency	73% @70 μA	75% @10 μA	60% @10 μA	68.8% @10 μA
Quiescent Current	45.4 μA	232.5 μA	1 μA	280 nA
$\triangle V_{RIPPLE}$	96 mV	5 mV	50 mV	28.9 mV
Diode	Off-chip	-	-	On-chip
Area	1.7 mm²	1.68 mm²	-	1.304 mm²

Table I presents the performance comparison. Compared with other published works, the proposed boost converter achieves higher efficiency under light-load conditions, lower quiescent current, and supports a wider output voltage range.

VI. SUMMARY

This paper presents an asynchronous Boost converter designed for applications with light-load and high-output-voltage requirements. To improve both light-load efficiency and output voltage accuracy, a dual-comparator burst control architecture is proposed. A current-limiting circuit ensures the operation of the converter in discontinuous conduction mode (DCM). With the $I_{LP,opt}$, the efficiency is further improved. Owing to its high light-load efficiency and precise output regulation, the proposed converter is particularly suitable for portable medical applications such as electroencephalographic (EEG) stimulation.

ACKNOWLEDGMENT

Supported by Laboratory Specialized Scientific Research Projects of Beijing Smart-chip Microelectronics Technology Co., Ltd.

REFERENCES

[1] J. Clerk Maxwell, L. Yao, J. Zhao, P. Li, R. -F. Xue, Y. P. Xu and M. Je, "A 20V-compliance implantable neural stimulator IC with closed-loop power control, active charge balancing, and electrode impedance check," 2014 IEEE Asian Solid-State Circuits Conference (A-SSCC), KaoHsiung, Taiwan, 2014, pp. 201-204.

[2] F. Reverter and M. Gasulla, "Optimal Inductor Current in Boost DC/DC Converters Operating in Burst Mode Under Light-Load Conditions," in IEEE Transactions on Power Electronics, vol. 31, no. 1, pp. 15-20, Jan. 2016.

[3] K. Lee et al., "An Asynchronous Boost Converter With Time-Based Dual-Mode Control for Wide Load Range and High Efficiency in SSD Applications," in IEEE Transactions on Industrial Electronics, vol. 67, no. 12, pp. 10520-10530, Dec. 2020.

[4] W. Hong and M. Lee, "A 10-MHz Current-Mode AOT Boost Converter With Dual-Ramp Modulation Scheme and Translinear Loop-Based Current Sensor for WiFi IoT Applications," in IEEE Journal of Solid-State Circuits, vol. 56, no. 8, pp. 2388-2401, Aug. 2021.

[5] TPS61099 Datasheet, " TPS61099x Synchronous Boost Converter with Ultra-Low Quiescent Current", Texas Instruments, Dallas, TX, USA, 2016.

A High-Efficiency Low-Ripple Buck Converter with Adaptive Load Frequency Control

Tao Ren[1,2], Xufeng Liao[1,2], Gefu Wang[1], Jiatong Wu[1], and Lianxi Liu*[1,2]

[1] Key Laboratory of Analog Integrated Circuits, Xidian University, Xi'an 710071, China.
[2] Chongqing Integrated Circuits Innovation Institute, Chongqing, 401331, China.

* Email: lxliu@mail.xidian.edu.cn

Abstract—A high-efficiency low-ripple buck converter is proposed in this letter to reduce ripples. An adaptive frequency control based on load variation is proposed by analyzing the relationship between the ripple and the output current. Thus, the output ripple is reduced with the increase of load current. The simulation results show that when the input voltage is 24V, the output voltage is 3.3V, and the load current is 5A, the output ripple of the Buck converter is only 1mV. When the input voltage is 7V and the load current is 0.6A, the converter achieves a peak efficiency of 95.2%.

Keywords—*DC-DC converter, high efficiency, low ripple, frequency control*

I. INTRODUCTION

Switch-mode power converters are indispensable in modern power electronics, enabling efficient energy conversion across diverse applications such as renewable energy systems, electric vehicles, consumer electronics, and industrial equipment. However, their fixed-frequency operation inherently introduces significant output voltage ripple, which can severely degrade performance in high-fidelity applications like audio amplifiers (5–24V), digital signal processors (typically 3.3V), and analog-to-digital conversion systems (several to tens of volts), compromising signal integrity [1].

Fixed-frequency PWM operation inherently generates significant output voltage ripple in buck converters under medium-to-heavy loads, especially at lower switching frequencies. Variable switching frequency (f_{SW}) techniques [2], [3] mitigate this issue by adjusting f_{SW} via load-dependent pulse width modulation (FPWM), typically inferring load current from switch-node voltage with multi-stage filtering. However, this introduces latency and causes frequency spikes during transients. This paper presents a buck converter with adaptive load-frequency control that reduces output ripple over wide load ranges by dynamically adjusting f_{SW} based on direct inductor current sensing, eliminating the need for voltage-based load inference.

The rest of this brief is organized as follows. Section II illustrates the operating principles of the adaptive load frequency control (ALFC) buck converter and presents theoretical ripple analysis under PWM mode. Circuit implementation is also discussed. The simulation results of the proposed ALFC-Buck are presented in Section III. The conclusion is provided in Section IV.

II. THE STRUCTURE OF THE PROPOSED BUCK CONVERTER

A. Adaptive Load Frequency Control Scheme

The conventional output ripple expression under fixed-frequency PWM mode is given by:

$$\Delta V_{OUT\_PWM} = \frac{V_{IN} - V_{OUT}}{L} \cdot D \cdot \frac{1}{f_{sw}} \left(R_{ESR} + \frac{1}{8 f_{sw} C_{OUT}} \right) \quad (1)$$

where ΔV_{OUT_PWM} is the output voltage ripple in PWM mode, L is the inductance, V_{IN} is the converter's input voltage, and V_{OUT} is the converter's output voltage. D is the duty ratio of the converter in CCM mode, C_{OUT} is the output filter capacitor, R_{ESR} is the equivalent series resistance (ESR) of the output filter capacitor, and f_{SW} is the operating frequency in PWM.

In PWM-modulated buck converters, output voltage ripple is inversely proportional to the switching frequency (f_{SW}). However, fixed-F_{SW} operation constrains ripple regulation, particularly at lower frequencies where ripple often exceeds 20mV. Considering the accumulation of inductor current by the output filter capacitor, the capacitive, the lower fixed switching frequency will produce a higher output filter capacitance, further increasing the output voltage ripple.

This paper proposes an adaptive load frequency control (ALFC) buck converter to eliminate the fixed-F_{SW} limitation in conventional buck converters. Referring to Fig. 1, the converter dynamically adjusts f_{SW} according to load variations under PWM mode, suppressing output ripple. The proposed ALFC buck converter employs peak current mode control (PCMC).

As shown in Fig.1, the ALFC loop consists of an oscillator (OSC), frequency-to-voltage converter (FVC), and frequency back protection (F_{BP}) circuit, enabling adaptive load frequency control under PWM mode. Frequency stabilization is achieved without the need for an external clock reference. Instead, it uses a sampling of the DC component of inductor current (I_{CS}) to generate a load-dependent clock frequency f_{CLK} for switching frequency regulation. The proposed loop converts f_{CLK} into voltage V_{af}, which superimposes on V_{FB} for PWM feedback control. The proposed frequency back protection (F_{BP}) ensures extended off-time during short-circuit conditions by increasing the switching period, thereby allowing sufficient inductor current decay.

The proposed adaptive load frequency control eliminates fixed-F_{SW} constraints by sampling the DC component of inductor current (I_{CS}) to generate a load-tracking clock f_{CLK}. This enhances reliability and ultimately minimizes output voltage ripple across medium-to-heavy load ranges.

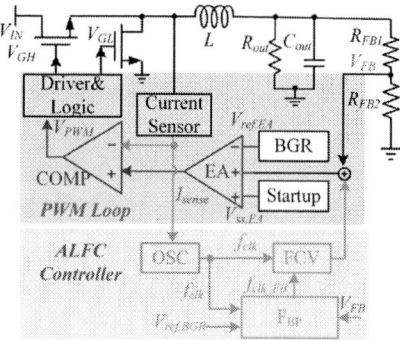

Fig. 1.   Structure of proposed ALFC-Buck converter.

B. Ripple Analysis in the Proposed Converter

Theoretical analysis of output voltage ripple for the proposed ALFC buck converter under PWM mode, considering the accumulation of inductor current by the output filter capacitor. The ripple is rigorously defined as comprising two components across medium-to-heavy loads: Equivalent series resistance (ESR)-induced ripple and Capacitor charge/discharge ripple. The ESR-induced ripple component is expressed as:

$$\Delta V_{ESR} = \Delta I_L \cdot R_{ESR} \qquad (2)$$

where R_{ESR} is the equivalent series resistance (ESR) of the output filter capacitor, for the capacitor charge/discharge ripple (ΔV_{cap}), the time-domain integral of inductor current represents the charge flowing through the inductor, equating to accumulated charge on the output capacitor. Thus, ΔV_{cap} is expressed as :

$$\Delta V_{cap} = \frac{\Delta Q}{C_{out}} = \frac{1}{C_{out}} \int_0^{T_S/2} i_L(t)dt \qquad (3)$$

Considering the inductor current waveform (i_L) in the buck converter, i_L is approximated as a triangular waveform expressed as $\Delta V_{cap} = \Delta I_L/8 f_{sw} C_{out}$.

In the proposed ALFC buck converter, the ALFC control loop dynamically adjusts f_{SW} via load current I_{load}. The DC component of inductor current (i_L) is sampled to generate sense current I_{sense}, which feeds into the oscillator (OSC). The OSC produces clock frequency f_{CLK} through dynamic charging/discharging. The charging current follows $I_{charge} = I_{sense} \propto I_{L,avg} = I_{load}$, with the charging duration.

$$t_{charge} = \frac{C_1 V_{ref}}{I_{charge}} = \frac{C_1 V_{ref}}{k_s I_{L,avg}} \qquad (4)$$

where k_s is the inductor current sampling coefficient set as $k_s = 1/2000$, C_1 is the charging capacitor in the dual-capacitor charge/discharge structure, V_O represents the charging voltage, and $I_{L,avg}$ is the DC component of inductor current satisfying $I_{L,avg} = I_{load}$. The output frequency is given by:

$$f_{clk} = \frac{1}{t_{charge}} = \frac{k_s I_{load}}{C_1 V_{ref}} \qquad (5)$$

That is $f_{clk} = k_1 I_{load}$, where $k_1 = k_s/C_1 V_{ref}$. Substituting $f_{clk} = k_1 I_{load}$ into ΔI_L gives:

$$\Delta I_L = \frac{(V_{in} - V_{out}) V_{out}}{V_{in} L (k_1 I_{load})} = \frac{K}{I_{load}} \qquad (6)$$

Thus, the output voltage ripple induced by adaptive load frequency control (ALFC) is expressed as:

$$\Delta V_{out\_ALFC} = \frac{K}{I_{load}} \left(R_{ESR} + \frac{1}{8 C_{out} k_1 I_{load}} \right)$$
$$= \frac{K R_{ESR}}{I_{load}} + \frac{K}{8 C_{out} k_1 I_{load}^2} \qquad (7)$$

where ΔV_{out_ALFC} is the output voltage ripple of the ALFC-Buck converter, L is the inductance, C_{out} is the output filter capacitor, and R_{ESR} is the output filter capacitor's equivalent series resistance (ESR). Here $K = (V_{in} - V_{out}) V_{out} C_1 V_{ref}/V_{in} L k_s$, C_1 is the capacitor in the dual-capacitor charge/discharge structure, V_{ref} is the reference voltage in the variable comparator, and k_s is the inductor current sampling coefficient. Critically, ΔV_{out_ALFC} exhibits an inverse relationship with I_{load}, as I_{load} increases, the output voltage ripple decreases proportionally.

Based on peak current mode control (PCMC), the proposed adaptive load frequency control (ALFC) establishes a functional relationship between load current I_{load} and switching frequency f_{sw}. This enables I_{load} based characterization of output voltage ripple, thereby eliminating the fixed frequency constraint inherent to conventional Buck converters. As the load transitions from medium to heavy levels, ΔV_{out_ALFC} progressively decreases. When dealing with varying loads, the zero point always remains within GBW. Conventional Type-II compensation is sufficient to achieve stability in PWM mode for ALFC-Buck converters, as demonstrated by the designed ALFC-Buck converter achieving loop stability with $R_c = 16.9k\Omega$ and $C_c = 4.7\mu F$.

C. Circuit Implementation of Adaptive Load Frequency Control

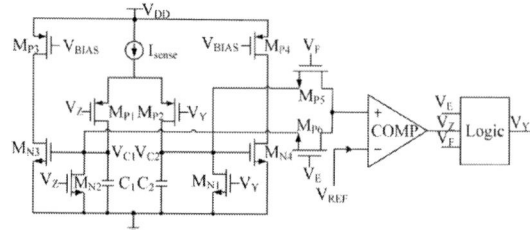

Fig. 2.   Structure of the oscillator.

The key enabler of the ALFC loop is the frequency generator from the load current. The oscillator circuit comprises a dual-capacitor charge/discharge structure and a variable comparator, with the frequency generation circuit under the PWM model shown in Fig. 2. Sampled inductor current I_{sense}, charges capacitors C_1 and C_2 alternately: during charging, only one capacitor path is active while the other is disabled. When V_Y is high and V_F is low, transistor M_{N1} turns on and M_{P2} off, discharging C_2 while disconnecting M_{P5} at the comparator input. Simultaneously, logic control sets V_E high and V_Z low, enabling I_{sense} to charge C_1 with the comparator input connected to C_1. Once V_{C1} reaches the reference level V_{REF}, the comparator toggles, switching V_E low and V_Z high. This discharges C_1 and connects the comparator input to C_2, allowing I_{sense} to charge C_2 until $V_{C2} = V_{REF}$, completing the self-oscillating cycle.

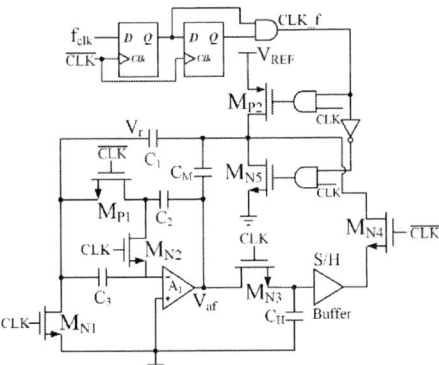

Fig. 3. Structure of the frequency to voltage converter.

Fig. 3 illustrates the circuit diagram of the proposed switched-capacitor frequency-to-voltage converter (FVC). Here, *CLK* and $\overline{CLK}$ denote non-overlapping complementary clock phases, while V_{REF} serves as the reference voltage. Operational amplifier A_1 forms an integrator with capacitors C_1 (input capacitor) and C_2 (integration capacitor). Capacitors C_3 and C_M ensure spike-free and offset-free operation. A unity-gain buffer and a capacitor C_H constitute the sample-and-hold (S/H) circuit.

Two *D* flip-flops detect rising edges of the input frequency signal f_{clk} synchronously with the *CLK* phase, generating a pulse train of width $1/2f_c$, where f_c is the dual-phase clock frequency. This pulse train activates the M_{p2} switch, charging C_1 to V_{REF}. During the subsequent *CLK* phase, the stored charge C_1V_{REF} transfers to C_2. Consequently, the integrator output voltage increases by $(C_1/C_2)V_{REF}$ per input pulse. Conversely, the S/H circuit feeds back the output voltage V_{af} to the integrator during each *CLK* phase. This extracts charge C_1V_{af} from C_2, reducing the integrator output by $(C_1/C_2)V_{af}$ per *CLK* cycle.

Assuming $C_1 \ll C_2$, the steady-state integrator output V_{af} remains approximately constant. The total charge extracted from C_2 over one input period equals $C_1V_{af}(f_c/f_{clk})$. Charge balance with the accumulated charge C_1V_r yields:

$$V_{af} = \left(V_r / f_c\right) f_{clk} \qquad (8)$$

As shown in Fig. 4, the proposed frequency back protection circuit (F_{BP}) operates such that when the V_{FB} voltage decreases from 0.8V to 0V, the oscillator frequency is progressively divided by factors of n (n=1,2,3,4). V_{FB} is compared against 200mV, 400mV, and 600mV reference levels. During operation, *ENV_CLK* remains high while EN_1 serves as the high-level enable signal.

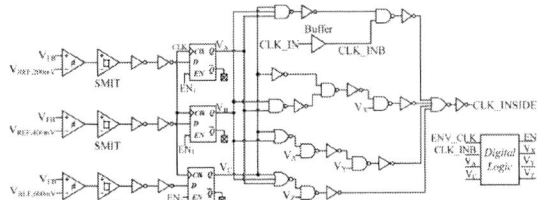

Fig. 4. Structure of the frequency back protection.

When V_{FB} maintains its nominal 0.8V level, V_C stays high and EN_2 remains low, turning off the frequency divider. Consequently, *CLK_INB* (generated by buffering *CLK_IN*) = *CLK_IN*. V_{FB} falls below 600mV but remains above 400mV, so V_C goes low to enable frequency division. With V_X becoming a 2:1 divided signal and all other inputs to the four-input NOR gate held low, *CLK_INSIDE* outputs half of *CLK_IN*'s frequency. Similarly, when V_{FB} drops below 400mV and 200mV, the frequency undergoes successive halving, thus achieving frequency feedback functionality.

The proposed frequency back protection circuit primarily serves dual purposes. Firstly, during the soft start phase, it implements frequency division to suppress inrush currents; on the other hand, under low output voltage or short-circuit conditions, where minimal duty cycle operation occurs to maintain inductor current below the peak current limit, the circuit extends the inductor current decay time through frequency division. This reduces the practical minimum duty cycle constraint imposed by the design-limited minimum on time of the high-side switch, thereby preventing inductor current from exceeding the peak current limit.

III. SIMULATION RESULT OF PROPOSED BUCK

Fig. 5 shows the layout of this circuit. The overall area is $2.46 \times 1.69 \text{mm}^2$. The proposed ALFC-Buck converter is designed using the 0.18 μm BCD process. Its input voltage range is 7V to 60V, and its output voltage is 3.3V. The output load current is 0–5A. A 7.2μH inductor and 141μF output capacitor were employed during all measurement processes mentioned in this manuscript.

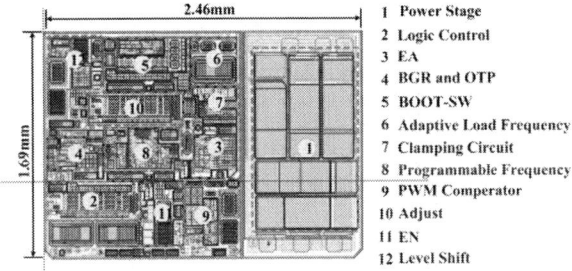

1	Power Stage
2	Logic Control
3	EA
4	BGR and OTP
5	BOOT-SW
6	Adaptive Load Frequency
7	Clamping Circuit
8	Programmable Frequency
9	PWM Comparator
10	Adjust
11	EN
12	Level Shift

Fig. 5. The layout of the proposed ALFC-Buck converter

As seen in Fig. 6, the ALFC-Buck converter operates at PWM mode under medium load (3A). The output voltage ripple measures 7mV, while the inductor current peak-to-peak ripple I_{RIPPLE} reaches 1.35A. Fig. 7 shows that the ALFC-Buck converter exhibits an output voltage ripple of merely 1mV under heavy load (5A) conditions. Adaptive load frequency control (ALFC) contributes significantly to ripple reduction.

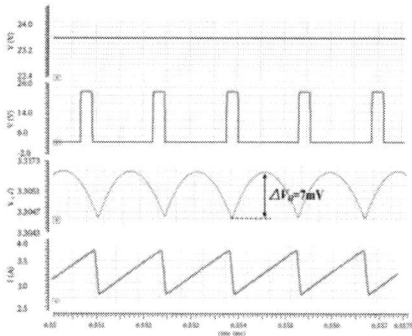

Fig. 6. Simulation waveforms of the proposed Buck converter at V_{in}=24V, V_{out}=3.3V, I_{load}=3A

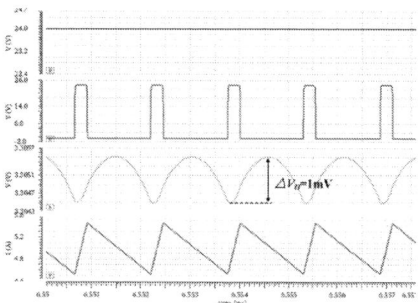

Fig. 7. Simulation waveforms of the proposed Buck converter at V_{in}=24V, V_{out}=3.3V, I_{load}=5A

Fig. 8 shows the simulated efficiency with several different I_{loads}. It demonstrates >70% power conversion efficiency above 0.4A load current, with peak efficiency exceeding 80% across all input conditions. Notably, 95.2% peak efficiency is achieved at 7V input voltage and 0.6A loads.

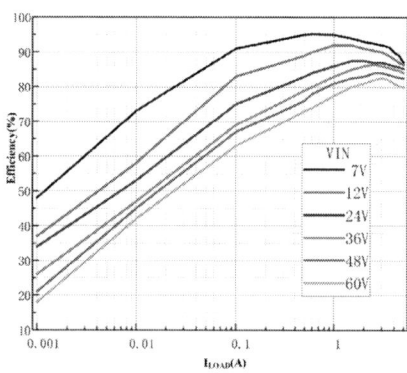

Fig. 8. The efficiency curve of the proposed ALFC-Buck converter

The performance of the proposed Buck converter is summarized in Table I and compared with other similar Buck converters. Table I shows that the adaptive load frequency control (ALFC) scheme achieves smaller output ripple across broader input voltage and full load ranges.

TABLE I.   PERFORMANCE COMPARISON

Parameter	[4]	[5]	[6]	This work
Technology(μm)	0.18	0.18	NA	0.18
Input Voltage (V)	5~30	5~32	4~40	7~60
Output Voltage (V)	3~15	3.3	5	3.3
Max .load(A)	2	1.4	3.5	5
Ripple @PWM(mV)	10	25	15	1
Peak efficiency(%)	94	91.7	92	95.2

CONCLUSIONS

A Buck converter using the 0.18μm BCD process and adaptive load frequency control (ALFC). Compared to a fixed frequency based Buck converter, the proposed design reduces output ripple while exhibiting decreasing voltage ripple across medium-to-heavy load variations. Thus, theoretical analysis validates the scheme's feasibility. Simulation results show that 1mV output ripples under a 5A load, and 95.2% peak efficiency is achieved under a 7V input and a 0.6A load.

ACKNOWLEDGMENT

This work was supported in part by the National Natural Science Foundation of China under Grant 62131010, 62204183, in part by the Chongqing Talents Program under Grant CQYC20210301367, and in part by the Fundamental Research Funds for the Central Universities under Grant YJSJ25013.

REFERENCES

[1] Z. Zhao, P. Luo, Z. Zhang, J. Fan, S. Zhen, and B. Zhang, "A VCO-Based Modulation Low Ripple PFM-PWM Buck Converter With Seamless Mode Transition," in IEEE Transactions on Circuits and Systems I: Regular Papers, vol. 72, no. 5, pp. 2456-2466, May 2025.J. Clerk Maxwell, A Treatise on Electricity and Magnetism, 3rd ed., vol. 2. Oxford: Clarendon, 1892, pp.68–73.

[2] TPS560430, Texas Instruments Inc. Accessed: Dec. 15, 2023. [Online]. Available: https://www.ti.com.

[3] B. Yuan, M. -X. Liu, W. T. Ng and X. -Q. Lai, "A Fast-Response RBAOT-Controlled Buck Converter With Pseudo fixed Switching Frequency and Enhanced Output Accuracy," in IEEE Journal of Emerging and Selected Topics in Power Electronics, vol. 9, no. 1, pp. 79-88, Feb. 2021.

[4] A. Besharati Rad, M. Kargaran, M. Meghdadi and A. Medi, "A Wide-Input-/Output-Voltage-Range Buck Converter With Adaptive Light-Load Efficiency Improvement and Seamless Mode Transition," in IEEE Transactions on Power Electronics, vol. 39, no. 2, pp. 2200-2212, Feb. 2024.

[5] P. Melillo, S. Zaffin, M. Leoncini, A. Brunero, A. Gasparini, S. Levantino, "A wide-input-range time-based buck converter with adaptive gain and continuous phase preset for seamless PFM/PWM transitions," IEEE Trans. Circuits Syst. I, Reg. Papers, vol. 71, no. 7, pp 3436–3447, Jul. 2024.

[6] LMR14030, Texas Instruments Inc. Accessed: Mar. 14, 2024. [Online]. Accessed: https://www.ti.com.cn/lit/ds/symlink/lmr14030.

A 400V High-speed Level-Shifting Gate Driver with Adaptive Signal-Path Disconnection for 278V/ns dv/dt Immunity in Soft-Switching Converters

Yile Xie [1], Hanyu, Shi [1], Ting Yi [1], Zhiliang Hong [1]

College of Integrated Circuits and Micro-Nano Electronics, Fudan University, Shanghai, China.
Email: 23212020024@m.fudan.edu.cn, zlhong@fudan.edu.cn

Abstract—High-voltage and high-frequency switch mode power converters commonly use pulse-triggered active coupling (PTAC) level shifters to drive high-side N-channel power transistor. However, high dv/dt at the switch node (V_{SW}) can mis-trigger the level shifter, causing erroneous conduction. To enhance dv/dt immunity, this paper presents a 400V high-speed level-shifting gate driver that adaptively disconnects signal paths during positive dV_{SW}/dt transitions and can be used in soft-switching converters. Implemented in an HV-BCD process, low- and high-side circuits of the gate driver occupy areas of $1.46\times 1.90mm^2$ and $1.40\times 2.00mm^2$, respectively. Post-layout simulations demonstrate dv/dt immunity up to 278V/ns. In a 300kHz/288W LLC converter, the driver achieves high-side gate drive delays below 11.32ns and level-shifter delays under 4.09ns.

Keywords—dv/dt immunity, level shifter, gate driver, soft-switching

I. INTRODUCTION

The AC/DC converter serves as the front end in many power-electronic systems, including server power supplies, laptop adapters, and mobile-phone chargers. Fig. 1 (a) shows a block diagram of a typical AC/DC converter. First, the power factor correction (PFC) stage rectifies and boosts 110/220V AC to 400V DC, then the intermediate-bus converter (IBC) stage steps 400V DC down to 48/24V DC. In a half-bridge shown in Fig. 1 (b), V_{SW} switches between GND and V_{IN} during low- and high-side power transistor conduction, causing hard switching loss $P_{loss, HS}$ proportional to V_{IN}^2:

$$P_{loss, HS} = 2 \cdot \frac{1}{2} C_{sw}V_{IN}^2 f_{sw} = C_{sw}V_{IN}^2 f_{sw} \qquad (1)$$

Here C_{sw} denotes the switch-node capacitance, and f_{sw} denotes the converter switching frequency. Equation (1) indicates that hard-switching loss becomes especially large under high V_{IN} and high f_{sw}. To eliminate hard-switching loss, the IBC stage typically uses a soft-switching converter that achieves zero-voltage switching (ZVS) or zero-current switching (ZCS). Fig. 2 shows the block diagram and operating signal waveforms of a typical soft-switching LLC converter. On each switching cycle, the resonant current I_{Lr} pre-charges V_{SW} to near V_{IN} before the high-side transistor M_H turns on and then discharges V_{SW} to near GND before the low-side transistor M_L conducts, thereby ensuring ZVS of the half-bridge switches. The proposed gate driver converts the control signals SWL/SWH into the gate drive signals G_L/G_H. It consists of a high-voltage (HV) level shifter followed by two driver stages. As the core component of the level-shifting gate driver, the HV level shifter must deliver fast, precise translation with minimal delay and low power consumption, while withstanding high dV_{SW}/dt slew rates.

The designs proposed by Moghe [1] and Cao [2] achieve delays of 2.4ns and 664ps at input voltages of 10V and 25V, respectively, but rely on V_{IN}-rated PDMOS devices, making them unsuitable for 400V applications. In [3], Yang demonstrates 25 MHz signal translation at 700 V, but he does not address dv/dt immunity. A 50V level shifter [4] uses a

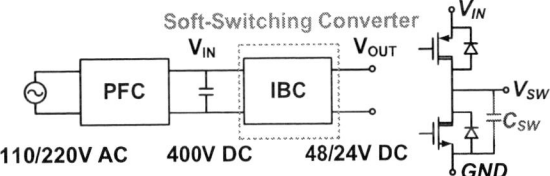

Fig. 1. (a) AC/DC converter block diagram (b) Half-Bridge

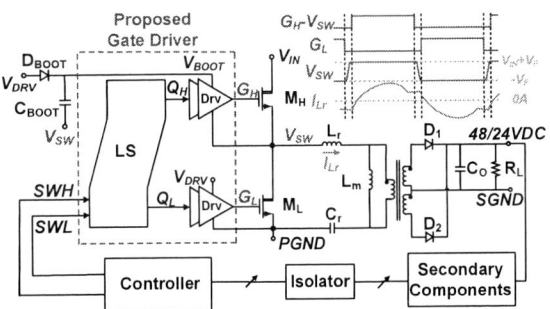

Fig. 2. Implementation of the proposed gate driver in a half-bridge LLC converter with operating signal waveforms

high-speed comparator on high-side to get 5ns delay and 20V/ns dv/dt immunity, though its 300µA static current (~0.12W at 400V) limits high-voltage use. A pulse-triggered active coupling (PTAC) design [5] delivers a 2ns level-shifter delay and uses a cross-coupled current-mirror to double dv/dt immunity at 40V. A capacitance-coupled level shifter [6] reaches 1.45ns delay at 50V, but need floating high-voltage capacitors rated for V_{IN} operation. Liu [7] achieves 200V/ns dv/dt immunity with 0.53ns delay at 50V by shunting dv/dt noise currents, but the design requires up to six HV devices and precise parasitic matching. A 20V level shifter [8] disconnects the high-side signal path after each Q_H transition, offering near-unlimited immunity, but its fixed timing sequence limits its use in soft-switching converters.

In summary, current shunting enhances dv/dt immunity by redirecting noise current to keep the latch input low, and signal-path disconnection cuts off the signal path to the latch inputs when a V_{SW} transition is expected. This work integrates both techniques and employs an adaptive signal-path disconnection synchronized to the V_{SW} transition, which boosts dv/dt immunity and ensures full compatibility with soft-switching converter operation.

II. PROPOSED GATE DRIVER

A. Proposed HV Level Shifter

The circuit of proposed level shifter is shown in Fig. 3, with the corresponding timing waveforms presented in Fig. 4, illustrating its operation in a soft-switching converter. The drains of the 30V NDMOS devices M_{1L} and M_{3L} serve respectively as the sources of the 600V depletion-mode JFETs M_{2L} and M_{4L} (threshold voltage = −29V), and other transistors

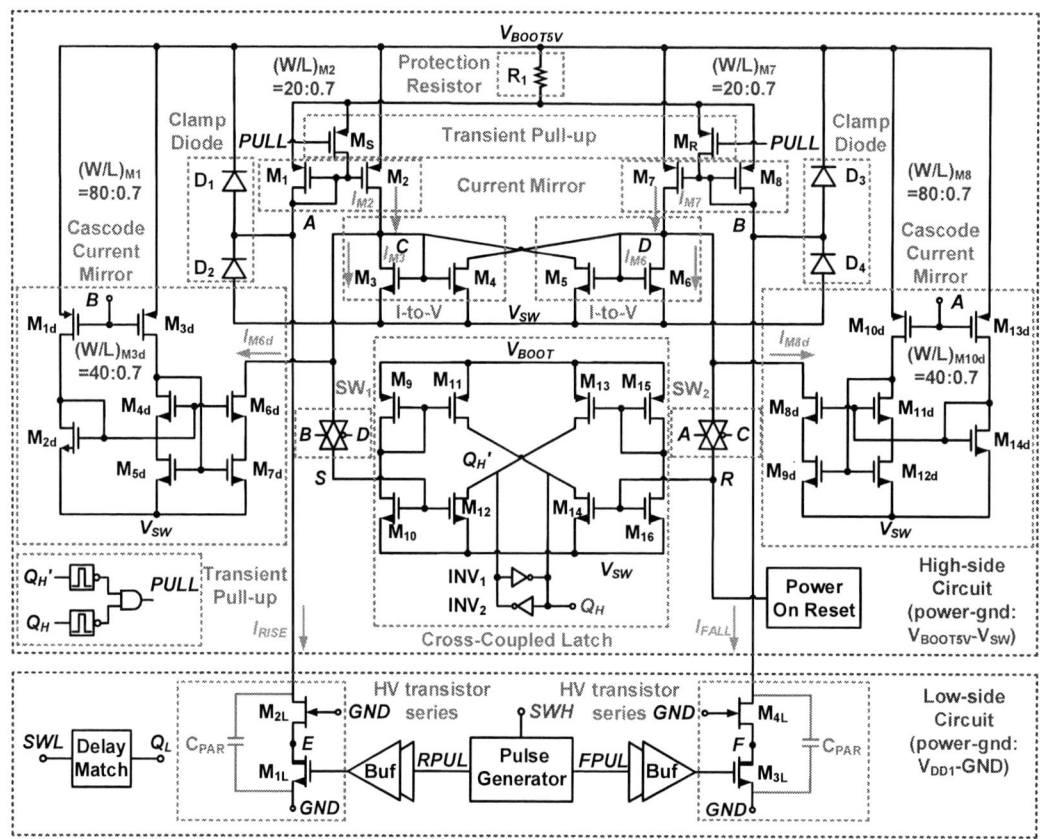

Fig. 3. Circuit Implementation of the proposed HV level shifter

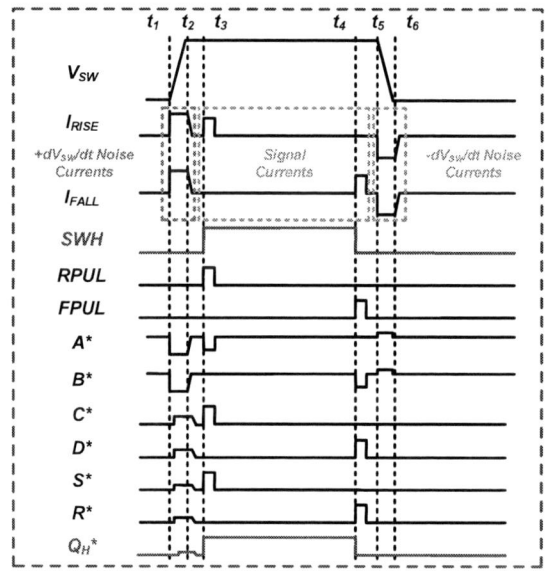

*: Signal A, B, C, D, S, R, Q_H relative to V_{SW}

Fig. 4. Signal waveforms of the proposed level shifter

in Fig. 3 are rated at 6 V. The low-side circuit has its substrate tied to GND, while the high-side circuit's substrate is tied to V_{SW}. The low-side circuit, powered by the 5 V V_{DD1} supply, includes a pulse generator that converts SWH rising edges into RPUL pulses and falling edges into FPUL pulses. It also includes two inverter-chain buffers and two HV transistor series M_{1L}–M_{2L} and M_{3L}–M_{4L} driven by the buffered pulses.

The high-side circuit, powered by the bootstrap voltage V_{BOOT5V}, consists of two current-mirror pairs, two I-to-V conversion stages, and a cross-coupled latch for signal transmission. It also employs two cascode current-mirror pairs with two transmission-gate switches to boost dv/dt immunity. Protection is ensured by four clamp diodes and a series resistor, while reliability is enhanced by an integrated power-on-reset and a transient pull-up circuit.

At t_1, V_{SW} begins its rising transition. During the interval from t_1 to t_2, parasitic capacitances C_{PAR} at the drains of M_{2L} and M_{4L} (nodes A and B) charge with I_{RISE} and I_{FALL}:

$$I_{RISE} = I_{FALL} = I_{PAR} = C_{PAR}\frac{dV_{SW}}{dt} \quad (2)$$

With C_{PAR} of 1pF and dV_{SW}/dt of 10V/ns, charge current I_{PAR} reaches approximately 10mA. These positive dV_{SW}/dt noise currents forward-bias clamp diodes D_2 and D_4, pulling nodes A and B to $-V_{D1}$ relative to V_{SW} (same applies to other high-side nodes), where V_{D1} is the diode forward-voltage drop. As shown by the sizing annotations, M_1 and M_8 are four times larger than M_2 and M_7, and twice the size of M_{3d} and M_{10d}. I_{RISE} and I_{FALL} are mirrored into I_{M2} and I_{M7} at a 4:1 ratio. The larger mirroring ratio of M_{3d} and M_{10d} and higher output impedance cause these cascode mirrors to shunt most of I_{M2} and I_{M7}. The remaining current is evenly shared between M_3–M_5 and M_4–M_6, significantly reducing the noise current coupled into I_{M3} and I_{M6}.

Since nodes C and D share the same voltage potential, the PMOS transistor in SW_1 (connecting C to S) is driven by D, whereas the PMOS transistor in SW_2 (connecting D to R) is driven by C. As long as the voltage difference between nodes C and D remains below $|V_{THP}|$ (the PMOS threshold voltage), the gates of the PMOS devices in SW_1 and SW_2 are held above

TABLE I. TRANSMISSION-GATE CONTROL SIGNALS AND STATES*

V_{SW} Status	Rising Edge	V_{IN}	Falling Edge	GND
A	L	H	H	H
B	L	H	H	H
C	H	L	L	L
D	H	L	L	L
SW$_1$	X	O	O	O
SW$_2$	X	O	O	O

*: L and H for low and high logic level, O and X for switch on and off state

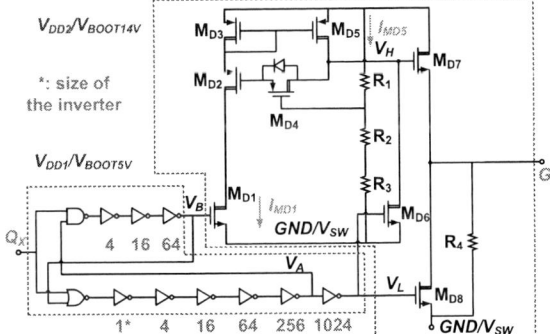

Fig. 5. Proposed Q_X to G_X driver stage

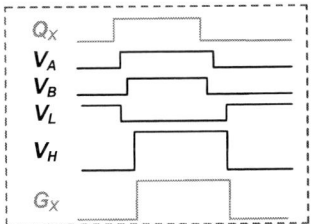

Fig. 6. Signal waveforms of the proposed driver stage

(C−|V_{THP}|) and (D−|V_{THP}|), respectively, corresponding to high-level (H) gate voltages that keep both PMOS devices off. Meanwhile, the −V_{D1} potentials at nodes A and B pull the NMOS gates of the two transmission-gates below V_{THN} (the NMOS threshold voltage), corresponding to low-level (L) gate voltages that turn off the NMOS devices. Therefore, both SW$_1$ and SW$_2$ remain disabled as shown in Table I. Only a small voltage couples through the transmission-gate parasitic capacitance into S and R, keeping Q_H undisturbed. After the V_{SW} rising transition completes at t_2, nodes A and B recharge to (V_{DD1}−|V_{THP}|) relative to V_{SW}, while nodes C and D discharge to V_{THN}. Under high dV_{SW}/dt conditions, current shunting prevents overvoltage on nodes C and D and helps discharging them after the transition. SW$_1$ and SW$_2$ are then enabled again, ready for signal transmission.

At t_3, the low-side pulse generator converts SWH rising edge to a 20ns RPUL pulse. After buffering, the pulse pulls node E from 29V to GND, generating the I_{RISE} signal current. When RPUL arrives, I_{M8d} and I_{M4} pulls node D to zero (relative to V_{SW}), fully turning off M$_5$. With FPUL held low, both I_{FALL} and I_{M6d} remain at zero. I_{M2} flows entirely into the I-to-V stage, I_{M3} equals to I_{M2}. At this point, node B is high and node D stays low, fully enabling SW$_1$. Node S follows node C and rises high. A current equal to I_{M3} is mirrored into I_{M10}, quickly setting the cross-coupled latch output Q_H with the assistance of INV$_1$ and INV$_2$. Once Q_H rises, the transient pull-up circuit detects the edge and generates a PULL pulse. The PULL pulse drives nodes A and B up to V_{DD1}, preventing subthreshold leakage of the current-mirrors that could slow FPUL transmission. At t_4, the FPUL pulse is transmitted to the

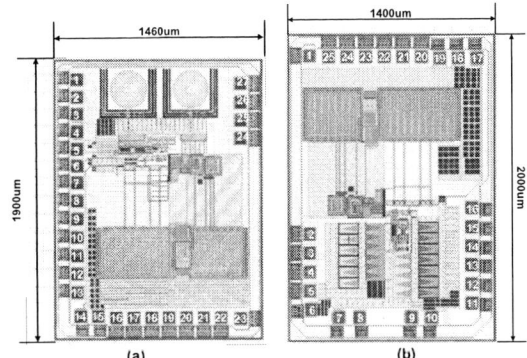

Fig. 7. (a) High-side circuit layout(b) Low-side circuit layout

high-side in a manner same as RPUL, resetting Q_H.

At t_5, V_{SW} begins its falling transition. During the interval from t_5 to t_6, negative dV_{SW}/dt noise currents from two parasitic capacitances discharge through clamp diodes D$_1$ and D$_3$. These currents are not mirrored by M$_1$ or M$_8$ and thus do not perturb Q_H. Due to the presence of the protection resistor R$_1$, the discharge of negative dV_{SW}/dt noise current through body diodes of M$_1$, M$_5$, M$_8$, and M$_R$ is suppressed.

By combining current shunting and adaptive signal-path disconnection, the level shifter keeps SW$_1$ enabled when dV_{SW}/dt is low, keeping nodes C, D, S, and R low without false Q_H triggers. When dV_{SW}/dt increases, the signal path is disconnected before nodes C and D can rise enough to perturb nodes S and R and false trigger Q_H, enabling ultra-high dV/dt immunity. Once the V_{SW} rising transition ends, the signal paths reconnect, ensuring no RPUL pulse is lost and preserving compatibility with soft-switching converters, particularly those employing adaptive dead-time control.

B. Q_X to G_X Driver Stage

Fig. 5 shows the Q_X to G_X driver stage, which converts the high- and low-side 5V logic signals Q_X (Q_L or Q_H) into the 13V gate drive signal G_X (G_L or G_H) for two power transistors in the half-bridge. Fig. 6 shows corresponding waveforms.

When Q_X is low, two scaled inverter chains pull V_L to V_{DD1}/V_{BOOT5V} and V_A, V_B, V_H to GND/V_{SW}, thus turning on M$_{D8}$ (D for isolated 30VLDMOS) and M$_{D6}$ while M$_{D1}$, M$_{D3}$, and M$_{D5}$ remain off, holding G_X low. R$_1$–R$_3$ bias M$_{D4}$'s gate V_D at (V_{DD2}−V_{DD1}), thereby setting M$_{D2}$'s gate V_C to (V_{DD2}−V_{DD1}+|V_{THP}|). On Q_X's rising edge, the upper chain's NAND gate ensures V_B rises after V_A goes high, creating dead time between V_H and V_L to prevent shoot-through. The falling V_L turns off M$_{D8}$ and M$_{D6}$ and then rising V_B turns on M$_{D1}$. Flowing through the deep linear region transistor M$_{D2}$, I_{MD1} is mirrored by M$_{D3}$ as I_{MD5}, which boosts V_H and turns on M$_{D7}$ to pull G_X high. Once V_H reaches V_{DD2}, M$_{D4}$'s body diode conducts with a forward voltage drop of V_{D2}, pulling V_C up to (V_H−V_{D2}), which disables M$_{D2}$ and cuts off static current. The NOR gate in the lower chain ensures V_B and V_L never go high at the same time. On Q_X's falling edge, V_L rises as G_X and V_H fall together, protecting M$_{D7}$'s gate oxide from overvoltage. Pull-down resistor R$_4$ pulls G_X to GND/V_{SW} by default.

III. RESULTS

The proposed gate driver is implemented in a HV-BCD process. Fig. 7(a) and (b) show the low- and high-side layouts, which occupy silicon areas of 1.46×1.90 mm² and 1.40×2.00 mm², respectively. Fig. 8(a) presents the post-layout simulation results of a 300 kHz/288 W half-bridge LLC converter driven by the proposed gate driver. Operating

979-8-3315-3918-4/25 $31.00 © 2025 IEEE

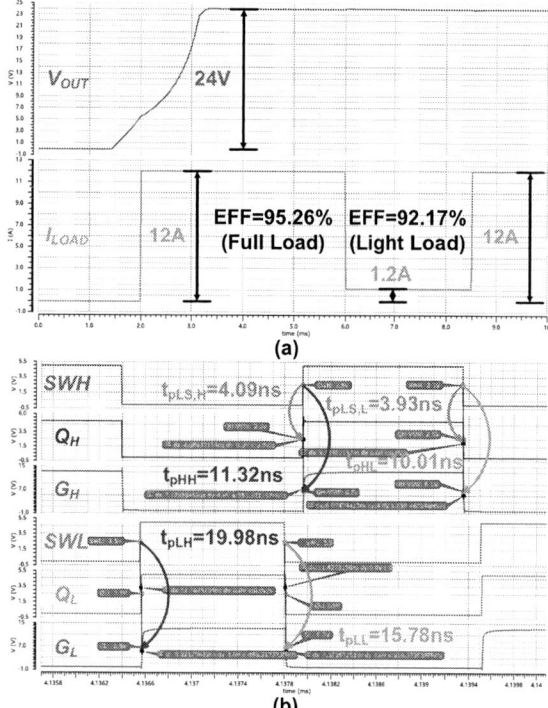

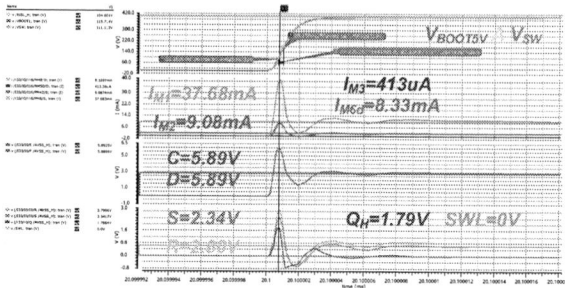

Fig. 9. Post-layout simulation results of dv/dt immunity

V_{THN} of 800mV, therefore Q_H returns to zero after the initial spike. Table II compares the results of this design with state-of-the-art designs. The figure of merit (FOM) of level shifter, defined as the delay divided by the product of the process minimum feature length and the high-to-low-side voltage difference, reaches a competitive value of 0.015 ns/(μm·V).

IV. CONCLUSION

A 400V level-shifting half-bridge driver with <5ns level-shifter delay and 278V/ns immunity is implemented in a HV-BCD process. The low- and high-side circuits occupy silicon areas of 1.46×1.90mm² and 1.40×2.00mm², respectively. In a 300kHz/288W LLC converter, the driver maintains soft-switching operation with high-side gate drive delays below 11ns and prevents false turn-on under extreme V_{SW} slew rates. A figure of merit of 0.015ns/(μm·V) highlights its superior trade-off between high-speed and high-voltage performance, making it suitable for high-frequency soft-switching converters.

REFERENCES

[1] Y. Moghe, T. Lehmann, T. Piessens, "Nanosecond Delay Floating High Voltage Level Shifters in a 0.35μm HV-CMOS Technology," IEEE Journal of Solid-State Circuits, vol. 46, no. 2, pp. 485-497, Feb 2011.

[2] J. Cao, Z. Zhou, Z. Wang, H. Tang, B. Zhang, "Design Techniques of Sub-ns Level Shifters With Ultrahigh dV/dt Immunity for Various Wide-Bandgap Applications," IEEE Transactions on Power Electronics, vol. 36, no. 9, pp. 10447-10460, Sep 2021.

[3] H. Yang, C. Chiu, S. Lai, J. Chen, C. Chang, C. Meng, K. Chen, C. Wey, Y. Lin, C. Lee, J. Lin, T. Tsai, H. Luo, "120V/ns output slew rate enhancement technique and high voltage clamping circuit in high integrated gate driver for power GaN FETs," ESSCIRC Conference 2015 - 41st European Solid-State Circuits Conference (ESSCIRC), pp. 291-294, Sep 2015.

[4] J. Wittmann, T. Rosahl, B. Wicht, "A 50V high-speed level shifter with high dv/dt immunity for multi-MHz DCDC converters," ESSCIRC 2014 - 40th European Solid State Circuits Conference (ESSCIRC), pp. 151-154, Sep 2014.

[5] Z. Liu, L. Cong, H. Lee, "Design of On-Chip Gate Drivers With Power-Efficient High-Speed Level Shifting and Dynamic Timing Control for High-Voltage Synchronous Switching Power Converters," IEEE Journal of Solid-State Circuits, vol. 50, no. 6, pp. 1463-1477, Jun 2015.

[6] D. Lutz, A. Seidel, B. Wicht, "A 50V, 1.45ns, 4.1pJ High-Speed Low-Power Level Shifter for High-Voltage DCDC Converters," ESSCIRC 2018 - IEEE 44th European Solid State Circuits Conference (ESSCIRC), pp. 126-129, Sep 2018.

[7] D. Liu, S.J. Hollis, B.H. Stark, "A New Design Technique for Sub-Nanosecond Delay and 200 V/ns Power Supply Slew-Tolerant Floating Voltage Level Shifters for GaN SMPS," IEEE Transactions on Circuits and Systems I: Regular Papers, vol. 66, no. 3, pp. 1280-1290, Mar 2019.

[8] Y. Yang, M. Huang, S. Du, R.P. Martins, Y. Lu, "A Level Shifter With Almost Full Immunity to Positive dv/dt for Buck Converters," IEEE Transactions on Circuits and Systems I: Regular Papers, vol. 70, no. 11, pp. 4595-4604, Nov 2023.

Fig. 8. Post-layout simulation results of proposed gate driver in LLC converter (a) V_{OUT} and I_{LOAD} (b) propagation delay

TABLE II. PERFORMANCE COMPARISON WITH SOTA

	[8]	[2]	[5]	[4]	[7]	This work
Voltage(V)	20	30	40	50	50	400
Delay(ns)	1.17	0.66	2.0	5	0.53	4.09
Process Node(μm)	0.18	0.5	0.5	0.18	0.18	0.7
dv/dt immunity(V/ns)	67	250*	-	20	200*	278*
FoM (ns/μm·V)	0.325	0.044	0.1	0.56	0.059	0.015

*: simulation result.

from soft-start to full-load (12 A), and down to 10% light-load, and back to full-load, the converter sustains a stable 24V DC output. Power efficiency under both full and light load conditions exceeds 92%, confirming the converter's soft-switching operation.

Propagation delays are measured from 50% input to 50% output thresholds. Fig. 8 (b) shows that the level-shifter delay from SWH to Q_H is 4.09ns on the rising-edge ($t_{pLS,H}$) and 3.93ns on the falling-edge ($t_{pLS,L}$). The high-side gate driver rising and falling delays are 11.32ns (t_{pHH}) and 10.01ns (t_{pHL}), while the low-side gate driver rising and falling delays are 19.98ns (t_{pLH}) and 15.78ns (t_{pLL}).

Fig. 9 shows dv/dt immunity simulation results. At the peak dV_{SW}/dt of 278V/ns, 37.68mA is sourced by M_1 and mirrored into 9.08mA in I_{M2}. I_{M6d} shunts 8.33mA and M_4 halves the remainder, leaving 413μA to M_3. Although nodes C and D still rise to 5.89V, the shunting network greatly reduces noise current and protects the 6V MOSFETs from overvoltage. With SW_1 and SW_2 off, only 2.80V and 2.34V couple into R and S, and Q_H peaks at 1.79V without false high-side turn-on. After the peak, I_{M1} falls to about 11.3mA, nodes C and D remain about 3V, but nodes S and R stay below

GaN-based complementary logic sawtooth generator for smart power ICs

Yutao Geng[1,2], Ji Shu[1,2], Tao Chen[1,2], Yan Cheng[1,2], Yat Hon Ng[1,2], and Kevin J. Chen*[1,2], *Fellow, IEEE*

[1] The Hong Kong University of Science and Technology Shenzhen Research Institute, Shenzhen, China

[2] The Department of Electronic and Computer Engineering, The Hong Kong University of Science and Technology, Hong Kong

*Email: eekjchen@ust.hk

Abstract—This work presents a GaN-based complementary logic sawtooth generator on a commercial *p*-GaN gate HEMT platform. The sawtooth generator circuit consists of a complementary logic comparator and a buffer inverter that employ GaN-based *n*/*p*-channel FETs. The complementary logic single-stage comparator with the buffer inverter achieves rail-to-rail operation with a 100-kHz triangular input. The sawtooth generator delivers a 54-kHz sawtooth signal with a voltage swing of 4.1 V. These demonstrations validate the potential of all-GaN complementary logic solutions for next-generation efficient power conversion systems.

Index Terms—GaN, complementary logic, comparator, sawtooth generator, *p*-FET, power integration.

I. INTRODUCTION

Emerging power-hungry applications like data centers, electric vehicles, and drones intensify the demand for high-efficient and high-power-density power conversion solutions. As a wide-bandgap semiconductor, gallium nitride (GaN) power devices demonstrate superior properties such as high switching frequencies, low on-resistance, and high critical electric field [1], which makes them as a crucial enabler for next-generation power conversion systems.

Monolithic integration of power devices and peripheral circuits further unlocks the potential of GaN power devices by substantially minimizing the parasitic effects caused by bond wires and package interconnects that limit the switching frequency of GaN-based power devices [2], [3]. Compared with direct coupled field effect transistor logic (DCFL), complementary logic exhibits suppressed static power consumption and is more favorable for constructing peripheral circuits [4]. All-GaN power integration with complementary logic circuits is becoming more attractive [5], [6], [7], [8].

The development of GaN complementary logic circuits has proved challenging because of the lack of high-performance *p*-FETs [9]. Nevertheless, substantial progress has been achieved in recent years. In 2020, a buried-channel *p*-FET demonstrated enhancement-mode operation and a high I_{ON}/I_{OFF} ratio of 10^7 [10]. Subsequently, *p*-FET stability has been significantly

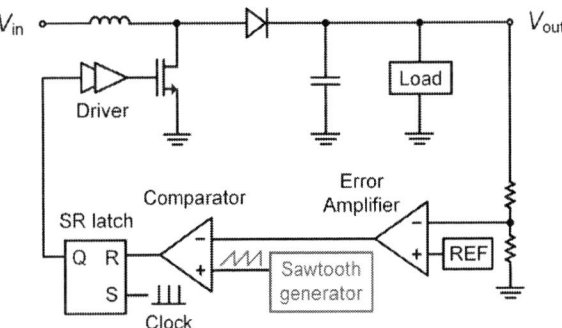

Fig. 1. Block diagram of a typical boost converter, where the sawtooth generator plays a vital role in the feedback control circuit.

enhanced through SiN_x/GaON gate stack [11], [12]. A family of GaN complementary logic gates and ring oscillators were successfully implemented, exhibiting rail-to-rail operation and suppressed power consumption [13], [14]. With these advancements in device engineering, it is essential and worthwhile to further explore the potential of the newly developed GaN complementary logic circuits in power conversion systems.

Fig. 1 depicts the block diagram of a typical boost converter, where the sawtooth generator provides a self-generated sawtooth wave in the feedback loop. A comparator then generates the PWM signal for power switch control by comparing the sawtooth signal and the output voltage sampled from the converter output. Following SR latch signal conditioning, a noise-immune gate control signal is delivered to the gate driver for power device switching.

In this work, a GaN complementary logic sawtooth generator featuring monolithically integrated *n*/*p*-FETs is demonstrated on a commercial *p*-GaN gate HEMT platform. As a core component of the sawtooth generator, the complementary logic two-stage comparator achieves a rail-to-rail operation under a triangular input waveform at 100 kHz. The sawtooth generator exhibits output waveforms of 54 kHz with a voltage swing of

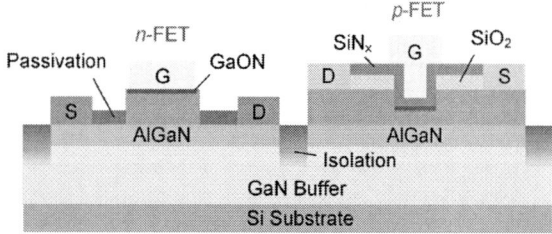

Fig. 2. Schematic cross-section of the monolithically integrated *n*-FET/*p*-FET.

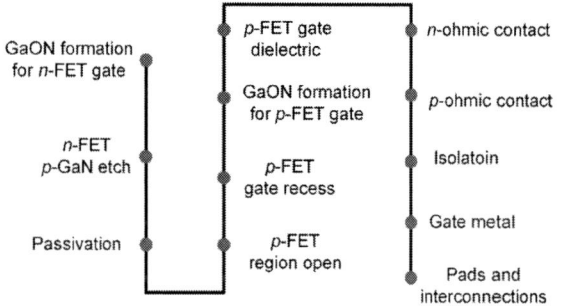

Fig. 3. Fabrication process flow of the GaN complementary logic circuit.

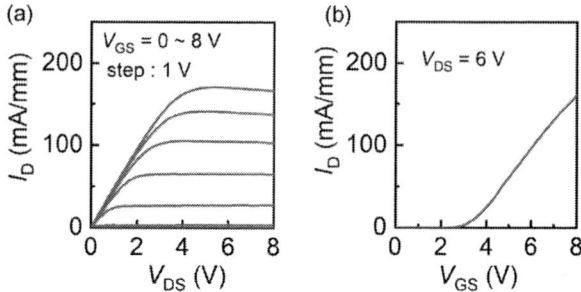

Fig. 4. (a) Output and (b) transfer characteristics of the *n*-FET with $L_{GS}/L_G/L_{GD} = 3/4/3$ μm.

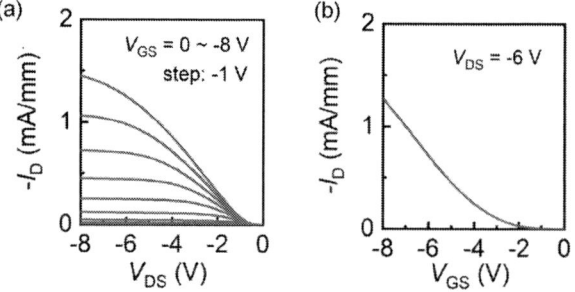

Fig. 5. (a) Output and (b) transfer characteristics of the *p*-FET with $L_{GS}/L_G/L_{GD} = 3/2/3$ μm.

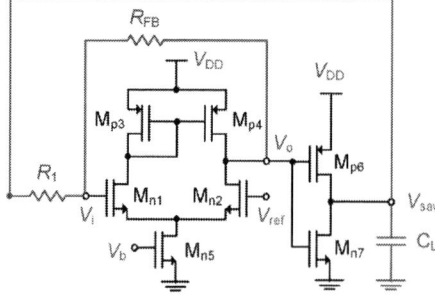

Fig. 6. Circuit schematic of the GaN complementary logic sawtooth generator with $W_{p\text{-}FET} = 200$ μm, $W_{n\text{-}FET} = 10$ μm.

4.1 V. These demonstrations further validate the viability of GaN complementary logic in smart power ICs.

II. DEVICE FABRICATION and CHARACTERISTICS

The GaN *n*/*p*-FETs were fabricated on a commercial GaN-on-Si wafer featuring an epitaxial stack consisting of 100 nm *p*-GaN, 15 nm AlGaN barrier, 0.8 nm AlN, 420 nm unintentionally doped GaN channel, and a 4.2 μm buffer layer. Fig. 2 shows the schematic cross-section of the monolithically integrated *n*/*p*-FETs. The fabrication process flow is illustrated in Fig. 3. The fabrication process began with GaON formation of *n*-FET for enhanced gate reliability, followed by *n*-FET gate etch and SiN$_x$/AlN passivation [15], [16]. After opening the *p*-FET region, the *p*-FET gate recess was conducted by ICP etching. Then, a SiN$_x$/GaON gate stack for *p*-FET stability enhancement was fabricated. Subsequently, *n*-ohmic and *p*-ohmic contacts were formed sequentially by opening the contact window, followed by e-beam evaporation. In the next step, planar device isolation was achieved by multi-energy fluorine ion implantation. Next, Ni/Au was evaporated as gate metals for both *n*-FETs and *p*-FETs. Finally, pads and interconnections were formed for device and circuit characterization.

Room-temperature characterization of the GaN *n*-FET ($L_{GS}/L_G/L_{GD} = 3/4/3$ μm) and *p*-FET ($L_{GS}/L_G/L_{GD} = 3/2/3$ μm) is presented in Fig. 4 and Fig. 5, respectively. Both devices demonstrate enhancement-mode operation: the *n*-FET exhibits $V_{TH} = +3.2$ V while the *p*-FET shows $V_{TH} = -3.8$ V, as extracted via the linear extrapolation method. Notably, the *p*-FET's current density is $\sim 10^{-2}$ lower than the *n*-FET's, primarily due to the inherent mobility disparity between electrons (~ 2000 cm²/V·s) and holes (~ 10 cm²/V·s) in GaN [17], [18].

III. CIRCUIT PERFORMANCES

Fig. 6 depicts the circuit schematic of the GaN complementary logic sawtooth generator, comprising two functional stages: a hysteresis comparator and a charging unit. Three off-chip components (two resistors R_1, R_{FB}, and one capacitor C_L, highlighted red) are connected to the circuit to enable periodic sawtooth generation. The circuit operation process is described as follows: When V_o is high, M$_{n7}$ is on to discharge C_L, causing V_{saw} to drop until reaching the lower threshold (V_{TL}) of the hysteresis comparator. This triggers the comparator output to toggle low, simultaneously turning on M$_{p6}$ while turning off M$_{n7}$. Consequently, C_L charges through M$_{p6}$ until V_{saw} reaches the upper threshold (V_{TH}), prompting the comparator to reset high and turn on M$_{n7}$ again to initiate a new discharge cycle. This autonomous charge-discharge oscillation continuously generates repetitive sawtooth waveforms through

979-8-3315-3918-4/25 $31.00 © 2025 IEEE

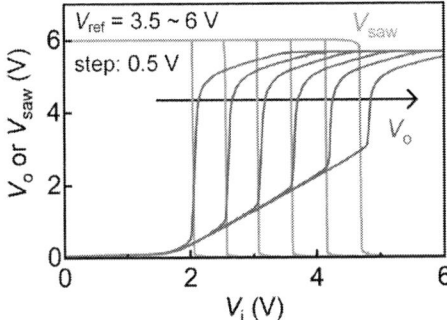

Fig. 7. Voltage transfer characteristics of the GaN complementary logic two-stage comparator with V_{DD} = 6 V.

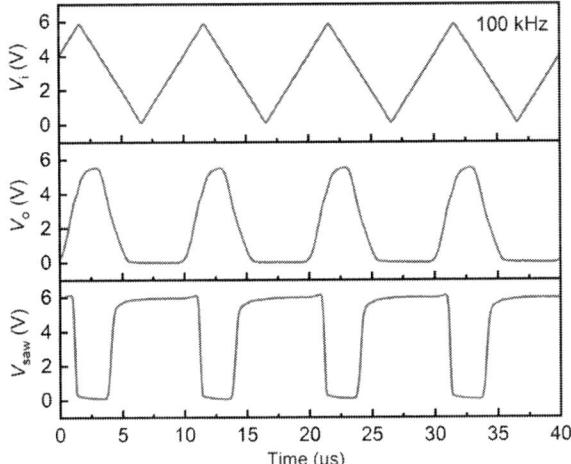

Fig. 8. Output waveforms of the GaN complementary logic two-stage comparator with a triangular wave input at 100 kHz (V_{DD} = 6 V and V_{ref} = 4.5 V).

the complementary transistor pair's alternating conduction states.

A. Comparator

The sawtooth generator is built based on a two-stage comparator composed of a five-transistor differential pair and a buffer stage (marked in black in Fig. 6) [19]. The gate widths of n/p-FETs in the circuit are 10 μm and 200 μm, respectively. The two-stage comparator (with all the capacitor and resistors disconnected) is characterized to evaluate the intrinsic circuit performance. Fig. 7 presents the transfer characteristics of the GaN complementary logic two-stage comparator with V_{DD} = 6 V. The output voltage shows a rail-to-rail operation and a sharp transition with V_{ref} of 3.5 ~ 6 V as V_i increases from 0 V to 6 V. The early increase in V_o before transition is due to the increasing current flowing through M_{n5}, which is an intrinsic characteristic of the five-transistor differential pair. It is also noted that after the transition, V_o increases slowly to a high level close to V_{DD}, which could be improved by optimizing the p-ohmic contact.

Fig. 8 presents the output waveforms of the GaN complementary logic two-stage comparator operating with a 100 kHz triangular wave input, under the condition that V_{DD} = 6 V and V_{ref} = 4.5 V. As the input voltage (V_i) approaches the switching threshold, the first stage output (V_o) initiates its

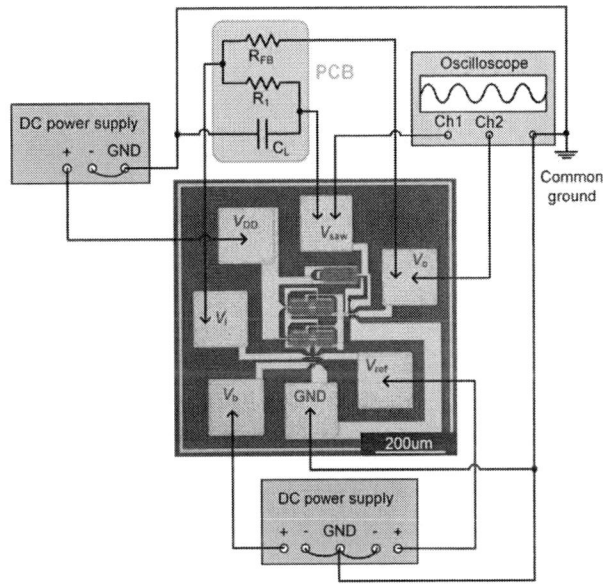

Fig. 9. Measurement setup of the GaN-based complementary logic sawtooth generator. The resistors and capacitor are soldered onto a PCB. Output waveforms are measured by high-impedance probes.

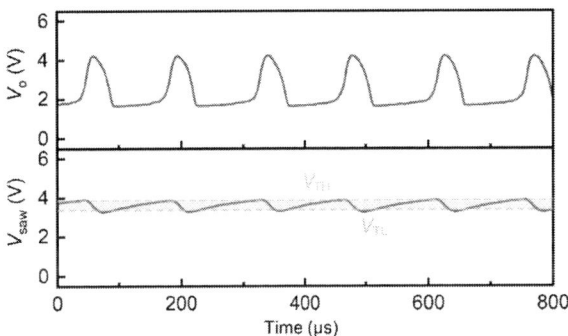

Fig. 10. Output waveforms of the GaN complementary logic sawtooth generator (f = 6 kHz with R_1/R_{FB} = 150 kΩ/390 kΩ and C_L = 4.7 nF).

transition. Subsequently, the buffer stage effectively conditions this V_o waveform, thus resulting in its output (V_{saw}) exhibiting a sharper transition and a rail-to-rail operation.

B. Sawtooth generator

The sawtooth generator is built based on the two-stage comparator, utilizing off-chip resistors and a capacitor. Fig. 9 depicts the measurement setup of the sawtooth generator. The off-chip RC components are soldered onto a PCB and connected to the circuit through probes. An oscilloscope is used to capture the output waveforms through high-impedance probes, while DC power supplies provide all other required voltages.

Fig. 10 presents the output waveforms of the sawtooth generator operating with R_1/R_{FB} = 150 kΩ/390 kΩ and C_L = 4.7 nF, yielding an oscillation frequency of 6 kHz. The load capacitor (C_L) charges through M_{p6} during the low state of V_o, while it discharges through M_{n7} during the high state of V_o. On the other hand, the output voltage V_o begins to decrease when the V_{saw} reaches the lower threshold (V_{TL}), while V_o starts to rise

979-8-3315-3918-4/25 $31.00 © 2025 IEEE

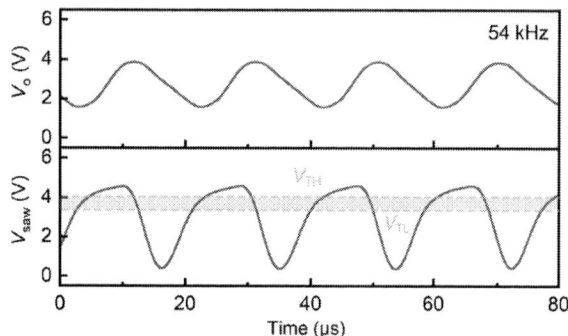

Fig. 11. Output waveforms of the GaN complementary logic sawtooth generator (f = 54 kHz with R_1/R_{FB} = 150 kΩ/390 kΩ and C_L = 82 pF).

once V_{saw} climbs to the upper threshold (V_{TH}). This cycle keeps repeating, where the first stage output switches the charging/discharging path, generating a repetitive 6-kHz sawtooth wave with a voltage swing of 0.7 V.

When the load capacitor is reduced to 82 pF, the oscillation frequency increases to 54 kHz. However, a significant delay appears in the V_{saw} waveform with an enlarged voltage swing of 4.1 V after V_{saw} reaches V_{TH} and V_{TL}. This delay occurs because the V_o charging/discharging process takes time. Specifically, V_{saw} only begins a ramp after V_o reaches the threshold voltage of the charging unit. The speed limitation primarily stems from the inherently low current density of p-FETs. Besides, off-chip interconnections introduce parasitic capacitance, directly constraining switching speed through increased RC time constants. Consequently, reducing the load capacitor alone cannot effectively increase the oscillation frequency. Monolithic integration of RC components is therefore essential for high-frequency operation.

IV. CONCLUSION

This work demonstrates a GaN-based complementary logic sawtooth generator on a commercial p-GaN HEMT platform, achieving a 54-kHz oscillation with a voltage swing of 4.1 V. While speed constraints are fundamentally rooted in the low current driving capability of p-FETs, monolithic RC integration is also essential for high-frequency operation. These results validate the viability of GaN multi-stage ICs for monolithic power integration.

ACKNOWLEDGMENT

This work was supported by the National Key Research and Development Program of China under Grant 2022YFB3604400.

REFERENCE

[1] K. J. Chen *et al.*, "GaN-on-Si Power Technology: Devices and Applications," *IEEE Transactions on Electron Devices*, vol. 64, no. 3, pp. 779–795, Mar. 2017, doi: 10.1109/TED.2017.2657579.

[2] H. Xu, G. Tang, J. Wei, Z. Zheng, and K. J. Chen, "Monolithic Integration of Gate Driver and Protection Modules With P-GaN Gate Power HEMTs," *IEEE Transactions on Industrial Electronics*, vol. 69, no. 7, pp. 6784–6793, Jul. 2022, doi: 10.1109/TIE.2021.3102387.

[3] G. Tang *et al.*, "Digital Integrated Circuits on an E-Mode GaN Power HEMT Platform," *IEEE Electron Device Lett.*, vol. 38, no. 9, pp. 1282–1285, Sep. 2017, doi: 10.1109/LED.2017.2725908.

[4] Z. Zheng, H. Xu, L. Zhang, and K. J. Chen, "On the operating speed and energy efficiency of GaN-based monolithic complementary logic circuits for integrated power conversion systems," *Fundamental Research*, vol. 1, no. 6, pp. 661–671, Nov. 2021, doi: 10.1016/j.fmre.2021.09.015.

[5] N. Chowdhury *et al.*, "p-Channel GaN Transistor Based on p-GaN/AlGaN/GaN on Si," *IEEE Electron Device Lett.*, vol. 40, no. 7, pp. 1036–1039, Jul. 2019, doi: 10.1109/LED.2019.2916253.

[6] J. Wei *et al.*, "GaN Power Integration Technology and Its Future Prospects," *IEEE Transactions on Electron Devices*, vol. 71, no. 3, pp. 1365–1382, Mar. 2024, doi: 10.1109/TED.2023.3341053.

[7] Q. Xie *et al.*, "Highly-Scaled Self-Aligned GaN Complementary Technology on a GaN-on-Si Platform," *IEDM*, 2022.

[8] J. Tang *et al.*, "Bipolar p-FET with Enhanced Conduction Capability on E-mode GaN-on-Si HEMT Platform," in *2023 International Electron Devices Meeting (IEDM)*, Dec. 2023, pp. 1–4. doi: 10.1109/IEDM45741.2023.10413728.

[9] S. J. Bader *et al.*, "Prospects for Wide Bandgap and Ultrawide Bandgap CMOS Devices," *IEEE Transactions on Electron Devices*, vol. 67, no. 10, pp. 4010–4020, Oct. 2020, doi: 10.1109/TED.2020.3010471.

[10] Z. Zheng, W. Song, L. Zhang, S. Yang, J. Wei, and K. J. Chen, "High ION and ION/IOFF Ratio Enhancement-Mode Buried p-Channel GaN MOSFETs on p-GaN Gate Power HEMT Platform," *IEEE Electron Device Lett.*, vol. 41, no. 1, pp. 26–29, Jan. 2020, doi: 10.1109/LED.2019.2954035.

[11] L. Zhang *et al.*, "SiN/in-situ-GaON Staggered Gate Stack on p-GaN for Enhanced Stability in Buried-Channel GaN p-FETs," in *2021 IEEE International Electron Devices Meeting (IEDM)*, San Francisco, CA, USA: IEEE, Dec. 2021, p. 5.3.1-5.3.4. doi: 10.1109/IEDM19574.2021.9720653.

[12] L. Zhang *et al.*, "Gate Leakage and Reliability of GaN -Channel FET With SiN$_x$/GaON Staggered Gate Stack," *IEEE Electron Device Letters*, vol. 43, no. 11, pp. 1822–1825, Nov. 2022, doi: 10.1109/LED.2022.3206470.

[13] Z. Zheng *et al.*, "Gallium nitride-based complementary logic integrated circuits," *Nat Electron*, vol. 4, no. 8, pp. 595–603, Aug. 2021, doi: 10.1038/s41928-021-00611-y.

[14] T. Li *et al.*, "Polarization Enhanced GaN Complementary Logic Circuits with Short Propagation Delay," *IEDM*, 2024.

[15] L. Zhang, Z. Zheng, S. Yang, W. Song, J. He, and K. J. Chen, "p-GaN Gate HEMT With Surface Reinforcement for Enhanced Gate Reliability," *IEEE Electron Device Letters*, vol. 42, no. 1, pp. 22–25, 2021, doi: 10.1109/LED.2020.3037186.

[16] J. Chen *et al.*, "Formation and Applications in Electronic Devices of Lattice-Aligned Gallium Oxynitride Nanolayer on Gallium Nitride," *Advanced Materials*, vol. 35, no. 12, p. 2208960, 2023, doi: 10.1002/adma.202208960.

[17] Y. H. Ng *et al.*, "Distribution and transport of holes in the p-GaN/AlGaN/GaN heterostructure," *Applied Physics Letters*, vol. 123, no. 14, p. 142106, Oct. 2023, doi: 10.1063/5.0172010.

[18] Y. H. Ng *et al.*, "p-GaN gate power HEMT heterostructure as a versatile platform for extremely wide-temperature-range (X-WTR) applications," *Applied Physics Letters*, vol. 124, no. 4, p. 043504, Jan. 2024, doi: 10.1063/5.0184784.

[19] Y. Geng *et al.*, "Monolithically Integrated GaN Comparator Based on Complementary Logic," *IEEE Transactions on Electron Devices*, vol. 72, no. 2, pp. 611–617, Feb. 2025, doi: 10.1109/TED.2024.3517603.

979-8-3315-3918-4/25 $31.00 © 2025 IEEE

Battery Charger Designs for Low-Voltage Energy Harvesting Based on the Return-on-Investment Concept (Invited)

W. Saito[1], A. Higuchi[1], T. Yamano[1], T. Tanzawa[2], *Fellow, IEEE*

[1] Shizuoka University, Hamamatsu, Japan, [2] Waseda University, Kitakyushu, Japan

Abstract— This paper studies battery charger charge pumps (CPs) for low-voltage energy harvesting based on the return-on-investment (RoI) concept. The charge transfer switches (CTSs) of the RoI CP are powered by a battery, V_{BAT}, providing a sufficient voltage for operating all the CTSs in linear regions (investment) whereas the power from an extremely low supply voltage of energy transducer, V_{IN}, is supplied into the battery (return). The battery charger RoI CP is designed to minimize V_{IN} under the condition where the return exceeds the investment, i.e., the battery is net charged with a positive net current. The circuit model is proposed to design this ROI battery charger charge pump. The charge pump was fabricated in 180 nm CMOS with 1.8 V transistors and MIM capacitors to evaluate the characteristics of the battery charger for a 2.3 V all-solid-state battery. The RoI CP was also compared with the RoI boost converter with respect to the minimum V_{IN}, total active area, and power efficiency.

Keywords—Battery charger, Charge pump, Energy harvesting, Low voltage, Return-on-Investment, Boost Converter

I. INTRODUCTION

When Internet of Things (IoT) sensing modules are deployed in the field, they communicate with one another or with cloud servers to collect environmental information and help maintain society safety with little or no human intervention [1], [2]. However, the batteries used in these sensor modules need to be replaced periodically. The cost of replacing old batteries with new ones has increased significantly as more sensor modules are deployed globally. To decrease the overall cost of sensor networks, technologies are needed to eliminate battery replacement. Energy harvesting is a key technology for achieving this. One solution is the use of an energy transducer (ET) to convert ambient energy sources, such as light, thermal flow, and vibration, into electrical power for the sensor modules, enabling battery-free operation [3], [4]. Since no battery is needed, battery replacement is unnecessary. However, if the modules require operation more frequently than the cycle time of the available ambient energy, a battery is needed as a backup [5] – [9]. The design of DC-DC boost converters operating at extremely low input voltages is a primal concern for battery charger applications. Various types of boost converters have been proposed, including switching boost converters using a chip inductor [10] – [13], fully integrated switched-capacitor boost converters, as known as charge pumps [14] – [16], hybrid converters that combine switching regulators and switched-capacitor converters [17] – [19], and boost circuits using a chip transformer [20] – [21]. Toward achieving a minimum-operating-voltage design, models for charge pumps have been discussed in the literature [22] – [24]. Hybrid power management circuits (PMCs) have both battery and energy transducer (ET) along with their power converters [6], [25] – [28]. Figs. 1 (a), (b) illustrate PMCs with parallel [6] and serial [25] connection of battery and thermoelectric generator (TEG). The battery in Fig. 1 (b) needs to invest power into the buck converter for every operation. Fig. 1 (c) illustrates a Return-on-Investment (RoI) battery charger based on the serially connected converter with battery and TEG [26].

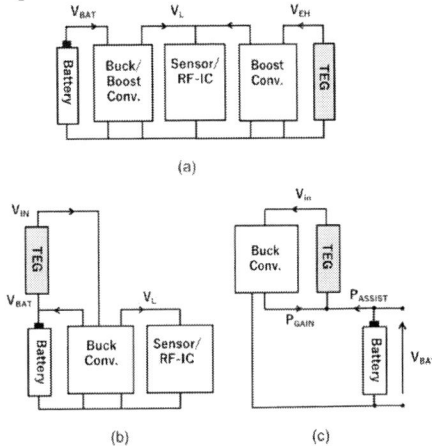

Fig. 1 PMCs with battery and ET connected in parallel (a) and in series (b). RoI battery charger based on the serially connected converter with battery and ET (c).

Fig. 2 shows standard (a) and RoI (b) battery charger boost converters [27]. The input voltage V_{IN} can be a terminal voltage of DC energy transducer such as TEG or a rectified voltage of AC vibration energy transducer (VET) [3]. VET requires AC-DC voltage-up or -down conversion depending on principles such as magnetostriction [29] or electrostatics [30], respectively. The minimum input voltage of a standard battery charger was limited by the control circuit. To eliminate the constraint that the controller must operate with V_{IN}, the supply voltage source was changed to the battery voltage V_{BAT}. The minimum V_{IN} is then limited by the condition that the power stage has the output current I_O which exceeds the input current I_{CKT}, i.e., the net current I_{NET} is positive. One of the drawbacks of the RoI boost converter is that it requires a chip inductor, which increases the size and cost of PMC.

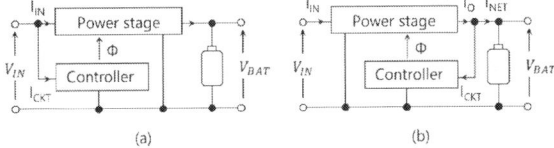

Fig. 2 Standard (a) and ROI (b) battery charger boost converters.

This paper focuses on the design of RoI battery charger charge pumps [28]. How the circuit can be designed optimally with respect to minimizing V_{IN} is formulated. Measured results on the minimum input voltage and peak power efficiency are compared with those of the RoI boost converter. The paper is organized as follows. Section II presents a model to design RoI battery charger charge pumps. Section III shows measured results. Comparison with the model calculations and SPICE simulations is made in Section IV.

II. DESIGN OF RoI BATTERY CHARGER CHARGE PUMP

Fig. 3 conceptually compares a standard charge pump (a) and the proposed RoI battery charger charge pump (b). φ_P and φ_G indicate the clock signals for the main capacitors and the gates of CTSs generated by a pulse generator (PG). Their voltage amplitudes are low input voltage V_{IN} in the standard CP. The switched-capacitor array (SC array) converts the input power from V_{IN} to charge the battery V_{BAT}. The minimum V_{IN} is limited by PG. On the other hand, the gate signal Φ_G is powered by V_{BAT} in the RoI boost converter. As a result, PG does not limit the minimum V_{IN}. Instead, the condition that $I_{NET} = I_O - I_{CKT} > 0$ determines the minimum V_{IN}.

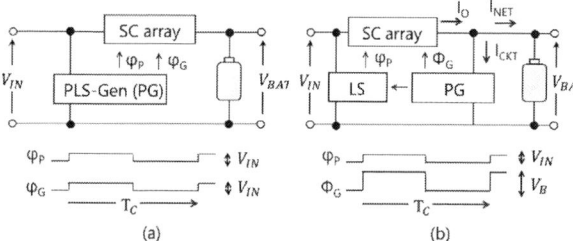

(a)   (b)

Fig. 3 Standard (a) and proposed RoI (b) battery charger CPs.

The initial guess on CP topology for low-voltage operation was the CMOS latched type [31], as illustrated in Fig. 4. Before the operation, all the internal capacitor nodes are discharged, as shown in Fig. 4 (a). With the initial clock edges, the gate-to-source voltage of the CTS becomes $2 \times V_{IN}$. Therefore, the minimum V_{IN} is limited to $V_{TH}/2$, where V_{TH} is the threshold voltage of the CTS transistors.

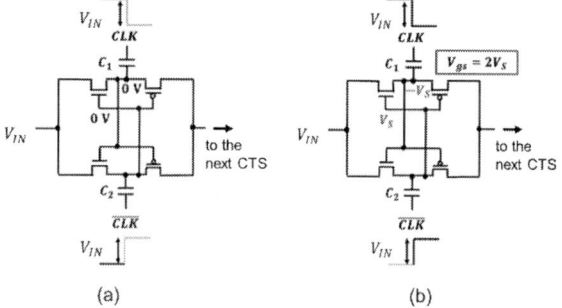

(a)   (b)

Fig. 4 States of a latched CMOS charge pump before the operation starts (a) and after a half cycle passes (b)

Therefore, the simplest CTSs were selected for demonstration of the RoI concept instead, as illustraetd in Fig. 5. The capacitor voltages in the former and latter halves of the SC array are lower and higher than $V_{BAT}/2$, respectively. As a result, NMOSFETs and PMOSFETs are used as CTSs in the former and latter halves like a simple CMOS inverter, as illustrated in Fig. 5. When $V_{BAT}/2$ is larger than V_{TH}, the transistors operate in linear regions, reulsintg in no voltage drops across CTSs.

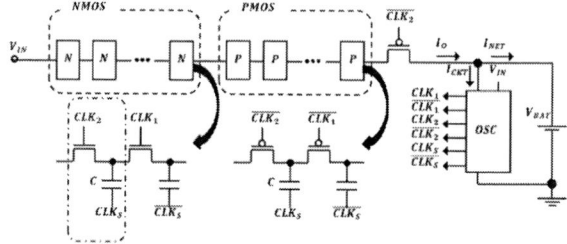

Fig. 5 An ROI battery charger charge pump

Fig. 6 shows the clock generator. Non-overlap clocks $CLK1, 2$ powered by V_{BAT} are level-shifted to CLK_S, $/CLK_S$ with amplitudes of V_{IN} by the circuits in Figs. 6 (c) and (d). Thus, the power from V_{IN} is input to the CP.

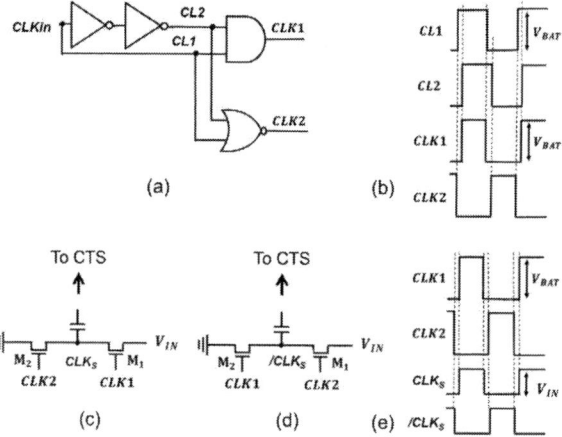

(a)   (b)

(c)   (d)   (e)

Fig. 6 CLK buffer for the RoI battery charger charge pump.

To design the battery charger CP, modeling is made as follows. Assuming I_{CKT} is given only by the switching of CTSs, (1) holds where C_P and M are the average gate capacitance of CTSs and the number of capacitors or stages, respectively. As noticed later, C_P should include the parasitic wiring capacitance as well, which can be more significant especially in the case that larger capacitors and longer wires to connect next neighbor CTSs are used.

$$I_{ckt} = f C_p M V_{BAT} \quad (1)$$

The relationship between the output current and output voltage is given by (2) [32], [33].

$$I_o = \frac{fC}{M}[(M+1)V_{IN} - V_{BAT}] \quad (2)$$

The relationship between I_{NET}, I_O, and I_{CKT} is given by (3).

$$I_{NET} = I_o - I_{CKT} \quad (3)$$

The condition of $I_{NET} = 0$ results in (4) based on $(1) - (3)$.

$$\frac{V_{IN}}{V_{BAT}} = \frac{\frac{C_p}{C}M^2 + 1}{M + 1} \quad (4)$$

The optimum M to minimize V_{IN} is determined by (5) in case of $M \gg 1$.

$$M^{opt} \sim \sqrt{\frac{C}{C_p}} \quad (5)$$

From (4) and (5), V_{IN}^{MIN} is determined to be (6).

$$\frac{v_{IN}^{min}}{V_{BAT}} \sim 2\sqrt{\frac{C_P}{C}} \qquad (6)$$

When the budget for the circuit area is constrained, or in other words, the total capacitance values are given to be C_{TOT}, one needs to fulfill (7), assuming the capacitors occupy the CP area.

$$C_{TOT} = CM \qquad (7)$$

Introducing a parameter α for C_P/C, M^{OPT}, C^{OPT}, and V_{IN}^{MIN} are simply given by (8), (9), and (10), respectively.

$$M^{opt} = 1/\sqrt{\alpha} \qquad (8)$$
$$C^{OPT} = \sqrt{\alpha}C_{TOT} \qquad (9)$$
$$V_{IN}^{MIN} = 2\sqrt{\alpha}V_{BAT} \qquad (10)$$

Power conversion efficiency was defined for RoI battery chargers as (11) [27].

$$\eta = \frac{V_{BAT} \cdot I_O}{V_{IN} \cdot I_{IN} + V_{BAT} \cdot I_{CKT}} \qquad (11)$$

A 180 nm CMOS technology with 1.8 V transistors and MIM capacitors available was used for validation of the design model as discussed above. A clock frequency of 5 MHz was selected to allow NMOSFETs with the minimum size and PMOSFETs with 2.5× wider channel for CTSs to minimize C_P, resulting in α of 2.8×10^{-4}. M^{OPT} and C^{OPT} are determined to be 60 and 60 pF for an available area C_{TOT} of 3.6 nF based on (8) and (9). From (10), V_{IN}^{MIN} was estimated to be 50 mV at V_{BAT} of 1.5 V, as shown by "1" in Fig. 7. Because the initial layout was not well done, additional wiring capacitances on the interconnection between next neighbor CTSs increased α to 4× larger, resulting in an increase in V_{IN}^{MIN} to 120 mV, as shown by "2" in Fig. 7.

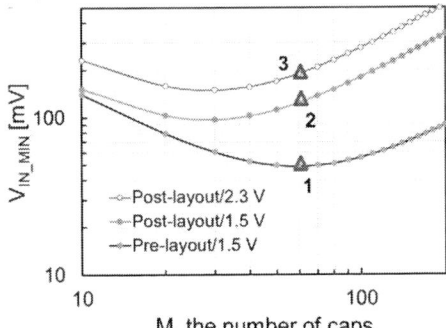

Fig. 7 M vs. V_{IN}^{MIN}

III. MEASUREMENT

To validate the design for RoI battery charger charge pump, a circuit was fabricated in 180 nm CMOS, as depicted in Fig. 8. The CLK buffer was placed next to the middle of the charge pump and the CLK signals were routed through the center between two halves. The total active silicon size was 3.1 mm².

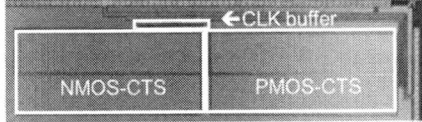

Fig. 8 Die photo.

Fig. 9 shows measured and simulated I_{NET}, I_O, and I_{CKT} across V_{IN} with V_{BAT} of 1.5 V. Even though I_O becomes positive at V_{IN} of 100 mV or higher, I_{NET} becomes positive at V_{IN} of 200 mV or higher because I_{CKT} was as high as 30 μA.

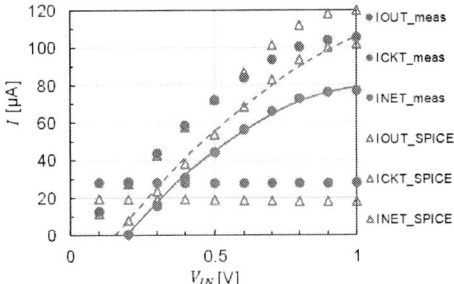

Fig. 9 Current components across V_{IN} for the RoI CP.

Fig. 10 shows the waveform for V_{BAT} of an all-solid-state battery with a size of 10.5×10.5 mm² [34] while the RoI charger is enabled with V_{IN} of 0.55 V. As a result, battery charger operation with the RoI charge pump was validated.

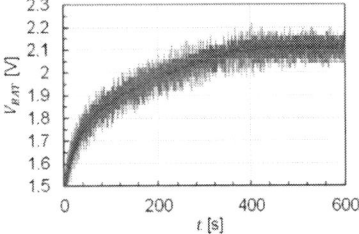

Fig. 10 Transient waveforms of V_{BAT} with the battery charger.

IV. COMPARISON BETWEEN CHARGE PUMP AND SWITCHING CONVERTER AS BATTERY CHARGERS

The current components of the RoI CP of Fig. 9 were compared with those of the RoI boost converter (BC) [27], as shown in Fig. 11. The device parameters were inductance of 64 μH, the series resisance of the incudtor of 0.7 Ω, the channel width of the low-side NMOSFET of 27 mm, and operation frequency of 9.3 kHz. Table I summarizes their performance comparison. The common conditions are 180 nm CMOS and 1.5 V of V_{BAT}. The RoI boost converter realized 20× lower V_{IN}^{MIN}, 10× smaller silicon area, and 4.8× higher peak power efficiency than the RoI charge pump with additional chip inductor, Schottky diode, and 500× slower operation frequency. The RoI charge pump could have V_{IN}^{MIN} lower by a factor of 4 with layout improvement with respect to wiring or smaller circuit area by a factor of 5 with denser MOS capacitors, but that could never outperform the RoI boost converter without any further improvement. On the other hand, the RoI charge pump has 40× smaller active area than the boost cpnverter.

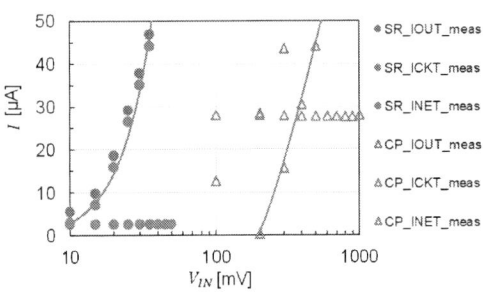

Fig. 11 Current components across V_{IN} for the RoI charge pump and boost converter.

Table I Performance comparison of the RoI battery chargers.

	BC [27]	CP
f	9 kHz	5 MHz
Si Area	0.3 mm²	3.1 mm²
V_{IN}^{MIN}	10 mV	200 mV
Peak η	53 % (at 20 kHz)	11 % (at 5 MHz)
Ext. elements	C_{IN}, L, SBD	C_{IN}
Total active area	160 mm²	3.6 mm²

V. CONCLUSION

RoI battery charger charge pumps were studied to minimize the input voltage from energy transducer for IoT application. The optimum number of capacitors can be determined by a technology dependent parameter as a ratio of the parasitic capacitance of the charge transfer switch to that of the pump capacitor. Due to additional wiring capacitance in the actual layout, the minimum input voltage was 200 mV higher by 150 mV from the pre-layout design in measurement of the circuit fabricated in 180 nm CMOS due to excessive wiring capacitance. The performance of the RoI charge pump was compared with the previously reported RoI boost converter. At a sacrifice in module size and cost due to additional external components of an inductor and a Schottky diode, the RoI boost converter showed much better performance with respect to the minimum input voltage and the circuit area for a battery charger whereas the RoI charge pump had 40× smaller active area than the boost cpnverter.

ACKNOWLEDGEMENT

This work is supported by The Iwatani Naoji Foundation, Zeon Corp., Maxell Corp., a Waseda University Grant for Special Research Projects (Project number: 2025C-528), dlab-VDEC, Synopsys Inc., Cadence Design Systems Inc., Siemens EDA.

REFERENCES

[1] A. Zanella, N. Bui, A. Castellani et al., "Internet of Things for smart cities," IEEE Internet of Things Journal, vol. 1, no. 1, pp. 22–32, Feb. 2014.
[2] A. Al-Fuqaha, M. Guizani, M. Mohammadi et al., "Internet of things: A survey on enabling technologies, protocols, and applications," IEEE Communications Surveys & Tutorials, vol. 17, no. 4, pp. 2347–2376, 2015.
[3] P. D. Mitcheson, E. M. Yeatman, G. K. Rao et al., "Energy harvesting from human and machine motion for wireless electronic devices," Proc. IEEE, vol. 96, no. 9, pp. 1457–1486, Sep. 2008.
[4] S. Sudevalayam and P. Kulkarni, "Energy harvesting sensor nodes: Survey and implications," IEEE Commun. Surveys Tuts., vol. 13, no. 3, pp. 443–461, 3rd Quater 2011.
[5] K. Kadirvel, Y. Ramadass, U. Lyles et al., "A 330nA Energy-Harvesting Charger with Battery Management for Solar and Thermoelectric Energy Harvesting," ISSCC, pp. 106 - 107, Feb. 2012.
[6] D. El-Damak and A. P. Chandrakasan, "A 10 nW–1 μW Power Management IC With Integrated Battery Management and Self-Startup for Energy Harvesting Applications," IEEE JSSC, vol. 51, no. 4, pp. 943 - 954, Apr. 2016.
[7] Q. Wan and P. K. T. Mok, "A 14-nA, highly efficient triple-output thermoelectric energy harvesting system based on a reconfigurable TEG array," IEEE JSSC, vol. 54, no. 6, pp. 1720 - 1732, Jun. 2019.
[8] Y. S. Noh, J. I. Seo, W. J. Choi et al., "A Reconfigurable DC-DC Converter for Maximum TEG Energy Harvesting in a Battery-Powered Wireless Sensor Node," ISSCC, pp. 266 - 267, Feb. 2021.
[9] S. Tanabe, Y. Sakamoto, H. Uchida et al., "A Hybrid Thermoelectric Generator–Battery Power Supply System Toward Replacement-Free Battery." In 2023 11th International Conference on Power Electronics and ECCE Asia (ICPE 2023-ECCE Asia), pp. 1817-1822. IEEE, May 2023.
[10] E. J. Carlson, K. Strunz, and B. P. Otis, "A 20 mV input boost converter with efficient digital control for thermoelectric energy harvesting," IEEE J. Solid-State Circuits, vol. 45, no. 4, pp. 741–750, Apr. 2010.
[11] Y. Ramadass, A. P. Chandrakasan, "A battery-less thermoelectric energy harvesting interface circuit with 35 mV startup voltage." IEEE Journal of Solid-State Circuits, vol. 46, no. 1, pp. 333-341, Jan. 2011.

[12] P. S. Weng, H. Y. Tang, P. C. Ku et al., "50 mV-Input Batteryless Boost Converter for Thermal Energy Harvesting," IEEE J. Solid-State Circuits, vol. 48, no. 4, pp. 1031 -1041, Apr. 2013.
[13] A. Shrivastava, N. E. Roberts, O. U. Khan et al., "A 10 mV-Input Boost Converter With Inductor Peak Current Control and Zero Detection for Thermoelectric and Solar Energy Harvesting With 220 mV Cold-Start and −14.5 dBm, 915 MHz RF Kick-Start," IEEE J. Solid-State Circuits, vol. 50, no. 8, pp. 1820 - 1832, Aug. 2015.
[14] P. H. Chen, K. Ishida, X. Zhang et al., "A 120-mV input, fully integrated dual-mode charge pump in 65-nm CMOS for thermoelectric energy harvester," 17th Asia and South Pacific Design Automation Conference, pp. 469-470, Jan. 2012.
[15] J. Goeppert, Y. Manoli, "Fully integrated startup at 70 mV of boost converters for thermoelectric energy harvesting," IEEE J. Solid-State Circuits, vol. 51, no. 7, pp. 1716 -1726, Jul. 2016.
[16] J. K. Yong, H. Ramiah, K. K. P. Churchill et al., "A 0.1-V VIN Subthreshold 3-Stage Dual-Branch Charge Pump With 43.4% Peak Power Conversion Efficiency Using Advanced Dynamic Gate-Bias," IEEE Transactions on Circuits and Systems II: Express Briefs, vol. 69, no. 9, pp. 3929 -3933, Sep. 2022.
[17] B. M. Lim, J. I, Seo, and S. G. Lee, "A Colpitts Oscillator-Based Self-Starting Boost Converter for Thermoelectric Energy Harvesting With 40-mV Startup Voltage and 75% Maximum Efficiency," IEEE J. Solid-State Circuits, vol. 53, no. 11, pp. 3293-3302, Nov. 2018.
[18] M. J. Chung, T. Hirose, T. Ono et al., "A 115× Conversion-Ratio Thermoelectric Energy-Harvesting Battery Charger for the Internet of Things," IEEE Transactions on Circuits and Systems I: Regular Papers, vol. 67, no. 11, pp. 4110-4121, Nov. 2020.
[19] S. Bose, T. Anand, and M. L. Johnston. "A 3.5-mV input single-inductor self-starting boost converter with loss-aware MPPT for efficient autonomous body-heat energy harvesting." IEEE journal of solid-state circuits, vol. 56, no. 6, pp. 1837-1848, Jun. 2021.
[20] LTC3108 Datasheet, Analog Devices. Inc. , Dec. 2011 [Online]. Available: https://www.analog.com/media/en/technical-documentation/data-sheets/LTC3108.pdf, [Accessed 28 Jun. 2025].
[21] Jong-Pil Im et al., "A 40 mV Transformer-Reuse Self-Startup Boost Converter With MPPT Control for Thermoelectric Energy Harvesting," IEEE JSSC, Volume: 47, Issue: 12, 3055 -3067, Dec.2012.
[22] S. Tokuda, T. Tanzawa. "Toward a minimum-operating-voltage design of DC-DC charge pump circuits for energy harvesting." IEEE International Symposium on Circuits and Systems (ISCAS), pp. 1-4, May 2019.
[23] A. Ballo, A. D. Grasso, G. Palumbo, "A High-Performance Charge Pump Topology for Very-Low-Voltage Applications," IEEE Transactions on Circuits and Systems II: Express Briefs, vol. 67, no. 7, pp. 1304-1308, Jul. 2020.
[24] A. Ballo, A. D. Grasso, G. Palumbo et al., "Charge pumps for ultra-low-power applications: Analysis, design, and new solutions." IEEE Transactions on Circuits and Systems II: Express Briefs, vol. 68, no. 8, pp. 2895-2901, Aug. 2021.
[25] S. Tanabe, Y. Sakamoto, et al., "A Hybrid Thermoelectric Generator – Battery Power Supply System Toward Replacement-Free Battery," International Conference on Power Electronics - ECCE Asia, May 2023.
[26] S. Tanabe, T. Tanzawa, "Battery-Assisted Battery Charger with Maximum Power Point Tracking for Thermoelectric Generator: Concept and Experimental Proof," Electronics (19), Sep. 2023.
[27] W. Saito, T. Tanzawa, "Design of a Battery Charger Boost Converter for Extremely Low Supply Voltages Below 10 mV in Energy Harvesting: A Return-On-Investment Approach", NEWCAS, Jun. 2025.
[28] T. Tanzawa, Chapter 5 Design of Battery Charger, "Fully-Integrated Power Management Circuits for Thermoelectric Energy Harvesting," Springer Nature (ISBN: 9783031597886), Jun. 2024.
[29] H. Kawauchi, T. Tanzawa, "A fully integrated clocked AC-DC charge pump for mignetostrictive vibration energy harvesting," Electronics, vol. 9, no. 12, Dec. 2020.
[30] N. Miyazaki, T. Tanzawa, "Design of Switched-Capacitor AC-DC Voltage Down-Converters for Micro-Watt Electrostatic Vibration Energy Harvesting," 23rd IEEE International NEWCAS Conference, Jun. 2025.
[31] R. Gariboldi, F. Pulvirenti, "A 70 mΩ Intelligent High Side Switch with Full Diagnostics," IEEE J. Solid-State Circuits, vol. 31, pp. 915–923, 1996.
[32] J. F. Dickson, "On-chip high-voltage generation in MNOS integrated circuits using an improved voltage multiplier technique," IEEE Journal of solid-state circuits, vol. 11, no. 3, pp. 374-378, 1976.
[33] T. Tanzawa, On-chip High-Voltage Generator Design, Springer, 2015.
[34] "Ceramic-packaged all-solid-state batteries & power supply module kit for evaluation", Maxell Ltd., [Online]. Available: https://biz.maxell.com/en/rechargeable_batteries/assb-spec-ceramicpackage.html [Accessed 28, Jun. 2025].

High-Efficiency Energy Extraction Interface for Piezoelectric Energy Harvesting

Chenghao Zhang [1], Junkai Chen [1], Jingjie Huang [2], Yue Shi [1,2]*, Zekun Zhou [1]*, Bo Zhang [1]

[1] State Key Laboratory of Electronic Thin Films and Integrated Devices, University of Electronic Science and Technology of China, Chengdu, China

[2] College of Communication Engineering, Chengdu University of Information Technology, Chengdu, China

*Email: october@cuit.edu.cn, zkzhou@uestc.edu.cn

Abstract—**Piezoelectric energy harvesting (PEH) offers significant advantages, including ubiquitous availability, structural flexibility, high integration density, and superior power conversion efficiency. These attributes render it a crucial enabling technology for addressing the energy supply challenges inherent in Internet of Things (IoT) networks and intelligent sensing systems, thereby holding substantial theoretical significance and practical value. However, the inherent alternating current (AC) characteristics of piezoelectric sources necessitate rectification via interface circuitry for practical energy harvesting. Conventional full-bridge rectifier topologies are fundamentally limited in achieving high-efficiency energy extraction due to significant charge loss incurred during voltage polarity reversal. Therefore, the implementation of specialized interface circuits is imperative for maximizing energy extraction efficiency. This paper comprehensively reviews state-of-the-art high-efficiency extraction techniques reported in the literature. Furthermore, these circuit design methodologies are systematically summarized and critically analyzed with respect to key performance metrics: extracted power capability, implementation cost, and system complexity. Finally, essential design considerations and practical implementation guidelines for piezoelectric energy harvesting systems are presented.**

Keywords—*Piezoelectric, energy harvesting, rectification stage, energy extraction*

I. INTRODUCTION

The advancement of next-generation information technologies, exemplified by the Internet of Things (IoT) and wireless sensor networks (WSNs), has underscored the limitations of conventional battery-powered solutions. These solutions struggle to meet the long-term operational requirements of distributed node systems due to finite energy capacity and high maintenance costs. Consequently, the demand for low-cost and stable power supplies has become increasingly critical [1-2]. Energy harvesting (EH) technology, which converts ambient renewable energy into electrical power, presents a viable alternative. It offers the potential for sustained and stable energy provisioning while significantly reducing the deployment and maintenance costs associated with traditional batteries [3-4]. Among various ambient energy sources, piezoelectric energy harvesting (PEH) stands out due to its ubiquitous and stable energy availability, design flexibility enabling diverse applications, high integration density facilitating miniaturization, and superior power conversion efficiency. These attributes confer significant theoretical importance and substantial practical value upon PEH systems [5-6].

Piezoelectric sources inherently generate alternating current (AC) output, necessitating rectification prior to energy utilization. The full-bridge rectifier (FBR), utilizing rectifying diodes for signal conditioning, was initially adopted for PEH interface circuits owing to its structural simplicity and ease of implementation [7-9]. However, the inherent two-diode forward voltage drop between the FBR's input and output terminals results in significant energy dissipation. To mitigate this limitation, research efforts have explored alternative techniques, including MOSFET-based active rectification, voltage double (VD), and switch-only rectifier (SOR). These approaches substantially reduce diode conduction losses, thereby enhancing rectification efficiency [10-12]. While effectively performing AC-to-DC conversion, these topologies suffer from a fundamental limitation: phase mismatch between the piezoelectric source voltage and current, caused by the inherent piezoelectric capacitance (C_P). This mismatch leads to substantial charge loss during voltage commutation events, severely degrading the system's energy extraction power.

To address this commutation loss, inductor-based rectification techniques, notably Synchronized Switch Harvesting on Inductor (SSHI) and Synchronous Electric Charge Extraction (SECE), have been proposed [13-15]. These methodologies leverage the energy storage characteristics of inductors to assist the commutation of C_P or directly act as an intermediate energy transfer bridge. Consequently, they effectively mitigate the charge loss stemming from voltage-current phase mismatch during commutation, enabling high-efficiency energy acquisition. Nevertheless, the energy conversion efficiency of inductor-based rectifiers is critically dependent on the inductor's quality factor (Q). The requirement for high-Q inductors, which are typically bulky and costly, presents a significant barrier to system miniaturization.

To circumvent the need for high-Q inductors, the Optimal Bias-Flip (OBF) technique has been developed [16-17]. This approach employs temporal and stepwise voltage flipping sequences, decomposing the single large-step resonant process inherent in traditional methods into multiple smaller steps. By reducing the peak inductor current during resonance, OBF achieves lower rectification losses and enables high-efficiency power extraction even with smaller, lower-Q inductors. An alternative strategy replaces the inductor with capacitors, embodied by the Synchronized Switch Harvesting on Capacitor (SSHC) technique [18-19]. The SSHC scheme utilizes flying capacitors to temporarily store energy from C_P before flipping their polarity and re-injecting the energy back into C_P, thereby achieving voltage inversion. Compared to inductor-based solutions, SSHC offers distinct advantages: it eliminates the requirement for high-Q inductors, benefits from the lower cost and higher integrability of capacitors, and simplifies control by avoiding the need for precise resonance period tuning. However, SSHC implementations still face challenges related to charge sharing losses and energy dissipation during C_P voltage reset at commutation, preventing ideal commutation efficiency.

This paper is dedicated to a comprehensive analysis of interface circuit implementations for piezoelectric energy harvesting. State-of-the-art PEH extraction techniques documented in the literature are systematically summarized, categorized, and critically analyzed. Concurrently, essential design considerations and guidelines for optimizing piezoelectric energy harvesting systems are presented.

II. REVIEW AND ANALYSIS INTERFACE OF PEH

Piezoelectric energy conversion leverages the fundamental property of piezoelectric materials wherein charge displacement occurs in response to applied mechanical stress. This phenomenon enables the transduction of ambient mechanical energy—including vibrations, impacts, and deformations—into electrical energy. Fig.1 illustrates the equivalent circuit model of a piezoelectric element operating under resonant conditions. A piezoelectric energy source can be accurately modeled as an AC source (I_P) connected in parallel with its inherent capacitance (C_P) and internal resistance (R_P). The resulting AC output exhibits characteristic high open-circuit voltage, low short-circuit current, and non-sinusoidal waveform characteristics. Given the inherent AC nature of the piezoelectric source, its output is incompatible with direct connection to subsequent DC-DC conversion stages. Consequently, a rectifying front-end interface circuit is essential for conditioning the generated AC power prior to its utilization or storage.

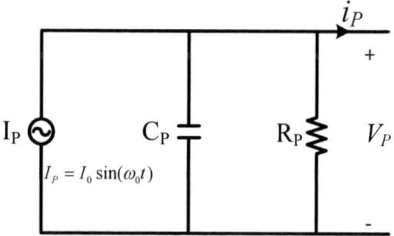

Fig.1 the equivalent circuit model of piezoelectric cell

A. Conventional FBR-based Interface Topology

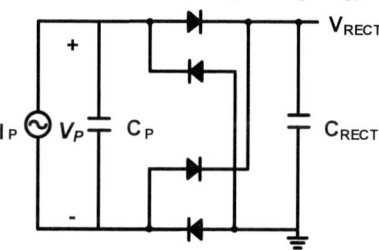

Fig.2 conventional FBR-based interface topology

Fig.2 depicts the conventional FBR-based interface topology. This approach directly couples the piezoelectric source to a standard diode bridge rectifier, employing a rectification capacitor (C_{RECT}) for energy storage. Fig.3 illustrates the voltage waveforms associated with this FBR interface circuit. Within one half-cycle of the piezoelectric current source (I_P), a single rectification cycle can be divided into two distinct phases: the piezoelectric capacitance (C_P) charging phase and the energy transfer phase. During charging phase immediately following a polarity reversal of I_P, the voltage across C_P (V_P) remains at the previous (negative) polarity. During this interval, I_P charges C_P until V_P reaches the voltage level necessary to enable forward conduction of the rectifier diodes and initiate energy transfer to the output. In transfer phase once V_P reaches the conduction threshold, I_P flows through the rectifier bridge, delivering energy to the output stage until the subsequent polarity reversal of I_P. The output power (P_{OUT}) delivered by the FBR-based interface can be expressed as:

$$P_{OUT,FBR} = 4C_P f_p V_{RECT}(V_{OC,P} - V_{RECT} - 2V_D) \quad (1)$$

Where f_P represents the vibration source frequency, V_{RECT} denotes the rectified output voltage, V_{OC} signifies the open-circuit voltage of the piezoelectric source, and V_D corresponds to the forward voltage drop of each rectifier diode.

Maximizing P_{OUT} is achieved by taking the derivative of Equation (1) with respect to V_{RECT} and equating it to zero:

$$\frac{dP_{OUT,FBR}}{dV_{RECT}} = 4C_P f_P(V_{OC} - 2V_D - 2V_{RECT}) = 0 \quad (2)$$

Solving Equation (2) yields the optimal rectified output voltage ($V_{RECT,OPT}$) for maximum power extraction:

$$V_{RECT,OPT} = \frac{V_{OC} - 2V_D}{2} \quad (3)$$

Substituting $V_{RECT,OPT}$ back into Equation (1) provides the maximum achievable output power:

$$P_{OUT,FBR,MAX} = C_P f_P(V_{OC} - 2V_D)^2 \quad (4)$$

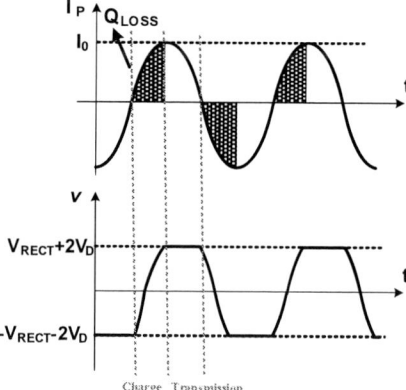

Fig.3 the voltage waveforms associated with FBR interface

Equation (4) reveals a fundamental limitation inherent in the full-bridge rectifier (FBR) topology: although capable of piezoelectric energy extraction, its maximum output power is severely constrained by the significant forward voltage drop (V_D) across the rectifier diodes. To mitigate this limitation, enhanced rectification techniques have been proposed. [10] replaces passive diodes with actively controlled MOSFETs, typically driven by comparators significantly reducing the effective voltage drop. [11-12] employ half-wave rectification utilizing voltage multiplication to achieve output power levels and drop reduction.

B. Inductor/Capacitor-based interface topology

The FBR-based interface topology facilitates the extraction of piezoelectric energy efficiently. However, due to the inherent capacitance C_P, the voltage and current cannot change in phase, leading to a situation where, after current reversal, the capacitance must be charged, resulting in a loss of nearly half of the charge. This reduces the overall power extraction efficiency. In [13], the SSHI technique was proposed to mitigate this issue by paralleling the inductor with the inherent capacitance C_P during current reversal. This configuration leverages an LC resonant circuit to assist in the voltage reversal of C_P, thus minimizing charge loss during the charging process. Fig.4 and Fig.5 respectively illustrate the SSHI circuit topology and voltage waveform. The operating principle of this structure is similar with that of the FBR structure, with the key difference being the addition of a "flip" stage during current reversal. During the current reversal, switch S_1 closes, and the inductor L forms an LC oscillatory circuit with the inherent capacitance C_P, which enables voltage reversal of C_P through half-cycle resonance. The output power P_{OUT} of the SSHI interface topology can be expressed as:

$$P_{OUT,SSHI} = 2C_P f_P V_{RECT}[2V_{OC} - (1-\eta)(V_{RECT} + 2V_D)] \quad (5)$$

Where η denotes the flip efficiency of the LC resonant circuit. Similarly, the output power can be derived to reach its maximum when $V_{RECT} = V_{OC}/(1-\eta) - V_D$. The maximum power can then be expressed as:

$$P_{OUT,SSHI,MAX} = 2C_P f_P (1-\eta)(\frac{V_{OC}}{1-\eta} - V_D)^2 \quad (6)$$

From Equation (6), it is evident that the piezoelectric extraction power in the SSHI structure is positively correlated with the flip efficiency during the flip phase. The higher the flip efficiency, the greater the extracted power. It is noteworthy that when $\eta = -1$ the flip phase voltage remains unchanged, and Equation (6) degenerates into Equation (4), indicating that the FBR structure can be considered as a simplified version of the SSHI structure. Compared to the FBR structure, the SSHI configuration utilizes inductor resonance during current reversal to assist the inherent capacitance C_P in voltage reversal, thereby reducing charge loss during the charging phase and ultimately enhancing the power extraction efficiency.

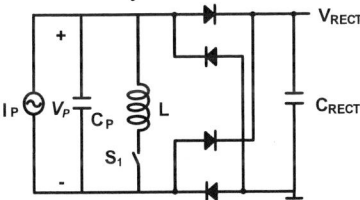

Fig.4 the SSHI interface topology

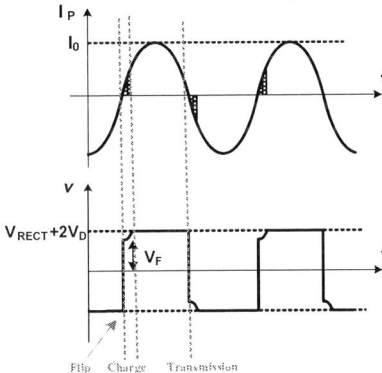

Fig.5 voltage waveform of SSHI interface

The inductor-based interface topology can leverage inductor resonance to accelerate the polarity reversal of the inherent capacitance, thereby reducing charge loss during the charging process. However, it requires high-Q inductive components and precise resonance cycle control to ensure

high flip efficiency. On one hand, high-Q inductors are expensive and bulky, which hampers the miniaturization of the system; on the other hand, precise resonance control necessitates the design of independent detection and control circuits, increasing the system's complexity. To overcome the limitations of the inductor-based rectifier interface topology, [18] proposed the SSHC technique, which uses flying capacitor arrays to replace inductor for the polarity reversal of C_P. Fig.6 and Fig.7 illustrate the SSHC circuit topology and voltage waveform, respectively.

The operating logic of this structure is similar with that of SSHI, with the addition of a flip phase during current reversal. The flip phase can be divided into three processes: In the first stage, switches Φ_N to Φ_1 sequentially turn on, connecting the flying capacitors C_N to C_1 in parallel with the inherent capacitance C_P. This allows charge sharing between the capacitors and stores the energy from C_P in the flying capacitors. In the second stage, switch Φ_0 turns on, resetting the voltage across C_P. In the third stage, switches Φ_{-1} to Φ_{-N} sequentially turn on, causing the flying capacitors C_1 to C_N to reverse polarity and be connected in parallel with C_P, thereby achieving the polarity reversal of the inherent capacitance.

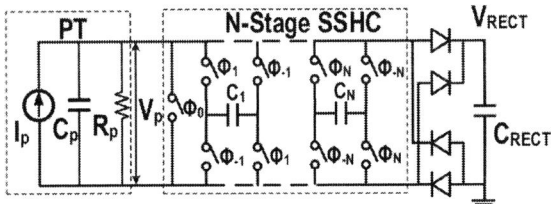

Fig.6 the SSHC interface topology

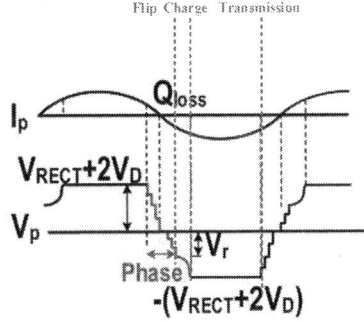

Fig.7 voltage waveform of SSHC interface

Similar with the SSHI structure, the voltage ratio of the inherent capacitance before and after the flip phase is defined as the flip efficiency η. The output power of the SSHC structure reaches its maximum when $V_{RECT} = V_{OC}/(1-\eta) - V_D$ and can be expressed as:

$$P_{OUT,SSHC,MAX} = 2C_P f_P (1-\eta)(\frac{V_{OC}}{1-\eta} - V_D)^2 \quad (7)$$

Compared to SSHI, SSHC uses multiple sets of flying capacitors to perform current reversal, thus avoiding the high cost and integration challenges associated with high-Q inductor. Additionally, capacitor charge sharing does not require precise resonance cycle control, reducing the complexity of the circuit design. However, the SSHC technique also has its drawbacks. Effective voltage reversal requires multiple stages of flying capacitors in the nF range,

979-8-3315-3918-4/25 $31.00 © 2025 IEEE

similar in magnitude to the inherent capacitance, making it difficult to achieve monolithic integration. Furthermore, although capacitor charge sharing does not require precise cycle control, the presence of multiple phases during the flip stage makes the design of switching timing for SSHC more complex.

III. DESIGN CONSIDERATIONS FOR ENERGY EXTRACTION INTERFACE FOR PEH

From the above analysis, several key factors must be considered when designing an extraction interface for PEH: high efficiency, low cost, and high integration.

For FBR-based interfaces, they offer significant advantages such as simple implementation, low cost, no reliance on external components, and ease of integration. Furthermore, these interface circuits do not require complex switching frequency control or precise phase synchronization algorithms, which results in high system reliability and operational stability. However, they cannot resolve the issue of voltage and current desynchronization caused by the inherent capacitance, leading to substantial charge loss during current reversal and, consequently, low overall power extraction.

For inductor/capacitor-based interfaces, the energy storage characteristics of inductors and capacitors are utilized during current reversal to assist with the switching of the inherent capacitance. This significantly reduces charge loss during the reversal process and greatly improves piezoelectric power harvesting efficiency. However, these interfaces also face challenges such as complex control, high circuit design difficulty, reliance on external components, and higher system costs, all of which limit their integration potential.

IV. CONCLUSION

A thorough review of extraction technologies for PEH is presented in this paper. The analysis primarily focuses on three distinct interface circuit topologies, detailing their respective operational principles and critically evaluating their comparative advantages and disadvantages. These design methodologies are systematically analyzed and summarized with respect to key performance metrics: power extraction efficiency, implementation cost, and integration complexity. FBR-based interfaces offer structural simplicity and eliminate the need for off-chip components. However, they are characterized by relatively low power extraction efficiency. Inductor/capacitor-based interfaces achieve superior power extraction efficiency. Nevertheless, this improvement necessitates the use of off-chip reactive components and entails increased design complexity. Given these persistent limitations across existing topologies, the development of high-performance PEH interface circuits remains an area demanding extensive ongoing research efforts.

V. ACKNOWLEDGENT

This work was supported in part by the National Natural Science Foundation of China under Grant 62074028, in part by the Sichuan Natural Science Foundation under Grant 23NSFSC0359, in part by the Chunhui Cooperative Research Program of the Ministry of Education of China under Grant HZKY20220583.

REFERENCES

[1] AL-FUQAHAA, GUIZANIM, MOHAMMADIM, etal. Internet of things: A survey on enabling technologies, protocols, and applications[J]. IEEE communications surveys & tutorials, 2015, 17(4): 2347-2376.

[2] TAN Y K, PANDA S K. Review of energy harvesting technologies for sustain able wireless sensor network[J]. Sustainable wireless sensor networks, 2010, 2010: 15-43.

[3] LU M, FUG, OSMANNB, etal. Green energy harvesting strategies on edge based urban computing in sustainable internet of things[J]. Sustainable cities and society, 2021, 75: 103349.

[4] LIU X, ANSARI N. Toward green iot: Energy solutions and key challenges[J]. IEEE Communications Magazine, 2019, 57(3): 104-110

[5] IKPEHAI A, ADEBISI B, RABIE K M, et al. Low-power wide area network technologies for internet-of-things: A comparative review[J]. IEEE Internet of Things Journal, 2018, 6(2): 2225-2240.

[6] KWON D, RINCÓN-MORA G A. A single-inductor 0.35 μm cmos energy investing piezoelectric harvester[J]. IEEE Journal of Solid-State Circuits, 2014, 49(10): 2277-2291.

[7] OTTMAN G K, HOFMANN H F, BHATT A C, et al. Adaptive piezoelectric energy harvesting circuit for wireless remote power supply[J]. IEEE Transactions on power electronics, 2002, 17(5): 669-676.

[8] JEON Y, SOOD R, JEONG J H, et al. Mems power generator with transverse mode thin film pzt [J]. Sensors and Actuators A: Physical, 2005, 122(1): 16-22.

[9] PETERSC, ORTMANNSM, MANOLIY. Low power high performance voltage rectifier for autonomous microsystems[C]//Proceedings of Power MEMS. 2007: 217-220.

[10] GUO S, LEE H. An efficiency-enhanced integrated cmos rectifier with comparator-controlled switches for transcutaneous powered implants[C]//2007 IEEE Custom Integrated Circuits Conference. IEEE, 2007: 385-388.

[11] LETT, HANJ, VONJOUANNEA, etal. Piezoelectric micro-power generation interface circuits[J]. IEEE journal of solid-state circuits, 2006, 41(6): 1411-1420.

[12] RAMADASSYK, CHANDRAKASANAP. An efficient piezoelectric energy harvesting interface circuit using a bias-flip rectifier and shared inductor[J]. IEEE journal of solid-state circuits, 2009, 45(1): 189-204.

[13] LEFEUVRE E, BADEL A, RICHARD C, et al. Piezoelectric energy harvest ing device optimization by synchronous electric charge extraction[J]. Journal of intelligent material systems and structures, 2005, 16(10): 865-876.

[14] GARBUIOL, LALLARTM, GUYOMARD, etal. Mechanical energy harvester with ultralow threshold rectification based on sshi nonlinear technique[J]. IEEE Transactions on Industrial Electronics, 2009, 56(4): 1048-1056.

[15] LALLART M, RICHARD C, GARBUIO L, et al. High efficiency, wide load bandwidth piezoelectric energy scavenging by a hybrid nonlinear approach[J]. Sensors and Actuators A: Physical, 2011, 165(2): 294-302.

[16] ZHAO Y, LIANG J. Synchronized triple bias-flip circuit for piezoelectric energy harvesting enhancement: Operation principle and experimental validation[C]// 2016 IEEE Energy Conversion Congress and Exposition (ECCE). IEEE, 2016: 1-6.

[17] LIANG J, ZHAO Y, ZHAO K. Synchronized triple bias-flip interface circuit for piezoelectric energy harvesting enhancement[J]. IEEE Transactions on Power Electronics, 2018, 34(1): 275-286.

[18] DUS, SESHIA A A. An inductor-less bias-flip rectifier for piezoelectric energy harvesting[J]. IEEE Journal of Solid-State Circuits, 2017, 52(10): 2746-2757.

[19] ÇIFTCI B, CHAMANIAN S, ULUŞAN H, et al. Low-cost fully autonomous piezoelectric energy harvesting interface circuit with up to 6.14 x power capacity gain[C]//2019 IEEE Custom Integrated Circuits Conference (CICC). Austin, TX, USA: IEEE, 2019: 1-4

PWM Scheme Selection Strategy for Fast Ramp-Up DC-DC Boost Converters in SSD Applications

Yuji Kanayama[1], Toru Tanzawa[2], *Fellow, IEEE*

[1] Shizuoka University, Hamamatsu, Japan, [2] Waseda University, Kitakyushu, Japan

Abstract— **How to control the pulse-width modulation (PWM) to minimize the ramp time for generating high programming voltages has not been discussed in detail. This paper therefore focuses on the design of the duty cycle control scheme to minimize the ramp time, particularly for NAND Flash in solid-state drive (SSD) applications. The PWM must be designed under conditions where the capacitive load (C_L) of the boost converters varies depending on the number of NAND dies selected for simultaneous programming, and the rated inductor current (I_R) varies depending on the inductor used in the boost converter. The ramp-up time is estimated using models for four PWM schemes: 1) maximizing output current at any output voltage (Io max), 2) maintaining the inductor current at I_R (I_R limit), 3) using a constant duty ratio (Const.D), and 4) operating in boundary current mode (BCM). An interesting result was that each of the four extreme cases, in terms of C_L and I_R, had a different optimal PWM scheme. The model results are compared with SPICE simulations and measured results using 24 V transistors in a 250 nm BCD process. To minimize ramp-up time across a wide range of load capacitances, PWM scheme selection strategy is proposed that selects one of two or three PWM schemes depending on the number of NAND chips being programmed simultaneously.**

Keywords— *High speed, Boost converter, Pulse width modulation, NAND Flash, SSD*

I. INTRODUCTION

The power consumption of datacenter globally has been increasing rapidly. AI datacenter accelerates this trend. Reduction in the power consumption of datacenter is one of the critical challenges for sustainability of the globe [1] – [4]. Because of its lower power and faster data access, Flash-based solid-state drives (SSDs) have been replacing traditional hard-disk drives (HDDs) [5] – [7]. NAND Flash memory is the densest non-volatile semiconductor memory [8], [9]. The cell structure has been changed from 2D to 3D. The data capacity per die has increased from 128 Gb [10] to 1 Tb [11] over ten years. Thus, the bit cost of SSDs has become competitive to that of HDDs.

To alter the data of NAND Flash, a high voltage of 20 V is required for the Flash cell to flow the tunneling current through the gate insulator. The initial applications of NAND Flash were to store the data of pictures taken by a digital camera and digital files with limited data sizes. As a result, a boost converter to generate the high voltage needed to be integrated into the same NAND die. Charge pumps using switches and capacitors meet such requirement with low area overhead [12] – [14]. Improvement of area efficiency was the priority to reduce the bit cost for single-die applications [15], [16]. After SSDs started replacing HDDs in personal computers and datacenters, the priority on reduction in power consumption became primal as well as the cost. Multiple NAND Flash chips are integrated into SSDs to store large volumes of data. Thus, more power-efficient switching boost converters using an inductor and switches, commonly used for multiple NAND chips in a single SSD, have been developed for reducing power consumption of SSDs [17] – [20]. The ramp-up time required to generate a high programming voltage is an overhead to program performance, because the programming voltage must be generated each time a program command is issued. Several design techniques have been proposed in the literature, but to the best authors' knowledge, how to control the pulse-width modulation (PWM) for minimizing the ramp time to generate the high programming voltages has not been discussed in detail. This paper therefore proposes PWM scheme selection strategy to minimize the ramp time. The discussion focuses on the transient state rather than the steady-state conditions of the boost converter. The outcome was that the optimum control scheme depends on the capacitive load. The number of NAND dies operating simultaneously can vary in SSDs. Thus, the duty cycle controller needs to function differently depending on the number of NAND dies. The paper is organized as below. Section II presents four schemes to be compared by load capacitance and inductor's rated current. Section III shows measured results and comparison with the model calculations and SPICE simulations.

II. FOUR DUTY CYCLE CONTROL SCHEMES

Fig. 1 (a) illustrates a simple circuit model of switching boost converters [17]. Duty cycle controller (DCC) generates a pulse Φ to drive the pull-down NMOSFET.

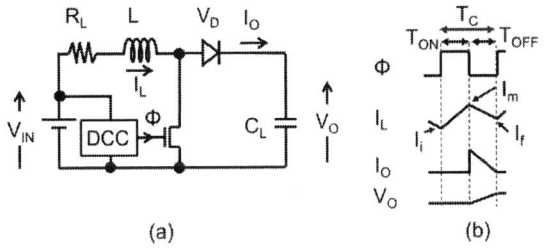

(a)   (b)

Fig. 1 Circuit model (a) and operation waveform (b) of a standard boost converter.

Fig. 1 (b) shows the waveforms of the inductor current I_L, the output current I_O, and the output volage V_O. I_L increases from I_i to I_m during T_{ON} and decreases from I_m to I_f during T_{OFF}. I_O is equal to I_L during T_{OFF}. T_C is a cycle time that is assumed to be a constant during ramp-up. R_L is the parasitic resistance of the inductor. For simplicity in the calculation, it is assumed that the ON resistance of the NMOSFET (R_M) is identical to that of the diode (R_D). V_D is a diode voltage drop, and C_L is a capacitive load.

A. General equations for the ramp-up operation

Equations (1) and (2) hold during T_{ON} and T_{OFF}, respectively, where $R_S = R_L + R_M = R_L + R_D$. It is assumed that the switching loss due to the parasitic capacitance of the MOSFET and diode is minor compared to the conduction loss, and V_O is considered constant in every cycle.

$$\frac{dI_L}{dt} = \frac{V_{IN} - R_S I_L}{L} \quad (1)$$

$$\frac{dI_L}{dt} = \frac{V_{IN} - V_O - V_D - R_S I_L}{L} \quad (2)$$

Equation (1) can be analytically solved as (3) using an initial current I_i.

$$I_L = \frac{V_{IN}}{R_S}\left(1 - e^{-\frac{R_S}{L}t}\right) + I_i e^{-\frac{R_S}{L}t} \quad (3)$$

The average output current I_O is calculated by (4) and can result in an analytical form (but not shown here). Then, V_O is slightly increased by ΔV_O, as shown in (5).

$$I_o = \frac{1}{T_C}\int_0^{T_C} I_o(t)dt = \frac{1}{T_C}\int_{T_{ON}}^{T_C} I_L(t)dt \quad (4)$$

$$\Delta V_O = \frac{I_o T_C}{C_L} \quad (5)$$

One can numerically predict the transient behavior of V_O using the above equations with any given duty ratio D, defined by T_{ON}/T_C. To obtain a simple equation to provide optimum control for D to minimize the ramp time, or in other words, to maximize I_O, a linear approximation for I_L is made. Using (6) and (7) instead of (1) and (2), respectively, I_L and I_O are analytically given by (8) and (9), respectively, where $I_f = I_i := I_L$ and $R_S T_C/L << 1$ are assumed; the former indicates that the inductor current stays at a constant current with a small ripple and the latter does that the inductor current changes linearly in T_{ON} and T_{OFF}.

$$L(I_m - I_i)/T_{ON} = V_{IN} - R_S I_i \quad (6)$$

$$L(I_f - I_m)/T_{OFF} = V_{IN} - V_O - V_D - R_S I_m \quad (7)$$

$$I_L = \{V_{IN} - (1 - D)(V_O + V_D)\}/R_S \quad (8)$$

$$I_O = \frac{T_{OFF}}{T_C} \times \frac{I_m + I_f}{2} = (1 - D) I_L \quad (9)$$

B. I_O max scheme

From (8), (9), I_O is maximized when D is controlled as in (10), where the first derivative of I_O with respect to D becomes zero.

$$D = 1 - V_{IN}/2(V_O + V_D) \quad (10)$$

The optimum I_L and I_O result in (11) and (12), respectively, with (10). Equation (11) indicates that half the input power is owned by L and another half is dissipated by heat via R_S.

$$I_{L\_OPT} = V_{IN}/2R_S \quad (11)$$

$$I_{O\_MAX} = V_{IN}^2/4R_S(V_O + V_D) \quad (12)$$

Thus, the "I$_O$ max" scheme is such an operation that the inductor current increases with $D = 1$ until I_L reaches I_{L_OPT}, and then D is controlled to meet (10) until V_O reaches the target voltage V_{PP}, as illustrated in Fig. 2 (a). The rise time for the former operation (T_{R1}) is determined by (3) and $I_i = 0$, resulting in (13).

$$T_{R1} = L \ln2/R_S \quad (13)$$

The rise time for the latter operation (T_{R2}) is calculated using (12) and (14), resulting in (15).

$$C_L dV_O/dt = I_{O\_MAX} \quad (14)$$

$$T_{R2} = 2C_L R_S\{V_{PP}^2 - V_{IN}^2 + 2V_D(V_{PP} - V_{IN})\}/V_{IN}^2 \quad (15)$$

The entire ramp time T_R is then given by (16).

$$T_R = T_{R1} + T_{R2} \quad (16)$$

The total input and output energies in T_R, E_{IN} and E_O, are calculated using the above equations to be (17) and (18), respectively. The average power efficiency η is given by (19).

$$E_{IN} = (T_{R1}^2/L + T_{R2}/R_S) V_{IN}^2/2 \quad (17)$$

$$E_O = C_L (V_{PP}^2 - V_{IN}^2)/2 \quad (18)$$

$$\eta = E_O/E_{IN} \quad (19)$$

It is assumed in (18) that the load capacitor is precharged to V_{IN} with a sufficiently short time.

C. I_R limit scheme

The Io max scheme is possible when $I_{O\_OPT} < I_R$, where I_R is the rated current of the inductor. When $I_{O\_OPT} > I_R$, I_L needs to be controlled at I_R or lower. This constrains D to meet (8) with $I_L = I_R$, resulting in (20).

$$D = 1 - (V_{IN} - R_S I_R)/(V_O + V_D) \quad (20)$$

The "I$_R$ limit" scheme is such an operation that the inductor current increases with $D = 1$ until I_L reaches I_R, and then D is controlled to be D with (20) until V_O reaches V_{PP}, as illustrated in Fig. 2 (a). T_{R2} and E_{IN} are calculated by (21) and (22), respectively. T_R and η are given by (13), (16), (21) and (18), (19), (22), respectively.

$$T_{R2} = C_L\{V_{PP}^2 - V_{IN}^2 + 2V_D(V_{PP} - V_{IN})\}/\{2I_R(V_{IN} - R_S I_R)\} \quad (21)$$

$$E_{IN} = V_{IN}^2 T_{R1}^2/2L + V_{IN} I_R T_{R2} \quad (22)$$

D. Const.D and BCM schemes

For comparison with the Io max and IR limit schemes, two other simple control schemes are also considered; Const.D and BCM schemes that operate the boost converter with a constant D and in boundary current mode, as illustrated in Fig. 2 (b) and (c), respectively. The BCM scheme is such an operation that 1) the inductor current increases with $D = 1$ until I_L reaches I_R in phase 1, then 2) the inductor current decreases with $D = 0$ until I_L reaches 0 in phase 2, and 3) the operations 1 and 2 are repeated until V_O reaches V_{PP}.

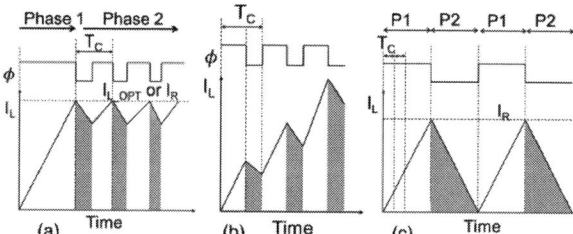

Fig. 2 Four PWM schemes; (a) Io max or IR limit, (b) Const.D, (c) BCM.

E. Comparison of the four schemes

In this paper, the design and device parameters shown in Table I are used for demonstration. I_{L_OPT} is 100 mA based on (10). Various conditions for C_L and I_R are considered to see if the optimum PWM scheme depends on those conditions.

Table I Design specifications and device parameters for demonstration.

	Param.	Value
Design spec.	V_{IN}	2.0 V
	V_{PP}	20 V
	T_C	10 μs
Device param.	V_D	0.3 V
	R_D	2.0 Ω
	L	1.0 mH
	R_L	8.0 Ω
	R_M	2.0 Ω
	C_L	10 nF, 100 nF, 1μF
	I_R	40, 80, 160 mA

Fig. 3 (a) shows T_R and its breakdown of T_{R1} and T_{R2} as a function of a steady inductor current I_{L_OP} at C_L of 100 nF, using the above model equations. This graph compares the Io max scheme at I_{L_OPT} = 100 mA with the I_R limit scheme at $I_{L\_OPT} \neq$ 100 mA. T_{R2} is minimized at I_{L_OPT} of 100 mA while T_{R1} is a monotonically increasing function. As a result, T_R is minimized when I_{L_OP} is slightly lower than I_{L_OPT}. As shown in Fig. 3(b), η is a monotonically decreasing function of I_{L_OP} because the conduction loss increases monotonically. Therefore, a practical design must take the trade-off into account. Fig. 3 (c) is based on the data points in Fig. 3 (b). I_{L_OP} of 80 mA minimizes T_R at a low η of 50% while I_{L_OP} of 40 mA increases T_R by 24% with a much higher η of 77%. In the following discussion, we will focus on the minimum T_R.

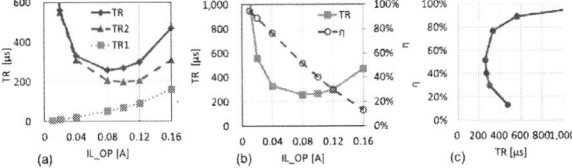

Fig. 3 (a) T_{R1}, T_{R2}, T_R vs. I_{L_OP}, (b) T_R and PCE vs. I_{L_OP}, (c) PCE vs. T_R, at C_L = 100 nF, using the models.

Fig. 4 shows transient behavior of D for I_O max (a), I_R limit (b), Const. D (c), BCM (d) at I_R = 160 mA and C_L = 100 nF. Fig. 5 (a) overlays V_O in the four schemes to see how the ramp speed is different among the schemes. Fig. 5 (b) summarizes the best scheme by $C_L - I_R$ condition. This result suggests that it is necessary to select one of two or three PWM schemes, depending on the number of NAND dies being programmed simultaneously, as discussed in Section II. Figs. 6 (a) and (b) compare the ramp time with different schemes at I_R = 40 mA and 160 mA, respectively. When I_R = 40 mA, the optimum scheme is the Const.D scheme for 10 nF and the I_R limit scheme for 100 nF and 1 μF. On the other hand, when I_R = 160 mA, the optimum scheme is the BCM scheme for 10 nF, the Const.D scheme for 100 nF, and the I_O max scheme for 1 μF. When an adaptive PWM scheme is implemented in an SSD controller with BCM for light load as 10 nF and Const.D for heavy load as 100 nF, the ramp-up time can be reduced by 30% in comparison with a non-adoptive PWM scheme, based on Fig. 6(b). Fig. 7 illustrates a proposed concept of PWM scheme selection strategy. NAND controller knows how many dies are going to execute program operation. Thus, selection signals S<1:0> enable the optimum PWM scheme. Table II compares T_R with the model calculations, normalized to SPICE simulation results. Among the thirty conditions, there was no condition where the error was larger than 24%.

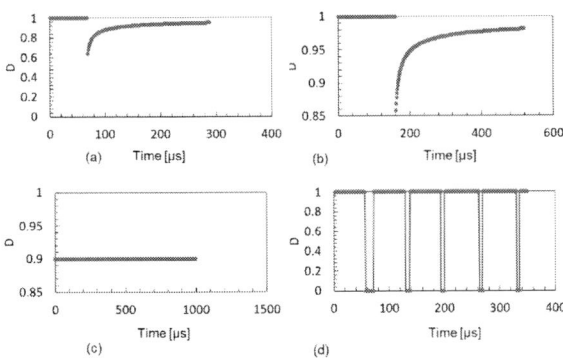

Fig. 4 Transient behavior of D for I_O max (a), I_R limit (b), Const. D (c), BCM (d) at I_R = 160 mA and C_L = 100 nF.

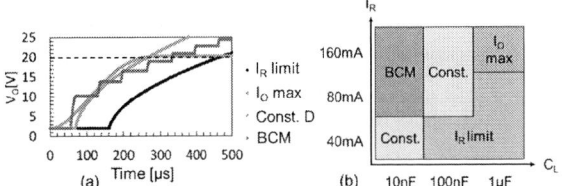

Fig. 5(a) Transient behavior for the four operation modes at I_R = 160 mA and C_L = 100 nF, and (b) the fastest operation mode by the condition on I_R and C_L.

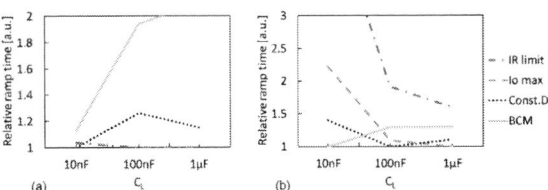

Fig. 6 Relative ramp time normalized by the fastest one at I_R = 40 mA (a) and 160 mA (b).

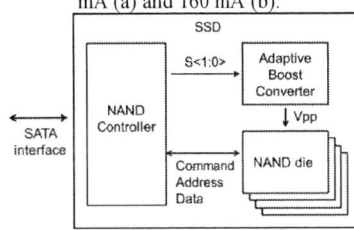

Fig. 7 Proposed concept of PWM scheme selection strategy

Table II Ratio of T_R with the model calculations to that with the SPICE simulation results.

I_R	Scheme	T_R (SPICE)			T_R (SPICE)/TR (Model)		
		10nF	100nF	1μF	10nF	100nF	1μF
40mA	I_R limit	59us	348us	3215us	105%	101%	99%
	I_O max						
	Const.D	59us	401us	3608us	109%	92%	97%
	BCM	62us	571us	5654us	102%	86%	83%
80mA	I_R limit	76us	287us	2370us	104%	106%	105%
	I_O max						
	Const.D	59us	284us	2433us	105%	105%	105%
	BCM	42us	370us	3406us	105%	100%	93%
160mA	I_R limit	201us	573us	4338us	104%	116%	124%
	I_O max	92us	296us	2366us	103%	105%	108%
	Const.D	59us	271us	2615us	105%	105%	108%
	BCM	42us	334us	3027us	105%	100%	106%

III. MEASUREMENT

To validate the four schemes, a boost converter was set up using 24 V transistors fabricated in 250 nm BCD process, a discrete Schottky diode (STPS5L25B-TR from STMicroelectronics), and an inductor (1410516C from Murata), as depicted in Fig. 8. A programmable pulse generator (PG) provides the gate signal Φ to the pull-down NMOS with a duty D as shown in Fig. 4.

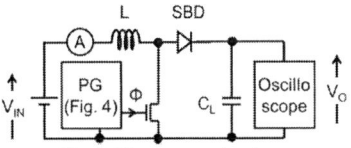

Fig. 8 Measurement set-up

Figs. 9 and 10 show the waveforms for V_{OUT} and I_L in the I_O max and I_R limit schemes.

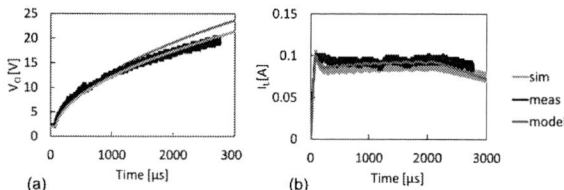

Fig. 9 Transient waveforms of V_{OUT} and I_L with I_O max scheme for I_R = 160 mA and C_L = 1 μF.

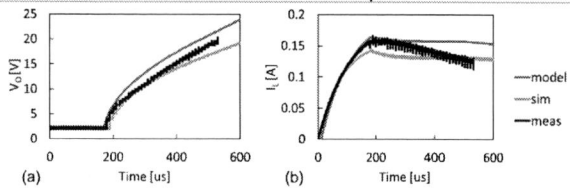

Fig. 10 Transient waveforms of V_{OUT} and I_L with I_R limit scheme for I_R = 160 mA and C_L = 100 nF.

Fig. 11 compares T_R between the model caluclations, SPICE simulation results and measured data in the I_O max and I_R limit schemes at C_L = 100 nF. The optimum I_{L_OP} was not matched between the model and measured results. However, the fastest operation scheme based on the measurement by the condition on I_R and C_L, as shown in Fig. 12, matched the model prediction shown in Fig. 5 (b) except for the condition of I_R = 160 mA and C_L = 100 nF. Therefore, the measured data confirmed that it is necessary to select one of two or three PWM schemes to minimize the ramp time, depending on the number of NAND dies being programmed simultaneously.

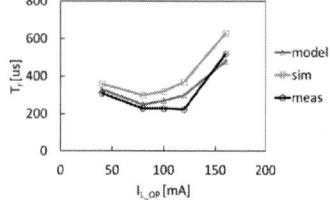

Fig. 11 T_R in the I_O max and I_R limit schemes at C_L = 100 nF.

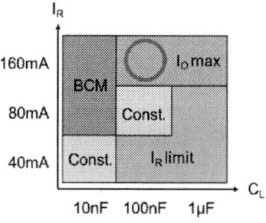

Fig. 12 Fastest operation scheme based on the measurement by the condition on I_R and C_L.

IV. CONCLUSION

Four PWM schemes were compared to identify the best option for minimizing the ramp time under a wide range of inductor rated current and load capacitance conditions. The model calculation results were compared with SPICE simulations and measured results. The model calculations showed good agreement with the SPICE results, with a maximum deviation of 24%. The measured data confirmed that an adaptive PWM scheme is effective for fast ramp-up operations, as it selects one of two or three PWM schemes depending on the number of NAND chips being programmed simultaneously.

ACKNOWLEDGEMENT

This work is supported by dlab-VDEC, Synopsys Inc., Cadence Design Systems Inc., Siemens EDA, TSMC Ltd., and TSMC Design Technology Japan, Inc.

REFERENCES

[1] Pelley, S., Meisner, D., Wenisch, T. F. et al., "Understanding and abstracting total data center power," In *Workshop on Energy-Efficient Design* (Vol. 11, pp. 1-6), Jun. 2009.
[2] Liu, L., Wang, H., Liu, X., et al., "GreenCloud: a new architecture for green data center," In *Proceedings of the 6th international conference industry session on Autonomic computing and communications industry session* (pp. 29-38), Jun. 2009.
[3] Zhang, Q., Meng, Z., Hong, X.,et al., "A survey on data center cooling systems: Technology, power consumption modeling and control strategy optimization, " *Journal of Systems Architecture, 119*, 102253, 2021.
[4] Zhang, Z., Zeng, Y., Liu, H., et al., "Smart DC: an AI and digital twin-based energy-saving solution for data centers," In *NOMS 2022-2022 IEEE/IFIP Network Operations and Management Symposium* (pp. 1-6), Apr. 2022.
[5] Carniel, A. C., Ciferri, R. R., & de Aguiar Ciferri, C. D. "Analyzing the performance of spatial indices on hard disk drives and flash-based solid state drives," *Journal of Information and Data Management, 8*(1), 34-34, 2017.
[6] Liang, S., Qiao, Z., Hochstetler et al., "Reliability characterization of solid state drives in a scalable production datacenter," In *2018 IEEE International Conference on Big Data (Big Data)* (pp. 3341-3349), Dec. 2018.
[7] K. Eshghi and R. Micheloni, "Chap. 2 SSD Architecture and PCI Express Interface" in "Inside solid state drives (SSDs)", Springer, 2018.
[8] Masuoka, F., Momodomi, M., Iwata, Y., & Shirota, R. "New ultra high density EPROM and flash EEPROM with NAND structure cell," In *1987 International Electron Devices Meeting* (pp. 552-555), Dec. 1987.
[9] Aritome S. "NAND flash memory technologies," John Wiley & Sons; Dec. 2015.
[10] K. Park, J. Han, D. Kim et al., "19.5 Three-dimensional 128Gb MLC vertical NAND Flash-memory with 24-WL stacked layers and 50MB/s high-speed programming," 2014 IEEE International Solid-State Circuits Conference Digest of Technical Papers (ISSCC), pp. 334-335, 2014.
[11] W. Jung, H. Kim, D. Kim et al., "13.3 A 280-Layer 1Tb 4b/cell 3D-NAND Flash Memory with a 28.5Gb/mm2 Areal Density and a 3.2GB/s High-Speed IO Rate," 2024 IEEE International Solid-State Circuits Conference (ISSCC), pp. 236-237, 2024.
[12] Dickson, J. F., "On-chip high-voltage generation in MNOS integrated circuits using an improved voltage multiplier technique," *IEEE Journal of solid-state circuits, 11*(3), 374-378, 1976.
[13] Gariboldi, R.; Pulvirenti, F. A 70 mΩ Intelligent High Side Switch with Full Diagnostics. IEEE J. Solid-State Circuits, 31, 915–923, 1996.
[14] Ballo, A., Grasso, A. D., Palumbo, G. et al., "Charge pumps for ultra-low-power applications: Analysis, design, and new solutions," *IEEE Transactions on Circuits and Systems II: Express Briefs, 68*(8), 2895-2901, 2021.
[15] Tanzawa, T., Tanaka, T., Takeuchi, K., & Nakamura, H., "Circuit techniques for a 1.8-V-only NAND flash memory," *IEEE Journal of Solid-State Circuits, 37*(1), 84-89, 2002.
[16] Tanzawa, T., Murakoshi, T., Kamijo, et al, "Design challenge in 3D NAND technology: A 4.8 X area-and 1.3 X power-efficient 20V charge pump using tier capacitors," In *2016 IEEE Asian Solid-State Circuits Conference (A-SSCC)* (pp. 165-168), Nov. 2016.
[17] Ishida, K., Yasufuku, T., Miyamoto, S et al., "1.8 V low-transient-energy adaptive program-voltage generator based on boost converter for 3D-integrated NAND flash SSD," *IEEE journal of solid-state circuits, 46*(6), 1478-1487, 2011.
[18] Hatanaka, T., Johguchi, K., & Takeuchi, K., "Experimental investigation of program voltage (20 V) generation with boost converter for 3-D-stacked NAND flash SSD," *IEEE Transactions on Components, Packaging and Manufacturing Technology, 5*(2), 188-193, . 2015).
[19] Lee, K., Kim, H., Yoon, J. et al., "An asynchronous boost converter with time-based dual-mode control for wide load range and high efficiency in SSD applications," *IEEE Transactions on Industrial Electronics, 67*(12), 10520-10530, 2020.
[20] Hasegawa, K., Aiba, Y., Li, X. et al., "Low Power and Thermal Throttling-less SSD with In-Package Boost Converter for 1000-WL Layer 3D Flash Memory," In *2023 IEEE International Memory Workshop (IMW)* (pp. 1-4), May 2023.

A 98.5% Efficiency Single-Mode Buck-Boost Converter with All-1.8-V-Switch and Non-Stopping Output Current Delivery

Qianhui Liu [1], Yuhua Chen [1], Yixing Wang [1], Yuming Zhang [1], Yimeng Zhang *[1]

[1] School of Microelectronics, Xidian University, Xian710000, China

* Email: zhangyimeng@xidian.edu.cn

Abstract—This paper proposes a high-efficiency single-mode buck-boost (SMBB) converter for lithium-ion battery-powered systems with a 3.3 V output. The proposed converter operates in a single mode without any mode transitions so that it regulates output voltage smoothly even when V_{in} varies at the boundary of buck and boost operation regions. This converter not only achieves lower average inductor current but also utilizes 1.8-V power switches that exhibit low on-resistance, thereby enhancing power efficiency. Furthermore, The SMBB converter features continuous output current delivery, with its transfer function containing only a pair of complex poles and no right-half-plane zero, enabling rapid load transient response. Implemented in a 0.18-μm BCD process using 1.8-V devices, the chip operates over an input voltage range of 2.7 V to 4.2 V. At a 1 MHz switching frequency, with a 4.7-μH inductor, two 10-μF flying capacitors, and a 10-μF output capacitor, it achieves a peak efficiency of 98.5%, demonstrating its potential for high-efficiency applications in portable electronics.

Keywords—*li-ion battery, SMBB, flying capacitor, average inductor current reduction, single-mode operation, fast transient response*

I. INTRODUCTION

In recent years, mobile devices have been evolving toward miniaturization and wearability. Lithium-ion (Li-ion) batteries, with advantages such as portability, high energy density, and user-friendly charging capabilities, have become the preferred power solution for such devices, typically offering an output voltage range of 2.7 V to 4.2 V. However, stringent physical size constraints limit battery capacity, while increasingly complex multifunctional circuits (e.g., adaptive display modules, heterogeneous computing units) generate dynamically varying load currents with instantaneous fluctuations spanning two orders of magnitude. Under such operating conditions, unpredictable voltage drops occur at the battery terminals, necessitating buck-boost converters to achieve millisecond-scale mode transitions to maintain a stable 3.3 V output.

Conventional four-switch Buck-Boost Converters (CBBC) face significant challenges: the average inductor current $\bar{I}_L=(M+1)\cdot I_{LOAD}$ (where M is the voltage conversion ratio) induces severe conduction losses under heavy loads. This issue is exacerbated when using compact inductors with large parasitic DC resistance (DCR), leading to drastic efficiency degradation. To address these limitations, numerous hybrid attempts incorporating flying capacitors have been explored. $\bar{I}_L$ However, the significantly different bias voltages of the flying capacitor in buck and boost modes make mode transitions extremely difficult at $M \approx 1$. The designs in [2] and [3] achieve reduced average inductor current through

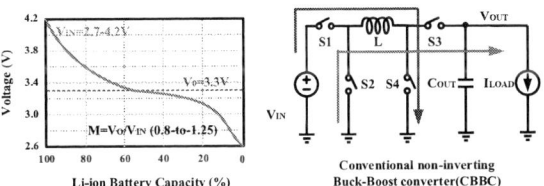

Fig. 1. Li-ion battery output voltage and CBBC structure

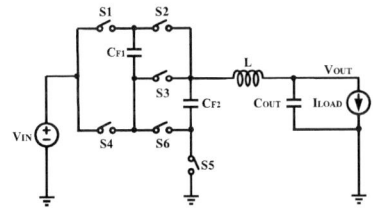

Fig. 2. Proposed SMBB converter topology

synergistic capacitor-inductor paths and operate in a single mode across the entire input voltage range. Nevertheless, they still employ two or three switches in the main current path, resulting in elevated conduction and switching losses. Recently, several articles have discussed enhancing power density in Buck and Boost converters for Li-ion battery-powered PMICs by utilizing lower-voltage-rated power switches, achieving higher performance at lower cost [4], [5], [6], [7]. However, research on improving Buck-Boost converter performance with such low-voltage-rated switches remains limited.

Based on the above background, this paper proposes a novel SMBB converter that operates with a single mode across the entire input voltage range, eliminating the complex mode transitions required in prior designs. The introduction of flying capacitor and stacked transistor structure reduces the switch voltage stress and average inductor current, resulting in low conduction losses. The entire chip is implemented using 1.8-V devices. Moreover, unlike CBBC, the proposed converter continuously delivers current to the load through the inductor during the entire switching cycle, enhancing transient response and output ripple performance.

II. SMBB CONVERTER

A. Topology and Operation Principle

The proposed SMBB converter topology is illustrated in Fig. 2, comprising of one inductor, one output capacitor, two flying capacitors, and six switches. Without any circuit configuration changes between the buck and boost modes, the converter performs the buck and boost transition in a single

979-8-3315-3918-4/25 $31.00 © 2025 IEEE

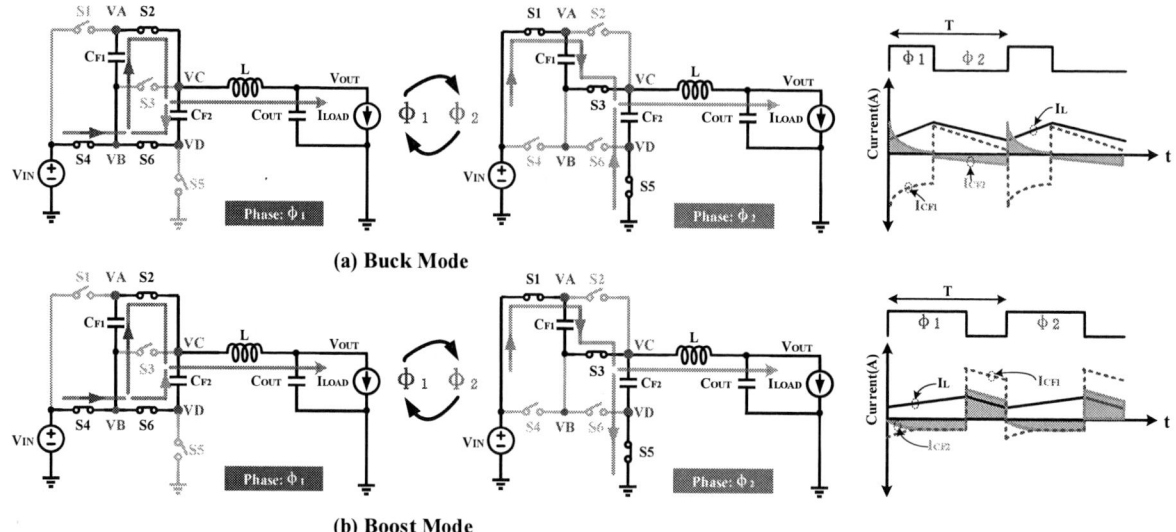

(a) Buck Mode

(b) Boost Mode

Fig. 3. Working principle and important waveforms of the proposed SMBB converter

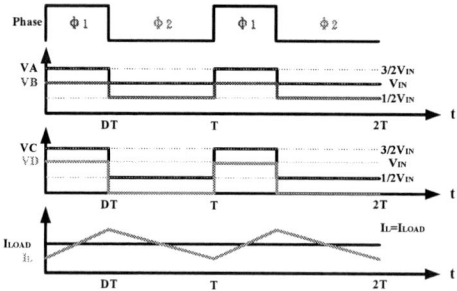

Fig. 4. SMBB converter operating waveforms

mode. This converter employs a simple two-phase operation: During phase $\phi1$, switches S1, S3, S5 are open while S2, S4, S6 are closed. Flying capacitors C_{F1} and C_{F2} are connected in parallel, ensuring identical voltages across them. The inductor is connected between the flying capacitors and the output, with a positive voltage applied across its terminals causing the inductor current to increase. Conversely, during phase $\phi2$, switches S1, S3, S5 close while S2, S4, S6 open. The flying capacitors C_{F1} and C_{F2} are in series and connected to the input V_{IN}, forming a capacitive voltage divider. The inductor is connected between the intermediate node of this divider and the output, experiencing a negative voltage that decreases the inductor current.

Small surface-mount inductors are essential for modern compact converters. However, the DCR of these inductors increases exponentially as their volume decreases, resulting in reduced efficiency and the heat generated also exacerbates thermal management. In the proposed SMBB converter, the inductor is directly connected to the load (as shown in Fig. 3), continuously supplying load current. Its average inductor current is given by:

$$\bar{I}_L = I_{LOAD} \tag{1}$$

Compared to CBBC, the average inductor current is significantly reduced. The conduction losses of the inductor DCR have significantly decreased, thereby alleviating the constraints on board space.

The periodic series-parallel reconfiguration of flying capacitors inherently establishes voltage balance. If $V_{C_{F1}}$ and $V_{C_{F2}}$ denote their steady-state voltages, the capacitor connection mechanism yields:

$$V_{C_{F1}} = V_{C_{F2}} = V_{IN}/2 \tag{2}$$

Consequently, during phase $\phi1$, the node voltages at VA, VB, VC and VD are $3V_{IN}/2$, V_{IN}, $3V_{IN}/2$ and V_{IN}, respectively. During phase $\phi2$, these voltages transition to V_{IN}, $V_{IN}/2$, $V_{IN}/2$ and 0, as illustrated in Fig. 4.

Based on the inductor volt-second balance principle, the voltage conversion ratio M can be derived as follows:

$$D(3V_{IN}/2 - V_{OUT}) + (1-D)(V_{IN}/2 - V_{OUT}) = 0 \tag{3}$$

$$M = \frac{V_{OUT}}{V_{IN}} = 0.5 + D \tag{4}$$

The duty cycle (D) is defined as the ratio of phase $\phi1$ duration to the switching period Ts. With D ranging from 0 to 1, the ideal M spans 0.5 to 1.5, encompassing both buck and boost operations. Consequently, distinct from other hybrid buck-boost converters, this topology achieves seamless regulation of V_{OUT} to 3.3V in a single operating mode, thereby suppressing output voltage ripple.

Applying charge balance principles to the flying capacitors C_{F1} and C_{F2}, the average currents through these capacitors during phases $\phi1$ and $\phi2$ are derived as:

$$\bar{I}_{C_{F1},\phi_1} = -\frac{1}{2D}\bar{I}_L = \frac{1}{2(0.5-M)}\bar{I}_L \tag{5}$$

$$\bar{I}_{C_{F1},\phi_2} = \frac{1}{2D-2}\bar{I}_L = \frac{1}{2(M-1.5)}\bar{I}_L \tag{6}$$

$$\bar{I}_{C_{F2},\phi_1} = \frac{1-2D}{2D}\bar{I}_L = \frac{1-M}{M-0.5}\bar{I}_L \tag{7}$$

$$\bar{I}_{C_{F2},\phi_2} = \frac{1-2D}{2-2D}\bar{I}_L = \frac{1-M}{1.5-M}\bar{I}_L \tag{8}$$

979-8-3315-3918-4/25 $31.00 © 2025 IEEE

As derived from (5)-(8), the current direction through flying capacitor C_{F2} reverses between buck and boost modes, as illustrated in Fig. 3. Nevertheless, the equalities presented in (1)-(4) remain valid across both operating regimes.

B. Power Stage Transfer Function

The control-to-output transfer function of the SMBB converter can be derived using standard state-space average modeling as follows:

$$G_{vd,SMBB}(s) = \frac{\hat{v}_{OUT}}{\hat{d}}$$
$$= \frac{V_{OUT}/0.5+D}{s^2 L C_{OUT} + s\,L/R_0 + 1} \quad (9)$$

From (9), it can be seen that the transfer function of the SMBB converter does not contain any right-half-plane zeros, thereby greatly improving the dynamic response.

III. IMPLEMENTATION OF THE SMBB CONVERTER

The detailed circuit of the proposed SMBB converter is shown in Fig. 5. The external components include two identical flying capacitors C_{F1} and C_{F2}, an output capacitor C_{OUT}, a hold capacitor C_{hold} and an inductor L.

A. Power Stage Circuits

As shown in Fig. 5, the power stage primarily consists of six power switches, corresponding gate drivers, and an LDO regulator, with the LDO regulator used to assist the gate drive of S5.

In the proposed SMBB converter, the voltage stress for the power switches S1, S2, S3, S4, and S6 is $V_{IN}/2$. The voltage stress for the power switch S5 is V_{IN}, but this can be halved using stacked transistors. In the 0.18-μm BCD process, the figure-of-merit (FoM) and Ron of 1.8-V MOSFETs are only 1/7 and 1/8 of those of 5-V MOSFETs, respectively. Clearly, low-voltage devices can provide better performance. To minimize conduction loss, all power switches employ 1.8-V transistors. S1 and S2 are implemented using PMOS, S3, S4, and S6 are implemented using NMOS, and S5 is implemented using two stacked NMOS.

In this design, the power switches do not require bootstrap circuits for driving, significantly saving area. Switches S1, S2, S3, and S4 can be directly driven by C_{F1}, switch S6 can be directly driven by C_{F2}, and the drive for switch S5 is achieved using an external capacitor C_{hold} (15 nF). As shown in Fig.5, C_{hold} can be charged via a small switch S7. As depicted in Fig. 6, C_{hold} charges through a small switch S7 during phase $\phi2$: when node VB settles at $V_{IN}/2$ and node VD at 0V, S7 closes to connect C_{hold} to VB, thereby charging it to $V_{IN}/2$. S7 is implemented using two 1.8-V PMOS transistors stacked together. To ensure that the voltage stress on both 1.8-V PMOS transistors is $V_{IN}/2$, a low-dropout regulator (LDO2) with an output voltage of $V_{IN}/2$ is used.

B. Control Stage Circuits

The control stage consists of a ramp generator, compensator, comparator, start module, logic control module, level shifter, and LDO1. To avoid the circuit design associated with inductor current detection required for current mode control, this work employs PWM voltage mode control with Type-III compensation. Since there is no RHP zero, Type-III compensation can achieve a large bandwidth and obtain a fast

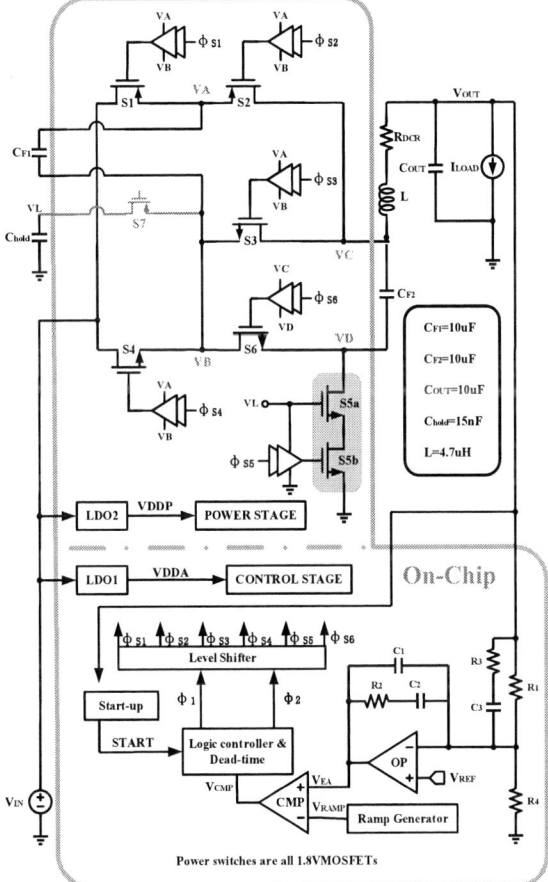

Fig. 5. Block diagram of the proposed SMBB converter

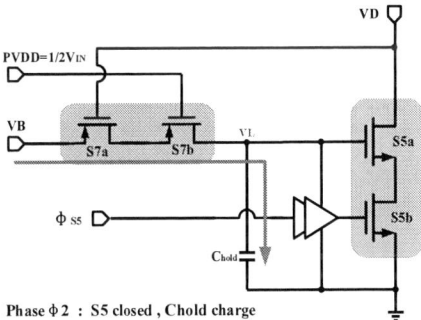

Phase $\phi2$: S5 closed , Chold charge

Fig. 6. Schematic of S5 driver

transient response. The SMBB converter maintains continuous conduction mode (CCM) operation across the full input range, even under light-load conditions. This is enabled by its reverse battery charging capability when inductor current crosses zero.

IV. SIMULATION RESULTS ANALYSIS

This SMBB converter is manufactured using 0.18-μm BCD process technology and uses only 1.8-V devices. The converter operates within an input voltage range of 2.7 V to 4.2 V, providing a stable 3.3 V output voltage with an output current of 0 to 1 A. External components include a 4.7 μH inductor (L), a 10 μF output capacitor C_{OUT}, two 10 μF flying

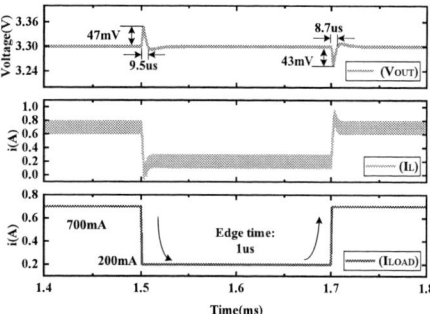

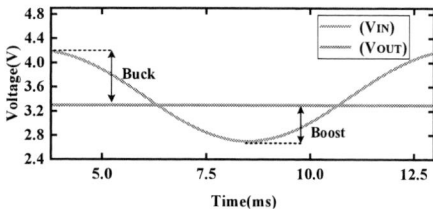

Fig. 7. Load transient response simulation results

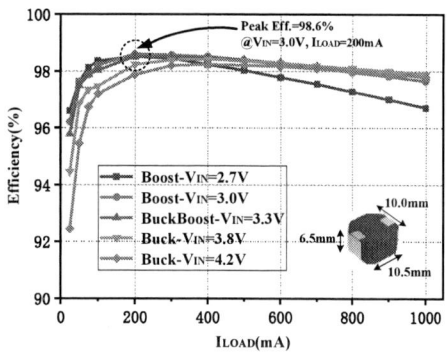

Fig. 8. Linear transient response results

the overshoot voltage is 47 mV, the overshoot recovery time is 9.5 us, the undershoot voltage is 43mV, and the undershoot recovery time is 8.7 us.

Fig. 8 shows the linear transient response. During simulation, V_{IN} uses a sine wave with a period of 10 ms and an amplitude ranging from 2.7 V to 4.2 V. Despite the significant fluctuations in V_{IN} the SMBB converter maintains a stable and perfect V_{OUT} at an input voltage of 3.3 V demonstrating its ability to seamlessly switch between buck and boost operations.

Fig. 9 shows the power conversion efficiency. Using a large-volume inductor with $DCR = 9m\Omega$, the peak efficiency reaches 98.5% when $V_{IN} = 3.0V$ and $I_{LOAD} = 300mA$. As shown in Fig. 10, even when using a small-volume chip inductor with a high $DCR = 250m\Omega$, a peak efficiency of up to 97.6% can be achieved. Compared to previous work, the efficiency has been significantly improved.

CONCLUSION

This paper presents an SMBB converter employing a single inductor and two flying capacitors. The proposed converter achieves a conversion ratio of $M = 0.5+D$, enabling both step-down and step-up operations solely through duty cycle control without requiring mode switching. The SMBB topology features reduced average inductor current, leading to high efficiency even with large inductor DCR. Utilizing 1.8-V rated devices, the converter confines all power switch voltage stress to $V_{IN}/2$ by virtue of its flying capacitor and stacked transistor structure, facilitating low-cost fabrication and minimizing conduction loss. Furthermore, the inductor continuously supplies current to the load, enabling fast transient response. The SMBB converter achieves a high peak efficiency of 98.5% with a 9-mΩ inductor DCR.

ACKNOWLEDGMENT)

This work was supported by the National Natural Science Foundation of China (Grant NO.62234010).

REFERENCES

[1] J. Jin et al., "A 98.6%-Peak-Efficiency 1.47A/mm2-Current-Density Buck-Boost Converter with Always Reduced Conduction Loss," ISSCC, pp. 448–449, Feb. 2023.

[2] D. Cho et al., "A high-efficiency single-mode dual-path buck-boost converter with reduced inductor current," IEEE J. Solid-State Circuits, vol. 58, no. 3, pp. 720–731, Mar. 2023.

[3] H. Shin et al., "A 96.6%-efficiency continuous-input-current hybrid dual-path buck-boost converter with single-mode operation and non-stoppingoutput current delivery," in Proc. Symp. VLSI Circuits, Jun. 2021.

[4] Y. Huh, S.-W. Hong, and G.-H. Cho, "A hybrid structure dual-path step-down converter with 96.2% peak efficiency using 250-m ω large-DCR inductor," IEEE J. Solid-State Circuits, vol. 54, no. 4, pp. 959–967, Apr. 2019.

[5] A. Abdulslam and P. P. Mercier, "A symmetric modified multilevel ladder PMIC for battery-connected applications," IEEE J. Solid-State Circuits, vol. 55, no. 3, pp. 767–780, Mar. 2020.

[6] P. Assem, W.-C. Liu, Y. Lei, P. K. Hanumolu, and R. C. N. Pilawa-Podgurski, "Hybrid Dickson switched-capacitor converter with wide con-version ratio in 65-nm CMOS," IEEE J. Solid-State Circuits, vol. 55, no. 9, pp. 2513–2528, Sep. 2020.

[7] S.-U. Shin et al., "A 95.2% efficiency dual-path DC-DC step-up converter with continuous output current delivery and low voltage ripple," in Proc. IEEE Int. Solid-State Circuits Conf., 2018, pp. 430–432.

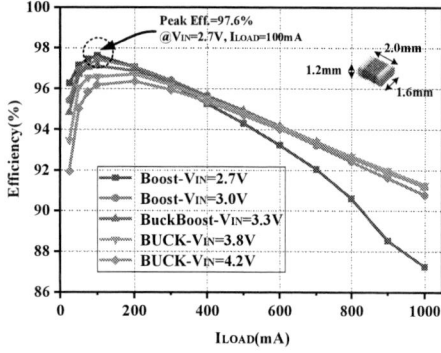

Fig. 9. Power efficiency with DCR=9mΩ

Fig. 10. Power efficiency with DCR=250mΩ

capacitors (C_{F1} and C_{F2}), and a 15 nF hold capacitor (C_{hold}), with a switching frequency of 1 MHz.

Fig. 7 shows the load transient response when $V_{IN} = 3.3V$. For a load transition from 200 mA to 700mA with $T_{edge} = 1us$,

Design of a Fully Integrated Low Dropout Linear Regulator with Bandgap Reference

Fan He [2,3], Jiao Liu [1], Yiyun Mao [1], Haoyuan Gao [1], Xianhui Wang [2,3], Yubing Zhang [2,3], Hao Xu [1], Na Yan*[1]

[1] State Key Laboratory of Integrated Chip and Systems, Fudan University, Shanghai 200433, China
[2] Beijing Smartchip Microelectronics Technology Co.,Ltd
[3] Beijing SmartChip Semiconductor Technology Co., Ltd

* Email: yanna@fudan.edu.cn

Abstract—A fully integrated Low-dropout regulator(LDO) with a low-temperature-coefficient bandgap reference(BGR) is proposed. This LDO incorporates a flipped voltage follower (FVF) buffer, a high-gain operational amplifier, and an additional auxiliary LDO. The proposed regulator demonstrates fast dynamic response speed and power supply rejection (PSR) capability in steady-state operation. The bandgap reference is designed with adjusted-temperature-curvature compensation method. The bandgap reference achieved a TC that varies from 7 to 16 ppm/°C from −40 °C to 125 °C. The simulation results indicate that under a supply voltage of 1.2V and an output current of 10mA, this LDO demonstrates a PSR greater than 80dB at low frequencies and over 40dB at 1MHz with a 200 pF on-chip output capacitor. The FVF LDO reaches 3 ns response time when I_{LOAD} changes between 100 µA and 20 mA with edge times less than 10 ps.

Keywords—flipped voltage follower (FVF), power supply rejection (PSR), fully-integrated low-dropout (LDO), bandgap reference(BGR), adjusted-temperature-curvature compensation

I. INTRODUCTION

With the continuous advancement of integrated circuit technology, there emerges a growing demand for low-dropout regulators that simultaneously achieve enhanced power supply rejection and fast transient response to meet the stringent requirements of mixed-signal systems. In the realm of power supply design for digital circuits, such as memory systems, switching voltage regulators are predominantly employed. This preference stems from the inherent characteristic of digital circuits to exhibit substantial noise immunity, which allows them to operate reliably even in the presence of moderate levels of power supply noise. Conversely, for high-performance analog-digital mixed-signal circuits—such as phase-locked loops (PLLs) and data converters—the design of analog low-dropout (LDO) regulators with high power supply rejection ratio (PSR) and broad bandwidth assumes paramount importance. These circuits are highly sensitive to power supply fluctuations due to their stringent requirements for low noise and precise signal integrity, necessitating LDO regulators that can effectively suppress power supply noise across a wide frequency range while maintaining stable voltage regulation.

Digital LDOs (DLDO) do not require extra frequency compensation capacitors and only need a comparator and a quantizer, making them more suitable for low-power voltage and small-area applications. However, due to the discontinuous quantization errors, digital LDOs are not suitable for circuits which are sensitive to noise and supply ripple [1]. Therefore, analog low dropout regulators (ALDO) are the foremost choice for oscillators due to the low-noise, ripple-free and fast-transient characteristics.

Researchers have contributed to improve PSR and transient response techniques. By utilizing a transient-enhanced buffer impedance attenuation technique, the architecture in [2] achieves <50mV voltage undershoot during 200mA load transients. However, this LDO exhibits a worst-case PSR of 45 dB at low frequencies. Furthermore, the existence of an external 1µF capacitor leads to a recovery time of approximately 10µs. The LDO in [3] introduces a feedforward ripple cancellation (FFRC) technique for the first time, demonstrating a worst-case power supply rejection (PSR) of 56 dB up to 10MHz. This implementation, however, requires a 4µF off-chip capacitor that fundamentally constrains its transient response. The capacitor-less LDO proposed in [4] requires only 128-pF on-chip output capacitance to achieve 6µs load transient recovery time. Its PSR enhancement circuitry maintains over 70 dB power supply rejection at 1 MHz while being robust against process and temperature variations. However, this architecture incurs additional cost due to the required auxiliary current amplifiers and buffers. Reference [5] employs adaptive body current injection technology to achieve 36 dB PSR at 10 GHz. The proposed LDO requires only 240-pF on-chip capacitance, demonstrating a 225-mV undershoot voltage during full-load step transients with voltage recovery within 100 nanoseconds.

II. CIRCUIT IMPLEMENTATION

A. LDO Circuit

Transistor-level implementation of the proposed LDO is shown in Fig. 1. The LDO consists of a folded cascode operational transconductance amplifier (OTA), a flipped voltage follower (FVF) buffer and an auxiliary LDO.

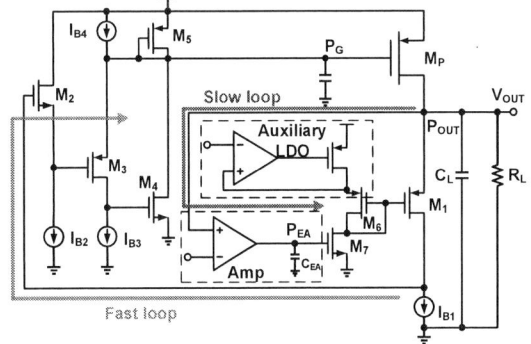

Fig. 1 Schematic of the proposed FVF LDO.

This LDO consists of a fast loop and a slow loop. The fast loop includes the power transistor M_P, the common-gate transistor M_1, and the super source follower (SSF) formed by M_2, M_3, and M_4. The common-gate transistor M_1 amplifies the variation in the output voltage ΔV_{OUT} and feeds it back to the gate of the power transistor. By introducing shunt feedback through the common-source transistor M4 within the SSF, the output resistance at the gate node of the power transistor is reduced, pushing the pole at the power transistor's gate (P_G) to a higher frequency.

B. Stability Analysis

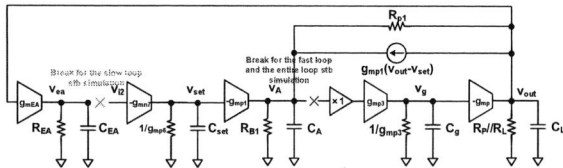

Fig. 2 Small-signal model for stability analysis.

The slow-regulation loop comprises an operational amplifier (op-amp), a diode-loaded common-source stage (M_6-M_7), and a common-gate stage (M_1). Within this loop architecture, the op-amp's output pole (P_{EA}) functions as the dominant pole, while the flipped voltage follower (FVF) loop's output pole (P_{OUT}) serves as the non-dominant pole. The incorporation of the op-amp enables the slow loop to achieve enhanced power supply rejection (PSR) performance. During steady-state operation, this loop continuously regulates power supply variations, thereby significantly improving the PSR. In contrast, the fast-response loop is specifically designed to handle large-signal load transients, ensuring robust dynamic performance while maintaining system stability. This dual-loop architecture effectively handles the static and dynamic regulation requirements, optimizing both steady-state precision and transient response. Equations (1)–(7) present the transfer function expressions of the entire loop.Fig. 3 shows the simulated loop stability results.

$$T(s) = \frac{v_{out}}{v_i} = \frac{-A_1 A_2 A_{mp}\left(1 + \frac{s}{A_2 p_{EA}}\right)}{\left(1 + \frac{s}{\frac{A_1 A_2 (R_P \parallel R_L)}{R_{n13}} p_{EA}}\right)\left(1 + \frac{s}{p_{OUT}}\right)\left(1 + \frac{s}{p_G}\right)} \quad (1)$$

$$A_1 = g_{m1} R_{B1} \quad (2)$$

$$A_2 = g_{mEA} R_{EA} \quad (3)$$

$$A_{mp} = g_{mp} R_L \quad (4)$$

$$p_{EA} = \frac{1}{2\pi R_{EA} C_{EA}} \quad (5)$$

$$p_{OUT} = \frac{1}{2\pi R_L C_L} \quad (6)$$

$$p_G = \frac{g_{mp3}}{2\pi C_g} \quad (7)$$

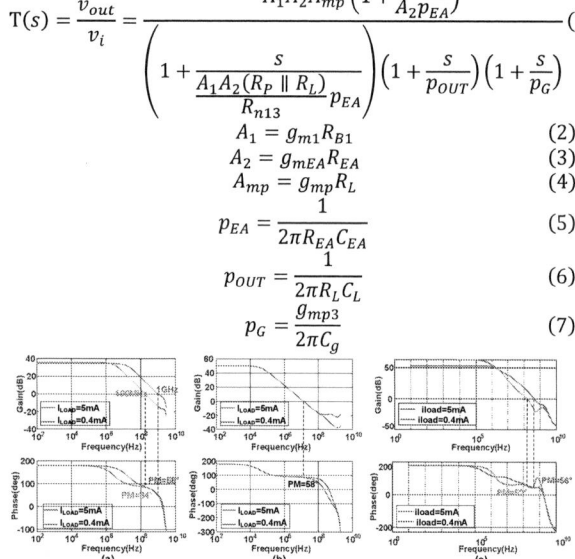

Fig. 3 Simulated Bode plot of (a) the fast loop,(b)the slow loop,(c)the entire loop.

C. Transient Analysis on FVF Stage

When the load current abruptly increases from low to high levels, the power supply must immediately deliver substantial charge. For high-speed circuits such as ADC and TDC references, the current transition time (typically tens of picoseconds) is significantly shorter than the loop bandwidth response time. Consequently, the decoupling capacitor initially discharges to supply the transient charge, resulting in a substantial voltage droop ΔV (highlighted in green in the figure).

The output voltage begins recovering only after the control loop responds, gradually increasing over time. As shown in the blue-shaded region, the settling time Δt_2 must span at least three time constants (3τ) of the loop to ensure the reference voltage error remains below 1 LSB for data converters[7]. Equations (8) and (9) provide the calculations for the maximum voltage drop ΔV and time constant τ, respectively.

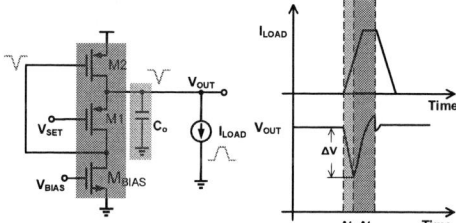

Fig. 4 Dynamic load response of the FVF buffer.

$$\Delta V = \frac{(I_{max} - I_{min})\Delta t_1}{C_O} \quad (8)$$

$$\tau = \frac{C_O}{g_{m2} g_{m1}(R_{BIAS} \parallel R_{O1})} \quad (9)$$

D. PSR Analysis

Fig. 5 shows the simplified schematic for PSR analysis, while Eqs. (10)-(12) represent the transfer functions from power supply noise to output. The analysis reveals that the PSR transfer function contains one zero and two poles.

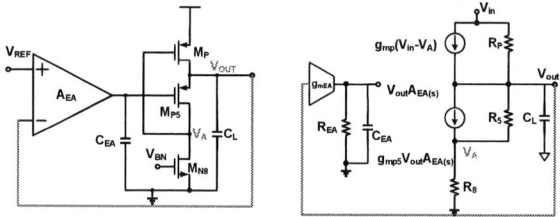

Fig. 5 Simplified PSR Analysis Diagram

$$\frac{v_{out}}{v_{in}} = \frac{(g_{mp}R_p + 1)(1 + \frac{s}{R_{EA}C_{EA}})}{1 + g_{mp}R_p A_{EA} g_{mp5}(R_5 R_8)\left(1 + \frac{s}{p_1}\right)\left(1 + \frac{s}{p_2}\right)} \quad (10)$$

$$p_2 = \frac{A_{EA} g_{mp5}(R_5 R_8)}{R_{EA} C_{EA}} \quad (11)$$

$$p_1 = \frac{g_{mp5} R_8 g_{mp} A_{EA}}{2\pi C_L} \quad (12)$$

The zero frequency coincides with the loop's dominant pole frequency (i.e., the operational amplifier's output pole frequency). This occurs because when the frequency reaches the op-amp's -3dB bandwidth, its gain begins to reduce, corresponding to PSR degradation, which explains the upward inflection in the PSR simulation results.

E. Bandgap Reference Circuit

The proposed bandgap reference (BGR) circuit architecture comprises five functional blocks: (i) the PTAT (proportional-to-absolute-temperature) current generation core, which establishes the fundamental temperature-dependent current reference; (ii) the CTAT (complementary-to-absolute-temperature) current generation module, providing the necessary temperature compensation; (iii) the nonlinear V_{BE} compensation module, which corrects higher-order temperature effects; (iv) an adaptive temperature coefficient compensation module for fine-tuning the thermal characteristics; and (v) a robust startup circuit ensuring reliable power-on initialization. Notably, the design employs a chopper-stabilized operational amplifier to mitigate low-frequency noise and offset voltage effects. Fig. 5 presents the detailed schematic of the PTAT and CTAT current generation circuit along with the startup circuit implementation.

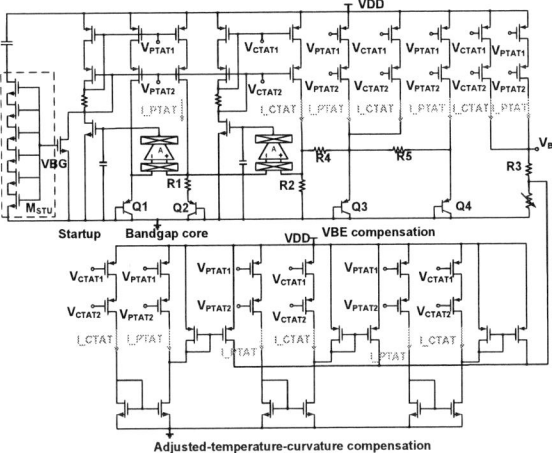

Fig. 6 Schematic of bandgap reference.

The start-up circuit comprises two NMOS transistors and one capacitor. Under abnormal startup conditions (when the output voltage remains at 0 V), the transistor M_{STU} connected to the output node remains in the off state, allowing the power supply to charge the capacitor. Once the voltage at the lower plate of the capacitor rises sufficiently to turn on the right NMOS transistor, the gate potential of the self-biased current mirror transistor M3 is pulled down, forcing M3 to conduct and draw current from the supply. This process ensures proper circuit startup.

Fig. 6 presents the pre-simulation results of the BGR output voltage using the piecewise adaptive temperature compensation method. Fig. 6(a) shows the temperature characteristics of a conventional uncompensated BGR output voltage, exhibiting a 5 mV variation across the −40 °C to 120 °C range, corresponding to a temperature coefficient of 25.6 ppm/°C. Fig. (b) displays the post-compensation simulation results, where the output voltage variation is reduced to 0.44 mV over the same temperature range, achieving a temperature coefficient of 2.2 ppm/°C. This demonstrates a 23 ppm/°C improvement in output precision after compensation.Fig. (c) illustrates the temperature characteristics at an output voltage of 0.7 V, where the temperature coefficient is approximately 4 ppm/°C. Fig.

(d) provides the simulated temperature-dependent behavior of the compensation current.

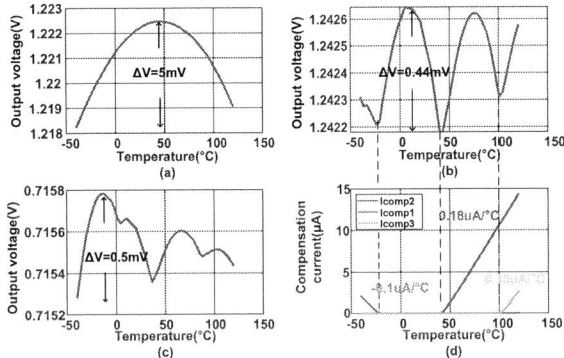

Fig. 7 Simulated results of output voltage and compensation current.

III. SIMULATION RESULTS

Fig. 8 shows the chip layout of the low-drift BGR cascaded with the FVF LDO, measuring approximately 380 μm in length and 130 μm in width, with a core area of 0.049 mm². The BGR operates from a 1.8 V supply, while the LDO uses a 1.2 V supply.

The operational amplifier employs a common-centroid matching layout to minimize the impact of input offset voltage on the BGR output voltage temperature coefficient. Current mirror transistors are placed in close proximity to reduce mismatch. To avoid additional IR drops caused by routing resistance, multilayer metal stacking is adopted for interconnects, albeit at the cost of increased parasitic capacitance. However, this design remains insensitive to parasitic effects due to the circuit's low operating frequency. Furthermore, power and ground lines utilize top-level metal layers to benefit from their significantly lower resistivity compared to lower metal layers.

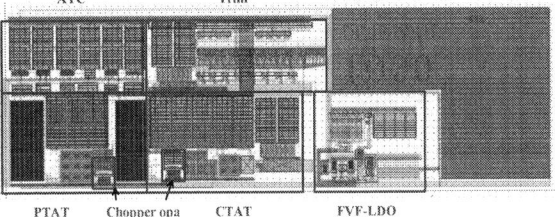

Fig. 8 Layout of FVF-LDO and BGR

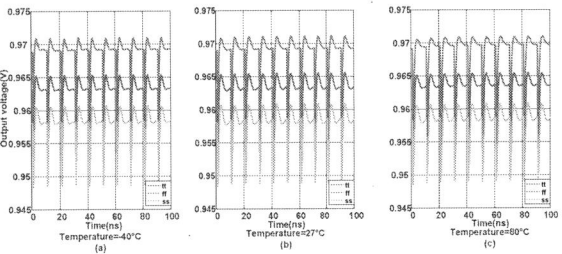

Fig. 9 Simulated results of transient output voltages.

Fig. 9 presents the simulated output voltage transient response under process corner variations (TT, SS, FF) at three temperature points (-40°C, 27°C, and 80°C), with the load current undergoing step changes over 10 cycles. For a 20 mA load current pulse with 10 ns period, the LDO exhibits a maximum output voltage droop of 10 mV and achieves recovery within approximately 3 ns. The steady-state voltage deviation originates from process-corner-induced variations in the BGR output voltage.

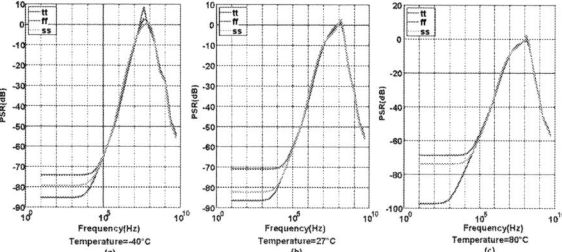

Fig. 10 Simulated results of power supply rejection ratio.

Fig. 10 presents the PSR simulation results, where subfigures (a) through (c) display the performance at -40°C, 27°C, and 80°C, respectively. Under TT process corner at 27°C, the measured PSR is 83 dB at DC and 39 dB at 1 MHz. The worst-case PSR of approximately 35 dB occurs near 1 MHz at 80°C. For the FF process corner at 80°C, the minimum low-frequency PSR degrades to 68 dB.

IV. CONCLUSION

This paper presents the design of a fast-transient low-dropout regulator (LDO) employing a flipped voltage follower (FVF) buffer and dual-loop architecture. The work systematically analyzes the LDO's loop transfer function, investigates its dynamic response characteristics, and examines the power supply rejection (PSR) performance. Furthermore, the design incorporates a low-temperature-drift bandgap reference (BGR) circuit to enhance voltage regulation accuracy. As summarized in Table 1, the proposed FVF-based LDO demonstrates better PSR performance compared to conventional architectures.

TABLE I. PERFORMANCE COMPARISON

Parameter	JSSC'[8]	JSSC'[6]	JSSC'[3]	This work
LDO type	Digital	Analog	Analog	Analog
$V_{OUT}(V)$	0.45-0.95	0.6	1	0.71-1.15
Dropout voltage(mV)	50	200	150	200
$I_q(\mu A)$	3.2	2	50	100
$I_{LOAD}(mA)$	12	10	25	10
C_L	0.1nF	1μF	4μF	0.1nF
PSR@DC(dB)	NAN	42.7	80	80
PSR@1MHz(dB)	NAN	NAN	60	40
$\Delta V_{OUT}(mV)/T_{edge}(ns)$	50/10	70/20	15/10	10/0.01
$T_R(ns)$	80	15	2400	3

This paper is supported by the Joint R&D Fund of Beijing Smartchip Microelectronics Technology Co.,Ltd(No.SGSCDT00TXQT2500108).

REFERENCES

[1] M. Huang, Y. Lu and R. P. Martins, "Review of Analog-Assisted-Digital and Digital-Assisted-Analog Low Dropout Regulators," in IEEE Transactions on Circuits and Systems II: Express Briefs, vol. 68, no. 1, pp. 24-29, Jan. 2021.

[2] M. Al-Shyoukh, H. Lee and R. Perez, "A Transient-Enhanced Low-Quiescent Current Low-Dropout Regulator With Buffer Impedance Attenuation," in IEEE Journal of Solid-State Circuits, vol. 42, no. 8, pp. 1732-1742, Aug. 2007

[3] M. El-Nozahi, A. Amer, J. Torres, K. Entesari and E. Sanchez-Sinencio, "High PSR Low Drop-Out Regulator With Feed-Forward Ripple Cancellation Technique," in IEEE Journal of Solid-State Circuits, vol. 45, no. 3, pp. 565-577, March 2010.

[4] C. -J. Park, M. Onabajo and J. Silva-Martinez, "External Capacitor-Less Low Drop-Out Regulator With 25 dB Superior Power Supply Rejection in the 0.4–4 MHz Range," in IEEE Journal of Solid-State Circuits, vol. 49, no. 2, pp. 486-501, Feb. 2014.

[5] Y. Lim, J. Lee, S. Park, Y. Jo and J. Choi, "An External Capacitorless Low-Dropout Regulator With High PSR at All Frequencies From 10 kHz to 1 GHz Using an Adaptive Supply-Ripple Cancellation Technique," in IEEE Journal of Solid-State Circuits, vol. 53, no. 9, pp. 2675-2685, Sept. 2018.

[6] N. Adorni, S. Stanzione and A. Boni, "A 10-mA LDO With 16-nA IQ and Operating From 800-mV Supply," in IEEE Journal of Solid-State Circuits, vol. 55, no. 2, pp. 404-413, Feb. 2020.

[7] C. -H. Chan et al., "60-dB SNDR 100-MS/s SAR ADCs With Threshold Reconfigurable Reference Error Calibration," in IEEE Journal of Solid-State Circuits, vol. 52, no. 10, pp. 2576-2588, Oct. 2017.

[8] M. Huang, Y. Lu, S. -P. U and R. P. Martins, "An Analog-Assisted Tri-Loop Digital Low-Dropout Regulator," in IEEE Journal of Solid-State Circuits, vol. 53, no. 1, pp. 20-34, Jan. 201

979-8-3315-3918-4/25 $31.00 © 2025 IEEE

A 455mV-Hysteresis, 120 nA, Bandgap less Power-on-Reset Circuit for IoT in 40nm CMOS

Mingzong Lin [1], Chaoran Chen [1], Jian Xu [2], Yue Lin [2], Wei Li [1], Hongtao Xu [1]

[1] State Key Laboratory of Integrated Chips and Systems, College of Integrated Circuits and Micro-Nano Electronics,
Fudan University, Shanghai, China
[2] ICLegend Micro, Shanghai, China

Email: 23112020020@m.fudan.edu.cn, hongtao@fudan.edu.cn

Abstract—This paper presents an ultra-low-power power-on-reset (POR) circuit with integrated brownout detection (BOD) for IoT devices. The resistor-less architecture, fabricated in 40nm CMOS, eliminates bandgap reference (BGR) while achieving 120nA quiescent current. The POR uses a sub-threshold current and NMOS switch to monitor the supply voltage in steady state, while the BOD circuit detects the time when power supply voltage drops through the cap charging. Measured results show 901 mV POR trip voltage and 446 mV BOD threshold. A novel state-transition mechanism enables 455 mV hysteresis voltage (V_{hyst}) with 3.4% σ/V_{hyst} simulated variation across PVT corners.

Keywords—Power-on-reset (POR), brownout detection (BOD), low voltage, ultra-low power, high hysteresis voltage

I. INTRODUCTION

The proliferation of IoT devices necessitates ultra-low-power IC designs, where power-on reset (POR) circuits face dual challenges: maintaining sub-μA quiescent current while ensuring robustness against supply fluctuations. On one hand, the POR cannot be completely shut down, as it is responsible for initializing digital circuits during power supply ramp-up [4]. On the other hand, low-voltage digital power supplies are commonly used in low-power systems. To ensure proper operation under low supply voltages, an accurate low-voltage POR is required. Additionally, high hysteresis voltage is essential to maintain digital circuit stability under fluctuating supply voltages.

In recent years, there are various types of low-power PORs proposed in previous works. Prior bandgap-based PORs [3] achieve accuracy but suffer from high static power. The bandgap generates an accurate reference voltage, and the hysteresis comparator compares the divided voltage generated from bandgap to obtain accurate trip voltage. However, bandgap-based POR is shorted of high-power-consumption and large area. Sub-threshold current designs [4]-[6] reduce I_q yet exhibit PVT sensitivity and inadequate hysteresis, making it sensitive to fluctuating supply voltages.

To eliminate resistors, a diode-connected transistor-based supply voltage detection circuit is adopted. To further reduce I_q, the circuit automatically shuts down after the POR process is completed. We propose a bandgap-free, resistor-less POR/BOD architecture achieving 120nA quiescent current and 470mV hysteresis voltage via synergistic state transition.

Modern Internet of Things (IoT) devices require PORs with sub-μW power consumption to extend battery lifetime. The proposed resistor-less architecture overcomes this limitation via synergistic POR/BOD integration. Specifically,

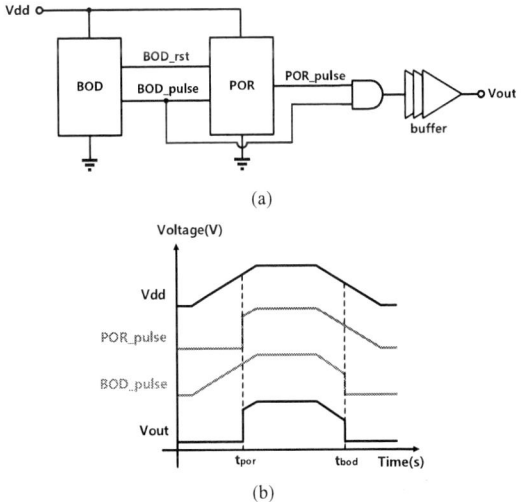

(a)

(b)

Fig. 1. (a) Proposed POR-BOD architecture (b) State transition timing diagram

the POR and BOD modules are reactivated by each other, achieving 120nA quiescent current without bandgap reference.

Fabricated in 40nm CMOS technology, the prototype occupies a 0.00726 mm^2 active area ($110\,\mu m \times 66\,\mu m$) while maintaining $120 \pm 80\,nA$ I_q across TT/FF/SS corners ($V_{dd} = 1.1\,V, 50°C$).

II. PROPOSED ARCHITECTURE

The proposed circuit addresses the challenges outlined in Section I, integrating power-on-reset (POR) and brown out detection (BOD) functions without bandgaps overhead (Fig.1). POR and BOD modules are provided by the sub-threshold current to get a low I_q. The hysteresis enhancement mechanism operates via two states between POR and BOD modules transitions:

1) POR→BOD transition: Post-reset completion disables POR via P1, activating BOD trough N_5

2) BOD → POR reset: Brownout event triggers Cres discharge, reactivating the POR module through P2

The POR pulse signal only detects power-on behavior, unable to determine the brownout behavior. Therefore, the POR would combine the POR pulse and the BOD pulse to generate output signal through AND logic. When the supply voltage powers off, POR output follows the supply voltage,

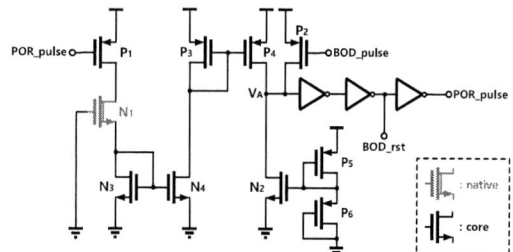

Fig. 2. Single threshold POR circuit

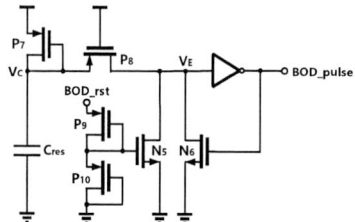

Fig. 3. Single threshold BOD circuit

and hence, it should be reset with the BOD module to prepare for the next power-on. When the BOD is activated to generate reset pulse, resetting digital circuit and POR module as well. After that, BOD shut off and POR is reactivated.

The cross-activation mechanism ensures only one module is active: (1) POR completion disables itself via P_1 while enabling BOD via N_5; (2) BOD-triggered C_{res} discharge reactivates POR via P_2. This duty-cycling reduces quiescent current to diode-connected transistor leakage levels.

III. CIRCUIT IMPLEMENTATION

A. POR module

The proposed POR module, shown in Fig. 2, utilizes N_1, a native NFET, to generate a sub-threshold current [1]:

$$I_{sub} = I_0 \cdot exp\left(\frac{V_{GS}}{nV_T}\right) \qquad (1)$$

Initially, this sub-threshold current charges node A via current mirror N_3, N_4, P_3, and P_6, pull its voltage up to V_{dd} until POR activation occurs. When V_{GS} of N_2 exceeds its threshold voltage, N_2 pulls node A down to V_{ss}, generating the POR reset pulse [2]. Moreover, POR sends a BOD reset pulse to activate BOD module. Upon POR completion, P_1 disables the detection circuit. Then the current mirror is shut down without current passing through. Consequently, the quiescent current of the module is only from the quiescent current of P_5 and P_6. Based on the above analysis, we determine the value of V_{POR} as followed:

$$V_{POR} = \frac{V_{GS,P5}+V_{GS,P6}}{V_{GS,P6}} \cdot V_{th,N2} \qquad (2)$$

Where $V_{th,N2}$ is the threshold voltage of N_2, $V_{GS,P5}$ and $V_{GS,P6}$ is the gate-source voltage of diode-connected transistors P_5 and P_6, respectively. This ratio stems from series-connected PMOS diodes (P_5-P_6) forming a voltage divider of V_{dd}. Since $V_{GS} = V_{th,p} + \sqrt{\frac{2I_q}{\mu C_{ox}(W/L)}}$, V_{POR} is tunable by adjusting PMOS ratio between the size (W/L) of P_5 and P_6 without changing circuit architecture. For 0.8V digital circuit core, increasing $(W/L)_{P5}$ by 2.6 × would lower V_{POR} values by ~150mV, ensuring POR generating reset pulse

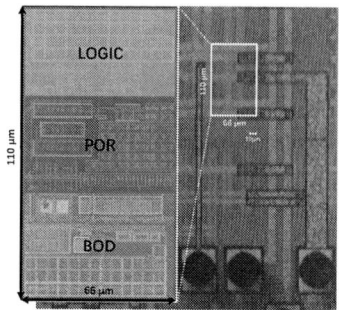

Fig. 4. Die photograph of the proposed circuit

supply voltage reaching 0.8V.

B. BOD module

The BOD circuit, shown in Fig. 3, operates in two states: active and inactive. In the active state, the BOD circuit detects the time when power supply voltage drops. When the BOD process is completed, the circuit remains inactive until subsequent POR triggering. BOD consists of N_3, N_4, P_5, P_6 and C_{res}. The capacitor C_{res} acts as a charge reservoir to maintain $V_D \approx V_{max}$ during brownout events. Initially, the BOD reset signal will pull V_E down to V_{ss} via N_5, forcing BOD following V_{dd} in the active state [2].

At the beginning of power off process, the gate voltage of P_8 is V_{dd}, turning P_8 off. V_D remains at V_{max} to the absence of a discharge path. When V_D exceeds $V_{dd} + |V_{th,P6}|$, P_6 turns on and attempts to charge node E. However, V_E could not be pulled up to V_{dd} because N_6 remains on, continuously discharging node E. When the supply voltage drops to $V_{th,N4}$, N4 turns off and allow the V_E raise to the V_{dd}, flipping the inverter. At this moment, BOD output turns to V_{ss} and the BOD process completes. Simultaneously, P_2 of POR turns on, resetting POR for the next power-up sequence (Fig. 3), as described before. After that, the quiescent current of the module is only from the diode-connected PMOS, same as the POR module. Hence, all the power consumption is formed by the quiescent current of diode-connected transistor. The value of V_{BOD} is given by:

$$V_{BOD} \approx V_{th,N6} \qquad (3)$$

Where $V_{th,N6}$ is the threshold voltage of N_6, making sure the robustness of V_{BOD}. Unlike V_{POR}, $V_{BOD} \approx V_{th,N6}$ lacks effective tunability due to fixed NMOS threshold. V_{BOD} exhibits $64\ mV(1\sigma)$ due to NMOS $V_{th,n}$ mismatch (Eq. 5). Assuming that $V_{th,N2} \approx V_{th,N6}$ And $\left(\frac{W}{L}\right)_{P5} = \left(\frac{W}{L}\right)_{P6}$, $V_{GS,P5} \approx V_{GS,P6}$, the hysteresis voltage is:

$$V_{hyst} = V_{POR} - V_{BOD} \approx V_{th,N} \qquad (4)$$

Eq. 4 promises the circuit achieving a high hysteresis voltage.

IV. EXPERIMENTAL RESULTS

The proposed POR has been fabricated in 40nm LP CMOS technology. The POR is design for the digital circuit at a 1.1-V supply. Fig. 4 shows a photograph of the proposed circuit. The chip occupies $110\ \mu m \times 66\ \mu m$ active area, where the POR, BOD, and digital buffer dominate the chip area. Post-layout simulations confirm a $120 \pm 30\ (1\sigma)\ nA$ quiescent current at 50°C, which is dominated by the quiescent current of diode-connected transistor from V_{dd} to V_{ss}.

979-8-3315-3918-4/25 $31.00 © 2025 IEEE

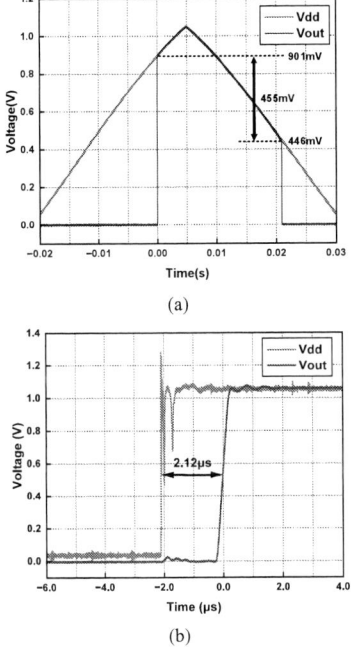

(a)

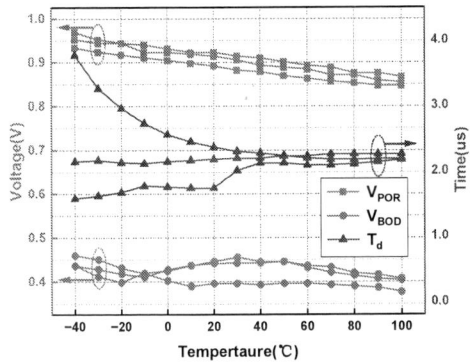

(b)

Fig. 5. Measured transient response with (a) power-up and brownout scenarios (b) fast power-up pulse

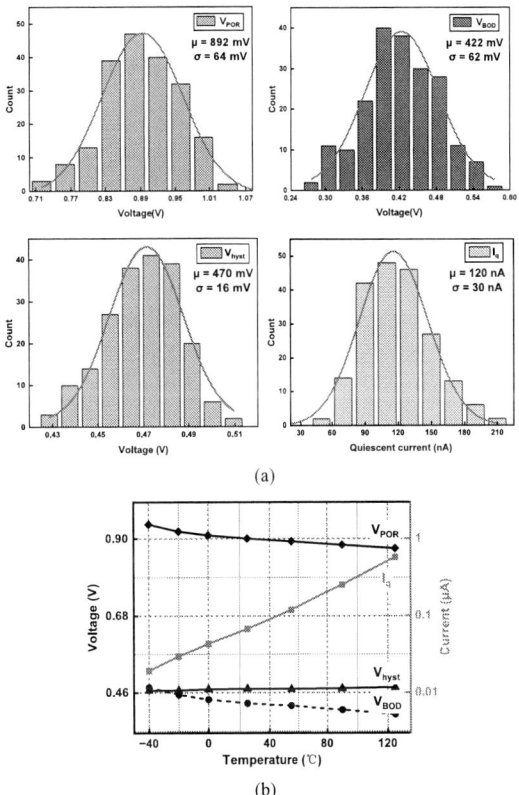

(a)

(b)

Fig. 7. Monte Carlo simulated result of proposed circuit at all process corners: (a) V_{POR}, V_{BOD}, V_{hyst} and I_q distributions at 50°C (b) the mean values of MC results versus temperature

TABLE I. PERFORMANCE SUMMARY AND COMPARISON

Parameter	This work	TCAS II [2]	TVLSI [3]	ESSCIRC [4]
Process	40nm	130nm	65nm	65nm
Supply voltage	1.1V	3.3V	1V	0.4V
Quiescent current	**120nA**	**1.1nA**	2.25uA	540nA
Hysteresis voltage	**470mV**	40mV	22mV	85mV
Area	0.007mm²	0.004mm²	0.014mm²	0.007mm²

$$V_{POR} = 892 \ mV \pm 64 \ mV (1\sigma)$$
$$V_{BOD} = 422 \ mV \pm 62 \ mV (1\sigma)$$
$$V_{hyst} = 470 \ mV \pm 16 \ mV (1\sigma)$$

Fig. 6. Measured results versus temperature (samples:3)

Fig.5 shows the measured V_{POR} trigger at $901 \ mV$ during power-up, where the rising and falling time of supply voltage is about 25ms. Measurements across 3 samples yield V_{POR}, V_{BOD}, and T_d respectively (Fig. 6). The test temperature ranges from -40°C to 100°C, with the step is 10°C/point. Due to the subthreshold voltage sensitive to the temperature, V_{POR} and V_{BOD} decrease as the temperature increases generally. The average measured V_{POR} variation coefficient exhibits $\Delta V/\Delta T = -95mV/140°C = -0.68 \ mV/°C$, aligning with threshold voltage variation of NMOS. The delay time of POR pulse is above $1.5\mu s$ across all temperatures and samples, ensuring the digital circuit has enough time to reset.

Post simulated Monte Carlo results (Fig. 7) across 200 samples at all process-voltage-temperature (PVT) corner (-40-125°C), included both global process variation and local mismatch, revealing 3.4% σ/V_{hyst} threshold variation and 12.3% current fluctuation (Fig.7(a)). At the 50°C , the quiescent current is $120 \pm 30 \ (1\sigma) \ nA$, and the threshold variations of POR are shown as (Fig.7(a)):

Local mismatch dominates $\sigma(V_{POR}) = 64mV$ due to uncorrelated $V_{th,p}$ variations in P5/P6 pair. Table. I show the summary of the POR performance and comparison with the state of the art. It is concluded that the proposed circuit consumed a lower quiescent current and a higher hysteresis voltage as compared with other works. A higher hysteresis voltage prevents false triggering due to supply noise or fluctuating power, which is crucial for IoT and battery-powered applications.

V. CONCLUSION

An on-chip POR integrated POR and BOD functions has been implemented for the 1.1 V digital circuit in IoT device.

979-8-3315-3918-4/25 $31.00 © 2025 IEEE

The prototype achieves $4.5\times$ lower I_q than prior resistor-less designs [4] and $20\times$ higher than bandgap-based PORs [3], enabling robust operation in noisy battery-supplied IoT nodes. Key innovations:

1) Synergistic POR/BOD state machine enabling low quiescent current and high hysteresis voltage.

2) V_{POR} tunability via PMOS ratio (Eq. 2) for multi-voltage domains.

3) Zero-bandgap architecture eliminating reference power.

ACKNOWLEDGEMENT

This work was supported by Special Funds for High-Quality Development (NICT) in Shanghai in 2024 (No. 2024-JCSS-01010)

REFERENCES

[1] H. You, J. Yuan, Z. Yu, and S. Qiao, "An Accurate Low-Power Power-on-Reset Circuit in 55-nm CMOS Technology", IEEE Trans. Circuits Syst. II Exp. Briefs, vol.69, no.8, pp.3361-3365, 2022.

[2] J. Guo et al., "An ultra-low quiescent current resistor-less power on reset circuit", IEEE Trans. Circuits Syst. II Exp. Briefs, vol. 68, no. 1, pp. 146-150, Jan. 2021.

[3] B. Zhou, Y. Jin and F. Zhao, "Sub-1-V BGR and POR Hybrid Circuit With 2.25-μA Current Dissipation and Low Complexity", IEEE Trans. VLSI, vol. 28, no. 10, pp. 2228-2232, Oct. 2020.

[4] O. Nechushtan, A. Feldman and J. Shor, "A 385mV 270nW Accurate Voltage Level Detector for IoT", Proc. ESSCIRC, pp. 369-372, Sep. 2022.

[5] H. You, D. Shi, D. Shang, Y. Zhou, and S. Qiao, "A 409mV, Sub-10nW Power-on Reset Circuit Using Adaptive Accuracy Adjustment for Low Voltage Applications", IEEE International Symposium on Circuits and Systems (ISCAS), pp.1-5, 2024.

[6] A. Antonov, M. Karpovich, and V. Vasilyev, "Power-On Reset Circuit in 180-nm CMOS With Brownout Detection Stable Switching Points Long Reset Pulse Duration and Resilience to Switching Noise", IEEE Trans. VLSI, vol. 30, no. 10, pp. 1373-1380, 2022.

An Anti-Single Particle Effect Over Temperature Protection Circuit Based on Dual Detectors

Ping Luo*[1], *Senior Member, IEEE*, Hong Zhao[1], Hao Wang[1], Fulin Yao[1,2], Kai Luo*[2]

[1] The State Key Laboratory of Electronic Thin Film and Integrated Devices, University of Electronic Science and Technology of China, Chengdu 611731, Sichuan, China
[2] the 24th Research Institute of China Electronics Technology Group Corporation, Chongqing, 400060, China

* Email: pingl@uestc.edu.cn, eeluok@163.com

Abstract— Over temperature protection (OPT) circuit is a key sub-circuit in power integrated circuits. This circuit can prevent the chip from being damaged due to excessively high operating temperatures. In space environment, OTP circuit is extremely prone to single-event transient flipping events, which may lead to the false shutdown of the chip. This paper proposes an OTP circuit based on dual detectors which adopts distinct voltage references. The proposed radiation hardened OTP circuit avoids the false triggering events caused by single-event transients affecting the reference module or the detector module. Compared with traditional radiation hardened redundant protection measures, the anti-single-event radiation OTP circuit proposed in this paper reduces the chip area while enhancing the radiation harden of the chip.

Keywords— Over temperature protection, radiation hardened, single particle effect, dual detector, distinct references

I. INTRODUCTION

With the advancement of aerospace technology, the application of integrated circuits (ICs) in the aerospace domain has progressively increased. However, the energy deposition of high-energy particles and radiation within devices in the space environment can induce various radiation effects, including total ionizing dose (TID) and single-event effects (SEEs), leading to device performance degradation or even failure. Research indicates that as IC feature sizes continue to scale down, logic gate counts increase, and operating voltages decrease, SEEs have become a predominant failure mechanism for spaceborne ICs [1].

Power integrated circuits (PICs) are indispensable components in space systems. When struck by high-energy particles in the space environment, PICs may experience SEEs, such as single-event transients (SETs), single-event upsets (SEUs), and single-event latchup (SEL). These effects can cause temporary operational disruptions or permanent damage to the circuit [2]. The over-temperature protection (OTP) circuit is a critical functional sub-circuit within PICs and is also highly sensitive to radiation. To mitigate susceptibility to SETs or SEUs, radiation hardening of this circuit is essential.

We conducted irradiation simulation tests on a low-dropout regulator (LDO) chip incorporating an OTP sub-circuit using the laser testing platform shown in Fig. 1, employing particles with a linear energy transfer (LET) of 37.5 MeV·cm²/mg. When the laser beam scanned over the OTP sub-circuit in the LDO chip, waveforms were observed on the oscilloscope as illustrated in Fig. 2. From top to bottom, the waveforms correspond to the power supply current I_{DD}, the LDO output voltage V_{OUT}, and external reference voltage V_{REF}.

This work was supported by project YG2404.

Fig. 1. Single particle laser simulation testing device.

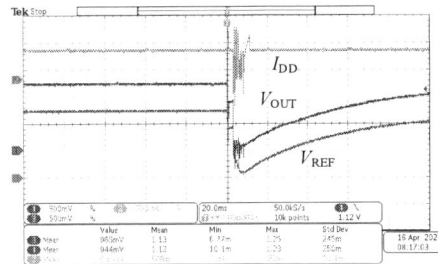

Fig. 2. Voltage waveforms of the LDO during a single-event strike on the OTP module in the chip.

It can be observed that following the single-event strike, both V_{OUT} and V_{REF} were pulled down. Upon the cessation of the single-event effect, output subsequently recovered, indicating the occurrence of a SET. Consequently, radiation hardening of the OTP module against SEEs is particularly crucial.

Currently, radiation harden techniques for digital IC against SEEs typically employ triple modular redundancy [3], while analog ICs often utilize RC filtering [4]. However, the OTP circuit represents a special type of mixed-signal circuit with analog inputs and digital outputs. Consequently, the aforementioned radiation hardening techniques cannot be directly applied to it. This paper proposes a dual detectors based OTP circuit with distinct voltage references, featuring a simple structure and small chip area.

II. TYPICAL OTP CIRCUIT AND ITS SINGLE-EVENT EFFECTS

A. Structure and Operating Principle of a Typical OTP Circuit

Figure 3 illustrates the structure of a typical hysteretic OTP circuit employed in PICs. The proportional to absolute temperature current I_{PTAT} generates a current proportional to temperature. By appropriate parameters design of M_{P1}, R_1, and R_2, the voltage V_A in Fig. 3 also exhibits a positive temperature coefficient. When the temperature rises sufficiently such that V_A equals the reference voltage V_{REF}, the OTP circuit triggers, driving the output V_{OTP} to a high logic level.

979-8-3315-3918-4/25 $31.00 © 2025 IEEE

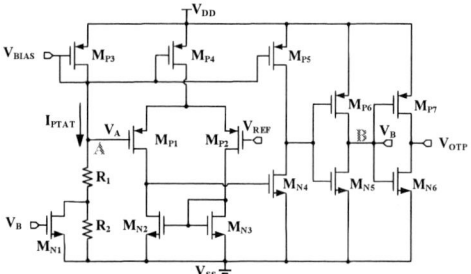

Fig. 3 Typical OTP circuit.

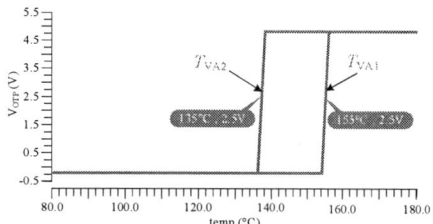

Fig. 4. Simulation results of OTP circuit in Fig.3.

Specifically, when temperature is low, V_A is less than V_{REF}, resulting in a high voltage of V_B, this turns on transistor M_{N1}. Consequently, the voltage at node A is given by equation (1)

$$V_A = V_{A1} = I_{PTAT} \times R_1 \qquad (1)$$

As the temperature increases, I_{PTAT} increases, and V_A begins to exceed V_{REF}. Hence, the output signal V_{OTP} transitions to a high logic level. Simultaneously, V_B toggles low, turning off M_{N1}. This action effectively places resistors R_1 and R_2 in series, altering the voltage at node A to

$$V_A = V_{A2} = I_{PTAT} \times (R_1 + R_2) \qquad (2)$$

When the temperature subsequently decreases from a high value, I_{PTAT} decreases, the voltage V_A shifts back to V_{A1}. Thus, the combination of transistor M_{N1} and resistor R_2 introduces hysteresis. This hysteresis prevents frequent toggling of the OTP output voltage near the protection threshold temperature, thereby avoiding disruption to the normal operation of the chip.

Set the power supply voltage V_{DD} in Fig. 3 to 5V. The temperature was swept from 80°C to 180°C during simulation. The simulation results depicted in Fig. 4 demonstrate that as temperature increases, the output signal V_{OTP} undergoes an abrupt transition from 0V to 5V when temperature reaches approximately T_{VA1} =155°C, triggering the OTP protection mechanism. Conversely, when temperature decreases from a high value, V_{OTP} initially remains at the high logic level and begins to fall to 0V only when temperature drops to approximately T_{VA2} =135°C. As evident from the figure, the designed OTP circuit functions correctly and exhibits a hysteresis window of approximately 15°C.

B. SEEs and False Triggering in OTP Circuit

When a PIC chip is struck by a single particle, a dense track of electron-hole pairs is formed along the particle trajectory. The charge induced by the single-event strike within the reverse-biased drain-substrate junction depletion region in MOS transistors is collected by the drain terminal. Concurrently, the inherent parasitic bipolar transistor effect in the circuit structure amplifies the transport of the single-event-induced charge. This amplification results in a transient current pulse generated by the single-particle strike [5].

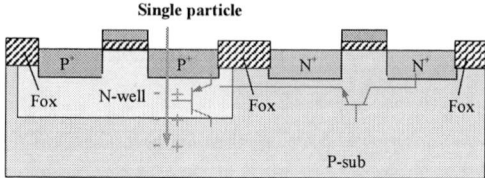

Fig. 5. Schematic diagram of MOS transistor subjected to single particle bombardment.

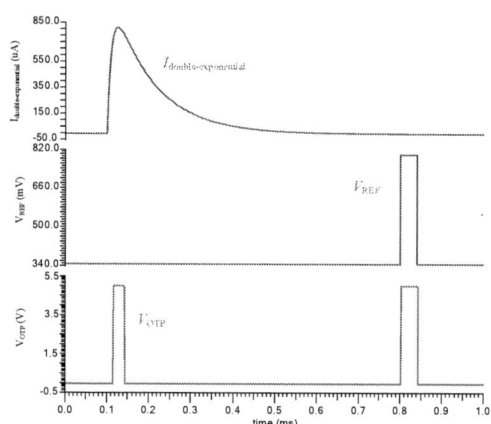

Fig. 6. Single particle simulation of non-hardened OTP.

Single-event strike simulations were performed on the LDO incorporating the non-hardened OTP circuit shown in Fig. 3. At $t = 100\mu s$, a double-exponential current source as [5] was applied to sensitive nodes within the internal circuitry. As observed, this induced false triggering of the OTP circuit. Furthermore, if the reference module experiences a single-event strike at $t = 800\mu s$ causing an overshoot in V_{REF}, false triggering of the OTP circuit also occurs. Thus, a single-event strike impacting either the internal OTP circuitry or the reference voltage V_{REF} can induce an erroneous transition in the final over-temperature signal, leading to unintended shutdown of the chip.

III. RADIATION HARDENED DESIGN FOR THE OTP CIRCUIT

A. Hardening Strategy

Considering that the voltage reference is also susceptible to single-event effects, this paper proposes a dual-detection OTP circuit based on distinct references shown in Fig.7. The fundamental strategy is to supply two separate OTP detection units with different voltage references. An AND-like logic block then ensures that the OTP circuit triggers only due to actual over-temperature conditions, rather than false triggering induced by a single-event strike.

B. Circuit Block Implementation

PICs typically incorporate a voltage reference block that generates a stable voltage independent of both the power supply voltage and temperature. The VREF1 block in Fig. 7 can utilize this inherent reference block within the PIC.

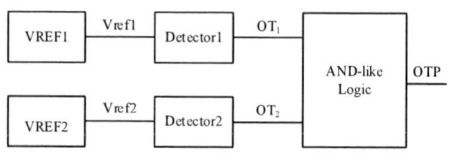

Fig. 7. Architecture of the dual detectors OTP circuit based on different voltage references.

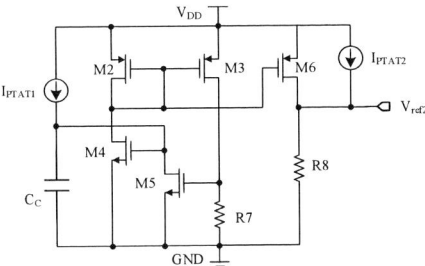

Fig. 8. Circuit diagram of voltage reference VREF2.

Furthermore, the inherent reference in a PIC often generates an I_{PTAT} current to bias other circuit blocks. Consequently, this I_{PTAT} current can be summed with a complementary-to-absolute-temperature (CTAT) current and flowed through a resistor to generate an additional temperature-independent reference voltage.

Figure 8 depicts the circuit structure designed for VREF2 in Fig. 7. This reference is an all-MOS reference, featuring a compact chip area. This block employs transistors M2, M3, M4, and M5 to form a negative feedback loop. Design M5 operating in subthreshold region, so that the feedback loop forces the gate-source voltage V_{GS5} of M5 to develop across resistor R7, generating a CTAT current I_{CTAT}. The I_{CTAT} and I_{PTAT} are summed and flowed through resistor R8. Transistors M2, M3, and M6 are designed with identical device sizes (W/L). The output voltage V_{ref2} can thus be expressed as:

$$V_{ref2} = (\frac{V_{GS5}}{R_7} + I_{PTAT2})R_8 \qquad (3)$$

While, V_{GS} of a MOS transistor operating in the subthreshold region can be expressed as the following equation [6]:

$$V_{GS}(T) = V_{TH}(T) + n(\frac{kT}{q})\ln[(\frac{L}{W})(\frac{q}{kT})^2(\frac{I_{DS}}{n\mu(T)C_{ox}})] \qquad (4)$$

where, T is the temperature, V_{TH}, C_{ox} are the threshold voltage and unit gate oxygen capacitance of a MOS transistors, respectively. The mobility $\mu(T)$ can be expressed as $\mu(T) = \mu(T_0)(T/T_0)^{-2}$.

Thus, the V_{GS} temperature coefficient of MOS transistor operating in the subthreshold region can be expressed as:

$$\frac{\partial V_{GS}}{\partial T} = \frac{\partial V_{TH}}{\partial T} + \frac{nk}{q}\ln[(\frac{L}{W})(\frac{q}{k})^2(\frac{I_{ds}T_0^2}{n\mu(T_0)C_{ox}})] \qquad (5)$$

where, V_{TH} decreases with increasing temperature, whereas the second term approaches zero. Consequently, the V_{GS} of a subthreshold region MOS exhibits a negative temperature coefficient (NTC). Therefore, the current flowing through resistor R7 possesses an NTC characteristic. By appropriately configuring the ratio of resistors R7 and R8, a temperature-independent reference voltage V_{ref2} can be obtained.

Detector1 and Detector2 in Fig. 7 form the dual detectors in the proposed OTP circuit, both of them adopt the same structure depicted in Fig. 9 for over temperature detection. This circuit leverages the NTC characteristic of the bipolar junction transistor (BJT)'s turn-on voltage $V_{BE,ON}$. At room temperature, $V_{BE,ON}$ is typically 700mV, with a temperature coefficient of approximately -2mV/°C. $V_{BE,ON}$ decreases as the temperature rises. When $V_{BE,ON}$ falls below V_{ref}, the detector

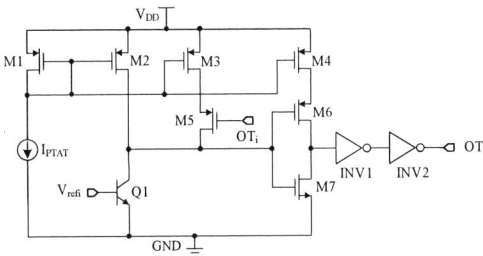

Fig. 9. Circuit structure of the over-temperature detector (where i=1,2).

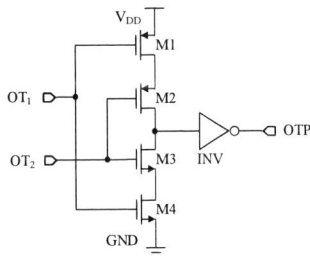

Fig. 10. Circuit of AND-like logic block.

output OT_i transitions to a high logic level, signaling an over-temperature condition. Consequently, the over-temperature threshold temperature T_{th} can be derived as follows:

$$T_{th} = \frac{700 - V_{refi}}{2} + 27 \qquad (6)$$

M3, M5, INV1 and INV2 in Fig. 9 form a hysteresis interval. When OT_i outputs a high level, M5 is disconnected, and the PTAT current flowing into Q1 changes, causing the turn-on voltage of Q1 to change. Thus forms a hysteresis interval to prevent the circuit from repeatedly switching.

Two parallel PMOS transistors in a normal AND gate circuit are prone to conducting due to SETs when both inputs are high, resulting in incorrect output flipping. Therefore, an AND-like logic block not AND logic circuit is utilized in the proposed OPT circuit in Fig.7 which is shown in Fig.10. The output logic signal OT_1 and OT_2 of dual detectors are the input signals of the AND-like logic block. OTP outputs a high level only when both OT_1 and OT_2 are high [7]. The truth table of the circuit shown in Fig. 10 is presented in Table 1.

TABLE I. TRUTH TABLE OF AND-LIKE LOGIC BLOCK

OT_1	OT_2	OTP
1	1	1
0	0	0
0	1	Retained
1	0	Retained

The inverter INV is used to ensure glitch-free signal propagation from the OTP detection units to the final output. That means, the over-temperature signal propagates to the output only when both Detector1 and Detector2 transition to a high logic level. Consequently, if a single-event strike affects one over-temperature detector, this logic circuit prevents the propagation of an erroneous over-temperature signal to the output. This design effectively avoids false shutdown of the chip induced by single-event effects within the OTP circuit.

IV. CIRCUIT SIMULATION

Figure 11 presents the simulation result of VREF2 block shown in Fig. 8. This reference achieves a temperature

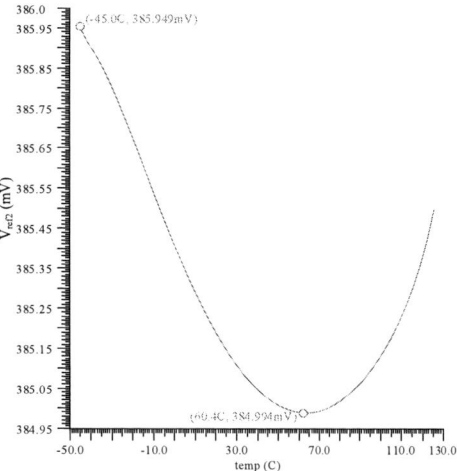

Fig. 11. Simulation waveforms of V_{ref2}.

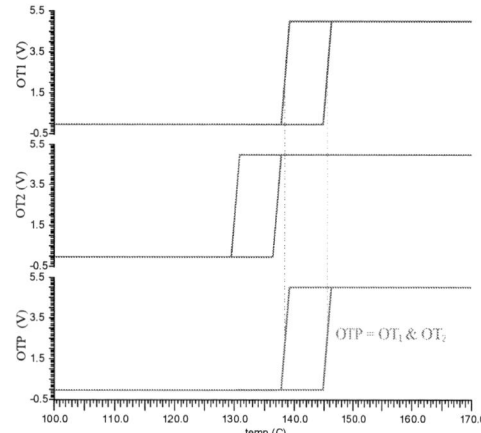

Fig. 12. Simulation waveforms of the over-temperature detector.

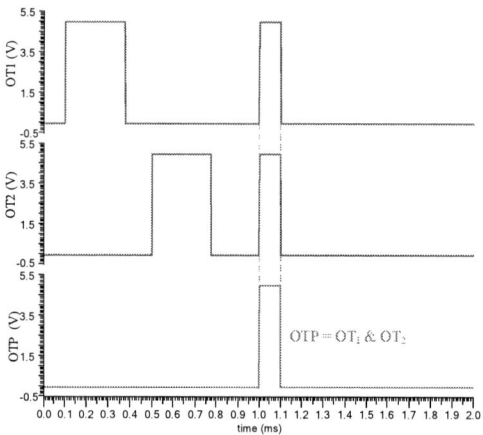

Fig. 13. Single particle simulation diagram of the proposed OTP.

V. CONCLUSION

This paper presents a radiation-hardened OTP circuit designed for PICs in aerospace applications. The proposed OTP circuit incorporates dual over-temperature detectors. Each detection path utilizes a distinct voltage reference: one leverages the inherent voltage reference module within the PIC, while the other generates a reference voltage by exploiting the negative temperature coefficient characteristic of the V_{GS} of MOS transistors operating in the subthreshold region, eliminating the need for BJTs or operational amplifiers. Furthermore, the over temperature detection blocks themselves are implemented without resistors or operational amplifiers. This design approach results in an OTP circuit characterized by structural simplicity and minimal silicon area overhead. Simulation results demonstrate that the proposed circuit effectively mitigates the probability of SEU induced false triggering within the OTP, significantly enhancing the overall radiation tolerance of the chip.

coefficient of 14.59 ppm/°C. The VREF2 is specially design for dual detectors in the anti-single particle effect OTP circuit flipping at high temperatures. The key design constraint of $V_{ref2} > V_{ref1}$ is maintained, the correct operation of the radiation-hardened OTP circuit can be ensured.

Figure 12 presents the simulation waveforms of the over-temperature detector shown in Fig. 9. It is shown that there is a temperature hysteresis of 15 °C for each over-temperature detector, and the temperature trip point of OT_2 in Fig. 7 is higher than that of OT_1, which is consistent with our preset $V_{ref2} > V_{ref1}$.

Figure 13 shows the single particle simulation results of the proposed OTP circuit. If single-event strikes impact either the Dtector1 or the Detector2, where encompasses a strike affecting either the reference voltage or the detection unit itself, causing an erroneous transition in the corresponding OT_1 or OT_2 signal. While the signal in the other detection remains unaffected and correctly indicates the temperature state, the final over-temperature signal OTP remains inactive. An erroneous OTP output occurs only if simultaneous single-event strikes affect both paths concurrently. Given the statistically improbable nature of two spatially distinct strikes occurring simultaneously, this design significantly enhances the single-event effect immunity of the over-temperature protection circuit.

REFERENCES

[1] V. Ferliet-Cavrois, L. W. Massengill, and P. Gouker. "Single event transients in digital CMOS − a review". IEEE Trans Transactions on Nuclear Science, 2013, 60 (3):1767-1790.

[2] Y F Zhao, S G Yue, and X Y Zhao, et al. "Single event soft error in advanced integrated circuit". Journal of Semiconductors, 2015, 36(11): 1-14.

[3] Methodology Ronald C. Lacoe. "Improving integrated circuit performance through the application of hardness-by-design methodology". IEEE Trans Transactions on Nuclear Science, 2008, 55 (4):1903-1925.

[4] Jingtian Liu, Dongsheng Wang, and Bin Liang, et al. "Soft error tolerant bandgap reference utilizing single-event transient filtering Technique". IEEE Transactions on Nuclear Science. 2024, 71(4): 895–901.

[5] D. A. Black, W. H. Robinson, and I. Z. Wilcox, et al. "Modeling of single event transients with dual double-exponential current sources: implications for logic cell characterization". IEEE Transactions on Nuclear Science, 2015, 62(4): 1540-1549.

[6] L H De Carvalho Ferreira and T C Pimenta. "A CMOS voltage reference based on threshold voltage for ultra low-voltage and ultra low-power". In 17th International Conference on Microelectronics, Islamabad, Pakistan, 2005, 10-12.

[7] M Zhang, Subhasish Mitra, and T. M. Mak, et al. "Sequential element design with built-in soft error resilience". IEEE Transactions on Very Large Scale Integration (VLSI) Systems, 2006, 14(12): 1368-1378.

A Fast Start-Up and Low-Power 32-kHz Crystal Oscillator for Real-Time Clock and Frequency Calibration

Jie Zheng[1], Qiang Li[2], Hao Min*[1]

[1] State Key Laboratory of ASIC and System(Fudan University), Shanghai 200433, CHINA
[2] Shanghai Quanray Electronics Co., Ltd.

* Email: 23212020033@m.fudan.edu.cn, hmin@fudan.edu.cn

Abstract—With the development of the Internet of Things (IOT), the requirements for clock signals have become increasingly stringent. As a commonly used clock unit in IoT nodes, the crystal oscillator's startup time, power consumption, and phase noise can impact the overall system performance. In this paper, a fast start-up and low-power 32-kHz crystal oscillator is proposed. Fabricated in a 55-nm CMOS process, the proposed XO's start-up time for real-time clock (RTC) and frequency calibration is less than 20ms and 80ms with the temperature range of -40 ℃ to 85 ℃ and its power consumption for low-power is less than 180nW. Also, it achieves a phase noise of -74.4dBc/Hz at 1.1Hz.

Keywords—fast start-up, low-power, crystal oscillator, phase noise

I. INTRODUCTION

Over the past few decades, the Internet of Things (IoT) has experienced rapid development, placing higher demands on various IoT sensor nodes. Generally, the equipment of each IoT sensor node typically consists of five core components: sensors, power management, clock, wireless communication, and central processing unit. For low-power applications, especially in duty-cycle devices [1], the power consumption of the clock module accounts for a significant portion of the entire system. This imposes dual requirements on clock design:(1) Start-up speed must be optimized to minimize power consumption during sleep modes;(2) Operational power consumption of the clock needs reduction as well.

Crystal Oscillators (XOs) are widely used as clocks in IoT nodes due to their low phase noise and high frequency stability [2]. Typically, an IoT system employs two reference clocks:(1) A low-frequency crystal oscillator (e.g., 32.768 kHz) for the always-on real-time clock (RTC);(2) A high-frequency crystal oscillator for frequency calibration to generate radio frequency (RF) transceiver clocks. As previously discussed, both start-up time and power consumption are critical metrics for oscillator circuits. Due to the exceptionally high quality factor (Q) of crystals, crystal oscillator circuits typically exhibit long start-up times. Consequently, designing XOs that achieve fast start-up while maintaining low power consumption has become a key research focus. Moreover, the need for two separate crystals imposes constraints on the area of individual IoT nodes. If a single XO could serve both purposes, significant improvements in power efficiency and area reduction would be realized

In this article, we present a fast start-up and low-power 32-kHz crystal oscillator operating at 0.8V for real-time clock and frequency calibration in order to solve the problem mentioned before. Fabricated in a 55-nm CMOS process, the proposed XO's start-up time for RTC and frequency calibration is less

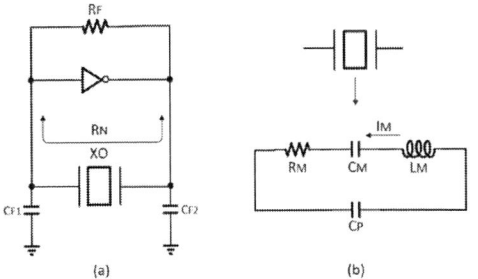

Fig. 1. (a) Pierce oscillator (b) Equivalent circuit model of quartz crystal

than 20ms and 80ms with the temperature range of -40 °C to 85 °C. Its low-power mode consumption is less than 180nW. Also, it achieves a phase noise of -74.4dBc/Hz at 1.1Hz.

This article is structured as follows. Section II analyzes methods for shortening start-up time and challenges encountered when using a 32-kHz crystal oscillator for frequency calibration. The proposed oscillator architecture is presented in Section III. Section IV shows the layout and simulation results of the crystal oscillator and Section V is the conclusion.

II. ANALYSIS OF XO

Fig. 1 shows the traditional Pierce crystal oscillator circuit architecture and the equivalent circuit of the quartz crystal. Pierce crystal oscillators are widely used because of their low power consumption and low phase noise [3]. As shown in the Fig. 1, The quartz resonator, which can be modeled as a series resonant circuit (R_M, C_M, L_M) in parallel with a capacitance (C_P), is connected between the input and output terminals of a resistive-feedback inverter. Two external capacitors are additionally connected from both sides of the resonator to ground. The resistive-feedback inverter and two external capacitances provide sufficient gain and phase shift for crystal oscillator startup. Based on the study of oscillator start-up time [4][5], the expression for the crystal oscillator start-up time can be derived from its equivalent circuit model in Fig. 1.

$$T_{startup} = -\frac{2L_M}{R_M - |R_N|} \ln\left(\frac{|i_{M,ss}|}{|i_M(0)|}\right) \quad (1)$$

where L_M and R_M are the motional inductance and motional resistance of the resonator, R_N is the negative resistance seen by the quartz resonator, i_M is the motional current through the series resonant circuit as shown in Fig. 1, $i_M(0)$ and $i_{M,ss}$ respectively represent the initial current and the steady-state current of i_M. R_N and $i_{M,ss}$ can be calculated as follows [4]:

$$R_N = \frac{g_m C_{F1} C_{F2}}{(g_m C_P)^2 + \omega_{XO}^2 (C_{F1} C_{F2} + C_{F1} C_P + C_{F2} C_P)^2} \quad (2)$$

979-8-3315-3918-4/25 $31.00 © 2025 IEEE

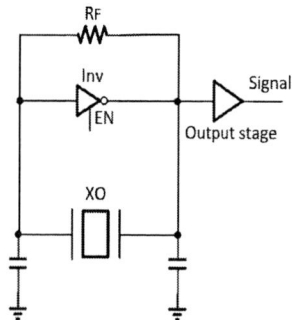

Fig. 2. Proposed architecture of the crystal oscillator

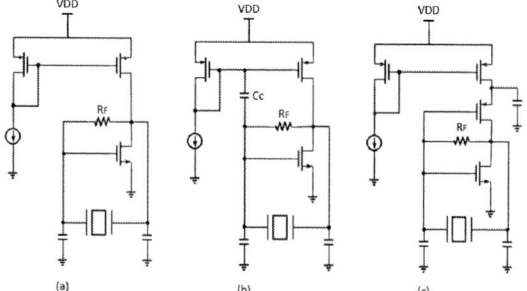

Fig. 3. Three traditional inverter architectures

$$|i_{M,ss}| \approx \omega_{XO} V_{XO} \left(\frac{C_{F1}C_{F2}}{C_{F1}+C_{F2}} + C_P \right) \quad (3)$$

where g_m is the transconductance of the inverter, ω_{xo} is the oscillation frequency of the crystal oscillator, V_{xo} is the voltage amplitude across the resonator, which can be approximated as the peak-to-peak voltage amplitude at a single crystal terminal, C_{f1} and C_{f2} are the external capacitance and C_P is the parallel capacitance shown in Fig. 1.

As can be seen from (1), the startup time of crystal oscillator depends only on R_n, $i_M(0)$ and V_{xo} because other parameters are inherent parameters determined by the crystal. However, it is challenging to reduce the startup time by increasing $i_M(0)$ and reducing V_{xo}. On the one hand, while pulse injection can enhance $i_M(0)$, the injected signal frequency must be near the crystal's fundamental frequency to effectively boost $i_M(0)$. This requires generating a low-frequency oscillation source around 32kHz in this work, which is difficult to achieve at low cost. On the other hand, since the oscillator circuit must also support frequency calibration, phase noise becomes a critical constraint. According to Hajimiri and Vittoz's theory [4][6], under large-signal conditions, the oscillator operates as a nonlinear time-varying system. Its phase noise power spectral density is inversely proportional to the maximum energy stored in reactive components at steady state. The maximum energy is proportional to the square of $i_{M,SS}$ and as established earlier, $i_{M,SS}$ is proportional to V_{xo}. Thus, the noise power spectral density is proportional to the square of V_{xo}. To achieve low phase noise, V_{xo} must oscillate at full swing, conflicting with the low V_{xo} required for fast startup. Therefore, increasing the negative resistance is the optimal approach to accelerate the startup time for 32kHz crystals. According to (2), it can be seen that the only variable parameter is g_m. We can calculate the maximum value of the negative resistance R_N as:

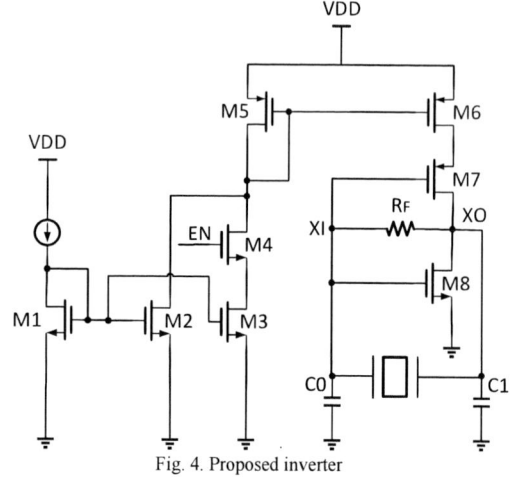

Fig. 4. Proposed inverter

$$R_{N,max} = \frac{C_{F1}C_{F2}}{2\omega_{XO}C_P(C_{F1}C_{F2}+C_{F1}C_P+C_{F2}C_P)} \quad (4)$$

when

$$g_{m,opt} = \frac{\omega_{XO}(C_{F1}C_{F2}+C_{F1}C_P+C_{F2}C_P)}{C_P} \quad (5)$$

Substituting into the calculation yields the required optimal inverter transconductance. Therefore, during the startup process, the transconductance should be maintained near this value to achieve the minimum startup time.

III. PROPOSED OSCILLATOR ARCHITECTURE

Fig. 2 shows the architecture of the crystal oscillator designed in this work. To simultaneously satisfy the requirements of low power consumption and fast startup, a dual-mode switching mechanism is inherently necessary. This paper achieves mode switching by adjusting the current magnitude of the tail current source, thereby modulating the transconductance g_m of the inverter. During fast startup mode, it maintains transconductance at the optimal value $g_{m,opt}$ to generate a square-wave signal for frequency calibration and RTC clock synchronization. Upon completion of frequency calibration, the circuit switches to a low-current sustain oscillation mode, producing only the RTC clock signal. The following sections detail the design of each modular circuit.

A. inverter

The oscillator in this design employs the Pierce oscillator. One critical aspect of Pierce oscillator design is the implementation of its inverter circuit. Typically, there are three variant inverter architectures [7], as illustrated in Fig. 3. In this work, Fig. 3. (c) is adopted. Compared to Fig. 3. (a), an additional PMOS is introduced in Fig. 3. (c). Under the small-signal model, the effective transconductance equals the sum of the PMOS and NMOS transconductances. This allows achieving higher transconductance with lower bias current, thereby reducing overall power consumption. Compared to Fig. 3. (b), the small-signal models are similar. However, as previously noted, the oscillation amplitude of our crystal circuit is exceptionally large. Due to the coupling capacitor C_c, the oscillating signal perturbs the gate voltage of the current source, increasing its average current and consequently elevating power dissipation. The final inverter architecture is shown in Fig. 4. The enable signal (EN), which indicates completion of frequency calibration, controls M4 to adjust the bias current, dynamically modulating the inverter's transconductance. Notice that M4 cannot be integrated within

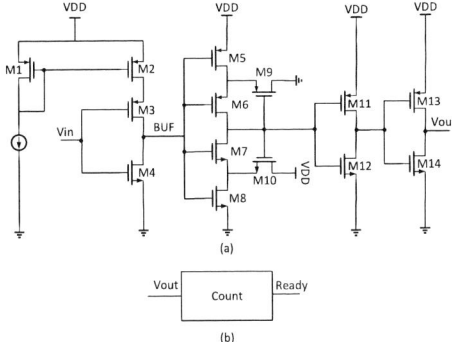

(a)

(b)

Fig. 5. (a) Square wave generator (b) Ready signal generator

the inverter branch composed of M7 and M8. Because at a supply voltage of 0.8 V, integrating M4 in this branch would violate voltage headroom requirements, causing the inverter to malfunction. Thus, a dedicated current source branch is necessary for biasing.

B. Output stage

Fig. 5 shows the basic structure of the output stage. It primarily consists of a high-gain amplifier, a Schmitt trigger, a counter, and a buffer. The structure of the high-gain amplifier is identical to the preceding inverter stage. This design choice eliminates the need for additional biasing circuit while simultaneously providing substantial gain. This is achieved because the preceding circuit, through resistive feedback, ensures that the input and output quiescent operating points of the inverter are biased at the same voltage. The subsequent stage employs a Schmitt trigger to convert the analog output into a square wave at the corresponding frequency. The positive and negative threshold voltage of a Schmitt trigger are given by

$$V_{T+} = \frac{\left(V_{DD} + \sqrt{\frac{(W/L)_{M8}}{(W/L)_{M10}}} V_{th,n}\right)}{1 + \sqrt{\frac{(W/L)_{M8}}{(W/L)_{M10}}}} \qquad (6)$$

$$V_{T-} = \frac{\sqrt{\frac{(W/L)_{M5}}{(W/L)_{M9}}}\left(V_{DD} + \sqrt{\frac{(W/L)_{M5}}{(W/L)_{M9}}} V_{th,p}\right)}{1 + \sqrt{\frac{(W/L)_{M5}}{(W/L)_{M9}}}} \qquad (7)$$

where $V_{th,n}$ and $V_{th,p}$ are the threshold voltages of the NMOS and PMOS transistors, respectively. As indicated by (6)(7), we can adjust the threshold voltage by modifying the width-to-length ratio (W/L) of the NMOS and PMOS transistors. This allows the circuit to directly output a square wave signal even if the oscillation amplitude is not full-swing. The frequency of this output signal is stable, but its duty cycle varies which means the phase of the signal shifts. If such a signal is directly used for frequency calibration, the phase variation would compromise calibration accuracy. Therefore, a ready signal for phase-stable indication is required. Typically, we only need to ensure that the signal reaching the Schmitt trigger attains full-swing amplitude to meet the phase-stable requirement. To achieve this, an additional counter module is necessary to record the time difference between the time when the square wave is generated and become phase-stable. Once the count reaches a predefined value, a 'ready' signal is outputted to indicate stability for frequency calibration. Finally, a driver stage is included to drive the subsequent load.

TABLE I. STARTUP TIME AND POWER CONSUMPTION IN DIFFERENT SITUATIONS

Corner	Startup time₁	Startup time₂	Power consumption
Tt(-40)	0.3ms	60ms	105nW
Tt(27)	5.6ms	65ms	114nW
Tt(85)	14.3ms	75ms	134nW
Ff(-40)	0.3ms	50ms	108nW
Ff(27)	6.3ms	60ms	125nW
Ff(85)	15.4ms	70ms	178nW
Ss(-40)	0.32ms	55ms	101nW
Ss(27)	5ms	65ms	108nW
Ss(85)	11.4ms	70ms	118nW
Fs(-40)	0.5ms	55ms	105nW
Fs(27)	5.4ms	60ms	114nW
Fs(85)	13.7ms	73ms	148nW
Sf(-40)	0.32ms	50ms	105nW
Sf(27)	5.4ms	60ms	114nW
Sf(85)	12.7ms	70ms	127nW

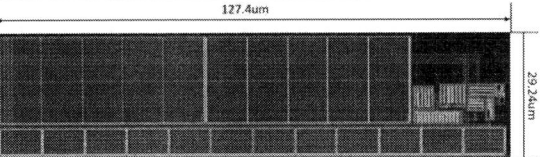

Fig. 6. Layout of the XO

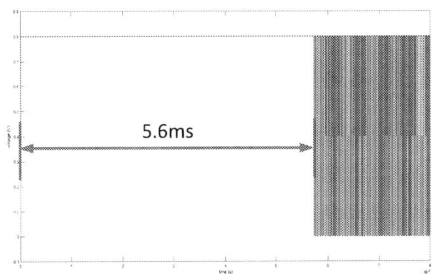

Fig. 7. Start-up time₁ at 27°C under the TT process corner.

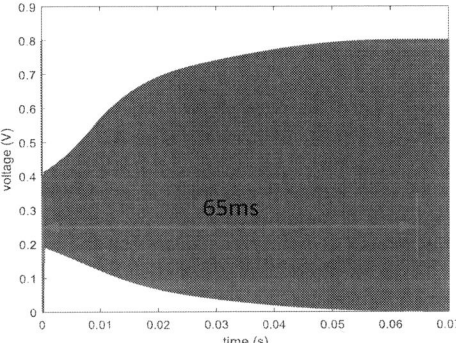

Fig. 8. Start-up time₂ at 27°C under the TT process corner.

IV. SIMULATION AND TEST RESULTS

The proposed oscillator is fabricated in 55nm CMOS technology. Its layout is shown in Fig. 6, occupying an area of $127.4 \times 29.24 \ \mu m^2$.

Table I shows the start-up times and the average power consumption in low-power mode under post-layout simulation. Start-up time₁ refers to the output timing of the V_{out} sig-

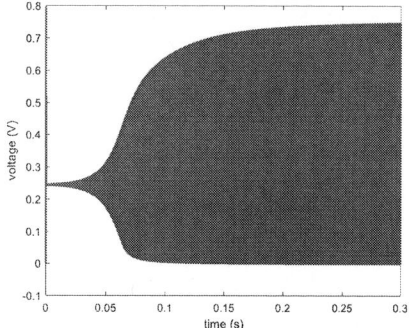

Fig. 9. Post-layout simulation result of XO

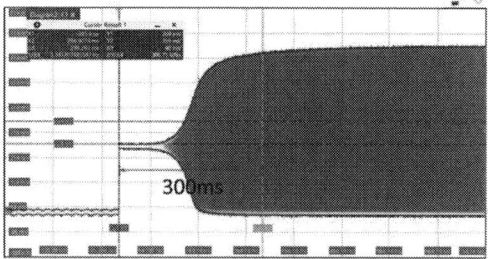

Fig. 10. test result of XO

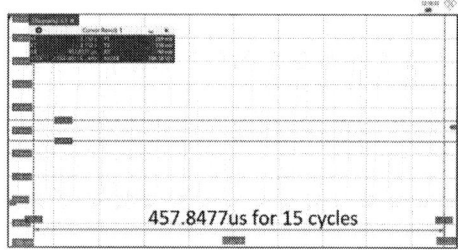

Fig. 11. Tested frequency of XO

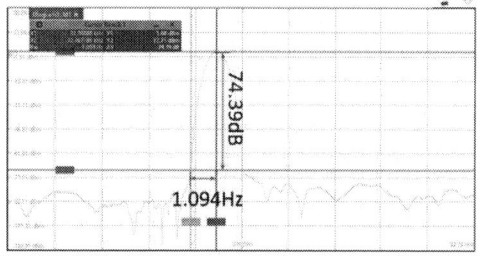

Fig. 12. Tested power spectrum of XO

nal in Fig. 5, indicating when the signal becomes usable for the RTC clock. Start-up time$_2$ refers to the timing when the BUF signal in Fig. 5 reaches full swing, marking when the signal becomes available for frequency calibration. For example, Fig. 7 and Fig. 8 displays both start-up times at 27°C under the TT process corner.

Due to the integration of the crystal oscillator within the entire chip, direct testing of the BUF and V_{out} terminals is not feasible. Instead, we verify the functionality of our module by monitoring signals at the XO terminal in Fig. 4. Fig. 9 and Fig. 10 shows the simulation result and test result of XO. It can be observed that both waveforms are highly consistent and reach nearly full swing at 300ms. The difference

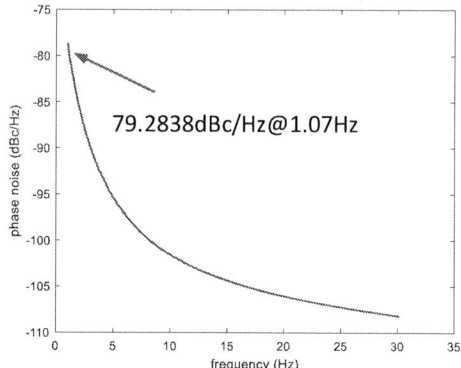

Fig. 13. Simulation results of XO phase noise

in peak-to-peak voltage is due to the actual VDD voltage being below 0.8V during chip operation.

Fig. 11 and Fig.12 show the frequency and power spectrum of XO respectively. Based on the data in the Fig. 11, the calculated crystal oscillator frequency is 32.762 kHz, which shows minimal deviation from the ideal value of 32.768kHz, confirming normal functionality of the oscillator. Also, it can be obtained the phase noise at 1.1Hz is -74.4dBc/Hz. Compared to the simulation result shown in Fig.13, the actual phase noise has degraded by 5dBc/Hz.

V. CONCLUSION

This paper proposed a fast start-up and low-power 32-kHz crystal oscillator for real-time clock and frequency calibration. Start-up time for RTC and frequency calibration is less than 20ms and 80ms with the temperature range of -40°C to 85 °C and its power consumption for low-power is less than 180nW as shown in Table I. Also, it achieves a phase noise of -74.4dBc/Hz at 1.1Hz.

VI. ACKNOWLEDGEMENTS

The paper is sponsored by National Science & Technology Major Project of China with No. 2023YFB4403600

[1] J. Gerard, J. A. Fraire, R. Ortigueira and S. Céspedes, "DtS-IoT Resource Allocation Analysis Framework: Assessing DQ and RESS-IoT," 2024 IEEE 10th World Forum on Internet of Things (WF-IoT), Ottawa, ON, Canada, 2024, pp. 672-677

[2] K. M. Megawer et al., "A Fast Startup CMOS Crystal Oscillator Using Two-Step Injection," in IEEE Journal of Solid-State Circuits, vol. 54, no. 12, pp. 3257-3268, Dec. 2019.

[3] S. Iguchi, H. Fuketa, T. Sakurai and M. Takamiya, "Variation-Tolerant Quick-Start-Up CMOS Crystal Oscillator With Chirp Injection and Negative Resistance Booster," in IEEE Journal of Solid-State Circuits, vol. 51, no. 2, pp. 496-508, Feb. 2016.

[4] Vittoz,Eric.Low-Power Crystal and MEMS Oscillators[M].Springer Netherlands,2010.

[5] A. Rusznyak, "Start-up time of CMOS oscillators," in IEEE Transactions on Circuits and Systems, vol. 34, no. 3, pp. 259-268, March 1987.

[6] A. Hajimiri and T. H. Lee, "A general theory of phase noise in electrical oscillators," in IEEE Journal of Solid-State Circuits, vol. 33, no. 2, pp. 179-194, Feb. 1998.

[7] W. Thommen, "An improved low power crystal oscillator," Proceedings of the 25th European Solid-State Circuits Conference, Duisburg, Germany, 1999.

A 0.067 mm^2 PNP-Based Temperature Sensor with ±0.6°C (3σ) Inaccuracy from −20°C to 80°C

Letian Li[1], Peilin Xiao[2], and Xuyang Lu[*1]

[1]Shanghai Jiao Tong University, Shanghai, China
[*]Email: lilt27@sjtu.edu.cn, xuyang.lu@sjtu.edu.cn

Abstract—This paper proposes a PNP-based temperature sensor that achieves a measured inaccuracy of ±0.6°C (3σ) across a temperature range of -20°C to 80°C. A novel combination of dynamic element matching (DEM) and finite current gain compensation is employed to effectively mitigate errors induced by mismatches. The temperature sensing module generates two temperature-dependent currents, I_{PTAT} and I_{CTAT}, which are subsequently sampled and quantized by a Σ-ΔADC. The chip has a 12-bit serial digital output and is manufactured using a 180-nm CMOS process, featuring a compact area of 0.067 mm^2, a power consumption of 39.6 μW, and a resolution FoM of 4.967 nJ·K^2. Additionally, the sensor demonstrates a remarkably low supply voltage sensitivity of 0.13 °C/V across a 1.6–2.0 V range, owing to the implementation of cascode current mirrors.

Keywords—*temperature sensor, BJT-based, dynamic element matching, finite current gain compensation, serial digital output.*

I. INTRODUCTION

Temperature sensors are widely used in System-on-Chips (SoC), Internet-of-Things (IoT) and wearable electronics. By continuously monitoring thermal conditions, they enable real-time system health assessment, anomaly detection, and dynamic adjustments such as dynamic voltage/frequency scaling (DVFS) in processors or thermal drift compensation in MEMS resonators. These sensors typically comprise a temperature-sensing element and a readout circuit that converts thermal information into digital codes or duty-cycle-modulated outputs. Among prevalent sensing technologies, resistor-based, MOS-based, and bipolar junction transistor (BJT)-based temperature sensors all exhibit distinct trade-offs in precision, energy efficiency, area, and calibration complexity [1]. Resistor-based temperature sensors offer simplicity and optimal energy efficiency but suffer from limited accuracy. MOS-based temperature sensors perform well under the constraints of low voltage, small area, and low power consumption, but are greatly affected by process deviations and nonlinearity. Both approaches necessitate two-point or multi-point calibration. Among these, a BJT-based temperature sensor is chosen in this paper for its high temperature linearity and good stability [2], [3].

Precision plays an important role in the design of temperature sensors. For MEMS resonators, ±0.5°C inaccuracy causes up to 12.5 ppm frequency error, degrading navigation precision by >10 m. Commonly used high-precision temperature sensing techniques include DEM technology, chopping, and

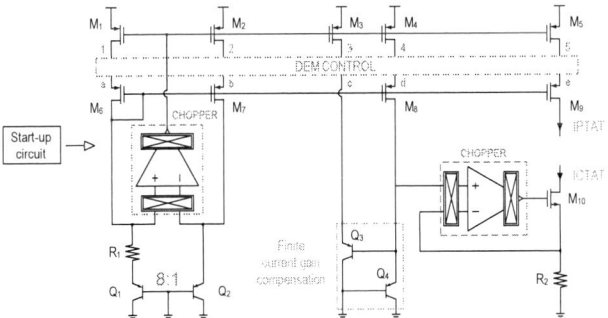

Fig. 1: Schematic of the proposed temperature sensing module.

finite-current gain compensation. Reference [4] proposed a low power NPN-based temperature sensor operating from −40 to 125°C, achieving an accuracy of ±0.85°C (3σ). Reference [5] proposed a wide sensing range temperature sensor with an inaccuracy of ±0.97°C.

This paper innovatively integrates DEM technology with finite current gain compensation to further enhance the robustness of the BJT's base-emitter voltage V_{BE}. Chopping is also used to improve accuracy and reduce the impact of noise. By properly using and combining these three methods, this paper achieves a PNP-based ±0.6°C (3σ) high precision temperature sensor. The rest of this paper is organized as follows: Section II details the architecture and operating principle of the proposed temperature sensor, Section III analyzes the source of errors and the effect of key high-precision technologies used in this design, Section IV shows the measurement results, and Section V summarizes this work.

II. DESIGN OF PROPOSED TEMPERATURE SENSOR

The temperature sensor chip designed in this paper is composed of a temperature sensing module, an Σ–ΔADC, and a digital readout circuit module. Temperature sensing module senses the temperature information in the environment and converts it into two temperature-related currents I_{PTAT} and I_{CTAT}. By using ADC to sample and quantify it, the temperature measurement results are obtained in digital form.

A. Temperature Sensing Module

The temperature sensing module is the core of the temperature sensor, and its basic working principle is shown in Fig.

979-8-3315-3918-4/25 $31.00 © 2025 IEEE

1. A dynamic start-up circuit ensures the system avoids operation at undesirable degeneration points. Once successfully activated, the start-up circuit itself consumes negligible power. As temperature sensing components, BJTs can separately generate complementary-to-absolute temperature (CTAT) voltage V_{BE} and proportional-to-absolute temperature (PTAT) voltage ΔV_{BE}. The base-emitter voltage V_{BE} is related to its bias current I_C:

$$V_{BE} = \frac{kT}{q}\ln\left(\frac{I_C}{I_S}\right), \qquad (1)$$

where k is the Boltzmann's constant, T is the absolute temperature, q is the electron charge, and I_S is the saturation current. The relationship between I_S and temperature can be written as:

$$I_S(T) = CT^{\eta}e^{-\frac{qV_{g0}}{kT}}, \qquad (2)$$

where V_{g0} is the bandgap voltage of silicon at 0K, η is a temperature constant that depends on the process, and it is equal to $4-\xi$. The constant ξ is derived from the temperature index of carrier mobility with temperature [6]. According to (1) and (2), V_{BE} can be described as:

$$V_{BE} = V_T\ln\left(\frac{I_C}{C}\right) - \eta V_T\ln(T) + V_{g0}. \qquad (3)$$

In (3), $V_T=kT/q$ is proportional-to-absolute temperature. Because of the term $-\eta V_T ln(T)$, V_{BE} has a high-order temperature item and a certain temperature non-linearity. Fortunately, this non-linearity can be effectively compensated by biasing PTAT current [7]. Considering only the first-order term, V_{BE} exhibits a negative linear dependence on temperature, defining it as the CTAT voltage V_{CTAT}.

According to (1), for two identical BJTs Q_1 and Q_2, if they have different collector currents I_{C1} and I_{C2}, they will have different base-emitter voltages V_{BE1} and V_{BE2}. This voltage difference can be expressed as:

$$\Delta V_{BE} = V_{BE1} - V_{BE2} = V_T\ln(N), \qquad (4)$$

where N is the ratio of the bias current I_{C1} and I_{C2}.

From (4), if we bias Q_1 and Q_2 proportionally, ΔV_{BE} will have an excellent positive linear relationship with temperature. By using two proportional resistors to convert V_{BE} and ΔV_{BE} into current domain, we can obtain a positive linear temperature current I_{PTAT} and a negative linear temperature current I_{CTAT} respectively.

B. Architecture of Proposed Temperature Sensor

Fig. 2 shows the architecture of the proposed temperature sensor. In the temperature sensing module, it separately generates two temperature dependent currents I_{CTAT} and I_{PTAT}. The Σ-ΔADC module converts these two currents into a bitstream output. The module mainly includes a comparator, a reference voltage V_{REF}, and an integrating capacitor C_{int}. It uses a single-bit feedback signal BS to control I_{PTAT} and I_{CTAT} to charge and discharge C_{int}. The ADC module has two different phases. When $V_{int}>V_{REF}$, BS=1, I_{CTAT} will discharge the capacitor, when $V_{int}<V_{REF}$, BS=0, I_{PTAT} will

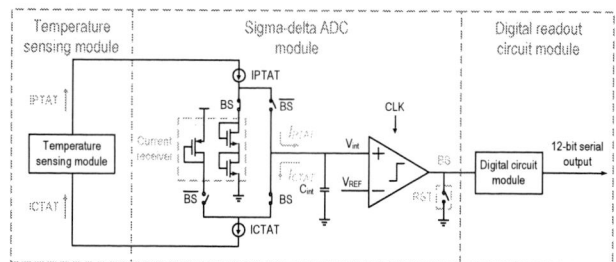

Fig. 2: Architecture of proposed temperature sensor.

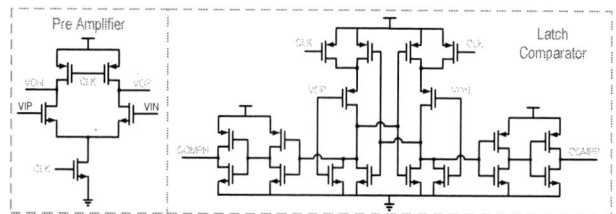

Fig. 3: Schematic of the comparator.

charge the capacitor. The current receiver provides a shunt path for unused current branches during ADC quantization cycles. When the charge-discharge cycles are enough, charge balance on C_{int} requires:

$$\mu I_{CTAT} = (1 - \mu) I_{PTAT}, \qquad (5)$$

where μ is the duty cycle of the ADC output bitstream (BS).

According to (5), μ can be expressed as:

$$\mu = \frac{I_{PTAT}}{I_{PTAT} + I_{CTAT}}. \qquad (6)$$

Given that I_{PTAT} and I_{CTAT} are respectively related to resistors R_1 and R_2, reasonably setting the ratio of resistors can make μ have a good temperature linearity.

Fig. 3 shows the comparator schematic, which combines a dynamic pre-amplifier with a latch comparator. At each rising edge of CLK, the comparator compares V_{int} with V_{REF}. The result is captured by an RS latch to generate a single-bit output BS, which then controls the current steering network at the subsequent falling edge. The dynamic comparator achieves a 0.5 ns decision time, significantly faster than the CLK period.

According to (6), μ is independent of V_{REF}. Since V_{REF} is only a reference voltage provided to the comparator and its absolute value does not affect the final measurement result, five identical PMOS transistors connected to the substrate and the source are connected end to end to generate a reference voltage $V_{REF} = 3/5V_{DD}$. A 1.3 pF capacitor is used to make the reference voltage stable.

The digital readout circuit module integrates a counter and a parallel-to-serial output (PISO) module. It counts the number of high levels in the bitstream BS over 4096 CLK cycles and outputs this count as a 12-bit serial stream, reducing the number of required chip output pins.

The final measured temperature T can be obtained from the following expression:

$$T = kD + b, \qquad (7)$$

979-8-3315-3918-4/25 $31.00 © 2025 IEEE

where D is the number corresponding to the 12-bit serial output, k and b are temperature-independent constants.

III. HIGH PRECISION TECHNOLOGY

A. Source of Errors

In the design of a temperature sensor, component mismatches often bring intolerable errors. In Fig. 1, the errors mainly come from the offset voltage V_{os} at the input end of the operational amplifier (op-amp), mismatches of the current mirror transistors M_1-M_9, resistor mismatches between R_1 and R_2 and non-ideal factors of the BJTs. The offset voltage V_{os} is directly applied to ΔV_{BE}, causing up to $10\,°C$ error.

Additionally, both the mismatches between the BJTs Q_1 and Q_2 and the process variation of Q_4 will bring some errors to the final temperature measurement results.

B. DEM Technology

Dynamic element matching (DEM) and chopping are indispensable technologies for high-precision temperature sensors, effectively reducing offset and mismatch errors [3].

The cascode current mirror in Fig. 1 enhances branch current matching accuracy, while DEM technology periodically swaps five current paths every $512\ \mu s$ to average mismatch-induced errors. We can analyze the effect of using DEM through calculation. Assume that due to current mirror transistor mismatches, five current branches have different currents $I_i = I + \delta I_i (i = 1, 2, 3, 4, 5)$. According to (4), before using DEM, the current mismatch will bring an error to ΔV_{BE}:

$$\Delta V_{BE,err} = V_T \ln \left(\frac{I + \delta I_2}{I + \delta I_1} \right) \tag{8}$$

On a full DEM cycle, average of ΔV_{BE} can be written as:

$$\Delta V_{BE,avg} = \frac{V_T}{5} \cdot \left(\sum_{i=1}^{4} \ln \left(\frac{NI_{i+1}}{I_i} \right) + \ln \left(\frac{NI_1}{I_5} \right) \right) \tag{9}$$
$$= V_T \ln (N)$$

where N is the current density ratio of Q_1 and Q_2.

Compare (8) and (9), DEM effectively eliminates the error of ΔV_{BE} caused by current mirror mismatches.

C. Finite Current Gain Compensation

In Fig. 1, the finite current gain β of the bipolar junction transistor Q_4 will bring some errors to the CTAT voltage V_{BE}. According to (1), V_{BE} is related to the collector current of Q_4, but due to the finite value of β, the relationship between I_{C4} and I_{PTAT} is:

$$I_{C4} = \frac{\beta}{1 + \beta} I_{PTAT}. \tag{10}$$

According to (10), voltage error of V_{BE} can be expressed as:

$$V_{BE,err1} = V_T \ln \left(1 + \frac{1}{\beta} \right) \tag{11}$$

From (11), the error increases with decreasing β and exhibits significant sensitivity to β variations. In the tt corner at $27\,°C$, $\beta \approx 2.88$. In the ff corner, $\beta \approx 3.3$, causing a voltage error of $0.863\,mV$. Using an identical BJT like Q_3

in Fig. 1 can effectively compensate for this error [8]. After compensation, the collector current of Q_4 can be written as:

$$I_{C4} = \frac{\beta}{(1 + \beta)} I_{E4} = \frac{\beta (2 + \beta)}{(1 + \beta)^2} I_{PTAT}. \tag{12}$$

Here, the voltage error can be described as:

$$V_{BE,err2} = V_T \ln \left(1 + \frac{1}{\beta (2 + \beta)} \right) \tag{13}$$

By comparing (11) and (13), it can be known that after adopting the finite current gain compensation, β^2 becomes the influence factor of error. If β changes from 2.88 to 3.3 due to process variation, it will cause a 0.34 mV error, achieving a 2.54× precision enhancement over the uncompensated topology.

D. Combination of DEM and Finite Current Gain Compensation

According to (9), applying DEM to the current paths driving Q_1 and Q_2 in Fig. 1 significantly reduces the ΔV_{BE} error caused by current mirror mismatch. To address the current mirror mismatch impeding Q_3's ability to adequately compensate Q_4 current, this paper combines DEM with the finite current gain compensation and extends DEM to the branches where Q_3 and Q_4 are located.

Taking the mismatch between current mirror transistors into account, I_{C4} can be rewritten as:

$$I_{C4,mis} = \frac{\beta}{(1 + \beta)} I_{E4,mis} = \frac{\beta I_3 + (\beta + \beta^2) I_4}{(1 + \beta)^2}. \tag{14}$$

Compare (12) with (14), the current ratio of I_{C4} in mismatch and no-mismatch scenarios can be written as:

$$\frac{I_{C4,mis}}{I_{C4,nomis}} = \frac{\beta I_3 + (\beta + \beta^2) I_4}{\beta (2 + \beta) I_4}. \tag{15}$$

We can assume that due to the mismatch, there is a relative variation $\delta I = (I_3 - I_4) / I_4$ between I_3 and I_4. The error of V_{BE} caused by the mismatch can be expressed as:

$$V_{BE,err3} = V_T \ln (1 + \delta I / (2 + \beta)). \tag{16}$$

Substituting the β value of 2.88 into (16), it can be calculated that if there is an occasional 10% mismatch between the current mirrors, the error is 0.525 mV, which is comparable to the error caused by not using finite current gain compensation. The combined use of DEM technology can effectively average out the mismatch-induced errors to avoid the occurrence of such extreme situations.

IV. MEASUREMENT RESULTS

The proposed BJT-based temperature sensor is fabricated in a 180-nm CMOS process. Fig. 4 shows the photograph and layout of the chip, with an area of $0.067\ mm^2$. It contains a temperature sensing module on the left side of the chip, an Σ–ΔADC module at the bottom, and a digital readout circuit module in the upper right corner. Four chips are packaged on the same printed circuit board (PCB) and placed in a

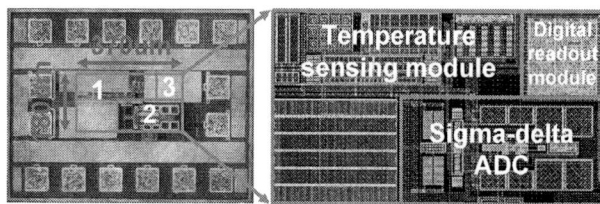

Fig. 4: Chip photograph and layout.

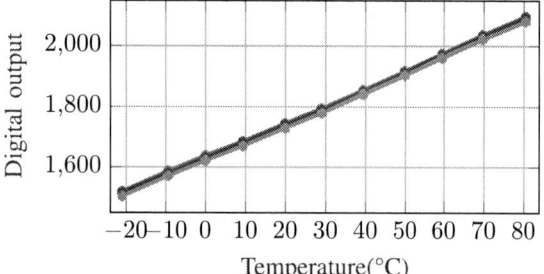

Fig. 5: Digital output measurement results of 4 chips.

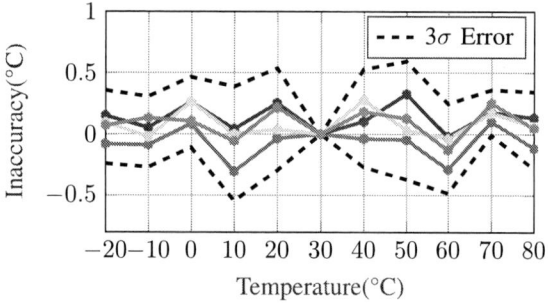

Fig. 6: Measured inaccuracy after one point calibration.

TABLE I: Temperature Sensor Summary and Comparison

	This Work	[5]	[9]	[10]
Technology(μm)	**0.18**	0.18	0.18	0.18
Calibration	**1-point**	1-point	2-point	2-point
Area (mm^2)	**0.067**	0.075	0.08	0.1225
Supply voltage (V)	**1.6-2.0**	1.8	1.8	2.7-3
Power (μW)	**39.6**	39.1	39.6	40
Temperature range ($^\circ$C)	**-20 to 80**	-40 to 120	-20 to 80	-20 to 80
Inaccuracy ($^\circ$C)	**±0.6**	±0.97	±1	±1.2
Conversion time (ms)	**4.096**	4.096	0.064	16
Resolution ($^\circ$C)	**0.175**	0.146	0.158	0.125
Resolution FoM (nJ·K^2)	**4.967**	3.414	0.063	10

resolution-FoM.

V. CONCLUSION

This paper shows a ±0.6°C (3σ) BJT-based temperature sensor which combines DEM technology and finite current gain compensation to achieve high accuracy. It occupies an area of 0.067 mm^2 and has a power consumption of 39.6 μW. It has 0.175°C resolution and 4.096 ms conversion time.

thermal chamber simultaneously for reliable testing. At each target temperature point, ten replicate measurements were performed. The arithmetic mean of these recorded values was subsequently designated as the final digital output representing the respective temperature point.

Fig. 5 shows the measured digital output result of 4 chips, the least significant bit (LSB) can be calculated as 0.175°C. Due to the non-uniform temperature of the chamber and the process deviation of the chip, their digital output results have a maximum temperature variation of ±1.2°C. One-point calibration is used to reduce this error. After calibration at a room temperature of 30°C, the measured temperature errors of the four chips are obtained in Fig. 6. The maximum measured inaccuracy is 0.33°C and the 3σ error is 0.59°C.

This chip can maintain a stable working status under different conditions, the supply voltage sensitivity is 0.13 $^\circ$C/V under different supply voltage ranges from 1.6 to 2.0 V. The power consumption of this chip is 39.6 μW, the conversion time is 4.096 ms, and the resolution FoM is 4.967 nJ·K^2. Table I summarizes this chip and compares it with prior works. Compared to [5], [9], [10], it achieves higher accuracy and fewer calibration points than [9] and [10]. Compared with [10], it has a 4× improvement in conversion time and better

REFERENCES

[1] A. Aprile, E. Bonizzoni, and P. Malcovati, "Temperature-to-digital converters' evolution, trends and techniques across the last two decades: A review," *Micromachines*, vol. 13, no. 11, p. 2025, 2022.

[2] N. G. Toth and A. K. Makinwa, "A β-compensated NPN-based temperature sensor with ± 0.1°C (3σ) inaccuracy from -55°C to 125°C and a 200 fJ·K^2 resolution FoM," in *2024 IEEE International Solid-State Circuits Conference (ISSCC)*, vol. 67, pp. 66–68, IEEE, 2024.

[3] N. G. Toth, Z. Tang, T. Someya, S. Pan, and K. A. Makinwa, "A BJT-based temperature sensor with ± 0.1°C (3σ) inaccuracy from -55°C to 125°C and a 0.85 pJ·K^2 resolution FoM using continuous-time readout," in *2023 IEEE International Solid-State Circuits Conference (ISSCC)*, pp. 358–360, IEEE, 2023.

[4] B. Wang, M.-K. Law, C.-Y. Tsui, and A. Bermak, "A 10.6 pJ·K^2 resolution FoM temperature sensor using astable multivibrator," *IEEE Transactions on Circuits and Systems II: Express Briefs*, vol. 65, no. 7, pp. 869–873, 2017.

[5] X. Lai, Z. Niu, B. Wang, and L. Li, "A 0.075 mm^2 BJT-based temperature sensor with a one-point trimmed 3σ inaccuracy of ±0.97°C from -40°C to 120°C," *Microelectronics Journal*, vol. 153, p. 106418, 2024.

[6] S. Lin and C. Salama, "ΔV/SUB be/(T) model with application to bandgap reference design," *IEEE Journal of Solid-State Circuits*, vol. 20, no. 6, pp. 1283–1285, 1985.

[7] B. Yousefzadeh, S. H. Shalmany, and K. A. Makinwa, "A BJT-based temperature-to-digital converter with ± 60 mK (3σ) inaccuracy from -55°C to +125°C in 0.16-μm CMOS," *IEEE Journal of Solid-State Circuits*, vol. 52, no. 4, pp. 1044–1052, 2017.

[8] G. Wang, A. Heidari, K. A. A. Makinwa, and G. C. M. Meijer, "An accurate BJT-based CMOS temperature sensor with duty-cycle-modulated output," *IEEE Transactions on Industrial Electronics*, vol. 64, no. 2, pp. 1572–1580, 2017.

[9] A. Aprile, M. Folz, D. Gardino, P. Malcovati, and E. Bonizzoni, "A BJT-based 0.08-mm^2 oversampling SAR temperature-to-digital converter for thermal drift compensation in MEMS inertial sensors," *IEEE Transactions on Instrumentation and Measurement*, vol. 73, pp. 1–11, 2024.

[10] S. Xie and A. J. P. Theuwissen, "On-chip smart temperature sensors for dark current compensation in CMOS image sensors," *IEEE Sensors Journal*, vol. 19, no. 18, pp. 7849–7860, 2019.

A Cryogenic Voltage Reference with Diode-Based Sensing and Substrate Resistor Compensation Compensation in 180-nm CMOS Process

Yixin Zhang, Hanze Liu, Jing Li*, Zhong Zhang, Ning Ning, Qi Yu

Ultra-Deep Submicron Integrated Circuits and Systems Laboratory, University of Electronic Science and Technology of China, Chengdu 610054, China

* Email: lijing686@uestc.edu.cn

Abstract—Conventional BJT-based voltage references face issues like abrupt current gain reduction and nonlinear temperature dependence of substrate resistance at cryogenic temperatures. This paper presents a diode-based voltage reference. Utilizing PN-junction diodes for temperature sensing—which are immune to current gain variations, the proposed design eliminates the impact of substrate resistance on the reference voltage and achieves exceptional temperature stability invariance. Implemented in 180-nm CMOS, it achieves 72 ppm/°C temperature coefficient, 1.3 mV/V line regulation at 1.8-3.3V, and 82uA current—enabling robust voltage references for cryogenic quantum systems.

Keywords—Cryogenic CMOS, resistor compensation, diode-based, voltage references.

I. INTRODUCTION

Applications such as space exploration (experiencing extreme temperature ranges, e.g., -230°C to +120°C on the Moon) and quantum computing/particle detection (requiring cryogenic temperatures as low as 100 mK) impose stringent thermal requirements on electronic devices that exceed standard operating ranges [1]. CMOS technology, recognized for its high integration density, wide supply voltage compatibility, broad temperature resilience, and excellent scalability, emerges as an ideal candidate for controlling electronics within these extreme cold environments. Within quantum interface System-on-Chip (SoC) architectures, which often incorporate numerous analog and RF circuits, reference circuits exhibiting robustness against temperature fluctuations, supply voltage variations, and process deviations are critically important [2]. Consequently, the development of precise voltage references for deep cryogenic temperatures has become a significant research challenge.

Due to significant differences in how CMOS, parasitic BJTs, and resistors behave at cryogenic temperatures compared to room temperature in standard CMOS processes, it is quite challenging to build precise voltage or current reference circuits under cryogenic temperatures. While parasitic BJT-based voltage references represent a classic room-temperature topology, their performance degrades severely below 77 K. This degradation manifests as a dramatic increase in base resistance and a sharp decline in current gain β, rendering the BJT's temperature-sensitive properties ineffective. Addressing this, literature [3] presented a family of sub-1V, all-CMOS voltage reference circuits utilizing weak

inversion MOSFETs as the core sensing element, achieving a temperature coefficient of 547 ppm/K from 300 K down to 4.2 K in a 40 nm CMOS process. Alternatively, Dynamic Threshold MOSFET (DTMOS) based references offer structural simplicity and stable performance, exhibiting only a 1.2% reference voltage variation from 4 K to 300 K. However, a primary challenge for DTMOS implementations lies in mitigating interface trap effects. Specific process technologies, such as Silicon-Germanium (SiGe) [4] and Fully Depleted Silicon-on-Insulator (FD-SOI) [5], incorporate BJTs and other devices exhibiting superior immunity to carrier freeze-out effects via bandgap engineering modifications and the inclusion of buried oxide layers, respectively. Nevertheless, these specialized processes often suffer from limited design flexibility and compatibility constraints.

As an alternative solution, this work proposes a voltage reference core based on diode. This topology utilizes diodes as the primary temperature-sensing element, enabling operation across an unprecedented wide temperature range from -250°C to +120°C. Crucially, a resistive compensation technique is employed to eliminate the detrimental impact of resistor temperature drift on the reference's overall temperature sensitivity. Diodes offer distinct advantages: they are readily integrable within standard CMOS processes for ease of implementation and provide a stable operating point even under extreme cryogenic conditions.

The rest of this paper is organized as follows: Section II provides a concise overview of behavioral changes in CMOS devices at cryogenic temperatures. Section III details the implementation of the proposed voltage reference architecture. Section IV presents post simulation results of the design. Finally, Section V offers the conclusions.

II. CRYOGENIC DEVICE PERFORMANCE

Conventional bandgap reference circuits typically employ bipolar junction transistors (BJTs) due to their superior stability, linear current-voltage characteristics, reduced process-corner sensitivity, high achievable accuracy (low temperature coefficient, TC), and favorable low-frequency noise performance [6]. The temperature-sensing mechanism of BJTs relies on their exponential current-voltage relationship.

$$I = I_S(e^{\frac{qV}{nkT}} - 1) \qquad (1)$$

Where I is device current, V is voltage drop, k is Boltzmann's constant, q is electron charge, T is absolute temperature; for a BJT, I_S is saturation current, $I = I_C$, $V = V_{BE}$, and $n = n_{BJT}$ is the effective emission coefficient. In standard CMOS

This work was supported in part by the National Natural Science Foundation of China under Grant 62004024 and 62204031, in part by the Science and Technology on Analog Integrated Circuit Laboratory under Grant JCKY2022210C009 and 2021JCJQLB0498..

979-8-3315-3918-4/25 $31.00 © 2025 IEEE

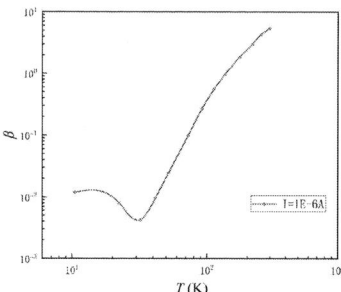

Fig. 1. Parasitic BJT current gain variation with temperature.

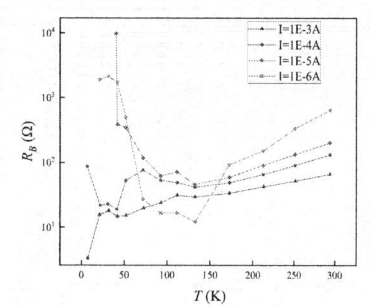

Fig. 2. Substrate resistance variation with temperature in parasitic BJTs.

processes, substrate PNP transistors can function as core temperature-sensing elements at temperatures significantly below typical military specifications (−55°C to +125°C). With room temperature down to 70 K [7], their performance degrades severely below 70 K owing to collapsing current gain (β) and increased parasitic base resistance R_B, as shown in Fig. 1 and Fig. 2.

Compared to BJTs, diodes, usually PN junction, offer simpler structures and exhibit similar temperature-dependent characteristics[8][9].

$$V_D = \frac{\eta kT}{q}\ln(\frac{I_D}{I_S}) \qquad (2)$$

Where I_D is the diode current, V_D is the voltage across the diode, and n is the ideality factor of the diode. Crucially, diodes are immune to the severe β variations that impair BJTs at cryogenic temperatures, making them promising sensing elements for ultra-low-temperature applications. At cryogenic temperatures, diodes affected by freeze-out effects can be simply modeled as an ideal pn-junction in series with substrate resistance. Although diode-based sensing remains susceptible to impedance variations induced by freeze-out, these effects can be mitigated through specific techniques. An optimal solution leverages identical diode structures and exploits two key physical properties[10]:

1) Logarithmic scaling of the intrinsic voltage with current;

2) Linear dependence of series resistance voltage drop on current.

For a typical silicon PN junction, carrier mobility follows classical Ohm's law, where resistance is approximately

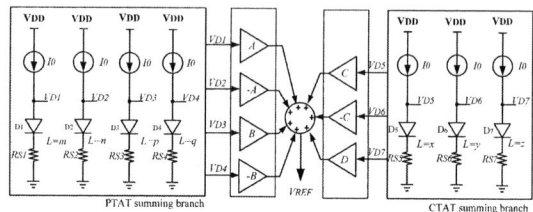

Fig. 3. Architecture of the proposed reference.

current-independent. When identical, precisely matched diodes are used, their substrate resistances (R_S) can be assumed equal. To eliminate impedance-induced errors in voltage expressions, we derive conditions for cancellation, as shown in Fig. 3 , only one diode-equivalent model unit is depicted per branch, where L represents the actual number of parallel-connected units in that branch, while A, B, C, and D denote the summing coefficients applied to V_{CTAT} and V_{PTAT} during the voltage summation process. Within each branch, the current is I_0.

For the CTAT (complementary to absolute temperature) voltage, resistive terms cancel when $1/x+1/y=1/z$

$$\begin{aligned}V_{CTAT} &= V_{D1} + V_{D2} - V_{D3} \\ &= V_T(\frac{zI_0}{xyI_S}) + \frac{I_0}{x}R_{S1} + \frac{I_0}{y}R_{S2} - \frac{I_0}{z}R_{S3}\end{aligned} \qquad (3)$$

For the PTAT (proportional to absolute temperature) voltage, resistive terms cancel when $1/m+1/p=1/n+1/q$

$$\begin{aligned}V_{PTAT} &= V_{D1} - V_{D2} + V_{D3} - V_{D4} \\ &= V_T\ln(\frac{nq}{mp}) + \frac{I_0}{m}R_{S1} - \frac{I_0}{n}R_{S2} + \frac{I_0}{p}R_{S3} - \frac{I_0}{q}R_{S4}\end{aligned} \qquad (4)$$

The output reference voltage V_{REF} is

$$\begin{aligned}V_{REF} &= V_{CTAT} + \alpha V_{PTAT} \\ &= V_T(\frac{zI_0}{xyI_S}) + \alpha V_T\ln(\frac{nq}{mp})\end{aligned} \qquad (5)$$

The scaling factor α for weighting V_{PTAT} and V_{CTAT} can be implemented via a current mirror configuration.

Conventionally, BJT-based voltage references exhibit unpredictable variations at cryogenic temperatures due to the nonlinear temperature dependence of base resistance (as depicted in Fig. 2). This architecture, however, achieves superior temperature stability by simultaneously nullifying the effects of current gain β and base resistance.

III. CIRCUIT DESIGN

Fig. 4 illustrates a bandgap reference circuit utilizing a switched-capacitor architecture. This structure comprises three primary sections: a CTAT voltage generation circuit, a PTAT voltage generation circuit, and a voltage summation circuit. The circuit branch containing diodes D_1-D_4 provides four diode voltage drops (V_D), which generates the V_{PTAT} and eliminates substrate parasitic resistance through capacitors C_1 and C_2. This V_{PTAT} voltage is then converted into a PTAT current via a regulator.The circuit branch containing diodes D_5-D_7, which generates the VCTAT voltage and eliminates substrate parasitic resistance through capacitors C_3 and C_4.

979-8-3315-3918-4/25 $31.00 © 2025 IEEE

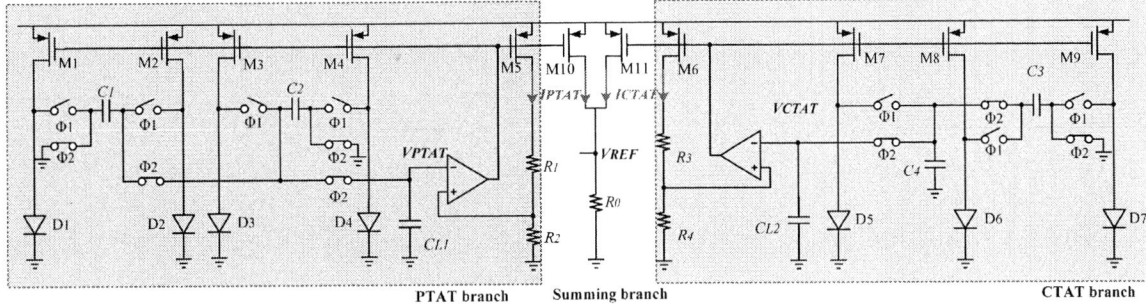

Fig. 4. Schematic of the cryogenic PTAT and CTAT generator circuits.

This V_{CTAT} voltage is then converted into a CTAT current via a regulator. The PTAT and CTAT currents are summed through transistors M_{10} and M_{11}, generating V_{REF} across resistor R0.

The circuit operates in two clock phases. During the initial sampling phase (Φ_1 high): In the PTAT path, capacitors C_1 and C_2 sample the voltage differences across diode pairs (D_1-D_2) and (D_3-D_4), respectively. In the CTAT path, capacitors C_3 and C_4 sample the voltage difference across diode pairs D_5 and (D_6-D_7). Load capacitors C_{L1} and C_{L2} sample the CTAT and PTAT voltages, respectively, at the inputs of the operational amplifier (op-amp). During the hold phase (Φ_2 high), all capacitors (C_1-C_4, C_{L1}, C_{L2}) are shorted to the op-amp input terminals. Charge redistribution occurs, establishing the CTAT and PTAT voltages at the op-amp inputs. For PTAT voltage generation,

$$V_{PTAT} = \frac{C_1(V_{D2}-V_{D1})+C_2(V_{D3}-V_{D4})+C_L V_{PTAT}}{C_1+C_2+C_{L1}} \quad (6)$$

The PTAT voltage generation circuit employs four diodes (D_1, D_2, D_3, D_4) with an area ratio of m:n:p:q.

$$\begin{aligned} V_{PTAT} &= \frac{C_1}{C_1+C_2}(V_{D2}-V_{D1}) + \frac{C_2}{C_1+C_2}(V_{D3}-V_{D4}) \\ &= V_T(\frac{C_1}{C_1+C_2}\ln(\frac{m}{n}) + \frac{C_2}{C_1+C_2}\ln(\frac{q}{p})) \\ &+ \frac{C_1}{C_1+C_2}I_0(\frac{R_{S2}}{n}+\frac{R_{S3}}{p}) - \frac{C_2}{C_1+C_2}I_0(\frac{R_{S1}}{m}+\frac{R_{S4}}{q}) \end{aligned}$$
$$(7)$$

Since the circuit employs matched diode devices and ensures identical layout geometries, the parasitic series substrate resistance can be considered identical across all devices. Therefore, the only requirement is:

$$C_1(\frac{1}{n}+\frac{1}{p}) = C_2(\frac{1}{m}+\frac{1}{q}) \quad (8)$$

In terms of component sizing, C_2 is set to $C_2 = 3C_1$, with the diode area ratios configured as $m = 36$, $n = 4$, $p = 3$, $q = 6$. This specific ratio ensures that the expression for the PTAT voltage (V_{PTAT}) excludes any dependency on the parasitic series substrate resistance.

For CTAT voltage generation, the circuit employs three diode-connected devices (D_5, D_6, D_7) with a multiplier ratio of $x : y : z$. According to charge redistribution,

$$V_{CTAT} = \frac{C_3 V_{D5}+C_4(V_{D6}-V_{D7})+C_L V_{CTAT}}{(C_3+C_4+C_{L2})} \quad (9)$$

When C_3 equals C_4,

$$V_{CTAT} = \frac{V_T}{2}\ln(\frac{z I_0}{xy I_S}) + \frac{1}{2x}I_0 R_{S5} + \frac{1}{2y}I_0 R_{S6} - \frac{1}{2z}I_0 R_{S7}$$
$$(10)$$

To eliminate the substrate resistance term and suppress the second-order component, the following condition must be satisfied:

$$\frac{1}{x} = \frac{1}{y} - \frac{1}{z} \quad (11)$$

Where the diode area ratios are configured as x= 1 , y= 2, z= 1. This topology ensures that the expressions for both V_{PTAT} and V_{CTAT} are inherently independent of the parasitic series substrate resistance. The expression for the reference voltage is therefore given by:

$$\begin{aligned} V_{REF} &= V_{CTAT} + \alpha V_{PTAT} \\ &= \frac{V_T}{2}\ln(\frac{I_0}{2I_S}) + \alpha(V_T(\frac{1}{4}\ln(\frac{m}{n}) + \frac{3}{4}\ln(\frac{q}{p}))) \end{aligned} \quad (12)$$

The generated PTAT and CTAT voltages undergo current summation via voltage-to-current converters. The summed current flows through resistor R_0 to produce the bandgap reference voltage V_{REF}. Crucially, non-overlapping clocks Φ_1 and Φ_2 to ensure accurate capacitor sampling.

In this design, capacitors are sized as $C_1 = C_3 = C_4 = 0.3$ pF, with $C_2 = 3C_1 = 0.9$ pF, mitigating charge injection effects from switches on the sampled voltage V_D. Switch dimensions are minimized (0.35 μm/0.18 μm, W/L) relative to the sampling capacitors to further suppress charge injection. As reported in Reference [9], leakage currents and non-idealities introduce voltage ripple in V_{REF} during steady-state operation. A time-interleaved replica branch (not shown in the figure) employs capacitors with identical values and switches operating with inverted clock phases but identical connectivity. This architecture significantly attenuates ripple amplitude.

Auxiliary circuits comprise the operational amplifier (op-amp), startup circuit, and bias circuit. To address increased threshold voltage and degraded transconductance (gm) at cryogenic temperatures [11] — which shift operating points — the design utilizes a basic OTA structure. The two-stage op-amp employs Miller compensation for enhanced stability while delivering 80 dB gain.

IV. SIMULATION RESULTS

Implemented in SMIC 180-nm CMOS technology with a 3.3 V supply (V_{DD}), Fig.5 shows that the proposed bandgap reference achieves a temperature coefficient (TC) of 72

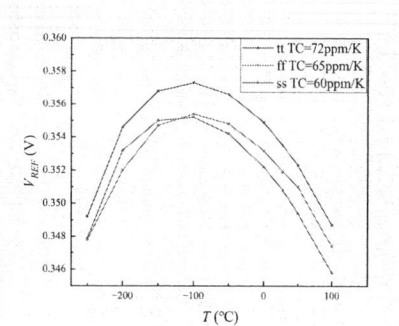

Fig. 5. Simulated reference voltage versus temperature and process variations at V_{DD} = 3.3V.

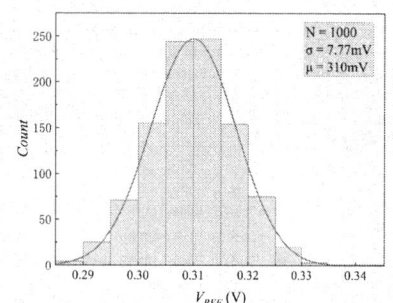

Fig. 6. Monte Carlo simulation results at V_{DD}=3.3V and 30°C.

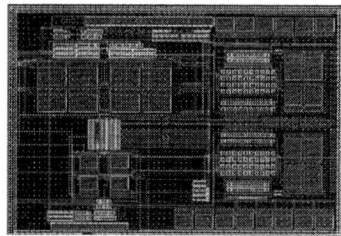

Fig. 7. Layout of the cryogenic voltage reference.

ppm/°C across −250°C to 120°C under typical-typical (tt) process corner, as per post-layout simulations. Monte Carlo analysis indicates an untrimmed output voltage variation of ±7.7 mV (1σ) in Fig. 6. Post-simulated output noise density measures 20 μV/ √ Hz at 1 Hz. Simulation results demonstrate: minimum V_{DD} = 2.5 V under cryogenic conditions due to operating point shifts from elevated diode substrate resistance and MOSFET threshold voltage (V_{TH}) increase. The circuit exhibits line regulation of 1.3 mV/V for V_{DD} = 1.8–3.3 V at 27°C. When injecting 1 Vpp AC ripple on the 3.3V DC supply, it achieves −32dB power supply rejection ratio (PSRR) at low frequencies. The core occupies 0.06 mm² (300 μm × 200 μm) in Fig. 7 and consumes 82μA quiescent current.

V. CONCLUSION

This work implements a voltage reference operational at cryogenic temperatures. The temperature-sensing core utilizes

TABLE I. SUMMARY AND COMPARISON WITH STATE OF THE ART

	This work[a]	[12]	[3]	[13]
Process	180nm	40nm	28 nm FDSOI	40nm
T range	-250~120℃	4~320K	4~320K	4~295K
Output(V)	0.310	0.96	0.485	0.699
Current(μA)	270	368	15.8	20.8
Area(um²)	60000	445	40920	725
TC(ppm/℃)	72	833	1214	744.3
PSR(dB)	-32	-23.1	-51	-12.25

[a.] Simulation results

diode-connected devices with substrate-resistance-variation compensation specifically designed for cryogenic conditions. The architecture achieves 72 ppm/°C temperature coefficient across an extended range of −250°C to 120°C.

REFERENCES

[1] J. D. Cressler and H. A. Mantooth, Extreme Environment Electronics. Boca Raton, FL, USA: CRC Press, 2013.

[2] C. Deng et al., "A Systematic Review of Voltage Reference Circuits: Spanning Room Temperature to Cryogenic Applications," in IEEE Transactions on Circuits and Systems I: Regular Papers, vol. 72, no. 4, pp. 1533-1546, April 2025.

[3] J. van Staveren et al., "Cryo-CMOS Voltage References for the Ultrawide Temperature Range From 300 K Down to 4.2 K," in IEEE Journal of Solid-State Circuits, vol. 59, no. 9, pp. 2884-2894, Sept. 2024.

[4] H. Bohuslavskyi et al., "28nm Fully-depleted SOI technology: Cryogenic control electronics for quantum computing," 2017 Silicon Nanoelectronics Workshop (SNW), Kyoto, Japan, 2017, pp. 143-144.

[5] A. A. Cherepanov, I. L. Novikov and V. Y. Vasiliev, "An Evaluation of SiGe HBT Operation at Cryogenic Temperatures," 2019 20th International Conference of Young Specialists on Micro/Nanotechnologies and Electron Devices (EDM), Erlagol, Russia, 2019, pp. 23-27.

[6] Gray P, Meyer R. Analysis and Design of Analog Integrated Circuits. 5th ed. Wiley; 2009.

[7] L. Song, H. Homulle, E. Charbon and F. Sebastiano, "Characterization of bipolar transistors for cryogenic temperature sensors in standard CMOS," 2016 IEEE SENSORS, Orlando, FL, USA, 2016, pp. 1-3.

[8] Pertijs, Micheal A. P., and J. H. Huijsing . Precision Temperature Sensors in CMOS Technology (Analog Circuits and Signal Processing). Springer-Verlag New York, Inc. 2006.

[9] S. M. Sze and K. K. Ng, Physics of Semiconductor Devices, 3rd ed. Hoboken, NJ, USA : Wiley, 2007.

[10] Y. Lin et al., "Characterization and Modeling of MOSFET Series Resistance Down to 4 K," in IEEE Journal of the Electron Devices Society, vol. 13, pp. 297-302, 2025.

[11] R. M. Incandela, L. Song, H. Homulle, E. Charbon, A. Vladimirescu and F. Sebastiano, "Characterization and Compact Modeling of Nanometer CMOS Transistors at Deep-Cryogenic Temperatures," in IEEE Journal of the Electron Devices Society, vol. 6, pp. 996-1006, 2018.

[12] H. Homulle, F. Sebastiano and E. Charbon, "Deep-Cryogenic Voltage References in 40-nm CMOS," in IEEE Solid-State Circuits Letters, vol. 1, no. 5, pp. 110-113, May 2018.

[13] M. Wen, K. G. McCarthy, I. O'Connel and G. M. Salgado, "Design of Low-Voltage and Low-Power Cryogenic CMOS Voltage Reference Circuits," 2024 IEEE 67th International Midwest Symposium on Circuits and Systems (MWSCAS), Springfield, MA, USA, 2024, pp. 1016-1020.

979-8-3315-3918-4/25 $31.00 © 2025 IEEE

77K Modeling and Implementation of a Cryogenic OTA for Infrared Sensors

Zhuokai Wang[1], Lei Deng[*1], Rui Yin[2,3], Jian Mei[1,3], Jiaming Zhang[4], Zhicheng Shi[4]

[1] National Integrated Circuit Innovation Center, Shanghai, China;
[2] College of Integrated Circuits and Micro-Nano Electronics, Fudan University, Shanghai, China;
[3] Jiashan Fudan Institute, Jiaxing, China;
[4] Beijing Institute of Space Mechanics and Electricity

* Email: wangzhuokai0520@163.com, lei.deng@shnicic.com

Abstract— This paper presents the design of a cryogenic operational transconductance amplifier (OTA) operating in the liquid nitrogen temperature (77K), targeting the low-temperature operational requirements of infrared imaging sensors. Guided by established 77K low-temperature models, a dual-mode (room temperature/cryogenic) OTA was fabricated and tested. This work details the modeling process, simulation results, and experimental data, validating the performance enhancement potential of low-temperature operation for circuits.

Keywords— Low-temperature semiconductor material modeling, MOSFET, Operational Amplifier, Infrared Sensor

I. BACKGROUND INTRODUCTION

Semiconductor materials are indisputably the core driving force of modern technological revolutions and are irreplaceable. They form the foundation of the information age—integrated circuits (e.g., CPUs, GPUs, memory)—supporting activities ranging from personal computing and artificial intelligence to information communication (e.g., 5G, optical fiber). They also play indispensable roles in energy conversion (e.g., solar panels, batteries), signal processing (e.g., various sensors), and cutting-edge fields like quantum computing. The advancement of semiconductor technology largely dictates the current level and application boundaries of technological development.

II. SPECIFICITY OF THE LOW-TEMPERATURE ENVIRONMENT AND RESEARCH MOTIVATION

A. Specificity of the Low-Temperature

For silicon materials, many physical properties, such as intrinsic carrier concentration, effective density of states, and Fermi level, undergo significant changes as temperature decreases.

When silicon operates at cryogenic temperatures (below 100K), the reduced thermal energy of free carriers leads to incomplete ionization of impurity atoms. As temperature further decreases, carrier concentration drops sharply, causing material conductivity to reach very low levels. Simultaneously, the low-temperature environment suppresses lattice vibrations (phonon scattering) in silicon crystals, thereby increasing carrier mobility. As conductivity decreases and mobility increases, conventional current mechanisms diminish drastically at low temperatures. The previously obscured tunneling effect which is less temperature-sensitive, gradually becomes dominant.

This work was supported by Beijing Engineering Research Center of Aerial Intelligent Remote Sensing Equipments Fund.

B. Research Motivation and Low-Temperature Application Scenarios

The unique advantages of cryogenic semiconductor technology are driving its adoption in multiple frontier and specialized fields. The core application scenario is quantum information processing, particularly the control and readout systems for superconducting quantum computers. To minimize thermal noise interference and prevent heat intrusion into quantum chips (operating at millikelvin levels), critical electronic circuits generating precise microwave control pulses and amplifying weak quantum state signals (e.g., Digital to Analog Converter, filters) must operate at 4K or lower. Furthermore, in mid-to-long-wave infrared high-performance imaging, cryogenic semiconductor technology is essential for achieving high-sensitivity, low-noise detection. Deep cooling to liquid nitrogen temperature (77K) or below significantly suppresses thermal noise and reduces background current by several orders of magnitude, enabling detectors to approach the theoretical sensitivity limit of background-limited detection. This study focuses on this scenario, aiming to develop a backend OTA adapted to the cryogenic environment of infrared sensors.

III. LOW-TEMPERATURE CMOS ELECTRICAL CHARACTERISTICS AND MODELING

A. Fundamental of Low-Temperature Physical Properties of Bulk Devices

1. Drastic reduction in intrinsic carrier concentration:

$$n_i = \sqrt{N_c N_v} \, exp\left(-\frac{E_g}{2k_B T}\right) \qquad (1)$$

2. Incomplete impurity ionization:

$$\frac{N_D^+}{N_D} = \left[1 + g_{CD}\, exp\left(\frac{\Delta E_D}{k_B T}\right) + \left(\frac{E_{fn} - E_c}{k_B T}\right)\right]^{-1} \qquad (2)$$

3. Reduced overall scattering and increased mobility

$$P_o \propto \frac{(\hbar\omega_1)^{3/2}}{(k_0 T)^{1/2}} \left[\frac{1}{exp\left(\frac{\hbar\omega_1}{k_0 T}\right) - 1}\right] \frac{1}{f\left(\frac{\hbar\omega_1}{k_0 T}\right)} \qquad (3)$$

B. DC Characteristics at Low Temperatures

Threshold Voltage (Vth) Increase: Due to the sharp drop in intrinsic carrier concentration, the Fermi potential rises, leading to an increase in threshold voltage at low temperatures.

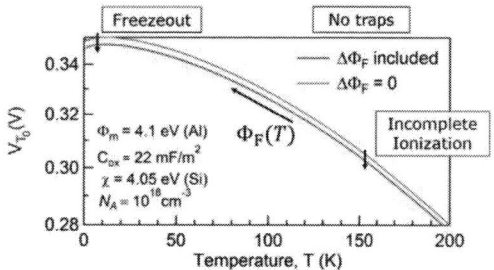

Fig. 1. Threshold voltage (Vth) increases as temperature drop [1].

Subthreshold Swing (SS) Behavior: Above 50K, SS decreases linearly with temperature. Below 50K, SS saturates and no longer decreases, exhibiting an anomalous phenomenon.

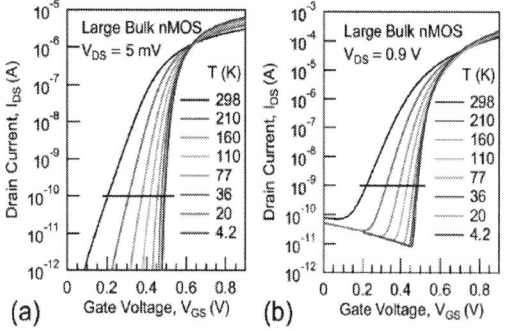

Fig. 2. Transconductance (gm) and Output Conductance (gds) [2].

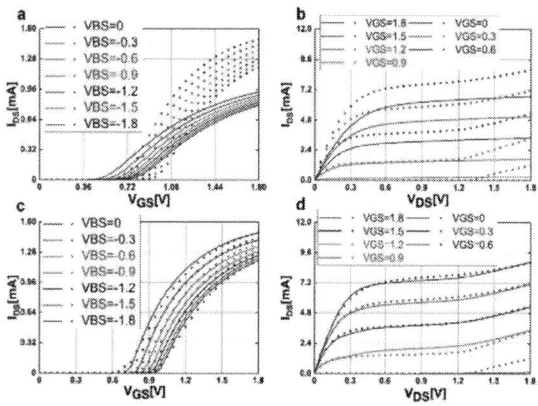

Fig. 3. I-V Characteristic Variation with Temperature [3].

Drift current increases at low temperatures due to enhanced mobility, while diffusion current decreases due to weakened diffusion.

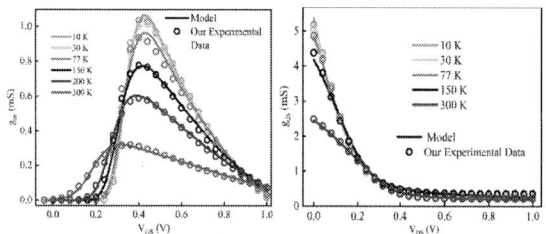

Fig. 4. Transconductance (gm) and Output Conductance (gds) Variation [4].

As Fig.4. shown, the peak transconductance (gm) increases as temperature decreases. The output conductance

(gds) in the linear region also increases with decreasing temperature.

C. 77K Low-Temperature Modeling

The low-temperature model was developed using Cadence Spectre (with its built-in BSIM4 model). The original model was retained, relevant parameters were modified, new parameters were obtained through fitting, and simulation results were generated via netlist input.

The BSIM model fitting process employed multiple devices for local fitting, aiming to preserve the physical meaning of individual parameters. Device dimensions (L & W) were varied to test short-channel and narrow-channel effects. Voltages were swept from low to high. Data was subsequently used for parameter tuning with Keysight's Device Modeling MBP.

D. Error Statistics

TABLE I 298K FIT RESULT

298K Data	RMS	MAX
Ids_Vds_Vgs@Vbs=-3.3	0.655%	2.834%
Ids_Vds_Vgs@Vbs=0	1.017%	3.144%
Ids_Vgs_Vbs@Vds=0.05	0.506%	1.894%
Ids_Vgs_Vbs@Vds=3.3	0.591%	2.693%
Isub_Vgs_Vds@Vbs=0	4.403%	24.553%

TABLE II 77K FIT RESULT

77K Data	RMS	MAX
Ids_Vds_Vgs@Vbs=-3.3	1.838%	4.465%
Ids_Vds_Vgs@Vbs=0	4.915%	10.195%
Ids_Vgs_Vbs@Vds=0.05	1.400%	5.177%
Ids_Vgs_Vbs@Vds=3.3	2.526%	9.897%
Isub_Vgs_Vds@Vbs=0	3.330%	20.191%

The fitting errors for the 77K data are relatively larger, likely due to the default geometric parameter values in the model's built-in calculation formulas not being fully optimized for the low-temperature environment.

IV. LOW-TEMPERATURE OTA DESIGN AND SIMULATION RESULTS

A. OTA layout design

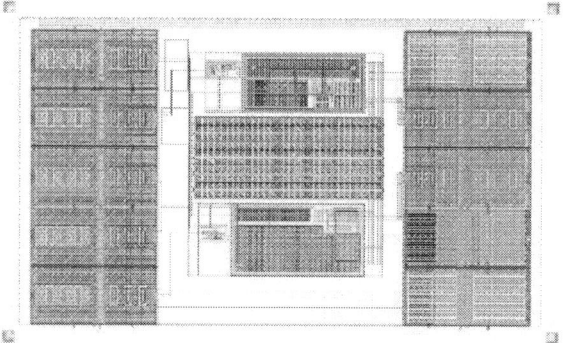

Fig. 5. Top Level layout.

B. OTA Cadence Spectre Simulation Results

TABLE III. 300K MODULE SIMULATION RESULTS (REFERENCE)

Item	Post-Layout result
Gain	92.8dB
GBW	30.7MHz
Phase Margin	69deg
Slew Rate	20.1V/us

TABLE IV. 77K MODULE SIMULATION RESULT

Item	77K	300K
Gain	103dB	90dB
GBW	283.6MHz	83.8MHz
Phase Margin	70.18deg	68.5deg
Slew Rate	60V/us	70V/us

TABLE V 77K STB SUMMARY

PM(deg)	@Freq(Hz)	GM(dB)	@Freq(Hz)
70.18	283.63<	14.388	881.22M

The simulations demonstrate significant improvements in gain and bandwidth at low temperatures. This aligns with the expected increase in transconductance (gm) due to enhanced carrier mobility. The boost in gm directly translates to substantial gains in performance metrics, proving the theoretical feasibility of low-temperature operation for optimizing infrared readout circuit performance.

C. OTA Physical Testing Results

1. Test Preparation

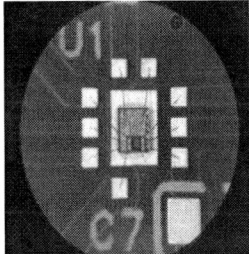 

Fig. 6. COB Solution, PCB, and Connection Diagram.

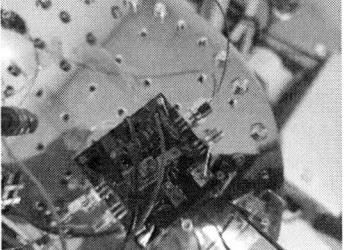 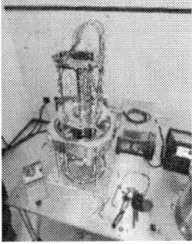

Fig. 7. Cryogenic Equipment and Connection.

2. Cryogenic Station Test Results (Low Temp 77K):

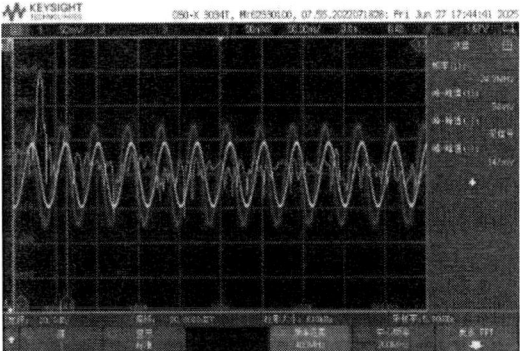

Fig. 8. 77K test result.

The measured waveforms at room and low temperatures meet expectations, confirming the OTA functions correctly at 77K. However, the measured gain and bandwidth were limited. This is attributed to constraints of the probe station setup: long transmission lines causing input signal phase shift, multiple port adapters connection which is inevitable, affecting impedance matching, and the oscilloscope's capacitive load (~12pF) reducing output swing. Additionally, the FFT analysis revealed a persistent second harmonic, likely caused by coupled noise from the cryogenic cooling system during operation.

V. CONCLUSION

This work, through 77K MOSFET modeling and OTA design, validates the key advantages of cryogenic technology for infrared sensor readout circuits: increased carrier mobility directly enhances gain/bandwidth, and thermal noise suppression potential is significant. Although measurements faced engineering challenges, this research establishes a crucial technical foundation for developing specialized cryogenic readout circuits for next-generation high-sensitivity, low-noise infrared imaging systems. Future optimization needs to address precise low-temperature modeling, improved test solutions and environments, and efficient cryogenic integration with infrared detectors.

REFERENCES

[1] A. Beckers, F. Jazaeri and C. Enz, "Cryogenic MOS Transistor Model," in IEEE Transactions on Electron Devices, vol. 65, no. 9, pp. 3617-3625, Sept. 2018, doi: 10.1109/TED.2018.2854701.

[2] A. Beckers, F. Jazaeri, A. Grill, S. Narasimhamoorthy, B. Parvais and C. Enz, "Physical Model of Low-Temperature to Cryogenic Threshold Voltage in MOSFETs," in IEEE Journal of the Electron Devices Society, vol. 8, pp. 780-788, 2020, doi: 10.1109/JEDS.2020.2989629.

[3] C. Luo et al., "0.18μm CMOS Ring Oscillator at Liquid Helium Temperature," 2019 IEEE 3rd International Conference on Circuits, Systems and Devices (ICCSD), Chengdu, China, 2019, pp. 97-100, doi: 10.1109/ICCSD.2019.8843292.

[4] W. Manzoor, A. K. Dutta, G. Pahwa, N. Manzoor, C. Hu and Y. Singh Chauhan, "Extending Standard BSIM-BULK Model to Cryogenic Temperatures," in IEEE Transactions on Electron Devices, vol. 71, no. 8, pp. 4510-4516, Aug. 2024, doi: 10.1109/TED.2024.3419783.

[5] Wang Zewei. Research on Cryogenic CMOS Devices and Circuits for High-Efficiency Computing and Quantum Computing [D]. University of Chinese Academy of Sciences, 2022. [in Chinese]

979-8-3315-3918-4/25 $31.00 © 2025 IEEE

Design and Analysis of PI Controller for Resonant Drive Circuits with AGC-PI Architecture

Yichen Lu[1,2], Tao Yin*[1,2], Ying Liu[3], Jian Liu[1,2], Nanjian Wu[1,2], Liyuan Liu[1,2]

[1] University of Chinese Academy of Sciences, Beijing, China
[2] State Key Laboratory of Semiconductor Physics and Chip Technologies, Institute of Semiconductors, Chinese Academy of Sciences, Beijing, China
[3] Beijing Information Science and Technology University Beijing, China

*Corresponding Author Email: yint@semi.ac.cn

Abstract—This paper analyzes the impact of PI controller on the environmental robustness of closed-loop resonant drive circuits. A comparative analysis of temperature-dependent variations in proportional (K_P) and integral (K_I) coefficients is conducted between open-loop and closed-loop PI controllers with the discussion of its influence on the oscillation amplitude stability. To mitigate thermal sensitivity, the AGC-PI drive circuit integrates an active closed-loop PI controller, enhancing system robustness and extending the operational temperature range. The drive circuit is designed and fabricated in a 0.18μm BCD process. It can drive a capacitive MEMS resonant pressure sensor at temperatures ranging from −55°C to 110 °C and supply voltages from 4 V to 5.5 V. Under a 5V power supply, the temperature coefficient of TIA output oscillation amplitude is 21 μV/°C, the phase noise of TIA output is −106 dBc/Hz at 10 kHz offset, and the bias instability of the output frequency is 0.0197 Hz.

Keywords—PI Controller, Closed-Loop Drive Circuit, Wide Temperature Range, MEMS

I. INTRODUCTION

With the increasing demand for high-temperature measurement and control in industrial, energy, and aerospace fields, the application of resonant sensors in high temperature environments has attracted increasing attention. In these application areas, pressure sensors are required to operate stably over a wide temperature range from −55 °C to above 100 °C, as well as within a wide power supply voltage range. This imposes stringent requirements on the robustness of the drive circuit. Automatic gain control (AGC) closed-loop drive circuits are widely used in resonant pressure sensors [1]. It constitutes an amplitude control loop with the microresonator, to stabilize the oscillation of the resonant pressure sensor. Consequently, the output signal amplitude is stabilized at a constant level. In addition, proportional-integral (PI) controllers are frequently used to improve the transient response characteristics of closed-loop drive circuits.

Conventional drive circuits often employ open-loop Gm-C PI controllers [2], which exhibit high susceptibility to temperature and supply voltage variations. This leads to significant gain fluctuations and large temperature-dependent drift in oscillation amplitude, severely limiting the operational temperature range of resonant sensors. To overcome this challenge, this paper analyzes the critical impact of the proportional and integral coefficients of the PI controller on the temperature dependence of oscillation amplitude in resonant drive systems.

This paper proposes an AGC-PI drive circuit with an active closed-loop PI controller, enabling precise control of sensor oscillation amplitude and enhanced tracking of resonant frequency variations. It can reduce environmental susceptibility and extend the operating temperature range.

II. SYSTEM ARCHITECTURE OF THE DRIVE CIRCUIT

A. System Design

The silicon micro-resonant pressure sensor employs a dual-resonator configuration which consists of a double-clamped resonant beam with drive and sense electrodes on both sides [3]. When pressure is applied to the diaphragm, two resonance frequencies are generated by the central beam (f_1) and side beam (f_2), respectively. The differential output frequency ($f_1 - f_2$) is then converted into the measured pressure. This MEMS sensor exhibits a resonant frequency of approximately 100 kHz, a quality factor up to 10,000, and a temperature coefficient of about 5000 ppm/°C.

An AGC-PI drive circuit is implemented for this sensor, as shown in Fig. 1. The system integrates an oscillation loop and an amplitude control loop. The oscillation loop consists of a transimpedance amplifier (TIA), a high-pass filter (HPF), a variable gain amplifier (VGA), and the microresonator. The TIA converts current from capacitance variations into voltage, the HPF compensates phase shifts, and the VGA amplifies the signal to drive the excitation electrodes. According to the Barkhausen criterion for self-excited oscillation [4], the HPF and VGA satisfy the phase and gain conditions respectively.

The AGC system comprises a band-pass filter (BPF), rectifier (REC), low-pass filter (LPF), PI controller and VGA. Signals pass through the HPF and BPF, then enter the rectifier and LPF for amplitude detection. The detected amplitude is compared with a reference voltage to generate an error signal. The PI controller processes this error signal to adjust VGA gain, thereby regulating excitation amplitude. Under disturbances such as temperature-induced gain drift, the PI controller dynamically adjusts VGA gain for compensation.

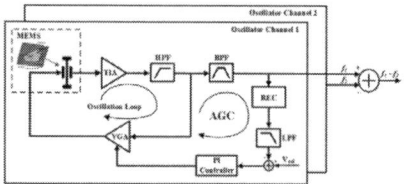

Fig. 1. Circuit structure of a resonant pressure sensor readout circuit.

B. Linearized Modeling and Analysis

The MEMS sensor and its interface circuit constitute a higher-order system with inherent nonlinear modules. To simplify stability analysis, a first-order linear model of the loop is established [5], allowing the stability assessment in the frequency domain. Within this approach, the dominant nonlinear elements, the sensor and VGA, are substituted with equivalent linear approximations.

The first-order linearized model of the microresonator is expressed as:

$$H_{amp}(s) = \frac{x_{amp}}{F_{amp}} = \frac{\omega_0}{2k} \cdot \frac{1}{s + \omega_0/2Q} \quad (1)$$

where $H_{amp}(s)$ is the first-order transfer function.

With high loop gain, the error signal becomes negligible. Under this condition, inputs to both rectifier and VGA stabilize at a constant value K_{VGA}.

Following the mentioned linearization method, the loop approximates a linear system as depicted in Fig. 2, where $K_{F/V}$, $K_{C/X}$, K_{TIA}, K_{hpf}, K_{bpf}, K_{REC}, K_{lpf}, are the gain converting drive voltage to electrostatic force on the resonant beam and translating sense plate displacement to capacitance, the gain of TIA, HPF, BPF, and rectifier, respectively. ω_{hpf}, ω_{bpf}, ω_{lpf} are the poles of the HPF, BPF and LPF. For the control subsystem, K_P and K_I are the proportional and integral coefficients of the PI controller, τ designates the integrator loss factor.

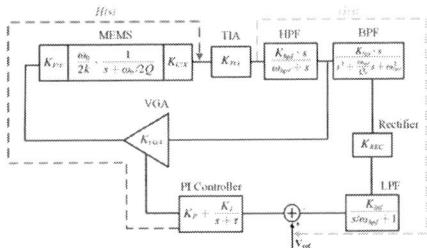

Fig. 2. Linearized model of the amplitude control loop in the AGC-PI architecture.

Amplitude variations at the TIA output under environmental changes indicate system robustness. Defining V_{ref} as the input and TIA output as the output, the forward path $H(s)$ encompasses from the PI controller to the TIA, while the feedback network $G(s)$ connects the TIA output back to the PI input. The resulting closed-loop transfer function is:

$$H_{cl}(s) = \frac{H(s)}{1 + G(s)H(s)} \quad (2)$$

$$H(s) = \left(K_P + \frac{K_I}{s+\tau}\right) \cdot \frac{\omega_0}{2k} \cdot \frac{1}{s + \frac{\omega_0}{2Q}} \cdot K_{\frac{F}{V}} \cdot K_{\frac{C}{X}} \cdot K_{VGA} \cdot K_{TIA} \quad (3)$$

$$G(s) = \frac{s}{\omega_{hpf}+s} \cdot \frac{s}{s^2 + \frac{\omega_{bpf}}{Q_0} + \omega_{bpf}^2} \cdot \frac{1}{\frac{s}{\omega_{lpf}}+1} \cdot K_{REC} \cdot K_{hpf} \cdot K_{bpf} \cdot K_{lpf} \quad (4)$$

where $H_{cl}(s)$ is the closed-loop transfer function. $H(s)$ is the open-loop transfer function. $G(s)$ is the feedback network transfer function.

From the open-loop transfer function $H(s)$, it can be observed that the PI controller introduces a zero at $-K_P / K_I$. This zero influences the root locus trajectory of the system,

ultimately causing the closed-loop poles to shift as K_P and K_I vary. When $K_P(T)$ and $K_I(T)$ decrease significantly with temperature, the loop gain is reduced. This leads to a shift in the open-loop zero and moves the closed-loop poles toward the left-half plane, resulting in slower system response.

According to negative feedback theory, variations in the analog front-end output amplitude are suppressed by approximately the reciprocal of the loop gain. Consequently, the amplitude control loop's suppression capability degrades with reduced loop gain, leading to larger TIA output deviations as K_P and K_I decrease. Since K_P and K_I exhibit temperature dependence, this degradation directly impacts the PI controller's ability to compensate for amplitude variations, particularly those induced by environmental factors such as temperature fluctuations.

Meanwhile, the gain coefficient K_{VGA} and K_{TIA}, the mechanical resonant frequency, quality factor (Q), and driving force constant $K_{F/V}$ of the oscillator can vary with temperature as well. These changes cause variations in the actual displacement amplitude, which in turn affects the detected voltage amplitude.

This paper focuses on the influence of the PI controller parameters, K_P and K_I, on the resonant drive circuit. These parameters play a critical role in maintaining amplitude stability across temperature variations.

III. DESIGN OF PI CONTROLLER

The PI controller plays a critical role in the AGC-PI closed-loop architecture by regulating the oscillation amplitude and enhancing system stability. It processes the amplitude error signal and adjusts the loop gain in real time, ensuring that the oscillation amplitude remains constant despite external disturbances such as temperature variations and supply voltage fluctuations.

A. Open-looped Gm-C PI Controller

In conventional AGC closed-loop circuits, a open-loop Gm-C PI controller is often used. Its transfer function is determined by the transconductance gm along with external passive components, including resistors and capacitors, as shown in Fig. 3(a). The transfer function of the implemented PI controller can be expressed as:

$$H_{open\_PI}(s) = g_m R_1 \cdot \frac{R_L}{R_1+R_L} + \frac{g_m}{C} \cdot \frac{[R_L/(R_1+R_L)]^2}{s+1/[C(R_1+R_L)]} \quad (5)$$

where $H_{open\_PI}(s)$ is the transfer function of open-loop Gm-C PI controller. R_L is the parallel combination of the OTA output resistance R_o and the resistor R_2. When $R_1 \ll R_2 \ll R_o$, the proportional and integral coefficients are given by:

$$\begin{cases} K_{I\_open} = g_m/C \\ K_{P\_open} = g_m R_1 \end{cases} \quad (6)$$

where g_m is the equivalent transconductance of the OTA, K_{P_open} is the proportional coefficient, and K_{I_open} is the integral coefficient of an open-loop Gm-C PI controller, and both K_{P_open} and K_{I_open} depend on g_m. The transconductance g_m is highly sensitive to temperature variations. In general, g_m decreases as the temperature increases, which leads to significant output drift with environmental temperature changes. This greatly limits the environmental robustness of the system.

979-8-3315-3918-4/25 $31.00 © 2025 IEEE

B. Closed-Loop Active PI Controller

In this work, a closed-loop active PI controller structure is adopted [6], as shown in Fig. 3(b). After full-wave rectification and low-pass filtering, a low-frequency signal is obtained. In the PI controller, this signal is compared with a reference voltage V_{ref} to generate an error signal e. Then the error signal is used to control the gain of the VGA.

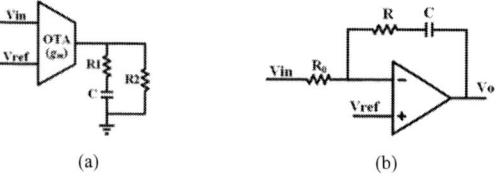

(a)   (b)

Fig. 3. (a) Gm-C-based PI controller architecture (b) Proposed closed-loop active PI controller architecture

The circuit transfer function is expressed as:

$$H_{closed\_PI}(s) = \frac{V_o}{V_{ref}-V_{in}}(s) = \frac{R}{R_0} + \frac{1}{sR_0C} = K_P + K_I\frac{1}{s} \quad (7)$$

where, $H_{closed\_PI}(s)$ is the transfer function of closed-loop Gm-C PI controller.

The proportional coefficient, integral coefficient, and integrator loss τ are given by:

$$\begin{cases} K_{I\_closed} = 1/(R_0C) \\ K_{P\_closed} = R/R_0 \end{cases} \quad (8)$$

where K_{P_closed} is the proportional coefficient, and K_{I_closed} is the integral coefficient. V_{in} is the input voltage signal, and V_{ref} is the reference voltage.

Adjusting the R/R_0 ratio sets the error signal amplification ratio. Varying capacitor C tunes the integral gain. In this closed-loop active PI controller, both K_{P_closed} and K_{I_closed} depend only on external passive components. Using low-temperature-coefficient resistors and capacitors minimizes their variations with temperature. Consequently, the PI controller exhibits negligible temperature sensitivity.

C. Simulation Results

Simulations characterize K_I and K_P variations across temperature (-55°C to 85°C) and supply voltages (4.5V, 5V, 5.5V). The temperature characteristics of K_I and K_P for the open-loop Gm-C PI controller and the closed-loop PI controller are shown in Fig. 4 and Fig. 5, respectively.

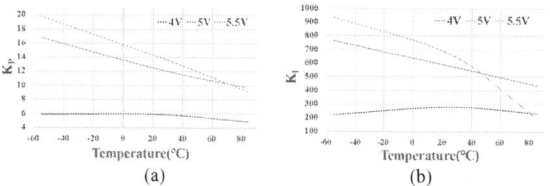

(a)   (b)

Fig. 4. Variation of (a) K_{I_open} and (b) K_{P_open} of the Gm-C PI controller with temperature and supply voltage.

TABLE I. compares the temperature and supply voltage characteristics of the two architectures. Under a 5V supply, the active closed-loop PI controller achieves a 10× lower temperature coefficient for K_P and 3.95× for K_I compared to the open-loop Gm-C PI controller design. At 25°C, it further improves voltage coefficient suppression by 76.7× (for K_P)

and 213× (for K_I). These reductions demonstrate significantly lower sensitivity of proportional and integral coefficient to temperature and supply voltage variations in the closed-loop controller. Consequently, circuit environmental robustness substantially improves.

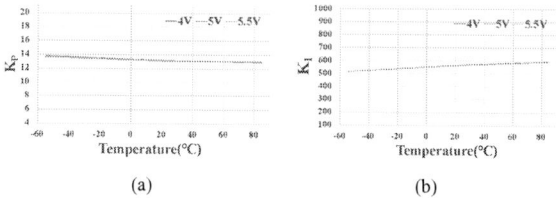

(a)   (b)

Fig. 5. Variation of (a) K_{I_closed} and (b) K_{P_closed} of the closed-loop PI controller with temperature and supply voltage.

TABLE I. TEMPERATURE AND VOLTAGE COEFFICIENTS COMPARISON BETWEEN GM-C AND ACTIVE CLOSED-LOOP PI CONTROLLERS

Structure	Pram.	Variation with Temperature @5V	Temperature Coefficient @5V	Variation with Supply Power @25°C	Supply Power Coefficient @25°C
Gm-C PI Controller	K_P	7.09	0.05/°C	8.05	-5.37/V
	K_I	303.33	2.17/°C	391.04	-260.69/V
Close-Loop PI Controller	K_P	0.72	0.005/°C	0.11	0.07/V
	K_I	77.71	0.55/°C	1.83	1.22/V

IV. CHIP TEST RESULTS

The circuit for the MEMS resonant pressure sensor proposed in this work was designed and fabricated using a 0.18 μm BCD process. The chip occupies an area of 1.29 mm × 0.82 mm, as shown in Fig. 6. It operates at a typical supply voltage of 5 V with a static current of 2.76 mA.

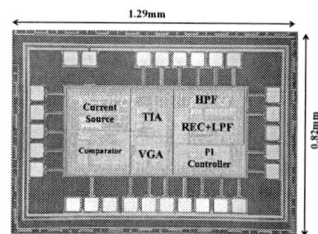

Fig. 6. Photograph of the closed-loop PI controller-based drive circuit.

This work compares two AGC-PI systems: one using a closed-loop PI controller and the other an open-loop Gm-C PI controller, while keeping all other modules identical, analyzing their different impacts on the oscillation amplitude temperature behavior. The proposed design with closed-loop PI controller extends the operating range from −55°C to 110°C (vs. conventional −55°C to 85°C) and achieves 0.45 pA/√Hz input-referred current noise (at 100 kHz). The system achieves a startup time of approximately 150 ms.

Fig. 7 compares the TIA output amplitude of both closed- and open-loop AGC-PI systems across temperatures (−55°C to 85°C) and supply voltages (4 V to 5.5 V), with key variations summarized in TABLE II.

In the AGC-PI driving circuit using the closed-loop PI controller, the temperature coefficient of the TIA output amplitude is approximately 21 μV/°C at a 5 V supply voltage. In contrast, the system using the open-loop Gm-C PI controller exhibits a temperature coefficient of 84 μV/°C at 5 V supply, which is about four times higher. At 25 °C, the TIA output resulting in a voltage coefficient of 0.83 mV/V. These results demonstrate that the AGC-PI structure with the closed-loop PI controller achieves better amplitude robustness under temperature variations.

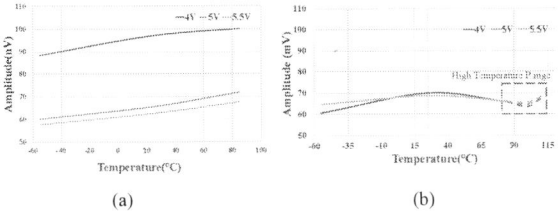

(a)   (b)

Fig. 7. TIA output amplitude variation with temperature in (a) the open-loop Gm-C PI controller and (b) the closed-loop PI controller under a supply voltage of 4–5.5 V.

TABLE II. COMPARISON OF TIA OUTPUT AMPLITUDE VARIATIONS WITH TEMPERATURE AND SUPPLY VOLTAGE FOR TWO PI CONTROLLERS

Structure	$\triangle TIA_{amp}$ @4V	$\triangle TIA_{amp}$ @5V	$\triangle TIA_{amp}$ @5.5V	Amplitude TC@5V
Gm-C PI Controller	11.75mV	11.75mV	10mV	84μV/°C
Closed-Loop PI Controller	9.5mV	4mV	4mV	21μV/°C

A pressure pump is used to provide the pressure input for the sensing element. The measured resonant frequency of the sensor is shown in Fig. 8(a). The resonant pressure sensor exhibits a sensitivity of approximately −44.66 Hz/kPa, with a nonlinearity of less than 0.11%. The Allan variance is shown in Fig. 8(b). The bias instability is approximately 0.0197 Hz at 7 s.

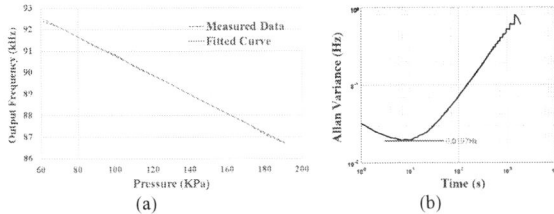

(a)   (b)

Fig. 8. (a) Relationship between input pressure and output frequency (b) Allan variance

At 25 °C, the phase noise of the TIA output oscillation signal is −106 dBc/Hz at 10 kHz offset. The measured result is shown in Fig. 9.

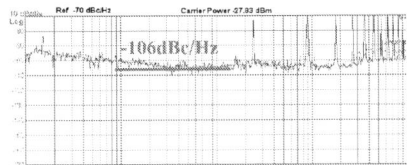

Fig. 9. Phase noise of the TIA output at 10kHz

TABLE III. summarizes the main performance metrics of the driving circuit and sensor system designed in this work. It also compares these results with those of the driving circuit based on the Gm-C PI controller architecture [7] and other reported circuits. The comparison shows that the proposed driving circuit achieves a wider operating temperature range and supply voltage range.

TABLE III. PERFORMANCE SUMMARY AND COMPARISON

	[8]	[9]	[7]	This Work
Technology (μm)	0.35	(Board)	0.18	0.18
Circuit Architecture	AGC + Phase Control Loop	AGC + Phase Control Loop	AGC-PI (Gm-C PI)	AGC-PI (Closed-Loop PI)
Supply Voltage (V)	5	5	4 V~5.5	4 ~5.5
Power Consumption (mW)	288	102.5	13.5	13.8
Temperature Range (°C)	-40~80	-40~70	-55~85	-55~110
Amplitude TC (V/°C)	/	/	84	21

V. CONCLUSION

This work presents a drive circuit for a resonant MEMS pressure sensor based on an AGC-PI architecture. To reduce the sensitivity of the PI controller parameters to temperature and supply voltage variations, a closed-loop PI controller is adopted. This design makes both the proportional and integral coefficients of the PI controller insensitive to temperature. It also enhances the ability to accurately correct amplitude deviations caused by environmental changes such as temperature. The circuit operates over a temperature range from −55 °C to 110 °C and a supply voltage range from 4 V to 5.5 V. The phase noise is around −106 dBc/Hz, and the bias instability is approximately 0.0197 Hz.

ACKNOWLEDGMENT

This work was supported by the National Key Research and Development Program of China (No. 2022YFB3204903).

REFERENCES

[1] J. Zhao, "A 0.23 μg bias instability and 1.6 μg/√Hz resolution silicon oscillating accelerometer with build-in Σ - Δ frequency-to-digital converter," IEEE Symposium on VLSI Circuits, Honolulu, HI, pp. 1-2, 2016.

[2] H. M. Wu, H. G. Yang, T. Yin, X. Y. Cheng, "An AGC-PI Based ClosedLoop Drive Circuit on Chip for Micro-Gyroscope," Nanotechnology and Precision Engineering, vol. 12, pp. 56-62, 2014.

[3] B. Xie, Y. H. Xing, Y. S. Wang, J. Chen, D. Y. Chen and J. B. Wang, "A Lateral Differential Resonant Pressure Microsensor Based on SOI-Glass Wafer-Level Vacuum Packaging," Sensors, vol. 15, no. 9, pp. 2425724268, Sep, 2015.

[4] J. S. Zhang, "Optimization of Measurement and Control Circuit for Silicon Oscillating Accelerometer." Nanjing University of Science and Technology, 2017.

[5] T. Lu, H. -M. Wu, T. Yin, L. -Y. Liu and W. Wang, "Loop Oscillation Analysis of MEMS Resonant Pressure Sensor Readout Circuit," 2023 IEEE 15th International Conference on ASIC (ASICON), Nanjing, China, 2023, pp. 1-4.

[6] X. Zhou, Research on Self-Excited Driving Technology for Silicon Micro-Resonant Pressure Sensors, M.S. thesis, Southeast University, Nanjing, China, 2020.

[7] Y. Liu, S. Yin, T. Yin, J. Liu, N. Wu and L. Liu, "An AGC-PI Based Readout Circuit for Resonant Pressure Microsensor," 2024 9th International Conference on Integrated Circuits and Microsystems (ICICM), Wuhan, China, 2024, pp. 538-542.

[8] J. Cui, Design of Self-Excited Drive Circuit for Resonant Pressure Sensors, M.S. thesis, Harbin Institute of Technology., Harbin, China, 2023.

[9] R. Liu, Parameter Testing of Silicon Micro-Resonant Pressure Sensor Resonators and Optimization of Measurement and Control Circuits, M.S. thesis, Southeast University, Nanjing, China, 2022.

Design of 11MHz Isolated Current Sense Amplifier Based on FDDA and Current Feedback Frequency Modulation Loop

Xinghong Chen[1], Jiahui Liu[1], Shaowei Zhen[1]*, Hongwei Shen[2], Wei Yang[2], Yongwang Ma[2], Bo Zhang[1]

[1]University of Electronic Science and Technology of China, Chengdu 611731, China

[2]Beijing Smartchip Microelectronics Technology Co., Ltd., Beijing 100089, China

*swzhen@uestc.edu.cn

Abstract—A high-bandwidth isolated current-sensing amplifier circuit was designed in a 0.18μm BCD process. A fully differential difference amplifier (FDDA) was employed as the analog front-end circuit for current sensing, achieving high sampling accuracy and bandwidth. A voltage controlled oscillator (VCO) and a frequency to current converter was used as the signal modulator loop, with the same frequency to current converter (FCC) serving as the demodulation circuit of receiver. This ensured that the signal can be quickly corrected by the loop when affected by non-ideal factors, and it guaranteed distortion-free transmission across the isolation barrier at high frequencies, significantly improving the bandwidth performance. The isolated current-sensing amplifier circuit could be applied in motor current-sensing systems. Simulation results indicated that when the input signal amplitude ranges from -10 mV to 10 mV, the amplifier exhibited a low-frequency fixed gain of 4 and a -3-dB bandwidth of 11 MHz.

Keywords—Fully differential difference amplifier (FDDA), voltage controlled oscillator (VCO) , frequency to current converter (FCC)

I. INTRODUCTION

With the development of power systems, more and more fields need to use isolated devices to reduce electrical interference between high-power, high-current circuits and high-precision control circuits. The isolated current sense amplifier (ICSA) is a common isolated device. Fig. 1. shows the diagram of ICSAs applied to motor current sense. As the operating frequency of circuit systems continues to increase, the bandwidth requirements for isolated devices are also rising. However, most commercially available ICSAs currently have a bandwidth in the hundreds of kilohertz range, with few reaching the megahertz range[1]. Therefore, it is of great significance to research high-bandwidth ICSA.

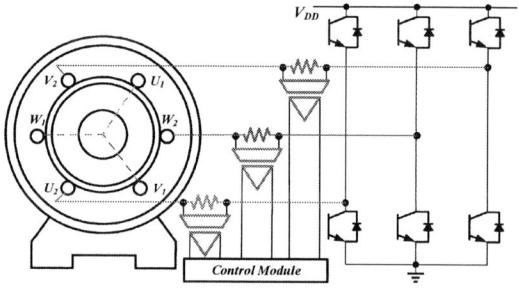

Fig. 1. Diagram of ICSAs applied to motor current sense.

Due to the limitations of power consumption and reliability, the resistance value of the current-sensing resistor is usually very small. This results in a very small voltage drop across the resistor when current flows through it. To

achieve high-precision detection of weak signals while providing a high bandwidth, a fully differential difference amplifier (FDDA) as the analog front-end circuit is presented. To maintain signal quality at high frequencies, a signal modulation loop composed of a voltage-controlled oscillator (VCO) and a frequency-to-current converter (FCC) are proposed. This allows for rapid correction of signals affected by non-ideal factors, thereby significantly enhancing the bandwidth of the isolated current-sensing amplifier.

This article is organized as follows. Section II introduces the principles and design of the high-bandwidth ICSA. Section III provides a detailed analysis of the modulation loop. Section IV presents the circuit simulation results, and Section V offers the conclusions.

II. ARCHITECTURE OF PROPOSED ICSA

The block diagram of the proposed high-bandwidth ICSA in this paper is shown in Fig. 2. It consists of a fully differential difference amplifier (FDDA)[2], a voltage controlled oscillator (VCO), a frequency to current converter (FCC), and three amplifiers for clamping and generating the common-mode signal.

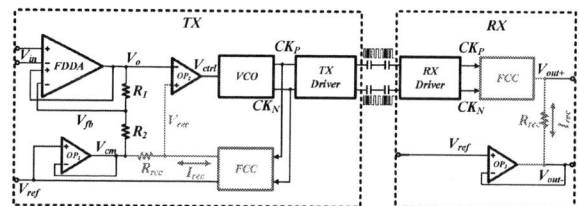

Fig. 2. The block diagram of the proposed high-bandwidth ICSA.

The FDDA possesses two pairs of input terminals. One pair is for signal detection, and the other pair, together with R_1 and R_2, forms a resistive feedback network to provide a fixed gain at low frequency.

The VCO, FCC, and clamping amplifier OP_2 constitute a modulation loop that functions as follows. When non-ideal factors cause V_{fb} to increase, V_{ctrl} is amplified by OP_2. This raises the frequency of the switch control signals (CK_P and CK_N) output by the VCO. Consequently, current I_{rec} flows into the FCC, and the voltage drop across R_{rec} reduces V_{rec}, creating a negative feedback that corrects signal offsets. Conversely, when V_{fb} decreases, I_{rec} flows out of the FCC, increasing V_{rec} to correct the signal offset.

The receiver circuit uses an identical FCC to demodulate the CK_P and CK_N signals that have crossed the isolation barrier, thereby reconstructing the sampled signal. As long as the parameters of the FCC, R_{rec}, and other components in the transmitter and receiver are matched, the output signal remains consistent with the clamped V_{rec} despite non-ideal factors. This ensures reliable signal transmission across the isolation barrier.

A. Fully Differential Difference Amplifier

The circuit diagram of the fully differential difference amplifier is shown in Fig. 3. The circuit is composed of two pairs of folded cascode and a floating Class AB. Compared to a conventional fully differential amplifier, the two pairs of input terminals offer more flexible connection options and can improve sampling accuracy to the µV level, making it suitable for current sense applications.

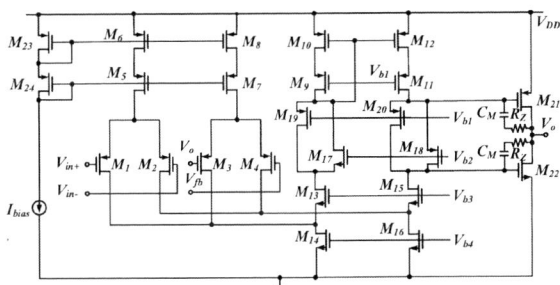

Fig. 3. Internal structure of FDDA.

When the FDDA is connected in the resistive feedback network configuration shown in Fig. 2. The fixed gain can be expressed as:

$$(V_o - V_{cm}) = \frac{R_2 + R_1}{R_1}(V_{in+} - V_{in-}) \qquad (1)$$

The dominant pole of the circuit is provided by the folded cascode output node, and the secondary pole is provided by the floating Class AB output node, the loop gain is:

$$G_{loop}(s) = \frac{g_{m21}g_{m1}R_1(g_{m11}r_{o11}r_{o12} \parallel g_{m15}r_{o15}r_{o16})}{(1+\frac{s}{w_1})(1+\frac{s}{w_2})}$$

$$w_1 = \frac{1}{C_M g_{m21}(R_1 + R_2)(g_{m11}r_{o11}r_{o12} \parallel g_{m15}r_{o15}r_{o16})} \qquad (2)$$

$$w_2 = \frac{1}{C_L(R_1 + R_2)}$$

Here, C_M is the Miller compensation capacitance, and C_L is the output capacitance of the FDDA. By adjusting R_Z, the right-half-plane zero caused by Miller compensation can be canceled out. Additionally, by tuning R_1 and R_2, the FDDA can be enabled to detect and transmit signals at the megahertz level.

B. Voltage Controlled Oscillator

The voltage-controlled oscillator consists of a voltage-to-current converter and cascaded inverters is shown in Fig. 4. The voltage-to-current converter transforms the control voltage V_{ctrl} into gate voltages V_P and V_N, ensuring that I_N is twice of I_P. The frequency of the output clocks CK_P and CK_N can be controlled by adjusting the inverter current.

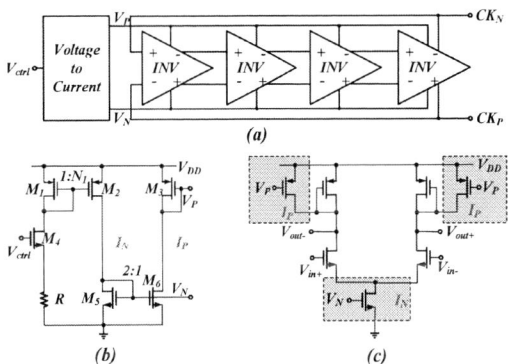

Fig. 4. (a) Block of VCO inverter cascade. (b)Internal structure of voltage-to-current converter. (c) Internal structure of inverter.

The output clock frequency f mainly depends on the ability to charge and discharge the load capacitance C_L. The relationship between f and I_N is:

$$f = \frac{I_N}{8C_L V_{DD}} \qquad (3)$$

Because the voltage-to-current converter uses a source-degeneration circuit, I_N is almost linearly related to the control voltage V_{ctrl}. Consequently, the output frequency f also exhibits a linear relationship with V_{ctrl}. The expression for this relationship is:

$$f = \frac{N_1 V_{ctrl}}{8RC_L V_{DD}} \qquad (4)$$

C. Frequency-to-Current Converter

The frequency-to-current converter is show in Fig. 5. This circuit modulates the current through transistor M7 by clock. A current mirror then replicates this change, altering the current through transistor M9. This mechanism controls the direction and magnitude of the output current Irec.

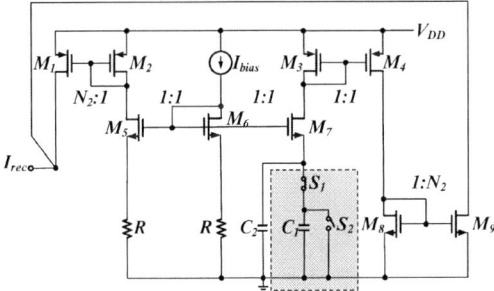

Fig. 5. Internal structure of FCC

The key part of this circuit is the switched-capacitor circuit highlighted in gray. Switches S_1 and S_2, controlled by clock signals, adjust the charging and discharging process of capacitor C_1. From the perspective of the source terminal of M_7, this process is equivalent to current flowing through a resistor. The equivalent resistance is given by Equation (5), where f is the clock frequency output by the VCO.

$$R_{eq} = \frac{1}{2\pi f C_1} \quad (5)$$

According to the current replication mechanism, the expression for Irec is obtained as follows:

$$I_{rec} = N_2(I_{bias} - I_7) \quad (6)$$

Where Ibias is the reference current, and I7 is the current through M7, expressed as:

$$I_7 = \frac{1}{2}\mu C_{ox}\frac{W}{L}(V_g - I_7 R_{eq} - V_{th}) \quad (7)$$

To derive the expression showing how I7 is affected by the clock frequency f, take the partial derivative of both sides of equation (7) with respect to Req:

$$\begin{aligned}
\frac{\partial I_7}{\partial R_{eq}} &= \mu C_{ox}\frac{W}{L}(V_g - I_7 R_{eq} - V_{th})(-I_7 - R_{eq}\frac{\partial I_7}{\partial R_{eq}}) \\
&= \frac{-\mu C_{ox}\frac{W}{L}(V_g - I_7 R_{eq} - V_{th})I_7}{1 + \mu C_{ox}\frac{W}{L}(V_g - I_7 R_{eq} - V_{th})R_{eq}} \\
&\approx -\frac{I_7}{R_{eq}}
\end{aligned} \quad (8)$$

From equations (5) and (8), the relationship between I7 and f can be further derived as:

$$\begin{aligned}
\frac{\partial I_7}{\partial f} &= \frac{\partial I_7}{\partial R_{eq}}\frac{\partial R_{eq}}{\partial f} = \frac{-I_7}{R_{eq}}\frac{-1}{2\pi C_1 f^2} = \frac{I_7}{f} \\
&\Rightarrow \frac{\partial I_7}{I_7} = \frac{\partial f}{f} \Rightarrow \ln I_7 = \ln(kf) \Rightarrow I_7 = kf
\end{aligned} \quad (9)$$

Where k is a gain factor obtained through derivation. Combining equations (6) and (9) yields the final relationship between Irec and the frequency f:

$$I_{rec} = N_2(I_{bias} - kf) \quad (10)$$

To ensure zero output current in common-mode operation, the following condition must be met at the common-mode frequency fcm:

$$k = \frac{I_{bias}}{f_{cm}} \quad (11)$$

From the above derivation, it is evident that the frequency-to-current converter can linearly transform frequency signals into current signals, thereby allowing the frequency at the VCO output to be reconstructed as a current.

III. MODULATION LOOP ANALYSIS

To effectively analyze the modulation loop, it is essential to first examine the poles and zeros of each circuit stage within the loop.

The clamping amplifier OP2, which employs a folded cascode configuration, provides a high-impedance node. Its transfer function is given by:

$$\begin{aligned}
H_1(s) &= \frac{A_0}{(1 + s/w_1)} \\
w_1 &= \frac{1}{(g_{mN}r_{oN1}r_{oN2} \| g_{mP}r_{oP1}r_{oP2})C_p}
\end{aligned} \quad (12)$$

The VCO involves voltage-to-frequency conversion. Considering the relationship between phase and frequency:

$$\frac{d\phi}{dt} = 2\pi f \quad (13)$$

From Equations (4) and (13), the VCO resembles an integrator, providing a pole at the origin. Additionally, the voltage-to-current converter circuit contributes another pole at the input terminal of M4. Thus, the transfer function of the VCO can be written as:

$$H_2(s) = \frac{\Phi(s)}{V(s)} = \frac{\frac{N_1\pi}{4RC_L V_{DD}}}{s(1 + RC_{gs4}s)} = \frac{K_{VCO}}{s(1 + s/w_2)} \quad (14)$$

The FCC converts frequency to current. Acting as a differentiator, it introduces a zero at the origin. The equivalent resistance R_{eq} is kept on the same order as the source degeneration resistance R, so the pole it creates is at a high frequency and can be neglected. The conversion of the output current I_{rec} to voltage through the large resistance R_{rec} results in a pole. Combining Equations (10) and (13), the transfer function is:

$$\begin{aligned}
H_3(s) &= \frac{V(s)}{\Phi(s)} = \frac{-\frac{k}{2\pi}N_2 R_{rec}s}{1 + R_{rec}(C_{gs1} + C_{gs9})s} \\
&= \frac{-K_{FCC}s}{1 + s/w_3}
\end{aligned} \quad (15)$$

From the above analysis, the loop can be simplified to the block diagram shown in Fig. 6. Given the three poles in the loop, a type-two compensation scheme using C_p and R_p is introduced to add a zero and enhance loop stability.

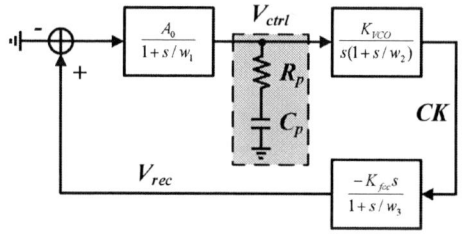

Fig. 6.   The diagram of modulation loop

In the loop, ω_1 has the highest impedance and thus is the dominant pole. The resistor R in the VCO is chosen to enhance linearity, but a very large R would reduce loop gain, so ω_2 is not at a low frequency and has minimal impact on stability. For the FCC, a small I_{rec} is preferred to minimize common-mode voltage impact. Therefore, R_{rec} must be large enough to provide sufficient feedback voltage even with a small I_{rec}, placing ω_3 at a relatively low frequency where it significantly affects stability. The overall loop gain expression is:

$$G_{loop}(s) = \frac{-A_0 K_{VCO} K_{FCC} (1 + s/w_z)}{(1 + s/w_d)(1 + s/w_{nd})}$$

$$w_z = \frac{1}{R_p C_p}$$

$$w_d = \frac{1}{(g_{mN} r_{oN1} r_{oN2} \| g_{mP} r_{oP1} r_{oP2}) C_p} \qquad (16)$$

$$w_{nd} = \frac{1}{R_{rec}(C_{gs1} + C_{gs9})}$$

By tuning C_p and R_p, it is possible to ensure high bandwidth while maintaining loop stability. This allows the modulation loop to quickly correct signal offsets during high-frequency signal transmission.

IV. SIMULATION RESULTS

The proposed high-bandwidth ICSA is designed by the 0.18 μm BCD process. Fig. 7 shows the gain Bode plot of the amplifier under an input voltage of -10 mV to 10 mV. The simulation shows that the low-frequency fixed gain is 12.08 dB (a 4× gain), and the -3 dB bandwidth reaches 11 MHz.

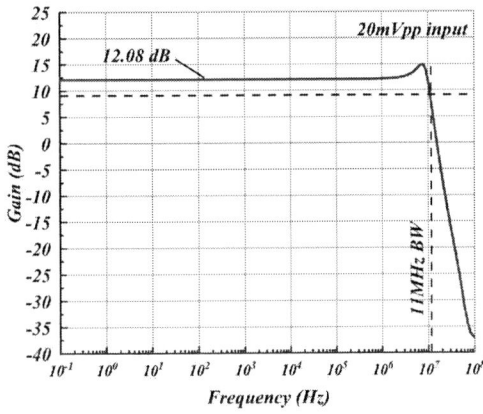

Fig. 7. The bandwidth diagram of the proposed ICSA

To verify the amplifier's transmission performance, a 1 MHz triangular wave with a peak-to-peak voltage of 60 mV is input at a common-mode level of 0 V. Fig. 8 shows the input and output waveforms. The output remains a 1 MHz triangular wave with a peak-to-peak voltage of 238 mV and a common-mode level of 900 mV, confirming the 4× gain and MHz-level signal transmission.

Fig. 8. Simulated results of the input voltage (V_{IN}) and output differential voltage (V_{OUT}) for a triangular wave input with a frequency of 1 MHz and a peak-to-peak amplitude of 60 mV

Table 1 compares the parameters of the proposed ICSA with those in the literature. Results show that the proposed circuit structure significantly improves signal bandwidth.

TABLE I. COMPARISON OF ICSAs PERFORMANCE

	This work	[1]	[3]	[4]	[5]	[6]
Tech	0.18μm BCD	0.18μm CMOS	-	-	-	-
Supply voltage [V]	1.8	5	3 to 20	3.3/5	3.3/5	3.3/5
Input range [mV]	±30	-	-	±250	±250	±250
Bandwidth [kHz]	11000	35200	1000	310	220	310
Gain	4	-	20	8.2	8.2	8.2

V. CONCLUSION

A high-bandwidth, high-precision ICSA realized with 0.18μm BCD process is presented. A FDDA is proposed as the analog front end for weak signal detection, and a modulation loop with a VCO and FCC is presented to correct high-frequency signals and suppress non-ideal effects, which significantly increases the transmission bandwidth. Simulation results indicate that within the -10 mV to 10 mV input range, the amplifier achieves a 4× low-frequency gain and an 11 MHz -3 dB bandwidth.

ACKNOWLEDGMENT

This work is supported by CIE-Smartchip research fund No.2023-006.

REFERENCES

[1] S. Takaya, H. Ishihara and K. Onizuka, "18.7 A DC to 35MHz Fully Integrated Single-Power-Supply Isolation Amplifier for Current- and Voltage-Sensing Front-Ends of Power Electronics," 2020 IEEE International Solid-State Circuits Conference - (ISSCC), San Francisco, CA, USA, 2020, pp. 298-300.

[2] Y. Zhang, "A High-Gain, Low-Power Fully Differential Amplifier with Common-Mode Feedback for High-Speed and High-Precision Analog-to-Digital Converters," 2024 IEEE 7th International Conference on Automation, Electronics and Electrical Engineering (AUTEEE), Shenyang, China, 2024, pp. 979-985.

[3] "TPA158 Datasheet," 3PEAK, 2024. [Online]. Available: https://static.3peak.com/res/doc/ds/Datasheet_TPA158.pdf.

[4] "AMC1400 Datasheet," Texas Instruments, July 2022. [Online]. Available: https://www.ti.com/lit/ds/symlink/amc1400.pdf?ts=1714484324325& ref_url=https%253A%252F%252Fwww.ti.com%252Fproduct%252F de-de%252FAMC1400.

[5] "NSI1400D Datasheet," Novosense, Rev. 1.0, Oct. 2023. [Online]. Available: https://www.novosns.com/Public/Uploads/uploadfile/files/20231206/ NSI1400_Datasheet_Rev1.0_EN.pdf.

[6] "CA-IS1300x Datasheet," Chipanalog Inc., Version 1.05, June 2024. [Online]. Available: https://www.chipanalog.com/web/bocupload/2024/06/11/ca-is1300x_datasheet_cn_version1.05.pdf

A Charge Pump Powered Current Sense Amplifier with -20 V to 40 V Input Common-Mode Range

Dejian Li[1], Hongwei Shen[1], Jinzhao Li[1], Jiahui Liu[2], Lixing Wang[2], Shaowei Zhen[2*], Bo Zhang[2]

[1]Beijing Smartchip Microelectronics Technology Co., Ltd., Beijing 100089, China

[2]University of Electronic Science and Technology of China, Chengdu 611731, China

*swzhen@uestc.edu.cn

Abstract—**This paper presents a high-performance current-sense amplifier (CSA) implemented in 0.18-μm BCD process, incorporating three key innovations. First, a dedicated charge pump supply circuit ensures stable CSA biasing across the wide operational voltage range. Second, a novel CSA architecture featuring an NPN-based input stage with current-sensing capability and an optimized feedback loop achieves both wide common-mode input range (-20 V to +40 V) and stable operation. Simulation results demonstrate a maximum output voltage variation of 3.19mV for a 100mV input signal under varying common-mode voltages. The proposed design enables robust performance in high-voltage applications.**

Keywords—current sense amplifier (CSA), charge pump

I. INTRODUCTION

Current measurement serves as a fundamental requirement in modern electronic systems, finding extensive applications in power management, battery monitoring, motor control, and industrial automation. The current sense amplifier (CSA), a key component in such systems, achieves high-accuracy current monitoring by amplifying the small voltage differential across a shunt resistor into a noise-resistant output signal. Growing demands for enhanced energy efficiency, system reliability, and real-time feedback control have established high-performance CSA design as a prominent research focus in both academic and industrial domains. In practical applications, as shown in Fig. 1, the common-mode voltage at the input of a current sense amplifier often undergoes wide-range fluctuations. When the common-mode voltage swings to extremely high or low levels, it may cause the sensing circuit to deviate from its normal operating condition or, in severe cases, damage circuit components. To address scenarios involving both excessively high and low common-mode voltages, the conventional approach employs dual sensing circuits on either side of the common-mode voltage range: a high-side and a low-side CSA[1]. When the common-mode voltage becomes too low, the low-side sensing circuit fails to operate due to compressed power rail headroom, while the high-side CSA remains functional with sufficient voltage margin. Conversely, during excessively high common-mode voltage conditions, only the low-side circuit maintains proper operation. This dual-circuit architecture enables current sensing across an extended common-mode voltage range. However, the low-side sensing circuit typically relies on the amplifier with PNP input transistors. In 0.18μm BCD (Bipolar-CMOS-DMOS) processes, PNP transistors generally cannot achieve performance parity with the NPN transistors. Consequently, the low-side circuit's performance often proves inferior to that of the high-side implementation.

To circumvent the use of PNP transistors in critical modules of the sensing circuit while maintaining operational capability under elevated common-mode voltages, this paper proposes a single-ended current sense amplifier powered by a charge pump, achieving the aforementioned objectives. The proposed amplifier demonstrates robust operation across a wide common-mode voltage range, particularly during high common-mode voltage conditions.

The second section will detail its key building blocks, including, the overall circuit architecture, the charge pump circuit design and the current sense amplifier circuit.

II. PROPOSED CHARGE PUMP POWERED CURRENT SENSE AMPLIFIER ARCHITECTURE

A. Top-Level Architecture

As shown in Fig. 1, the current-sense amplifier circuit uses the sense line as its own ground. If the power supply is a fixed voltage, variations in the common-mode voltage will cause the supply rails of the current-sense amplifier to fluctuate, significantly limiting its ability to operate under high common-mode voltages. By introducing a charge pump circuit to power the current-sense amplifier, the charge pump can continue supplying power to the main circuit through capacitive charge transfer when the common-mode voltage changes. This ensures a stable voltage difference between the supply voltage of the main circuit and the common-mode voltage, allowing proper circuit operation.

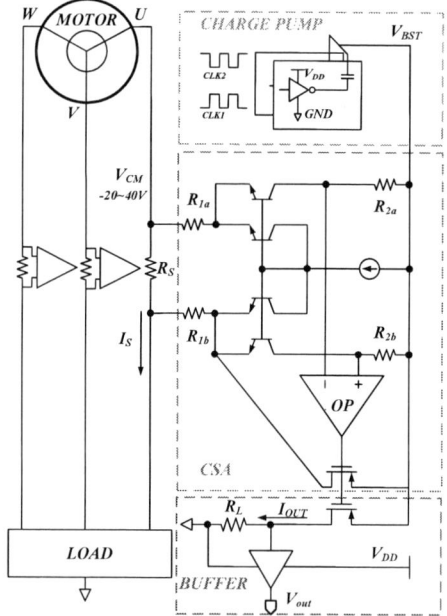

Fig. 1.   The application and the top architecture of the proposed CSA.

979-8-3315-3918-4/25 $31.00 © 2025 IEEE

Since the main circuit operates in a floating voltage domain, conventional charge pump methods[2] are limited in accommodating common-mode voltage variations, especially when the common-mode voltage rises. Because their output voltages are referenced to a fixed ground at a high potential. In practical applications, the common-mode voltage can often reach very high levels, making such fixed high-voltage supply schemes insufficient.

This paper proposes a novel charge pump architecture capable of supporting higher common-mode voltages. The design of the sense amplifier circuit includes an NPN-based current-input amplifier, followed by a differential amplifier that detects current differences between two paths and feeds back to the input to form a control loop. The output is then buffered and sent to a back-end analog-to-digital converter (ADC) module for quantization.

B. Proposed charge pump circuit

Fig. 2(a) illustrates the architecture of the proposed charge pump circuit, where the two-phase non-overlapping clocks CLK1 and CLK2 are generated by the circuit shown in Fig. 2(b). The non-overlapping clock scheme is employed to prevent both clocks from being high simultaneously, thereby avoiding large voltage spikes in the bootstrap voltage.

The working principle of the charge pump can be briefly analyzed as follows: In the initial state when the common-mode voltage (V_{CM}) is near ground (GND) potential, the supply voltage V_{DD} directly powers the main current-sense amplifier circuit while the charge pump remains idle. As V_{CM} increases, the charge pump activates to deliver charge to the main circuit. Specifically, when CLK1 is high and V_{CM} is higher than GND, the common-mode voltage charges capacitor Cf1 through a pair of cross coupled NMOS, establishing a pumped voltage V_{CP} that is given

$$V_{CP} = V_{CM} \tag{1}$$

In the subsequent phase when CLK1 goes low, the inverter elevates the lower plate of Cf1 to V_{DD}, consequently raising the upper plate of the capacitor to $V_{DD} + V_{CP}$. This boosted voltage then supplies the bootstrap power V_{BST} for the main circuit through the diode and the cross coupled PMOS. Therefore, V_{BST} can be expressed as

$$V_{BST} = V_{DD} + V_{CM} - V_F \tag{2}$$

V_F is the diode forward voltage drop of D1. Since CLK2 and CLK1 operate in complementary phases, the bootstrap power supply for the main circuit is alternately maintained by both charge pump circuits, thereby keeping V_{BST} relatively stable. As evident from Equation (2), the actual bootstrap voltage falls slightly short of a full V_{DD}. However, given that the diode forward voltage drop (V_F) is significantly smaller than the supply voltage V_{DD}, and considering that the main circuit is typically designed to operate across a wide voltage range, this power delivery scheme adequately meets the operational requirements of the main circuit.

Remarkably, this charge pump architecture demonstrates excellent simplicity, requiring only two inverters, a pair of capacitors, a diode and two pairs of cross coupled MOS to reliably provide power rails for normal circuit operation under

high common-mode voltages while maintaining low power consumption.

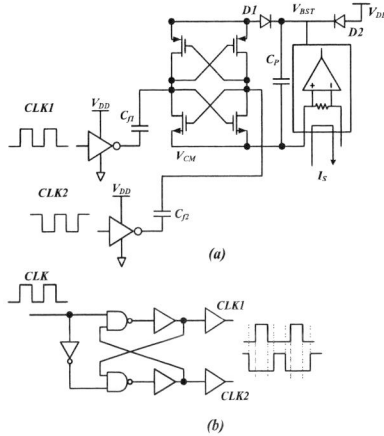

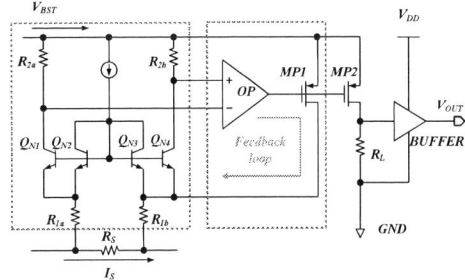

Fig. 2. (a)Structure of proposed charge pump circuit.(b)The circuit that generates the two-phase non-overlapping clocks CLK1 and CLK2.

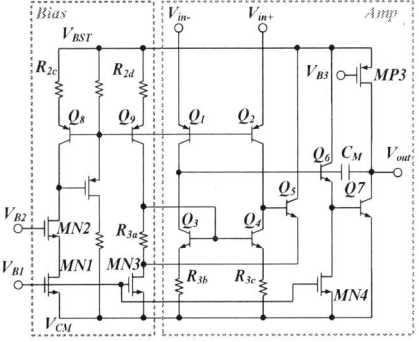

Fig. 3. Architecture of the overall CSA circuit.

C. Architecture of the current-sense amplifier

The current-sense amplifier circuit is illustrated in Fig. 3. Within the blue dashed box, the circuit represents a current input amplifier with resistive loads. Here, the input transistors employ NPN configurations to achieve high transconductance. The circuit samples currents from both the high-side and low-side of Rs and converts them into a voltage difference across load resistors R_{2a} and R_{2b}. This differential voltage is then amplified by OP. In the red dashed box, The amplified output voltage is converted back to current through transistor MP1, which injects the feedback current into the low-side of R_S, thereby closing the feedback loop. Simultaneously, MP2 mirrors this injection current and delivers it to the subsequent BUFFER circuit, whose output is ultimately quantized by the backend ADC unit.

Fig. 4. The details of the OP circuit.

Notably, this top-level architecture employs a feedback structure that significantly enhances the circuit's robustness against PVT (Process, Voltage, Temperature) variations. The loop gain is contributed by three primary components: OP, the current-input amplifier, and MP1. The loop contains only two high-impedance nodes in the OP. Through simple Miller compensation applied to the amplifier circuit itself, the system effectively maintains loop stability while achieving optimal performance characteristics.

Fig. 4 details the internal circuit of the OP. The blue dashed box encloses the biasing network, while the red dashed box contains the core amplifier circuit. In the main circuit, the emitters of input transistors Q_1 and Q_2 connect to resistors R_{2a} and R_{2b}, respectively. A symmetric design is implemented using Q_8, Q_9, R_{2c}, and R_{2d} to ensure proper biasing for Q_1 and Q_2. The differential current signal is mirrored and summed at the base of Q_6 through current mirrors formed by Q_3 and Q_4. Here, Q_5 essentially functions as a β-helper transistor. The voltage drop across resistor R3a precisely regulates the potential difference between the base voltages of Q_5 and Q_4/Q_3, preventing excessive voltage headroom consumption. By carefully designing the circuit, the voltage drops across R_{3a}, R_{3b}, and R_{3c} are set to match the combined V_{BE} of Q_3, Q_4, Q_5, and Q_6, as well as the V_{DS} of MN3 and MN4, satisfying

$$V_{DS} = V_{BE} = IR \qquad (3)$$

This ensures optimal biasing for this stage. The current I in each branch is supplied by the external biasing circuit, while the primary branch currents in the BJT amplifier heavily depend on the matching accuracy of the aforementioned biasing scheme. Fig. 3 illustrates the operational principle where the output voltage of OP serves dual functions: it is first converted into a current through MP1 and fed back to the input of the sensing amplifier to establish the feedback loop, while simultaneously being mirrored by MP2 to generate a proportional current that is transformed into a voltage via a resistor before being delivered to the output through the BUFFER circuit.

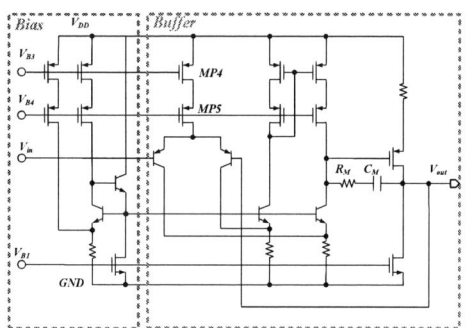

Fig. 5.   The details of the output buffer circuit.

Fig. 5 presents the detailed architecture of this BUFFER circuit, with the biasing network enclosed in the blue dashed box and the core amplifier circuit in the red dashed box. The amplifier employs a sophisticated two-stage topology combining a folded-cascode input stage with a common-source output stage. The design deliberately utilizes PNP transistors as the input pair to capitalize on their inherent high transconductance (g_m) advantages, while also providing an extended input voltage range expressed by the equation(4)

$$IR + |V_{CE}| - |V_{BE}| \le V_+ \le V_{DD} - 2|V_{ov}| - |V_{BE}| \quad (4)$$

where V_{CE} and V_{BE} represent the collector-emitter and base-emitter voltages of the PNP transistors (both negative values), and V_{GS} denotes the gate-source voltage of the PMOS bias current source (also negative). Notably, the upper voltage limit only subtracts $2|V_{OV}|$ term due to the implementation of a low-voltage cascode current mirror , shown in Fig. 6, a configuration that optimally preserves output headroom. The biasing circuit within the blue box incorporates not only a β-helper structure to address base current errors but also carefully replicates the common-base NPN environment of the folded-cascode stage to ensure superior matching characteristics.

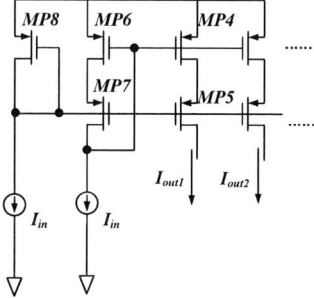

Fig. 6.   The bias circuit for MP4, MP5.and other PMOS.

III. SIMULATION RESULT OF PROPOSED CSA

The proposed CSA circuit with charge pump supply was implemented in a 0.18-μm BCD process. Fig. 7 presents the magnitude and phase frequency responses of the control loop in the proposed CSA, which consists of an NPN amplifier for current input, OP, and feedback current transistor MP1. The results demonstrate a loop gain of 101.8dB, a bandwidth of 1.3MHz, and a phase margin of 65.5°. The CSA circuit consumes 59μA when there is no input.

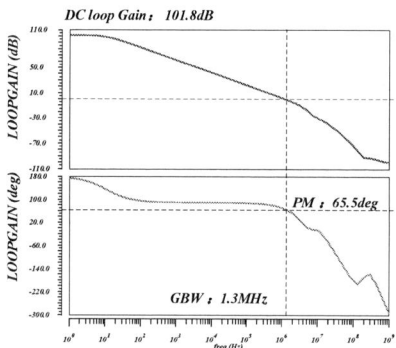

Fig. 7.   The Bolt plot of the proposed CSA.

To verify that the proposed charge pump circuit enables the CSA to operate reliably across a wide common-mode voltage range, the common-mode voltage is varied from −20 V to +40 V while maintaining a sensed input voltage difference of 100 mV. The output voltage difference is then measured, as shown in Fig. 8(a). In this situation, the charge pump circuit consumes 211μA. Fig. 8(b) illustrates the variation curve of the bootstrap supply V_{BST} of CSA with respect to the common mode voltage V_{CM}. The designed CSA exhibits an 8× gain between its output voltage and input voltage. In the Monte Carlo simulation, the input voltage

979-8-3315-3918-4/25 $31.00 © 2025 IEEE

difference detected by the current-sense amplifier was configured at 100 mV. The distribution of the output voltage is shown in Fig. 9, with a calculated offset of 660uV. The gain error is about 0.08% as the gain is set to 8 shown in Fig. 10.

A performance comparison between the proposed CSA and prior designs is summarized in Table I. The proposed architecture demonstrates superior tolerance to common-mode voltage variations.

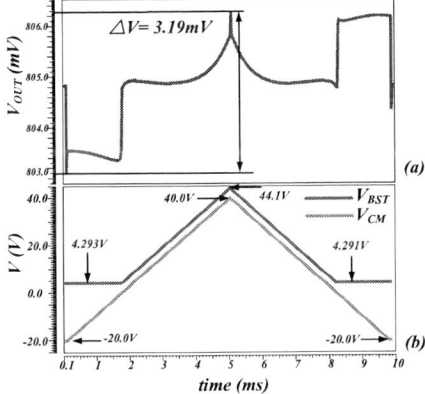

Fig. 8. (a)Simulation result of the output voltage defference as the common-mode voltage was varied from −20 V to +40 V while maintaining a sensed input voltage difference of 100 mV.(b) The variation curve of the bootstrap supply V_{BST} of CSA with respect to the common mode voltage V_{CM}.

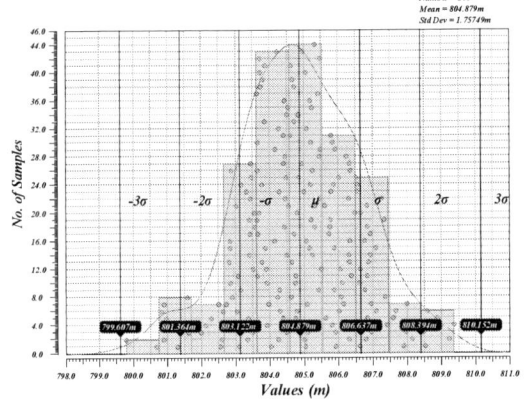

Fig. 9. The distribution of the output voltage for the input voltage difference is 100 mV.

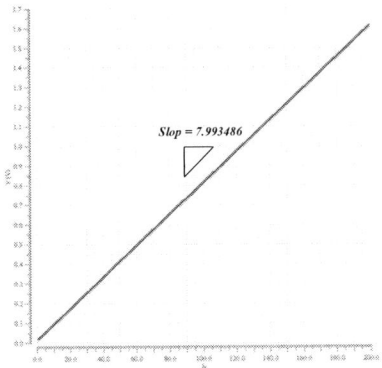

Fig. 10. Gain of the CSA.

IV. CONCLUSION

A high-performance current-sense amplifier is implemented in 0.18-μm BCD technology. A dedicated charge pump supply circuit is proposed which is specifically designed to maintain stable biasing for the CSA across the entire operational voltage range. A novel CSA architecture is presented, which achieves both wide common-mode input range and stable operation through an NPN-based input stage with current-sensing capability and an optimized feedback loop incorporating a precision operational amplifier. Simulation results indicate a maximum output voltage variation of 3.19mV with a 100 mV input signal under common-mode voltage variations ranging from -20 V to +40 V.

TABLE I. COMPARISON OF PROPOSED CSA PERFORMANCE

	This work	[3]	[4]	[5]	[6]
Tech	0.18μm BCD	0.8μm BICMOS	-	0.5μm BCD	0.8μm BCD
Input common-mode range [V]	-20-40	1.9-30	-6-30	-0.2-30	0-30
Supply voltage [V]	4-24.3	2.8-5.5	3.5-12	2.5-30	2.7-30
I_{CSA} [μA]	59	650	200	57	-
Gain error	0.08%	0.1%	-	0.05%	0.05%

ACKNOWLEDGMENT

This work is supported by CIE-Smartchip research fund No.2023-006.

REFERENCES

[1] "INA197 Datasheet," Texas Instruments, January 2015. [Online]. Available: https://www.ti.com.cn/cn/lit/ds/symlink/ina197.pdf?ts=175237555640 0&ref_url=https%253A%252F%252Fwww.ti.com.cn%252Fproduct %252Fcn%252FINA197.

[2] TianRui Ying, Wing-Hung Ki and M. Chan, "Area-efficient CMOS charge pumps for LCD drivers," in IEEE Journal of Solid-State Circuits, vol. 38, no. 10, pp. 1721-1725, Oct. 2003.

[3] J. F. Witte, J. H. Huijsing and K. A. A. Makinwa, "A Current-Feedback Instrumentation Amplifier with 5μV Offset for Bidirectional High-Side Current-Sensing," 2008 IEEE International Solid-State Circuits Conference - Digest of Technical Papers, San Francisco, CA, USA, 2008, pp. 74-596.

[4] AD8203 Datasheet," Analog Devices, [Online]. Available: https://www.analog.com/media/en/technical-documentation/data-sheets/AD8203.pdf.

[5] R. Puşcaşu, P. Brînzoi, L. Creoşteanu and G. Brezeanu, "High accuracy current sense amplifier with extended input common mode range," 2014 10th Conference on Ph.D. Research in Microelectronics and Electronics (PRIME), Grenoble, France, 2014, pp. 1-4.

[6] M. Zhang, X. Wang, X. Li, X. Qi, H. Liu and S. Wang, "A High Accuracy and Wide Input Voltage Range Current Sense Amplifier for Bidirectional High-Side and Low-Side Current-Sensing," 2022 IEEE 2nd International Conference on Electronic Technology, Communication and Information (ICETCI), Changchun, China, 2022, pp. 245-249.

The Sequency Domain: a new Approach for Radio Frequency Front End

François Rivet, Pierre Ferrer, Maxandre Fellmann, Nathalie Deltimple,
Hervé Lapuyade, Eric Kerhervé, Yann Deval

University of Bordeaux, CNRS, Bordeaux INP, IMS, UMR 5218, F-33400 Talence, France

Abstract—This paper presents the first experimental demonstration of a fully Walsh-domain Radio Frequency Front-End (RFFE) architecture implemented in 28nm FDSOI CMOS technology. The system integrates a Walsh-based RF digital-to-analog converter (WDAC) and a Walsh-based RF analog-to-digital converter (WADC) for digital predistortion (DPD) purpose. The WDAC achieves an instantaneous bandwidth (BW) of 4.6875 GHz with an energy efficiency of 0.38 pJ/bit. The WADC supports a 2.5 GHz BW and consumes only 4 mW. This experimental demonstration showcases wideband and power-efficient signal processing and highlights the potential operating entirely in the Walsh basis as a compelling approach for future low-power, high-throughput wireless systems, combining both energy efficiency and wideband performances.

Index Terms—Walsh Transform, Radio Frequency Front-End, Digital Predistortion, RF DAC/ADC

I. INTRODUCTION

BY 2040, the number of smart robots and autonomous systems, including UxVs, is projected to grow significantly, potentially surpassing many human-operated systems. They will need to communicate without any human intervention. The current connectivity solutions have limited capabilities. They use RF (Radio Frequency) transceivers with several RF Front-Ends (RFFE), traditionally integrated in expensive non-CMOS technologies to deal with relatively high data rates at high reliability and fairly low latency while balancing a constrained energy budget. The telecommunication industry is constrained to utilize the available frequencies below 100GHz by aggregating a fragmented spectrum. The design of these transceivers is highly complex, as are the efforts to design them in a power-efficient manner to handle large bandwidths and multiple frequency bands. Traditional IQ-based RFFE architectures presented in Fig. 1a rely on low-frequency DACs followed by quadrature upconversion using a local oscillator LO and mixers. While this approach provides excellent linearity, it is inherently narrowband, as the maximum signal bandwidth (BW) is constrained by the DAC sampling rate around hundreds of MHz [1]. Meanwhile, the Power Amplifier (PA) requires a Digital Pre-Distortion (DPD) feedback loop, with on the top of Local Oscillator (LO), mixers, and associated digital circuitry increase both complexity and power consumption.

This paper proposes a novel approach to address the critical challenges of complexity, efficiency, and linearity, offering solutions at the system level to achieve low power consumption, high power efficiency, and wideband operation, with

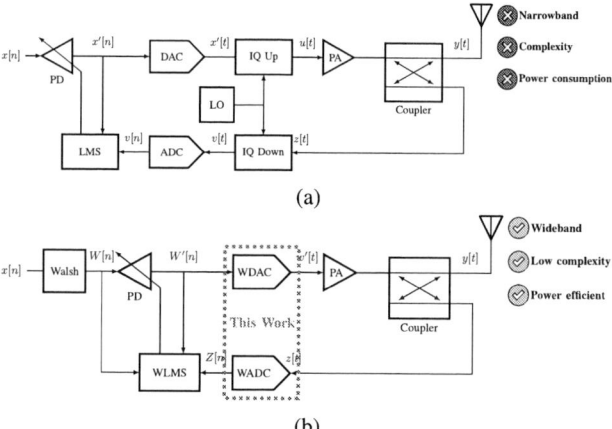

Fig. 1: (a) Standard IQ RFFE architecture. (b) Walsh-based RFFE architecture.

enhanced spectrum access and linearization to reduce costs tenfold compared to conventional methodologies. At the heart of it is the application of the Walsh theory over the whole RF chain to perform signal processing, from digital to analog, from baseband to RF. The chain operates fully in the sequency domain and demonstrates:

- **diversity**: the wideband conversion captures information thanks to the orthogonality of the Walsh domain,
- **sparsity**: the inherent sequency-domain properties release strong energy-efficient performances,
- **binary**: the digital nature of the Walsh signal processing is aligned with CMOS technology.

This choice allows the system to perform signal generation, processing, and linearization simultaneously, all within a unified framework. In this paper, we demonstrate experimentally the RFFE based on the Walsh signal processing [2] with a wideband and direct conversion from digital-to-analog (DAC) and analog-to-digital conversion (ADC) framework for an energy efficient RF signal generation and linearization. The RFFE integrates a Walsh-based RF DAC (WDAC) [3] and a Walsh-based RF ADC (WADC) for digital predistortion (DPD) as recommended in [4]. Section II provides an overview of Walsh-based signal processing. Section III and IV detail respectively the WDAC and the WADC with their measurement results. Finally, Section V concludes on the results.

II. Walsh Theory and Sequency Domain for RF

Joseph Walsh developed his "Walsh" transform as part of his broader work on harmonic analysis looking for alternatives to Fourier transforms which was not tailored for digital applications. He aimed to simplify the mathematical tools for signal analysis requiring rapid, computationally efficient transformations. He introduced his system of functions in 1923 [5], organizing them based on "sequence" (cf. Fig. 2a) as a definition of a measure of the number of zero crossings or sign changes per unit interval (cf. Fig. 2b). The Walsh functions are binary, making them inherently digital-friendly. Their properties demonstrated a relevant application to digital signal processing, communication theory, and data compression offering advantages in speed, simplicity, and noise resistance [6]–[8].

- Spectroscopy uses Walsh [9] for decoding the output of a two-beam interferometer with rapid spectral analysis.
- Medical Applications with electrocardiograms (ECG) and electroencephalograms (EEG) analysis with ECG waveforms and patterns recognition [10].
- Speech processing with dynamic spectral analysis [11] by adapting to changes in pitch and frequency to reduce bandwidth and improve recognition rates.
- Seismology with rapid and reliable event detection [12] such as the Goforth-Herrin algorithm [13].
- Communications with multiplexing, coding, improving signal clarity, and supporting error management. For instance, Sequency-Division Multiplexing (SDM) uses Walsh functions as carriers, allowing for efficient, low-interference transmission [14].

The sequency domain simultaneously captures characteristics across multiple temporal and frequency scales within a single set of coefficients. The Walsh functions, forming an orthonormal basis, allow for any usual function to be projected onto this set. It offers efficient signal processing with both diversity for RF signals (frequencies, modulations) and sparsity for computation (operations, algorithms).

The RFFE depicted in Fig. 1b eliminates LO and mixers through direct RF conversion. The WDAC generates the RF signal, and the WADC captures the PA output signal for DPD processing. Both integrate the Walsh transform, enabling a wideband conversion [7] while improving energy efficiency due to its sequential nature [6]. The Walsh sequences encode the sequency-domain information using Walsh coefficients. The Walsh sequences are high-frequency binary-valued signals and, thus, are compliant with an integration in CMOS technologies. The Walsh coefficients carry the signal information with both time and frequency properties and are converted as follows:

- The WDAC allocates Walsh sequences to each DAC similarly to frequency-interleaved architectures [15]. Walsh coefficients are processed digitally to weight the Walsh sequences using DACs whose outputs are combined in current to build up the RF signal as detailed in [16].
- The WADC samples the PA output and performs the Walsh transform using charge sharing. The Walsh coefficients display a compact data representation to improve DPD algorithm performances and to provide wideband properties for the PA linearization as detailed in [17].

Two integrated circuits (IC) are designed as a proof-of-concept (PoC) of a Walsh-based RFFE to demonstrate the benefits of its RF signal generation and its associated digital signal processing in the Walsh domain.

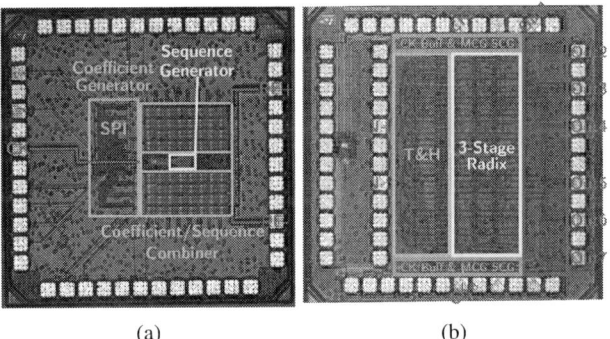

(a)   (b)

Fig. 3: Die photographies of (a) the WDAC (b) the WADC.

III. Walsh-based RF DAC

The WDAC has an instantaneous BW of $f_{WDAC} = 5\,GHz$ as detailed in [3]. The architecture is an order-6 inverse Fast Walsh transform (IFWT) and generates an RF signal based on 64 8-bit Walsh coefficients stored in a Coefficient Generator implemented as a 512-bit SPI. A Sequence Generator displays 64 Walsh sequences clocked at f_{WDAC}, weighted by each corresponding Walsh coefficient and summed in a Sequences and Coefficients Combiner (cf. Fig. 3a). The WDAC circuit has been fabricated with a core area of $0.339\,mm \times 0.415\,mm = 0.14\,mm^2$ and characterized up to 5 GHz.

The 64 8-bit Walsh coefficients are set to display a continuous wave (CW). f_{WDAC} is set at 5 GHz. The minimum and maximum CW frequencies of WDAC are evaluated:

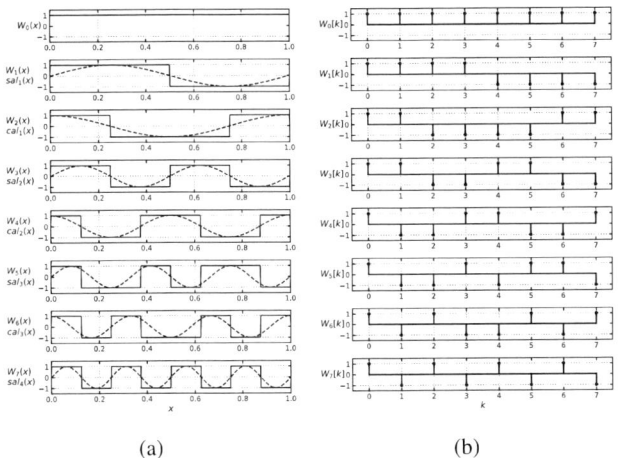

(a)   (b)

Fig. 2: (a) Set of 8 Walsh functions and their associated sinusoids. (b) Set of 8 discrete Walsh functions.

979-8-3315-3918-4/25 $31.00 © 2025 IEEE

- The minimum frequency is $f_{WDAC}/20 = 250\ MHz$ with the RF signal at Fig. 4a and its spectrum at Fig. 4b. It depicts a Spurious Free Dynamic Range (SFDR) of 10.01 dB with an output power of −3.9 dBm, associated with a power consumption of 44.8 mW and an energy efficiency of 0.346 pJ/bit.
- The maximum frequency is $f_{WDAC} = 5\ GHz$ with the RF signal at Fig. 4c and its spectrum at Fig. 4d. It depicts an SFDR of 10.33 dB with an output power of −3.28 dBm, associated with a power consumption of 45.9 mW and an energy efficiency of 0.38 pJ/bit.

TABLE I: State-of-the-Art comparison

Paper	[18]	[19]	[20]	[21]	**This Work**
Architecture	TI-DEM	Current steering	Current steering	TI-Current steering	**Walsh current steering**
CMOS node (nm)	7	28	5	40	**28**
Area (mm²)	0.5	0.28	0.57	0.045	**0.140**
Maximum operating frequency (GHz)	7.8	4.95	7.1	13.34	**5**
Instantaneous BW (GHz)	-	-	0.094	-	**4.6875**
Consumption (mW)	509	957	486	103	**49.8**
Quantization bits	14	16	12	16	**8**
SFDR$_{1CW}$ (dB)	54	70.1	58	34.6	**10.33**
Efficiency (pJ/bit)	2.27	5.98	2.53	0.61	**0.38**

To the best knowledge of the authors, the proposed WDAC has the lowest power consumption, the best energy efficiency and the largest instantaneous BW of the State of the Art as exhibited in Table I. Nevertheless, the linearity is challenged and can be improved thanks to lessons learnt by the experimental demonstration.

IV. WALSH-BASED RF ADC

The WADC has an instantaneous BW of $f_{WADC}/2 = 2.5\ GHz$ as detailed in [17]. The architecture is an order-3 analog Fast Walsh transform (AFWT) to convert 8 discrete signal samples into 8 Walsh coefficients using analog signal processing of [22] with only addition or subtraction operators. The AFWT block diagram is depicted in Fig. 6. Pipeline operations are performed thanks to a time-interleaved Track-and-Hold (T&H) composed of 32 branches, 3 stages of 16 radix units and a main clock generation (MCG) operating at f_{WADC}. The Walsh coefficients are converted at a rate of 1.25 GHz.

Analog signal processing uses the sampling capacitor of the T&H circuits as the charge-sharing capacitor to reduce the number of capacitors and the digital control signals. 2 duplicated architectures A and B sample and process AFWT continuously, with architecture A performing the analog signal processing when architecture B collects new samples, and vice-versa. For each architecture, the input signal is sampled by an 8-branch T&H at $f_{WADC}/16$ with a duty cycle of 12.5% and a phase shift of $f_{WADC}/2$. The WADC circuit has been fabricated with a core area of 0.74 mm × 0.96 mm = 0.71 mm² and characterized up to 5 GHz (cf. Fig. 3b).

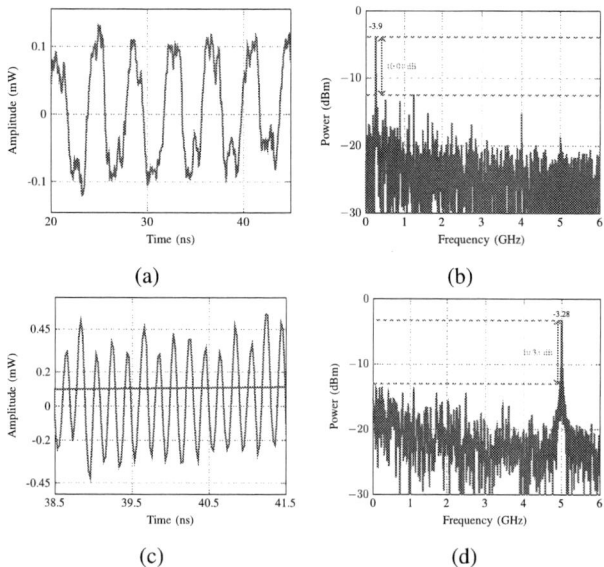

(a)

(b)

(c)

(d)

Fig. 4: (a) Time domain 250 MHz. (b) Frequency domain 250 MHz. (c) Time domain 5 GHz. (d) Frequency domain 5 GHz.

The wideband property of the WDAC is showcased in Fig. 5 with a 2-tone signal made out of 2 CW at 312.5 MHz and 5 GHz. The output power is −2.85 dBm and −3.83 dBm at 312.5 MHz and 5 GHz respectively, with an amplitude drop of 0.98 dB over a 4.6875 GHz BW. The measured SFDR is 5.02 dB with a power consumption of 49.8 mW and an energy efficiency of 0.38 pJ/bit.

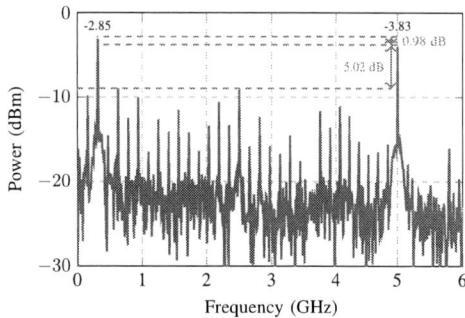

Fig. 5: 2-tone frequency domain reconstructed signal (312.5 MHz - 5 GHz).

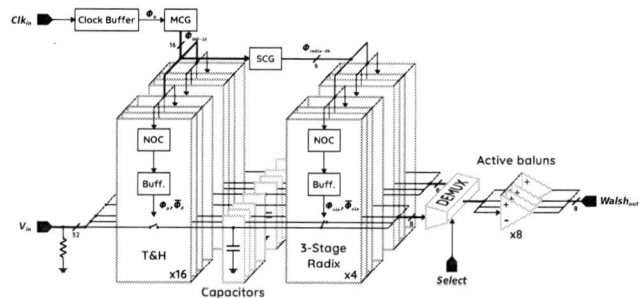

Fig. 6: Charge-reuse Analog Walsh Transform system.

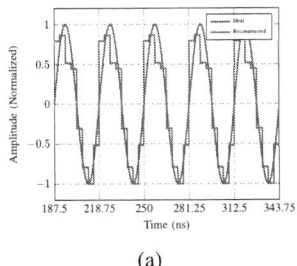
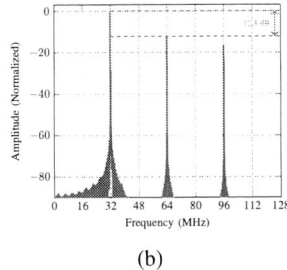

Fig. 7: (a) Time domain reconstructed signal. (b) Frequency domain reconstructed signal.

The AFWT is demonstrated at $f_{WADC} = 128\ MHz$ with an input CW frequency of $f_{WADC}/4 = 32\ MHz$ due to measurement capabilities limitations because of the charge sharing. The output Walsh coefficients are converted in the time domain using an inverse Walsh transform. Fig. 7a shows the properly reconstructed CW. Fig. 7b exhibits its FFT with an SFDR of 12.3 dB. The measured power consumption is 4 mW. The SFDR by from the 3rd harmonics but it does not affect being out of band of the digital signal processing for DPD. As detailed in [17], the Walsh Least Mean Squares (WLMS) algorithm can be performed without any complex calculation and is compared to recursive least mean squares (RLS) algorithm. The energy efficiency is up to 20 times better while achieving RLS-like performances in Adjacent Channel Leakage Power Ratio (ACLR) and Error Vector Magnitude (EVM).

V. CONCLUSION

This work experimentally demonstrated a Walsh-domain RFFE architecture fabricated in 28nm FDSOI CMOS technology, showcasing its potential for the next-generation of wireless systems. By integrating a WDAC and a WADC for DPD, the proposed system enables energy-efficient and wideband RF signal processing directly in the Walsh domain. The WDAC achieves up to 4.6875 GHz instantaneous BW with an energy efficiency of 0.38 pJ/bit, demonstrating the feasibility of Walsh-based RF signal generation. The WADC carries out a charge-reuse analog Walsh transform to efficiently extract Walsh coefficients from a 2.5 GHz-BW signal with only 4 mW power consumption and to improve by 20 times the energy efficiency in the DPD feedback loop. These results validate the effectiveness of Walsh-domain architectures for RF transmission and linearization. The experimental demonstration provides a scalable and power-efficient alternative to conventional RFFE, paving the way to industrial versions for future wireless transceivers in 6G.

ACKNOWLEDGMENT

The project HERMES has received funding from the European Union's Horizon 2020 research and innovation program under grant agreement No 964246. Publication reflects only the author's view, the Commission is not responsible for any use that may be made of the information it contains.

REFERENCES

[1] B. Jann, G. Chance, A. G. Roy, A. Balakrishnan, Karandikar *et al.*, "21.5 A 5G Sub-6GHz Zero-IF and mm-Wave IF Transceiver with MIMO and Carrier Aggregation," *2019 IEEE International Solid- State Circuits Conference - (ISSCC)*, pp. 352–354, 2019.

[2] J. Manz, "A sequency-ordered fast walsh transform," *IEEE Transactions on Audio and Electroacoustics*, vol. 20, no. 3, pp. 204–205, 1972.

[3] P. Ferrer, F. Rivet, H. Lapuyade, and Y. Deval, "A Walsh-Based Arbitrary Waveform Generator for 5G Applications in 28nm FD-SOI CMOS Technology," *IEEE Access*, vol. 11, pp. 117 434–117 442, 2023.

[4] R. Quéheille, F. Rivet, N. Deltimple, Y. Deval, and E. Kerhervé, "Demonstration of a Walsh-based Arbitrary Waveform Generator using Components Off-The-Shelf," *2021 28th IEEE International Conference on Electronics, Circuits, and Systems (ICECS)*, pp. 1–4, 2021.

[5] J. L. Walsh, "A Closed Set of Normal Orthogonal Functions," *American Journal of Mathematics*, vol. 45, pp. 5–24, 1923.

[6] H. F. Harmuth and N. Ahmed, "Sequency theory: foundations and applications," *IEEE Transactions on Systems, Man, and Cybernetics*, vol. 9, no. 5, pp. 312–312, 1979.

[7] K. G. Beauchamp, "Applications of walsh and related functions, with an introduction to sequency theory," *(No Title)*, 1984.

[8] A. Deb, S. K. Sen, and A. K. Datta, "Walsh functions and their applications: a review," *IETE Technical Review*, vol. 9, no. 3, pp. 238–252, 1992.

[9] H. Gebbie, "Walsh functions and the experimental spectroscopist," *Proc. Sympos. Applic. Walsh Functions*, pp. 99–100, 1970.

[10] R. Srivastva and Y. N. Singh, "Ecg biometric analysis using walsh–hadamard transform," in *Advances in Data and Information Sciences*, M. L. Kolhe, M. C. Trivedi, S. Tiwari, and V. K. Singh, Eds. Springer Singapore, 2018, pp. 201–210.

[11] J. Tyler, "Speech recognition system using walsh analysis and dynamic programming," *Microprocessors and Microsystems*, vol. 10, no. 8, pp. 427–433, 1986.

[12] S. K. Fletcher, "Walsh transforms in seismic-event detection," *IEEE Transactions on Electromagnetic Compatibility*, no. 3, pp. 367–369, 1983.

[13] T. Goforth and E. Herrin, "An automatic seismic signal detection algorithm based on the walsh transform," *Bulletin of the Seismological Society of America*, vol. 71, no. 4, pp. 1351–1360, 1981.

[14] H. Zhang and D. Rutkowski, "A sequency multiplexing technique for mobile communication systems," in *Proceedings of IEEE Singapore International Conference on Networks/International Conference on Information Engineering'93*, vol. 1. IEEE, 1993, pp. 226–230.

[15] G. Ding, C. Dehollain, M. Declercq, and K. Azadet, "Frequency-interleaving technique for high-speed A/D conversion," *Proceedings of the 2003 International Symposium on Circuits and Systems, 2003. ISCAS '03.*, vol. 1, pp. I–I, 2003.

[16] P. Ferrer, F. Rivet, H. Lapuyade, and Y. Deval, "A Charge Injection Neutralization Technique for Wide-Band Interleaved DAC," *2024 31st IEEE International Conference on Electronics, Circuits and Systems (ICECS)*, pp. 1–2, 2024.

[17] M. Fellmann and et al., "A Block-Based LMS Using the Walsh Transform for Digital Predistortion of Power Amplifiers," *IEEE Transactions on Communications*, vol. 71, no. 10, pp. 6074–6087, 2023.

[18] W.-H. Tseng and et al., "17.3 A 14b 16GS/s Time-Interleaved Direct-RF Synthesis DAC with T-DEM Achieving -70dBc IM3 up to 7.8GHz in 7nm," *2023 IEEE International Solid- State Circuits Conference (ISSCC)*, pp. 268–270, 2023.

[19] C. Huang, K. Ma, S. Chen, J. Fan, N. Sun, H. Yang, and X. Li, "A 16-bit 10-GS/s Calibration-Free DAC Achieving <-77dBc IM3 up to 4.95GHz in 28nm CMOS," *2024 IEEE Custom Integrated Circuits Conference (CICC)*, pp. 1–2, 2024.

[20] B. Koo and et al., "A 12-bit 16GS/s Single-channel RF-DAC with Hybrid Segmentation for Digital Backoff and Code-dependent Free Switch Driver achieving -85dBc IMD3 in 5nm FinFET," *2024 IEEE Symposium on VLSI Technology and Circuits*, pp. 1–2, 2024.

[21] W.-C. Kim, D.-s. Jo, Y.-J. Roh, Y.-D. Kim, and S.-T. Ryu, "A 6b 28GS/s Four-channel Time-interleaved Current-Steering DAC with Background Clock Phase Calibration," *2019 Symposium on VLSI Circuits*, pp. C138–C139, 2019.

[22] B. Sadhu, "Circuit techniques for cognitive radio receiver front-ends," 2012, retrieved from the University Digital Conservancy. [Online]. Available: https://hdl.handle.net/11299/165309

300-GHz Phased-Array Transceiver in 40-nm CMOS with Interpolated Feeding and OTA Metrics

Minoru Fujishima

Hiroshima University, Japan

Email: fuji@hiroshima-u.ac.jp

Abstract— This paper presents a fully integrated 300-GHz phased-array transceiver in 40-nm CMOS, addressing key circuit- and system-level challenges in sub-terahertz wireless communication. To overcome CMOS limitations at 300 GHz—such as poor gain, limited oscillator capability, and inefficient power amplification—a mixer-centric signal architecture is adopted: a mixer-last transmitter performs frequency conversion near the antenna after intermediate-frequency (IF) signal generation, while a mixer-first receiver performs immediate downconversion starting from near the antenna. An interpolated antenna feeding technique is introduced to reduce the number of active RF chains by directly driving only a subset of antenna elements, while passively connecting auxiliary elements at their feed points, thereby enabling compact 2D beamforming. A 2×2 prototype demonstrates ±30° beam steering in both E- and H-planes and achieves 40 Gb/s wireless data transmission using 16QAM at 275 GHz. To evaluate receiver performance under realistic over-the-air (OTA) conditions, two system-level metrics—effective isotropic conversion gain (EICG) and effective isotropic noise figure (EINF)—are introduced and experimentally validated. These capture antenna-circuit interactions and spatial effects, offering a practical alternative to conventional port-based metrics. The combination of mixer-centric signal flow, interpolated feeding, and OTA-based evaluation provides a scalable and energy-efficient framework for CMOS-based sub-terahertz transceivers targeting future 6G and beyond systems.

Keywords— 300 GHz, CMOS transceiver, phased-array, sub-terahertz wireless, mixer-first receiver, mixer-last transmitter, interpolated feeding, beam-steering, EICG, EINF, OTA evaluation, 6G

I. INTRODUCTION

The growing demand for ultra-high-speed wireless communication, fueled by the emergence of sixth-generation (6G) systems, is pushing the limits of current millimeter-wave (mmWave) technologies. Existing commercial mmWave bands, such as 28 GHz and 60 GHz, offer only several gigahertz of bandwidth, which is insufficient to meet the requirements of terabit-per-second (Tb/s) data rates, broadband beamforming, and other key enablers for immersive and high-density applications. To overcome these limitations, attention has increasingly turned to the sub-terahertz (sub-THz) frequency range, where significantly wider bandwidths are available. Among the candidate bands, the 252–296 GHz spectrum has emerged as particularly promising. As shown in Fig. 1, this band includes a contiguous 44 GHz allocation for fixed and mobile wireless systems, as designated by the World Radiocommunication Conference in 2019 (WRC-2019) [1]. When including additional non-contiguous segments, the total available bandwidth in this range reaches approximately 155 GHz. Compared to the D-band (~31.8 GHz) [2] and the 28 GHz band, the 252–296 GHz allocation offers over 5× and 100× the bandwidth, respectively, making it a strong candidate for future ultra-broadband wireless links and phased-array architectures.

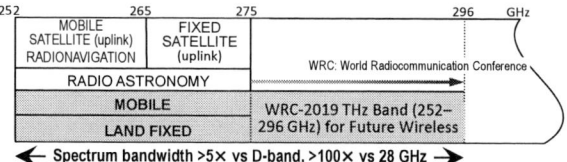

Fig. 1. Frequency allocation in the 252-296 GHz band based on WRC-2019 decisions. The spectrum includes 44 GHz of contiguous bandwidth for fixed and mobile wireless systems, offering over 5× the bandwidth of the D-band and more than 100× that of the 28 GHz band. This allocation enables new opportunities for future sub-terahertz wireless communication.

Despite its spectral advantages, realizing practical wireless systems at 300 GHz remains a substantial technological challenge. Photonic solutions, such as optical heterodyning using uni-traveling carrier photodiodes (UTC-PDs), offer excellent spectral purity and frequency stability [3], but require bulky components and complex optical setups, making them unsuitable for compact and cost-effective integration. On the other hand, CMOS technology enables highly integrated, low-cost solutions that combine RF, baseband, and control circuitry on a single die. However, CMOS suffers from inherent performance degradation near 300 GHz due to limited transistor gain, low maximum oscillation frequency, and poor power amplifier efficiency [4]. To bridge this gap between spectrum availability and circuit feasibility, we propose a 300-GHz transceiver architecture implemented in 40-nm CMOS that introduces three key innovations. First, we employ a mixer-centric signal architecture—specifically, a mixer-last configuration on the transmitter (TX) side and a mixer-first configuration on the receiver (RX) side—to relocate essential signal processing tasks such as frequency conversion and amplification to frequency bands where CMOS device performance is more favorable [5]. Second, we implement an interpolated antenna feeding scheme, which drives only a subset of array elements while passively coupling to adjacent elements. This approach reduces the number of required RF chains while preserving full two-dimensional beamforming capability, enabling scalable phased-array integration without excessive area or complexity [6]. Third, we introduce two novel over-the-air (OTA) system-level figures of merit—effective isotropic conversion gain (EICG) and effective isotropic noise figure (EINF)—which provide a realistic and comprehensive evaluation of receiver performance under practical operating conditions where conventional port-based measurements are inadequate [7].

These innovations collectively establish a practical and scalable foundation for CMOS-compatible transceivers operating in the 300 GHz band. By addressing both circuit-level and system-level challenges, the proposed approach offers a path toward compact, high-capacity, and beam-steerable wireless systems suitable for next-generation communication platforms such as 6G and beyond.

II. System Architecture

CMOS technology offers a promising platform for large-scale integration of sub-terahertz (sub-THz) transceivers due to its cost-effectiveness and compatibility with digital processing. However, operation near 300 GHz exposes fundamental device-level limitations: transistors exhibit drastically reduced gain, power amplifiers (PAs) lack sufficient output power, oscillators become unreliable, and low-noise amplifiers (LNAs) suffer from poor sensitivity. These constraints render conventional mmWave architectures—which rely on high-gain PAs, on-chip LOs, and cascaded amplification—unsuitable for this frequency range (Fig. 2). To address these issues, we adopt a mixer-centric signal architecture that relocates frequency-critical operations to lower bands where CMOS devices remain effective. In the transmitter (TX) path, intermediate-frequency (IF) signal generation and modulation occur at frequencies well below 300 GHz, followed by upconversion near the antenna, thus avoiding the need for 300 GHz on-chip LO generation or high-power amplification. In the receiver (RX) path, a mixer placed at the antenna interface immediately downconverts the incoming 300 GHz signal to baseband. This mixer-first configuration allows all subsequent processing—amplification, filtering, demodulation—to occur at lower frequencies, where CMOS offers better gain and noise performance. The viability of this architecture was demonstrated in prior single-element CMOS prototypes [5], which achieved 80 Gb/s communication at 265.68 GHz using 16QAM, without directly generating or amplifying 300 GHz signals on-chip.

While the single-element approach validates the architectural concept, scaling to a two-dimensional (2D) phased array introduces physical integration challenges. At 300 GHz, the free-space wavelength is approximately 1 mm, limiting the inter-element spacing to about 0.5 mm ($\lambda_0/2$). If every antenna were equipped with a dedicated RF chain—including mixer, LO, and amplifier—the required circuit area per element would exceed this pitch, making dense array integration impractical. To overcome this, we adopt an interpolated antenna feeding technique (Fig. 3), where only a subset of elements (main antennas) are actively driven by the transmitter circuitry. The remaining auxiliary antennas are passively excited through feed-point connections to adjacent main antennas and are not independently driven. When the phase difference $\Delta\theta$ between main elements is small, each auxiliary element assumes an intermediate phase of approximately $\Delta\theta/2$ with matched amplitude, preserving aperture coherence and enabling full 2D beamforming in both azimuth and elevation. The effectiveness of this approach has been confirmed through full-wave simulations and experimental validation.

To further support array scalability, we implement orthogonal routing of LO and IF signals: LO signals are distributed along one axis of the array, while IF signals are routed orthogonally. This matrix-style topology minimizes routing congestion, supports modular extension, and maintains beamforming flexibility without increasing layout complexity. Collectively, these architectural strategies—a mixer-centric signal flow that confines high-frequency operation to the antenna interface, interpolated antenna feeding that minimizes active circuitry, and orthogonal signal routing that supports planar scaling—provide a practical and scalable framework for realizing 300-GHz phased-array transceivers in CMOS. They directly address the hardware limitations highlighted in Fig. 2 and serve as the basis for the integrated system described in the following sections.

III. Experimental Demonstration

To validate the proposed 300-GHz phased-array transceiver architecture described in Section II, a 2×2 prototype system was fabricated using a 40-nm CMOS process. The implementation includes both a transmitter (TX) and a receiver (RX), each adopting the mixer-centric signal processing strategy and interpolated antenna feeding technique. This section describes the design methodology, physical implementation, and comprehensive over-the-air (OTA) evaluation of the system.

A. Transmitter Implementation and Characterization

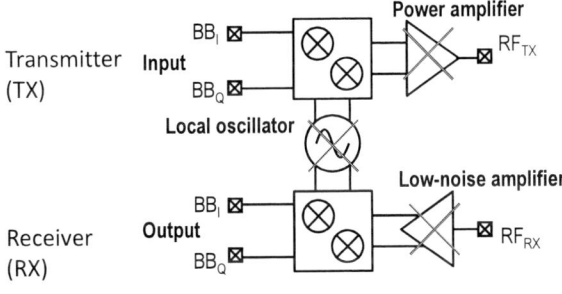

Fig. 2. Limitations of 300-GHz transceiver integration in CMOS. Power amplifiers lack sufficient gain, oscillators cannot reliably operate at 300 GHz, and low-noise amplifiers offer poor sensitivity. These limitations motivate the adoption of mixer-centric architectures for signal upconversion and downconversion.

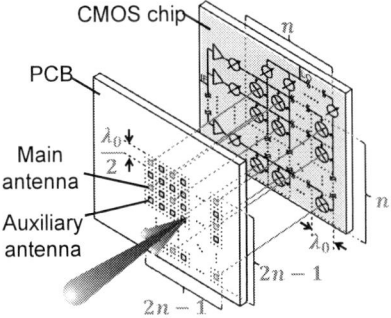

Fig. 3. Concept of interpolated antenna feeding in a phased array. Main antennas are actively driven, while auxiliary antennas are passively excited through near-field coupling. When the inter-element spacing is sufficiently small, the interpolated elements maintain phase and amplitude coherence across the aperture.

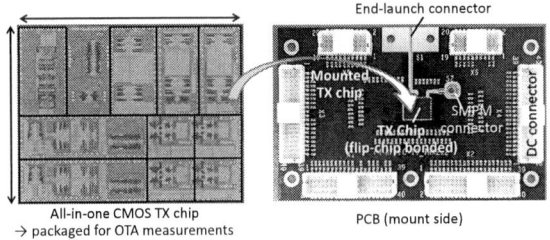

Fig. 4. Die photo and evaluation board of the 2×2 CMOS phased-array transmitter. The left image shows the chip layout including 75 GHz and 150 GHz LO chains, IF stages, and four TX elements. The right image shows the flip-chip bonded chip mounted on a multilayer PCB with SMPM, DC, and end-launch connectors for OTA evaluation.

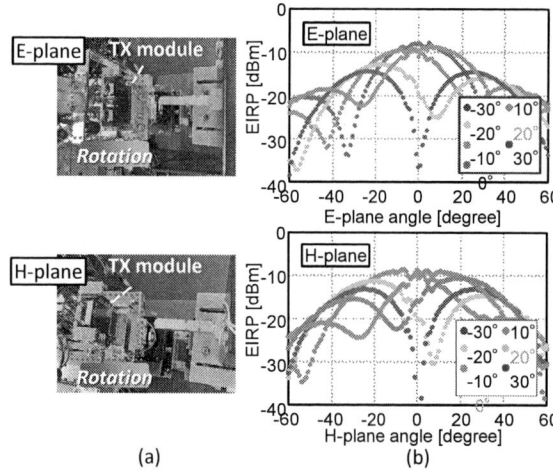

(a)   (b)

Fig. 5. (a) Over-the-air beam measurement setup for the 300-GHz transmitter. A flip-chip TX module is mounted on a motorized linear stage and rotation platform for 2D beam scanning. (b) Measured E-plane and H-plane beam patterns at 275 GHz, showing ±30° steering capability with consistent radiated power, validating the interpolated antenna feeding scheme.

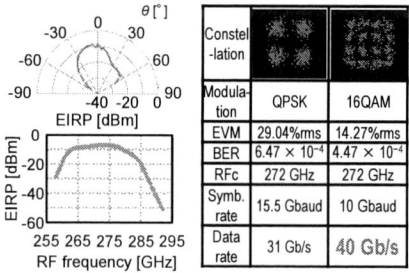

Fig. 6. 16QAM constellation diagrams captured at 275 GHz under beam-steering. Results at θ = 0° (E-plane) and θ = -10° (H-plane) show minimal degradation in modulation fidelity, confirming robust high-speed transmission using the phased-array system.

A 2×2 phased-array TX chip was developed to demonstrate the effectiveness of mixer-last upconversion and interpolated beamforming. Each of the four transmitter paths (TX1–TX4) comprises a two-stage IF chain, a frequency doubler-based local oscillator (LO) chain operating at 75 GHz and 150 GHz, an upconversion mixer, and a driver amplifier. Per-path amplitude and phase control are supported to enable beam steering. The die photo in Fig. 4 (left) shows the integrated layout. The chip was flip-chip bonded to a custom multilayer PCB, shown in Fig. 4 (right), which includes SMPM connectors for RF output, end-launch connectors for IF and LO inputs, and DC pads for biasing. This flip-chip integration minimizes parasitics and enables stable high-frequency performance required for OTA testing.

Beamforming capability was assessed using the measurement setup shown in Fig. 5(a). The TX module was mounted on a motorized linear and rotary positioning system inside an anechoic chamber to enable 2D scanning of the radiation pattern. A 300-GHz downconverter and vector signal analyzer (VSA) were used to measure the radiated signals. As illustrated in Fig. 5(b), the measured E-plane and H-plane beam patterns exhibit ±30° steering with smooth and symmetric lobes, validating the performance of the

interpolated antenna feeding scheme. To evaluate high-speed wireless transmission, 16QAM signals were modulated at 20 Gbaud (corresponding to 40 Gb/s) and transmitted using the TX array. Fig. 6 presents constellation diagrams at steering angles θ = 0° and θ = −10°, confirming minimal degradation in modulation fidelity. The results demonstrate that the TX array supports robust dynamic beam steering while maintaining signal integrity.

B. Receiver Implementation and OTA Evaluation

A 2×2 RX chip was implemented to complement the TX prototype, employing the mixer-first topology for immediate downconversion at the antenna interface. The RX die, measuring 6.23 × 4.73 mm², integrates four mixer cores with IF amplification stages and is designed to operate with an interpolated reception array. As in the TX, only the main antennas are actively driven by RF circuitry, while auxiliary antennas are connected at the feed point to extend the aperture. The RX module was assembled by flip-chip bonding the chip to a multilayer PCB exposing a 3×3 antenna array with $\lambda_0/2$ spacing on the bottom surface, as shown in Fig. 7. To assess array performance under OTA conditions, two new system-level figures of merit were introduced: effective isotropic conversion gain (EICG) and effective isotropic noise figure (EINF). EICG is defined as the ratio of baseband output power to the incident free-space RF power, capturing the overall signal processing efficiency of the RX array. EINF, derived from cold-source measurements, reflects the total noise including contributions from antennas, mixers, and IF circuitry. These metrics offer meaningful performance assessment beyond traditional port-based definitions, which are inapplicable in fully integrated systems.

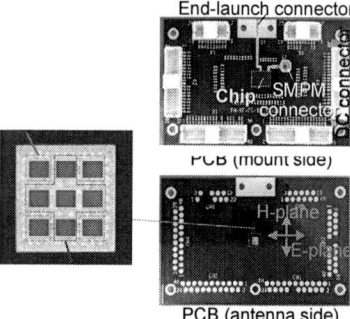

Fig. 7. Assembled 300-GHz CMOS receiver module. The 2×2 RX chip is flip-chip bonded to a multilayer PCB exposing a 3×3 antenna array with $\lambda_0/2$ spacing. Auxiliary antennas are passively excited to extend the effective aperture under interpolated reception.

Beam-steering performance of the RX array was characterized by sweeping phase codes while illuminating the array from varying angles. As shown in Fig. 8, the RX system successfully steered the receive beam across −30° to +30° at 270 GHz with well-defined main lobes, confirming beam control capability consistent with TX-side performance. End-to-end wireless communication was then demonstrated by transmitting modulated signals from the TX array and receiving them with the RX module. Fig. 9 summarizes OTA communication results, including demodulation of QPSK and 16QAM signals at data rates up to 36 Gb/s. The measured constellation diagrams, error vector magnitudes (EVM), and bit error rates (BER) validate the transceiver system's ability to support high-speed wireless links.

979-8-3315-3918-4/25 $31.00 © 2025 IEEE

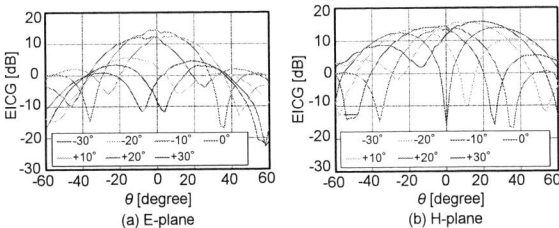

Fig. 8. Measured beam patterns of the 2×2 RX array for different beam-steering angles from −30° to +30° at 270 GHz. The plots confirm effective beam-steering with distinct main lobes corresponding to the intended angles.

E-plane (θ = 0°)			H-plane (θ = −10°)	
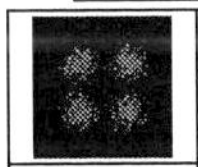		Constellation		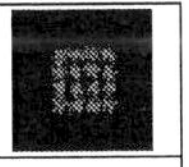
QPSK	16QAM	Modulation	QPSK	16QAM
30.52	13.31	EVM [% rms]	31.01	14.42
5.24	2.93	BER [×10^{-4}]	6.30	4.89
268	274	RFc [GHz]	270	270
15	6	Symbol Rate [Gbaud]	18	7
30	24	Data Rate [Gb/s]	36	28

Fig. 9. OTA digital communication performance using the 2×2 RX array. QPSK and 16QAM demodulation results are shown, including constellation diagrams, EVM, BER, RF center frequency, symbol rate, and data rate.

Lastly, frequency sweeps across 270–280 GHz confirmed consistent EICG and EINF characteristics with minimal variation across the band. Angular measurements showed that EICG remains stable across ±20° in the E-plane and extends beyond ±30° in the H-plane. These results affirm the scalability and robustness of the receiver design and demonstrate the practical utility of the proposed OTA figures of merit.

IV. CONCLUSION AND OUTLOOK

This work has demonstrated the feasibility of implementing compact, beam-steerable, and high-speed wireless links at 300 GHz using standard 40-nm CMOS technology. To overcome the fundamental limitations of CMOS devices in the sub-terahertz regime—namely, diminished transistor gain, limited oscillator performance, and inefficient power amplification—a mixer-centric transceiver architecture was adopted. By shifting frequency conversion, LO generation, and amplification to lower-frequency domains through mixer-last transmission and mixer-first reception, the proposed approach enables reliable signal processing without relying on high-performance mmWave blocks. To address the physical integration challenges of 300-GHz phased arrays—particularly the sub-millimeter spacing between antenna elements—an interpolated antenna feeding technique was introduced. This method directly drives only a subset of antennas (main elements) with active circuits, while passively exciting the auxiliary elements via feed-point connections. This reduces the number of RF chains, enabling two-dimensional beamforming with minimal circuit complexity and silicon area. The implemented 2×2 CMOS prototype achieved ±30° beam steering in both E- and H-planes and demonstrated 40 Gb/s wireless communication using 16QAM at 275 GHz, validating the architecture in over-the-air (OTA) conditions.

In addition to hardware-level innovations, this study introduced two OTA-based system-level evaluation metrics—effective isotropic conversion gain (EICG) and effective isotropic noise figure (EINF)—to assess receiver performance in fully integrated arrays where antenna-circuit interactions are inseparable. These metrics reflect the aggregate effects of the antenna, front-end circuit, and spatial propagation environment, offering a more realistic alternative to conventional port-based definitions. EICG and EINF were experimentally shown to remain stable over wide frequency ranges (270–280 GHz) and beam-steering angles (±30°), confirming the robustness and scalability of the receiver design as well as the practical utility of the proposed metrics for evaluating integrated sub-THz systems.

The combined architectural contributions—mixer-centric signal paths, interpolated antenna feeding, and OTA-centric performance metrics—establish a scalable, CMOS-compatible foundation for future sub-terahertz transceivers. Looking ahead, several development directions are envisioned. First, expanding to fully driven arrays could enhance spatial resolution and beamforming flexibility, albeit with increased circuit density and power consumption. Second, continued miniaturization and optimization of mixers, LO distribution, and antenna interfaces will be essential for pushing operation into higher bands such as 400–500 GHz. Third, extended OTA evaluations over longer distances and under dynamic channel conditions will be needed to verify link robustness in practical deployment scenarios. Finally, integration with advanced packaging and interposer technologies, including hybridization with photonic components, may further enhance signal quality, linearity, and system-level functionality. In summary, this work presents a viable path toward practical 300-GHz phased-array transceivers using deeply scaled CMOS, laying the groundwork for compact, energy-efficient, and high-capacity wireless systems for 6G and beyond.

ACKNOWLEDGMENT

This work was supported in part by the Ministry of Internal Affairs and Communications (MIC), Japan, under Grant JPJ000254, and in part by JSPS KAKENHI under Grant Number 25H00744. The author would like to express sincere gratitude to all collaborators and supporting institutions for their invaluable contributions to the research and development presented in this work.

REFERENCES

[1] *Sharing and Compatibility Studies Between Land-Mobile, Fixed and Passive Services in the Frequency Range 275–450 GHz*, ITU-Rec. SM.2450-0, Int. Telecommun. Union, Geneva, Switzerland, Jun. 2019.

[2] (Electron. Commun. Committee, Copenhagen, Denmark) *ECC Recommendation (18)01*. 2018.

[3] D. Guillaume and T. Nagatsuma, "Wireless communications in the THz range," in Fundamentals of Terahertz Devices and Applications. Hoboken, NJ, USA: Wiley, 2021, pp. 479–510.

[4] M. Fujishima, "Challenges and Innovations in CMOS-Based 300-GHz Transceivers for High-Speed Wireless Communication," in *IEEE Open Journal of the Solid-State Circuits Society*, vol. 5, pp. 21-32, 2025.

[5] S. Lee et al., "An 80 Gb/s 300 GHz-band single-chip CMOS transceiver," *IEEE J. Solid-State Circuits*, vol. 54, no. 12, pp. 3577–3588, Dec. 2019.

[6] K. Takano et al., "A 300-GHz-Band 40-Gb/s 2D phased-array CMOS transmitter with near-half-wave antenna pitch," in *Proc. IEEE Radio Frequency Integr. Circuits Symp. (RFIC)*, 2024, pp. 335–338.

[7] S. Tanaka et al., "A 300-GHz-band 36-Gb/s scalable 2 × 2 2D phased-array CMOS receiver," in *Proc. A-SSCC*, 2024, pp. 1–3.

979-8-3315-3918-4/25 $31.00 © 2025 IEEE

A Polar-Modulation OFDM Backscatter System for Passive IoT Communication

Qijing Xiao , Xin Hu , Weixiao Wang, Yuxuan Luo , Bo Zhao*

College of Integrated Circuits, Zhejiang University, Hangzhou, China.
*Corresponding Author: zhaobo@zju.edu.cn

Abstract—**This paper presents a polar-modulation-based OFDM backscatter system for passive internet of things (IoT) communication. The design separates amplitude and phase control by decomposing the IFFT output into magnitude and sign components, enabling polar-domain backscatter modulation without complex DACs. A digital pre-distortion compensation linearizes the nonlinear reflection coefficient control realized by a multi-bit switch array. Simulation results show a significantly reduced error vector magnitude (EVM) of 4.87%, an improved spurious-free dynamic range (SFDR) of 20 dB, and a data rate of 14.4 Mbps with 6 bps/Hz spectral efficiency. These results demonstrate the effectiveness of the proposed architecture in enhancing modulation accuracy and spectral purity for backscatter communication.**

Keywords—Backscatter communication, OFDM, polar modulation, pre-distortion compensation

I. INTRODUCTION

With the rapid proliferation of IoT devices, ensuring reliable wireless communication under ultra-low power constraints has become a significant research focus. Due to the high-power consumption associated with power-hungry components such as power amplifiers and frequency synthesizers, conventional active communication is unsuitable for battery-free or long-lifetime deployments [1-2]. In contrast, backscatter communication achieves signal modulation by reflecting ambient continuous waves, thereby obviating the need for active RF transmission and significantly reducing power consumption. This makes it a promising solution for energy-constrained or battery-free systems [3-5]. Compared to conventional single-carrier backscatter, orthogonal frequency-division multiplexing (OFDM) provides high spectral efficiency and robustness against multipath fading, and has been increasingly adopted in backscatter systems to improve data rate and communication reliability [6-8].

Existing OFDM backscatter systems can be broadly categorized into two architectures. The first employs analog baseband modulation, as illustrated in Fig. 1(a), where DACs and vector-modulated reflection switches are used to backscatter OFDM signal [6]. This approach, however, suffers from high power consumption and circuit complexity, and it is difficult to ensure a linear mapping between the DAC output and the resulting reflection coefficient, which degrades modulation accuracy. The second adopts an all-digital architecture, which generates time-domain OFDM signals through a switch array. While more power-efficient, the limited number of switch elements restricts the modulation resolution. To overcome this, delta-sigma modulation (DSM) is often introduced to shape the quantization noise, as shown in Fig. 1(b). Nonetheless, DSM-based designs typically demand higher clock rates, leading to increased power consumption, and the shaped bitstream may still suffer from

nonlinear control over the reflection coefficient, further limiting error vector magnitude (EVM) performance [8].

To address these issues, this paper proposes a novel polar-modulation-based OFDM backscatter architecture (Fig. 1(c)). By separating amplitude and phase, the system enables polar-domain backscatter modulation. Furthermore, a pre-distortion compensation mechanism is introduced to linearize the mapping between the digital control code and the resulting reflection coefficient. As a result, the EVM is significantly reduced to 4.87%. This paper details the system architecture and pre-distortion method, and validates its effectiveness through spectral and constellation diagram analysis.

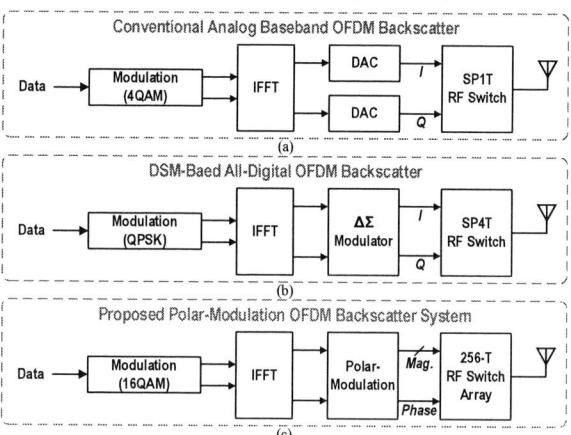

Fig. 1. Comparison of OFDM backscatter architectures: (a) conventional analog baseband OFDM backscatter; (b) DSM-enhanced all-digital OFDM backscatter; (c) proposed polar-modulation OFDM backscatter system.

II. POLAR-MODULATION BACKSCATTER TECHNIQUE

The proposed polar-modulation-based OFDM backscatter architecture is illustrated in Fig. 2. The system adopts OFDM modulation comprising 120 subcarriers, where each subcarrier carries data symbols modulated using 64QAM. The resulting frequency-domain symbols are processed through an IFFT to generate the time-domain OFDM waveform, which is then separated into I and Q branches for further processing.

In the I branch, the IFFT output is first processed to extract the sign information, denoted as I_{sgn}, This signal is fed into the phase modulator, where it is mixed with an intermediate frequency (IF) signal, IF_0, to generate the phase-modulated IF signal $IF_{OUT,I}$. Simultaneously, the original I-path IFFT signal is sent to the amplitude modulator. After passing through an absolute value extractor to obtain the magnitude information I_{abs}, the signal is processed by a digital pre-distortion that compensates for the nonlinearity in the reflection coefficient of the switch array. The resulting signal controls the on/off states of 256 backscatter switch units, which in turn

979-8-3315-3918-4/25 $31.00 © 2025 IEEE

determines the amplitude of the backscatter signal. The $IF_{OUT,I}$ signal drives the switch array, which reflects the phased modulated signal based on the incident continuous-wave (CW) carrier.

The Q branch follows the same processing structure. The sign bit Q_{sgn} is extracted from the IFFT output and mixed with a quadrature-phase IF signal IF_{90} to produce $IF_{OUT,Q}$. The magnitude component is similarly passed through an absolute value module and the digital pre-distortion to generate the control signal for the Q-path backscatter switches.

Finally, the reflected signals from both I and Q branches are combined through an off-chip power combiner to generate the final polar-modulated OFDM backscatter signal.

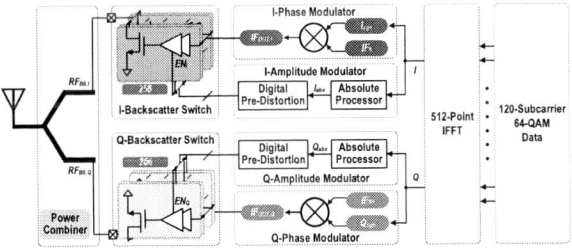

Fig. 2.   Proposed polar-modulation OFDM backscatter system

The following sections present the polar-modulation backscatter and pre-distortion compensation technique in detail.

A. Polar-Modulation Backscatter

To enable efficient backscatter OFDM modulation, a polar-modulation Backscatter method based on coordinate decomposition is proposed. It separates the real-valued IFFT time-domain symbols into amplitude and sign components, which are independently mapped to reflection amplitude and phase modulation paths.

Consider the real-valued I or Q component of the IFFT output, denoted as $x[n]$, which is normalized to the range $[-256, 256]$ to facilitate subsequent amplitude mapping. The signal can be decomposed as:

$$x[n]=|x[n]| \cdot sgn(x[n]) \tag{1}$$

Where $|x[n]| \in [0, 256]$ represents the instantaneous amplitude, and $sgn(x[n]) \in \{-1, +1\}$ denotes the sign bit. By reducing the phase to a single-bit sign, the polar decomposition simplifies intermediate-frequency modulation and allows independent and low-complexity amplitude and phase mapping.

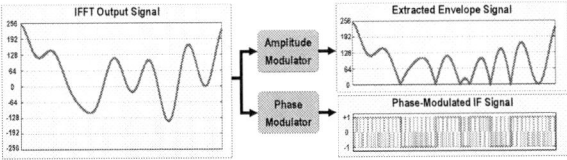

Fig. 3. Polar-domain decomposition of IFFT waveform for backscatter modulation, including amplitude extraction and sign-based phase modulation.

As illustrated in Fig. 3, The amplitude modulator outputs amplitude components $|x[n]|$, while the phase modulation is implemented by mixing the sign bit $sgn(x[n])$ with an IF tone

$cos(2\pi f_{IF} nt)$. The resulting phase-modulated waveform is expressed as:

$$X_{IF}[t]=sgn(x[n]) \cdot cos(2\pi f_{IF} t) \tag{2}$$

Both amplitude and phase components are then upconverted via backscatter by mixing with the incident RF carrier $cos(2\pi f_{RF} t)$. The resulting backscattered signal is given by:

$$S_{RF}[t]=x[n] \cdot cos(2\pi (f_{RF}+f_{IF})t) \tag{3}$$

This effectively realizes $\pm 180°$ binary phase shift keying (BPSK) at the IF frequency, where the sign bit encodes phase information. The amplitude control signal and the phase-modulated waveform are applied to the backscatter switch array, enabling polar-domain OFDM backscatter modulation with reduced hardware complexity.

B. Pre-Distortion Compensation

In OFDM systems employing high-order modulation schemes such as 64-QAM, accurate amplitude control is critical to maintain modulation fidelity. In backscatter systems, the amplitude of the reflected signal is determined by the load impedance, which is modulated by digitally controlling the conduction states of multiple reflection switches. However, the mapping between the number of activated switches and the resulting reflection coefficient is highly nonlinear, thereby degrading amplitude modulation accuracy without proper compensation.

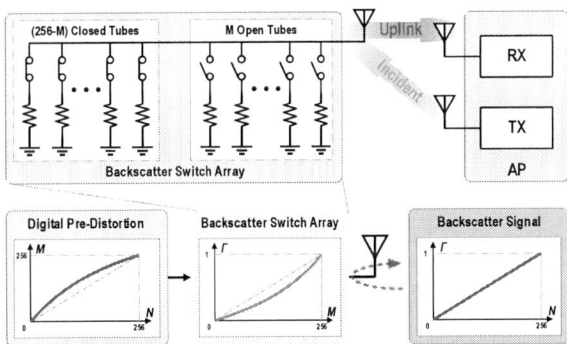

Fig. 4. Backscatter switch array diagram and key curves illustrating nonlinear reflection coefficient versus switch count, digital pre-distortion mapping, and the resulting linearized reflection control.

Specifically, the backscatter array consists of 256 reflection switches, as depicted in the top of Fig. 4, where each switch has a closed-state impedance of $R_{unit}=256 \times R_s$. Here, R_s is the source impedance. When M switches are turned off (i.e., M open tubes as depicted), the remaining $256-M$ switches form the equivalent load impedance:

$$R_L=\frac{R_{unit}}{256-M}=\frac{256 \times R_s}{256-M} \tag{4}$$

From transmission line theory, the reflection coefficient can then be expressed as:

$$\Gamma=\frac{R_L-R_s}{R_L+R_s}=\frac{M}{512-M} \tag{5}$$

979-8-3315-3918-4/25 $31.00 © 2025 IEEE

This indicates a nonlinear relationship between the switch control quantity M and the reflection coefficient Γ, which would introduce amplitude distortion if the input control word N is directly mapped to M, To address this issue, a digital pre-distortion technique based on a lookup table (LUT) is proposed to linearize the mapping.

Assuming a target linear reflection coefficient defined as:

$$\Gamma = \frac{N}{256} \tag{6}$$

The corresponding switch control quantity M can be derived by inverting (5), resulting in the pre-distortion mapping:

$$M = \frac{512N}{256+N} \tag{7}$$

As shown in Fig. 4, the digital pre-distortion aligns the nonlinear hardware behavior to the desired linear reflection response. three key characteristics are captured: the intrinsic nonlinear mapping between the backscatter switch control M and reflection coefficient Γ, the pre-distortion transformation from desired amplitude index N to hardware control M, and the resulting linearized Γ-to-N response of the backscattered signal. The proposed pre-distortion scheme effectively linearizes the amplitude modulation, thereby providing enhanced signal quality for subsequent OFDM demodulation.

III. SIMULATION RESULTS

To further verify the proposed architecture, Fig. 5 shows simulated time-domain waveforms at key points along the signal path. Baseband time-domain waveforms on the I and Q branches, labeled $IFFT_{OUT,I}$ and $IFFT_{OUT,Q}$, are generated by the IFFT module. These signals correspond to the orthogonal components of the 64QAM-OFDM baseband.

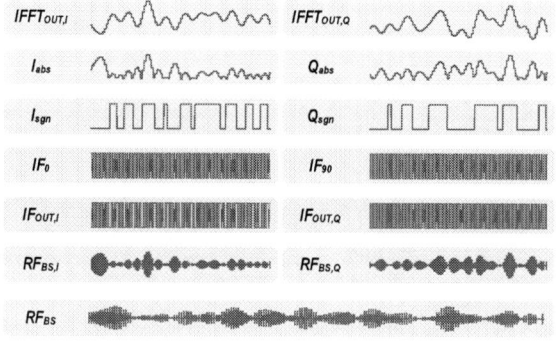

Fig. 5. Simulated time-domain waveform of the polar-modulated backscatter system.

Subsequently, each branch undergoes amplitude and phase modulation. The amplitude extraction yields I_{abs} and Q_{abs}, capturing the envelope variation of the respective signals. Concurrently, the sign bits I_{sgn} and Q_{sgn} are derived to represent the polarity of the IFFT outputs. These sign sequences are mixed with orthogonal IF carriers (IF_0 and IF_{90}), resulting in modulated intermediate-frequency outputs $IF_{OUT,I}$ and $IF_{OUT,Q}$. Finally, the polar components are reflected and modulated onto the incident carrier by backscatter switch arrays, generating $RF_{BS,I}$ and $RF_{BS,Q}$. These components are then coherently combined to form the composite backscattered signal RF_{BS}. The waveform transitions across

stages clearly demonstrate the separate amplitude and phase paths and confirm the feasibility of polar-domain OFDM modulation in the time domain.

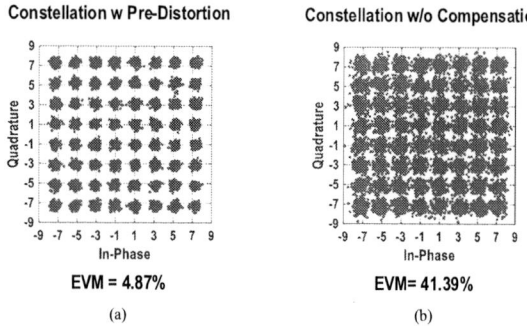

Fig. 6. Simulated backscattered signal constellations: (a) with pre-distortion compensation; (b) without pre-distortion compensation.

As illustrated in Fig. 6, the constellation diagrams of the backscattered signal are presented with and without pre-distortion compensation. With compensation applied, the constellation points are tightly clustered, resulting in a low EVM of 4.87%. In contrast, the uncompensated case exhibits severe constellation spreading, with the EVM increasing to 41.39%. These results demonstrate the effectiveness of the proposed compensation method in enhancing modulation accuracy.

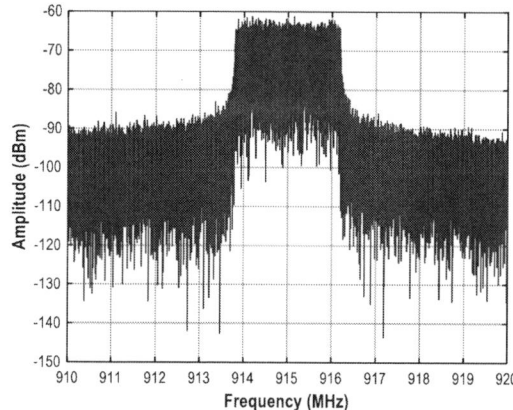

Fig. 7. Spectrum of the polar-modulated OFDM 64QAM backscattered signal.

The OFDM backscatter system is simulated using 64-QAM modulation with 120 subcarriers and a subcarrier spacing of 20kHz, resulting in a symbol rate of 20kSym/s per subcarrier. The intermediate frequency is set to 10.24 MHz, which also serves as the IFFT sampling clock. A 904.75MHz CW, emitted by the access point (AP), is used as the incident signal.

Fig. 7 shows the simulated spectrum of the backscattered OFDM signal, centered at 915MHz with a 2.4MHz bandwidth ranging from 913.8MHz to 916.2MHz. The results confirm the effectiveness of the proposed scheme in frequency translation and bandwidth control. Under this configuration, a data rate of 14.4Mbps is achieved, corresponding to a spectral efficiency of 6bps/Hz, demonstrating the bandwidth

979-8-3315-3918-4/25 $31.00 © 2025 IEEE

efficiency of the proposed polar-modulated backscatter scheme.

120-Subcarrier Band Spectrum

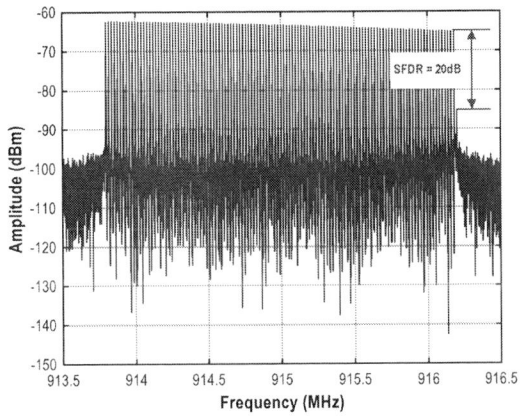

Fig. 8. Spectrum of OFDM backscattered signal with 120 unmodulated subcarriers (tones) for SFDR analysis.

To evaluate the spectral quality of the proposed polar-modulated backscatter signal, Fig. 8 presents the spectrum of the backscattered signal with all 120 subcarriers modulated by fixed symbols ('111111'), effectively acting as unmodulated carriers (tones). This configuration is ideal for evaluating the Spurious-Free Dynamic Range (SFDR) across the signal band. Following the methodology described in [8], the SFDR is defined as the ratio between the weakest subcarrier power and the strongest spur or noise component within the spectral band that encompasses all subcarriers:

$$SFDR = P_{min,subcarrier} - P_{max,spur} \quad \text{(dB)} \quad (8)$$

Using this definition, the proposed design achieves an SFDR of 20 dB, which represents a significant improvement compared to the 12.4 dB reported in [8] for DSM-based OFDM modulation. The enhanced SFDR benefits from the linear control of the reflection coefficient, the use of a multi-bit switch array, and the suppression of quantization artifacts through pre-distortion. Moreover, the frequency spectrum in Fig. 8 clearly shows 120 distinct unmodulated carrier tones with minimal spectral leakage, further validating the effectiveness of the proposed polar modulation in maintaining spectral purity.

IV. SUMMARY

This paper presents a novel polar-modulation-based OFDM backscatter architecture that effectively separates amplitude and phase modulation to enable energy-efficient, hardware-friendly modulation for battery-free IoT devices. By employing a digital pre-distortion scheme, the proposed design significantly improves linearity and reduces EVM to 4.87%. Simulation results demonstrate enhanced spectral purity with an SFDR of 20 dB and a data rate of 14.4 Mbps at 6 bps/Hz spectral efficiency. These findings validate the proposed architecture as a promising solution for high-speed, low-power backscatter communication systems for passive IoT.

ACKNOWLEDGMENT

This work was supported by the National Key R&D Program of China under Grant 2024YFE0203500.

REFERENCES

[1] T. Wang et al., "A Fully Integrated Digital Polar Transmitter With Single-Ended Doherty PA and DLL-Based Three-Segment Hybrid DTC in 28 nm CMOS, " in IEEE Journal of Solid-State Circuits, vol. 59, no. 2, pp. 388-399, Feb. 2024.

[2] M. Beikmirza, Y. Shen, L. C. N. de Vreede and M. S. Alavi, "A Wideband Energy-Efficient Multi-Mode CMOS Digital Transmitter, " in IEEE Journal of Solid-State Circuits, vol. 58, no. 3, pp. 677-690, March 2023.

[3] Z. Chang, Q. Xiao, W. Wang, Y. Luo and B. Zhao, "A Passive Bidirectional BLE Tag Demonstrating Battery-Free Communication in Tablet/Smartphone-to-Tag, Tag-to-Tablet/Smartphone, and Tag-to-Tag Modes, " 2023 IEEE International Solid-State Circuits Conference (ISSCC), San Francisco, CA, USA, 2023, pp. 468-470.

[4] Z. Chang et al., "A Passive Crystal-Less Wi-Fi-to-BLE Tag Demonstrating Battery-Free FDD Communication with Smartphones, " 2024 IEEE International Solid-State Circuits Conference (ISSCC), San Francisco, CA, USA, 2024, pp. 404-406.

[5] T. Wu, Y. Zhao, X. Peng, J. Feng and H. Min, "A 0.037-mm2, 65.8-nW Temperature and Capacitance Sensor With Analog Pulse-Width-Modulation Backscatter, " in IEEE Journal of Radio Frequency Identification.

[6] V. Ranganathan, S. Gupta, J. Lester, J. R. Smith, and D. Tan, "RF bandaid: A fully-analog and passive wireless interface for wearable sensors," Proc. ACM Interact., Mobile, Wearable Ubiquitous Technol., vol. 2, no. 2, pp. 1–21, Jul. 2018.

[7] J. D. Rosenthal and M. S. Reynolds, "Hardware-Efficient All-Digital Architectures for OFDM Backscatter Modulators, " in IEEE Transactions on Microwave Theory and Techniques, vol. 69, no. 1, pp. 803-811, Jan. 2021.

[8] J. D. Rosenthal and M. S. Reynolds, "Single Sideband Noise Shaping for All-Digital Delta-Sigma OFDM Backscatter Modulators, " in IEEE Journal of Radio Frequency Identification, vol. 8, pp. 270-276, 2024.

A Compact Q/V Band Bidirectional Phase Shifter with 0.32° Phase Error

Congrui Li, Yan Wang, Lei Zhang*

School of Integrated Circuits, Tsinghua University, Beijing 100084, China

Email: zhang.lei@tsinghua.edu.cn

Abstract—This letter presents a Q/V band passive 7 bits bidirectional vector modulated phase shifter. Based on the insertion loss tunable (ILT) baluns, more states are introduced to choose from to calibrate for a certain state, the minimum Root-Mean-Square (RMS) phase error is reduced to 0.32° at 50GHz. While the average peak gain of the phase shifter is -17.7dB, the average 1dB bandwidth covers from 44-55GHz, the RMS amplitude error is 0.7dB at 50GHz. The core area is 0.078mm². The simulation result shows the phase shifter has a robustness of the process.

Index Terms—Q/V Band, Phase Shifter, Vector Modulated, Insertion Loss Tunable, Phase Error.

I. INTRODUCTION

In Q/V band, the phased array beamforming technique has been widely used to overcome the pathloss in the free space and to ensure the precise detection. The phase shifter (PS) is a critical block in a phased-array system, it determines the resolution of the phased array. The PS can be roughly categorized into active PSs and passive PSs.

Active PSs utilizing vector modulation exhibit high gain, but they often cost more power. In an active PS, two variable gain amplifiers (VGA) are used, it costs twice DC power, but the gain level is same as a single VGA, in multi-channel phased array systems, the active PSs will cost more DC power.

Passive PSs can be categorized into switched-type PS (STPS), reflective-type PS (RTPS), and passive vector-modulated PS (VMPS)[1]. Generally, they consume no power and have relatively good linearity. What's more, the passive VMPS can provide bidirectional phase shift operations. However, compared to the active PSs, the resolution and the phase error of the passive PSs is not as outstanding as the active PSs. Previous works for the passive PSs indicate that Root-Mean-Square (RMS) phase error do not exhibit a considerable advantage, which are typically larger than 1.5°[2].This letter utilizes the Insertion-Loss Tunable (ILT) balun to induce more superfluous states to choose from, and provide a scheme to choose the proper states, so the RMS phase error is minimized.

This letter presents a compact broadband 7 bits bidirectional passive VMPS with a -17.7dB average peak gain, a 11GHz 1dB bandwidth, the RMS phase error is minimized to 0.32° at 50GHz, while the amplitude error is minimized to 0.71dB at 50GHz. The core area of the proposed VMPS is 0.078mm².

II. DETAILED OF THE PROPOSED VMPS

A. The Architecture of the VMPS

The proposed 7-bits passive phase shifter combines four parts, the input match balun, two 5 bits X-Type attenuators two 3bits ILT baluns, and the quadrature signal generator (QSG). Fig. 1 shows the architecture of the proposed VMPS.

Each attenuator is implemented as some binary-weighted vector modulated cells connected in parallel to work as an attenuator. The 3 bits ILT balun in I/Q path can serve as redundant bits of the VMPS.

The VMPS can offer two modes: the forward phase shifting mode and the backward phase shifting mode, which founded on the bidirectionality of the attenuator, the balun and the quadrature signal generator in the circuit.

For the forward phase shifting mode, the input signal is first split by the input match balun, next the signal in each I/Q path is weighted by the 5 bits attenuator, the 3 bits ILT balun can realize the impedance matching and can attenuate the signal in each path independently, finally the signal will be summed up by the quadrature signal generator at the output node. For the backward phase shifting mode, the signal is split by the quadrature signal generator and then matched by the 3 bits ILT balun and weighted by the 5 bits passive attenuator, finally the signal in I/Q path will be combined by the input match balun. To keep the attenuator in a correct DC operating point, a proper bias voltage 0.6V is needed at the input match balun. The bias voltage can be adjusted to alleviate the PVT variation.

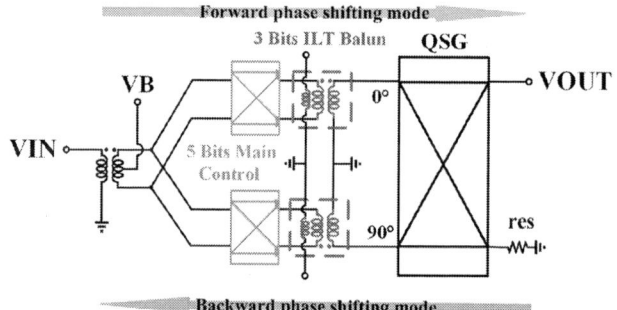

Fig. 1. The schematic of the proposed bidirectional VMPS.

B. The 5 bits X-Type Attenuator

The main issue of the attenuator in I/Q path is to provide different level of insertion loss. The Fig. 2(a) shows the schematic of the 5 bits attenuator. It consists 6 cells in parallel, the basic cell (1×) consists 4 transistors and control by a pair of complementary signals. For the cell weight of 2, there are two basic cells in parallel, etc. Especially, the 8× cell consists two 4× cells in parallel, and the size of the 0.5× cell is the same as that of the basic cell, but a pair of transistors are controlled by the supply voltage directly so it can offer smaller steps. In the layout of the attenuator, each metal line in the input node and the output node will introduce extra parasitic inductance, which will make the impedance matching difficult. The attenuators provides 32dB range for gain control, while its peak

979-8-3315-3918-4/25 $31.00 © 2025 IEEE

insertion loss is 4.4dB at 50GHz and the 1-dB bandwidth can cover 40-60GHz, so the bandwidth of the whole VMPS can be ensured. And the simulation result shows that the forward insertion loss and the backward insertion loss is nearly the same.

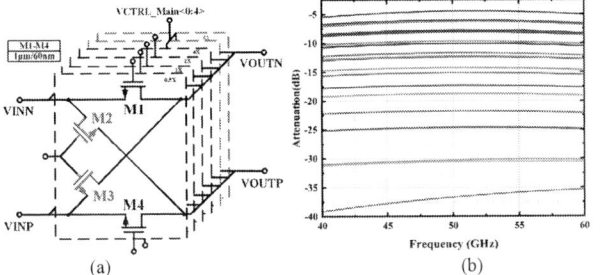

(a)   (b)

Fig. 2. The schematic of the proposed X-Type attenuator and the attenuation of the X-Type attenuator.

C. The 3 bits ILT Transformer

The 3D model of the proposed ILT transformer is shown in Fig. 3(a), which contains a conventional octagonal transformer implemented by M8, M9 and AP, an additional inductor M1 and three additional transistors. The direction of the magnetically induced small current in M1 will result a small magnetic field opposed to the conventional transformer, so the coupling coefficient of the whole balun will decrease as the three transistors turn on in turns, which introduces the change of the insertion loss. Fig.3(b) shows the basic parameter of the ILT balun. As Fig.3(c) shows, each ILT balun will offer 2.6dB tuning range and the phase variation is less than 5°.

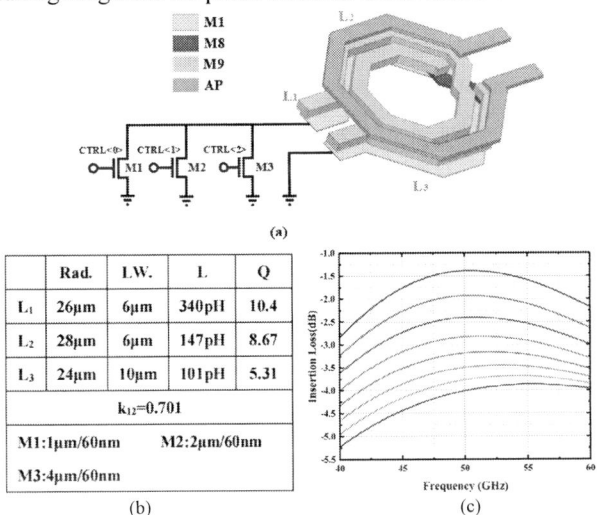

(b)   (c)

Fig. 3. (a)The 3D model of the proposed ILT transformer. (b) The basic parameter of the ILT transformer. (c)The insertion loss of the ILT transformer.

D. The Quadrature Signal Generator

The traditional QSG is based on a transformer, the primary coil and the secondary coil of the quadrature signal generator will use one single kind of thick metal such as M8 and M9, but M9 is much thicker than M8, which will induce the imbalance between the primary coil and the secondary coil. In this letter, the QSG is based on balanced transformer to solve this problem. The layer AP and M7 are introduced to serve as the short distance bridge. There are same length of M9 and M8 in

each path, the balance of the QSG is improved. Fig. 4(a) shows the 3D model and the parameter of the QSG. The radius of the inner coil is 52μm, the radius of the outer coil is 76μm, while the linewidth is 6μm.

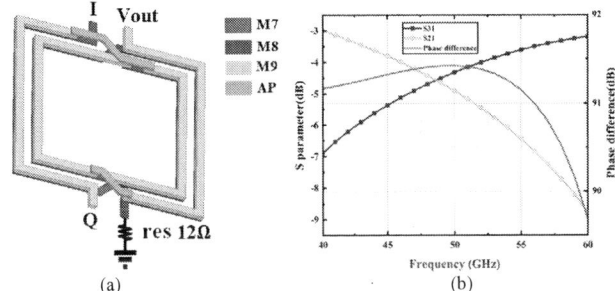

(a)   (b)

Fig. 4. (a)The 3D model of the proposed QSG. (b) The S parameter and the phase difference of the QSG.

E. The Scheme to Choose Proper States

The 5 bits X-Type attenuators in I/Q path offer only 1024 states. To reach the improvement of the phase resolution and minimize the RMS phase error, more states need to be introduced to choose from. The proposed 3 bits ILT balun can attenuate the signal in I/Q path. As the Fig. 5. shown, to calibrate a certain phase state, first let the ILT balun off, only sweep the 10 main control bits in I/Q path to find an approximate state. Next, sweep the 6 auxiliary bits of the ILT balun in I/Q path, another redundant 63 states near the approximate state will be introduced. The final calibrated phase state is chosen from all of the 64 states, so the phase will be calibrated while the amplitudes of these states are approximately same.

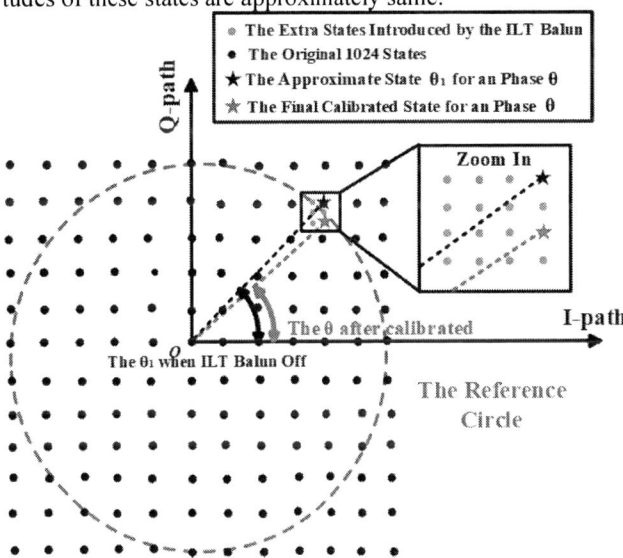

Fig. 5. The principle of the phase calibration.

III. SIMULATION RESULTS AND ANALYSIS

The full layout of the proposed bidirectional VMPS is implemented in TSMC 65nm CMOS process. As the Fig. 6. shown, the core area is 0.078mm².

979-8-3315-3918-4/25 $31.00 © 2025 IEEE

TABLE I COMPARISON WITH STATE-OF-THE-ARTS

Reference	This work*	IMS 23[2]	TCASI 22[3]	TCASII 21[4]	TVLSI 24[5]	MWCL18[6]
Technology	65nm CMOS	40nm CMOS	28nm FDSOI	55nm CMOS	40nm CMOS	65nm CMOS
Peak Average Gain(dB)	-17.7	-19	-17.5	-13.6[c]	-14	-14.5[c]
Frequency (GHz)	44-55[a]	26-32[b]	32-40	30-36	26-32[a]	52-57
Phase Range/Resolution	360/7	360/6	360/7	360/7	360/6	360/6
RMS Phase Error(deg)	0.32-2.2	1.8-2.6	0.45-1.6	0.69-1.1	0.4-1.3	2.8-3.76
RMS Amplitude Error(dB)	0.71-1.1	0.8-1.6	0.2-0.36	1.31-2.07	0.5-0.9	2.07-2.23
Core Area(mm²)	0.078	0.15	0.14	0.24	0.16	0.32
P_{DC}(mW)	0	0	0	0	0	14.3

* Post-layout Simulation Results   [a] 1 dB Bandwidth   [b] 3 dB Bandwidth   [c] Estimated from Figure

The full layout simulation result is shown in Fig. 7(a)-(d).

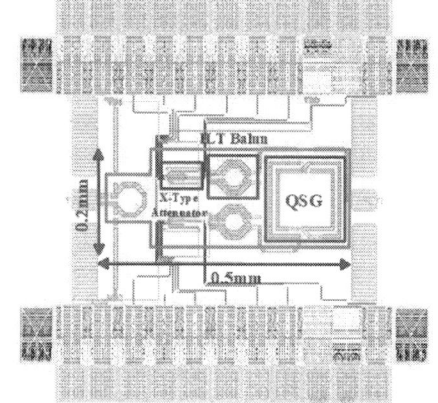

Fig. 6. The full layout of the proposed VMPS.

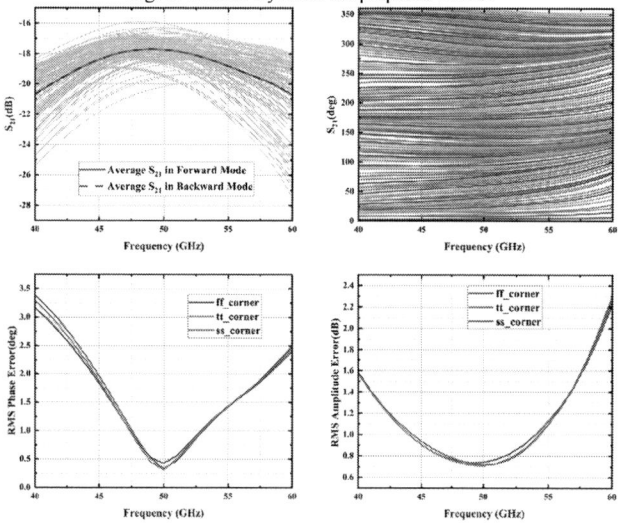

Fig. 7. (a)The gain of the chosen 128 states of the VMPS. (b) The phase of the chosen 128 states. (c)The RMS phase error in different corner (forward mode).(d) The RMS amplitude error in different corner (forward mode).

From Fig. 7(a)-(d), the average peak gain is -17.7dB, the 1dB bandwidth covers 44-55GHz.As indicated, the 360° tuning range is achieved with 2.8° phase step. At the typical corner, the RMS gain error of 0.71dB-1.1dB, and the RMS phase error of 0.31°-2.2°.At the fast corner, the peak gain is -17.6dB, the RMS amplitude/phase error is 0.74dB-1.12dB/0.42°-2.16°,the bias voltage is 0.65V.While at the slow corner, the peak gain is -17.9dB, the RMS amplitude/phase error is 0.73dB-1.09dB/0.33°-2.31°,the bias voltage is 0.55V. Table I summarizes the performance of this work and compares it with other PSs. Benefitting from the proposed 3 bits ILT balun and the footprint of the layout, our work achieves the lowest RMS phase error, core area and the largest bandwidth. Meanwhile, this work demonstrates competitive performance in terms of peak gain, the RMS amplitude error.

IV. CONCLUSION

A compact 44-55GHz 7 bits passive bidirectional phase shifter with minimum 0.32° RMS phase error is presented. A pair of ILT baluns is proposed in the phase shifter to introduce another 63 states for a certain phase state to calibrate. Only one balun is used in the input match stage to make the layout of the proposed phase shifter compact. And the simulation result shows that it has the robustness of process.

ACKNOWLEDGMENT

This work was supported by the National Key Research and Development Program of China 2024YFF1400200.

REFERENCES

[1] P. Gu and D. Zhao, "Geometric analysis and systematic design of a reflective-type phase shifter with full 360° phase shift range and minimal loss variation," *IEEE Trans. Microw. Theory Techn.*, vol. 67, no. 10, pp. 4156-4166, Oct. 2019.
[2] Y. Tian *et al.*, "A 26-32GHz 6-bit Bidirectional Passive Phase Shifter with 14dBm IP1dB and 2.6° RMS Phase Error for Phased Array System in 40nm CMOS," in Proc. *IEEE/MTT-S Int. Microw. Symp. (IMS)*, San Diego, CA, USA, 2023, pp. 195-198.
[3] Y. Li, Z. Duan, Y. Fang, X. Li, B. Deng, Y. Dai, L. Sun, and H. Gao, "A 32-40GHz 7-bit Bi-Directional Phase Shifter With 0.36 dB/1.6° RMS Magnitude/Phase Errors for Phased Array Systems," *IEEE Trans. Circuits and Syst. I: Reg. Papers*, vol. 69, no. 10, pp. 4000-4013, 2022.
[4] X. Li, B. Liu, H. Fu and K. Ma, "A 30-36 GHz Passive Hybrid Phase Shifter With a Transformer-Based High-Resolution Reflect-Type Phase Shifting Technique," *IEEE Trans. Circuits Syst. II: Exp. Briefs*, vol. 68, no. 7, pp. 2419-2423, July 2021.
[5] Y. Tian *et al.*, "Design and Analysis of a 26-32-GHz 6-bit Passive Vector Modulation Phase Shifter for CMOS Bidirectional Transceiver," in *IEEE Trans. Very Large Scale Integr. (VLSI) Syst.* doi: 10.1109/TVLSI.2024.
[6] X. Quan *et al.*, "A 52-57 GHz 6-Bit Phase Shifter With Hybrid of Passive and Active Structures," *IEEE Microw. Wireless Compon. Lett.*, vol. 28, no. 3, pp. 236-238, Mar. 2018.

A 300GHz Coherent Radiator Array with Multi-functional Antenna in 65nm CMOS

Houyi Yan[1], Kaizhe Guo[1]*

[1] National Mobile Communication Research Lab, Southeast University, Nanjing 210096, China

* Email: 220230869@seu.edu.cn, kaizhe.guo@hotmail.com

Abstract—This paper presents a high-efficiency, high effective isotropic radiated power (EIRP) terahertz (THz) scalable radiation source array implemented in 65nm CMOS technology. To address the inherent limitation of the maximum oscillation frequency (fmax) in CMOS and enable signal generation in the terahertz band, a push-push second harmonic oscillator was designed based on a differential colpitts oscillator topology, achieving a significant improvement in DC-to-THz efficiency. Furthermore, a scalable oscillator array coupling technique was employed to enhance the output power and EIRP of the terahertz source. A compact layout was identified as a critical factor for achieving high-performance metrics of radiator array. To this end, a multi-functional on-chip folded monopole antenna was proposed, which not only facilitated antenna miniaturization and harmonic selective radiation but also resolved the challenging issue of DC power distribution in large-scale radiation arrays. A 36-unit scalable coupled oscillator radiator array was fabricated within a core circuit area of 0.81 mm². Simulation results demonstrated a maximum radiated power of 14 dBm and an EIRP of 40 dBm. Additionally, the design achieved a DC-to-THz efficiency of 2.3% and a frequency tuning range of 0.85% (301.9-304.5GHz), showcasing its potential for high-performance terahertz applications.

Keywords—*CMOS, harmonic oscillator, multi-functional antenna, terahertz (THz) source.*

I. INTRODUCTION

Silicon-based terahertz integrated systems demonstrate significant application potential in sensing and communication, particularly excelling in high-resolution imaging, high-speed wireless data transmission, and molecular detection. In recent years, with the growing demand for terahertz band, the design and implementation of high-power terahertz signal sources have emerged as a critical technological challenge. Although CMOS devices hold a prominent position in the terahertz domain due to low cost, high integration density, and mature fabrication processes, the inherent maximum oscillation frequency (fmax) limitations pose substantial technical barriers to generating terahertz signals.

In existing technologies, harmonic oscillator design is commonly employed to overcome the low fmax of CMOS devices, aiming to extract higher-order harmonic frequencies. However, when operating close to the device fmax, harmonic oscillators exhibit a significant decline in output power, thereby limiting the overall performance of terahertz signal sources. Mutually coupled harmonic oscillator arrays are commonly employed to generate high-power terahertz signals [1-6]. However, the challenges of low DC-to-THz efficiency and the difficulty of DC power supply in large-scale arrays remain to be addressed.

This paper presents a scalable terahertz radiation source array integrated with multi-functional on-chip folded monopole antenna. By efficiently generating terahertz signals using a second-harmonic oscillator based on a differential Colpitts prototype, 36 oscillator units are coupled together, significantly enhancing the radiated power and effective isotropic radiated power (EIRP) of the array. The proposed multi-functional on-chip antenna achieves a compact size while selectively radiating harmonic signals generated by the oscillators. Additionally, its unique antenna structure provides a solution to the DC power supply challenge in large-scale radiation array.

The outline of the paper is as follows. Section II shows the circuit design of the proposed radiator array. The simulation results are presented in section III. At last, some conclusions are drawn in section IV.

II. CIRCUIT DESIGN

A. Array Architecture and Oscillator Core Design

The topology of the proposed coupled oscillator array with 36 elements is shown in Figure 1, which can be scaled up to a larger size. The differential unit cells are coupled out-of-phase with the adjacent unit cells using the coupling structure at the fundamental frequency. Therefore, the second harmonics generated by each oscillator are radiated from antenna and combine in-phase in free space.

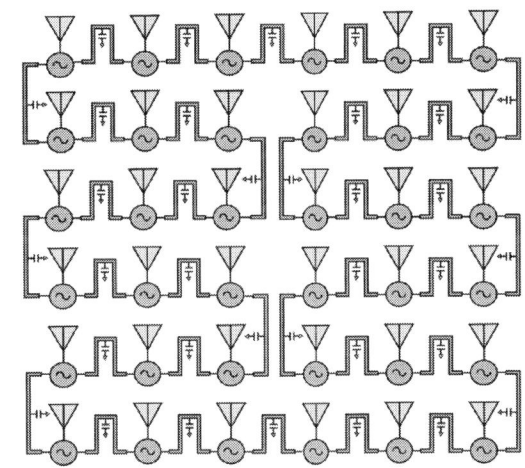

Fig. 1. Architecture of the 300GHz radiator array

The coupling structure are shown in Figure. 2, including a parallel capacitor in the middle and two 6um-wide microstrip lines on both sides connected to the gates of transistors of adjacent oscillator cores. The capacitor is implemented as parallel-plate capacitor formed between the M4 layer metal and M3 layer metal (ground) to suppress even-mode oscillation with no undesired effects for the differential oscillation. Under differential-mode excitation, a virtual ground is created at center of the microstrip, the structure acts as a 45.7pH inductor in parallel with a 1.85kΩ resistor at f0, slightly shifting frequency without disrupting core operation. Under common-mode excitation, the capacitor is transformed to an impedance equivalent to a 1.82Ω resistor parallel to a 5.52pH inductor. This parallel resistor stops any common-mode oscillation.

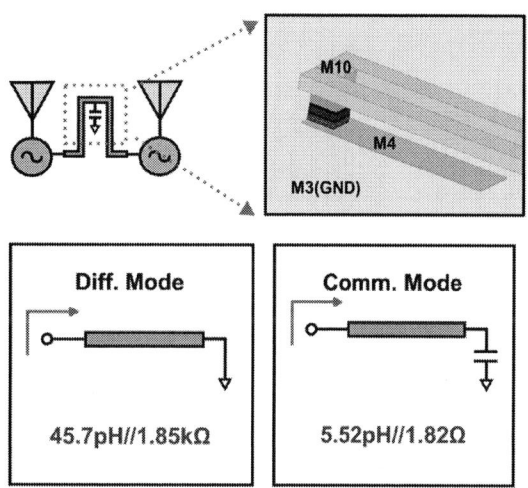

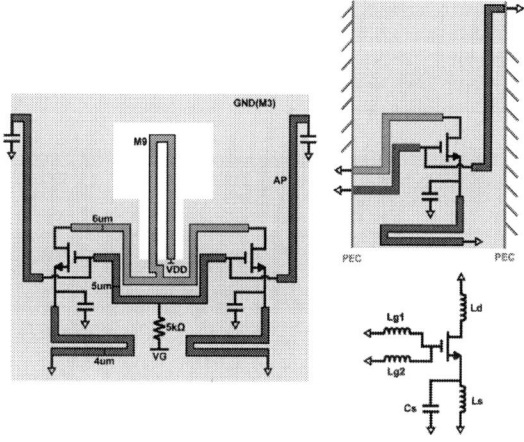

Fig. 2. Mechanizm of the couple structure

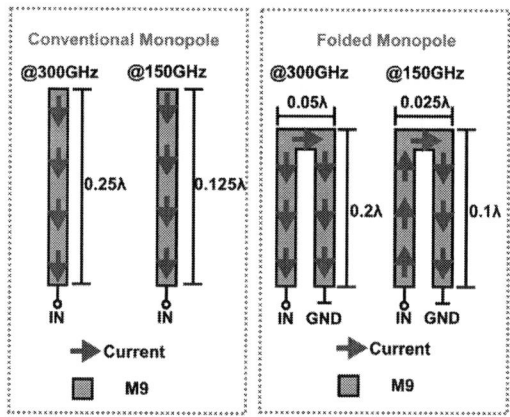

Fig. 3. Structure of the oscillator core

Figure.3 shows the detailed structure of the differential oscillator core. Once the oscillator cores are correctly coupled, the PEC boundaries are formed as the red line shown. At fundamental frequency, the equivalent half-circuit of the oscillator unit is shown in the lower right corner of Figure 3. The gate inductor can be separated into two parallel inductors (Lg1 and Lg2). Capacitor Cs provides the source capacitive impedance required for the Colpitts oscillator. The inductor Ls implemented using a quarter-wavelength transmission line provides a DC current return path and exhibits high impedance at fundamental frequency, thereby ensure minimal impact on the oscillator. The proposed on-chip antenna is connected to the virtual ground at the drain terminal, providing the required second-harmonic impedance without affecting the fundamental frequency impedance. enabling more second harmonic power is generated and delivered to the antenna for radiation.

B. Design of Multi-functional Antenna and Novel Array Power Supply Method

The multi-functional antenna is derived from the conventional monopole antenna. The proposed monopole antenna is extended, folded, and connected to ground as shown in Figure 4 right. In comparison to the conventional dipole antenna shown on the left side of Figure 4, the proposed multi-functional antenna exhibits harmonic-selective radiation capabilities: At second harmonic frequency, the standing wave on the folded metal line reverses its current direction at a point located λ/4 from the ground connection. This results in the currents on both halves of the folded metal line being in phase, enabling effective radiation. Conversely, at fundamental frequency, the length of the folded metal line is less than λ/4, causing the currents on the two halves to be out of phase, which leads to the cancellation of radiation. Additionally, by folding the antenna, its electrical size is reduced by approximately 20%, making it more compatible with the compact layout requirements of radiator arrays.

Fig. 4. Layout of the multi-functional antenna

In terahertz integrated circuit design, the utilization of wide and thickened metal layers for DC power routing is a common practice. However, in the design of radiator arrays, the requisite number of DC power supply lines escalates with the expansion of array elements, thereby imposing significant complexity on the layout process. Furthermore, the dense arrangement of metallic interconnects can detrimentally impact the performance of high-frequency core circuit. As such, the challenge of delivering DC power to each individual radiating element emerges as a pivotal issue that must be resolved to facilitate the scalability of the array. Building upon the proposed multi-functional on-chip antenna architecture, this paper presents an innovative and efficient DC power distribution scheme for the array, as depicted in Figure 5. This method utilizes the bottom three metal layers for DC power distribution. The gate bias voltage VG of the transistor is provided by the M1 metal layer, which is connected to the virtual ground point of gate in each oscillator unit through big resistor. The drain voltage VDD is supplied by the M2 metal layer, achieved by connecting the ground terminal of the on-chip antenna to the M2 metal plane. Due to the large area of the DC power supply metal planes, each layer effectively functions as a large capacitance, forming an ac short to the ground plane (M3 metal layer). This DC supply network offers two advantages. Firstly, the three DC power supply metal planes are located directly beneath each oscillator unit, enabling the DC paths to connect directly to the corresponding transistor ports from below. This significantly reduces the complexity of DC routing. Then, by utilizing the ac ground point of the on-chip antenna to deliver VDD, eliminates the area and power loss incurred by quarter-wavelength transmission line which always used in single-ended circuit power supply.

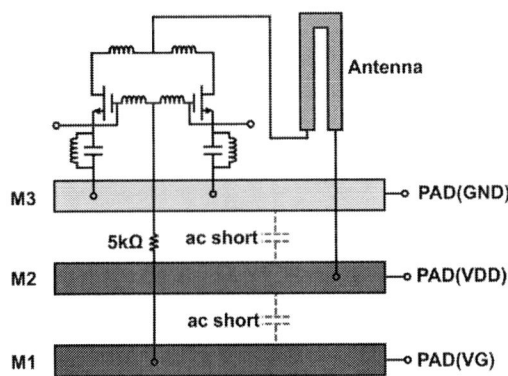

Fig. 5. The innovative DC power distribution scheme

III. SIMULATION RESULT

The proposed radiator array is implemented in 65nm CMOS technology. The layout is shown in Figure 6. The area is 0.81mm² without pad. The total power consumption is 1.008W from 1V power supply.

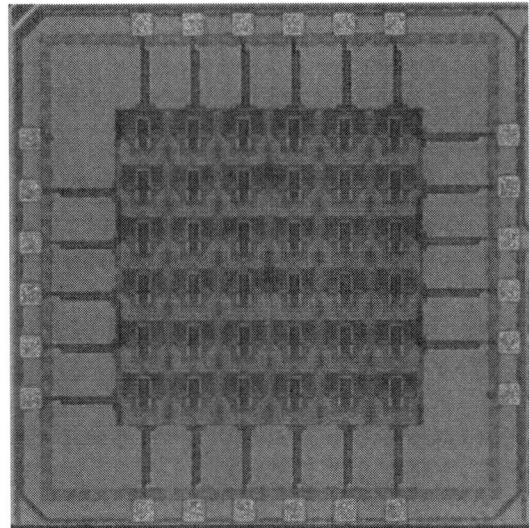

Fig. 6. The layout of the radiator array

By adjusting the gate bias voltage of the transistor in oscillator cores (VG), frequency tunability is achieved. Figure

7 illustrates the simulated second harmonic output power and frequency of single unit for different VG. The frequency turning range of the radiator array is 301.9-304.5GHz. The maximum radiated power of single unit is -1.43dBm at 302GHz.

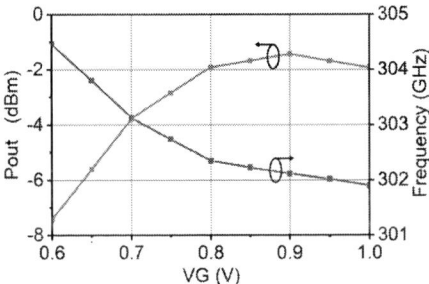

Fig. 7. Simulated output power and frequency for different VG

The simulation results of the on-chip multi-functional antenna are shown in Figure 8. The backside of the chip is attached onto a hemispherical (radius = 5mm) high-resistivity silicon lens for backside radiation. The simulated radiation efficiency of single antenna is shown in Figure 8 lower right. At 300 GHz, the antenna achieves a radiation efficiency of 35%, significantly higher than that at 150 GHz. The simulated radiation patterns of the 6×6 antenna array at 300 GHz and 150 GHz are shown in the upper right of Figure 8. The array achieves a gain of 26 dB at 300 GHz, with a radiation suppression ratio of 21 dB compared to 150 GHz. Based on the aforementioned data, the proposed radiator array achieves a maximum EIRP of 40 dBm and a DC-to-THz efficiency of 2.3%.

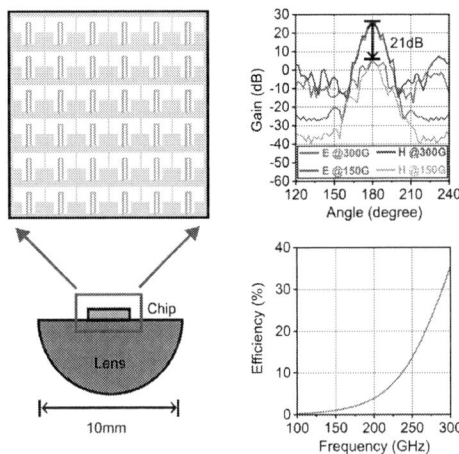

Fig. 8. Simulation result of the multi-functional antenna

IV. CONCLUSION

In this paper, a terahertz radiator array with high EIRP and high DC-to-THz efficiency is proposed. The array integrates an on-chip multi-functional antenna, enabling harmonic-selective radiation and compact antenna dimensions. Additionally, an efficient DC power supply scheme for large-scale arrays is introduced based on a novel on-chip antenna structure. The array achieves a radiated power of 14 dBm, an EIRP of 40 dBm, and the frequency turning range of 0.85%. Table I compares this work with the prior-art silicon-based signal sources with similar output frequencies，demonstrating that this work achieves the highest DC-to-THz efficiency, radiated power and EIRP.

TABLE I.   COMPARISON OF THE RADIATOR ARRAYS

	[2] 2012JSSC	[3] 2016RFIC	[5] 2015JSSC	This Work
Frequency (GHz)	280	280	338	302
Turning Range (%)	3.2	4.11	2.1	0.85
EIRP (dBm)	9.4	24.1	17	40
Radiated Power (dBm)	-7.2	9	-0.9	14
DC Power (W)	0.81	0.421	1.54	1.008
DC-to-THz Efficiency (%)	0.0235	1.88	0.053	2.3
Radiating Element	DAR + Subs. Thin.	Loop Ant.	Patch	Folded monopole
Array Size	4×4	5×6	4×4	6×6
Technology	45nm SOI CMOS	65nm CMOS	65nm CMOS	65nm CMOS

REFERENCES

[1] R. Han and E. Afshari, "A CMOS high-power broadband 260-GHz radiator array for spectroscopy," in IEEE Journal of Solid-State Circuits, vol. 48, no. 12, pp. 3090–3104, Dec. 2013.

[2] K. Sengupta and A. Hajimiri, "A 0.28 THz Power-Generation and Beam-Steering Array in CMOS Based on Distributed Active Radiators," in IEEE Journal of Solid-State Circuits, vol. 47, no. 12, pp. 3013–3031, Dec. 2012.

[3] N. Buadana, S. Jameson and E. Socher, "A 280GHz +9dBm TRP Dense 2D multi-port radiator in 65nm CMOS," 2018 IEEE Radio Frequency Integrated Circuits Symposium (RFIC), Philadelphia, USA, 2018, pp. 248-251.

[4] S. Jameson, E. Halpern and E. Socher, "A 300GHz wirelessly locked 2x3 array radiating 5.4dBm with 5.1% DC-to-RF efficiency in 65nm CMOS," 2016 IEEE International Solid-State Circuits Conference (ISSCC), San Francisco, CA, 2016, pp. 348-349.

[5] Y. Tousi and E. Afshari, "A high-power and scalable 2-D phased array for terahertz CMOS integrated systems," in IEEE Journal of Solid-State Circuits, vol. 50, no. 2, pp. 597–609, Feb. 2015.

[6] H. Jalili and O. Momeni, "A 0.34-THz wideband wide-angle 2-D steering phased array in 0.13-µ m SiGe BiCMOS," in IEEE Journal of Solid-State Circuits, vol. 54, no. 9, pp. 2449–2461, Sep. 2019.

Design of a 300GHz Wideband On-Chip Antenna in 28nm CMOS

Jinghao Zhang and Chen Jiang *

State Key Laboratory of Integrated Chips and Systems (SKLICS), Fudan University, Shanghai, China

* Email: cjiang@fudan.edu.cn

Abstract—**Terahertz (THz) waves (0.1–10 THz) offer high frequencies and wide bandwidths, enabling promising applications in future technologies. A high-performance on-chip antenna (OCA) is crucial for bridging integrated circuits with the external environment at these frequencies. This work designs a microstrip patch antenna targeting 300 GHz using the TSMC 28 nm CMOS process. The performance of OCAs using M9 and AP metals as patch layers is compared. To enhance bandwidth, a U-slot structure is introduced. Simulations show that the AP-metal OCA achieves a peak directivity of 7.2 dB, a gain of 2.3 dBi, 31.9% radiation efficiency, and a −10 dB bandwidth of 6.9 GHz (297.3–304.2 GHz). With the U-slot, the antenna achieves a significantly wider −10 dB bandwidth of 27.7 GHz (294.0–321.7 GHz), with a directivity of 6.9 dB, gain of −1.0 dBi, and 17.2% efficiency.**

Keywords—Terahertz Chip, interface, OCA, bandwidth

I. INTRODUCTION

Despite ongoing advancements in chip technology, conventional integrated circuits are increasingly constrained by inherent limitations. For instance, the narrow bandwidth of traditional RF chips makes it difficult to meet the demands of high-resolution sensing and high-bandwidth communication applications. Moreover, their limited imaging resolution restricts their use in high-precision fields such as medical imaging. The rapid development of THz chips has opened up new possibilities. Thanks to their high frequency, high resolution, strong penetration, large bandwidth, and low power consumption, THz chips offer promising solutions to overcome these challenges. Notably, THz chips are expected to play a vital role in advanced secure communication systems, 6G wideband communication systems, high resolution terahertz radar and so on, highlighting vast potential.

The OCA, serving as the physical interface between THz chips and the external world, plays a dual role: it acts as a bridge while also potentially becoming a performance bottleneck. If the OCA lacks sufficient bandwidth or radiation efficiency, it limits the data rate and overall communication performance of the chip. Therefore, OCAs with desirable characteristics—such as wide bandwidth and high radiation efficiency—are critical for a THz chip.

Various types of OCAs have been explored by domestic and international scholars, including dipole, Yagi-Uda antennas, slot antennas, and microstrip antennas. Among them, the microstrip patch antenna stands out due to its relatively simple structure, intuitive operating principle, and ease of implementation, making it a suitable candidate for design. This antenna typically consists of a single metal patch, with the patch length designed to be approximately half the wavelength. This configuration induces electromagnetic resonance in the cavity beneath the patch, allowing electromagnetic waves to leak out from the edges of the resonant cavity and radiate into free space, thereby realizing antenna functionality. It is worth noting that the ground plane of a microstrip antenna should be positioned as far as possible from the patch to maximize the quality factor (Q) and bandwidth.

In this work, a microstrip patch antenna is designed targeting a resonant frequency of 300 GHz and implemented using a TSMC 28 nm process. During design, layout of the OCA is firstly done in Virtuoso, then transferred to a 3D model in HFSS, and finally simulated in HFSS. OCAs using M9 metal and AP metal as patch layers are simulated and compared. The better-performing design is further optimized using the U-slot technique to enhance its bandwidth. Before optimization, the OCA achieves a peak directivity of 7.2 dB, a peak gain of 2.3 dBi, an efficiency of 31.9%, and a −10 dB impedance bandwidth of 6.9 GHz (297.3 GHz–304.2 GHz). After optimization, the OCA exhibits a peak directivity of 6.9 dB, a peak gain of −1.0 dBi, an efficiency of 17.2%, and a significantly broadened bandwidth of 27.7 GHz (294.0 GHz–321.7 GHz).

II. INTRODUCTION OF TSMC 28NM PROCESS

The TSMC 28 nm process is the last technology node that utilizes conventional planar transistor structures. It features ten metal layers, including copper layers (M1-9) and an aluminum layer (AP). These metal layers are interconnected through vias. The thickness and relative positions of the metal layers are shown in Fig 1.

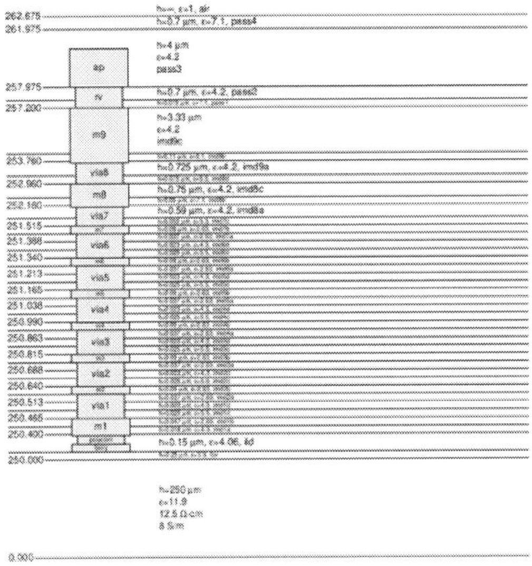

Fig. 1. Thickness and positions of the metal layers in 28 nm CMOS

For OCA design, the patch layer can be implemented using either M9 or AP, as these two layers are the farthest from the ground plane. When M9 is used, the material is copper, which has a lower electrical resistivity than the aluminum used in the AP layer. This results in reduced ohmic losses and improved efficiency. On the other hand, when AP is selected, the vertical distance from the AP layer to the M1 layer is 7.51 μm, which is more than twice the distance compared to the case where M9 is used (3.235 μm). This greater separation from the ground plane is beneficial for enhancing the antenna bandwidth.

III. DESIGN OF ON-CHIP TRANSMISSION LINE

In this work, the OCA adopts edge feeding via an on-chip transmission line. A microstrip line structure is chosen for the on-chip transmission line, as microstrip lines are among the most commonly used types of transmission lines. They can be fabricated using simple photolithographic processes and are easily integrable with other passive or active components [1]. A schematic cross-sectional view of the microstrip line is shown in Fig 2.

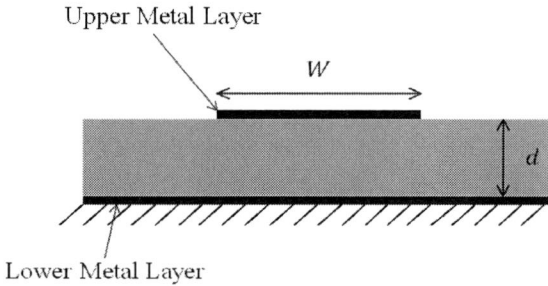

Fig. 2. The cross-sectional view of the microstrip line

Under TSMC 28 nm process, the vertical spacing between the M1 and M9 metal layers is only 3.235 μm, allowing the transmission line to satisfy the quasi-TEM condition. The microstrip line satisfies the following relationships[1]:

$$Z_0 = \begin{cases} \dfrac{60}{\sqrt{\epsilon_e}} \ln\left(\dfrac{8d}{W} + \dfrac{W}{4d}\right), & W/d \leq 1 \\[3mm] \dfrac{120\pi/\sqrt{\epsilon_e}}{\dfrac{W}{d} + 1.393 + 0.667\ln\left(\dfrac{W}{d} + 1.444\right)}, & W/d \geq 1 \end{cases} \quad (1)$$

$$\epsilon_e = \frac{\epsilon_r + 1}{2} + \frac{\epsilon_r - 1}{2}\frac{1}{\sqrt{1 + 12d/W}} \quad (2)$$

where Z_0 denotes the characteristic impedance, ϵ_e is the effective dielectric constant, ϵ_r is the relative permittivity of the dielectric, W is the width of the top metal layer of the microstrip line, and d is the dielectric thickness.

The layout structure of the on-chip transmission line is shown in Fig 3. The design goal is to achieve a characteristic impedance of $50\,\Omega$, in alignment with common industry standards. After importing the layout into HFSS, parametric sweeps and adjustments were conducted within HFSS to observe the variation in characteristic impedance and to tune it to the target value of 50 Ω.

The design variables include the width of the M9 metal line, the CPW (coplanar waveguide) channel width, and the outer wall width of the M1–M9 metal stack. Through controlled parameter variation experiments, the final design values were determined as follows: M9 physical width of 1.8

μm, CPW channel width of 15.83 μm, and metallic sidewall width of 13.5 μm. The HFSS simulation results are presented in Fig 4, indicating that the characteristic impedance of the on-chip transmission line is 49.91 Ω, which is within an acceptable margin of error.

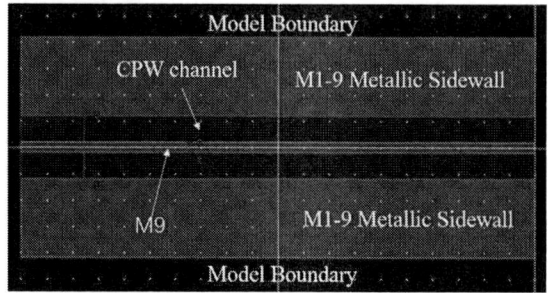

Fig. 3. The layout of the on-chip transmission line structure

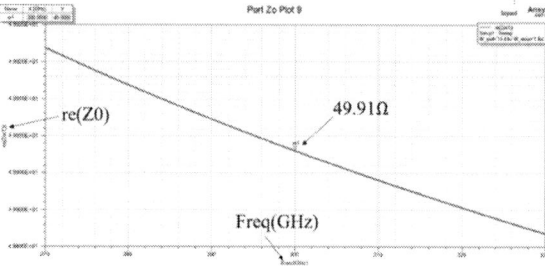

Fig. 4. Simulation results of the characteristic impedance

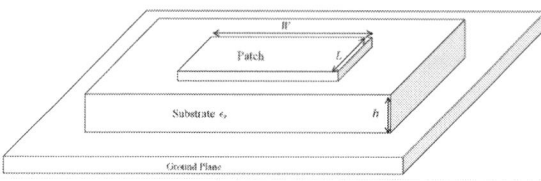

Fig. 5. The typical structure of an on-chip microstrip patch antenna

IV. DESIGN OF MICROSTRIP PATCH ANTENNA

The typical structure and parameters of an on-chip microstrip patch antenna is illustrated in Fig 5. This type of antenna offers several advantages: it features a simple structure composed of a single metal patch, making it easy to implement; it is a well-studied design with good compatibility for integration with IC processes; and a wide range of mature bandwidth enhancement techniques are available for optimization. The primary parameter relationships satisfy the following [2]:

$$L = \frac{c}{2f\sqrt{\epsilon_e}} \quad (3)$$

$$W = \frac{c}{2f\sqrt{2/(\epsilon_r + 1)}} \quad (4)$$

$$\Delta L = \frac{0.412(\epsilon_e + 0.3)\left(\dfrac{W}{h} + 0.264\right)h}{(\epsilon_e - 0.258)\left(\dfrac{W}{h} + 0.8\right)} \quad (5)$$

$$L_e = L - 2\Delta L \quad (6)$$

where c denotes the speed of light in vacuum, f represents the resonant frequency, and L_e refers to the effective (or corrected) length of the patch element.

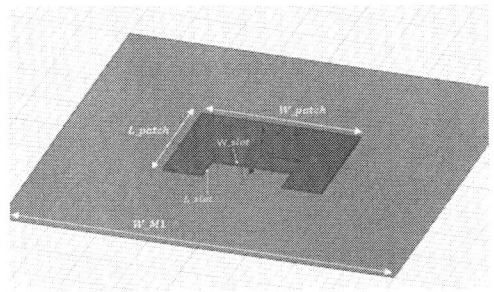

Fig. 6. The structure of OCA and corresponding design variables in HFSS

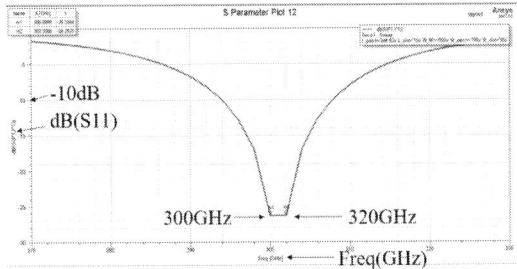

(a) The simulated S11

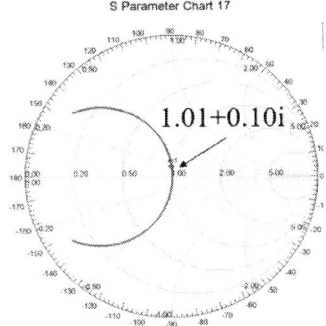

(b) The simulated S11 on a Smith Chart

Fig. 7. The simulated S-parameter curve and Smith Chart results (M9)

A. M9 used as the patch layer

When M9 is used as the patch element, the structure of the OCA and the corresponding design variables in HFSS are shown in Fig 6. A slot extension is introduced to the antenna because impedance matching cannot be easily achieved by adjusting only the patch's length and width.

The design objective is to achieve resonance at 300 GHz with proper impedance matching. In HFSS, five parameters are swept and adjusted: the edge length of the M1 ground plane (W_M1), the patch length (L_patch), the patch width (W_patch), the slot length (L_slot), and the slot width (W_slot). The final optimized values are as follows: W_M1 (800 μm), W_patch (100 μm), L_patch (244.8 μm), W_slot (30 μm), and L_slot (10 μm)

The simulated S11 and the corresponding curve on a Smith Chart are shown in Fig 7, respectively. The results meet the design expectations.

B. AP used as the patch layer

When AP is used as the patch element, the OCA structure in HFSS is similar to the previous design; however, no slot

extension is required in this case, as impedance matching can be more easily achieved by adjusting the patch width.

The design targets antenna resonance at 300 GHz with proper impedance matching. In HFSS, three parameters are optimized: the M1 ground plane side length, patch length, and patch width. The final values are 500 μm, 231.3 μm, and 370 μm, respectively. The simulated S11 and corresponding Smith chart in Fig. 8 confirm resonance at 300 GHz with satisfactory impedance matching.

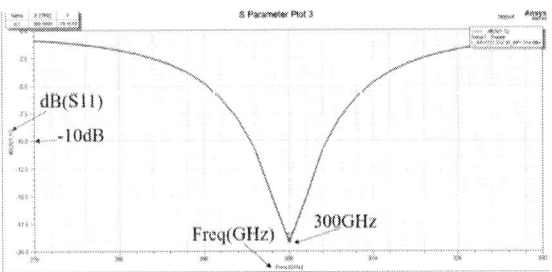

(a) The simulated S11

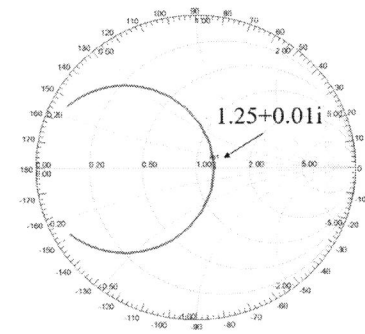

(b) The simulated S11 on a Smith Chart

Fig. 8. The simulated S-parameter curve and Smith Chart results (AP)

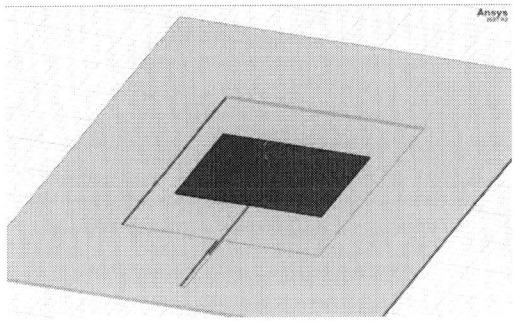

Fig. 9. The structure of the complete OCA in HFSS

C. Comparation

Table 1 compares the antenna performance of the two designs. When M9 is used as the patch layer, both radiation efficiency and peak gain are critically low, primarily due to the minimal separation between M9 and the M1 ground plane. This results in an overly compact structure with narrow patch edges, causing most electromagnetic energy to remain confined within the cavity rather than radiating into free space. In contrast, the AP-based design achieves performance metrics within a practical range, making it a more effective and suitable choice for terahertz applications.

979-8-3315-3918-4/25 $31.00 © 2025 IEEE

TABLE I. COMPARISON OF ANTENNA PERFORMANCES (M9 AND AP)

Antenna Parameters	M9 as Patch Layer	AP as Patch Layer
Peak Directivity (dB)	6.1	7.2
Peak Gain (dBi)	-10	1.1
Efficiency	2.7%	25%

D. Complete OCA design with integrated feed line

The microstrip patch antenna using AP as the patch element is integrated with the on-chip transmission line. The structure implemented in HFSS is shown in Fig 9. In this design, the feed point position is introduced as a new variable and is swept and optimized. The final feed location is determined to be at 40 μm. In this case, the matching performance remains as good as in the previous design.

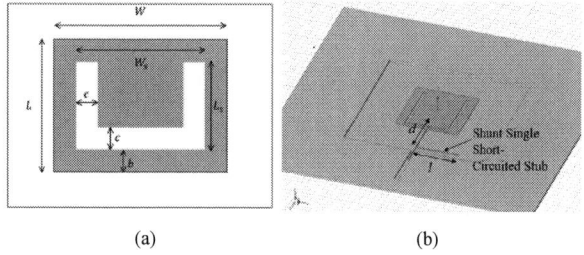

(a)   (b)

Fig. 10. (a) U-slot modification (b) The complete 3D structure of OCA

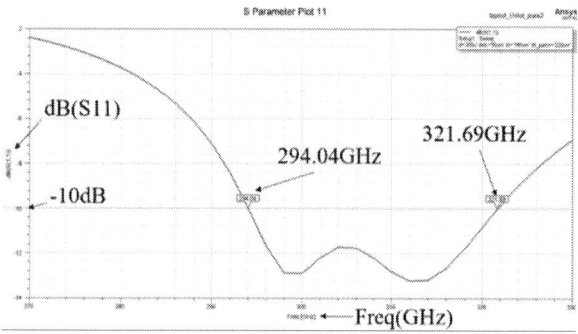

Fig. 11. The simulated S11 (U-slot)

V. BANDWIDTH OPTIMIZATION OF THE OCA

The U-slot technique is employed to enhance bandwidth, as it is a well-established broadband method based on a single-layer, single-patch structure. It requires only modifying the patch element without altering the antenna structure or feed mechanism, offering ease of implementation (Fig. 10 (a)).

As described by Huynh and Lee [3], the U-slot improves bandwidth through two primary mechanisms: Firstly, it introduces additional capacitance that compensates for the feed inductance, effectively lowering the system Q-factor via LC compensation. Secondly, it generates a secondary resonant mode that couples with the fundamental mode, resulting in broader impedance bandwidth.

This dual-resonance effect leads to a circular S11 trajectory around the origin on the Smith chart, indicating improved impedance matching across a wider frequency range. The relative contribution of each mode is mainly influenced by the feed point location, while the U-slot's

geometry and dimensions also play critical roles in bandwidth tuning

Furthermore, a single short-circuited stub impedance matching technique is employed to further optimize impedance matching. This involves adding a shunt open-circuited stub along the transmission line. The complete 3D structure of the optimized OCA is shown in Fig 10 (b).

Based on Fig 10, the following design variables can be identified, swept and designed: Side slot length of the U-slot (L_s, 200 μm), Opening width of the U-slot (W_s, 160 μm), Position of the short-circuited stub (d, 200 μm), Stub length (l_0, 180 μm) and Feed point position ($delL$, 95 μm).

The final simulated S-parameter curve is shown in Fig 11.

TABLE II. COMPARISON OF ANTENNA PARAMETERS (U-SLOT OR NOT)

Parameters	With U-slot	Without U-slot
Frequency (GHz)	300	300
Process	28nm CMOS	28nm CMOS
Peak Directivity (dB)	6.9	7.2
Peak Gain (dBi)	-1.0	2.3
Efficiency	17.2%	31.9%
Bandwidth (GHz)	27.7	6.9

VI. SIMULATION RESULTS AND PERFORMANCE COMPARISON

Table 2 compares the antenna performance before and after bandwidth optimization. Compared to the original OCA, the U-slot-based design shows a slight reduction in key metrics: peak directivity decreases by 0.3 dB, gain by 3.3 dB, and efficiency by 46.1%. Despite these reductions, all values remain within acceptable ranges for terahertz antenna applications. In contrast, the impedance bandwidth is significantly improved to 27.7GHz (294.0–321.7 GHz), indicating excellent broadband performance.

VII. CONCLUSION

The OCA employing M9 as the patch layer demonstrates extremely low radiation efficiency and gain, making it unsuitable for practical applications. In contrast, the AP-based OCA shows significantly improved efficiency and peak gain. Incorporating a U-slot into the AP-based design results in a slight 0.3 dB reduction in peak directivity, a 3.3 dB drop in peak gain, and a 46.1% decrease in efficiency, while achieving a fourfold increase in bandwidth (294.0–321.7 GHz). This trade-off leads to substantially enhanced broadband performance in the THz range, with all other key metrics remaining within practical limits. Thus, the proposed design offers a promising and well-balanced interface solution for terahertz applications.

ACKNOWLEDGMENT

This work is partially supported by the National Natural Science Foundation of China under Grant 62341409 and 92473101.

REFERENCES

[1] POZAR D M. Microwave engineering[M]. 4th Edition. Chichester: Wiley, 2012, pp.117-118.

[2] STUTZMAN W L, THIELE, GARY A. Antenna theory and design[M]. 3rd Edition. Hoboken, NJ: Wiley, 2013, pp.197-199.

[3] Chen, Z. N., Liu, Duixian. editor, Nakano, Hisamatsu. editor, Qing, Xianming. editor, Zwick, Thomas. editor, & SpringerLink. Handbook of Antenna Technologies[M]. 2016, pp.804.

ATSim: A Fast and Accurate Simulation Framework for 2.5D/3D Chiplet Thermal Design Optimization

(Invited Paper)

Qipan Wang[1,2], Tianxiang Zhu[1], Jiajia Cui[1], Yicheng Wei[1], Linxiao Shen[1], Zhe Cheng[1], Runsheng Wang[1,3,4*], Ru Huang[1,3,4], Yibo Lin[1,3,4*]

[1]School of Integrated Circuits, Peking University   [2]Academy for Advanced Interdisciplinary Studies, Peking University
[3]Institute of Electronic Design Automation, Peking University, Wuxi, China
[4]Beijing Advanced Innovation Center for Integrated Circuits, Beijing, China
Email: {qpwang,yibolin}@pku.edu.cn

Abstract—**This paper reviews the thermal challenges in 2.5D/3D chiplet integration systems and introduces `ATSim`, a simulation framework with applications to chiplet thermal optimization. `ATSim` enables fast and accurate thermal simulation for both steady-state and transient conditions. It supports nonlinear, heterogeneous, and anisotropic materials. The framework features a multilevel grid generation scheme based on a novel hybrid tree structure. Compared to mainstream academic and commercial tools, `ATSim` achieves high accuracy and efficiency, making it a powerful tool for evaluating and improving thermal designs, including applications like thermal-aware placement.**

I. INTRODUCTION

Thermal management has become a critical challenge in modern integrated circuit (IC) design, especially with the rise of aggressive integration techniques such as 2.5D, 3D, and 3.5D-ICs [1], [2]. These technologies enhance performance by densely integrating multiple dies in compact footprints, but they also exacerbate thermal dissipation difficulties. High temperatures, in turn, degrade chip performance, increase power consumption, and lead to reliability and failure issues. Accurate and efficient thermal simulation is thus indispensable for thermal sign-off and optimization.

Numerical thermal simulation methods can be roughly categorized into three types: 1) finite element method (FEM); 2) finite difference method (FDM); 3) finite volume method (FVM). Popular thermal simulators fall into these categories, including FEM-based platforms such as COMSOL [3], Celsius [4], and MTA [5], as well as FDM/FVM-based solvers like ANSYS ICEPAK [6], HotSpot [7], and 3D-ICE [8]. Table I summarizes the methods and features of representative simulators from both industry and academia. Despite the unique characteristics of each tool, we can see that they fall short in addressing the following three challenges raised by next-generation ICs.

(1) Multi-scale modeling complexity: The thermal behavior of modern ICs spans a wide range of scales — from centimeter-sized heat sinks to sub-micron features like TSVs and microbumps. Capturing this range accurately requires multi-scale modeling, which is difficult for traditional mesh-based solvers. Uniform meshing either loses detail at small scales or leads to excessive computational costs at larger scales.

(2) Complex material properties: IC materials often have anisotropic, nonlinear, and heterogeneous thermal conductivities.

For example, epoxy and silicon dioxide show different thermal conductivities in the in-plane and through-plane directions [9]. Additionally, materials like silicon and copper exhibit nonlinear behavior at high temperatures, as demonstrated in previous studies [10], [11]. These properties complicate the development of robust, general-purpose thermal simulators.

(3) Design workflow integration: Many thermal simulators operate in isolation, making it hard to incorporate thermal analysis into the early design stages. However, integrating thermal modeling with chip design tools is essential for thermal-aware optimization and power-thermal co-simulation, especially when accounting for temperature-dependent power behavior.

In this work, we introduce `ATSim`[1], a thermal simulation framework specifically designed for multiscale 2.5D/3D chiplet systems, with extensions to 3.5D-ICs. As shown in Table I, `ATSim` addresses the challenges above through adaptive meshing, nonlinear solver techniques, and seamless integration with existing EDA workflows. Our contributions are summarized as follows.

- A scalable thermal simulation framework addressing multiscale challenges with an adaptive multilevel grid generation scheme based on a novel hybrid tree structure.
- Holistic strategies to handle nonlinear, anisotropic, and heterogeneous materials.
- Seamless integration with chip design tools for power-thermal co-simulation and thermal-aware placement.

The accuracy and efficiency of `ATSim` are verified against both commercial and academic tools using a set of representative benchmarks. It achieves an average acceleration of $80\times$ and $40\times$ compared to ICEPAK and COMSOL, respectively, with a relative error of $< 3\%$.

II. THE `ATSim` FRAMEWORK

A. Overview

The overall workflow of `ATSim` is illustrated in Fig. 1, where input data including geometrical configuration (.xml file of 3D layer-stacking description and .csv files for the floorplan of each layer), material parameters, power library, and simulation settings are processed to generate accurate temperature maps for post-process like visualization. The geometrical configuration describes

---

*Corresponding author.

[1]Compiled binary available at https://github.com/Brilight/ATSim_pub.

TABLE I: Comparison of common thermal simulators' features.

Simulator	Algorithm	Efficiency	Material Properties			3.5D-IC	Power-Thermal Co-simulation
			Nonlinear	Anisotropy	Heterogeneous		
COMSOL [3]	FEM	Low	✓	✓	✓	✓	×
Celsius [4]		High	×	×	✓	✓	✓
MTA [5]		High	✓	×	✓	✓	×
HotSpot [7]	FDM	Low	×	×	✓	×	×
PACT [12]		High	×	×	✓	×	✓
3D-ICE [8]		Medium	×	×	✓	×	×
ICEPAK [6]	FVM	Medium	✓	✓	✓	✓	✓
ATSim		High	✓	✓	✓	✓	✓

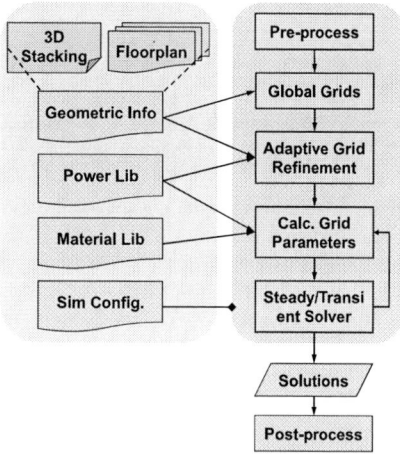

Fig. 1: The overall flow of ATSim.

the 3D layer-stacking structure of the system, including the positions, materials, and detail description files of different layers or components. The floorplan files provide detailed information about the layout of each component in the system, including the dimensions and positions of units or modules inside. The material parameters include thermal conductivity, heat capacity, and other relevant properties used. The power library contains steady or transient power profiles for each source, which includes the internal power, switching power, and leakage power. These profiles can be either fixed or temperature-dependent, allowing for flexible modeling of the thermal behavior of the system. The simulation settings specify the type of analysis to be performed, such as steady-state or transient, and include parameters like convergence criteria and time step sizes.

B. Multilevel Grid Generation

The proposed meshing algorithm begins by partitioning the whole package into a set of global grids, distinguishing between heating and cooling components. Local refinement is then applied based on the structural characteristics and power distribution. For example, chiplets like HBM can be coarsely partitioned, while those with non-uniform power maps—such as CPUs and ASICs—require finer meshing at hotspots. Each grid is further subdivided both horizontally and vertically. Vertical subdivisions are determined by the number of layers and number of tiers to partition for each layer, while horizontal divisions follow a quadtree structure guided by material properties and power density. The refinement process

terminates when either the minimum grid size (resolution limit) is reached or the maximum depth level is achieved.

After the grid generation, the whole system can be described with a hybrid tree structure, where the root nodes correspond to the coarsest partitions and leaf nodes represent the finest elements, which is different from the hybrid octree in [13] and the quadtree in [14]. Each node maintains metadata about its position, size, material composition, and power profile. This hierarchical representation facilitates efficient traversal with the depth-first search, recording the level and information of each grid during the solving phase. This strategy avoids the imbalance of grid density observed in tools like ICEPAK and HotSpot and prevents excessive fine meshing that increases computational cost. Additionally, it supports modeling of temperature-dependent cooling effects.

C. Complex Material Properties

In 3.5D-IC systems, materials such as silicon, silicon dioxide, copper, and epoxy exhibit significant variations in thermal conductivity, often being anisotropic and temperature-dependent. In ATSim3D [10], nonlinear thermal conductivity is addressed via Kirchhoff's transformation. ATSim employs Newton iteration to iteratively update the temperature-dependent thermal conductivity until convergence is achieved.

For anisotropic materials, the governing equation is discretized using finite volume methods with anisotropic gradient terms. A 3x3 diagonal thermal conductivity matrix is constructed for each grid, allowing for directional heat transfer modeling. When heterogeneous or nonlinear materials exist within a single grid, an equivalent thermal model is derived by averaging the conductivities of subgrids following [15], [16], considering both material properties and volume ratios. The thermal conductivity matrix is updated at each iteration based on the current temperature distribution, ensuring accurate modeling of nonlinear and anisotropic thermal behaviors. This enables precise simulation of the temperature distribution across the system. Additionally, we propose an analytical model [11] to more accurately extract the equivalent thermal conductivity of the back end of line (BEOL) layers.

D. Steady-state and Transient Solver

ATSim supports both steady-state and transient thermal analysis for the nonlinear heat diffusion PDE, which reads:

$$C_v(\mathbf{r}, T)\frac{\partial T}{\partial t} = \nabla \cdot (\kappa(\mathbf{r}, T)\nabla T(\mathbf{r})) + \mathbf{P}(\mathbf{r}, T), \qquad (1)$$

where $C_v(\mathbf{r}, T)$ is the volumetric heat capacity, $\kappa(\mathbf{r}, T)$ is the thermal conductivity, and $\mathbf{P}(\mathbf{r}, T)$ is the power density. Various types of boundary conditions are supported, including fixed temperature (Dirichlet), heat flux (Neumann), convective cooling (Robin), and radiation boundary conditions.

For the steady-state solver, the Eq.(1) reads after discretization:

$$\mathbf{K}(T)T = \mathbf{b}(T)(T), \tag{2}$$

where $\mathbf{K}(T)$ and $\mathbf{b}(T)$ are the temperature-dependent conductivity matrix, and source vector, respectively. Our solver supports linear and nonlinear variants, all accelerated via multigrid techniques. Linear solvers include conjugate gradient, LU decomposition, and so on, while the nonlinear variant uses Full Approximation Scheme Multigrid (FAS-MG) [17] for robust convergence.

The transient solver currently assumes linear thermal behavior due to the lack of clarity regarding the underlying mechanisms of temperature-dependent thermal properties in time-domain simulations. The transient problem is modeled using the following ODE:

$$\mathbf{C}\frac{dT}{dt} + \mathbf{K}T = \mathbf{b}(t, T), \tag{3}$$

where $\mathbf{C}$ is the heat capacity matrix, and the source term $\mathbf{b}(t, T)$ is calculated based on the current temperature distribution and updated at each time step. Time integration is performed using both explicit and implicit schemes depending on stability requirements. The explicit method is used for small time steps, while the implicit method is used for larger time steps to ensure stability.

E. Parallelization

To improve computational efficiency, `ATSim` leveraging the multilevel grid structure to distribute the computational workload across multiple threads or processes. This allows parallel computation of the temperature field across the system, significantly reducing runtime without compromising accuracy.

F. Integration with Chip Design Tools

A key feature of `ATSim` is its seamless integration with EDA tools, supporting power-thermal co-simulation and thermal-aware placement. Layout and power information generated by the standard EDA flow can be imported into `ATSim` to generate high resolution temperature distributions, which is then fed back to the EDA flow to decide the temperature corners of the standard cell libraries, resulting in a iterative co-simulation procedure. The temperature results can also be used to construct compact thermal models to feed back into the design flow for iterative optimization, exemplified by the analytical thermal-aware chiplet placement framework in [18].

III. EXPERIMENTAL RESULTS

To verify our simulator, we arrange experiments on several systems, including 2D-ICs Intel Multi-core CPU [12] (**2DIC**), a 2.5D-IC of the Nvidia V100 GPU [16] (**V100**), a 3D-IC of 7-layer Mono3D-IC with 2 active layers [12] (**Mono3D**), and a 3.5D-IC of 3D near-data-processor (NDP) chiplets forming 2×2 array (**3.5D-NDPs**) [16]. The floorplan and power maps are adopted from relevant works. Appropriate material properties and boundary conditions are also configured based on typical values reported in the literature. To ensure a fair comparison, appropriate mesh resolution is implemented in each experiment to guarantee convergence of the solution and to maintain a balance between accuracy and efficiency.

TABLE II: Performance of different simulators on linear steady state simulation.

Simulator	V100				3.5D-NDPs			
	MARE/%	MaxE/°C	MAE/°C	Time/s	MARE/%	MaxE/°C	MAE/°C	Time/s
ICEPAK (gloden)	0	0	0	315	0			299
HotSpot†	3.00	2.66	1.46	40	Cannot Handle			
MTA	1.15	3.08	0.57	12	0.67	1.17	0.29	9
ATSim	**0.82**	**1.36**	**0.41**	**5**	**0.34**	**0.95**	**0.15**	**3**
	2DIC				**Mono3D**			
	MARE/%	MaxE/°C	MAE/°C	Time/s	MARE/%	MaxE/°C	MAE/°C	Time/s
COMSOL (gloden)	0	0	0	60	0	0	0	1410
PACT	2.40	1.81	0.22	55	**0.53**	**0.71**	**0.21**	255
ATSim	**2.07**	**1.62**	**0.20**	**8**	0.54	0.78	0.22	**18**

TABLE III: Performance of different simulators on non-linear steady state simulation.

Simulator	V100				3.5D-NDPs			
	MARE/%	MaxE/°C	MAE/°C	Time/s	MARE/%	MaxE/°C	MAE/°C	Time/s
COMSOL (golden)	0	0	0	301	0	0	0	509
MTA‡	1.23	3.98	0.66	26	0.54	0.56	0.23	17
ATSim	**0.83**	**1.30**	**0.44**	**20**	**0.49**	**0.53**	**0.21**	**16**
	2D-IC				**Mono3D**			
	MARE/%	MaxE/°C	MAE/°C	Time/s	MARE/%	MaxE/°C	MAE/°C	Time/s
COMSOL (golden)	0	0	0	117	0	0	0	5541
ATSim	**2.69**	**1.54**	**0.18**	**25**	**1.06**	**1.15**	**0.41**	**53**

‡ The released binary of MTA supports nonlinear conductivity only.

A. Evaluation of the Steady Solver

We first evaluate simulators under the linear configuration. The results are shown in Table II, here the accuracy is evaluated by the mean absolute relative error (MARE), maximum error (MaxE), and mean absolute error (MAE). Notably, our simulator achieves the lowest error for most of the cases (MAE of 0.20 °C for **2DIC**, 0.41 °C for **V100**, and 0.14 °C for **3.5D-NDPs**), along with the highest efficiency of 57× acceleration compared to the golden simulator on average thanks to the multilevel grid strategy.

For the nonlinear simulation, as **ICEPAK** can only support nonlinear leakage and not nonlinear thermal conductivity, we use COMSOL as the golden simulator instead. The released binary of MTA supports nonlinear conductivity only without nonlinear leakage, so we modify the leakage power according to temperature results, when running MTA. As shown in Table III, our simulator exhibits high accuracy ($< 3\%$ MARE and $< 2 °C$ MaxE) and efficiency for all cases. The runtime is much larger than that under linear configuration as the number of grids is enlarged to tackle the nonlinear conductivity. `ATSim`'s errors arise mostly from two sources. The first is the coarse-grained error, which can be reduced by refining the global grids. The second concentrates in the high-temperature areas, where the leakage power and conductivity are updated differently between FEM-based COMSOL and our FVM-based simulator. Future efforts will focus on optimizing these errors.

B. Evaluation of the Transient Solver

To validate the accuracy of `ATSim` concerning transient simulation, we simulate the **3.5DNDP** case with a fixed power map as in the steady simulation. The MTA work as golden simulator and the results are summarized in Table IV with the MAE and MARE averaged during the whole 50s. The overall MAE, MARE are 0.23K, and 0.16%, respectively, and the maximum error is 1.0K. Our simulator is 7× faster than MTA. Fig. 2 shows the temperature traces for two points of minimum and maximum temperatures. We can observe a close matching, implying the accuracy of `ATSim`.

TABLE IV: Performance of simulators on transient simulation.

Simulator	MARE/%	MAE/°C	MaxE/°C	Time/s
MTA (golden)	0	0	0	1790
ATSim	0.16	0.23	1.03	231

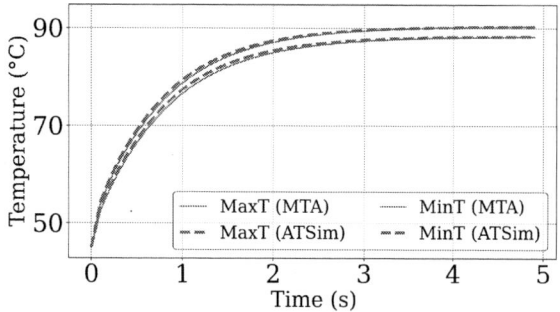

Fig. 2: The transient temperatures of MTA and `ATSim` on **3.5DNDP**.

IV. CONCLUSION AND FUTURE WORK

This paper presents `ATSim`, a unified thermal simulation framework designed for 2.5D/3D chiplet-integrated systems, with natural extensibility to 3.5D-ICs. `ATSim` effectively addresses key challenges such as multiscale modeling, nonlinear thermal effects, and material anisotropy and heterogeneity by leveraging multilevel grid generation and equivalent thermal parameter extraction techniques. Compared to both commercial and academic simulation tools, the framework achieves significant speedup while maintaining high accuracy. In future work, we plan to extend `ATSim` to support system-technology co-optimization (STCO) and large-scale thermo-mechanical co-simulation [19]. We will also perform temperature measurements on real-world chips [20] to calibrate and validate the simulation results. Our long-term goal is to develop a comprehensive multi-physics platform that enables integrated simulation and co-optimization of electrical, thermal, and mechanical interactions in advanced packaging systems.

ACKNOWLEDGEMENT

This work was supported in part by the National Science Foundations of China (Grant No. 62125401, 62034007), the Natural Science Foundation of Beijing, China (Grant No. Z230002), Grant QYJS-2023-2303-B, Beijing Outstanding Young Scientist Program (JWZQ20240101004), and the 111 project (B18001).

REFERENCES

[1] J. H. Lau, "Recent advances and trends in advanced packaging," *IEEE Transactions on Components, Packaging and Manufacturing Technology*, vol. 12, no. 2, pp. 228–252, 2022.

[2] C. S. Mandalapu, C. Buch, P. Shah, R. Topacio, P. Cheng, L. Wang, R. Swaminathan, A. Smith, J. Wuu, K. Mysore *et al.*, "3.5 d advanced packaging enabling heterogenous integration of hpc and ai accelerators," in *2024 IEEE 74th Electronic Components and Technology Conference (ECTC)*. IEEE, 2024, pp. 798–802.

[3] "Comsol Multiphysics."

[4] "Cadence Celsius," http://www.cadence.com.

[5] S. Ladenheim, Y.-C. Chen, M. Mihajlović, and V. F. Pavlidis, "The mta: An advanced and versatile thermal simulator for integrated systems," *IEEE Transactions on Computer-Aided Design of Integrated Circuits and Systems*, vol. 37, no. 12, pp. 3123–3136, 2018.

[6] "Ansys ICEPAK," http://www.ansys.com.

[7] M. R. Stan, K. Skadron, M. Barcella, W. Huang, K. Sankaranarayanan, and S. Velusamy, "Hotspot: A dynamic compact thermal model at the processor-architecture level," *Microelectronics Journal*, vol. 34, no. 12, pp. 1153–1165, 2003.

[8] F. Terraneo, A. Leva, W. Fornaciari, M. Zapater, and D. Atienza, "3d-ice 3.0: efficient nonlinear mpsoc thermal simulation with pluggable heat sink models," *IEEE Transactions on Computer-Aided Design of Integrated Circuits and Systems*, vol. 41, no. 4, pp. 1062–1075, 2021.

[9] C. Wang, Q. Xu, C. Nie, H. Cao, J. Liu, D. Zhang, and Z. Li, "A multiscale anisotropic thermal model of chiplet heterogeneous integration system," *IEEE Transactions on Very Large Scale Integration (VLSI) Systems*, vol. 32, no. 1, pp. 178–189, 2024.

[10] Q. Wang, T. Zhu, Y. Lin, R. Wang, and R. Huang, "Atsim3d: Towards accurate thermal simulator for heterogeneous 3d-ic systems considering nonlinear leakage and conductivity," in *2024 2nd International Symposium of Electronics Design Automation (ISEDA)*. IEEE, 2024, pp. 618–623.

[11] T. Zhu, Q. Wang, Y. Lin, and R. Wang, "High-resolution full-chip thermal resistance extraction of beol interconnects in 3-d ics considering detailed via connectivity," in *2025 IEEE/ACM International Conference on Computer-Aided Design (ICCAD)*. IEEE, 2025, pp. 1–8.

[12] Z. Yuan, P. Shukla, S. Chetoui, S. Nemtzow, S. Reda, and A. K. Coskun, "Pact: An extensible parallel thermal simulator for emerging integration and cooling technologies," *IEEE Transactions on Computer-Aided Design of Integrated Circuits and Systems*, vol. 41, no. 4, pp. 1048–1061, 2021.

[13] Y. Yang, Z. Gu, C. Zhu, R. P. Dick, and L. Shang, "Isac: Integrated space-and-time-adaptive chip-package thermal analysis," *IEEE Transactions on Computer-Aided Design of Integrated Circuits and Systems*, vol. 26, no. 1, pp. 86–99, 2006.

[14] T. Smy, D. Walkey, and S. Dew, "Transient 3d heat flow analysis for integrated circuit devices using the transmission line matrix method on a quad tree mesh," *Solid-State Electronics*, vol. 45, no. 7, pp. 1137–1148, 2001.

[15] D. Stefaniuk and M. Kachanov, "Voigt-reuss and hashin-shtrikman bounds revisited," *International Journal of Engineering Science*, vol. 191, p. 103903, 2023.

[16] Q. Wang, T. Zhu, Y. Lin, R. Wang, and R. Huang, "Atsim3.5d: A multiscale thermal simulator for 3.5d-ic systems based on nonlinear multigrid method," in *2025 3rd International Symposium of Electronics Design Automation (ISEDA)*. IEEE, 2025.

[17] V. Henson *et al.*, "Multigrid methods nonlinear problems: an overview," *Computational imaging*, vol. 5016, pp. 36–48, 2003.

[18] Q. Wang, X. Li, T. Jia, Y. Lin, R. Wang, and R. Huang, "Atplace2.5d: Analytical thermal-aware chiplet placement framework for large-scale 2.5d-ic," in *2024 IEEE/ACM International Conference on Computer-Aided Design (ICCAD)*. IEEE, 2024, pp. 1–8.

[19] T. Zhu, Q. Wang, Y. Lin, R. Wang, and R. Huang, "More-stress: Model order reduction based efficient numerical algorithm for thermal stress simulation of tsv arrays in 2.5 d/3d ic," in *Design, Automation & Test in Europe Conference & Exhibition (DATE), 2025*. IEEE, 2025.

[20] L. Shen, J. Cui, and G. Li, "Cmos image sensor with 3×3 grid interpolation readout based on spatial similarity," Patent, april, 2025, patent pending in China.

Radio Frequency Integrated Circuits Generated by AI-based Design Automation

Ruoyu Wang[1,2], Meijun Hou[2], Jun Wu[2], Hongtao Xu[1], Ye Lu*[1]

[1] State Key Laboratory of Integrated Chips and Systems, Fudan University, Shanghai, China
[2] IC Prophet Microelectronics, Shanghai, China

* Email: lu_ye@fudan.edu.cn

Abstract—This paper presents an AI-enabled automated design flow of radio frequency integrated circuits (RFICs) for end-to-end synthesis from circuit parameter optimization to automated layout generation. The proposed flow integrates (i) a collaborative engine of multiple optimization algorithms for parameter optimization; and (ii) a reinforcement learning (RL) based placement and routing engine for automated layouts synthesis. (iii) The methodology is validated with two low-noise amplifiers (LNAs) in 40-nm CMOS technology: a 2.4 GHz differential cascode LNA and a 5.5 GHz two-stage differential common-source (CS) LNA. The results highlight the feasibility and efficiency of this automated design flow, offering a promising direction for RFIC design automation.

Keywords—radio frequency integrated circuits (RFIC), design automation, black-box optimization, reinforcement learning (RL), layout synthesis, low-noise amplifier (LNA)

I. INTRODUCTION

Radio frequency (RF) integrated circuits (ICs) have become increasingly critical in modern electronic systems, driven by the rapid growth of wireless communication technologies (5G/6G), the Internet of Things (IoT), and advanced radar systems. With the desire for lower power consumption, higher integration, and enhanced performance, the complexity of RFICs has escalated correspondingly [1].

The design of RFICs is considered to be one of the most challenging areas in IC design due to the frequency dependent parasitic effects and time-consuming simulations, particularly the electromagnetic (EM) simulations. Passive components, such as inductors and transformers, play a crucial role in RF circuits as their performance directly impacts specifications including gain, noise figure, and linearity [2]. Moreover, RFICs design remains heavily reliant on the expertise and intuition of experienced designers, requiring numerous iterative tuning and manual optimizations due to the nonlinear interactions between active and passive circuits. Conventional design flows, illustrated in the top half of Fig. 1, tend to be time-consuming and inefficient. Consequently, exploring efficient and automated methodologies to streamlining the design process while ensuring optimal performance has becoming a key focus of research and industry [3].

Driven by escalating design complexity, electronic design automation (EDA) of RFICs has attracted increasing attention in recent years [4]. Several optimization-based approaches and computer-aided design (CAD) tools have been reported. Bayesian Optimization (BO) efficiently optimizes expensive black-box functions by surrogate model and acquisition function. [5]. J. Zhou et al. proposed an AI-enabled algorithmic flow based on reinforcement learning (RL) [6] for

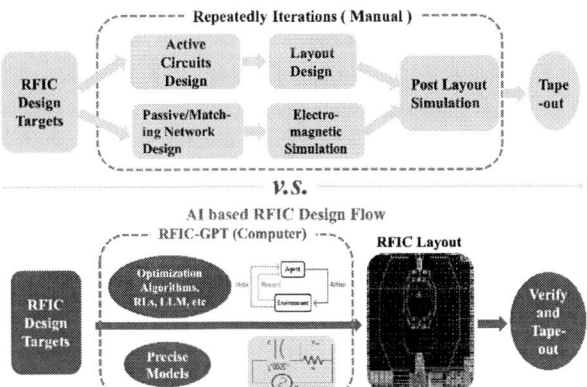

Fig. 1. Conventional RFIC design v.s. AI based RFIC design.

architecture selection, circuit topology, and parameter optimization, including inverse EM synthesis. A stand-alone CMOS LNA synthesis tool [7] enabled fast sizing with embedded foundry models and a validated 0.25-μm 900-MHz prototype, but remained single-block and lacked broader EM coupling treatment. AIDA (Analog IC Design Automation) [8] extended automation to three 2.4 GHz front-end blocks (LNA, mixer, LC-VCO) using NSGA-II multi-objective sizing under fixed topologies.

In this work, we developed an AI-enabled automated design flow of end-to-end synthesis based on RFIC-GPT framework, integrating multiple optimization algorithms and precise device models. As shown in the bottom half of Fig. 1, given predefined circuit specifications and topology, this process enables automated circuit parameter optimization and end-to-end synthesis of a DRC/LVS clean layout, including placement and routing. Compared to traditional manual design flows that require repeated iterations between circuit design, layout, and EM simulation, the proposed approach enables efficient exploration of the extensive design space, which is one of the most significant challenges in design automation[9]. We illustrate the automated design flow with two example low-noise amplifiers (LNAs) in 40-nm CMOS technology: a differential cascode LNA at 2.4 GHz and a two-stage differential CS LNA at 5.5 GHz. The synthesized layouts closely satisfy the circuit specifications without manual iteration, which verifies the feasibility and robustness of the automated flow across diverse circuit topologies.

II. AUTOMATED DESIGN FLOW

The overall framework of the proposed automated RFIC design flow is depicted in Fig. 2. This methodology is organized into three stages: circuit topology selection and specification definition, parameter optimization, and layout

979-8-3315-3918-4/25 $31.00 © 2025 IEEE

synthesis. Each stage is tightly integrated within the flow, employing systematic approaches to enable efficient exploration and high-performance RFIC design.

A. Circuit topology selection and Specification definition

The proposed automated flow begins with the selection of an appropriate circuit topology and the definition of key performance specifications. The target specifications and circuit topology should meet the functional requirements.

The choice of topology is influenced by factors such as technology node, available passive and active devices, and integration constraints. Once the architecture is determined, the next step is to define the target specifications, which serve as the optimization objectives and constraints. Accurate and proper specification definition is crucial to ensure that the synthesized circuits can fulfill both performance and manufacturability criteria.

For automation, the specifications are formalized into quantifiable targets and boundaries, which systematically guide the parameter optimization process and enables thorough exploration of the solution space, ensuring that the circuit satisfies all required standards.

B. Parameter optimization

The second stage of the automated flow is circuit parameter optimization based on the collaboration of multiple optimization algorithms, which includes various black-box optimization approaches.

A black-box problem refers to an optimization scenario where the internal structure of the objective function is unknown and only its output for given inputs can be observed. Black-box optimization algorithms are designed to efficiently optimize such functions, especially when evaluations are costly, by adaptively selecting evaluation points. These algorithms are widely used for tuning system parameters, such as hyperparameters in machine learning or settings in engineering applications [10].

RFIC design inherently involves strongly coupled, nonlinear, multi-objective trade-offs (e.g., NF, gain, matching, linearity, power, and area) over a high-dimensional design space. These characteristics make RFIC design a typical black-box optimization problem, well-suited for advanced algorithms such as BO [11], genetic algorithms (GA) [12], particle swarm optimization (PSO), simulated annealing [13], and RL. A batch multi-objective Bayesian optimization (MOBO) approach, which ensembles multiple acquisition functions to iteratively approximate the Pareto front, is adopted to obtain well-balanced candidate solutions [14]. This method is particularly effective, as it enables efficient exploration of the design space while progressively refining a diverse set of near-optimal solutions.

The implementation begins with the definition of the design space, which consists of all tunable circuit parameters, such as transistor dimensions, bias voltages, and passive component values. Each parameter is assigned a feasible range based on technology constraints and initial design considerations. The dimensionality of the design space is a critical factor; as the number of parameters increases, the optimization problem becomes significantly more challenging due to the curse of dimensionality. Therefore, careful selection and reduction of variables may be necessary to ensure tractable optimization.

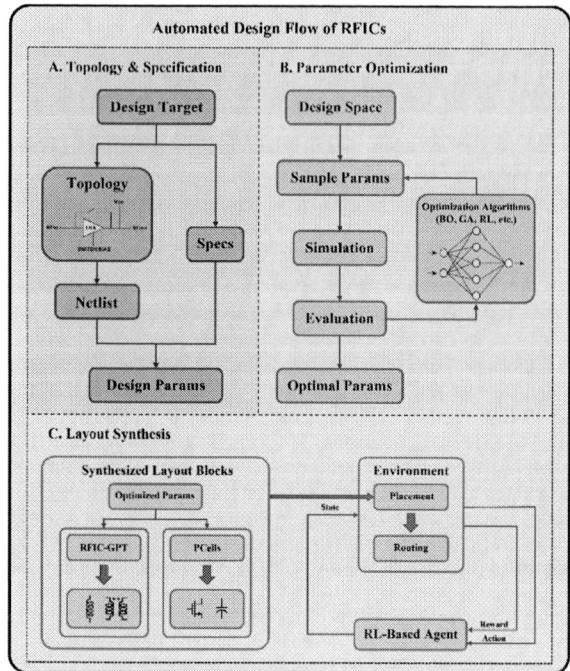

Fig. 2. Automated Design Flow of RFICs.

Optimization objectives are derived from the predefined specifications. The optimization algorithms interact with a circuit simulator, proposing new parameter sets and evaluating their performance. Simulation results are fed back to guide subsequent searches. This iterative process continues until a set of optimal solutions is obtained, representing the best trade-offs among the target circuit performance specifications.

By integrating simulation feedback into the optimization loop, the automated flow adapts to real circuit behavior, effectively navigating complex and high-dimensional design spaces to identify optimal parameter combinations.

C. Layout synthesis

The final stage of the automated design flow of RFICs focuses on layout synthesis. Once the circuit parameters are optimized, the corresponding schematic is automatically translated into a physical layout. Passive components are synthesized using the RFIC-GPT tool [15]. Active devices are instantiated using foundry parameterized cells (PCells) in conjunction to the optimization results.

Placement and routing are subsequently performed within a RL-based Actor-Critic proximal policy optimization (PPO) framework [16], where the state is defined by the position and orientation of each device, the action correspond to the movement direction and distance for the next placement step, the reward function is designed to optimize key layout metrics such as area utilization and density. Once the placement is finished, routing is performed by algorithm that efficiently determines the shortest path for signal wires while avoiding layout rule violations. The detailed algorithms of place and route will be presented in the future work.

Overall, the automated synthesis flow of RFIC enabled by multiple AI-enabled algorithms markedly shortens the RFIC design cycle relative to conventional manual efforts.

III. EXPERIMENTS

The design of LNA involves navigating trade-offs among several design goals (NF, gain, power, linearity, impedance matching, etc.). Typical design strategies tend to minimize NF, while others prioritize power reduction or linearity enhancement [17].

To demonstrate the viability and effectiveness of the proposed automated design flow, it is applied to two different LNAs in 40-nm CMOS technology: a 2.4 GHz differential cascode LNA and a 5.5 GHz two-stage differential CS LNA.

A. A Differential Cascode LNA at 2.4 GHz

The schematic of the proposed 2.4 GHz differential cascode LNA is presented in Fig. 2, where two transformers are used for input and output matching networks. The cross-couple capacitor structure is introduced to neutralize C_{gd}, enhance gate-drain isolation and reduce nonlinear distortion.

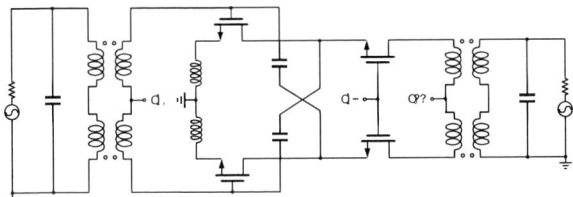

Fig. 3. Schematic of the proposed differential cascode LNA with cross-coupled capacitors.

Based on the given circuit topology, the target circuit specifications at 2.4 GHz are as follows: NF < 1.8 dB, S_{21} > 20 dB, S_{11} < -10 dB, S_{22} < -10 dB, S_{12} < -25dB, P_{dc} < 30 mW, and IIP3 > -6 dBm, and the supply voltage is 1.1 V.

This example features a relatively large design space with 18 design variables and 7 optimization objectives. By applying the proposed automated design flow, circuit parameter optimization and end-to-end layout (DRC/LVS clean) synthesis are accomplished within a few hours. Fig. 3 shows the synthesized layout of the proposed differential cascode LNA, which occupies a die area of 0.38×0.94 mm^2.

Within the proposed design flow, the schematic and the layou are automatically synthesized. Subsequent verification involves (i) pre-simulatiom, (ii) EM simulation of passive elements, (iii) parasitic extraction for the active devices, and (iv) post- simulation. Fig. 4 illustrates the post-simulated NF and S-parameters as well as the pre-simulated results, all specifications are satisfied. The post-simulated S_{21} of the proposed LNA shows a 3-dB bandwith of 2 GHz to 2.7 GHz. Table I compares the post-simulation performance with the pre-simulation performance of the presented LNA, revealing slight differences.

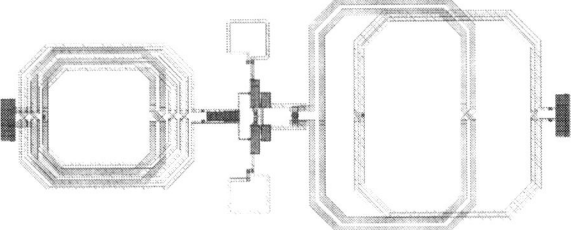

Fig. 4. Synthesized layout of the proposed differential cascode LNA with cross-coupled capacitors.

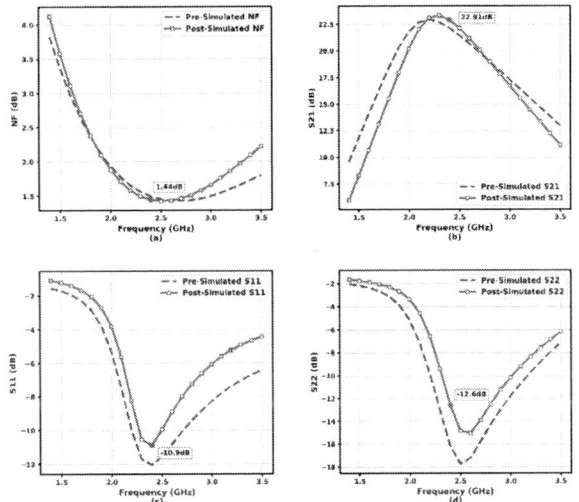

Fig. 5. Pre-simulated and post-simulated NF and S-parameters.

TABLE I. PERFORMANCE OF THE DIFFERENTIAL CASCODE LNA

Reference	Spec	Pre-Simulation	Post-Simulation
Technology	40-nm CMOS	40-nm CMOS	40-nm CMOS
Frequency (GHz)	2.4	2.4	2.4
V_{dd} (V)	1.1	1.1	1.1
NF (dB)	< 1.8	1.49	1.44
Gain (dB)	> 20	22.14	22.91
S_{11} (dB)	< -10	-12.06	-10.89
S_{22} (dB)	< -10	-16.19	-12.61
S_{12} (dB)	< -25	-34.25	-34.89
P_{dc} (mW)	< 30	28.27	28.47
IIP3 (dBm)	> -6	-2.317	-0.02

B. A Two-Stage Differential CS LNA at 5.5 GHz

The schematic of the proposed 5.5 GHz two-stage differential CS LNA is presented in Fig. 2, in which three transformers implement the input, interstage and output matching networks while the cross-couple capacitor structure is applied to each CS stage.

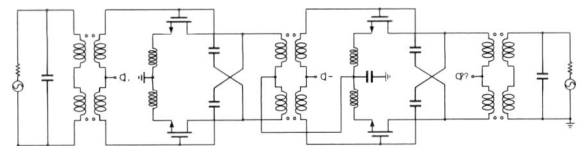

Fig. 6. Schematic of the proposed two-stage differential CS LNA with cross-coupled capacitors.

Given the specified circuit topology, the target circuit specifications at 5.5 GHz are as follows: NF < 2 dB, S_{21} > 27 dB, S_{11} < -7 dB, S_{22} < -7 dB, S_{12} < -50dB, P_{dc} < 20 mW, and IIP3 > -8 dBm, and the supply voltage is 1.2 V.

This architecture compared with the first topology introduces approximately ten additional design variables (26 in total), substantially expanding the design space and increasing optimization complexity. Fig. 6 shows the synthesized layout of the proposed differential two-stage CS LNA, which occupies a die area of 0.27×0.8 mm².

As shown in Fig. 7, both post-simulated and pre-simulated NF and S-parameters meet the target specifications. The post-simulated S_{11} of the proposed LNA is below -9 dB from 5 GHz to 6 GHz. The gain flatness is only 0.5 dB within a $\pm 10\%$ relative bandwidth. Table II shows close agreement between the post-simulation performance and the pre-simulation performance of the presented LNA.

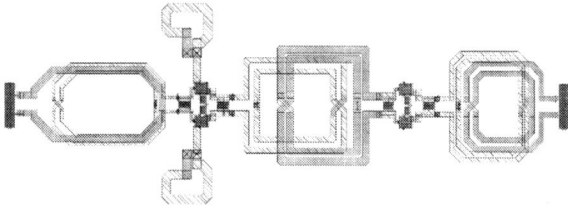

Fig. 7. Synthesized layout of the proposed two-stage differential CS LNA with cross-coupled capacitors.

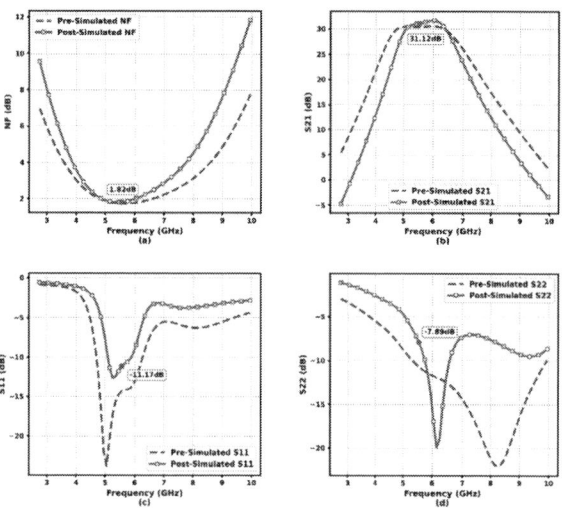

Fig. 8. Pre-simulated and post-simulated NF and S-parameters.

TABLE II. PERFORMANCE OF THE TWO-STAGE DIFFERENTIAL CS LNA

Reference	Spec	Pre-Simulation	Post-Simulation
Technology	40-nm CMOS	40-nm CMOS	40-nm CMOS
Frequency (GHz)	5.5	5.5	5.5
V_{dd} (V)	1.2	1.2	1.2
NF (dB)	< 2	1.73	1.81
Gain (dB)	> 27	30.41	31.12
S_{11} (dB)	< -7	-14.77	-11.75
S_{22} (dB)	< -7	-10.67	-7.89
S_{12} (dB)	< -50	-61.53	-57.44
P_{dc} (mW)	< 20	20.92	19.63
IIP3 (dBm)	> -8	-6.68	-8.01

IV. CONCLUSION

In this paper, an AI-enabled automated design flow of RFICs based on optimization algorithms is presented. This work achieves precise and efficient circuit parameter optimization as well as end-to-end layout synthesis. The effectiveness, robustness, and potential of the proposed methodology are validated through two example designs, highlighting its promise for the future of RFIC design automation.

REFERENCES

[1] L. E. Larson, "Integrated circuit technology options for RFICs-present status and future directions," *IEEE Journal of Solid-State Circuits*, vol. 33, no. 3, pp. 387–399, 2002.

[2] R. Gupta, B. M. Ballweber, and D. J. Allstot, "Design and optimization of CMOS RF power amplifiers," *IEEE Journal of Solid-State Circuits*, vol. 36, no. 2, pp. 166–175, 2002.

[3] S. E. Sorkhabi and L. Zhang, "Automated topology synthesis of analog and RF integrated circuits: A survey," *Integration*, vol. 56, pp. 128–138, 2017.

[4] R. A. Rutenbar, G. G. Gielen, and J. Roychowdhury, "Hierarchical modeling, optimization, and synthesis for system-level analog and RF designs," *Proceedings of the IEEE*, vol. 95, no. 3, pp. 640–669, 2007.

[5] S. Maji, A. F. Budak, S. Poddar, and D. Z. Pan, "Toward end-to-end analog design automation with ML and data-driven approaches," in *2024 29th Asia and South Pacific Design Automation Conference (ASP-DAC)*, 2024: IEEE, pp. 657–664.

[6] J. Zhou, E. A. Karahan, S. Ghozzy, Z. Liu, H. Jalili, and K. Sengupta, "25.3 AI-Enabled Design Space Discovery and End-to-End Synthesis for RFICs with Reinforcement Learning and Inverse Methods Demonstrating mm-Wave/sub-THz PAs Between 30 and 120GHz," in *2025 IEEE International Solid-State Circuits Conference (ISSCC)*, 2025, vol. 68: IEEE, pp. 1–3.

[7] G. Tulunay and S. Balkir, "A synthesis tool for CMOS RF low-noise amplifiers," *IEEE Transactions on Computer-Aided Design of Integrated Circuits and Systems*, vol. 27, no. 5, pp. 977–982, 2008.

[8] R. Povoa, I. Bastos, N. Lourenço, and N. Horta, "Automatic synthesis of RF front-end blocks using multi-objective evolutionary techniques," *Integration*, vol. 52, pp. 243–252, 2016.

[9] Y. Li, Y. Wang, Y. Li, R. Zhou, and Z. Lin, "An artificial neural network assisted optimization system for analog design space exploration," *IEEE Transactions on Computer-Aided Design of Integrated Circuits and Systems*, vol. 39, no. 10, pp. 2640–2653, 2019.

[10] D. Golovin, B. Solnik, S. Moitra, G. Kochanski, J. Karro, and D. Sculley, "Google vizier: A service for black-box optimization," in *Proceedings of the 23rd ACM SIGKDD international conference on knowledge discovery and data mining*, 2017, pp. 1487–1495.

[11] P. I. Frazier, "A tutorial on Bayesian optimization," *arXiv preprint arXiv:1807.02811*, 2018.

[12] J. B. Grimbleby, "Automatic analogue circuit synthesis using genetic algorithms," *IEE Proceedings-Circuits, Devices and Systems*, vol. 147, no. 6, pp. 319–323, 2000.

[13] D. Joshi, S. Dash, S. Reddy, R. Manigilla, and G. Trivedi, "Multi-objective hybrid particle swarm optimization and its application to analog and RF circuit optimization," *Circuits, Systems, and Signal Processing*, vol. 42, no. 8, pp. 4443–4469, 2023.

[14] W. Lyu, F. Yang, C. Yan, D. Zhou, and X. Zeng, "Batch Bayesian optimization via multi-objective acquisition ensemble for automated analog circuit design," in *International conference on machine learning*, 2018: PMLR, pp. 3306–3314.

[15] RFIC GPT tool: https://rfic-gpt.com/

[16] J. Schulman, F. Wolski, P. Dhariwal, A. Radford, and O. Klimov, "Proximal policy optimization algorithms," *arXiv preprint arXiv:1707.06347*, 2017.

[17] J.-S. Goo, H.-T. Ahn, D. J. Ladwig, Z. Yu, T. H. Lee, and R. W. Dutton, "A noise optimization technique for integrated low-noise amplifiers," *IEEE Journal of Solid-State Circuits*, vol. 37, no. 8, pp. 994–1002, 2002.

Advancing Sparse Matrix Solvers via Exploring More Parallelism and Random Sketching

Wenjian Yu, Jiawen Cheng, and Baiyu Chen

Department of Computer Science and Technology, BNRist, Tsinghua University, Beijing, China

Abstract—Sparse matrix solver plays a vital role in circuit simulation, as it is the core of mathematics for accurately predicting the dynamical behaviors of analog circuits. In this invited paper, we present two novel techniques to develop more efficient sparse matrix solvers for circuit simulation. The first one is the parallel LU factorization with a novel task scheduling approach. It originates from the nested dissection approach for matrix reordering. Through partitioning the separator tree derived from nested dissection ordering, a less-synchronization scheme for parallelizing LU factorization is derived. This enables significant improvements over parallel direct solvers PARDISO and CKTSO without sacrificing robustness. The other technique is the randomized generalized minimal residual (GMRES) algorithm. With a highly efficient sketched least-squares solver for the Gram-Schmidt process, the randomized Arnoldi process for orthogonalizing Krylov subspace basis is proposed. Combining the skills of estimating residual error during the iterative process, we present the practical randomized GMRES algorithm which has theoretically-supported stability and higher efficiency than the GMRES algorithm on various circuit simulation problems.

Index Terms—Circuit Simulation, Sparse Matrix Solver, Parallel LU Factorization, Random Sketching, GMRES Algorithm.

I. INTRODUCTION

Circuit simulation tools such as SPICE play a crucial role in modern electronic design automation by analyzing the behavior of integrated circuits. The major tasks of circuit simulation include DC analysis, AC analysis, transient analysis, RF simulation, etc. The core problem in them is the solution of linear equations $Ax = b$. Due to the nature of circuit topology, the matrix A is usually very sparse. Solving such a large-scale sparse linear system dominates the total runtime of circuit simulation. So, how to accelerate it is of concern.

Methods for solving a sparse linear equation system can be divided into two categories: direct solver and iterative solver. The direct solver is based on matrix factorization such as LU factorization, and usually accurate and robust. However, it could be very inefficient or infeasible for very large problems, due to the excessive fill-ins generated during the factorization and memory bottleneck. On the other hand, the iterative solver, such as the preconditioned generalized minimal residual (GMRES) algorithm [1], does not affect the sparsity of coefficient matrix, and is thus more scalable to large-scale circuit simulation problems [2], [3]. However, it is less robust and not efficient for repeatedly solving the fixed-structure equations in small and medium size.

In this paper, we present the recent advancements on developing efficient sparse matrix solvers for circuit simulation. One

is a parallel direct solver named **SubtreeLU** run on shared-memory computer, which is suitable for the transient analysis solving a number of fixed-structure equations [4]. The other revolutionizes the classical GMRES algorithm, reducing the cost of Arnoldi process via random sketching based techniques and benefiting other circuit simulation problems [5].

II. PRELIMINARIES

A. Sparse LU Factorization for Transient Simulation

The procedure of solving a linear system with sparse LU factorization is composed of three phases: symbolic analysis, matrix factorization, and triangular solving (substitutions). Fig. 1 shows a typical flow of transient simulation incorporating sparse LU factorization, where the symbolic analysis is conducted only once if the sparsity pattern of A is not changed. The symbolic analysis involves the static pivoting for numerical stability and the matrix reordering to minimize fill-ins during LU factorization, etc. LU factorization is conducted on A to obtain lower (L) and upper (U) triangular matrices. Finally, the substitutions are conducted through solving the resulting triangular systems $Ly = b$ and $Ux = y$ in turn. These steps repeat within each Newton-Raphson (NR) iteration until convergence is achieved. The simulation then proceeds to the next time point in the whole time integration process.

Compared to the substitutions, LU factorization consumes much more time. Existing direct solvers for circuit simulation such as KLU [6], NICSLU [7] and CKTSO [8] adopt the Gilbert-Peierls (G-P) algorithm for LU factorization [9] as the computational kernel. They all support two modes of LU factorization: **factorization with pivoting** and **refactorization without pivoting**. The former dynamically selects large matrix entries as pivots to ensure numerical stability. This process

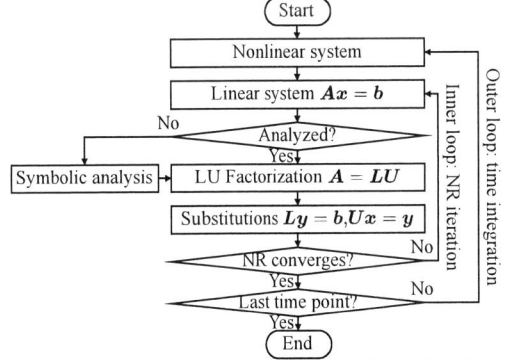

Fig. 1: A typical flow of circuit transient simulation.

requires element exchanges, which directly affect the sparsity pattern of the LU factors. Consequently, the symbolic factorization determining the sparsity pattern of the LU factors must be tightly interleaved with numerical computation. This significantly increases the computational cost of factorization. In contrast, refactorization without pivoting reuses the pivoting order and the sparsity pattern of LU factors obtained from a previous factorization. This allows symbolic factorization to be skipped entirely. However, refactorization may lead to numerical instability if the matrix values vary significantly.

The G-P algorithm is also known as sparse up-looking LU factorization algorithm (Alg. 1). It proceeds row by row. For each row i, a working row vector x is first initialized as $A_{i,:}$. Symbolic factorization identifies the sparsity pattern of the i-th row of the LU factors through a depth-first search on the previously processed rows [9]. Numerical update then modifies the working vector using the previously computed LU factors. Partial pivoting ensures numerical stability by selecting the largest entry within the row as pivot if the diagonal is too small. Finally, the updated row is written into the LU factors. For the refactorization, line 3 and lines 7-10 of Algorithm 1 are no longer needed.

Algorithm 1 Sparse up-looking LU factorization

Input: Sparse matrix $A \in \mathbb{R}^{n \times n}$, pivoting tolerance θ
Output: LU factors L and U ▷ $l_{i,j}$ and $u_{i,j}$ are their elements
1: **for** $i = 1, \ldots, n$ **do**
2:    $x \leftarrow a_{i,:}$ ▷ *the i-th row of A*
3:    Determine the sparsity pattern of the i-th row of L and U
4:    **for** each k in non-zero indices of L's i-th row **do**
5:      **for** each j in non-zero indices of U's k-th row **do**
6:        $x_j \leftarrow x_j - x_k u_{k,j}$ ▷ *update the j-th element of x*
7:    $m \leftarrow \arg\max_{i+1 \leq k \leq n} |x_k|$ ▷ *partial pivoting*
8:    **if** $|x_i| < \theta |x_m|$ **then**
9:      Swap x_i and x_m, and record the permutation
10:    Prune the previous rows   ▷ *quicken symbolic factorization*
11:    $l_{i,1:i} \leftarrow x_{1:i}$, $u_{i,i:n} = x_{i:n}/x_i$

B. Generalized Minimal Residual (GMRES) Algorithm

The GMRES algorithm aims at minimizing the residual (initially $r_0 = b - Ax_0$) to obtain an approximate solution within the Krylov subspace $\mathbb{K}_m = span\{r_0, \cdots, A^{m-1}r_0\}$. Let V_m denote the matrix formed by the basis vectors of $\mathbb{K}_m$. The approximate solution can be written as $x_m = x_0 + V_m y_m$, where y_m includes the approximation coefficients corresponding to V_m. The residual $r_m = b - Ax_m$ is minimized via solving the least-squares problem: $AV_m y_m \approx r_0$. The Arnoldi process with modified Gram-Schmidt (MGS) process makes V_m including the orthonormal basis vectors of $\mathbb{K}_m$, ensuring $AV_m = V_{m+1}\overline{H}_m$, where upper Hessenberg matrix $\overline{H}_m$ includes the projection coefficients in orthonormalization.

Now, the least-squares problem can be easily solved as

$$V_{m+1}\overline{H}_m y_m \approx r_0 \Longleftrightarrow \overline{H}_m y_m \approx V_{m+1}^T r_0 = \|r_0\|e_1 , \quad (1)$$

where e_1 denotes the first column of the identity matrix. The converted least-squares problem can be efficiently solved by applying Givens transformations. Furthermore, both the

---

Algorithm 2 The restarted GMRES algorithm

Input: $A \in \mathbb{R}^{n \times n}$, $b \in \mathbb{R}^n$, $x_0 \in \mathbb{R}^n$, restart parameter $m_{\max}$, tolerance of relative residual error tol.
Output: solution x_0
1: **while** $\gamma > tol \cdot \|b\|$ **do**
2:    $r \leftarrow b - Ax_0$, $g \leftarrow \|r\|$, $v_1 \leftarrow r/g$
3:    **for** $j = 1, 2, \cdots, m_{\max}$ **do**
4:      $w_j \leftarrow Av_j$, update $g \in \mathbb{R}^{j+1}$ by appending a zero
5:      Run MGS process to obtain $h_{1:j,j}$ and update w_j
6:      $h_{j+1,j} \leftarrow \|w_j\|$ ▷ *element of matrix $\overline{H}_m$*
7:      $v_{j+1} \leftarrow w_j/h_{j+1,j}$
8:      Apply the Givens transformations $G_1, \cdots, G_{j-1}$ to the column vector $h_{1:j,j}$
9:      Compute the Givens matrix G_j to make $h_{j+1,j} = 0$
10:      Apply the Givens transformation G_j to vector g
11:      $\gamma \leftarrow |g_{j+1}|$ ▷ *last element of vector g*
12:      **if** $\gamma \leq tol \cdot \|b\|$ **then**
13:        **Break**
14:    $m \leftarrow j$
15:    Solve y_m with the transformed matrix $\overline{H}_m$ and g
16:    $x_0 \leftarrow x_0 + V_m y_m$, with $V_m = [v_1, \cdots, v_m]$

Arnoldi process and the solution of (1) can be run in an incremental manner as the dimensionality of the Krylov subspace increases. Combining the above derivation and the restarting strategy, one obtains the restarted GMRES algorithm as Alg. 2, where the residual error of the approximate solution is

$$\|r_m\| = \|b - Ax_m\| = \|r_0 - V_{m+1}\overline{H}_m y_m\|$$
$$= \|V_{m+1}(\|r_0\|e_1 - \overline{H}_m y_m)\| \quad (2)$$
$$= \|\|r_0\|e_1 - \overline{H}_m y_m\| = \gamma ,$$

where γ is the absolute value of the $(m+1)$-th element of the right-hand side (originally $\|r_0\|e_1$) of the least-squares equation after transforming $\overline{H}_m$ to an upper triangular matrix with the Givens transformations. With (2), one can easily determine whether the iteration in GMRES algorithm should be terminated (see line 12 of Alg. 2). In practice, the preconditioning technique is also applied to fasten the convergence. With the preconditioner generated in advance, one just applies the preconditioner in line 4 of Alg. 2.

III. PARALLEL SPARSE LU FACTORIZATION BASED ON NESTED DISSECTION

The up-looking LU factorization (Alg. 1) exhibits potential for parallelization, where the computation of a row is treated

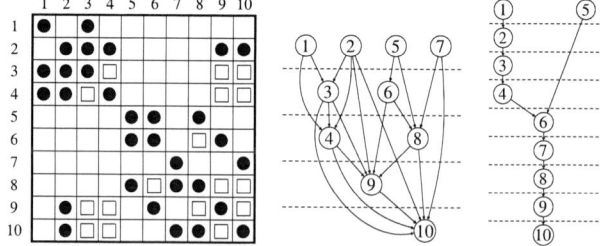

(a) LU's sparsity pattern   (b) Dependency DAG (c) The ETree
Fig. 2: An example result of sparse up-looking LU factorization and the corresponding dependency DAG and the ETree. In (a), solid circles represent the nonzero elements in A, and hollow squares represent the fill-ins.

as a basic task (referred to by row number) [7], [8]. Task i depends on task $k < i$ if and only if $l_{ik} \neq 0$. And, the parallelism exists among the tasks with no dependencies. A directed acyclic graph (DAG) can be constructed from the sparsity pattern of L to represent the dependencies. In this dependency DAG, a directed edge from node k to node i implies that task i depends on task k (see Fig. 2). It can be used to schedule the parallel refactorization without pivoting. However, the sparsity pattern of L cannot be determined prior to factorization with pivoting since the symbolic factorization is interleaved with numerical computation. To address this issue, the elimination tree (ETree) of AA^T is introduced as an approximation by considering all possible pivoting choices and fill-ins [10]. This **severely overestimates true dependencies**, leading to conservative scheduling for parallel factorization. With a dependency graph (either the DAG or ETree), levelization-based scheduling can be applied, where the nodes (i.e., rows) in the same level can be processed in parallel. Both NICSLU and CKTSO implement a dual-mode scheduling scheme based on this levelization [7], [8]. For the top levels with many independent nodes, they are processed level by level with parallel threads and a global barrier is set after all tasks in a level are finished. This is the so-called **cluster mode**. For the bottom levels with less nodes, **pipeline mode** is employed by applying fine-grained dynamic scheduling. The use of global barriers in cluster

mode introduces **synchronization overhead and idle time**, particularly when workload imbalance exists across threads.

Nested dissection is a graph partitioning technique used to improve the performance of sparse matrix factorization by minimizing fill-ins [11]. It recursively partitions the graph associated with the matrix into two balanced subgraphs by identifying the vertex separator. This process yields a hierarchical decomposition of the matrix into independent subdomains and their connecting separators, until the limits of size and recursion depth are met. The result of nested dissection can be represented by a separator tree, where each non-leaf node corresponds to a separator at a particular level of recursion, and its children represent the subdomains separated by it. The leaves represent the smallest subdomains. Fig. 3 shows a matrix reordered using nested dissection and its associated separator tree. Note that **the separator tree also captures the inter-dependencies between matrix blocks, valuable for guiding parallel task scheduling in LU factorization**.

In practice, we apply the nested dissection on the undirected graph corresponding to $A + A^T$, where METIS [12] is used to compute the vertex separator. We set the minimum recursion depth to ensure there are at least P subdomains (P is the number of threads). So, each subdomain is assigned to a **private task queue** for a thread. This largely reduces the synchronization cost. The remaining tasks are processed with the pipeline mode as [8]. Compared to the cluster-pipeline scheduling adopted by CKTSO, our approach **avoids severe dependency overestimation** by bypassing the ETree while partial pivoting is still enabled. It also **eliminates the multiple global barriers** in cluster mode by processing the subdomains independently. And, by **supporting supernode detection in private mode**, our approach further improves the efficiency.

With this approach, we developed a parallel sparse matrix solver in C++, named **SubtreeLU**. It is compared with CKTSO [8], Intel MKL PARDISO [13] and MUMPS. CKTSO is a state-of-the-art parallel sparse direct solver for circuit simulations, while PARDISO and MUMPS are general-

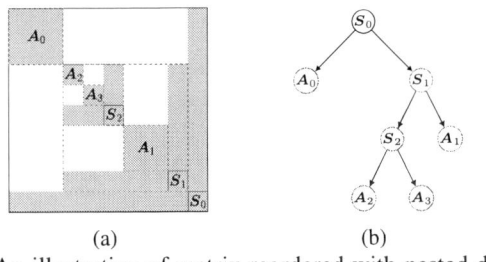

Fig. 3: An illustration of matrix reordered with nested dissection (a) and its corresponding separator tree (b) [4].

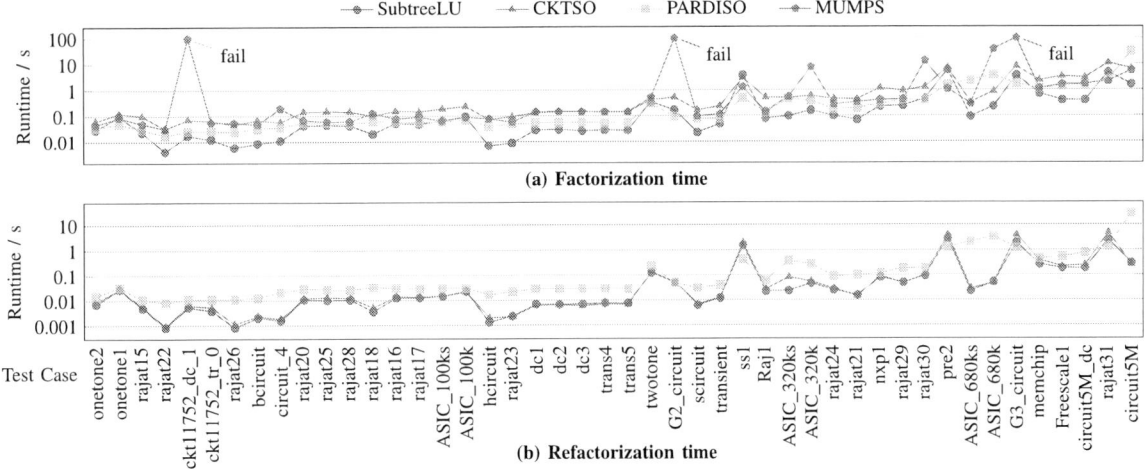

Fig. 4: Comparison on the 100-run average time for factorization and refactorization (16 threads). "fail" means wrong result.

purpose solvers exhibiting superior performance. The pivoting tolerance is set to 0.001 for SubtreeLU, CKTSO and MUMPS. The test cases are the 46 circuit matrices with the size larger than 30,000 (up to 5,558,326), obtained from SuiteSparse Matrix Collection. The results run on Intel Xeon Gold 6230R 26-core CPU are shown in Fig. 4. From it we see that our SubtreeLU outperforms the others on most cases. In terms of geometric mean of speedup ratio, the SubtreeLU achieves a **3.99**×, **1.85**×, and **3.30**× average speedup over CKTSO, PARDISO, and MUMPS respectively for factorization. The maximum speedups over CKTSO and PARDISO are **11**× on case rajat23 and **25**× on case ASIC_680ks, respectively. For the refactorization, SubtreeLU also exhibits the fastest refactorization across most matrices. It achieves the average speedup of **1.21**× over CKTSO and **3.68**× over PARDISO, with the maximum speedups **3.51**× on case ASIC_320ks and **94**× on case ASIC_680ks, respectively.

IV. A PRACTICAL RANDOMOIZED GMRES ALGORITHM

A drawback of the GMRES is the $O(nj)$ time complexity of the orthogonalization process in each iteration, which can be very expensive if n is large and a large number of iterations occur. A randomized Gram-Schmidt process was recently proposed to reduce the cost of orthogonalization process [14] and naturally applies to the Arnoldi process in GMRES algorithm. It means we replace line 4-7 of Alg. 2 with the following:

4:	$w_j \leftarrow Av_j$, $p_j \leftarrow \Theta w_j$	▷ *random sketching*
5a:	Solve $z_j = \arg\min_z \|S_j z - p_j\|$, with $S_j = [s_1, \cdots, s_j]$	
5b:	$v_{j+1} \leftarrow w_j - V_j z_j$, with $V_j = [v_1, \cdots, v_j]$	
5c:	$s_{j+1} \leftarrow \Theta v_{j+1}$	▷ *random sketching*
6:	$h_{1:j,j} \leftarrow z_j$, $h_{j+1,j} \leftarrow \|s_{j+1}\|$	
7:	$v_{j+1} \leftarrow v_{j+1}/\|s_{j+1}\|$, $s_{j+1} \leftarrow s_{j+1}/\|s_{j+1}\|$ ▷ *normalize* s_{j+1}	

where Θ is a $k \times n$ random matrix with $k \ll n$. This is inspired by that the MGS orthogonalization can be approximately computed by solving the sketched least-squares problem:

$$(\Theta V_j)z \approx \Theta w_j, \quad \text{i.e.} \quad \min_z \|(\Theta V_j)z - \Theta w_j\|. \quad (3)$$

In order to make the revised GMRES algorithm efficient, we need overcome two drawbacks: 1) the random sketching should run fast while keeping good accuracy; 2) Eq. (2) does not hold as $\{v_j\}$ are not orthogonormalized, which means terminating the GMRES iteration for a given error tolerance could be expensive. We address the first drawback by proposing a random sketching approach which is derived from a $\xi \times n$ Rademacher matrix ($\xi \ll k$, e.g. $\xi = 4$) and does not explicitly generate the $k \times n$ matrix. The setup of this sketching and applying it to get Θx are both of $O(n)$ time complexity [5]. Then, we realize that the problem of Eq. (2) can be solved by enforcing the normalization on $\{v_j\}$ since they are approximately orthogonal vectors with good sketching applied. So, we exchange the roles of s_j and v_j in terms of normalization while keeping $s_j = \Theta v_j$. Now, the revised GMRES has an approximate residual error indicator based on (2). To ensure the returned solution fulfills the user-specified tolerance on residual error, we accurately compute and check it in the outer loop of GMRES. So far, we obtain a

TABLE I: The computational results of the GMRES and randomized GMRES algorithms for three industrial cases.

Case	n	nnz	GMRES		RGMRES [14]		PRGMRES			
			Iter	T (s)	Iter	T (s)	Iter	T (s)	Sp_1	Sp_2
industry1	1.4E6	7.6E6	117	21.1	120	16.6	118	15.5	1.36	1.07
industry2	9.0E5	5.0E6	161	23.6	170	17.1	162	15.3	1.54	1.12
industry3	1.4E6	7.6E6	115	24.5	120	19.1	116	18.5	1.33	1.04
Average	-	-	-	-	-	-	-	-	**1.41**	**1.08**

practical randomized GMRES algorithm named **PRGMRES**.

We have implemented the GMRES algorithm and the randomized GMRES algorithms in C++, based on Intel MKL for optimizing sparse matrix-vector multiplication and other matrix/vector operations. Their performance is compared with serial computing on three industrial cases. The tolerance for relative residual is set to 1E-7, and the ILU(1) preconditioner is applied which enables fast convergence of GMRES. The information of the matrices and computational results are listed in Table I. From it we can see, our PRGMRES achieves an average speedup of **1.41**× to GMRES. More experimental results can be found in [5], which shows on matrices from power grid analysis and circuit simulation PGRMRES runs **1.5**× faster than GMRES on average, with comparable memory cost. All experiments also validate the robustness of PRGMRES, whose convergence behavior is similar to GMRES. And, in the future we will extends its application and explore its parallelization.

V. CONCLUSION

We have presented two advancements for fast sparse matrix solver: the parallel **SubtreeLU** for LU factorization and a practical randomized GMRES (PRGMRES) algorithm. In the future, the parallelization of the PRGMRES can be explored.

REFERENCES

[1] Y. Saad and M. Schultz, "GMRES: A generalized minimal residual algorithm for solving nonsymmetric linear systems," *SIAM J. Sci. Stat. Comp.*, vol. 7, no. 3, pp. 856–869, 1986.

[2] S. Zeng, W. Yu, X. Hong, and et al., "Efficient power network analysis with modeling of inductive effects," *IEICE Trans. Fundamentals*, 2010.

[3] X. Zhao, L. Han, and Z. Feng, "A performance-guided graph sparsification approach to scalable and robust SPICE-accurate integrated circuit simulations," *IEEE Trans. CAD*, vol. 34, no. 10, pp. 1639–1651, 2015.

[4] J. Cheng, Y. Zhang, and W. Yu, "SubtreeLU: High-performance parallel sparse LU factorization for circuit simulation," in *Proc. ICCAD*, 2025.

[5] B. Chen, J. Cheng, and W. Yu, "A practical randomized GMRES algorithm for solving linear equation system in circuit simulation," in *Proc. ASP-DAC*, 2025, pp. 183–189.

[6] T. Davis and E. Palamadai, "Algorithm 907: KLU, a direct sparse solver for circuit simulation problems," *ACM Trans. Math. Softw.*, 2010.

[7] X. Chen, Y. Wang, and H. Yang, *Parallel Sparse Direct Solver for Integrated Circuit Simulation*. Springer, 2017.

[8] X. Chen, "CKTSO: High-performance parallel sparse linear solver for general circuit simulations," *IEEE Trans. CAD*, 2024.

[9] J. Gilbert and T. Peierls, "Sparse partial pivoting in time proportional to arithmetic operations," *SIAM J. Sci. Stat. Comp.*, 1988.

[10] J. W. Liu, "The role of elimination trees in sparse factorization," *SIAM J. Matrix Ana. Appl.*, vol. 11, pp. 134–172, 1990.

[11] M. Khaira, G. Miller, and T. Sheffler, *Nested Dissection: A Survey and Comparision of Various Nested Dissection Algorithms*, 1992.

[12] G. Karypis and V. Kumar, "A fast and high quality multilevel scheme for partitioning irregular graphs," *SIAM J. Sci. Comp.*, pp. 359–392, 1998.

[13] O. Schenk and K. Gärtner, "Solving unsymmetric sparse systems of linear equations with pardiso," *Future Generation Computer Systems*, vol. 20, no. 3, pp. 475–487, 2004.

[14] O. Balabanov and L. Grigori, "Randomized Gram–Schmidt process with application to GMRES," *SIAM J. Sci. Comp.*, p. A1450–A1474, 2022.

Snow Ablation Optimizer Accelerator Based on High Level Synthesis

Maoshuo He[*1], Renjing Hou[*2], Zirui Li[2], Kang Zhao[†2],

[1]Xidian University   [2]Beijing University of Posts and Telecommunications

maoshuohe@stu.xidian.edu.cn, {hourj, lzr_official, zhaokang}@bupt.edu.cn

Abstract—How to reduce the execution time in the snow ablation optimizer (SAO) is a key problem. However, it is very hard because the space for achieving time optimization through code modification is limited. To resolve this issue, this paper utilizes the Vitis HLS tool to deploy the SAO onto the FPGA and optimize the SAO's execution time. Vitis HLS is a key tool in high-level synthesis (HLS). Until now, there have been few attempts to optimize the execution time of this type of algorithm through HLS. This paper proposes a pragma integration model combined with a modified linear feedback shift register (LFSR) algorithm to reduce execution time. The experimental results show that this approach reduces the latency to 44.50% while maintaining an acceptable convergence error, compared to previous optimizations.

Index Terms—High-Level Synthesis, Snow Ablation Optimizer, FPGA, Accelerators, Linear Feedback Shift Register

I. INTRODUCTION

In the field of very large-scale integration (VLSI), high-level synthesis (HLS) [1] has emerged as a crucial technology and is now extensively adopted by VLSI design companies. When contrasted with the traditional register transfer level (RTL) design approach, HLS exhibits remarkable technical advantages in VLSI development. Its core value lies in the elevation of the design abstraction level: by leveraging high-level languages such as C/C++/SystemC to depict algorithmic behavior, HLS redirects the design emphasis from the microcontrol of sequence and circuit structure to the efficient implementation of functional logic. This can significantly abbreviate the development cycle and reduce the threshold for hardware development.

At the optimization stage, the HLS tool deploys strategies including pipeline, loop unroll, and array partition to generate a pareto-optimal architecture with respect to throughput, latency, and hardware resource consumption. For example, as illustrated in Fig. 1, in this case, the HLS tool initiates the next operation before the current one is fully completed. Suppose the major cycle consists of three cycles and the initiation interval is set to 1. Once one cycle is finished, the next major cycle commences. This approach can effectively reduce latency. Through synthesis directives, other structures like arrays can also be manipulated. Arrays can be synthesized into register-based random access memory (RAM) or fully

---

*Co-first authors with equal contribution.

†Corresponding author.

This work is supported in part by National Key R&D Program of China (2022YFB2901100) and Beijing Natural Science Foundation under Grant 4244107.

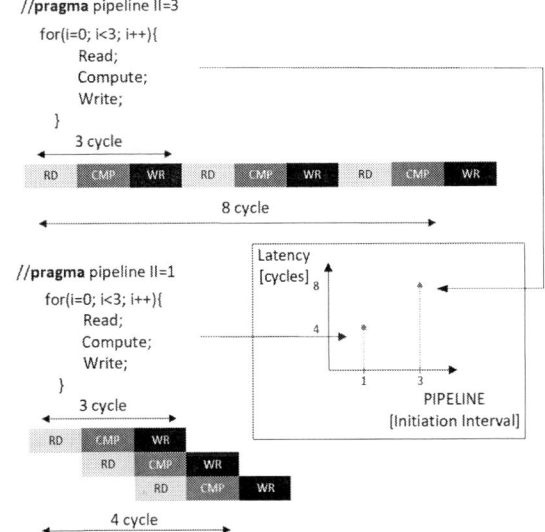

Fig. 1 The pragma of pipeline on hardware circuit generated by HLS

expanded into independent trigger circuits. Now there are numerous cases [2], [3] have demonstrated that algorithms implemented on FPGAs can be efficiently accelerated using Vitis HLS.

The snow ablation optimizer (SAO) [4] is a newly proposed innovative meta−heuristic algorithm. In comparison with other heuristic algorithms, it demonstrates a superior ability to balance the exploration and exploitation processes. This enables it to effectively avoid premature convergence and achieve more favorable outcomes. Moreover, it exhibits remarkable global search capabilities and strong general applicability. Nevertheless, the SAO has a relatively high time complexity because it uses dual−population mechanism and the dependence on random number generation for convergence. When addressing high−dimensional or large−scale optimization problems, it consumes the longer processing time for handling such problems because it has the substantial number of iterations and the extended convergence time. Consequently, the reduction of the SAO's execution time has emerged as an urgent issue.

In this work, we have proposed an effective optimization method to achieve better performance. The main contributions of this paper include:

979-8-3315-3918-4/25 $31.00 © 2025 IEEE

- A pragma integration model designed based on the latency of the objective function.
- A random number generation model based on HLS for LFSR combined with the pragma integration model
- Based on the optimization of the previous two points, the experiment shows that the latency has decreased by 44.50%.

The rest of the paper is organized as follows: Section II introduces the preliminary knowledge. Section III presents the proposed methodology. Section IV gives the experimental results. Section V concludes the paper.

II. PRELIMINARIES

A. Vitis HLS

To shorten the execution time of SAO, we contemplated implementing SAO on an FPGA. In contrast to the conventional RTL design approach, HLS enables rapid algorithm deployment. Consequently, we decide to utilize HLS for this deployment. There are numerous HLS tools available. ROCCC HLS tool can transform C programs into hardware accelerators. GAUT can convert C programs into a pipeline structure that satisfies specific constraints. LegUp HLS can take C programs as input and automatically convert them into a hybrid system. Vitis HLS is capable of automatically converting C/C++/OpenCL code into an optimized RTL implementation. Through the comprehensive comparative analysis, we determined that Vitis HLS provides a more extensive library set in comparison to other HLS tools. Its framework integrates C simulation, C/RTL co-simulation, and waveform analysis within a unified IDE, enabling accelerated hardware implementation.

Vitis HLS, an advanced HLS tool developed by Xilinx, supersedes Vivado HLS and features enhanced IDE integrations. Its architecture comprises a front-end and a back-end. The open-source front-end, available on GitHub, supports custom optimization plugins and deep integration with the Vitis AI and Vitis vision libraries. The back-end implements FPGA device-specific resource and timing path optimizations. The operation process of the Vitis HLS tool is shown in Fig. 2.

B. Snow Ablation Optimizer

The Snow Ablation Optimizer (SAO) is a novel meta-heuristic algorithm inspired by snow melting phenomena. It simulates two physical processes: snow formation and melting, as shown in Fig. 3. During the exploration phase, SAO employs Brownian motion to model the stochastic diffusion of water vapor. During exploitation, it utilizes the degree-day method [5] to update individual positions relative to the swarm centroid. It incorporates a dual-population mechanism to maintain a dynamic equilibrium between global exploration and local exploitation throughout the optimization process.

The position update equation of the entire SAO algorithm is shown as Equation (1):

$$Z_i(t+1) = \begin{cases} \text{Elite}(t) + BM_i(t) \otimes [\theta_1(G - Z_i) \\ + (1 - \theta_1)(\bar{Z} - Z_i)], i \in index_a \\ M \times G(t) + BM_i(t) \otimes [\theta_2(G - Z_i) \\ + (1 - \theta_2)(\bar{Z} - Z_i)], i \in index_b \end{cases} \quad (1)$$

where $index_a$ and $index_b$ denote a set of indexes that include the line numbers of individuals in P_a and P_b throughout the position matrix. θ_1 and θ_2 denote a number randomly produced in $[0, 1]$. $\text{Elite}(t)$ denotes the elite pool. $Z_i(t)$ denotes the ith individual during the tth iteration. $\bar{Z}(t)$ denotes the centroid position of the whole swarm.

Although SAO exhibits exceptional global search capabilities and strong universality, it has high time complexity, as shown in Equation (2). It leads to scalability challenges. When applied to high-dimensional or large-scale optimization problems, the algorithm requires extensive iterations and prolonged convergence time. Consequently, the efficient hardware acceleration of SAO becomes imperative.

$$O\left(N * Dim + N * t_{\max} * (\log N + Dim + 1)\right) \quad (2)$$

where N denotes the number of search agents. Dim denotes the dimension.

C. Linear Feedback Shift Register

The convergence of SAO depends on random number generation. Compared to alternatives like the linear congruential generator [6], mersenne twister [7], and wichmann-hill [8] algorithms, the linear feedback shift register (LFSR) offers a key FPGA implementation advantage. It requires only shift registers and XOR gates. This eliminates the necessity for complex arithmetic units such as multipliers and dividers, thereby significantly reducing resource consumption. Therefore, we employ LFSR as the random number generation algorithm for the implementation of SAO.

The linear feedback shift register (LFSR) generates pseudorandom numbers. It generates random numbers from initial seeds via linear feedback and tap-selected XOR operations. At the same time, it will create a new seed for future use. This feedback can shifts the register states, as formalized in Equation (3). This architecture enhances the algorithm's performance longevity while maintaining superior optimization efficiency on FPGAs.

$$s_n = s_1 \oplus s_2 \oplus ... \oplus s_m \quad (3)$$

where s denotes the data of each bit. $\oplus$ denotes the XOR operation.

III. METHODOLOGY

A. Pragma Integration Model

The SAO implementation on FPGA achieves accelerated execution and enhanced resource efficiency via pragma-directed HLS optimizations. However, algorithm-specific

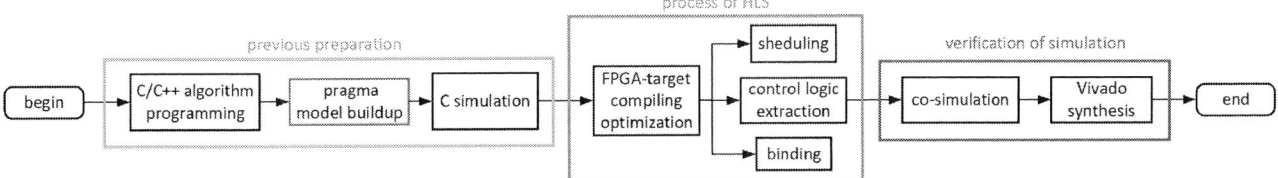

Fig. 2 The process of Vitis HLS

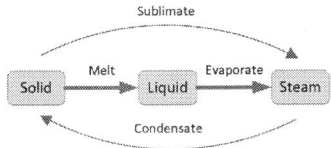

Fig. 3 Three-state change diagram

characteristics dominate pragma optimization outcomes. Systematic exploration of pragma interdependencies is therefore critical for optimizing SAO hardware performance.

In Vitis HLS, diverse pragmas optimize algorithm implementations. Directly applying dataflow pragmas to the main function of SAO is infeasible, because SAO has the prevalent imperfect loop nesting and its subsequent loop iterations depend on prior computations. Although temporary arrays can decouple dependencies, they incur prohibitive resource overhead. To resolve this, we propose a hierarchical pragma integration model that can significantly reduce latency in SAO.

Taking init_loop as an example, the init_loop needs to call the objective function during its operation. Each objective function has the different running cycles. For high-dimensional or large-scale optimization problems, this variability leads to significant latency overhead. The entire process of the init_loop algorithm is summarized in Algorithm 1.

Algorithm 1 init_loop process

Require: $function, current\_best\_score, current\_best\_pos$
1: **for** $i = 0 \rightarrow N - 1$ **do**
2:     $obj\_val = objective\_function(X[i])$
3:     $objective\_function[i] = obj\_val$
4:     **if** $obj\_val < current\_best\_score$ **then**
5:         $current\_best\_score = obj\_val$
6:         **for** $j = 0 \rightarrow dim$ **do**
7:             $current\_best\_pos[i]=X[i][j]$
8:         **end for**
9:     **end if**
10: **end for**

Based on the above pseudo-code, we significantly accelerate the overall running speed through the micro-combination of pipeline and unroll. Taking the three-dimensional sphere function as the objective function as an example, analysis shows that the clock cycle for calling the objective function once is shown in Equation (4)

$$L_y = L_m + \log_2(\text{dim}) \cdot L_a \qquad (4)$$

where L_y represents the total cycles, L_m represents the cycle of each level of multiplication, L_a represents the cycle of each level of addition, dim represents the dimension of the function.

In this example, we use the pragma directive inline to specify inlining of the target function and use the pragma directive unroll to expand the entire for loop. We obtained the total latency for one cycle is 3. In the SAO algorithm, multiple data need to be read simultaneously. Since each block of RAM has only two data ports at most, the loading operation cannot be completed within a single cycle. We propose a model that can minimize the initiation interval of the pipeline in SAO. This model can significantly reduce the latency. Taking the three-dimensional sphere function as an example, in the init_loop, we set the initiation interval of pipeline to 3, which is the total latency number. The unroll factor should be divisible by the number of search agents. We control throughput at 1.6 data per cycle. In thie case, this approach partitions combinational logic depth to prevent single-cycle path timing violations and reduce critical path delay. It can also avoid the explosion of fully expanded resources, resulting in a significant decrease in latency.

B. LFSR Based on HLS

The SAO algorithm mainly reduces the solution space through random generation. It generates random numbers to search for the approximate path and conducts detailed search through population segmentation. The entire process can consume a large amount of latency during the generation of random numbers. Therefore, the processing of random number generation is crucial for reducing latency.

The initial implementation of the LFSR algorithm generates integers ranging from 0 to 2^{16} through a linear feedback shift register and then divides by 2^{16} to obtain decimal numbers between 0 and 1. However, this method consumes a large amount of multipliers during repeated calls. It will significantly reduce the operational efficiency.

To further reduce latency, we optimize the LFSR algorithm. The optimization is summarized in Fig. 4. To reduce the risk of overflow and truncation, we generate a 16-bit fixed-point number as the required random number. We initially generate a 32-bit binary number using LFSR, where the first 16 bits represent the integer part and the last 16 bits represent the fractional part. Setting the upper 16 bits to zero, we obtain a random number in the range [0,1]. Based on the

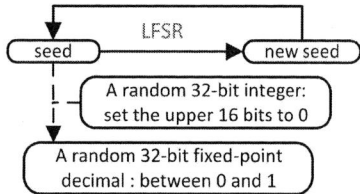

Fig. 4 The process of LFSR

analysis, the operation of generating random numbers can be completed in just 3 clock cycles. To adapt LFSR for HLS, we employ inline directives to enable function inlining. Through the analysis of the LFSR function implementation, we have developed a random number generation algorithm that can be executed within a single cycle in Vitis HLS. This algorithm exclusively employs operations such as shifting and concatenation. These operations are inherently free from additional latency. It does not involve multiplication or division operations that are unsuitable for FPGA implementation. By integrating this algorithm with the pragma integration model, we have achieved a significant reduction in latency.

IV. EVALUATION

A. Experimental Environment

We implemented the FPGA deployment of the SAO algorithm based on Vitis HLS, and achieved another FPGA acceleration by modifying its code and fine-tuning pragmas. All test cases are written in C/C++. All experiments were conducted on an Ubuntu 20.04.6 LTS system, with the CPU being an Intel (R) Core (TM) i5-12500H @ 3.1GHz.

B. Experimental Setting

We use the Ackley function [9] as a test case. The Ackley function is a multi-modal test function often used to evaluate the performance of optimization algorithms. It has one global minimum point and multiple local minimum points, which is shown in Equation (5)

$$f(\mathbf{x}) = -a \cdot e^{-b\sqrt{\frac{1}{n}\sum_{i=1}^{n} x_i^2}} - e^{\frac{1}{n}\sum_{i=1}^{n}\cos(c \cdot x_i)} + a + e \quad (5)$$

where a,b,c denotes the constant, n denotes dimension, x_i denotes input parameter.

Because the global minimum point of the arkley function is at the origin (0, 0, 0), we use the distance from the optimal convergence point to the origin as the evaluation criterion, which is shown in Equation (6)

$$score = x^2 + y^2 + z^2 \quad (6)$$

where x, y, z denotes the coordinate of the outcome.

C. Experimental Result

We set the dimension of the SAO algorithm to 3, the number of search agents to 50, and the maximum number of iterations to 2000. To minimize data precision loss, all data are represented as 32-bit fixed-point numbers. The convergence results obtained before and after optimization are shown in Fig. 5. The final result is shown in TABLE I.

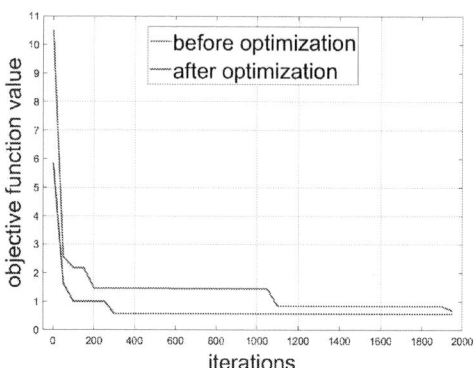

Fig. 5 The result of the snow ablation optimizer

TABLE I result of accelerator

Result	Best Score	Latency(s)	Improvement (%)
before optimization	0.57	0.1827	—
after optimization	0.65	0.1014	44.5%(Latency ↓)

According to TABLE I, the error of its best score is 14.04%, which is within the acceptable range. Compared with previous optimizations, the optimization method can reduce latency by 44. 50%.

V. CONCLUSION

In this work, we propose an implementation and optimization of the SAO algorithm based on HLS, which will be beneficial for the future optimization of algorithms involving random number generation using the Vitis HLS tool. By proposing the pragma integration model and optimizing the LFSR function, we achieve a reduction in its latency.

REFERENCES

[1] P. Coussy, D. D. Gajski, M. Meredith, and A. Takach, "An introduction to high-level synthesis," *IEEE Design & Test of Computers*, vol. 26, no. 4, pp. 8–17, 2009.

[2] M. A. Elhewehy, K. O. Abbass, and O. A. Nasr, "Hardware-software co-design implementation of fixed-point googlenet on soc using xilinx vitis," in *5th Novel Intelligent and Leading Emerging Sciences Conference (NILES)*, 2023, pp. 274–278.

[3] M. Zou, L. Ma, and J. Li, "Implementation and optimization of polyphase channelization using vitis hls," in *IEEE 12th International Conference on Information, Communication and Networks (ICICN)*, 2024, pp. 42–46.

[4] L. Deng and S. Liu, "Snow ablation optimizer: A novel metaheuristic technique for numerical optimization and engineering design," *Journal of Expert Systems with Applications*, vol. 225, p. 120069, 2023.

[5] G. Zhou, M. Cui, J. Wan, and S. Zhang, "A review on snowmelt models: progress and prospect," *Sustainability*, vol. 13, no. 20, p. 11485, 2021.

[6] I. Borosh and H. Niederreiter, "Optimal multipliers for pseudo-random number generation by the linear congruential method," *BIT Numerical Mathematics*, vol. 23, no. 1, pp. 65–74, 1983.

[7] M. Matsumoto and T. Nishimura, "Mersenne twister: a 623-dimensionally equidistributed uniform pseudo-random number generator," *ACM Transactions on Modeling and Computer Simulation (TOMACS)*, vol. 8, no. 1, pp. 3–30, 1998.

[8] B. McCullough, "Microsoft excel's 'not the wichmann–hill' random number generators," *Computational Statistics & Data Analysis*, vol. 52, no. 10, pp. 4587–4593, 2008.

[9] D. Ackley, *A connectionist machine for genetic hillclimbing.* Springer science & business media, 2012, vol. 28.

An MLIR-Based Framework for Efficient Dynamic Circuits Generation

Yuxuan Guan, Jiangnan Li, Lingli Wang*

State Key Laboratory of Integrated Chips and Systems, Fudan University, Shanghai, China

* Email: 22307130004@m.fudan.edu.cn, llwang@fudan.edu.cn

Abstract—We introduce an **MLIR-based framework that transforms C/C++ kernels into dynamically scheduled accelerators, with hardware modules designed in Chisel. Beginning with Polygeist-lowered affine and SCF dialects, our flow applies loop extraction, operator fusion, dependency analysis, and affine memory banking to produce a hardware-oriented intermediate representation (IR) featuring explicit ready/valid handshakes. The generated RTL employs elastic operators and multi-bank memory controllers to support fine-grained out-of-order execution. On an FPGA, our design reduces area to roughly one-quarter of a conventional dataflow baseline and achieves up to 3× lower latency across diverse benchmarks.**

Keywords—MLIR, High-Level Synthesis, Dynamic Scheduling, Hardware Accelerators

I. INTRODUCTION

High-level synthesis (HLS) has revolutionized custom hardware design by automatically translating high-level code into synthesizable hardware descriptions. Since its inception in the 1990s, most HLS tools have relied on static scheduling, where operations are assigned fixed execution cycles and coordinated by finite-state machines that periodically activate data path components [1][2]. This approach effectively uncovers instruction-level parallelism in regular, loop-centric kernels—much like very long instruction word (VLIW) processors—but often struggles with irregular control flow, unpredictable memory accesses, and data dependencies that cannot be fully resolved at compile time.

To address these limitations, recent research has revisited the concept of dynamically scheduled dataflow circuits [3]. In such architectures, individual operators are interconnected through handshake protocols and fire as soon as their input operands become valid, eliminating a central global controller. This fine-grained activation model offers the promise of matching out-of-order superscalar processors in exploiting parallelism at runtime [4]. However, conventional dataflow generators impose a strict ordering on repeated operation instances—particularly within loops—which prevents true dynamic reordering across iterations. As a result, they cannot leverage hardware features such as non-blocking caches or variable-latency functional units to the same extent as modern CPUs [5].

In this work[1], we aim to bridge the semantic gap between static HLS and fully out-of-order execution by enabling selective dynamic scheduling within dataflow accelerators. Our compiler framework automatically identifies independent operation instances and schedules them opportunistically, allowing later iterations or subsequent operations to proceed ahead of earlier, slower tasks. This approach dramatically increases pipeline utilization and overall throughput, while maintaining a modest area overhead and compatibility with existing HLS toolchains.

We introduce a framework that leverages MLIR to insert fine-grained dynamic scheduling into dataflow accelerators, narrowing the gap between static HLS and fully out-of-order execution. Our flow automatically lifts C/C++ kernels through affine and SCF dialects into a unified dataflow IR, from which it derives elastic-handshake circuits and multi-bank memory controllers without manual annotations. We also provide a set of parameterized hardware primitives, each with built-in ready/valid logic and local FSMs, that can opportunistically fire independent operations—even across loop iterations—to boost pipeline utilization and throughput with minimal area impact.

II. BACKGROUND

A. High-Level Synthesis

High-Level Synthesis (HLS) raises the abstraction of hardware design by allowing developers to describe functionality in C-like languages rather than at the register-transfer level (RTL). An HLS compiler [6] automates the transformation from high-level source code to hardware description languages (HDLs) through three principal stages: allocation, scheduling, and binding. During allocation, the compiler determines the types and quantities of compute and storage resources required. Scheduling assigns each operation to a specific clock cycle, balancing data dependencies against resource constraints. Finally, binding maps scheduled operations onto the allocated hardware units.

Modern HLS tools support two contrasting scheduling paradigms. Static scheduling computes a fixed execution timeline at compile time. A typical static flow consists of [7]: (1) software compilation into an intermediate representation, (2) a scheduling algorithm that assigns cycles to operations, (3) allocation and binding that exploit resource sharing, and (4) finite-state machine generation to control data path activation. Static scheduling excels on regular code—such as loops with predictable bounds—enabling deep pipelining where successive iterations overlap, improving throughput by reducing the initiation interval (II) [8].

In contrast, dynamic scheduling [9] constructs a dataflow circuit in which operators communicate via handshake protocols (ready/valid signals) and execute as soon as their operands become available. This loose coupling allows run-time resolution of dependencies, which benefits programs with irregular control flow or memory access patterns. However, purely dynamic dataflow implementations often incur significant area and control overhead. Hybrid scheduling approaches combine static and dynamic methods: they statically pipeline outer loops for coarse control while

---

[1] This work is supported by the National Natural Science Foundation of China under grant 62174035 and Fudan's Undergraduate Research Opportunities Program (24940).

dynamically scheduling inner kernels to adapt to run-time conditions.

B. The MLIR Compilation Framework

MLIR (Multi-Level Intermediate Representation)[2] is a flexible compiler infrastructure that supports building and transforming nested IRs at varying abstraction levels. At its core lies an SSA-based IR [10] in which the fundamental unit is an Operation. Each operation declares a fixed number of Operands and Results, both of which carry static types. The flow of data is captured by wiring a result from one operation into an operand of another, while Attributes supply compile-time constants or metadata that parameterize operation behavior.

Operations are grouped into Blocks, which form the nodes of a Control-Flow Graph. One or more blocks are enclosed within a Region, and regions in turn nest inside operations. This hierarchy allows MLIR to express everything from straight-line arithmetic sequences to complex control constructs. A top-level function is itself an operation that owns a primary region containing its entry block and any subsequent blocks.

To illustrate these concepts, Fig. 1 shows a simple IR fragment. It includes both structured control-flow operations from the affine and scf dialects (e.g., affine.for, scf.if) and their equivalent unstructured branches (br, cond_br) after lowering. Types have been elided for clarity.

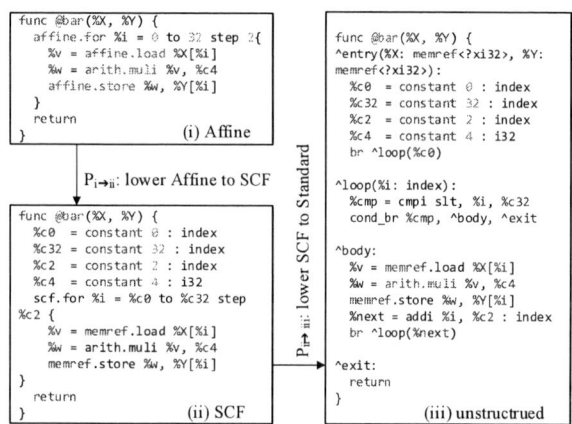

Fig. 1. Example MLIR fragment showcasing structured loops and conditional operations, and their lowered unstructured form.

A key feature of MLIR is its dialect mechanism: each dialect defines a namespace of related operations, types, and attributes. Built-in dialects such as arith, memref, affine, and scf cover common functionality—arithmetic, memory accesses, affine loop analysis, and structured control flow— while users can introduce custom dialects for domain-specific abstractions. Passes, which include both Transforms (intra-dialect optimizations) and Conversions (cross-dialect lowerings), systematically rewrite the IR. Within this ecosystem, lowering refers to the controlled reduction of higher-level constructs into simpler or more target-specific representations, making MLIR an ideal foundation for multi-stage HLS compilers.

III. THE PROPOSED FRAMEWORK

We present an end-to-end flow that transforms C/C++ kernels into dynamically scheduled dataflow accelerators. First, the front-end lowers source code into MLIR and performs dataflow-centric optimizations to generate a richly annotated control/dataflow graph. Next, each graph node is mapped to a parameterized elastic hardware module implementing ready/valid handshakes and local FSMs. Finally, these modules are wired according to the graph to produce synthesizable RTL supporting fine-grained out-of-order execution with predictable timing and resource utilization.

A. Compilation Flow

The hardware generation flow begins by lowering C/C++ kernels into MLIR (via Polygeist [11]) and then applying three compiler stages on top of the open-source DataFlowGen framework[3]:

1) Hierarchical Dataflow Construction. We identify loop nests and control constructs, flatten them into hardware-scheduling regions, rewrite φ nodes into merge operations, lower branches to simple selects, and fuse scalar operator chains with register insertion to shorten the loop initiation interval.

2) Dependency Analysis and Control/ Dataflow Graph Generation. We traverse the optimized IR, instantiate each operation as an adaptive elastic unit (e.g., ComputeNode, LSNode, MergeNode, StateBranchNode) with ready/valid ports and control masks, and wire data channels under the ready/valid protocol. Loop startup, back-edge iteration, and exit are encoded as token-based control edges, while merge operations carry inter-iteration values, producing a complete hardware control/dataflow graph.

3) Memory Access Analysis & Allocation. In the final pass, each affine.load/store is analyzed via MLIR's affine API to extract linear address parameters per

$$Addr = BaseAddr + \sum_{i=1}^{N} DimScale_i \times Index_i \quad (1)$$

Regular patterns are mapped to LSNode instances with built-in stride generators; irregular accesses fall back to chained multiply-add trees. The compiler then analyzes intra-task access patterns to instantiate multi-bank MemoryNode controllers and connect them via dynamic request-response interfaces, ensuring high throughput with minimal stalls.

After these passes, the IR is ready for RTL emission, with all control/dataflow and memory interfaces explicitly modeled for backend code generation.

B. Hardware Modules

Our dataflow accelerator is composed of modular nodes, each providing a specific data path function and synchronized via ready/valid handshakes:

Argument & Loop Control Nodes initialize the kernel by broadcasting inputs and a start token, simply forwarding their DataBundle when enabled.

Execution Block Node groups operations into an atomic region. Its FSM waits for a block-enable token and then issues

---

[2] https://mlir.llvm.org/

[3] https://github.com/jiangnan7/DataFlowGen

979-8-3315-3918-4/25 $31.00 © 2025 IEEE

enable tokens in parallel to all child nodes, preventing early firing.

ComputeNode executes a two-operand ALU operation. It latches inputs and a control token, computes the result, asserts output valid, and resets upon downstream readiness.

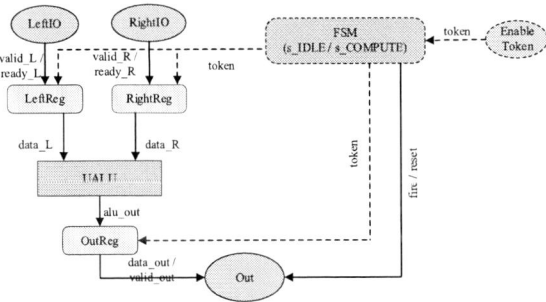

Fig. 2. Internal structure of the ComputeNode showing input registers, ALU computation, output register, and finite-state machine.

LSNode manages on-chip memory accesses. It captures address and control tokens, arbitrates ports if needed, interacts with SyncReadMem, and buffers read data until consumption. Write operations join address and data before issuing a memory write.

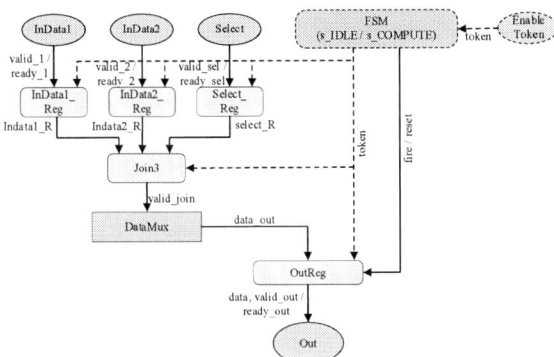

Fig. 3. Internal structure of the LSNode showing address capture, arbitration, synchronous memory interface, and response buffering.

SelectNode implements a two-input multiplexer under handshake. It latches both operands and a select signal, then forwards the selected DataBundle once the FSM transitions on valid handshakes.

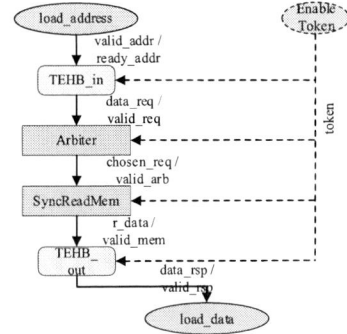

Fig. 4. Internal structure of the SelectNode showing input registers, three-way join, data multiplexer, and output register.

MergeNode performs φ-style selection among N inputs using a one-hot mask. It latches inputs and mask bits, decodes the mask to an index, and broadcasts the chosen DataBundle, coordinating valid and reset via its FSM.

Additional nodes (e.g., GepNode for address computation, ConstNode for constants, ReturnNode for completion) follow the same pattern of input latching, simple combinational logic, and standard handshake interfaces.

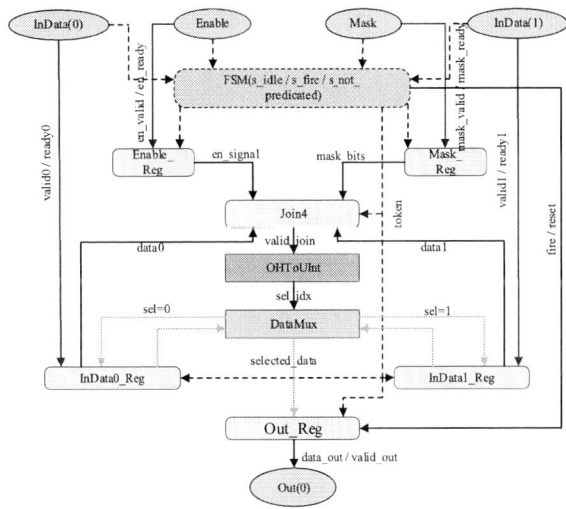

Fig. 5. Internal structure of the MergeNode showing multiple input registers, mask register, one-hot-to-index conversion, data multiplexer, and output register.

C. Overall Hardware Execution Flow Graph

In this final design stage, we compose the individual elastic units into a full end-to-end hardware execution flow graph (H-CDFG), hiding each node's internal implementation behind a simple black-box interface. The graph begins with an ArgCallNode that injects input arguments and an enable token, proceeds through one or more LoopNode and ExecutionBlockNode pairs to manage iterative and block-level control, and interconnects data-path units—ComputeNode, LSNode, SelectNode, MergeNode—via ready/valid handshakes. Upon completion, a ReturnNode collects final outputs and issues the termination token.

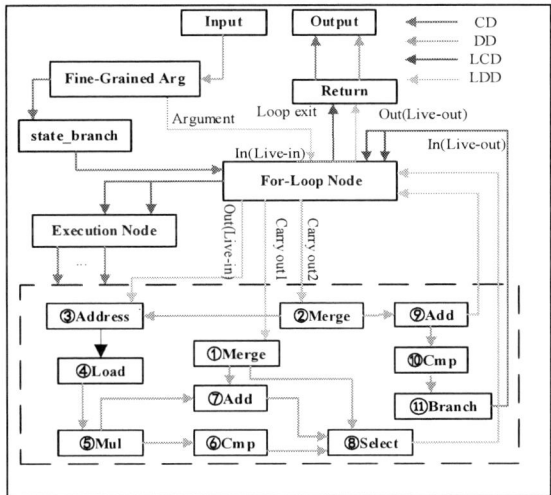

Fig. 6. A full end-to-end H-CDFG depicting the accelerator structure of the kernel. The CD, DD, LCD, and LDD represent control dependency, data dependency, loop-dependent control dependency, and loop-dependent data dependency, respectively.

IV. Experimental Evaluation

A. Evaluation Setup

To validate our compiler and hardware modules, we compare against the µIR framework n a set of compute kernels drawn from its public benchmark suite. All designs target the Xilinx xc7vx485t FPGA. We use Polygeist [11] to lower C++ kernels into MLIR, then invoke our modified DataFlowGen flow to generate Verilog. Cycle-accurate simulation is performed via Verilator, and synthesis/routing reports are obtained from Vivado 2022.1. Experiments run on a server equipped with an Intel® Xeon® Gold 6148 CPU.

B. Results and Discussion

TABLE I. and TABLE II. summarize the area and performance comparison between µIR and our design across the eight benchmarks.

TABLE I.   Resource Utilization Comparison

Benchmark	Slices		LUTs		FFs		DSPs	
	µIR	Ours	µIR	Ours	µIR	Ours	µIR	Ours
Loop with condition	375	77	949	175	1190	259	3	0
Sumi3 memory	367	141	967	274	1135	490	6	4
FIR	465	122	1079	241	1373	443	3	0
GenTanh	521	96	1396	206	1544	322	12	0
Dropout	556	125	1428	320	1844	386	0	0
DoitgenTriple	786	162	2115	449	2572	542	6	0
Histogram	489	75	1293	178	1625	256	0	0
Matrix add	1140	325	2982	959	3650	1100	0	0
Normalized mean	1	0.24	1	0.22	1	0.22	1	0.14

TABLE II.   Performance Comparison

Benchmark	Cycles		Period(ns)		Execution Time(ns)	
	µIR	Ours	µIR	Ours	µIR	Ours
Loop with condition	1906	713	7.24	3.38	13.80	2.41
Sumi3 memory	6722	1413	7.76	5.10	52.17	7.20
FIR	9180	915	7.12	3.07	65.37	2.81
GenTanh	2706	706	8.13	3.89	21.99	2.74
Dropout	49157	10256	4.52	4.25	222.24	43.57
DoitgenTriple	19667	3118	7.13	4.02	140.21	12.53
Histogram	2634	699	4.66	3.33	61.69	2.33
Matrix add	10326	459	5.15	4.414	53.19	2.03
Normalized mean	1	0.20	1.00	0.64	1.00	0.12

1) Area Reduction. Our design reduces slice usage to 24 % and LUT usage to 22 % of the µIR baseline, with FFs down to 22 % and DSPs to 14 %, due to streamlined handshake logic, operator fusion, and bit-width optimizations.

2) Performance Improvement. Dynamic scheduling and operator fusion shrink cycle counts to 20 % of µIR's, and shorter critical paths cut clock period to 64 %. Overall execution time averages 12 % of the baseline, achieving up to 3× lower latency.

These results confirm that our MLIR-based dynamic dataflow framework delivers significant area savings and

substantial performance gains compared to the µIR[12] approach.

V. Conclusion

This paper introduces a compiler framework that leverages MLIR to generate dynamically scheduled dataflow accelerators from C/C++ kernels. By extending the open-source DataFlowGen, we propose composable hardware modules with built-in handshake logic and state machines, and apply loop fusion, dependency analysis, and affine memory banking. FPGA experiments demonstrate that our approach reduces area to roughly one-quarter of µIR [12] and cuts latency by up to 3×. These results confirm the effectiveness of combining MLIR-driven compilation with dynamic dataflow modules for high-efficiency accelerator design.

References

[1] J. Cong, B. Liu, S. Neuendorffer, J. Noguera, K. Vissers, and Z. Zhang. High-level synthesis for FPGAs: From prototyping to deployment. *IEEE Transactions on Computer-Aided Design of Integrated Circuits and Systems*, 30(4):473–91, Apr. 2011.

[2] J. Cong and Z. Zhang. An efficient and versatile scheduling algorithm based on SDC formulation. In *Proceedings of the 43rd Design Automation Conference*, pages 433–38, San Francisco, Calif., July 2006.

[3] M. Budiu, P. V. Artigas, and S. C. Goldstein. Dataflow: A complement to superscalar. In *Proceedings of the IEEE International Symposium on Performance Analysis of Systems and Software*, pages 177–86, Austin, Tex., Mar. 2005.

[4] L. Josipović, A. Guerrieri, and P. Ienne. Synthesizing general-purpose code into dynamically scheduled circuits. *IEEE Circuits and Systems Magazine*, 21(2):97–118, Second quarter 2021.

[5] K. I. Farkas and N. P. Jouppi. Complexity/performance tradeoffs with nonblocking loads. In *Proceedings of the 21st Annual International Symposium on Computer Architecture*, pages 211–22, Chicago, Ill., Apr. 1994.

[6] Andrew Canis, Jongsok Choi, Mark Aldham, Victor Zhang, Ahmed Kammoona, Jason H. Anderson, Stephen Brown, and Tomasz Czajkowski. 2011. LegUp: High-Level Synthesis for FPGA-Based Processor/Accelerator Systems. In *Proceedings of the 19th ACM/SIGDA International Symposium on Field Programmable Gate Arrays* (Monterey, CA, USA) *(FPGA '11)*.

[7] Philippe Coussy, Daniel D. Gajski, Michael Meredith, and Andres Takach. 2009. An Introduction to High-Level Synthesis. IEEE Design Test of Computers 26, 4 (2009).

[8] Yu-Chin Hsu and Yuang-Long Jeang. 1993. Pipeline scheduling techniques in high-level synthesis. In *Sixth Annual IEEE International ASIC Conference and Exhibit*.

[9] Jianyi Cheng, Lana Josipovic, George A. Constantinides, Paolo Ienne, and John Wickerson. 2020. Combining Dynamic & Static Scheduling in High-Level Synthesis. In *Proceedings of the 2020 ACM/SIGDA International Symposium on Field-Programmable Gate Arrays* (Seaside, CA, USA)

[10] Ron Cytron, Jeanne Ferrante, Barry K Rosen, Mark N Wegman, and F Kenneth Zadeck. 1991. Efficiently computing static single assignment form and the control dependence graph. *ACM Transactions on Programming Languages and Systems* (TOPLAS) 13, 4 (1991), 451–490.

[11] William S. Moses, Lorenzo Chelini, Ruizhe Zhao, and Oleksandr Zinenko. Polygeist: Raising C to Polyhedral MLIR. In *2021 30th International Conference on Parallel Architectures and Compilation Techniques (PACT)*, pages 45–59, 2021

[12] A. Sharifian, R. Hojabr, N. Rahimi, S. Liu, A. Guha, T. Nowatzki, and A. Shriraman, "µIR -An intermediate representation for transforming and optimizing the microarchitecture of application accelerators," in *Proceedings of the 52nd Annual IEEE/ACM International Symposium on Microarchitecture*, ser. MICRO '52, 2019, p. 940–953.

HybridEPP: Hybrid Numerical and Symbolic Error Probability Propagation in Logic Network

Gaopeng Shen, Chang Wu*

School of Microelectronics, Fudan University, Shanghai, China

* Email: 23212020013@m.fudan.edu.cn, wuchang@fudan.edu.cn

Abstract—**Error probability propagation (EPP) is an important problem in soft error mitigation in radiation environment. Due to the high cost of full Triple Modular Redundancy (TMR) design, people proposed to perform partial TMR based on soft error rate. Numerical error propagation method, though very fast, suffers low accuracy due to reconvergent paths and signal correlation in digital circuits. Full symbolic method, on the other hand, suffers long runtime problem. In this paper, we propose a hybrid numerical and symbolic EPP algorithm. Our results show that we can improve the error rate estimation accuracy by up to 56% over numerical method and 8× and 48× faster than full symbolic and simulation methods, respectively.**

Keywords—error probability propagation, reconvergence, symbolic calculation, probabilistic graph

I. Introduction

Single-Event Upset (SEU) is a serious problem in radiation environment [1]. Triple Modular Redundancy (TMR) is a widely used technology to mitigate SEUs [2], however, suffers high-cost problem of 3~5x logic duplications. Recently, people proposed partial TMR technology based on error rates of different logic errors on the circuit output [3, 4]. The main idea of partial TMR is to only perform logic duplication for logics or gates which have a high probability of causing output errors. This requires an efficient and accurate Error Probability Propagation (EPP) method.

Previous works on EPP analysis approaches can be categorized as simulation (with fault injection) [1,5,6] and error propagation method.

There are many algorithms tackling the EPP of logic network. G. Fey et al. proposed to use formal verification to provide lower and upper bounds of EPP [7]. But their approach cannot compute the exact EPP of a logic network. N. Miskov-Zivanov and D. Marculescu proposed a symbolic method to compute logic error rate [8], which presents higher accuracy at the cost of exponential runtime complexity. These approaches have two major obstacles on error masking effect and reconvergent path handling which lead to inaccuracy on error rate estimation. A numerical method [9] turned the EPP problem into a probabilistic graph to overcome the first obstacle but with inaccuracy problem on reconvergent path. W. Xiao and W. Qian proposed a mixed symbolic and numerical probability calculation algorithm that efficiently overcomes the second obstacle [10]. But this algorithm did not take the former obstacle into account.

In this paper, we propose a new mixed numerical and symbolic error probability propagation approach to compute accurate logic error rate efficiently for SEU mitigation. Symbolic error rate propagation is used on reconvergent paths for accuracy consideration and numeric propagation is used in other parts of the circuit for efficiency. The probabilistic graph model and Bayesian network are used for logic error rate propagation. Reconvergence significance analysis (RSA) including reconvergent path reduction and symbol elimination are used in our symbolic error rate propagation for efficiency enhancement. Our experimental results show that our method improves the error rate accuracy by up to 56% over numerical method, while is 48× faster than the simulation method.

The rest of the paper is organized as follows. Section 2 presents our logic error rate propagation model and symbolic formula handling on reconvergent paths. Our HybridEPP algorithm is presented in Section 3 with experimental results in section 4. Our conclusion is in Section 5.

II. Methodological Foundations and Derivation

This section illustrates our theoretical model and its derivation.

A. Probabilistic Gate Model

In the probabilistic gate model (PGM) of EPP, we consider 4 types of signals with their propagation probabilities P_0, P_1, P_0^e and P_1^e [9], which stand for the probabilities of being 0, 1, 0^e and 1^e, where 0^e means the signal should be logic 1 but becomes 0. Likewise, 1^e means that 0 becomes 1 in the circuit.

Fig. 1 shows the truth table of a 2-input AND gate in PGM of EPP. It is worth noticing that entries $(1^e, 0^e)$ and $(0^e, 1^e)$ are 0. It means error inputs may cancel each other and produce correct output. This error masking effect actually creates complexity on error propagation computation.

Based on this expanded truth table, the probabilities of the output can be derived, e.g., P_0^e equals the sum of probabilities of inputs being 10^e, 0^e1, 0^e0^e.

B. Probabilistic Graph Model of Logic Network

Based on the PGM, probabilistic graph model of EPP is realized as the substitution of probabilities for signals in the logic network.

Because the logic network is a directed acyclic graph (DAG), the corresponding probabilistic graph is a Bayesian Network, in which the signal probabilities satisfy the Bayes formula. Fig. 2 shows a logic network and its probabilistic graph. We assume that the primary inputs are independent. The probabilities of the graph are expressed as:

$$P(a, b, m_0, m_1, m_2, out) = P(a) \times P(b) \times P(m_0|a, b)$$

$$\times P(m_1|a, m_0) \times P(out|m_1) \quad (1)$$

$$P(out) = \sum_a \sum_b \sum_{m_0} \sum_{m_1} P(a, b, m_0, m_1, m_2, out) \quad (2)$$

979-8-3315-3918-4/25 $31.00 © 2025 IEEE

AND	0	1	0^e	1^e
0	0	0	0	0
1	0	1	0^e	1^e
0^e	0	0^e	0^e	0
1^e	0	1^e	0	1^e

Fig. 1. The expanded truth table of a 2-input AND gate

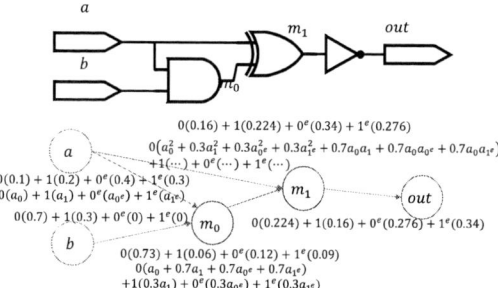

Fig. 2. A logic network and its corresponding probabilistic graph with numerical and symbolic probabilities

$P(a)$ and $P(b)$ are the probabilities of primary inputs while $P(m_0|a,b)$, $P(m_1|a,m_0)$ and $P(out|m_1)$ are conditional probabilities dependent on the types of logic gates and . As a result, the edge probability $P(out)$ is a sum of a set of $P(a)P(b)$. In other words, regarding every probability of a and b as first-order term, $P_0(out)$, $P_1(out)$, $P_0^e(out)$ and $P_1^e(out)$ are all quadratic homogeneous polynomials. Then, it is easy to prove that the probabilities of every node are all homogeneous polynomials of which each monomial is the product of one probability of each input. This inference is generalized to any logic network and its sub-network.

C. Reconvergent Path

A reconvergent path (RP) is defined as a set of multiple paths between two nodes in the DAG. Neglect of it can lead to significant error, because RPs will induce the appearance of the emergence of higher-order terms in the signal probability propagation calculation.

To simplify the illustration, we use a simplified signal probability to demonstrate how the RPs influences the accuracy where $0(a_0)+1(a_1)+0^e(a_0^e)+1^e(a_1^e)$ means that the probabilities of a signal being 0, 1, 0^e and 1^e are equal to a_0, a_1, a_0^e, a_1^e. The probability propagation progress of the logic network is shown in Fig. 2 with all numerical and selected symbolic probability results of each node. It is evident that the symbolic expression of $P_0(m_1)$ is inconsistent with the inference (2) we obtain before. The reason is that there is a reconvergent path from a to m_1 which consists of two paths a, m_1 and a, m_0, m_1. As a result, each term in the expression at the endpoint of the reconvergent path contains the symbols of a twice which gives rise to the appearance of estimation error.

To address the issue, we introduce a new calculation operation in symbolic computation called *Quasi-Multiplication*. There are three symbolic values f, x and y, where f is equivalent to the quasi-multiplication result of x and y. Then f is defined as:

$$f = x * y = \begin{cases} xy & node(x) \neq node(y) \\ x & x=y \\ 0 & otherwise \end{cases} \quad (3)$$

in which $node(x)$ is the node that the symbolic probability x belongs to. And the operation satisfies the laws of associativity and commutativity. The operation inherently prevents potential logical conflicts. After the substitution of quasi-multiplication for multiplication, the expression of $P_0(m_1)$ is:

$$P_0(m_1) = a_0 + 0.3a_1 + 0.3a_0^e + 0.3a_1^e \quad (4)$$

After elimination of symbols, it equals 0.37, which suggests considerable error that reconvergent paths cause compared to numerical result 0.16. This result could be verified by the method in Section II(B). Therefore, we substitute quasi-multiplication for multiplication to minimize the error that reconvergent paths cause in the symbolic calculation.

In summary, it is feasible that the combination of symbolic calculation and quasi-multiplication can eliminate the error reconvergent paths cause. However, symbolic calculation is extremely time-consuming. Thus, symbolic calculation only performs on the probabilities of nodes related to reconvergent paths and the remaining part of logic network is determined numerically for acceleration. The appropriate combination of them achieves a balance in accuracy and complexity.

III. HYBRIDEPP ALGORITHM

In this section, HybridEPP algorithm is presented based on the derivation of Section II. Our target is to minimize the error of reconvergence by symbolic calculation and accelerate the program by numerical calculation. Meanwhile, RSA provides a heuristic approach for accelerating symbolic computation with acceptable precision loss.

A. Reconvergent Paths Collection

The first step of the algorithm is to collect the reconvergent paths in the logic network as presented in Algorithm 1. Our basic idea is to propagate the significance values from parent nodes to child nodes. That the significance of a node v merges at another node u and the reconvergent significance is more than the threshold T ($0<T<1$) mean there is a reconvergent path from v to u. Then by Depth-First Search (DFS) we can find the multiple paths between two nodes. In our algorithm, the significance of node v is defined as the degrees of ancestor nodes reconvergence on node v ranging between zero and one where the significance being one indicates all the fanout reconverging at a node namely dominator and a node without fanout results in significance begin zero. *v.significance* stores all the ancestor nodes and the corresponding reconvergence significance. *RPlist* contains all the reconvergent paths. There is a trick that there is no need for DFS to search the entire graph but the nodes in the significance map.

B. Edge Classification

The second step is to classify the types of probabilities of each node. Instead of explicitly inferring them, edge types are utilized to implicitly describe them. Edge types are organized into VALUE, SYMBOL and EXPRESSION.

Algorithm 1: The procedure *GetReconvergentPaths* for obtaining all the reconvergent paths in the logic network

Input: A network $G = (V, E)$ and a threshold T
Output: A list of Reconvergent paths *RPlist*

1: $RPlist \leftarrow []$
2: **foreach** *node v of G* **do**
3:      $v.significance \leftarrow \{\}$
4: **foreach** *node v of G in topologic order* **do**
5:      **if** v.outdegree=0 **then**
6:          **continue**
7:      **if** *v.outdegree*=1 **then**
8:          $v.significance[v] \leftarrow 0$
9:      **else** $v.significance[v] \leftarrow 1$
10:     **foreach** *node child in v.ChildrenNode* **do**
11:         **foreach** *s in v.significance.keys* **do**
12:             $value = v.significance / v.outdegree$
13:             $child.significance[s] += value$
14:             **if** $child.significance[s] > T$ **then**
15:                 $starts.append(s)$
16:         **foreach** *node s in starts* **do**
17:             $RPlist.append(DFS(G,s,child))$
18: **return** *RPlist, significance*

1) VALUE: The propagation of probabilities through this edge is based on values, which indicates the target node is out of every reconvergent path.

2) SYMBOL: The source delivers new symbols to the target. It happens when the source node is the start of the reconvergent path that the edge belongs to.

3) EXPRESSION: When the edge is in one reconvergent path but not the start, the edge is EXPRESSION which implies the expression-based propagation.

After traversing every edge on the reconvergent paths, we can determine the type of edges based on the rules.

C. Reconvergence Significance Analysis

Algorithm 2: The procedure *GetProbability* for appending the probability for current node *cv* into probability list *prob*

Input: A network $G = (V, E)$, an edge type map *Edgetype, significance,* current node *cv*, Threshold *K*, probability list *prob*

1: $cv.parentprobability \leftarrow []$
2: $cv.symbols \leftarrow \{\}$
3: **foreach** *node v of cv.parents* **do**
4:      $p \leftarrow getinputprob(Edgetype, v, cv)$
5:      $cv.parentprobability.append(p)$
6:      $cv.symbols.append(p.symbols)$
7: **if** $cv.symbols.size < K$ **then**
8:      $prob[v] \leftarrow PGM(v.parentprobability, v.gatetype)$
9:      **return**
10: $symbols \leftarrow sort(v.symbols, significance)$
11: $newparprob \leftarrow []$
12: **foreach** *p in v.parentprobability* **do**
13:     $newparprob.append(p.eliminate(symbols, K))$
14: $prob[v] \leftarrow PGM(newparprob, v.gatetype)$

RSA is a heuristic approach to accelerate symbolic computation. When the expression of a node probability contains too many symbols potentially, elimination is performed based on the significance of the node with priority given to eliminating symbols corresponding to smaller values. The larger the value of significance, the more reconvergences occur at that node. This implies that, if converted into numerical computation, a greater error would be introduced.

The implementation of RSA is fulfilled by obtaining the symbols of parent nodes and then performing elimination before the probability calculation as shown in Algorithm 2. The function *eliminate* is to get the probability expression after eliminating symbols after *K-th* element in the *symbols*.

D. Flow of HybridEPP

The flow of HybridEPP is shown in Algorithm 3. Given the logic network G, two thresholds T and K, probabilities of input nodes, finally we can get the EPP of every node.

In Lines 2 and 3, it acquires reconvergent paths and significance maps in the graph, as well as determining the edge types. Line 4 initializes the error-free probabilities which is calculated in Lines 6, 7 and after that we define result to store the EPP result. In Lines 8-13, the EPP is computed by successively injecting errors into different nodes and then computing the EPP progress of the subsequent nodes.

The worst computation complexity of Algorithm 2 is $O(2^{0.5K+1} + 2^K)$ because of comprising two symbol elimination and one computation. It is a constant value, denoted as H subsequently. Thus, the computation complexity for each error propagation progress is $O(HN)$ where N is the number of affected nodes. Therefore, the overall time complexity of Algorithm 3 is $O(H|V|^2)$ where $|V|$ is the number of nodes in the network.

Algorithm 3: The procedure *GetEPP* for obtaining the EPP results of all potential fault nodes

Input: A network $G = (V, E)$, thresholds T, K, input signal probabilities *inprob*
Output: The error propagation probability map *result*

1: $rplist, sig \leftarrow GetReconvergentPaths(G,T)$
2: $Etype \leftarrow GetEdgeTypes(G,rplist)$
3: $prob \leftarrow inprob$
4: $result \leftarrow \{\}$
5: **foreach** *node v of G in topologic order* **do**
6:      $GetProbability(G,Etype,sig,u,K,errprob)$
7: **foreach** *node v of G in topologic order* **do**
8:      $epp \leftarrow prob$
9:      $seterrorprob(epp,v)$
10:         **foreach** *node u of G in topologic order after v* **do**
11:             $GetProbability(G,Etype,sig,u,K,epp)$
12:     $result[v] \leftarrow epp$
13: **return** *result*

IV. EXPERIMENTAL RESULTS

A. Experimental Setup

We choose a set of designs from LGSynth91 for our tests, which are listed in Table I, where columns *#PIs*, *#POs* and #Nodes are the number of primary inputs, outputs and logic gates of the circuits.

TABLE I. STATISTICS OF CIRCUITS

Circuits	#POs	#PIs	#Nodes
9symml	9	1	201
alu2	10	6	348
apex7	49	37	167
c1355	41	32	183
c1908	33	25	222
c2670	233	140	480
c499	41	32	182
c880	60	26	257
average	59.5	37.4	255.0

Our algorithm is implemented in C++ and runs on an Intel i7-12700H computer with 16GB RAM. A Monte-Carlo based error injection simulation is used in test as a comparison basis with HybridEPP. For circuits with less than 20 PIs, 10000 random test patterns are generated for each simulation experiment, while 100000 patterns are for the rest. HybridEPP is our method with threshold argument values T=0.15 and K=6. The numerical method is based on [9] while the symbolic method is the full symbolic version of our algorithm.

The initial probabilities of the primary inputs (PIs) are set to 0(0.8)+1(0.2) in our tests. The fault model for each node is configured as a functional inversion, e.g., the function of an AND gate node is turned into the function of a NAND gate.

B. Performance of HybridEPP

The experimental results are presented in Table II. The average error (Ave), max error (Max) and run time per error injection are compared.

Compared to the numerical algorithm, our method has improved the average accuracy by 31% on average, with a maximum increase of 48%. The maximum error has decreased by an average of 19%, with the largest reduction reaching 56%. Our algorithm is about 8× faster than the symbolic method while suffering from nearly no precision loss. Our algorithm's speed has improved by up to 1685×, with an average improvement of 48× to simulation method.

V. CONCLUSION

In this work, a novel error probability propagation algorithm for logic networks named HybridEPP is proposed.

Symbolic calculation is utilized to eliminate the error caused by reconvergence, while it consists of two acceleration techniques: numerical calculation for non- reconvergence nodes and symbol elimination according to RSA. HybridEPP achieves higher accuracy than the conventional numerical algorithm and much shorter runtime than the full symbolic and simulation methods. In the future, we will research reliability estimation and error mitigation methods, including partial TMR designs based on HybridEPP.

REFERENCES

[1] S. Xu, Q. Liu, T. Li, and H. Fan, "IC security evaluation against fault injection attack based on FPGA emulation," in Proc. 2016 Int. Conf. Field-Programmable Technol. (FPT), Xi'an, China, 2016, pp. 285-288.

[2] F. L. Kastensmidt, L. Sterpone, L. Carro and M. S. Reorda, "On the optimal design of triple modular redundancy logic for SRAM-based FPGAs," in Proc. Design, Automation and Test in Europe (DATE), Munich, Germany, 2005, pp. 1290-1295 Vol. 2.

[3] P. K. Samudrala, J. Ramos and S. Katkoori, "Selective triple modular redundancy (STMR) based single-event upset (SEU) tolerant synthesis for FPGAs," IEEE Trans. Nucl. Sci., vol. 51, no. 5, pp. 2957-2969, Oct. 2004.

[4] O. Ruano, J. A. Maestro and P. Reviriego, "A Methodology for Automatic Insertion of Selective TMR in Digital Circuits Affected by SEUs," IEEE Trans. Nucl. Sci., vol. 56, no. 4, pp. 2091-2102, Aug. 2009.

[5] R. Zhang, L. Xiao, J. Li, X. Cao, and C. Qi, "A fault injection platform supporting both SEU and multiple SEUs for SRAM-based FPGA," IEEE Trans. Device Mater. Rel., vol. 18, no. 4, pp. 599-605, Dec. 2018.

[6] X. Du, C. He, S. Liu, Y. Zhang, Y. Li, C. Xiong, and P. Tan, "Soft error evaluation and vulnerability analysis in Xilinx Zynq-7010 system-on-chip," Nuclear Instrum. Methods Phys. Res. A, vol. 831, pp. 344-348, Apr. 2016.

[7] G. Fey, A. Sulflow, and R. Drechsler, "Computing bounds for fault tolerance using formal techniques," in Proc. 46th ACM/IEEE Design Automation Conference (DAC), San Francisco, CA, USA, 2009, pp. 190-195.

[8] N. Miskov-Zivanov and D. Marculescu, "Circuit reliability analysis using symbolic techniques," IEEE Trans. Comput.-Aided Des. Integr. Circuits Syst., vol. 25, no. 12, pp. 2638-2649, Dec. 2006.

[9] G. Asadi and M. B. Tahoori, "Soft error rate estimation and mitigation for SRAM-based FPGAs," in Proc. 13th Int. Symp. Field-Programmable Gate Arrays (FPGA), New York, NY, USA, 2005, pp. 149-160.

[10] W. Xiao and W. Qian, "ASPPLN: Accelerated Symbolic Probability Propagation in Logic Network," in Proc. 2022 IEEE/ACM International Conference on Computer-Aided Design (ICCAD), San Diego, CA, USA, 2022, pp. 1-9.

TABLE II. COMPARISON OF HYBRIDEPP WITH OTHER METHOD AND SIMULATION

Circuit	Symbolic			Numerical			HybridEPP			Simulation
	Max	Ave	Time(s)	Max	Ave	Time(s)	Max	Ave	Time(s)	Time(s)
9symml	0.1995	0.1225	1.689	0.2526	0.0978	0.007	0.1608	0.0505	0.837	28.870
alu2	0.1728	0.0286	9.966	0.1651	0.0209	0.022	0.1587	0.0170	1.194	25.133
apex7	0.1420	0.0231	1.483	0.1894	0.0256	0.007	0.2063	0.0207	0.024	40.447
c1355	0.0428	0.0073	13.647	0.0428	0.0048	0.020	0.0428	0.0048	0.382	46.323
c1908	0.1265	0.0133	10.989	0.1408	0.0131	0.016	0.1408	0.0118	2.845	76.400
c2670	0.1966	0.0087	9.684	0.3291	0.0354	0.034	0.2023	0.0286	2.477	128.880
c499	0.0206	0.0073	12.921	0.0205	0.0048	0.019	0.0206	0.0047	0.413	33.733
c880	0.1209	0.0160	5.877	0.1623	0.0125	0.018	0.1211	0.0107	0.324	41.237
average	0.1277	0.0283	8.282	0.1628	0.0269	0.018	0.1317	0.0186	1.062	52.628

979-8-3315-3918-4/25 $31.00 © 2025 IEEE

Success-Rate Improvement of Analog Circuit Topology Generation by Large Reasoning Model

Koutaro Hachiya*[1], Kentaro Yoshikawa[1], Atsushi Kurokawa[2]

[1] Teikyo Heisei University, Tokyo, Japan
[2] Hirosaki University, Aomori, Japan

* Email: k.hachiya@thu.ac.jp

Abstract— Analog circuit topology generation remains a challenging task due to the vast design space and stringent performance constraints. While prior approaches such as AnalogCoder have demonstrated the potential of large language models (LLMs) for automating topology synthesis, their success rates remain limited due to invalid netlists, simulation failures, and unmet specifications. This work addresses these limitations by introducing a Large Reasoning Model (LRM) that structures the design process into interpretable subproblems and incorporates simulation-guided feedback during iterative refinement. Experimental results show a significant improvement in topology-generation success rate compared with baseline methods. The paper also discusses broader opportunities for LLMs and vision-language models in analog design automation.

Keywords—analog circuit, topology generation, generative artificial intelligence, large language model, reasoning model

I. INTRODUCTION

While the share of analog circuitry in modern systems has shrunk with the rise of sophisticated digital signal processing [1], analog blocks remain indispensable wherever electronics interface with humans or the physical world and in high-speed communication front ends. Even within predominantly digital SoCs, power transfer and regulation, clock synthesis and distribution, biasing, and sensing rely on analog circuits that determine overall system robustness, efficiency, and performance.

Compared to digital design automation, progress in analog automation has lagged due to several intrinsic challenges. The search space couples discrete topology choices with continuous, highly nonlinear sizing, objectives are multi-criteria and often conflicting, and performance must be met across process, voltage, and temperature corners. Reliable evaluation requires expensive simulation loops that may fail to converge, labeled design data are scarce, and successful practice still encodes tacit expert heuristics that are hard to formalize.

Recent advances in AI, especially transformer-based large language models (LLMs), have enabled autonomous execution of tasks previously accessible only to human experts, including code synthesis, tool orchestration, and multi-step problem solving. Their application to electronic design automation is growing [2]: adoption began in digital flows and is steadily extending to analog tasks such as specification interpretation, constraint formulation, sizing, and verification. Among these efforts, topology generation stands out as particularly difficult because it demands both symbolic structural reasoning and quantitative performance awareness.

AnalogCoder is a representative prior approach that applies LLMs to analog circuit topology generation [3]. It demonstrated that language-driven synthesis can discover feasible graphs, yet its overall success rate, defined as the fraction of generated topologies that both simulate successfully and satisfy target specifications, remains limited due to invalid netlists, SPICE non-convergence, and unmet constraints. These limitations point to the need for stronger, explicitly structured reasoning and tighter feedback from verification to guide generation.

This work addresses the limitations of existing LLM-based topology generation by replacing the LLM in AnalogCoder with a Large Reasoning Model (LRM) and evaluating its impact on generation success rates. We focus on assessing whether the reasoning capabilities of LRMs can improve the generation of complex analog circuit topologies compared to conventional LLMs.

The remainder of the paper is organized as follows. Section II reviews related work on LLM-assisted analog design. Section III presents our methodology and experimental results on improving topology generation success rate. Section IV discusses broader opportunities for LLMs and vision-language models in analog design workflows. Section V concludes.

II. LARGE LANGUAGE MODELS IN ANALOG CIRCUIT DESIGN

A. Design Process of Analog Circuits

The design process of an analog circuit can be summarized as follows:

1) Determine the required specifications.
2) Select an appropriate circuit architecture that meets the specification and decompose it into functional blocks.
3) Decide on the components to be used and their interconnections, then create a circuit schematic.
4) Perform circuit simulations considering manufacturing variations to optimize design parameters and verify the circuit.
5) Conduct layout design, determining the physical placement of components and wiring.
6) Perform circuit simulations that account for parasitic elements introduced by the layout and conduct the final verification.

In previous applications of LLMs, only a subset of the above design steps has typically been addressed. Few approaches have attempted to cover multiple steps in an integrated manner. Therefore, in the following subsections, we introduce representative examples of LLM applications in analog circuit

This work was supported by JSPS KAKENHI Grant Number 23K22730.

design, organized sequentially according to the steps outlined above.

B. Application Examples of LLMs

1) Specification and Architecture Determination: In the initial step of determining the design requirements, no concrete applications of LLMs have been identified thus far. However, setting realistic specifications—while considering trade-offs among the required PPA (Power, Performance, and Area), cost, and delivery schedule—is not an easy task, and AI-based support is therefore desirable. Examples supporting Step 2 are rare, with only a simple implementation found in Atelier [4]. There are also cases where circuit diagrams and design information are extracted from datasheets or textbooks to build databases or to train LLMs, which could potentially be leveraged. DocEDA combines LLMs with computer vision techniques to read circuit diagrams, parameters, and descriptive text from datasheets, enabling the construction of a database of SPICE netlists [5]. AMSnet similarly extracts circuit diagrams and related information from textbooks to create datasets containing SPICE netlists and their descriptions [6][7]. Masala-CHAI likewise builds datasets of SPICE netlists and descriptive text from textbooks, and furthermore uses these datasets to train LLMs and evaluate the success rate of SPICE netlist generation [8].

2) Topology and Schematic Generation: Circuit topology generation for Step 3 is currently one of the hottest research areas. Existing approaches can be broadly classified into three categories:

- those that explore new circuit topologies,

- those that use general-purpose LLMs without additional training to generate circuit topologies, and

- those that incorporate domain knowledge via retrieval-augmented generation (RAG) or supervised fine-tuning (SFT) before generating circuit topologies.

LaMAGIC randomly generates power converter circuit topologies, runs circuit simulations, and fine-tunes an LLM using the simulation results. After SFT, the LLM can generate an optimal circuit topology when given specifications [9]. AnalogGenie employs a sequence-based graph representation that expresses devices at the pin level and trains on a dataset of 3,350 real circuits, successfully generating high-performance circuits not present in the dataset. It uses a custom tokenizer and a decoder-only transformer for pretraining; therefore, strictly speaking, it is not an LLM [10].

AnalogCoder uses one-shot prompting to have an LLM generate PySpice circuit descriptions, then verifies the results through circuit simulation. If an error occurs during verification, the error message is fed back to the LLM for regeneration, repeating this loop [3]. It also defines a benchmark consisting of 24 circuits. SPICEPilot follows a similar approach: it verifies PySpice circuit descriptions generated by an LLM, feeds detailed error information back to the LLM, and iterates the regeneration loop. The generated circuit descriptions are intended for use as training datasets for LLMs. A benchmark of 60 circuits is also defined [11]. AnalogXpert also uses a general-purpose LLM with a feedback loop, but its prompting and feedback strategies are closer to those used by analog design experts [12].

As an example of incorporating domain knowledge into an LLM using RAG, Atelier adopts this approach [4]. Additionally, Masala-CHAI, mentioned earlier, improves topology generation success rates by training on SPICE netlists through SFT [8].

There are also examples of applying LLMs to the conversion from netlists to schematics. Schemato enables netlist-to-schematic conversion by fine-tuning an LLM on a dataset of LTSpice netlists and schematic files (.asc) [13].

3) Parameter Optimization: Several studies have explored the use of LLMs for device parameter optimization in Step 4. ADO-LLM is a framework that integrates Bayesian Optimization (BO) with LLMs for analog circuit parameter optimization. By combining the exploration capability of BO with the knowledge-driven design suggestions of LLMs, it improves both the efficiency of design space exploration and the quality of the resulting designs [14]. AnalogCoder-Pro extends AnalogCoder by not only generating topologies but also assisting in extracting parameters to be optimized and supporting their optimization through BO with the help of an LLM [15].

4) Layout Design: There are still few efforts to leverage LLMs for layout design in Step 5. Reference [16] uses an LLM to create layout constraints for layout synthesis based on BO. LayoutCopilot enables natural language interaction with layout tools, automatically breaking down abstract instructions into detailed operations, reducing the need for manual GUI manipulation [17].

5) Test Bench Generation: Here we present an example of using LLMs for test bench generation, which is required in Step 6. AnalogTester extracts experimental conditions and evaluation metrics from research papers and builds an experimental scheme template along with a TED function database, thereby enabling test bench code generation using RAG [18].

III. SUCCESS RATE IMPROVEMENT IN ANALOGCODER

AnalogCoder uses a general-purpose LLM to generate analog circuit topologies without applying SFT or RAG; however, its success rate was low when generating complex circuits. To address this, in this work we employ a Large Reasoning Model (LRM) trained to autonomously perform Chain-of-Thought (CoT) reasoning for analog circuit topology generation and evaluate its success rate.

A. AnalogCoder Benchmark

Table 1 lists the circuits that make up the benchmark used in AnalogCoder [3]. The benchmark consists of 24 circuits, classified as follows: 1–8 are easy, 9–13 are medium, and 14–24 are hard. In addition, circuits 1–15 are basic circuits, while 16–24 are composite circuits. For composite circuit generation, the prompt encourages the use of sub-circuits registered in the circuit tool library. The success rate of benchmark circuit generation by AnalogCoder using GPT-4o [3] is shown in the second column of Table 2.

TABLE I.   ANALOGCODER BENCHMARK [3]

Id	Type	Circuit Description	Id	Type	Circuit Description
1	Amplifier	Common-source amp. with R load	13	Opamp	Common-source op-amp with R loads
2	Amplifier	3-stage common-source amplifier with R loads	14	Opamp	2-stage op-amp with active loads
3	Amplifier	Common-drain amp. with R load	15	Opamp	Cascode op-amp with cascode loads
4	Amplifier	Common-gate amp. with R load	16	Oscillator	RC Shift oscillator
5	Amplifier	Cascode amp. with R load	17	Oscillator	Wien Bridge oscillator
6	Inverter	NMOS inverter with R load	18	Integrator	Op-amp integrator
7	Inverter	Logical inverter with NMOS and PMOS	19	Differentiator	Op-amp differentiator
8	Current Mirror	NMOS constant current source with R load	20	Adder	Op-amp adder
9	Amplifier	Common-source amp. with diode-connected load	21	Subtractor	Op-amp subtractor
10	Amplifier	2-stage amplifier with Miller compensation C	22	Schmitt trigger	Non-inverting Schmitt trigger
11	Opamp	Op-amp with active current mirror loads	23	VCO	Voltage-Controlled Oscillator
12	Current Mirror	Cascode current mirror	24	PLL	Phase-Locked Loop

B. Success Rate Improvement Methods

General-purpose LLMs often lack sufficient knowledge about analog circuits and SPICE/PySpice descriptions. Adding such domain knowledge through SFT or RAG can improve the success rate. For example, additional training via SFT using a dataset of analog circuits with SPICE netlists created by Masala-CHAI improved the topology generation success rate of AnalogCoder [3].

Another approach is to provide reasoning processes in the prompts and apply Chain of Thought (CoT), Tree of Thoughts (ToT), or Graph of Thoughts (GoT) [19], which is expected to enhance the success rate of topology generation. However, creating a universal prompt applicable to various analog circuits is not easy. Therefore, this paper applies a Large Reasoning Model (LRM) [20] that has been fine-tuned on diverse reasoning processes.

There are also attempts to improve topology generation success rates using Vision-Language Models (VLMs). For instance, AnalogCoder-Pro incorporates VLM feedback by recognizing circuit simulation waveform images and feeding this information back into the circuit generation loop [15].

Moreover, many recently released LLMs feature tool-calling capabilities, allowing them to act as agents that autonomously invoke functions or external tools [21]. By enabling an LLM to call external tools during the reasoning process to search for relevant information or evaluate the circuit under generation, it is expected that the success rate of generation can be further improved.

C. Improvement by Large Reasoning Models

Table 2 shows the success rate (Pass@1) of benchmark circuit generation when replacing the LLM used in AnalogCoder with various models. Each circuit was generated 20 times to calculate the success rate. We evaluated representative (LLM, LRM) pairs from major providers: OpenAI (GPT-4o mini, o4 mini high), Google (Gemini 2.5 Flash, Gemini 2.5 Pro), DeepSeek (DeepSeek-V3 0324, DeepSeek-R1 0528), and Alibaba (Qwen3 235B A22B 2507, Qwen3 235B A22B Thinking 2507). For combinations involving an LRM, the paired LLM was selected as the base model for that LRM, except for the OpenAI case. For example, DeepSeek's LRM, DeepSeek-R1 0528, was developed based on its LLM, DeepSeek-V3 0324.

When comparing the success rates of LLMs and LRMs for each provider individually, the average success rate for composite circuits (Id = 16–24) improved by a factor of 1.47–1.91 with LRMs compared to LLMs excluding OpenAI's models. In contrast, for basic circuits (Id = 1–15), there was no significant difference between LLMs and LRMs.

IV. OPPORTUNITIES FOR FUTURE RESEARCH

As seen in Section II-B, current research efforts are primarily focused on creating training datasets for analog circuits and on topology generation, with generation systems gradually integrating device parameter optimization. The application of LLMs and VLMs to layout design automation remains limited and is an area of future interest. Similarly, the use of LLMs for analog circuit verification and testing is still scarce. Furthermore, promising directions include the generation of Built-In Self-Test (BIST) circuits and improving defect coverage in analog testing.

V. CONCLUSIONS

In this paper, we reviewed research on leveraging LLMs in analog circuit design. Furthermore, we demonstrated that applying an LRM to topology generation significantly improves the success rate for composite circuits. Finally, we highlighted areas where automation is expected to advance through the application of LLMs.

REFERENCES

[1] P. Toledo, R. Rubino, F. Musolino and P. Crovetti, "Re-Thinking Analog Integrated Circuits in Digital Terms: A New Design Concept for the IoT Era," IEEE Trans. on Circuits and Systems II: Express Briefs, vol. 68, no. 3, pp. 816–822, March 2021.

[2] J. Pan, G. Zhou, C.-C. Chang, I. Jacobson, J. Hu, and Y. Chen, "A Survey of Research in Large Language Models for Electronic Design Automation," ACM Trans. Des. Autom. Electron. Syst. 30, 3, pp. 1–21, May 2025.

[3] Y. Lai, et al., "AnalogCoder: Analog Circuit Design via Training-Free Code Generation," The Thirty-Ninth AAAI Conference on Artificial Intelligence (AAAI-25), Feb. 2025, pp. 379–387.

[4] J. Shen, et al., "Atelier: An Automated Analog Circuit Design Framework via Multiple Large Language Model-Based Agents," IEEE Transactions on Computer-Aided Design of Integrated Circuits and Systems, May 2025.

[5] H. C. Chen, et al., "DocEDA: Automated Extraction and Design of Analog Circuits from Documents with Large Language Model," arXiv:2412.05301 [cs.AR], Nov. 2024.

[6] Z. Tao, et al., "AMSNet: Netlist Dataset for AMS Circuits," 2024 IEEE LLM Aided Design Workshop (LAD), San Jose, CA, USA, 2024, pp. 1–5.

[7] Y. Shi, et al., "AMSnet 2.0: A Large AMS Database with AI Segmentation for Net Detection," arXiv:2505.09155 [cs.CV], May 2025.

[8] J. Bhandari, V.P. Bhat, Y. He, S. Garg, H. Rahmani and R. Karri, "Masala-CHAI: A Large-Scale SPICE Netlist Dataset for Analog Circuits by Harnessing AI," arXiv:2411.14299 [cs.AR], Mar. 2025.

[9] C. C. Chang, Y. Shen, S. Fan, J. Li, S. Zhang, N. Cao, Y. Chen and X. Zhang, "LaMAGIC: language-model-based topology generation for analog integrated circuits," Proceedings of the 41st International Conference on Machine Learning (ICML'24), Vol. 235. July 2024, pp. 6253–6262.

TABLE II. SUCCESS RATE (PASS@1) OF THE BENCHMARK CIRCUIT GENERATION

Id	GPT-4o [3]	GPT-4o mini	o4 mini high	Gemini 2.5 Flash	Gemini 2.5 Pro	DeepSeek -V3 0324	DeepSeek -R1 0528	Qwen3[a]	Qwen3 Thinking[b]
1	100	85	100	100	100	100	100	100	100
2	100	100	90	95	100	100	100	100	100
3	100	90	100	100	95	100	100	100	100
4	100	95	100	100	100	100	100	100	100
5	100	95	90	100	80	100	95	100	100
6	100	100	100	100	100	100	100	100	100
7	100	90	100	100	100	100	100	100	100
8	100	90	60	45	0	100	70	0	20
9	100	15	95	0	25	100	75	85	100
10	100	45	100	100	100	100	100	100	100
11	100	0	0	0	0	0	0	0	0
12	13.3	0	25	10	10	60	15	35	25
13	100	0	0	0	0	0	0	0	0
14	73.3	0	0	0	0	0	0	0	0
15	13.3	0	0	0	0	0	0	0	0
16	6.7	0	70	70	10	0	60	25	85
17	0	5	60	25	90	10	45	10	50
18	100	0	0	0	0	0	0	0	0
19	60	0	0	0	0	0	0	0	0
20	100	0	90	5	95	35	95	65	100
21	20	0	95	90	100	100	100	80	100
22	0	0	5	0	0	5	5	5	0
23	0	5	90	50	95	80	80	45	90
24	0	0	5	25	0	30	15	0	15
Avg. (1–15)	86.7	53.7	57.3	56.7	54.0	70.7	63.7	61.3	63.0
Avg. (16–24)	31.9	1.10	46.1	29.4	43.3	28.9	44.4	25.6	48.9

[a.] Qwen3 235B A22B 2507

[b.] Qwen3 235B A22B Thinking 2507

[10] J. Gao, W. Cao, J. Yang, X. Zhang, "AnalogGenie: A Generative Engine for Automatic Discovery of Analog Circuit Topologies," The Thirteenth International Conference on Learning Representations (ICLR), April 2025.

[11] D. Vungarala, S. Alam, A. Ghosh and S. Angizi, "SPICEPilot: Navigating SPICE Code Generation and Simulation with AI Guidance," IEEE International Conference on Rebooting Computing (ICRC), 2024, pp. 1–6.

[12] H. Zhang, S. Sun, Y. Lin, R. Wang, J. Bian, "AnalogXpert: Automating Analog Topology Synthesis by Incorporating Circuit Design Expertise into Large Language Models," arXiv:2412.19824 [cs.AR], June 2025.

[13] R. Matsuo, et al., "Schemato – An LLM for Netlist-to-Schematic Conversion," arXiv:2411.13899 [cs.LG], June 2025.

[14] Y. Yin, Y. Wang, B. Xu, P. Li., "ADO-LLM: Analog Design Bayesian Optimization with In-Context Learning of Large Language Models," The 43rd IEEE/ACM International Conference on Computer-Aided Design. April 2025, pp. 1–9.

[15] Y. Lai, et al., "AnalogCoder-Pro: Unifying Analog Circuit Generation and Optimization via Multi-modal LLMs," arXiv:2508.02518 [cs.LG], Aug. 2025.

[16] G. Chen, "LLM-Enhanced Bayesian Optimization for Efficient Analog Layout Constraint Generation," arXiv:2406.05250 [cs.AI], Dec. 2024.

[17] B. Liu, et al., "LayoutCopilot: An LLM-Powered Multiagent Collaborative Framework for Interactive Analog Layout Design," IEEE Transactions on Computer-Aided Design of Integrated Circuits and Systems, vol. 44, no. 8, pp. 3126–3139, Aug. 2025.

[18] W. Chen, et al., "AnalogTester: A Large Language Model-Based Framework for Automatic Testbench Generation in Analog Circuit Design," 2025 International Symposium of Electronics Design Automation (ISEDA), May 2025, pp. 201–207.

[19] M. Besta et al., "Graph of thoughts: solving elaborate problems with large language models," The Thirty-Eighth AAAI Conference on Artificial Intelligence and Thirty-Sixth Conference on Innovative Applications of Artificial Intelligence and Fourteenth Symposium on Educational Advances in Artificial Intelligence (AAAI'24/IAAI'24/EAAI'24), Vol. 38. AAAI Press, Article 1972, pp. 17682–17690, 2024.

[20] Z.-Z. Li et al., "From System 1 to System 2: A Survey of Reasoning Large Language Models," arXiv:2502.17419 [cs.AI], June 2025.

[21] H. Xu et al., "Alignment for Efficient Tool Calling of Large Language Models," arXiv:2503.06708 [cs.CL], March 2025.

Fast Thermal-driven 3D Fixed-outline Floorplanning By Learning-based Thermal Analysis

Yikai Liu, Jindong Zhou, Jiayi Li, Pingqiang Zhou

School of Information Science and Technology, ShanghaiTech University

Shanghai, China

{liuyk2023,zhoujd,lijy22023,zhoupq}@shanghaitech.edu.cn

Abstract—Higher integration density in 3D ICs brings severe thermal issues which potentially degrades the performance and reliability of a 3D chip. To address this problem, thermal-aware optimization is necessary at early floorplanning stage. In this work, we proposed a novel thermal-driven 3D fixed-outline floorplanning algorithm with learning-based fast thermal evaluation method. To integrate the neural network model into the floorplanning framework, we also proposed an adaptive cost function. Experimental results shows that compared with thermal simulation tool *HotSpot* [1], our method can speed up the thermal analysis by about 3700× with high accuracy. Leveraging the accuracy and efficiency of deep neural network, the floorplanner can effectively reduce the peak temperature of the chip.

Index Terms—3D-IC, Floorplan, Thermal Analysis, Deep Neural Networks

I. INTRODUCTION

Three Dimensional (3D) IC technology develops rapidly recently with the maturity of process and manufacturing [2]. While enjoying the advantages of small package area and high bandwidth, the 3D systems are suffering from severe thermal problems brought by the die stacking [3]–[5], as power density increases linearly with the number of stacked tiers.

Many investigations have been conducted to alleviate the thermal problems of Integrated Circuits (IC) [6] in physical design, by arranging the layout of the circuit modules to reduce the peak temperature. For 3D circuits, thermal-aware floorplanning and placement have also been explored [7]–[9]. The general idea is to adjust the locations of modules or macros according to the results of thermal analysis after each iteration.

In our work, we focus on 3D floorplanning. During the iterative process of 3D fixed-outline floorplanning, *it is complex and time consuming to do 3D thermal analysis.* Accurate evaluation methods like Finite Element Method (FEM) and classical thermal analysis tool *HotSpot* [1] involve complex matrix solving, which consumes massive amount of time. When it comes to a multi-tier 3D circuit, the analysis time is even longer. However, typical optimization process requires thousands of iterations, thus makes accurate thermal analysis impractical. To address this problem, existing methods usually use simplified approximation thermal models to guide the optimization, including green function method [10], power-blurring method (PB) [11], [12], or directly using the power

This work is supported by the Science and Technology Commission of Shanghai Municipality (STCSM) under Grant 24JD1402500.

features of each block like vertical-heat-flow method (VHF) [13]–[15] as an indicator of the thermal performance.

In recent years, deep learning has shown its efficiency in fitting complex functions with powerful feature extraction and regression capabilities. Many researches have adopted deep learning to accelerate the thermal evaluation [16]–[19] of a 2D or 3D IC. Results show that a well-trained model can greatly accelerate the expensive thermal analysis process and accurately predict a temperature distribution.

In our work, we accelerate the thermal-driven 3D floorplanning process by incorporating learning-based thermal analysis method. The contribuitons of this work include:

- We develop a thermal-aware fixed-outline floorplan framework for 3D-IC, leveraging sequence pair as the floorplan representation and simulated annealing as the optimization method.
- We use deep neural networks to estimate the final steady-state temperature distribution from 3D-IC layout and power distribution in each iteration of 3D floorplanning.
- We evaluate the two-tier 3D-IC on GSRC benchmarks. The results show that our method can reduce the maximum temperature by 11.26K on the largest case. The learning-based model can accelerate the thermal evaluation by about 3700× compared with *HotSpot* simulation.

II. OUR WORK

A. Learning-based Thermal Analysis

To avoid the huge time cost of traditional thermal evaluation methods like *HotSpot* or FEM while achieving high accuracy, we propose a novel neural network based regression model to evaluate the steady-state temperature distribution of a 3D chip. In order to capture the relationship between power distribution and temperature effectively, we choose U-Net [20] architecture as the backbone of our model.

1) Temperature distribution prediction: For each tier, the input floorplan is first converted into a power map P. We divide the floorplan region into $N \times N$ grids, and the power at grid (i, j) is calculated by

$$p_{ij} = \sum_{b \in B} Overlap_{ij}(b) \times PowerDensity(b)$$

where B is the set of all blocks, $Overlap_{ij}(b)$ is the overlap area between grid (i, j) and the block b. Then, the power map is fed into the trained model and the final output temperature

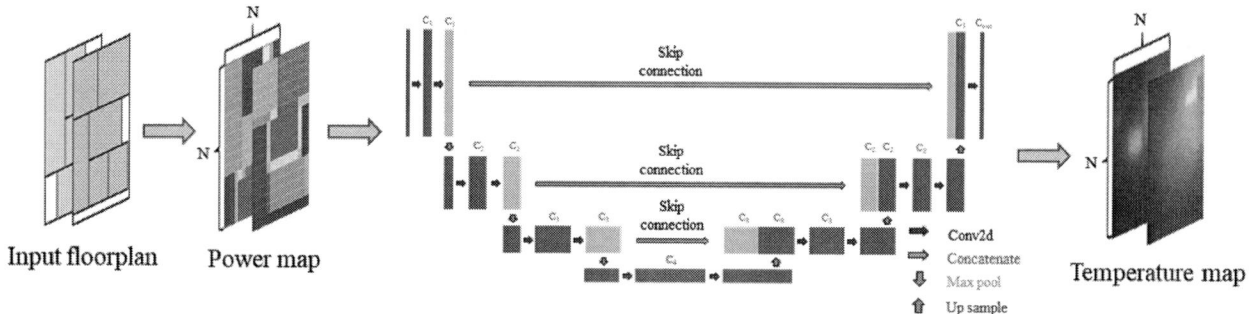

Fig. 1. The proposed thermal analysis flow and the U-Net network architecture .

map size is $N \times N$. The proposed flow and the network structure of our U-Net is shown in Fig. 1. It's worth noting that the model size is determined by the intermediate channel number $[C_1, C_2, C_3, C_4]$. To enable the thermal evaluation in the optimization process, a trade-off is made between the size and accuracy of the model which will be discussed in Section III.

2) Dataset Generation: In order to train the U-Net model for thermal analysis, we employ *HotSpot* to generate the groundtruth temperature distribution of a given 3D chip. The chip structure is shown in Fig. 2. In this work, we modeled

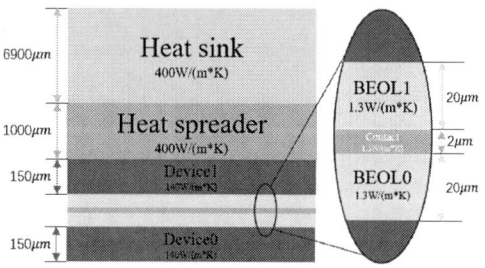

Fig. 2. 2-tier face-to-face chip structure.

a 2-tier face-to-face 3D system. The whole structure has seven layers: Device0, BEOL0, Contact, BEOL1, Device1, heat spreader and a heat sink.

One key procedure in dataset generation is to produce random legal floorplans. We develop an algorithm based on the slicing tree representation [21], where the floorplan can be viewed as the result of recursively applying horizontal and vertical cuts to a rectangle. The detail of the recursive process is described by algorithm 1, where the strategy set S = {CutLeft, CutRight, CutTop, CutBottom, BisectHorizontal, BisectVertical}, CutX means discard the corresponding part of current node. Fig. 3 gives an example of how these strategies are applied to a node. min_ratio controls the minimal block area and k controls the average block size. A larger k will cause the bisection stop earlier and lead to a larger average block size.

Algorithm 1 recursive_bisection

Input: node, S;

1: area = node.width × node.height
2: ratio = area / total_chip_area
3: prob = $\max(\frac{\text{ratio-min\_ratio}}{\text{1-min\_ratio}}, 0)^k$
4: **if** random() < prob **then**
5:     strategy = random_choice(S)
6:     new_nodes = apply_strategy(node, strategy)
7:     **for** new_node in new_nodes **do**
8:         recursive_bisection(new_node, S)
9:         node.children.append(new_node)

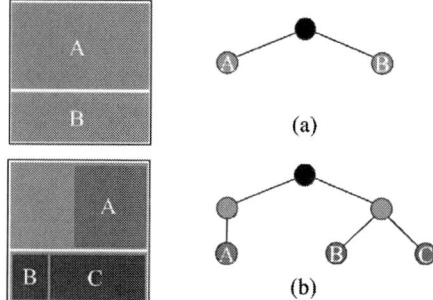

Fig. 3. An example of floorplan generation. (a) The original floorplan and tree representation. (b) The floorplan after applying CutLeft strategy on tree node A and BisectHorizontal on tree node B.

B. Thermal-driven Fixed-outline Floorplanning

The thermal-aware fixed-outline floorplanning problem can be formulated as follows. Given a set of M blocks $B = \{b_i | 1 \le i \le M\}$ with width w_i and height h_i, the goal is to find the position x_i, y_i and the orientation o_i of each block, such that $0 \le x_i \le W - w_i$, $0 \le y_i \le H - h_i$ and the blocks should not overlap with each other. Here W, H are the width and height of the chip. In the meanwhile, the wirelength and peak temperature should be optimized.

In this work, we use a simulated-annealing based method with multi-tier sequence pair representation to solve the problem. The cost function of simulated-annealing method can be written as

$$cost = WL + \lambda C_T + \mu C_A$$

where WL is the total HPWL of the circuit, C_T is the thermal cost and C_A is the area cost. Here, we use the total exceeding width and height as the penalty to the fixed-outline constraint, i.e. $C_A = \max(0, W_f - W) + \max(0, H_f - H)$. W_f and H_f are the width and height of the bounding rectangle of the floorplan. λ, μ control the penalty strength.

However, special caution should be taken when adding the temperature cost C_T. In our learning-based method, the peak temperature can only be obtained when the floorplan satisfies the fixed-outline constraint. However, this constraint is often violated during the early stage of the annealing process. Thus, the actual cost function should be:

$$cost = \begin{cases} WL + \lambda C_T & \text{if constraint is met} \\ WL + \lambda C_{T0} + \mu C_A & \text{otherwise} \end{cases} \quad (1)$$

As a result, a proper value of the constant C_{T0} needs to be assigned before getting the first legal floorplan.

In fact, the value of C_{T0} cannot be chosen arbitrarily. As shown in Fig. 4, a high C_{T0} will enforce the solver to search only within feasible solutions, while a low C_{T0} may fail to find a solution if the peak temperature of feasible floorplan is higher. This is mainly caused by the "discontinuity" of the cost function in the vicinity of feasible region boundary. To address the problem, we use an adaptive cost function to alleviate the discontinuity:

$$cost = \begin{cases} WL + \lambda (C_T - C_{init}), & \text{if constraint is met} \\ WL + \mu C_A, & \text{otherwise} \end{cases}$$
$$(2)$$

Instead of adding λC_{T0} to the cost of unsatisfied case, we decrease the cost of the satisfied case by λC_{init}, where C_{init} is the peak temperature of the first feasible solution during the optimization process.

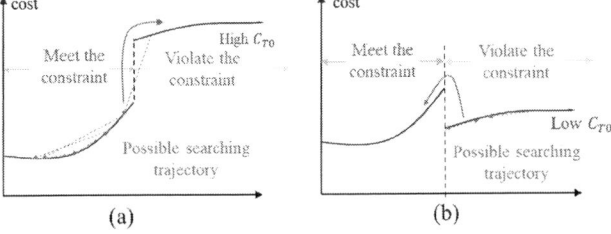

Fig. 4. Improper C_{T0} choice. (a) A high C_{T0} will cause the solver to only search within feasible solutions. (b) A low C_{T0} might fail to find a feasible solution.

III. EXPERIMENTAL RESULTS

A. Experiment setup

We implement the proposed floorplanning algorithm in C/C++ on a Linux environment using an Intel i9-13900K processor. The neural network model is trained with PyTorch on an RTX 4070 GPU. We evaluate our algorithm using the GSRC [22] benchmark suite. For each benchmark, we reserve

10% of the total block area as whitespace. The width and height of the 2-tier chip are calculated as:

$$W = H = \sqrt{\frac{\text{total\_block\_area} \times (1 + 0.1)}{2}}.$$

The grid size used for both the power map and temperature map is 64×64. The power of each block ranges from 0.05W to 1W.

B. Accuracy and Runtime of the Model

To train the neural network model, we generate 10,000 random legal floorplans as the dataset. The dataset is split into a training set with 7,000 samples, a validation set with 1,500 samples, and a test set with 1,500 samples. To balance runtime and accuracy, we fine-tune the number of intermediate channels in the model to $[C_1, C_2, C_3, C_4] = [4, 8, 16, 24]$, as this configuration achieves the best performance on the validation set with relatively few parameters. The model is evaluated on the test set using root mean square error (RMSE) and mean absolute error (MAE). The test results and runtime are shown in Table I. Compared with *HotSpot* simulation, our model achieves approximately 3700× speedup with acceptable accuracy loss. More importantly, the significant runtime reduction enables peak temperature evaluation during the floorplanning iteration process. Fig. 5 shows a comparison between the thermal map predicted by our U-Net model and the groundtruth thermal map generated by *HotSpot*.

TABLE I
MODEL ACCURACY AND RUNTIME COMPARISON

Metrics	Our model	HotSpot 7.0 [1]
RMSE	0.56K	
MAE	0.39K	
RMSE (peak temperature)	1.34K	groundtruth
MAE (peak temperature)	0.91K	
runtime	0.87ms	3208ms
(speed up)	(3687.4x)	(1x)

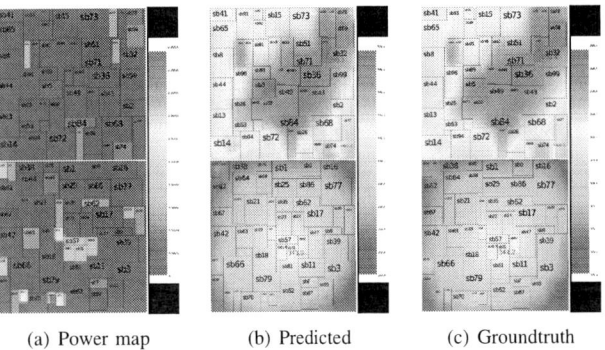

(a) Power map   (b) Predicted   (c) Groundtruth

Fig. 5. Comparison between the predicted temperature map and groundtruth temperature map of n100.

C. Result of our Thermal-driven 3D Floorplanning Framework

To demonstrate the effectiveness of our algorithm, we compare the 3D floorplan results with 1) a basic wirelength-driven method and 2) the vertical-heat-flow (VHF) method [14], [15]. The VHF method does not perform full thermal analysis; instead, it estimates peak temperature by leveraging the power map and thermal resistance. While this approach significantly reduces runtime, it may compromise accuracy. For each method, we run 10 trials and report the average to mitigate the randomness inherent in the simulated annealing algorithm. The results are presented in Table II.

TABLE II
AVERAGE WIRELENGTH, PEAK TEMPERATURE AND RUNTIME OF WIRELENGTH-DRIVEN, VERTICAL-HEAT-FLOW(VHF) AND OUR METHOD.

Cases		n30	n50	n100	n200	n300
Wirelength Driven	WL	68292	106814	177384	361051	550347
	Peak T	333.50K	341.06K	355.93K	392.71K	436.28K
	RT	1.38s	10.52s	22.17s	104.05s	151.68s
VHF	WL	73157 (+4865)	108719 (+1878)	181094 (+3710)	367146 (+6095)	558405 (+8058)
	Peak T	330.74K (-2.76K)	337.36K (-3.70K)	353.30K (-2.63K)	388.63K (-4.08K)	429.63K (-6.65K)
	RT	1.72s	12.27s	24.11s	108.20s	157.93s
Our method	WL	71756 (+3464)	108429 (+1615)	181628 (+4244)	368015 (+6964)	560186 (+9839)
	Peak T	330.83K (-2.67K)	337.27K (-3.79K)	350.76K (-5.17K)	384.26K (-8.45K)	425.02K (-11.26K)
	RT	11.62s	62.94s	69.87s	207.32s	279.26s

Our learning-based method effectively reduces peak temperature with minimal wirelength overhead. For small cases, the VHF method can achieve similar temperature reduction as our method. While in larger cases such as n100, n200 and n300, due to the lack of horizontal thermal coupling consideration, the VHF method fails to separate high power density blocks from each other, thereby degrading the solution quality. In contrast, our method demonstrates superior performance, as the neural network effectively captures thermal coupling effects via convolution operations in the U-Net architecture.

Regarding overall runtime, neural network inference accounts for a significant portion. However, the inference overhead remains constant as circuit size increases. Thus, for larger circuits, the relative runtime overhead becomes smaller (e.g., $8.42\times$ for n30, $1.84\times$ for n300).

IV. CONCLUSION

This paper proposed a novel fixed-outline thermal-aware floorplanning algorithm. By introducing a nerual network based thermal evaluation method, accurate thermal map is available for the floorplanner in each iteration, which greatly improves the solution quality. To incorporate this evaluation method into the floorplanner with fixed-outline constraint, we presented an adaptive cost function to help the convergence of the stochastic searching. Experiment result shows that our method can effectively reduce the peak temperature with little wirelength overhead.

REFERENCES

[1] J.-H. Han, X. Guo, K. Skadron, and M. R. Stan, "From 2.5D to 3D chiplet systems: Investigation of thermal implications with HotSpot 7.0," in *IEEE Intersociety Conference on Thermal and Thermomechanical Phenomena in Electronic Systems (iTherm)*, 2022, pp. 1–6.

[2] Y. Zhao, L. Zou, and B. Yu, "Physical design for advanced 3D ICs: Challenges and solutions," in *Proceedings of the International Symposium on Physical Design (ISPD)*, 2025, pp. 209–216.

[3] S. S. Sapatnekar, "Addressing thermal and power delivery bottlenecks in 3d circuits," in *Proceedings of Asia and South Pacific Design Automation Conference (ASP-DAC)*, 2009, pp. 423–428.

[4] T. Lu, C. Serafy, Z. Yang, S. K. Samal, S. K. Lim, and A. Srivastava, "TSV-based 3-D ICs: design methods and tools," *IEEE Transactions on Computer-Aided Design of Integrated Circuits and Systems (TCAD)*, vol. 36, no. 10, pp. 1593–1619, 2017.

[5] A. Todri-Sanial and C. S. Tan, *Physical Design for 3D Integrated Circuits*. CRC Press, 2017.

[6] H. Sultan, A. Chauhan, and S. R. Sarangi, "A survey of chip-level thermal simulators," *ACM Comput. Surv.*, vol. 52, no. 2, 2019.

[7] S. K. Samal, S. Panth, K. Samadi, M. Saeidi, Y. Du, and S. K. Lim, "Adaptive regression-based thermal modeling and optimization for monolithic 3-D ICs," *IEEE Transactions on Computer-Aided Design of Integrated Circuits and Systems (TCAD)*, vol. 35, no. 10, pp. 1707–1720, 2016.

[8] J.-M. Lin, W.-Y. Chang, H.-Y. Hsieh, Y.-T. Shyu, Y.-J. Chang, and J.-M. Lu, "Thermal-aware floorplanning and tsv-planning for mixed-type modules in a fixed-outline 3-D IC," *IEEE Transactions on Very Large Scale Integration Systems (TVLSI)*, vol. 29, no. 9, pp. 1652–1664, 2021.

[9] P. Zhou, Y. Ma, Z. Li, R. P. Dick, L. Shang, H. Zhou, X. Hong, and Q. Zhou, "3D-STAF: scalable temperature and leakage aware floorplanning for three-dimensional integrated circuits," in *Proceedings of International Conference on Computer-Aided Design (ICCAD)*, 2007, pp. 590–597.

[10] S. S.-Y. Liu, R.-G. Luo, S. Aroonsantidecha, C.-Y. Chin, and H.-M. Chen, "Fast thermal aware placement with accurate thermal analysis based on green function," *IEEE Transactions on Very Large Scale Integration Systems (TVLSI)*, vol. 22, no. 6, pp. 1404–1415, 2014.

[11] W. Guan, X. Tang, H. Lu, Y. Zhang, and Y. Zhang, "Thermal-aware fixed-outline 3-D IC floorplanning: An end-to-end learning-based approach," *IEEE Transactions on Very Large Scale Integration Systems (TVLSI)*, vol. 31, no. 12, pp. 1882–1895, 2023.

[12] A. Ziabari, J.-H. Park, E. K. Ardestani, J. Renau, S.-M. Kang, and A. Shakouri, "Power blurring: Fast static and transient thermal analysis method for packaged integrated circuits and power devices," *IEEE Transactions on Very Large Scale Integration Systems (TVLSI)*, vol. 22, no. 11, pp. 2366–2379, 2014.

[13] G. Luo, Y. Shi, and J. Cong, "An analytical placement framework for 3-D ICs and its extension on thermal awareness," *IEEE Transactions on Computer-Aided Design of Integrated Circuits and Systems (TCAD)*, vol. 32, no. 4, pp. 510–523, 2013.

[14] J. Cong, G. Luo, J. Wei, and Y. Zhang, "Thermal-aware 3D IC placement via transformation," in *Proceedings of the Asia and South Pacific Design Automation Conference (ASP-DAC)*, 2007, pp. 780–785.

[15] L. Xiao, S. Sinha, J. Xu, and E. F. Young, "Fixed-outline thermal-aware 3d floorplanning," in *Proceedings of the Asia and South Pacific Design Automation Conference (ASP-DAC)*, 2010, pp. 561–567.

[16] L. Chen, J. Lu, W. Jin, and S. X.-D. Tan, "Fast full-chip parametric thermal analysis based on enhanced physics enforced neural networks," in *Proceedings of the International Conference on Computer Aided Design (ICCAD)*, 2023, pp. 1–8.

[17] Z. Liu, Y. Li, J. Hu, X. Yu, S. Shiau, X. Ai, Z. Zeng, and Z. Zhang, "DeepOHeat: Operator learning-based ultra-fast thermal simulation in 3D-IC design," in *Proceedings of the Design Automation Conference (DAC)*, 2023, pp. 1–6.

[18] L. Chen, W. Jin, and S. X.-D. Tan, "Fast thermal analysis for chiplet design based on graph convolution networks," in *Proceedings of the Asia and South Pacific Design Automation Conference (ASP-DAC)*, 2022, pp. 485–492.

[19] A. Sridhar, A. Vincenzi, M. Ruggiero, and D. Atienza, "Neural network-based thermal simulation of integrated circuits on gpus," *IEEE Transactions on Computer-Aided Design of Integrated Circuits and Systems (TCAD)*, vol. 31, no. 1, pp. 23–36, 2012.

[20] O. Ronneberger, P. Fischer, and T. Brox, "U-Net: Convolutional networks for biomedical image segmentation," *CoRR*, vol. abs/1505.04597, 2015.

[21] R. Otten, "Automatic floorplan design," in *Proceedings of the Design Automation Conference (DAC)*, 1982, pp. 261–267.

[22] http://vlsicad.eecs.umich.edu/BK/GSRCbench.

979-8-3315-3918-4/25 $31.00 © 2025 IEEE

Systematic design for coupled heterogeneous accelerators

Tim Todman, Wayne Luk

Department of Computing, Imperial College London, London, UK

{timothy.todman, w.luk}@imperial.ac.uk

Abstract—This paper explores a systematic approach for developing designs for heterogeneous accelerators coupled together so that the workload can be partitioned appropriately to suit the capability and resources of each accelerator. To model the potential performance gains, we introduce an analytic performance model supporting multiple distributed heterogeneous accelerators coupled together. Based on the performance model, we develop schemes to partition the work between accelerators to reduce overall system runtime. We illustrate our approach by targeting a system with a Groq AI accelerator coupled to a U250 FPGA. The strengths and drawbacks of this approach are evaluated based on a number of common workloads.

I. INTRODUCTION

Reconfigurable hardware such as FPGAs have often been used for compute-intensive linear algebra applications such as deep learning. Modern FPGAs include dedicated multiply and digital signal processing (DSP) blocks, and even dedicated hardware for neural network inference.

In this work, we explore the potential of accelerating FPGA designs using loosely-coupled, off-chip accelerators, such as those designed for artificial intelligence (AI). Conversely, we accelerate AI accelerators using finer-grain, memory-attached computation in FPGAs.

Computer systems are increasingly heterogeneous. Modern datacentres may include not just CPUs, but also GPUs, FPGAs, and lately more specialized accelerators for AI. Examples of the latter include Cerebras, Groq, Google TPU.

Recently there has been a large variety of new architectures for AI acceleration and related computations from Groq, Google, Cerebras, and various startups. We observe that FPGAs and AI accelerators, which we call NPUs (Neural Processing Units), have complementary strengths. NPUs are good at matrix-matrix multiply, matrix-vector multiply, and the non-linear parts of neural pipeline (e.g. SoftMax, ReLU). FPGAs by contrast are good at finer-grain bit- and word-twiddling. Both architectures have common strengths: allowing performance to be predicted statically, low latency processing.

FPGAs themselves are increasingly heterogeneous (hardened multipliers, memories, floating-point, even some AI acceleration in Xilinx Versal devices), but are unlikely to include all capabilities of NPUs.

There has been a rapid development of AI accelerators for deep learning, including multiple kinds of NPUs and FPGAs. We propose a model for developing systems based on such architectures that can be scaled to multiple cards and racks, within data centres.

We target *coupled heterogeneous accelerators (CHA)*: two or more different kinds of accelerators used closely together. Historically, accelerators have been used to offload from general-purpose computers; here, we offload from general purpose accelerators, such as FPGAs, to special-purpose or fixed-function accelerators, such as NPUs.

While NPUs are best suited to dense linear algebra and neural networks, FPGAs could help them to scale to problems with variable precision across the matrix, in space or time — unlike CPUs, GPUs and NPUs, FPGAs are not limited to small set of data types. They can be used to support data structures customised for specific applications.

The contributions of this paper are:

1) An approach to map computations to loosely-coupled accelerators in multiple hosts on a network;
2) An analytical performance model taking communication overhead into account, since this has a significant effect on CHAs;
3) A partitioning scheme based on the performance model;
4) Evaluation of the proposed approach based on matrix multiplication.

The rest of this paper is organized as follows. Section II describes related work. Section III introduces a performance model. Section IV presents a partitioning scheme. Section V evaluates the approach on matrix multiplication. Finally, section VI concludes and suggests future work.

II. BACKGROUND

Potted history: the earliest FPGAs included just lookup tables (LUTs) and flip-flops (FFs); later FPGAs included embedded memories, hard multipliers, digital signal processing (DSP) units, IO interfaces, and recent FPGAs include some artificial intelligence (AI) blocks.

Partitioning schemes: traditional approaches to partitioning act at low level, operating on graphs of bit-level operations. While this might give scope for the best partitions, the search space is large. We adopt a matrix-level partitioning scheme extending work on parallel computing.

Heterogeneous hardware: FPGAs are increasingly heterogeneous and include hardened cores for floating-point, direct network and bus attachment. Some FPGAs such as Xilinx Versal and Altera Stratix-10 NX include dedicated hardware to accelerate AI inferencing.

Matrix calculations: matrix computations have long been optimised for FPGAs, including both dense and sparse matrices.

979-8-3315-3918-4/25 $31.00 © 2025 IEEE

Recent work includes systolic arrays for matrix multiplication [1], by three-level tiling using the AI accelerators on Xilinx Versal FPGAs in CHARM [2] [3], sparse matrix multiply [4], and QR decomposition [5]. Compared to CHARM, our work uses loosely-coupled off-chip rather than on-chip accelerators.

III. PERFORMANCE MODEL

This section shows our performance model. A simple but effective model is adopted to cover various kinds of accelerators including FPGAs and NPUs.

The target architectures of the performance model are multiple machines, with one or more accelerators each. Initially we consider two kinds of accelerators, but the approach is not limited to them.

We consider both *fixed-function* and *fully configurable* accelerators, examples of which include respectively NPUs and FPGAs. Although many NPUs can be programmed, the set of functional units is fixed, so we call them fixed-function. By contrast, FPGAs can be configured to implement any function that will fit in the device.

Much work involves capturing performance using high-level models such as Amdahl's law. Such models result in a lower bound on performance, since parts of the system may have to wait for other parts and communication delays are not well modelled, if they are addressed at all. Our approach, in contrast, includes modelling communication delays to predict the performance of statically-scheduled accelerators.

The proposed performance model captures NPU and FPGA performance based on clock frequency, number of cycles run, bandwidth between host CPU and accelerator, and between accelerator and local memory.

Figure 1 illustrates our architecture and model parameters. The general-purpose network connecting hosts has bandwidth B_N, while there is bandwidth B_{HA1} (A1) and B_{HA2} (A2) between accelerators (A1 and A2) and their respective hosts. Both accelerators additionally have local memory connected via buses of bandwidth B_{A1M} and B_{A2M} respectively. Each machine could have more than one accelerator. All bandwidths are measured in bytes per second. We do not currently consider accelerators without hosts, or memory shared between host and accelerator.

Our network model is deliberately simple. General-purpose networks such as Ethernet have complex, non-deterministic behaviour. We assume that since B_N will in practice be much less than B_{HA1} or B_{HA2}, we can conservatively model it as a fixed bandwidth which will not affect the partitioning.

Our model adapts previous work modelling run-time reconfigurable FPGAs [6]. While we do not consider run-time reconfiguration in this work, this could be added to our model as an extension. The model considers a *baseline* design with a single accelerator, and compares with an accelerator enhanced by off-chip communication to other accelerators. The speedup S is the factor of improvement of the enhanced design over the baseline design:

$$S = \frac{T_b N_b}{T_c N_c + T_e N_e}$$

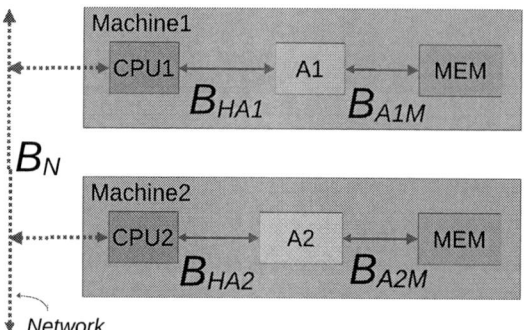

Fig. 1. Target architecture parameters, with accelerators A1 and A2: network bandwidth B_N, host-A1 bandwidth B_{HA1} and host-A2 bandwidth B_{HA2}; both A1 and A2 are also connected to their own local memory with bandwidths B_{A1M} and B_{A2M} respectively. Each host could have more than one accelerator, and further hosts could join the network.

where the baseline design runs for N_b cycles with cycle time T_b, and the enhanced design runs for N_e cycles with cycle time T_e, and communicates for N_c cycles with cycle time T_c. Hence $T_c N_c$ represents the *communication overhead* of using the extra accelerator. For the off-chip acceleration to be profitable, we require $S > 1$, however designs with $S \leq 1$ may still be useful if they use less power or energy than the baseline design. For most accelerators, $T_c = T_e$. Note that communication time is added to computation time, as in most accelerators they may be partially overlapped but not parallel.

The above models communication with a single off-chip accelerator. In general, there may be more than one accelerator connected to each other. In this case, the speedup becomes:

$$S = \frac{T_b N_b}{\max_j (T_{c,j} N_{c,j} + T_{e,j} N_{e,j})} \qquad (1)$$

where multiple accelerators connect to each other and accelerator j executes for $N_{e,j}$ cycles, with cycle time $T_{e,j}$, and communicates for $N_{c,j}$ cycles with cycle time $T_{e,j}$. Communication time $T_{c,j} N_{c,j}$ is taken the maximum of network time N_N/B_N and bus transfer to the accelerator N_{HAj}/B_{HAj}, where N_N items are transferred over the network, and N_{HAj} items are transferred over the bus with bandwidth B_{HAj}. The total time is bounded by the slowest accelerator.

Accelerators can also be memory-bound, so the accelerator performance becomes:

$$T_{A1} = \max(T_{comp}, T_{bus}, T_{mem}, T_{net})$$
$$= \max \left(N_{A1} T_{CA1}, \frac{N_{HA1}}{B_{HA1}}, \frac{N_{A1M}}{B_{A1M}}, \frac{N_{M1N}}{B_N} \right) \quad (2)$$

Respectively these are computation, bus, memory and network times for A1 running for N_{A1} cycles with cycle time T_{CA1}, bus time (N_{HA1} items across bandwidth B_{HA1}), memory access (N_{A1M} items across bandwidth B_{A1M}), and network (N_{M1N} items over bandwidth B_N). Not all accelerators have local memory, in which case $N_{A1M} = 0$. Equation 2 is used in section IV to predict performance of different partition options.

Since all accelerators are statically scheduled and all links are also statically determinate, all these quantities can be calculated statically, and are reported by vendor tools.

IV. PARTITIONING SCHEME

This section details our partitioning scheme. The key idea is to partition to use off-chip accelerators. Given two accelerators A1 and A2, part of the computation is mapped to A1, and part to A2.

As an example, consider the following:
- application: matrix computation;
- target architecture: number of A1s, number of A2s, links characterised by latency and throughput.

Given this input, the output is the application partitioned into CPU, A1 and A2 parts, making use of network connections.

Our partitioning scheme works at array, matrix or tensor levels, rather than on the low-level data path. We aim to keep large-scale operators on the same device rather than finding inter-device partitions. The smallest unit is a subarray —— we trade fine grain solutions for rapid design space exploration

The advantages of high-level partitioning include: we can rapidly explore design space since graph will be much smaller: each node is an entire vector, matrix or tensor computation. We avoid tricky partitions of matrices between devices which may give only marginal gains. High-level partitioning also has disadvantages, eliminating many partitions between devices.

We expect that overall runtime will be dominated by matrix calculations, particularly matrix-matrix multiply. Hence, our partitioning scheme builds on standard parallel matrix multiply partitioning schemes, adapted to heterogeneous hardware. Most parallel computers use a single kind of processing node. We consider three kinds: CPUs, possibly with multiple cores, A1 accelerators, and A2 accelerators, with A1 and A2 instances being attached to CPUs.

Figure 2 shows the partitioning options include:
1) Homogeneous partitioning onto one machine with n CPU cores;
2) Heterogeneous partitioning onto one machine with n CPU cores and one A1 accelerator, or multiple machines, each with n cores and an A1 or A2 accelerator.

Each partition is evaluated using equation 2 to predict performance, and the best-performing partition is chosen.

Note that there is no need for the number of A1s and A2s to be identical. Accelerators A1 and A2 could be on the same or separate machines, the only difference being the bandwidth between them.

V. EVALUATION

We evaluate our approach using a case study on Groq and Maxeler systems. The configuration is shown in figure 1.
- Accelerator A1 is a GroqCard: a fixed-function NPU intended for deep learning applications, containing dedicated hardware for matrix-matrix and matrix-vector multiply;
- Accelerator A2 is an AMD U250 FPGA.

Experimental setup: each accelerator is hosted in a separate machine. The machines are connected via a standard gigabit

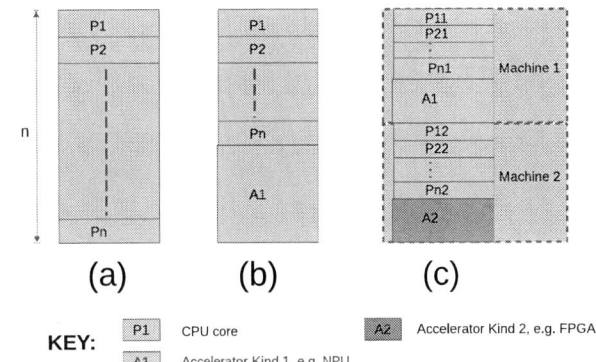

KEY: P1 CPU core   A2 Accelerator Kind 2, e.g. FPGA
      A1 Accelerator Kind 1, e.g. NPU

Fig. 2. Partitioning schemes: horizontal striping. (a) striping of matrix height N onto n CPU cores; (b) striping of same matrix onto n CPU cores and one accelerator, in this case an NPU; (c) striping onto two machines, each with n CPU cores and one accelerator; a machine could have more than one accelerator.

Size	Bitwidth	Logic%	DSP%	BRAM%	URAM%	Speed
1536	32	22.26	25.10	66.15	100.0	150
1664	32	23.97	27.08	88.99	100.0	100
1024	16	22.17	24.98	56.99	48.75	150

TABLE I

FPGA DESIGNS FOR $N \times N$ MATRIX MULTIPLY, SHOWING (TOP) THE RESOURCES USED FOR THE LARGEST 32-BIT DESIGN TO FIT AT 150MHZ ($N = 1536$) AND 100MHZ ($N = 1664$); (BOTTOM) THE LARGEST 16-BIT DESIGN TO FIT AT 150MHZ.

ethernet network. The NPU host contains one 900MHz Groq-Card with GroqWare 0.10.0; the FPGA host contains an AMD U250 FPGA.

We use Maxeler MaxCompiler 2023.1 to develop our FPGA designs with 16- and 32-bit floating point data elements; 32-bit designs are based on Maxeler MaxPower Tiled Matrix Multiply, while 16-bit designs use our own design based on dot product. Both designs take N^2 processing clock cycles for an $N \times N$ matrix, with memory and bus access in parallel to the processing.

Groq designs also use two bit sizes: 32-bit uses Groq compiler flags to disable the automatic conversion to 16-bit floating point internally, making the results more comparable to those from the FPGA, while 16-bit results allow this conversion.

To coordinate and run on multiple machines we use Unix shell scripts to synchronize machines, and distribute data.

FPGA designs: Table I shows the FPGA design, parametric in the matrix size. The largest design (at 150MHz) has a matrix size of 1536×1536. The limiting factor is the on-chip memory, not the DSP resources. Designs are built for a target clock frequency of 150MHz. A larger matrix size of 1664×1664 requires the clock frequency to reduce to 100Mhz. This is likely due to the larger design crossing between SLR (Super Logic Region) boundaries: bandwidth between SLRs is limited.

Groq designs: the largest 32-bit matrix size that will fit on the Groq device is 2656×2656; for 16-bit data, the largest is 3200×3200. Larger designs will not fit the Groq device.

Accelerator	Tile Size N	Bitwidth	Flops
FPGA	1536	32	2.14e11
Groq	2656	32	1.46e12
Groq	3200	16	3.91e12
Groq serving FPGA	2656(Groq)+1536(FPGA)	32	2.81e11
FPGA serving Groq	1536(FPGA)+2656(Groq)	32	1.50e12

TABLE II
RESULTS FOR GROQ AND FPGA ACCELERATORS, ALONE AND IN PARALLEL; THE TILE SIZE IS THE LARGEST MATRIX ($N \times N$) THAT WILL FIT IN EACH ACCELERATOR AT ONCE.

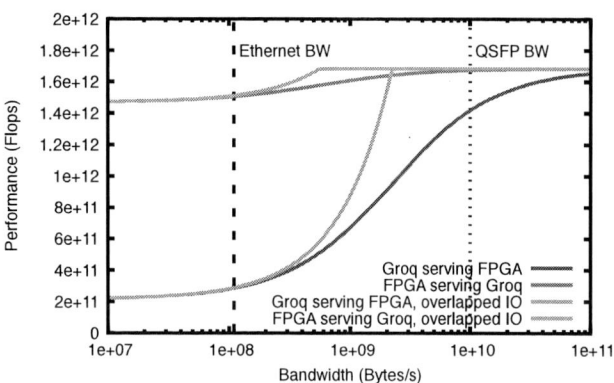

Fig. 3. Design-space exploration showing effect on increasing network bandwidth on performance.

Table II summarises results for one and two accelerators. Performance is measured over 1000 runs of the largest matrix tile size that will fit on each device. Peak FPGA performance occurs for a tile size of 1536; a larger tile of 1664 will fit but can only be clocked to 100MHz. Groq designs can fit larger tile sizes than the FPGA and have a higher throughput.

Our *parallel FPGA and NPU designs* use the Groq NPU as an off-chip accelerator for the FPGA. The FPGA host sends part of its work over the network to the Groq host, via file transfer. We measure network speed $B_N = 109.6$MB/s.

Table II also shows estimates for Groq and FPGA working together: (a) Groq serving FPGA, where the Groq is used as a networked accelerator for the FPGA system, and (b) FPGA serving Groq. The Groq 32-bit design computes 1000 matrix multiplies in 25.8 seconds, giving 1.458e12 Flops locally; remotely, file (58.5s) and network access (480s) increase the time to 575 seconds, a rate of 6.509e10. In parallel, we can run the FPGA 16,900 times, giving an overall rate of 2.81e11 Flops – about $1.31\times$ faster than the FPGA alone. The FPGA can be used in the same way to accelerate the Groq system, but network access times mean this is only slightly faster.

Figure 3 shows a design-space exploration of improving network speeds, plotted using the denominator of equation 1. The ethernet speed of 109.6MB/s restricts speeds; increasing the network speed to Quad Small Form-factor Pluggable (QSFP28) (about 10GB/s) improves speeds considerably. Overlapped Compute and IO also improves speeds until the design becomes compute-bound.

VI. CONCLUSION

We show methods for using coupled heterogeneous accelerators, including a performance model to guide application partitioning, taking into account limited off-chip bandwidth. Evaluation shows that a Groq NPU can be used to offload some matrix computations from an FPGA.

Our work assumes static scheduling of the accelerators. Future work could include dynamic scheduling, which has been applied to FPGAs [7]; some NPUs can also be dynamically scheduled, as are GPUs.

We use a straightforward model for performance. More sophisticated estimation models, such as Gaussian process regression [8] could be used with our work to give a better estimate of performance.

This work could also be extended to related applications such as sparse matrix multiply. For example, given a partially-sparse matrix, FPGAs might be better suited to the sparse parts, while fixed-function NPUs could process the dense parts.

Future systems are likely to involve more heterogeneous, off-chip accelerators. Designers of such systems would also benefit from this approach, which would also guide the development of novel tools for design automation.

Acknowledgement. The support of the United Kingdom EPSRC (grant number UKRI256, EP/V028251/1, EP/S030069/1, and EP/X036006/1), KIAT, Groq and Xilinx is gratefully acknowledged. We thank Dr Tobias Becker for his help.

REFERENCES

[1] J. Wang, L. Guo, and J. Cong, "AutoSA: A polyhedral compiler for high-performance systolic arrays on FPGA," in *The 2021 ACM/SIGDA International Symposium on Field-Programmable Gate Arrays*, ser. FPGA '21. New York, NY, USA: Association for Computing Machinery, 2021, p. 93–104.

[2] J. Zhuang, J. Lau, H. Ye, Z. Yang, Y. Du, J. Lo, K. Denolf, S. Neuendorffer, A. Jones, J. Hu, D. Chen, J. Cong, and P. Zhou, "CHARM: Composing heterogeneous accelerators for matrix multiply on versal ACAP architecture," in *Proceedings of the 2023 ACM/SIGDA International Symposium on Field Programmable Gate Arrays*, ser. FPGA '23. New York, NY, USA: Association for Computing Machinery, 2023, p. 153–164.

[3] J. Zhuang, J. Lau, H. Ye, Z. Yang, S. Ji, J. Lo, K. Denolf, S. Neuendorffer, A. Jones, J. Hu, Y. Shi, D. Chen, J. Cong, and P. Zhou, "CHARM 2.0: Composing heterogeneous accelerators for deep learning on Versal ACAP architecture," *ACM Trans. Reconfigurable Technol. Syst.*, vol. 17, no. 3, Sep. 2024. [Online]. Available: https://doi.org/10.1145/3686163

[4] G. Gerogiannis, S. Aananthakrishnan, J. Torrellas, and I. Hur, " HotTiles: Accelerating SpMM with Heterogeneous Accelerator Architectures ," in *2024 IEEE International Symposium on High-Performance Computer Architecture (HPCA)*. Los Alamitos, CA, USA: IEEE Computer Society, Mar. 2024, pp. 1012–1028.

[5] M. Langhammer and B. Pasca, "High-performance QR decomposition for FPGAs," in *Proceedings of the 2018 ACM/SIGDA International Symposium on Field-Programmable Gate Arrays*, ser. FPGA '18. New York, NY, USA: Association for Computing Machinery, 2018, p. 183–188.

[6] W. Luk, "Analysing reconfigurable computing systems," in *Transforming Reconfigurable Systems*. World Scientific, 2015, pp. 101–115.

[7] A. Vaishnav, K. D. Pham, and D. Koch, "Heterogeneous resource-elastic scheduling for CPU+FPGA architectures," in *Proceedings of the 10th International Symposium on Highly-Efficient Accelerators and Reconfigurable Technologies*, ser. HEART '19. New York, NY, USA: Association for Computing Machinery, 2019.

[8] M. Ferianc, H. Fan, D. Manocha, H. Zhou, S. Liu, X. Niu, and W. Luk, "Improving performance estimation for design space exploration for convolutional neural network accelerators," *Electronics*, vol. 10, no. 4, p. 520, 2021.

TorchLitho 2.0: Differentiable Lithography Simulation Engine for Large-Scale Layouts

Shuo Yin   Su Zheng   Ziyang Yu   Bei Yu
The Chinese University of Hong Kong

Abstract—As circuit complexity rapidly evolves and technology nodes continue to shrink, the manufacturing process becomes increasingly critical. Lithography is one of the most critical steps in the semiconductor manufacturing process, as it is responsible for transferring circuit patterns onto silicon wafers. In this paper, we introduce TorchLitho 2.0, an open-sourced differentiable lithography framework built on a differentiable programming toolkit with GPU acceleration, designed for large-scale layout simulation. TorchLitho 2.0 supports both Abbe and Hopkins imaging models and leverages source-level parallelism to accelerate the simulation process for large-scale layouts. Furthermore, TorchLitho 2.0 offers a comprehensive set of APIs that allow users to define custom lithography process parameters. We believe that TorchLitho 2.0 will be a valuable tool for optimizing the manufacturing process of advanced technology nodes and will pave the way for future research in full-chip-scale lithography simulation and optimization. The source code of TorchLitho 2.0 is available at https://github.com/OpenOPC/TorchLitho-Lite.

I. INTRODUCTION

In recent years, optical proximity correction (OPC) algorithms [1]–[15] have been widely developed to address the challenges associated with advanced lithographic processes. However, there remains a significant gap between academic research and industrial applications. Most studies are evaluated using the ICCAD13 contest [16] benchmarks, where the patterns are even smaller than a standard cell, making them unrealistic for industry-level layouts. Furthermore, full-chip scale OPC faces additional critical challenges, such as boundary errors caused by optical diffraction and the substantial memory footprint required during optimization.

Recently, some studies have started addressing the challenges of full-chip scale OPC to bridge the gap between academic research and industrial needs. For example, DAMO [17] and AdaOPC [18] introduced methods to improve the efficiency of full-chip mask optimization by incorporating a novel partitioning approach and adaptive solver selection. However, both methods still depend on the ICCAD13 simulator, which functions as a black-box kernel and only supports fixed input sizes. This simulator restricts users from modifying physical parameters, limiting its scalability and applicability to other benchmarks. FuILT [19] is the first work to propose a distributed full-chip mask optimization engine that addresses boundary healing. Unfortunately, the simulator used in FuILT is a commercial tool which is not publicly available.

These limitations in existing works underscore the need for new simulators that are better aligned with real-world,

The project is supported in part by Research Grants Council of Hong Kong SAR (No. RFS2425-4S02). (*Corresponding authors: Ziyang Yu, Bei Yu*)

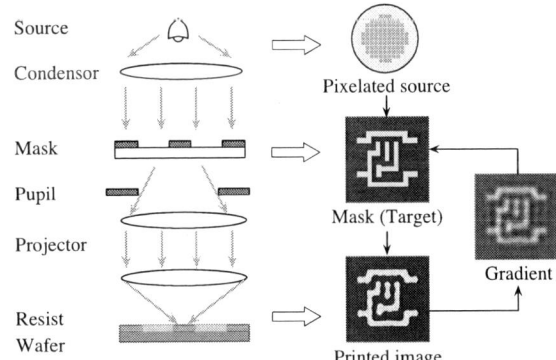

Fig. 1 Schematic diagram of differentiable lithography system.

industrial-level lithography. A simplified illustration of the lithography process is shown in Fig. 1. TorchLitho [20] is the first open-source, differentiable lithography simulator built on PyTorch [21] with GPU acceleration. It provides a comprehensive set of APIs, enabling users to define custom lithography process parameters. However, TorchLitho is primarily designed for small-scale layouts and lacks support for large-scale layouts due to memory constraints.

In this paper, we present TorchLitho 2.0, an enhanced version of TorchLitho that supports large-scale layout simulation. TorchLitho 2.0 integrates both Abbe and Hopkins imaging models and utilizes source-level parallelism to accelerate the simulation process for large-scale layouts. We believe that TorchLitho 2.0 will serve as a valuable tool for optimizing the manufacturing processes of advanced technology nodes and will pave the way for future research in full-chip-scale lithography simulation and optimization.

II. FUNDAMENTAL COMPONENTS OF AN OPTICAL SYSTEM

Lithography Formulation. The Fourier Transform is a fundamental technique in signal processing, enabling efficient computation and analysis of frequency-domain information. In computational lithography simulation, the optical system is modeled in the frequency domain using Fourier optics and scalar diffraction theory.

The aerial image of lithography optical systems $I(x, y)$ can be described by the Equation (1), which is derived from the convolution of the illumination and the mask, followed by a Fourier transform. $\bar{O}(f, g)$ is the object spectrum (Fourier transform of the mask transmission $O(x, y)$), $\tilde{H}(f, g)$ is the projection lens transfer function, and $\tilde{J}(f, g)$ is the effective mutual intensity distribution due to the source illumination.

$$I(x,y) = \iiiiint \tilde{J}(f,g)\tilde{H}(f+f',g+g')\tilde{H}^*(f+f'',g+g'')$$
$$\tilde{O}(f',g')\tilde{O}(f'',g'')e^{2\pi i[(f'-f'')x+(g'-g'')y]}$$
$$df\,dg\,df'\,dg'\,df''\,dg''. \tag{1}$$

Here (x,y) and (f,g) represent spatial and frequency coordinates, respectively.

Source Definition. The source illuminates the mask at varying angles, determined by its shape. The simplest source is circular, characterized by its radius σ, which defines source coherence. In lithography, the source is temporally coherent due to its monochromatic nature, and the overall image is an incoherent sum of contributions from each source point. The source image is projected onto the projection lens's entrance pupil. Illumination at non-zero angles to the optical axis shifts the diffraction pattern without changing the diffraction orders' amplitude. The frequency shift Δf for an oblique angle θ is given by:

$$\Delta f = \frac{\sigma \sin(\theta)}{\lambda}, \tag{2}$$

where λ is the wavelength of the light source.

Projection Lens Definition. The mask's diffraction pattern is focused at the entrance pupil of the lens, which, according to Fourier optics, performs an inverse Fourier transform of the field at the pupil and projects it onto the image plane.

The projection lens acts as a low-pass filter for the mask's Fourier transform. Low frequencies pass through the lens aperture unperturbed and are recombined before being projected onto the substrate. However, for frequencies larger than the projection lens cutoff frequency, the signal is lost. The cutoff frequency, f_{cut}, is defined as $f_{cut} = \frac{NA}{\lambda}$. This frequency behavior is determined by the numerical aperture (NA) of the projection lens. Mathematically, the effect of the projection lens is described by a circular transfer function, as follows:

$$\tilde{H}(f,g) = \begin{cases} 1, & \text{if } \sqrt{f^2+g^2} \leq f_{\text{cut}}; \\ 0, & \text{otherwise.} \end{cases} \tag{3}$$

III. TORCHLITHO 2.0 FRAMEWORK

Abbe Imaging. The Abbe imaging model performs source point integration directly in the frequency domain. The aerial image is obtained as the incoherent sum of the contributions from each source point to the overall image. The equation can be rewritten as Equation (4):

$$I(x,y) = \iint \tilde{J}(f,g) \left[\left| \iint \tilde{H}(f+f',g+g') \right. \right.$$
$$\left. \left. \tilde{O}(f',g')e^{2\pi i[f'x+g'y]}\,df'\,dg' \right|^2 \right] df\,dg. \tag{4}$$

Due to the necessity of enumerating all source points during computation, the computational complexity becomes exceedingly high. To address this challenge, the TorchLitho 2.0 framework introduces an innovative approach to efficiently mitigate this complexity by employing parallel computing techniques. The pseudocode is presented in Algorithm 1.

Algorithm 1 Parallel Abbe Imaging

Input: Mask $\tilde{O}(f,g)$ with shape $[H,W]$, source points $\tilde{J}(f,g)$
Output: Aerial image I.
1: $\tilde{F} \leftarrow \sqrt{f_x \tilde{f}_x + f_y \tilde{f}_y}$;
2: $pupils \leftarrow [\ ]$;
3: **for** $s \in \tilde{J}$ **do**
4: $\quad \tilde{F}' \leftarrow \tilde{F} - s$;
5: $\quad \tilde{H} \leftarrow (\tilde{F} \leq \frac{NA}{\lambda}) \times \mathbf{1} + (\tilde{F} > \frac{NA}{\lambda}) \times \mathbf{0}$; $\quad \triangleright$ Equation (3)
6: $\quad pupils \leftarrow pupils \cup \tilde{H}$;
7: **end for**
8: $\tilde{H}' \leftarrow \texttt{concat}(pupils, \text{dim=0})$;
9: $\tilde{O}' \leftarrow \tilde{O} \odot \tilde{H}'$;
10: $I \leftarrow \texttt{mean}(\|\texttt{ifft}(\tilde{O}')\|^2, \text{dim=0})$;

We modify the computation order of Equation (4). Specifically, we first compute the pupil transfer matrices with the source illumination. Subsequently, the mask integration is performed using element-wise matrix multiplication.

Hopkins Imaging. The Hopkins model is a standard framework for describing how mask patterns are transferred onto the wafer through a projection system under partially coherent illumination. It reformulates the aerial image intensity as a bilinear functional of the mask spectrum, with the illumination and optical system jointly represented by a kernel function known as the Transmission Cross Coefficient (TCC). Mathematically, the aerial image intensity on the wafer is expressed as

$$I(x,y) = \iint \tilde{O}(f',g')\,\tilde{O}^*(f'',g'')\,TCC(f',g',f'',g'')$$
$$e^{i2\pi[(f'-f'')x+(g'-g'')y]}\,df'\,df''\,dg'\,dg'', \tag{5}$$

where the TCC is defined as

$$TCC(f',g',f'',g'') = \iint \tilde{J}(f,g)\,H(f'+f,g'+g)$$
$$H^*(f''+f,g''+g)\,df\,dg. \tag{6}$$

The TCC therefore encodes the combined effects of the illumination source and the optical system on the interference of diffracted orders from the mask.

While the Hopkins model provides rigorous accuracy, its computational cost is substantial due to the quadratic scaling with the number of spatial frequency components. To overcome this limitation, eigen-decomposition of the TCC is commonly employed, enabling the bilinear form to be approximated as a sum of coherent imaging systems, thereby significantly reducing computational complexity. Formally, the TCC can be decomposed as:

$$TCC(f',g',f'',g'') = \sum_k \lambda_k \, \Phi_k(f',g') \, \Phi_k^*(f'',g''), \tag{7}$$

where λ_k are the non-negative eigenvalues and $\Phi_k(f,g)$ are the corresponding orthonormal eigenfunctions of the TCC operator.

Based on the decomposition, the formation of aerial image can be rewritten as

$$I(x,y) = \sum_k \lambda_k \left| \phi_k \otimes O(x,y) \right|^2, \tag{8}$$

where $\phi_k = \mathcal{F}^{-1}[\Phi_k]$. This representation is known as the sum of coherent systems (SOCS) method. Each coherent mode corresponds to an eigenfunction of the TCC, with its contribution to the aerial image weighted by the associated eigenvalue. By

TABLE I Parameter descriptions and defaults for the simulation.

Param	Type	Description	Default
Pixel	int	Pixel size of the mask (unit: nm).	14
Sigma	float	Source broadening parameter σ.	0.05
NA	float	Numerical Aperture.	1.35
Wavelength	int	Illumination wavelength (unit: nm).	193
Defocus	List[int]	List of defocus values.	None
Parallel	bool	Enable parallel mode for Abbe imaging simulation (performance only; no physical change).	False
Batch	bool	Enable a batch dimension to simulate multiple masks at once (inputs must include a batch axis).	False

---

Algorithm 2 TCC Generation

Input: Pupil function $\tilde{H}$, source function $\tilde{J}$, and threshold λ_{th}
Output: Decomposed TCC kernels $\phi = \{\phi_1, \phi_2, \cdots\}$, and weights $\boldsymbol{\lambda} = \{\lambda_1, \lambda_2, \cdots\}$
1: $\tilde{h} \leftarrow \mathcal{F}[\tilde{H}]$;
2: $\tilde{J}_0 \leftarrow \mathcal{F}[\tilde{J}]$;
3: $\boldsymbol{W}(f', g', f'', g'') \leftarrow \tilde{J}_0(f' - f'', g' - g'')\tilde{h}(f', g')\tilde{h}^*(f'', g'')$;
4: Call randomized SVD to get $\boldsymbol{W} \leftarrow \sum_{k=1}^{K} \lambda_k \phi_k$;
5: $\phi \leftarrow [\phi_k$ for $k \in \{1, 2, \cdots, K\}$ if $\lambda_k < \lambda_{th}]$;
6: $\boldsymbol{\lambda} \leftarrow [\lambda_k$ for $k \in \{1, 2, \cdots, K\}$ if $\lambda_k < \lambda_{th}]$;
7: **return** $\phi, \boldsymbol{\lambda}$;

---

retaining only the dominant modes with the largest eigenvalues, the SOCS method achieves an effective balance between accuracy and computational efficiency, forming the foundation of modern large-scale lithography simulation frameworks.

We generate the TCC via Algorithm 2. In practice, the construction of matrix $\boldsymbol{W}(f', g', f'', g'')$ can take too much memory when the target layout is large or the pixel size is very small. To mitigate this for small pixel sizes (e.g., 4×4 nm^2), we first use a coarser pixel size (e.g., 16×16 nm^2) to decompose the TCC into $\phi = \phi_1, \phi_2, \ldots$ and $\boldsymbol{\lambda} = \lambda_1, \lambda_2, \ldots$. We then upsample the kernels by zero-padding in the frequency domain: $\phi_k = \mathcal{F}^{-1}\left[\text{pad}\left(\mathcal{F}[\phi_k]\right)\right]$. Here, $\text{pad}(\cdot)$ zero-pads the spectrum so that the resulting kernels match the desired TCC kernel size. For large layout sizes, we apply an analogous zero-padding procedure in the spatial domain.

API Design. The API related to our lithography simulation model is listed in TABLE I. TorchLitho 2.0 supports dynamic layout sizes based on user requirements. TorchLitho 2.0 also supports batch processing, with the mask represented such that the first axis denotes the batch dimension.

IV. EXPERIMENTAL RESULTS

All experiments were conducted on a server equipped with an NVIDIA 3090 GPU featuring 24GB of on-chip memory.

A. Visualization of Lithography Process

We demonstrate the differentiable lithography process of TorchLitho 2.0 by simulating the lithography process on two benchmark datasets: the ICCAD-2013 contest dataset [16] and the full-chip mask for the GCD design.

The ICCAD-2013 contest dataset consists of 10 mask patterns, each with a size of 2048×2048 pixels. The pixel size is set to 1 nm, and the lithography parameters are configured to the default values listed in TABLE I. The simulation results are presented in Fig. 2. As shown in the figure, the aerial images and gradient maps are successfully generated from the input mask patterns, demonstrating the effectiveness of TorchLitho 2.0 in simulating the lithography process.

Additionally, we simulate the full-chip mask of the GCD design, optimized using FuILT [19], which includes both metal and via layers. The layout size is 3.5×3.5 μm^2, with a pixel size of 14 nm. The printed images are generated through direct simulation without any partitioning. he target intensity in the resist model is set to 0.225. The simulation results are shown in Fig. 3, demonstrating the scalability of TorchLitho 2.0 in handling large-scale layouts.

B. Parallel Mode of Abbe Imaging Model

To evaluate the efficiency of the parallel mode in the Abbe imaging model, we compare the performance of its serial and parallel implementations. We test various pixel sizes, including 2, 4, and 8 nm, as well as different layout sizes: 512×512, 1024×1024, 2048×2048, and 4096×4096. The experimental results are presented in Fig. 4. As shown in the figure, the parallel implementation consistently outperforms the serial implementation across all configurations for both forward and backward passes. This highlights the efficiency of the parallel mode in accelerating the Abbe imaging model.

V. CONCLUSION

In this work, we introduce TorchLitho 2.0, an ultrafast, differentiable lithography simulation engine built on PyTorch for full-chip designs. TorchLitho 2.0 integrates advanced lithography models, including the Abbe imaging model with a parallel mode and the Hopkins imaging model, to improve simulation accuracy. To support further research in lithography simulation and optimization, TorchLitho 2.0 is open-sourced.

REFERENCES

[1] G. Chen, Z. Yu, H. Liu, Y. Ma, and B. Yu, "DevelSet: Deep neural level set for instant mask optimization," in *Proc. ICCAD*, 2021.
[2] Z. Yu, G. Chen, Y. Ma, and B. Yu, "A GPU-enabled level set method for mask optimization," in *Proc. DATE*, 2021.
[3] B. Jiang, L. Liu, Y. Ma, H. Zhang, E. F. Y. Young, and B. Yu, "Neural-ILT: Migrating ILT to nerual networks for mask printability and complexity co-optimizaton"," in *Proc. ICCAD*, 2020.
[4] H. Yang, Z. Li, K. Sastry, S. Mukhopadhyay, M. Kilgard, A. Anandkumar, B. Khailany, V. Singh, and H. Ren, "Generic lithography modeling with dual-band optics-inspired neural networks," *arXiv preprint arXiv:2203.08616*, 2022.

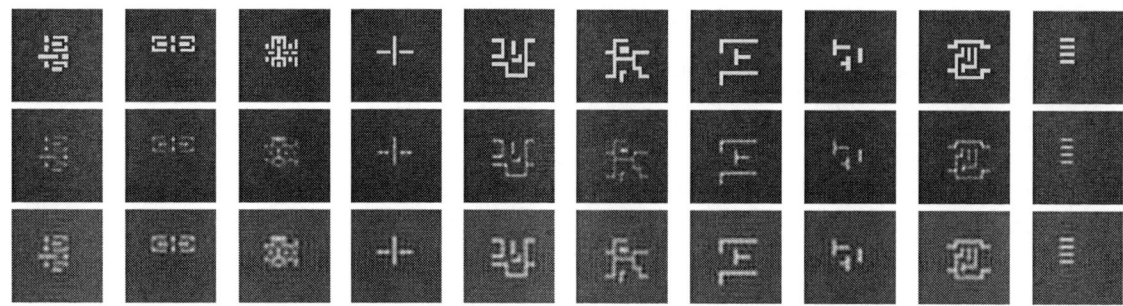

Fig. 2 Experimental results on the ICCAD-2013 contest benchmark. The first row shows the input mask patterns, the second row shows the corresponding aerial images, and the third row shows the gradient map of the lithography process.

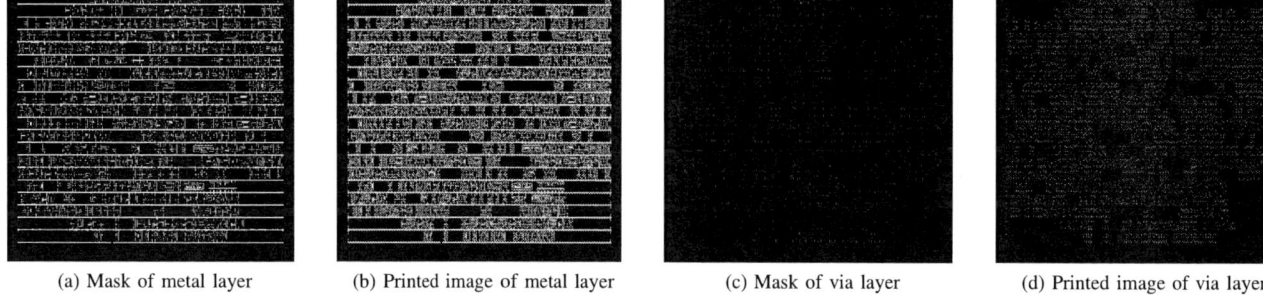

(a) Mask of metal layer   (b) Printed image of metal layer   (c) Mask of via layer   (d) Printed image of via layer

Fig. 3 Experimental results on the full-chip mask of the GCD design, where the mask pattern is optimized using FuILT [19].

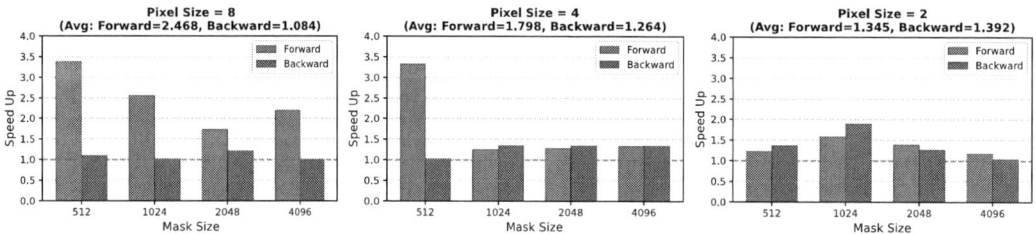

Fig. 4 Performance comparison between serial and parallel implementations of the Abbe imaging model.

[5] H. Yang, S. Li, Z. Deng, Y. Ma, B. Yu, and E. F. Y. Young, "GAN-OPC: Mask optimization with lithography-guided generative adversarial nets," *IEEE TCAD*, 2020.

[6] X. Zhang, S. Zheng, G. Chen, B. Zhu, H. Xu, and B. Yu, "Fracturing-aware curvilinear ILT via circular e-beam mask writer," in *Proc. DAC*, 2024, pp. 1–6.

[7] B. Zhu, S. Zheng, Z. Yu, G. Chen, Y. Ma, F. Yang, B. Yu, and M. D. Wong, "L2O-ILT: Learning to optimize inverse lithography techniques," *IEEE TCAD*, vol. 43, no. 3, pp. 944–955, 2023.

[8] S. Zheng, B. Yu, and M. Wong, "OpenILT: An open source inverse lithography technique framework," in *Proc. ASICON*, 2023, pp. 1–4.

[9] G. Chen, H. Yang, H. M. Ren, B. Yu, and D. Z. Pan, "Differentiable edge-based OPC," in *Proc. ICCAD*, 2024, pp. 1–9.

[10] S. Zheng, X. Liang, Z. Yu, Y. Ma, B. Yu, and M. Wong, "Curvilinear Optical Proximity Correction via Cardinal Spline," in *Proc. DAC*, 2025, pp. 1–7.

[11] Z. Yu, S. Zheng, W. Zhao, S. Yin, X. Liang, G. Chen, Y. Ma, B. Yu, and M. D. Wong, "Rulelearner: Opc rule extraction from inverse lithography technique engine," *IEEE TCAD*.

[12] G. Chen, H. He, P. Xu, H. Geng, and B. Yu, "Efficient bilevel source mask optimization," in *Proc. DAC*, 2024, pp. 1–6.

[13] S. Zheng, Y. Ma, B. Yu, and M. D. Wong, "Emogen: Enhancing mask optimization via pattern generation," in *Proc. DAC*, 2024, pp. 1–6.

[14] Z. Yu, P. Liao, Y. Ma, B. Yu, and M. D. Wong, "Ctm-sraf: Continuous transmission mask-based constraint-aware subresolution assist feature generation," *IEEE TCAD*, vol. 42, no. 10, pp. 3402–3411, 2023.

[15] J.-R. Gao, X. Xu, B. Yu, and D. Z. Pan, "MOSAIC: Mask optimizing solution with process window aware inverse correction," in *Proc. DAC*, 2014, pp. 52:1–52:6.

[16] S. Banerjee, Z. Li, and S. R. Nassif, "ICCAD-2013 CAD contest in mask optimization and benchmark suite," in *Proc. ICCAD*, 2013, pp. 271–274.

[17] G. Chen, W. Chen, Y. Ma, H. Yang, and B. Yu, "DAMO: Deep agile mask optimization for full chip scale," in *Proc. ICCAD*, 2020.

[18] W. Zhao, X. Yao, Z. Yu, G. Chen, Y. Ma, B. Yu, and M. D. Wong, "Adaopc: A self-adaptive mask optimization framework for real design patterns," in *Proc. ICCAD*, 2022.

[19] S. Yin, W. Zhao, L. Xie, H. Chen, Y. Ma, T.-Y. Ho, and B. Yu, "FuILT: Full chip ilt system with boundary healing," in *Proc. ISPD*, 2024, pp. 13–20.

[20] G. Chen, H. Geng, B. Yu, and D. Z. Pan, "Open-source differentiable lithography imaging framework," in *DTCO and Computational Patterning III*, vol. 12954. SPIE, 2024, pp. 118–127.

[21] A. Paszke, S. Gross, F. Massa, A. Lerer, J. Bradbury, G. Chanan, T. Killeen, Z. Lin, N. Gimelshein, L. Antiga *et al.*, "Pytorch: An imperative style, high-performance deep learning library," *Proc. NeurIPS*, vol. 32, 2019.

Hierarchical Residual Fitting for Enhanced S-Parameter Accuracy in Devices Exhibiting Complex Delay

Jiaxin Wei[1,2] Haonan Wang[3], Ting-Jung Lin*[3] Lei He[3]

[1]Shanghai Jiao Tong University, Shanghai, China

[2]BTD.Tech Inc., Ningbo, China

[3]Ningbo Institute of Digital Twin, Eastern Institute of Technology, Ningbo, China

Email: 81210201153371@sjtu.edu.cn[1], tlin@idt.eitech.edu.cn[3]

Abstract—Modern distributed devices often exhibit complex time-delay effects. However, existing S-parameter fitting algorithms struggle to capture their frequency-domain behavior accurately and efficiently. This paper proposes a novel Hierarchical Residual Fitting (HRF) algorithm that decomposes the device response into multiple residual layers, each addressing progressively finer delay-induced features. By adaptively fitting and subtracting residual components, HRF achieves substantial improvements in both fitting precision and computational efficiency compared to conventional approaches. Numerical experiments on representative high-speed circuit models demonstrate that HRF reduces the mean fitting error by up to 40% and accelerates convergence by an order of magnitude.

Index Terms—vector fitting, s-parameter, delay

I. INTRODUCTION

Modern high-speed electronic systems require higher data rates, lower voltage swings, and denser multilayer PCB interconnects, which intensify parasitic effects and degrade signal integrity (SI)—a key measure of transmission fidelity [1]. In practice, SI degradation results from reflections, crosstalk, electromagnetic interference, and power-integrity issues, which distort signal waveforms. These impairments can cause data misinterpretation at the receiver, making accurate modeling and mitigation essential for robust system performance.

A common SI analysis strategy models each interconnect segment as a transmission line characterized by its S-parameters, which capture frequency-dependent reflection and transmission characteristics [1]. There are two mainstream approaches for time-domain simulation. The first applies inverse Fourier transform to raw S-parameters, which yields impulse responses for convolution. This method provides higher fidelity at the cost of long runtime and truncation errors. The second method is macromodeling, which converts S-parameters into compact and SPICE-compatible equivalent circuits. This delivers higher stability while significantly improving simulation efficiency.

This work by Jiaxin Wei was supported by the National Natural Science Foundation of China (Grant No. 92473206), the "Science and Technology Innovation in Yongjiang 2035" Key Technology Project (Grant No. 2024Z283), and research support from BTD Inc.

Rational-function fitting (RF) is pivotal to macromodeling, as it represents sampled S-parameter data as a sum of poles and residues. Vector Fitting (VF) and its variants (e.g., Relaxed VF [2], VFAS [3], Delayed VF [4]) have been widely adopted for their robust convergence and causality guarantees. However, existing RF algorithms often struggle with multiple complex time-delay effects or highly oscillatory spectra. These challenges require manual pole tuning, which may hinder timely convergence or the desired accuracy.

This paper presents **Hierarchical Residual Fitting (HRF)**, a delay-aware RF algorithm tailored for devices with complex delay profiles. HRF hierarchically decomposes the frequency response into residual layers, adaptively fitting each to capture delay-induced oscillations while maintaining stability and causality. The proposed method integrates delay extraction with VFAS and automatically generates SPICE-compatible models featuring embedded lossless transmission-line sections. Simulation results on representative S-parameter datasets demonstrate that HRF achieves over 40% RMSE reduction and an order-of-magnitude speedup in convergence compared to state-of-the-art techniques. Patents corresponding to this work are under review.

The paper is organized as follows. Section II reviews RF methods and equivalent-circuit synthesis. Section III details the HRF algorithm with hierarchical residual decomposition and delay-adaptive fitting. Section IV presents case studies and performance comparisons. Section V concludes the paper.

II. BACKGROUND

This section reviews key existing VF algorithms, including the basic VF method [5], the improved VF with Adding and Skimming (VFAS) [3], and the Delayed VF (DVF) algorithm [4]. Their strengths and limitations are discussed, which motivate our proposed HRF algorithm in Section III.

A. The Basic Vector Fitting Algorithm (VF)

VF is an iterative rational function approximation technique. It converts the nonlinear fitting problem into a linear least-squares problem via pole relocation. This method efficiently extracts stable poles and residues from frequency-domain samples. For simplicity, we consider a single-port network as an example. Suppose a set of frequency-domain sampled data $\{s_k, \breve{H}_k\}, k = 1, 2, \ldots, K$ is available. The transfer function $H(s)$ is approximated by a rational function,

$$\hat{H}(s) = c_0 + \sum_{j=1}^{N} \frac{c_j}{s - p_j} \tag{1}$$

with unknown poles $\{p_j\}$ and residues $\{c_j\}$. Standard VF is an iterative algorithm that refines an initial estimate $\{q_j^0\}$ of the N dominant poles of the structure.

Denote the poles at iteration i as $\{q_j^i\}$, we can define the weight function as

$$\sigma^i(s) = 1 + \sum_{j=1}^{N} \frac{d_j^i}{s - q_j^i} = \frac{\prod_{j=1}^{N}(s - z_j^i)}{\prod_{j=1}^{N}(s - q_j^i)} \tag{2}$$

with unknown residues $\{d_j^i\}$. Then, the least-squares approach can be employed to obtain the optimal approximation of the unknown parameters, as expressed in the equation,

$$\sigma^i(s)H(s) = c_0^i + \sum_{j=1}^{N} \frac{c_j^i}{s - q_j^i}. \tag{3}$$

The current fitting function employs predefined poles $\{q_j^i\}$, which do not match the true poles $\{p_j^i\}$. Therefore, the current poles need to be updated using the zeros $\{p_j^i\}$ of the weight function σ.

$$\{q_j^i\} \leftarrow \{p_j^i\} \tag{4}$$

Once these poles are determined, a second linear least-squares solution to (1) yields the residues $\{r_n\}$. Consequently, the rational approximation in (1) is obtained through a multi-stage linear least-squares procedure, thereby eliminating the need for potentially unstable nonlinear optimization techniques. This algorithm can be readily extended to multi-port cases, with the detailed methodology presented in [6].

Although the VF algorithm performs efficiently for smooth data, its accuracy degrades when the data exhibits numerous peaks. This limitation arises due to the restricted number of poles and their constrained mobility at high frequencies.

B. Vector Fitting with Adding and Skimming (VFAS)

The classical VF algorithm employs fixed poles, which lack flexibility and can cause underfitting or overfitting. Too few poles result in poor approximations, while too many increase computational cost and instability risk due to closely spaced poles, creating ill-conditioned least-squares problems. Selecting the proper pole count requires significant expertise, which can reduce adaptability in practice.

The Vector Fitting with Adding and Skimming (VFAS) algorithm was developed in 2006 to address these limitations. Specifically, VFAS adaptively adjusts both the number and positions of poles during iterations. After preset iterations, it inserts new poles at the peaks of approximation errors to enhance local accuracy; simultaneously, it merges closely spaced poles based on a threshold distance to prevent ill-conditioning. Additionally, poles with negligible residues are pruned to maintain model compactness. These adaptive operations significantly improve the computational robustness and flexibility of the algorithm, enabling automatic parameter adjustments.

VFAS consists of initialization, iterative pole relocation, adaptive pole management, and convergence evaluation. The initial pole placements reflect the global trend of frequency-domain S-parameters. Iterations refine pole positions to capture system dynamics better. Convergence is measured by root-mean-square error and pole count criteria, ensuring accurate and compact models. Although VFAS performs well for smooth responses, it requires many poles to fit periodic oscillations in devices with significant delay, increasing complexity, and degrading convergence and accuracy. Thus, VFAS has limitations when modeling data with pronounced delay characteristics.

C. Delayed Vector Fitting (DVF)

For systems with significant delays, frequency-domain responses show strong oscillations, limiting conventional VF. Existing methods include structure segmentation [6], frequency partitioning [7], and DVF [4]. DVF adds delay terms into VF, improving oscillatory response modeling while balancing accuracy, efficiency, and implementation. This section presents the DVF formulation and procedure.

To enhance the ability of VF to handle systems with long delays, DVF introduces delay coefficients into the fitting process, modifying the target function as:

$$H(s) = \sum_{m} Q_m(s)e^{-s\tau_m}. \tag{5}$$

Before fitting, delay values must be estimated. Techniques such as Gabor analysis [8] are widely used for this purpose. Once delays are determined, the rational approximation takes the form:

$$H(s) \simeq \sum_{m=1}^{\bar{m}} \frac{\sum_{n=0}^{\bar{n}} c_{mn}\varphi_n(s)}{\sum_{n=0}^{\bar{n}} d_n\varphi_n(s)} e^{-s\tau_m} \tag{6}$$

with basis functions defined as

$$\varphi_n(s) = \begin{cases} 1, & n = 0 \\ \frac{1}{s - q_n}, & n = 1, \ldots, \bar{n}. \end{cases} \tag{7}$$

The fitting error is minimized by the following RMS criterion

$$\varepsilon = \sqrt{\frac{1}{\bar{k}} \sum_{k=1}^{\bar{k}} |\mathcal{E}_k|^2} \tag{8}$$

where the residual at the k-th frequency sample is

$$\mathcal{E}_k = H_k - \frac{\sum_{n=0}^{\bar{n}} \sum_{m=1}^{\bar{m}} c_{mn}\varphi_n(j\omega_k)e^{-j\omega_k\tau_m}}{\sum_{n=0}^{\bar{n}} d_n\varphi_n(j\omega_k)}. \tag{9}$$

To preserve linearity with respect to unknowns in the current iteration, a weight function is defined using poles from iterations i and $i-1$:

$$W_k^{(i)} = \frac{\sum_{n=0}^{\bar{n}} d_n^{(i)}\varphi_n(j\omega_k)}{\sum_{n=0}^{\bar{n}} d_n^{(i-1)}\varphi_n(j\omega_k)} \tag{10}$$

979-8-3315-3918-4/25 $31.00 © 2025 IEEE

Substituting this into (9), the updated linear residual becomes

$$\mathcal{E}_k^{(i)} = \frac{H_k \sum_n^{\bar{n}} d_n^{(i)} \varphi_n(j\omega_k) - \sum_n^{\bar{n}} \sum_m^{\bar{m}} c_{mn}^{(i)} \varphi_n(j\omega_k) e^{-j\omega_k \tau_m}}{\sum_n^{\bar{n}} d_n^{(i-1)} \varphi_n(j\omega_k)}.$$

(11)

This enables least-squares optimization and pole updating until convergence, yielding the final fitted result shown below.

$$\hat{H}(s) = \sum_{m=1}^{\bar{m}} \left(\sum_{n=0}^{\bar{n}} \frac{c_{mn}}{s - p_n} + c_{m0} \right) e^{-s\tau_m}.$$

(12)

The DVF algorithm has five steps: initialization, delay estimation, weight computation, pole relocation, and error evaluation. It adaptively refines pole positions and adds delay terms to better model oscillatory responses. However, in complex multi-delay scenarios, DVF adds all delays simultaneously using shared poles, which results in redundant poles, higher computational costs, and near-singular matrices that degrade convergence and accuracy. These issues limit DVF's performance with complex delay structures.

III. HIERARCHICAL RESIDUAL FITTING (HRF)

To balance automation and modeling accuracy, this work proposes the HRF algorithm. HRF combines VFAS for smooth components and delay compensation for oscillatory residuals caused by transmission delays. By hierarchically decomposing the frequency response into smooth low-frequency and oscillatory high-frequency parts via adaptive partitioning, HRF enables accurate modeling with reduced complexity. Smooth components are fitted using pole-residue models, while oscillatory parts are captured via time-domain kernels, avoiding the high dimensionality of global methods.

The target model structure of HRF resembles that of DVF:

$$H(s) = \sum_m Q_m(s) e^{-s\tau_m}$$

(13)

where

$$Q_m(s) = c_{m0} + \sum_{j=1}^{n} \frac{c_{mj}}{s - p_j}.$$

(14)

The HRF algorithm extracts transmission delays and fits residuals to improve stability and accuracy. It estimates the dominant delay and smooths the response to reduce high-order oscillations. VFAS fits the smoothed data. The model is updated with the delay, and the residual is computed for the next iteration. This repeats until convergence or the maximum iteration is reached.

The hierarchical design confines high-frequency jitter to lower layers with bounded matrix sizes, avoiding dimensionality issues. Upper layers capture smooth trends with fewer poles. Adaptive pole relocation and pruning improve convergence and model compactness. All parameters are constrained to be real [6].

In multi-port vector fitting, many S-parameter elements share similar frequency traits. We propose a peak-based delay-grouping scheme that normalizes impulse responses and extracts local peaks of above-average magnitude. Elements

---

Algorithm 1: Hierarchical Residual Fitting (HRF)

Input: Spectral data $\{s_i, H(s_i)\}$, max iterations $N_{\max}$, error threshold ε
Output: Hierarchically fitted model $\hat{H}(s)$

1   $H_{\text{total}} \leftarrow 0$, $r \leftarrow H$, $n \leftarrow 0$;
2   **while** $n < N_{\max}$ and error $> \varepsilon$ **do**
3     $\{\tau_j\} \leftarrow \text{DetectDelay}(r)$;
4     $r_{\text{nd}} \leftarrow \text{RemoveDelays}(r, \{\tau_j\})$;
5     $r_{\text{sm}} \leftarrow \text{Smooth}(r_{\text{nd}})$;
6     $M(s) \leftarrow \text{VFAS}(r_{\text{sm}}, \{s_i\})$;
7     $M_d(s) \leftarrow \text{ApplyDelays}(M(s), \{\tau_j\})$;
8     $H_{\text{total}} \leftarrow H_{\text{total}} + M_d(s)$;
9     $r \leftarrow H - H_{\text{total}}$;
10   error $\leftarrow \text{Evaluate}(r)$;
11    $n \leftarrow n + 1$;
12   **return** H_{total}

---

with similar peak patterns are grouped and jointly modeled, enabling efficient pole reuse. This approach is suitable for delay-dominant systems and has a linear time complexity of $O(n)$, outperforming Euclidean or cross-correlation methods. Experiments show over 30% reduction in fitting time with minimal accuracy loss. The grouping improves scalability, enhancing the generation and simulation of equivalent circuits for large, high-speed interconnect systems. The grouping algorithm is summarized in Algorithm 2.

IV. EXPERIMENTS AND DISCUSSIONS

This section evaluates the proposed HRF algorithm on a high-speed interconnect case. It assesses delay extraction accuracy, robustness for coupled S-parameters, and performance in real transmission-line scenarios. Results are benchmarked against VFAS and DVF. Full-band root-mean-square error

---

Algorithm 2: Delay-Based Data Grouping

Input: Spectral data $\{s_i, H(s_i)\}$, port count n
Output: Grouped indices $\{A_{ij}\}$

1   delays $\leftarrow \text{DetectDelays}(\{H(s_i)\})$;
2   mask $\leftarrow [\textbf{false}] \times n^2$, Groups $\leftarrow \emptyset$;
3   **for** $i \leftarrow 0$ **to** $n^2 - 1$ **do**
4   **if** mask$[i]$ **then**
5   **continue**;
6   group $\leftarrow \{i\}$;
7   mask$[i] \leftarrow \textbf{true}$;
8   **for** $j \leftarrow i + 1$ **to** $n^2 - 1$ **do**
9   **if** $\neg$mask$[j]$ **and** $\text{Match}(\text{delays}[i], \text{delays}[j])$ **then**
10   group $\leftarrow$ group $\cup \{j\}$;
11   mask$[j] \leftarrow \textbf{true}$;
12   Groups $\leftarrow$ Groups $\cup \{$group$\}$;
13   **return** Groups

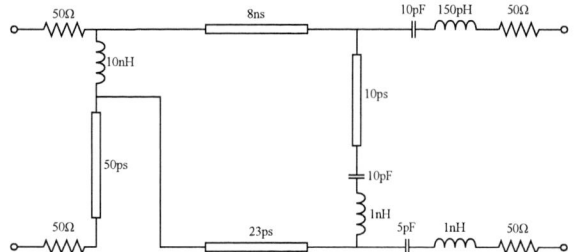

Fig. 1: Schematic of the Crosstalk Delay Network

TABLE I: Performance Comparison: Fitting Time and RMSE (S_{22} Data)

Algorithms	Time (s)	RMSE ($\times 10^{-2}$)
VFAS	382.52	2.5740
DVF	138.10	2.8166
HRF (Proposed)	73.09	0.5987

(RMSE) shown below is used to quantify modeling accuracy and convergence.

$$\text{RMSE} = \sqrt{\frac{1}{N}\sum_{k=1}^{N}(|\tilde{S}_{ij,k} - \hat{S}_{ij,k}|)^2}. \quad (15)$$

Our experimental case models a crosstalk delay network shown in Fig. 1. It is composed of multiple lossless transmission lines coupled through resistors, inductors, and capacitors, with impedance matching at the terminations to emulate parallel interconnect crosstalk. LC paths are formed by inserting mid-network resistors and inductors, inducing capacitive and inductive coupling. The goal is to assess HRF's performance on S-parameters with complex interactions.

Fitting results are shown in Fig. 2. For the diagonal S-parameters (Fig. 2a), HRF achieves an RMSE of 5.987×10^{-3}, outperforming VFAS and DVF. For off-diagonal terms (Fig. 2b), HRF yields 3.114×10^{-3}, again superior. As summarized in Table I, HRF also reduces runtime significantly, improving efficiency by $5.23\times$ and $1.89\times$ compared to VFAS and DVF. These results demonstrate that HRF ensures high accuracy with faster convergence and high robustness when modeling multiport crosstalk systems.

V. CONCLUSIONS

This paper proposes an efficient fitting method called Hierarchical Residual Fitting (HRF) to model systems with complex delays accurately. HRF effectively decomposes delay-induced oscillations, improving fitting accuracy with high computational efficiency. The hierarchical approach ensures robust convergence and stability in challenging multi-delay scenarios. Extensive evaluations confirm that HRF's performance surpasses that of existing methods. Patents corresponding to this work are under review. Overall, HRF provides a practical solution for modeling high-fidelity delay-dominant systems.

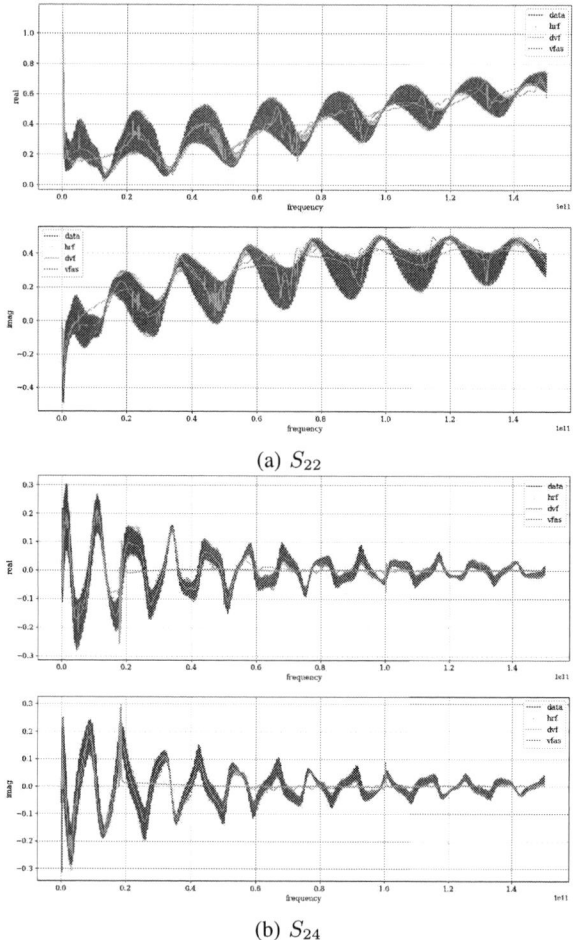

(a) S_{22}

(b) S_{24}

Fig. 2: Fitting Results

REFERENCES

[1] P. J. Pupalaikis, *S-Parameters for Signal Integrity*. University Printing House, Cambridge CB2 8BS, United Kingdom: Cambridge University Press, 2020, printed in the United Kingdom by TJ International Ltd.

[2] B. Gustavsen, "Improving the pole relocating properties of vector fitting," *IEEE Transactions on Power Delivery*, vol. 21, no. 3, pp. 1587–1592, 2006.

[3] S. Grivet-Talocia and M. Bandinu, "Improving the convergence of vector fitting for equivalent circuit extraction from noisy frequency responses," *IEEE Transactions on Electromagnetic Compatibility*, vol. 48, no. 1, pp. 104–120, 2006.

[4] A. Chinea, P. Triverio, and S. Grivet-Talocia, "Compact macromodeling of electrically long interconnects," in *2008 IEEE-EPEP Electrical Performance of Electronic Packaging*, 2008, pp. 199–202.

[5] B. Gustavsen and A. Semlyen, "Rational approximation of frequency domain responses by vector fitting," *IEEE Transactions on Power Delivery*, vol. 14, no. 3, pp. 1052–1061, 1999.

[6] S. Grivet-Talocia and B. rn Gustavsen, *Passive Macromodeling*, ser. Wiley Series in Microwave and Optical Engineering. Hoboken, NJ, USA: John Wiley & Sons, Inc., 2016.

[7] A. Charest, M. S. Nakhla, R. Achar, D. Saraswat, N. Soveiko, and I. Erdin, "Time domain delay extraction-based macromodeling algorithm for long-delay networks," *IEEE Transactions on Advanced Packaging*, vol. 33, no. 1, pp. 219–235, 2010.

[8] S. Grivet-Talocia, "Delay-based macromodels for long interconnects via time-frequency decompositions," in *2006 IEEE Electrical Performane of Electronic Packaging*, 2006, pp. 199–202.

Hybrid Model-Based Hardware Acceleration for Diesel Engine NOx Emission Prediction

Xinlei Su [1], Shanqiang Yang [1], Tianliang Xu [1], Xiaozhen Yan [1], Jianfeng Li [1], Tian Rong [1], Chenxu Wang*[1,2], Yuhang Wang [1], Zhiwei Han [3]

[1] Harbin Institute of Technology, Weihai 264209, China
[2] Key Laboratory of Cross-Domain Synergy and Comprehensive Support for Unmanned Marine Systems, Ministry of Industry and Information Technology, Weihai 264209, China
[3] Shandong Huayi Micro-Electronics Technology Co, Ltd, Jinan, 250001, Shandong, China

* Email: 23S030117@stu.hit.edu.cn, wangchenxu@hit.edu.cn

Abstract—Nitrogen oxides (NOx) from diesel engines pose significant environmental and public health challenges. To comply with stringent emission standards, this study proposes a hybrid CNN-LSTM model for real-time prediction of engine-out NOx emissions. Using real-vehicle operating data, the CNN extracts spatial features from engine parameters while the LSTM captures temporal dependencies in emission sequences. The model achieves a mean absolute error (MAE) of 28.05 ppm, reducing errors by 15.71% and 19.97% compared to standalone CNN and LSTM models, respectively. In the context of in-vehicle deployment, INT16 quantization limits MAE degradation to 8.3% while enabling FPGA acceleration. The customized hardware accelerator leverages parallel computing and on-chip memory to optimize convolution and LSTM operations via time-division multiplexing. Implemented on a Kintex-7 FPGA at 100 MHz, it achieves a latency of 0.12ms with a power consumption of 0.71 W. This solution offers a high-precision, low-latency deployment option for real-time monitoring of diesel engine emissions.

Keywords—NOx prediction, CNN-LSTM, low power, FPGA implementation

I. INTRODUCTION

The accelerated development of the Chinese industry has led to a substantial increase in demand for diesel vehicles used for heavy-duty transportation. However, nitrogen oxide (NOx) emissions from diesel engines pose severe environmental and public health challenges[1]. In recent years, nations worldwide have intensified efforts to mitigate the environmental impact of NOx, implementing increasingly stringent emission standards, such as the Euro 7 in the European Union and China VIb[2]. Among NOx control technologies, selective catalytic reduction (SCR) systems represent the mainstream approach for medium- and heavy-duty diesel engines to meet these rigorous standards[3]. Effective closed-loop control of SCR systems requires rapid and accurate measurement of engine-out NOx emissions.

Direct NOx measurement using physical sensors faces limitations in both cost and accuracy[4]. Traditional emission models exhibit significant latency and signal drift during complex transient operating conditions. To address these challenges, this paper proposes a hybrid architecture integrating convolutional neural networks (CNN) and long short-term memory (LSTM) networks, utilizing on-board operating data from a China VIb-compliant diesel engine. The CNN component extracts spatial correlations from engine operating parameters, while the LSTM captures long-term temporal dependencies in emission sequences. This dual approach aims to enhance the prediction accuracy of raw NOx

emissions during transient conditions. Furthermore, to enable real-time, low-power operation in vehicular environments, we designed an optimized hardware architecture. We implemented it on a field-programmable gate array (FPGA) platform for inference validation.

II. DATASET AND DATA PREPROCESSING

The study utilizes on-board operational data from a China VIb-compliant six-cylinder 12.52L diesel engine, comprising over 200 parameters (vehicle speed, torque, etc.)

To screen out input variables that are highly correlated with NOx emissions, the key parameters are preliminarily screened. This is based on the NOx generation mechanism and engine operating conditions. Secondly, the data cleaning process is initiated to identify and remove any outliers or invalid data. Finally, Pearson correlation coefficient analysis is employed to quantify the strength of the correlation between each parameter and NOx emissions[5]. The Pearson's correlation coefficient formula is shown in (1):

$$r = \frac{\sum_{i=1}^{n}(X_i - \overline{X})(Y_i - \overline{Y})}{\sqrt{\sum_{i=1}^{n}(X_i - \overline{X})^2}\sqrt{\sum_{i=1}^{n}(Y_i - \overline{Y})^2}} \,. \tag{1}$$

Where n represents the total number of samples, X_i and Y_i represent the tested variable and NOx emission value, respectively, and $\overline{X}$ and $\overline{Y}$ represent the means of these two variables.

The closer the absolute value of the correlation coefficient is to 1, the stronger the linear relationship between the parameters. Based on the results of the Pearson correlation analysis, six variables were selected as input features for the model. The Pearson coefficients are detailed in TABLE I. .

TABLE I. CORRELATION ANALYSIS RESULTS

Variable	Value
Output torque	0.832902
Lambda value	-0.832085
Final fuel injection volume	0.814188
Intake pressure	0.681085
Exhaust flow rate	0.635757
Manifold inlet airflow value	0.623756

979-8-3315-3918-4/25 $31.00 © 2025 IEEE

Given the determined input and output feature dimensions of the model, the original data is linearly transformed using Min-Max Scaling:

$$X' = \frac{X - X_{min}}{X_{max} - X_{min}} \qquad (2)$$

Mapping each feature value to the unit interval [0, 1] significantly improves the convergence speed and stability of the gradient descent algorithm, as it eliminates differences in feature dimension and deviations in feature scale.

III. CNN-LSTM MODEL CONSTRUCTION

A. Convolution Neural Network Model

Convolutional neural networks are a type of deep learning model that is inspired by the way the human visual system works. The core concept is the efficient extraction of spatial or temporal features from input data through local perception (i.e., the sliding scan of a convolution kernel) and parameter sharing (i.e., the reuse of the same convolution kernel). In 1D-CNN, the convolutional kernel slides along a one-dimensional direction (primarily along the time sequence), and it is widely applied in feature extraction for non-image sequence data (such as time series, sensor signals, audio spectra, and text)[6].

B. Long Short-Term Memory Network Model

The Long Short-Term Memory network is a specialized variant of the recurrent neural network (RNN)[7] architecture. Its core innovation lies in replacing traditional RNN hidden layer nodes with purpose-designed LSTM units to overcome the vanishing or exploding gradient problem inherent in standard RNN during the propagation of extended temporal information. The LSTM structure comprises four fundamental components: the input gate, the forget gate, the memory cell, and the output gate[8]. Its neural network architecture is illustrated in Fig. 1.

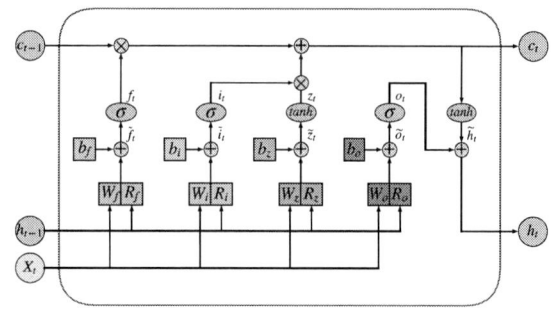

Fig. 1. The architecture of a typical LSTM cell.

Input gate, forget gate, storage unit, and output gate as shown in (3).

$$
\begin{aligned}
f_t &= sigmoid\left(\tilde{f}_t\right), \tilde{f}_t = W_f X_t + R_f h_{t-1} + b_f \\
i_t &= sigmoid\left(\tilde{i}_t\right), \tilde{i}_t = W_i X_t + R_i h_{t-1} + b_i \\
z_t &= tanh\left(\tilde{z}_t\right), \tilde{z}_t = W_z X_t + R_z h_{t-1} + b_z \\
o_t &= sigmoid\left(\tilde{o}_t\right), \tilde{o}_t = W_o X_t + R_o h_{t-1} + b_o
\end{aligned}
\qquad (3)
$$

Where the weight parameters W_f, W_i, W_z, and W_o correspond to the input weight vectors between the input X_t and the forget gate, input gate, cell input gate, and output gate,

respectively. The weights R_f, R_i, R_z, and R_o correspond to the recurrent weights between the hidden state h_{t-1} and the forget gate, input gate, cell input gate, and output gate, respectively, with b_z, b_i, b_f, and b_o being their corresponding biases. *Sigmoid* and *tanh* denote the activation functions.

The LSTM state update is described by (4).

$$
\begin{aligned}
c_t &= f_t \otimes c_{t-1} + i_t \otimes g_t \\
h_t &= o_t \otimes tanh\left(c_t\right)
\end{aligned}
\qquad (4)
$$

Where c_t denotes the cell state update, h_t denotes the hidden state update.

C. CNN-LSTM Model

In the hybrid architecture, the CNN extracts spatial features from preprocessed parameters, which serve as inputs to the LSTM module. The LSTM network possesses the capability to maintain long-term dependencies and effectively captures temporal characteristics within sequential data. The detailed model structure is illustrated in Fig. 2.

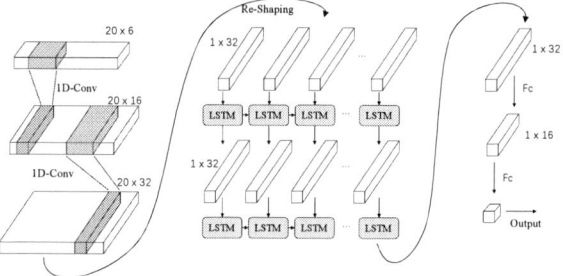

Fig. 2. CNN-LSTM neural network structure

D. Evaluation criteria

The evaluation of prediction models typically employs mean squared error (MSE) and mean absolute error (MAE). These metrics comprehensively quantify model performance. The mathematical formulations are given in (5) and (6):

$$MSE = \frac{1}{n}\sum_{i=1}^{n}\left(y_i - \hat{y}_i\right)^2 \qquad (5)$$

$$MAE = \frac{1}{n}\sum_{i=1}^{n}\left|y_i - \hat{y}_i\right| \qquad (6)$$

Where n denotes the sample size, y_i represents the actual value, and $\hat{y}_i$ indicates the predicted value.

E. Comparative Model Performance Evaluation

To validate the efficacy of the hybrid CNN-LSTM model, standalone CNN and LSTM models were constructed as benchmarks. The CNN model retained the dual-convolutional-layer architecture of the hybrid framework, adopting identical kernel dimensions, activation functions (Relu), and terminal fully connected layers. Similarly, the LSTM model replicated the LSTM layer and fully connected layer configurations from the hybrid structure. All models maintained consistent input/output dimensions and identical training parameters. As shown TABLE II. , the hybrid model achieved a 22.65% reduction in MSE compared to the standalone CNN and a 33.99% reduction compared to the standalone LSTM, along with MAE reductions of 15.71% and 19.97%, respectively.

TABLE II.   MODEL PERFORMANCE COMPARISON

Model	MSE	MAE
CNN	3796.930608	33.283507
LSTM	4449.380882	35.055104
CNN-LSTM	2936.989624	28.053213

IV. HARDWARE ACCELERATOR ARCHITECTURE DESIGN

A. System Architecture

The hardware accelerator architecture illustrated in Fig. 3. The PE array handles convolutional operations and matrix multiplications for both LSTM and fully connected (FC) layers. The Relu unit computes activation functions for convolutional and FC layers, while the LSTM state updata computation unit manages LSTM gate activations (e.g., *sigmoid/tanh*) and sequential state updates. The control unit orchestrates dataflow scheduling and neural network state transitions. For intermediate data buffering, the Temp Memory subsystem utilizes ping-pong buffering to temporarily cache inter-layer results.

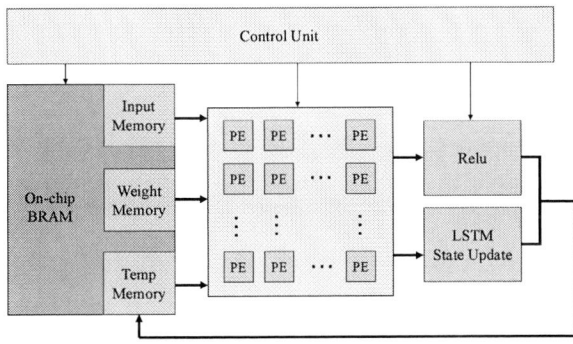

Fig. 3. System architecture diagram

B. PE Computation Array

The computation array serves as the computational core of the accelerator, with its overall architecture illustrated in Fig. 4. Each processing element (PE) unit is implemented using the FPGA's embedded DSP48E1 blocks. This two-dimensional arrangement optimizes FPGA routing resource utilization. Targeting the operational characteristics of CNN and LSTM, both convolutional operations and matrix multiplications are decomposed into parallel multiply-accumulate computations. By strategically allocating input data and weights, the PE array enables parallel acceleration for both CNN convolutions and LSTM matrix multiplications. The designed 32×4 PE array supports concurrent processing of up to 32 convolutional kernels or matrix multiplication operations with weight matrices containing ≤32 rows.

When computing a 1D-CNN, the input data, weight data, and bias data are loaded into the cache register group of the PE computing array. They are then sequentially multiplied and accumulated into register D. The temporary results in register D can be returned to the accumulation input via a data selector to participate in further computations. The input data and weight data are sequentially input into the PE array to complete the computation of a convolution kernel. Each row of the PE computation array corresponds to the computation of a convolution kernel, thereby achieving parallel acceleration. The final computation result is input into the Relu unit for activation processing and stored in the Temp

Memory as input data for the next layer to continue participating in the computation.

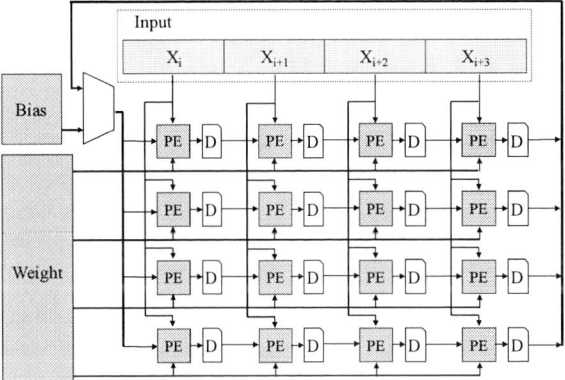

Fig. 4. Schematic of PE computation array

When calculating the matrix multiplication in the LSTM layer, based on the characteristics of matrix multiplication in the LSTM layer, the input weight matrix W_x and the hidden state weight matrix W_h are concatenated column-wise to form the weight matrix. The current input, X_t, is concatenated with the hidden state, h_{t-1}, from the previous time step to form a one-dimensional input vector. Each row of the weight matrix is assigned to a different PE row for parallel computation of the multiplication and accumulation with the input vector, thereby achieving efficient matrix multiplication.

C. LSTM State Update

The LSTM state update is completed in four sequential stages through time-division multiplexing hardware modules, as shown in Fig. 5. In the first stage, the *sigmoid* activation function is used to generate the input gate i_t from $W_iX_t + R_ih_{t-1}$; in the second stage, the *tanh* activation function is applied to $W_gX_t + R_gh_{t-1}$ to generate g_t, which is then multiplied by i_t; in the third stage, the sigmoid function is used to calculate the forget gate f_t from $W_fX_t + R_fh_{t-1}$, which is multiplied by c_{t-1} and added to the result from the second stage to obtain c_t, followed by a *tanh* transformation on c_t; In the fourth stage, the output gate o_t is generated by calculating $W_oX_t+R_oh_{t-1}$ using the *sigmoid* activation function, and finally multiplied by $tanh(c_t)$ to output h_t. The module adopts a time-division multiplexing strategy, and in actual hardware, there is only one *sigmoid* module, two *tanh* modules, one multiplication module, and one addition module.

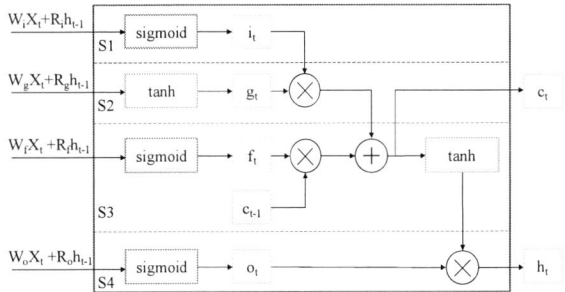

Fig. 5. Translation of LSTM state update schematic

V. EXPERIMENTAL RESULTS

A. Experimental Setup

The trained model weights and test dataset underwent 16-bit fixed-point quantization (1 sign bit, four integer bits, 11 fractional bits) and were deployed on a Kintex-7 XC7K325T-2FFG676 FPGA operating at 100 MHz. Weight and input data were transmitted to the FPGA via Universal Asynchronous Receiver/Transmitter (UART), with inference results returned to the PC for statistical analysis.

For power consumption and performance comparison, experiments were conducted on a CPU platform (Intel Core i9-11390HX, 2.2 GHz base clock frequency, CPU power measured using HWINFO during inference) and a GPU platform (NVIDIA GeForce RTX 4060 Laptop GPU, 1785 MHz boost clock frequency, GPU power monitored via the pynvml library).

B. Experimental Results

Fig. 6 compares the actual values, hybrid neural network predictions, and FPGA predictions. Statistical analysis reveals that the quantized neural network achieves a mean squared error (MSE) of 3042.529698 (3.59% increase) and a mean absolute error (MAE) of 30.38034 (8.3% increase), both of which remain within the tolerance thresholds of SCR closed-loop control.

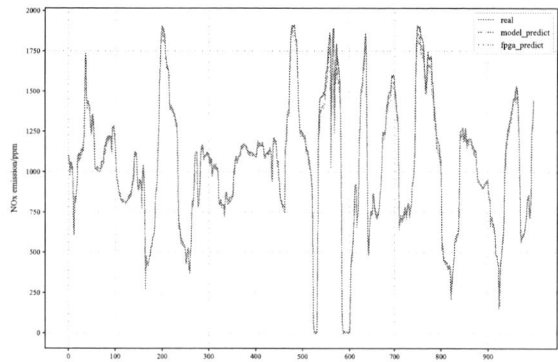

Fig. 6. Prediction results comparison.

TABLE III. presents the resource utilization of the accelerator on the Kintex-7 XC7K325T platform. The generalized PE array design effectively reduces resource occupancy, achieving 15.48% DSP utilization and 18.76% BRAM utilization while maintaining a compact footprint.

TABLE III. THE RESOURCE UTILIZATION OF ACCELERATOR

Resource	LUT	DSP	BRAM	FF
Used	13005	130	83.50	37075
Total	203800	840	445	407600
Utilization	6.38%	15.48%	18.76%	9.10%

TABLE IV. compares performance and power consumption across platforms. The proposed FPGA accelerator achieves a throughput of 6.16 GOP/s and an energy efficiency ratio of 8.7 GOPs/W. Compared to the CPU, the FPGA accelerator delivers a 335× speedup while consuming only 2% of the CPU's power. Against the GPU, it achieves 76× faster inference with 12% of the GPU's power consumption. This architecture provides a real-time, low-power, and high-reliability NOx prediction solution for diesel SCR systems, balancing computational density and energy efficiency.

TABLE IV. THE PERFORMANCE OF EACH PLATFORM

Platform	Time(ms)	Frequency(Hz)	Power(W)
i9-113900HX	40.20	2.2G	34.95
RTX4060	9.13	700M	5.93
This work	0.12	100M	0.71

VI. CONCLUSION

This study proposes a real-time NOx prediction method for diesel engines based on a CNN-LSTM hybrid model, combined with an FPGA hardware accelerator to achieve efficient and low-power deployment. The model achieves a mean absolute error of 28.05 ppm, reducing errors by 15.71%–19.97% compared to single CNN and LSTM models. The designed accelerator achieved ultra-low latency of 0.12 ms and power consumption of 0.71 W when deployed on an FPGA. This achievement provides high-precision, low-latency technical support for real-time closed-loop control of diesel engine SCR systems.

ACKNOWLEDGMENT

This work was mainly supported by Major scientific and technological innovation projects of Shandong Province of China, with Grant No.2022ZLGX04. The research presented in this paper is also partially supported by the NSF project of China with granted No. U2106202 and Shandong Provincial Natural Science Foundation with Grant ZR2023MA074.

REFERENCES

[1] H. Sun, G. Li, J. Li, Z. Zheng, Q. Tang, and M. Yao, "Development of an LSTM-CCF-MA Model for Predicting NOx Emission and Exhaust Temperature of a Diesel Engine," Int.J Automot. Technol., vol. 26, no. 2, pp. 437–450, Apr. 2025.

[2] M. Pan, X. Cao, C. Fu, S. Liao, X. Zhou, and W. Guan, "Emission prediction and optimization of methanol/diesel dual-fuel engines based on ITransformer-BiGRU and NSGA-III," Energy and AI, vol. 19, p. 100466, Jan. 2025.

[3] M. K. A. Wardana, K. Oh, Y. J. Lee, Y. M. Woo, and O. Lim, "Effects of Urea Injection Timing on Predicting NoX Conversion In SCR Systems," Int.J Automot. Technol., vol. 21, no. 1, pp. 137–145, Feb. 2020

[4] O. Y. Odufuwa, L. K. Tartibu, and K. Kusakana, "Artificial neural network modelling for predicting efficiency and emissions in mini-diesel engines: Key performance indicators and environmental impact analysis," Fuel, vol. 387, p. 134294, May 2025.

[5] J. J. Park, S. Lee, S. Shin, M. Kim, and J. Park, "Development of a Light and Accurate Nox Prediction Model for Diesel Engines Using Machine Learning and Xai Methods," Int.J Automot. Technol., vol. 24, no. 2, pp. 559–571, Apr. 2023.

[6] Y. Yang, F. Ge, D. Qiu, X. Yue, Z. Li, and F. Zhou, "Implementation of Reconfigurable CNN-LSTM Accelerator Based on FPGA," in 2021 IEEE 21st International Conference on Communication Technology (ICCT), Tianjin, China: IEEE, 2021, pp. 1026–1030.

[7] B. Wu, X. Wu, P. Li, Y. Gao, J. Si, and N. Al-Dhahir, "Efficient FPGA Implementation of Convolutional Neural Networks and Long Short-Term Memory for Radar Emitter Signal Recognition," Sensors, vol. 24, no. 3, p. 889, Jan. 2024.

[8] Z. Gao, W. Xiao, W. Zhou, and Z. Yang, "FPGA Implementation of CNN-LSTM Classifier in Speech Emotion Recognition System," in 2023 International Conference on High Performance Big Data and Intelligent Systems (HDIS), Macau, China: IEEE, 2023, pp. 47–52.

Extending Straight-Through Estimation for Robust Neural Networks on Analog CIM Hardware

Yuannuo Feng[†1,3], Wenyong Zhou[†2,3], Yuexi Lyu[‡3], Yixiang Zhang[3], Zhengwu Liu[*2], Ngai Wong[*2] and Wang Kang[*1]

[1]School of Integrated Circuit Science and Engineering, Beihang University, Beijing, China
[2]Department of Electrical and Electronic Engineering, The University of Hong Kong, Hong Kong
[3]Zhicun Research Lab, Beijing, China
[†]: Equal Contribution. [‡]: Project Leader. [*]: Corresponding Author(s).

Abstract—**Analog Compute-In-Memory (CIM) architectures promise significant energy efficiency gains for neural network inference, but suffer from complex hardware-induced noise that poses major challenges for deployment. While noise-aware training methods have been proposed to address this issue, they typically rely on idealized and differentiable noise models that fail to capture the full complexity of analog CIM hardware variations. Motivated by the Straight-Through Estimator (STE) framework in quantization, we decouple forward noise simulation from backward gradient computation, enabling noise-aware training with more accurate but computationally intractable noise modeling in analog CIM systems. We provide theoretical analysis demonstrating that our approach preserves essential gradient directional information while maintaining computational tractability and optimization stability. Extensive experiments show that our extended STE framework achieves up to 5.3% accuracy improvement on image classification, 0.72 perplexity reduction on text generation, 2.2× speedup in training time, and 37.9% lower peak memory usage compared to standard noise-aware training methods.**

Index Terms—**Analog Compute-In-Memory, Straight-Through Estimator, Hardware-Aware Training**

I. INTRODUCTION

The exponential growth of neural network applications has intensified demand for energy-efficient computing solutions, particularly for edge devices with severe power and computational constraints [1], [2]. Analog Compute-In-Memory (CIM) architectures address these challenges by performing matrix-vector multiplications directly within memory arrays, eliminating energy-intensive data movement and achieving orders of magnitude energy efficiency improvements over traditional von Neumann architectures through analog weight storage and physical law-based computation [3], [4].

However, the analog nature introduces significant deployment challenges, as analog CIM hardware suffers from various noise sources that severely degrade inference accuracy [5]. These include device-level variations from manufacturing and aging, circuit-level noise from voltage drops and current leakage, and peripheral circuit imperfections such as ADC quantization errors and amplifier non-linearities [6], [7]. The cumulative effect of these noise sources can lead to substantial accuracy degradation, limiting practical neural network deployment on analog CIM hardware [8], [9].

To address this challenge, researchers have developed noise-aware training methods that expose neural networks to simulated hardware noise during training, enabling them to develop robustness against deployment conditions [10], [11]. However, existing approaches face a fundamental limitation: they rely on simplified, differentiable noise models that can be easily integrated into standard gradient-based optimization frameworks. While Gaussian noise injection or uniform quantization can be readily handled by automatic differentiation, they fail to capture the full complexity of real analog CIM hardware variations, as many critical hardware effects are inherently non-differentiable, exhibit complex spatial and temporal correlations, or involve computationally expensive physical simulations that make direct gradient computation impractical [6]. This limitation creates a significant gap between the simplified noise models used during training and the complex noise characteristics encountered during deployment, where networks trained with oversimplified noise models often fail to generalize to real hardware conditions, resulting in substantial accuracy drops when deployed on actual analog CIM systems [12].

We extend the Straight-Through Estimator (STE) framework, originally for quantization, to address complex noise in analog CIM hardware [13], [14]. By decoupling forward noise simulation from backward gradient computation, we enable realistic noise models in forward passes and simplified gradients in backpropagation, achieving efficient noise-aware training with high-fidelity hardware models. This paper provides three key contributions:

- We formalize the extension of STE to complex noise environments, providing a general framework that can accommodate various types of hardware noise beyond simple quantization.
- We develop a STE-based gradient approximation strategy and provide theoretical understanding of its effectiveness from the gradient perspective.
- Experiments demonstrate that our extended STE framework achieves up to 5.3% higher accuracy on image classification, 0.72 lower perplexity on text generation, 2.2× faster training, and 37.9% less peak memory usage than standard noise-aware methods.

979-8-3315-3918-4/25 $31.00 © 2025 IEEE

II. Preliminaries

The STE was originally developed to enable gradient-based training of neural networks containing non-differentiable functions, particularly quantization operations [15]. In the standard quantization scenario, the forward pass applies a quantization function $Q(\cdot)$ to the weights or activations:

$$y = Q(x) \tag{1}$$

where $Q(\cdot)$ is typically a step function that maps continuous values to discrete levels.

The challenge arises during backpropagation, as the gradient of $Q(\cdot)$ is zero almost everywhere, preventing effective learning. The STE addresses this by using different functions for forward and backward passes. During the forward pass, the actual quantization function is applied, while during the backward pass, the gradient flows through as if the quantization operation were the identity function:

$$\frac{\partial L}{\partial x} = \frac{\partial L}{\partial y} \cdot 1 \tag{2}$$

where 1 represents the identity operation [16].

III. Methodology

A. Problem Formulation

We consider training neural networks that operate robustly under complex deployment noise. Let $\mathbf{W} \in \mathbb{R}^d$ denote the clean weights and $\mathcal{N}(\mathbf{W}, \boldsymbol{\xi})$ represent a complex noise function, where $\boldsymbol{\xi}$ encapsulates noise parameters. The noisy weights are:

$$\tilde{\mathbf{W}} = \mathcal{N}(\mathbf{W}, \boldsymbol{\xi}) \tag{3}$$

The training objective minimizes the expected loss:

$$\mathbb{E}_{\boldsymbol{\xi}}[L] = \mathbb{E}_{\boldsymbol{\xi}}[\ell(f(\tilde{\mathbf{W}}, \mathbf{x}), \mathbf{y})] \tag{4}$$

The challenge arises when computing gradients $\frac{\partial L}{\partial \mathbf{W}}$ through $\mathcal{N}(\mathbf{W}, \boldsymbol{\xi})$, which may be non-differentiable, have zero gradients almost everywhere, or involve computationally expensive stochastic processes that make standard backpropagation infeasible.

Algorithm 1 demonstrates our STE approach for a linear layer. During the forward pass, inputs and weights are quantized to int8 precision, converted to appropriate formats for the noise model, and processed to generate noisy outputs. The difference between the noisy and clean outputs is computed as a delta term, which is then added to the clean output using gradient detachment to preserve backpropagation through the original linear computation while incorporating the effects of quantization noise.

B. Theoretical Analysis

To understand how the STE can be extended to complex noise environments, we analyze the approach from the perspective of gradient direction and magnitude. The effectiveness of STE can be understood by decomposing any gradient vector g into its components:

$$\mathbf{g} = \|\mathbf{g}\| \cdot \frac{\mathbf{g}}{\|\mathbf{g}\|} \tag{5}$$

Algorithm 1 Noisy Linear Layer with Quantization and Noise

1: **function** LINEAR(x, *linear*, *noise_level*, *with_noise*)
2:      **if** not *with_noise* **then**
3:          **return** Linear(x, *linear.weight*, *linear.bias*)
4:      **end if**
5:      y_{clean} = Linear(x, *linear.weight*, *linear.bias*)
6:      *noise_model* = NoiseModelNNTrain(*noise_level*)
7:      **with** torch.no_grad():
8:          $x_{q8}, scale_x$ = Quant$_{int8}$(x)
9:          $w_{q8}, scale_w$ = Quant$_{int8}$(*linear.weight*)
10:         x_{uint8}, w_{split} = SplitInput(x_{q8}, w_{q8})
11:         $bias_q$ = QuantBias(*linear.bias*, $scale_x$, $scale_w$)
12:         g = ComputeGain($scale_x$, $scale_w$, y_{clean})
13:         *noise_model*.program($x_{uint8}, w_{split}, bias_q, g$)
14:         y_{noisy} = *noise_model*.infer(x_{uint8})
15:         y_{scaled} = Rescale($y_{noisy}, g, scale_x, scale_w$)
16:         $\delta = y_{scaled} - y_{clean}$.detach()
17:      **return** $y_{clean} + \delta$.detach()
18: **end function**

---

where $\|\mathbf{g}\|$ represents the gradient norm (step size information) and $\frac{\mathbf{g}}{\|\mathbf{g}\|}$ represents the gradient direction (update direction).

Consider a neural network layer computation affected by complex hardware noise. The forward computation can be expressed as:

$$\mathbf{y} = \tilde{\mathbf{W}}\mathbf{x} = \mathcal{N}(\mathbf{W}, \boldsymbol{\xi})\mathbf{x} \tag{6}$$

where $\mathcal{N}(\cdot, \cdot)$ represents the complex noise function that transforms the clean weights $\mathbf{W}$ into noisy weights $\tilde{\mathbf{W}}$.

The ground truth gradient that properly accounts for the noise characteristics would be:

$$\mathbf{g}^* = \frac{\partial L}{\partial \mathbf{W}} = \frac{\partial L}{\partial \mathbf{y}} \frac{\partial \mathcal{N}(\mathbf{W}, \boldsymbol{\xi})}{\partial \mathbf{W}} \mathbf{x} \tag{7}$$

However, computing this gradient exactly is often intractable due to the complexity and non-differentiable nature of $\mathcal{N}(\cdot, \cdot)$. The STE approach approximates this ground truth gradient using a simplified gradient that assumes $\frac{\partial \mathcal{N}(\mathbf{W}, \boldsymbol{\xi})}{\partial \mathbf{W}} \approx \mathbf{I}$:

$$\tilde{\mathbf{g}} = \frac{\partial L}{\partial \mathbf{y}} \mathbf{x} \tag{8}$$

From a statistical perspective, we can view the relationship between the ground truth gradient and the STE gradient as:

$$\mathbf{g}^* = \tilde{\mathbf{g}} + \boldsymbol{\delta} \tag{9}$$

where $\boldsymbol{\delta} = \frac{\partial L}{\partial \mathbf{y}} \left(\frac{\partial \mathcal{N}(\mathbf{W}, \boldsymbol{\xi})}{\partial \mathbf{W}} - \mathbf{I} \right) \mathbf{x}$ represents the bias term introduced by the noise function.

Direction Analysis: The STE gradient $\tilde{\mathbf{g}}$ represents the full-precision gradient direction, which captures the correct optimization trajectory for the clean network. To quantify the directional reliability, consider the cosine similarity between the STE gradient and the ground truth gradient:

$$\cos(\theta) = \frac{\langle \tilde{\mathbf{g}}, \mathbf{g}^* \rangle}{\|\tilde{\mathbf{g}}\| \|\mathbf{g}^*\|} = \frac{\langle \tilde{\mathbf{g}}, \tilde{\mathbf{g}} + \boldsymbol{\delta} \rangle}{\|\tilde{\mathbf{g}}\| \|\tilde{\mathbf{g}} + \boldsymbol{\delta}\|} \tag{10}$$

Expanding this expression:

$$\cos(\theta) = \frac{\|\tilde{\mathbf{g}}\|^2 + \langle \tilde{\mathbf{g}}, \delta \rangle}{\|\tilde{\mathbf{g}}\|\|\tilde{\mathbf{g}} + \delta\|} \tag{11}$$

When the noise-induced bias δ is uncorrelated with the clean gradient $\tilde{\mathbf{g}}$ (i.e., $\mathbb{E}[\langle \tilde{\mathbf{g}}, \delta \rangle] \approx 0$), the cosine similarity approaches:

$$\mathbb{E}[\cos(\theta)] \approx \frac{\|\tilde{\mathbf{g}}\|}{\|\tilde{\mathbf{g}} + \delta\|} \approx \frac{\|\tilde{\mathbf{g}}\|}{\sqrt{\|\tilde{\mathbf{g}}\|^2 + \|\delta\|^2}} \tag{12}$$

This shows that the directional alignment deteriorates as the noise bias magnitude $\|\delta\|$ increases relative to the clean gradient magnitude $\|\tilde{\mathbf{g}}\|$. The variance of the ground truth gradient can be expressed as:

$$\mathrm{Var}[\mathbf{g}^*] = \mathrm{Var}[\delta] = \mathbb{E}[\|\delta\|^2] - \|\mathbb{E}[\delta]\|^2 \tag{13}$$

Since $\tilde{\mathbf{g}}$ is deterministic given the clean weights, it exhibits zero variance, making it statistically more reliable for consistent optimization direction.

The key insight is that $\tilde{\mathbf{g}}$ provides a projection of the optimization direction onto the clean parameter space. Mathematically, this can be viewed as:

$$\tilde{\mathbf{g}} = \mathbb{E}_\xi[\mathbf{g}^*] \quad \text{when} \quad \mathbb{E}_\xi[\delta] = \mathbf{0} \tag{14}$$

When noise is not zero-mean, $\tilde{\mathbf{g}}$ provides a simplified approximation of the optimization direction by ignoring the systematic bias introduced by the noise in $\mathbf{g}^*$.

Magnitude Analysis: While the magnitude of $\tilde{\mathbf{g}}$ may differ from $\mathbf{g}^*$, this discrepancy is less critical for optimization success. The magnitude relationship can be expressed as:

$$\|\mathbf{g}^*\|^2 = \|\tilde{\mathbf{g}} + \delta\|^2 = \|\tilde{\mathbf{g}}\|^2 + 2\langle \tilde{\mathbf{g}}, \delta \rangle + \|\delta\|^2 \tag{15}$$

The relative magnitude difference is:

$$\frac{\|\mathbf{g}^*\|^2 - \|\tilde{\mathbf{g}}\|^2}{\|\tilde{\mathbf{g}}\|^2} = \frac{2\langle \tilde{\mathbf{g}}, \delta \rangle + \|\delta\|^2}{\|\tilde{\mathbf{g}}\|^2} \tag{16}$$

Modern optimizers such as Adam effectively normalize gradients using adaptive scaling:

$$\mathbf{W}_{t+1} = \mathbf{W}_t - \alpha \frac{\hat{\mathbf{m}}_t}{\sqrt{\hat{\mathbf{v}}_t} + \epsilon} \tag{17}$$

where $\hat{\mathbf{m}}_t$ and $\hat{\mathbf{v}}_t$ are bias-corrected momentum estimates. This adaptive mechanism makes the optimization relatively insensitive to the exact gradient magnitude, as the effective step size is determined by the gradient history rather than instantaneous magnitude.

IV. EXPERIMENTS

A. Experiment Setup

We develop a comprehensive noise simulator modeling realistic analog CIM accelerator conditions with both I/O and tile-level non-idealities detailed in Table I. I/O non-idealities include ADC/DAC quantization noise and device non-linearity effects. Tile-level non-idealities encompass programming noise from weight fabrication variations, cycle-by-cycle read variance, thermal-induced parameter drift,

TABLE I
HARDWARE NON-IDEALITIES MODELED IN CIM NOISE SIMULATOR.

Category	Noise Source	Type
I/O non-idealities	ADC noise	Quantization noise
	DAC noise	Quantization noise
	Device non-linearity	System non-linearity
Tile non-idealities	Programming noise	Weight fabrication non-ideality
	Cycle-by-cycle read variance	Computational consistency
	Thermal variation	Temperature-dependent drift
	Retention	Memory degradation
	IR-drop	Wire resistance non-ideality

TABLE II
PERFORMANCE COMPARISON OF MODELS ACROSS DIFFERENT
LOGNORMAL NOISE LEVELS.

Dataset	Model	Performance under Noise					
		$level = 1.0$		$level = 2.0$		$level = 3.0$	
		Value	Δ	Value	Δ	Value	Δ
CIFAR-10	VGG-8	90.14%	+1.43%	89.56%	+1.99%	89.38%	+2.75%
	VGG-11	89.90%	+3.20%	89.56%	+4.50%	89.90%	+5.30%
	ResNet-20	90.77%	+2.57%	90.66%	+2.65%	90.44%	+3.02%
	ResNet-18	94.00%	+3.58%	93.59%	+1.98%	93.78%	+2.56%
Shakespeare	BERT	3.10	-0.31	3.11	-0.33	3.16	-0.24
	GPT	4.65	-0.51	4.69	-0.50	4.69	-0.72

TABLE III
TRAINING PHASE HARDWARE COST COMPARISON OF DIFFERENT
GRADIENT MANAGEMENT METHODS.

Method	Time (s)	Memory (MB)		GPU Utilization (%)	
		Peak	Average	Compute	Memory
Baseline	0.136	1501	342	64.9	4.14
Full Gradient	1.172	4019	343	96.9	7.94
`torch.no_grad`	0.608	2496	342	94.5	8.44
`.detach()`	0.616	3124	342	94.3	8.40

retention-based memory degradation, and IR-drop effects from wire resistance causing voltage variations across crossbar arrays.

B. Experiment Result

Table II presents the performance comparison of various models across different lognormal noise levels, demonstrating that our method significantly improves performance under different noise conditions compared to baseline models. For CIFAR-10 classification tasks, all models show substantial accuracy increases, with improvements ranging from 1.43% to 5.3%. VGG-11 exhibits the most remarkable gains, achieving up to 5.3% accuracy increase at noise level 3.0. For the Shakespeare language modeling task, both BERT and GPT models experience slight perplexity decreases, with reductions ranging from 0.24 to 0.72.

Table III presents a comprehensive comparison of training phase hardware costs across different gradient management

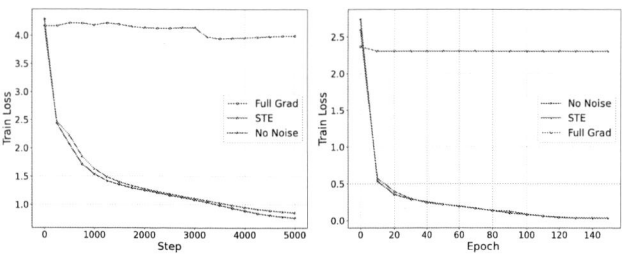

Fig. 1. The training loss reduction curves for GPT-2 on the Shakespeare dataset and ResNet on the CIFAR-10 dataset.

methods. The baseline method, which represents standard training without complex noise considerations, demonstrates the most efficient resource utilization with the lowest time consumption (0.136s), minimal memory usage (1501MB peak, 342MB average). In contrast, the full gradient approach exhibits significantly higher computational overhead, requiring 1.172s per iteration and consuming substantially more memory (4019MB peak, 343MB average). The gradient management techniques `torch.no_grad` and `.detach()` show similar performance characteristics, with execution times of 0.608s and 0.616s respectively, and comparable memory consumption patterns (2496MB and 3124MB peak usage, both with 342MB average). Our proposed methods effectively save memory and accelerate training speed compared to the full gradient approach, with `torch.no_grad` emerging as the superior solution due to its optimal balance between computational efficiency and resource utilization.

The left panel of Fig. 1 shows the training loss curves for GPT-2 models trained on the Shakespeare dataset using three different methods: baseline (No Noise), noise with gradients (Full Grad), and STE. Both the baseline and our STE method exhibit rapid convergence during the initial 1000 training steps, dropping from approximately 4.0 to below 2.0 and reaching final losses around 0.8-1.0. In contrast, the noise gradient method shows minimal improvement throughout training, maintaining consistently high loss around 4.0 for both training and validation sets, suggesting that directly propagating gradients through noisy operations significantly impairs the optimization process. Similarly, the right panel shows ResNet's performance on CIFAR-10 for image classification, highlighting a significant advantage of our STE method over full gradient training.

V. CONCLUSION

This work addresses the gap between realistic analog CIM noise modeling and practical training by extending the STE framework to handle complex, non-differentiable hardware variations. By decoupling forward noise simulation from backward gradient computation, our approach enables high-fidelity noise modeling while maintaining computational tractability. Our work improves image classification accuracy by 5.3%, cuts text generation perplexity by 0.72, speeds training 2.2×, and reduces peak memory use by 37.9% versus standard noise-aware methods.

ACKNOWLEDGEMENT

This work was supported by ACCESS – AI Chip Center for Emerging Smart Systems under the InnoHK initiative of the Innovation and Technology Commission, Hong Kong SAR Government; the Research Grants Council of Hong Kong through TRS project T45-701/22-R and GRF project 17203224; the National Natural Science Foundation of China (62404187, 62274008); the Beijing Nova Program (20250484807); the Zhejiang Provincial Key R&D Program (2025C01071); and the Research Funding of Hangzhou International Innovation Institute of Beihang University (2024KQ157).

REFERENCES

[1] Z. Guan *et al.*, "A hardware-aware neural architecture search pareto front exploration for in-memory computing," in *2022 IEEE 16th International Conference on Solid-State & Integrated Circuit Technology (ICSICT)*, pp. 1–4, 2022.

[2] W. Zhou *et al.*, "A time- and energy-efficient cnn with dense connections on memristor-based chips," in *2023 IEEE 15th International Conference on ASIC (ASICON)*, pp. 1–4, 2023.

[3] A. Shafiee *et al.*, "ISAAC: A Convolutional Neural Network Accelerator with In-Situ Analog Arithmetic in Crossbars," in *2016 ACM/IEEE 43rd Annual International Symposium on Computer Architecture (ISCA)*, pp. 14–26, 2016.

[4] P. Chi *et al.*, "PRIME: A Novel Processing-in-Memory Architecture for Neural Network Computation in ReRAM-Based Main Memory," in *2016 ACM/IEEE 43rd Annual International Symposium on Computer Architecture (ISCA)*, pp. 27–39, 2016.

[5] W. Mao *et al.*, "Hyimc: Analog-digital hybrid in-memory computing soc for high-quality low-latency speech enhancement," in *2025 Design, Automation & Test in Europe Conference (DATE)*, pp. 1–2, IEEE, 2025.

[6] G. Charan *et al.*, "Accurate Inference with Inaccurate RRAM Devices: Statistical Data, Model Transfer, and On-line Adaptation," in *ACM/IEEE Design Automation Conference (DAC)*, 2020.

[7] N. a. Ye, "BayesFT: Bayesian Optimization for Fault Tolerant Neural Network Architecture," in *2021 58th ACM/IEEE Design Automation Conference (DAC)*, pp. 487–492, 2021.

[8] Y. Chen *et al.*, "Device characteristic-aware quantization for eflash-based in-memory computing soc," in *2024 IEEE International Conference on Integrated Circuits, Technologies and Applications (ICTA)*, pp. 70–71, IEEE, 2024.

[9] T. Bai *et al.*, "An end-to-end in-memory computing system based on a 40-nm eflash-based imc soc: Circuits, toolchains, and systems co-design framework," *IEEE Transactions on Computer-Aided Design of Integrated Circuits and Systems*, vol. 43, no. 6, pp. 1729–1740, 2024.

[10] W. Zhou *et al.*, "Towards Robust RRAM-based Vision Transformer Models with Noise-aware Knowledge Distillation," in *Design, Automation & Test in Europe Conference & Exhibition (DATE)*, pp. 1–2, 2025.

[11] W. Zhou *et al.*, "Enhancing robustness of implicit neural representations against weight perturbations," in *ICASSP 2025 - 2025 IEEE International Conference on Acoustics, Speech and Signal Processing (ICASSP)*, pp. 1–5, 2025.

[12] G. Jung *et al.*, "Cost- and Dataset-free Stuck-at Fault Mitigation for ReRAM-based Deep Learning Accelerators," in *Design, Automation & Test in Europe Conference & Exhibition (DATE)*, 2021.

[13] Z. Liu *et al.*, "Nonuniform-to-uniform quantization: Towards accurate quantization via generalized straight-through estimation," in *2022 IEEE/CVF Conference on Computer Vision and Pattern Recognition (CVPR)*, pp. 4932–4942, 2022.

[14] Z. Liu *et al.*, "Bi-real net: Enhancing the performance of 1-bit cnns with improved representational capability and advanced training algorithm," (Berlin, Heidelberg), p. 747–763, Springer-Verlag, 2018.

[15] R. Gong *et al.*, "Differentiable soft quantization: Bridging full-precision and low-bit neural networks," in *2019 IEEE/CVF International Conference on Computer Vision (ICCV)*, pp. 4851–4860, 2019.

[16] H. Qin *et al.*, "Forward and backward information retention for accurate binary neural networks," in *2020 IEEE/CVF Conference on Computer Vision and Pattern Recognition (CVPR)*, pp. 2247–2256, 2020.

979-8-3315-3918-4/25 $31.00 © 2025 IEEE

A Parallel Level-Set Based Approach for Etching Topography Simulation in Process Emulation

Yin Cheang Ng[1], Xin Wen[2], Boyuan Yu[1], Wenjian Yu[1,*]

[1]Department of Computer Science and Technology, BNRist, Tsinghua University, Beijing, China
[2]Hubei NineCube Microelectronics Co. Ltd

Abstract—In process emulation, 3D etching topography simulation is a computationally challenging task. Based on the level-set method, we employ high-order spatial discretization and advanced time integration schemes to tackle this challenge. It leads to a software prototype named pLSM-etching, which includes third-order Runge-Kutta (RK3) and third-order spatial reconstruction schemes and a parallel-computing framework based on sparse matrix operations and OpenMP parallelization. The accuracy of pLSM-etching is meansured with quantitative metrics including Hausdorff distance, area difference, perimeter ratio, shape context, and Hu moments. The experiments demonstrate its strong robustness in capturing topological changes, reaching 97.74% similarity with industrial standards SEMulator3D, along with a 11.2X parallel speedup with 16-thread computing.

Index Terms—Level-Set Method, Hamilton-Jacobi Equation, High-Order Numerical Scheme, Parallel Computation

I. INTRODUCTION

Semiconductor etching is a critical manufacturing process in which materials are selectively removed to create intricate microstructure. Etching topography simulation is an important step in process emulation, which saves a lot of cost in process development. The etching simulation involves accurately tracking the evolution of complex geometries, handling intricate boundary conditions, and modeling interactions between multiple materials. Numerically evolving interfaces in semiconductor processes present significant challenges because of their topological changes, sharp corner formation, and extreme velocity variations. Traditional approaches exhibit fundamental limitations: *marker/string methods* suffer from swallowtail instabilities in discontinuities and struggle with topological changes [1]–[3]; *cell-based methods* compromise geometric accuracy despite handling topology well [4]; *characteristic methods* encounter stability issues in 3D implementations [3].

The level set method (LSM), introduced by Osher and Sethian [2], revolutionized interface tracking by representing surfaces as zero-level sets of higher-dimensional functions, naturally accommodating topology changes while providing weak solutions satisfying entropy to the Hamilton-Jacobi equation [3], [5]. Subsequent advances include: 1) *fast marching methods* [6], 2) *narrow band techniques* handling orientation-dependent etching [7]. Although LSM has been proven effective for semiconductor applications including directional etching [1], exiting work only employ first-order numerical scheme while efficient parallel implementation for large-scale problem lacks.

In this work, we present a parallel level-set based approach for 3D etching simulation. It employs third-order numerical scheme to achieve high accuracy, and exploits efficient multi-thread parallel computing to boost the performance. Experiments with industrial test cases have validated the accuracy of our approach and the parallel algorithm exhibits 11.2X speedup with 16-thread computing. For reproducibility, we share the codes of the proposed algorithm and experimental data on https://github.com/Yin169/pLSM-etching.

II. PRELIMINARIES

The level set method represents an etching front $\Gamma(t)$ as the zero level set of a higher-dimensional function $\phi(\mathbf{r}, t)$:

$$\Gamma(t) = \{ \mathbf{r} \mid \phi(\mathbf{r}, t) = 0 \} \quad (1)$$

where $\phi(\mathbf{r}, t)$ is defined as a signed distance function with negative values inside the geometry and positive values in the etched regions. This implicit representation automatically handles topological changes [4]. The evolution of $\phi(\mathbf{r}, t)$ is governed by:

$$\frac{\partial \phi}{\partial t} + \mathbf{U} \cdot \nabla \phi = 0 \quad (2)$$

and its Hamilton-Jacobi form:

$$\frac{\partial \phi}{\partial t} + F \|\nabla \phi\| = 0 \quad (3)$$

where the velocity field $\mathbf{U}$ determines the velocity of interface movement and F represents the normal component of $\mathbf{U}$ to the interface. The initial condition for $\phi(\mathbf{r}, t)$ is given by signed distance function:

$$\phi(\mathbf{r}, 0) = d(\mathbf{r}, \Gamma_0) \quad (4)$$

where $d(\mathbf{r}, \Gamma_0)$ denotes the signed distance to the initial interface Γ_0. In order to solve equation (3), finite difference method with first-order upwind scheme was proposed [7]. A three-dimensional spatial grid is imposed in the simulation region, and ϕ_{ijk}^n denotes the value of ϕ at the (i, j, k)-index grid point and the n-th time step. The first-order upwind scheme converts equation (3) to the following form:

$$\phi_{ijk}^{n+1} = \phi_{ijk}^n - \Delta t \left[\max(F_{ijk}, 0) \nabla^+ + \min(F_{ijk}, 0) \nabla^- \right] \quad (5)$$

where

$$\nabla^+ = \Big[\max(D_{ijk}^{-x} \phi, 0)^2 + \min(D_{ijk}^{+x} \phi, 0)^2 \\ + \max(D_{ijk}^{-y} \phi, 0)^2 + \min(D_{ijk}^{+y} \phi, 0)^2 \\ + \max(D_{ijk}^{-z} \phi, 0)^2 + \min(D_{ijk}^{+z} \phi, 0)^2 \Big]^{1/2} \quad (6)$$

$$\nabla^- = \left[\max(D_{ijk}^{+x}\phi, 0)^2 + \min(D_{ijk}^{-x}\phi, 0)^2 \right.$$
$$+ \max(D_{ijk}^{+y}\phi, 0)^2 + \min(D_{ijk}^{-y}\phi, 0)^2 \tag{7}$$
$$\left. + \max(D_{ijk}^{+z}\phi, 0)^2 + \min(D_{ijk}^{-z}\phi, 0)^2 \right]^{1/2}$$

and $D_{ijk}^v (v = x, y, z)$ denotes difference operator along the corresponding direction v. The method in [7] performs the time integration with first-order forward Euler method. Note that most published methods are of first-order accuracy.

III. A PARALLEL LEVEL-SET BASED APPROACH FOR ETCHING SIMULATION

A. Tracking Etching Front by Numerically Solving PDE

Fig. 1 demonstrates the geometry considered in the etching simulation. The problem is to obtain the etched contour after a specified time, which is the zero level set of $\phi(\mathbf{r}, t)$ in equation (3). In this case, $\mathbf{U}$ represents the motion of the etching front $\Gamma(t)$ and has the following form:

$$\mathbf{U}(\mathbf{r}) = [\alpha R_m, \alpha R_m, R_m]^T \tag{8}$$

where R_m is the vertical etching rate for material m, and α is a factor less than 1 that defining the lateral etching rate. For the etching process of complex structures, the first-order methods are no longer sufficient. Therefore, a higher-order method is urgently needed to be developed. Because of the specialty of hyperbolic conservation law, finite volume method with Riemann solver and third-order reconstruction approach has been employed and combined with the high-order Runge-Kutta method to obtain a more precise modeling result. To better capture discontinuity and benefit by inherent conversation of the finite volume method, we adopted semi-discretizations of conversation law as follows:

$$\frac{\partial \phi}{\partial t} + \frac{f\left(\phi_{i+\frac{1}{2}}\right) - f\left(\phi_{i-\frac{1}{2}}\right)}{\Delta x} = 0 \tag{9}$$

where

$$f(\phi) = \mathbf{U} \cdot \nabla \phi \tag{10}$$

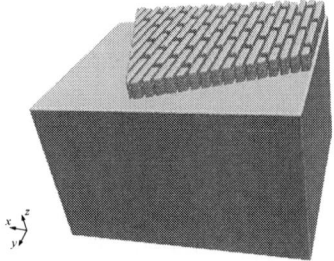

Fig. 1. The geometry considered in the etching simulation.

The convective term $\mathbf{U} \cdot \nabla \phi$ poses significant numerical challenges due to its hyperbolic characteristics. The Roe scheme, introduced by [8], is a linearized approximate Riemann solver used to compute numerical fluxes in the context of hyperbolic conservation laws. It is widely used for its ability to accurately capture discontinuities while maintaining numerical stability.

The numerical flux at the interface between cells i and $i+1$ is defined as:

$$f_{i+1/2} = \frac{1}{2} \left[U\phi_L + U\phi_R - |U|(\phi_R - \phi_L) \right] \tag{11}$$

The scheme offers a good balance between accuracy and efficiency. To enhance spatial accuracy beyond first order, the Roe scheme can be combined with a weighted interpolation of three neighboring cell values to achieve third-order accuracy in smooth regions.

$$\phi_L = \begin{cases} \frac{6\phi_{i-1} + 3\phi_i - \phi_{i-2}}{8} & U_f \geq 0 \\ \frac{6\phi_i + 3\phi_{i-1} - \phi_{i+1}}{8} & U_f < 0 \end{cases}$$
$$\phi_R = \begin{cases} \frac{6\phi_i + 3\phi_{i+1} - \phi_{i-1}}{8} & U_f \geq 0 \\ \frac{6\phi_{i+1} + 3\phi_i - \phi_{i+2}}{8} & U_f < 0 \end{cases} \tag{12}$$

However, higher-order schemes are prone to spurious oscillations near sharp gradients. To mitigate this, flux limiters such as the Van Leer limiter are introduced to enforce monotonicity:

$$\phi(r) = \frac{r + |r|}{1 + |r|}, \quad r = \frac{\phi_i - \phi_{i-1}}{\phi_{i+1} - \phi_i + \epsilon}, \tag{13}$$

where r is the ratio of successive gradients and ϵ is a small number to prevent division by zero. The limiter smoothly transitions between first-order upwind and higher-order schemes based on the local solution profile, ensuring Total Variation Diminishing behavior. This combination yields a robust and accurate method suitable for capturing both smooth features and sharp discontinuities.

For three-dimensional implementations, all schemes are extended dimension-by-dimension using operator splitting. Grid spacings $\Delta x, \Delta y, \Delta z$ may be anisotropic to accommodate semiconductor feature geometries, with special attention to boundary conditions:

- Dirichlet conditions: $\phi = \phi_{\text{specified}}$
- Neumann conditions: $\partial \phi / \partial n = 0$

In the beginning of our work, we employ **Backward Euler Scheme** as time integrator to gain better numerical stability:

$$\frac{\phi^{n+1} - \phi^n}{\Delta t} = -\mathbf{U} \cdot \nabla \phi^{n+1} \tag{14}$$

The linear system after spatial scheme has been employed then turns into:

$$(I + \Delta t A)\phi^{n+1} = \phi^n \tag{15}$$

where A is the convection operator matrix. The stability of the method makes it robust, but introduces $\mathcal{O}(\Delta t)$ dissipation. To further enhance our model, **Third Order Runge-Kutta Scheme** [9] was adopted:

$$\begin{cases} \phi^{(1)} = \phi^n + \Delta t L(\phi^n) \\ \phi^{(2)} = \frac{3}{4}\phi^n + \frac{1}{4}\phi^{(1)} + \frac{1}{4}\Delta t L(\phi^{(1)}) \\ \phi^{n+1} = \frac{1}{3}\phi^n + \frac{2}{3}\phi^{(2)} + \frac{2}{3}\Delta t L(\phi^{(2)}) \end{cases} \tag{16}$$

where $L(\phi) = -\mathbf{U} \cdot \nabla \phi$. The method requires CFL condition $\Delta t \leq C \frac{\min(\Delta x, \Delta y, \Delta z)}{\max |\mathbf{U}|}$ where the dimensionless number C is called the Courant number which less than or equal to 1.

B. Efficient Numerical Solution and Parallelization

During evolution, ϕ drifts from the signed distance property ($\|\nabla\phi\| \neq 1$). For every 5-10 physical steps, we periodically solve the re-initialization equation with first-order forward Euler scheme and central difference discretization:

$$\phi_\tau = \text{sign}(\phi_0)(1 - \|\nabla\phi\|) \tag{17}$$

using smoothed sign function:

$$\text{sign}(\phi_0) = \frac{\phi_0}{\sqrt{\phi_0^2 + \|\nabla\phi_0\|^2 \epsilon^2}}, \quad \epsilon = 0.5\Delta x \tag{18}$$

terminates when $\left| \|\nabla\phi\| - 1 \right| < 0.01$ is met. The completed process of the algorithm is shown in Algorithm 1.

To manage the computational demands of 3D semiconductor etching simulations, we adopt a high-performance computing technique that combines sparse linear algebra optimization with advanced solver configuration. Efficient solution of large sparse systems arising from implicit temporal discretization is achieved through parallel matrix assembly techniques and parallelism of the BiCGSTAB solver [10]. This includes the use of Triplet storage with thread-local buffers, pre-computation of sparsity patterns to eliminate dynamic allocation during assembly, and Lock-free insertion strategies using OpenMP parallel. To enhance memory efficiency and performance, matrices are stored as compressed row storage (CRS) format employing blocked CRS layouts for improved cache locality, and utilize SIMD-optimized packing to increase arithmetic throughput. For solving the discretized Hamilton-Jacobi equations, we configure an accelerated BiCGSTAB solver based on Eigen's vectorized implementation. The solver is further enhanced with diagonal pre-conditioning, together with a strict convergence tolerance of 10^{-8} to guarantee the accuracy of the results. Meanwhile, OpenMP enable the scaling of explicit method by parallelizing stencil operator over fixed grid.

Algorithm 1 High-Accuracy Level-Set Based Approach for Etching Simulation

---

Input: Δt, $maxSteps$, reinitFreq, geometry, material Prorperties

Output: surfaceMesh, DF-ISE file

1: Initialize ϕ as a signed distance function by equation (4)
2: Assign etching rates to each region by equation (8)
3: **for** $step = 1$ to $maxSteps$ **do**
4:     Apply Dirichlet/Neumann boundary conditions
5:     Parallel compute $L(\phi)$ using equations (11) and (12)
6:     Apply Van Leer limiter using equation (13)
7:     Update ϕ using Runge-Kutta scheme in equation (16)
8:     **if** $step$ is a multiple of reinitFreq **then**
9:         Reinitialize ϕ to maintain signed distance property by equation (17)
10:     **end if**
11: **end for**
12: Extract $\phi = 0$ surface using Marching Cubes [11]
13: Convert surface to DF-ISE format

IV. NUMERICAL RESULTS

The proposed approach is implemented in C++. Numerical experiments are carried out with the 3D structure shown in Fig. 1. With the proposed approach, a $600 \times 600 \times 600$ spatial grid and fixed time step $\Delta t = 1s$ are enforced. The simulated results at $t = 60s$ are compared with those obtained with SEMulator3D [12]. Our program is run on a Linux Server with Intel (R) Xeon (R) Gold 6230R CPU @ 2.10GHz.

A. Metric for Evaluating Accuracy

Comparison of the simulated contour with the reference solution is performed by comparing multiple 2D cross-sectional contours. The following metrics: Hausdorff distance [13], Area Distance quantifying the enclosed area consistency between shapes S_1 and S_2

$$\Delta_A = \left| \iint_{S_1} dx\, dy - \iint_{S_2} dx\, dy \right|, \tag{19}$$

Perimeter Ratio which compares boundary lengths of shapes S_1 and S_2 by

$$R_P = \frac{\oint_{\partial S_1} ds}{\oint_{\partial S_2} ds}, \tag{20}$$

Shape Context Distance [14] and Hu Moments [15] are considered to evaluate the similarity between the simulated 2D contour and the reference structure.

For each metric excluding Perimeter Ratio, we normalize its value with

$$s_i = \exp(-\lambda d_i), i = 1, 2, 4, 5, \tag{21}$$

where d_i is the raw value of the metric. λ is a calibrated decay factor such that $d_i = 0$ yields $s_i = 1$, while d_i become larger and eventually to reach maximum tolerable deviation yields s_i approach to zero. The Perimeter Ratio (s_3) is adopted directly in final calculation, since it has been scaled between 0 and 1. Subsequently, we calculate an overall similarity score by

$$S = \sum_{i=1}^{5} w_i \cdot s_i \tag{22}$$

The weight of each component is defined as $w_1 = 0.2, w_2 = 0.15, w_3 = 0.15, w_4 = 0.35$ and $w_5 = 0.15$.

B. Accuracy and Runtime Comparison

In this section, we benchmark the backward Euler and first-order upwind scheme against the third-order Runge-Kutta and scheme (12) by comparing the corresponding execution time and contour precision with industrial references obtained from SEMulator3D [12]. From Table I, third-order Runge-Kutta

TABLE I
RUNTIMES AND PARALLEL SPEEDUPS

#Thread	Backward Euler (s)	Speedup	Runger-Kutta3 (s)	Speedup
1	5126	-	12340	-
2	3670	1.39X	6406	1.92X
4	3764	1.36X	3414	3.61X
8	2683	1.91X	1834	6.73X
16	2684	1.90X	1098	11.22X

scales significantly better with increasing threads, achieves 11.22X speedup when the number of threads is increased to 16 while Backward Euler method demonstrates scaling limitation, with minimal gain beyond 8 threads. SEMulator3D's runtime for simulating this case is 940 seconds with 8 threads on a computer with AMD Rvzen9 5900HX CPU@3.30GHz. Although SEMulator3D involves 4,614,538 triangles in spatial discretizations and undisclosed techniques for the etching simulation, we can see the proposed algorithm already exhibits comparable efficiency to SEMulator3D.

TABLE II
SIMILARITY COMPARISON WITH SEMULATOR3D AT TWO Y-SLICES

Method	Score S at $y = -184$	Score S at $y = 254$
Backward Euler	0.9880	0.9837
Runge-Kutta3	0.9774	0.9770

Fig. 2. The simulated 2D contours at $y = -184$, with the figures of stacked contours shown at the bottom row.

The similarity scores of the sliced contours of our simulated results at positions y = -184 and 254, in comparison with those obtained from SEMulator3D, are listed in Table II. It demonstrates incredibly high consistency with respect to industrial reference. Fig 2 and Fig 3 show the relevant 2D contours, which indicate that Runge-Kutta3's contour is more accurate on capturing the evolving of rectangular fin and sharp geometric details.

V. CONCLUSION

In this work, we present a parallel level-set based framework for semiconductor etching which achieves high-order accuracy and efficient parallel performance. Using third-order spatial reconstruction and Runge-Kutta integration, it preserves curvature features and handles topological changes, matching SEMulator3D's results with 97.74% similarity. By leveraging optimized linear algebra operation and OpenMP parallelization, our Runge-Kutta3 scheme achieves a 11.2X

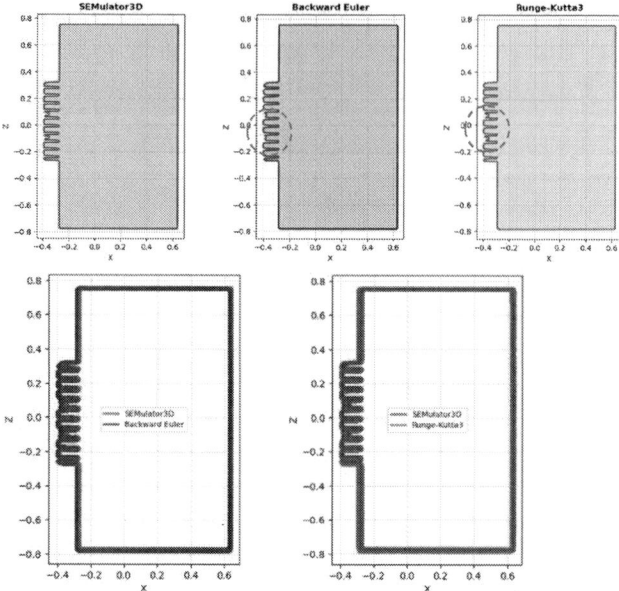

Fig. 3. The simulated 2D contours at $y = 254$, with the figures of stacked contours shown at the bottom row.

speedup on 16 threads, reducing the runtime from 12,340s to 1,098s. This makes the framework well-suited for industrial-scale simulations, offering a strong balance between accuracy and computational efficiency.

REFERENCES

[1] J. Helmsen, "A comparison of three-dimensional photolithography development methods," Ph.D. dissertation, Ph. D. thesis, UC Berkeley, 1994.

[2] S. Osher and J. A. Sethian, "Fronts propagating with curvature-dependent speed: Algorithms based on hamilton-jacobi formulations," *Journal of computational physics*, vol. 79, no. 1, pp. 12–49, 1988.

[3] J. A. Sethian, "Curvature and the evolution of fronts," *Communications in Mathematical Physics*, vol. 101, pp. 487–499, 1985.

[4] J. A. Sethian and D. Adalsteinsson, "An overview of level set methods for etching, deposition, and lithography development," *ieee transactions on semiconductor manufacturing*, vol. 10, no. 1, pp. 167–184, 2002.

[5] J. Sethian, "An analysis of flame propagation [ph. d. dissertation]," *Department of Mathematics, University of California, USA*, 1982.

[6] J. A. Sethian, "A fast marching level set method for monotonically advancing fronts." *proceedings of the National Academy of Sciences*, vol. 93, no. 4, pp. 1591–1595, 1996.

[7] D. Adalsteinsson and J. Sethian, "A level set approach to a unified model for etching, deposition, and lithography," *Journal of computational physics*, vol. 138, no. 1, pp. 193–223, 1997.

[8] P. L. Roe, "Characteristic-based schemes for the euler equations," *Annual review of fluid mechanics*, vol. 18, no. 1, pp. 337–365, 1986.

[9] W. Yu, *Numerical Analysis and Algorithms (the 3rd version)*. Tsinghua University Press, 2020.

[10] (2025) A C++ template library for linear algebra. [Online]. Available: https://eigen.tuxfamily.org/

[11] W. E. Lorensen and H. E. Cline, "Marching cubes: A high resolution 3d surface construction algorithm," in *Seminal graphics: pioneering efforts that shaped the field*, 1998, pp. 347–353.

[12] (2025) SEMulator3D. [Online]. Available: https://www.lamresearch.com/products/semulator3d/

[13] R. T. Rockafellar and R. J.-B. Wets, *Variational analysis*. Springer Science & Business Media, 2009, vol. 317.

[14] S. Belongie, G. Mori, and J. Malik, "Matching with shape contexts," *Statistics and analysis of shapes*, pp. 81–105, 2006.

[15] M.-K. Hu, "Visual pattern recognition by moment invariants," *IRE transactions on information theory*, vol. 8, no. 2, pp. 179–187, 1962.

Some Signal Processing Techniques for Testing Wireless Communication LSIs

(*Invited Paper*)

Koji Asami

Systems Design Lab, School of Engineering, The University of Tokyo, Tokyo, 113-0032 Japan

Email: asami@vdec.u-tokyo.ac.jp

DSP-based testing method has been used for around 50 years. Testing a Large Scale Integration circuit (LSI) examines whether the LSI is correctly manufactured as designed. In analog LSI testing, such as RF and mixed-signal devices, performance measurements were traditionally conducted using dedicated instruments. Since around the 2nd half of 1970s, DSP-based testing has become mainstream, which can perform various measurements with the identical measurement system by using analog-to-digital converters (ADCs) and digital-to-analog converters (DACs) in combination with digital signal processing (DSP) on the automatic test equipment (ATE) [1].

In the 1990s, the emergence of second-generation (2G) cellular networks introduced digital modulation, which required more advanced analysis techniques for DSP. A key metric for evaluating digital modulation performance is Error Vector Magnitude (EVM), which quantifies the deviation between ideal and actual transmitted signal vectors. EVM was initially employed to assess the communication quality of transmitters and receivers, but it has since been adopted for testing semiconductor devices in wireless systems [2].

Subsequently, as communication bandwidth continues to expand, the analysis of frequency-dependent characteristics that impact EVM, such as I/Q imbalance, has gained attention as a test application. In response, DSP techniques are being employed to improve the performance of ATE hardware.

This presentation introduces signal processing techniques used on the measurement side for testing wideband modulated signals, focusing on three key aspects: test applications, hardware performance improvements, and measurement techniques.

I. Evaluation Technique of Frequency-Dependent I/Q Imbalances

As an example of a test application for wideband modulators, a method for identifying frequency-dependent I/Q imbalances, which significantly affect EVM values, is introduced [3]. EVM is defined according to the calculation procedures specified by each communication standard, and it differs slightly from a pure physical quantity due to intentional correction processes such as channel compensation. However, I/Q imbalances are not corrected in these calculations, as they are not channel characteristics but rather imperfections inherent to the device under test. With the recent trend toward wideband communications, these imbalances have become frequency-dependent. In this method, time-shifted multitone signals are employed to separate and identify I/Q imbalances and carrier phase offset.

II. Digital Compensation for Time-Interleaved ADCs

As an example of hardware performance enhancement in ATE, a compensation algorithm for time-interleaved analog-to-digital converters (TI-ADCs) is introduced. To measure increasingly broadband signals, ATE systems require high-precision, wideband waveform digitizers. However, there exists an inherent dilemma in that electronic measurement instruments and ATE are constructed using currently available devices, yet are required to evaluate next-generation, cutting-edge devices that emerge subsequently. To address this challenge, innovative circuit design techniques and measurement methodologies have been developed, with DSP playing a key role. Although the TI-ADCs is a representative example, it is well known that inter-channel mismatches pose a significant challenge [4]. In ATE systems, DSP is one of the effective approaches to address this issue [5].

III. Local Sweep Digitizing Method

Furthermore, in anticipation of continued broadband expansion, a novel approach called the Local Sweep Digitizer method [6] is introduced. This DSP-based measurement technique utilizes the synchronization system of the ATE and is implemented through an effective integration of hardware and DSP. In this method, wideband signals are divided into subbands, captured using a single ADC, and recombined through DSP. To ensure accurate reconstruction, a technique for aligning phase uncertainty between subbands is essential.

References

[1] M. Mahoney,"DSP-BASED TESTING of Analog and Mixed-Signal Circuits," Computer Society Press of the IEEE, 1987.

[2] K. B. Schaub and J. Kelly, Production Testing for RF and System-on-a-Chip Devices for Wireless Communications, Artech House Publishers, 2004.

[3] K. Asami, T. Kurihara, and Y. Inada,"Evaluation Techniques of Frequency-Dependent I/Q Imbalances in Wideband Quadrature Mixers," Proc. International Test Conference, pp.1-7, Nov. 2010.

[4] N. Kurosawa, H. Kobayashi, K. Maruyama, H. Sugawara and K. Kobayashi, "Explicit Analysis of Channel Mismatch Effects in Time-Interleaved ADC Systems," IEEE Trans. Circuits and Systems I, vol. 48, No. 3, Mar. 2001.

[5] K. Asami, "An Algorithm to Improve the Performance of M-Channel Time-Interleaved A-D Converters," IEICE Transactions on Fundamentals of Electronics, Communications and Computer Sciences, vol. E90-A, no.12, pp.2846-2852, December 2007.

[6] K. Asami, K. Kusunoki, N. Shimizu, Y. Aoki, "Ultra-Wideband Modulation Signal Measurement Using Local Sweep Digitizing Method," The 38th IEEE VLSI Test Symposium (VTS'20), session RP10-1, April 5-8, 2020 (San Diego).

Voltage-Domain vs. Time-Domain: Trade-offs in High-Speed Applications

Haoyu Li, Sai-Weng Sin, Rui P. Martins, Mingqiang Guo*

University of Macau, Macau

* Email: mqguo@um.edu.mo

Abstract— **In recent years, the demand for high-speed analog-to-digital converters (ADCs) has grown significantly. For both conventional successive approximation register (SAR) ADCs and emerging time-domain ADCs, numerous technological innovations have been proposed to enhance their performance. This paper will review cutting-edge research, discuss the advantages and drawbacks of various high-speed architectures, and explore solutions to key challenges: breaking the speed limitations of SAR ADCs, addressing the fine time-to-digital converter (TDC) and process-voltage-temperature (PVT) issues in time-domain ADCs, and further improving the speed and energy efficiency of time-domain and voltage-domain hybrid ADCs.**

Keywords—SAR ADC, time-domain ADC, time-interleaving, time-to-digital converter, high-speed ADC.

I. INTRODUCTION

The rapid advancement of artificial intelligence (AI), big data analytics, and autonomous vehicle technologies has escalated demands for ADCs with high-speed operation and moderate resolution in modern wireline or wireless communication systems. Flash ADCs provide rapid single-conversion speeds and low latency, but their complexity scales exponentially with resolution, leading to excessive kickback noise, power consumption, and area overhead. Benefiting from process scaling and technological innovations [1]-[5], SAR ADCs exhibit exceptional energy efficiency. However, their sequential loop architecture fundamentally limits conversion speed, presenting a critical bottleneck for higher-throughput applications.

Time-domain (TD) ADCs [6]-[8] have emerged as an attractive option for high-speed applications. Compared to conventional SAR ADCs, TD ADCs offer significant speed advantages, enabling a reduced number of TI factors. The voltage-to-time converter (VTC) isolates kickback noise from the sampling function by decoupling the sampling capacitor from the quantizer, which achieves lower input loading. However, to achieve high speed and moderate resolution simultaneously, TD ADCs need to generate time steps with sub-gate delay, resulting in a jitter-sensitive and complex fine TDC with large power consumption.

A hybrid ADC [9]-[11] employs a TD ADC as the first stage and a SAR ADC as the second stage. The first stage TD ADC achieves wide input bandwidth by small input capacitance and fast coarse conversion; the second stage SAR ADC quantizes the residue with low power consumption, improving energy efficiency. Compared to SAR ADCs, the time/voltage-domain hybrid ADC demonstrates faster conversion speed and wider input bandwidth. Compared to conventional TD ADCs, the hybrid ADC exhibits jitter insensitive, simpler structures and better energy efficiency.

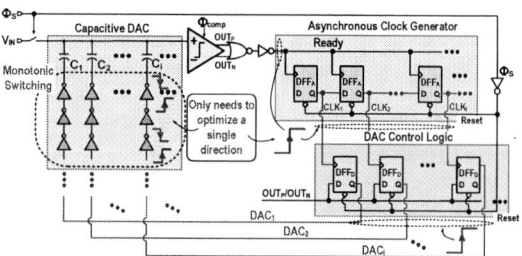

Fig. 1 The unidirectional data transmission path in the monotonic switching SAR ADC.

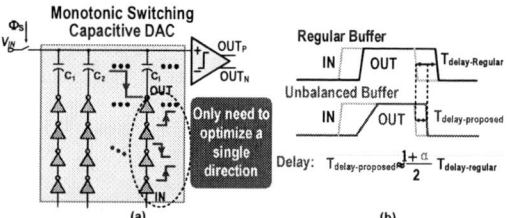

Fig. 2 (a)The DAC buffers in the monotonic switching SAR ADC, and (b) the comparison of using the regular and the unbalanced buffer.

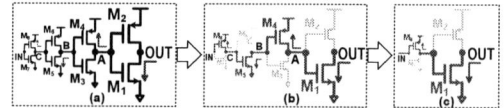

Fig. 3 (a) 4-stage balanced inverters, (b) 4-stage unbalanced inverters, and (c) 2-stage unbalanced inverters.

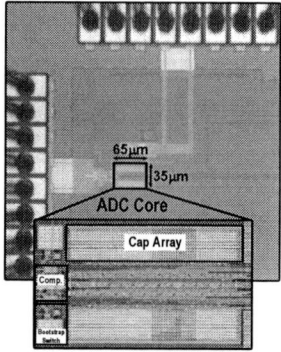

Fig. 4. Die microphotograph of the prototype SAR ADC. [5]

However, in [9] [10], the time-to-voltage converter (TVC) consumes extra power, and the SAR conversion speed is limited due to the sub-ranging architecture and no interstage amplification. [11] proposes a pipelined hybrid architecture with residual time-voltage amplification, which further improves the performance.

979-8-3315-3918-4/25 $31.00 © 2025 IEEE

This paper is organized as follows. Section II introduces the architecture of the conventional SAR ADC and advanced power-delay-optimized techniques. Section III presents the TD ADCs and the Vernier-based multipath Flash (VMF) TDC. Section IV discusses the time/voltage-domain hybrid ADCs. Section V draws conclusions.

II. HIGH-SPEED SAR ADCs

A. SAR ADC

Fig. 1 illustrates the architecture of a conventional 1-bit/cycle SAR ADC, comprising a sample-and-hold (S/H) circuit, a comparator, a capacitive digital-to-analog converter (CDAC), and SAR logic. By leveraging a binary-search quantization algorithm and low-complexity circuitry, the SAR ADC achieves high power efficiency. Over the past two decades, technological innovations—such as asynchronous SAR logic [1], monotonic switching [2], and dynamic logic circuits [3][4]—have significantly improved SAR ADC performance. Nevertheless, an N-bit SAR ADC typically requires N conversion cycles, with each cycle encompassing comparator decision time, logic propagation delay, and DAC settling time. Consequently, conversion speed remains the fundamental bottleneck of SAR ADC.

B. Power-Delay-Optimized Unbalanced N/P-MOS Sizing Technique

In [5], a power-delay-optimized SAR loop architecture (Fig. 2–3) leverages asymmetric NMOS/PMOS transistor sizing to overcome the speed limitations of single-channel SAR ADCs. By exploiting the unidirectional switching behavior intrinsic to monotonic SAR ADCs, the design minimizes comparator-to-DAC latency through unilateral device scaling along critical timing paths, including monotonic DAC drivers, DAC switch control logic, and asynchronous control logic. This approach accelerates critical-edge transitions while downsizing transistors in non-critical directions, thereby simultaneously enhancing conversion speed, reducing power consumption, and minimizing area overhead—effectively breaking the conventional power-delay tradeoff. Silicon measurements demonstrate a 9-bit ENOB at 700 MS/s in a 28-nm CMOS process, consuming only 2.02 mW (Fig. 4).

III. TIME-DOMAIN HIGH-SPEED ADCs

B. Two-stage TD ADC Architecture

Fig. 5 presents the basic architecture and operational principle of a two-stage TD ADC [6] [7] [8]. Firstly, the VTC charges/discharges the sampling capacitor and detects the ramp voltage to converter the input voltage to time signals. The information of the input signal is converted from the voltage difference into the time difference between the rising edges of the time signals. Then, the first stage TDC coarsely quantizes the time signals with time steps T_D, producing a time residue for fine quantization by the second stage TDC with time steps T_{LSB}. Unlike SAR ADCs using a CDAC composed of capacitor array as the sampling capacitor, TD ADCs utilizes a single capacitor for sampling. Additionally, the VTC isolates comparator kickback noise, which relaxes the requirement of the input capacitor and enhances input bandwidth. The most circuits of the TD ADC are digital, which is easy to scale down. Since the conversion time of the TD ADC is determined by the T_{LSB}, how to generate a fine time step with sub-gate delay is the key challenge. Thanks to [12], the PVT tracking between VTC and TDC is achieved. However, the inter-stage

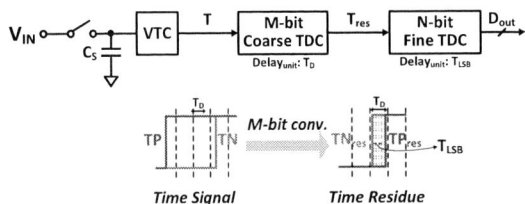

Fig. 5. Block diagram of TD ADCs and the signal path.

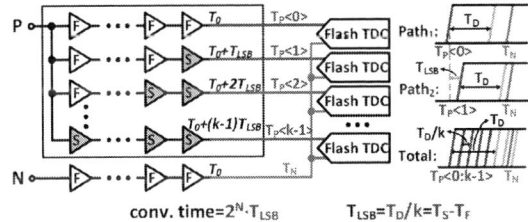

Fig. 6. Block diagram of the VMF TDC and time signals of each path and total paths.

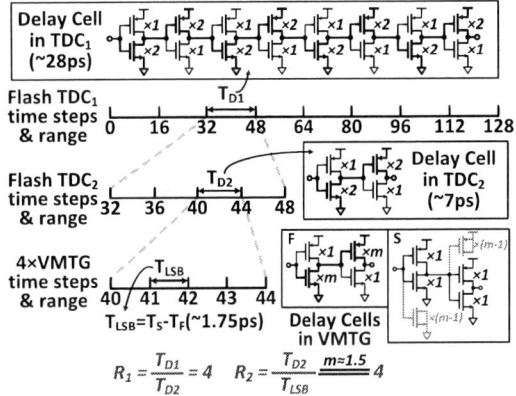

Fig. 7. Circuit implementation of delay cells and time steps and range of the VMF TDC.

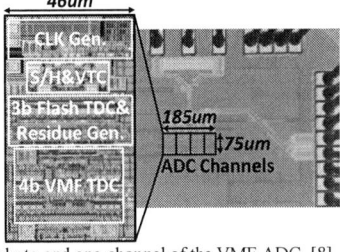

Fig. 8. Die photo and one-channel of the VMF ADC. [8]

gain varies as PVT variation due to the different structures of the coarse TDC and the fine TDC.

C. A Vernier-based Multipath Flash TDC

The conventional Vernier TDC quantizes the time signals by one path with slow delay cells and one path with fast delay cells. As a result, it achieves a fine time step, which is the time difference between the slow delay and fast delay, but long conversion time due to the slow path. The conventional Flash TDC achieves fast conversion speed, but its time steps are the delay of a buffer, which is coarse.

[8] proposes a VMF TDC, which utilizes the Venier TDC to generate fine time steps and the Flash TDC to increase conversion speed, taking advantage of these two structures. Fig. 6 displays the architecture of the VMF TDC, which consists of a k× Vernier-based Multipath TLSB Generator (VMTG) and multipath Flash TDCs. There are k× paths with different delay time in the VMTG and the time difference between each path is (T_S-T_F), which is set to T_{LSB}. By the VMTG, the signal P is converted into multipath signals $T_P<0:k-1>$ and the rising edge of each path differs by T_{LSB}. For signal N, it is delayed by T_0 to match the latency of P side. After that, the k× Flash TDCs quantize the signal $T_P<0:k-1>$ and T_N in parallel. For one path, the Flash TDC does quantization with coarse time steps T_D. For total paths, the multipath Flash TDC achieves fine time steps T_{LSB}. Setting $T_{LSB} = T_D/k$ for the matching between VMTG and Flash TDCs.

Since the information of time signals is the time difference between rising edges, the speed-critical edges are certain in the delay cell. By increasing the size of certain transistors, the transmission of the time signal in a buffer could be accelerated. As shown in Fig. 7, the slow delay cell in the VMTG is a buffer with regular size. For the fast delay cell, to achieve a shorter delay time, a buffer with unbalanced N/P-MOS size [5] is implemented. Dummy transistors are added in slow buffers to keep the loading of every node consistent with that in fast buffers. In that case, the topologies of the fast and slow delay cells in the VMTG are not changed, that are same as the delay cells in Flash TDCs. Therefore, the time steps T_{LSB} in the VMTG and time steps TD in Flash TDCs have the same response as PVT varies, achieving PVT-robust time steps ratios and solving the interstage gain issues of two-stage TD ADCs.

Fig. 8 presents the die photograph of the two-step TD ADC implementing the VMF architecture; the single channel achieves 4 GS/s sampling rate with 7-bit resolution in a 28-nm CMOS process.

IV. TIME-DOMAIN AND VOLTAGE-DOMAIN HYBRID ADCs

For the two-step TD ADC, the fine time steps, that are sensitive to jitter and mismatch, highly rely on advanced process and cause the fine TDC to become the most power-hungry block. To improve energy efficiency, time/voltage-domain hybrid ADCs are proposed in [9]-[11] to replace the fine TDC with a low power SAR ADC.

B. Conventional Sub-ranging Hybrid ADC

Fig. 9 shows the architecture of the sub-ranging hybrid ADC [9] [10] and the voltage signal on the CDAC. Firstly, a VTC converts the input voltages to time signals. Then, the time signals are quantized by a M-bit TDC and control the TVC to rebuild the input voltages on the CDAC, simultaneously. Next, the CDAC generates a voltage residue via M-bit codes feedback. Last, a TI-SAR replaces conventional fine TDCs and performs LSB quantization. This hybrid architecture combines the high-speed conversion of TD ADCs and the energy efficiency of SAR ADCs. However, the absence of interstage amplification restricts the voltage residue swing to merely 100 mVpp[9] /50 mVpp[10], slowing down the SAR conversion speed. And the time-to-voltage conversion introduces additional large power consumption, since the loading of the TVC is 90fF [9] /100fF [10]. Furthermore, the sub-ranging architecture forces the fast M-

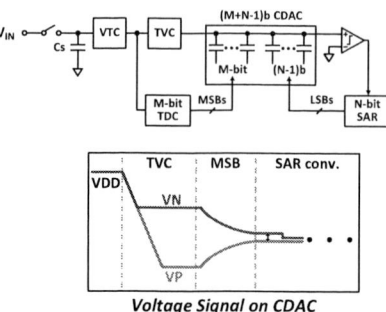

Fig. 9. Block diagram of the time/voltage domain hybrid sub-ranging ADC.

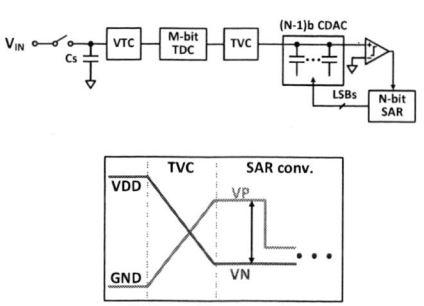

Fig. 10. Block diagram of the time/voltage-domain hybrid pipelined ADC.

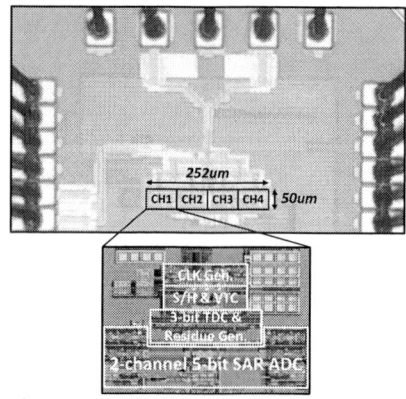

Fig. 11. Die photo of the hybrid pipelined ADC. [11]

bit TDC to enter a standby period until the slower N-bit SAR completes its conversion cycle.

C. Pipelined Hybrid Architecture with Residual Time-Voltage Amplification

To address these limitations, [11] propose a power-efficient pipelined hybrid TD-SAR ADC incorporating residual time-voltage amplification (Fig. 10). In this architecture, the TDC quantizes the time signals on the signal path and generates a time residue. Then a large output swing TVC converts the time residue to a voltage residue with 16× amplification. This amplified residue on the CDAC is subsequently quantized by the 5-bit SAR ADC. Compared to the conventional sub-ranging hybrid ADCs, this pipelined approach offers following advantages: 1) The residue is generated in TD rather than do subtraction in VD, which reduces the CDAC to 15 fF, lowering power consumed by

TABLE I Performance Summary and Comparison With State-of-the Art ADCs

	CICC-25 H. Li	ISSCC-24 A. Whitcombe	CICC-24 W. Zhang	VLSI-25 H. Li	ISSCC-20 M. Zhang	ISSCC-22 J. Liu	VLSI-21 E. Martens	ISSCC-25 J. Shen
Architecture	Hybrid TD and VD			Time-Domain			Voltage-Domain	
	Pipelined	Sub-ranging		VMF TDC	Interpolation TDC	SAR TDC	SAR	
Technology	28nm CMOS	22nm FinFET	28nm CMOS	28nm CMOS	65nm CMOS	14nm FinFET	16nm FinFET	16nm FinFET
f_s (GS/s)	12.5	40	20	16	10	10	8	12
No. of channels	4/8	12/48	4/16	4	4	2	8	16
Resolution (bits)	7	7	8	7	8	8	8	9
Supply (V)	1	0.85	0.95	0.9	1	0.8	0.85	0.8
SNDR@Nyq.(dB)	36.7	32.3	38.9	35.3	40.1	37.2	42.4	48.1
SFDR@Nyq.(dB)	53.4	45.9	51.9	51.2	52.8	50.7	46	60.8
Power (mW)	14.7	66.1	42.1	19.6	50.8	14.8	26	160
FoMw@Nyq. (fJ/conv-step)	21.1	49.1	29.1	25.7	61.5	24.8	30.2	64.2
Active area (mm²)	0.0126	0.103	0.042	0.0139	0.095	0.00285	0.023	0.25

TVC conversion and SAR operation. 2) Benefitting from the 16× interstage gain and large output swing TVC, the SAR input swing is significantly increased to 1400 mVpp, accelerating SAR conversion. 3) Due to the pipelined architecture, the linearity requirement of the TVC is relaxed to 5-bit, which is easy to implement.

Fig. 11 presents the die photograph of the time/voltage domain hybrid pipelined ADC; the single channel achieves 3.125 GS/s sampling rate with 7-bit resolution in a 28-nm CMOS process.

Table I summarizes the performance comparison of advanced high-speed ADCs. It presents that [8] with a VMF TDC achieves fast conversion speed and competitive energy efficiency. Although [11] with a pipelined hybrid architecture is slower than [8], it achieves best energy efficiency among >5GS/s ADCs [15].

V. CONCLUSION

This paper reviews published high-speed ADCs and explored three techniques to overcome limitations and improve performance: 1. For SAR ADCs, a monotonic switching with unbalanced N/P-MOS sizing minimizes the latency. 2. For TD ADCs, a VMF TDC achieves fast conversion, fine time steps and stable time steps ratios. 3. For TD and VD hybrid ADCs, a pipelined TD-SAR architecture with residual amplification significantly improves the energy efficiency.

ACKNOWLEDGMENT

This work was supported by Macau FDCT under Grant 0010/2023/RIA1, 0100/2022/A, 0038/2025/RIB1 and 004/2023/SKL; by NSFC of China under Grant 62204002; by UM under Grant MYRG-GRG2023-00178-IME, MYRG-GRG2024-00118-IME and SRG2022-00056-IME.

REFERENCES

[1] S. -W. M. Chen and R. W. Brodersen, "A 6-bit 600-MS/s 5.3-mW Asynchronous ADC in 0.13- μm CMOS," IEEE J. of Solid-State Circuits, vol. 41, no. 12, pp. 2669-2680, Dec. 2006.

[2] C. -C. Liu, S. -J. Chang, G. -Y. Huang and Y. -Z. Lin, "A 10-bit 50-MS/s SAR ADC With a Monotonic Capacitor Switching Procedure," IEEE J. Solid-State Circuits, vol. 45, no. 4, pp. 731-740, Apr. 2010.

[3] Y.-Z. Lin, C.-H. Tsai, S.-C. Tsou, and C.-H. Lu, "A 8.2-mW 10-b 1.6-GS/s 4x TI SAR ADC with fast reference charge neutralization and background timing-skew calibration in 16-nm CMOS," in Proc. IEEE Symp. VLSI Circuits, Jun. 2016, pp. 1–2.

[4] M. Zhan, L. Jie, Y. Zhong and N. Sun, "A 10-mW 10-ENoB 1-GS/s Ring-Amp-Based Pipelined TI-SAR ADC With Split MDAC and Switched Reference Decoupling Capacitor," IEEE J. Solid-State Circuits, vol. 58, no. 12, pp. 3576-3585, Dec. 2023.

[5] M. Guo, L. Qi, W. Zhao, G. Xiao, R. P. Martins and S. -W. Sin, "A 10b 700 MS/s Single-Channel 1b/Cycle SAR ADC Using a Monotonic-Specific Feedback SAR Logic With Power-Delay-Optimized Unbalanced N/P-MOS Sizing," IEEE Trans. Circuits Syst. I: Reg. Papers, vol. 70, no. 12, pp. 4767-4780, Dec. 2023

[6] M. Zhang, Y. Zhu, C. -H. Chan and R. P. Martins, "An 8-Bit 10-GS/s 16× Interpolation-Based Time-Domain ADC With <1.5-ps Uncalibrated Quantization Steps," IEEE J. Solid-State Circuits, vol. 55, no. 12, pp. 3225-3235, Dec. 2020.

[7] J. Liu, M. Hassanpourghadi and M. S. -W. Chen, "A 10-GS/s 8-bit 2850-μm² Two-Step Time-Domain ADC with Speed and Efficiency Enhanced by the Delay-Tracking Pipelined-SAR TDC," IEEE J. Solid-State Circuits, vol. 57, no. 12, pp. 3757-3767, Dec. 2022.

[8] H. Li et al., " A PVT-Robust 16GS/s 4×TI Time-Domain ADC with Vernier-based Multipath Flash TDC achieving 25.7fJ/c-s FoM in 28nm CMOS, " in Proc. Symp. VLSI Circuits, Jun. 2025, pp. 1-3.

[9] A. Whitcombe et al., "22.3 A 76mW 40GS/s 7b Time-Interleaved Hybrid Voltage/Time-Domain ADC with Common-Mode Input Tracking," in IEEE Int. Solid-State Circuits Conf. (ISSCC) Dig. Tech. Papers, Feb. 2024, pp. 392-394.

[10] W. Zhang, M. Zhang, Y. Zhu, R. P. Martins and C. -H. Chan, "A PVT-Robust 8b 20GS/s Time-Interleaved SAR ADC with Quantization-Embedded Current-Mode Buffer and Differ-Based Dither Timing Skew Calibration," in Proc. Custom Integer. Circuits Conf. (CICC), Apr. 2024, pp. 1-2.

[11] H. Li et al., "A 12.5GS/s 14.7mW 4×TI Pipelined Hybrid TD-SAR ADC with Residual Time-Voltage Amplification," in Proc. Custom Integer. Circuits Conf. (CICC), Apr. 2025, pp. 1-3.

[12] M. Zhang, C. -H. Chan, Y. Zhu and R. P. Martins, "A 0.6-V 13-bit 20-MS/s Two-Step TDC-Assisted SAR ADC With PVT Tracking and Speed-Enhanced Techniques," IEEE J. Solid-State Circuits, vol. 54, no. 12, pp. 3396-3409, Dec. 2019.

[13] E. Martens, D. Dermit, M. Shrivas, S. Nagata and J. Craninckx, "A Compact 8-bit, 8 GS/s 8×TI SAR ADC in 16nm with 45dB SNDR and 5 GHz ERBW," in Proc. Symp. VLSI Circuits, Jun. 2021, pp. 1-2.

[14] J. Shen et al., "24.8 A 12GS/s 9b 16× Time-Interleaved SAR ADC in 16nm FinFET," in *IEEE Int. Solid-State Circuits Conf. (ISSCC) Dig. Tech. Papers*, Feb. 2025, pp. 442-444.

[15] B. Murmann, *ADC Performance Survey 1997-2025*. Accessed: 2025. [Online]. Available: https://github.com/bmurmann/ADC-survey.

A Pitch-Matched Transceiver ASIC with Element ADC and Continuous-Time Gain Compensation for 3D Ultrasound Probes

Jing Li, Tianci Zhang, Li Dai, Yingchen Liu, Jinlai Fu, Zhongshan Wang, Penghao Jiang, Yihu Yu,
Zhong Zhang, Kejun Wu, Ning Ning* and Qi Yu
State Key Laboratory of Electronic Thin Films and Integrated Devices
University of Eletronic Science and Technology of China, Chengdu, China
Email: lijing686@uestc.edu.cn, ning_ning@uestc.edu.cn

Abstract—This paper introduces a pitch-matched transceiver ASIC for 3D Ultrasound Probes. In the receiver(RX), an analog-front-end(AFE) with hybrid continuous-time (CT) gain compensation architecture is proposed: a low-noise amplifier (LNA) firstly divides the compensation range by 6dB/step in 6 steps, and then the time gain compensation(TGC) smooths the gain in each step continuously. This hybrid combination realizes continuous-time gain compensation with a 36dB range, ±0.9dB gain error, low hardware cost and low power consumption. A C2C SAR ADC with compact size and low power consumption is proposed, achieving a SNR of 54.7dB. The prototype was fabricated in a 130 nm BCD process with a 250 μm pitch. The RX achieves 1.43 mW/channel of power and an 84dB DR.

Keywords—Ultrasound ASIC, Transceiver, ADC, TGC, Continuous-Time

I. INTRODUCTION

Three-dimensional (3D) intra-cardiac echography (ICE) has emerged as a leading technique for diagnosing cardiovascular diseases. Advanced integrated circuits (ICs) embedded in miniature probes are critical for improving signal quality and minimizing cable count by processing data from 2D transducer arrays [1-6]. While integrating an analog-to-digital converter (ADC) within the probe enhances signal transmission and enables digital beamforming, existed solutions face challenges related to ADC size, power consumption, and limited access to individual transducer elements due to shared ADCs among subarrays [2,4,5]. Traditional discrete-time gain compensation methods offer superior noise performance and dynamic range with minimal hardware and power requirements, but at the cost of image artifacts. Reducing step sizes to alleviate these artifacts increases hardware complexity. Continuous-time compensation approaches using transimpedance amplifiers (TIAs) have faced integration difficulties due to their size and power demands [2,3]. This paper introduces a transceiver ultrasound ASIC that integrates element-level ADC and continuous-time gain compensation for the first time.

II. CIRCUIT DESIGN

A. System Architecture

Fig.1 illustrates the proposed ultrasound ASIC architecture, featuring 36 channels: 12 transmitting (TX) channels that generate maximum ±30V pulses to stimulate the transducer, employing digital beamforming with 15.4 ns time resolution and 3.9 μs time range. The 24 receiving (RX)

This work is funded by National Natural Science Foundation of China under Grant 62004024, National Key Laboratory of Integrated Circuits and Microsystems under Grant JCKY2023210C003.

channels handle signal compensation and digitization. To optimize ADC design and address ultrasound attenuation, the RX path integrates an AFE and ADC. The proposed AFE adopts a hybrid approach to realize continuous-time gain compensation, enhancing gain accuracy with minimal hardware compared to previous approaches [2,3]. The proposed ADC features a C2C SAR architecture, providing high-speed sampling in a compact size, effectively integrated into each element.

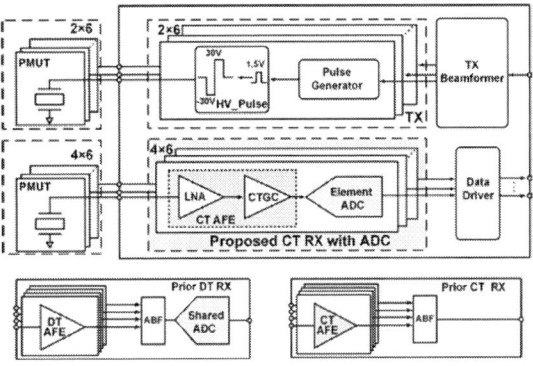

Fig. 1. Overall block diagram of the proposed US imaging system.

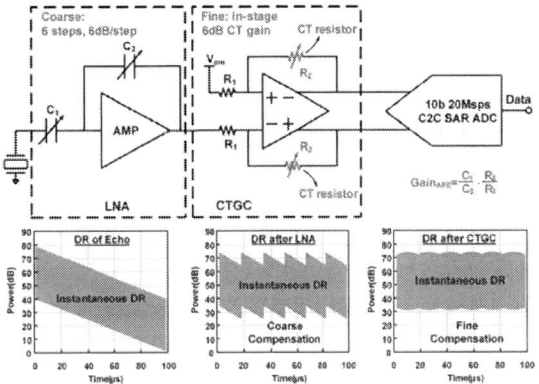

Fig. 2. Circuit diagrams of the proposed RX and evolution of DR as a function of time.

B. AFE Design

Fig. 2 presents the CT RX circuit comprising a low-noise amplifier (LNA), a continuous-time gain compensation (CTGC) amplifier, and a C2C SAR ADC. The LNA behaves as a discrete-time gain compensation, dividing the compensation range by 6 steps from -6 dB to 24 dB. The CTGC provides in-stage continuous-time gain from 6 dB to

12 dB, smoothing discrete steps in LNA. This hybrid combination realizes continuous-time gain compensation with a wide range of 36dB, high accuracy, low hardware cost and power consumption.

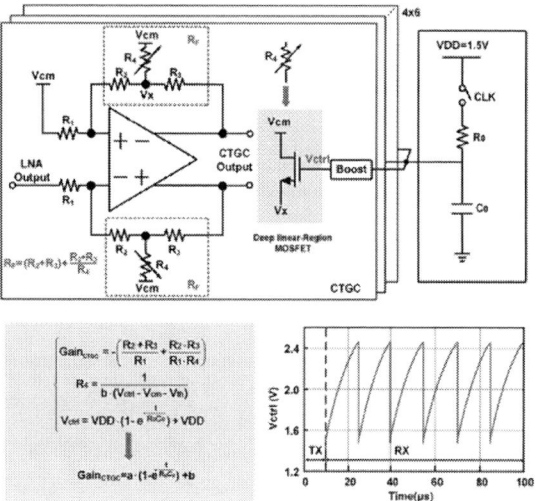

Fig. 3. Circuit diagrams of the CTGC

Traditional continuous-time compensation approaches using transimpedance amplifiers (TIAs) have faced integration difficulties due to their size and power demands [2,3]. This paper proposed a CTGC amplifier, depicted in Fig. 3, adopting a resistor-ratio amplifying architecture and realizing the continuous-time gain by continuously tuning the feedback resistor. The feedback resistor ladder consists of R_1, R_2, and R_3 for basic gain of $(R_2+R_3)/R_1$. R_4, composed by an NMOS operating in the linear region, provides the variable gain of $(R_2 \cdot R_3)/(R_1+R_4)$. The NMOS is biased by an exponential voltage, contributing its resistance exponentially dependent on time, achieving dB-linear and continuous-time gain. The exponential voltage is generated by a simple RC network by periodically charging along with LNA steps. It is further boosted by a VDD to enhance the NMOS working in linear region. To minimize the hardware cost, the RC network is shared by the RX array. Compared to traditional continuous-time compensation approaches, the proposed CTGC is compact, low power and easy to integrate in each channel.

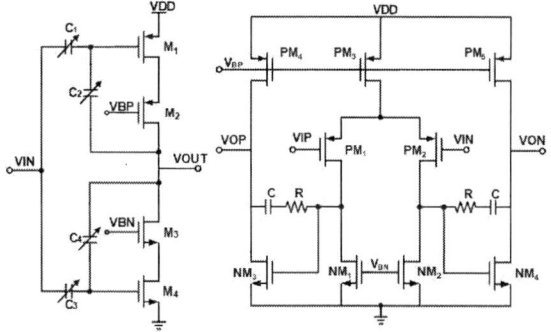

Fig. 4. Circuits of LNA(left) and amplifier in TGC(right)

The details of LNA and TGC amplifier are shown in Fig.4. The LNA features a single-ended class-AB amplifier to conserve power and provides discrete gain steps through capacitor ratios. The CTGC converts the single-ended signal

to differential format for improved common-mode noise rejection. A symmetric two-stage amplifier is adopted in the CTGC.

C. ADC Design

Due the small pitch and restricted heat, ADC integrated in each element faces serious difficulties in area and power consumption. This work proposed a C2C SAR ADC whose capacitor DAC (CDAC) is segmented into four parts, reducing the total number of capacitors from 1024 to 42, making the ADC minimal area, enhancing conversion speed by 95.9%, achieving resolution of 10 bits and conversion speed of 20 MS/s. The MSB capacitors sample the input signal, and a capacitor ladder at the MSB minimizes ADC gain errors. The calibration algorithm, which utilizes a split-switch mode for MSB quantization and single-ended switching for LSB quantization, improves area efficiency, reduces power consumption, and maintains high speed. Despite parasitic capacitance and resolution challenges, the C2C SAR ADC's resolution is improved using a foreground calibration algorithm. In calibration mode, dither is injected at MSB, and quantization are performed twice before and after dither injection. By comparing the two results, ADC's weight values are iteratively refined using the LMS algorithm. Also, the ADC operates with asynchronous timing, clocked by a 260 MHz off-chip clock, delivering serial digital outputs.

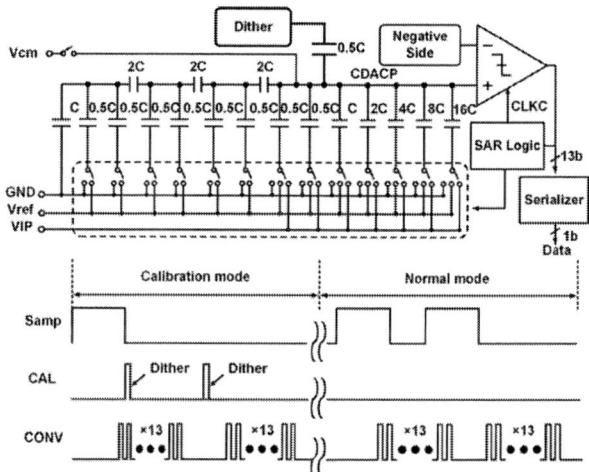

Fig. 5. Circuit diagrams and timing diagrams of the C2C SAR ADC

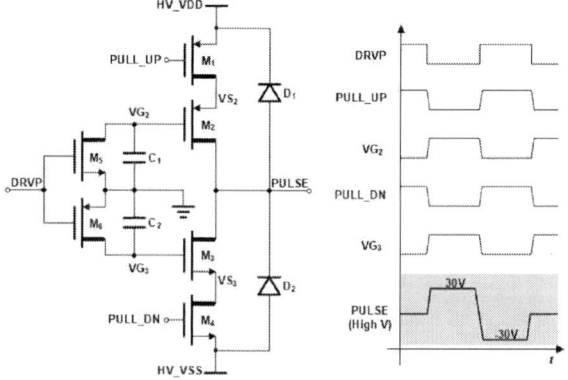

Fig. 6. Overall block diagram of the proposed US imaging system

D. TX Design

The bipolar high-voltage pulser in TX is shown in Fig. 6, which adopts a push-pull structure to reduce static power consumption, while using a floating-gate bootstrap transistor for isolation to ensure that the pulser does not break down at high output voltage. It is combined with a level shifter and a logic circuit in the front to form a high-voltage ultrasonic transmitter circuit with fewer high-voltage transistors.

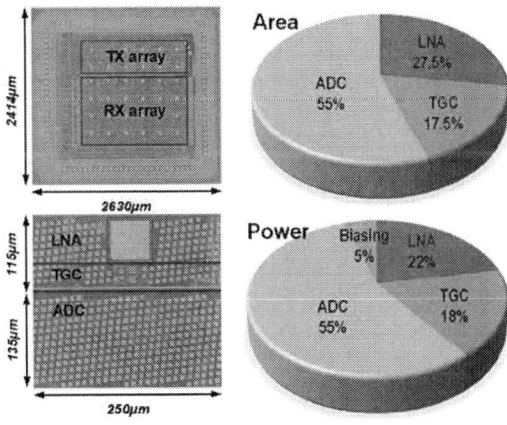

Fig. 7. Distribution of area and power

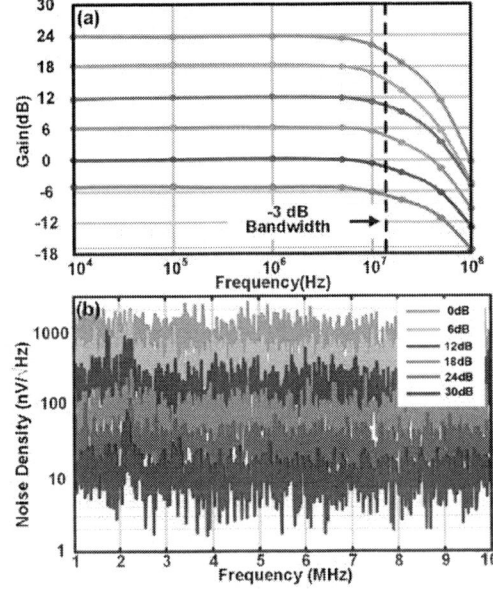

Fig. 8. (a)LNA Gain and Bandwidth (b)Input-referred noise

III. EXPERIMENT RESULT

The prototype (see Fig. 7), fabricated using a 130 nm BCD process, shows that the each element of RX and TX occupy 0.0625 mm² with a 250 µm pitch. The AFE and ADC occupy 0.02875 mm² and 0.03375 mm², respectively. The total RX power consumption is 1.43 mW, with 0.64 mW for the AFE and 0.79 mW for the ADC. Fig. 8-10 presents measurement results. Fig. 8(a) illustrates the transfer function of LNA with six precise gain settings around the center frequency of 5 MHz. Fig. 8(b) shows the input referred noise of AFE, whose best value of 11.4nV/ √ Hz is obtained at the highest setting. Fig. 9(a) depicts the total input dynamic

range of 84dB. Fig. 9(b) presents the transient output signal of the AFE with attenuated echo input. The signal is partitioned into six segments with 15 µs time steps by the LNA and continuously compensated by the CTGC. AFE gain increases dB-linearly across a 36-dB range, with maximum and minimum gain errors of ±0.9 dB, respectively, as shown in Fig. 10(a). The ADC output spectrum in Fig. 10(b) reveals that the initial SNDR is limited to 50.7 dB due to parasitic capacitance and capacitor mismatch. After calibration, the SNDR improves to 54.7 dB.

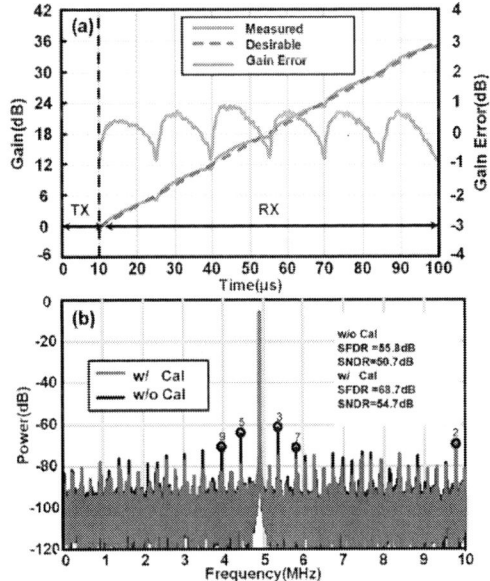

Fig. 9. (a)AFE Gain and Gain Error (b)Single Channel Power Spectrum

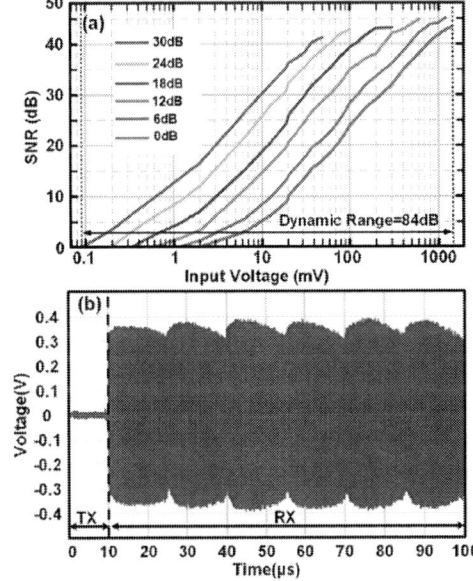

Fig. 10. (a)Dynamic Range (b)AFE Output Voltage

TABLE I compares this work with previous 3D ultrasound imaging solutions. This work integrates element-level ADC digitization and continuous-time gain compensation in an ultrasound ASIC, which is the first reported. The proposed CT AFE and C2C ADC are area-efficient and power-efficient, providing competitive power

TABLE I. PERFORMANCE SUMMARY AND COMPARISON

	This Work	ISSCC'18[1]	ISSCC'20[2]	ISSCC'22[3]	ISSCC'22[4]	ISSCC'21[5]	VLSI'19[6]
Process	**130nm BCD**	180nm	180nm BCD	180nm BCD	180nm BCD	130nm BCD	180nm
Transducer	**PMUT**	PZT	CMUT	PZT	PZT	CMUT	PZT
Center Freq.	**5MHz**	5MHz	5MHz	10MHz	6MHZ	1-10MHz	5MHz
Element Pitch	**250μm**	150μm	400μm	100μm	160μm	208μm	150μm
TX Beamforming	**√ (digital)**	×	×	×	×	√ (digital)	×
RX Beamforming	**√ (digital)**	√(analog)	×	√ (analog)	√ (analog)	√ (digital)	√ (analog)
Gain Control	**CT**	DT	CT	CT	DT	DT	DT
Gain Error	**±0.9dB**	/	±1dB	±0.4dB	±6dB	/	±6dB
Gain Range	**36dB**	48dB	36dB	36dB	54dB	/	/
Element ADC	**√**	Shared ADC	×	×	Shared ADC	Shared ADC	√
Input Referred Noise	**11.4nV/√Hz @5MHz**	6.38nV/√Hz @5MHz	2.0pA/√Hz @5MHz	1.31pA/√Hz @10MHz	12.7nV/√Hz @6MHz	/	/
Input DR	**84dB**	/	/	82dB	91dB	/	/
Peak SNR	**54.7dB**	51.8dB	/	54dB	52.3dB	54.1dB	49.8dB
RX Power/EL.	**1.43mW**	0.91mW	4.9mW	1.17mW	1.23mW	0.66mW	1.54mW

consumption, a large dynamic range, minimal gain error, and small pitch size, making this design a promising solution for next-generation imaging catheters.

REFERENCES

[1] C. Chen et al., "A 0.91mW/element pitch-matched front-end ASIC with integrated subarray beamforming ADC for miniature 3D ultrasound probes," ISSCC, pp. 186-188, Feb. 2018.

[2] E. Kang et al., "A 2pA/√Hz Transimpedance Amplifier for Miniature Ultrasound Probes with 36dB Continuous-Time Gain Compensation," ISSCC, pp. 354-356, Feb. 2020.

[3] Peng Guo et al., " A 1.2mW/channel 100μm-Pitch-Matched Transceiver ASIC with Boxcar-Integration-Based RX Micro-Beamformer for High-Resolution 3D Ultrasound Imaging," ISSCC, pp. 496-497, Feb. 2022.K. Elissa, "Title of paper if known," unpublished.

[4] Yannick Hopf et al., "A Pitch-Matched ASIC with Integrated 65V TX and Shared Hybrid Beamforming ADC for Catheter-Based High-FrameRate 3D Ultrasound Probes," ISSCC, pp. 494-495, Feb. 2022.

[5] Nevada Sanchez et al., "An 8960-Element Ultrasound-on-Chip for Point-of-Care Ultrasound," ISSCC, pp. 480-481, Feb. 2021.

[6] J. Li, et al., "A 1.54mW/Element 150μm-Pitch-Matched Receiver ASIC with ElementLevel SAR/Shared-Single-Slope Hybrid ADCs for Miniature 3D Ultrasound Probes," IEEE Symp. VLSI Circuits, pp. C220-C221, 2019.

A Low-Power-Consumption Capacitance to Digital Converter with Novel Calibration Technology

Xiwen Zhu*[1], Yufeng Zhang[1], Xiaoming Teng[1], Yihan Wang[1]

[1] MEMS Center, Harbin Institute of Technology, Harbin 150001, China

* Email: zhuxiwen@hit.edu.cn

Abstract—Currently, capacitive sensors are extensively applied in environmental monitoring, medical equipment, automotive electronics, and so on. However, conventional capacitive sensor readout circuits based on analog to digital converter (ADC) architecture have a high power consumption and lack the self-calibration function. This paper presents a low power consumption capacitance to digital converter (CDC) circuit with calibration. The circuit can achieve a high-precision monitoring for capacitive sensors with low-frequency variation. The detectable capacitance range is 0pF to 5pF. The simulated maximum absolute error of CDC is 26.3fF. The application specific integrated circuit (ASIC) is fabricated in 180nm CMOS process. The chip area is 1.4mm×1.05mm. Experimental results demonstrate that the measured absolute error is less than 0.1pF.

Keywords—CDC, Calibration, Sensor readout, Low power consumption

I. INTRODUCTION

Capacitive sensors are essential in diverse fields, including environmental monitoring, medical equipment, and automotive electronics, etc[1]. The sensitive front-end structures of high-precision instruments such as pressure meter[2], hygrometer[3], altimeter[4], or vibrometer[5]are capacitive. The signal from sensitive front-end needs to convert into digital quantity for further processing or display. Consequently, readout circuits are necessary. Conventional readout circuits, typically based on Analog-to-Digital Converter (ADC) modules, often exhibit complex architectures and a high power consumption[6]. This characteristic renders them unsuitable for applications demanding ultra-low power, such as wireless sensor network (WSN) nodes. Recently, capacitance to digital converter (CDC) circuits have been proposed, which convert the capacitance to measurable periods[7]. In this paper, we present a novel low-power consumption CDC circuit with novel calibration technology for capacitive sensors to eliminate the influence of parasitic capacitance.

II. IMPLEMENTATION OF CAPACITANCE TO DIGITAL CONVERTER

A. Description of Proposed System architecture

The proposed system architecture is depicted in Fig. 1. It comprises primarily a multiplexer (MUX), a capacitance to voltage converter, a relaxation oscillator, and frequency dividers. To eliminate the adverse effects of the drift of the transmitted parameters, an automatic calibration circuit was adopted. The MUX circuit was used to connect the selected input capacitor to the capacitance to voltage converter. Besides, a relaxation oscillator was used to output the periodic voltage signal of the capacitance to voltage converter. Finally, a frequency division unit was used as a filter. This frequency reduction facilitates more convenient

measurement of the waveform periods. Fig.2 and Fig.3 show the block diagram of the interface and its input and output signals, respectively.

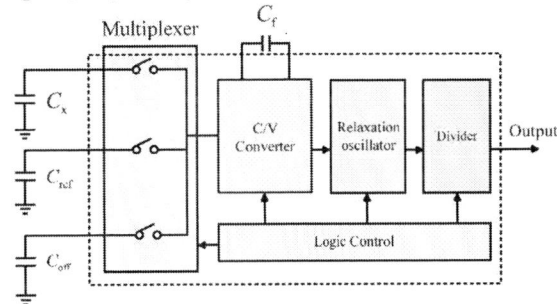

Fig. 1. The system architecture of proposed CDC

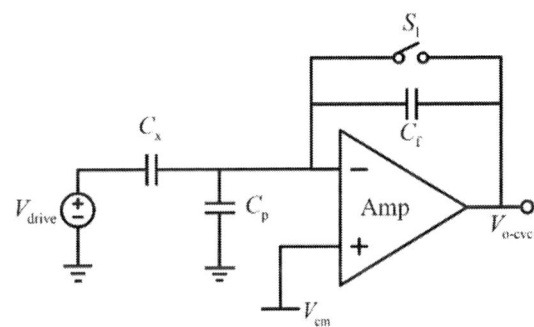

Fig. 2. The circuit diagram of capacitor to voltage converter

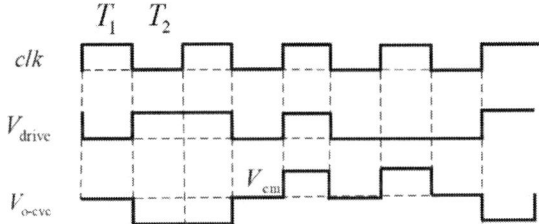

Fig. 3. The timing diagram of capacitor-voltage converter

The driving voltage level of the capacitor to voltage converter is 0 to V_{dd}. The timing diagram is shown in the Fig. 3. To prevent unnecessary charge loss, switch *S1* should operates in a break-before-make mode. During the sampling phase of the non-overlapping clock, the sensor capacitor is closed and connected between the drive voltage and the common mode reference voltage for sampling, with the value being the magnitude of the common mode reference voltage. In the subsequent transfer phase T_2, the capacitive sensor is

979-8-3315-3918-4/25 $31.00 © 2025 IEEE

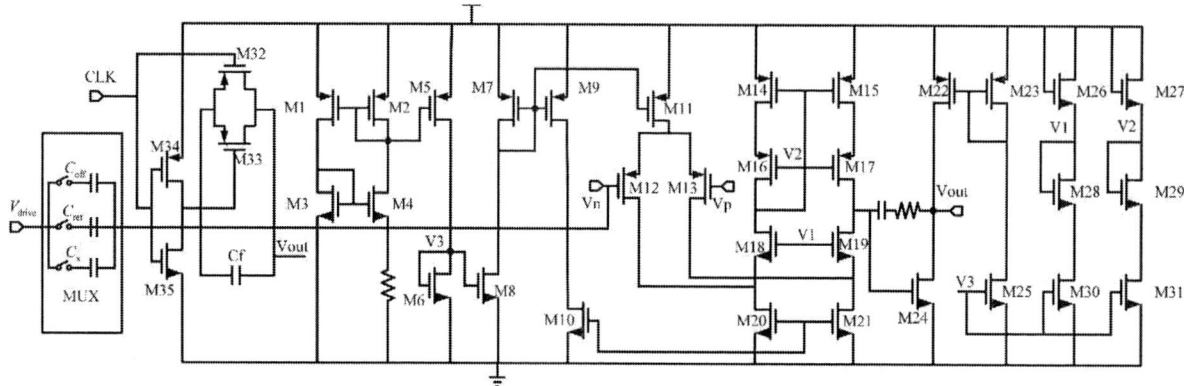

Fig. 4. The circuit diagram of capacitor to voltage amplifier

connected between the power supply voltage and the common mode reference voltage. Thus, the charge charged on the sensor capacitor is transferred to the integrating capacitor, causing a change in the voltage at the output voltage terminal. The voltage change can be written as equation (1).

$$V_x = C_x V_{dd} / C_f \tag{1}$$

the excitation voltages of the capacitor are namely 0, V_{cm} and V_{dd}. When using an off-chip capacitor, the setting range of the input capacitor is very wide, and the maximum setting can be 220pF. The detailed circuit is shown in the Fig. 4. The operational amplifier uses a cascode structure, which can achieve a larger output swing amplitude. When the multiplexer at the left end of the circuit selects a capacitor to connect to the capacitor-voltage converter under the control of the driving voltage. Transistors M_{32} and M_{33} act as switches under the control of the clock signal. The capacitor is selected by the multiplexer and transfer the charge to the two ends of the integrating capacitor in a certain stage, causing periodic changes in the output terminal voltage.

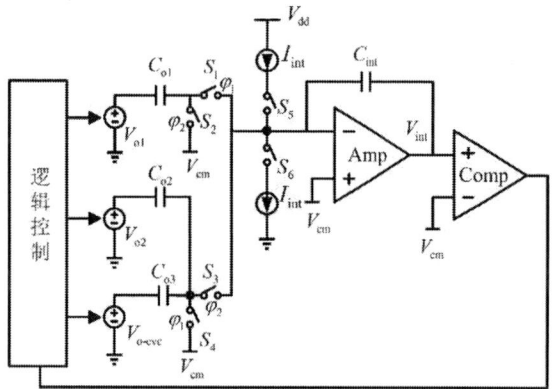

Fig. 5. The circuit diagram of Relaxation Oscillator

V_{o1} and V_{o2} are square wave signals with the level of 0-V_{dd}. During the period of T_1, the charge of C_{o1} is transferred to the integrating capacitor, causing the output voltage of the integrator to rise. Then, the charge is removed by a constant integrating current I_{int}. The comparator at the output end of the integrator will detect the change in voltage. During this period, the charge is transferred to the integrating capacitor. Similarly, this charge is also removed by the integrating current. The time consumed in this stage is. By repeating the above process,

periodic oscillations are generated at the output end. And a complete period can be obtained as equation (2):

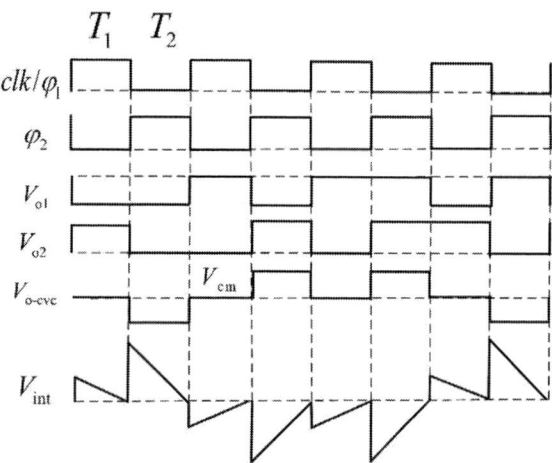

Fig. 6. The timing diagram of Relaxation Oscillator

$$
\begin{aligned}
T_{msm} &= 4(T_1 + T_2) \\
&= 4\frac{V_{dd}(C_{o1} + C_{o2}) + V_x C_{o3}}{I_{int}}
\end{aligned} \tag{2}
$$

III. SELF-CALIBRATION TECHNOLOGY AND SIMULATION

In this paper, we proposed an automatic calibration technology by measuring the C_{off}, C_{ref}, and C_x. The coefficients in the equation (1) are actually affected by process characteristics and non-ideal characteristics of various circuits, such as comparator delay, changes in temperature characteristics, and unstable power supply voltage, etc. We need automatic calibration technology to eliminate the unstable influence on the coefficients. We employed a multiplexer to measure the offset capacitance, reference capacitance and sensor capacitance respectively. Three different measurements were made using three capacitors, and their values were linearly converted to the time domain, generating three corresponding different periods. and. In order to specifically identify these three cycles, the logical control of the output signal needs to generate two relatively short time interval. The timing diagram of control signals are shown in Fig. 7.

979-8-3315-3918-4/25 $31.00 © 2025 IEEE

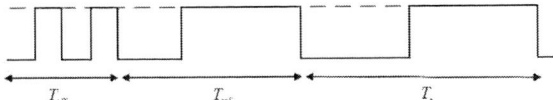

Fig. 7. The timing diagram of control signal

As is shown in Fig. 8, a multiplexer is employed to select offset capacitors, reference capacitors, and sensor capacitors respectively at different time periods for automatic calibration. Both PLUS-V1 and PLUS-V2 are square wave signals, and the high and low levels are V_{dd} and 0 respectively. The pulse time and delay time can be set by oneself to control and select the length of the capacitance time through the switch, and then connected to the next stage for capacitor-voltage conversion. M7-M12 and M13-M18 are switches that conduct the circuits of the selected capacitors at different time periods.

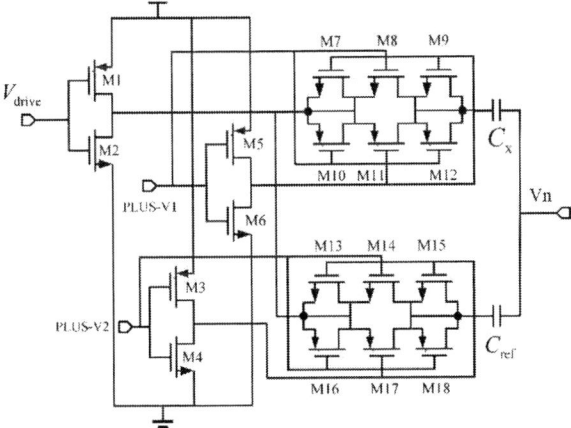

Fig. 8. The circuit diagram of Multiplexer

We introduce linearity M to measure C_x. The expression of M is as follows: can be obtained as:

$$M = \frac{T_x - T_{off}}{T_{ref} - T_{off}} = \frac{C_x - C_{off}}{C_{ref} - C_{off}} \tag{3}$$

From the equation (3), it can be concluded that the measured values of linearity M and are independent of the parameters sum. To achieve good resolution, the difference between the offset capacitor and the reference capacitor in the above formula should be large enough. Then, we simulated the circuit. The power supply voltage of the circuit is 1.8V. We set the offset capacitor and the reference capacitor to 4pF and 2pF respectively (which can be interchanged for measurement), and set the pulse duration of the square wave PLUS-V1 to 2.5ms, with no delay time. Set the pulse time length of the square wave PLUS-V2 to 5, the delay time to 5, and the total time length to 10ms. The current source in the relaxation oscillator has a magnitude of 1uA. The transient simulation results of the circuit are shown in Fig. 9. By measuring the waveform periods of 0-2.5ms, 2.5-5ms, and 5-10ms respectively, the following periods can be obtained: 347.68, 515.96, and 477.11us. According to the calculation formula of linearity (the above formula), M can be obtained as 0.767. Furthermore, we simulated the absolute error of CDC circuit as Fig. 10. The results show that the maximum error is 26.3fF when input capacitance is 1.5pF.

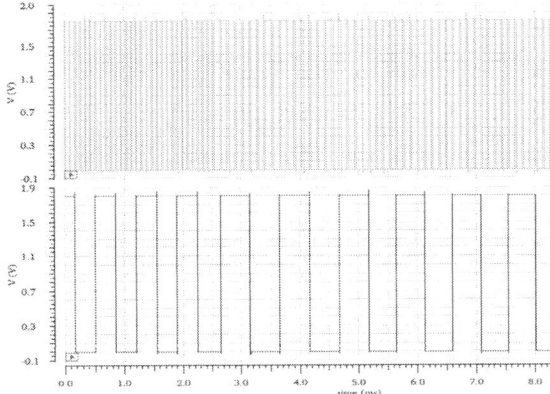

Fig. 9. The output waveform of CDC before dividers

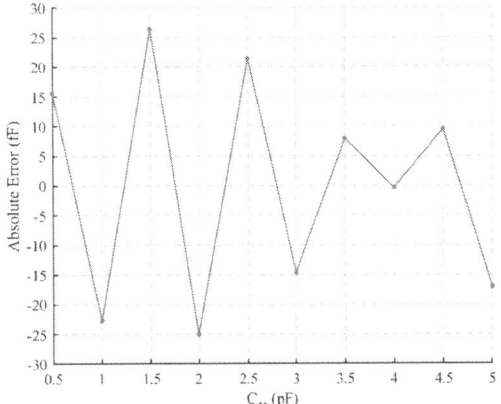

Fig. 10. The simulated absolute error of CDC

IV. MEASUREMENT RESULTS

The low-power-consumption CDC with calibration ASIC was implemented and fabricated in 180nm CMOS process. The die micrograph is shown in Fig. 11. The chip occupies an effective area of 1.05mm×1.4mm. Fig. 12 shows the output frequency signals. When reference capacitance is 4pF, the frequency of output signal is 78.678Hz. When circuit was set to tested capacitance phase, the frequency of output signal is 63.155Hz. When the circuit was set to no input capacitance phase, the frequency of output signal is 107.68Hz. After calculation, the value of the capacitor to be measured is 2.09pF. the absolute error is less than 0.1pF.

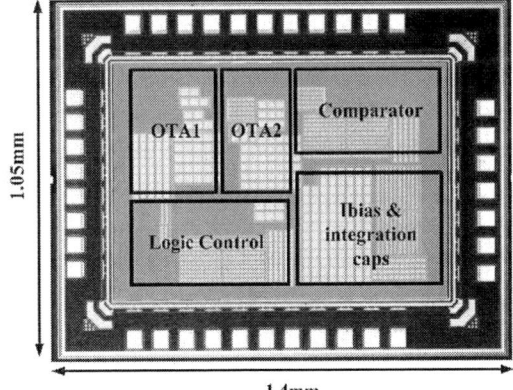

Fig. 11. The die photograph of proposed CDC

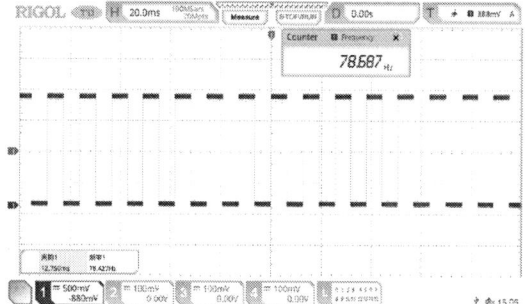

(a) The output waveform with the reference capacitance (4pF)

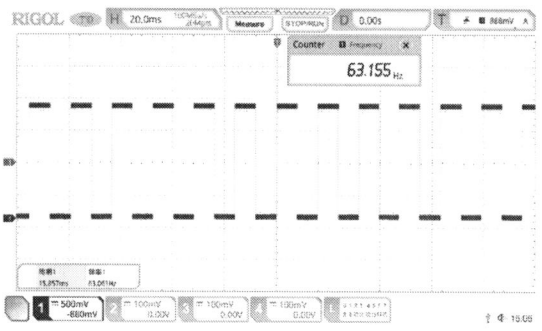

(b) The output waveform with tested capacitance (2pF)

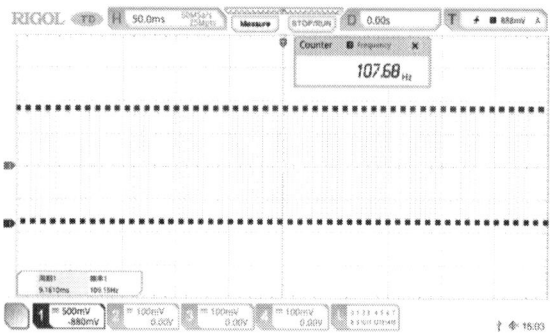

(c) The output waveform with no input capacitance (0pF)

Fig. 12. The output waveform of CDC under different input capacitance

V. CONCLUSION

This paper presents a low-power consumption CDC circuit. The proposed circuit employs a calibration technology to improve measurement accuracy with parasitic capacitance. The circuit includes a multiplexer, a capacitance-voltage converter, a relaxation oscillator, and frequency dividers. The chip was fabricated in 180nm CMOS process. The simulation results show that the maximum error is 26.3fF. Besides, the experimental results demonstrate that the absolute error is less than 0.1pF.

REFERENCES

[1] Klein, H.W. et al. (2019). Advanced Capacitive Sensing for Mobile Devices. In: Makinwa, K., Baschirotto, A., Harpe, P. (eds) Low-Power Analog Techniques, Sensors for Mobile Devices, and Energy Efficient Amplifiers . Springer, Cham.

[2] S. Oh et al., "A Dual-Slope Capacitance-to-Digital Converter Integrated in an Implantable Pressure-Sensing System," in IEEE Journal of Solid-State Circuits, vol. 50, no. 7, pp. 1581-1591, July 2015, doi: 10.1109/JSSC.2015.2435736.

[3] B. Yousefzadeh, W. Wu, B. Buter, K. Makinwa and M. Pertijs, "A compact sensor readout circuit with combined temperature, capacitance and voltage sensing functionality," 2017 Symposium on VLSI Circuits, Kyoto, Japan, 2017, pp. C78-C79, doi: 10.23919/VLSIC.2017.8008555.

[4] P. Maniraman and L. Chitra, "Comparitive analysis of capacitive type MEMS pressure sensor for altitude sensing," 2014 IEEE National Conference on Emerging Trends In New & Renewable Energy Sources And Energy Management, Chennai, India, 2014, pp. 195-199, doi: 10.1109/NCETNRESEM.2014.7088766.

[5] N. E. Wu, R. N. Miles and J. Huang, "Feasibility study of a low-cost feedback damping scheme for a micromachined capacitive microphone," Proceedings of the 2010 American Control Conference, Baltimore, MD, USA, 2010, pp. 3415-3422, doi: 10.1109/ACC.2010.5531112.

[6] J. -C. Liu, Y. -S. Hsiung and M. S. . -C. Lu, "A CMOS Micromachined Capacitive Sensor Array for Fingerprint Detection," in IEEE Sensors Journal, vol. 12, no. 5, pp. 1004-1010, May 2012, doi: 10.1109/JSEN.2011.2167748.

[7] R. Nojdelov and S. Nihtianov, "Capacitive-Sensor Interface With High Accuracy and Stability," in IEEE Transactions on Instrumentation and Measurement, vol. 58, no. 5, pp. 1633-1639, May 2009, doi: 10.1109/TIM.2009.2012957.

A 10-bit 4 GS/s 67.79-dBc SFDR Switched-Capacitor DAC with Reservoir Capacitor-based Reference Generation

Yitao Wang, Meng Xu, Qiang Pan, Jize Liu, Yuekang Guo*, Jing jin

School of Integrated Circuits, Shanghai Jiao Tong University, Shanghai, 200240, China
*Email: guoyuekang@sjtu.edu.cn

Abstract—This paper presents a 4 GS/s 10-bit switched-capacitor digital-to-analog converter (DAC) in a 40-nm CMOS process. A reference generation method is proposed, which employs reservoir capacitors to isolate data correlated spurs from main signal paths with considerable power efficiency. This method operates under a two-phase clock and can efficiently eliminate spurs at high speed. This paper also analyzes the generation mechanism and optimization methods of output signal nonlinearity in inverter-based output buffers. Simulations show that the DAC exhibits an SFDR of 67.79 dBc under an input signal at 1.96 GHz. The DAC consumes a total power of 51 mW, which demonstrates competitive power efficiency.

Index Terms—switch-capacitor, digital-to-analog converter (DAC), reservoir capacitor

I. INTRODUCTION

High-speed digital-to-analog converters (DACs) are widely employed in RF systems. Nowadays single-core DACs capable of achieving 1 GS/s sampling rates are no longer uncommon, among which current-steering architecture remains the most prevalent architecture.

However, as technology scales down, lower supply voltages increase current mirror design challenges in current-steering DACs [1], particularly for cascode structures. The architecture's constant current paths also lead to worse power efficiency. Moreover, process variations complicate precise current-cell matching [2], while background calibration remains difficult to implement at high speeds.

In contrast, switched-capacitor DACs show better scaling compatibility as their signal chain remains digital, avoiding low-voltage limitations while reducing power [3]. Capacitors' inherent matching properties also improve linearity with fewer process variations.

However, switched-capacitor DACs face two challenges: limited driving capability and digital interference. The driving capability limitation originates from their fundamental voltage-mode output. Digital interference arises from parasitic coupling of high-speed switching noise and clock feedthrough.

This paper proposes a switched-capacitor DAC based on reservoir capacitors, which isolates data-dependent spurs from main signal paths. Different from the reservoir capacitors operating under three-phase clocking in successive-approximation

register analog-to-digital converters (SAR ADCs), the proposed design requires only two clock phases in DAC applications. Additionally, this work analyzes the nonlinearity mechanisms in the output buffer and introduces source-degeneration to improve linearity.

The remaining sections of this paper are organized as follows: Section II introduces the isolation reservoir capacitance principle proposed in this paper and provides a nonlinear analysis of the output buffer based on active inductance. Section III presents the implementation of the DAC core circuit. Section IV displays and analyzes the simulated results of the DAC, and Section V summarizes the entire paper.

II. PROPOSED LINEARITY IMPROVING METHODS

DAC nonlinearity mainly stems from the digital-to-analog signal interface such as switches and the conversion circuitry, for example current sources in current-steering DACs or capacitors in capacitor-based DACs. This section analyzes the two key nonlinearity sources, capacitors and the output buffer, and proposes techniques to improve them.

A. Reservoir Capacitors Array

In DACs, digital and clock signals from the front-end can couple into the analog output through various parasitic paths. The digital signal coupling is more concerning as it generates input-dependent spurious components, particularly harmonic distortions such as third-order distortion.

The most common sources of spurious signals are switches, because they carry both data and clock signals. When placed in the signal path, it causes data-related glitches at the output called Code-Dependent Switching Transient (CDST), which has been widely researched in current-steering DACs [4].

In recent years, reservoir capacitors have been employed in SAR ADCs to mitigate interference from input digital signals to the output analog signal [5] [6].

In this work, reservoir capacitors are introduced to alleviate CDST problem. As shown in Fig. 1(a), the DAC conversion process operates across two clock phases. During the charging phase ϕ_{Charge}, two reservoir capacitors are charged to reference voltage in opposite polarities, while the DAC capacitor is reset. In the converting phase ϕ_{Convert}, depending on the digital input

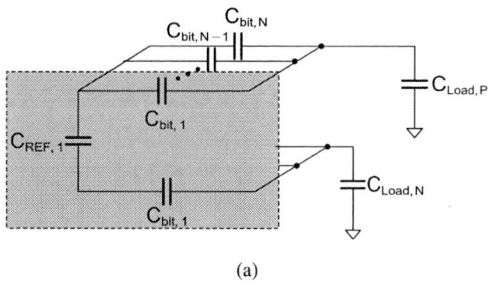

(a)

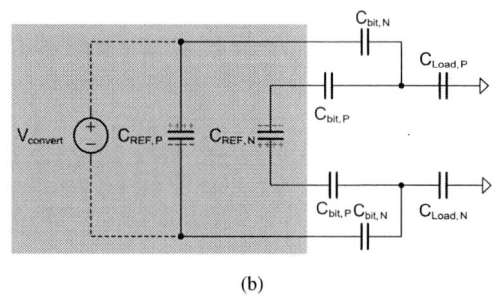

(b)

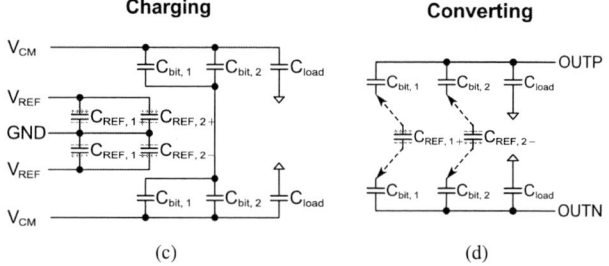

(c)　　　　　　　(d)

Fig. 1: Two-phase operation mechanism of capacitor arrays (a) capacitors array structure (b) simplified charging model (c) charging phase (d) converting phase

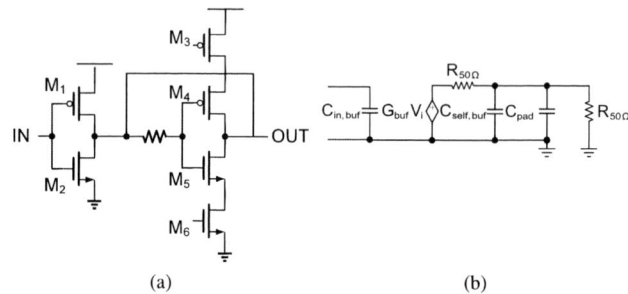

(a)　　　　　　　(b)

Fig. 2: Inverter-based output buffer (a) schematic (b) small signal model

$$V_{\mathrm{P}} = s \cdot C_{\mathrm{LSB}} \cdot V_{\mathrm{REF}} \sum_{i=1}^{N} 2^{i-1} b_i \tag{1}$$

$$V_{\mathrm{N}} = s \cdot C_{\mathrm{LSB}} \cdot V_{\mathrm{REF}}(2^{N} - 1 - \sum_{i=1}^{N} 2^{i-1} b_i) \tag{2}$$

$$V_{\mathrm{convert}} = V_{\mathrm{P}} - V_{\mathrm{N}} = s \cdot C_{\mathrm{LSB}} \cdot V_{\mathrm{REF}}(\sum_{i=1}^{N} 2^{i} \cdot b_i + 1 - 2^{N}) \tag{3}$$

If the reservoir capacitor is designed to be proportional to the bit capacitor, it will not introduce any quantization step variation.

B. Output Buffer

Due to insufficient driving capability of the voltage signal from the capacitor array, a capacitive DAC requires an output buffer, which has always been a critical design challenge in CDAC implementations.

Inverter-based active inductors are widely used in high-speed continuous-time linear equalizers (CTLEs) [7] to enhance output drive capability. An output buffer based on inverters was proposed for use in CDACs [3].

The basic circuit structure, as illustrated in Fig. 2(a), consists of a transconductance unit (g_{m}) and a closed-loop transconductance load unit ($1/g_{\mathrm{m}}$). The g_{m} unit is implemented using an inverter biased at VDD/2, which not only performs voltage-to-current conversion but also provides signal amplification. The $1/g_{\mathrm{m}}$ load unit also serves as an impedance-matching network.

This structure offers significant advantages. During operation, the inverter is placed in the amplification region where both the NMOS and PMOS operate in saturation. Consequently, its small-signal model is equivalent to that of a common-source amplifier. It can thus be inferred that its dominant pole frequency is on the same order as the intrinsic cutoff frequency of the MOS transistors, indicating that its bandwidth can nearly reach the process limit.

However, this structure introduces extra nonlinearity in high-speed applications. The gain of output buffer contains

code, one of the two reservoir capacitors is connected to the capacitor array to complete the charge transfer.

It can be observed that during the phase when the digital-to-analog conversion actually generates the output, the input signal and the output signal path are disconnected, making it impossible for the input to have any influence.

In conventional capacitive DAC designs, an attenuation capacitor is often required to adjust the output voltage swing. However, in this work, the reservoir capacitors inherently introduce a voltage division effect when charging the DAC capacitor array. By appropriately sizing the reservoir capacitors, the need for an attenuation capacitor can be eliminated.

Previous work has mentioned the issue of quantization step variation caused by the reservoir capacitor during charging phase [5]. This topic is also discussed here. As shown in Fig. 1(b), during data converting, the circuit essentially consists of N voltage sources with values of s (the Laplace variable) multiplied by $2^{i-1} C_{\mathrm{LSB}} V_{\mathrm{REF}}$ connected either forward or reverse between the positive and negative capacitor groups, so the total voltage source is:

* *The actual circuit is of a differential structure*

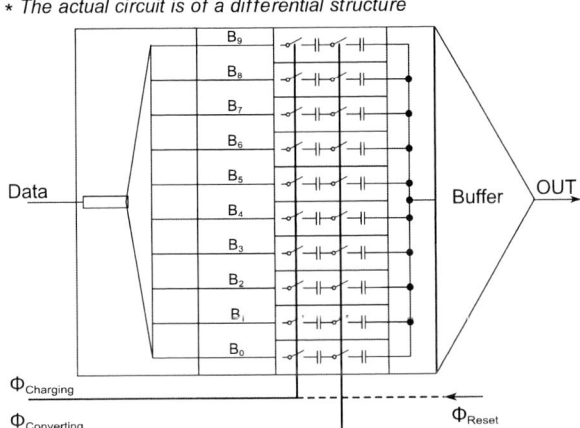

Fig. 3: Structure of the DAC core

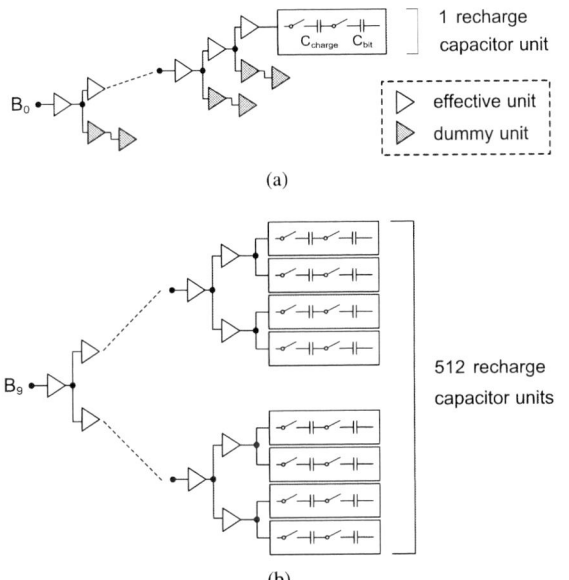

Fig. 4: Digital decoder circuit structure (a) B_0 decoder chain (b) B_9 decoder tree

quadratic and higher-order components of the signal. Since differential circuits can cancel out even-order harmonics, the third-order harmonic introduced by the output buffer consequently becomes the primary concern in the circuit.

To deliver output swings over 300 mV with single-digit gain, the buffer must handle proportionally large input signals, thereby exposing its gain nonlinearity due to changes in DC operating points.

III. CIRCUIT IMPLEMENTATION

A. DAC Structure

As shown in Fig. 3, the switched-capacitor DAC presented in this work employs a charge-redistribution architecture. Each DAC bit of the 10-bit switched-capacitor array is implemented

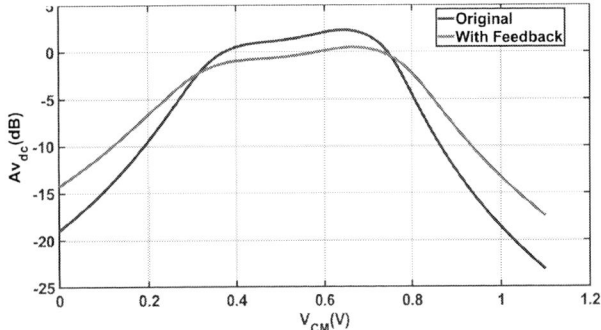

Fig. 5: Simulated relationship between gain of output buffer and the common-mode bias voltage

using identical unit capacitors. This approach increases matching between every bits, thus improving DAC output linearity.

The digital input first undergoes decoding through a digital decoder circuit, and then are used to control the charging phase of the reservoir capacitors. During the converting phase, charge redistribution happens between reservoir capacitors and bit capacitors, generating two opposite DAC output signals. These signals are subsequently processed by output buffers.

B. Decoding of the Capacitor Array

The switched capacitor DAC proposed in this work adopts an 10 bit binary-segmented architecture. Each binary-weighted bit consists of parallel-connected LSB capacitors and corresponding recharging capacitors, ensuring consistent matching accuracy and improved linearity.

As shown in Fig. 4, this work implements a fan-out-of-2 decoder stages to drive the 2^N switched-capacitors array. For the most significant bit, a complete tree structure is used. For lower bits, the optimized architecture, illustrated in Fig. 4(a), preserves two stages of dummy decoding units while pruning redundant circuitry. These dummy units play a critical role in load balancing and delay matching across all signal paths.

C. Output Buffer

Based on the previously described circuit, the output buffer incorporates source degeneration in both the transconductance and load stages to improve linearity. Although this design compromises bandwidth, it remains sufficient for the speed requirements of the recharge operation. Furthermore, with the addition of feedback resistors, the buffer's bandwidth can exceed 10 GHz.

This work simulates the frequency response of the buffer under different DC bias conditions as Fig. 5 shows. The comparison reveals that to confine the gain variation induced by DC offset within 1 dB, the allowable range of inverter transconductance is 200 mV. However, with the introduction of source degeneration, this range expands to 450 mV.

IV. SIMULATION RESULTS

A 10-bit DAC core circuit is designed in 40-nm CMOS process. For testing purposes, the DAC is simulated with

TABLE I: Performance Comparison with State-of-the-Art DACs

	Murmann [3] JSSC 2021	Duncan [8] JSSC 2017	Wang [9] ASICON 2021	*Xing [1] ASICON 2021	*THIS WORK
Resolution (bits)	8	10	6	10	10
f_{sample} (GS/s)	28	3.35	1	1.6	4
V_{supply} (V)	0.8	1.5/3.5	1.2/3.3	1.1/2.2	1.1
SFDR (dB)	37	48	46.6	64	67.79
Power (mW)	88	1910	–	54	51
Technology (nm)	16 FinFET	130 SiGe	130	40	40
Interleaving Factor	2	1	1	1	1

*Results are obtained from simulation.

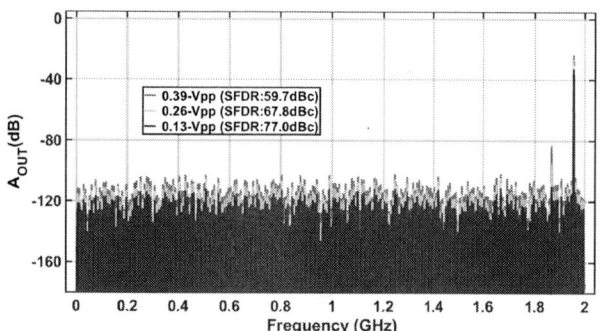

Fig. 6: The output spectrum under different output swings

different output swings of 0.13-Vpp, 0.26-Vpp and 0.39-Vpp, terminated on a 50 Ω equivalent differential load resistance.

Fig. 6 presents the simulated spectrum of the output signal at the end of the output buffer. All these results are measured when DAC works at a sampling frequency of 4 GHz, and the input signal is 1.956 GHz. When the output is 0.26-Vpp, the chip consumes a total power of 51 mW and exhibits an SFDR of 67.79 dBc.

As shown in Fig. 6, the simulated output spectrum reveals key insights into DAC nonlinearity. While the capacitor array maintains consistent dynamic performance across input amplitudes, the output buffer exhibits significant third-harmonic distortion variations. Notably, this distortion becomes negligible for output swings below 0.13-Vpp, confirming its origin in the buffer's large-signal gain nonlinearity. This finding provides clear optimization directions for future DAC designs.

Further more, from the nonlinearity mechanism of the output buffer, it can be inferred that variations in the power supply voltage will also cause the DC operating point to shift, thereby introducing nonlinearity. This poses challenges for the practical application of the circuit.

Table I presents a performance summary and comparison with prior works. The power consumption of our design is on par with the state of the art, while achieving significantly improved linearity despite a speed trade-off, but still outperforms comparable CS architectures in speed.

V. CONCLUSION

This paper presents a 10-bit 4 GS/s switched-capacitor DAC with reservoir capacitors. Simulation results achieve 67.79 dBc SFDR with a 1.95 GHz input signal while delivering 0.26-Vpp

output swing at a power consumption of 51 mW. The proposed two-phase charge redistribution method can isolate input digital signal from output signal path. Source degeneration is introduced in the output buffer aiming at nonlinearity caused by large input signal swing, and can significantly broaden the input linear range. Measurement results demonstrate that the third harmonic has been significantly attenuated. However, it remains the dominant non-ideality source. Further analysis reveals that power supply voltage deviations also severely degrade performance, providing critical guidance for improving future capacitor-based DAC designs.

ACKNOWLEDGMENT

This work was supported in part by the National Natural Science Foundation of China under Grant 62401360 and Grant 62431016, and in part by the Shuguang Program of Shanghai Education Development Foundation and Shanghai Municipal Education Commission under Grant 23SG10.

REFERENCES

[1] Y. Zhang and X. Xing, "A 10bit 1.6GS/s current-steering DAC in 40nm CMOS," in *IEEE International Conference on ASIC (ASICON)*, Kunming, China, 2021, pp. 1-4.

[2] Z. Dongmei, F. Dongbing, S. Jiangang and L. Kaicheng, "Digital static calibration technology used for 16-bit DAC," in *IEEE International Conference on ASIC (ASICON)*, Changsha, China, 2009, pp. 1081-1084.

[3] P. Caragiulo, O. E. Mattia, A. Arbabian and B. Murmann, "A 2× time-interleaved 28-GS/s 8-bit 0.03-mm² switched-capacitor DAC in 16-nm FinFET CMOS," *IEEE Journal of Solid-State Circuits*, vol. 56, no. 8, pp. 2335-2346, Aug. 2021.

[4] W. -H. Tseng, C. -W. Fan and J. -T. Wu, "A 12-bit 1.25-GS/s DAC in 90 nm CMOS with >70 dB SFDR up to 500 MHz," *IEEE Journal of Solid-State Circuits*, vol. 46, no. 12, pp. 2845-2856, Dec. 2011.

[5] J. Shen et al., "A 16-bit 16-MS/s SAR ADC with on-chip calibration in 55-nm CMOS," *IEEE Journal of Solid-State Circuits*, vol. 53, no. 4, pp. 1149-1160, April 2018.

[6] C. Yuan et al., "A compact low-power 16 b SAR ADC using reservoir-charge-redistributed DAC and configurable FIA-based comparator," *IEEE Journal of Solid-State Circuits*.

[7] K. Zheng, Y. Frans, K. Chang and B. Murmann, "A 56 Gb/s 6 mW 300 um² inverter-based CTLE for short-reach PAM2 applications in 16 nm CMOS," in *IEEE Custom Integrated Circuits Conference (CICC)*, San Diego, CA, USA, 2018.

[8] L. Duncan et al., "A 10-bit DC-20-GHz multiple-return-to-zero DAC with >48-dB SFDR," *IEEE Journal of Solid-State Circuits*, vol. 52, no. 12, pp. 3262-3275, Dec. 2017.

[9] X. Zhang et al., "A 6-bit, 1GS/s digital to analog converter for automotive ethernet PHY," in *IEEE International Conference on ASIC (ASICON)*, Kunming, China, 2021, pp. 1-4.

A 16-channel Neural Signal Acquisition Analog Front-End with Foreground Calibration for High-Precision Backend SAR ADC

Chun Feng, Junfeng Tang, Longhao Chen, Songping Mai, Xian Tang

Shenzhen International Graduate School, Tsinghua University

* Email: feng-c23@mails.tsinghua.edu.cn, tang.xian@sz.tsinghua.edu.cn

Abstract—This paper presents a 16-channel neural signal acquisition analog front-end (AFE) circuit implemented in 180 nm CMOS technology, designed for low-noise amplification and high-resolution digitization of weak neural signals. The system integrates **programmable-gain** capacitively-coupled chopper-stabilized low-noise amplifier (LNA) and a 16-channel-shared **16-bit resolution** successive approximation register (SAR) analog-to-digital converter (ADC). To address the capacitor mismatch in SAR ADC's capacitive digital-to-analog converter (CDAC), a FFT-based calibration algorithm is used to weight calibration. At 1.8-V supply voltage, the LNA achieves a power consumption of 2.16 μW, a total harmonic distortion (THD) of −63 dB, maintaining a noise efficiency factor (NEF) of 2.43 in 0.3-200 Hz frequency range and 2.02 in 0.2 k-10k Hz frequency range, with an input impedance of 750 MΩ. The backend SAR ADC operates at a sampling rate of 320 kS/s and achieves an ENOB of 15.06 bits with 214 μW power consumption.

Keywords—*Neural signal acquisition, AFE, chopper-stabilized LNA, SAR ADC, capacitor mismatch calibration.*

I. INTRODUCTION

The increasing demand for wearable and implantable biomedical devices has driven the need for high-performance neural signal acquisition systems. These systems must feature low noise, low power consumption, high linearity, and scalability to support high-density recording. Neural signals originate from a widely distributed network of neurons, making multi-channel acquisition crucial to extract inter-neuron correlations such as synchronization, inhibition, or excitatory interactions.

The neural signals of interest are typically categorized into two types: local field potentials (LFPs) and action potentials (APs). LFPs reflect the collective activity of nearby neurons and exhibit frequency components between 1 Hz and 300 Hz with amplitudes ranging from 1 μV to 1 mV. APs, on the other hand, are brief high-frequency events (300 Hz to 10 kHz) generated by single-neuron firing [1]. Capturing both types simultaneously poses significant challenges to the design of analog front-end (AFE), particularly in terms of noise, bandwidth, power and resolution.

In response to these challenges, this work proposes a 16-channel AFE system. The architecture is illustrated in Fig. 1. Each channel integrates a dedicated programmable-gain capacitively-coupled chopper-stabilized low-noise amplifier (LNA) that amplifies neural signals acquired by electrode. A multiplexer (MUX) selects the amplified analog signal to be sampled and quantified by the backend 16-channel-shared 16-bit resolution successive approximation register (SAR) analog-to-digital converter (ADC).

The AFE achieves input impedance matching for high-resistance electrodes and minimizes noise through a subthreshold-biased complementary input amplifier topology.

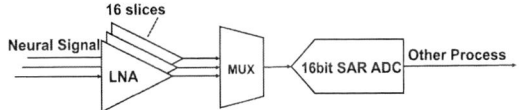

Fig. 1. The block diagram of the proposed architecture

The weight calibration of capacitive digital-to-analog converter (CDAC) constitutes a critical hurdle in achieving reliable performance in SAR ADCs targeting resolution above 12 bits. Due to the capacitor mismatch, parasitic and some other problem in the circuit, the weight of CDAC will no longer be a strict binary relationship. The purpose of calibration is to accurately obtain the weight of each digital output code. In [2], CDAC mismatch is calibrated using a least-mean-square (LMS) algorithm with pre-stored mismatch weights in a ROM. However, this approach often suffers from a trade-off between convergence speed and accuracy, and requires a high-precision reference. Another foreground digital calibration proposed in [3] uses the lower bit weights to estimate the higher bit weights. Errors in the reference bits may affect other bits and lower the accuracy. In [4], a pseudorandom noise (PN) injection background calibration scheme is proposed. This method requires additional PN injection conversion time, and the correlation-based computation leads to slow convergence. This paper uses a foreground calibration method – FFT based calibration algorithm. This method is derived from [5], where the algorithm is applied to compute interstage filter coefficients in continuous-time pipeline ADC to reconstruct the digital outputs across stages. We have made an extension so it can be applied to the weight mismatch calibration of SAR ADC.

The rest of this paper is organized as follows: Section II introduces the designed LNA, Section III shows the circuit details of the 16-bit high-precision SAR ADC and its capacitor mismatch calibration algorithm; Section IV shows the simulation results, and Section V summarizes the entire work.

II. LNA IMPLEMENTATION

A. Details of capacitively-coupled chopper-stabilized LNA

The overall structure of the proposed capacitively-coupled chopper-stabilized LNA is illustrated in Fig. 2. The structure is mainly composed of two-stage Miller compensation main amplifier, chopper switches, capacitor feedback loop, DC-servo Loop (DSL) and impedance boosting loop.

The closed-loop gain is determined by the ratio of the capacitors C_{in} and C_{fb} shown in Fig. 2. When the open-loop gain A_o of the main operational amplifier is not infinite, the closed-loop gain A_{CL} can be calculated as:

$$A_{CL} = \frac{A_o}{1 + A_o \frac{C_{fb}}{C_{in}}} = \frac{C_{in}}{C_{fb}} \frac{A_o}{\frac{C_{in}}{C_{fb}} + A_o} \qquad (1)$$

The proposed design implements a programmable-gain LNA by selectively switching C_{in} shown in Fig. 2 under external digital control. This enables the LNA gain to be configured among 20 dB, 32 dB, and 40 dB, offering flexibility for different signal amplitudes to avoid amplifier saturation.

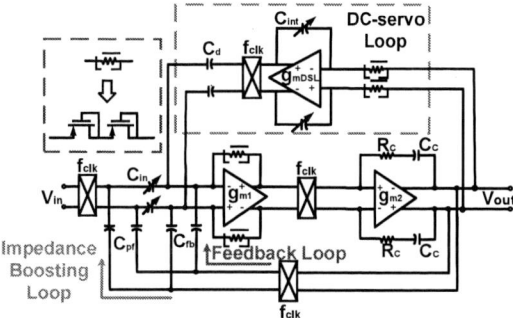

Fig. 2. Structure of the proposed capacitively-coupled chopper LNA

To balance between output swing and open-loop gain, a fully differential two-stage operational amplifier with Miller compensation is adopted which is shown in Fig. 3(a). Both stages employ the common-mode feedback (CMFB) circuit shown in Fig. 3(b) to regulate the output common-mode voltage and ensure the stability of the differential amplifier. The first stage employs a complementary-input, current-reuse operational amplifier, which effectively boosts the overall transconductance to $g_{mp} + g_{mn}$ without incurring additional bias current, thereby significantly improving the noise efficiency factor (NEF) [6].

Given the stringent low-noise and low-voltage design constraints, most of transistors are dimensioned with long channel lengths and relatively small widths, and the input differential pair is biased in the subthreshold region to minimize flicker noise and enhance output impedance.

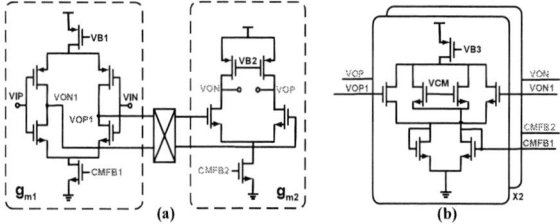

Fig. 3. (a) Schematic of the fully differential two-stage amplifier with Miller compensation labeled as g_{m1} and g_{m2} in Fig. 2. (b) Schematic of the CMFB used for g_{m1} and g_{m2}.

B. DSL and Impedance Boosting Loop Design

DSL is designed to suppress the electro-dc-offset and ensure stable operation of the amplifier without saturation. It consists of an integrator whose output signal is fed back to the main amplifier input. Since the high-pass corner frequency is low, a very small bandwidth integrator is required with large values of its integrating resistor and capacitor, which will

cause a large area consumption. In this design, pseudo-resistors are employed to realize the required extremely large resistance. However, due to the inherent nonlinearity and strong sensitivity of pseudo-resistors to process-voltage-temperature (PVT) variations, a capacitor trimming scheme is implemented to stabilize the high-pass corner frequency to about 0.3 Hz.

In the design of this article, Cin is set to 4 pF and f_{clk} is set to 20 kHz. The equivalent input impedance is calculated to be 6.25 MΩ. Since the electrode source impedance of neural signals is usually high (hundreds of kΩ to several MΩ), if the front-end input impedance is too low, it will lead to severe signal voltage division, energy loss and reduced signal-to-noise ratio. Therefore, this design adds an impedance boosting circuit which is shown in Fig. 2 to significantly increase the AC impedance of the input node to 750MΩ.

III. SAR ADC IMPLEMENTATION

A. 16-bit SAR ADC overall architecture

To reduce the overall capacitor array area, the SAR ADC adopts a two-stage bridged CDAC architecture, where only the MSB array participates in the sampling phase. This configuration minimizes the input capacitance and thereby alleviates the loading on the preceding LNA. As illustrated in Fig. 4, the MSB array comprises 7 bits, while the LSB array comprises 9 bits, with two redundant bits inserted at the 4th and 13th positions respectively. The implement of these redundant bits enhances the robustness of the conversion by correcting comparator decision errors.

The offset of the comparator will introduce a fixed DC offset voltage in the quantization stage, and the offset calibration technology is required to compensate for the offset voltage. As shown in Fig. 4, this design adopts the output offset storage technology (OOS) to stores the offset voltage on the output coupling capacitor. The working timing diagram illustrated in the lower left corner of Fig. 4. The OOS comparator is composed of four-stage pre-amplifier and one-stage LATCH implemented by strong arm architecture.

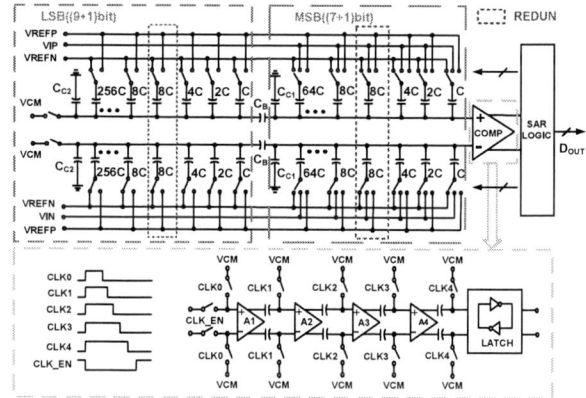

Fig. 4. The overall structure of the proposed 16-bit SAR ADC.

The offset voltage (V_{os}) of the comparator without and with OOS was simulated 200 times by Monte Carlo simulation. The simulation results are shown in Fig. 5. The standard deviation (std) of offset voltage before OOS calibration is 2.71 mV, while after calibration is 6.73 μV. After OOS, the std of offset voltage is much smaller than half of LSB of the 16-bit SAR ADC at power supply voltage of 1.8V.

979-8-3315-3918-4/25 $31.00 © 2025 IEEE

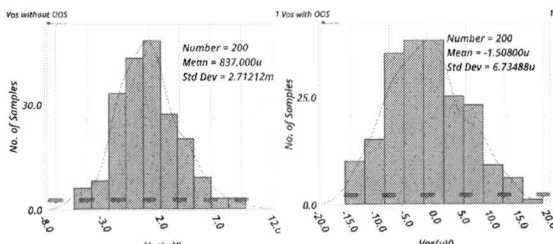

Fig. 5. Comparison of simulation results of Vos without and with OOS.

B. Bootstrap switch with DNW

To meet the stringent linearity requirements of high-resolution ADC design, a bootstrapped switch is employed in the sampling network. During the sampling phase, this technique ensures a constant gate-to-source voltage (V_{GS}) across the sampling switch, thereby maintaining a stable on-resistance (R_{on}) regardless of the input voltage amplitude.

Furthermore, as shown in Fig. 6, a deep N-well (DNW) NMOS device is used for the sampling switch. During the sampling phase, the body of M_1 is dynamically biased to the input voltage V_{IN}, which suppresses body effect and further enhances the linearity of the switch.

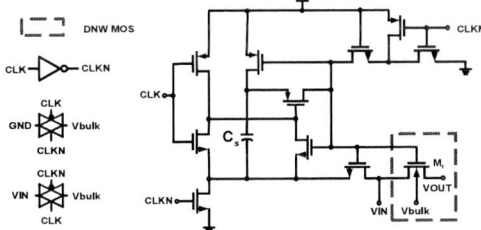

Fig. 6. Detailed schematic of bootstrap with DNW.

C. FFT-based Weight Calibration algorithm

Inspired by the methodology presented in [5], this work uses a simplified calibration algorithm that requires no complex computations, no long-time convergence and no architectural modifications to SAR ADC. These advantages make it well-suited for integration in low-power wearable analog front-end systems. The core principle of the algorithm is when the result of the combination of each digital code has the minimum total power at a frequency other than the input frequency, the noise energy in the corresponding bandwidth is the minimum. The detailed operation flow of the FFT-based algorithm is described as follows.

Considering an output weight model of any ADC, when we perform a single tone test, the output is $y = xw$, where y is the output of the test, $x = [x_1 x_2 \dots x_{M-1} x_M]$ is the data that needs to be weighted, and $w^T = [w_1 w_2 \dots w_{M-1} 1]$ are their weights. After N-point FFT, the output in the frequency domain can be written as

$$[X_1 \ X_2 \dots X_{M-1}][w_1 \ w_2 \dots w_{M-1}]^T + X_M = Y \quad (2)$$

where $Y^T = [Y_1 \ Y_2 \dots Y_N]$ and $X_i^T = [X_{i\_1} \ X_{i\_2} \dots X_{i\_N}]$ are the N-point FFT of y and x_i, $i=1\sim M$ respectively. Then we remove the input bins in $X_1\sim X_M$ and Y to obtain $X_{1O}\sim X_{MO}$ and Y_O, (2) can be written as

$$[X_{1O} \ X_{2O} \dots X_{(M-1)O}][w_1 \ w_2 \dots w_{M-1}]^T + X_{MO} = Y_O \quad (3)$$

so Y_O should only contain noise components from frequency points 1 to N. The optimal weights $w' = [w_1 \ w_2 \dots w_{M-1}]^T$ will minimize the 2-norm of Y_O, which means N-dimensional vectors $X_{1O}\sim X_{(M-1)O}$ should approximate $-X_{MO}$, and w' is their optimal coordinate. So next, we solve the following equation

$$Vw' = V_b \quad (4)$$

where $V = [X_{1O} \ X_{2O} \dots X_{(M-1)O}]$, $V_b = -X_{MO}$. When V^H is the conjugate transpose of V, the solution to w' is

$$w' = \mathrm{Re}(V^H V)^{-1} \mathrm{Re}(V^H V_b) \quad (5)$$

In summary, the process of this calibration method is as follows and is stressed in Fig. 7.

a. Excite the ADC with a known sinusoidal tone. Collect signals $x = [x_1 \ x_2 \dots x_M]$ which require weight calibration.

b. Compute the FFTs $X_1\sim X_M$ of $x_1\sim x_M$. Then remove the input bins in frequency domain. This step can directly set the input bins to 0.

c. Choose the frequency range we care about to form V and V_b. In this article, we hope that the noise of the entire frequency band is minimized. The frequency points are chosen from 1 to N, so the number of rows for V and V_b is N.

d. Calculate w' through (5). $\mathrm{Re}(V^H V)$ must be full rank, meaning that signals in x are uncorrelated.

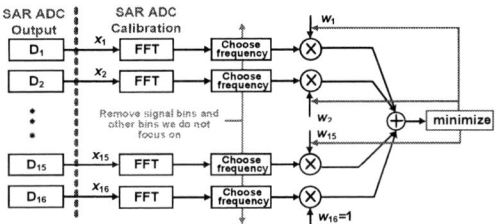

Fig. 7. Data processing flow of the calibration algorithm.

We excite SAR with input sine at f_{IN}, and choose the output $D_1[n]\sim D_{16}[n]$ as $x = [x_1 \ x_2 \dots x_{16}]$. Next, we obtain FFTs of these 16 signals and set the input bins to 0. Then the wight vector $w' = [w_1 \ w_2 \dots w_{15}]^T$ can be computed by (5). Based on the weight vector, the digital outputs of SAR can be calibrated as

$$D_{out} = \sum_{i=1}^{15} w_i D_i[n] + D_{16}[n] \quad (6)$$

This section introduces a simple foreground calibration algorithm. There is no additional test process because single tone test is an inherent process for any ADC.

IV. SIMULATION RESULT

The proposed 16-channel AFE is designed using a 180-nm technology. The core building blocks include the programmable-gain capacitively coupled chopper-stabilized LNA and a 16-bit high-resolution SAR ADC, designed for amplifying and digitizing weak neural signals.

Fig. 8(a) shows the simulated gain profiles of the LNA, which supports three gain settings: 20 dB, 32 dB, and 40 dB. Simulation results under SS, TT, and FF process corners demonstrate that the amplifier maintains a high-pass cutoff frequency of 0.3 Hz and a low-pass cutoff frequency greater than 10 kHz, effectively amplifying neural signals while suppressing DC offsets.

At the highest gain setting of 40 dB, the amplifier consumes current of 1.2 μA. Driven with a 5-mV$_{pp}$ input, it achieves a total harmonic distortion (THD) below -63 dB shown in Fig. 8(b) and a common-mode rejection ratio (CMRR) greater than 100 dB. The integrated input-referred noise is below 0.82 μV in the 0.3 Hz–200 Hz band and below 4.77 μV in the 200 Hz–10 kHz band，resulting NEF of 2.43 and 2.02 respectively.

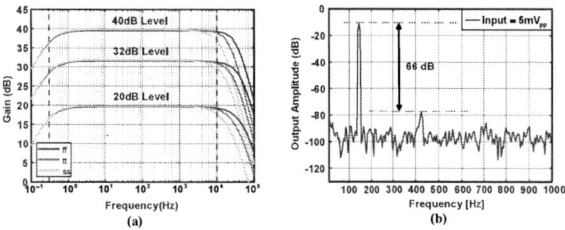

Fig. 8. (a) The PAC simulation results-gain of the LNA. (b) THD of the LNA for a 5-mV$_{pp}$ input.

To balance noise performance, capacitor area, and mismatch, the SAR ADC employs MIM capacitors with a unit capacitance of 164 fF whose Monte Carlo simulations indicate a typical mismatch of 0.2%. After incorporating the capacitor mismatch into the transistor-level simulation circuit, the calibration algorithm is applied to correct the resulting weight errors. As shown in Fig. 9(a), after calibration, the noise floor is significantly suppressed and harmonic suppression is improved. As shown in Fig. 9(b), this calibration significantly improves the ADC's performance without complex computations and architectural modifications, making it well suited for resource-constrained AFE systems.

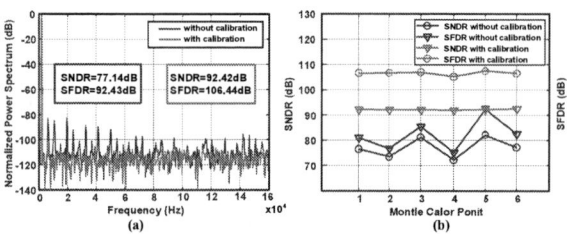

Fig. 9. (a) Power spectrum comparison before and after calibration. (b) Calibration-performance with different Montle Calor Point.

Table I summarizes a performance comparison between this work and several state-of-the-art AFE designs. The proposed AFE achieves a favorable noise-power trade-off, enabling high-performance amplification and quantization of both LFPs and APs signals. Benefiting from the effectiveness of the SAR ADC calibration algorithm, ADC performance remains robust against capacitor mismatch and parasitic bridging capacitance, achieving a FoM$_s$ of 181 dB.

V. CONCLUSION

This paper presents a 16-channel neural signal acquisition AFE integrating chopper amplifiers featuring low-noise，low-power, high-linearity and a 16-bit SAR ADC which used FFT-based noise-minimized weight calibration algorithm to overcome the impact of capacitor mismatch. The proposed solution offers a promising strategy for future high-resolution, low-power neural recording systems.

TABLE I.   PERFORMANCE AND COMPARISON OF THE PROPOSED AFE

	LNA Performance			
	JSSC-18[7]	JSSC-19[6]	TCSI-24[8]	This work[a]
Technology	180nm	180nm	180nm	180nm
Topology	CCIA	Chop.amp	CCIA	Chop.amp
VDD(V)	1	1.8	1.3/1.8	1.8
Gain(dB)	25.4	40	40	20/32/40
Power(uW)	0.25	3.24	10.8	2.16
IRnoise(V$_{rms}$)	5.5u (10-10kHz)	0.65u (0.3-200Hz) 2.14u (200-5kHz)	0.93 (0.5-500Hz)	0.82u (0.3-200Hz) 4.77u (200-10kHz)
NEF	1.07	2.37 (0.3-200Hz) 1.56 (200-5kHz)	3.09	2.43 (0.3-200Hz) 2.02 (200-10kHz)
THD(dB)	—	-61	-64.3	-63
	SAR ADC Performance			
	BioCAS[2]	TCSI-20[3]	TCSI-22[9]	This work[a]
Technology	180nm	130nm	180nm	180nm
VDD(V)	1.5	1.2/3.3	1.8/3.3	1.8
Resolution(bit)	16	14	16	16
Fs(kHz)	1000	200	1000	320
SNDR(dB)	91.33	71.89	83	92.42
Power(uW)	800	57	1050	214
FOM$_s$(dB)	179.3	164.3	169.8	181.2

[a.] Simulation results [b.] NEF=$V_{ni,rms}*\sqrt{(2*I_{tot}/\pi*U_T*4KT*BW)}$

[c.] FOMs(dB) = SNDR + 10×log(BW/Power)

ACKNOWLEDGMENT

This work was supported by Shenzhen Science and Technology Program under project number: JCYJ20220818101001003.

REFERENCES

[1] R. Muller, S. Gambini and J. M. Rabaey, "A 0.013 mm², 5 μW, DC-Coupled Neural Signal Acquisition IC With 0.5 V Supply," in IEEE Journal of Solid-State Circuits, vol. 47, no. 1, pp. 232-243, Jan. 2012.

[2] Y. Wang et al., "A Closed-Loop Neuromodulation Chipset With 2-Level Classification Achieving 1.5-Vpp CM Interference Tolerance, 35-dB Stimulation Artifact Rejection in 0.5ms and 97.8%-Sensitivity Seizure Detection," in IEEE Transactions on Biomedical Circuits and Systems, vol. 15, no. 4, pp. 802-819, Aug. 2021.

[3] Q. Zhang, N. Ning, J. Li, Q. Yu, Z. Zhang and K. Wu, "A High Area-Efficiency 14-bit SAR ADC With Hybrid Capacitor DAC for Array Sensors," in IEEE Transactions on Circuits and Systems I: Regular Papers, vol. 67, no. 12, pp. 4396-4408, Dec. 2020.

[4] L. Zhang, P. Wang, J. Sun and J. Wu, "Correlation-Based Background Calibration of Bit Weight in SAR ADCs Using DAS Algorithm," in IEEE Transactions on Circuits and Systems II: Express Briefs, vol. 68, no. 4, pp. 1063-1067, April 2021.

[5] N. Basavaraj, S. Manivannan and S. Pavan, "Simplified Simulation and Measurement of the Signal Transfer Function of a Continuous-Time Pipelined Analog-to-Digital Converter," in IEEE Transactions on Circuits and Systems II: Express Briefs, vol. 69, no. 10, pp. 3993-3997, Oct. 2022.

[6] D. Luo, M. Zhang and Z. Wang, "A Low-Noise Chopper Amplifier Designed for Multi-Channel Neural Signal Acquisition," in IEEE Journal of Solid-State Circuits, vol. 54, no. 8, pp. 2255-2265, Aug. 2019.

[7] L. Shen, N. Lu and N. Sun, "A 1-V 0.25-μW Inverter Stacking Amplifier With 1.07 Noise Efficiency Factor," in IEEE Journal of Solid-State Circuits, vol. 53, no. 3, pp. 896-905, March 2018.

[8] K. Wen, S. Liu, L. Zhong, Y. Shen and Z. Zhu, "A −64.3 dB THD, 26 nV/√ Hz Bio-Potential Readout Analog-Front-End Amplifier With a Gm-C Integrator-Implanted DC Servo Loop, and a Bulk-Driven Ripple Reduction Loop," in IEEE Transactions on Circuits and Systems I: Regular Papers, vol. 71, no. 2, pp. 537-547, Feb. 2024.

[9] Y. -H. Chung, C. -H. Tien and Q. -F. Zeng, "A 16-Bit Calibration-Free SAR ADC With Binary-Window and Capacitor-Swapping DAC Switching Schemes," in IEEE Transactions on Circuits and Systems I: Regular Papers, vol. 69, no. 1, pp. 88-99, Jan. 2022.

An 18-bit 1MS/s SAR ADC with Weight-Fitting Digital Calibration and High-Linearity Capacitor Array Design

Baoyi Zheng [1], Guoao Wang [2], Zongmin Wang*[2], Jin Qian [2], Bosen Liu [3], Zhaohang Bing [2], Tieliang Zhang [2]

[1] School of Integrated Circuit, Tsinghua University, Beijing, China
[2] Beijing Microelectronics Technology Institute, Beijing, China
[3] School of Integrated Circuit Science and Engineering, Beihang University, Beijing, China

* Email: zby23@mails.tsinghua.edu.cn, wzongmin@sohu.com

Abstract—This paper presents an 18-bit 1MS/s successive approximation register (SAR) analog-to-digital converter (ADC) that employs a high-linearity capacitor array and digital foreground calibration. A three-segment bridged binary-weighted capacitor array is adopted to reduce capacitance area, while a split high-segment capacitor is applied to enhance sampling linearity. To address capacitor mismatch, a weight-fitting algorithm based on an extended calibration matrix is implemented. The proposed ADC is fabricated in a 0.25μm CMOS process, occupying an area of 0.93mm². It achieves a signal-to-noise-and-distortion ratio (SNDR) of 92.8dB and a spurious-free dynamic range (SFDR) of 113.3dB with a 1kHz input at 1MS/s, representing improvements of 33.3dB and 43.8dB, respectively, through weight-fitting calibration. From a 2.5V supply and a 5V reference voltage, the ADC consumes only 7.3mW.

Keywords—Analog-to-Digital Converter, Capacitor Mismatch, Digital Foreground Calibration

I. INTRODUCTION

Successive approximation register analog-to-digital converters (SAR ADCs) are widely regarded for their simple architecture and excellent tradeoff between energy efficiency and performance, making them broadly applicable in sensor interfaces, biomedical instrumentation, and other low-power mixed signal systems. However, designing a high-linearity capacitor array becomes increasingly challenging when the target resolution exceeds 16bits and the sampling rate reaches the megahertz level. In such applications, the impact of non-idealities such as capacitor mismatch, comparator offset, and noise becomes more pronounced and significantly degrades overall performance. Although digital calibration techniques have been proposed to compensate for these factors, implementing them fully on-chip is often limited by the complexity of the required logic and the associated increase in circuit area and power consumption [1].

To overcome the above issues, a three-segment bridged binary-weighted capacitor array combined with weight-

fitting digital foreground calibration is applied to an 18-bit 1MS/s SAR ADC. Fabricated in a 0.25μm CMOS process, the proposed ADC demonstrates a signal-to-noise and distortion ratio (SNDR) of 92.8dB and a spurious-free dynamic range (SFDR) of 113.3dB with a 1kHz input signal at 1MS/s after calibration.

II. ARCHITECTURE OF THE CAPACITOR ARRAY

A. Bridged Binary-Weighted Structure

The linearity of the SAR ADC is largely determined by its capacitive digital-to-analog converter (CDAC). For ADCs with resolutions below 12bits, a simple binary-weighted CDAC with redundancy [2] is commonly used. However, the area of the binary-weighted CDAC grows exponentially with increasing resolution, making the CDAC impractical for resolutions beyond 14bits. Therefore, the bridged binary-weighted CDAC [3] is a strong candidate for meeting the demands of high-resolution applications.

Fig. 1 shows the bridged binary-weighted CDAC used in the 18-bit SAR ADC. The capacitor array is divided into high-, mid-, and low-segment by C_{B1} and C_{B2}, respectively, and the value of C_{B1} is given by:

$$C_{B1} = \frac{C_L}{C_L - C_u},\qquad(1)$$

where C_L is the total capacitance of the low-segment, and C_u (170fF) is the unit capacitor. It should be noted that C_{B1} is a fractional value, which makes it difficult to fabricate an accurate capacitance based on C_u. Therefore, additional capacitors ($C_{ext1}=30C_u$, $C_{ext2}=13C_u$) are added to each sub-segment, enabling C_{B1} and C_{B2} to be integer multiples of C_u. The total resolution of the capacitor array is the sum of the resolutions of each sub-segment (7+5+6). Furthermore, redundancy bits and 5 least significant bits (LSBs) repeats [4] are employed to enhance fault tolerance, as illustrated in Fig. 3, resulting in a total of 28 comparison cycles. Compared

Fig. 1. The bridged binary-weighted capacitor array (single-ended) with three segments for the 18-bit SAR ADC.

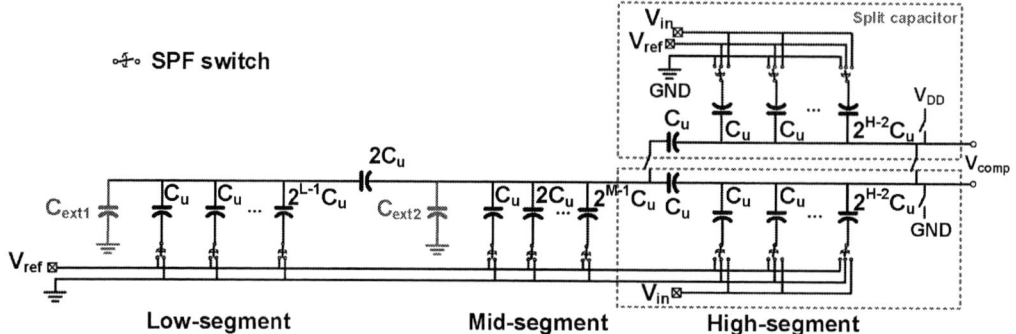

SPF switch

Fig. 2. Simplified diagram of the proposed capacitor array (single-ended) with high-segment split.

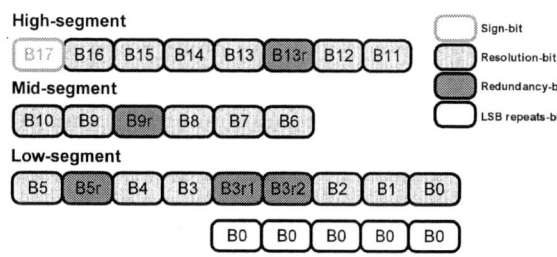

Fig. 3. Comparison cycle diagram of the proposed ADC.

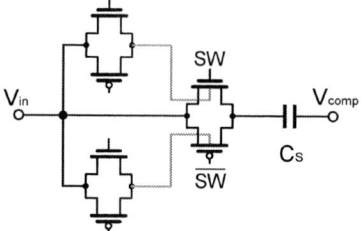

Fig. 4. Schematic diagram of the SPF switch.

with a conventional binary-weighted CDAC, the total capacitance of the proposed bridged binary-weighted CDAC is only 78.3pF, while maintaining high resolution, which contributes to reduced power consumption and compact circuit implementation.

B. High-Segment Splitting

However, the voltage variation (ΔV) across a capacitor can induce a corresponding change in capacitance, which can be modeled as:

$$C(V) = C_0(1 + c_1\Delta V + c_2(\Delta V)^2), \qquad (2)$$

where c_1 and c_2 are the first- and second-order voltage coefficients, respectively. During the sampling phase, this voltage-dependent capacitance can introduce harmonic distortion to the input signal, thereby degrading ADC performance. In the proposed architecture, the high-segment capacitor (C_H=44.5pF) serves as the sampling capacitor. To mitigate the impact of capacitance variation, a split capacitor array [5] is adopted in the high-segment, where each sub-segment capacitor is designed to be half the value of the original segment capacitor, as illustrated in Fig. 2.

Due to the use of opposite sampling voltages (V_{DD} and GND), the ΔV across each split sub-capacitor in the high segment is also opposite in polarity. Accordingly, the voltage-dependent capacitance variation can be re-expressed as:

$$\begin{aligned} C(V) &= 0.5C_0(1 + c_1\Delta V + c_2(\Delta V)^2) + \\ &\quad 0.5C_0(1 - c_1\Delta V + c_2(\Delta V)^2 \\ &= C_0(1 + c_2(\Delta V)^2). \end{aligned} \qquad (3)$$

The c_1 of split capacitors cancel each other, effectively eliminating the first-order nonlinearity of the overall high-segment capacitor array. As shown in Fig. 4, CMOS sampling switches with substrate potential following (SPF) technique are employed to further enhance sampling linearity

by minimizing switch-induced nonlinearity. After the sampling phase, the inputs of the comparator are shorted to generate a common-mode voltage (V_{CM}), allowing for reduced power consumption without requiring an additional V_{CM} generation circuit.

III. DIGITAL FOREGROUND CALIBRATION

A. Impact of Capacitor Mismatch

In practical circuits, components with identical design values often exhibit mismatch due to process variations. Among these, capacitor mismatch is a primary contributor to performance degradation in high-resolution SAR ADC. Two types of code transition discrepancies arise in SAR ADC due to capacitor mismatch. The first, known as "deviation due to lost analog signal," refers to situations where the analog input falls within a range that no longer maps to a valid digital output code. The second, referred to as "deviation due to lost digital codes," describes cases where certain digital codes never appear in the output, despite corresponding analog inputs existing.

As illustrated in Fig. 5, lost analog signal typically occurs when the actual capacitance of a given bit exceeds the total capacitance of all lower-order bits. This causes the decision threshold to shift beyond its intended range, rendering some input voltages unresolvable. In contrast, lost digital codes occur when the actual capacitance of a bit is smaller than the sum of the lower-order capacitors, leading to overlapping decision regions or missing output codes.

When such mismatch occurs, the SAR ADC may still perform the quantization procedure correctly; however, the resulting digital output no longer corresponds to ideal binary weighting. Interpreting these outputs under the assumption of ideal binary weighting results in a shifted transfer curve, elevated noise floor, and increased nonlinear harmonic distortion.

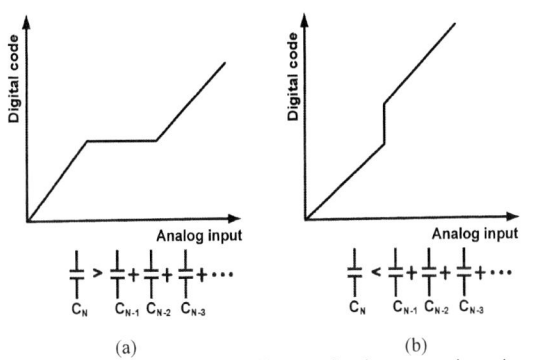

(a)   (b)

Fig. 5. Two types of code transition discrepancies due to capacitor mismatch: (a) lost analog signal; (b) lost digital codes.

B. Weight-Fitting Digital Calibration

Assume the digital output of the ADC is A, CDAC weight is x, and the corresponding analog input is b; then the relationship is given by:

$$
\begin{pmatrix} b_1 \\ b_2 \\ \vdots \\ b_M \end{pmatrix} = \begin{pmatrix} A_{11} & A_{12} & \cdots & A_{1N} \\ A_{21} & A_{22} & \cdots & A_{2N} \\ & & \vdots & \\ A_{M1} & A_{M2} & & A_{MN} \end{pmatrix} \begin{pmatrix} x_1 \\ x_2 \\ \vdots \\ x_N \end{pmatrix} \tag{4}
$$

The weight of each bit is predetermined by the capacitor array. By analyzing the relationship between the input-output data pairs, the actual contribution of each bit to the analog output is extracted. The corrected bit weights are then applied in the digital domain to reconstruct the accurate analog value, effectively completing the calibration of the SAR ADC without modifying the analog circuitry.

C. Extended Calibration Matrix

In addition to capacitor mismatch, SAR ADC errors also encompass other factors. Accordingly, the modified conversion equation for the calibration is given by:

$$
b = Ax + e, \tag{5}
$$

e denotes the error introduced during the conversion process. Given that the influence of noise on each column of the conversion result matrix A, corresponding to individual comparison bits, statistically diminishes as the number of conversions increases, it is reasonable to assume that:

$$
A^T e = 0. \tag{6}
$$

Further:

$$
x = (A^T A)^{-1} A^T b. \tag{7}
$$

Solving the above equation constitutes a least squares problem, which can be addressed using regression techniques. Provided that the matrix satisfies the full column rank condition, the solution can be efficiently obtained via QR decomposition, which is widely used in least squares estimation due to its numerical stability.

Employing a periodic sine wave as the input to the SAR ADC, the matrix A is constructed from the digital conversion results, while the vector b contains the corresponding analog input values. In a practical testing environment, the data in vector b may be affected by limitations in measurement instrument accuracy. Furthermore, for the equation (7), each row of matrix A must precisely correspond to the associated analog value in vector b, requiring strict time alignment and zero time difference between the two sets of signals, which is difficult to guarantee in practice.

To address this issue, it is necessary to extend the calibration matrix. Let Δt represent the time difference between the recorded analog signal and the converted digital code, and T denotes the conversion time of the ADC. Then, b can be re-written as:

$$
\begin{pmatrix} b_1 \\ b_2 \\ \vdots \\ b_M \end{pmatrix} = \begin{pmatrix} \sin(T + \Delta t) \\ \sin(2T + \Delta t) \\ \vdots \\ \sin(MT + \Delta t) \end{pmatrix}
$$

$$
= \begin{pmatrix} \sin(T)\cos(\Delta t) + \cos(T)\sin(\Delta t) \\ \sin(2T)\cos(\Delta t) + \cos(2T)\sin(\Delta t) \\ \vdots \\ \sin(MT)\cos(\Delta t) + \cos(MT)\sin(\Delta t) \end{pmatrix} . \tag{8}
$$

Since both $\cos(\Delta t)$ and $\sin(\Delta t)$ are constant, the impact of phase difference can be mitigated by inserting cosine vector c orthogonal to the matrix b and adjusting w_C components accordingly. Moreover, the calibration matrix can be further extended by accounting for the offset present in the actual conversion process. This is achieved by appending a column vector of ones to matrix A, while simultaneously introducing an offset adjustment parameter δ_{offset} in the solution vector x. In practical scenarios, the input signal source may also contain harmonic components. Therefore, adjustment coefficients δ_{hs} and δ_{hc}, corresponding to the sine and cosine components of the input signal harmonics are incorporated into vector b. The final extended matrix is given by:

$$
\begin{pmatrix} b_1 \\ b_2 \\ \vdots \\ b_M \end{pmatrix} = \begin{pmatrix} A_{11} & A_{12} & \cdots & A_{1N} & c_1 & 1 & h_{s1} & h_{c1} \\ A_{21} & A_{22} & \cdots & A_{2N} & c_2 & 1 & h_{s2} & h_{c2} \\ & & \vdots & & & & & \\ A_{M1} & A_{M2} & & A_{MN} & c_M & 1 & h_{sM} & h_{cM} \end{pmatrix} \begin{pmatrix} x_1 \\ x_2 \\ \vdots \\ x_N \\ w_c \\ \delta_{\text{offset}} \\ \delta_{hs} \\ \delta_{hc} \end{pmatrix} \tag{9}
$$

The number of columns in the extended matrix is $N+2+2*K$, where N represents the number of bits in the CDAC, which is typically larger than the nominal resolution due to redundancy. K represents the number of considered harmonics, with $2K$ terms included for sine and cosine calibration components. To ensure a valid least-squares solution, it is critical that the extended calibration matrix maintains full column rank.

IV. MEASUREMENT RESULT

The proposed 18-bit 1MS/s SAR ADC was fabricated in a 0.25μm CMOS process, occupying an area of 0.93mm². It consumes 7.3mW from a 2.5V power supply and a 5V reference voltage (V_{REF}).

979-8-3315-3918-4/25 $31.00 © 2025 IEEE

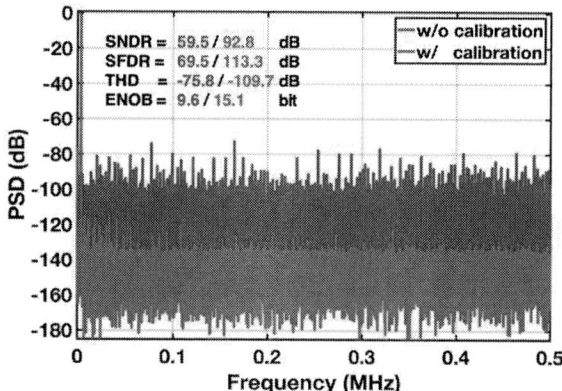

Fig. 6. BS spectra (262144fft) of the proposed ADC with and without digital calibration.

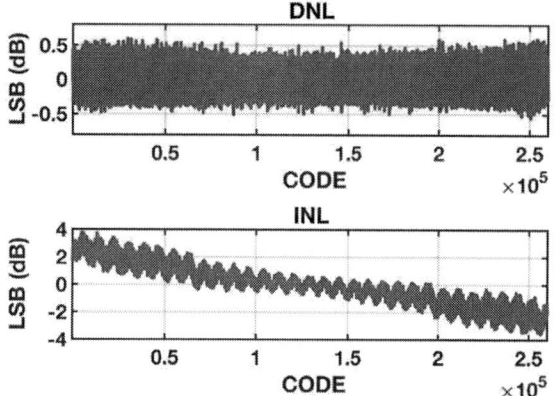

Fig. 7. DNL and INL of the proposed 18-bit ADC.

A. Measurement Performance

Fig. 6 shows the dynamic performance of the proposed SAR ADC with and without digital foreground calibration. At 1MHz sampling rate, the ADC achieves an SNDR of 92.8dB with a 1kHz input signal of -0.9dBFS, compared to 59.5dB without digital calibration, corresponding to an improvement of 5.5bits in resolution. Additionally, the SFDR improves by 43.8dB following calibration. The derived Schreier figure-of-merit (FoM$_S$) of the ADC is 171dB.

As shown in Fig. 7, the static performance after digital calibration of the SAR ADC is evaluated based on the code density method. The measured differential nonlinearity (DNL) is +0.59/-0.55LSB, and the integral nonlinearity (INL) is +3.76/-3.55LSB.

B. Comparison to Previous Work

Table I summarizes the measurement performance of the proposed ADC and compares it with state-of-the-art SAR and SAR-hybrid ADCs featuring resolution above 16bits and sampling rates in the megahertz range. Despite being implemented in a more mature 0.25μm CMOS process, the proposed ADC occupies a smaller area than other 18-bit designs. Moreover, it exhibits lower power while maintaining a respectable SNDR of 92.8dB and a competitive FoM$_S$ of 171dB.

V. CONCLUSIONS

This paper presents a high-linearity capacitor array design along with a digital foreground calibration technique,

TABLE I. PERFORMANCE SUMMARY AND COMPARISON

	This work	[6]	[7]	[8]
Architecture	SAR	P-SAR	SAR	SAR
Technology (nm)	250	180	180	180
Area (mm²)	0.93	3.9	5.74	0.81
Resolution (bit)	18	18	18	16
Supply/V$_{REF}$ (V)	2.5/5	1.8/5	1.8/5	1.8/3
V$_{FS}$ (V)	10	10	10	6
F$_S$ (MS/s)	1	2	5	4
SNDR (dB)	92.8 @1kHz	100 @100kHz	98.6 @1kHz	87.5 @15kHz
Power (mW)	7.3	11.35	30.52	10.1
FoM$_S$[1] (dB)	171	179.9	179.3	170.5

[1]FoM$_S$=SNDR+10log$_{10}$(BW/Power)

targeting an 18-bit 1MS/s SAR ADC in a 0.25μm CMOS process. The proposed capacitor array employs a three-segment bridged architecture for the reduced capacitance, combined with high-segment split. Measurement results demonstrate that the ADC achieves an SNDR of 92.8dB and an SFDR of 113.3dB with a 1kHz input signal, which improve by 33.3dB and 43.8dB, respectively, after applying the weight-fitting digital calibration. The ADC consumes only 7.3mW from a 2.5V supply and a 5V V$_{REF}$. These results validate the effectiveness of the proposed architecture and calibration approach in achieving high-resolution and medium-to-high-speed SAR ADC applications.

ACKNOWLEDGMENT

This work was supported by Beijing Microelectronics Technology Institute. Corresponding author: Zongmin Wang.

REFERENCES

[1] Y. Zhou, B. Xu and Y. Chiu, "A 12 bit 160 MS/s Two-Step SAR ADC With Background Bit-Weight Calibration Using a Time-Domain Proximity Detector," in *IEEE Journal of Solid-State Circuits*, vol. 50, no. 4, pp. 920-931, April 2015.

[2] C. -C. Liu, S. -J. Chang, G. -Y. Huang and Y. -Z. Lin, "A 10-bit 50-MS/s SAR ADC With a Monotonic Capacitor Switching Procedure," in *IEEE Journal of Solid-State Circuits*, vol. 45, no. 4, pp. 731-740, April 2010.

[3] Z. Du, B. Yao, W. Xu, X. Wang, H. Hu and L. Qiu, "Capacitor Mismatch Calibration of a 16-Bit SAR ADC Using Optimized Segmentation and Shuffling Scheme," in *IEEE Transactions on Circuits and Systems II: Express Briefs*, vol. 70, no. 8, pp. 2789-2793, Aug. 2023.

[4] Z. Jiao, H. Luo, J. Zhang, X. Wang, L. Chen and H. Zhang, "An 84dB-SNDR 1-0 Quasi-MASH NS SAR with LSB Repeating and 12-bit Bridge-Crossing Segmented CDAC," *2023 IEEE Custom Integrated Circuits Conference (CICC)*, San Antonio, TX, USA, 2023, pp. 1-2.

[5] Y. Zhu, C. Chan, U. Chio, S. Sin, S. U and R.Martins "A 10-bit 100-MS/s Reference-Free SAR ADC in 90 nm CMOS," in *IEEE Journal of Solid-State Circuits*, vol. 45, no. 6, pp. 1111-1121, June 2010.

[6] D. Hummerston and P. Hurrell, "An 18-bit 2MS/s pipelined SAR ADC utilizing a sampling distortion cancellation circuit with −107dB THD at 100kHz," *2017 Symposium on VLSI Circuits*, Kyoto, Japan, 2017, pp. C280-C281.

[7] A. Bannon, C. P. Hurrell, D. Hummerston and C. Lyden, "An 18 b 5 MS/s SAR ADC with 100.2 dB dynamic range," *2014 Symposium on VLSI Circuits Digest of Technical Papers*, Honolulu, HI, USA, 2014, pp. 1-2.

[8] Y. Chen, Q. Huang, Y. Fan, Q. Zhao, S. Huang and J. Yuan, "A 16-bit 4-MS/s SAR ADC With Dual-Segmental Bit Weight Self-Calibration," in *IEEE Transactions on Circuits and Systems I: Regular Papers*, vol. 71, no. 9, pp. 3961-3974, Sept. 2024.

AUTHOR INDEX

Arafune, T. 310
Asami, K. 465
Bai, B. 614
Bao, L. 521
Bao, W. 13
Bendra, M. 211
Bi, H. 586
Bi, X. 769
Bing, Z. 487
Bo, P. 702
Cai, C. 793
Cai, D. 219
Cai, R. 834
Cai, Y. 521
Cao, J. 706
Cao, S. 156, 267
Cao, X. 858
Cao, Y. 566
Chai, Z. 1
Chakrabarti, C. 566
Chan, M. 32, 100, 637
Che, H. 682
Chen, B. 28, 199, 239, 417, 622, 706, 793
Chen, C. 351, 503, 814, 898
Chen, C.-C. 255
Chen, G. 144, 649, 667, 801, 838, 894
Chen, H. 698
Chen, J. 160, 164, 168, 175, 187, 199, 215, 231, 335, 511, 578, 630, 749, 753, 757, 917
Chen, K. 16, 530, 594
Chen, K.J. 327
Chen, L. 219, 223, 259, 483, 674
Chen, M. 874
Chen, N. 511, 630
Chen, S. 128, 550, 890
Chen, T. 92, 327, 674
Chen, X. 378, 814, 909
Chen, Y. 148, 207, 311, 343, 491, 519, 521, 606, 789
Chen, Z. 92, 207, 219, 239, 898
Cheng, J. 417, 530
Cheng, P. 660
Cheng, R. 28, 88, 199
Cheng, X. 507, 582
Cheng, Y. 327
Cheng, Z. 409
Chi, Y. 749, 753, 797
Chu, B. 866
Chu, Y. 5

Coutinho, J.G.F. 132
Cui, H. 519
Cui, J. 409, 850
Cui, K. 602
Cui, W. 64
Dai, L. 471, 499
Dai, S.-Q. 902
Dai, Y. 60, 271, 678
Deltimple, N. 386
Demirsoy, S. 132
Deng, H. 44
Deng, J. 534
Deng, K. 781
Deng, L. 271, 371, 942
Deval, Y. 84, 386
Di Hu, H. 195
Di Zhang, W. 183
Ding, D. 633
Ding, F. 626
Ding, J. 491
Ding, Q. 72
Ding, X. 777
Ding, Y. 160, 164, 199, 223, 235, 578
Dong, G. 310, 315
Dong, H. 168, 578, 781
Dong, J. 36
Dong, S. 785
Dong, Y. 44
Dong, Z. 219, 586, 925
Dou, X. 215
Du, J. 657
Du, P. 28
Du, Y. 674
Evans, R. 525
Fan, Y. 645, 745
Fan, Z.-C. 267
Fellmann, M. 386
Feng, A. 805
Feng, C. 483
Feng, H. 781
Feng, J. 706, 805
Feng, L. 834
Feng, P. 511, 630, 657
Feng, Q. 44
Feng, Y. 187, 457, 554
Feng, Z. 711
Ferrer, P. 386
Freitas, J.R.D. 132

Fu, B.	777
Fu, C.	263
Fu, G.	515
Fu, H.	40
Fu, J.	471, 499
Fu, R.	124, 846
Fu, X.	299
Fujishima, M.	390
Fukui, H.	664
Furuta, T.	664
Gai, W.	633, 769
Gan, Y.	68
Gao, D.	88
Gao, H.	36, 347
Gao, S.	649, 667
Gao, Y.	215, 745
Ge, Y.	48
Geng, Y.	327
Goes, W.	211
Gong, Y.	299, 917
Gu, H.	80, 96, 136, 203
Gu, S.	259
Guan, Y.	425
Guillot, M.	84
Guo, A.	882
Guo, C.	116, 132
Guo, K.	401
Guo, M.	466
Guo, P.	160, 164
Guo, Y.	72, 295, 479, 626, 757
Hachiya, K.	433
Han, D.	36
Han, G.	28, 199
Han, J.	152, 263, 275, 822
Han, M.	519
Han, S.	287
Han, W.	44
Han, X.	777
Han, Y.	525
Han, Z.	453, 866
Hao, J.	949, 973
Hao, K.	582
Hao, W.	562
Haque, E.	566
He, C.	913
He, F.	347
He, J.	718, 874
He, K.	72

He, L.	449
He, M.	421
He, Q.	279
He, S.	818
He, Y.	215, 574, 660, 842, 913
Higuchi, A.	331
Hong, X.	92
Hong, Y.	653
Hong, Z.	323, 789
Hou, M.	413
Hou, R.	421
Hou, Y.	215
Hou, Z.	745
Hu, A.	259, 850
Hu, C.	797
Hu, D.	68
Hu, J.	698
Hu, L.	215
Hu, Q.	515
Hu, S.	965, 969
Hu, X.	195, 394, 618, 961
Hu, X.Y.	183
Hu, Y.	175, 215, 761
Hu, Z.	602
Huang, G.	570
Huang, H.	938
Huang, J.	72, 335, 530, 586
Huang, Q.	9
Huang, R.	9, 409
Huang, T.	259
Huang, W.	124, 846
Huang, Y.	582, 789, 965, 969
Huang, Y.-S.	255
Huang, Z.	906
Huo, R.	637
Huo, Y.	223
Inoue, T.	664
J.Wan	20
Jayarajan, J.	76
Ji, Z.	574, 590, 913
Jia, S.	898
Jian, J.	1
Jiang, A.	191, 195
Jiang, A.Q.	183
Jiang, B.	694
Jiang, C.	405, 858
Jiang, H.	882, 957
Jiang, P.	471

Jiang, Q. 88
Jiang, Y. 92, 874
Jiang, Z. 156
Jiao, S. 574, 913
Jin, C. 199, 243
Jin, J. 479, 626, 757
Jin, X. 315, 598, 946
Jin, Y. 251, 674
Jing, C. 953
Jing, M. 645
Jing, N. 136
Jou, S.-J. 255
Jørstad, N.P. 211
Kanayama, Y. 339
Kang, J. 179, 934
Kang, Q. 602
Kang, W. 457, 554
Kang, X. 965, 969
Kang, Y. 550
Katayama, S. 310
Ke, M. 28
Kerhervé, E. 386
Khatami, R. 310
Kim, I. 730
Kim, N. 64
Kishine, K. 664
Kobayashi, H. 310
Kobori, Y. 310
Kong, M. 44
Kong, Z. 179
Kuai, R. 614
Kuang, X. 737
Kurokawa, A. 433
Kuwana, A. 310
Lai, R. 60
Lapuyade, H. 84, 386
Le Wu, W. 140
Lee, C. 207
Lei, Y. 902
Li, A. 259, 271, 678, 854
Li, B. 307
Li, C. 398, 534
Li, D. 28, 315, 382, 499, 682
Li, G. 247, 546, 761
Li, H. 40, 124, 168, 235, 295, 466, 578, 846, 850, 934
Li, J. 40, 219, 303, 367, 382, 425, 437, 453, 471, 499, 866, 965, 969
Li, K. 850

Li, L. 363, 586, 653
Li, M. 5, 112
Li, N. 715, 854
Li, Q. 120, 359, 558, 830
Li, R. 72, 160, 164, 283
Li, W. 351, 503, 674, 785, 793
Li, X. 187, 215, 562, 641, 930
Li, Y. 24, 574, 582
Li, Z. 140, 287, 291, 421, 586, 682, 698, 862
Lian, J. 645
Liang, B. 749, 753, 797
Liang, F. 682
Liang, J. 64, 870
Liang, L. 521
Liao, J. 231
Liao, X. 315, 319
Liao, Y. 88
Lin, F. 52, 56
Lin, M. 351, 503
Lin, T.-J. 449
Lin, Y. 351, 409, 503, 610, 622, 793, 965, 969
Lin, Z. 854
Liu, B. 487, 797
Liu, C. 287, 291, 578, 618, 745, 749, 753, 862
Liu, D. 88, 259, 850, 953
Liu, E. 649, 667
Liu, F. 491
Liu, G. 890
Liu, H. 28, 367, 554, 749, 753, 949
Liu, J. 108, 263, 347, 374, 378, 382, 479, 511, 614, 630, 645, 657, 757, 878, 882, 886
Liu, K. 722, 726, 733
Liu, L. 108, 315, 319, 374, 511, 630, 657, 874
Liu, M. 745
Liu, Q. 311, 343, 810
Liu, S. 68, 602, 722, 726, 733, 761, 814, 961
Liu, W. 52, 56
Liu, X. 24, 179, 315, 530, 757, 870, 921
Liu, Y. ... 5, 28, 72, 88, 179, 199, 231, 374, 437, 471, 499, 598, 674, 781
Liu, Z. 457, 542, 554, 574, 870, 913, 957
Lu, D. 247
Lu, G. 906
Lu, H. 711
Lu, J. 40, 259, 307, 789, 842, 850
Lu, X. 363
Lu, Y. 243, 374, 413
Luan, Y. 495

Luk, W. 132, 441
Luo, D. 749, 753
Luo, H. 602, 610
Luo, K. 355
Luo, M. 283
Luo, P. 355
Luo, Q. 814
Luo, Y. 394
Lv, F. 614, 618
Lv, J.H. 934
Lv, S. 606
Lv, Z. 866
Lyu, Y. 457, 554
Ma, B. 92
Ma, C. 52, 56
Ma, J. 80, 96, 203
Ma, K. 76
Ma, L. 144, 838
Ma, R.C. 826
Ma, X. 251
Ma, Y. 378, 546, 789
Ma, Z. 530
Mahalingam, N. 76
Mai, S. 483
Mao, J. 60
Mao, K. 698
Mao, Y. 347, 737
Marium, S.M. 128
Martins, R.P. 466
Mei, J. 271, 371, 678, 715, 854
Mei, S. 942
Meng, F. 76
Miao, R. 562
Min, H. 359, 765
Min, T. 1
Miu, L. 886
Mo, Y. 686
Mou, C. 749, 753
Mu, C. 814
Nalla, P.S. 566
Nan, L. 674
Ng, W.T. 64
Ng, Y.C. 461
Ng, Y.H. 327
Nie, H. 715
Nie, X. 279
Ning, N. 72, 367, 471, 499
Niu, Q. 120

Ou, S. 637
Pan, Q. 479, 660
Pan, Z. 562
Pang, Z. 614
Peng, L. 72
Peng, P. 949
Pi, C. 637
Pruckner, B. 211
Pu, Y. 818
Qi, H. 711
Qi, J. 227
Qi, M. 686
Qian, H. 199
Qian, J. 487, 641
Qian, L. 842
Qiao, G.C. 826
Qiao, M. 52, 56, 72, 930
Qin, C. 745
Qin, Y. 730, 961
Qin, Z. 862
Qing, Y. 60
Qiu, H. 307
Qiu, T. 657
Qiu, X. 590
Qu, X. 546
Qu, Y. 175, 207, 223, 570
Quan, S. 104
Que, Z. 132
Rahardja, S. 267
Ren, H. 694
Ren, P. 92
Ren, S. 239, 251
Ren, T. 140, 319
Rivet, F. 84, 386
Rong, T. 453
RS.He, B. 20
Ruan, A. 124, 846
S.Cristoloveanu, Y. 20
Saito, W. 331
Sang, P. 160, 164, 168, 175, 187, 215, 231, 578
Sang, W. 674
Sapatnekar, S.S. 566
Sarfraz, K. 100
Sawan, M. 737
Selberherr, S. 211
Shangqian, C. 961
Shao, J. 797
Shen, A. 934

Shen, B. 80, 195
Shen, C. 88
Shen, G. 429
Shen, H. 378, 382
Shen, L. 283, 409
Shen, R. 199
Shen, T. 538, 606
Shi, C. 606, 874
Shi, F. 890
Shi, H. 323
Shi, L. 953
Shi, Q. 965, 969
Shi, X. 152
Shi, Y. 60, 116, 307, 335, 495, 722, 733
Shi, Z. 271, 371, 854
Shu, J. 327
Shu, Z. 152
Si, X. 542
Sin, S.-W. 466
Skafidas, E. 525
Song, C. 810
Song, R. 797
Song, X. 909
Su, H. 76
Su, R. 810
Su, X. 92, 283, 453
Su, Y. 92, 235, 287, 291
Sui, Z. 172, 534
Sun, C. 36, 761
Sun, H. 805
Sun, J. 307
Sun, Q. 16, 80, 96, 203
Sun, W. 602
Sun, X. 726
Sun, Y. 148, 310, 538, 542, 637, 822, 842, 957
Sverdlov, V. 211
Tan, Z. 737
Tang, H. 626, 973
Tang, J. 483
Tang, K. 749, 753
Tang, X. 483
Tang, Y. 542
Tang, Z. 702
Tanzawa, T. 331, 339
Tao, M. 749, 753
Tao, Q. 215
Tao, R. 88
Teng, X. 303, 475

Thangarasu, B.K. 76
Tian, J. 649, 667
Tiancong, W. 671
Todman, T. 441
Tsuchiya, A. 664
Tsukiji, N. 310
Tu, C.-L. 255
Unnithan, R.R. 525
Wan, C. 519
Wan, J. 13
Wan, X. 690
Wang, A. 562
Wang, C. 16, 283, 453, 594, 866, 917, 938
Wang, F. 215
Wang, G. 319, 487, 511, 630
Wang, H. 355, 449, 906, 909
Wang, J. 495, 653
Wang, K. 9, 641
Wang, L. 382, 425, 558, 718, 858
Wang, M. 702
Wang, N. 694
Wang, P. 239, 243, 247, 830
Wang, Q. 24, 104, 409, 930
Wang, R. 5, 112, 409, 413
Wang, S. 283, 949
Wang, W. 394
Wang, X. 160, 164, 168, 347, 542, 578, 925, 957
Wang, Y. ... 16, 72, 92, 299, 311, 343, 398, 453, 475, 479, 519,
    530, 578, 590, 594, 653, 797, 810, 846, 866, 909, 913
Wang, Z. 13, 16, 371, 471, 487, 521, 594, 641
Wei, C. 299
Wei, J. 32, 310, 449, 682
Wei, W. 16, 594
Wei, X. 231
Wei, Y. 92, 409
Wen, L. 558
Wen, X. 461
Wen, Y. 108
Weng, Z. 219, 223, 227
Wong, N. 457, 554
Wu, B. 156
Wu, C. 80, 96, 203, 429
Wu, D. 542
Wu, F. 818
Wu, H. 5, 641, 737
Wu, J. 80, 96, 160, 164, 168, 175, 187, 203, 215, 231, 319,
    413, 491, 578, 757
Wu, J.H. 826

Wu, K.	471, 626
Wu, L.	140, 698, 741, 834
Wu, N.	108, 374, 511, 630, 657
Wu, Q.	24, 235, 582, 906
Wu, T.	965, 969
Wu, W.	68, 227
Wu, X.	144, 586, 801, 838
Wu, Y.	76, 88, 116, 172, 175, 785, 793, 886
Wu, Z.	622, 741
Xia, H.	906
Xia, J.	586
Xia, Y.	80, 96, 203
Xiang, X.	108
Xiao, J.	749, 753
Xiao, L.	618
Xiao, P.	363
Xiao, Q.	394
Xiao, S.	315
Xiao, Y.	862, 909
Xiao, Z.	938
Xiaoqiang, L.	671
Xie, T.	112
Xie, X.	227
Xie, Y.	323
Xing, H.	231
Xiong, L.	538, 606
Xiong, Q.	930
Xiong, S.	279, 641
Xu, C.	917
Xu, G.	104
Xu, H.	347, 351, 413, 503, 610, 622, 761, 785, 793
Xu, J.	351, 503, 610, 618
Xu, K.	88, 275
Xu, L.	136
Xu, M.	295, 479
Xu, P.	52, 56, 586
Xu, T.	283, 453, 538, 866, 917
Xu, W.	894
Xu, X.	534
Xuan, Z.	550
Xue, X.	148, 538, 606
Yamano, T.	331
Yan, B.-P.	902
Yan, C.	570
Yan, H.	235, 291, 401
Yan, N.	347, 761
Yan, X.	453
Yan, Z.	60

Yang, C.	850
Yang, G.	160, 164, 660
Yang, J.	104, 534, 737
Yang, L.	108
Yang, S.	283, 453, 866, 917
Yang, W.	378
Yang, X.	199, 243
Yang, Y.	140, 745, 946
Yang, Z.	16, 594, 618, 913
Yao, F.	355, 965, 969
Yao, R.	870
Yao, S.	718
Yao, Y.	120, 586
Ye, B.	633, 769
Ye, F.	690, 925
Ye, G.	921
Ye, H.	247, 706, 805
Ye, R.	602
Ye, S.	495
Ye, W.	602
Ye, Y.	737
Yeo, K.S.	76
Yi, H.	534
Yi, T.	36, 323, 495
Yin, J.	878
Yin, M.	953
Yin, P.	275
Yin, R.	271, 371, 614, 678, 715, 854, 942
Yin, S.	445
Yin, T.	374
Yin, W.	144, 838
Yin, X.	44
Yin, Y.	711
Yoshikawa, K.	433
You, J.	44
You, X.	890
You, Z.W.	826
Yu, B.	445, 461, 793
Yu, C.	534
Yu, F.	28
Yu, H.	68
Yu, Q.	367, 471, 499
Yu, S.	68, 108, 973
Yu, W.	417, 461
Yu, X.	28, 199, 822
Yu, Y.	80, 96, 203, 471
Yu, Z.	445, 530
Yuan, L.	614, 618

Yuan, X. 1
Yuan, Y. 761
Yue, Z. 562
Yueng, C.Y.A. 64
Zeng, J. 765
Zeng, L. 702
Zeng, X. 148, 538, 606, 618, 645
Zeng, Z. 72
Zhai, D. 40
Zhai, Q. 909
Zhan, W. 582
Zhan, X. 160, 164, 168, 175, 187, 215, 231, 578
Zhang, B. 52, 56, 60, 72, 335, 378, 382, 507, 722, 733, 930
Zhang, C. 335
Zhang, D.W. 16, 80, 96, 203, 594
Zhang, F. 574
Zhang, G. 614, 711
Zhang, H. 104, 199, 287, 542, 637, 830, 862
Zhang, J. 36, 371, 405, 566, 657, 854, 878, 882, 886
Zhang, J.F. 1
Zhang, K. 773
Zhang, L. 68, 398, 793
Zhang, P. 530
Zhang, Q. 40, 818
Zhang, R. 148, 921
Zhang, S. 295, 718
Zhang, T. 471, 487, 499
Zhang, W. 1, 191, 195, 822
Zhang, W.J. 48, 64
Zhang, X. 36, 140, 172, 530, 534, 698, 711, 722, 733, 741, 953
Zhang, Y. 100, 104, 235, 287, 291, 295, 303, 311, 343, 347, 367, 457, 475, 515, 542, 558, 618, 711, 866, 930, 961, 965, 969
Zhang, Z. 16, 112, 263, 279, 367, 471, 499, 594, 894, 938, 949
Zhao, B. 394
Zhao, C. 120, 267, 660, 818
Zhao, G. 890
Zhao, H. 136, 355
Zhao, J. 726
Zhao, K. 36, 279, 421
Zhao, W. 120
Zhao, X. 773
Zhao, Y. 40, 207, 219, 223, 227, 279, 570, 653, 822
Zhao, Z. 144, 801
Zhen, S. 378, 382
Zheng, B. 487

Zheng, C. 789
Zheng, J. 359
Zheng, L. 842
Zheng, S. 445
Zheng, X. 168, 534, 578
Zheng, Z. 530
Zhong, Z. 874
Zhou, C. 637
Zhou, J. 437, 626, 649, 667
Zhou, P. 13, 437
Zhou, Q. 148
Zhou, R. 530
Zhou, T. 546
Zhou, W. 457, 550, 554, 894
Zhou, X. 1, 507, 801
Zhou, Y. 160, 164, 578
Zhou, Z. 60, 251, 291, 295, 335, 521, 546, 781
Zhu, C. 124, 949, 973
Zhu, K. 88, 718, 870
Zhu, R. 645
Zhu, T. 409, 649
Zhu, X. 475, 801, 838, 949, 973
Zhu, Y. 830, 878, 894
Zhu, Z. 515, 834, 842
Zhuang, Q. 307
Zhuge, F. 660
Zou, Z. 842

2025 IEEE 16th International Conference on ASIC (ASICON 2025)

Kunming, China
21-24 October 2025

Pages 491–976

IEEE Catalog Number: CFP25442-POD
ISBN: 979-8-3315-3918-4

Copyright © 2025, IEEE

All Rights Reserved

Copyright and Reprint Permissions:

Abstracting is permitted with credit to the source. Libraries are permitted to photocopy beyond the limit of U.S. copyright law for private use of patrons those articles in this volume that carry a code at the bottom of the first page, provided the per-copy fee indicated in the code is paid through Copyright Clearance Center, 222 Rosewood Drive, Danvers, MA 01923.

For other copying, reprint or republication permission, write to IEEE Copyrights Manager, IEEE Service Center, 445 Hoes Lane, Piscataway, NJ 08854. All rights reserved.

*** This is a print representation of what appears in the IEEE Digital Library. Some format issues inherent in the e-media version may also appear in this print version.

IEEE Catalog Number:	CFP25442-POD
ISBN (Print-On-Demand):	979-8-3315-3918-4
ISBN (Online):	979-8-3315-3917-7
ISSN:	2162-7541

Additional Copies of This Publication Are Available From:

Curran Associates, Inc
57 Morehouse Lane
Red Hook, NY 12571 USA

Phone:	(845) 758-0400
Fax:	(845) 758-2633
E-mail:	curran@proceedings.com
Web:	www.proceedings.com

TABLE OF CONTENTS

Generation, Modulation and Application of Spintronic Markov Chain Signal 1
    Xihui Yuan, Jiajia Jian, Zheng Chai, Xue Zhou, Weidong Zhang, Jian Fu Zhang, Tai Min

Opportunities for Advanced Logic Technology with Dual-sided Integrations: From Lateral to Vertical
Transistors 5
    Yanbang Chu, Yu Liu, Runsheng Wang, Ming Li, Heng Wu

Si Hybrid Tunnel FET-CMOS Foundry Platform for Ultra-low-Power Circuit Applications 9
    Qianqian Huang, Kaifeng Wang, Ru Huang

Si-MoS₂ Heterogeneous CFET for Ultra-low Power Logic Technology Scaling 13
    Zehua Wang, Wenzhong Bao, Peng Zhou, Jing Wan

Impact of Off-state Stress on the Reliability of 14nm nFinFETs 16
    Wendi Wei, Kun Chen, Chen Wang, Yaolin Wang, Zhao Yang, Zhiteng Zhang, Zhuming Wang
    Qingqing Sun, David Wei Zhang

Performance Comparison Between Bulk-Si and FDSOI Nanosheet GAAFETs 20
    RS.He, BX.Gan, S.Cristoloveanu, Y.Xu, J.Wan

Improving EUV Patterning Fidelity and Aberration Control through Source-Mask Co-Optimization 24
    Qi Wang, Qiang Wu, Ying Li, Xianhe Liu, Yanli Li

Enhancement of HfO₂-Based Ferroelectric Thin Film Performance via Interface and Defect Engineering 28
    Xiao Yu, Peiyuan Du, Huan Liu, Dongya Li, Fei Yu, Bing Chen, Ran Cheng, Mengnan Ke, Yan Liu
    Genquan Han

Threshold Voltage Swing Caused by Intense Phonon-Electron Interaction in High-k Dielectrics 32
    Jinchen Wei, Mansun Chan

A Ti/ITO Bilayer Gate Electrode Strategy for Improving Subthreshold Swing of Oxide Transistors 36
    Chuanlin Sun, Tingchen Yi, Han Gao, Jiakang Zhang, Junchen Dong, Kai Zhao, Dedong Han, Xing Zhang

Fabrication of High-Performance β-Ga₂O₃ MOSFETs via Ohmic Contact Optimization 40
    Hui Li, Qihao Zhang, Haodong Fu, Jianguo Li, Dongyuan Zhai, Yi Zhao, Jiwu Lu

Experimental Study on 1.2kV/40mΩ SiC MOSFET with Integrated JBS Diode 44
    Moufu Kong, Qizhi Feng, Hongfei Deng, Yufeng Dong, Wei Han, Xuequan Yin, Jiakai You

A Review of Active Gate Drivers for SiC Power MOSFETs 48
    Yuchu Ge, Wei Jia Zhang

A Quadruple RESURF LDMOS with Enhanced Hot-Carrier-Induced Degradation Immunity 52
    Wenliang Liu, Ming Qiao, Penglong Xu, Chunxia Ma, Feng Lin, Bo Zhang

Investigation of Dual-Mode R_{on} Degradation Mechanisms in LOCOS-Based LDMOS 56
    Wenliang Liu, Ming Qiao, Penglong Xu, Chunxia Ma, Feng Lin, Bo Zhang

Advanced Gate Driver Solutions for Fast-Switching SiC Power Device Applications 60
    Yu Qing, Zhihao Yan, Zekun Zhou, Jiaxing Mao, Zijun Zhou, Yun Dai, Rongxing Lai, Yue Shi, Bo Zhang

Design Considerations for Smart Gate Drivers 64
  Wai Tung Ng, Jingyuan Liang, Wentao Cui, Chun Yin Au Yueng, Namjee Kim, Wei Jia Zhang

Investigation of Threshold Voltage Instability in GaN HEMTs Using Rapid Ramp Sweeping Technique 68
  Diangang Hu, Yutian Gan, Shufu Yu, Sai Liu, Lirong Zhang, Weijing Wu, Hongyu Yu

Design and Kirk Effect Improvement of 30V NLDMOS Base on 0.18µm BCD Platform 72
  *Qi Ding, Ning Ning, Renxiong Li, Jun Huang, Yutuo Guo, Yu Wang, Kunqin He, Yaxin Liu, Ziyi Zeng
  Ming Qiao, Lulu Peng, Bo Zhang*

Device Modeling Based on Residual Neural Network with Ensemble-Based Active Learning 76
  *Hongfei Su, Yutong Wu, Jithish Jayarajan, Bharatha Kumar Thangarasu, Nagarajan Mahalingam
  Fanyi Meng, Kaixue Ma, Kiat Seng Yeo*

Performance Benchmark of Gate-All-Around Nanosheets Transistors Based on DTCO Simulation 80
  *Chunlei Wu, Jian Ma, Hanzhi Gu, Yueyuan Yu, Yiming Xia, Jiayi Wu, Boqian Shen, Qingqing Sun
  David Wei Zhang*

The impact of Back-Gate biasing and layout on temperature sensitivity of transistors in FD-SOI CMOS
technology 84
  Yann Deval, Maxime Guillot, Hervé Lapuyade, François Rivet

Considerations for Low Temperature High-Performance Computing: Design Technology
Co-optimization, Device Variation, and Hot Carrier Injection 88
  *Yuanxi Liao, Keyang Zhu, Ran Tao, Yongyu Wu, Qinglan Jiang, Yaxiong Liu, Kai Xu, Chen Shen, Dong Liu
  Dawei Gao, Ran Cheng*

Feature clustering-driven data augmentation in multi-level hotspot detection for integrated circuits
based on GAN 92
  *Pengyu Ren, Bojie Ma, Yajuan Su, Xiaojing Su, Xin Hong, Yuqin Wang, Yujie Jiang, Zhanzi Chen
  Tianao Chen, Yayi Wei*

A New TCAD Simulation Framework for Strain-Aware Quantum Tunneling Current Modeling 96
  Jian Ma, Chunlei Wu, Hanzhi Gu, Yueyuan Yu, Yiming Xia, Jiayi Wu, Qingqing Sun, David Wei Zhang

DTCO-based Hybrid Rail 8T Complementary FET SRAM Design towards advanced node 100
  Yutian Zhang, Khawar Sarfraz, Mansun Chan

Design of VCM Motor Coil Based on Five Factor Integration 104
  Shengxian Quan, Huihong Zhang, Yuejun Zhang, Qiang Wang, Guanglong Xu, Jinsheng Yang

Full-spiking Bio-inspired Target Detection Vision Algorithm based on Gating Attention Prediction for
DVS and SPAD Sensors 108
  Lengjun Yang, Xingyu Xiang, Yiyao Wen, Jian Liu, Nanjian Wu, Liyuan Liu, Shuangming Yu

The Quest for Reliable AI Accelerators: Cross-Layer Evaluation and Design Optimization 112
  Meng Li, Tong Xie, Zuodong Zhang, Runsheng Wang

A Lightweight Hardware Defense Against DSE-Based Trojans in NN Accelerators 116
  Yujing Wu, Chao Guo, Youhua Shi

SpykSim: A Cycle-Level Full-System Simulator for Systolic SCNN Accelerators 120
  Wanwan Zhao, Yichu Yao, Qiang Niu, Qian Li, Chen Zhao

A Data-Efficient Deep Reinforcement Learning Algorithm and FPGA Accelerator for Real-Time Robot Motion Control Applications 124
Wenhao Huang, Rao Fu, Aiwu Ruan, Huiyun Li, Chongyang Zhu

GraphFlow-PIM: Annotated Execution Graphs of DNN Workloads across Diverse PIM Configurations 128
Syeda Munazza Marium, Song Chen

Optimizing LLM inference for FPGAs 132
Jorge R De Freitas, Jose G. F. Coutinho, Ce Guo, Suleyman Demirsoy, Wayne Luk, Zhiqiang Que

Fine-Grained Layer Scheduling and Mapping for Chiplet-Based LLM Inference 136
Hongyang Gu, Lei Xu, Haochen Zhao, Naifeng Jing

A 16×16 High-Utilization Systolic Array Hardware Accelerator for Long-Sequence Flash-Attention Computation in Transformer 140
Zhenkun Li, Liji Wu, Yi Yang, Tianling Ren, Le Wu, Xiangmin Zhang

Sparse Approximation of Softmax: Hardware-Efficient Acceleration for Long Sequence Inference 144
Lanqi Ma, Zifeng Zhao, Xiaoxing Wu, Gengsheng Chen, Wenbo Yin

A Hybrid Processing-in-Memory and Computing-in-Memory Architecture for Large Language Model Inference in Edge Devices 148
Yujia Sun, Ruicong Zhang, Yuanfeng Chen, Qiang Zhou, Xiaoyong Xue, Xiaoyang Zeng

MCDC: A Memory-efficient and Computation-efficient Architecture for Deformable Convolutions 152
Zhiyi Shu, Xinhua Shi, Jun Han

Hardware-Efficient Lightweight Feature Map Compression for Convolutional Neural Networks 156
Bing Wu, Shan Cao, Zhiyuan Jiang

A Study on Dwell Time Impacts in Charge-trapping 3D NAND Flash Memory 160
Yining Zhou, Ruidong Li, Xuepeng Zhan, Guangkuo Yang, Yujiao Ding, Xinghao Wang, Pengpeng Sang Peng Guo, Jixuan Wu, Jiezhi Chen

Access Mode Impacts on 3D Charge-trapping (CT) QLC (4bit/cell) Raw NAND Chip 164
Guangkuo Yang, Ruidong Li, Yining Zhou, Yujiao Ding, Xinghao Wang, Pengpeng Sang, Peng Guo Xuepeng Zhan, Jixuan Wu, Jiezhi Chen

The Influence of Radiation on Reliability of Cold Data in 3D CT NAND Flash Memory 168
Haitao Dong, Xinghao Wang, Haotian Li, Xuesong Zheng, Pengpeng Sang, Xuepeng Zhan, Jixuan Wu Jiezhi Chen

Process Co-Optimization of Void Suppression in ULK Dielectric Layers for 28 nm RRAM Arrays Towards High-density Integration 172
Zhenchao Sui, Yanqing Wu, Xing Zhang

A Study on Performance Enhancement of TiO_2/HfO_2 Memristors through Rapid Thermal Annealing 175
Yifan Wu, Yuzhe Hu, Yuewei Qu, Pengpeng Sang, Jixuan Wu, Xuepeng Zhan, Jiezhi Chen

Investigation of Self-Heating Effects in InGaZnO Vertical Channel Transistors for DRAM Application 179
Zhuoran Kong, Yizhan Liu, Jinfeng Kang, Xiaoyan Liu

Broadband Characterization of Ferroelectric Domain Switching Hysteresis Loops in $TiN/Hf_{0.5}Zr_{0.5}O_2/TiN$ Thin-film Capacitors 183
Wen Di Zhang, Xian Yu Hu, An Quan Jiang

On the Reliability of Sub-10nm Ultra-thin Ferroelectric HZO Thin Film 187
Xiaopeng Li, Yang Feng, Pengpeng Sang, Xuepeng Zhan, Jixuan Wu, Jiezhi Chen

High Dielectric Permittivity in Size-scaled $Hf_{0.5}Zr_{0.5}O_2$ Thin-film Capacitors 191
Wendi Zhang, Anquan Jiang

1.2 V Operation of $Hf_{0.5}Zr_{0.5}O_2$ Ferroelectric Thinfilm Capacitors for Low-Power ASICs after
Interfacial-layer Engineering 195
Xianyu Hu, Di Hu, Wendi Zhang, Bowen Shen, Anquan Jiang

Reliability Enhancement of $Hf_{0.5}Zr_{0.5}O_2$-Based Ferroelectric Capacitors via Argon Plasma Treatment 199
Hongrui Zhang, Xiu Yang, Rongzong Shen, Yian Ding, Haoji Qian, Jiajia Chen, Chengji Jin, Ran Cheng
Bing Chen, Xiao Yu, Yan Liu, Genquan Han

Cryogenic Ferroelectricity (10-298 K) of Superlattice and Solid-solution HZO Films 203
Yiming Xia, Chunlei Wu, Jian Ma, Hanzhi Gu, Yueyuan Yu, Jiayi Wu, Qingqing Sun, David Wei Zhang

Systematic Review of Write Reliability in Spin-Transfer Torque Magnetic Random-Access Memory 207
Yuhao Chen, Yiming Qu, Ziyuan Chen, Choonghyun Lee, Yi Zhao

Emerging Magnetoresistive Memories 211
Viktor Sverdlov, Nils Petter Jørstad, Bernhard Pruckner, Mario Bendra, Wolfgang Goes
Siegfried Selberherr

Comprehensive Characterizations of Polarization Switching Dynamics in HfO_2-based FRAM across a
Broad Temperature Spectrum 215
Yilin Hou, Jixuan Wu, Xiaopeng Li, Xiaoyu Dou, Yaoyu He, Pengpeng Sang, Xuepeng Zhan, Yuqi Gao
Linhui Hu, Feng Wang, Yushi Hu, Qian Tao, Jiezhi Chen

WO_x interlayer employed to improve the imprint effect on $HfZrO_2$ ferroelectric capacitors 219
Zibo Dong, Zeping Weng, Jianguo Li, Lijian Chen, Ziyuan Chen, Yi Zhao, Daolin Cai

Understanding the Physical Mechanism of Endurance Cycling in Antiferroelectric Memories 223
Y. Qu, Y. Huo, Y. Ding, L. Chen, Z. Weng, Y. Zhao

Accelerated polarization switching speed and durable endurance enabled by confined domain size and
solid defect migration barrier in FE/AFE multilayer stacked $Hf_xZr_{1-x}O_2$ ferroelectric capacitor 227
Wenhao Wu, Xinyu Xie, Zeping Weng, Yi Zhao, Jiabin Qi

A High-Speed Dual-Entropy Sources True Random Number Generator Implemented on FPGA 231
Yizhi Liu, Jierui Liao, Hao Xing, Pengpeng Sang, Jixuan Wu, Jiezhi Chen, Xuepeng Zhan, Xiangye Wei

A Lightweight Arbiter PUF Design Based on Threshold Loss in Transmission Gates 235
Haoxuan Yan, Yitian Su, Qiwen Wu, Yong Ding, Hui Li, Yuejun Zhang

Destruction-Free Soft PUF Architecture: Merging Security and Efficiency in 4T2M TCAM Without
Data Migration 239
Shimao Ren, Pengjun Wang, Bo Chen, Zhenhong Chen

An RRAM-Based Soft PUF Achieving Near-Zero BER through Skewed Voltage Masking 243
Xinrong Yang, Pengjun Wang, Cailong Jin, Yixin Lu

A Performance Enhancement Strategy for Strong PUF Circuits to Improve IoT Authentication Security 247
Dong Lu, Pengjun Wang, Gang Li, Hao Ye

A Sequential Obfuscation PUF Resistant to Machine Learning Attacks Based on AES Key Expansion 251
Xuejiao Ma, Yimeng Jin, Shuyang Ren, Ziyu Zhou

Hardware-Efficient Doppler Estimation and Compensation in PDSCH for 5G Non-Terrestrial Networks 255
Chih-Chen Chen, Yi-Shan Huang, Chung-Lun Tu, Shyh-Jye Jou

FPGA Bitstream Modification Attacks on CRYSTALS Kyber 259
Lei Chen, Jiahao Lu, Tianze Huang, Aobo Li, Shengfei Gu, Ang Hu, Dongsheng Liu

BIND: A Batch Cache-Invalidation Framework Based on Doorbell Mechanism 263
Jialin Liu, Zhiyuan Zhang, Chao Fu, Jun Han

High-Throughput Multiplier-Free FPGA Implementation for Pure-Number Discrete Fractional Complex
Hadamard Transform 267
Chengqi Zhao, Zi-Chen Fan, Shan Cao, Susanto Rahardja

Design and Implementation of a Bilateral Filtering Accelerator Based on RISC-V 271
Zhengyao Shi, Yushan Dai, Angyang Li, Jian Mei, Lei Deng, Rui Yin

Skip-Zero Strategy: A Latency and Power Optimization for SRT Divider 275
Ke Xu, Ping Yin, Jun Han

A High-precision Stochastic Computing Multiplier with Co-optimization of Area and Latency 279
Qiang He, Yudi Zhao, Zhihuai Zhang, Xiaofei Nie, Shisheng Xiong, Kai Zhao

A Low-Overhead Fault-Tolerant Design for Quantized CNN Accelerators 283
Shanqiang Yang, Chenxu Wang, Lexiang Shen, Xinlei Su, Min Luo, Tianliang Xu, Ruoshi Li, Siyuan Wang

A Real-Time and Reconfigurable Pre-Driver Design for ABS Solenoid Valve Applications 287
Zhinan Li, Yitian Su, Shaochen Han, Huihong Zhang, Yuejun Zhang, Cang Liu

Real-Time Highly Flexible Wheel Speed Sensing Interface IP Design 291
Yitian Su, Zhinan Li, Haoxuan Yan, Zhenkai Zhou, Yuejun Zhang, Cang Liu

Design of Secure Storage Circuit Based on Reversible Logic XOR-Toffoli Gate 295
Yiting Guo, Yuejun Zhang, Shutong Zhang, Mengfan Xu, Zhenkai Zhou, Hui Li

A Hierarchical Approximate Floating Point MAC Unit with Precision-Adaptive Self-Configuration 299
Xianghui Fu, Yike Wang, Chaojie Wei, Yu Gong

High-Performance Radiation-Hardened Flip-flop for Reliable Systems 303
Jie Li, Xiaoming Teng, Yufeng Zhang

Analysis and Design of Regulating Rectifier with Multiple Outputs for Wirelessly Powered Biomedical
Devices 307
Quanrong Zhuang, Junyi Sun, Bo Li, Jie Lu, Yi Shi, Hao Qiu

Noise Notch Frequency Design for EMI Mitigation in DC-DC Converters Using Digital-to-Time
Converter 310
Yasunori Kobori, Yifei Sun, Guiyi Dong, Nobukazu Tsukiji, Ramin Khatami, Takuya Arafune
Shogo Katayama, Anna Kuwana, Jianglin Wei, Haruo Kobayashi

A 240nA-1μA Quiescent SIMO Converter Featuring 3mV Undershoot under 30mA/μs Transients 311
Yuhua Chen, Qianhui Liu, Yixing Wang, Yimeng Zhang, Yuming Zhang

A 280-nA, 85.8% Efficiency Boost Converter with Optimal Inductor Current in Burst Mode for Brain Stimulation 315
Dejian Li, Xin Jin, LianXi Liu, Gang Dong, Xufeng Liao, Shihao Xiao, Xincai Liu

A High-Efficiency Low-Ripple Buck Converter with Adaptive Load Frequency Control 319
Tao Ren, Xufeng Liao, Gefu Wang, Jiatong Wu, Lianxi Liu

A 400V High-speed Level-Shifting Gate Driver with Adaptive Signal-Path Disconnection for 278V/ns dv/dt Immunity in Soft-Switching Converters 323
Yile Xie, Hanyu Shi, Ting Yi, Zhiliang Hong

GaN-based complementary logic sawtooth generator for smart power ICs 327
Yutao Geng, Ji Shu, Tao Chen, Yan Cheng, Yat Hon Ng, Kevin J. Chen

Battery Charger Designs for Low-Voltage Energy Harvesting Based on the Return-on-Investment Concept (Invited) 331
W. Saito, A. Higuchi, T. Yamano, T. Tanzawa

High-Efficiency Energy Extraction Interface for Piezoelectric Energy Harvesting 335
Chenghao Zhang, Junkai Chen, Jingjie Huang, Yue Shi, Zekun Zhou, Bo Zhang

PWM Scheme Selection Strategy for Fast Ramp-Up DC-DC Boost Converters in SSD Applications 339
Yuji Kanayama, Toru Tanzawa

A 98.5% Efficiency Single-Mode Buck-Boost Converter with All-1.8-V-Switch and Non-Stopping Output Current Delivery 343
Qianhui Liu, Yuhua Chen, Yixing Wang, Yuming Zhang, Yimeng Zhang

Design of a Fully Integrated Low Dropout Linear Regulator with Bandgap Reference 347
Fan He, Jiao Liu, Yiyun Mao, Haoyuan Gao, Xianhui Wang, Yubing Zhang, Hao Xu, Na Yan

A 455mV-Hysteresis, 120 nA, Bandgap less Power-on-Reset Circuit for IoT in 40nm CMOS 351
Mingzong Lin, Chaoran Chen, Jian Xu, Yue Lin, Wei Li, Hongtao Xu

An Anti-Single Particle Effect Over Temperature Protection Circuit Based on Dual Detectors 355
Ping Luo, Hong Zhao, Hao Wang, Fulin Yao, Kai Luo

A Fast Start-Up and Low-Power 32-kHz Crystal Oscillator for Real-Time Clock and Frequency Calibration 359
Jie Zheng, Qiang Li, Hao Min

A 0.067 mm² PNP-Based Temperature Sensor with $\pm 0.6°C$ (3σ) Inaccuracy from $-20°C$ to $80°C$ 363
Letian Li, Peilin Xiao, Xuyang Lu

A Cryogenic Voltage Reference with Diode-Based Sensing and Substrate Resistor Compensation Compensation in 180-nm CMOS Process 367
Yixin Zhang, Hanze Liu, Jing Li, Zhong Zhang, Ning Ning, Qi Yu

77K Modeling and Implementation of a Cryogenic OTA for Infrared Sensors 371
Zhuokai Wang, Lei Deng, Rui Yin, Jian Mei, Jiaming Zhang, Zhicheng Shi

Design and Analysis of PI Controller for Resonant Drive Circuits with AGC-PI Architecture 374
Yichen Lu, Tao Yin, Ying Liu, Jian Liu, Nanjian Wu, Liyuan Liu

Design of 11MHz Isolated Current Sense Amplifier Based on FDDA and Current Feedback Frequency Modulation Loop 378
Xinghong Chen, Jiahui Liu, Shaowei Zhen, Hongwei Shen, Wei Yang, Yongwang Ma, Bo Zhang

A Charge Pump Powered Current Sense Amplifier with -20 V to 40 V Input Common-Mode Range 382
Dejian Li, Hongwei Shen, Jinzhao Li, Jiahui Liu, Lixing Wang, Shaowei Zhen, Bo Zhang

The Sequency Domain: a new Approach for Radio Frequency Front End 386
François Rivet, Pierre Ferrer, Maxandre Fellmann, Nathalie Deltimple, Hervé Lapuyade, Eric Kerhervé Yann Deval

300-GHz Phased-Array Transceiver in 40-nm CMOS with Interpolated Feeding and OTA Metrics 390
Minoru Fujishima

A Polar-Modulation OFDM Backscatter System for Passive IoT Communication 394
Qijing Xiao, Xin Hu, Weixiao Wang, Yuxuan Luo, Bo Zhao

A Compact Q/V Band Bidirectional Phase Shifter with 0.32° Phase Error 398
Congrui Li, Yan Wang, Lei Zhang

A 300GHz Coherent Radiator Array with Multi-functional Antenna in 65nm CMOS 401
Houyi Yan, Kaizhe Guo

Design of a 300GHz Wideband On-Chip Antenna in 28nm CMOS 405
Jinghao Zhang, Chen Jiang

ATSim: A Fast and Accurate Simulation Framework for 2.5D/3D Chiplet Thermal Design Optimization 409
Qipan Wang, Tianxiang Zhu, Jiajia Cui, Yicheng Wei, Linxiao Shen, Zhe Cheng, Runsheng Wang Ru Huang, Yibo Lin

Radio Frequency Integrated Circuits Generated by AI-based Design Automation 413
Ruoyu Wang, Meijun Hou, Jun Wu, Hongtao Xu, Ye Lu

Advancing Sparse Matrix Solvers via Exploring More Parallelism and Random Sketching 417
Wenjian Yu, Jiawen Cheng, Baiyu Chen

Snow Ablation Optimizer Accelerator Based on High Level Synthesis 421
Maoshuo He, Renjing Hou, Zirui Li, Kang Zhao

An MLIR-Based Framework for Efficient Dynamic Circuits Generation 425
Yuxuan Guan, Jiangnan Li, Lingli Wang

HybridEPP: Hybrid Numerical and Symbolic Error Probability Propagation in Logic Network 429
Gaopeng Shen, Chang Wu

Success-Rate Improvement of Analog Circuit Topology Generation by Large Reasoning Model 433
Koutaro Hachiya, Kentaro Yoshikawa, Atsushi Kurokawa

Fast Thermal-driven 3D Fixed-outline Floorplanning By Learning-based Thermal Analysis 437
Yikai Liu, Jindong Zhou, Jiayi Li, Pingqiang Zhou

Systematic design for coupled heterogeneous accelerators 441
Tim Todman, Wayne Luk

TorchLitho 2.0: Differentiable Lithography Simulation Engine for Large-Scale Layouts 445
Shuo Yin, Su Zheng, Ziyang Yu, Bei Yu

Hierarchical Residual Fitting for Enhanced S-Parameter Accuracy in Devices Exhibiting Complex Delay 449
Jiaxin Wei, Haonan Wang, Ting-Jung Lin, Lei He

Hybrid Model-Based Hardware Acceleration for Diesel Engine NOx Emission Prediction 453
Xinlei Su, Shanqiang Yang, Tianliang Xu, Xiaozhen Yan, Jianfeng Li, Tian Rong, Chenxu Wang Yuhang Wang, Zhiwei Han

Extending Straight-Through Estimation for Robust Neural Networks on Analog CIM Hardware 457
Yuannuo Feng, Wenyong Zhou, Yuexi Lyu, Yixiang Zhang, Zhengwu Liu, Ngai Wong, Wang Kang

A Parallel Level-Set Based Approach for Etching Topography Simulation in Process Emulation 461
Yin Cheang Ng, Xin Wen, Boyuan Yu, Wenjian Yu

Some Signal Processing Techniques for Testing Wireless Communication LSIs 465
Koji Asami

Voltage-Domain vs. Time-Domain: Trade-offs in High-Speed Applications 466
Haoyu Li, Sai-Weng Sin, Rui P. Martins, Mingqiang Guo

A Pitch-Matched Transceiver ASIC with Element ADC and Continuous-Time Gain Compensation for 3D Ultrasound Probes 471
Jing Li, Tianci Zhang, Li Dai, Yingchen Liu, Jinlai Fu, Zhongshan Wang, Penghao Jiang, Yihu Yu Zhong Zhang, Kejun Wu, Ning Ning, Qi Yu

A Low-Power-Consumption Capacitance to Digital Converter with Novel Calibration Technology 475
Xiwen Zhu, Yufeng Zhang, Xiaoming Teng, Yihan Wang

A 10-bit 4 GS/s 67.79-dBc SFDR Switched-Capacitor DAC with Reservoir Capacitor-based Reference Generation 479
Yitao Wang, Meng Xu, Qiang Pan, Jize Liu, Yuekang Guo, Jing Jin

A 16-channel Neural Signal Acquisition Analog Front-End with Foreground Calibration for High-Precision Backend SAR ADC 483
Chun Feng, Junfeng Tang, Longhao Chen, Songping Mai, Xian Tang

An 18-bit 1MS/s SAR ADC with Weight-Fitting Digital Calibration and High-Linearity Capacitor Array Design 487
Baoyi Zheng, Guoao Wang, Zongmin Wang, Jin Qian, Bosen Liu, Zhaohang Bing, Tieliang Zhang

A digital front-end self-calibration algorithm for SAR ADC 491
Fuming Liu, Jie Ding, Jiangfeng Wu, Yongzhen Chen

Bitwise Bayesian Optimization for SAR ADC Calibration 495
Yu Shi, Shen Ye, Yihang Luan, Jiahao Wang, Ting Yi

An Area-Efficient C2C SAR ADC with Hybrid Switching Mode for Ultrasound Miniature Probes 499
Tianci Zhang, Jinlai Fu, Li Dai, Dongxu Li, Yingchen Liu, Jing Li, Zhong Zhang, Ning Ning, Qi Yu

A Reconfigurable 9-to-14b 15MS/s 4th-Order NS-SAR ADC with Self-Calibrated Open-loop FIA 503
Chaoran Chen, Mingzong Lin, Jian Xu, Yue Lin, Wei Li, Hongtao Xu

A 16-bit 4-MS/s Deadlock-free Asynchronous SAR ADC Using High-level First Transmission Gate 507
Xiaokun Zhou, Baijie Zhang, Xu Cheng

A Differential SAR-SS ADC with Gain-Scaled Ramp Quantization for High-Speed CMOS Image Sensors 511
Nanbo Chen, Jingyang Chen, Gang Wang, Peng Feng, Jian Liu, Nanjian Wu, Liyuan Liu

A 79.2dB-SNDR 12.5MHz-BW Pipelined SAR ADC with Analog-Domain Gain Error Shaping 515
Qiaoyu Hu, Guolong Fu, Yanbo Zhang, Zhangming Zhu

A Deep Reservoir Computing System based on IGZO Electrical-Double-Layer Transistors 519
M. Han, Y. Chen, H. Cui, Y. Wang, C. Wan

High-density and High-reliability (H²DR) RRAM for Energy-efficient AI Computing 521
Yimao Cai, Yiyun Chen, Lin Bao, Ling Liang, Zheng Zhou, Zongwei Wang

The Digital Coupled Ring Oscillator Ising Machine 525
Yue Han, Ranjith R Unnithan, Robin Evans, Efstratios Skafidas

Nanocrystal-Si Flash Memory-based Engergy-efficient Multi-bit Compute-in-Memory Design for Edge Neural Networks 530
Xianping Liu, Jian Huang, Zihan Zheng, Xinrui Zhang, Ruibin Zhou, Zhiyi Yu, Zhongyuan Ma, Kunji Chen Yuhan Wang, Jian Cheng, Peng Zhang

A Multi-level RRAM-based Ising Machine for Solving Combinatorial Optimization Problems 534
Zhenchao Sui, Xiaoxin Xu, Chengshuo Yu, Jingxin Deng, Xu Zheng, Chengyue Li, Hailan Yi, Jianguo Yang Xing Zhang

DRAM-Centric Near-Data Processing: A Survey of Architectures, Technologies, and Trends 538
Taoran Shen, Yujia Sun, Tingyi Xu, Li Xiong, Xiaoyong Xue, Xiaoyang Zeng

Challenges and Trends of SRAM based Floating Point Computing-in-Memory Circuits 542
Yuchen Tang, Yanqi Zhang, Zhichao Liu, Xing Wang, Defa Wu, Huaiwen Zhang, Yeqi Sun, Xin Si

Mapping of Graph Convolution Network on Sparse-Aware Computing-In-Memory Macros 546
Guoxiang Li, Tianhang Zhou, Xinyu Qu, Zecheng Zhou, Yufei Ma

CDCC: A High-Efficiency SRAM-Based Charge-Domain Compute-in-Memory Macro with Complement Compensation Design for AI Applications 550
Wanting Zhou, Zihao Xuan, Song Chen, Yi Kang

HPD: Hybrid Projection Decomposition for Robust State Space Models on Analog CIM Hardware 554
Yuannuo Feng, Wenyong Zhou, Yuexi Lyu, Hanjie Liu, Zhengwu Liu, Ngai Wong, Wang Kang

ADC-Free RRAM-Based XNOR-Bitcount Architecture for Hand Gesture Recognition 558
Lixun Wang, Yuejun Zhang, Qikang Li, Liang Wen

ESD Reliability Roadmap Considerations for 3D Heterogeneous Integration Microsystems (Invited) 562
Zijin Pan, Xunyu Li, Weiquan Hao, Runyu Miao, Zijian Yue, Albert Wang

Tiny Chiplets Enabled by Packaging Scaling: Opportunities in ESD Protection and Signal Integrity 566
Emad Haque, Pragnya Sudershan Nalla, Jeff Zhang, Sachin S. Sapatnekar, Chaitali Chakrabarti, Yu Cao

Time-dependent Dielectric Breakdown in Advanced MOSFET: From Theoretical Models to Experimental Findings 570
Chu Yan, GuoQiXin Huang, Yiming Qu, Yi Zhao

Mechanical Stress Induced by Temperature Cycling: Impact of MOSFET Placement on Bandgap Reference Voltage Offset 574
Fengbo Zhang, Yancong He, Zhinong Liu, Shuang Jiao, Yang Li, Zhigang Ji

Experimental and Theoretical Study of Single Event Latchup in a 3D TLC NAND Flash Memory Under Heavy Ion Irradiation 578
Xinghao Wang, Haitao Dong, Yujiao Ding, Yining Zhou, Haotian Li, Xuesong Zheng, Yuhang Wang
Pengpeng Sang, Jixuan Wu, Xuepeng Zhan, Chaoming Liu, Jiezhi Chen

A Data Hierarchy-Based Adaptive Testing Method for Integrated Circuit Parameter Sets 582
Kaiming Hao, Yan Li, Xu Cheng, Qiong Wu, Wenfa Zhan, Yujie Huang

Microstructural Evolution and Reliability Analysis of RDL Copper Interconnects under High-Temperature Conditions 586
Peng Xu, Lan Li, Jialu Huang, Yu Yao, Hengchang Bi, Jiang Xia, Zongyi Li, Zuoyuan Dong, Xing Wu

Reliability Screening for Yield Improvement in IC Design Industry: Progress, Challenges and Prospects 590
Yixian Wang, Xiaoxiao Qiu, Zhigang Ji

Impact of Thermal Shock on the Threshold Voltage and Transconductance of FinFET I/O Devices 594
Yaolin Wang, Kun Chen, Wendi Wei, Zhao Yang, Zhiteng Zhang, Zhuming Wang, Chen Wang
David Wei Zhang

Effects of Total Ionizing Dose on ESD Performance in High-Voltage SCR with Double Snapback Characteristics 598
Yujie Liu, Xiangliang Jin

A New Surge Protection Circuit with Low Dynamic Leakage Current 602
Zhiqiang Hu, Ran Ye, Qiao Kang, Ke Cui, Hao Luo, Weipeng Ye, Siyang Liu, Weifeng Sun

Reliability Enhancement in HfO$_2$-Based FeRAM: Circuit-Level Solutions for Insufficient Polarization and Memory Window Degradation 606
Changnan Shi, Taoran Shen, Li Xiong, Shuyang Lv, Yuanfeng Chen, Xiaoyong Xue, Xiaoyang Zeng

A PVT-Tolerant Quick Startup CMOS Crystal Oscillator With Chirp-Assisted Fixed Injection 610
Hao Luo, Yue Lin, Jian Xu, Hongtao Xu

A 20 Gb/s/Wire Short-Reach Simultaneous Bi-Directional Transceiver with DuoBinary Coding for Die-to-Die Interface in 28 nm CMOS 614
Bohui Bai, Fangxu Lv, Zhengbin Pang, Geng Zhang, Ruixiao Kuai, Liangyong Yuan, Ruotian Yin
Jiliang Liu

A 56Gb/s PAM4 Transceiver Based on BSS-LMS Algorithm With 3-Taps Adaptive TX FFE 618
Xianchao Zeng, Fangxu Lv, Liquan Xiao, Jiaqing Xu, Zhouhao Yang, Liangyong Yuan, Cewen Liu
Xiaoyue Hu, Yingjie Zhang

A Low-Power Gm-Boosted VCO with Multi-Transformer in 40nm CMOS 622
Zilong Wu, Bowen Chen, Yue Lin, Hongtao Xu

A 64 Gbps 10 mW 0.0081 mm^2 Inverter-Based CTLE Employing Power-Efficient Split Biasing Topology in 40 nm CMOS 626
Fang Ding, Huzhi Tang, Ke Wu, Yuekang Guo, Jing Jin, Jianjun Zhou

A Low-Power Area-Efficient Serializer for CMOS Image Sensors 630
Jingyang Chen, Nanbo Chen, Gang Wang, Peng Feng, Jian Liu, Nanjian Wu, Liyuan Liu

A 6b 14GHz Phase Interpolator with 2-Stage Injection-Locked Ring Oscillators in 28nm CMOS 633
Danqi Ding, Bingyi Ye, Weixin Gai

Boosting Self-Powered Properties of 2D Material-Based Photodetectors via Asymmetry Engineering 637
Ran Huo, Han Zhang, Yihong Sun, Shijun Ou, Changming Pi, Mansun Chan, Changjian Zhou

Interfacial adhesion enhancement enabled mechanically durable flexible organic optoelectronics 641
Ziqi Wang, Xiangzhe Li, Huimin Wu, Kai Wang, Sixing Xiong, Jin Qian

Image Flare Removal via Stable Diffusion Framework 645
Jiazheng Lian, Ruoxi Zhu, Jiaming Liu, Ming'e Jing, Xiaoyang Zeng, Yibo Fan

AI-Assisted Droplet Splitting on a Parallel-Plate Optoelectrowetting Chip 649
Junyan Tian, Shang Gao, Tengpu Zhu, Enqing Liu, Gaifang Chen, Jia Zhou

A Sub-1mV Voltage-Variation Pixel Power Supply Architecture with Radiation-Hardened Built-In
LDO for Pixel Readout ASIC 653
Lei Li, Jinxiang Wang, Yini Hong, Yuxiao Zhao, Yongsheng Wang

A CMOS Pixel with Gradient-Doped PPD and LOFIC for 1.7 ns Charge Transfer Time and 92 dB
Dynamic Range 657
Tianjing Qiu, Jinglei Du, Junli Zhang, Peng Feng, Jian Liu, Nanjian Wu, Liyuan Liu

Stacked 2D materials Nanopore Sensors 660
Candong Zhao, Qinjie Pan, Guangyi Yang, Peng Cheng, Fuwei Zhuge, Yuhui He

On-chip Contact Angle Sensor Using Coplanar Capacitors for Digital Microfluidic Systems 664
Akira Tsuchiya, Hayato Fukui, Tsubasa Furuta, Toshiyuki Inoue, Keiji Kishine

Selective manipulations of droplets on photo-driven microfluidic chip with virtual electrowetting
channels 667
Gaifang Chen, Enqing Liu, Junyan Tian, Shang Gao, Jia Zhou

A MEMS Rectenna for RF energy harvesting around 2.4GHz 671
Liu Xiaoqiang, Wang Tiancong

Design and Implementation of Shared Storage Communication Architecture for MCCSIP-RAA 674
Longmei Nan, Yu Jin, Yiran Du, Tao Chen, Lin Chen, Yanjiang Liu, Wei Li, Weiquan Sang

A Fully Quantized LeNet-5 accelerator for Edge Computing with Quantization-Aware Training 678
Yushan Dai, Angyang Li, Jian Mei, Rui Yin

A High-Voltage And High-Precision Operational Amplifier 682
Juan Wei, ZongLin Li, HongRui Che, FuMei Liang, DaGang Li

A 12-bit 1MS/s SAR ADC design for high-temperature MEMS accelerometers 686
Yanlin Mo, Min Qi

A Low-Noise Ultrasound Analog Front-End with Low Gain Error Time-Gain Compensation 690
Xiangchen Wan, Fan Ye

Adaptive Frequency Modulation Buck Converter Based on Valley Current Mode ACOT Control 694
Bowen Jiang, Hong Ren, Ningning Wang

A High-Voltage Level Shifter for BMS Chip in EV with 0-80V Input Range 698
Kunning Mao, Liji Wu, Jing Hu, Zhiwei Li, Haifeng Chen, Xiangmin Zhang

A New Circuit for Generating Half of VDD 702
    Li Zeng, Ming Wang, Peng Bo, Zhangwen Tang

A Boost DC-DC Converter with Low Power and High Efficiency for Portable Device Applications 706
    Jing Cao, Bingjie Chen, Hongfei Ye, Jianhua Feng

A Hybrid Complex-Filtering Scheme with High Image Rejection and Efficient Channel Selection for
Low-IF Receivers 711
    Yue Yin, Guanlin Zhang, Haobo Qi, Haodong Lu, Xinbing Zhang, Ziting Feng, Ye Zhang

Design of a Nonlinear Temperature Compensated Bandgap Reference in 55nm Process 715
    Hezhuang Nie, Ningning Li, Jian Mei, Rui Yin

A Novel Light-load Control Method For Switching Converters in Portable Devices 718
    Jie He, Shuyu Zhang, Langyuan Wang, Suyi Yao, Kejia Zhu

Adaptive on-time Control Buck Converter Based on Phase-Locked-loop and Dynamic Calibration of
DC Offset 722
    Xinyu Zhang, Sujuan Liu, Kun Liu, Bingxue Zhang, Yahua Shi

Design of a Fast Transient Response LDO Circuit Based on Transient Enhancement Structure 726
    Xudong Sun, Sujuan Liu, Kun Liu, Junchao Zhao

A 32-MHz FLL-Based RC Oscillator with PVT Compensation Using Frequency Tripler 730
    Ikhwan Kim, Yajie Qin

A Constant On-Time Buck Converter with VCO-based DC Offset Calibration Technique 733
    Bingxue Zhang, Sujuan Liu, Kun Liu, Xinyu Zhang, Yahua Shi

A Low-Power High-Precision Impedance Measurement Circuit Using DC Servo Loop for Closed-Loop
DBS Systems 737
    Ziqi Tan, Yijun Ye, Yutao Mao, Hui Wu, Xiaofei Kuang, Jie Yang, Mohamad Sawan

A Low-Power BJT-Based Thermal Shutdown Circuit with Hysteresis for BMS chip in EV 741
    Zonghuan Wu, Xiangmin Zhang, Liji Wu

Design of CRFF-B Loop Filter Architecture for Wideband Continuous Time Sigma-Delta Modulators
in CMOS 28 nm 745
    Zhihao Hou, Yuqi Fan, Yifei Gao, Chuan Liu, Chuan Qin, Maliang Liu, Yintang Yang

An Open-Loop Residue Amplifier with SSF Structure Achieving 69dBc SFDR for High-Speed and
High-Precision PSAR ADCs 749
    Chengjun Liu, Deng Luo, Hanbing Liu, Chengchao Mou, Bin Liang, Yaqing Chi, Jianjun Chen, Kai Tang
    Jing Xiao, Ming Tao

A 14-bit R-2R DAC with All-Digital Foreground Calibration based on Redundant LSB 753
    Hanbing Liu, Deng Luo, Chengjun Liu, Chengchao Mou, Bin Liang, Yaqing Chi, Jianjun Chen, Kai Tang
    Jing Xiao, Ming Tao

A Multi-Channel Reconfiguration and Combination Technique for Timing Mismatch Calibration in
Time-Interleaved ADCs 757
    Jize Liu, Jinwei Wu, Jiayi Chen, Xinqi Liu, Yuekang Guo, Jing Jin

A 12-bit 620 MS/s Pipelined-SAR ADC with Feedforward Compensation Closed-loop Residual Amplifier in 28 nm CMOS 761

Shuai Liu, Yi Hu, Guoyu Li, Congyang Sun, Yidong Yuan, Hao Xu, Na Yan

A novel RA architecture and digital calibration method for SAR-assisted pipeline ADCs 765

Jieqiong Zeng, Hao Min

An 8-bit 0.4-mW 740-µm² DS Digital-to-Analog Converter in 28nm CMOS with 60.89-dBc SFDR 769

Xiongfeng Bi, Bingyi Ye, Weixin Gai

A K-Band CMOS Switched-Type Attenuator with Temperature Compensation Technique 773

Xiaodong Zhao, Kai Zhang

An Ultra-Wideband 1.5–18.5 GHz MMIC Phase Shifter in 0.25-µm GaAs Technology 777

Bo Fu, Xuan Ding, Xuesong Han, Xiao Ding

A 0.2–7.3-GHz Compact LNA with Super Linearity for 5G NR in 22-nm CMOS Technology 781

Kaiyun Deng, Zan Zhou, Yingqi Liu, Haoyu Dong, Haigang Feng

An Area-Efficient Bi-directional Cascode PA-LNA For 5G NR in 28-nm CMOS 785

Yue Wu, Wei Li, Shijiao Dong, Hongtao Xu

A 223M-235MHz Fully-Integrated Differential Class-E Power Amplifier with 45.5% PAE and 22.8dBm 789

Chaoyang Zheng, Yanxiang Chen, Jianhua Lu, Yan Ma, Zhiliang Hong, Yumei Huang

A 16~46-GHz, >77-dB IRR, Low-Amplitude and Phase-Error IQ Generator with Self-Adaptive I/Q Calibration in 28-nm CMOS 793

Lijiang Zhang, Wei Li, Bowen Yu, Chengzhang Cai, Yue Wu, Bowen Chen, Yue Lin, Hongtao Xu

Impact of Process Parameter Variations on the Random Values of SRAM-Based PUFs 797

JinJin Shao, Ruiqiang Song, Chunmei Hu, Biwei Liu, Bin Liang, Yaqing Chi, Yaohua Wang

MIVO: Operator-Level On-Chip Memory System with Dynamic Bank Scheduling for Many-Core Neural Processing Unit 801

Xinghao Zhu, Zifeng Zhao, Xiaoxing Wu, Gengsheng Chen, Xiaofang Zhou

DyQRA: A Deadlock-free Routing Algorithm for Large-Scale Mesh NoCs 805

Haoxiang Sun, Aoyun Feng, Hongfei Ye, Jianhua Feng

An Enterprise Solid-State Drive Controller Supporting Spin-transfer Torque Magnetoresistive Random Access Memory 810

Chao Song, Qihao Liu, Yunzhe Wang, Rufa Su

An IO Die with Collective-Aware Routing and In-Situ Processing for Data Synchronization in Multi-Chiplet Systems 814

Qi Luo, Chen Mu, Chixiao Chen, Xin Chen, Shiwei Liu

An STT-MRAM Last Level Cache Management Method Based on Write Intensity Prediction for GPUs 818

Yujie Pu, Qiaoran Zhang, Shitong He, Fanchen Wu, Chen Zhao

Towards Scalable and High-Throughput NTT Acceleration On Hybrid-Bonding Architecture 822

Wenxuan Zhang, Yi Sun, Xinglong Yu, Yifan Zhao, Jun Han

A Low-Cost Multiplier-Free Accelerator for Binary Neural Network 826
  Z. W. You, J. H. Wu, R. C. Ma, G. C. Qiao

Design of a MobileNetV2 FPGA Accelerator for Low-Power Real-Time Identification of Plant
Nematodes 830
  Ying Zhu, Pengjun Wang, Qikang Li, Huihong Zhang

A 65nm Analog-Computing Chip With Reconfigurable Charge-Pump-Based Adders for
5.26nJ/Decision Retrainless Keyword-Spotting 834
  Lichen Feng, Rundong Cai, Lin Wu, Zhangming Zhu

HAMP: Head-Aware Mixed-Precision Token Pruning and Quantization for Efficient ASR 838
  Xiaoxing Wu, Xinghao Zhu, Lanqi Ma, Gengsheng Chen, Wenbo Yin

A Compressed Sensing Spiking Neural Network System for Radar-Based HGR 842
  Liyu Qian, Zikai Zhu, Yuhan He, Jie Lu, Yaojie Sun, Lirong Zheng, Zhuo Zou

FlexiCore-DNN: A Configurable and Templated Architecture for End-to-End FPGA Acceleration of
Deep Neural Networks 846
  Rao Fu, Wenhao Huang, Aiwu Ruan, Huiyun Li, Yongqing Wang

A 7-bit 6.25-GHz Low Power High Linearity DPC for CDR Applications 850
  Jingsong Cui, Kai Li, Chengyu Yang, Jiahao Lu, Hao Li, Ang Hu, Dongsheng Liu

Design of Low-Voltage Differential Signaling Driver for Image Sensor 854
  Zhongwei Lin, Ningning Li, Angyang Li, Jian Mei, Rui Yin, Jiaming Zhang, Zhicheng Shi

Exploring The Further Fracturability of Intel ALM 858
  Chenyu Jiang, Xianfeng Cao, Lingli Wang

A General and Modular FPGA Hardware Architecture for Enhanced Scalability and Flexibility 862
  ZiRui Qin, ZhiNan Li, YaBo Xiao, Hui Zhang, Cang Liu

Pipelined Parallel Design of SIFT Algorithm on FPGA 866
  Yuanhao Zhang, Tianliang Xu, Jianfeng Li, Zhenbin Lv, Shanqiang Yang, Chenxu Wang, Yuhang Wang
  Bo Chu, Zhiwei Han

Design and Implementation of an FPGA-based MIPI DSI Interface for Micro-LED Displays 870
  Runfeng Yao, Xinyi Liu, Kaisong Zhu, Jinbo Liang, Zhaojun Liu

A Sub-100μs-Latency Visual-Cortex-Mimicking Heterogeneous Multi-Core Edge Neuromorphic
Processor Enabling On-Chip High-Accuracy Learning 874
  Junxian He, Ying Jiang, Zhengqing Zhong, Mingju Chen, Liyuan Liu, Cong Shi

An Energy-Optimized FPGA Implementation for Convolutional Neural Networks Accelerator 878
  Yujie Zhu, Jianxuan Yin, Jingjing Liu, Jianhua Zhang

A Lightweight Low-Latency Hardware Architecture for Dual Attention Super-Resolution Network 882
  Haocan Jiang, Aiying Guo, Jianhua Zhang, Jingjing Liu

A Scalable Channel-Parallel Accelerator for Spiking Neural Network 886
  Yuchun Wu, Lingling Miu, Jingjing Liu, Jianhua Zhang

A precise current-controlled resistor and its applications in zero-pole tracking frequency compensation for LDO 890

Guanting Liu, Guijuan Zhao, Feng Shi, Xiaohuan You, Shuhai Chen

ASSVD: A Self-Supervised Surgical Video Desmoking Network with Sparse Attention 894

Yinna Zhu, Wanyi Zhou, Zijing Zhang, Gengsheng Chen, Wei Xu

A 71 TOPS/W 24.2 TOPS/mm² 14nm SRAM CIM Macro with a Capacitor-less ADC for Edge AI 898

Zexing Chen, Siyao Jia, Chixiao Chen

Data-Centric Automatic Design Migration of Low Voltage CMOS Bandgap Reference Circuit 902

Shun-Qi Dai, Yuan Lei, Bei-Ping Yan

Innovative Detection Capacitor Utilization in ESD Power Clamp Circuits for HBM Residual Voltage Suppression 906

Zelong Huang, Guangyi Lu, Haoyu Xia, Qi Wu, Haiming Wang

High Efficient Efuse Full Process Burning Solution Based on ATE 909

Qian Zhai, Yichen Xiao, Xin Song, Haobin Wang, Yuyuan Wang, Xuxin Chen

Study of Reliability Screening Method to Improve the DPPM of IC Products 913

Yancong He, Zhiyong Yang, Zhinong Liu, Shuang Jiao, Chuyuan He, Yixian Wang, Zhigang Ji

Weight Bit Sensitivity Analysis and FPRH-Based Hardening Strategy for CNN Accelerators 917

Jinghao Chen, Shanqiang Yang, Tianliang Xu, Congan Xu, Yuehong Gong, Chenxu Wang

An effective method for low-contrast high-noise lithography SEM image contour extraction 921

Ruirui Zhang, Gongyan Ye, Xianhe Liu

Design of A Dual-Mode Analog Front-End Circuit Applied in the Voice Activity Detection System 925

Zirui Dong, Xuhaohan Wang, Fan Ye

Research on Radiation-Hardened High-Voltage Gate Driver Circuit Based on 0.8μm 1200V Bulk Silicon BCD Process 930

Xiaohui Li, Yi Zhang, Qiang Wang, Qiankun Xiong, Bo Zhang, Ming Qiao

Parameter identification of single-phase inverter digital twin system 934

Ao Shen, Hui Li, Jie Kang, Jia Hao Lv

Optimization of Three-dimensional High-k Superjunction under Non-Punch-Through Mode: Theoretical Modeling and Comparison 938

Zhentao Xiao, Chenxing Wang, Zonghao Zhang, Haimeng Huang

Smart Adaptive Perception for High-Precision Lightweight Infrared UAV Detection and Tracking 942

Shiyu Mei, Lei Deng, Rui Yin

Design and validation of fluorescence lifetime solving algorithm for fiber-optic temperature sensor 946

Yuxuan Yang, Xiangliang Jin

A Multi-Cycle Pulse Transfer Timing Scheme for Enhancing Charge Efficiency in CMOS Image Sensors 949

Zhenhao Zhang, Chiang Zhu, Haiyang Liu, Peng Peng, Sikai Wang, Junjie Hao, Xiaona Zhu

Design of RF Microsystem Based on Silicon-based Stereoscopic Integration Technology 953

Xiaoqing Zhang, Lei Shi, Mengmeng Yin, Cui Jing, Dexi Liu

A Novel Pretreatment Approach to High-quality SiO$_2$ Surface Applied for C2W Cu/SiO$_2$ Hybrid Bonding 957
Han Jiang, Xianlong Wang, Ziyu Liu, Yabin Sun

Approximately Timed Scalable DSP Model Based on SystemC 961
Yongwang Qin, Sheng Liu, Yang Zhang, Xing Hu, Chen Shangqian

Microscopic Mechanisms of Bias Temperature Instability Induced by Defects in Si/SiO$_2$/HfO$_2$ Gate Stacks: A DFT and NEGF Study 965
Yantao Huang, Yunzhi Lin, Yixin Zhang, Junlong Li, Xiaoxu Kang, Fengying Yao, Shaojian Hu, Qing Shi Tao Wu

Mechanism of Leakage Current Enhancement Induced by La Doping in HfO$_2$ Gate Stacks: A DFT Investigation 969
Yunzhi Lin, Yantao Huang, Yixin Zhang, Qing Shi, Fengying Yao, Junlong Li, Shaojian Hu, Xiaoxu Kang Tao Wu

Layout-Aware Performance Analysis of the CFET based NAND2 constructed Ring Oscillator 973
Junjie Hao, Chiang Zhu, Huawei Tang, Xiaona Zhu, Shaofeng Yu

A digital front-end self-calibration algorithm for SAR ADC

Fuming Liu [1], Jie Ding [1], Jiangfeng Wu [1], Yongzhen Chen *[1]

[1] School of Electronics and Information Engineering, Tongji University, Shanghai 201804, China.

Email: 2111144@tongji.edu.cn，*Email: yzchen@tongji.edu.cn

Abstract —Capacitance mismatch is the key factor leading to the reduction of conversion accuracy of a Successive Approximation Register Analog-to-Digital Converter (SAR ADC). This paper proposes an improved digital self-calibration technique, the uncalibrated high capacitor is set to the suspended state first, and the low capacitor is used to convert the high capacitance mismatch error cyclically to achieve the mismatch calibration of the high capacitor, so as to solve the capacitance mismatch induced accuracy degradation problem, Based on the digital self-verification algorithm, a 16-bit SAR ADC chip is designed and fabricated by 180 nm CMOS process. The tape-out test results show that the overall conversion accuracy and performance are high. The Effective Number of Bits (ENOB) of the ADC has increased by 3.55 bits, the signal-to-noise ratio (SNDR) and spurious-free dynamic range (SFDR) are increased by 19dB and 42.5dB, respectively, proving that the designed digital self-calibration technology is effective.

Keywords - SAR ADC, Capacitance mismatch, self-calibration.

I. INTRODUCTION

SAR ADC has the advantages of simple structure, low power consumption and high accuracy, and has attracted much attention due to its excellent power efficiency and good compatibility with advanced CMOS technology. However, the existence of non-ideal factors such as capacitance mismatch, charge injection, and clock jitter affects the accuracy of the overall ADC performance, and capacitance mismatch is the key factor. In the case of severe mismatch, the SAR ADC digital domain signal will be significantly distorted and cannot accurately reflect the true value of the input voltage signal, degrading the performance of the overall electronic system.

To control the capacitor mismatch, optimizing the manufacturing process, circuit compensation, and calibration technology are often taken to reduce the impact of ADC error on SAR ADC performance. There are two types of calibration technology, analog calibration and digital calibration. The analog calibration technology is to compensate for the DAC error in analog domain, and the digital calibration technology is to correct the conversion results in digital domain and find out the error rules and correct it through the analysis and processing of a large number of conversion data.

In recent years, several practical calibration techniques have been proposed. Lee pioneered the front-end calibration technique for capacitor mismatch in SAR ADCs. This technique quantifies the mismatch error by measuring additional resistance and incorporates a calibration DAC to address the mismatch in the capacitor DAC [1]. McNeil used digital calibration to mitigate the effects of capacitor mismatch in SAR ADCs [2]. Ku proposed to use residual integral noise averaging and digital domain capacitance error correction techniques to rectify capacitor mismatch issues [3].

Self-calibration based on bottom–up using LSB capacitors are an effective way to address capacitance mismatch errors. This technique involves digitally calibrating the capacitor array layer by layer, initiating from the least significant bit and progressively detecting and compensating for each capacitor bit, thereby enhancing the overall performance of the SAR ADC. However, calibration inaccuracies arising from comparator static offset and low-level capacitor mismatch can diminish the precision of self-calibration.

In view of the limitations of self-calibration based on bottom–up using LSB capacitors, this paper proposes an improved front-end digital self-calibration method suitable for high-precision SAR ADCs. The basic idea is to use capacitor suspension operation to prevent saturation of sub-analog-to-digital conversion during calibration. The averaging technology and the low-level capacitance cyclic quantification of high-level capacitance calibration technology are used to reduce the random mismatch of the low-level capacitance.

II. DIGITAL SELF-CALIBRATION TECHNOLOGY

A. Technical principle

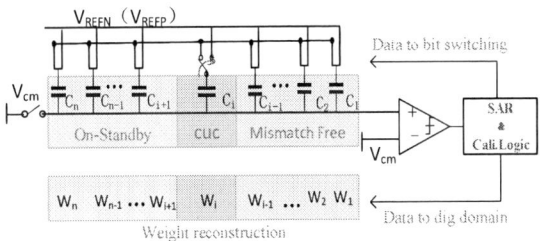

Fig. 1. Self-calibration based on bottom–up using LSB capacitors

The self-calibration technology is mainly designed for the problem of CDAC capacitance mismatch, and its core principle is to use the existing structural resources inside the SAR ADC to automatically detect and perform calibration operations through specific circuit structure and calibration logic to eliminate the influence of offset error, gain error, nonlinear error.

Figure 1 shows the correction of the capacitance mismatch using the digital self-calibration method. The high capacitance using CDAC is firstly quantified by a sub-ADC consisting of a low capacitor, comparator, and SAR logic. Subsequently, the quantization results are used to reconstruct the weights of the calibrated capacitors.

B. Calibration procedure

The digital self-calibration method assumes that the low-level capacitance meets the requirements of matching by accuracy and is an ideal capacitor bank that conforms to binary weighting, and the calibration is performed from the Capacitor Under Calibration (CUC) C_i.

979-8-3315-3918-4/25 $31.00 © 2025 IEEE

The calibration procedure is as follows:

Reset: The input of the comparator is reset to V_{CM}, the capacitance to be calibrated (CUC) C_i and the low capacitor are reset to V_{CM}, and the higher uncalibrated capacitor $C_{i+1}{\sim}C_n$ is kept at V_{REFN}.

Flip: First, to avoid charge injection errors, disconnect the comparator from the VCM, and the successive approximation of the register logic blocks forces Ci to switch from V_{CM} to V_{REFN}, resulting in a differential voltage at the comparator input, as shown below:

$$V_{e,i} = -\frac{C_i}{C_{tot}} \cdot V_{CM} \qquad (1)$$

Quantization: Quantization $V_{e,j}$ by the built-in comparator, successive approximation register logic module, and $(i-1)$ bit CDAC composed of $C_i {\sim} C_{i-1}$, the quantization process table is as follows:

$$\frac{C_i}{C_{tot}} \cdot V_{CM} = \frac{\sum_{j=1}^{i-1} D_j C_j}{C_{tot}} \cdot V_{CM} \qquad (2)$$

At the same time, the corresponding weights of the quantization process in the numerical domain can be calculated as follows:

$$W_{cail,i} = \sum_{j=1}^{i-1} D_j W_j \qquad (3)$$

$D_{i-1}{\sim}D_1$ represents the quantized output code of CDAC, and $W_{i-1}{\sim}W_1$ represents the capacitance weight value. The C_i calibration is complete, and its weight is updated to the calibration values in the numeric field, and then the C_{i+1} is calibrated. In this way, the calibration of all high capacitors is completed by analogy, and the corresponding calibration weights $W_{cali, n}{\sim}W_{cali,i}$ are obtained.

C. The limitations of digital self-calibration algorithms

From the perspective of circuit implementation, the above digital self-calibration method is relatively simple, but it also has its own limitations.

The static offset and calibration error of the comparator caused by low-level capacitor mismatch will reduce the self-calibration accuracy. Although, multiple continuous calibration and averaging can effectively reduce the sampling noise and comparator noise in the calibration process. But the simple average method cannot eliminate the static dissonance of the comparator. On the other hand, in the Self-calibration based on bottom–up using LSB capacitors, it is assumed that the low-level capacitor is an ideal capacitor bank and does not need to be calibrated, but in the actual application scenario, the low-level capacitor will also have mismatch, which cannot be eliminated by averaging, and the calibration error of the low-level capacitor will accumulate exponentially with the calibration.

Ding used the method of increasing the low-level capacitance area to lower the calibration error caused by capacitor mismatch，which significantly improved the accuracy of digital calibration [4].

III. DESIGN AND IMPLEMENTATION OF IMPROVED SELF-CALIBRATION TECHNOLOGY

The improved self-calibration algorithm proposed in this paper, which uses the capacitor suspension，operation to

prevent the saturation phenomenon of the sub-analog-to-digital conversion during the calibration process，and fuses the Forward and Reverse Switching techniques of the Ci flip stage of the capacitor, to offset the static offset error of the comparator. At the same time, the averaging technology and the low-level capacitance cyclic quantification of high-level capacitance calibration technology are used to reduce the random mismatch of the low-level capacitance and improve the accuracy of digital calibration.

The improved digital self-calibration technology is used to realize a 16-bit resolution SAR ADC with sampling rate of 1 MS/s based on 180 nm process, as shown in Figure 2.

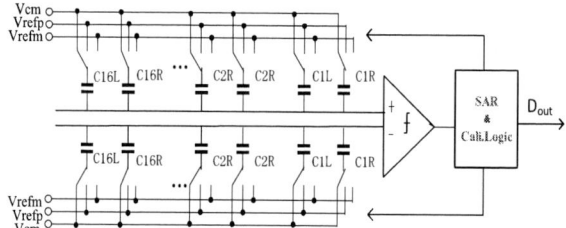

Fig. 2. Block diagram of SAR ADC with digital self-calibration

A. Capacitor suspension calibration

This calibration technique is based on the principle of conservation of charge, and the mismatch error of the SAR ADC capacitor is calibrated by introducing additional suspended capacitance. During the implementation phase of the calibration step, the input signal is first grounded to initialize the capacitor array; Then, according to the calibration algorithm, the suspended capacitors are connected to the array in a specific order, and the charge distribution of the array is changed, and the output of the comparator changes accordingly. By detecting the output produced by the comparator, the capacitor mismatch calibration is completed by adjusting the access state of the suspended capacitor and optimizing the charge distribution until the array output is close to the ideal value. The high capacitor to be calibrated is set to float, and Forward and Reverse Switching techniques of the C_i flip stage of the capacitor to be calibrated are used to offset the static offset error of the comparator.

The capacitance C_i to be calibrated is switched from V_{CM} to V_{REFN} in the flipping stage, and the quantification process is as follows:

$$-C_i + \frac{V_{OS}}{V_{CM}} \cdot C_{tot} = \sum_{j=1}^{i-1} D_j C_j \qquad (4)$$

The reverse switch switches in the direction of switching from V_{CM} to V_{REFP}, and the quantification process of $C_{(i-1)}$ CDAC is as follows:

$$C_i + \frac{V_{OS}}{V_{CM}} \cdot C_{tot} = \sum_{k=1}^{i-1} D_k C_k \qquad (5)$$

The calibration value of the capacitance C_i can be subtracted from equations (4) and (5) to:

$$C_{i,cali} = (\sum_{k=1}^{i-1} D_k C_k - \sum_{j=1}^{i-1} D_j C_j) / 2 \qquad (6)$$

The effect of comparator static offset is cancelled by subtraction.

B. Low Capacitance Cyclic Quantification High Capacitance Calibration

In SAR ADC capacitor arrays, the high capacitance mismatch error has a significant impact on the conversion accuracy, while the low capacitor has higher manufacturing accuracy due to its small capacitance value. The improved calibration technology proposed in this paper utilizes the high-precision characteristics of low-level capacitors to calibrate the mismatch error of high-level capacitors by designing calibration circuits and algorithms. This method transfers the high-precision information of low-level capacitors to high-level capacitors.

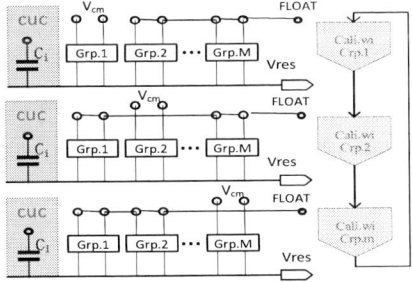

Fig. 3. Diagram of a self-calibrating SAR ADC for low-level capacitance cycling

As shown in Figure 3, several low-level capacitors are selected to form a calibration DAC. Assuming the calibration DAC has ideal binary weighting, it is used to quantify the real weighting of the capacitors to be calibrated. During the calibration phase, low-level capacitors generate accurate reference voltages, which are compared with the relevant voltages of high-level capacitors. The states of high-level capacitors are adjusted based on the comparison results. Then, the calibrated low-level capacitors are continuously added to the calibration DAC to form a new calibration DAC. The above procedure is iterated for high-level capacitors to complete the entire calibration and achieve mismatch error calibration.

In this paper, combined with the principle of averaging, a method is designed to quantify the high capacitance cyclically by multiple groups of low capacitances, to reduce the calibration error caused by the mismatch of the lowest significant bit capacitance. N repeated calibrations of the capacitor Ci and N times averaging of its calibration values to remove noise means that the equivalent calibration error is minimized after averaging the Ci from a probabilistic point of view, thus reducing the effect of error accumulation.

C. Simulation verification

The simulation model considers a CDAC with the same capacitance value and random mismatch as the actual circuit and uses the lower 4 bits of the CDAC to calibrate the high capacitance. Each bit is averaged 32 times to ensure that the effects of noise and mismatched least significant bit (LSB) capacitance mismatch.

To verify the calibration algorithm and test its improvement effect on the effect of capacitor mismatch, five simulation environments were established to simulate the ADC performance. Figure 4 illustrates the spectrum of different calibration schemes with the same CDAC mismatch and the static offset of 30 LSBs (1.65 mV) in the comparator.

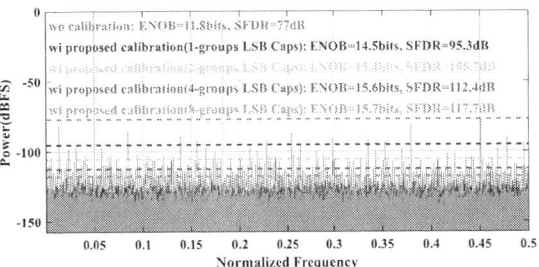

Fig. 4. Spectrum with different calibration methods under same mismatch

Uncalibrated, the ENOB is 11.8-bit and the SFDR is 77-dB SFDR. The performance of the calibration scheme was compared by sampling different LSB groups separately by using the capacitor suspension operation cyclic of the least significant bit (LSB) capacitor bank. Tests show that ADC performance improves as the number of LSB groups increases. Further improved calibration using eight sets of least significant bit (LSB) capacitance cycling techniques, the ADC demonstrated improved performance with an ENOB of 15.7bits and an SFDR of up to 117.7dB.

D. Implementation of digital system

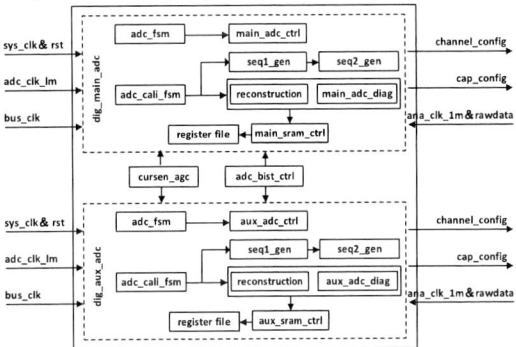

Fig. 5. Digital Systems Framework

Based on the improved self-calibration algorithm, the design of the digital circuit considers the requirements of area power consumption, and the RTL code is completed, and the code structure is shown in Figure 5. The state machine module and the reconstruction module are the core of the entire digital system, which respectively control the calibration mode of the whole system and the timing and reconstruction of normal ADC measurements.

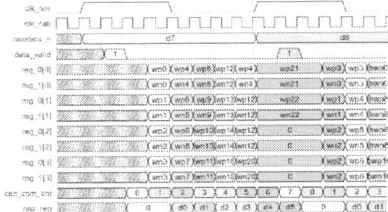

Fig.6. Calculate the timing diagram.

The state machine module flexibly selects the corresponding digital calibration mode while realizing the switch between calibration and normal measurement work without causing disorder in the working state. The reconstruction module is used to reconstruct the output of the ADC. In this design, the calculation logic of the normal

979-8-3315-3918-4/25 $31.00 © 2025 IEEE

measurement mode and the calibration mode are shared to reduce the overhead, and the partial weight storage method is used to reduce the cost of the registers. At the same time, the logic of parallel computing is used to improve work efficiency. Taking normal working mode as an example, the calculation timing is shown in Figure 6.

IV. TAPE-OUT TEST RESULTS

The improved digital calibration scheme proposed in this paper is applied to a 1MS/s, 16-bit SAR ADC system on 180nm CMOS process, and its micrographs are shown in Figure 7.

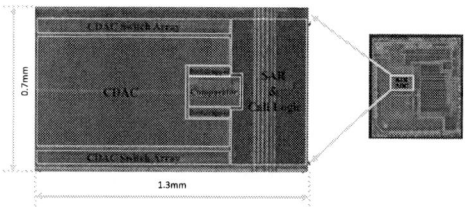

Fig. 7. 16-bit SAR ADC Chip micrograph.

All tape-out measurements were obtained at room temperature. Fig. 8 shows that the nominal weights of CDAC are used before calibration, and the maximum values of INL are +35.4/−34.3 LSB, and the maximum values of DNL are +5.9/−1.0 LSB. After calibration, the maximum INL improved to +1.6/−1.2 LSB, the maximum DNL improved to +0.86/−0.85 LSB. The comparison shows that after using the improved calibration method, the true weight of the capacitance is reconstructed in the digital domain, and the linearity of the ADC is significantly improved.

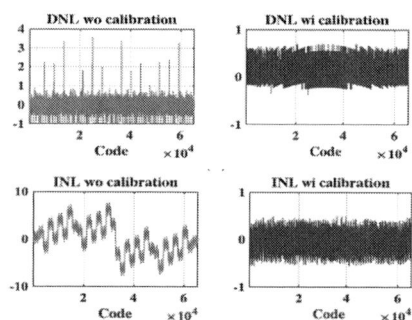

Fig.8. DNL/INL before/after the proposed calibration

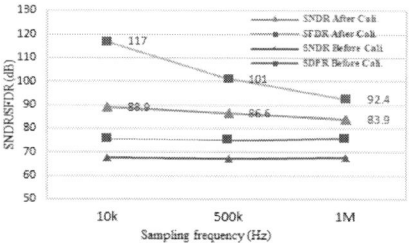

Fig.9. Measured Dynamic performance.

Figure 9 shows dynamic performance as a function of the sampling frequency at the 1 kHz sine wave input state. At sampling frequencies up to 1 MS/s, the SNDR are 83.9-dB，

SFDR are 92.4-dB.The calibrated analog-to-digital converter achieves a spurious-free dynamic range of 117-dB at a sampling rate of 10KS/s, which is 42.5-dB higher than the uncalibrated case.

This paper compares the performance of digital calibration algorithms adopted in other domestic and international literature in recent years, as shown in Table 2. Through comparison with similar studies, it is found that the ADC performance indicators proposed in this paper are close to or exceed the reference sample values, demonstrating that the proposed digital calibration technology can achieve high-precision SAR ADC calibration and realize extremely high-performance improvement.

TABLE I. PERFORMANCE COMPARISON

	This work	Zhang [5]	Wang [6]	Wang [7]	Fan [8]
Supply voltage(V)	1.8	1.8	2.5	1.8	1.8
Technology(nm)	180	180	180	180	180
Fs (MS/s)	1	1	1		0.1
Resolution(bits)	16	14	16	16	14
ENOB (bit)	14.5	12.81	14.21	15.82	11.29
SFDR (dB)	117	93.43	101.31	106.4	83.77
SNDR (dB)	88.9	78.88	87.31	97.80	69.75

V. CONCLUSION

An improved digital self-calibration technique is proposed in this paper. The basic principle is to set the uncalibrated high capacitor in the floating state and use the low capacitor to periodically convert the high capacitor mismatch error, so as to realize the mismatch calibration of the high capacitor. This method effectively solves the problem of accuracy degradation caused by capacitor mismatch in DAC.

REFERENCES

[1] H.-S. Lee, D. A. Hodges, P. Gray, "A self-Calibrating 15-bit CMOS A/D Converter," IEEE J. Solid-State Circuits, Dec. 1984, Vol.19. 813-819.

[2] J. McNeill, K. Y. Chan, M. C. W. Coln, "All-digital background calibration of a successive approximation ADC using the 'Split ADC' Architecture," IEEE Trans. Circuits and Systems I, Oct.2011, 58(10). 2355-2365.

[3] Hwan S K, Seun Guam C, Jae Y S, "A 87.5-dB-SNDR Residue-integrated SAR ADC with a Digital-domain Capacitor Mismatch Calibration," Journal of Semiconductor Technology and Science, 2021,21 (2) ,143-151.

[4] J. Ding et al., "A 16-bit 1-MS/s SAR ADC With Capacitor Mismatch Self-Calibration," in IEEE Transactions on Very Large-Scale Integration (VLSI) Systems, vol. 33, no. 1, pp. 10-20, Jan. 2025.

[5] Chang Zhang, Design of 14-bit SAR_ADC based on deterministic digital calibration [D]. Xi'an University of Posts and Telecommunications, 2024.

[6] Hanfeng Wang, Research and design of a high-precision SAR_ADC with self-calibration technology[D]. University of Electronic Science and Technology of China, 2024.

[7] G. Wang, Z. Wang, Z. Gao, Z. Bing, J. Zhao and T. Zhang, "A Design of SAR ADC Calibration Technology based on Weight Fitting," 2024 9th International Conference on Electronic Technology and Information Science (ICETIS), Hangzhou, China, 2024, pp. 6-10.

[8] Hua Fan, Zhuorui Chen,"14-Bit SAR ADC with on-Chip Digital Bubble Sorting Calibration Technology" , Chinese Journal of Electronics vol. 34, no. 1, pp. 125-136, January,2025.

Bitwise Bayesian Optimization for SAR ADC Calibration

Yu Shi [1], Shen Ye[1], Yihang Luan[1], Jiahao Wang[1], Ting Yi*[1]

[1] State Key Laboratory of Integrated Chips and Systems, Fudan University

* Email : shiy23@m.fudan.edu.cn, yiting@fudan.edu.cn

Abstract—This paper proposes a bitwise Bayesian Optimization method for the digital calibration of SAR ADCs. By leveraging Bayesian Optimization, the linearity of the SAR ADC is efficiently improved without requiring prior knowledge. The proposed method requires only a single set of ADC outputs and prevents overfitting by optimizing the decoder instead of directly altering the output. Experiment results show that the SNDR of a modeled 12-bit SAR ADC (with DAC mismatch) in MATLAB has improved from 50.41 dB to 71.11 dB, while the SNDR of a 12-bit SAR ADC chip has been improved from 61.47 dB to 70.24 dB.

Keywords—ADCs, calibration, machine learning, Bayesian Optimization

I. INTRODUCTION

The growing demand for high-resolution and high-speed data acquisition systems has underscored the need for precise Analog-to-Digital Converters (ADCs) in a wide range of applications, such as instrumentation, radar imaging, wireless communications, and biomedical signal processing. Among the various ADC architectures, the successive approximation register (SAR) ADC has attracted significant attention due to its favorable trade-off among power efficiency, resolution, and scalability. However, as ADC resolution increases, non-idealities like capacitor mismatch and clock jitter become increasingly detrimental to the dynamic performance. Therefore, many recent works have incorporated calibration techniques to address nonlinearity issues in ADCs [1]-[6].

Traditional calibration approaches [2] [3] [4] (Fig. 1(a)) for SAR ADCs often rely on additional analog circuits, requiring a detailed understanding of the system's physical characteristics and device-level modeling. These methods demand extensive prior knowledge about the system and are typically time-consuming and labor-intensive. Furthermore, the integration of additional hardware may not be feasible in resource-constrained or low-power applications.

Recently, machine learning (ML)-based calibration techniques [5] [6] (Fig. 1(b)) have emerged as promising alternatives. These methods aim to overcome the limitations of traditional approaches by learning from the system's behavior without relying on exhaustive prior modeling. However, existing ML-based solutions still face significant challenges, such as the need for additional reference channels, large amounts of calibration data, and complex neural network architectures. These systems may also struggle with high-dimensional search spaces and the impact of hardware variations, complicating their convergence and limiting their effectiveness in practical applications.

In this paper, we propose a bitwise Bayesian Optimization approach for the digital calibration of SAR ADCs. As illustrated in Fig. 1(c), the proposed method iteratively adjusts the digital decoding weights to enhance ADC performance without introducing any additional analog hardware, complex modeling procedures, or prior circuit knowledge. Compared to conventional ML-based approaches, this framework is

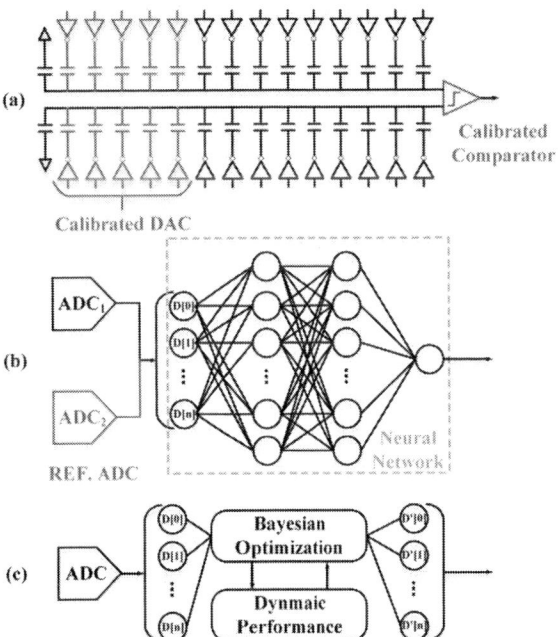

Fig. 1. ADC calibration framework of (a) traditional analog intensive scheme, (b) previous ML-based scheme, and (c) the proposed Bayesian Optimization approach

sample-efficient, requiring only a single set of ADC outputs. It avoids overfitting by optimizing weights rather than directly modifying the digital output. The method exhibits rapid convergence and maintains robustness when applied to ADC chips. Experimental results verify its effectiveness, showing improvements of 8.77 dB in Signal-to-Noise-and-Distortion Ratio (SNDR) and 10.59 dB in Spurious-Free Dynamic Range (SFDR) on a fabricated 12-bit SAR ADC.

This paper is organized as follows. Section II describes the proposed Bayesian optimization for SAR ADC calibration. Section III presents the experimental results, including both simulation-based evaluations and measurement results. Section IV concludes the work.

II. PROPOSED BAYESIAN OPTIMIZATION METHOD FOR SAR ADC CALIBRATION

Capacitor mismatch in SAR ADCs is a significant source of nonlinearity, leading to distortion and errors in the digital output. Traditional solutions, such as analog and digital calibration, are often limited in effectiveness and require hardware overheads. To address these challenges, we propose an optimized decoder architecture based on bitwise Bayesian Optimization. As shown in Fig. 2, this approach leverages a Gaussian process surrogate model to iteratively calibrate the decoding weights, improving dynamic performance without the need for extensive hardware modifications. Once the optimal weights are determined, they ensure performance improvements across the entire Nyquist frequency range.

979-8-3315-3918-4/25 $31.00 © 2025 IEEE

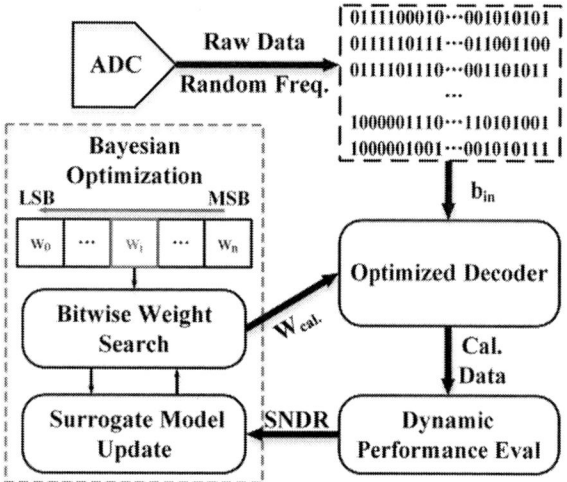

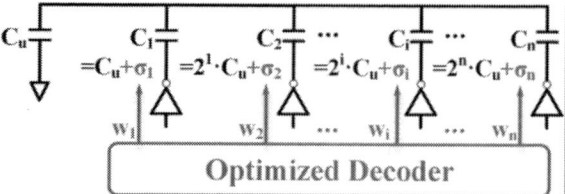

Fig. 2. Bitwise Bayesian optimization of decoder weights

Fig. 3. Capacitor mismatch model and decoder-level correction

for additional reference ADC and large datasets, and offers a more efficient digital solution.

Importantly, although the optimization is performed using a representative waveform at a single frequency, the calibrated weights exhibit strong generalization across the entire Nyquist frequency range. Moreover, the calibration frequency can be randomly selected, and the output variation across different calibration frequencies is less than 0.7%, which demonstrates the robustness of the proposed algorithm. This is because capacitor mismatch is a static issue that does not vary with the input signal frequency. As a result, the optimized decoder compensates for these fixed structural nonidealities, enabling broadband calibration with minimal measurement cost and robust performance across varying frequencies.

B. Bitwise Bayesian Optimization Framework

To compensate for the capacitor mismatch in SAR ADCs, we focus on optimizing the digital decoder weights to correct the mismatch-induced nonlinearity. The challenge lies in efficiently determining the optimal decoder weights that can effectively address these distortions. To solve this problem, we propose a bitwise Bayesian Optimization framework. Bayesian Optimization offers distinct advantages, particularly its ability to optimize complex objective functions with limited data [7].

As a first step, we formulate the digital reconstruction process to identify the key parameters subject to optimization, the decoder weights associated with each bit output.

$$D_{out} = \sum_{i=1}^{n} w_i \cdot bin_i \qquad (4)$$

where w_i is the corresponding decoder weight. For each bit w_i, we define a constrained search space centered around its ideal binary weight, bounded within $2^{i-1} \pm 10\%$. This constraint prevents instability from extreme values and preserves the exponential structure of the decoder.

The proposed Bayesian Optimization is carried out bit by bit, from the most significant bit (MSB) to the least significant bit (LSB), as higher-order bits contribute more suggestively to quantization error. After the higher-order bits are fixed, the optimization proceeds with the lower-order bits. Each bit has undergone the same optimization steps as Algorithm 1:

A Gaussian Process (GP) serves as the surrogate model to approximate the relationship between the weight w_i and the objective function metric, which will be introduced in section C. Specifically, the GP models the objective function as a distribution over possible functions, parameterized by a kernel function like Matérn kernel ($\nu = 2.5$). Given a set of sampled weight values $\{w_i^{(1)}, w_i^{(2)}, \dots\}$ and their corresponding measured objective function, the GP produces a posterior distribution that provides both a mean prediction and a variance estimate at any unsampled point in the domain.

A. Optimizing Decoder Weights for Mismatch

Capacitor mismatch is a main nonlinearity source in SAR ADCs, arising from manufacturing variations and not easily addressable through analog design techniques. As shown in Fig. 3, we focus on the digital decoding phase, proposing a correction that adapts to the mismatched behavior without requiring invasive hardware changes.

To better understand its impact, we provide a quantitative analysis of the mismatch effect by examining the capacitive Digital-to-Analog Converter (DAC), a key component of SAR ADCs. The ideal voltage is given by:

$$V_{DAC,ideal} = \sum_{i=1}^{n} \frac{2^i \cdot C_u}{2^n \cdot C_u} \cdot bin_i \cdot V_{ref} = \sum_{i=1}^{n} 2^i \cdot bin_i \cdot \frac{V_{ref}}{2^n} \quad (1)$$

where C_u is the unit capacitor, V_{ref} is the reference voltage, and $bin_i \in \{0,1\}$ denotes the comparator output of the i-th bit. In the digital domain, this corresponds to a digital output code, which represents the weighted sum of each bit:

$$D_{out,ideal} = \frac{V_{DAC,ideal}}{V_{ref} / 2^n} = \sum_{i=1}^{n} 2^i \cdot bin_i \qquad (2)$$

However, due to capacitor mismatch, the actual capacitors deviate from their nominal values, resulting in a distorted voltage:

$$V_{DAC,real} = \sum_{i=1}^{n} \frac{2^i \cdot C_u + \sigma_i}{2^n \cdot C_u} \cdot bin_i \cdot V_{ref} = \sum_{i=1}^{n} w_i \cdot bin_i \cdot \frac{V_{ref}}{2^n} (3)$$

where σ_i represents the mismatch associated with the i-th capacitor. These deviations distort the DAC output and introduce quantization errors. To compensate for this effect, updated decoding weights are required.

To address this issue, we propose a new calibrated decoding strategy that uses bitwise Bayesian Optimization to directly optimize the decoder weights based on performance metrics such as SNDR. Instead of analytically modeling the mismatch or altering the capacitor array, this method iteratively refines the decoding weights, allowing the decoder to adapt to the true behavior of the DAC and compensate for the static mismatch effects. This approach eliminates the need

Algorithm 1 Bayesian Optimization of Decoder Weights

1: **Input:** ADC output **x**(randomly select),resolution **n**
2: **Output:** Optimized weight vector $\mathbf{w} = [w_1, \ldots, w_n]$

3: Initialize decoder weights $\mathbf{w}_{\text{ideal}} = [2^{n-1}, \ldots, 1]$
4: **for** $i = 1$ to n **do**
5:   Define search range: $w_i \in [0.9 \cdot w_i^{\text{ideal}}, 1.1 \cdot w_i^{\text{ideal}}]$
6:   Generate random samples of w_i
7:   Fit Gaussian Process (GP) surrogate model on $[w_i, \text{SNDR}_{w_i}]$ pairs
8:   **for** epoch $= 1$ to max_epochs **do**
9:     Compute Expected Improvement (EI) acquisition function
10:     Select $w_i^{\text{new}} = \arg\max \text{EI}(w_i)$
11:     Compute decoded signal: $\hat{\mathbf{x}}_{\text{new}} = \mathbf{x} \cdot \mathbf{w}_{\text{new}}^{(i)}$
         where $\mathbf{w}_{\text{new}}^{(i)} = [\ldots, w_{i-1}^*, w_i^{\text{new}}, w_{i+1}^{\text{ideal}}, \ldots]$
12:     Evaluate $\text{SNDR}(\hat{\mathbf{x}}_{\text{new}})$
13:     Update GP with new sample
14:   **end for**
15:   Set $w_i^* = \arg\max_{w_i} \text{SNDR}(w_i)$ and fix it
16: **end for**

This predictive uncertainty is then leveraged by the Expected Improvement (EI) acquisition function, which selects the next candidate weight to evaluate. EI quantifies the expected improvement over the current best observation, considering both the predicted mean and uncertainty from the GP model. By design, EI balances the trade-off between exploration (prioritizing regions of high predictive uncertainty that may yield better performance) and exploitation (refining search within regions already known to exhibit favorable objective function).

In each epoch, the candidate weight that maximizes the EI is selected and evaluated via the objective function. The resulting objective function is then used to update the GP surrogate model. This iterative process continues for 20 optimization steps per bit, following 15 initial random evaluations, yielding a total of 35 evaluations per bit. This sampling strategy ensures efficient convergence while maintaining robustness to noise and local optima.

C. Objective Function

Bayesian Optimization is employed to iteratively optimize the decoding weights by constructing a probabilistic model of the system's performance. This model is updated as new data is collected, guiding the search for the optimal weight configuration. A critical aspect of applying Bayesian Optimization in our framework is selecting an appropriate objective function. In this case, we use SNDR, a widely recognized metric that reflects the overall dynamic performance of the ADC. SNDR accounts for both random noise and harmonic distortion, making it a comprehensive indicator of signal fidelity in the digital output.

By maximizing SNDR, we ensure that the optimization process reduces unwanted distortions by effectively guiding the decoder weights toward values that approximate the capacitor ratios and compensate for mismatch. This approach does not require knowledge of internal mismatch sources and remains effective across various input conditions, ensuring that the calibrated weights yield high-resolution, low-distortion conversion results.

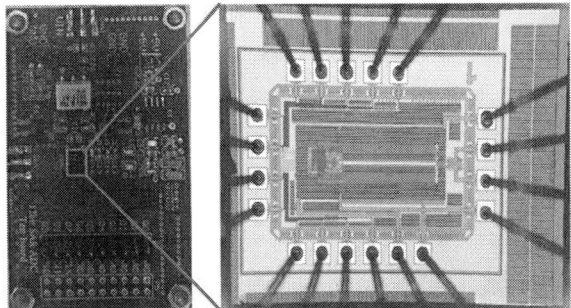

Fig. 4. Test board and micrograph of the SAR ADC

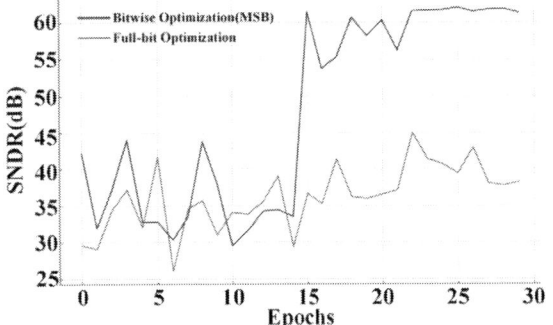

Fig. 5. Performance comparison of Bitwise and Full-bit Optimization over epochs.

III. EXPERIMENT RESULTS

To validate the effectiveness and generalizability of the proposed method, we conducted a series of experiments covering multiple aspects of the decoder calibration process. Specifically, we evaluate the convergence efficiency of the bitwise Bayesian Optimization strategy, compare the calibration consistency under different input waveforms, and examine its performance on both a modeled SAR ADC and a SAR ADC chip. In the latter case, the test board and ADC layout are shown in Fig. 4.

The bitwise optimization strategy ensures efficient convergence by focusing on the most impactful parts of the decoder weights. As shown in Fig. 5, this approach is efficient, converging typically in 15 epochs. The results, indicate that our method significantly outperforms traditional optimization techniques, which attempt to jointly optimize the full decoding vector, both in terms of computational efficiency and performance improvement.

To enable comparison with other works that employ simulation-based ADC calibration techniques, we developed a behavioral model of a 12-bit, 20MHz SAR ADC in MATLAB. This model incorporates key non-idealities, including comparator offset, noise, and capacitor mismatch. The mismatch among the unit capacitors is modeled as independent and normally distributed, with a 3% standard deviation (1σ) relative to the nominal unit capacitance. The calibration results are presented in Fig. 6(a), where the SNDR improves by 20.7 dB and the SFDR increases by 37.94 dB. To further evaluate robustness and effectiveness, we simulate additional mismatch levels of 1% and 5%. As shown in Fig. 7, the calibration consistently restores SNDR to approximately 71 dB under 1% and 3% mismatch, confirming effective correction of capacitor mismatch. Even under the extreme 5% case, SNDR improves significantly from 48.54 dB to

979-8-3315-3918-4/25 $31.00 © 2025 IEEE

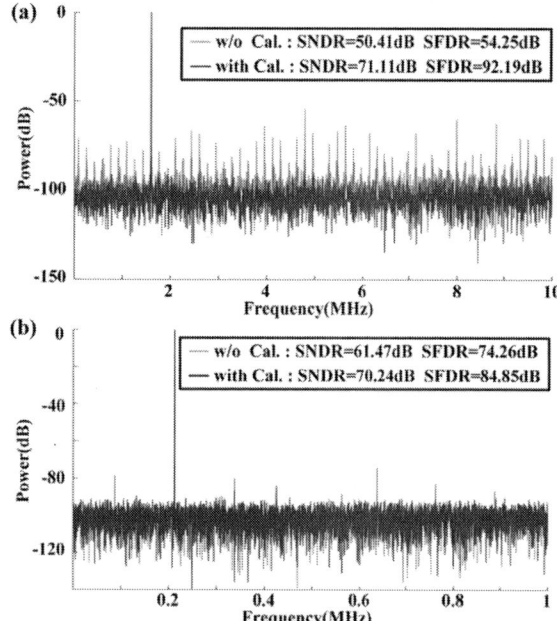

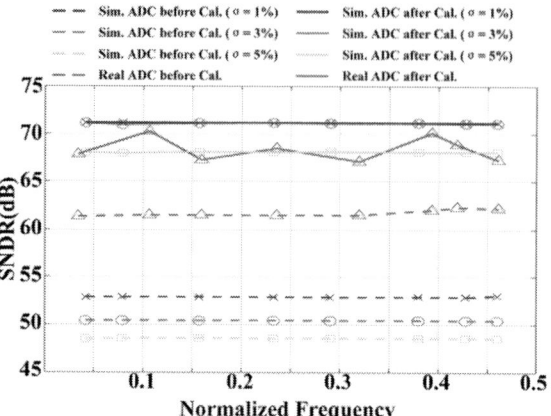

Fig. 7. Measured SNDR versus input frequency

Fig. 6. Power spectrum comparison before and after calibration based on (a) a SAR ADC model(σ=3%), (b) a SAR ADC chip

68.03 dB, demonstrating strong resilience against severe non-idealities.

As shown in Fig. 6(b), we also carried out calibration on measurement data obtained from a fabricated 12-bit 2 MHz SAR ADC (Fig. 4) to assess the robustness and effectiveness of our approach. Experimental results demonstrate that our method successfully enhanced the performance of the tested ADCs, achieving a SNDR improvement of 8.77 dB, a SFDR improvement of 10.59 dB demonstrating its effectiveness in real-world applications. The performance gap can be attributed to additional non-idealities in the physical chip, such as layout parasitic, measurement noise and process variations.

of these results with those from other works is provided in Table I. As shown in Fig. 7, our calibration algorithm can be directly applied across the full bandwidth without the need for re-calibration, whether for modeled ADCs or ADC chips.

IV. CONCLUSION

In this work, we introduce a novel approach for calibrating SAR ADCs by combining Bayesian Optimization with bitwise decoding weight adjustment. Without relying on prior knowledge or circuit-level changes, this method enhances ADC accuracy purely through digital decoding optimization.

The innovation lies in using Bayesian Optimization for efficient, iterative calibration, coupled with bitwise optimization to refine each bit's weight. We demonstrated the proposed approach in a 12-bit SAR ADC chip, which has achieved an 8.77 dB SNDR improvement with only 4096 data points to converge. This solution offers an efficient, scalable, and low-cost solution for SAR ADC calibration, ensuring high accuracy and broad applicability in real-world systems.

TABLE I. WORK COMPARISON

	This Work		ESSCIRC-2022[6]	ICCS-2023[2]	ICSICT-2024[4]
ADC	Sim.	Real	Real	Sim.	Sim.
Resolution	12	12	6	12	12
Applicable Bandwidth	0-0.5fs		0.10-0.11fs	0-0.5fs	0-0.5fs
Calibration Method	Bayesian Optimization		Neural Network	H2L & L2H	EKF & LMS
Required Points	4096	4096	16384000	8192	5120
SNDR(dB) (w/o--w/)	50.41-71.11	62.47-70.24	26.68-34.80	50.7-68.9	50.9-70.1
SNDR(dB) Improvement	20.7	8.77	8.12	18.2	19.2

Sim.: **Output from simulated ADC model**
Real: **Measured data from fabricated ADC chip**

The results confirm that our calibration method remains effective across ADC chips and diverse simulation-based models, requiring only 4096 calibration points. A comparison

REFERENCES

[1] M. Bagheri, F. Schembari, N. Pourmousavian, H. Zare-Hoseini, D. Hasko and R. B. Staszewski, "A Mismatch Calibration Technique for SAR ADCs Based on Deterministic Self-Calibration and Stochastic Quantization," in IEEE Transactions on Circuits and Systems I: Regular Papers, vol. 67, no. 9, pp. 2883-2896, Sept. 2020.

[2] X. Xia, J. Sun and W. Liu, "Analysis and Design of Calibration Technique for Capacitor Mismatch in SAR ADCs," 2023 5th International Conference on Circuits and Systems (ICCS), Huzhou, China, 2023, pp. 108-113.

[3] C. Chen, Z. Yuan, P. Cao, J. Xu and Z. Hong, "A 71.5-dB SNDR 475-MS/s Ringamp-Based Pipelined SAR ADC with On-Chip Bit-Weight Calibration," 2024 IEEE Symposium on VLSI Technology and Circuits, Honolulu, HI, USA, 2024, pp. 1-2.

[4] D. Zhou, Y. Xiang, J. Ren and F. Ye, "A Digital Foreground Calibration Method for Pipeline SAR ADCs Using Extended Kalman Filter," 2024 IEEE 17th International Conference on Solid-State & Integrated Circuit Technology (ICSICT), Zhuhai, China, 2024, pp. 1-3.

[5] T. Zhang, Y. Cao, S. Zhang, C. Chen, F. Ye and J. Ren, "Machine Learning Based Prior-Knowledge-Free Calibration for Split Pipelined-SAR ADCs with Open-Loop Amplifiers Achieving 93.7-dB SFDR," ESSCIRC 2019 - IEEE 45th European Solid State Circuits Conference (ESSCIRC), Cracow, Poland, 2019, pp. 189-192.

[6] E. Ware, J. Correll, S. Lee and M. Flynn, "6GS/s 8-channel CIC SAR TI-ADC with Neural Network Calibration," ESSCIRC 2022- IEEE 48th European Solid State Circuits Conference (ESSCIRC), Milan, Italy, 2022, pp. 325-328.

[7] B. Shahriari, K. Swersky, Z. Wang, R. P. Adams and N. de Freitas, "Taking the Human Out of the Loop: A Review of Bayesian Optimization," in Proceedings of the IEEE, vol. 104, no. 1, pp. 148-175, Jan. 2016.

An Area-Efficient C2C SAR ADC with Hybrid Switching Mode for Ultrasound Miniature Probes

Tianci Zhang[1], Jinlai Fu[1], Li Dai[1], Dongxu Li[1], Yingchen Liu[1],
Jing Li*[1], Zhong Zhang[1], Ning Ning[1], Qi Yu[1]

[1] StateKey Laboratory of Electronic Thin Films and Integrated Devices, University of Electronic Science and
Technology of China, Chengdu 610054, China

* Email: ztc17@qq.com, lijing686@uestc.edu.cn

Abstract—This paper presents an area-efficient 10-bit hybrid C2C successive approximation register analog-digital converter (SAR ADC) specifically designed for three-dimensional (3D) intracardiac echography (ICE) ultrasound ASIC. The proposed a C2C SAR ADC embedded with hybrid switching method, achieving 95% capacitor area reduction compared to the conventional SAR ADC. To mitigate the effect of parasitic capacitor, we implement hybrid switching mode for the MSB and LSB to cooperate with the LMS correction algorithm. Fabricated in 130nm BCD technology, the ADC occupies 0.034mm². Operating at 20 MS/s sampling rate, it achieves 54.7dB SNDR and 68.7dB SFDR for ultrasound center frequency at 5MHz, and consumes 0.79 mW with a 1.5-V supply, resulting in a figure of merit (FoM) of 38.6 fJ/conversion step.

Keywords—ultrasound, SAR ADC, C2C array, capacitor array, low area.

I. INTRODUCTION

Three-dimensional (3D) intracardiac echography (ICE) has emerged as a pivotal technology for cardiovascular disease diagnosis. Miniature probes embedded with advanced integrated circuits (ICs) play a critical role in enhancing signal quality and reducing cable complexity by processing data from 2D transducer arrays. Integrating analog-to-digital converters (ADCs) into the probe enables digital beamforming and enhances signal transmission fidelity. However, for arrays with a large number of elements, integrating ADCs faces challenges in terms of area and power consumption. Fig.1 show the block diagram of traditional ultrasound front-end, generally, the echo signals from transducer need to be quantized by the array readout architecture with ADCs. With the steady increase in array counts, frame rate, and dynamic range of digital outputs, hundreds of ADCs are adopted in the array sensors. Therefore, a trade-off in size, reliability, and power consumption is critical to each ADC.

Several types of ADCs have been utilized in array sensors, such as Single-slope (SS) ADCs and Successive approximation register analog-to-digital converters (SAR ADCs). SAR ADCs are heavily utilized in energy limited applications with moderate resolutions and sampling rates. However, SAR ADC require a capacitor array with a large area to generate the reference voltages for comparison operation. SS ADCs have excellent linearity and minimal occupation. Nevertheless, a disadvantage of Single-slope ADC is its relatively slow conversion speed. To overcome the drawback of the two ADCs, the hybrid SS SAR ADC has been used [1], [2], [3]. Each ADC still requires an extra ramp generator with an additional capacitor array and encoder logic.

This work was supported by National Key Laboratory of Integrated Circuits and Microsystems under Grant YG2407-3.

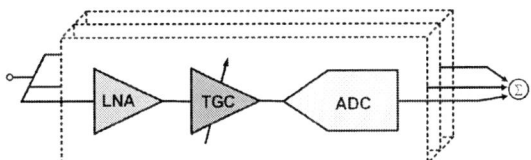

Fig. 1 Block Diagram of Traditional Ultrasound Front-End

In this work, A C2C ladder network is utilized to overcome large area of capacitor array. Compared with the traditional parallel capacitor structure, the C2C array can significantly reduce the number of unit capacitors due to its series architecture. However, the series connection introduces weight variations caused by parasitic effects, which restricts the application of C2C arrays in higher-precision ADCs. At the terminal of the parallel capacitor structure, introducing a partial series configuration can strike a balance between parallel and series capacitor arrays, achieving both a small array area and moderate precision. This may still lead to changes in the least significant bits (LSBs). Fortunately, we adopt a split switching approach, where each capacitor is split into two for switching [4]. Combined with single-side switching [5], this allows us to use only one capacitor for single-side switching in LSB switching, while using the other capacitor as redundancy. This approach strengthens the power advantage of SAR ADC and reduces the area overhead. Additionally, building upon the split-based switching method, we propose a latch with four outputs that individually control the four capacitors required for split switching, thereby reducing the path delay in signal transmission. Furthermore, to read out the actual weight values, we employ the LMS algorithm with dither injection for foreground calibration. A prototype ADC is implemented in 130-nm BCD. It achieves a 54.7-dB signal to noise-plus-distortion ratio (SNDR) and a 68.7-dB SFDR at a sampling rate of 20 MS/s. Notably, the design consumes only 0.79 mW of power under a 1.5-V supply voltage, occupying a compact area of 0.034 mm².

This paper is organized as follows. Section II describes the principle and overall architecture. Section III describes a latch with four outputs and detailed circuits design. The experimental results measured from the prototype are presented in Section IV. Section V concludes this brief.

II. PROPOSED HYBRID C2C SAR ADC

Fig. 2 illustrates the concept of the proposed SAR ADC. It consists of a hybrid-structured capacitor array, a comparator, a SAR logic, a dither generation circuit, and a parallel-to-serial converter (P/S) circuit. The hybrid-structured capacitor array

979-8-3315-3918-4/25 $31.00 © 2025 IEEE

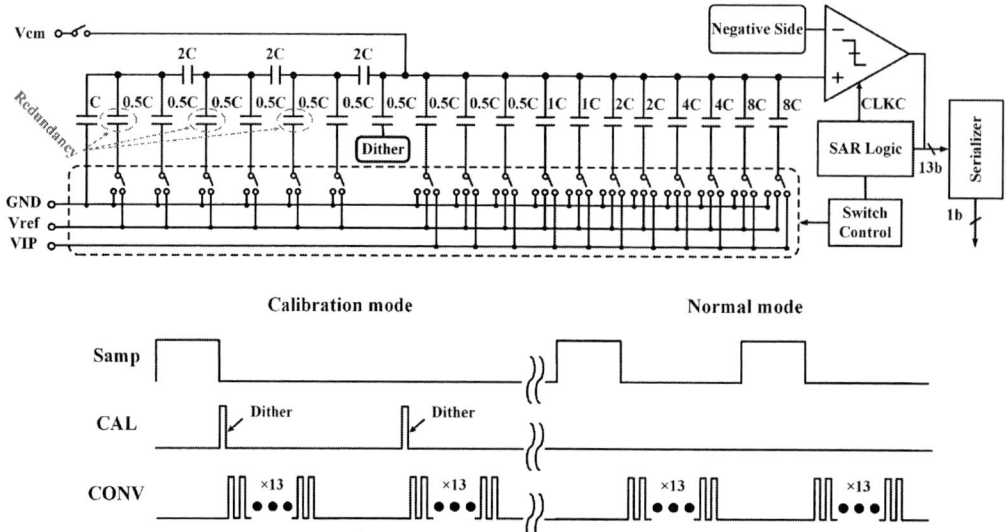

Fig. 2 Circuit schematic and timing diagram of the proposed SAR ADC

is composed of a parallel 5-bit capacitor array and a 3-bit C2C capacitor-string. All capacitors are split into two for hybrid switching operation. MSB DACs utilize binary-weighted parallel capacitors, all of which are split into two to implement split switching. The terminal capacitors are also split into two, one is dedicated to quantization using single-side switching while the other is employed for dither injection.

Dither injection is enabled exclusively during the foreground calibration phase, as illustrated in the timing diagram of Fig. 2 During calibration phase, the ADC quantizes the same sampled signal twice, with opposite dither injection applied in each quantization. To avoid errors caused by sampling clock skew between MSB and LSB, only the MSB capacitors including the terminal ones participate in sampling. The terminal capacitors are fully utilized: they not only contribute to sampling to mitigate gain errors but also serve as a transition between the MSB split-switching scheme and the LSB single-side switching scheme. Additionally, they function as dither injection capacitors. Although the LSB capacitors do not participate in sampling, they are leveraged post-sampling to elevate the common-mode voltage on the top plates of the MSB capacitors. Specifically, during the sampling phase, all bottom plates of the LSB capacitors are connected to GND. After sampling, half of these bottom plates are switched to VDD. This approach effectively raises the comparator's input common-mode voltage, thereby achieving a faster decision speed.

The quantization of LSBs is achieved by switching the C2C capacitor string. Due to the parasitic capacitance on the bridge capacitor plates, LSB accuracy is compromised. To address this parasitic-induced quantization error, we employ single-side switching technology in the C2C capacitor string. As previously mentioned, each LSB capacitor is split into two, unlike the split capacitor switching where both capacitors are switched simultaneously, the two capacitors of LSBs here are switched independently, with one of them serving as a redundant capacitor for switching. As shown in Fig. 3, split switching is used for MSBs while single-side switching is adopted for LSBs, with the lowest three bits each having redundancy.

Redundancy is applied between the two switching mode without introducing additional capacitors. However, single-side switching inevitably causes changes in the comparator's input common-mode voltage, leading to variations in offset, noise, power consumption, and speed, which degrade the overall performance of the ADC [5]. To address this, our single-side switching scheme alternates between positive and negative capacitor arrays. Thanks to the combination of split capacitors and single-side switching, the comparator's common-mode voltage ultimately returns to its initial value, as shown in Fig. 4.

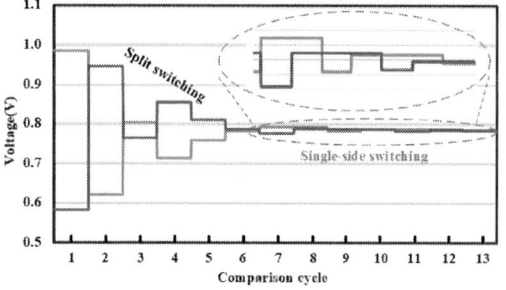

Fig. 3 Timing diagram of the proposed SAR ADC conversion

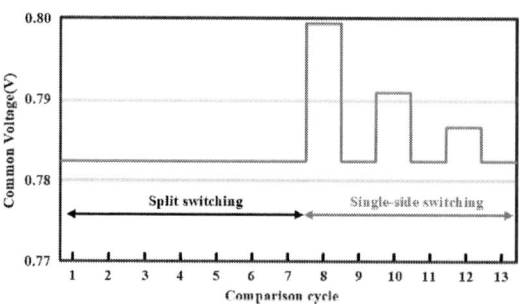

Fig .4 Common voltage variation for comparation cycle

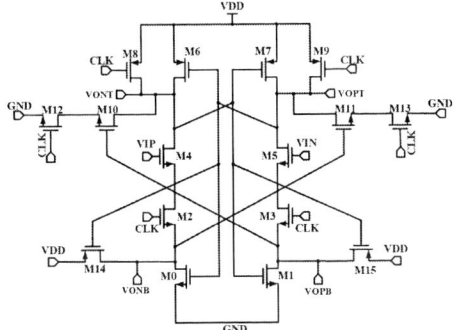

Fig. 5. Proposed quad output latch

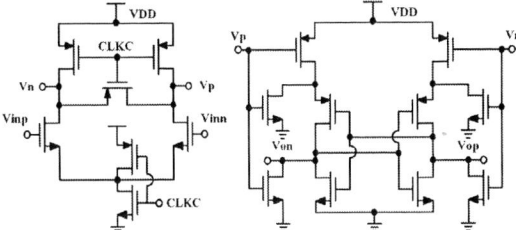

Fig. 6 The schematic of comparator

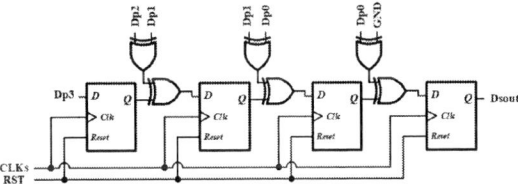

Fig 7 3b parallel to serial converter

III. CIRCUIT IMPLEMENTATIONS

A. Quad Output Latch

After each comparison in the split capacitor array, at least four control bits are required to regulate the switching of four capacitors. However, traditional latches can only provide two output states per operation, which necessitates a larger logic circuit scale and longer operation delay. Additionally, input-triggered latches require complex clock logic to shut down input channels after latching is completed, which increases hardware and power consumption overhead.

To address these issues, we propose a quad-output latch as shown in Fig. 5 which eliminates the need for complex clock signals. Within one conversion cycle, its clock is enabled before the comparator completes its decision, latches the first input signal received, and then holds the output steady regardless of subsequent input changes. During the sampling phase of ADC, the clock follows the sampling clock to disable, resetting the quad-output latch.

The four outputs of the latch always change in pairs. During the ADC sampling phase, *CLK* is disabled, resetting *VOPT* and *VONT* to *VDD*, while *VOPB* and *VONB* are reset to *GND*. Subsequently, before the comparator's comparison, *CLK* is enabled. Depending on the comparator's output, either *VOPT* and *VONB* (or *VONT* and *VOPB*) are inverted. After the latch operation is completed and the comparator reset finalizes, the *CLK* of the subsequent latch is enabled. This process repeats until the final comparison is concluded.

B. Comparator

Due to the extremely small capacitor array, the kickback noise from the comparator's input differential pair becomes significant. Therefore, a dynamic amplifier is employed as a preamplifier for the latch. The comparator used is shown in Fig. 6. Instead of requiring an additional clock control for resetting the latch, the output of the preamplifier is utilized. The preamplifier and the latch are connected via a single NMOS transistor to further enhance the gain. Since the sampling capacitors in this design are relatively small, additional reset transistors are added to the drain and source terminals of the input differential pair to mitigate the impact of parasitic capacitances from the input pair on the small capacitor array. These extra reset transistors ensure that the input differential pair of the comparator starts each comparison cycle in the same state.

C. Parallel-to-Serial Converter

Constrained by the number of cables, the RX outputs of numerous channels cannot be transmitted in parallel. Therefore, a parallel-to-serial converter (P/S) must be integrated after each ADC to serialize the ADC data. Fig. 7 shows a 3-bit parallel-to-serial circuit, serving as an illustration for the 13-bit parallel-to-serial converter used in this design. CLKs as the 260MHz main clock of the ADC provides the clock signal for the parallel-to-serial converter, while RST serves as the reset signal. The ADC and the parallel-to-serial converter share the 260MHz main clock, with the ADC internally dividing this clock to 20MHz as the sampling clock. By sharing the main clock, the converter generates a 13-bit serial output within one quantization cycle (50ns) of the ADC.

This parallel-to-serial converter is based on XOR gates and shift registers, eliminating the need for additional circuits or control signals. The XOR gate has the following properties:

$$A \oplus B \oplus A = B \qquad (1)$$

$$A \oplus 0 = A \qquad (2)$$

During the reset phase, the outputs of all D flip-flops are set to 0, while the input of the last D flip-flop is equivalent to $Dp0$. Therefore, when the first clock edge arrives, the MSB code $Dp0$ of the ADC is output to $Dsout$. Meanwhile, the third D flip-flop outputs the result of $Dp0 \oplus Dp1$ to the final stage. Since $Dp0 \oplus Dp1 \oplus Dp0$ equals $Dp1$, the input of the last D flip-flop becomes $Dp1$. Therefore, when the second clock edge arrives, the MSB-1 code $Dp1$ of the ADC is output to $Dsout$. Following this pattern, the LSB code $Dp13$ is output to $Dsout$ on the 13th clock edge.

IV. MEASUREMENT RESULTS

The prototype (see Fig. 8), fabricated using a 130 nm BCD process, shows that the RX sections each occupy 0.0625 mm² with a 250 μm pitch. The prototype ADC occupy 0.034 mm², respectively. The total ADC power consumption is 0.79 mW. Fig. 9 shows the distribution of area and power. Thanks to the C2C structure, the capacitor array of the SAR ADC is significantly minimized, with DAC power consumption

979-8-3315-3918-4/25 $31.00 © 2025 IEEE

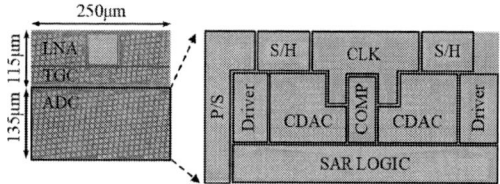

Fig. 8 Die micrograph

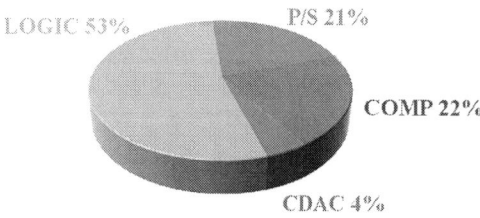

Fig. 9 Power breakdown of proposed ADC

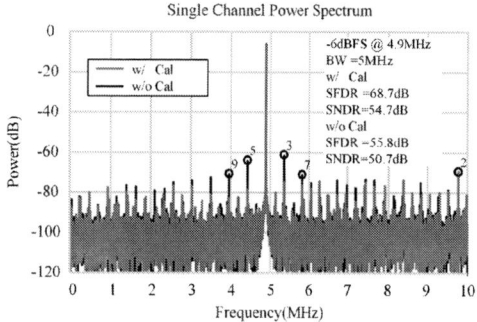

Fig. 10 Power breakdown of proposed ADC

	This work	[1]	[7]	[8]
Process	130nm BCD	130nm BCD	180nm CMOS	180nm CMOS
Pitch	250μm	150μm	-	150μm
Transducer	PZT	PZT	-	PZT
Architecture	C2C SAR	SAR SS	BF SAR	ΔΣ
Sampling Rate	20MS/s	30MS/s	20MS/s	-
Center Frequency	5MHz	5MHz	5MHz	5MHz
SNDR	54.7dB	49.8dB	60.1dB	47.0dB
Power consumption	0.79mW	1.5mW	2.6mW	0.8mW
area	0.034mm²	0.026 mm²	0.525 mm²	0.025 mm²

TABLE I.   TABLE TYPE STYLES

V. CONCLUSION

In this work, a hybrid C2C SAR ADC for 3-D ICE is proposed. It can solve the problem of too many 3-D transesophageal catheters. At the same time, the introduction of the hybrid C2C SAR ADC architecture can effectively solve the area and power consumption requirements of the ICE system. The proposal of the hybrid structure reduces the requirement of the ADC for the driving capability of the AFE when the AFE is directly connected to the ADC. The proposed structure SNDR of 54.7dB is achieved within 100% bandwidth of the transducer center frequency of 5MHz.

REFERENCES

[1] J. Li *et al.*, "A 1.54mW/Element 150μm-Pitch-Matched Receiver ASIC with Element-Level SAR/Shared-Single-Slope Hybrid ADCs for Miniature 3D Ultrasound Probes," *2019 Symposium on VLSI Circuits*, Kyoto, Japan, 2019, pp. C220-C221.

[2] Y. M. Hopf et al., "A Pitch-Matched Transceiver ASIC With Shared Hybrid Beamforming ADC for High-Frame-Rate 3-D Intracardiac Echocardiography," in IEEE Journal of Solid-State Circuits, vol. 57, no. 11, pp. 3228-3242.

[3] Y. M. Hopf et al., "A Pitch-Matched High-Frame-Rate Ultrasound Imaging ASIC for Catheter-Based 3-D Probes," in IEEE Journal of Solid-State Circuits, vol. 59, no. 2, pp. 476-491.

[4] B. P. Ginsburg and A. P. Chandrakasan, "500-MS/s 5-bit ADC in 65-nm CMOS With Split Capacitor Array DAC," in IEEE Journal of Solid-State Circuits, vol. 42, no. 4, pp. 739-747.

[5] C. -C. Liu, S. -J. Chang, G. -Y. Huang and Y. -Z. Lin, "A 10-bit 50-MS/s SAR ADC With a Monotonic Capacitor Switching Procedure," in IEEE Journal of Solid-State Circuits, vol. 45, no. 4, pp. 731-740.

[6] C. Liu, J. Liu and N. Sun, "Mitigating Sampling Noise for Energy-Efficient ADCs: A Tutorial Brief," in IEEE Transactions on Circuits and Systems II: Express Briefs, vol. 71, no. 3, pp. 1638-1643.

[7] T. Kim, S. Shin and S. Kim, "An 80.2 dB DR 23.25 mW/channel 8 channel ultrasound receiver with a beamforming embedded SAR ADC," in IEEE Transactions on Circuits and Systems II: Express Briefs, vol. 66, no. 9, pp. 1487-1491.

[8] M. D'Urbino et al., "An Element-Matched Electromechanical ΔΣ ADC for Ultrasound Imaging," in IEEE Journal of Solid-State Circuits, vol. 53, no. 10, pp. 2795-2805.

accounting for only 4% of the total power consumption. The echo received by the ultrasonic ASIC applied in this design has a center frequency of 5 MHz, and the SNDR of proposed hybrid C2C SAR ADC measured at this test frequency is 50.7 dB. The ADC output spectrum in Fig. 10 reveals that the initial SNDR is limited to 50.7 dB due to parasitic capacitance and capacitor mismatch. After calibration, the SNDR improves to 54.7 dB.

A Reconfigurable 9-to-14b 15MS/s 4th-Order NS-SAR ADC with Self-Calibrated Open-loop FIA

Chaoran Chen [1], Mingzong Lin [1], Jian Xu [2], Yue Lin [2], Wei Li [1], Hongtao Xu [1]

[1] State Key Laboratory of Integrated Chips and Systems, College of Integrated Circuits and Micro-Nano Electronics,
Fudan University, Shanghai, China
[2] ICLegend Micro, Shanghai, China

Email: 22112020005@m.fudan.edu.cn, hongtao@fudan.edu.cn

Abstract—This paper presents a reconfigurable noise-shaping SAR (NS-SAR) ADC for multi-scenario smart sensor system-on-chip (SoC) application. In this work, an optimal 4th-order EF-CIFF-CIFF architecture is implemented based on the proposed optimization model of cascaded NS structures. With a floating inverter amplifier (FIA) gain self-calibration scheme and a CMOS thyristor-based auto-timer, the proposed ADC is trim-less and widely reconfigurable on sampling rate, bandwidth, resolution and power consumption. Simulations results shows the proposed ADC achieves 85.84 dB SNDR, 1.25MHz bandwidth and Schreier figure of merit (FoMs) of 180.9 dB. Furthermore, it exhibits a flat FoMs around 180 dB with sampling clock scaling from 1MS/s to 15MS/s.

Keywords—*Noise-shaping, NS-SAR ADC, floating inverter amplifier, FIA gain self-calibration*

I. INTRODUCTION

In recent years, the rapid development of Artificial Intelligence of Things (AIoT) has been fueling the interest of highly integrated smart sensor system-on-chips (SoCs). The varying sensing scenarios and operation modes demand for high performance Analog-to-digital converters (ADCs) to be reconfigurable within a wide range of bandwidth around sub-mega-hertz and resolutions between 10-14 bit, while maintaining power efficiency and flat figure-of-merit.

Conventional ADC architectures exhibits limited speed and resolution to be reconfigured [1], [2]. In recent research, noise shaping SAR (NS-SAR) ADC features reconfigurable f_s within several hundred kHz [3]. The scarcity of medium-to-high precision reconfigurable ADC motivates investigation into higher order of NS-SAR ADCs.

In this work, a reconfigurable NS-SAR ADC with an optimized noise-shaping structure and floating inverter amplifier (FIA) self-calibration technique is proposed, which is highly reconfigurable on sampling rate, bandwidth, resolution and power.

II. NOISE-SHAPING ARCHITECTURE OPTIMIZATION

A generalized NS-SAR frame incorporating both error feedback (EF) and cascaded integrator feed-forward (CIFF) loop is depicted in Fig. 1, with their noise transfer functions (NTF) in which E_Q, E_{N1}, E_{N2} are quantization noise, input referred noise introduced in EF and CIFF loop filter respectively. In a simple 1st-order case, loop filters H_{EF} and H_{CIFF} take the term of z^{-1} and $z^{-1}/(1- z^{-1})$. It reveals that E_Q is shaped by both EF and CIFF loop, and E_{N1} is not shaped (since H_{EF} is low-pass), while E_{N2} is shaped by EF loop (through 1-H_{EF}), making it advantageous to combine EF-CIFF loop filters. It is worth noting that the sampling noise (E_S) and distortion

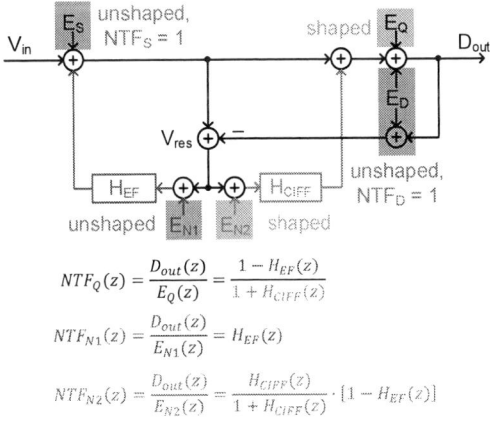

Fig. 1. A generalized NS-SAR incorporating both EF and CIFF loop filters and transfer function of non-idealities.

$$NTF_Q(z) = \frac{D_{out}(z)}{E_Q(z)} = \frac{1 - H_{EF}(z)}{1 + H_{CIFF}(z)}$$

$$NTF_{N1}(z) = \frac{D_{out}(z)}{E_{N1}(z)} = H_{EF}(z)$$

$$NTF_{N2}(z) = \frac{D_{out}(z)}{E_{N2}(z)} = \frac{H_{CIFF}(z)}{1 + H_{CIFF}(z)} \cdot [1 - H_{EF}(z)]$$

(E_D) are not suppressed, due to their unity noise transfer functions (NTF).

To be more comprehensive, a cascaded NS-SAR model is illustrated in Fig. 2. The overall NTF of E_Q is the product of each cascade stage [4], and $E_{N1,i}$ or $E_{N2,i}$ (i denotes a certain stage) are first shaped by current stage, and then by the following stages, as formulated in Fig. 2. To determine the optimum architecture for our nominal design point (i.e. SNDR>86 dB, BW = 1.25 MHz), an estimation model is developed, in which the integral noise of each non-idealities is presented to determine the signal-to-noise-ratio (SNR). Furthermore, with the estimated power of SAR core and loop filter, the model yields a FoMs of a certain cascaded structure. Traversing all possible structure, an optimized NS architecture under certain constrains is determined by ranking FoMs.

Before we start, based on the loop filter property, a prediction could be made: An ideal NS-SAR architecture features inner loop filter(s) with relaxed noise requirement, because its EN will be shaped by the subsequent stages, and outer loop filters, in which noise from CIFF filter is shaped, remaining only outer EF filter noise significant

The estimation model is initiated with the constrains:

- Total order up to 4, with EF in each stage up to 2, and CIFF up to 4.

- SAR core bits ranges from 6-bit to 12-bit.

- Bandwidth fixed at 1.25 MHz, while over-sampling ratio (OSR) ranges between 6 and 12.

979-8-3315-3918-4/25 $31.00 © 2025 IEEE

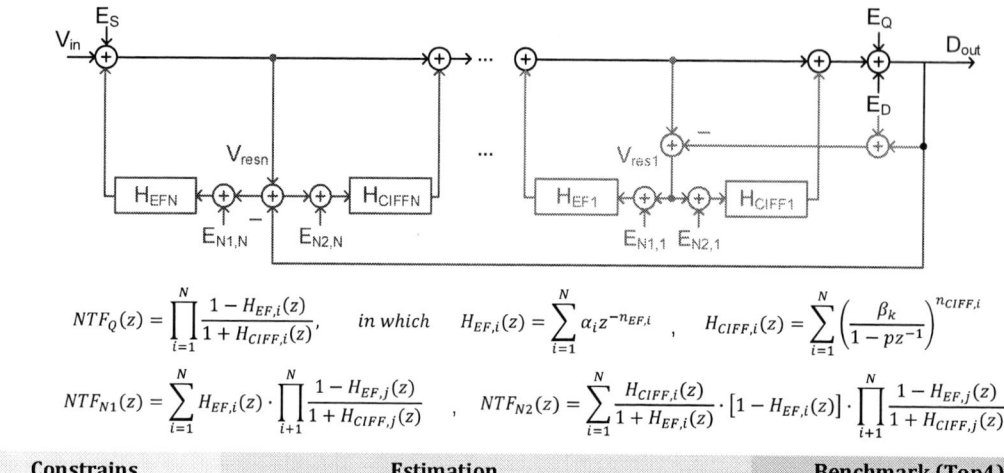

$$NTF_Q(z) = \prod_{i=1}^{N} \frac{1 - H_{EF,i}(z)}{1 + H_{CIFF,i}(z)}, \quad in\ which \quad H_{EF,i}(z) = \sum_{i=1}^{N} \alpha_i z^{-n_{EF,i}}, \quad H_{CIFF,i}(z) = \sum_{i=1}^{N} \left(\frac{\beta_k}{1 - pz^{-1}}\right)^{n_{CIFF,i}}$$

$$NTF_{N1}(z) = \sum_{i=1}^{N} H_{EF,i}(z) \cdot \prod_{i+1}^{N} \frac{1 - H_{EF,j}(z)}{1 + H_{CIFF,j}(z)}, \quad NTF_{N2}(z) = \sum_{i=1}^{N} \frac{H_{CIFF,i}(z)}{1 + H_{EF,i}(z)} \cdot [1 - H_{EF,i}(z)] \cdot \prod_{i+1}^{N} \frac{1 - H_{EF,j}(z)}{1 + H_{CIFF,j}(z)}$$

Constrains	Estimation	Benchmark (Top4)				

Constrains	**Estimation**			**Benchmark (Top4)**		
• Order: total=[1,4]; EF=[0,2], CIFF=[0,4] • SAR core bits L : [6,12] • BW=1.25M; OSR=[4,16] • E_S=33µ; E_{N1}=E_{N2}=10µ • K_1=1.6×10^{-15}; K_2=4×10^{-22}	$P_Q = \frac{E_Q^2}{\pi} \int_0^{\frac{\pi}{OSR}} \|NTF_Q(j\omega)\|^2 d\omega \quad P_{sig} = \frac{V_{FS}^2}{8}, \quad P_S = \frac{E_S^2}{OSR}$ $P_N = \frac{E_{N1}^2}{\pi} \int_0^{\frac{\pi}{OSR}} \|NTF_{N1}(j\omega)\|^2 d\omega + \frac{E_{N2}^2}{\pi} \int_0^{\frac{\pi}{OSR}} \|NTF_{N2}(j\omega)\|^2 d\omega$ $P_{tot} = P_{core} + P_{filter} = BW \cdot OSR \cdot \left(\frac{K_1}{E_Q} + \frac{K_2}{P_N}\right) \quad E_Q = \frac{2^{-L}}{\sqrt{12}}$	**Stage2 order**	**Stage1 order**	**SAR bits**	**OSR**	**FoM**
		EF=2 CIFF=1	EF=0 CIFF=1	9	6	183.20
		EF=2 CIFF=1	EF=1 CIFF=0	9	6	183.08
		EF=0 CIFF=1	EF=2 CIFF=1	9	6	182.98
Aim: $SNR = 10\log_{10} \frac{P_{sig}}{P_S + P_Q + P_N} > 86dB, FoM_S = SNR + 10\log_{10}\frac{BW}{P_{tot}}\ maximized$			EF=2 CIFF=1	EF=0 CIFF=1	10	5	182.92

Fig. 2. Estimation model for a general cascaded NS-SAR, and the configurations of the top-4 structure.

- Sampling noise fixed at 33 uV, i.e. a fixed CDAC area.
- Estimated power relates to, noise, OSR and survey-process-based factors K_1, K_2.

By ranking the yielding FoMs, configurations of the top-4 structure are presented in Fig. 2. The optimum structure is in line with our prediction: As depicted in Fig. 3, it features a 1st-order CIFF inner loop, and outer loop with 2nd-order EF and 1st-order CIFF, forming an EF-CIFF-CIFF structure in general. The proposed architecture has several advantages: As mentioned before, the noise introduced in inner CIFF is shaped by outer stage, and the outer CIFF shaped by EF loop. Furthermore, as illustrated in Fig. 3 and [5], the comparator offset is suppressed by the outer CIFF path, if the outer loop is implemented by a merged op-amp manner.

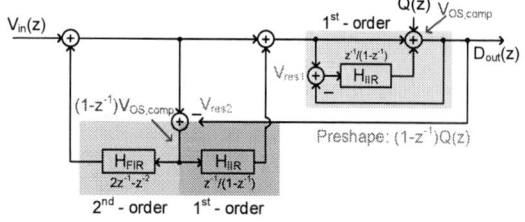

Fig. 3. The proposed NS-SAR architecture.

III. CIRCUIT IMPLEMENTATION

The top-level schematic of the proposed 4th-order NS-SAR ADC is illustrated in Fig. 5. According to the optimization

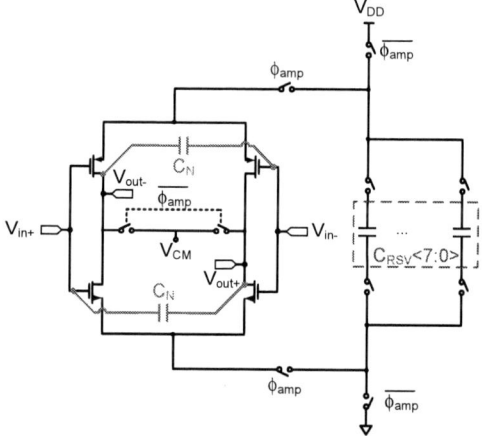

Fig. 4. FIA with gain controlled by C_{RSV}.

results in section II, a SAR core of 9-bit with 1-bit redundancy is adopted. The proposed NS-SAR structure utilizes 2 op-amp for 4th-order noise-shaping with optimized zeros, and incorporates a sampling-noise-cancellation (SNC) scheme to allow for smaller CDAC area [5]. In this section, an open-loop floating-invert amplifier (FIA) self-calibration technique and a corresponding auto-timer are presented.

A. Self-Calibrated Open-loop FIA

In this work, a floating-inverter amplifier (FIA) is employed as the loop filter op-amp for its high current efficiency and stable output common mode. Furthermore, the open-loop gain features time and reservoir relating property,

979-8-3315-3918-4/25 $31.00 © 2025 IEEE

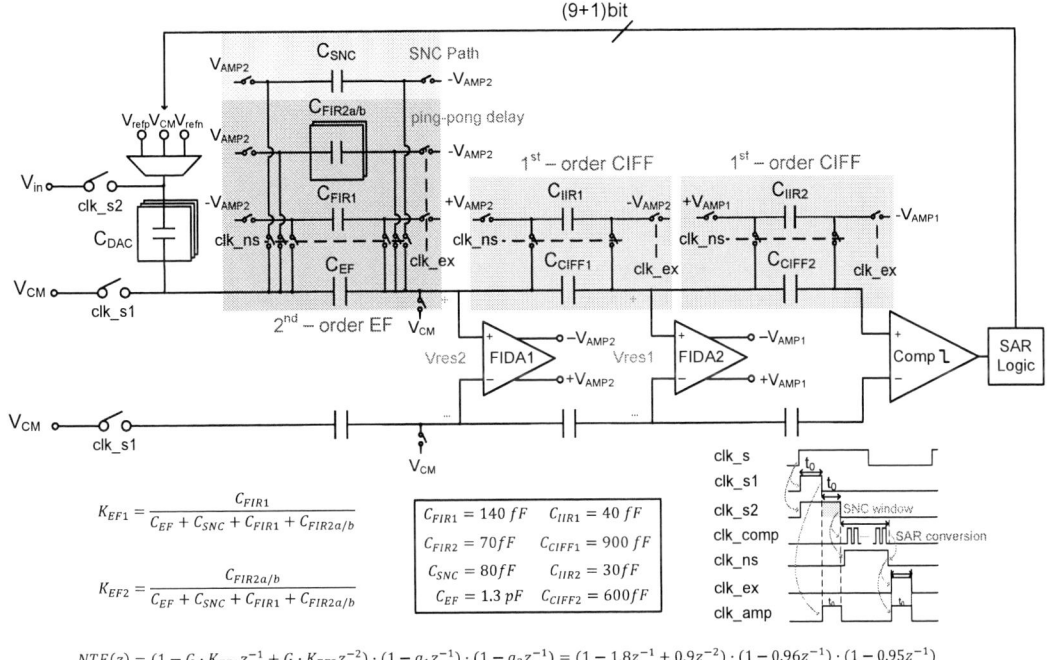

$$K_{EF1} = \frac{C_{FIR1}}{C_{EF} + C_{SNC} + C_{FIR1} + C_{FIR2a/b}}$$

$$K_{EF2} = \frac{C_{FIR2a/b}}{C_{EF} + C_{SNC} + C_{FIR1} + C_{FIR2a/b}}$$

$C_{FIR1} = 140\,fF$	$C_{IIR1} = 40\,fF$
$C_{FIR2} = 70\,fF$	$C_{CIFF1} = 900\,fF$
$C_{SNC} = 80\,fF$	$C_{IIR2} = 30\,fF$
$C_{EF} = 1.3\,pF$	$C_{CIFF2} = 600\,fF$

$$NTF(z) = (1 - G \cdot K_{EF1} z^{-1} + G \cdot K_{EF2} z^{-2}) \cdot (1 - a_1 z^{-1}) \cdot (1 - a_2 z^{-1}) = (1 - 1.8 z^{-1} + 0.9 z^{-2}) \cdot (1 - 0.96 z^{-1}) \cdot (1 - 0.95 z^{-1})$$

Fig. 5. Top-level schematic of the proposed 4th-order NS-SAR ADC.

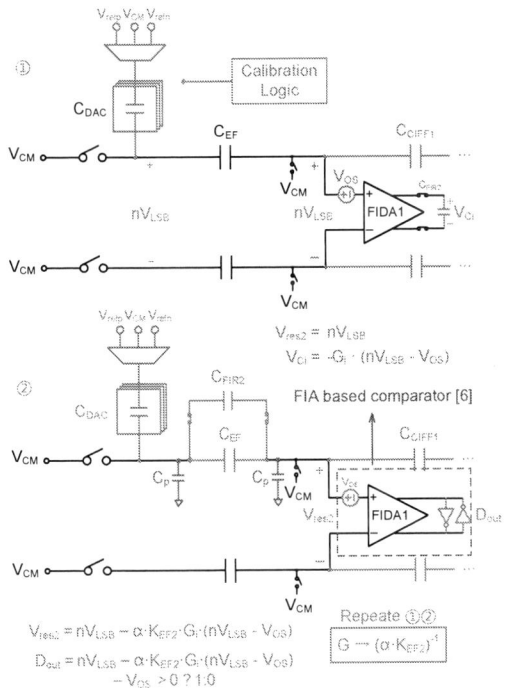

Fig. 6. FIA gain self-calibration scheme.

making it possible to control the residue gain by controlling a C_{RSV} array [5]:

$$G \approx \frac{\alpha C_{RSV}}{C_L} ln \left(1 + \frac{t}{\tau}\right), \quad \tau = \frac{2n V_t C_{RSV}}{I_{AMP}(0^+)} \quad (1)$$

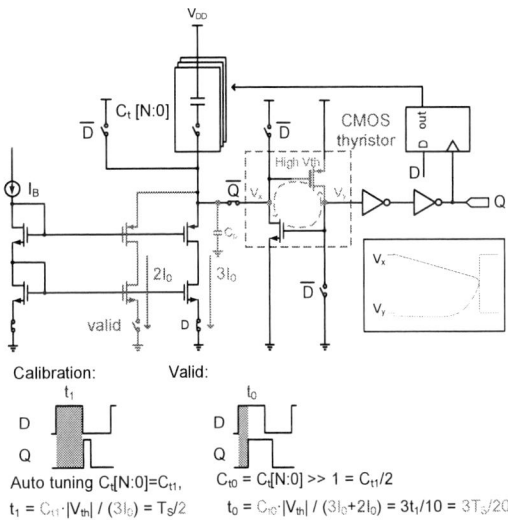

Fig. 7. CMOS thyristor-based auto-timer with calibration scheme.

As presented in Fig. 6, the FIA self-calibration scheme works as follow: First, the CDAC array generates a moderate voltage of several LSB, then FIA extracts and amplifies the nV_{LSB} along with offset V_{OS}. The capacitor storing the FIA output (C_{FIR2}) is then charge-shared with C_{EF}. With a parasitic attenuation factor α, the voltage at FIA input is $nV_{LSB} - \alpha \cdot G_i \cdot K_{EF2} \cdot (nV_{LSB} - V_{OS})$. The FIA is then configured as an FIA based comparator by adding a strong-arm latch at FIA output, as illustrated in [6]. By auto controlling $C_{RSV}<n:0>$ and comparing, $nV_{LSB} - \alpha \cdot G_i \cdot K_{EF2} \cdot (nV_{LSB} - V_{OS}) \to 0$, thus $G \to 1 / (\alpha K_{EF2})$, i.e. the calibrated gain is element-ratioed, and free from parasitic attenuation and FIA offset.

B. CMOS Thyristor-Based Auto-Timer

In this work, a timer featuring a delay (t_0) which is adaptive to f_S and regulates the FIA working time is essential for SNC and residue extraction. As depicted in Fig. 7, an auto-timer with CMOS thyristor-based delay element [7] is proposed. It is similar to a simple delay-lock-loop (DLL). When the calibration is done, the clock-to-Q delay is settled at a fraction of the system clock. It's worth noting that when a clock with precise duty-cycle is not available, the calibration scheme can be changed to single edge triggered. Furthermore, t_0 need not to be accurate because of the FIA self-calibration technique, and even need not to cover all the f_S, because the saturation of the timer will avoid the saturation of FIA in the cases where f_S is slow. The auto-timer generates a 10ns delay and consumes 1.5uW at f_S = 15MHz, with enhanced power and jitter performance compared with RC-type and inverter-chain-type delay elements [7].

IV. SIMULATION RESULTS

The proposed NS-SAR ADC is implemented in 40nm CMOS process, and simulation results in Fig. 8(a) illustrate that it achieves 85.84 dB SNDR at nominal condition (i.e. f_S = 15MS/s, OSR = 6, BW = 1.25 MHz). Employing a fully asynchronous self-timed logic, the proposed ADC is clocked by system clock f_S only. With f_S scaling within 1-15MHz, the proposed NS-SAR ADC exhibits a flat FoMs of around 180-dB, as Fig. 8(b) shows. Fig. 8(c) illustrates that the proposed ADC is power-scalable by NS-order, until scaled down to a conventional SAR ADC with 9-bit ENOB and 7.5MHz bandwidth. Fig.8 (d) compares this work with other NS-SAR ADCs above 80 dB SNDR. The proposed ADC exhibits a wide reconfigurable range in both bandwidth and resolution: It performs towards lower BW and resolution by lowering NS order or f_S, and towards higher resolution by increasing OSR.

TABLE I. PERFORMANCE SUMMARY AND COMPARISON

	ISSCC 21	ISSCC 22	CICC 17	*This work**
Process [nm]	65	65	65	40
Architecture	EF-CIFF	EF-CRFF	CIFF	EF-CIFF-CIFF
Reconfigurable	no	no	Yes	**Yes**
f_S [MS/s]	10	5	25**	15**
BW [kHz]	625	500	625**	**1250****
OSR	8	5	20	6**
SNDR [dB]	84.8	84.1	80.4**	85.84**
Power [µW]	119	133.8	630.2**	338.2**
FoMs [dB]	182	180	170.4	180.9

*Simulation results. **Reconfigurable parameters.

V. CONCLUSION

In this paper, a reconfigurable NS-SAR ADC with an optimized noise-shaping structure is presented. Employing FIA gain self-calibration and timer auto-tuning techniques, the proposed ADC exhibits a wide reconfigurable range on

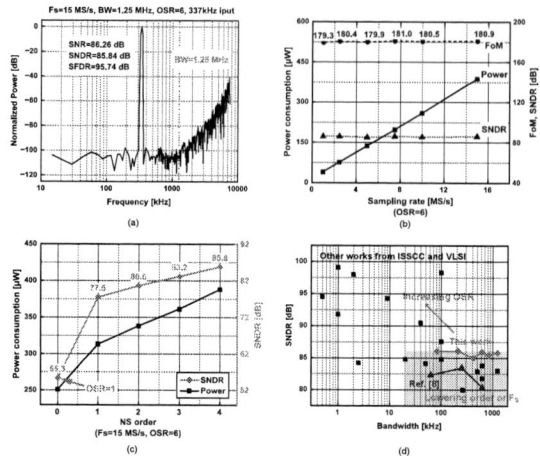

Fig. 8. Simulation results of the proposed NS-SAR ADC. (a)FFT plot under nominal condition. (b)Sampling frequency vs. power consumption, FoMs and SNDR. (c)Power-scalability with NS order. (d) Comparison with other NS-SAR ADCs over 80dB SNDR.

sampling rate, bandwidth, resolution and power, making it appropriate for multi-scenario smart sensor applications.

ACKNOWLEDGMENT

This work was supported by Special Funds for High-Quality Development (NICT) in Shanghai in 2024 （ No. 2024-JCSS-01010).

REFERENCES

[1] M. Taherzadeh-Sani and A. A. Hamoui, "A reconfigurable and power-scalable 10–12 bit 0.4–44 MS/s pipelined ADC with 0.35–0.5 pJ/step in 1.2 V 90 nm digital CMOS," IEEE Trans. Circuits Syst. I, Reg. Papers, vol. 60, no. 1, pp. 74–83, Jan. 2012.

[2] M. Yip and A. P. Chandrakasan, "A resolution-reconfigurable 5-to-10-bit 0.4-to-1 V power scalable SAR ADC for sensor applications," IEEE J. Solid-State Circuits, vol. 48, no. 6, pp. 1453–1464, Jun. 2013.

[3] R. Karim, M. Grassi, and P. Malcovati, "An event-triggered asynchronous incremental NS-SAR ADC featuring sampling-rate reconfigurability with power-scalability and enabling AFE-ADC co-design approach," IEEE Access, early access, 2024.

[4] L. Jie et al., "An Overview of Noise-Shaping SAR ADC: From Fundamentals to the Frontier," in IEEE Open Journal of the Solid-State Circuits Society, vol. 1, pp. 149-161, 2021.

[5] T.-H. Wang, R. Wu, V. Gupta, and S. Li, "27.3 A 13.8-ENOB 0.4pF-CIN 3rd -order noise-shaping SAR in a single-amplifier EF-CIFF structure with fully dynamic hardware-reusing kT/C noise cancellation," in Proc. IEEE Int. Solid- State Circuits Conf. (ISSCC), 2021, pp. 374–376.

[6] X. Tang et al., "An energy-efficient comparator with dynamic floating inverter amplifier," IEEE J. Solid-State Circuits, vol. 55, no. 4, pp. 1011–1022, Apr. 2020.

[7] G. Kim, Min-Kyu Kim, Byoung-Soo Chang and W. Kim, "A low-voltage, low-power CMOS delay element," in IEEE Journal of Solid-State Circuits, vol. 31, no. 7, pp.966-971, July 1996.

[8] M. Miyahara and A. Matsuzawa, "An 84-dB dynamic range 62.5–625 kHz bandwidth clock-scalable noise-shaping SAR ADC with open-loop integrator using dynamic amplifier," in Proc. IEEE Custom Integr. Circuits Conf., 2017, pp. 1-4.

A 16-bit 4-MS/s Deadlock-free Asynchronous SAR ADC Using High-level First Transmission Gate

Xiaokun Zhou, Baijie Zhang, Xu Cheng*

State Key Laboratory of Integrated Chips and Systems, Fudan University, Shanghai 200433, China

* Email: 23212020215@m.fudan.edu.cn, chengxu@fudan.edu.cn

Abstract—A Successive-approximation-register analog-to-digital converter (SAR ADC) using bottom-plate sampling and a split capacitor digital-to-analog converter (CDAC) may suffer from the nonlinearity caused by the negative voltage of the top plate in the redistribution phase. Therefore, this paper proposes a high-level first transmission gate (HFTG) that uses a dedicated delay cell to avoid the negative voltage. Besides, using an inverter-based preamplifier without initialization in an asynchronous SAR ADC possibly results in a deadlock. Therefore, a deadlock-free common-mode feedback (CMFB) timing sequence with an extra initialization phase is also proposed. Employing these two techniques, an asynchronous SAR ADC is designed in 28 nm standard CMOS process. Simulation shows that the SAR ADC achieves an SFDR of 101.7 dB, an SNDR of 92.9 dB and an ENOB of 15.13 bits, while HFTG achieving an improvement of 14.2 dB SFDR and 7.2 dB SNDR.

Keywords—Successive-approximation-register (SAR), analog-to-digital converter (ADC), asynchronous, high-level first transmission gate (HFTG), deadlock-free.

I. INTRODUCTION

Successive-approximation-register (SAR) analog-to-digital converter (ADC) mainly holds the advantages of low power and simple structure. However, the decreasing supply voltage accompanied by the scaling down of the CMOS feature size challenges the design of high-resolution SAR ADCs.

Therefore, inverter-based preamplifiers are proposed [1-4]. Since this structure is pseudo-differential, a common-mode feedback (CMFB) circuit is necessary. In [4], the common-mode feedback only is activated after a conversion of the SAR ADC, so it takes a lot of cycles for the preamplifier to output a stable common-mode voltage. Before that, the comparator may generate no output results due to an inappropriate common-mode voltage.

In this case, the synchronous SAR ADC will not stop and the common-mode feedback continues operating. In contrast, this mechanism may not work in an asynchronous SAR ADC. For an asynchronous logic, if no result of comparison is sent to the logic module, the operation of SAR ADC will stop, which causes the common-mode feedback procedure also to stop. This phenomenon is called "deadlock" [16]. Therefore, this paper proposes a deadlock-free CMFB timing sequence to avoid the deadlock by adding initialization phase.

In terms of the capacitor digital-to-analog converter (CDAC), The split CDAC [5] which fixes the common-mode voltage of the top plate is employed. However, since the top plate is connected to ground in sampling phase, the range of input voltage is limited. Connecting the top-plate of the CDAC in [5] to V_{CM} (reference common-mode voltage) when

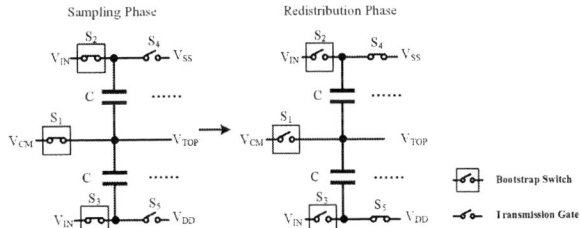

Fig. 1. The employed split capacitor digital-to-analog converter.

sampling solves this problem. This structure is employed as shown in Fig.1.

In redistribution phase in Fig.1, the level of the top plate V_{TOP} is pulled up by S_5 and is pulled down by S_4. However, if a weak V_{DD} is transferred first, the switch of the top plate S_1 may generate the leakage current, leading to nonlinearity in sampling. It is noted that the same problem also exists in [5]. Therefore, this paper proposes a high-level first transmission gate (HFTG) that makes the PMOS in the transmission gate conduct before the NMOS by specifically introducing a delay cell.

This paper is organized as follows. Section II presents the proposed techniques and theories. Section III shows the implementation of the designed SAR ADC. Then, simulation results are exhibited in Section IV. Finally, Section V draws the conclusion.

II. PROPOSED TECHNIQUES

A. High-level First Transmission Gate (HFTG)

In Fig.1, according to the law of conservation of charge, the top-plate voltage V_{TOP} in the redistribution phase can be expressed as

$$V_{TOP} = V_{CM} + \frac{V_{DD}}{2} - V_{IN}, \tag{1}$$

where V_{IN} is the input signal level.

For ease of understanding, t_p and t_n are defined to describe the moment when the PMOS conducts and the moment when the NMOS conducts. Then $t_d = t_n - t_p$ is defined to describe the time when the PMOS conducts before the NMOS in a transmission gate.

- If $t_d < 0$, the NMOS of the transmission gate conducts before the PMOS as shown in Fig. 2(a) and Fig. 2(b). Since the NMOS can only transfer a weak V_{DD} due to the threshold voltage $V_{TH,N}$, V_{TOP} before the PMOS is conducted can be expressed as

979-8-3315-3918-4/25 $31.00 © 2025 IEEE

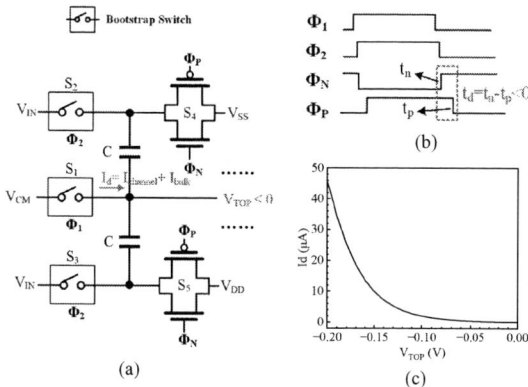

(a)

(b)

(c)

Fig. 2. (a) Diagram of the generation of leakage current. (b) Timing sequence of the generation of leakage current. (c) Relationship of the leakage current and the top-plate voltage.

$$V_{TOP} = V_{CM} + \frac{V_{DD} - V_{TH,N}}{2} - V_{IN}. \qquad (2)$$

When $V_{IN} > V_{CM} + \frac{V_{DD}-V_{TH,N}}{2}$, V_{TOP} will be pulled down below 0. For the sampling NMOS in the bootstrap switch S_1 [14], the subthreshold leakage current $I_{channel}$ is generated caused by the gate-to-source voltage $V_{GS} > 0$, and the forward diode leakage current [15] I_{bulk} is generated caused by the bulk-to-source voltage $V_{BS} > 0$. The total leakage current I_d consists of the $I_{channel}$ and the I_{bulk} shown in Fig. 2(a), injecting charges related to the input signal into the floating node V_{TOP}.

- If $t_d = 0$, the NMOS and the PMOS are turned on simultaneously. V_{TOP} may be pulled down faster than being pulled up because of the stronger driving ability of the NMOS, leading to the generation of leakage current.

- If $t_d > 0$, the PMOS is conducted first and transfer a weak V_{SS}. V_{TOP} will be always over 0, avoiding the leakage current mentioned above.

According to the analysis above, only $t_d > 0$ is acceptable in a high-resolution SAR ADC. It should be noted that the control signals Φ_N and Φ_P of the transmission gate in Fig.2 are generated by an inner asynchronous SAR logic. If t_d is not set dedicatedly between them, it will be uncertain caused by layout parasitics and process, which is uncontrollable in actual circuits.

Simulation shows the relationship between I_d and V_{TOP} as Fig. 2(c). It verifies the generation of I_d caused by a negative V_{TOP}. This problem also occurs in [5], deteriorating the resolution of ADCs.

Thus, Fig. 3(a) shows the proposed structure of HFTG used for S_4/S_5. The HFTG topology differs from the traditional transmission gate, in that a delay cell is added into the control path, thereby making sure that t_d is over 0.

Fig. 3(b) and Fig .3(c) compare the simulation results of SFDR and SNDR with different input signal amplitudes and t_d. It can be clearly seen that compared with the results of a

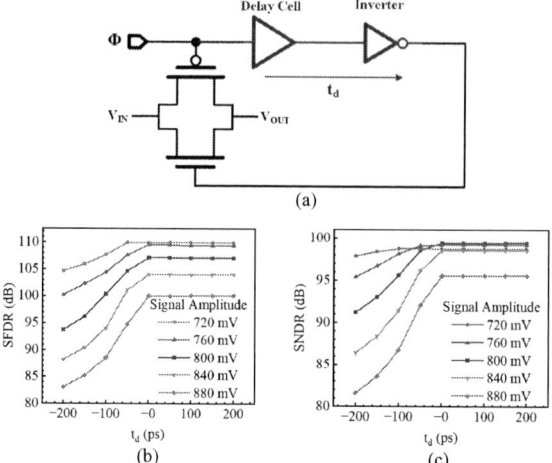

(a)

(b)

(c)

Fig. 3. (a) Schematic of the proposed high-level first transmission gate. (b) SFDR and (c) SNDR in redistribution phase with different input signal amplitudes and delay times if transmission gates.

negative t_d, HFTG has a higher sampling accuracy, and as the amplitude of the input signal increases, the difference will become more obvious. This conclusion is consistent with the above analysis. Therefore, it is clear to see that HFTG performs a significant improvement of sampling accuracy and robustness.

B. Deadlock-free CMFB Timing Sequence

The deadlock is caused by an inappropriate common-mode voltage of the comparator output. Fig.4. shows the schematic of the inverter-based comparator in [4]. The common-mode (CM) comparator compares the output common-mode voltage with V_{CM}. Based on the comparison result, the CMFB circuit performs charging or discharging operations on the common-mode bias capacitors C_{fi}, and then adjusts the common-mode bias of the preamplifier.

It should be noted that the latch with NMOS as the input transistor is employed in this paper. If the output common-mode voltage of the preamplifier is too low, the latch may generate no output results. Fig. 5(a) shows the deadlock caused by this reason in an asynchronous SAR ADC. Ideally, asynchronous SAR logic reads the output result of the comparator and drives the next bit cycle. And then drive common-mode feedback once when the conversion is completed. If the comparator has no output result available for asynchronous SAR logic to read due to incorrect common-mode bias, SAR logic will keep waiting in the current state and fall into a deadlock state. In Fig. 5, *INIT* is the initialization signal, *SAMPLE* is the sampling control signal, *Latch* is the latch control signal, V_{OUTP} and V_{OUTN} are the output results of the comparator, *CONV_RDY* is the signal indicating the completion of ADC conversion, and *EN_CMFB* is the enable signal for common-mode feedback.

To solve this problem, a deadlock-free timing sequence for common-mode feedback is proposed as shown in Fig. 5(b). When powered on, the common-mode bias is quickly established by inputting INIT and an external clock signal to generate EN_CMFB. In addition, the SAR ADC refreshes the common-mode bias once when a conversion is finished. Compared with the method in [4], the mechanism proposed ensures that there will be no deadlock due to the improper bias of the comparator in an asynchronous SAR ADC.

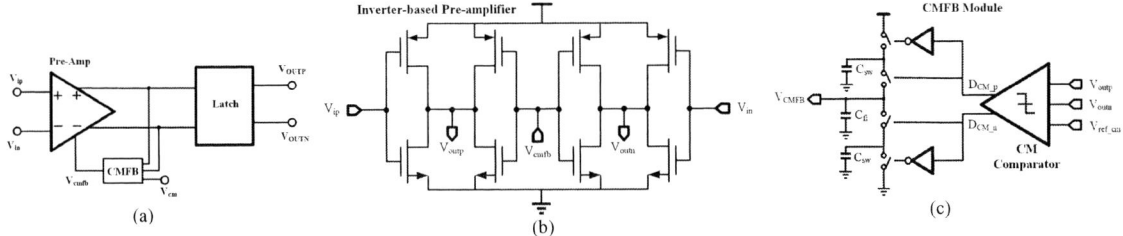

Fig. 4. Schematic of (a) the comparator, (b) the Inverter-based preamplifier and (c) the common-mode feedback circuit in [4].

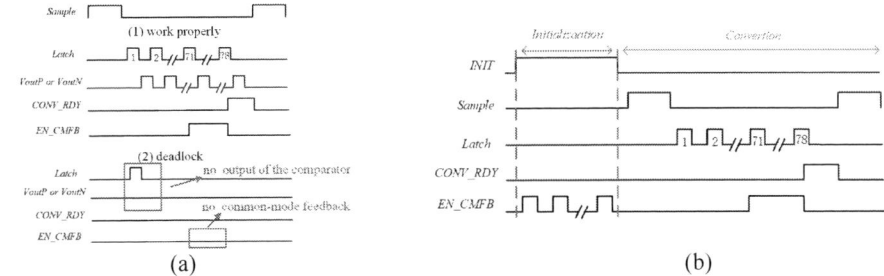

Fig. 5. Waveform diagram of (a) the deadlock in asynchronous SAR ADC. (b) the proposed deadlock-free common-mode feedback timing.

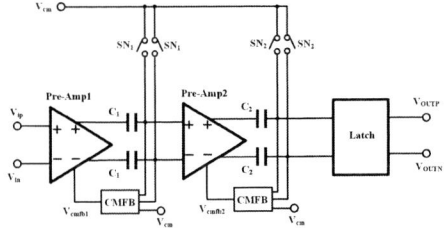

Fig. 6. Schematic of the two-stage comparator with OOS.

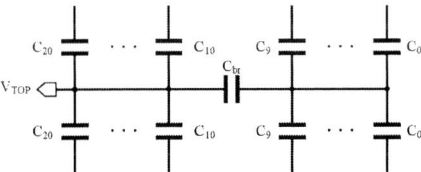

Fig. 7. One of the two CDACs of the designed SAR ADC.

III. IMPLEMENT OF THE SAR ADC

The asynchronous SAR ADC in this paper is designed in 28 nm standard CMOS process and the supply voltage is 0.9 V. Then, some significant components of the ADC are discussed.

A. Two-stage Comparator with OOS

Single-stage inverter for preamplification in [4] cannot meet the requirement of a high-resolution SAR ADC. Therefore, in this paper, a two-stage comparator is employed as the structure for preamplification, and the output offset storage (OOS) technology is used to improve the accuracy as shown in Fig.6.

B. Splitting CDAC with redundancy

Fig. 7 shows one of the two CDACs in the proposed differential 16-bit SAR ADC. In order to reduce the size of CDAC, the structure of bridge capacitor is employed. In addition, to reduce the establishment time and facilitate background digital calibration, redundancy is utilized in the CDAC in this paper. The weights of C_{20} to C_{10} are respectively {3000, 2000, 1250, 750, 488, 289, 170, 100, 60, 35, 20}, and the weights of C_9 to C_0 are {24, 14, 8, 5, 3, 2, 1, 1, 1, 1}. Besides, the weight of the bridge capacitor C_{br} is taken as 120 to avoid a fraction and the unit capacitance is set to $4\,fF$ to meet the KT/C noise requirement of 16-bit resolution.

C. Asynchronous SAR Logic

After the sampling is completed, a small amount of time is reserved for the CDAC to flip and establish the signal voltage.

Subsequently, SAR logic outputs the first latch signal to the comparator, and the comparator outputs the comparison result to SAR logic. After SAR logic obtains the comparison result, it drives the CDAC to switch and outputs the second latch signal to the comparator. The asynchronous SAR ADC repeatedly implements such operations until all bits of the output digital codes are determined. It is noted that the last seven bits in an ADC conversion are decided by 9-time majority voting of the comparator to reduce the equivalent noise [10].

IV. SIMULATION RESULTS

To evaluate the influence of HFTG on the resolution of ADC, a spectrum analysis on the ADC conversion results is conducted. Using the two techniques proposed in Section II, under the conditions of $t_d = \pm100\,ps$, the simulation results are shown as Fig. 8.

According to the theoretical analysis in Section II, the leakage current exhibits a strong dependency on the amplitude of the input signal. Specifically, negligible leakage current is observed under small-signal conditions, while the leakage charge increases significantly with higher input signal levels. These results demonstrate a direct correlation between the input signal amplitude and the generated leakage charge.

Fig. 8 verifies the analysis very well. HFTG avoids generating high odd harmonics caused by the leakage current that causes the obvious deterioration of SFDR (the even harmonics have been eliminated by the differential structure). The proposed HFTG achieves an improvement of 14.2 dB SFDR and 7.2 dB SNDR compared with the normal topology,

979-8-3315-3918-4/25 $31.00 © 2025 IEEE

TABLE I.   PERFORMANCE SUMMARY AND COMPARISON

	This work	[4]	[11]	[12]	[13]
Process [nm]	28	28	55	130	180
Stage of Results	Pre-layout Simulation	Post-layout Simulation	Measurement	Post-layout Simulation	Pre-layout Simulation
Resolution [bit]	16	16	16	12	12
Supply [V]	0.9	1	3.3/1.2	3.3	1.8
Speed [MS/s]	4	200	16	16	20
Synchronous/Asynchronous	Asynchronous	Synchronous	Synchronous	Asynchronous	Asynchronous
SFDR [dB]	101.7	86	98	73	-
SNDR [dB]	92.9	80.1	78	70.8	70.48
ENOB [bit]	15.13	13.01	12.66	11.46	11.42

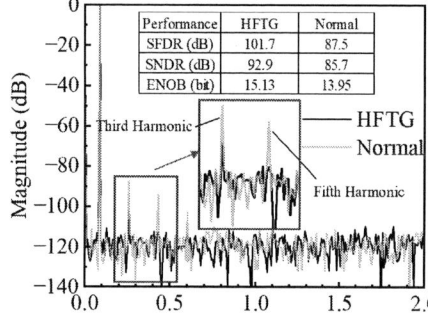

Fig. 8.   Measured ADC output spectra (512 sampling points, FFT).

significantly improving the resolution and robustness of the SAR ADC.

Table I summarizes the performance of the proposed asynchronous SAR ADC and comparisons with other advanced SAR ADCs. It can be seen that this design achieves higher resolution under a lower voltage supply. Therefore, this work adapts to the trend of CMOS development and the requirements of high-resolution SAR ADC quite well.

V.   CONCLUSION

The proposed high-level first transmission gate improves SFDR of SAR ADC by 14.2 dB. It is verified that HFTG significantly improves the sampling accuracy, and then increases the resolution of the ADC. In addition, the deadlock-free CMFB timing sequence avoid the potential deadlock in an asynchronous SAR logic. Both proposed techniques significantly strengthen the robustness of the ADC. Therefore, the proposed techniques are suitable for high-resolution and high-reliability asynchronous SAR ADC with a low voltage supply.

REFERENCES

[1] B. Verbruggen, K. Deguchi, B. Malki and J. Craninckx, "A 70 dB SNDR 200 MS/s 2.3 mW dynamic pipelined SAR ADC in 28nm digital CMOS," 2014 Symposium on VLSI Circuits Digest of Technical Papers, Honolulu, HI, USA, 2014, pp. 1-2.

[2] D. Luu et al., "A 12-bit 300-MS/s SAR ADC With Inverter-Based Preamplifier and Common-Mode-Regulation DAC in 14-nm CMOS FinFET," in IEEE Journal of Solid-State Circuits, vol. 53, no. 11, pp. 3268-3279, Nov. 2018.

[3] X. Tang et al., "An Energy-Efficient Comparator With Dynamic Floating Inverter Amplifier," in IEEE Journal of Solid-State Circuits, vol. 55, no. 4, pp. 1011-1022, April. 2020.

[4] L. Qiu, T. Meng, B. Yao, Z. Du and X. Yuan, "A High-Speed Low-Noise Comparator With Auxiliary-Inverter-Based Common Mode-Self-Regulation for Low-Supply-Voltage SAR ADCs," in IEEE Transactions on Very Large Scale Integration (VLSI) Systems, vol. 31, no. 1, pp. 152-156, Jan. 2023.

[5] B. P. Ginsburg and A. P. Chandrakasan, "500-MS/s 5-bit ADC in 65-nm CMOS With Split Capacitor Array DAC," in IEEE Journal of Solid-State Circuits, vol. 42, no. 4, pp. 739-747, April 2007.

[6] Y. Zhu et al., "A 10-bit 100-MS/s Reference-Free SAR ADC in 90 nm CMOS," in IEEE Journal of Solid-State Circuits, vol. 45, no. 6, pp. 1111-1121, June 2010.

[7] Chun-cheng Liu, Design of High-Speed Energy-Efficient Successive-Approximation Analog-to-Digital Converters [D], National Cheng Kung University, 2010.

[8] P. Harpe, E. Cantatore and A. van Roermund, "A 10b/12b 40 kS/s SAR ADC With Data-Driven Noise Reduction Achieving up to 10.1b ENOB at 2.2 fJ/Conversion-Step," in IEEE Journal of Solid-State Circuits, vol. 48, no. 12, pp. 3011-3018, Dec. 2013.

[9] M. Ahmadi and W. Namgoong, "Asynchronous SAR logic design using majority vote comparison for configurable SAR ADCs," 2015 IEEE Dallas Circuits and Systems Conference (DCAS), Dallas, TX, USA, 2015, pp. 1-4.

[10] K. Yoshioka, "VCO-Based Comparator: A Fully Adaptive Noise Scaling Comparator for High-Precision and Low-Power SAR ADCs," in IEEE Transactions on Very Large Scale Integration (VLSI) Systems, vol. 29, no. 12, pp. 2143-2152, Dec. 2021.

[11] J. Shen et al., "A 16-bit 16-MS/s SAR ADC With On-Chip Calibration in 55-nm CMOS," in IEEE Journal of Solid-State Circuits, vol. 53, no. 4, pp. 1149-1160, April 2018.

[12] L. Xie, X. Han, H. Zhang and X. Jin, "A 12bit 16MS/s Asynchronous SAR ADC with Speed-Enhanced Comparator and TSPC Latch," 2019 IEEE 4th International Conference on Integrated Circuits and Microsystems (ICICM), Beijing, China, 2019, pp. 104-108.

[13] N. Jiang, L. Meng, M. Zhao and Z. Tan, "A 12-bit 20MS/s Asynchronous SAR ADC," 2022 IEEE 16th International Conference on Solid-State & Integrated Circuit Technology (ICSICT), Nanjing, China, 2022, pp. 1-3.

[14] K. Cornelissens and M. Steyaert, "A Novel Bootstrapped Switch Design, Applied in a 400 MHz Clocked ΔΣ ADC," 2006 13th IEEE International Conference on Electronics, Circuits and Systems, Nice, France, 2006, pp. 1156-1159.

[15] N. H. E. Weste and D. M. Harris, CMOS VLSI Design: A Circuits and Systems Perspective, 4th ed. Boston, MA, USA: Addison-Wesley, 2011.

[16] B. Hershberg et al., "3.6 A 6-to-600MS/s Fully Dynamic Ringamp Pipelined ADC with Asynchronous Event-Driven Clocking in 16nm," 2019 IEEE International Solid-State Circuits Conference - (ISSCC), San Francisco, CA, USA, 2019, pp. 68-70.

979-8-3315-3918-4/25 $31.00 © 2025 IEEE

A Differential SAR-SS ADC with Gain-Scaled Ramp Quantization for High-Speed CMOS Image Sensors

Nanbo Chen[1,2], Jingyang Chen[1,2], Gang Wang[2], Peng Feng[1,3*], Jian Liu[1,3], Nanjian Wu[1,3], Liyuan Liu[1,3]

[1] State Key Laboratory of Semiconductor Physics and Chip Technologies, Institute of Semiconductors, Chinese Academy of Sciences, Beijing 100083, China

[2] School of Physical Science and Technology, Ningbo University, Ningbo 315211, China

[3] Center of Materials Science and Optoelectronics Engineering, University of Chinese Academy of Sciences, Beijing 100049, China

* Email: fengpeng06@semi.ac.cn

Abstract—This paper presents a column-parallel 12-bit differential Successive Approximation Register and Single-Slope analog-to-digital converter (SAR-SS ADC) with gain-scaled ramp quantization, designed for high-speed CMOS image sensors (CIS). The single-slope (SS) quantization stage utilizes a novel gain scaled mechanism, switching from high to low gain during the residual voltage conversion. Compared to conventional designs, the proposed architecture achieves faster quantization. The ADC has been designed in a 180nm 1P6M CMOS process, powered by 3.3 V/1.8 V supplies. Post-layout simulation results show that at a sampling rate of 526 kS/s with a 49.85 kHz input, the ADC achieves an SNDR of 68.59 dB, SFDR of 76.56 dB, THD of 72.90 dB, and an ENOB of 11.10 bits.

Keywords—Gain-Scaled, SAR ADC, SS ADC, CMOS image sensor

I. INTRODUCTION

The growing demands of autonomous driving and machine vision have raised the need for high-speed, high-resolution image sensors. Analog-to-digital Converter (ADC) plays a key role in determining image sensor performance [1].

Among various architectures, column-parallel ADCs are widely used in high-speed CMOS image sensors (CIS) due to their balance of speed, area, and cost. Typical column-level architectures include single-slope (SS) ADC, successive approximation register (SAR) ADC, and cyclic ADC. SS ADC provides high accuracy but suffer from long conversion time (2^n cycles). SAR ADC is faster and more power-efficient, but area grows exponentially with resolution [2]. Two-step SAR-SS ADC combines the advantages of both architectures [3]. For SAR-SS ADC, coarse quantization is performed by a fast SAR stage, followed by a SS stage for fine quantization, enabling high speed and compact layout. However, the sampling speed has still need to be improved to satisfy the improved resolution and frame rate of the CIS.

This paper presents a two-step SAR-SS ADC that employs gain-scaled ramp quantization to accelerate the fine conversion process. During the quantization of the SAR residual voltage, the ramp gain is dynamically adjusted to further improve speed. Compared to conventional architectures, this approach introduces only a pair of switches for ramp gain control along with timing signals, yet it doubles the overall quantization speed, reduces the number of

comparisons required for residual voltage, and lowers the dynamic power consumption.

II. ARCHITECTURE AND CIRCUIT IMPLEMENTATION

A. Architecture Overview

The proposed 12-bit differential SAR-SS ADC with gain-scaled ramp quantization is shown in Fig. 1. It consists of bootstrap switches, 7-bit capacitor DAC, comparator, SAR logic, and a globally shared ramp generator.

Compared to the traditional architecture, the proposed design introduces a pair of switches at the top plate of the capacitor array to enable ramp gain switching. When the differential input signals *Vin* and *Vip* are applied, the bootstrap switches sample and hold the signals. The SAR ADC then performs coarse quantization by adjusting the bottom plate voltages of the capacitor array through SAR logic, allowing *Vxn* and *Vxp* to gradually converge. The comparator output is recorded to obtain the upper 7-bit result. Then, the ramp generator connects the ramp signal to the unit capacitor through switches S3 and S4, enabling fine 5-bit quantization. During this phase, S1 and S2 control the ramp slope to implement gain-scaled quantization, effectively reducing both the number of comparisons and fine quantization time. In this ADC architecture, the ramp generator is globally shared, and the comparator is reused for both the SAR and SS stages. The detailed operation and circuit implementation are described in the following sections.

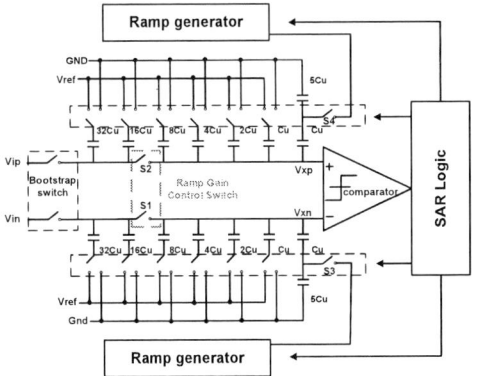

Fig. 1. Architecture of the proposed 12-bit SAR-SS ADC with gain-scaled ramp quantization.

This work is supported by the National Key Research and Development Program of China (2022YFB2804402), National Natural Science Foundation of China (62134004).

B. Operating Principle

The timing diagram of the proposed SAR-SS ADC is illustrated in Fig. 2. During phase T_1, switches S1 and S2 are switched on to reset digital logic. In phase T_2, the bootstrap switches sample the input signal. The asynchronous 7-bit SAR conversion performs in phase T_3. In phase T_4, ramp voltage is injected by switching on S3 and S4 to perform residual quantization. During the residual quantization, taking the case where Vxn is lower than Vxp as an example: the top-plate switch S1 is switched off, while S2 on the Vxp side remains switched on. Each step of the ramp generator is set to 64 LSB. Due to capacitive voltage division, the resulting comparator top-plate voltages are:

$$\Delta Vxp = 64\,\text{LSB} \times \frac{C_u}{64C_u} = 1\,\text{LSB} \qquad (1)$$

$$\Delta Vxn = 64\,\text{LSB} \times \frac{C_u}{16C_u} = 4\,\text{LSB} \qquad (2)$$

The residual voltage is thus quantized in steps of 3 LSB. When Vxn exceeds Vxp, the 5-bit ramp counter value at that moment is stored as SS_DATA1.

In phase T_5, S3 is switched off to disconnect the ramp from the Vxn node, holding its voltage constant. Meanwhile, the ramp remains connected to the Vxp side and continues quantization with a step size of 1 LSB. When Vxp exceeds Vxn, a second 5-bit counter value is recorded as SS_DATA2. The final residual code is calculated as:

$$RES\_CODE = 4 \times SS\_DATA1 - SS\_DATA2 \qquad (3)$$

As shown in Fig. 3, take the residual voltage is 20.5 LSB as a example, and the ramp clock is 20 MHz, the conventional architecture requires 21 ramp cycles to complete the residual quantization. In contrast, the proposed architecture performs coarse quantization in the 7th cycle, switches the ramp gain, and completes fine quantization in the 8th cycle, thus finishing the residual conversion.

$$T_{con} \approx T_{SAR}(200\,ns) + 21 \times T_{SS}(50ns) = 1250ns \qquad (4)$$

$$T_{proposed} \approx T_{SAR}(200\,ns) + 8 \times T_{SS}(50ns) = 650ns \qquad (5)$$

Here, T_{SAR} refers to the quantization time of the 7-bit asynchronous SAR quantization step, which is less than 200 ns. T_{SS} denotes the comparison time of a single SS quantization step, with a ramp clock of 20 MHz, each ramp step takes 50 ns. When the residual voltage is 20.5 LSB, the proposed architecture reduces the total quantization time by approximately 50% compared to the conventional structure.

In two-step ADCs, parasitic effects and offset errors may introduce inaccuracies at the transition point between coarse and fine quantization. To mitigate this effect, redundancy is typically added. In the proposed design, the residual voltage after 7-bit SAR quantization is theoretically less than 32 LSB, so the ramp quantization range is extended by 50% to 48 LSB, providing sufficient margin to prevent missing codes.

As shown in Fig. 4, for a quantization range of 48 LSB, the conventional architecture requires at least 48 ramp cycles for residual quantization, while the proposed design requires only 19 cycles, which is 40% of the conventional case. Since the SS quantization step is the main speed bottleneck in SAR-

SS architectures, this design effectively doubles the overall quantization speed and halves the number of comparator operations, significantly reducing dynamic power consumption.

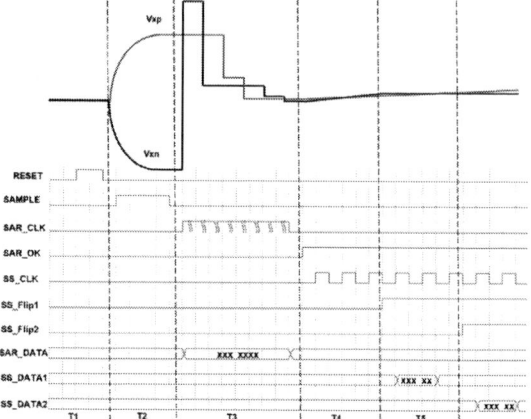

Fig. 2. Timing diagram of the proposed SAR-SS ADC.

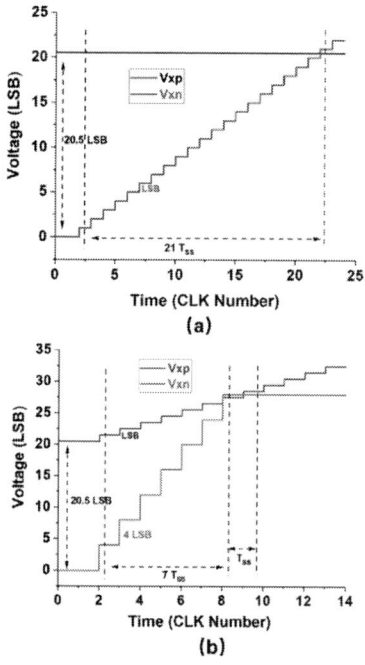

Fig. 3. Ramp quantization process: (a) Conventional architecture, (b) Proposed architecture.

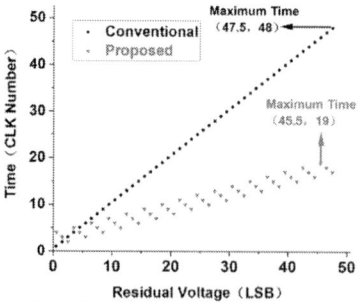

Fig. 4. Comparison of residual quantization cycles between conventional and gain-scaled architectures.

C. Gain-Scaling Control Signal Generation

The control signal generation circuit for gain-scaled ramp quantization is shown in Fig. 5(a). The RESET signal serves as a global reset. After the SAR conversion step is completed, SAR_OK signal is asserted.

P_SAR<7> and N_SAR<7> represent the seventh-bit outputs from the positive and negtive sides during the SAR conversion step, respectively, and are also used to control the ramp gain by switching S1 and S2.

The ramp flip indication signal SS_Change is generated from positive and negative outputs of the comparator OUTP and OUTN. In the example of Fig. 5, SS_Change equals with OUTN. The signal SS_Flip1 records the timestamps when Vxn exceeds Vxp, and the signal SS_Flip2 records the timestamps when Vxp exceeds Vxn. Therefore, two 5-bit values can be obtained, which can be used to reconstruct the actual 5-bit fine quantization result.

In the example of Fig. 5, P_RAMP and N_RAMP are the control signals for ramp injection switches S3 and S4 in Fig.1. When SAR_OK is asserted, both S3 and S4 are switched on, enabling ramp input to both sides for SS conversion. After coarse SS conversion step is completed, S3 is switched off to hold the ramp signal, which is used in the fine SS conversion step.

III. SIMULATION RESULTS

The proposed ADC is implemented using a 180 nm 1P6M CMOS process, with supply voltages of 3.3 V (analog) and 1.8 V (digital). A 20 MHz master clock is used. The input quantization range is -1 V to +1 V, and the total conversion time is 1.9 μs. When the input signal frequency is 49.85 kHz, pre-layout simulation results is shown in Fig. 6. The performances of the SNDR, SFDR, THD and ENOB are 71.3 dB, 83.30 dB, 80.40 dB and 11.55-bit respectively.

The layout of a single column is shown in Fig. 7, occupying an area of 80 μm × 660 μm. In addition, circuits outside the column-parallel path, such as the ramp generator and clock control logic, share an area of 220 μm × 400 μm. The post-layout simulation results is shown in Fig. 8. The performances of the SNDR, SFDR, THD and ENOB are 68.59 dB, 76.56 dB, 72.90 dB and 11.10-bit respectively.

Table I. summarizes the performance of the proposed ADC and compares it with recent designs. Compared to conventional SAR-SS ADC architectures, the proposed ADC achieves a significantly higher sampling rate while maintaining a high ENOB.

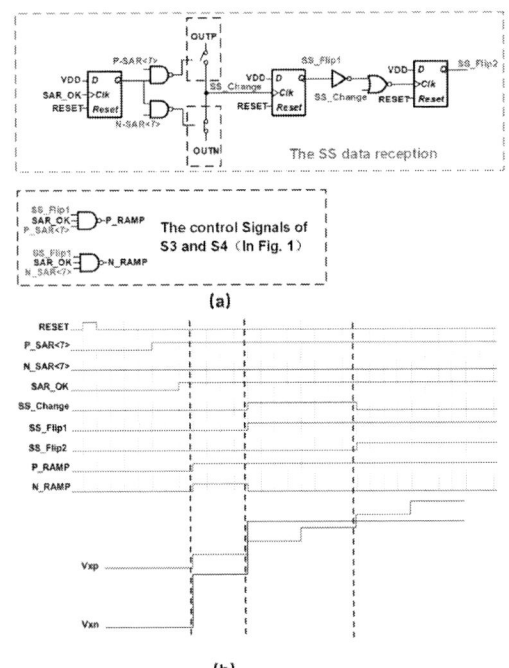

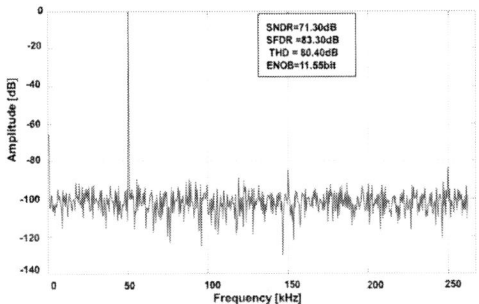

Fig. 5. (a) Control signal generation circuit，(b) Timing diagram of the corresponding control signals.

Fig. 6. Pre-layout dynamic performance results of the ADC.

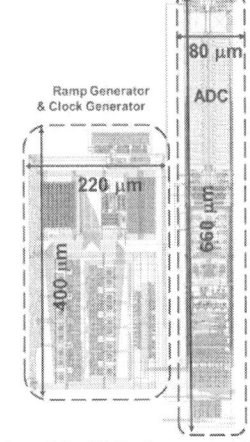

Fig. 7. Layout view of the ADC column and globally shared ramp and clock generator circuits.

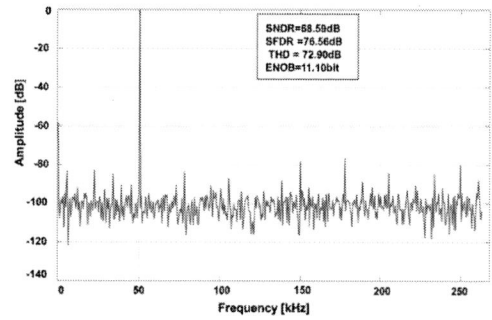

Fig. 8 Post-layout dynamic performance results of the ADC.

TABLE I.   SUMMARIZED COMPARISON WITH OTHER TYPES OF ADC WITH COLUMN-PARALLEL ARCHITECTURES

	[4]	[5]	[6]	[7]	This Work
Process	130 nm CMOS	180 nm CMOS	180 nm CMOS	110 nm CMOS	180 nm CMOS
Architecture	SS	SS/SAR	SAR/SS	SAR/SS	SAR/SS
ADC Area (μm^2)	7.5×675	14.5×856	70×1100	94×528	80×660
Sampling Rate (kS/s)	100	240	83.3	357	526
Resolution (bits)	12	11	11	12	12
ENOB (bits)	9.8	8.3	-	11.17	11.10

IV. CONCLUSION

This paper presents a 12-bit gain-scaled SAR-SS ADC for high-speed CMOS image sensors. A novel gain-scaled ramp quantization technique is applied to the SS conversion step, achieving faster quantization with reduced comparision times and lower dynamic power. The ADC covers ±1 V input, operates at 526 kS/s, and achieves a resolution of 488.3 µV, the ENOB reaches 11.10 bits, demonstrating the effectiveness of the proposed architecture.

REFERENCES

[1] F. Morishita, W. Saito, Y. Iizuka, N. Kato, R. Otake and M. Ito, "A 30.2-µ Vrms Horizontal Streak Noise 8.3-Mpixel 60-Frames/s CMOS Image Sensor With Skew-Relaxation ADC and On-Chip Testable Ramp Generator for Surveillance Camera," in IEEE Journal of Solid-State Circuits, vol. 57, no. 10, pp. 3103-3113, Oct. 2022.

[2] X. Tang et al., "Low-Power SAR ADC Design: Overview and Survey of State-of-the-Art Techniques," in IEEE Transactions on Circuits and Systems I: Regular Papers, vol. 69, no. 6, pp. 2249-2262, June 2022.

[3] S. Dai, K. Hu and J. K. Rosenstein, "A Segmented SAR/SS ADC with Digital Error Correction and Programmable Resolution for Column-Parallel Sensor Arrays," 2020 IEEE International Symposium on Circuits and Systems (ISCAS), Seville, Spain, 2020, pp. 1-5.

[4] Q. Zhang, N. Ning, Z. Zhang, J. Li, K. Wu and Q. Yu, "A 12-Bit Two-Step Single-Slope ADC With a Constant Input-Common-Mode Level Resistor Ramp Generator," in IEEE Transactions on Very Large Scale Integration (VLSI) Systems, vol. 30, no. 5, pp. 644-655, May 2022.

[5] S.-J. Byun, J.-T. Seo, T.-H. Kim, J.-H. Lee, Y.-K. Kim, and K.-H. Baek, "An 11 Bit single slope/successive approximation register analog to digital converters with on chip fine step range calibration for CMOS image sensors," Electronics, vol. 14, no. 1, Art. 83, 2025.

[6] F. Tang, D. G. Chen, B. Wang and A. Bermak, "Low-Power CMOS Image Sensor Based on Column-Parallel Single-Slope/SAR Quantization Scheme," in IEEE Transactions on Electron Devices, vol. 60, no. 8, pp. 2561-2566, Aug. 2013.

[7] H. Zhang, Z. Fang, N. Yu, N. Lv and Z. Guo, "A Column-parallel SAR/SS ADC with Multi-column Shared Capacitor DAC for CMOS Image Sensor," 2022 IEEE 16th International Conference on Solid-State & Integrated Circuit Technology (ICSICT), Nangjing, China, 2022, pp. 1-3.

A 79.2dB-SNDR 12.5MHz-BW Pipelined SAR ADC with Analog-Domain Gain Error Shaping

Qiaoyu Hu [1], Guolong Fu [1], Yanbo Zhang*[1], Zhangming Zhu [1]

[1] Key Laboratory of Analog Integrated Circuits and Systems (Xidian University), Ministry of Education, School of Integrated Circuits, Xidian University, Xi'an 710071, China

* Email: huqy@stu.xidian.edu.cn, zhybor@163.com

Abstract—This article proposed a pipelined successive approximation register (SAR) analog-to-digital converter (ADC) with a second-order analog-domain gain error shaping (AD-GES) technique that effectively suppresses the in-band quantization leakage error caused by interstage gain error. The AD-GES is implemented through a subrange structure in the first stage, where the coarse ADC performs the noise shaping (NS) with minimal overhead and transfers its decision to directly generate the shaped quantization noise on the top plate of the fine DAC. Through analog-domain NS, the proposed AD-GES eliminates the digital-domain prediction error inherent in traditional GES techniques, thereby enhancing the GES efficiency. Verified by simulation in 28-nm CMOS process, the prototype achieves a 79.2-dB SNDR over 12.5-MHz BW while operating at 200MS/s and consuming 1.54mW. It exhibits a 178.3-dB Schreier FoM, and the SNDR deviates less than 3 dB within −28% to +39% gain error.

Keywords—Analog-to-digital converter (ADC), pipelined successive-approximation-register (SAR), noise shaping (NS), gain error shaping (GES)

I. Introduction

The pipelined successive approximation register (SAR) analog-to-digital converter (ADC) offers high resolution, fast speed, and excellent power efficiency, making it a popular architecture in recent years. However, inaccurate interstage gain degrades the overall ADC's quantization accuracy.

Gain Error Shaping (GES) has emerged as a promising solution to address the gain error problem [1][2][3]. Existing GES methods typically rely on digital prediction-based approaches, where the first-stage quantization error (Q_1) is estimated in the digital domain and fed back to the first and second stages with opposite polarity, achieving an outstanding GES ability. However, the traditional GES technique [1] suffers from the truncation error and gain error, degrading the GES efficiency. The digital error feedback (DEF) technique was introduced in [2] to mitigate the truncation error. However, the positive gain error leads to degradation in GES ability. Subsequently, the QPU-GES was proposed in [3] to eliminate the degradation of GES efficiency caused by gain error. Nevertheless, the limitations of finite-bit quantization-based prediction still constrain the GES efficiency.

In this paper, we propose an analog-domain GES (AD-GES) technique. By employing a subrange structure in the first stage, the coarse SAR ADC with noise shaping (NS) can directly generate a shaped Q_1 on the top plate of fine DAC without relying on prediction, thereby effectively addressing the performance degradation due to inaccurate Q_1 prediction in digital domain, which is caused by finite-bit quantization and gain error. Simulated in a 28-nm CMOS process, the ADC prototype runs at 200 MS/s with 12.5-MHz bandwidth (BW)

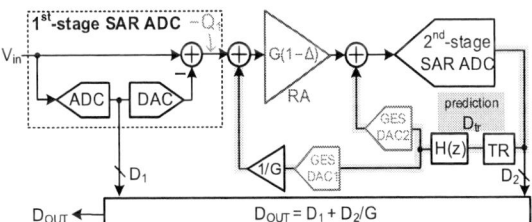

Fig. 1. Block diagram of the two-stage pipelined SAR ADC equipped with traditional GES technique.

and 79.2-dB SNDR, consuming a total of 1.54-mW power from a 1-V supply. It delivers a 178.3-dB Schreier FoM and demonstrates a −28% to +39% 3-dB-SNDR gain error tolerance ability at OSR = 8.

II. Proposed Analog-Domain GES

A. Prior Art

Fig. 1 shows the block diagram of the two-stage pipelined SAR ADC equipped with the traditional GES technique. V_{in}, D_1, D_2, and D_{OUT} represent the input signal, the first stage output, the second stage output, and the overall output, respectively. Assuming the actual gain of the residue amplifier (RA) is $G(1-\Delta)$, where G represents the ideal gain and Δ denotes the gain error. It is well known that the gain error introduces quantization noise leakage (ΔQ_1) and harms the overall ADC performance.

The traditional GES [1] is implemented based on the digital domain prediction, where D_2/G is used as the predicted value of Q_1 by feeding D_2 back to a GES DAC that is ratioed $1/G$ to the first-stage DAC. Taking the first-order GES as an example, as long as $Q_1 - Q_1 z^{-1}$ is realized at the input of the RA and Q_1 is reconstructed at the input of the second-stage ADC, ΔQ_1 will be shaped by the noise transfer function (NTF) of $1 - z^{-1}$. The ideal transfer function is

$$D_{OUT} = V_{in} + \Delta Q_1 (1 - z^{-1}) + Q_2/G. \tag{1}$$

However, this implementation results in two limitations. First, D_2 must undergo a truncation (TR) operation for a feasible GES DAC, where only the most significant bits (MSBs) D_{tr} are used for prediction, introducing the truncation error Q_{tr}, Secondly, in the presence of Δ, D_2/G actually represents $Q_1 - \Delta Q_1 (1 - z^{-1})$ rather than Q_1, consequently degrading the GES performance. Therefore, the actual transfer function is

$$D_{OUT} = V_{in} + \Delta Q_1 \frac{1 - z^{-1}}{1 - \Delta z^{-1}} + \frac{Q_2 - \Delta Q_{tr} z^{-1}}{G(1 - \Delta z^{-1})} \tag{2}$$

Evidently, the prediction error caused by Q_{tr} and Δ degrades the GES ability.

The DEF technique is proposed in [2] to suppress the Q_{tr}. However, the positive Δ introduces a positive pole in the NTF as shown in (2), still degrading the GES ability. The QPU-GES method [3] employs additional digital codes from the first-stage auxiliary ADC to predict Q_1. Since this prediction is based on the first-stage digital codes rather than the second-stage outputs, it eliminates the influence of Δ on the NTF. However, the QPU-GES method remains constrained by finite-bit quantization, resulting in degraded shaping ability.

B. Proposed AD-GES Technique

Fig. 2 illustrates the block diagram of the proposed AD-GES technique in a pipelined SAR ADC. Unlike the traditional GES, the first-stage SAR ADC is replaced with an NS SAR ADC, and both the forward and backward feedback paths from D_2 to the RA are eliminated. The transfer function of the first stage is represented as follows:

$$D_1 = V_{in} + \text{NTF} \times Q_1 \qquad (3)$$

The difference between AD-GES and traditional GES is the realization method of NTF·Q_1. Due to the NS, the residue voltage V_{res} can be expressed as

$$V_{res} = V_{in} - D_1 = -\text{NTF} \times Q_1 \qquad (4)$$

which precisely equals $-$NTF·Q_1 and can be directly obtained from the DAC's top plate. Therefore, the NS in SAR ADC determines the shaping order and the NTF form of the AD-GES. In this work, to improve the first-stage conversion speed while relaxing the NS design requirements (analyzed in detail in Section III-A), a subrange structure is adopted in the first stage where the NS is integrated into the coarse SAR ADC. Under the subrange mechanism, the coarse NS SAR ADC generates D_1 and copies D_1 to the fine DAC to directly create the residue voltage V_{res} (equal to $-$NTF·Q_1) at the top plate.

The AD-GES technique implements NTF·Q_1 directly in the analog domain through the coarse NS SAR ADC, departing from subtraction-based generation methods. By eliminating both the digital-domain prediction and GES DAC feedback paths, this approach simultaneously resolves the prediction error caused by Q_{tr} and Δ.

Fig. 3 shows the signal flowchart and the spectral view of the two-stage pipelined SAR ADC equipped with the proposed AD-GES technique. As shown in Fig. 3(a), the input signal V_{in} represents a sinusoidal waveform. Fig. 3(b) indicates that the top-plate voltage V_{res} of the fine DAC is equal to $-$NTF·Q_1. With a residue amplifier gain of $G(1-\Delta)$, the amplified output voltage $V_{out,RA}$ is obtained after processing V_{res}. Fig. 3(c) reveals that $V_{out,RA}$ contains two high-pass components, expressed as

$$V_{out,RA} = -\text{NTF} \times GQ_1 + \text{NTF} \times \Delta GQ_1 \qquad (5)$$

The second-stage SAR ADC subsequently samples $V_{out,RA}$ and completes the remaining quantization to generate D_2. The final output D_{OUT} is then digitally reconstructed by combining D_1 and D_2, with its spectral view shown in Fig. 3(d). The overall transfer function can be expressed as

$$D_{OUT} = V_{in} + \text{NTF} \times \Delta Q_1 + Q_2/G \qquad (6)$$

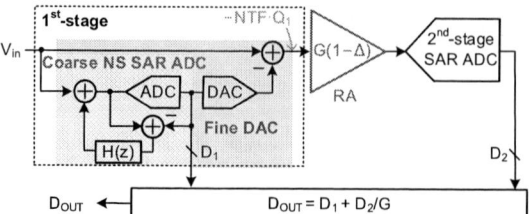

Fig. 2. Block diagram of the two-stage pipelined SAR ADC equipped with proposed AD-GES technique.

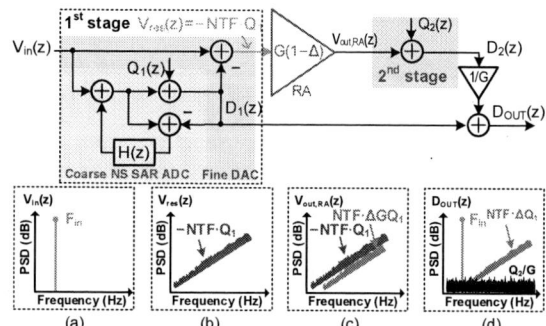

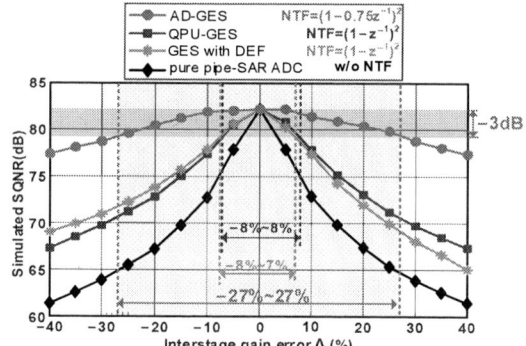

Fig. 3. Signals flowchart and the spectral view of the two-stage pipelined SAR ADC equipped with the proposed AD-GES technique.

Fig. 4. Simulated SNDR versus interstage gain error for a two-stage pipelined SAR ADC equipped with different GES techniques.

The transfer function demonstrates that ΔQ_1 is high-pass shaped, thereby suppressing the in-band quantization noise leakage error induced by the gain error.

To demonstrate the capability of the proposed AD-GES, we compare it with the GES with DEF and QPU-GES under the same architecture configuration. Both the first and second stages quantize 7-bit, with 2-bit redundancy. The NTFs of both GES with DEF and QPU-GES architectures are characterized by $(1 - z^{-1})^2$, while the AD-GES exhibits an NTF of $(1 - 0.75z^{-1})^2$ to align with the implementation. Fig. 4 illustrates that the SQNR varies with the gain error.

As observed, all architectures show SQNR degradation as $|\Delta|$ increases. However, under the same gain error, the ADC with the AD-GES technique achieves higher SQNR than other GES techniques. The finite-bit quantization error limits the GES ability of GES with DEF and QPU-GES, causing a more obvious decrease in SQNR. In addition, the positive Δ deteriorates the GES capability of GES with DEF, further reducing the error tolerance range. Simulation results

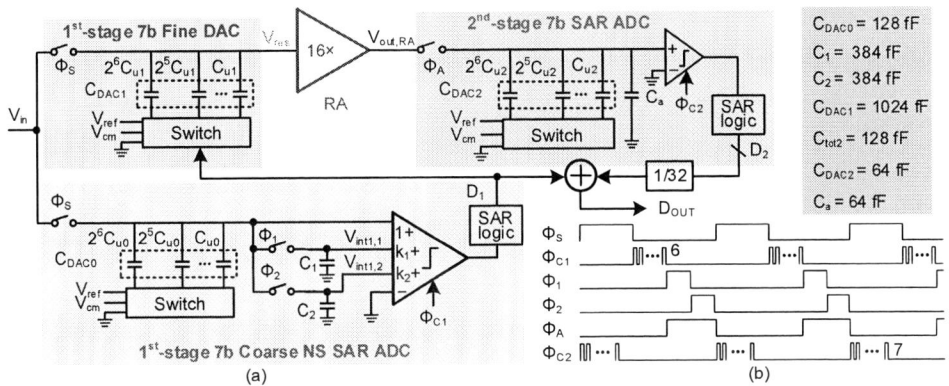

Fig. 5. Top-level schematic and timing diagram of the prototype ADC.

demonstrate that the pipelined SAR ADC employing AD-GES maintains SQNR degradation within 3 dB across a wide gain error range of −27% to +27%. In comparison, the −3 dB ranges for GES with DEF and QPU-GES are only −8% to +7% and −8% to +8%, respectively. Although AD-GES adopts a more moderate NTF of $(1 - 0.75z^{-1})^2$, it still performs better than other GES techniques. In summary, the proposed AD-GES significantly enhances the GES efficiency and improves the overall quantization accuracy.

III. CIRCUIT IMPLEMENTATION OF THE ADC

This chapter performs the nonideality analysis first, then presents the detailed schematic and its corresponding clock timing (Fig. 5) according to the error specifications.

A. Nonideality Analysis of the Proposed Architecture

Fig. 6 displays the major sources of nonideality in the two-stage pipelined SAR ADC with AD-GES, where $n_{s1,c}$ and $n_{s1,f}$ represent the sampling thermal noise of the first-stage coarse NS SAR ADC and fine DAC, respectively, $e_{mis1,c}$, $e_{mis1,f}$ and e_{mis2} denote the DAC mismatch error in the coarse NS SAR ADC, fine DAC, and second-stage SAR ADC, respectively, n_{amp} and $n_{NS,c}$ are added to model the noise from the RA and the loop filter for NS. The comparator noises are characterized by $n_{comp1,c}$ and n_{comp2} for the coarse NS SAR ADC and second-stage SAR ADC, respectively. When accounting for these nonidealities (noise, mismatch, etc.), the ADC's transfer function becomes:

$$D_{OUT} = V_{in} + (1-\Delta) \times \left(e_{mis,f} + n_{s1,f} + n_{amp}\right)$$
$$+ \Delta \times \left[NTF \cdot \left(Q_1 + n_{comp1,c}\right) + H(z) \cdot n_{NS,c} + n_{s1,c} + e_{mis,c}\right] \quad (7)$$
$$+ \frac{Q_2 + n_{comp2} + e_{mis2}}{G}$$

Since $H(z) \cdot n_{NS,c}$, $n_{s1,c}$, and $e_{mis,c}$ undergo the same transfer function as $NTF \cdot Q_1$, the requirements of noise caused by NS and thermal noise and mismatch error in coarse ADC need only satisfy the resolution of the coarse NS SAR ADC. Therefore, by implementing NS in the coarse SAR ADC based on a subrange architecture, the design requirements for the NS can be relaxed, decreasing the power consumption and hardware overhead of the NS. However, the $e_{mis,f}$, $n_{s1,f}$, and n_{amp} appear at the output directly without any suppression. Therefore, the noise requirement from the sampler and the RA is fulfilled by budgeting large enough sampling and loading capacitors, respectively. The Q_2 and other nonideal noise from the second stage are attenuated by the interstage gain.

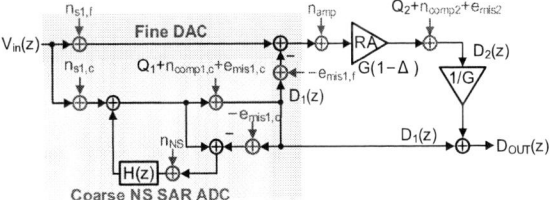

Fig. 6. Signal flowchart of the Coarse NS SAR ADC.

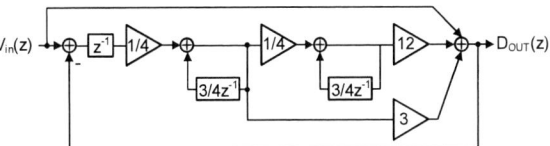

Fig. 7. Signal flowchart of the proposed architecture with the nonideality.

Furthermore, since AD-GES eliminates the traditional GES DAC, a larger interstage gain is achieved, significantly relaxing the design requirements for the second-stage SAR ADC, which allows a low-power design in second stage.

B. Detailed Schematic of the Proposed Architecture

Fig. 5 shows the top-level schematic and timing diagram of the proposed pipelined SAR ADC with the AD-GES technique. It consists of a 7-bit coarse NS SAR ADC, a 7-bit fine DAC, an RA with 16× gain, and a 7-bit SAR ADC in the second stage with 2-bit redundancy. Both stages adopt top-plate sampling and V_{cm}-based DAC switching. To ensure robustness, the coarse NS SAR ADC employs a second-order passive FF NS scheme. The signal flow diagram of the coarse NS SAR ADC is illustrated in Fig. 7. The NTF of the NS SAR ADC is derived as $(1 - 0.75z^{-1})^2$. The integration capacitors, C_1 and C_2, are directly connected to the DAC array, where $C_1 = C_2 = 3C_{DAC}$. Since the zeros of the NTF are solely determined by the capacitor ratio, they exhibit low sensitivity to PVT variations. Furthermore, the NS SAR ADC utilizes a three-input comparator [4] to provide relative gain, with $k_1 = 3$ and $k_2 = 12$, for the integrated voltages $V_{int1,1}$ and $V_{int1,2}$. The mismatch error of the input pairs in the coarse comparator can be carefully designed to control the gain ratio and thus maintain a stable NS efficiency [4]. Capacitor C_a acts as the attenuation capacitor to scale down the second-stage reference voltage. Therefore, an inter-stage gain of 16× is implemented to ensure sufficient RA's linearity.

Fig. 5(b) shows the timing diagram of the ADC. During

Φ_S, both the coarse NS SAR ADC and fine DAC sample the input signal simultaneously, after which the coarse NS SAR ADC performs quantization to obtain D_1. These digital codes are transferred to the fine DAC, which switches according to these codes, generating V_{res} on the DAC's top plate with a value of $-NTF \cdot Q_1$. Next, V_{res} is amplified by the RA to obtain the voltage $V_{out,RA}$ during Φ_A. At the same time, the coarse DAC (C_{DAC0}) sequentially performs charge sharing with C_1 and C_2 during Φ_1 and Φ_2, generating integrated voltages $V_{int1,1}$ and $V_{int1,2}$ for the next quantization cycle. It allows the residue process to operate in parallel with the amplification without extra time overhead. After amplification, the second-stage SAR ADC then performs conversion during Φ_{C2} to obtain D_2. The total capacitance of the fine DAC is 1024 fF for enough kT/C noise budget. Since the coarse NS SAR ADC does not need to produce a full-resolution residue, its DAC is just 7-bit, and the total single-ended capacitance is only 128 fF. The second-stage total capacitance is 128 fF, which is the sum of C_{DAC2} and C_a capacitances (both are 64 fF).

IV. SIMULATION RESULTS

The proposed ADC was implemented in 28-nm CMOS process for verifying specific performance. Fig. 8 shows the ADC's output spectrum with OSR = 8. The proposed ADC achieves SFDR and SNDR of 102.1 dB and 79.2 dB, respectively, with the input frequency of 0.82 MHz, encompassing more than 15 harmonics. Thanks to its excellent shaping capability, the SNDR drops only 3 dB under +39% gain error. At a sampling rate of 200 MS/s, the total power consumption is 1.54 mW with a 1-V power supply. Fig. 9 shows the simulated SNR and SNDR versus the input amplitude, indicating a dynamic range (DR) of 80.7 dB. The simulated SNDR and SQNR versus gain error are reported in Fig. 10. The simulation results show that the chip maintains less than 3-dB SNDR degradation (from the nominal value) within −28 % to +39 % gain error. Table I summarizes the ADC's performance and compares it with prior art. Under the 3-dB deviation of SNDR, the gain error tolerance range is from −28% to +39%, which greatly relaxes the gain accuracy requirement of the RA.

V. CONCLUSION

This paper presents a second-order AD-GES in a pipelined SAR ADC to attain better gain error tolerance ability than prior techniques. A subrange structure is adopted in the first stage, speeding up the conversion and relaxing the design requirements for the NS. Thanks to analog-domain NS in the coarse NS SAR ADC, $-NTF \cdot Q_1$ can be obtained directly on the fine DAC's top plate, removing the limitations caused by the prediction error of digital-domain GES techniques. The −3dB SNDR gain error range is expanded to −28% to +39% at the OSR of 8, demonstrating exceptional GES capability.

ACKNOWLEDGMENT

This work was funded by the National Key R&D Program of China (2023YFB4405002), the National Natural Science Foundation of China (62574155, 62361166671), the Fundamental Research Funds for the Central Universities (YJSJ25013), and the Innovation Fund of Xidian University. (Corresponding author: Yanbo Zhang.)

REFERENCES

[1] C. -K. Hsu, T. R. Andeen and N. Sun, "A Pipeline SAR ADC With Second-Order Interstage Gain Error Shaping," in IEEE Journal of Solid-State Circuits, vol. 55, no. 4, pp. 1032-1042, April 2020.

TABLE I.   PERFORMANCE COMPARISON

Specifications	JSSC 2020[1][a]	JSSC 2021[2][a]	JSSC 2023[3][a]	This Work[b]
Process (nm)	40	40	28	28
Architecture	PipeSAR	PipeSAR	TI-PipeSAR	PipeSAR
Supply (V)	1	1	1	1
GES order	2	2	2	2
F_S (MHz)	100	100	400	200
OSR	4	8	8	8
-3dB SNDR Gain Error (%)	+5	+25	−18 to +24	−28 to +39
BW (MHz)	12.5	6.25	25	12.5
SNDR (dB)	75.8	77.1	77.2	79.2
Power (mW)	1.54	1.38	2.03	1.54
FoM$_S$[c] (dB)	174.9	173.7	178.1	178.3

[a] Mesurement results [b] Simulation results [c] FoM$_S$=SNDR+10log$_{10}$(BW/Power)

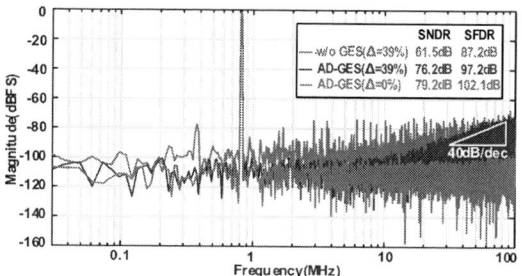

Fig. 8. The ADC's output spectrum.

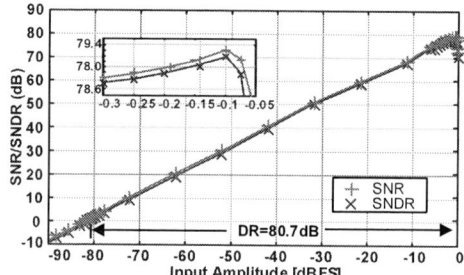

Fig. 9. Measured SNR and SNDR versus input amplitude.

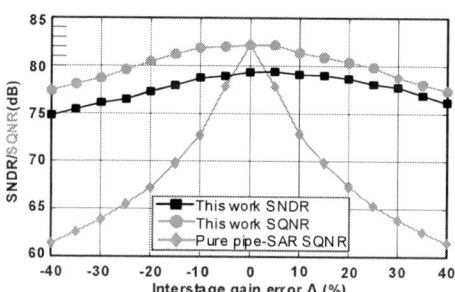

Fig. 10. Gain error tolerance performance of the ADC.

[2] C. -K. Hsu et al., "A 77.1-dB-SNDR 6.25-MHz-BW Pipeline SAR ADC With Enhanced Interstage Gain Error Shaping and Quantization Noise Shaping," in IEEE Journal of Solid-State Circuits, vol. 56, no. 3, pp. 739-749, March 2021.

[3] H. Zhang, Y. Zhu, R. P. Martins and C. -H. Chan, "A Second-Order NS Pipelined SAR ADC With Quantization-Prediction-Unrolled Gain Error Shaping and Fully Passive Integrator," in IEEE Journal of Solid-State Circuits, vol. 58, no. 12, pp. 3565-3575, Dec. 2023.

[4] J. Liu, S. Li, W. Guo, G. Wen and N. Sun, "A 0.029-mm² 17-fJ/Conversion-Step Third-Order CT ΔΣ ADC With a Single OTA and Second-Order Noise-Shaping SAR Quantizer," in IEEE Journal of Solid-State Circuits, vol. 54, no. 2, pp. 428-440, Feb. 2019.

979-8-3315-3918-4/25 $31.00 © 2025 IEEE

A Deep Reservoir Computing System based on IGZO Electrical-Double-Layer Transistors

M. Han [1†], Y. Chen [1†], H. Cui [1], Y. Wang [1], C. Wan*[1]

[1] School of Electronic Science and Engineering, Nanjing University, Nanjing, 210023, China.

†These authors contributed equally: M. Han, Y. Chen; *Email: cjwan@nju.edu.cn

Abstract—Conventional physical reservoir computing typically relies on time-division multiplexing to generate diverse reservoir states. In this work, a deep reservoir computing architecture based on layer-wise coupling is proposed based on IGZO electrical-double-layer transistor, enabling the generation of rich reservoir dynamics without time multiplexing. By leveraging the nonlinear capacitive characteristics of electrical-double-layer transistors, this system supports multiple temporal signal classification tasks, and achieves accuracies of 96% and 82% on speech recognition and human action recognition, respectively. Our results indicate a energy-efficient computing framework for edge AI scenarios.

Keywords—Deep reservoir computing, Electrical-double-layer transistors, IGZO TFTs

I. INTRODUCTION

Reservoir computing (RC) has offered a hardware-friendly framework for temporal data recognition. However, conventional implementations often rely on time-division multiplexing to enhance reservoir dynamics, which increases latency and complexity. To address this issue, a deep RC architecture based on electrical-double-layer (EDL) transistors is proposed, utilizing their nonlinear and memory properties to enable multi-layer state transformation without time multiplexing. This design supports efficient, low-latency processing for temporal tasks within a compact in-sensor computing system.

II. DEEP RESERVOIR COMPUTING SYSTEM

Fig. 1 illustrates the architectures of conventional and deep RC. In conventional RC, the input is processed through time-division multiplexing using a random mask matrix before entering a single-layer reservoir [1]. The resulting reservoir states are combined with output weights trained via linear regression to generate predictions. While time multiplexing enriches reservoir dynamics, it also introduces additional latency and computational overhead.

In contrast, the proposed deep RC adopts a layer-wise coupling architecture, where multi-layer state propagation produces diverse reservoir states without the need for time multiplexing. This structure reduces resource consumption while preserving computational performance.

III. EDL TRANSISTOR CHARACTERISTICS FOR DEEP RC

A. Transistor Structure and Fabrication

Fig. 2 shows the schematic structure and image of the fabricated transistor. The transistor consists of a four-layer stack: a silicon substrate serving as the gate electrode, a SiO_2 gate dielectric deposited by PECVD, an IGZO channel layer formed by RF magnetron sputtering, and aluminum source/drain electrodes deposited by thermal evaporation.

B. Memory Characteristics of the Transistor

Fig. 3 shows the transfer and output characteristics of the transistor. The transfer curve exhibits a hysteresis window, indicating the memory behavior of the device. An excitatory postsynaptic current (EPSC) is induced by a gate voltage pulse applied to the EDL transistor, as shown in Fig. 4. The EPSC response consists of two phases: Process I, a rapid increase caused by proton accumulation at the SiO_2/IGZO interface; and Process II, a gradual decay as the protons diffuse back to equilibrium. When the pulse amplitude increases from 0.6 V to 1.6 V, the EPSC intensity correspondingly increases, indicating a tunable synaptic response under different input strengths. The EPSC behavior is described by (1) [2], where a is 2.84, b is 1.27, τ_{DL} is 346.7, τ is 13.01, β is 0.64.

$$\Delta I_{DS} = \begin{cases} \left(a \times V_{GS} - b\right) \cdot \left[1 - \exp\left(-\dfrac{t}{\tau_{DL}}\right)\right] & (t \le T) \\ \left(a \times V_{GS} - b\right) \cdot \left[1 - \exp\left(-\dfrac{T}{\tau_{DL}}\right)\right] \cdot \exp\left[-\left(\dfrac{t-T}{\tau}\right)^{\beta}\right] & (t > T) \end{cases} \tag{1}$$

Memory capacity and Mackey–Glass sequence prediction are widely adopted benchmarks for evaluating reservoir computing systems. As shown in Fig. 5, the proposed deep RC system demonstrates competitive performance on both tasks."

IV. MULTI-TASK PROCESSING BY DEEP RC SYSTEM

The proposed deep RC system was evaluated on two classification tasks. On the NIST-46 spoken digit dataset ("zero" to "nine"), it achieved 96% accuracy (Fig. 6). On the UTD-MHAD human action dataset, the accuracy reached 82%, which is approaching to the previously reported 92% [3], demonstrating comparable performance. Fig. 7b shows reservoir states across five layers. Fig. 7c illustrates how accuracy varies with layer number, showing a sub-exponential increase. To control for data volume, a single-layer RC with replicated reservoir states was used for comparison. The improvement is not due to data size, indicating a structural benefit from depth.

ACKNOWLEDGMENT

This work was supported by the National Natural Science Foundation of China (Grant no. 62174082).

REFERENCES

[1] Y. Zhong, et al., "Dynamic memristor-based reservoir computing for high-efficiency temporal signal processing," Nat Commun, vol. 12, no. 1, pp. 408, Jan. 2021.

[2] C. J. Wan, et al., "Organic/inorganic hybrid synaptic transistors gated by proton conducting methylcellulose films," Appl. Phys. Lett., vol. 108, no. 4, Jan. 2016.

[3] Cui, H., Xiao, Y., Yang, Y. et al. A bioinspired in-materia analog photoelectronic reservoir computing for human action processing. Nat Commun 16, 2263 (2025).

979-8-3315-3918-4/25 $31.00 © 2025 IEEE

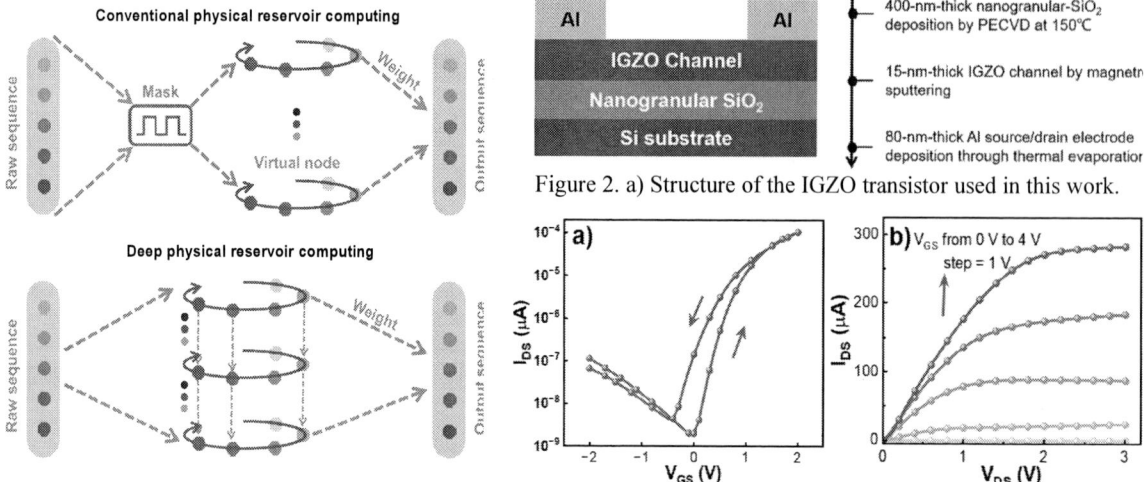

Figure 1. Structures of conventional and deep physical reservoir computing. The deep model eliminates the need for input masking by using a multilayer reservoir structure.

Figure 2. a) Structure of the IGZO transistor used in this work.

Figure 3. a) Transfer curve showing hysteresis and memory effect. b) Output curve under varying gate-drain voltage combinations.

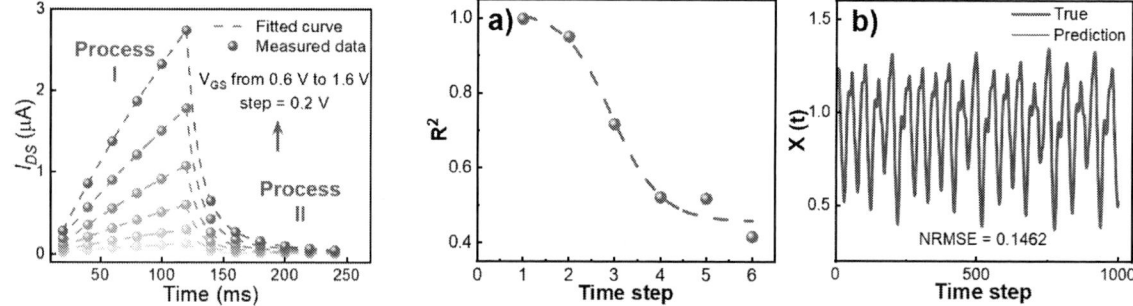

Figure 4. EPSC responses under different stimulus intensities, illustrating synaptic-like behavior that supports neuromorphic network implementation.

Figure 5. a) Memory capacity of the deep RC system, quantified by the area under the curve. b) Prediction of the classical Mackey–Glass chaotic sequence, indicating that the deep RC system is capable of retaining temporal information effectively.

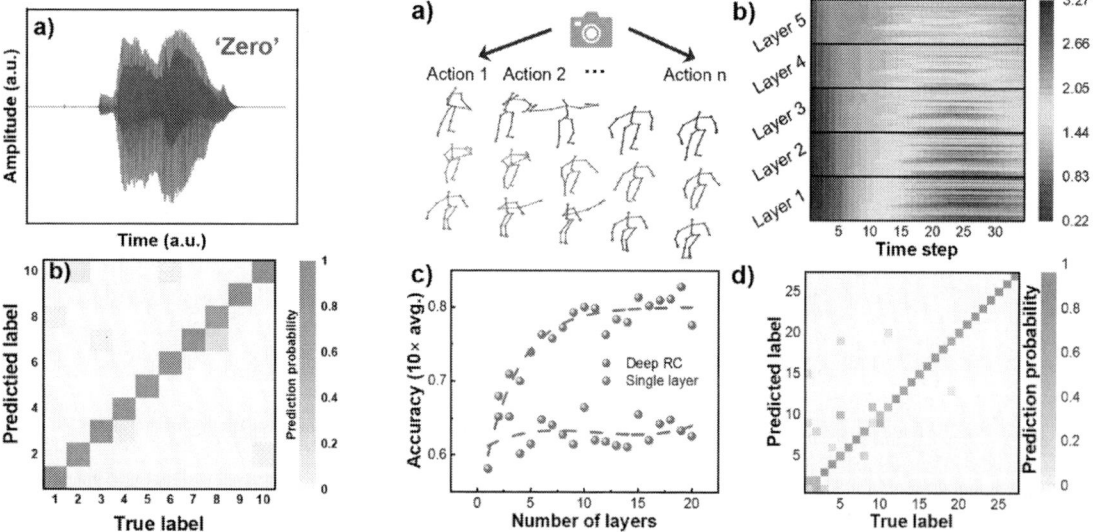

Figure 6. a) Waveform of the input speech signal. b) Confusion matrix for speech recognition using 10-layers deep RC system.

Figure 7. a) Input skeleton sequences. b) Reservoir states from each layer of the deep RC system (example: 5-layer). c) Accuracy comparison between deep and single-layer RC. d) Confusion matrix for human action recognition using the 10-layers deep RC system.

979-8-3315-3918-4/25 $31.00 © 2025 IEEE

High-density and High-reliability (H^2DR) RRAM for Energy-efficient AI Computing

Yimao Cai[1*], Yiyun Chen[1], Lin Bao[2], Ling Liang[1], Zheng Zhou[1], Zongwei Wang[1]

[1] School of Integrated Circuits, Beijing Advanced Innovation Center for Integrated Circuits, Peking University, Beijing 100871, P. R. China
[2] State Key Laboratory of Information Photonics and Optical Communications, Beijing University of Posts and Telecommunications, Beijing 100876, P. R. China
* Email: caiyimao@pku.edu.cn

Abstract — **The rapid evolution of embodied intelligence has spurred an escalating demand for ultra-energy-efficient computing paradigms. Resistive random-access memory (RRAM)-based in-memory computing (IMC) has positioned itself as a leading candidate, delivering unparalleled memory density and energy efficiency. This paper presents a comprehensive review of our research on high-density and high-reliability (H^2DR) RRAM technology and H^2DR RRAM-based IMC systems, spanning device innovation, architectural optimization, and full-chip integration. We elucidate both the transformative potential and persistent challenges in realizing this technology for next-generation AI applications.**

Keywords — RRAM, In-memory computing, M3D Integration

I. INTRODUCTION

Embodied intelligence (EI), an emerging interdisciplinary domain bridging artificial intelligence and robotics, has advanced significantly in recent years by enabling autonomous learning in intelligent agents through environmental interaction. The EI systems demand rapid processing of multimodal sensory data (e.g., images, point clouds) while maintaining stringent requirements for response latency and energy efficiency. Traditional von Neumann architectures, with their physically separated memory and compute units, inherently suffer from the "memory wall" bottleneck, leading to prohibitive data transfer overheads in multimodal processing. Resistive random-access memory (RRAM)-based in-memory computing (IMC) architectures have emerged as a transformative solution by natively integrating computation with memory [1], thereby drastically reducing data movement and enabling energy-efficient, real-time processing of complex sensory inputs (Fig. 1). This paradigm shift addresses critical performance limitations in EI systems, paving the way for next-generation autonomous agents capable of operating in complex environments.

This paper presents a holistic approach to advancing RRAM-based IMC systems, encompassing device innovation, architectural optimization, and chip-scale implementation. At the device level, we developed high-density high-reliability (H^2DR) RRAM arrays and a high-bandwidth monolithic 3D (M3D) integration process. Architecturally, we introduced the hybrid-domain polynomial transformation (HDPT) framework for distortion calibration, alongside sparsity-aware IMC architectures for efficient image recognition and heterogeneous processing pipelines tailored for point cloud analysis. These architectural innovations were realized through dedicated chip with H^2DR RRAM array, enabling experimental validation across intelligent applications. Empirical results demonstrate superior performance and energy efficiency in tasks spanning image calibration, Bayesian decision-making, and point cloud processing, addressing critical bottlenecks in von Neumann-based chips and paving a way for energy-efficient EI systems.

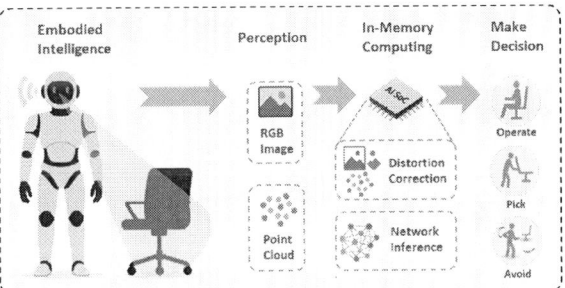

Fig. 1. IMC system for embodied intelligent application scenario.

II. HIGH-DENSITY AND HIGH-RELIABILITY RRAM

H^2DR RRAM serves as the foundational technology for energy-efficient IMC systems. Through systematic optimization of device fabrication, array architecture, and endurance/retention characterization, we have achieved significant advancements in RRAM integration density and operational reliability. By combining this H^2DR RRAM technology with M3D integration methodologies, we establish a scalable technological framework enabling next-generation 3D IMC architectures with unprecedented computational density and energy efficiency for EI applications.

A. High-reliability RRAM Cell based on SPS Technology

In RRAM devices, the reliability is profoundly influenced by the dynamics of oxygen vacancy (V_O) distribution and migration. To address this issue, we introduce an innovative self-passivation sidewall (SPS) process designed to enhance V_O migration uniformity [2]. Fig. 2(a) depicts the device architecture alongside its transmission electron microscope (TEM) image, illustrating the SPS technology's integration. This approach employs a self-formed passivation sidewall — generated during room-temperature etching — to spatially confine V_O migration, thereby minimizing exchange between the switching region and adjacent areas. By eliminating high-temperature annealing, the process achieves seamless compatibility with CMOS back-end-of-line (BEOL) fabrication. Electrical characterization in Fig. 2(b) & (c) reveals robust DC performance, with cumulative distribution function (CDF) analysis demonstrating a 10× memory window and exceptional cycle-to-cycle consistency, underscoring the SPS methodology's efficacy in stabilizing V_O behavior.

B. High-density RRAM array with nTmR Layout Design

At the memory cell level, the one-transistor-one-resistive-RRAM (1T1R) cell is extensively utilized to mitigate sneak-path currents and write disturbances in RRAM arrays. However, the cell area of 1T1R configurations is fundamentally constrained by the access transistor's dimensions, as further downscaling risks compromising the

979-8-3315-3918-4/25 $31.00 © 2025 IEEE

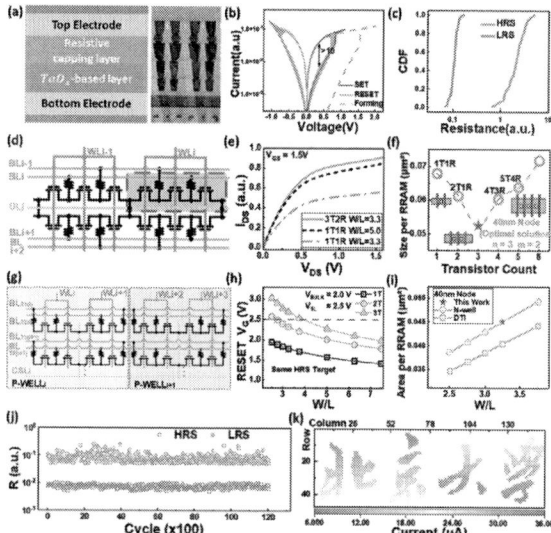

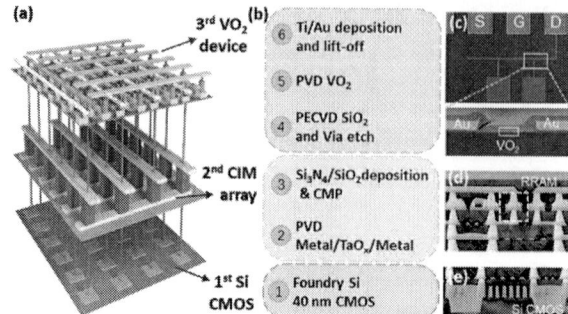

Fig. 3. (a) Schematic of three-tier M3D IMC system. (b) The fabrication process of the chip. (c) SEM image and TEM image of the VO_x-based IMT device (d) TEM image of the H²DR RRAM-based IMC layer. (e) TEM image of logic circuit layers. Adopted from Ref. [5].

Fig. 2. (a) The schematic and TEM image of SPS-based H²DR RRAM device. (b) Typical I-V characteristics. (c) CDF of the HRS and LRS. (d) Schematics of 3T2R array with a common SL. (e) Driving capability comparisons between 3T2R and 1T1R cells. (f) Relationship between different transistor counts and nTmR cell area. (g) Schematic of the proposed array with adjacent T1 and T2 cells. (h) Relationship between RESET V_G and the W/L ratio with the p-well biased at $V_{P\text{-well}}$. (i) Trends in unit area reduction as a function of W/L under different p-well isolation schemes. (j) Endurance performance of H²DR RRAM. (k) A 5-level pattern was programmed into four 4 kb test chips with a 100% yield. Adopted from Ref. [2] & [3] & [4].

drive current and voltage margins essential for reliable resistive switching. To address this issue, we present an *n*-transistors-*m*-RRAM (*n*T*m*R) array topology enabling high-density RRAM integration [3]. This design leverages multiple narrow-channel transistors to reduce cell footprint while preserving adequate drive capability. Fig. 2(d) details the 3T2R array schematic with shared source-line (SL) configuration. Driving capability comparisons in Fig. 2(e) demonstrate that the 3T2R cell outperforms conventional 1T1R counterparts in drive strength at equivalent area. Area scaling analysis for variable *m*/*n* configurations in Fig. 2(f) reveals that the 3T2R configuration achieves the minimum cell area of 0.0525 μm^2 at the 40 nm technology node, enabling unprecedented RRAM integration densities.

At the array topology level, source-line contacts impose significant area penalties that degrade integration density. To mitigate this, we introduce a high-density NAND-gate array architecture employing localized series-connected architectures to minimize contact overhead (Fig. 2(g)) [4]. While increasing local series elements enhances density, it elevates operating gate voltages, compromising transistor reliability. A substrate bias scheme is implemented to reduce RESET operation gate voltage (Fig. 2(h)), with a triple-well process isolating the array's p-well from the substrate to limit substrate bias impact. Deep trench isolation (DTI) technology further enables aggressive isolation scaling. Fig. 2(i) demonstrates how transistor width/length (W/L) ratios and isolation schemes collectively optimize cell area, achieving minimal footprint through synergistic design of series topology, bias engineering, and advanced isolation.

C. Wafer-scale Measurement Results of H²DR RRAM

Based on the aforementioned device and array innovations, a test wafer incorporating H²DR RRAM technology was fabricated for comprehensive evaluation. Endurance testing

under optimal drive current conditions (Fig. 2(j)) demonstrated reliable operation exceeding 12,000 cycles, maintaining a distinct current differential of ~20 µA at 0.2 V. Leveraging dynamic gate voltage modulation, we achieved 32 discrete conductance states with exceptional cycle-to-cycle uniformity on 4 kb arrays [2], while simultaneous programming of a 5-level pattern across four test chips yielded 100% operational success (Fig. 2(k)). Accelerated retention testing via high-temperature baking (150°C) projected data stability exceeding a decade through extrapolation of post-baking read margins [2]. These results — combining record-high integration density, sub-10-year retention at aggressive nodes, and precise multilevel programming — position H²DR RRAM as a compelling solution for energy-efficient in-memory AI acceleration.

D. H²DR RRAM in M3D IMC Chips

Combine the H²DR RRAM with monolithic 3D (M3D) integration technology offers a pathway to significantly enhance the integration density and data bandwidth of IMC systems. We demonstrate a M3D IMC chip (Fig. 3(a)) comprising three functional layers: a bottom CMOS circuit layer, a middle TaO_x-based H²DR RRAM array, and a top VO_x-based insulator-metal-transition (IMT) device layer. The fabrication process is shown in Fig. 3(b). Scanning electron microscopy (SEM) and cross-sectional TEM images (Fig. 3(c)) confirm the high quality of the VO₂ IMT devices, while Fig. 3(d) & (e) illustrate the layer-specific RRAM and logic circuit integration. Experimental validation demonstrates that this M3D architecture achieves 2.1× operational speed enhancement alongside reduced energy consumption and hardware overhead compared to planar implementations, underscoring its viability for AI computing applications [5].

III. HIGH ENERGY-EFFICIENT IMC ARCHITECTURES

High-performance IMC architectures are pivotal for realizing AI systems. We have developed a suite of energy-optimized IMC frameworks tailored for embodied intelligence applications, enabling efficient distortion calibration, feature extraction, and intelligent recognition across image and point cloud data modalities. By systematically optimizing parameter mapping, circuit-level precision, and dataflow orchestration, these architectures provide a unified computational substrate for low-latency, multimodal data processing in resource-constrained environments.

A. Hybrid-domain Architecture for Distortion Calibration

Accurate environmental perception is essential for intelligent agents to make reliable decisions, yet pure AI

979-8-3315-3918-4/25 $31.00 © 2025 IEEE

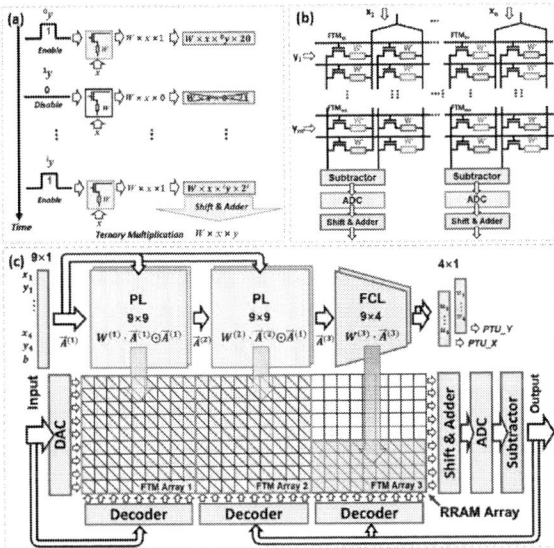

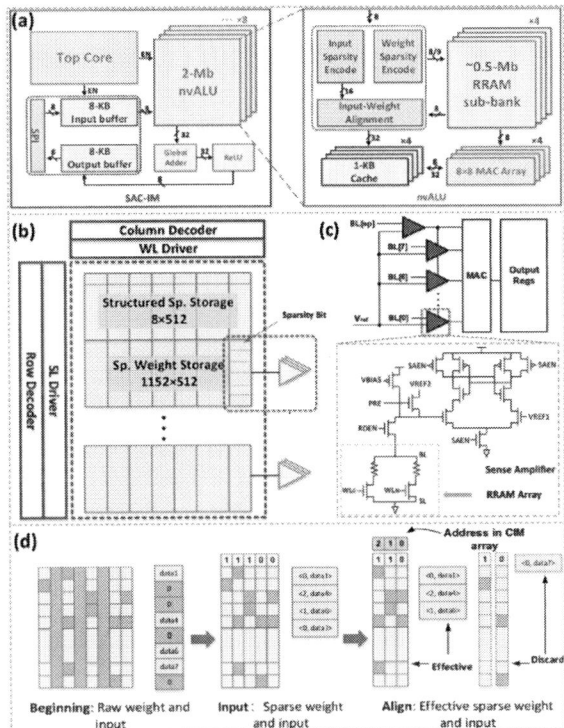

Fig. 4. (a) Schematic of hybrid-domain computing of TM operation. (b) The schematic of RRAM based FTM array and corresponding circuit topology. (c) The parallel polynomial transformation model and the corresponding PTU. Adopted from Ref. [6].

systems inherently struggle with data distortion (e.g., images, point clouds) that degrades fidelity. To address this, we propose a hybrid-domain polynomial transformation (HDPT) architecture leveraging synergies across time, analog, and digital domains for accelerated distortion calibration [6]. Fig. 4(a) illustrates the core unsigned three-operand multiplication (TM) operator in polynomial transformation algorithm, realized via an expandable ternary multiplier (ETM) where a time-domain control signal (input y) modulates the multiplication of analog input x and digital weight W, followed by accumulation through a shift-and-adder stage. For signed operands, Fig. 4(b) introduces a four-quadrant ternary multiplier (FTM) array enabling TM operations with signed x/W and unsigned y. The polynomial transformation unit (PTU) in Fig. 4(d) integrates two polynomial layers (PLs) and a fully connected layer (FCL) to support third-order polynomial algorithms with massively parallel computation, ensuring scalable, energy-efficient distortion correction while maintaining compatibility with advanced process nodes.

B. Sparsity-aware Architecture for Image Processing

Intelligent image processing algorithms often exhibit high parameter sparsity, leading to excessive hardware overhead and computational waste when directly implemented on IMC arrays. To mitigate this, we propose a sparsity-aware compute-in-memory (SA-CIM) architecture tailored for efficient image processing [7]. Fig. 5(a) outlines the SA-CIM framework, featuring non-volatile arithmetic logic units (nvALUs) governed by a Top Core controller that enables in-situ data programming, verification, and parallel readout during computation. Each nvALU integrates a sparsity-aware RRAM sub-bank (Fig. 5(b)) with a dedicated "sparsity bit" per word-line, flagging unstructured sparsity patterns. The sub-bank architecture couples sense amplifiers (SAs), multiply-accumulate (MAC) units, and output registers to streamline dataflow. As detailed in Fig. 5(c), SA-CIM first performs in-situ sparsity detection and encoding on raw input data and weight matrices. During computation, input vectors are dynamically aligned with pre-encoded sparse weights,

Fig. 5. (a) Overall architecture of SA-CIM and the architecture of NvALU. (b) Internal structure of the sparsity-aware RRAM sub-bank. (c) The connection of CIM cell, MAC and output registers and the schematic of SA and bias circuit. (d) The compute process of SA-CIM. Adopted from Ref. [7].

automatically discarding mismatched input-weight pairs. This two-stage compression reduces MAC operations by over 80% while maintaining algorithmic accuracy, achieving 1.4× energy efficiency improvement compared to conventional IMC designs.

C. Heterogeneous Architecture for Point Cloud Processing

Point cloud processing, which provides three-dimensional object representations for precise system decision-making, demands efficient acceleration of Euclidean distance (L2D) calculations and matrix-vector multiplications (MVM). Existing IMC solutions face challenges in high-precision L2D implementation and suffer from inter-stage data transfer bottlenecks. To address this issue, we propose a heterogeneous IMC architecture for point cloud neural networks (PNNs) that integrates 2T0C DRAM-based IMC array for high-accuracy L2D operations and RRAM-based IMC array for MVM acceleration (Fig. 6(a)) [8]. The architecture introduces a novel in-memory L2D scheme (Fig. 6(b)) that reorders computational priorities: unlike conventional methods requiring complex pre-alignment and exponent shifting operations, our approach prioritizes mantissa multiplication while deferring exponent adjustments to post-processing stages, significantly reducing hardware complexity. To compensate for quantization-induced accuracy loss, we developed a hybrid IMC operator combining analog and digital domains for dynamic data conversion and precision calibration (Fig. 6(c)). The design eliminates conventional coarse-grained dataflow limitations by directly coupling hybrid operators with MVM units, bypassing multilevel caching and enabling parallel L2D/MVM execution (Fig. 6(d)). This pipeline optimization

979-8-3315-3918-4/25 $31.00 © 2025 IEEE

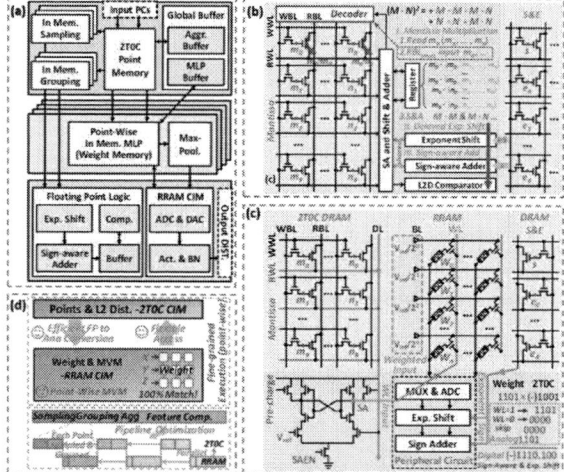

Fig. 6. (a) The H-M3D-based point cloud processing architecture. (b) L2D computation scheme utilizing the 2T0C array. (c) Schematic of hybrid operator for feature computation, supporting high-bit quantization for the first layer MLP. (d) Optimized dataflow pipeline. Adopted from Ref. [8].

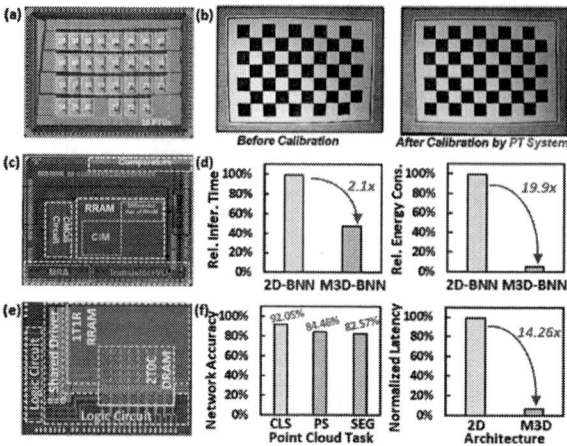

Fig. 7. (a) The microscopy image of multi-core IMC chip. (b) The comparison between the images w/ and w/o calibration. (c) The microscopy image of M3D-BNN chip. (d) The comparison of inference time and energy consumption between 2D-BNN and M3D-BNN system. (e) The microscopy image of point cloud processing chip. (f) Simulated accuracy and speed enhancement. Adopted from Ref. [5] & [6] & [8].

achieves 1.5× throughput improvement and 31% latency reduction compared to state-of-the-art point cloud accelerators.

IV. H²DR RRAM-BASED ENERGY-EFFICIENT IMC CHIP

A. Multi-core IMC Chip for Distortion Calibration

The multi-core IMC chip, based on the HDPT architecture, is designed for image calibration tasks. As depicted in Fig. 7(a), the chip's microscopy image reveals 30 polynomial transformation units (PTUs) integrated with a H²DR RRAM-based FTM array, enabling parallel computation of distortion calibration. Fig. 7(b) shows the chip's processing effect on distorted images. Evaluation results show that the chip's throughput can reach 158M pixels/s and its energy efficiency can reach 3.81G pixels/W.

B. M3D IMC Chip for Intelligent Decision-making

The high-bandwidth M3D IMC chip is designed for intelligent decision-making applications. As shown in the microscopy image in Fig. 7(c), the chip integrates CMOS circuitry, a H²DR RRAM-based IMC array, and VO₂-based IMT devices, all working in unison to implement the core operations of Bayesian neural networks (BNNs). Fig. 7(d) compares inference latency and energy efficiency between conventional planar BNN implementations and the proposed M3D-BNN system, revealing a 2.1× reduction in latency and a 19.9× improvement in energy efficiency for the M3D architecture relative to its planar counterpart.

C. Hybrid M3D IMC Chip for Point Cloud Processing

The hybrid M3D IMC chip, tailored for three-dimensional point cloud processing, integrates a high-precision 2T0C DRAM-IMC module optimized for Euclidean distance computations with an RRAM-IMC array dedicated to matrix-vector multiplications, collectively enabling efficient feature extraction through dataflow optimization at the granularity of individual points (Fig. 7(e)). Fig. 7(f) shows the chip performance evaluation results based on experimental data, which demonstrates a 14× reduction in point cloud processing latency, underscoring the architecture's capacity to balance precision and computational efficiency in 3D spatial tasks.

V. CONCLUSION

RRAM-based IMC technology demonstrates enormous potential to deliver solutions for highly energy-efficient AI computing. This paper highlights our comprehensive efforts in H²DR RRAM-based IMC systems, including high-performance device, energy-efficient IMC architectures and high-performance chips, providing new ideas for the design of next-generation AI processors.

ACKNOWLEDGMENT

This work was supported by National Natural Science Foundation of China under Grant 62025401, Grant 62322401, and Grant 62341407, Beijing Nova Program under Grant 20220484113, and in part by "111" Project under grant B18001.

REFERENCES

[1] Y. Cai et al., 2023 IEEE 15th International Conference on ASIC (ASICON), Nanjing, China, 2023, pp. 1-4.

[2] Q. Wang et al., "A logic-process compatible RRAM with 15.43 Mb/mm2 density and 10years@150°C retention using STI-less dynamic-gate and self-passivation sidewall," 2023 International Electron Devices Meeting (IEDM), San Francisco, CA, USA, 2023, pp. 1-4.

[3] S. Bao et al., "Design technology co-optimization of high-density RRAM array for advanced CMOS technology node," in IEEE Transactions on Electron Devices, vol. 72, no. 5, pp. 2278-2284, May 2025.

[4] S. Bao et al., "Advancing toward 4F² 1T1R RRAM with local NAND-gate and isolation scheme," in IEEE Transactions on Electron Devices, vol. 72, no. 5, pp. 2327-2333, May 2025.

[5] L. Shan et al., "Monolithically 3D integrated memristive bayesian neural network for intelligent motion planning," 2024 IEEE International Electron Devices Meeting (IEDM), San Francisco, CA, USA, 2024, pp. 1-4.

[6] L. Bao et al., "Hybrid-domain in-memory polynomial acceleration based on 40nm RRAM multi-core chip for machine vision calibration," 2023 International Electron Devices Meeting (IEDM), San Francisco, CA, USA, 2023, pp. 1-4.

[7] H. Ding et al., "SA-CIM: a 28nm 16Mb RRAM-based sparsity-aware compute-in-memory macro for edge AI algorithm processing," 2025 IEEE International Symposium on Circuits and Systems (ISCAS), London, United Kingdom, 2025, pp. 1-5.

[8] Y. Gao et al., "A hybrid monolithic 3D integration of 2T0C DRAM and RRAM chip for high-precision in-memory point cloud acceleration with ultra-fine-grained dataflow," 2025 Symposium on VLSI Technology and Circuits (VLSI Technology and Circuits), Kyoto, Japan, 2025, pp.1-3

The Digital Coupled Ring Oscillator Ising Machine

Yue Han [1], Ranjith R Unnithan [2], Robin Evans [3], Efstratios Skafidas*[1]

The University of Melbourne, Australia
hany6@student.unimleb.edu.au, sskaf@unimelb.edu.au

Abstract—Ising machines offer a high-performance approach to solving combinatorial optimization problems by evolving toward the ground state of the system Hamiltonian. Among these, oscillator-based Ising machines have gained attention due to their ultra-low power consumption and efficiency in reaching optimal configurations. However, existing implementations are often constrained by specific graph topologies and problem constraints, limiting their scalability and versatility. In this work, we propose a mixed-signal oscillator-based Ising machine that utilizes digital coupling with programmable weights, enabling support for problems with arbitrary graph connections and weight values. By eliminating the need for direct analog coupling, our design significantly reduces hardware complexity while expanding the problem space to all-to-all connectivity. We present a 128-node digital-coupled ring oscillator Ising machine to solve weighted Max-Cut problems, demonstrating its optimality, efficiency, and computational robustness advantages. Comprehensive simulation studies address key challenges, including sampling frequency, resolution, coupling strength, and second harmonic injection schemes. Our results show that the proposed system achieves high success probabilities in finding near-optimal solutions efficiently, paving the way for scalable, reconfigurable Ising-based computing architectures.

Keywords—*Ising machine, combinatorial optimization, oscillator-based computing, simulated annealing, mixed-signal circuits*

I. INTRODUCTION

General-purpose processors (GPPs) have been the backbone of computational systems for over fifty years due to their ease of programming and extensive software support. They power applications from small embedded devices to large-scale data centres. However, advances in performance and energy efficiency of GPPs have slowed, creating a bottleneck for solving increasingly complex computational tasks, particularly in artificial intelligence and optimisation applications [1].

Combinatorial optimization problems (COPs), such as Max-Cut, graph coloring, and the travelling salesman problem, are NP-hard and widely encountered in IC design, communication networks, computer vision, and machine learning [2]. Traditional approaches like Monte Carlo simulations and simulated annealing offer approximate solutions but suffer from long execution times, limiting their scalability, optimality and efficiency [3]. Quantum computing promises speedups via quantum superposition and entanglement [4], but its practical development is restricted by stringent requirements, including absolute zero temperatures, ultra-high vacuum, and electromagnetic shielding, which make quantum systems unsuitable for incorporation in portable and low-power embedded systems.

The Ising model, initially developed for ferromagnetic systems, represents a problem using binary spins with configurable coupling strengths and external fields. The Ising Hamiltonian is given by:

$$H(s) = -\sum_{i,j} J_{ij} s_i s_j - \sum_i h_i s_i \tag{1}$$

where J_{ij} is the coupling strength, $s_i \in \{\pm 1\}$ is the Ising spin, and h_i is the external field acting on the spin. By minimizing (1) through annealing techniques, Ising machines efficiently compute optimal solutions to NP-hard problems [5]. As illustrated in Fig. 1, the Ising machine workflow involves a pre-processing stage that formulates the problem using the Ising model directly or via quadratic unconstrained binary optimization (QUBO) to extract J_{ij} and h_i for configuring the Ising machine and post-processing stage that reads out the final spin states and maps to the problem's solution. With the growing adoption of Ising formulations for NP problems [6], Ising machines are emerging as promising low-power and high-efficiency solvers for COPs.

Various Ising machine implementations have been explored, including quantum dots [7], photonic annealers [8], digital annealers [9], optical oscillators [10], [11] and electrical oscillators [12], [13], [14]. Among these, oscillator-based Ising machines have gained significant attention due to its potential for ultra-low-power, CMOS-compatible scalable computing architectures [15]. This permits oscillator-based Ising machine solvers to be built alongside advanced general-purpose processors on the same die, allowing systems to recruit the optimal compute architecture for the problem that needs to be solved.

The oscillator-based Ising approach uses coupled nonlinear oscillators to emulate the Ising model, operating based on the Kuramoto model:

$$\frac{d}{dt}\phi_i(t) = -K \sum_{j=1, j\neq i}^{n} J_{ij} \sin\left(\phi_i(t) - \phi_j(t)\right) \tag{2}$$

where ϕ_i is the oscillator phase, K is the overall coupling strength, J_{ij} is the coupling strength between the oscillators, and

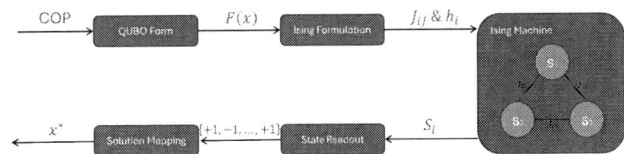

Fig. 1. The workflow of COP solving using an Ising machine.

979-8-3315-3918-4/25 $31.00 © 2025 IEEE

n is the total number of spins. The system represents coupled oscillators as Ising spins, encoding each oscillator's phase as a corresponding spin state. It evolves toward the lowest-energy configuration, which corresponds to the Ising Hamiltonian's ground state when the oscillator's phase is restricted to either in-phase or out-of-phase.

Recent oscillator-based Ising machines have explored various types of oscillators and coupling methods to improve performance by scaling up the spin count and expanding the coupling range for solving more complex COPs [12], [13], [14], [16], [17]. However, physical interconnections and analog couplings like resistive or capacitive coupling networks between the oscillators limit the scalability and flexibility. They cannot instantiate the arbitrary coupling values, which restricts applicability to a narrow class of problems defined at chip fabrication.

This work, for the first time, proposes a digital-coupled ring oscillator Ising machine with a reconfigurable all-to-all connectivity architecture and fully programmable coupling weights. The architecture supports up to 128 nodes with coupling weights ranging from -128 to +127, enabling broader problem coverage and enhanced flexibility. Here, we show that this approach is feasible, after significant evaluation to show performance and convergence when the coupling weights are implemented using a mixed signal system. The proposed machine has been evaluated on randomly generated Max-Cut problems, achieving 100% success probability for complex instances with up to 128 nodes.

The paper is organized as follows: Section II introduces the top-level architecture, highlighting the fully programmable coupling weights and reconfigurable network topology. Section III describes the circuit-level implementation and simulation setup used to evaluate system performance. Section IV presents simulation results for solving randomly generated max-cut problems, analyzing the impact of key design parameters.

II. DIGITAL-COUPLED RING OSCILLATOR ISING MACHINE

Fig. 2 shows the top-level architecture of the digital-coupled ring oscillator Ising machine. The system consists of three primary modules: the analog ring oscillator array representing Ising spins, the digital coupling module that samples oscillators with analog-to-digital converters (ADCs) and performs coupling interactions for each oscillator with a coupling unit, and the digital-to-analog converters (DACs) injecting the overall coupling signals back into the oscillators. The system operates as a real-time mixed-signal loop. Initially, oscillators run freely while coupling strengths are stored in a symmetrical matrix. In each clock cycle, ADCs digitise oscillator amplitudes, either 0 or 1, forming a column vector which multiplies with the coupling matrix for interconnecting oscillators with the corresponding coupling strength:

$$J * A + A_r * H = \begin{bmatrix} 0 & J_{12} & J_{13} & \cdots & J_{1N} \\ J_{12} & 0 & J_{23} & \cdots & J_{2N} \\ \vdots & \vdots & \vdots & \ddots & \vdots \\ J_{1N} & J_{2N} & J_{3N} & \cdots & 0 \end{bmatrix} \begin{bmatrix} A_1 \\ A_2 \\ \vdots \\ A_N \end{bmatrix} + A_r \begin{bmatrix} h_1 \\ h_2 \\ \vdots \\ h_N \end{bmatrix}$$

$$= \begin{bmatrix} A_2 J_{12} + A_3 J_{13} + \cdots + A_N J_{1N} + A_r h_1 \\ A_1 J_{12} + A_3 J_{23} + \cdots + A_N J_{2N} + A_r h_2 \\ \vdots \\ A_1 J_{1N} + A_2 J_{2N} + \cdots + A_{N-1} J_{(N-1)N} + A_r h_N \end{bmatrix} \quad (3)$$

where N is the number of oscillators, J is an N-by-N coupling matrix, A is an N-by-1 amplitude matrix, A_r is the amplitude of the reference oscillator with a fixed phase of zero to account for self-biasing effects induced by external fields, and H is an N-by-1 external fields matrix. The resulting product matrix in (3) represents cumulative coupling signals for each oscillator from all other oscillators in the network and self-biasing contributions. The coupling signals are then injected back into the oscillators through DACs, prompting the phases to shift according to the Kuramoto model. This iterative process continues until the system reaches a steady state, where the oscillators exhibit synchronization behaviour based on their coupling dynamics.

Compared to state-of-the-art oscillator-based Ising machines, which rely on fixed or programmable analog resistive coupling for cross-interactions, the proposed digital coupling module performs cross-coupling through matrix multiplication in the digital domain. This approach significantly reduces hardware complexity by eliminating the need for direct analog interconnections between oscillators. The system allows the integration of a large number of spins with unrestricted connections, overcoming scalability constraints imposed by the physical limitations of analog coupling. By avoiding up to 128^2 physical interconnections, the system achieves all-to-all connection for 128 nodes, which is more than twice that of the leading oscillator-based Ising machine implementation [18], significantly enhancing both scalability and architectural flexibility.

Furthermore, digital coupling also enables programmable 8-bit coupling weights, offering 256 discrete coupling levels and fully reconfigurable topologies by setting unused connections to zero. This flexibility supports a wider range of COPs and allows the system to operate in parallel by partitioning the oscillator network into independent subgroups through selective disconnection of spins. Such parallelism significantly improves throughput and enhances solution accuracy on small-scale problems by enabling majority-vote mechanisms across oscillator groups.

III. SYSTEM IMPLEMENTATION

The 128-node Ising machine is implemented using TSMC 65nm process, incorporating 128 ring oscillators, a second-

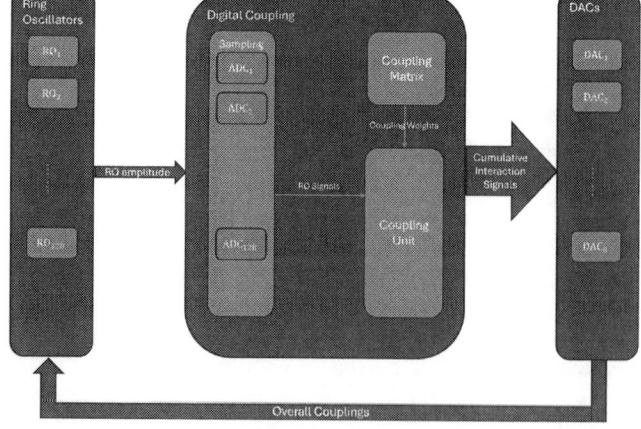

Fig. 2.   Top-level architecture of digital-coupled ring oscillator Ising machine.

979-8-3315-3918-4/25 $31.00 © 2025 IEEE

harmonic injection locking (SHIL) module, a digital coupling unit, and high-speed 8-bit current steering DACs.

The ring oscillator is the core analog computing element, with its phase encoding the Ising spin state. As shown in Fig. 3, each oscillator consists of a 9-stage CMOS inverter loop with a measured natural frequency of 25 MHz. The digital coupling module comprises an SRAM that stores the upper triangle of the symmetric coupling matrix J, 128 double flip-flops for sampling and synchronizing the ring oscillator outputs, and 128 parallel processing elements (PEs) that compute coupling interactions. The PEs accumulate relevant coupling weights based on the sampled spin states and output the resulting coupling signals to the DACs. As a critical module in the mixed-signal Ising machine, the DAC directly impacts the coupling quality and computation accuracy. To achieve high speed and resolution, a segmented current steering DAC architecture is designed. It combines a 5-bit unary-weighted segment and a 3-bit binary-weighted segment to improve linearity while maintaining a compact area and low power consumption [19]. The unary segment comprises 32 unit current sinks, each delivering a current of I_{unit}, controlled by a thermometer code, ensuring excellent monotonicity by activating one sink per asserted bit. The binary segment, handling the 3 least significant bits, includes current sinks scaled to $I_{unit}/2$, $I_{unit}/4$, and $I_{unit}/8$, where the mismatch-induced error has minimal effect on overall DAC performance. To reduce area, each DAC is time-multiplexed across a group of ring oscillators, sequentially delivering the computed coupling signals during each sampling cycle. Compared to voltage injection, current injection simplifies integration by removing the need for passive components and enables precise control of the overall coupling strength K via the digital coupling module.

System performance is evaluated by solving random weighted Max-Cut problems. The problem is first formulated in Ising form, and the corresponding coupling matrix is stored in SRAM to configure the Ising machine. After a defined convergence period, the solution is extracted from the stabilized phase

configuration. Simultaneously, the same problem is solved using the Tabu search algorithm, and its optimal solution is recorded as the ground truth. The Ising machine output is then classified as follows:

- Optimal: if the configuration matches the ground truth.

- Sub-optimal: if it differs but achieves 95% of the optimal max-cut value.

- Non-optimal: if the normalized max-cut value is below 95%.

IV. RESULTS AND ANALYSIS

A. Sampling Frequency and Resolution

Sampling frequency and resolution are critical parameters in mixed-signal systems, directly affecting the accuracy and computational efficiency of the proposed Ising machine. The sampling frequency influences both the bandwidth and signal precision. As shown in Fig. 4, low sampling frequencies introduce aliasing, distorting signals and degrading accuracy, while higher sampling frequencies improve success probability, which is defined as the percentage of optimal and sub-optimal solutions, until saturation occurs around 15 times the oscillator's natural frequency. However, computational time increases with the sampling frequency due to the higher data throughput required for convergence.

Resolution impacts the precision of the overall coupling signals injected into the oscillators. While higher resolution reduces quantization error, it also increases noise sensitivity, power consumption, and circuit complexity. In the proposed digital-coupled Ising machine, an 8-bit resolution ensures sufficient amplitude granularity for accurately converting cumulative coupling signals. A higher resolution for exact precision is not essential for convergence, as long as the relative order of the cumulative coupling values is maintained.

B. Problem Complexity and SHIL

The simulation results demonstrate the effectiveness of the digital-coupled architecture in mapping and solving complex

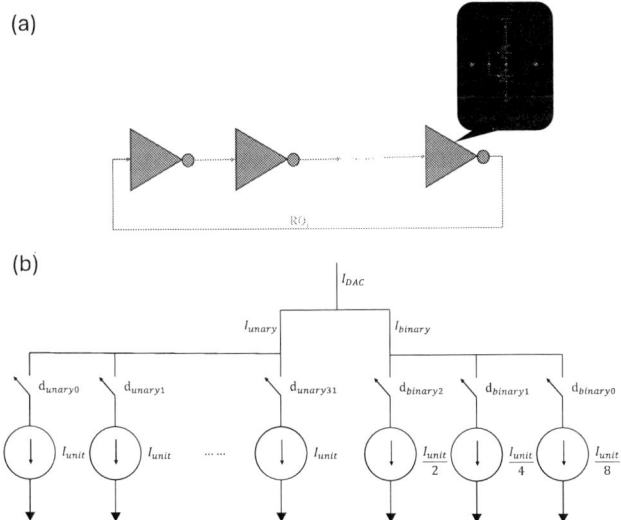

Fig. 3. (a) 9-stage inverter-based ring oscillator schematic. (b) Simplified segmented current steering DAC schematic.

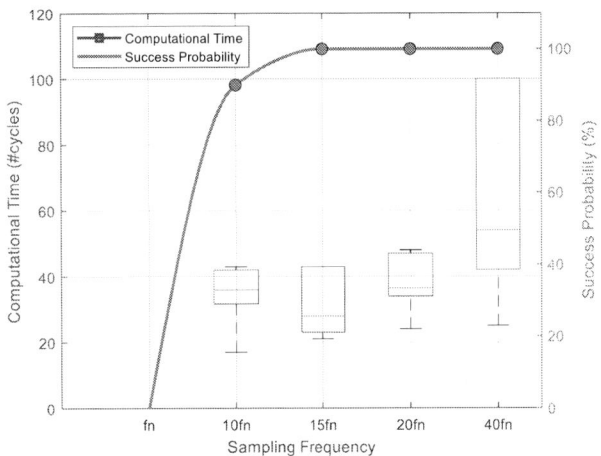

Fig. 4. Success probability and box plot of computational time in solving randomly generated max-cut problems with increasing sampling frequencies, where f_n is the oscillator's natural frequency.

COPs. Fig. 5 illustrates a piecewise linearly increasing computational time while maintaining a 100% success probability when solving problems with exponentially increasing complexity, characterized by $O(m2^n)$, where n is the number of variables and m is the connectivity. As the problem size increases, the search space expands exponentially, and the number of feasible solutions grows rapidly with rising connectivity. The system consistently finds optimal solutions for low-complexity problems, but yield near-optimal configurations once the complexity exceeds $O(16 \times 2^{16})$, corresponding to 16-node all-to-all connectivity. Notably, for extremely high-complexity problems, such as 128-node all-to-all connected problems, the computational time saturates and even decreases.

This behaviour aligns with the stability analysis of the coupled oscillator system with SHIL, given by:

$$\frac{d}{dt}\phi_i(t) = -K \sum_{j=1, j \neq i}^{n} J_{ij} \sin\left(\phi_i(t) - \phi_j(t)\right)$$
$$-K_s \sin\left(2\phi_i(t)\right) \qquad (4)$$

where K_s is the SHIL strength. A configuration is considered stable if all Lyapunov exponents, determined by eigenvalues of the Jacobian matrix of (4) at equilibrium points, are negative [20]. This analysis highlights the critical role of SHIL in oscillator-based Ising machines, as the Lyapunov exponents scale proportionally with the ratio K/K_s. With weak SHIL, even the optimal configuration with the lowest Hamiltonian may become unstable, hindering the system from converging to the global optimal solution. In contrast, strong SHIL suppresses Lyapunov exponents across all configurations and stabilizes multiple configurations, including both the global optimal and near-optimal configurations. These stable configurations act as equilibrium points that attract the system within their attraction domains [21]. For complex problems, rising SHIL strength leads to a significant increase of stable near-optimal configurations with small energy difference, leading the system to converge towards these configurations. As a result, the system tends to converge faster but more often settles into one of the sub-optimal configurations rather than the global optimum.

To address stability issues and further enhance performance, a stepped increasing amplitude-modulated SHIL scheme is proposed to reduce the number of stable near-optimal configurations. This approach is inspired by the adiabatic theorem, which states that a physical system remains in its instantaneous eigenstate if a given perturbation is applied slowly enough and a gap exists between its eigenvalue and the rest of the Hamiltonian spectrum [22]. Adiabatic control is widely applied in quantum computing, enabling gradual evolution from a known ground state to the optimal solution under a slowly changing perturbation. Similarly, in the proposed Ising machine, the slowly increasing SHIL amplitudes guide the system toward lower-energy states while preserving stability. The stepped amplitude increments provide a certain convergence period for oscillators at each SHIL strength. It allows the system to gradually settle into the nearest stable configuration with a lower Hamiltonian. As shown in Fig. 6, this SHIL scheme significantly increases the frequency of global optimal solutions for low-complexity problems and improves solution quality for high-complexity problems, with the normalized max-cut values approaching 100%, indicating that sub-optimal configurations closely approximate the global optimum.

V. CONCLUSION

The proposed digital-coupled ring oscillator Ising machine achieves over 90% success probability for global optimal solutions in low-complexity problems and 100% success probability for near-optimal solutions in high-complexity problems, with the cost value within 97% of the global minimum under the stepped amplitude-modulated SHIL scheme. Compared to traditional oscillator-based Ising machines, the digital coupling architecture supports fully programmable couplings and reconfigurable topologies, enabling the system to address a significantly broader class of COPs with diverse constraints. Simulation results demonstrate the system's ability to solve complex problems with up to 128-node all-to-all connection, achieving 100% success probability within less than 250 oscillation cycles. This represents a substantial improvement in both maximum connectivity and coupling flexibility over the current oscillator-based Ising machine.

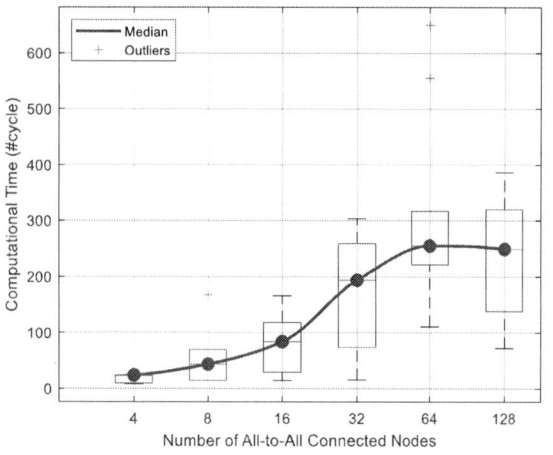

Fig. 5. Box plot of Computational time for randomly generated max-cut problems with exponentially increasing complexities in terms of $O(m2^n)$, where $m = n = \# nodes$.

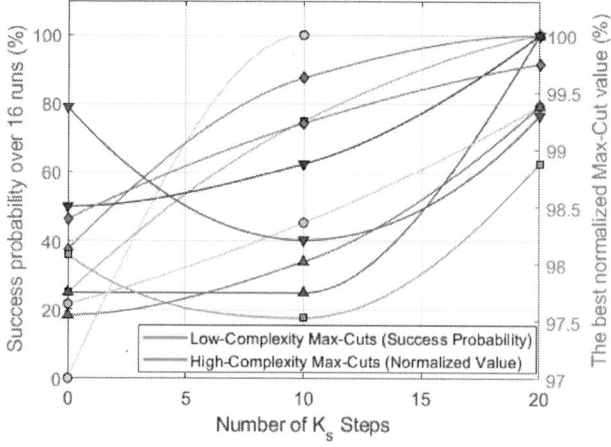

Fig. 6. Success probability and normalized max-cut value for solving randomly generated max-cut problems with low and high complexity when enabling slower stepped-increasing amplitude-modulated SHIL.

979-8-3315-3918-4/25 $31.00 © 2025 IEEE

REFERENCES

[1] W. J. Dally, Y. Turakhia, and S. Han, "Domain-specific hardware accelerators," Communications of the ACM, vol. 63, no. 7, pp. 48-57, 2020.

[2] X. Gao, H. Du, and M. Han, "Combinatorial optimization and applications," Proc. COCOA, p. 10628, 2017.

[3] Y.-C. Lin, C.-C. Wang, C.-H. Tu, and S.-H. Hung, "Towards Optimizations of Quantum Circuit Simulation for Solving Max-Cut Problems with QAOA," arXiv preprint arXiv:2312.03019, 2023.

[4] A. Steane, "Quantum computing," Reports on Progress in Physics, vol. 61, no. 2, p. 117, 1998.

[5] A. Lucas, "Ising formulations of many NP problems," Frontiers in physics, vol. 2, p. 5, 2014.

[6] N. Mohseni, P. L. McMahon, and T. Byrnes, "Ising machines as hardware solvers of combinatorial optimization problems," Nature Reviews Physics, vol. 4, no. 6, pp. 363-379, 2022.

[7] S. Sarkar and S. Bhanja, "Synthesizing energy minimizing quantum-dot cellular automata circuits for vision computing," in 5th IEEE Conference on Nanotechnology, 2005., 2005: IEEE, pp. 541-544.

[8] D. Pierangeli, G. Marcucci, and C. Conti, "Large-scale photonic Ising machine by spatial light modulation," Physical review letters, vol. 122, no. 21, p. 213902, 2019.

[9] S. Matsubara et al., "Digital annealer for high-speed solving of combinatorial optimization problems and its applications," in 2020 25th Asia and South Pacific Design Automation Conference (ASP-DAC), 2020: IEEE, pp. 667-672.

[10] T. Inagaki et al., "A coherent Ising machine for 2000-node optimization problems," Science, vol. 354, no. 6312, pp. 603-606, 2016.

[11] P. L. McMahon et al., "A fully programmable 100-spin coherent Ising machine with all-to-all connections," Science, vol. 354, no. 6312, pp. 614-617, 2016.

[12] T. Wang and J. Roychowdhury, "OIM: Oscillator-based Ising machines for solving combinatorial optimisation problems," in Unconventional Computation and Natural Computation: 18th International Conference, UCNC 2019, Tokyo, Japan, June 3–7, 2019, Proceedings 18, 2019: Springer, pp. 232-256.

[13] J. Chou, S. Bramhavar, S. Ghosh, and W. Herzog, "Analog coupled oscillator based weighted Ising machine," Scientific reports, vol. 9, no. 1, p. 14786, 2019.

[14] W. Moy, I. Ahmed, P.-w. Chiu, J. Moy, S. S. Sapatnekar, and C. H. Kim, "A 1,968-node coupled ring oscillator circuit for combinatorial optimization problem solving," Nature Electronics, vol. 5, no. 5, pp. 310-317, 2022.

[15] R. Lyon, "A computational model of filtering, detection, and compression in the cochlea," in ICASSP'82. IEEE International Conference on Acoustics, Speech, and Signal Processing, 1982, vol. 7: IEEE, pp. 1282-1285.

[16] J. Vaidya, R. Surya Kanthi, and N. Shukla, "Creating electronic oscillator-based Ising machines without external injection locking," Scientific Reports, vol. 12, no. 1, p. 981, 2022.

[17] H. Lo, W. Moy, H. Yu, S. Sapatnekar, and C. H. Kim, "An Ising solver chip based on coupled ring oscillators with a 48-node all-to-all connected array architecture," Nature Electronics, vol. 6, no. 10, pp. 771-778, 2023.

[18] H. Cılasun et al., "A coupled-oscillator-based Ising chip for combinatorial optimization," Nature Electronics, pp. 1-10, 2025.

[19] X. Li and L. Zhou, "A survey of high-speed high-resolution current steering DACs," Journal of Semiconductors, vol. 41, no. 11, p. 111404, 2020.

[20] M. K. Bashar, Z. Lin, and N. Shukla, "Stability of Oscillator Ising Machines: Not All Solutions Are Created Equal," arXiv preprint arXiv:2301.07601, 2023.

[21] Y. Cheng, M. Khairul Bashar, N. Shukla, and Z. Lin, "A control theoretic analysis of oscillator Ising machines," Chaos: An Interdisciplinary Journal of Nonlinear Science, vol. 34, no. 7, 2024.

[22] M. Born and V. Fock, "Beweis des Adiabatensatzes," Zeitschrift für Physik, vol. 51, no. 3, pp. 165-180, 1928/03/01 1928, doi: 10.1007/BF01343193.

Nanocrystal-Si Flash Memory-based Engergy-efficient Multi-bit Compute-in-Memory Design for Edge Neural Networks

Xianping Liu[1,2], Jian Huang[1,*], Zihan Zheng[1], Xinrui Zhang[1], Ruibin Zhou[1], Zhiyi Yu[1], Zhongyuan Ma[3], Kunji Chen[3], Yuhan Wang[1], Jian Cheng[2], Peng Zhang[2]

[1]School of Microelectronics Science and Technology, Sun Yat-sen University, Zhuhai, 510275, China
[2]Peng Cheng Laboratory, Shenzhen, 518055, China
[3]School of Electronic Science and Engineering, Nanjing University, Nanjing, 518055, China
*E-mail: huangj573@mail.sysu.edu.cn

Abstract—With the rapid advancement of artificial intelligence, the computational demands of neural network models have grown significantly. The traditional Von Neumann architecture separates memory and computation units and incurs substantial power overhead. Computing-in-Memory (CIM) has emerged as a promising paradigm to overcome this problem. However, most existing CIM systems still suffer from high memory cell operating currents and substantial power overhead from peripheral circuits. This work presents a novel CIM macro based on nanocrystal silicon (nc-Si) Flash memory. Leveraging the low-power characteristics of nc-Si Flash memory, we propose a multi-bit storage CIM macro in which digital-to-analog conversion for multi-bit inputs is performed within the nc-Si Flash CIM unit itself. Experimental results show that the proposed nc-Si Flash CIM array achieves an energy efficiency of 193.1TOPS/W. The classification accuracies of the LeNet-5 network on the MNIST and Letters datasets are 98.5% and 90.77%, respectively. The quantized VGG-9 achieves a classification accuracy of 76.2% on the CIFAR-10 dataset.

I. INTRODUCTION

As the number of parameters in neural networks continues to increase, the bottleneck of computing hardware based on the Von Neumann architecture is becoming severe. In recent years, numerous studies have shown that Compute-in-Memory(CIM) solutions provide an effective path to overcome this limitation. A variety of memory devices, such as SRAM, RRAM, MRAM, FeRAM and Flash memory have been widely used to implement different CIM designs. Flash-based CIM presents a promising solution, as Flash memory offers a small cell footprint, nonvolatility, and excellent CMOS compatibility when employed as a data storage element. These features enable seamless on-chip integration and process scalability, thereby reducing overall system complexity and manufacturing costs [1].

Among the various types of Flash memory, nanocrystal-based non-volatile memories exhibit several advantages compared to traditional floating-gate and nitride-type devices, owing to their discrete charge storage mechanism. In these devices, the storage of charge within isolated nanocrystals significantly mitigates the risk of charge loss due to defects in the tunnel oxide layer. The spatially distributed nature of charge storage

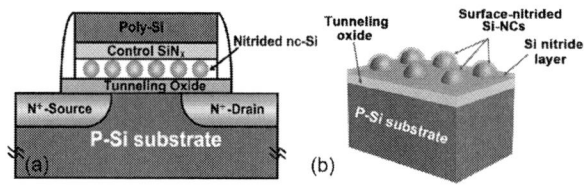

Fig. 1. (a) Cross-sectional structure of the proposed nc-Si Flash memory cell. (b) discrete surface-nitrided nc-Si dots, demonstrating uniform distribution for reliable charge storage.

effectively diminishes lateral leakage currents and enables the implementation of thinner tunnel barriers, thereby facilitating reduced programming voltages and superior scalability. Moreover, nanocrystal memories demonstrate robust radiation tolerance, making them well-suited for deployment in harsh environments [2]. Based on the nanocrystal silicon (nc-Si) Flash memory fabricated using a standard 0.13 μm CMOS, we investigated its potential for CIM design.

The simulation results demonstrate that that the proposed nc-Si Flash CIM array achieves an energy efficiency of 193.1TOPS/W. For recognition tasks, the classification accuracies of the LeNet-5 network on the MNIST and Letters datasets are 98.5% and 90.77%, respectively. The quantized VGG-9 achieves a classification accuracy of 76.2% on the CIFAR-10 dataset.

II. CIM NC-SI FLASH MEMORY WEIGHT STORAGE

This section introduces the device fundamentals of nc-Si Flash memory and details the proposed 3 bit neural weight storage strategy.

A. Basics of Nc-Si Flash Memory

As shown in Fig. 1(a), the structure of nc-Si Flash memory is identical to that of a MOS transistor with an additional floating layer. The tunneling oxide layer is specifically designed

979-8-3315-3918-4/25 $31.00 © 2025 IEEE

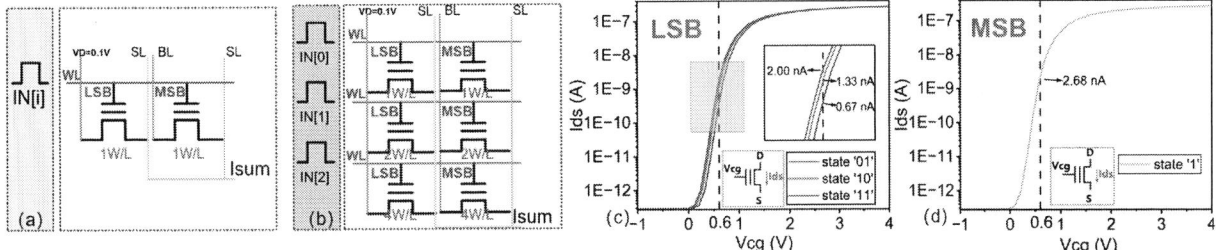

Fig. 2. (a) An example of single weight storage using two memory cells: the MSB cells store 1-bit and LSB cells store 2-bit of the weight. (b) Structure of the NFC unit consisting of three pairs of 2T nc-Si Flash cells. (c) The working current levels for the LSB cells under a fixed input voltage. (d) The working current levels for the MSB cells under a fixed input voltage.

to enable quantum tunneling, allowing electrons to move into and out of the nitrided nc-Si floating gate. The control SiNx layer, located above the floating gate, is made of silicon nitride and serves as an insulating barrier between the control and floating gates. The nc-Si floating gate, which consists of individual nanoscale silicon particles, is insulated on both sides. As a result, it can retain stored electrons even when the Flash memory is powered off, Fig. 1(b) shows the uniform distribution charge.

B. CIM Weight Storage in Nc-Si Flash Memory

Flash memory stores digital information by modulating the threshold voltage (V_{th}) of floating gate transistors, where each cell retains charge in its floating gate to represent specific logic levels.

In this work, we utilize two nc-Si Flash cells to store 3-bit neural network weights. By adding an additional dimension of summation to the multi-threshold voltage programming, 3-bit data can be represented using 5 threshold voltages in 2 cells, rather than 8 threshold voltages in TLC. As illustrated in Fig. 2(a), using a 1-bit input and a unsigned 3-bit weight as an example, Nc-Si Flash memory-based CIM unit denoted as Cell MSB and Cell LSB, can function as analog memory, with their source terminals connected together. To achieve a low-power design, the drain voltage (VD) was set to 0.1V, which is the same as the fabricated nc-Si Flash memory test. Under the condition of an input control gate voltage (Vcg) of 600mV, the data and the corresponding operating current are shown in Fig. 2(c-d).

III. NC-SI FLASH MEMORY BASED CIM MACRO

This section presents a multi-bit CIM macro based on nc-Si Flash memory. In this macro, two Flash cells are used to store 3-bit weight data, and an input bit-slicing strategy is further employed.

A. Nc-Si Flash Memory Based CIM unit

In quantized neural networks, with 3-bit input and 3-bit weight quantization used as an example, multiply-accumulate (MAC) operation can be formally expressed as (1).

$$R = (IN_2 \cdot 2^2 + IN_1 \cdot 2^1 + IN_0 \cdot 2^0)$$
$$\times (W_2 \cdot 2^2 + W_1 \cdot 2^1 + W_0 \cdot 2^0) \quad (1)$$

Fig. 2(b) illustrates the circuit structure of the proposed nc-Si Flash multiply CIM (NFC) unit. A 3-bit input is used in this CIM macro. The NFC unit comprises three pairs of 2T nc-Si Flash cells, each pair capable of storing 3-bit quantized weights. And each pair is independently controlled by its corresponding input signal through wordline(WL). Leveraging the non-volatile nature of nc-Si Flash memory, the 3-bit weights are pre-programmed into the three cell pairs prior to the computing operation, enabling efficient parallel computation. To further improve energy efficiency, we program 2 bits into the LSB cell and only 1 bit into the MSB cell, as shown in Fig. 2, this configuration results in a smaller computational summation current, thereby enhancing energy efficiency.

The implementation details are as follows: the bitline (BL) voltage is set to 0.1V to ensure consistency with the actual test conditions of the nc-Si Flash memory cells. The input signal is applied through the WL, where a high voltage of 600mV represents logic '1', and a voltage of 0V represents logic '0'. For the input bit slicing strategy, taking an example of a 3-bit input with all bits set to '1', **IN[0]**: corresponds to the activation of the first wordline (WL0), during which an input control gate voltage V_{cg} is applied to drive the first pair of 2T nc-Si multi-bit Flash memory cells. Under a fixed weight configuration, the resulting operating current is denoted as I_0. **IN[1]**: Corresponds to the activation of the second wordline (WL1), during which an operating V_{cg} is applied to drive the second pair of 2T nc-Si multi-bit Flash memory cells. The width-to-length (W/L) ratio of this set is configured to be twice that of the first set. Under the same stored weight, the resulting operating current is denoted as I_1, where $I_1 = 2I_0$. **IN[2]**: Corresponds to the activation of the third wordline (WL2), during which an operating control gate voltage V_{cg} is applied to drive the third pair of 2T nc-Si multi-bit Flash memory cells. The W/L ratio of the third pair is configured to be four times that of the first pair. Under the same stored weight as in the first and second pairs, the resulting operating current is denoted as I_2, where $I_2 = 2I_1 = 4I_0$.

Through parallel activation of wordlines corresponding to individual input bits, the analog current output encodes the weighted dot product of input and stored data. This bitwise accumulation forms the basis of the MAC operation.

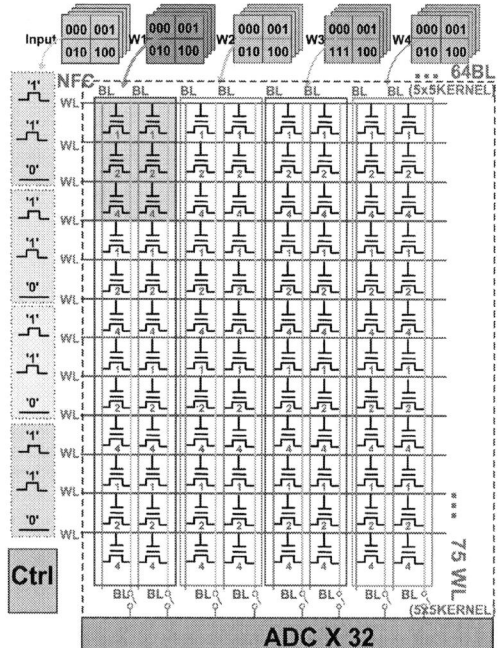

Fig. 3. Structure of proposed nc-Si Flash memory-based macro which contains 75 WLs and 64 BLs.

B. Nc-Si Flash based CIM Macro

As shown in Fig. 3, the proposed CIM macro consists of a 25×32 array of NFC units, 75 word lines (WLs), 64 bit lines (BLs), and 32 ADCs. The architecture is compatible with both 5×5 and 3×3 convolution kernel sizes.

In the proposed NFC CIM macro, the input matrix is first unfolded along the column direction and then applied to the array via WLs in the form of voltage pulses. Owing to the replicated storage of weight data across three adjacent word lines in the NFC unit, our design enables the parallel loading of all input matrix elements. In the quantized computation described by (1), the corresponding output vector is formed by sensing currents on two adjacent source lines. And in this macro, neural network weights are preprogrammed into the NFC array. Since our design primarily targets edge-side inference scenarios, frequent weight reprogramming is unnecessary, thus contributing to reduced power consumption and extended device reliability. The aggregated current generated during computation is sampled and digitized using the high-side current-sensing ADC architecture proposed in [3], as it offers compatibility with the targeted current range of the CIM operation.

IV. PERFORMANCE EVALUATION

This section presents the simulation results of the proposed nc-Si Flash memory-based CIM macro.

A. Evaluation Workflow

For the circuit simulation, we developed a compact nc-Si Flash memory model based on PSP model from NXP [9]. And we adopted a co-simulation approach. Specifically, based on the hardware MAC operation simulation results in Cadence, we simulated the entire inference process. Meanwhile, to evaluate the system accuracy, we developed a hardware simulation platform, as outlined in Algorithm 1.

Algorithm 1 Quantized Training & Hardware Deployment

Stage 1: Online Training

Forward:
    $x \leftarrow$ data each quantized layer L (QConv2d/QLinear)
    $W_{\text{fp}} \leftarrow$ floating-point weights of layer L
    $W_q \leftarrow$ Quantize($W_{\text{fp}}, 3$) {Weight quantization}
    $x \leftarrow$ Conv/Linear(x, W_q) + BatchNorm
    $x \leftarrow$ Quantize($x, 4$) {Activation quantization}
    Compute loss via cross-entropy on x (logits)

Backward:
    Zero gradients; backpropagate loss (loss.backward())
    Use STE in Quantize.backward
    Update floating-point parameters with lr

Stage 2: Offline Quantization
Quantize all trained weights to signed 4-bit uniformly

Stage 3: Hardware Mapping
Map quantized parameters to 4-bit weight(3-bit integer) matrices in PSP model for hardware simulation

---

B. Accuracy Evaluation

In this section we conducted precision experiments with weights of different bit widths. As shown in Fig. 4, the LeNet-5 model achieved 98.45% accuracy on the MNIST dataset and 90.77% on the Letters dataset. Additionally, leveraging the multi-bit storage capability of our nanocrystal silicon devices and the design of our multi-bit computational cells, the increase in weight bit width from 3 bits to 4 bits led to a substantial improvement in accuracy, the sign bit is computed by separate macros. Specifically, for the MNIST dataset, the hardware recognition accuracy was enhanced by 9.11%, and for the Letters dataset, the hardware recognition accuracy was elevated by 26.27%. For the VGG9 network, on the CIFAR-10 dataset, where the software and hardware accuracies remained consistent, and the hardware accuracy improved by 7.7% when increasing the weight precision. While the quantization results in a slight accuracy degradation relative to state-of-the-art baselines, the proposed macro achieves nearly twice the energy efficiency compared to other CIM designs.

C. Engergy Efficiency Evaluation

In the computing phase of our nc-Si Flash CIM macro, although the DAC has been omitted, the ADC remains an essential component. Our study utilizes the high-side current sensing ADC architecture developed in [3].

TABLE I
COMPARISON OF CIM DESIGNS USING VARIOUS MEMORY TECHNOLOGIES

	DAC 2024 [4]	TCAD 2024 [5]	ISSCC 2023 [6]	JSSC 2022 [7]	JSSC 2021 [8]	This Work
Memory Type	MRAM	RRAM	SRAM	SRAM	Poly-Si Flash	Nc-Si Flash
Technology	14nm	40nm	12nm	65nm	65nm	130nm
Computing Type	Full Digital	Current Mode	Charge Sharing Mode	Voltage Mode	Current Mode	Current Mode
Cell Type	1T	1T1RRAM	(4×8T)4C	8T	1T3poly-Si Flash	6 nc-Si Flash
Array Size	8×32	128×128	256×8	128×128	64×320	75×64
Device Model	14nm MTJ PDK	40nm RRAM PDK	CMOS PDK	CMOS PDK	3T eFlash Model	PSP-modified
V_{DD}	0.5-0.8V	0.9	0.5/0.85V	0.45/0.8V	1.2V/2.5V	0.1V
In/Weight Width Bit	8/8	1,2,4/8	8/8	1/1-5	8/8	3/4
Out Width Bit	24	3	8	1	16	4
Energy Efficiency (TOPS/W)	56.72	7.01-56.1	70.85/86.27	15.8-490	97.9	193.1
Accuracy (MNIST/CIFAR-10)	96.65% / 82.7%	98.52/86.12%	–	96.2% / 85.6%	98.42% / 81.01%	98.45% / 76.2%

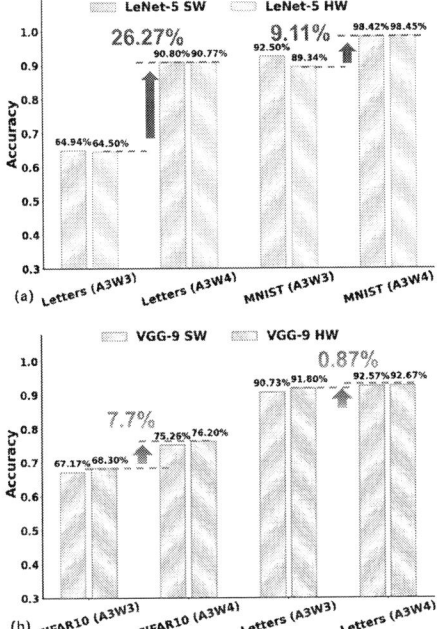

Fig. 4. Software(SW) and hardware(HW) inference accuracy across different quantization bit widths.

The energy efficiency evaluation is conducted with an input bit-width of 3 bits and a weight bit-width of 4 bits. When the ADC resolution to 4 bits raises the total ADC energy consumption to 349.44 fJ, with a conversion delay of 5 ns, resulting in a reduced energy efficiency of 193.1 TOPS/W. As shown in Table I, a comparison with other CIM schemes reveals that the proposed design exhibits significant advantages in energy efficiency. This advantage mainly arises from the lower supply voltage, reduced operating current, and innovative macro design, even though it is implemented using a larger process node than the reference design.

V. CONCLUSION

In summary, this paper presents a multi-bit CIM solution based on nc-Si Flash memory. Due to its lower operating current under low supply voltage, the analog matrix computation achieves higher energy efficiency. Moreover, the digital-to-analog conversion for multi-bit inputs is performed internally within the nc-Si Flash CIM unit, further reducing the energy consumption. In addition, in NFC unit, the two nc-Si Flash cells are used to store multi-bit weights, resulting in improved inference accuracy. The results shows that our nc-Si Flash based CIM design has a strong potential for enabling high ennergy efficient edge intelligence applications.

ACKNOWLEDGEMENTS

This work was supported by the Key-Area Research and Development Program of Guangdong Province under Grant 2023B0303030004 and the 100 Talents Program of Sun Yat-sen University (Grant No. 76220-12255002).

REFERENCES

[1] A. F. Laguna, M. M. Sharifi, A. Kazemi, X. Yin, M. Niemier, and X. S. Hu, "Hardware-software co-design of an in-memory transformer network accelerator," *Frontiers in Electronics*, vol. 3, p. 847069, 2022.

[2] W. K. Chim, "Germanium nanocrystal non-volatile memory: Fabrication, charge storage mechanism and characterization," *Nanoscale*, 2025.

[3] T. Soliman, F. Müller, T. Kirchner, T. Hoffmann, H. Ganem, E. Karimov, T. Ali, M. Lederer, C. Sudarshan, T. Kämpfe *et al.*, "Ultra-low power flexible precision fefet based analog in-memory computing," in *2020 IEEE International Electron Devices Meeting (IEDM)*. IEEE, 2020, pp. 29–2.

[4] J. Wang, Z. Wang, B. Zhang, Z. Gu, Y. Chen, W. Zhao, and Y. Zhang, "Frm-cim: Full-digital recursive mac computing in memory system based on mram for neural network applications," in *Proceedings of the 61st ACM/IEEE Design Automation Conference*, 2024, pp. 1–6.

[5] W. Li, S. Huang, X. Sun, H. Jiang, and S. Yu, "Secure-rram: A 40nm 16kb compute-in-memory macro with reconfigurability, sparsity control, and embedded security," in *2021 IEEE Custom Integrated Circuits Conference (CICC)*. IEEE, 2021, pp. 1–2.

[6] S.-E. Hsieh, C.-H. Wei, C.-X. Xue, H.-W. Lin, W.-H. Tu, E.-J. Chang, K.-T. Yang, P.-H. Chen, W.-N. Liao, L. L. Low *et al.*, "7.6 a 70.85-86.27 tops/w pvt-insensitive 8b word-wise acim with post-processing relaxation," in *2023 IEEE International Solid-State Circuits Conference (ISSCC)*. IEEE, 2023, pp. 136–138.

[7] C. Yu, T. Yoo, K. T. C. Chai, T. T.-H. Kim, and B. Kim, "A 65-nm 8t sram compute-in-memory macro with column adcs for processing neural networks," *IEEE Journal of Solid-State Circuits*, vol. 57, no. 11, pp. 3466–3476, 2022.

[8] M. Kim, M. Liu, L. R. Everson, and C. H. Kim, "An embedded nand flash-based compute-in-memory array demonstrated in a standard logic process," *IEEE Journal of Solid-State Circuits*, vol. 57, no. 2, pp. 625–638, 2021.

[9] G. Gildenblat, X. Li, W. Wu, H. Wang, A. Jha, R. Van Langevelde, G. D. Smit, A. J. Scholten, and D. B. Klaassen, "Psp: An advanced surface-potential-based mosfet model for circuit simulation," *IEEE Transactions on Electron Devices*, vol. 53, no. 9, pp. 1979–1993, 2006.

979-8-3315-3918-4/25 $31.00 © 2025 IEEE

A Multi-level RRAM-based Ising Machine for Solving Combinatorial Optimization Problems

Zhenchao Sui[1,4], Xiaoxin Xu[2], Chengshuo Yu[3], Jingxin Deng[3], Xu Zheng[2], Chengyue Li[2], Hailan Yi[3],
Jianguo Yang*[2,3], Xing Zhang[1*]

[1] School of Software and Microelectronics, Peking University, Beijing, China
[2] Institute of Microelectronics of Chinese Academy of Sciences, Beijing, China
[3] Zhangjiang Laboratory, Shangha, China
[4] Semiconductor Manufacturing Beijing Corporation, Beijing, China

* * Email: yangjianguo@ime.ac.cn, zhx@pku.edu.cn

Abstract—This work presents a high-density Ising machine based on multi-level Resistive RAM (RRAM) that combines precise interaction modeling with energy-efficient analog in-memory computing. The architecture supports compact spin encoding and enables interactions with up to eight neighbors at 7-level coefficient precision, allowing for more expressive problem representations. Implemented in 28nm CMOS, the system's performance and resilience are validated through detailed simulations that incorporate measured RRAM variability. To demonstrate practical utility, we use software-based models to solve combinatorial optimization problems, such as Max-Cut, while explicitly accounting for analog and device-level non-idealities. The proposed spin unit occupies just 23.6 µm², achieving a 28.6% to 95.8% area reduction compared to recent state-of-the-art designs, normalized by feature size.

Keywords—*Multi-level RRAM, Compute-in-memory, Ising Machine, Combinatorial optimization problem, Max-Cut problem*

I. INTRODUCTION

Combinatorial optimization problems (COPs) are widely encountered in real-world applications. As the problem size increases, however, the number of possible combinations grows exponentially, posing significant computational challenges. The Ising model provides an effective framework for addressing such problems. For example, a Max-Cut problem as illustrated in Fig. 1(a), by mapping them onto a network of binary spins taking values of +1 or −1. The system evolves to find the global minimum of its energy space, described by the Ising Hamiltonian, which is expressed as:

$$H = -\sum_{i,j} J_{ij}\sigma_i\sigma_j - \sum_i h_i\sigma_i \quad (1)$$

where J_{ij} is the interaction coefficient between spins σ_i and σ_j, and h_i represents an external magnetic bias. Fig. 1(b) shows the corresponding mapped Ising model and the update principle for each spin. In the absence of an external field, the update value of σ_i is determined by its local Hamiltonian, formulated as:

$$H_{\sigma_i} = -\sum_j J_{ij}\sigma_j \quad (2)$$

To maintain a negative local Hamiltonian, σ_i is flipped to -1 if H_{σ_i} is positive, and to +1 if H_{σ_i} is negative. Notably, when H_{σ_i} equals 0, the spin is updated randomly to +1 or -1 with equal probability. As shown in Fig. 1(c), this approach enables a gradual reduction in the system's global energy and introduces stochasticity to escape local minima, ultimately ensuring convergence to the global optimum. Finally, the spin

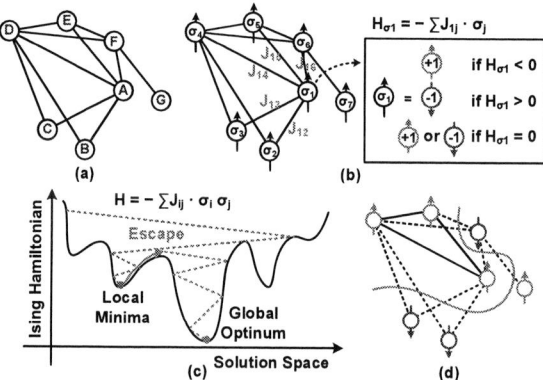

Fig. 1. (a) Max-cut problem. (b) Ising model and spin update. (c) Ising Hamiltonian. (d) Solution to the Max-Cut problem.

configuration is translated back to the solution of the original COP, as shown in Fig. 1(d).

The state-of-the-art CMOS Ising machines can be categorized into discrete-time and continuous-time approaches. Discrete-time Ising machines are updated cycle by cycle and are compatible with in-memory computing, resulting in better scalability and higher coefficient precision. Fully digital workflows have been adopted to improve the scalability and reconfigurability of these machines [1], while analog and mixed-signal in-memory computing approaches achieve higher energy efficiencies [2][3][4]. Additionally, fully connected Ising machines have been proposed to enhance connectivity [5][6]. In contrast, continuous-time Ising machines behave like dynamic energy systems in which spins evolve naturally toward a minimum energy state. Once the spins are interconnected, they begin to interact simultaneously and continuously. Ring oscillator-based Ising machines use the phase of oscillators to represent spins [7][8][9]. These works progressively improve the coefficient precision and connectivity of the Ising machine. Others employ binary-phase circuits as spins, such as latches and inverter chains [10][11][12]. However, it is still necessary to enhance the scalability and problem-solving capabilities of the Ising machine.

In this work, we present a high-density Ising machine leveraging multi-level RRAM based compute-in-memory. This work features programmable RRAM devices with three distinct resistance states enabled through precision voltage tuning. Each spin unit executes dot-product operations between 7-level interaction coefficients and adjacent binary spin values. This ultra-compact solution of Ising machine delivers high computational precision and massively scalability, efficiently addressing large-scale complex combinatorial optimization problems.

II. PROPOSED ISING MACHINE

A. Multi-level RRAM

Fig. 2(a) illustrates the layout of the eRRAM macro. The memory cells are integrated on a 28 nm standard logic platform. The RRAM cells are constructed between Metal 4 (M4) and Metal 5 (M5), requiring two additional mask layers to pattern the bottom via and the RRAM cell, as shown in Fig. 2(b). Fig. 2(c) is the enlarged view of the cell structure, which reveals the detailed TMO (Transition Metal Oxide) stack, comprising the Bottom Electrode (BE), Top Electrode (TE), and TMO layer. First, the Bottom Electrode (BE) via with dimensions of 80 nm × 80 nm is patterned using lithography and etching processes. This via is filled with TaN via Physical Vapor Deposition (PVD), followed by Chemical Mechanical Polishing (CMP). The 60-nm-thick BE via serves to prevent copper diffusion from Metal 4 (M4). Subsequently, the MOx (Metal Oxide) storage layer and the TaOx barrier layer are sequentially deposited onto the BE via via PVD. The Top Electrode (TE) is then deposited via sputtering. Following the completion of the RRAM cell fabrication, the logic Back-End-of-Line (BEOL) metals are deposited according to the standard logic process flow.

Fig. 3(a) shows the typical I-V curve of the RRAM, demonstrating its bipolar switching behavior. During the voltage sweep from 0 V to 1.2 V, the device resistance switches from the high-resistance state (HRS) to the low-resistance state (LRS). Conversely, when swept from 0 V to -1.5 V, the device transitions back to the HRS. Fig. 3(b) presents the switching characteristics under pulse operation. The set operation requires a voltage of 1.3 V with a 50-ns pulse width, while the reset operation is achieved using a -1.5 V pulse (50-ns width). The programming speed is 50 ns, and the read speed can be as fast as 20 ns, highlighting the device's high-speed and low-power consumption characteristics. Fig. 3(c) demonstrates the robust endurance characteristics of the device, which maintains stable operation for over 5000 programming cycles. To investigate the statistical properties of the RRAM, a 1 Kb array was tested. The statistical distributions of the programming voltage and resistance (Fig. 3(d)) indicate excellent uniformity across the RRAM array.

The proposed RRAM device can also achieve three distinct resistance states (with an approximate ratio of 1:2:4) by adjusting the programming voltage. This feature enables

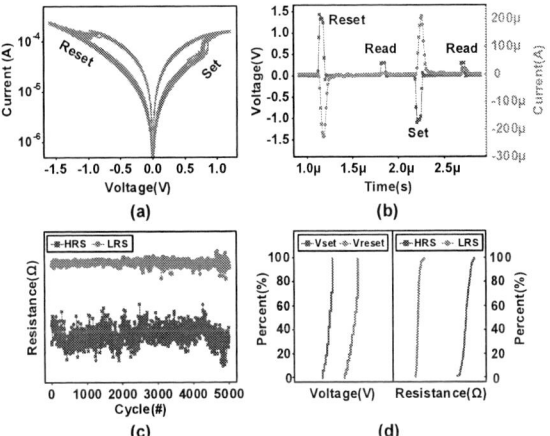

Fig. 3. Basic characteristics of the RRAM array. (a) DC I-V characteristic curve. (b) Pulse I-V characteristic curve. (c) Endurance characteristics of the RRAM device. (d) Statistical distribution of voltage and resistance values for the RRAM array.

efficient implementation of high precision computing, thereby enhancing the computational capability of the Ising machine. Subsequent sections demonstrate the resistance distribution characteristics for each of these three resistance states.

B. Spin with Analog Compute-in-memory Operation

Fig. 4 illustrates the architecture of the processing element (PE), highlighting the mechanism by which the compute-in-memory unit, built on the proposed multi-level RRAM, executes dot-product operations between spin states and their corresponding interaction coefficients. The multiplication results govern the magnitude of current injected into capacitors C_P and C_N effectively encoding computational outcomes in the resulting voltage levels. A sense amplifier then detects the voltage differences on bit-lines BL_P and BL_N, determining the next spin state, which is latched using a D flip-flop. This new spin value is distributed to all eight neighboring PEs in subsequent clock cycles. Once the spin network reaches convergence, the final spin states across each row are output sequentially through a shift register mechanism.

The circuit employs dual 1T1R RRAM cells to execute multiplication operation between adjacent spins σ_i (binary) and interaction coefficients J_{ij} (7-level precision from eight directions: N, NW, W, SW, S, SE, E, NE). As depicted in Fig. 5 (left), complementary spin states drive the word line (WL) and complementary word line (WLB). The spin '+1' activates WL while disabling WLB, establishing the upper charging pathways. Distinct resistance combinations represent

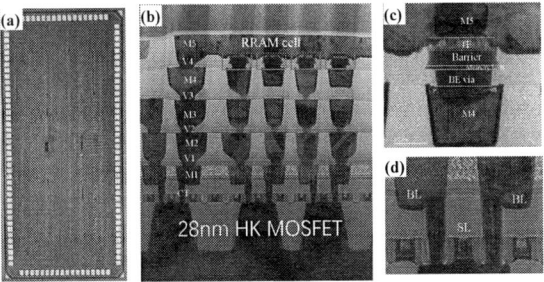

Fig. 2. (a) Optical micrograph of the chip. (b) Cross-sectional TEM image of the RRAM array. (c) Magnified views of the RRAM cell. (d) Magnified views of the transistor.

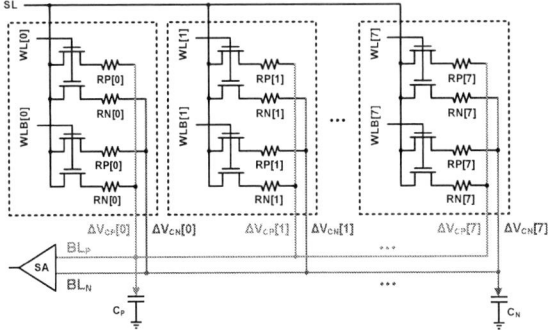

Fig. 4. Proposed spin (process element) with King's graph topology.

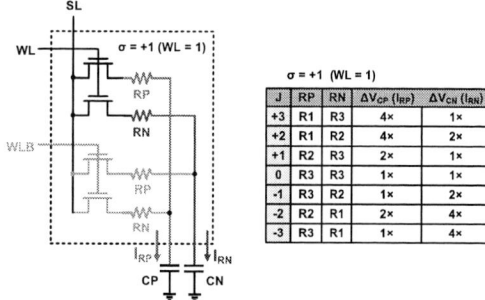

Fig. 5. Precise interaction coefficient J (7-level) and binary spin state.

programmable interaction strengths J_{ij} across seven discrete levels (ranging from -3 to +3 with a step of one). These resistance configurations produce corresponding distinction in discharge current magnitude, as demonstrated in Fig. 5 (right).

Fig. 6 shows four exemplary multiplication operations implemented with the proposed multi-level RRAM compute-in-memory module. Different RRAM resistance combinations correspond to distinct interaction strength coefficients J_{ij}, which generate corresponding discharge currents flowing into capacitors C_p and C_n. The resulting voltage difference (V_{CP}-V_{CN}) from $-3\times\Delta V$ to $+3\times\Delta V$ directly represents the multiplication result.

III. MEASUREMENT AND SIMULATION RESULTS

A. Hardware Variation and Nonlinearity

Fig. 7 presents experimental and simulation results for the multi-level RRAM implementation. Fig.7 (top) displays resistance distribution data obtained from 5,000 device measurements per resistance state. Fig.7 (bottom) shows simulation results for dot-product linearity. While global linearity is suboptimal due to resistance variations and inherent nonlinear characteristics across the three resistance states, the critical operating region near the switching threshold maintains acceptable linearity. Error rate analysis based on measured resistance distributions reveals a 3% deviation for two critical dot-product operations (+1 and -1). This error margin has been incorporated into subsequent algorithmic simulations.

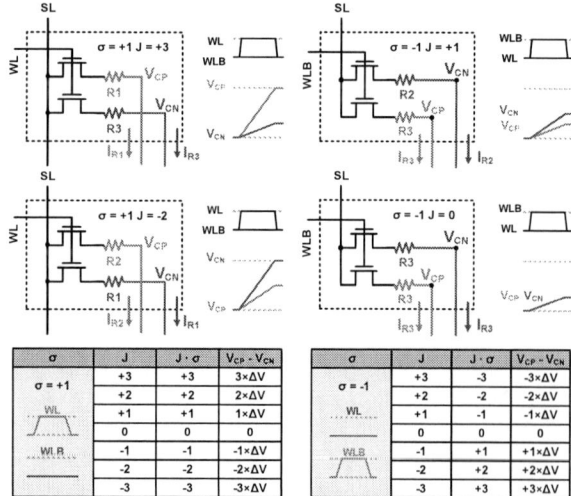

Fig. 6. Analog compute-in-memory based MAC operation.

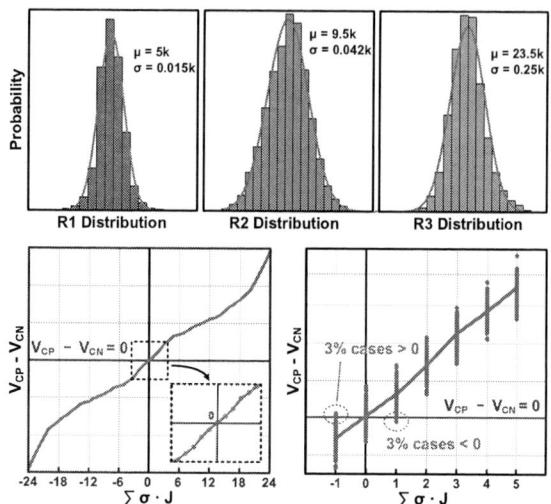

Fig. 7. Measured multi-level RRAM resistance distribution (top), and simulated voltage difference (V_{CP}-V_{CN}) linearity with error rates for critical dot-product results (bottom).

B. Max-Cut Problem Demonstration

Fig. 8 demonstrates the evolution toward the energy ground state in a 20×20 spin array configured with the predefined "1234 ABC" pattern. Sequential snapshots of spin configurations across annealing cycles reveal progressive alignment of spins, while the corresponding Ising Hamiltonian exhibits monotonic convergence. The final state achieves a well-resolved representation of the target pattern with minimal defects. Statistical validation through 100 independent simulation runs under 3% linearity error constraints for critical dot-product results (+1 and -1) confirms robust performance, with the mean final Hamiltonian reaching 99.4% of the theoretical ground state energy.

IV. CONCLUSION

This work demonstrates a high-density Ising machine implementing a 20×20 spin array in 28nm technology. The core innovation leverages multi-level RRAM devices with three programmable resistance states (approximate ratio 1:2:4) enabled by precision voltage tuning. Software simulations incorporating device variations and nonlinearity confirm the system achieves solutions reaching 99.4% of theoretical ground-state energy for Max-Cut problems. As benchmarked in Table I against state-of-the-art implementations, our design achieves a 23.6 µm²/spin, representing a 28.6% to 95.8% area reduction versus recent works after feature size normalization.

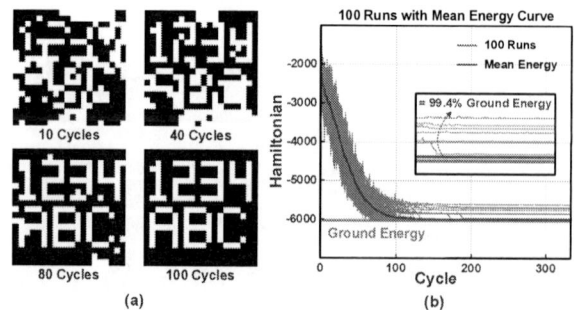

Fig. 8. Max-Cut problem demonstration. (a) Spin maps for solving a Max-Cut problem. (b) Hamiltonian iterations of the proposed method vs. ground energy.

This compact compute-in-memory scheme delivers both high computational precision and exceptional scalability.

TABLE I.

COMPARISON WITH STATE-OF-THE ART ISING MACHINES

	VLSI'20 [7]	JSSC'22 [3]	CICC'23 [10]	This Work
Technology (nm)	65	65	65	28
Topology	Hexagonal	King's Graph	Lattice Graph	King's Graph
Core Voltage (V)	1	1	0.6-1.2	0.4-0.9
Spin Circuit	Ring Oscillator (Analog)	eDRAM (Digital)	Inverter-Chain (Analog)	Register (Digital)
Coefficient	SRAM	N/A	SRAM	Multibit-RRAM
Coefficient Bit-Width	2b	1-4b	2b	7-level
Hamiltonian Computing	ROSC Coupling	Volatge Accu.& Charge Sharing	Inverter-Chain Coupling	Voltage Accumulation
# of Spins	560	6400	1920	400
Spin Area (µm²) (Norm. Area)	946 (23.8×)	1b J: 48 (1.8×) 2b J: 216 (4.1×)	224 (8.5×)	23.6 (1×)

*Normalized Area = (Spin Area) / (# of Coef Bit-Width) * (Connectivity Degree) * (Feature Size)²

ACKNOWLEDGMENT

This research was supported in part by Strategic Priority Research Program of the Chinese Academy of Sciences under Grant No. XDA0330100 and National Natural Science Foundation of China under Grant No. 92164204, 62222119, 62322412, and in part by Youth innovation Promotion Association CAS.

REFERENCES

[1] Y. Su, J. Mu, H. Kim and B. Kim, "A Scalable CMOS Ising Computer Featuring Sparse and Reconfigurable Spin Interconnects for Solving Combinatorial Optimization Problems," in IEEE Journal of Solid-State Circuits, vol. 57, no. 3, pp. 858-868, March 2022

[2] J. Mu, Y. Su and B. Kim, "A 20x28 Spins Hybrid In-Memory Annealing Computer Featuring Voltage-Mode Analog Spin Operator for Solving

Combinatorial Optimization Problems," 2021 Symposium on VLSI Technology, Kyoto, Japan, 2021, pp. 1-2.

[3] S. Xie, S. R. S. Raman, C. Ni, M. Wang, M. Yang and J. P. Kulkarni, "Ising-CIM: A Reconfigurable and Scalable Compute Within Memory Analog Ising Accelerator for Solving Combinatorial Optimization Problems," in IEEE Journal of Solid-State Circuits, vol. 57, no. 11, pp. 3453-3465, Nov. 2022.

[4] Deng, K. Zhou, H. Yang, C. Yu and J. Yang, "A High-Density RRAM-Based Ising Machine with Analog In-Memory Operation for Solving Combinatorial Optimization Problems," *2025 IEEE International Symposium on Circuits and Systems (ISCAS)*, London, United Kingdom, 2025

[5] Y. Liu et al., "A 1024-Spin Scalable Ising Machine With Capacitive Coupling and Progressive Annealing Method for Combination Optimization Problems," in IEEE Transactions on Circuits and Systems II: Express Briefs, vol. 71, no. 12, pp. 5009-5013, Dec. 2024

[6] Yue, W., Zhang, T., Jing, Z. et al. "A scalable universal Ising machine based on interaction-centric storage and compute-in-memory," Nat Electron 7, 904–913 (2024).

[7] I. Ahmed, P. -W. Chiu and C. H. Kim, "A Probabilistic Self-Annealing Compute Fabric Based on 560 Hexagonally Coupled Ring Oscillators for Solving Combinatorial Optimization Problems," 2020 IEEE Symposium on VLSI Circuits, Honolulu, HI, USA, 2020.

[8] Moy, W., Ahmed, I., Chiu, Pw. et al. A 1,968-node coupled ring oscillator circuit for combinatorial optimization problem solving. Nat Electron 5, 310–317 (2022).

[9] Lo, H., Moy, W., Yu, H. et al. An Ising solver chip based on coupled ring oscillators with a 48-node all-to-all connected array architecture. Nat Electron 6, 771–778 (2023).

[10] C. Yu, J. Mu, K. Chai, T. Kim and B. Kim, "A Continuous-Time Ising Machine using Coupled Inverter Chains Featuring Fully-Parallel One-Shot Spin Updates," 2023 IEEE Custom Integrated Circuits Conference (CICC), San Antonio, TX, USA, 2023.

[11] J. Bae, W. Oh, J. Koo, C. Yu and B. Kim, "CTLE-Ising: A Continuous-Time Latch-Based Ising Machine Featuring One-Shot Fully Parallel Spin Updates and Equalization of Spin States," in IEEE Journal of Solid-State Circuits, vol. 59, no. 1, pp. 173-183, Jan. 2024.

[12] J. Bae, C. Shim and B. Kim, "15.6 e-Chimera: A Scalable SRAM-Based Ising Macro with Enhanced-Chimera Topology for Solving Combinatorial Optimization Problems Within Memory," 2024 IEEE International Solid-State Circuits Conference (ISSCC), San Francisco, CA, USA, 2024.

979-8-3315-3918-4/25 $31.00 © 2025 IEEE

DRAM-Centric Near-Data Processing: A Survey of Architectures, Technologies, and Trends

Taoran Shen[1], Yujia Sun[1], Tingyi Xu[1], Li Xiong[3], Xiaoyong Xue*[1,2], Xiaoyang Zeng[1,2]

[1] State Key Lab of Integrated Chips and Systems, College of Integrated Circuits and Micro-Nano Electronics, Fudan University, Shanghai, China
[2] School of Microelectronics, Fudan University, Shanghai, China
[3] School of Physics and Electromechanical Engineering, Hexi University, Zhangye, 734000, China

* Email: xuexiaoyong@fudan.edu.cn

Abstract—To alleviate the performance bottleneck caused by the memory wall, DRAM-centric Near-Data Processing (NDP) has become a promising solution by minimizing data movement and exploiting memory-level parallelism. This paper provides a structured survey of recent advances in DRAM-based NDP, focusing on Processing-In-Memory (PIM) and Processing-Near-Memory (PNM) architectures. We summarize representative efforts from both academia and industry, compare their design trade-offs in terms of performance, energy efficiency, and interface compatibility, and analyze system-level integration strategies. Application scenarios including AI inference, graph processing, and data analytics are discussed to demonstrate the practical value of NDP. Finally, we outline key challenges and future directions for standardization and deployment.

Keywords—DRAM, Near-Data Processing, Processing-in-Memory, Processing-near-Memory

I. INTRODUCTION

The explosive growth of data-intensive applications, including artificial intelligence (AI), large-scale graph analytics, and high-performance computing (HPC), has exacerbated the memory wall problem in conventional von Neumann architectures. Frequent data shuttling between physically separated processing and memory units dominates system energy consumption while contributing substantially to operational latency [1]. While caching, prefetching, and (Non-Uniform Memory Access) NUMA-aware scheduling techniques have been proposed [2], they fail to address the fundamental bottleneck caused by the physical separation of compute and memory units [3].

Near-Data Processing (NDP) has emerged as a compelling architectural paradigm. By moving computation closer to memory, NDP minimizes data movement and exploits memory-level parallelism, resulting in significant improvements in performance and energy efficiency. DRAM-centric NDP designs stand out for their compatibility with existing memory infrastructure and ability to leverage mature DRAM fabrication standards. Notable advancements include the GDDR6-based AiM of SK Hynix [4-7] and Aquabolt-XL HBM2-PIM of Samsung [8], demonstrating the feasibility of embedding compute logic directly into memory modules.

DRAM-NDP can be classified into Processing-In-Memory (PIM), where computation is placed within DRAM banks, and Processing-Near-Memory (PNM), where compute units reside close to memory with greater design flexibility[3], as shown in Fig.1. The ecosystem surrounding DRAM-based NDP is rapidly evolving, with JEDEC specifications for LPDDR6-PIM and CXL-based memory-mapped compute models paving the way for widespread deployment.

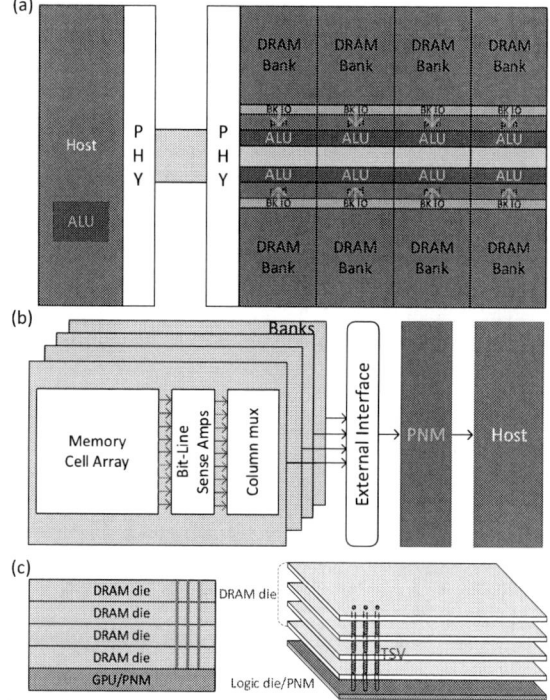

Fig.1. Different types of NDP architectures: (a) Typical PIM architecture (b) Memory controller based PNM architecture (c) 3D-stacked PNM architectures

II. ARCHITECTURAL CLASSIFICATION OF DRAM NDP

A. Taxonomy of NDP: PIM vs PNM

Near-Data Processing architectures are primarily divided based on compute placement: PIM places compute elements directly within DRAM banks, providing highest potential bandwidth by leveraging DRAM's internal architecture. PNM positions compute units at higher memory hierarchy levels, typically near memory controllers or within 3D-stacked logic dies, maintaining compatibility with standard interfaces [12-15]. While PNM offers reduced bandwidth compared to PIM, it provides superior compatibility with existing memory standards.

B. Processing-In-Memory (PIM) Approaches

PIM integrates compute logic directly within DRAM banks to minimize data transfers and exploit high internal bandwidth, yielding substantial performance and energy efficiency gains in AI inference, graph analytics, and database filtering.

Early commercial designs are dominated by fixed-function accelerators. SK Hynix proposed the GDDR6-based

979-8-3315-3918-4/25 $31.00 © 2025 IEEE

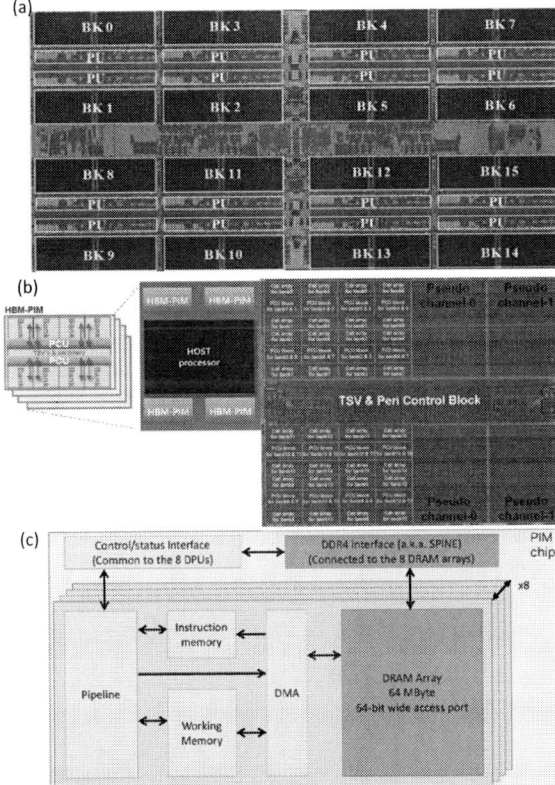

Fig.2. Representative industrial PIM implementations: (a) GDDR6-based AiM of SK Hynix (b) HBM2-PIM of Samsung (c) DDR4-based PIM of UPMEM

AiM embedding multiply-accumulate units with bank-wide mantissa shift scheme, delivering 1 TFLOPS throughput and 7.5×-10.5× speedups for GEMV and MNIST inference while maintaining JEDEC compatibility, as shown in Fig.2a [4-6]. Samsung proposed the Aquabolt-XL HBM2-PIM integrating FP16 SIMD units into HBM2 logic layers, reaching 1.2 TFLOPS per cube with 3.5× and 8.9× speedups for speech recognition and GEMV respectively, achieving over 60% system-level energy reduction, as illustrated in Fig.2b [8].

Beyond fixed-function designs, as shown in Fig.2c, UPMEM proposed DDR4-based PIM embedding thousands of RISC-V cores (DPUs) with 64KB local memory, enabling C-programmable task offloading with over 20× speedups on TPC-H queries while reducing off-chip bandwidth usage[13], [14]. AttAcc! implements adaptive precision (INT4/FP8) using bit-decomposed weights, reducing data movement by 3.2× and achieving 188.8 GOPS/W energy efficiency[15].

Overall, PIM architectures demonstrate significant potential in addressing the memory wall through diverse design approaches ranging from fixed-function accelerators to programmable and adaptive-precision solutions. While these implementations achieve substantial performance gains and energy efficiency improvements, challenges in thermal management, interface standardization, and workload generality remain key considerations for broader deployment in heterogeneous computing environments.

C. Processing-Near-Memory (PNM) Approaches

PNM architectures strategically position computational units adjacent to DRAM modules to mitigate data movement bottlenecks[16]. This paradigm exploits physical proximity to

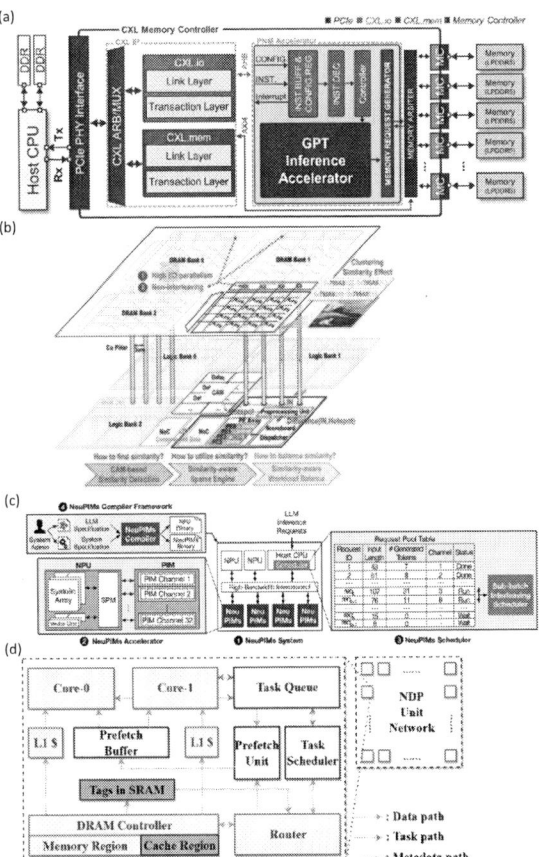

Fig.3. Representative PNM implementations: (a) LPDDR-based CXL-PNM platform of Samsung (b) Hybrid bonding PNM architecture (c) NPU-PIM Heterogeneous accelerator (d) ABNDP architecture

significantly enhance memory bandwidth utilization, reduce access latency, and improve energy efficiency for data-intensive workloads, as evidenced by contemporary research in interconnect scalability[11], [12], three-dimensional integration, and workload-specific optimizations[17].

CXL-enabled architectures leverage Compute Express Link protocol for cache-coherent, high-bandwidth interconnects. LPDDR5X-based CXL-PNM platforms construct scalable memory pools that deliver 512GB capacity and 1.1TB/s bandwidth, reducing large language model inference latency by 23% with 31% throughput improvement, as shown in Fig.3a [11].

Three-dimensional integration employing hybrid bonding techniques enables ultra-high bandwidth between processing logic and memory dies. Alibaba's implementation achieves 64Mb/mm² on-chip memory density through hybrid wafer-to-wafer bonding, delivering >200× better energy efficiency compared to off-chip memory solutions for recommendation systems[10]. Complementary research on vision AI models exploits this architectural approach to optimize feature map similarity detection and parallel processing within the memory layer through novel SRAM-based Content Addressable Memory techniques, as shown in Fig.2c [18].

Workload-specific optimizations demonstrate how tailored data management strategies enhance PNM efficacy. Stream-based data placement techniques dynamically map contiguous access patterns to adjacent DRAM banks, integrating memory controller prefetch logic and adopting stream models to optimize data locality for graph traversal and other irregular workloads[19].As shown in Fig.3c and Fig.4a, NeuPIMs and pSyncPIM frameworks address neural network training constraints through mixed-precision gradient reduction and hardware synchronization primitives respectively [20], [21]. The ABNDP architecture is shown in Fig.3d which employs a mechanism with skewed data mapping and hybrid scheduling to minimize remote accesses while dispersing hotspots, achieving 1.68× performance improvement for data-intensive algorithms[22].

The evolution of PNM architectures from domain-specific accelerators toward generalized computational frameworks reflects significant advancements in interconnect standardization and three-dimensional integration technologies. PNM's fundamental capacity to address memory-centric bottlenecks establishes its critical role in contemporary computing infrastructure [16].

III. SYSTEM-LEVEL INTEGRATION STRATEGIES

A. Support from memory controller and task scheduling mechanisms

System-level integration requires sophisticated memory controller enhancements and task scheduling mechanisms. PIM-MMU introduces hardware/software co-design integrating dedicated data copy engine, PIM-aware memory scheduler and heterogeneity-aware memory mapping. This approach accelerates DRAM↔PIM transfers by 4.1× in throughput and energy efficiency, yielding 2.2× end-to-end speedup [1].

For irregular sparse workloads, pSyncPIM addresses limitations of traditional all-bank execution through partially synchronous execution model, enabling better handling of sparse matrix operations like SpMV and SpTRSV[21]. DEAR-PIM proposes disaggregated execution approach incorporating disaggregated command queue (DCQ) that buffers all-bank commands, as shown in Fig.4c [23].

These architectural innovations enable PIM and near-memory architectures to handle diverse workloads efficiently while maintaining memory coherence and compatibility with conventional host systems.

B. Host interaction models (memory-mapped execution, offloading interfaces)

Effective host interaction models are critical for real-world deployment. UM-PIM addresses early PIM challenges by introducing unified memory abstraction supporting both CPU and PIM cores within shared physical memory space, eliminating separate PIM-specific address regions [24].

Compiler and runtime frameworks bridge high-level applications and PIM hardware. PIMFlow provides end-to-end software stack for CNN workloads, performing operator-level partitioning and automatically mapping compute-intensive layers to PIM units, achieving 82% reduction in end-to-end latency and 26% energy savings [25]. UniNDP offers unified compilation and simulation framework supporting multiple near-DRAM platforms, achieving up to 3.4× speedup over static mapping approaches [17].These developments represent progression toward unified, scalable, and software-transparent host interaction models that significantly lower adoption barriers for DRAM-centric NDP deployment.

C. Compatibility with standard memory interfaces

Ensuring compatibility with standard memory interfaces is critical for widespread NDP adoption. AiM GDDR6 embeds MAC units while maintaining mechanical and electrical compatibility with standard GDDR6 [5], [6]. UPMEM DDR4 integrates DPUs into standard DDR4-2400 DIMMs with full interface compatibility [13], [14].

Samsung's HBM-PIM solutions preserve external HBM2 interfaces achieving full pin-compatibility with existing processors. Aquabolt-XL uses aliased PIM instructions to backward-compatible column commands, allowing existing HBM2 systems to utilize PIM capabilities without host memory controller modifications [8, 26, 27].Standardization efforts advance interface compatibility through ongoing JEDEC standardization for LPDDR-PIM, targeting mobile applications with implicit mode changes and support for FP16 and INT8 operations [28]. Additionally, CXL-based PNM platforms enable memory-mapped processing through CXL.mem protocol without disrupting existing server infrastructures [11].

These developments demonstrate successful integration of NDP technologies while preserving standard interfaces, facilitating broader deployment across diverse computing environments.

IV. CHALLENGES AND FUTURE DIRECTIONS

DRAM-centric NDP architectures provide a promising approach to overcoming the von Neumann bottleneck, with strong potential in AI inference, edge computing, and HPC. Key challenges hinder large-scale deployment. PIM architectures face thermal issues from increased switching activity and heat concentration in 3D-stacked configurations, potentially degrading reliability and performance. Interface constraints and coherence management between host processors and in-memory compute units require specialized memory controllers and custom command sets, complicating system integration. Workload generality beyond domain-specific designs is limited, as programming complexity increases when developers must explicitly manage data placement and synchronization. The absence of standardized interfaces further slows adoption, with companies reluctant to commit to proprietary solutions.

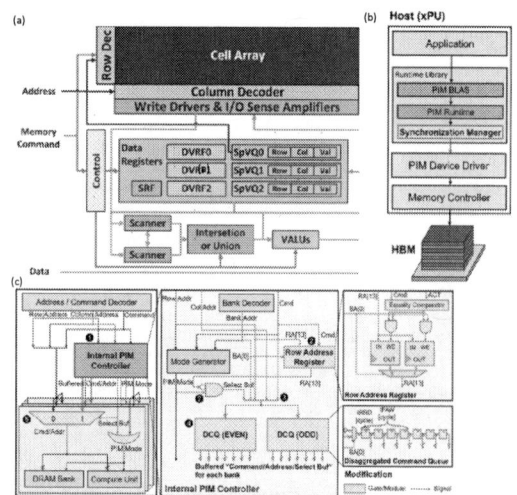

Fig.4. (a) Architecture of a pSyncPIM processing unit for each bank (b) DEAR-PIM software stack (c) DEAR-PIM Microarchitecture

979-8-3315-3918-4/25 $31.00 © 2025 IEEE

Although JEDEC specifications for LPDDR6-PIM are advancing, broader vendor support is essential. The software ecosystem remains immature, with compiler toolchains and debugging support still emerging. Future DRAM-NDP systems are expected to employ chiplet architectures and advanced packaging to achieve higher bandwidth, while rising AI-driven DRAM demand—projected to grow over 50% annually—will promote the use of machine learning for intelligent data placement. Continued memory scaling toward sub-10 nanometer nodes by 2030 will enable deeper compute integration at competitive cost, positioning DRAM-centric NDP as a cornerstone of next-generation data-intensive systems.

ACKNOWLEDGMENT

This work was supported by STI 2030-Major Projects (2022ZD0209200), in part by the National Natural Science Foundation of China (62274038), the Science and Technology Commission of Shanghai Municipality (24JD1400200), and ZTE Industry-University-Institute Cooperation Funds under Grant (IA20241120003).

REFERENCES

[1] D. Lee, B. Hyun, T. Kim, and M. Rhu, "PIM-MMU: A Memory Management Unit for Accelerating Data Transfers in Commercial PIM Systems," in *2024 57th IEEE/ACM International Symposium on Microarchitecture (MICRO)*, Austin, TX, USA: IEEE, Nov. 2024, pp. 627–642. doi: 10.1109/MICRO61859.2024.00053.

[2] M. Dashti *et al.*, "Traffic management : A Holistic Approach to Memory Placement on NUMA Systems".

[3] Y. Gu *et al.*, "PIM Is All You Need: A CXL-Enabled GPU-Free System for Large Language Model Inference," in *Proceedings of the 30th ACM International Conference on Architectural Support for Programming Languages and Operating Systems, Volume 2*, Mar. 2025, pp. 862–881. doi: 10.1145/3676641.3716267.

[4] M. He *et al.*, "Newton: A DRAM-maker's Accelerator-in-Memory (AiM) Architecture for Machine Learning," in *2020 53rd Annual IEEE/ACM International Symposium on Microarchitecture (MICRO)*, Athens, Greece: IEEE, Oct. 2020, pp. 372–385. doi: 10.1109/MICRO50266.2020.00040.

[5] S. Lee *et al.*, "A 1ynm 1.25V 8Gb, 16Gb/s/pin GDDR6-based Accelerator-in-Memory supporting 1TFLOPS MAC Operation and Various Activation Functions for Deep-Learning Applications," in *2022 IEEE International Solid- State Circuits Conference (ISSCC)*, San Francisco, CA, USA: IEEE, Feb. 2022, pp. 1–3. doi: 10.1109/ISSCC42614.2022.9731711.

[6] D. Kwon *et al.*, "A 1ynm 1.25V 8Gb 16Gb/s/Pin GDDR6-Based Accelerator-in-Memory Supporting 1TFLOPS MAC Operation and Various Activation Functions for Deep Learning Application," *IEEE J. Solid-State Circuits*, vol. 58, no. 1, pp. 291–302, Jan. 2023, doi: 10.1109/JSSC.2022.3200718.

[7] H. Choi *et al.*, "AiMX: Accelerator-in Memory Based Accelerator for Cost-effective Large Language Model Inference (Invited)," in *2024 IEEE International Electron Devices Meeting (IEDM)*, San Francisco, CA, USA: IEEE, Dec. 2024, pp. 1–4. doi: 10.1109/IEDM50854.2024.10873583.

[8] J. H. Kim *et al.*, "Aquabolt-XL HBM2-PIM, LPDDR5-PIM With In-Memory Processing, and AXDIMM With Acceleration Buffer," *IEEE Micro*, vol. 42, no. 3, pp. 20–30, May 2022, doi: 10.1109/MM.2022.3164651.

[9] H. Ham *et al.*, "Low-overhead General-purpose Near-Data Processing in CXL Memory Expanders," Sept. 23, 2024, *arXiv*: arXiv:2404.19381. doi: 10.48550/arXiv.2404.19381.

[10] D. Niu *et al.*, "184QPS/W 64Mb/mm^2 3D Logic-to-DRAM Hybrid Bonding with Process-Near-Memory Engine for Recommendation System," in *2022 IEEE International Solid- State Circuits Conference (ISSCC)*, San Francisco, CA, USA: IEEE, Feb. 2022, pp. 1–3. doi: 10.1109/ISSCC42614.2022.9731694.

[11] S.-S. Park *et al.*, "An LPDDR-based CXL-PNM Platform for TCO-efficient Inference of Transformer-based Large Language Models," in *2024 IEEE International Symposium on High-Performance Computer Architecture (HPCA)*, Edinburgh, United Kingdom: IEEE, Mar. 2024, pp. 970–982. doi: 10.1109/HPCA57654.2024.00078.

[12] S. Yun *et al.*, "CLAY: CXL-based Scalable NDP Architecture Accelerating Embedding Layers," in *Proceedings of the 38th ACM International Conference on Supercomputing*, Kyoto Japan: ACM, May 2024, pp. 338–351. doi: 10.1145/3650200.3656595.

[13] "The true Processing In Memory accelerator," 2019.

[14] B. Friesel, M. Lütke Dreimann, and O. Spinczyk, "A Full-System Perspective on UPMEM Performance," in *Proceedings of the 1st Workshop on Disruptive Memory Systems*, Koblenz Germany: ACM, Oct. 2023, pp. 1–7. doi: 10.1145/3609308.3625266.

[15] J. Park *et al.*, "AttAcc! Unleashing the Power of PIM for Batched Transformer-based Generative Model Inference," in *Proceedings of the 29th ACM International Conference on Architectural Support for Programming Languages and Operating Systems, Volume 2*, La Jolla CA USA: ACM, Apr. 2024, pp. 103–119. doi: 10.1145/3620665.3640422.

[16] O. Mutlu, A. Olgun, G. F. Oliveira, and I. E. Yuksel, "Memory-Centric Computing: Recent Advances in Processing-in-DRAM (Invited)," in *2024 IEEE International Electron Devices Meeting (IEDM)*, San Francisco, CA, USA: IEEE, Dec. 2024, pp. 1–4. doi: 10.1109/IEDM50854.2024.10873410.

[17] T. Xie *et al.*, "UniNDP: A Unified Compilation and Simulation Tool for Near DRAM Processing Architectures," in *2025 IEEE International Symposium on High Performance Computer Architecture (HPCA)*, Las Vegas, NV, USA: IEEE, Mar. 2025, pp. 624–640. doi: 10.1109/HPCA61900.2025.00054.

[18] Z. Yue *et al.*, "Exploiting Similarity Opportunities of Emerging Vision AI Models on Hybrid Bonding Architecture," in *2024 ACM/IEEE 51st Annual International Symposium on Computer Architecture (ISCA)*, Buenos Aires, Argentina: IEEE, June 2024, pp. 396–409. doi: 10.1109/ISCA59077.2024.00037.

[19] Y. Li, B. Tian, Y. Ren, and M. Gao, "Stream-Based Data Placement for Near-Data Processing with Extended Memory," in *2024 57th IEEE/ACM International Symposium on Microarchitecture (MICRO)*, Austin, TX, USA: IEEE, Nov. 2024, pp. 1648–1662. doi: 10.1109/MICRO61859.2024.00120.

[20] G. Heo *et al.*, "NeuPIMs: NPU-PIM Heterogeneous Acceleration for Batched LLM Inferencing," in *Proceedings of the 29th ACM International Conference on Architectural Support for Programming Languages and Operating Systems, Volume 3*, La Jolla CA USA: ACM, Apr. 2024, pp. 722–737. doi: 10.1145/3620666.3651380.

[21] D. Baek, S. Hwang, and J. Huh, "pSyncPIM: Partially Synchronous Execution of Sparse Matrix Operations for All-Bank PIM Architectures," in *2024 ACM/IEEE 51st Annual International Symposium on Computer Architecture (ISCA)*, Buenos Aires, Argentina: IEEE, June 2024, pp. 354–367. doi: 10.1109/ISCA59077.2024.00034.

[22] B. Tian, Q. Chen, and M. Gao, "ABNDP: Co-optimizing Data Access and Load Balance in Near-Data Processing," in *Proceedings of the 28th ACM International Conference on Architectural Support for Programming Languages and Operating Systems, Volume 3*, Vancouver BC Canada: ACM, Mar. 2023, pp. 3–17. doi: 10.1145/3582016.3582026.

[23] J. Hyun, M. Seo, S. Jeong, H.-J. Lee, and X. T. Nguyen, "DEAR-PIM: Processing-in-Memory Architecture with Disaggregated Execution of All-bank Requests," in *2025 Design, Automation & Test in Europe Conference (DATE)*, Lyon, France: IEEE, Mar. 2025, pp. 1–7. doi: 10.23919/DATE64628.2025.10992691.

[24] Y. Zhao *et al.*, "UM-PIM: DRAM-based PIM with Uniform & Shared Memory Space," in *2024 ACM/IEEE 51st Annual International Symposium on Computer Architecture (ISCA)*, Buenos Aires, Argentina: IEEE, June 2024, pp. 644–659. doi: 10.1109/ISCA59077.2024.00053.

[25] Y. Shin, J. Park, S. Cho, and H. Sung, "PIMFlow: Compiler and Runtime Support for CNN Models on Processing-in-Memory DRAM," in *Proceedings of the 21st ACM/IEEE International Symposium on Code Generation and Optimization*, Montréal QC Canada: ACM, Feb. 2023, pp. 249–262. doi: 10.1145/3579990.3580009.

[26] Y.-C. Kwon *et al.*, "25.4 A 20nm 6GB Function-In-Memory DRAM, Based on HBM2 with a 1.2TFLOPS Programmable Computing Unit Using Bank-Level Parallelism, for Machine Learning Applications," in *2021 IEEE International Solid- State Circuits Conference (ISSCC)*, San Francisco, CA, USA: IEEE, Feb. 2021, pp. 350–352. doi: 10.1109/ISSCC42613.2021.9365862.

[27] J. H. Kim *et al.*, "Samsung PIM/PNM for Transformer based AI".

[28] B. Kim *et al.*, "The Breakthrough Memory Solutions for Improved Performance on LLM Inference," *IEEE Micro*, vol. 44, no. 3, pp. 40–48, May 2024, doi: 10.1109/MM.2024.3375352.

Challenges and Trends of SRAM based Floating Point Computing-in-Memory Circuits

Yuchen Tang [12], Yanqi Zhang [1], Zhichao Liu [1], Xing Wang [1], Defa Wu [1], Huaiwen Zhang [1], Yeqi Sun [1], Xin Si*[1]

[1] Southeast University
[2] National Center of Technology Innovation for EDA

* Email: xinsi@seu.edu.cn

Abstract—This paper reviews the challenges and trends in floating-point (FP) computing-in-memory (CIM) circuits based on SRAM technology. Three exponent alignment schemes—ShareFloat, Input Alignment, and Product Alignment—are analyzed in terms of architecture, computation flow, and hardware overhead. Experimental evaluations under Normal, Uniform, and Pareto data distributions reveal significant precision variations, with Product Alignment achieving the highest accuracy in most cases and Input Alignment having unique advantages in Pareto distributions. The study highlights trade-offs between precision, energy efficiency, and circuit complexity, emphasizing the impact of data distribution on alignment performance. Future directions include distribution-aware alignment strategies, hardware–software co-design, and support for emerging low-bit-width FP formats (e.g., FP8, FP4, MX). This work provides design insights for efficient and precise FP CIM architectures targeting edge AI and large-scale generative model acceleration.

Keywords—Computing-in-Memory (CIM), SRAM, Floating-Point (FP) Arithmetic, Exponent Alignment, Data Distribution, Edge AI

I. INTRODUCTION

SRAM-based CIM technology integrates computation units directly within memory arrays, effectively reducing the communication overhead associated with data transfer between storage and computation units, thereby significantly improving energy efficiency. This architecture is well-suited for large-scale matrix computation workloads due to the regular and structured circuit topology of the compute cells, analogous to conventional SRAM. Edge-deployed large-scale generative models, such as Large Language Models (LLMs) and Diffusion Models, are typically power-sensitive while requiring extensive matrix operations, rendering SRAM-based CIM highly advantageous in such applications. Prior studies have shown that, for the same data bit-width, floating-point formats can markedly improve inference accuracy over integer (INT) formats [1–3]. For FP arithmetic, the exponent alignment constitutes one of the most critical operations. Thus, implementing floating-point computation in SRAM-based CIM requires additional circuits to support alignment. Various shift-and-alignment schemes have been proposed and extensively investigated to enable efficient alignment operations in CIM architectures. This paper focuses on the research progress of shift-and-alignment techniques in SRAM-based FP CIM circuits, providing a systematic review of the key challenges and trends in SRAM-based FP CIM.

II. EXPONENT ALIGNMENT SCHEME

A. ShareFloat

In SRAM CIM, the main obstacle for FP Multiply-Accumulate (MAC) is the inefficiency in in-array exponent alignment and parallel shifting, which causes large area and power overhead. The ShareFloat scheme addresses this by sharing the same exponent within a data group (e.g., the same input channel), storing only the sign bit and aligned mantissa in the array, and isolating the exponent by handling it in peripheral logic. This removes per-element exponent storage and eliminates costly in-array shifting. The computation flow includes the following steps:

- Find the exponent maximum E_{max} as the shared exponents from inputs and weights separately:

$$E_{max\_in}, E_{max\_w} \quad (1)$$

- Compute the shift values ΔE_{in} and ΔE_w from the inputs and weights separately:

$$\Delta E_{in}=E_{max\_in}-E_{in,i} \quad (2)$$
$$\Delta E_w=E_{max\_w}-E_{w,i} \quad (3)$$

- Align the mantissas of the inputs and weights separately:

$$M_{in\_aligned,i}=M_{in,i}<< \Delta E_{in} \quad (4)$$
$$M_{w\_aligned,i}=M_{w,i}<< \Delta E_w \quad (5)$$

- Compute the MAC value:

$$MACV=\sum_i M_{in\_aligned,i} \times M_{w\_aligned,i} \quad (6)$$

- Compute exponent sum:

$$E_{max\_sum}=E_{max\_in}+ E_{max\_w} \quad (7)$$

- Normalize MACV and E_{max_sum} to a standard FP format.

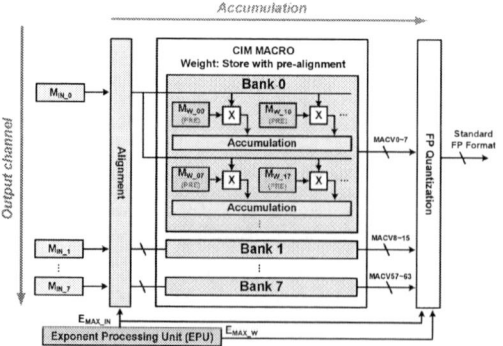

Fig. 1. ShareFloat Scheme

Guo et al. [4] implemented the first ShareFloat CIM macro using a voltage-coupled Float-Compute-Cell and a four-stage pipelined accumulator, achieving FP64-level precision with 18.8 TFLOPS/W and 73.11% accuracy on VGG-16@CIFAR-100. Tu et al. [5] extended the design with an offline pre-alignment scheme and Bitwise in-Memory Booth Multiplication (BM²), enabling a unified FP/INT CIM macro (BF16/FP32/INT8/INT16) with a peak efficiency of 29.2 TFLOPS/W in BF16 mode. Guo et al. [6] further proposed a DBcell-based hybrid structure integrating high-bit full-precision multiply cells (HFMC) and low-bit approximate multiply cells (LAMC), achieving double-bit parallel MAC and up to 31.6 TFLOPS/W in BF16 mode.

B. Input Alignment

In the ShareFloat scheme, mantissa input (M_{in}) and mantissa weight (M_w) are aligned based on exponent input (E_{in}) and exponent weight (E_w) respectively. This scheme can cause non-negligible computation errors in certain data distributions because M_{in} and M_w are truncated before multiplication and accumulation, and the truncation errors are unrecoverable. The input alignment scheme is proposed to solve this problem [7-9], and the computation flow includes the following steps:

- Compute exponent sum:

$$E_{sum,i} = E_{in,i} + E_{w,i} \quad (8)$$

- Find the maximum E_{max_sum} and compute ΔE_{in}:

$$\Delta E_{in} = E_{max\_sum} - E_{sum,i} \quad (9)$$

- Align $M_{in,i}$ based on ΔE_{max_sum}:

$$M_{in\_aligned,i} = M_{in,i} << \Delta E_{in} \quad (10)$$

- Compute the MAC value:

$$MACV = \sum_i M_{in\_aligned,i} \times M_{w,i} \quad (11)$$

- Normalize MACV and E_{max_sum} to a standard FP format.

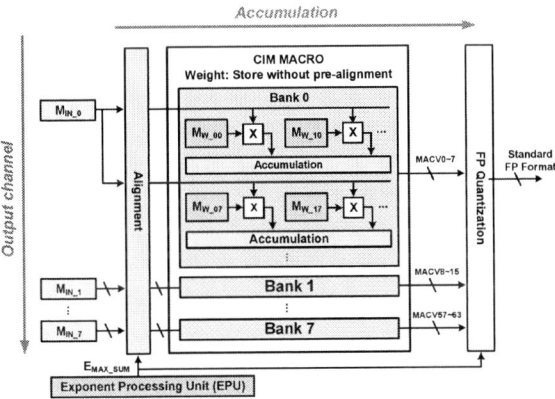

Fig. 2. Input Alignment Scheme

The input alignment scheme dynamically obtains the shift amount based on input and weight distributions and shifts M_{in} by computing the sum of E_{in} and E_w.

Wu et al. [7] introduced the first input-alignment FP in-memory-computing macro combining time-domain, digital, and analog-voltage processing for high energy efficiency. Khwa et al. [8] further proposed an INT/FP dual-mode CIM macro, enhancing energy and area efficiency for INT8 and BF16 MAC operations. These works adopted peripheral alignment circuits to implement mantissa alignment, inducing large area overhead and compromising the input reusability. To overcome these limitations, Wang et al. [9] proposed a serial alignment scheme, embedding a light local converter into the SRAM array to reduce area overhead and enhance input-reusability.

C. Product Alignment

The product alignment scheme is the approach most closely resembling standard FP computation. After summing the exponent input ($E_{in,i}$) and exponent weight ($E_{w,i}$), the mantissas are first multiplied, and the resulting product is then shifted according to the exponent difference. This effectively avoids the accumulation of truncation errors caused by pre-multiplication mantissa alignment. The computation flow includes the following steps:

- Compute exponent sum:

$$E_{sum,i} = E_{in,i} + E_{w,i} \quad (12)$$

- Compute mantissa product:

$$M_{p,i} = M_{in,i} \times M_{w,i} \quad (13)$$

- Find the maximum E_{max_sum} and compute ΔE_p:

$$\Delta E_p = E_{max\_sum} - E_{sum,i} \quad (14)$$

- Align the mantissa product according to ΔE_p:

$$M_{p\_aligned,i} = M_{p,i} << \Delta E_p \quad (15)$$

- Compute the MAC value:

$$MACV = \sum_i M_{p\_aligned,i} \quad (16)$$

- Normalize MACV and E_{max_sum} to a standard FP format.

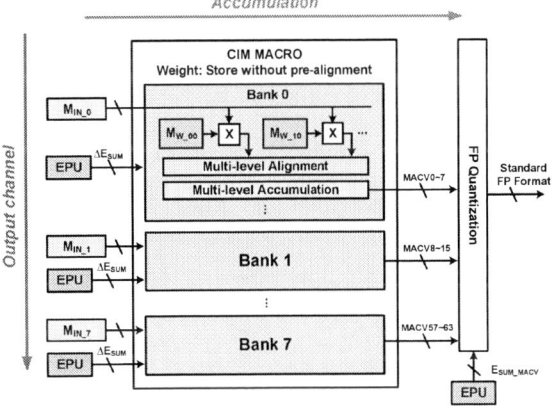

Fig. 3. Product Alignment Scheme

Since the mantissa multiplication is performed before the alignment shift, the product alignment Scheme achieves higher precision while reducing dependency on shift accuracy, thereby outperforming the input alignment scheme in truncation-sensitive data distributions. It should be noted that the preserved bit-width of the mantissa product significantly impacts both accuracy and circuit implementation.

Jyotishman et al. [10] presented a full-precision FP8 CIM implementation at ESSCIRC 2023. In this implementation,

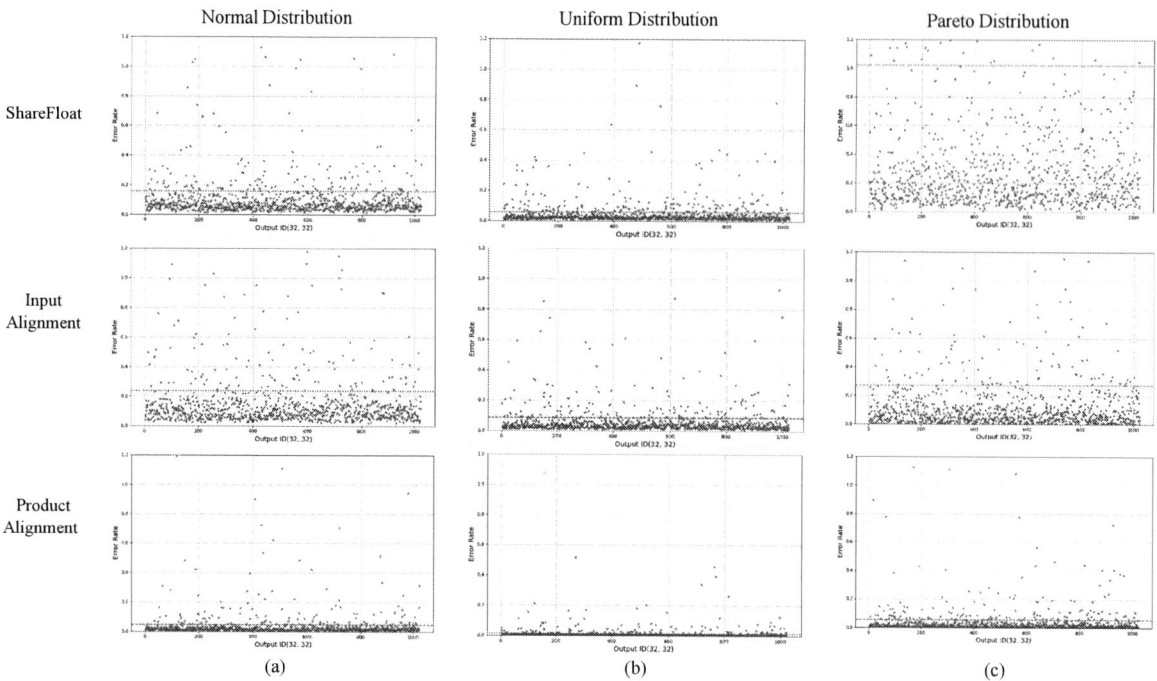

Fig. 4. Accuracy Variations of Different Alignment Schemes under Different Data Distributions

FP operations were performed pairwise on the data, followed by accumulation using a FP adder tree. Due to the large adder tree, this approach incurs considerable hardware cost.

Yue et al. [11] adopted an approach in which the exponent and mantissa are computed in parallel, followed by a single alignment operation for a group of numbers. The alignment is postponed until after partial product generation, leveraging the wider product bit-width to reduce truncation loss. Compared to the two alignment schemes, this method improves adaptability to high-dynamic-range data but introduces larger circuit overhead.

III. EXPERIMENT RESULTS

This section systematically evaluates the accuracy performance of different exponent alignment schemes under various data distributions and the accuracy loss when using different schemes under the same distribution. Three representative data distributions(Fig.5)—Normal, Uniform and Pareto—are selected to analyze the precision variation of three typical alignment schemes (ShareFloat, input alignment, and product alignment), thus analyzing the relationship between data distributions, alignment schemes, and the dynamic range of input data.

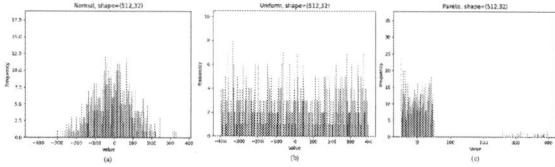

Fig. 5. Data Distributions: (a) Normal Distribution; (b) Uniform Distribution; (c) Pareto Distribution

A. Experimental Setup

Random matrices A(32×512) and B(512×32) are generated using Python according to the target distributions,

and then mapped to A' and B',which introduce truncation errors similar to those caused by hardware alignment. The shift retention bit-width is set to 8 bits for all cases.

Matrix multiplications A×B and $A' \times B'$ are performed to obtain output matrices O (32×32) and O', respectively. The error rate is computed as:

$$\text{Error Rate} = \frac{ABS(O - O')}{ABS(O)} \quad (17)$$

For each data distribution, 1000 random simulations are conducted, and element-wise averaging of error Rate is performed to obtain the mean error distribution, observing the distribution of error rates in the output matrix.

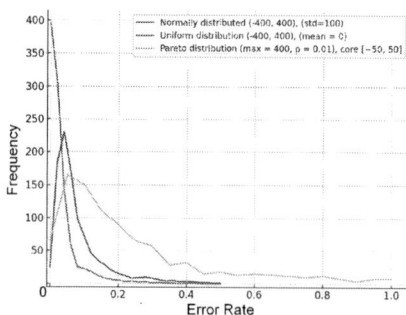

Fig. 6. Accuracy Variation of ShareFloat under Different Data Distributions

B. Impact of Different Data Distributions on Accuracy

As shown in Fig.6, under the ShareFloat alignment scheme, the error rate varies significantly across data distributions. The uniform distribution yields the lowest average error rate, followed by the normal distribution, while the Pareto distribution produces the highest error rate. The uniform distribution has a larger proportion of data near the maximum value, leading to smaller shift amounts, thereby reducing the truncation errors caused by alignment shifting.

In the normal distribution, although values are centered around the mean, alignment shifting is determined by the maximum exponent, and the relatively low proportion of large values increases exponent differences and truncation loss compared to the uniform case. The Pareto distribution contains a small number of extreme values, significantly increasing exponent differences and requiring large shifts, which discard a substantial amount of significant information during truncation, thus accumulating more accuracy loss. Overall, the uniform distribution shows the most significant precision advantage, followed by the normal distribution, and the Pareto distribution performs the worst. These results suggest that a more uniform distribution can effectively mitigate precision degradation. Through software-hardware co-design, the above strategy can be efficiently implemented. For example, Tender [12] offers an innovative approach to enhance data distribution handling. This method applies bias and data grouping techniques, which effectively reduce inter-group differences, making the data distribution more uniform and minimizing precision loss caused by alignment shifting, thereby improving the system's efficiency in maintaining data accuracy. At the same time, handling outlier values separately, thereby leaving more uniform data for FP CIM, is also a wise optimization direction.

C. Impact of Different Alignment Schemes on Accuracy

As shown in Fig.4, the x-axis represents the ID of the output matrix, and the y-axis represents the error rate of each output value in the output matrix. The scatter points in the figure clearly reflect the distribution of error rates in the output matrix. As illustrated, the three exponent alignment schemes exhibit significant differences in precision. The product alignment scheme achieves the highest accuracy. Product alignment first performs mantissa multiplication to accumulate partial products, then computes the exponents of each partial product by summing the exponents. By utilizing both the mantissa and exponent information, Product Alignment can identify more important values and truncate only the smaller values, thereby minimizing the impact on the final precision. Input alignment, on the other hand, aligns mantissas based on the sum of exponents from both the input and weight. However, due to the broader distribution of the exponent sum, the alignment increases sparsity in the input, resulting in additional precision loss. Consequently, the accuracy of Input Alignment is slightly lower than that of ShareFloat.

However, under Pareto distribution, input alignment exhibits significantly higher precision than ShareFloat. In this random number model, both the input and weights follow a Pareto distribution, containing a certain amount of extreme values. By prioritizing the summation of exponents, input alignment alleviates the impact of the extreme exponent distribution in both the input and weight to some extent, thus showing a distinct advantage in the context of Pareto distribution.

This study concludes that selecting an appropriate alignment scheme is crucial for precision optimization, with Input Alignment showing advantages in handling extreme or long-tailed distributions like Pareto.

IV. CONCLUSION AND FUTURE WORK

This paper discusses three exponent alignment schemes for FP CIM circuits and their impact on data precision under different data distributions. While the product alignment

scheme provides high precision, it increases circuit overhead and reduces energy efficiency. The ShareFloat scheme, although more hardware-friendly, causes significant precision loss in some applications, limiting its practicality. Data distribution significantly affects precision, with uniform distributions enabling more efficient utilization of CIM's advantages. Balancing precision and circuit overhead is key for future FP CIM optimization. Future directions include software-hardware co-design for better data distribution handling, new low-bit-width data formats like FP8 and FP4 for edge AI, and optimizing shift retention bit-width to improve overall system performance.

ACKNOWLEDGMENT

This work is funded with National Science and Technology Major Project under Grant 2022ZD0118902, National Natural Science Foundation of China under Grant 6232B2022, 92264203, 62204036, 92464302, Jiangsu Provincial Key Research and Development Program under Grant BE20230201 and Xiaomi. All the authors are with the School of Integrated Circuits, Southeast University, Nanjing, 210096, China.

REFERENCES

[1] S. Liu, Z. Liu, X. Huang, *et al.*, "LLM-FP4: 4-bit floating-point quantized transformers," *arXiv preprint* arXiv:2310.16836, 2023.

[2] Y. Zhang, L. Zhao, S. Cao, *et al.*, "Integer or floating point? New outlooks for low-bit quantization on large language models," in *Proc. IEEE Int. Conf. Multimedia and Expo (ICME)*, 2024, pp. 1–6.

[3] B. D. Rouhani, R. Zhao, A. More, *et al.*, "Microscaling data formats for deep learning," *arXiv preprint* arXiv:2310.10537, 2023.

[4] A. Guo, Y. Zhou, B. Wang, *et al.*, "ShareFloat CIM: A compute-in-memory architecture with floating-point multiply-and-accumulate operations," in *Proc. IEEE Int. Symp. Circuits and Systems (ISCAS)*, 2022, pp. 2276–2280.

[5] F. Tu, Y. Wang, Z. Wu, *et al.*, "A 28 nm 29.2 TFLOPS/W BF16 and 36.5 TOPS/W INT8 reconfigurable digital CIM processor with unified FP/INT pipeline and bitwise in-memory Booth multiplication for cloud deep learning acceleration," in *Proc. IEEE Int. Solid-State Circuits Conf. (ISSCC)*, vol. 65, 2022, pp. 1–3.

[6] A. Guo, C. Xi, F. Dong, *et al.*, "A 28-nm 64-kb 31.6-TFLOPS/W digital-domain floating-point-computing-unit and double-bit 6T-SRAM computing-in-memory macro for floating-point CNNs," *IEEE J. Solid-State Circuits*, vol. 59, no. 9, pp. 3032–3044, 2024.

[7] P. C. Wu, J. W. Su, L. Y. Hong, *et al.*, "A 22 nm 832 Kb hybrid-domain floating-point SRAM in-memory-compute macro with 16.2–70.2 TFLOPS/W for high-accuracy AI-edge devices," in *Proc. IEEE Int. Solid-State Circuits Conf. (ISSCC)*, 2023, pp. 126–128.

[8] W. S. Khwa, P. C. Wu, J. J. Wu, *et al.*, "A 16 nm 96 Kb integer/floating-point dual-mode-gain-cell-computing-in-memory macro achieving 73.3–163.3 TOPS/W and 33.2–91.2 TFLOPS/W for AI-edge devices," in *Proc. IEEE Int. Solid-State Circuits Conf. (ISSCC)*, vol. 67, 2024, pp. 568–570.

[9] X. Wang, T. Jiao, Y. Yang, *et al.*, "A 28 nm 17.83-to-62.84 TFLOPS/W broadcast-alignment floating-point CIM macro with non-two's-complement MAC for CNNs and transformers," in *Proc. IEEE Int. Solid-State Circuits Conf. (ISSCC)*, vol. 68, 2025, pp. 254–256.

[10] J. Saikia, A. Sridharan, I. Yeo, *et al.*, "FP-IMC: A 28 nm all-digital configurable floating-point in-memory computing macro," in *Proc. IEEE 49th Eur. Solid State Circuits Conf. (ESSCIRC)*, 2023, pp. 405–408.

[11] Z. Yue, X. Xiang, Y. Wang, *et al.*, "A 51.6 TFLOPS/W full-datapath CIM macro approaching sparsity bound and <2–30 loss for compound AI," in *Proc. IEEE Int. Solid-State Circuits Conf. (ISSCC)*, vol. 68, 2025, pp. 1–3.

[12] J. Lee, W. Lee, and J. Sim, "Tender: Accelerating large language models via tensor decomposition and runtime requantization," in *Proc. ACM/IEEE 51st Annu. Int. Symp. Comput. Archit. (ISCA)*, 2024, pp. 1048–1062.

Mapping of Graph Convolution Network on Sparse-Aware Computing-In-Memory Macros

Guoxiang Li[1,2], Tianhang Zhou[3], Xinyu Qu[1,2], Zecheng Zhou[1,2], Yufei Ma[*1,2]

[1]Institute for Artificial Intelligence, Peking University, Beijing, China
[2]School of Integrated Circuits, Peking University, Beijing, China
[3]School of Integrated Circuits, Anhui University, Beijing, China
[*]Email: yufei.ma@pku.edu.cn

Abstract—Graph Convolution Network(GCN) is an important model in the field of graph data processing, requiring extensive matrix computations. Computing-In-Memory (CIM) is a computational architecture that has gained significant attention in recent years, with the potential to overcome the limitations of the traditional von Neumann architecture. The deployment of GCN on CIM architectures is worth studying and optimizing. This paper investigates the mapping of GCN on a Sparse-Aware CIM. We considered the changes in computational load due to sparse processing and investigated the impact of the order of the two main matrix multiplications in GCN on the overall computation. Furthermore, we explored the deployment of matrix multiplication on a multi-CIM macro architecture based on loop tiling, loop interchange, and loop unrolling techniques.

Index Terms—Mapping, GCN, Sparse, Computing-In-Memory

I. INTRODUCTION

In recent years, neural network models have emerged as a pivotal research focus in the realm of artificial intelligence. Graph Convolutional Networks (GCNs) [1], have established themselves as a hallmark model within the domain of graph-related tasks. GCNs can directly operate on graph-structured data consisting of nodes and edges. By aggregating information from neighboring nodes, they realize feature propagation and updating, thereby efficiently capturing the latent topological relationships and node association patterns inherent in the graph. This inherent trait endows them with significant advantages in fields including social network analysis, molecular structure modeling, and recommendation systems, leading to their extensive application in addressing complex problems in practical scenarios. [2]–[4]

Graph Convolutional Networks (GCNs) involve a large number of inherent sparse computations in their operations. To address the computational challenges posed by GCNs, some works [5]–[7] have been proposed to enhance the computational efficiency of GCNs. Although these works have achieved significant improvements over CPUs and GPUs, they still suffer from the limitations of the von Neumann architecture due to their adoption of the traditional compute-memory

This work was supported in part by National Natural Science Foundation of China (No. 62204003) and in part by Beijing Natural Science Foundation (No. L257017). *(Guoxiang Li and Tianhang Zhou are co-first authors.) (Corresponding author: Yufei Ma.)*

separation structure. Computing-In-Memory is one of the most promising solutions to break through this bottleneck. Under the compute-in-memory architecture, data can be processed within the memory or near memory cells, leading to a significant reduction in data movement overhead. Some previous works [8], [9] have made attempts to apply the CIM architecture to GCN accelerators.

While traditional GCN accelerators have each conducted analyses on GCN mapping, research on CIM-based GCN accelerators still lacks a systematic exploration of this mapping process. Some studies have investigated neural network mapping onto CIM, but the majority focus on convolutional neural networks (CNNs) rather than GCNs [10]–[13].

To address this gap, this paper performs modeling and analysis of GCN mapping in CIM systems and develops a corresponding tool. Given the extremely high sparsity inherent to GCNs, a specialized sparse CIM architecture is adopted. By integrating a multi-address structure, the flexibility of the CIM is enhanced, enabling effective leveraging of GCN sparsity to a certain extent and thereby improving overall computational efficiency.

II. BACKGROUND

A. Basics of Graph Convolutional Networks

Graphs are fundamental data structures used to model relationships between entities, formally defined as $G = (V, E)$ where V is the set of vertices (nodes) and E is the set of edges.Each node is associated with a feature vector capturing its attributes.The connectivity of the graph is described by an adjacency matrix A, which is of dimension $|V| \times |V|$ with $|V|$ denoting the number of nodes.

A typical GCN is composed of multiple graph convolutional layers stacked in sequence. Each layer transforms node features through two core stages—Aggregation and Combination—by aggregating information from neighboring nodes. The Aggregation stage aggregates feature information from a node's neighbors and itself using the adjacency matrix, while the Combination stage applies a linear transformation (via a weight matrix W) and a non-linear activation function (e.g., $ReLU$, $Softmax$) to the aggregated features. The core computation of a GCN layer can be formulated as:

$$X^{(l+1)} = \sigma\left(\hat{A}X^{(l)}W^{(l)}\right),$$

where $X^{(l)}$ and $W^{(l)}$ denote the input feature matrix and the trainable weight matrix of the l-th layer, and $\hat{A}$ is the normalized adjacency matrix to avoid feature scaling issues. $\sigma()$ is an activation function. $X^{(l+1)}$ is the output feature of layer l.

B. Conventional Computing-In-Memory Architecture

The structure of the conventional Compute-In-Memory (CIM) macro is depicted in Fig.1(a). Subarrays dedicated to weight data storage are arranged in a two-dimensional configuration. Within each subarray, computational gates are positioned in close proximity to memory cells. The multiplication results of each column are accumulated via an adder tree, thereby generating partial outputs (Y). Notably, the computational logic gates in each row share a common input line, which implies that an input vector can synchronously perform multiplication operations with the weights of all columns. Furthermore, this primitive array structure typically employs a single-address mechanism, where all rows share a common address line. In other words, the indices of the activated word lines (WLs) for each row remain consistent. Such a design largely constrains the flexibility of the CIM macro.

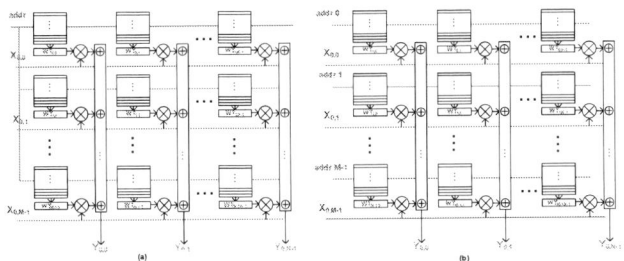

Fig. 1. (a)Traditional single address CIM architecture; (b)Multi-address CIM architecture.

C. Sparse-Aware CIM Structure and Data reorder

To address the requirements of sparse computation, this paper proposes the architecture of a multi-address Compute-In-Memory (CIM) macro. As illustrated in Figure (b), the indices of the selected word lines (WLs) for each row are no longer identical; consequently, the memory cells read in the subarray no longer correspond to the same offset addresses. This improvement enables the CIM to support more flexible weight reading operations, thereby eliminating unnecessary zero computations.

In common sparse computation scenarios, for a given matrix multiplication $A \times B$, it is often necessary to process each row of matrix A through the following steps:

Step 1: Splitting: Given the input channel count M of the CIM, a specific row of matrix A needs to be split into multiple

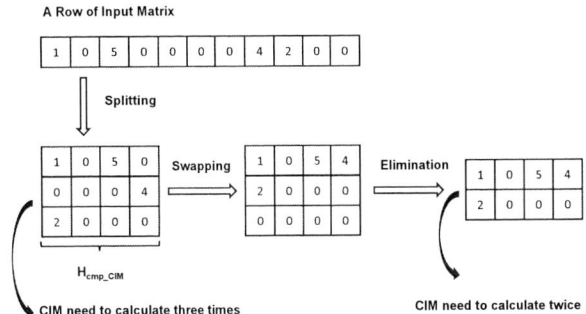

Fig. 2. Steps for sparse-aware data reorder.

blocks, where each block has a length of M, and the last block can be any positive integer less than or equal to N. For instance, when a row of A contains 11 elements and $M = 4$, the splitting method is shown in Fig.2.

Step 2: Swapping: Benefiting from the support of the multi-address CIM for sparse computation, data at the same position across all groups (i.e., the same column in the table) can be swapped arbitrarily. This allows all non-zero values to be clustered in the preceding blocks as much as possible.

Step 3: Elimination: All-zero blocks do not require any computation and thus can be skipped directly.

Ultimately, through the aforementioned data rearrangement, a large number of unnecessary zero operations are eliminated.

III. GCN COMPUTATIONAL MODELING

A. Execution Order

The overall number of operations is affected by the execution order of (AX)W and A(XW). Despite the fact that the impact of the execution order of AXW on the computational load have been analyzed by the previously-cited GCN accelerators (AWB-GCN, GCNAX), the analyses are based on digital architectures, without considering sparsity. Therefore, based on the sparse CIM, this paper re-analyzes the computational order of AXW. The sparse-sware used in this paper need to pre-process the input matrix. For example, when calculating the computational load of (AX)W, sparse processing is first performed on matrix A. The processed matrix A is then multiplied by matrix X to obtain the (AX) matrix. After the (AX) matrix undergoes sparse processing, it is multiplied by matrix W, and the computational load is counted. Results of the calculations indicate that the computational load of A(XW) is usually significantly lower than that of (AX)W.

B. Matrix Multiplication Analysis

The two matrix multiplications required under the computational order A(XW) are both implemented via CIM units. Specifically, for the computation of (XW), the weight matrix W is stored in the CIM as fixed weight, while the feature matrix X is input to the CIM after undergoing sparse processing. For the computation of A(XW), the intermediate result (XW) is stored in CIM as weight, and the adjacency matrix A completes the computation after sparse processing.

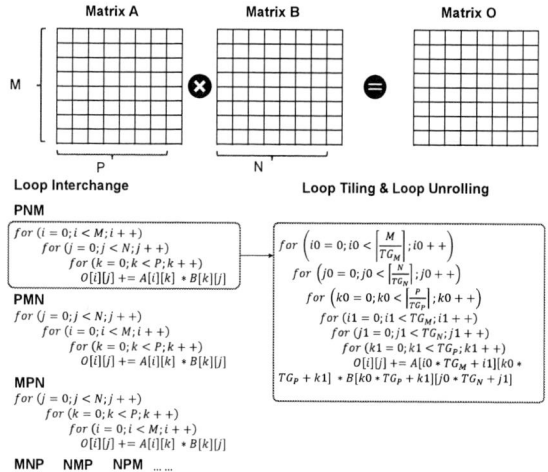

Fig. 3. Matrix multiplication.

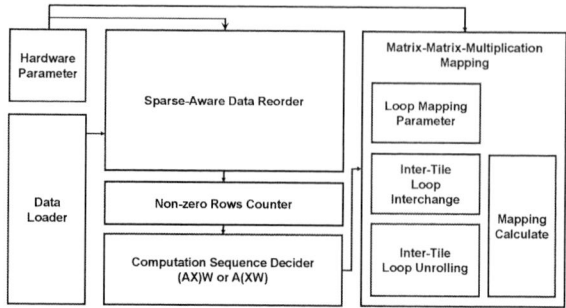

Fig. 4. Python-based mapping tool.

To clarify hardware constraints and the computational model, key parameters of the CIM architecture are defined as follows: N_{CIM} denotes the number of compute-in-memory macros; H_{mem_CIM} and W_{mem_CIM} represent the height and width of the memory cells, respectively; the computational dimensions of the CIM macro are denoted as H_{cmp_CIM} and W_{cmp_CIM}.

Let the three dimensions of matrix multiplication be M, N, and P, with corresponding indices i, j, and k specifically, the output dimensions correspond to M and N, and the inner product dimension corresponds to P.

The mapping of these tiles to the CIM architecture is divided into two steps: inter-tile mapping and intra-tile mapping. The former focuses on how to distribute the partitioned tiles across multiple CIM macros, while the later concerns the specific computation process for a single tile within a CIM macro. We define one tile as the load for one CIM macro. As for the multi CIM macros architecture, the load of all the CIM macros is defined as tile group.

1) Inter-Tile:

a) Inter-Tile Loop interchange: A matrix multiplication can be decomposed into multiple tile groups(TG). The tile group has a maximum of six loop orders, corresponding to the six loop orders for matrix calculation. Set the size of the tile group of matrix A to $TG_M \times TG_P$, and the size of the tile group of matrix B to $TG_P \times TG_N$. Set the size of matrix A to S_M, S_P, and the size of matrix B to S_P, S_N. The priority of the three dimensions will bring about different impacts. When prioritizing the P dimension (e.g., in PMN orders), the final result is accumulated directly, requiring only 1 copy of psum to be stored. However, the loading times of both X and W reach $\frac{S_P}{TG_P} \times \frac{S_M}{TG_M} \times \frac{S_N}{TG_N}$, as each P-cycle necessitates reloading tiles in the $M \times N$ dimension, resulting in more frequent data movement. When prioritizing the M dimension (e.g., in MPN order), the W matrix can be reused, but $\frac{S_M}{TG_M}$ copies of psum need to be cached (partial sums in the M dimension). When

prioritizing the N dimension (e.g., in NMP order), the X matrix can be reused across N-cycles, reducing the loading times to $\frac{S_M}{TG_M} \times \frac{S_P}{TG_P}$, while the number of psum copies increases to $\frac{S_M}{TG_M} \times \frac{S_N}{TG_N}$.

b) Inter-Tile Unrolling: Tile unrolling refers to partitioning matrix multiplication across three dimensions (including the inner product dimension) and enabling parallel computation of multiple tiles by multiple CIMs, thereby fully utilizing the parallelism among CIM macros. Let the unrolling parameters be denoted as P_M, P_N, and P_P, respectively. Unrolling strategies for the three dimensions yield distinct benefits. These parameters have the following constraints.

$$N_{CIM} = P_M \times P_N \times P_P \quad (1)$$

$$TG_M = P_M \quad (2)$$

$$TG_P = P_P \times H_{mem\_CIM} \quad (3)$$

$$TG_N = P_N \times W_{mem\_CIM} \quad (4)$$

2) Intra-Tile: For CIM, the hardware-native computation mode naturally couples the inner-product operations along the P-dimension with the column-wise parallelism along the N-dimension within the CIM macro. In this way, the parallel computing capability of the array can be fully leveraged without additional loop scheduling or data reorganization.

IV. EXPERIMENTS

We used Python to develop a mapping tool based on the analysis in the previous section as in Fig. 4. The tool takes hardware parameters and dataset-related data as input, reorder

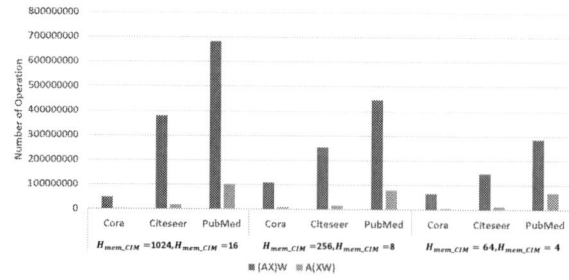

Fig. 5. Number of operations using the two execution orders.

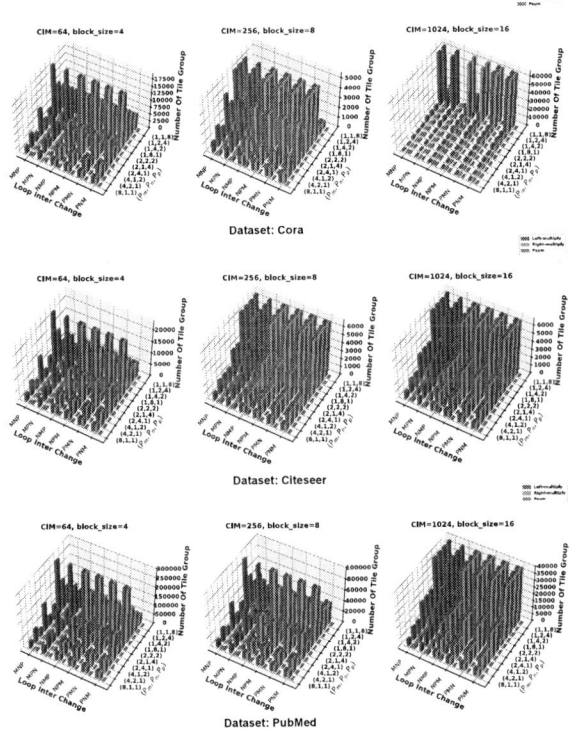

Fig. 6. Number of data movements and psum of varied desighn options.

the left-multiplied matrix as sparsity processing, and counts the number of non-zero rows. Based on the computational load, it determines the execution order as either A(XW) or (AX)W. Subsequently, it calculates the data movement and PSUM storage counts after loop tiling, interchange, and unrolling.

A. Number of Operations of Two Execution Order

We have statistically analyzed the number of blocks to be calculated in the AXW matrix multiplication operation after sparse processing. The results are in Fig. 5. Typically, choosing A(XW) results in less computational load.

B. Loop Mapping

We used the mapping tool to evaluate the impact of different hardware configurations and different mapping options for matrix multiplication. The blue bars represents the number of tile groups the input matrix for CIM needs to load, the orange bars represents the number of tile groups the weight matrix for CIM needs to load, and the green bars represents the total number of tile groups that need to store PSUM. To simplify, we set the number of DIM, N_{CIM}, as 8. We explored the effects of CIM size and mapping options (loop interchange and loop unrolling) on these results, calculating the data loading counts and PSUM storage amounts for different design options, as in Fig. 6. In the actual development of GCN accelerator, these results can assist in determining the mapping method.

V. CONCLUSION

This paper mainly discusses the GCN mapping problem on the Sparse-Aware CIM. First, we pre-process the data of GCN to fit for the sparse-aware CIM. Second, we examine the effect of the two execution orderings, (AX)W and A(XW), on total number of operations, and demonstrates through experiments that the A(XW) ordering is more effective in minimizing computation in most situations. Finally, by leveraging loop tiling, loop interchange, and loop unrolling, we calculate the data movements and number of psum. In our GCN accelerator design, this analytical data plays a supportive role in selecting the mapping of GCN on the sparse-aware CIM system.

REFERENCES

[1] Thomas N. Kipf and Max Welling. Semi-supervised classification with graph convolutional networks. In International Conference on Learning Representations (ICLR), 2017.

[2] K. Do, T. Tran, and S. Venkatesh, "Graph transformation policy network for chemical reaction prediction," in KDD, 2019, pp. 750–760.

[3] A. Fout, J. Byrd, B. Shariat, and A. Ben-Hur, "Protein interface prediction using graph convolutional networks," NeurIPS, vol. 30, 2017.

[4] R. Ying, R. He, K. Chen, P. Eksombatchai, W. L. Hamilton, and J. Leskovec, "Graph convolutional neural networks for web-scale recommender systems," in KDD, 2018, pp. 974–983.

[5] M. Yan et al., "HyGCN: A GCN Accelerator with Hybrid Architecture," Jan. 06, 2020, arXiv: arXiv:2001.02514. Accessed: May 23, 2024. [Online]. Available: http://arxiv.org/abs/2001.02514

[6] T. Geng et al., "AWB-GCN: A Graph Convolutional Network Accelerator with Runtime Workload Rebalancing," in 2020 53rd Annual IEEE/ACM International Symposium on Microarchitecture (MICRO), Athens, Greece: IEEE, Oct. 2020, pp. 922–936. doi: 10.1109/MICRO50266.2020.00079.

[7] J. Li, A. Louri, A. Karanth, and R. Bunescu, "GCNAX: A Flexible and Energy-efficient Accelerator for Graph Convolutional Neural Networks," in 2021 IEEE International Symposium on High-Performance Computer Architecture (HPCA), Seoul, Korea (South): IEEE, Feb. 2021, pp. 775–788. doi: 10.1109/HPCA51647.2021.00070.

[8] T. Yang et al., "PIMGCN: A ReRAM-Based PIM Design for Graph Convolutional Network Acceleration," 2021 58th ACM/IEEE Design Automation Conference (DAC), San Francisco, CA, USA, 2021, pp. 583-588, doi: 10.1109/DAC18074.2021.9586231.

[9] Y. Ma et al., "DCIM-GCN: Digital Computing-in-Memory Accelerator for Graph Convolutional Network," in IEEE Transactions on Circuits and Systems I: Regular Papers, vol. 71, no. 6, pp. 2735-2748, June 2024, doi: 10.1109/TCSI.2024.3384748.

[10] S. Wang, F. Liang, Q. Cao, Y. Wang, H. Li, and J. Liang, "A Weight Mapping Strategy for More Fully Exploiting Data in CIM-Based CNN Accelerator," IEEE Transactions on Circuits and Systems II: Express Briefs, vol. 71, no. 4, pp. 2324–2328, Apr. 2024, doi: 10.1109/TCSII.2023.3336299.

[11] X. Peng, R. Liu, and S. Yu, "Optimizing Weight Mapping and Data Flow for Convolutional Neural Networks on Processing-in-Memory Architectures," IEEE Transactions on Circuits and Systems I: Regular Papers, vol. 67, no. 4, pp. 1333–1343, Apr. 2020, doi: 10.1109/TCSI.2019.2958568.

[12] Y. Wang and X. Fong, "Benchmarking DNN Mapping Methods for the in-Memory Computing Accelerators," IEEE Journal on Emerging and Selected Topics in Circuits and Systems, vol. 13, no. 4, pp. 1040–1051, Dec. 2023, doi: 10.1109/JETCAS.2023.3328864.

[13] K. E. Jeon, J. Rhe, H. Bang, and J. H. Ko, "Weight-Aware Activation Mapping for Energy-Efficient Convolution on PIM Arrays," in 2023 IEEE/ACM International Symposium on Low Power Electronics and Design (ISLPED), Aug. 2023, pp. 1–6. doi: 10.1109/ISLPED58423.2023.10244618.

979-8-3315-3918-4/25 $31.00 © 2025 IEEE

CDCC: A High-Efficiency SRAM-Based Charge-Domain Compute-in-Memory Macro with Complement Compensation Design for AI Applications

Wanting Zhou [1], Zihao Xuan *[2], Song Chen [1], Yi Kang [1]

[1] School of Microelectronics, University of Science and Technology of China
Hefei, 230026, China
[2] The Hong Kong University of Science and Technology, Hong Kong SAR, China

* Email: zihaoxuan@ust.hk

Abstract—Computing-in-Memory (CIM) is an efficient architecture for AI applications. This paper presents an SRAM-based charge-domain Computing-in-Memory (CIM) macro called CDCC. CDCC uses charge-sharing of capacitors and performs multiplication and accumulation computing (MAC) operations entirely in analog domain to achieve high energy efficiency. Charge shares between rows and columns in the array, traditional shift-and-add circuits are replaced in a CDCC macro. To compensate the loss of analog computing and sign-magnitude data transforming, we propose a new coding scheme called negative offset-encoded sign-magnitude code, and design a complement compensation mechanism. For architecture optimization, we proposed a ping-pong buffer like architecture to improve efficiency. Simulations show the proposed CIM arrays achieves 4740 TOPS/W running a CNN model and BERT respectively, which are better than SOTA baselines.

Keywords—Analog Computing, Compute-in-memory, Charge Sharing, Complement Compensation

I. INTRODUCTION

Deep Neural Networks (DNNs) show strong vitality in realizing complex artificial intelligence (AI) applications. For high performance DNNs accelerators, Computing-in-Memory (CIM) was proposed [1] because CIM addresses the problem of memory wall based on traditional von Neumann architectures. CIM macros perform in-situ MAC operations, which minimizes energy overhead from external memory access. Among CIM technologies, SRAM-based CIM is often used due to its proximity to CPUs, high read & write speed, low energy consumption, and mature fabrication.

Previous SRAM-based analog CIM designs [2] often rely on bit-serial inputs. Each bit input produces an analog computing result which is converted by ADC. The Results of different bits do shift-and-add operations in digital domain. This process has expensive ADC cost and require extra digital circuits and latency. Among those design, Charge-domain CIM [3] offers superior linearity compared to current-domain CIM, which shows promising potential for multi-bit parallel computing. However, due to the charge-sharing feature of capacitors, unsigned accumulation can be easily implemented, signed arithmetic and subtraction operations remain challenges. Previous work [4] proposed a sign-separated computing scheme based on the sign-magnitude representation. Due to the existence of positive and negative zeros and the limited numerical range, sign-separated

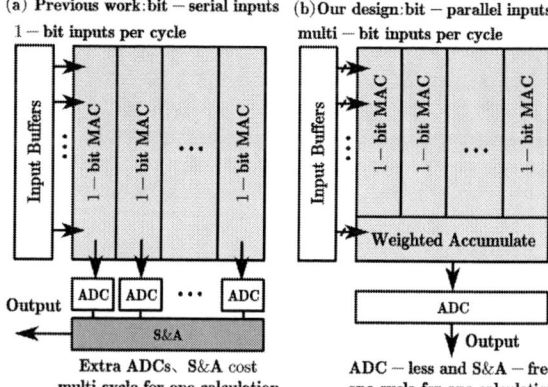

Fig. 1. Comparison of bit-serial design and our bit-parallel design

computing schemes introduce computational errors. These errors are critical to large language models which require higher precision in computing.

To overcome these limitations, we propose a Charge-domain Complement Compensation (CDCC) SRAM-CIM macro in this paper. Our contribution is as follows: (1) Complement compensation for signed arithmetic: Signed MAC operations are performed by separating positive and negative multiplications and computing positive and negative absolute values, then the absolute values are subtracted by differential-ADC. To compensate precision loss, a new code: negative offset-encoded-sign-magnitude (OESM) code is introduced. (2) Ping-pong buffer like architecture: we use a ping-pong buffer like architecture to shares ADCs between adjacent arrays and improve the efficiency. (3) Fully parallel multi-bit signed MAC with signed DAC-less input: Signed input is converted by charge sharing along rows without DAC. The signed multi-bit MAC is implemented by analog circuits, carrying out accumulation operations along single column and weighted accumulations between columns by charge-sharing.

II. CDCC MACRO

The proposed CDCC macro is illustrated in Fig. 2(a), comprises 4 CIM banks, each integrating the following components: (1) Input/Output buffers: Facilitate data transfer between external interfaces and internal computing units. (2) Complement-to-OESM-Converter (COC) circuit: Converts INT4 two's complement inputs into OESM code. (3) 64×8

979-8-3315-3918-4/25 $31.00 © 2025 IEEE

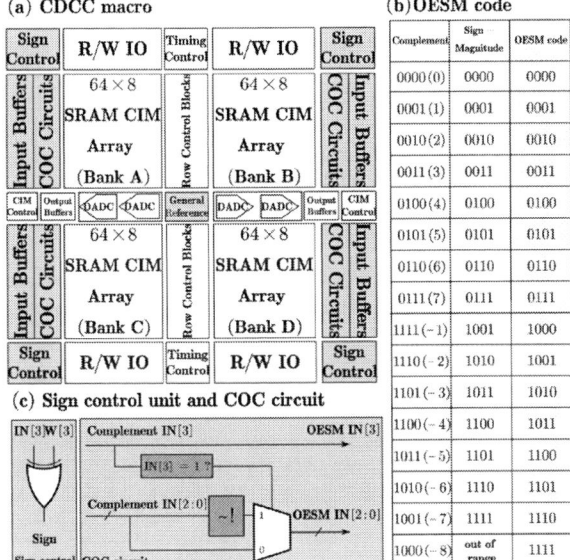

(a) CDCC macro

(b) OESM code

Complement	Sign Magnitude	OESM code
0000(0)	0000	0000
0001(1)	0001	0001
0010(2)	0010	0010
0011(3)	0011	0011
0100(4)	0100	0100
0101(5)	0101	0101
0110(6)	0110	0110
0111(7)	0111	0111
1111(−1)	1001	1000
1110(−2)	1010	1001
1101(−3)	1011	1010
1100(−4)	1100	1011
1011(−5)	1101	1100
1010(−6)	1110	1101
1001(−7)	1111	1110
1000(−8)	out of range	1111

(c) Sign control unit and COC circuit

Fig. 2. (a) CDCC macro (b) OESM Code (c) Sign control unit and COC circuit

CDCC SRAM-CIM array: Implements the charge-domain MAC operations detailed in Section 3, with pre-stored weights formatted in the OESM scheme. (4) Sign control Module: Performs operations on the sign bits of inputs and weights to enable polarity-specific calculation units (positive/negative paths).

The macro further integrates two 8-bit differential-ADC(DADC), two CIM control modules, a general reference block for ADC voltage calibration, and standard SRAM peripherals including row control buffers, R/W IO banks, and timing controllers. Each 64×8 CIM array follows the hierarchical structure detailed in Section 3, supporting parallel charge-domain computation, produces 2 pairs of positive and negative MAC results through two output paths, connected to two differential-ADC respectively. Each differential-ADC can do subtraction of positive and negative MAC results and AD-convert at the same time due to differential characteristic.

A. OESM code

When quantizing both inputs and weights into 4-bit original sign-magnitude representations, the theoretical computation separate positive and negative MAC results in absolute value scheme. Normal CPU use two's complement code, we need complement-to-sign-magnitude conversion to get absolute value before computing. However, direct complement-to-sign-magnitude conversion introduces inherent errors: for INT4 operands, the two's complement range (-8 to 7) mismatches the sign-magnitude range (-7 to 7).

To solve this, we propose an negative offset-encoded sign-magnitude (OESM) code tailored for our signed dual-path (positive and negative) architecture. The OESM encoding rules are shown in Fig .2(c). Positive values maintain their original sign-magnitude form, negative values are mapped as: -1 to 1000, -2 to 1001, …, -8 to 1111.This encoding aligns with standard sign-magnitude for positives but offsets negatives by -1 compared to their absolute values. Consequently, during the MAC process, positive input and weight remain unchanged, but negative input and weight inherently require a +1 offset compensation. This compensation can be realized in the analog process easily,

detailed in section 3. After compensation, the error between complement and absolute value disappears. The loss of data accuracy brought by signed dual-path absolute value calculation is minimized in this way. COC circuit is shown in Fig. 2(c). The logic of the circuit is: The output sign bit matches the input sign bit. If the sign bit (IN[3]) is 1, the data bits (IN[2:0]) are output after bitwise inversion; otherwise, the data bits are output unchanged.

B. Ping-pong buffer like architecture

Specially, 4 CIM bank are arranged in a 2×2 grid, where horizontally adjacent banks share row control blocks to minimize routing cost, while vertically paired banks share the same ADC block via working like ping-pong buffers mode: While one array performs charge-domain computation, the ADC concurrently digitizes the prior result from the neighboring array. This macro operates either as a conventional 512-bit SRAM or as a low-energy CIM accelerator for multi-bit vector-matrix multiplication (VMM), which can complete multi-bit parallel computation in only one operational cycle.

III. SRAM-BASED CHARGE-DOMAIN CIM ARRAY

For a CDCC macro, CIM array is the key part. In the following context, a CIM array used in CDCC macros is described assuming that the array is of size 64×8, stores 128 4-bit weights, performs 64 parallel 4-bit inputs 4-bit weights vector multiplications and generates outputs through two independent channels.

A. Sign Calculation Unit

In the CIM array, each SRAM cell interfaces with 2 calculation units (CU) that operate under complementary activation conditions. One CU is enabled only when the single multiplication result of 4-bit input and weight is positive, otherwise maintaining zero output, referred to as positive calculation unit (PCU). The other CU is enabled for negative multiplication results, referred to as negative calculation unit (NCU). Overall array is shown in Fig .3(a).

Each CU comprises three switches (S_{sign}, S_0 and S_{out}), one capacitor(C_0), and two MOSFETs (M_0, M_1), shown in Fig .3(b). The control signals for S_{sign} and S_0 are generated by the sign control unit, which produce an XOR signal of the

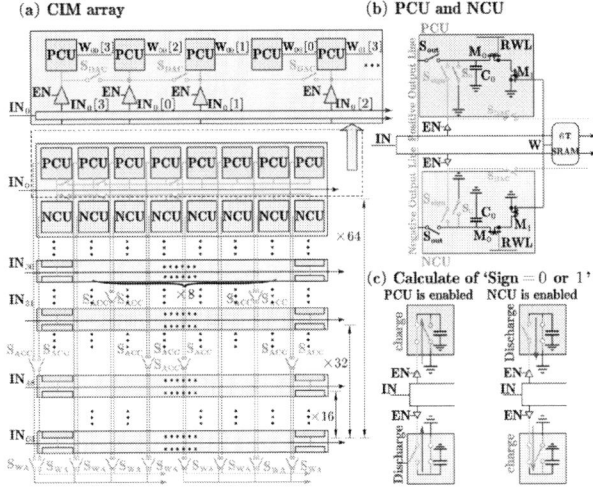

Fig. 3. (a) CIM array. (b) PCU and NCU. (c) Calculate of 'Sign = 0 or 1'.

MSBs of input and weight, detailed in section 2. When XOR result is 0, input and weight have same sign, multiplication result is positive, S_{sign} of PCU closes to enable charge-sharing, S_0 remain open, S_{sign} of NCU turned off either, S_0 of NCU closes to reset the voltage of C_0 to GND, in this way PCU is enabled. When XOR result is 1, input and weight have different sign, multiplication result is negative. S_{sign} of NCU and S_0 of PCU close, S_0 of NCU and S_{sign} of PCU open to enable NCU and reset PCU. PCUs and NCUs are connected to dedicated output lines corresponding to positive and negative MAC results. The whole process is shown in Fig .3(c).

B. DAC-Free Input

During the data input phase, the row-wise capacitors are used to achieve DAC-free input. As Fig. 4(a), after EN is enabled, S_{sign} closes based on XOR result, the input bits IN[0], IN[1], and IN[2](excluding the sign bit IN[3]) are configured in parallel groups of 1,2, and 4 CU respectively. Depending on the single-bit logic state, the designated capacitors are selectively charged or discharged. The sign bit IN[3], connected to a single CU, serves as a complement compensation component, which is elaborated in section 3.D. Then, S_{DAC} closes to initiate charge sharing among the 8 row-aligned capacitors, as illustrated in Fig. 4(b). S_{sign} opens when voltage on the capacitors become stable, charge is stored individually on different capacitors, as shown in Fig. 4(c). Following this phase, the stabilized voltage of the capacitor array can be mathematically expressed as:

$$V_{in} = \frac{IN[3]+IN[0]+2\times IN[1]+4\times IN[2]}{8} VDD \qquad (1)$$

C. Multi-bit Weight MAC

Fig. 4(d). and Fig. 4(e) shows the process of MAC operations. After input data conversion, capacitor C_0 stores quantized charge individually. The Read Word Line (RWL) is activated to perform keeping charge or discharging operations based on the 1-bit weight data stored in the SRAM cell. The gate of M_1 is connected to $\overline{W}$ stored in SRAM cell. Specifically, if the SRAM stores logic "1"(corresponding to $\overline{W}$=0), the capacitor keeps its charge. Otherwise, if logic "0" is stored, the capacitor discharges to GND. This mechanism inherently implements in-situ 1-bit multiplication, as Fig. 4(d). Subsequently, S_{out} is enabled to initiate charge sharing across capacitors of CU along the same column, accumulating charge along single column without weighting. When stable, voltage on the column capacitors can be mathematically expressed as:

$$V_{out}^j = \frac{\sum_{i=1}^{64} v_{in}^i \times w_{ij}}{64} \qquad (2)$$

V_{in}^i represents the analog input voltage from row i, and W_{ij} represents the 1-bit weight stored in the SRAM cell at row i and column j. To enable weighted accumulation across columns, switches S_{ACC} and S_{WA} are incorporated to configure the number of capacitors of each column involved in final charge sharing. For a 4-bit weight precision, weighted accumulation requires proportional contributions from four columns. Specifically, S_{ACC} switches are strategically placed along the output line to establish a binary-scaled capacitor ratio between columns. After 1-bit MAC operation within each column, S_{ACC} switches are opened to freeze the weighted charge distribution. Then S_{WA} switches are closed (following the disconnection of S_{ACC} switches) to finalize the weighted

analog accumulation across columns. For a 4-bit weight implementation, columns corresponding to weight bit W[0], W[1], W[2] integrate 16, 32 and 64 capacitors respectively, the sign bit W[3] incorporates 16 capacitors to provide complement compensation, whose operational details will be elaborated in section 3.D. The whole process is shown in Fig. 4(e). After the multi-bit weighted charge sharing, the final output voltage can be mathematically expressed as:

$$V_{out} = \frac{V_{out}^3 + V_{out}^0 + 2\times V_{out}^1 + 4\times V_{out}^2}{8} \qquad (3)$$

Finally, we get the analog voltage represents positive and negative MAC results on positive and negative output lines.

D. Complement Compensation of Input and Weight

OESM code need a +1 offset compensation for negatives. We compensate inputs and weights respectively. Complement compensation mechanism is shown in Fig. 4(f). For input compensation: During row-wise charge sharing, the input sign bit IN [3] is connected to a single unit capacitor. For positive inputs (IN[3] = 0), this capacitor remains zero that has no influence on the input conversion result. For negative inputs (IN[3] = 1), it contributes an additional unit charge to offset the OESM mapping. For weight compensation: The weighted sign bit W[3] is assigned to a column with 16 CU capacitors while the non-sign bits W[0], W[1], and W[2] scale with 16, 32 and 64 capacitors respectively. By equating the sign bit's weight (16 capacitors) to the LSB column (W[0]), we compensate for -1 offset when processing negative weights.

IV. EVALUATION AND DISCUSSION

A. Methodology

To validate the proposed architecture, we conducted a co-simulation of analog and digital subsystems under 28nm CMOS process with commercial EDA tools and open-source tool MICsim [5]. The analog subsystem is comprised of the whole CIM array, is first verified by circuit level simulations, then layout of PCU and NCU and post-layout simulation is carried out subsequently. The digital subsystem is compromise of COC circuits and sign control part and is designed on RTL. The RTL-level design is synthesized and

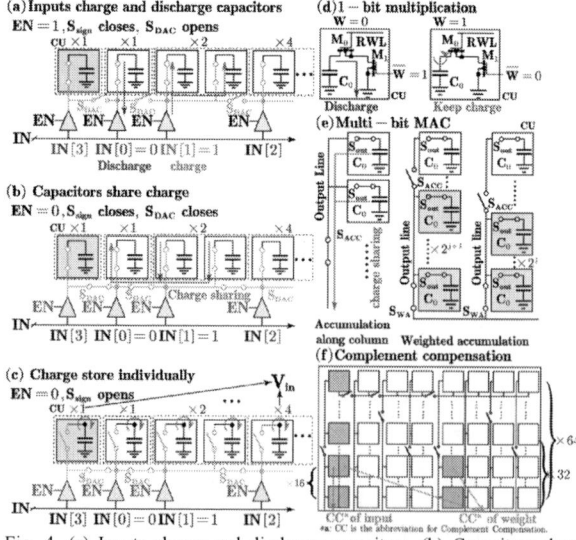

Fig. 4. (a) Inputs charge and discharge capacitors. (b) Capacitors share charge. (c) Charge store individually. (d) 1-bit multiplication. (e) Multi-bit MAC. (f) Complement compensation.

verified with gate-level netlists. Then MICsim is used to co-simulate the analog and digital subsystems for a CNN and a Transformer. For CNN VGG-8[6] is used, and Q8-BERT[7] is used for Transformer.

B. Accuracy and Precision

To evaluate the accuracy of the CIM arrays, its inputs and weights are swept separately over their possible range to get two transfer-curves shown in Fig. 5(a) and Fig. 5(b). From the transfer-curves the analog output voltage linearity defined as MAC error is shown in Fig. 5(c) and Fig. 5(d), showing MAC error of input sweep and weight sweep can be seen within ± 2 LSB, ± 1 LSB in most cases. We also performed 200-run Monte Carlo simulations on PCU across FF (Fig. 7), SS (Fig. 8), and TT (Fig. 9). Output voltage across all corners in Fig. 7, Fig. 8 and Fig. 9 remain within 3σ and with >90% of samples confined to 1σ. Absolute deviations are consistently under 1 LSB for all process corners.

C. Area, Power and Delay

The analog subsystems simulation shows the analog compute array achieves 6 ns computation latency per array operation with a 9 μW array power consumption, yielding an energy efficiency of 4740 TOPS/W.

D. System verification and evaluation

The proposed CIM macro was validated on different kind of neural network models, including CNN and Transformer in MICsim. We achieve 96.3% inference accuracy on VGG-8 on CIFAR10 and 91.6% inference accuracy on Q8-BERT on GLUE. The comparison with SOTA baselines is listed in Table 1, which shows that CDCC has better energy efficiency while having in par performance on accuracy.

TABLE I. FEATURE SUMMARY AND COMPARISON TO PRIOR WORKS

	This Work	JSSC'23 [3]	JSSC'25 [8]	JSSC'24 [9]
Technology	28nm	22nm	28nm	65nm
MAC operation	charge domain	charge domain	Analog digital	charge domain
Input precision	4	8	1-8	4-8
Weight precision	4	8	1-8	4/6/8
Output precision	8	8	1-8	-
VMM time (ns)	6	6.9	3.49-20	-
Energy Efficiency (TOPS/W)	4740	2061	206-1158	4094
Inference Accuracy	96.3%[a] 91.6%[b]	96.85%[a]	96.16%[a]	91.7%[a] 95.8%[b]

*a: CNN mode, b: Transformer mode

V. CONCLUSION

In this work, we propose a high efficiency charge domain SRAM-based CIM macro that enables high-precision signed matric-vector multiplication through a designed OESM code. The proposed architecture adopts a ping-pong buffer like mode to improve the efficiency. We achieve 4740 TOPS/W energy efficiency, 96.3% inference accuracy on CNN and 91.6% inference accuracy on transformers.

REFERENCES

[1] C.-J. Jhang, C.-X. Xue, J.-M. Hung, F.-C. Chang, and M.-F. Chang, "Challenges and Trends of SRAM-Based Computing-In-Memory for AI Edge Devices," IEEE Transactions on Circuits and Systems I: Regular Papers, vol. 68, no. 5, pp. 1773-1786, 2021.

[2] B. Yan et al., "A 1.041-Mb/mm2 27.38-TOPS/W Signed-INT8 Dynamic-Logic-Based ADC-less SRAM Compute-in-Memory Macro in 28nm with Reconfigurable Bitwise Operation for AI and Embedded Applications," 2022 IEEE International Solid-State Circuits Conference (ISSCC), San Francisco, CA, USA, 2022, pp. 188-190.

[3] H. Wang, R. Liu, R. Dorrance, D. Dasalukunte, D. Lake and B. Carlton, "A Charge Domain SRAM Compute-in-Memory Macro With C-2C Ladder-Based 8-Bit MAC Unit in 22-nm FinFET Process for Edge Inference," in IEEE Journal of Solid-State Circuits, vol. 58, no. 4, pp. 1037-1050, April 2023.

[4] H. Zhang, W. Yin, S. He, Y. Du and L. Du, "An Efficient Two-Stage Pipelined Compute-in-Memory Macro for Accelerating Transformer Feed-Forward Networks," in IEEE Transactions on Very Large Scale Integration (VLSI) Systems, vol. 32, no. 10, pp. 1889-1899, Oct. 2024.

[5] C. Wang, Z. Chen, S. Huang, "MICSim: A Modular Simulator for Mixed-signal Compute-in-Memory based AI Accelerator," arXiv preprint, arXiv:2409.14838, 2024.

[6] K. Simonyan, A. Zisserman, "Very Deep Convolutional Networks for Large-Scale Image Recognition," arXiv preprint, arXiv:1409.1556, 2014.

[7] O. Zafrir, G. Boudoukh, P. Izsak and M. Wasserblat, "Q8BERT: Quantized 8Bit BERT," 2019 Fifth Workshop on Energy Efficient Machine Learning and Cognitive Computing - NeurIPS Edition (EMC2-NIPS), Vancouver, BC, Canada, 2019, pp. 36-39.

[8] s. Yue et al., "CV-CIM: A Hybrid Domain Xor-Derived Similarity-Aware Computation-in-Memory Supporting Cost-Volume Construction," in IEEE Journal of Solid-State Circuits, vol. 60, no. 2, pp. 719-733, Feb. 2025.

[9] K. Yoshioka, "A 818–4094 TOPS/W Capacitor-Reconfigured Analog CIM for Unified Acceleration of CNNs and Transformers," in IEEE Journal of Solid-State Circuits, vol. 60, no. 5, pp. 1844-1855, May 2025.

[10] Z. Xuan, Y. Yang, W. Xuan, Z. Su, S. Chen, and Y. Kang, "YOCO: A hybrid in-memory computing architecture with 8-bit sub-petaops/W in-situ multiply arithmetic for large-scale AI," arXiv preprint, arXiv:2312.11836, 2023.

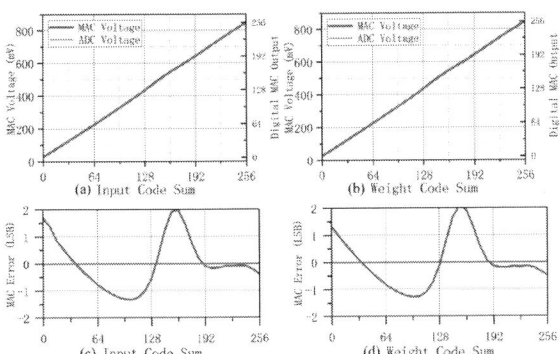

Fig. 5. Transfer curve and MAC error

Fig. 6. Layout of PCU and NCU

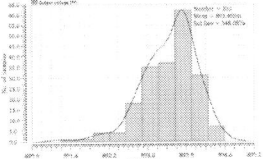

Fig. 7. Monte Carlo across FF

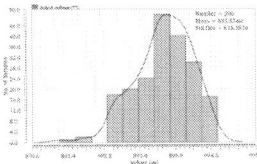

Fig. 8. Monte Carlo across SS

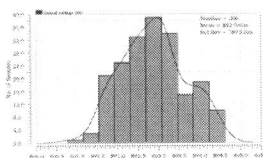

Fig. 9. Monte Carlo across TT

979-8-3315-3918-4/25 $31.00 © 2025 IEEE

HPD: **H**ybrid **P**rojection **D**ecomposition for Robust State Space Models on Analog CIM Hardware

Yuannuo Feng[†1,3], Wenyong Zhou[†2,3], Yuexi Lyu[‡3], Hanjie Liu[3], Zhengwu Liu[*2], Ngai Wong[*2], Wang Kang[*1]

[1]School of Integrated Circuit Science and Engineering, Beihang University, Beijing, China
[2]Department of Electrical and Electronic Engineering, The University of Hong Kong, Hong Kong
[3]Zhicun Research Lab, Beijing, China
[†]: Equal Contribution. [‡]: Project Leader. [*]: Corresponding Author(s).

Abstract—State Space Models (SSMs) are efficient alternatives to traditional sequence models, excelling at processing long sequences with lower computational complexity. Their reliance on matrix multiplications makes them ideal for compute-in-memory (CIM) architectures, which improve energy efficiency by computing within memory arrays. However, device non-idealities in CIM introduce weight perturbations that can degrade inference accuracy. In this paper, we systematically analyze the robustness of SSMs under noisy conditions, identifying that the final block and output projection layers are more susceptible to perturbations compared to other components. Building on these insights, we propose HPD, a Hybrid Projection Decomposition strategy for the last output projection layer. We replace the original weight matrix with the multiplication of U and Σ in its SVD to ensure compatibility with existing hardware architectures, while offloading $V^\top$ to digital hardware for precise and robust correction. Comprehensive tests on Mamba models show that our method reduces perplexity by up to 99.57% under various noise conditions compared to baseline models, with accuracy gains of up to 96.67% on the PIQA benchmark for commonsense reasoning.

Index Terms—State Space Models, Compute-in-Memory

I. INTRODUCTION

State Space Models (SSMs) have emerged as efficient alternatives to traditional sequence modeling architectures, with recent variants like Mamba [1], [2] achieving Transformer-competitive performance [3] while using substantially fewer computational resources for long sequences, as shown in Fig. 1. SSMs achieve their efficiency through structured parameterization that enables linear scaling with sequence length, relying heavily on matrix multiplication operations that are ideally suited for analog compute-in-memory (CIM) acceleration [4]–[7].

CIM architectures implement matrix operations directly within memory arrays, eliminating energy-intensive data movement and offering order of magnitude efficiency improvements over conventional digital systems [8]. However, analog computing elements suffer from inherent device non-idealities that introduce weight perturbations, which can severely degrade neural network inference accuracy and potentially offset the hardware acceleration benefits [9]–[12].

The impact of weight perturbations on SSMs is particularly concerning due to their recurrent nature and the importance of precisely tuned parameters for maintaining stable dynamics. Perturbations in SSM parameters can compound across the

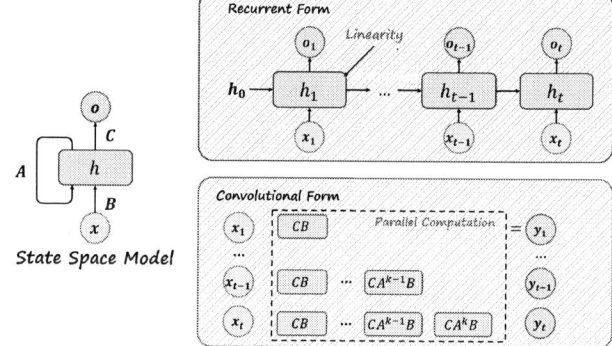

Fig. 1. An illustration of SSM architectures, combining recurrent and convolutional computations via their linear nature. SSMs enable recurrent inference and parallel training, leveraging RNN and Transformer strengths.

sequence length, potentially causing significant performance degradation [13]. Despite the growing interest in deploying SSMs on efficient hardware, the robustness of these models to weight perturbations remains largely unexplored [14].

To bridge this gap, we analyze the vulnerability of SSM components to weight perturbations, identifying that the final block and output projection layers are particularly susceptible. Based on these findings, we propose a lightweight output correction strategy that factorizes the output projection matrix into two components: one implemented on CIM hardware to ensure compatibility with existing hardware architectures, and the other offloaded to digital hardware for precise calibration. In summary, this paper makes the following key contributions:

- We systematically investigate the robustness of SSMs under noisy CIM conditions, revealing that the final block and output projection layers are more vulnerable to perturbations compared to other components.
- We propose a novel strategy that factorizes the last output projection matrix into two components: $U\Sigma$ implemented on CIM hardware to ensure compatibility with existing hardware architectures, and $V^\top$ offloaded to digital hardware for correction.
- Tests on Mamba models show our approach cuts perplexity by up to 99.57% under noise and boosts PIQA accuracy by up to 96.67% compared to baselines.

979-8-3315-3918-4/25 $31.00 © 2025 IEEE

II. PRELIMINARIES

SSMs represent a class of sequence modeling architectures that parameterize sequence transformations using linear time-invariant (LTI) systems. An SSM transforms an input sequence $\mathbf{x}(t) \in \mathbb{R}$ into an output sequence $\mathbf{y}(t) \in \mathbb{R}$ through a hidden state $\mathbf{h}(t) \in \mathbb{R}^N$ according to the following continuous-time system:

$$\frac{d\mathbf{h}(t)}{dt} = \mathbf{A}\mathbf{h}(t) + \mathbf{B}\mathbf{x}(t) \quad (1)$$

$$\mathbf{y}(t) = \mathbf{C}\mathbf{h}(t) + \mathbf{D}\mathbf{x}(t) \quad (2)$$

where $\mathbf{A} \in \mathbb{R}^{N \times N}$, $\mathbf{B} \in \mathbb{R}^{N \times 1}$, $\mathbf{C} \in \mathbb{R}^{1 \times N}$, and $\mathbf{D} \in \mathbb{R}$ are learnable parameters. For discrete inputs with time step Δ, this system is converted to its discrete-time equivalent:

$$\mathbf{h}_t = \bar{\mathbf{A}}\mathbf{h}_{t-1} + \bar{\mathbf{B}}\mathbf{x}_t \quad (3)$$

$$\mathbf{y}_t = \mathbf{C}\mathbf{h}_t + \mathbf{D}\mathbf{x}_t \quad (4)$$

where $\bar{\mathbf{A}} = e^{\mathbf{A}\Delta}$ and $\bar{\mathbf{B}} = (\mathbf{A}^{-1}(e^{\mathbf{A}\Delta} - \mathbf{I}))\mathbf{B}$. This discretization allows SSMs to process sequence data efficiently.

Recent SSM variants have improved upon the basic formulation. The S4 model [15] introduced structured parameterizations of $\mathbf{A}$ using the HiPPO matrix [16], enabling efficient modeling of long-range dependencies. S4 computes the convolution between input $\mathbf{x}$ and a structured SSM kernel $\mathbf{K}$ using fast Fourier transforms:

$$\mathbf{y} = \mathbf{x} * \mathbf{K}, \quad \mathbf{K} = (\mathbf{C}\bar{\mathbf{B}}, \mathbf{C}\bar{\mathbf{A}}\bar{\mathbf{B}}, \mathbf{C}\bar{\mathbf{A}}^2\bar{\mathbf{B}}, \ldots) \quad (5)$$

Mamba [1] introduced selective state space models, which make the SSM parameters input-dependent:

$$\mathbf{A}(\mathbf{x}), \mathbf{B}(\mathbf{x}), \mathbf{C}(\mathbf{x}) \quad (6)$$

This input-dependent parameterization allows the model to selectively process different parts of the input sequence with varying dynamics. In practice, Mamba computes:

$$\bar{\mathbf{B}}, \bar{\mathbf{C}}, \Delta = \text{Projections}(\mathbf{x}) \quad (7)$$

$$\bar{\mathbf{A}} = \exp(\Delta \cdot \text{diag}(\mathbf{A})) \quad (8)$$

$$\mathbf{h}_t = \bar{\mathbf{A}} \odot \mathbf{h}_{t-1} + \bar{\mathbf{B}} \odot \mathbf{x}_t \quad (9)$$

$$\mathbf{y}_t = \bar{\mathbf{C}} \odot \mathbf{h}_t \quad (10)$$

where $\odot$ represents element-wise multiplication, and the time step Δ is computed from the input, enabling selective forgetting and retention of information based on the input content.

III. METHODOLOGY

A. Vulnerability of SSMs under Weight Perturbation

Fig. 2 illustrates how weight perturbation affects the performance of Mamba models of varying sizes. As noise standard deviation increases from 0 to 0.05, we observe a clear relationship between model size and robustness. The smallest model (Mamba-130M) exhibits the highest baseline perplexity and shows the most dramatic degradation under noise, with perplexity increasing from approximately 36 to 45. In contrast, larger models demonstrate significantly better performance and greater resilience to noise perturbations. A similar trend can

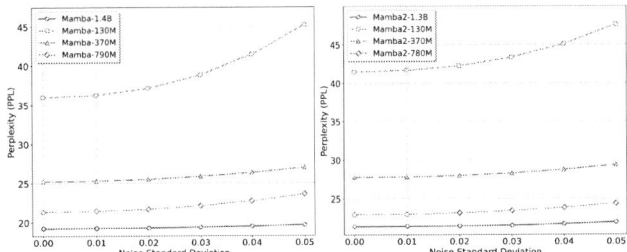

Fig. 2. Perplexity versus noise standard deviation for Mamba (left) and Mamba2 (right) models of varying sizes. Both model families show increased perplexity with higher noise levels, with smaller models (130M, orange) being most sensitive to noise injection.

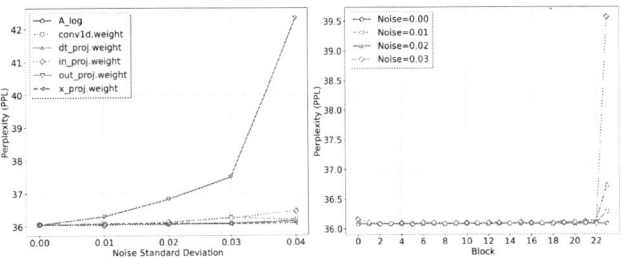

Fig. 3. (Left) Perplexity versus noise standard deviation for different layer types. (Right) Per-block sensitivity analysis across the 24-layer Mamba-130M model, demonstrating that last blocks exhibit higher sensitivity to noise perturbations.

be observed in the Mamba-2 series in the right panel. This pattern suggests that parameter count serves as a buffer against noise interference, with larger models maintaining more stable internal representations despite increasing noise levels.

Fig. 3 illustrates the robustness of different components within the Mamba architecture when subjected to noise perturbations. The left panel reveals that while the matrix A in SSM mechanism, convolution layer, dt projection, x projection and input projection layer maintain relatively stable perplexity under increasing noise levels, the output projection layer exhibits significant vulnerability, with perplexity rising dramatically at noise standard deviations of 0.03 and 0.04. The right panel demonstrates how noise affects different blocks of the model, showing consistent performance across most blocks but a sharp performance degradation at the final blocks. This block-wise analysis confirms that later stages of the model's computation are particularly susceptible to noise interference.

B. Hybrid Projection Decomposition

Based on our vulnerability analysis, we identified that the output projection layer is particularly susceptible to weight perturbations in CIM architectures. To address this issue, we propose Hybrid Projection Decomposition (HPD), a novel approach that strategically decomposes the weight matrix to mitigate the effects of hardware non-idealities while maintaining computational efficiency.

In SSMs, the output projection layer transforms the state representation into the final output space through a linear

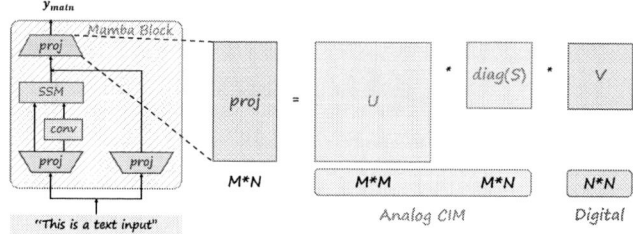

Fig. 4. An illustration of SSM architectures, combining recurrent and convolutional computations via their linear nature. SSMs enable recurrent inference and parallel training, leveraging RNN and Transformer strengths.

transformation:

$$\mathbf{y} = \mathbf{W}_{\text{out}}\mathbf{h} + \mathbf{b} \tag{11}$$

where $\mathbf{W}_{\text{out}} \in \mathbb{R}^{d_{\text{model}} \times d_{\text{vocab}}}$ is the weight matrix, $\mathbf{h} \in \mathbb{R}^{d_{\text{model}}}$ is the hidden state, and $\mathbf{b} \in \mathbb{R}^{d_{\text{vocab}}}$ is the bias term.

Our approach leverages Singular Value Decomposition (SVD) to factorize the weight matrix $\mathbf{W}_{\text{out}}$ as:

$$\mathbf{W}_{\text{out}} = \mathbf{U}\mathbf{\Sigma}\mathbf{V}^{\top} \tag{12}$$

where $\mathbf{U} \in \mathbb{R}^{d_{\text{model}} \times r}$ contains the left singular vectors, $\mathbf{\Sigma} \in \mathbb{R}^{r \times r}$ is a diagonal matrix with singular values, $\mathbf{V} \in \mathbb{R}^{d_{\text{vocab}} \times r}$ contains the right singular vectors, and r is the rank of the decomposition.

The key innovation of HPD lies in the strategic distribution of computation between CIM and digital hardware:

$$\mathbf{y} = (\mathbf{V}^{\top})^{\top}[(\mathbf{U}\mathbf{\Sigma})\mathbf{h}] + \mathbf{b} \tag{13}$$

We implement this as a two-stage process:

$$\mathbf{z} = (\mathbf{U}\mathbf{\Sigma})\mathbf{h} \tag{14}$$

$$\mathbf{y} = \mathbf{V}\mathbf{z} + \mathbf{b} \tag{15}$$

The first computation $(\mathbf{U}\mathbf{\Sigma})\mathbf{h}$ is performed on CIM hardware, where we precompute $\mathbf{W}_{\text{CIM}} = \mathbf{U}\mathbf{\Sigma}$ to maintain full compatibility with existing CIM array architectures. This is possible because the product $\mathbf{U}\mathbf{\Sigma}$ preserves the same dimensionality and matrix multiplication structure as the original weight matrix, requiring no modifications to the underlying CIM hardware design or control circuitry. The second stage $(\mathbf{V}\mathbf{z})$ is executed on digital hardware, which provides higher precision and immunity to analog noise.

IV. EXPERIMENTS

A. Experiment Setup

All models are implemented using PyTorch 2.2.0 and tested on an NVIDIA L20 GPU. To evaluate performance, we compare our model to its vanilla counterparts under various noise conditions, including Gaussian noise and lognormal noise, due to the lack of relevant research in this area.

TABLE I
PERPLEXITY (PPL) COMPARISON OF MAMBA AND MAMBA2 MODELS ACROSS DIFFERENT GAUSSIAN NOISE LEVELS.

Model	Size	PPL under Gaussian Noise		
		$\sigma = 0.01$	$\sigma = 0.03$	$\sigma = 0.05$
Mamba	130M	33.64 (+84.93%)	34.35 (+88.12%)	35.78 (+90.34%)
	370M	24.58 (+57.13%)	24.89 (+12.60%)	25.60 (+7.37%)
	790M	21.90 (+38.29%)	22.25 (+57.06%)	22.95 (+63.71%)
	1.4B	19.74 (+36.38%)	19.83 (+11.32%)	20.04 (+5.56%)
Mamba2	130M	37.21 (+99.57%)	37.70 (+35.14%)	39.04 (+25.69%)
	370M	26.27 (+51.81%)	26.70 (+32.44%)	27.57 (+25.10%)
	780M	22.43 (+85.51%)	22.84 (+23.17%)	23.69 (+22.85%)
	1.3B	21.19 (+91.53%)	21.19 (+97.18%)	21.45 (+78.49%)

TABLE II
PERPLEXITY (PPL) COMPARISON OF MAMBA AND MAMBA2 MODELS ACROSS DIFFERENT LOGNORMAL NOISE LEVELS.

Model	Size	PPL under lognormal Noise		
		$\sigma = 0.01$	$\sigma = 0.03$	$\sigma = 0.05$
Mamba	130M	33.65 (+85.64%)	34.37 (+88.30%)	35.78 (+90.61%)
	370M	24.60 (+58.38%)	24.90 (+13.22%)	25.60 (+7.95%)
	790M	21.91 (+37.36%)	22.27 (+56.47%)	22.97 (+62.78%)
	1.4B	19.74 (+35.58%)	19.83 (+11.16%)	20.05 (+5.56%)
Mamba2	130M	37.21 (+90.50%)	37.72 (+33.92%)	39.11 (+24.34%)
	370M	26.33 (+87.73%)	26.95 (+50.41%)	28.08 (+22.24%)
	780M	22.43 (+85.49%)	22.84 (+25.00%)	23.68 (+38.83%)
	1.3B	21.19 (+91.65%)	21.18 (+91.39%)	21.49 (+85.28%)

B. Experiment Result

The results in Tables I and II demonstrate that our method achieves consistent improvements over the vanilla Mamba models under both Gaussian and lognormal noise on the Wikitext dataset. The primary metric is perplexity (PPL), and the robustness ratio is defined as:

$$\text{Ratio} = 1 - \frac{\text{PPL}_{\text{noise, ours}} - \text{PPL}_{\text{no noise, original}}}{\text{PPL}_{\text{noise, original}} - \text{PPL}_{\text{no noise, original}}} \tag{16}$$

where $\text{PPL}_{\text{no noise, original}}$ represents the PPL of the original model under no noise conditions, $\text{PPL}_{\text{noise, original}}$ is the PPL of the original model with noise, and $\text{PPL}_{\text{noise, ours}}$ is the PPL of our method under noise. This ratio quantifies how much our method reduces the noise-induced performance degradation relative to the original model.

Under Gaussian noise, as shown in Table I, our method consistently improves robustness across both Mamba and Mamba2 models, as evidenced by lower perplexity (PPL) increases compared to their no-noise conditions. For smaller models (e.g., 130M), our method demonstrates significant robustness improvement, achieving PPL increases of 99.57% and 84.93% at $\sigma = 0.01$ for Mamba2 and Mamba models, respectively, compared to substantially higher increases in baseline models. As the model size increases (e.g., 1.3B), our method maintains consistent effectiveness for Mamba2, achieving a PPL increase of 78.49% compared to 90.34% for the Mamba model at $\sigma = 0.05$. These results highlight

TABLE III
EVALUATION OF ACCURACY FOR MAMBA MODELS UNDER GAUSSIAN NOISE.

Model	Size	Accuracy (%)		
		ARC-e	PIQA	LAMBADA
Mamba	130M	47.73 (+77.78%)	64.24 (+96.67%)	42.39 (+57.22%)
	370M	54.55 (+37.50%)	69.22 (+66.67%)	53.65 (+27.50%)
	790M	60.98 (+59.09%)	72.36 (+50.00%)	61.23 (+66.98%)
	1.4B	65.26 (+9.52%)	75.06 (+33.33%)	65.08 (+36.51%)

the consistent ability of our method to reduce noise-induced degradation across different model architectures and sizes, with the most pronounced benefits observed in smaller models where inherent noise resilience is typically limited.

For lognormal noise, as shown in Table II, our method similarly retains its generalization ability across different statistical noise distributions. For smaller models (*e.g.*, 130M), our method achieves PPL increases of 90.50% and 85.64% for Mamba2 and Mamba models, respectively, at $\sigma = 0.01$, demonstrating robust performance even under the heavy-tailed characteristics of lognormal perturbations. For larger models (*e.g.*, 1.3B), PPL increases of 85.28% and 90.61% are observed for Mamba2 and Mamba models, respectively, at $\sigma = 0.05$, indicating that the method's effectiveness scales appropriately with model capacity. The consistent improvements across both noise distributions provide strong evidence for the fundamental robustness mechanisms introduced by our approach, rather than overfitting to specific noise characteristics.

To verify the generalization ability of our method, we evaluate Mamba models of varying sizes under Gaussian noise perturbations on commonsense reasoning datasets (Table III). Our approach demonstrates significant improvements in robustness across all model sizes, with particularly substantial gains observed in the ARC-e benchmark: 77.78% improvement for the 130M model, 37.50% for the 370M model, and 59.09% for the 790M model. The PIQA benchmark shows consistent improvements ranging from 33.33% to 96.67%, demonstrating enhanced physical commonsense reasoning capabilities under noise conditions. Similarly, the LAMBADA benchmark exhibits improvements from 36.51% to 66.98%, confirming the method's effectiveness across diverse reasoning tasks and linguistic contexts.

V. CONCLUSION

This work presents the first systematic study of weight perturbations in SSMs on analog CIM hardware, pinpointing the output projection layers in the final block as highly noise-sensitive. Our HPD method splits the output projection matrix into CIM-compatible and digitally calibrated components for better precision. Tests on Mamba models show HPD cuts perplexity by up to 99.57% under noise and boosts PIQA accuracy by up to 96.67% for commonsense reasoning, compared to baselines.

ACKNOWLEDGEMENT

This work was supported by ACCESS – AI Chip Center for Emerging Smart Systems under the InnoHK initiative of the Innovation and Technology Commission, Hong Kong SAR Government; the Research Grants Council of Hong Kong through TRS project T45-701/22-R and GRF project 17203224; the National Natural Science Foundation of China (62404187, 62274008); the Beijing Nova Program (20250484807); the Zhejiang Provincial Key R&D Program (2025C01071); and the Research Funding of Hangzhou International Innovation Institute of Beihang University (2024KQ157).

REFERENCES

[1] A. Gu and T. Dao, "Mamba: Linear-time sequence modeling with selective state spaces," in *arXiv preprint arXiv:2312.00752*, 2023.

[2] T. Dao and A. Gu, "Transformers are SSMs: Generalized models and efficient algorithms through structured state space duality," in *International Conference on Machine Learning (ICML)*, 2024.

[3] T. Brown *et al.*, "Language Models Are Few-Shot Learners," in *Advances in neural information processing systems*, vol. 33, pp. 1877–1901, 2020.

[4] Z. Guan *et al.*, "A hardware-aware neural architecture search pareto front exploration for in-memory computing," in *2022 IEEE 16th International Conference on Solid-State & Integrated Circuit Technology (ICSICT)*, pp. 1–4, 2022.

[5] Y. Chen *et al.*, "Device characteristic-aware quantization for eflash-based in-memory computing soc," in *2024 IEEE International Conference on Integrated Circuits, Technologies and Applications (ICTA)*, pp. 70–71, IEEE, 2024.

[6] T. Bai *et al.*, "An end-to-end in-memory computing system based on a 40-nm eflash-based imc soc: Circuits, toolchains, and systems co-design framework," *IEEE Transactions on Computer-Aided Design of Integrated Circuits and Systems*, vol. 43, no. 6, pp. 1729–1740, 2024.

[7] W. Zhou *et al.*, "A time- and energy-efficient cnn with dense connections on memristor-based chips," in *2023 IEEE 15th International Conference on ASIC (ASICON)*, pp. 1–4, 2023.

[8] A. Shafiee *et al.*, "ISAAC: A Convolutional Neural Network Accelerator with In-Situ Analog Arithmetic in Crossbars," in *2016 ACM/IEEE 43rd Annual International Symposium on Computer Architecture (ISCA)*, pp. 14–26, 2016.

[9] W. Zhou *et al.*, "Enhancing robustness of implicit neural representations against weight perturbations," in *ICASSP 2025 - 2025 IEEE International Conference on Acoustics, Speech and Signal Processing (ICASSP)*, pp. 1–5, 2025.

[10] W. Mao *et al.*, "Hyimc: Analog-digital hybrid in-memory computing soc for high-quality low-latency speech enhancement," in *2025 Design, Automation & Test in Europe Conference (DATE)*, pp. 1–2, IEEE, 2025.

[11] G. Wang *et al.*, "A 40nm 5-16tops/w@ int8 eflash in-memory computing soc chip with noise suppression and compensation techniques to improve the accuracy," in *2023 IEEE International Conference on Integrated Circuits, Technologies and Applications (ICTA)*, pp. 128–129, IEEE, 2023.

[12] B. Pan *et al.*, "A mini tutorial of processing in memory: From principles, devices to prototypes," *IEEE Transactions on Circuits and Systems II: Express Briefs*, vol. 69, no. 7, pp. 3044–3050, 2022.

[13] W. Zhou *et al.*, "Towards Robust RRAM-based Vision Transformer Models with Noise-aware Knowledge Distillation," in *Design, Automation & Test in Europe Conference & Exhibition (DATE)*, pp. 1–2, 2025.

[14] Y. Cho *et al.*, "Ptq4vm: Post-training quantization for visual mamba," in *2025 IEEE/CVF Winter Conference on Applications of Computer Vision (WACV)*, pp. 1176–1185, 2025.

[15] A. Gu *et al.*, "Efficiently Modeling Long Sequences with Structured State Spaces," in *International Conference on Learning Representations*, 2022.

[16] A. Gu *et al.*, "Hippo: Recurrent Memory with Optimal Polynomial Projections," in *Advances in neural information processing systems*, vol. 33, pp. 1474–1487, 2020.

ADC-Free RRAM-Based XNOR-Bitcount Architecture for Hand Gesture Recognition

Lixun Wang[1], Yuejun Zhang*[1], Qikang Li[1], Liang Wen[2]

[1] Faculty of Electrical Engineering and Computer Science, Ningbo University, Ningbo 315211, China
[2] Department of Electronic Technology, China Coast Guard Academy, Ningbo, 315801, China

* Email: 2301100038@nbu.edu.cn, zhangyuejun@nbu.edu.cn

Abstract—**With the growing demand for contactless human machine interaction, hand gesture recognition has become an increasingly important input modality in applications such as wearable devices and virtual reality. However, as the depth of convolutional neural networks (CNNs) increases, the accumulation of convolutional kernels leads to a super-linear growth in computational requirements. The resulting high computational complexity and memory overhead make it challenging to deploy such models on resource-constrained edge devices. To address this issue, this paper proposes a compute-in-memory XNOR-bitcount convolution accelerator based on resistive random-access memory (RRAM), tailored for efficient deployment of binary neural networks in gesture recognition tasks. The proposed architecture employs 2T2R cells to construct XNOR computing arrays, and integrates a structured input mechanism with a task-specific preprocessing strategy to enhance robustness against common distortions such as boundary blur and pose variation in gesture images. Simulation results based on the Hand Gesture Recognition Database demonstrate that the four-layer binary convolution network achieves nearly 100% recognition accuracy and reaches a peak throughput of 320 GOPS under a 65 nm CMOS process.**

Keywords—RRAM, ADC-Free, Compute-in-Memory, Binary Neural Network

I. INTRODUCTION

With the rapid advancement of human machine interaction (HMI), the demand for contactless HMI systems is surging, particularly in scenarios such as wearable electronics, virtual reality, and in-vehicle control [1,2]. As a natural and intuitive form of interaction, hand gesture recognition is gradually becoming a standard interface for edge devices. However, conventional gesture recognition models based on CNNs suffer from high computational complexity and significant memory demands, making real-time execution on low-power embedded systems challenging [3].

To address these limitations, compute-in-memory (CIM) architectures have emerged in recent years as a promising solution for in-situ data processing [4,5]. In particular, resistive random-access memory (RRAM), a representative non-volatile memory technology, offers excellent compatibility with standard CMOS processes and inherent support for parallel in-memory computation [6,7].

This work was supported in part by the National Natural Science Foundation of China under Grant 62474100; in part by the Major Special Project of China Innovation Challenge (Ningbo) from Ningbo Science and Technology Program under Grant 2024T016; in part by the Ningbo University "Double First-Class" Cooperation Special Directional Entrusted Scientific and Technological Cooperation Project under Grant HX2024000574, HX2025000106; in part by the Cixi Science and Technology Program under Grant CZ2025006; in part by the Graduate Student Scientific Research and Innovation Project of Ningbo University under Grant IF2025030.

Simultaneously, Binary Convolutional Neural Networks (BCNNs) have been introduced to replace traditional multiply–accumulate operations with lightweight bitwise XNOR and bitcount operations. This approach performs approximate dot-product computation by applying bitwise XNOR between input vectors and kernel weights, followed by counting the number of resulting 1s. Due to its extremely compact representation of weights and activations, BCNNs significantly reduce the computational and memory overhead, making them highly suitable for deployment on resource-constrained edge platforms [8,9].

Building upon our previously developed RRAM-based binary convolutional neural network macro [10], this work reconstructs its architecture to adapt specifically to hand gesture recognition tasks. We propose a gesture-oriented neural array architecture with XNOR logic as the computing core, leveraging the non-volatility and high parallelism of RRAM devices to enable efficient matching and recognition of gesture templates. To address challenges such as boundary blur and frequent pose variations inherent to gesture inputs, a structured input mechanism and task-specific preprocessing strategy are introduced. Based on these components, the RRAM-XNOR cell enable ultra-low-power binary inference without requiring any ADCs or external processors. As a dedicated hardware accelerator for gesture recognition, the proposed architecture integrates binary convolution and bitcount operations to efficiently support feature extraction and classification within edge AI systems.

II. BCNN FULL HARDWARE ARCHITECTURE

A. RRAM Device

To construct a high-efficiency and highly integrated accelerator array for hand gesture recognition, this work adopts RRAM as the core storage and computation unit. A typical RRAM cell consists of a top electrode (TE), a bottom electrode (BE), and a metal oxide switching layer in between. When a specific voltage pulse is applied across the device, a conductive filament can be formed or ruptured within the oxide layer, enabling a reversible transition between a high resistance state (HRS) and a low resistance state (LRS), which are used to represent logical "0" and "1", respectively. Thanks to its non-volatility, the resistance state can be retained even without power, making RRAM inherently suitable for data storage and weight mapping in binary neural networks.

B. RRAM-Based XNOR Array Architecture

To efficiently support hardware implementation of XNOR-bitcount convolution, as shown in Fig. 1，this work proposes a compute-in-memory XNOR array architecture based on RRAM technology. The architecture is constructed

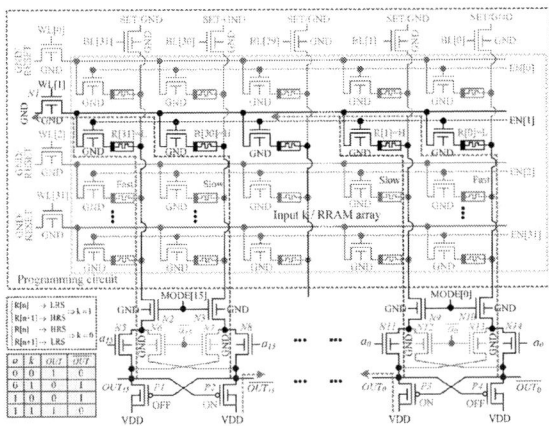

Fig. 1. 32×32 XNOR computational subarray with 2T2R structure.

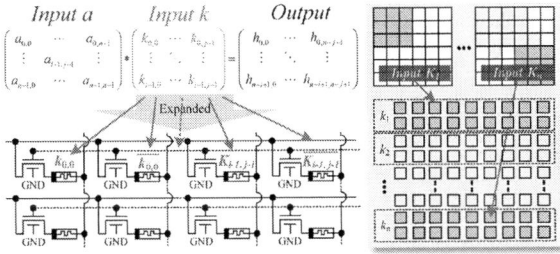

Fig. 2. Convolutional kernel unfolding by bits, and its complementary coded mapping inside the array.

using 2T2R memory-compute units as the fundamental building blocks. Each subarray consists of 32×32 cells, where each column corresponds to a specific bit position in the weight matrix, and each row stores the bitwise-expanded vector of a different convolutional kernel. As shown in Fig. 2, during the preprocessing stage, each kernel is flattened and encoded into complementary resistance states and programmed into the corresponding 2T2R cells. During inference, input data is applied as voltage signals and interacts with the stored weights within the same array, enabling large-scale parallel logic computation.

In the kernel programming phase (MODE=0), the RRAM tuning path is electrically decoupled from the cross-coupled feedback structure. In this mode, a row-wise enable signal activates the target cells, and appropriate voltages are applied to the word lines and bit lines to selectively turn on the tuning transistors. By applying either VDD or ground to the transistor source terminals, a voltage drop is created across each RRAM device to trigger a SET or RESET operation, depending on the required polarity and magnitude. This scheme supports either full-row or partial-row parallel programming, providing flexibility to balance write throughput and individual cell control. Each convolutional kernel weight k is encoded into the corresponding 2T2R memory cell in the form of complementary resistive states, i.e., R_n=LRS, R_{n+1}=HRS means k=1; and vice versa R_n=HRS, R_{n+1}=LRS means k=0).

During the computation phase, the array switches to XNOR mode, where each input bit a is logically compared with the stored weight k through in-situ bitwise XNOR operations. In this mode, the tuning transistors are fully

disabled, while only the cross-coupled feedback structure and discharge paths remain active. The computation row is selected via an enable signal, and the input vector a is applied as voltage signals to determine the discharge path of each cell. Specifically, due to the distinct conduction speeds of different resistance states, the discharge path with an LRS device conducts significantly faster than that with an HRS. This creates a voltage difference at the gates of the cross-coupled PMOS transistors, pulling one output node to VDD and realizing the XNOR operation.

For example, as shown in Fig. 1(left), when a=1 and k=0 (i.e., R[30]=HRS, R[31]=LRS), transistors N5 and N8 are turned on, while N6 and N7 remain off. The low resistance of R[31] enables a faster discharge path, quickly lowering the voltage at the OUT node and turning on PMOS P2, which pulls the output to VDD. Consequently, OUT outputs a logical "0", representing XNOR(1,0)=0. Conversely, as shown in Fig. 1(right), for a=1 and k=1 (R[0]=LRS, R[1]=HRS), the output is driven high, indicating a logic "1". Notably, this structure allows multiple cells to operate in parallel, further improving the computational throughput and array-level parallelism.

C. Hand Gesture Recognition BCNN Circuit Design

To efficiently support the inference execution of BCNN for hand gesture recognition tasks, this work addresses the challenges commonly encountered in gesture image acquisition, such as boundary blur, occlusion, and pose variations. A structured input mechanism and task-aware preprocessing flow are designed at the network input stage. First, to mitigate the boundary blur typically seen in gesture images, a local edge enhancement filter is applied for spatial preprocessing, which strengthens the contour information and suppresses background noise. Second, to accommodate the shape variations resulting from different gesture poses, a geometric transformation-based structural template is constructed, projecting the original image onto a set of pose-normalized feature spaces, ensuring key shape alignment and representation consistency. Additionally, considering that the gesture region typically occupies the center of the image, a center-weighted encoding strategy is introduced. This strategy prioritizes the binarization of the central region, improving perception accuracy while maintaining computation resource efficiency.

After preprocessing, all images are uniformly encoded into binary input vectors and mapped to the recognition circuit. The overall system is built upon a 64×128 RRAM array, as shown in Fig. 3(a), composed of 8 individual 32×32 subarray modules. The array units are based on a 2T2R structure, providing in-situ hardware support for the bitwise XNOR and bitcount operations in the BCNN. During the computation mode, row control signals WL<63:0> are used to activate the designated computation path, while the input data vector Input<127:0> is applied as voltage signals to the array. Subarrays are interconnected via row-column multiplexing strategies and buffer isolation, ensuring signal transmission integrity and cross-block communication. Dedicated buffer units are inserted between every two 32×32 modules to alleviate signal attenuation and timing offsets caused by long-range communication. The array output is directly connected to the bitcount module, enabling near-value reconstruction without the need for an ADC. All computation results are processed within the array, where logical decision-making

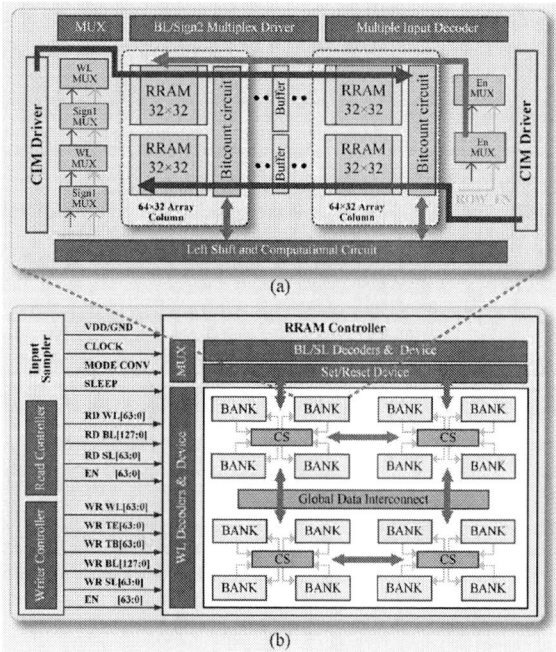

(a)

(b)

Fig. 3. Convolutional kernel unfolding by bits, and its complementary coded mapping inside the array.

and accumulation are carried out. The final convolution results are produced by shifting and binary encoding the output. As shown in Fig. 3(b), the entire computational flow consists of a CIM driver, input selection module, row activation circuits, RRAM XNOR array, bitcount logic, and output processing path, forming a complete BCNN binary convolution hardware flow.

III. SIMULATION RESULTS

A. Functional Simulation Waveform

To verify the functionality of the proposed RRAM-XNOR array structure, functional-level simulation tests were conducted based on the array circuit shown in Fig. 1. This test targets multiple computation units within the same row, each of which stores different weight values k to validate the XNOR operation between the input vector a and the stored weights. As shown in Fig. 4, upon activating the enable signal EN1, the units on the left side of the array are pre-set to the resistance states corresponding to logical "0", while those on the right side correspond to logical "1". The simulation waveform demonstrates the dynamic current response of the RRAM cells as the resistance states change after loading the input vector. Additionally, the waveform shows the stable output results generated by the array after completing the XNOR operation in a single cycle for various input combinations.

The verification results indicate that, for different input-weight matching relations, the output signal accurately reflects the expected XNOR logic characteristics. The difference in current conduction paths clearly illustrates the role of RRAM high and low resistance states in the computation process, providing a reliable functional foundation for subsequent large-scale array mapping and bitcount accumulation.

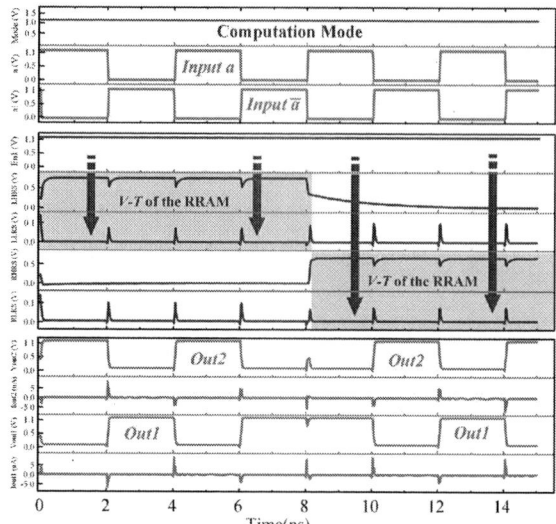

Fig. 4. Function test waveform of RRAM-based XNOR array with different inputs.

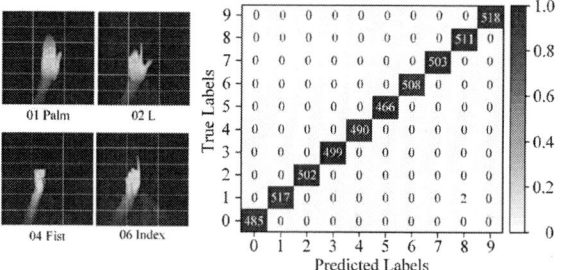

Fig. 5. Gesture recognition accuracy confusion matrix.

B. Recognition Accuracy

To evaluate the recognition performance of the proposed RRAM-XNOR array in practical applications, as shown in Fig. 4, simulation tests were conducted using the Hand Gesture Recognition Database. The network structure used in the tests consists of four binary convolution layers, with input images mapped to the hardware array for inference computation after structured preprocessing. The experimental results demonstrate that, for the trained gesture categories, the accelerator achieves near 100% recognition accuracy, thoroughly validating the feasibility and robustness of the proposed architecture for edge-based hand gesture recognition tasks.

C. Comparative Analysis

Table I presents a comparison with recent related works. This paper, based on a 65nm process and utilizing 2T2R compute-in-memory units, ensures write independence. Additionally, the fully binary input/output design reduces interface complexity and avoids the power-hungry ADC modules. In terms of overall performance, although the proposed design has a smaller array size compared to some high-density implementations, it achieves a peak throughput of 320 GOPS by leveraging the lightweight input strategy and ADC-free inference path. Figure 6 compares the performance of the proposed system with related works on the CIFAR-10 and MNIST datasets, demonstrating its potential for edge recognition tasks.

TABLE I.   COMPARISON WITH RECENT CIM WORKS

	VLSI 2020 [5]	CICC 2021 [7]	TCAS-I 2022 [8]	TCAS-II 2023 [6]	TCAS-I 2023 [9]	This work
Technology	130nm	40nm	40nm	180nm	40nm	**65nm**
Supply Voltage	0.27-0.4V	0.9V	1V	0.5-1.9V	0.9V	**1.2V**
Memory	2T2R	1T1R	1T2R	3T2R	1T1R	**2T2R**
Array Size	256×256	128×128	256×128	0.8k	256×256	**64×128**
I/O Precision	2bit/2bit	1bit/3bit	8bit/18bit	Binary	Analog	**Binary**
Weight Precision	Ternary	8bit	9bit	Binary	Binary	**Binary**
Operation Mode	Voltage	Current	Voltage	Voltage	Voltage	**Voltage**
Without ADC	NO	NO	NO	YES	YES	**YES**
MAX Throughput [GOPS]	N/A	20.96	N/A	200	222.88	**320**

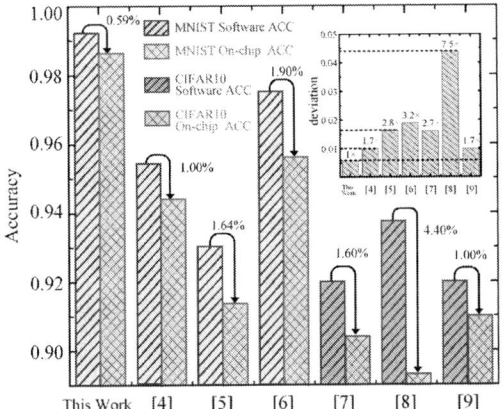

Fig. 6.   Gesture recognition accuracy confusion matrix.

IV. CONCLUSIONS

This paper presents a low-power XNOR-bitcount convolution accelerator for hand gesture recognition, based on a 2T2R RRAM array that implements a compute-in-memory design. The architecture performs in-situ binary convolution within the array without the need for ADCs or external processors, while incorporating a structured input mechanism to enhance robustness against gesture features such as boundary blur and pose variation. Simulation results on the Hand Gesture Recognition Database show that the proposed design achieves nearly 100% recognition accuracy for trained categories and reaches a peak throughput of 320 GOPS under a 65nm process, demonstrating its potential for low-power edge AI applications.

REFERENCES

[1] P. J. Lin, C. H. Shih and T. H. Weng, "Contactless and real-time hand gesture recognition using inductive proximity technique for wrist-worn wearables," IEEE Sens. J., vol. 25, no. 11, pp. 20474-20485, June, 2025.

[2] Z. Chen, X. Qiao, S. Liang, T. Yan and Z. Chen, "sEMG-based gesture recognition via multi-feature fusion network," IEEE J. Biomed. Health., vol. 29, no. 4, pp. 2570-2580, April 2025.

[3] X. Wang, C. Dong, P. Zhou, S. Nandi, S. Nath, R. Elliman, H. Iu, S. Kang and J. Eshraghian, "Low-variance memristor-based multi-level ternary combinational logic," IEEE Trans. Circuits Syst. I, Reg. Papers, vol. 69, no. 6, pp. 2423-2434, June 2022.

[4] Q. Liu, B. Gao, P. Yao, D. Wu, J. Chen, Y. Pang, W. Zhang, Y. Liao, C. X. Xue, W. H. Chen, J. Tang, Y. Wang, M. F. Chang, H. Qian and H. Wu, "A fully integrated analog ReRAM based 78.4tops/w compute-in-memory chip with fully parallel mac computing," IEEE Int. Solid-State Circuits Conf. (ISSCC) Dig. Tech. Papers, San Francisco, CA, USA, 2020, pp. 500-502.

[5] W. Wan, R. Kubendran, B. Gao, S. Joshi, P. Raina, H. Wu, G. Cauwenberghs and H. S. P. Wong, "A voltage-mode sensing scheme with differential-row weight mapping for energy-efficient RRAM-based inmemory computing," IEEE Symp. VLSI Technol., Honolulu, HI, USA, 2020, pp. 1-2.

[6] Y. Li, J. Chen, L. Wang, W. Zhang, Z. Guo, J. Wang, Y. Han, Z. Li, F. Wang, C. Dou, X. Xu, J. Yang, Z. Wang and D. Shang, "An ADC-less RRAM-based computing-in-memory macro with binary CNN for efficient edge AI," IEEE Trans. Circuits Syst. II, Exp. Briefs, vol. 70, no. 6, pp. 1871-1875, June 2023.

[7] W. Li, S. Huang, X. Sun, H. Jiang and S. Yu, "Secure-RRAM: A 40 nm 16kb compute-in-memory macro with reconfigurability, sparsity control, and embedded security," IEEE Custom Integr. Circuits Conf. (CICC), Austin, TX, USA, 2021, pp. 1-2.

[8] Z. Jing, B. Yan, Y. Yang and R. Huang, "VSDCA: A voltage sensing differential column architecture based on 1T2R RRAM array for computing-in-memory accelerators," IEEE Trans. Circuits Syst. I, Reg. Papers, vol. 69, no. 10, pp. 4028-4041, October 2022.

[9] H. Jiang, S. Huang, W. Li and S. Yu, "ENNA: An efficient neural network accelerator design based on ADC-free compute-in-memory subarrays," IEEE Trans. Circuits Syst. I, Reg. Papers, vol. 70, no. 1, pp. 353-363, January 2023.

[10] L. Wang, Y. Zhang, P. Wang, J. Yang, H. Zhang, G. Li and Q. Li, "A 578-Tops/w RRAM-based binary convolutional neural network macro for tiny AI edge devices," IEEE Trans. Very Large Scale Integr. (VLSI) Syst., vol. 33, no. 2, pp. 371-383, February 2025.

979-8-3315-3918-4/25 $31.00 © 2025 IEEE

ESD Reliability Roadmap Considerations for 3D Heterogeneous Integration Microsystems (Invited)

Zijin Pan, Xunyu Li, Weiquan Hao, Runyu Miao, Zijian Yue and Albert Wang

Dept. of Electrical and Computer Engineering, University of California, Riverside, CA, USA, aw@ece.ucr.edu

Abstract— **Ending of Moore's Law and incoming societal transition from information technology (IT) to internet of everything (IoET) call for 3-dimensional (3D) heterogeneous integration (HI) microsystem chips, featuring systems-on-integrated-chiplets (SoIC) and advanced micro-scale packaging (μ-packaging). Reliability plays a key role in fully unlocking the native, full potentials of heterogeneous components in HI forms, where electrostatic discharge (ESD) is a major reliability problem. This paper provides a roadmap view to highlight emerging challenges (e.g., complex interfaces, life cycle reliability, new ESD phenomena) for and future research directions (e.g., novel low-overhead ESD protection, holistic ESD protection, co-design, co-simulation and artificial intelligence) in ESD protection for 3D SoICs with μ-packaging.**

I. HETEROGENEOUS INTEGRATION

Ending of Moore's Law hits a brake on the otherwise never-ending advances in IC scaling. Yet the demands for higher performance of microsystem chips continue, which plays a vital role in enabling the merging societal transition from the IT age to IoET era. Fortunately, heterogeneous integration seems to open a new pathway towards smart future chips of more functionalities and higher performance at affordable costs [1]. HI technologies hetero-integrate different devices made in dissimilar materials using various technologies at multiple nodes into SoIC microsystem chips to achieve functionality diversity and performance supremacy. It is recognized that reliability must be ensured in order to unlock the full potentials of heterogeneous devices utilizing their native properties in SoIC formats in terms of functionality, performance and product costs. Among various reliability problems, ESD failure is a major challenge in developing 3D HI-based SoIC microsystem chips [2]. From the technology roadmap angle, this paper discusses emerging ESD reliability challenges and highlights potential future research directions in developing robust ESD protection solutions for 3D HI-based SoIC microsystems utilizing advanced μ-packaging.

II. EMERGING ESD CHALLENGES FOR 3D SoICs

ESD phenomena involve transient exchange of electrostatic charges between objects of different electric potentials, producing strong, fast voltage and current pulses, and easily damaging integrated circuits and microsystems [3]. ESD failure is a life-long reliability problem to any microelectronic products occurring both in manufacturing and usage phases.

ESD protection is hence required for all ICs. Many industry ESD test standards exist to model ESD events of different origins and to characterize ESD reliability of products, such as human body model (HBM), machine model (MM), human metal model (HMM) and IEC, etc. [4-7]. These ESD test standards share one common nature that is they model external-oriented, from-external-to-internal ESD discharging events where the ESD pulse (charges) comes externally to an IC chip. Accordingly, a pad-based on-chip ESD protection strategy (Fig. 1) is commonly used to discharge an incident ESD transient into a ground (GND) via a low-resistance ESD protection device at an IC pad, hence avoiding any internal ESD failures to a chip [3]. The main ESD protection design task includes accurate design of ESD-Critical Parameters (e.g., ESD triggering voltage, V_{t1}), compliance of ESD Design Window (Fig. 2, set by supply voltage and breakdown voltage, BV), and minimization of ESD-induced Design Overhead (e.g., parasitic capacitance, C_{ESD}, and ESD device size) [3].

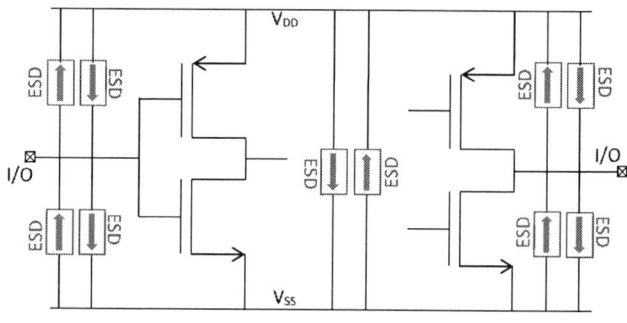

Fig. 1 Full-chip ESD protection scheme.

The main challenge of ESD protection reliability for 3D HI-based SoICs in advanced μ-packaging originates from the heterogeneity of heterogeneous integration microsystem chips, which is the core feature delivered by HI technologies. The pursuit for vast functionalities and ultra performance of smart chips critically leverages different devices made in different materials fabricated by different technologies to deliver the best native device functions. Such heterogeneity complexity means greater diversity in materials, fabrication technologies, chip architectures and functional domains, which leads to very sophisticated interfaces and interactions within SoIC chips,

979-8-3315-3918-4/25 $31.00 © 2025 IEEE

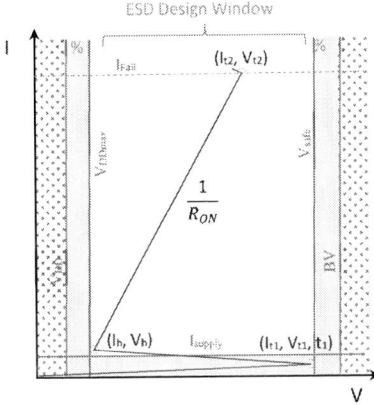

Fig. 2 Concept of ESD-critical parameters and ESD protection design window.

will be very important in handling ESD reliability problems for SoIC chips, particularly, one must recognize that ESD reliability is a problem throughout the entire life cycle of any products and ESD risk is an everyday concern primarily related to end-users. Therefore, while adequate ESD Control in the making phase is important, warranting some reduction in ESD protection targets at advanced technology nodes to accommodate IC performance [11, 12], the real-world ESD risks never decrease in the end-users' hands throughout the entire product life cycle. In fact, ESD reliability is a much bigger problem for SoICs and hence requires more robust ESD protection, because devices made in advanced technologies (e.g., 2nm node) are much more vulnerable to even lower ESD stresses, while the costs of ESD failure of advanced SoIC chips are becoming unbearable.

substantially complicating the ESD reliability problem. While continuous evolving, the major emerging ESD protection challenges for 3D chiplets-based SoIC chips, very different from ESD protection for single-die ICs, are highlighted below [8, 9]: 1) Obviously, vast functional and structural heterogeneity makes ESD reliability extremely challenging for 3D SoICs over their single-chip IC counterparts, this is because ESD charging/discharging involves the whole chip (single-die IC or chiplets-based SoIC); therefore, microsystem-level holistic ESD protection is essential for 3D HI SoIC chips, which must be cross-layer, cross-domain and cross-chiplet in nature. 2) Accordingly, a new holistic ESD protection strategy is highly needed to simultaneously achieve both best chip performance and highest ESD robustness for SoICs, which requires ESD-technology co-development, ESD-IC co-design and ESD-SoIC so-simulation. 3) Die-to-die-package interfaces must be thoroughly considered in SoIC ESD protection design and qualification, including chip-package interactions (CPI) and chip-package-board interactions (CPBI). 4) New modelling and CAD techniques, including algorithms, software tools and design verification procedures, will be essential to developing any good SoIC ESD protection solutions. 5) New ESD phenomena may appear for 3D SoIC chips, which may blur the boundaries of existing industry ESD test standards, e.g., HBM ESD model versus charged device model (CDM) [10]. 6) New ESD testing techniques and standards will be needed to accurately characterize and qualify ESD protection solutions for SoIC chips. 7) It will be very beneficial to think about ESD protection non-traditionally for SoICs, both to explore disruptive ESD protection concepts (e.g., new materials and devices) and to leverage unique features of SoIC/μ-packaging (e.g., utilizing interposers and TSV). 8) Artificial intelligence (AI) can be leveraged to investigate ESD phenomena and to develop robust ESD protection solutions. 9) New industry SoIC ESD compliance standards are needed in order to establish a new supply chain ecosystem (from chiplets to packaging to qualification); for example, to ensure chiplet interoperability for developing new SoIC products. 10) A new strategic mindset

III. FUTURE DIRECTIONS FOR SoIC ESD PROTECTION

The prior discussions on emerging ESD reliability challenges naturally lead to several important future research directions on ESD protection for 3D HI-based SoIC chips as outlined below: First, efforts must be given to develop holistic ESD protection solutions for 3D chiplets-based SoIC microsystem chips as a whole, not simply relying on summation of ESD features of individual dies within a SoIC chip. A good SoIC-scale ESD protection solution should take a global view of this entire packaged chip, i.e., the entire ESD protection network across an SoIC chip, and pay special attention to overall ESD discharging routes, globally and locally. Second, holistic ESD protection requires accurate design of ESD-Critical Parameters and compliance of ESD Design Window at SoIC level (as opposed to single-chip), which calls for new design techniques to estimate SoIC-equivalent ESD-critical parameters, ESD design windows and ESD discharging paths. Novel ESD design concepts are needed to facilitate flexible SoIC construction using chiplets (multi-domains, multi-supplies, multi-technologies and multi-dies) from different vendors using different technologies, e.g., utilizing field-programmable ESD protection [13]. Third, research is needed to thoroughly investigate SoIC heterogeneity-induced complex interfaces and interactions, such as CPI and CPBI interactions within SoICs, which is extremely difficult due to many different factors involved (materials, devices, processes, dies, packages, substrate carriers, boards, etc.). For example, construction of SoICs involve interposers, TSV/TGV, embedded Si bridges, micro bumps, RDL, which may affect ESD discharging paths, hot spots and ESD weak points, etc., which is much more complicated than in single-chip ICs. Fourth, new ESD-SoIC co-design methodology should be developed to balance the design to achieve best ESD protection and SoIC performance simultaneously. For example, die-level ESD protection may be reduced (following strict ESD Control in die-making phases) without affecting SoIC-level ESD robustness by using add-on ESD protection in an interposer. This is particularly important for advanced SoICs featuring extreme data throughputs (e.g., for GPU for AI and autonomous driving) where any ESD-

979-8-3315-3918-4/25 $31.00 © 2025 IEEE

induced parasitic capacitance can seriously affect SoIC performance. ESD innovation is therefore highly needed to balance ESD reliability and chip specifications; for example, field-dispensable ESD concept may be a potential solution to ensure both robust ESD protection and ultra SoIC performance (e.g., for AI data centers) [14]. Fifth, major research is needed to develop SoIC-level modelling and holistic ESD co-simulation techniques to support SoIC ESD protection design optimization, prediction and verification. For example, new models are needed to tackle the complicated physical and thermal boundaries within SoIC that can seriously affect transient thermal boundary conditions, heat flow and dissipation, and hot spot formation. ESD co-simulation requires new CAD algorithms and software to conduct atom-to-system holistic simulation that will involve both numerical and schematic simulation at SoIC/package scale. More importantly, whole-SoIC/package ESD design verification will be required to ensure first-design success in making SoIC chips (avoiding unbearable costs and shortening time-to-market due to design iterations). Such new CAD techniques must be able to auto-extract ESD devices, equivalent ESD-critical parameters, ESD netlists and ESD discharging paths from given SoIC physical design data, and conduct ESD-function-based ESD design verification at SoIC chip level. It is noteworthy that such new whole-SoIC/package ESD CAD frameworks must be physical-design-based, because schematic-only CAD methods (e.g., PERC) is incapable of dealing with the complex physical interfaces and functional interactions inside 3D HI SoICs. Fig. 3 depicts a new ESD CAD framework with desired features to facilitate whole-SoIC/package ESD design verification (throughout chiplets, SoIC and package; and addressing CPI/CPBI interactions) [15]. Sixth, non-traditional ESD protection design concepts should be explored, particularly being suitable for SoIC ESD protection. For example, novel phase-changing materials may be synthesized to make simple resistive ESD switch, as opposed to conventional PN-based active ESD devices, which may be hetero-integrated into SoICs. Seventh, efforts should be given to leverage any unique features for SoIC for novel ESD protection solutions. For example, as shown in Fig. 4, ESD protection (individual ESD devices or entire ESD protection network) may be placed in an interposer in SoICs, which can be naturally suitable for making 3D holistic ESD protection at SoIC level, e.g., a diode or resistive ESD device can be placed inside a TSV/TGV hole in an interposer, both to save ESD device area and to achieve low-R ESD discharging vertically and locally (hence, minimizing overheating). Eighth, research is needed to discover any new ESD phenomena unique to SoICs. For example, it is recently reported that classic pad-pad CDM ESD protection approach may be fundamentally wrong since CDM ESD is an internal-oriented, from-internal-to-external event, entirely different from HBM ESD events. Hence, a new non-pad-based internal-distributed CDM ESD protection technique was proposed [16]. The CDM-type ESD phenomenon may be much more complicated in SoICs for several reasons: 1) Heterogeneity

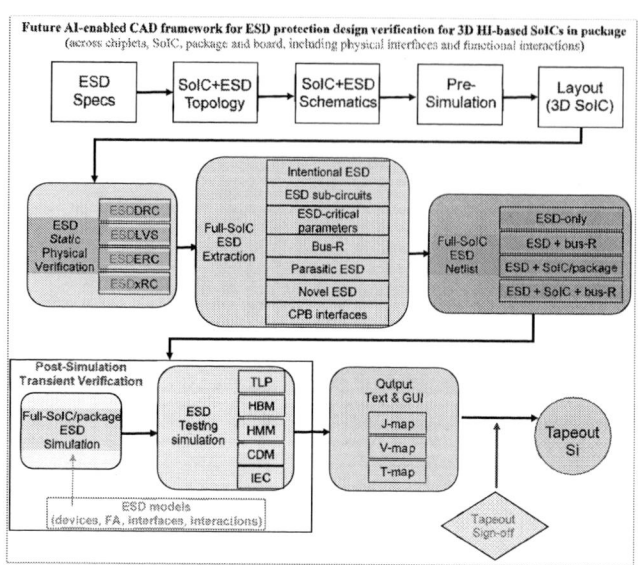

Fig. 3 Proposed AI-enabled whole-SoIC/package ESD CAD framework for holistic SoIC ESD protection design verification covering any physical interfaces and functional interactions.

leads to complexity in charge generation, storage and distribution inside an SoIC chip, 2) SoIC structure makes internal ESD discharge routing much more complicated; both may increase internal ESD vulnerability and ESD failure randomness within SoIC chips. Similar internal-distributed ESD protection may be a potential solution for SoICs, which is however different from single-chip ICs in that, to each individual chiplet, its "CDM" ESD discharging seems to be from-external-to-internal (hence, pad-based ESD protection may be used); however, for the entire SoIC chip, the "CDM" ESD event is from-internal-to-external in nature that cannot rely on pad-based ESD protection. To this point, an internal-distributed CDM-like ESD protection may also be used for SoIC where in-TSV ESD devices can be readily used in an interposer (Fig. 4) [17, 18]. Overall, this seems to be a new hybrid CDM/HBM ESD phenomenon that must be further investigated for SoIC chips. Ninth, AI-for-ESD will be an emerging research topic, particularly for SoICs. On one hand, ESD reliability is very knowledge-centric and data-driven, yet full of unknowns; hence, it will be very beneficial to combine many brains together globally and develop AI/ML algorithms for statistical data analysis, e.g., ESD device modelling by ML [19]. On the other hand, AI can enable otherwise impossible design tasks, for example, allowing smart partitioning of an SoIC chip into pieces to understand random internal charge distribution, internal ESD discharge routing and transient local hot spot formation, etc. ESD-specific large language models (LLM) can be developed that will be very beneficial for practical ESD protection designs addressing data complexity and uncertainty. Further, a global ESD reliability ecosystem, e.g., AI-for-ESD Forum, can serve this purpose ideally [20].

More importantly, combining bottom-up approach (physics-

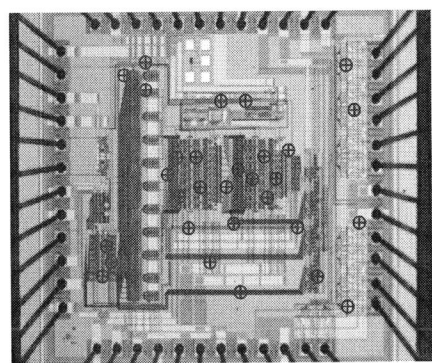

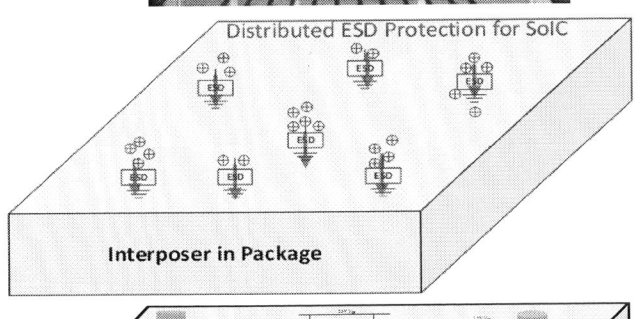

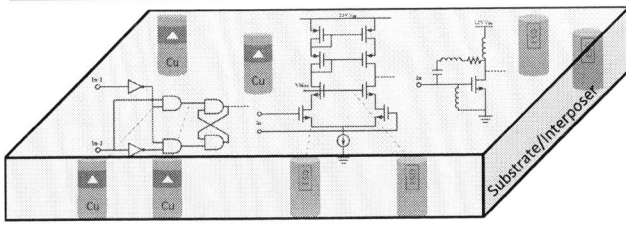

Fig. 4 Internal-distributed CDM-like ESD protection scheme implemented in an interposer featuring in-TSV/TGV ESD switching devices for efficient local ESD discharging for SoICs.

based) and top-down approach (data-driven, AI-enabled) will lead to more powerful and efficient ESD protection design methodologies for complex SoICs. Tenth, research is needed to develop new ESD testing standards for 3D HI-based SoICs facilitating meaningful and accurate characterization of ESD protection designs within SoICs, e.g., stress testing for hybrid HBM/CDM ESD phenomena and identifying transient hot spots throughout a packaged SoIC chip. Eleventh, industry ESD qualification and compliance standards are desired in production, which supports a smooth global supply chain where chiplet interoperability is highly needed for making new SoIC products in terms of time-to-market and costs.

IV. CONCLUSIONS

Heterogeneous integration opens a pathway towards smart future SoIC chips comprising different devices made in dissimilar materials using various technologies to achieve rich functionality and high performance at low costs. Reliability is critical to unlocking full potentials of heterogeneous devices and chiplets in their native forms to deliver smart SoIC microsystem chips. ESD reliability is an emerging challenge for smart future SoICs. This paper discusses key merging ESD reliability challenges and outlines important future research directions on ESD protection for SoICs in advanced packages, offering a roadmap view in the ESD-SoIC field.

REFERENCES

[1] *HIR: Heterogeneous Integration Roadmap*, IEEE Electronics Packaging Society, https://eps.ieee.org/technology/heterogeneous-integration-roadmap/hir-specialsessions-at-estc-2024.html.

[2] Z. Pan, X. Li, W. Hao, R. Miao, Z. Yue and A. Wang, "Challenges: ESD Protection for Heterogeneously Integrated SoICs in Advanced Packaging", *Electronics 13, No. 12: 2341*, 2024.

[3] A. Wang, Practical ESD Protection Design; Wiley-IEEE Press: New York, NY, USA, 2022; ISBN 978-1119850403.

[4] ANSI/ESDA/JEDEC JS-001-2017; For Electrostatic Discharge Sensitivity Testing—Human Body Model (HBM)—Component Level, The ESD Association; JEDEC, 2017.

[5] ESD STM5.2-2012, Electrostatic Discharge Sensitivity Testing: Machine Model - Component Level, the ESD Association, 2012.

[6] ESD SP5.6-2019, ESD Association Standard Practice for Electrostatic Discharge Sensitivity Testing – Human Metal Model (HMM) Component Level, the ESD Association, 2019.

[7] IEC 61000-4-2, "Electromagnetic Compatibility, Part 4: Testing and Measurement Techniques, Section 2: Electrostatic Dis-charge Immunity Test", the International Electrotechnical Com-mission (IEC), 2008.

[8] X. Li, Z. Pan, W. Hao, R. Miao, Z. Yue and A. Wang, "Interposer-Based ESD Protection: A Potential Solution for □μ-Packaging Reliability of 3D Chips", *Micromachines, 16(4)*, pp. 488, 2025.

[9] Z. Pan, X. Li, W. Hao, R. Miao, Z. Yue and A. Wang, "Challenges: ESD Protection for Heterogeneously Integrated SoICs in Advanced Packaging", *Electronics 13, No. 12: 2341*, 2024.

[10] ANSI/ESDA/JEDEC JS-002-2018; For Electrostatic Discharge Sensitivity Testing—Charged Device Model (CDM)—Device Level, the ESD Association; JEDEC, 2018.

[11] Industry Council on ESD Target Levels. White Paper 1: A Case for Lowering Component Level HBM ESD Specifications and Requirements, Industry Council on ESD Target Levels, Rev. 4.0; Industry Council on ESD Target Levels: Online, 2018.

[12] Industry Council on ESD Target Levels. White Paper 2: A Case for Lowering Component-Level CDM ESD Specifications and Requirements, Industry Council on ESD Target Levels, Rev. 3.0; Industry Council on ESD Target Levels: Online, 2021.

[13] Z. Shi, X. Wang, J. Liu, L. Lin, H. Zhao, Q. Fang, L. Wang, C. Zhang, S. Fan, H. Tang, B. Li, A. Wang, J. Liu and Y. Cheng, "Programmable on-Chip ESD Protection Using Nano Crystal Dots Mechanism and Structures", IEEE Trans. Nanotechnology, Vol. 11, Issue 5, pp. 884-889, September 2012.

[14] C. Zhang, Z. Dong, F. Lu, R. Ma, L. Wang, H. Zhao, X. Wang, X. S. Wang, H. Tang and A. Wang, "Fuse-Based Field-Dispensable ESD Protection for Ultra-High-Speed ICs", IEEE Electron Device Letters. Vol. 35, No. 3, pp.381-383, March 2014.

[15] Z. Pan, X. Li, W. Hao, R. Miao and A. Wang, "On-Chip ESD Protection Design Methodologies by CAD Simulation", ACM Transactions on Design Automation of Electronic Systems, (TODAES), Vol. 29, Issue 1, Article No. 4, pp. 1-41, November 2023.

[16] M. Di, C. Li, Z. Pan and A. Wang, "Non-Pad-Based in Situ in-Operando CDM ESD Protection Using Internally Distributed Network", *IEEE J-EDS*, pp. 1248, 2021.

[17] A. Wang, "ESD protection structures and local grounding using through-silicon-vias (TSV) for ICs", U. S. Patent App. No. 62/385,770, 2016; UC Case No. 2016-973, 2015.

[18] A. Wang, "Interposer-based ESD Protection Structures", U.S. Patent App. No. 62/412,105, 2016; UC Case No. 2016-985-1, 2015.

[19] W. Liang, X. Yang, M. Miao, A. Loiseau, S. Mitra, and R. Gauthier, "Novel ESD compact modeling methodology using machine learning techniques for snapback and non-snapback ESD devices," IEEE Trans. Device and Materials Reliability, vol. 21, no. 4, pp. 455–464, Dec. 2021.

[20] W. Hao, Z. Pan, X. Li, R. Miao, Z. Yue and A. Wang, "On-Chip ESD Protection: Design Innovation", IEEE Electron Devices Reviews, Vol. 2, pp. 32-51, December 2024.

Tiny Chiplets Enabled by Packaging Scaling: Opportunities in ESD Protection and Signal Integrity

Emad Haque*[1], Pragnya Sudershan Nalla[2], Jeff Zhang[1], Sachin S. Sapatnekar[2],
Chaitali Chakrabarti[1] and Yu Cao*[2]

[1] School of Electrical, Computer and Energy Engineering, Arizona State University, Tempe, AZ 85282, USA
[2] Department of Electrical and Computer Engineering, University of Minnesota, Minneapolis, MN 55455, USA
* Email: ehaque5@asu.edu, yucao@umn.edu

Abstract—The scaling of advanced packaging technologies provides abundant interconnection resources for 2.5D/3D heterogeneous integration (HI), thereby enabling the construction of larger-scale VLSI systems with higher energy efficiency in data movement. However, conventional I/O circuitry, including electrostatic discharge (ESD) protection and signaling, introduces significant area overhead. Prior studies have identified this overhead as a major constraint in reducing chiplet size below 100 mm². In this study, we revisit reliability requirements from the perspective of chiplet interface design. Through parasitic extraction and SPICE simulations, we demonstrate that ESD protection and inter-chiplet signaling can be substantially simplified in future 2.5D/3D packaging technologies. Such simplification, in turn, paves the road for further chiplet miniaturization and improves the composability and reusability of tiny chiplets.

Keywords—Advanced packaging, Heterogeneous integration, ESD, Signal integrity, Chiplet

I. INTRODUCTION

Heterogeneous integration (HI) systems, which leverage 2.5D and 3D packaging technologies to integrate multiple chiplets on an advanced packaging substrate, have emerged as a key enabler in modern VLSI design [1, 2]. These systems offer high flexibility, scalability, and high energy efficiency, particularly for data-intensive workloads such as high-performance computing, autonomous vehicles and AI tasks. By disaggregating a large monolithic design into smaller chiplets and interconnecting them through 2.5D and 3D integration, these systems effectively reduce design and fabrication costs, improve overall yield, provide higher bandwidth and lower energy consumption in data movement, and achieve better system reconfigurability [3, 4].

However, previous studies have identified the I/O interface of chiplets, such as electrostatic discharge (ESD) protection, clock and data synchronization, and related area cost as a key limitation in scaling chiplet sizes below 100 mm² [5]. For instance, an implementation of the Advanced Interface Bus (AIB) [6], a widely adopted I/O module for 2.5D integration, occupies several mm² at 22nm, larger than many design IP blocks (such as CPUs, DSPs, FFT accelerators, and systolic arrays) [7]. These overheads restrict the reusability and composability of chiplets in heterogeneous integration [7].

In this study, we revisit the reliability requirements of chiplet interfaces, with a focus on advanced packaging technologies. As packaging continues to scale, with finer pitch, shorter inter-chiplet spacing, and lower electrical parasitics, we explore how these advancements can significantly reduce the overhead of ESD protection and inter-chiplet signaling, thereby eliminating conventional I/O bottlenecks and enabling future scaling of chiplet sizes (i.e., tiny chiplets).

II. SCALING OF 2.5D/3D PACKAGING

A heterogeneous system typically comprises three primary components [1, 2]: the chiplets, which serve as the functional units for computing, memory, control and other tasks; the interconnects, either horizontal or vertical, that connect the chiplets together to form a complete system; and the substrate, which can be silicon-, organic- or glass-based, providing the foundation for hosting both the interconnects and chiplets. Figure 1 illustrates such a heterogeneous system with multiple stacks of 2.5D and 3D chiplets on a common substrate, highlighting key features and parameters of wires between chiplets. Figure 2 presents a more detailed view of an individual 2.5D chiplet, with µbump arrays on each side for delivering signals, clocks, and power supply.

As packaging technologies advance, several physical features of HI systems continue to scale down. Table I outlines important geometric parameters related to the chiplet size and interconnect dimensions. Based on HI roadmaps [1, 2], Tables II and III summarize the scaling trends of µbumps and hybrid bonds. These two structures form the critical interface between chiplets and the substrate or between chiplets themselves. Their dimensions directly influence the electrical properties relevant to ESD and signal integrity analysis.

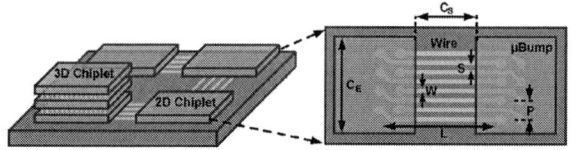

Fig. 1. Overview of a 2.5D/3D heterogeneous system.

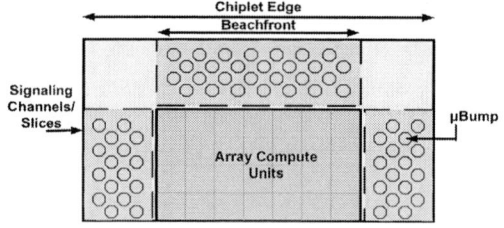

Fig. 2. Top view of a 2.5D chiplet with compute units and µbump arrays as the inter-chiplet interface [1, 2].

TABLE I. PHYSICAL GEOMETRY IN A HI SYSTEM.

Parameter	Description	Parameter	Description
C_E	Chiplet edge length	S	Wire spacing
C_S	Chiplet spacing	H	Wire height
L	Wire length	T	Wire thickness
W	Wire width	P	µbump pitch

979-8-3315-3918-4/25 $31.00 © 2025 IEEE

TABLE II. SCALING OF μBUMP TECHNOLOGIES [1, 2].

Generation	0	1	2	3	4	5
L (mm)	4	2	1.6	0.9	0.4	0.15
W = S (μm)	2.5	2	1.5	1	0.5	0.25
T = H (μm)	5	4	3	2	1	0.5
P (μm)	70	55	40	30	20	10

TABLE III. SCALING OF HYBRID BONDING TECHNOLOGIES [1, 2].

Generation	0	1	2	3	4
L (μm)	150	100	75	50	25
W = S (μm)	0.25	0.20	0.15	0.10	0.05
T = H (μm)	0.5	0.4	0.3	0.2	.0.1
P (μm)	10	5	2.5	1	0.5

III. ELECTROSTATIC DISCHARGE PROTECTION

One of the key cost factors in chiplet-based design is electrostatic discharge protection for the I/O interface [5, 8]. ESD pose a critical concern for the reliability of on-chip transistors, and its adverse impact becomes even more pronounced as CMOS technology continues to scale. At the chiplet level, only the interfaces, such as μbumps and hybrid bonds, are exposed to potential ESD from the external environment and therefore, require protection.

ESD protection typically involves large diodes within I/O cells that clamp the ESD voltage below the damage threshold. However, this approach inevitably introduces significant area overhead and additional capacitive load on the signal channel, thereby limiting the usable chip area and reducing the data rate of inter-chiplet communication. Given the controlled clean room environment in advanced packaging, the JEDEC roadmap recommends reducing ESD protection targets from 250V today to 125V for scaled packaging technologies, and even down to as low as 5V for chiplet I/Os using hybrid bonding in 2.5D/3D systems [8].

To meet the JEDEC requirements of ESD protection, it is essential to determine the minimum diode size, which is affected by many factors, such as the parasitic resistance (R), inductance (L), and capacitance (C) along the I/O path. In this study, we use the circuit schematic for ESD validation, as shown in Fig. 3 [9, 10]. We perform SPICE simulations to evaluate the diode size needed for ESD protection under the Charged Device Model (CDM) across generations. This generic schematic includes both the interconnect component and the pad structure, such as μbumps and hybrid bonds used in 2.5D/3D integration. Based on the dimensions listed in Table II, we adopt compact models to calculate the corresponding RLC parameters for each generation [11]. As an example, Table IV summarizes the parameters for μbumps.

Using the values in Table IV, we simulate the gate voltage waveforms with μbumps to search for the required diode size. It is determined by applying the target voltage to the package capacitance and checking whether the gate voltage is below the gate oxide breakdown voltage. Table V lists the minimum size necessary to keep the gate voltage below the gate-oxide breakdown voltage (3.8V for the 28 nm process node). Figure 4 presents the gate voltage waveforms at the minimum diode

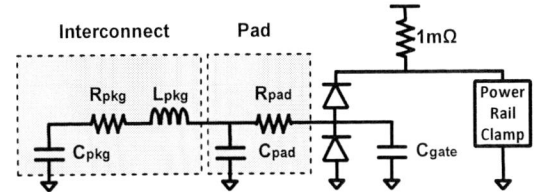

Fig. 3. SPICE simulation model to evaluate the required ESD diode size for CDM protection [9, 10].

TABLE IV. PARAMETERS FOR μBUMPS IN SPICE SIMULATIONS.

Parameter	C_{pkg} (fF)	R_{pkg} (Ω)	L_{pkg} (nH)	C_{pad} (fF)	R_{pad} (mΩ)
Gen 0	1141.04	7.040	5.978	5.911	4.574
Gen 1	423.11	11.00	2.801	4.645	5.822
Gen 2	427.89	7.333	2.262	3.378	8.004
Gen 3	233.26	9.899	1.242	2.533	10.673
Gen 4	103.67	17.599	0.542	1.689	16.009
Gen 5	38.87	26.400	0.195	0.844	32.018

size, confirming that all voltages remain below the threshold. As observed in Table V, the diode area of μbumps decreases across generations, because dimension leads to a lower impedance path, thereby improving the protection for downstream transistors. Nevertheless, for μbump technologies, a significant area must still be allocated for the diode to drive the interconnect in 2.5D integration.

The situation improves substantially for hybrid bonding with shorter 2.5D inter-chiplet spacing. The RLC parameters are calculated from the dimensions in Table III. Due to the much smaller feature sizes here as compared to μbumps, the need for diodes is eliminated. Figure 5 presents the voltage waveforms for a 10V ESD event (above the 5V JEDEC threshold). Even without any ESD protection diodes, the voltage remains below 3.8V. This indicates that 2.5D tiny chiplets with hybrid bonding can eliminate the expensive on-chiplet ESD protection.

TABLE V. REQUIRED ESD DIODE AREAS AT 28NM WITH μBUMPS.

ESD Target Voltage (V)	Total Required Diode Area (μm^2)					
	Gen 0	Gen 1	Gen 2	Gen 3	Gen 4	Gen 5
10	6.46	6.11	6.38	6.15	6.15	6.02
30	20.4	18.2	20.1	18.7	18.7	17.9
50	34.3	30.5	34.2	31.2	31.2	29.8
125	51.8	46.6	51.6	47.9	47.9	45.4

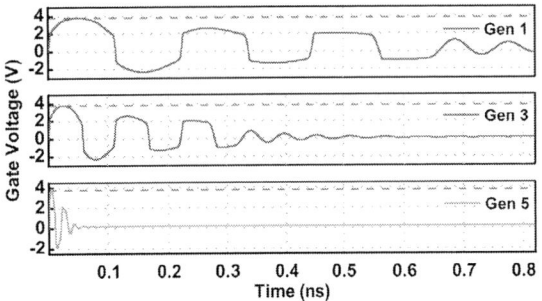

Fig. 4. Simulated gate voltage for a 125V ESD event for μbumps at Gen 1, Gen 3, and Gen 5. Gate-oxide breakdown voltage is marked in red.

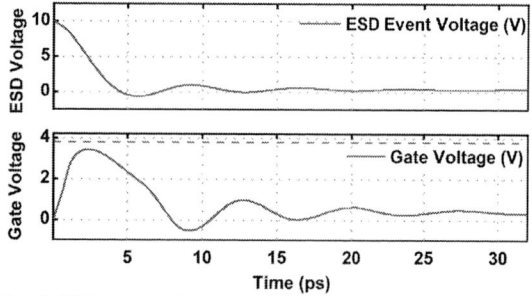

Fig. 5. ESD event voltage and gate voltage for hybrid bonding at Generation 4, without using any ESD diode for protection. Gate-oxide breakdown voltage is marked in red.

IV. INTER-CHIPLET SIGNALING

For 2.5D inter-chiplet communication, multiple I/O protocol standards have been proposed, including AIB [6], Universal Chiplet Interconnect Express (UCIe) [12] and Bunch of Wires (BoW) [13]. These protocols typically need to drive interconnects spanning several millimeters, as shown in Table II. As a result, they are often adapted from conventional high-speed I/O designs, such as DDR and PCIe, and then customized for the specific packaging wire load. Like DDR and PCIe, these modules are implemented as IP blocks integrated at the chiplet edges, but they still involve considerable design complexity and area cost.

In this study, we explore direct signaling link (DSL), a much simpler I/O approach that resembles multiple parallel data buses within a chiplet. It is inspired by the observation that inter-chiplet spacing is approaching the typical length of on-chip interconnects. With simplified transceivers and receivers, essentially on-chip buffers, DSL offers a lightweight signaling structure that is particularly appealing for tiny chiplets. Figure 6 presents the DSL schematic used in our SPICE simulations. The CMOS driver is determined by the required bandwidth and channel parasitics to ensure signal integrity. The receiver is the same size as the driver.

To assess the signal integrity of DSL in inter-chiplet communication, we examine eye diagrams under various configurations for a fixed driver size. Figure 7 presents the eye diagrams at 1GHz for four generations of μbumps. As packaging technology scales, our analysis shows continuous improvement in signal quality, making DSL increasingly viable. Figure 8 further studies eye diagrams for Generation 5 with varying channel length. As spacing decreases, DSL demonstrates better quality in signal transmission. Therefore, in advanced packaging where inter-chiplet distances for adjacent chiplets decrease to hundreds of micrometers, DSL emerges as a suitable I/O solution for tiny chiplets. However, for longer channel lengths in the order of millimeters, special

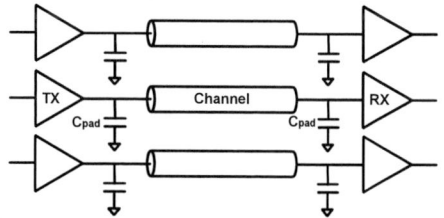

Fig. 6. SPICE models to simulate Direct Signaling Links, with three parallel channels shown as an example. The two outer channels act as aggressors to evaluate crosstalk noise. (TX: transceiver; RX: receiver)

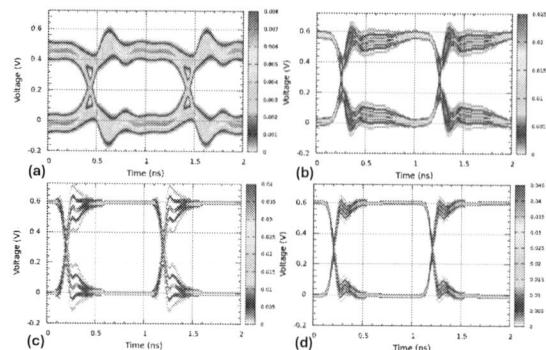

Fig. 7. Eye diagrams for DSL for a fixed driver size, for (a) Gen 0, (b) Gen 1, (c) Gen 3, (d) Gen 5.

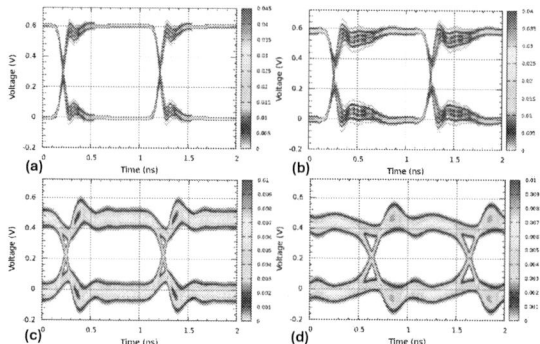

Fig. 8. Eye diagrams of DSL for packaging Gen 5, with channel length of (a) 150 μm, (b) 750 μm, (c) 2 mm, and (d) 4 mm.

protocols, such as AIB or UCIe, remain necessary. Note that when DSL is used for signaling across a short distance, clock synchronization can also be relaxed by leveraging globally asynchronous locally synchronous (GALS) architectures or similar techniques.

V. IMPACT ON TINY CHIPLETS

As highlighted in previous studies, the cost of ESD and other circuitry in I/O cells has long been the key barrier to chiplet scaling [1, 2, 5]. This limitation further restricts the flexibility and composable of chiplets, as reusable, low-cost hardened IPs across a broad range of applications.

Based on our simulations in Sections III and IV, we demonstrate that such I/O costs are expected to decrease significantly as packaging technologies continue to scale down, with shorter and narrower wires and smaller interface pads. For μbumps, Fig. 9(a) summarizes the area of ESD protection diode per pad. Use of hybrid bonding along with new I/O structures and cleaner assembly environments, helps eliminate the need for on-chip ESD protection, dramatically simplifying I/O design and planning.

Furthermore, Fig. 9(b) compares the total area of I/O bump arrays for both AIB and the proposed DSL scheme. At shorter distances, DSL not only ensures the integrity in inter-chiplet signaling, but also reduces the area cost by more than 8× compared to AIB. At longer distances, the special I/O protocols are necessary for signal integrity. The AIB I/O array area saturates above a 4 mm edge length due to its architectural limit of 24 channels per column. With a fixed bump pitch (10 μm in this study), this channel limit becomes the dominant constraint, and increasing the edge length no

979-8-3315-3918-4/25 $31.00 © 2025 IEEE

longer improves I/O density. Overall, DSL delivers a compact and cost-effective I/O solution than current 2.5D I/O modules for future tiny chiplets.

2.5D/3D advanced packaging offers a key advantage over monolithic design by improving I/O data rates. This improvement stems from the increased number of pads and the use of narrower wires to boost channel density. For both AIB and DSL, the implementations are aligned along the chiplet edge, which inherently limits the number of available pads and the achievable data bandwidth. In this scenario, the I/O bandwidth scales linearly with the chiplet edge length, while the computing capability and its associated data demands increase quadratically with chiplet size. This mismatch implies that as chiplet size increases, performance will eventually be limited by I/O bandwidth.

To illustrate this, we set up the general chiplet structure, following Fig. 2, consisting of a systolic array of multiply-accumulate (MAC) units for compute and µbump arrays for inter-chiplet signaling [14]. Based on the number of available channels and compute units, we derive both the maximum supported bandwidth, as well as the data rate required to fully utilize the compute resources. Figure 10(a) shows that for an older technology node, both AIB and DSL can meet the data demands of compute units up to a chiplet size of approximately 10 mm × 10 mm (i.e., 100 mm²). However, for more advanced nodes in Fig. 10(b), AIB increasingly struggles to meet the data requirements due to its larger area overhead. In contrast, the more compact design of DSL continues to support higher data bandwidths, enabling tiny chiplets down to 2 mm × 2 mm (i.e., 4 mm²). This confirms our earlier observations on the benefits of chiplet scaling.

VI. CONCLUSION

With lower packaging parasitics, the need of ESD protection is significantly reduced or even eliminated in future HI systems. High-speed inter-chiplet signaling is further supported by simpler circuits, such as direct signaling links, than 2.5D I/O modules today (AIB, UCIe and BoW). Such a trend enables the scaling of tiny chiplets down to 4 mm², leading to better composability and IP reusability.

ACKNOWLEDGMENT

This work is supported in part by COCOSYS, one of six centers in JUMP 2.0, a Semiconductor Research Corporation

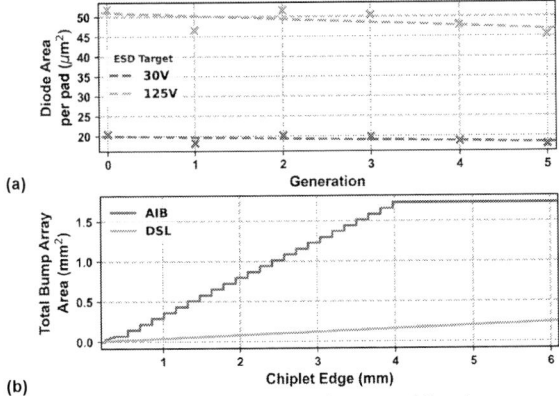

Fig. 9. Area overhead in (a) ESD protection per pad for µbumps across generations; and (b) total bump array of AIB and DSL vs. chiplet edge size.

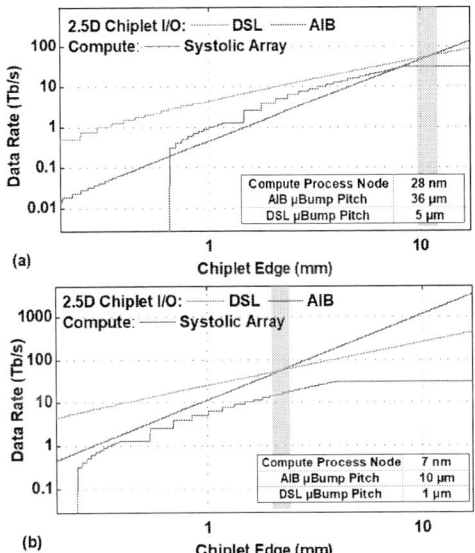

Fig. 10. Effect of chiplet size on the maximum achievable data rate for AIB and DSL interfaces per chiplet. DSL is able to support the high data rate required for smaller chiplet sizes (< 4 mm²) through packaging scaling. Array compute is the theoretical maximum data rate when all array units in the chiplet transfer data in parallel.

(SRC) program sponsored by DARPA. This work is also partially supported by National Science Foundation under grant CCF-2403408 and CCF-2403409.

REFERENCES

[1] "Heterogeneous Integration Roadmap (HIR)," IEEE Electronics Packaging Society, 2024.

[2] "Microelectronics and Advanced Packaging Technologies Roadmap (MAPT)," Semiconductor Research Corporation, 2025.

[3] S. Naffziger, K. Lepak, M. Paraschou, M. Subramony, "AMD chiplet architecture for high-performance server and desktop products," ISSCC, pp. 44-45, 2020.

[4] S. R. Srinivasa, et al., "A 300MB SRAM, 20Tb/s bandwidth scalable heterogeneous 2.5D system inferencing simultaneous streams across 20 chiplets with workload-dependent configurations," ISSCC, pp. 50-51, 2025.

[5] A. Graening, S. Pal, P. Gupta, "Chiplets: How small is too small?" pp. 1-6, DAC, 2023.

[6] "Advanced Interface Bus Specification," Intel, 2019.

[7] P. S. Nalla, E. Haque, Y. Liu. S. Sapatnekar, J. Zhang, C. Chakrabarti, Y. Cao, "CLAIRE: Composable chiplet libraries for AI inference," DATE, pp. 1-7, 2025.

[8] "JEDEC – JEP157A: Recommended ESE-CDM target levels," JEDEC, 2022.

[9] "An introduction to IBIS (I/O Buffer Information Specification) modeling," Texas Instruments, 1998.

[10] "Elements of an IBIS model," Intel, 2003.

[11] W. Zhao, Y. Cao, "New generation of predictive technology model for sub-45nm early design exploration," IEEE TED, vol. 53, no. 11, pp. 2816-2823, November 2006.

[12] D. D. Sharma, G. Pasdast, Z. Qian, K. Aygun, "Universal Chiplet Interconnect Express (UCIe): An open industry standard for innovations with chiplets at package level," IEEE TCPMTB, vol. 12, no. 9, pp 123-1431, September 2022.

[13] S. Ardalan, R. Farjadrad, M. Kuemerle, K. Poulton, S. Subramaniam, B. Vinnakota, "An open inter-chiplet communication link: Bunch of Wires (BoW)," IEEE Micro, vol. 41, no. 1, pp. 54-60, 2021.

[14] Z. Wang, et al., "HISIM: Analytical performance modeling and design exploration of 2.5D/3D heterogeneous integration for AI computing," IEEE TCAD, 2025.

Time-dependent Dielectric Breakdown in Advanced MOSFET: From Theoretical Models to Experimental Findings

Chu Yan [1], GuoQiXin Huang [3], Yiming Qu [*2], Yi Zhao [2]

[1] College of Information Science and Electronic Engineering, Zhejiang University, Hangzhou 310027, China
[2] Research Center of Integrated Circuits, Huada Semiconductor, Shanghai, 201210, China
[3] School of Integrated Circuits, East China Normal University, Shanghai 200241, China

* Email: quym@hdsu.com.cn

Abstract—In this article, we present a comprehensive summary of the current understanding and experimental observations of time-dependent dielectric breakdown (TDDB) for the gate dielectrics in advanced MOSFETs. The evolution and fundamental principles of breakdown statistics and the percolation model are critically reviewed. A detailed exploration of trap generation mechanisms and the voltage dependency model of TDDB is summarized, offering insights into the underlying physics of dielectric degradation. Furthermore, the latest advancements in AC TDDB research are highlighted, with a focus on its frequency-dependent behavior and its implications for device reliability. This review aims to bridge the gap between theoretical models and experimental findings, providing a holistic perspective on TDDB in modern MOSFETs.

Keywords—Time-dependent dielectric breakdown (TDDB), dielectrics, high-k, percolation model

I. INTRODUCTION

As the dimensions of modern MOSFET aggressively scale down, ultra-thin gate dielectric layers are required to maintain gate control capability. However, the ultra-thin gate dielectric layers not only lead to the problem of increased gate leakage, but also make the time-dependent dielectric breakdown (TDDB) one of the most serious reliability concerns [1]. TDDB refers to gradual degradation and eventual failure of the gate dielectric under prolonged electrical stress, which is primarily driven by the formation of a conductive path through trap generation within the dielectric bulk. Since the occurrence of breakdown (BD) is intrinsically linked to the random generation of defects within the dielectric, the evaluation and description of TDDB require statistical methods and models. Although BD statistics of conventional devices has been well described and predicted by percolation model [1-5], its applicability has gradually decreased with the emergence of advanced devices such as high-*k* metal-gate (HKMG) transistor [7]. In addition, the specific defect-generating behavior is still under debate. More recently, AC TDDB, which mimics actual circuit operating conditions, has gradually become the mainstream of TDDB research [7-13]. The physical mechanism of TDDB frequency dependence has been extensively studied.

In this article, the fundamentals and development of breakdown statistics and percolation model are reviewed in section II. The trap generation related acceleration models are

This work was supported by the National Key Research and Development Program of China (2020AAA0109001), the Chenguang Program of Shanghai Education Development Foundation and Shanghai Municipal Education Commission (23CGA35) and the ECNU/HDSC Integrated Circuit Engineering Technology Joint Laboratory.

summarized in section III. Additionally, section IV presents the state-of-art methodological and physical pictures related to AC TDDB research. Finally, the summary of the article is presented in section V.

II. BREAKDOWN STATISTICS AND PERCOLATION MODEL

The breakdown of the gate dielectric layer is a stochastic process, and therefore standard reliability assessment methods are based on the statistics of the time-to-failure (TTF) and the physical properties of the defect generation. The Weibull distribution has been shown to be the statistical function that most appropriately describes the stochastic features of the TDDB in the form of cumulative density function given by (1)

$$F(t) = 1 - \exp\left[-(t/T_{63})^{\beta}\right] \quad (1)$$

$$W_{BD}(t) = \ln\left[1 - \ln\left(F(t)\right)\right] = \beta\left[\ln(t) - \ln(T_{63})\right] \quad (2)$$

where t the statistical variable (TTF for TDDB), T_{63} is the scale factor and β is the shape factor (also known as Weibull slope). This function is commonly shown in Weibull plot as shown in (2), where W_{BD} is known as Weibit.

At the early stage of TDDB research, experimental data showed that the Weibull distribution was the most suitable statistical function for dielectrics BD [15], but the physical mechanism behind it was not clear until the proposal of percolation theory [2]. The core argument of percolation theory is that defects can be randomly generated in dielectric

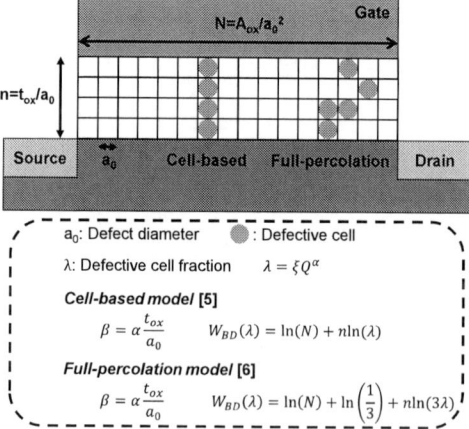

Fig. 1 Schematic and equations of the current cell-based 2-D percolation model for the dielectric layer. Defects are assumed to be randomly generated in the dielectric layer and λ is used to represent the fraction of defective cells. A measurable stress variable (Q) is usually related to λ to deal with experimental data [22].

under stress and these defects accumulate and form a percolation path, which leads to BD, as shown in Fig. 1. Based on this theory, the first statistical TDDB model for the distribution of TTF was proposed by Suñé in 1990 [2], which is widely known as percolation model. Subsequently, a mathematical framework was developed [3], [16] and an analytical cell-based model was further proposed [5], which derives the relationship between β and the gate dielectric thickness in a simple form. Shortly thereafter, the cell-based model was extended to the full-percolation model by accounting for the nearest neighbor cells [6].

As shown in Fig. 1, current percolation models for TDDB predominantly assume the oxide defect generation to be random and uniform across the channel region. However, due to the introduction of the HKMG devices, the mechanism of TDDB becomes more complicated as compared to the conventional SiO$_2$ devices [17]. One of the most significant differences is a nontraditional TTF distribution with the bimodal Weibull shape parameter which has been experimentally observed [18], as show in Fig. 2(a). The authors of [7] state that the bimodal Weibull slope is attributed to the different rates of defect generation in the high-k (HK) and interfacial (IL) layers and it is verified by kinetic Monte-Carlo (kMC) simulations as shown in Fig. 2(b). Another explanation for the bimodal Weibull slope is the higher defect generation rate in grain boundary (GB) of the HK layer [19]. Both explanations suggest that there are multiple defects generation rates in the gate stack of HKMG. Under these circumstances, the conventional analytic percolation is thought to be invalid. Great efforts have been made by

researchers to modify the analytic percolation model [7], [19-21]. However, such modulations are usually specialized for certain cases and not universal. More recently, a matrix version of the analytic percolation model has been proposed [22], which describes the various defect generation rates in the gate dielectric through a matrix.

III. DEFECT GENERATION AND VOLTAGE DEPENDENCE

Since the characterization of TDDB is performed under accelerated voltage conditions, understanding the voltage dependence of TTF is naturally a key focus in evaluating device reliability. Several defect generation based voltage dependence models have been proposed, including the Thermo-Chemical (TC) model [23], [24], the Anode Hole Injection (AHI) model [25], [26], the Anode Hydrogen Release (AHR) Model [27], [28], and combined models [29] (Fig. 3). Significant differences exist in TDDB lifetime predictions using different acceleration models on the same data, making model selection controversial in academia. This section briefly reviews the current acceleration models, their underlying principles, and application scope in the TDDB research field.

The TC model is mainly based on the thermal chemical process of the chemical bond breaking in the gate dielectric layer. It connects the externally applied electric field E_{ox} to the energy required to break the chemical bonds ΔH. Under stress voltage, the positions of atoms will shift relatively, causing changes in the angles between chemical bonds and thereby affecting the energy required for the breaking of chemical bonds. Based on this, the time to breakdown T_{BD} is modeled by (3), where ΔH_0 is the energy required to break the chemical bonds without an external field, k_B is the Boltzmann constant, T is the temperature, α is acceleration factor of the electric field, and A_0 is a constant. This model provides physical explanations for some TDDB experiments, but some studies have shown its inaccuracy under low stress voltage. And it fails to offer a reasonable explanation for the relationship of TTF as well as stress polarity.

$$T_{BD} = A_0 \exp\left[\frac{\Delta H_0}{k_B T} - \alpha E_{ox}\right] \quad (3)$$

The AHI model is the first proposed current-driven acceleration model. Under stress voltages, electrons tunneling from the cathode to the anode will undergo collisional ionization and then excite electron-hole pairs, where electrons flow into the anode while holes tunnel back into the dielectric layer, causing damage to it. This model in (4) is as a function positively correlated with $1/E$ and is essentially electric-driven, where B is a temperature-dependent constant, and γ is

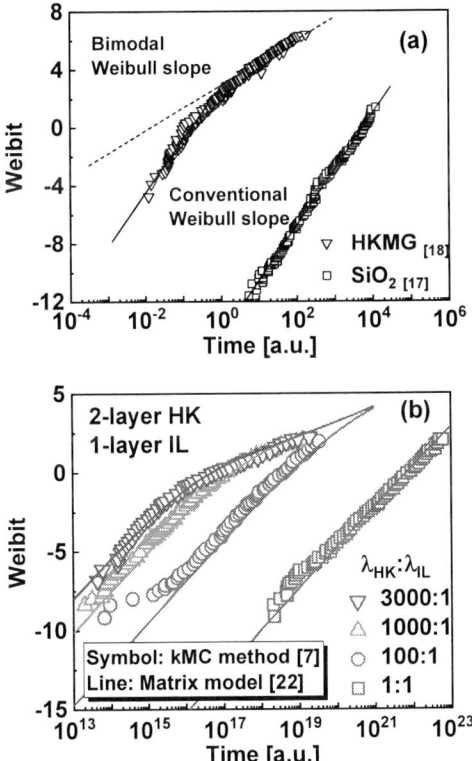

Fig. 2 (a) Comparison of the conventional Weibull slope in SiO$_2$ devices [17] and the bimodal Weibull slope in HKMG devices [18]. (b) Weibull distribution with different ratios of HK and IL defect generation rates simulated by kMC method [7] and matrix percolation model [22].

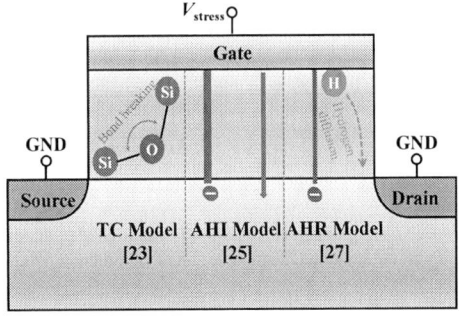

Fig. 3 Schematic diagram of mechanism for the different TDDB models.

acceleration factor. A more obvious problem of is that it has limitations for ultra-thin gate dielectric layers.

$$T_{BD} = B \exp\left[\frac{\gamma}{E_{ox}}\right] \quad (3)$$

The AHR model depicts the TDDB mechanism as hydrogen release from the Si-H bond at the interface of the channel and gate stack and focuses on the defect generation. With the concept of critical breakdown defect density N_{BD} in the percolation model, the defect generation efficiency ξ is established. Moreover, the gate current is included and described as gate tunneling current $J(V_G, t_{ox})$, which is determined by the stress voltage V_G and oxide thickness t_{ox}. And this model of (5) has also been verified by a large amount of experimental data, especially for ultra-thin gate dielectric layers.

$$T_{BD} = \frac{q t_{ox} N_{BD}}{J(V_G, t_{ox}) \xi(V_G)} \quad (5)$$

IV. FREQUENCY DEPENDENCE OF TDDB

With the rapid development of devices, the concept of circuit-like reliability characterization methodologies has evolved in tandem. AC test is one of the industry recognized circuit-like reliability characterization methodologies and has been documented in JEDEC [30]. For advanced MOSFET devices, their operation frequency typically exceeds GHz, and the stress voltage required for TDDB testing is relatively high. Consequently, applying stress voltage accurately to devices at such frequencies poses a significant challenge. Table I summarizes the AC stress provider and the corresponding upper frequency limits used in recent AC TDDB studies.

TABLE I.   AC STRESS PROVIDER AND FREQUENCY USED IN RECENT AC TDDB STUDIES

AC Stress provider	Frequency	Ref.
Pulse generator	100 kHz	[8]
Keysight 1530 WGFMU	100 kHz	[9]
Keysight 1530 WGFMU	1 MHz	[10]
Pulse generator	500 MHz	[31]
Keysight 1530 WGFMU	1 MHz	[12]
-	1 MHz	[13]
Pulse generator	500 MHz	[14]
Built-in RO	300 MHz	[32]
Embedded RO	GHz	[11]
Embedded RO	GHz	[33]

In the current study for AC TDDB, stress providers can be divided into three main categories. The first is the commercial semiconductor analyzer mentioned earlier which is the easiest one to use but the frequency range is limited. The second choice is the pulse generator or arbitrary waveform generator. With the help of impedance matching probes [14], this is the choice for most high-frequency AC TDDB studies. The third choice is using embedded ring oscillator (RO) as the stress provider which can reach very high frequency. However, this method requires an advanced specially designed layout and is not suitable for most cases.

The AC TDDB property of MOSFET devices has been extensively investigated and the TTF deterioration under low-frequency AC stress and gain under high-frequency AC stress

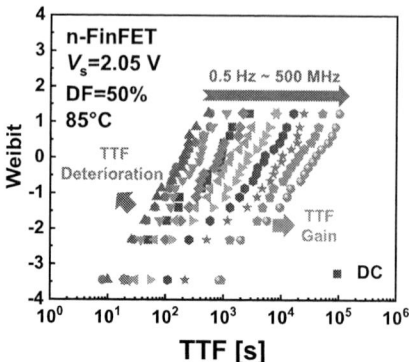

Fig. 4. **Fig. 4** AC TDDB TTF deterioration under low-frequency AC stress and gain under high-frequency AC stress [36].

are widely reported in both planar devices and FinFETs [9], [10], [13], [34]. Fig. 4 gives an example of previous experimental data [36]. Researchers have analyzed the impact of stress frequency on defect generation in the gate dielectric layer through electrical characterization of devices during the TDDB stress.

The physical mechanisms underlying TTF variations under both low-frequency and high-frequency conditions have been summarized. For low-frequency TDDB lifetime degradation, electrical characterization results reveal that low-frequency AC stress signals induce additional defects within the gate dielectric layer [35]. This extra degradation was attributed to the Maxwell–Wagner (MW) instability of the bilayer gate stack [36], as shown in Fig. 5(a). For the phenomenon of high-frequency lifetime gain, experimental results indicate that high-frequency AC stress signals significantly suppress defect generation, which is attributed to the nonlinear defect generation rate in a single stress pulse [14], [33]. As shown in Fig. 5(b), the defect generation rate starts

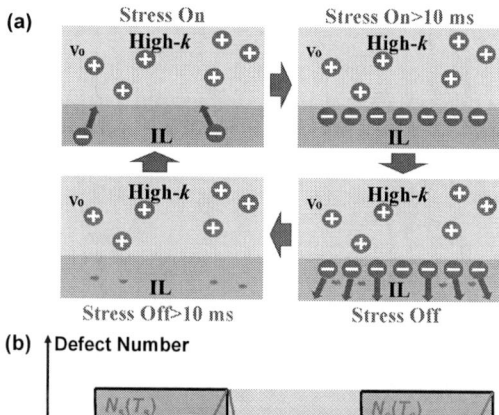

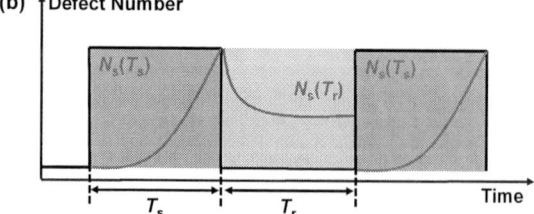

Fig. 5 (a) Schematic describes the triggering of the physical mechanism responsible for low-frequency TTF deterioration [36]. (b) Schematic of nonlinear defect generation rate in a single stress pulse [34].

from a lower value, gradually increases, and eventually saturates at the DC stress level. Additionally, the defect generation rate resets to its initial state at the onset of the next stress pulse. When high-frequency AC stress is applied to the device, the defect generation rate remains consistently at a low level. As a result, defect generation is suppressed compared to the DC stress condition and the TTF is extended.

V. CONCLUSIONS

This article comprehensively reviews experimental observations, physical mechanisms, and model developments in TDDB including the evolution of the percolation model and trap generation-based acceleration models. Besides，current understandings of AC TDDB are also reviewed including low and high frequency range. However, unresolved discrepancies remain, such as the physics mechanism of bimodal Weibull slope. Further research into the mechanism of BD requires more accurate and detailed characterization.

REFERENCES

[1] S. Kim et al., "Reliability Assessment of 3nm GAA Logic Technology Featuring Multi-Bridge-Channel FETs," in 2023 IEEE International Reliability Physics Symposium (IRPS), Mar. 2023, pp. 1–8.

[2] J. Suñé, I. Placencia, N. Barniol, E. Farrés, F. Martín, and X. Aymerich, "On the breakdown statistics of very thin SiO2 films," Thin Solid Films, vol. 185, no. 2, pp. 347–362, Mar. 1990.

[3] R. Degraeve et al., "New insights in the relation between electron trap generation and the statistical properties of oxide breakdown," IEEE Transactions on Electron Devices, vol. 45, no. 4, pp. 904–911, Apr. 1998.

[4] M. Houssa, P. W. Mertens, and M. M. Heyns, "Relation between stress-induced leakage current and time-dependent dielectric breakdown in ultra-thin gate oxides," Semicond. Sci. Technol., vol. 14, no. 10, pp. 892, Oct. 1999.

[5] J. Sune, "New physics-based analytic approach to the thin-oxide breakdown statistics," IEEE Electron Device Lett., vol. 22, no. 6, pp. 296–298, Jun. 2001.

[6] A. T. Krishnan and P. E. Nicollian, "Analytic Extension of the Cell-Based Oxide Breakdown Model to Full Percolation and its Implications," in 2007 IEEE International Reliability Physics Symposium Proceedings. 45th Annual, Apr. 2007, pp. 232–239.

[7] T. Nigam, A. Kerber, and P. Peumans, "Accurate model for time-dependent dielectric breakdown of high-k metal gate stacks," in 2009 IEEE International Reliability Physics Symposium, Apr. 2009, pp. 523–530.

[8] K. T. Lee et al., "Frequency dependent TDDB behaviors and its reliability qualification in 32nm high-k/metal gate CMOSFETs," in 2011 International Reliability Physics Symposium, IEEE, 2011, pp. 2A – 3.

[9] R. Ranjan, Y. Liu, T. Nigam, A. Kerber, and B. Parameshwaran, "Impact of AC voltage stress on core NMOSFETs TDDB in FinFET and planar technologies," in 2017 IEEE International Reliability Physics Symposium (IRPS), Monterey, CA, USA: IEEE, Apr. 2017, pp. DG-10.1-DG-10.5.

[10] M. Rafik, A. P. Nguyen, X. Garros, M. Arabi, X. Federspiel, and C. Diouf, "AC TDDB extensive study for an enlargement of its impact and benefit on circuit lifetime assessment," in 2018 IEEE International Reliability Physics Symposium (IRPS), Mar. 2018, pp. 4A.3-1-4A.3-6.

[11] M. Arabi et al., "New Insights on device level TDDB at GHz speed in advanced CMOS nodes," in 2018 International Integrated Reliability Workshop (IIRW), Oct. 2018, pp. 1–5.

[12] X. Liu, Y. Sun, J. Huang, X. Shang, and M. Yao, "Study of voltage margin of Gate oxide TDDB between AC and DC stress," in 2023 IEEE International Symposium on the Physical and Failure Analysis of Integrated Circuits (IPFA), Jul. 2023, pp. 1–4

[13] P. S. Chen et al., "AC TDDB Analysis for HK/IL Gate Stack Breakdown and Frequency-dependent Oxygen Vacancy Trap Generation in Advanced nodes FinFET Devices by SILC Spectrum Methodology," in 2022 IEEE International Reliability Physics Symposium (IRPS), Mar. 2022, pp. 11A.4-1–11A.4-6.

[14] X. Yu, C. Yan, Y. Ding, Y. Qu, and Y. Zhao, "GHz AC to DC TDDB Modeling with Defect Accumulation Efficiency Model," in 2023 IEEE International Reliability Physics Symposium (IRPS), Mar. 2023, pp. 1–6.

[15] J. H. Stathis, "Physical and predictive models of ultrathin oxide reliability in CMOS devices and circuits," IEEE Trans. Device Mater. Rel., vol. 1, no. 1, pp. 43–59, 2001.

[16] J. H. Stathis, "Percolation models for gate oxide breakdown," Journal of Applied Physics, vol. 86, no. 10, pp. 5757–5766, Nov. 1999.

[17] E. Y. Wu, "Facts and myths of dielectric breakdown processes—part I: statistics, experimental, and physical acceleration models," IEEE Trans. Electron Devices, vol. 66, no. 11, pp. 4523–4534, Nov. 2019.

[18] A. Kerber, E. Cartier, B. P. Linder, S. A. Krishnan, and T. Nigam, "TDDB failure distribution of metal gate/high-k CMOS devices on SOI substrates," in 2009 IEEE International Reliability Physics Symposium, Apr. 2009, pp. 505–509.

[19] N. Raghavan, K. L. Pey, K. Shubhakar, and M. Bosman, "Modified Percolation Model for Polycrystalline High- \kappa Gate Stack With Grain Boundary Defects," IEEE Electron Device Letters, vol. 32, no. 1, pp. 78–80, Jan. 2011.

[20] J. Sune, S. Tous, and E. Y. Wu, "Analytical Cell-Based Model for the Breakdown Statistics of Multilayer Insulator Stacks," IEEE Electron Device Letters, vol. 30, no. 12, pp. 1359–1361, Dec. 2009.

[21] S. Choi and Y. J. Park, "Physical modeling of time dependent dielectric breakdown (TDDB) of BEOL oxide using Monte Carlo particle simulation," in 2014 International Conference on Simulation of Semiconductor Processes and Devices (SISPAD), Sep. 2014, pp. 157–160.

[22] C. Yan and Y. Zhao, "A new matrix percolation model for dielectric breakdown with nonuniform defect generation," IEEE Electron Device Lett., vol. 46, no. 7, pp. 1027–1030, Jul. 2025.

[23] J. McPherson and R. Khamankar, "Molecular model for intrinsic time-dependent dielectric breakdown inSiO2 dielectrics and the reliability implications for hyper-thin gateoxide," Semicond. Sci. Technol., vol. 15, no. 5, pp. 462, 2000.

[24] J. McPherson and H. Mogul, "Disturbed bonding states in SiO/sub 2/thin-films and their impact on time-dependent dielectric breakdown," in 1998 IEEE International Reliability Physics Symposium Proceedings. 36th Annual (cat. No. 98CH36173), IEEE, 1998, pp. 47–56.

[25] I.-C. Chen, S. E. Holland, and C. Hu, "Electrical breakdown in thin gate and tunneling oxides," IEEE J. Solid-State Circuits, vol. 20, no. 1, pp. 333–342, 1985.

[26] K. F. Schuegraf and C. Hu, "Metal-oxide-semiconductor field-effect-transistor substrate current during fowler–nordheim tunneling stress and silicon dioxide reliability," J. Appl. Phys., vol. 76, no. 6, pp. 3695–3700, 1994.

[27] M. Rohner, A. Kerber, and M. Kerber, "Voltage acceleration of TBD and its correlation to post breakdown conductivity of N-and P-channel MOSFETs," in 2006 IEEE International Reliability Physics Symposium Proceedings, IEEE, 2006, pp. 76–81.

[28] E. Wu et al., "Interplay of voltage and temperature acceleration of oxide breakdown for ultra-thin gate oxides," Solid-State Electron., vol. 46, no. 11, pp. 1787–1798, 2002.

[29] J. W. McPherson, Reliability physics and engineering: time-to-failure modeling. Springer, 2018.

[30] JEDEC, | JESD263.

[31] M. Arabi et al., "New Insights on Device Level TDDB at GHz Speed in Advanced CMOS Nodes," IEEE Transactions on Device and Materials Reliability, vol. 19, no. 2, pp. 255–261, Jun. 2019.

[32] T.-Y. Yew et al., "The impacts of inverter-like transitions on AC TDDB in a fast switching logic circuit," in 2014 IEEE International Integrated Reliability Workshop Final Report (IIRW), Oct. 2014, pp. 47–50.

[33] M. Arabi et al., "Frequency dependant gate oxide TDDB model," in 2022 IEEE International Reliability Physics Symposium (IRPS), Dallas, TX, USA: IEEE, Mar. 2022, pp. P25-1-P25-5.

[34] I. K. Chen et al., "The physical mechanism investigation of off-state drain bias TDDB and its implication in advance HK/MG FinFETs," in 2018 IEEE International Reliability Physics Symposium (IRPS), Mar. 2018, pp. 4A.2-1-4A.2-6.

[35] A. Bezza et al., "Physical understanding of low frequency degradation of NMOS TDDB in High-k metal gate stack-based technology. Implication on lifetime assessment," in 2015 IEEE International Reliability Physics Symposium, Apr. 2015, pp. 5A.5.1-5A.5.5.

[36] C. Yan, Y. Qu, and Y. Zhao, "Physical study of low-frequency TDDB lifetime deterioration in advanced FinFETs," in 2024 IEEE International Reliability Physics Symposium (IRPS), Grapevine, TX, USA: IEEE, Apr. 2024, pp. 1–6.

979-8-3315-3918-4/25 $31.00 © 2025 IEEE

Mechanical Stress Induced by Temperature Cycling: Impact of MOSFET Placement on Bandgap Reference Voltage Offset

Fengbo Zhang [1], Yancong He *[1], Zhinong Liu [1], Shuang Jiao [1], Yang Li [1], Zhigang Ji [2]

[1] UNISOC(Shanghai)Technologies Co., Ltd, Shanghai 201203, China
[2] Shanghai Jiao Tong University, Shanghai 200240, China

* Email: Fengbo.Zhang@unisoc.com

Abstract—Mechanical stress significantly impacts the voltage offset of bandgap reference circuits (BGRs). This study investigates the impact of mechanical stress induced by temperature cycling on BGR voltages, with a focus on the placement of Metal-Oxide-Semiconductor Field-Effect Transistors (MOSFETs). Simulations and packaging reliability Design of Experiments (DOEs) were conducted using WLCSP (Wafer-Level Chip-Scale Package) chips. Results show how different MOSFET placements affect the BGR voltage offset, with practical implications for chip design.

Keyword—bandgap reference, mechanical stress, voltage offset, MOSFET placement, temperature cycling

I. INTRODUCTION

Bandgap Reference (BGR) voltage sources play a crucial and fundamental role in integrated circuits. As voltage and current references, BGR counteract the effects of thermal stress on circuit voltage and current. However, in actual semiconductor chips, in addition to thermal stress, mechanical stress (such as tension or compression) can also affect the BGR voltage[1]. Mechanical stress primarily originates from packaging stress, which can lead to physical effects such as changes in semiconductor carrier behavior and alterations in band structure. Mechanical stress may also change the base-emitter voltage V_{BE} of bipolar junction transistors (BJTs) through the piezojunction effect, causing deviations in the BGR voltage. It may also affect MOSFET mobility and threshold voltage through the piezoresistive effect, thereby impacting circuit performance[2]. In minor cases, it can lead to accuracy degradation, such as abnormal voltage output, incorrect sensor readings, or communication desynchronization. In severe cases, it can trigger failures like gate oxide breakdown or device burnout [3].

In the packaging process, several mature methods are used to avoid the impact of mechanical stress, such as selecting packaging materials with matched thermal expansion coefficients, and eliminating internal residual stress through baking during the post-mold cure (PMC) stage of back-end-of-line (EOL) processes. These methods are often effective for "fresh" chips just off the production line. However, during the chip's lifecycle, packaging stress caused by soldering, storage, and long-term operation can influence and ultimately cause the BGR voltage to shift. This paper investigated and discussed the impact of packaging stress on BGRs through package reliability tests (primarily temperature cycling) and provides solutions.

II. BGR PRINCIPLES AND STRESS-INDUCED MECHANISMS

A simplified BGR circuit providing a reference voltage is shown in Fig. 1. The ΔV_{BE} drop falls across R1, generating a positive temperature coefficient (PTAT) current. V_{BE} has a negative temperature coefficient (CTAT) through R2/R3.

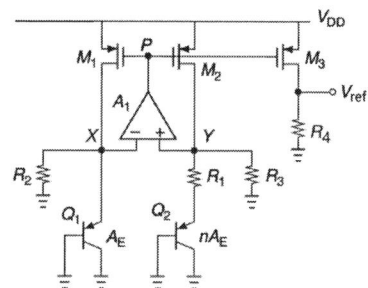

Fig. 1 Simplified Bandgap Reference Circuit

After resistance adjustment, the current through M1/M2 can achieve ZTAT. Applying such a ZTAT current to resistor R_4 generates a ZTAT voltage V_{ref}.

$$V_{ref} = \frac{R_4}{R_2}\left(|V_{BE}| + \frac{R_2}{R_1}V_T \ln n\right) \qquad (1)$$

Mechanical stress primarily affects the output voltage of bandgap reference circuits by altering the electrical characteristics of key components (BJTs, resistors, and MOSFETs). The following detailed the impact mechanisms from the perspectives of BJTs, resistors, and MOSFETs.

A. BJT

Mechanical stress applied to BJTs changes the mobility μ of minority carriers and the intrinsic carrier concentration n_i in the silicon material, thereby altering the saturation current I_s, and consequently changing V_{BE} and the PTAT voltage ΔV_{BE}.

$$\Delta V_{BE} = -\frac{kT}{q} \cdot \frac{\Delta I_s}{I_s} \qquad (2)$$

979-8-3315-3918-4/25 $31.00 © 2025 IEEE

where $\frac{\Delta I_S}{I_S} = \pi \cdot \sigma$, π is the piezo-junction coefficient, σ is the stress, k is Boltzmann's constant, T is the absolute temperature, and q is the elementary charge.

Studies have shown that in 0.13 μm BiCMOS process, packaging stress (-200 to +200 MPa) causes V_{BE} to change by approximately 0.3 mV, leading to an output voltage drift of ±0.6%[4]. Thermomechanical stress (temperature cycling -55°C to 125°C) causes a change in β of BJT, thus affecting current matching and V_{ref} stability [6].

B. Resistor

Polysilicon resistors undergo resistivity changes under mechanical stress. The piezoresistive effect in silicon material causes the resistance value to vary with stress direction and crystal orientation.

$$\frac{\Delta R}{R} = \pi_l \sigma_l + \pi_t \sigma_t \tag{3}$$

where π_l and π_t are the longitudinal and transverse piezoresistive coefficients, σ_l and σ_t are the corresponding stresses. For p-type silicon, π_l is approximately $71.8 \times 10^{-11} Pa^{-1}$, π_t is approximately $-66.3 \times 10^{-11} Pa^{-1}$. Studies indicate that in 0.18 μm CMOS technology, p-type silicon resistors change by approximately ±8% under packaging stress (-150 to +150 MPa), leading to a bandgap output voltage drift of approximately 0.5%-1%[5].

C. MOSFET

Mechanical stress changes the carrier mobility μ of MOSFETs through the piezoresistive effect, directly affecting channel resistance.

$$\frac{\Delta \mu}{\mu} = -\pi_l \sigma_l - \pi_t \sigma_t \tag{4}$$

Studies have shown that in 7nm FinFET process, local stress can increase PMOSFET mobility by up to 70%, significantly reducing resistance. STI stress causes MOSFET resistance to change by ±10%[7].

At the same time, mechanical stress can also affect the threshold voltage (V_{th}) by changing the bandgap energy and interface state density, thereby influencing the reference voltage.

$$\Delta V_{th} = k\sigma \tag{5}$$

where k is the stress-related coefficient, which depends on the device structure and stress type. The sensitivity of typical circuits is generally $k = 5\mu V/MPa$ [8].

This paper focused on the impact of mechanical stress on MOSFETs, and conclusions are drawn through simulation and experimental analysis.

III. SIMULATION EXAMPLES

To verify the effect of mechanical stress on V_{ref} of BGR circuits when applied to MOSFETs, this study used a chip equipped with multiple BGR circuits. By changing the placement of the MOSFETs, the trend of the reference voltage change was investigated.

This paper chose a WLCSP packaged chip, which has no substrate or molding, and bumps are grown directly on the

die. By designing different MOSFET placement, they were laid out either (a) underneath the bump or (b) free from the bump. Since there is no substrate to transfer stress, the mechanical stress caused by the bump will be directly transmitted to the MOSFET. As shown in Fig. 2, stress simulation indicated that after 700 cycles of Temperature Cycling (-55~125C), the stress on the MOSFETs below the bump was 30.93MPa, while the stress on BJTs and MOSFETs free from the bump area was 20.52MPa. The stress in the resistor area was 19.906MPa.

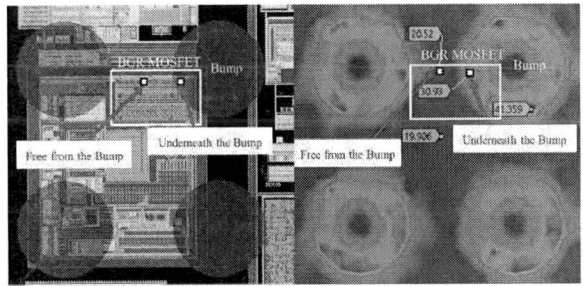

Fig. 2 BGR circuit placement on layout and simulation results

Next, we calculated the V_{ref} change based on the simulated stress values:

A. BJT

For NMOS, the transistor sensitivity σ under tensile stress is 0.0194mV/MPa, leading to $\Delta V_{BE} = 0.398mV$. Under compressive stress, the sensitivity is 0.0097mV/MPa, leading to $\Delta V_{BE} = -0.199mV$. Considering that the basic bandgap voltage is generally around 1.2V, and the bandgap voltage of the experimental chip used in this paper is 3.3V, the amplification factor M is approximately 2.75. Therefore, when the stress is 20.52MPa, the V_{ref} change is as shown in TABLE I.

TABLE I. ΔV_{ref_BJT} under Different Stress Conditions

Stress (MPa)	Tension Stress		Compression Stress	
	ΔV_{ref_BJT} (mV)	ΔV_{ref_BJT} (%)	ΔV_{ref_BJT} (mV)	ΔV_{ref_BJT} (%)
20.52	1.095	0.0332	-0.547	-0.0166

B. Resistor

From Equation (1), V_{ref} is affected by R4/R2 and R2/R1. Since the resistors in the BGR circuit are in the same area and subjected to stress in the same direction. $\Delta(R_4/R_2)/(R_4/R_2) = 0$, $\Delta(R_2/R_1)/(R_2/R_1) = 0$. Therefore, the contribution of resistance change to V_{ref} is 0.

C. MOSFET

The effects of mechanical stress on MOSFET μ and threshold voltage V_{th} are calculated separately.

Mobility change affects the current mirror current ($I \propto \mu$). The PTAT current change is:

$$\frac{\Delta I}{I} \approx \frac{\Delta \mu}{\mu} \tag{6}$$

V_{PTAT} change is:

$$\Delta V_{PTAT} \approx V_{PTAT} \cdot \frac{\Delta\mu}{\mu} \qquad (7)$$

Considering that V_{PTAT} generally accounts for 40% of the V_{ref}, so $V_{PTAT} \approx 0.48V$, and the amplification factor M≈2.75. Then

$$\Delta V_{ref\_\mu} = M \cdot V_{PTAT} \cdot (-\pi_l \cdot \sigma) = -2.606 \times 10^{-3} \cdot \sigma \quad (8)$$

V_{ref} change for two stress states on MOSFETs is calculated as shown in TABLE II.

TABLE II. ΔV_{ref_μ} under Different Stress Conditions

Stress (MPa)	Tension Stress		Compression Stress	
	ΔV_{ref_μ} (mV)	ΔV_{ref_μ} (%)	ΔV_{ref_μ} (mV)	ΔV_{ref_μ} (%)
20.52	-19.448	-0.589	19.448	0.589
30.93	-29.331	-0.889	29.331	0.889

The influence of V_{th} on ΔV_{ref} refers to (5). Based on the low sensitivity of normal circuits, k=5μV/MPa, and the circuit gain is approximately 1, i.e., $\Delta V_{ref} = \Delta V_{th}$. The calculation is shown in TABLE III.

From the above analysis, in this case, the largest impact on the ΔV_{ref} is from MOSFET mobility offset, while the change in MOSFET threshold voltage and the effect on BJT's V_{BE} are relatively small.

TABLE III. $\Delta V_{ref\_V_{th}}$ under Different Stress Conditions

Stress (MPa)	Tension Stress		Compression Stress	
	$\Delta V_{ref\_V_{th}}$ (mV)	$\Delta V_{ref\_V_{th}}$ (%)	$\Delta V_{ref\_V_{th}}$ (mV)	$\Delta V_{ref\_V_{th}}$ (%)
20.52	0.1026	0.0031	-0.1026	-0.0031
30.93	0.1547	0.0047	-0.1547	-0.0047

IV. DOES VERIFICATION

80 chip samples were subjected to temperature cycling (Condition B, -55~125C). 8 sets of bandgap output voltage $V_{BG1} \sim V_{BG8}$, were recorded at Duration=0 (T0) and Duration=700 cycles (TCB700). V_{BG} for each group of bandgap showed consistency and conformed to a normal distribution, as exemplified by V_{BG1} in Fig. 3.

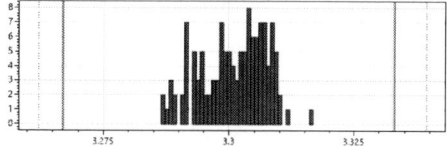

(a) V_{BG1} at T0

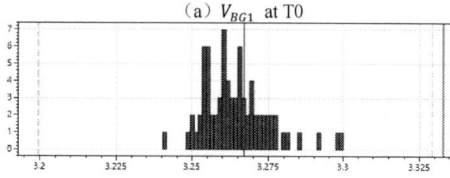

(b) V_{BG1} at TCB700

Fig. 3 Distribution of V_{BG1} at (a) T0 and (b) TCB700

The change rates for the 8 groups of V_{BG} before and after TCB700 were processed, in which 4 groups free from the bump and 4 groups underneath the bump.

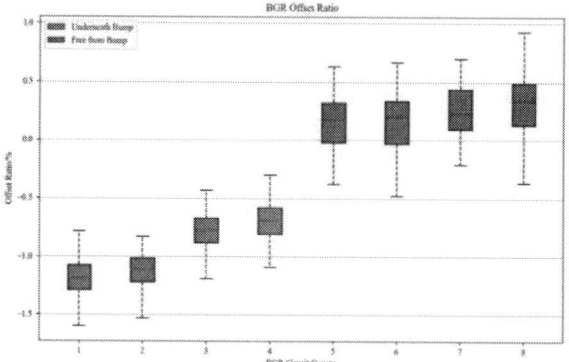

Fig. 4 V_{BG} offset ratio of Different BGR Circuits

As shown in Fig. 4, it can be seen that for BGR MOSFETs free from the bump, V_{ref} increased by 0.17%, 0.18%, 0.25%, and 0.32% after TCB700. For underneath the bump, V_{ref} decreased by 1.18%, 1.1%, 0.77%, and 0.67% after TCB700.

From the measured data, it can be observed that when the MOSFET is not pressed by the bump, the reference voltage consistently shows an increasing trend, which is consistent with the theoretical trend of compressive stress. Therefore, it can be concluded that when the BGR is free from the bump, it experiences compressive stress, and the theoretical $\Delta V_{ref}/V_{ref} \approx 0.57\%$. When the MOSFET is pressed by the bump, the reference voltage consistently shows a decreasing trend, which is consistent with the theoretical trend of tensile stress. Therefore, it can be concluded that when the BGR is underneath the bump, it experiences tensile stress, and the theoretical $\Delta V_{ref}/V_{ref} \approx -0.9\%$. Considering the complexity of circuit design and the error in mesh division during simulation, the measured reference voltage changes for different bandgap reference circuits will fluctuate.

To verify the validity of the results, this paper replaced the four MOSFETs underneath the bump to be free from the bump, meaning that the BGR circuit design remained unchanged, only the placement on the layout was altered. After measurement, as shown in TABLE IV, for the same BGR circuit, merely changing the MOSFET placement on the layout caused the reference voltage to change from decreasing to increasing, and the offset rate was similar to that of other BGR reference voltages located away from the bump.

TABLE IV. ΔV_{BG} of the Same BGR Circuit After Changing MOSFET Placement at TCB700

BGR Offset	Underneath the bump	Free from the bump
ΔV_{BG1}	-1.18%	0.17%
ΔV_{BG2}	-1.10%	0.04%
ΔV_{BG3}	-0.77%	0.21%
ΔV_{BG4}	-0.67%	0.12%

The offset of V_{BG1} is shown in Fig. 5, where the reference voltage change after TCB700 went from -1.18% to

979-8-3315-3918-4/25 $31.00 © 2025 IEEE

0.17%, which is consistent with the results derived above.

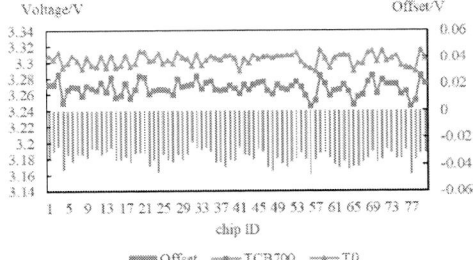

(a) MOSFET underneath the bump

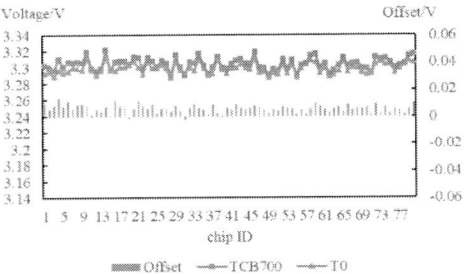

(b) MOSFET free from the bump

Fig. 5 V_{BG1} offset ratio with different MOSFET placement

V. CONCLUSION

Mechanical stress affects BJTs, resistors, and MOSFETs in BGR circuits, leading to reference voltage offset through piezoresistive or piezo-junction effects. This paper designed a WLCSP chip for simulation and experimental verification, demonstrating that

1) BGR circuits not directly pressed by a bump experience tensile stress, causing the reference voltage to increase.

2) BGR circuits directly pressed by a bump experience compressive stress, causing the reference voltage to decrease.

3) Different BGR circuits exhibit different reference voltage offset rates.

Therefore, during chip design, placing the BGR circuit, especially the MOSFETs, free from the bump can reduce the reference voltage offset rate.

REFERENCES

[1] F. Fruett, G. C. M. Meijer, and A. Bakker, " Minimization of the mechanical-stress-induced inaccuracy in bandgap voltage references, " IEEE Journal of Solid-State Circuits, vol. 38, no. 7, pp. 1288-1291, 2003.

[2] Bonev, Nikolay et al. "Mechanical stress impact on CMOS low suppl y voltage bangap reference circuit." 2016 International Semiconducto r Conference (CAS) , 133-136,2016.

[3] G. A. Rincon-Mora, "Impact of packaging on bandgap references," IEEE Transactions on Circuits and Systems II: Analog and Digital Signal Processing, vol. 48, no. 5, pp. 504-507, May 2001.

[4] F. Fruett, G. Wang, and G. C. M. Meijer, "The piezojunction effect in NPN and PNP vertical transistors and its influence on silicon temper ature sensors," Sensors and Actuators A: Physical, vol. 85, no. 1– 3, pp. 70–74, Aug. 2000.

[5] P. Qu, Z. Xiao, Y. Zhao, and K. Yousef, "A high accuracy bandgap reference with adaptive mechanical stress compensation," AEU - International Journal of Electronics and Communications, 2025.

[6] K. Chen, Y. Zhang, and X. Liu, "Package stress effects on circuit perf ormance," IEEE Journal of Solid-State Circuits, vol. 55, no. 11, pp. 2978–2986, Nov. 2020.

[7] S. Li, J. Zhang, and Y. Wang, "Bandgap and stress engineering in 7n m FinFET CMOS design," IEEE Transactions on Electron Devices, v ol. 61, no. 5, pp. 1391–1397, May 2014.

[8] M. Motz, U. Ausserlechner and M. Holliber, "Compensation of Mechanical Stress-Induced Drift of Bandgap References With On-Chip Stress Sensor," in IEEE Sensors Journal, vol. 15, no. 9, pp. 5115-5121, Sept. 2015.

Experimental and Theoretical Study of Single Event Latchup in a 3D TLC NAND Flash Memory Under Heavy Ion Irradiation

Xinghao Wang [1], Haitao Dong[1], Yujiao Ding[1], Yining Zhou[1], Haotian Li[1], Xuesong Zheng [2,3], Yuhang Wang [2], Pengpeng Sang[1], Jixuan Wu[1], Xuepeng Zhan*[1], Chaoming Liu*[2], and Jiezhi Chen[1]

[1] School of Information Science and Engineering, Shandong University, Qingdao, P. R. China
[2] School of Astronautics, Harbin Institute of Technology，Harbin, China
[3] China Aerospace Components Engineering Center, Beijing, China

* Email: zhanxuepeng@sdu.edu.cn; cmliu@hit.edu.cn

Abstract—As a key solid-state mass storage technology for space applications, the reliability of 3D NAND flash memory is susceptible to significant impacts from complex irradiation environments. Focusing on the Single Event Latchup (SEL) effect induced by heavy ion irradiation, this study systematically evaluates its impact on the reliability parameters of 3D NAND flash memory, including read disturb (RD), degradation from program/erase (P/E) cycles, and data retention characterization through measurement and simulation. The results of the experiment showed that the raw bit error rate (RBER) of the RD cycle of the flash memory improves by up to 30.4% under the irradiation effect for different PE cycles conditions. The irradiation effect improves the downshift error in all states, while increases the fail bit count (FBC) in the lower states. In addition, it is shown that the SEL effect can lead to the improvement of up to 21.9% in the data retention characteristics of the memory. The irradiation effect reduces the threshold voltage offset by 27.7% at the simulation level. Our findings provide a useful reference for improving the reliability design and radiation-tolerant 3D NAND flash memory in radiation environments.

Keywords—3D NAND, Heavy Ion Irradiation, Reliability Characterizations, Single Event Latchup

I. INTRODUCTION

As one of the most commonly used data storage devices for many space missions, 3D NAND flash memory plays a critical role in the commercial market for semiconductor memory for space applications due to its advantages of fast access speeds, high storage densities, large capacity, and low power consumption [1-2]. It has been shown that the radiation environment in space poses a great risk to the reliability of 3D NAND flash memory [3]. With the decreasing feature size of the device, the error cross section caused by the Single Event Effect (SEE) should not be underestimated [4-5]. Heavy ions will interact with nuclei or electrons as they pass through the material leading to transient, localized energy deposition. When the incident energy of the ions is sufficiently high or the irradiation time is sufficiently long, the Single Event Latchup (SEL) effect may be triggered. This can form a short circuit locally and lead to the reversal of the device state, which results in the loss of data in the NAND flash array [6-8]. Therefore, it is particularly important for the study of such short-time irradiation effects. It is worth noting that when considering the effects of irradiation on flash memory, mainstream work has normally focused on threshold voltage

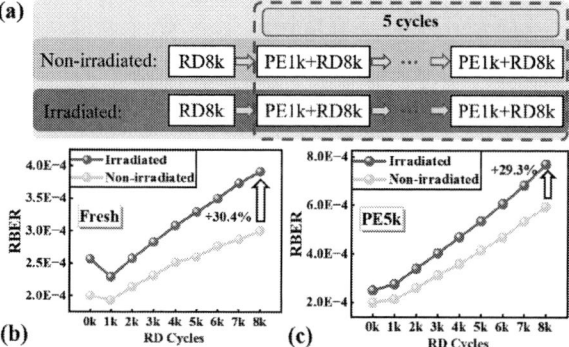

Fig. 1. (a) Flowchart for irradiated or non-irradiated testing. The relationship between RD cycles and RBER under irradiated or non-irradiated conditions at (b) Fresh state and (c) 5k PE cycles.

(V_{th}) shifts [9-10]. However, few works have been reported on the reliability of data under high-frequency interference especially after SEL effect.

In this work, the impact of SEL on the reliability of 3D charge-trap (CT) TLC NAND flash memory is mainly evaluated in terms of the raw bit error rate (RBER) and fail bit count (FBC) by applying variations of read disturb (RD), degradation from program/erase (P/E) cycles, and other parameters. In addition, this paper shows that the data retention characteristics of irradiated 3D NAND flash memory are improved to 21.9%. It gives the validation and internal mechanism explanation via systematical simulations. Our results contribute to the understanding the change of SEL on the reliability of 3D NAND memory. Moreover, the trend of the data retention characteristics provides valuable insights for future radiation-tolerant 3D NAND flash memory.

II. EXPERIMENTAL METHOD

A 128 Gb TLC NAND flash memory is adopted to evaluate the radiation effects. For simulating the irradiation effect on electronic components in the space environment, the memory chip is irradiated with Ar ions. The ion energy is 12.00 MeV/u, which corresponds to the Linear energy transfer (LET) value of 15.50 MeV/(mg/cm²) to evaluate the SEE. In the experiments, a current surge of up to 0.095 A is detected as a criterion by the programmable power supply connected to the chip. It exceeds the normal current value by more than 14

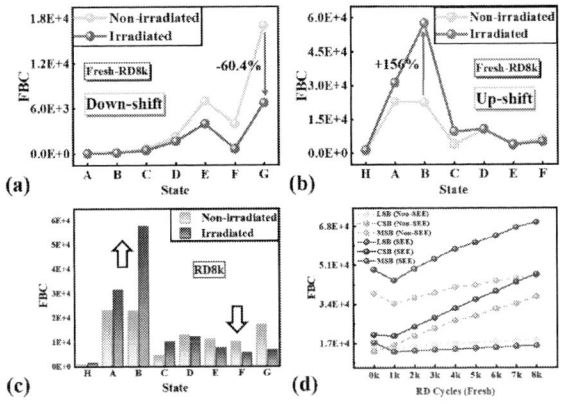

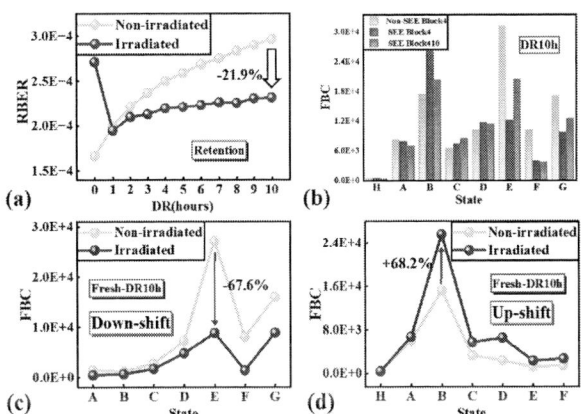

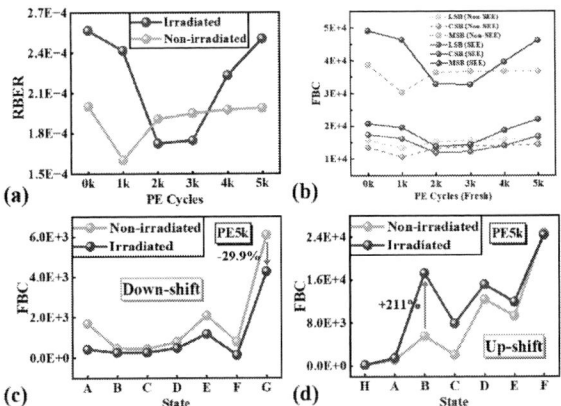

Fig. 2. Error bit distribution patterns under irradiated or non-irradiated conditions at 8k RD cycles. (a) Down-shift. (b) Up-shift. (c) All the error counts of each state. (d) Trend of FBC with RD cycles at fresh state for different page types.

Fig. 4. (a) The relationship between data retention characteristics and RBER under irradiated or non-irradiated conditions. (b) FBC of each state at DR 10h. Error bit distribution patterns under irradiated or non-irradiated conditions at DR 10h with (c) Down-shift and (d) Up-shift.

Fig. 3. The relationship between PE cycles and (a) RBER (b) Trend of FBC for different page types under irradiated or non-irradiated conditions in the fresh state. Error bit distribution patterns under irradiated or non-irradiated conditions at 5k PE cycles with (c) Down-shift and (d) Up-shift.

each node with corresponding RBER analyzed. The results show that the RBER of the chips with the irradiation generally increases by about 30% overall under different P/E cycles. With the increase in the number of RD cycles, the RBER difference is higher as compared to the one without the irradiation. The results of RD cycles performed in the fresh state are shown in Fig. 1(b) and the results of P/E 5k cycles are plotted in Fig. 1(c). In addition, the increase in RBER becomes smaller at higher P/E cycles. For example, the maximum increase at 5k cycles is 29.3%, which is smaller than 30.4% in the Fresh state.

By analyzing the FBC for different states, it is found that the FBC can be reduced under the SEL effect for down-shift errors in all states under a certain number of RD cycles, as shown in Fig. 2(a). The higher state reduction accounts for the more pronounced one. For the up-shift errors, the FBC are raised under the irradiation effect, especially in the B-state errors, as shown in Fig. 2(b). All the error counts of each state are displayed in Fig. 2(c). The FBC of the lower states are generally increased, while the FBC of the higher states show a slight downward shift after irradiation. In addition, the impact of irradiation effect on pages under RD cycles are also discussed. The trend of FBC for different page types are shown in Fig. 2(d). It is clear that the FBC of Most Significant Bit (MSB) and Center Significant Bit (CSB) increase after irradiation with more significantly for MSB. The FBC of Least Significant Bit (LSB) decreased under a certain number of RD cycles. All the pages showed consistent trends under the condition of different number of P/E cycles.

B. Effects of Program Erase Cycles

To study the impact of the SEL effect under P/E cycles separately, the experiments also evaluated the trend associated with increasing the number of P/E cycles. The data are separately extracted for each node in the process of the previous section. The trend of RBER of the irradiated chip is not as strong regular as that of the chip without the irradiation as the number of P/E cycles increases, as shown in Fig. 3(a). The RBER of the chip after irradiation is even lower when the P/E cycles are from 2k to 3k. The SEL effect has the same trend on pages under the PE cycles, it can be seen in Fig. 3(b) that the trend of the MSB is the most significant. In addition, the FBC of different states are extracted for analysis with the results under P/E 5k cycles plotted in Fig. 3(c-d). It can be

times, which determine whether it reaches the SEL effect. At this point the total injection of ions reaches 8.9×10^5 ions/cm², which in turn stops the beam supply for subsequent reliability tests. The commercial platform NplusT automated test equipment is utilized for the experimental measurements [11-12]. All tests are conducted at the room temperature with random numbers as programming data. The main difference in the experiment was whether or not irradiation is performed. The chip performance is evaluated by comparing write data with read data to ensure consistency of the experiments. The only different condition is with or without irradiation. In the simulations, three adjacent cells are constructed with the middle cell as the target cell. The displacement damage is increased on the basis of ionization, which simulates the irradiation by heavy ions [13].

III. RESULTS AND DISCUSSIONS

A. Effects of Read Disturb Cycles

In order to ensure the uniqueness of the variables in the experiments, a unified test procedure is performed for the same block of the same chip to analyze the impact of the SEE on the RBER. The flow is illustrated in Fig. 1(a) to evaluate different nodes under cumulative P/E cycles ranging from Fresh to 5k cycles. Then 8k RD cycles are performed under

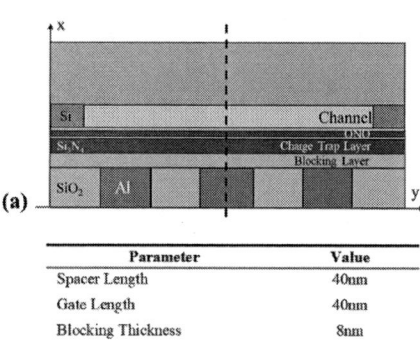

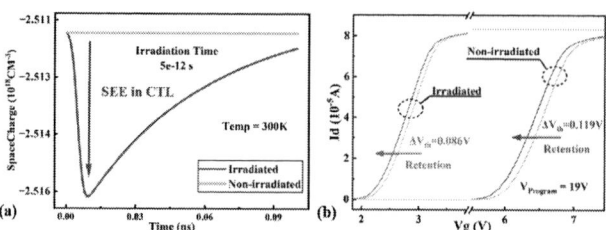

Fig. 6. (a) Ion implantation induces a surge of space charge in the CT layer within a short period of time. (b) I_d-V_g curves with and without the irradiation before and after performing the data retention process.

Parameter	Value
Spacer Length	40nm
Gate Length	40nm
Blocking Thickness	8nm
CT Thickness	8nm
Channel Thickness	10nm
Bandgap in CT Layer	5eV
Electron trap density in CT Layer	3e19 cm^{-3}
Electron trap volume in CT Layer	1e-15 μm^3

Fig. 5. (a) Schematic diagram of memory structure by sectional view. (b) Relevant parameter settings for the simulation.

found that after a certain number of P/E cycles, the SEL effect still tends to improve for down-shift errors, which is similar to the previous section. For up-shift errors, the FBC improve more after irradiation, and the change in the B-state errors remains significant.

C. Effects of Data Retention

The effects of the SEL effect on the data retention characteristics are evaluated. The irradiated and non-irradiated chips were subjected to a 10 hours data retention at room temperature and the RBER for each hour is plotted in Fig. 4(a). It shows a certain number of retention hours cause the RBER to rise over time. The RBER of the irradiated chips decreased from the first hour compared with the non-irradiated group. This phenomenon is related to the ion implantation into the CT layer, which effectively reduces the charge spreading during the subsequent data retention period. A mechanistic explanation at the simulation level is in the next section.

It is found that the data retention characteristics are more consistent across time nodes. When analyzing the FBC of different states, the 10h data retention is selected for further analysis. Different blocks are tested for horizontal comparison as shown in Fig. 4(b). All the states still satisfy the phenomenon of increasing bit counts in the lower states and decreasing bit counts in the higher states. Except for a small reduction in the bit counts of the A state, there is an increasing in the bit counts of the D state. Fig. 4(c-d) demonstrates the trends between the FBC for down-shift and up-shift errors under the 10 hours point. The results show that after the irradiation, the down-shift errors of the higher states show a greater decrease in data retention, while the increase in the FBC of the lower states in the up-shift errors is more significant.

D. Simulation Verification on Irradiation Effect

In order to further explore and verify the phenomenon that the irradiation effect leads to better data retention characterization, the Technology Computer Aided Design (TCAD) simulation model is used to simulate this process. The cross-sectional structure and some of the parameters of the created Charge-trap 3D NAND flash memory are shown in Fig. 5. This model matches the infrastructure of the chip

used for testing. It has three adjacent cells, which center cell is designated as the target cell. For the CT layer, Si_3N_4 is used as the material in the simulation as well as the Pool Frenkel emission model and the Nonlocal tunneling model. The irradiation effects on the tunneling layer (TNL) and material interface is neglected to simplify the simulations [14-15]. The relevant parameter settings used for some of the simulations are listed in Fig. 5(b).

In the simulation process, the program voltage of the target cell is firstly set to 19 V and the side cells are kept in the erased state. The temperature remains at 300 K. Then the CT layer is irradiated. In order to map the actual test, the LET value is set to 15.50 MeV/(mg/cm²). The irradiation time is set to 5×10^{-12} s. The change curve of the space charge concentration in the CT layer with the time is shown in Fig. 6(a). As shown in the Fig. 6(a), the space charge concentration in the CT layer surges instantaneously at SEE, proving that the incident ions introduce a large amount of additional charge in the CT layer. After performing the irradiation, the V_{th} after programming and reading is shifted in a negative direction. The shift leads to an increase in the FBC of the chip, which is consistent with the actual test. In order to enable better mapping of cell-size simulations to array-level tests, data retention characterization is chosen for simulation verification. A data retention for 10^6 s is set up on it, which is used to verify the relevant content in the test. A non-irradiated group without irradiation effect is established.

Fig. 6(b) shows the variation of the I_d-V_g curves with and without the irradiation as well as whether performing the data retention process or not. From the offset of the two threshold voltages, it can be found that the threshold voltage offset of 0.086 V after irradiation, which is smaller than the non-irradiated threshold voltage offset of 0.119 V. The phenomenon is the same in the control group with different program voltages. This suggests that the irradiation effect inhibits the lateral spreading of charge and further reduces the impact of the threshold voltage offset.

IV. CONCLUSIONS

This study systematically analyzes the impact of SEL effect on the reliability of 3D NAND flash memory, focusing on the mechanism and its influence on RD, degradation from PE cycles, and data retention characteristics. The results of the experiment show that under the irradiation effect, the RBER of the chip's RD cycles with the above parameters is increased by up to 30.4%. The primary manifestation is the increase in the chip's lower state FBC under the above parameters, while the FBC of the higher states is reduced. Moreover, with respect to the shift errors of each state, the irradiation effect has the most significant suppression effect on down-shift error, with a maximum improvement of 67.6%. This phenomenon is accompanied by a rise in up-shift error. It is further found that the irradiation effect can enhance the data retention characteristics of the chip under certain conditions, which characteristics can be improved by up to 21.9%. It is verified by simulation and rationally explained in the context of the internal mechanism at the device level.

ACKNOWLEDGMENT

This work was supported by National Natural Science Foundation of China (U2441248, U23B2040), Natural Science Foundation of Shandong Province (ZR2023LZH007, ZR2023QF054, tsqn202306059), and MIND project (MINDXZ202407).

REFERENCES

[1] M. Fabiano and G. Furano, "NAND flash storage technology for mission-critical space applications," in IEEE Aerospace and Electronic Systems Magazine, vol. 28, no. 9, pp. 30-36, Sept. 2013.

[2] S. W. Samwel, E. A. El-Aziz, H. B. Garrett, A. A. Hady, M. Ibrahim, and M. Y. Amin, "Space radiation impact on smallsats during maximum and minimum solar activity," Adv. Space Res. 64, 239–251 (2019).

[3] D. Chen et al., "Heavy Ion Irradiation Fluence Dependence for Single-Event Upsets in a NAND Flash Memory," in IEEE Transactions on Nuclear Science, vol. 64, no. 1, pp. 332-337, Jan. 2017.

[4] M. Bagatin, S. Gerardin, A. Paccagnella and A. Visconti, "Impact of Technology Scaling on the Heavy-Ion Upset Cross Section of Multi-Level Floating Gate Cells," in IEEE Transactions on Nuclear Science, vol. 58, no. 3, pp. 969-974, June 2011.

[5] E. P. Wilcox and M. J. Campola, "A TID and SEE Characterization of Multi-Terabit COTS 3D NAND Flash," 2019 IEEE Radiation Effects Data Workshop, San Antonio, TX, USA, 2019, pp. 1-7.

[6] M. Bagatin, G. Cellere, S. Gerardin, A. Paccagnella, A. Visconti and S. Beltrami, "TID Sensitivity of NAND Flash Memory Building Blocks,"

in IEEE Transactions on Nuclear Science, vol. 56, no. 4, pp. 1909-1913, Aug. 2009.

[7] E. R. Benton and E. V. Benton, 'Space radiation dosimetry in low-Earth orbit and beyond', Nuclear Instruments and Methods in Physics Research Section B: Beam Interactions with Materials and Atoms, vol. 184, no. 1–2, pp. 255–294, Sep. 2001.

[8] E. P. Wilcox, "Observation of low-energy proton direct ionization in a 72-layer 3-D NAND flash memory," IEEE Trans. Nucl. Sci., vol. 68, no. 5, pp. 835–841, May 2021.

[9] T. R. Oldham, "SEE and TID characterization of an advanced commercial 2 Gbit NAND flash nonvolatile memory," IEEE Trans. Nucl. Sci., vol. 53, no. 6, pp. 3217–3222, Dec. 2006.

[10] L. D. Edmonds, F. Irom and G. R. Allen, "Total Ionizing Dose Influence on the Single Event Effect Sensitivity in Samsung 8Gb NAND Flash Memories," in IEEE Transactions on Nuclear Science, vol. 64, no. 8, pp. 2046-2053, Aug. 2017.

[11] X. Fang, "High-precision short-term lifetime prediction in TLC 3-D NAND flash memory as hot-data storage," in Proc. Int. Conf. Compil., Archit., Synth. Embed. Syst., vol. 42, no. 10, pp. 3224–3235, Oct. 2023.

[12] Z. Chen, Y. Pan, M. Gong, H. Zhang, M. Zhang, and Z. Liu, "A NAND flash endurance prediction scheme with FPGA-based memory controller system," in Proc. 32nd IEEE Int. Syst. Chip Conf. (SOCC), 2019, pp. 68–73.

[13] M. G. Esposito, "Investigating Heavy-Ion Effects on 14-nm Process FinFETs: Displacement Damage Versus Total Ionizing Dose," in IEEE Transactions on Nuclear Science, vol. 68, no. 5, pp. 724–732.

[14] X. Wei, "Enhanced Total Ionizing Dose Response of 16 nm n-FinFETs With a Single Fin," in IEEE Electron Device Letters, vol. 44, no. 12, pp. 1931–1934, Dec. 2023.

[15] R. D. Schrimpf, D. M. Fleetwood, M. L. Alles, R. A. Reed, G. Lucovsky, and S. T. Pantelides, "Radiation effects in new materials for nano-devices," Microelectron. Eng., vol. 88, no. 7, pp. 1259–1264, Jul. 2011.

A Data Hierarchy-Based Adaptive Testing Method for Integrated Circuit Parameter Sets

Kaiming Hao [1], Yan Li [2], Xu Cheng [2], Qiong Wu [1], Wenfa Zhan [1], Yujie Huang *[2]

[1] School of Mathematics and Physcis, Anqing Normal University, Anqing, China
[2] the state key laboratory of integrated chips and systems, Fudan University
Shanghai, China

* Email: yujiehuang@fudan.edu.cn

Abstract—To address the escalating costs of wafer testing, the adaptive testing field has proposed various data-driven quality prediction methods. Conventional adaptive testing approaches dynamically adjust the testing process by real-time monitoring of test data distribution variations, thereby effectively reducing overall testing costs. This paper proposes a novel low-cost testing method that introduces a quartile-based data stratification technique based on the statistical distribution characteristics of test data itself, enabling efficient organization and feature selection of test data. Furthermore, this study combines two distinct machine learning algorithms to collaboratively screen test items, which not only preserves both linear and non-linear relationships within the test data but also significantly enhances the representativeness and discriminative power of the selected parameter items. Experimental results demonstrate that the stratified data achieves zero levels in both test escape rate and test yield loss, while reducing the number of test items to merely 13.2% of the original set. This method provides an efficient and scalable low-cost testing solution for quality model training in large-scale integrated circuits.

Keywords—quartile-based stratification, adaptive testing, quality predictions

I. INTRODUCTION

Integrated circuits (ICs) have evolved from simple logic devices to highly complex SoC architectures, serving as the foundational hardware for information technology and artificial intelligence (AI) advancements. This exponential growth in design complexity—characterized by billion-transistor integration and heterogeneous 3D packaging—has fundamentally transformed testing requirements, rendering conventional methods increasingly inadequate. Significant progress has been made in the field of integrated circuit testing and optimization[1][2][3][4].

Although existing methods have achieved the goals of low-cost testing and reduced test escape rates by leveraging diverse machine learning models, they exhibit significant limitations in handling data variability—in the field of machine learning, even subtle differences in data can substantially impact model training. Since raw datasets may contain noise interference, outliers, or imbalanced class distributions, direct use for model training can easily lead to performance degradation. Therefore, scientifically and rationally partitioning the dataset is of critical importance: it not only effectively suppresses overfitting but also significantly enhances the model's generalization ability and robustness.

This study proposes an Non-redundant Low-cost Testing(NLT) adaptive testing solution aimed at eliminating redundant test items to reduce testing costs in large-scale chip manufacturing. The constructed testing framework trains quality prediction models using specified parameter test results and employs a hierarchical strategy to process parameter test outcomes, thereby ensuring synergistic optimization of stability and accuracy across different predictive models. The purpose of parameter set screening is to obtain a test suite that not only meets the requirements for low test escape but also demonstrates cost-effectiveness. Meanwhile, parameter selection (which essentially constitutes variable selection in statistics and machine learning) represents a critical aspect of model construction. It plays a pivotal role in determining the model's predictive power, generalization capability, computational efficiency, and interpretability. Product quality prediction relies on in-depth mining of historical manufacturing data to uncover the complex mapping relationships between quality-influencing factors and product performance. This enables the assessment of production quality levels and facilitates the proactive planning of countermeasures to prevent quality issues before they occur.

The main contributions of this paper can be summarized as follows:

- **Multi-method collaborative optimization of parameter sets**:

  Integrates multiple machine learning techniques to select optimal parameter combinations, significantly reducing testing costs while ensuring test quality.

- **LLP-rule-based data stratification mechanism**:

  Innovatively designs the Layered Learning Partitioning(LLP) rule for data stratification, enhancing machine learning models' adaptability to data distribution variations.

- **Standardized Training Framework**:

  By establishing a unified training paradigm across different data levels, it significantly reduces the technical barriers and implementation complexity of the training process while ensuring model performance.

- **Lightweight software implementation**:

  The solution is entirely deployed on a software architecture, requiring no additional hardware resources, thus demonstrating significant engineering application value.

II. ADAPTIVE TESTING SOLUTION

This section provides a detailed description of the proposed NLT adaptive testing methodology, which consists of three sequential stages as illustrated in the figure: (1) parameter selection, (2) data stratification, and (3) quality model training, as shown in Fig. 1.

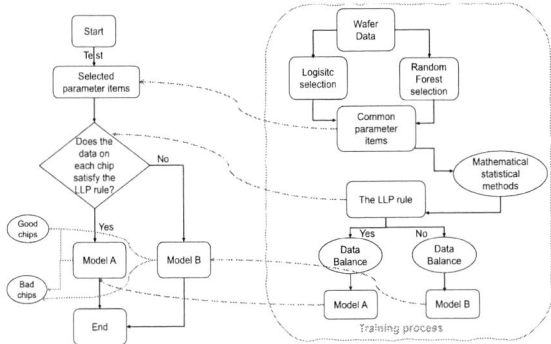

Fig. 1. Non-redundant Low-cost adaptive testing

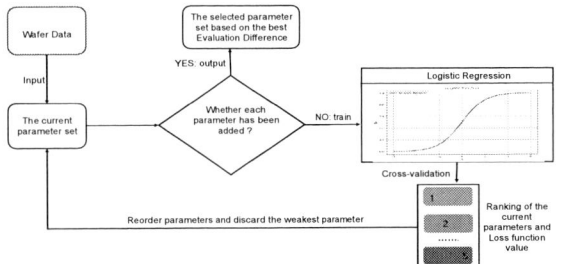

Fig. 2. Parameter Optimization Process of Logistic Regression

In Stage (1), parameter subsets are screened based on historical wafer data to identify high-efficiency parameter sets that can preserve sufficient information about:Relationships among different parameters, and Correlations between parameters and test outcomes. Building upon the selected parameters from Stage (1), Stage (2) applies the Layered Learning Partitioning(LLP) rule for data stratification. For the stratified data, Stage (3) employs distinct model training strategies to:Thoroughly capture parameter interdependencies, and Achieve accurate classification and prediction of test chip quality.

A. Parameter Selection Process

The logistic regression model [5] inherently possesses strong interpretability, and the parameter selection process further enhances this advantage. By retaining only the parameters that significantly influence the prediction outcome, it becomes easier to identify the key factors driving classification decisions.

As shown in Fig. 2, logistic regression is used as the classifier, with 5-fold cross-validation (where k = 5) and StratifiedKFold as the splitting strategy to preserve class distribution proportions. During the parameter addition process, model performance is evaluated based on the rate of change in the negative log-likelihood loss (NLL). By minimizing NLL, the model optimizes its parameters, improving both predictive accuracy and generalization capability. Scoring Criterion Calculation Formula as in (1), Here, $score\ newx_i$ is the score of the i - th newly added parameter. $curloss$ is the value of the loss function of the parameter set after the current parameter is added. $preloss$ refers to the value of the loss function of the parameter set when the previous parameter was added. $prenum$ is the number of parameters in the optimal parameter set before adding new parameters. When $\frac{score\ newx_{i-1}}{2} < score\ newx_i$, this parameter is added to the optimal parameter set; otherwise, it is excluded. The purpose here is to find the parameters that can enable the parameter set to reach the minimum value of the loss function more quickly.

$$score\ newx_i = \frac{curloss - preloss}{prenum + 1} \qquad (1)$$

Random Forest (RF) is an ensemble learning method that utilizes multiple decision trees as base models, employing the bagging (Bootstrap Aggregating) technique to process training data[6]. During the bagging process, N samples are randomly drawn with replacement from the original training set to construct the training dataset for each individual tree. This study adopts the following steps to select features:

Step 1: Model Training with Random Forest

A Random Forest algorithm is employed to train the dataset. Random Forest consists of multiple decision trees, each constructed based on distinct random sampling and feature subsets.

Step 2: Calculation of Gini Importance

For each feature, its Gini importance within the Random Forest is computed. The Gini importance values across all trees are averaged to derive the final importance score for each feature.

Step 3: Feature Ranking Based on Importance

Features are ranked according to their computed Gini importance scores. Higher-importance features are prioritized, as they contribute more significantly to the model's classification performance.

Step 4: Selection of Final Feature Subset

A threshold is applied to the ranked Gini importance scores to extract the most critical features, forming the final feature subset.

B. The LLP rule

Excessive data volume may lead to inefficiency in the training and prediction of machine learning models. Large datasets increase computational and storage overhead, often requiring more resources and time for model training. Furthermore, large datasets may introduce noise and unnecessary complexity, negatively impacting the model's generalization ability and predictive performance. To address this issue, the LLP rule proposed in this paper is designed based on quartiles, employing mathematical rules to partition the dataset into smaller subsets, thereby mitigating the adverse effects of large datasets on model training.

Quartiles are statistical measures that describe the distribution of a dataset by dividing it into four equal parts, each containing approximately 25% of the data. The four quartiles, in ascending order, are the first quartile (Q1), the median (Q2), the third quartile (Q3), and the interquartile range (IQR). The first quartile (Q1) indicates that 25% of the observations in the dataset are less than or equal to this value;

the median (Q2) indicates that 50% of the observations are less than or equal to this value; the third quartile (Q3) indicates that 75% of the observations are less than or equal to this value. The specific workflow of the LLP rule is outlined as follows:

Step 1: Calculate the first quartile ($q1_j$) and third quartile ($q3_j$) for each parameter j.

Specifically, for the j-th parameter, compute the 25th percentile ($q1_j$) and the 75th percentile ($q3_j$) of the test data observations.

Step 2: Compute the feature value $ValueX_{i,j}$

$$ValueX_{i,j} = \begin{cases} \dfrac{|X_{i,j} - q1_j|}{3\sigma_j}, & if X_{i,j} \leq q1_j \\ \dfrac{X_{i,j} - q3_j}{3\sigma_j}, & if X_{i,j} \geq q3_j \\ 0, & Other\ cases \end{cases} \quad (2)$$

$X_{i,j}$ refers to the test data of the j-th parameter item of the i-th sample, σ_j refers to the standard deviation of the test data of the j-th parameter item, and $ValueX_{i,j}$ is the feature value. Through calculation, the value of $X_{i,j}$ is transformed into data about quartiles, making the information contained in the data more abundant, as in (2).

Step 3: Data partitioning based on quartiles.

The advantages of data partitioning lie in its ability to facilitate a more comprehensive understanding of the characteristics and distribution of a dataset. By calculating the relative position of a feature - value with respect to the mean, we can divide the dataset into two parts, which respectively represent the samples relatively higher and lower than the average level. This partitioning is beneficial for identifying and comprehending outliers and anomalous data points within the data, offering a more in - depth perspective on the data. Moreover, the partitioned data can be utilized for specific analyses, modeling, or processing tailored to different data subsets. This approach is conducive to optimizing the modeling process, enhancing model performance, and more effectively recognizing and leveraging the information within the dataset, as in (3).

$$X_i \in \begin{cases} above_{data}\ if \sum_{j=1}^{m} ValueX_{i,j} > \dfrac{\sum_{i=1}^{n}\sum_{j=1}^{m} ValueX_{i,j}}{n} \\ below_{data}\ if \sum_{j=1}^{m} ValueX_{i,j} \leq \dfrac{\sum_{i=1}^{n}\sum_{j=1}^{m} ValueX_{i,j}}{n} \\ i \in 1,2,3 \dots n \end{cases} \quad (3)$$

C. Quality Model Training Method

Data preprocessing is one of the critical stages in data mining for improving the quality of a dataset. This study addresses imbalanced data in chip datasets. The algorithm employed to handle imbalanced data in the dataset is Synthetic Minority Oversampling Technique (SMOTE). SMOTE is one of the most widely used oversampling methods for addressing data distribution imbalance in machine learning modeling. SMOTE [7] balances class distributions by increasing the quantity of minority-class samples, as in (4), x' denotes the newly generated data point; x^i represents a sample from the minority class; x^j is a value randomly selected from the k-nearest neighbors of x^i; μ is a randomly chosen value within the range of 0 to 1.

$$x' = x^i + (x^j - x^i) * \mu \quad (4)$$

Data preprocessing significantly enhances model training efficacy. In this study, SMOTE is applied to all stratified (layered) subsets of the dataset to achieve balanced class distributions.

XGBoost (eXtreme Gradient Boosting), a scalable and high-performance machine learning framework, has been extensively adopted for classification, regression, and ranking tasks [8]. To identify the optimal hyperparameter configuration, we implemented a grid search methodology integrated with k-fold cross-validation (CV). This systematic approach enables exhaustive exploration of the hyperparameter space while leveraging CV to ensure statistically robust performance evaluation. The optimization process was specifically tailored to address the unique characteristics of IC testing datasets, thereby maximizing the model's adaptability and predictive fidelity for semiconductor applications. In this context, the objective function for the parameter combination is defined as in (5).

$$score = TER + TYL \quad (5)$$

This formulation ensures that during model training, simultaneous attention is given to both Test Escape Rate(TER) and Test Yield Loss(TYL), enabling the overall architecture to leverage their respective strengths and enhance performance in testing scenarios.

III. EXPERIMENTAL RESULTS

The experimental platform was configured with an AMD Ryzen 5 5600H processor featuring Radeon Graphics, operating at a base clock frequency of 3.30 GHz, and equipped with 16 GB of DDR4 3200 MHz high-bandwidth memory to ensure efficient data processing. The simulation environment was established using Python 3.9 as the primary runtime framework, integrating key computational libraries including NumPy for numerical operations, Pandas for data manipulation, and Scikit-learn for machine learning algorithm implementation.

Regarding the dataset, IC test data typically involve proprietary commercial intellectual property. The experimental data were obtained from ICND2263 chips produced by Chizhou Huayu Semiconductor Technology Co., Ltd. The test parameters encompassed critical electrical characteristics, including supply voltage, current consumption, output signal voltage amplitude, and offset values. The dataset comprised 10,911 IC samples, each characterized by 151 test parameters. Among these, 10,875 samples were classified as qualified products (passing all test criteria), while the remaining 36 samples were nonconforming.

The Test Escape Rate (TER) quantifies the proportion of defective chips incorrectly predicted as qualified (false negatives) relative to the total number of chips subjected to testing. Mathematically, TER is defined as the ratio of the number of defective chips escaping detection (N_E) to the total number of tested chips (s), where actual faulty chips are identified exclusively through comprehensive (100% coverage) post-production testing. A lower TER value indicates higher test quality and greater manufacturing reliability, as in (6).

$$TER = \frac{N_E}{s} * 100\% \quad (6)$$

Test Yield Loss(TYL) quantifies the proportion of good chips incorrectly predicted as defective (false positives) relative to the total number of chips tested. It serves as a critical metric for evaluating the economic and operational impact of over-rejection in semiconductor manufacturing. Mathematically, TYL is defined as the ratio of the number of good chips misclassified as defective (N_L) to the total number of tested chips (s), where ground truth classifications are determined through comprehensive physical testing. A lower TYL value indicates higher test precision and reduced manufacturing waste, as in (7).

$$TYL = \frac{N_L}{s} * 100\% \tag{7}$$

Test Implementation Rate (TIR) is a metric quantifying test cost reduction efficiency by evaluating the proportion of actual test items deployed relative to the standard test item baseline, It serves as a critical indicator for optimizing semiconductor test processes, balancing cost-efficiency with defect detection coverage, as in(8).

$$TIR = \frac{N_{actual}}{N_{standard}} * 100\% \tag{8}$$

We performed a comparative analysis among the non-stratified data approach inTABLE. I, the methodology proposed in 2021[9], and the technique introduced in 2025[10]. Upon contrasting our stratified methodology with the non-stratified data ,The method which, within our framework, refrains from performing the data - stratification operation, it was evident that our approach yielded substantial improvements. Specifically, our model achieved zero test escapes and a perfect test yield of 100%, underscoring the efficacy of our stratification strategy. In aggregate, despite a marginal uptick in test escapes, our method achieved a noteworthy 0.09% reduction in test yield loss. Furthermore, when benchmarked against the 2021 approach, our technique realized a reduction in test yield loss exceeding 80%. In comparison to the 2025 method, our approach resulted in a 4% decrease in test reduction rate and a 0.18% decline in test escapes.

TABLE. I Comparative Performance Analysis of Diverse Algorithmic Approaches

Method		Evaluation Metrics		
		TIR	TYL	TES
A non-data stratification method		13.2%	0.23%	0.09%
	2021[9]	33.8%	83.5%	0.04%
	2025[10]	17.2%	0	0.32%
Our method	above	13.2%	0.33%	0
	below	13.2%	0	0.23%
	general	13.2%	0.14%	0.14%

IV. CONCLUSION

This paper presents an ensemble learning-based adaptive testing methodology specifically designed for high-yield integrated circuits. The proposed approach achieves cost reduction in semiconductor testing while maintaining minimal test escape rates. By focusing on data-driven optimization, we introduce an enhanced quartile-based stratification technique that significantly improves testing quality-cost efficiency. The developed methodology not only optimizes the trade-off between testing costs and quality assurance but also establishes a novel framework for quality prediction in high-reliability semiconductor manufacturing processes.

ACKNOWLEDGMENT

This work was supported by the Graduate Education Qual ity Engineering Project of Anqing Normal University(2023cx cysj137)

REFERENCES

[1] M. Abadir, "Economics modeling of multichip modules testing strategies," in IEEE Transactions on Components, Packaging, and Manufacturing Technology: Part B, vol. 21, no. 4, pp. 360-370, Nov. 1998J. Clerk Maxwell, A Treatise on Electricity and Magnetism, 3rd ed., vol. 2. Oxford: Clarendon, 1892, pp.68–73.

[2] Li B, Agrawal V D. Applications of mixed-signal technology in digital testing[J]. Journal of Electronic Testing, 2016, 32: 209-225.

[3] Yohannes D, Kirichenko A, Sarwana S, et al. Parametric testing of HYPRES superconducting integrated circuit fabrication processes[J]. IEEE transactions on applied superconductivity, 2007, 17(2): 181-186.

[4] PAN R, ZHANG Z, LI X, CHAKRABARTY K, GU X. Black-box test-cost reduction based on Bayesian network models[J]. IEEE Transactions on Computer-Aided Design of Integrated Circuits and Systems, 2020, 40(2): 386-399..

[5] Shen, S., Zhang, Z., Jin, R., & Deng, X. Efficient estimation and selection for regularized dynamic logistic regression. IISE Transactions, 2024, 57(6), 639–654.

[6] Q. Fang, "Research on Remote Monitoring and Fault Diagnosis System of New Energy Vehicles Based on Internet of Things," 2024 IEEE 6th International Conference on Civil Aviation Safety and Information Technology (ICCASIT), Hangzhou, China, 2024, pp. 1742-1746.

[7] . Ahmad et al., "Vehicle Recognition using Multi-Layer Perceptron and SMOTE Technique," 2022 2nd International Conference of Smart Systems and Emerging Technologies (SMARTTECH), Riyadh, Saudi Arabia, 2022, pp. 190-193.

[8] P. Rani, P. Sharma and I. Gupta, "Class-Balanced by SMOTE & Filtering Mechanism Combined with XGBoost Algorithm for Classifying Imbalanced Data," 2023 Second International Conference On Smart Technologies For Smart Nation (SmartTechCon), Singapore, Singapore, 2023, pp. 55-61.

[9] X. Chen, Y. Zhao, H. Lü, X. Shao, C. Chen and Y. Huang, "A Machine Learning-based Approach for Failure Prediction at Cell Level based on Wafer Acceptance Test Parameters," 2021 IEEE Microelectronics Design & Test Symposium (MDTS), Albany, NY, USA, 2021, pp. 1-5.

[10] Pan Y, Liang H, Li J, et al. Low test cost adaptive testing method for high yield IC products[J]. Integration, 2025, 103: 102401.

979-8-3315-3918-4/25 $31.00 © 2025 IEEE

Microstructural Evolution and Reliability Analysis of RDL Copper Interconnects under High-Temperature Conditions

Peng Xu[1], Lan Li[1], Jialu Huang[1], Yu Yao[1], Hengchang Bi[1], Jiang Xia[3], Zongyi Li[3], Zuoyuan Dong[1,2]*, Xing Wu[1]

[1] In Situ Devices Center, School of Integrated Circuits, East China Normal University, Shanghai, 200241, China
[2] School of Integrated Circuits, Peking University, Beijing, 100871, China
[3] JCET Semiconductor Integration (Shaoxing) CO., LTD., Shaoxing, 312000, China

* Email: zydong@pku.edu.cn

Abstract—In this work, we conduct an in situ thermal experiment focused on redistribution layers (RDLs) copper interconnect structures in integrated circuits (ICs), utilizing advanced characterization techniques such as focused ion beam (FIB) and transmission electron microscopy (TEM). The experiment reveals the formation of dislocations and void defects within the copper during heating, indicating underlying microstructural failure mechanisms induced by thermal stress. These findings highlight critical reliability concerns associated with copper interconnects under high-temperature conditions in IC applications, providing valuable insights for the design of advanced packaging.

Keywords—*in situ TEM, FIB, interconnect, defects, reliability*

I. INTRODUCTION

With the continuous advancement of semiconductor technology and the increasing demand for integrated circuit (IC) device miniaturization and functional integration, the reliability of interconnect structures in advanced packaging has become a critical concern [1-5]. Among various interconnect materials, copper (Cu) has been widely adopted in redistribution layers (RDLs) due to its excellent electrical conductivity and compatibility with back-end-of-line (BEOL) processes [6, 7]. However, under high-temperature operating conditions—such as those encountered in high-performance computing or automotive electronics—copper interconnects are susceptible to thermal degradation, which may compromise the overall reliability of ICs [8]. Understanding the failure mechanisms of copper interconnects at elevated temperatures is therefore essential for improving the thermal stability and service life of advanced packaging systems [9]. In particular, microstructural defects such as dislocation motion and void formation can significantly impact the electrical and mechanical performance of interconnects, especially in densely packed RDL architectures [10, 11]. Kwon et al. [12] conducted a systematic evaluation of electromigration-induced failure behaviors in copper RDLs with line widths of 2 μm and 10 μm within high-density fan-out packaging, under varying current densities (7.5–12.5×10^5 A/cm^2) and temperature conditions (157–194 °C). In their work, focused ion beam (FIB) and scanning electron microscopy (SEM) were employed for precise sample preparation and microstructural characterization. The investigation revealed multiple failure phenomena, including void

nucleation and growth, Cu/PI interfacial delamination, copper oxidation, and copper diffusion into the PI layer. These findings offer critical experimental insights for the reliability-oriented design of high-density interconnect structures. Based on previous work, although SEM/in situ SEM can observe the morphological evolution of materials under thermal, electrical, or mechanical loading at relatively large scales, its limited spatial resolution hinders the revelation of the intrinsic mechanisms of microstructural and interfacial failures. In contrast, in situ transmission electron microscopy (TEM) combines atomic-level resolution with real-time loading capabilities, enabling direct observation of defect evolution, interfacial reactions, and void nucleation, among other microscopic failure behaviors [13, 14]. This provides a more in-depth and precise experimental approach for comprehensively understanding the structural stability and failure mechanisms of copper RDL structures under thermal environments. There is currently a lack of systematic research on in situ thermal experiments of copper RDL.

In this work, we employ FIB techniques to isolate and transfer the copper region of an RDL sample onto an in situ thermal chip platform for in situ TEM observation. This setup enables real-time observation of the microstructural evolution of copper interconnects at a controlled temperature exceeding 500 °C. The experimental results reveal the emergence of dislocation activity and the formation of void defects during heating, indicating pronounced thermal-stress-induced degradation in the copper structure. These findings highlight the critical importance of assessing copper interconnect reliability under thermal loading, particularly in the context of ensuring the long-term stability and performance of IC devices. This work not only provides fundamental insights into the high-temperature behaviour of copper in RDLs but also contributes to the development of design guidelines for thermally robust interconnect architectures in advanced packaging.

II. EXPERIMENTAL SETUP AND SAMPLE PREPARATION

Fig. 1 illustrates a typical interconnect structure used in advanced packaging technologies, comprising Controlled Collapse Chip Connection (C4) bumps, micro bumps (μbumps), and an RDL built on an RDL interposer. Among these, the RDL serves as a critical interconnect layer that reroutes the internal chip pads to align with the μbumps. It

979-8-3315-3918-4/25 $31.00 © 2025 IEEE

is typically constructed through alternating stacks of copper metallization and dielectric layers. The design of the RDL must simultaneously meet requirements for high-density routing capability, electrical performance, and thermo-mechanical reliability. As such, the RDL plays a pivotal role in enhancing both system-level performance and the overall integration density of advanced packaging architectures [15-18].

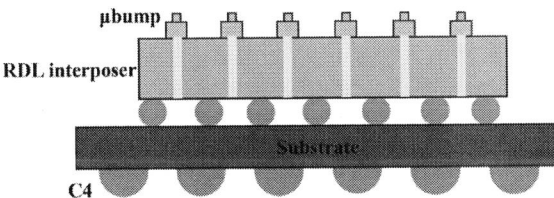

Fig. 1 A typical interconnect structure used in advanced packaging technologies.

Fig. 2 presents the optical microscopy and SEM image of the sample region targeted for observation in this work. As shown in Fig. 2 (a), after cross-sectional cutting, the sample's side exposes a relatively flat copper metal layer. The corresponding SEM image of this region is shown in Fig. 2 (b), where polyimide (PI) is distributed around the copper interconnect lines. PI in RDL structures primarily functions as an insulating and stress-buffering layer, ensuring electrical isolation of the copper lines and mitigating thermomechanical stress [19]. In this work, the region of interest is confined exclusively to the copper metal area.

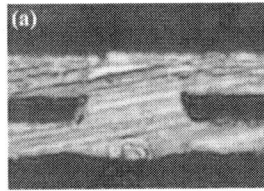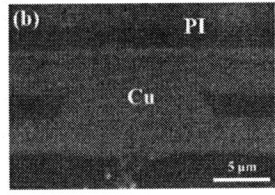

Fig. 2 (a) Optical microscopy image of the Cu RDL; (b) Corresponding SEM image of the Cu RDL.

To gain an in-depth understanding of the failure mechanisms of Cu RDL structures under high-temperature environments, it is worthwhile to employ an in situ TEM experiment to investigate the thermal reliability. Through in situ heating, the structural evolution of Cu RDL under thermal loading can be directly observed in real time, providing visual evidence for elucidating thermally induced failure mechanisms. To enable high-resolution structural characterization and precise selection of the region of interest, FIB technology is utilized for cross-sectional sample preparation and precise transfer of the target region, ensuring the accuracy of in situ TEM observations. FIB technology plays a pivotal role in device failure analysis and reliability assessment. Its primary advantage lies in its ability to precisely remove and deposit materials at the nanoscale, enabling accurate cross-sectional preparation, defect localization, and structural inspection in targeted regions. In failure analysis, FIB can be employed to directly mill critical areas of the device, thereby exposing hidden defects such as contact failure, metal migration, and interfacial delamination. When integrated with SEM and TEM, FIB facilitates the multiscale characterization of structural and compositional properties [20-25]. The following section presents the sample preparation procedure for in situ thermal chip specimens using FIB technology as employed in this work.

Fig. 3 illustrates the preparation process of the sample used for in situ thermal TEM using FIB technology in this work. As shown in Fig. 3 (a), the sample is initially mounted on a 0° stage, and the region of interest is identified through simultaneous observation at the electron and ion beam crossover point in a dual-beam system. Subsequently, as depicted in Fig. 3 (b)&(c), trenches are milled on both the upper and lower sides of the target region, followed by undercutting at the base. The desired sample is then attached to a nanomanipulator via gas injection systems (GIS) and lifted out.

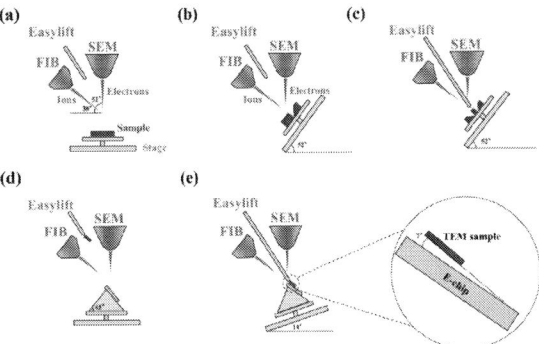

Fig. 3 Schematic of the FIB sample preparation process for in situ TEM experiments: (a) Schematic illustration of the FIB setup; (b) Trench milling of the sample; (c) Sample lift-out using a nanomanipulator; (d) Mounting of the in situ thermal chip; (e) Attachment of the extracted sample onto the surface of the in situ thermal chip (with an inset showing a magnified view of the contact interface between the chip and the sample).

As shown in Fig. 3 (d), the initial in situ thermal chip is mounted on a 45° stage, and a nanomanipulator is prepared to transfer the extracted specimen onto the surface of the in situ thermal chip, where it is fixed using GIS deposition, followed by subsequent thinning processes. During this orientation, tilting the stage by an additional 7° aligns the ion beam parallel to the surface of the in situ thermal chip, allowing for post-attachment milling of the sample. However, to mitigate ion beam-induced damage to the Si_3N_4 membrane on the in situ thermal chip during milling, a slight intentional tilt is introduced between the sample and the chip surface. As shown in Fig. 3 (e), tilting the stage to 14° results in a 7° angle between the sample and the chip surface. This configuration effectively reduces ion beam damage to the Si_3N_4 membrane during subsequent milling while preserving high imaging quality in TEM analysis.

Fig. 4 (a)&(b) show SEM images of the initial in situ thermal chip and the chip with the sample mounted on the Si_3N_4 observation window, respectively. The resistive heater on the in situ thermal chip is confined to a microscale region, enabling rapid heating, minimal thermal drift, and controllable cooling without the need to heat the entire chip [26]. However, due to the limited size of the observation window, precise alignment and transfer of the sample to the target area are required. The circular Si_3N_4 observation window has a diameter of approximately 10 μm. During the sample transfer process, the region of interest must be simultaneously observed under both electron and ion beams to ensure precise alignment with the Si₃N₄ window. Meanwhile, two additional details must be carefully considered during the sample preparation process. Firstly, when employing the GIS to deposit a platinum precursor for securing the specimen onto the in situ thermal chip, only a minimal amount of platinum should be used—just sufficient to achieve reliable adhesion. Excessive deposition may lead to unintended coverage of the Si₃N₄ observation window, thereby compromising the imaging quality during subsequent TEM analysis. Secondly, during the thinning process, it is crucial to avoid ion beam-induced damage to the Si₃N₄ observation window, as such damage can compromise the structural integrity and hinder accurate microstructural characterization. Fig. 4 (c) shows the final milled sample, with the observation region correctly positioned above the Si₃N₄ membrane, thereby ensuring suitability for subsequent in situ TEM analysis.

Fig. 4 SEM images of the sample prepared for in situ thermal TEM: (a) Pristine in situ thermal chip; (b) In situ thermal chip with the target sample mounted; (c) Milled in situ thermal chip after final sample preparation.

III. IN SITU THERMAL EXPERIMENT

Fig. 5 shows the results of the in situ thermal experiment. The system of in situ thermal experiment is linearly heated from the ambient temperature of 25 °C to 400 °C, with isothermal holding periods of approximately 30 seconds at key temperature points such as 100 °C and 200 °C. Fig. 5 (e) presents the temperature–time profile recorded during the in situ thermal TEM process of the copper region. In situ TEM observations reveal pronounced dislocation activity in the copper during the heating process. Dislocations are first observed at a temperature of 150 °C, and this dislocation behaviour becomes increasingly prominent as the temperature is maintained at 200 °C and 300 °C, as shown in Fig. 5 (a-d).

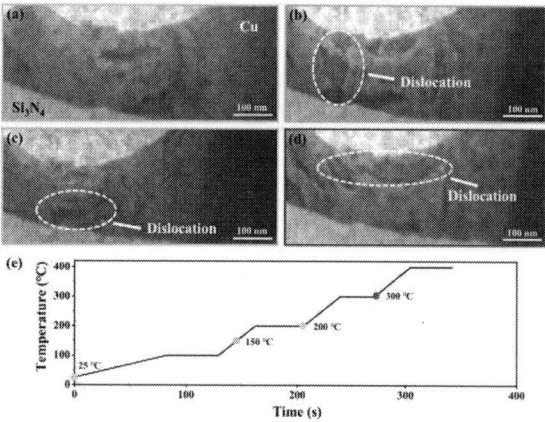

Fig. 5 (a–d) TEM images of the copper region during the in situ thermal TEM process; (e) Temperature–time profile recorded during the in situ thermal process.

This phenomenon can be attributed to the enhanced atomic vibrations within the copper crystal lattice during thermal activation, which effectively reduce the energy barrier for dislocation nucleation and lower the resistance to dislocation glide. As the temperature rises beyond a certain threshold, the accumulation of thermally induced stress facilitates the activation of dislocation sources and promotes the mobility of dislocation lines. Moreover, the elevated temperature may also lead to the activation of pre-existing lattice defects, thereby increasing the likelihood of dislocation generation from multiple sources and intensifying their mutual interactions. These effects collectively manifest as a significant increase in dislocation density observed in the TEM images. The accumulation of dislocations may lead to localized stress concentration, which accelerates void nucleation and crack propagation, thereby compromising the reliability of copper interconnect structures.

Based on the aforementioned observations, the temperature is further increased while continuously recording the temperature–time profile. As the temperature rises, thermal effects intensify, leading to the formation of void defects within the copper region. As shown in Fig. 6 (a), when the temperature reaches 400 °C, an increase in image contrast in the area indicated by the arrow suggests a reduction in local copper thickness, accompanied by the emergence of voids. With continued heating, the void progressively enlarges. As shown in Fig. 6 (b-d), during the temperature increase from 400 °C to 500 °C, including the holding period at 500 °C, the void area continuously enlarges until complete melting and breakage of the copper metal occur.

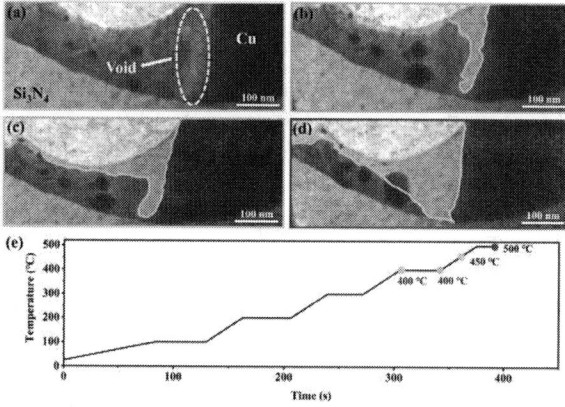

Fig. 6 (a–d) TEM images of the copper region during the in situ thermal TEM process; (e) Temperature–time profile recorded during the in situ thermal TEM process.

These results can be attributed to the significantly enhanced atomic diffusion rates under high-temperature conditions, which promote the formation of voids within the copper matrix. As the temperature continues to rise, thermally induced stress and surface energy gradients further intensify localized material transport, exacerbating the non-uniform thinning of the copper layer. Ultimately, the progressive growth of voids leads to the rupture of the conductive pathway, manifesting as catastrophic melting-induced failure of the copper interconnect. These findings indicate that when the temperature exceeds 400 °C, the structural stability of copper interconnects deteriorates markedly, accompanied by irreversible microstructural evolution and substantial reliability degradation. Therefore, thermal diffusion effects and associated failure mechanisms must be carefully considered in the thermal design of microelectronic devices.

IV. CONCLUSION

In summary, in situ investigation of the high-temperature reliability of copper interconnects within RDL is conducted using advanced materials characterization techniques, including FIB and TEM. Through precise fabrication and transfer processes, the copper region of the RDL sample is successfully mounted onto an in situ thermal chip platform, enabling real-time observation under elevated temperatures within the TEM. Experimental results reveal the progressive emergence of pronounced dislocation motion and void defect evolution in copper as the temperature increases, indicating microstructural instability mechanisms driven by thermal stress. These defect evolution processes pose significant threats to the electrical conductivity and mechanical integrity of RDL structures, underscoring the critical role of copper interconnect reliability in advanced packaging and IC applications. The findings provide essential experimental evidence and theoretical insight for understanding and enhancing the thermal stability of interconnect structures in semiconductor packaging.

REFERENCES

[1] Y. L. Liu et al., "Hyper RDL (HRDL) Interposer by Layer Transfer Technology for 3D IC and Advanced Packaging," in 2024 IEEE International Electron Devices Meeting (IEDM), 2024, pp. 1-4.

[2] Z. Choi et al., "Effect of pre-existing void in sub-30nm Cu interconnect reliability," in 2010 IEEE International Reliability Physics Symposium (IRPS), 2010, pp. 903-905.

[3] F. Xia et al., "Characterization and challenge of TDDB reliability in Cu/low K dielectric interconnect," in 2011 International Reliability Physics Symposium, 2011, pp. 2C.1.1-2C.1.4.

[4] M. Nandy et al., "Reliability Testing by Mechanical and Electrical Characterization of Flexible and Stretchable Interconnect Materials," in 2021 IEEE 71st Electronic Components and Technology Conference (ECTC), 2021, pp. 1744-1748.

[5] R. Kasim et al., "Reliability Modeling of Middle-Of-Line Interconnect Dielectrics in Advanced process nodes," in 2023 IEEE International Reliability Physics Symposium (IRPS), 2023, pp. 1-8.

[6] H. Lu et al., "Advances in Panel Scalable Planarization and High Throughput Differential Seed Layer Etching Processes for Multilayer RDL at 20 Micron I/O Pitch for 2.5D Glass Interposers," in 2016 IEEE 66th Electronic Components and Technology Conference (ECTC), 2016, pp. 2210-2215.

[7] F. Liu et al., "Low Cost Panel-Based 1-2 Micron RDL Technologies with Lower Resistance than Si BEOL for Large Packages," in 2018 IEEE 68th Electronic Components and Technology Conference (ECTC), 2018, pp. 613-618.

[8] Y. C. Huang et al., "Void Migration Kinetics in Fine Line Cu RDL under Electric Current Stressing and the Improvement of Electromigration Reliability by Polyimide Passivation," in 2024 IEEE 74th Electronic Components and Technology Conference (ECTC), 2024, pp. 576-580.

[9] Z. Hongchao et al., "A 2D Clock Interconnect Electromigration-Thermal Coupling Simulation Method Based on COMSOL," in 2023 IEEE 15th International Conference on ASIC (ASICON), 2023, pp. 1-4.

[10] C. L. Liang et al., "Direct Observation of Void Nucleation and Growth in a 2-μm-Wide Cu Redistribution Line During In Situ Electromigration," in 2024 International Conference on Electronics Packaging (ICEP), 2024, pp. 1-2.

[11] A. Chaudhuri et al., "Influence of Passivation Layer on Electromigration Lifetime of Fine-Pitch Cu RDL," in 2025 IEEE 75th Electronic Components and Technology Conference (ECTC), 2025, pp. 1126-1133.

[12] J. Kwon et al., "Electromigration Performance of Fine-Line Cu Redistribution Layer (RDL) for High-Density Fan-Out Packaging," in 2023 IEEE 73rd Electronic Components and Technology Conference (ECTC), 2023, pp. 1297-1302.

[13] C. Luo et al., "In Situ Transmission Electron Microscopy Characterization and Manipulation of Two-Dimensional Layered Materials beyond Graphene," Small, vol. 13, no. 35, p. 1604259, 2017.

[14] Z. Wang et al., "Benchmarking Heterogeneous Integration with 2.5D/3D Interconnect Modeling," in 2023 IEEE 15th International Conference on ASIC (ASICON), 2023, pp. 1-4.

[15] C. F. Yu et al., "Warpage Assessment of System in Wafer-level Package Technology with RDL Process through Theoretical Approach and Experimental Validation," in 2023 International VLSI Symposium on Technology, Systems and Applications (VLSI-TSA/VLSI-DAT), 2023, pp. 1-2.

[16] W. S. Liao et al., "A high-performance low-cost chip-on-Wafer package with sub-μm pitch Cu RDL," in 2014 Symposium on VLSI Technology (VLSI-Technology): Digest of Technical Papers, 2014, pp. 1-2.

[17] K. L. Suk et al., "Low Cost Si-Less RDL Interposer Package for High Performance Computing Applications," in 2018 IEEE 68th Electronic Components and Technology Conference (ECTC), 2018, pp. 64-69.

[18] C. K. Hsiung et al., "Glass Panel Process Integrated Low Stress Organic Dielectric RDL Structure," in 2024 IEEE 74th Electronic Components and Technology Conference (ECTC), 2024, pp. 1896-1899.

[19] X. Lai et al., "Influence of material property of photosensitive polyimide on typical structural stress in redistribution layer of fan-out package," in 2022 23rd International Conference on Electronic Packaging Technology (ICEPT), 2022, pp. 1-5.

[20] X. Wu et al., "Probing and Manipulating the Interfacial Defects of InGaAs Dual-Layer Metal Oxides at the Atomic Scale," Adv. Mater., vol. 30, no. 2, p. 1703025, 2018.

[21] X. Wu et al., "Atomic Scale Modulation of Self-Rectifying Resistive Switching by Interfacial Defects," Advanced Science, vol. 5, no. 6, p. 1800096, 2018.

[22] Z. Zhang et al., "The Trends of In Situ Focused Ion Beam Technology: Toward Preparing Transmission Electron Microscopy Lamella and Devices at the Atomic Scale," Advanced Electronic Materials, vol. 8, no. 9, p. 2101401, 2022.

[23] C. Luo et al., "Tailoring the phase transition of silver selenide at the atomistic scale," Nanoscale, 10.1039/D2NR04248G vol. 14, no. 43, pp. 16077-16084, 2022.

[24] Z. Dong et al., "Catching the Missing EM Consequence in Soft Breakdown Reliability in Advanced FinFETs: Impacts of Self-heating, On-State TDDB, and Layout Dependence," in 2023 IEEE Symposium on VLSI Technology and Circuits (VLSI Technology and Circuits), 2023, pp. 1-2.

[25] Z. Dong et al., "Towards Understanding Cryogenic Reliability in FinFETs Under Hot Carrier Stress: New Findings on Ge Migration, and Impacts of Tail States Evolution," in 2025 Symposium on VLSI Technology and Circuits (VLSI Technology and Circuits), 2025, pp. 1-3.

[26] T. Zhao et al., "On-chip gas reaction nanolab for in situ TEM observation," Lab on a Chip, 10.1039/D3LC00184A vol. 23, no. 17, pp. 3768-3777, 2023.

979-8-3315-3918-4/25 $31.00 © 2025 IEEE

Reliability Screening for Yield Improvement in IC Design Industry: Progress, Challenges and Prospects

Yixian Wang [1], Xiaoxiao Qiu [2], Zhigang Ji*[1]

[1] *Shanghai Jiaotong University, Shanghai, China*

* Email: zhigangji@sjtu.edu.cn

Abstract—Reliability screening is pivotal for yield improvement in IC design, evolving alongside semiconductor manufacturing advancements. Parametric strategies progressed from early statistical methods to AI models like Random Forest, XGBoost, and DNNs, leveraging electrical data but considering spatial correlations in a limited way. Spatial strategies, particularly GDBN, advanced from fixed windows to AI-enhanced frameworks with CNNs and transformers, excelling in cluster detection but suffering from noise sensitivity. Standalone approaches face limitations: parametric methods over-rely on distributional assumptions, while spatial-only GDBN lacks interpretability. Fusion of both, integrating parametric precision and spatial context via AI, addresses these gaps, enhancing robustness and reducing false positives for advanced nodes.

Keywords—Reliability Screening, IC Yield, Parametric Strategy, Spatial Strategy, GDBN, AI Integration, Fusion Approach

I. INTRODUCTION

Yield, the proportion of functional devices from a manufacturing batch, directly determines profitability. Reliability screening is a cornerstone process in the integrated circuit (IC) design industry, with its core objective firmly rooted in improving yield. To achieve this, different stakeholders adopt distinct strategies: fabrication facilities (FABs) focus on optimizing manufacturing processes to reduce inherent defects, while fabless design houses, which lack in-house manufacturing capabilities, rely heavily on reliability screening algorithms to identify and eliminate potentially faulty units. For fabless companies, these algorithms are critical tools to maximize the number of marketable devices, as they cannot directly intervene in the fabrication process. Traditionally, such screening has centered on electrical parametric testing (e.g., measuring voltage and current) to detect abnormal units, using statistical methods to flag outliers—an approach that laid the groundwork for ensuring product quality in earlier technology generations.

The evolution of reliability screening has been driven by the need to address increasingly complex manufacturing challenges. Early methods[1], built on statistical analysis of electrical parameters, such as identifying outliers through threshold-based tests, effectively served earlier technology generations. However, semiconductor manufacturing's shift to advanced nodes (e.g., 3nm, 7nm) has intensified defect detection challenges: shrinking feature sizes increase pattern complexity, noise sensitivity, and the need for robust methods to distinguish critical defects from process variations. These advancements render traditional standalone parametric or spatial approaches insufficient, making it difficult for early statistical methods to keep pace with intricate defect mechanisms. The rise of artificial intelligence (AI) has since revolutionized the field[2-10]: AI algorithms, capable of processing vast datasets, uncover subtle patterns in both parametric data and spatial information.

This review focuses on two complementary AI-driven strategies: parametric-based and spatial-information-based strategies. Section II traces the development of parametric-based strategy, from early statistical approaches to modern AI models, and their evolving characteristics. Section III explores the evolution of GDBN strategy, from traditional fixed-window strategies to AI-enhanced frameworks integrated with deep learning and transformers, highlighting their strengths in spatial defect detection. Section IV analyzes the challenges of reliability screening: it first compares the two strategies, then elaborates on the drawbacks of parametric-only approaches (with early methods ignoring spatial correlations, later AI models considering them limitedly, over-reliance on distributional assumptions, and scalability issues) and spatial-only GDBN (noise sensitivity, poor interpretability, and data scarcity). Section V presents a perspective on fusion-based reliability screening, discussing the necessity of integrating parametric and spatial information, early exploration efforts, and exploratory experimental results. Section VI concludes with a summary and prospects. From a historical perspective, advanced manufacturing challenges reveal critical gaps in standalone parametric and spatial methods, highlighting the need for their fusion, which is inevitable. Our work advocates for deep integration of parametric precision with spatial context, aiming to establish a robust, next-generation reliability screening framework that addresses the multifaceted challenges of cutting-edge semiconductor manufacturing.

II. PARAMETRIC-BASED RELIABILITY SCREENING STRATEGY

A. Historical evolution

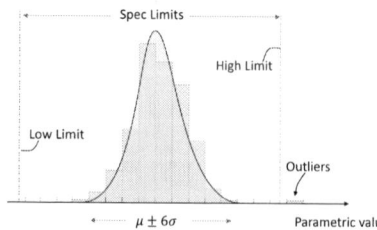

Fig. 1   Illustration of PAT for a specific parametric test result: screening out die outside defined ranges based on the parametric value distribution(High Limit, Low Limit, $\mu \pm 6\sigma$).

Thank UNISOC (Shanghai) Technologies for the support of this work.

As introduced above, parametric-based reliability screening parametric-based reliability screening (R.S.) relies on electrical metrics to identify at-risk dies, which originally began with statistical process control (SPC) and evolved into machine learning (ML) models.

As introduced in [1], early methods (SPAT) relied on fixed thresholds, limiting flexibility for diverse defect modes. DPAT introduced dynamic range adjustments but remained vulnerable to outliers skewing μ/σ estimates. Subsequent advances (AEC DPAT, RDPAT) unified solutions to these gaps: AEC DPAT adapted to process shifts, while RDPAT integrated Grubbs' algorithm to robustly filter outliers—collectively enhancing resilience to noise and variability. The core idea of SPC-based approaches (especially PAT) is illustrated in Fig. 1.

B. ML techniques

Tree-based models were adopted to predict failure probabilities, which advanced semiconductor reliability screening by learning correlations between parametric features and field failures, evolving from Random Forest (RF) to XGBoost for increasingly complex challenges[7, 11]. RF pioneered ensemble learning for high-dimensional parametric data. By aggregating decorrelated decision trees, it balanced class imbalance (rare defects vs. good die) and resisted overfitting, as shown in Fig. 2. XGBoost refined ensemble learning for complex parametric patterns. By leveraging gradient boosting and iterative tree refinement, it prioritized misclassified samples and mitigated overfitting via regularization.

Compared to tree models, DNNs elevated parametric screening by modeling complex spatial relationships in electrical data. Though their consideration of spatial cluster information remains limited, Chuang's CNN for IPDs (structure in Fig. 3, simplified schematic from [9]) predicts breakdown voltage using image-like wafer probing data (capacitance, leakage current, etc.) with neighboring dies, achieving 65.8% hit rate and 10.9% overkill rate on 360,000 samples—outperforming traditional methods.

III. SPATIAL-BASED RELIABILITY SCREENING STRATEGY

A. Evolution of GDBN

Traditional GDBN (as shown in Fig. 4), which evolved for almost two decades from 2001 to 2018[12-14], assumes that a die in bad clusters is more likely to fail in the future than one far from any cluster. It relied on rigid fixed windows yet struggled to adapt to irregular defect patterns.

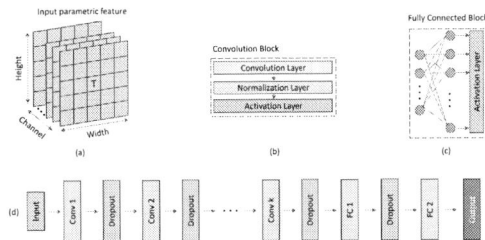

Fig. 2. Mechanism of DNN operation based on parametric test results. (a) Dies' parametric values are reshaped into a 5x5 window, aligned with spatial positions, with channels representing different parametric tests. (b) Internal structure of convolution blocks (Conv1, Conv2...) in (d). (c) Structure of fully connected blocks: a fully connected layer + an activation layer. (d) Overall CNN structure, with details of Conv 1-Conv k and FC1-FC2 shown in (b) and (c).

Post-2020, a paradigm shift to ML-enabled GDBN introduced data-driven pattern recognition[8, 9]. Subsequent deep learning revolutionized GDBN, with AI-enhanced methods integrating CNNs and transformers to model intricate spatial patterns. 2023 advancements[4] added spatial attention and MobileNet for holistic wafer-level analysis. By 2024[2, 3], Transformer-driven variants like MetaFormer-GDBN extended to full-wafer observation, modeling global dependencies, reducing false positives via denoising and multi-DUI insertion, and outperforming prior methods.

B. Advantages of GDBN

GDBN (Good Die in a Bad Neighborhood) brings distinct advantages to semiconductor reliability screening, addressing critical gaps left by parametric-only methods. Its core strength lies in spatial defect detection. In semiconductor manufacturing, defects frequently stem from localized process variations that create clustered anomalies[1, 15]. GDBN excels here: by learning "normal" spatial distributions of good dies (e.g., randomness in their arrangement), it flags deviations like clusters of adjacent dies with subtle, in-spec parametric shifts.

Another key advantage is early latent risk identification. For example, pre-failure defects caused by nano-particles don't trigger parametric test failures initially but form spatial clusters. GDBN detects such clusters before they cause outright failures, preventing costly field returns.

GDBN also offers synergy with advanced processes. In shrinking nodes, where spatial patterns grow more complex, nanoscale defects like line edge roughness become critical[5]. GDBN's spatial sensitivity allows it to spot these subtle, clustered defects—filling gaps left by parametric methods, which struggle with such patterns due to their focus on individual die metrics. By integrating with modern AI

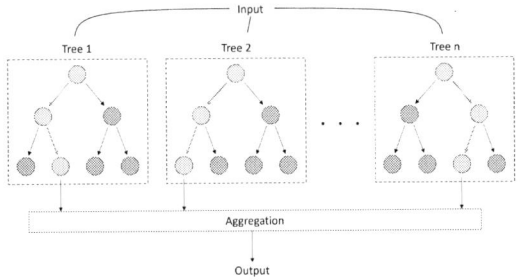

Fig. 3 The principle of Random Forest: aggregating decorrelated decision trees to balance class imbalance (rare defects vs. good die) and resist overfitting.

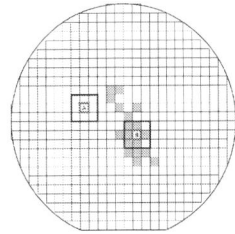

Fig. 4 Traditional GDBN principle: Based on fixed window (usually a 3x3 or 5x5 window), it assumes dies in bad clusters (i.e. die B) are more prone to failure than die A, thus screening potential defects.

enhancements (e.g., spatial attention), GDBN adapts to the complexities of advanced manufacturing, making it a valuable tool for next-gen semiconductor quality control.

IV. CHALLENGES FOR RELIABILITY SCREENING

As manufacturing progresses to finer nodes, the complexity of defect mechanisms and process variations has surged[16], these limitations (parametric/spatial gaps) become more pronounced. While parametric-based and spatial-based reliability screening methods each play irreplaceable roles, their inherent limitations have increasingly become bottlenecks in ensuring screening accuracy and efficiency.

A. Drawbacks of Parametric-Only Strategy

Parametric-based reliability screening methods, despite their foundational role and continued relevance, face significant drawbacks that limit their efficacy in modern semiconductor manufacturing. Chief among these is their limited or inadequate consideration of spatial context. In semiconductor fabrication, dies on a wafer are not isolated entities; they are part of a larger spatial system where process variations and defects can exhibit clustering behavior. Parametric methods, which typically analyze each die's electrical metrics (such as leakage current, voltage levels) in isolation, fail to account for these spatial relationships.

Moreover, these methods often suffer from over-reliance on distributional assumptions. Many traditional parametric techniques, like early statistical process control (SPC) or Part Average Testing (PAT), assume that parametric data follows well-behaved distributions (e.g., Gaussian distributions). However, in advanced semiconductor nodes[16], the nature of defects and process variations has become far more complex. Defect mechanisms are no longer neatly captured by simple statistical distributions, leading to situations where these assumptions break down. As a result, parametric-only methods may either fail to detect genuine defects that do not conform to these assumed distributions or incorrectly flag good dies as defective, leading to unnecessary yield loss.

Additionally, scalability challenges pose a significant hurdle. With the continuous shrinkage of semiconductor feature sizes and the corresponding increase in the number of parametric test metrics (e.g., hundreds or even thousands of electrical parameters per die in advanced nodes), parametric-only approaches struggle to keep pace. Machine learning models employed in parametric screening, such as DNNs, require substantial computational resources (e.g., GPUs) and large volumes of training data. This makes real-time or near-real-time screening in high-volume manufacturing fabs a formidable task, as the computational complexity can bottleneck the production process, hindering efficient quality control.

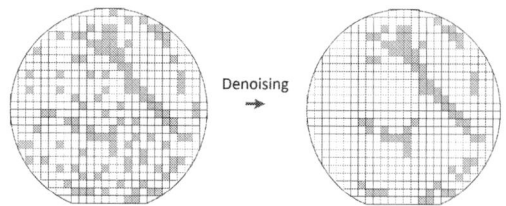

Fig. 6 Illustrative wafermaps of spatial distributions before denoising (left) and after denoising (right), a preprocessing step for AI-assisted GDBN.

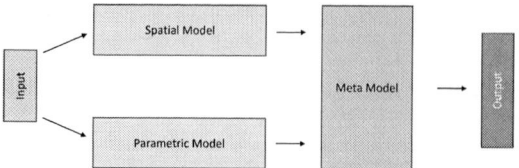

Fig. 5 Framework of fusion strategy: Parametric and spatial information processed by their respective AI models, fused via a stacking structure with learnable weights.

In essence, parametric-based methods provide valuable individual die-level insights but suffer from three critical gaps: limited consideration of spatial context, over-reliance on simplistic statistical assumptions, and scalability limits in advanced nodes.

B. Drawbacks of Spatial-Only Strategy

Spatial-based methods (particularly approaches related to the GDBN theory), excel at capturing clustered defects but suffer from their own set of limitations when applied in isolation.

A major issue is noise sensitivity. Fig. 5 shows the noisy spatial distributions and the resulting wafermap after denoising, which is usually used by AI-assisted GDBN methods as preprocessing technique. Wafer test data inherently contains random variations, causing noisy maps[17]. GDBN is sensitive to noise in wafer data; fixed-window methods struggle to distinguish noise from genuine clusters (see Fig. 5), leading to false positives and unnecessary yield loss.

Interpretability challenges also plague spatial-only GDBN. Deep GDBN models, especially those with complex architectures, act as "black boxes." Engineers find it hard to trace why a cluster is flagged, limiting root-cause analysis. This opacity is problematic in industries like automotive, where traceability and explainability are crucial.

Additionally, data scarcity hampers spatial-only GDBN. High-quality, labeled datasets of spatial defects are rare—manufacturing defects are inherently infrequent. Without sufficient training data, GDBN models struggle to generalize, reducing their accuracy in real-world screening.

These drawbacks highlight why GDBN benefits from integration with other data sources (e.g., parametric metrics) for robust reliability screening.

V. PERSPECTIVE: FUSION-BASED RELIABILITY SCREENING

A. Rationale for Fusion

In semiconductor manufacturing processes, defects often tend to cluster[8]. As Section IV highlights, parametric methods miss clustered defects, while spatial methods lack electrical validation—their weaknesses are reciprocal, making fusion imperative.

A simple linear fusion of parametric and spatial information—such as flagging a die as defective if either model does so—is flawed. It fails to leverage their complementarity, instead amplifying individual limitations, leading to severe overkill: many sound dies are misidentified, reducing yield and forfeiting fusion's core benefit of cutting false positives through cross-validation of parametric and spatial cues.

979-8-3315-3918-4/25 $31.00 © 2025 IEEE

TABLE I. Three Strategies: Key Attributes

	Parametric	Spatial	Fusion
Interpretability	√		√
Spatial Correlations	√	√	√
Scalability		√	√
Robustness			√
Detection of clustered latent risks		√	√

To enhance the effectiveness of reliability screening, a much deeper merging of parametric and spatial information is required. By integrating parametric data with spatial context, we can potentially develop more sophisticated models that can not only identify the location of potentially defective dies but also understand the nature of the defects based on their electrical characteristics. For instance, if a die in a spatially-flagged cluster also shows abnormal parametric values related to leakage current and resistance, it becomes more likely that there is a real defect present, such as a short circuit or a broken interconnect. That is, a die that is flagged as potentially defective by a spatial-based method can be further evaluated using parametric data to confirm or refute the suspicion.

Conversely, parametric-based anomaly detection can be complemented by spatial analysis to determine if the anomaly is part of a larger, spatially-related defect pattern. In essence, the fusion of parametric and spatial information has the potential to provide a more complete picture of the semiconductor device's quality and reliability, making it an essential direction for future research and development in the field of semiconductor testing.

B. Exploratory Results

To validate the feasibility of deep fusion between parametric and spatial information, we propose an comprehensive framework[18] that addresses the limitations of standalone methods and avoids the pitfalls of linear fusion (as illustrated in Fig. 6). This framework processes parametric and spatial information through their respective AI models, fused via a stacking structure with learnable weights—avoiding linear fusion pitfalls and leveraging their unique strengths, which directly addresses these complexities, where neither parametric nor spatial methods alone can keep pace. Their distinct characteristics, along with the advantages of the fusion strategy, are systematically compared in TABLE I.

VI. CONCLUSIONS

Semiconductor reliability screening stands at a critical evolutionary juncture, driven by the growing complexities of advanced node manufacturing. From early statistical parametric methods to AI-enhanced models, and from traditional fixed-window GDBN to Transformer-driven spatial frameworks, a clear reality emerges: standalone parametric strategies are both inadequate for cutting-edge fabrication.

As detailed in Section V.A, fusing these paradigms resolves their reciprocal limitations: parametric precision grounds spatial analysis, while spatial context refines parametric thresholds. Empowered by AI, this integration boosts robustness, reduces false positives, and uncovers subtle clustered latent defects critical to advanced nodes.

Looking ahead, dismantling information silos between parametric and spatial domains is key. Only through such deep fusion can the industry transcend current bottlenecks, moving toward the "zero-defect" vision vital for next-generation semiconductor ecosystems, where yield, quality, and efficiency converge.

REFERENCES

[1] M. J. Moreno-Lizaranzu and F. Cuesta, "Improving Electronic Sensor Reliability by Robust Outlier Screening," Sensors, vol. 13, pp. 13521-13542, Oct 2013.

[2] C. C. Lu, C. C. Chang, C. H. Yen, S. W. Chang, Y. H. Chu, K. C. Wu, et al., "Transformer and Its Variants for Identifying Good Dice in Bad Neighborhoods," 42nd VLSI Test Symposium (VTS), April 2024

[3] S. W. Li, C. H. Yen, S. W. Chang, Y. H. Chu, K. C. Wu, and M. C. T. Chao, "Wafer-View Defect-Pattern-Prominent GDBN Method Using MetaFormer Variant," 2024 IEEE International Test Conference (ITC), November 2024

[4] C. M. Liu, C. H. Yen, S. W. Lee, K. C. Wu, and M. C. T. Chao, "Enhancing Good-Die-in-Bad-Neighborhood Methodology with Wafer-Level Defect Pattern Information," 2023 IEEE International Test Conference (ITC), pp. 357-366, October 2023

[5] P. Lenhard, A. Kovalenko, and R. Lenhard, "Die level predictive modeling to reduce latent reliability defect escapes," Microelectronics Reliability, vol. 148, September 2023.

[6] S. Jayaram, H. J. Lee, D. Kim, S. Choi, S. Hong, S. Lee, et al., "Automotive Process Reliability Prediction for 5,7nm using ML," 2023 34th Annual SEMI Advanced Semiconductor Manufacturing Conference, May 2023

[7] S. Y. Lin, P. Y. Tan, C. W. Wu, M. D. Shieh, C. H. Chuang, and G. Liao, "Weak Die Screening by Feature Prioritized Random Forest for Improving Semiconductor Quality and Reliability," 6th IEEE International Test Conference in Asia (ITC-Asia), pp. 25-30, August 2022

[8] C. Xanthopoulos, A. Neckermann, P. List, K. P. Tschernay, P. Sarson, and Y. Makris, "Automated Die Inking," IEEE Transactions on Device and Materials Reliability, vol. 20, pp. 295-307, Jun 2020.

[9] C. H. Chuang, K. W. Hou, C. W. Wu, M. Lee, C. H. Tsai, H. Chen, et al., "A Deep Learning-Based Screening Method for Improving the Quality and Reliability of Integrated Passive Devices," 4th IEEE International Test Conference in Asia (ITC-Asia), pp. 13-18, September 2020.

[10] N. Sumikawa, J. Tikkanen, L. C. Wang, L. Winemberg, and M. S. Abadir, "Screening Customer Returns With Multivariate Test Analysis," 2012 IEEE International Test Conference, November 2012

[11] S. Wang and Y. Chen, "Improved Yield Prediction and Failure Analysis in Semiconductor Manufacturing with XGBoost and Shapley Additive exPlanations Models," 2024 IEEE International Symposium on the Physical and Failure Analysis of Integrated Circuits (IPFA), July 2024

[12] H. Chen, H. C. Lin, and M. J. Wang, "Smart GDBC Screening for High Quality IPD," E-Manufacturing and Design Collaboration Symposium (eMDC), September 2018

[13] T. S. Barnett, M. Grady, K. Purdy, and A. D. Singh, "Exploiting prediction defect clustering for yield and reliability," Iee Proceedings-Computers and Digital Techniques, vol. 152, pp. 407-413, May 2005.

[14] R. B. Miller and W. C. Riordan, "Unit level predicted yield: a method of identifying high defect density die at wafer sort," International Test Conference, pp. 1118-1127, November 2001

[15] M. Lee, C. T. Lu, C. H. Tsai, H. Chen, and M. J. Wang, "Site-aware Anomaly Detection with Machine Learning for Circuit Probing to Prevent Overkill," 4th IEEE International Test Conference in Asia (ITC-Asia), September 2020

[16] Y. S. Ma, F. Wang, Q. Xie, L. Hong, J. Mellmann, Y. Y. Sun, et al., "Machine Learning Based Wafer Defect Detection," Conference on Design-Process-Technology Co-Optimization for Manufacturability XIII, vol. 10962, February 2019

[17] K. S. M. Li, P. Y. Y. Liao, L. Chou, K. C. C. Chen, A. Y. A. Huang, S. J. Wang, et al., "PWS: Potential Wafermap Scratch Defect Pattern Recognition with Machine Learning Techniques," 25th IEEE European Test Symposium (ETS), May 2020

[18] Y. Wang and Z. Ji, "Fusion strategy for reliability screening," to be published.

Impact of Thermal Shock on the Threshold Voltage and Transconductance of FinFET I/O Devices

Yaolin Wang[1], Kun Chen[1,2], Wendi Wei[1], Zhao Yang[1], Zhiteng Zhang[1],
Zhuming Wang[1], Chen Wang*[1,2] and David Wei Zhang[1,2]

[1] School of Microelectronics, Fudan University, Shanghai 200433, China
[2] National Integrated Circuit Innovation Center, Shanghai 201203, China

* Email: yaolinwang24@m.fudan.edu.cn, chen_w@fudan.edu.cn

Abstract—The research on device reliability of transistors under high or low temperatures is quite extensive. However, the impact of rapid temperature changes on device performance has rarely been investigated. We propose a thermal shock experiment to study the degradation of device parameters induced by rapid switching between high and low temperature environments. The results show that after thermal shock, the threshold voltage of the device decreases while the transconductance increases. The magnitude of these changes increases linearly with the number of shock cycles. We characterize the traps of the device before and after thermal shock using stress-induced leakage current and charge pumping tests and discuss the mechanism by which thermal shock affects device performance.

Keywords—FinFET, thermal shock, thermal mechanical stress

I. INTRODUCTION

In modern electronic systems, integrated circuits are exposed to complex and highly variable environmental conditions. In particular, applications like automobiles and aerospace frequently encounter thermal shock - rapid and extreme temperature excursions[1-2]. Such abrupt temperature transitions can induce significant thermal mechanical stress within the CMOS devices, thereby affecting their electrical performance and reliability[3].

The effects of mechanical stress on CMOS devices have been extensively researched, leading to the development of theories such as the deformation potential theory. This theory enables the calculation of strain-induced changes in the band structure. Simplified models, including the piezoresistance model and the drift-diffusion approach, are widely used in TCAD simulations for analyzing band structure and carrier mobility[4-5]. Most strain related research efforts have focused on the application of strain engineering techniques[6]. In current FinFET technology, applying specific types of strain to different devices has become standard practice. Silicon-germanium (SiGe) is utilized in pMOS devices to provide compressive stress, while silicon carbide (SiC) is applied in nMOS devices to provide tensile stress.

Although some studies have examined the impact of thermal mechanical stress induced by the self-heating effect of devices on their performance[7-8]. Research into the influence of the accumulated thermal mechanical stress after multiple thermal shocks on the performance of the device is still relatively limited. When devices experience large, rapid temperature changes in a short period, thermal mechanical stress arises within the devices due to the differences in the thermal expansion coefficients of various materials. The accumulation of this thermal mechanical stress can significantly alter the structure of the device, potentially leading to the degradation of the device and even the device failure. Therefore, investigating the impact of the thermal shock on the performance of the FinFET devices holds significant value.

In this paper, a systematic thermal shock experiment was designed and conducted on the FinFET I/O devices. Electrical performance parameters were recorded before and after each thermal shock cycle, and we find out that the thermal shock reduces the absolute value of the threshold voltage (V_{TH}), and enhances the transconductance and the channel carrier mobility. A relationship model between the thermal shock cycles and the performance of the devices was established. Additionally, the underlying mechanisms by which the thermal shock causes the device degradation and eventual failure were analyzed.

II. EXPERIMENT

The experimental samples in this study are 14 nm FinFET I/O devices from a commercial foundry, which employ the high-κ metal gate (HKMG) and strained silicon technologies. As shown in Fig. 1(a), the gate oxide thickness of the devices is 3 nm, and the fins exhibit a pitch of 50 nm, with uniform dimensions of 10 nm in width and 50 nm in height, though slight sidewall tapering is observed. The use of the silicon germanium source and drain strain technology in the pFinFET I/O devices can be seen with the energy dispersive X-ray spectroscopy (EDS) (Fig. 1(b)).

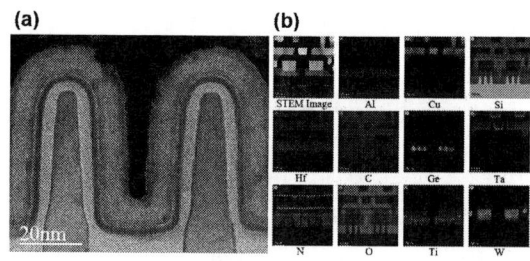

Fig. 1. (a) A cross-sectional transmission electron microscope (TEM) image of the FinFET I/O Device. (b) EDS images demonstrating the strain technology implementation of the pFinFET with strained SiGe in source/drain regions.

To simulate the thermal shock in the aerospace environment on the performance of the FinFET I/O devices, a thermal shock testing procedure is designed in this paper. The specific steps are as follows: each thermal shock cycle consists of two stages: low temperature (77 K or 273 K) and high temperature (273 K or 573 K). Initially, the devices are placed

979-8-3315-3918-4/25 $31.00 © 2025 IEEE

in the low-temperature environment for 15 minutes to ensure that the devices are thorough cooling down and thermal equilibrium[9]. Subsequently, the devices are then transferred to the high-temperature environment within five seconds, which would achieve a heating rate as high as 100 K/s. Subsequently the devices remain in the high-temperature environment for 15 minutes. Afterward, the devices are returned to a room-temperature environment for slow cooling. The electrical performance tests are conducted before and after each thermal shock cycle using Keysight B1500A semiconductor parameter analyzer, and the key electrical performance parameters of the devices, including the threshold voltage, the subthreshold swing (SS), and the maximum transconductance (gm_{max}), are extracted.

III. RESULT AND DISCUSSION

As shown in Fig. 2, the absolute value of the threshold voltage of the pFinFET I/O devices decreases as the number of the thermal shock cycles increases. While the maximum transconductance increases, the subthreshold swing shows minimal change and no distinct trend with the increasing thermal shock cycles. For comparison, the electrical characteristics of the nFinFET I/O devices on the same wafer were also measured after the same thermal shock exposure. Unlike the pFinFET I/O devices, the threshold voltage, the maximum transconductance, and the subthreshold swing of the nFinFET I/O devices did not show any significant trends with the increasing number of the thermal shock cycles, even after ten cycles. Given that the back-end-of-line (BEOL) interconnect processes for the nFinFET and the pFinFET devices on the same wafer are identical, the substantial changes observed only in the performance of the pFinFET I/O devices indicate that these alterations are not related to the BEOL processes.

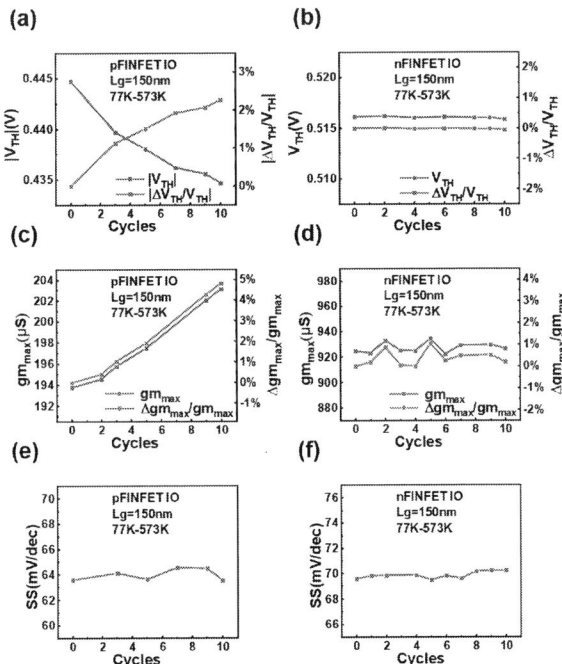

Fig. 2. (a) pFinFET I/O $|V_{TH}|$, (c) pFinFET I/O gm_{max}, (e) pFinFET I/O SS over thermal shock cycles, respectively. (b) nFinFET I/O $|V_{TH}|$, (d) nFinFET I/O gm_{max}, (f) nFinFET I/O SS over thermal shock cycles, respectively. With increasing thermal shock cycles, pFinFET I/O exhibits a decrease in the absolute value of V_{TH} and an increase in gm_{max}, whereas nFinFET I/O shows

negligible variations in both parameters. Notably, SS remains stable for both device types.

To further clarify the threshold voltage and the transconductance changes after the thermal shock on the pFinFET I/O devices, we systematically characterized devices subjected to the thermal shock cycling across various temperature ranges to further investigate the impact of thermal shock amplitude on their electrical performance. As depicted in Fig. 3(a), regardless of the temperature range, all the devices exhibited the consistent performance trends: a reduction in the threshold voltage, coupled with the increases in the transconductance and the channel carrier mobility (μ), while the subthreshold swing remained largely unaltered. What's more, Fig. 3(b) further demonstrates that the extent of the electrical parameter degradation directly correlates with the thermal shock amplitude. Devices exposed to wider temperature ranges exhibited more pronounced performance shifts, including a greater V_{TH} reduction and a higher Gm enhancement, compared to those subjected to narrower temperature ranges.

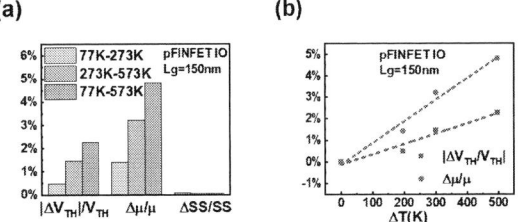

Fig. 3. pFinFET I/O (a) performance difference with different temperature ranges. SS remains almost invariant across all the temperature conditions. (b) $|\Delta V_{TH}/V_{TH}|$ and $\Delta\mu/\mu$ versus ΔT. The change of the performance scales linearly with the temperature swing amplitude of the thermal shock.

It's well-established that threshold voltage shifts can be attributed to either interface traps or gate oxide traps[10]. Consequently, we quantified both interface trap density and gate oxide trap density in this study to investigate their potential correlation with threshold voltage shifts induced by thermal shock.

Among various characterization techniques, charge pumping is an efficient and sensitive method for measuring interface trap density[11]. By applying periodic pulses to the gate and utilizing the injection-trapping process of interface traps, the substrate current or source-drain current generated by recombination can be measured. For this study, measurements were performed at ambient room temperature with the substrate grounded. Periodic gate pulses with varying baseline voltages were applied while the source and drain terminals were short-circuited, allowing the resulting source-drain current to be measured.

By applying the formula (1) for calculating the charge pumping current, we can accurately compare the interface trap density before and after thermal shock. Where N_{it} is the interface trap density, I_{cp} is the charge pumping current, f is gate pulse frequency, A_G is gate oxide area, and q is the electron charge.

$$N_{it} = I_{cp} / (f \cdot A_G \cdot q) \qquad (1)$$

Substituting the gate pulse frequency of 100 MHz, the gate length of 150 nm, and the gate width of 2200 nm into the formula (1), we calculated the interface trap density before and after the thermal shock. The N_{it} before the thermal shock was 4.46E10cm^{-2}, and after 10 thermal shock cycles, it was 4.48E10cm^{-2}, resulting in a ΔN_{it} of 0.4%. It can be seen that compared to the change in the threshold voltage of the device, the interface trap density did not change significantly before and after 10 thermal shock cycles, this illustrates that the thermal shock did not generate a significant number of new interface traps in the devices. This indicates that the threshold voltage shift of the devices is not primarily caused by the interface trap.

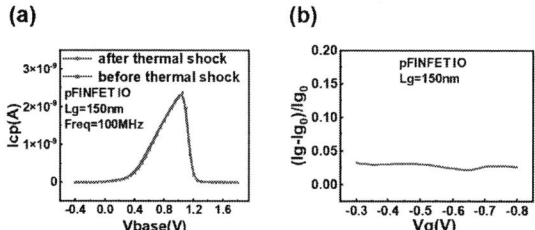

Fig. 4. (a) Comparison of charge pumping I_{cp}-V_{base} curve of pFinFET I/O device before and after 10 thermal shock cycles, the interface trap density remains largely unchanged. (b) SILC test of pFinFET I/O devices, the result shows the gate oxide trap density shows no significant changes after 10 thermal shock cycles.

Stress-induced leakage current (SILC) is a simple and effective means of characterizing gate oxide traps[12]. According to the SILC formula (2), where I_g represents the gate current after thermal shock and I_{g0} is the gate current before thermal shock, the SILC component (I_g-I_{g0}) directly corresponds to the current generated by the trap assisted tunneling of the gate oxide layer. Thus, the type of the oxide layer trap can be identified by the SILC spectrum, and the magnitude of SILC quantifies the change in the number of gate oxide layer traps[10].

$$SILC = \frac{I_g - I_{g0}}{I_{g0}} \qquad (2)$$

As shown in Fig. 4 (b), after 10 thermal shock cycles, no significant peaks appeared in the SILC spectrum. Compared to the initial state, the gate current of the device increases by approximately 2.5%, showing that thermal shock generated only a limited number of new oxide traps in the gate oxide layer. Moreover, the measured gate oxide trap density increases slightly with the number of thermal shock cycles. This finding contradicts the observed decrease in the absolute value of the threshold voltage for pFinFET in this study. It is well-established that gate oxide traps in pFinFET capture holes, which would typically increase the absolute value of the threshold voltage. Therefore, these findings suggest that the observed threshold voltage shift is not primarily caused by charge capture in the gate oxide traps.

An additional experiment aimed to verify whether the initial trap density level influences the device's response to thermal shock has been conducted. Another device with a lower initial interface trap density has been tested for comparison, the measured initial N_{it} was 1.11E10cm^{-2} before the thermal shock, which changed to 1.10E10cm^{-2} after 10 thermal shock cycles, resulting in a ΔN_{it} of 0.9%. The charge pumping result shows that even with a lower initial interface trap density, there are minimal changes in interface trap

density after thermal shock, reinforcing the conclusion that thermal shock does not significantly alter the interface trap density. Moreover, as shown in Fig. 5, the trend of the electrical characteristic changes in this low interface trap density device is consistent that observed in the device with a higher initial trap density. The threshold voltage of the pFinFET I/O device decreases by about 2.2%, while the carrier mobility increases by about 5.0% after 10 thermal shock cycles, which indicates that the performance changes of the pFinFET I/O devices are not related to the difference of the initial interface trap density.

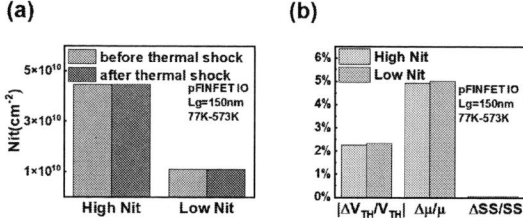

Fig. 5. (a) Interface trap density difference before and after the thermal shock for two different pFinFET I/O devices with initial high and low Nit. (b) Performance impact versus initial trap densities. The initial difference in interface trap density does not affect the performance changes caused by the thermal shock in the devices.

In conjunction with the trend of the noticeable increasing transconductance, the channel carrier mobility increases with the increasing number of thermal shock cycles. Given that silicon-germanium is incorporated into this 14 nm pFinFET device to provide compressive stress and enhance channel carrier mobility, we hypothesize that thermal shock introduces additional compressive stress into this pFinFET. This hypothesis is supported by the significant difference in the linear thermal expansion coefficients of germanium and silicon, with germanium's coefficient being nearly twice that of silicon. Under thermal shock, which involves drastic temperature changes, this difference in thermal expansion coefficients may lead to the higher compressive mechanical stress in the channel. Such stress can alter the band structure of the channel[13], thereby changing the threshold voltage and carrier mobility of pFinFET.

Moreover, the measurements indicate that changes in the pFinFET performance are directly proportional to the amplitude of thermal shock. A broader temperature range leads to more significant shifts in threshold voltage and enhancements in channel carrier mobility in the devices when compared to narrower temperature ranges. It is well established that a larger temperature change causes a greater mismatch in thermal expansion, thereby generating higher thermal mechanical stress. The observed relationship between the performance changes and the thermal shock amplitude further corroborates that the thermal mechanical stress is the primary factor that drives the performance changes of the device under the thermal shock conditions.

IV. SUMMARY

This paper systematically investigates the impact of the thermal shock on the electrical performance of the FinFET I/O devices. By comparing the performance changes of the nFinFET I/O and pFinFET I/O devices on the same wafer before and after the thermal shock, we find that the performance changes of the devices induced by the thermal shock are not related to the back end interconnect processes.

Furthermore, through the charge pumping tests and the stress-induced leakage current measurements, it is explicitly demonstrated that the change in the threshold voltage and the transconductance induced by thermal shock is unrelated to the interface traps and the gate oxide traps. Besides, we find out that the initial interface trap density in the devices doesn't influence the trends of the electrical characteristic changes. Based on these observations, we conclude that the thermal shock introduces new thermal mechanical stress in the pFinFET, leading to the changes in the band structure, a decrease in the absolute value of the threshold voltage, and an increase in the carrier mobility.

ACKNOWLEDGMENT

The authors would like to thank the National Integrated Circuit Innovation Center for its invaluable support and for providing access to its IC common-technology R&D and testing platform, both of which were essential to the successful completion of this research.

REFERENCES

[1] J. H. L. Pang and T. H. Low, "Modeling thermal cycling and thermal shock tests for FCOB," ITherm 2002. Eighth Intersociety Conference on Thermal and Thermomechanical Phenomena in Electronic Systems (Cat. No.02CH37258), San Diego, CA, USA, 2002, pp. 987-992.

[2] M.A. Belaïd and A. Nahhas, "Experimental and numerical studies on the power RF N-LDMOS transistor under cold and hot thermal shock tests based aging mechanism," Results in Engineering, Volume 17, 2023.

[3] L. Anoldo et al., "Study of the Thermomechanical Strain Induced by Current Pulses in SiC-Based Power MOSFET," in IEEE Electron Device Letters, vol. 42, no. 7, pp. 1089-1092, July 2021.

[4] Kuan-Ting Chen, Ren-Yu He, Yun-Fang Chung, Min-Hsin Hsieh, Shu-Tong Chang, "Mobility model based on piezoresistance coefficients for Ge 3D transistor, " in Solid State Electronics Letters, Volume 1, Issue 2, 2019, Pages 92-97.

[5] M. Koganemaru, K. Yoshida, T. Ikeda, N. Miyazaki and H. Tomokage, "Device Simulation for Evaluating Effects of Inplane Biaxial Mechanical Stress on n-Type Silicon Semiconductor Devices," in IEEE Transactions on Electron Devices, vol. 58, no. 8, pp. 2525-2536.

[6] Weimin Zhang and J. G. Fossum, "On the threshold Voltage of strained-Si-Si/sub 1-x/Ge/sub x/ MOSFETs," in IEEE Transactions on Electron Devices, vol. 52, no. 2, pp. 263-268, Feb. 2005.

[7] H. Duan, E. Li, W. Zang, Q. Huang, Y. Wang and W. Chen, "Theoretical Study of Self-Heating-Induced Thermal Stress Effects on Quantum Transport in p-Type Ultrathin Body-FinFET by Multiphysics Simulation," in IEEE Transactions on Electron Devices, vol. 70, no. 8, pp. 4001-4007, Aug. 2023.

[8] Y. Wang et al., "An Artificial Neural Network Model for Electro-Thermal Effect Affected Hot Carrier Injection Reliability in 14-nm FinFETs," in IEEE Transactions on Microwave Theory and Techniques, vol. 70, no. 11, pp. 4827-4834, Nov. 2022.

[9] S. Zhou and J. Wang, "An RF Stress-Based Thermal Shock Test Method for a CMOS Power Amplifier," in IEEE Journal of the Electron Devices Society, vol. 9, pp. 1024-1029, 2021.

[10] Z. Yu, J. Zhang, R. Wang, S. Guo, C. Liu and R. Huang, "New insights into the hot carrier degradation (HCD) in FinFET: New observations, unified compact model, and impacts on circuit reliability," 2017 IEEE International Electron Devices Meeting (IEDM), San Francisco, CA, USA, 2017, pp. 7.2.1-7.2.4.

[11] D. Veksler, G. Bersuker, A. Koudymov, C. D. Young, M. Liehr and B. Taylor, "Comprehensive analysis of charge pumping data for trap identification," 2011 International Reliability Physics Symposium, Monterey, CA, USA, 2011, pp. GD.4.1-GD.4.5.

[12] L. Pantisano and K. P. Cheung, "Stress-induced leakage current (SILC) and oxide breakdown: are they from the same oxide traps?" in IEEE Transactions on Device and Materials Reliability, vol. 1, no. 2, pp. 109-112, June 2001.

[13] L. Yu, W. -Y. Chang, K. Zuo, J. Wang, D. Yu and D. Boning, "Methodology for analysis of TSV stress induced transistor variation and circuit performance," Thirteenth International Symposium on Quality Electronic Design (ISQED), Santa Clara, CA, USA, 2012, pp. 216-222.

979-8-3315-3918-4/25 $31.00 © 2025 IEEE

Effects of Total Ionizing Dose on ESD Performance in High-Voltage SCR with Double Snapback Characteristics

Yujie Liu [1,2], and Xiangliang Jin *[1,2]

[1] School of Physics and Electronics, Hunan Normal University, Changsha 410081, China
[2] Key Laboratory of Physics and Devices in Post-Moore Era, College of Hunan Province, Changsha 410081, China

* Email: Yujie Liu@727805321@qq.com, Xiangliang Jin@jinxl@hunnu.edu.cn

Abstract—A high-voltage silicon-controlled rectifier (HVSCR) for on-chip electrostatic discharge (ESD) protection is developed using a 0.18 μm bipolar CMOS DMOS (BCD) process. With the extra shunt n-p-n bipolar junction transistor (BJT) path added parallelly to the main SCR path, this device exhibits superior ESD characteristics like double snapback, high holding voltage and holding current. Total-ionizing-dose (TID) irradiation of the devices using ^{60}Co γ-rays, followed by Transmission Line Pulse (TLP) testing. Experimental results demonstrate that the device exhibits a TID tolerance exceeding 150 krad(Si), indicating its potential as a robust ESD protection solution for space applications. Furthermore, two-dimensional TCAD simulations were conducted to investigate the device operation mechanism and radiation effects to reveal the impact of radiation-induced interface traps on the device's ESD performance.

Keywords—silicon-controlled rectifier (SCR), electrostatic discharge (ESD), total-ionizing-dose (TID)

I. INTRODUCTION

In the current era, integrated circuits (ICs) are threatened by electrostatic discharge (ESD) from humans [1], [2], machines [3], [4], and charged devices [5]. Consequently, on-chip ESD protection is critical to the reliability of silicon-based integrated circuit products. SCR devices offer superior performance in terms of discharge area efficiency and reduced parasitic effects [6], [7]. However, traditional SCRs are prone to latch-up risks due to their low holding voltage [8]. Currently, extensive research focuses on increasing the holding voltage of SCRs to meet the latch-up immunity requirements for high-voltage applications. These approaches include widening the base region of the parasitic bipolar junction transistor (BJT) [9], segmenting the emitter of the parasitic BJT [10], [11] and stacking unit devices [12]. However, all of these methods inevitably reduce the current handling capability of SCRs or significantly increase the layout area, thereby decreasing their discharge efficiency per unit area. Furthermore, ICs designed for space or terrestrial nuclear environments require careful consideration of radiation effects on silicon devices. The dominant radiation-induced electrical phenomena include transient single-event effects (SEEs) and cumulative total ionizing dose (TID) effects [13]. Research has shown that ESD devices themselves exhibit some level of immunity to single-event effects [14]. The single event strike induced high current

transient along the incident path can be handled by ESD devices for its discharge ability for excessive charges. It has been reported that Transmission Line Pulse (TLP) testing can be used to evaluate the single-event burnout threshold [15]. Therefore, the radiation effects that should be specially emphasized is TID effects for ESD devices. The trapped charge in the oxide layer and interface traps generated by the TID effect can shift the ESD characteristics of the device beyond the design window, reducing I_{t2} and compromising the protection functionality of the ESD device [16]. This requires a focused investigation into the mechanisms of TID degradation in radiation-hardened ESD protection solutions.

In order to address the limitations of traditional SCRs, we designed a high-voltage SCR (HVSCR) structure [17]. In this structure, we embedded an additional BJT shunting path on the surface of the device. This shunting path enables the HVSCR to exhibit double snapback characteristics, along with relatively high holding voltage and holding current. This paper systematically investigates the ESD characteristics of HVSCR through a combined experimental and technology computer-aided design (TCAD) simulation approach. The results mechanistically elucidate how radiation-induced interfacial traps affect ESD performance and provide important insights for the development of radiation-resistant ESD protection devices.

II. EXPERIMENT DETAILS

The HVSCR device in this work is fabricated using a standard 0.18 μm BCD process for high voltage ESD protection of 12 V ICs. The device structure and equivalent circuit are shown in Fig. 1. It can be seen that the HVSCR has more discharge paths provided by Q_{NPN2} than a conventional SCR device. When the electrostatic pulse arrives, the HVSCR initially activates the Q_{NPN1} through the avalanche effect at the NW/PW reverse-biased PN junction. This corresponds to the first snapback event. As the current increases, Q_{PNP} and Q_{NPN2} gradually begin to turn-on. The electron current flowing through Q_{NPN1} provides base current to the Q_{PNP}, creating a positive feedback effect that results in a second snapback. The inherent SCR path, composed of Q_{PNP} and Q_{NPN1}, along with Q_{NPN2}, forms a current shunting mechanism. During the current discharge process, some minority carriers in the base region of Q_{NPN1} are shunted by Q_{NPN2}, leading to a sharp reduction in the current gain (β) of Q_{NPN1}. As a result of this surface current path shunting effect, the HVSCR structure requires a higher voltage to maintain conduction in the SCR path.

This work was supported in part by the National Natural Science Foundation of China under Grant 62174052, in part by the Postgraduate Scientific Research Innovation Project of Hunan Province under Grant CX20240532.

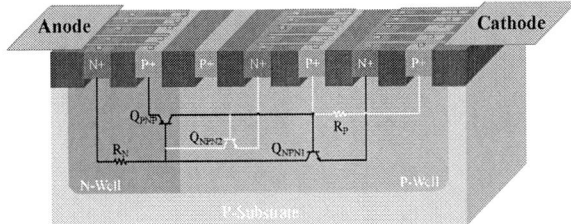

Fig. 1. Cross-sectional view and equivalent circuit of the HVSCR device.

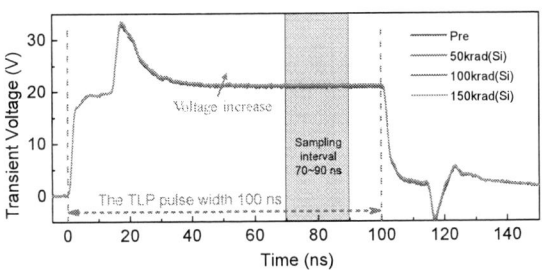

Fig. 2. TLP test results of HVSCR at different total doses of radiation.

Fig. 3. Measured transient voltage curves of the HVSCR device under a 20 V TLP pulse at different doses.

The devices were exposed to a ^{60}Co γ-ray radiation source with a dose rate of 100 rad(Si)/s. A total of four device samples were used: one unirradiated sample for pre-irradiation measurements (Pre), and three irradiated samples subjected to doses of 50, 100, and 150 krad(Si) under a bias voltage of 13.2 V. Immediately following the irradiation experiments, TLP testing was performed on all irradiated devices. TLP testing was conducted immediately after irradiation for each irradiated device. The rise time of the TLP pulse was 10 ns, with a pulse width of 100 ns. During the test, the bias voltage was set to 13.2 V to monitor the leakage current.

III. RESULTS AND DISCUSSION

A. TLP Test Results

The TLP test results for the HVSCR device, before and after exposure to different total ionizing doses, are shown in

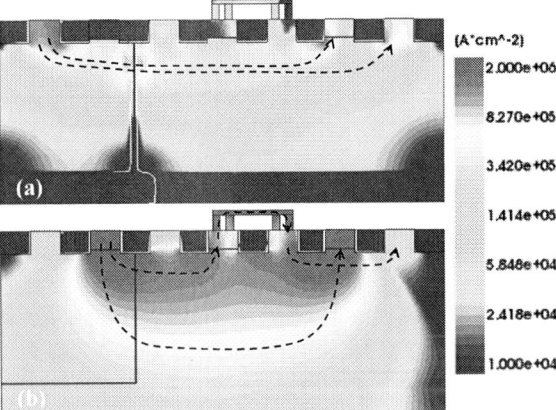

Fig. 4. The TCAD simulated current density distribution of the HVSCR under (a) 0.5 mA and (b) 0.1 A current pulses before TID irradiation.

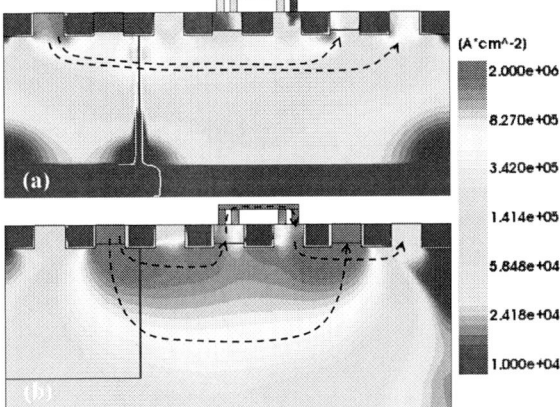

Fig. 5. The TCAD simulated current density distribution of the HVSCR under (a) 0.5 mA and (b) 0.1 A current pulses after TID irradiation.

Fig. 2. The TLP test curves exhibit a double snapback characteristic. As depicted in the equivalent circuit in Fig. 1, there are two ESD charge flow paths. The main SCR path is composed of Q_{PNP} and Q_{NPN1}, similar to conventional SCR devices. An additional path is formed by Q_{PNP} and Q_{NPN2}, which accounts for the observed double snapback characteristic. When the ESD current arrives, Q_{NPN1} is triggered earlier due to its higher base-emitter parasitic resistance, and this path contributes to the first snapback. As the ESD amplitude increases, Q_{PNP} and Q_{NPN2} gradually turn on. Since Q_{NPN2} diverts some of the base current from Q_{NPN1}, the current gain of Q_{NPN1} decreases. The triggering condition for the intrinsic SCR, $\beta_{QPNP} \times \beta_{QNPN1} > 1$, is only met when the current is larger, resulting in the second snapback. This mechanism helps the HVSCR achieve higher holding voltage and holding current.

Furthermore, we observe that the HVSCR maintains a high level of ESD robustness after 150 krad (Si) irradiation. The holding voltage increases slightly with increasing total dose. Fig. 3 shows the measured transient voltage curves of the HVSCR device under a 20 V TLP pulse. These curves nearly overlap, exhibiting a similar shape, with a slight increase in the clamping voltage. This behavior corresponds to the trend observed in the variation of the holding voltage. These results show that HVSCR has good ESD robustness and stability in radiation environments. Table I summarizes

TABLE I. COMPREHENSIVE PERFORMANCE COMPARISON

	Ref. [19]	*Ref. [20]*	*Ref. [21]*	*This work*
V_{t1} (V)	42.3	30	14.97	**22.94**
V_h (V)	8.28	17	5.37	**16.65**
I_{t2} (A)	10.70	9	4.53	**5.81**
FoM (mA/μm²)	0.33	1.52	-	**1.88**
TID Tolerance krad(Si)	100	~200	100	**>150**

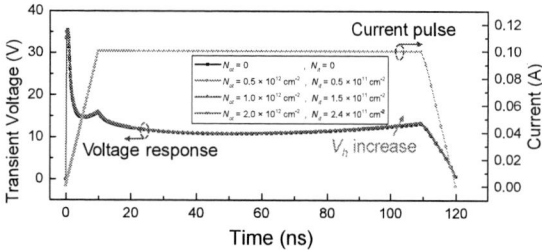

Fig. 6. The TCAD simulated transient voltage response curves of the HVSCR for different TID radiation with an ESD current pulse of 0.1A.

the performance of recently proposed SCR-based ESD protection devices. The results show that the proposed HVSCR achieves the highest figure of merit (FoM) of 1.88. After exposure to 150 krad (Si) irradiation, no decrease in I_{t2} was observed. These findings highlight the significant potential of this device for ESD protection in 12 V ICs used in space applications.

B. ESD Current Path Simulation

The effects of radiation on the ESD characteristics of the HVSCR device is studied through TCAD Sentaurus tools. The physical models considered include collision ionization, Fermi-Dirac statistical distribution, hydrodynamic carrier transport model, the Shockley-Read-Hall (SRH) model, Auger recombination, and bandgap narrowing. To simulate the electrical characteristics of the HVSCR device under transient pulses at different current levels, the anode is subjected to pulses of 0.5 mA and 0.1 A. The rise and fall times of the pulses are 10 ns, with a pulse width of 100 ns. Fig. 4 shows the current density distribution of the TCAD-simulated pre-irradiated HVSCR under two current pulses, corresponding to the current path of the device during initial conduction and after full conduction, respectively. The initial current discharge path of the HVSCR is primarily governed by the Q_{NPN} structure and the reverse-biased N-Well/P-Well junction. The anode current is mainly concentrated in the N+ region, while the cathode current density is significantly distributed across both the N+ and P+ regions. Due to the larger base-emitter parallel resistance of Q_{NPN1} compared to Q_{NPN2}, the avalanche carriers initially cause a positive bias at the emitter junction of Q_{NPN1}, leading to its conduction first. This is the cause of the first snapback occurrence. When the current increases to 0.1 A, as shown in Fig. 4(b), the inherent SCR path of the HVSCR has formed and becomes the primary discharge path. The current distribution becomes wider and more uniform, penetrating deeper into the device. The peak anode current density shifts toward the P+ region, while the cathode current density is primarily distributed in the N+ region. In the inherent SCR path, the base and

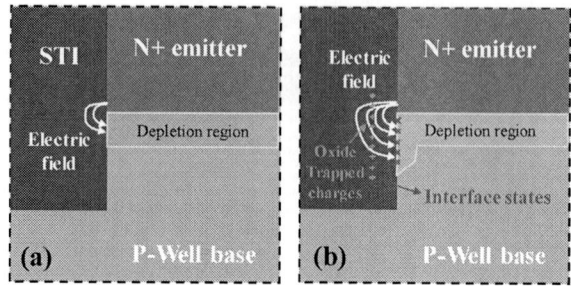

Fig. 7. Magnified cross-section of base-emitter of the QNPN1 (a) before and (b) after TID irradiation.

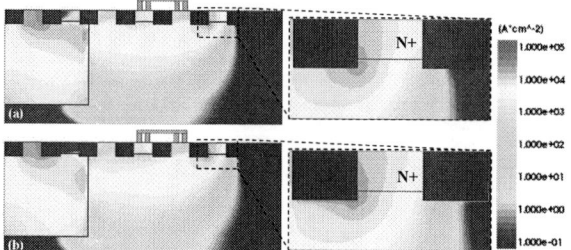

Fig. 8. Simulation results of the electron current density (a) before and (b) after radiation for the HVSCR device under a 0.5mA ESD current.

collector of Q_{PNP} and Q_{NPN1} are coupled to provide a driving current, which initiates a positive feedback loop that reliably releases the ESD current. This led to the second snapback. The above simulation results verify the working mechanism of HVSCR.

To further investigate the effect of irradiation on the HVSCR, a TCAD simulation was conducted to examine the TID effect on the HVSCR device. This simulation incorporated fixed positive charge (N_{ot}) in the SiO₂ layer and interface trap charges (N_{it}) between the SiO₂ and Si layers. The N_{ot} values are set to 0, 0.5, 1.0, and 2.0 × 10¹² cm⁻², while the N_{it} values are set to 0, 0.5, 1.5, and 2.4 × 10¹¹ cm⁻² [18]. The current density distribution of the HVSCR device after irradiation is shown in Fig. 5. Since the main current path of the HVSCR is deep inside the device, the surface shallow trench isolation (STI) interface trap has little effect on the ESD characteristics of the HVSCR. Especially under high current conditions, in which the main current flows deep and bypasses the STI, the current distribution of the HVSCR remains almost constant.

C. TID effect on holding voltage simulation

Fig. 6 shows the simulated transient voltage response of the HVSCR under different TID irradiations, with an ESD current pulse of 0.1 A. The transient voltage increases slightly with increasing N_{ot} and N_{it}, which coincides with the test results in Fig. 3. Fig. 7(a) and (b) show the amplified cross-sections of the base-emitter junction of Q_{NPN1} before and after irradiation. N_{ot} cause the depletion region to widen in the P-Well junction area, also modulating the surface potential. This modulation increases the interface SRH recombination rate per unit area.

Additionally, the increase in N_{it} exposes more recombination sites that contribute to the base recombination current. Both effects contribute to the increase in the base recombination current near the edge of the STI/P-Well interface in the emitter region. The simulation results of the electron current density before and after radiation of the

979-8-3315-3918-4/25 $31.00 © 2025 IEEE

HVSCR device at 0.5 mA ESD current are shown in Fig. 8. After ionizing radiation, the electron current in the P-Well below the STI of the device, particularly in the base region of Q_{NPN1} near the emitter junction, shows a slight increase. Since the collector current remains nearly constant, the increase in base current results in a decrease in the amplification factor β_{QNPN1}. Simulation results indicate that the slight increase in the device's holding voltage after total dose radiation is due to a slight reduction in β_{QNPN1} and a slight weakening of the positive feedback effect of the SCR. Consequently, the holding voltage of the device increases slightly.

IV. CONCLUSION

A HVSCR for on-chip ESD protection has been designed and validated in the standard 0.18 μm BCD process. Benefit to the additional surface n-p-n path, the device exhibits double snapback characteristics and a high holding voltage. TID radiation experiments and TLP testing reveal that the HVSCR device's TID tolerance exceeds 150 krad(Si), with a slight increase in holding voltage after radiation exposure. Furthermore, TCAD simulations confirmed the operational mechanism of the HVSCR, showing that the increase in holding voltage is due to radiation-induced charge trapping and the interface traps, which increase the base recombination current of Q_{NPN1}.

REFERENCES

[1] H Yi-Jie Huang, Ming-Dou Ker, Investigation of Human-Body-Model and Machine-Model ESD Robustness on Stacked Low-Voltage Field-Oxide Devices for High-Voltage Applications. IEEE Transactions on Electron Devices, 2016, 63(8): 3193-3198.

[2] Jian-Hsing Lee, Natarajan Mahadeva Iyer, Analytical Model of Correlation Factor for Human-Body Model to Transmission-Line Pulse ESD Testing. IEEE Electron Device Letters, 2017, 38(7): 952-954.

[3] Vadim Kuznetsov, HBM, MM, and CBM ESD Ratings Correlation Hypothesis. IEEE Transactions on Electromagnetic Compatibility, 2018, 60(1): 107-114.

[4] Ming-Dou Ker, Hsin-Chyh Hsu, Jeng-Jie Peng, Electrostatic discharge implantation to improve machine-model ESD robustness of stacked NMOS in mixed I/O interface circuits. Fourth International Symposium on Quality Electronic Design, 2003. Proceedings, 2003, San Jose, USA.

[5] Yi-Chun Huang, Ming-Dou Ker, Investigation of CDM ESD Protection Capability Among Power-Rail ESD Clamp Circuits in CMOS ICs With Decoupling Capacitors. IEEE Journal of the Electron Devices Society, 2022, 11: 84-94.

[6] Milova Paul, B. Sampath Kumar, Kranthi Karmel Nagothu, Pulkit Singhal, Harald Gossner, Mayank Shrivastava, Drain-Extended FinFET with Embedded SCR (DeFinFET-SCR) for High-Voltage ESD Protection and Self-Protected Designs. IEEE Transactions on Electron Devices, 2019, 66(12): 5072-5079.

[7] Rong-Kun Chang, Bo-Wei Peng, Ming-Dou Ker, Schottky-Embedded Silicon-Controlled Rectifier with High Holding Voltage Realized in a 0.18-μm Low-Voltage CMOS Process. IEEE Transactions on Electron Devices, 2021, 68(4): 1764-1771.

[8] Wenqiang Song, Ruibo Chen, Zhuang Tong, Fei Hou, Feibo Du, Zhiwei Liu, Hongxia Liu, Robust Silicon-Controlled Rectifier with High-Holding Voltage for On-Chip Electrostatic Protection. IEEE Transactions on Electron Devices, 2022, 69(2): 696-703.

[9] Song S, Du F, Hou F, Song W, Liu, Z, and Liu J, "A New dual directional SCR with high holding voltage for High Voltage ESD protection," 2019 IEEE International Conference on Electron Devices and Solid-State Circuits (EDSSC). IEEE, 2019: 1-2.

[10] Du F, Hou F, Liu Z, Liu J, and Liou J J, "Bidirectional silicon‐controlled rectifier for advanced ESD protection applications," Electronics Letters, 2019, 55(2): 112-114. doi: 10.1049/el.2018.6686.

[11] Liu Z, Liou J J and Vinson J, "Novel silicon-controlled rectifier (SCR) for high-voltage electrostatic discharge (ESD) applications," IEEE electron device letters, 2008, 29(7): 753-755.

[12] Liu Z W, Liou J J, Dong S R, Han Y, "Silicon-controlled rectifier stacking structure for high-voltage ESD protection applications," IEEE Electron Device Letters, 2010, 31(8): 845-847.

[13] Zhengxuan Zhang and Shichang Zou, Radiation hardening technology in SOI materials and devices. Kexue Tongbao/Chinese Sci. Bull., 2017, 62(10): 1004-1017.

[14] Moon-Kyu Cho, Ickhyun Song, Fleetwood E. Zachary, Ani Khachatrian, Jeffrey H. Warner, Stephen P. Buchner, Dale McMorrow, Pauline Paki, John D. Cressler, "Best Practices for Using Electrostatic Discharge Protection Techniques for Single-Event Transient Mitigation," IEEE Transactions on Nuclear Science, vol. 66. no. 1, pp. 240-247, Jan. 2019.

[15] M. Hamlyn, P. L. Hower, K. Warren, R. C. Baumann, "Transmission Line Pulse Test Method for Estimating SEB Performance of n-Channel Lateral DMOS Power Transistors," IEEE Transactions on Nuclear Science, vol. 65, no. 1, pp. 249-255, Jan. 2018.

[16] Lin Zhongyu, Zhong Daohong, Ma Zhen, "Breakdown Voltage Shift Mechanism Analysis of Bi-directional ESD Protection Device After Total Ionizing Dose Test," in 2022 International EOS/ESD Symposium on Design and System (IEDS), Chengdu, China, Nov. 2022, pp. 1-5.

[17] Yujie Liu, Yang Wang, Xiangliang Jin, Jian Yang, Yan Peng, Jun Luo, A novel robust SCR with high holding voltage for on-chip ESD protection of industry level bus. Solid State Electronics, 2023, 208: 108762.

[18] Chao Peng, Zhiyuan Hu, Zhengxuan Zhang, Huixiang Huang, Bingxu Ning, Dawei Bi, "Total ionizing dose effect in 0.2 lm PDSOI NMOSFETs with shallow trench isolation," Microelectronics Reliability, vol. 54, pp. 730-737, 2015.

[19] Wu M, Lu W, Zhang C, et al. The impact of radiation and temperature effects on dual-direction SCR devices for on-chip ESD protections[J]. Semiconductor Science and Technology, 2020, 35(4): 045016.

[20] Duan Y, Li X, Lu P, et al. High-voltage ESD protection devices with high robustness of 13 kV and strong radiation tolerance up to 200 krad (Si)[J]. IEEE Transactions on Electron Devices, 2024, 71(6): 3518-3524.

[21] Zhuojun Chen, Zhiqiang Wu, Ming Wu, Wei Peng, Yun Zeng, Xiangliang Jin, Binhong Li, Bo Li, "A Comprehensive Study of a Bidirectional ESD Protection Device Under Harsh Environment," in 2019 12th International Workshop on the Electromagnetic Compatibility of Integrated Circuits (EMC Compo), Hangzhou, China, Dec. 2019, pp. 111-113.

A New Surge Protection Circuit with Low Dynamic Leakage Current

Zhiqiang Hu, Ran Ye*, Qiao Kang, Ke Cui, Hao Luo, Weipeng Ye, Siyang Liu*, Weifeng Sun

National ASIC System Engineering Research Center, Southeast University, Nanjing, China

*Email: 101300309@seu.edu.cn, liusy2017@seu.edu.cn

Abstract—This study proposes a new MOSFET-based power clamp circuit that has robust surge protection capacity while achieving dramatic suppression of dynamic leakage current during rapid power-on process. The design has been experimentally validated. A dynamic leakage suppression circuit at the gate of the discharge field-effect transistor ensures it off under rapid power-on events, effectively reducing the dynamic leakage current. Practical applications and performance comparisons with traditional circuits are also included.

Keywords—Clamp circuit, surge, dynamic leakage current, rapid power-on process, dynamic leakage suppression circuit

I. INTRODUCTION

Surge, as a high-energy form of electrical over-stress (EOS), poses a serious threat to semiconductor devices and integrated circuits[1]. It may cause irreversible thermal damage, fusing, or dielectric breakdown, leading to abnormal device operation or even permanent failure. To address this potential risk, integrating surge protection circuits between sensitive circuits and surge sources (such as power ports or signal interfaces) is particularly important.

The prevailing approach for surge protection utilizes integrated diode-triggered clamping circuits, which generally consist of a voltage sensing module, a gate driver, and a protection field-effect transistor (FET)[2]. The voltage sensing module is used to set the clamp breakdown voltage (V_{BR}). When the input voltage (V_{IN}) is lower than V_{BR}, the gate driver and the protection FET are in an off state, and the circuit has no static power consumption current. However, when the input voltage (V_{IN}) exceeds the breakdown voltage (V_{BR}), the gate driver is activated, thereby initiating conduction in the integrated protection field-effect transistor and creating a low-impedance discharge path for surge currents originating from the V_{DD}/IO pins[3], [4]. Despite these advantages, conventional diode-triggered clamp circuits suffer from a critical limitation: during the rapid power-on process, the clamp circuits have significant dynamic leakage current due to unintended gate overdrive in the protection field-effect transistor. The lack of a dedicated suppression circuit for this phenomenon leads to a substantial increase in dynamic power consumption and may cause damage to the protected circuit and instability of the system.

This paper proposes a novel surge protection circuit design with low dynamic leakage current characteristics. Simulation and experimental results show that this circuit not only effectively achieves the surge protection function but also significantly reduces dynamic leakage current during the rapid power-on process of the system, thereby effectively controlling dynamic power consumption.

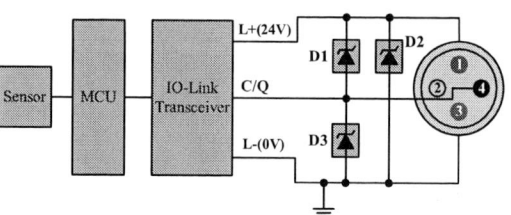

Fig.1. Schematic of the Proposed Surge Protection Circuit Integrated into IO-Link Transceiver.

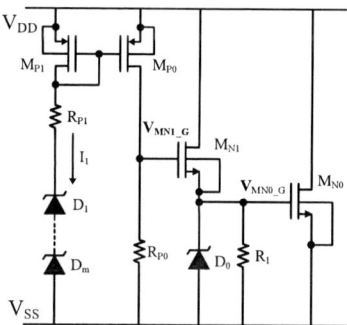

Fig.2. Schematic of traditional surge protection circuit.

II. CONVENTIONAL SURGE PROTECTION CIRCUIT

Fig.1 shows the topology of the surge protection circuits integrated into the IO-Link transceiver[5]. Clamp circuits is deployed between the 24 V power supply line (L+) and the signal line (C/Q) to resist possible high-voltage transient impacts. Fig.2 illustrates a traditional surge protection circuit, whose working principle is based on the detection of the supply voltage (V_{DD})[6]. A voltage detector is formed by a resistor, Zener diodes, M_{P0} and M_{P1}. When V_{DD} exceeds the trigger voltage of the protection circuit, the mirror circuit is turned on and the M_{N1} is activated. The M_{N1} raises the gate voltage of the M_{N0}. M_{N0} turns on and discharges the surge current from the source voltage to ground. The trigger voltage V_{t1} can be expressed as:

$$V_{t1} = mV_Z + \left| V_{th\_MP0} \right| + I_1 R_0 \quad (1)$$

where n is the number of Zener diodes, V_Z is the reverse Zener voltage of the Zener diode, and V_{th_MP0} is the threshold voltage of M_{P0}. I_1 is the current flowing through resistor R_0. The number of Zener diodes m is chosen as five.

However, during the rapid power-on process, a significant dynamic leakage current exists within the circuit. When the internal circuit is powered on, V_{DD} rises from 0 to 24 V, with a rise time of 2 μs. The normal power-up

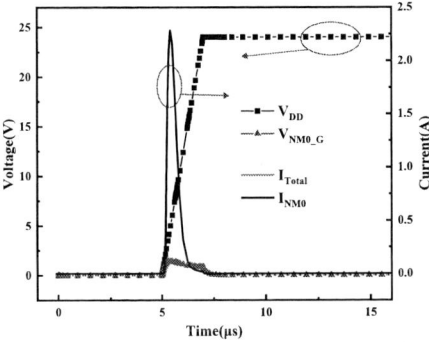

Fig.3. Simulated voltage waveform on the nodes and current waveform of the traditional diode-triggered clamp circuit under the rapid power-on event.

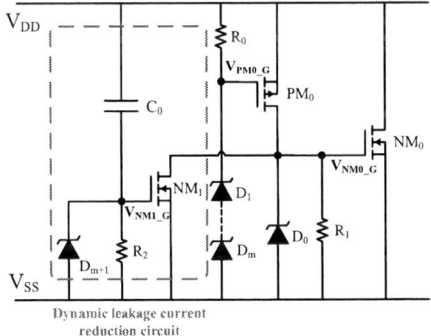

Fig.4. Schematic of proposed clamp circuit.

simulation results of the traditional surge protection circuit are shown in Fig.3. It can be observed that during the V_{DD} ramp-up process, the gate voltage of M_{N0} (V_{MN0_G}) gradually rises to its threshold voltage (0.96 V), triggering NM_0 turn-on. In this state, M_{N0} operates in the linear region, showing a low channel resistance. As a result, the applied V_{DD} across the drain-source terminals generates a transient current surge through M_{N0}, with the measured peak current reaching 2.28 A. The large dynamic leakage current will cause additional dynamic power consumption and may even pose a threat to the protected circuit.

III. NEW PROPOSED CLAMP CIRCUIT

A. Structure of Proposed Circuit

The schematic diagram of the surge protection power clamp circuit proposed for reducing dynamic leakage current is shown in Fig.4. Different from the traditional surge protection circuit, the proposed power clamp circuit includes an additional dynamic leakage current suppression circuit composed of C_0, R_2, D_{m+1} and NM_1. NM_1 is a regular transistor with a threshold voltage (V_{th_NM1}) of approximately 0.96 V. C_0 is a MIM capacitor, D_{m+1} is a Zener diode with a V_Z of 6.3V and R_2 is a large resistance. m is chosen as five.

During the rapid power-on process, the circuit charges capacitor C_0, generating a voltage drop V_{NM1_G} across resistor R_2. When this voltage exceeds V_{th_NM1}, NM_1 turns on, pulling down the gate voltage of NM_0 (V_{NM0_G}) to a level below its threshold voltage, thereby maintaining NM_0 in an off state and effectively suppressing leakage current caused by unintended conduction of NM_0. Meanwhile, during a surge event, the gate voltage of NM_0 must be increased

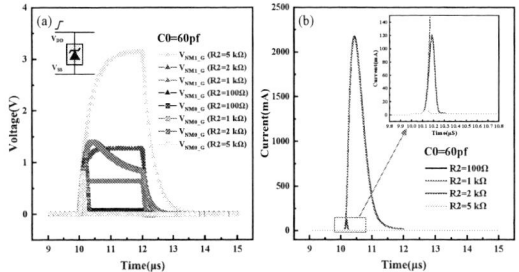

Fig.5. Simulated voltage waveform on the nodes and (b) current waveform of the proposed clamp under the rapid power-on event with different R_2.

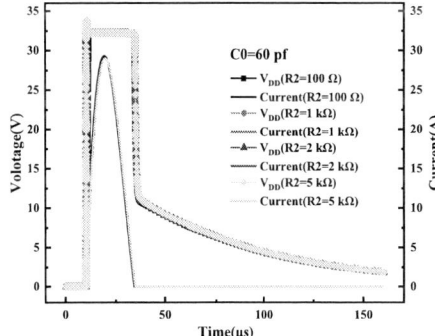

Fig.6. Simulated voltage waveform on the nodes and current waveform of the proposed clamp under the surge event with different R_2.

beyond its threshold voltage to activate the device, thereby establishing a low-impedance discharge path. To achieve this objective, it is critical to ensure that V_{NM1_G} remains below V_{th_NM1}, thereby keeping NM_1 in the off state. Therefore, the RC time constant in the dynamic leakage current suppression circuit needs to be precisely designed to achieve effective surge protection while maintaining a low dynamic leakage current during the rapid power-on process.

B. Parameter Configurations

To investigate the impact of C_0 and R_2 on the key performance of the current suppression circuit, the parameter scanning method was employed to conduct multiple Power-On and Surge simulation analyses. When C_0 is set to 60 pF, the simulation results of the proposed circuit during the rapid power-on process under different R_2 parameter configurations are presented in Fig.5(a)(b). To simulate a rapid power-on event, a voltage ramp with a rise time of 2 μs and an amplitude of 24 V was applied to the V_{DD}. As the resistance of R_2 increases from 100 Ω to 5 kΩ, the gate voltage V_{NM1_G} of NM_1 correspondingly rises from 63 mV to 3.13 V. When R_2 is set to 100 Ω and 1 kΩ respectively, the gate voltage V_{NM1_G} remains below V_{th_NM1}, resulting in NM_1 remaining in the off state and failing to effectively pull down the gate potential of NM_0. Consequently, the gate voltage of NM_0 exceeds its threshold voltage V_{th_NM0}, causing the device to conduct and generating a leakage current as high as 2.17 A. In contrast, when R_2 is configured to 2 kΩ and 5 kΩ respectively, V_{NM1_G} surpasses V_{th_NM1}, thereby activating NM_1 and pulling V_{NM0_G} down to near zero, forcing NM_0 into the off state and significantly suppressing the leakage current. Specifically, when R_2 is set to 2 kΩ, the peak leakage current is reduced to 121 mA; further increasing R_2 to 5 kΩ results in a peak

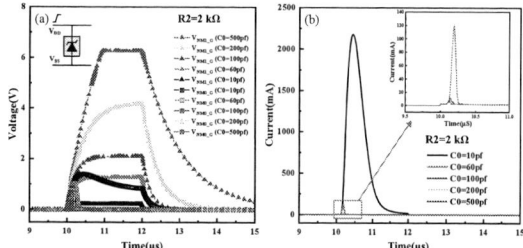

Fig.7. (a) Simulated voltage waveform on the nodes and (b) current waveform of the proposed clamp under the rapid power-on event with different C_0.

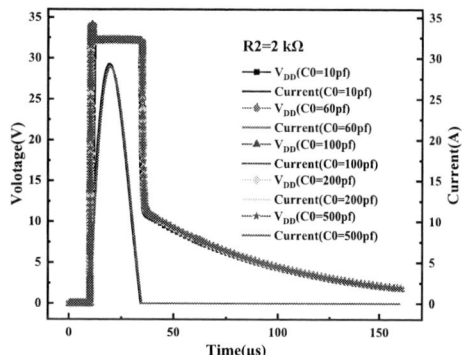

Fig.8. Simulated voltage waveform on the nodes and current waveform of the proposed clamp under the surge event with different C_0.

leakage current as low as 7.3 mA. To ensure effective suppression of the dynamic leakage current, with C_0 fixed at 60 pF, the resistance of R_2 should be at least 2 kΩ.

Fig.6 shows the simulation results of the proposed circuit during a surge event. The calibrated short-circuit current waveform at a 100 V surge source voltage meets the requirements of the 8/20 μs short-circuit current waveform defined by the IEC 61000-4-5 standard[7]. All R_2 values can maintain a stable clamp voltage of 32.2 V, and the current decay time remains within the range of 24.3-24.7 μs, indicating that the variation of the R_2 parameter has no significant impact on the surge withstand capability.

When R_2 is set to 2 kΩ, Fig.7(a)(b) shows the simulation results of the proposed circuit during the rapid power-on process under different C_0 parameter configurations. As C_0 increases from 10 pF to 500 pF, V_{NM1_G} rises from 0.2 V to 6.26 V. When $C_0 = 10$ pF, $V_{NM1\_G} < V_{th\_NM1}$, causing NM_1 to turn off and resulting in a leakage current of 2.17 A. However, when $C_0 \geq 60$ pF, $V_{NM1\_G} > V_{th\_NM1}$, NM_0 is effectively turned off, and the leakage current shows an exponential decay trend: from 121 mA ($C_0 = 60$ pF) to 11.7 mA ($C_0 = 100$ pF), and further decreases to approximately 8 mA ($C_0 = 200$ pF and 500 pF).

Fig.8 shows the simulation results of the proposed circuit during a surge event, further verifying that the C_0 parameter is insensitive to the surge response characteristics. Results indicate that the clamp voltage remains stable at 32.2V under different C_0 values; meanwhile, the decay time of the current injected into the clamping circuit shows good consistency, ranging from 24.3 μs ($C_0 = 10$ pF) to 24.7 μs ($C0 > 10$ pF). The simulation results indicate the presence of distinct parameter boundary conditions for suppressing dynamic leakage current: when C_0 is set to 60 pF, R_2 must

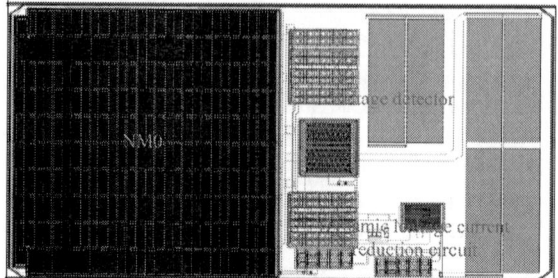

Fig.9. The layout of proposed clamp circuit.

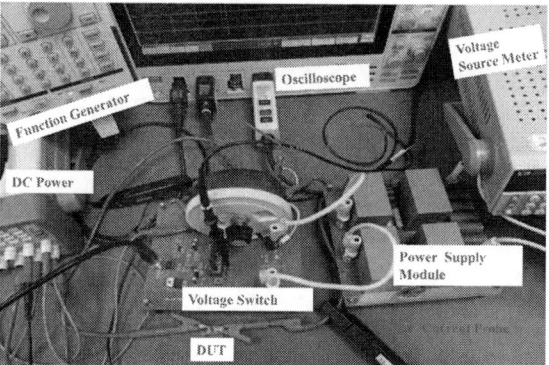

Fig.10. The test environment for rapid power-on process.

be at least 2 kΩ; conversely, with R_2 fixed 2 kΩ, C_0 should not fall below 100 pF. Furthermore, surge protection performance remains consistently stable across all tested configurations, thereby confirming the robustness of the circuit under various combinations of C_0 and R_2.

C. Discussion

For different power-on times, different R_2 and C_0 can be set to reduce the leakage current during the power-on process. Based on the investigation into the dynamic leakage current suppression mechanism, in the case of a power-on time of 2 μs, the optimal configuration of $C_0 = 100$ pF and $R_2 = 2$ kΩ was ultimately determined. Under this configuration, during rapid power-on process, the gate voltage V_{NM1_G} of NM_1 reaches 2.1 V, which is 2.18 times V_{th_NM1}. As a result, NM_1 turns on and actively pulls down the gate voltage of NM_0 to 3 mV, thereby suppressing the dynamic leakage current to 11.7 mA, which represents a 90.3% reduction compared to the $C_0 = 60$ pF configuration (121 mA). In the surge test, the circuit maintained a stable clamping voltage of 32.2 V under the 8/20 μs surge waveform specified by the IEC 61000-4-5 standard, with a current decay time of 24.7 μs[7]. Compared to traditional clamp circuits, which exhibit a peak dynamic leakage current of 2170 mA, the proposed design achieves a key technical breakthrough with a dynamic leakage current suppression rate of 99.4%, while maintaining equivalent surge protection performance.

IV. EXPERIMENTAL RESULTS

Fig.9 shows the full layout of the proposed circuit. Fig.10 shows the test environment for rapid power-on, including the power supply module, the switch module and the proposed surge protection circuit. The surge waveform used for the surge test complies with the 8/20 μs surge specification of IEC 61000-4-5 standard[7].

979-8-3315-3918-4/25 $31.00 © 2025 IEEE

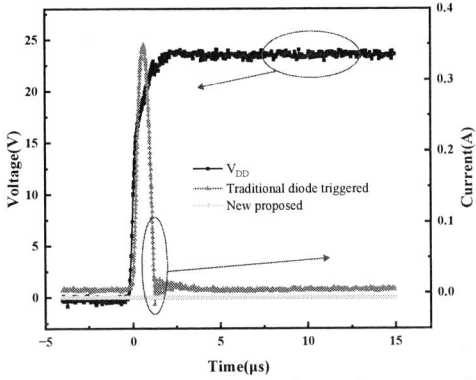

Fig.11. Measured transient voltage waveform and current waveform of the proposed circuits and traditional circuit under the rapid power-on event.

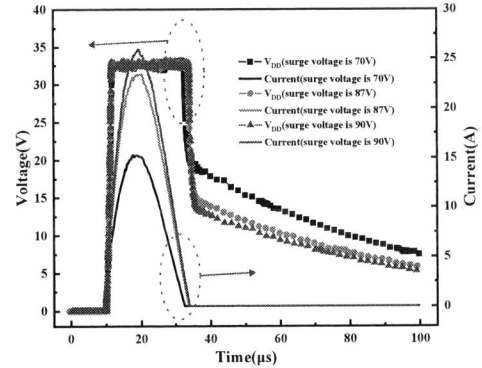

Fig.12. Measured transient V_{DD} voltage waveform of the proposed clamp circuit with different surge source voltages.

A. Rapid Power-on Tests

A rapid power-on test was conducted on the proposed circuit, with a voltage ramp signal of 24 V and a rise time of 2 μs applied at the V_{DD} terminal. Fig.11 shows the test results of the traditional surge protection circuit and the proposed circuit under rapid power-on conditions. The test indicates that during the power-on process, the leakage current generated by the traditional surge protection circuit is 350 mA, while that of the proposed circuit is only 6 mA, achieving a dynamic leakage current suppression rate of 98.3%. The experimental results confirm that the proposed circuit can effectively suppress the leakage current generated during rapid power-on, demonstrating significant performance.

B. Surge Tests

In the three-stage surge voltage tests (70 V/87 V/90 V), the proposed circuit demonstrated outstanding voltage clamp stability. The test results are shown in Fig.12. The results indicate that when the surge voltage increased from 70 V to 90 V, the clamping voltage remained consistently stable at 32.8 V, verifying the robustness of its overvoltage protection mechanism. The surge discharge current showed a linear growth trend with the increase of test voltage, with peak currents of 15.4 A (surge voltage of 70 V), 23.5 A (surge voltage of 87 V), and 25.8 A (surge voltage of 90 V). The current decay time fluctuated within a narrow range of 24.7-25.0 μs (with a maximum deviation of only 1.2%). This characteristic proves that the circuit maintains a stable energy discharge capacity even in extreme surge events (> 3 times the working voltage).

V. CONCLUSION

In this article, a new power clamp circuit with low dynamic leakage current for surge protection is proposed and experimentally verified. The proposed circuit has a special RC network, which can ensure that the discharge FET remains in an off state during the rapid power-on process. Compared with the traditional surge protection circuit, the proposed power clamp circuit reduces the leakage current generated during the rapid power-on process by 98.3%.

REFERENCES

[1] Huang J S, Olson T, Isip E. Human-body-model electrostatic-discharge and electrical-overstress studies of buried-heterostructure semiconductor lasers[J]. IEEE Transactions on Device and Materials Reliability, 2007, 7(3): 453-461.

[2] Choi J Y. A comparison study of input ESD protection schemes utilizing NMOS, thyristor, and diode devices[J]. Communications and Network, 2010, 2(1): 11-25.

[3] Cao Y, Glaser U. Novel active ESD clamps for high-voltage applications[J]. IEEE Transactions on Device and Materials Reliability, 2013, 13(2): 388-397.

[4] Altolaguirre F A, Ker M D. Power-rail ESD clamp circuit with diode-string ESD detection to overcome the gate leakage current in a 40-nm CMOS process[J]. IEEE transactions on electron devices, 2013, 60(10): 3500-3507.

[5] "Semtech," 2023. [Online]. Available: http://www.semtech.com

[6] Wang D, Roland S O N. Precision surge clamp with constant clamping voltage and near-zero dynamic resistance under various thermal, power and current levels: U.S. Patent 10,014,682[P]. 2018-7-3.

[7] International Electrotechnical Commission. Electromagnetic Compatibility (EMC)—Part 4-5: Testing and Measurement Techniques—Surge Immunity Test[J]. IEC: Geneva, Switzerland, 2017.

Reliability Enhancement in HfO_2-Based FeRAM: Circuit-Level Solutions for Insufficient Polarization and Memory Window Degradation

Changnan Shi [1], Taoran Shen[1], Li Xiong [1,4], Shuyang Lv[1], Yuanfeng Chen[3], Xiaoyong Xue*[1,2,3], Xiaoyang Zeng[1]

[1] State Key Lab of Integrated Chips and Systems, College of Integrated Circuits and Micro-Nano Electronics, Fudan University, Shanghai, China
[2] School of Microelectronics, Fudan University, Shanghai, China
[3] TRANSCPUTING Technology LTD, Shanghai 201203, China
[4] School of Physics and Electromechanical Engineering, Hexi University, Zhangye, 734000, China

* Email: xuexiaoyong@fudan.edu.cn, xl-427814@163.com

Abstract—With the booming development of portable electronic devices, Ferroelectric RAM (FeRAM) has garnered significant attention as a cutting-edge NVM, thanks to its advantages of ultra-low static power consumption, fast access speed, and high endurance. This paper addresses the critical reliability challenges in FeRAM operations: during writing, the threshold voltage loss of NMOS transistors conflicts with the demand for low supply voltage; during reading, the memory window (MW) shrinks due to degraded remanent polarization (Pr) after multiple cycles. To tackle these issues, two innovative schemes are proposed: a Word Line (WL) Boost circuit for writing operations, which enhances the voltage across the ferroelectric capacitor to compensate for NMOS threshold loss; and a novel Voltage Sense Amplifier (VSA) for reading operations, which amplifies the input voltage difference to expand the MW. Fabricated using the CSMC 180nm CMOS process, the prototype demonstrates promising results: the WL Boost scheme nearly eliminates NMOS threshold loss at a driven voltage of 2.4V, while the proposed VSA increases the MW by 38.9% after 10^4 cycles at 2.2V, significantly improving FeRAM's operational reliability.

Keywords—FeRAM reliability, WL Boost, voltage sense amplifier, memory window.

I. INTRODUCTION

The industry's demand for the continuous reduction of the cost and power consumption and the improvement of system reliability in memories has become increasingly intense with the explosive growth of the mobile phone and other portable electronic device markets. Non-Volatile Memories (NVM) have attracted much attention due to their ability to prevent data loss in case of power failure, which proves to be a shining point compared to conventional RAMs (such as DRAM, SRAM) in terms of power saving and improving data reliability. Among NVMs, FeRAM stands out for the advantage of ultra-low static power consumption, fast access speed, long endurance of 10^{12} cycles, and strong radiation resistance, which makes FeRAM suitable for electronic device applications and enables further expansion to the aviation field.

The ferroelectric capacitor employed in this work features a sandwich-like structure composed of $TiN-HfO_2-TiN$. The working mechanism of FeRAM is fundamentally rooted in the hysteresis loop characteristics exhibited by ferroelectric materials, which can be depicted through the polarization-voltage curve (P-V curve). As illustrated in Figure 1, the coordinates (-Vc, 0) and (+Vc, 0) represent the minimum voltage required to cause a polarization reversal in the ferroelectric capacitor. Even if the applied voltage is removed, this polarization state can still be maintained until a sufficient voltage is applied across the ferroelectric capacitor again, triggering a new polarization reversal. The coordinates (0, +Pr) and (0, -Pr) denote two opposite remanent polarization states, which are the polarization remaining in the ferroelectric material when the applied voltage is reduced to zero. These two remanent polarization states serve as the physical basis for data storage in FeRAM, with (0, +Pr) representing the stored data "0" and (0, -Pr) corresponding to the stored data "1". This clear binary mapping between the polarization states and digital information is what enables FeRAM to reliably store and retrieve data, leveraging the inherent stability and reversibility of the ferroelectric polarization process.

This paper is structured as follows: In Section II, we demonstrate the challenges related to data reliability that arise during the data operation process in the design of FeRAMs. The proposed operating scheme in peripheral circuits corresponding to these challenges are explained in detail in Section III. Simulation results and effect verification are presented in Section IV, followed by work conclusion in Section V.

This work was supported by STI 2030-Major Projects (2022ZD0209200), in part by the National Natural Science Foundation of China (62274038), the Science and Technology Commission of Shanghai Municipality (24JD1400200), and ZTE Industry-University-Institute Cooperation Funds under Grant (IA20241120003).

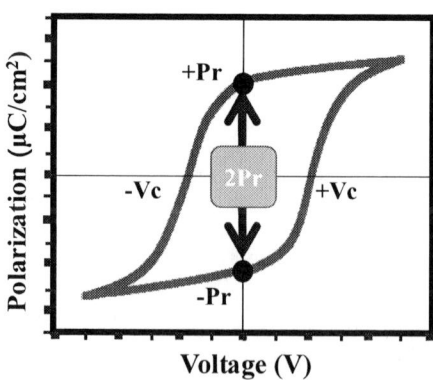

Figure 1. Polarization-voltage (P-V) loop of ferroelectric capacitor

II. CHALLENGES IN FeRAM DESIGN

The storage capacity of ferroelectric memory relies on the polarization characteristics. The magnitude of remanent polarization (Pr) and the voltage applied across the ferroelectric capacitor have a crucial impact on the data writing and reading functions, which is related to the memory's reliability directly.

A. Challenge in Writing Operations

During the writing operation, as Figure 2 shows, a sufficient voltage is needed considering whether the data can be written into the ferroelectric capacitor successfully. Meanwhile, as the demand for chips with lower power consumption has been increasing continuously, the supplied voltage in memories tends to be low. Thus, there is a contradiction between the requirement for data-writing reliability and the need for low power consumption of the chip.

Due to the inevitable trend of reduced supply voltage in consideration of reducing power consumption, the existing work has focused on reducing the remanent polarization of ferroelectric capacitors to adapt to reduced supply voltage[1]. However, this solution brings about a new problem that the difference in polarization charge quantity of the ferroelectric capacitor when storing data "1" and data "0" decreases, which also leads to a reduction in the voltage margin during data reading operations.

In order to achieve a sufficient voltage across the ferroelectric capacitor while maintaining a relatively high remanent polarization, we need to avoid any unnecessary voltage transmission loss during writing operations. When reviewing the structure of 1T1C storage units, it can be observed that the NMOS transistor will cause a threshold voltage loss when writing data "1". Usually, the power supplied voltage can be as low as 1.8V, while the threshold voltage of NMOS in this work is generally about 0.6V, which is around one third of the supplied voltage. Therefore, how to reduce or avoid this voltage loss becomes a challenge in memory design for data reliability.

In this work, we proposed a Word-line (WL) Boost writing scheme to raise the voltage of WL by at least one NMOS threshold voltage, which can enable full BL voltage transmission to the ferroelectric capacitor without threshold loss while writing data "1".

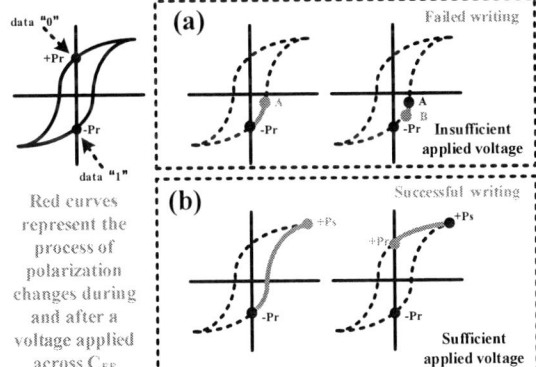

Figure 2. Diagram of the polarization inversion process in the case of (a) insufficient applied voltage (<Vc) which causes data-writing failure (b) sufficient applied voltage (>Vc) which makes data "0" written in successfully.

B. Challenge in Reading Operations

The read timing diagram and its operation principle is shown in Figure 3. Data reading can be divided into two stages: charge sharing and subsequent voltage amplification. During the first stage, as the WL turns on, the current channel is activated, the ferroelectric capacitor (C_{FE}) charges the parasitic capacitance (C_{BL}) on the BL, resulting in a change in the voltage on the BL depending on whether the stored data is "0" or "1".

However, as the operation cycles increase, the remanent polarization of ferroelectric capacitor degrades, which results in a reduction in the difference of charging amount between stored data "0" and stored data "1" during reading, in other words, degraded Pr reduces the voltage difference on Bit-line (BL), bringing about a reduction in memory window (MW). As a critical parameter, MW is regarded as an indicator for measuring the reliability of memory. The MW should be large enough to provide a sufficient read margin and therefore, protect the stored data in memory from noise and signal disturbance.

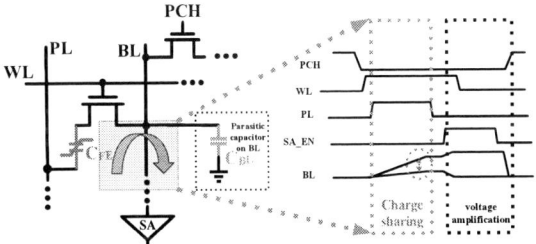

Figure 3. Operation principle and timing diagram of read operation

The conventional solutions focused on how to enhance the endurance of ferroelectric capacitors to maintain the original remanent polarization for as long as possible. For example, the non-destructive read scheme can reduce the cycling times[2], thus slowing down the degradation of Pr, and the previously proposed new Ru/HZO/Ru sandwich structure of ferroelectric capacitor improves the endurance of memory through material innovation[3]. However, these solutions either increase the area cost due to the increase of peripheral read circuit complexity or raise the process cost caused by material update.

Unlike the emphasis of the above-mentioned work, which primarily focuses on mitigating the degradation of remanent polarization in ferroelectric capacitors, our research adopts a fundamentally different approach. Rather than attempting to preserve the intrinsic polarization properties of the ferroelectric material itself, we direct our efforts toward optimizing the peripheral read circuitry, specifically by enhancing the sensitivity of the voltage sense amplifier (VSA), whose input is electrically connected to BL. This strategic shift in focus stems from the recognition that Pr degradation is an inevitable consequence of repeated operational cycles, and efforts to delay this process often incur significant trade-offs. In contrast, through the proposed VSA, we can achieve a relatively high read-out data accuracy and a high memory reliability by boosting the VSA's ability to detect minute voltage differences on the BL without considering the issue of deterioration in remanent polarization of the P-V curve.

979-8-3315-3918-4/25 $31.00 © 2025 IEEE

III. PROPOSED OPERATING SCHEME FOR HIGH-RELIABILITY

A. Proposed WL Boost Writing Scheme

Figure 4 provides a detailed illustration of the proposed WL Boost circuit, which is specifically engineered to mitigate the threshold voltage loss inherent in NMOS transistors during writing operations. This innovative circuit design addresses the critical challenge mentioned in the previous section: ensuring sufficient voltage reaches the ferroelectric capacitor to achieve reliable polarization switching. The operational sequence of the WL Boost circuit unfolds as follows: upon the initiation of a write operation, the input control signal WLEN undergoes a controlled transition from the ground potential (VSS) to the supply voltage (VDD), generating a rising edge that serves as a critical trigger for subsequent circuit activation. The rising edge can be sampled by a positive-edge (posedge) detection circuit, which is optimized to sense rapid voltage changes with high fidelity. Upon capturing the rising edge, the posedge detection circuit will generate a pulse voltage that can turn on the pass gate of the memory cell to allow the WLEN voltage to propagate to the WL in a fixed duration. Once the pulse expires, the pass gate is deactivated, returning to its off state. This action electrically isolates the WL from the WLEN signal, leaving the WL in a floating state. Concurrently, the voltage on the negative plate of the coupling capacitor (C-) is switched from VSS to VDD. Under the effect of capacitive coupling, this voltage transition on C- induces a corresponding increase in the voltage of the WL, which is electrically connected to the positive plate of C-. This coupling-induced voltage elevation effectively compensates for the threshold voltage loss of the NMOS transistor, ensuring that the final voltage applied to the ferroelectric capacitor meets the critical threshold required for reliable polarization switching, even under low supply voltage conditions.

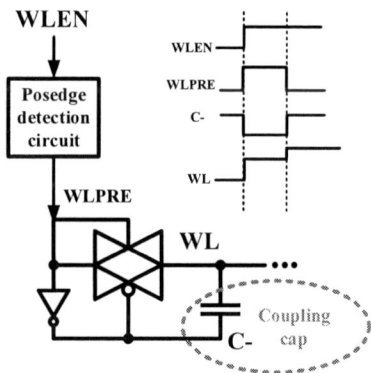

Figure 4. The proposed WL Boost writing scheme

In essence, the WL Boost circuit leverages edge detection and capacitive coupling to dynamically enhance the Word line voltage, reconciling the conflicting demands of low-power operation and robust data writing in FeRAM designs.

B. Proposed VSA Circuit in Reading Scheme

To address the inevitable problem of reduced memory window that arises due to the increased memory operation cycles, a novel VSA in reading scheme is proposed, which can ensure the accuracy of read-out data under the condition of a decreasing input voltage difference, thus the reliability of data reading is enhanced. As shown in Figure 5, the complete read operation process can be divided into two phases.

- During the first phase, the control signal CNTL is at a high voltage, the SAIN signal is transmitted to the SAPRE terminal, while the SABIN signal is transmitted to the SAPREB terminal. Meanwhile, the negative plates of the coupling capacitors C1 and C2 are connected to SABIN and SAIN signals, respectively. At this point, a voltage difference of $V_{SAIN}-V_{SABIN}$ is formed across C1, and that of $-(V_{SAIN}-V_{SABIN})$ is formed across C2.

- The second phase is about amplifying the input voltage difference of the sense amplifier. At this stage, the control signal CNTL changes to a low voltage. At this time, the two passgates are closed, leaving two terminals, SAPRE and SAPREB, in a floating state. Meanwhile, the negative plate voltage of C1 switches to SAIN, and that of C2 switches to SABIN. Due to the voltage switch from the first phase to the second phase, the voltage variation of the negative plate of C1 is $V_{SAIN}-V_{SABIN}$, while that of C2 is $-(V_{SAIN}-V_{SABIN})$. According to the coupling effect of capacitors, a voltage change will occur on the other plate of the capacitor. Accumulating with the voltage generated in the first phase, the voltage difference between the two input terminals SAPRE and SAPREB of the VSA can achieve $3\times(V_{SAIN}-V_{SABIN})$, which is three times the original input voltage difference between SAIN and SABIN. Therefore, the VSA circuit proposed in this paper can theoretically amplify the sensing memory window by three times compared to the conventional VSA.

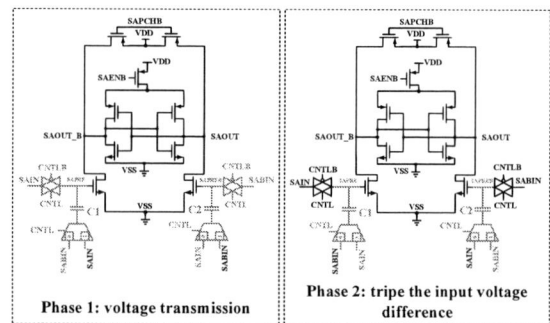

Figure 5. The proposed VSA reading scheme

IV. SIMULATION AND RESULTS

To comprehensively verify the performance metrics of the proposed schemes and to test the high-reliability characteristics targeted in this work, a ferroelectric random-access memory (FeRAM) prototype has been meticulously designed and fabricated using the CSMC 180nm CMOS standard process.

By leveraging this process, the prototype not only embodies the core innovations of the proposed schemes but also allows for precise measurement of key performance indicators such as polarization switching speed, memory window stability, endurance (number of reliable read-write cycles), and data retention time—all of which are critical to validating the efficacy of the design. Moreover, the standard process ensures that the reliability characteristics, including resistance to environmental variations (e.g., temperature fluctuations and voltage noise) and long-term operational stability, can be evaluated under conditions that closely align with industrial deployment requirements. This approach thus provides a robust experimental foundation to confirm whether the proposed schemes meet the targeted performance benchmarks and deliver the high-reliability attributes necessary for practical FeRAM applications.

A. The Enhancement in Memory Window

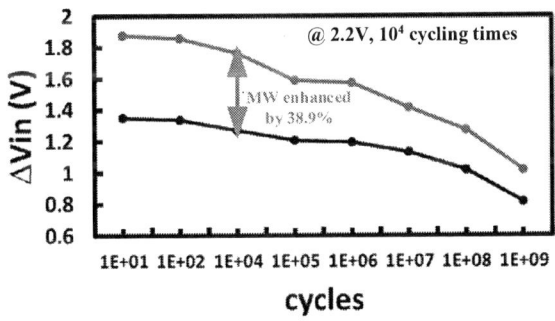

Figure 6. Comparison between conventional VSA and the proposed VSA of differential input voltage and the improvement of memory window

As shown in Figure 6, we conducted a detailed comparison between the differential input voltage of a typical VSA and the proposed VSA design. The gap between the two ΔVin curves can reflect the enhancement of the memory window in ferroelectric memory. This can be explained that the proposed VSA can pre-amplify the differential voltage read from BL (V_{BL}), then send the processed voltage to a latch-up structure, which performs as a conventional VSA. The two-stage read scheme can thus improve the sensitivity of differential V_{BL} and the reliability of data reading. It can be seen from the test results that the memory window can be enhanced by 38.9% when the operating voltage is 2.2V after 10^4 operation cycles.

B. The Increase of Voltage across The Ferroelectric Capacitor

Figure 7 presents a comprehensive visualization of the experimental test results, specifically focusing on the comprehensive analysis of the voltage magnitude achieved by the proposed Word Line Boost mechanism versus the voltage magnitude of the original Word Line configuration. These results are derived from a series of systematic measurements conducted under identical operating conditions, including consistent input power levels, ambient temperature, and memory cell loading, to ensure the validity and fairness of the comparison.

As shown in Figure 7, it can be noticed that as the driven voltage applied to the Word Line increases, the enhancing effect of the proposed WL Boost mechanism becomes increasingly pronounced and measurable. This trend can be attributed to the inherent working principle of the boost circuit:

when the input driven voltage is relatively low, the energy stored in the coupling capacitors is limited, resulting in a moderate voltage elevation that does not exceed the NMOS threshold voltage. However, as the driven voltage continues to rise beyond a certain threshold around 2.4V, as is shown, the WL Boost scheme can almost eliminate the impact caused by NMOS threshold loss.

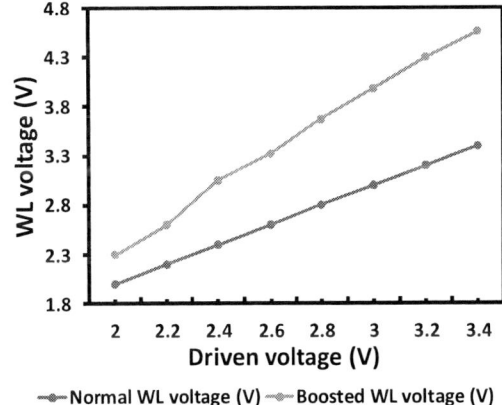

Figure 7. Comparison of Word Line voltage with and without Boost scheme

V. CONCLUSION

This paper focuses on the challenges pertaining to circuit reliability encountered by FeRAM, currently a cutting-edge non-volatile memory technology in its practical applications. We innovatively proposed a WL Boost scheme for the challenge of threshold voltage loss during writing operation and a novel voltage sense amplifier structure to address the challenge caused by the reduction of memory window due to the deterioration of remanent polarization while reading.

Our test results show that the proposed WL Boost scheme can achieve an almost complete compensation of NMOS threshold voltage loss (0.6~0.7V) while the driven voltage is as low as 2.4V. In a read operation, our proposed VSA can enlarge the memory window by 38.9% after 10^4 operation cycles at an operating voltage of 2.2V, thus improving the sensitivity of input differential V_{BL} and reliability of read-out data.

REFERENCES

[1] S. Guo *et al.*, "Low Operation Voltage, High-Temperature Reliable, and High-Yield BEOL Integrated $Hf_{0.5}Zr_{0.5}O_2$ Ferroelectric Memory Arrays," IEEE Transactions on Electron Devices, vol. 71, no. 6, pp. 3645-3650, 2024, doi: 10.1109/TED.2024.3394460.

[2] S. Mukherjee et al., "Capacitive Memory Window with Non-Destructive Read in Ferroelectric Capacitors," IEEE Electron Device Letters, vol. 44, no. 7, pp. 1092-1095, 2023, doi: 10.1109/LED.2023.3278599.

[3] R. Cao et al., "Improvement of Endurance in HZO-Based Ferroelectric Capacitor Using Ru Electrode," IEEE Electron Device Letters, vol. 40, no. 11, pp. 1744-1747, 2019, doi: 10.1109/LED.2019.2944960.

A PVT-Tolerant Quick Startup CMOS Crystal Oscillator With Chirp-Assisted Fixed Injection

Hao Luo [1], Yue Lin [2], Jian Xu [2], Hongtao Xu [1]

[1] State Key Laboratory of Integrated Chips and Systems, College of Integrated Circuits and Micro-Nano Electronics, Fudan University, Shanghai, China
[2] ICLegend Micro, Shanghai, China

Email: hluo23@m.fudan.edu.cn

Abstract—**This paper presents a quick startup technique for 39.57MHz Crystal Oscillator(XO). PVT-tolerant XO startup time is achieved using chirp-assisted fixied frequency injection in 1.2V 40nm CMOS process. The startup circuit pre-energizes the crystal with a chirp signal, and the crystal acts as a reference to assist a fixed frequency injection .The proposed XO startup circuit has a startup time of 48.63us and startup energy of 127.6nJ. The startup time variation across temperature variation from -40℃~125℃ is ±4.25%.**

Keywords—Crystal oscillator(XO), chirp injection, fixed frequency injection, quick startup, variation tolerant

I. INTRODUCTION

To extend the battery lifetime of IoT devices, minimizing system power consumption is essential. Low-power wireless communication nodes operate intermittently to achieve energy efficiency. In such systems, crystal oscillators (XOs) are commonly employed as reference clock sources due to their high frequency stability, low temperature drift, and low phase noise. However, due to its high quality factor ($Q \sim 100000$), megahertz-XOs typically exhibit startup times of 1–4 milliseconds [1],[2], which degrades the efficiency of duty-cycled operations.

Startup time(T_S) is defined as the time taken for the XO output amplitude to reach steady state when starting from off state. T_S is derived in (1) using the electrical model [3] of the Pierce oscillator shown in Fig. 1.

$$T_S = -\frac{2L_m}{R_m - |R_N|} \ln\left(\frac{0.9 w_{XO} C_T V_{XO}}{|i_m(0)|}\right) \quad (1)$$

L_m and C_m are the motional branch's inductor and capacitor, R_m is the resistive loss of the crystal, C_T is the equivalent parallel capacitance of the resonator ($C_T = C_1 C_2/(C_1+C_2)+C_P$). C_P is the parasitic capacitance across the the to terminals of the crystal. w_{XO} is the XO oscillation frequency($w_{XO} = w_m/(1+C_m/2C_T)$). ω_m is the motional branch fundamental frequency ($\omega_m = 1/(L_m C_m)^{1/2}$). R_N is the negative resistance seen by the motional branch. V_{XO} is the steady-state XO amplitude. Motional branch current (i_m) is the current flowing through the $R_m L_m C_m$ series branch and $i_m(0)$ is the initial motional branch current amplitude.

The motional branch current reprensents energy stored in the crystal, and the aim of XO fast startup techniques is to speed up the growth of i_m envelope. In previous arts, startup time was reduced using negative resistance boost or energy injection techniques. In [4], the startup time was reduced by increasing the transconductance (g_m) of the amplifier temporarily, thereby enhancing the negative resistance.

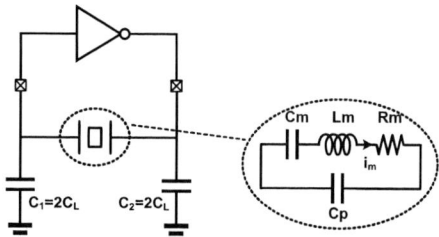

Fig. 1. Electrical model of the Pierce Oscillator

However, this approach increases power consumption during startup, resulting in no startup energy reduction.

Energy injection methods can significantly improve startup energy efficiency. However, to be effective, the injected signal must be in phase with the current within the crystal resonator, imposing stringent frequency accuracy requirements on the auxiliary injection source. The two-step injection method in [5] relaxes the frequency accuracy requirement but still requires a frequency offset within 5000 ppm for the first injection. In [6] the chirp injection technique further lowers the requirement by sweeping across oscillation frequency, but energy is wasted outside the oscillation frequency band.

To optimize startup energy efficiency and to relax the accuracy requirment of the injection source, a chirp-assisted fixed injection is demonstrated in this work. The article is arranged as follows: In section II, the implementation of the circuit is explained. In section III, the simulation results of the proposed circuit are introduced, and in section IV, the conclusion of the article is given.

II. SYSTEM IMPLEMENTATION

Despite its high energy efficiency, the fixed-frequency injection method is sensitive to injection source inaccuracy. For the startup of a large-amplitude XO, a longer injection time is needed. This makes the system more susceptible to phase errors caused by frequency offset accumulation, which in turn degenerates energy in the crystal. When a differential square wavform with fixed frequency is applied to the crystal nodes, the i_m envelope is a function of both injection time (T_{inj}) and frequency deviation of the injector [7], defined by (2).

$$i_{m,env}(T_{inj}) = \frac{T_{inj} V_{inj}}{2L_m} \left| \text{sinc}\left(\frac{T_{inj}}{2}\Delta w\right) \right| e^{-\frac{T_{inj}}{\tau}} \quad (2)$$

where w_{inj} is the injection source frequency and Δw is the frequency deviation between w_{inj} and w_m. V_{inj} is the amplitude of injection waveform. $\tau = 2Q/w_m$ is the time constant of the injection circuit.

TABLE I. PIERCE OSCILLATOR PARAMETERS

L_m	C_m	C_p	R_m	C_L	V_{inj}	V_{XO}
5.12mH	3.16fF	820fF	12.42Ω	12pF	1.2V	0.5V

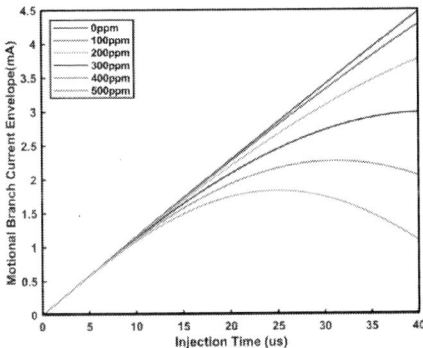

Fig. 2. i_m envelope under different injection frequecny offsets

$$\frac{\Delta w_{worst}}{w_m} = \pm \frac{4V_{inj}}{\pi V_{XO}} \frac{C_m}{C_T} \qquad (3)$$

Based on the oscillator parameters from this work listed in Table I, Fig. 2 shows the envelope of i_m over injection time under different frequency offsets. In the presence of frequency offsets, the i_m envelope decays after a certain period of injection. The maximum attainable crystal energy determines the worst-case tolerable frequency deviation, given by (3). To achieve a V_{XO} of 0.5V, the tolerable frequency offset range is ±753.2 ppm. Due to VCO temperature drift and process variations, meeting this injection accuracy requirement is challenging.

Chirp injection readily pre-energizes the crystal but struggles to sustain amplitude growth due to its imprecise frequency sweeping method. After chirp injection, although the energy within the crystal oscillator is relatively low, it still provides a clock with sufficient frequency accuracy, making it suitable for calibrating the auxiliary injection source. The two injection methods are combined by using a PLL to calibrate the auxiliary injection source to XO frequency after chirp pre-energization.

A. System Overview

Fig. 3 illustrates the operation flow of the system. It consists of four stages: In the first stage, a slope signal controls the VCO and pre-energizes the crystal oscillator, as shown in Fig. 3(a). The tuning range of the designed VCO is sufficient to cover the oscillation frequency of the crystal oscillator over supply voltage and temperature variations. When the frequency is swept across the VCO tunning range, the crystal is pre-energized, inducing a small amplitude oscillation. In the second stage, the oscillation is amplified and used as the reference clock for a PLL to calibrate the VCO, as shown in Fig. 3(b). After the PLL is locked, the control voltage for the VCO to oscillate at XO frequency is stored on the capacitor of the low-pass filter (LPF). In the third stage, the PLL loop is disconnected, and the voltage on the LPF directly controls the VCO for fixed frequency injection, as shown is Fig. 3(c). The XO is configured in the Pierce oscillator topology after steady state is reached, as shown in Fig. 3(d).

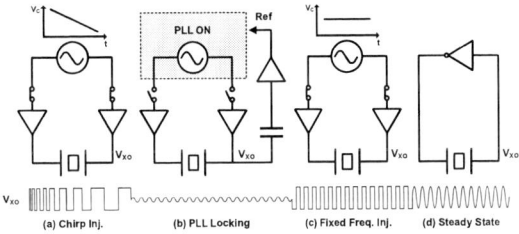

Fig. 3. XO startup operation flow

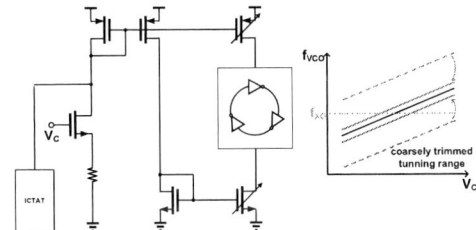

Fig. 4. Auxilary injection VCO

B. Chirp generator

The chirp signal is obtained by discharging the capacitor with a current source. When the XO is enabled, a current source discharges the capacitor and the control voltage Vc decreases steadily. According to [6], the current injected into the crystal oscillator is given by (4)

$$|i_m|_{CI} = \frac{4w_m C_m V_{inj}}{\sqrt{2\pi S}} \qquad (4)$$

where S is the normalized frequency slope factor, defined in (5).

$$S = \frac{w_2 - w_1}{t_{CI}} \times L_m C_m \qquad (5)$$

$|i_m|_{CI}$ is the current in the motional branch after chirping injection. S is determined by the VCO's Kvco and the slope of voltage Vc. $|i_m|_{CI}$ is insensitive to S variation, considering that it is inversely proportional to the square root of S. In this work, the VCO tuning range is designed to be 20MHz to 60MHz, and chirp injection time is chosen to be 13us. Thus, $|i_m|_{CI}$ =85uA is obtained. Even when S is deviated by 20% due to the capacitor process variation, $|i_m|_{CI}$ has a minimum of 77.6uA, which is sufficient to drive a voltage amplifier, showing its robustness againt process variation.

C. Auxilary Injection VCO

A current starved ring oscillator is used in this work. The structure of the VCO is shown in Fig. 4. V_C controls the bias current and thus controls the oscillation frequency. A 4-bit code is used to trim the VCO against process variations. Since it is sufficient to pre-energize the crystal as long as the VCO tuning range covers the crystal oscilation frequency, the accuracy requirments for trimming is significantly relaxed.

979-8-3315-3918-4/25 $31.00 © 2025 IEEE

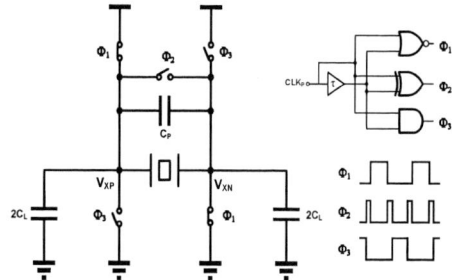

Fig. 5. Injection buffer scheme during fixed frequency injection

TABLE II. SUPPLY ENERGY CONSUMPTION IN HALF A PERIOD

	E_{SE}	E_{DIFF}
Conventional	$2C_L V_{DD}^2$	$2C_P V_{DD}^2$
Buffer in [8]	$C_L V_{DD}^2$	$C_P V_{DD}^2$
Buffer in this work	$(C_L + 0.5C_P)V_{DD}^2$	$C_P V_{DD}^2$

The frequency of the ring oscillator exhibits significant temperature drift due to variations in carrier mobility and MOSFET threshold voltage. A contrast to abusolute temperature (CTAT) current source is added to compensate for tuning range deviation under different temperatures. Therefore only a single-point calibration is required.

D. Energy Efficient Injection Buffer

Regardless of the techniques used in energy injection startup circuits, the differential nodes of the crystal are periodically charged and discharged. Part of the energy is stored in the crystal, but most of the energy is wasted in driving the load capacitor C_L and the parasitic capacitance C_P. In [8], energy loss is reduced by disconnecting the external capacitor during startup. However, energy dissipation through C_P remains unavoidable. In [9], a step-charging technique is proposed, where a buffer capacitor is used to gradually charge and discharge the load capacitor. This approach reduces power consumption by half, at the cost of requiring a huge buffer capacitor array C_{buffer} ($C_{buffer} \gg C_L$).

In this work, considering that the load capacitance C_L is significantly larger than the cross-coupled parasitic capacitance C_P, load capacitor C_L at the opposing node is repurposed as the buffer capacitor. The injection buffer scheme during fixed frequency injection is shown in Fig. 5. During chirp injection stage, Φ_2 is disabled, and the switches are driven directly by the differential clocks from the VCO, alternately connecting the crystal terminals to supply voltage and ground. During fixed-frequency injection, since the injection period is predetermined, the charge sharing control signal Φ_2 can be obtained with a delay cell briefly. During half the injection cycle, Φ_1 is first enabled to pull V_{XP} to supply voltage and V_{XN} to ground; Then, Φ_2 is enabled to short C_P and makes the charge on X_P node flow into the load capacitor on X_N ; Finally, Φ_3 is enabled to charge both the C_L on the X_N node and C_P. Among direct injection, 2-stepwise injection with C_{buffer}, and the method used in this work, a comparison of supply energy consumption in every half injection period is made in Table II. E_{SE} and E_{DIFF} denote the energy consumed in charging and discharging single-ended and differential capacitors, respectively. Method used in this work costs more energy because a portion of charge flows into C_P instead of C_L

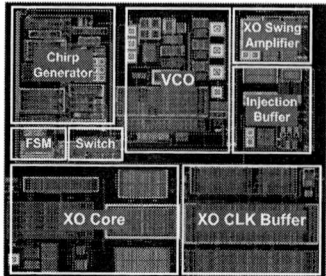

Fig. 6. Layout of the proposed XO

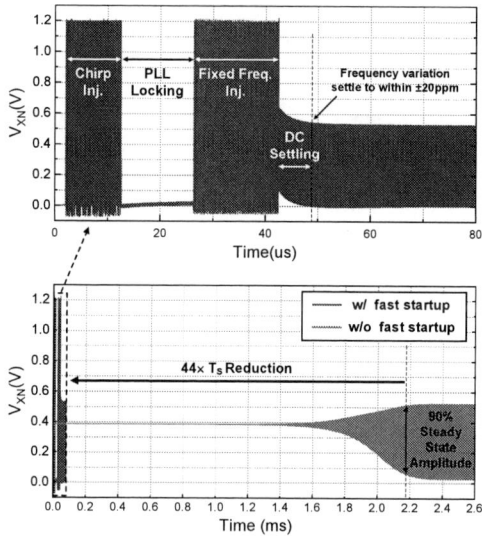

Fig. 7. XO startup waveforms with and without the proposed technique

when Φ_2 is enabled. Thus degenerating the energy efficiency slightly.

E. PLL specification

A type-II charge-pump PLL is used to calibrate VCO frequency to XO frequency. Considering that the VCO has a wide frequency tuning range to ensure that it can cover the crystal resonant frequency under PVT, its Kvco is high. Thus the VCO output frequency is sensitive to control voltage V_C. A large off-chip capacitor (140pF) is chosen in LPF. A high threshold MOS switch is used to lock the VCO control voltage, preventing current leakage from degenerating to VCO frequency accuacy. The bandwidth of PLL is designed to be 1MHz to balance locking time while preventing ripples on V_C. The system reserves 13us time for PLL to ensure that it locks successfully under PVT.

III. SIMULATION RESULTS

Using 40nm process, the layout area shown in Fig. 6 is 192um × 240um. Post-layout simulation is performed under PVT variations, where the temperature ranges from -40°C to 125°C, and a ±5% deviation on the 1.2V power supply voltage is considered. Fig. 7 shows the comparison between startup waveforms with and without using the proposed technique. In the regular pierce crystal oscillator configuration without fast startup, the oscilation needs 2.17ms to reach steady state,

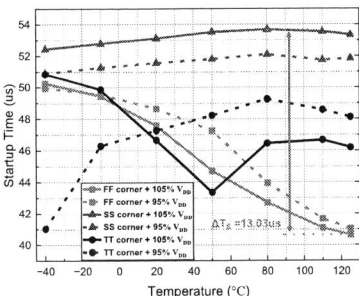

Fig. 8. T_S variation over temperature

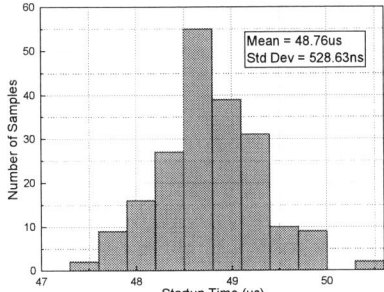

Fig. 9. T_S distribution under mismatch-only Monte-Carlo simulation

TABLE III. COMPARISON WITH PORIOR ARTS

	This Work	[5] JSSC' 2019	[6] JSSC' 2016
CMOS process (nm)	40	65	180
XO frequency (MHz)	39.57	54	39.25
Suply voltage, V_{dd} (V)	1.2	1	1.5
Steady State, Vxo(V)	0.5	0.7	1.4
Load capacitor, C_L (pF)	12	6	6
Startup time, T_S (us)	48.63	19	158
Startup cycles	1924	1026	6201
Startup energy, E_S (nJ)	127.6	34.9	349
Startup technique	Chirp+Fixed Injection	2-tep Injection	Chirp+RNB
T_S variation over Temp.	+-4.25% -40°C~125°C	+-1.25% -40°C~85°C	+-7% -30°C~125°C
PVT-tolerant Injection	Yes	No	Yes

while 48.63us is needed with chirp assisted fixed injection, leading to a 44× *Ts* reduction.

Fig. 8 shows the startup time variation over temperature. The temperature dependence results primarily from the varied DC settling process after injection. It affects the time required for the frequency to stabilize within ±20 ppm deviation. The startup time change under typical corner is 4.2us (±5.76%). Under PVT combinations, T_S variation is 13.03us (±13.39%). The startup time remains below 54 μs under worst-case scenarios.

To validate the design robustness to device mismatch, a 200-point Monte-Carlo simulation is performed. As shown in Fig. 9, the average startup time of the circuit is 48.76us, and 3σ deviation is 1.58us. MC simulation indicates that the design is insensitve to mismatch. Mismatch between the switch and its dummy connected to VCO control voltage can inject charge into the LPF and potentially degenerates frequency accuracy. Yet, V_C exhibits only minimal variation due to the large capacitance in the LPF, resulting in negligible injection frequency deviation.

In Table III, comparison is made between fast startup using energy injection techniques. Compared to chirp injections, the startup time is reduced due to high efficiency of the fixed injection. Compared to 2-step injections, the required accuracy for the injection source is significantly relaxed, reducing the difficulty of calibration.

IV. CONCLUSION

In this paper, a fast startup XO technique is reported. Chirp injection and fixed frequency injection are combined. During chirp injection, the injection frequency only needs to sweep across the XO frequency, therefore it mitigates the difficulty of calibrating the injection source. VCO frequency is then calibrated to XO frequency and boosts energy in crystal rapidly with fixed frequency injection. An injection buffer using step-wise charging is introduced to save energy. Prototyped with 40nm CMOS, this 39.57MHz XO achieves a startup time of 48.63us. The startup time variation over temperature is ±4.25%, and ±13.39% over PVT combinations, showing PVT-tolerant performace.

ACKNOWLEDGMENT

This research was supported by Special Funds for High-Quality Development (NICT) in Shanghai in 2024（No. 2024-JCSS-01010)

REFERENCES

[1] Y. Chang, J. Leete, Z. Zhou, M. Vadipour, Y. T. Chang, and H. Darabi, "A Differential Digitally Controlled Crystal Oscillator With a 14-Bit Tuning Resolution and Sine Wave Outputs for Cellular Applications," IEEE Journal of Solid-State Circuits, vol. 47, no. 2, pp. 421-434, 2012

[2] S. Farahvash, C. Quek, and M. Mak, "A Temperature-Compensated Digitally-Controlled Crystal Pierce Oscillator for Wireless Applications," in *2008 IEEE International Solid-State Circuits Conference - Digest of Technical Papers*, 3-7 Feb. 2008 2008, pp. 352-619

[3] A. Rusznyak, "Start-up time of CMOS oscillators," *IEEE Transactions on Circuits and Systems,* vol. 34, no. 3, pp. 259-268, 1987

[4] M. Miyahara, Y. Endo, K. Okada, and A. Matsuzawa, "A 64μs Start-Up 26/40MHz Crystal Oscillator with Negative Resistance Boosting Technique Using Reconfigurable Multi-Stage Akemplifier," in *2018 IEEE Symposium on VLSI Circuits*, 18-22 June 2018 2018, pp. 115-116

[5] K. M. Megawer *et al.*, "A Fast Startup CMOS Crystal Oscillator Using Two-Step Injection," *IEEE Journal of Solid-State Circuits,* vol. 54, no. 12, pp. 3257-3268, 2019

[6] S. Iguchi, H. Fuketa, T. Sakurai, and M. Takamiya, "Variation-Tolerant Quick-Start-Up CMOS Crystal Oscillator With Chirp Injection and Negative Resistance Booster," *IEEE Journal of Solid-State Circuits,* vol. 51, no. 2, pp. 496-508, 2016

[7] H. Esmaeelzadeh, and S. Pamarti, "A Quick Startup Technique for High-Q Oscillators Using Precisely Timed Energy Injection," *IEEE Journal of Solid-State Circuits,* vol. 53, no. 3, pp. 692-702, 2018

[8] A. Karimi-Bidhendi, H. Pu, and P. Heydari, "Study and Design of a Fast Start-Up Crystal Oscillator Using Precise Dithered Injection and Active Inductance," *IEEE Journal of Solid-State Circuits,* vol. 54, no. 9, pp. 2543-2554, 2019

[9] J. B. Lechevallier, H. S. Bindra, R. A. R. v. d. Zee, and B. Nauta, "Energy Efficient Startup of Crystal Oscillators Using Stepwise Charging," IEEE Journal of Solid-State Circuits, vol. 56, no. 8, pp. 2427-2437, 2021

A 20 Gb/s/Wire Short-Reach Simultaneous Bi-Directional Transceiver with DuoBinary Coding for Die-to-Die Interface in 28 nm CMOS

Bohui Bai, Fangxu Lv, Zhengbin Pang*, Geng Zhang, Ruixiao Kuai,
Liangyong Yuan, Ruotian Yin, Jiliang Liu

College of Computer Science and Technology, National University of Defense Technology, Changsha, China

* Email: zhengbinpang@nudt.edu.cn

Abstract—This article presents a short-reach simultaneous bidirectional (SBD) transceiver that employs DuoBinary coding, targeting the resolution of signal integrity challenges and enhancement of pin efficiency in high-throughput die-to-die (D2D) interconnection systems. DuoBinary coding mitigates channel insertion loss while preserving signal integrity. When transmitting DuoBinary signals, the horizontal margin under the worst-case bathtub curve (BER<10^{-7}) reaches 17.5% unit interval (UI), demonstrating a 55.5% improvement in margin over PAM4 coding and no worse than NRZ performance. By adopting an improved N-over-N low-power driver design, the transceiver enables low-swing three-level modulation, achieving an 8.92% reduction in power consumption. Meanwhile, the simultaneous bidirectional transceiving architecture doubles the data rate and improves pin efficiency, boosting data throughput to four times that of conventional unidirectional NRZ transmission.

Keywords—Inter-Chip Interface(D2D), DuoBinary, Low-Power Transceiver, Simultaneous Bidirectional Transmission

I. INTRODUCTION

With the rapid development of high-performance computing and artificial intelligence (AI) technologies, traditional single-chip architectures are facing challenges in implementing large-scale complex system-on-chips (SOCs)[1]. As shown in Fig.1, the decline in cost-performance ratio brought about by process improvements and the problem of shrinking markets have driven the integrated circuit industry to shift towards multi-chip solutions[2].

Heterogeneous integration (Chiplet) technology offers a novel approach to low-cost high-performance chips implementation by decomposing complex large chips into functional-specific small chips (Dies) and realizing die interconnection through packaging technology[4],[4]. The multi-chip architecture imposes high-throughput requirements on the die-to-die (D2D) links within the package. Traditional solutions achieve this by increasing the data rate of a single channel (such as PAM4 technology) or increasing the channel density[5]. However, PAM4 reduces the signal-to-noise ratio (SNR), while the increase in channel density leads to insufficient pin efficiency, making it difficult to simultaneously meet the requirements of signal integrity and high-density interconnection.

II. SIGNAL SCHEME

The Nyquist frequency of DuoBinary coding is only half that of NRZ signals, that enables DB signals to exhibit significc-

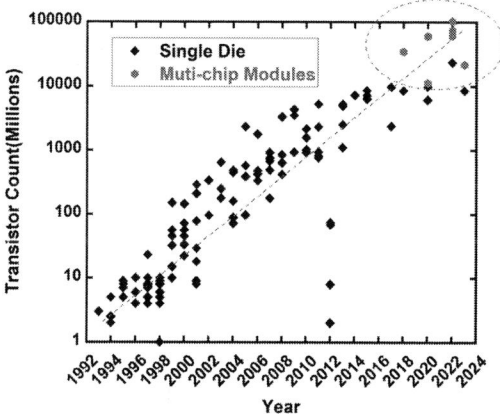

Fig. 1. Chip complexity trends in ISSCC.

ant transmission loss advantages over communication channels. Taking a 56 Gb/s signal as an example, its loss at the Nyquist frequency is only 20.9 dB, far lower than 70 dB for NRZ and 36.16 dB for PAM4[6]. As shown in Fig.2 (a) and (b), at a data rate of 20 Gb/s, the equalization strength required for DB in a 20 cm FR4 channel is 2.8 dB and 8.9 dB lower than that for PAM4 and NRZ, respectively. In a 40 cm Rogers channel, this gap further expands to 4.0 dB and 6.8 dB[7].

Additionally, compared with the complex circuit design of PAM4 that needs to process four signal levels, DuoBinary coding does not require multi-level processing, and the circuit implementation and signal processing algorithms are more concise. It can also incorporate channel loss into the overall response, relaxes the design requirements for equalizers, reduces the hardware design difficulty and system power consumption. Meanwhile, it has the advantages of large signal-to-noise ratio tolerance, maximum eye height, etc., which improves the transmission reliability.

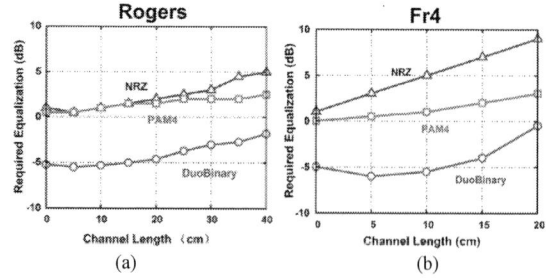

Fig. 2. Required boost at Nyquist frequency. (Data rate 20 Gb/s.)

The DuoBinary coding, as shown in Fig.3, consists of three levels, "1", "0", and "-1".

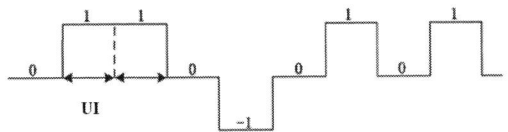

Fig. 3. DuoBinary coding.

The coding formula of DB signal is shown in (1) and (2)

$$b(n)=2a(n)-1 \qquad (1)$$

$$c(n)=1/2[b(n)+b(n-1)] \qquad (2)$$

First, the original binary code $a(n)$ with "0" and "1" is converted into a bipolar code $b(n)$ with "-1" and "+1", and then the code values at the previous and current moments are added to obtain the DB signal $c(n)$. The DB signal is not a conventional three-level signal, and the signal sequence exhibits correlation. Specifically, a zero level must appear between opposite polarities, meaning that sequences like "-1,1" and "1,-1" do not exist. If a zero level occurs between two same polarities, it must be followed by two consecutive zero levels, so sequences such as "-1,0,-1" and "1,0,1" are nonexistent. Thus, there is no direct level transition from low to high, which not only reduces dynamic power consumption but also ensures signal integrity. Meanwhile, due to the correlation of the signal sequence, a bit error is prone to cause consecutive errors in subsequent sequences, a phenomenon known as error propagation. Therefore, a precoding process must be added before binary encoding. As shown in (3)-(5), where '^' denotes XOR operation, $b(n)$ denotes the bipolar code, $a(n)$ the original code, $c(n)$ the DuoBinary code value, and $d(n)$ the precoded value.

$$d(n)=a(n)\wedge d(n-1) \qquad (3)$$

$$b(n)=2d(n)-1 \qquad (4)$$

$$c(n)=1/2[b(n)+b(n-1)]=d(n)+d(n-1)-1 \qquad (5)$$

For the decoding logic, when the original code is "0", it is encoded as "1" and "-1"; when the original code is "1", it is encoded as "0".

III. ARCHITECTURE AND CIRCUIT DESIGN

A. Overall Framework

The overall system framework is illustrated in Fig.4. Chip A and Chip B have completely identical and symmetric structures, both of which can be used as transceivers simultaneously. The half-rate clock signal, after being modulated by DCC and QEC, serves as the homologous clock signal for both the transmitter and receiver. At the transmitter side, the PRBS generators at both Chip A and Chip B generate NRZ signals, which are then encoded by DuoBinary encoders and fed into N-over-N drivers. These drivers convert the DuoBinary signals into the required output format before transmission over the channel, where the signals from both terminals are superimposed. At the receiver side, the hybrid signals from the channel undergo level restoration and sampl-

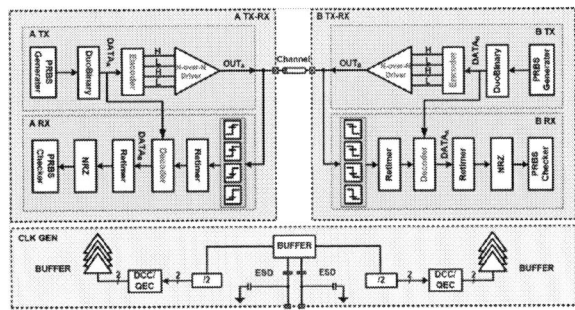

Fig. 4. Top block diagram.

ing through comparators and a retimer module. Subsequently, the locally generated DB signal from the DuoBinary encoder is subtracted from the hybrid signals to isolate the counter-directional transmission, thereby achieving simultaneous bidirectional signals separation.

B. Simultaneous Bidirectional (SBD)

The operational principle of the simultaneous bidirectional transceiver is illustrated in Fig.5, where DATA$_A$ and DATA$_B$ represent the transmission data from Terminal A and Terminal B respectively, while TX$_A$ and TX$_B$ denote their corresponding transmitter circuits. As both sides of the channel transmit data simultaneously, bidirectional signals are coupled on this channel. In the reception process, the system implements a key signal processing operation by subtracting the local transmission data from this hybrid signal, which effectively extracts the remote transmission data[8]. This architecture achieves communication through simultaneous bidirectional transmission, demonstrating superior spectral efficiency compared to conventional half-duplex systems. The design's core innovation lies in its ability to isolate and recover the desired signal while canceling out the strong self-interference from the local transmitter, enabling reliable data reception during concurrent transmission.

C. Three-level Transmitter

For the driver and precoding module, the signal mapping follows a three-level logic: when the input is at the high level, the outputs H and L are driven to high and low states, respectively; for a low-level input, H and L assume low and high states, correspondingly. Notably, an intermediate input level results in both H and L outputs being pulled low.

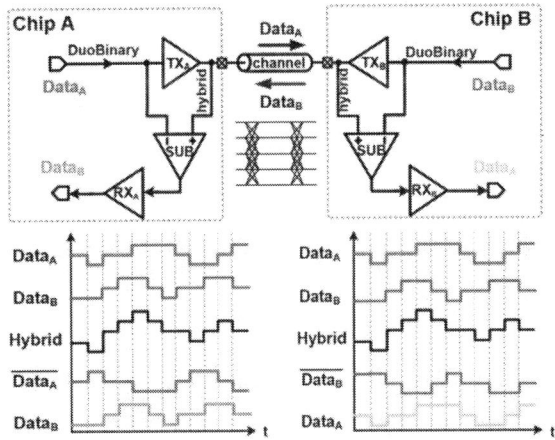

Fig. 5. Example of Simultaneous Bidirectional Interface.

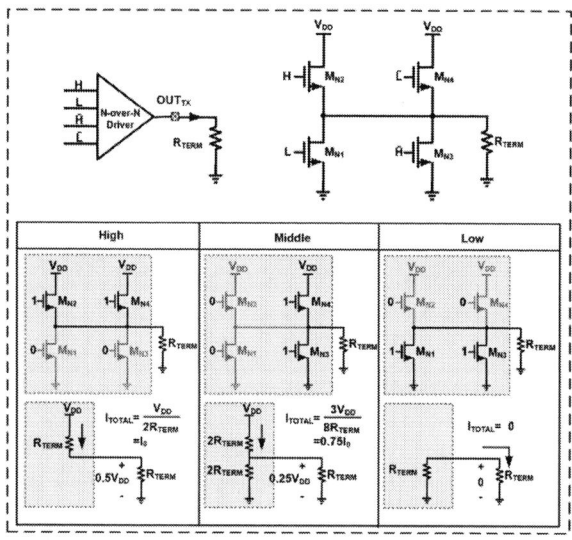

Fig. 6. N-over-N Driver schematic and operation.

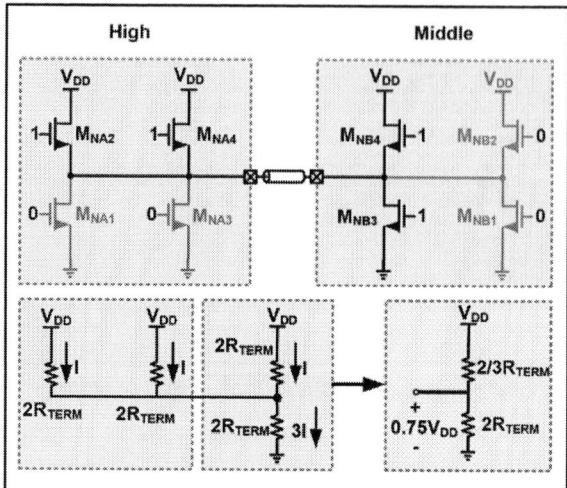

Fig. 7. Hybrid signal coding.

The output driver adopted in previous studies generates a significant transient short-circuit current at the intermediate level, as it only uses two N-type metal-oxide-semiconductor (NMOS) transistors to perform pull-up and pull-down functions[9]. Fig.6 shows the modified schematic of the driver, which consists of MN_1-MN_4. Three different input cases are used for the three-level modulation. To reduce the short current, two additional branches, MN_3 and MN_4, are added for the middle-level modulation. Each NMOS is constructed to modulate the three output levels, namely, 0.5 V_{DD}, 0.25 V_{DD}, and zero, with a ground termination resistor R_{TERM}. In the case of a high level, both MN_2 and MN_4 are turned on with a series resistance of R_{TERM}, generating 0.5 V_{DD}. As for the middle level, MN_3 and MN_4 are turned on with the series resistance of $2R_{TERM}$.

TABLE I.   ENCODING TABLE

Encoded Data(Odd)		Encoded Data(Even)		Output$_{TX}$ (Odd+Even)	Current	
H_O	L_O	H_E	L_E		Odd	Even
1	0	1	0		I_0	I_0
1	0	0	0		I_0	$0.75I_0$
1	0	0	1		I_0	0
0	0	1	0		$0.75I_0$	I_0
0	0	0	0		$0.75I_0$	$0.75I_0$
0	0	0	1		$0.75I_0$	0
0	1	1	0		0	I_0
0	1	0	0		0	$0.75I_0$
0	1	0	1		0	0

These resistances are set for the 0.25 VDD output level and for impedance matching with RTERM. In this case, the total current is 3 VDD/8RTERM, which is 0.75 times lower than that for a high level. In the case of a low level, both MN1 and MN3 are turned on with a series resistance of RTERM, generating zero value[10]. As shown in Table I, for three-level signals, there are nine possible level transitions. Since DB coding does not allow direct transitions from low to high or from high to low levels, two of these nine transitions are eliminated. By using this coding scheme, the total power consumption of the output driver is reduced by 8.92% compared to traditional approaches (which utilize all three-level transitions), while further ensuring signal integrity.

Taking the example of Terminal A transmitting a high level and Terminal B transmitting an intermediate level, the encoding of the hybrid signal on the channel is as follow. As shown in Fig.7, when M_{NA2} and M_{NA4} at Terminal A are turned on, and M_{NB3} and M_{NB4} at Terminal B are turned on, a voltage of $0.75V_{DD}$ is obtained on the channel through series-parallel voltage division. Similarly, when Terminal A transmits an intermediate level and Terminal B transmits a high level, the same principle applies. Other cases can be deduced by analogy.

IV. EXPERIMENT RESULTS

The system was simulated and tested in a 28 nm CMOS technology, achieving a data transmission rate of 20 Gb/s/Wire. In order to fairly compare the signal integrity, we employ the same simultaneous bi-directional transceiver architecture to transmit three different signals. We operate the three transceivers with the same supply voltage of 1 V and examine the eye opening at the receiving end. As shown in Fig.8, it can be clearly shown that the duobinary signal presents the largest eye opening (87.5mV), whereas the PAM4 signal exhibits the smallest (56.3 mV opening).

As shown in Fig.9, when the bit error rate (BER) is below 10^{-7}, the horizontal margin of the DuoBinary signal's bathtub curve under the worst-case scenario reaches 17.5% unit interval (UI), representing a 55.5% improvement over the 11.25% UI of PAM4 signals, and not inferior to the 16.67%

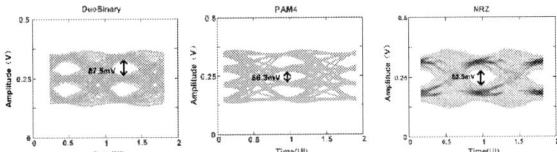

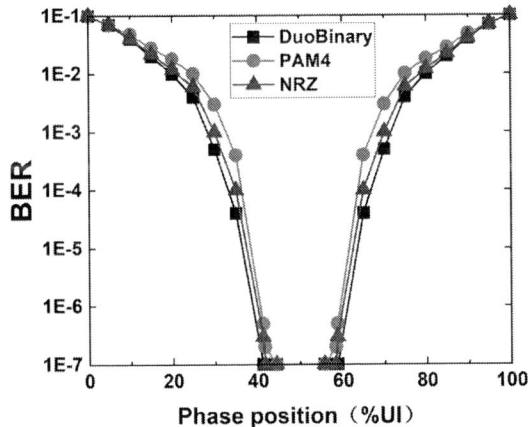

Fig. 8. Comparison of received eye diagrams.

Fig. 9. Measured bathtub curves at RX.

V. CONCLUSION

This paper proposes a bi-directional simultaneous transmission structure based on DuoBinary coding, which achieves a short-distance high-speed chip interconnection at 20 Gb/s/Wire in 28 nm CMOS technology. The scheme optimizes channel loss through DuoBinary coding and achieves low-power three-level modulations with an N-over-N driver, addressing the contradiction between high throughput and pin efficiency in multi-chip architectures. Experimental results show that the system outperforms traditional solutions in signal integrity, power consumption, and data throughput, providing a feasible technical path for next generation high-speed inter-chip interfaces. Future work can further explore the application of this technology in 3D packaging and scenarios with higher data rates.

UI of NRZ signals, verifying its advantages in transmission. In full-duplex mode, the data throughput is quadrupled compared to traditional unidirectional NRZ transmission, meeting the low-latency and high-bandwidth requirements of high-performance computing.

Table II compares the performance of these three signals with prior art.

TABLE II. PERFORMANCE COMPARISON WITH PRIOR ART

Reference	[11]	This work		
Data Format	**DuoBinary**	**DuoBinary**	**PAM4**	**NRZ**
Technology	**90nm**	**28nm**	**28nm**	**28nm**
Data Rate	12Gb/s	20Gb/s	**20Gb/s**	**20Gb/s**
BER (2^{31}-1 PRBS)	**N/A**	<10^{-7}	<10^{-7}	<10^{-7}
Eye Opening	**73.5mV**	87.5mV	56.3mV	83.5mV
Driver	**N-over-N**	**Enhanced N-over-N**	**SST**	**SST**
Transmission Mode	**UD**	**SBD**	**SBD**	**SBD**

REFERENCE

[1] P. Vivet et al., "IntAct: A 96-core processor with six chiplets 3D-stacked on an active interposer with distributed interconnects and integrated power management," IEEE J. Solid-State Circuits, vol. 56, no. 1,pp. 79-97, Jan. 2021.

[2] K. Seong et al., "A 4nm 32Gb/s 8Tb/s/mm Die-to-Die Chiplet Using NRZ Single-Ended Transceiver With Equalization Schemes And Training Techniques," ISSCC, pp. 114-116, Feb. 2023.

[3] J. Gu et al., "A 32Gb/s 0.36pJ/bit 3nm Chiplet IO using 2.5D CoWoS Package with Real-Time and Per-Lane CDR and Bathtub Monitoring," IEEE Symp. VLSI Technology, pp. C19-3, Jun. 2024.

[4] Li, T., Hou, J., Yan, J., Liu, R., Yang, H., & Sun, Z. (2020). Chiplet heterogeneous integration technology – status and challenges. Electronics (Switzerland), 9(4).

[5] S. Kim et al., "A 0.458-pJ/bit 24-Gb/s/pin Capacitively Driven PAM-4 Transceiver With PAM-Based Crosstalk Cancellation for High-Density Die-to-Die Interfaces," in IEEE Journal of Solid-State Circuits, vol. 59, no. 11, pp. 3730-3740, Nov. 2024.

[6] Z. Tang, F. Lv, J. Shi, J. Zhang, Z. Wang and P. Li, "112Gbps High-speed SerDes Transmitter Based on Duo-Binary Pam4 Encoding," 2021 6th International Conference on Integrated Circuits and Microsystems (ICICM), Nanjing, China, 2021, pp. 73-76, doi: 10.1109/ICICM54364.2021.9660352.

[7] J. Lee, M. -S. Chen and H. -D. Wang, "Design and Comparison of Three 20-Gb/s Backplane Transceivers for Duobinary, PAM4, and NRZ Data," in IEEE Journal of Solid-State Circuits, vol. 43, no. 9, pp. 2120-2133, Sept. 2008, doi: 10.1109/JSSC.2008.2001934.

[8] Y. Nishi et al., "A 0.297-pJ/Bit 50.4-Gb/s/Wire Inverter-Based Short-Reach Simultaneous Bi-Directional Transceiver for Die-to-Die Interface in 5-nm CMOS," in IEEE Journal of Solid-State Circuits, vol. 58, no. 4, pp. 1062-1073, April 2023.

[9] H. Park, J. Song, Y. Lee, J. Sim, J. Choi and C. Kim, "23.3 A 3-bit/2UI 27Gb/s PAM-3 Single-Ended Transceiver Using One-Tap DFE for Next-Generation Memory Interface," 2019 IEEE International Solid-State Circuits Conference - (ISSCC), San Francisco, CA, USA, 2019, pp.

[10] H. Park et al., "30-Gb/s 1.11-pJ/bit Single-Ended PAM-3 Transceiver for High-Speed Memory Links," in IEEE Journal of Solid-State Circuits, vol. 56, no. 2, pp. 581-590, Feb. 2021.

[11] K. Yamaguchi et al., "12 Gb/s Duobinary Signaling with ×2 Over-sampled Edge Equalization," in IEEE International Solid-State Circuits Conference (ISSCC) Digest of Technical Papers, 2005, pp. 70-71.

A 56Gb/s PAM4 Transceiver Based on BSS-LMS Algorithm With 3-Taps Adaptive TX FFE

Xianchao Zeng , Fangxu Lv*, Liquan Xiao , Jiaqing Xu , Zhouhao Yang , Liangyong Yuan , Cewen Liu , Xiaoyue Hu , Yingjie Zhang

College of Computer Science and Technology, National University of Defense Technology, Changsha, China

* Email: zxc_17550311598@163.com, lvfangxu1988@nudt.edu.cn

Abstract—This paper proposes a block sign-sign least mean square (BSS-LMS) algorithm based on a dual transceiver (TRX) architecture. The algorithm facilitates real-time main channel feedback via the back channel, enabling closed-loop adaptive tuning of the transmitter's 3-tap FFE to reduce the bit error rate (BER). Simulation results show that, in a 56 Gbps 4-pulse amplitude modulation (PAM4) transmission, the proposed solution can effectively compensate for 15–30 dB insertion loss. The BSS-LMS reduces the BER by two orders of magnitude compared to conventional TX FFE with fixed coefficient.

Keywords—Back-channel, BSS-LMS, FFE, Adaptive equalization

I. INTRODUCTION

With the growing demand for high-performance computing and data center resources driven by artificial intelligence applications, high-speed wireline communication systems are facing increasing bandwidth requirements and more stringent signal integrity challenges. While high-speed SerDes interfaces enable increased data rates, they also introduce signal quality issues such as dielectric loss, skin effect, and impedance discontinuities, which in turn lead to inter-symbol interference (ISI). To this end, equalization techniques such as continuous-time linear equalizers (CTLE), and decision feedback equalizers (DFE) are commonly employed at the receiver, while pre-emphasis using feed-forward equalization is applied at the transmitter (TX) to compensate for channel-induced high-frequency losses [1][2]. However, since the transmitter typically lacks access to the channel information at the receiver side, its equalization parameters are difficult to adjust dynamically, making it challenging to cope with variations caused by changing environmental conditions.

To resolve this problem, transmitter adaptation techniques have emerged, enabling the transmitter to dynamically adjust equalizer parameters in real time, thereby enhancing signal integrity and system stability under complex channel conditions. Peak voltage swing constraints in transmitter pre-emphasis cause unknown equalization reference levels, a problem addressed by dual-loop adaptive algorithms [3]. For updating transmitter equalizer tap coefficients, sign-sign group LMS adaptive algorithms have been proposed [4]. Furthermore, a low-cost S-ZF algorithm leverages data edge correlation to automatically adjust tap weights in transmitter feed-forward equalizer (FFE) [5]. However, the implementation of existing algorithms primarily focuses on lower-rate application scenarios. Adaptive equalization algorithms for transmitters in high-rate application lack effective solutions.

National key research and development program(2021YFB2206600)

To overcome these high-rate limitations. We propose a transmitter adaptive equalization algorithm based on BSS-LMS. Leveraging a parallel architecture for high-speed serial links, the algorithm overcomes the inherent limitations of conventional TX FFE with fixed coefficients, which struggle to adapt to dynamic channel variations. Compared to conventional fixed coefficient approaches, the proposed algorithm achieves a two order of magnitude reduction in BER. Experimental validation confirms the feasibility of the adaptive equalization algorithm for 56 Gb/s PAM4 signals and demonstrates its potential for higher-speed applications.

II. CRUCIAL TECHNOLOGIES

A. Overall Architecture

To overcome the challenges of non-dynamic adjustment of transmitter equalizer parameters and insufficient responsiveness to environmental variations, this paper adopts a dual-TRX-based adaptive equalization architecture for high-speed serial links, as illustrated in Fig. 1. In this architecture, the main channel is used for high-speed forward transmission of data signals, while the back-channel enables low-speed feedback of equalization parameters derived from receiver-side, thereby facilitating closed-loop adaptive adjustment of the transmitter's FFE tap weights.

TRXI serves as the transmitter and performs pre-equalization on the generated PAM4 digital signal using a 3-tap FFE, proactively compensating for potential high-frequency attenuation and ISI introduced by the transmission channel. Subsequently, the pre-equalized digital signal is converted into an analog signal by a digital-to-analog converter (DAC) and transmitted through the main channel for forward transmission.

TRXII serves as the receiver and first processes the analog signal transmitted through the main channel using an analog front-end (AFE). The AFE integrates a CTLE and a variable gain amplifier (VGA), where the CTLE compensates for high-frequency signal loss, and the VGA dynamically adjusts the signal amplitude to match the input range of the analog-to-digital converter (ADC) while suppressing out-of-band noise. The analog signal processed by the AFE is sampled by an ADC into 64 parallel digital lanes, which are then fed into the digital signal processors (DSP) module. The DSP module performs fine-grained channel distortion compensation using a 16-tap FFE, followed by a 1-tap DFE to mitigate residual ISI, thereby collaboratively enhancing signal quality. The Error module calculates the signal error, which is then fed into the LMS algorithm module to adaptively adjust the tap coefficients of both the FFE and DFE. Meanwhile, the BSS-LMS algorithm is introduced to compute the tap coefficients for the transmitter-side FFE.

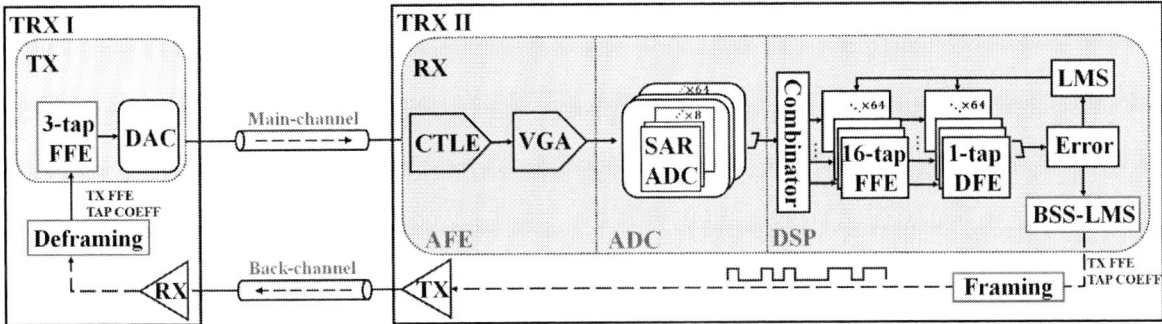

Fig. 1. Overall Architecture

The computed tap coefficients are processed through framing and then fed back to TRXI through the back-channel by the transmitter module of TRXII. TRXI receives the back-channel data and extracts the FFE tap coefficients through deframing, enabling real-time updates of the transmitter's 3-tap FFE. Ultimately, a closed-loop adaptive adjustment mechanism is established.

B. Transmitter FFE Adaptive Equalization Algorithm

The transmitter FFE and receiver FFE essentially function as finite impulse response (FIR) filters. In practical applications, the receiver can estimate the channel response and feed it back to the transmitter through the back-channel. However, if the back-channel is constrained, the transmitter FFE's ability to adapt to the end-to-end channel response will also be limited, resulting in inferior adaptive performance compared to the receiver FFE.

Due to the lack of knowledge about the actual channel characteristics, the transmitter equalizer parameters can only be initialized based on empirical values, which are clearly suboptimal. Therefore, employing an efficient adaptive algorithm for parameter optimization becomes particularly important.

To address this issue, this paper proposes a BSS-LMS algorithm tailored for transmitter FFE, which reduces computational complexity and back-channel overhead through parallel processing, enabling efficient adaptive adjustment of tap weights.

The LMS algorithm, based on the method of steepest descent, is widely used in the field of adaptive filtering due to its simple structure, good stability, and ease of implementation. The core idea of the LMS algorithm is to iteratively minimize the error between the input signal and the desired signal, thereby gradually optimizing the filter coefficients until they converge to the optimal solution.

Specifically, the current input signal vector be defined as:

$$X(n) = [x(n), x(n-1), \dots, x(n-M+1)]^T \quad (1)$$

the current filter coefficient vector be defined as:

$$W(n) = [w_0(n), w_1(n), \dots, w_{M-1}(n)]^T \quad (2)$$

the desired output at this time is $d(n)$, and the actual output can be written as follows:

$$y(n) = W^T(n)X(n) \quad (3)$$

The error signal is defined as:

$$e(n) = d(n) - y(n) = d(n) - W^T(n)X(n) \quad (4)$$

The objective of the LMS algorithm is to minimize the mean squared error (MSE), it can be expressed as follow:

$$J(W) = E[e^2(n)] = E[(d(n) - W^T X(n))^2] \quad (5)$$

Optimization is performed using the gradient descent method, the results are shown below:

$$W(n+1) = W(n) - \mu \nabla J(W) \quad (6)$$

Here, μ is the learning rate or step-size factor, and $\nabla J(W)$ is the gradient of the cost function with respect to the coefficient vector.

The gradient is computed as:

$$\nabla J(W) = -2E[e(n)X(n)] \quad (7)$$

Since the expected value $E[e(n)X(n)]$ is not known in practical environments, the LMS algorithm replaces it with the instantaneous value. The most basic update equation is derived as follows:

$$W(n+1) = W(n) + \mu e(n)X(n) \quad (8)$$

To further reduce computational and implementation costs, the output of the DFE slicer is used as the ideal reference input for the LMS algorithm in practical applications, thereby enabling a blind adaptive equalization strategy. In this case, the weight update equation becomes can be expressed as follows:

$$W(n+1) = W(n) + \mu e(n)Y(n) \quad (9)$$

Where $Y(n)$ is the output signal after slicer decision.

To further reduce the complexity of transmitting the transmitter's equalizer coefficients, the Sign-Sign LMS (SS-LMS) algorithm is adopted, in which both the error $e(n)$ and the input vector $X(n)$ are replaced by their respective signs can be expressed as:

$$W(n+1) = W(n) + \mu \cdot sign(e(n)) \cdot sign(Y(n)) \quad (10)$$

Where $sign(e(n))$ is the sign of the error, and $sign(Y(n))$ is the sign of the decision signal.

Considering the transmission rate limitations of the back-channel, the Block LMS (BLMS) algorithm is introduced. In contrast to the traditional LMS algorithm, which updates the filter weights after each received sample. In contrast, the BLMS algorithm updates the filter weights only after collecting a block of L samples. To simplify algorithm complexity, the block size in this design is set to L = 64,

979-8-3315-3918-4/25 $31.00 © 2025 IEEE   619

matching the number of parallel outputs from the DFE. The weight update equation is given by:

$$W(n+1) = W(n) + \mu \sum_{i=0}^{L-1} e(n+i)Y(n+i) \quad (11)$$

Building upon the BLMS and SS-LMS algorithms, this paper proposes a novel adaptive algorithm tailored for transmitter feed-forward equalizers, namely the BSS-LMS algorithm. The formula is shown below:

$$W(n+1) = W(n) + \\ \mu \sum_{i=0}^{L-1} sign\big(e(n+i)\big) \cdot sign(Y(n+i)) \quad (12)$$

This algorithm combines the advantages of block processing and sign-based operations.

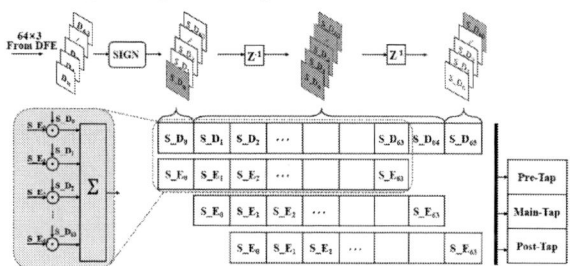

Fig. 2. Design of 64-Lane Parallel BSS-LMS Algorithm

Fig. 2 presents the BSS-LMS algorithm implemented on a 64-channel ADC-DSP architecture. First, the multi-lane data from the DFE is sign-processed and reorganized through a delay module to enable XNOR and accumulation operations. This allows for the separate computation of weight updates for the pre-tap, main-tap, and post-tap, which are then used to optimize the transmitter FFE tap coefficients in real time.

C. Framing and Deframing

To enable efficient and reliable transmission of equalization coefficients over the back-channel, the system incorporates a framing and deframing mechanism based on a fixed frame structure. As shown in Fig. 3, each complete transmission frame consists of three parts: a frame header, a data field, and a frame tail, with a total length of 32 bytes.

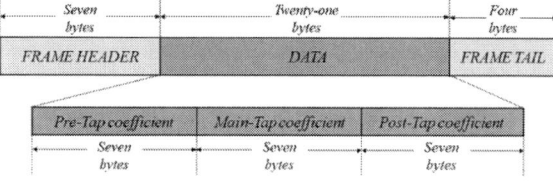

Fig. 3. Data Frame Structure

The frame header occupies 7 bytes and is primarily used for synchronization; the data field spans 21 bytes and carries the core equalization coefficient data; the frame tail takes up 4 bytes and is used to indicate the end of the frame.

The data field is further divided into three sections: Pre-Tap Coefficient, Main-Tap Coefficient, and Post-Tap Coefficient. Each type of coefficient occupies 7 bytes, allowing for the representation of precision-weighted values across multiple taps. This ensures that the FFE can effectively compensate for varying channel conditions.

III. EXPERIMENT RESULTS

In this section, system-level verification of the proposed high-speed adaptive equalization architecture is conducted on a mixed digital-analog simulation platform. A complete data link simulation environment was constructed, and system performance was evaluated under various channel attenuation conditions. Fig. 4 illustrates the selected channels with insertion losses of 15dB, 20dB, 25dB and 30dB @14GHz respectively.

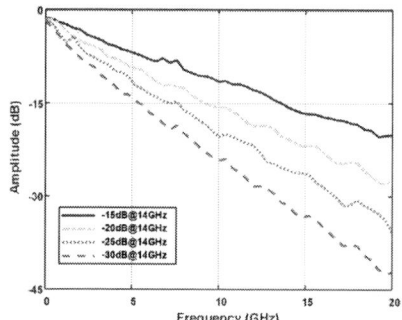

Fig. 4. Insertion Loss of Channels

At the transmitter side, a signal generator produces a 56 Gb/s PAM4 signal. The signal first undergoes pre-encoding, followed by pre-emphasis through a fixed coefficient FFE, and is then converted to an analog signal by a DAC before being transmitted through the channel.

The signal undergoes initial gain adjustment and high-frequency compensation through the CTLE and VGA modules in the receiver's AFE. Based on this, the optimal sampling point for the ADC is further determined. However, as shown in Fig. 5(a), After CTLE equalization, the distribution of sampling points remains highly disordered, with no clear boundaries between the four PAM4 levels. The eye diagram is severely closed, indicating that analog equalization alone cannot recover a valid signal.

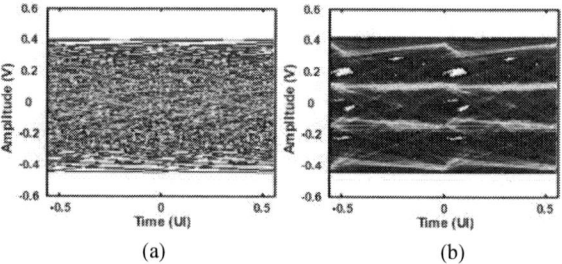

(a)   (b)

Fig. 5. Eye diagrams at different phases. (a) After ADC (b) TX FFE with fixed coefficient

The data output from the ADC is fed into the DSP module, where it undergoes combined equalization processing through FFE and DFE stages. As shown in Fig. 5(b), although the signal exhibits some level of decision capability after digital equalization, the four PAM4 levels remain unevenly distributed, and the eye opening is still insufficient. The root cause lies in the transmitter FFE parameters being empirically set, making them poorly adapted to the current channel characteristics. Additionally, the equalization capability of the receiver-side FFE and DFE is inherently limited, making it difficult to achieve full compensation.

979-8-3315-3918-4/25 $31.00 © 2025 IEEE   620

On this basis, the transmitter adaptive equalization algorithm proposed in this paper is introduced, and a back-channel is established to provide real-time feedback of receiver information. The final converged signal waveform is shown in Fig. 6(a), where the four PAM4 levels are clearly and evenly distributed, and the eye diagram is significantly open. The BER is below 1e-12, indicating that the system has achieved effective equalization convergence.

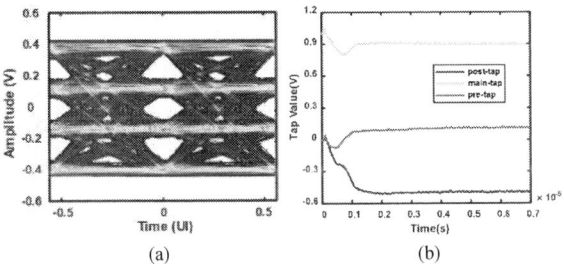

Fig. 6. Eye diagrams and coefficient curve. (a) TX FFE with BSS-LMS. (b) TX FFE coefficient.

Fig. 6(b) also presents the variation curves of the receiver-side FFE and DFE tap coefficients, as well as the convergence process of the transmitter's adaptive FFE tap weights. These results demonstrate the system's ability to effectively sense channel conditions and achieve rapid self-adjustment, thereby significantly improving signal integrity.

Fig. 7 shows the eye diagram of the back-channel. In this link, although the transmitter of the back-channel has not undergone equalization adjustment, it is still capable of directly transmitting low-rate control signals using its original driver. As shown in the figure, the eye diagram exhibits a vertical opening of approximately 200 mV and the BER is below 1e-14, with clearly defined zero-crossing points and low jitter in the horizontal time domain, indicating good signal integrity. This back link is capable of stable and reliable reception without the need for an equalizer.

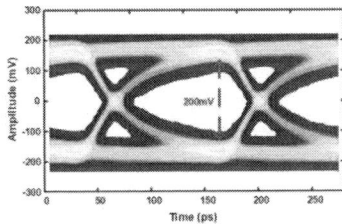

Fig. 7. The transmission eye diagram of a back-channel without any equalization.

Fig. 8 compares the BER performance of transceivers utilizing BSS-LMS adaptive equalization and those employing conventional fixed coefficient equalization under varying channel attenuation conditions. The results indicate that both equalization methods achieve a BER better than 1e-12 when the channel attenuation is below 20 dB. However, as the channel attenuation exceeds 20 dB, the BSS-LMS adaptive equalization increasingly demonstrates a significant

performance advantage. At a channel attenuation of 30 dB, the BER of the system with BSS-LMS adaptive equalization remains below 1e-12, whereas the BER of the fixed coefficient equalization degrades significantly to approximately 1e-8. This demonstrates that the BSS-LMS adaptive equalization method exhibits superior BER performance in high-channel-attenuation environments.

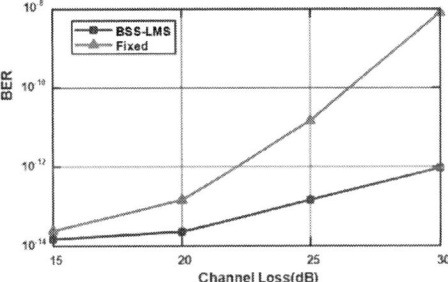

Fig. 8. BER of BSS-LMS and fixed coefficient in different channels

IV. CONCLUSION

This paper presents a dual-TRX-based adaptive equalization architecture combined with the BSS-LMS algorithm. The proposed design leverages a cooperative mechanism between the main channel and the back-channel to enable real-time updates of the transmitter FFE tap weights. It demonstrates the capability to continuously track environmental disturbances such as channel attenuation and temperature drift. In the 56 Gbps PAM4 transmission scenario, the proposed BSS-LMS algorithm employs a block-level sign-based computation strategy, which effectively reduces the computational complexity of implementation and significantly lowers the bandwidth overhead of the back-channel. While ensuring rapid tracking of time-varying channels, the back-channel is capable of delivering stable and reliable data feedback without the need for equalization. Simulation results show that under channel conditions with insertion loss as high as 15-30 dB, and the BER is below 1e-12. The back-channel achieves a vertical eye opening of 200 mV, and the BER is below 1e-12.

[1] J. Im, "A 40-to-56 Gb/s PAM-4 receiver with ten-tap direct decision-feedback equalization in 16-nm FinFET," IEEE Journal of Solid State Circuits, vol. 52, no. 12, pp. 814–827, 2017.

[2] D. Wang et al., "A 56-Gbps PAM-4 wireline receiver with 4-tap direct DFE employing dynamic CML comparators in 65 nm CMOS," IEEE Transactions on Circuits and Systems I: Regular Papers (TCAS-I), vol. 69, no. 3, pp. 1027–1040, 2022.

[3] V. Stojanovic, "Autonomous dual-mode (PAM2/4) serial link transceiver with adaptive equalization and data recovery," IEEE J. Solid-State Circuits, vol. 40, no. 4, pp. 1012–1026, Apr. 2005.

[4] K. Elissa, J. T. Stonick, G.-Y. Wei, J. L. Sonntag and D. K. Weinlader, " An adaptive PAM-4 5-Gb/s backplane transceiver in 0.25-μm CMOS ", IEEE J. Solid-State Circuits, vol. 38, no. 3, pp. 436-443, Mar. 2003.

[5] X. Zheng et al., "A 40-Gb/s quarter-rate SerDes transmitter and receiver chipset in 65-nm CMOS," IEEE J. Solid-State Circuits, vol. 52, no. 11, pp. 2963–2978, Nov. 2017.

A Low-Power Gm-Boosted VCO with Multi-Transformer in 40nm CMOS

Zilong Wu[1], Bowen Chen[2], Yue Lin[2], Hongtao Xu[1]

[1] State Key Laboratory of Integrated Chips and Systems, College of Integrated Circuits and Micro-Nano Electronics, Fudan University, Shanghai, China
[2] ICLegend Micro, Shanghai, China

Email: 22112020124@m.fudan.edu.cn

Abstract—**This paper proposes a low power broadband continuously tuning voltage-controlled oscillator (VCO) designed using TSMC's 40nm CMOS process, operating in the 8 to 9GHz frequency range. The design employs a Gm-boosted active NMOS pair based on capacitive division and a transformer-based resonant tank with capacitive linear compensation, achieving a single-band continuously frequency tuning range of 1GHz and a medium Kvco of 790MHz/V. The phase noise performance reaches -115dBc/Hz at 1MHz offset from an 8GHz carrier, while maintaining low power consumption of 3mW, resulting in a figure of merit (FoMA) as low as 199dBc/Hz.**

Keywords—VCO, Gm-boost, Low power, Capacitive division

INTRODUCTION

Since frequency-modulated continuous-wave (FMCW) radar technology offers significant advantages including exceptionally high range resolution capabilities, virtually minimal range blind spot limitations, and reduced transmission power requirements, it has experienced remarkable and continuous technological development across diverse automotive and consumer electronic applications. The voltage-controlled oscillator (VCO) serves as the fundamental core component of the sophisticated phase-locked loop (PLL) frequency generation system architecture in advanced FMCW radar implementations, necessitating the critical capability to provide a precisely controllable and continuously tunable frequency band operation within a well-defined specified range, while simultaneously maintaining sufficiently low phase noise characteristics and achieving optimal low power consumption performance for demanding specific applications. This work presents an innovative low-power consumption and low phase-noise VCO design based on advanced Gm-boosted amplification and varactor compensation techniques specifically optimized for fractional - N phase-locked loop applications operating within the near-8GHz frequency spectrum.

CIRCUIT DESIGN

A. Gm-boost technology

For achieving optimal low-power voltage-controlled oscillator design implementations, reduced bias current levels are fundamentally required, which consequently introduces the challenging problem of significantly diminished transconductance (Gm) characteristics of the active differential pair configuration at this operating point. The conventional design approach involves increasing the physical geometric size of the core transistors to effectively enhance the overall transconductance performance, however, when the dimensions of the MOS transistor devices are enlarged beyond a certain critical threshold under specific process conditions, the parasitic capacitance components of core transistors begin to become increasingly non-negligible, which will dramatically and detrimentally reduce the effective transconductance performance at elevated high-frequency operating conditions. Lee et al. previously proposed an innovative circuit structure topology that can significantly enhance the cross-coupling transconductance characteristics under identical transistor sizing constraints [1].

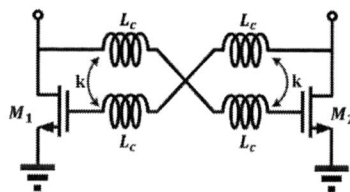

Fig. 1. Gm-boosted active pair

This work introduces a transformer-based coupling element positioned at the gate terminals of the active differential pair to achieve Gm-boosted performance as shown in Figure 1. Actually, the designed inductive impedance introduced by the transformer coupling at the gate terminals is utilized to compensate for the unexpected parasitic capacitance effects, thereby obtaining larger transconductance values within the targeted operational frequency band. Under these optimized design conditions, the complex impedance characteristics of the active circuit region can be calculated based on the small-signal equivalent circuit shown in Figure 2.

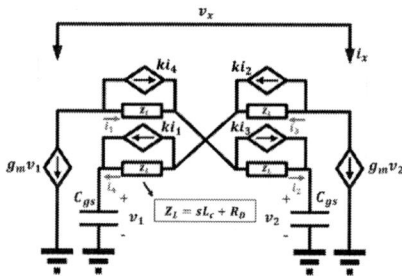

Fig. 2. Small-signal equivalent circuit of the Gm-boosted active pair

The specific expression is as follows:

$$Y_{in} = \frac{i_x}{v_x} = \frac{\left(\frac{g_m}{sC_{gs}} - 1\right)(k-1)}{2\left[\frac{1-k}{sC_{gs}} + 2(sL_c + R_D)\right]} \quad (1)$$

$$Re[Y_{in}] = \frac{(-1+k)\left(-g_m(-1+k+2\omega^2 C_{gs}L_c) - 2\omega^2 C_{gs}^2 R_D\right)}{2\left(-1+k+2\omega^2 C_{gs}L_c\right)^2 + 8\omega^2 C_{gs}^2 R_D^2} \quad (2)$$

$$Im[Y_{in}] = \frac{(-1+k)\omega C_{gs}(-1+k+2\omega^2 C_{gs}L_c - 2g_m R_D)}{2(-1+k+2\omega^2 C_{gs}L_c)^2 + 8\omega^2 C_{gs}^2 R_D^2} \quad (3)$$

Where Lc represents the inductance value of the transformer, Rd represents the equivalent resistance characteristics of the circuit configuration, k signifies the magnetic coupling coefficient of the transformer, gm represents to the small-signal transconductance gain of the active transistor device, and Cgs represents the intrinsic gate-to-source parasitic capacitance of the MOS transistor. In direct comparison with the conventional cross-coupled differential pair configuration, the innovative active differential pair incorporating the advanced gm-boosted structural enhancement demonstrates larger transconductance gain performance characteristics.

The Y-parameter characteristic curves of both the conventional cross-coupled differential pair and the enhanced Gm-boosted cross-coupled differential pair configuration are illustrated in Figure 3. The simulation results clearly demonstrate that through transformer-based compensation techniques, the transconductance performance within the targeted operational frequency band of 8 - 9 GHz is significantly and measurably higher than that achieved by the conventional cross-coupled differential pair implementation.

As can be distinctly observed from Figure 3, the introduction of the transformer coupling element fundamentally alters the transconductance frequency response characteristics of the traditional cross-coupled differential pair configuration. The transconductance performance of the cross-coupled differential pair without transformer enhancement exhibits a nearly linear degradation relationship with increasing frequency, whereas the transconductance characteristic curve with the introduction of the transformer coupling demonstrates enhanced transconductance values at elevated higher frequency operations. Through careful optimization and precise setting of both the inductance value and magnetic coupling coefficient parameters of the transformer element, higher transconductance gain performance can be achieved within the targeted operational frequency band. The simulation results make it clear that within the specific frequency band range of 8 to 9 GHz, the transconductance value of the Gm-boosted configuration is demonstrably at least 50% higher than that of the conventional cross-coupled differential pair implementation.

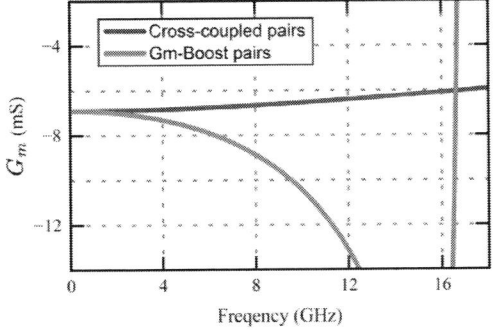

Fig. 3. Comparison of Gm curves using Gm-boosted and conventional cross-coupling pairs

B. Varactor Bias Compensation

In order to meet the stringent continuous frequency modulation operational requirements of advanced FMCW radar systems and effectively avoid the detrimental noise folding problem that is typically caused by voltage-controlled oscillator gain nonlinearity characteristics (particularly problematic in fractional-N frequency division phase-locked loop implementations), a substantially larger linear frequency tuning range needs to be achieved within a single operational frequency band, which fundamentally means obtaining a flatter and more uniform Kvco curve. A transformer configuration with a positioned center tap connection is introduced in this design implementation to provide a secondary bias voltage source that effectively compensates for the inherent nonlinearity characteristics of the varactor diode under conventional single bias voltage operation, as illustrated in the detailed schematic presented in Figure 5. The implementation of this additional secondary bias voltage will increase the linear operational range of the Vtune.

The two selected bias voltage levels implemented in this design are precisely set to vdd = 0.6 V and vb = 1.8 V respectively. It should be particularly noted and emphasized that the vdd supply voltage level also determines the operational characteristics and physical dimensions of the active circuit region, therefore the compensation of the varactor diode linearity and the optimal sizing of the active circuit area need to be simultaneously considered and carefully balanced during the design optimization.

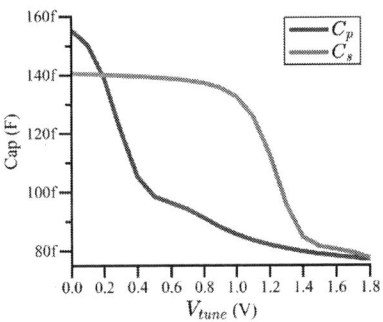

Fig. 4. C-V curve of the varactor with additional bias compensation

According to Figure 4, the capacitance-voltage (C-V) characteristic curves of the primary and secondary varactor elements provide complementary linearity compensation across the complete tuning voltage range.

After the transformer is introduced, the equivalent circuit model of the resonant tank is shown in Figure 5.

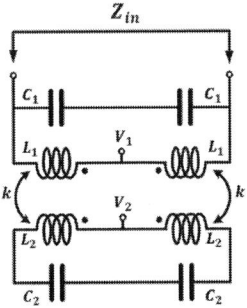

Fig. 5. Schematic diagram of the resonant tank with transformer

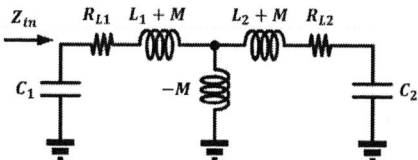

Fig. 6. Simplified model of a resonant tank with a transformer

As illustrated in the simplified circuit diagram of the transformer-coupled resonant tank shown in Figure 6, the resonant circuit exhibits two distinct resonance frequencies, with the specific expressions given as follows:

$$Z_{in,1}(j\omega) = \frac{[-C_2L_1L_2(1-k^2)\omega^3 + L_1\omega]i}{L_1L_2C_1C_2(1-k^2)\omega^4 - (L_1C_1 + L_2C_2)\omega^2 + 1} \quad (4)$$

When resonance occurs, the denominator of the expression (4) approaches zero, and at this point:

$$L_1L_2C_1C_2(1-k^2)\omega^4 - (L_1C_1 + L_2C_2)\omega^2 + 1 = 0 \quad (5)$$

Solving equation (5) yields:

$$\omega_1 = \frac{1}{\sqrt{2}} \cdot \frac{1}{\sqrt{1-k^2}} \cdot \sqrt{\frac{1}{L_2C_2} + \frac{1}{L_1C_1} - \frac{\sqrt{X}}{L_1L_2C_1C_2}} \quad (6)$$

$$\omega_2 = \frac{1}{\sqrt{2}} \cdot \frac{1}{\sqrt{1-k^2}} \cdot \sqrt{\frac{1}{L_2C_2} + \frac{1}{L_1C_1} + \frac{\sqrt{X}}{L_1L_2C_1C_2}} \quad (7)$$

The expression of X is as follows:

$$X = L_2^2C_2^2 + L_1^2C_1^2 + (4k^2 - 2)L_1L_2C_1C_2 \quad (8)$$

Note that the above solution ignores the parasitic resistance that may exist in the transformer inductance, the parasitic capacitance to ground, and the inter-layer parasitic capacitance.

Using this method requires determining through calculation or simulation that the main gain peak of the VCO occurs at the desired frequency.

C. Summary for Circuit Design

As shown in Figure 7, The active circuit section is composed of stacked NMOS differential pairs, with the entire VCO operating in Class-B mode. The transistor sizing ratio of M1 to M2 is configured as 1:2.

The passive circuit section includes the Gm-boost transformer and the resonator transformer. the transformer parameters are precisely obtained through EM electromagnetic field simulation analysis. Specifically, the primary parameters are Lp = 186pH, Ls = 232pH, with coupling coefficient k = 0.77. While the Gm-boosted section exhibits Lc = 73pH with coupling coefficient k = 0.57. The transformers utilized for both Gm enhancement and varactor compensation are fabricated using the topmost metal layer M8, with the effective magnetic induction area of each coil designed to be as large as practically possible. The main transformer coil implemented in this work features two turns for both the primary coil and secondary coil, thereby maximizing the quality factor of the passive circuit to the greatest extent possible. The overall layout is shown in Figure 8.

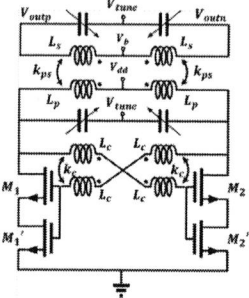

Fig. 7. Proposed VCO schematic with Lp=186pH, Ls=232pH, k=0.77, Lc = 73pH, k = 0.57

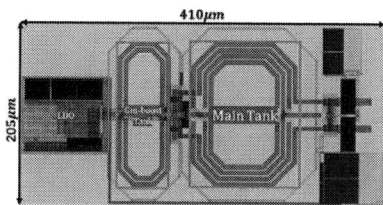

Fig. 8. Layout of proposed VCO

LAYOUT AND POST-SIMULATED RESULTS

The post-layout simulation results demonstrate single-band continuous frequency tuning operation within the 8-9GHz frequency range. The phase noise achieves -115dBc/Hz at 1MHz offset from an 8GHz carrier with an overall power consumption of 3mW. This design exhibits superior performance compared to similar low-power VCO implementations.

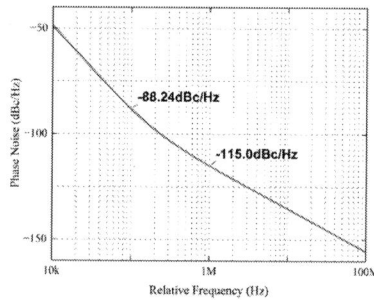

Fig. 9. Phase Noise at 8 GHz

As shown in Figure 9, at 8GHz, the phase noise measures -115dBc/Hz at 1MHz offset and -88.24dBc/Hz at 100kHz offset, representing the best phase noise performance across the entire tuning range.

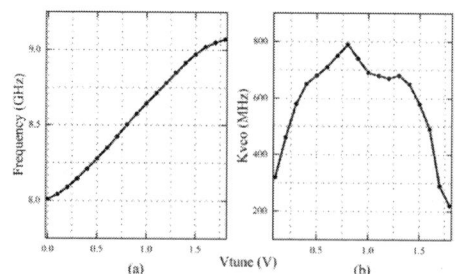

Fig. 10. (a) Tuning range of VCO; (b) K_{vco} of VCO

979-8-3315-3918-4/25 $31.00 © 2025 IEEE

As illustrated in Figure 10(a), the VCO achieves continuous frequency tuning from 8 to 9GHz over the control voltage range of Vtune = 0 to 1.8V. Figure 10(b) shows that Kvco ranges from 320 to 790MHz/V with great linearity over the Vtune range of 0.3 to 1.5V.

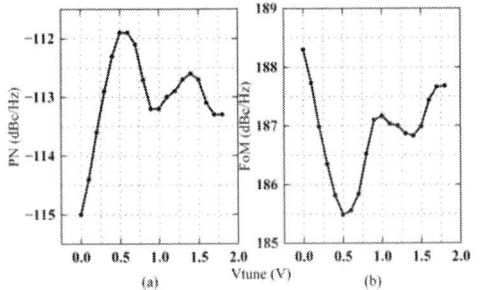

Fig. 11.   (a) Phase Noise of VCO; (b) FoM of VCO

As shown in Figures 11(a) and (b), the VCO achieves phase noise performance of -112 to -115dBc/Hz and FoM values of 185.5 to 188.3dBc/Hz across the tuning range.

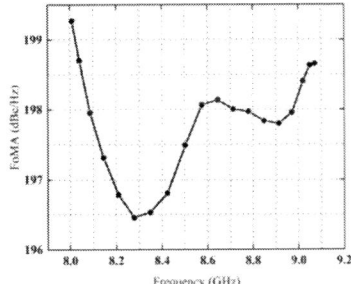

Fig. 12. FoM_A of VCO

As shown in Figure 12, the FoMA achieves an optimal value of 199dBc/Hz within the tuning range, which benefits from the transformer-based design that improves area efficiency.

$$FoM^1=|PN| + 20\log\left[\frac{f_{osc}}{f_{offset}}\right] - 10\log\left[\frac{P_{dc}}{1mW}\right], FoM_T{}^2 = FoM + 20\log\left[\frac{TR(10\%)}{10}\right]$$

$$FoM_A{}^3 = FoM + 10\log\left[\frac{Area(mm^2)}{1mm^2}\right]$$

The performance comparison with state-of-the-art LC VCO designs is presented in TABLE I. The results clearly demonstrate that among the LC VCOs designed for millimeter-wave applications, this work achieves excellent FoMA performance indicators while maintaining extremely low power consumption characteristics. The mathematical definitions of each FoM value presented in the comparison table are provided in the aforementioned formulas.

CONCLUSION

This research work presents a novel single-band continuously frequency-modulated 8-9GHz voltage-controlled oscillator design that incorporates advanced Gm-boost technology specifically optimized for low-power operation. The final implementation successfully achieves an outstanding figure of merit value of 187dBc/Hz in the average case, while maintaining power consumption at only 3mW, which represents exceptional performance levels among comparable research works in the field.

ACKNOWLEDGMENT

This work is supported by National Key R&D Program Special Projects of China (No.2023YFB4403800).

REFERENCES

[1] H. S. Lee, D. M. Kang, S. J. Cho, C. W. Byeon and C. S. Park, "Low-Power, Low-Phase-Noise Gm-Boosted 10-GHz VCO With Center-Tap Transformer and Stacked Transistor," in IEEE Transactions on Circuits and Systems II: Express Briefs, vol. 67, no. 10, pp. 1710-1714, Oct. 2020

[2] C. Wan, T. Xu, X. Yi and Q. Xue, "A Current-Reused VCO With Inductive-Transformer Feedback Technique," in IEEE Transactions on Microwave Theory and Techniques, vol. 70, no. 5, pp. 2680-2689, May 2022

[3] A. Franceschin, P. Andreani, F. Padovan, M. Bassi and A. Bevilacqua, "A 19.5-GHz 28-nm Class-C CMOS VCO, With a Reasonably Rigorous Result on 1/f Noise Upconversion Caused by Short-Channel Effects," in IEEE Journal of Solid-State Circuits, vol. 55, no. 7, pp. 1842-1853, July 2020

[4] Y. Hu, T. Siriburanon and R. B. Staszewski, "A Low-Flicker-Noise 30-GHz Class-F23 Oscillator in 28-nm CMOS Using Implicit Resonance and Explicit Common-Mode Return Path," in IEEE Journal of Solid-State Circuits, vol. 53, no. 7, pp. 1977-1987, July 2018

[5] M. Haghi Kashani, R. Molavi and S. Mirabbasi, "A 2.3-mW 26.3-GHz G_{m}-Boosted Differential Colpitts VCO With 20% Tuning Range in 65-nm CMOS," in IEEE Transactions on Microwave Theory and Techniques, vol. 67, no. 4, pp. 1556-1565, April 2019

TABLE I.   COMPARISON OF THE STATE-OF-THE-ART ON LC VCO

	This work	T-MTT [2]	JSSC [3]	JSSC [4]	T-MTT [5]
Technology	40nm	65nm	28nm	28nm	65nm
Topology	Gm-Boost	Current-Reuse	Class-C	Class-F23	Gm-Boost+Colpitts
TR(GHz)	8 to 9	5.18 to 9.62	17.6 to 21.84	27.3 to 31.2	23.7 to 28.9
TR(%)	11.76	60	24	13.33	20
Power(mW)	3	7.7	20.7	12 to 22	2.3
Core Area(mm²)	0.08	0.06	0.07	0.15	0.22
Phase Noise @1MHz(dBc/Hz)	-113*	-115.35	-112	-105*	-98.3
FoM¹(dBc/Hz)	187*	183.11	185	181	183
FoMT²(dBc/Hz)	189*	192.64	193	183	189
FoMA³(dBc/Hz)	198*	195.33	197	189	190

*Average value over the entire frequency tuning range

A 64 Gbps 10 mW 0.0081 mm^2 Inverter-Based CTLE Employing Power-Efficient Split Biasing Topology in 40 nm CMOS

Fang Ding, Huzhi Tang, Ke Wu, Yuekang Guo*, Jing Jin, Jianjun Zhou

School of Integrated Circuits, Shanghai Jiao Tong University, Shanghai, 200240, China

guoyuekang@sjtu.edu.cn

Abstract—This paper presents an inverter-based continuous time linear equalizer (CTLE) for high-speed compact wireline interface. A novel inverter topology is proposed, which employs resistors and tunable currents to split the DC bias voltages of the NMOS and the PMOS, achieving a good trade-off between tunable peaking gain range and power efficiency. Post-layout simulation results show that the proposed CTLE demonstrates an 8 dB tunable peak gain range with a peak frequency of 16 GHz. The CTLE consumes 10 mW power, resulting in an energy efficiency of 0.15 pJ/b.

Keywords—*Split Bias, Inverter, CTLE, High-Speed Wireline, Tunable Gain, Power Efficiency*

I. INTRODUCTION

The rapid growth of 5G communication and data center interconnects has significantly increased the demand for high-speed wired communication systems. As data rates continue to rise, channel attenuation and inter-symbol interference (ISI) have become critical bottlenecks that limit the performance of these systems [1]. Among the various components in a high-speed serializer/deserializer (SerDes) system, the continuous-time linear equalizer (CTLE) plays a pivotal role in compensating for channel loss and improving signal integrity at the receiver front-end.

Traditional CTLE implementations based on current-mode logic (CML) architectures have been widely used in high-speed communication systems. However, these conventional approaches face several limitations in advanced technology nodes. The inductive peaking technique commonly employed in CML CTLEs requires large area-consuming passive components [2], which becomes prohibitive in highly integrated multi-lane SerDes implementations.

Moreover, the reduced supply voltages in advanced CMOS processes make it increasingly challenging for CML-based CTLEs to achieve sufficient gain and bandwidth while maintaining acceptable linearity and power efficiency.

To address these challenges, this paper presents an innovative CTLE design implemented in a 40 nm CMOS process. Unlike conventional CML-based approaches, the proposed CTLE utilizes standard CMOS inverter cells as the fundamental gain elements. This architectural shift offers several advantages, including reduced area consumption, improved power efficiency, and enhanced scalability in advanced technology

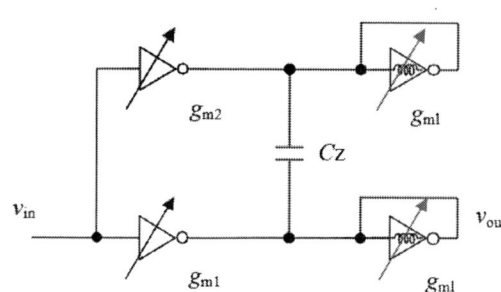

Fig. 1: Schematic diagram of the proposed CTLE architecture.

nodes [3]. The proposed inverter-based CTLE achieves a compact core area of 90 μm × 90 μm, which is approximately 4 times smaller than traditional CML implementations [2]. Most critically, the design incorporates advanced biasing techniques and a novel active inductive peaking method to extend the bandwidth while maintaining low power consumption.

II. INVERTER-BASED CTLE CIRCUIT TOPOLOGY

Fig. 1 shows the single-ended circuit of the proposed fully differential CTLE. The proposed CTLE features an innovative inverter-based architecture, incorporating two stages of equalization for enhanced performance. This design hinges on a hybrid circuit topology that adeptly merges low-frequency and high-frequency gain paths. This topology is meticulously crafted from core and tunable slices, optimizing the equalizer's functionality. To further broaden the bandwidth of the CTLE, active inductors are integrated, each equipped with tunable biasing circuits. The transfer function characterizing the operation of the proposed CTLE can be succinctly expressed as follows:

$$\frac{v_o}{v_i} = -\frac{g_{m1}}{g_{ml}} \frac{1 + s\frac{g_{m1}+g_{m2}}{g_{m1}}\frac{C}{g_{m1}}}{1 + s\frac{2C}{g_{ml}}} \cdot P(s), \qquad (1)$$

where g_{m1} and g_{m2} are the transconductances of the low-frequency and high-frequency gain paths, respectively. g_{ml} represents the transconductance of the active inductive load,

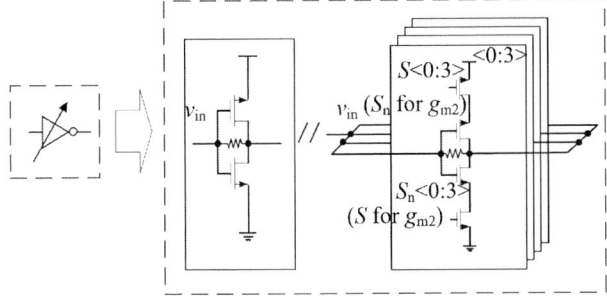

Fig. 2: Core circuit schematic of the proposed CTLE.

which is tunable with the current biasing circuits. This tunability is essential for extending the bandwidth of the CTLE. To maintain a stable peak gain while adjusting the DC gain, the tunable slices of the g_{m1} and g_{m2} cells employ opposite switches $S[3:0]$ and $S_n[3:0]$. This configuration ensures that the sum of g_{m1} and g_{m2} remains constant [4]. The capacitors C are utilized to determine the zero location of the transfer function. Additionally, $P(s)$ accounts for the additional zeros and poles introduced by the active inductive load, as well as other parasitic capacitors and resistors.

A. Tunable active inductors

As shown in Fig. 3, a crucial aspect of this innovative design lies in the introduction of an extra current source biasing technique specifically tailored for g_{ml}. By employing this sophisticated method, the design enables highly accurate regulation of the operating point of the active inductive load. As a result, the performance of the CTLE is markedly improved. Notably, this enhancement manifests as a substantial increase in the bandwidth, reaching up to 2 GHz, all while incurring a minimal additional power consumption of merely 1 mW. This elegant solution effectively strikes a balance between performance optimization and power efficiency, making it a standout feature of the design.

The additional current source biasing applied to g_{ml} provides several advantages. By carefully adjusting the bias current, the transconductance g_{ml} can be dynamically tuned. This tunability translates into enhanced flexibility in shaping the frequency response of the CTLE. The active inductive peaking effect is significantly improved, enabling the CTLE to achieve a higher peak gain while maintaining a wide bandwidth. This is particularly beneficial for compensating severe channel loss at high data rates.

The additional current source biasing also helps in optimizing the linearity and power efficiency of the CTLE. By maintaining the active inductive load in its optimal operating region, the circuit can achieve better linearity, which is crucial for handling multi-level modulation schemes such as four-level pulse amplitude modulation (PAM4). The ability to precisely control the bias current enables the CTLE to operate efficiently across a range of process, voltage, and temperature (PVT) variations.

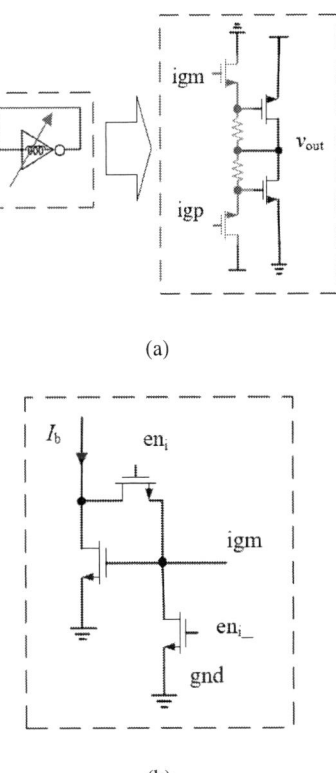

(a)

(b)

Fig. 3: Schematic of (a) active inductor and (b) pared current mirror

Overall, the additional current source biasing technique plays a pivotal role in enhancing the performance of the proposed CTLE. It provides the necessary flexibility and adaptability to meet the demanding requirements of high-speed link communication systems.

The tunable inverter in Fig. 3 (a) uses switched current mirrors to adjust V_{bias}:

$$g_{ml} = k_\mu C_{ox} \frac{W}{L}(I_b R_l - V_{th}), \qquad (2)$$

where I_b is the input current and R_l is the feedback resistor of the inverter-based active inductor. A pair of current mirrors separately controls the current through NMOS and PMOS, which can increase the V_{gs} of the inverter sell to expand the band width of the proposed CTLE, meanwhile enabling 6 - 14 dB gain variation while maintaining $V_{DS} > V_{dsat}/2$ for PAM4 linearity.

B. T-coil Implementation

The proposed CTLE architecture integrates T-coil technology to enhance impedance matching and signal integrity. The T-coil is strategically positioned in the input stage of the CTLE to achieve optimal impedance matching with the transmission channel. This design effectively reduces reflection losses and significantly improves the overall performance of the receiver

979-8-3315-3918-4/25 $31.00 © 2025 IEEE

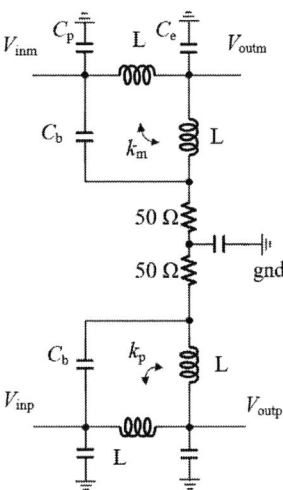

Fig. 4: T-coil in CTLE.

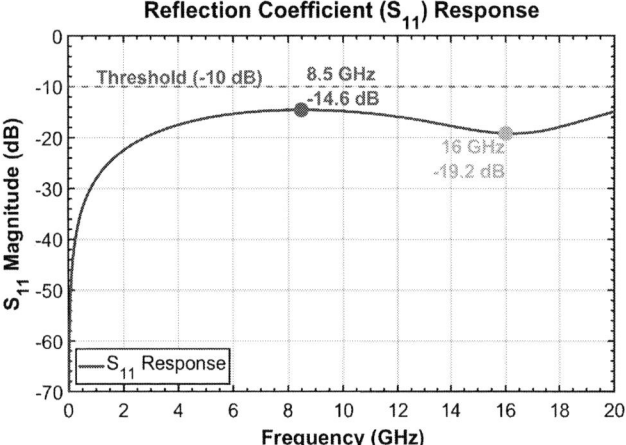

Fig. 5: S_{11} Curve.

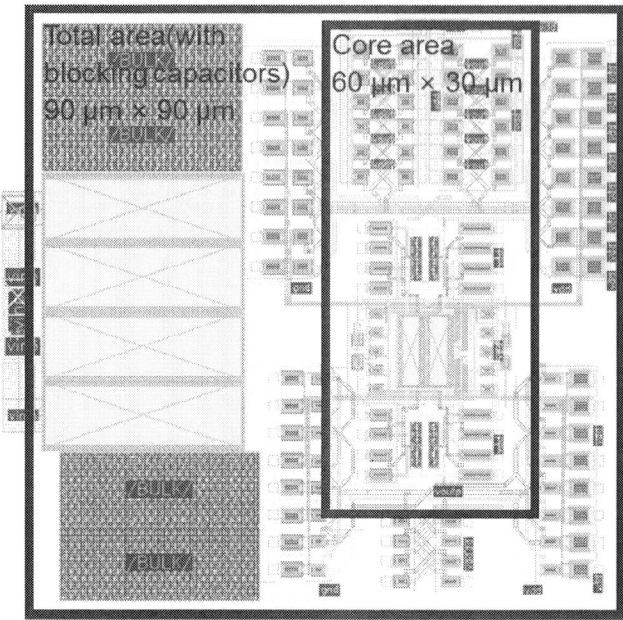

Fig. 6: Layout of proposed CTLE with blocking capacitors and feedback resistors.

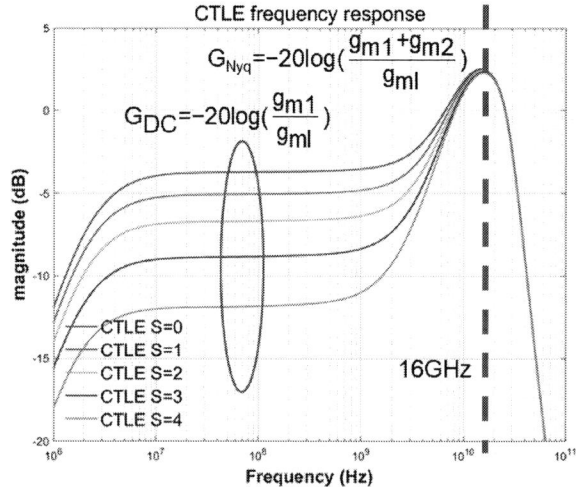

Fig. 7: AC simulation frequency response of the proposed CTLE

front-end. Subsequent simulations confirmed that the input impedance matching remains stable across the nyquist bandwidth, as shown in Fig. 5, with S_{11} maintained below -14 dB, ensuring minimal reflection loss.

III. LAYOUT AND SIMULATION RESULTS

As shown in Fig. 6, the layout of the CTLE is composed of coupling capacitors and the main circuit. It is a fully differential structure with a total size of 90 μm × 90 μm. The core part in the middle is 60 μm × 30 μm, including the current biasing circuit. The inverter gain cells are arranged in a centroid symmetric manner, which greatly reduces the impact of parasitic resistance and capacitance. Post-layout simulations in 40 nm CMOS process demonstrate that the CTLE can compensate for 8 dB channel loss while maintaining the integrity of a 64 Gbps PAM4 signal. This is made possible through dynamic split biasing, which extends the bandwidth of the proposed CTLE to handle high-frequency signals more effectively. Additionally, active inductive peaking is employed

to flatten the gain across the frequency spectrum, ensuring a more uniform signal response and reducing distortion.

The AC response demonstrates 8 dB peaking gain at 16 GHz with 6–14 dB programmable range. Bandwidth extends to 15.9 GHz through active peaking showing a 17% improvement over the same topology without the proposed split biasing designs.

Transient analysis with PRBS31 patterns reveals the compensation effectiveness of the proposed CTLE. As quantified in Table I, the equalized eyes show average 25 mV vertical openings and 0.29 UI horizontal margins. The clear separation between all three PAM4 levels shown in Fig. 8 confirms the design's linearity, with RLM exceeding 92%.

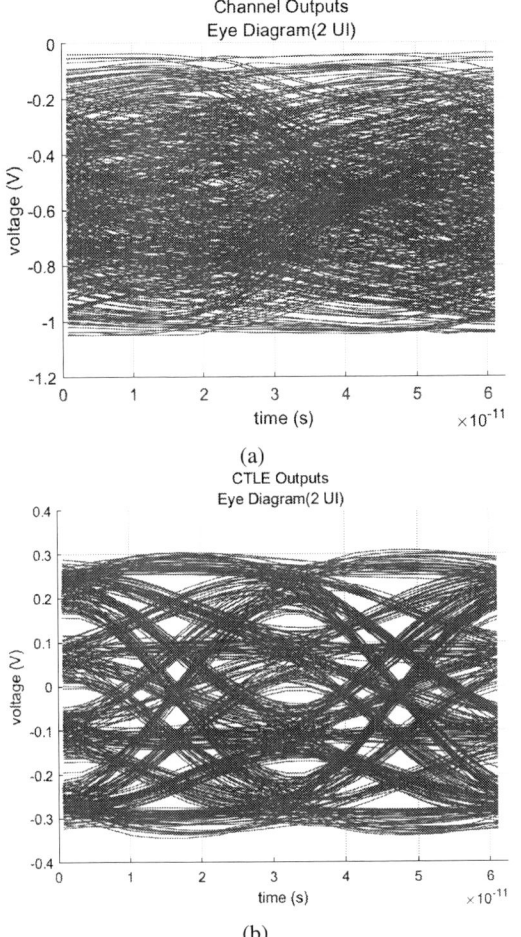

Channel Outputs
Eye Diagram(2 UI)

(a)

CTLE Outputs
Eye Diagram(2 UI)

(b)

Fig. 8: Post-equalization 64 Gbps PAM4 eye diagram of (a) channel output (b) CTLE output.

TABLE I: Eye Diagram Measurement Results

Metric	Threshold (mV)	Height (mV)	Width (UI)
Upper Eye	-183.7	22.49	0.28
Middle Eye	-18.68	28.93	0.30
Lower Eye	147.7	22.72	0.30

The combined results verify that the inverter-based architecture achieves competitive performance (0.15 pJ/b efficiency, 0.0081 mm^2 core area) while overcoming traditional CML CTLE limitations in PAM4 applications. The dynamic biasing technique proves particularly effective for maintaining consistent equalization across process corners.

IV. CONCLUSION

This work demonstrates that tunable biasing enables inverter-based CTLEs to meet 64 Gbps PAM4 requirements. The design overcomes the linearity limitations of previous fixed-bias methods and achieves a 4× reduction in area compared to [2] while using the same technology. Additionally,

TABLE II: Performance Comparison of CTLE Architectures

Specification	[2]	[5]	[6]	This Work
Technology	40 nm	16 nm	28 nm	40 nm
Architecture	CML	Inverter	Cherry-Hooper	Inverter
Modulation	PAM4	PAM4	NRZ	PAM4
Data rate (Gbps)	64	56	32	64
Gain@f_{Nyq} (dB)	22.5	14 [a]	9.5	6–14
Efficiency (pJ/b)	0.20	0.61	0.32	0.15
Area (μm × μm)	325 × 110 [b]	56 × 75	50 × 85	90 × 90

[a] Estimated from the CTLE frequency response figure

[b] Estimated from the layout figure

it achieves a 4× improvement in energy efficiency compared to [5]. The proposed CTLE achieves a compact core area of 90 μm × 90 μm and consumes only 10 mW of power while delivering excellent performance in compensating for channel loss and improving signal integrity. The innovative inverter-based CTLE architecture offers a promising alternative to traditional CML-based designs, paving the way for advanced SerDes implementations that can meet the ever-increasing demands of modern communication technologies.

ACKNOWLEDGMENT

This work was supported by the National Natural Science Foundation of China under Grant 62401360 and Grant 62431016.

REFERENCES

[1] J. Chen et al., "A 32-GS/s 7-bit TI-SAR ADC in 28-nm for 32-gbps ADC-based SerDes receiver," in *Proc. IEEE Int. Conf. ASIC (ASICON)*, Nanjing, China, Oct. 2023, pp. 1–4.

[2] G. Wang and Z. Zhang, "A 64-gbps 0.33-pJ/bit PAM4 receiver analog front-end with a single-stage triple-peaking CTLE achieving 22.5-dB boost in 40-nm CMOS process," *Integr. Circuits Syst.*, vol. 1, no. 2, pp. 103–108, May-Jun. 2024.

[3] K. Zheng et al., "A 56 gbps 6 mW 300 um2 Inverter-Based CTLE for Short-Reach PAM2 Applications in 16 nm CMOS," *2018 IEEE Custom Integrated Circuits Conference (CICC)*, pp. 1-4, Sep. 2018.

[4] A. Ensinger et al., "Minimum Power Point Design of Inverter Based Continuous Time Linear Equalizer (CTLE)," *2024 IEEE International Symposium on Circuits and Systems (ISCAS)*, pp. 1-5, May 2024.

[5] K. Zheng et al., "An Inverter-Based Analog Front-End for a 56-gbps PAM-4 Wireline Transceiver in 16-nm CMOS," *IEEE Solid-State Circuits Letters*, vol. 1, no. 12, pp. 249-252, Dec. 2018.

[6] S. Lee et al., "Feedforward Cherry-Hooper continuous-time linear equalizer in 28-nm CMOS," in *Proc. Int. Tech. Conf. Circuits/Syst., Comput. Commun. (ITC-CSCC)*, Phuket, Thailand, Jul. 2022, pp. 507–510.

979-8-3315-3918-4/25 $31.00 © 2025 IEEE

A Low-Power Area-Efficient Serializer for CMOS Image Sensors

Jingyang Chen[1,2], Nanbo Chen[1,2], Gang Wang[2], Peng Feng[1,3*], Jian Liu[1,3], Nanjian Wu[1,3], Liyuan Liu[1,3]

[1] State Key Laboratory of Semiconductor Physics and Chip Technologies, Institute of Semiconductors, Chinese Academy of Sciences, Beijing 100083, China

[2] School of Physical Science and Technology, Ningbo University, Ningbo 315211, China

[3] Center of Materials Science and Optoelectronics Engineering, University of Chinese Academy of Sciences, Beijing 100049, China

* Email: fengpeng06@semi.ac.cn

Abstract—**This paper proposes a low-power, area-efficient serializer circuit for column-parallel output CMOS image sensors (CIS). The circuit replaces conventional D flip-flops (DFFs) with set-reset true single-phase clocking (TSPC) DFFs and employs low-frequency multiphase clocks instead of high-frequency clocks. This circuit has been designed in a 1P6M 180 nm CMOS process with layout area of 1470 μm x 390 μm. The post-layout simulation results show that it can achieve a maximum transmission rate of 1 Gbps with low-voltage differential signaling (LVDS) driver. The power consumption of the serializer and the LVDS driver is 2.21 mW and 12.28 mW.**

Keywords—*CMOS image sensor (CIS), Serializer, Multiphase clock, Low voltage differential signaling (LVDS)*

I. INTRODUCTION

With the continuous performance improvement in CMOS image sensor (CIS), the data amount of CIS has increased significantly, imposing higher demands on data transmission. The use of serializer [1] and low-voltage differential signaling (LVDS) drivers [2] in CIS offers advantages such as high transmission speed, high reliability and simple connectivity, making them ideal choices for high-speed data transfer. As chip integration level continues to increase, there is a growing requirement to reduce the power consumption and area of the readout circuit while maintaining high-speed data transmission [3].

This paper proposes a low-power, area-efficient serializer for 64-column-parallel 12-bit analogue-to-digital converter (ADC) in CIS, which converts the low-speed parallel data from the ADC into high-speed serial data and transmits data off-chip via an LVDS driver at a maximum rate of 1 Gbps for a single channel.

II. PROPOSED CIRCUIT IMPLEMENTATION

A. Architecture of the Serializer

To meet the data transmission requirements of the column parallel CIS, this paper adopts a hybrid parallel-tree structure in the serializer. The overall architecture is shown in Fig. 1. This serializer is composed of two stages. In the 1st stage, the 12-bit parallel digital signals from the ADC columns are input into corresponding 12:1 serializers. In 12:1 serializer, two group of 6-bit parallel data are input into two 6:1 serializers and then serialized by a 2:1 serializer for transmission to the next stage. As a result, the ADC data is converted into 64-channel medium-speed serial data by 64 parallel 12:1 serializers. A global multiphase clock generator provides the required clock signals, generating different clock phases to control the 6:1 serializers. In 2nd stage, the tree-structured 64:1 serializer, composed of two stages of 8:1 serializers, further converts the 64-channel medium-speed data into a single-channel high-speed serial data, which is finally output off-chip through the LVDS driver.

B. Global Multiphase Clock Generator

The first stage of the serializer consists of 64 12:1 serializers and a global multiphase clock generator, achieving low-power and area-efficient. The structure and timing of the global multiphase clock generator are shown in Fig. 2(a) and (b). It contains ring shift registers by six set-reset D flip-flops (DFFs) connected in series. Under the control of a reset signal, CLK5 is set high, while CLK4-CLK0 are reset low. Driven by the CLK which is divided by the frequency divider, the states of CLK5-CLK0 sequentially transition from: 100000 → 010000 → 001000 → 000100 → 000010 → 000001.

In the DFF, traditional architecture typically consists of multiple NAND and NOR gates. In this paper, a true single-phase clocking (TSPC) DFF [4] is used, whose structure is shown in Fig. 3. It contains only 18 transistors, significantly reducing power and area. The TSPC DFF operates as follows: When reset is high, output Q is reset to logic "0"; When set is high, Q is set to logic "1", thus realizing the set-reset function. If reset and set are both low: For D = logic "1": When CLK is

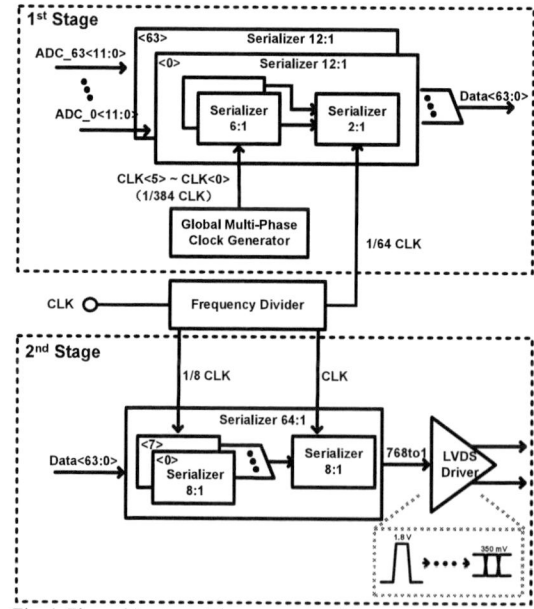

Fig. 1. The architecture of the proposed serializer.

This work is supported by the National Key Research and Development Program of China (2022YFB2804402), National Natural Science Foundation of China (62134004).

979-8-3315-3918-4/25 $31.00 © 2025 IEEE

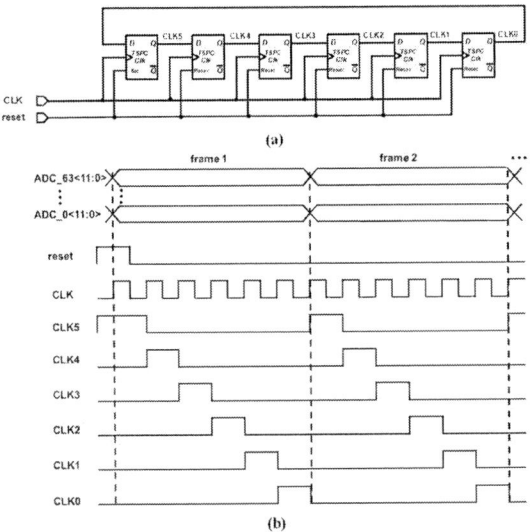

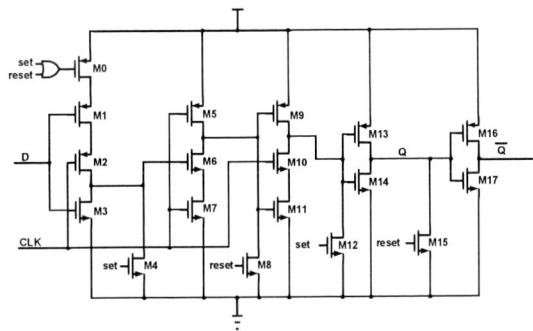

Fig. 2. (a) The structure of the global multiphase clock generator, (b) The timing diagram of the global multiphase clock generator.

Fig. 3. The schematic of the TSPC DFF.

TABLE I. TRUTH TABLE OF TSPC DFF

reset	set	D	CLK	Q	$\overline{Q}$
1	0	-	-	0	1
0	1	-	-	1	0
0	0	1	0/1	hold	hold
0	0	1	0→1	1	0
0	0	0	0/1	hold	hold
0	0	0	0→1	0	1

low, M2 and M5 turn on, pulling M3's drain low, and M6's drain is precharged to a high by M5. At this time, M10 turns off, the M9/M11 inverter can't pass a valid signal to the next stage, with a level-holding characteristic. When the CLK jumps to a high level, M2 and M5 turn off, M3's drain stays low, and M6's drain still stays high. At this time, M10 turns on under the CLK control, pulling its drain low and through the M13/M14 inverter setting the output Q to logic "1". For D = "0": When the CLK jumps to a high level, the output Q is set to logic "0". The truth table of TSPC DFF is shown in Table I.

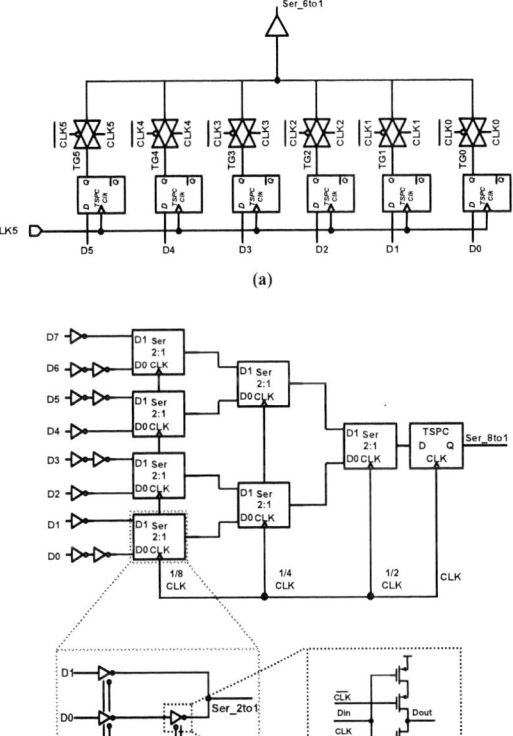

Fig. 4. (a) The structure of the 6:1 serializer, (b) The architecture of the 8:1 serializer.

C. Serializer Cell

The 12:1 serializer consists of two 6:1 serializers based on a multiphase clock generator and a 2:1 serializer based on clock-controlled inverter, resulting in optimized area and power consumption. The architecture of the 6:1 serializer is shown in Fig. 4(a). The DFFs align the phases of input data D5-D0 with the clocks from the global multiphase clock generator. Under CLK5-CLK0 control, transmission gates TG5-TG0 are sequentially enabled, converting 6-channel parallel data to 1-channel serial data.

The 64:1 serializer consists of two stages of 8:1 serializers, and the structure of the 8:1 serializer is shown in Fig. 4(b). The 8:1 serializer comprises multiple 2:1 serializers based on clock-controlled inverter. Each 2:1 serializer uses only three clock-controlled inverters, and output $\overline{D1}$ when CLK is high, and output D0 when CLK is low, as a result one bit data can be transmitted per half clock cycle. To output correct data, the number of inverters in different paths must be matched by adding extra inverters. Since only one extra inverter stage is required at the input, area and power are minimized.

D. LVDS Driver

The LVDS driver's output differentially transmits to an off-chip receiver capable of driving 10 pF loads. As shown in Fig. 5, the driver employs 3.5 mA constant-current sources and converts 3.3 V differential inputs into ±350 mV LVDS outputs by controlling M1–M4. The dual-current-source architecture incorporates a common-mode feedback circuit to

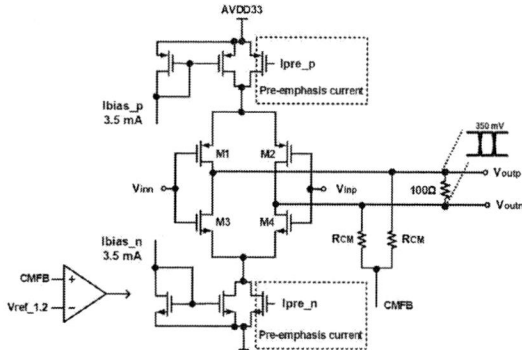

Fig. 5. The schematic of the LVDS driver.

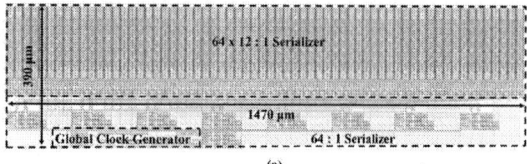

(a)

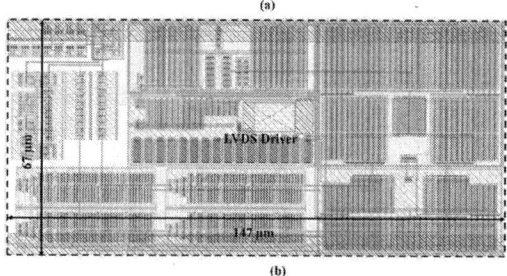

(b)

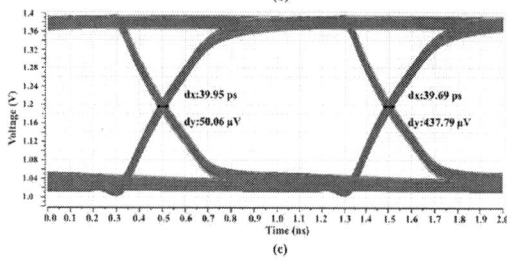

(c)

Fig. 6. (a) Layout of the proposed serializer, (b) Layout of the proposed LVDS driver, (c) The eye diagram of the output data at 1 Gbps.

stabilize output common-mode voltage. Since transmission line parasitics cause high-frequency attenuation, the pre-emphasis circuit injects additional current during signal transitions, compensating high-frequency losses.

III. SIMULATION RESULTS AND COMPARATION

In this paper, the circuit is designed in a 180 nm 1P6M CMOS process. The serializer circuit layout is shown in Fig. 6(a) with an area of 1470 μm x 390 μm. The LVDS driver circuit layout is shown in Fig. 6(b) with an area of 147 μm x 67 μm. And Fig. 6(c) shows the eye diagram of the output LVDS data at 1 Gbps, with a jitter of 39.95 ps. The power consumption of the serializer and the LVDS driver is 2.21 mW and 12.28 mW, respectively. Comparison with other literature results is demonstrated in Table II. The results show that the designed circuit has significant advantages in power efficiency and area.

TABLE II.   PERFORMANCE COMPARISON

	This Work	[5]	[6]	[7]
Technology	180 nm CMOS	180 nm CMOS	150 nm CMOS	130 nm CMOS
Data Rate (Gbps)	1	1.2	1.25	1.25
Jitter (ps)	39.9	40	60	25.6
Load (pF)	10	12	10	N/A
Serializer Energy Efficiency (mW/Gbps)	2.21	N/A	2.88	11.6
Serializer Area (μm²/per channel)	746	N/A	1654	6006
LVDS Power (mW)	12.28	12.7	N/A	N/A
LVDS Area (mm²)	0.0099	0.067	0.023	N/A

IV. CONCLUSION

This paper presents a readout circuit for column-parallel output CIS, comprising a 768:1 serializer and an LVDS driver. The design converts 64-column 12-bit parallel digital signals from the ADC columns into a single-channel 1 Gbps high-speed data, achieving superior power and area efficiency. The layout areas of them are 1470 μm × 390 μm and 147 μm × 67 μm, with power consumption of 2.21 mW and 12.28 mW, respectively.

REFERENCES

[1] H. Lu, C. Su and C. -N. J. Liu, A Tree-Topology Multiplexer for Multiphase Clock System, in IEEE Transactions on Circuits and Systems I: Regular Papers, vol. 56, no. 1, pp. 124-131, Jan. 2009.

[2] J. Park, J. -H. Chae, Y. -U. Jeong, J. -W. Lee and S. Kim, A 2.1-Gb/s 12-Channel Transmitter with Phase Emphasis Embedded Serializer for 55-in UHD Intra-Panel Interface, in IEEE Journal of Solid-State Circuits, vol. 53, no. 10, pp. 2878-2888, Oct. 2018.

[3] M. Furuta, Y. Nishikawa, T. Inoue and S. Kawahito, A High-Speed, High-Sensitivity Digital CMOS Image Sensor With a Global Shutter and 12-bit Column-Parallel Cyclic A/D Converters, in IEEE Journal of Solid-State Circuits, vol. 42, no. 4, pp. 766-774, April 2007.

[4] A. K. Mishra, U. Chopra and D. Vaithiyanathan, A Partially Static High Frequency 18T Hybrid Topological Flip-Flop Design for Low Power Application, in IEEE Transactions on Circuits and Systems II: Express Briefs, vol. 69, no. 3, pp. 1592-1596, March 2022.

[5] W. Fan, Z. Li, J. Xi, L. He, K. Sun and N. Xie, A 1.2 Gbps failsafe low jitter LVDS transmitter-receiver applied in CMOS image sensor, 2018 7th International Conference on Modern Circuits and Systems Technologies, Thessaloniki, Greece, 2018, pp. 1-4.

[6] F. Zhan et al., Invited Paper: Area Efficient Low Power Data Transmission Circuit with Analog Differential Data Selector for CMOS Image Sensor, 2024 IEEE International Conference on Integrated Circuits, Technologies and Applications, Hangzhou, China, 2024, pp. 128-131.

[7] X. Niu, W. Han, W. Zhou and C. Zhao, Design and Simulation of HiGBt, a 5 Gb/s SerDes for Heavy-Ion Physics Experiments, in IEEE Transactions on Nuclear Science, vol. 70, no. 6, pp. 1083-1089, June 2023.

A 6b 14GHz Phase Interpolator with 2-Stage Injection-Locked Ring Oscillators in 28nm CMOS

Danqi Ding[1] , Bingyi Ye[*2] , Weixin Gai[*1, 3]

[1] Peking University, Beijing, China
[2] East China Normal University, Shanghai, China
[3] Beijing Advanced Innovation Center for Integrated Circuits, Beijing, China

*byye@ic.ecnu.edu.cn, wgai@pku.edu.cn

Abstract—This paper presents a 6-bit digitally controlled 8-phase 14 GHz phase interpolator for 224 Gb/s PAM-4 receivers, based on a 2-stage injection-locked ring oscillator(ILRO) architecture. The first stage ILRO employs two-phase injection signals to generate initial 8-phase clocks. The 2nd-stage ILRO suppresses phase errors to within ±1.5°. The phase interpolator, implemented in 28nm CMOS, shows a peak-to-peak INL of 1.26 LSB, a peak-to-peak DNL of 0.16 LSB, and an RMS jitter of 61.8 fs integrated from 10 kHz to 100 MHz offset under TT process corner at 27° C.

Keywords—injection-locked ring oscillator, multi-phase injection, phase interpolator

I. INTRODUCTION

Next-generation AI workloads require highly interconnected systems to enable distributed and efficient computing. To support this demand for distributed computing, Ethernet data rates continue to increase rapidly. The recent 2024 standard targets 200 Gb/s, representing a doubling in less than 3 years [1] . To meet these bandwidth requirements, wireline receivers beyond 100 Gb/s increasingly require 8-way interleaving for SAR ADC implementations. Furthermore, newer receiver designs are advancing towards 8-way [2] and even 16-way interleaving [3]. Conventional phase rotation solutions for 224 Gb/s PAM-4 receivers requiring 8-phase clocks face significant challenges due to prohibitively high-cost high-frequency clock generation and inefficient operation. The number of PIs scales with the number of sampling phases, greatly adding to the power and area cost for these many-phase ADCs. [1].

ILROs have emerged as a promising alternative for multi-phase clock generation due to their capability to operate at high frequencies, favorable phase noise performance, and low power consumption [4]. However, multiphase injection requires a number of injection circuits corresponding to the number of phases, significantly increasing power consumption and area overhead [5]. While employing only two-phase injection reduces this overhead, it introduces critical drawbacks such as a reduced locking range and significant phase misalignment among the output phases.

II. INJECTION SIGNAL GENERATION CIRCUIT

This paper proposes a phase interpolator based on 2-stage ILROs architecture, as illustrated in Fig. 1 [6] .

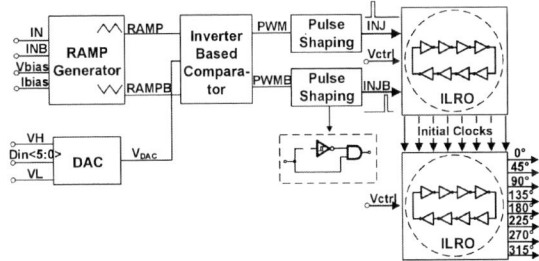

Fig.1 Architecture of the Phase Interpolator with 2-Stage ILRO

The operation of the injection signal generation circuit is further illustrated by the timing diagram of its critical nodes shown in Fig. 2, which demonstrates how the phase shift is implemented. The ramp generator employs 14GHz differential clock signals IN to produce differential ramp signals RAMP. A 6-bit input code, DIN<5:0>, controls the DAC to generate 64 discrete DC levels V_{DAC}. These levels set the DC bias voltage for the differential ramp signals. A low-power, high-speed inverter-based comparator compares the DC-biased ramp signals against the threshold voltage of the inverters, generating PWM outputs with duty cycles that vary according to the DAC input code. The pulse shaping circuit converts PWM into differential injection signals INJ with identical, fixed duty cycles. These signals are injected into the first-stage ILRO to produce initial 8-phase clocks [6] . Due to the use of only two-phase injection in the first stage, these initial clocks exhibit phase misalignment. These initial clocks are then injected into the 2nd-stage ILRO to compensate for phase misalignment.

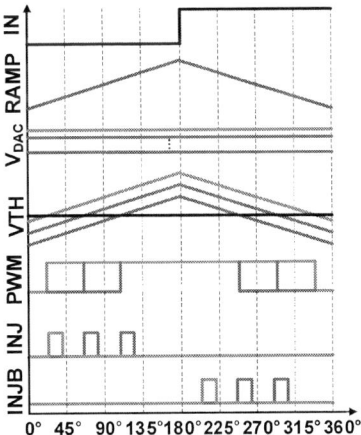

Fig.2 Timing Diagram of Critical Nodes

This work was supported in part by the National Key R&D Program of China (Grant No. 2022YFB2803301)

979-8-3315-3918-4/25 $31.00 © 2025 IEEE

Compared to prior work[5] , the ramp generator proposed in this work utilizes common-mode feedback (CMFB) to set the DC bias voltage for ramp signals to overcome matching-dependent deviations inherent in replica biasing methods and thereby enhancing accuracy. The CMFB's dynamic regulation mechanism further suppresses nonlinearity from input signal coupling and process/temperature variations, while its real-time compensation capability ensures robust operation under 250-mV low-swing conditions—critical for high-frequency, low-power designs. The design reuses the ring oscillators structure to employ a two-stage ILRO configuration to correct phase offset.

A. Ramp Generator Circuit

Figure 3 shows a detailed circuit diagram of the ramp generator. This circuit integrates current onto capacitors based on the 14GHz differential input clocks to generate differential ramp signals RAMP and RAMPB. When IN is high (and INB is low), the NMOS transistor M6 turns on, discharging capacitor C1 and causing the voltage at RAMP to decrease. Simultaneously, the PMOS transistor M7 turns on, charging capacitor C2 and causing the voltage at RAMPB to increase.

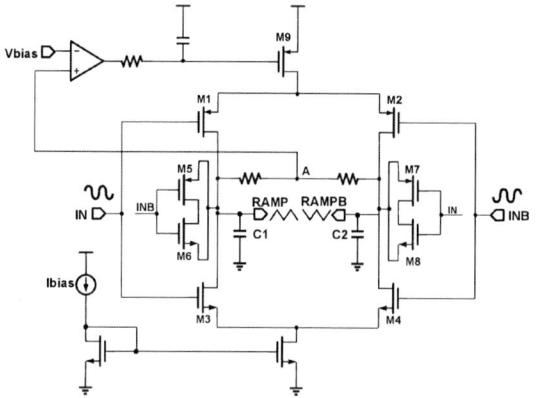

Fig.3 detailed circuit diagram of the ramp generator

Compensation transistors (e.g., M5-M8 in Fig. 3) are added to suppress clock feedthrough effects. CMFB stabilizes the DC level of node V_A at V_{bias}. A dedicated bias circuit provides the bias current Ibias. To minimize dynamic power consumption, the Ibias and capacitor values are chosen to generate differential ramp signals (RAMP and RAMPB) with a low swing of 250 mV. An RC low-pass filter (LPF) with a 21-MHz bandwidth is incorporated at the output of the operational transconductance amplifier (OTA). This LPF, leveraging the voltage-stabilizing capacitor in its RC network, confines the fluctuation at the gate of M9 caused by coupled input signal variations to less than 0.09%. This enhances the stability of the feedback loop while reducing high-frequency noise contributed by the CMFB loop.

B. 6-bit binary-decoded resistor-string DAC

A 6-bit resistor-string DAC, implemented using a series resistor voltage divider, employs binary-weighted multiplexers (MUXes) to control the output level, as depicted in Fig. 5. The DAC generates a linear output voltage, V_{DAC}, which sets the DC bias level of the differential ramp signals (RAMP and RAMPB). The output low-pass filter (LPF),

with a bandwidth of 24.9 MHz, confines the V_{DAC} ripple to less than 10 µV.

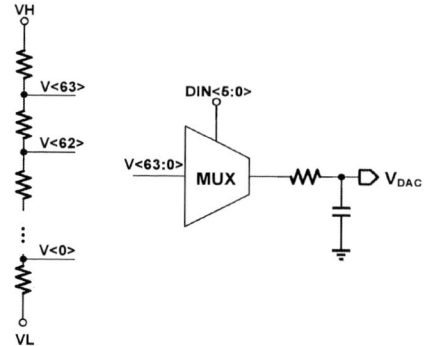

Fig.5 6-bit binary-decoded resistor-string DAC

C. Inverter-Based Comparator and Pulse Shaping Circuit

Conventional static op-amp-based comparators are difficult to operate at high speed. Therefore, we employ an inverter-based comparator (shown in Fig. 6), which operates at frequencies up to 14 GHz and offers the advantage of low power consumption [5] . This comparator generates signals PWM and PWMB by comparing one of the DC-biased ramp signals against the threshold voltage of the inverter stages.

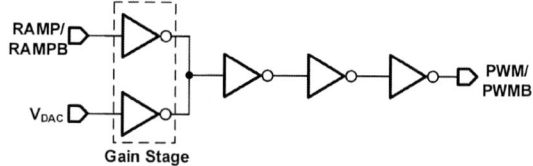

Fig.6 Circuit of inverter-based Comparator

Fig. 7 depicts the DC-biased ramp signals and the PWM signals generated by the comparator under varying V_{DAC} voltages. The varying duty cycles of PWM and PWMB would introduce nonlinearity if these signals were directly injected into the ILRO[5] . Consequently, a pulse-shaping circuit proposed by [5] is added to reshape the PWM signals (shown in Fig. 1). This produces the injection signals INJ and INJB with an identical duty cycles, thereby reducing nonlinearity.

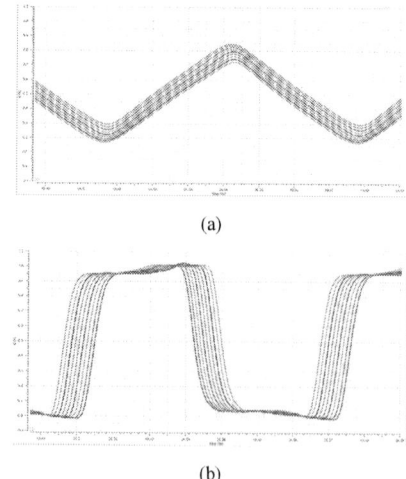

(a)

(b)

Fig.7 (a) DC-biased ramp signals with varying V_{DAC} voltages. (b) PWM signals with varying duty cycles generated by the comparator.

D. The 2-stage ILROs

The first-stage ILRO employs four stages of inverter-based pseudo-differential delay cells to generate the required 8-phase clocks, as shown in Fig. 8(a). Cross-coupled inverters I_x0-I_x7, sized at half the width of the main inverters I0-I7, are integrated to eliminate common-mode gain, thereby preventing the oscillator from latching into a DC state. The injection stage I_x8-I_x11 are sized identically to the cross-coupled inverters. The injection path incorporates a first inverter stage I_x9 and I_x11(see Fig. 8), consisting of a level-shifting AC-coupled self-biased inverter, proposed in[8] . A second inverter stage I_x8 and I_x10 determines the injection strength. The phase shift introduced at the two injection point results in significant phase misalignment, causing deviations from the ideal 45° spacing between the eight output phases. Therefore, the output phases from the first-stage ILRO are injected into the 2nd-stage ILRO for phase compensation.

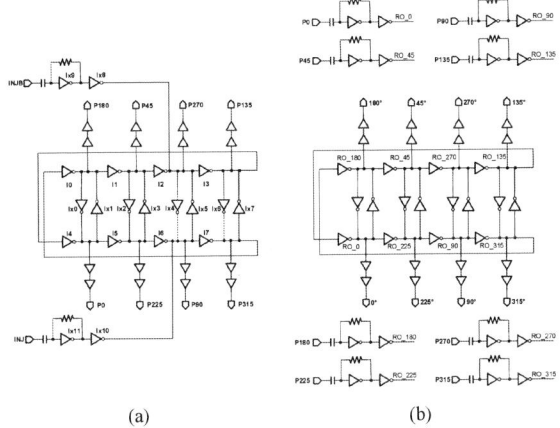

(a)   (b)

Fig.8 Detailed circuit diagrams of the 2-Stage ILROs: (a)

the first-stage ILRO. (b) the 2nd-stage ILRO.

Multiphase injection techniques exhibit the capability to broaden the locking range of injection-locked ring oscillators (ILROs) [9]. Multiphase injection across all stages can restore clock phase symmetry. As depicted in Fig. 8, the initial 8-phase clocks from the first stage are injected into the corresponding phases of the 2nd-stage ILRO. The oscillation frequency of 2-stage ILROs is controlled by supply voltage Vctrl. In particular, multiphase injection-locked ring oscillators (MPIL-ROSCs) have demonstrated the ability to generate highly accurate multiphase clocks even from poorly matched injection signals [8]. Although inaccuracies in the injection timing of individual phases can cause phase lead or lag in the oscillator output, the ring feedback inherently enforces a total loop phase shift of 2π, suppressing these deviations [9] . This stabilization arises because any deviation from the exact 2π phase accumulation around the loop violates the Barkhausen oscillation criterion, forcing the oscillator to dynamically adjust its frequency until the cumulative phase shift realigns to precisely 2π.

III. SIMULATION RESULTS

Figure 9 shows the PI output phase variation versus input code. The simulated integral nonlinearity (INL_{pp}) and differential nonlinearity (DNL_p) are 1.26 LSB and 0.16 LSB,

respectively. Additionally, the simulated RMS jitter integrated from 10 kHz to 100 MHz offset is 61.8 fs under TT process corner at 27° C. Notably, by injecting the initial 8-phase clocks (with phase error up to $\pm 4.5°$) from the first ILRO into the 2nd-stage ILRO (which has a similar structure), the phase error is suppressed to within $\pm 1.5°$. This demonstrates the effectiveness of the two-stage injection-locked architecture in correcting phase errors.

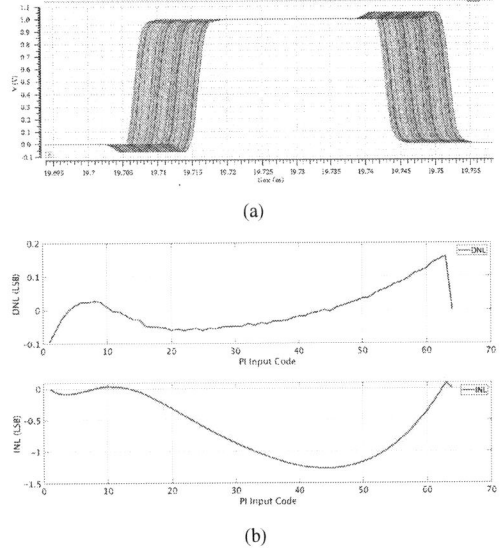

(a)

(b)

Fig.9 (a) the PI output phase variation versus input code. (b)DNL and INL vs. phase code for the proposed PI. a

The entire circuit consumes 28.394 mW of power, with the power breakdown detailed in figure 10. Table I compares the performance of circuits featuring multi-phase clock generation and phase interpolation. Specifically, the 2-stage ILRO dominates 79.7% of the total power, while the high-speed comparator achieves low power consumption at 14.1%. Notably, the dynamic power consumption of the ramp generator is negligible; combined with the DAC, it contributes merely 1.3%. As summarized in Table I, the proposed circuit demonstrates excellent DNL, competitive INL and jitter, and low power consumption.

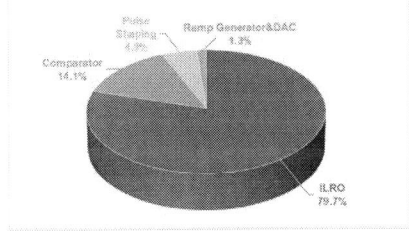

Fig.10 Power Breakdown

979-8-3315-3918-4/25 $31.00 © 2025 IEEE

TABLE I. PERFORMANCE SUMMARY AND COMPARISON

Metrics	Comparison and Summary				
	This work	Chen[6]	E.Monca [10]	Wang[11]	Huang [12]
architecture	Two-stage ILRO	QLL	QLL	MPI-ILO	MP-ILO
Process	28nm CMOS	7nm FinFET	28nm FD SOI	65nm CMOS	16nm FinFET
Freq[GHz]	14	16	11	7	16
Phases	8	2	2	2	8
Resolutions[bit]	9	7	7	7	8
INL_{pp}[LSB]	1.26	1.74	1.1	1.72	1.14
DNL_p[LSB]	0.16	0.87	0.5	1.13	1.31
Jitter[fs RMS]	61.8	80	139	84	254
Supply[V]	1	1.2	1	1.2	1.2
Power[mW]	28.39	22.34	18.48	22.95	11.4

IV. CONCLUSION

This paper presents a phase interpolator in 224 Gb/s PAM-4 wireline receivers. Employing a two-stage ILRO architecture with differential injection for initial multiphase generation and multiphase injection for calibration, the proposed PI synthesizes 8-phase 14 GHz output clocks. The prototype achieves a peak-to-peak INL of 1.26 LSB, a peak-to-peak DNL of 0.16 LSB, and an RMS jitter of 61.8 fs integrated from 10 kHz to 100 MHz offset under TT process corner at 27℃.

REFERENCES

[1] B. Zhou and B. Nikolić, "Design Techniques for a Multi-Phase Injection-Based Eight-Phase 17-GHz Clock Generator for Multi-Phase Wireline Receivers," in IEEE Transactions on Circuits and Systems I: Regular Papers, doi: 10.1109/TCSI.2025.3563517.

[2] H. Lin et al., "A 4×112 Gb/s ADC-DSP Based Multistandard Receiver in 7nm FinFET," 2020 IEEE Symposium on VLSI Circuits, Honolulu, HI, USA, 2020, pp. 1-2, doi: 10.1109/VLSICircuits18222.2020.9162802.

[3] Y. Segal et al., "A 1.41pJ/b 224Gb/s PAM-4 SerDes receiver with 31dB loss compensation," in IEEE Int. Solid-State Circuits Conf.(ISSCC) Dig. Tech. Papers, vol. 65, Feb. 2022, pp. 114−116, doi:10.1109/ISSCC42614.2022.9731794.

[4] Lin Zhang, D. Karasiewicz, B. Cifctioglu and Hui Wu, "A 1.6-to-3.2/4.8 GHz dual-modulus injection-locked frequency multiplier in 0.18μm digital CMOS," 2008 IEEE Radio Frequency Integrated Circuits Symposium, Atlanta, GA, 2008, pp. 427-430, doi: 10.1109/RFIC.2008.4561469.

[5] S. Mohapatra, E. Afshar, Z. Zhou and D. Heo, "7.9 An 8b 6-12GHz 0.18mW/GHz DC Modulated Ramp-Based Phase Interpolator in 65nm CMOS Process," 2024 IEEE International Solid-State Circuits Conference (ISSCC), San Francisco, CA, USA, 2024, pp. 140-142, doi: 10.1109/ISSCC49657.2024.10454438.

[6] S. Chen et al., "A 4-to-16GHz inverter-based injection-locked quadrature clock generator with phase interpolators for multi-standard I/Os in 7nm FinFET," 2018 IEEE International Solid-State Circuits Conference - (ISSCC), San Francisco, CA, USA, 2018, pp. 390-392, doi: 10.1109/ISSCC.2018.8310348.

[7] J. -C. Chien and L. -H. Lu, "Analysis and Design of Wideband Injection-Locked Ring Oscillators With Multiple-Input Injection," in IEEE Journal of Solid-State Circuits, vol. 42, no. 9, pp. 1906-1915, Sept. 2007, doi: 10.1109/JSSC.2007.903058.

[8] P. Kinget, R. Melville, D. Long and V. Gopinathan, "An injection-locking scheme for precision quadrature generation," in IEEE Journal of Solid-State Circuits, vol. 37, no. 7, pp. 845-851, July 2002, doi: 10.1109/JSSC.2002.1015681.

[9] Z. Wang, Y. Zhang, Y. Onizuka and P. R. Kinget, "Multi-Phase Clock Generation for Phase Interpolation With a Multi-Phase, Injection-Locked Ring Oscillator and a Quadrature DLL," in IEEE Journal of Solid-State Circuits, vol. 57, no. 6, pp. 1776-1787, June 2022, doi: 10.1109/JSSC.2021.3124486.

[10] E. Monaco, G. Anzalone, G. Albasini, S. Erba, M. Bassi and A. Mazzanti, "A 2–11 GHz 7-Bit High-Linearity Phase Rotator Based on Wideband Injection-Locking Multi-Phase Generation for High-Speed Serial Links in 28-nm CMOS FDSOI," in IEEE Journal of Solid-State Circuits, vol. 52, no. 7, pp. 1739-1752, July 2017, doi: 10.1109/JSSC.2017.2702742.

[11] Z. Wang, Y. Zhang, Y. Onizuka and P. R. Kinget, "11.4 A High-Accuracy Multi-Phase Injection-Locked 8-Phase 7GHz Clock Generator in 65nm with 7b Phase Interpolators for High-Speed Data Links," 2021 IEEE International Solid-State Circuits Conference (ISSCC), San Francisco, CA, USA, 2021, pp. 186-188, doi: 10.1109/ISSCC42613.2021.9365800.

[12] Y. -C. Huang and B. -J. Chen, "30.7 An 8b Injection-Locked Phase Rotator with Dynamic Multiphase Injection for 28/56/112Gb/s Serdes Application," 2019 IEEE International Solid-State Circuits Conference - (ISSCC), San Francisco, CA, USA, 2019, pp. 486-488, doi: 10.1109/ISSCC.2019.8662292.

Boosting Self-Powered Properties of 2D Material-Based Photodetectors via Asymmetry Engineering

Ran Huo[1], Han Zhang[1], Yihong Sun[1], Shijun Ou[1], Changming Pi[2], Mansun Chan[3], Changjian Zhou*[1]

[1]School of Microelectronics, South China University of Technology, Guangzhou, China
[2]College of Integrated Circuits and Micro-Nano Electronics, Fudan University
[3]Department of Electronic and Computer Engineering, The Hong Kong University of Science and Technology,
Kowloon, Hong Kong, China
*Email: zhoucj@scut.edu.cn

Abstract—Self-powered photodetectors (SPPDs) based on two-dimensional materials (2DMs) have attracted significant interest as promising candidates for next-generation wearable electronics and intelligent sensing systems, owing to their intrinsic advantages, such as low-power consumption characteristics, broadband spectral response, and ultrathin mechanical flexibility. This review provides a comprehensive summary of recent progress in the development of self-powered PDs promoted by different types of symmetry-breaking strategies, including structural designs, material selections, and process innovations. Furthermore, representative applications in imaging and recognition, optical communication, and health monitoring are presented to highlight the practical utility of these devices. Finally, critical challenges and prospective research directions are discussed, with emphasis on scalable manufacturing techniques, integration into device arrays, and the establishment of standardized performance benchmarking frameworks to support sustainable development and real-world deployment of 2D material-based photodetectors.

Keywords—Self-powered photodetector, 2D material, asymmetry engineering

I. INTRODUCTION

With the rapid advancement of the electronics industry, there is a growing demand for ultrathin, flexible, and wearable photodetectors (PDs) with low-power consumption across diverse fields such as biomedicine and outdoor sensing [1]. Two-dimensional materials (2DMs) cover a broad range of band gaps, emerging as competitive candidates for the next-generation optoelectronic devices. 2DM self-powered photodetectors (SPPDs) exhibit unique advantages such as device miniaturization, mechanical portability, fast response, and zero external energy consumption, thereby demonstrating great potential for complementing or even outperforming traditional optoelectronic materials. The devices typically rely on intrinsic material characteristics or engineered asymmetry (e.g., structural or process-induced) that facilitates directional separation of photoinduced carriers to form photocurrent (I_{ph}). Representative mechanisms include bulk photovoltaic effect, photoconductive effect, photothermoelectric effect (PTE), and photogating effect [2]. In light of these developments, this review highlights recent progress in 2DM SPPDs with asymmetric structural designs, multi-effect synergy, and process modification. Their emerging applications (e.g., imaging and recognition, optical communication, health monitoring) and challenges in manufacturing and integration are discussed. Furthermore, the necessity and importance of establishing standardized performance benchmarks are also emphasized.

This work is supported by Guangdong Provincial Key Field Research and Development Program (2022B0701180002) and GRF (16201223) from the Research Grant Council of Hong Kong.

II. RESEARCH PROGRESS

A. Symmetry-Breaking Structure

Using asymmetric structures is one of the most widely adopted strategies to induce self-powered properties of 2DM SPPDs. With different engineering methods in contact materials, electrode thicknesses, junction types, and contact areas, the Schottky barrier height and carrier transport mechanisms are modulated to facilitate efficient separation of photoinduced carriers under illumination. Zhou et al. designed a Gr/WSe$_2$/Au asymmetric SPPD (Fig. 1. (a, b)) [3]. Utilizing dry-transfer methods to form van der Waals contacts, the device successfully avoided Fermi-level pinning effects and achieved an ultralow dark current of approximately 10^{-14} A, a high zero-bias responsivity of 7.55 A/W, and a high linear dynamic range approaching 60 dB for light power density. Ma et al. introduced a strategy leveraging transmittance contrast by fabricating GaN/MXene electrodes of varying thicknesses through spin-coating and drop-casting techniques (Fig. 1. (c)) [4]. The PD exhibited a transmittance contrast up to 84%, contributing to an on/off ratio of 1.33×10^6, and a high specific detectivity of approximately 7.57×10^{12} Jones in the ultraviolet (UV) band (340 nm), which surpassed the devices with fully transparent electrodes. Pan et al. reported a Gr/ReSe$_2$/SnSe$_2$ heterojunction PD which possessed two kinds of contacts with the same metal electrodes (Fig. 1. (d)) [5]. The broken-gap band alignment of the heterojunction for ReSe$_2$ and SnSe$_2$ enabled tunneling of photoinduced carriers under different bias conditions, rather than conventionally relying on thermal excitation. The configuration achieved an outstanding signal-to-noise ratio of 10^5 and an anisotropic I_{ph} ratio (I_{max}/I_{min}) of 13.27 which was approximately three times higher than that of ReSe$_2$-based photodetectors. Abnavi et al. fabricated a flexible lateral MoS$_2$-Cr/Au Schottky PD on a PET substrate with a pronounced contact area asymmetry with a ratio about 8:1 (Fig. 1. (e)) [6]. The device exhibited a high rectification ratio up to 10^5 without illumination and retained 94.4% of its I_{ph} and 88.2% of its power conversion efficiency (PCE) after 5000 bending cycles, demonstrating comparable performance with devices on SiO$_2$/Si substrates. Furthermore, the self-powered properties of 2DM PDs are influenced by channel layer thicknesses, attributed to their tunable bandgaps and associated electronic properties [7]. These studies indicate the key role of asymmetric structures in tuning the optoelectronic behavior for exploring multi-effect synergy in 2DM SPPDs.

B. Multi-Effect Synergy

Integrating materials with unique intrinsic properties (e.g., ferroelectricity, thermoelectricity, piezoelectricity, and electromagnetism) into devices offers promising avenues to enhance device functionality and broaden application scopes [8][9]. Jiang et al. developed a series of ferroelectric field-

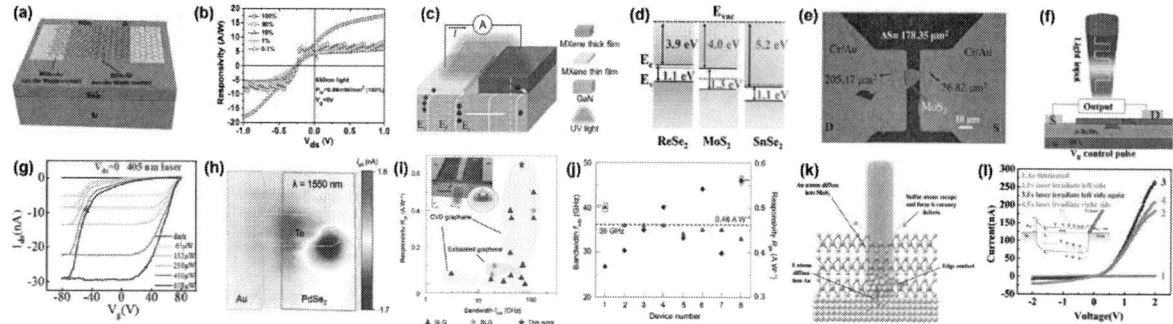

Fig. 1. Research advancements of 2DM SPPDs. (a, b) Schematic of a Gr/WSe$_2$/Au PD and the dependence of measured responsivity on light power density. (c) Schematic of a GaN/MXene PD with electrodes of varying thicknesses. (d) Band diagrams of ReSe$_2$, MoS$_2$, and SnSe$_2$ before contact. (e) Optical microscopy image of a MoS$_2$-Cr/Au PD. (f, g) Schematic illustration of a Fe-FED based on WSe$_2$/MoS$_2$ and the hysteretic photovoltaic effect based on different light intensities. (h) The scanning I$_{ph}$ mappings of Te/PdSe$_2$ heterojunction under zero bias for the incident wavelength of 1,550 nm. (i, j) Schematic illustration of a tBLG PD and comprehensive comparisons of 3-dB bandwidth and photoresponse. (k, l) Schematic illustration of the effect of fs laser on the interface and the energy band diagram of an Au/MoS$_2$/fs-Au PD with/without one/two times fs laser irradiation. (a, b) Reproduced from Ref. [3]. © 2020, Springer Nature. (c) Reproduced from Ref. [4]. © 2024, Wiley-VCH. (d) Reproduced from Ref. [5]. © 2024, Wiley-VCH. (e) Reproduced from Ref. [6]. © 2022, Wiley-VCH. (f, g) Reproduced from Ref. [10]. © 2025, Springer Nature. (h) Reproduced from Ref. [11]. © 2024, Wiley-VCH. (i, j) Reproduced from Ref. [12]. © 2024, Springer Nature. (k, l) Reproduced from Ref. [14]. © 2023, Elsevier Ltd.

effect diodes (Fe-FED) based on WSe$_2$/MoS$_2$ van der Waals heterojunctions (vdWHs) utilizing several non-volatile and stable ferroelectric gate materials (e.g., α-In$_2$Se$_3$, PVDF-TrFE, and CuInP$_2$S$_6$) (Fig. 1. (f, g)) [10]. These ferroelectric gates show tunable polarization behaviors, with hysteresis windows up to 30 V induced from a bias of ±80 V, enabling ambipolar carrier transport and achieving significant enhancements in photovoltaic performance. The open-circuit voltage (V$_{oc}$), short-circuit current (I$_{sc}$), and PCE were improved 4.3 times (0.492 V), 7.1 times (5.78 nA), and 29 times (7.25%), respectively. Moreover, these devices showed multifunctional capabilities, including photodetection, photomemory, and photologic operations, indicating the potential of reducing hardware complexity in computing architectures. Wang et al. used solution-growth, spin-coating, exfoliation, and dry-transfer methods to form PDs with Te/PdSe$_2$ vdWHs possessing ultra-broadband detection (405 ~ 4000 nm) and anisotropy ratios of 1.48 at 405 nm, 3.56 at 1550 nm, and 1.62 at 4000 nm, which was the most excellent device (Fig. 1. (h)) [11]. Photons with different energies were absorbed by either Te or PdSe$_2$ due to the bandgap difference, and localized thermal effect led to Seebeck coefficient difference, which dominated reversed I$_{ph}$ under 1550 nm short-wavelength infrared (SWIR) light. These advances highlight the ideas of incorporating multifunctional materials into SPPDs to couple multi-physical effects to enhance self-powered characteristics.

C. Process Modification

Several categories of innovative fabrication strategies have been reported to show great promise for future integrated applications. Traditional methods, such as dry transfer, can effectively reduce interfacial defects and suppress carrier scattering. However, these methods suffer from relatively low fabrication efficiency. At the same time, emerging process techniques on different parts of PDs can modulate material properties (e.g., carrier transport mechanisms) and enhance photoelectric performance, which offers more margin for processes such as chemical vapor deposition (CVD). Wu et al. boosted the light absorption of twisted bilayer graphene (tBLG) by three times through adjusting the intensive incident electrical field by waveguide modulation (Fig. 1. (i, j)) [12]. They also realized controllable twist-angle from CVD growth to form moiré superlattices, contributing to a fourfold

enhancement in I$_{ph}$ under 1550 nm light. The single PD exhibited a responsivity of 0.65 A/W and an ultra-high 3 dB bandwidth exceeding 65 GHz, while the 8-PD array exhibited an average responsivity of 0.46 A/W and an average bandwidth of 36 GHz, which indicated the potential for large-scale integration in optical communication systems. Guo et al. utilized high-temperature sintering and thermal evaporation to tune the composition ratio of Te$_{1-x}$Se$_x$, thereby successfully achieving continuous bandgap modulation and extended photoresponse into SWIR up to 1550 nm, while still maintaining low noise power density without any bias [13]. Peng et al. introduced localized femtosecond (fs) laser irradiation to break the symmetry of Schottky contacts (Fig. 1. (k, l)) [14]. If one of the MoS$_2$/Au contacts is treated, localized thermal effects would facilitate Au atom diffusion into the MoS$_2$ layer, and reduce about 50% Schottky barrier height. Meanwhile, additional hole traps would be generated and induce a local electric field that enhances carrier separation. As a result, the PD could exhibit a high external quantum efficiency of 341.9%, with significant improvements in I$_{sc}$ (approx. 39 times), V$_{oc}$ (approx. 17 times), and maximum output electrical power (approx. 1000 times), respectively. Notably, the enhancements could accumulate with continued fs laser exposure, demonstrating effectiveness in tuning device performance. These process techniques show significant potential for tailoring the optoelectronic properties of SPPDs and pave the way for scalable, high-performance devices suitable for future photonic integration.

III. ADVANCED APPLICATION

A. Imaging and Recognition

Image reconstruction systems play a pivotal role in optical communication. So far, there have been primarily two distinct operating principles. One method relies on mask modulation, which employs masks to project patterned illumination onto PDs. The induced I$_{ph}$ signals are then processed with algorithms to reconstruct the original images. Che et al. designed a PtSe$_2$/MoSe$_2$ heterojunction PD capable of broadband detection (532~1550 nm) (Fig. 2. (a, b)) [15]. Under 635 nm illumination and a moving mask for scanning, the I$_{ph}$ image showed clear reconstruction of the "CIOMP" pattern both at zero and +3 V bias. He et al. also designed an

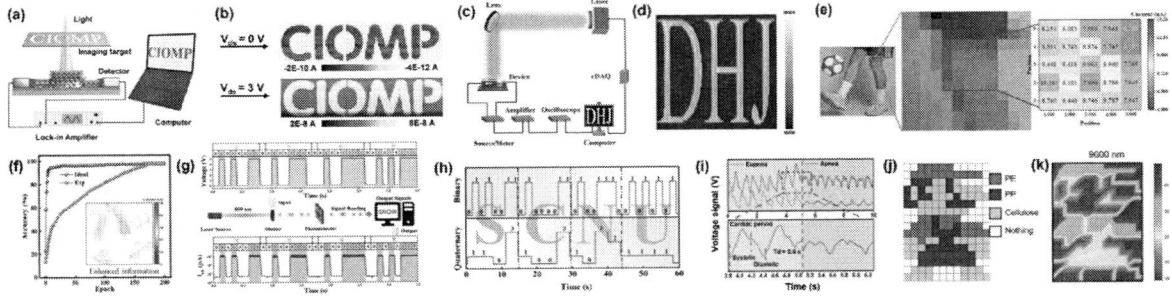

Fig. 2. Advanced applications of 2DM SPPDs. (a, b) Schematic of a mask-based imaging system and imaging results under 635 nm light. (c, d) Schematic of a laser-intensity-based imaging system and imaging result under 638 nm light. (e, f) Image-irradiation array of 25 pixels, the enhanced imaging result, and the comparison of experimental and ideal simulation with software weight updates for recognition accuracy. (g) Schematic of a typical optical communication system (here is a system for transmitting ASCII codes "SNOW"). (h) Binary and polarization-coded quaternary technology used in optical communication systems. (i, j, k) Monitored PPG waveforms of eupnea and apnea, a composite "Super Mario" image using different materials, and the resulting I_{ph} intensities due to the varying penetration abilities of IR light through different materials. (a, b) Reproduced from Ref. [15]. © 2024, American Chemical Society. (c, d) Reproduced from Ref. [17]. © 2023, American Chemical Society. (e, f) Reproduced from Ref. [18]. © 2025, American Chemical Society. (g) Reproduced from Ref. [19]. © 2023, Wiley-VCH. (h) Reproduced from Ref. [5]. © 2024, Wiley-VCH. (i, j, k) Reproduced from Ref. [20]. © 2025, Wiley-VCH.

all-2D Gr/WSe$_2$/NbSe$_2$ heterostructure PD with ultrafast response times of 80 µs (rise) and 72 µs (fall) without bias [16]. The device successfully reconstructed high-contrast images under 405 nm near-UV light with rapid response. The other method is based on laser-intensity modulation, where grayscale pixel values from target images are used to control the laser source power. The spatially encoded light produces distinguishable I_{ph} of PDs, enabling pixel-wise image reconstruction. Zheng et al. designed a WSe$_2$/Ta$_2$NiSe$_5$/WSe$_2$ dual sandwich-stacked heterojunction PD with excellent polarization sensitivity (>10) under near-infrared (NIR) light and successfully reconstructed a recognizable image with 442×174 pixels (Fig. 2. (c, d)) [17]. Yang's team designed a reconfigurable MoS$_2$ PD with Ag and Pt electrodes. The Ar plasma-treated surface induced localized electron doping with numerous Sulphur vacancies (Fig. 2. (e, f)) [18]. The PD demonstrated a fast response of 224/293 µs and non-volatility. Using a convolutional neural network, a 5×5 PD array scanned by only 64×63 grayscale image pixels could achieve 96% classification accuracy in 10000 images after 200 training cycles. These findings show the viability of 2DM SPPDs for future intelligent optical communication systems.

B. Optical Communication

Currently, optical communication systems utilizing 2DM SPPDs are generally implemented by encoding signals before optical transmission. Li et al. developed high-speed PDs (180/80 µs fast response) based on all-2D NbSe$_2$/MoSe$_2$ heterostructures (Fig. 2. (g)) [19]. By transforming target information into ASCII code and sequentially mapping it to laser source voltages, the device demonstrated distinguishable output signal recognition. The PD designed by Pan et al. possessed excellent polarization sensitivity (Fig. 2. (h)) [5]. The device could rapidly, accurately, and steadily recognize quaternary optical communication signals when receiving polarized light at 0°, 30°, 60°, and 90°. The standardized n-nary coding strategy provides new possibilities for integrated applications, such as polarized imaging and coding in optical communication systems. These progresses also indicate future demands of security issues in the short/medium distance.

C. Health Monitoring

Health signal monitoring is of huge research interest and drives demand for flexible and portable devices. Tang et al. pioneered the wafer-scale fabrication of 1D GaN nanorods/2D

MoS$_2$/PEDOT:PSS heterojunction [8]. Their PDs achieved a fast response of 54/71 µs recorded under zero bias with 5 kHz UV light pulse, surpassing state-of-the-art flexible devices. The -0.78% strain increased 68% responsivity (2.47 A/W), showing outstanding enhancement from the piezoelectric effect. The hand-detection system they designed could induce varying I_{ph} under different UV intensities sensitively, which indicated its suitability as a portable and wearable system for real-time prevention of skin cancer. Ling et al. designed a 2D WS$_2$/monodisperse hexagonally stacked (MHS) 3D PdTe$_2$/Si mixed-dimensional PD (Fig. 2. (i, j, k)) [20]. When applied to a dual-wavelength photoplethysmography sensor, it could reflect the pulse and blood oxygen saturation level of subjects non-invasively and accurately. Furthermore, leveraging the light penetration difference for different materials, the sensor possessed promising potential in IR-based safety and health monitoring scenarios. These practices show the values of health monitoring systems of 2DM PDs, laying a critical paradigm for developing new smart healthcare technologies.

IV. OUTLOOK

A. Process for manufacturing

The widely adopted fabrication method for 2DM SPPDs to break symmetry is mechanical exfoliation. However, shape randomness and limited efficiency cause trouble in experiment reproducibility and the feasibility of large-scale manufacturing. With the continued development of fabrication processes, CVD, epitaxy, atomic layer deposition, and sputtering have provided promising solutions for controllable wafer-scale production of PDs and PD arrays [21]. Notably, processes compatible with Si-based complementary MOS (CMOS) are still actively studied. For example, Ye's team reported a fully Si-CMOS-compatible method for PtSe$_2$ thin films [22]. They synthesized PtSe$_2$ films at a temperature similar to Si-CMOS on naturally oxidized SiO$_2$/Si substrates via thermally assisted conversion, CVD, and evaporation, and then electrodes (Cr/Au/InGa) are deposited using magnetron sputtering. Based on this, they successfully fabricated a 9×9 PD array with excellent optoelectrical response, showing potential in reproducibility and broadband detection of UV-visible-NIR light. However, significant improvements are still needed to address process challenges such as aspect ratio limits in anisotropic etching, epitaxial selectivity, and defects from deposition/thermal processes.

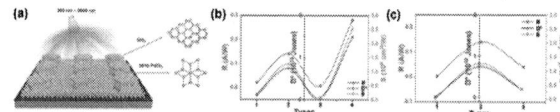

Fig. 3. Structure evolution of 2DM SPPDs. (a) Schematic of WS₂/MHS PdTe₂/Si periodic array protrusions. (b, c) The performance of the arrays with different WS₂ or PdTe₂ growth times at 808 nm. (a, b, c) Reproduced from Ref. [20]. © 2025, Wiley-VCH.

B. Structure Evolution

Besides using 2D materials with the potential of CMOS compatibility, structural upgrading should also be taken into account for device integration. Drawing inspiration from the development of planar field-effect transistors (FETs), where FinFET structure was employed to increase gate control area, similar innovations may also provide ideas for SPPDs to boost light absorption and spectral response. Surface engineering like periodic array protrusions with sputtered particles can increase surface area and roughness, and boost absorption from enhanced diffuse reflection (Fig. 3. (a)). This structure makes more use of vertical space on planar SiO_2/Si or flexible substrates due to light re-absorption through internal reflection between adjacent nanostructures within the array plane. It was also proved that the performance of this SPPD array exhibits excellent tolerance to the growth time of these micro-features (Fig. 3. (b, c)) [20]. Recent reports indicate that structural strategies are likely to provide promising pathways for PD array integrations in the near future.

C. Benchmark and Framework

The lack of standardized benchmarks and frameworks for SPPDs has been a persistent challenge in the optoelectronic field for a long time. Many studies are not using unified sets of criteria, and different testing equipment and environments will result in difficulties reproducing and fairly comparing published results. Moreover, inherent trade-offs between key figures of merit often lead to selective reports in papers. To address these issues, it's essential to propose authoritative organizations, such as the International Roadmap for Devices and Systems, to establish comprehensive multi-dimensional evaluation workflows and frameworks. Standardizing test procedures for performance metrics will help facilitate fair performance evaluation, guide technology development, and fuel sustainable transitions from academy to industry.

V. SUMMARY

In summary, this review offers a comprehensive overview of the recent progress in 2DM SPPDs, focusing on typical strategies such as asymmetric structures, multi-effect synergy, and process modifications. Promising applications in imaging and recognition, optical communication, and health monitoring are presented to demonstrate the versatility and practical potential of 2DM SPPDs across diverse fields. Moreover, the review provides prospective development directions in the future, emphasizes the importance of compatible and scalable fabrication methods, innovative structures, and the essential need for the establishment of standardized benchmarking protocols. The development of unified evaluation criteria and testing procedures will be vital to promoting the sustainable and regulated advancement of SPPD technologies.

REFERENCES

[1] S. Chang et al., "Flexible and Stretchable Light-Emitting Diodes and Photodetectors for Human-Centric Optoelectronics," *Chemical Reviews*, vol. 124, pp. 768-859, February 2024.

[2] H. Zeng et al., "Recent developments in CVD growth and applications of 2D transition metal dichalcogenides," *Frontiers of Physics*, vol. 18, 53603, May 2023.

[3] C. Zhou et al., "Self-driven WSe₂ photodetectors enabled with asymmetrical van der Waals contact interfaces," *npj 2D Materials and Applications*, vol. 4, 46, December 2020.

[4] H. Ma et al., "Transmittance contrast-induced photocurrent: A general strategy for self-powered photodetectors based on MXene electrodes," *InfoMat*, vol. 6, e12540, May 2024.

[5] Y. Pan et al., "High-Performance Photoinduced Tunneling Self-Driven Photodetector for Polarized Imaging and Polarization-Coded Optical Communication based on Broken-Gap ReSe₂/SnSe₂ van der Waals Heterojunction," *Small*, vol. 20, 2311606, August 2024.

[6] A. Abnavi et al., "Flexible High-Performance Photovoltaic Devices based on 2D MoS₂ Diodes with Geometrically Asymmetric Contact Areas," *Advanced Functional Materials*, vol. 33, 2210619, February 2023.

[7] Y. Liu et al., "Approaching the Schottky-Mott limit in van der Waals metal-semiconductor junctions," *Nature*, vol. 557, pp. 696-700, May 2018.

[8] X. Tang et al., "Wafer-Scale Vertical 1D GaN Nanorods/2D MoS₂/PEDOT:PSS for Piezophototronic Effect-Enhanced Self-Powered Flexible Photodetectors," *Nano-Micro Letters*, vol. 17, 56, November 2024.

[9] Q. Wang, C. Zhou, and Y. Chai, "Breaking symmetry in device design for self-driven 2D material based photodetectors," *Nanoscale*, vol. 12, pp. 8109-8118, March 2020.

[10] Y. Jiang et al. "Ferroelectric Gate Enabled Programmable Photovoltaics in vdWHs for Self-Powered In-Memory Logics," *ACS Photonics*, vol. 11, pp. 1557-1564, March 2024.

[11] Pu. Wang et al., "Anisotropic Te/PdSe₂ Van Der Waals Heterojunction for Self-Powered Broadband and Polarization-Sensitive Photodetection," *Small*, vol. 20, 2401216, April 2024.

[12] Q. Wu et al., "Waveguide-integrated twisted bilayer graphene photodetectors," *Nature Communications*, vol. 15, 3688, May 2024.

[13] Z. Guo et al., "Self-driven Te₀.₆₅Se₀.₃₅/GaAs SWIR photodiode with spectral response to 1.55 μm for broadband imaging and optical communication," *Nano Energy*, vol. 133, 110452, January 2025.

[14] J. Peng et al., "Asymmetric Schottky contacts induced via localized ultrafast laser irradiation for ultrasensitive, self-powered, 2D photodetectors," *Nano Energy*, vol. 117, 108891, December 2023.

[15] M. Che et al., "High-Efficiency Self-Powered Broadband Photodetector Based on PtSe₂/MoSe₂ Heterojunction," *ACS Photonics*, vol. 11, pp. 1693-1702, March 2024.

[16] S. He et al., "All-2D asymmetric self-powered photodetectors with ultra-fast photoresponse based on Gr/WSe₂/NbSe₂ van der Waals heterostructure," *Journal of Materials Science & Technology*, vol. 219, pp. 205-212, June 2025.

[17] T. Zheng et al., "Self-Powered Photodetector with High Efficiency and Polarization Sensitivity Enabled by WSe₂/Ta₂NiSe₅/WSe₂ van der Waals Dual Heterojunction," *ACS Applied Materials & Interfaces*, vol. 15, pp. 29363–29374, June 2023.

[18] C. Zhou et al., "Self-Powered and Reconfigurable Double-Terminal MoS₂ Photodetector for Image Recognition," *Nano Letters*, vol. 25, pp. 3515–3523, February 2025.

[19] C. Li et al., "Self-Powered Photodetector with High Performance Based on All-2D NbSe₂/MoSe₂ van der Waals Heterostructure," *Advanced Optical Materials*, vol. 11, 2300905, July 2023.

[20] C. Ling et al., "WS₂/MHS PdTe₂/Si Mixed-Dimensional Heterojunction as Ultra-Broadband Photodetector for Health and Safety Monitoring," *Advanced Healthcare Materials*, vol. 14, 2402507, January 2025.

[21] X. Guan et al., "New paradigms of 2D layered material self-driven photodetectors," *Nanoscale*, vol. 16, pp. 20811-20841, October 2024.

[22] P. Ye et al., "Si-CMOS-compatible 2D PtSe₂-based self-driven photodetector with ultrahigh responsivity and specific detectivity," *Science China Materials*, vol. 66, pp. 193-201, July 2022.

Interfacial adhesion enhancement enabled mechanically durable flexible organic optoelectronics

Ziqi Wang[1], Xiangzhe Li[1], Huimin Wu[1], Kai Wang[1], Sixing Xiong*[1], Jin Qian*[1]

[1] Huanjiang Laboratory, School of Aeronautics and Astronautics, Zhejiang University, Zhuji, ZJ 311800, China

* Email: sixing_xiong@hust.edu.cn; jqian@zju.edu.cn

Abstract—Flexible organic optoelectronics play a significant role in health monitoring and intelligent robotics due to their advantages of lightness and high adaptability. The functionality of these devices stems from the integration of multiple layer components. However, mechanical mismatches among flexible substrates, metal/metal oxide layer, and organic semiconductor composed of different materials often lead to interfacial failure under load. Consequently, maintaining the stability of the multi-layered structure during deformation has become a critical technical challenge. This study presents our progress in mitigating interface failure in flexible multi-layered optoelectronics under mechanical deformation. By employing interface adhesion enhancement engineering, the mechanical durability of flexible electronics can be effectively improved. This enhancement is achieved through the thermo-mechanical coupling, offering great potential for improving device stability under mechanical deformation or load.

Keywords—flexible organic optoelectronics, multi-layer structure, mechanical deformation, interface failure, adhesion enhancement

I. INTRODUCTION (*HEADING 1*)

In recent years, flexible electronics technology has garnered increasing attention and found a wide range of applications in wearable electronics, biomedical sensing, soft robotics, and implantable devices [1-5]. This is because flexible electronic devices not only provide intelligent response properties comparable to those of conventional rigid electronic devices but also possess mechanical deformation capabilities such as stretchability and bendability. These features enable them to completely conform to and seamlessly attach to irregular surfaces, including human skin and soft tissues [6-8].

Flexible organic optoelectronic devices typically rely on the photoelectric effect of organic semiconductor to achieve the conversion of photoelectric signals. Organic semiconductor materials undertake a key role in this photoelectric conversion process [9]. These materials are excellent candidates for wearable electronics due to their unique optoelectronic properties and outstanding biocompatibility with humans. Additionally, organic semiconductor materials exhibit superior mechanical performance compared to inorganic semiconductors, enabling flexible organic optoelectronic devices to achieve physiologically adaptable optoelectronic signal acquisition without motion interference [10]. Furthermore, the organic semiconductor materials also offer low manufacturing costs, excellent optoelectronic tunability, and ease of large-scale

production [11, 12], which significantly facilitate the commercialization of flexible organic optoelectronic devices.

Although flexible organic optoelectronic devices have shown tremendous application potential in the era of the Internet of Things (IoT), their large-scale and commercial application still faces significant challenges. These devices typically require the lamination of thin films with diverse properties, which inevitably creates different interfaces. The flexible multi-layered device structure is prone to interface failure under mechanical deformation or load. For instance, flexible organic optoelectronic devices generally consist of a flexible substrate, a transparent electrode, a photosensitive layer, an electron transporting layer, a hole transporting layer, and an opaque electrode [13-15]. However, organic semiconductor materials usually possess low Young's moduli, whereas the metal electrodes and metal oxide transporting layers exhibit high Young's moduli. Due to this modulus mismatch between different materials, the interfaces within the multi-layer structure become unstable under mechanical deformation loads [16]. The differing mechanical properties between different materials will lead to interface delamination.

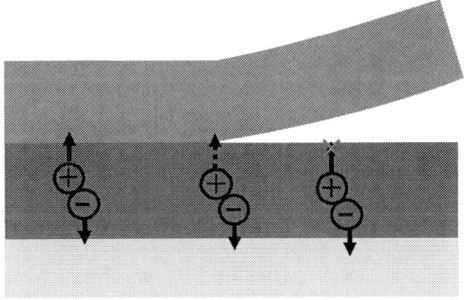

Fig. 1 Schematic diagram of a multi-layered device structure and charge carriers transportation under partial interface delamination.

The charge transfer efficiency at the interface is closely related to the adhesive condition of the adjacent layers. When the interface delamination occurs under mechanical deformation or load, it adversely affects the device performance [17]. As shown in Fig. 1, interface delamination in the device will suppress the transportation of charge carriers. Therefore, the interfacial failure in flexible organic optoelectronic devices under deformation or load poses a significant challenge to their reliability in practical applications.

In this study, mechanical simulation was employed to analyze the strength of interface adhesion. The results revealed that the interface between the MoO_x metal oxide hole transporting layer and the active layer exhibited the weakest adhesion. To address this issue, we introduced flexible organic electronics incorporating the in-situ growth of an AgO_x hole transporting layer, which significantly enhances stability under mechanical stress by reinforcing the adhesion between the hole transporting layer and the active layer. The in-situ growth process involves directly depositing Ag onto the active layers, followed by annealing in air. Consequently, the designed flexible organic optoelectronic devices retained over 95% of their initial performance after 5,000 compression–stretching cycles with 30% compression, demonstrating superior mechanical stability compared to devices without enhanced interface adhesion. These features enable our flexible organic optoelectronics structure well-suited for wearable applications.

II. RESULT AND DISCUSSION

A. Simulation analysis

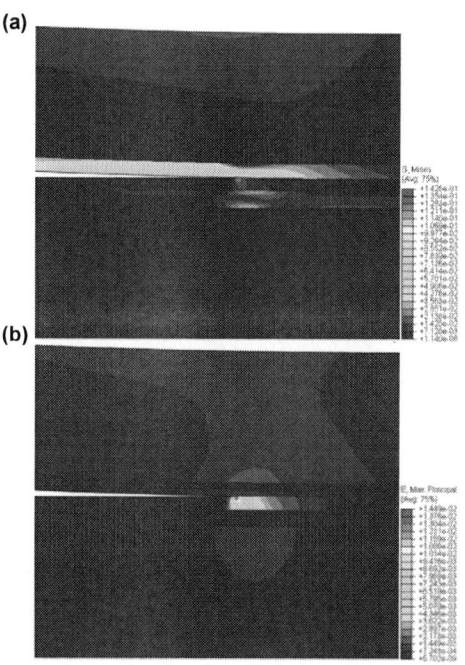

Fig. 2 Finite element simulation of (a) Mises stress and (b) maximum principal strain distribution for interface failure

To identify the location of interfacial failure in multi-layered flexible electronic devices under deformation stress, the commercial finite element software ABAQUS was used to analyze the stress and strain distribution of the devices. The device structure consisting of PI/ITO/PEI-Zn/organic active layer/MoO_x/Ag was selected as the simulation target. This device structure was reported in our previous work [11]. For the stress/strain analysis of multi-layered device structure, Mises stress can be used to assess stress concentration and the tendency of yield failure, while the maximum principal strain (or stress) focuses on the risk of brittle cracking or fracture failure. The combination of the two can provide a key basis for optimizing the mechanical properties of the structure (such as layer thickness design and material matching). As

illustrated in Fig. 2a, the delamination occurs at the interface between the MoO_x hole transporting layer (significant stress concentration) and the organic active layer when the flexible optoelectronic device is subjected to external load. The device strain indicates the risk of cracking at the interface (on the side close to the organic active layer) near separation position (as shown in Fig. 2b). Therefore, to improve the mechanical stability of flexible electronic devices, it is necessary to suppress interfacial failure between the MoO_x hole transport layer and the organic semiconductor active layer.

B. Interface adhesion enhancement

We noticed that the weakest interface (between MoOx and organic active layer) in abovementioned structure, indicating that this interface is most prone to interface failure under mechanical deformation or load. To improve the interface adhesion, we employed an in-situ growth method to form a new hole-transporting layer, thereby strengthening the adhesion between the active layer and the hole transporting layer. The device structure with the new AgO_x hole transporting layer is shown in Fig. 3a. The devices were fabricated on flexible a transparent polyimide (PI) substrate. Before spin-coating the PI substrate precursor, the glass supporting substrates were cleaned by plasma treatment. Subsequently, the glass surface was fluorinated by using polymer (Novec). After fluorinating, the flexible optoelectronic device could be easily peeled-off from the glass. Then, the PI precursor was spin-coated on the glass substrate at 3000 rpm to form an approximately 1.4 µm-thick PI layer. Following deposition, the samples were transferred to an N_2-filled oven and annealed at 250 °C for 8 hours. Here, we utilized indium tin oxide (ITO) as a transparent electrode. To form a uniform and transparent electrode, the ITO target materials were sputtered onto the PI substrate. The ITO electrode was patterned by photolithography to create the desired shape. The PEI-Zn (ethoxylated polyethyleneimine (PEIE) chelated with Zn2+) electron transporting layer was spin-coated at 3500 rpm onto the ITO electrode after oxygen plasma. The PEI-Zn precursor solution was prepared as follows: 70 mg of zinc acetate dehydrate was dissolved in 1 ml 1 wt% PEIE in 2-methoxyethanol solution. Then, the PEI-Zn film was thermal annealed at 180 °C for 30 min in air to form a uniform film approximately 10 nm thick. Subsequently, the active layer was deposited on the substrate. In this study, the poly[(2,6-(4,8-bis(5-(2-ethylhexyl-3-fluoro)thiophen-2-yl)-benzo[1,2-b:4,5b']dithiophene))-alt-(5,5-(1',3'-di-2-thienyl-5',7'-bis(2-ethylhexyl)benzo[1',2'-c:4',5'-c']dithiophene-4,8-dione)]:2,2'-[[12,13-Bis(2-ethylhexyl)−12,13-dihydro-3,9-diundecylbisthieno[2',3':4',5']thieno[2',3':4,5]pyrrolo[3,2-e:2',3'-g][2,1,3]benzothiadiazole-2,10-diyl]bis[methylidyne(5,6-difluoro-3-oxo-1H-indene2,1(3H)-diylidene)]]bis[propane dinitrile] (PM6:Y6) was selected as the efficient organic active layer. The PM6:Y6 solution (7 mg : 9 mg in 1 ml mixed solvent of chloroform : 1-chloronaphthalene) was spin-coated in the glovebox at 3500 rpm and then annealed at 110 °C for 10 min [18]. The thickness of the active layer is approximately 120 nm. Next, a 100-nm-thick Ag electrode was deposited onto the active layer. To form the AgO_x hole transporting layer, the devices were annealed in the air. The annealing temperature and time should be further optimized according to the performance of the device. Finally, the flexible organic optoelectronic devices were peeled off from the supporting glass substrate. The effective area of the device is defined by the overlapping area of the opaque metal electrode and the transparent electrode.

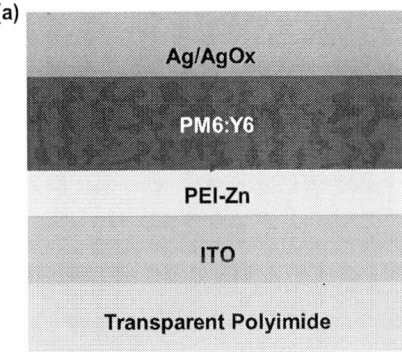

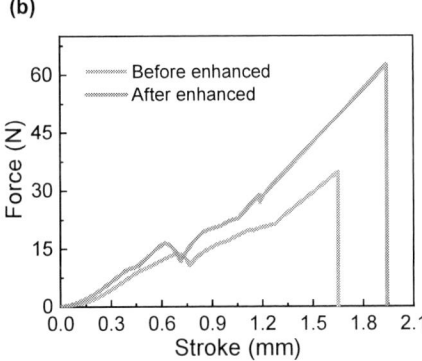

Fig. 3 (a) Flexible organic optoelectronic device structure after enhancing interface adhesion. (b) Force-stroke curves of samples before and after interface adhesion enhancement.

To quantitatively compare the interface adhesion of different devices, a high-precision tensile testing machine was used for the tensile test. A stud was bonded to the Ag electrode using an adhesive tape. The sample was fixed on the tensile tester. Then, the holder of the stretching machine clamped the stud and gradually applied pull force until the samples were completely delaminated from the stud. The tensile force at separation was recorded to evaluate the interface adhesion. As shown in Fig. 3b, the tensile force between AgO_x and the active layer was approximately 62 N, whereas the force between MoO_x and the active layer was only approximately 31 N. These results indicate that the weakest interface adhesion strength in the flexible organic optoelectronic devices was enhanced by two times after thermal annealing. The thermo-mechanical coupling strategy significantly improved the interface adhesion in flexible devices, which provides a feasible solution for the manufacture of firm flexible electronic.

C. Mechanical durability

To evaluate the mechanical stability of the flexible device, we tested the performance of the device under cyclic compressing–stretching process. The flexible device was attached to a pre-stretched elastomer. When the pre-stretched elastomer was released to its initial state, the flexible organic

optoelectronic devices were compressed [19]. Fig. 4a shows the optical images of the flexible organic optoelectronic device in the initial state and compressed state. The deformation was calculated based on the initial and compressed length of the elastomer [20, 21].

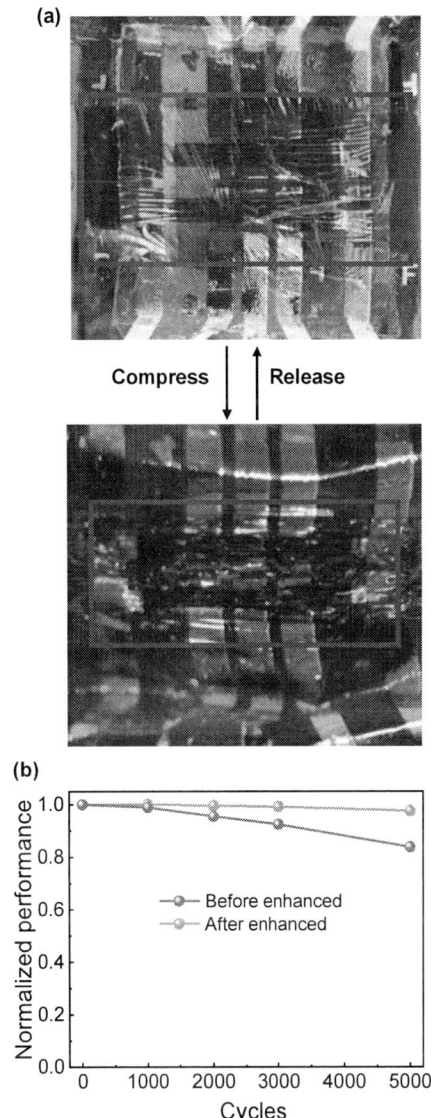

Fig. 4 (a) Schematic illustrations of pre-stretching and compressing. (b) Evolution of power conversion efficiency (PCE) of the flexible organic optoelectronic devices under cyclic compressing–stretching deformation with 30% compression.

Fig. 4b illustrates the performance evaluation of the flexible device under cyclic compressing-stretching deformation with 30% compression. After 1,000 deformation cycles, there is only a negligible difference in the performance before and after interface adhesion enhancement, indicating the excellent mechanical stability of both fabricated devices. As the number of deformation cycles increases, the performance of the device with the MoO_x hole transporting layer showed a significant decrease, whereas the performance

979-8-3315-3918-4/25 $31.00 © 2025 IEEE

of the device with the AgO$_x$ layer, which enhanced the interface adhesion, still maintained over 95% of its initial efficiency. These results demonstrate that the mechanical durability of the flexible organic optoelectronic device with a multi-layered structure is significantly improved by enhancing the interface adhesion.

III. SUMMARY

We simulated and analyzed the position of interface failure in a widely used multi-layered flexible organic optoelectronic device structure under deformation stress. We then proposed an interface adhesion enhancement by thermo-mechanical coupling strategy to suppress the interface delamination, thereby improving the mechanical durability of flexible organic optoelectronic devices. After applying this adhesion enhancement, the flexible organic optoelectronic device demonstrated excellent mechanical durability, maintaining over 95% of its initial performance after 5,000 compressing–stretching cycles with 30% compression. These findings are highly significant for the development of wearable multi-layered structure electronic products.

IV. ACKNOWLEDGEMENT

The authors thank Prof. T. Someya, Prof. T. Yokota of The University of Tokyo, Prof. K. Fukuda of Osaka University, and Dr. S. H. Lee, Dr. S. Y. Lee of Center for Emergent Matter Science, RIKEN for their support.

V. REFERENCES

[1] P. Le Floch, S. Zhao, R. Liu, N. Molinari, E. Medina, H. Shen *et al.*, 3D spatiotemporally scalable in vivo neural probes based on fluorinated elastomers. *Nat. Nanotechnol.*, 19, 319-329, (2024).

[2] X. He, X.-L. Shi, X. Wu, C. Li, W.-D. Liu, H. Zhang *et al.*, Three-dimensional flexible thermoelectric fabrics for smart wearables. *Nat. Commun.*, 16, 2523, (2025).

[3] Y. Yi, H. Liao, E. Liu, Y. Ye, Z. Cai, X. Li *et al.*, Review of flexible biomedical sensors: design, application, and challenge. *IEEE Sens. J.*, 24, 2321-2328, (2024).

[4] S. Xiong, K. Fukuda and T. Someya, "Waterproof and wearable power sources," *2024 IEEE 17th International Conference on Solid-State & Integrated Circuit Technology (ICSICT)*, doi: 10.1109/ICSICT62049.2024.10831890.

[5] M. H. Malekoshoaraie, B. Wu, D. D. Krahe, Z. Ahmed, S. Pupa, V. Jain *et al.*, Fully flexible implantable neural probes for electrophysiology recording and controlled neurochemical modulation. *Microsyst. Nanoeng.*, 10, 91, (2024).

[6] R. Dahiya, D. Akinwande and J. S. Chang, Flexible electronic skin: from humanoids to humans. *Proc. IEEE.*, 107, 2011-2015, (2019).

[7] Y. Kim, J. M. Suh, J. Shin, Y. Liu, H. Yeon, K. Qiao *et al.*, Chip-less wireless electronic skins by remote epitaxial freestanding compound semiconductors. *Science*, 377, 859-864, (2022).

[8] Y. Liu, J. Li, S. Song, J. Kang, Y. Tsao, S. Chen *et al.*, Morphing electronics enable neuromodulation in growing tissue. *Nat. Biotechnol.*, 38, 1031-1036, (2020).

[9] J. Yi, G. Zhang, H. Yu and H. Yan, Advantages, challenges and molecular design of different material types used in organic solar cells. *Nat. Rev. Mater.*, 9, 46-62, (2024).

[10] Y. Bian, K. Liu, Y. Ran, Y. Li, Y. Gao, Z. Zhao *et al.*, Spatially nanoconfined N-type polymer semiconductors for stretchable ultrasensitive X-ray detection. *Nat. Commun.*, 13, 7163, (2022).

[11] S. Xiong, K. Fukuda, S. Lee, K. Nakano, X. Dong, T. Yokota *et al.*, Ultrathin and efficient organic photovoltaics with enhanced air stability by suppression of zinc element diffusion. *Adv. Sci.*, 9, 2105288, (2022).

[12] C. Zhao, J. Park, S. E. Root and Z. Bao, Skin-inspired soft bioelectronic materials, devices and systems. *Nat. Rev. Bioeng.*, 2, 671-690, (2024).

[13] S. Xiong, K. Fukuda and T. Someya, "Ultra-flexible organic photovoltaics for powering wearable electronics," *2023 IEEE 15th International Conference on ASIC (ASICON)*, doi: 10.1109/ASICON58565.2023.10396389.

[14] H. Tang, Y. Bai, H. Zhao, X. Qin, Z. Hu, C. Zhou *et al.*, Interface engineering for highly efficient organic solar cells. *Adv. Mater.*, 36, 2212236, (2024).

[15] S. Najam and B. Kumar, "Organic solar cell: operating principle, performance parameters, structures and its advantages," *2018 5th IEEE Uttar Pradesh Section International Conference on Electrical, Electronics and Computer Engineering (UPCON)*, doi: 10.1109/UPCON.2018.8597120.

[16] S. Cheng, Z. Lou, L. Zhang, H. Guo, Z. Wang, C. Guo *et al.*, Ultrathin hydrogel films toward breathable skin-integrated electronics., *Adv. Mater.*, 35, e2206793, (2023).

[17] S. Xiong, K. Fukuda, K. Nakano, S. Lee, Y. Sumi, M. Takakuwa *et al.*, Waterproof and ultraflexible organic photovoltaics with improved interface adhesion. *Nat. Commun.*, 15, 681, (2024).

[18] J. Yuan, Y. Zhang, L. Zhou, G. Zhang, H.-L. Yip, T.-K. Lau *et al.*, Single-junction organic solar cell with over 15% efficiency using fused-ring acceptor with electron-deficient core. *Joule*, 3, 1140-1151, (2019).

[19] M. Kaltenbrunner, M. S. White, E. D. Głowacki, T. Sekitani, T. Someya, N. S. Sariciftci *et al.*, Ultrathin and lightweight organic solar cells with high flexibility. *Nat. Commun.*, 3, 770, (2012).

[20] M. Kaltenbrunner, T. Sekitani, J. Reeder, T. Yokota, K. Kuribara, T. Tokuhara *et al.*, An ultra-lightweight design for imperceptible plastic electronics. *Nature*, 499, 458-63, (2013).

[21] B. Du, S. Xiong, L. Sun, Y. Tagawa, D. Inoue, D. Hashizume *et al.*, A water-resistant, ultrathin, conformable organic photodetector for vital sign monitoring. *Sci. Adv.*, 10, eadp2679, (2024).

Image Flare Removal via Stable Diffusion Framework

Jiazheng Lian [1,2], Ruoxi Zhu [2], Jiaming Liu [2], Ming'e Jing [2], Xiaoyang Zeng [2], Yibo Fan [2]

[1] College of Intelligent Robotics and Advanced Manufacturing, Fudan University
[2] State Key Laboratory of Integrated Chips and Systems, Fudan University

* Email: jzlian20@fudan.edu.cn, fanyibo@fudan.edu.cn

Abstract—**Flare artifacts, commonly caused by strong light sources in mobile photography, can significantly degrade image quality by reducing contrast and introducing undesired brightness in localized regions. These distortions not only obscure scene content but also disrupt overall luminance consistency, posing challenges for accurate image restoration. To address this problem, we propose a flare removal framework based on ControlNet, a conditional diffusion architecture that enables precise spatial control during image generation. The proposed method incorporates spatially aligned flare priors to guide the diffusion process, providing explicit structural cues that help localize and suppress flare-affected regions. By leveraging both the generative capability of diffusion models and the structural guidance from flare priors, the framework effectively removes light artifacts while preserving natural textures and fine details. Extensive experiments on both synthetic and real-world flare datasets demonstrate that the proposed approach achieves superior performance compared to existing methods in terms of perceptual quality and quantitative accuracy. The framework also exhibits strong generalization ability, making it suitable for practical deployment in mobile imaging scenarios.**

Keywords—*Image Restoration, Flare Removal, Diffusion Model*

I. INTRODUCTION

Lens flare is a prevalent optical degradation that arises when intense light scatters or reflects within a camera lens system, resulting in undesired artifacts such as radial streaks, polygonal reflections, or bright spots superimposed on captured images [1]. These artifacts typically originate from strong light sources—such as the sun, street lamps, or vehicle headlights—and are especially pronounced in low-light or night-time conditions, where reflections and scattering effects are exacerbated by surface contaminants like dust, scratches, or fingerprints on the lens elements. Two primary types of lens flare are commonly observed: Reflective Flare (RF), caused by internal lens reflections and often manifested as geometric patterns, and Scattering Flare (SF), which produces diffused streaks due to irregular surface scattering [2]. These flare patterns severely degrade image quality by reducing contrast, washing out dark regions, and obscuring semantic details in critical areas near light sources. Such degradation not only affects human visual perception but also impairs the performance of downstream computer vision tasks, including object detection and autonomous navigation. Despite advances in optical design and image post-processing, effectively mitigating lens flare remains an open and

challenging problem, particularly in uncontrolled, real-world environments.

Conventional flare removal techniques primarily rely on handcrafted priors, heuristic rules, or classical image filtering. While these methods can suppress simple flare patterns, they often fail to generalize to complex or highly entangled artifacts, especially when flares overlap with semantically important regions. Recent learning-based approaches have shown promise by leveraging data-driven representations [3], [4]. However, they typically require large-scale paired training data [1], [5], which is difficult to obtain due to the unpredictable nature of real-world flare artifacts. Moreover, existing methods often lack spatial controllability, making it challenging to selectively suppress flare without compromising scene integrity.

To address flare removal challenges, we propose a Uni-ControlNet-based diffusion framework that introduces explicit flare structure priors as control signals for spatially-aware denoising. This enables targeted suppression of flares while preserving surrounding details. Unlike prior unconditional or image-to-image methods, our approach offers fine-grained, structure-aware guidance. Experiments on synthetic and real-world datasets show superior performance in perceptual quality, structural fidelity, and robustness under diverse nighttime conditions.

II. RELATED WORKS

A. Lens Flare Removal

Previous studies have predominantly addressed flare removal under daytime conditions [6], [7], as daytime flares often exhibit scattering patterns that closely resemble those of the background, making them more amenable to general image restoration techniques. Methods such as net [8], Uformer [9], and Restormer [10] have demonstrated notable effectiveness in suppressing these flares. In particular, transformer-based architectures have achieved state-of-the-art performance not only in daytime flare removal but also across a wide range of image restoration tasks. [5], [11].

B. Incorporating Priors for Flare Removal

Compared with existing methods, our approach introduces a detail-aware generative framework that explicitly incorporates handcrafted priors to guide flare removal. Previous works largely focused on dataset construction or adopted GAN-based architectures, which often lacked precise spatial control and tended to produce over-smoothed results [3]. More recent latent diffusion-based [12], rarely addressed the recovery of fine structural details underneath.

This work was supported in part by the China NSF under Grant 62427801, in part by the National Key R&D Program of China (2023YFB4502802), in part by the China NSF under Grant 62031009, in part by Fudan-ZTE joint lab, in part by Alibaba Research Fellow (ARF) Program.

Corresponding author: Yibo Fan

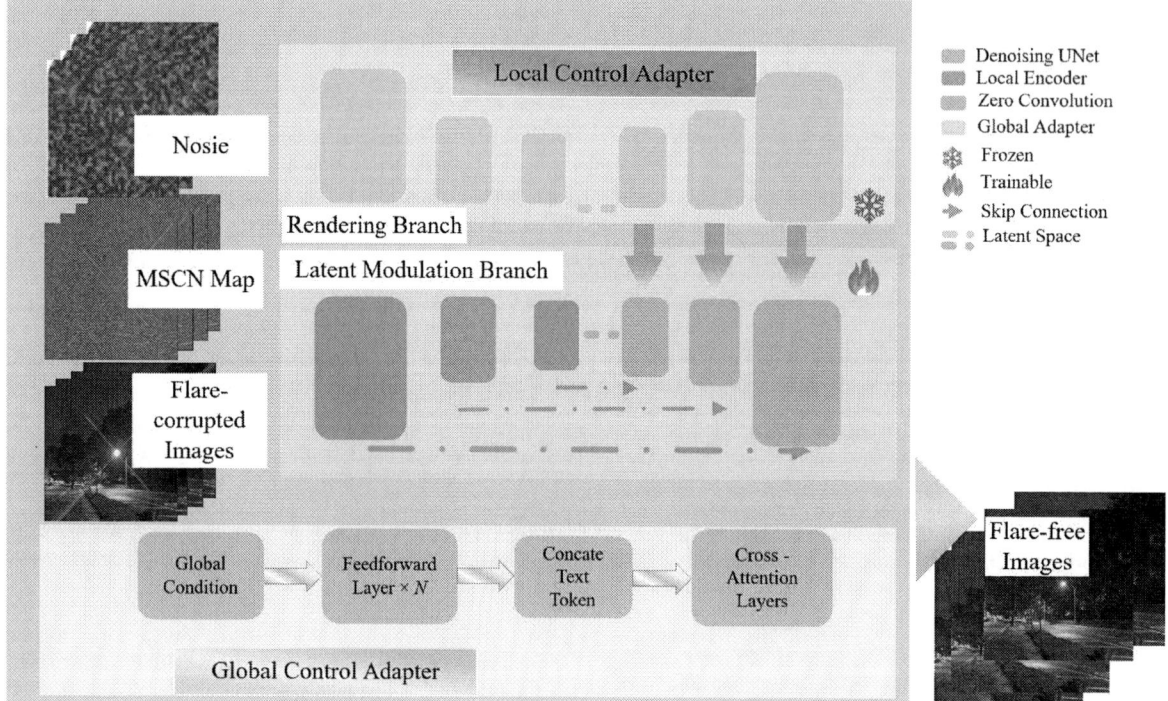

Figure 1 Illustration of the proposed global-local collaborative framework for flare removal. The framework integrates a Global Control Adapter and a Local Control Adapter to guide the diffusion model. The Global Adapter extracts global semantic conditions from the input image and injects them as tokens into cross-attention layers. The Local Adapter takes both the flare-corrupted image and its MSCN map, extracting local features through two branches: a Rendering Branch and a Latent Modulation Branch. Together, these modules enable the model to produce flare-free images with consistent global appearance and enhanced local details

In contrast, our method emphasizes structural fidelity by integrating a prior derived from Mean Subtracted Contrast Normalization (MSCN) [13], which captures local contrast degradation indicative of flare. This prior provides statistical guidance that enables the model to focus on precise detail reconstruction. In addition, we leverage the structural consistency between LDR and HDR representations to enforce semantic alignment in the latent space, promoting coherent texture across both flare-affected and clean regions. These innovations together allow our model to better restore high-frequency details and achieve superior perceptual quality, particularly in challenging night-time and low-light scenarios.

III. METHOD

Diffusion models, such as Stable Diffusion (SD) [14], have emerged as powerful frameworks for image generation and restoration due to their ability to model complex data distributions through iterative denoising. SD employs a U-Net-based denoising backbone, which consists of a multi-scale encoder-decoder architecture with residual blocks and attention layers, making it highly adaptable to diverse visual tasks.

Building on this foundation, as shown in Figure 1, we extend SD with Uni-ControlNet to enable diverse conditional controls tailored for flare removal [15]. Specifically, we use both the flare-corrupted image and its corresponding MSCN map as condition inputs. Prior studies have shown that handcrafted structural priors—such as MSCN—can

effectively guide restoration models, even under zero-shot settings, by highlighting local contrast degradations indicative of artifacts [16]. We propose a dual-branch architecture based on the Uni-Control framework for effective night-time flare removal. By integrating handcrafted priors through Local and Global Adapter branches, our unified control mechanism guides the denoising process, enabling accurate recovery of fine structures obscured by flares.

A. Input Representation

To better capture the statistical irregularities introduced by night flares, we use two types of inputs: the raw flare-contaminated image and its corresponding Mean Subtracted Contrast Normalized (MSCN) image. The MSCN transformation helps highlight local structural inconsistencies caused by flares, enabling the model to better isolate and remove such artifacts [16].

B. Local Adapter Branch

The local branch focuses on fine-grained, spatially localized flare removal. We design a Local Adapter that takes the concatenated raw and MSCN images as input. This input is passed through a modified control module, which follows the structure of the Uni-Control framework but is adapted to enhance its responsiveness to local statistical anomalies introduced by flares. The output of this branch retains detailed texture information while suppressing localized flare patterns.

The MSCN image is concatenated with the original image to form the input to the Local Adapter Branch. The MSCN image $\hat{I}(i,j)$ is concatenated with the original image to form the input to the Local Adapter Branch, as shown in (1).

$$\hat{I}(i,j) = \frac{I(i,j) - \mu(i,j)}{\sigma(i,j) + C} \quad (1)$$

where $I(i,j)$ denotes the intensity of the original image, and $\mu(i,j)$ is the local mean, as shown in (2).

$$\mu(i,j) = \sum_{k,j} \omega(k,l) \cdot I(i+k, j+l) \quad (2)$$

Besides, $\sigma(i,j)$ is the local standard deviation computed using a Gaussian-weighted window centered at (i,j), as shown in (3).

$$\sigma(i,j) = \sqrt{\sum_{k,l} \omega(k,l) \cdot \left[I(i+k, j+l) - \mu(i,j)\right]^2} \quad (3)$$

Where $\omega(k,l)$ is a normalized 2D Gaussian kernel, used to smooth the local neighborhood , typically with a kernel size of 5×5 or 7×7, and C is a small constant (e.g., 10^{-3}) added to prevent division by zero. The MSCN representation enhances the visibility of local contrast anomalies, making it well-suited for tasks such as flare removal.

C. Global Adapter Branch

To ensure that local flare correction is globally consistent, we introduce a Global Adapter Branch that complements the local guidance. While the local branch focuses on fine-grained, region-specific corrections, the global branch aggregates holistic contextual information from the flare-contaminated image. Specifically, it is designed to capture large-scale luminance shifts and scene-level illumination patterns that may be distorted by flare artifacts. By providing a broader view of the scene, this branch enables the model to reason about long-range dependencies and lighting consistency across the image. Structurally, it follows the modified Uni-Control paradigm and operates in parallel with the local branch, enabling joint optimization of both local precision and global coherence.The local branch focuses on fine-grained, spatially localized flare removal. We design a Local Adapter that takes the concatenated raw and MSCN images as input. This input is passed through a modified control module, which follows the structure of the UNI-Control framework but is adapted to enhance its responsiveness to local statistical anomalies introduced by flares. The output of this branch retains detailed texture information while suppressing localized flare patterns.

D. Fusion and Output

For inference strategy, the outputs of the model of the both local and global adapter branches are manually integrated to produce the final restored image. Specifically, we assign fixed weights to the global branch's output based on empirical observations, prioritizing local details in regions with dense flare artifacts, while emphasizing global consistency in areas affected by broader luminance distortion. This manual fusion strategy enables us to balance fine-grained correction with holistic restoration without introducing additional trainable parameters.

IV. EXPERIMENTS

A. Experiment Settings

As shown in Table I and Table II, our framework consists of a memory-efficient Uni-ControlNet backbone (859M), a pre-trained Autoencoder KL (83.7M), a frozen CLIP encoder (123M), and a 411M Local Adapter for injecting handcrafted priors. Of the total 1.515B parameters, 1.309B are trainable and 206M are frozen, enabling precise flare removal while preserving semantic stability. The design remains efficient for training on a single NVIDIA A6000 GPU.

| MSCN Map | Flare- corrupted | DCP [17] | UFormer [9] | Flare7kpp [1] | Zhou et al. [18] | Ours | Ground Truth |

Figure 2 Qualitative results on the Flare7K RFR dataset show that our model better preserves scene details around light sources in both indoor and outdoor settings.

TABLE I. MODEL STRUCTURE AND PARAMETER SUMMARY

Name	Type	Function	Params
Model	Diffusion Wrapper	UNet Backbone and Interface	859 M
First Stage Model	Autoencoder KL	Encode and Decode	83.7 M
Condition Stage Model	Frozen Clip Embedder	Semantic Control	123 M
Local Adaper	Local Adapter	Local Hinter	411 M

TABLE II. THE OVERLOAD SUMMARY

Type	Params
Trainable Params	1.309 B
Non trainable Params	0.206 B
Total Params	1.515 B

B. Results and Analysis

Our method demonstrates superior performance in removing flare artifacts, as shown in Figure 2 particularly in preserving structural details around strong light sources in both indoor and outdoor scenes. Compared to existing methods such as DCP [17], UFormer [9], and Flare7K++ [1], which fail to eliminate streak-like flares near light sources, and Zhou *et al.*, which struggle to localize flare regions accurately, our approach produces cleaner and more visually consistent results.

Table III reports the quantitative comparison of our method against state-of-the-art flare removal approaches on the Flare7K dataset, including both Real Flare Removal (RFR) and Synthetic Flare Removal (SFR) settings. Across all three metrics—PSNR, SSIM [19] , and LPIPS [20], our method demonstrates competitive performance.

For the SFR setting, our method achieves the best LPIPS score (0.012), indicating superior perceptual quality and fidelity. It also attains the highest PSNR (31.10), surpassing all baselines including Uformer (30.47) and FFFormer (30.88), and ranks second in SSIM (0.950), slightly below FFFormer (0.969). This suggests that our global-local control strategy effectively removes synthetic flare while preserving structural similarity.

TABLE III.   RESULTS ON THE DATASETS FROM FLARE7K [5] . RED INDICATES THE BEST RESULT, AND BLUE INDICATES THE SECOND-BEST RESULT.

Method	RFR			SFR		
	PSNR	SSIM	LPIPS	PSNR	SSIM	LPIPS
Input	22.56	0.8557	0.078	22.77	0.921	0.060
Wu [6]	24.61	0.871	0.060	27.88	0.952	0.031
UNet[8]	26.11	0.879	0.055	29..07	0.958	0.022
HINet [21]	26.74	0.882	0.048	29.97	0.959	0.021
Zhou et al [18]	25.18	0.872	0.055	28.779	0.939	0.0287
Restormer [10]	26.38	0.883	0.054	29.45	0.950	0.025
Uformer [9]	26.98	0.890	0.047	30.47	0.965	0.017
FFFormer [7]	27.35	0.901	0.044	30.88	0.969	0.019
Ours	27.12	0.899	0.039	31.10	0.950	0.012

In the more challenging RFR setting, our method achieves a PSNR of 27.12 and SSIM of 0.899, which are very close to the best results (FFFormer: 27.35 / 0.901), and still maintains a competitive LPIPS score (0.050). Notably, while FFFormer slightly outperforms us in RFR metrics, our model shows better consistency across both real and synthetic domains.

Overall, these results confirm that our method achieves a strong balance between structural restoration and perceptual quality, especially excelling in synthetic and real flare removal scenarios.

REFERENCES

[1] Y. Dai, C. Li, S. Zhou, R. Feng, Y. Luo, and C. C. Loy, "Flare7K++: Mixing synthetic and real datasets for nighttime flare removal and beyond," *IEEE Transactions on Pattern Analysis and Machine Intelligence*, vol. 46, no. 11, pp. 7041–7055, 2024, doi: 10.1109/TPAMI.2024.3406821.

[2] F. Lan and C. W. Chen, "Tackling Scattering and Reflective Flare in Mobile Camera Systems: A Raw Image Dataset for Enhanced Flare Removal," Jul. 26, 2023, *arXiv*: arXiv:2307.14180. Accessed: Jun. 06, 2024. [Online]. Available: http://arxiv.org/abs/2307.14180

[3] X. Qiao, G. P. Hancke, and R. W. H. Lau, "Light Source Guided Single-Image Flare Removal from Unpaired Data," in *2021 IEEE/CVF International Conference on Computer Vision (ICCV)*, Montreal, QC, Canada: IEEE, Oct. 2021, pp. 4157–4165. doi: 10.1109/ICCV48922.2021.00414.

[4] Y. Kotp and M. Torki, "Flare-free vision: Empowering uformer with depth insights," in *2024 IEEE international conference on acoustics, speech and signal processing*, 2024, pp. 2565–2569. doi: 10.1109/ICASSP48485.2024.10446006.

[5] Y. Dai, C. Li, S. Zhou, R. Feng, and C. C. Loy, "Flare7k: A phenomenological nighttime flare removal dataset," *Advances in Neural Information Processing Systems*, vol. 35, pp. 3926–3937, 2022.

[6] Y. Wu *et al.*, "How to Train Neural Networks for Flare Removal," Oct. 07, 2021, *arXiv*: arXiv:2011.12485. Accessed: Jan. 13, 2023. [Online]. Available: http://arxiv.org/abs/2011.12485

[7] D. Zhang, J. Ouyang, G. Liu, X. Wang, X. Kong, and Z. Jin, "FF-Former: Swin Fourier Transformer for Nighttime Flare Removal".

[8] O. Ronneberger, P. Fischer, and T. Brox, "U-net: Convolutional networks for biomedical image segmentation," in *Medical image computing and computer-assisted intervention – MICCAI 2015*, N. Navab, J. Hornegger, W. M. Wells, and A. F. Frangi, Eds., Cham: Springer International Publishing, 2015, pp. 234–241.

[9] Z. Wang, X. Cun, J. Bao, W. Zhou, J. Liu, and H. Li, "Uformer: A General U-Shaped Transformer for Image Restoration," in *2022 IEEE/CVF Conference on Computer Vision and Pattern Recognition (CVPR)*, Jun. 2022, pp. 17662–17672. doi: 10.1109/CVPR52688.2022.01716.

[10] S. W. Zamir, A. Arora, S. Khan, M. Hayat, F. S. Khan, and M.-H. Yang, "Restormer: Efficient transformer for high-resolution image restoration," in *Proceedings of the IEEE/CVF conference on computer vision and pattern recognition*, 2022, pp. 5728–5739.

[11] Y. Dai, Y. Luo, S. Zhou, C. Li, and C. C. Loy, "Nighttime Smartphone Reflective Flare Removal Using Optical Center Symmetry Prior," Mar. 27, 2023, *arXiv*: arXiv:2303.15046. Accessed: Apr. 07, 2023. [Online]. Available: http://arxiv.org/abs/2303.15046

[12] J. Lian, Z. Liu, M. Jing, J. Zhou, and Y. Fan, "Improving nighttime flare removal with subspace basic projection," in *2024 IEEE international conference on signal, information and data processing (ICSIDP)*, IEEE, 2024, pp. 1–6.

[13] A. Mittal, A. K. Moorthy, and A. C. Bovik, "No-reference image quality assessment in the spatial domain," *IEEE Transactions on Image Processing*, vol. 21, no. 12, pp. 4695–4708, 2012, doi: 10.1109/TIP.2012.2214050.

[14] R. Rombach, A. Blattmann, D. Lorenz, P. Esser, and B. Ommer, "High-resolution image synthesis with latent diffusion models," in *2022 IEEE/CVF conference on computer vision and pattern recognition (CVPR)*, 2022, pp. 10674–10685. doi: 10.1109/CVPR52688.2022.01042.

[15] S. Zhao *et al.*, "Uni-ControlNet: All-in-One Control to Text-to-Image Diffusion Models," Oct. 29, 2023, *arXiv*: arXiv:2305.16322. Accessed: Jul. 20, 2024. [Online]. Available: http://arxiv.org/abs/2305.16322

[16] R. Zhu *et al.*, "Zero-Shot Structure-Preserving Diffusion Model for High Dynamic Range Tone Mapping".

[17] K. He, J. Sun, and X. Tang, "Single image haze removal using dark channel prior," in *2009 IEEE conference on computer vision and pattern recognition*, 2009, pp. 1956–1963. doi: 10.1109/CVPR.2009.5206515.

[18] Y. Zhou, D. Liang, S. Chen, S.-J. Huang, S. Yang, and C. Li, "Improving Lens Flare Removal with General-Purpose Pipeline and Multiple Light Sources Recovery," presented at the Proceedings of the IEEE/CVF International Conference on Computer Vision, 2023, pp. 12969–12979. Accessed: Nov. 03, 2023. [Online]. Available: https://openaccess.thecvf.com/content/ICCV2023/html/Zhou_Improving_Lens_Flare_Removal_with_General-Purpose_Pipeline_and_Multiple_Light_ICCV_2023_paper.html

[19] Z. Wang, A. C. Bovik, H. R. Sheikh, and E. P. Simoncelli, "Image quality assessment: from error visibility to structural similarity," *IEEE transactions on image processing*, vol. 13, no. 4, pp. 600–612, 2004.

[20] R. Zhang, P. Isola, A. A. Efros, E. Shechtman, and O. Wang, "The unreasonable effectiveness of deep features as a perceptual metric," in *Proceedings of the IEEE conference on computer vision and pattern recognition*, 2018, pp. 586–595.

[21] L. Chen, X. Lu, J. Zhang, X. Chu, and C. Chen, "Hinet: Half instance normalization network for image restoration," in *Proceedings of the IEEE/CVF conference on computer vision and pattern recognition*, 2021, pp. 182–192.

979-8-3315-3918-4/25 $31.00 © 2025 IEEE

AI-Assisted Droplet Splitting on a Parallel-Plate Optoelectrowetting Chip

Junyan Tian, Shang Gao, Tengpu Zhu, Enqing Liu, Gaifang Chen, Jia Zhou*

State Key Laboratory of Integrated Chips and Systems, College of Integrated Circuits and Micro-Nano Electronics,
Fudan University, Shanghai 200433, China
School of Microelectronics, Fudan University, Shanghai 200433, China

* Email: jia.zhou@fudan.edu.cn

Abstract—This paper presents an AI-assisted droplet generation system integrated with a parallel-plate optoelectrowetting (OEW) chip. The system enables real-time identification of droplets with varying colors, sizes, and quantities, achieving an average detection confidence exceeding 85%. By leveraging AI, the system dynamically generates and projects light patterns onto the OEW chip, facilitating precise droplet deformation and splitting. Experimental studies examine the effects of light intensity and plate spacing on contact angle and splitting time. The results demonstrate that stronger illumination enhances the optoelectrowetting effect on the chip surface, while increased electrode spacing prolongs splitting time or hinders droplet division completely. This non-contact, AI-assisted approach to droplet manipulation represents a promising advancement in digital microfluidics, offering a powerful strategy for future applications in lab-on-a-chip technologies.

Keywords—optoelectrowetting, AI, droplet splitting

I. INTRODUCTION

Optoelectrowetting (OEW) chips represent a light-actuated droplet manipulation technology, in which light irradiation modulates the electrowetting effect on the chip surface to enable precise control of droplet movement and behavior [1,2]. Unlike conventional electrowetting-on-dielectric (EWOD) chips that rely on pre-patterned electrodes [3], OEW utilizes dynamically reconfigurable "virtual electrodes" generated by localized illumination on photosensitive materials such as amorphous silicon (α-Si) [4-6]. This mechanism eliminates the need for predesigned electrode arrays, simplifies device fabrication, and effectively addresses the challenges associated with complex electrode wiring, thereby significantly enhancing chip integration and design flexibility [7].

Despite these advantages, OEW chip still faces certain limitations in terms of droplet generation and splitting precision. Accurate positioning of virtual electrodes typically relies on mechanical components such as stepper motors or mirror arrays [8,9]. These devices often suffer from limited spatial resolution, slow response times, and bulky form factors, all of which constrain the speed, accuracy, and programmability of droplet operations—particularly during dynamic processes such as droplet generation [10].

To overcome these challenges, this study proposes an AI-assisted parallel-plate OEW system. In this system, a deep learning model is employed to accurately detect droplet morphology and position [11,12], while a commercially available digital projector is used to generate adaptive light patterns to induce virtual electrodes. Through this approach, precise and programmable control over droplet generation is achieved without the need for mechanical motion components, thereby improving the overall manipulation accuracy and enabling full automation of the OEW process.

II. EXPERIMENTAL

A. Working Principle

In an OEW chip, controlled droplet splitting is governed by a localized electrowetting effect induced by optical illumination. The electrowetting behavior follows the classical Lippmann–Young equation [13]:

$$cos\theta = cos\theta_0 + \frac{C}{2\gamma}V^2 \tag{1}$$

where θ_0 represents the initial contact angle of the droplet in the absence of an applied voltage, while θ denotes the contact angle under an applied voltage. C is the capacitance per unit area, γ is the interfacial tension between the liquid and the surrounding medium, and V is the voltage applied across the locally illuminated region of the photoconductive layer.

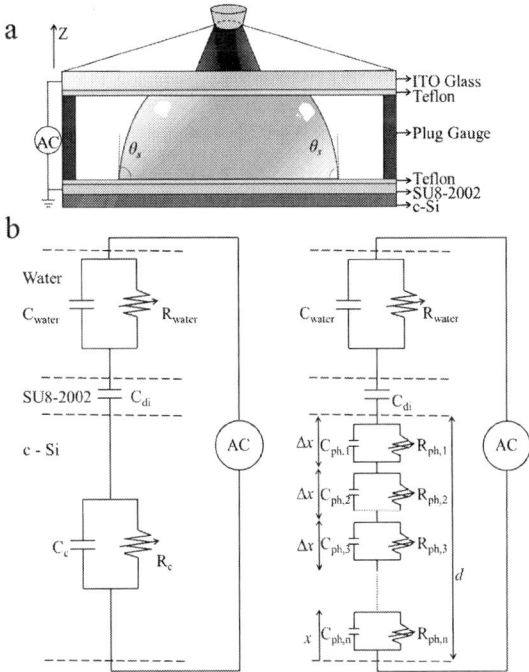

Fig.1. Structure of the parallel-plate OEW chip. (a) Composition of the parallel-plate OEW chip. (b) Equivalent circuit of the OEW chip.

The OEW chip employs single-crystalline silicon as the photoconductive layer. As a semiconductor material, it exhibits a significant change in surface conductivity under varying light intensities. By modulating the spatial distribution of illumination on the chip surface, the local potential of the photoconductive layer can be precisely controlled, thereby inducing localized electrowetting effects, as illustrated in Fig. 1(a).

In this system, changes in the applied voltage V lead to variations in the contact angle. According to the equivalent circuit diagram shown in Fig. 1(b), V can be expressed as:

$$V = U \frac{Z_{di}}{Z_{di} + Z_{c,total} + Z_{water}} \quad (2)$$

where U denotes the externally applied AC voltage, Z_{di} is the impedance of the dielectric layer, $Z_{c,total}$ represents the total impedance per unit area of the single-crystalline silicon layer, and Z_{water} is the impedance of the droplet. These parameters can be calculated using the following equations:

$$Z_{di} = \frac{1}{j\omega C_{di}} \quad (3)$$

$$Z_{c,total} = \int_0^d Z_{\Delta x}(x)dx \quad (4)$$

$$Z_{water} = (\frac{1}{R_{water}} + j\omega C_{water})^{-1} \quad (5)$$

where ω is the angular frequency of the applied AC signal, C_{di} denotes the capacitance of the dielectric layer, and R_{water} and C_{water} represent the resistance and capacitance of the droplet, respectively. These parameters are influenced by factors such as the plate spacing and the type of droplet used. The impedance per unit area element, $Z_{\Delta x}(x)$ consists of a resistance $R_{ph}(x)=\rho(x)\Delta x$ and a capacitance $C_{ph}(x)=\varepsilon_{ph}\varepsilon_0/\Delta x$, and can be expressed as [14]:

$$Z_{\Delta x}(x) = \frac{\rho(x)\Delta x}{1 + j\omega\varepsilon_{ph}\varepsilon_0\rho(x)} \quad (6)$$

where $\rho(x)$ denotes the resistivity of the single-crystalline silicon at position x, ε_0 is the permittivity of free space, and ε_{ph} represents the relative permittivity of the single-crystalline silicon. According to the relationship between light intensity and resistivity, the resistivity at position x, $\rho(x)$ can be determined by the following equation:

$$\rho(x) = \frac{1}{q\mu_n\tau_n + q\mu_p\tau_p} \frac{hv\Delta x}{I_0(e^{-\alpha x} - e^{-\alpha(x+\Delta x)})} \quad (7)$$

where α is the absorption coefficient of the photoconductor. The absorption coefficient of silicon varies depending on the wavelength (color) of the incident light. h, v, τ_n and τ_p represent Planck's constant, optical frequency, electron lifetime, and hole lifetime, respectively. μ_n and μ_p denote the electron and hole mobilities, respectively.

In this study, a dynamic control strategy based on symmetric virtual electrodes was employed to achieve droplet generation, as illustrated in Fig. 2(b). At the initial stage, two light patterns fully overlap at the droplet center, forming a symmetric electrowetting excitation region. As the light patterns move outward from the droplet center at equal velocities, the droplet ends are exposed to illumination while the central region remains dark, creating a contact angle gradient. Under this condition, electrowetting driving forces F act on both ends of the droplet along the X-axis, whereas the central region retains its original contact angle due to the

absence of illumination, leading to the formation of a necking structure in the middle. The electrowetting driving force exerted on the droplet can be described by the following equation:

$$F = \frac{1}{2}\frac{\varepsilon_{di}\varepsilon_0}{t_{di}}(V_{di,light}^2 - V_{di,dark}^2) \quad (8)$$

where ε_{di}, t_{di} are the relative permittivity and thickness of the dielectric layer, respectively. $V_{di,light}$ and $V_{di,dark}$ are the voltage dropped across the dielectric layer, calculated from equation(2), when the photoconductor is illuminated with and without light, respectively.

When the droplet is stretched beyond its critical deformation threshold, the neck undergoes rupture due to the combined effects of Laplace pressure and electrowetting forces, ultimately resulting in the successful formation of two daughter droplets.

B. Fabrication of the OEW System

The structure and process flow of the OEW system employed in this study are illustrated in Fig. 2. The system primarily consists of three components: a Thorlabs CCD camera for image acquisition, an AI-based intelligent recognition module, and a projector for light pattern projection, as depicted in Fig. 2(a). The Thorlabs CCD camera captures real-time images of droplets on the chip and transmits the image data to a computer. The computer runs a deep learning model based on the You Only Look Once (YOLO) v11 algorithm [15], which accurately detects the spatial position and boundaries of droplets. Based on this detection, corresponding dual light patterns are generated to excite virtual electrodes that drive droplet deformation.

To ensure precise droplet position recognition during the OEW droplet splitting process, a lightweight object detection model was constructed and trained in this work. The training dataset comprises 700 manually annotated droplet images, covering three droplet types—colorless, red, and blue—and various droplet counts and volumes, enhancing the model's generalization capability for practical applications. Finally, the commercial projector projects the generated light patterns onto the OEW chip, enabling dynamic control of droplet behavior.

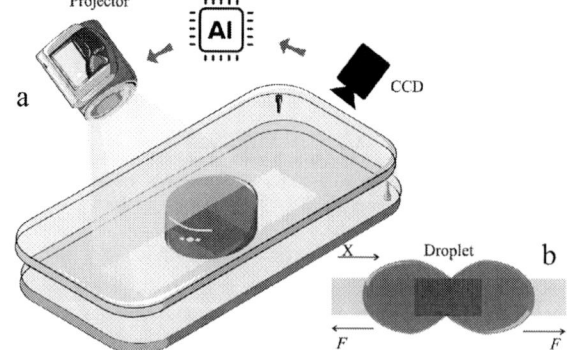

Fig. 2. Design of a parallel-plate OEW chip system. (a) Schematic diagram of the system as a whole. (b) Optical virtual electrode and droplet driving direction.

The structural composition of the OEW chip is shown in Fig. 1(a). Along the Z-axis (from bottom to top), the lower substrate consists of a 500 μm-thick single-crystalline silicon wafer, a spin-coated SU8-2002 layer (purchased from

MicroChem) serving as the dielectric layer, and a surface hydrophobic coating of Teflon® AF2400. The upper plate is formed by spin-coating Teflon® AF2400 onto an indium tin oxide (ITO)-coated glass substrate, which provides both hydrophobicity and electrical grounding. The gap between the upper and lower plates is controlled using plug gauge. During the experiments, a sinusoidal AC signal is applied across the two plates.

III. RESULTS AND DISCUSSION

This section presents an analysis from two perspectives: system functionality and droplet generation performance. Firstly, the AI-based droplet generation system was developed and validated, with a focus on evaluating its accuracy. Subsequently, a systematic investigation was conducted on the electrowetting behavior of the chip surface under different illumination conditions, as well as the effect of plate spacing on droplet splitting dynamics. These studies clarified the role of these parameters in influencing the efficiency of droplet splitting.

A. Verification of AI system

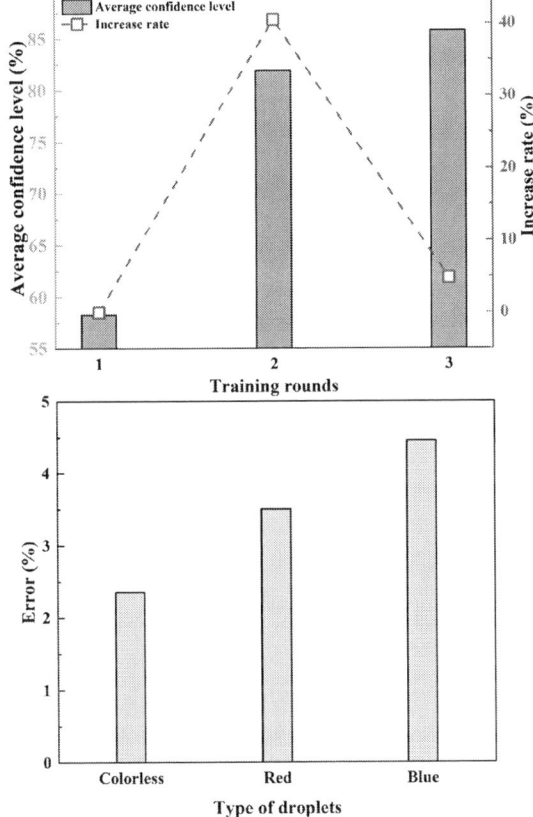

Fig. 3. Demonstration of AI model performance. (a) Confidence scores and improvement across three training rounds. (b) Classification performance of the trained model on different droplet types.

As shown in Fig. 3(a), the model underwent three rounds of training. In the second round, the number of training epochs was increased based on the first round, which significantly improved the consistency and stability of detection. The third round further incorporated a small set of extreme cases—such as low-contrast and irregularly shaped droplets—to enhance the model's robustness under complex conditions. After the

final training phase, the model achieved an average detection confidence exceeding 85%. Although this is slightly lower than that of some recent YOLO variants, it is sufficient to meet the real-time and accuracy requirements of the present study. With further expansion of the training dataset, the model performance is expected to improve significantly.

The classification accuracy of the model is shown in Fig. 3(b) Due to the relatively small number of red and blue droplet samples, the classification error for these categories is slightly higher compared to colorless droplets. Nevertheless, the overall classification error remained below 5%, demonstrating strong multi-class recognition capabilities.

Fig. 4 demonstrates the model's classification results in practical tests, where droplets of different colors were correctly identified and their volumes estimated with high confidence.

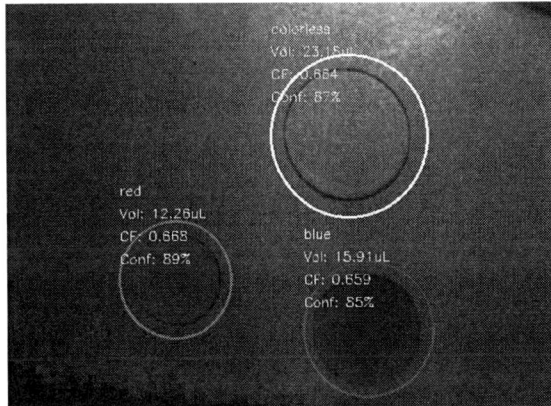

Fig. 4. Actual classification performance of the AI model.

B. Investigation of droplet splitting behavior

To investigate the influence of illumination intensity on electrowetting, grayscale light patterns were projected onto the chip surface, and contact angle changes of fixed-volume droplets were measured under constant electrical conditions. Results showed that higher illumination led to stronger wettability, confirming that the electrowetting force can be effectively tuned by adjusting light intensity.

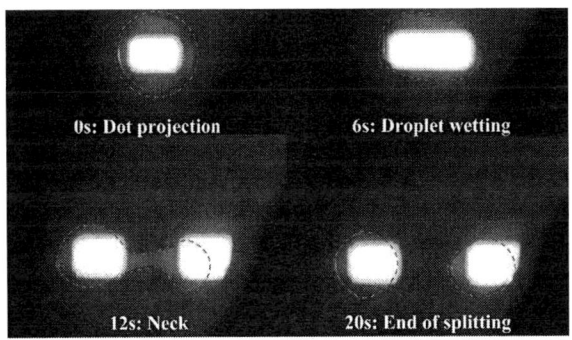

Fig. 5. Schematic diagram of the 12μL droplet splitting process under 300Vpp and 300μm plate spacing.

The electrowetting force acting on the droplet is not only dependent on the illumination intensity, but also influenced by the geometric parameters of the system. Among those, the spacing between the upper and lower plates is one of the key factors. It affects the equivalent capacitance between the electrodes and the droplet, thereby altering the voltage distribution and the strength of the electrowetting effect.

Consequently, it impacts the force dynamics and response time during the droplet splitting process. To elucidate this relationship, we further investigated the changes in droplet splitting behavior under different plate spacings.

Initially, the droplet splitting process was observed under a plate spacing of 300 μm and an applied voltage of 300 Vpp, using a 12 μL droplet driven by dual light patterns , as shown in Fig. 5 The droplet underwent three typical stages: initiation of wetting, neck formation, and final breakup. Subsequently, the plate spacing was set to 100, 200, 300, 400 and 500 μm in separate experiments, while keeping the voltage and the top-view droplet area constant to eliminate the influence of factors such as contact line length.

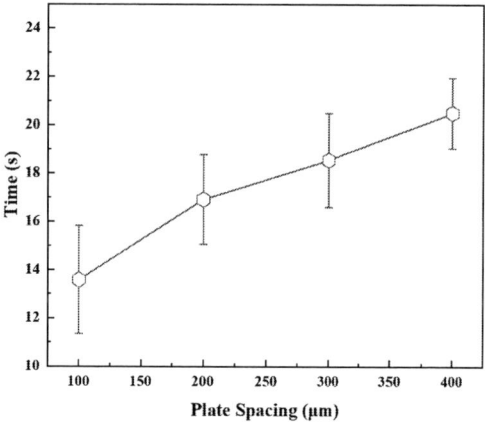

Fig. 6. Droplet splitting time versus plate spacing.

The experimental results are presented in Fig. 6. As the plate spacing increased, the droplet splitting time was prolonged. Notably, under a 500 μm spacing, although significant necking was observed under light stimulation, the droplet failed to complete the splitting and instead stabilized in a stretched state (see Fig. 7). This indicates that the system did not provide sufficient driving force to overcome surface tension and induce droplet rupture. These observations suggest the existence of a threshold electrowetting voltage for droplet splitting. When the combination of applied voltage and system parameters fails to generate adequate electrowetting force, the droplet can be stretched but cannot be divided. Future work will focus on systematically modeling the threshold voltage for splitting and exploring optimization strategies to enable more efficient and programmable droplet manipulation

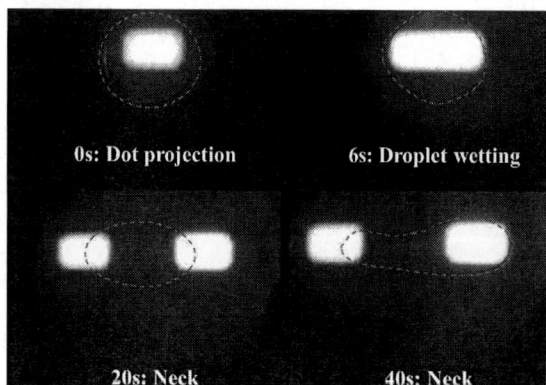

Fig. 7. Schematic diagram of the 20μL droplet splitting process under 500Vpp and 300μm plate spacing. The droplet is eventually pulled toward one side.

IV. .CONCLUSION

This work presents an AI-assisted parallel-plate OEW system for programmable droplet splitting. By combining AI-based real-time detection with dynamic light projection, the system achieves accurate and autonomous control of droplet motion. Experimental results confirm that light intensity and electrode spacing significantly affect electrowetting efficiency and splitting behavior. The system demonstrates reliable performance and offers a promising platform for intelligent, non-contact microfluidic manipulation.

ACKNOWLEDGMENT

This work was supported by the National Natural Science Foundation of China with Grant No. 62274039.

REFERENCES

[1] P. Y. Chiou, H. Moon, H. Toshiyoshi, C.-J. Kim, and M. C. Wu, "Light actuation of liquid by optoelectrowetting," *Sensors and actuators A: physical*, vol. 104, no. 3, pp. 222-228, 2003.

[2] P.-Y. Chiou, Z. Chang, and M. C. Wu, "Droplet manipulation with light on optoelectrowetting device," *Journal of Microelectromechanical Systems*, vol. 17, no. 1, pp. 133-138, Feb 2008.

[3] G. J. Shah, A. T. Ohta, E. P. Y. Chiou, M. C. Wu, and C.-J. Kim, "EWOD-driven droplet microfluidic device integrated with optoelectronic tweezers as an automated platform for cellular isolation and analysis," *Lab on a Chip*, vol. 9, no. 12, pp. 1732-1739, 2009 2009.

[4] S. K. Thio and S.-Y. Park, "A review of optoelectrowetting (OEW): from fundamentals to lab-on-a-smartphone (LOS) applications to environmental sensors," *Lab on a Chip*, Review vol. 22, no. 21, pp. 3987-4006, Oct 25 2022.

[5] S. Gao, J. Tian, J. Yao, H. Zheng, E. Liu, and J. Zhou, "Study on droplet splitting in single-plate OEW chips," *Journal of Electrostatics*, vol. 136, Aug 2025, Art no. 104111.

[6] S.-Y. Park, S. Kalim, C. Callahan, M. A. Teitell, and E. P. Y. Chiou, "A light-induced dielectrophoretic droplet manipulation platform," *Lab on a Chip*, vol. 9, no. 22, pp. 3228-3235, 2009 2009.

[7] Z. Hayat and A. I. El Abed, "High-Throughput Optofluidic Acquisition of Microdroplets in Microfluidic Systems," *Micromachines*, Review vol. 9, no. 4, Apr 2018, Art no. 183.

[8] C. Doering, J. Strassner, and H. Fouckhardt, "Lithography-Free Technology for the Preparation of Digital Microfluidic (DMF) Lab-Chips with Droplet Actuation by Optoelectrowetting (OEW)," *International Journal of Analytical Chemistry*, vol. 2022, May 29 2022, Art no. 2011170.

[9] M. P. Kremer and A. Tortschanoff, "Thermally induced light-driven Microfluidics using a MOEMS-based Laser Scanner for Particle Manipulation," in *Conference on Microfluidics, BioMEMS, and Medical Microsystems XII*, San Francisco, CA, 2014 Feb 02-04 2014, vol. 8976, in Proceedings of SPIE, 2014

[10] S.-Y. Park, "Optofluidic devices and their applications," in *Optical MEMS, Nanophotonics, and Their Applications*: CRC Press, 2017, pp. 347-376.

[11] E. Liu, C. Wang, L. Du, S. Li, A. Riaud, and J. Zhou, "AI-powered modular and general-purpose droplet processing system based on single-sided continuous optoelectrowetting chip," *Sensors and Actuators B-Chemical*, vol. 420, Dec 1 2024, Art no. 136445.

[12] T. Wu *et al.*, "Investigation into the optoelectrowetting droplet transport mechanism," *Electrophoresis*, vol. 45, no. 15-16, pp. 1428-1442, Aug 2024.

[13] F. Mugele and J. C. Baret, "Electrowetting: From basics to applications," *Journal of Physics-Condensed Matter*, Review vol. 17, no. 28, pp. R705-R774, Jul 20 2005.

[14] S. N. Pei, J. K. Valley, Y.-L. Wang, and M. C. Wu, "Distributed Circuit Model for Multi-Color Light-Actuated Opto-Electrowetting Microfluidic Device," *Journal of Lightwave Technology*, vol. 33, no. 16, pp. 3486-3493, Aug 15 2015.

[15] J. Redmon, S. Divvala, R. Girshick, A. Farhadi, and Ieee, "You Only Look Once: Unified, Real-Time Object Detection," in *2016 IEEE Conference on Computer Vision and Pattern Recognition (CVPR)*, Seattle, WA, 2016 Jun 27-30 2016, in IEEE Conference on Computer Vision and Pattern Recognition, 2016, pp. 779-788.

979-8-3315-3918-4/25 $31.00 © 2025 IEEE

A Sub-1mV Voltage-Variation Pixel Power Supply Architecture with Radiation-Hardened Built-In LDO for Pixel Readout ASIC

Lei Li [1], Jinxiang Wang [1], Yini Hong [1], Yuxiao Zhao [2], Yongsheng Wang *[1]

[1] Department of Microelectronics, Harbin Institute of Technology, Harbin 150001, China
[2] State Key Laboratory of Integrated Chips and Systems, Fudan University, Shanghai 200433, China

* Email: 18846446949@163.com, yswang@hit.edu.cn

Abstract—As the number of pixels in the hybrid pixel detector readout ASIC increases, the variation among pixel supply voltage becomes more significant due to the voltage drop problem. This problem will lead to the distribution of the gain, and noise of each pixel, which will deteriorate the performance of the pixel detector readout ASIC. The pixel power supply architecture using built-in LDO can greatly reduce the variation of pixel power voltage. Additionally, pixel readout ASICs need enough radiation tolerance to ensure normal circuit function when exposed to radiation in high-radiation environments. This paper presents a pixel power supply architecture with radiation-hardened built-in LDO for pixel readout ASIC. To avoid the impact of radiation hardening structures on chip area and performance, this paper proposed a novel radiation hardening structure for built-in capacitor-less LDO, which can quickly discharge the charges generated by radiation at the circuit nodes with $34.5 \mu m^2$/pixel and 2.5nA/pixel. The LDO is designed in CMOS 180nm process, and post-simulation show that this power supply architecture can reduce the variation among pixel supply voltage to sub-1mV. The area of the built-in LDO is 404.5 μm^2/pixel. The current of this LDO is $1.1 \mu A$/pixel under typical conditions. The output voltage of LDO varies within 32.04 mV, the recovery time is less than $1.5 \mu s$, and no significant degradation is observed in other performance of the built-in LDO, when the radiation intensity is LET < 50MeV·cm²/mg.

Keywords—hybrid pixel detector, readout ASIC, built-in LDO, radiation hardening

I. INTRODUCTION

Hybrid pixel detectors (HPD) are constructed of a separate sensor array and pixel readout Application Specific Integrated Circuits (ASIC), fabricated independently and interconnected via bump bonding, as shown in Fig.1. The separate production allows independent optimization of readout ASIC and sensor, which can obtain excellent detection efficiency and readout efficiency. Nowadays, hybrid pixel detectors are widely used in various fields such as high-energy physics experiments, medical imaging, and deep-space exploration[1].

Currently, hybrid pixel detectors typically have thousands of pixels. To achieve higher spatial resolution and a larger detector area, the pixel scale is still increasing. However, voltage drop problems become more serious as the number of pixels of the hybrid pixel detector continues to grow and their pixel functions become complex[2]. Additionally, in hybrid pixel detectors readout ASIC, the power pads and the input/output pads are generally placed on one side of the readout ASIC, as shown in Fig.1. When readout ASICs are tiled to expand the pixel detector area, this placement strategy minimizes the dead area and enhances the scalability of the

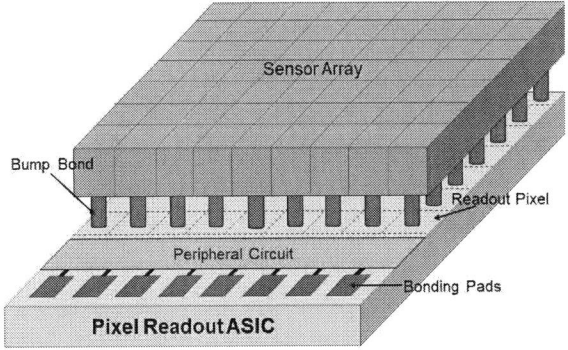

Fig. 1. Typical geometry of hybrid pixel detector

detector chip. This placement strategy gives rise to a problem, that is, the power voltage drops of the pixels close to the pad side are relatively small, while the power voltage drops of the pixels far from the pad side are relatively large. Therefore, the increase in pixel scale and the single-sided arrangement architecture of power pads have made the voltage variation of pixels more serious. This problem will not only lead to the nonuniformity of the power voltage of each pixel but also the distribution of the gain, and noise of each pixel, which will deteriorate the performance of the pixel detector readout ASIC[3].

To address this issue, many power supply architectures have been used, but most of them are not effective. In [4], a traditional column power supply architecture is adopted, which results that the voltage variation of pixels is 92mV. Recently, a power supply architecture based on built-in Low-Dropout Regulator (LDO) has been proposed in [2] and [5], which can significantly mitigate the voltage variation of pixels. However, the current consumption is increased by the built-in LDO, which is 8.31uA/pixel in [2] and 1.89uA/pixel in [5]. Thus, an LDO-based power supply architecture with lower current and area consumption is necessary.

Additionally, pixel readout ASICs must have enough radiation tolerance to ensure normal circuit function when exposed to radiation in high-radiation environments, such as high-energy physics experiments and deep-space exploration and so all [1]. For analog circuits, the hardening methods require either a special process [6] or a complex structure [7], which will inreoduce extra area and power consumption. In this paper, a pixel power supply architecture with radiation-hardened built-in LDO is proposed to address the voltage variation of pixel and radiation effect for hybrid pixel detector readout ASIC.

979-8-3315-3918-4/25 $31.00 © 2025 IEEE

The rest of this brief is structured as follows: Section II shows the power supply architecture for pixel readout ASIC. Section III introduces the design of the radiation-hardened built-in LDO. Section IV presents the post-simulation results. Finally, Section V concludes this brief.

II. POWER SUPPLY ARCHITECTURE FOR PIXEL READOUT ASIC

In the design process of this power supply architecture, the first step is to evaluate the voltage drop using calculation and modeling to obtain the parameter for the LDO design. Then, a built-in LDO could be design for the pixel readout ASIC to obtain better performance.

A traditional power supply architecture for pixel readout ASIC is the column power supply architecture [4], as shown in Fig.2(a). Some power supply architectures are evolved from this architecture [8]. The expression for the voltage drops of this power supply architecture is as follows, where i and j is the row and column coordinates of pixels, respectively; the Rcol and Rrow is the parasitic resistance of the routing; the Ipixel is the current of one pixel.

$$\Delta V_{ij} = \left\{ \frac{m \times [n+(n-j+1)] \times j}{2} R_{col} + \frac{[m+(m-i+1) \times i]}{2} R_{row} \right\} \times I_{pixel} \quad (1)$$

Another power supply architecture is the mesh-based power supply architecture, as shown in Fig.2(b). Compared with the column power supply architecture, this architecture has a lower voltage drop, but its power routing becomes more complex. This power supply architecture is adopted in [9]. It is quite complex to get the expression for the voltage drop of the mesh-based power supply architecture. Thus, we can calculate the voltage drop of each pixel node through the method of modeling.

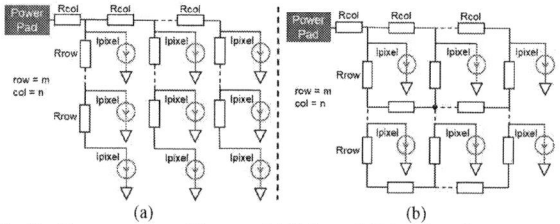

Fig. 2. Power supply architecture. (a) Column (b) Mesh-based

The following is an example to compare the voltage variation of the two power supply architectures. It is assumed that the scale of the pixel readout ASIC is a 128×128 array, Ipixel = 15μA, power supply = 3.3V and the number of power pads is 1. The routing resistance parameters are referred to in the layout of XPIXR2B in [9]. In the column power supply architecture, Rcol = 0.1Ω, Rrow = 1.5 Ω. In the mesh-based power supply architecture, Rcol = 0.75 Ω, Rrow = 1.5 Ω.

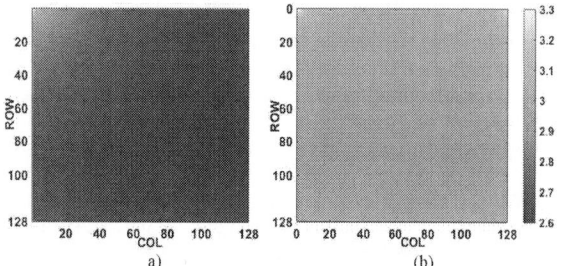

Fig. 3. Power voltage of the readout ASIC employing mesh-based power supply architecture. (a) one power pad. (b) two power pads

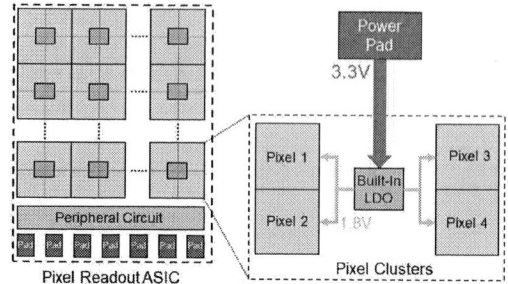

Fig. 4. LDO-based power supply architecture

Calculated using Eq. (1) and simulated through modeling, the variation among pixel supply voltage under the column power supply architecture can reach 1.77V, while the mesh power supply architecture is 0.61V as shown in Fig.4(a). Obviously, the voltage variation of pixels can be mitigated by increasing the number of power pads. As shown in Fig.4(b), the power voltage of each pixel in the mesh-based power supply architecture with two power pads is presented, and the maximum voltage variation is 0.31V. Since the power consumption is only 8 μW/pixel in [4] and the pixel array scale is only 40×50 in [9], these two power supply architectures are acceptable in [4] and [9].

The LDO-based power supply architecture is shown in Fig.4. The pixel pitch is 100μm. The power pad supplies power to the pixel clusters through built-in LDO. A pixel cluster is composed of four pixels. When the power supply voltage is 3.3V and the output of the pixel-level LDO is 1.8V, the tolerable voltage drop can reach 1.3V or even lower. In theory, this architecture can reduce the variation among pixel supply voltage to zero.

Of course, the LDO in the architecture shouldn't introduce much extra power consumption, while ensuring correct functionality in radiation environment. Therefore, a low-power and radiation-hardened LDO is essential.

III. PROPOSED RADIATION-HARDENED BUILT-IN LDO

A. The Proposed LDO Radiation-Hardening Method

Research indicates that the threshold voltage drift caused by TID can be neglected when the gate oxide thickness is less than 10nm. In the 180nm CMOS process, the gate oxide thickness is already less than 3.8nm[10]. Thus, more attention is paid to SET in design. It is generally agreed that the change amplitude caused by SET in the output voltage <5% and the transient pulse width is <10μs is acceptable[11].

One hardening method is the RC filter structure (RCF) to reduce voltage fluctuations at sensitive nodes [11], as shown in Fig.5(a). Although RCF has a simple structure, it introduces big capacitance to circuit nodes, which may impact circuit performance. Another approach is to accelerate the discharge of radiation-induced charge pulses by using a path to ground (PTG)[6], as shown in Fig.5(b). PTG affects neither the dc nor ac performance. However, this method requires additional layout structures and special techniques.

To achieve radiation hardening without significantly affecting circuit performance in CMOS process, this paper proposes a novel radiation-hardening structure. The novel radiation-hardening structure is shown in Fig.5(c). This structure can fast discharge the radiation-induced charge at the node，named Fast Discharge Structure (FDC). Its working

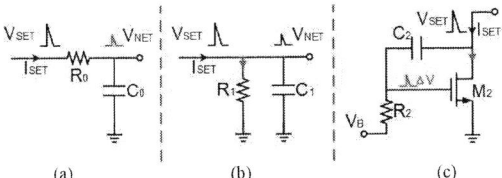

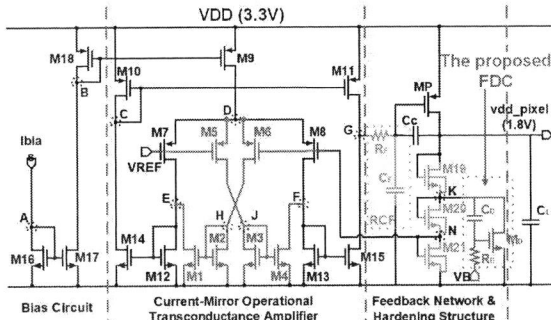

Fig. 5. Topological Diagram of Hardening. (a) RCF. (b) PTG. (c) This work

process is described as follows. When radiation-induced charges are not generated at the circuit node, the gate voltage of M_2 is V_B, which is relatively low compared to its threshold voltage Vth. Therefore, M_2 operates in the cut-off region, drawing only a few nA of leakage current. M_2 can be regarded as a high-impedance state, so it can be considered that the impact on the circuit node is very small. And C_2 is typically around 20fF, so its impact is also small for circuit nodes that are not sensitive to capacitance. In summary, the impact of this structure on circuit nodes is very small when it is not triggered.

When radiation-induced charges are generated at the circuit node, they cause a voltage pulse at the node, whose amplitude is related to the charge generated by radiation. This voltage pulse is coupled to the gate of M_2 through C_2, inducing a voltage change of ΔV at the gate of M_2. Easily, $V_B + \Delta V$ will exceed Vth of M_2, which turns M_2 on. In this state, we call this structure triggered. During this triggered state, M_2 rapidly discharges the radiation-induced charges to achieve radiation hardening.

Because the coupled voltage ΔV appearing at the gate of M_2 is only related to C_2 and the gate capacitance C_{gg}, and C_{gg} is small, thus, C_2 generally doesn't need a large value . In this design, a coupling capacitor of 20fF can achieve good results. For V_B, a fixed voltage generated by peripheral circuits (such as a bandgap reference) can be used. However, this method leads to significant variations in the leakage current of M_2 in the non-triggered state under PVT corners. Therefore, V_B can be implemented using a current mirror biasing scheme, which minimizes variations in the non-triggered leakage current of M_2 under PVT corners. The leakage current of M_2 in the non-triggered state can be determined by the current mirror. In this structure, R_2 is used to block V_B from the gate of M_2, preventing V_B from clamping the gate voltage of M_2.

B. The Proposed Radiation-Hardened Built-In LDO

The radiation-hardened built-in capacitor-less LDO proposed in this paper is shown in Fig.6. The LDO is composed of the Error Amplifier (EA), power transistor ($M13$), feedback network, Miller compensation capacitor (Cc), and radiation hardening structure.

The EA employs a current-mirror operational transconductance amplifier (OTA) structure, and a self-biased shunt current mirror structure[5] is added to achieve higher gain. This self-biased shunt current mirror structure aims to increase both transconductance and output resistance. The self-biased shunt current mirror structure is composed of $M1 \sim M6$, which can increase the transconductance and the output resistance. Assuming that the size ratio of transistors $M2 : M1$ is $1 : k$, $M3 : M4$ is $1:k$, $M12 : M14$ is $m : n$, and $M13 : M15$ is $m : n$, the gain of the OTA can be expressed as:

$$A_v = G_m R_{out} = \frac{m+2k}{m} g_{m7,8}\left(r_{11} // r_{15}\right) \quad (2)$$

where $g_{m7,8}$ is the transconductance of M7 or M8; r_{11} and r_{15} is the equivalent resistance of M11 and M15.

Fig. 6. The proposed radiation-hardened built-in capacitor-less LDO.

The feedback network is realized using diode-connected n-type transistors to minimize area. Additionally, DNW (Deep N-Well) devices are employed to prevent threshold voltage variations due to the body effect.

The radiation hardening structure consists of an RCF and an FDC. The RCF, composed of R_F and C_F, is used for radiation hardening of node G. The FDC, composed of C_D, R_D, and M_D, is employed for radiation hardening of node K. Not all nodes require radiation hardening. Whether a node needs radiation hardening can be determined through simulation. The double-exponential current source model is commonly used to simulate the charge pulses generated by SET. The time constants of the double-exponential current source are set to τ_1 = 10ps and τ_2 = 100ps, and the peak current of the double-exponential current source is set to I_p = 1mA, to simulate radiation with $LET < 50$MeV·cm²/mg. When radiation occurs at nodes A, B, C, D, H, and N, the variation of vdd_pixel does not exceed 3% of 1.8V, which is 54mV, so radiation hardening is not required. For nodes E, F, and J, although radiation-induced variations in vdd_pixel exceed 54mV, indirectly influence vdd_pixel by perturbing the gate of power transistor MP. Therefore, it is sufficient to harden node G, and nodes E, F, and J do not need to be hardened again.

For the gate of power transistor MP, even a slight variation can cause significant fluctuations in vdd_pixel. Therefore, node G needs to be hardened by RCF. Since this node is the main pole of the negative feedback loop, increasing the equivalent capacitance via RCF does not degrade loop stability. For node K , applying RCF would influence the dc of vdd_pixel. Therefore, FDC is used for this node. This structure allows the hardening design for node K to be achieved with a small impact on the dc operating point. The specific values of the components in the radiation hardening structure are as follows: R_F = 155kΩ, C_F = 215fF, C_D = 22.6fF, R_D = 31.1kΩ.

IV. POST-SIMULATION AND DISCUSSION

This power supply architecture is implemented using CMOS 180nm process. The total area of the radiation-hardened built-in LDO is 1618 μm², of which the radiation hardening structure RCF consumes 552 μm² and FDC consumes 138μm². Since the built-in LDO in this power supply architecture powers a pixel cluster composed of four pixels, the area of the built-in LDO is 404.6 μm²/ pixel. For a pixel with a pitch of 100μm, the built-in LDO consumes 4.1% of the area. Post-simulation shows that in typical condition, the total current consumed by the built-in LDO is 4.3 μA, which is 1.1 μA / pixel. The RCF consumes zero current, and the FDC consumes 10 nA, which is 2.5 nA/pixel.

979-8-3315-3918-4/25 $31.00 © 2025 IEEE

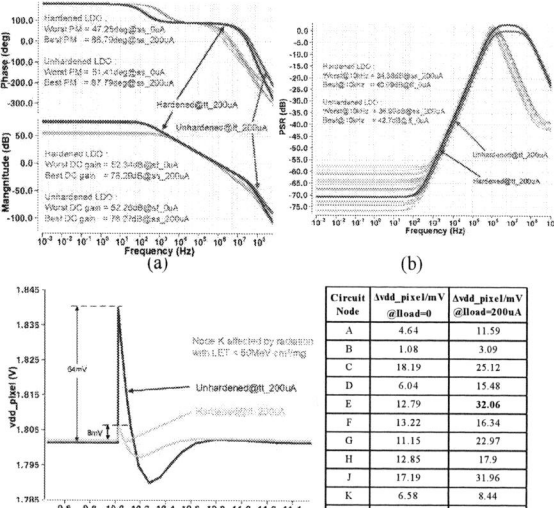

Fig. 7. The performance of the proposed LDO. (a) loop stability. (b) PSR. (c) transient of vdd_pixel. (d) the variation of vdd_pixel in radiation

Circuit Node	Δvdd_pixel/mV @Iload=0	Δvdd_pixel/mV @Iload=200uA
A	4.64	11.59
B	1.08	3.09
C	18.19	25.12
D	6.04	15.48
E	12.79	32.06
F	13.22	16.34
G	11.15	22.97
H	12.85	17.9
J	17.19	31.96
K	6.58	8.44
N	5.43	7.72

within the range of 1.7982V to 1.7989V (3-sigma range), so the voltage variation of pixels using this power supply architecture is sub-1mV.

Table I summarizes and compares some similar works. First of all, in a large array scale, the architecture in this paper can significantly reduce the worst variation of pixel. Besides, compared with other LDO-based power supply architectures, this paper introduces a lower extra current consumption and a low area consumption. Furthermore, the architecture in this paper is radiation hardened to ensure normal circuit functionality when hybrid pixel detector readout ASIC exposed to radiation in high-radiation environments.

V. CONCLUSION

In this paper, a power supply architecture with radiation-hardened built-in capacitor-less LDO for hybrid pixel detector readout ASIC is designed in CMOS 180nm process. In this architecture, the LDO powers a pixel cluster composed of four pixels. The radiation hardening structures adopt RCF and the proposed FDC structures, which consume lower area and power. Post-simulation shows that the voltage variation of pixels in this power supply architecture is sub-1mV. The built-in LDO consumes an area of 404.5μm^2/pixel and a current of 1.1μA/pixel. The output voltage variation of the LDO is less than 32.06mV under the radiation of LET < 50MeV·cm²/mg. The results prove that this power supply architecture can be benefit to the mitigation of the voltage variation of pixels and have enough radiation tolerance to ensure normal circuit functionality when exposed to radiation environments.

Fig.7(a) compares the loop stability of the built-in LDO before and after hardening. In typical condition, the loop gain changes from 55.53dB to 55.59dB, and the phase margin (PM) decreases from 66.42 deg to 62.43 deg, showing no significant variation. Across different process corners, the loop stability also shows negligible variation. Fig. 7(b) compares the PSRR performance of the LDO. In typical condition, the PSRR at 10 kHz changes from 39.86 dB to 37.25 dB, showing negligible variation. Moreover, across different process corners, the PSRR also shows negligible variation. Fig. 7(c) shows the transient waveform of vdd_pixel when node K is affected by radiation in typical condition. Simulation results indicate that the radiation hardening structure FDC can reduce the voltage variation of vdd_pixel caused by radiation at node K from 64 mV to 8 mV, with a recovery time of less than 1.5μs. Fig. 7(d) shows that the output voltage variation of the LDO is less than 32.06mV, when affected by radiation with LET < 50MeV·cm²/mg.

As mentioned above, the LDO-based power supply architecture can reduce the voltage variation of pixels to zero under ideal conditions. In practice, it is influenced by routing and other non-ideal factors. In order to minimize the impact of routing, the approach of making the routing from the LDO output to each pixel as same as possible is adopted during layout design. To evaluate the impact of the LDO's non-ideal factors on t the voltage variation of pixels, we performed 2000 times Monte Carlo simulations. The Monte Carlo simulation results show that the output voltage of this LDO fluctuates

REFERENCES

[1] M. Garcia-Sciveres, "Hybrid pixel readout integrated circuits," Nuclear Instruments and Methods in Physics Research Section A: Accelerators, Spectrometers, Detectors and Associated Equipment, vol. 1057, 2023.

[2] M. Li et al., "A charge-integration pixel readout chip features IR-drop effect mitigation by distributed LDOs," Journal of Instrumentation, vol. 17, no. 09, 2022.

[3] R. Szczygiel, P. Grybos, and P. Maj, "Design of Pixel Readout Integrated Circuits in Submicron Technology to Minimize the Mismatch Effects," (in English), Mixdes 2009: Proceedings of the 16th International Conference Mixed Design of Integrated Circuits and Systems, pp. 51-54, 2009.

[4] X. Llopart, M. Campbell, D. S. Segundo, E. Pernigotti, and R. Dinapoli, "Medipix2, a 64k pixel read out chip with 55 /spl mu/m square elements working in single photon counting mode," in 2001 IEEE Nuclear Science Symposium Conference Record (Cat. No.01CH37310), 2001, vol. 3, pp. 1484-1488 vol.3.

[5] J. Cheng, W. Gao, C. Yu, and X. Wu, "Design of a Pixel Readout ASIC Using Super Pixel Circuits With a Built-In LDO Regulator for Hybrid X-Ray Imaging Detectors," IEEE Transactions on Circuits and Systems II: Express Briefs, vol. 71, no. 8, pp. 3680-3684, 2024.

[6] A. K. Sutton et al., "An Evaluation of Transistor-Layout RHBD Techniques for SEE Mitigation in SiGe HBTs," IEEE Transactions on Nuclear Science, vol. 54, no. 6, pp. 2044-2052, 2007.

[7] M. Um, D. Ro, I. J. Chang, and H. M. Lee, "A Radiation-Hardened Readout Integrated Circuits for Sensor Systems," in 2020 IEEE International Conference on Consumer Electronics - Asia (ICCE-Asia), 2020, pp. 1-4.

[8] E. Monteil et al., "RD53A: a large scale prototype for HL-LHC silicon pixel detector phase 2 upgrades," in PoS, 2019, p. 157.

[9] Y. Wang et al., "Design of a Readout Chip for Pixel Silicon Detector With Event-Driven Readout Method," IEEE Transactions on Nuclear Science, vol. 72, no. 3, pp. 559-566, 2025.

[10] H. L. Hughes and J. M. Benedetto, "Radiation effects and hardening of MOS technology: devices and circuits," IEEE Transactions on Nuclear Science, vol. 50, no. 3, pp. 500-521, 2003.

[11] R. Yao et al., "Design-technology-co-hardening for voltage reference and linear voltage regulator based on bipolar technology," Microelectronics Reliability, vol. 147, 2023.

TABLE I. THE OVERALL OF COMPARISON OF THE READOUT ASICs

Readout ASIC	Medipix2 [4]	J.Inst 2022 [2]	TCASII 2024 [5]	This work
power supply schemes	Column	LDO-based	LDO-based	LDO-based
Process	CMOS 130nm	CMOS 130nm	CMOS 180nm	CMOS 180nm
Array scale	256×256	16×24	16×16	128×128
Pixel pitch	55μm	100μm	100μm	100μm
Worst variation among pixel supply voltage	92mV	NA	80μV	sub-1mV
Power consumption/pixel	8uW	34uW	54uW	28.98uW
Current consumption for power supply schemes	NA	8.31μA/pixel	1.89μA/pixel	1.1μA/pixel
Area consumption for power supply schemes	NA	NA	462μm²/pixel	404.5μm²/pixel
Radiation hardened	NO	NO	NO	YES

A CMOS Pixel with Gradient-Doped PPD and LOFIC for 1.7 ns Charge Transfer Time and 92 dB Dynamic Range

Tianjing Qiu [1,2], Jinglei Du [1,2], Junli Zhang [1], Peng Feng [2,3*], Jian Liu [2,3], Nanjian Wu [2,3], Liyuan Liu [2,3]

[1] School of Physical Science and Technology, Lanzhou University, Lanzhou 730000, P. R. China
[2] State Key Laboratory of Semiconductor Physics and Chip Technologies, Institute of Semiconductors, Chinese Academy of Sciences, Beijing 100083, China
3 Center of Materials Science and Optoelectronics Engineering, University of Chinese Academy of Sciences, Beijing 100049, China

* fengpeng06@semi.ac.cn

Abstract—This paper proposes a CMOS pixel with gradient-doped PPD and Lateral Overflow Integration Capacitor (LOFIC). The LOFIC architecture is used to improve the FWC of the pixel without increasing the PD doping concentration, which can further improve the charge transfer speed with gradient-doped PPD. The proposed pixel has been designed in a 0.18 μm CMOS process with layout size of $12\mu m^H \times 9\mu m^V$. The simulation results show that 1.7 ns charge transfer time and 92 dB dynamic range can be achieved.

Keywords—*CMOS Image Sensor, Lateral Overflow Integration Capacitor, gradient-profile doping, dynamic range, high-speed*

I. INTRODUCTION

With the accelerated development of industrial intelligence, medical precision and premium consumer electronics, the application area of CMOS image sensors (CIS) continues to expand. However, conventional high-speed CIS typically offer a dynamic range limited to 70 dB [1], which cannot meet the demand for high-dynamic-range (HDR) and high-speed imaging in high-end scenarios such as machine vision, automotive electronics, scientific instruments, and absorption imaging [2]. In CIS, Pinned-Photodiode (PPD) is widely used as the photon-sensitive device.

Fig. 1 illustrates the charge-transfer process within the PPD:

- (1): N-layer fringe charge is transferred into the channel by the built-in field at the edge of PPD and transfer gate through drift mechanism.

- (2)-(3) :Remaining charge is transferred into channel by diffusion mechanism.

Especially during stage (3), the charge gather at the point which has the lowest electrostatic potential. This charge must overcome a higher energy barrier. When the pin voltage of the PPD is too high, this barrier increases significantly, trapping more residual electrons within the photodiode. This phenomenon physically limits the full-well capacity (FWC), thereby constraining the dynamic range.

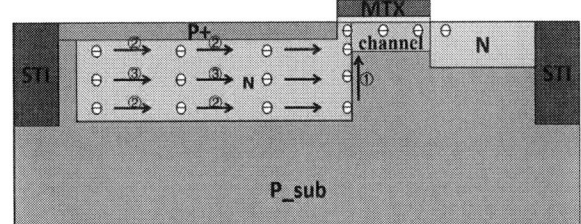

Fig. 1. Charge-transfer process within the PPD

In addition, the frame rate of CIS is limited by multiple factors. For burst high-speed CIS, the analog pixel signal is firstly saved in the on-chip memories and then read out, which breaks the speed limitation of the readout circuits [3]. So the transfer time of charge within the photodiode becomes one of the key limitation for achieving higher frame rates. However, the diffusion-dominated charge transport mechanism with slow speed in conventional large PPD designed for high-speed CIS severely limits the frame rate. If an electric field is built within the PPD by a gradient-doped method, the transfer speed of the charge can be significantly enhanced [4]; however, the FWC will be limited because of the low doping concentration.

This work proposed a novel pixel which combines the gradient-doped PPD structure with Lateral Overflow Integration Capacitor (LOFIC) structure. The gradient-doped PPD creates a lateral electric field through non-uniform doping to enable high-speed transfer of charge. The LOFIC structure expands dynamic range via overflow paths while improves the FWC of the pixel [5].

II. PROPOSED PIXEL DEVICE

A. Proposed LOFIC structure

Fig. 2 compares the conventional 4T pixel (a) with the proposed LOFIC pixel (b). The key enhancements include:

- Source-Follower (SF) capacitive boosting: In proposed pixel, the M_{SEL} is designed between the power and the Source-Follower (M_{SF}), when it is turned on, the gate-drain capacitance of the M_{SF} can couple to the floating diffusion (FD) node, raising the reset voltage at FD and expanding its voltage swing, thereby enhancing dynamic range.

This work is supported by the National Key Research and Development Program of China (2022YFB2804402), National Natural Science Foundation of China (62134004).

979-8-3315-3918-4/25 $31.00 © 2025 IEEE

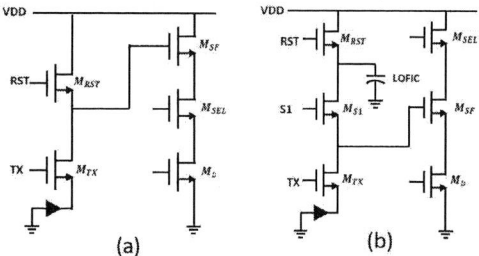

Fig. 2. (a)The conventional 4T pixel (b)The proposed LOFIC pixel

- The conversion gain (CG) and FWC can be designed independently: a LOFIC capacitor is integrated between the reset transistor M_{RST} and the transfer transistor M_{S1}. Switch transistor M_{S1} couples FD to LOFIC, enabling independent optimization of CG and FWC [6].

During exposure, the overflow charge moves from the saturated PD to FD through the transfer transistor M_{TX}. If FD saturates, excess charge flow into the LOFIC via the switch transistor M_{S1}. This allows the pixel to keep integrating after PD saturation, and significantly extends the effective dynamic range.

Fig. 3 and Fig. 4 present the operational timing diagram and potential diagram for the proposed pixel.

Fig. 3 shows the timing control process as follows:

- Simultaneously reset PD, FD, and LOFIC, then sample the low-gain (LG) reset signal (stage t1).

- Open M_{S1} and sample the high-gain (HG) reset signal (stage t2).

- Enable M_{TX} to transfer photo-generated charge and sample the HG pixel signal (stage t3).

- Simultaneously activate M_{TX} and M_{S1} to transfer residual charge, then sample the LG pixel signal (stage t4).

Under high-illumination conditions, pixel performance is dominated by shot noise, whereas in low-light environments, the input-referred noise at the floating diffusion (FD) node dominates. By relaxing the required FWC of the PD, the LOFIC structure reduces dark current and the pin voltage. Consequently, the energy barrier that photo-generated charge must overcome is lowered, shortening the charge transfer time, and improving charge-transfer efficiency. Meanwhile, the electric field produced by gradient-doped method accelerates charge transfer process, thereby eliminating image lag.

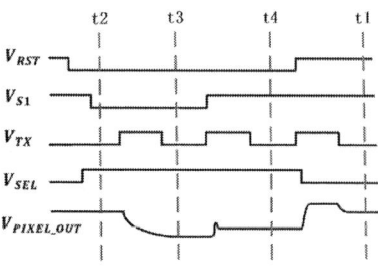

Fig. 3. The operational timing diagram

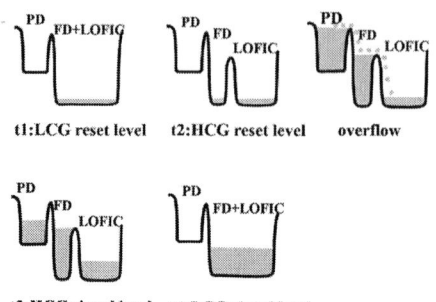

Fig. 4. The potential state diagram

B. Gradient doping

Fig. 5(a) shows the layout design of the proposed gradient-doped pixel, while Fig. 5(b) shows its cross-sectional view. Two ion implant steps create an N^+/N^- non-uniform doping profile, this builds a continuous potential gradient along the charge transfer path. The lateral electric field points from high-concentration to low-concentration regions. It rapidly drives electrons via drift mechanism.

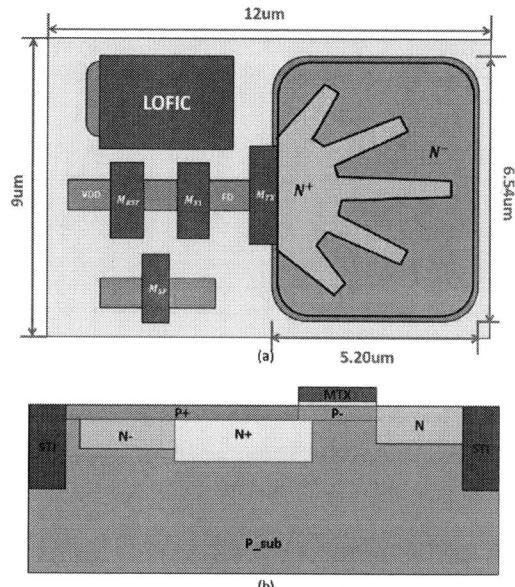

Fig. 5. (a) Pixel layout (b)Cross-sectional view of the gradient-doped PPD

III. SIMULATION RESULTS

This pixel has been designed and simulated in a 0.18 μm CMOS process. Fig. 6 presents the FWC and electron transfer time of the proposed pixel. The photodiode achieves a FWC of 45.7 ke$^-$. With the gradient-doped design, the electron transfer time of PPD is reduced to 1.7 ns while the residual charge is less than 0.01e$^-$, which significantly enhances charge transfer efficiency.

In this pixel, the capacitor is implemented by a MOS transistor. Fig. 7 shows the capacitance simulation of the MOS capacitor. Moreover, the key aspect of this pixel lies in the design of the overflow path, which requires precise gate-voltage control.

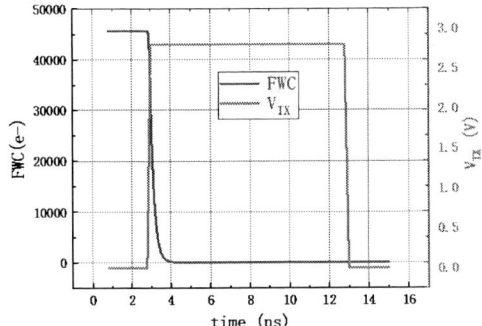

Fig. 6. The full-well capacity and transfer time of Pixel

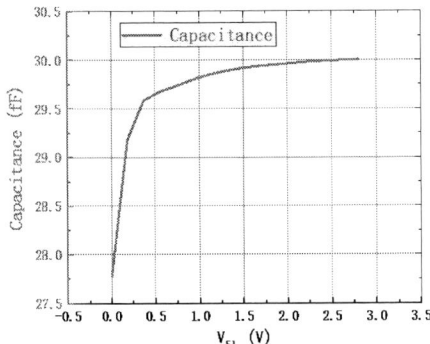

Fig. 7. Capacitance of the MOS capacitor

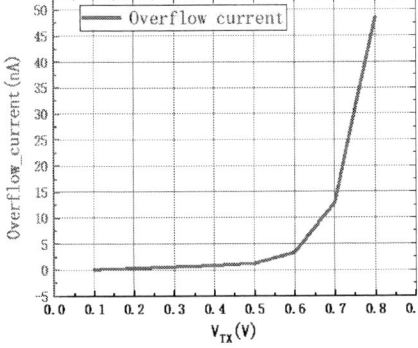

Fig. 8. The dependence of overflow current on TX gate voltage

Fig. 8 shows the dependence of overflow current on M_{TX} voltage. The overflow path designed in pixel guarantees that all photo-generated electrons are accumulated synchronously during integration. The bias voltage of M_{TX} and M_{S1} are designed to be 0.6V and 0.9V, to balance the saturation characteristics and overflow efficiency of PD and FD.

Table I summarizes the performance comparison between this work and several other image sensors.

TABLE I. COMPARISON OF SPECIFICATIONS AND PERFORMANCES BETWEEN SEVERAL HDR CIS

		This work	*[1]*	*[6]*
Process technology		0.18μm CMOS with PPD(FSI)	65 nm CMOS with PPD(BSI)	0.18μm CMOS with PPD(FSI)
Power supply voltage (V)		3.3	2.8	3.3
Pixel size		$12\mu m^H \times 9\mu m^V$	2.8μm	16μm
Fill factor		31.5%	*N/A*	52.8%
Charge Transfer Time(ns)		1.7	*N/A*	*N/A*
Conversion Gain (μV/e^-)	HCG	61.6	160	*N/A*
	LCG	5.38	10	*N/A*
FWC (e^-)	FD	29.2k	7k	17.8k
	LOFIC1	334k	120k	509k
	LOFIC2	*N/A*	*N/A*	11.4M
Readout noise FD (e^-_{rms})		8.64	*N/A*	3.5
Dynamic range (dB)		92	>100	>120

IV. CONCLUSION

The proposed high-speed LOFIC CMOS image sensor achieves 92 dB dynamic range with the charge transfer time of only 1.7ns, through integration of gradient-doped PPD and LOFIC technologies. The sensor shows promising applications in various fields such as machine vision, automotive, analytical instruments and absorption imaging.

REFERENCES

[1] Takayanagi I, Miyauchi K, Okura S, Mori K, Nakamura J, Sugawa S. A 120-ke- Full-Well Capacity 160-μV/e- Conversion Gain 2.8-μm Backside-Illuminated Pixel with a Lateral Overflow Integration Capacitor. Sensors. 2019; 19(24):5572.

[2] S. Iida et al. "A 0.68e-RMS random-noise 121dB dynamic-range subpixel architecture CMOS image sensor with LED flicker mitigation, " in IEDM Tech. dig. dec. 2018, pp. 10.2.1-10.2.4.

[3] Yue X, Fossum E R. Design and characterization of a burst mode 20 mfps low noise cmos image sensor[J]. Sensors, 2023, 23(14): 6356.

[4] Zhongxiang C, Quanliang L. High-speed CMOS image sensor and its application[J]. Journal of Tianjin University (Science and Technology), 2021, 54(4): 426-434.

[5] S. Sugawa et al. "A 100dB dynamic range CMOS image sensor using a lateral over flow integration capacitor," in IEEE ISSCC Dig. Tech. Papers, 2005, pp. 352-353.

[6] Y. Fujihara, M. Murata, S. Nakayama, R. Kuroda and S. Sugawa, "An Over 120 dB Single Exposure Wide Dynamic Range CMOS Image Sensor With Two-Stage Lateral Overflow Integration Capacitor," in IEEE Transactions on Electron Devices, vol. 68, no. 1, pp. 152-157, Jan. 2021, doi: 10.1109/TED.2020. 3038621.

Stacked 2D materials Nanopore Sensors

Candong Zhao[1,2], Qinjie Pan[1], Guangyi Yang[1], Peng Cheng[1], Fuwei Zhuge[3], Yuhui He*[1,2]

[1]School of Integrated Circuits, Huazhong University of Science and Technology, Wuhan, China
[2]Shenzhen Loop Area Institute, Shenzhen, China
[3]School of Materials Science & Technology, Huazhong University of Science and Technology, Wuhan, China

* Email: zcd@hust.edu.cn, heyuhui@hust.edu.cn

Abstract—This paper presents a solid-state nanopore sensor based on stacked two-dimensional (2D) materials, designed to enhance the sensing accuracy for DNA and protein detection compared to conventional solid-state nanopores. Traditional solid-state nanopores utilize an applied voltage across the nanopore to provide an electrophoretic driving force for target biomolecules while simultaneously measuring the resulting translocation current signals. By incorporating the stacked 2D material nanopore design, this work introduces an additional thermophoretic driving force acting on the target biomolecules. This innovation provides the solid-state nanopore sensor with an extra manipulation mechanism, thereby enhancing the signal-to-noise ratio (SNR) of the sensing signals and enabling the sensor to acquire multi-modal sensing data.

Keywords—Solid-state nanopores, Stacked 2D materials, Thermophoresis

I. INTRODUCTION

Nanopore sensors possess characteristic dimensions comparable to biological molecules such as DNA, proteins, and viruses, enabling the analysis of individual biomolecules. The sensing principle is illustrated in Figure 1: when a biomolecule translocate through the nanopore, it causes a physical blockade of the pore. Each biomolecule passing through the nanopore generates a distinct blockade signal. Consequently, nanopore sensors achieve an exceptionally low detection limit, reaching the attomolar (aM) level. Solid-state nanopore sensors offer advantages including compatibility with complementary metal-oxide-semiconductor (CMOS) fabrication processes, high stability, and low cost. They also hold potential for fabrication into high-density electronic arrays.

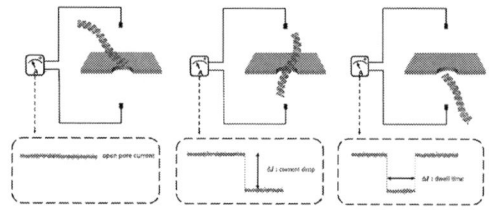

Fig. 1. Schematic of Nanopore Sensor Operating Principle

DNA sequencers based on biological nanopores, utilizing the same fundamental sensing principle, have achieved commercialization. However, solid-state nanopores face two critical challenges on the path towards practical application: 1. achieving spatial resolution down to several bases, and 2. reducing the translocation velocity of DNA molecules to improve the signal-to-noise ratio (SNR). This work introduces a thermoregulation layer onto a conventional silicon nitride (SiN) nanopore. This layer serves to establish a temperature gradient within the nanopore region. The thermoregulation layer controls the translocation velocity of biomolecules and provides an additional control parameter. Constructed from stacked two-dimensional (2D) materials, the thermoregulation layer exploits their ultrathin nature and thermal isolation properties. This enables the creation of extremely high temperature gradients within a minute volume. Consequently, the thermophoretic driving force becomes comparable to and competes with the electrophoretic driving force, collectively determining the motion of the biomolecules.

II. SYSTEM OVERVIEW

Conventional solid-state nanopore systems employ an electrical driving mechanism to propel target biomolecules through the nanopore[1]. Most biomolecules acquire charge in solution due to hydrolysis (e.g., DNA carries a negative charge). By applying a voltage across the nanopore, an electric field is established within the solution. This electric field drives charged biomolecules to move from one side of the nanopore to the other. When a biomolecule passes through the nanopore, its physical blockade of the pore alters the overall electrical resistance of the system. This manifests as a measurable blockade current. The nanopore system infers the properties of the biomolecule by analyzing changes in this current.

Traditional nanopore systems achieve both the driving of biomolecules through the nanopore and the acquisition of translocation current signals by applying a voltage across the nanopore and measuring the resulting current. However, current solid-state nanopore systems require biomolecules to translocate as slowly as possible to enable the use of lower current sampling rates (which inherently generate less noise). They also require the blockade current signal to be as large as possible to improve the signal-to-noise ratio (SNR). Nevertheless, increasing the voltage across the nanopore strengthens the electric field. This larger electrophoretic driving force accelerates the motion of the biomolecules, which is detrimental to acquiring data with higher SNR. To address this, we propose the use of stacked two-dimensional (2D) materials as a thermoregulation layer to independently modulate the translocation velocity of biomolecules.

A. System Structure

The operational system of the stacked two-dimensional (2D) material nanopore sensor is illustrated in Figure 2. Firstly, consistent with conventional solid-state nanopore sensors, a voltage is applied across the nanopore to measure the ionic current. This simultaneously applies an electrophoretic force on the target biomolecules and acquires the translocation signal generated as they pass through the pore. Secondly, we direct a laser beam onto the nanopore to induce localized heating while simultaneously monitoring the nanopore's temperature. In the exemplary configuration, the nanopore structure consists of graphene/molybdenum disulfide

(MoS$_2$)/tungsten diselenide (WSe$_2$)/silicon nitride (SiN). We utilize a Raman spectrometer to direct a red laser beam, selectively heating the graphene layer. The stacked 2D material architecture enables the creation of a sufficient temperature gradient within the few-layer stack. Concurrently, the temperature at the nanopore is measured by analyzing the Raman spectral signal.

Biomolecules in solution are subjected to electrophoretic and thermophoretic forces. The electrophoretic force originates from an externally applied voltage and exhibits a fixed direction. The thermophoretic force arises from temperature gradients: a positive temperature gradient (increasing temperature) extends from the upper chamber to the nanopore, while a negative temperature gradient (decreasing temperature) spans from the nanopore to the lower chamber. Owing to the design of stacked 2D materials and laser heating, the temperature gradient distribution—like the electric field—is highly nonuniform, with significantly stronger temperature gradients and electric field intensities near the nanopore than in other regions. For DNA, which is negatively charged in solution and possesses a negative Soret coefficient (Fig.2), the molecule experiences an electrophoretic driving force F_E and a thermophoretic resistive force F_T when approaching the nanopore. After translocation through the pore, the thermophoretic force reverses direction and cooperates with the electrophoretic force to propel DNA away from the nanopore.

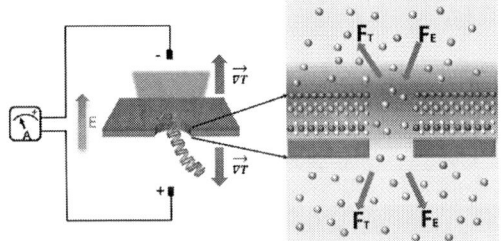

Fig. 2.   Schematic of the Stacked 2D Material Nanopore

The fabrication process flow for the stacked two-dimensional (2D) material nanopore device is illustrated in Figure 3. First, the window region is prepared via backside ultraviolet (UV) photolithography followed by reactive ion etching (RIE) to remove the silicon nitride (SiN) layer, exposing the underlying silicon substrate. Subsequently, potassium hydroxide (KOH) wet etching is employed to etch the silicon, thereby exposing the thin SiN membrane window. Next, two-dimensional (2D) materials are transferred layer-by-layer using polydimethylsiloxane (PDMS) stamping to construct the stacked 2D material layer, which serves as the thermoregulation layer. Finally, the nanopore pattern is defined using electron beam lithography (EBL), and the nanopore is etched through the stack via RIE. Following photoresist removal, the stacked 2D material nanopore chip is obtained.

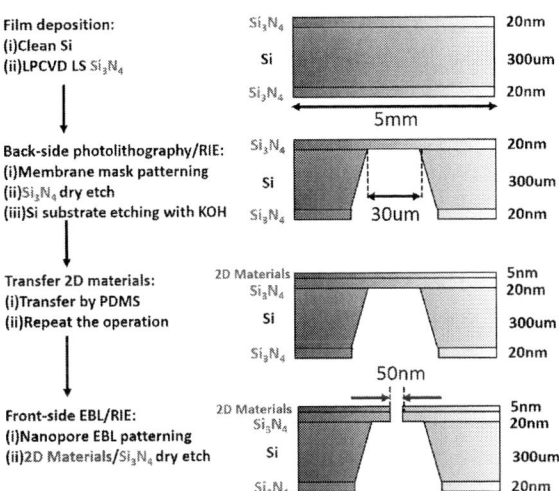

Fig. 3.   Stacked 2D Material Nanopore Fabrication Flow

B. Temperature Gradient Construction and Functionality

Compared to conventional solid-state nanopore sensors, our designed stacked two-dimensional (2D) material nanopore sensor incorporates an additional input control parameter. This enables simultaneous modulation of biomolecule motion within the solution environment by adjusting both the voltage applied across the nanopore and the laser heating power. Due to the nanopore's structure, the electrical resistance within the nanopore region significantly exceeds that of the adjacent chambers. This results in a substantial voltage drop near the nanopore, generating a sufficiently strong electrophoretic driving force to capture biomolecules. To ensure the temperature field exerts a comparable influence to the electric field, an equally substantial temperature gradient is required. We implemented two key design strategies to establish a sufficiently large temperature gradient near the nanopore: firstly, selective heating of 2D materials – graphene possesses advantageous properties including a zero bandgap, ultrathin structure, and stable laser absorption efficiency; secondly, exploitation of low interlayer thermal conductivity – the stacked 2D layers are bonded via van der Waals forces, which drastically suppress interlayer heat transport.

Based on prior research[2], the heat source within the stacked 2D structure is a monolayer 2D material, leveraging its atomic-scale thickness and stable interaction with laser irradiation. For instance, a red light laser can selectively heat a single graphene layer. Due to the monolayer's minuscule volume, minimal laser power is required to elevate it to a high temperature. Furthermore, heat dissipation from a monolayer is highly constrained. The extremely small solid-state heat source suppresses blackbody radiation and convective heat transfer. Most heat transfers to adjacent 2D layers and the surrounding solution via phonon-mediated heat transfer. However, the atomically smooth surfaces and absence of chemical bonding between layers create a very high interfacial thermal resistance[3], severely impeding phonon transport. Consequently, the stacked 2D structure enables the creation of temperature gradients reaching tens of degrees Celsius across nanometer-scale interlayer distances, fulfilling the nanopore sensor's requirement for a significant thermophoretic force.

To accurately monitor the nanopore temperature, we utilize the Raman signal to determine the temperature of the

2D materials. The Raman laser serves the dual purpose of heating the 2D material and eliciting its Raman spectrum. By analyzing this Raman spectral signal, we can precisely determine the temperature of the irradiated 2D material layer. Moreover, leveraging the distinct spectral signatures of different 2D materials, we can measure temperatures across individual layers within the stack, thereby directly mapping the temperature gradient within the nanopore region. The Raman spectrum of monolayer molybdenum disulfide (MoS_2) is presented in Figure 4. The material's temperature rises significantly with increasing laser power.

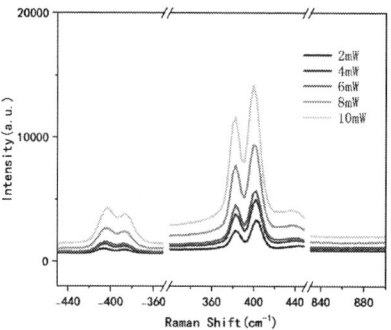

Fig. 4. Power-Dependent Raman Spectra of Single-Layer MoS₂

As the heating laser power or ambient temperature changes, the temperature of the two-dimensional (2D) material correspondingly varies. This is reflected in the Raman spectrum as a shift in the peak positions of the Raman modes. For instance, the Raman peak positions of monolayer molybdenum disulfide (MoS_2) shift as the temperature increases, as depicted in Figure 5. Conversely, we can utilize this shift in the Raman peaks to monitor the temperature variation of the monolayer MoS_2 material. This allows us to infer the temperature at the nanopore, thereby enabling precise thermal control of the nanopore to the desired temperature.

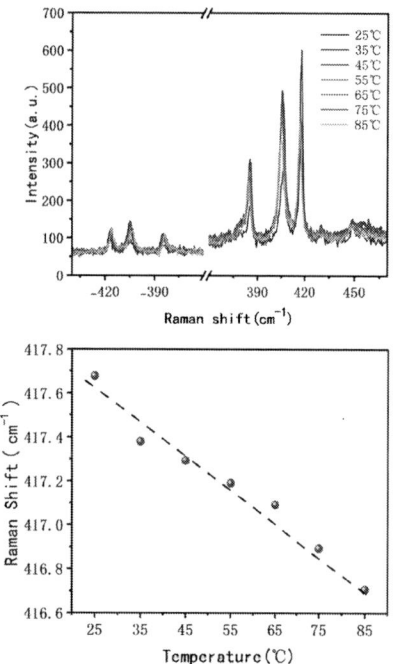

Fig. 5. Temperature-Dependent Raman Spectra of Single-Layer MoS₂

To extract temperature information at the nanopore from the Raman signal, we employ temperature-dependent Raman experiments to establish the relationship between the Raman peak positions of the 2D material and temperature. The material's temperature is determined by both the ambient temperature and the laser power. To obtain more accurate material temperature readings, we utilize low-power laser excitation during signal acquisition to minimize laser-induced heating. The entire solid-state nanopore chip is uniformly heated using a temperature-controlled stage. At each stabilized temperature point, Raman signals are acquired using long integration times. By processing the acquired Raman spectra, the correlation between Raman peak positions and material temperature is established. In subsequent heating experiments, the measured Raman signal under specific heating laser powers provides the corresponding Raman peak positions. These positions are then used to deduce the material's temperature, thereby obtaining temperature information within the nanopore region.

III. TWO-DIMENSIONAL MATERIAL NANOPORE SENSORS

Compared to biological nanopores, which have achieved commercial sequencing, solid-state nanopores, particularly those based on two-dimensional (2D) materials, can now be fabricated with dimensions comparable to their biological counterparts. However, the absence of motor proteins – the protein structures responsible for regulating translocation velocity – results in DNA translocation speeds through solid-state nanopores being nearly three orders of magnitude faster than through biological nanopores[5]. This excessive translocation velocity drastically reduces the dwell time of each nucleotide within the nanopore. Consequently, extremely high sampling frequencies are required to ensure signal acquisition for each nucleotide as it traverses the pore. Unfortunately, such high sampling frequencies introduce significant additional noise, thereby diminishing the signal-to-noise ratio (SNR) of the nucleotide sensing signal. This degradation ultimately hinders the accurate reconstruction of the nucleotide sequence from the recorded ionic current trace. As illustrated in Fig. 6, the characteristic signal from solid-state nanopores exhibits a low SNR due to the rapid molecular translocation, rendering it currently impossible to resolve individual nucleotides (e.g., DNA, RNA).

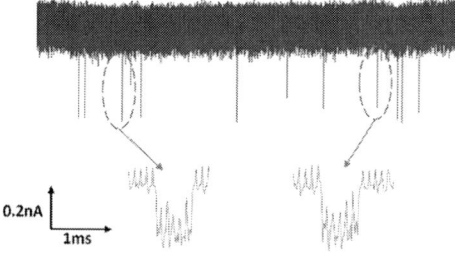

Fig. 6. Schematic of Solid-State Nanopore Translocation

Biological nanopores, the foundation of next-generation commercial sequencing technology, also utilize an electric field to drive DNA translocation and read the signal[6]. The reason biological nanopores do not suffer from excessively rapid DNA movement is that they typically incorporate a DNA-handling protein called a motor protein at the cis entrance of the pore. This motor protein actively slows down

979-8-3315-3918-4/25 $31.00 © 2025 IEEE

the velocity imparted to the DNA by the strong electric field force, enabling DNA to translocate through the pore at a significantly reduced speed. However, integrating such additional control structures onto solid-state nanopores to regulate biomolecular motion presents extreme fabrication challenges due to the stringent dimensional and thickness requirements of the nanopore itself. Furthermore, approaches that increase the nanopore aperture or thickness to accommodate control structures inevitably compromise the spatial resolution and overall sensing capabilities of the nanopore.

Following a comparative analysis of several external forces, we have selected thermophoresis to counteract the excessive electrophoretic force[7]. This approach aims to enhance the translocation signal amplitude while simultaneously reducing the translocation velocity of biomolecules. Two-dimensional (2D) material nanopores inherently represent a leading contender for ultrathin nanopore platforms, eliminating the need for introducing additional complex structures. By simply incorporating laser heating of the 2D material during testing, an intense, highly localized temperature gradient can be generated near the nanopore. This gradient induces a thermophoretic force on biomolecules that is sufficient to counteract the electrophoretic force. Consequently, the nanopore sensor transitions from a two-terminal device controlled solely by the voltage bias across the pore to a three-terminal device governed by both the transmembrane voltage and the laser heating power. The introduction of this additional control terminal not only partially emulates the function of motor proteins in biological nanopores by regulating biomolecular motion but also unlocks the potential for acquiring richer sensing signals. For instance, the motion of DNA within the temperature field can be characterized by its Soret coefficient. Crucially, the Soret coefficient can be modulated by adjusting the solution composition, enabling control over both the magnitude and the direction of the thermophoretic force. By strategically manipulating the electric field strength, laser heating power, and solution composition, a significantly broader spectrum of biomolecular sensing information becomes accessible.

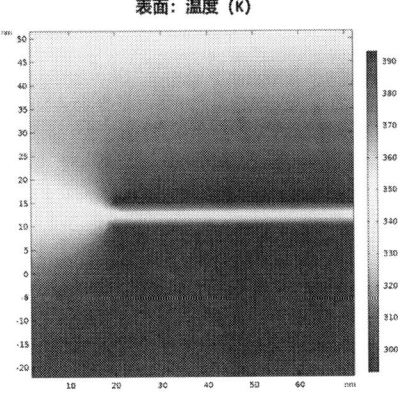

Fig.7. Schematic of Solid-State Nanopore Translocation

Previous studies have demonstrated that adjacent layers of two-dimensional (2D) materials in air can develop temperature differences on the order of tens of degrees Celsius under selective laser heating[8]. In an aqueous solution environment, the thermal conductivity of the solution is higher than that of air, leading to a slight reduction in the temperature difference between adjacent 2D material layers. However, due to the high thermal conductivity of the substrate, which accounts for the dominant heat dissipation pathway, and the highly localized heat source provided by the single-layer 2D material resulting in minimal lateral heat diffusion, a substantial temperature gradient persists within the nanopore region. Temperature simulations of the nanopore region, presented in Fig. 7, confirm the existence of a significant temperature gradient. This gradient remains sufficiently large to exert a measurable influence on the motion of biomolecules.

IV. CONCLUSION

This paper presents a novel nanopore sensor based on stacked two-dimensional (2D) materials. The sensor innovatively proposes the concept of utilizing stacked 2D materials to achieve an extremely high localized temperature gradient. This approach evolves the conventional solid-state nanopore sensor, which is controlled solely by a two-terminal voltage bias, into a three-terminal device regulated by both a temperature gradient and an electric field. This dual-control mechanism enables the simultaneous reduction of the translocation velocity of target biomolecules and the enhancement of the detection signal amplitude, thereby significantly improving the sensor's signal-to-noise ratio (SNR).

ACKNOWLEDGMENT

This work was supported by the National Key Research a nd Development Program of China (No. 2023YFB4502200), Natural Science Foundation of China (Nos. 92164204 and 6 2374063) and HUST 2024JCYJ008.

REFERENCES

[1] M. Graf, M.Lihter et al. "Fabrication and practical applications of molybdenum disulfide nanopores" Nature Protocols, vol.4, pp.1130-1168,2019.

[2] S. Vaziri, E.Yalon et al. "Ultrahigh thermal isolation across heterogeneously layered two-dimensional materials" Science Advances, vol.5, no.8, eaax1325,2019.

[3] S.E.Kim, F.Mujid et al. "Extremely anisotropic van der Waals thermal conductors" Nature, vol.597, pp. 660-665,2019.

[4] M.Kuball, J.W.Pomeroy et al. "A Review of Raman Thermography for Electronic and Opto-Electronic Device Measurement With Submicron Spatial and Nanosecond Temporal Resolution" IEEE Transactions on Device and Materials Reliability, vol.16, no.4, 2016.

[5] S.Carson, M.Wanunu. "Challenges in DNA motion control and sequence readout using nanopore devices" Nanotechnology, vol.26, no.7,074004 ,2015.

[6] Y.Wang, Y.Zhao et al. "Nanopore sequencing technology, bioinformatics and applications" Nature Biotechnology, vol. 39,pp.1348-1365 ,2021.

[7] J.Li, L.Lin et al. "Opto-Thermophoretic Tweezers and Assembly" Journal of Micro and Nano-Manufacturing, vol.6, no.4,2018.

[8] R.Zhang, L.Gan et al. "Extreme Thermal Insulation and Tradeoff of Thermal Transport Mechanisms between Graphene and WS2 Monolayers" Advanced Materials, vol.36, no.21, 2313753,2024.

On-chip Contact Angle Sensor Using Coplanar Capacitors for Digital Microfluidic Systems

Akira Tsuchiya, Hayato Fukui, Tsubasa Furuta, Toshiyuki Inoue, Keiji Kishine

Dept. Electronic Systems Engineering, The University of Shiga Prefecture

Hikone-shi, Shiga, Japan

tsuchiya.a@e.usp.ac.jp

Abstract—This paper proposes a capacitive sensor for digital microfluidic systems. In digital microfluidic systems, EWOD (Electro-Wetting On Dielectric) is used to move liquid droplets. The contact angle of liquids and the substrate is essential for EWOD, but it is difficult to measure the contact angle. We developed a contact-angle estimation method using two coplanar capacitors. The proposed method is verified by electromagnetic simulations and measurement with scaled models.

Index Terms—microfluidics, capacitive sensor, contact angle

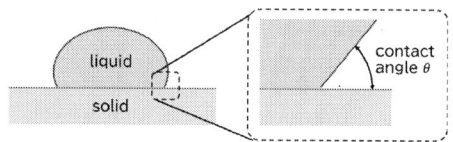

Fig. 1. Contact angle of droplet.

I. INTRODUCTION

Digital microfluidics is a system to control (move, divide, mix) tiny liquid droplets on small substrates or chips [1]. It is also called "Lab-on-a-Chip", and automated experiments with tiny amount of test reagents is highly expected especially in chemical, biomedical, and so on. Many digital microfluidic systems use EWOD (Electro-Wetting On Dielectric) to move droplets [2], [3]. By applying electric field to a droplet on the substrate, the shape of droplet changes. The shape change generates moving force. So EWOD can control moving forces by electrical signals. It means that digital microfluidics requires enough change of shape to move droplets. If the change is too small, they cannot move droplets. The major figure of the change of shape is contact angle. To realize higher contact angle, digital microfluidics employs water-repelling material or coating. However, almost one way to measure the contact angle is imaging analysis with camera [4]. If small sensor is developed, it contributes small and low-cost digital microfluidics.

To realize an integrated droplet shape measurement system, the shape sensor should be small enough to integrate, and flat shaped not to disturb droplet movement. To satisfy these requirements, we propose to use coplanar capacitor. It can be small and flat shape. Sensing liquid by capacitance is widely used, for example water-depth sensor [5]. We extend this idea to estimate contact angle between droplet and substrate by two coplanar capacitors. Capacitive water-depth sensor can sense water existence but cannot sense the shape of water. We utilize estimation error depending on the contact angle and the shape of electrode. The proposed method estimates the contact angle from two capacitors with different electrode widths. The proposed sensor is verified by electromagnetic simulations and measurements of scaled model. The experimental results show

This work was supported by JSPS KAKENHI (23K03959).

that the proposed sensor can estimate the contact angle from 50 degrees to 130 degrees with about 10-degree error. The accuracy is lower compared with imaging analysis by camera; however this sensor contributes to realize integrated contact-angle measurement in digital microfluidic systems.

II. CONTACT ANGLE ESTIMATION BY CAPACITORS

As a measure of the shape of droplet, we focus on contact angle shown in Fig. 1. Contact angle is the angle between the liquid and the substrate which is determined by the surface tension. EWOD lowers the contact angle by electric field, so the contact angle should be large. The contact angle depends on the surface tension of the droplet, so the contact angle can change during experiments by mixing with test reagents. Also, if water-repelling coating is used to have higher contact angle, the coating can degrade during experiments. If the contact angle becomes lower, contamination can occur due to incomplete move or divide. Then, the experiment fails. Therefore, contact angle monitoring is desired in digital microfluidic systems.

Capacitance is widely used for electrical sensing for liquid, because water has high relative permittivity, about 80. Fig. 2 shows a mechanism of a capacitive water depth sensor. Water makes capacitance larger, so this sensor can estimate the edge of water from the capacitance value. The water depth sensing assumes 90-degree angle between the water and the electrodes. If the contact angle is not 90 degrees, the position of the water edge estimated from the capacitance changes as shown in Fig. 3. We utilize this estimation error to estimate the contact angle. The error between the actual edge and the estimated edge depends on the width of the electrodes. Thus, we can estimate the contact angle from two different estimated positions as shown in Fig. 4. Fig. 5 shows the structure of two coplanar capacitors.

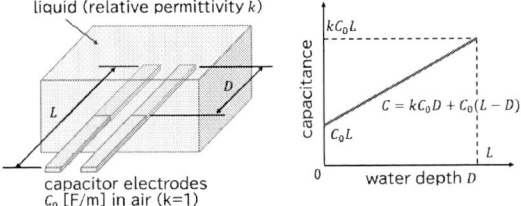

Fig. 2. Water depth sensing by capacitor.

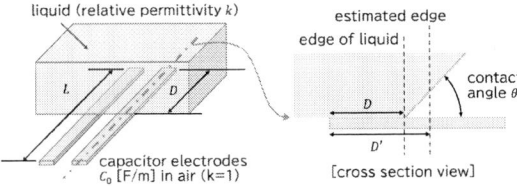

Fig. 3. Impact of contact angle on estimated edge position.

III. VERIFICATION

This section shows some experimental results. First, verification by electromagnetic simulation is shown. Then, measurement results with scaled model are demonstrated.

A. Electromagnetic Simulation

We evaluate the proposed sensor by a 3D electromagnetic simulator [6]. For on-chip integration, we assume an 180-nm CMOS process and the cross section is shown in Fig. 6. Since the thickness of each layer is fixed by the fabrication process, the design parameters are the width and the spacing of the small capacitor and the large capacitor. Fig. 7 shows the actual contact angle versus the estimated contact angle. The estimation accuracy depends on the size of electrodes

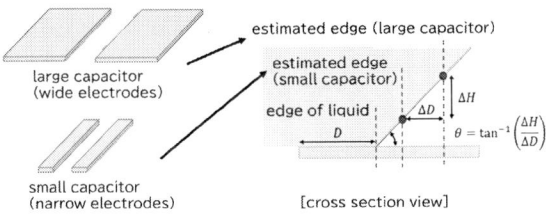

Fig. 4. Estimation of the contact angle by two capacitors.

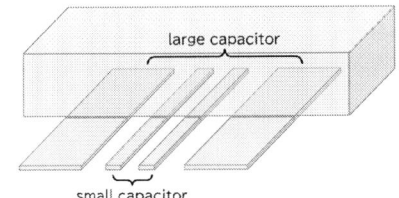

Fig. 5. Structure of two capacitors for contact angle estimation.

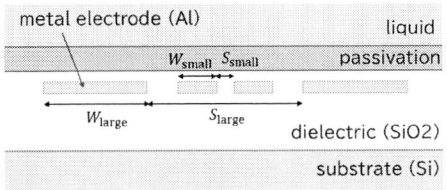

Fig. 6. Structure of on-chip capacitors (cross section view).

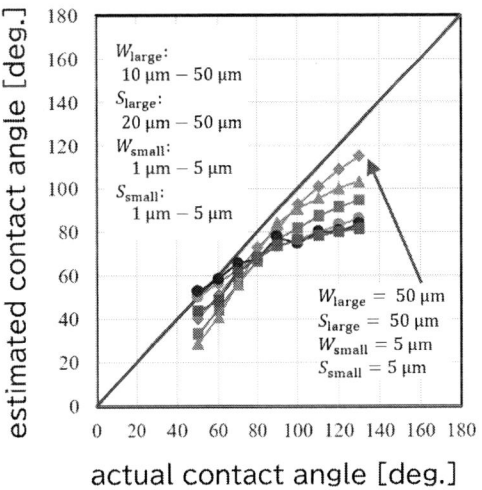

Fig. 7. Simulation results with various electrode sizes.

and the thickness of electrode, dielectric, and passivation. The estimation error is about 10 degrees in the best case. As shown in Fig. 7, the estimated contact angle changes monotonically against the actual contact angle. Thus, the proposed sensor can estimate the contact angle electrically.

B. Measurement of Scaled Model

It is difficult to measure the proposed sensor in IC, because it is hard to prepare droplets with various contact angle. So, we fabricated a scaled model by a PCB (Printed Circuit Board). To change the contact angle, dedicated water containers are created by a 3D printer. As shown in Fig. 8, the containers with various angled walls are made to reproduce a certain contact angle. The capacitance is measured by an LCR meter as shown in Fig. 9. Fig. 10 shows the measurement results. The results show that the proposed method can estimate the contact angle within about 10-degree error. We also tested a low permittivity liquid (food oil). As shown in Fig. 10, the proposed method is valid for low permittivity liquid.

Please note that the estimation accuracy of this measurement is higher than that of electromagnetic simulation because the dielectric covering the electrode is thin. So, the sensitivity of capacitance against the liquid shape is higher than electromagnetic simulation for on-chip electrodes.

979-8-3315-3918-4/25 $31.00 © 2025 IEEE

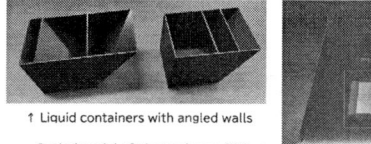

↑ Liquid containers with angled walls

Scaled model of electrodes on PCB →

Fig. 8. Liquid containers and scaled model on PCB.

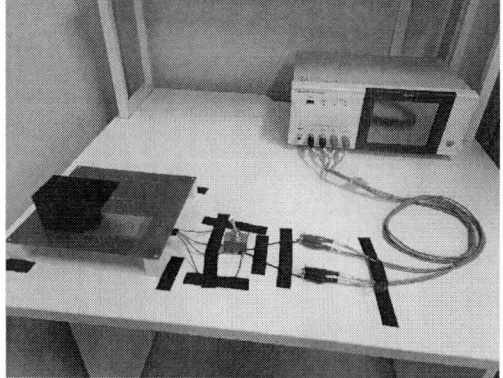

Fig. 9. Measurement equipment.

IV. CONCLUSION

This paper proposes a contact-angle sensor by two coplanar capacitors. Though the contact angle of droplets is an essential factor for digital microfluidic systems, there is no electrical sensor to monitor the contact angle. We proposed a capacitive contact angle sensor. The proposed sensor is small and flat shaped, so it can be integrated into digital microfluidic systems. The proposed sensor was verified by electromagnetic simulation and measurement by a scaled model. The experimental results show that the estimation error is about 10 degrees from 50 degrees to 130 degrees. The accuracy is lower

than imaging analysis using camera; however this sensor is the first contact angle sensor which can be integrated in standard CMOS ICs.

As a future work, we fabricate the sensor with capacitor sensing circuits and demonstrate the proposed sensor integrated in IC.

REFERENCES

[1] R. Fair, "Digital microfluidic chips for chemical and biological applications," in *2009 Annual International Conference of the IEEE Engineering in Medicine and Biology Society*, 2009, pp. 6560–6564.

[2] Y.-C. Lin, K.-C. Chuang, T.-T. Wang, C.-P. Chiu, and S.-K. Fan, "Integrated digital and analog microfluidics by EWOD and LDEP," in *2006 1st IEEE International Conference on Nano/Micro Engineered and Molecular Systems*, 2006, pp. 1414–1417.

[3] A. Tröls, E. K. Reichel, and B. Jakoby, "FEM modeling and capillary wave analysis of electrowetting induced droplet oscillations," in *2018 IEEE SENSORS*, 2018, pp. 1–4.

[4] L. Li, W. Kang, and D. Ye, "A contact angle measurement method for the droplets in EWOD-based chips," in *2007 2nd IEEE International Conference on Nano/Micro Engineered and Molecular Systems*, 2007, pp. 1071–1075.

[5] Texas Instruments Incorporated, "Capacitive-based liquid level sensing sensor reference design," TI Designs, TIDU736A, 2015.

[6] Ansys, *Q3D extractor*.

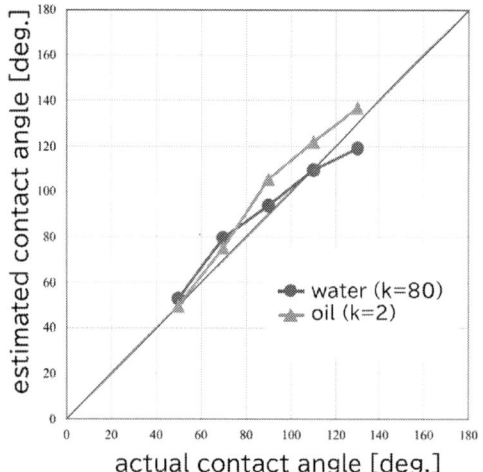

Fig. 10. Measurement results.

Selective manipulations of droplets on photo-driven microfluidic chip with virtual electrowetting channels

Gaifang Chen[1†*], Enqing Liu[1,2†*], Junyan Tian[1], Shang Gao[1], Jia Zhou[1*]

1 State Key Laboratory of Integrated Chips and Systems, College of Integrated Circuits and Micro-Nano Electronics, Fudan University, Shanghai 200433, China
School of Microelectronics, Fudan University, Shanghai 200433, China
2 Physics of Complex Fluids, MESA+ Institute, University of Twente, PO box 217, 7500AE, Enschede, the Netherlands
[†] these authors contribute equally

* Email: 20112020020@fudan.edu.cn, jia.zhou@fudan.edu.cn

Abstract—**This paper achieves selective manipulation of droplets, by modifying the surface potential configuration of the photo-driven microfluidic chip. We use a commercial projector to shine programmable patterns on the chip, realizing the local modulation of contact angles of droplets. This approach achieves high selectivity in sample manipulation for lab-on-a-chip applications, and reduces the risk of cross-contamination between samples, representing a significant technical improvement and optimization in OEW.**

Keywords—*optoelectrowetting, selective droplet manipulation, photo-driven microfluidics*

I. INTRODUCTION

In the modern microfluidics field, digital microfluidics has gained attention for its precise control of discrete droplets, with electrowetting (EW) and optoelectrowetting (OEW) being key driving methods [1]. Initially, EW was prominent due to its fast response time, while its flexibility is restricted by the nonreconfigurable electrode layout, which make the target droplets' sizes and positions limited as well. OEW solves this problem by generating a localized electric field through light exposure to the photosensitive material, which enables the real time designing/programming of droplet driving paths [2,3]. The photosensitive layer is transparent and compatible with optical detection, making it suitable for applications, such as single-cell analysis. Currently, photoelectrowetting is becoming a key driver in advancing digital microfluidics technology in areas like biomedicine and drug screening [4].

Since Chiou et al. introduced the first light-driven droplet mechanism [5], OEW has integrated photoconductive materials (such as amorphous silicon) between the bias

electrodes. Current OEW systems focus on manipulating a small numbers of droplet samples on the same path, although research on OEW devices and systems has already gone deeply into combining AI for real-time droplet detection and path planning [6-8]. In practical applications, various types of samples often need to be driven selectively in different paths to achieve their specific functions, for example, reactions with different agents. Based on our previous work of driving droplets in any direction using Z-shaped light patterns [9], we have further proposed selective droplet driving. In this study, we change the contrast between the driving dark stripe and the background to control the potential distribution underneath the droplets to locally modify the droplet contact angles and then selective droplet manipulation. This approach allows for arbitrary control of the light path, is easy to integrate, and avoids cross-contamination between droplets in biochemical applications, representing a significant technical improvement and optimization in OEW. It enhances the precision of microscale droplet manipulation, making it more promising for practical applications.

II. EXPERIMENTAL

A. Working Principle

The chip adopts a single-plane structure. Firstly, an amorphous silicon (α-Si:H 500 nm) film is deposited on the glass, and then Cr/Au electrodes are deposited at both ends of the chip. Followed by, SU8 2000.5 and Teflon are spin-coated as the dielectric layer and the hydrophobic layer, respectively, as shown in Fig. 1(a). Fig. 1(b) shows the virtual channel generated by Pygame which is often used to create animation games in Python®.

As illustrated in Fig. 1, when the dark stripes are projected onto the chip, the resistance of α-Si at that location increases, resulting in a voltage drop across the SU8 and Teflon layers above it. This voltage drop will drive the droplet to move. Specifically, due to the photoconductivity effect of α-Si, the stronger light, the lower the resistivity. Therefore, the resistance of R3 at the dark stripe is greater than that of R2 at the bright location. So that the voltage on C3 is very high, which results in a higher voltage on the right side compared to the left. This leads to a smaller contact angle on the right side according to the Lippman-Young equation [10]:

$$cos\theta = cos\theta_0 + \frac{C}{2\gamma}V^2 \qquad (1)$$

where θ_0 represents the initial contact angle at zero potential, and C stands for the capacitance of the dielectric layer per unit area. V denotes the voltage on the dielectric capacitor, while γ represents the surface tension between the droplet and the surrounding medium. Therefore, the droplets on the light regions under a dark stripe shining will experience an asymmetric capillary force due to electrowetting, while the droplets on the dark regions won't be activated because of the zero contrast between the dark stripe and the background.

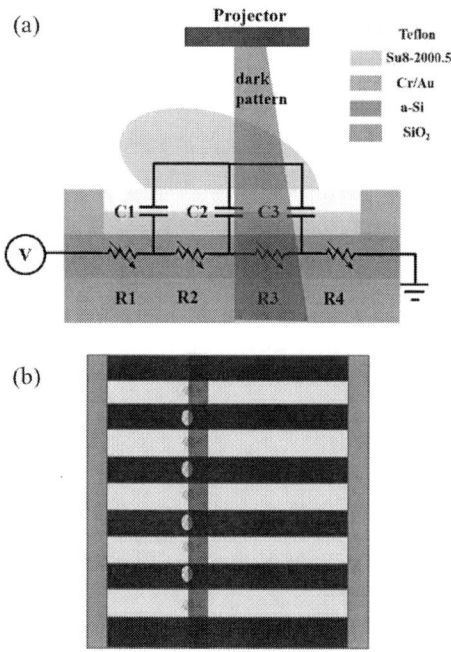

Fig.1. Working principle of (a)structure and schematic diagram of the OEW chip, (b)AI-generated light virtual channels for selective driving.

B. OEW System integration

The experimental setup for testing is shown in Fig. 2, consisting of a computer(a), a projector(b), a camera(c), a signal generator(e), a signal amplifier(f) and the OEW chip(d), with the projector serving as the shining pattern source.

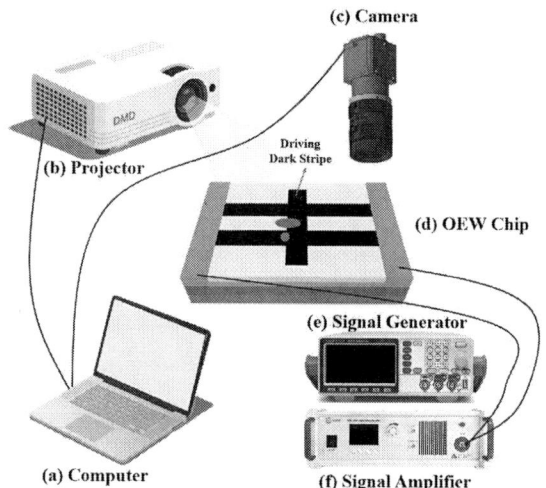

Fig.2. Schematic diagram of the OEW chip experiment system: (a)computer, (b)projector, (c)camera, (d)the OEW chip, (e)signal generator, and (f)signal amplifier.

III. RESULTS AND DISCUSSION

The key parameter of the driving force is the voltage drop in the dielectric layer underneath the droplet which is determined by the photo-dark conductivity ratio α. The test results of photoconductivity are shown in Fig. 3(a). In the Pygame language, (0,0,0) represents full-black and (255,255,255) is the full-white. Therefore, we set six shades of light and dark from 0 to 255 as the back ground, and the dark stripes are always full-dark. The light intensities of different background are also measured and shown in Fig.3(a). The biggest contrast between full-dark and full-white generates a significant voltage drop under dark stripes while no pronounced voltage drop will happen if the background is the same as the dark stripe. Fig. 3(b) shows the changes in contact angle under different brightness conditions for six brightness levels, with different voltages applied. The contact angle decreases as the background brightness increases under the same bias voltage. As the voltage increases, the difference in contact angle changes become more pronounced between the droplet in bright and dark areas, where the contact angles drop 50°and 10° under 250 volts, respectively. This suggests that only the droplets on the bright regions could be driven by moving the dark stripe while the droplets on dark regions will keep staying, enabling the selective manipulation of multiple droplet of interest.

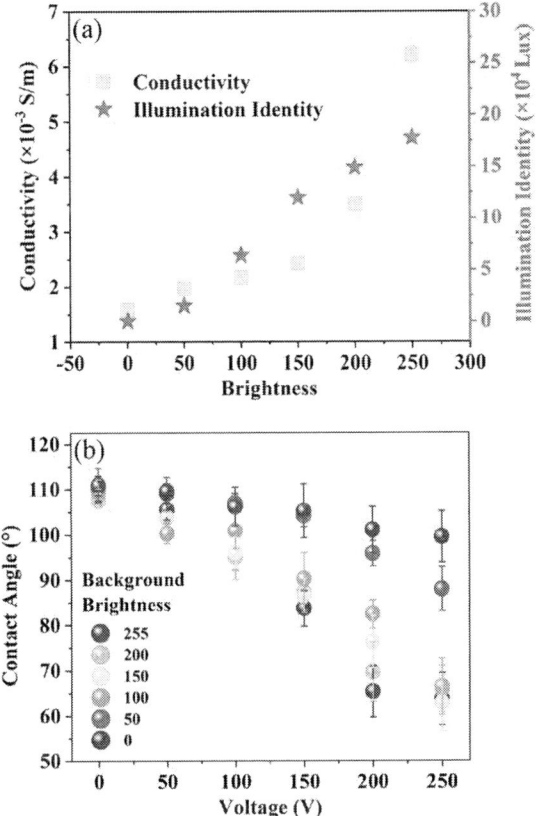

Fig.3. Experimental results under different brightness of (a)light intensity and conductivity of the chip and (b)contact angle of the droplet under different voltage.

We also simulated the potential distributions under five different background lights using Comsol Multiphysics 6.3®. Different background brightness are reflected by different conductivity, i.e., the higher the conductivity, the greater the light intensity. $\sigma = 1 \times 10^{-8}$ S/m is set to the condition of total darkness. As can be seen from Fig. 4(a), the color shows the voltage drop across the dielectric layer, the maximum voltage drop is at the position where the right boundary of the droplet touches the dark driving stripe for both droplets, while it is much larger when the droplet is on the bright region. In Fig. 4(b), we calculated the contact angles contour a droplet contact line with a full-dark pattern and different background, the contact angle drops show the same trend as we measured in the experiments in Fig 3(b).

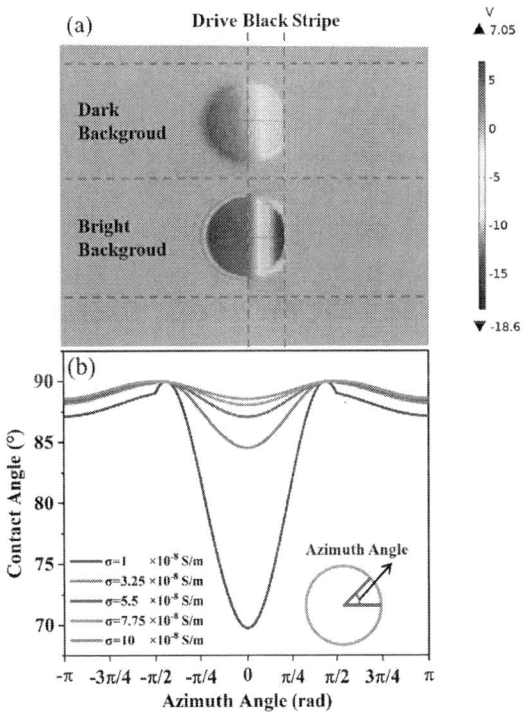

Fig.4. Simulation results under different background light, of (a) the potential of droplets, (b) of contact angles at different locations.

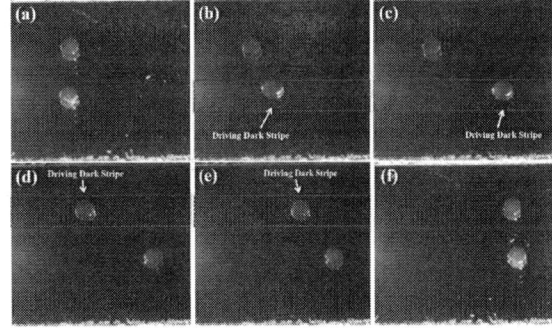

Fig.5. Demonstration of droplet selective drive.

Fig. 5 shows how the droplets are driven by dark stripes in a selective manner. Fig. 5(a-c) show the cyan droplet in the bright channel moves to follow of dark stripes, while the orange droplet remains in the dark. Since the dark background stripes do not cause a voltage drop, the orange droplet is not activated. Similarly, when the channels position shift to the orange droplet, it can be activated. Fig. 5(d-f) demonstrate the movement of the orange droplet.

IV. CONCLUSION

In this paper, we proposed a method for droplet movement on light defined virtual channels and explained its mechanism. We then tested the chip's conductivity and contact angle changes under various light and dark backgound. Comparisons between the experiment and simulation results confirm the feasibility of dark stripe-driven droplet movement. Finally, the selective and programmable manipulation of droplets is achieved. More precise and selective operation of droplets will make the chip laboratory have greater application prospects and practical feasibility, which is a technological progress in the field of OEW.

ACKNOWLEDGMENT

This work was supported by the National Natural Science Foundation of China with Grant No. 62274039 and China Scholarship Council No.202306100244.

REFERENCES

[1] S. K. Thio and S. Y. Park, "A review of optoelectrowetting (OEW): from fundamentals to lab-on-a-smartphone (LOS) applications to environmental sensors," Lab on a Chip, vol. 22, no. 21, pp. 3987-4006, 2022.

[2] D. Jiang and S. Y. Park, "Light-driven 3D droplet manipulation on flexible optoelectrowetting devices fabricated by a simple spin-coating method," Lab on a Chip, vol. 16, no. 10, pp. 1831-1839, 2016.

[3] S. Y. Park, S. Kalim, C. Callahan, M. A. Teitell, and E. P. Y. Chiou, "A light-induced dielectrophoretic droplet manipulation platform," Lab on a Chip, vol. 9, no. 22, pp. 3228-3235, 2009.

[4] U. A. Gurkan, D. K. Wood, D. Carranza, L. H. Herbertson, S. L. Diamond, E. Du, S. Guha, et al., "Next generation microfluidics: fulfilling the promise of lab-on-a-chip technologies," Lab on a Chip, vol. 24, no. 7, pp. 1867-1874, 2024.

[5] P. Y. Chiou, H. Moon, H. Toshiyoshi, C.-J. Kim, and M. C. Wu, "Light actuation of liquid by optoelectrowetting," *Sensors and actuators A: physical,* vol. 104, no. 3, pp. 222-228, 2003.

[6] W. Yang, X. Li, M. Li, and Z. Hao, "Droplet deposition characteristics detection method based on deep learning," Computers and Electronics in Agriculture, vol. 198, 2022, Art no. 107038.

[7] K. Deng, J. Ji, J. Zhou, C. Chang, J. Ding, Z. Jia, et al., "A Hybrid Closed-Loop Droplet Detection for Digital Microfluidics," IEEE Sensors Journal, vol. 25, no. 9, pp. 15809-15821, 2025.

[8] J. Zhu, Y. Meng, W. Gao, S. Yang, W. Zhu, X. Ji, et al., "AI-driven high-throughput droplet screening of cell-free gene expression," Nature Communications, vol. 16, no. 1, 2025, Art no. 2720.

[9] E. Liu, C. Wang, H. Zheng, S. Song, A. Riaud, and J. Zhou, "Two-dimensional manipulation of droplets on a single-sided continuous optoelectrowetting digital microfluidic chip," Sensors and Actuators B: Chemical, vol. 368, 2022, Art no. 132231.

[10] S. Gao, J. Tian, J. Yao, H. Zheng, E. Liu, and J. Zhou, "Study on droplet splitting in single-plate OEW chips," Journal of Electrostatics, vol. 136, 2025, Art no.104111.

A MEMS Rectenna for RF energy harvesting around 2.4GHz

Liu Xiaoqiang *, Wang Tiancong

Microelectronics Center, Harbin Institute of Technology, Harbin 150001, China

* Email: xiaoqiang_liu@hit.edu.cn

Abstract—**This article implements a rectenna on a silicon wafer. The rectenna is fabricated by etching conductive copper foil and affixed to the silicon wafer. The overall dimensions of the rectenna are 50mm × 50mm × 0.8mm. The proposed antenna has an S11 of -18dB and a gain of 4.96dBi at 2.4GHz. The maximum rectification efficiency of the rectifier circuit reaches 31.96% when the input power is 8.4dBm.**

Keywords—RF energy harvesting, antenna design, rectifier

I. INTRODUCTION

The growth of applications such as the internet of things (IoT), smart home and wireless sensor networks are significantly changing our daily lives. The conventional power sources of these low-power devices usually depend on batteries, which leads to frequent battery replacements, high maintenance costs, and environmental pollution[1]. As an eco-friendly alternative, radio frequency energy harvesting (RF-EH) technology shows excellent potential in alleviating these problems. A crucial component in ambient RF energy harvesting systems is the rectenna, which could scavenge ambient RF signals emitted by various sources like Wi-Fi routing, broadcast towers, and consumer electronics[2]. With receiving antennas, impedance matching networks, and rectifier circuits as its key components, the performance of a rectenna determines its ability to output direct current (DC) power and the amount of power it can generate. Consequently, the rectenna has become hotpots for researchers[3].

In recent study, conventional substrates such as Rogers, RT/Duroid, and FR4 are widely adopted in the design of rectennas due to their low cost and easy availability[4]. These rectennas are typically implemented using standard printed circuit boards (PCBs). However, these substrates have inherent shortcomings: due to their weak mechanical strength, porous structure, and reliance on the performance layer of the material covering their surface, they exhibit poor performance in terms of structural stability and mechanical properties. Therefore, Silicon substrate provides a viable choice for rectenna. Meanwhile, it is notable that the silicon substrate is common used for various micro-electro-mechanical-systems devices (MEMS), which facilitates the integration of RF-EH technologies with MEMS devices[5].

In this paper, we proposed a novel RF rectangular microstrip rectenna fabricated on a silicon substrate. It bridges the gap between ambient energy availability and the power requirements of modern low power MEMS devices. An overview of typical application of the proposed rectenna is shown in Figure 1. It contains of three blocks. The first is radio frequency source. The second is the free space as the medium to propagate the signal from RF source to the rectenna. The third is the proposed rectenna, which captures the RF energy to drive low power MEMS devices.

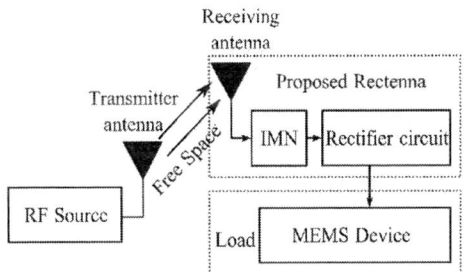

Fig. 1. The typical application schematic of the proposed rectenna.

The remainder of this paper is organized as follows: Section II details antenna design. Section III shows the rectifier design. Rectenna integration is presented in Section IV. Section V concludes the paper.

II. DESIGN OF ANTENNA

In the field of RF-EH, antennas, as the core components for both energy reception and transmission, play a crucial role in rectenna's performance. Rectangular microstrip antennas have become the preferred solution for many application scenarios due to their advantages such as compact structure, easy integration and low cost. This article will design an energy harvesting antenna based on this structure for 2.4GHz radio frequency signals. The antenna is simulated using the professional antenna design software High Frequency Structure Simulator (HFSS). The antenna structure and parameters are shown in Fig. 2 and Table1.

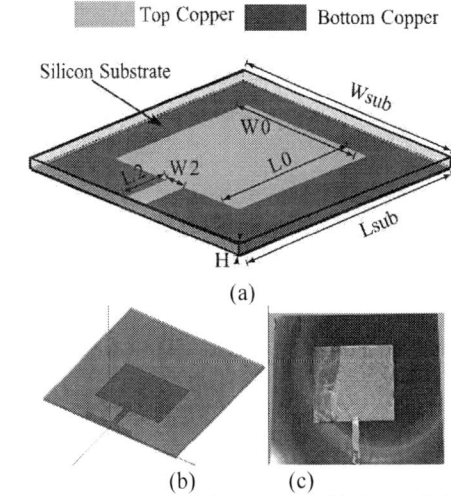

Fig. 2 (a) The structure of the antenna. (b) the model of antenna in HFSS. (c) the photograph of designed antenna.

Table1 Geometric parameters of the antenna

Parameter	Wsub	Lsub	L0	L2
Value (mm)	50	50	17.7	5
Parameter	W0	W2	H	
Value (mm)	24.1	2.12	0.8	

In this design, in order to matching the radiation path and the rectifier, the length of the microstrip line serving as the feed line is determined to be a quarter-wavelength. Due to the high dielectric constant of the silicon substrate, when the impedance of the microstrip line is set to 50 ohms, its width becomes excessively narrow, making it extremely challenging to fabricate. To address this issue, this paper employs a non-standard 50-Ω matching technique. Specifically, during the design process, the impedance value at the antenna radiation patch port is obtained through simulation, as shown in Fig. 2, at a frequency of 2.42 GHz, the impedance of antenna is 214 + 0.5j Ω. Based on the real impedance of 214 Ω, a microstrip line with the width of 2.12mm is used to achieve matching the impedance to 3 Ω at the input of rectifier.

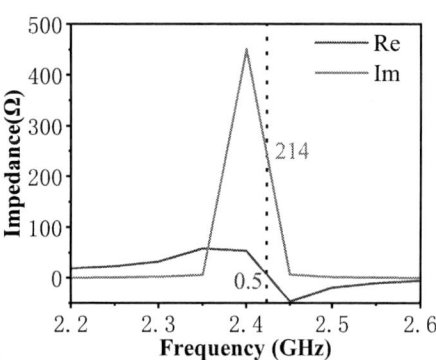

Fig. 3. Simulated impedance of the antenna

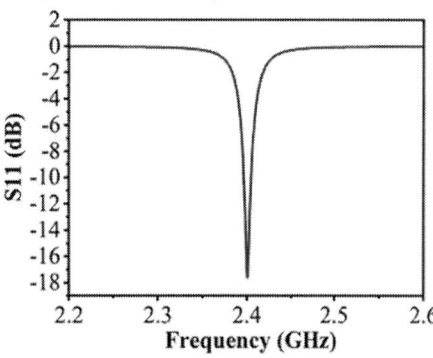

Fig. 4. The S11 of designed antenna

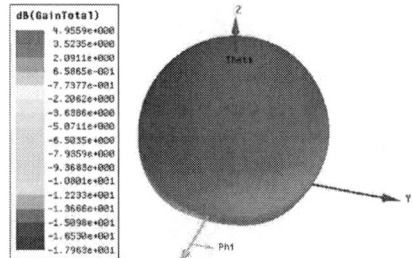

Fig.5. 3D Radiation pattern of the Antenna

Based on the simulation results, the total dimensions of the designed antenna are determined to be 50mm * 50mm. Fig. 4 shows the S11 of the proposed antenna, it reaches -18 dB at

the central resonant frequency of 2.4 GHz. And the antenna's max gain reaches 4.96 dBi, as shown in Fig. 5.

III. RECTIFIER DESIGN

The rectifier is the key component to convert high-frequency RF signals into DC signals (RF-DC), it consists of impedance matching networks and rectifier circuit. To minimize the transmission loss, the IMN of rectifier is designed with lumped component, and a two-stage Dickson charge pump is adopted as the rectifier circuit. Fig.4. depicts the structure of rectifier.

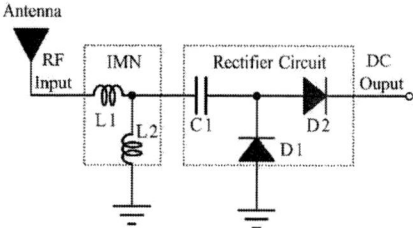

Fig. 6. Rectifier Structure

To improve the energy conversion efficiency of the rectifier, we modeled and simulated the rectifier in Advanced Design System (ADS). High-frequency capacitors and inductors from Murata and diode from Broadcom are used for the simulation. the diode is SMS7630, its SPICE model (as shown in Fig. 7) is established with reference to its datasheet.

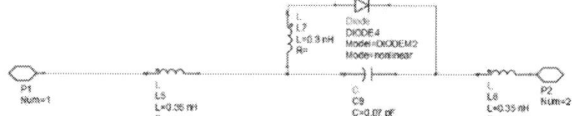

Fig. 7. SPICE model of SMS7630.

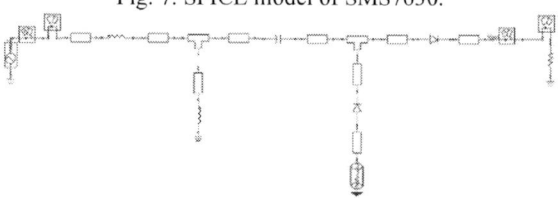

Fig.8. the simulation model of Rectifier in ADS

Based on the SPICE model of SMS7630, the whole model of Rectifier is established in ADS(as shown in Fig. 8), After simulation optimization, the rectification efficiency of the rectifier is shown in Fig. 9. At an input power of 9 dBm, the conversion efficiency reaches 30%. The final determined component parameter values are as follows: L1 is 2 nH, L2 is 4.7 nH, C1 is 5.1 pF, and both D1 and D2 are SMS7630.

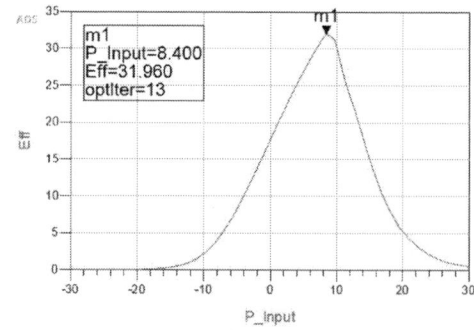

Fig. 9. Simulated energy conversion efficiency of the rectifier.

IV. RECTENNA INTEGRATION

In the previous work, the parameters of the rectenna were determined. In this section, we will integrate the antenna. First, we used a cutting machine to fabricate the radiation patch and the rectifier circuit pads. The cutting platform, as shown in Figure 10, consists of a control computer and a high-precision cutting machine. We created the cutting layout for the antenna radiation patch and the rectifier circuit pads according to Figure 11a. The fabrication was carried out by cutting a 0.1-mm-thick copper foil, and the finished product is shown in Figure 11b. Finally, the Radiation patch and the rectifier circuit pads are integrated into the silicon substrate, and the photograph is shown in Fig. 12.

Fig. 10. Rdiation patch and rectifier pads fabrication platform

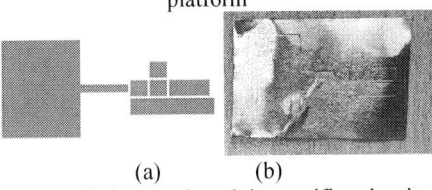

(a)   (b)

Fig.11 (a) Radiation patch and the rectifier circuit pads schematic diagram. (b) The product of radiation patch and the rectifier circuit pads.

Fig. 12 The photograph of proposed rectenna.

V. CONCLUSION

In this paper, a rectenna based on a silicon substrate is proposed. Utilizing the conductive copper foil etching technique, the simulated antenna and rectifier patterns can be etched and then attached to the silicon wafer to form the rectenna. The MEMS rectenna design method proposed in this paper enables rapid validation of MEMS RF energy harvesters.

REFERENCES

[1] N. Shinohara, "History and innovation of wireless power transfer via microwaves," *IEEE journal of microwaves*, vol. 1, no. 1, pp. 218-228, 2021.

[2] C. Song, Y. Ding, A. Eid, J. G. Hester, X. He, R. Bahr, A. Georgiadis, G. Goussetis, and M. M. Tentzeris, "Advances in wirelessly powered backscatter communications: From antenna/rf circuitry design to printed flexible electronics," *Proceedings of the IEEE,* vol. 110, no. 1, pp. 171-192, 2021.

[3] M. Wagih, A. S. Weddell, and S. Beeby, "Rectennas for Radio-Frequency Energy Harvesting and Wireless Power Transfer: A Review of Antenna Design [Antenna Applications Corner]," *IEEE Antennas and Propagation Magazine,* vol. 62, no. 5, pp. 95-107, Oct, 2020.

[4] S. Muhammad, J. J. Tiang, S. K. Wong, A. Smida, M. I. Waly, and A. Iqbal, "Efficient quad-band RF energy harvesting rectifier for wireless power communications," *Aeu-International Journal of Electronics and Communications,* vol. 139, Sep, 2021.

[5] N. H. Mohd Yunus, J. Sampe, J. Yunas, A. Pawi, and Z. A. J. M. t. Rhazali, "MEMS based antenna of energy harvester for wireless sensor node," vol. 26, pp. 2785-2792, 2020.

Design and Implementation of Shared Storage Communication Architecture for MCCSIP-RAA

Longmei Nan*[1], Yu Jin[1], Yiran Du[1], Tao Chen[1], Lin Chen[1], Yanjiang Liu[1], Wei Li[1], Weiquan Sang[1]

[1]Information Engineering University, Zhengzhou 450001, China

* Email: lnan13@fudan.edu.cn

Abstract—**The Multi-core cryptographic specific instruction processor with reconfigurable accelerated array (MCCSIP-RAA) combines flexibility, efficiency, and resource utilization, but its high computational throughput also puts enormous pressure on communication structures. Based on the analysis of MCCSIP-RAA architecture and multi-core processor communication systems, combined with the data interaction characteristics of MCCSIP-RAA, a shared storage communication architecture based on mailbox for MCCSIP-RAA was designed. A communication structure based on register implementation for fast exchange and shared storage was designed for small batch and large bit width intermediate result data, named the Shared Cross Memory structure. A shared storage communication structure based on RAM implementation was designed for the interaction of wheel key data with large amounts of data that need to be repeatedly and continuously used, named the shared key pool RAM structure. A shared storage communication structure based on FIFO implementation was designed for continuous and large-scale streaming data processed by different inter core links, named Link Data Buffer FIFO. Furthermore, a synchronization mechanism based on mailbox was designed to resolve the issues of data exchange correctness and storage consistency. The experimental results show that shared storage communication architecture can effectively alleviate the contradiction between cryptographic processing flexibility, efficiency, and resource utilization in the target architecture MCCSIP-RAA. Adding smaller resources to cryptographic algorithm processing can achieve processing performance of 3-7 times for different cryptographic algorithms, effectively improving resource utilization.**

Keywords—*MCCSIP-RAA, Shared Storage, Communication Architecture, Cross Memory, Mailbox, Shared Key Pool, Shared Link FIFO*

I. INTRODUCTION

The implementation of cryptographic algorithms plays an important role in information security. It is effective to design a flexible and efficient MCCSIP-RAA. So designing flexible and efficient communication structure that meets the communication characteristics of cryptographic processing for MCCSIP-RAA is of great significance.

II. ANALYSIS OF MCCSIP-RAA DATA INTERACTION CHARACTERISTICS AND COMMUNICATION STRUCTURE

A. Analysis of MCCSIP-RAA Architecture

The communication structure designed in this paper is oriented towards to the MCCSIP-RAA, so it is necessary to analyze the MCCSIP-RAA, firstly.

MCCSIP-RAA is proposed based on the fusion analysis of multi-core cryptographic processors and acceleration arrays,

---

This work is supported by Natural Science Foundation of Henan (No. 232300421393).

and on the analysis of cryptographic parallel processing structural parameters such as computing granularity, parallelism, and number of computing cores. It contains four cryptographic processors and a shared acceleration array, data storage, input/output microprogram controller, as shown in Fig. 1. The four cryptographic processors, named as SPcore1-SPcore4 respectively, are designed in an isomorphic manner and are the core computing components of task level parallel processing. The reconfigurable acceleration array is the main computational component for nonlinear operations in stream cipher algorithms, named as Array. It is shared by four instruction processors in a shared form and can also work independently.

Efficient communication structures need to be established between the four instruction stream cryptographic processors, as well as between the four instruction stream cryptographic processors and the reconfigurable acceleration array. It can be seen that designing efficient communication structures that meet communication requirements is an important module of MCCSIP-RAA, and researching efficient communication structures for MCCSIP-RAA is of great significance.

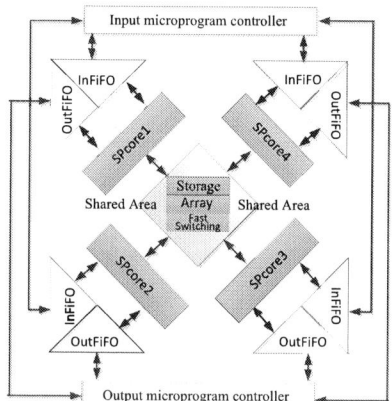

Fig. 1. The structure of MCCSIP-RAA

B. Analysis of MCCSIP-RAA Data Interaction Characteristics

To design an efficient communication structure, it is necessary to analyze the interactive characteristics of cryptographic algorithm data streams. (a) Cryptographic processing is usually a type of data operation with a large bit width, typically 128 bits and 256 bits, and some can even reach 512 bits and 1024 bits or more. The data has strong parallelism and correlation, making it a parallel and correlation intensive operation. (b) When mapping cryptographic algorithms, in order to reduce communication overhead, operations with strong data correlation in the cryptographic algorithm are often mapped to the same

---

979-8-3315-3918-4/25 $31.00 © 2025 IEEE

processor, or the cryptographic algorithm is divided into stages and mapped to different processors. For block cipher algorithms, the sub key generation process is mapped to one processor, while the round operation process is mapped to another processor. The two processors need to exchange wheel key data, which is relatively large and needs to be continuously and repeatedly used multiple times. For stream cipher algorithms based on linear operations, the initialization process is often mapped to one processor, and the normal key generation process needs to be mapped to another processor. The data that needs to be exchanged between the two processors is the state sequence of the shift register, which has a relatively small amount of data. For hash algorithms, the data processing bit width is relatively large, the round operation is relatively complex, and the message block needs to be filled. Often, data filling occupies one processor, and the complex round operation decomposition maps to different processors. The data exchanged between different processor cores is the filled data and the intermediate result data of the round operation, which has a relatively small amount of data.

C. Analysis of Multi-core Processor Communication Systems

To design a more reasonable communication structure for MCCSIP-RAA , this paper further analyzes the typical multi-core processor communication structure [1-3]. (a) Bus sharing structure, where various computing processors, input/output ports, memory, etc. are all connected to the same bus for data exchange through a shared bus. Its characteristics are simple protocol and flexible setting of bus priority; The existing problems are poor scalability, a small number of processors, and low communication parallelism, which makes it impossible to support communication for more than one pair of users simultaneously, that is, broadcast communication cannot be well supported. The various arithmetic processors can only occupy the bus sequentially in series, so when the number of processors increases, arbitration becomes complex. (b) Shared storage structure, where each computing processor shares the same memory bank and uses read-write shared memory to achieve data exchange communication. Its characteristics are simple programming model, high parallelism in data exchange compared to buses, and the ability to support the transmission of large continuous data blocks; The problem is poor scalability, and the probability of access conflicts increases as the number of processors increases. (c) The message passing method is mainly used in multi-core processors of on-chip networks, which has good communication bandwidth and scalability. However, the problem is that when transmitting a large amount of data, it can cause congestion and result in high communication latency and low efficiency.

After analyzing the characteristics of various communication structures and combining them with the data exchange characteristics of cryptographic processing, it can be concluded that when continuous data needs to be transmitted and the amount of data is large, a shared storage inter core communication structure is suitable, which only requires fast access between the processor and the shared storage unit for data exchange without affecting communication between other processors. Therefore, a communication structure based on shared storage is designed.

III. DESIGN OF SHARED STORAGE COMMUNICATION STRUCTURE BASED ON MAILBOX FOR MCCSIP-RAA

A. Design of Shared Storage Communication Architecture for MCCSIP-RAA

Based on the above analysis, a shared Cross Memory structure as shown in Fig. 2 was designed, mainly consisting of register stacking storage unit (the main carrier of data storage), flag register unit, output selection unit, and synchronization unit. The synchronization mechanism consists of a gating unit (which cooperates with a flag register to complete "locking") and a clearing unit (which cooperates with a flag register to complete "unlocking"). Its main function is to ensure the correctness and consistency of data exchange, which will be introduced together with the mailbox synchronization mechanism in the next section.

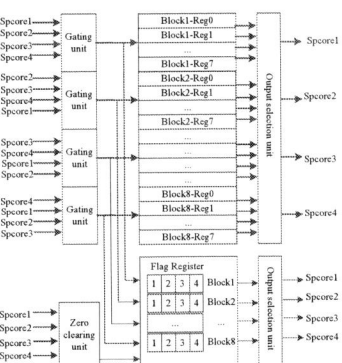

Fig. 2. The structure of Cross Memory for MCCSIP-RAA

The register stacking storage unit is the core of the entire shared Cross Memory structure, mainly used to store data. Due to the large bit width of the data exchanged during password processing, the storage capacity of each block in the register storage unit is set to 256 bits, which is 8x32 bits, divided into 8 blocks. Each block can be accessed by any processor to temporarily store the data exchanged with other processors. For each Block, set it to have 4 write ports, each with 32 bits, supporting 128 bits data writing, and 8 read ports, each with 32 bits, supporting 256 bits data reading. This enables parallel transmission of multiple words and further designs its access methods for arbitrary writing and fixed reading to improve access efficiency. For each processor, in addition to its own private general-purpose register file, sub key shadow register, and dedicated bit register, there are also 8x8 32-bit shared registers that can be used as source or destination registers by each processor. In addition, there are 4-bit flag bits set for each block, strictly corresponding to each processor, denoted as S3 S2 S1 S0, where flag bit S0 corresponds to the programmable symmetric cryptographic processor Spore1, flag bit S1 corresponds to Spore2, flag bit S2 corresponds to Spore3, and flag bit S3 corresponds to Spore4. When a block has a read task from a processor, the corresponding flag bit is set. Other processors are not allowed to perform write operations on the block until the corresponding processor completes the read operation, and then the corresponding flag bit is reset. Only then can the block perform write operations. It can be concluded that only when all 4-bit S3, S2, S1, and S0 corresponding to the block are in the reset state, can the block be written by the processor. After writing, the status bit corresponding to the processor to be read needs to be set. For each block, it supports multiple processors to perform read operations simultaneously, that is, it supports a "one to many" data broadcast communication interaction mode, but only one processor can perform write operations, and the gating unit determines which processor can perform the write operation. When a 4-bit block has all 0 flags and multiple processors want to write access to the block, conflicts can be resolved by setting write access priority policies.

A shared key pool RAM structure based on RAM implementation was designed for the wheel key data exchange in various cryptographic processors, which has a large amount of data and needs to be repeatedly and continuously used. This approach can effectively reduce the area consumed by storage resources compared to register implementation. The RAM structure of the shared key pool is shown in Fig. 3 (a), which is divided into 4 sub blocks. Each sub block of RAM is called Bank, with a capacity of 128x32 bits, totaling 512x32 bits. Except for the RAM based implementation of the memory bank, which differs from the register implementation of Cross Memory, its block based approach, write/read logic, and block based memory bank access priority setting are all similar to the register implementation of Cross Memory. A shared link data buffer FIFO based on FIFO implementation was designed for processing continuous and large batches of streaming data between different processors, as shown in Fig.3 (b). This method does not require consideration of read and write addresses during the continuous reading and writing of large amounts of data. Compared to register and RAM implementation methods, the read and write logic is very simple and easy to control. Its access priority is Spcore1 Spcore2、Spcore3、Spcore4，The synchronization method also adopts a combination of mailbox and status identification register, which will not be repeated here.

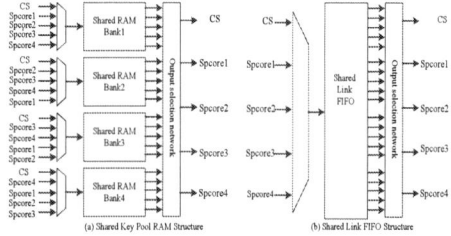

Fig. 3.   Shared Key Pool RAM Structure and Shared Link FIFO Structure

B. Shared Storage Mailbox Synchronization Mechanism for MCCSIP-RAA

This article uses mailbox synchronization mechanism to solve the important storage consistency problem in the shared Cross Memory structure, namely the correctness and consistency of data exchange. Each processor has a structurally similar mailbox, and here we take the internal structure of the Spcore1 mailbox as an example to illustrate. As shown in Fig.4, the Spcore1 mailbox structure consists of six 8-bit registers, grouped in pairs and corresponding to a fixed data source. Spcore2 corresponds to register 20 and register 21, Spcore3 corresponds to register 30 and register 31, and Spcore4 corresponds to register 40 and register 41. The purpose of setting up two register structures is to support ping-pong operations between two arithmetic processors for writing and reading data. After the data source processor completes the write to the storage, it needs to inform the destination processor of the completion of the write and the data type of the communication. Therefore, it is necessary to

provide operation instructions for the corresponding bits of the destination processor mailbox that it owns. Before starting mailbox communication, each processor needs to agree on the relevant email register information rules and the meaning of information transmission. Therefore, the email register is set to 7 bits and can work in three modes: number-setting, reset, and accumulation. The data value of the email register corresponds one-to-one with the transmission data type. For example, writing 1 in the email register indicates that the initial IV vector is being transmitted, and writing 2 in the email register indicates that the initial sub key is being transmitted.

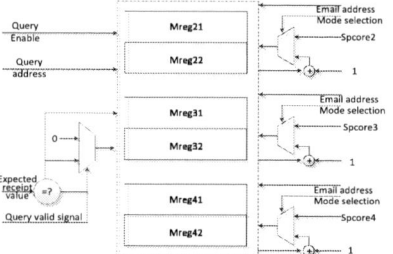

Fig. 4.   Internal Structure of Spcore1 mailbox Module

Based on the mailbox structure and tag bits, the inter core communication process using shared storage units as communication intermediaries consists of five steps. Here, taking Spcore1 sending data to Spcore3 as an example, the communication process is shown in Fig.5

(a) The Spcore1 processor queries the flag bits of a block to be written to the Cross Memory storage. When all 4-bit flag bits of the block are 0, it indicates that the block can perform a write operation, and the Spcore1 processor can perform the write operation. Otherwise, if all 4-bit flag bits are not 0, the Spcore1 processor waits and continues the query until it can be written. (b) After writing data into the corresponding storage block, the Spcore1 processor places Spcore3 at the corresponding identification position of the block to avoid other processors writing data that may overwrite the valid data to be transmitted. (c) The Spcore1 processor sends the agreed upon email data to the Spcore3 processor, notifying them of the data and data type to be read from the corresponding block in the shared storage of Spcore3. The Spcore3 processor queries the content of local email data and compares it with pre agreed data to obtain synchronization information such as data type. After successful query, the Spcore3 processor will reset the email register corresponding to the local email Spcore1 processor to zero. (d) The Spcore3 processor reads the valid data in the corresponding block of the shared memory, transfers it to its corresponding register, and after reading, resets and clears the flag bit of the corresponding block of the shared memory, releasing the corresponding block resources of the shared memory for a new round of data transmission of the block. It can be seen that the entire communication process takes 5 clock cycles to complete up to 8 32-bit parallel transmissions, and when the program mapping is ideal, the synchronization of the query flag and query mailbox can be ignored in other operation instructions. Therefore, up to 8 32-bit data can be transmitted within 2 clock cycles.

The inter core communication structure based on Cross Memory adopts a register design method based on Block, which not only supports one-to-one correspondence transmission, but also supports "one to many" data broadcast

979-8-3315-3918-4/25 $31.00 © 2025 IEEE   676

transmission, which can greatly increase communication efficiency and data interaction capability. This will not be repeated here.

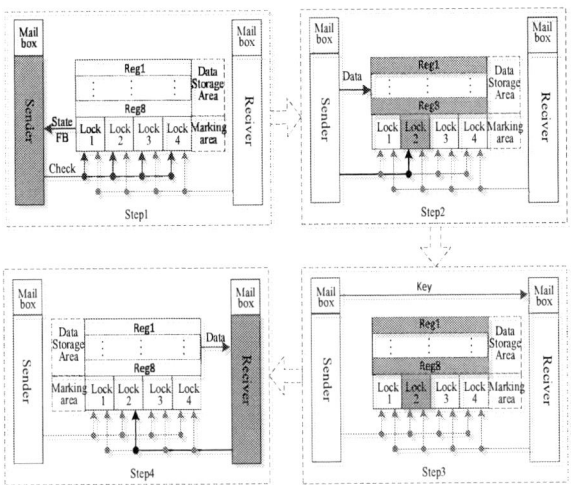

Fig. 5. Inter core communication mechanism based on shared storage

IV. MAPPING AND PERFORMANCE EVALUATION

Taking the communication mapping between Spcore1 and Spcore2, Spcore3 as an example, it can be mapped into four processes: write CrossMem shared storage by four Spcore process, send mailbox process, check received mailbox process, read CrossMem process, as shown in Table 1.

TABLE I. COMMUNICATION MAPPING OF SPCORE1 AND SPCORE2/3

Instructions
step 1. Spcore1 uses the write fast swap memory instruction to write the data from RgA0 to Rc0 in Block0 of the fast swap memory
Write_CrossMemA1 Block0, Rc0, RgA0, Spcore1
Step 2. Spcore1 sends an email to Spcore2, notifying them that the write operation to the fast swap storage area has been completed
Send_mailbox ClusterA0.Spcore1ToSpcore2, Mbox0
Step 3. Spcore1 sends an email to Spcore3, notifying them that the write operation to the fast swap storage area has been completed
Send_mailbox ClusterA0.Spcore1ToSpcore3, Mbox0
Step 4. Spcore2 queries the email to confirm that Spcore1 has completed the write operation for the quick swap storage area
Checkreceive_mailbox Spcore1.Mbox0 # imm8 #addr_imm11
Step 5. Spcore3 queries the email to confirm that Spcore1 has completed the write operation for the quick swap storage area
Checkreceive_mailbox Spcore1.Mbox0 # imm8 #addr_imm11
Step 6. Spcore2 reads the data from the corresponding register in the fast swap storage area to its general register RgA0:
Read_CrossMemA1 RgA0, Block0, Rc0
Step 7. Spcore3 reads the data from the corresponding register in the fast swap storage area to its general register RgA0
Read_CrossMemA1 RgA0, Block0, Rc0

In protocols such as IPSec, SSL/TLS, etc., block algorithms (such as SM4) are first used to encrypt the data to be transmitted. Then, in order to ensure the integrity and authenticity of the data, hash algorithms (such as SM3) are used to calculate the hash value of the relevant data. Therefore, block algorithm can be mapped to one processor, and hash algorithm can be mapped to another processor, resulting in data exchange between the two. Here, the mapping of SM3 algorithm is taken as an example, as shown in Table 2.

The executive performance of typical algorithms in each platforms[4-6] and this work is shown in Table 3. It can be

seen that with the support of the hierarchical storage structure designed in this article, MCCSIP-RAA can efficiently and flexibly implement different cryptographic algorithms, and its processing performance is significantly better than other platforms.

TABLE II. SM3 ALGORITHM MAPPIN

Instructions
Parallel injection main program
Write_CrossMemA1 Block0, RgC0,SPCore1   //Cromem Spcore1- ->Spcore2 ··· Send_mailbox ClusterA1.SPCore1ToSPCore2, Mbox1,#1 Write_CrossMemA1 Block3, RgC1,PCore3   //Cromem Spcore3- ->Spcore4 ··· Send_mailbox ClusterA1.SPCore3ToSPCore4, Mbox0,#1
Message Block Calculation main program
IROL32 RgA10, RgA1, #12 | IROL32 RgB13, RgC3, j → TMODADD32 RgC14, RgA10, RgB13, RgA2 → IROL32 RgD10, RgA1, #7 TL RgB11, RgA1, RgB1, RgC1, immed8 | XOR RgA11, RgA10, RgD10 → TMODADD32 RgC11, RgA11, RgD1, RgB3 | TL RgD11, RgA2, RgB2, RgC2, immed8   MODADD32 RgA12, RgB11, RgC11 | TMODADD32 RgB12, RgB3 RgD10, RgD2 | IROL32 RgC12, RgB1, #9 | IROL32 RgD12, , RgB1, #19   MODADD32 RgA13, RgB12, RgD11 → IROL32 RgB13, RgA13, #9 | IROL32 RgC13, RgA13, #17 → TL RgD13, RgA13, RgB13, RgC13, immed8
CCO encrypted data output program
Task1:oMovi_1 Rig2, #0x8 oCmpz Rig2 oWait.Valid PCore0.unempty ··· oLoad Rig0, PCore0 oGe OUTFIFO ··· oStore.OutLength Rig3

TABLE III. PERFORMANCE COMPARISON ANALYSIS OF MCCSIP-RAA

Design	Frequency (MHz)	Area (mm²)	Process (nm)	Algorithm name	Throughput rate (Mbps)
[4]	1000	16	65	AES	2160
				IDEA/SM4	700/670
[5]	500	1.89	65	AES	2181
				DES/ IDEA	491/563
[6]	350	18	180	AES	1587
				DES/ IDEA	460/400
This Work	300	17	65	SM4/AES	4978
				DES /IDEA	3079
				SM3/ Shink	2214/4756
				A5/W7	394
				Grain/ LILI	300
				TRIVIUM	240
				Toyocrypt/E0	1200

V. CONCLUSIONS

In summary, based on the typical communication structures of multi-core system, and combined with the encrypted data exchange characteristics of MCCSIP-RAA, this article designs a mailbox based shared storage communication structure that can effectively support efficient processing of MCCSIP-RAA and provide reference for other multi-core systems.

EFERENCES

[1] Quan Heng, Research on the Design of Multi core Processor Computing and Inter core Communication Modules [D]. Shanghai: Fudan University, 2012.

[2] Chen Fan, Research and Design of Inter core Communication Mechanism for Password Multi core Processor [D]. Zhengzhou: University of Information Engineering, 2014.

[3] Li Wei, Heterogeneous multi-core information security processor [D]. Shanghai: Fudan University,2017

[4] Fengxiao, Reconfigurable Asymmetrical Multi-core Architecture for Block Cipher[J]. Chinese Journal of Electronics, In press, 2016.

[5] Li Gongli, Research on Key Technologies of Block Cipher Stream Processor [D], Zhengzhou: Information Engineering University, 2018.

[6] Li Wei, A reconfigurable block cryptographic processor based on VLIW architecture[J]. China Communications, 2016, 13(1):91-99.

A Fully Quantized LeNet-5 accelerator for Edge Computing with Quantization-Aware Training

Yushan Dai [1], Angyang Li [2], Jian Mei [1,2], Rui Yin* [1,3]

[1] College of Integrated Circuits and Micro-Nano Electronics, Fudan University, Shanghai, China
[2] National Integrated Circuit Innovation Center, Shanghai, China
[3] Jiashan Fudan Institute, Jiaxing, China

* Email: ysdai24@m.fudan.edu.cn, yinrui@fudan.edu.cn

Abstract—This paper presents a fully quantized LeNet-5 accelerator optimized for edge-based handwritten digit recognition. Using pseudo-quantization and quantization-aware training, the model adopts integer arithmetic (INT8/INT9/INT12/INT14) to maintain high accuracy (98.6%-99.3%) while reducing computational overhead. Implemented in 180nm CMOS, the accelerator integrates pipelined computation units and SRAM-based data reuse to minimize area and power consumption, supporting dual operating modes: a 100MHz high-speed mode delivers 863 FPS for real-time tasks, while a 10MHz low-power mode achieves 45mW consumption with 101 FPS, ideal for battery-powered devices. Post-simulation results validate its ability to balance performance, efficiency, and accuracy, providing a flexible edge solution for handwritten digit recognition.

Keywords—CNN; quantization; quantization-aware training; hardware accelerator; edge computing

I. INTRODUCTION

In recent years, with the rapid development of AI, people have begun to explore its applications in various fields. Through edge deployment, AI-specific chips are being used to advance research progress across disciplines. In the latest studies, numerous efforts have focused on implementing CNNs through edge deployment to meet diverse domain-specific needs. For instance, some research integrates CNN chips with transformers to build hybrid models for semantic segmentation[1], while others utilize integrated CNN chips for 3D gesture recognition[2]. Additionally, there are initiatives to design autonomous driving processor chips based on CNNs[3]. These works have opened up new research avenues for the hardware deployment of CNNs.

LeNet-5, a classic CNN architecture designed for digit recognition[4], has been widely studied for its simplicity and effectiveness. However, conventional implementations of LeNet-5 using floating-point arithmetic are unsuitable for edge devices due to high computational latency and power consumption. To address this, model quantization—converting floating-point weights and activations to integer types[5,6] (e.g., INT8)—has emerged as a critical technique to reduce memory usage and computational complexity, enabling efficient hardware acceleration.

In addition, the quantization-aware training (QAT) proposed by Google in 2018 can train CNN models with integer data while maintaining accuracy[7]. Compared with floating-point operations, it is more amenable to hardware implementation and can significantly improve the computational accuracy of quantized models on the hardware side.

II. LENET-5 MODEL TRAINED VIA QUANTIZATION AWARE TRAINING

Given that our work aims to quantize the CNN model into INT-type data for ASIC implementation, we adopt a pseudo-quantization method during model training to ensure the accuracy of handwritten digit recognition after quantization. This approach simulates the precision loss that occurs in hardware circuits, thereby maintaining the recognition accuracy of the designed ASIC for handwritten digits. Consequently, we have made certain modifications to the classic LeNet-5 model architecture.

A. Quantized Model Architecture

The specific model structure is illustrated in Figure 1. The input image needs to be adjusted to a single-channel 32×32 size. To improve the operational efficiency of the hardware circuit, the ReLU function is adopted as the activation function. The convolution kernel size is 5×5 with a stride of 1. For the first convolutional layer, both the input weights and image data are quantized to the INT8 type to

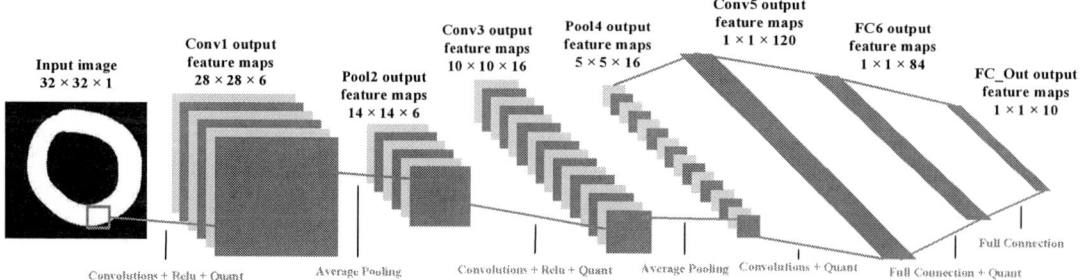

Fig. 1. The proposed architecture of quantized model.

This work is supported by Shanghai Pudong New Area Science and Technology Development Fund Industry-Academia-Research Collaboration Special Funding PKX2024-D06.

simulate hardware. The output feature maps are computed using integer-type data and extended by one bit to INT9 for output. The second pooling layer uses average pooling with no change in data bit width. To further compensate for the precision loss caused by quantization, the third convolutional layer extends the output feature maps to 12 bits (INT12). The fourth layer is still an average pooling layer, maintaining the output bit width of 12 bits. The fifth convolutional layer convolves the three-dimensional feature maps into a one-dimensional vector, keeping the bit width at 12 bits. The subsequent two fully connected layers also use weights and biases quantized to INT8 for computation. After these two fully connected layers, 10 integer data items with a bit width of 14 are finally output, representing the probabilities of the 10 digits from 0 to 9, respectively. So far, the overall model has been introduced. The specific parameters of each layer can be seen in Table Ⅰ.

TABLE I. QUANTIZED MODEL LAYER PARAMETERS

Layer	Type	Kernel/ Stride	Input shape	Output shape
Input	Input Layer	-	-	-
Conv1	Convolutional	5×5/1	1×32×32_int8	6×28×28_int9
Pool2	Average Pooling	2×2/2	6×28×28_int9	6×14×14_int9
Conv3	Convolutional	5×5/1	6×14×14_int9	16×10×10_ int12
Pool4	Average Pooling	2×2/2	16×10×10_ int12	16×5×5_int12
Conv5	Convolutional	5×5/1	16×5×5_int12	120×1×1_int12
Fc_6	Fully Connected	-	120×1×1_int12	84×1×1_int12
Fc_out	Fully Connected	-	84×1×1_int12	10×1×1_int14

B. Quantization Method and Quantization Aware Training

In this work, since INT-type data is required for subsequent chip design operations, we need to perform quantization processing on the model in advance at the software end. The pixels of grayscale images can usually be stored using INT8, that is, 0-255. In the normal model training process, weight values are often stored using FLOAT32 floating-point data. To achieve INT8 quantization for all weights and biases here, we need to scale the trained model weights proportionally, that is, convert the weights from the original 0-1 floating-point to 0-255 integer. Equation (1) shows the convolution method of our unquantized model.

$$out = \sum\nolimits_{kernel} in * weight + bias \qquad (1)$$

Correspondingly, we need to quantize the input, weights, and bias to integer types. Equation (2) represents the convolution calculation formula after quantization.

$$out = \left[\left(\sum\nolimits_{kernel} in * (q\_weight - Z) + \frac{bias*gain}{scale \gg shift} \right) * scale \right] \gg shift \qquad (2)$$

In the equation, each parameter is an integer type. Here, q_weight represents the quantized integer weights. Since the weights are quantized to INT8, $gain$ is set to 256. Z is the zero_point, indicating the integer value corresponding to 0 in the floating-point weights after quantization. $scale$ is an integer scaling factor. Given that real scaling factors are often decimals and cannot be stored as integer data, the integer shift parameter $shift$ is introduced to assist the hardware implementation. Therefore, the actual scaling factor is $scale \gg shift$. The specific calculations for $scale$ and $shift$ are shown in Equation (3) and the calculation of Z is defined in Equation (4).

$$scale \gg shift = (weight_{max} - weight_{min})/256 \qquad (3)$$

$$Z = (weight_{min} - 0)/(scale \gg shift) \qquad (4)$$

Combining Equations (3) and (4), the quantized integer q_weight in Equation (2) is computed as:

$$q\_weight = \frac{weigh}{scale \gg shift} + Z \qquad (5)$$

In the above, we introduced the specific quantization process for weights, which will also be applied in the subsequent circuit implementation. However, during the model training process, we need to consider not only the quantization of weights but also the quantization of the output of each layer, which serves as the input to the next layer. Therefore, during the training process, we adopt quantization-aware training (QAT) to simulate the precision loss caused by the quantization process on the hardware side. The specific operation method is to quantize the weights and outputs of each layer according to Equation (5), then dequantize them back to floating-point numbers, and use the dequantized data as the input for training. Meanwhile, according to the quantization bit width of each layer in the hardware circuit, the output of each layer can be truncated to the maximum value corresponding to the bit width. This process of first quantizing, then dequantizing, and finally truncating the data according to the bit width can effectively simulate the precision loss caused by quantization in each layer on the hardware side. Thus, the trained model can also have robustness against quantization loss in the subsequently designed hardware circuit.

III. HARDWARE CIRCUIT IMPLEMENTATION OF QUANTIZED MODEL

Based on the quantized LeNet-5 model trained through quantization-aware training in the previous section, we need to realize the edge deployment of the model through actual circuit implementation. In the operation module, we design operators using quantized integer data, which effectively reduces the power consumption and area of the chip. Meanwhile, a 5-stage pipeline design method is adopted for the multiplier to improve the computing speed of the chip. In terms of data storage, to reduce area and power consumption, we use 4 SRAM modules for the storage and reuse of weights and data respectively, which significantly reduces the overall area and power consumption of the chip. The overall system structure of the chip is shown in the figure.

The overall structure is composed of the model calculation part, data storage part, control module part and interface part. The model calculation part mainly implements the LeNet-5 model trained in the second section, including three convolutional layers, two pooling layers and two fully connected layers. The data storage part includes four SRAM storage modules, which are used for storing convolutional layer weights, input image data, fully connected layer weights and output feature images of each layer respectively. The control module part is an SRAM control module, which is used to realize the data calling and storage between the overall calculation part and the data storage module. In terms of interface, the SPI module is adopted. The SPI module is used to realize the input of all weights, biases and image data, and also to realize the output of final calculation results.

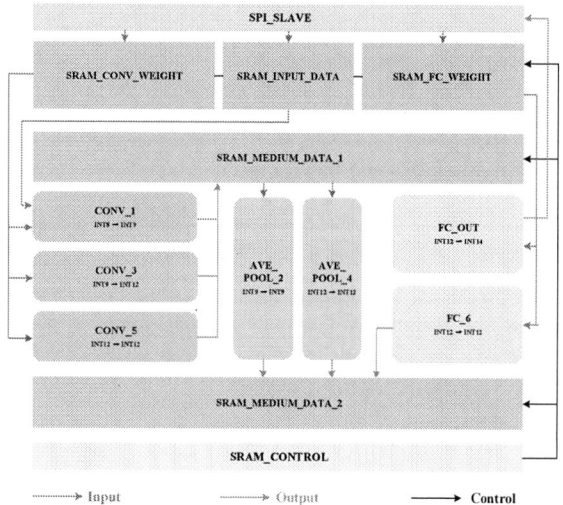

Fig. 2. Overall architecture diagram of the chip.

A. Model Calculation

In this section, the architecture consists of three convolutional layers, two average pooling layers, and two fully connected layers. To minimize area overhead in both convolutional and fully connected layers, instead of storing all weights and data in flip-flops, a specialized computation scheduling strategy is employed. This strategy prioritizes convolving the same set of weight data with an entire channel of image data before updating the convolution kernel values stored in flip-flops, thereby optimizing area through data reuse.

For image data reuse, a shift-based update mechanism is implemented. Upon completing computations for the current row of image data, the flip-flop contents are shifted left, and a new row of data is loaded from SRAM. This approach leverages SRAM storage in place of flip-flops, significantly reducing both chip area and power consumption by minimizing redundant data storage and leveraging efficient data movement patterns.

The computational modules in each convolutional and fully connected layer are implemented using a 5-stage pipelined multiplier architecture. For example, in the convolutional layer, since the convolution kernel of this module is 5×5 in size, the convolution of one convolution kernel with the feature image requires 25 multiplications and 24 additions. To achieve high computational efficiency while reducing area and power consumption, this design implements these 25 multiplications via five multipliers in a 5-stage pipeline, with the 24 additions realized through an adder tree structure.

B. Data Storage

In the data storage module, we utilize four single-ended SRAM blocks. Among them, SRAM_CONV_WEIGHT is used to store the weights and biases of all convolutional layers, while SRAM_FC_WEIGHT is employed to store the weights and biases of the fully connected layers.

SRAM_INPUT_DATA stores the input digital images. SRAM_MEDIUM_DATA_1/2 are respectively used to store the feature maps output by different layers. Since the output of each layer serves as the input of the next layer, and a single-ended SRAM cannot support simultaneous input and output operations, we adopt two SRAM blocks to achieve timely data reuse.

C. Control Module

The Control Module (SRAM_CONTROL), acting as the central nervous system for data management in the quantized LeNet-5 hardware implementation, bridges computational units (convolutional, pooling, fully connected layers) and four single-ended SRAM modules (SRAM_CONV_WEIGHT, SRAM_FC_WEIGHT, SRAM_INPUT_DATA, SRAM_MEDIUM_DATA_1/2) to orchestrate data flow, resolve memory access conflicts, and ensure efficient resource utilization—critical for low-power, high-speed edge deployment; it employs a hierarchical scheduling algorithm to manage bidirectional data transfers between layers and memory, dynamically generating addresses and control signals based on layer-specific parameters and computation phase, coordinating weight fetching from SRAM_CONV_WEIGHT and feature map access from SRAM_INPUT_DATA or SRAM_MEDIUM_DATA_1/2 for convolutional layers while adapting access patterns for pooling and fully connected layers to match their computational needs.

A key challenge in the design is managing the limitations of single-ended SRAMs, which cannot support simultaneous read and write operations. The SRAM_CONTROL module addresses this through two complementary strategies: time-division multiplexing and dual-buffer switching. Time-division multiplexing ensures that each SRAM is accessed in non-overlapping cycles, with read operations preceding writes to prevent bus contention. For intermediate data storage, the module employs a ping-pong buffering scheme using SRAM_MEDIUM_DATA_1/2. While one buffer is being read by the current layer, the other is written to by the previous layer, enabling continuous data flow without stalls. This approach effectively doubles the available bandwidth for intermediate data, critical for maintaining pipeline efficiency.

D. Interface

The interface module (SPI_SLAVE) serves as the communication interface between the quantized LeNet-5 accelerator and external host devices (such as microcontrollers and processors), responsible for data interaction in edge deployment scenarios. Implemented in slave mode, this module adheres to the Serial Peripheral Interface (SPI) protocol and enables bidirectional data transmission via a standard four-wire interface (clock signal SCK, chip select signal CS, master-out-slave-in line MOSI, and master-in-slave-out line MISO). Its main functions include: receiving configuration information (e.g., operation parameters of each layer, quantization-related parameters) sent by external hosts, loading quantized weights and biases of the model into corresponding SRAM storage modules, and transmitting inference results back to the host. By parsing SPI commands to access internal control registers, the module ensures efficient collaboration between the accelerator and external control units, acting as a core component for stable communication between the system hardware and external devices.

IV. LAYOUT AND POST SIMULATION RESULTS

Based on the hardware circuit structure described in the previous section, after undergoing front-end simulation, logic synthesis, and physical design processes (including placement and routing), the circuit diagram of the fully quantized LeNet-5 accelerator implemented using a 180nm process is finally obtained, as shown in Figure 3. The overall chip size is 5mm × 5mm, and the standard voltage for this tape-out is 1.8V. This integrated design includes all optimized core modules, such as computational units like convolutional layers, pooling layers, and fully connected layers, as well as SRAM storage modules, the SRAM_CONTROL module, and the SPI_SLAVE interface module. It completely presents a hardware implementation scheme tailored to the constraints of the 180nm process node and oriented towards efficient edge deployment.

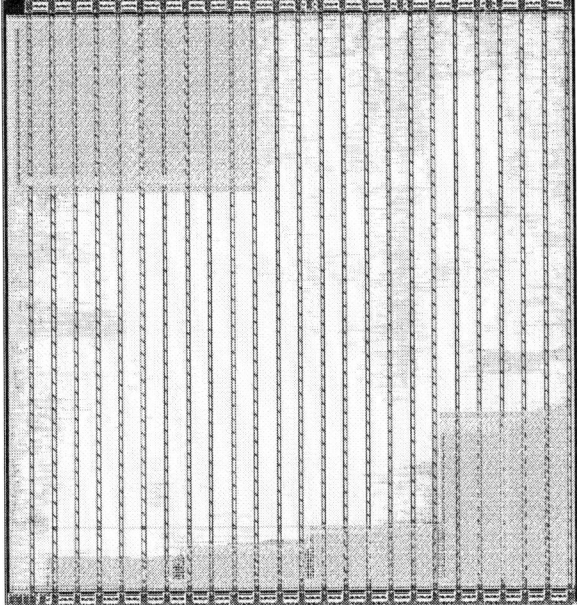

Fig. 3. The layout of the quantized LeNet-5.

In the post simulation of the hardware circuit, 1,000 handwritten digit images were tested at a clock frequency of 100 MHz. The final output achieved an accuracy rate of 98.6%. At this 100 MHz frequency, the recognition frame rate for 32×32 digit images reached a maximum of 863 FPS. Detailed results are presented in Table Ⅱ.

TABLE II. CHIP PARAMETERS

Operating Mode	High-Speed Processing Mode	Low-Power Mode
Chip Process	180nm CMOS	180nm CMOS
Standard Voltage	1.8V	1.8V
Power	517mW	45mW
Area	5mm*5mm	5mm*5mm
Clock Frequency	100Mhz	10Mhz
Interface Frequency	100Mhz	100Mhz
Image Processing Rate	863FPS	101FPS
Recognition Accuracy	98.6%	99.3%

To reduce power consumption, we also tested 1,000 images in a low-power mode with a clock frequency of 10 MHz and an interface frequency of 100 MHz. In this mode, the overall accuracy increased to 99.3%, the total power consumption decreased to 45 mW, and the recognition rate remained at 101 FPS. These two operation modes enable the accelerator to switch between high-frame-rate recognition and low-power consumption scenarios based on application requirements.

V. CONCLUSION

This work presents a fully quantized LeNet-5 accelerator optimized for edge-based handwritten digit recognition, addressing the need for efficient deployment on resource-constrained hardware: by combining pseudo-quantization with quantization-aware training, the model retains 98.6% – 99.3% accuracy using integer arithmetic (INT8/INT9/INT12/INT14), eliminating floating-point operations; implemented in 180nm CMOS with pipelined units and SRAM-based reuse to minimize area and power, it supports dual modes—100MHz high-speed (863 FPS for 32 × 32 images) for real-time tasks and 10MHz low-power (45mW, 101 FPS) for battery-powered devices—validating its balance of performance, efficiency, and accuracy as a robust edge solution, with future work focusing on scaling to larger models, advanced process nodes, and dynamic optimization for better power-performance trade-offs.

REFERENCES

[1] P. Dong et al., "A 28nm 0.22 μ J/Token Memory-Compute-Intensity-Aware CNN-Transformer Accelerator with Hybrid-Attention-Based Layer-Fusion and Cascaded Pruning for Semantic-Segmentation," 2025 IEEE International Solid-State Circuits Conference (ISSCC), San Francisco, CA, USA, 2025, pp. 01-03, doi: 10.1109/ISSCC49661.2025.10904499.

[2] J. Sim, J. -S. Park, M. Kim, D. Bae, Y. Choi and L. -S. Kim, "14.6 A 1.42TOPS/W deep convolutional neural network recognition processor for intelligent IoE systems," 2016 IEEE International Solid-State Circuits Conference (ISSCC), San Francisco, CA, USA, 2016, pp. 264-265, doi: 10.1109/ISSCC.2016.7418008.

[3] K. Matsubara et al., "A 12-nm Autonomous Driving Processor With 60.4 TOPS, 13.8 TOPS/W CNN Executed by Task-Separated ASIL D Control," in IEEE Journal of Solid-State Circuits, vol. 57, no. 1, pp. 115-126, Jan. 2022, doi: 10.1109/JSSC.2021.3120191.

[4] Y. Lecun, L. Bottou, Y. Bengio and P. Haffner, "Gradient-based learning applied to document recognition," in Proceedings of the IEEE, vol. 86, no. 11, pp. 2278-2324, Nov. 1998, doi: 10.1109/5.726791.

[5] Young S I, Zhe W, Taubman D, et al. Transform quantization for CNN compression[J]. IEEE Transactions on Pattern Analysis and Machine Intelligence, 2021, 44(9): 5700-5714.

[6] Cheng J, Wu J, Leng C, et al. Quantized CNN: A unified approach to accelerate and compress convolutional networks[J]. IEEE transactions on neural networks and learning systems, 2017, 29(10): 4730-4743.

[7] B. Jacob et al., "Quantization and Training of Neural Networks for Efficient Integer-Arithmetic-Only Inference," 2018 IEEE/CVF Conference on Computer Vision and Pattern Recognition, Salt Lake City, UT, USA, 2018, pp. 2704-2713, doi: 10.1109/CVPR.2018.00286.

A High-Voltage And High-Precision Operational Amplifier

Juan Wei[1], ZongLin Li[1], HongRui Che[1], FuMei Liang[1], DaGang Li[*1]

[1]Chengdu Sino Microelectronics Technology Co., Ltd., Chengdu 610000, China
[*]Email: dagang@csmsc.com

Abstract—This paper proposes an operational amplifier design suitable for high voltage applications. The device operates at a ± 15 V power supply voltage . The circuit adopts a three-stage cascade architecture: the first stage is a folded common source and common gate circuit and a gain bootstrap structure to achieve high gain; the second stage is a common source amplifier to provide a wide output swing; the third stage is a common drain amplifier circuit, which is connected to the first stage op amp output cross-stage Miller capacitor to achieve flexible zero-pole compensation. The simulation results after layout show that the open-loop DC gain is greater than 150 dB, the phase margin is greater than 60° when driving a 50 pF capacitive load and a 5 kΩ resistive load , and the input offset 3σ is less than 15 μV. This design takes into account high voltage output, ultra-high linearity and extremely low offset, and is suitable for industrial testing, precision signal conditioning, high voltage data acquisition and other fields.

Index Terms—High voltage operational amplifier, folded cascode amplifier circuit, gain bootstrapping, multi-stage Miller compensation, low offset

I. INTRODUCTION

With the increasing requirements for analog front-end performance in fields such as industrial testing, precision signal conditioning, and high-voltage data acquisition, high-voltage operational amplifiers (OPAs) play a key role in applications such as power management, isolation amplification, sensor interfaces, and high-voltage drive. In typical application scenarios, OPAs not only need to provide wide-swing outputs at power supply voltages of ±15 V or higher, but also must have extremely high open-loop gain and bandwidth to ensure the linearity and bandwidth requirements of closed-loop systems. In addition, low input offset and extremely low 1/f noise are essential for the precision amplification of microvolt-level signals.

The traditional two-stage OPA has a simple structure and low power consumption, but its limited open-loop gain and bandwidth, as well as the pair of zero poles introduced by the Miller capacitor, are not easy to adjust flexibly, and it is often difficult to strike a balance between high voltage, wide bandwidth and ultra-high linearity. In order to resolve this contradiction, some studies have changed the second-stage load to an active current mirror load to increase the gain [1], but the pole configuration is limited, making it difficult to simultaneously meet the high performance requirements of gain ≥ 150 dB, phase margin ≥ 60° and wide output swing.

*Corresponding author: dagang@csmsc.com

To address the above challenges, this paper proposes a high-voltage operational amplifier design based on a three-stage cascade architecture:

The first stage adopts a folded common source and common gate amplifier circuit and a gain bootstrap structure to greatly improve the primary gain;

The second stage uses a common source amplifier to provide sufficient swing margin for the output stage;

The third stage introduces a common drain amplifier circuit and connects the Miller compensation capacitor between the first and third stage outputs to achieve flexible arrangement of poles and zeros.

After layout simulation verification, the design achieves open-loop gain ≥ 150 dB , phase margin ≥ 60° , and input offset 3σ is reduced to ≤ 15 μV under ±15 V power supply. This paper will also introduce the circuit structure, compensation mechanism and simulation results in detail, and compare and analyze with the existing high-voltage OPA to prove the advantages of this solution in high-voltage, broadband, high-linearity and high-precision applications.

II. DESIGN METHODOLOGY

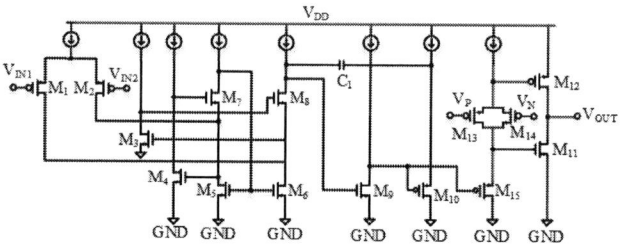

Fig. 1: Overall op amp architecture

As shown in Figure 1, this figure is the overall architecture diagram of the operational amplifier. The first stage adopts a folded common source and common gate structure plus gain bootstrapping, the second stage adopts a common source amplifier circuit, and the third stage adopts a common drain amplifier circuit. This structure has the following innovations.

1.Add a gain bootstrap circuit in the first stage to increase the overall gain of the op amp to ≥ 150 dB;

2.The drain of M7 is introduced into the negative feedback network to avoid large voltage changes due to process corner and temperature deviation, which will affect the operating point of the op amp.

979-8-3315-3918-4/25 $31.00 © 2025 IEEE

3.By introducing the third-stage common-drain amplifier circuit and connecting the Miller capacitor between the first-stage output and the third-stage output, the zero and pole can be flexibly adjusted.

This structure op amp is used in a DAC with a power supply voltage of ±15V. The power supply operating range of the op amp is also ±15V. First, at the beginning of the op amp design, its gain requirements are evaluated as shown below:

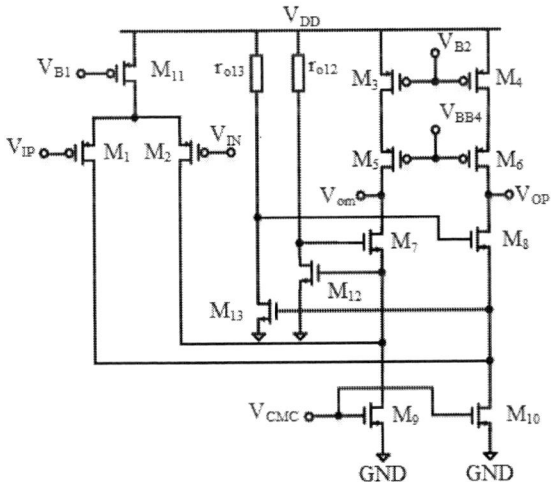

Fig. 2: Folded cascode + gain bootstrapping

As shown in Figure 2, this is the first-stage structure of this op amp, and its gain expression is as follows:

$$A_{V2} = g_{m1}(g_{m8}r_{o8}r_{o10}//g_{m6}r_{o6}r_{o4})g_{m13}\frac{r_{o13}}{2} \quad (1)$$

Here we assume that all gm and ro are equal, which can be obtained from the above equation:

$$A_{V2} = \frac{g_m^3 r_o^3}{4} \quad (2)$$

Considering that the output swing of the second-stage op amp needs to be in a larger range, the second-stage op amp is temporarily assumed to be a common-source amplifier circuit, and its overall gain is:

$$A_V = \frac{g_m^4 r_o^4}{8} \geq 10^6 \quad (3)$$

After calculation, its intrinsic gain needs to satisfy:

$$g_m r_o \geq 53.18 \quad (4)$$

The intrinsic gain requirement for the tube is not high, and there are many processes to choose from , so the first-stage structure op amp is determined to be a folded cascode + gain bootstrap. Considering that the second-stage op amp needs to provide a large output swing, a common source amplifier circuit will be used.

In the second innovation, the drain of M7 is connected to the gate of M5. In order to avoid M7 entering the linear region

due to large changes in the drain voltage of M7 when the process angle and temperature are biased, the drain of M7 is introduced into the overall negative feedback network to make its voltage relatively stable.

The third innovation is the introduction of a third-stage common-drain amplifier circuit with a gain of approximately 1, which makes its poles and zeros flexibly adjustable. In the traditional circuit, the Miller capacitor is connected across the first-stage output and the second-stage output. When adjusting its phase margin, the capacitor size and the size of the second-stage op amp can be changed. By only changing the capacitor size, considering the slew rate problem, the zero-pole adjustment range is limited. If the size of the second-stage op amp is adjusted, the overall gain will change. Its simplified structure is shown below.

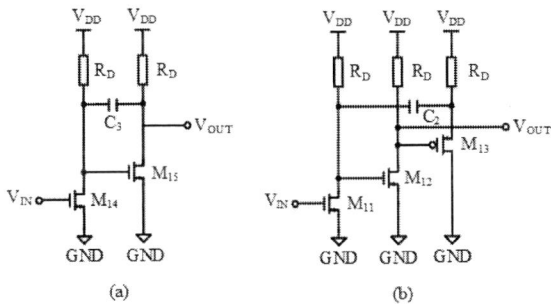

Fig. 3: (a) Traditional Miller structure; (b) Improved structure;

The first and second stage op amps are traditional common source stage circuits, and the third stage is a common drain amplifier circuit, that is, a source follower. Figure 3 (a) is a traditional circuit, and Figure 3 (b) is an improved structure.

For the sake of simplicity, the following analysis method uses Miller approximation and introduces a general formula for op amp gain, as follows:

$$A_V = A_{V0}\frac{(1 + \frac{s}{z})}{(1 + \frac{s}{\omega_1})(1 + \frac{s}{\omega_2})} \quad (5)$$

In the above formula, z represents the zero point, ω represents the pole, and A_V0 represents the low-frequency gain.

Without considering parasitic capacitance, the main pole of the circuit is:

$$p_1 = \frac{1}{r_{ds4}(1 + g_{m15}(r_{ds15}//R_D)C_3)} \quad (6)$$

the secondary pole is:

$$p_2 = \frac{1}{r_{ds15}(1 + \frac{1}{g_{m15}(r_{ds15}//R_D)}C_3)} \quad (7)$$

The solution of the zero point is as follows. According to the definition of the zero point, when $s = s_z$, $V_{OUT}(s_z) = 0$, we can get:

$$V_{G15}S_ZC_3 = g_{m15}V_{G15} \quad (8)$$

The zero point is:

$$S_Z = \frac{g_{m15}}{C_3} \quad (9)$$

The final gain formula is:

$$A_V = A_{V14}A_{V15}\frac{1+\frac{s}{z}}{(1+\frac{s}{\omega_1})(1+\frac{s}{\omega_2})}$$

$$= A_{V14}A_{V15}\frac{(1+\frac{s}{\frac{g_{m3}}{C_3}})}{1+\frac{s}{\frac{1}{r_{ds14}(1+g_{m15}(r_{ds15}//R_D)C_3)}}}$$

$$*\frac{1}{(1+\frac{s}{\frac{C_3}{r_{ds15}(1+g_{m15}(r_{ds15}//R_D))}})} \quad (10)$$

Among them, A_{V14} and A_{V15} represent the low-frequency gain of the first-stage op amp and the second-stage op amp respectively. According to the effect of Miller capacitance, the primary and secondary poles are separated. While moving the primary pole to low frequency, a zero point is introduced at the second stage. The zero point can compensate the secondary pole by adjusting the position, thereby improving the phase margin of the circuit. According to the above formula, the way to adjust the zero point is to directly adjust C_3 or to adjust g_{m15}. In the actual circuit, if you want to move the zero point to low frequency and compensate the secondary pole, you can increase C_3, but the capacitance cannot be increased all the time. Increasing the capacitance will bring about the disadvantages of reduced slew rate and reduced op amp speed. If the zero point position of g_{m15} is adjusted by lowering, the gain of the second-stage op amp will decrease, thereby affecting the overall circuit gain. To solve the above problem, the circuit of Figure 3 (b) is introduced, and the gain expression of Figure 3 (b) is written using the same method.

The poles of Figure3(b) are as follows: the main pole:

$$\omega_1 = \frac{1}{r_{ds11}(1+g_{m12}(r_{ds12}//R_D)\frac{g_{m13}R_D}{1+g_{m13}R_D})C_2} \quad (11)$$

and there are two secondary poles, namely:

$$\omega_2 = \frac{1}{r_{ds12}C_{DS12}} \quad (12)$$

$$\omega_3 = \frac{1}{r_{ds13}(1+\frac{1}{(1+g_{m12}(r_{ds12}//R_D)\frac{g_{m13}R_D}{1+g_{m13}R_D})C_2})} \quad (13)$$

The zero point solution is as follows:

$$V_{G12}S_ZC_2 = g_{m13}V_{G13} \quad (14)$$

$$V_{G12}A_{V12} = V_{G13} \quad (15)$$

Available: $S_Z = \frac{A_{V12}g_{m13}}{C_2}$;
Substituting the above formula into the following formula:

$$A_V = A_{V11}A_{V12}\frac{(1+\frac{s}{z})}{(1+\frac{s}{\omega_1})(1+\frac{s}{\omega_2})(1+\frac{s}{\omega_3})} \quad (16)$$

It can be seen that the low-frequency gain of the above formula is mainly affected by A_{V11} and A_{V12}. According to the zero-point formula, the zero-point position is affected by A_{V12}, g_{m13}, and C_2. Compared with the zero-point formula in Figure 3 (a), the phase margin can be adjusted by adjusting g_{m13} and C_2 without affecting the size of the low-frequency gain.

III. LAYOUT AND SIMULATION

As shown in the figure below, this is the layout of this op amp.

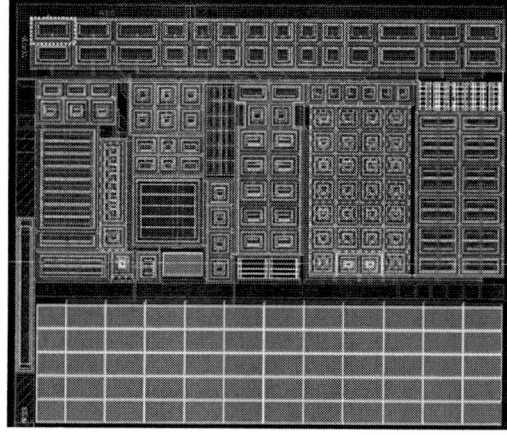

Fig. 4: Op amp layout

Under the TT process angle, the gain and phase margin of different common mode voltages are shown in the following table.

TABLE 1: Gain and Phase Margin

	Corner	Temperature	GAIN	PM
VIN=-5V	TT	-55°C	163.5dB	87.86°
		25°C	160.9dB	92.73°
		125°C	153.4dB	96.97°
VIN=0V	TT	-55°C	163.8dB	88.61°
		25°C	161.5dB	92.87°
		125°C	159dB	96.68°
VIN=5V	TT	-55°C	164dB	91.57°
		25°C	161.7dB	95.66°
		125°C	159.3dB	99.45°

As shown in the figure below, this is the simulation diagram of the op amp after input offset.

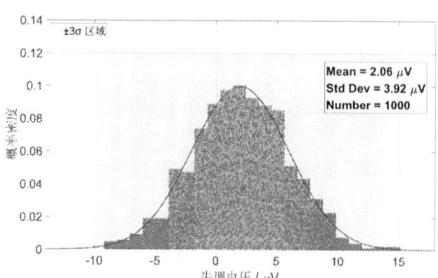

Fig. 5: Monte Carlo simulation of op amp input offset

The op amp works under ±15V power supply, $5\,\text{k}\Omega$ load, 50pF, and its performance indicators are as follows:

TABLE 2: Op Amp Indicators

index	scope
Gain	≥ 150 dB
Phase Margin	$\geq 60°$
Common Mode Rejection Ratio	-220dB
Power Supply Rejection Ratio	-187dB
Gain BandWidth	1.7MHZ
Slew rate	$3V/\mu s$
Offset voltage	$2.06\mu V$
Load Capacitance	50pF
Load resistance	$5\,\text{k}\Omega$

TABLE 3: Performance index comparison

	Stanescu[2]	Zhang[3]	Fan [4]	This article
V_{DD}	4V~24V	5/40V	20V	15V
V_{SS}	0V	0V	0V	-15V
C_L	100pF	10pF	70pF	50pF
R_L	$10\,\text{k}\Omega$	-	-	$5\,\text{k}\Omega$
$Gain$	134dB	119.5dB	78dB	150dB
P_M	$\geq 62°$	$\geq 90.8°$	$\geq 80°$	$\geq 60°$
GBW	1.8MHZ	1.9MHZ	0.8MHZ	1.7MHZ
SR	$1.6V/\mu s$	$1.3V/\mu s$	-	$3V/\mu s$
V_{OFFSET}	$10\mu V$	$0.78\mu V$	$3\mu V$	$2.06\mu V$

IV. CONCLUSION

This op amp is a high-voltage op amp designed based on the $0.6\mu m$ BCD process. The op amp uses gain bootstrapping to increase the gain of the op amp, introduces a third-stage common-drain amplifier circuit to flexibly adjust the gain of the op amp, and uses a common-source class AB output at the output stage to increase the output range. In the previous section and the comparison table of high-voltage op amps in other papers, the advantage of the op amp in this article is that it can achieve greater gain and greater slew rate under positive and negative power supplies. The op amp of this structure is suitable for high-voltage drive circuits with high precision requirements.

ACKNOWLEDGMENT

This work was completed by the in-house team at Chengdu Sino Microelectronics Technology Co., Ltd. We especially thank the company and its departments for their technical support and resource provisions, which enabled the successful development and validation of the high-voltage, high-precision operational amplifier project.

REFERENCES

[1] Jiayuan Zhang, "Two-Stage Operational Amplifier Design by Using Direct and Indirect Feedback Compensations," AIP Advances , vol. 11, no. 2, 020040, Feb. 2021.

[2] Stanescu C, Dinca C, Mcdonald D, et al. A 24 V chopper offset-stabilized operational amplifier with symmetrical RC notch filters[C]//2017 International Semiconductor Conference (CAS). IEEE, 2017: 167-170.

[3] Zhang Junan , Zhang Chuandao, Yang Faming, et al. A 40 V high voltage output auto-zero operational amplifier based on 0.6 μm BCD process[J]. Microelectronics, 2023, 53(5): 786-793.

[4] FAN Q, HUIJSING JH, MAKINWA KA A. A capacitively-coupled chopper operational amplifier with 3 μV offset and outside-the-rail capability [C] //ESSCIRC. Bordeaux, France. 2012: 73-76.

A 12-bit 1MS/s SAR ADC design for high-temperature MEMS accelerometers

Yanlin Mo [1], Min Qi*[1]

[1] Institute of Acoustics, Chinese Academy of Sciences

* Email: moyanlin23@mails.ucas.ac.cn, min1983@mail.ioa.ac.cn

Abstract—**This paper presents a 12-bit 1MS/s successive approximation register analog-to-digital converter (SAR ADC) designed for high-temperature MEMS accelerometers. The circuit is implemented using X-FAB's 0.18μm XH018 high-temperature, high-voltage CMOS process, targeting an operational temperature range of -40°C to 175°C. To ensure stable performance at 175°C, a trimming structure is proposed, which supports both manual and automatic adjustments to maintain ADC functionality across the specified temperature range. The analog and digital circuits operate at 5V and 3.3V supplies, respectively. Pre-layout simulation achieves 12-bit resolution, while post-layout simulation yields 10.65-bit resolution, with a total power consumption of 7.36mW.**

Keywords—*Analog-to-digital converter, high-temperature SAR ADC, pre-amplified latch, internally generated clock*

I. INTRODUCTION

In recent years, With the continuous advancement of integrated circuit technologies, analog-to-digital converters, serving as critical interface components for signal conversion between analog and digital domains have become indispensable in modern signal acquisition systems. In cutting-edge applications such as petroleum exploration, where ADC performance directly determines the accuracy of signal acquisition and control systems in precision microelectronic platforms, operational environments often demand extreme temperature tolerance up to 175°C. Conventional MEMS accelerometer servo ASIC circuits typically operate within 125°C, proving inadequate for oil drilling environments. Among various ADC architectures, successive approximation register ADCs (SAR ADCs) have gained significant adoption due to their superior power efficiency, compact footprint, and balanced precision-speed characteristics. This paper presents a high-temperature resilient SAR ADC [1]-[3] chip design specifically engineered for such specialized industrial applications, addressing the critical need for robust signal conversion under extreme thermal conditions.

This paper introduces a 12-bit, 1Ms/s SAR ADC with a trimming structure, including the DAC circuit, preamplifier comparator circuit, asynchronous clock generation circuit schematic, SAR logic circuit design, and I/O port design. The structure of the article is as follows: Part II discusses the ADC architecture and circuit implementation, Part III covers the ADC layout design, and Part IV presents the simulation results and summary.

II. ADC ARCHITECTURE AND CIRCUITS IMPLEEMENTATION

A. DAC switch design and switching strategy

The fully differential 12-bit DAC architecture presented in this work employs a capacitive DAC (CDAC) with a V_{CM}-based switching strategy [4].

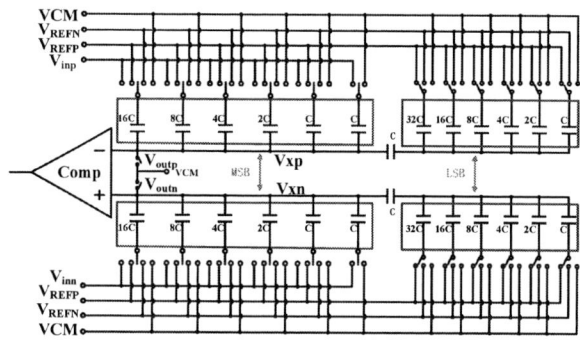

Fig. 1. DAC capacitor array architecture.

During the sampling phase, the top plates of the capacitor array are connected to the common-mode voltage (V_{CM}), while the bottom plates of the MSB capacitors are connected to the differential input signals V_{in} and V_{ip}. The bottom plates of the LSB capacitors remain connected to the V_{CM} throughout this phase.

In the hold phase, the top plates are disconnected from V_{CM}, and the bottom plates of both the MSB and LSB capacitors are switched to V_{CM}. The voltage established at the top plates during this phase serves as the input for the initial comparator comparison. Based on the comparator decision ($V_{xn} > V_{xp}$ or $V_{xn} < V_{xp}$), the bottom plates of the corresponding MSB capacitors are switched to the reference voltages: the highest MSB capacitor in the V_{xp} path is connected to V_{ref} (4 V) and the corresponding capacitor in the V_{xn} path is connected to V_{nref} (0 V) if $V_{xn} > V_{xp}$; conversely, if $V_{xn} < V_{xp}$, the second-highest MSB capacitor in the V_{xn} path is connected to V_{ref} and the corresponding capacitor in the V_{xp} path is connected to V_{nref}. This switching logic is iterated sequentially for subsequent bits. Notably, the bottom plate of the final MSB capacitor C is connected exclusively to the input signals during the sampling phase and remains connected to V_{CM} during all subsequent operational phases.

At the end of the sampling phase (when SWb transitions from 8 V to 0V), ENb goes high while Q_{in} remains reset to 0. During this period, V_{CM} is connected to the output. When the comparison for a specific bit is completed, Q_{in} transitions to a high state, thereby disconnecting V_{CM} from the output. Simultaneously, the reference voltage(either V_{ref} = 4V or V_{nref} = 0V) is connected to the output based on the comparator decision result. Notably, the bottom-plate switch of the final MSB capacitor C remains permanently connected to V_{CM} after the sampling phase. This configuration ensures stable common-mode voltage maintenance while enabling precise reference switching according to the successive approximation logic.

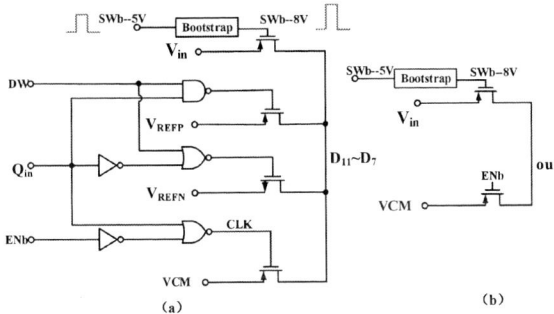

Fig. 2. (a) Bottom-plate switches for the MSB segment 16C-C, and(b) Bottom-plate switch for the final MSB capacitor C.

TABLE I

Port Name	Definition
V_{REFP}/V_{REFN}	Positive/Negative Reference Voltage
V_{CM}	Common-mode Voltage
SWb	Bottom Plate Sampling Switch Clock
DW	Comparator Comparison Result
Q_{in}	Comparator Comparison Complete Signal
ENb	Sampling Start/End Signal

At the instant sampling concludes, Q_{in} remains in its reset state 0, and the common-mode voltage is connected to the output. Upon completion of the corresponding bit comparison, Q_{in} transitions to a high state, thereby disconnecting V_{CM}. The reference voltage V_{ref} or V_{nref} is then connected to the output based on the comparator decision result (DW value).

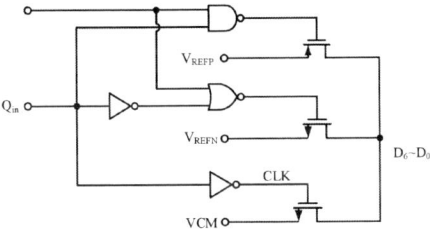

Fig. 3. Bottom Plate Switch.

At the moment sampling ends, Q_{in} remains reset to 0, and V_{CM} stays connected to the output. Once the comparison for the corresponding bit is completed, Q_{in} goes high, thereby disconnecting V_{CM}. Depending on the comparison result (i.e., the value of DW), either V_{ref} or V_{nref} is then connected to the output.

B. Pre-Amplified Auto-Zeroing Comparator

The comparator employs a two-stage preamplifier dynamic comparator with auto-zeroing technology. the

comparator module is powered by a 0V to 5V supply, with bias voltages generated via a constant-Gm bias circuit. In this design, the reference voltage provided to the ADC is 4 V, and the value of 1 LSB can be determined as follows:

$$1LSB = \frac{V_{ref}}{2^N} = \frac{4}{2^{12}} = 0.98mV \quad (1)$$

The preamplifier needs to be able to recognize 0.5LSB = 0.49mV; the minimum recognition voltage of the latch is designed to be 30mV, so the gain of the preamplifier should be:

$$Av = \frac{30mV}{0.49mV} = 61.22 = 35dB \quad (2)$$

Design the preamplifier to settle to 90% within 40 ns, i.e.:

$$e^{-2\pi f_0 t} < 10\% \Rightarrow f_0 > 9.16MHz \quad (3)$$

Therefore, the dominant pole frequency f_0 should be at least 9.16 MHz.

To ensure the proper operation of the comparator's pre-amplifier across process corners and temperature ranges, a trimming circuit for the amplifier's bias current is designed. The current in the M2 branch serves as the bias current for the comparator, while IB, generated by the bias circuit (a), is the total bias current. This total current is distributed through four shunt branches M3, M5, M7, M9 to maintain the comparator's bias current within the normal range under varying temperatures. The switches M4, M6, M8, and M10 control the shunt branches, ensuring the output bias current remains 25μA across different conditions. As shown in Figure 6, it is the input and output waveform diagram of the comparator.

By dynamically adjusting the combination of shunt branches via the switches, this trimming structure achieves precise compensation, stabilizing the comparator's bias current at 25μA despite temperature variations. The core mechanism relies on binary-weighted shunt branches and switch state encoding to flexibly adapt to current changes under diverse environmental conditions.

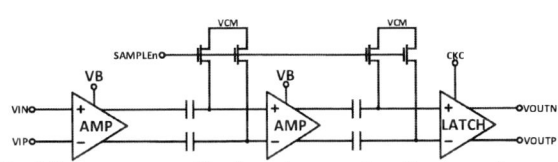

Fig. 4. Two-stage preamplifier dynamic comparator with auto-zeroing technology.

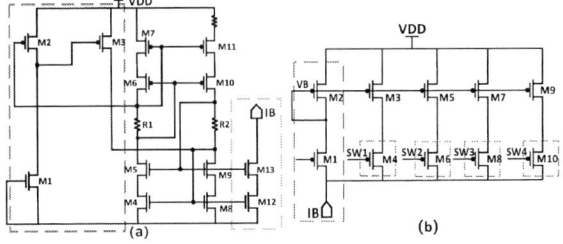

Fig. 5. (a)Bias Circuit, and (b)Current trimming Circuit.

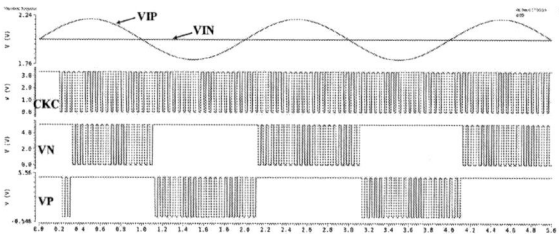

Fig. 6. Output waveform diagram of the comparator.

Fig. 8. (a)Negative Pulse Generation, and (b) Single Negative Pulse Generation.

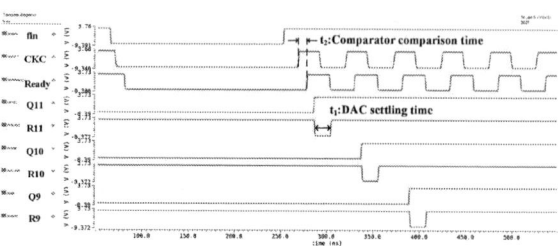

Fig. 9. Timing diagram of asynchronous clock circuit.

C. Design of Digital Logic Control Circuit

VP and VN are the two output voltages of the comparator in Fig. 7. During comparison, both VP and VN are reset to high level, while Ready is set to low level. When the comparator completes the comparison, the comparison completion signal Ready transitions to high level. After a brief delay, Q_{11} becomes high, and D_{11} selects whether to connect the lower plate to V_{REFP} or V_{REFN} based on the comparator result. The content shown in Fig. 8 are Negative Pulse Generation and Single Negative Pulse Generation, this rising edge triggers R_{11} to generate a negative pulse with a fixed delay of t_1 shown in Fig. 9. During this delay, the DAC completes its settling.

Since R_{11} generates a negative pulse with a fixed delay of t_1, CKC correspondingly goes low, causing the latch output to be set to 1, thereby pulling Ready back to low level. After the t_1 delay, CKC transitions to high again due to R_{11}, producing a second rising edge, and the process repeats from step 10 until all bits ($D_{10} \sim D_0$) are determined. The system then waits for the next rising edge of Clkin, and the same process repeats to complete the output of the previous 12-bit digital code ($D_{11} \sim D_0$), followed by sampling, holding, and conversion.

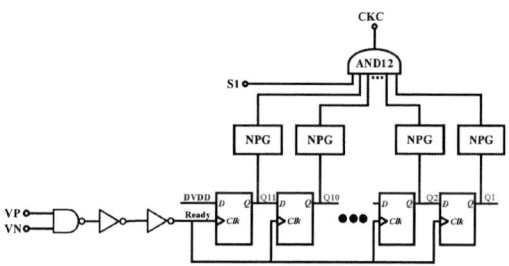

Fig. 7. Asynchronous Clock Generation Circuit.

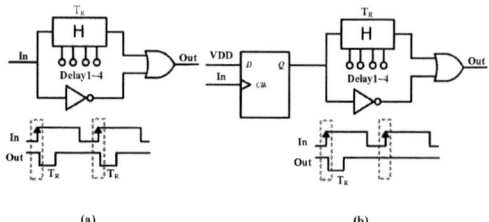

D. Trimming Delay Circuit Design

Fig. 10. shows the delay module with trimming structure designed in this work. Here, capacitor 2C is permanently connected in the circuit. The external inputs Delay1~4 can adjust the delay time of the module. When all inputs are at high level, capacitor 8C is connected to the circuit, increasing the total capacitance and thus prolonging the delay time to ensure sufficient DAC settling. When all inputs are at low level, capacitor 8C is disconnected from the circuit, reducing the total capacitance and shortening the delay time to guarantee ADC conversion completion within the specified period. The appropriate delay setting can be selected according to the operating temperature of the chip to maintain consistent delay across the entire temperature range.

The Delay1~4 inputs can be manually adjusted by external switches to determine whether they are set to high or low level. Alternatively, a separate external circuit can be designed to automatically adjust Delay1~4 based on the chip's operating state, enabling trimming functionality.

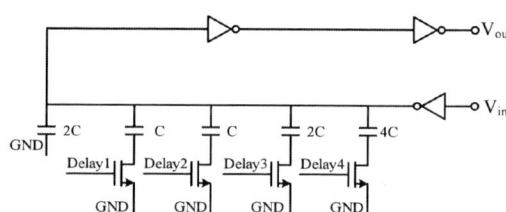

Fig. 10. The delay circuit with a trimming structure.

III. SIMULATION RESULTS

The proposed SAR ADC was designed in a 180nm CMOS high-temperature process. In the layout of the DAC capacitor array, the parasitics at both ends of the bridge capacitor are crucial factors affecting the weight. To reduce the impact of parasitic capacitance on the overall weight of the capacitor array, thereby improving the output accuracy of the ADC. Fig. 11 shows the overall layout structure of the SAR ADC, including the layout of the SAR ADC core circuit and the I/O port circuit.

As show in Fig. 12, with a 1MS/s rate input, the measured SNDR, SFDR, and ENOB are 67.34dB, 81.25dB, and 10.65 bit.Under the conditions of an analog power supply voltage of 5V, a digital power supply voltage of 3.3V, and an ambient temperature of 175°C, the measured power consumption is 7.36mW. The power consumption distribution is as follows: DAC which including the sample-

979-8-3315-3918-4/25 $31.00 © 2025 IEEE

and-hold circuit accounts for 27%, digital circuits account for 24%, and the comparator account for 49%, as shown in Fig. 13. Table II summarizes the performance of the proposed SAR ADC and compares it to the other three designs, and features a temperature range of -40°C to 175°C, outperforming other designs in high-temperature tolerance while maintaining comparable resolution and dynamic performance.

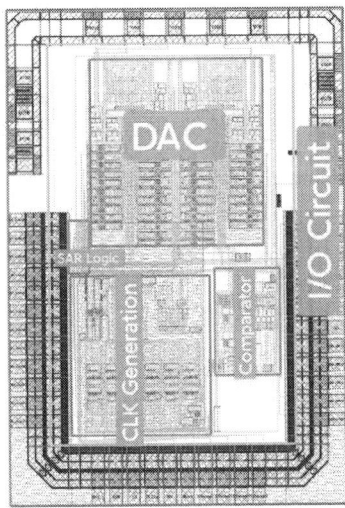

Fig. 11. The layout of the proposed ADC.

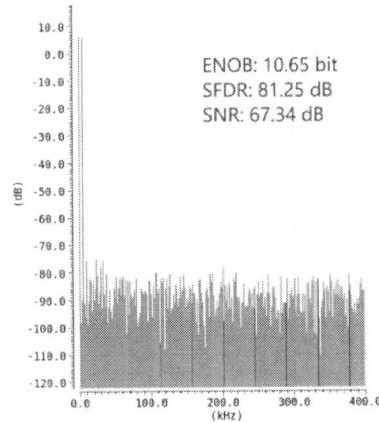

ENOB: 10.65 bit
SFDR: 81.25 dB
SNR: 67.34 dB

Fig. 12. FFT spectrum at 1 MS/s.

Fig. 13. Power dissipation.

TABLE II.

	[5]	[6]	[7]	This work
Technology(nm)	28	28	180	180
Active Area(mm²)	0.525	\	0.057	1.467
Resolution(bits)	9	12	9	12
Sampling Rate(MS/s)	500	1000	0.8	1
SFDR(dB)	60.4	62	\	80
ENOB(bits)	7.8	>10	8.5	10.63
Temperature range(°C)	-30~80	-40~125	-50~110	-40~175

IV. CONCLUSION

This paper presents a 12-bit 1MS/s SAR ADC tailored for high-temperature MEMS accelerometers, implemented in 0.18μm CMOS process to withstand -40°C to 175°C. Key innovations include a manual-mode trimming structure to stabilize performance across high temperature variations, a fully differential CDAC with a VCM-based switching strategy for precise charge redistribution, and a two-stage pre-amplified auto-zeroing comparator with bias current trimming to to reduce offset voltage and improve comparator accuracy. The asynchronous clock generation and delay trimming circuits optimize conversion timing, enable the ADC to operate stably and normally within a wide temperature range without compromising performance while the common-centroid layout of the DAC capacitor array mitigates parasitic effects. This design can address the critical need for robust signal conversion in extreme-temperature industrial applications like petroleum exploration, demonstrating reliable performance and power efficiency in high-temperature MEMS systems.

REFERENCES

[1] Babayan-Mashhadi S, Jahangiri-Khah M. A low-power, signal-specific SAR ADC for neural sensing applications[J]. Journal of Circuits, Systems and Computers, 2018, 27(14):1850230.

[2] Q. Zhao, Q. Huang, Y. Chen, Y. Fan, S. Huang and J. Yuan, "A 16-bit 1-MS/s SAR ADC With Asynchronous LSB Averaging Achieving 95.1-dB SNDR and 98.1-dB DR," in IEEE Transactions on Circuits and Systems I: Regular Papers, vol. 71, no. 12, pp. 6447-6458, Dec. 2024.

[3] J. Yang, J. Du, Z. Ma and K. Wang, "A BJT-Based Amplifier-Less Temperature Sensor with 12-bit SAR ADC Readout and ±0.6°C (3σ) Inaccuracy from -40°C to 120°C," in IEEE Transactions on Circuits and Systems II: Express Briefs.

[4] C. -C. Liu, S. -J. Chang, G. -Y. Huang and Y. -Z. Lin, "A 10-bit 50-MS/s SAR ADC With a Monotonic Capacitor Switching Procedure," in IEEE Journal of Solid-State Circuits, vol. 45, no. 4, pp. 731-740, April 2010.

[5] H. Kim, S. Lee and J. Kim, "A 9-Bit 500-ms/s 4-Stage Pipelined SAR ADC With Wide Input Common-Mode Range Using Replica-Biased Dynamic Residue Amplifiers," in IEEE Access, vol. 11, pp. 22531-22541, 2023.

[6] J. Goes, B. Tardivel, J. de Melo and J. Marques, "A Temperature-Compensated Class-AB Parametric Residue Amplifier for SAR-Assisted Pipeline ADCs," 2020 IEEE International Symposium on Circuits and Systems (ISCAS), Seville, Spain, 2020, pp. 1-5.

[7] A. Aprile, M. Folz, D. Gardino, P. Malcovati and E. Bonizzoni, "An Area-Efficient Smart Temperature Sensor Based on a Fully Current Processing Error-Feedback Noise-Shaping SAR ADC in 180-nm CMOS," in IEEE Journal of Solid-State Circuits, vol. 59, no. 3, pp. 716-727, March 2024.

A Low-Noise Ultrasound Analog Front-End with Low Gain Error Time-Gain Compensation

Xiangchen Wan, Fan Ye*

The State Key Lab of Integrated Chips and Systems, Fudan University.

* Email: Xiangchen Wan 22212020031@m.fudan.edu.cn, Fan Ye fanye@fudan.edu.cn

Abstract—This paper presents a low-noise ultrasound analog front-end (AFE) circuit with low gain error time-gain compensation (TGC). The AFE consists of a low noise amplifier (LNA), a TGC circuit and an anti-aliasing filter (AAF). The LNA uses a split varied input stage to ensure uniform bandwidth and reduce power consumption without sacrificing the noise performance. The TGC adopts an interpolating structure to achieve a monotonically smooth linear-in-dB gain with low gain error. The AAF is designed as a second-order low-pass filter, which determines the whole AFE's bandwidth. Simulation results show the AFE achieves a gain range of 36 dB with an input-referred noise of 1.6 nV/√Hz at 5 MHz and a low gain error below ± 0.3 dB, and the bandwidth is 15/30 MHz.

Keywords—Medical ultrasound imaging, Analog front-end (AFE), Time gain compensation (TGC).

I. INTRODUCTION

In recent years, with its numerous advantages such as high resolution, non-invasiveness, real-time imaging, and cost-effectiveness, ultrasound imaging has become an important method of medical diagnosis and has been widely used in various applications including vascular assessments, tumor localization and echocardiography. In an ultrasound imaging system, the ultrasound signal is transmitted by the transducer and propagates through the tissues and organs in human body, suffering from attenuation and reflection, and the reflected echo signal will be received and processed by the ultrasound receiver, which plays an important role in determining the whole system's performance.

Fig.1 shows the system diagram of a typical ultrasound receiver, the echo signal needs to be pre-processed by the analog front-end (AFE) circuit first, and then converted to digital signal by the analog-to-digital converter (ADC), and finally sent to digital system for further processing. The AFE usually consists of a low noise amplifier (LNA), a time-gain compensation (TGC) circuit and an anti-aliasing filter (AAF). The LNA is usually used to amplify hint signal while keeping the noise floor enough low, and the AAF is needed to filter out out-of-band spurious signal to prevent aliasing. In addition, when the ultrasound signal propagates into the human body, the signal undergoes attenuation, following a linear-in-dB trend [1]. Due to this, the echo signal from deep distance will arrive later and suffer more attenuation, which results in the exponentially decreasing amplitude of the echo signal, as shown in Fig.2. Meanwhile, as illustrated in Fig.3, considering the instantaneous dynamic range (DR) required at any depth for enough image resolution, the overall DR's requirement will be too strict to meet if the echo signal is sent to ADC directly, so a TGC circuit is required to compensate for the attenuation by adjusting the gain of the AFE as a function of time. Through the TGC, the output will have a constant amplitude and the overall DR will be reduced to the instantaneous DR, which can be handled easier by the ADC.

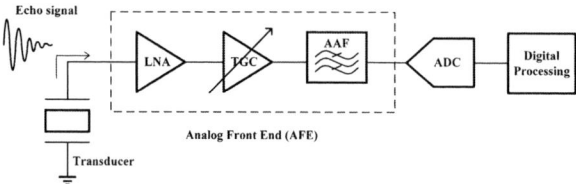

Fig. 1. Diagram of a typical ultrasound receiver.

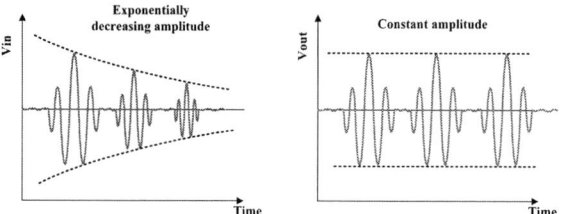

Fig. 2. Input and output signals of the TGC.

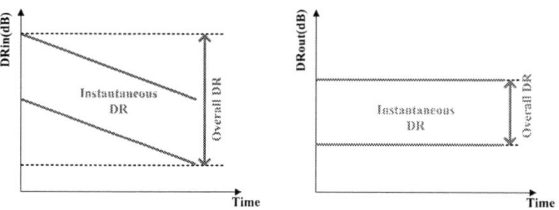

Fig. 3. Input and output DR as a function of time with ideal TGC.

In this brief, a low-noise ultrasound AFE with low gain error continuous TGC is proposed. This AFE can achieve linear-in-dB gain with low input-referred noise (IRN) and low gain error, keeping enough high bandwidth for high frequency ultrasound applications. This brief is organized as follows: in section II, the AFE's overall requirements and specifications are introduced and discussed, and then the specifications will be allocated to each module. Section III introduces each module's architecture and analyze some key points of each module in detail. Section IV presents some simulation results of the proposed AFE. Finally, section V concludes the paper.

II. THE OVERALL SYSTEM'S REQUIREMENTS

For the ultrasound AFE, a key challenge is to maintain sufficient DR for handling both large and small signals. For the ultrasound signal, it penetrates most human tissues with an attenuation rate ranging from 0.3 to 0.6 dB·MHz^{-1}·cm^{-1} [2]. Taking the average attenuation rate of 0.45 dB·MHz^{-1}·cm^{-1} as an example, an ultrasound signal with 5 MHz center frequency will suffer around 36 dB attenuation to penetrate 8 cm depth. Adding a minimum instantaneous DR of 50 dB for obtaining acceptable image display resolution [3], the overall DR of the whole ultrasound AFE must be over 86 dB. To achieve this, the AFE's noise, TGC's gain range and the ADC's signal-to-noise ratio (SNR) must be all considered. As illustrated in

979-8-3315-3918-4/25 $31.00 © 2025 IEEE

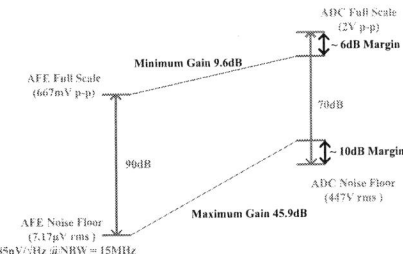

Fig. 4. AFE's gain range and noise requirements.

Fig.4, when the signal is reflected from the surface of the human tissues, the input signal has a relatively large amplitude, so the AFE must have enough full scale to receive it and low gain to avoid saturating the signal and exceeding the ADC's full scale. On the other hand, if the signal is reflected from deep layers of the human tissues, the signal will be very week. Consequently, the AFE is required to have a very low noise floor and relatively high gain to compensate the attenuation, and the AFE's output noise should be higher than ADC's noise floor to avoid the output's SNR being degraded.

In this design, the AFE's input peak-to-peak full scale is 667 mV, and the ADC's peak-to-peak full scale is 2 V, so the minimum gain of the whole AFE can be obtained by (1), which is 9.6 dB. In order to achieve the DR of 90 dB, AFE's noise floor should be less than 7.17 µV. Based on this, the AFE's IRN can be calculated as 1.85 nV/$\sqrt{\text{Hz}}$ if the noise bandwidth is 15 MHz. Considering the mainstream high-end ultrasound receivers employ 12-bit resolution ADC with 70 dB SNR [4], [5], so the ADC's noise floor can be calculated as 447 µV. Similarly, the maximum gain of the AFE can be obtained by (2), which is 45.9 dB, so the TGC's gain range is about 36 dB. In addition, the LNA's gain can be determined by the LNA's input and output full scale, and the TGC's gain and the AAF's gain can be allocated according to the gain requirements of the whole AFE.

$$\text{Min Gain} = \frac{\text{ADC Full-Scale Input}}{\text{AFE Full-Scale Input}} - \text{Top\_Margin} \quad (1)$$

$$\text{Max Gain} = \frac{\text{ADC Noise Floor}}{\text{AFE Noise Floor}} - \text{Bottom\_Margin} \quad (2)$$

To receive and process high-frequency ultrasound signals, the AFE's bandwidth is set as 15 MHz or 30 MHz which is determined by the AAF, and the bandwidth of the LNA and the TGC should be much larger than AAF to avoid the reduction in overall bandwidth. For AFE's noise, although the overall IRN will become significant when the AFE's gain is relatively low due to the TGC's noise contribution, the input signal's amplitude is also large, so the SNR will not be affected. On the contrary, when the AFE has maximum gain, the SNR is worst assuming the output's amplitude is constant, so the noise allocation of each module can be completed according to the noise requirements in this case.

III. CIRCUIT IMPLEMENTATION

A. Low Noise Amplifier

As the first stage of the AFE, LNA's performance is crucial to the whole system. In this design, the core amplifier of the LNA adopts a current feedback topology [6] as shown in Fig.5. The input stage adopts source degeneration twice, for one at input transistor M_1/M_2 aims to improve linearity while the

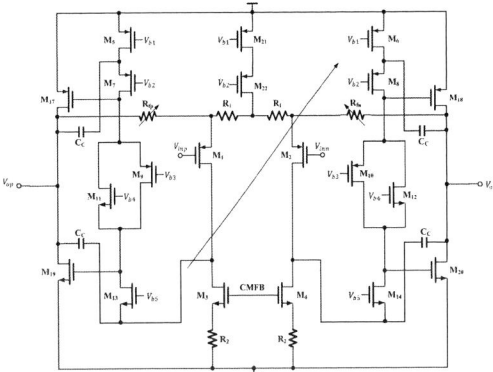

Fig. 5. Schematic of the core amplifier of the LNA.

other at current source transistor M_3/M_4 is used to reduce the noise of M_3/M_4. Although the noise of M_3/M_4 and R_2 can be decreased as the resistance of R_2 increases, it is limited by the voltage dropout on R_2 to ensure a reasonable DC operation point. The output stage employs the class AB structure to save static power because the LNA is needed to drive the TGC's input network, a low value resistive load, which will be introduced in the next part.

The closed-loop gain of the LNA is determined by the ratio between feedback resistor and source degeneration resistor. In order to adapt to various ultrasound application scenarios, the LNA usually has multiple gain settings, so the feedback resistor is varied, which can be realized by digitally-controlled resistor arrays. Correspondingly, the input stage is split into multiple parts, controlled by the same signal as the gain setting, to achieve varied input transconductance. This design has two main advantages: for one thing, when the LNA's gain varies, the feedback factor will also change. If the input stage transconductance follows the change in gain, then the gain bandwidth (GBW) product of the loop gain can keep constant at different gain settings. For another thing, when the LNA's gain is low, the input signal has relatively large amplitude, so the noise requirement of the LNA becomes more relaxed. Consequently, the input stage transconductance and bias current can be decreased to save power.

In addition, the echo signal is AC-coupled to the LNA input, so the input common mode voltage of the LNA needs to be set separately. To avoid static current across the feedback resistor, the source voltage of the input transistors should be the same as the output common mode voltage, so an extra input common mode voltage generation circuit with feedback is employed to ensure the DC operating point of the LNA stable even if the process, voltage, temperature (PVT), and gain setting varies.

B. Time Gain Compensation

The TGC circuit in ultrasound application is designed to achieve linear-in-dB gain to compensate the echo signal with exponentially decaying amplitude. In general, it can be divided into three categories: discrete programmable gain amplifiers (PGA), amplifiers with an approximately exponential transfer function and interpolating variable-gain amplifiers. The discrete PGA may lead to imaging artifacts when switching between gain steps, while the amplifiers with an approximately exponential transfer function usually use open-loop structures so that they are sensitive to PVT variations. Consequently, the interpolating variable-gain amplifier is chosen in this design, which makes a good tradeoff between the two structures mentioned before.

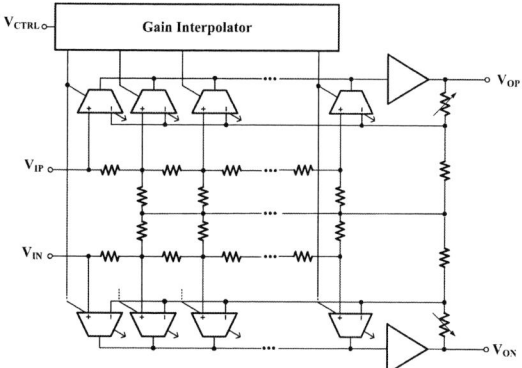

Fig. 6. Schematic of the TGC.

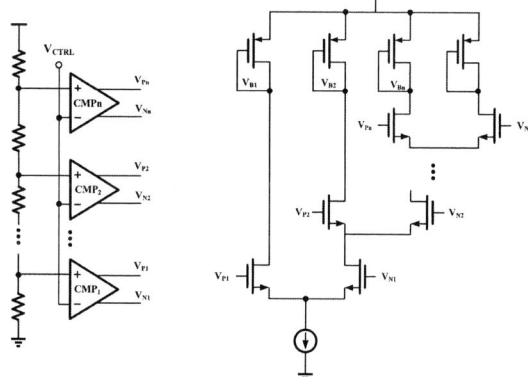

Fig. 7. Schematic of the gain interpolation control signal generation circuit.

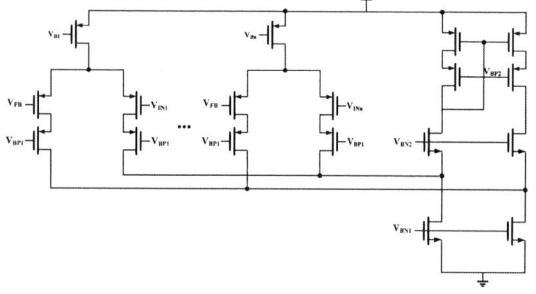

Fig. 8. Schematic of the input stages of the TGC.

As shown in Fig.6, the TGC circuit is composed of a tapped attenuator and a PGA [7]. The attenuator is a 12-stage differential recursive resistor ladder with 3 dB per tap, so the total gain range of the TGC is 36 dB. To avoid introducing much noise, the input resistance per side is 200 Ω, so the LNA uses class AB output stage to drive this small resistance load. The input stages of the PGA are distributed along the ladder, and the control signal V_{CTRL} determines the input tap point as well as gain through the gain interpolator. The gain interpolator is illustrated in Fig.7, since the interpolation has 12 stages, the voltage difference between adjacent reference voltages is relatively small, only 50 mV. Therefore, a set of open-loop comparators are first used to compare the control voltage with the reference voltage to achieve pre-amplification. The resulting voltage controls the current-steering circuit to generate a set of signals to control the bias current of the PGA's input stages. It should be noted that the comparators cannot have a very steep transfer function; otherwise, the total gain curve will exhibit a sawtooth pattern like that of discrete gain control method. Instead, through setting an appropriate gain of

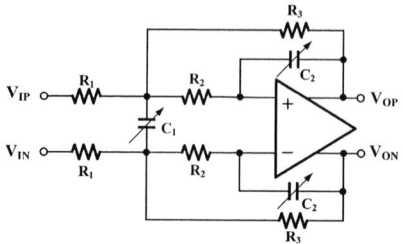

Fig. 9. Schematic of the input stages of the TGC.

comparators, it can generate overlapping bias currents, so the signal from successive taps merge to provide a monotonically smooth gain range, which results in relatively low gain error. Fig.8 shows the input stages of the PGA, the cascode transistors are used to reduce feedthrough from the signal taps to the output and ensure excellent frequency response uniformity across the gain range. In addition, the tapped attenuator only includes passive elements and the PGA employs negative feedback to determine closed-loop gain, so the total gain is relatively precise and the linearity is excellent.

C. Anti-aliasing Filter

The AAF is a second-order low-pass filter which uses multiple feedback structure as depicted in Fig.9. The closed-loop gain is set by the ratio of R_3 and R_1, while the cut-off frequency is determined by R_2, R_3, C_1, and C_2. The cut-off frequency is adjusted by changing the capacitance of C_1 and C_2, so the gain and bandwidth can be adjusted independently.

IV. SIMULATION RESULTS

The proposed AFE is designed in 28 nm CMOS technology. To adapt to different ultrasound applications, the LNA and PGA are designed with multiple gain settings, in which the LNA's gain varies from 12 dB to 18 dB, while the PGA provides a gain range from 24 dB to 30 dB, and the gain step is 3 dB, which is shown in Fig.10.

In the proposed AFE, the TGC is required to compensate the attenuated ultrasound echo signal. To verify the TGC's function, a sinusoidal signal with exponentially decaying amplitude envelop is used as the input, which is illustrated in Fig.11. When the TGC is applied a linear ramp-up control voltage with an appropriate speed, the output sinusoidal signal shows an almost constant amplitude, which is in accord with expectation. Fig.12 shows the TGC's varied gain is an approximately dB-linear function of the control signal, and the gain error can be obtained by subtracting from the ideal linear-in-dB gain, which is below ± 0.3 dB within a 36.2 dB gain range, which well meets the compensation requirements of the TGC.

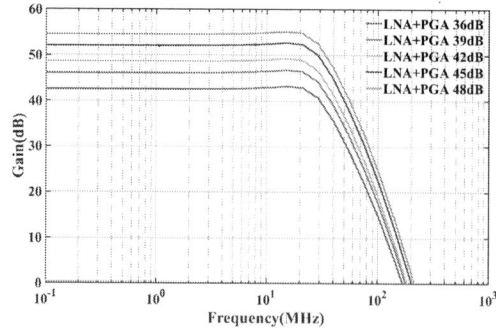

Fig. 10. The gain range of the LNA and PGA.

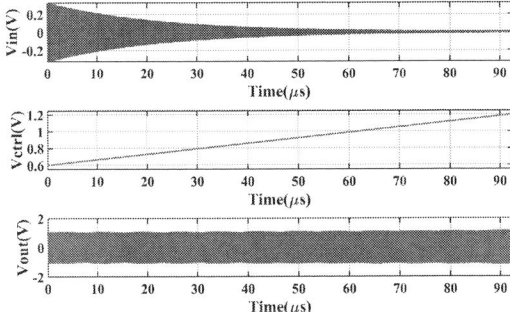

Fig. 11. The transient waveform of the TGC's function verification.

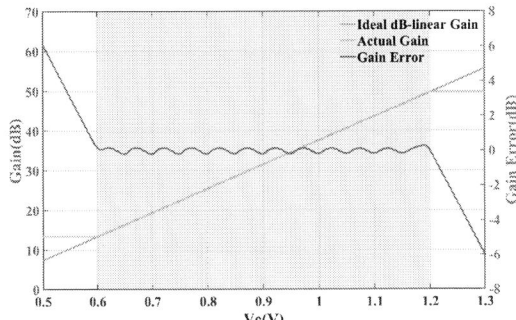

Fig. 12. The TGC's gain range and gain error as a function of Vc.

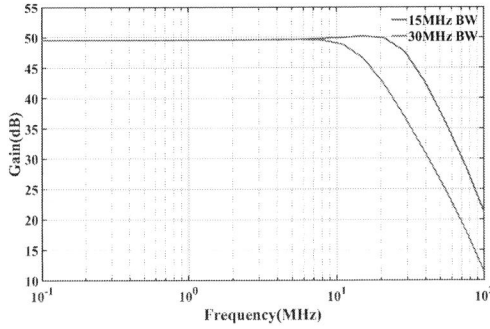

Fig. 13. The bandwidth of the AFE.

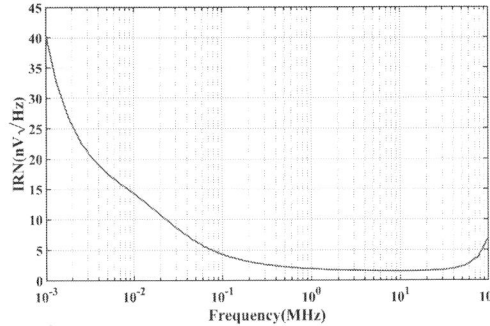

Fig. 14. The input noise density of the AFE.

The AAF in the AFE is designed in two bandwidth modes for ultrasound applications of different frequencies, and the simulated bandwidths are about 30 MHz and 15 MHz respectively which is shown in Fig.13. Fig.14 shows the input noise spectrum of the AFE, and the IRN at 5 MHz frequency is about 1.6 nV/√Hz, which meets the noise specification of the AFE.

Table I gives the performances of this AFE compared with some other ultrasound AFEs, which illustrates the AFE in this work has a very low noise and low gain error. Additionally, it can provide a relatively large tunable bandwidth for different ultrasound applications.

TABLE I.   PERFORMANCE COMPARISON

	This Work	[8]	[9]	[10]	[11]
Process	28 nm CMOS	28 nm CMOS	180 nm CMOS	40 nm CMOS	180 nm BCDMOS
Supply Voltage (V)	2.5	2.5	1.8	1.1	±0.9
Bandwidth (MHz)	15/30	15/30	3.1	3-18	7
TGC Type	Continuous	Continuous	Continuous	Discrete	Continuous
TGC Range (dB)	36	42	37	24	33
Gain Error (dB)	±0.3	±1	±1.4	-	±1
IRN [nV/√Hz]	1.6	2.38	8.6	3.9	3.6

V. CONCLUSION

This paper presents a low-noise AFE with low gain error continuous TGC. The LNA adopts a current feedback topology to achieve a very low input referred noise, and the split varied input stage compensates the bandwidth variation and reduces the power consumption. The TGC uses a differential resistor ladder and PGA to achieve linear-in-dB gain with low gain error, and it also provides excellent frequency response uniformity and linearity. Simulation results show the AFE achieves a gain range of 36 dB with a low IRN of 1.6 nV/√Hz at 5 MHz and a low gain error below ± 0.3 dB. The AFE also has a tunable bandwidth and gain setting, allowing it to suit different ultrasound application scenarios.

REFERENCES

[1] T. L. Szabo, *Diagnostic Ultrasound Imaging: Inside Out*, 2nd ed. Boston, MA, USA: Academic, 2014.

[2] P. R. Hoskins, K. Martin, and A. Thrush, *Diagnostic Ultrasound: Physics and Equipment*, 3nd ed. Cambridge, U.K.: CRC Press, 2019.

[3] H. B. Meire, Basic Ultrasound. Hoboken, NJ, USA: Wiley-Blackwell, 1995.

[4] Texas Instruments, "Fully-integrated, 8-channel analog front-end for ultrasound, 0.85nV/√Hz, 12-bit, 50 MSPS, 122 mW/Channel", AFE5805 Datasheet, 2008.

[5] Analog Devices, Inc., "Octal LNA/VGA/AAF/ADC and Crosspoint Switch", AD9271 Datasheet, 2007.

[6] S. -J. Jung, S. -K. Hong and O. -K. Kwon, "Low-Power Low-Noise Amplifier Using Attenuation-Adaptive Noise Control for Ultrasound Imaging Systems," in *IEEE Transactions on Biomedical Circuits and Systems*, vol. 11, no. 1, pp. 108-116, Feb. 2017.

[7] B. Gilbert, "A low-noise wideband variable-gain amplifier using an interpolated ladder attenuator," in *IEEE Int. Solid-State Circuits Conf.(ISSCC) Dig. Tech. Papers*, 1998, pp. 280–281.

[8] X. Yu et al., "28-nm CMOS Ultrasound AFE With Split Attenuation for Optimizing Gain-Range, Noise, and Area," in *IEEE Transactions on Circuits and Systems I: Regular Papers*, vol. 70, no. 12, pp. 4742-4754, Dec. 2023.

[9] J. -Y. Um, "A Compact Variable Gain Amplifier With Continuous Time-Gain Compensation Using Systematic Predistorted Gain Control," in IEEE Transactions on Circuits and Systems II: Express Briefs, vol. 69, no. 2, pp. 274-278, Feb. 2022.

[10] M. Zhou, S. Ouzounov, E. Cantatore and P. Harpe, "An RX AFE With Programmable BP Filter and Digitization for Ultrasound Harmonic Imaging," in *IEEE Transactions on Biomedical Circuits and Systems*, vol. 15, no. 6, pp. 1430-1440, Dec. 2021.

[11] E. Kang et al., "A Variable-Gain Low-Noise Transimpedance Amplifier for Miniature Ultrasound Probes," in *IEEE Journal of Solid-State Circuits*, vol. 55, no. 12, pp. 3157-3168, Dec. 2020.

Adaptive Frequency Modulation Buck Converter Based on Valley Current Mode ACOT Control

Bowen Jiang[1], Hong Ren[1], Ningning Wang*[1]

[1] the College of Electronics and Information, Hangzhou Dianzi University, Hangzhou 310018, China

* Email:231040020@hdu.edu.cn, ning.wang@hdu.edu.cn

Abstract—**This paper proposes a valley current-mode adaptive-constant-on-time (ACOT) based adaptive frequency modulation circuit for buck converters. The circuit aims to address the limitations of traditional ACOT control concerning duty cycle range and input voltage adaptability. The design employs an adaptive frequency modulation technique, which dynamically lowers the switching frequency near the maximum duty cycle. This effectively extends the input voltage range, enabling stable converter operation across a wide input voltage range from 3.5V to 20V and supporting duty cycles up to 96%. The approach is implemented in a buck converter design fabricated using 180nm BCD process. Test results demonstrate that the circuit can adaptively scales its frequency from the nominal 1MHz down to 250KHz and recovers to the original frequency as duty cycle decreases, validating its dynamic regulation capability. Compared to similar solutions, the proposed design features a wider input voltage range, higher maximum duty cycle, and multiphase compatibility.**

Keywords—Buck Converter, Valley Current Mode Constant-On-Time Control, Adaptive Frequency Modulation, Wide Input Voltage, Maximum Duty Cycle Extension

I. INTRODUCTION

Buck converters provide precise voltage regulation for individual devices within various circuits. Their performance critically impacts overall system functionality, even determining whether a circuit operates correctly. Consequently, research into power supply technologies capable of operating under wide input voltage conditions is essential for enhanced adaptability. In Buck converters, the absolute input voltage range is fundamentally constrained by the maximum voltage tolerance of the fabrication process and the minimum voltage required for the operation of internal circuitry. However, beyond these fundamental limits, the practical input voltage range is also restricted by the switching frequency and the minimum on-time and minimum off-time. The enforcement of minimum on-time and off-time is crucial for converter stability and to prevent noise-induced false triggering. Nevertheless, with a fixed switching frequency, the minimum on-time imposes a limitation on the maximum allowable input voltage, while the minimum off-time imposes a limitation on the minimum allowable input voltage.

Among various control modes, COT control is currently widely adopted due to its intrinsic advantages: relatively simple circuit implementation, fast transient response to load steps, extended off-time at light loads for maintaining high efficiency, and minimal constraints on duty cycle and input voltage range. However, its switching frequency is inherently variable, making synchronization with an external clock challenging. Furthermore, its stability is highly dependent on the output voltage ripple[1][2]. Subsequently, frequency-constant ACOT control emerged, incorporating techniques

such as Discontinuous Conduction Mode (DCM) at light loads to enhance light-load efficiency[3] and internal ripple compensation to improve stability[4]. However, these advancements fail to address the fundamental limitations imposed by the fixed switching frequency, minimum on-time, and minimum off-time, which restrict both the achievable duty cycle range and the practical input voltage range.

This work proposes a valley current mode ACOT with adaptive frequency modulation capability to overcome the aforementioned limitations. The valley current mode ACOT architecture inherently enables precise phase current balancing[5], ensures robust phase margin[6], and supports fixed-frequency operation with multiphase synchronization capability. Additionally, a duty cycle detection circuit is utilized to dynamically reduce the switching frequency as the system approaches maximum duty cycle, effectively expending the practical input voltage range.

II. SYSTEM ARCHITECTURE AND OPERATING PRINCIPLE

A. Valley Current Mode ACOT System Architecture

Fig. 1. illustrates a simplified block diagram of the Valley-Current Adaptive Constant On-Time (ACOT) control scheme. The primary components include an error amplifier and a Pulse Width Modulation (PWM). The later consists of a comparator, a valley-current sampling circuit, an on-time calculation circuit, logic and driver circuits, and the power stage (switches). The error amplifier amplifies the difference between the feedback voltage (V_{FB}) and the reference voltage (V_{REF}). This amplified error signal is then compared against

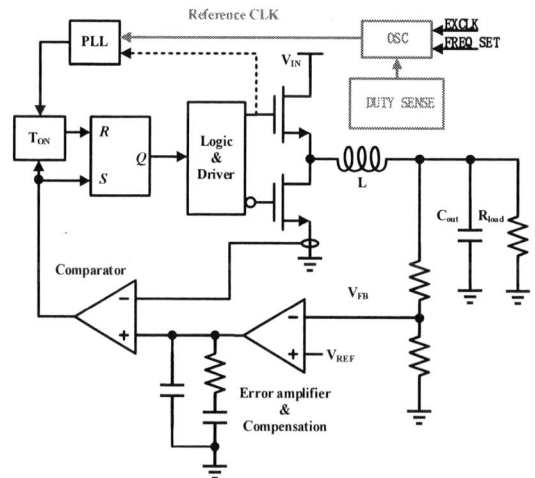

Fig. 1. Valley Current Mode ACOT System Block

979-8-3315-3918-4/25 $31.00 © 2025 IEEE

the sampled inductor valley current to determine whether to turn on the control switch. Since the turn-on decision is governed by the valley current detection and its comparison result, a prolonged minimum off-time is essential to avoid erroneous triggering. The turn-off timing of the switch is determined by the on-time calculation circuit. The calculated on-time (T_{ON}) is inversely proportional to the input voltage and directly proportional to the output voltage, as defined by the duty cycle relationship given in Eq. (1). Consequently, the switching period remains nominally fixed (neglecting parasitic effects). Furthermore, a Phase-Locked Loop (PLL) dynamically adjusts T_{ON} to synchronize the switching frequency with a reference clock signal, thereby enhancing switching frequency stability and enabling multi-phase operation.

B. Operating Principle

Imposing minimum and maximum on-time limits on Eq. (1) yields the inequality in Eq. (2). It is evident from Eq. (2) that the constraints on V_{IN} imposed by the minimum and maximum duty cycle become less restrictive as the switching period increases. As depicted in Fig. 2., for a nominal period of 1 μs with a minimum off-time of 110 ns, extending the period to 1.5 μs increases the maximum achievable duty cycle from 89% to 93%. This effectively lowers the minimum allowable input voltage closer to the output voltage. This method of extending the input voltage range via adaptive frequency variation is also applicable to circuits limited by the minimum on-time. However, within the operational conditions of the circuit presented in this work, minimum on-time does not constrain the input voltage range; therefore, the adaptive frequency modulation discussed pertains solely to extending the range at the maximum duty cycle limit. The underlying principle can be extended to Boost and Buck-Boost converter topologies to enhance their input voltage range[7].

$$T = \frac{V_{IN}}{V_{OUT}} T_{ON} \quad (1)$$

$$V_{OUT} \times \frac{T}{T - T_{MINOFF}} < V_{IN} < V_{OUT} \times \frac{T}{T - T_{MINON}} \quad (2)$$

Fig. 2. Schematic illustrating the mechanism of duty cycle limitation

III. CORE IMPLEMENTATION CIRCUITRY

The adaptive frequency modulation circuitry is depicted in Fig. 3. The operating frequency of the power supply is configured by an external resistor, which converts the frequency setting information into a current signal to control the default oscillation frequency of the oscillator. The frequency setting signal, after passing through a current mirror, can be further modulated by current i1. This enables both external clock synchronization and adaptive frequency reduction during duty cycle limitation. Current i2 represents

the final setting current flowing into the oscillator. It linearly controls the oscillation frequency by regulating the charge/discharge rate of the oscillator's internal capacitor. Transistors M1, M2, and M3 form the matched input stage of the comparator. Transistor M4 acts as the switch enabling the adaptive frequency reduction functionality. The voltage signal V_{DUTY} applied to the M2 input contains duty cycle information. When the duty cycle is not approaching its maximum limit, V_{DUTY} remains fixed at the bandgap reference voltage (V_{BG}), and M4 remains turned off. When the duty cycle approaches its maximum value, M4 is turned on. Simultaneously, V_{DUTY} decreases proportionally with increasing duty cycle. This causes current i1 to increase and current i2 to decrease. Consequently, the oscillator frequency is reduced, achieving the adaptive frequency modulation function.

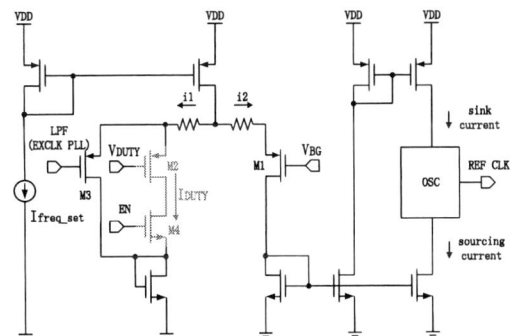

Fig. 3. Adaptive Frequency Modulation Implementation Circuitry

In the adaptive frequency modulation circuit, the function of converting the duty cycle into a corresponding voltage is implemented by the charge pump circuit, as shown in Fig. 4. In the figure, EN is the enable signal for the adaptive frequency modulation circuitry; BG is the low-side (synchronous rectifier) gate drive signal; BG_MIN_ON is a fixed-pulse-width signal generated from the BG signal, serving as a flag indicating proximity to the minimum on-time; AG is the high-side (control switch) gate drive signal; and REF CLK is the reference clock signal.

When the EN signal is low, the charge pump circuit is disabled by transistors PM1, NM1, and NM5 (shown in Fig. 4.), and V_{DUTY} is connected to V_{BG} via NM5. When EN is high, the BG and BG_MIN_ON signals control the charging and discharging of capacitor C1, respectively. Consequently, the voltage across capacitor C1 (which is V_{DUTY}) decreases as the pulse width of the BG signal decreases.

Fig. 5. illustrates the decision logic for enabling and disabling the adaptive frequency modulation function. The adaptive frequency modulation mode is activated when the pulse width of the BG signal is less than the BG_MIN_ON width for eight consecutive cycles. To prevent the circuit from repeatedly entering and exiting the frequency modulation mode, sufficient hysteresis (overlap region) is designed between the entry and exit decision thresholds. Furthermore, since the fixed pulse width derived from BG after frequency reduction no longer accurately reflects the actual duty cycle, to prevent erroneous exit decisions, the criterion for exiting the frequency modulation mode is based on the duty cycle of the AG signal. The circuit automatically exits the frequency-scaled operating mode when the duty cycle of AG falls below 75% (the maximum operating frequency is 2 MHz).

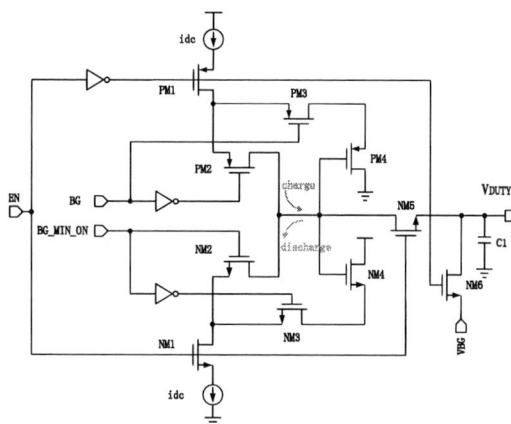

Fig. 4. Charge pump circuit for generating the V_{DUTY} signal

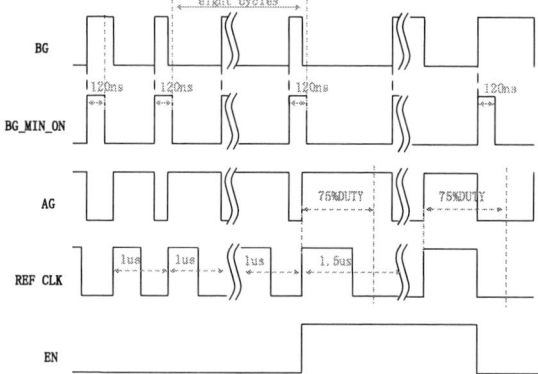

Fig. 5. Schematic Diagram of the Enabling and Disabling Logic for Adaptive Frequency Modulation

IV. TEST RESULTS AND ANALYSIS

This paper implements the proposed BUCK converter circuit with adaptive downclocking capability in extreme modes using 180nm BCD process. Fig. 6. displays the die shot of the BUCK converter, where the annotated circuit represents the oscillator module generating the reference clock, incorporating both duty cycle detection circuit and adaptive frequency modulation circuit.

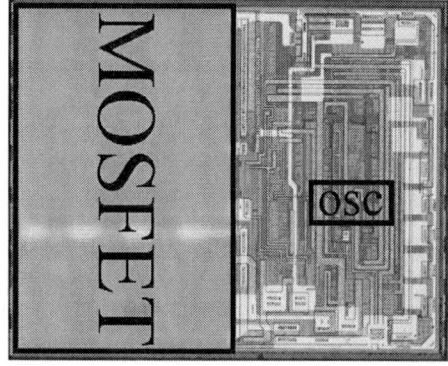

Fig. 6. Die shot

Fig. 7. shows the test platform. The test conditions were set as follows: output voltage 5V, output current 2A, and default switching frequency 1MHz. The input voltage was controlled to decrease continuously until the adaptive frequency modulation mode was triggered. Then, the input voltage was increased to observe whether the switching frequency could undergo controlled reduction normally and return to normal operating frequency.

Fig. 7. Test board

Fig. 8. presents the test results of the adaptive frequency modulation function. In Graph A, with an input voltage of 16V, the switching frequency remains fixed at 1 MHz while the circuit operates normally. Graph B demonstrates successful entry into the adaptive frequency modulation mode at an input voltage of 5.4V, where the operating frequency drops to 250 kHz. When the input voltage increases to 6.2V (Graph C), the operating frequency correspondingly rises to 580 kHz. This behavior verifies the proper operation of the charge pump circuit shown in Figure 4, confirming that the switching frequency is stably controlled by the low-side switch on-time. In Graph D, with the input voltage elevated to 6.81V, the duty cycle reaches 75%, causing the operating frequency to revert to 1 MHz. This response complies with the design specifications, indicating normal circuit operation. Post-test analysis confirms an achievable maximum duty cycle of 96% for this circuit implementation.

The comparative performance parameters of this BUCK converter versus other step-down converter implementations are summarized in Table 1.

TABLE 1 PERFORMANCE PARAMETERS COMPARISON TABLE

Performance	TPS54325[8]	2019[9]	2024[10]	This work
Input voltage range	4.5V-18V	1.8V	5V-36V	3.5V-20V
Switching frequency	700Khz	10Mhz	2.2Mhz	1M-2Mhz
Maximum duty cycle	90%	93%	95%	96%
Maximum output current	3A	0.6A	0.3A	5A
Multi-phase synchronization	Not supported	Not supported	Not supported	supported
Peak efficiency	91%	95%	92%	96%
DCM mode	supported	Not supported	Not supported	supported

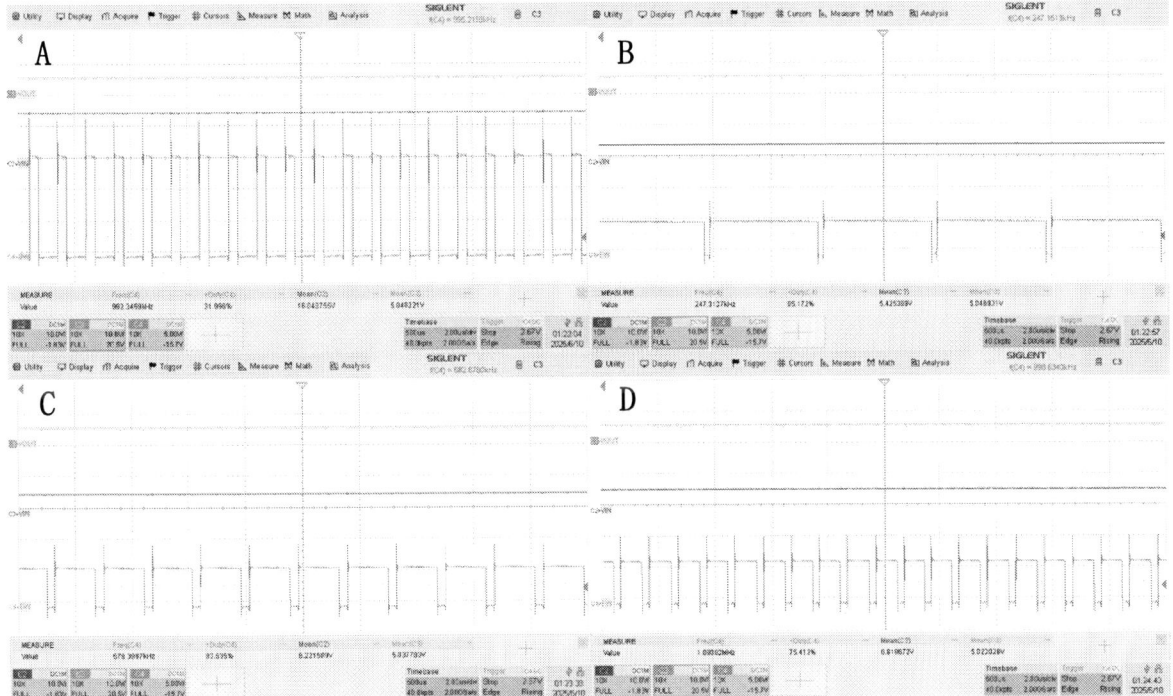

Fig. 8. adaptive frequency modulation Function Test

V. CONCLUSIONS

This paper proposes a valley current-mode constant-on-time (COT) based adaptive frequency modulation circuit for buck converters, enabling maximum duty cycles up to 96%. By implementing a simplified oscillator control circuit, the design effectively broadens the input voltage operating range, ensuring reliable circuit operation across various operating conditions. The designed state decision circuit has no interference with normal operational functions. Additionally, the circuit architecture demonstrates substantial potential for functional improvement, such as transient response enhancement through frequency-varying circuits[11] and electromagnetic interference (EMI) mitigation[12].

REFERENCES

[1] K. -Y. B. Cheng, F. C. Lee and P. Mattavelli, "Adaptive ripple-based constant on-time control with internal ramp compensations for buck converters," 2014 IEEE Applied Power Electronics Conference and Exposition - APEC 2014, Fort Worth, TX, USA, 2014, pp. 440-446.

[2] Y. -C. Lin, C. -J. Chen, D. Chen and B. Wang, "A novel ripple-based constant on-time control with virtual inductance and offset cancellation for DC power converters," 2011 IEEE Energy Conversion Congress and Exposition, Phoenix, AZ, USA, 2011, pp. 1244-1250.

[3] I-Chieh Wei, Dan Chen, Yu-Cheng Lin and Ching-Jan Chen, "The stability modeling of ripple-based constant on-time control schemes used in the converters operating in DCM," 2012 International Conference on Renewable Energy Research and Applications (ICRERA), Nagasaki, 2012, pp. 1-8.

[4] Texas Instruments, "D-CAP2TM Frequency Response Model based on frequency domain analysis of Fixed On-Time with Bottom Detection having Ripple Injection," Application Report SLVA546, January.2013.

[5] Y. -J. Chen, D. Chen, Y. -C. Lin, C. -J. Chen and C. -H. Wang, "A novel constant on-time current-mode control scheme to achieve adaptive voltage positioning for DC power converters," IECON 2012 - 38th Annual Conference on IEEE Industrial Electronics Society, Montreal, QC, Canada, 2012, pp. 104-109.

[6] J. Li and F. C. Lee, "New Modeling Approach for Current-Mode Control," 2009 Twenty-Fourth Annual IEEE Applied Power Electronics Conference and Exposition, Washington, DC, USA, 2009, pp. 305-311.

[7] Z. Xiao, G. Zhao, Y. Wang, H. Wang, and W. Hu, "Adaptive frequency adjustment technique in current mode DC–DC converters for input voltage range extension, " IET Power Electronics, vol. 12, no. 3, pp. 557–566.

[8] Texas Instruments, "TPS54325 4.5-V to 18-V, 3-A Output Synchronous Step Down Switcher with Integrated FET," TPS54325 datasheet, November 2014.

[9] J. -G. Kang, J. Park, M. -G. Jeong and C. Yoo, "A Time-Domain-Controlled Current-Mode Buck Converter With Wide Output Voltage Range," in IEEE Journal of Solid-State Circuits, vol. 54, no. 3, pp. 865-873, March 2019.

[10] Z. Zhao, P. Luo, Z. Zhang, J. Fan, B. Zhang and X. Chen, "A Peak-Valley Current-Mode Buck Converter With 3% to 95% Duty Cycle," in IEEE Transactions on Circuits and Systems II: Express Briefs, vol. 72, no. 1, pp. 328-332.

[11] S. Bari, Q. Li and F. C. Lee, "A New Fast Adaptive On-Time Control for Transient Response Improvement in Constant On-Time Control," in IEEE Transactions on Power Electronics, vol. 33, no. 3, pp. 2680-2689, March 2018.

[12] N. Miki, N. Tsukiji, K. Asaishi, Y. Kobori, N. Takai and H. Kobayashi, "EMI reduction technique with noise spread spectrum using swept frequency modulation for hysteretic DC-DC converters," 2017 International Symposium on Intelligent Signal Processing and Communication Systems (ISPACS), Xiamen, China, 2017, pp. 889-894.

A High-Voltage Level Shifter for BMS Chip in EV with 0-80V Input Range

Kunning Mao[1], Liji Wu*[2], Jing Hu*[1], Zhiwei Li[2], Haifeng Chen[2], Xiangmin Zhang[2]

[1] School of Electronic Engineering, Heilongjiang University, Harbin, China
[2] School of Integrated Circuits Tsinghua University, Beijing National Research Center for Information Science and Technology, Beijing, China

* Email: lijiwu@mail.tsinghua.edu.cn , hjlyh@126.com

Abstract—Battery Management System (BMS) plays a crucial role in electric vehicles (EVs). The stacked batteries employed in EVs and numerous energy storage systems operate in a high-voltage domain after being connected in series through multiple stages, while the signal processing circuits within the chip operate in a low-voltage domain. The high-voltage characteristics of batteries pose a significant challenge to signal processing circuits in EVs and energy storage systems. Currently, the cross-coupled structure of level shifters is predominantly utilized for low-voltage conversion between I/O and core circuit voltage domains, with limited application in processing dynamic high voltages. In the domain of automotive electronics, particularly in battery management systems where high-voltage applications are prevalent, level shifters capable of handling voltages above 70V while meeting the performance requirements of automotive-grade chips are relatively rare. Therefore, this paper proposes a level shifter designed to operate under high-voltage conditions, characterized by a wide common-mode input range. The design presented in this paper is a level shifter that operates under high-voltage conditions, fabricated using the 180 nm high-voltage BCD (bipolar CMOS DMOS) process. This level shifter is characterized by a wide common-mode input range, small error, and the ability to provide a stable and high-precision voltage output. The proposed level shifter operates within a temperature range of -40°C to 150°C and features a wide input voltage range of 0-80V. It has a static input current of 10uA and an input signal frequency of 1kHz. The level shifter can stably output a common-mode voltage of 2.5V，with the maximum error of ±3.72mV. This level shifter effectively addresses the compatibility issue between high-voltage battery signals and low-voltage signal processing circuits. The high-precision, stable low-voltage common-mode signal provided by the converter can significantly enhance the detection accuracy of the front-end circuit in the battery management system (BMS).

Keywords—electric vehicle, Battery Management System, level shifter

I. INTRODUCTION

Lithium-ion batteries used in electric vehicles (EVs) and energy storage systems typically employ a stacked IC structure to meet the demand for high voltage and high current output. The voltage range of a single cell is 3 V to 4.2 V. When 16 cells are connected in series by the battery management system, the total voltage can reach 48 V to 67.2 V. The design of this work necessarily allows for an overvoltage margin of about 10 V to enhance the circuit's ability to handle high voltage signals. Therefore, the input voltage design specification is set at 0 to 80 V. In contrast, the operating voltage of most microcontrollers, analog-to-digital converters (ADC), and other signal processing circuits is usually around 1.8V or 3.3V. These low-voltage circuits are incapable of directly handling high-voltage signals, which may cause damage to the circuits. Due to the high voltage, most integrated circuit (IC) technologies cannot be directly interfaced with the entire battery pack. This presents a significant challenge in designing a robust and cost-effective battery management system.

Incorporating a level shifter in the BMS can address the compatibility issue between high-voltage signals and low-voltage processing circuits. Moreover, the level shifter can provide electrical isolation, preventing interference from high-voltage signals to low-voltage circuits, thereby enhancing the reliability and stability of the system. The BMS requires precise measurement of the voltage and current of each battery cell to ensure the health and safe operation of the battery. The level shifter is capable of accurately converting high-voltage signals into low-voltage signals, thereby improving the measurement accuracy of the ADC. Without a level shifter, designing circuits that can directly handle high-voltage signals would be extremely complex and costly. The function of the level shifter in the BMS system is shown in Figure 1.

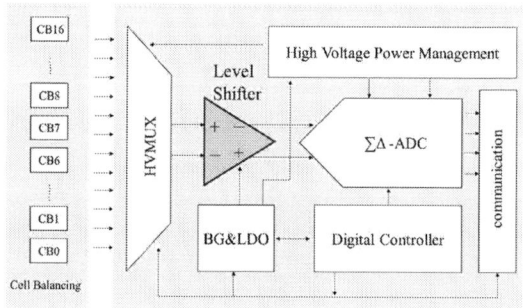

Fig. 1. The overall architecture of the battery management system of our research group

By multiplexing the signals from multiple battery cells to a shared level shifter and ADC, the circuit can be simplified, resulting in reduced chip area and lower cost.

This work is divided into five sections. Section 2 provides an overview of the level shifter circuit architecture. Section 3 presents the specific implementation of the voltage translation circuit. Section 4 demonstrates the experimental layout and post-simulation results and compares this work with the published literature. Finally, Section 5 draws the conclusion.

II. A HIGH-VOLTAGE LEVEL SHIFTER

The primary function of the level shifter circuit is to convert the high-voltage signal ranging from 0 to 80 V on the battery side into a stable low-voltage signal of 2.5 V, thereby matching the input voltage range of the on-chip signal processing circuit. The overall architecture of the circuit consists of three key sub-modules, a step-down sampling circuit, a transimpedance amplifier, and an output buffer, respectively. The circuit architecture is shown in Figure 2.

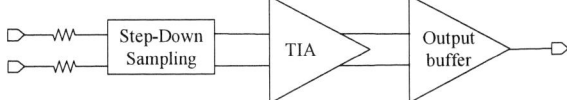

Fig. 2. Circuit Architecture

The common-mode voltage sampling circuit serves as the front-end module, primarily responsible for detecting the common-mode level of the input signal. By employing a resistor divider network, it linearly scales down the input voltage from 0-80 V to a common-mode voltage signal ranging from 0 to 3.2 V, with an attenuation factor of 0.04. This module essentially functions as a voltage-controlled voltage source, providing a stable common-mode voltage reference for the subsequent stages.

As a key part of this design, the transimpedance amplifier converts the voltage signal after buck sampling into a current signal, completing the switch from a high-voltage variable common-mode input signal to a constant common-mode signal. This amplifier exhibits excellent linear conversion characteristics, ensuring that the output common-mode current signal remains stable as the input voltage signal varies within the 0-80 V range. Consequently, the output common-mode voltage signal is maintained at a constant 2.5 V. This feature is crucial for maintaining the stability of the entire circuit, effectively preventing signal distortion caused by large fluctuations in the input voltage.

The output buffer is implemented using a source follower structure, which leverages its high input impedance and low output impedance. The high input impedance ensures minimal loading effect on the preceding stages, avoiding interference with their normal operation.

III. CIRCUIT DESIGN

A. Step-Down Sampling Circuit

In the context of fully differential signal processing architecture, the design of the step-down sampling circuit is of paramount significance. Given that the signal processing circuit primarily adopts a fully differential structure, the input terminals of the voltage processing circuit in this design are selected as V_A and V_B, with the specific circuit structure depicted in Figure 3. The common-mode input voltage ranges from 0 to 80V, and through the common-mode feedback resistor network constituted by resistors R_1 to R_6, a linear voltage drop is achieved at node VICM, resulting in an output ranging from 0 to 3.2V, with an attenuation coefficient of 0.04. This process can be described by the following equations. Among them, V_A and V_B are the differential input high-voltage signals:

$$V_1 = R_2/(R_1 + R_2) \times V_A \qquad (1)$$

$$V_2 = R_4/(R_3 + R_4) \times V_B \qquad (2)$$

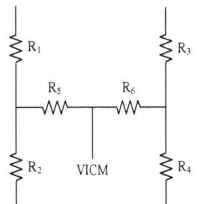

Fig. 3. Step-Down Sampling Circuit

This design endows the module with the functionality of a voltage-controlled voltage source, enabling the conversion of differential signals with high common-mode voltage into single-ended signals with low common-mode voltage. Figure 4 validates the accuracy and effectiveness of this linear scaling process.

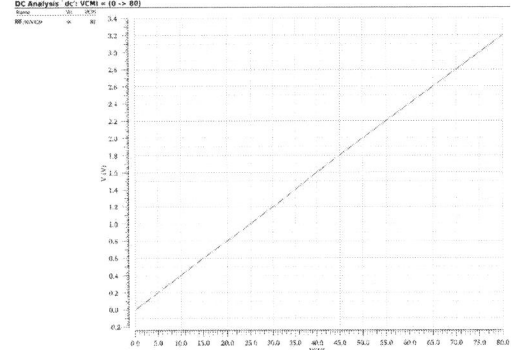

Fig. 4. 0~80V high voltage linear voltage reduction

B. Transimpedance Amplifier

The most critical part of this design is the transition from a variable wide-range input signal to a constant low-voltage common-mode signal. Considering the high common-mode voltage at the battery input terminal, a transimpedance amplifier (TIA) is designed to maintain the common-mode voltage at the amplifier input around 2.5 V. There are key nodes V_{o1} and V_{o2} between the step-down sampling module and the TIA, and the constant voltage point can be determined through a clever structural design, as shown in Figure 5.

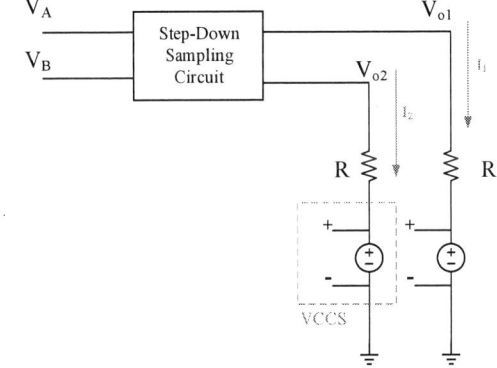

Fig. 5. Transimpedance Amplifier

The red part in Figure 5 is actually a voltage-controlled current source (VCCS) structure, the specific structure of which is given in Figure 6.

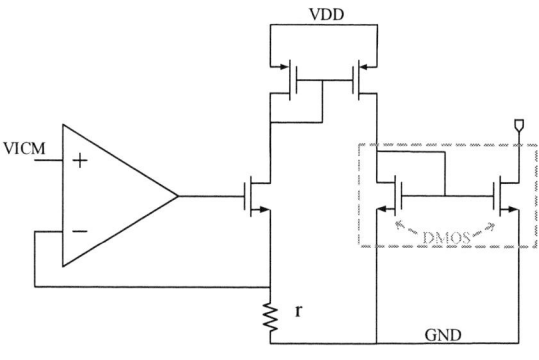

Fig. 6.   VCCS

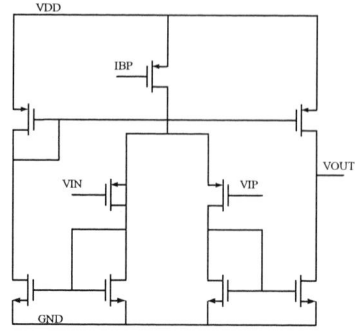

Fig. 7.   Operation Amplifier

The amplifier uses the negative feedback characteristic to replicate the common-mode voltage signal VICM to the resistor R, and the function of determining the common-mode point is described by the following equation:

$$I_1 = VICM/r \qquad (3)$$

$$VICM = V_A \cdot R_2/(R_1 + R_2) \qquad (4)$$

$$V_{o1} = V_A - I_1 \cdot R \qquad (5)$$

$$(dVo_1)/(dV_A) = 1 - (R_2 \cdot R)/(r(R_1 + R_2)) \qquad (6)$$

The differential of V_B clearly shows a linear relationship, and V_{o1} is a constant determined solely by the resistance value. This indicates that the circuit design precisely controls the gain output through the resistor network. In addition, a transistor with a threshold voltage ($V_{th}<0$) is used at the amplifier output to ensure that the output common-mode voltage remains operational over a wide input range. This design ensures that although the input voltage V_A (V_B) dynamically varies between 0 V and 80 V, the potential at the key node V_{o1} remains unchanged, and the same is true for V_{o2}. The input resistor is used to absorb the differential voltage between the battery input voltage and the amplifier common-mode voltage. The power supply current mirror in the blue part of the amplifier in Figure 6 uses the high-voltage-resistant DMOS devices unique to the BCD process to isolate the circuit and prevent breakdown, ensuring normal operation over a wide input range of 0 to 80 V.

The symmetric OTA comprises one differential pair and three current mirrors. The load of the input differential pair consists of two identical current mirrors, which provide a current gain. This can also be interpreted as load compensation for the OTA, given that the two loads are identical.

The output resistance *Rout* at the VOUT node exhibits a high impedance, making it the sole high-impedance node in the circuit. In contrast, the resistances at all other nodes are approximately 1/*gm*. This node features a relatively high gain and a large output swing, ultimately forming the dominant pole of the circuit. Compared with a conventional two-stage Miller OTA, the structure adopted in this paper does not require additional compensation structures. This design ensures high performance while reducing circuit complexity. The lightweight and efficient operational amplifier structure is shown in Figure 7.

C. Output Buffer Circuit

The output buffer stage employs a negative feedback buffer structure, which stabilizes the output common mode voltage signal VOCM at 2.5V through the negative feedback loop, improving accuracy and reducing errors. The negative feedback operational amplifier converts the high-impedance input signal into a low-impedance output signal, thereby achieving impedance matching. This is crucial for driving low-impedance loads, as it minimizes signal loss during transmission and effectively isolates the preceding stage circuit from the load.

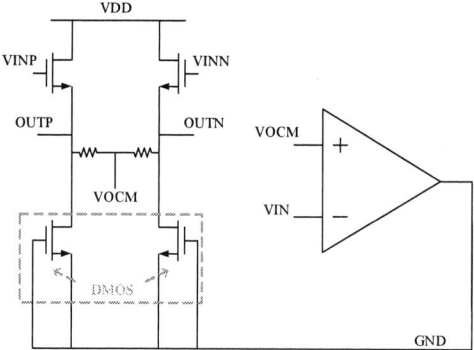

Fig. 8.   Output buffer

IV.   VERIFICATION

When the input common-mode voltage parameter VCMI varies within the range of 0 to 80V, the common-mode value of the output signal VOCM is maintained at 2.5V, with an error not exceeding 145uV.

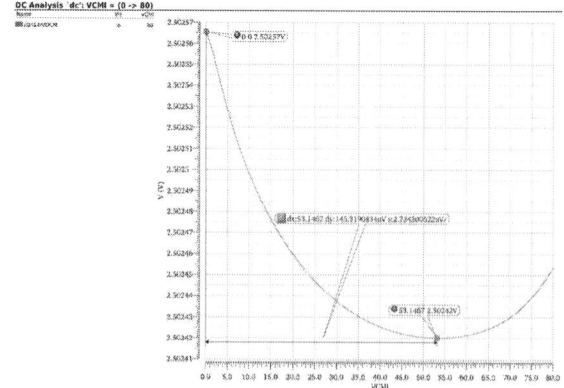

Fig. 9.   Simulation of Common Mode Output Signal

979-8-3315-3918-4/25 $31.00 © 2025 IEEE

The layout design is completed as shown in Figure 10.

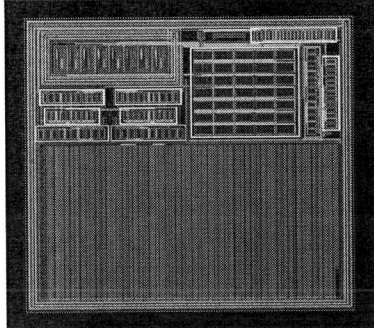

Fig. 10. Overall layout of level shifting circuit

The post-simulation results are shown in Figures 11 and 12. To verify the compliance with the automotive-grade temperature range of -40°C to 150°C, 30 sets of corner validation simulations were performed. The average power consumption of the circuit was 2.414 mW and the maximum error of the output signal was ±3.7 mV under the conditions of an input signal frequency of 1 kHz and an operating voltage of 5 V.

Output	Min	Max	Mean	Median	Std Dev
(abs(average(IT("/I0/VDD"))) * 5)	1.446m	11.82m	5.033m	4.64m	2.414m

Fig. 11. -40~ 150 ℃ Power consumption corner simulation

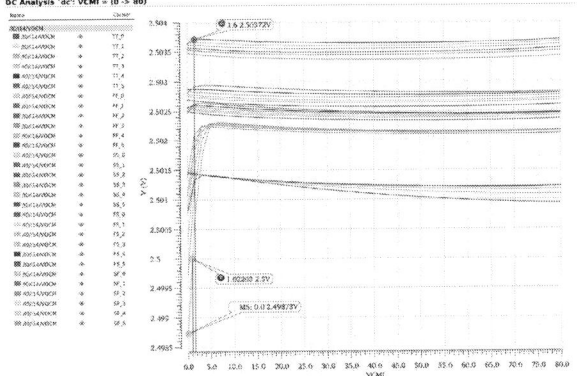

Fig. 12. Output corner simulation at -40~150 ℃

Table 1 summarizes the parameters of the high-voltage level shifter and compares it with other state-of-the-art designs in both academia and industry. The proposed circuit design achieves a wider measurement voltage range, a broader operating temperature range, and excellent measurement accuracy. The low-error and low-power design ensures that the subsequent ADC circuit in the BMS system can perform high-precision current or voltage detection.

TABLE I.   COMPARISON OF PERFORMANCES

properties	this work	[2]	[3]	[4]	[5]
Technology (um)	0.18	0.25	0.35	0.18	\
Temperature (℃)	-40~150	25	25	25	-40~105
Input voltage (V)	0~80	-16~32	0~33.6	0~70	6~72
Output voltage (V)	2.5	0~2.5	\	7.5	3
Error (mV)	±3.7	\	32	\	±60

V.   CONCLUSION

The design employs a 180 nm high-voltage BCD (bipolar CMOS DMOS) process to integrate DMOS devices, enabling the circuit to operate in higher voltage domains and making it suitable for applications with high voltage requirements。 The high-voltage level shifter circuit is powered by a 5 V supply and operates stably within the automotive-grade temperature range of -40°C to 150°C. It converts any high voltage signal in the range of 0 to 80 V into a constant static low voltage signal of 2.5 V. With an input signal frequency of 1 kHz, the average power consumption across various process corners is 2.4mW. The maximum common-mode output error is ±3.7 mV, and under the typical tt process corner at 25°C, the error is only 145 uV. Therefore, the proposed level shifter is a wide-input-range, low-error, and low-power voltage level shifter.

REFERENCES

[1] B. Ragchaa, X. He, L. Wu and X. Zhang, "A high precision voltage reference circuit for battery management system chip of new energy electric vehicle," 2022 IEEE 16th International Conference on Solid-State & Integrated Circuit Technology (ICSICT), Nanjing, China, 2022, pp. 1-3

[2] C. -L. Chen, D. -S. Wang, J. -J. Li and C. -C. Wang, " A Voltage Monitoring IC with HV Multiplexer and HV Transceiver for Battery Management Systems," in IEEE Transactions on Very Large-Scale Integration (VLSI) Systems, vol. 23, no. 2, pp. 244-253, Feb. 2015.

[3] D. G. Muratore, E. Bonizzoni, S. Verri and F. Maloberti, "High-Resolution Time-Interleaved Eight-Channel ADC for Li-Ion Battery Stacks," in IEEE Transactions on Circuits and Systems II: Express Briefs, vol. 64, no. 6, pp. 620-624, June 2017

[4] J. -K. Lee et al., "ASIL-D Compliant Battery Monitoring IC with High Measurement Accuracy and Robust Communication," 2023 IEEE International Solid-State Circuits Conference (ISSCC), San Francisco, CA, USA, 2023, pp. 322-324

[5] MAX1106812Channel, High Voltage Sensor, SmartData-Acquisition Interface. Maxim Incorporated, Sunnyvale, CA, USA.

979-8-3315-3918-4/25 $31.00 © 2025 IEEE

A New Circuit for Generating Half of VDD

Li Zeng*, Ming Wang, Peng Bo, Zhangwen Tang*

State Key Laboratory of ASIC & System, Fudan University, Shanghai 200433, China

* Email: 19112020034@fudan.edu.cn, zwtang@fudan.edu.cn

Abstract—**The paper presents a new circuit for generating half of VDD. The highlight of this work is verifying through theoretical calculations and SPICE simulations that the proposed circuit can generate a stable VDD/2. Meanwhile, the output noise is decreased by adding the filter capacitor. At the same power consumption, the proposed circuit exhibits stronger driving capability compared with the traditional Vcm generation circuit. Furthermore, under the same settling accuracy condition, the proposed circuit demonstrates faster settling speed, meeting the application requirements for high-speed SAR ADCs.**

Keywords—VDD/2, driving capability, settling speed, high-speed SAR ADCs

I. INTRODUCTION

Successive Approximation Register Analog-to-Digital Converter based on common voltage method (Vcm-based SAR ADC) has emerged as an attractive architecture [1-3]. The Vcm serves as a critical reference during the conversation process. In sampling stage, the analog input voltage is sampled on capacitor array by using the Vcm as the voltage benchmark. In conversion phase, the comparator compares the estimated voltage with the input signal by regarding the Vcm as the common reference input, and the SAR logic will be guided to refine the precise quantified results bit by bit. By applying the Vcm as the reference voltage, the switching number between high-power reference voltage (e.g., VDD and VSS) can be decreased significantly, which leads the Vcm-based SAR ADC to a lower power consumption architecture [4-6]. Moreover, the Vcm-based SAR ADC relies on relative voltage comparisons rather than high absolute reference voltages, which makes the Vcm-based SAR ADC feasible and effective in sub-1.2V systems [7-9]. As a result, the Vcm is a strategic design element that enables SAR ADCs to be employed in low-power and low-voltage applications.

The linearity of the Vcm-based SAR ADC largely depends on the accuracy of Vcm. Therefore, a stable Vcm generation circuit is crucial for SAR ADCs. Firstly, the noise cancellation of Vcm is important for the ADC input stages, and the high-frequency noise should be eliminated by using R-C filter. Furthermore, after each conversation cycle, the sampling capacitors are charged or discharged by Vcm within the limited sampling time, which means the driving requirements of the Vcm design are very high. However, there are rare studies attempt to improve the performance of Vcm generation circuits.

Herein, a novel Vcm generation circuit named half-VDD are firstly proposed. Secondly, the theoretical analysis and derivation of half-VDD are displayed. Thirdly, the noise characteristics of half-VDD are determined. Lastly, the transient settling behavior of half-VDD are discussed by comparing with another traditional Vcm generation circuit.

II. PROPOSED HALF-VDD GENERATION CIRCUIT

A. Circuit Architecture

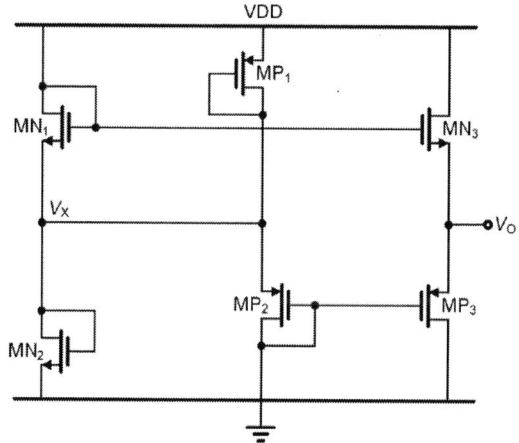

Figure 1. Architecture of the proposed Half-VDD

Fig. 1 displays the circuit configuration of Half-VDD. For the three PMOS and three NMOS transistors, $(W_1/L_1)_P = (W_2/L_2)_P = (W_3/L_3)_P$, $(W_1/L_1)_N = (W_2/L_2)_N = (W_3/L_3)_N$. Moreover, the substrates of these six MOS transistors are connected to the sources for suppressing the body effect. According to

$$V_{TH} = V_{TH0} + \gamma \left(\sqrt{2\phi_F + V_{SB}} - \sqrt{2\phi_F} \right), \qquad (1)$$

it can be seen that the V_{TH} the of these six MOS transistors remain relatively stable (V_{TH0}, γ, ϕ_F, and V_{SB} are constants). As a result, the current of MN1 and MN2 can be written as

$$I_{MN1} = \frac{1}{2} \mu_N C_{ox} \left(\frac{W}{L} \right)_N \left(\text{VDD} - V_X - V_{THN} \right)^2, \qquad (2)$$

$$I_{MN2} = \frac{1}{2} \mu_N C_{ox} \left(\frac{W}{L} \right)_N \left(V_X - 0 - V_{THN} \right)^2. \qquad (3)$$

Due to $I_{MN1} = I_{MN2}$, it is clear that

$$V_X = \frac{\text{VDD}}{2}, \qquad (4)$$

If $I_{MN1} = I_{MP1}$, (2) becomes

$$\frac{1}{2}\mu_N C_{ox}\left(\frac{W}{L}\right)_N \left(\text{VDD} - V_X - V_{THN}\right)^2 \\ = \frac{1}{2}\mu_P C_{ox}\left(\frac{W}{L}\right)_P \left(V_X - 0 - |V_{THP}|\right)^2 \qquad (5)$$

The current of MN3 and MP3 can be written as

$$\frac{1}{2}\mu_N C_{ox}\left(\frac{W}{L}\right)_N \left(\text{VDD} - V_O - V_{THN}\right)^2 \\ = \frac{1}{2}\mu_P C_{ox}\left(\frac{W}{L}\right)_P \left(V_O - 0 - |V_{THP}|\right)^2, \qquad (6)$$

and substituting (4) and (5) into (6), V_O can be written as

$$V_O = V_X = \frac{\text{VDD}}{2}. \qquad (7)$$

B. Noise Analysis and Optimization

In consideration of the noise problem, it is necessary to add the filter capacitor. As shown in Fig. 2, C is applied to reduce the output noise. As demonstrated clearly in Fig. 3, larger filter capacitor results in smaller output noise, but scaling the filter capacitor from 600fF to 1000fF results in only around 1 nV^2 reduction in output noise, suggesting negligible benefit beyond optimal capacitance value. Therefore, considering area consumption, the subsequent simulations are all based on the condition that $C = 600$fF.

III. COMPARISON WITH TRADITIONAL VCM CIRCUIT

In this section, the performance of Half-VDD is compared with that of the traditional two-stage transconductance. As depicted in Fig. 4, the first stage (AMP) is a folded-cascode amplifier, which primarily provides high gain. And a negative feedback structure can be formed by a source follower as the second stage, which ensures low output impedance and high

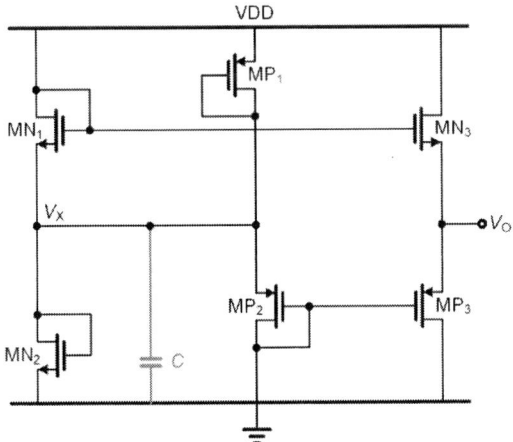

Figure 2. Architecture of the proposed Half-VDD with the filter capacitor

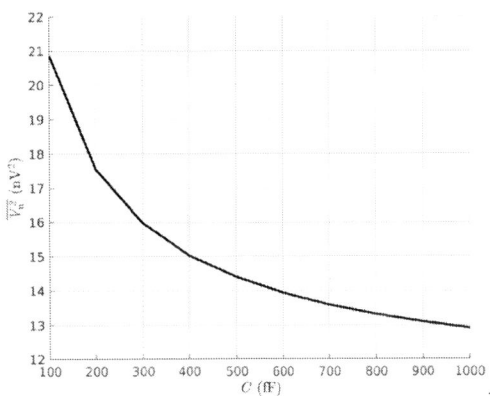

Figure 3. Output noise cancellation by the filter capacitor

slew rate. For Fig. 4, because the input voltage is VDD/2, the output voltage V_O is theoretically expected to be exactly VDD/2.

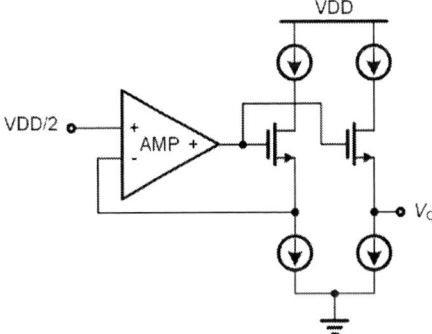

Figure 4. The framework of the two-stage-amp

A. Comparison of Driving Current

As discussed in section I, the circuit of Half-VDD is specifically designed to provide Vcm for the capacitive DAC, comparator, and amplifier in SAR ADCs. Thus, the current driving capability of this circuit is of significant importance. In Fig. 5, when the circuit is in quiescent condition, $I_O = 0$. In the case of the voltage fluctuation ΔV at V_O, current of MN$_3$ and MP$_3$ can be written as

$$\Delta I_{MN3} = \frac{1}{2}\mu_N C_{ox}\left(\frac{W}{L}\right)_N \\ \left[\Delta V^2 + 2\Delta V\left(V_O - \text{VDD} + V_{THN}\right)\right], \qquad (8)$$

$$\Delta I_{MP3} = \frac{1}{2}\mu_P C_{ox}\left(\frac{W}{L}\right)_P \\ \left[\Delta V^2 + 2\Delta V\left(V_O - 0 + |V_{THP}|\right)\right], \qquad (9)$$

Due to driving current $\Delta I_O = \Delta I_{MN3} - \Delta I_{MP3}$, ΔI_O can be

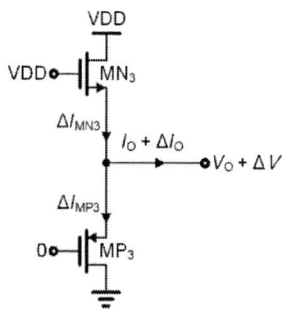

Figure 5. Driving current of Half-VDD

calculated as

$$\Delta I_O = \left[\mu_N C_{ox} \left(\frac{W}{L} \right)_N - \mu_P C_{ox} \left(\frac{W}{L} \right)_P \right] \Delta V^2 \atop +2\Delta V \left(2V_O - \text{VDD} + V_{TH} + \left| V_{THP} \right| \right), \quad (10)$$

As demonstrated in (10), Half-VDD shows strong current driving capability for ΔV variation.

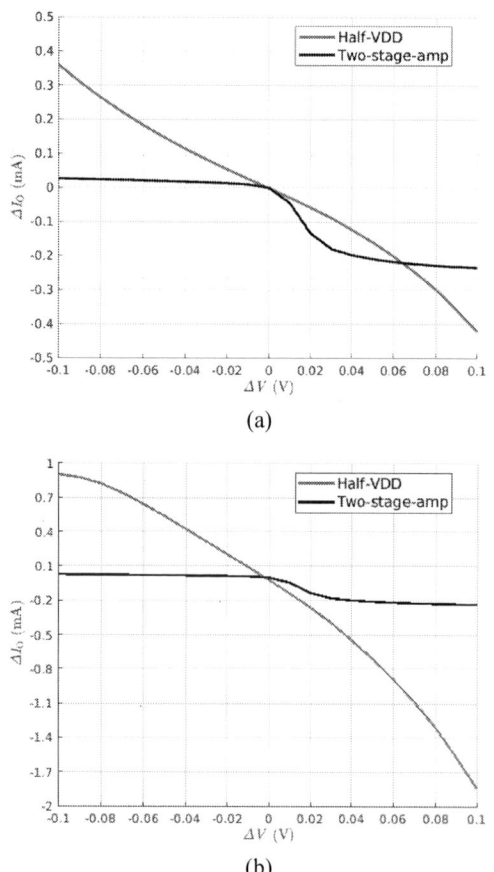

Figure 6. Driving current of original circuits (a) and same power-consumption circuits (b)

Let $\Delta V = \pm 0.1\text{V}$, the driving current of Half-VDD and traditional two-stage-amp are compared in Fig. 6(a). It can be readily observed that when the ΔV is negative, the driving current of Half-VDD is larger than that of two-stage-amp. This characteristic suggests that Half-VDD maintains effective driving capability over a wide range of external voltages. Furthermore, it is found that two-stage-amp consumes twice the power of Half-VDD. By adding numbers of MN_3 and MP_3 in Fig. 5, the two circuits achieve the same power consumption. Under this condition, Fig. 6(b) shows the driving capability of Half-VDD is significantly superior to that of two-stage-amp.

B. Comparison of Settling Time

For SAR ADCs, the sampling time of the input capacitors should not exceed half of sampling period. In other words, as a module for generating Vcm, Half-VDD is required to restore the output to VDD/2 within an ultra-short time in high-speed SAR ADCs.

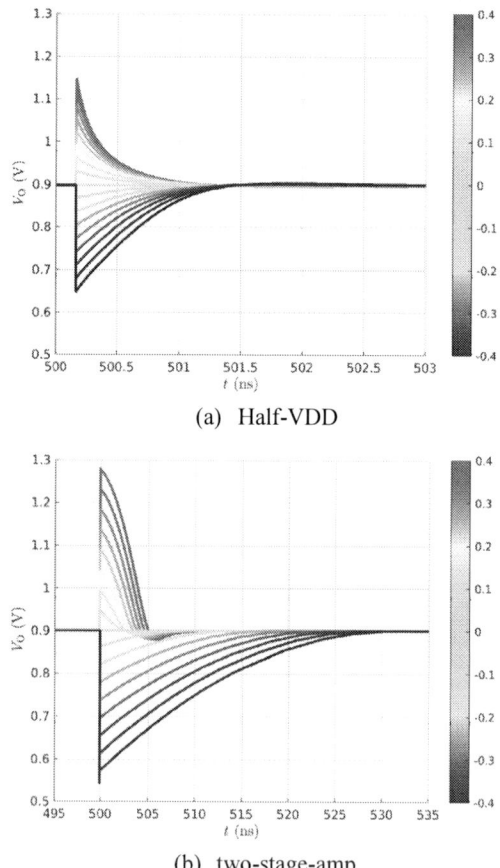

(a) Half-VDD

(b) two-stage-amp

Figure 7. Voltage setting capability

In the following simulation, the VDD and load capacitance are set to 1.8V and 2pF, respectively. As shown in Fig. 7, there is a voltage difference ($\Delta V = \pm 0.4\text{V}$) between the initial voltage and target voltage (0.9V) across the load capacitor at $t = 500\text{ns}$. In Fig. 7(a), as expected, Half-VDD exhibits relatively balanced pullback capability for both positive and negative voltage errors, and can settle V_O back to around 0.9V within 1.5ns. However, in Fig. 7(b), two-stage-amp shows unbalanced settling for ± 0.4V voltage difference, and its required settling time is at least 10 times that of Half-VDD.

According to Fig. 7, Fig. 8 displays the settling times of Half-VDD and two-stage-amp when the settling error is constrained within 1% (0.891V < V_O < 0.909V). Fig. 8 clearly illustrates the two-stage-amp fails to achieve fast settling of ΔV, especially negative ΔV. Fortunately, within the ± 0.4V voltage difference range, Half-VDD settles the desired output voltage (V_O = 0.9V ± 0.01V) within 2 ns, meeting the 250 MS/s ADC specification.

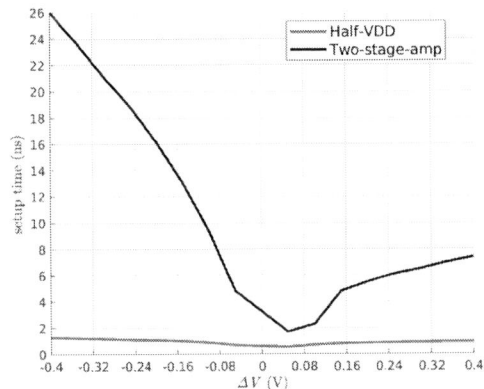

Figure 8. Required setup time of two circuits

IV. CONCLUSION

In summary, a novel circuit for generating half of VDD (Half-VDD) is proposed. Theoretical derivation based on the structure of Half-VDD suggests that the output voltage is accurate VDD/2. The noise analysis results show that the filter capacitor can effectively reduce the output noise. By comparing with traditional two-stage amplifier, Half-VDD not only demonstrates strong current driving capability but also achieves high settling accuracy within a short time, indicating that Half-VDD can be applied to high-speed and high-precision SAR ADCs.

REFERENCES

[1] Y. Wu, X. Cheng and X. Zeng, "A split-capacitor vcm-based capacitor-switching scheme for low-power SAR ADCs," 2013 IEEE International Symposium on Circuits and Systems (ISCAS), Beijing, China, 2013, pp. 2014-2017.

[2] M. B. S. Carvalho, T. C. dos Santos, R. D. P. de Oliveira, A. G. Girardi and P. C. C. de Aguirre, "A Low-Power 10-Bit VCM-Based SAR ADC with 15.4-fJ/conv in 65-NM CMOS," 2024 Argentine Conference on Electronics (CAE), Bahía Blanca, Argentina, 2024, pp. 109-114.

[3] Z. -H. Chen, Y. -C. Zhang, W. -G. Lu and Z. -J. Chen, "A 12-bit 1-MS/s SAR ADC Using Vcm-based Split MSB Switching and Segmented CDAC," 2024 IEEE 17th International Conference on Solid-State & Integrated Circuit Technology (ICSICT), Zhuhai, China, 2024, pp. 1-3.

[4] W. C. Lai, "A 10-bit 40 MS/s successive approximation register analog-to-digital converter with Vcm-based method for wireless communications," 2015 IEEE 11th International Conference on ASIC (ASICON), Chengdu, China, 2015, pp. 1-5.

[5] Z. Fu, X. Tang, D. Li, J. Wang, D. Basak and K. -P. Pun, "A 10-bit 2 MS/s SAR ADC using reverse VCM-based switching scheme," 2016 IEEE International Symposium on Circuits and Systems (ISCAS), Montreal, QC, Canada, 2016, pp. 1030-1033.

[6] D. -J. Kim, Y. -O. Kim and G. -C. Ahn, "A 12-bit 40-kS/s VCM-based switching C-C SAR ADC," 2015 International SoC Design Conference (ISOCC), Gyeongju, Korea (South), 2015, pp. 83-84.

[7] L. Huang, J. Li, X. Jiang and J. Wu, "A 2.1-fJ/Conversion-Step 10-bit 125-KS/s SAR ADC with Vcm-based Bidirectional Single-side Switching Scheme," 2023 12th International Conference on Modern Circuits and Systems Technologies (MOCAST), Athens, Greece, 2023, pp. 1-4.

[8] X. Ma, Q. Duan, S. Huang and P. Li, "Design and verification of a 10-bit asynchronous logic SAR ADC," 2023 10th International Forum on Electrical Engineering and Automation (IFEEA), Nanjing, China, 2023, pp. 1041-1044.

[9] B. Xie, Q. Lei, Z. Zhang, Q. Liu, Y. Yang and S. Feng, "A 12Bits segment SAR ADC with low power switching method," 2023 5th International Conference on Electronic Engineering and Informatics (EEI), Wuhan, China, 2023, pp. 37-40.

A Boost DC-DC Converter with Low Power and High Efficiency for Portable Device Applications

Jing Cao
School of Software and
Microelectronics
Peking University
Beijing, PR China
caoj@stu.pku.edu.cn

Bingjie Chen
School of Software and
Microelectronics
Peking University
Beijing, PR China
henry.chen@stu.pku.edu.cn

Hongfei Ye
School of Integrated Circuits
Peking University
Beijing, PR China
hfye@ime.pku.edu.cn

Jianhua Feng*
School of Integrated Circuits
Peking University
Beijing, PR China
fengjh@pku.edu.cn

Abstract— This paper presents a high-efficiency, low-power peak-current-mode boost DC-DC converter designed for portable devices. The chip operates with an input voltage range of 1.8V–4.5V and delivers an adjustable output voltage of 2.5V–5V (including a fixed 3.3V option). It achieves a 500mA load current at 1.8V input and 3.3V output while maintaining a quiescent current as low as 19µA and peak efficiency up to 93%. Key features include a zero-crossing detection (ZCD) circuit to prevent reverse current flow and a burst-mode control scheme for enhanced light-load efficiency. The design exhibits excellent transient response (load regulation: 0.57%/A, line regulation: 0.04%/V) and integrates comprehensive protection functions (soft-start, UVLO, OCP, OVP, thermal shutdown). Implemented in a 90nm BCD process, the converter demonstrates robust performance across load conditions, making it suitable for battery-powered portable electronics requiring extended operational life and high reliability.

Keywords—Boost DC-DC converter, conversion efficiency, linear transient response, low power

I. INTRODUCTION

In portable consumer products, achieving a lightweight design often necessitates the use of a single battery or a limited number of batteries for power supply, which can result in potentially low voltage levels at the battery terminals. However, many mixed-signal systems still use 5V standard CMOS technology, which cannot be directly powered by a single battery. Consequently, a boost converter is required to elevate the supply voltage to meet the system's requirements[1]. These electronic devices must support multiple operating modes, such as low-power standby mode and high-performance operating mode, and there are significant differences in power consumption between different modes. A key challenge is maintaining high conversion efficiency across these different operational modes. Additionally, as the battery voltage gradually decreases during use, minimizing the converter's input voltage threshold can further extend the battery's lifespan. Therefore, to enhance the battery life of electronic products, it is crucial to develop a boost DC-DC converter that offers high reliability, low power consumption, low input voltage capability, and fast mode-switching performance.

The proliferation of battery-powered portable electronics (wearables, IoT sensors, medical devices) intensifies demands for power management ICs that extend operational life while maintaining performance[2]. As these systems increasingly employ single-cell batteries (1.8–3.6V) yet require higher supply rails (e.g., 3.3V/5V), efficient boost DC-DC converters become critical. However, reconciling three key requirements remains challenging:

1. Ultra-Low Input Voltage Operation: Functionality down to deep discharge levels (≤ 1.8V) to maximize battery utilization.

2. High Efficiency Across Loads: Maintaining >90% efficiency from heavy load (500mA) to standby (µA-range), despite dominant switching/control losses at light loads.

3. Dynamic Response: Fast transient performance for modern multi-mode devices switching abruptly between sleep/active states.

Conventional PWM converters suffer significant efficiency degradation under light loads due to fixed-frequency switching losses. While PFM and burst modes offer improvements, they often compromise transient response or introduce excessive output ripple. Furthermore, synchronous rectification—essential for efficiency—risks reverse inductor current flow in Discontinuous Conduction Mode (DCM), wasting energy unless precisely controlled.

This paper presents a monolithic peak-current-mode boost DC-DC converter addressing these challenges:

• Operates at 1MHz with an input range of 1.8–4.5V, delivering 2.5–5V outputs (including a fixed 3.3V option).

• Integrates a zero-crossing detection (ZCD) circuit to eliminate reverse current losses during DCM operation.

• Implements an adaptive burst-mode control scheme, reducing quiescent current to 19µA during light/no-load conditions while minimizing output ripple.

• Achieves 93% peak efficiency and sustains 500mA load at 1.8V input via optimized current-mode control with Type II compensation.

Section II details the system architecture. Section III describes the ZCD and burst-mode efficiency optimizations. Section IV presents a low-voltage bandgap reference. Section V validates performance through simulation, and Section VI concludes.

979-8-3315-3918-4/25 $31.00 © 2025 IEEE

II. SYSTEM DESIGN

The schematic diagram of the proposed Boost DC-DC system is presented in Fig.1, which includes components such as a bandgap reference, a current sampling module, an error amplifier, and a zero-crossing detection circuit. The chip adopts peak current mode control, operates at a frequency of 1MHz, and can functions effectively within a input voltage range of 1.8V~4.5V and an output voltage range of 2.5-5V. The chip can achieve a load current of 500mA under the conditions of 1.8V input voltage and 3.3V output voltage. The peak efficiency is above 90%, and the static current is below 20μA. To enhance the accuracy of the output voltage and the stability of the circuit, a type II compensation circuit is implemented to improve loop gain and phase margin. Additionally, the chip is equipped with output short-circuit protection, overvoltage protection, and thermal shutdown protection functions to ensure its reliability.

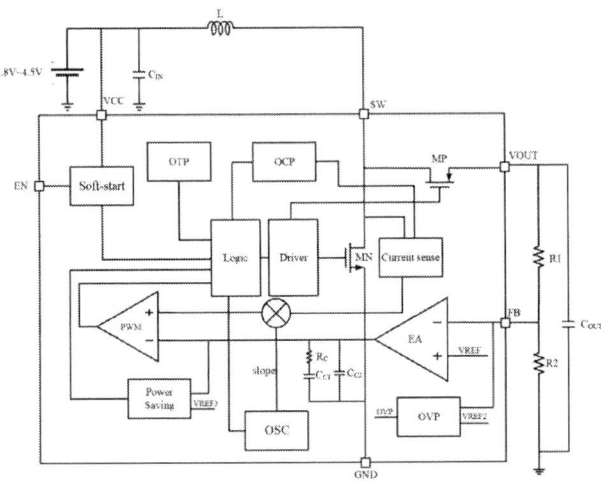

Fig. 1. System block diagram.

Its basic working principle can be summarized as follows: first, the bandgap reference module generates a reference voltage V_{REF}, the output voltage V_{OUT} generates a feedback voltage V_{FB} by voltage division. And then the error amplifier amplifies the difference between V_{REF} and V_{FB} to obtain the control signal V_C. The inductor current sampling circuit samples the sampling current I_{SENSE}, which is superimposed with the fixed current I_{REF} provided by the bias circuit and the slope compensation current I_{SLOPE} generated by the slope compensation circuit. The voltage signal V_S is obtained when the current passing through the sampling resistor R_{SENSE}. In normal operation, the oscillator generates a clock signal that initiates the switching cycle, causing the switching transistor to conduct. The comparison of the control signals V_C and V_S generates a PWM signal that regulates the switching transistor's turn-off. If the inductor current does not decrease to zero during this cycle, the circuit operates in Continuous Conduction Mode (CCM). Conversely, if the inductor current falls to zero, the zero crossing detection circuit need to output a ZCD signal to control the rectifier, turning it off to prevent current backflow. At this time, it will work in discontinuous conduction mode (DCM). When the load is lighter, due to the insufficient energy consumed by the load in one cycle to deplete the energy stored in the inductor, the output voltage V_{OUT} will continue to rise, and

the feedback voltage V_{FB} will also increase accordingly. As a result, the output signal V_C of the error amplifier will progressively decrease. When it falls below the set threshold voltage, the system will enter power-saving mode and deactivate modules such as the switch tube, oscillator, and PWM comparator. This significantly reduces power consumption.

III. EFFICIENCY OPTIMIZATION

In order to improve the efficiency of DC-DC converters within the full load current range, the following design techniques were adopted in this paper:

A. ZERO CURRENT DETECTOR

As the system transitions from heavy load to light load, the average inductor current gradually decreases with the load. When the average inductor current falls to half of the ripple current, it will reach zero within the cycle, as shown in Fig.2. However, in synchronous boost converters, the MOS transistor lacks the unidirectional conduction properties of a diode. If the inductor current continues to decrease into negative values, it may lead to reverse current flow from the output capacitor back to the input through the inductor, resulting in significant energy loss. Therefore, implementing a zero current detection circuit is essential[3]. This circuit will turn off the circuit breaker when the inductor current reaches zero, thereby preventing current backflow.

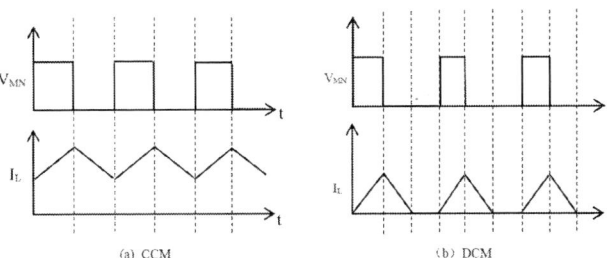

Fig. 2. The main waveforms of the CCM and DCM operating modes.

The circuit structure of the zero crossing detection circuit is shown in Fig.3. PM1 to PM7 are the same transistors. Its basic principle is to compare the output voltage V_{OUT} and the voltage V_{SW} of the switching node. When V_{OUT} is equal to V_{SW}, the output ZCD signal will turn off the synchronous freewheeling transistor. The specific working process is as follows: when the system is in the charging state, the switch transistor opens and the freewheeling transistor closes. At this state, DRVH is high, and DRINH is opposite to DRVH. Therefore, PM8 is turned off, SW is not connected to the circuit. And NM15 is conducting, the current flowing through PM1, PM2, and PM3 is greater than the current flowing through PM4, PM5, PM6, and PM7. Thus, V_1 is less than V_2, and the output ZCD signal is low. After the system enters the freewheeling state, the switch transistor is closed and the freewheeling transistor is opened. At this time, DRVH is low, the SW voltage is connected to the circuit. DRINH is high, NM15 is turned off, then the current flowing through PM1, PM2, and PM3 is equal to the current flowing through PM5, PM6, and PM7. During the freewheeling process, the output voltage V_{OUT} continuously increases. when V_{OUT} is about to exceed the SW point voltage, V_1 will rise to be greater than V_2, the output ZCD signal will flip from low to high, which will turn off the freewheeling transistor, effectively preventing the occurrence of current backflow phenomenon.

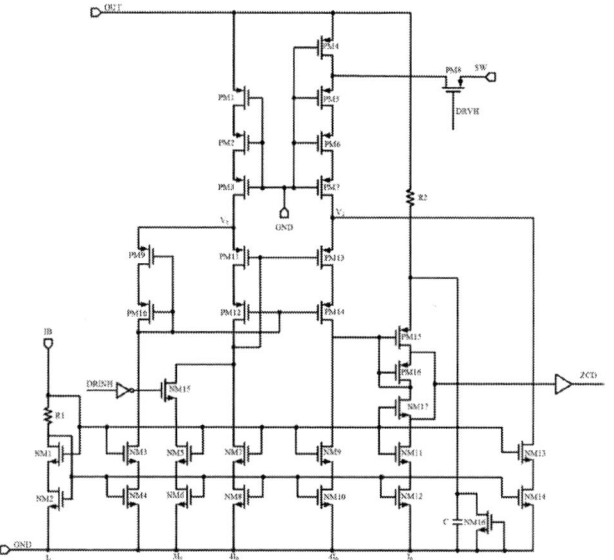

Fig. 3. The zero crossing detection circuit.

B. BURST MODE

Under light or no-load conditions, continuing to operate DC-DC converters at high frequencies can significantly reduce efficiency. This reduction occurs because the proportion of switch losses and other fixed losses relative to the output power increases. Such inefficiency is unacceptable in many power-sensitive applications, such as portable devices. Consequently, during light or no-load conditions, the converter mitigates switching losses by reducing the switching frequency. When the load demand is low, the switching frequency decreases markedly, and the converter enters an intermittent 'pulse' operating state, also referred to as Burst working mode, thereby lowering overall power consumption[4].

In the current-mode controlled boost DC-DC converter, the output voltage of the EA is directly related to the load current during stable operation. In the case of a fixed frequency current control method, the stable output voltage of the operational amplifier is described by equation (1).

$$V_{EA} = (\langle I_L \rangle + \frac{V_{IN}}{2L}T_{on}) \times R_{sen} + V_{slope(max)} \quad (1)$$

where V_{slope} is the slope compensation voltage, T_{on} is the conduction time, R_{sen} is the equivalent sampling resistance. And the relationship between the average current of the inductor and the output load current is

$$\langle I_L \rangle = \frac{I_o}{1-D} \quad (2)$$

In conclusion, the output voltage of the operational amplifier is proportional to the load current. Therefore, whether to enter burst mode can be determined by the stable output voltage of the operational amplifier. Set a lower threshold for the output of the operational amplifier. It can be identified that when the output voltage falls below this threshold, the system is in a light load

state. At this point, the boost DC-DC converter stops switching, and both the high side switch and the low side switch are in the off state. Due to the high side switch being turned off, the output load needs to extract charge from the load capacitor, resulting in a decrease in V_{OUT} and V_{FB}. The decrease in V_{FB} will cause the output voltage of the EA to rise and exceed the lower threshold, then the boost DC-DC converter will turn on the switch again to oscillate charging and discharging, thereby completing a large charging and discharging cycle. Like the waveform diagram shown in Fig4. This method achieves flexible switching between PWM mode and PSM mode.

The burst mode is typically implemented using a hysteresis comparator. The circuit structure of the hysteresis comparator designed in this article is illustrated in Fig.5. The hysteresis function can be implemented with or without a resistor connection. In burst mode, when the output voltage reaches a specific value and the output of the error amplifier falls below the threshold, the system will cease operation for an extended period until the output voltage decreases to a predetermined level, at which point it will resume operation. Although the longer shutdown duration in burst mode significantly reduces losses associated with power transistor switching operations, it may also result in considerable output ripple.

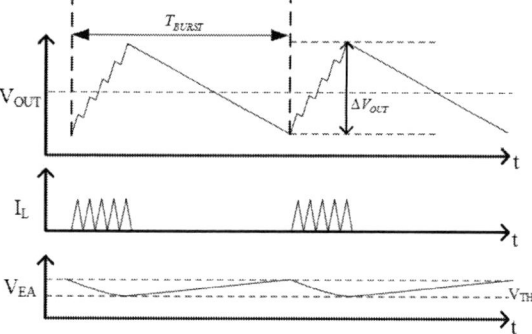

Fig. 4. Schematic diagram of working waveform in burst mode.

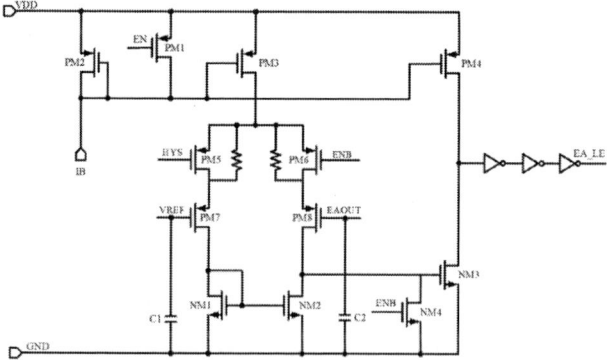

Fig. 5. The circuit structure of the hysteresis comparator.

IV. DESIGN OF THE BANDGAP REFERENCE

Due to the low input voltage of the system, it is crucial to utilize low threshold transistors and native transistors when designing bandgap reference circuits to ensure adequate voltage margins. Fig.6 shows the bandgap reference circuit designed in this paper. Considering the layout design, we have designed two branch ratios of 8:1, which means that the ratio of Q1 to Q3 is 1/8. To address the leakage issue of the transistor at elevated temperatures, we have incorporated Q2 to counteract the leakage from the two branches. The m value of Q2 is set at 7. The error amplifier designed in this article consists of Q4, Q5, NNM3, NNM4, PM5, PM6, PM7, and PM8. The potentials of V1 and V2 are stable in an equal state due to control and adjustment by the internal loop of the amplifier:

$$V_1 = V_2 = V_{BE3} \qquad (3)$$

The voltage at both ends of R4 is

$$V_{R4} = V_{BE1} - V_1 = V_{BE1} - V_{BE3} = \Delta V_{BE} = V_T \ln N \qquad (4)$$

And R4 and R5 select resistors of the same size and type, so the reference voltage V_{REF} is obtained as

$$V_{REF} = 2 \times I_{R4}(R_3 + R_6) + V_{BE1} \qquad (5)$$

So

$$V_{RFE} = \frac{2 \times V_T \ln N}{R_4}(R_3 + R_6) + V_{BE1} \qquad (6)$$

From this equation, it can be seen that by cleverly adjusting the resistance values of R3, R4, and R6 to precisely cancel out the positive and negative temperature coefficients, a zero temperature drift reference voltage that is almost unaffected by temperature can be achieved.

Given the potential for degeneracy during power-on, it is essential to design a startup circuit that ensures the entire system can quickly and reliably achieve normal operating conditions. The startup process of this circuit can be outlined as follows: upon powering on, the enable signal (EN) is high while ENB is low, resulting in both PMOS PM1 and PM2 being in a conductive state. The native NMOS NNM1 also enters a conductive state, raising the source potential of NNM1. Subsequently, NNM2 becomes conductive, allowing the circuit to transition out of the zero current state. As the reference voltage is established, the pull-down current (ID) gradually develops, leading to a decrease in the gate potential of NNM1, which ultimately results in the closure of NNM1. At this point, the startup circuit no longer influences the operation of the circuit.

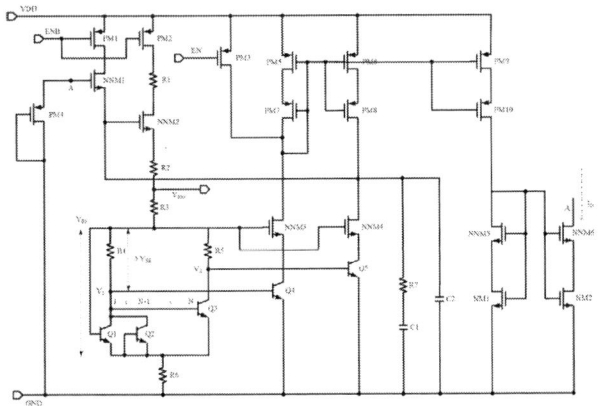

Fig. 6. The circuit structure of the bandgap.

V. SIMULATION RESULTS AND DISCUSSIONS

The proposed DC-DC is implemented using a 90nm Bipolar-CMOS-DMOS (BCD) process. The chip adopts peak current mode control, with a working frequency of 1MHz, and can operate normally within the input voltage range of 1.8V~4.5V. The output voltage is 2.5~5V, and it can achieve a load current of 500mA under the conditions of 1.8V input voltage and 3.3V output voltage.

At an input voltage of 1.8V, Fig.7 and Fig.8 illustrate the transient responses under heavy and no-load conditions, respectively. From Fig.7, it is evident that the circuit operates in Continuous Conduction Mode (CCM), where the inductor current does not decrease to zero. The output voltage remains stable at approximately 3.3V, with an output ripple of about 16mV. Under no-load conditions, as shown in Fig.8, the circuit transitions into Burst Mode, allowing the inductor current to drop to zero. The output voltage remains stable at around 3.3V, with an output ripple of about 12mV. Table I compares the DC-DC proposed in this paper with several recent publications. The results show that our work has high peak efficiency and low static current.

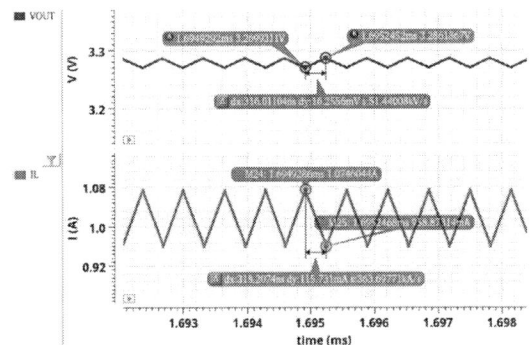

Fig. 7. The transient responses under heavy load.

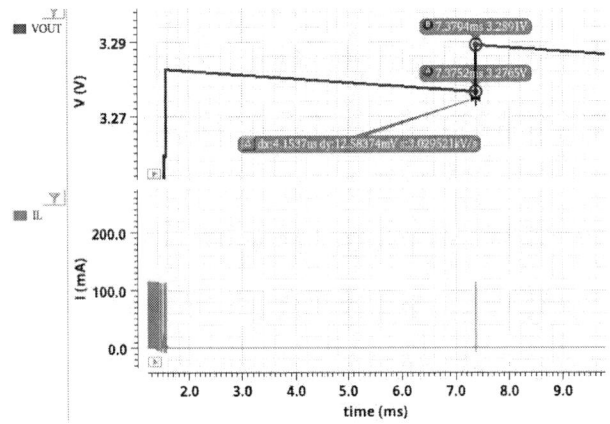

Fig. 8. The transient responses under no-load condition.

Table I PERFORMANCE SUMMARY OF THE PROPOSED DC-DC

	[3]	[5]	[6]	This work
Process	180nm	500nm	130nm	90nm
Switching frequency	10MHz	10MHz	3.2MHz	1MHz

Input Voltage	2.5~5V	3.3V-5.5V	1.8V-2.4V	1.8V~4.5V
Output Voltage	NA	3.3V	3.0V-3.6V	2.5~5V
Quiescent Current	37uA	NA	40μA	19uA
Peak efficiency	NA	92%	91.6%	93%

VI. CONCLUSIONS

This paper has presented a high-efficiency, low-power boost DC-DC converter designed specifically for portable battery-powered devices. Implemented in a 90nm BCD process, the converter achieves an input voltage range of 1.8–4.5 V and delivers an adjustable output voltage of 2.5–5 V (including a fixed 3.3-V option). Key innovations include: (1) A zero-crossing detection (ZCD) circuit effectively prevents reverse inductor current flow, minimizing conduction losses during discontinuous conduction mode (DCM). (2) A burst-mode control scheme significantly reduces switching losses and quiescent current (down to 19 μA) under light/no-load conditions while maintaining output regulation. (3) The peak-current-mode control architecture with Type II compensation ensures stable operation across load transients, achieving excellent load regulation (0.57%/A) and line regulation (0.04%/V). Measurement results demonstrate a peak efficiency of 93% and the capability to deliver 500 mA at 1.8-V input and 3.3-V output. Comprehensive protection features (soft-start, UVLO, OCP, OVP, thermal shutdown) ensure high reliability. The combination of wide input range, ultra-low quiescent current, high efficiency across load conditions, and robust transient response makes this converter highly suitable for extending battery life in portable electronic systems such as IoT devices, wearables, and handheld electronics.

ACKNOWLEDGEMENT

This work is supported by the National Science Foundation of China (No. 62174002) and the National Key R&D Program of China under grant (2022YFF1202302).

REFERENCES

[1] Mao Fangyu, Lu Yan, Lin Jie, et al. "A Single-Stage Current-Mode Active Rectifier with Accurate Output-Current Regulation for IoT," 2018 IEEE International Symposium on Circuits and Systems (ISCAS), 2018: 1-4.

[2] Alevoor S, Nayaket R D, Talele B, et al. "A 95.2% Efficiency DC-DC Boost Converter Using Peak Current Fast Feedback Control (PFFC) for Improved Load Transient Response," IEEE Transactions on Circuits and Systems I: Regular Papers, vol.70, no. 3, pp.1097-1109, 2023.

[3] X. He, P. Huang, Z. Wu, Y. Yan, J. Yang and Y. Zheng, "Synchronous coarse-fine comparators based zero-current detector for DC-DC converter operating in switching frequency beyond 10 MHz," 2017 International Conference on Electron Devices and Solid-State Circuits (EDSSC), Hsinchu, Taiwan, 2017, pp. 1-2.

[4] Wang Y, Li P and Lai S. "Robust and Efficient Transistor-Level Envelope-Following Analysis of PWM/PFM/PSM DC-DC Converters," IEEE Transactions on Computer-Aided Design of Integrated Circuits and Systems, vol.35, no. 11, pp. 1836-1847, 2016.

[5] Q. Khan et al., "A 3.3V 500mA digital Buck-Boost converter with 92% peak efficiency using constant ON/OFF time delta-sigma fractional-N control," 2011 Proceedings of the ESSCIRC (ESSCIRC), Helsinki, Finland, 2011, pp. 439-442.

[6] X. Jing, P. K. T. Mok and M. C. Lee, "A Wide-Load-Range Constant-Charge-Auto-Hopping Control Single-Inductor-Dual-Output Boost Regulator With Minimized Cross-Regulation," IEEE Journal of Solid-State Circuits, vol. 46, no. 10, pp. 2350-2362, 2011.

A Hybrid Complex-Filtering Scheme with High Image Rejection and Efficient Channel Selection for Low-IF Receivers

Yue Yin[1], Guanlin Zhang[1], Haobo Qi[1], Haodong Lu[1], Xinbing Zhang[1], Ziting Feng*[1], Ye Zhang*[2]

[1]School of Microelectronics, Northwestern Polytechnical University (NPU), Xi'an 710072, China

[2]China Electronic Product Reliability and Environmental Testing Research Institute, Guangzhou 511370, China

* Email: yinyue@nwpu.edu.cn, ztfeng@mail.nwpu.edu.cn, z18675872121@sina.com

Abstract—In this paper, a new hybrid filtering scheme is proposed for low-intermediate frequency (IF) receivers, implemented in a 0.18-μm CMOS process. The design adopts a compact two-stage architecture: the first stage integrates a polyphase filter (PPF) with a biquad complex band-pass filter (CBPF) to improve image rejection ratio (*IRR*) with reduced power overhead, while the second stage enhances aliasing rejection by combining a PPF with a biquad low-pass filter (LPF). Unlike conventional approaches, the PPF functions as the input transconductance stage, eliminating redundant OTAs, which reduces power and area. Besides, the current-mode signal path allows direct I/Q summation, while good anti-aliasing is achieved by the cascading biquad LPF. Simulation results demonstrate that the hybrid filter achieves over 60 dB *IRR* from 3 MHz to 5 MHz, with an aliasing rejection ratio (*ARR*) of 30 dB at 10 MHz and a total power consumption of 0.75mW.

Keywords—*Hybrid filtering scheme, image rejection, out-of-band rejection, low power.*

I. INTRODUCTION

With the rapid development of wireless communication, portable receivers impose stringent requirements on power consumption and cost. The low-intermediate frequency (IF) architecture, widely adopted in GNSS, Bluetooth and Zigbee receivers, offers high integration and good comprehensive performance but remains vulnerable to image interference that degrades signal-to-noise ratio (*SNR*). So to achieve image rejection, complex filters are proposed, which can be divided into three types normally. Passive RC polyphase filter (PPF) achieves high single-stage image rejection ratio (*IRR*) but requires external buffers with additional power and area[1]. Tuning is required due to process and temperature variations, but capacitor-array-based tuning is discrete and leads to increased chip area. Active PPF[2] eliminates the need for buffers and enables continuous frequency adjustment via transconductance control. However, bandwidth extension demands multistage networks and band-pass filters (BPFs), resulting in increased power consumption and design complexity. By contrast, active complex band-pass filter (CBPF) provides simultaneous image and out-of-band rejection within a single stage, but suffers from lower *IRR* compared to PPF[3].Moreover, *IRR* improvement is constrained in higher-order CBPFs by non-idealities such as I/Q mismatch, limiting its performance despite increased power [4].

This paper proposes a hybrid complex filter topology implemented in a 0.18-μm CMOS process. The design integrates a PPF and a biquad CBPF within a single circuit, forming the first stage, where the PPF functions as the input transconductance stage of the CBPF. This configuration not only reduces power consumption by eliminating the input transconductor in traditional CBPF, but also enhances *IRR* by

the advantage of the PPF. The second stage is comprised of a PPF and a biquad low-pass filter (LPF). It directly sums the I/Q output currents from the PPF, removing the need for a conventional I/Q combination circuit. The combined signal is then passed through the biquad LPF, which provides further out-of-band rejection, while maintaining low power and area. So the hybrid topology offers an efficient solution for low-power, high-*IRR* applications in modern low-IF receivers.

II. THE HYBRID COMPLEX FILTERING SCHEME AND ITS PERFORMANCE REQUIREMENTS

Designing a single complex filter to simultaneously achieve high *IRR*, wide bandwidth, good anti-aliasing, high linearity, low noise, low power and small chip area is impractical, so trade-offs are required. The proposed hybrid complex filter prioritizes *IRR* and power while still achieving adequate bandwidth and out-of-band rejection. Performance is quantified using the "carrier-to-noise density" ratio (C/N_0) as the metric, which normalizes *SNR* to a 1 Hz bandwidth and thus accounts for both image-band and aliasing interference[5].

$$\frac{C}{N_0} = \left(\frac{S}{N}\right) BW \tag{1}$$

where S/N is *SNR*, *BW* is signal bandwidth, and N_0 is noise power density.

A. Image rejection

Image rejection is the core function of any complex filter. If *IRR* denotes the ratio of desired-band to image-band gain, then the output C/N_0 degrades according to

$$\left(\frac{C}{N_0}\right)_{OUT} = \frac{C}{N_0 + N_{image}} = \left(\frac{C}{N_0}\right)_{IN} \times \left(\frac{1}{1 + \frac{1}{IRR}}\right) \tag{2}$$

and the corresponding loss in decibels is

$$\left(\frac{C}{N_0}\right)_{LOSS} = \left(\frac{C}{N_0}\right)_{IN} [dB] - \left(\frac{C}{N_0}\right)_{OUT} [dB]$$
$$= 10 \log \left(1 + \frac{1}{IRR}\right) \tag{3}$$

As illustrated in Fig. 1, maintaining $IRR \geq 20$ dB limits the C/N_0 loss to under 0.05 dB, which is negligible for most applications. In order to provide adequate margin against weak desired-signal conditions and front-end variability, a design target of $IRR \geq 60$ dB is adopted.

B. Bandwidth and out-of-band rejection

Bandwidth is another requisite performance of complex filter, which limits the applications of the complex filter. For a low-IF receiver with center frequency f_{IF} and bandwidth *BW*, and for an ADC sampling rate f_S satisfying $f_S > 2f_{IF} + BW$, the first alias occurs at $f_S - f_{IF}$.By requiring that the rejection at the half-sampling frequency ($f_S/2$) match our *IRR* target as

979-8-3315-3918-4/25 $31.00 © 2025 IEEE

shown by the solid line in Fig. 2, the aliasing rejection ratio (*ARR*) of ≥ 30 dB at $f_S/2$ is specified, ensuring effective anti-aliasing without unnecessary overdesign.

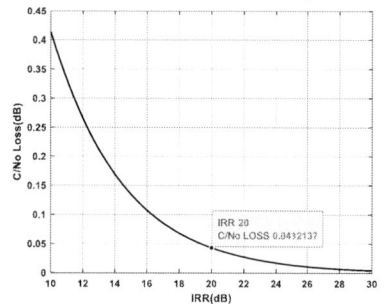

Fig. 1. Effect of *IRR* on C/N_0 performance.

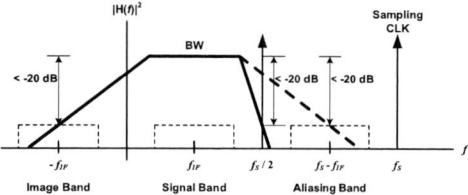

Fig. 2. Requirements of channel selectivity.

C. Comparison of PPF, CBPF and the proposed

In a representative low-IF system (f_{IF} = 4 MHz, *BW* = 2 MHz, f_S = 10 MHz), the above requirements translate to *IRR* ≥ 60 dB over 3–5 MHz and *ARR* ≥ 30 dB at 10 MHz. Fig. 3 compares the channel selectivity responses of three architectures under these targets. The PPF+BPF comprises three cascaded polyphase stages followed by a 5th-order BPF (2nd-order HPF and 8th-order LPF). The CBPF achieves 4th-order selectivity by cascading 2 stage biquads. The hybrid filter is that a core circuit (one stage PPF and CBPF) and an additional hybrid polyphase-lowpass filter (one stage PPF and a biquad LPF) in cascade. Block-diagram representations are provided in Fig. 4.

Under ideal conditions, the hybrid scheme attains an 80 dB *IRR*, exceeding both PPF and CBPF, while its *ARR* remains above 40 dB. Fig. 4 shows that the PPF with BPF requires a total of 38 OTAs, the CBPF requires 24 OTAs, while the proposed hybrid filter uses only 22 OTAs. Therefore, by eliminating standalone operational transconductance amplifiers (OTAs), the hybrid design matches CBPF in resource consumption and outperforms the PPF cascade. Uniform pole angular frequency and $Q = 0.7071$ achieve the flattest pass-band response, as well as improved matching, linearity and noise performance.

In practice, I/Q mismatch degrades *IRR* in higher-order CBPFs. According to the research of Jirayuth Mahattanakul, *IRR* falls to approximately 50 dB when the order of complex filter exceeds 3[4]. By limiting the effective complex order to three or less, the proposed hybrid filter preserves high *IRR* in comparison with CBPFs.

III. CIRCUIT DESIGN

The complete hybrid filter is organized into two cascaded stages, Hybrid Complex Filter (HCF) and Hybrid Poly-phase-Lowpass Filter (HPLF), as shown in Fig. 5. This scheme isolates image rejection and anti-aliasing functions for analysis and tuning.

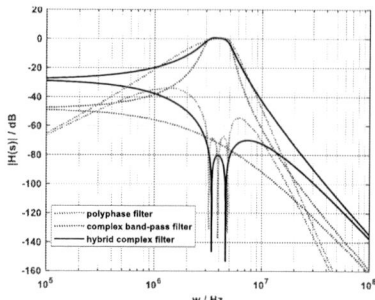

Fig. 3. Channel selectivity of PPF with 5th-order BPF, CBPF and the proposed with 2nd-order LPF.

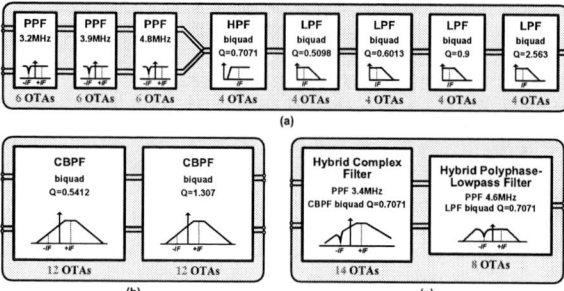

Fig. 4. Block diagram of (a) PPF with BPF, (b) CBPF, (c) the proposed with LPF.

A. Hybrid Complex Filter (HCF)

The core stage merges a PPF cell and a CBPF cell into a single current-mode core circuit, achieving high image rejection and band-pass selectivity without external buffers or standalone OTAs.

According to the general diagram, PPF contains 1st-order LPFs and HPFs, and output of the filter cells need to add in quadrature. Fig. 6 shows the circuit of high-pass trans-conductor Gm_H and low-pass transconductor Gm_L in PPF cell. Since the output conductance g_{d0} of current source M_0 is negligible, their pole frequency satisfies the relation as follows, ensuring perfect quadrature outputs and high *IRR*.

$$\omega_0 = \frac{g_m}{2C_H} \qquad (4)$$

Where g_m and C_H are transconductance of transistor M_1 and capacitance, respectively.

Summing the low-pass and high-pass currents with a +90° phase shift on the high-pass branch, the quadrature branches can be combined into a single PPF output current expression

$$i_{out,\,PPF} = i_L + j i_H = v_{in}\, g_m \left(\frac{\omega_0}{s+\omega_0} + j \frac{s}{s+\omega_0} \right) \qquad (5)$$

Then, the quadrature current signal feeds into the CBPF cell. Generally, CBPF is transformed from LPF, adding a linear frequency transformation path. A five-transistor (5-T) OTA-C low-pass filter is converted into a one-pole complex band-pass by a linear frequency-shift operation. Fig. 7 shows the circuit of OTA.

The input-to-output relation of 5-T OTA can be written as

$$i_{out} = g_{m1} \cdot v_{in} \qquad (6)$$

where g_{m1} is the transconductance value of the whole OTA circuit, which is equal to the transconductance of the transistor M_1 in Fig. 7.

Therefore, the pole frequency in the OTA-C filter is

$$\omega_p = \frac{g_{m1}}{C} \qquad (7)$$

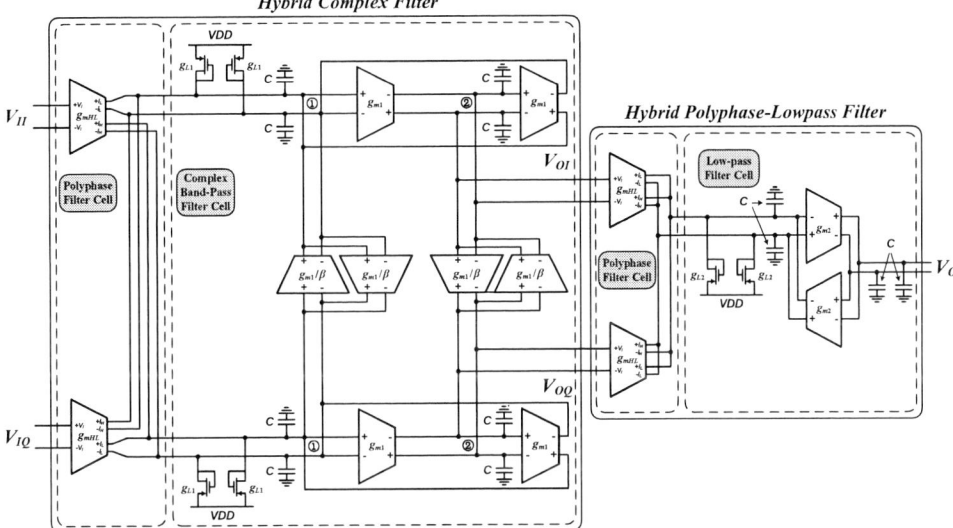

Fig. 5. Complete circuit of the hybrid complex filtering scheme.

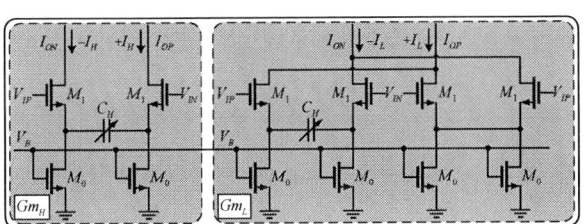

Fig. 6. Circuit of Gm_H and Gm_L in complex notch filter cell.

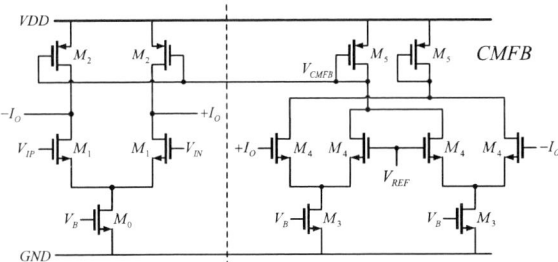

Fig. 7. Circuit of differential OTA.

After shifting by $j \cdot g_{m1}/\beta C$, the input-to-output relation of the CBPF cell can be written as

$$v_{out} = \frac{i_{in,\,CBPF}}{g_{L1}} \frac{\left(\frac{g_{m1}}{C}\right)^2}{\left(s + j\frac{g_{m1}}{\beta C}\right)^2 + \frac{g_{L1}}{C}\left(s + j\frac{g_{m1}}{\beta C}\right) + \left(\frac{g_{m1}}{C}\right)^2} \quad (8)$$

Since the PPF output current $i_{out,\,PPF}$ equals the CBPF input $i_{in,\,CBPF}$, the transfer function of HCF is

$$H_{\mathrm{HCF}}(s) = \frac{g_m}{g_{m1}}\left(\frac{\frac{g_m}{2C_H}}{s + \frac{g_m}{2C_H}} + j\frac{s}{s + \frac{g_m}{2C_H}}\right) \cdot$$
$$\frac{\left(\frac{g_{m1}}{C}\right)^2}{\left(s + j\frac{g_{m1}}{\beta C}\right)^2 + \frac{g_{L1}}{C}\left(s + j\frac{g_{m1}}{\beta C}\right) + \left(\frac{g_{m1}}{C}\right)^2} \quad (9)$$

From equation (9), the filter performance parameters of HCF can be derived as follows

$$A_{V1} = \frac{g_m}{g_{m1}} \quad (10)$$

$$\omega_c = \frac{g_{m1}}{\beta C} \quad (11)$$

$$BW = \frac{2g_{m1}}{C} \quad (12)$$

$$Q_1 = \frac{g_{m1}}{g_{L1}} \quad (13)$$

where A_{V1} is pass-band gain, ω_c is center frequency, BW is bandwidth, and Q_1 is quality factor.

Actually, as the value g_{m1}/C determines the BW by determining the cut-off frequency of LPF, it should be set up

first. Then, by adjusting g_m, β, and g_{L1}, the parameters A_{V1}, ω_c and Q_1 can be independently tuned.

B. Hybrid Polyphase-Lowpass Filter (HPLF)

The second stage repeats the PPF cell as the input transconductance stage, and the combined I/Q output currents are fed into a biquad LPF to provide enhanced out-of-band rejection for effective aliasing suppression.

The input-to-output function of the biquad LPF is given by

$$v_O = \frac{i_{in,LPF}}{g_{m2}} \frac{\left(\frac{g_{m2}}{C}\right)^2}{s^2 + \frac{g_{L2}}{C}s + \left(\frac{g_{m2}}{C}\right)^2} \quad (14)$$

Since the PPF output current equals the LPF input current $i_{in,LPF}$, the transfer function of HPLF is

$$H_{\mathrm{HPLF}}(s) = \frac{g_m}{g_{m2}}\left(\frac{\frac{g_m}{2C_H'}}{s + \frac{g_m}{2C_H'}} + j\frac{s}{s + \frac{g_m}{2C_H'}}\right) \cdot$$
$$\frac{\left(\frac{g_{m2}}{C}\right)^2}{s^2 + \frac{g_{L2}}{C}s + \left(\frac{g_{m2}}{C}\right)^2} \quad (15)$$

From equation (15), the filter performance parameters of HPLF can also be derived as follows

$$A_{V2} = \frac{g_m}{g_{m2}} \quad (16)$$

$$\omega_0 = \frac{g_{m2}}{C} \quad (17)$$

$$Q_2 = \frac{g_{m2}}{g_{L2}} \quad (18)$$

where ω_0 is the cut-off frequency of the biquad LPF.

C. The complete hybrid filter

Based on the previous analysis, the transfer function of the complete hybrid filter can be expressed as

$$H(s) = H_{\text{HCF}}(s) \cdot H_{\text{HPLF}}(s) \qquad (19)$$

In the proposed circuit design, while satisfying the *IRR* requirement, the poles of the two-stage PPF are adjusted to closely but different frequencies to achieve a wider bandwidth. In addition, to achieve better out-of-band rejection, the pole of the second-stage LPF is set equal to the frequency-shifted pole of the first-stage CBPF, satisfying the following relationship.

$$\frac{g_{m2}}{C} = \frac{g_{m1}}{C} + \frac{g_{m1}}{\beta C} \qquad (20)$$

IV. SIMULATION RESULTS

The proposed Hybrid Complex Filter was implemented in 0.18-μm CMOS process for verifying specific performance. Fig. 8 and Fig. 9 shows the frequency response and *IRR*. The overall circuit achieves an *IRR* in excess of 60 dB across 3–5 MHz, while also delivering better *ARR* of 30 dB at 10 MHz, enhancing the system's anti-aliasing capability. The input 3rd order intercept point (IIP3) and input referred noise (IRN) are -6.2 dBm and 70 μV, respectively. The performance comparison with previous PPF and CBPF is given in TABLE I, from which it is evident that the proposed design achieves better integrated performance.

V. CONCLUSION

Implemented in a 1.2 V supply, the design achieves a low static current of 0.625 mA. By matching the poles of the PPFs and CBPF, image-signal rejection is enhanced while preserving the desired signal over a wide bandwidth, thus ensuring better overall performance. Additionally, a cascaded LPF provides high-frequency attenuation beyond the CBPF, enhancing anti-aliasing rejection. By eliminating redundant OTAs, the hybrid filter achieves low power while achieving $IRR > 60\text{dB}$, *ARR* of 30dB within a single circuit. These attributes make the proposed filter especially attractive for low-IF receiver applications, matching current industry needs and future development trends.

TABLE I. PERFORMANCE COMPARISON

	This work	2019 MWSCAS[6]	2019 TMTT[7]	2021 TCAS-I[8]
Technology (nm)	180	350	130	90
Supply (V)	1.2	2.7	1.2	1.2
Topology	Hybrid	Active-RC PPF	Passive-RC PPF+LPF	Active-RC CBPF
Gain (dB)	3	17	0	47.5
Bandwidth/ Center frequency (MHz)	2/4	2.6/4.1	2/1.4	1.57/1.33
IRR (dB)	>-60	>-42	-36	-20.5
Power dissipation (mW)	0.75	2.7	0.24	0.48
FoM [dB(J⁻¹)]	206.138	188.064	–	171.325

$*IMFDR_3 = \left(\frac{IIP3}{IRN}\right)^{4/3}$

$**\text{FoM} = 10\log_{10}\left(\frac{\sqrt{f_c + \frac{BW^2}{2}} \cdot IMFDR_3}{IRR \cdot \frac{P_\omega}{N}}\right)$

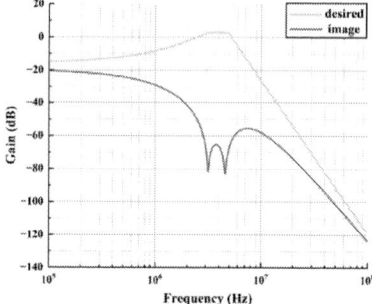

Fig. 8. The frequency response of the complete circuit.

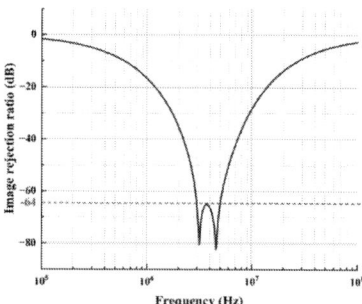

Fig. 9. The *IRR* of the complete circuit.

ACKNOWLEDGMENT

This work was supported by the Key Project of Industrial Technology Infrastructure Platform of Ministry of Industry and Information Technology of China under Grant 000645-23ZB0717/01.

REFERENCES

[1] Tran MT, Kushita N, Kuwana A, Kobayashi H. Pass-Band Gain Improvement Technique for Passive RC Polyphase Filter in Bluetooth Low-IF Receiver Using Two RC Band-Stop Filters. AEF 2020;38:192–205.

[2] Yin Y, Wang S, Ma Y,et al.The design of large image rejection and wideband CMOS active polyphase filter for BeiDou RF receiver[J].IEICE Electronics Express, 2020, 17(12).

[3] A. A. Emira and E. Sanchez-Sinencio, "A pseudo differential complex filter for Bluetooth with frequency tuning," in IEEE Transactions on Circuits and Systems II: Analog and Digital Signal Processing, vol. 50, no. 10, pp. 742-754, Oct. 2003.

[4] J. Mahattanakul, "The effect of I/Q imbalance and complex filter component mismatch in low-IF receivers," in IEEE Transactions on Circuits and Systems I: Regular Papers, vol. 53, no. 2, pp. 247-253, Feb. 2006.

[5] Jinho Ko, Jongmoon Kim, Sanghyun Cho and Kwyro Lee, "A 19-mW 2.6-mm/sup 2/ L1/L2 dual-band CMOS GPS receiver," in IEEE Journal of Solid-State Circuits, vol. 40, no. 7, pp. 1414-1425, July 2005.

[6] S. Delshadpour, "A 2.6 MHz Bandwidth, 3rd/5th Order Active-RC Polyphase Filter with Quadrature Offset Cancellation for Low-IF GPS Radio," 2019 IEEE 62nd International Midwest Symposium on Circuits and Systems (MWSCAS), Dallas, TX, USA, 2019, pp. 1017-1020.

[7] M. Silva-Pereira, J. T. de Sousa, J. Costa Freire and J. Caldinhas Vaz, "A 1.7-mW−92-dBm Sensitivity Low-IF Receiver in 0.13-μm CMOS for Bluetooth LE Applications," in IEEE Transactions on Microwave Theory and Techniques, vol. 67, no. 1, pp. 332-346, Jan. 2019.

[8] M. Cavallaro and G. Nicollini, "A Complex Band-Pass Filter for Low-Power and High-Performance Transceivers," in IEEE Transactions on Circuits and Systems I: Regular Papers, vol. 68, no. 12, pp. 5018-5028, Dec. 2021.

Design of a Nonlinear Temperature Compensated Bandgap Reference in 55nm Process

Hezhuang Nie [1], Ningning Li [2], Jian Mei [1,3], Rui Yin*[1,2]

[1] College of Integrated Circuits and Micro-Nano Electronics, Fudan University, Shanghai, China
[2] Jiashan Fudan Institute, Jiaxing, China
[3] National Integrated Circuit Innovation Center, Shanghai, China

* Email: hznie25@m.fudan.edu.cn, yinrui@fudan.edu.cn

Abstract—**In analog integrated circuits, the bandgap reference circuit is a typical component, which provides a stable reference voltage for ADC and determines the accuracy and performance of the temperature sensor system. In this paper, the bandgap voltage of silicon is extracted by employing a nonlinear temperature compensation technique, and the trimming technique is used to modify the reference voltage. The bandgap reference circuit is based on 55nm HLMC process. By adopting this circuit topology, the temperature coefficient (TC) of 6.9 ppm/°C is achieved within the temperature range of -40°C to 125°C, and the power supply rejection (PSR) is -53 dB at 10 Hz.**

Keywords—reference voltage, temperature coefficient, bandgap, current mode, trimming

I. INTRODUCTION

Natural phenomena like sound and temperature inherently exist as analog signals. Since computers exclusively process digital information, these continuous analog signals must undergo conversion into discrete digital formats. Within industrial control applications, temperature sensor systems are extensively employed. As science and technology advance rapidly, demands on temperature sensor systems within the industrial automation landscape are becoming increasingly significant. Consequently, there is a growing need to design and implement high-performance temperature sensor systems.

The proposed chip architecture of temperature sensor system is shown in Fig. 1, comprising a high-precision bandgap reference, DAC, clock, ADC, and digital processing. The sensor system must achieve high performance metrics, including exceptional accuracy, robustness in harsh environments, low power consumption. The reference voltage of the bandgap is widely used in the temperature sensor system. Temperature changes are an important factor affecting the performance of electronic devices. Reference sources with minimal temperature drift are essential for keeping a consistent and robust output voltage across varying temperatures, which is vital for sensor networks and high-precision equipment. With the increasing demand for high-performance circuits, the need for reference sources with a lower temperature coefficient is also on the rise.

The reference voltage source significantly influences the circuit's overall stability and is a fundamental part of the circuit system. For the bandgap reference with ordinary first-order compensation, its accuracy in practical circuits often cannot meet the expected goals. Thus, we employ a nonlinear compensation technique to improve its precision. The temperature coefficient of the reference source needed to be less than 10 ppm/°C in -45 °C to 125 °C temperature range.

In this paper, a high-order nonlinear compensation technique is incorporated into the conventional current mode bandgap reference[1]. The bandgap reference achieves low temperature coefficient. In addition, trimming techniques are used to improve the deviations caused by process changes in the production of chips.

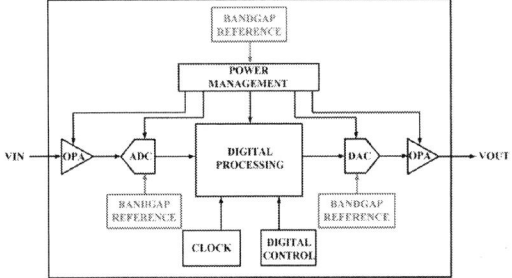

Fig. 1. Temperature sensor system block diagram.

The following is a brief description of the structure of this paper: Section II specifically introduces the bandgap reference theory and the proposed bandgap structure of this paper, containing the choose of circuit parameters. Section III details the circuit's simulation outcomes. Lastly, Section IV provides a conclusion of this paper and makes comparison with other works.

II. STRUCTURE OF BANDGAP REFERENCE

The conventional approach to create a bandgap reference (BGR) involves developing a circuit that forms a linear combination to produce both the base-emitter voltage VBE of a bipolar junction transistor (BJT) and the voltage proportional to absolute temperature (PTAT). However, in addition to linear term, VBE also contains nonlinear curvature Tln(T) term. Therefore, it is necessary to compensate for the nonlinear term to obtain a high-precision BGR. This work compensates for the linear term in the transistor voltage VBE on the basis of the conventional circuit and eliminates the Tln(T) term in VBE by introducing a nonlinear term.

A. The Theory of Bandgap Reference

An important characteristic of bandgap reference is the temperature stability of its output voltage. As is shown in Fig. 2, the PTAT voltage can be obtained by the use of an operational amplify, cancelling with the VBE of a transistor. The Vref of conventional BGR can be defined as

---

This work is supported by Shanghai Pudong New Area Science and Technology Development Fund Industry-Academia-Research Collaboration Special Funding PKX2024-D06.

$$V_{ref} = V_{BE2} + [V_T \ln N]\left[1 + \frac{R_3}{R_1}\right] \qquad (1)$$

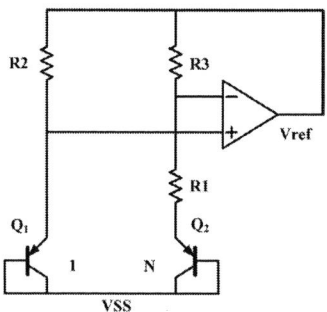

Fig. 2. conventional bandgap reference.

B. Proposed Structure of Bandgap Reference

This design circuit compensates for the linear term in the voltage VBE of the transistor and introduces a nonlinear term to achieve high-order compensation, which is shown in Fig. 3. The Vref can be defined as

$$V_{ref} = \frac{R_4}{R_2}\left[V_{BE1} + \frac{R_2}{R_1}V_T \ln N + \frac{R_2}{R_3}\ln\left(\frac{T}{T_r}\right)V_T\right] \qquad (2)$$

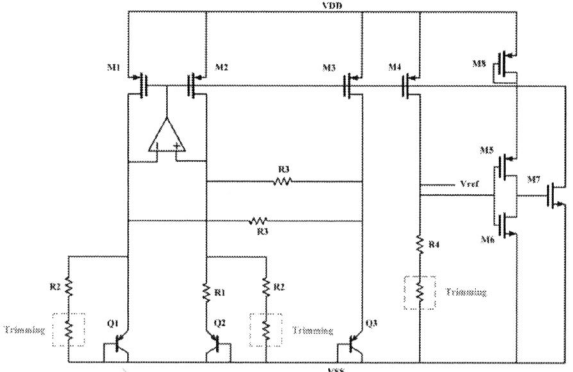

Fig. 3. Schematic of the proposed BGR.

The current flowing through M1 and M2 is approximately zero temperature coefficient. The current of zero temperature coefficient is copied to M3 through a current mirror, and high-order temperature compensation can be achieved through the utilization of M3, R3, and Q3. By changing the value of R4, the magnitude of the voltage can be changed.

The ratio of R2 and R1 affects the effectiveness of first-order compensation, while the ratio of R3 and R2 is significant to nonlinear compensation.

MOSFET M5, M6, M7, and M8 constitute the startup circuit of the BGR. When the circuit output is at a low level, the inverter formed by M5 and M6 outputs a high-level signal, turning on M7. This pulls down the gate terminal of the PMOS current mirror, enabling current generation in the bandgap core. Once normal operation is established, the inverter shuts off. The gate-drain shorted M8 minimizes power consumption in the startup path.

The temperature curve of conventional BGR is illustrated in Fig 4(a). Fig 4(b) demonstrates the temperature curve with nonlinear temperature compensation. After

nonlinear compensation, the temperature curve of its reference voltage is no longer parabolic in shape, but presents two extreme points. Voltage changes are smoother than in Fig 4(a).

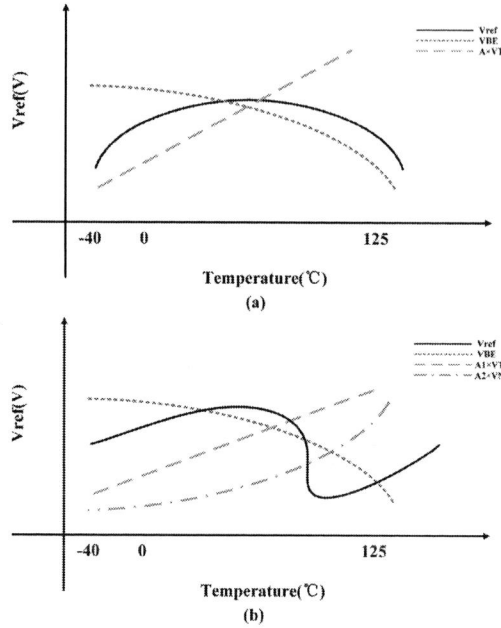

Fig. 4. (a) first-order temperature compensation (b) with the addition of nonlinear compensation.

In this circuit design, the ratio of transistors Q1, Q2, and Q3 is 1:5:3, forming a 3 × 3 common center matching pattern on the layout. The ratio of resistors R2 to R3 is 8.5, and the ratio of resistors R3 to R2 is approximately 4.1.

In the actual fabrication process of integrated circuits, inherent process variations inevitably occur. These manufacturing inconsistencies introduce deviations in the electrical characteristics of circuit components, including the resistors within a bandgap reference circuit. To be exact, these process variations can cause significant fluctuations in the output voltage. Therefore, trimming techniques are employed. In this circuit , trimming is performed on R2 and R4. By precisely altering the resistance of R2, the PTAT current can be changed. Thus, an accurate reference voltage can be obtained.

III. LAYOUT DESIGN AND SIMULATION RESULT

The bandgap reference is based on 55 nm HLMC process. With a supply voltage of 2.5V. The circuit layout is illustrated in Fig. 5. The total area of the layout is 0.024 mm².

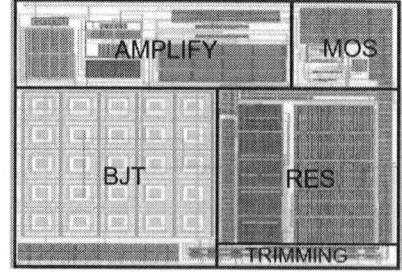

Fig. 5. Layout of the proposed BGR.

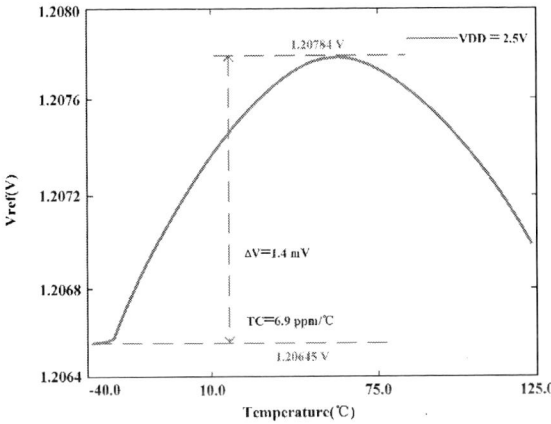

Fig. 6. Simulation results of temperature coefficient.

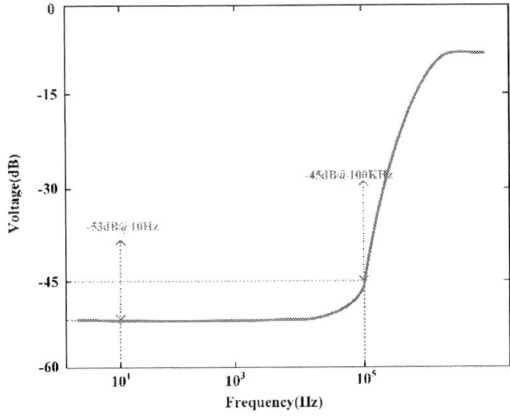

Fig. 7. Simulation results of PSR.

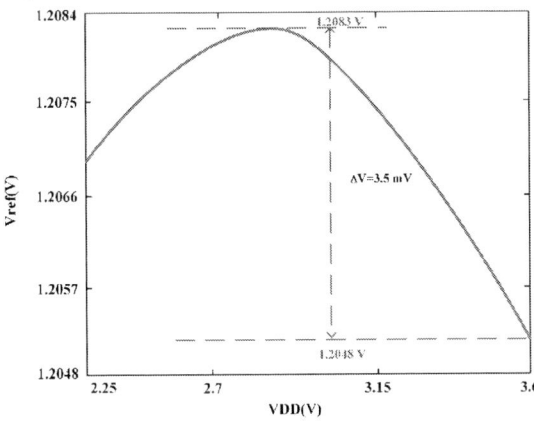

Fig. 8. Simulation results of Vref as a function of supply voltage.

Fig. 6 illustrates the Vref as a function of temperature. Within the temperature range of-40 °C to 125 °C, Vmax is 1.20784 V, Vmin is 1.20645 V, and ΔV is 1.4 mV. The average output voltage is 1.207 V. Based on these values, the temperature coefficient can be calculated is 6.9 ppm/°C.

Fig. 7 illustrates the PSR of the circuit. The PSR was -53 dB at DC frequency and -45 dB at 100 KHz, exhibiting a good frequency stability.

Fig. 8 illustrates the variation of Vref with the power supply voltage, where VDD ranges from 2.25V to 3.6V. The average output voltage is 1.2071V and the output voltage fluctuation range remains below 3.5mV.

IV. CONCLUSION

Based on the traditional first-order compensation, this paper introduces a nonlinear term through an additional branch to compensate for VBE. The designed bandgap reference has better performance and meets the fulfils the design requirements of the temperature sensor system. The value of temperature coefficient is 6.9 ppm/°C and the PSR is -53 dB at 10 Hz.

TABLE I.   COMPARISON RESULTS WITH OTHER WORKS

Parameter	This Work	[2]	[3]	[4]
Process (nm)	55	45	350	600
Output Voltage(V)	1.207	0.5	1.18	0.603
Supply Voltage(V)	2.5	1.05	1.4-3.0	0.98
Temperature Range(°C)	-40~125	-40~125	-10~110	0~100
Temperature Coefficient(ppm/°C)	6.9	24.4	13	15
PSR(dB)	-53	-60	-53	-

ACKNOWLEDGMENT

This work was supported by Shanghai Pudong New Area Science and Technology Development Fund Industry-Academia-Research Collaboration Special Funding PKX2024-D06.

REFERENCES

[1] Liu, Nanqi, Randall L. Geiger, and Degang Chen. "Sub-ppm/° C bandgap references with natural basis expansion for curvature cancellation." IEEE Transactions on Circuits and Systems I: Regular Papers 68.9 (2021): 3551-3561.

[2] R. Nagulapalli, R. K. Palani and S. Bhagavatula, "A 24.4 ppm/°C Voltage Mode Bandgap Reference With a 1.05V Supply," in IEEE Transactions on Circuits and Systems II: Express Briefs, vol. 68, no. 4, pp. 1088-1092, April 2021.

[3] J.M.Leeet al., "A 29nW bandgap reference circuit," in IEEE Int. Solid-State Circuits Conf. (ISSCC) Dig. Tech. Papers, Feb. 2015, pp. 100–101.

[4] Ka Nang Leung and P. K. T. Mok, "A sub-1-V 15-ppm//spl deg/C CMOS bandgap voltage reference without requiring low threshold voltage device," in IEEE Journal of Solid-State Circuits, vol. 37, no. 4, pp. 526-530, April 2002.

A Novel Light-load Control Method For Switching Converters in Portable Devices

Jie He[1], Shuyu Zhang[1*], Langyuan Wang[1], Suyi Yao[1], Kejia Zhu[1]

[1] Common Mode Semiconductor Technology (Suzhou) Co., Ltd

* Email: hj@manbasemi.com, zsy@manbasemi.com

Abstract—This paper proposes a peak-current PWM control method of DC-DC switching converters specifically for light load conditions. Conventional switching converters exhibit significant variations in critical current threshold during light-load-to-heavy-load switching, which not only affects chip efficiency but also leads to system instability and complicates circuit design. To address these issues, the proposed design optimizes the circuit structure by incorporating a secondary slope compensation block and an anti-interference logic block. These enhancements eliminate critical peak inductor current threshold variations while improving system stability. The proposed control method is verified by an integrated converter implemented in 180 nm CMOS process and characterized experimentally. Measurement results show stable operations with fixed peak inductor current threshold under different light loads and I/O voltage conditions, which demonstrates significant application value for portable devices with stringent low-consumption and stability requirements, particularly in automotive and medical applications.

Keywords— light load; switching converter; control method; slope compensation; inductor current

I. INTRODUCTION

With the rapid development of consumer and industrial electronics, such as healthcare devices [1], wearable technology [2], and automotive electronics [3], there is an increasing demand for power supplies to maintain low power consumption and high performance even under light loads or standby conditions. Conventional switching converters face multiple challenges during load switching, including significant differences in critical current thresholds that affect power efficiency and may even cause system instability, thereby increasing circuit design complexity [4-5]. Therefore, developing a novel control method that effectively addresses these issues under different light loads and input/output (I/O) voltages shows significant practical importance.

The rest of the paper is organized as follows: Section II describes the conventional light-load control strategy of a DC-DC converter and its drawbacks of variable inductor current thresholds. Section III presents an improved control method achieving fixed peak inductor current threshold. In Section IV, the measurement results presented. Section V concludes the paper.

II. CONVENTIONAL LIGHT-LOAD CONTROL STRATEGY FOR SWITCHING CONVERTERS

A. Circuit Structure and Operational Principles

The structure of the existing conventional peak-current pulse-width modulation (PWM) controlled switching converter is shown in Fig. 1 [6].

Conventional switching converter consists of key components including error amplifier (EA), Vc buffer (BUF), PWM comparator (CMP1), light load comparator (CMP2), slope compensator (SLOPE1), current sensor (Current Sense), clock oscillator (OSC), zero current switching detector (ZCS), control logic and driver block, high-side power transistor (Mhpwr), and low-side power transistor (Mlpwr).

The operational principles of the conventional switching converter are as follows:

The feedback voltage is compared and amplified with the reference voltage V_{ref1} through FB pin to generate the error voltage V_c. V_c is buffered by BUF to generate the voltage V_{c_buf}. After superimposing the compensation slope, the sensed peak inductor current is compared with V_{c_buf} to generate the PWM signal V_{pwm}. Then V_{pwm} can turn off the high-side power transistor Mhpwr and turn on the low-side power transistor Mlpwr.

In heavy-load mode, the low-power control signal, V_{sleep}, remains low, the turn-on activation of Mhpwr and the turn-off activation of Mlpwr are determined by the clock signal (CLK) falling edge. In light-load mode, power consumption is reduced by disabling OSC, CMP1 and SLOPE1. A fixed light-load reference voltage V_{ref2} is compared with V_c to determine V_{sleep}, which can turn on Mhpwr and turn off Mlpwr when triggered.

B. Problems Existing in Conventional Control Strategy

As shown in Fig. 1, the error voltage V_c is enhanced by BUF to generate the buffered voltage V_{c_buf}. There will be a voltage offset V_{offset} between V_c and V_{c_buf} due to process, temperature, supply voltage and so on.

The timing waveforms of the conventional control strategy are shown in Fig. 2. The relationships among V_c, V_{c_buf} and V_{offset} can be calculated as follows:

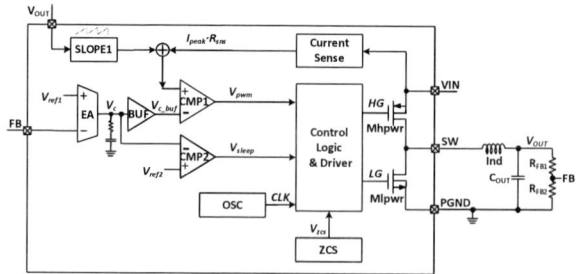

Fig. 1. The structure of conventional peak-current PWM controlled switching converter.

979-8-3315-3918-4/25 $31.00 © 2025 IEEE

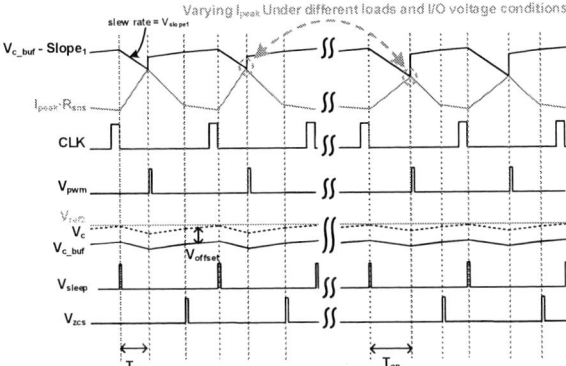

Fig. 2. The timing waveforms of the conventional control strategy.

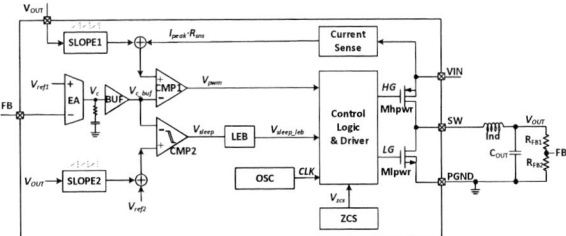

Fig. 3. The structure of the switching converter using the proposed control method.

$$V_c = V_{ref2} \tag{1}$$

$$V_{c\_buf} = I_{peak} \cdot R_{sns} + V_{slope1} \cdot T_{on} \tag{2}$$

$$V_{offset} = V_c - V_{c\_buf} \tag{3}$$

where R_{sns} is the equivalent current sensing resistance, V_{slope1} is the slew rate of compensation slope voltage, T_{on} is the on-time of high-side power transistor Mhpwr and I_{peak} is the critical peak-current threshold for light and heavy load mode switching. Then I_{peak} can be derived as:

$$I_{peak} = \frac{V_{ref2} - V_{slope1} \cdot T_{on} - V_{offset}}{R_{sns}} \tag{4}$$

Different input voltage V_{IN}, output voltage V_{OUT} and switching frequency F_{SW} will affect the on-time of Mhpwr, T_{on}, in each cycle. Combined with the effect of offset voltage V_{offset}, the critical current threshold I_{peak} in different operating conditions will be greatly different.

The current threshold deviation of switching between light and heavy load modes will worsen the efficiency of the switching converter. It can even cause system instability, which affects the reliability of the converter, and greatly increase the difficulty of circuit design.

In order to avoid mode switching problems, there is one solution that the circuit can work with a fixed switching frequency under light load. Therefore, blocks consuming high power (such as OSC, CMP1 and SLOPE1) need to be always enabled, resulting in high static power consumption and low system efficiency under light load.

At the same time, in the light load mode without synchronous clock OSC, asynchronous signal V_{sleep} is easy to be affected by noise and other interference, resulting in instability and affecting system robustness.

III. IMPROVED SWITCHING CONVERTER AND CONTROL METHOD UNDER LIGHT LOAD

A. Circuit Structure and Operation Principles of The Proposed Control Method

A novel control method is proposed for peak-current controlled DC-DC converter in this paper with fixed inductor current threshold fixed I_{peak} under different loads and I/O conditions, which optimize the light load operation as shown in Fig. 3.

Apart from the same blocks as the conventional structure, a secondary slope compensation block SLOPE2 related to the first slope compensation block SLOPE1 and another adder are added to the proposed design.

The operational principles of the proposed control method are as follows:

During light-load-to-heavy-load switching operation, the proposed control method incorporates a compensation slope with a slew rate of V_{slope2} by block SLOPE2 on the basis of V_{ref2}. V_{slope2} can equal to the voltage slope V_{slope1}. Simultaneously, the error voltage V_c is buffered through the buffer (BUF) to generate voltage V_{c_buf}, which is used for both CMP1 and CMP2 to produce V_{pwm} and V_{sleep}, correspondingly. Oppositely, V_{c_buf} is only used for V_{pwm} and V_c is used for V_{sleep} in the prior conventional design. These approaches eliminate the impact of varied input voltage V_{IN}, output voltage V_{OUT}, switching frequency F_{SW}, and offset voltage V_{offset} on the critical current threshold I_{peak} during different loads and I/O conditions, thereby enhancing the system performance under light-load conditions.

To further enhance the stability of converter under light load conditions, a leading-edge blanking logic block (LEB) is added after V_{sleep} for anti-interference. The circuit design employs hysteresis and blanking mechanisms to filter transient spikes in the V_{sleep} signal, ensuring stable signal V_{sleep_leb} transmitted to the down-stream components. This approach prevents asynchronous V_{sleep} signal from mistakenly triggering the heavy/light-load mode switching, thereby improving overall system stability during low-power operation.

B. Timing Waveforms and Mathematical Analysis

The timing waveforms of the DC-DC converter using the proposed control method are shown in Fig. 4. V_{c_buf} and I_{peak} can be calculated as follows:

$$V_{c\_buf} = V_{ref2} + V_{slope2} \cdot T_{on} \tag{5}$$

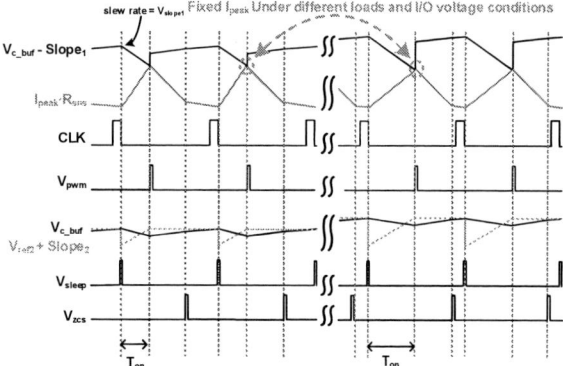

Fig. 4. The timing waveforms of the proposed control method.

$$I_{peak} = \frac{V_{c\_buf} - V_{slope1} \cdot T_{on}}{R_{sns}} \quad (6)$$

The relationship between V_{slope2} and V_{slope1} is given by

$$V_{slope2} = V_{slope1} \quad (7)$$

Then the critical peak-current threshold I_{peak} can be derived from Equation (5-7) as:

$$I_{peak} = \frac{V_{ref2} + V_{slope2} \cdot T_{on} - V_{slope1} \cdot T_{on}}{R_{sns}} = \frac{V_{ref2}}{R_{sns}} \quad (8)$$

It can be obtained from the Equation (8) that when the reference voltage V_{ref2} is determined, the critical peak-current threshold for light and heavy load mode switching (I_{peak}) is unchanged even under different input voltage V_{IN}, output voltage V_{OUT} and switching frequency F_{SW} and load currents, which greatly reduces the difficulty of circuit design and improves the reliability of the system.

In this way, the critical current threshold I_{peak} can be effectively guaranteed to remain stable under different working conditions, so as to improve the performance of system under light load and avoid the problems such as reduced efficiency and system instability caused by huge varations in I_{peak}.

C. Optimized Specifications

The proposed control method for the switching converter has the following beneficial effects.

- The proposed system enhances the driving capability of the error voltage V_c through the buffer to generate the buffered voltage V_{c_buf}. V_{c_buf} is simultaneously used in CMP1 and CMP2 to generate V_{pwm} and V_{sleep}. It eliminates the impact of offset voltage V_{offset} on the critical peak-current threshold I_{peak}, effectively resolving the I_{peak} variation issue caused by V_{offset} in prior conventional designs.

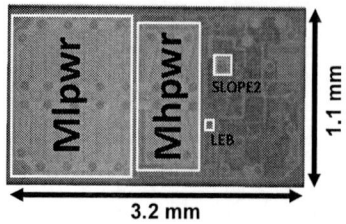

Fig. 5. Microphotograph of the chip.

- By incorporating a secondary compensation slope SLOPE1 into V_{ref2}, the system ensures stable I_{peak} values across varying loads and I/O conditions. It significantly reduces circuit complexity, enhances chip reliability, and effectively prevents performance degradation or system instability caused by drastic I_{peak} variations. Furthermore, it improves system performance under light load conditions.

- By adding the LEB anti-interference logic block based on hysteresis and blanking mechanisms after the mode switching signal V_{sleep}, the proposed design prevents asynchronous V_{sleep} signal from triggering erroneously. Then it can ensure stable V_{sleep_leb} for downstream components, significantly enhancing system stability under light load conditions while improving robustness. These improvements enable the chip to better adapt to various complex operational environments.

IV. MEASUREMENT RESULTS

The proposed control method is verified by a monolithic integrated buck converter fabricated in a 180nm CMOS process. As shown in Fig. 5, the active area of chip occupies 3.2×1.1 mm^2. The proposed converter was measured with an off-chip inductor L of 220 nH and an output capacitor C_O of 44 µF.

Fig. 6 and Fig. 7 compare the measured waveforms of V_{OUT}, V_{sw} and I_L at different light-load conditions, which are 50 /100mA. The waveforms in Fig. 6 correspond to the case when $V_{IN} = 2.7V$, $V_{OUT} = 0.9V$, while the waveforms in Fig. 7 correspond to the case when $V_{IN} = 3.3V$, $V_{OUT} = 0.8V$. Measurement results show stable operations with fixed peak inductor current threshold I_{peak} of 1.471 A under different loads and I/O voltage conditions, which demonstrates significant application value for portable devices with stringent low-consumption and stability requirements.

V. CONCLUSION

A novel light-load control method for DC-DC converter is proposed in this paper, which effectively addresses variations in critical current threshold during light-load-to-heavy-load switching encountered in conventional designs. The approach significantly enhances system performance and stability under light loads without complex circuit design, improving chip robustness and applicability. The proposed method demonstrates broad application potential and shows significant practical importance.

979-8-3315-3918-4/25 $31.00 © 2025 IEEE

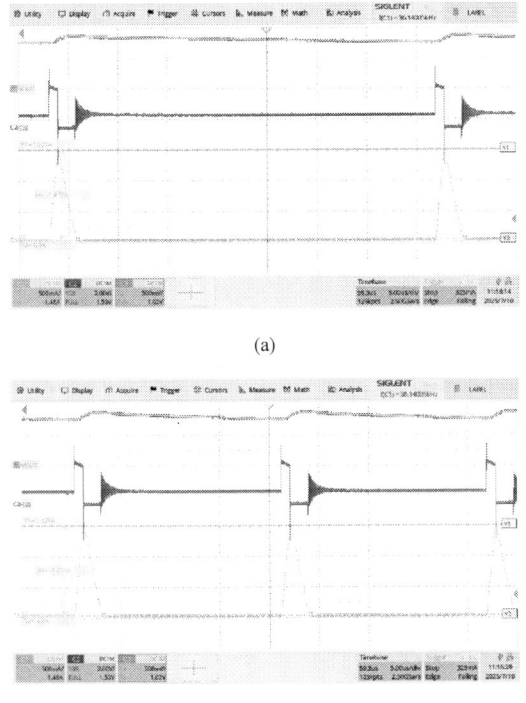

(a)

(b)

Fig. 6. Measured light-load waveform when V_{IN} = 2.7V, V_{OUT} = 0.9V. (a) Load = 50mA. (b) Load = 100mA.

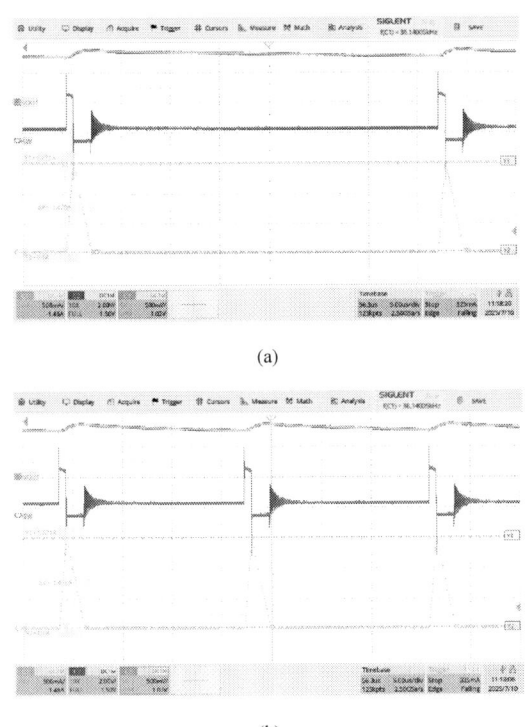

(a)

(b)

Fig. 7. Measured light-load waveform when V_{IN} = 3.3V, V_{OUT} = 0.8V. (a) Load = 50mA. (b) Load = 100mA.

ACKNOWLEDGMENT

Thanks to the sponsorship from Common Mode Semiconductor Technology (Suzhou) Co. Ltd.

REFERENCES

[1] W. Park *et al.*, "A 94% Peak Efficiency Dual Mode Buck Converter With Fully Integrated On-Time-Based Mode Control for Implantable Medical Devices," *IEEE Trans. Circuits Syst. II: Express Briefs*, vol. 69, no. 11, pp. 4458-4462, Nov. 2022.

[2] S. Zhang, M. Zhao, X. Bai, Y. Yao and X. Wu, "A 6A, 2.5MHz integrated dual-phase DC-DC buck converter with low quiescent consumption for mobile devices," *43rd Annual Conference of the IEEE Industrial Electronics Society (IECON), 2017*, pp. 497-502.

[3] F. Santoro *et al.*, "A Hysteretic Buck Converter With 92.1% Maximum Efficiency Designed for Ultra-Low Power and Fast Wake-Up SoC Applications," *IEEE J. Solid-State Circuits*, vol. 53, no. 6, pp. 1856-1868, Jun. 2018.

[4] J. -S. Kim, J. -O. Yoon and B. -D. Choi, "A High-Light-Load-Efficiency Low-Ripple-Voltage PFM Buck Converter for IoT Applications," *IEEE Trans. Power Electron.*, vol. 37, no. 5, pp. 5763-5772, May 2022.

[5] A. Besharati Rad, M. Kargaran, M. Meghdadi and A. Medi, "A Wide-Input-/Output-Voltage-Range Buck Converter With Adaptive Light-Load Efficiency Improvement and Seamless Mode Transition," *IEEE Trans. Power Electron.*, vol. 39, no. 2, pp. 2200-2212, Feb. 2024.

[6] Linear Technology Corporation. (Milpitas, CA), "Circuits and methods for adjustable peak inductor current and hysteresis for burst mode in switching regulators", US Patent *US20080030178A1*.

Adaptive on-time Control Buck Converter Based on Phase-locked-loop and Dynamic Calibration of DC Offset

Xinyu Zhang [1], Sujuan Liu*[1], Kun Liu [1], Bingxue Zhang [1], Yahua Shi [1]

[1] School of Information Science and Technology, Beijing University of Technology, Beijing, 100124, China

* Email:liusujuan@bjut.edu.cn

Abstract—Constant on-time (COT) control is widely adopted in BUCK converters owing to its advantages of fast transient response and high light-load efficiency. However, the inherent principle of conventional COT control leads to limitations, including low output voltage regulation accuracy and significant switching frequency variation. Addressing the demand for high-precision output voltage regulation and stable switching frequency, this paper proposes a high-precision adaptive on-time control method utilizing phase-locked loop (PLL) and dynamic calibration techniques. The proposed method enhances output voltage accuracy by dynamically detecting the DC level of the output voltage and reducing its offset via a digital-to-analog converter (DAC). Concurrently, switching frequency stability is achieved by embedding the BUCK converter within a PLL to adaptively adjust the on-time. The circuit architecture, implemented in a 0.18 μm BCD process, demonstrates that under a 3 A load current and a 3.57 V output voltage, the output voltage DC offset is reduced to 0.767 mV (0.02%), while the switching frequency remains stable at 2 MHz.

Keywords—constant on time (COT), BUCK converter, DC offset, phase-locked loop (PLL).

I. INTRODUCTION

In recent years, with the rapid development of the electronics industry, the demand for power supply output accuracy, fast transient response, and efficiency in equipment performance has been increasing, which is critical to maintain voltage stability under various operating conditions, constant on-time (COT) controlled buck converters are widely used due to their simple control and high efficiency[1]. However, there are still two problems with COT controlled buck converters: output voltage offset caused by the control mechanism and compensation ripple[2], and switching frequency variation caused by input/output voltage and load current[3]. To improve the accuracy of output voltage, an internal ripple compensation method is proposed. By injecting the virtual inductor current ripple and its DC component into comparator for valley comparison, but the output voltage still has DC offset due to the DC error between the virtual inductor current ripple and its DC component[4][5].Q. ul Ain et al.[6] employed a differential adder to sum the inductor current ripple and the feedback voltage V_{FB}, yielding V_{SUM}. Then a subtractor removes the DC offset voltage V_{OFFSET} from V_{SUM}, producing a new feedback voltage V_{FB2}. For the problem of switching frequency drift, Y. -L. Chao et al.[7]proposed a V^2 adaptive on-time control strategy. This approach eliminates loop delay via a V²-loop, making the system frequency depend only on the duty cycle and achieving pseudo-constant switching frequency. However, the introduced error amplifier requires an exceptionally high gain-bandwidth product, and

This work was supported in part by Beijing Natural Science Foundation under Grant No.4232063 and the National Natural Science Foundation of China under Grant 62074010.

the need for numerous passive components for compensation complicates loop stability design.

This paper introduces a method of adaptive on-time control based on phase-locked loop (PLL) and dynamic calibration. This technique feeds back a new reference voltage through an integrator circuit and a digital-to-analog converter (DAC) module. Since the DC offset in the output voltage is directly calibrated, the loop stability of the converter remains unaffected. By incorporating a PLL, the whole converter is embedded in the phase-locked loop. The system frequency is locked by comparing it to a reference clock, and the on-time is dynamically adjusted to achieve constant frequency under input/output different voltages. As this mechanism operates outside the main control loop, it does not affect the transient response speed of the converter.

The structure of this paper is as follows: Section II describes the system design and DC offset calibration techniques; Section III describes the implementation of key circuits; Sections IV and V present simulation verification results and conclusions to support effectiveness.

II. SYSTEM ANALYSIS AND DESIGN

This section first analyzes the sources of output voltage offset and frequency drift in constant on-time control. Subsequently, a self-calibration structure for eliminating the DC offset and the use of a PLL to lock down the switching frequency are presented.

A. Output DC offset and switching frequency drift in the Traditional Method

Fig. 1 (a) illustrates COT control circuit structure diagram based on a comparator (Comp), on-time timer (T_{ON} generator), and driver block (Driver logic). When analyzing the offset voltage, the parasitic inductance (ESL) is ignored since the output voltage ripple is primarily determined by the equivalent series resistance (ESR) of the output filter capacitor. If the peak-to-peak output voltage ripple amplitude is $2\Delta V_{OUT}$,the expression for the output dc level V_{OUT_DC} is

$$V_{OUT\_DC} = K_R \cdot V_{REF} + \Delta V_{OUT} \qquad (1)$$

Where K_R is the voltage divider resistance coefficient $(R_1 + R_2)/R_2$, V_{REF} is the reference level provided by bandgap. Consequently, the magnitude of the ripple directly affects the output DC voltage. In the converter, the peak-to-peak inductor current ripple is given as

$$2\Delta i_L = (V_{IN} - V_{OUT})T_{ON}/L \qquad (2)$$

therefore，the ripple caused by the filtering capacitor's ESR can be expressed as

$$2\Delta V_{OUT} = 2\Delta i_L \cdot ESR \qquad (3)$$

combining (1) and (3), V_{OUT_DC} can be obtained

$$V_{OUT\_DC} = K_R V_{REF} + \frac{V_{IN} - V_{OUT}}{2L} \frac{K V_{OUT}}{V_{IN}} \cdot ESR. \qquad (4)$$

Equation (4) clearly demonstrates the impact of input voltage V_{IN}, output voltage V_{OUT}, and parasitic ESR on the output DC level. As shown in Fig. 1 (b), with the reference voltage fixed, the feedback ripple magnitude directly affects the DC level of the output voltage. Therefore, variations in output voltage ripple produce DC offset.

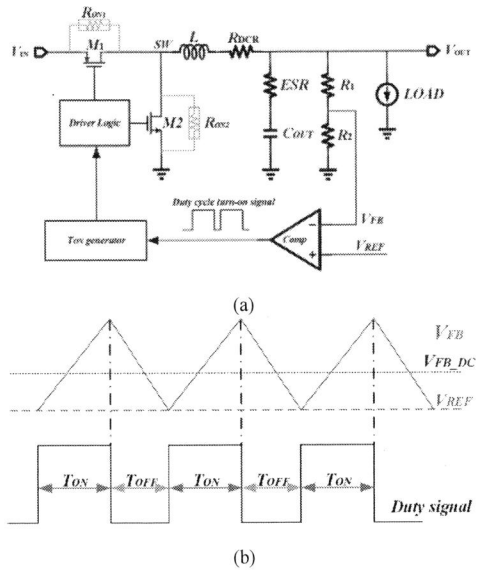

(a)

(b)

Fig. 1 Conceptual diagram of DC offset (a) Traditional COT buck converter (b) The key waveforms of dc output

As mentioned above, switching frequency instability may arise from variations in input voltage, output voltage, and load current. During operation, the output voltage V_{OUT} is fed back to the comparator through a voltage divider as V_{FB}. This feedback signal V_{FB} is then compared with the reference voltage V_{REF}. As shown in Fig. 1 (b), when V_{FB} falls below V_{REF}, a new on-time cycle begins. Due to the parasitic resistances of the power switch and the inductor, the SW node voltages—high-level voltage V_{SW_H} and low-level voltage V_{SW_L} can be expressed as

$$V_{SW\_H} = V_{IN} - I_L(R_{ON1} + R_{DCR}) \qquad (5)$$

$$V_{SW\_L} = -I_L(R_{ON2} + R_{DCR}). \qquad (6)$$

Since the on-time is fixed, the relationship between the system frequency F_{SW}, the duty cycle D, the input voltage V_{IN}, and the output voltage V_{OUT} can be expressed as

$$T_{ON}/T_S = D = T_{ON} \cdot F_{SW} \qquad (7)$$

$$V_{OUT} = D \cdot V_{SW\_H} - (1 - D) \cdot V_{SW\_L} \qquad (8)$$

$$F_{SW} = \frac{V_{OUT} - V_{SW\_L}}{V_{SW\_H} - V_{SW\_L}} \cdot \frac{1}{T_{ON}} = \frac{V_{OUT} + I_L(R_{ON2} + R_{DCR})}{V_{IN} - I_L(R_{ON1} - R_{ON2})} \cdot \frac{1}{T_{ON}}. (9)$$

Equation (9) explicitly demonstrates the dependence of the switching frequency F_{SW} on the input voltage, output voltage and inductor current I_L. Specifically, the switching frequency F_{SW} increases with I_L.

B. The Proposed Self-calibration Structure

To enhance output voltage accuracy in COT controlled converters, a DC calibration technique utilizing a DAC is proposed. As shown in Fig. 2, the self-calibration structure integrated into the adaptive on time (AOT) control loop comprises an integrator, dynamic comparator (DCOMP), digital counter (COUNTER), and DAC. Unlike conventional COT control, V_{REF} is not directly compared to the feedback voltage V_{FB}. Instead, the self-calibration system generates a calibrated voltage V_{COMP}, which is compared to both the ripple component and valley point of the compensated feedback signal to determine the on-time. Since the self-calibration structure only detects the DC component of V_{FB}, it preserves the transient response characteristics of the converter.

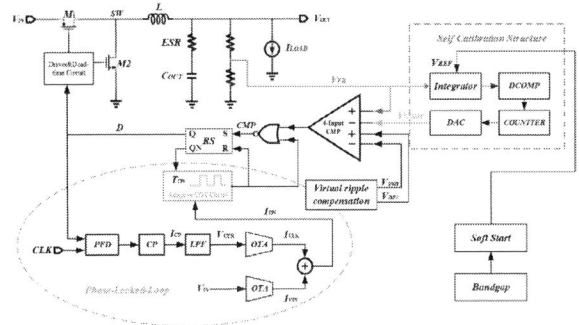

Fig. 2 Structure of proposed method in AOT buck converter

C. Loop Structure Embedded in PLL

In COT control, the factors influencing switching frequency deviation are derived from equation (9). To enhance the high stability of chip operating frequency, an AOT controlled converter embedded with PLL is proposed, as shown in Fig. 2. Compared with traditional PLL, this design eliminates the voltage-controlled oscillator (VCO) and integrates the on-time timer and RS latch into the loop. Crucially, the combined action of the on-time generator and latch module functionally emulates a VCO by regulating output frequency. Consequently, the system achieves full frequency and phase locking. In this paper, the PLL samples the frequency error between F_{SW} and CLK. It then generates a corrective current to the on-time module, ensuring precise frequency locking to CLK. At high operating frequencies, the switching frequency is prescaled before PLL sampling. The design addresses the inherent loop delay Δt in the converter, as frequency increases, the switching period T_S decreases, amplifying the relative delay error $\Delta t / T_S$. Frequency division effectively minimizes this error.

III. CIRCUIT REALIZATION

A. DC Offset Calibration Circuit

The calibration system is illustrated in Fig. 3, comprises a fully differential RC integrator (Integrator), a dynamic comparator (DCOMP), an 8bit counter (COUNTER), a Dead Time Logic and Drive Circuit (Dead-time), an Adder/Subtractor circuit (AS Logic), a CLK GEN and a DAC. The integrator module integrates the feedback voltage V_{FB} and reference signal V_{REF} over 16 cycles, filtering out high-frequency ripple components to extract the DC level V_{FB_DC} and V_{REF_DC}, which is then fed to the comparator. The counter generates an increment/decrement control signal based on the output of DCOMP. The DAC quantizes this error, and subtracts the quantization result from V_{REF} to produce the calibrated reference voltage V_{COMP}. This calibration is performed bit by bit during each integration cycle.

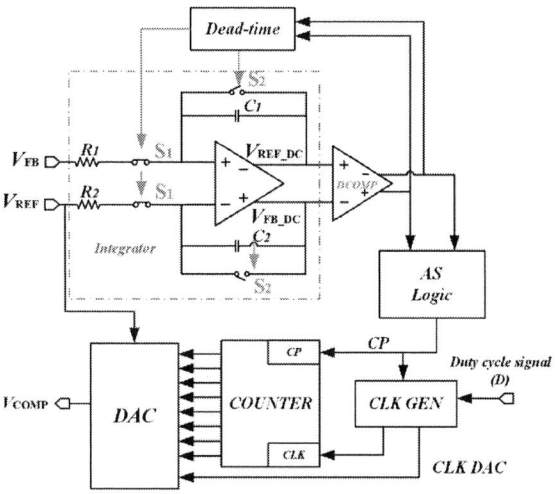

Fig. 3 system structure of the proposed dc calibration

During calibration, the dead-time control block monitors the comparator output signal and produces a pair of non-overlapping clock signals. This sequence first turn-off switch S_1 (disconnecting the integration resistor R_1/R_2 from the operational amplifier), then turn-off switch S_2 to discharge the integration capacitor C_1/C_2, thereby resetting its stored charge for the next integration cycle. Prior to calibration termination, based on the judgment result of DCOMP, the AS Logic module generates a count signal CP that controls the increment or decrement of the counter. The CLK GEN monitors CP and produces the clock signal for the counter. When the DC error is reduced below 1 LSB, the CLK GEN shuts off the clock of the counter, thereby freezing the output of DAC and finalizing the precision calibration. To enhance integration significance, the calibration frequency must be lower than the switching frequency of system. Moreover, higher DAC bit counts yield smaller LSB values, thereby improving calibration accuracy.

B. Switching frequency calibration circuit

Fig. 4 shows a simplified schematic of a module that adjusts the switching frequency by modulating the power switch conduction time (T_{ON}) through PLL. Switch S_1 resets the circuit when triggered by the rising edge of the RS latch, discharging capacitor C_1. Current source I_{ON} charges C_1, causing its voltage to ramp up linearly. When this voltage exceeds the comparator's positive input threshold, it generates a reset signal for the latch. The expression for T_{ON} can be derived as follows

$$I_{ON} \cdot T_{ON} = C \cdot k V_{OUT} \qquad (10)$$

$$I_{ON} = I_{CLK} + I_{VIN} \qquad (11)$$

$$T_{ON} = C \cdot k V_{OUT} \cdot \left(\frac{1}{I_{ON}}\right) = \frac{C \cdot k V_{OUT}}{I_{VIN} + I_{CLK}} \qquad (12)$$

where k is the proportionality coefficient, I_{ON} is a current that contains frequency information, from equation (9), the switching frequency can be expressed

$$F_{SW} = \frac{V_{OUT} + I_L(R_{ON2} + R_{DCR})}{V_{IN} - I_L(R_{ON1} - R_{ON2})} \cdot \frac{I_{VIN} + I_{CLK}}{C \cdot k \cdot V_{OUT}}. \qquad (13)$$

The switching frequency remains constant during stable operation, locked to the external clock CLK. This synchronization ensures switching frequency stability under adaptive adjustment, matching the CLK frequency. Notably, the GBW of the PLL must be designed at approximately 1/4 of the GBW of the converter loop. This ratio ensures the converter responds faster than the PLL, thereby preserving the fast transient response characteristic of the COT control.

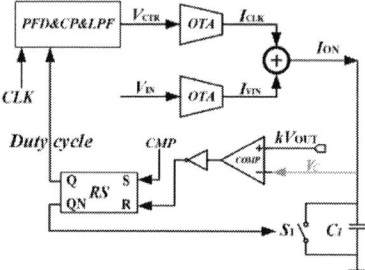

Fig. 4 system structure of the proposed frequency calibration

IV. Verification Of The Proposed Method

In order to verify the DC offset calibration adaptive conduction time control method based on PLL, the whole system circuit is realized by 0.18μm BCD process. The chip layout is shown in Fig. 5, and the total chip area including ESD is 1267 μm×2352 μm.

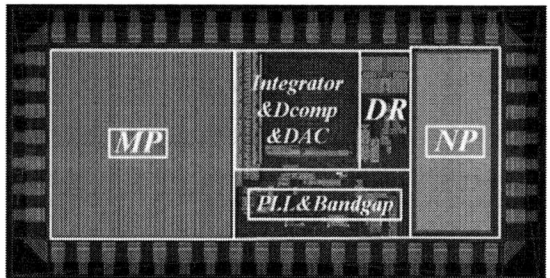

Fig. 5 The diagram of the chip layout

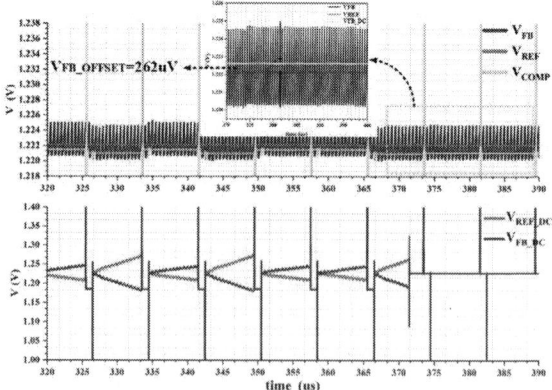

Fig. 6 Key nodes waveform of DC calibration module under output voltage V_{OUT}=3.57V, V_{IN}=4.2V.

Fig. 6 presents the waveforms at critical nodes within the DC calibration module under the operating conditions of V_{IN}=4.2V, V_{OUT}=3.57V, and I_{LOAD}=3A. As shown in Fig. 6, the calibration structure accurately integrates the DC levels of V_{FB} and V_{REF}, generating a comparison voltage V_{COMP} that is

979-8-3315-3918-4/25 $31.00 © 2025 IEEE

fed back into the system loop for dynamic adjustment. The calibration process terminates when the DC error falls below 1 LSB, yielding a measured DC offset of 0.767 mV (0.02%). These results validate the significant improvement in output voltage accuracy achieved by the proposed method.

Fig. 7 demonstrates the PLL calibration results of the switching frequency F_{SW} under varying load conditions, along with its response to load transients. Under different load currents, the PLL circuit locks the F_{SW} to 2MHz by regulating the on-time, which eliminates the impact of I_L on F_{SW} in equation (9). During load transients, the control voltage V_{CTR} does not change abruptly, which maintains near-constant on-time and enables timely modulation of the off-time to rapidly relock the switching frequency.

Fig. 8 shows the load transient response with an input of 4.2 V and an output of 3.57 V. For the load step from 1 to 3 A, it can be seen that the overshoot and undershoot deviations of 19.2 mV and 13.9 mV respectively, with a recovery time under 4.1 μs.

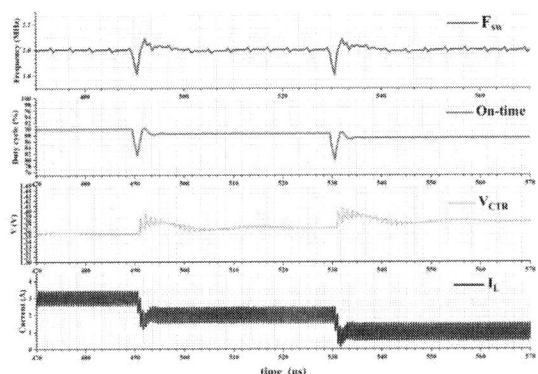

Fig. 7 Key waveforms of the PLL circuit

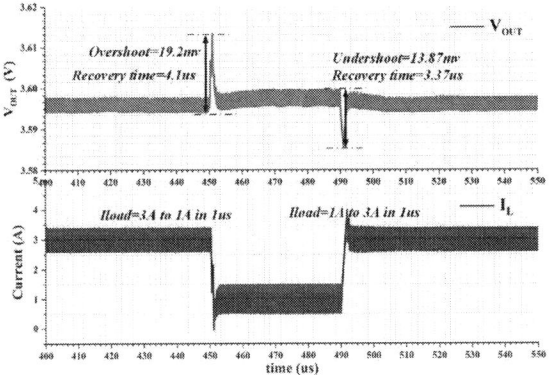

Fig. 8 Load transient response under output voltage V_{OUT} = 3.57V, V_{IN} =4.2V, the load changes between 1A and 3A

The simulated performance metrics and comparisons with prior works are shown in Table I, compared with prior work, the proposed method enhances output voltage accuracy to 0.767mV (0.02%) while confining transient recovery within 4.1 μs. Simulation results confirm that the proposed control methodology achieves high-precision output voltage regulation, stable switching frequency, and fast load transient response.

TABLE I. COMPARISONS WITH THE PRIOR WORK

References	2020[2] (Measure)	2021[3] (Measure)	2024[5] (Simulation)	This work (Simulation)
Process	BCD 0.18 μm	BCD 0.25 μm	BCD 0.18 μm	BCD 0.18 μm
Input voltage	3.3 V	7 V	8 to 24 V	4 to 5.5 V
Output voltage	1.03 V	5 V	5 V	3.57 V
Inductor	10 μH	4.7 μH	5.6 μH	240 nH
Capacitor	10 μF	NA	30 μF	22 μF
Load Current	1 A	6 A	0-2 A	0-3 A
Switching Frequency	NA	0.1 to 1 MHz	1 MHz	2 MHz
Offset Voltage	8 mV	NA	6 mV	0.767 mV
Peak Efficiency	93.4%	98%	NA	94.7%
Recovery Time	30 μs (0.1 to 1A)	20 μs (1.2 to 4.8A)	27 μs (0.1 to 2A)	4.1 μs (1 to 3A)

V. CONCLUSION

This paper presents an AOT control method based on a PLL with dynamically calibrated DC offset. By embedding the buck converter into the PLL loop, this method achieves stable system frequency across different input/output voltages and load currents to lock the switching frequency at 2MHz. The integrator and DAC are employed to tune the reference signal, thereby eliminating the DC offset. This successfully enhances the output voltage accuracy to within 0.767 mV (0.02%). Since the calibration circuitry does not participate in loop stability compensation, it does not affect the output voltage feedback ripple or transient response speed, the Recovery Time is 4.1 μs.

VI. REFERENCES

[1] S. Bari, Q. Li and F. C. Lee, "A New Fast Adaptive On-Time Control for Transient Response Improvement in Constant On-Time Control," IEEE Trans. Power Electron, vol.33, no. 3, pp. 2680-2689, March.2018.

[2] R. C. -H. Chang, W. -C. Chen and J. K. -S. Huang, "A 93.4% Efficiency 8-mV Offset Voltage Constant On-Time Buck Converter With an Offset Cancellation Technique," IEEE Trans Circuits Syst II vol. 67, no. 10, pp. 2069-2073, Oct.2020.

[3] J. Zhao, Q. Ye and X. Lai, "A Frequency Stable On-Time Control Buck Converter With Reference and Frequency Compensation Technique Using Low ESR Output Capacitor," IEEE Trans. Ind. Electron., vol.69, no.4, pp. 35363545, April.2022.

[4] X. Ming, Y. -L. Xin, T. -S. Li, H. Liang, Z. -J. Li and B. Zhang, "A Constant On-Time Control With Internal Active Ripple Compensation Strategy for Buck Converter With Ceramic Capacitors," IEEE Trans Power Electron, vol. 34, no. 9, pp. 9263-9278, Sept.2019.

[5] J. Lv, L. Huang, C. Wang and D. Sun, "A Novel Internal Ripple Compensation Technology With DC Offset Cancellation Based on COT Buck Converter,"2024 IEEE China Int Youth Conference on Electrical Engineering (CIYCEE), Wuhan, China, 2024, pp.15.

[6] Q. ul Ain et al., "A High-Efficiency Fast Transient COT Control DC–DC Buck Converter With Current Reused Current Sensor," IEEE Trans on Power Electronics, vol. 36, no. 8, pp. 9521-9535, Aug. 2021.

[7] Y. -L. Chao, C. -J. Tsai, Y. -R. Huang, W. -C. Liu, S. -H. Ma and C. -J. Chen, "A 4-MHz Ultra-Fast Transient Response Capacitor Current Adaptive On-Time (CCAOT) Controlled Buck Converter With Passive Ramp Compensation," IEEE Trans on Industry Applications, vol.60, no.2, pp.3397-3410, March-April 2024.

Design of a Fast Transient Response LDO Circuit Based on Transient Enhancement Structure

Xudong Sun [1], Sujuan Liu*[1], Kun Liu [1], Junchao Zhao [1]

[1] School of Information Science and Technology, Beijing University of Technology, Beijing, 100124, China

* Email: liusujuan@bjut.edu.cn

Abstract—This paper proposes a frequency-compensated, capacitor-less low-dropout linear regulator (LDO) that achieves high accuracy and fast transient response. The design employs a buffer stage composed of a super source follower and a flipped voltage follower, effectively extending the bandwidth by pushing the pole of the power transistor gate to higher frequencies. To further provide frequency compensation, this design utilizes Feedback Amplifier Miller Compensation (FAMC), introducing a compensating zero to ensure loop stability. A transient enhancement structure is developed based on three key mechanisms: charging and discharging of the power transistor gate, discharge at the output node, and injection current into the error amplifier. These enhancements significantly improve the transient response of the regulator. The proposed LDO is implemented in a 0.18 μm BCD CMOS process, occupying a silicon area of 0.033 mm². It operates over an input voltage range of 4–5 V and regulates the output voltage at 3.3 V with a load capacitance of 80 pF. Simulation results demonstrate that the LDO maintains a phase margin exceeding 60° across the full load range, with a transient settling time of less than 400 ns.

Keywords—power management, low-dropout linear regulator (LDO), frequency compensation, fast transient response

I. INTRODUCTION

With the rapid advancement of integrated circuit technology, the integration density of System on Chip (SoC) designs continues to increase. As a critical component within SoCs, the Power Management Integrated Circuit (PMIC) faces both new opportunities and challenges. Among PMIC components, low-dropout linear regulators (LDO) have gained significant attention due to their high integrability, fast transient response, and excellent noise suppression capabilities. Based on the location of the dominant pole in the regulation loop [1], LDOs can generally be categorized into two types: off-chip capacitor-based LDOs and capacitor-less LDOs. The former requires a large external capacitor at the output to ensure loop stability, which makes it unsuitable for full on-chip integration. The work in [2] implemented in a 65 nm process achieves an ultra-low static power consumption of only 5 nA, but relies on a 1 μF output capacitor. The 10-mA LDO in [3] employs a zero compensation structure and achieves 16 nA quiescent current under light load, but operates within a limited load current range of 100 μA to 10 mA. Currently, capacitor-less LDOs have become a major focus of research. An adaptive frequency compensation technique, incorporating a novel transistor degeneration frequency compensation method to ensure stability under ultralow static bias conditions, is presented in [4]. The local positive-feedback loop technique can be used in a single-stage structure to enhance the general performance of CL-LDO [5]. The LDO in [6] employs a feed-forward ripple cancellation circuit and a negative capacitance circuit to improve power

supply rejection (PSR), while a voltage damper is utilized to enhance transient response.

This paper presents an LDO circuit implemented in a 0.18 μm BCD CMOS process, occupying a silicon area of 0.033 mm². The design employs a source follower structure that combines a flipped voltage follower (FVF) with a super source follower (SSF) to achieve pole separation. Feedback Amplifier Miller Compensation (FAMC) is utilized to introduce a compensating zero. To enhance transient response, the circuit incorporates techniques that address three key aspects: charging and discharging of the power transistor gate, discharge at the output node, and injection current into the error amplifier. The proposed LDO operates over an input voltage range of 4–5 V, delivers a regulated output of 3.3 V, and supports a maximum output current of 30 mA.

II. PROPOSED LDO STRUCTURE

A. Stability analysis

Fig. 1 illustrates the structure of the proposed LDO. It consists of an error amplifier (EA), a buffer stage implemented using a source follower structure, a PMOS power transistor (M_P), an output capacitor (C_L), an impedance feedback network consisting of R_1 and R_2, compensation capacitor (C_f), and a frequency compensation loop composed of a feedback amplifier (G_Z) and a capacitor (C_Z). In addition, a transient enhancement structure is integrated to improve the response speed of the circuit.

Fig. 2 shows the small-signal model of the LDO. Since the capacitors in the transient enhancement circuit are very small, the transient enhancement structure can be neglected when constructing the small-signal model. In this model, the transconductance of the error amplifier, buffer, and PMOS power transistor are represented by g_{mEA}, g_{mB} and g_{mP}, respectively. $R_{O,EA}$, $R_{O,B}$ and $C_{O,EA}$ denote the output resistance of the error amplifier and the buffer stage, and the output capacitance of the error amplifier, respectively. C_{PASS} is the parasitic capacitance at the gate of the power transistor. C_{Out} is the output capacitance of the LDO. The feedback error amplifier has a transconductance denoted by g_{mZ} and an equivalent output impedance denoted by R_Z. R_1 and R_2 are the feedback resistors, and C_f is the compensation capacitor used to introduce a zero-pole pair for stability.

The open-loop transfer function of the LDO can be derived based on the analysis of Fig. 2:

$$A_V(s) = \frac{A_{dc}(1+sR_1C_f)(1+sR_zC_Z)}{(1+sR_{O,EA}R_{O,B}R_{OUT}g_{mB}g_{mP}C_Z)[1+s(R_1\|R_2)C_f]\left(1+s\frac{g_{mZ}C_{PASS}}{g_{mB}g_{mP}}+s^2\frac{C_{PASS}C_{OUT}}{g_{mB}g_{mP}}\right)}. \quad (1)$$

Here, $A_{dc}(=-\beta g_{mEA}g_{mB}g_{mP}R_{O,EA}R_{O,B}R_{OUT})$ represents the DC gain of the LDO. The proposed LDO includes four poles and two zeros, which are attributed to the following: the output

This work was supported in part by Beijing Natural Science Foundation under Grant No.4232063 and the National Natural Science Foundation of China under Grant 62074010.

979-8-3315-3918-4/25 $31.00 © 2025 IEEE

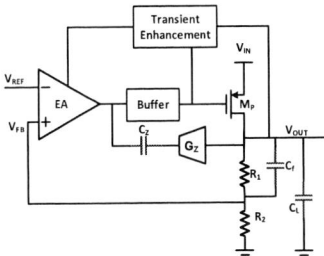

Fig. 1. Block diagram of the proposed LDO

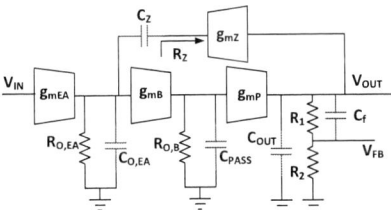

Fig. 2. Small-signal model of the proposed LDO

pole of the error amplifier, the gate pole of the power transistor, the output pole, the pole-zero pair generated by the resistive feedback network, and the zero introduced by the feedback amplifier. Their expressions are given as follows:

$$P_1 = \frac{1}{R_{O,EA} R_{O,B} R_{OUT} g_{mB} g_{mP} C_Z} \qquad (2)$$

$$P_2 = \frac{g_{mB} g_{mP}}{g_{mZ} C_{PASS}} \qquad (3)$$

$$P_3 = \frac{g_{mZ}}{C_{OUT}} \qquad (4)$$

$$P_4 = \frac{1}{(R_1 || R_2) C_f} \qquad (5)$$

$$Z_1 = \frac{1}{R_1 C_f} \qquad (6)$$

$$Z_2 = \frac{1}{R_z C_Z}. \qquad (7)$$

The detailed implementation of the proposed LDO is shown in Fig. 3. The error amplifier (M_1–M_{11}) employs a folded cascode structure, which provides a wide input common-mode range and high gain, thereby enhancing the accuracy of the LDO. In two-stage LDO architectures, the error amplifier typically presents a high output impedance, and the power transistor gate exhibits significant parasitic capacitance—both of which severely degrade the phase margin. To mitigate this, a source follower is inserted between the error amplifier and the power transistor to shift the gate pole to a higher frequency. In Fig. 3, transistors M_{12}–M_{17} form the buffer stage. Specifically, M_{12}, M_{16}, and M_{17} constitute the super source follower (SSF), while M_{12}, M_{13}, and M_{15} form the flipped voltage follower (FVF), which further reduces the buffer's output impedance. The equivalent output impedance of the buffer stage is given by:

$$R_{O,B} = \frac{1}{g_{m12} (g_{m13} + g_{m17}) r_{o12}}. \qquad (8)$$

By configuring the bias and buffer DC operating points such that M_{15} and M_{42} have same size and conduct equal currents, the source voltage of M_{15} closely approximates V_{B3}, enabling precise control of the DC current through M_{17}. Moreover, when the LDO output voltage exhibits an overshoot or undershoot, M_{13} and M_{17} can respectively charge or discharge the gate of the power transistor, thereby improving the transient response of the LDO.

The feedback amplifier is implemented using transistors M_{18}–M_{25}, with C_Z serving as the Miller compensation capacitor. This combination forms the feedback amplifier miller compensation (FAMC) structure. The compensation loop not only provides capacitive loading at the output of the error amplifier to push the dominant pole to a lower frequency, but also introduces a zero into the frequency response to compensate for the output pole, especially under light-load conditions. Fig. 4 illustrates the frequency response with and without the FAMC structure, where the phase response exhibits a rise near 100 kHz due to the left-half-plane zero introduced by the FAMC.

Resistors R_1 and R_2 form the feedback network. Together with the resistance feedback network capacitance (RFNC) C_f, they generate a pole P_4 and a zero Z_1. In LDO circuits with low feedback voltage (such as those using a bandgap reference with a low output voltage), the resistance of R_2 is much smaller than that of R_1, which leads to a wide separation between the introduced pole and zero. This allows for the value of C_f to be selected such that the generated zero Z_1 effectively compensates one of the internal poles of the LDO. In this design, the introduced pole-zero pair improves the frequency response and extends the circuit bandwidth. As shown in Fig. 5, with RFNC compensation, the phase margin increases from 68° to 79°, and the bandwidth improves from 1.18 MHz to 2.71 MHz.

B. Transient response analysis

When the LDO circuit experiences a sudden change in load, the variation in output voltage can be expressed as:

$$\Delta V_{OUT} = \frac{\Delta I_L T_R}{C_L} \qquad (9)$$

where T_R is the response time of the LDO circuit, and it can be expressed as:

$$T_R = \frac{1}{BW} + C_{PASS} \frac{\Delta V_G}{I_G} \qquad (10)$$

where BW is the loop bandwidth of the circuit, ΔV_G is the gate voltage variation of the power transistor under different load conditions, and I_G is the charging or discharging current at the power transistor gate. As indicated by (10), increasing the loop bandwidth and enhancing the gate slew current of the power transistor can accelerate the transient response of the LDO.

The transient enhancement structure (TES) shown in Fig. 3 consists of overshoot and undershoot compensation circuits. The overshoot compensation includes a gate charging module for the power transistor and an output current discharge module. The power transistor gate charging module consists of a coupling capacitor C_2, a bias circuit formed by M_{26} and M_{27}, a push-pull structure formed by M_{28} and M_{29}, and a

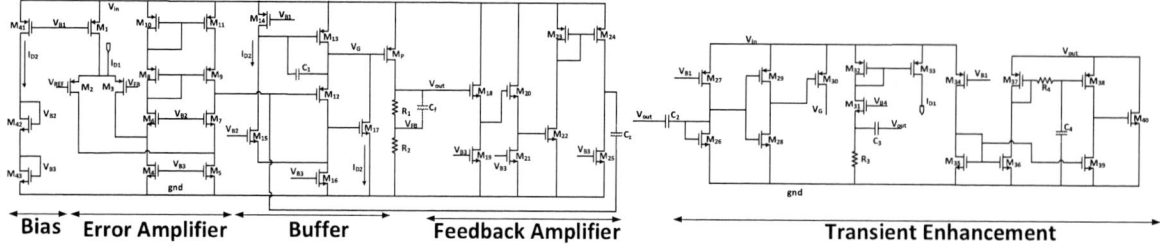

Fig. 3. Schematic of the proposed LDO

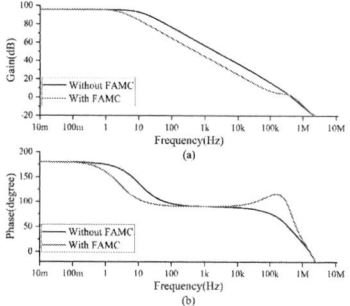

Fig. 4. Frequency response with and without the proposed FAMC (a) Gain curve (b) Phase curve

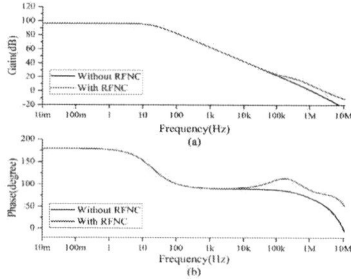

Fig. 5. Frequency response with and without the proposed RFNC (a) Gain curve (b) Phase curve

PMOS transistor M_{30}. In steady-state operation, M_{26} biases M_{29} in the triode region, keeping M_{30} in the subthreshold region so that the static power consumption of the module remains low. When the load suddenly decreases, the limited bandwidth of the circuit causes the gate voltage of the power transistor to respond slowly, leading to an increase in the LDO output voltage. This output voltage change is coupled through C_2 to the input of the push-pull structure, which lowers the gate voltage of M_{30}. As a result, M_{30} enters saturation and generates a large current to charge the gate capacitance of the power transistor. This accelerates the response of the power transistor gate voltage and enables the LDO output voltage to stabilize quickly.

The output current discharge module consists of a bias circuit formed by M_{34}–M_{39} and an NMOS transistor M_{40}. In steady-state operation, the current through M_{38} is small, resulting in a low gate voltage for M_{40} that keeps it in the cutoff region. When the load suddenly decreases, the source voltage of M_{38} experiences an overshoot, while the gate voltage of M_{38} remains unchanged. This causes the drain current of M_{38} to increase, which in turn raises the gate voltage of M_{40}. As a result, M_{40} enters saturation and discharges the

excess current generated by the power transistor, allowing the LDO output voltage to stabilize quickly.

The undershoot compensation structure consists of M_{31}–M_{33} and the coupling capacitor C_3. In steady-state operation, M_{31} remains in the cutoff region, resulting in no static power consumption. When the output voltage experiences an undershoot, the voltage change is coupled through C_3 to the source of M_{31}, causing M_{31} to enter saturation. M_{33} mirrors the current generated by M_{31} and injects it into the error amplifier. Fig. 6 shows the transient response with and without the transient enhancement structure, where Fig. 6(a) and Fig. 6(b) illustrate the time-domain output voltage response during a sudden decrease and increase in load, respectively. With the transient enhancement structure, the response time for overshoot is reduced from 21.8 µs to 972 ns, and the response time for undershoot is reduced from 788 ns to 396 ns. The magnitudes of both the overshoot and undershoot are also significantly reduced.

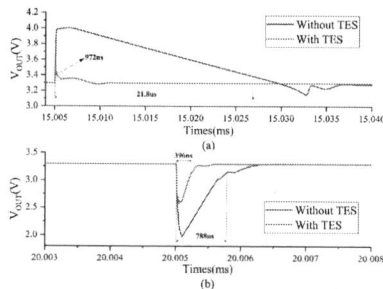

Fig. 6. Transient response with and without the proposed TES (a) step-down load transient (b) step-up load transient

III. SIMULATION RESULTS

The circuit is implemented using a 0.18 µm BCD CMOS process, and its layout is shown in Fig. 7. The total area is 0.033 mm². The overall circuit operates with an input voltage of 4-5 V and provides an output voltage of 3.3 V.

Fig. 8 shows the simulated frequency response of the LDO under different load current conditions. The LDO output capacitance is 80 pF. It can be seen from the simulation results that the DC gain is greater than 90 dB, and the phase margin exceeds 60° across the entire load range.

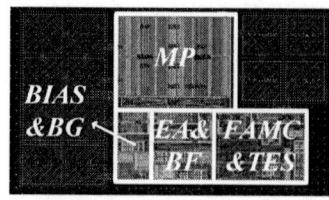

Fig. 7. Layout of the proposed LDO

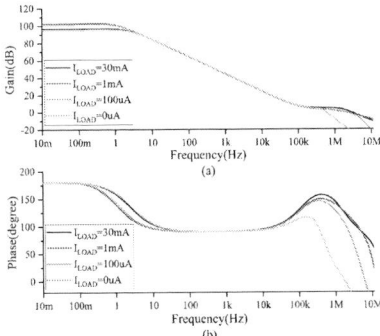

Fig. 8. Frequency response of the proposed LDO with different load current (a) Gain curve (b) Phase curve

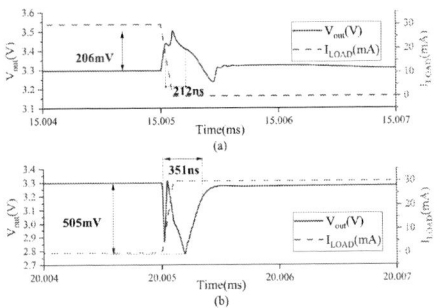

Fig. 9. Transient response of the proposed LDO (a) step-down load transient (b) step-up load transient

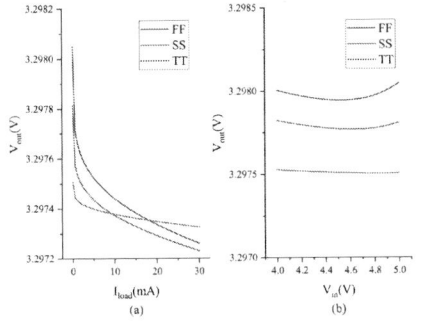

Fig. 10. (a) Load regulation and (b) Line regulation of the proposed LDO

Fig. 9 shows the simulated transient response of the LDO when the load current steps from 0 to 30 mA within 100 ns. The time required for the output to settle to the DC value with an error within 3% is defined as the settling time of the LDO. The settling times for the LDO's overshoot and undershoot are 212 ns and 351 ns, respectively, with corresponding voltages of 206 mV and 505 mV.

Fig. 10 shows the load regulation and line regulation of the LDO. When the load current varies from 0 mA to 30 mA and the input voltage varies from 4 V to 5 V, the load regulation and line regulation under the TT corner are 0.019 mV/mA and 0.116 mV/V, respectively.

Comparison with existing references in Table I demonstrates that the proposed circuit exhibits superior transient response performance.

TABLE I.   COMPARISONS WITH THE PRIOR WORK

Ref	*This work*[b]	*[4]*[c]	*[5]*[c]	*[6]*[c]	*[7]*[b]
Tech. (nm)	180	180	180	65	130
Area (mm^2)	0.033	0.055	0.033	0.039	N/A
V_{OUT} (V)	3.3	1	1.06	1.8	1.2
$\Delta V_{OUT}/V_{OUT}$ (mV/V)	153	117	273.58	118	81.67
$I_{LOAD, MAX}$ (mA)	30	100	50	50	20
$I_{Q, MAX}$ (uA)	37.4	245	138	110	7.2
Current efficiency (%)	99.88	99.76	99.72	99.78	99.96
Load Regulation (mV/mA)	0.019	0.077	0.476	N/A	0.011
Line Regulation (mV/V)	0.116	0.283	3.484	N/A	N/A
C_L (pF)	80	100	100	10	100
T_R (ns)	351	1560	270	1100	1500
FOM[a] (ns)	0.436	3.822	0.745	2.42	0.54

a. $FOM = T_R \frac{I_{Q.MAX}}{I_{LOAD.MAX}}$ b. Simulation results   c. Measure results

IV. CONCLUSION

This paper presents a capacitor-less LDO circuit with fast transient response based on transient enhancement structure. A source follower structure combining a flipped voltage follower (FVF) and a super source follower (SSF) is employed to achieve pole separation, while feedback amplifier Miller compensation is introduced to maintain stability across the entire load range. Power transistor gate charging, output node current discharge, and error amplifier current injection modules are designed to enhance the response speed of the LDO. The circuit is implemented in a 0.18 μm BCD CMOS process, with a layout area of 0.033 mm². Simulation results show that the settling time of the LDO does not exceed 400 ns.

REFERENCES

[1] S. Bhuiyan, M. A. Hossain, K. N. Minhad, F. Haque, M. S. K. Hemel, O. Md Dawi, M. B. Ibne Reaz, and K. J. A. Ooi, "CMOS low-dropout voltage regulator design trends: An overview," Electronics, vol. 11, no. 2, p. 193, Jan. 2022.

[2] Z. Yang, Q. Chen, S. Zhen, and M. Huang, "A LDO with 5-nA quiescent current and improved transient response within a 50-mA load current range," in Proc. IEEE Int. Symp. Circuits Syst. (ISCAS), Singapore, 2024, pp. 1–5.

[3] N. Adorni, S. Stanzione, and A. Boni, "A 10-mA LDO with 16-nA IQ and operating from 800-mV supply," IEEE J. Solid-State Circuits, vol. 55, no. 2, pp. 404–413, Feb. 2020.

[4] Y. Jiang, D. Wang, and P. K. Chan, "A quiescent 407-nA output-capacitorless low-dropout regulator with 0–100-mA load current range," IEEE Trans. Very Large Scale Integr. (VLSI) Syst., vol. 27, no. 5, pp. 1093–1104, May 2019.

[5] P. Li, X. Zhao, X. Zhang, M. Li, and B. Wen, "General performance enhancement for capacitor-less low-dropout regulator using local positive-feedback technique," IEEE Trans. Power Electron., vol. 40, no. 1, pp. 1498–1507, Jan. 2025.

[6] X. Cheng et al., "A fast transient response capless LDO regulator achieving −78 dB of PSR up to 2 MHz," in Proc. IEEE Int. Symp. Circuits Syst. (ISCAS), Singapore, 2024, pp. 1–5

[7] J. Tang, L. Ouyang, C. Dai, Y. Wang, and W. Zou, "A low-power fast-transient output-capacitorless LDO," in Proc. Int. Symp. Next Gener. Electron. (ISNE), Changsha, China, 2021, pp. 1–4.

A 32-MHz FLL-Based RC Oscillator with PVT Compensation Using Frequency Tripler

Ikhwan Kim, Yajie Qin*

School of Information Science and Technology, Fudan University, Shanghai 200433, China
* Email: 20110720122@fudan.edu.cn, yajieqin@fudan.edu.cn

Abstract—This paper presents a compact and low-power 32-MHz RC oscillator implemented in a 110 nm CMOS process. The proposed design employs a frequency-locked loop (FLL) architecture to mitigate supply voltage variation and integrates a chopper-stabilized integrator to suppress offset and flicker noise. To enhance robustness against process variation, a frequency tripler is used to dynamically adjust the divide ratio of the switched-capacitor resistor (SCR) control clocks, to compensate for feedback voltage variations caused by process drift. Temperature dependence is addressed by combining poly and diffusion resistors with complementary temperature coefficients (TC) in the RC time constant network. This design achieves a frequency stability of 44.4 ppm/°C and a maximum frequency error of ±1041 ppm across a temperature range of -40 °C to 150 °C. The design occupies 0.028 mm² and consumes 81μA, making it suitable for area- and power-constrained SoC applications.

Keywords—RC oscillator, Frequency-locked loop, PVT compensation, low-power oscillator

I. INTRODUCTION

With the growing demands of mobile, automotive, and IoT applications, precision frequency references are essential for accurate sensing and communication systems. While crystal oscillators offer high accuracy and stability, they cannot be fully integrated on-chip and are sensitive to PVT variations. LC oscillators are also accurate and reliable but are not suitable for low-frequency integration due to the area constraints of on-chip inductors [1]. As an alternative, RC oscillators have been proposed [2], offering advantages such as compact area and low power consumption, making them well suited for full integration. However, maintaining frequency stability remains a significant challenge due to variations in the RC time constant caused by PVT fluctuations.

Fig. 1 illustrates a conventional FLL-based RC oscillator. The main advantage of the FLL-based architecture lies in its ability to minimize output frequency sensitivity to supply variation while achieving high stability through closed-loop operation.

In this work, we propose an FLL-based RC oscillator architecture that not only inherits the robustness against supply voltage and temperature variations characteristic of conventional FLL-based designs but also incorporates additional techniques to minimize the impact of process variation.

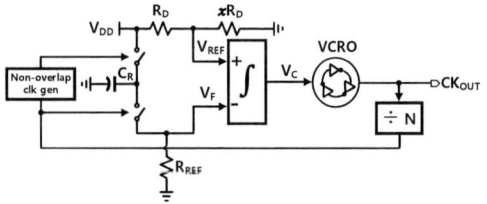

Fig. 1. Conventional FLL-based RC oscillator.

This work was supported by the National Key Research and Development Program of China (2023YFB4704000).

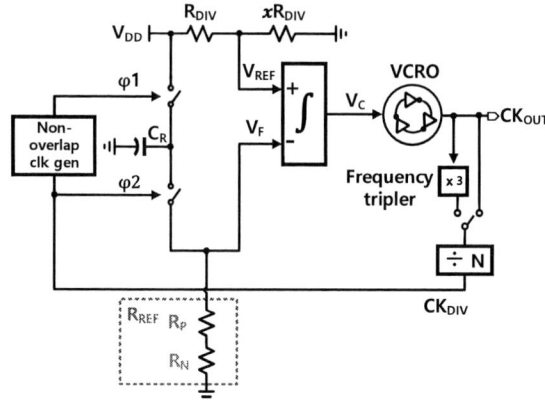

Fig. 2. Proposed architecture.

II. SYSTEM ARCHITECTURE

Fig. 2 shows the architecture of the proposed RC oscillator. While the overall structure follows that of a conventional FLL-based design, two types of resistors are employed in the R_{REF} to compensate for temperature-induced variations in the output clock frequency. By optimizing the combination of a poly resistor with a negative TC and a diffusion resistor with a positive TC, the temperature dependency of the RC time constant is significantly reduced.

To control the SCR, a non-overlapping clock generator is used to produce the internal control signals $\varphi1$ and $\varphi2$. However, due to the non-idealities of this circuit, process variation leads to changes in the feedback voltage V_F, which in turn introduces errors in the output clock frequency. Specifically, at the FF process corner, V_F decreases to approximately 70% of the TT corner value, whereas at the SS corner, V_F increases to around 130% of the TT value.

To mitigate these effects, a frequency tripler is introduced, enabling dynamic adjustment of the CK_{DIV} divide ratio. As a result, the RC time constant remains stable across process corners, while the output frequency variation is effectively minimized.

A. FLL Operation

To analyze the system, consider the reference voltage V_{REF} generated using matched resistors. Assuming identical resistor types and dimensions, the effects of process and temperature variations on V_{REF} cancel out:

$$V_{REF} = \frac{x\, R_{DIV}}{(1+x)\, R_{DIV}} \cdot V_{DD} \qquad (1)$$

The feedback voltage V_F is determined by the V_{DD}, the SCR, and the R_{REF}. Since the resistance of the SCR is given by $1/(C_R f_{DIV})$, V_F can be expressed as follows:

Assuming high DC gain and zero offset in the integrator, $V_F = V_{REF}$, leading to the output frequency:

$$F_{zero\_os} = \frac{x\,N}{C_R\,R_{REF}} \qquad (3)$$

To ensure independence from the supply voltage, V_{REF} and V_F must be matched precisely. However, integrator offset V_{OS} due to process mismatch causes deviation:

$$V_F = V_{REF} + V_{OS} \Rightarrow F_{os} = \frac{F_{zero\_os}}{x} \cdot \frac{x\,V_{DD} + (1+x)\,V_{OS}}{V_{DD} - (1+x)\,V_{OS}} \qquad (4)$$

To mitigate this, a chopper-stabilized integrator, as shown in Fig. 3, is introduced. This technique effectively reduces offset and $1/f$ noise, thereby enhancing the overall stability and accuracy of the system.

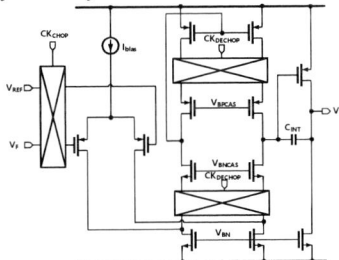

Fig. 3. Integrator with chopper stabilizer.

B. Compensation for Process Variation

In this system, the V_F is a critical factor in determining the output frequency. From the perspective of process variation, two primary contributors to fluctuations in V_F must be considered: the variation in SCR and the variation in R_{REF}.

While the SCR ideally has resistance $R_{SCR} = 1/(C_R f_{DIV})$, practical implementations include a non-overlapping time t_{gap} between clock phases. This modifies the effective switching frequency, resulting in:

$$R_{SCR,eff} = \frac{T_{CKDIV} - 2t_{gap}}{C_R} \qquad (5)$$

In the proposed design, with $C_R = 1\,pF$, and t_{gap} values of 4 ns for TT, 6.21 ns for SS, and 2.76 ns for FF, the change in SCR resistance is negligible. Hence, the dominant variation in V_F is due to shifts in R_{REF}.

While one potential solution is to trim R_{REF} directly, this approach must simultaneously account for both temperature and process variations, thereby increasing the complexity of the trimming mechanism and overall circuit implementation. As an alternative, the proposed design employs a frequency tripler, as shown in Fig. 4, to simplify the trimming strategy. This enables effective compensation for process-induced variations while preserving the desired TC of R_{REF}.

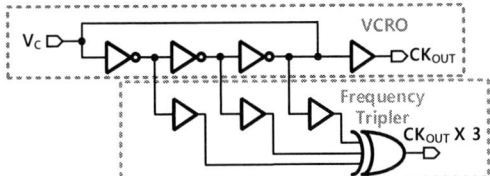

Fig. 4. Frequency tripler.

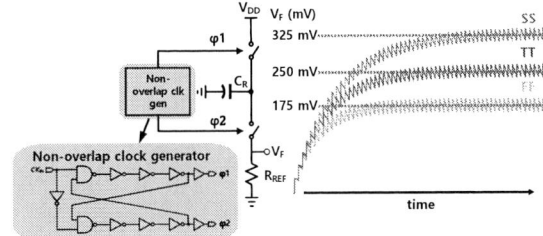

Fig. 5. SCR circuit and changes in V_F due to process variation.

As shown in Fig. 5, the process-induced variation causes V_F at the SS corner to increase to approximately 1.3 times its TT value, while at the FF corner, it decreases to about 0.7 times the TT value. To account for the process dependence of V_F, a scaling factor α is introduced into the expression derived from (2), resulting in:

$$V_F = \frac{V_{DD}}{1 + \dfrac{N}{R_{REF}\,C_R\,F_{OUT}}} \cdot \alpha \qquad (6)$$

The value of α corresponds to the deviation caused by process variation in V_F. Based on the results in Fig. 5, $\alpha = 1$ for the TT corner, $\alpha = 1.3$ for SS, and $\alpha = 0.7$ for FF.

To compensate for the variation in α, the proposed design adjusts the frequency divide ratio N using the frequency tripler circuit shown in Fig. 4. A conventional frequency divider supports only power-of-two values for N, which makes it difficult to precisely counteract the variation in α. Therefore, the proposed design sets $N = 32$ for the TT corner, and $N = 64/3$ and $N = 128/3$ for the FF and SS corners, repectively, by employing a tripled 96 MHz clock. This compensation method corrects process-induced error while minimizing changes to the TC of V_F, enabling a simple and cost-effective solution.

III. SIMULATION RESULTS

The system is implemented using a standard 110 nm CMOS process. All simulation results presented herein are based on post-layout simulations. The average power consumption of the system is 81 μA, and the layout occupies an area of 0.028 mm². Fig. 6 shows the layout of the proposed RC oscillator.

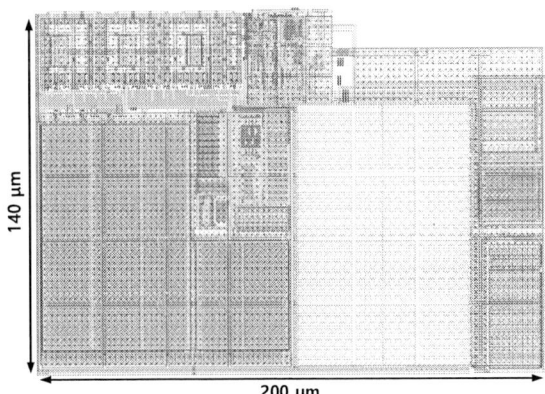

Fig. 6. Layout of the RC oscillator.

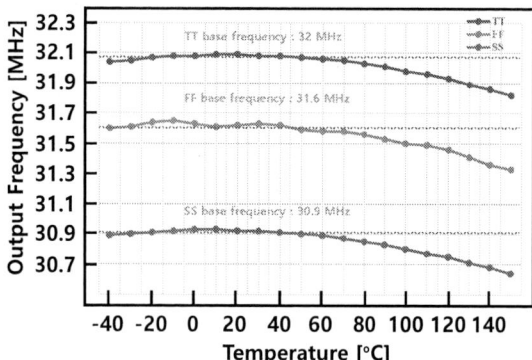

Fig. 7. Output frequency variation across process corner.

The variation of the output frequency with respect to temperature is illustrated in Fig. 7. Due to a combination of imperfect matching between the scaling factor α, which accounts for process variation, and the corresponding divide ratio N, as well as changes in internal delays of logic circuits, the base output frequency decreases slightly under process variations, reaching 31.6 MHz at the FF corner and 30.9 MHz at the SS corner. Across the temperature range of -40 °C to 150 °C, the output frequency variation is measured to be 44.4 ppm/°C at the TT corner, 53.3 ppm/°C at the FF corner, and 49.4 ppm/°C at the SS corner.

Since the adjustment of the clock divide ratio N and the delay variations in digital logic due to process variation have rarely affect the TC of the RC time constant, the TC of the output frequency remains relatively consistent across process corners.

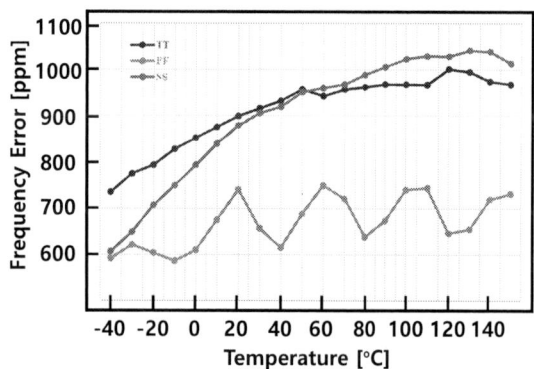

Fig. 8. Output frequency accuracy.

Fig. 8 illustrates the output frequency accuracy of the proposed system over the temperature range of -40 °C to 150 °C. At the TT corner, the minimum frequency error is 734 ppm at -40 °C, and the maximum error is 1001 ppm at 120 °C. For the FF corner, the minimum and maximum errors are 586 ppm at -10 °C, and 748 ppm at 60 °C, respectively. At the SS corner, the frequency error ranges from a minimum of 606 ppm at -40 °C to a maximum of 1041 ppm at 130 °C. Therefore, under worst-case conditions, the maximum frequency error of the proposed system is ±1041 ppm.

TABLE I. PERFORMANCE COMPARISON

	JSSC2020 [3]	JSSC2022 [4]	ISSCC2023 [5]	This work
Process	180	65	180	110
Frequency [MHz]	10.5	32	10	32
Power [μW]	219.8	34	85	121
Power efficiency [μW/MHz]	20.9	1.06	8.5	3.78
Temp. range	-45~125	-40~85	-45~125	-40~150
TC [ppm/°C]	137	8.4	31.5	44.4
Max. Freq. error [ppm]	-	±400	±2800	±1041
Supply Voltage [V]	1.4~2.2	1.1~2.3	1.5~1.8	1.4~1.7
Area [mm²]	0.015	0.18	0.01	0.028

Table I presents a comparison between the proposed design and previously reported state-of-the-art works. The proposed oscillator achieves a relatively low frequency error and low TC over a wide temperature range from -40 °C to 150 °C, while also maintaining low power consumption and occupying a small area.

IV. CONCLUSION

This paper proposes a compact and low-power FLL based RC oscillator capable of compensating for PVT variations. An FLL-based architecture is adopted to mitigate the effects of supply voltage variation. The RC time constant's sensitivity to temperature is reduced by combining a poly resistor with a negative TC and a diffusion resistor with a positive TC. In addition, a frequency tripler is employed to compensate for feedback voltage V_F variations caused by process variation. This approach effectively minimizes the error in the output frequency due to process-induced changes. By applying these techniques, the proposed design achieves a maximum frequency error of ±1041 ppm and a TC of 44.4 ppm/°C across the -40 °C to 150 °C temperature range, all within a compact area of 0.028 mm².

REFERENCES

[1] E. O. Ates, A. Ergul, and D. Y. Aksin, "Fully Integrated Frequency Reference With 1.7 ppm Temperature Accuracy Within 0–80°C," IEEE J. Solid-State Circuits, vol. 48, no. 11, pp. 2850–2859, Nov. 2013.

[2] G. Zhang, K. Yayama, A. Katsushima, and T. Miki, "A 3.2 ppm/°C Second-Order Temperature Compensated CMOS On-Chip Oscillator Using Voltage Ratio Adjusting Technique," IEEE J. Solid-State Circuits, vol. 53, no. 4, pp. 1184–1191, Apr. 2018.

[3] J. Lee, A. K. George, and M. Je, "An Ultra-Low-Noise Swing-Boosted Differential Relaxation Oscillator in 0.18-μm CMOS," IEEE J. Solid-State Circuits, vol. 55, no. 9, pp. 2489–2497, Sep. 2020.

[4] A. Khashaba, J. Zhu, N. Pal, M. G. Ahmed, and P. K. Hanumolu, "A 32-MHz, 34-μW Temperature-Compensated RC Oscillator Using Pulse Density Modulated Resistors," IEEE J. Solid-State Circuits, vol. 57, no. 5, pp. 1470–1479, May 2022.

[5] X. An, S. Pan, H. Jiang, and K. A. A. Makinwa, "A 0.01 mm2 10MHz RC Frequency Reference with a 1-Point On-Chip-Trimmed Inaccuracy of ±0.28% from -45° C to 125° C in 0.18μm CMOS," in 2023 IEEE International Solid-State Circuits Conference (ISSCC), San Francisco, CA, USA: IEEE, Feb. 2023, pp. 60–62.

979-8-3315-3918-4/25 $31.00 © 2025 IEEE

A Constant On-Time Buck Converter with VCO-based DC Offset Calibration Technique

Bingxue Zhang [1], Sujuan Liu*[1], Kun Liu [1], Xinyu Zhang [1], Yahua Shi [1]

[1] School of Information Science and Technology, Beijing University of Technology, Beijing, 100124, China

* Email: liusujuan@bjut.edu.cn

Abstract—The constant on-time (COT) buck converters have the advantage of simple control mechanism, excellent load transient response, and high efficiency, making them well-suited to meet the power requirements of modern electronic devices. However, the COT control mode has inherent limitations in output voltage accuracy, and load variations further affect this precision, which restricts its application in scenarios with stringent voltage accuracy demands. This paper presents a novel DC offset calibration circuit to enhance output voltage precision. The circuit integrates the DC error between the feedback signal V_{FB} and the reference signal V_{REF} using the straightforward voltage-controlled oscillator (VCO) and Digital Logic module. The integration result is then digitized and encoded by the Digital Logic module, and a digital-to-analog converter (DAC) generates the output offset calibration result V_R based on this encoded value, effectively eliminating the output DC offset. The converter is implemented using a 0.18μm BCD process, with an input voltage range of 4.2 V to 5.5 V and an output voltage of 3.52 V. The output DC offset is effectively reduced to 0.545 mV, ensuring improved voltage accuracy.

Keywords—*constant on-time (COT), buck converter, output DC offset, efficiency*

I. INTRODUCTION

Compared to pulse width modulation (PWM) technology, the constant on-time (COT) buck converters have garnered increasing attention and research due to their simple control mechanism, excellent load transient response, and high efficiency. The COT converters have been widely adopted in applications such as computers, telecommunication equipment, and others. However, the operating principle of traditional COT buck converters leads to inherent limitations, including poor output voltage accuracy. This inherent issue with output precision [1] has long restricted the use of COT buck converters, especially in applications requiring precise voltage regulation, such as high-performance processors and memory devices. The output DC offset of the COT converters primarily arises from the output capacitor's equivalent series resistance (ESR) and the ripple generated by ripple compensation techniques. In addition, the output DC offset of COT converters is proportional to the amplitude of the introduced ripple.

A common approach to eliminating DC offset is to improve the ripple compensation circuit [2]. This is done by shifting the DC level of the ripple signal within the ripple compensation circuit so that the ripple's valley reaches zero, and then superimposing this adjusted ripple signal onto the feedback voltage V_{FB}, resulting in a ripple whose valley voltage equals V_{FB}. When the valley voltage of the ripple signal equals the reference voltage V_{REF}, the output DC offset is effectively eliminated. A fixed-coefficient calibration technique has been proposed [3], but its drawback is that the calibration coefficient remains constant. When the duty cycle

of the COT Buck converter varies, this leads to DC offset. An improved ripple compensation circuit can also detect the actual valley of the ripple to precisely eliminate the output offset caused by the ripple. In recent years, several ripple valley detection circuits have been proposed. The D-CAP3 technology by Texas Instruments detects the ripple valley using a sample-and-hold circuit [4]. However, clock feedthrough and channel charge injection in the sampling switch can adversely affect sampling accuracy. To resolve this, a new calibration technique has been proposed that improves both sampling speed and accuracy [5]. Nevertheless, ripple valley detection techniques can only mitigate the impact of ripple introduced by the ripple compensation circuit. They do not resolve output DC offset caused by the ripple induced by the actual ESR. Another commonly used approach to eliminating DC offset is to lower the reference voltage V_{REF}. This method is similar to the constant on-time V^2 control strategy [6]. However, the main drawback is the increased complexity of the control loop.

This paper presents the DC offset calibration module for eliminating DC offset in the COT buck converter. The DC offset calibration module eliminates DC offset caused by the ripple introduced by the ripple compensation circuit, as well as DC offset resulting from the ESR voltage ripple, even when using low-ESR ceramic capacitors. The COT buck converter forms a fast control loop through the interaction of current ripple and the comparator, enabling rapid transient response. In parallel, the DC offset calibration module operates as a slow control loop, gradually adjusting the DC accuracy of the output voltage. A four-input comparator superimposes the difference between the ripple voltage V_{SAC} and its DC component V_{SDC} onto the feedback voltage V_{FB}, enhancing the ripple signal. This approach improves the output voltage accuracy and ensures the stability of the control loop.

II. DC OFFSET PRINCIPLE AND SYSTEM DESIGN

The concept of output DC offset in a COT converter is shown in Fig. 1. Fig. 1 (a) shows the COT control architecture, while Fig. 1 (b) presents the waveforms of key nodes. Ignoring other non-ideal factors, this control mode results in the average value of the feedback voltage $V_{FB(dc)}$ consistently being higher than the reference voltage V_{REF} by half of the ripple peak-to-peak value.

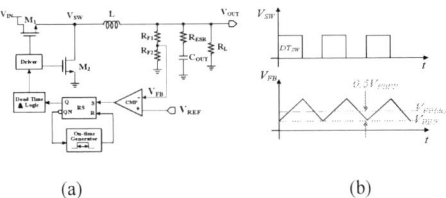

(a)   (b)

Fig. 1 Conceptual illustration of output DC offset (a) The COT control architecture (b) Key node waveforms

---

This work was supported in part by Beijing Natural Science Foundation under Grant No.4232063 and the National Natural Science Foundation of China under Grant 62074010.

979-8-3315-3918-4/25 $31.00 © 2025 IEEE

To maintain the stability of the control loop, the ripple caused by ESR must be substantially greater than the ripple generated by the output capacitance. As a result, only the ESR ripple voltage is considered, and the output capacitor ripple voltage is neglected. Under this assumption, the relationship between the average feedback voltage $V_{FB(dc)}$ and the valley voltage $V_{FB(valley)}$ can be approximated as

$$V_{FB(valley)} = V_{FB(dc)} - \frac{V_{OUT}(1-D)T_{SW}R_{ESR}R_{F2}}{2L(R_{F1}+R_{F2})} \quad (1)$$

where D represents the duty cycle and T_{SW} denotes the switching period. Based on the divider ratio of the feedback resistors R_{F1} and R_{F2}, the expression for the output DC offset V_{OUT_OFFSET} can be derived as

$$V_{OUT\_OFFSET} = \frac{V_{OUT}(1-D)T_{SW}R_{ESR}}{2L}. \quad (2)$$

The system architecture of the proposed COT converter is given in Fig. 2. To ensure the stability of the control loop, the system utilizes a Ripple Compensation Circuit to filter the switching node voltage V_{SW}, generating a ripple voltage V_{SAC} that is in phase with the inductor current, along with its corresponding DC component V_{SDC}. A four-input comparator adds the difference between V_{SAC} and V_{SDC} to the feedback voltage V_{FB}, thereby enhancing the ripple signal. The COT buck converter also incorporates the DC offset calibration module to improve output voltage accuracy. In addition, the system includes a PSM module for light-load operation, a protection circuit, and a Soft-Start module.

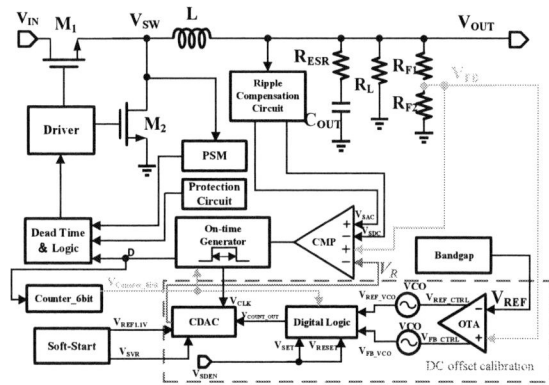

Fig. 2 System architecture of the proposed COT converter

The architecture of the DC offset calibration module is shown in the dashed box in Fig. 2. It consists of a transconductance amplifier OTA, two VCOs, a Digital Logic module, and a CDAC. This DC offset calibration module dynamically adjusts the reference voltage V_R based on the output voltage to correct the DC offset. The duty cycle signal D serves as the clock input for a 6-bit counter Counter_6bit. When the Counter_6bit reaches 64, it generates a pulse signal $V_{Counter\_6bit}$ and then resets, defining one quantization cycle as 64 switching cycles. During each quantization cycle, the DC error between the feedback voltage V_{FB} and the reference voltage V_{REF} is integrated using the VCO and the Digital Logic module. The Digital Logic module also compares the integration result. At the rising edge of $V_{Counter\_6bit}$, which marks the end of the quantization cycle, the Digital Logic

module converts the comparison results into a digital output code V_{COUNT_OUT}. Then the CDAC uses this output code V_{COUNT_OUT} to generate the calibrated reference voltage V_R for that quantization cycle. During the high level of $V_{Counter\_6bit}$, the integration state in the Digital Logic module is reset in preparation for the next cycle.

The DC offset calibration module eliminates DC offset caused by the ripple introduced by the Ripple Compensation Circuit, as well as DC offset resulting from the ESR voltage ripple, even when using low-ESR ceramic capacitors. The COT buck converter forms a fast control loop through the interaction of current ripple and the comparator, enabling rapid transient response. In parallel, the DC offset calibration module operates as a slow control loop, gradually adjusting the DC accuracy of the output voltage. When the DC offset of the output voltage is sufficiently small, the Digital Logic module engages a lock mechanism to lock the output code V_{COUNT_OUT}. If the system is required to operate in a low-power mode, the DC offset calibration module can be powered down. Since the CDAC input is a digital code generated by the Digital Logic module, the COT buck converter offers improved compatibility with digital power management systems.

III. CIRCUIT IMPLEMENTATION

The following section presents the key circuit implementation of the DC offset calibration module.

A. VCO

Since the VCO used in the DC offset calibration module is applied within power management applications, it does not require a high oscillation frequency, nor does it impose stringent phase noise performance. Therefore, the ring VCO is selected as the practical circuit. As shown in Fig. 3, the ring VCO primarily consists of four differential delay cells DIFF.

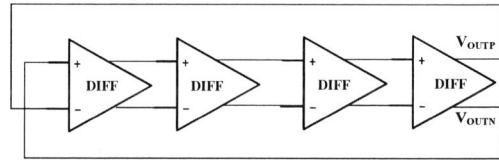

Fig. 3 Main schematic diagram of the ring VCO

VCO can be considered as voltage V_{in} to phase Φ_{VCO} integrator with a transfer function, which can be expressed as follows

$$\Phi_{VCO} = K_{VCO} \int V_{in}\, dt \quad (3)$$

where, K_{VCO} represents the gain of VCO. The expression in (3) is then transformed into the following

$$\frac{\Phi_{VCO}}{V_{in}}(s) = \frac{K_{VCO}}{s}. \quad (4)$$

The VCO uses tail current-controlled delay cells DIFF, as shown in Fig. 4. As the input voltage V_N gradually decreases, the drain voltage of transistor M_1 increases. The supply voltage VDD charges the load capacitor through the transistor M_3, providing a high current path for the load capacitor, which allows the output swing to reach a maximum of approximately VDD. Transistors M_3 and M_4 extend the maximum output swing of the DIFF, improving phase noise and linearity.

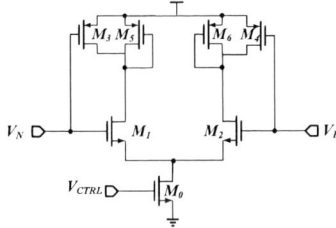

Fig. 4 The tail current-controlled delay cell DIFF

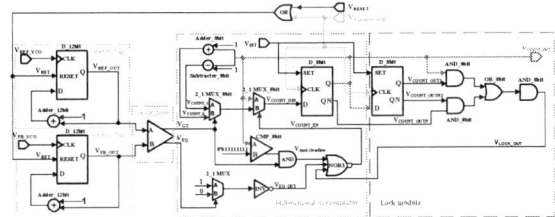

Fig. 5 Block diagram of the Digital Logic module

B. Digital Logic module

The Digital Logic module primarily consists of the counter Counter_V_{FB}, counter Counter_V_{REF}, comparator CMP_12bit, Bidirectional Accumulator, and Lock module, as shown in Fig. 5.

The reset signal V_{RESET} and the pulse signal $V_{Counter\_6bit}$ from the Counter_6bit in Fig. 2 control the reset signal $\overline{V_{RST}}$ of both the Counter_V_{FB} and Counter_V_{REF} through an OR gate. The Counter_V_{FB} uses the output signal V_{FB_VCO} of VCO as its clock. During a quantization cycle, when the signal $V_{Counter\_6bit}$ is low, the Counter_V_{FB} counts the rising edges of the signal V_{FB_VCO}, with the count result being V_{FB_OUT}. The Counter_V_{REF} operates in the same manner as Counter_V_{FB}, with the count result being V_{REF_OUT}. The count results V_{FB_OUT} and V_{REF_OUT} are then provided to the two inputs of the CMP_12bit. The CMP_12bit has two outputs: the comparison output V_{GT} (A>B or A< B) and the equality output V_{EQ} (A=B). When the rising edge of the pulse signal $V_{Counter\_6bit}$ occurs at the end of a quantization cycle, the comparator CMP_12bit outputs the V_{GT} and V_{EQ} based on the integration result, which are then sent to the Bidirectional Accumulator. The Bidirectional Accumulator increments, decrements, or holds the value of the output code V_{COUNT_OUT} based on the CMP_12bit's outputs V_{GT} and V_{EQ}. When V_{GT} is 1 (A>B), V_{COUNT_OUT} is incremented by 1; when V_{GT} is 0 (A<B), V_{COUNT_OUT} is decremented by 1; and when V_{EQ} is 1 (A = B), V_{COUNT_OUT} remains unchanged.

The Bidirectional Accumulator also includes an overflow protection feature. When V_{COUNT_OUT} approaches its maximum value, it performs a hold operation to prevent overflow. Fig. 5 also includes a Lock module. When the output DC offset is sufficiently small, the Lock module outputs a high-level signal V_{LOCK_OUT} to lock the value of V_{COUNT_OUT}.

C. CDAC

Fig. 6 shows the 8bit CDAC, where A_0-A_7 are reset switches and D_0-D_7 are conversion switches. V_{SVR} and $V_{REF1.1V}$ are the reference voltage and reset voltage

respectively, generated by the Soft-Start module. One side of the reset switches A_0-A_7 is connected to the bottom plate of the capacitors, while the other side is connected to the reference voltage V_{SVR}. The conversion switches D_0-D_7 are connected across the capacitors. The clock signal V_{CLK} passes through the LOG_DEAD module to generate two non-overlapping clocks, V_{NRESET} and $V_{NRESULT}$, which control the conversion switches D_0-D_7 and the reset switches A_0-A_7, respectively. The switches corresponding to D_0-D_7 and A_0-A_7 are not on at the same time. $V_{NRESULT}$ and the code V_{COUNT_OUT} are combined through NAND_8bit to generate the switch signal A[7:0], which is then passed through NOR_8bit with V_{NRESET} to produce the switch signal D[7:0]. The capacitors are connected according to the corresponding switch signals.

The operation of the CDAC consists of two phases. During the charging phase, V_{RESET} is high and switch S_0 is turned on, setting the top plates of the capacitors to the reset voltage $V_{REF1.1V}$. V_{NRESET} is low and $V_{NRESULT}$ is high, and the capacitors are connected according to the digital code V_{COUNT_OUT}. During the sharing phase, all reset switches A_0-A_7 are turned on and all conversion switches D_0-D_7 are turned off. The bottom plates of all capacitors are connected to the reference voltage V_{SVR}, and the top plates of the capacitors generate the DC offset calibration result V_R.

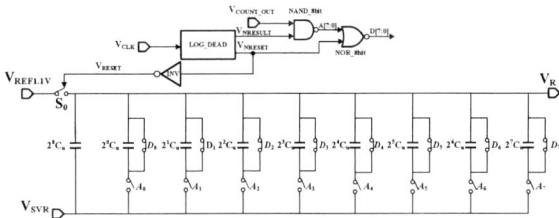

Fig. 6 8bit CDAC

IV. VALIDATION OF THE PROPOSED METHOD

To validate the proposed COT buck converter with the DC offset calibration module, the entire circuit was implemented using 0.18μm BCD process. Fig. 7 shows the chip layout, which occupies an area of 2.4 × 1.4mm².

Fig. 7 The diagram of the chip layout

Fig. 8 shows the waveforms of the feedback voltage V_{FB}, the DC offset calibration result V_R, and the reference voltage V_{REF} under load current of 3 A. As shown in Fig. 8, the DC offset of V_{FB} is 186μV. When the difference between V_{FB} and V_{REF} becomes sufficiently small, the output code V_{COUNT_OUT} is locked, the DC offset calibration result V_R remains constant, and the feedback voltage V_{FB} stabilizes, thereby achieving high output accuracy. Based on the resistor divider ratio, the output DC offset is 545 μV.

979-8-3315-3918-4/25 $31.00 © 2025 IEEE

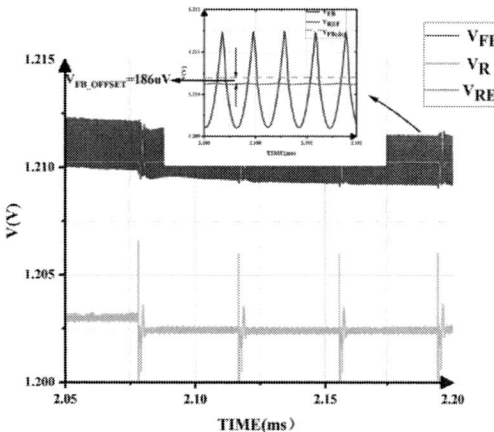

Fig. 8 Waveforms of V_{FB}, V_{REF}, and V_R under 3 A load current

Fig. 9 shows the load transient response at input voltage of 4.2 V and output voltage of 3.52 V. When the load steps from 1.5 A to 3 A, the output voltage is settled with 31.24 μs, and the undershoot voltage is 81.44 mV. When the load steps from 3 A to 1.5 A, the output voltage is settled with 28.45 μs, and the overshoot voltage is 87.09 mV.

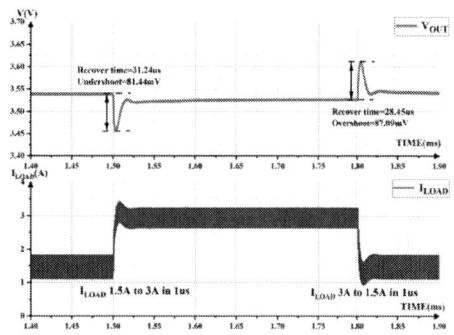

Fig. 9 Load transient response under output voltage V_{OUT}=3.52 V, V_{IN}=4.2 V, load changes between 1.5 and 3 A.

Fig. 10 shows the efficiency of the proposed COT converter at input voltage of 4.2 V and output voltage of 3.52 V under different process corners. As shown in Fig. 10, the converter achieves the peak efficiency of 96.82%.

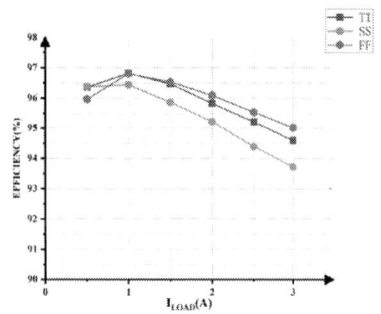

Fig. 10 Efficiency of the COT buck Converter

Table I summarizes the simulated performance results of this work and compares them with those of advanced COT converters. The proposed COT converter achieves high output

voltage accuracy, fast load transient response, and quasi-constant switching frequency. Among the compared designs, this work exhibits the lowest output DC offset, which is 545 μV.

TABLE I.   COMPARISONS WITH THE PRIOR WORK

References	*This work*[a]	*[5]*[b]	*[3]*[a]	*[7]*[a]
Control	COT	COT	AOT	COT
Process (μm)	0.18	0.5	0.18	0.18
V_{IN} (V)	4.2-5.5	12	5-18	8-24
V_{OUT} (V)	3.52	NA	0.8-5	5
F_{SW} (MHz)	2	0.4	1	1
Range of I_{LOAD} (A)	0-3	1.2-4	0-2	0-2
L (μH)	0.47	2.2	1.5	5.6
C_{OUT} (μF)	22	88	10	30
Offset (mV)	0.545	2	1	6
Recovery time T_R (μs)	31.24	28	NA	27
Highest efficiency	96.82%	95%	93%	NA

a. Simulation results   b. Measure results

V.   CONCLUSION

This paper presents the DC offset calibration module designed to improve output voltage accuracy. The COT buck converter utilizes current ripple and a comparator to form a fast feedback loop, enabling rapid transient response. In contrast, the DC offset calibration module operates as a slow feedback loop, gradually adjusting output voltage accuracy. Simulation results demonstrate that the proposed approach effectively enhances output voltage precision while maintaining loop stability across a wide range of load condition and input voltage. The proposed method improves output voltage accuracy, with the output DC offset of 545 μV.

REFERENCES

[1] Xin Zhou, Jiwei Fan and A. Huang, "Monolithic dc offset self-calibration method for adaptive on-time control buck converter," 2009 IEEE Energy Conversion Congress and Exposition, San Jose, CA, USA, 2009, pp. 655-658.

[2] W. -H. Yang et al., "A Constant-on-Time Control DC–DC Buck Converter With the Pseudowave Tracking Technique for Regulation Accuracy and Load Transient Enhancement," in IEEE Transactions on Power Electronics, vol. 33, no. 7, pp. 6187-6198, July 2018.

[3] B. Yuan, M. -X. Liu, W. T. Ng and X. -Q. Lai, "A Fast-Response RBAOT-Controlled Buck Converter With Pseudofixed Switching Frequency and Enhanced Output Accuracy," in IEEE Journal of Emerging and Selected Topics in Power Electronics, vol. 9, no. 1, pp. 79-88, Feb. 2021.

[4] Song Guo. Accuracy-Enhanced Ramp-Generation Design for D-CAP3 Modulation [EB/OL]. (2016-4) [2022-03-03]. https://www.ti.com.cn/cn/lit/pdf/slva762.

[5] X. Ming, Y. -L. Xin, T. -S. Li, H. Liang, Z. -J. Li and B. Zhang, "A Constant On-Time Control With Internal Active Ripple Compensation Strategy for Buck Converter With Ceramic Capacitors," in IEEE Transactions on Power Electronics, vol. 34, no. 9, pp. 9263-9278, Sept. 2019.

[6] S. Tian, F. C. Lee, P. Mattavelli, K. -Y. Cheng and Y. Yan, "Small-Signal Analysis and Optimal Design of External Ramp for Constant On-Time V^2 Control With Multilayer Ceramic Caps," in IEEE Transactions on Power Electronics, vol. 29, no. 8, pp. 4450-4460, Aug. 2014.

[7] J. Lv, L. Huang, C. Wang and D. Sun, "A Novel Internal Ripple Compensation Technology With DC Offset Cancellation Based on COT Buck Converter," 2024 IEEE China International Youth Conference on Electrical Engineering (CIYCEE), Wuhan, China, 2024, pp. 1-5.

A Low-Power High-Precision Impedance Measurement Circuit Using DC Servo Loop for Closed-Loop DBS Systems

Ziqi Tan[1,2,3,4], Yijun Ye[2,3,4], Yutao Mao[1,2,3,4], Hui Wu[3,4], Xiaofei Kuang[1],
Jie Yang[2,3,4], *Senior Member, IEEE*, and Mohamad Sawan[2,3,4], *Life Fellow, IEEE*
[1]College of Electronics Information, Hangzhou Dianzi University, Hangzhou, Zhejiang 310018, China
[2]Westlake Institute for Optoelectronics, Hangzhou 310024, Zhejiang, China
[3]CenBRAIN Neurotech, School of Engineering, Westlake University, Hangzhou 310024, Zhejiang, China
[4]Integrated-on-Chips Brain-Computer Interfaces Zhejiang Engineering Research Center, Hangzhou 310058, Zhejiang, China
Email: kuangxiaofei@hdu.edu.cn, {yangjie,sawan}@westlake.edu.cn

Abstract—This paper presents a multifunctional impedance detection system for neural modulation chips. Circuit reuse and topology optimization, the system achieves triple-mode coexistence of 16-channel signal acquisition, electrical stimulation, and impedance measurement with only 1 μA additional quiescent current. A differential step-current excitation strategy is implemented, leveraging the programmable gain amplifier (PGA) in the signal path for amplification. A DC Servo Loop (DSL) is incorporated to suppress DC drift errors caused by leakage currents in pseudo-resistors. Fabricated in 40 nm CMOS technology, the system achieves an impedance detection range of 10 Ω–100 kΩ, with a measurement error of 2.3% at 10 kΩ. The additional power consumption is only 1.2 μW, and the chip occupies 0.97 mm^2. Notably, the impedance detection module incurs no additional hardware overhead. Experimental results confirm its capability for real-time monitoring of electrode-tissue contact conditions, offering an ultra-low-power on-chip solution for high-precision impedance monitoring in closed-loop neural modulation systems.

Index Terms—Bio-impedance measurement, closed-loop deep brain stimulation, implantable medical systems, step current, low power, pseudo-resistors

I. INTRODUCTION

Closed-loop neural modulation technology offers precise treatment methods for neurological disorders such as epilepsy and Parkinson's disease by actively monitoring neural electrical activity and applying adaptive electrical stimulation [1] [2]. The real-time detection of the electrode-tissue interface impedance is crucial: a sudden increase in contact impedance due to electrode detachment may lead to abnormal distribution of stimulation currents, decreasing the efficiency of target region stimulation [3], and potentially causing overheating in non-target tissues, which can trigger inflammatory responses. Moreover, dynamic impedance changes can reveal high-frequency oscillatory activities or pathological states during epileptic seizures by correlating fluctuations in extracellular matrix ion concentrations with changes in neuronal membrane capacitance. Typically, implantable SoC electrode impedance measurement circuits must ensure low power consumption, a simple architecture, and seamless integration with front-end biosignal acquisition modules.

To support low-power portable applications, impedance measurement systems face two major challenges: precision and system integration. In [4], a phase-calibrated impedance detection circuit based on a DAC to inject current into brain tissue, followed by splitting the instrumentation amplifier (IA) output into two paths for quadrature demodulation. One path is used to detect the resistive component of the impedance, while the other is used to detect the capacitive component. However, this design suffers from measurement errors due to signal delays caused by the voltage drop across the measured impedance from the DAC-generated current. Moreover, the DAC itself requires significant power consumption and occupies a large area, which hinders system integration.

In [5], an innovative electrode-tissue contact impedance measurement circuit is proposed for implantable medical devices, integrated within a System-on-Chip (SoC) architecture. The design operates by injecting a differential step current into the electrode-tissue interface, generating a measurable voltage that facilitates the extraction of both real and imaginary impedance components. To ensure stability, a common-mode feedback (CMFB) bias circuit maintains the common-mode voltage level. Notably, the implementation partially reuses the SoC's existing biosignal acquisition circuitry, enhancing integration efficiency. Nevertheless, this approach exhibits several limitations: (1) a constrained output dynamic range, and (2) the necessity of dedicated bias circuits per channel, which escalates both silicon area overhead and power consumption.

In this study, we propose an impedance detection scheme integrated into a neural modulation chip. To address measurement errors caused by pseudo-resistor leakage and MOS transistor mismatch in traditional methods, this design incorporates a DC servo loop (DSL) in the amplifier, effectively reducing DC offset errors to below 500 μV. The scheme fully reutilizes the system's existing stimulator, 11-bit SAR ADC, and programmable gain amplifier (PGA), achieving complete impedance detection functionality with an addition

979-8-3315-3918-4/25 $31.00 © 2025 IEEE

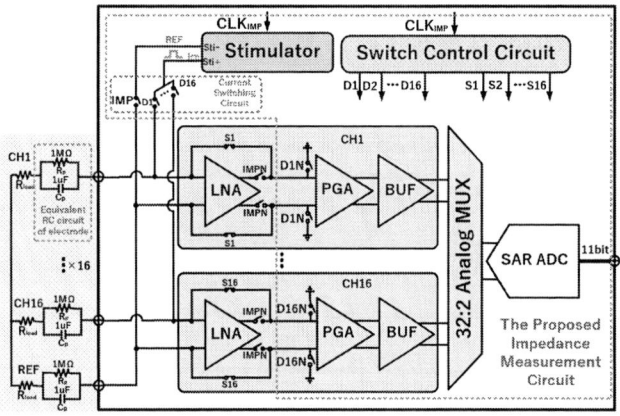

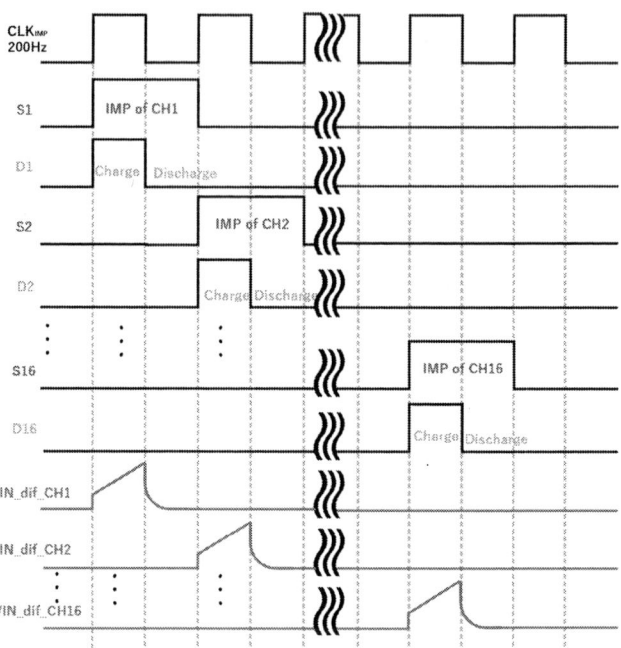

Fig. 1. Block diagram of the overall system circuit, illustrating the structural relationships and functional integration among the impedance electrode model, current excitation module, signal acquisition module, and timing control module.

Fig. 2. Timing diagram of the operating clock generated by the switch control circuit and PGA output waveforms.

of only $0.0005\,\mathrm{mm}^2$ to the chip area. Test results based on the $40\,\mathrm{nm}$ CMOS process indicate that the design achieves a maximum measurement error of 2.3% within the $10\,\Omega$–$100\mathrm{k}\,\Omega$ impedance range.

II. CIRCUIT AND OPERATION

The proposed multi-channel impedance measurement system architecture is illustrated in Figure 1. The system employs a 16-channel shared single 11-bit successive approximation register analog-to-digital converter (SAR ADC) to achieve high integration density. Each measurement channel reuses the negative terminal of the neural stimulator (Sti-) as the reference potential (REF), while a digital control logic dynamically allocates the positive terminal (Sti+) to sequentially inject step excitation currents (I_{STI}) into individual channels for time-multiplexed impedance detection. The core signal chain comprises four key modules:

- A PGA for signal amplification,
- A buffer (BUFFER) to enhance driving capability,
- A multiplexer (MUX) for channel selection,
- An 11-bit SAR ADC for analog-to-digital conversion.

For electrode-tissue interface modeling, the system incorporates the double-layer effect. The deep brain stimulation (DBS) electrode impedances of Channel 1 (CH1) and Channel 16 (CH16) are modeled as a parallel combination of charge transfer resistance (R_p) and double-layer capacitance (C_p), further connected in series with tissue resistance (R_{load}). When the system operates in impedance detection mode, IMPN is disconnected, and the switch array (S1–S16) is sequentially activated under clock control. During this phase, the low-noise amplifier (LNA), which does not participate in signal acquisition, is disabled to significantly reduce dynamic power consumption.

Throughout the detection cycle, the negative terminal of the stimulator (Sti-) remains connected to the common reference (REF) of all 16 channels. The positive terminal (Sti+) delivers time-multiplexed $1\,\mu\mathrm{A}$ step currents (I_{STI}) via a high-precision

current-switching matrix. The switching sequence is generated by a switch control circuit, which employs a 16-stage D-flip-flop-based ring counter operating at 200Hz to ensure strict timing synchronization. S1–S16 select measurement channels, while D1–D16 control stimulation current delivery.

Figure 2 illustrates the clock timing diagram and differential output waveforms of different channels. At any given moment, only one channel's measurement path is activated. The operational sequence is as follows:

1) **Charging Phase:** On the rising edge of CLK, S1 is activated to enable signal amplification, while D1 goes high. The stimulator delivers a $1\,\mu\mathrm{A}$ current to charge the electrode load for 2.5 ms.

2) **Discharging Phase:** On the falling edge of CLK, D1 resets to low and its complementary signal (D1N) activates, grounding the electrode to form a discharge path. This eliminates residual charges from the double-layer capacitance.

The impedance measurement system employs a capacitive-feedback PGA architecture to process the voltage drop of the electrode-tissue interface induced by current excitation. As illustrated in Figure 3(a), compared to the traditional resistive feedback architecture, this design offers two main advantages: significant reduction in power consumption and higher gain accuracy [6]. Its transfer function can be expressed as:

$$H(j\omega) = \frac{A}{\left(1 - j\frac{1}{\omega R_p C_{fb}}\right)}, A = \frac{C_{in}}{C_{fb}} \qquad (1)$$

Where the C_{in} is the input capacitance, C_{fb} is the feedback capacitance, and R_p is the pseudo-resistor. From the above

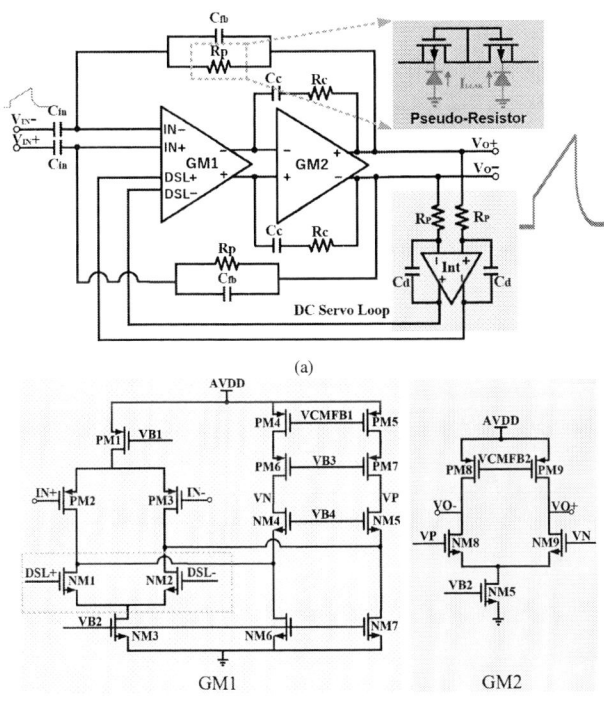

(a)

GM1   GM2

(b)

Fig. 3. (a) Overall Architecture of the PGA. (b)Detailed Circuit Implementation of Transconductance Units GM1 and GM2.

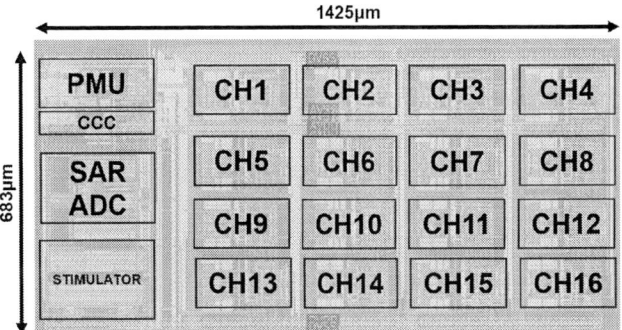

Fig. 4. Chip Layout: Highly Integrated 16-Channel Acquisition Circuits, Stimulators, ADC, and Power Management Unit (PMU).

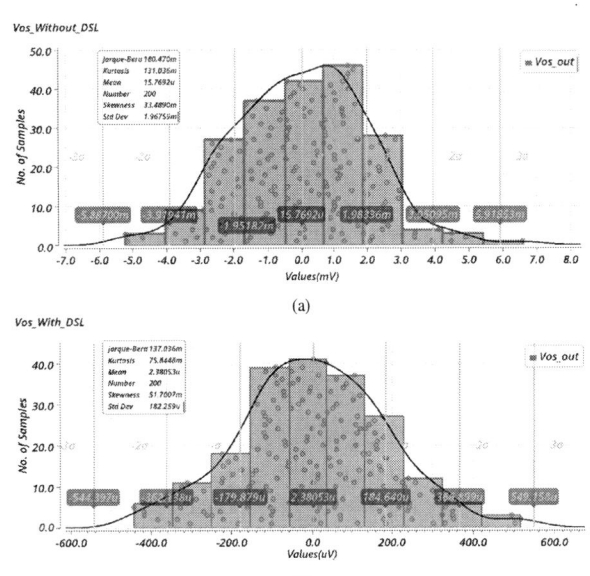

(a)

(b)

Fig. 5. Offset voltage comparison: (a) with DSL and (b) without DSL. The DSL reduces the offset from 5 mV to below 500 μV.

equations, it can be seen that the gain of the circuit is C_{in}/C_{fb}, and the high-pass cutoff frequency is $f_{hp} = 1/(C_{fb} \cdot R_p)$.

To achieve a wide dynamic range measurement, a PGA is designed with selectable gains of 4, 16, 40, and 80. For the acquisition of low-frequency neural (LFP) signals, a 200 fF feedback capacitor C_{fb} is implemented to constrain the high-frequency cut-off to approximately 0.5 Hz, requiring TΩ-level pseudo-resistor R_p. However, substrate leakage currents I_{LEAK} in ultra-high-resistance pseudo-resistors [7] [8] can generate significant DC voltage drops throughout R_p, causing output offset voltages or even amplifier saturation.

To address this issue, a DSL compensation mechanism is integrated into the PGA circuit (Fig 3(a)). The DSL employs an integrator architecture: when output DC offsets are detected, NMOS transistors NM1/NM2 convert the offset voltage to current signals, which are amplified by a cascode stage to generate inverse compensation voltages injected at the PGA input node, dynamically nullifying DC offsets (Fig 3(b)). To prevent interference with the main circuit's high-pass cutoff frequency, the DSL loop bandwidth must satisfy:

$$f_{\text{UGB}} = \frac{1}{2\pi R_p C_D} < 0.5\,\text{Hz} \quad (2)$$

A 2 pF integrating capacitor C_{D} is selected with matched R_p values to ensure loop stability and compensation efficiency.

III. EXPERIMENTAL RESULTS

This study presents a deep brain stimulation closed-loop neural regulation system integrated with an electrode impedance measurement circuit, manufactured using a

40nm CMOS process. The entire system includes 16-channel impedance measurement, ADC, stimulator, and power management modules, with a chip size of only 1425 μm × 683 μm. As shown in Figure 4, this SoC demonstrates a highly integrated and miniaturized design, suitable for deep brain stimulation applications.

To minimize the undesirable error caused by PGA's DC offset in impedance measurement, the design incorporates a DSL circuit. Figure 5 shows the post-simulation results before and after introducing DSL. As observed, without DSL, the PGA's output offset can reach up to 5 mV, potentially leading to over 5% measurement error. However, with DSL, the PGA's offset is reduced to only 0.5 mV, affecting the measurement accuracy by no more than 0.5%, which is negligible.

Figure 6 illustrates the impedance measurement waveforms of four channels. These waveforms are generated by the voltage drop across the electrode and load, amplified 40× by the PGA. In the electrode equivalent model, $R_p = 1\,\text{M}\Omega$, $C_p = 1\,\mu\text{F}$, and the load resistance is $R_{\text{load}} = 2\,\text{k}\Omega$. The

979-8-3315-3918-4/25 $31.00 © 2025 IEEE

739

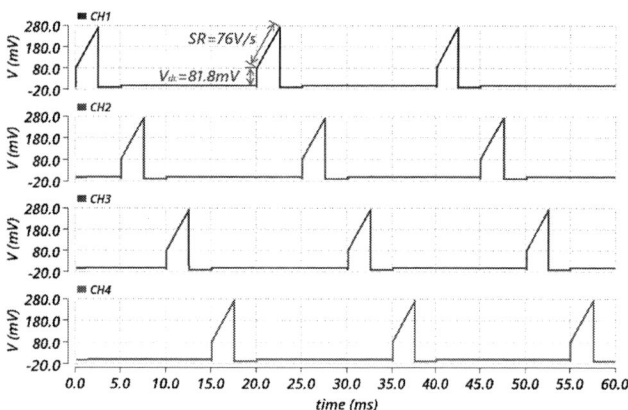

Fig. 6. Four-Channel Waveforms of PGA-Amplified Voltage Between Electrodes.

TABLE I. Comparison of impedance measurement systems in terms of key performance metrics.

Parameter	This Work	TBioCAS'23 [9]	JSSC'24 [10]
Technology	40 nm CMOS	180 nm CMOS	180 nm CMOS
Supply Voltage(V)	1.2	1.8	1.2/0.9
Measurement Frequency(Hz)	200	125	1k-200k
Injection Current (μA)	1	0.26	3-100
Extra Power Consumption(μW)	1.2	2.65	4.3
Modulation Type	Square	Square	Sinusoid
Max. Error	2.3% ($R_{load} = 2k\Omega$)	8.3% ($R_{load} = 2.4k\Omega$)	NA
Extra Area(mm^2)	0.0005	0.047	NA
Measured Impedance (kΩ))	0.01-100	N/A	0.01-37

amplifier output waveform consists of two distinct phases:

Step Response: The amplifier output transitions from 0 to a DC voltage V_{dc}, representing the real part of the impedance. The measured voltage aligns with the expression:

$$V_{dc} = I_{STI} \times R_{load} \times A_v \tag{3}$$

where I_{STI} is the step current amplitude, R_{load} is the load resistance, and A_v is the PGA's voltage gain. From the waveform, $V_{dc} = 81.8\,mV$, $I_{STI} = 1\,\mu A$, and the calculated resistance is $R_{load} = 2.045\,k\Omega$, yielding a measurement error of 2.3%.

Ramp Waveform: The ramp phase results from C_p charging. The slope SR (slew rate) is used to calculate the capacitance via:

$$SR = 2 \times C_p \times A_v \times I_{STI} \tag{4}$$

where A_v is the voltage gain. In this example, $SR = 76$ V/s, $I_{STI} = 1\,\mu A$, and $A_v = 40$. Substituting these values yields $C_p = 0.95\,\mu F$, with a measurement error of 5%.

This precision meets the requirements of closed-loop neuromodulation systems and enables detection of circuit short-circuit or open-circuit conditions.

CONCLUSION

This paper presents a low-power impedance detection system for closed-loop neuromodulation. By integrating a DSL to suppress leakage-induced drift and reusing existing circuits (PGA, SAR ADC, stimulator), the design achieves 2% measurement error and a wide dynamic range (0.01–100 kΩ) with only 1.2 μW additional power consumption and 0.0005 mm^2 chip area. Compared to prior works [6], [7], the proposed system reduces extra power consumption by 54.7% and shrinks the chip area by 98.9% . The 0.01 kΩ detection limit enables short-circuit monitoring. Furthermore, the square-wave modulation strategy balances accuracy and hardware simplicity, providing robust real-time electrode-tissue interface monitoring for closed-loop neural feedback. Future work will focus on in vivo validation of the system for clinical applications.

REFERENCES

[1] L. Rossi, S. Marceglia, G. Foffani, F. Cogiamanian, F. Tamma, P. Rampini, S. Barbieri, F. Bracchi, and A. Priori, "Subthalamic local field potential oscillations during ongoing deep brain stimulation in parkinson's disease," *Brain research bulletin*, vol. 76, no. 5, pp. 512–521, 2008.

[2] H. Wu, J. Chen, X. Liu, W. Zou, J. Yang, and M. Sawan, "An energy-efficient small-area configurable analog front-end interface for diverse biosignals recording," *IEEE Transactions on Biomedical Circuits and Systems*, vol. 17, no. 4, pp. 818–830, 2023.

[3] R. F. Yazicioglu, S. Kim, T. Torfs, H. Kim, and C. Van Hoof, "A 30 μ w analog signal processor asic for portable biopotential signal monitoring," *IEEE Journal of Solid-State Circuits*, vol. 46, no. 1, pp. 209–223, 2011.

[4] S.-I. Cheon, S.-J. Kweon, Y. Kim, J. Koo, S. Ha, and M. Je, "A polar-demodulation-based impedance-measurement ic using frequency-shift technique with low power consumption and wide frequency range," *IEEE Transactions on Biomedical Circuits and Systems*, vol. 15, no. 6, pp. 1210–1220, 2021.

[5] C.-W. Huang, C.-H. Chung, R.-S. Syu, and C.-Y. Wu, "The design of cmos electrode-tissue impedance measurement circuit using differential current switch with cmfb bias for implantable neuro-modulation socs," in *2019 IEEE Biomedical Circuits and Systems Conference (BioCAS)*. IEEE, 2019, pp. 1–4.

[6] R. R. Harrison and C. Charles, "A low-power low-noise cmos amplifier for neural recording applications," *IEEE Journal of solid-state circuits*, vol. 38, no. 6, pp. 958–965, 2003.

[7] H. Wu, Z. Tan, X. Liu, J. Chen, W. Zou, Q. Hou, S. Lin, Y. Mao, X. Kuang, J. Yang *et al.*, "Efficient self-adaptive pseudo-resistor with rapid settling and high linearity for neurorecording front-end circuits," in *2025 IEEE International Symposium on Circuits and Systems (ISCAS)*. IEEE, 2025, pp. 1–5.

[8] H. Wu, X. Liu, J. Yang, and M. Sawan, "A power-efficient source-follower based tunable pseudo-rc low-pass filter for wearable biomedical applications," pp. 178–182, 2022.

[9] C.-W. Huang, C.-K. Lai, C.-C. Hung, C.-Y. Wu, and M.-D. Ker, "A cmos synchronized sample-and-hold artifact blanking analog front-end local field potential acquisition unit with ±3.6-v stimulation artifact tolerance and monopolar electrode-tissue impedance measurement circuit for closed-loop deep brain stimulation socs," *IEEE Transactions on Circuits and Systems I: Regular Papers*, vol. 70, no. 6, pp. 2257–2270, 2023.

[10] H. Choi, S.-I. Cheon, G. Yun, S. Oh, J.-H. Suh, S. Ha, and M. Je, "A bio-impedance readout ic with complex-domain noise-correlated baseline cancellation," *IEEE Journal of Solid-State Circuits*, vol. 59, no. 11, pp. 3538–3548, 2024.

A Low-Power BJT-Based Thermal Shutdown Circuit with Hysteresis for BMS chip in EV

Zonghuan Wu[1], Xiangmin Zhang[3], Liji Wu[2,*]

[1,2,3] School of Integrated Circuits, Tsinghua University, Beijing National Research Center for Information Science and Technology, Beijing, China

Email: [1] wu-zh20@tsinghua.org.cn, [2,*] lijiwu@mail.tsinghua.edu.cn

Abstract— **This paper presents a thermal shutdown circuit using an NPN BJT's temperature-dependent collector current for a Battery Management System (BMS) in Electric Vehicle (EV). The circuit is designed in a 0.18μm Bipolar CMOS DMOS (BCD) process, but the design procedure is applicable to other processes. A set of 20 thermal shutdown modules, each implementing a transistor with unique gain type, emitter area, and voltage tolerance, is tuned to set the output when temperature rises above 140°C (the shutdown temperature) and to clear below 120°C (the release temperature), as the design requirements for the BMS chip of our lab. Pre-simulation results show positive correlation between emitter area and process stability in terms of standard deviation of the shutdown temperature in Monte Carlo analysis. The module with the best process stability has the layout drawn. Post-simulation results demonstrate robustness against process variations and mismatch. The shutdown temperature and hysteresis temperature within ±3σ meet design specification. The circuit operates correctly from -50°C to 160°C while consuming less than 120nA at room temperature, making it suitable for power sensitive automotive grade integrated circuits, such as the BMS chip of our lab.**

Keywords— thermal shutdown, 0.18μm BCD process, BMS, temperature characteristics of BJT

I. INTRODUCTION

Electric vehicles (EVs) have gained significant market and research attention for the last decade as economic and environmental benefits of using electricity surpasses those of burning fossil fuel. The power battery stores electric energy and is an essential component of EVs. To safely monitor and control the energy in the battery, a Battery Management System (BMS) is required. The BMS is responsible for the safety of the battery and must operate within its designed temperature range to avoid unexpected behavior. For this purpose, a thermal shutdown circuit is required.

In integrated circuits, Bipolar Junction Transistors (BJTs) are widely used as temperature sensing elements due to their excellent temperature characteristics, for example, the Proportional-To-Absolute-Temperature (PTAT) voltage drop across the base-emitter node. Understanding these characteristics is crucial for designing reliable temperature sensing modules, such as thermal shutdown modules.

A typical overtemperature protection circuit, such as [1], compares the voltage drop across a diode-connected BJT with a reference voltage using comparators. However, comparators consume much static current, which is not suitable for power sensitive devices such as BMS chips. This paper aims to propose an overtemperature sensing circuit that consumes little current, takes up little area, and provides acceptable sensing accuracy for the BMS chip of our lab.

II. TEMPERATURE CHARACTERISTICS OF BIPOLAR TRANSISTORS

For an NPN BJT in forward-active region, the base-emitter junction is forward biased and the base-collector junction is reversely biased. Since the emitter region is heavily doped compared to the base region, and the base width W_B is much smaller, significantly more electrons than holes are ready to recombine in the base-emitter junction. The accumulated electrons diffuse across the base region to the base-collector junction and recombine with holes from the collector region, which is observed as the macroscopic collector current [2].

In this case, the collector current can be described by:

$$I_C = \frac{kTAn_i^2\overline{\mu_n}}{W_B N_a} \exp\frac{qV_{BE}}{kT} \tag{1}$$

where k is the Boltzmann constant, T is temperature, A is emitter area, n_i is the intrinsic carrier concentration, $\overline{\mu_n}$ is the effective electron mobility, N_a is the acceptor concentration in the base.

The intrinsic carrier concentration and effective electron mobility are temperature dependent described in (2) and (3), respectively, where n is a constant [3]. The acceptor concentration is temperature independent because the dopant atoms are already fully ionized at a very low temperature.

$$n_i^2(T) \propto T^3 \, exp\left(-\frac{q(V_{g_0}-\alpha T)}{kT}\right) \tag{2}$$

$$\overline{\mu_n}(T) \propto T^{-n} \tag{3}$$

Regardless of nonideal effects such as base width modulation, the collector current can therefore be described by:

$$I_C(V_{BE}, T) = CT^\eta \exp\frac{q(V_{BE}-V_{g_0})}{kT} \tag{4}$$

where C is the combination of all temperature independent terms, $\eta = 4 - n \geq 1.4$ [4], and V_{g_0} is the extrapolated bandgap voltage at 0K.

When $V_{BE} - V_{g0}$ is sufficiently small or negative, I_C increases with temperature. If V_{BE} varies with temperature, e.g. $V_{BE} = V_{BE_0} + \lambda T$, then, with V_{BE_0} being the extrapolated V_{BE} at 0K, regrouping the elements in (4) gives:

$$I_C(T) = CT^\eta \exp\frac{q(V_{BE_0}-V_{g_0})}{kT} \tag{5}$$

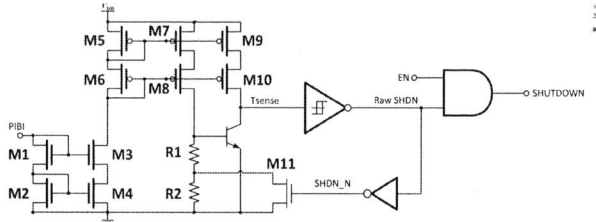

Fig. 1. Schematic of Thermal Shutdown Module

TABLE 1. DESIGN SPECIFICATIONS

Spec.	Min	Typ	Max	Unit
V_{DD}	4.5	5.0	5.5	V
T_{SHDN}	130	140	150	$°C$
T_{HYST}	15	20	25	$°C$
I_Q		100		nA

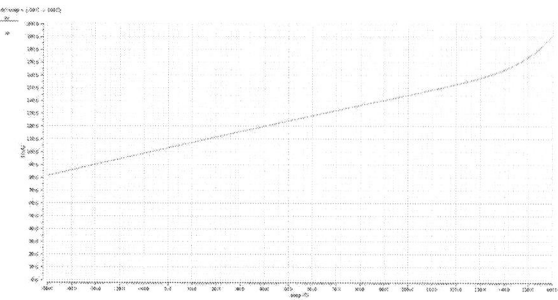

Fig. 2. Reference current source temperature characteristic from -50°C to 160°C

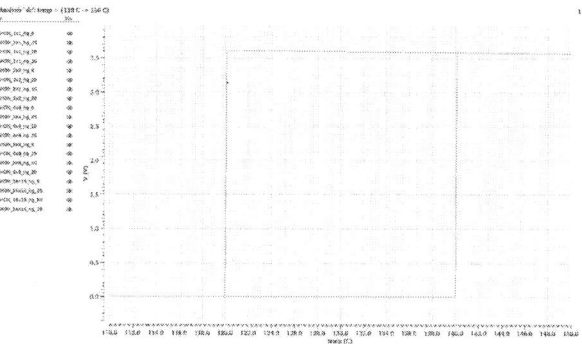

Fig. 3. DC analysis with temperature hysteresis sweep. All signals overlap on one trace.

III. DESIGN OF A THERMAL SHUTDOWN CIRCUIT

The circuit in this paper is designed in a 0.18μm BCD process from DongBu HiTek.

A. Circuit structure

Based on (5), a general overtemperature detection circuit is created by connecting a current source to the collector and a voltage source to the base. The collector voltage will suddenly drop to about 0.2V when the collector saturation current exceeds the current source current as the temperature rises.

In the proposed design, the collector current and base voltage are sourced from a current mirror and the voltage drop of another mirrored current through a resistor, respectively, as shown in Fig. 1. This thermal shutdown module has two power ports (VDD and GND), one enable port (EN), one reference current input port (PIBI), and one output port (SHUTDOWN).

When raw SHDN is low, V_{BE} equals the voltage drop across R1. As temperature rises, V_{BE} increases with reference current in polynomial order while the collector current, according to (5), increases exponentially. As the temperature rises across the shutdown temperature where the collector saturation current equals the current mirror output, the rapidly falling collector voltage is detected by a Schmitt trigger to prevent glitching. The raw SHDN signal is buffered through an AND gate if the module is enabled.

To achieve a hysteresis between the shutdown temperature and the release temperature, positive feedback is added to the circuit. The raw SHDN signal after the Schmitt trigger is inverted to drive an N-FET which shorts the lower resistor when the output is not set. Once the temperature increases above the shutdown temperature, the raw output rises, and the lower resistor comes in series and pushes the base voltage slightly higher. The adjusted base voltage, according to (4), causes the collector saturation current to increase so that the temperature must fall a few kelvins below the shutdown temperature to clear the output signal.

B. MOSFET ratio selection

In practice, it is likely that the input reference current is given and the shutdown temperature, T_{SHDN}, and release temperature, T_{RLS}, are to be satisfied with power and area constrains. For the BMS chip of our lab, the design goal is shown in table 1, where $T_{RLS} = T_{SHDN} - T_{HYST}$.

Generally, any strictly temperature positive monotonic current source can be used as reference. The input reference current in this design is generated by a bandgap voltage source whose output current varies with temperature as shown in Fig. 2, but any PTAT current source would work just fine. To keep the quiescent current no higher than the desired value of 100nA during normal operation, the size ratios of M2:M4 and M5:M7:M9 are 2:1 and 2:2:1, respectively.

Since the proposed circuit will be integrated in an automotive grade BMS chip, it is important that it works properly across the entire temperature range of, in this case, -40°C to 150°C. Based on (5), the circuit clearly never outputs high level on low temperature. This is verified by simulation but not shown in the Results section as the focus of this work is the behavior at high temperature.

C. BJT selection

Different BJT results in different yield distribution as shown in the Results section. All BJTs can be used for the typical condition case by only recalculating the resistance value.

D. Resistance calculation

Here is the complete procedure to calculate the value of R1 and R2:

- Find the lower switching threshold, V_{THL}, at T_{SHDN} and the higher switching threshold, V_{THH}, at T_{RLS} via simulation.

- Find the current mirror output I_{D7} and I_{D9} at T_{SHDN} and I'_{D7} and I'_{D9} at T_{RL} via simulation.

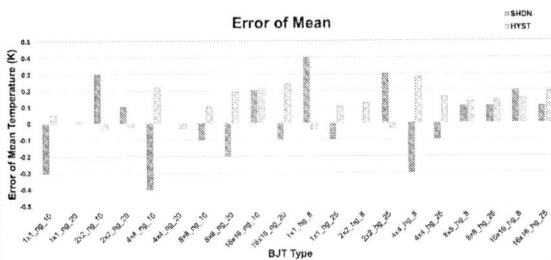

Fig. 4. Thermal Shutdown module with different BJT Monte Carlo pre-layout simulation mean temperature error.

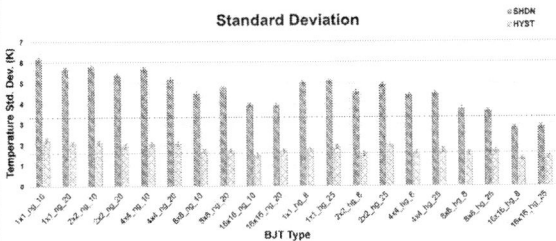

Fig. 5. Thermal Shutdown module with different BJT Monte Carlo pre-layout simulation standard deviation.

- Find V_B and I_B when $V_{CE} = V_{THL}$ and $I_C = I_{D9}$ at T_{SHDN}, and V_B' and I_B' when $V_{CE} = V_{THH}$ and $I_C = I_{D9}'$ at T_{RLS}.

- the resistance can be calculated:

$$R_1 @ T_{SHDN} = \frac{V_B}{I_{D7} - I_B} \tag{6}$$

$$(R_1 + R_2) @ T_{RLS} = \frac{V_B'}{I_{D7}' - I_B'} \tag{7}$$

IV. RESULTS

A. Pre-simulation

To find the most reliable design, 20 thermal shutdown modules each with a unique transistor of a high-gain or low-gain, low-voltage-tolerance or high-voltage-tolerance with emitter area of $1\mu m \times 1\mu m$, $2\mu m \times 2\mu m$, $4\mu m \times 4\mu m$, $8\mu m \times 8\mu m$, or $16\mu m \times 16\mu m$, have their resistance tuned to behave the same on temperature sweeps. A DC analysis is carried out with temperature hysteresis sweep between 110°C and 150°C with typical conditions to verify its functionality. Then, a Monte Carlo analysis for supply voltage of 4.5V, 5.0V, and 5.5V each with more than 100 sampling points is conducted to evaluate the process stability and mismatch robustness of each module.

The DC analysis results of the modules are presented in Fig. 3. The output rises at $140 \pm 0.5°C$ when temperature increases and falls at $120 \pm 0.5°C$ when temperature decreases.

The Monte Carlo analysis results show a normal distribution of T_{SHDN} and T_{HYST}. The errors of mean temperatures are shown in Fig. 4, and the standard deviation in Fig. 5. The results indicate that the larger the emitter area and the higher gain of transistor, the more robust the module is against process and mismatch. Although T_{SHDN} and T_{HYST} still are strongly affected by process and mismatch, a standard deviation of less than $3.3°C$ for T_{SHDN} and $1.6°C$ for T_{HYST} is

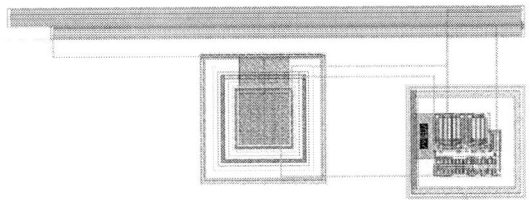

Fig. 6. Thermal Shutdown module test layout.
ISO rings are added since the devices used are isolated MOSFETs (bottom right) and BJT with buried layer (bottom left).

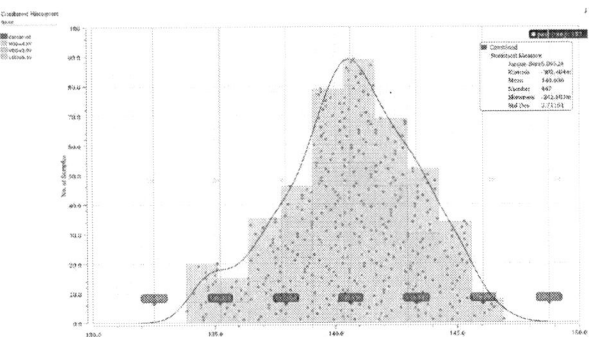

Fig. 7. Thermal Shutdown module shutdown temperature Monte Carlo post-layout simulation result.

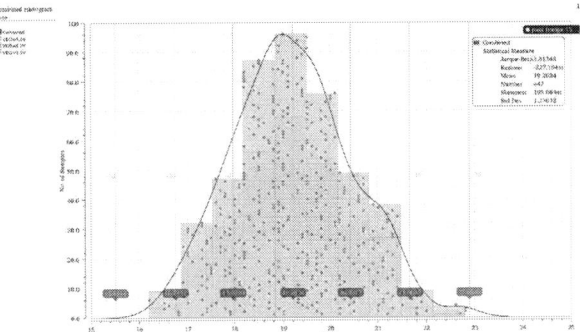

Fig. 8. Thermal Shutdown module temperature hysteresis Monte Carlo post-layout simulation result.

tolerable and that at least 99.7% of the results satisfies the design goal.

As the module using high-gain 8V-tolerable transistor with emitter area of $16\mu m \times 16\mu m$ outperforms the rest, it is chosen to draw a layout for post-simulation to test its robustness against parasitic effects.

B. Layout and post-simulation

Fig. 6 presents the layout for parasitic extraction. Fig. 7 and Fig. 8 show the distribution of T_{SHDN} and T_{HYST}. The results indicate negligible effect of parasitic on the distribution of T_{SHDN} and T_{HYST}. The total current consumption at 300K is 120nA.

V. CONCLUSION

This paper presented a comprehensive analysis of the temperature characteristics of NPN BJTs and their application in a thermal shutdown circuit with thorough design procedures. The derived equations for collector current as a function of temperature and base-emitter voltage formed the foundation for the circuit design, which is also applicable in any process with any BJT. The proposed design provides a complete solution of a thermal shutdown module with low quiescent current and small device count in a 0.18μm BCD process.

Simulation results confirmed the circuit's robustness, with acceptable deviation in shutdown and release temperatures across process variations and supply voltage fluctuations within design specification. The design is submitted to the chief engineer for integration in the BMS chip.

The proposed design is also suitable for power-sensitive and area-constrained chips and may find further application in power storage systems and mobile devices.

Future work could optimize for insensitivity on input reference current and resistor type, or explore tradeoffs in increased collector current for better process stability.

[1] J. Dang et al., "A Novel Reliability-Enhanced Dual Over-Temperature Protection Circuit With Delayed Thermal Restart for Power ICs," in IEEE Transactions on Circuits and Systems II: Express Briefs, vol. 71, no. 3, pp. 1471-1475, March 2024, doi: 10.1109/TCSII.2023.3322472.

[2] M. Alawein (2020, August 16). 8 bipolar transistor. Academia.edu. https://www.academia.edu/43872860/8_Bipolar_Transistor

[3] M. Pertijs, J. Huijsing. Precision temperature sensors in cmos technology[M/OL]. 2006. DOI: 10.1007/1-4020-5258-8.

[4] J. W. Slotboom and H. C. de Graaff, "Bandgap narrowing in silicon bipolar transistors," in IEEE Transactions on Electron Devices, vol. 24, no. 8, pp. 1123-1125, Aug. 1977, doi: 10.1109/T-ED.1977.18889.

Design of CRFF-B Loop Filter Architecture for Wideband Continuous Time Sigma-Delta Modulators in CMOS 28 nm

Zhihao Hou [1], Yuqi Fan [1], Yifei Gao [1], Chuan Liu[1], Chuan Qin[1], Maliang Liu*[1], Yintang Yang[1]

[1] School of Microelectronics, Xidian University, Xi'an 710071 China

* Email: houzhihao2022@163.com, mlliu@xidian.edu.cn

Abstract—This paper presents a cascade of resonator feedforward-feedback (CRFF-B) type loop filter for high-performance continuous-time (CT) sigma-delta modulators (SDM). The proposed architecture synergistically combines merits of conventional continuous-time cascade of integrator feedforward (CIFF) and feedback (CIFB) topologies. A resonator is introduced to optimize the zero positioning of the noise transfer function (NTF), effectively suppressing in-band quantization noise. Moreover, to reduce hardware area, this design utilizes proportional-integrator (PI) for excess loop delay (ELD) compensation. For better power efficiency, the implementation utilizes third-order feedforward amplifiers. Post-layout simulation results demonstrate 79.8dB dynamic range (DR) at low-frequency inputs. For input frequencies of 15.3 and 93.1 MHz, the prototype achieves 73.3 and 70.1 dB SNDR, respectively. The implemented loop filter consumes merely 12 mW power while occupying 0.039 mm² core area. The CRFF-B loop filter circuit was fabricated as part of a chip in 28 nm CMOS technology.

Keywords—*Analog-to-digital converter (ADC), continuous-time sigma-delta modulator (CT SDM), excess loop delay (ELD) compensation, amplifier.*

I. INTRODUCTION

The escalating demands for signal bandwidth and dynamic range (DR) in wired/wireless communications (e.g., digital modulation, LTE-Advanced), medical imaging, and ultra-high-definition video processing are driving continuous evolution of analog-to-digital converters (ADCs) [1][2]. Emerging multi-channel communication technologies such as 5G require the development of ADCs with nearly bandwidth of 100 MHz and DR of 70dB[3].

Although Nyquist switched-capacitor ADCs offer high dynamic range and bandwidth, the KT/C thermal noise constraint necessitates the use of extremely large sampling capacitors. Additionally, anti-aliasing filters and input buffers must be integrated when these ADCs are deployed in transceivers, leading to increased system complexity, power consumption, and area. Conventional high-performance Nyquist ADCs, such as pipeline, TI-SAR are inherently characterized by high power consumption and reliance on complex calibration algorithms. Continuous-time (CT) sigma-delta modulator (SDM) ADCs, in contrast, are recognized as preferred architectures for RF direct-sampling transceivers. Their intrinsic anti-aliasing properties and resistive input impedance enable front-end circuit complexity to be significantly reduced, while power-hungry ADC drivers and anti-aliasing filters are eliminated. Through CT signal

processing and noise-shaping techniques, high bandwidth and DR are achieved by CT SDM ADCs with lower power consumption compared to Nyquist counterparts. Therefore, the research and design of CT SDM ADCs are of significant value.

Within sigma-delta ADCs, the loop filter is identified as the core circuit defining modulator performance. This paper proposes a cascade of resonator feedforward-feedback (CRFF-B) structured circuit serving as the loop filter for a high-speed single-loop CT SDM ADC. Operating at 4-GS/s sampling rate, the design achieves over 79.8dB DR within 100-MHz signal bandwidth. The loop filter employs proportional-integral (PI) elements to compensate for loop excess delay while utilizing resonator zero-positioning techniques to enhance DR under fixed oversampling ratio (OSR) and system order. A third-order feedforward operational amplifier constitutes the active-RC integrator for improved power efficiency.

The reminder of this paper is organized as follows. Section II reviews basic principles of SDM and presents the proposed loop filter architecture for CT SDM. Section III presents the detailed circuit implementation of the loop filter. Section IV reports the post-layout simulation results, and Section V presents the conclusion.

II. PROPOSED CRFF-B LOOP FILTER WITH PI

A. CIFF-B Type

The design of CT SDM builds upon well-established discrete-time (DT) system methodologies. As illustrated in Fig. 1, the simplified model comprises the noise transfer function (NTF) and signal transfer function (STF), which correspond to the forward and feedback transmission paths of the loop filter, respectively. Leveraging the superposition theorem of linear circuit theory, the signal and noise components satisfy the relationship:

$$STF(Z) = \frac{L_0(Z)}{1 + L_1(Z)} \quad (1)$$

$$NTF(Z) = \frac{1}{1 + L_1(Z)} \quad (2)$$

In the implementation of SDM, priority is given to the selection of the NTF, as it dictates the theoretically achievable optimal performance of the system. The NTF is used as a high- can be employed to straightforwardly derive

979-8-3315-3918-4/25 $31.00 © 2025 IEEE

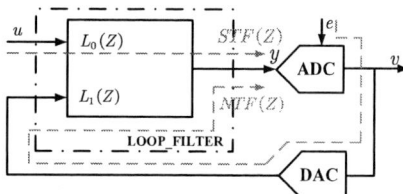

Fig 1. Model of SDM.

an NTF that meets predefined specifications. Based on Eq. (2), the loop filter transfer function $L_i(Z)$ $_{i=0,1}$ is determined, thereby finalizing the modulator architecture. The system obtained at this stage remains DT. The critical step in transitioning from a DT SDM to a CT SDM lies in the impulse invariance method, which transforms the DT loop filter $L_1(Z)$ into a CT counterpart $L_1(S)$ of equivalent order. Subsequently, the coefficients K_1, K_2, and K_3 derived from $L_1(S)$ are used to synthesize the continuous-time filter.

Common third-order CT filter architectures are illustrated in Fig. 2(a) and (b). While they share the same NTF, their characteristics differ significantly. Fig. 2(a) demonstrates the cascade of integrator feedback (CIFB) topology. Its STF exhibits an optimal third-order high-frequency roll-off. This property is critical, as input signals with substantial out-of-band interference may be received by the modulator in certain applications (e.g., wireless transceivers). Such inherent anti-aliasing characteristics can inherently simplify the design of front-end filters.

Under stable deep negative feedback, the input DC level of any integrator must theoretically be zero. This implies that the output swing of the first integrator approximates the large-signal range of a_2*v. After dynamic scaling, the first integrator attains a reduced unit-gain frequency, thereby amplifying non-ideal effects (e.g., noise and distortion). Concurrently, this necessitates larger capacitor values, resulting in an increase in the area of the chip. In contrast, the cascade of integrator feedforward (CIFF) topology shown in Fig. 3(b) inherently avoids aforementioned issue. Its first integrator exhibits minimal output swing. Post dynamic scaling, the unit-gain bandwidth of the first integrator is extended. The resultant in-band gain enhancement significantly suppresses input-referred noise and distortion. Smaller capacitors and fewer feedback DACs further alleviate area constraints in highly integrated designs. However, this comes at the cost of suboptimal first-order roll-off (lower attenuation slope), which reduces system tolerance to signal peaks[4]. Additionally, CIFF couples low-order fast feedback paths with high-order slow feedback loops, complicating circuit implementation compared to CIFB architectures.

$$L_1(Z) = \mathcal{Z}\left\{\mathcal{L}^{-1}\left[L_1(S)\right]_{t=\frac{n}{fs}}\right\}$$
$$= \mathcal{Z}\left\{\mathcal{L}^{-1}\left[K_3 S^{-3} + K_2 S^{-2} + K_1 S^{-1}\right]_{t=\frac{n}{fs}}\right\} \quad (3)$$

The proposed architecture in this work adopts the cascade of integrators with feedforward and balanced feedback (CIFF-B) topology shown in Fig. 3 (c), which hybridizes the characteristics of both aforementioned filter architectures. In this configuration, the second-order feedback loop leverages

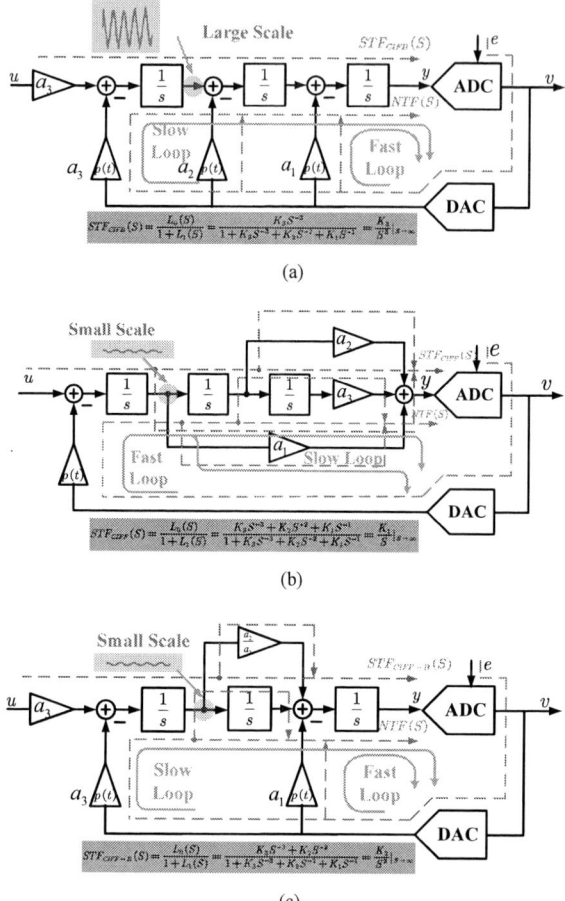

Fig. 2 Common third-order CT filter architectures (a) CIFB (b) CIFF (c) CIFF-B.

feedforward paths, while the first- and third-order loops employ feedback paths. Inheriting the advantages of the CIFF structure, the first integrator inherently avoids handling large-swing signals from feedback cancellation. Simultaneously, inspired by the CIFB architecture, the fast and slow feedback loops are functionally decoupled to optimize dynamic stability. The resultant STF achieves a balanced second-order high-frequency roll-off. By synergistically integrating the merits of both topologies, the CIFF-B architecture ensures systematic optimization across noise suppression, linearity, and area efficiency while mitigating the limitations of individual structures.

B. PI-Element method for ELD Compensation

Excess loop delay (ELD), a well-known nonideality in CT SDM, arises from the non-instantaneous feedback signal propagation caused by the finite switching times of transistors in the ADC and DAC circuitry. This delay is further exacerbated when linearization techniques such as dynamic element matching (DEM) are applied to the DAC. To mitigate ELD effects, the modulator's loop filter coefficients must be retuned to compensate for the delayed feedback. Specifically, for ELD values exceeding one clock cycle (i.e., feedback pulses delayed beyond the unit interval), compensation via a zero-order feedback path becomes mandatory. The retuned coefficients must satisfy :

$$L_1(Z) = \mathcal{Z}\left\{ \mathcal{L}^{-1}\left[K_3^{''} S^{-3} + K_3^{'} S^{-3} + K_3^{'} S^{-3} + K_0^{'} \right] \times P(S)e^{-\tau_{da}S} \right\} \quad (4)$$

As illustrated in Fig. 3(a), the conventional approach to implementing the compensation path involves inserting an additional DAC feedback path directly preceding the ADC. However, this method necessitates an active summing amplifier, significantly increasing hardware complexity. The inclusion of a direct feedback DAC approximately doubles the output swing requirement of the third integrator[5]. In the topology of Fig. 2(c), the output swing is limited to approximately the full-scale range, whereas in Fig. 3(a), it escalates to $(1+a_0)$ times the full-scale swing. To address these limitations, an alternative architecture utilizing a PI element is proposed, as shown in Fig. 3(b).

The direct compensation path shares a unified feedback DAC with the first-order compensation path. The active summing amplifier and third integrator shares a single amplifier. By employing the PI, the system achieves significant reductions in power consumption and area overhead. Furthermore, the PI maintains the same output swing as the uncompensated case, thereby relaxing the drive strength requirements of the final integrator stage.

C. Zero Optimization of NTF with Resonator

In prior discussions, all zeros of the NTF are located at DC. To achieve higher DR, it is standard practice to optimize the zero locations of the NTF. Table I summarizes the optimal zero locations for NTF systems of various orders.

Zero tuning is accomplished via a second-order resonator formed by the r_z, a feedback loop, and the first two integrators. Upon integrating the resonator, the feedback transfer function is expressed as:

$$L_1^{'}(S) = \frac{a_3 a_0 a_1 S^3 + (a_3 + a_0 a_2 a_3^{-1})S^2 +}{(S^2 + r_z)S} +$$
$$\frac{(a_1 a_2 a_3^{-1} + a_0 + a_0 a_1^{-1} a_3 r_z)S + a_1 a_3 r_z}{(S^2 + r_z)S} \quad (5)$$

The zero location yields $r_z = \sqrt{\omega_z}$. By multiplying the denominator of the right-hand side to the left-hand side and dividing both sides by S^3. Approximating the newly obtained left-hand term $(1 + r_z S^{-2})L_1(S)$ as a third-order polynomial in S^{-1}, the coefficients, $K_1^{''}$, $K_2^{''}$ and $K_3^{''}$ are related to the system parameters as follows Eq.(6).

$$\begin{cases} K_3^{''} = a_3 a_0 a_1^{-1} \\ K_2^{''} = a_3 + a_0 a_2 a_3^{-1} \\ K_1^{''} = a_1 a_2 a_3^{-1} + a_0 a_1^{-1} a_3 r_z + a_0 \\ K_0^{''} = a_1 a_3 r_z \end{cases} \quad (6)$$

TABLE I Zero Locations

Order	Zero Locations Relative to Band Edges	SQNR Improvement(dB)
1	0	0
2	± 0.577	3.5
3	$0, \pm 0.775$	8
4	$\pm 0.340, \pm 0.861$	13

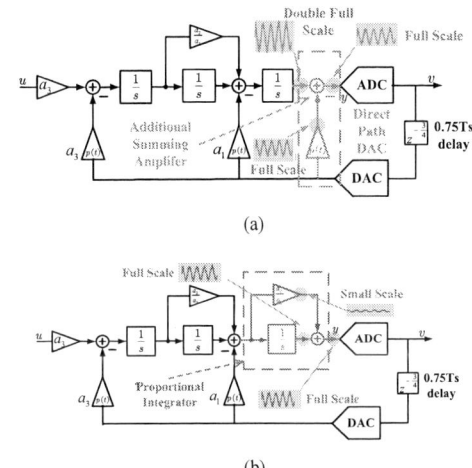

(a)

(b)

Fig 3 The compensation scheme of ELD (a) ELD compensation using a direct DAC path. (b) ELD compensation using a PI-element.

III. CIRCUIT IMPLEMENTATION

This work employs an active-RC integrator for CT SDM. To meet the system's linearity and noise requirements, the operational amplifier must achieve a gain of over 40dB[6]. Employing multistage feedforward op-amps with higher-order roll-off characteristics enables improved power efficiency. Additionally, for a continuous-time input system, the zero-induced effects on signal settling in the feedforward amplifier are negligible. Fig. 6 illustrates the detailed op-amp circuit. The two zeros generated by the parallel connection of two low-order paths and a third-order path compensate the amplifier into a pseudo-first-order system.

The first stage of the third-order path $G_{M,1}$ is designed as a cascade structure. This configuration optimizes the amplifier's noise performance. The cascade transistor suppresses the input parasitic capacitance induced by the gate-drain parasitic capacitance of the input transistor via the Miller effect. $G_{M,2a}$ and $G_{M,2b}$ share a single current source, further enhancing power efficiency. The fast path ($G_{M,c}$) and the output stage ($G_{M,3}$) of the third-order path are both implemented as push-pull amplifiers. The AC-coupled configuration, which combines the fast path and high-order path, isolates the low output impedance of $G_{M,c}$ from Gm3, thereby ensuring low-frequency gain. Furthermore, the large input parasitic capacitance introduced by the parallel-connected $G_{M,c}$ and the high-gain path due to the Miller effect is mitigated by the AC-coupling arrangement. The other two amplifiers in this work share the same architecture. Special attention is paid to the settling speed of the output-stage integrator during design, as its output exhibits relatively significant jitter.

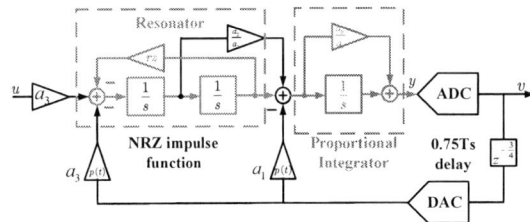

Fig. 4 Block diagram of proposed loop filter.

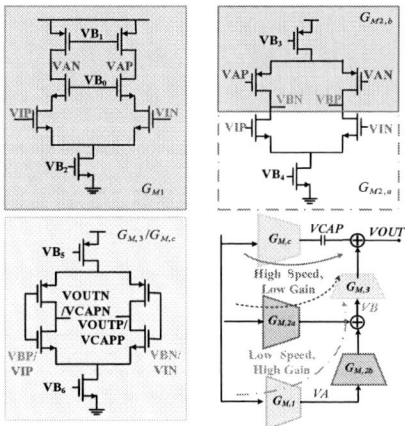

Fig. 5 The circuit diagram of the proposed amplifier.

IV. POST-LAYOUT SIMULATION RESULTS

To demonstrate the effectiveness of the proposed loop filter, a third-order CTSDM employing this architecture is simulated and fabricated in 28 nm CMOS technology. The die photograph is shown in Fig.6, with the loop filter occupying a silicon area of 130μm × 300μm. Fig.7(a) presents the simulated STF and NTF of the loop filter, demonstrating third-order noise shaping with an out-of-band NTF gain capped at 5 dB to ensure stability. The optimized STF exhibits less than 3 dB peaking to suppress out-of-band interference, while featuring first-order attenuation for high-frequency signals as discussed.

Fig.7(b) plots the SNDR versus input amplitude, revealing a peak SNDR of 73.3 dB and a DR of 79.8 dB, with SNDR degradation observed at higher input levels. Output spectra for -3dBFS input tones at 15.3 MHz and 93.1 MHz are shown in Fig. 7(c) and (d), achieving SNDR values of 73.3 dB and 70.1 dB, respectively. The complete loop filter consumes 12mW, predominantly attributed to three operational amplifiers.

V. CONCLUSION

This paper presents a CRFF-B loop filter architecture for high-performance CTSDMs, merging the benefits of CIFF and CIFB topologies. By replacing the conventional DAC-based direct feedback path with PI for ELD compensation, the design reduces power and area. A resonator is introduced to enhance the achievable SNDR of the system. The loop filter incorporates a third-order multistage multipath feedforward amplifier with low flicker noise and high bandwidth. The implemented CTSDM based on this

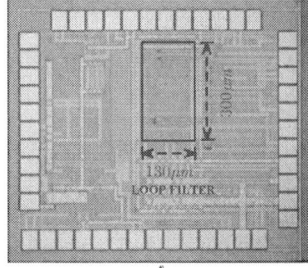

Fig. 6 Die photo of loop filter.

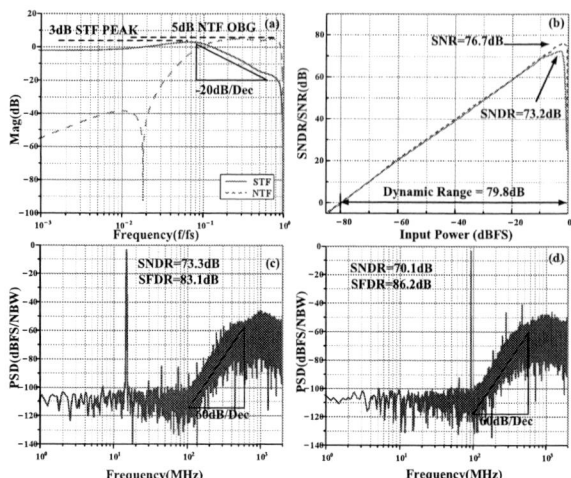

Fig.7 SDM's result (a) STF and NTF magnitude response. (b) SNDR vs input amplitude for an input at 15.3 MHz. (c) spectrum at 15.3 MHz and (d) spectrum at 93.1 MHz.

architecture achieves a DR of 79.8 dB. For input frequencies of 15.3 and 93.1MHz, the prototype achieves 73.3 and 70.1 dB SNDR, respectively. The implemented loop filter consumes merely 12 mW power while occupying 0.039 mm² core area.

ACKNOWLEDGMENT

This work was supported in part by the Fundamental Research Funds for the Central Universities under Grant KYFZ25008, in part by the Natural Science Foundation of China under Grant 8091B02042301, in part by Shaanxi Provincial Key Research and Development Program under Grant 2024CY2GJHX34 and in part by National Science and Technology Major Project under Grant 2024ZD0302600. (Corresponding author: Maliang Liu.)

REFERENCES

[1] M. Bolatkale, L. J. Breems, R. Rutten and K. A. A. Makinwa, "A 4 GHz Continuous-Time ΔΣ ADC With 70 dB DR and −74 dBFS THD in 125 MHz BW," in IEEE Journal of Solid-State Circuits, vol. 46, no. 12, pp. 2857-2868, Dec. 2011, doi: 10.1109/JSSC.2011.2164963.

[2] S. -H. Wu, T. -K. Kao, Z. -M. Lee, P. Chen and J. -Y. Tsai, "A 160MHz-BW 72dB-DR 40mW continuous-time ΔΣ modulator in 16nm CMOS with analog ISI-reduction technique," 2016 IEEE International Solid-State Circuits Conference (ISSCC), San Francisco, CA, USA, 2016, pp. 280-281, doi: 10.1109/ISSCC.2016.7418016.

[3] M. B. Dayanik, D. Weyer and M. P. Flynn, "A 5GS/s 156MHz BW 70dB DR continuous-time sigma-delta modulator with time-interleaved reference data-weighted averaging," 2017 Symposium on VLSI Circuits, Kyoto, Japan, 2017, pp. C38-C39, doi: 10.23919/VLSIC.2017.8008539.

[4] L. Breems et al., "A 2.2 GHz Continuous-Time ΔΣ ADC With −102 dBc THD and 25 MHz Bandwidth," in IEEE Journal of Solid-State Circuits, vol. 51, no. 12, pp. 2906-2916, Dec. 2016, doi: 10.1109/JSSC.2016.2591826.

[5] C. -Y. Ho, C. Liu, C. -L. Lo, H. -C. Tsai, T. -C. Wang and Y. -H. Lin, "A 4.5 mW CT Self-Coupled ΔΣ Modulator With 2.2 MHz BW and 90.4 dB SNDR Using Residual ELD Compensation," in IEEE Journal of Solid-State Circuits, vol. 50, no. 12, pp. 2870-2879, Dec. 2015, doi: 10.1109/JSSC.2015.2475160.

[6] T. Caldwell, D. Alldred and Z. Li, "A Reconfigurable ΔΣ ADC With Up to 100 MHz Bandwidth Using Flash Reference Shuffling," in IEEE Transactions on Circuits and Systems I: Regular Papers, vol. 61, no. 8, pp. 2263-2271, Aug. 2014, doi: 10.1109/TCSI.2014

An Open-Loop Residue Amplifier with SSF Structure Achieving 69dBc SFDR for High-Speed and High-Precision PSAR ADCs

Chengjun Liu [1], Deng Luo [2,3], Hanbing Liu [1], Chengchao Mou [1], Bin Liang [2,3], Yaqing Chi [2,3]
Jianjun Chen [2,3], Kai Tang [1], Jing Xiao [1], Ming Tao*[1]

[1] College of Electrical and Information Engineering, Hunan University, China
[2] College of Computer Science and Technology, National University of Defense Technology, Changsha 410073, China
[3] Key Laboratory of Advanced Microprocessor Chips and Systems, National University of Defense Technology, Changsha, 410073, China

* Email: liucj2001@163.com, tming@hnu.edu.cn

Abstract—This paper proposes an open-loop residue amplifier (RA) based on a super source follower (SSF) structure, designed for 14-bit high-speed pipelined-SAR (PSAR) analog-to-digital converters (ADCs). The gain is determined purely by resistor ratios, significantly improving linearity by minimizing dependence on transistor transconductance nonlinearity. Designed in a 28nm CMOS process, the proposed RA achieves an 8× gain within 300ps, with a total harmonic distortion (THD) below −67dB and a spurious-free dynamic range (SFDR) exceeding 69dBc across −40°C to 125°C. Compared to existing open-loop RAs, the proposed design shows superior linearity and robustness, making it highly suitable for high-performance PSAR ADCs.

Keywords—Residue amplifier, pipelined-SAR ADC, super source follower, open-loop, high linearity, low THD.

I. INTRODUCTION

High-speed and high-resolution analog-to-digital converters (ADCs) have become essential components in modern communication systems. Among various architectures, the pipelined-SAR ADC (PSAR ADC) combines the high speed and resolution of pipelined ADCs with the low power of SAR ADCs, making it an attractive choice for high-performance applications [1].

In PSAR ADCs, the residue amplifier (RA) is a critical module that directly impacts overall conversion speed and resolution. The architectures of RA are typically categorized into closed-loop and open-loop structures. Closed-loop RAs offer high precision but suffer from limited speed [2]-[3], while open-loop RAs are preferred for high-speed PSAR ADCs due to their high speed and low power [4]-[7]. However, traditional Gm-C and Gm-R open-loop RAs exhibit poor linearity due to gain dependency on transconductance, capacitance/resistance values, and amplification time, all of which are nonlinear with respect to device characteristics.

To address this problem, various techniques have been proposed. In [4], an inverter-based RA with a harmonic-injecting cross-coupled pair (HXCP) is introduced, achieving 8x gain with THD below −52dB under a 70mV$_{pp}$ input. Similarly, [5] proposed a Gm-R-based RA with a differential flipped voltage follower (DFVF), achieving 8× gain and −48dB THD under the same conditions. However, under an input swing of 70mV$_{pp}$, these RAs cannot meet the linearity

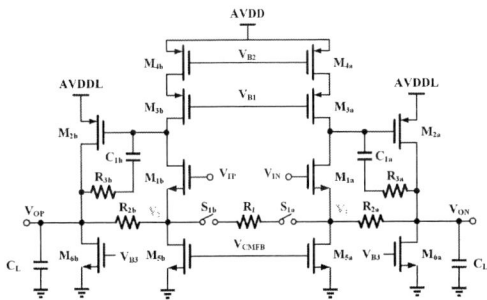

Fig. 1. Schematic diagram of the proposed open-loop residual amplifier.

requirements of 14-bit ADCs, whose THD is around −50dB in the worst-case scenario.

To overcome these limitations, this paper presents a novel open-loop RA based on a super source follower (SSF) structure designed in 28nm CMOS. In the proposed design, gain is determined strongly by the ratio of passive resistors, greatly improving linearity by eliminating the influence of transistor nonlinearity. Under 70mV$_{pp}$ input and 300ps amplification time, the RA achieves an 8× gain, THD < −67dB, and SFDR > 69dBc, suitable for 14-bit PSAR ADCs operating at up to 625MSPS.

II. PROPOSED RESIDUAL AMPLIFIER DESIGN

A. Circuit Design

As illustrated in Fig. 1, the proposed open-loop residue amplifier (RA) adopts a fully differential symmetric structure, where signal amplification is realized through a super source follower (SSF) topology and feedback resistors. During amplification, switches S_{1a} and S_{1b} are closed. Transistors M_{3a}, M_{4a}, M_{5a}, M_{6a}, M_{3b}, M_{4b}, M_{5b}, and M_{6b} serve as bias transistors, supplying constant currents via bias voltages. Among them, M_{5a} and M_{5b} are regulated by a common-mode feedback (CMFB) voltage to stabilize the output common-mode level. Differential input transistors M_{1a} and M_{1b} make nodes V_1 and V_2 track input signals from V_{IN} and V_{IP}, generating a voltage difference across resistor R_1, which induces current flow through R_1. Since the bias transistors maintain constant current, the differential current must flow through resistors R_{2a} and R_{2b}, which are equal in value. This

979-8-3315-3918-4/25 $31.00 © 2025 IEEE

creates a differential output voltage at nodes V_{OP} and V_{ON}, with the gain determined by the resistor ratio of R_{2a} and R_1.

For example, assuming a differential input voltage V_i, with $+V_i/2$ and $-V_i/2$ at nodes V_{IP} and V_{IN} respectively, the voltage across R_1 becomes v_i, resulting in a current I. Setting the resistance values of R_1 and R_{2a} respectively to R and αR, I is equal to V_i/R. This current flows through R_{2a} toward node V_{ON}, generating an output voltage V_{on} which is $-(\alpha+1/2)*V_i$. By symmetry, the positive output V_{op} is $(\alpha+1/2)*V_i$, yielding a differential gain G of $2\alpha+1$. This gain depends on passive resistor ratios, and is therefore independent of transistor transconductance, which mitigates nonlinear effects and enhances the linearity of the amplifier.

When S_{1a} and S_{1b} are open, the RA enters tracking mode. Transistors M_{1a}, M_{2a}, and M_{6a} form an SSF structure, as do M_{1b}, M_{2b}, and M_{6b}. In this mode, the gain is close to unity, and the output tracks the input voltage directly.

In addition to improving linearity, the circuit also addresses bandwidth and stability. To meet high-speed requirements, all pole frequencies are designed to be high. However, since the dominant pole at M_{2a} and M_{2b} gates is close to the output pole, insufficient phase margin could result in instability. To address this, a Miller compensation network comprising C_{1a}, C_{1b}, R_{3a}, and R_{3b} is employed to separate the dominant and secondary poles, thereby improving phase margin and ensuring stable operation.

The dominant pole is located at the gate of M_{2a} and M_{2b}. Taking M_{2a} as an example, the pole frequency is approximately:

$$\omega_{p1} = \frac{1}{g_{m2}R_{out}C_1r_{eq}} \tag{1}$$

Where r_{eq} is the equivalent resistance to ground seen from the gate, and R_{out} is the output resistance at node V_{ON}. Compared to transistor output resistances, R_1 and R_{2a} are small, so both r_{eq} and R_{out} are relatively low, allowing the dominant pole to be pushed to a high frequency. The bandwidth, limited by this dominant pole, reaches more than 4GHz, which meets the high-speed requirements of PSAR ADCs.

Moreover, voltage overstress protection is implemented. In a 28 nm CMOS process, the core device breakdown voltage is around 1.09V. With a 0.6V common-mode output voltage and a 0.25V signal amplitude, the output may reach 0.35V. If only supplied with 1.8V, the drain-source voltage of M_{2a} and M_{2b} could rise to 1.45V, potentially damaging the transistors. To mitigate this, the sources of M_{4a} and M_{4b} are connected to the 1.8V AVDD, while the sources of M_{2a} and M_{2b} are tied to a 1.3V AVDDL generated by an on-chip LDO.

B. Gain Calculation and Analysis

To further analyze gain characteristics, the RA is simplified into a single-ended equivalent model as shown in Fig. 2(a). Taking the negative input side as an example, assume resistance values of R_1 and R_{2a} respectively to R and αR, and for AC signals the center node of R_1 is at virtual ground. M_{3a} and M_{4a} are modeled as a current source I_3 with output impedance $g_{m3}r_{o3}r_{o4}$, and similarly for M_{5a} and M_{6a} with r_{o5} and r_{o6}.

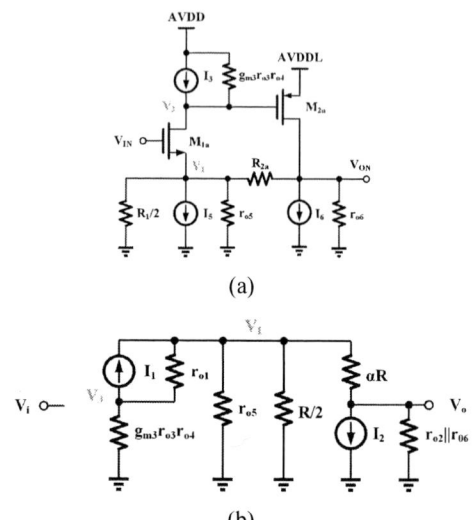

Fig. 2. (a) Single-ended equivalent circuit of the residual amplifier; (b) Small-signal equivalent circuit of the residual amplifier.

The small-signal equivalent circuit, shown in Fig. 2(b), is used to derive the output gain of the RA. Kirchhoff's Current Law (KCL) is applied to nodes V_1:

$$\frac{V_1}{r_{o5} \parallel \frac{R}{2}} + \frac{V_3}{g_{m3}r_{o3}r_{o4}} + \frac{V_o - V_1}{\alpha R} = 0 \tag{2}$$

Among them, I_2 is equal to:

$$I_2 = g_{m2}V_3 \tag{3}$$

Then KCL is applied to nodes V_1:

$$I_2 + \frac{V_o}{r_{o2} \parallel r_{o6}} + \frac{V_o - V_1}{AR} = 0 \tag{4}$$

Combining equations (2), (3), and (4) yields:

$$V_3 = \frac{\dfrac{1}{r_{o5} \parallel \dfrac{R}{2}} + \dfrac{1}{\alpha R + r_{o2}}}{\dfrac{\alpha g_{m2}R}{\alpha R + r_{o2}} - g_{m2} - \dfrac{1}{g_{m3}r_{o3}r_{o4}}} \cdot V_1 = \beta \cdot V_1 \tag{5}$$

According to the KCL equation for the V_3 node:

$$g_{m1}(V_i - V_1) + \frac{V_3 - V_1}{r_{o1}} + \frac{V_3}{g_{m3}r_{o3}r_{o4}} = 0 \tag{6}$$

Substituting (5) into (6) yields the following.

$$V_i = \frac{[g_{m1} + \dfrac{1}{r_{o1}} - \beta(\dfrac{1}{r_{o1}} + \dfrac{1}{g_{m3}r_{o3}r_{o4}})]\cdot V_1}{g_{m1}} = \frac{\gamma \cdot V_1}{g_{m1}} \tag{7}$$

From (4), (5) and (7), Vo can be calculated as follows.

$$V_o = \frac{r_{o2} + \beta \cdot \alpha g_{m2} r_{o2} R}{\alpha R + r_{o2}} \cdot \frac{g_{m1}}{\gamma} \cdot V_i \qquad (8)$$

Assuming that the output impedances of the bias transistors are significantly higher than those of other components, (8) can be greatly simplified.

$$V_o = \frac{2\alpha + 1}{1 + \dfrac{1}{g_{m1} r_{o1}} + \dfrac{2\alpha + 1 + \dfrac{2r_{o2}}{R}}{g_{m1} r_{o1} g_{m2} r_{o2}}} \cdot V_i \qquad (9)$$

When $g_{m1} r_{o1}$ and $g_{m2} r_{o2}$ are very large, then the output gain G is as follows.

$$G = \frac{V_o}{V_i} = 2\alpha + 1 \qquad (10)$$

The analysis reveals that, under the condition of high output impedance and stable bias current, and with sufficient loop gain, the voltage gain of the RA is predominantly determined by the resistor ratio. Specifically, the differential gain approaches $2\alpha + 1$.

This confirms that the proposed RA architecture achieves high linearity by decoupling gain from transistor transconductance nonlinearity. Since the gain depends solely on resistor ratios, the design effectively mitigates distortion introduced by device variations. However, this also imposes stringent requirements on resistor matching. Mismatches among R_1, R_{2a} and R_{2b} may result in noticeable gain errors, underscoring the importance of precise layout and calibration strategies.

C. Adjustable Resistor Array for Gain Calibration

In practical implementations, due to variations in process and temperature, the ratio between R_{2a} and R_1 may deviate from the intended design value of α :1. These deviations can lead to significant fluctuations in the overall gain of the RA, potentially degrading system performance. To mitigate this issue and ensure accurate gain calibration, this work replaces R_1 with an adjustable resistor array, as illustrated in Fig. 3.

The resistor array consists of five parallel-connected resistors with a binary-weighted ratio of 8:4:2:1:1. The first four branches are gated by NMOS switches, which are controlled by a 4-bit digital calibration code D<3:0>. By adjusting the digital code, the effective resistance can be tuned, allowing fine-grained control of the gain of RA.

III. SIMULATION RESULT

The proposed open-loop residue amplifier based on the SSF structure was designed in a 28nm CMOS process. The layout of the RA is shown in Fig. 4, occupying an area of 1369μm². The circuit operates with a 1.8V power supply, while a 1.3V bias voltage is generated by an on-chip LDO. Post-layout simulations were performed under a single-ended load capacitance of 250fF, an amplification time of 300ps, and across process-voltage-temperature (PVT) corners, covering a temperature range from −40°C to 125°C.

Fig. 5 and Fig. 6 illustrate the simulated SFDR and THD versus input swing under both nominal and worst-case conditions. With a differential input swing of 70mV$_{pp}$ the RA

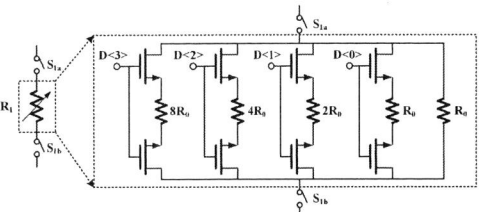

Fig. 3. Resistor array circuit for gain calibration.

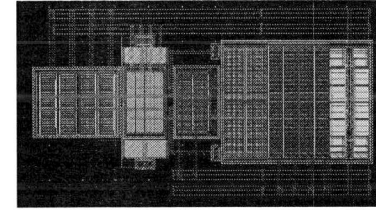

Fig. 4. Layout of the proposed residual amplifier.

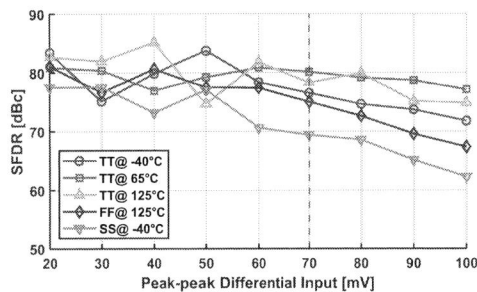

Fig. 5. Simulated SFDR versus input swing under different temperatures and process corners.

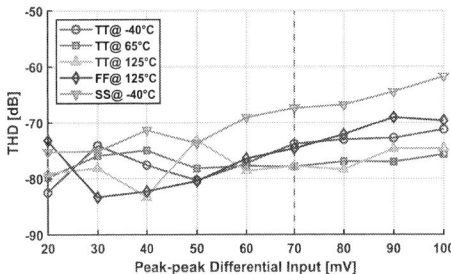

Fig. 6. Simulated THD versus input swing under different temperatures and process corners.

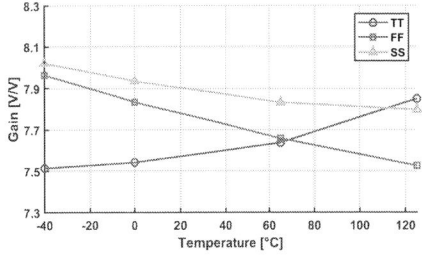

Fig. 7. Simulated gain versus temperature under TT, FF, and SS process corners.

achieves a SFDR of 80dBc and a THD of −78dB under typical conditions. The SFDR remains above 69dBc and THD stays below −67dB, even under the worst-case PVT

979-8-3315-3918-4/25 $31.00 © 2025 IEEE

corner, demonstrating excellent linearity and robustness across a wide operating range.

Moreover, Fig. 7 and Fig. 8 illustrate the gain and bandwidth characteristics of the proposed RA. Across the temperature range from –40°C to 125°C, the RA maintains a gain around 8×, which satisfies the requirements for PSAR ADC applications. Under the typical process corner (TT) and standard operating temperature of 65°C, the –3dB bandwidth reaches 4.48GHz, demonstrating the capability to support high-precision amplification within 300ps.

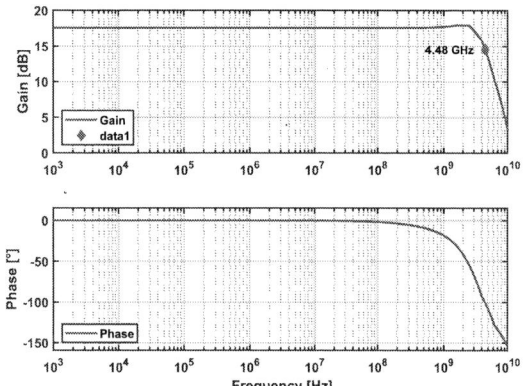

Fig. 8. Simulated AC Bode plot of the residual amplifier under typical conditions.

TABLE I. summarizes the key performance metrics of the proposed RA and compares it with state-of-the-art designs. In [4] and [5], the worst-case THD achieved under similar input swing conditions is –52dB and –48dB, respectively, with an operating temperature range limited to 0°C to 80°C. In [6], a two-stage Gm-R based RA with built-in linearity compensation achieves –52dB THD over a range of –20°C to 80°C. In contrast, the proposed RA achieves approximately 15dB improvement in THD over a wider temperature range, from –40°C to 125°C, making it highly suitable for robust, high-performance ADC systems.

In terms of power consumption, the RA dissipates a maximum of 18mW, excluding peripheral circuits, which is acceptable for high-speed and high-resolution PSAR ADC applications.

TABLE I.   RA PERFORMANCE SUMMARY AND COMPARISON

	This work	[4]	[5]	[6]
Process	28nm	28nm	28nm	65nm
Supply Voltage(V)	1.8	0.9	1	1.2
Nominal Gain(V/V)	8	8	8	5
Temperature Range(°C)	–40~125	0~80	0~80	–20~80
THD(dB)	–67	–52	–48	–52

IV. CONCLUSION

This paper presents a high-linearity open-loop residue amplifier based on a Super Source Follower architecture, designed for use in 14-bit high-speed Pipelined-SAR ADCs. The amplifier leverages a resistor-ratio-based gain structure, effectively decoupling gain performance from transistor nonlinearity and thereby significantly improving linearity.

Designed in a 28nm CMOS process and powered by a 1.8V supply, the proposed RA achieves an 8× voltage gain within an amplification time of only 300ps. Under a $70mV_{pp}$ input swing, the RA demonstrates excellent linearity with a THD lower than –67dB and a SFDR exceeding 69dBc. Furthermore, the RA maintains robust performance across a wide temperature range of –40°C to 125°C, confirming its suitability for high-resolution, high-speed PSAR ADC applications. Compared with state-of-the-art open-loop RA designs, the proposed amplifier offers superior linearity and thermal robustness while maintaining moderate power consumption of 18mW. These attributes make it a promising solution for future low-power, high-performance ADC systems in advanced communication and signal processing platforms.

ACKNOWLEDGMENT

This work was supported in part by the National Natural Science Foundation of China under Grant 62304258, in part by Shandong Provincial Natural Science Foundation Innovation and Development Joint Fund under Grant ZR2023LZH005, and in part by Hunan Provincial Natural Science Foundation of China under Grant 2023JJ40176.

REFERENCES

[1] S. Palermo, S. Hoyos, S. Cai, S. Kiran and Y. Zhu, "Analog-to-Digital Converter-Based Serial Links: An Overview," in IEEE Solid-State Circuits Magazine, vol. 10, no. 3, pp. 35-47, Summer 2018.

[2] C. C. Lee and M. P. Flynn, "A 12b 50MS/s 3.5mW SAR assisted 2-stage pipeline ADC," 2010 Symposium on VLSI Circuits, Honolulu, HI, USA, 2010, pp. 239-240.

[3] Y. Zhu, C. -H. Chan, S. -W. Sin, S. -P. U and R. P. Martins, "A 34fJ 10b 500 MS/s partial-interleaving pipelined SAR ADC," 2012 Symposium on VLSI Circuits (VLSIC), Honolulu, HI, USA, 2012, pp. 90-91.

[4] L. Fang, T. Fu, X. Wen and P. Gui, "A 12-b 1-GS/s 61-dB SNDR Pipelined-SAR ADC With Inverter-Based Residual Amplifier and Tunable Harmonic-Injecting Cross-Coupled-Pair for Distortion Cancelation Achieving 6.3 fJ/conv-step," in IEEE Solid-State Circuits Letters, vol. 5, pp. 194-197, 2022.

[5] W. Jiang, Y. Zhu, M. Zhang, C. -H. Chan and R. P. Martins, "A Temperature-Stabilized Single-Channel 1-GS/s 60-dB SNDR SAR-Assisted Pipelined ADC With Dynamic Gm-R-Based Amplifier," in IEEE Journal of Solid-State Circuits, vol. 55, no. 2, pp. 322-332, Feb. 2020.

[6] N. Li et al., "A 10-Bit 500-MS/s Pipelined SAR ADC With Nonlinearity-Compensated Open-Loop Amplifier and Parallel Conversion Through Comparator Reusing," in IEEE Transactions on Circuits and Systems II: Express Briefs, vol. 72, no. 2, pp. 354-358, Feb. 2025.

[7] Y. Zhang, M. Zhang, Z. Wu, Y. Zhu, R. P. Martins and C. -H. Chan, "24.5 A 72GS/s 9b Time-Interleaved Pipeline-SAR ADC Achieving 55.3/49.3dB SFDR at 20GHz/Nyquist Inputs in 16nm FinFET," 2025 IEEE International Solid-State Circuits Conference (ISSCC), San Francisco, CA, USA, 2025, pp. 436-438.

A 14-bit R-2R DAC with All-Digital Foreground Calibration based on Redundant LSB

Hanbing Liu [1], Deng Luo [2,3], Chengjun Liu [1], Chengchao Mou [1], Bin Liang [2,3], Yaqing Chi [2,3]
Jianjun Chen [2,3], Kai Tang [1], Jing Xiao [1], Ming Tao*[1]

[1] College of Electrical and Information Engineering, Hunan University, China
[2] College of Computer Science and Technology, National University of Defense Technology, Changsha 410073, China
[3] Key Laboratory of Advanced Microprocessor Chips and Systems, National University of Defense Technology, Changsha, 410073, China

* Email: lhb2023@hnu.edu.cn, tming@hnu.edu.cn

Abstract—This work implements a 14-bit R-2R DAC at the transistor level in standard 55-nm CMOS technology. An all-digital foreground calibration technique based on redundant LSB enhances linearity, with 4 additional redundant bits preventing calibration overflow. Compared to existing solutions requiring auxiliary DAC, the proposed scheme utilizes only these redundant bits and the DAC's intrinsic LSB for calibration and compensation. Transistor-level simulation results demonstrate that the application of the proposed calibration technique increases the effective number of bit (ENOB) by 1.9 bit, while the Spurious-Free Dynamic Range (SFDR) improves by 10 dB. Furthermore, the Differential Nonlinearity (DNL) and Integral Nonlinearity (INL) are reduced by factors of 6.34 and 6.27, respectively.

Keywords—Digital-to-analog convert, calibration technique, resistor mismatch.

I. INTRODUCTION

The R-2R DAC has gained widespread adoption in many high accuracy applications, particularly in audio systems and industrial control systems, owing to its simple structure, compact area, and low power consumption. However, the design of R-2R DACs presents several technical challenges, with resistor mismatch emerging as the most critical limitation. While the ideal implementation requires all resistors to maintain precise values of either R or 2R, but manufacturing variations inevitably introduce mismatches that severely degrade DAC linearity. In this case, a few techniques have been proposed to reduce the negative impact of mismatch errors. [1] employed Dynamic Element Matching (DEM) to transform harmonic distortion into white noise at the cost of elevated noise floor. Alternative approaches using analog calibration techniques [2]-[3] achieve higher ENOB, while the problem is the additional circuit complexity, noise, and spurs. The solution proposed in [4] utilizes an auxiliary DAC to provides compensation currents without introducing additional noise or spurs. However, this method suffers from inherent matching challenges between the main and calibration DACs during physical implementation, which ultimately limits the achievable performance.

To overcome the excessive analog overhead inherent in conventional calibration techniques, this work presents an all-digital foreground calibration technique based on redundant LSB for a 14-bit R-2R DAC. The proposed technique eliminates the need for an auxiliary DAC by leveraging the DAC's native LSB for both calibration and compensation, thereby substantially reducing analog overhead. Transistor-level simulation results verify the

calibration effectiveness, showing measured improvements of +1.9 bit in ENOB and +20 dB in SFDR after calibration implementation.

The remainder of this paper is organized as follows. Section II presents the DAC architecture. Section III presents the proposed calibration technique and the circuit implementation. Section IV presents transistor-level simulation results, followed by the conclusions in Section V.

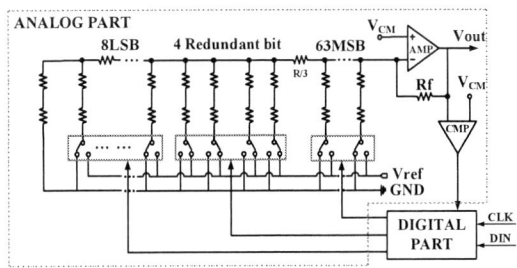

Fig. 1. The overall architecture of the proposed 14-bit R-2R DAC

II. DAC ARCHITECTURE

The overall architecture of the proposed 14-bit R-2R DAC is depicted in Fig. 1. The DAC employs a 6-8 segmentation scheme, comprising a 6-bit most significant bit (MSB) segment and an 8-bit least significant bit (LSB) segment. The MSB segment is thermometer coded, which supports both data-weighted averaging (DWA) and DEM techniques. The LSB segment is binary coded. Four additional redundant bits are incorporated to provide an extended quantization range, thereby preventing calibration overflow (the specific calibration procedure and metho-dology are detailed in Section III). These redundant bits possess a weight of 128 LSB, equivalent to the weight of the DAC's 8th bit (counting from the LSB).

Based on the parameters provided by the foundry, the random mismatch standard deviation (σ) for the MSB resistors is approximately 0.5%, corresponding to 1.28 LSB. In the worst-case scenario, the cumulative 3σ error for the 63 MSB unit elements amounts to approximately 242 LSB. The maximum compensation range provided by the four additional redundant bits is ±256 LSB, which is sufficient to accommodate a 6σ variation. The MSB and LSB segments are interconnected via a bridge resistor with a value of R/3. An operational amplifier incorporating a feedback resistor is connected after the resistor array. This configuration enhances the DAC's current-driving capability while

979-8-3315-3918-4/25 $31.00 © 2025 IEEE

simultaneously acting as an isolation buffer, effectively minimizing the influence of subsequent circuit stages on the resistor array, functionally equivalent to treating each resistor as a current source. The DAC output voltage is given by

$$Vout = \frac{32 * Rf}{R} * (V_{CM} - D * LSB) + V_{CM} \quad (1)$$

where D is the decimal equivalent of the input code, R is the unit resistance within the array, Rf is the feedback resistor with a value of $R/32$, and V_{CM} is the reference common-mode voltage.

During calibration cycles, the comparator performs a comparison between the DAC output voltage and V_{CM}. The resulting error signal is then processed by the successive approximation register (SAR) logic within the digital section. Through iterative adjustments, the SAR logic progressively minimizes this error, driving the difference between the $Vout$ and V_{CM} towards zero.

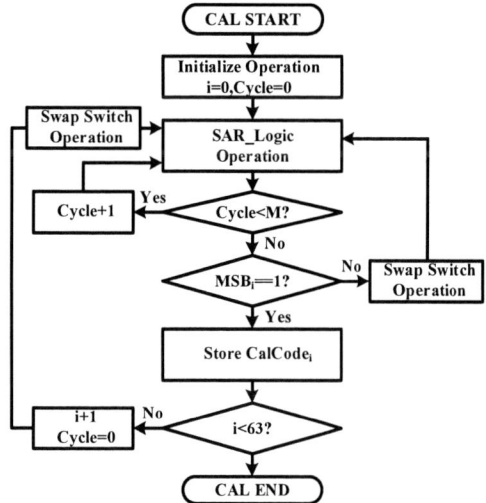

Fig. 2. The sequential calibration flowchart.

III. CALIBRATION TECHNIQUE

A. Redundant LSB calibration technique

Owing to the mathematically dominant weighting of the MSB segment, the INL and DNL performance of the R-2R DAC are predominantly governed by resistor mismatch within this segment. Consequently, the proposed foreground calibration technique selectively targets the 63-bit MSB segment. This calibration is executed prior to normal DAC operation; the sequential calibration flow is detailed in Fig. 2. To elucidate the calibration principle, the procedure for the first MSB (MSB<0>) serves as an exemplar. When calibrating MSB<0>, its switch is configured to the positive reference terminal, as illustrated in Fig. 3(a). Concurrently, the switches corresponding to the remaining 62 MSBs are partitioned equally between positive reference terminal and negative reference terminal, maintaining this fixed configuration throughout MSB<0> calibration. All LSB segment switches are connected to the negative reference terminal. Furthermore, the redundant bit switches are equally distributed between P and N. Under this specific switch configuration, the DAC output voltage is expressed as follows:

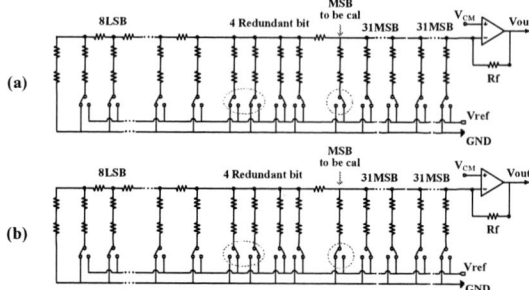

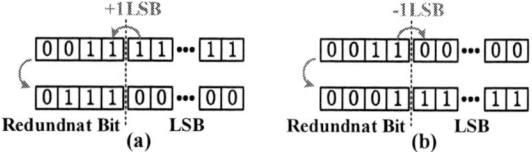

Fig. 3. The switch configuration acquiring (a) the calibration code at terminal P, (b) the calibration code at terminal N.

Fig. 4. During the calibration process, cases of LSB (a) carry , (b) borrow.

$$Vout = (\sum_{i=33}^{62} IMSB_i - \sum_{i=1}^{32} IMSB_i - IMSB_0 + IRB_{init} + ILSB_{init}) \times Rf + V_{CM} \quad (2)$$

$$IRB_{init} + ILSB_{init} = \sum_{i=0}^{7} ILSB_i + \sum_{i=0}^{1} IRB_i + \sum_{i=2}^{3} IRB_i \quad (3)$$

where $IMSBi$ denotes the current contributed by the i-th MSB to the output node, $ILSB_{init}$ and IRB_{init} represent the initial currents contributed by the LSB segments and the redundant bits to the output node, respectively, during the initialization phase. Following initialization, the calibration commences. The two inputs of the comparator are connected to $Vout$ and V_{CM}, respectively. The SAR logic controller adjusts the digital codes of the LSB segments and redundant bits over M adjustment cycles, driving the difference between $Vout$ and V_{CM} toward zero. This calibration procedure is mathematically expressed as

$$(\sum_{i=33}^{62} IMSB_i - \sum_{i=1}^{32} IMSB_i - IMSB_0 + IRB_{over} + ILSB_{over}) \times Rf - Vos = 0 \quad (4)$$

Where LSB_{over} and IRB_{over} represent the currents contributed by the LSB segments and the redundant bits, respectively, to the output node after the adjustment phase is completed. V_{OS} denotes the comparator offset voltage. The calibration code $Ical_P$ obtained when the target MSB is connected to terminal P during calibration is given by

$$Ical_P = IRB_{over} + ILSB_{over} - IRB_{init} + ILSB_{init} - Vos \quad (5)$$

During the SAR logic adjustment process, potential carry-over or borrow events in the LSB segments necessitate the inclusion of 4 redundant bits to prevent calibration overflow. As illustrated in Fig. 4(a), when the LSB segments are at an "all-1s" digital code value, an increment operation triggers a carry-over to the redundant bits. This transitions the redundant bits from 0011 to 0111, while resetting the LSB segments to "all-0s". Conversely, Fig. 4(b) depicts the borrow scenario: when the LSB segments hold an "all-0s" code, a decrement operation initiates a borrow from the redundant bits. This changes the redundant bits from 0011 to 0001, simultaneously setting the LSB segments to "all-1s".

After acquiring the calibration code $Ical_P$ at terminal P, the switch configuration is updated to Fig. 3(b). The switch

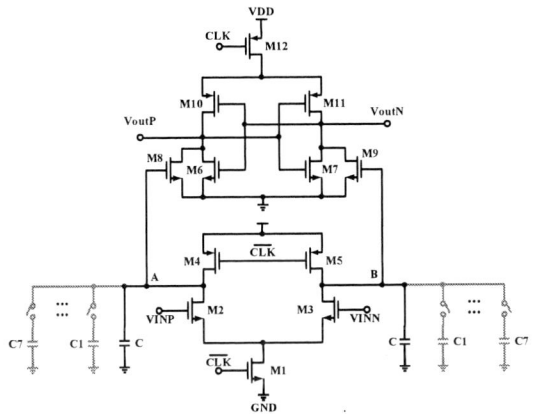

Fig. 5. The comparator circuit proposed in this paper.

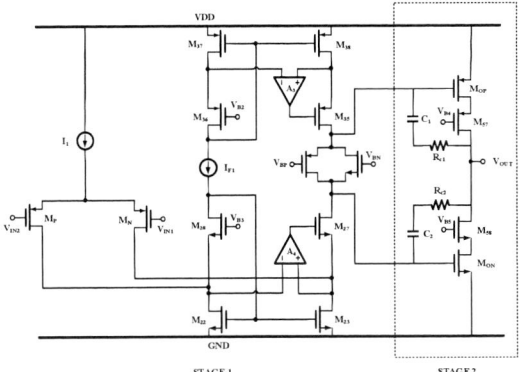

Fig. 6. The operational amplifier circuit proposed in this paper.

corresponding to the target MSB under calibration is now connected to terminal N. Simultaneously, all four redundant bit switches are set to terminal P, while all LSB segment switches are connected to terminal N. Under this configuration, the DAC output voltage is expressed as

$$Vout = (\sum_{i=33}^{62} IMSB_i - \sum_{i=1}^{32} IMSB_i + IMSB_0 + IRB_{init} + ILSB_{init}) \times Rf + V_{CM} \quad (6)$$

$$IRB_{init} + ILSB_{init} = \sum_{i=0}^{7} ILSB_i - \sum_{i=0}^{3} IRB_i \quad (7)$$

Following this, the identical SAR logic adjustment process is then executed over M cycles to acquire the terminal-N calibration code $Ical_N$ for the target MSB. Finally, subtracting $Ical_P$ from $Ical_N$ yields the compensated code $Ical$ for the first MSB, computed as

$$Ical = IRB_1 + IRB_0 - IMSB_0 \quad (8)$$

The aforementioned calibration procedure is sequentially performed for each of the remaining 62 MSBs, yielding individual calibration codes $Ical$ for every MSB. Equation (8) reveals that the final calibration code represents the difference between the MSB current and the combined contributions of the redundant bits. Consequently, the fundamental principle of this calibration technique employs the redundant bits as a reference to generate corresponding calibration codes for each MSB. During normal DAC operation, these calibrated codes are utilized to compensate the MSB currents, thereby enhancing overall linearity performance.

B. Comparator offset calibration

In this work, the typical LSB current of the DAC is 24.2 nA. The redundant bits provide a maximum compensation current of 256 LSB (equivalent to 6.25 μA). When this current flows through the 2500 Ω feedback resistor Rf, it generates a maximum compensation voltage of 15.625 μV. While the calibration technique proposed in Part A can cancel comparator offset voltages, this capability is contingent upon the offset being smaller than the maximum compensation voltage. Should the comparator offset exceed this compensation range, calibration becomes infeasible. To prevent calibration failure due to excessive equivalent offset Critically, comparator offset calibration precedes MSB calibration to further mitigate its impact on calibration accuracy. A binary-weighted capacitor array (C_1–C_7) is connected to nodes A and B in Fig. 5. This array decays the discharge rate at nodes A and B, with capacitance values scaling binarily from C_1 (smallest unit capacitor) to C_7

(largest unit capacitor). The capacitor switches are governed by the DAC's digital control logic. During offset calibration initialization, all capacitor switches connect to node N, resistor array switches preserve their pre-MSB calibration configuration.

The SAR logic then executes iterative comparisons while progressively adjusting capacitor connections between nodes P and N. This process systematically minimizes the input-referred offset voltage between the comparator's differential inputs. Upon convergence, the capacitor switch states remain fixed for subsequent operations.

C. Amplifier designs

The operational amplifier (op amp) within the DAC must meet stringent requirements: its input common-mode voltage is fixed at the reference VREFCM, while its output must swing across the full range of analog voltages generated by the DAC decoder. Additionally, the op amp must drive a heavy 1 Ω resistive load, necessitating high dc gain. Given the feedback configuration employed, stability is a paramount design consideration.

As shown in Fig. 6, a two-stage op amp topology is implemented. The first stage utilizes a folded-cascode amplifier providing high gain. The second stage employs a class AB output buffer to achieve wide voltage swing and robust output drive capability. The gain of the first stage is enhanced by actively driving the cascode gates. To ensure stability, Miller compensation with nulling resistors (C1/Rc1, C2/Rc2) is adopted. This technique splits the dominant and non-dominant poles. Critically, the series resistors Rc1 and Rc2 reposition the right-half-plane zero (RHPZ) inherent to Miller compensation into the left-half plane. This left-half-plane zero (LHPZ) is then placed to cancel the non-dominant pole, achieving a phase margin >60° for robust operation.

IV. SIMULATION

The proposed DAC was designed in standard 55-nm CMOS technology. During simulations, random mismatch parameters were incorporated for every analog component. To comprehensively validate the calibration efficacy, separate dynamic and static performance simulations were conducted.

A. Dynamic performance simulations

Dynamic performance simulations employed a 1-MSPS sampling rate with a 1.4-kHz digital sine-wave input. Monte

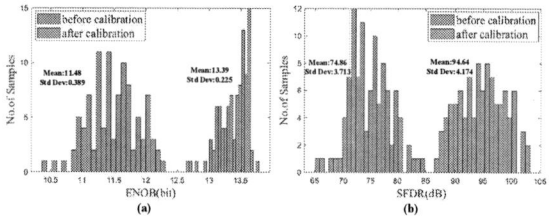

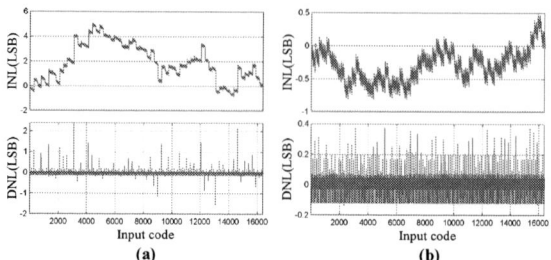

Fig. 7. The dynamic performance with 100 Monte Carlo simulation.

Fig. 9. Simulation results of INL and DNL (a) before calibration (b) after calibration.

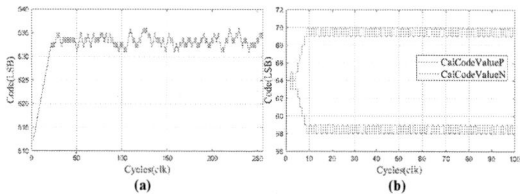

Fig. 8. Calibration code convergence in (a) MSB calibration (b) comparator offset convergence.

TABLE I. PERFORMANCE COMPARISON

	This work	[5]	[6]	[7]
Technique	55nm	65nm	180nm	180nm
Architecture	R-2R	R-2R	R-2R	R-2R
Resolution (bit)	14	16	13	16
Sample rate (MS/s)	1	4	0.01	1
SFDR(dB)	94.64	65	87.39	101.5
ENOB(bit)	13.39	N/A	12.87	15.34
DNL/(LSB)	±0.38	±0.04	±0.34	±0.5
INL(LSB)	±0.81	±0.26	±0.58	±1

Carlo simulations incorporating random device mismatches revealed severe ENOB degradation in the uncalibrated 14-bit R-2R DAC, averaging merely 11.48 bit. This confirms significant performance deterioration due to resistor mismatches. Fig. 7 compares the dynamic performance with 100 Monte Carlo simulation before and after calibration, demonstrating that the proposed technique effectively compensates MSB mismatches. Post-calibration simulations show the mean of ENOB improvement from 11.48 bit to 13.39 bit and the mean of SFDR enhancement from 74.86 dB to 94.64 dB. Fig. 8 illustrates the convergence behavior observed in simulation for the MSB calibration and the comparator offset calibration.

B. Static performance simulations

Static performance simulations employed identical test conditions to the dynamic analysis, utilizing a full-scale ramp input from all-zeros to all-ones. Fig. 9 contrasts INL and DNL characteristics before and after calibration. Pre-calibration simulations exhibited ±5.08 LSB INL and ±2.41 LSB DNL. Implementation of the foreground calibration technique reduced these errors to ±0.81 LSB INL and ±0.38 LSB DNL, demonstrating significant improvement in static linearity performance. The comparison with other publications is shown in Table I.

V. CONCLUSION

This paper presents a transistor-level 14-bit R-2R DAC implemented in 55nm CMOS process, featuring an all-digital foreground calibration based on redundant LSB to enhance linearity. Simulation results demonstrate that the proposed calibration method effectively compensates for nonlinearity induced by resistor mismatch. Furthermore, the calibration technique exhibits notable extensibility, it can be readily adapted to other DAC architectures such as current-steering DAC, indicating broad application potential.

ACKNOWLEDGMENT

This work was supported in part by the National Natural Science Foundation of China under Grant 62304258, in part by Shandong Provincial Natural Science Foundation Innovation and Development Joint Fund under Grant ZR2023LZH005, and in part by Hunan Provincial Natural Science Foundation of China under Grant 2023JJ40176.

REFERENCES

[1] J. Remple, A. Panigada and I. Galton, "An ISI Scrambling Technique for Dynamic Element Matching Current-Steering DACs," in IEEE Journal of Solid-State Circuits, vol. 57, no. 2, pp. 465-479, Feb. 2022.

[2] M. Clara, W. Klatzer, B. Seger, A. D. Giandomenico, and L. Gori, "A 1.5 V 200 MS/s 13b 25 mW DAC with randomized nested background calibration in 0.13 μm CMOS," in Proc. IEEE Int. Solid-State Circuits Conf., Dig. Tech. Papers, vol. 50, 2007, pp. 250-251.

[3] C. -H. Lin et al., "A 16b 6GS/S nyquist DAC with IMD <-90dBc up to 1.9GHz in 16nm CMOS," 2018 IEEE International Solid-State Circuits Conference - (ISSCC), San Francisco, CA, USA, 2018, pp. 360-362.

[4] Kong, K. Rivas-Rivera and I. Galton, "A 600-MS/s DAC With Over 87-dB SFDR and 77-dB Peak SNDR Enabled by Adaptive Cancellation of Static and Dynamic Mismatch Error," in IEEE Journal of Solid-State Circuits, vol. 54, no. 8, pp. 2219-2229, Aug. 2019.

[5] A. A. Noorwali, S. M. Qasim, A. S. Doost and A. Huynh, "A 16-bit 4 MSPS DAC for lock-in amplifier in 65nm CMOS," 2016 IEEE 13th International Conference on Networking, Sensing, and Control (ICNSC), Mexico City, Mexico, 2016, pp. 1-5.

[6] M. Wang, Q. Li and J. Li, "A Low-power High-precision Differential Output 13bit R-2R DAC Pathway in 180nm CMOS," 2024 9th International Conference on Integrated Circuits and Microsystems (ICICM), Wuhan, China, 2024, pp. 852-856.

[7] K. Wu, Y. Liu, Y. Hu, Y. He, Z. Yu and N. Ning, "An Area-Efficient 16-Bit Four-Channel R-2R DAC Based on Switching On-Resistance Adaptive Calibration Technique," 2024 IEEE 17th International Conference on Solid-State & Integrated Circuit Technology (ICSICT), Zhuhai, China, 2024, pp. 1-3.

A Multi-Channel Reconfiguration and Combination Technique for Timing Mismatch Calibration in Time-Interleaved ADCs

Jize Liu [1,2], Jinwei Wu [1], Jiayi Chen [1], Xinqi Liu [1], Yuekang Guo [1], Jing Jin *[1]

[1]School of Integrated Circuits, Shanghai Jiao Tong University, Shanghai, 200240, China
[2]Zhiyuan College, Shanghai Jiao Tong University, Shanghai, 200240, China

* Email: aurora0611@sjtu.edu.cn, jinjing@sjtu.edu.cn

Abstract—Time-interleaved (TI) ADCs enable high sampling rates but suffer from timing mismatches that introduce interleaving spurs, which limit the SNDR of the ADC. This paper presents a multi-channel reconstruction and combination (MCRC) technique for real-time calibration, which forms high-speed sub-ADCs to improve the effective Nyquist rate. A novel channel alignment strategy calibrates all channels to the average timing error, enhancing robustness against reference ADC failures. Simulation results show that MCRC effectively restores SFDR, SNR, and ENOB, and achieves fast, accurate convergence, making it well-suited for dynamic, high-speed ADC applications.

Keywords—Time-Interleaved ADC, Timing Mismatch, Real-Time Calibration, Spline Interpolation, Channel Combination, Multi-channel

I. INTRODUCTION

Time-interleaved (TI) techniques are crucial for increasing sampling rates in modern RF ADCs, addressing the limitations of single ADCs in high-speed, wide-bandwidth applications. By using multiple ADC modules in parallel, an M-channel TI-ADC system can achieve a sampling rate M times that of a single channel. However, mismatches such as gain, offset, and timing errors, along with signal noise, degrade signal accuracy, reducing SFDR and SNDR. To improve performance in ultra-high-speed TI-ADC systems, it is essential to compensate for these mismatched errors, with timing mismatch being the most complex and critical to calibrate.

In recent years, numerous digital calibration methods have been proposed to address timing mismatches in TI-ADC systems. Methods based on correlation, such as those in [1] and [2], are effective but struggle with high-frequency inputs due to model inaccuracies and clock jitter. Neural network and genetic algorithm-based approach-es, like those in [3] and [4], offer global optimization but come with high computational complexity and the need for dedicated calibration signals. [5] introduced a wideband compensation method, but its reliance on offline training and high computational demands limits its real-time applicability. Reference-assisted background calibration for real-time operation, as proposed in [6], offers a solution but introduces additional hardware overhead and is sensitive to clock jitter. Finally, [7][8] explored reference ADC-based correlation methods, which reduce complexity but still require a reference ADC and are constrained by input frequency limitations.

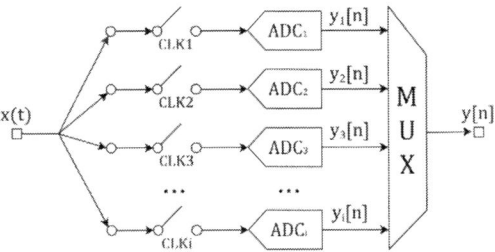

Fig. 1.   Time-interleaved ADC system diagram.

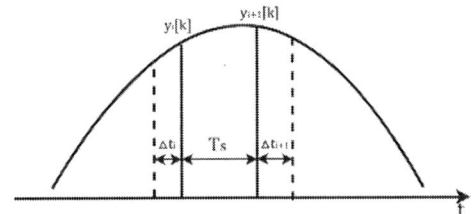

Fig. 2.   Timing mismatch schematic sampling waveform.

This paper proposes a multi-channel reconfiguration and combination technique (MCRC) for channel merging to address timing mismatches. In the proposed calibration structure, a multi-channel TI-ADC can be reconstructed into a reduced number of channels, thereby preventing calibration failure caused by reference ADC malfunctions. Section II introduces the M-channel Time-Interleaved ADC model and presents the multi-channel reconfiguration and combination technique accordingly. Numerical simulation results are displayed in Section III and Section IV concludes this paper.

II. THE PROPOSED CALIBRATION STRUCTURE

A. M-channel Time-Interleaved ADC Model

As shown in Fig. 1, an M-channel TI-ADC consists of M sub-ADCs with a sampling rate of Fs/M, which sample the signal at different phase points. After merging the outputs, the system behaves as an ADC with a sampling rate of Fs. Assuming the sampling period for the M-channel TI-ADC is Ts, the sampling period for each sub-ADC is M·Ts. The output of the i-th channel can be expressed as:

$$y_i[n] = \sum_{n=0}^{\infty} \delta(t - (nM + i - 1)T_s) \cdot x(t) \qquad (1)$$

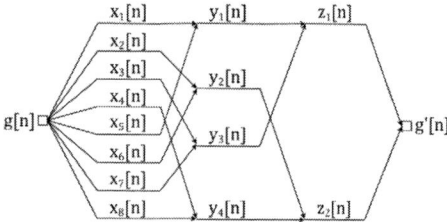

Fig. 3. Signal combination schematic diagram.

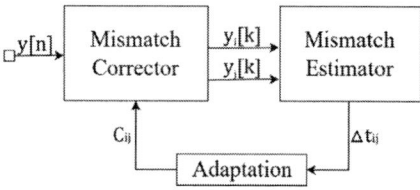

Fig. 4. System calibration algorithm block diagram.

Considering timing mismatch, the sampling instants of the sub-ADCs deviate by Δt_i due to timing errors. The output of the i-th channel with timing mismatch can be expressed as:

$$y_i[n] = \sum_{n=0}^{\infty} \delta(t-(nM+i-1+\Delta t_i)T_s) \cdot x(t) \qquad (2)$$

where i is the channel index($i=1,2,...,M$). According to (2), different sub-ADCs have different amounts of mismatch, and the merging of channels can improve the minimum Nyquist frequency. However, after merging, the ADC performance, such as SNDR and SFDR, deteriorates due to the timing mismatch. A calibration structure is required to correct the timing errors during the merging process to improve the performance of the merged ADC.

Fig. 3 illustrates the proposed calibration structure. In a multi-channel TI-ADC system, to perform channel merging, the i-th channel is first merged with the $|M/2+i|$-th channel. This is because the sampling instants between the i-th and $|M/2+i|$-th channels differ by $M/2 \cdot Ts$. In an ideal case, the sampling period after merging is $M/2 \cdot Ts$, and the sampling rate is $2 \cdot Fs$. However, the merging process must account for the timing mismatch, so each merging step involves a calibration to minimize performance degradation. After merging, the number of TI-ADC channels is reduced to half, while the minimum Nyquist frequency increases to twice the original value. By repeatedly merging the channels, the number of channels in the TI-ADC can be reduced to 1/2n, with the Nyquist frequency increasing to 2n times the original value.

B. Correlation-Based Mismatch Estimator

The calibration loop during the merging process is shown in Fig. 4, consisting of two stages: timing mismatch detection and timing mismatch calibration. In the timing mismatch detection stage, the time-interleaved output signal is first divided into two sub-channels, and the correlation between them is computed to identify the mismatch. Let the input signal be x(t) with sampling frequency Fs. For N samples, the sampled signals are:

$$y_1[n] = x(t_1 + \Delta T_1) \qquad (3)$$

$$y_2[n] = x(t_2 + \Delta T_2) \qquad (4)$$

Where ΔT_1 and ΔT_2 are the timing errors for the first and second sub-ADCs, respectively. The timing misalignment between $y_1[n]$ and $y_2[n]$ is detected by calculating their product and averaging the difference:

$$\Delta T = mean(P_{12} - P_{21}) \qquad (5)$$

where:

$$P_{12} = y_1[n] \cdot y_2[n-1] \qquad (6)$$

$$P_{21} = y_2[n-1] \cdot y_1[n-1] \qquad (7)$$

This detection formula estimates the timing error by comparing the signal differences between the two channels and outputs a value ΔT, which is used for subsequent error compensation.

C. Spline Mismatch Corrector

In order to correct timing mismatches in time-interleaved ADC systems, spline interpolation is utilized to recover signal values at the desired, uniformly aligned time instants based on nonuniformly sampled data. Let the distorted sampled signal be denoted as

$$\tilde{y}[n] = x(t_n + \Delta t_n) \qquad (8)$$

where Δt_n represents the channel-specific timing skew. The goal is to estimate the signal at the corrected time point $t_n + \Delta t$, where Δt denotes the average skew calculated across all channels.

To achieve this, a cubic spline interpolation function S(t) is constructed over a local neighborhood of samples surrounding t_n, typically including four consecutive samples such as y[n−1],y[n],y[n+1],y[n+2], with their corresponding time indices. Within each subinterval $[t_i,t_{i+1}]$, the spline is defined by a piecewise cubic polynomial of the form

$$S_i(t) = a_i + b_i(t-t_i) + c_i(t-t_i)^2 + d_i(t-t_i)^3 \qquad (9)$$

where the coefficients ai, bi, ci, di are determined by imposing continuity of the function and its first and second derivatives across the interval boundaries. Specifically, the interpolation conditions $S_i(t_i)=y[i]$ and $S_i(t_{i+1})=y[i+1]$, as well as derivative continuity conditions $S_i'(t_{i+1})=S_{i+1}'(t_{i+1})$ and $S_i''(t_{i+1})=S_{i+1}''(t_{i+1})$, ensure smooth transitions between adjacent spline segments. At the boundaries, natural spline conditions are typically applied, setting the second derivatives at the endpoints to zero unless otherwise constrained.

Once the spline has been constructed, the corrected value at the target time is obtained by evaluating the spline function at the adjusted instant $t_n+\Delta t - \Delta t_n$, yielding

$$y_{corrected}[n] = S(t_n + \Delta \bar{t} - \Delta t_n). \qquad (10)$$

Cubic spline interpolation offers a high degree of smoothness and low interpolation error, making it well-suited for timing correction in high-speed ADC applications.

Compared to lower-order methods, it provides improved approximation of the underlying continuous-time signal, leading to more accurate and robust compensation of timing skew.

D. Adaptation Module

To achieve adaptive adjustment of time mismatch correction coefficients in time-interleaved ADC systems, a hierarchical pairwise calibration framework is adopted. This framework organizes the entire N-channel ADC system into $\log_2(N)$ layers, with each layer corresponding to progressively refined channel grouping for calibration purposes. In each calibration window, the adjustment of mismatch correction coefficients starts from coarse-grained grouping and gradually proceeds to more fine-grained pairwise comparisons.

In each layer l, all channels are divided into $2^{(l-1)}$ groups, with each group containing two sub-channels: one reference channel and one target channel. Time mismatch correction coefficients are adaptively updated based on the feedback of relative alignment between paired channels. For a given pair of channels, the correction coefficient of the target channel is updated adaptively according to the following formula:

$$\delta_{new} = \delta_{old} + \eta \cdot \Delta \qquad (11)$$

where η denotes the learning rate of the current layer, and δ represents the cumulative correction applied to the target channel.

The iterative adjustment process continues until convergence criteria are met. After completing coefficient adjustments for all channel pairs in the given layer, the cumulative corrections are globally applied to the corresponding channels through spline-based resampling. These corrections are inherited and further optimized in subsequent layers to achieve finer channel alignment.

To accommodate time-varying non-idealities in the system, this framework supports dynamic adjustment of correction coefficients across calibration windows. By maintaining a cumulative correction matrix across layers and channels, the system can respond to changes in the distribution of time mismatches, ensuring robust performance even under non-stationary conditions. This hierarchical adaptive coefficient adjustment method provides scalable calibration capabilities for large ADC systems while reducing computational complexity and enabling efficient hardware or real-time implementation.

E. Theoretical Analysis of Channel Reconfiguration and Calibration Technique

The proposed calibration method provides several significant theoretical advantages. By aligning each channel to the average timing skew $\Delta \bar{t}$ rather than to a single reference channel, the method avoids the inherent risks associated with reference ADC failures or large offsets. This symmetry ensures consistent calibration across all channels and enhances system robustness.

In terms of interpolation accuracy, cubic spline interpolation achieves superior performance. Its residual error can be analytically bounded by:

$$\acute{U}_{spline} = \frac{1}{384} \left| h^4 max \right| \left\| x^{(4)}(t) \right\| \qquad (12)$$

where h is the sampling interval. This error is significantly lower, enabling more precise signal reconstruction.

III. SIMULATION RESULTS

We conducted a simulation study in the MATLAB environment to evaluate the performance of the proposed timing calibration algorithm for a 16-channel time-interleaved analog-to-digital converter (TI-ADC) system with a resolution of 10 bits and a sampling rate of 16 GHz. During the simulation, nominal timing mismatches were introduced into the behavioral model with a deviation set to 100 femtoseconds (fs). The test input signal was a combination of sinusoidal signals at 0.1 GHz, 0.15 GHz, and 0.2 GHz. The simulation results demonstrated that the proposed calibration algorithm effectively mitigates the impact of timing mismatches on signal quality.

Figure 5 compares the spectra of the input signal affected by timing mismatches (red curve) and the signal after calibration (blue curve). Without calibration, timing mismatches introduced significant spurious tones in the signal spectrum, leading to a substantial degradation in spectral purity. After applying the calibration algorithm, these spurious tones were successfully suppressed, resulting in a notable improvement in spectral purity. The key performance metrics of the system after calibration were: a spurious-free dynamic range (SFDR) of 75.69 dB, a signal-to-noise ratio (SNR) of 63.16 dB, and an effective number of bits (ENOB) of 10.20 bits. These metrics clearly demonstrate the effectiveness of the calibration algorithm in eliminating timing mismatches and significantly enhancing signal quality.

To further validate the applicability of the calibration algorithm in dynamic environments, we applied the calibration system to real-time input signal streams and designed a dynamic testing scheme. During signal processing, calibration was performed every 1024 signal samples, and new timing mismatches were introduced during the 10th and 20th calibration cycles to evaluate the algorithm's ability to adapt to changing timing errors. Experimental results showed that the calibration algorithm quickly adjusted and converged to the correct output state after timing mismatches changed. The algorithm exhibited high adaptability and robustness, with rapid convergence ensuring the stability of system operation. Figure 6 illustrates the calibration performance under dynamic conditions, verifying the reliability of the algorithm in handling varying timing mismatches.

Additionally, to assess the calibration algorithm's effectiveness across all channels, we recorded the convergence trends of timing mismatches for all 16 channels during the calibration cycles, as shown in Figure 7. The results reveal that timing mismatches for all channels converged rapidly to below 0.01 fs within five calibration cycles. This demonstrates that the algorithm not only achieves high-precision calibration but also offers high convergence efficiency, completing calibration in a minimal number of iterations. These findings further confirm the superior performance of the algorithm in practical applications.

In summary, the simulation results strongly demonstrate that the proposed calibration algorithm effectively mitigates timing mismatches, significantly enhances signal quality, and

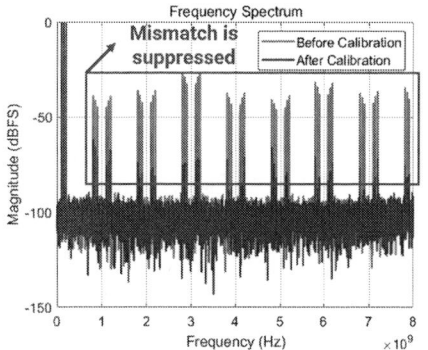

Fig. 5. Frequency spectrum comparison of uncalibrated and calibrated signals.

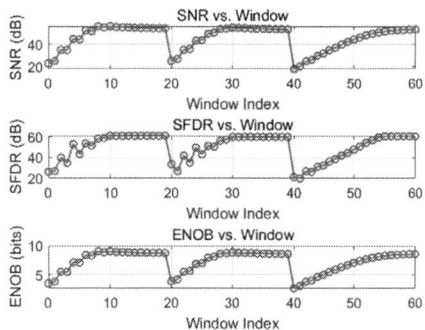

Fig. 6. Timing mismatch reduction (Δt) over calibration iterations.

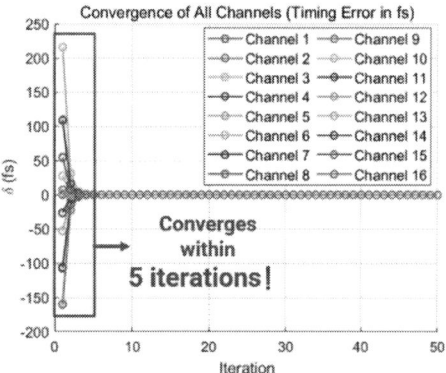

Fig. 7. Improvement in SNDR, SFDR, and ENOB after calibration.

maintains stable performance under dynamic conditions. The algorithm shows great potential for application in high-speed, wideband TI-ADC systems, offering promising prospects for future development. Furthermore, as shown in Table I, the proposed MCRC algorithm eliminates the need for a digital derivative filter and reference ADC, while employing a low-complexity computational approach to reduce performance overhead. This ensures effective calibration of high-speed ADC systems and enhances their fault tolerance and reliability.

IV. CONCLUSION

This paper presents a novel multi-channel reconfiguration and combination (MCRC) technique for timing mismatch calibration in time-interleaved ADCs.

TABLE I. HORIZONTAL COMPARISON OF CALIBRATION ALGORITHMS

	D. Stepanovic JSSC-13	M. Guo JSSC-20	M. El-Chammas JSSC-11	J. Song JSSC-18	H. Wei JSSC-14	N. L. Dortz ISSCC-14	This Work
Ref. Channel	2×SAR	Sub-Channel	Comparator	Window Detector	Sub-Channel	None	None
Detection Methods	LMS-based	LMS-based	Auto correlation-based	Variance-based	Auto correlation-based	Auto correlation-based	Auto correlation-based
Correction Types	Analog Capacitor DAC	Digital	Analog Capacitor DAC	Analog Capacitor DAC	Analog Capacitor DAC	Digital	Digital
Require Digital Derivative Filter	No	Yes	No	No	No	Yes	No

The proposed method effectively addresses the limitations of conventional approaches by enhancing robustness, scalability, and adaptability in dynamic high-speed ADC systems. Through innovative channel merging and spline-based interpolation, the technique achieves precise timing alignment, significantly improving key performance metrics such as SFDR, SNR, and ENOB while maintaining rapid convergence. Simulation results validate the effectiveness of the algorithm in mitigating timing mismatches. The MCRC technique demonstrates substantial potential for advancing the performance and reliability of ultra-high-speed TI-ADC systems.

ACKNOWLEDGMENT

This work was supported in part by the Hui-Chun Chin and Tsung-Dao Lee Chinese Undergraduate Research Endowment (CURE), the Zhiyuan Future Scholar Program under Grant ZIRC2024-24, and in part by the Shuguang Program of Shanghai Education Development Foundation and Shanghai Municipal Education Commission under Grant 23SG10.

REFERENCES

[1] Y. Qiu, J. Zhou, Y. Liu and Y. Huangfu, "A Novel Calibration Method of Gain and Time-skew Mismatches for Time-interleaved ADCs Based on Neural Network," 2019 IEEE MTT-S International Wireless Symposium (IWS), Guangzhou, China, 2019, pp. 1-3.

[2] M. Ni, X. Wang, F. Li, W. Rhee and Z. Wang, "A 13-Bit 2-GS/s Time-Interleaved ADC With Improved Correlation-Based Timing Skew Calibration Strategy," in IEEE Transactions on Circuits and Systems I: Regular Papers, vol. 69, no. 2, pp. 481-494, Feb. 2022.

[3] M. Ni et al., "A Correlation-based Timing Skew Calibration Strategy Using a Time-Interleaved Reference ADC," 2020 IEEE 63rd International Midwest Symposium on Circuits and Systems (MWSCAS), Springfield, MA, USA, 2020, pp. 345-348.

[4] J. Qin, W. Zhong, Y. Cao, J. Li, Z. Cao and L. Zhao, "Machine-Learning-Based Mismatch Calibration for Time-Interleaved ADCs," in IEEE Transactions on Nuclear Science, vol. 71, no. 8, pp. 2012-2019, Aug. 2024.

[5] C. K. Su, P. J. Hurst and S. H. Lewis, "A Time-Interleaved SAR ADC With Signal-Independent Background Timing Calibration," in IEEE Transactions on Circuits and Systems I: Regular Papers, vol. 69, no. 2, pp. 620-633, Feb. 2022.

[6] S. Liu, L. Zhao, Z. Deng and Z. Zhang, "A Low-Complexity Timing Mismatch Calibration Method for Four-Channel Time-Interleaved ADCs Based on Cross Correlation," 2020 IEEE 15th International Conference on Solid-State & Integrated Circuit Technology (ICSICT), Kunming, China, 2020, pp. 1-3.

[7] W. Xu, B. Yao, Q. Cheng, Z. Du and L. Qiu, "A Reference Assisted Background Calibration Technique With Constant Input Impedance for Time-Interleaved ADCs," in IEEE Transactions on Circuits and Systems II: Express Briefs, vol. 71, no. 10, pp. 4437-4441, Oct. 2024.

[8] Y. A. Tavares and M. Lee, "A Foreground Calibration for M-Channel Time-Interleaved Analog-to-Digital Converters Based on Genetic Algorithm," in IEEE Transactions on Circuits and Systems I: Regular Papers, vol. 68, no. 4, pp. 1444-1457, April

A 12-bit 620 MS/s Pipelined-SAR ADC with Feed-forward Compensation Closed-loop Residual Amplifier in 28 nm CMOS

Shuai Liu[1], Yi Hu[2,3], Guoyu Li[1], Congyang Sun[1], Yidong Yuan[2,3], Hao Xu*[1], Na Yan[1]

[1] State Key Laboratory of Integrated Chip and Systems, Fudan University, Shanghai
[2] Beijing Smartchip Microelectronics Technology Co., Ltd
[3] Beijing Smartchip Semiconductor Technology Co., Ltd

* Email: haoxu@fudan.edu.cn

Abstract—This paper presents a 12-bit three-stage 620MS/s pipelined-SAR ADC in 28 nm CMOS process, where the inter-stage residue amplification employs a feed-forward compensation charge-redistribution MDAC to achieve accurate gain and high energy efficiency. With one-time bit weight calibration, the 620 MS/s 12-bit pipelined SAR ADC achieves 53.1 dB SNDR at 2 MHz input frequency and 44.2 dB SNDR at Nyquist frequency while dissipating 20.6 mW power and occupying 450 μm×130 μm area.

Keywords—ADC, pipelined SAR, MDAC, closed-loop amplifier

I. INTRODUCTION

In recent years, the advancement of wireless communication technologies has significantly expanded receiver operating bands, imposing stringent demands on analog-to-digital converters (ADCs) regarding both conversion rate and resolution.

Successive-Approximation Register (SAR) ADCs based on charge-redistribution capacitive digital-to-analog converters (CDACs) are widely adopted for high linearity and energy efficiency. However, the conversion rate is inherently limited by the sequential comparison process. To enhance the sampling rate, two primary approaches exist: time-interleaving and pipeline-assisted architectures. Time-interleaving (TI) techniques require calibration to mitigate inter-channel mismatches (e.g., offset, gain, and timing skew), while significantly increasing the design complexity of the sampling frontend and multi-phase clock distribution network[1]. On the other hand, pipelined-SAR ADCs face the primary challenge of designing a high-speed, high-precision residue amplifier.

Open-loop amplifiers offer fast response and low power consumption[2,3], but suffer from inaccurate gain, often requiring calibration that introduces additional hardware overhead. Inverter-based ring amplifiers can improve the speed of traditional closed-loop amplifiers, but their performance varies significantly with PVT variation[4,5], necessitating bias tuning for proper operation. The feed-forward compensation amplifier enhances stability by introducing a feedforward zero to cancel one pole in the two-stage operational amplifier, thereby reducing power consumption in the second stage while achieving precise

closed-loop gain. This paper explores the application of feedforward-compensated amplifiers in pipelined-SAR ADCs.

II. ARCHITECTURE

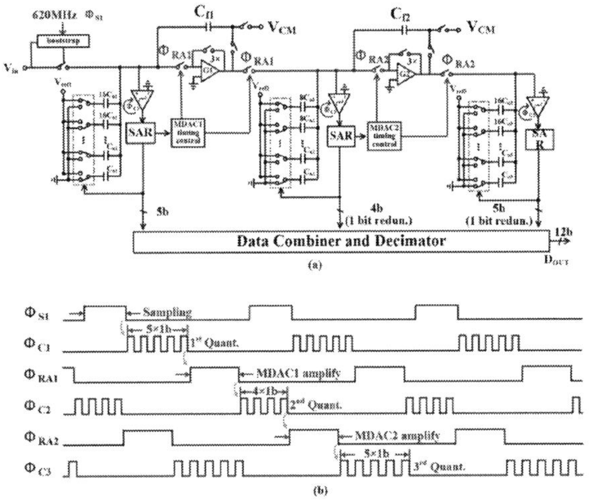

Fig. 1. (a) Block diagram of the implemented 3-stage pipelined SAR. (single-ended shown) (b) Applied asynchronous timing sequence.

The single-ended block diagram of the proposed pipelined-SAR ADC is shown in Fig. 1(a). 5bit-4bit-5bit SAR quantizer with 2-bit redundancy for comparator offset tolerance and 2 charge-redistribution MDAC are applied. The MDAC adopts a flip-around charge-redistribution structure to achieve closed-loop 3-times amplification, thereby increasing bandwidth, reducing power consumption, and improving capacitor utilization efficiency.

The applied asynchronous timing sequence is shown in Fig. 1(b). The enable signal of MDACs are triggered by the end of SAR quantization, eliminating the stand-by time between quantization and amplification in synchronous timing schemes, thereby increasing the sampling rate.

III. CIRCUIT IMPLEMENTATION

A. Feed-forward Compensation Amplifier

The closed-loop amplifier provides stable and precise gain, eliminating the need for complex calibration. The two-stage feed-forward amplifier is employed to introduce an in-band zero for frequency compensation, eliminating the power consumption typically required to push the secondary pole as shown in Fig. 2(a). By sharing the bias current between the feedforward stage and the second stage, the design achieves significant power savings.

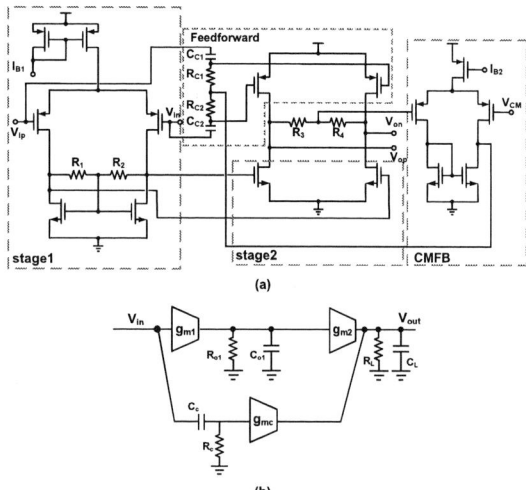

(a)

(b)

Fig. 2. (a) Schematic of feed-forward compensation amplifier and (b) single-ended differential equivalent model.

The equivalent model of the feed-forward amplifier is shown in Fig. 2(b), and the transfer function is

$$H(s) = \frac{G_{m1}G_{m2}R_{o1}R_L}{\left(1+\frac{s}{\omega_{p1}}\right)\left(1+\frac{s}{\omega_{p2}}\right)} + \frac{\frac{s}{\omega_c}}{1+\frac{s}{\omega_c}}\frac{G_{mc}R_L}{1+\frac{s}{\omega_{p2}}}$$

$$= \frac{G_{m1}G_{m2}R_{o1}R_L + \frac{s}{\omega_c}(G_{m1}G_{m2}R_{o1}R_L + G_{mc}R_L) + \frac{s^2}{\omega_c\omega_{p1}}G_{mc}R_L}{\left(1+\frac{s}{\omega_{p1}}\right)\left(1+\frac{s}{\omega_{p2}}\right)(1+\frac{s}{\omega_c})}$$

The zero and pole introduced by AC coupling occur at the same frequency, while the additional zero from the feedforward path ω_{z1} is located at a frequency of

$$\omega_{z1} \approx \omega_{z1} + \omega_{z2} = -\frac{G_{m1}G_{m2}R_{o1}R_L + G_{mc}R_L}{G_{mc}R_L}\omega_{p1} = \frac{G_{m1}G_{m2}}{G_{mc}C_{o1}}$$

The feed-forward zero ω_{z1} is utilized to compensate the output pole of the first stage ω_{p1}:

$$\frac{G_{m1}G_{m2}}{G_{mc}C_{o1}} = \frac{1}{R_{o1}C_{o1}}$$

The GBW of the feed-forward compensation amplifier can be expressed as:

$$GBW = \frac{A_v\omega_{p1}}{2\pi} = \frac{G_{m1}G_{m2}R_{o1}R_L}{2\pi R_{o1}C_{o1}} = \frac{G_{mc}}{2\pi C_L}$$

The feed-forward compensation amplifier achieves 3–5 times wider bandwidth compared to Miller-compensated topologies with the same power consumption.

For 250 ps settling time for MDAC and <1/2 LSB settling error, the DC gain and GBW requirement for 1st stage amplifier are 51.7 dB and 9.3 GHz, and for 2nd stage amplifier are 33.7 dB and 5.3 GHz. Figure 3 shows the simulation results with the extracted layout at different PVT conditions. At 100°C, the transconductance of the differential input pair degrades, resulting in insufficient open-loop

gain and GBW. Under all other conditions, the two-stage amplifier meets or exceeds the designed open-loop gain and GBW specifications. The power consumption of the 1st stage amplifier and 2nd amplifier is 5.2 mW and 3.1 mW, respectively.

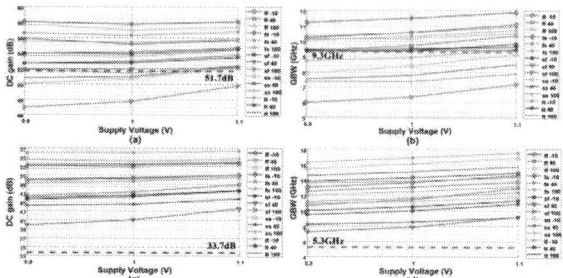

Fig. 3. Simulation result with extracted layout of (a) DC gain of the 1st stage amplifier, (b) GBW of the 1st stage amplifier, (c) DC gain of the 2nd stage amplifier, and (d) GBW of the 2nd stage amplifier with PVT variation.

B. Comparator

Dynamic comparators are favored by designers in high-speed ADCs due to their fast response and low power consumption. The two most common architectures are the Strong-Arm comparator and the double-tail comparator.

The Strong-Arm latch comparator offers advantages of simple structure, low power consumption, and fast response speed, but it imposes strict requirements on input common-mode voltage range and supply voltage, and its regeneration current is limited by the input common-mode voltage. This design employs three double-tail comparators as shown in Fig. 4. To ensure the comparator offset remains within the redundant range, varactor-based capacitor arrays are integrated into the output latch stages of both the first-stage and second-stage SAR comparators for offset calibration.

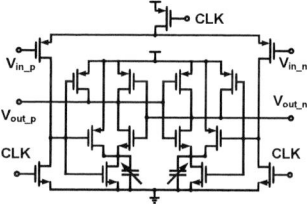

Fig. 4. Schematic of the double-tail comparator.

Figure 5 shows the simulated results of the comparators of the three stages. Due to inter-stage redundancy, only the final-stage comparator impacts the overall SNR. The noise from the first two stages merely needs to stay within the redundant range. With the third-stage comparator exhibiting 400 μV input-referred noise, and after two 3× MDACs gain suppression, the equivalent input-referred noise is 18 nV². For 1V input amplitude, this enables an SNR of 68 dB.

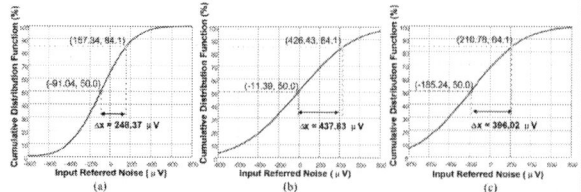

Fig. 5. Simulated results of input referred noise of comparators with extracted layout of (a) the 1st stage, (b) the 2nd stage, and (c) the 3rd stage.

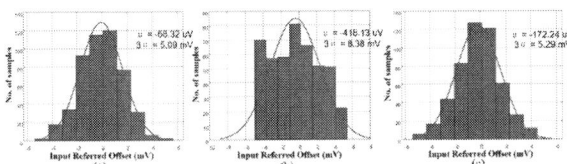

Fig. 6. 500-run Monte Carlo simulated comparator offset of (a) the 1st stage, (b) the 2nd stage, and (c) the 3rd stage.

Figure 6 presents 500-run Monte Carlo simulated results of the comparators of the three stages. The offset calibration range of the 1st and the 2nd comparators are 9 mV, covering 3-σ offset range.

C. Clock and timing distribution

A differential sinusoidal clock at two times the sampling frequency f_s is fed into an input buffer to generate near-square-wave signals CLKP and CLKN. These signals then pass through a frequency divider to produce four-phase clocks (CLK0, CLK1, CLK2, CLK3) at frequency f_s. By combining two 90° phase-shifted clocks via a NAND logic circuit, the final 25% duty cycle sampling clock CLKS is generated. Figure 8 shows the simulated sampling jitter with PVT variation. The sampling jitter is less than 45 fs at all PVT conditions, and SNR is >80 dB at 350 MHz input frequency, satisfying the 12-bit resolution requirement.

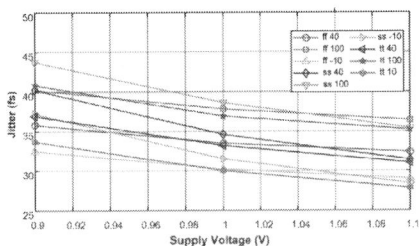

Fig. 7. Simulated sampling jitter with PVT variation.

The timing logic generators for the 1st MDAC and 2nd MDAC are identical as shown in Fig. 8. When the quantization of the 1st SAR is completed, the Ready1 signal is pulled high. CLK2 enables MDAC1 and the sampling switch of the second stage. After the amplification is completed, the sampling switch of the second stage is first turned off, followed by the reset of the 1st SAR logic and residue voltage. The residue voltage reset should occur after MDAC1 completes amplification and before the sampling of the next cycle of the first-stage SAR ADC. Therefore, the residue reset signal is generated by performing a logical NOR operation between the logic reset signal rstn and the first-stage sampling clock CLKS.

To achieve adjustable amplification time under different sampling rates, a 4-bit switched-capacitor-based adjustable delay chain is implemented, controlled by a thermometer code. Each control bit provides approximately 100 ps of delay.

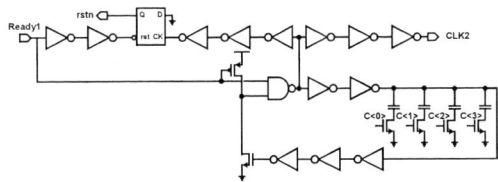

Fig. 8. MDAC asynchronous clock generator.

IV. MEASUREMENT

The prototype pipelined-SAR ADC is fabricated in 28-nm CMOS process, occupying 450 μm×130 μm core area as shown in Fig. 9.

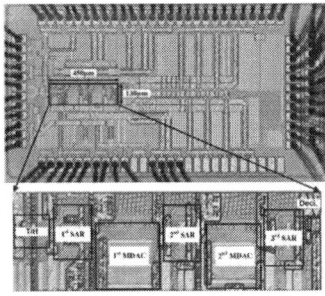

Fig. 9. Micrograph of the prototype pipelined SAR ADC

The prototype chip was measured according to the setup shown in Fig. 10 and the photograph is shown in Fig. 11. A voltage source, Keysight E36311A, provides a 5 V power supply to the PCB and the LDO on the power board. The clock signal is generated by the signal source SMA100B and connected to the clock interface of the PCB through the balun BAL0067. The input signal is provided by the signal source Ceyear 1465L, filtered by a bandpass filter to remove harmonics. The signal source and clock source are synchronized using a 10 MHz reference clock to avoid spectral leakage. The 14-bit digital codeword output of the Pipelined-SAR ADC is captured by the logic analyzer Agilent 16806A and subjected to spectral analysis on PC.

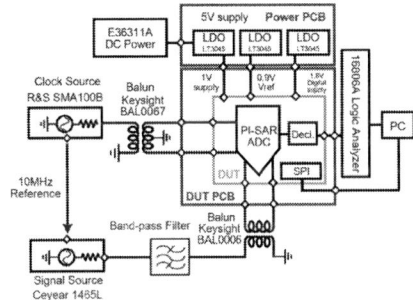

Fig. 10. Schematic of the measurement testbench and (b) the photograph of the testbench.

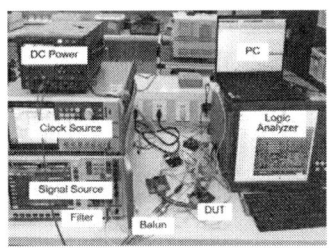

Fig. 11. The photograph of the testbench.

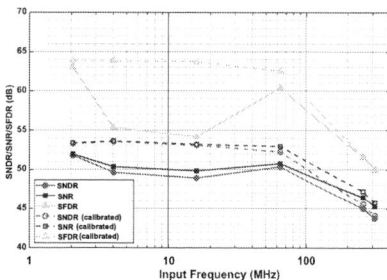

Fig. 12. Measured SNDR, SNR and SFDR versus input frequency at 620MS/s sampling rate and 1V full-scale input amplitude with and without bit-weight calibration.

Figure 12 shows the measured SNDR, SNR and SFDR versus input frequency at 620 MS/s sampling rate. One-time bit-weight calibration is applied to compensate the mismatch of the capacitor array of CDACs and gain variation of MDACs and SARs. When the input frequency is below 80 MHz, the SNDR remains at 53 dB and is primarily limited by the SNR. For 310 MHz Nyquist input frequency, SNDR drops to 44.2 dB and SFDR drops to 50 dB. This is caused by insufficient settling of the MDAC. The non-reset voltage of the second-stage CDAC affects the subsequent comparison, introducing input-dependent nonlinearity. The spectrum exhibits a series of high-order harmonics induced by nonlinearity as shown in Fig. 13.

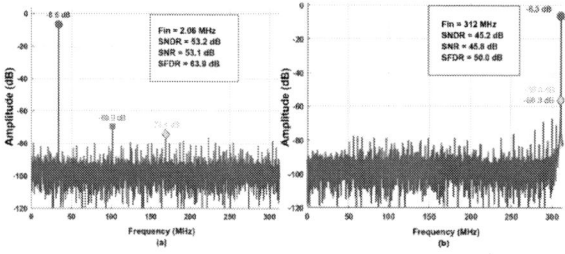

Fig. 13. Measured spectra at (a) 2 MHz input frequency and (b) 310MHz input frequency with 620 MS/s sampling rate and 1V full-scale input amplitude with bit-weight calibration.

The total power consumption is 20.6 mW at 620 MS/s sampling rate, with the power distribution of each component shown in Fig. 14. The sampling clock generation circuit consumes 2.4 mW, the reference voltage contributes 3 mW of power, the two residual amplifiers collectively consume 8.7 mW, and the remaining parts of the circuit—including digital logic, comparators, and bootstrap switch circuits—consume a total of 6.5 mW. Table I summarizes the measured performance of the Pipelined-SAR ADC chip designed in this work.

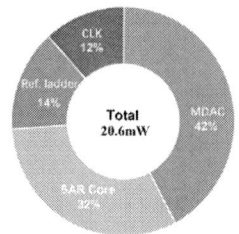

Fig. 14. Power breakdown at 620MS/s sampling rate.

TABLE I
PERFORMANCE SUMMARY

Process	28 nm
Resolution	12 bit
Sampling rate	620 MS/s
Architecture	Pipelined-SAR
Amplifier structure	Close-loop
Supply	1 V
SFDR @ low fin	63.9 dB
SFDR @ Nyq. fin	50.0 dB
SNDR @ low fin	53.1 dB
SNDR @ Nyq. fin	44.2 dB
Power	20.6 mW
Area	0.059 mm²
FoM$_W$	201 fJ/c-s
FoM$_S$	146 dB

V. CONCLUSION

This paper presents a 12-bit three-stage Pipelined-SAR ADC implemented in 28 nm CMOS technology. An AC-coupled feed-forward compensation amplifier is employed to achieve a high-gain, wide-bandwidth residue amplifier, reducing reliance on inter-stage calibration circuits. The chip occupies core area of 450 μm×130 μm. Measurement results demonstrate that at a sampling rate of 620 MS/s with a 2 MHz low-frequency input, the SNDR reaches 53.1 dB and SFDR reaches 63.9 dB. Under Nyquist frequency input, the SNDR is 44.2 dB and SFDR is 50 dB.

ACKNOWLEDGMENT

This paper is supported by the Joint R&D Fund of Beijing Smartchip Microelectronics Technology Co., Ltd （No.SGSCDT00MNMM2500121）. Shuai Liu and Yi Hu contributed equally to this work.

REFERENCES

[1] E. Martens, D. Dermit, M. Shrivas, S. Nagata, and J. Craninckx, "A compact 8-bit, 8 GS/s 8×TI SAR ADC in 16 nm with 45dB SNDR and 5 GHz ERBW," in Proc. Symp. VLSI Circuits, Jun. 2021, pp. 1–2.

[2] L. Wei et al., "A 12-bit 1GS/s ADC With Background Distortion and Split-ADC-Like Gain Calibration," in IEEE Transactions on Circuits and Systems I: Regular Papers, vol. 70, no. 12, pp. 4679-4691, Dec. 2023.

[3] W. Jiang, Y. Zhu, M. Zhang, C. -H. Chan and R. P. Martins, "A Temperature-Stabilized Single-Channel 1-GS/s 60-dB SNDR SAR-Assisted Pipelined ADC With Dynamic Gm-R-Based Amplifier," in IEEE Journal of Solid-State Circuits, vol. 55, no. 2, pp. 322-332, Feb. 2020.

[4] J. Lagos, B. Hershberg, E. Martens, P. Wambacq and J. Craninckx, "A Single-Channel, 600-MS/s, 12-b, Ringamp-Based Pipelined ADC in 28-nm CMOS," in IEEE Journal of Solid-State Circuits, vol. 54, no. 2, pp. 403-416, Feb. 2019.

[5] B. Hershberg, N. Markulić, J. Lagos, E. Martens, D. Dermit and J. Craninckx, "A 1-MS/s to 1-GS/s Ringamp-Based Pipelined ADC With Fully Dynamic Reference Regulation and Stochastic Scope-on-Chip Background Monitoring in 16 nm," in IEEE Journal of Solid-State Circuits, vol. 56, no. 4, pp. 1227-1240, April 2021.

A novel RA architecture and digital calibration method for SAR-assisted pipeline ADCs

Jieqiong Zeng[1,2], Hao Min[1]*

[1] College of Integrated Circuits and Micro-Nano Electronics, Fudan University, Shanghai, China
[2] CRM ICBG (wuxi) Co., ltd, Wuxi, China

* Email: hmin@fudan.edu.cn

Abstract—This paper presents a novel residue amplification (RA) architecture and a digital calibration method for SAR-assisted pipeline ADCs. The RA architecture achieves non-attenuated passive residue transfer and enables the two time-consuming tasks (the first-stage sampling and RA) to be executed in parallel rather than in series, thereby improving the overall conversion rate. Furthermore, the proposed digital calibration method effectively mitigates ADC precision degradation caused by three key factors: capacitor mismatch in the first-stage SAR ADC, inter-stage gain errors, and parasitic-induced errors. Theoretical analysis demonstrates that the proposed calibration improves the mean ENOB from 10.6 bits to 12.26 bits under conditions of 1% unit capacitor mismatch and 1% inter-stage gain error. Circuit simulations show an ENOB enhancement from 9.61 bits to 12.43 bits after calibration under the same conditions.

Keywords—residue amplification, digital calibration, SAR-assisted pipeline ADC

I. INTRODUCTION

The SAR-assisted pipeline ADC [1] has emerged as an increasingly popular hybrid architecture due to its potential to simultaneously provide high speed, high precision, and excellent power efficiency.

Fig. 1 depicts a typical architecture and timing sequence of the two-stage SAR-assisted pipeline ADC. The first-stage performs three serial phases: sampling, SAR conversion and residue amplification (RA), whereas the second-stage performs sampling during the RA phase and then executes its SAR conversion. The conversion times of the two SAR ADCs can be considered nearly equal, as the conventional balanced bit allocation between pipeline stages can help to achieve the shortest clock period. The sampling accuracy requirement inherently constrains the first-stage sampling time, necessitating a significantly extended duration for high precision requirements. This indicates that the first-stage sampling time, SAR conversion, and RA form the speed bottleneck. As shown in Fig. 1, the conversion cycle Tconv can be given by:

$$Tconv = Tra + Ts1 + Tc1 \qquad (1)$$

Where Tra denotes the RA time, Ts1 and Tc1 denote the sampling and conversion time of the first-stage ADC, respectively.

A perspective to alleviate the speed bottleneck is to remove the RA timing from the timing budget of the first-stage, so that the two time-consuming tasks: the first-stage sampling and RA, can be executed in parallel rather than in series. In [2], a transfer capacitor is employed to sample and hold the first-stage residue voltage. This allows the first-stage to be immediately released for sampling the next input, while the residue voltage retained on the transfer capacitor is amplified. However, the signal attenuation due to the charge sharing will dramatically impact the overall noise budget of the ADC.

Additionally, the performance of SAR-assisted pipeline ADCs is constrained by various non-ideal factors, primarily due to process variations and parasitic effects. Among these constraints, the capacitor mismatch of the first-stage SAR ADC and the gain error of the inter-stage emerge as the predominant error sources, necessitating calibration techniques to preserve linearity specifications.

In this paper, a SAR-assisted pipeline ADC with a novel RA architecture is proposed. This RA architecture achieves non-attenuated passive residue transfer while enabling first-stage release during RA operations. Consequently, the proposed architecture substantially improves the ADC's overall conversion rate and allows for the relaxation of the bandwidth requirements for the residue amplifier by allocating additional time for the RA process. Moreover, a digital calibration technique is presented to measure and compensate for both capacitor mismatch errors in the first-stage SAR ADC and inter-stage gain errors. This approach effectively mitigates performance degradation caused by three key factors: capacitor mismatch in the first-stage SAR ADC, inter-stage gain errors, and parasitic-induced errors, as validated by simulation results.

II. PROPOSED TWO-STAGE SAR-ASSISTED PIPELINE ADC

Fig. 2 illustrates the architecture and timing of the proposed SAR-assisted pipeline ADC consisting of a 7-bit first-stage SAR ADC and an 8-bit second-stage SAR ADC. The ADC achieves a 14-bit resolution after digital correction with one-bit inter-stage redundancy. The Vcm-based switching technique [3] is adopted to achieve good energy efficiency and good common-mode control.

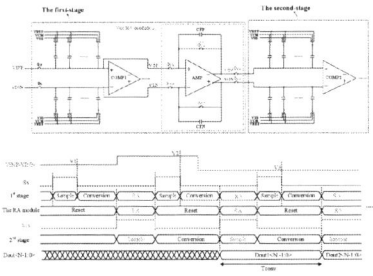

Fig. 1. A typical architecture of the two-stage SAR-assisted pipeline ADC and its operation timing.

979-8-3315-3918-4/25 $31.00 © 2025 IEEE

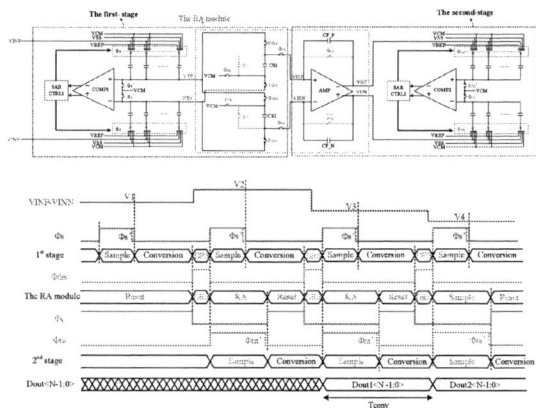

Fig. 2. The architecture and timing of the proposed SAR-assisted pipeline ADC.

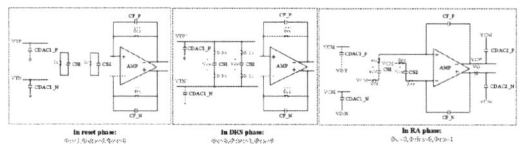

Fig. 3. The operational principle of the RA module proposed.

A. The proposed RA architecture

As shown in Fig. 2, a novel three-phase RA module is incorporated in the proposed ADC. This RA architecture inserts a differential residue sampling (DRS) phase during which two dedicated capacitors (CS1 and CS2) sample the residue voltage of the first-stage ADC. This enables the first-stage CDAC array to be released for the next sampling during the RA operation, meaning the two time-consuming tasks: the first-stage sampling and RA, can be executed in parallel rather than in series. So the speed bottleneck is now formed by the first-stage sampling time, SAR conversion, and DRS. The conversion cycle Tconv of the proposed ADC can be expressed as:

$$Tconv=Tdrs+Ts1+Tc1 \tag{2}$$

Where Tdrs denotes the DRS time, Ts1 and Tc1 are the same with (1). Compared to the conventional design shown in Fig. 1 and (1), this work achieves a significant reduction in the conversion cycle by replacing Tra with Tdrs, which is much shorter than the former. This improvement stems from the fact that the passive charge-sharing process employed by DRS is significantly faster than the time-consuming active amplification process performed in the RA operation.

Fig. 3 illustrates the operational principle of the three-phase RA module proposed. During the reset phase ($\Phi c=1$, $\Phi drs=0$, $\Phi ra=0$), the two plates of sampling capacitors (CS1 and CS2) are shorted together to fully discharge residual charges from the previous cycle. Following the completion of the first-stage SAR conversion, the residue voltage Vres (=VTN-VTP) is generated across the top plates of P/N arrays (CDAC1_P and CDAC1_N). Then, the module enters the DRS phase ($\Phi c=0$, $\Phi drs=1$, $\Phi ra=0$), where the switches controlled by Φdrs turn on, enabling a capacitive charge sharing process between CS1, CS2, CDAC1_P, and CDAC1_N.

The voltage stored across CS1 and CS2 can be derived as:

$$Vcs = \frac{-Vres*CDAC1}{CDAC1+4*CS} \tag{3}$$

Where CDAC1 denotes the total capacitance value of one side of the CDAC array (CDAC1=CDAC1_P=CDAC1_N), CS denotes the sampling capacitance value (CS=CS1=CS2).

In this design, $CS = \frac{1}{4}*CDAC1$ is implemented, simplifying the equation to:

$$Vcs =-\frac{1}{2}*Vres \tag{4}$$

Subsequently, the module enters the RA phase ($\Phi c=0$, $\Phi drs=0$, $\Phi ra=1$). During this phase, the voltages stored in CS1 and CS2 are stacked with the common-mode voltage Vcm and applied differentially to the amplifier input. It is worth noting that with the DRS operation and capacitor stacking, the differential input of the amplifier equals Vres without any signal loss. Simultaneously, through the feedback capacitors CF_P and CF_N, the residue voltage is amplified, and the second stage samples the amplified residue output (VOP-VON) through the capacitor arrays CDAC2_P and CDAC2_N, which act as the amplifier load. The amplified residue output can be expressed by:

$$VOP - VON = Vres * \frac{CS}{CF} \tag{5}$$

Where CF represents the feedback capacitance value (CF =CF_P=CF_N). In this work, the inter-stage gain is configured as 64× by setting CS/CF=64.

With the proposed three-phase RA architecture, the two time-consuming tasks: the first-stage sampling and RA, can be executed in parallel rather than in series. This proposed architectural innovation significantly improves the conversion rate of the ADC.

B. The proposed digital calibration method

The first-stage SAR ADC adopts a binary-scaled recombination redundant CDAC array architecture [4], as illustrated in Fig.4. The capacitor weighting ratios from C7P/N to C0P/N are configured as 48, 32, 24, 12, 4, 4, 2, 1, where the capacitor pair C0P/N is utilized to generate the residue voltage. A unit capacitor (Cu) is employed as the termination capacitor (CdP/N), which combines with the weighted capacitors to form a total capacitance of 128Cu per CDAC branch. During each conversion cycle, eight comparison operations are performed to generate an 8-bit redundant digital code D1r[7:0]. This code is then converted into a 7-bit binary output D1[6:0] via a dedicated redundant-to-binary conversion logic, according to the capacitor weighting ratios. This redundant architecture allows for relaxed comparator noise requirements and inherent error tolerance. However, a strict capacitor mismatch specification remains critical for maintaining the CDAC array's linearity performance, which directly affects the overall ADC accuracy. Furthermore, inter-stage gain errors can significantly degrade the linearity of the entire ADC. The study in [5] employs a startup calibration method that measures nonlinear code jumps in ADC output to correct capacitor mismatches in the CDAC. However, for SAR-assisted pipeline ADCs, since capacitor mismatch errors and inter-stage gain errors are correlated, mutual calibration between these error sources becomes essential to achieve accurate conversion results. To address these issues, a digital

979-8-3315-3918-4/25 $31.00 © 2025 IEEE

calibration method is proposed to measure and compensate for both capacitor mismatch errors in the first-stage SAR ADC and inter-stage gain errors.

The core concept involves utilizing the second-stage SAR ADC to quantify the mismatch information of each capacitor pair in the first-stage CDAC array during calibration. Each differential capacitor pair consists of P- and N-branch capacitors. For example, the MSB capacitor pair comprises C7P and C7N, each with an ideal capacitance value of 48Cu. Fig. 5 presents the calibration flowchart. During the dedicated test mode, the calibration sequence is triggered. First, the digital values corresponding to each capacitor pair in the first-stage CDAC array are acquired through the second-stage 8-bit ADC. These include Datad (associated with the termination capacitor pair CdP/N) and Data0~Data7 (corresponding to the capacitor pairs C0P/N~C7P/N, respectively).

The measurement process for the 8-bit data Datad corresponding to the termination capacitor pair CdP/N is detailed as follows: During the sampling phase of the first-stage ADC, connect CdP's bottom plate to VREF and CdN's bottom plate to VSS, while all remaining capacitors' bottom plates are connected to VCM. Simultaneously, connect all capacitor top plates to VCM. After sampling time, control the top plates of all capacitors disconnect from VCM, and the bottom plates of all capacitors connect to VCM. At this point, a differential voltage (VTPd-VTNd) across P/N arrays is generated. Subsequently, a standard conversion cycle is executed by the RA stage and the second-stage ADC to obtain the 8-bit output data Datad. For this case, the input differential voltage is equal to the reference voltage VREF multiplied by the average of the actual ratios of CdP and CdN. So, the final data Datad reflects the actual ratio of CdP and CdN.

The 8-bit data (Data0) corresponding to the capacitor pair C0P/N can be obtained through a similar operational procedure, which involves connecting the bottom plates of C0P/N to VREF/VSS (rather than CdP/CdN) during the sampling phase of the first-stage ADC. However, for the data Data1~Data7 corresponding to the capacitor pairs C1P/N~C7P/N, the process becomes more complex. This is because the differential voltage generated across the P/N arrays in these cases cannot be directly amplified by the RA stage, as this would cause an over-range condition. Therefore, in these cases, the first-stage ADC must perform a partial standard conversion cycle to mitigate the amplitude of the differential voltage before transmitting it to the RA stage, as illustrated in Fig.6. To prevent noise interference, each final data can be obtained by averaging multiple measurements.

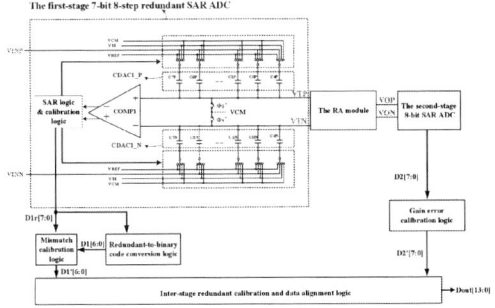

Fig. 4. Block diagram of the proposed two-stage SAR-assisted pipeline ADC incorporating calibration circuitry

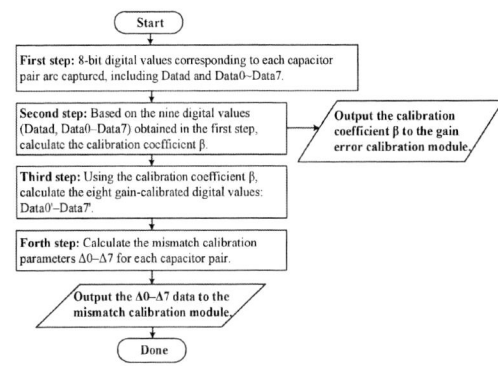

Fig. 5. The proposed calibration flowchart.

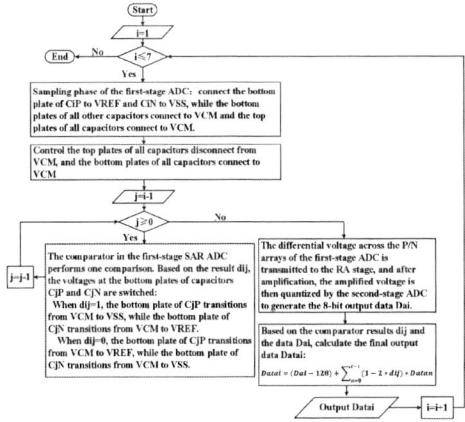

Fig. 6. The measurement flowchart for Data1~Data7.

With the nine data values (Datad, Data0~Data7) obtained in the first step, the calibration coefficient β is calculated as follows: First, compute the true ratio DIV0 of the capacitor pair C0P/N, expressed as:

$$DIV0 = \frac{Data0}{Datad + \sum_{i=0}^{7} Datai} \tag{6}$$

The actual differential voltage value for the residue amplifier during Data0 measurement can be derived as VTP0–VTN0=-DIV0*VREF. The theoretical value Data0' can be calculated for the ideal case where this differential voltage is amplified by -64 times and then quantized by the second-stage ADC. So, the true value of the inter-stage gain G can be derived as:

$$G = \frac{Data0 - 128}{Data0' - 128} * 64 = \beta * 64 \tag{7}$$

Where $\beta = \frac{Data0 - 128}{Data0' - 128}$ is defined as the calibration coefficient.

Using the calibration coefficient β, the eight gain-calibrated digital values (Data0'–Data7') can be calculated as follows:

$$Datai' = 128 + \frac{Datai - 128}{\beta}, \ i = 0 \sim 7 \tag{8}$$

Then, the mismatch compensation parameter Δi (i=0~7) for each capacitor pair is calculated as follows:

$$\Delta i = Datai' - DWAi * Da \tag{9}$$

Where DWAi is the ideal capacitor weighting ratio of CiP/N, and Da=$\frac{\sum_{i=0}^{7} \text{Datai}'}{128}$, for instance, DWA7=48.

The above calculated calibration coefficient β and mismatch compensation parameters Δ0–Δ7 are stored in system memory, then transmitted to the gain and mismatch calibration module, where they are utilized for real-time data calibration during normal operation. The calibration process proceeds as follows:

- Both D1r[7:0] and D1[6:0] obtained by the first-stage 7-bit 8-step redundant architecture SAR ADC are fed into the mismatch calibration module to compute the calibrated output D1'[6:0] , defined as: D1'[6:0] = D1[6:0] + Δ, where Δ=$\sum_{i=0}^{7} (2 * D1r[i] - 1) * \Delta i$.

- D2[7:0] obtained by the second-stage 8-bit SAR ADC is processed by a gain calibration module to produce D2'[7:0], calculated as:

  D2'[7:0] = 128 + (D2[7:0] - 128)/β.

- Finally, D1'[6:0] and D2'[7:0] are combined in the inter-stage redundancy calibration and data alignment module to generate the final 14-bit quantized output Dout [13:0].

Notably, the digital hardware overhead is small for this calibration, as most computations are performed off-chip.

III. SIMULATION RESULTS

The proposed 14-bit SAR-assisted pipeline ADC is first simulated in MATLAB. Fig. 7 displays the results of a 1000-run Monte Carlo simulation under 1% unit capacitor mismatch and 1% inter-stage gain error for both pre-calibration and post-calibration cases. It shows that the mean ENOB value improves from 10.6 bits to 12.26 bits after applying the proposed calibration technique. Furthermore, the prototype has been implemented in a 180nm BCD process for ultrasonic echo signal detection applications with a 4MSPS conversion rate and 5V power supply. The layout view is shown in Fig.8, and the circuit simulation results is presented in TABLE I. For the simulations without layout parasitics, the ENOB improves from 9.61 bits to 12.43 bits after calibration, indicating consistency with the MATLAB simulation outcomes. The result of 12.43 bits falls below the ideal case value of 13.31 bits, primarily due to truncation errors introduced during the digital implementation of the calibration algorithm. When layout parasitics are included, the results demonstrate that the ENOB degrades to 11.27 bits due to layout-induced parasitics in the CDAC array and RA module, even in the absence of artificially added mismatch and inter-stage errors. Through application of the proposed calibration technique, the ENOB improves to 12.53 bits, confirming the method's effectiveness against parasitic-induced errors.

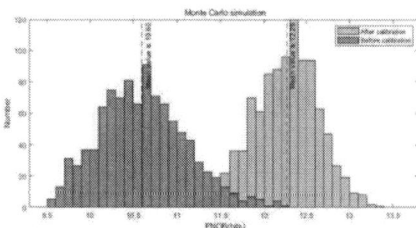

Fig. 7. The ENOB distribution before and after calibration.

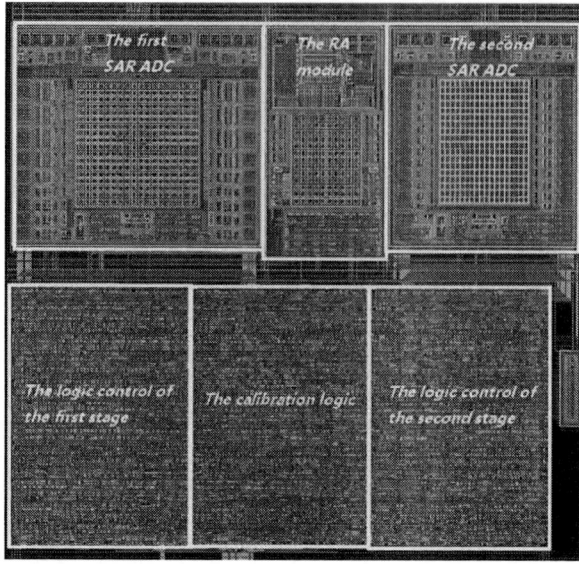

Fig. 8. The layout view of the proposed ADC.

TABLE I. THE CIRCUIT SIMULATION RESULTS

Simulation Condition			SFDR (dB)	SNDR (dB)	ENOB (bits)
Without layout parasitics	Ideal case, no mismatch and inter-stage error added		93.38	81.94	13.31
	Add 1% random mismatch and 1% inter-stage error	Before calibration	63.73	59.65	9.61
		After calibration	82.72	76.6	12.43
With layout parasitics	No mismatch and inter-stage error added	Before calibration	79.23	69.62	11.27
		After calibration	87.06	77.22	12.53

IV. CONCLUSION

In this work, a SAR-assisted pipeline ADC with a novel RA architecture and digital calibration is presented. The simulation results illustrate that the proposed calibration technique effectively mitigates overall ADC precision degradation caused by three key factors: capacitor mismatch in the first-stage SAR ADC , inter-stage gain errors, and parasitic-induced errors.

REFERENCES

[1] C. C. Lee and M. P. Flynn, "A SAR-assisted two-stage pipeline ADC," IEEE J. Solid-State Circuits, vol. 46, no. 4, pp. 859–869, Apr. 2011.

[2] H. Huang, H. Xu, B. Elies, and Y. Chiu, "A non-interleaved 12-b 330-MS/s pipelined-SAR ADC with PVT-stabilized dynamic amplifier achieving sub-1-dB SNDR variation," IEEE J. Solid-State Circuits, vol. 52, no. 12, pp. 3235–3247, Dec. 2017.

[3] Y. Zhu et al., "A 10-bit 100-MS/s reference-free SAR ADC in 90 nm CMOS," IEEE J. Solid-State Circuits, vol. 45, no. 6, pp. 1111–1121,Jun. 2010.

[4] C. C. Liu, C. H. Kuo, Y. Z. Lin. "A 10 bit 320MS/s low-cost SAR ADC for IEEE 802.11ac applications in 20 nm CMOS". IEEE J. Solid-State Circuits, vol. 50, no. 11, pp. 2645–2654,Nov. 2015.

[5] C. C. Lee, C. -Y. Lu, R. Narayanaswamy, et al. "A 12b 70MS/s SAR ADC with digital startup calibration in 14nm CMOS". 2015 Symposium on VLSI Circuits (VLSI Circuits), Kyoto, Japan, 2015, pp. C62-C63, doi: 10.1109/VLSIC.2015.7231328.

An 8-bit 0.4-mW 740-μm² DS Digital-to-Analog Converter in 28nm CMOS with 60.89-dBc SFDR

Xiongfeng Bi[1], Bingyi Ye *[2], Weixin Gai*[1,3]

[1] Peking University, Beijing, China
[2] East China Normal University, Shanghai, China
[3] Beijing Advanced Innovation Center for Integrated Circuits, Beijing, China

*byye@ic.ecnu.edu.cn, wgai@pku.edu.cn

Abstract—This paper presents an ultra-compact 8-bit DAC for receiver threshold generation that combines a 5-bit resistor-string DAC with a 3-bit ΔΣ modulator to achieve extreme area-efficiency. The hybrid architecture leverages monotonicity of the resistor-string DAC and noise-shaping characteristics of the ΔΣ modulator, eliminating bulky current sources while enhancing linearity. Measured results demonstrate 60.89-dBc Nyquist SFDR at 10-MS/s operation. The core occupies 740 μm² and consumes 0.4 mW at 0.9-V supply.

Keywords—DAC, delta-sigma modulator，digital-to-analog converter, receiver，segmented DAC

I. INTRODUCTION

Over the past decades, the explosive growth in network bandwidth demands driven by data centers, AI computing, and high-definition media transmission has continuously pushed the data rates of electrical and optical interconnect transceivers to exponentially higher levels. High-speed serial communication circuits usually employ mixed-signal PAM-4 receivers. These receivers process PAM-4 signals directly through analog circuits combined with mixed-signal processing, aiming to compensate for channel loss, eliminate intersymbol interference (ISI), and perform data decision in real time—all without relying on digital signal processing (DSP). In mixed-signal PAM-4 receivers, digital-to-analog converters (DACs) are usually employed to provide the precise threshold voltages required for data slicers and error slicers. Due to multi-channel interleaving combined with PAM4's requirement for multiple voltage level comparisons, and critically, the distinct offset voltages exhibited by individual comparators, these DACs simultaneously perform dual functions: threshold generation and calibration. This results in 112G/224G RX designs requiring 64-256 DACs. Consequently, even though each DAC consumes minimal power and occupies small area, the aggregate cost of the DAC array becomes non-negligible [1-3]. Fig. 1 illustrates the circuit architecture of a 224Gb/s extra-short reach (XSR) receiver and its integrated DAC array within it [1]. For threshold DACs, the core challenge lies in achieving sufficient precision within stringent area and power budgets. Conventional current-steering DACs [4-5] necessitate large bias currents to reduce relative noise and mitigate matching errors to meet precision requirements, leading to significant power overhead. Pure resistor-string DACs offer monotonicity but suffer from exponentially growing resistor network area at high resolutions and speed limitations [6-7].

To meet the design objectives of low power, small area, and low noise, this paper proposes a segmented 8-bit DAC architecture. The 5 most significant bits (MSBs) are implemented using a binary-weighted resistor string structure, while the 3 least significant bits (LSBs) are realized using a

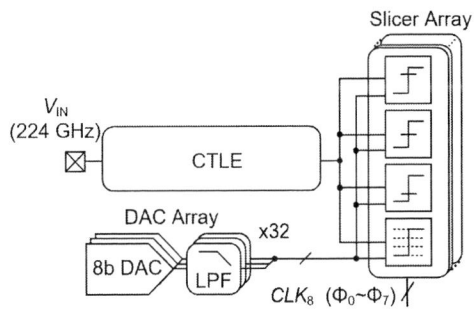

Fig. 1. A 224Gb/s XSR receiver architecture

first-order Delta-Sigma modulator (DSM). This combination harnesses the advantages of both techniques: 1) Area minimization: The resistor-string eliminates large current-source arrays and complex decoding logic, while the DSM requires minimal digital components to achieve high linearity conversion through oversampling and noise shaping. 2) *SFDR enhancement*: The DSM shapes the quantization noise of the LSBs to higher frequencies, significantly reducing in-band noise power within the signal baseband and thereby enhancing overall SFDR. Concurrently, the resistor-string architecture ensures inherent monotonicity and superior DNL performance for the MSBs. 3) *Reduced matching sensitivity*: The architecture minimizes dependence on precise current-source matching, instead leveraging resistor ratios and switching characteristics to enhance process robustness. This preference stems from fundamental physical advantages: resistor matching primarily depends on geometric dimensions, while current-source matching is compromised by variations in carrier mobility, threshold voltage, and oxide thickness—factors exhibiting 3 to 5 times stronger process sensitivity in nanoscale CMOS nodes.

Post-layout simulation results based on a 28nm CMOS process demonstrate that at 10 MS/s, the segmented DAC achieves an SFDR of 63.28 dBc for an input signal at 488 KHz, and a Nyquist SFDR of 60.89 dBc for a 4.99 MHz input signal. The proposed DAC occupies an area of 740 μm² and consumes 0.4 mW of power.

II. CIRCUIT ARCHITECTURE

The block diagram of the proposed DAC is illustrated in Fig. 2. In the figure, the 5 MSBs of the input control a 5-bit binary resistor-string DAC, while the 3 LSBs control the DSM. The 5-bit resistor-string DAC comprises a global resistor string and two 32-to-1 multiplexers (MUXs). The resistor string generates 32 reference voltages V_{REF} [31:0],

This work was supported in part by the National Key R&D Program of China (Grant No. 2022YFB2803301).

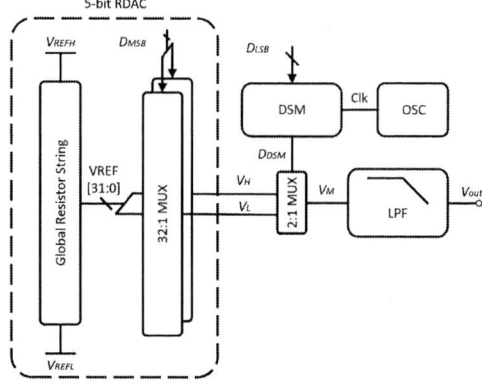

Fig. 2. Proposed segmented DAC architecture

which are selected by the two 32:1 MUXs to produce two adjacent voltages, V_H and V_L. An internal ring oscillator (OSC) generates a 5 GHz clock signal to oversample the DSM's input digital signal. The DSM output V_{DS} is a single-bit digital signal. The two output signals from the resistor-string DAC are connected to the data input ports of a 2:1 MUX. The single-bit digital output of the DSM (D_{DSM}) is connected to the select control port of this same 2:1 MUX. The output logic of the 2:1 MUX is defined as follows:

- When $D_{DSM} = 1$: V_H is selected as output voltage V_M.
- When $D_{DSM} = 0$: V_L is selected as output voltage V_M.

A passive low-pass filter (LPF) is applied to remove the high-frequency quantization noise and glitches from V_M, yielding the final output V_{out}. The expression of the final output voltage is:

$$V_{out} = V_{REFL} + \left[\frac{dec(D_{MSB})}{32} + \frac{dec(D_{LSB})}{256} + \varepsilon_{DSM}\right] * (V_{REFH} - V_{REFL}) \quad (1)$$

where N denotes the number of oversampling cycles, ε_{DSM} is the quantization error term and will be filtered out by the LPF.

III. CIRCUIT IMPLEMENTATION

A. Resistor-String DAC

Fig. 3 depicts the 5-bit resistor-string DAC. The resistor-string DAC is composed of a global resistor string and two 32:1 MUXs. The global resistor string generates 32 reference voltages between V_{REFL} and V_{REFH}. The two 32:1 MUXs select two adjacent voltages, V_H and V_L, from these 32 reference voltages based on the 5 MSBs of the DAC's digital input signal, satisfying the condition:

$$V_H - V_L = \frac{V_{REFH} - V_{REFL}}{2^5} = V_{LSB} \quad (2)$$

A single MUX requires 62 switch elements. The two MUXs share a single global resistor string, thereby reducing the overall area requirement.

B. Ring Oscillator

The internal sampling clock is generated by a 5-stage ring oscillator. The output frequency of the ring oscillator depends on the oversampling ratio (OSR) of the DSM. A key benefit of DSM is noise shaping, which pushes a significant portion of this quantization noise power from the baseband signal band into higher frequencies. The effectiveness of noise shaping and the resulting in-band quantization noise reduction is directly proportional to the OSR, defined as:

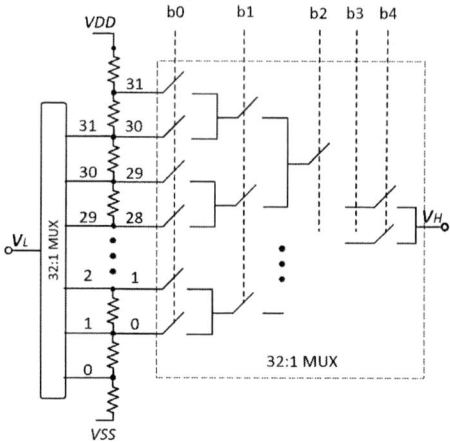

Fig. 3. Proposed 5-bit resistor-string DAC

$$OSR = \frac{f_s}{2 * f_B} \quad (3)$$

Where f_s is the sampling clock frequency and f_B is the signal bandwidth of interest (the Nyquist frequency of the entire 8-bit DAC). For this DAC operating at an overall conversion rate of 10 MS/s, the Nyquist frequency f_B is 5 MHz.

The theoretical SQNR improvement due to oversampling and noise shaping is approximately given by:

$$SQNR \approx 6.02 * N - 3.41 + (20 * L + 10) * log_{10}(OSR) \quad (4)$$

Where N is the number of bits in the quantizer (effectively 1-bit here, N=1), and L is the order of the modulator (L=1). While SFDR relates to spurious tones and SNR/SQNR to noise, achieving good SFDR at high frequencies also relies on sufficient quantization noise suppression within the signal band. Targeting an effective SQNR > 60 dB for the LSBs within the 5 MHz band necessitates a high OSR. An OSR of 128 would theoretically provide ~60 dB SQNR improvement for L=1. However, the implemented 5 GHz clock corresponds to an OSR of 500. This provides a substantial theoretical SQNR improvement margin, ensuring the quantization noise of the LSBs is suppressed far below the thermal and resistor noise floor, contributing significantly to the achieved high SFDR even at Nyquist input frequencies.

Two common methods exist to achieve the target frequency: One involves adding capacitors after each inverter stage and tuning the capacitance to adjust the output frequency. The other directly modifies the inverter sizes to obtain the target frequency. To minimize power consumption, this work employs the inverter sizing method to generate the 5 GHz clock signal.

C. Delta-Sigma Modulator

The first-order DSM employs a digitally implemented error-feedback structure (Fig. 4) for minimal area. Its operation is defined by two critical design aspects: The Σ adder acts as a unit-gain integrator (accumulator). This choice ensures unconditional stability for a first-order modulator while eliminating multiplier hardware. The register stores the accumulated value $S[n]$, updated as:

979-8-3315-3918-4/25 $31.00 © 2025 IEEE

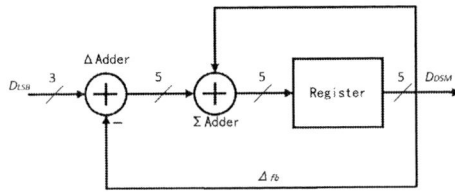

Fig. 4. Proposed Delta-Sigma Modulator

$$S[n] = S[n-1] + \Delta[n] \qquad (5)$$

where $\Delta[n]$ is the output of the Δ adder. The absence of scaling coefficients ($\alpha=1$) simplifies digital implementation but limits noise shaping optimization – a deliberate trade-off for area reduction.

The DSM's signal transfer function (STF) and noise transfer function (NTF) in the z-domain are:

$$STF(z) = z^{-1}, NTF(z) = 1 - z^{-1} \qquad (6)$$

This NTF exhibits a first-order high-pass characteristic for quantization noise. The magnitude response is:

$$|NTF(f)| = 2\left|\sin\left(\frac{\pi f}{f_s}\right)\right| \qquad (7)$$

where f_s = 5 GHz is the sampling frequency. Within the baseband ($f \ll f_s$), $NTF\ |\ NTF(f)\ |\approx 2\pi f/f\ s$, providing 9 dB/octave (or 30 dB/decade) of quantization noise attenuation. This suppresses LSB errors in the critical low-frequency region where receiver thresholds operate.

The 3-bit unsigned input D_{LSB} is converted to a signed 5-bit representation $X[n]$ by sign-extension and offset subtraction ($X[n] = \{0,0,D_{LSB}\} - 4$ in decimal), centering the input range around the integrator's zero. The feedback value Δ_{fb} is generated as:

$$\Delta_{fb} = \begin{cases} 0 & if\ D_{DSM} = 0 \\ -8 & if\ D_{DSM} = 1 \end{cases} \qquad (8)$$

This asymmetric feedback compensates for the input offset, ensuring the modulator's output duty cycle linearly represents the 3 LSBs. The Δ adder computes the error term:

$$\Delta[n] = X[n] - \Delta_{fb} \qquad (9)$$

driving the integrator. The DSM output D_{DSM} is the MSB of the accumulator $S[n]$, providing the 1-bit stream controlling the output MUX.

D. Low-Pass Filter

The proposed DAC employs a third-order passive RC filter to suppress high-frequency quantization noise introduced by the DSM. Identical RC values are adopted for all stages *(R1=R2=R3=R* and *C1=C2=C3=C)* to achieve the flattest possible passband response. The third-order topology increases the roll-off slope to 60 dB/dec and attenuate the out-of-band noise while relaxing the cutoff frequency specification. The LPF's cutoff frequency was determined through simulation experiments. Simulation results indicate that setting the filter cutoff frequency to 119 MHz achieves >60-dBc SFDR with compact area implementation. The magnitude-frequency response of the LPF is illustrated in Fig. 5.

To achieve reasonable area allocation, set R=3 kΩ and

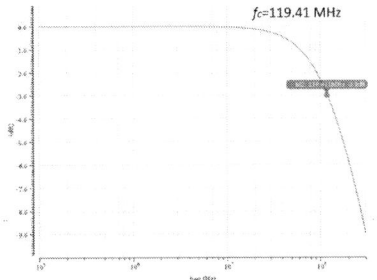

Fig. 5. Magnitude-frequency response of the proposed LPF

C=86 fF so that the resistor and capacitor occupy comparable area on the chip.

IV. SIMULATION RESULTS

Fig. 6 presents the layout photograph of the DAC integrated in 28nm CMOS technology, featuring a core area of 740 µm². Key modules are explicitly annotated in the layout.

Simulated output spectrum is shown in Fig. 7. Post-layout results confirm that at 10 MS/s, the segmented DAC achieves an SFDR of 63.28 dBc for a 488 KHz input signal, and a Nyquist SFDR of 60.89 dBc for a 4.99 MHz input signal, a total power consumption of 0.4 mW. The simulation results are summarized in the Table I along with comparison with other reported works.

V. CONCLUSION

This work implements an ultra-low-power hybrid 8-bit DAC for high-speed transceivers, combining a 5-bit resistor-string DAC (MSBs) with a 3-bit $\Delta\Sigma$ modulator (LSBs) in 28nm CMOS. The design achieves 740µm² core area and 0.4 mW. Integrated with a passive 3rd-order RC filter exhibiting zero static power, the DAC delivers 60.89 dBc Nyquist SFDR at 10 MS/s. This provides a low-area solution for low- power transceivers.

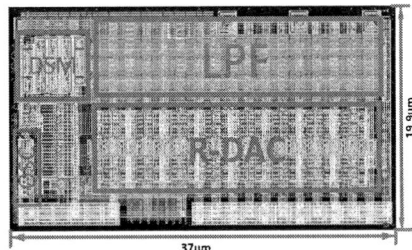

Fig. 6. Core layout of proposed DAC

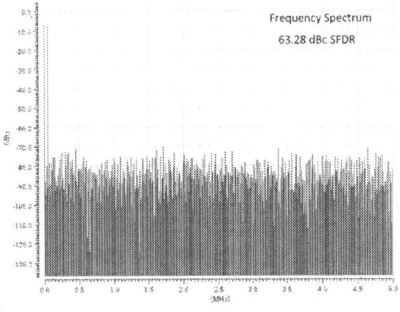

(a)

979-8-3315-3918-4/25 $31.00 © 2025 IEEE

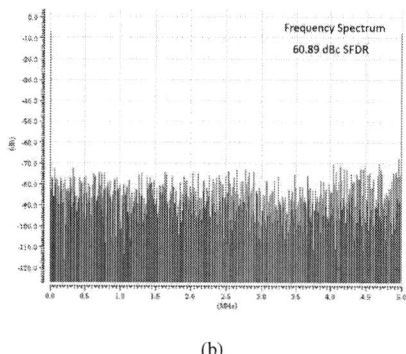

(b)

Fig. 7. (a) Output Spectrum and SFDR under Low-Frequency Input (b) Output Spectrum and Dynamic Range at Nyquist Frequency Input

TABLE I.   SIMULATED PERFORMANCE COMPARISON WITH PRIOR WORKS

Parameters	This Work	[8]	[9]
Architecture	Segmented DAC	C.DAC	R.DAC
CMOS Technology(nm)	28	28	180
Resolution(bits)	8	8	10
Power Supply(V)	0.9	0.9	1.8
Power Dissipation(mW)	0.4	0.85	0.54
SFDR at Nyquist Signal (dBc)	60.89	58	NA
Area(μm²)	740	NA	46000

REFERENCES

[1] R. Shivnaraine et al., "11.2 A 26.5625-to-106.25Gb/s XSR SerDes with 1.55pJ/b Efficiency in 7nm CMOS," 2021 IEEE International Solid-State Circuits Conference (ISSCC), San Francisco, CA, USA, 2021, pp. 181-183.

[2] C. F. Poon et al., "A 1.24-pJ/b 112-Gb/s (870 Gb/s/Mm) Transceiver for In-Package Links in 7-nm FinFET," in IEEE Journal of Solid-State Circuits, vol. 57, no. 4, pp. 1199-1210, April 2022

[3] B. Ye et al., "A 1.11pJ/b 224Gb/s XSR Receiver with Slice-Based CTLE and PI-Based Clock Generator in 12nm CMOS," 2025 IEEE International Solid-State Circuits Conference (ISSCC), San Francisco, CA, USA, 2025, pp. 140-142

[4] J. Bastos, A. M. Marques, M. S. J. Steyaert and W. Sansen, "A 12-bit intrinsic accuracy high-speed CMOS DAC," in IEEE Journal of Solid-State Circuits, vol. 33, no. 12, pp. 1959-1969, Dec. 1998.

[5] V. Kommangunta, K. Shehzad, D. Verma and K. -Y. Lee, "Low Power 10-bit 100 MSPS Segmented Current Steering DAC with > 78 dB SFDR," 2021 International Conference on Electronics, Information, and Communication (ICEIC), Jeju, Korea (South), 2021.

[6] H. -C. Seol, S. -K. Hong and O. -K. Kwon, "An Area-Efficient High-Resolution Resistor-String DAC with Reverse Ordering Scheme for Active Matrix Flat-Panel Display Data Driver ICs," in Journal of Display Technology, vol. 12, no. 8, pp. 828-834, Aug. 2016.

[7] C. -W. Lu, P. -Y. Yin, C. -M. Hsiao, M. -C. F. Chang and Y. -S. Lin, "A 10-bit Resistor-Floating-Resistor-String DAC (RFR-DAC) for High Color-Depth LCD Driver ICs," in IEEE Journal of Solid-State Circuits, vol. 47, no. 10, pp. 2454-2466, Oct. 2012.

[8] V. Kommangunta, K. Shehzad, D. Verma and K. -Y. Lee, "Low Power 10-bit 100 MSPS Segmented Current Steering DAC with > 78 dB SFDR," 2021 International Conference on Electronics, Information, and Communication (ICEIC), Jeju, Korea (South), 2021, pp. 1-3.

[9] C. -W. Lu, C. -C. Shen and W. -C. Chen, "An Area-Efficient Fully R-DAC-Based TFT-LCD Column Driver," in IEEE Transactions on Circuits and Systems I: Regular Papers, vol. 57, no. 10, pp. 2588-2601, Oct. 2010.

A K-Band CMOS Switched-Type Attenuator with Temperature Compensation Technique

Xiaodong Zhao *, Kai Zhang

Southwest China Institute of Electronic Technology, Chengdu, China

* Email: nanod@qq.com

Abstract—This paper presents a concise temperature compensation method for CMOS digitally controlled attenuators. By utilizing the negative temperature coefficient of diode junction resistance, a variable resistor with a negative temperature coefficient is constructed and applied to compensate for the attenuation errors caused by resistance changes in traditional switched-type attenuators under varying ambient temperatures. Based on this method, a K-band 4-bit temperature-compensated attenuator was designed using a 65nm CMOS process. Simulation results show that, under the temperature range of -40°C to 100°C, the RMS attenuation error of the attenuator is less than 0.18dB, and the RMS phase error is less than 1.6° within the frequency range of 15GHz to 25GHz. The RMS amplitude error is reduced by 0.17 dB compared to the attenuator without temperature compensation, achieving high-precision broadband attenuation with low phase error.

Keywords—broadband high-precision attenuation, CMOS digitally controlled attenuator, low phase error, temperature compensation

I. INTRODUCTION

With the rapid development of terrestrial and satellite mobile communications, phased array systems have seen unprecedented applications. The digitally controlled attenuators, as the amplitude control component of the phased array transceiver, serve to compensate for gain imbalance between array elements, or to perform amplitude weighting on the transmitted and received signals to reduce sidelobe level of the array [1], [2], thereby reducing interference and achieving optimal signal transmission. For this purpose, an amplitude control circuit with wide bandwidth, high precision, and a resolution of at least 0.5 dB is required [2]. The switched T-type structure is a good choice for implementing a circuit with such low attenuation, as shown in Fig. 1(a), where C_{comp} is the phase compensation capacitor. This structure is simple and compact, and it has minimal impact on the impedance of the signal path. Since the parallel resistor R_p is much larger than the series resistor R_s in the low-attenuation T-type circuit, the series resistor R_s and the series MOS transistor switch M_1 can be omitted, with negligible impact on port impedance matching [1]-[3]. Therefore, the 0.5 dB and 1 dB attenuation units can adopt a simplified T-type structure, as shown in Fig. 1(b). For attenuation units with attenuation values of 2 dB or larger, the Π-type structure is generally used, as shown in Fig. 1(c).

The attenuation of passive attenuators is sensitive to changes in ambient temperature. As the ambient temperature varies, the value of the resistor in the attenuator can change significantly, resulting in large attenuation errors. For example, in low-temperature environments, the parallel resistance R_p in T-type and Π-type structures decreases, leading to a larger attenuation compared to that at room temperature. Reported temperature compensation measures include: adding an extra

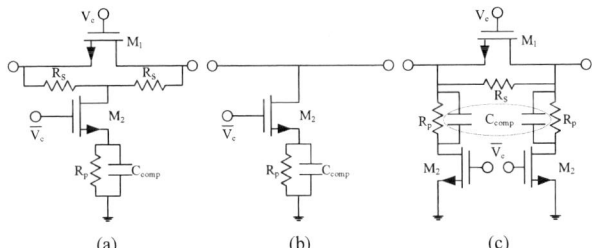

Fig. 1. Schematic of the (a) T-type, (b) simplified T-type and (c) Π-type attenuators.

unit with a smaller attenuation, such as a 0.25dB attenuation unit, which can be turned on or off via digital control when the attenuation error reaches a certain level [4]. The drawback of this approach is that it requires additional chip area and digital control bits, as well as more complex control logic. Moreover, the 0.25dB unit itself is also affected by temperature changes. Other temperature compensation measures include: using a proportional to absolute temperature (PTAT) current source to design a negative temperature coefficient temperature-dependent voltage source to control the ON-state resistance (R_{ON}) of MOSFETs for temperature compensation of attenuation [5]; or using a temperature-dependent voltage source to control the gate of an amplifier for gain compensation [6]. The problem with these circuits is that the design of the temperature-dependent voltage source is complex and requires additional auxiliary circuits such as PTAT current sources and operational amplifiers, and the DC power consumption is also significant [6].

Hangai et al. [7] proposed a simple temperature-dependent voltage control circuit that utilizes the negative temperature coefficient characteristic of the Schottky diode under forward bias condition. This diode is connected in series with a resistor to form a voltage divider, and the voltage across the diode is applied to the gate of the FET in the bypass resonator, achieving temperature compensation for both attenuation and phase. But the bandwidth of this compensation circuit is narrow, and the compensation effect is not significant either. When the ambient temperature changes by 75°C, the attenuation variation exceeds 10dB, and the phase variation exceeds 10 degrees.

This paper proposes a new temperature compensation circuit for passive attenuators, which consists of a CMOS on-chip diode connected in series with a polysilicon resistor, and a forward bias is applied to the diode to form a temperature-dependent voltage output circuit with a negative temperature coefficient. Unlike Hangai et al. [7], the nMOS FET controlled by the diode terminal voltage is part of the parallel resistor R_p of the simplified T-type circuit, rather than part of the bypass resonator. A 65-nm CMOS K-band digitally controlled passive attenuator is designed for validation purposes. Simulation results show that this 4-bit passive

979-8-3315-3918-4/25 $31.00 © 2025 IEEE

attenuator with a 7.5dB attenuation range and 0.5dB step size achieves high attenuation accuracy (i.e., <0.18 dB RMS amplitude error), low phase error (i.e., <1.6° RMS phase error), and wideband operation (15GHz~25GHz) within the ambient temperature range of -40°C to 100°C. Compared to the circuit without temperature compensation, both the RMS attenuation error and RMS phase error are significantly reduced.

II. CIRCUIT DESIGN AND VERIFICATION

A. Temperature compensation circuit design

The influence of ambient temperature changes on the simplified T-type attenuation structure is more significant. This is because in the Π-type structure, the series resistor R_s, also varies with temperature, can partially compensate for the attenuation error caused by the parallel resistor R_p. The 0.5dB and 1dB simplified T-type attenuation units are key components in the attenuator, determining the attenuation resolution of the circuit. Therefore, this paper proposes a temperature compensation circuit for the simplified T-type attenuation structure, as shown in Fig. 2. The diode and polysilicon resistor R_1 are connected in series to form a voltage divider circuit. When the control voltage V_c equals the supply voltage (e.g., 1V), the diode is forward-biased. Due to the negative temperature coefficient characteristic of its junction resistance, a temperature-dependent output voltage V_A is generated at point A. V_A controls the R_{ON} of the nMOS FET M_c through the bias resistor R_{bc} to achieve temperature compensation for the parallel resistor R_p, thereby compensating for the attenuation in high and low-temperature environments. The RF port is connected to the signal path, M_2 is the switch transistor, R_b is the gate bias resistor of M_2, and C_{de} is the bypass capacitor.

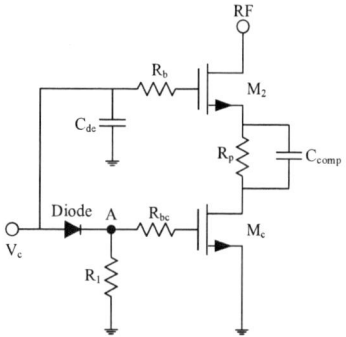

Fig. 2. The proposed temperature compensation circuit.

The component values for the 0.5dB and 1dB temperature compensation circuits are shown in Table 1. The values of the diode and R_1 are the same in both the 0.5dB and 1dB units, so the generated A_V are the same for both units. Fig. 3 shows the simulated curve of V_A varying with temperature in the range of -40°C to 100°C. It can be seen that V_A varies between 0.39V and 0.64V. To ensure the temperature compensation circuit functions within this voltage range, a low-threshold voltage transistor M_c is used. The size of M_c is related to the value of R_p. The 1dB unit has a smaller R_p, and correspondingly, its M_c has a larger size to obtain a smaller R_{ON} to matches the R_p. The simulated R_{ON} values of M_c for the two units are shown in Fig. 3. It can be seen that R_{ON} decreases with the increase in ambient temperature, which is opposite to the trend of R_p with temperature, thus providing temperature compensation for R_p. Since the resistance values of the diode and R_1 are very large, the maximum current generated by the 1V control voltage at

different ambient temperatures is less than 10uA, resulting in negligible power consumption.

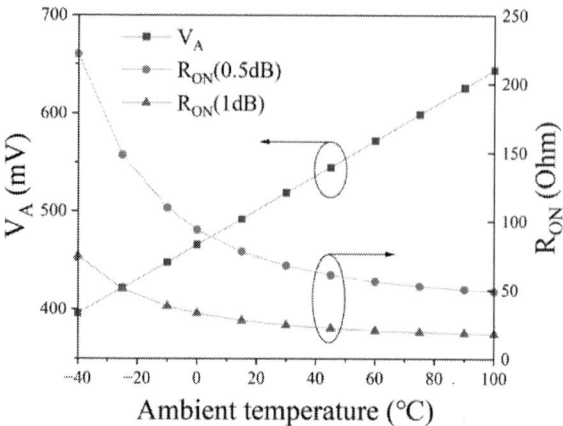

Fig. 3. The relationship curves of V_A and R_{ON} with respect to ambient temperature.

TABLE I. COMPONENT VALUES OF THE TEMPERATURE COMPENSATION CIRCUIT

Component	0.5dB unit	1dB unit
M_c(W/L)	12μm/60nm	32μm/60nm
Diode(W×L)	10μm×10μm	10μm×10μm
R_1(Ohm)	68.5K	68.5K
R_{bc}(Ohm)	10K	20K

B. Circuit Verification

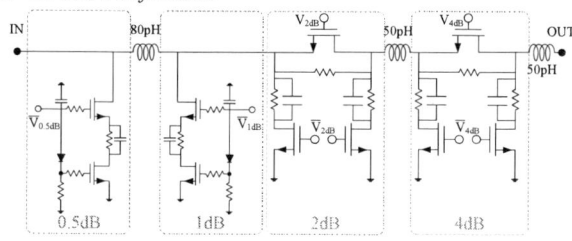

Attenuation	M₁(W/L)	M₂(W/L)	R_s(Ohm)	R_p(Ohm)	C_{comp}(fF)
0.5dB	/	6μm/60nm	/	308	0
1dB	/	6μm/60nm	/	129	5
2dB	20μm/60nm	5μm/60nm	16.5	270	19
4dB	20μm/60nm	5μm/60nm	40	141	19

(a)

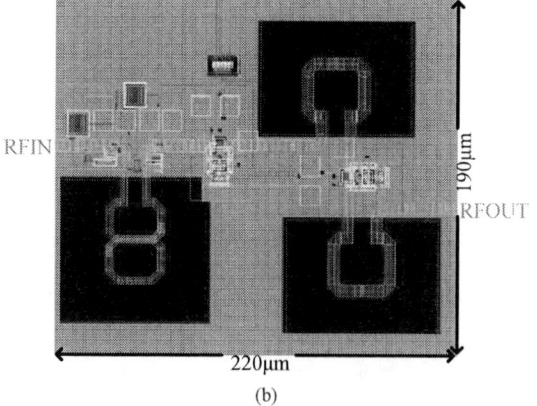

(b)

Fig. 4. (a) schematic and (b) layout of the 4-bit digital step attenuator

The complete schematic of the 4-bit digital step attenuator is shown in Fig. 4(a). The attenuation values of the four attenuation units are 0.5, 1, 2, and 4 dB, respectively, covering an amplitude adjustment range of 7.5 dB with a step of 0.5 dB. The 0.5 dB and 1 dB units use the aforementioned simplified T-type structure with temperature compensation, while the 2 dB and 4 dB units use the Π-type structure without temperature compensation measures. The series inductors between the units serve the purpose of impedance matching to obtain better input and output return loss. According to the methods introduced in [1] and [2], the values of the components in each unit can be calculated, as shown in the table in Fig. 4(a). Fig. 4(b) shows the circuit layout of the attenuator. Excluding the pads, the core area of the chip is only 0.042 mm². To accurately calculate the circuit's performance, full wave electromagnetic simulation was performed to extract the S-parameters of the passive structure of the complete layout, and parasitic parameters of circuit devices (i.e. resistors, FET transistors, etc.) were extracted for post-simulation. Fig. 5(a) shows the simulated S-parameter curves of the attenuator in the reference state at an ambient temperature of 45°C. Results show that the insertion loss is less than 2.3 dB in the range of 15 GHz to 25 GHz. Fig. 5(b) shows the simulated attenuation of all 16 attenuation states relative to the reference state at 45°C. Fig. 5(c) and 5(d) show the S_{11} and S_{22} curves of the 16 attenuation states at ambient temperatures of -40°C, 45°C, and 100°C, respectively. It can be seen that the S_{11} and S_{22} of the attenuator are both less than -14.5 dB, which meets the requirements of most phased array applications.

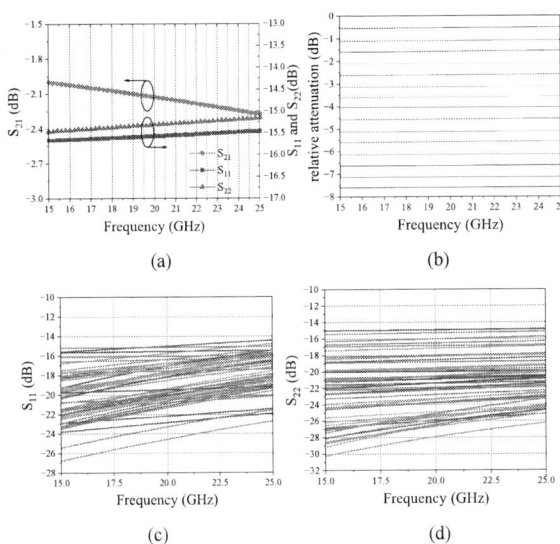

Fig. 5. Simulated (a)S-parameters at the reference state, (b) relative attenuation, (c) S_{11} and (d) S_{22} of the temperature compensated attenuator.

To verify the effectiveness of the temperature compensation circuit, a 4-bit attenuator without temperature compensation measures was also designed. Except for the removal of the temperature compensation components in the 0.5 dB and 1 dB units (their R_p and C_{comp} values were fine-tuned to ensure attenuation accuracy at 45°C), the rest of the circuit are identical to those of the temperature compensation attenuator. Fig. 6 shows the RMS attenuation error and phase error of the two attenuators in the frequency range of 15 GHz to 25 GHz, under ambient temperatures of -40°C, 45°C, and 100°C. The simulation results indicate that the temperature

compensation attenuator has an RMS amplitude error < 0.18 dB and an RMS phase error < 1.6°, while the attenuator without temperature compensation has an RMS amplitude error < 0.35 dB and an RMS phase error < 1.4°. The temperature compensation circuits reduce the RMS amplitude error of the attenuator by 0.17 dB, which is a significant accuracy improvement for an attenuator with a resolution of 0.5 dB.

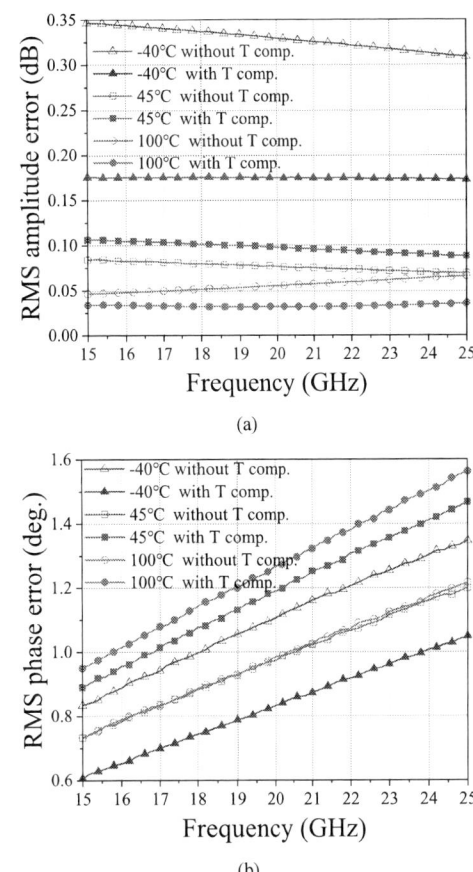

Fig. 6. Simulated (a) RMS amplitude error and (b) RMS phase error of the attenuators at different ambient temperatures.

Large-signal simulation results show that at 20 GHz, over the entire temperature range, the input 1 dB compression point (IP_{1dB}) of the temperature compensation attenuator is larger than 13 dBm. For the attenuator without temperature compensation, the IP_{1dB} is larger than 12.9 dBm, demonstrating that the temperature compensation circuit does not reduce the overall linearity of the circuit. Table II summarizes the performance of the proposed attenuator and compares it with previous designs. It can be seen that the attenuator proposed here occupies the smallest chip area and has low insertion loss and good impedance matching. Additionally, the temperature compensation attenuator exhibits low RMS amplitude and phase error in high and low-temperature environments, within the 15 GHz to 25 GHz range. The temperature compensation structure proposed here can also be applied to Π-type attenuation circuits.

III. CONCLUSION

Based on a 65 nm CMOS process, a high-precision, wideband switched-type passive attenuator with automatic temperature compensation is proposed. To address the large attenuation errors under high and low-temperature environments, a temperature compensation design is introduced to the simplified T-type circuit, effectively reducing the RMS amplitude error and enhancing the robustness of the circuit design. Post-simulation results show that within the ambient temperature range of -40°C to 100°C, the temperature compensation attenuator exhibits a low RMS amplitude error (<0.18 dB) and low RMS phase error (<1.6°) in the frequency range of 15 GHz to 25 GHz, showing a significant performance improvement compared to the attenuator without temperature compensation. Additionally, the core chip size is only 0.042 mm². Therefore, this attenuator is highly suitable for K-band phased array systems, and the temperature compensation design method presented here can be widely applied to switch-type passive attenuators.

TABLE II.   COMPARISON OF THE PRIOR-ART ATTENUATORS AND VGAS

Reference		This work*	[4]*	[5]	[6]
Technology		65-nm CMOS	40-nm CMOS	0.13-μm SiGe BiCMOS	40-nm CMOS
Topology		Switched T-/Π-type	Switched T-type	Switched T-/Π-type	2-stage VGA
Bandwidth (GHz)		15-25	27.5-31	19-24	6-15.3
Attenuation range/step (dB)		7.5/0.5	31.5/0.5	31.5/0.5	15.5/0.5
Insertion loss (dB)		2.3	8.76	N/A	-3.2~-1.7 ***
Return loss (dB)		>14.5	N/A	>15**	>10
RMS amplitude error (dB)	Room Temp.	<0.11	<0.11	N/A	<0.3
	High-low Temp.	<0.18	<0.33	0.5@−55°C~125°C	N/A
RMS phase error (deg.)	Room Temp.	<1.5	<1.4	N/A	<2.95
	High-low Temp.	<1.6	<4.8	4.1@−55°C~125°C	N/A
IP_{1dB} (dBm)		>13 @20GHz	15.35 @29.5GHz	N/A	>-1.07
P_{DC} (mW)		<0.01	0	N/A	18.4-29.6
Core chip area (mm²)		0.042	0.447	0.514	0.075

*: Simulated results.

**: Deduced from measurements photo.

***: Insertion Gain.

REFERENCES

[1] X. Li, "Low phase variation digital controlled attenuator with amplitude calibration function," Application of Electronic Technique, vol. 49(2), pp. 26-31, 2023.

[2] P. Gu, D. Zhao, and X. You, "A DC-50 GHz CMOS switched-type attenuator with capacitive compensation technique," IEEE Trans. Circuits Syst. I, Reg. Papers, vol. 67, no. 10, pp. 3389-3399, Oct. 2020.

[3] M. Cho, I. Song, Z. E. Fleetwood, and J. D. Cressler, "A SiGe-BiCMOS wideband active bidirectional digital step attenuator with bandwidth tuning and equalization," IEEE Trans. Microw. Theory Techn., vol. 66(8), pp. 3866-3876, Aug. 2018.

[4] T. Zhao, Q. Li, L. Xu and S. Wang, "A Ka-band 7-bit digital attenuator with temperature compensation technique in 40nm CMOS," 2021 International Conference on Microwave and Millimeter Wave Technology (ICMMT), Nanjing, China, May 2021, pp. 1-3.

[5] Y. Yuan, S. Mu, and Y. Guo, "6-bit step attenuators for phased array system with temperature compensation technique," IEEE Microw. Wireless Compon. Lett., vol. 28, no. 8, pp. 690–692, Aug. 2018.

[6] B. Chen, Z. Li, Z. Xia, Z. Fang and D. Zhou, "A 0.075-mm² 6-15.3 GHz active digital step attenuator with novel current-tuning topology for phased-array radar system," IEEE Trans. Circuits Syst., II, Exp. Briefs, vol. 71, no. 9, pp. 4116–4120, Sep. 2024.

[7] M. Hangai, H. Asao, M. Hieda, M. Yamaguchi, and M. Miyazaki, "Amplitude/phase temperature compensation attenuators with variable-FET resonators," IEEE Trans. Microw. Theory Techn., vol. 56(12), pp. 3058-3065, Dec. 2008.

An Ultra-Wideband 1.5–18.5 GHz MMIC Phase Shifter in 0.25-μm GaAs Technology

Bo Fu [1], Xuan Ding [2], Xuesong Han [3], Xiao Ding*[1], *Senior Member, IEEE*

[1] University of Electronic Science and Technology of China, China
[2] Georgia Institute of Technology, USA
[3] Chengdu Huaxing Dadi Technology Co., Ltd, China,

* Email: b.fu@std.uestc.edu.cn, xding@uestc.edu.cn

Abstract — This paper presents a six-bit digital phase shifter operating from 1.5 to 18.5 GHz (12.3:1 bandwidth) using a 0.25-μm GaAs pHEMT process. A magnetically coupled all-pass network (MCAPN) is proposed as the core topology, where scattering parameters and all-pass response conditions are derived through rigorous odd-even mode analysis. The MCAPN enables broadband return loss and insertion loss performance. Three types of switch network architectures are implemented: (1) Single-network switched-capacitor cells (for 5.625°, 11.25°, 22.5° bits), (2) Dual-network path-select switches (for 45°, 90° bits), and (3) A two-stage cascaded MCAPN structure for the 180° bit, trading area for enhanced bandwidth. The measurement results demonstrate the RMS phase error < 8.5°, insertion loss < 19 dB, and input/output return loss < –9 dB across the band. The proposed chip size is 4.1 × 1.6 mm².

Keywords—wideband, phase shifter，magnetic-coupled all-pass network（MCAPN）

I. INTRODUCTION

Phase shifters are indispensable components in advanced radio frequency systems, underpinning critical functionalities within phased array antennas, radar systems, microwave instrumentation, and wireless communications [1]. As core elements enabling beamforming and signal modulation, their performance fundamentally governs system-level characteristics, including dynamic range, response speed, and interference resilience.

Compared with analog phase shifters, digital phase shifters have obvious advantages in structural complexity and power consumption, etc. Traditional phase shifters mostly use silicon-based substrates. Silicon-based digital phase shifters have advantages such as low cost and easy integration. However, as the operating frequency increases, the high parasitic capacitance of MOS transistors will lead to increased insertion loss and limited phase accuracy of passive digital phase shifters[2][3]. Although silicon-based active vector phase shifters can achieve high gain and high-precision phase shifters, they introduce additional power consumption [4].

On the contrary, digital phase shifters based on GaAs pHEMT technology have demonstrated great potential in phased array antennas, satellite communications, and electronic countermeasure systems due to their advantages such as fast response speed and high phase control resolution. Traditional GaAs pHEMT passive digital phase shifters have various implementation methods, including PIN diode structure phase shifters, high and low pass structures, and reflective phase shifters. However, the bandwidth of phase shifters using the above structures is limited. Therefore, how to achieve low phase error, high linearity, and low power consumption circuit design within a wide frequency band is currently a research difficulty.

This study addresses the above challenges by proposing a six-bit digital phase shifter based on GaAs pHEMT technology. To expand the bandwidth, the phase shifter adopts MCAPN . Through the optimization of the circuit structure, a bandwidth of 170% has been successfully achieved. Section II provides a detailed analysis of the MCAPN phase shifter designed in this study. To verify the experimental structure, Section III presents the measured results of the ultra-wideband MCAPN phase shifter. Finally, in Section IV, we will draw a conclusion.

II. CIRCUIT DESIGN

The MCAPN has been proven to be applicable in the design of digital phase shifters [4]. The phase shifter adopting the MCAPN structure can achieve a wideband frequency response. Moreover, the MCAPN phase shifter can further expand the phase shift bandwidth through multi-stage design at the expense of circuit area and insertion loss. In Section II of this paper, the scattering parameters of the MCAPN are derived first by using the odd-even mode analysis method. Then, three types of topologies are analyzed respectively, and finally the overall phase shifter design structure is obtained.

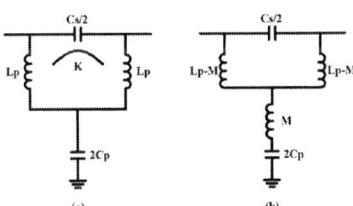

Fig. 1. (a)MCAPN   (b)Equivalent circuit

A. Analysis of Magnetic Coupling All-Pass Filter Network （MCAPN）

The topology and equivalent circuit of the MCAPN are shown in Fig. 1. In this network, a coupling exists between the two inductors. And the coupling coefficient is denoted by k, with its value ranging from -1 to 1. The absolute value of k indicates the coupling strength, and the sign represents the relative winding polarity of the two coils. The mutual inductance M between the two inductors can be expressed as:

$$M = kL_p \qquad (1)$$

Define the normalized angular frequency as follows:

$$\bar{\omega} = \frac{\omega}{\omega_T} \quad (2)$$

Where ω_T is the transition frequency.

Normalized reactance and normalized susceptance are expressed as:

$$x = \frac{\omega L_p}{Z_0} \quad (3)$$

$$b_s = \omega C_s Z_0 \quad (4)$$

$$b_p = \omega C_p Z_0 \quad (5)$$

Next, as shown in Fig. 2, the series and shunt branches of the network are divided into two, forming a symmetrical plane for odd-even mode analysis. The even-mode impedance and odd-mode admittance can be easily obtained:

$$Z_{even} = j\omega(L_p + M) + \frac{1}{j\omega C_p} \quad (6)$$

$$Y_{odd} = j\omega C_s + \frac{1}{j\omega(L_p - M)} \quad (7)$$

the reflection coefficients of odd and even modes were derived and expressed as follow:

$$\Gamma_{even} = \frac{Z_{even} - Z_0}{Z_{even} + Z_0} \quad (8)$$

$$\Gamma_{odd} = \frac{Y_{odd} - Y_0}{Y_{odd} + Y_0} \quad (9)$$

Ultimately, the S parameters of the MCAPN are derived:

$$S_{11} = \frac{1}{2}(\Gamma_{even} + \Gamma_{odd}) \quad (10)$$

$$S_{21} = \frac{1}{2}(\Gamma_{even} - \Gamma_{odd}) \quad (11)$$

Ideally:

$$S_{11} = 0 \quad (12)$$

Therefore, through observation and calculation, it can be calculated that:

$$b_s = (1+k)x \quad (13)$$

$$b_p = (1-k)x \quad (14)$$

The phase expression of the is obtained through calculation as:

$$\phi = -\pi - 2\tan^{-1}\left[(1+k)x - \frac{1}{(1+k)x}\right] \quad (15)$$

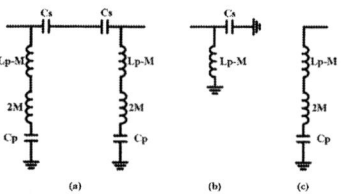

Fig. 2. (a) Schematic for even-odd mode analysis of MCAPN (b) odd-mode and (c) even-odd half-circuits.

B. Single Network With Internal Switched Capacitors

The block diagram of the topology is shown in Fig 3. As shown in Fig. 3, this structure typically adopts a design with a smaller phase shift, which has the advantages of lower loss and smaller circuit area occupation.

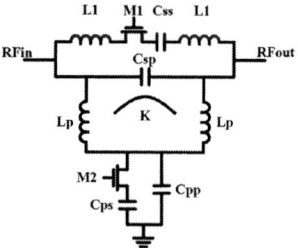

Fig. 3. Single Network With Internal Switched Capacitors of the phase shifter

The variable capacitor is composed of a switch tube and a fixed capacitor, as shown in Fig 4，By controlling the voltage to regulate the on and off states of the switch tube, the equivalent capacitance value of the entire circuit can be changed, thereby achieving the conversion of the phase shift circuit between the basic state and the phase shift state.

In this study, the 5.625°, 11.25° and 22.5° phase shift circuits are all designed using this structure. As shown in Fig. 3, by introducing high-Q value inductors on both sides of the series capacitor in the variable capacitor structure, the odd and even mode impedances of the circuit can be compensated, achieving frequency compensation for high-frequency phase and obtaining lower parasitic amplitude modulation across the entire frequency band.

C. Dual Networks With External Path-Select Switches.

The block diagram is shown in Fig 5，By controlling the single-pole double-throw (SPDT) switches, the signal path can be switched from one network to another. The two networks are both MCAPNs, but with different transition frequencies. The phase difference between the two branches is the phase shift of the circuit. Although this type of topology can achieve a large phase shift range, it is accompanied by significant insertion loss and occupies a large circuit area, so it is mainly suitable for 45° and 90° phase shift applications. As shown in Fig .5., by connecting capacitors in series at both ends of the phase shift port of MCAPN, the low-frequency working bandwidth can be effectively expanded and phase shift accuracy can be improved.

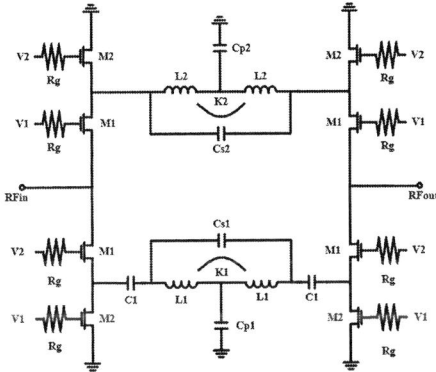

Fig. 4.   Dual Networks With External Path-Select Switches.

As shown in Fig. 6, in the 180° phase design, a series structure of two-stage magnetically coupled all-pass filter networks is adopted. The design effectively expands the bandwidth by appropriately increasing the circuit area, achieving a large phase shift characteristic under broadband matching.

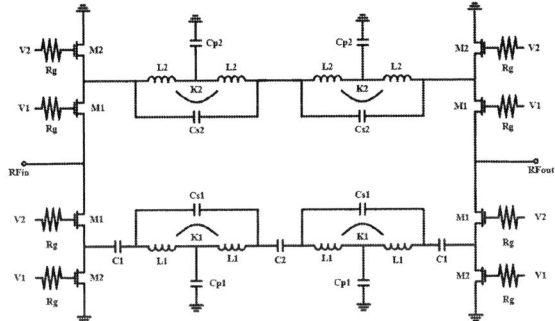

Fig. 5.   The design phase shifter structure of 180°

D. Design of a Six-Digital Phase Shifter

Fig 7 illustrates the overall architecture diagram of the six-bit digital phase shifter. The design adopts a topology structure with alternating large-phase units and small-phase units, effectively reducing phase shift errors. By configuring the 5.625° unit and the 45° unit with excellent standing wave characteristics at the input and output ports respectively, the phase shifter ensures superior return loss performance overall.

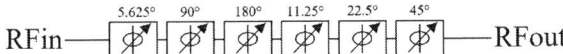

Fig. 6.   Block diagram of Phase Shifter

III. MEASUREMENT RESULTS

This research presents an ultra-wideband digital phase shifter designed based on GaAs 0.25um E/D mode pHEMT process，operating with a frequency range from 1.5GHz to 18.5GHz.The physical dimensions of the chip measure 4.1mm*1.6mm.As illustrated in Fig. 8 , the physical layout of the phase shifter is displayed. The control voltage of the device is 0V and -5V, and no additional power supply voltage is required.

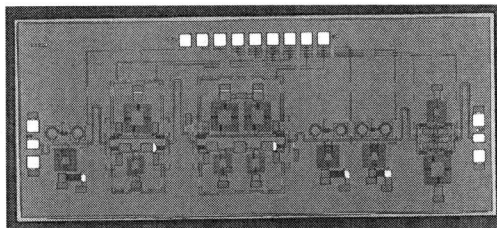

Fig. 7.   Physical layout of the phase shifter

Fig. 9 demonstrates the test date of the 6-bit digital phase shifter across 64 full-state phase shifter conditions. The test results in Fig. 10 indicate that the root mean square (RMS) phase error of the phase shifter is less than 8.5°. As shown in Fig. 11, through the measurement of insertion loss across 64 states, the full-state insertion loss of the phase shifter is consistently below 19dB. Furthermore, Fig. 12 and 13 illustrate the input and output return loss of the phase shifters are less than -9dB, respectively.

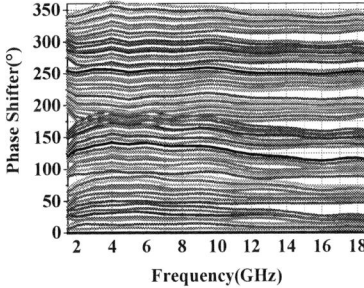

Fig. 8.   Phase Shift in all 64 States

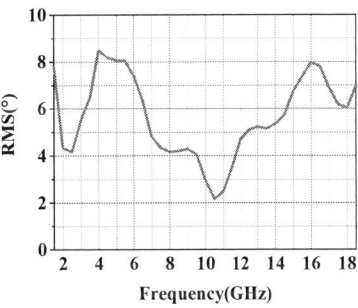

Fig. 9.   RMS phase error of the phase shifter

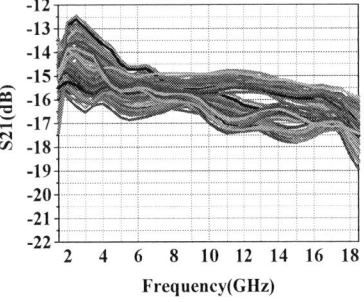

Fig. 10. Insertion loss in all 64 States

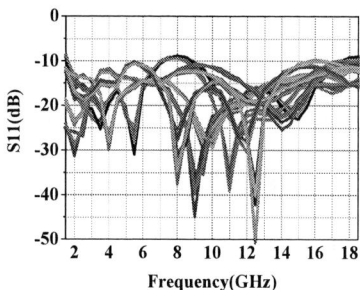

Fig. 11. Input return in all 64 States

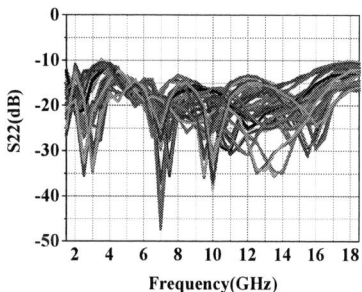

Fig. 12. Output return in all 64 States

TABLE I.   PERFORMANCE COMPARISON OF PHASE SHIFTERS

Ref	[6]	[7]	[8]	[9]	This work
Frequency (GHz)	6-18	4-19	12-18	5-20	1.5-18.5
Phase Resolution (GHz)	6	6	5	5	6
Insertion Loss(dB)	9	18	9.4	13.5	19
RMS(deg)	8	5	7	5	8.5
Return Loss(dB)	-10	-10	-8.8	-11	-9.5
Size(mm2)	2.9*1.87	5*3.5	4.27*3.17	4.2*2.98	4.1*1.6
Technology	GaAs pHEMT	GaAs pHEMT	GaAs pHEMT	GaAs pHEMT	GaAs pHEMT

Table 1 shows the comparison between the phase shifter in this paper and the state-of-the-art designs. Through the comparison, it can be found that the phase shifter designed in this paper has obvious advantages in bandwidth and area.

IV. CONCLUSION

This work demonstrates a novel six-bit digitally controlled phase shifter achieving 170% fractional bandwidth (1.5-18.5 GHz) through magnetic-coupled all-pass network (MCAPN) architectures in 0.25-μm GaAs pHEMT technology. The stratified design approach—implementing switched-capacitor cells for 5.625°-22.5° bits, path-select switches for 45°-90° bits, and cascaded dual-MCAPN for 180° bit—resolves fundamental bandwidth-area tradeoffs. Measurement results confirm breakthrough performance: 64-bits operation with <8.5° RMS phase error, <-9 dB return loss, and <19 dB insertion loss across the band.

ACKNOWLEDGMENT

This work was supported in part by the Shenzhen Science and Technology Program (Grant NO. KJZD20240903102730039).

REFERENCES.

[1] R. V. Garver, "Broadband Diode Phase Shifters," *1971 IEEE GMTT International Microwave Symposium Digest*, Washington, DC, USA, 1971, pp. 178-179, doi: 10.1109/GMTT.1971.1122955.

[2] M. Meghdadi, M. Azizi, M. Kiani, A. Medi and M. Atarodi, "A 6-Bit CMOS Phase Shifter for S-Band," in *IEEE Transactions on Microwave Theory and Techniques*, vol. 58, no. 12, pp. 3519-3526, Dec. 2010, doi: 10.1109/TMTT.2010.2086310.

[3] Z. Duan, Y. Wang, W. Lv, Y. Dai and F. Lin, "A 6-bit CMOS Active Phase Shifter for Ku-Band Phased Arrays," in *IEEE Microwave and Wireless Components Letters*, vol. 28, no. 7, pp. 615-617, July 2018, doi: 10.1109/LMWC.2018.2837885.

[4] D. Wei *et al.*, "Analysis and Design of a 35-GHz Hybrid π-Network High-Gain Phase Shifter With 360° Continuous Phase Shifting," in *IEEE Access*, vol. 9, pp. 11943-11953, 2021, doi: 10.1109/ACCESS.2021.3051246

[5] H. -Y. Li and J. -S. Fu, "Analysis of Magnetically Coupled All-Pass Network for Phase-Shifter Design," in *IEEE Transactions on Microwave Theory and Techniques*, vol. 62, no. 9, pp. 2025-2037, Sept. 2014, doi: 10.1109/TMTT.2014.2334065

[6] C. Wang, Y. Jiang, H. Ji, H. Yin, X. Wu and Y. Liu, "A Wideband 6-bit Phase Shifter with Low Insertion Loss," 2024 6th International Conference on Circuits and Systems (ICCS), Chengdu, China, 2024, pp. 279-283, doi: 10.1109/ICCS62517.2024.10846514.

[7] C. Shireesha, K. Y. Varma, P. Mohan and N. Simplice Rufin, "Broadband Monolithic 6-Bit Digital Phase Shifter," *2019 IEEE MTT-S International Microwave and RF Conference (IMARC)*, Mumbai, India, 2019, pp. 1-4, doi: 10.1109/IMaRC45935.2019.9118653

[8] K. Miyaguchi et al., "A 6-18 GHz 5-Bit Phase Shifter MMIC Using Series/Parallel LC Circuit," 2002 32nd European Microwave Conference, Milan, Italy, 2002, pp. 1-4, doi: 10.1109/EUMA.2002.339224.

[9] Dai Yongsheng *et al.*, "A novel multi-octave five-bit monolithic phase shifter," *ICMMT 2000. 2000 2nd International Conference on Microwave and Millimeter Wave Technology Proceedings (Cat. No.00EX364)*, Beijing, China, 2000, pp. 215-218, doi: 10.1109/ICMMT.2000.895660.

979-8-3315-3918-4/25 $31.00 © 2025 IEEE

A 0.2–7.3-GHz Compact LNA with Super Linearity for 5G NR in 22-nm CMOS Technology

Kaiyun Deng*[1], Zan Zhou [1], Yingqi Liu [1], Haoyu Dong [1] and Haigang Feng*[1]

[1] Shenzhen International Graduate School, Tsinghua University, Shenzhen 518055, China.

* Email: dkj23@mails.tsinghua.edu.cn, feng.haigang@sz.tsinghua.edu.cn

Abstract—This paper presents a broadband low-noise amplifier (LNA) with 6.7dBm third-order input intercept point (IIP3), 2.83 dB noise figure (NF) and 5.5mW power consumption. The LNA is suitable for 5G New Radio (NR) applications. To address the issue of low IIP3 caused by the reduced supply voltage in advanced processes, the derivative superposition (DS) technique is employed to enhance the IIP3. Simultaneously, the noise canceling (NC) method is applied to mitigate the trade-off between NF and wideband input impedance matching. The LNA is designed under the 22-nm CMOS process. Post-simulation results indicate that the proposed LNA achieves a peak gain of 13.3 dB and a -3dB bandwidth of 0.2-7.3 GHz. The FoM reaches 85.57 with die area of 0.0016 mm^2.

Keywords—CMOS LNA, broadband, current reuse, noise canceling, linearity improvement

I. INTRODUCTION

As the foundation of next-generation mobile systems, 5G wireless communication demands ultra-high capacity, data throughput, and spectral efficiency [1]. These requirements necessitate highly flexible and reconfigurable RF front-end architectures capable of supporting multiple frequency bands and communication standards. In this context, the design of broadband low-noise amplifiers (LNAs) has become a key enabler for software-defined radios (SDRs) and multiband receivers, which are essential for modern wireless platforms. 5G New Radio (NR), standardized by 3GPP, defines two frequency ranges, with Frequency Range 1 (FR1) extending the conventional sub-6 GHz spectrum up to 7125 MHz. This expanded frequency coverage imposes stringent requirements on the LNA in terms of bandwidth, gain flatness, noise figure, and linearity. Moreover, since the LNA directly influences the receiver's sensitivity and overall power consumption, its performance is critical to meeting 5G NR system-level specifications. As a result, broadband LNA design has attracted substantial attention in both academia and industry, as it plays a pivotal role in enabling scalable, power-efficient, and standard-compliant 5G receiver architectures.

Conventional approaches to extending the bandwidth of LNA typically rely on inductive peaking techniques, which require the integration of bulky passive inductors, leading to increased chip area and reduced design flexibility [2]–[7]. With the scaling of CMOS technologies, however, the intrinsic parasitic capacitance of MOS transistors has significantly decreased, enabling the realization of wideband LNAs without the need for inductors. This inductor-less design paradigm is particularly advantageous for broadband and multi-standard receivers, where compact area and integration are critical. Nonetheless, modern advanced processes also impose new challenges: the reduced supply voltage inherently limits the available voltage headroom,

which degrades the linearity of the LNA and consequently constrains the overall dynamic range of voltage-mode receiver front-ends. Previously reported LNAs in advanced processes have shown linearity below -2 dBm [3]–[5], [8]–[11], underscoring the need for new design strategies that can maintain both wideband operation and high linearity under low-voltage constraints. Moreover, noise cancellation (NC) techniques are widely adopted in broadband LNA designs to suppress the excessive noise contributed by the input matching transistors [3]–[5], [7]–[11].

In order to solve the problem of poor linearity of LNA under low power supply voltage in advanced processes, this paper designs a broadband LNA based on NC complementary pMOS-nMOS configurations. The LNA not only has excellent linearity in the entire 5G NR FR1, but also has a good trade-off in noise figure (NF), gain, and power consumption. The structure of this paper is organized as follows. Section II introduces the design of the proposed broadband LNA. Section III discusses the post-layout simulation results. Finally, section IV draws the conclusions.

II. PROPOSED BROADBAND LNA

The proposed LNA is shown in Fig. 1. Unlike the design in [4], which employs a voltage-shunt negative feedback resistor (R_F) to improve input matching and linearity, the R_F is intentionally removed in this work to eliminate the associated noise and stability concerns. The LNA covers overall 5G NR bands without inductors, with complementary pMOS-nMOS configurations in each stage to enhance linearity and reduce power consumption. In addition, the NC structure is applied to mitigate the trade-off between NF and wideband input impedance matching of the input common gate (CG) transistors. The common source (CS) NC stage includes an additional bias and is isolated by an ac coupling capacitor.

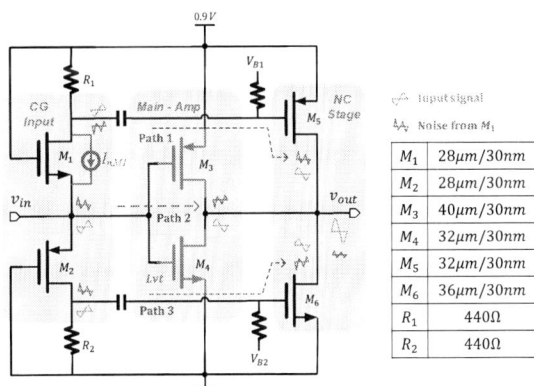

Fig. 1. Schematic of the proposed LNA.

979-8-3315-3918-4/25 $31.00 © 2025 IEEE

A. Wideband Impedance Matching

With small parasitic capacitance in advanced processes, the LNA can achieve direct matching via the CG stage and load resistors R_1 and R_2 without voltage shunt negative feedback resistors, and high linearity can be further achieved using linearization techniques. The input impedance of this LNA is:

$$Z_{in} = \frac{R_1 + r_{o1}}{(1 + g_{m1} + g_{mb1})r_{o1}} // \frac{R_2 + r_{o2}}{(1 + g_{m2} + g_{mb2})r_{o2}} // \frac{1}{sC_X} \quad (1)$$

The output resistors r_{o1}, r_{o2} of M_1, M_2 and the parasitic capacitance C_X at the input node play key roles in impedance matching. The input impedance matching remains stable throughout the entire 5G NR FR1 without using inductors, thanks to advanced processes with higher eigen-frequency f_T, and small C_X ensures good high-frequency matching. The LNA's output impedance, determined by four parallel drain impedances, offers flexibility for adjustment. For instance, [1] uses this design as a low-noise transconductance amplifiers in current-mode receivers, achieving higher output impedance and improved linearity by tuning M_3~M_6. Therefore, through appropriate design, the proposed LNA supports impedance matching and efficient 50 Ω load driving without requiring an additional buffer stage. This not only simplifies the signal path and reduces power consumption, but also minimizes noise contribution and potential bandwidth limitations associated with conventional output buffers.

B. Analysis of Triple-Path Noise Canceling

Fig. 1 illustrates how the proposed LNA eliminates transistor channel thermal noise through three signal paths. Taking the thermal noise of M_1 as an example, the noise generates two opposite-phase voltages at its source and drain. Assuming the source noise is positive-phase and the drain noise is negative-phase, the thermal noise passes through Path 1 and produces positive-phase noise at the output. Meanwhile, the source noise of M_1 is processed by main amplifier and M_2&M_6 auxiliary path, which produces negative-phase noise at the output. Unlike traditional CG noise cancellation, the thermal noise in Path 1 is canceled by Paths 2 and 3. The following discussion presents a calculation-based explanation. Consider only the thermal current noise generated by M_1. According to the noise cancellation theory, the ratio of the resulting noise voltage at the source and drain of M_1 is given by

$$\frac{V_{n,D}}{V_{n,S}} = \frac{R_1}{R_S'} \quad (2)$$

Let R_S be the impedance looking into the source of M_1. If the transconductances of M_1 and M_2 are approximately equal, and their parallel combination matches the source impedance $1/R_S$, then the following condition holds:

$$\frac{V_{n,D}}{V_{n,S}} = \frac{R_1}{R_S} \approx \frac{R_1}{R_S //(1/g_{m1})} = 3g_{m1}R_1 \quad (3)$$

Without incorporating any auxiliary cancellation branches, and assuming the transconductances of M_3 and M_4, as well as M_5 and M_6, are equal respectively, the condition for noise cancellation becomes:

$$V_{n,D} \cdot g_{m5} = V_{n,S} \cdot 2g_{m3} \quad (4)$$

$$3g_{m1}R_1g_{m5} = 2g_{m3} \quad (5)$$

However, when auxiliary cancellation branches are introduced, and assuming $g_{m1}R_1 = g_{m2}R_2$, the new noise cancellation condition is given by:

$$3g_{m1}R_1g_{m5} = 2g_{m3} + g_{m2}R_2g_{m6} \quad (7)$$

$$2g_{m1}R_1g_{m5} = 2g_{m3} \quad (7)$$

By comparing equations (5) and (7), it is evident that the inclusion of auxiliary branches reduces the required g_{m3} by approximately 30%, which lowers the overall current consumption. This also allows for smaller device sizing in the common-source stage, resulting in improved impedance matching and enhanced high-frequency performance.

This reduces the current requirement of the main amplifier stage, lowering power consumption. Additionally, it reduces the size of the CG stage, improving high-frequency matching. The NC mechanism applies similarly to other transistors. On the other hand, the useful signals passing through these three paths are amplified with the same polarity and superposed at the output. Fig. 2 shows the noise contribution of each component. It can be seen that the noise from M_1 and M_2 is only 0.46%, which is relatively small due to the high NC ratio of triple-path structure. And this is consistent with the above analysis.

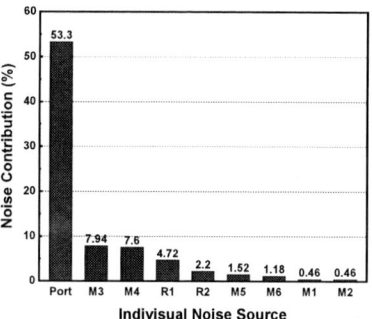

Fig. 2. Simulated relative noise contributions at 2 GHz.

C. Derivative superposition Linearization

Operating under a low supply voltage of 0.9 V inherently limits the voltage headroom and makes achieving high linearity particularly challenging. To mitigate this limitation, the proposed LNA adopts the derivative superposition (DS) technique, which effectively suppresses nonlinearity by engineering the transconductance (g_m) characteristics of the input stage [12]. Specifically, as demonstrated in [3] and [4], this technique utilizes carefully biased complementary pMOS–nMOS transistor pairs to cancel the first- and second-order derivatives of the transconductance, thereby minimizing third-order intermodulation distortion. In the proposed design, optimized biasing ensures that the nonlinear components of the overall transconductance curve are suppressed around the operating point, significantly improving linearity without incurring additional power or area overhead. This makes the DS method particularly well-suited for low-voltage broadband LNAs targeting modern wireless systems where both linearity and power efficiency are critical. The input CG transistors M_1 and M_2 generate nonlinear currents, but their gain is low, contributing mainly to g_m for input matching rather than linearity. In contrast, the complementary output pair M_5 and M_6 dominate the LNA's linearity. Their transconductance and nonlinear components, modeled as a nonlinear current source

979-8-3315-3918-4/25 $31.00 © 2025 IEEE

using Volterra series expansion, determine the total output current. Considering third-order nonlinear distortions, the total output current of the M_5 & M_6 pairs is equal to

$$i_{total} = i_{dsN} + i_{dsP} \tag{7}$$
$$= (g_{m6} + g_{m5})v_{gs} + (g'_{m6} - g'_{m5})v^2_{gs} + (g''_{m6} + g''_{m5})v^3_{gs}$$

It is obvious that the total transconductance (g_{m56}), subtraction of first-order (g'_{m56}), and sum of second-order derivatives (g''_{m56}) determine the output current and linearity through DS. Since the ac input signal for NMOS/PMOS are out of phase, g_{m56} increases in the triple-path NC structure, while g'_{m56} and g''_{m56} decrease at optimal bias. Fig. 3 shows g'_{m56} and g''_{m56} varying with gate bias voltage ($V_{B1} = V_{B2} = V_B$). At specific bias points, g'_{m56} and g''_{m56} drop to nearly zero. This demonstrates effective cancellation of second- and third-order nonlinear currents, aligning with theoretical analysis. By reducing nonlinear currents mixed with inputs via C_{gd} feedback, both second-order input intercept point (IIP2) and third-order input intercept point (IIP3) can significantly improve, with IIP3 peaking at the optimal bias. Consequently, this DS technique solves LNA linearity challenges under advanced processes, enabling high linearity at a low supply voltage.

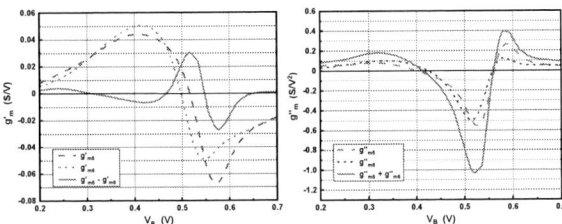

Fig. 3. (a) First-order derivatives of the g_m of the output NC stage. (b)Second-order derivatives of the g_m of the output NC stage.

III. POST-LAYOUT SIMULATION RESULT

The proposed LNA is implemented using 22-nm CMOS technology. All design parameters are shown in Fig. 1. The parasitic capacitance of 100 fF on pads and the inductance of 0.8 nH on the bonding wires are considered. The layout of the proposed LNA is shown in Fig. 4, with a core area of 40 μm × 40 μm. By carefully designing the MOS transistor of output stage, it does not require an additional buffer to drive the output load, which is beneficial for the power consumption, gain, and NF of the entire circuit.

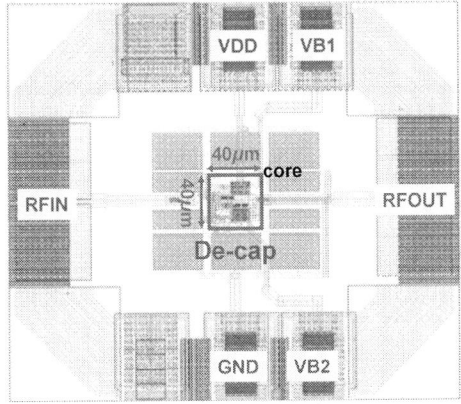

Fig. 4. Layout of the proposed LNA.

A. Monte-Carlo Simulation Results

Process variations and device mismatches can significantly degrade the effectiveness of noise-cancellation techniques by disrupting the precise conditions required for constructive signal summation and destructive noise suppression. To quantitatively evaluate their impact on the noise performance, extensive Monte Carlo simulations were conducted under two scenarios: mismatch-only and full process variation. As shown in Fig. 5, at 1 GHz, the standard deviation of the NF is 0.005 dB due to device mismatch and 0.0854 dB under full process variation. At 7 GHz, the standard deviation increases slightly to 0.0102 dB and 0.0944 dB, respectively. These results demonstrate that while mismatch introduces only minor variation, process fluctuations have a more pronounced effect, particularly at higher frequencies. Nevertheless, the overall NF variation remains within a narrow range, indicating that the proposed architecture maintains strong robustness against both mismatch and process-induced deviations, making it well-suited for wideband and process-tolerant RF applications.

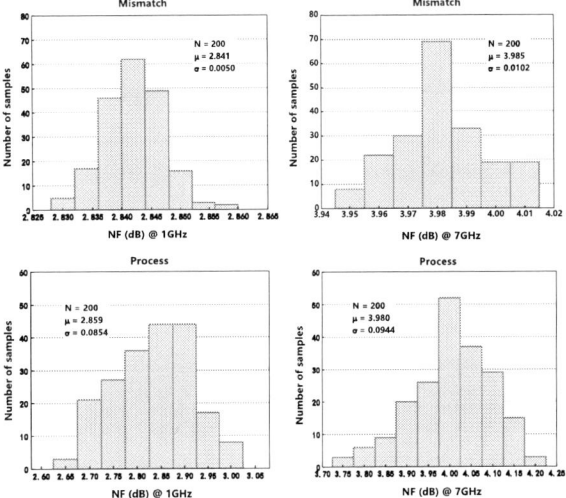

Fig. 5. Monte Carlo simulation results for NF.

B. S- parameter, NF and Linearity Simulation Results

Fig. 6 presents the simulation results of the S-parameter analysis and NF. IIP3 at 5.5 GHz and the entire frequency band are shown in Fig. 7. P1dB at 7 GHz and the entire frequency band are shown in Fig. 8. The proposed LNA matches well in the 0.2-7.3 GHz, covering the entire bandwidth of 5G NR FR1. Within the 3-dB bandwidth, the peak gain is 13.3 dB and the minimum NF is 2.83 dB. The IIP3 of LNA is obtained through a two-tone test, reaching up to 6.7 dBm across the entire operating range. The proposed LNA operates at a low supply voltage of 0.9 V, with a total static current of 6.1 mA and a total power consumption of only 5.5 mW. Table I benchmarks the performance with previously published broadband LNAs. Compared to previous designs, the proposed LNA demonstrates excellent linearity at a low supply. Although the LNA in [4] also achieves IIP3 of 5.8 dBm, it operates at a power supply voltage of 1.6V. In contrast, the proposed LNA offers a good trade-off across other performance metrics, with its FoM reaching a very high level. This makes it a viable solution for overcoming the linearity limitations of LNAs in receiver at a low supply using advanced processes.

979-8-3315-3918-4/25 $31.00 © 2025 IEEE

TABLE I
Summary of Typographical Settings

Ref.	CMOS Tech.	BW (GHz)	Gain (dB)	NF$_{min}$ (dB)	IIP3 (dBm)	Supply (V)	Power (mW)	Num. of inductors	Area (mm^2)	FoM$_1$	FoM$_2$
This Work*	22 nm	0.2-7.3	13.3	2.83	6.7	0.9	5.5	0	0.0016	29.65	85.57
[4] TMTT'20	65 nm	1-20	12.8	3.3	5.8	1.6	20.3	3	0.096	22.70	43.06
[5] JSSC'21	28 nm	0.02-4.5	15.2	2.09	-4.63	1	4.5	1	0.03	10.08	40.54
[6] TMTT'21	40 nm	1-11	17	3.5	-2.8	1.2	9	2	0.061	10.46	34.75
[9] ASICON'23*	40 nm	0.3-4.3	21.7	3.2	-12.4**	1.1	3.6	0	NR	2.53	/
[10] TCAS I'24	28 nm	0.2-2.85	20	2.9	-12.3	0.6	1.74	0	0.0048	-0.5	45.88

*Post-layout simulated results   ****calculate from IIP3=P1dB+9.6dB**   **NR: Not report**

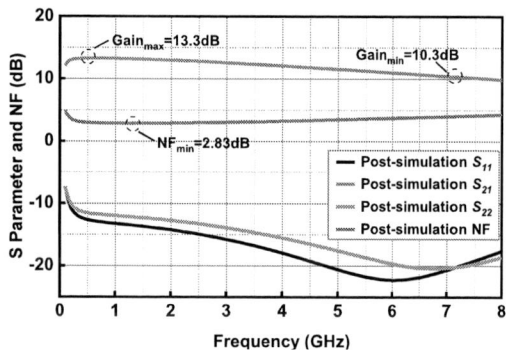

Fig. 6. Post-layout simulated results of S-parameter and NF.

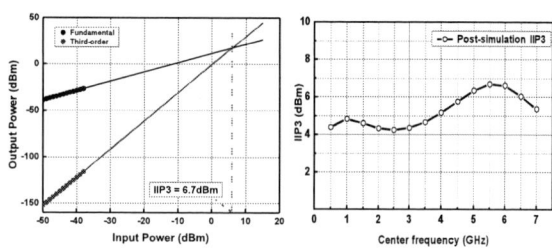

Fig. 7. Post-layout simulated results of IIP3. (a) IIP3 at 5.5 GHz. (b) IIP3 of the whole band.

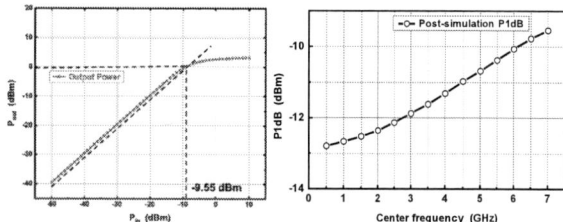

Fig. 8. Post-layout simulated results of P1dB. (a) P1dB at 7 GHz. (b) P1dB of the whole band.

$$FoM_1 = 20log_{10}\left(\frac{BW[GHz] \cdot Gain[abs] \cdot IIP_3[mW]}{P_{DC}[mW] \cdot (F_{min}-1)}\right) \quad (8)$$

$$FoM_2 = 20log_{10}\left(\frac{BW[GHz] \cdot Gain[abs] \cdot IIP_3[mW]}{P_{DC}[mW] \cdot (F_{min}-1) \cdot A[mm^2]}\right) \quad (9)$$

IV. CONCLUSION

This paper presents a broadband LNA topology to solve the problem of poor linearity caused by low supply voltage in advanced processes for 5G NR applications. A novel derivative superposition technique is utilized to enhance IIP3. The proposed LNA features a compact design with extremely small area and covers the full 5G NR bandwidth without inductors, which is well-suited for high-linearity MIMO 5G mobile communication receivers.

REFERENCES

[1] Z. Luo et al., "A 0.4–6-GHz Blocker-Tolerant Receiver in 65-nm CMOS With Bandwidth-Extended Technologies for Future V2X Applications," IEEE Trans. Circuits Syst. II, Exp. Briefs, vol. 71, no. 5, pp. 2634-2638, May 2024.

[2] S. Shekhar, J. S. Walling, and D. J. Allstot, "Bandwidth extension techniques for CMOS amplifiers," IEEE J. Solid-State Circuits, vol. 41, no. 11, pp. 2424–2438, Nov. 2006.

[3] A. Bozorg and R. B. Staszewski, "A 0.02–4.5-GHz LN(T)A in 28-nm CMOS for 5G exploiting noise reduction and current reuse," IEEE J. Solid-State Circuits, vol. 56, no. 2, pp. 404–415, Feb. 2021.

[4] H. Yu, Y. Chen, C. C. Boon, P.-I. Mak, and R. P. Martins, "A 0.096-mm2 1–20-GHz triple-path noise-canceling common-gate common-source LNA with dual complementary pMOS-nMOS configuration," IEEE Trans. Microw. Theory Techn., vol. 68, no. 1, pp. 144–159, Jan. 2020.

[5] Z. Liu, C. C. Boon, X. Yu, C. Li, K. Yang, Y. Liang, "A 0.061-mm2 1–11-GHz noise-canceling low-noise amplifier employing active feedforward with simultaneous current and noise reduction", IEEE Trans. Microw. Theory Techn., vol. 69, no. 6, pp. 3093–3106, Jun. 2021.

[6] G. Ma et al., "A 0.3-7.7GHz Noise-Canceling Low-Noise Amplifier in 55nm CMOS," 2023 IEEE MTT-S International Wireless Symposium (IWS), 2023, pp. 1-4.

[7] M. A. Karami, M. Lee, R. Mirzavand and K. Moez, "A 0.1–20.1-GHz Wideband Noise-Canceling gm-Boosted CMOS LNA With Gain Reuse," IEEE Trans. Microw. Theory Techn., vol. 72, no. 5, pp. 2990-3000, May 2024.

[8] Z. Liu, C. C. Boon, et al., "A 0.0078 mm2 3.4 mW Wideband Positive feedback-Based Noise-Cancelling LNA in 28nm CMOS Exploiting Gm Boosting," in 2022 IEEE International Solid-State Circuits Conference (ISSCC), vol. 65, pp. 1–3, IEEE, 2022.

[9] Z. Yang, J. Jin, Y. Chen, J. Zhou and X. Liu, "A Wideband Inductorless LNA Employing Dual-Loop Feedback for Low-Power Applications," 2023 IEEE 15th International Conference on ASIC (ASICON), Nanjing, China, 2023, pp. 1-4.

[10] Z. Liu, C. Chye Boon and Y. Dong, "A 0.6 V, 1.74 mW, 2.9 dB NF Inductorless Wideband LNA in 28-nm CMOS Exploiting Noise Cancellation and Current Reuse," in IEEE Transactions on Circuits and Systems I: Regular Papers, vol. 71, no. 8, pp. 3561-3572, Aug. 2024.

[11] H. Dong, K. Deng, B. Qiu, Shaohui and H. Feng, "A 0.6-8.1 GHz, 2 dB NF inductorless LNTA in 22 nm CMOS with novel bandwidth extension technique," 2024 IEEE International Conference on Integrated Circuits, Technologies and Applications (ICTA), Hangzhou, China, 2024, pp. 154-155.

[12] H. Zhang and E. Sánchez-Sinencio, "Linearization techniques for CMOS low noise amplifiers: A tutorial," IEEE Trans. Circuits Syst. I, Reg. Papers, vol. 58, no. 1, pp. 22–36, Jan. 2011.

An Area-Efficient Bi-directional Cascode PA-LNA For 5G NR in 28-nm CMOS

Yue Wu, Wei Li*, Shijiao Dong, Hongtao Xu

State Key Laboratory of Integrated Chips and Systems, College of Integrated Circuits and Micro-Nano Electronics,
Fudan University, Shanghai, China

* Email: w-li@fudan.edu.cn

Abstract—This paper presents a bi-directional cascode power amplifier-low-noise amplifier (PA-LNA) in 28-nm CMOS technology. In the proposed PA-LNA, the cascode PA achieves high output power and efficiency by variable source voltage, while the LNA realizes lower noise figure and high gain with low power consumption by combining current reuse topology and the negative coupled transformer. Furthermore, the transformer based matching network is fully shared by PA and LNA for a compact area. In PA mode, the proposed PA-LNA achieves a stimulated peak gain of 20.4 dB across 22.3-30.7GHz, a saturated output power (P_{sat}) of 16.9dBm with a peak power-added-efficiency (PAE) of 17.3% at 27GHz. In LNA mode, the proposed PA-LNA demonstrates a peak gain of 12.7dB across 24.6-30.7GHz, a noise figure (NF) of 6.7-7.0dB across the frequency band and a power consumption of 43mW. The core area of the PA-LNA is 0.19mm^2.

Keywords—5G, Bi-directional, power amplifier-low noise amplifier (PA-LNA), millimeter-wave (mm-wave)

I. INTRODUCTION

With the demand for high data rate transmissions, the millimeter-wave FR2 band 5G new radio is being rapidly developed. As the first globally coordinated 5G millimeter wave spectrum resources in the millimeter wave band, the 24.25-29.5GHz band (n257/n258) has a continuous bandwidth of up to 1GHz. Compared with Sub-6GHz band, this band faces the propagation challenge of an increase in free-space path loss which is to be compensated with large scaled multiple-input multiple-output (MIMO) arrays. As one of the crucial components of millimeter wave phased-array system, conventional RF front-end (FE) consisting of low-noise amplifiers (LNAs), power amplifiers (PAs), and T/R switches consume a large chip area due to separate matching networks (MNs). Therefore, a compact RF front-end solution is required.

In recent years, bi-directional PA-LNAs, are introduced where most of the MNs are fully shared by PA and LNA to reduce die area [1]-[4], the core schematic of the traditional bi-directional PA-LNA is shown in Fig.1(a). There are two key points in the design of bi-directional PA-LNA: one is to simultaneously achieve noise matching of LNA and load matching of PA over a shared MN. The other is to prevent the large output signal swing of the PA from turning on the LNA transistors. In previously reported mm-wave bi-directional PA-LNA[2]-[4][6], transformers are often used to achieve both impedance and noise matchings, but it is still very challenging to achieve large output power of PA and low noise of LNA simultaneously. The Fig.1(b) shows that for the large transient amplitude signal in PA mode, the V_{GS} of M_{LNA} has to exceed the threshold voltage V_T, which will definitely cause the transistors of LNA to turn on, consequently worsen the PA

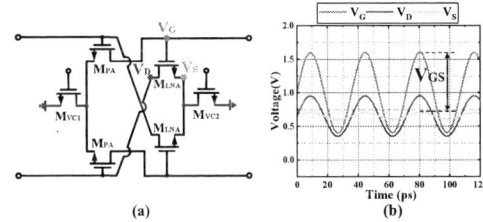

Fig. 1. Conventional bi-directional PA-LNA[1] (a) Core schematic. (b) Large transient output signal waveform in PA mode.

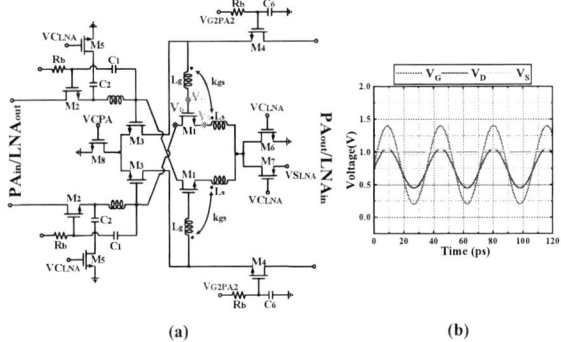

Fig. 2. Proposed bi-directional cascode PA-LNA (a) Core schematic. (b) Large transient output signal waveform in PA mode.

performance. The methods in previous studies [2]-[4] tried to solve such problem by means of lowering PA supply voltage, hybrid P/N-MOS, bias separation, gate switching, etc., but deteriorated the PA performance or consumed more power in LNA mode due to the parasitic capacitance introduced by the large size PA transistors, and caused rapidly decreased gain of LNA at high frequency.

In order to address the above challenges, a cascode PA-LNA in 28-nm CMOS is proposed in this paper. The current reuse topology [5] is adopted to realize high gain and low power consumption in LNA mode, in which the weak coupling of the transformer is applied to tune the value of the inductor and optimizes the noise figure (NF). A cascode PA with variable source voltage is proposed to prevent LNA transistors from turning on at high gain and output power in PA mode. Section II presents the detailed design of the proposed PA-LNA. Section III shows the post-simulation results and followed by the conclusions in Section IV.

II. CIRCUIT DESIGN

Fig.3 shows the entire schematic of the proposed bi-directional PA-LNA. It consists of a driving amplifier, a cascode power/low-noise amplifier, and transformer-based MNs. The MNs are fully shared in both PA and LNA modes for a compact area. In order to improve the matching

---

This paper is supported by National Key R&D Program of China under Grant 2023YFB4403802, and by the Fundamental Research Funds for Central Universities in China.

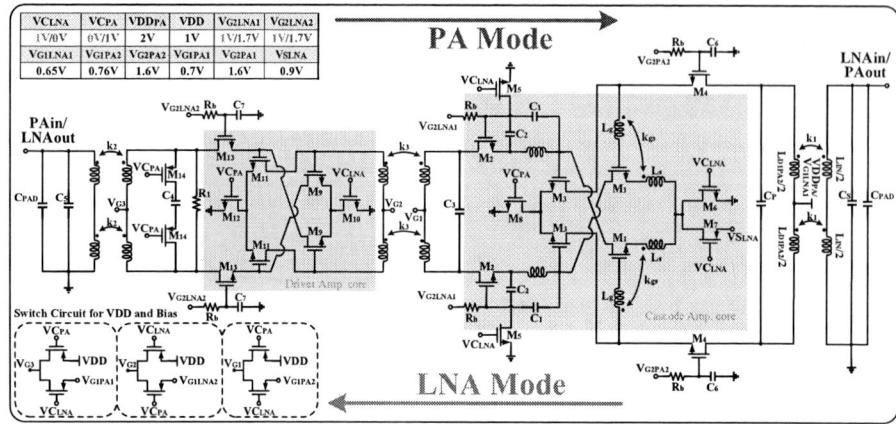

Fig. 3. Entire schematic of the proposed two stage bi-directional cascode PA-LNA with current reuse topology.

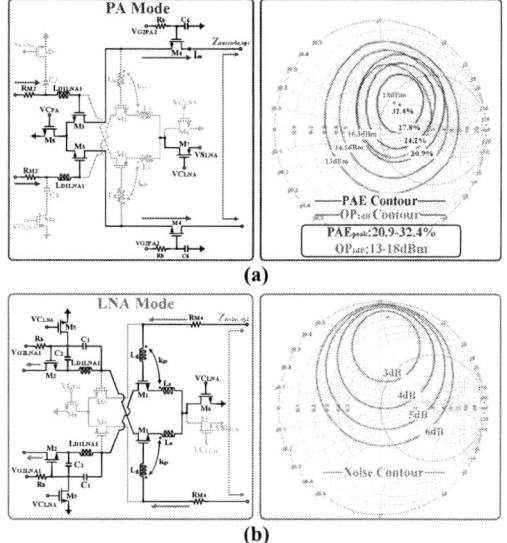

Fig. 4. Schematic of the core stage of proposed PA-LNA (a) in PA mode with load-pull simulation results at 28GHz, and (b) in LNA mode with optimum noise factor simulation results at 28GHz.

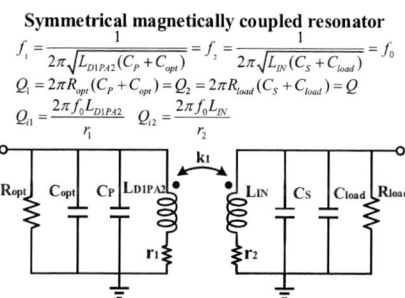

Fig. 5. The schematic of symmetrical magnetically coupled resonator.

performance, resistor R1 and capacitor C4 are utilized and controlled by switch M14, which is turned on in LNA mode and turned off in PA mode. The bias and VDD of the PA-LNA are controlled by the switch circuits shown in Fig.3.

The cascode PA-LNA, as depicted in Fig.2 (a), is the core module in the proposed bi-directional PA-LNA, and will be focused on more detail design descriptions and discussions in this session.

In PA design, due to the large swing of the signal at PA output node, conventional stacked PA tends to cause the transistors M_1 of the LNA to turn on, resulting in the deterioration of the output power and efficiency. However, in our proposal as shown in Fig.2(a), the gate of the common source (CS) amplifier M_1 of LNA is connected to the middle node of the cascode PA, which not only ensures the large supply voltage of the PA in output node, but also prevents M_1 from turning on by adjusting the sizes of M_3 and M_4 and their bias V_{G1PA2} and V_{G2PA2} to control the voltage of the middle node. Moreover, we apply a dc voltage V_{SLNA} to the source of M_1 in PA mode, which can further prevent M_1 from turning on.

In PA mode, the transistor M_2 in the common gate (CG) stage of LNA operates as a switch by adjusting bias V_{G2LNA1}. The traditional CS PA has $Z_{cs,opt} = \frac{V_m - V_{knee}}{I_m}$, which is much smaller than the $Z_{noise,opt}$, due to the large I_m of PA we required.

Therefore, it is difficult to simultaneously realize the load matching and noise matching by a shared MN. In our design, the proposed cascode PA has a $Z_{cascode,opt} = \frac{2V_m - V_{knee}}{I_m} \approx 2Z_{cs,opt}$ is similar to the $Z_{noise,opt}$ we required with M_3 transistor size of 2×72um. The evidences are the load-pull simulation result of PA and the noise-source simulation result of LNA at 28GHz that are shown in Fig.4(a) and (b) respectively. According to the simulation results of Smith chart, Z_{opt}^* is selected as 79Ω//45fF. Meanwhile, the symmetrical magnetically coupled resonator (MCR) is adopted to realize the matching between the Z_{load} and Z_{opt}^*, where Z_{load}=50Ω//36fF. The schematic of symmetrical MCR is shown in Fig.5. Since the transformer is a lossy model, the maximum efficiency of the transformer can be derived as [9]

$$\eta_{max} = \cfrac{1}{1 + \cfrac{2}{Q_{i1}Q_{i2}k_1^2} + 2\sqrt{\cfrac{1}{Q_{i1}Q_{i2}k_1^2}\left(1 + \cfrac{1}{Q_{i1}Q_{i2}k_1^2}\right)}} \quad (1)$$

Where, Q_{i1} and Q_{i2} are the quality factor of the coupling inductor L_{D1PA2} and L_{IN} respectively. Due to the process limitation, in order to realize the maximum efficiency, the coupling coefficient k_1 can only be designed up to 0.7 at 28GHz. According to the condition of minor gain ripple of transformer $k_1^2(1 + Q^2) = 1$ [10], the resonant quality factor Q can be calculated as 1 for both Q_{i1} and Q_{i2}. According to the

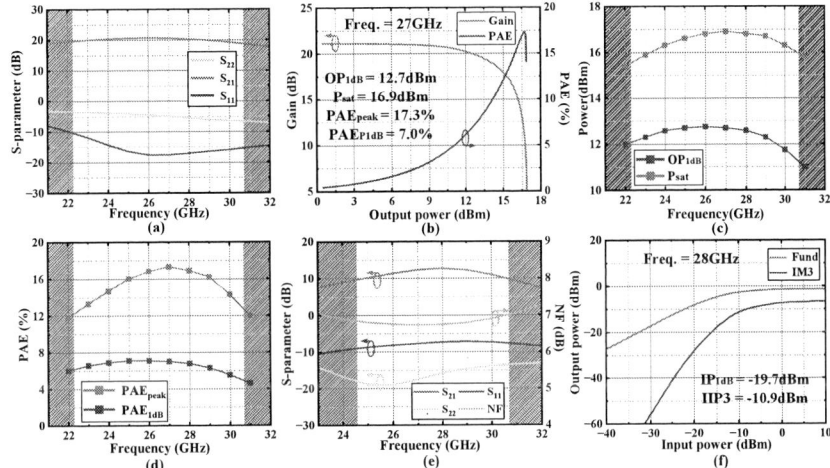

Fig. 8. Post-stimulation results of (a) S-parameters, (b) OP_{1dB} and PAE at 27GHz, (c) OP_{1dB} and P_{sat} over the frequency range, and (d) PAE over the frequency in PA mode. (e) S-parameters and NF, and (f) IP_{1dB} and IIP3 at 28GHz in LNA mode.

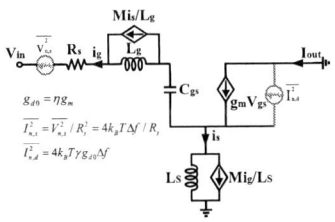

Fig. 6. Equivalent circuit of the input matching network used for noise analysis in LNA mode.

equation $Q_1 = 2\pi R_{opt}(C_P + C_{opt})$, we can get the value of the additional parallel capacitance C_P, and as a result of which the value of L_{D1PA2} can be calculated by the equation $L_{D1PA2} = \frac{1}{2\pi f_0^2 (C_P + C_{opt})}$. Based on the calculations combined with the simulation results, the L_{D1PA2}, L_{IN} and C_P are designed as 380pH, 260pH and 30fF respectively at 28GHz.

Generally, in LNA design, the CS amplifier is always adopted in the input stage to reduce noise, while cascode amplifier is adopted to provide high gain. In the mm-wave band, the parasitic capacitance of the transistor cannot be ignored, which will have a large impact on the input impedance and high frequency gain. In order to deal with the parasitic capacitance introduced by large-sized transistors in PA mode, the current reuse topology, which consists of M_1 and M_2, is adopted to increase gain and reduce power consumption, as is shown in Fig.4(b). In LNA mode, M_4 in the CG stage of PA operates as a switch, due to the small on-resistance R_{M4}, it has a weak effect on the NF of LNA. Meanwhile, the inductor L_g and L_s are coupled with the coupling coefficient k_{gs} to optimize the noise figure [6]. The small signal analysis of the input matching network of LNA is shown in Fig.6. The transconductance G_m of the input stage can be calculated as follows

$$G_m = \frac{I_{out}}{V_{in}} = \frac{g_m}{1 + s(L_s g_m + c_{gs} R_s + M g_m) + s^2 C_{gs}(L_g + L_s + 2M)} \quad (2)$$

Here, $M = k_{gs}\sqrt{L_g L_s}$, we make $A = 1 + s(L_s g_m + c_{gs} R_s + M g_m) + s^2 C_{gs}(L_g + L_s + 2M)$, the mean-squared output current noise of the source resistance $\overline{I_{o,ns}^2}$ and the channel thermal noise $\overline{I_{o,nd}^2}$ can be respectively derived as

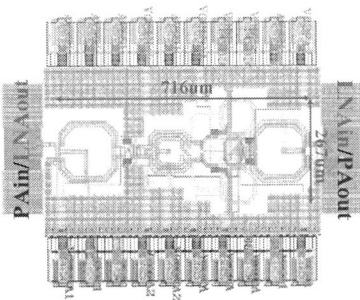

Fig. 7. The layout of the proposed PA-LNA.

$$\overline{I_{o,ns}^2} = \left| \frac{g_m}{A} R_s \right|^2 \overline{I_{n,s}^2} \quad (3)$$

$$\overline{I_{o,nd}^2} = \left| \frac{1 + s c_{gs} R_s + s^2 C_{gs}(L_g + L_s)}{A} R_s \right|^2 \overline{I_{n,d}^2} \quad (4)$$

According to equation (3) and (4), the noise factor with magnetic coupling feedback can be derived as

$$F = \frac{|I_{o,total}|^2}{|I_{o,ns}|^2} = 1 + \frac{\eta\gamma \left(1 + s c_{gs} R_s + s^2 C_{gs}(L_g + L_s + M)\right)^2}{g_m R_s} \quad (5)$$

Where, γ is the channel thermal noise coefficient and η is the ratio of the channel conductance g_{d0} under zero-bias condition to the transconductance g_m. According to (5), the NF can be optimized by the negative value of coupling coefficient k_{gs}. The L_g and L_s are designed as 196pH and 40pH respectively with the negative coupling coefficient k_{gs}=-0.3 at 28GHz.

III. SIMULATION RESULTS

The proposed cascode PA-LNA was implemented in 28-nm CMOS. Fig.7 shows the layout of the VG-LNA with a core size of 0.19 mm². The PA-LNA consumes 43mW in LNA mode from benefit of current reuse topology. Fig.8(a)-(f) show the post-simulated results of PA and LNA respectively. In PA mode, the peak gain is 20.4dB at 26GHz, and the input return loss stays below -10dB for input matching over the 22.3-30.7GHz. The OP_{1dB} and PAE_{peak} are 12.7dBm and 17.3% at 27GHz, respectively. The proposed PA-LNA achieves the

979-8-3315-3918-4/25 $31.00 © 2025 IEEE

TABLE I. PERFORMANCE SUMMARY AND COMPARISON WITH STATE-OF-THE-ART WORKS

		This work*	SSCL 2023[7]	TMTT 2024[4]	RFIC 2024[2]*	ISSCC 2022[3]	JSSC 2020[1]	RFIT 2020[8]
	Technology	28-nm CMOS	65-nm CMOS	40-nm CMOS	28-nm CMOS	45-nm CMOS SOI	65-nm CMOS	65-nm CMOS
	Architecture	MNs shared	MNs separated	MNs shared	MNs shared	MNs shared	MNs shared	MNs separated
PA	Freq. (GHz)	22.3-30.7	26.5-29.5	52-67	27.3-35.4	25.3-42	22.0-34.0	27
	Frac.BW (%)	31.6%	10.7%	25.2%	25.8%	49.6%	42.8%	-
	Peak Gain (dB)	20.4	18.0	8.02	20.4	18.9	15	35
	P_{sat}/OP_{1dB} (dBm)	16.9/12.7	16.3/N.A	15.2/12.8	16.3/15.0	19.4/17.8	15.1/11.3	16/N.A
	PAE_{peak}/PAE_{1dB} (%)	17.3/7.0	21/N.A	9.35/N.A	14.9/9	42.9/34.9	20.0/14.0	23.2/14.5
	Supply (V)	2.0	1.0	2.1	3.0	2.0	N.A	1.0
LNA	Freq. (GHz)	24.6-30.7	26.5-29.5	52-66	28.0-36.0	27.0-38.0	23.0-34.0	27
	Frac.BW (%)	22%	10.7%	23.7%	25%	28.5%	38.5%	-
	Peak Gain (dB)	12.7	15.0	11.29	17.3	17.6	15	25.5
	NF (dB)	6.73-7.01	6.2-7.2	>6.87*	5.3-7.4	5.2-7.8	4.2-5.0	>5.9
	IP_{1dB}/IIP3 (dBm)	-19/-10.3	N.A	N.A	-10.3/0	-8.6/0.9	-14.6/-5	N.A
	Supply (V)	1.0	1.0	1.5	1.0	1.1	N.A	1.0
	P_{DC} (mW)	43	33	78	63	66	31	50
	Core Area (mm^2)	0.19	0.33	0.08	0.1	0.19	N.A	0.26

* post-simulated results.

performance of P_{sat}>15.6dBm and OP_{1dB}>11.2 dBm with PAE_{peak}>12% in the target frequency band. The LNA operates in the frequency band of 24.6-30.7GHz, and the peak gain is 12.7dB at 28GHz, the NF remains ≤ 7.01dB over the entire frequency range. The input P_{1dB} (IP_{1dB}) and input third-order intercept point (IIP3) are -19.7dBm and -10.9dBm at 28GHz respectively.

The summary of the proposed bi-directional cascode PA-LNA performance and the state-of-arts is shown in Table I. It is demonstrated that the current reuse technique is effective in reducing power consumption with lower NF in LNA mode, and high output power and efficiency are realized over wider bandwidth in PA mode by applying variable source voltage. Shared MNs helps significantly reduce area.

IV. CONCLUSIONS

This paper presents a mm-wave bi-directional cascode PA-LNA with fully shared MNs in 28-nm CMOS process. A cascode PA with variable source voltage is proposed to improve matching performance for a large output power and efficiency. The current reuse topology is adopted to achieve high gain with low power consumption in LNA mode. In PA mode the proposed PA-LNA has a peak gain of 20.4 dB over 22.3-30.7 GHz with OP_{1dB} of 12.7dBm and PAE_{peak} of 17.3% at 27GHz. In LNA mode, the peak gain is 12.7dB over 24.6-30.7GHz, the NF is 6.73-7.01 dB across the frequency band and the power consumption is 43mW.

REFERENCES

[1] J. Pang et al., "A 28-GHz CMOS Phased-Array Beamformer Utilizing Neutralized Bi-Directional Technique Supporting Dual-Polarized MIMO for 5G NR," in IEEE Journal of Solid-State Circuits, vol. 55, no. 9, pp. 2371-2386, Sept. 2020.

[2] J. Hwang and B. -W. Min, "A Compact Ka-Band Bi-Directional PA-LNA with 17.4-dBm Psat Using Three-Stack Power Amplifier in 28-nm CMOS," 2024 IEEE Radio Frequency Integrated Circuits Symposium (RFIC), Washington, DC, USA, 2024, pp. 63-66.

[3] J. Park and H. Wang, "A 26-to-39GHz Broadband Ultra-Compact High-Linearity Switchless Hybrid N/PMOS Bi-Directional PA/LNA Front-End for Multi-Band 5G Large-Scaled MIMO System," 2022

IEEE International Solid-State Circuits Conference (ISSCC), San Francisco, CA, USA, 2022, pp. 322-324.

[4] H. Jia, L. Jiang, X. Y. Zhang, Y. Wang and A. Zhu, "An Ultracompact Bidirectional CMOS Gate-Switching Cascode Amplifier for Millimeter-Wave Transceiver Front End," in IEEE Transactions on Microwave Theory and Techniques, doi: 10.1109/TMTT.2024.3518207.

[5] R. -M. Weng, C. -Y. Liu and P. -C. Lin, "A Low-Power Full-Band Low-Noise Amplifier for Ultra-Wideband Receivers," in IEEE Transactions on Microwave Theory and Techniques, vol. 58, no. 8, pp. 2077-2083, Aug. 2010.

[6] R. Wang, C. Li and Y. Wang, "A Broadband Variable-Gain Low-Noise Amplifier With Low NF and Dual Phase Compensation," in IEEE Transactions on Circuits and Systems II: Express Briefs, vol. 71, no. 9, pp. 4086-4090, Sept. 2024.

[7] J. Pang et al., "A Compact 28 GHz Bi-Directional Power-Combined Antenna Interface in WLCSP for 5G and B5G Transceivers," in IEEE Solid-State Circuits Letters, vol. 6, pp. 149-152, 2023.

[8] J. Wang, W. Zhu and Y. Wang, "A 24.25-27.5 GHz Front-End Module with Transformer-Based T/R Switch for 5-G communications," 2020 IEEE International Symposium on Radio-Frequency Integration Technology (RFIT), Hiroshima, Japan, 2020, pp. 205-207.

[9] G. Liu, P. Haldi, T. -J. K. Liu and A. M. Niknejad, "Fully Integrated CMOS Power Amplifier With Efficiency Enhancement at Power Back-Off," in IEEE Journal of Solid-State Circuits, vol. 43, no. 3, pp. 600-609, March 2008.

[10] H. Jia, L. Jiang, X. Y. Zhang, Y. Wang and A. Zhu, "An Ultracompact Bidirectional CMOS Gate-Switching Cascode Amplifier for Millimeter-Wave Transceiver Front End," in IEEE Transactions on Microwave Theory and Techniques, doi: 10.1109/TMTT.2024.3518207.

A 223M-235MHz Fully-Integrated Differential Class-E Power Amplifier with 45.5% PAE and 22.8dBm

Chaoyang Zheng[1], Yanxiang Chen[1], Jianhua Lu[2,3], Yan Ma[2,3], Zhiliang Hong[1], Yumei Huang[1*]

[1]College of Integrated Circuits & Micro-Nano Electronic, Fudan University, Shanghai 200433, China
[2] Beijing Smartchip Microelectronics Technology Co., Ltd
[3]Beijing Smartchip Semiconductor Technology Co., Ltd
Email: 22212020197@m.fudan.edu.cn, *yumeihuang@fudan.edu.cn

Abstract—RF power amplifiers (PAs) are essential components in wireless transceivers. To achieve high-efficiency performance, a differential Class E architecture based on the finite DC-feed inductance is employed. Additionally, on-chip transformers are used for power combination and impedance conversion in the differential PA. This design further optimizes the circuit structure and increases performance insensitivity to transformer parameters. The chip area is approximately 2.9mm², based on 55nm CMOS technology, operating at a supply voltage of 1.5V and a frequency range of 223MHz to 235MHz. According to post-simulation analysis, it achieved a peak power-added efficiency (PAE) of 45.5% and an output power of 22.8dBm within the frequency band on a 50ohms load.

Keywords—PA, Transformer, Class-E, Full integrated, finite DC-feed inductance

I. INTRODUCTION

RF PAs are generally considered the main contributors to power consumption in RF transmission systems. Nonlinear PAs, typically used in constant-envelope modulation schemes, are driven by input signals that continuously switch the transistors between on and off states, offering theoretical energy efficiency up to 100%. However, practical implementations suffer from switching and harmonic losses, which limit the achievable efficiency.

Illustrated in the Fig.1 is the simple topology of a switching-mode Class-E PA[1].

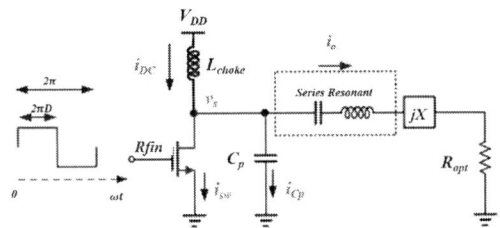

Fig. 1. Conventional Class-E power amplifier structure

In Class-E PAs, the drain waveform v_s is designed to satisfy the Zero Voltage Switching (ZVS) and Zero Derivative Switching (ZDS) conditions, as defined in equations (1) and (2), to optimize the drain load design [2].

$$\text{ZVS: } v_s(\omega t = 2\pi) = 0 \qquad (1)$$

$$\text{ZDS: } \frac{dv_s}{dt}(\omega t = 2\pi) = 0 \qquad (2)$$

Owing to their topology and operating principles, Class-

Supported by the Joint R&D Fund of Beijing Smartchip Microelectronics Technology Co., Ltd. SGTYHT/21-JS-223.

E PAs are more suitable for CMOS integration compared to other switching-mode architectures. Firstly, the large parasitic drain-to-source capacitance inherent in CMOS transistors can be fully absorbed into the drain load network. Secondly, the harmonic-controlled load design in Class-E PAs reduces the frequency sensitivity.

However, the reduced breakdown voltage in advanced CMOS technologies, along with the increasing demand for low-voltage applications, imposes limitations on the maximum achievable output power of Class-E PAs. To address this challenge, designers have explored the integration of on-chip transformers and inductors to enhance PA performance. In [3], a phase-staggered Class-E PA incorporating a transformer-based passive enhancement network was presented. Fabricated in a 45 nm CMOS process, the design achieved an output power of 29.5 dBm and a peak PAE of 46.76%. In [4], a choke-less Class-DE PA was proposed, utilizing a single center-tapped inductor for power delivery and resonant filtering at the output. This design, implemented in 40 nm CMOS and operating at 1 GHz, achieved 21dBm output power with a PAE of 41.1%.

This paper presents a differential Class-E PA with an on-chip transformer. This design aims to realize a fully integrated power amplifier operating in the 223 MHz to 235 MHz frequency range, achieving an output power above 20dBm and PAE exceeding 40%. The work primarily focuses on the matching network between the switching transistors and the 50 ohms load.

The remainder of this paper is organized as follows: Section II details the design of the Class-E PA incorporating a finite DC-feed inductance and introduces the circuit implementation, with particular emphasis on the equivalent modeling and functionality of the on-chip transformer. Section III presents the post-layout simulation results Section IV compares the performance metrics with prior state-of-the-art works.

II. CIRCUIT DESIGN

A. Class-E PA with the Finite DC-feed Inductance

In conventional Class-E PA structures, a large choke inductance is typically required to suppress current ripple, which presents significant challenges for on-chip integration in CMOS processes and constrains the flexibility of parameter tuning. As illustrated in Fig. 2, adopting a finite-value feed inductance not only enables full integration but also introduces an additional degree of freedom for design, allowing more flexible optimization of device parameters [5].

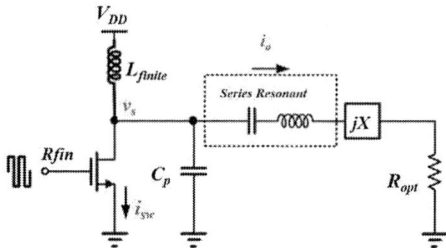

Fig. 2. Class-E power amplifier with the finite feed inductance

In CMOS implementations of this circuit, the capacitance C_p is lower-bounded by the parasitic capacitance of the switching transistor. Moreover, the large impedance transformation ratio between the 50 ohms antenna and the optimal load impedance increases the frequency sensitivity of the matching network. Reference [6] shows that, under fixed values of parasitic capacitance and optimal load impedance, the product of output power and operating frequency is maximized when:

$$\frac{1}{\omega\sqrt{L_{finite}C_p}} = 1.442 \qquad (3)$$

Under this condition, the circuit design parameters satisfy the following relationships:

$$0.669R_{opt} = \omega L_{finite} \qquad (4)$$

$$0.693 = \omega C_p R_{opt} \qquad (5)$$

$$1.423V_{DD}^2 = P_{out}R_{opt} \qquad (6)$$

$$-0.086R_{opt} = X \qquad (7)$$

B. Proposed Power Amplifier

The proposed differential Class-E power amplifier, illustrated in Fig. 3, incorporates finite-valued feed inductors and utilizes an on-chip transformer to realize the drain load network for the switching transistors. The differential outputs are combined and matched to a single-ended antenna.

One key advantage of the differential structure is the reduction of peak drain voltage on each switching transistor to approximately half, thereby improving device reliability. Additionally, the series resonant circuit does not need to suppress the second harmonic component, which reduces the inductor area and associated losses.

Using a cascode configuration for the switch transistor offers several advantages. Firstly, it reduces the voltage stress on the transistor's gate-drain voltage (the maximum voltage at the source node is approximately 3.6×V_DD). Additionally, it minimizes the impact of dynamic changes in the capacitance's parasitic transistor during the switching process on C_p. Furthermore, the output of the driver stage must maintain a low DC level to ensure proper turn-off of the switch. Otherwise, during power-up, if the node voltage v_s starts from a low voltage, excessive voltage swing and oscillation may occur, potentially leading to transistor breakdown.

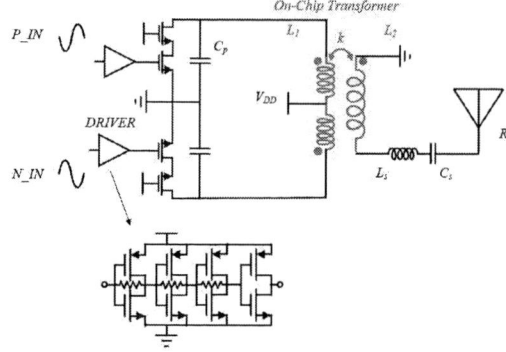

Fig. 3. Proposed differential Class-E power amplifier

The transformer's primary winding serves as the feed inductor and facilitates impedance transformation across a broad frequency band. Its equivalent circuit model is employed in the design of the drain-side matching network. The on-chip transformer cleverly integrates the finite DC-feed inductance into the transformer, as depicted in Fig.4, through an effective equivalent process.

Fig. 4(a) shows the load seen by the drain of the switch, including the actual transformer, drain-source shunt capacitor, series resonant supplementary circuit, and load.

$$n = \frac{\sqrt{L_1/L_2}}{k} \qquad (8)$$

The first step of the equivalent process is demonstrated in Fig. 4(b). The actual transformer can be equivalent to an ideal transformer with a turns ratio of n (n is defined by (8)) and two inductors distributed by the coupling coefficient k.

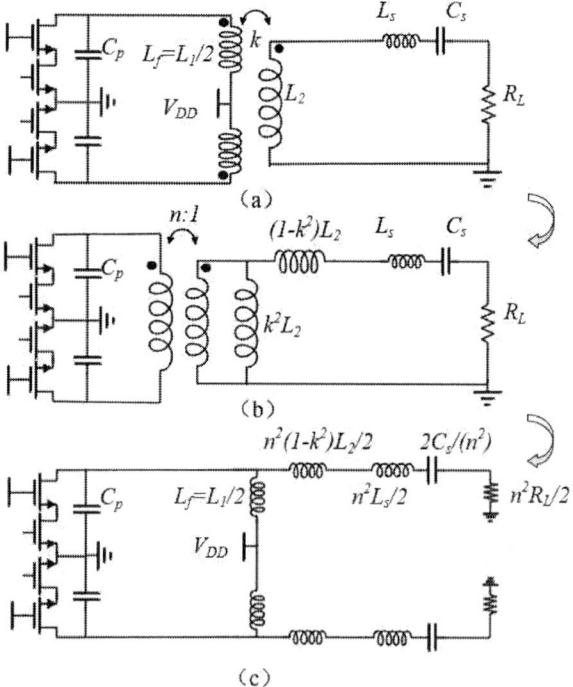

Fig. 4. Equivalent process of transformer matching load network.

Fig. 4(c) illustrates the equivalent result, after transformation by factor n and the conversion of the single-ended load to differential, resulting in the typical structure of the Class-E PA with the finite DC-feed inductance. The relationship between the parameters of this equivalent structure and the schematic diagram is depicted in Fig. 4(c).

C. Transformer Implementation

By analyzing Fig. 2 and Fig. 4(c) in conjunction with the design parameter equations, it can be deduced that the transformer should satisfy the following condition:

$$L_2 = \frac{R_L}{2k^2}\frac{L_1}{R_{opt}} \qquad (9)$$

In the transformer layout, interleaved winding is used to implement the primary and secondary coils on the top metal layer. According to the design equations, the rated output power is inversely proportional to the optimal load resistance. Based on this relationship, a turns ratio of 2:5 is selected to reduce the minimum load impedance seen by the circuit. To minimize the required value of L_2, the transformer's inner diameter is maximized. The PGS is designed using POLY silicon to reduce the coupling between the coils and the substrate, thereby increasing the quality factor Q. At the resonant frequency, the coupling coefficient is denoted as 0.82, and the quality factors of the primary and secondary windings are 6.2 and 8.2, respectively. The resonant frequency of the transformer is about 1.52GHz.

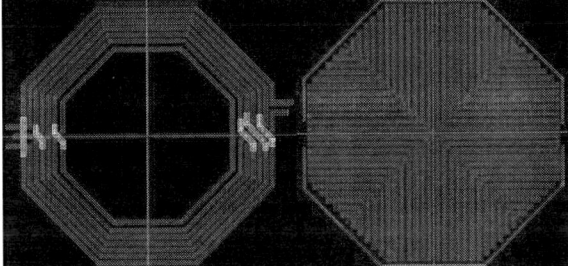

Fig. 5. Layout view of the transformer and PGS.

III. SIMULATION RESULTS

As illustrated in Fig. 6, the circuit layout based on the SMIC 55nm process is shown. The layout clearly presents the transformer, driver circuit, and core passive components, occupying an area of approximately 2.88 mm².

Figs. 7 and 8 respectively illustrate the variation of PAE and Pout across the frequency band under various process corners. Post-layout simulation results show that across all process corners and over the full frequency range, the output power exceeds 21dBm and the power-added efficiency exceeds 40%. As shown in Fig. 9 , the driver circuit enables the switching transistors to operate in their optimal region when the input power is -30dBm. Figs. 10 and 11 demonstrate that the designed PA's efficiency and output power both exhibit strong robustness against variations in temperature and supply voltage.

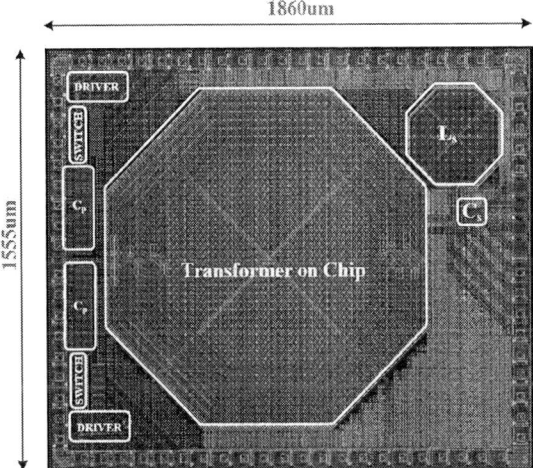

Fig. 6. Layout view of the designed PA.

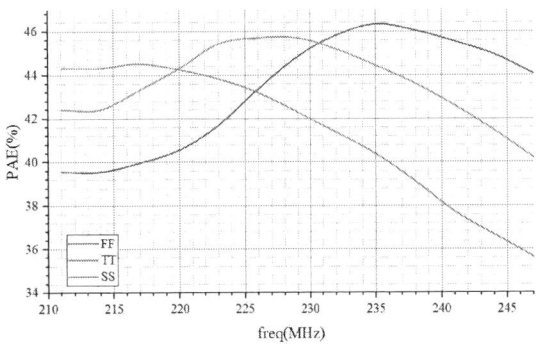

Fig. 7. The post-simulated PAE with supply voltage of 1.5V and input frequency of 229MHz under different process corners.

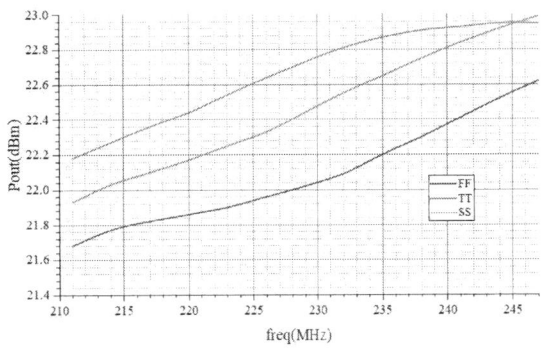

Fig. 8. The post-simulated output power with supply voltage of 1.5V and input frequency of 229MHz under different process corners.

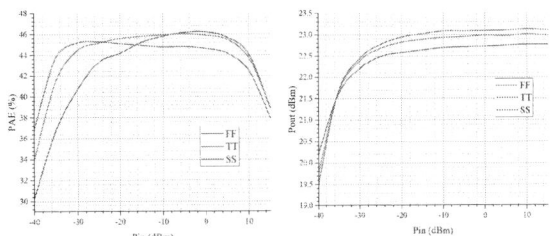

Fig. 9. The post-simulated PAE and Pout with input power.

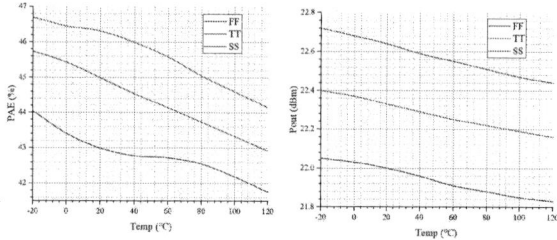

Fig. 10. The post-simulated PAE and Pout with temperature.

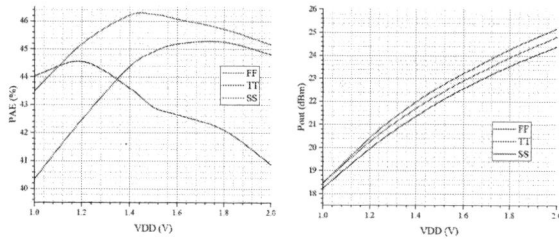

Fig. 11. The post-simulated PAE and Pout with supply voltage.

IV. CONCLUSION

A Class-E power amplifier featuring a finite DC-feed inductance has been successfully implemented, with an on-chip transformer facilitating impedance matching and power combining through the integration of a tuning capacitor. By carefully designing the parameters of the on-chip transformer, the equivalent finite DC-feed inductance has been effectively realized, thereby enhancing the level of integration. As summarized in Table I, a comparison with prior works demonstrates the advantages of the proposed design. The amplifier delivers the targeted output power with relatively high PAE across the chip. While inductive components tend to occupy a larger portion of the chip area at lower operating frequencies, this does not significantly impact the overall design efficiency.

TABLE I. PERFORMANCES COMPARISON

Parameter	Ref[7]	Ref[8]	Ref[9]	This work
VDD (V)	1.2	3.3	1.5	1.5
Frequency (GHz)	3-5	1.8	2.45	0.223-0.235
Area (mm²)	-	1.53	-	2.88
Technology	65nm	350nm	130nm	55nm
Pout (dBm)	19.9	29	23	22.8
PAE (%)	38.5	38.7	29	45.5

REFERENCES

[1] F. Raab, "Idealized operation of the class E tuned power amplifier," in IEEE Transactions on Circuits and Systems, vol. 24, no. 12, pp. 725-735, December 1977, doi: 10.1109/TCS.1977.1084296.

[2] Kazimierczuk M K. RF power amplifiers[M]. John Wiley & Sons, 2014.

[3] Banerjee A, Hezar R, Ding L, et al. A 29.5 dBm class-E outphasing RF power amplifier with efficiency and output power enhancement circuits in 45nm CMOS[J]. IEEE Transactions on Circuits and Systems I: Regular Papers, 2017, 64(8): 1977-1988.

[4] Singh G D, Nallam N. An RF choke-less class E power amplifier[J]. IEEE Transactions on Circuits and Systems II: Express Briefs, 2020, 67(11): 2422-2426.

[5] R. Zulinski and J. Steadman, "Class E Power Amplifiers and Frequency Multipliers with finite DC-Feed Inductance," in IEEE Transactions on Circuits and Systems, vol. 34, no. 9, pp. 1074-1087, September 1987, doi: 10.1109/TCS.1987.1086268.

[6] Acar M, Annema A J, Nauta B. Generalized design equations for class-E power amplifiers with finite DC feed inductance[C]//2006 European Microwave Conference. IEEE, 2006: 1308-1311.

[7] H. Ruan, T. Yan and Y. Huang, "A high-efficiency class e power amplifier with integrated finite DC feed inductance," 2017 IEEE 12th International Conference on ASIC (ASICON), Guiyang, China, 2017, pp. 128-131.

[8] Zhai, Chenxi, and Kwok-Keung M. Cheng. "Fully-integrated CMOS differential class-E Power Amplifier with combined waveform-shaping network and transformer-based balun." 2014 Asia-Pacific Microwave Conference. IEEE, 2014.

[9] P. Reynaert and M. S. J. Steyaert, "A 2.45-GHz 0.13-μm CMOS PA With Parallel Amplification," in IEEE Journal of Solid-State Circuits, vol. 42, no. 3, pp. 551-562, March 2007.

A 16~46-GHz, >77-dB IRR, Low-Amplitude and Phase-Error IQ Generator with Self-Adaptive I/Q Calibration in 28-nm CMOS

Lijiang Zhang [1], Wei Li *[1], Bowen Yu [1], Chengzhang Cai [1], Yue Wu [1], Bowen Chen [2], Yue Lin [2], Hongtao Xu [1]

[1] State Key Laboratory of Integrated Chips and Systems, College of Integrated Circuits and Micro-Nano Electronics, Fudan University, Shanghai, China
[2] ICLegend Micro, Shanghai, China

* Email: w-li@fudan.edu.cn

Abstract—This work proposes a broadband IQ generator based on a 1-stage Type-I RC-PPF (Poly Phase Filter), integrated with a digital-controlled 6-bit resistor array and a closed-loop self-adaptive I/Q calibration scheme. The proposed method achieves a wide frequency range of 16~46-GHz, enabling a 110% fractional bandwidth. Post-simulation results demonstrate excellent image rejection ratio (IRR) exceeding 77-dB with less than 0.1 dB amplitude imbalances and less than 1° phase imbalance by applying the self-adaptive I/Q calibration, and insertion loss around 4.6~6.1-dB while phase noise at 1MHz offset remains under -160-dBc/Hz. The entire design occupies a compact silicon area of 0.003 mm² and operates under a 1.1 V supply with 3 mW power consumption. These results validate the effectiveness of the self-adaptive I/Q calibration scheme in enhancing broadband IRR performance, making it suitable for mm-wave phased-array transceivers.

Keywords—Millimeter-wave (mm-wave), RC polyphase filter (RC-PPF), self-adaptive calibration

I. INTRODUCTION

In-phase and quadrature (IQ) signal generation is a key enabler in modern wireless communication systems, underpinning RF front-ends, modulators/demodulators, radar systems, software-defined radios, and high-frequency mixers. By decomposing signals into I and Q components, it facilitates advanced modulation schemes (e.g., QAM), thereby enhancing spectral efficiency and overall system performance. On the receiver side (RX), IQ demodulation enables synchronous extraction of modulation information while suppressing image signals. On the transmitter side(TX), accurate IQ generation is essential for image rejection, wideband modulation, and high-fidelity spectral synthesis.

Among various types of quadrature networks, hybrid couplers (HCs) [1] and quadrature all-pass filters (QAFs) [1],[2],[3],[5] have been widely adopted due to their low insertion loss at millimeter-wave (mm-wave) frequencies. However, both rely on transmission lines and/or inductors, resulting in larger chip area overhead. Another drawback lies in their limited bandwidth with respect to phase accuracy, which diminishes their suitability for wideband or dual-band applications. Although efforts have been made to extend the bandwidth of QAFs—often at the cost of increased insertion loss—QAFs remain sensitive to the imaginary part of the

load impedance (i.e., the input impedance of programmable weights), especially at mm-wave frequencies.

In contrast, the RC polyphase filter (RC-PPF)[6] I/Q network exhibits phase behavior that is insensitive to load impedance due to its inherent symmetry. Moreover, to meet a specific bandwidth requirement, two RC-PPF stages can be cascaded. In most practical scenarios, a two-stage RC-PPF suffices to fulfill application demands.

Nevertheless, a major limitation of multi-stage RC-PPFs, compared to HCs, QAFs, and even single-stage RC-PPFs, is their higher insertion loss [7]. To reduce insertion loss, these works[6],[8] had to insert amplifiers or buffers between multiple PPF stages to boost the gain, which consequently resulted in significant power consumption. Furthermore, multi-stage RC-PPFs can achieve both amplitude and phase balance only at their designed pole frequency. It means within the operating bandwidth, they can guarantee either amplitude balance or phase balance depending on the filter type, i.e. Type-I RC-PPFs ensure phase balance, while Type-II RC-PPFs ensure amplitude balance. This leads to a deterioration in the in-band IRR performance.

In this work, a broadband IQ generator with self-adaptive feedback network to extend the fractional bandwidth while enhance IRR performance is proposed. Instead of other refs [4],[6],[7],[8] using multi-stage RC-PPFs to extend bandwidth with penalty of high insertion loss, this design employs a 1-stage RC type-I PPF topology with digital-controlled self-adaptive feedback to successfully achieve 16~46-GHz wide bandwidth with > 77-dB IRR, and low power consumption without deteriorating signal phase noise. This paper is organized as follows. Section II discusses the circuit design of the proposed IQ generator. Section III shows the post-simulation results and a comparison with state-of-the-art works. Conclusion is given in Section IV.

II. CIRCUIT DESIGN

A. Proposed Type-I PPF with Self-Adaptive I/Q Calibration

The proposed self-adaptive PPF, illustrated in Fig. 1(a), consists of four main components: a Type-I 1-stage RC-PPF integrated with a 6-bit resistor array (RA), an envelope detector (ED), a comparator, and a binary search controller. The operation principle shown in Fig. 1(b) is as following.

This paper is supported by National Key R&D Program of China under Grant 2023YFB4403803, and by the Fundamental Research Funds for Central Universities in China.

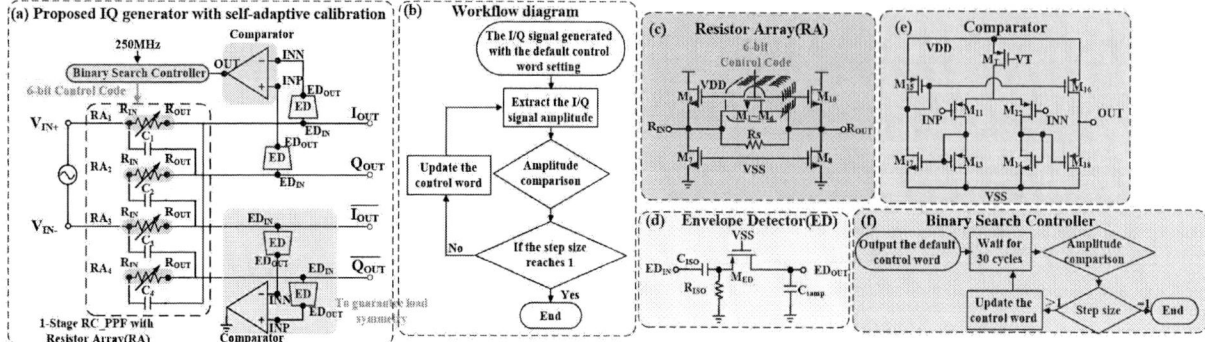

Fig. 1. The block diagram of (a) the proposed 1-stage RC-PPF I/Q generator with self-adaptive I/Q calibration and (b) its workflow, and the schematics of (c) 6-bit resistor array, (d) envelope detector, (e) comparator and (f) binary search controllr

In the initial state, the envelope detector monitors the amplitudes of the I/Q output signals from the Type-I RC-PPF under the default control code of RA, and converts it into a DC voltage. This voltage is then fed into a differential comparator, which outputs either a logic low (0) or high (VDD) based on the amplitude comparison, i.e. the amplitude imbalance. The binary search controller uses this result to perform a binary search algorithm, updating the control signals applied to the resistor array accordingly. After five iterations of the binary search, the self-calibration process "concludes", and the output control word remains fixed.

B. Resistor Array

For a conventional 1-stage Type-I RC polyphase filter (RC-PPF), as illustrated in Fig. 2(a), the amplitude ratio of the output I/Q signals can be expressed by (1). It can be observed that the phase balance of the I/Q outputs is independent of the operating frequency, whereas both amplitude and phase balance can only be achieved simultaneously when the frequency satisfies $\omega=1/RC$. As a result, when applied to broadband scenarios, amplitude imbalance tends to increase significantly.

$$\frac{H_{I, conventional}(s)}{H_{Q, conventional}(s)} = \frac{(1/sC)/(1/sC + R)}{R/(1/sC + R)} = \frac{1}{sRC} \quad (1)$$

To address the issue in the conventional 1-stage Type-I RC-PPF, this work proposes an approach: by making the resistance R tunable, it becomes possible to identify a suitable resistance value for each target frequency, thereby enabling broadband I/Q signal generation. Since the pole frequency ω of the RC-PPF is inversely proportional to the resistance R, this work adopts a parallel-weighted resistor array (RA) architecture as depicted in Fig. 1(c).

When the capacitance $C_n=200fF$ ($n=1,2,3,4$) in the RC-PPF remains constant, the extent to which the pole can be tuned is determined by the range of variation in the admittance of the resistor array. Since the admittance values introduced by the resistor array are discrete rather than continuous, the number of distinct admittance levels—i.e., the resolution—within the same bandwidth can significantly affect the quality of the I/Q output signals. In order to achieve high resolution while maintaining a 30 GHz bandwidth. a 6-bit control word is employed, yielding 63 distinct admittance levels. Additionally, a fixed resistor R_S is introduced in parallel to constrain the maximum achievable pole frequency without affecting the location of the lowest-frequency pole, thereby enhancing the overall resolution of

the system. When the resistor $R_S =66\Omega$ and the admittance of the resistor array ranges from 16mS to 65mS, the proposed 1-stage Type-I RC-PPF can achieve a tunable frequency range from 16 to 46-GHz.

Each resistor branch is controlled by a cluster switches ($M_1 \sim M_6$), where the effective resistance is determined by the ON-state resistance of the NMOS transistors. Given that the ON-resistance of an NMOS transistor is approximately proportional to the ratio L/W, the channel lengths for $M_1 \sim M_6$ are chosen as $350nm \times 2^n$, where n represents the corresponding switch transistor controlled by each bit of the control word. Therefore, the conduction admittance of the switch transistors increases by a factor of two for each successive bit from the least significant bit (LSB) to the most significant bit (MSB).

When taking into account the effects of the load capacitance C_L and the parasitic capacitance C_P in parallel with the resistors in a 1-stage RC-PPF, which is plotted in Fig. 2(b) , the amplitude ratio of the I-path and Q-path signals can be expressed by (2) and (3), respectively. From these equations, it can be observed that the phase balance characteristic of the Type-I RC-PPF is independent of the load magnitude but is significantly influenced by the parasitic capacitance C_P in parallel with the resistor. To minimize the drain-source parasitic capacitance C_P when the NMOS switch ($M_1 \sim M_6$) are turned on, all transistors ($M_1 \sim M_6$) used in the switch network shown in Fig. 1(a) are designed with a minimum channel length of 30nm. The curves showing the variation of the resistor array's equivalent admittance and parasitic capacitance as functions of the digital control word are plotted in Fig. 3.

$$\frac{H_{I, with C_L}(s)}{H_{Q, with C_L}(s)} = \frac{1/(1 + sR(C + C_L))}{sRC/(1 + sR(C + C_L))} = \frac{1}{sRC} \quad (2)$$

$$\frac{H_{I, with C_P}(s)}{H_{Q, with C_P}(s)} = \frac{1/sC}{R/(1 + sRC_P)} = \frac{1 + sRC_P}{sRC} \quad (3)$$

The NMOS transistors $M_7 \sim M_{10}$ are used to set the DC operating point of the output node. M_7 and M_8 are permanently turned on to provide a DC path to ground, while M_9 and M_{10} remain off to maintain a high-impedance state.

C. Envelope Detector

To compare the amplitude of sinusoidal signals with identical frequencies but different phases, an envelope detector, as shown in Fig. 1 (d), is employed to measure the

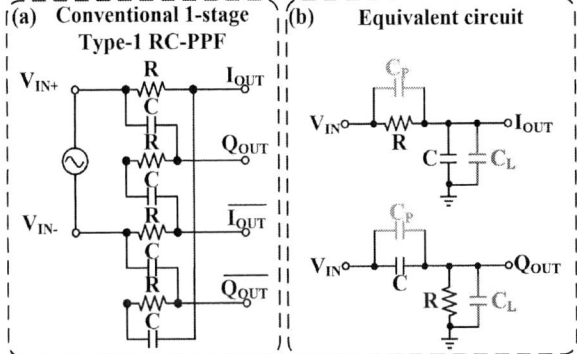

Fig. 2. Schematics of (a) conventional 1-stage type-1 RC-PPF and (b) its Equivalent circuit, which takes into account both the load capacitance C_L and parasitic capacitance C_P.

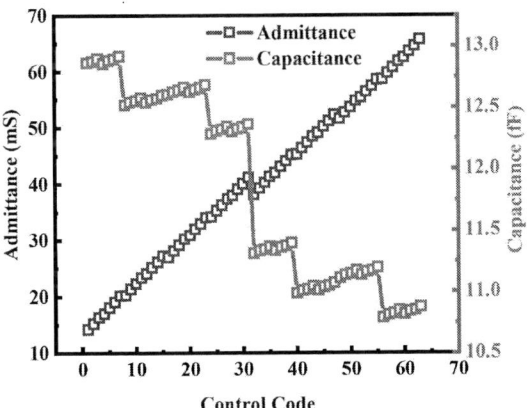

Fig. 3. Admittance and parasitic capacitance characteristics of the resistor array as a function of control code

magnitude of the I/Q signals. While the proposed envelope detector exhibits a relatively longer response time (100 ns) compared to prior designs, it benefits from zero static power consumption and a straightforward architecture, making it highly suitable for low-power and compact applications. An AC coupling capacitor C_{ISO}=100fF and a resistor R_{ISO}=600Ω to ground are inserted before the envelope detector to establish a well-set DC operating point and enhance input isolation, while maintaining the integrity of the high-frequency signal. When the input voltage ED_{IN} exceeds the voltage across the capacitor C_{samp} (=50fF), the PMOS M_{ED} conducts, allowing the capacitor C_{samp} to track the rising edge of the input. Once the input falls below the capacitor voltage, the diode turns off, and the capacitor C_{samp} gradually discharges through the load resistor, realizing a track-and-hold behavior in ED_{OUT} for the input envelope.

D. Comparator and Binary Search Controller

After the envelope detector converts the amplitude of the I/Q signals into a DC voltage, a comparator is used to generate a binary output signal (either 0 or VDD) based on the relative magnitudes of the two inputs. The schematic of the comparator is shown in Fig. 1(e). The comparison stage is implemented using a two-stage amplifier based on current mirror topology. The input voltages INP and INN are applied to the gates of the first-stage PMOS transistors M_{11} and M_{12}, and the voltage difference is translated into a corresponding voltage at node OUT through the current mirror structure.

Based on the output of the comparator, the binary search controller updates the digital control word using a binary search algorithm shown in Fig. 1(f). The initial control word is set to 32, and the adjustable step size begins at 16, halving in each subsequent iteration. Given the response time of the envelope detector (~100ns), the control unit samples the comparator output and updates the output control word every 30 clock cycles at a 250MHz clock frequency. When the step size reaches 1, the search process terminates, and the controller holds the final control word constant. This iterative process lasts approximately 400 ns.

III. SIMULATION RESULTS

The proposed self-adaptive IQ generator was implemented in 28-nm CMOS. Fig. 4(a) shows the layout of the IQ Generator with core size of 0.003 mm^2. The IQ generator operates from 16 to 46-GHz in 1.1V power supply.

Fig. 4(b) shows the post-simulated image rejection ratio (IRR), amplitude imbalance and phase imbalance as a function of the frequency. Within the 16 to 46-GHz range, the phase mismatch is less than 1°, and the amplitude mismatch is less than 0.1-dB. The phase mismatch primarily results from the parasitic capacitance introduced by the resistor array and the load asymmetry which is caused by layout-induced imbalances. The IRR remains higher than 77-dB within the operating band. Fig. 4(c) shows the post-simulated insertion loss and phase noise under 50fF load. The insertion loss varies from 4.5dB to 6.1dB while phase noise at 1MHz offset remains under -160-dBc/Hz.

A summary of the IQ generator performance and comparison is shown in Table I. Compared with existing designs, the proposed architecture achieves wide bandwidth and high IRR while maintaining low power consumption and compact area. It is demonstrated that the proposed IQ generator with self-adaptive calibration scheme achieves an excellent IRR>77-dB within 110% fractional bandwidth from 16 to 46-GHz. The core area is only 0.003mm^2 and the total power consumption is as low as 3mW. It is verified by the above results that the proposed self-adaptive structure can successfully enhance IRR performance, which will be applied in the future work.

IV. CONCLUSIONS

In this paper, we propose a IQ generator that achieves high IRR performance by self-adaptive I/Q calibration. The designed IQ generator employs a 1-stage type-I PPF with digital-controlled 6-bit resistor array to achieve a fractional bandwidth of 110% from 16 to 46-GHz. The post-simulation results show the insertion loss from 4.5 to 6.1-dB and lower than -160-dBc/Hz dB phase noise at 1MHz offset over the operating frequency band. With a core area of only 0.003mm^2 and power consumption of 3mW, amplitude imbalance less than 0.1-dBm, phase imbalance less than 1° and IRR higher than 77-dB are achieved. These results confirm that the proposed self-adaptive structure is an efficient approach to enhance IRR performance and to extend bandwidth without introducing high insertion loss compared with multi-stage PPF counterparts.

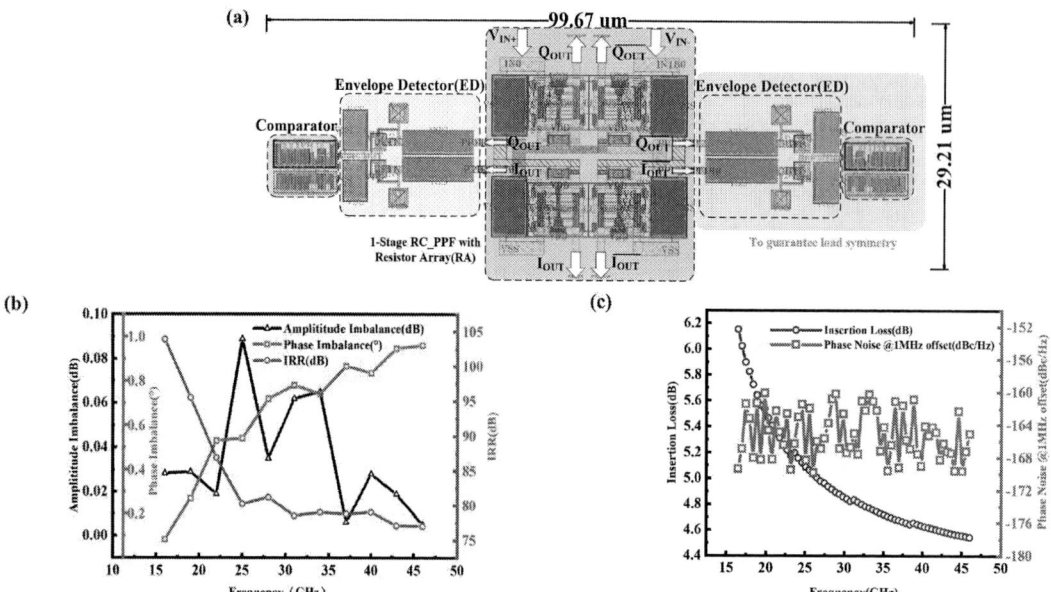

Fig. 4. (a) The layout of the proposed IQ Generator and its post simulation results of (b) image rejection ratio (IRR), amplitude imbalance and phase imbalance, and (c) insertion loss under 50fF load.

TABLE I.   PERFORMANCE SUMMARY & COMPARISON WITH STATE-OF-THE-ART CMOS IQ GENERATORS

	This work*	TMTT'24 [4]	TMTT'25 [5]	TMTT'23 [6]	RWS'23 [7]	ICICM'24 [8]
Technology	28-nm CMOS	40-nm CMOS	65-nm CMOS	22-nm FDSOI	22-nm CMOS	130-nm BiCMOS
Topology	Self-adaptive +1-stage RC PPF	Coupler +1-stage RC PPF	Improved QAF	gmC +3-stage RC PPF	Gain Boosting +2-stage RC PPF	AMP +6-stage RC PPF
Freq.(GHz)	16~46 FBW=110%	24~44 FBW=61%	21.5~35 FBW=61%	19~25 FBW=27%	24~40 FBW=51%	0.4~8 FBW=424%
Insertion Loss(dB)	4.5~6.1	4.4~5.8	-1~-3	-3	1~10	-13
Amplitude Imbalance(dB)	0.1	0.1	1	0.2	0.2	0.005
Phase Imbalance(°)	1	2.4	0.5	2	2.5	0.3
IRR(dB)	77	60.5	43.6	61	58	97
VDD(V)	1.1	0.9	3	0.85	0	1.2
Power Consumption(mW)	3	0	33.8	82.3	0	667.9
Core Area(mm²)	0.003	0.054	1.085	0.02	0.079	1.324

*post-simulated results

REFERENCES

[1] K. Kwang-Jin and G. M. Rebeiz, "0.13-mu m cmos phase shifters for x-, ku-, and k-band phased arrays," IEEE Journal of Solid State Circuits, vol. 4, 2007, pp. 2535–2546.

[2] S. Y. Kim, D.-W. Kang, K.-J. Koh, and G. M. Rebeiz, "An improved wideband all-pass i/q network for millimeter-wave phase shifters," IEEE transactions on microwave theory and techniques, vol. 60, no. 11, 2012, pp. 3431–3439.

[3] A. Hirai, T. Fujiwara, M. Tsuru, K. Mori, and M. Shimozawa, "Vectorsum phase shifter using a tunable active g m-c polyphase filter," IEEE Transactions on Microwave Theory and Techniques, vol. 68, no. 10, 2020, pp. 4091–4102.

[4] P. Gu, D. Zhao and X. You, "A Wideband Vector-Modulated Variable Gain Phase Shifter for 5G NR FR2 in 40-nm CMOS," in IEEE Transactions on Microwave Theory and Techniques, vol. 72, no. 9, 2024, pp. 5274-5284.

[5] T. Xu et al., "A Novel Wideband I / Q Network With Multiple Phase Error Zeros for Millimeter-Wave High Image Rejection Mixer," in IEEE Transactions on Microwave Theory and Techniques, vol. 73, no. 5, 2025, pp. 2640-2652.

[6] V. Åberg, C. Fager, R. Hou and L. Svensson, "Ultrawideband RF-IQ Modulator Using Segmented Nonlinearly Scaled RF-DACs and Nonoverlapping LO Signals," in IEEE Transactions on Microwave Theory and Techniques, vol. 71, no. 5, 2023, pp. 1899-1910.

[7] M. G. Bardeh, N. Naseh, J. Fu, J. Paramesh and K. Entesari, "A mm-wave RC PPF Quadrature Network with Gain Boosting in 22nm CMOS FDSOI," 2023 IEEE Radio and Wireless Symposium (RWS), Las Vegas, NV, USA, 2023, pp. 108-110.

[8] J. Liu, Q. Li and C. Zhu, "A High-Precision Broadband Quadrature Signal Generation Circuit Based on 130nm BiCMOS," 2024 9th International Conference on Integrated Circuits and Microsystems (ICICM), Wuhan, China, 2024, pp. 898-902

Impact of Process Parameter Variations on the Random Values of SRAM-Based PUFs

JinJin Shao,[1,2] Ruiqiang Song,[1,2] Chunmei Hu,[1,2] Biwei Liu,[1,2] Bin Liang,[1,2] Yaqing Chi,[1,2] Yaohua Wang*[1,2]

[1] College of Computer Science and Technology, National University of Defense Technology, Changsha, China, 410073
[2] Key Laboratory of Advanced Microprocessor Chips and Systems, Changsha, China, 410073

* Email: yaowangeth@gmail.com

Abstract—SRAM physically unclonable functions (PUFs) are low-cost cryptographic primitives implemented in secret key generation and device authentication strategies. Process parameter variations change the random values of SRAM PUF cells, which ultimately produce unique keys. This paper presents the random value of the SRAM PUF cell with multiple process parameters using a three-dimensional technology computer-aided design (3D-TCAD) simulation tool. Simulated results show that process parameters have different effects on the bias of the SRAM PUF cell. These parameters can be divided into no-impact parameters, relative-impact parameters, and absolute-impact parameters. The reason for causing the above phenomenon is analyzed. Simulated results demonstrate that sufficient process parameter variations break the current balance between transistors in the cross-coupled inverter. This is the main mechanism affecting the random values of SRAM PUF after power-up.

Keywords—Physical unclonable functions, SRAM, Process parameter variations, voltage bias

I. INTRODUCTION

Physical unclonable functions (PUFs) have recently become an innovative primitive that is being used for authentication, identification, and secret key storage [1], [2]. PUFs utilize the inherent transistor parameter variations to generate a random key. For instance, SRAM PUFs use the randomness of SRAM PUF cells at power-up to generate a unique digital code [3], [4]. The principle of SRAM PUFs is based on the incomplete symmetry of SRAM PUF cells. During the power-up stage, the random values of SRAM PUF cells are toward "0" or "1". During the read cycle, the power-up random values of SRAM PUF cells are output and combined to form a digital code. Each SRAM PUF cell has a preferred bias state at power-up due to variations in process parameters. All bias states of SRAM PUF cells form a unique data pattern that can act as a "fingerprint" of integrated circuits.

Based on the principle of SRAM PUFs, the random value of each SRAM PUF cell is strongly dependent on the variations in process parameters [5]. Different process parameters may lead to changes in the bias state of SRAM PUF cells, which will affect the final data pattern and even impact the uniformity and randomness of SRAM PUFs. Due to the uncontrollable variations of process parameters, it is difficult to fully investigate the effect of process parameter variations on the bias of SRAM PUF cells using experimental methods. Only few measurable process parameters have been investigated [6].

Recently, a three-dimensional technology computer-aided design (3D-TCAD) device-level simulation tool provides an effective method to investigate the effect of process parameter variations on SRAM PUFs. The 3D-TCAD device process simulation tool can realistically simulate the manufacturing process of integrated circuit devices. It also allows for convenient adjustments to process parameters, which is useful for simulating process parameter variations in integrated circuits. In our previous works, the 3D-TCAD device-level simulation tool proved to be a valuable method for investigating the physical mechanisms of circuit responses [7]. In this paper, it is used to investigate the effect of process parameter variations on the random value of SRAM PUF. Several process parameters are selected to change the initial values during TCAD simulations. These changes are used to simulate process parameter variations to investigate their effect on the random value of SRAM PUFs. The simulated results are statistical, and the mechanisms that cause changes in SRAM PUFs are analyzed.

II. SIMULATION SETUP

A. TCAD model and simulation conditions

A simple 6T SRAM PUF cell is used as the basic cell of SRAM PUFs. The basic schematic of an SRAM PUF cell is illustrated in Fig. 1. The cross-coupled inverters (P1, P2, N1 and N2) are depicted using the TCAD model, while the remaining transistors (T1 and T2) utilize the corresponding SPICE model. The TCAD model is calibrated to align with the electrical characteristics obtained from standard compact models for planar bulk 28nm CMOS technology. The sizes of all transistors are consistent with the SRAM PUF cell layout. The supply voltage (VDD) is set to 0.9V. The write line (WR) and bit line (Bit and Bit_n) voltages are set to 0. The power-up duration is set to 100 milliseconds. All transistor voltages and currents are recorded during TCAD simulations.

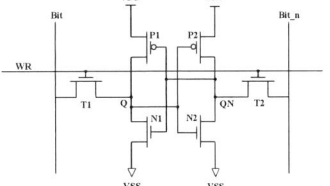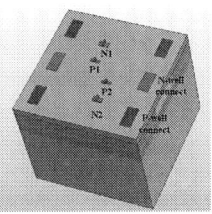

Fig.1. The basic schematic and TCAD model of 6T SRAM PUF cell

B. Process parameter selection

In order to thoroughly investigate the impact of process parameter variations on the random value of SRAM PUFs, 20 process parameters have been selected, as detailed in Table 1. These parameters are associated with transistor manufacturing, such as size parameters and doping parameters required to form a transistor. The locations of transistor manufacturing process in the TCAD model are illustrated in Fig. 2.

Table.1. The selected process parameters

Process parameter types	Process parameter name
Transistor process parameters	PMOS/NMOS_PolyGateDoping
	PMOS/NMOS_GateOxideThickness
	PMOS/NMOS_SourceDoping
	PMOS/NMOS_DrainDoping
	PMOS/NMOS_SourceLddDoping
	PMOS/NMOS_DrainLddDoping
	PMOS/NMOS_SourceHaloDoping
	PMOS/NMOS_DrainHaloDoping
	PMOS/NMOS_ThresholdDoping
	PMOS/NMOS_LeakageDoping

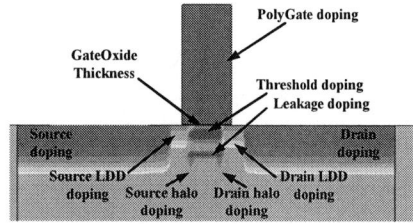

Fig.2. The locations of the selected process parameters

Two simulation strategies are used during TCAD simulations. The first simulation strategy only selects one process parameter. Subsequently, a total of 50 TCAD simulations are conducted with variations in the chosen process parameter. The change in the selected process parameter is random for each TCAD simulation, with the maximum change amplitude not exceeding ±15%. Following each TCAD simulation, the power-up random value (0.0V, 0.9V, or 0.45V) of the SRAM PUF cell is recorded. The probability of each bias state is calculated after 50 TCAD simulations. The first simulation strategy is primarily utilized to quantitatively assess the impact of a single process parameter on the random value of the SRAM PUF cell.

The second simulation strategy involves selecting more than two process parameters simultaneously. A total of 50 TCAD simulations are conducted with the chosen process parameter variations. The change in value of the selected process parameters is randomized during each TCAD simulation. The maximum change amplitude of each selected process parameter does not exceed ±15%. Following each TCAD simulation, the power-up random value (0.0V, 0.9V, or 0.45V) of the SRAM PUF cell is recorded. The probability of each bias state is calculated after 50 TCAD simulations. This simulation strategy is mainly used to assess the collective impact of multiple process parameter variations on the random value of the SRAM PUF cell.

III. SIMULATION RESULTS

A. Single parameter variation-induced SRAM PUF bias

The simulated results with a single basic process parameter variation are shown in Fig. 3. It is evident that each process parameter has a distinct effect on the random value of the SRAM PUF cell. For instance, the source/drain region doping density and the leakage-suppressing doping density do not affect the random value of the SRAM PUF cell. In the absence of changes in all process parameters, the cross-coupled inverter in the SRAM PUF cell exhibits perfect symmetry. Upon powering up the SRAM PUF cell, the node voltages Q and QN bias to a metastable value (0.45V). The simulated results

indicate that these process parameters are incapable of altering the bias of the SRAM PUF cell. These process parameters can be classified as no-impact parameters.

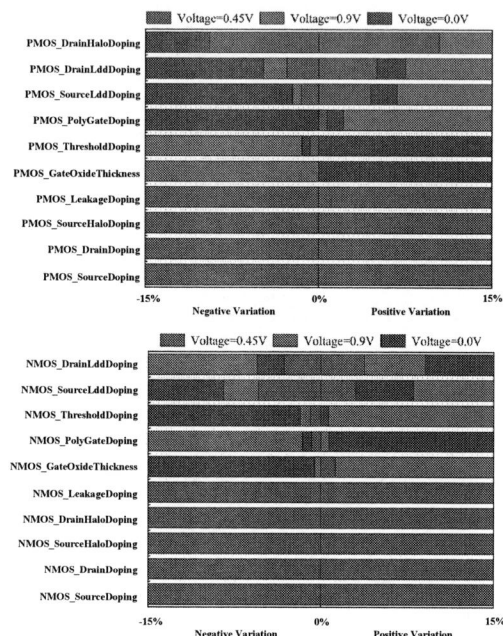

Fig.3. The simulation results of transistor process parameter variations

However, the light doping drain (LDD) density has an obvious effect on the random value of the SRAM PUF cell. For instance, the random value of the SRAM PUF cell is more likely towards 0.0V when the PMOS LDD density has a negative variation. On the contrary, when the PMOS LDD density has a positive variation, the random value of the SRAM PUF is more likely towards 0.9V. It is worth noting that not all changes in the PMOS LDD density can lead to a value change of the SRAM PUF cell. When the PMOS LDD density changes slightly, the SRAM PUF cell still maintains a metastable value (0.45V). With an increased variation in the PMOS LDD density, the SRAM PUF cell shows different values. Simulated results indicate that a sufficient variation in the LDD density affects the random value of the SRAM PUF cells. These process parameters can be referred to as the relative-impact parameters.

The other parameters, such as the threshold doping density, also impact the random value of the SRAM PUF cell. Unlike the LDD density, these parameters significantly change the random value of the SRAM PUF cell even if they only have slight variations. These parameters can be referred to as absolute-impact parameters.

B. Muti-parameter variations-induced SRAM PUF bias

Based on the single-parameter variation simulated results, process parameters have different effects on the random value of the SRAM PUF cell. These process parameters can be divided into no-impact parameters, relative-impact parameters, and absolute-impact parameters. To effectively simulate the effect of multiple process parameter variations on the random value of the SRAM PUF cell, two or more process parameters are selected from the above three types and constituted as a group of parameters. The group of parameters changes their values simultaneously during TCAD simulation, aiming to investigate the effect of multiple process parameter variations on the random value of the SRAM PUF cell.

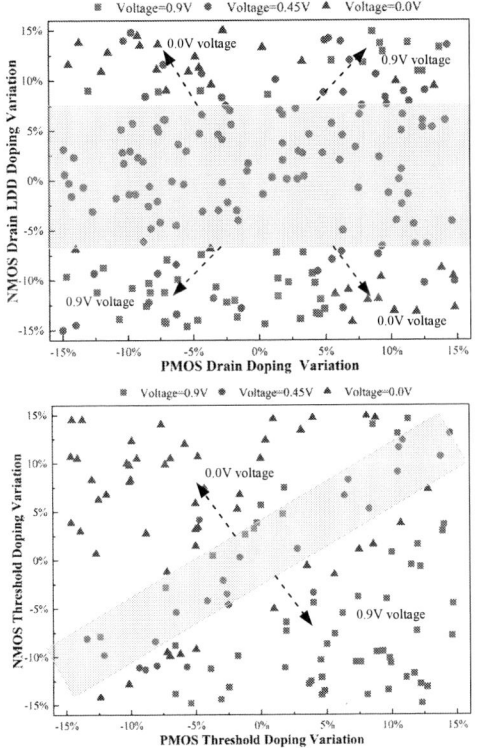

Fig.4. The simulated results of muti-process parameter variations

Fig. 4 illustrates the impact of two distinct groups on the random value of the SRAM PUF cell. The first group includes the related-bias parameter of the NMOS drain LDD density and the no-impact parameter of the PMOS drain doping density. Simulated results indicate a significant metastable region in the random value distribution. A slight change in the NMOS drain LDD density does not influence the random value of the SRAM PUF cell, even if the other parameter changes significantly. However, a substantial change in the NMOS drain LDD density determines the random value of the SRAM PUF cell. When both parameters vary in the same direction, they lead to the value of the SRAM PUF cell towards 0.9V. Conversely, when the two parameters vary in different directions, they cause the value of the SRAM PUF cell towards 0.0V.

The second group consists of two absolute-impact parameters PMOS threshold doping density and NMOS threshold doping density. The simulated bias distribution differs from the previous results. The simulation results indicate that the random value distribution includes a small metastable region. Any changes in the PMOS threshold doping density or NMOS threshold doping density immediately change the random value of the SRAM PUF cell.

IV. DISCUSSION

A. The mechanism for parameter variation-changed SRAM PUF values

Based on the simulated results, most process parameters can influence the random value of the SRAM PUF cell. How-

ever, the positive and negative variations of each process parameter do not yield the same direction for the SRAM PUF cell. To investigate the mechanism of process parameter variations on the random value of the SRAM PUF cell, the voltage and current curves of each transistor in the TCAD model are characterized during the power-up duration.

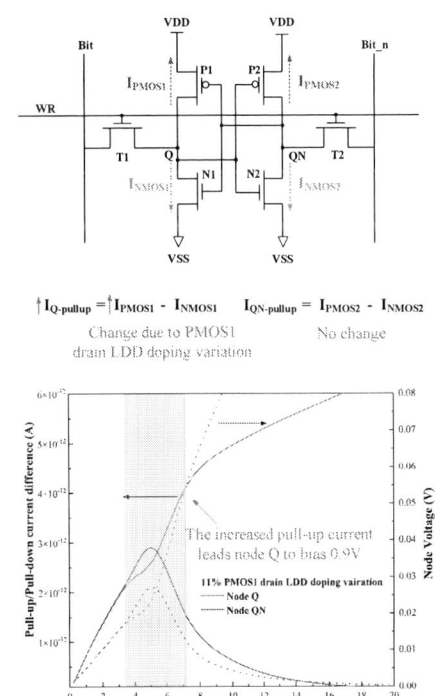

Fig.5. The mechanism for parameter variation-changed SRAM PUF values

Fig. 5 illustrates the current curves of each transistor alongside the voltage curves of the corresponding nodes Q and QN. When the SRAM PUF cell is powered up, each transistor in the cross-coupled inverter generates a corresponding drive current. For the PMOS transistors (P1 and P2), the generated drive current biases the node voltage towards 0.9V. Conversely, for the NMOS transistors (N1 and N2), the drive current biases the node voltage towards 0.0V. The final bias state of nodes Q and QN is determined by the drive currents of the PMOS and NMOS transistors. In the absence of process parameter variations, all PMOS and NMOS transistors in the cross-coupled inverter exhibit identical parameters. Consequently, the discrepancy in driving currents is consistent for both nodes Q and QN. As a result, nodes Q and QN maintain the same voltage, causing the SRAM PUF cell to reach a metastable state (0.45V).

However, when considering variations in process parameters, such as an 11% positive variation in the PMOS1 drain LDD density, the P1 transistor is impacted. When the SRAM PUF cell is powered up, the drive currents of the N1, P2, and N2 transistors remain unaffected by these variations. In contrast, the driving current of the P1 transistor increases with the PMOS1 drain LDD density variation. Consequently, this variation increases the pull-up current of node Q, resulting in the pull-up current of node Q exceeding that of node QN. The higher pull-up current makes it easier for node Q to be pulled up, causing the random value of the SRAM PUF cell towards 0.9V. Simulated results demonstrate that process parameter variations disrupt the current balance between the transistors

979-8-3315-3918-4/25 $31.00 © 2025 IEEE

in the cross-coupled inverter, which is the primary mechanism affecting the random value of the SRAM PUF cell.

B. Parameter variation-induced relative impact

Based on the previous simulated results, process parameters can be categorized into three types: no-impact parameters, relative-impact parameters, and absolute-impact parameters. To investigate the mechanisms that cause the variations in these parameter types, a specific process parameter is selected, and the voltage and current curves of each transistor are extracted during the power-up duration.

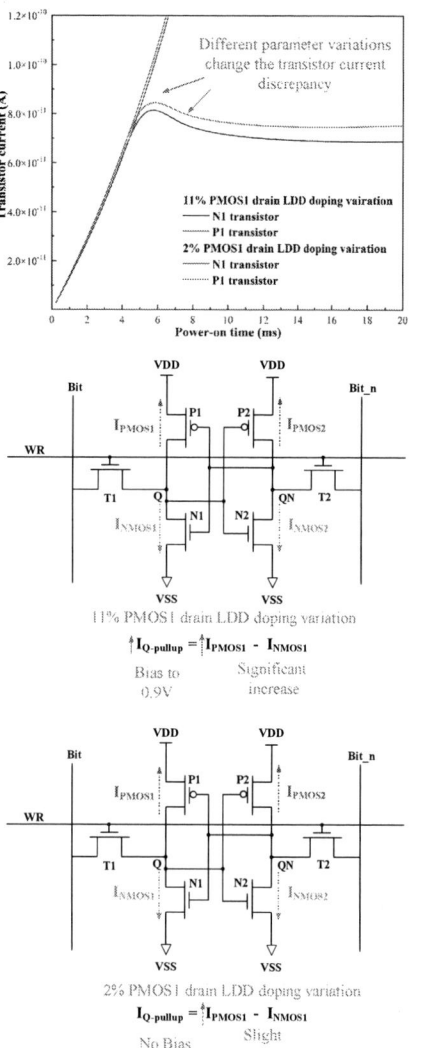

Fig.6. Parameter variation-induced relative impact

Fig. 6 illustrates the current curves of the N1 and P1 transistors with different PMOS1 drain LDD density variations. When the PMOS1 drain LDD density only produces a 2% positive variation, it slightly affects the drive current curve of the P1 transistor. The increase drive current of the P1 transistor is not enough to significantly increase the pull-up current at node Q. When the SRAM PUF cell is powered up, the current discrepancy is not enough to pull up node Q to 0.9V. Therefore, the SRAM PUF cell is still in a metastable value. However, when the PMOS1 drain LDD density produces a large positive

variation, the current discrepancy at node Q is sufficient to pull up. Thus, the current balance is broken and the SRAM PUF cell is easier to towards 0.9V. Simulated results demonstrate some process parameters with slight variations are not enough to significantly break the current balance between transistors in the cross-coupled inverter. It is the main mechanism to produce the relative-impact parameters.

V. CONCLUSIONS

SRAM PUFs use the process parameter variations of SRAM PUF cells to generate a unique digital code. Different process parameter values change the random value of SRAM PUF cells, which finally produce unique digital codes. 3D-TCAD simulation tool is used to investigate several process parameter variations on the random of the SRAM PUF cell. Simulated results show that process parameters have different effects on the SRAM PUF cell. These parameters can be divided into no-impact parameters, relative-impact parameters, and absolute-impact parameters. The main mechanism to change the random value of the SRAM PUF cell is analyzed. Simulated results demonstrate the process parameter variations break the current balance between transistors in the cross-coupled inverter. The main mechanism that causing different types of parameters is also analyzed. Simulated results demonstrate some process parameters with slight variations are insufficient to significantly break the current balance. The above studies fully clarify the effect of process parameters on the SRAM PUF cells. They can play a theoretical support for deeply understanding the random values generation of SRAM PUFs.

ACKNOWLEDGMENT

This work is supported in part by the National Natural Science Foundation of China (NSFC) under Grant Nos. 62272477 and 62174180.

REFERENCES

[1] G. E. Suh and S. Devadas, "Physical Unclonable Functions for Device Authentication and Secret Key Generation", 2007 44th ACM/IEEE Design Automation Conference, San Diego, CA, USA, pp:9-14, 2007.

[2] R. L. Sembiring, R. R. Pahlevi and P. Sukarno, "Randomness, Uniqueness, and Steadiness Evaluation of Physical Unclonable Functions", 2021 9th International Conference on Information and Communication Technology (ICoICT), Yogyakarta, Indonesia, pp:429-433, 2021.

[3] S. Xu, K. Liu, Y. Tang, R. Zhang et. al., "Effect of Quadruple Size Transistor on SRAM Physically Unclonable Function Stabilized by Hot Carrier Injection", 35th International Conference on Microelectronic Test Structure (ICMTS), Tokyo, Japan, pp:1-6, 2023.

[4] Y. Zheng, A. Bystrov and A. Yakovlev, "A Rapid Reset 8-Transistor Physically Unclonable Function Utilising Power Gating", Design, Automation & Test in Europe Conference & Exhibition (DATE), Antwerp, Belgium, pp:1-2, 2023.

[5] S. Masoumian et al., Modeling and Analysis of SRAM PUF Bias Patterns in 14nm and 7nm FinFET Technology Nodes, IFIP/IEEE 31st International Conference on Very Large-Scale Integration (VLSI-SoC), Dubai, United Arab Emirates, pp:1-6, 2023.

[6] M. R. Faragalla, M. A. Ewais, H. F. Ragai, et al., "Impact of process variability on FinFET 6T SRAM cells for physical unclonable functions (PUFs)", 12th International Conference on Computer Engineering and Systems (ICCES), Cairo, Egypt, pp:31-36, 2017.

[7] Jinjin Shao, Ruiqiang Song, Shaoqing Li, Yang Guo, Yaqing Chi, Bin Liang, Jianjun Chen, Yaohua Wang, Heavy ion-induced microdose effects on the reliability of planar and FinFET-based SRAM physical unclonable functions[J], in Appl. Phys. Lett. 124, 262103 (2024).

MIVO: Operator-Level On-Chip Memory System with Dynamic Bank Scheduling for Many-Core Neural Processing Unit

Xinghao Zhu [1], Zifeng Zhao [1], Xiaoxing Wu [1], Gengsheng Chen [1,2], Xiaofang Zhou*[1]

[1] College of Integrated Circuits and Micro-Nano Electronics, Fudan University, Shanghai, China
[2] Jiashan Fudan Institute, Jiaxing, Zhejiang Province, China

* Email: zhuxh23@m.fudan.edu.cn, xiaofangzhou@fudan.edu.cn

Abstract—With the development of artificial intelligence models, many-core Neural Processing Units (NPUs) featuring a collaborative design of parallel compute cores and hierarchical memory have shown significant advantages. However, existing on-chip memory systems encounter challenges such as inadequate utilization of storage resources and poor dynamic adaptability, limiting the processing efficiency. In this paper, we propose MIVO, an operator-level hardware-assisted memory management framework for many-core NPUs. MIVO adopts a multi-bank on-chip memory with real-time fused bank scheduling and virtual-physical bank mapping. The bank scheduling is supported by space exploration mechanism and memory fragmentation collection, improving the utilization of on-chip memory and the performance of NPU. Experiment results indicate that MIVO decreases required on-chip memory size by at most 48% compared to the baseline. In terms of performance, MIVO has a 3.3× performance boost over baseline with only 3% additional area overhead and 5% extra power consumption. Compared to the state-of-the-art schemes, MIVO achieves a performance enhancement from 1.2× to 2.7×.

Keywords—On-Chip Memory Management, Multi-bank Memory, Many-Core, Neural Processing Unit

I. INTRODUCTION

In recent years, hardware acceleration for neural network models has advanced significantly. Neural Processing Units (NPUs) tailored for deep learning tasks are evolving towards many-core architectures, providing performance benefits with large-scale parallel compute cores and hierarchical memory [1]. However, frequent off-chip DRAM accesses cause latency and bandwidth wastage in shared bus or on-chip network (NoC). Inefficient memory system has emerged as a critical factor constraining NPU performance.

To reduce off-chip accesses, high-speed on-chip memory is used for data buffering [2]. This hierarchical storage approach promotes data reuse to improve performance [2-4]. Modern NPUs organize compute cores in a cluster, which further enhances performance [5]. The single memory structure faces limitations in port number, causing address conflicts that hinder meeting multi-stream access needs. To ensure parallelism, researchers employ multi-bank memory that allows simultaneous accesses to different banks [6-8]. Nevertheless, these studies adopt static memory partitioning strategies by pre-allocating fixed banks for different data types (inputs, weights, outputs), leading to underutilization of banks and limited adaptability for various applications.

To guarantee the uninterrupted execution of calculation and processing, it is important to allocate sufficient space for every data block. However, Data blocks often come in varying

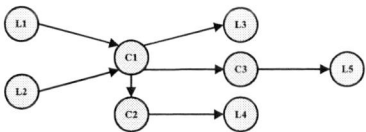

Figure 1: A part of neural network with producer-consumer relationship.

sizes with unpredictable expiration times, which inevitably give rise to memory fragmentation issues, reducing the utilization of on-chip memory. Existing works often optimize on-chip buffering from the perspective of data reuse, but lacks fragmentation collection for high utilization of memory itself [2-4] [6-8]. To the best of our knowledge, there is currently no real-time fused bank allocation supporting memory fragmentation collection mechanism.

To address these issues, we introduce MIVO (Memory System Maintaining Block Information, Block Virtual-Physical Mapping and Bank Occupancy). MIVO is an operator-level on-chip memory system for cluster-based many-core NPUs. Within each compute node, MIVO integrates a multi-bank on-chip memory (MIVO Memory, IVOM) and a controller system (MIVO Controller, IVOC). Our work makes the following contributions:

1. We design IVOM, a multi-bank on-chip memory with fused bank allocation. IVOM ensures parallelism and high utilization. We also enable virtual-physical bank mapping to guarantee IVOM's access efficiency.

2. We design IVOC, an efficient memory controller with dynamic bank scheduling, which involves space exploration and memory fragmentation collection mechanism to further improve the utilization of IVOM.

3. We implement IVOC using three lightweight tables and read/write channel controllers for operator-level memory management to enhance hardware efficiency.

4. IVOC and IVOM constitute MIVO. Experiment results indicate that MIVO reduces required on-chip memory size by at most 48% and increases performance to at least 3.3× with only 3% additional area and 5% extra power. Compared to state-of-the-art solutions, MIVO shows an improvement of 1.2×-2.7× in performance.

II. BACKGROUND AND MOTIVATIONS

A. NPU Operators and Producer-Consumer Relationships

Typical NPU supports computational operators such as convolution and matrix multiplication, along with data movement operators like load and store. We establish producer-consumer relationships among operators according to data dependencies. Figure 1(a) illustrates a part of a neural network, comprising three computational operators (C1-C3) and five data movement operators (L1-L5). Arrows represent data dependency relationships, indicating that data blocks produced by producers are used by consumers.

B. Baseline many-core NPU

The emergence of systolic arrays [9] has significantly enhanced computation efficiency. As the needs for parallel computation grow, a single systolic array falls short. Many-core NPUs boost computation parallelism. Nonetheless, the introduction of numerous data movement operators for inter-node data transmissions can lead to a decrease in processing efficiency. The cluster-based many-core NPU architecture strikes a balance between parallelism and processing delay. Figure 2 demonstrates our baseline NPU. Within a compute node, the compute engine is a cluster composed of multiple cores. The compute core is based on a Processing Unit (PE) array. We utilize Local DMA for handling data transmissions between IVOM and compute cores, and Remote DMA for IVOM and Network Interface (NI).

C. Motivations

a) Inefficient Bank Allocation Strategy: NPU requires simultaneous accesses to multiple data blocks for uninterrupted processing. Figure 3(a) demonstrates a common method involving using distinct memory modules for various data types [10][11]. While simple to implement, it repeatedly stores some data blocks to be reused. An improved approach is utilizing multi-bank memory shared among all compute cores and pre-allocating banks separately for different data types [6-8], as Figure 3(b) shows. Such solution enhances access parallelism, but lacks adaptability when dealing with different applications. Figure 3(c) illustrates our method. MIVO adopts a real-time fused bank allocation approach, which dynamically schedules banks for data blocks to reduce bank wastage.

b) Memory Fragmentation: Memory fragmentation occurs due to varying data block sizes, hindering optimal on-chip memory utilization. Figure 4 demonstrates such phenomenon. In the 10-bank memory shown in Figure 4(a), data blocks A, B, and C occupy 3, 2, and 3 banks respectively. Then data block B leaves and its banks are released, as shown in Figure 4(b). Subsequently, a new data block D requiring 4 banks needs to be written. Even though there are enough remaining banks, D cannot be processed due to the absence of fragmentation collection mechanism. To this end, MIVO enables efficient space exploration and memory fragmentation collection, thus allowing data blocks to be processed without relying on a continuous segment of address, which maximizes on-chip memory utilization.

III. MIVO

A. Architecture Overview

MIVO comprises IVOM and IVOC. IVOM is a multi-bank memory featuring adaptive fused bank allocation and virtual-physical bank mapping. IVOC consists of three lightweight tables including the Block Information Table (BIT), the Block Virtual Table (BVT) and the Bank Occupancy Table (BOT), along with read/write channel controllers. The microarchitecture of MIVO is demonstrated in Figure 5. Each bank of IVOM is a dual-port SRAM. IVOC tables are also based on SRAM modules. MIVO can be deployed in the baseline NPU shown in Figure 2 by replacing

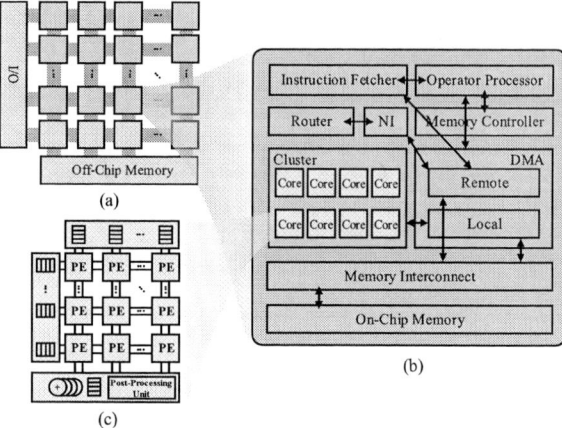

(a)

(b)

(c)

Figure 2: Baseline NPU architecture. (a) Top-level many-core architecture with compute nodes (green), mem nodes and I/O ports (blue). All the nodes are connected to a 2-D mesh NoC. (b) The architecture of compute node. Inter-node communication is accomplished by NI and Router. (c) The architecture of compute core, which is based on processing element (PE) array.

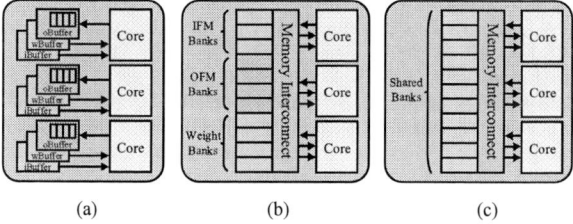

(a)   (b)   (c)

Figure 3: Several on-chip memory allocation strategies.

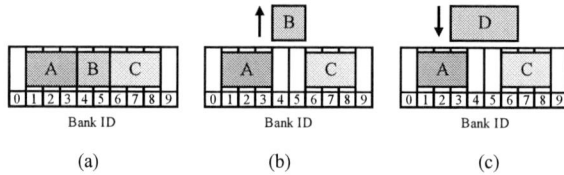

(a)   (b)   (c)

Figure 4: Memory fragmentation problem. (a) Initial situation. (b) Data block B leaves. (c) Even though there are enough remaining banks, but they are memory fragmentations.

the memory controller with IVOC, substituting the on-chip memory and memory interconnect with IVOM, and adding some additional logic circuits.

The IVOC tables, as shown in Figure 6(a), maintain important information for the efficient function of MIVO. Regardless of the type, each operator at most produces only one data block. Therefore, we assign an ID to each operator. Correspondingly, any data block produced by an operator inherits its ID. To constrain the size of tables, we establish *LOCAL_TH* as the threshold for the number of data blocks stored in IVOM per compute core and set it as 4. BIT is responsible for maintaining fundamental details regarding the data blocks generated within this node. Moreover, we define several states for data blocks, detailed in Table I, with BIT also recording these states. BOT is responsible for tracking each bank is occupied by which data block. BVT records the virtual-physical bank mappings of the data blocks in the IVOM of this node. The capacity (maximum entries) of BOT is equal to the number of banks. BIT and BVT must satisfy a scenario where the number of data blocks in IVOM from all compute cores reaches the threshold. Hence, their capacity satisfies the following formula:

$$Capacity = \min \{LOCAL\_TH \times Cores\_per\_cluser, \#Banks\}$$

TABLE I.   DATA BLOCK STATES

State	Description
USELESS	This block has been fully consumed by all consumers. No further read is required and it can be overwritten.
UNREADY	This block has not been fully produced yet and is in an indeterminate state. It cannot be read or overwritten.
ONLINE	This block is in IVOM, not fully consumed by all consumers yet. It is readable and cannot be overwritten.

TABLE II.   COMPUTATIONAL MIVO OPERATORS

Name	Description
MISO	Multiple-in-single-out computation
SISO	Single-in-single-out computation

TABLE III.   DATA MOVEMENT MIVO OPERATORS

Name	Description
GLOAD	Receive a data block from a remote node to this node
GSEND	Send a data block to a remote node
GSTORE	Send a data block to a memory node

B. High-Level Idea

To maximize the utilization of IVOM while ensuring the processing efficiency, we design an adaptive bank scheduling mechanism for MIVO, which is capable of space exploration and memory fragmentation collection. For the collaboration with NPU, MIVO supports a set of MIVO operators for describing and handling NPU behaviors. We have computational operators in Table II and data movement operators in Table III. The function of MIVO is developed based on these operators.

For a write operation, the Operator Processor sends a write request to IVOC carrying the type of operator, the number of required banks, the data block's ID, and the number of consumers. Upon receiving the request, IVOC explores the IVOM and attempts to allocate banks for the data block. As shown in Figure 6(b), BOT has two pointers (head and tail). Initially, both pointers are in the same position. During the stage of exploring space, the head pointer traverses along the BOT. Whenever it encounters an idle bank, the bank ID is recorded. Once the recorded number of banks matches the requirement, a response message containing the sequence of recorded banks is generated and sent. Simultaneously, the tail pointer marks the corresponding entries in the BOT as occupied, updates the information of this data block in BIT, and sets the state to UNREADY. Finally, the bank sequence is written to BVT as the virtual-physical list. After the operator is finished, the state is changed to ONLINE. If the two pointers overlap but recorded banks are not enough, IVOC will return a response indicating insufficient space. During this process, IVOC collects all the memory fragmentations it meets.

For a read operation, the Operator Processor sends a read request to IVOC carrying the ID of the data block to be read and the type of operator. The read channel controller then searches BIT and BVT according to the data block ID and returns the virtual-physical list as well as the block's state.

C. Virtual-Physical Mapping

To ensure the efficient reading and writing of data blocks across non-contiguous banks, MIVO employs a virtual-physical bank mapping mechanism. The IVOM address is divided into virtual segment and physical segment. The virtual segment corresponds to the virtual ID of the bank, and the

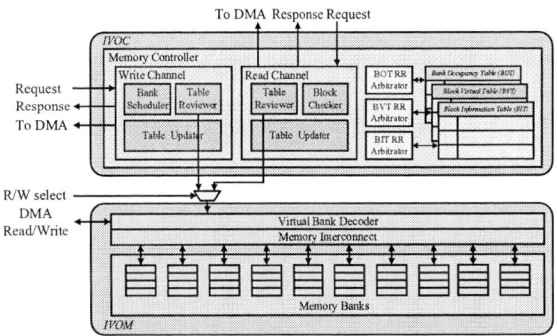

Figure 5: Microarchitecture of MIVO system.

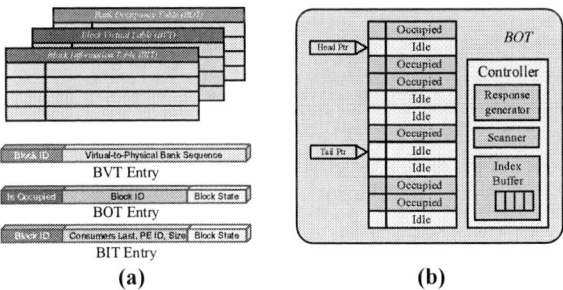

Figure 6: IVOC tables and the microarchitecture of BOT. (a) IVOC table and their entries. (b) The micro architecture of BOT. Response Generator is responsible for generating the read/write response. Scanner is responsible for controlling the two pointers. Index Buffer is responsible for recording the allocated bank IDs.

physical segment represents the address within the bank. Due to our bank scheduling strategy, a data block can occupy one or multiple banks, which may be contiguous or non-contiguous. With the virtual-physical address mapping, the virtual address segment of the read/write addresses issued by the DMA can directly be contiguous addresses from zero to the block's size. The virtual address and the virtual-physical bank mapping sequence are fed into the Virtual Bank Decoder to generate the physical address.

IV. EVALUATION

A. Experiment Setup

We present RTL implementation of MIVO on a cluster-based NPU comprised of 16 compute nodes. Each cluster contains 8 compute cores, which are based on 16×16 16-bit systolic arrays. The nodes are interconnected via a 2-D mesh NoC utilizing XY routing with a buffer depth of 8. The size of each bank of IVOM is set to 32KB, and the number of banks will be adjusted according to the specific needs of the following experiments. Our benchmarks include MobileNet v2 [12] and ResNet 50 [13].

B. Memory Utilization

To measure memory utilization, we use the Stall-Free Threshold (SFT) as the evaluation metric, which represents the minimum memory size required for the model to run without stalling due to insufficient memory space. The tests compare three configurations: fixed allocation without memory fragmentation collection (fixed_alloc), fused allocation without memory fragmentation collection (fused_alloc), and fused allocation with memory fragmentation collection (fused_alloc + mem_frag). The results are shown in Figure 7. The fused bank allocation strategy reduces the probability of bank wastage and improves

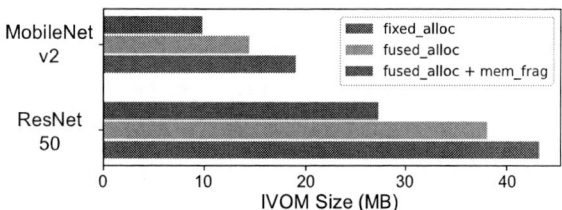

Figure 7: SFT (Stall-Free Threshold) of different schemes.

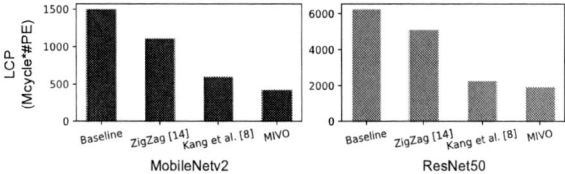

Figure 8: LCT (Latency-Compute Produce) of different schemes.

memory utilization. Experiment results indicate that such allocation decreases the required memory size by at most 24%. The memory fragmentation collection strategy further reduces the likelihood of bank wastage, thus reducing SFT by at most 32%. Overall, compared to the basic scenario, MIVO provides a maximum optimization of 48%.

C. Runtime Performance

We conduct tests for runtime performance with 8MB on-chip memory. Our baseline architecture is a basic cluster-based many-core NPU with fixed bank allocation. We compare MIVO with baseline and two state-of-the-art schemes. For fairness in compute resources, we use Latency-Compute Product (LCP) as the measurement criterion, which is the product of processing delay and the number of PEs. The results are illustrated in Figure 8. Benefiting from the collaborative design of IVOC and IVOM, MIVO has an improvement of at least $3.3\times$ compared baseline architecture and a $1.2\times\text{-}2.7\times$ improvement over the state-of-the-art schemes.

D. Area and Power

We synthesize and implement MIVO as well as baseline using a commercial 28nm CMOS process at 500MHz to evaluate area and power. During the back-end flow, we set the on-chip memory size of both MIVO and the baseline to 8MB. The results are summarized in Table IV. Due to the lightweight implementation of IVOC, MIVO only incurs additional 3% area overhead and 5% power consumption, showcasing high hardware efficiency for achieving the minimum $3.3\times$ performance improvement mentioned in IV-C.

TABLE IV. AREA AND POWER OF MIVO AND BASELINE

Schemes	Area (mm²)	Power (W)
Baseline	5.547079	1.3000733
MIVO	5.702082	1.3689666

V. CONCLUSION

In this paper, we propose MIVO, a novel on-chip memory system for cluster-based many-core NPUs. MIVO is an operator-level hardware framework consisting of a dynamic fused multi-bank on-chip memory named IVOM and an efficient memory controller named IVOC. The fused bank allocation ensures high access parallelism and adaptability. We introduce space exploration and memory fragmentation collection mechanism to maximize the utilization of IVOM and improve processing performance. Additionally, we enable the virtual-physical bank mapping mechanism for high efficiency in IVOM access. The experiment results indicate that MIVO reduces required on-chip memory size by at most 48%. In terms of processing performance, MIVO exhibits an improvement of at least $3.3\times$ over baseline with only 3% extra area and 5% extra power. Compared to state-of-the-art schemes, MIVO has a $1.2\times\text{-}2.7\times$ performance improvement.

REFERENCES

[1] Norm Jouppi et al., "TPU v4: An Optically Reconfigurable Supercomputer for Machine Learning with Hardware Support for Embeddings," 2023 50th Annual International Symposium on Computer Architecture (ISCA), pp. 1–14.

[2] D. T. Nguyen, H. Je, T. N. Nguyen, S. Ryu, K. Lee and H. -J. Lee, "ShortcutFusion: From Tensorflow to FPGA-Based Accelerator With a Reuse-Aware Memory Allocation for Shortcut Data," in IEEE Transactions on Circuits and Systems I: Regular Papers, vol. 69, no. 6, pp. 2477-2489, June 2022.

[3] H. Zhou et al., "RISC-V based Fully-Parallel SRAM Computing-in-Memory Accelerator with High Hardware Utilization and Data Reuse Rate," 2023 5th International Conference on Artificial Intelligence Circuits and Systems (AICAS), Hangzhou, China, 2023, pp. 1-5.

[4] A. Marchisio, M. A. Hanif and M. Shafique, "CapsAcc: An Efficient Hardware Accelerator for CapsuleNets with Data Reuse," 2019 Design, Automation & Test in Europe Conference & Exhibition (DATE), Florence, Italy, 2019, pp. 964-967.

[5] Y. -H. Chen, T. -J. Yang, J. Emer and V. Sze, "Eyeriss v2: A Flexible Accelerator for Emerging Deep Neural Networks on Mobile Devices," in IEEE Journal on Emerging and Selected Topics in Circuits and Systems, vol. 9, no. 2, pp. 292-308, June 2019.

[6] M. Shi et al., "CMDS: Cross-layer Dataflow Optimization for DNN Accelerators Exploiting Multi-bank Memories," 2023 24th International Symposium on Quality Electronic Design (ISQED), San Francisco, CA, USA, 2023, pp. 1-8.

[7] J. Song et al., "7.1 An 11.5TOPS/W 1024-MAC Butterfly Structure Dual-Core Sparsity-Aware Neural Processing Unit in 8nm Flagship Mobile SoC," 2019 IEEE International Solid-State Circuits Conference (ISSCC), San Francisco, CA, USA, 2019, pp. 130-132.

[8] D. Kang et al., "Multi-Bank On-Chip Memory Management Techniques for CNN Accelerators," in IEEE Transactions on Computers, vol. 71, no. 5, pp. 1181-1193, 1 May 2022.

[9] Kung, "Why systolic architectures?," in Computer, vol. 15, no. 1, pp. 37-46, Jan. 1982.

[10] D. Kang, D. Kang and S. Ha, "HierArch: A Cluster-Based DNN Accelerator with Hierarchical Buses for Design Space Exploration," 2023 IEEE 36th International System-on-Chip Conference (SOCC), Santa Clara, CA, USA, 2023, pp. 1-6.

[11] Z. Li, W. Mao, S. Zhang, Q. Dong and Z. Wang, "An Efficient Sparse Hardware Accelerator for Spike-Driven Transformer," 2024 IEEE Asia-Pacific Conference on Applied Electromagnetics (APACE), Langkawi, Kedah, Malaysia, 2024, pp. 250-253.

[12] M. Sandler, A. Howard, M. Zhu, A. Zhmoginov and L. -C. Chen, "MobileNetV2: Inverted Residuals and Linear Bottlenecks," 2018 IEEE/CVF Conference on Computer Vision and Pattern Recognition (CVPR), Salt Lake City, UT, USA, 2018, pp. 4510-4520.

[13] K. He, X. Zhang, S. Ren and J. Sun, "Deep Residual Learning for Image Recognition," 2016 IEEE Conference on Computer Vision and Pattern Recognition (CVPR), Las Vegas, NV, USA, 2016, pp. 770-778.

[14] L. Mei, P. Houshmand, V. Jain, S. Giraldo and M. Verhelst, "ZigZag: Enlarging Joint Architecture-Mapping Design Space Exploration for DNN Accelerators," in IEEE Transactions on Computers, vol. 70, no. 8, pp. 1160-1174.

DyQRA: A Deadlock-free Routing Algorithm for Large-Scale Mesh NoCs

Haoxiang Sun
School of Software and Microelectronics
Peking University
Beijing, China
sunhx@stu.pku.edu.cn

Aoyun Feng
School of Software and Microelectronics
Peking University
Beijing, China
1832903742@qq.com

Hongfei Ye
School of Integrated Circuits
Peking University
Beijing, China
hfye@ime.pku.edu.cn

Jianhua Feng
School of Integrated Circuits
Peking University
Beijing, China
fengjh@pku.edu.cn

Abstract—Adaptive routing using Q-learning in Network-on-Chips (NoCs) often fails to outperform static algorithms like XY or DyXY due to delayed Q-table updates causing inflexible routing during convergence. This paper presented DyQRA (Dynamic Q-learning Routing Algorithm), a novel deadlock-free routing algorithm designed for large-scale mesh NoCs. DyQRA synergistically combines the congestion awareness of DyXY with the adaptability of Q-learning. It dynamically adjusts routing paths based on real-time network stress by utilizing backward credit flits to update the Q-table, prioritizing less congested routes. A key innovation is the strategic use of DyXY during the initial Q-table stabilization phase, seamlessly transitioning to optimized Q-learning-based routing once trained. DyQRA incorporates a deadlock-free turning model to ensure operational safety. Evaluated using the Gem5 simulator, DyQRA demonstrates significant performance improvements: it reduces average flit latency by 40-50% compared to XY routing under congestion and outperforms DyXY by 3.79% in full-system simulations using the PARSEC benchmark suite, achieving up to an 11.86% reduction in simulation time for specific workloads. DyQRA provides enhanced congestion handling and lower latency without substantial resource overhead.

Keywords—Q-learning, Network-on-Chip, adaptive routing, deadlock-free, congestion

I. INTRODUCTION

With the growing complexity of many-core processors, efficient and reliable on-chip communication has become critical [1]. Static routing algorithms like XY and DyXY, while simple and deadlock-free, are inefficient in dynamic traffic conditions. Adaptive algorithms such as Q-learning offer potential but face challenges in convergence speed and reliability, especially during initial routing phases. This paper introduces DyQRA, a hybrid solution leveraging the fast-decision-making capability of DyXY with Q-learning's adaptability to traffic congestion. This approach ensures deadlock-free operation and improves congestion handling without significantly increasing resource overhead.

In chip design, various IPs are interconnected in a two-dimensional layout. While numerous flexible neural network routing algorithms have been proposed to enhance performance under complex traffic patterns [2][3], their adoption by IC

designers has been limited due to concerns about power consumption and area overhead. The traditional XY routing algorithm [4] routes packets to their destination in a fixed dimension, but it struggles with adaptability in dynamic traffic scenarios. To overcome this limitation, the Dynamic XY (DyXY) routing algorithm [5] has emerged, which builds on the traditional XY framework by implementing congestion-aware routers to eliminate certain constraints, thereby better accommodating dynamically changing traffic patterns.

Despite its potential as an adaptive routing algorithm, Q-learning has not demonstrated significant advantages over static algorithms like XY or DyXY, mainly due to the delays associated with updating the Q-table. In the initial phases of network operation, the time required for the Q-table to converge leads to inflexible routing decisions, which can negatively impact overall performance. The Q-learning algorithm relies on a routing table that registers the expected rewards for each action, but its slow convergence in high-congestion scenarios limits its effectiveness.

This paper presents DyQRA, a novel routing algorithm that combines the strengths of Dynamic XY and Q-learning with a deadlock-free turning model. DyQRA utilizes a dedicated routing table in each routing unit, enabling real-time decision-making for the next step every time a flit enters the router. This approach allows DyQRA to effectively avoid severe traffic stalls while maintaining a deadlock-free operation, all without imposing substantial resource overhead.

II. DyQRA AND ROUTER ARCHITECTURE

A. Q-learning Routing Algorithm

The Q-learning Routing Algorithm (QRA) is an adaptive routing method designed for on-chip networks, based on the Q-learning model as adopted by Farahnakian et al. [6]. QRA aims to reduce congestion caused by complex traffic patterns, offering a lightweight alternative to deep learning routing algorithms. It can be implemented using a routing table and a few multiplexers, as its quantification methods differ from more resource-intensive approaches. In QRA, latency is quantified as rewards, where higher latency corresponds to lower rewards. The primary goal of the Q-learning model is to

979-8-3315-3918-4/25 $31.00 © 2025 IEEE

develop an accurate estimation of each path's latency to the destination node. The core mechanism of QRA involves storing estimated Q-values for every destination node in the routing table, allowing each router to determine the next move for flits in order to maximize its Q-value. The Q-table is updated based on the true reward received from the downstream router, using the Bellman equation, as illustrated in:

$$Q(s,a) \leftarrow Q(s,a) + \alpha \left[R(a) + \gamma \max_{a'} Q(s',a') - Q(s,a) \right]$$

In the equation, s stands for the state of a router, a means the routing decision it selects, s' represents all possible states of this router, and a' represents the related action that leads to state s'. $Q(s,a)$ represents the estimated value, $R(a)$ represents the actual reward from action a. Additionally there are two parameters controlling the model, α is the learning rate, lesser α results in slower learning, γ is the decay parameter which indicates how much the model values the future Q value. The decision-making process is based on epsilon-greedy strategy, allowing the model to deviate from the action that leads to a higher Q value to explore all possible routes' latency. Of all the elements, Q values are registered with a Q-table as Table I shows.

TABLE I.   Q-TABLE EXAMPLE IN MESH NETWORK

Destination Node	Q-table at node X			
	East	*South*	*West*	*North*
0	Q(0,E)	Q(0,S)	Q(0,W)	Q(0,N)
1	Q(1,E)	Q(1,S)	Q(1,W)	Q(1,N)
2	Q(2,E)	Q(2,S)	Q(2,W)	Q(2,N)
...	...	...	...	...

a. Each destination node is maintained as a Q-table entry

b. Unreachable directions are marked as -1

To implement Q-learning within the mesh network, we utilize a dedicated routing table in each routing unit. This routing table records all destination nodes and their corresponding four possible directions. When a flit arrives at a router, it consults the routing table to select the direction with the highest expected reward based on historical experience. The routing table is updated each time a flit exits to the next routing unit, integrating feedback from credit flits that return to the previous router.

In this paper, we adopt routers using virtual channels [7]. As Fig. 1 illustrates, the router architecture comprises input units, output units, a virtual channel allocator, a switch allocator, a crossbar switch, a routing compute unit, and a stress bank. The routing algorithm is executed within the routing compute unit, which utilizes stress information from the stress bank. The stress bank receives stress values from link credit flits sent by the upstream router [8]. Each stress counter accumulates stress information, which is conveyed to the downstream node through link credit flits, indicating the congestion level. As shown in Table II, stress values are categorized into four classes. Higher input virtual channel occupancy correlates with increased stress levels, resulting in

reduced rewards for congested routes. This reward classification can be customized based on the specific requirements of the network, and a rigorous quantization of rewards can significantly enhance overall system performance.

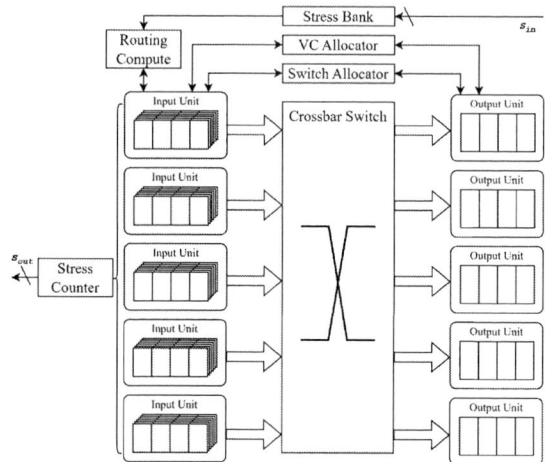

Fig. 1.   Router architecture with stress bank.

TABLE II.   STRESS VALUE CATEGORIZATION USING VC OCCUPATION

Input VC Occupation(%)	Stress Class	Reward
0~25%	Low	1
25~50%	Middle	0.7
50~75%	High	0.5
75~100%	Extreme	0

B. Proposed DyQRA

DyXY efficiently routes packets toward their destinations using the minimum number of steps, thus avoiding unnecessary detours and mitigating potential deadlocks [9]. However, since the Q-table in Q-learning requires time to initialize and stabilize, DyQRA strategically employs DyXY during this initialization phase, leveraging its rapid decision-making capabilities. DyQRA seamlessly transitions to QRA once the Q-table has had sufficient time to train and provide effective routing decisions.

The routing algorithm's decision-making process is illustrated in Table III, where a threshold parameter governs the choice between DyXY and QRA. A higher threshold increases the possibility of using DyXY, serving as a conservative measure in fluctuating traffic conditions.

Deadlock-free routing is achieved through a turning model that prohibits certain turns as depicted in Fig. 2. According to [10], routing in a 2D mesh can be categorized into eight turning patterns, and a routing algorithm is considered deadlock-free if it guarantees no traffic loops are formed. In

DyQRA, the turning restrictions are implemented as Table III shows to maintain this property.

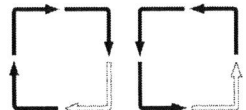

Fig. 2. Router architecture with stress bank.

TABLE III. PSEUDO CODE OF ALGORITHM DECISION

1. Candidate_Direction: All output directions to be selected.
2. N_d: Destination node.
3. Q(N, a): Q-table entry for node N.
4. FOR dirn IN Candidate_Direction:
5.　　use_QRA = FALSE;
6.　　IF $Q(N_d, dirn)$ - average_q_value > threshold;
7.　　　　use_QRA = TRUE;
8.　　　　EXIT;

TABLE IV. ROUTING RULES IN DYQRA

1. No turn-around
2. If inport direction is West, forbid North
3. If inport direction is North, forbid West
4. If N_d is at straight South, forbid East and North
5. If N_d is at straight East, forbid West and South
6. If N_d is at South West/North West/North East/Straight West/Straight North, forbid East and South

Routing unit masks all forbidden directions when a flit enters the router before making QRA decisions. During QRA, a Q-table lookup is made to check if the best action's q value is higher enough than the average value of this q-table entry, This step allows DyQRA to differentiate the most optimal action from others; if the Q-value is insufficient, DyXY is applied.

We use Fig. 3 as an example to illustrate how a DyQRA routing decision is made. A flit is injected from node 9 with a destination at node 7. Assuming router 9 used a greedy decision rather than following the Q-table, the flit may be routed to router 8. As Fig. 3(b) shows, node 8 references the Q-table and finds two potential directions, ultimately choosing to move north toward node 12. At node 12, if the Q-value for routing to node 7 is deemed insufficient, DyXY will take precedence, although only the east direction remains available. As shown in Fig. 3(d), router 13 applies the turning restrictions to ensure that the flit cannot move west, and there are two directions left to choose from. Whether QRA or DyXY is made, the flit might move to router 14 or router 9. Ideally, the flit should move east; however, if it moves south instead, it could create a round trip back to router 9. To prevent this, our routing scheme enforces rule 3, which forbids the flit from moving west. This mechanism effectively breaks any potential traffic loops, ensuring that no deadlocks occur.

C. Q-table Customization

In DyQRA, each routing unit maintains a dedicated Q-table for every destination node, allowing for fine-tuned traffic management. This customization enables us to adapt the routing decisions based on specific traffic patterns and priorities. In Fig. 4, when implementing DyQRA in a 2D mesh network with memory controllers positioned along the bottom edge, it is crucial to prioritize transactions from high-speed Direct Memory Access (DMA) units. To achieve this, the Q-tables of adjacent routers can be pre-programmed to avoid routing flits toward the DMA node, ensuring that high-priority transactions are efficiently handled. For example, at router 10, the Q-table entry for node 2 can be configured to -1 to prohibit routing to the south, thereby preventing any flits from traveling to router 2 through router 6. This strategic customization allows DyQRA to meet complex requirements such as fault tolerance, power efficiency, and enhanced hardware security.

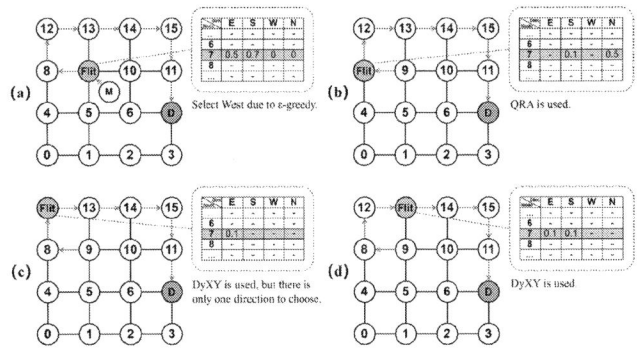

Fig. 3. Routing Example using DyQRA.

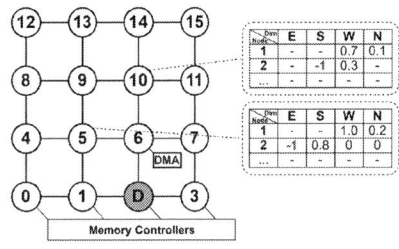

Fig. 4. Customizing Q-table to set preserved zone.

By tailoring the Q-tables to reflect the specific operational needs of the network, DyQRA can optimize routing decisions, thereby improving overall performance and responsiveness in various scenarios.

III. EXPERIMENTS AND RESULTS

To evaluate the performance of DyQRA, we used the Gem5 simulator, which provides an accurate environment for full-system simulations in computer architecture research. Gem5 is particularly well-suited for validating Network-on-Chip (NoC) routing algorithms. We conducted our experiments using synthetic traffic patterns and full-system benchmarks.

A. Synthetic Traffic Experiment

In our first experiment, we aimed to assess DyQRA's performance under conditions of random partial congestion. We set up a 16-core system, where each router is connected to a pseudo CPU and a directory controller. We utilized uniform random traffic pattern with a 0.1 injection rate, randomly designating some routers as congested by artificially increasing their latency.

The results, illustrated in Fig. 5, demonstrate that DyQRA significantly outperforms both DyXY and QRA in terms of average flit latency across varying levels of congestion. Initially, with no congested nodes present, all algorithms performed comparably. However, as we increased the number of congested routers to 12%, DyQRA exhibited a latency reduction of 40-50% compared to the static XY routing algorithm. When compared to QRA and DyXY, DyQRA achieved a latency reduction of 7.8-15.9%, showcasing its effectiveness in managing congestion and maintaining performance as the network approached saturation.

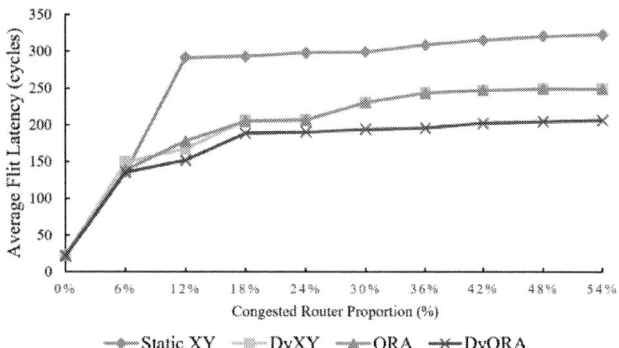

Fig. 5. Average flit latency of four algorithms at different congestion level.

B. Full-System Simulation

To further validate DyQRA's performance in a realistic environment, we implemented the algorithm alongside DyXY in a full-system simulation within Gem5. We modeled a 16-core system with four DDR4 memory controllers located at the corners of a 2D mesh network. The PARSEC benchmark suite was selected for this phase of testing, as it includes a range of emerging applications relevant to recognition, mining, synthesis, and system-level performance evaluation. The input size was SimSmall for all workloads. Table V outlines other detailed parameters for the full-system experiments.

TABLE V.   PARAMETERS IN FULL-SYSTEM EXPERIMENT

Parameter	Description
Network Topology	2D Mesh with Cornered Directory
Network Size	4×4
VCs per Vnet	4
CPU Type	X86KvmCPU × 16
CPU Frequency	2GHz
Memory Type	DualChannel DDR4 2400 × 4
Coherence Protocol	MOESI_CMP_Directory
Kernel	X86_64 Linux Kernel 4.19.83

Fig. 6 demonstrates DyQRA's consistent performance advantage over DyXY across the PARSEC benchmark suite under SimSmall input sizes. Notably, DyQRA achieves its most significant improvement in the Canneal workload, reducing simulation time by 11.86%. This substantial gain can be attributed to Canneal's irregular communication patterns and high memory contention, where DyQRA's congestion-aware routing dynamically avoids hotspots more effectively than the reactive DyXY mechanism. Significant improvements are also observed in Fluidanimate (7.2%) and Blackscholes (5.1%), workloads characterized by bursty traffic and varying computation-memory ratios. These results underscore DyQRA's strength in adapting to dynamically congested paths through real-time Q-table updates. Conversely, workloads with more uniform traffic patterns like Bodytrack show marginal differences (0.8%), indicating that DyXY's static optimization suffices under low network stress. Critically, DyQRA never underperforms DyXY, achieving an average simulation time reduction of 3.79% across all benchmarks. This performance gap directly correlates with DyQRA's hybrid strategy: its ability to leverage Q-learning's long-term congestion avoidance while relying on DyXY's deadlock-free guarantees during transient periods ensures optimal path selection under diverse traffic loads, validating its efficacy in real-world heterogeneous many-core systems.

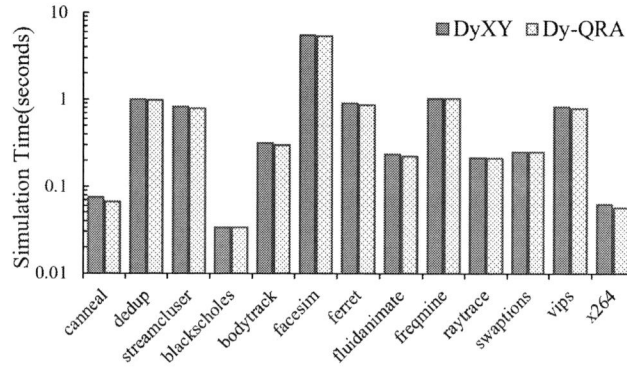

Fig. 6. Simulation time of workload in SimSmall input size.

IV. CONCLUSION

This paper presented DyQRA (Dynamic Q-learning Routing Algorithm), a novel deadlock-free routing algorithm designed to address the limitations of existing static and adaptive routing schemes in large-scale mesh Network-on-Chips (NoCs). DyQRA uniquely integrates the congestion-awareness and fast decision-making of the DyXY algorithm with the adaptability of Q-learning, while ensuring deadlock freedom through a structured turning model.

The key innovation of DyQRA lies in its hybrid routing strategy: it strategically employs DyXY during the critical Q-table initialization and stabilization phase to avoid the performance pitfalls of untrained Q-learning, and seamlessly transitions to optimized Q-learning-based routing once the Q-table converges. This leverages the strengths of both approaches. Furthermore, DyQRA utilizes backward credit flits to dynamically update Q-tables with real-time network stress information, enabling congestion-aware path selection. The dedicated Q-table structure also allows for customizable routing policies to meet specific requirements like priority handling or fault tolerance.

Experimental evaluation using the Gem5 simulator demonstrated DyQRA's significant performance advantages: (1) Under synthetic traffic with partial congestion (12% congested nodes), DyQRA achieved a 40-50% reduction in average flit latency compared to static XY routing, and a 7.8-15.9% reduction compared to both standard Q-learning (QRA) and DyXY. (2) In full-system simulations using the PARSEC benchmark suite (SimSmall input), DyQRA consistently outperformed DyXY, achieving an average 3.79% reduction in simulation time across all workloads. Notably, it delivered an 11.86% reduction for the Canneal workload. These results validate DyQRA's effectiveness in mitigating congestion and reducing communication latency, particularly under dynamic and stressful network conditions, without imposing significant resource overhead.

ACKNOWLEDGMENT

This work is supported by the National Natural Science Foundation of China (No. 62174002) and the National Key R&D Program of China under grant (2022YFF1202302).

REFERENCES

[1] D. Wu, B. M. Al-Hashimi, and M. T. Schmitz, "Improving Routing Efficiency for Network-on-Chip through Contention-Aware Input Selection", in Proc. of 11th Asia and South Pacific Design Automation Conference, 2006, pp. 36-41.

[2] M. F. Reza, "Deep Reinforcement Learning Enabled Self-Configurable Networks-on-Chip for High-Performance and Energy-Efficient Computing Systems," IEEE Access, vol. 10, pp. 65339-65354, 2022.

[3] K. Wang and A. Louri, "CURE: A High-Performance, "Low-Power, and Reliable Network-on-Chip Design Using Reinforcement Learning," IEEE Transactions on Parallel and Distributed Systems, vol. 31, no. 9, pp. 2125-2138, 2020.

[4] Dally W , Towles B .Principles and Practices of Interconnection Networks. 2004.

[5] M. Li, Q. Zeng, and W. Jone, "DyXY-a proximity congestion-aware deadlock-free dynamic routing method for network on chip," in Proc. 43rd annual Design Automation Conference (DAC 06), 2006, pp. 849-852.

[6] Farahnakian F , Ebrahimi M , Daneshtalab M ,et al. "Q-learning based congestion-aware routing algorithm for on-chip network," in Proc. of the 2nd IEEE International Conference on Networked Embedded Systems for Enterprise Applications, 2011, pp.8-9.

[7] Dally, William J ."Virtual-Channel Flow Control," Acm Sigarch Computer Architecture News, vol. 18, no. 3, pp. 60-68,1990.

[8] Y. Liu, R. Guo, C. Xu, X. Weng and Y. Yang, "A Q-Learning-Based Fault-Tolerant and Congestion-Aware Adaptive Routing Algorithm for Networks-on-Chip," IEEE Embedded Systems Letters, vol. 14, no. 4, pp. 203-206, 2022.

[9] Y. Xiang, J. Meng and D. Ma, "A Q-routing based self-regulated routing scheme for network-on-chip," in IEEE 9th International Conference on Communication Software and Networks (ICCSN), 2017, pp. 177-181.

[10] Glass C J , Ni L M. "The Turn Model for Adaptive Routing," Acm Sigarch Computer Architecture News,1994.

An Enterprise Solid-State Drive Controller Supporting Spin-transfer Torque Magnetoresistive Random Access Memory

Chao Song *[1], Qihao Liu [2], Yunzhe Wang [2], Rufa Su [2]

[1] Shandong Yunhai Guochuang Innovative Technology Co., Ltd, Jinan, China
[2] Shandong SinoChip Semiconductors Co., Ltd, Jinan, China

* Email: songchao05@inspur.com

Abstract—Flash-based solid state drives (SSDs) incorporate DRAM for data cache, which significantly optimize performance. Due to the volatile nature of DARM, the essential metadata information and address mapping table are easy to be lost, hence it is difficult to guarantee the reliability of the information. One promising solution is replacing DRAM with Spin-transfer Torque Magnetoresistive Random Access Memory (STT-MRAM) which is a non-volatile memory with low latency. This paper proposed a SSD controller architecture of employing STT-MRAM and DRAM residing on the DDR4 bus. The DDR4 controller contained in the SSD controller was optimized for support handling differing timing, command and addressing requirements for STT-MRAM. Furthermore, a system level error correcting code (ECC) was applied to the STT-MRAM interface to correct bit errors occurring inside the STT-MRAM chip. This SSD controller was silicon-proven working well, and presented the expected read & write speed of STT-MRAM.

Keywords—*Enterprise SSD, DDR4, STT-MRAM, System-Level ECC*

I. INTRODUCTION

Flash-based solid state drives are used more and more widely in data centers, especially in the recent years, thanks to those new NAND flash technologies including triple-level cell (TLC) technology, quad-level cell (QLC) technology, and 3-D stack technology[1-3]. Those new technologies have together contributed a sustained drop in bit costs and higher performance. Despite all the new technology, there are still many drawbacks existing in the application process: unbalanced write/read latency, low write endurance, and high write energy.

Traditionally, DRAM-based SSD caching solutions have sought to avoid these issues by incorporating DRAM for write buffering, which significantly reduce write latency issues and extend the service life of SSDs. The essential metadata information and address mapping table are stored in DRAM. Due to the volatile nature of DRAM, those in-the-fly data and essential information are vulnerable to be lost, hence it is difficult to guarantee the reliability of the information. Meanwhile, DRAM requires periodically refreshing the data due to the leakage static power, which means Additional energy consumption. To protect in-the-fly data and map table, enterprise SSDs integrate capacitors as back-up power source for internal write buffer and data cache. In the event of an unexpected power loss, there is enough energy to support critical data migration tasks, so that critical data from DRAM can be promptly written to NAND flash to safeguard in-the-fly data security. But the aging of power backup unites might

be accelerated by high temperature in the most enterprise usage scenarios, which greatly reduces the lifetime of enterprise SSDs. As a promising solution of the above problems, storage class non-volatile semiconductor memories (SCM) such as spin-transfer torque magnetoresistive random access memory (STT-MRAM), phase-change memory (PCM), XL-Flash, and ferroelectric memory (FeRAM) have enabled technological innovations of SSDs.[4-7] Among the many SCMs, STT-MRAM has the advantages of the most mature offerings, which makes it a good candidate for cache memory. [8-10]

Considering that the comprehensive cost of STT-MRAM chips is relatively, adding STT-MRAM to complement the volatile DRAM on the SSD controller's DDR bus is more suitable. However, it's difficult to handle differences between two different memory types in the degree of command, timing and addressing parameters in a single DDR controller.[11, 12]

In this paper, an enterprise SSD controller was introduced, featuring two parallel DDR 4.0 controllers interfaced with DRAM and STT-MRAM chips respectively. Furthermore, considering STT-MRAM chips contain some intrinsic defects, a system-level hard error correct code (ECC) is developed to provide further data integrity protection. The chip was fabricated and functionally tested using a 12 nm process node.

II. HARDWARE ARCHITECTURE OF SSD CONTROLLER

The enterprise SSD controller encompasses essential components such as the MCU subsystem, PCIe Gen4 interface, NVMe command parser, FTL mapping manager, NAND Flash controller, and DRAM controller, which are requisite for conventional enterprise SSD controllers. As shown in Fig. 1, it also incorporates a DDR4 MRAM controller and a System-Level ECC, which enable compatibility with STT-MRAM chips. All these units are interconnected via the AMBA bus.

For enterprise SSD applications, system-level ECC is essential to lower the Bit Error Rate (BER) of STT-MRAM, achieving acceptable failure in time (FIT) rates or high mean time between failures (MTBF). As ECC integrated into the STT-MRAM chip may be insufficient for error situations like SEC, DED, or triple-bit error detection, external system-level ECC implementation is necessary.[13-15] This ensures data integrity, prevents premature failures, and helps attain system reliability goals.

The DDR4 MRAM Controller realizes steady transmission of signal between the AXI4 bus and the PHY via a DDR4 MRAM PHY interface (DFI) connection, and

979-8-3315-3918-4/25 $31.00 © 2025 IEEE

supports all signals, signal relationships and timing parameters required to transfer control and data information between AXI4 bus and the commercial DDR4 PHY. Meanwhile, the controller interfaces with three AXI4 controller core interface modules and one APB register port. Each AXI4 controller core interface contains five separate channels of traffic to/from the AXI4 bus: read command, write command, read data, write data, and write response. The single APB register port converts the APB register addresses to DDR4 Controller core register addresses. As another important component of the DDR4 MRAM controller, the commercial high-speed DDR MRAM PHY implements the functionality as connection between the commercial DDR4 controller and the STT-MRAM chip. The PHY generates and manages the command and clock signals for STT-MRAM chips, including data strobes and control clocks.

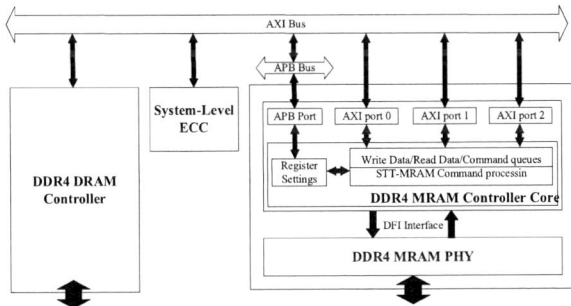

Fig. 1. Part of SSD internal architecture with DDR4 MRAM controller.

III. SYSTEM-LEVEL ECC DEISG

In enterprise SSD applications, the system-level ECC plays a critical role in reducing the Bit Error Rate (BER) of STT-MRAM, thereby achieving acceptable Failure In Time (FIT) rates and high Mean Time Between Failures (MTBF). While STT-MRAM chips incorporate on-chip ECC, such mechanisms are often inadequate for handling error scenarios like Single Error Correction (SEC), Double Error Detection (DED), or triple-bit errors. Thus, the implementation of an external system-level ECC becomes necessary to safeguard data integrity, prevent premature failures, and meet the stringent reliability requirements of enterprise systems.

To balance the kind of uniting the function, power consumption and area, a coding and decoding scheme of single-error correction-double-error detection (SEC-DED) Hamming code is given out, which can enable a higher speed and save certain area of the chip. The relationship between the original data bits and extra parity bits Hamming codes can be described simply as a mathematical formula:

$$2^m \geq n + m + 1 \qquad (1)$$

To ensure effective error correction and prevent error propagation in STT-MRAM chips, the system-level ECC must be orthogonal to the on-chip ECC word. Orthogonality enables the system-level ECC to isolate the impact of errors within the chip. For instance, if triple-bit errors occur in a vertical column of the on-chip ECC word, an orthogonal system-level ECC (implemented horizontally) ensures that only a single-bit error is observable on the external DQ lane when data is clocked out, which can then be corrected by the system-level ECC logic. In contrast, non-orthogonality can lead to multi-bit errors damaging both data bits and system-level ECC bits, causing the system-level ECC to malfunction.

A. Algorithmic Design

For reading process of DDR4 MRAM Controller, the Controller core transfers the data to the port read data FIFO in a memory data width of 64 bits. According to equation (1), for case where number of origin data bits (n) = 64, necessary value of ECC bits (m) is 7 for single-bit error correction (SEC). In addition to the SEC above, one more bit is required for double-bit error detection. Hence the total number of ECC (parity) bits will be 8. In other words, this ECC Hamming code for SEC-DED can be written as (72, 64) where total block code size is 72 bits with 64 data bits and 8 ECC (parity) bits. The 8 ECC bits is calculated based on equations (2) - (9).

$$
\begin{aligned}
P_0 = &D_0 \oplus D_1 \oplus D_2 \oplus D_4 \oplus D_5 \oplus D_5 \oplus D_{10} \oplus \\
&D_{11} \oplus D_{12} \oplus D_{14} \oplus D_{17} \oplus D_{21} \oplus D_{22} \oplus D_{24} \oplus \\
&D_{27} \oplus D_{31} \oplus D_{34} \oplus D_{37} \oplus D_{38} \oplus D_{39} \oplus D_{41} \oplus \\
&D_{44} \oplus D_{49} \oplus D_{49} \oplus D_{55} \oplus D_{62}
\end{aligned} \qquad (2)
$$

$$
\begin{aligned}
P_1 = &D_0 \oplus D_1 \oplus D_3 \oplus D_4 \oplus D_6 \oplus D_8 \oplus D_{10} \oplus \\
&D_{11} \oplus D_{13} \oplus D_{15} \oplus D_{18} \oplus D_{21} \oplus D_{23} \oplus D_{25} \oplus \\
&D_{28} \oplus D_{32} \oplus D_{34} \oplus D_{37} \oplus D_{38} \oplus D_{40} \oplus D_{42} \oplus \\
&D_{44} \oplus D_{46} \oplus D_{50} \oplus D_{54} \oplus D_{56} \oplus D_{60}
\end{aligned} \qquad (3)
$$

$$
\begin{aligned}
P_2 = &D_0 \oplus D_2 \oplus D_3 \oplus D_5 \oplus D_6 \oplus D_9 \oplus D_{10} \oplus \\
&D_{12} \oplus D_{13} \oplus D_{16} \oplus D_{19} \oplus D_{22} \oplus D_{23} \oplus D_{26} \oplus \\
&D_{29} \oplus D_{33} \oplus D_{34} \oplus D_{37} \oplus D_{39} \oplus D_{40} \oplus D_{43} \oplus \\
&D_{44} \oplus D_{47} \oplus D_{51} \oplus D_{54} \oplus D_{57} \oplus D_{62}
\end{aligned} \qquad (4)
$$

$$
\begin{aligned}
P_3 = &D_1 \oplus D_2 \oplus D_3 \oplus D_7 \oplus D_8 \oplus D_9 \oplus D_{10} \oplus \\
&D_{14} \oplus D_{15} \oplus D_{16} \oplus D_{20} \oplus D_{24} \oplus D_{22} \oplus D_{25} \oplus \\
&D_{26} \oplus D_{30} \oplus D_{35} \oplus D_{37} \oplus D_{41} \oplus D_{42} \oplus D_{43} \oplus \\
&D_{44} \oplus D_{48} \oplus D_{52} \oplus D_{58} \oplus D_{60} \oplus D_{63}
\end{aligned} \qquad (5)
$$

$$
\begin{aligned}
P_4 = &D_4 \oplus D_5 \oplus D_6 \oplus D_7 \oplus D_8 \oplus D_9 \oplus D_{10} \oplus \\
&D_{17} \oplus D_{18} \oplus D_{19} \oplus D_{20} \oplus D_{27} \oplus D_{28} \oplus D_{29} \oplus \\
&D_{30} \oplus D_{36} \oplus D_{37} \oplus D_{45} \oplus D_{46} \oplus D_{47} \oplus D_{48} \oplus \\
&D_{53} \oplus D_{54} \oplus D_{59} \oplus D_{60} \oplus D_{63}
\end{aligned} \qquad (6)
$$

$$
\begin{aligned}
P_5 = &D_{11} \oplus D_{12} \oplus D_{13} \oplus D_{14} \oplus D_{15} \oplus D_{16} \oplus \\
&D_{17} \oplus D_{18} \oplus D_{19} \oplus D_{20} \oplus D_{31} \oplus D_{32} \oplus D_{33} \oplus \\
&D_{34} \oplus D_{35} \oplus D_{36} \oplus D_{37} \oplus D_{49} \oplus D_{50} \oplus D_{51} \oplus \\
&D_{52} \oplus D_{53} \oplus D_{54} \oplus D_{61} \oplus D_{62} \oplus D_{63}
\end{aligned} \qquad (7)
$$

$$
\begin{aligned}
P_6 = &D_{21} \oplus D_{22} \oplus D_{23} \oplus D_{24} \oplus D_{25} \oplus D_{26} \oplus \\
&D_{27} \oplus D_{28} \oplus D_{29} \oplus D_{30} \oplus D_{31} \oplus D_{32} \oplus D_{33} \oplus \\
&D_{34} \oplus D_{35} \oplus D_{36} \oplus D_{37} \oplus D_{55} \oplus D_{56} \oplus D_{57} \oplus \\
&D_{58} \oplus D_{59} \oplus D_{60} \oplus D_{61} \oplus D_{62} \oplus D_{63}
\end{aligned} \qquad (8)
$$

$$
\begin{aligned}
P_7 = &D_{38} \oplus D_{39} \oplus D_{40} \oplus D_{41} \oplus D_{42} \oplus D_{43} \oplus \\
&D_{44} \oplus D_{45} \oplus D_{46} \oplus D_{47} \oplus D_{48} \oplus D_{49} \oplus D_{50} \oplus \\
&D_{51} \oplus D_{52} \oplus D_{53} \oplus D_{54} \oplus D_{55} \oplus D_{56} \oplus D_{57} \oplus \\
&D_{58} \oplus D_{59} \oplus D_{60} \oplus D_{61} \oplus D_{62} \oplus D_{63}
\end{aligned} \qquad (9)
$$

B. Hardware Design

Fig. 2 illustrates the hardware architecture of system-level ECC. It mainly comprises AXI for bus communication, the Write Address (WA) and Read Address (RA) modules for managing read/write addressing, and Write Data (WD) and Read Data (RD) modules for write and read operations, respectively.

For write operations, the WA module receives write commands and collaborates with the WD module. In non-full-width write operations, a data merge operation is involved.

979-8-3315-3918-4/25 $31.00 © 2025 IEEE

The data first enters the cache's FIFO buffer. The WA module generates corresponding ECC read commands and data read commands based on the received write instructions. The ECC read command is issued first via RA module, with the received data temporarily stored in the ECC cache. This data is then sequentially combined with data read back via the subsequent data read command and fed into the DEC)module for verificationThe read-back data is merged with the write data temporarily stored in the FIFO, forming full-width transmission data. This data passes through a Multiplexer (MUX) to the ENC module for ECC encoding. The regular data is sent via the MUX, while the ECC code is temporarily stored in the ECC cache. Once data transmission is complete, an ECC write operation is initiated to write the ECC code to the corresponding address space, completing the current write operation. The WA module then readies itself for new write instructions. In full-width write operations, the write data passes through the MUX to the ENC module for ECC encoding. The regular data is sent via the MUX, while the ECC code is temporarily stored in the ECC cache. Once data transmission is complete, an ECC write operation is initiated to write the ECC code to the corresponding address space, completing the current write operation. The WA module then readies itself for new write instructions.

For read operations, the RA module, upon receiving a read command, generates and issues an ECC read instruction. The received ECC code is stored in the ECC cache of the RD module. Once ECC code collection is complete, the RA issues the received read command normally. The read-back data and temporarily stored ECC code are sequentially fed into the DEC module for verification. After verification, the read-back data is returned to the upstream AXI bus. If verification fails, an error response is returned via the RRESP channel or an error flag is set. Once the read operation is complete, the RA readies itself to accept new read commands.

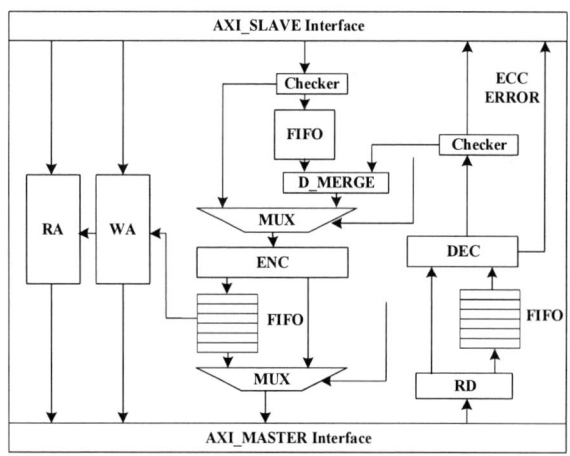

Fig. 2. Hardware architecture of System-level ECC

IV. IMPLEMENTATION

The enterprise SSD controller was synthesized in 12 nm process. As shown in Fig. 3 and Fig. 4, the layout area of MRAM Controller is about 1824.384×2603.04 μm^2 and the layout area of DRAM Controller is about 4125.264×2130.24 μm^2. And the area proportion of the System-Level ECC module is far lower than that of the MRAM controller. Given the significant disparity in area scales between the two, the area overhead of the System-Level ECC module can be excluded from the comparative analysis. Its negligible proportion means including it would not alter the core conclusions of the area comparison.

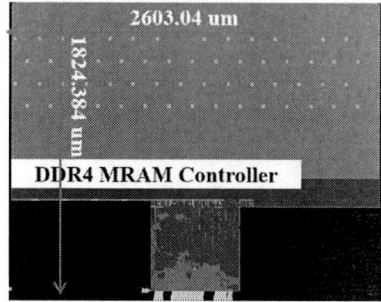

Fig. 3. Layout of the DDR4 MRAM Controller

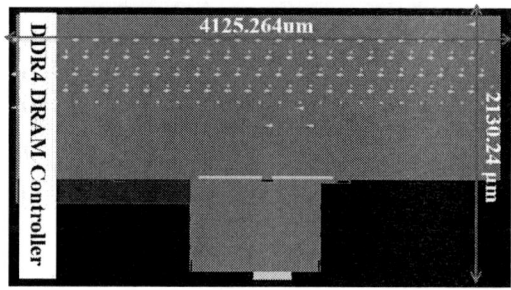

Fig. 4. Layout of the DDR4 DRAM Controller

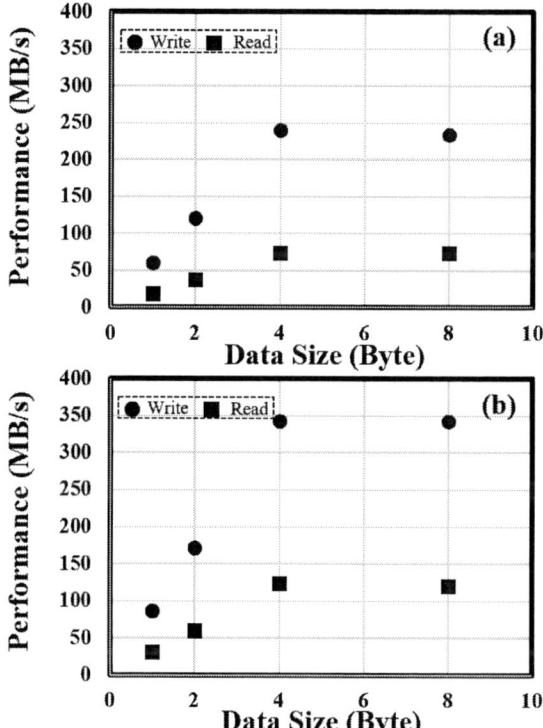

Fig. 5. The read and write performances of (a) STT-MRAM and (b) DRAM;

As illustrated in Figure 5, a comprehensive evaluation of the read and write performance of STT-MRAM and DRAM under cache miss conditions was conducted. Each test was repeated 1000 times, with the results presented as averages. The findings indicate that as the data byte size increased, the performance of both memory types improved. Specifically, STT-MRAM's read speed rose from 18.46MB/s at 1 byte to 73.28MB/s at 8 bytes, while its write speed increased from 60MB/s to 234.15MB/s. For DRAM, the read performance escalated from 30.77MB/s at 1 byte to 120MB/s at 8 bytes, and the write performance climbed from 85.71MB/s to 342.86MB/s. Despite the relatively lower read and write speeds of STT-MRAM compared to DRAM, STT-MRAM's non-volatile and high - durability characteristics make it suitable for specific applications in SSDs, where it can function effectively.

V. CONCLUSION

We have developed an enterprise SSD controller supporting STT-MRAM chips by adding a DDR4 MRAM controller and system-level ECC in conventional enterprise SSD controller architecture. At the 12nm process node, the area only increases by approximately 4.75 mm². The SSD controller exhibited remarkable read and write performance for STT-MRAM chips, akin to that of DRAM chips. When Cache miss occurred, the write performance reached 234.15 MB/s, and the read performance attained 73.28 MB/s.

ACKNOWLEDGMENT

This work was supported in part by the Shandong Provincial Natural Science Foundation (ZR2024QF162).

REFERENCES

[1] M. Fukuchi, Y. Sakaki, C. Matsui, K. Takeuchi, and Ieee, "20% System-performance Gain of 3D Charge-trap TLC NAND Flash over 2D Floating-gate MLC NAND Flash for SCM/NAND Flash Hybrid SSD," in *IEEE International Symposium on Circuits and Systems (ISCAS)*, Florence, ITALY, 2018, NEW YORK: Ieee, 2018.

[2] Y. Takai, M. Fukuchi, R. Kinoshita, C. Matsui, and K. Takeuchi, "Analysis on Heterogeneous SSD Configuration with Quadruple-Level Cell (QLC) NAND Flash Memory," in 2019 IEEE 11th International Memory Workshop (IMW), May 2019, pp. 1–4.

[3] S. Hachiya, T. Onagi, S. Y. Ning, K. Takeuchi, and Ieee, "Comprehensive Comparison of 3D-TSV Integrated Solid-State Drives (SSDs) with Storage Class Memory and NAND Flash Memory," in

IEEE International 3D Systems Integration Conference (3DIC), Sendai, JAPAN, 2015, NEW YORK: Ieee, 2015.

[4] Z. Chen, "A Novel SSD Buffer Design Based on STT-MRAM," in 2024 5th International Seminar on Artificial Intelligence, Networking and Information Technology (AINIT).

[5] X. Zhang, X. Duan, J. Yang, and J. Wang, "ARW: Efficient Replacement Policies for Phase Change Memory and NAND Flash," IEICE Transactions on Information and Systems, vol. E100.D, no. 1, pp. 79-90, 2017.

[6] C. Matsui, C. Sun, and K. Takeuchi, "Design of Hybrid SSDs With Storage Class Memory and NAND Flash Memory," Proceedings of the IEEE, vol. 105, no. 9, pp. 1812-1821, 2017.

[7] S. Mittal and J. S. Vetter, "A Survey of Software Techniques for Using Non-Volatile Memories for Storage and Main Memory Systems," (in English), Ieee Transactions on Parallel and Distributed Systems, Article vol. 27, no. 5, pp. 1537-1550, May 2016.

[8] C. Zambelli, G. Navarro, V. Sousa, I. L. Prejbeanu, and L. Perniola, "Phase Change and Magnetic Memories for Solid-State Drive Applications," Proceedings of the IEEE, vol. 105, no. 9, pp. 1790-1811, 2017.

[9] J. Choe, "Recent Technology Insights on STT-MRAM: Structure, Materials, and Process Integration," presented at the 2023 IEEE International Memory Workshop (IMW), 2023.

[10] Z. Li *et al.*, "A Software-Defined Fusion Storage System for PCM and NAND Flash," in *IEEE Non-Volatile Memory Systems and Applications Symposium (NVMSA)*, Hong Kong, HONG KONG, 2015, NEW YORK: Ieee, 2015

[11] B. Oh, N. Abeyratne, N. S. Kim, J. Ahn, R. G. Dreslinski, and T. Mudge, "Rethinking DRAM's Page Mode With STT-MRAM," IEEE Transactions on Computers, vol. 72, no. 5, pp. 1503-1517, 2023.

[12] S. Vartanian et al., "Single Event Effects Characterization of the ST-DDR4 Spin-transfer Torque Magnetoresistive Random Access Memory (STT-MRAM)," presented at the 2021 IEEE Nuclear and Space Radiation Effects Conference (NSREC), 2021.

[13] D. C. Worledge, "Write-error-rate of Spin-Transfer-Torque MRAM (Invited)," presented at the 2023 IEEE International Reliability Physics Symposium (IRPS), 2023.

[14] S. Javed, U. U. Fayyaz, and T. Mahmood, "Polar Code and Symbol Mapping Design for Multi-Level Cell Spin-Torque Transfer Magnetic Random Access Memory," IEEE Transactions on Magnetics, vol. 58, no. 2, pp. 1-5, 2022.

[15] J. Liu and P. Chen, "LDPC Joint Decoding Scheme for STT MRAM Storage," presented at the 2021 IEEE 32nd Magnetic Recording Conference (TMRC), 2021.

An IO Die with Collective-Aware Routing and In-Situ Processing for Data Synchronization in Multi-Chiplet Systems

Qi Luo[1], Chen Mu[1], Chixiao Chen[1,2], Xin Chen[3] Shiwei Liu*[2]

[1] State Key Laboratory of Integrated Chips and Systems, Fudan University, Shanghai, China

[2] Fudan Shaoxin Laboratory, Zhejiang, China

[3] Nanjing University of Aeronautics and Astronautics, Jiangsu, China

* Email: luoq24@m.fudan.edu.cn, liusw18@fudan.edu.cn

Abstract—Chiplet integration removes the yield and cost barriers of large monolithic SoCs, but it shifts the performance bottleneck to inter-chiplet communication. Conventional die-to-die (D2D) links offer only point-to-point transfers and are poorly suited to modern HPC/AI workloads, which issue frequent data synchronization (broadcast, gather, all-reduce). To solve this challenge, we present a scalable IO Die that turns the point-to-pint interconnection to centralized IO network. This IO die embeds a virtual-channel router with a collective-aware scheduler and reorder buffer enforces deadlock-free context exchange. Additionally, it integrates a lightweight In-Situ Data Processing Engine (IDPE) for on-the-fly vector arithmetic and reduction. These two techniques eliminate repetitive D2D data transfers and satisfy the requirement for dedicated collective communication hardware accelerator in multi-chiplet system. Tested in a 32-chiplet system for LLAMA-70B training task, the prototype in 28 nm CMOS achieves an average throughput of 1.8Tb/s at 500MHz with a burst length of 256, resulting in only 13% lantency overhead within the training process. The IO die reduces data synchronization overhead by up to 86% compared to prior point-to-point interconnect designs.

Index Terms—IO Die, collective communication, chiplet

I. INTRODUCTION

The exponential growth in compute and storage demands of high-performance computing (HPC) has pushed traditional monolithic chip designs to their limit. Monolithic chip designs are increasingly hindered by fundamental challenges, including lithography constraints and reduced manufacturing yields [1]. To sustain the progress of Moore's Law, the research and industry have rapidly shifted toward chiplet-based architectures, breaking down large system-on-chip (SoC) designs into smaller, modular dies using advanced packaging and high-speed wireline technologies [2]. This strategy enhances manufacturing yield, reduces cost, and provides flexibility to integrate heterogeneous technologies within a single package.

Nevertheless, the shift from monolithic chips to multi-chiplet systems introduces a critical bottleneck, ***inter-chiplet communication***. Although chiplets successfully mitigate die-size constraints, overall system performance increasingly depends on efficient data exchange between chiplets. Existing D2D interfaces [3] primarily support simple point-to-point transmissions, where data is unicast from one chiplet to another. However, this is insufficient for modern parallel workloads, which require frequent and complex communication patterns such as broadcast, gather, and all-reduce operations. The use of conventional D2D interfaces for these complex communication patterns introduces substantial inefficiencies, placing a heavy burden on NPUs/GPUs to manage data synchronization and aggregation. Previous studies [4] have shown that D2D data communication accounts for over 70% of the total latency, leading to underutilization of GPUs.

To enable efficient data synchronization in multi-chiplet systems, we propose a centralized IO die equipped with a deadlock-free virtual channel router and an In-Situ Data Processing Engine (IDPE). Our primary contribution lies in (1) the router controller manages unicast and collective communication (multicast and broadcast). It features a high-throughput, non-blocking crossbar switch for low-latency traffic and employs a collective-aware scheduler to orchestrate complex operations, guaranteeing deadlock-free data exchange across multiple chiplets. (2) a lightweight IDPE within the IO die performs vector arithmetic, including addition and normalization. the IDPE extends traditional point-to-point communication by integrating hardware-accelerated collective communication primitives. Executing these operations directly within the communication pathway enables asynchronous communication, significantly reducing synchronization overhead and freeing up valuable computational resources. (3) evaluated on a 32-chiplet system for an LLAMA-70B training workload, our proposed IO die delivers 40.5x the throughput of the recent chip-to-chip (C2C) link [5], 125x that of the SerDes-based interconnect [6], and a 284.5x increase over the AIB standard. In terms of energy efficiency, it consumes only 2.78 pJ/bit, achieving a 5.4× reduction compared to the C2C link [5]. These experimental results show that the proposed IO die holds significant potential for AI and HPC applications.

II. BACKGROUND AND MOTIVATION

A. Primer on Chiplet

In modern semiconductor design, chiplet architectures integrate multiple specialized dies into a single logical device using advanced heterogeneous packaging. This approach overcomes the scaling and photomask limitations of monolithic chips by dividing large designs into smaller, independently

979-8-3315-3918-4/25 $31.00 © 2025 IEEE

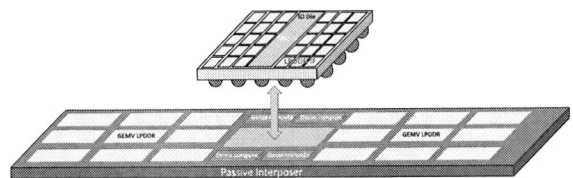

Fig. 1: A Domain-Specific AI Accelerator Architecture.

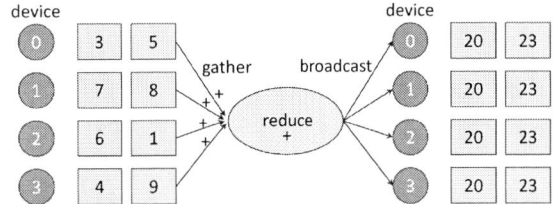

Fig. 2: Commonly Used Collective Communication Primitives.

manufactured blocks that connect via high-density substrates.

Chiplet designs improve manufacturing yield and modularity at the cost of complex packaging and expensive D2D interconnects, which suffer from higher power/area overhead and lower bandwidth than monolithic wiring. This creates a fundamental trade-off for architects: balancing chiplet granularity against communication overhead to optimize for performance, power, and cost. The chiplet-based system in Fig. 1 employs a high-density interposer with a low-latency fabric to connect chiplets on both sides. These chiplets are specialized for either dense matrix multiplication (GEMM) or memory-centric computation (GEMV with LPDDR). A dedicated IO die provides a uniform D2D interface for low-overhead, asynchronous synchronization across all components.

B. Data Synchronization in Multi-Chiplet System

Despite the advantages of chiplet-based systems, interchiplet communication has emerged as a major performance bottleneck in modern HPC workloads. These parallel and distributed workloads exhibit complex data communication and synchronization patterns that go beyond simple point-to-point unicast transfers, including broadcast, gather, and all-reduce operations. For instance, distributed AI training depends on efficient gradient aggregation [7] [8], typically implemented as an all-reduce operation, during each training iteration. Edge-deployed large model inference tasks, which operate using tensor, data, and pipeline parallelism under strict memory and storage constraints [9] [10], similarly rely on effective interchip synchronization. These data synchronization operations shown in Fig. 2 serve as fundamental primitives for distributed processing in multi-chiplet systems.

Broadcast: An one-to-all operation, where a single source chiplet sends identical data to all other chiplets.

Gather: An all-to-one operation, where each chiplet transmits unique data to a root chiplet, aggregating the data typically by concatenation.

All-reduce: A many-to-many operation that aggregates data from all chiplets and broadcasts the final reduced result. Gradient aggregation during distributed AI training is a prime example of this operation.

Previous works [11] provided limited support for these complex D2D data synchronization patterns, resulting in a latency overhead of 76%. To improve the efficiency of multi-chiplet systems, we propose a centralized IO die equipped with a deadlock-free virtual channel router and an IDPE.

III. CENTRALIZED ARCHITECTURE

A. IO Die Architecture Overview

We first present the overall IO-die-based architecture targeting at the 2.5D chiplet system as shown in Fig. 3. The IO die serves as a control hub to direct the data com-

munication among the surrounding compute dies. The IO die comprises four primary modules, including several input buffers (implemented as virtual channels), an IDPE block, a central controller, and a data-distribution crossbar. The virtual channels are divided into four VC queues, each buffering data packets from the compute dies according to their request priority. The IO die preferentially processes high-priority data communications, while lower-priority data and requests are buffered in their respective VC queues to prevent deadlock. The IDPE block is a single-instruction-multiple-data (SIMD) unit that performs arithmetic operations, such as accumulation and averaging during data synchronization, and handles nonlinear normalization required in LLMs. The controller manages the necessary D2D protocol and translates logical addresses from each compute die into a unified global address space maintained within the IO die. With its built-in scheduler and dispatch units, the system also accelerates collective-communication operations directly in hardware. The non-blocking crossbar serves as the physical data path for distributing data among the compute dies.

B. Deadlock-free Virtual Channel Router

Data transmission deadlocks can be classified into two types: (1) D2D interface deadlocks and (2) priority deadlocks. The first type of deadlock occurs when the D2D interface fails to handshake properly due to issues such as channel noise or crosstalk. As a result, the source die continues to send data requests, causing the data link to be blocked. Priority deadlock occurs when the IO die is processing a high-priority request, while a low-priority request with the same destination is issued by a compute die. In this case, the compute die is blocked, waiting for the IO die to become available.

To resolve interface deadlock, we introduce two optimizations based on the D2D interface [12]. As shown in Fig. 4, we redesign the data packet format to include explicit Source and Destination IDs, thereby enabling arbitrary D2D communication across chiplets. This modification transforms static point-to-point physical links into dynamic communication channels compatible with the non-blocking crossbar in Fig 3. Second, we implement a two-fold fault-handling mechanism to prevent deadlocks. A timeout counter measures the duration until an acknowledgment (ACK) or negative acknowledgment (NACK) is received. Additionally, a retry limit counter tracks consecutive NACKs when the interface fails to complete the handshake. If any counter exceeds a predefined threshold, an interrupt is issued to the host to handle the error. Subsequently, both the IO die and the compute die release the D2D interface, allowing it to serve other requests.

To resolve priority deadlocks, the router employs a coordinated approach involving the virtual-channel, VC allocation,

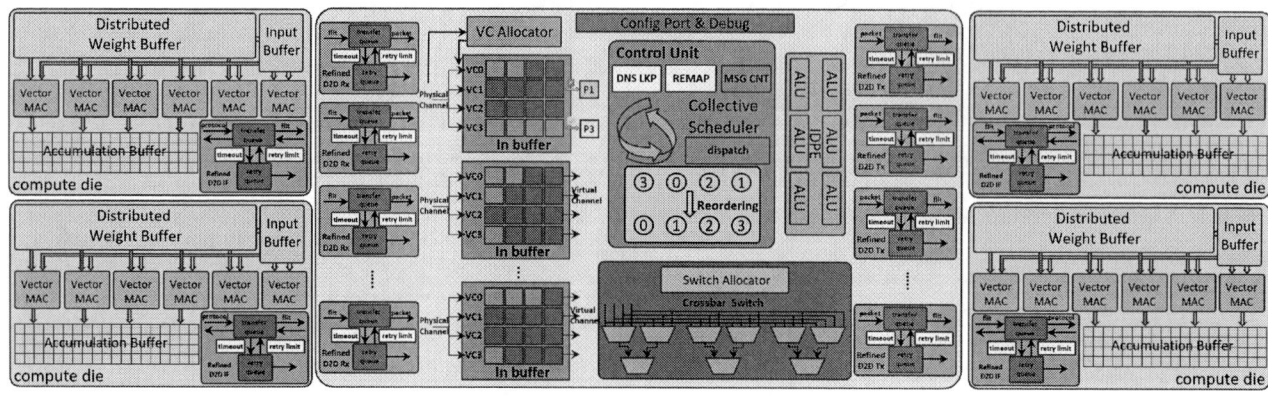

Fig. 3: Scalable IO Die Architecture Overview.

Packet Head	SRC & DEST	Collective Pattern & ID	Packet ID	CRC Code	Address Infomation	Related Data	Priority ID	Related Control
4/8/16bits	Depends on device num	P2P/Broadcast/ Gather/allreduce etc.	4/8/16bits	16bits	Depends on protocol address bitwidth	Depends on protocol data	2bits	Depends on protocol

Fig. 4: Package Format.

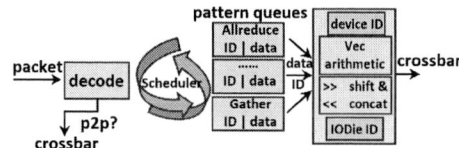

Fig. 5: Operation For Collective Communication.

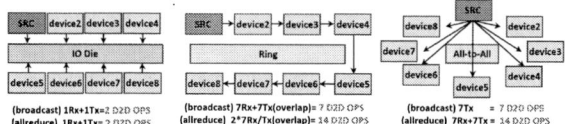

Fig. 6: Communication Steps Comparison.

Switch allocation and the collective-aware scheduler in the controller. Each IO port connects to a Compute Die, and each port features an independent buffer partitioned into per-priority VC queues. Incoming packets are first steered to the appropriate queue based on their priority. If a high-priority packet stalls, the router drains the remaining queues so that lower-priority packets can still advance. This mechanism eliminates head-of-line blocking and prevents priority deadlocks.

In addition, the modern D2D interface supports interleave and out-of-order transmission to improve bandwidth utilization. To enable arbitrary point-to-point communication, packets are forwarded from the VC queues to the crossbar switch for a second arbitration round. Switch allocation then routes them to the corresponding output buffers. To maintain correct data synchronization, the router utilizes the non-blocking crossbar, collective scheduler and dispatch unit to distribute the data extracted from packets tagged with the same collective pattern and ID to the IDPE for processing. This enables accurate all-reduce operations on partial results of the same matrix originating from different compute dies.

C. In-Situ Data Processing Engine

The IO die incorporates a lightweight but efficient IDPE to handle various Collective communication pattern, such as broadcast, gather and all-reduce. Fig. 5 illustrates the operation performed on data packet. Due to space constraints, we describe the all-reduce operation as an example. Other operations follow a similar process. As described in Sec. III-B, the IO die first extracts the data packets with the same collective pattern and ID from the VC queues for accurate all-reduce. The IDPE first accumulates the data segments to generate the synchronized result. It then appends the IO Die's ID as source ID to initiate the broadcast to all compute dies. In addition,

instead of generating individual data packets for each compute die, the IDPE assigns a unified broadcast collective pattern to signal a broadcast operation. This approach significantly reduces memory overhead for data packet construction in the IO die. Fig. 6 compares the IO die with all-to-all and ring-based topologies for D2D communication in multi-chiplet system. For ring-based broadcast, overlapping 7 Rx and 7 Tx operations incurs 7 D2D operations, whereas an all-to-all broadcast uses 7 Tx operations, likewise totaling 7 D2D operations. In contrast, the I/O die requires only 1 Rx and 1 Tx for broadcast. Similarly, for all-reduce, both ring-based and all-to-all topologies incur 14 D2D operations (7 Rx and 7 Tx), while the I/O die completes the all-reduce with just 1 Rx and 1 Tx. The IO die reduces D2D operations by 71% for broadcast and by 86% for all-reduce. It minimizes D2D communication during collective synchronization.

IV. EXPERIMENTAL RESULTS

The IO Die prototype was implemented in Chisel, then compiled to Verilog, and synthesized with Synopsys Design Compiler in 28 nm CMOS for power and area estimation. The prototype runs at 500MHz and 0.9V. We configure the number of compute die and the bit-width of D2D interface to validate the scalability of the IO-die-based architecture. The D2D protocol follows Liao et al. [12]. The compute die utilizes the design of previous work [13]. In our implementation, flit size is defined by the number of physical IO pads.

Fig.7(a) shows latency as a function of flit size and compute die count. For 256-bit packets, 16/32-bit flits minimize latency by reducing serialization overhead. Moreover, increasing the device count has a negligible impact on overall latency. Fig.7(b) and Fig.7(c) present synthesis area and power for various configurations. By contrast, area and power consumption scale primarily with port count and protocol data width. For a given data width, increasing the port count from 8 to 32 causes roughly a 3x rise in both metrics. Furthermore, adjusting flit size or other logic level parameters adds almost no additional

TABLE I: Comparison With Prior D2D Links

	C2C [5]	D2D [6]	AIB [14]	This Work	
				2.5D	3D
Freq. (MHz)	N/A	300	1000	500	500
PHY	silicon photonic	SerDes	Parallel-IO	Parallel-IO	Parallel-IO
Latency (cycles)[1]	N/A	105–132	<5	81	65
Throughput (Gb/s)[2]	45	14.56	6.4	1024.9	1820.7
Energy (pJ/bit)[2]	15	N/A	0.44	8.97	2.78

[1] Latency measured with burst length = 1
[2] Throughput and Energy measured with burst length = 256

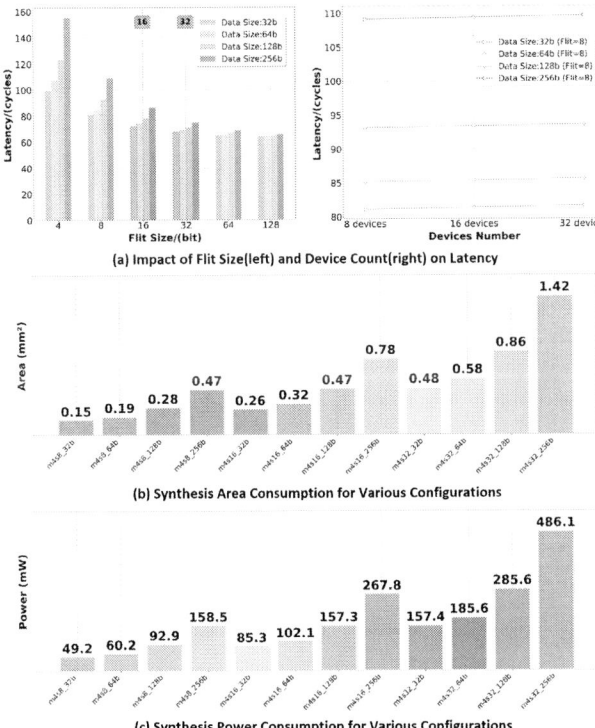

(a) Impact of Flit Size(left) and Device Count(right) on Latency

(b) Synthesis Area Consumption for Various Configurations

(c) Synthesis Power Consumption for Various Configurations

Fig. 7: Impact Of Various Configurations.

area or power cost.

As shown in Table I, our IO Die is configured for 2.5D/3D integration. In the 2.5D configuration, the IO die sends 32-bit protocol data continuously via eight pads; in the 3D configuration, the 256-bit protocol data is sent via 128 pads. The results show that the proposed IO Die, in its 2.5D configuration, delivers approximately 70.4x higher throughput and a 1.3x to 1.6x reduction in latency when compared to the SerDes-based interconnect [6]. In addition, the 3D configuration achieves over 40.5x the throughput of the recent C2C work [5] and 125x that of the SerDes-based interconnect [6]. When benchmarked against the AIB standard, our 3D integration provides a staggering 284.5x improvement in throughput. In terms of energy efficiency, the 2.5D configuration consumes 8.97 pJ/bit, a 1.7× better than the C2C link [5]. In the 3D configuration, consumption reduces to 2.78 pJ/bit. In comparison, The AIB link achieves an energy-efficient of 0.44 pJ/bit.

V. CONCLUSION

This paper proposes a centralized IO die that transforms point-to-point links into a fabric capable of natively handling complex communication. Compared to prior interconnect architectures, executing collective communications directly within this fabric substantially reduces synchronization overhead, dramatically improving both latency and effective bandwidth. This approach delivers a scalable, cost-effective solution for multi-chiplet HPC and AI systems.

ACKNOWLEDGMENTS

This work was supported by the Research on Peng Cheng Laboratory and China Mobile (PCL-CMCC) Foundation for Science and Innovation under Grant 2024ZY2B0070 and Heterogeneous Storage and Computing Architecture Integrating Different Storage Media under Grant 92464204.

REFERENCES

[1] IEEE International Roadmap for Devices and Systems, "2023 International Roadmap for Devices and Systems (IRDS): Lithography," IEEE, 2023. [Online]. Available: https://irds.ieee.org/images/files/pdf/2023/2023IRDS_Litho.pdf. [Accessed: Jul. 11, 2025].
[2] Chester Liu, Jacob Botimer, and Zhengya Zhang. 2021. A 256Gb/s/Mm-Shoreline AIB-Compatible 16nm FinFET CMOS Chiplet for 2.5D Integration with Stratix 10 FPGA on EMIB and Tiling on Silicon Interposer. In 2021 IEEE Custom Integrated Circuits Conference (CICC). IEEE, Austin, TX, USA, 1-2. https://doi.org/10.1109/CICC51472.2021.9431555
[3] Xiaohan Ma, Ying Wang, Yujie Wang, Xuyi Cai, and Yinhe Han. 2022. Survey on Chiplets: Interface, Interconnect and Integration Methodology. CCF Transactions on High Performance Computing 4, 1 (March 2022), 43–52. https://doi.org/10.1007/s42514-022-00093-0
[4] Kashyap A, Lu X. NVMe-oAF: Towards adaptive NVMe-oF for IO-intensive workloads on HPC cloud[C]//Proceedings of the 31st International Symposium on High-Performance Parallel and Distributed Computing. 2022: 56-70.
[5] Sun, Chen, et al. "A monolithically-integrated chip-to-chip optical link in bulk CMOS." IEEE Journal of Solid-State Circuits 50.4 (2015): 828-844.
[6] W. Liao, Y. Guo, S. Xiao, and Z. Yu, "A low-cost and high-throughput noc-aware chip-to-chip interconnection," in 2020 IEEE International Symposium on Circuits and Systems (ISCAS), 2020, pp. 1–5.
[7] Shoeybi, Mohammad, et al. "Megatron-lm: Training multi-billion parameter language models using model parallelism." arXiv preprint arXiv:1909.08053 (2019).
[8] Fang, Jin, et al. "GRID: Gradient routing with in-network aggregation for distributed training." IEEE/ACM Transactions on Networking 31.5 (2023): 2267-2280.
[9] Pan, Xiurui, et al. "Instinfer: In-storage attention offloading for cost-effective long-context llm inference." arXiv preprint arXiv:2409.04992 (2024).
[10] Du, Hongchao, et al. "FlexInfer: Breaking Memory Constraint via Flexible and Efficient Offloading for On-Device LLM Inference." Proceedings of the 5th Workshop on Machine Learning and Systems. 2025.
[11] Ardalan, Shahab, et al. "Bunch of wires: An open die-to-die interface." 2020 IEEE Symposium on High-Performance Interconnects (HOTI). IEEE, 2020.
[12] J. Liao et al., "A Scalable Die-to-Die Interconnect with Replay and Repair Schemes for 2.5D/3D Integration," in 2023 IEEE International Symposium on Circuits and Systems (ISCAS), Monterey, CA, USA, 2023, pp. 1-5.
[13] C. Mu et al., "A 28-nm RRAM/SRAM Collaborative CIM Accelerator Supporting RRAM-Endurance-Latency Awareness for Edge Fine-Tuning," in IEEE Journal of Solid-State Circuits, doi: 10.1109/JSSC.2025.3577335.
[14] David Kehlet. [n.d.]. Accelerating Innovation Through a Standard Chiplet Interface: The Advanced Interface Bus (AIB). https://www.intel.com/content/dam/www/public/us/en/documents/white-papers/accelerating-innovationthrough-aib-whitepaper.pdf.

979-8-3315-3918-4/25 $31.00 © 2025 IEEE

An STT-MRAM Last Level Cache Management Method Based on Write Intensity Prediction for GPUs

Yujie Pu, Qiaoran Zhang, Shitong He, Fanchen Wu, Chen Zhao

College of Computer Science

Northwestern Polytechnical University

Xian, China

Email: yujiepu@mail.nwpu.edu.cn, chenzhao@nwpu.edu.cn

Abstract—Spin-transfer torque random access memory (STT-MRAM) has emerged as an ideal candidate for last level cache (LLC) in core-intensive GPUs due to its high density and low leakage power. However, its high write energy consumption and latency remain critical challenges. In this paper, we introduce WRIP, a write intensity prediction-based LLC management method that leverages the inherent write characteristics of STT-MRAM and reduces the write current for cache blocks with high write intensity. This effectively mitigates the write energy consumption and latency of STT-MRAM. Experimental results demonstrate that compared to a traditional SRAM-based L2 cache of the same die area, WRIP achieves a 13% average improvement in IPC (Instructions Per Cycle) and a 51% reduction in average total power consumption. Furthermore, exploiting STT-MRAM's high density, systems using STT-MRAM L2 cache achieve 39% lower average DRAM power versus those with SRAM-based L2 cache.

Keywords—GPU, GPGPU Application, STT-MRAM, Cache Management

I. INTRODUCTION

Over the past decade, graphics processing units (GPUs), equipped with hundreds of simple cores, have emerged as a mainstream technology for many-core high-performance systems. In the early GPU memory architectures, the capacity of register files was typically larger than that of L1 caches/shared memories, which in turn were larger than L2 cache. However, in recent years, the complexity of artificial intelligence (AI), high-performance computing (HPC), and data analytics has increased exponentially, imposing higher requirements on GPU memory performance. Consequently, starting with the Ampere architecture, NVIDIA has substantially increased L2 cache capacity, reaching 50 MB in the latest Hopper architecture[1].

Many large-scale GPU applications generate various irregular memory accesses. Studies have shown that the latency consumed by off-chip memory requests accounts for an average of 75% of the total execution time of GPU devices, and off-chip data access I/O services account for 71% of the total GPU energy consumption[2]. A straightforward solution is to use a larger L2 cache to reduce off-chip memory requests. However, simply expanding the L2 cache severely reduces on-chip space. Meanwhile, the power consumption brought by a larger L2 cache is also a non-negligible issue.

In the deep nanometer technology era, leakage current increases by 10 times for each generation of process node

shrinkage, imposing severe scalability and power consumption constraints on SRAM arrays[3]. Spin-transfer torque random-access memory (STT-MRAM)[4] emerges as a promising candidate to replace SRAM. Early studies on STT-MRAM mainly focused on last level cache (LLC) in CPUs, and there have been numerous research studies on STT-MRAM for CPUs. In GPUs, research on STT-MRAM has primarily centered on register files, with very few studies addressing last level cache. The storage cells of STT-MRAM exhibit almost no leakage power and have a density approximately four times that of SRAM cells, making them ideal for constructing future GPU L2 cache. However, writing to STT-MRAM cells requires prolonged injection of large currents, leading to access latency and energy consumption far higher than those of SRAM.

To address the high write latency and energy consumption of STT-MRAM, we propose a write intensity prediction-based STT-MRAM last level cache management method (WRIP), with the following main contributions: 1) Through targeted design based on the write characteristics of STT-MRAM and the write behaviors of GPGPU applications, we reduce the high write latency and energy consumption of STT-MRAM by designing a write intensity predictor and a write access controller. 2) We model and evaluate our scheme in GPGPU-SIM[5]. Compared with traditional L2 cache of the same on-chip area, WRIP achieves an average 13% improvement in instructions per cycle (IPC) and a 51% reduction in average total power consumption.

II. MOTIVATION

A. Asymmetric latency of STT-MRAM write operations

STT-MRAM employs magnetic tunnel junctions (MTJs) as storage cells, as illustrated in Figure 1. An MTJ consists of two ferromagnetic layers and an oxide barrier (MgO): one ferromagnetic layer serves as the reference layer with a fixed magnetic orientation, while the other is the free layer whose magnetic orientation can be altered during write operations. Based on the magnetic orientation of the free layer, an MTJ exhibits two resistance states: parallel and anti-parallel. The transition latency from the parallel state to the anti-parallel state is 9.7 ns, which is significantly higher than the 3.9 ns transition latency from the anti-parallel state to the parallel state[6]. Therefore, for cache blocks in the L2 cache that are frequently written, we preset the MTJ cells to the anti-parallel state before each write operation. This approach ensures that write operations primarily involve transitions from the anti-parallel to the parallel state, effectively reducing the high write latency inherent in STT-MRAM.

979-8-3315-3918-4/25 $31.00 © 2025 IEEE

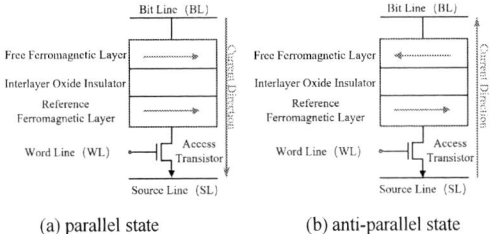

(a) parallel state	(b) anti-parallel state

Fig. 1: STT-MRAM structure. Figure 1a is the parallel state (when the free layer's orientation matches the reference layer's), representing the value 0, and Figure 1b is the anti-parallel state (opposite orientations), representing the value 1.

B. GPGPU application characterization

To analyze the write intensity of each cache block in the L2 cache, we analyzed the write access patterns of cache blocks from various data processing benchmarks[7]-[9], as shown in Figure 2. Based on the write frequency of each cache block, cache blocks are classified into two types: i) high write intensity (WH) and ii) low write intensity (WL). WH blocks require frequent data updates, while WL blocks are rarely updated after an initial write. Since WL blocks retain data for longer periods, they are not our optimization targets. In this work, the write intensity threshold is set to 3 based on the write intensity of all tested workloads and references to the study in [10].

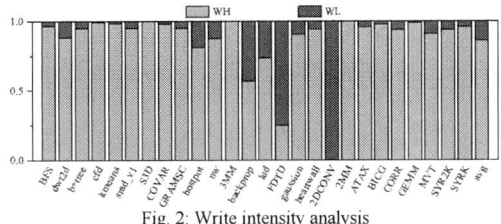

Fig. 2: Write intensity analysis

Table 1 lists the retention time (R.T), write energy (W.E), and write latency (W.L) corresponding to different magnetization stability height (△). STT-MRAM cells with a 10-year retention time exhibit the highest write energy and latency due to high data stability requirements. However, studies show that most data blocks in applications are rewritten within 100μs, and a 40ms retention time can meet the write requirements of cache blocks with high write intensity[10]. By reducing the retention time, we can decrease the required write current, thereby lowering energy consumption and further reducing latency.

TABLE I.   STT-MRAM PARAMETERS FOR DIFFERENT DATA RETENTION TIMES.

△	R.T	W.L(ns)	W.E(nJ)	Refreshing(block)
40.29	10 year	10	1.158	0
17.50	40ms	2	0.161	1867
11.51	100us	0.8	0.070	2750755

III. CACHE MANAGEMENT METHOD BASED ON WRITE INTENSITY PREDICTION

A. Overview

We propose a write intensity prediction-based STT-MRAM last level cache management method (WRIP) for GPUs, which consists of a write intensity predictor and a write access controller. As shown in Figure 3, the write intensity predictor includes a cache request sampler and a write intensity prediction table. The cache request sampler is used to sample the access requests of the L2 cache in the

GPU and extract the features of access patterns for the write intensity prediction table. Based on the historical access pattern information of each sampled cache block, the write intensity prediction table statistically analyzes the write intensity of each cache block, and adds marks to the cache blocks predicted to have high write intensity. Before each write operation to the L2 cache, the write access controller checks the mark bit of the corresponding cache block in the write intensity predictor according to the cache block address. If the block to be written is a high write intensity block, it will perform a special write operation on the block.

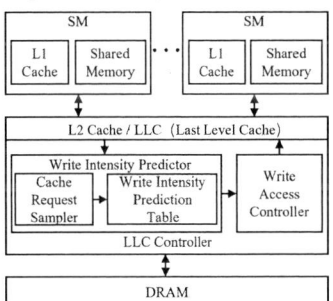

Fig. 3: Overview of WRIP architecture

B. Write intensity predictor

Figure 4 illustrates the detailed architectural information of our predictor. The sampler analyzes memory requests to the L2 cache, identifies block addresses to be accessed, access types (read/write), hit/miss status, and then filters these messages to collect only write hit information, which is sent to the write intensity prediction table for updating.

The write intensity prediction table is updated based on the sampler's results. For example, if a write hit occurs for the block with address 3222357120 in the L2 cache, the sampler collects the write information and sends it to the table. If this block enters the table for the first time, a new row is allocated to store its address, the write hit count is incremented by 1, and the write intensity flag is initialized to 0 (indicating low write intensity). A timestamp records the last write access time of cache blocks to provide a reference for subsequent replacements. When the write hit count of a block exceeds the threshold, the write intensity flag is automatically set to 1 (indicating high write intensity). During program execution, there may be blocks with infrequent data updates. Take block address 3222160768 as an example: its write hit count consistently remains below the threshold, resulting in it being continuously marked as having low write intensity.

Initially, all blocks are marked as low write intensity (WL). As the program executes, the write intensity predictor dynamically identifies and marks cache blocks with updated write intensities in real-time, providing references for subsequent decision-making by the write access controller. When the write intensity prediction table becomes full, we adopt a replacement algorithm that combines write intensity and LRU (Least Recently Used). This algorithm gives priority to replacing the cache blocks with low write intensity that have been least recently used. In extreme cases, if all the cache blocks with low write intensity in the prediction table have been replaced, then the cache blocks with high write intensity that have been least recently used will be replaced. This maximizes the retention of cache blocks with high write intensity in the table.

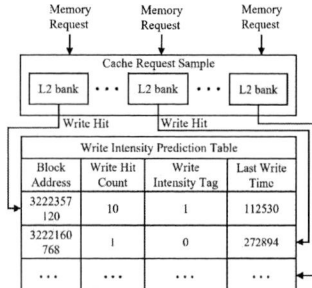

Fig. 4: Write intensity predictor

C. Write access controller

The write access controller leverages the unique writing characteristics of STT-MRAM to reduce write latency and power consumption. Based on the information from the write intensity predictor, it presets the STT-MRAM cells of WH cache blocks to the anti-parallel state before the next write operation, so that writing new values only requires a transition to the parallel state. Additionally, using a 10-year retention current for WH blocks is unnecessary; a 40ms retention current suffices to meet their write requirements.

Figure 5 illustrates the write operation flow of the write access controller in the LLC. The controller first determines whether the cache access request is a write or read operation. For read requests, it proceeds with normal read access. For write requests, it checks for a write hit: if it is a write miss, a write-allocate policy is employed for handling. If it is a write hit, the controller matches the block address with the write intensity prediction table. If the block address exists in the table and its write intensity flag is 1, the STT-MRAM cells of the cache block are set to the anti-parallel state, and the write current is reduced for fast writing. If the block address is not in the table or the flag is 0, a slow writing operation is performed.

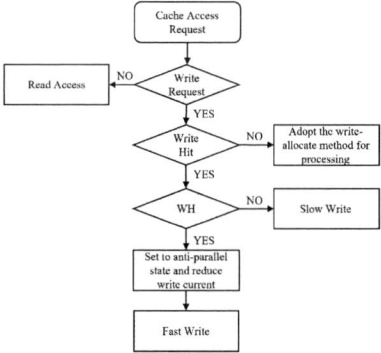

Fig. 5: Cache management with the proposed WRIP scheme

IV. METHODOLOGY AND EXPERIMENT SETUP

A. Experimental environment and configuration

We implemented the WRIP on the GPGPU-Sim 4.2.1 simulator [5], which can simulate the behaviors of SMs, L2 cache, interconnection network, and off-chip main memory. The power consumption of the simulated GPU system is evaluated using GPUWattch[11]. Two configurations were evaluated: 1) STT: Construct an STT-MRAM L2 cache with four times the capacity under the same area. 2) WRIP: Our proposed scheme is implemented on top of the STT configuration. The experimental configuration is detailed in

Table 2. The read/write energy parameters of SRAM and STT-MRAM are configured based on [10]-[12].

TABLE II. SYSTEM CONFIGURATIONS

SM config	80SMs,1132MHZ
L1D cache	32KB/SM
Shared Memory	96KB/SM
Write intensity predictor	256KB
Baseline LLC(SRAM)	96KB/Memory sub partition, 6MB in total
STT LLC(STT-MRAM)	384KB/Memory sub partition, 24MB in total, Read:1cycle, Write:5cycles
WRIP LLC(STT-MRAM)	384KB/Memory sub partition, 24MB in total, Read:1cycle, Fast Write:1cycle, Slow Write:5cycles
SRAM Read/Write Energy	1.175/1.175 nJ/access
STT-MRAM Read/Fast Write/Slow Write Energy	0.293/1.354/2.351 nJ/access

B. Testbench

We evaluated a large number of workloads from the PolyBench[7], Rodinia[8], and SHOC[9] benchmark suites. According to [13], GPGPU applications can be classified into two categories based on cache behavior: 1) cache capacity-insensitive; 2) cache capacity-sensitive. Increasing the cache capacity can improve the performance of some applications, while others are less affected.

V. RESULT

Figure 6 illustrates the IPC values of two configurations evaluated using workloads, normalized to Baseline LLC. We observe that compared to the baseline, STT exhibits a 7% average IPC drop, yet significant variability exists across different workloads. In cache capacity-sensitive workloads (e.g., cfd, kmeans), STT achieves 1.9× and 2.6× IPC improvements, respectively, attributed to its larger cache capacity from its high density. The cache capacity gain outweighs the performance degradation caused by STT's high write latency. Conversely, in write-intensive workloads (e.g., 2MM, 3MM), STT's performance degrades by 69% and 65%, respectively. This is because the marginal cache capacity benefit is overshadowed by the substantial write operations (tens of millions). After applying WRIP, the IPC improves by 13% on average compared to the baseline and 20% compared to STT. In write-intensive workloads (e.g., 2MM, 3MM), WRIP achieves 64% and 58% improvements compared to STT, respectively, leveraging their high write intensity and frequent updates.

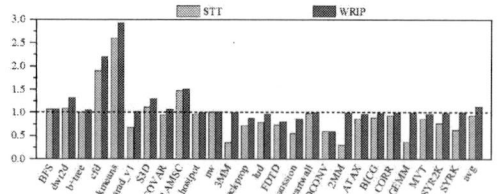

Fig. 6: IPC normalized to Baseline LLC

Figure 7 illustrates the L2 cache dynamic power of two configurations, normalized to Baseline LLC. While STT exhibits higher write power but lower read power than SRAM, the program's read/write characteristics significantly influence L2 cache power. For write-intensive workloads (e.g., 2MM, 3MM), massive write operations drive dynamic power 84% and 87% higher than baseline, respectively. The high proportion of high write intensity blocks enables significant power reduction via write current optimization. In read-intensive workloads (e.g., hotspot, SYR2K), far more read operations lead to 55% and 40% lower dynamic power than baseline. Overall, STT's high write energy causes a 10% higher average dynamic power than baseline. WRIP mitigates this, achieving an average 31% reduction in dynamic power compared to the baseline.

As shown in Figure 8, due to negligible leakage power in magnetic cells, STT and WRIP achieve 20% and 51% lower average total L2 cache power than the Baseline LLC, respectively. Notably, while STT-MRAM has low leakage power, its energy consumption still exceeds the baseline for extremely write-intensive workloads (2MM, 3MM, GEMM, SYRK) due to high write demands.

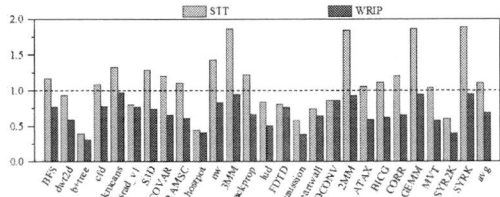

Figure. 7: Dynamic power normalized to Baseline LLC

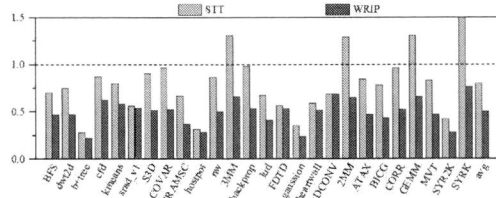

Figure. 8: Total power normalized to Baseline LLC

We also investigated the L2 access hit rates of Baseline LLC and WRIP, as shown in Figure 9. An important rationale for analyzing L2 hit rates is to observe DRAM power consumption changes. Each DRAM access consumes significantly more energy than SRAM access. A higher L2 hit rate reduces DRAM accesses, and since DRAM power constitutes a major portion of GPU system consumption, reducing DRAM power lowers overall system power.

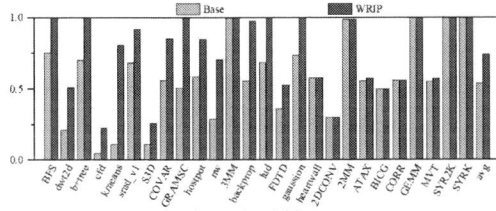

Figure. 9: L2 hit rate

Figure 10 illustrates the DRAM power consumption of WRIP, normalized to Baseline LLC. Increased L2 hit rates reduce DRAM accesses, resulting in significantly lower DRAM power for cache capacity-sensitive workloads. Overall, WRIP achieves an average 39% DRAM power reduction compared to the baseline.

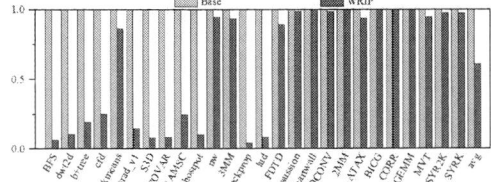

Figure. 10: DRAM power normalized to Baseline LLC

VI. Conclusion

In this paper, we propose WRIP, a write intensity prediction-based STT-MRAM last level cache management method for GPUs. By leveraging the unique writing characteristics of STT and the access patterns of workloads to the L2 cache, WRIP specifically targets the reduction of high write latency and write energy in STT-MRAM. Extensive evaluations show that compared with traditional SRAM-based methods, WRIP achieves an average 13% improvement in IPC and an average 51% reduction in power consumption. Additionally, benefiting from the high density of STT-MRAM, WRIP reduces DRAM power consumption by an average of 39%.

Acknowledgment

This work was supported in part by the STI2030-Major Project under Grant 2022ZD0208805; in part by the National Key Research and Development Program of China under Grant 2024YFB4303700; in part by the Fundamental Research Funds for the Central Universities under Grant 24GH0201339 and Grant WH00001161.

References

[1] NVIDIA, Whitepaper: NVIDIA H100 Tensor Core GPU Architecture, 2022.

[2] J. Zhang, M. Jung, M. Kandemir, "FUSE: Fusing STT-MRAM into GPUs to Alleviate Off-Chip Memory Access Overheads," in: 2019 IEEE International Symposium on High Performance Computer Architecture (HPCA), 2019, pp. 426-439.

[3] ITRS, Emerging Research Devices, Nov. 2011.

[4] X. Dong, X. Wu, G. Sun, Y. Xie, H. Li, Y. Chen, "Circuit and Microarchitecture Evaluation of 3D Stacking Magnetic RAM (MRAM) as a Universal Memory Replacement," in: 2008 45th ACM/IEEE Design Automation Conference (DAC), 2008, pp. 554-559.

[5] A. Bakhoda, G. L. Yuan, W. W. L. Fung, H. Wong, T. M. Aamodt, "Analyzing CUDA Workloads Using a Detailed GPU Simulator," in: 2009 IEEE International Symposium on Performance Analysis of Systems and Software, 2009, pp. 163-174.

[6] R. Bishnoi, M. Ebrahimi, F. Oboril, M. B. Tahoori, "Improving Write Performance for STT-MRAM," IEEE Transactions on Magnetics, vol. 52, no. 8, pp. 1-11, Aug. 2016.

[7] S. Grauer-Gray, L. Xu, R. Searles, S. Ayalasomayajula, J. Cavazos, "Auto-tuning a High-Level Language Targeted to GPU Codes," in: 2012 Innovative Parallel Computing (InPar), 2012, pp. 1-10.

[8] S. Che et al., "Rodinia: A Benchmark Suite for Heterogeneous Computing," in: 2009 IEEE International Symposium on Workload Characterization (IISWC), 2009, pp. 44-54.

[9] A. Danalis et al., "The Scalable Heterogeneous Computing (SHOC) Benchmark Suite," in: 2010 Proceedings of the 3rd Workshop on General-Purpose Computation on Graphics Processing Units (GPGPU-3), 2010, pp. 63-74.

[10] M. H. Samavatian, H. Abbasitabar, M. Arjomand, H. Sarbazi-Azad, "An Efficient STT-RAM Last Level Cache Architecture for GPUs," in: 2014 51st ACM/EDAC/IEEE Design Automation Conference (DAC), 2014, pp. 1-6.

[11] J. Leng et al., "GPUWattch: Enabling Energy Optimizations in GPGPUs," SIGARCH Computer Architecture News, vol. 41, no. 3, pp. 487-498, Jun. 2013.

[12] X. Liu, M. Mao, X. Bi, H. Li, Y. Chen, "Exploring Applications of STT-RAM in GPU Architectures," IEEE Transactions on Circuits and Systems I: Regular Papers, vol. 68, no. 1, pp. 238-249, Jan. 2021.

[13] J. Lee, H. Kim, "TAP: A TLP-aware Cache Management Policy for a CPU-GPU Heterogeneous Architecture," in: 2012 IEEE International Symposium on High Performance Computer Architecture (HPCA), 2012, pp. 1-12.

Towards Scalable and High-Throughput NTT Acceleration On Hybrid-Bonding Architecture

Wenxuan Zhang[1], Yi Sun[1], Xinglong Yu[1], Yifan Zhao[1], Jun Han*[1]

[1] State Key Laboratory of Integrated Chips and Systems, Fudan University, Shanghai, China

* Email: wenxuanzhang25@m.fudan.edu.cn, junhan@fudan.edu.cn

Abstract—Fully Homomorphic Encryption (FHE) enables computation on encrypted data but suffers from severe performance bottlenecks due to high computational and memory overhead. In this paper, we present a scalable and high-throughput NTT acceleration scheme leveraging a hybrid-bonding (HB) 3D memory-compute architecture, which introduces an optimized NTT unit that integrates a Swizzling memory scheme to ensure conflict-free memory access under varying parallelism levels, a unified butterfly unit, and a tailored modular reduction scheme enabling over 3× overall performance improvement for specific moduli. These optimizations are integrated into a vertically stacked 3D HB-based memory-compute architecture, featuring node-level scalability. Through hardware-software co-simulation, we explore optimal configurations under various transform sizes and application scenarios. Experimental results demonstrate that our design can achieve a throughput of up to 10^7 NTT/s and a bandwidth of up to 10^3 GB/s, and a throughput gain of up to 93.8% when integrated with an external NTT engine.

Keywords—*Fully Homomorphic Encryption(FHE), Number Theoretic Transform(NTT), Swizzling memory scheme, Hybrid-Bonding(HB), 3D Stacked Embedded DRAM(3D SEDRAM)*

I. INTRODUCTION

As a pivotal technology in privacy computing, Fully Homomorphic Encryption (FHE) allows arbitrary computations on ciphertexts without decryption. However, the practical deployment of FHE remains constrained by performance bottlenecks, especially in scenarios involving large-scale data processing or high-depth computations. These bottlenecks primarily stem from the Number Theoretic Transform (NTT), which are computational and memory intensive operations at the core of FHE schemes based on the Ring Learning with Errors (RLWE) problem. While combining NTT with hardware acceleration has improved FHE performance, the frequent access to large-scale polynomial coefficients, twiddle factors, and ciphertexts imposes heavy pressure on memory systems. Conventional solutions often rely on large on-chip SRAM and high-bandwidth memory (HBM) to alleviate data transfer bottlenecks [1-2]. However, due to fundamental physical and scalability limitations, the capacity of SRAM and the bandwidth of HBM cannot increase infinitely, rendering them increasingly inadequate for supporting the growing problem sizes and higher degrees of parallelism in modern FHE workloads.

Moreover, most existing NTT accelerator designs [2-3] are tailored to specific configurations, lacking the architectural flexibility to support scalability in parallelism. In particular, inefficient on-chip coefficient addressing strategies can lead to potential memory bank conflicts under high parallelism. Additionally, prior works have predominantly focused on isolated accelerator designs, with little attention to scalable architectures that enable multi-node computation. These limitations highlight the need for a more integrated,

scalable, and memory-efficient solution to support NTT computation under demanding FHE workloads.

To overcome these limitations, this work proposes an NTT accelerator utilizing Stacked Embedded DRAM (SEDRAM) based on a hybrid-bonding (HB) architecture. By tightly integrating logic and memory layers through vertical interconnects, the HB-based 3D memory system enables low-latency, high-bandwidth data access tailored for highly parallel NTT computation. Furthermore, we redesign the NTT processing unit, introducing architectural optimizations to both its arithmetic and memory modules. This design enables a high-throughput NTT accelerator capable of significantly improving the efficiency of FHE operations. The main contributions of this paper are summarized as follows:

1) We design an optimized NTT unit featuring a unified butterfly unit, a tailored modular reduction circuit optimized for specific moduli, and a Swizzling-based memory access strategy that enables conflict-free memory access across arbitrary degrees of parallelism, allowing the architecture to scale seamlessly with diverse application scenarios.

2) We construct a 3D-stacked SEDRAM based on hybrid-bonding architecture that tightly integrates logic and memory layers through high-density vertical interconnects, achieving high bandwidth and low latency tailored for NTT acceleration and supporting flexible node expansion.

3) We establish a joint hardware-software co-simulation framework that can explore NTT unit configurations under different design trade-offs and derive the optimal architecture across different application scenarios. Besides, leveraging the scalability of the proposed architecture, we achieve 10^7 NTT/s throughput and over 10^3 GB/s memory bandwidth, while observing a maximum 93.8% per-node throughput improvement when integrating a referenced NTT engine.

II. BACKGROUND & MOTIVATION

A. Number Theoretic Transform

The Number-Theoretic Transform (NTT) is a finite-field analogue of the Discrete Fourier Transform (DFT), operating over the polynomial ring $R_q = \mathbb{Z}_q[x]/(x^N + 1)$ where q is a prime satisfying $q \equiv 1 \bmod N$. For a given polynomial, the NTT maps it from coefficient form to evaluation form following Eq.(1), where the variable ω_N is the N-th primitive root of unity in $\mathbb{Z}_q$. The powers of ω_N are called twiddle factors.

$$\hat{a}_k = \sum_{i=0}^{N-1} a_i \omega_N^{ik} \bmod q \tag{1}$$

In the NTT domain, polynomial multiplication can be efficiently performed as a component-wise product of transformed coefficients, leveraging the convolution theorem under the negacyclic ring. This method requires that the input polynomials need to be pre-processed by multiplying each coefficient with powers of a 2N-th primitive root of unity ζ_{2N},

---

This work was supported by the National Natural Science Foundation of China under Grant 61934002 and 62234008.

979-8-3315-3918-4/25 $31.00 © 2025 IEEE

ensuring compatibility with the negacyclic structure. After performing point-wise multiplication and applying the inverse NTT, a symmetric post-processing step is required to recover the correct result by reversing the twiddle multiplication [4].

Two algorithms are commonly used to reduce the computational complexity from $O(N^2)$ to $O(N\log N)$, namely decimation in time (DIT) and decimation in frequency (DIF). DIT generally adopts the Cooley–Tukey (CT) approach, while DIF is based on the Gentleman–Sande (GS) algorithm. NTT hardware implementations often exploit parallelism by instantiating multiple processing elements (PEs) to compute concurrently, whose number is denoted n_{PE}.

B. Motivations for using HB 3D SEDRAM

Traditional 2D and 2.5D DRAM architectures face critical limitations in sustaining the high bandwidth and low-latency requirements of modern, massively parallel workloads. Specifically, the limited I/O parallelism of planar DRAM and the coarse granularity of memory access in interleaved configurations hinder fine-grained, high-throughput data retrieval. As computational density increases, these constraints form a severe bandwidth bottleneck and degrade overall system efficiency, motivating the exploration of novel architectures.

Hybrid Bonding (HB) technology has emerged as a promising 3D integration solution that enables direct metal-to-metal wafer bonding at sub-10 µm pitch [5]. Compared to traditional Through-Silicon Via (TSV)-based 2.5D stacking, HB offers dramatically increased vertical interconnect density, lower parasitic resistance and capacitance, and reduced power consumption, facilitating tightly coupled compute-memory pairs with local, low-latency data access. Furthermore, its natural support for fine-grained, modular integration allows scalable multi-node expansion, where both DRAM capacity and computational resources within each node can be independently configured. This architectural flexibility enables the system to scale bandwidth and throughput according to varying workload demands, making it highly suitable for domain-specific accelerators.

III. PROPOSED NTT UNIT DESIGN

A. Overall Architecture

Fig. 1 illustrates the proposed NTT unit which adopts a modular architecture composed of four primary components: a memory interface, an address generator, a butterfly processing unit (BFU) array, and a controller. These components are hierarchically organized to form a pipelined structure that enables efficient and high-throughput polynomial transformation. The memory interface handles the access to input coefficients and twiddle factors, while the address generator produces conflict-free addresses based on the computation stage and data layout. The BFU array contains n_{PE} BFUs that execute modular arithmetic operations in parallel, while the controller coordinates control signals and data flow throughout the computation process.

B. Swizzling Memory Scheme

Since each butterfly unit simultaneously requires two input coefficients and one twiddle factor per operation, the increase of the degree of parallelism often leads to memory access conflicts when multiple operands reside in the same SRAM bank. To address this, we employ a Swizzling-based coefficient mapping scheme that remaps coefficient indices to ensure conflict-free memory access, detailed in Algorithm 1.

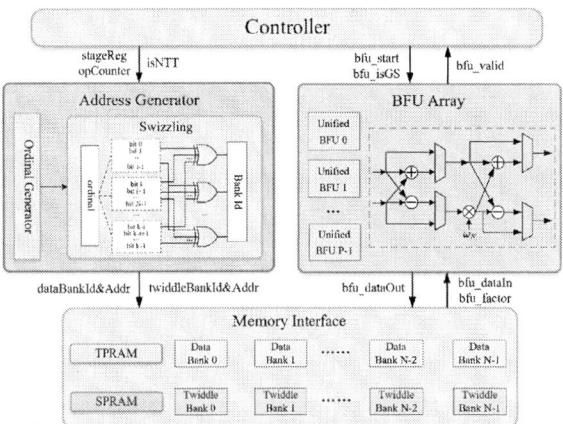

Fig. 1.   Overall architecture of the proposed NTT Unit

Each coefficient index is partitioned into fixed-width groups according to the number of bank address bits. The bitwise XOR result of these groups determines the target SRAM bank, while the intra-bank address is obtained via shift operation. This method guarantees that data required by different butterfly units in the same clock cycle are distributed across distinct banks in all computation stages, enabling conflict-free memory access across arbitrary degrees of parallelism, thus sustaining high NTT throughput with minimal logic overhead.

Based on the mapping scheme, to guarantee conflict-free access for n_{PE} butterfly units, $2n_{PE}$ memory banks are required to store the coefficients. When $N = 16$ and $n_{PE} = 2$, the coefficient distribution across banks is illustrated in Fig. 2. At each computation stage, the input coefficients for each clock cycle are carefully arranged to enable efficient and correct NTT execution. During the first stage, coefficients $(a_0,\ a_8)$ and $(a_1,\ a_9)$ are accessed simultaneously by the two butterfly units in the first cycle, while $(a_0,\ a_4)$ and $(a_2,\ a_6)$ are accessed in the first cycle of the second stage, ensuring no bank conflicts. As is demonstrated, NTT computation can be sustained without introducing additional memory stalls with the Swizzling memory scheme and proper scheduling strategy.

Algorithm 1 XOR-Swizzling Algorithm

Input: *ordinal, bankNum, Width*
Output: *bank, bankAddr*

1 $bankBits \leftarrow \lceil \log_2(bankNum) \rceil$
2 **if** $bankNum = 1$ **then**
3     $bank \leftarrow 0$
4     $bankAddr \leftarrow ordinal$
5 **else**
6     $groupWidth \leftarrow \lceil Width/bankBits \rceil$
7   **for** $i \leftarrow 0$ **to** $bankBits - 1$ **do**
8        $x \leftarrow 0$
9   **for** $j \leftarrow 0$ **to** $groupWidth - 1$ **do**
10          $bitIdx \leftarrow i + j \times bankBits$
11   **if** $bitIdx < Width$ **then**
12             $x \leftarrow x \oplus ordinal[bitIdx]$
13   **end if**
14   **end for**
15        $bank[i] \leftarrow x$
16   **end for**
17     $addr \leftarrow ordinal \gg bankBits$
18 **end if**
19 **return** $\{bank, bankAddr\}$

C. Unified Butterfly Unit

To eliminate the need for an explicit bit-reversal stage, our design adopts a mixed computation scheme in which the forward NTT is performed using the CT butterfly, while the inverse NTT is carried out using the GS butterfly, avoiding costly data permutations. Although CT and GS differ in their

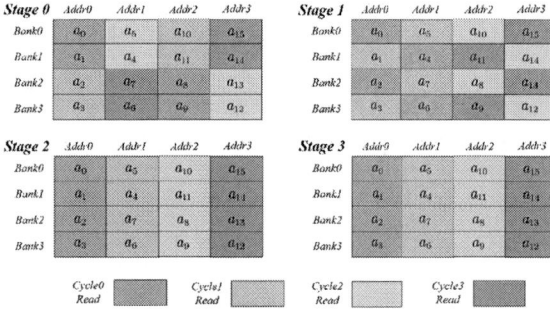

Fig. 2.   Swizzling Scheme for $N=16$, $n_{PE}=2$ NTT

operand and twiddle factor usage, they both require one modular multiplication, one modular addition, and one modular subtraction. Leveraging this similarity, we design a Unified BFU capable of supporting both CT and GS operations by introducing a mode-select signal and using carefully placed multiplexers to switch operand roles, enabling the same hardware datapath to be reused for both algorithms.

D. Unified Modular Multiplication Unit

Our design features a unified modular multiplication unit that integrates multiple reduction strategies to accommodate different moduli. For 32-bit multiplication, the unit employs a Karatsuba-based [6] structure, where operand decomposition and parallel computation of partial terms enable full multiplication to complete within a single clock cycle. For the reduction, we incorporate both the Montgomery and Barrett reduction schemes as the baseline implementations. On top of these, we further integrate a tailored reduction based on the K-Red algorithm [4], which targets moduli of the form $Q=2^u\pm 2^v+1$, leveraging the congruence property $2^u\equiv -2^v-1\ mod\ Q$. Based on this, the intermediate product is segmented into fixed-width chunks, and the reduction is performed iteratively by folding higher chunks into lower ones using the derived congruence, eliminating the need for full-width multiplication.

IV. PROPOSED HB 3D SEDRAM DESIGN

Building on the characteristics of HB technology, this work proposes a 3D SEDRAM architecture that vertically stacks logic and DRAM dies in a fine-grained, one-to-one mapping configuration. As depicted in Fig. 3(shown for $n=4$), the top layers consist of n DRAM dies partitioned into n^2 equally sized memory blocks. Each block includes 4 bank groups, with 8 banks per group, yielding a capacity of 768 Mb per die and a total of 6 GB across the stack. The bottom layer is a monolithic logic die, also partitioned into $n\times n$ homogeneous tiles. Each tile integrates a complete NTT processing unit and is vertically aligned to its corresponding DRAM block using high-density micro-TSVs, ensuring direct and exclusive memory access without cross-tile contention.

During computation, each logic tile captures memory traces emulating SRAM-based behavior and dynamically maps them to DRAM-compatible addresses, coordinated by a lightweight memory controller through the TSV interface. As illustrated in the right part of Fig. 3, our design introduces a fine-grained scheduling strategy that leverages the aggregated intra-node area margins of expanded nodes to place additional processing elements (PEs), enabling adaptive and flexible resource allocation under different computational demands

without increasing per-node footprint. Overall, the proposed architecture establishes an efficient coupling between memory and computation, where the DRAM layer serves as a high-bandwidth, parallel-access memory backend, while the logic layer is responsible for arithmetic processing, dataflow management, and memory coordination—jointly enabling a scalable and high-throughput acceleration platform.

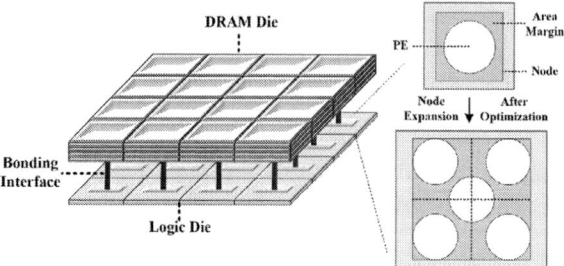

Fig. 3.   Proposed HB 3D SEDRAM design scheme

V. EXPERIMENTAL RESULTS

In this work, the proposed NTT hardware units are evaluated using a Verilator-based simulation flow for functional correctness and execution latency. Area and power estimates for the arithmetic cores are obtained using Synopsys Design Compiler under the TSMC 28nm technology node, while SRAM area is derived from Memory Compiler. All modular multipliers are tested using a 30-bit prime modulus that satisfies the structural requirements of all the reduction algorithms described in Section III. For the cycle-accurate DRAM analysis, we extend Ramulator 2.0 [7] to simulate the proposed HB 3D SEDRAM architecture. The configuration of DRAM capacity has been explained in Section IV, while timing parameters are detailed in Table I.

TABLE I.   TIMING PARAMETERS OF DRAM TILES

Parameter	Value	Parameter	Value	Parameter	Value
tRCD	16.4ns	tRP	11.7ns	tRAS	33.6ns
tCL	1.7ns	tRC	45.3ns	tREF	32ms

A. Design Space Exploration for NTT Unit Configuration

To explore the optimal hardware configurations of the proposed NTT unit across diverse application scenarios, we conduct a design space exploration. Based on measurement results, we first select the tailored modular reduction method for subsequent discussions, as it yields 3× overall performance improvement compared to baseline implementations. Three indicators—latency, area, and power—are normalized, and a composite PPA score is calculated using a weighted model with configurable parameters (α, β, γ) representing the relative importance of performance, area, and power. Inspired by [8], this framework supports flexible tuning toward high-performance (0.6, 0.2, 0.2), balanced (1/3, 1/3, 1/3), or edge-oriented (0.2, 0.4, 0.4) deployment scenarios through the formulation defined in Eq. (2).

$$Score=\alpha\cdot Perf+\beta\cdot Area+\gamma\cdot Power\ \ s.t.\ \alpha+\beta+\gamma=1\ (2)$$

Fig. 4 shows that the optimal degree of parallelism is highly scenario-dependent. At $N=4096$, the best PPA scores are achieved with $n_{PE}=16$ for high-performance use, and $n_{PE}=4$ for both balanced and edge-oriented scenarios. For $N=16384$, optimal choices shift to $n_{PE}=32$, $n_{PE}=16$, and $n_{PE}=4$ respectively.

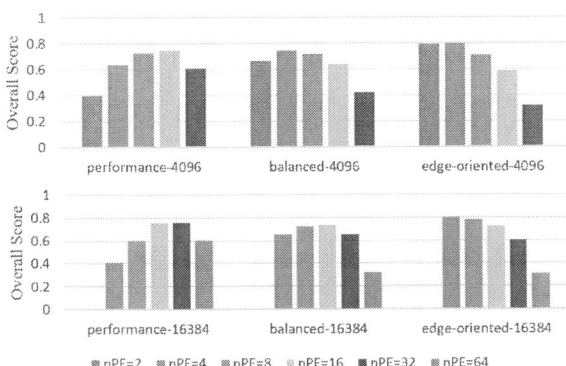

Fig. 4.   Results of DSE for NTT Unit Configuration

B. Node Scalability Exploration of HB-Based 3D SEDRAM

Based on the HB 3D SEDRAM model established in Section IV, we perform cycle-accurate simulations to evaluate the achievable throughput and external memory bandwidth under various node configurations, subject to the node-level area constraint of 12 mm^2 per logic tile defined in [9]. For each NTT size, we identify the optimal parallelism level that maximizes throughput and memory bandwidth within the logic tile constraints. Detailed results are presented in Table II.

TABLE II.   RESULTS OF DSE FOR NODE SCALABILITY

N	#Node	n_{PE}	#NTTU	Throughput (NTT/s^a)	Bandwidth (GB/s)
4096	1	16	7	354897	189.5
	4	16	29	1470289	785.2
	9	16	65	3295477	1760.1
	16	16	116	5881159	3141.0
16384	1	32	2	43794	105.7
	4	32	11	240869	581.5
	9	32	25	547429	1321.6
	16	32	45	985372	2379.0
65536	1	32	1	4834	53.9
	4	64	3	28998	320.7
	9	64	7	67664	748.3
	16	64	12	115996	1282.8

a. The number of N-point NTT operations that can be computed per second

At small transform sizes ($N = 4096$), although higher parallel levels can enhance performance, increasing parallelism beyond the sweet spot leads to disproportionate growth in area, resulting in suboptimal gains in throughput. As the transform size increases ($N = 16384$), the resource footprint of each NTT unit increases, resulting in noticeable area margins in a single node, which can be effectively leveraged to accommodate additional compute units in multi-node configurations. As a result, increasing the number of nodes under these conditions leads to superlinear gains in aggregate throughput and bandwidth. At large scales ($N = 65536$), individual tile constraints limit parallelism per node, but multi-node expansion enables higher configuration flexibility, supporting wider datapaths and higher memory utilization per node. These findings demonstrate that the HB 3D SEDRAM architecture not only sustains compute-memory balance under tight constraints, but also offers scalable performance benefits across diverse workload granularities.

To further validate the scalability of the proposed framework, we integrate an NTT engine design from [10] and conduct node-level scalability experiments under different per-node area constraints. As shown in Table III, the architecture consistently demonstrates improved per-node throughput with increasing node counts across all area settings, yielding up to a 93.8% throughput improvement under the 8mm^2 constraint. Exploration results underscore the framework's strong scalability and ability to flexibly accommodate diverse NTT designs while maintaining efficient memory-compute balance through its high-bandwidth, low-latency DRAM integration.

TABLE III.   RESULTS OF DSE UNDER VARYING AREA CONSTRAINTS

Area Constraint (per node)	6mm^2	8mm^2	12mm^2
Improvement of Throughput	↑25.0%	↑93.8%	↑14.6%

VI. CONCLUSION

We propose a highly scalable and efficient NTT acceleration framework leveraging HB architecture. We innovate in NTT unit design by introducing a unified butterfly unit, a tailored modular reduction scheme, and a Swizzling-based memory strategy that enables conflict-free access under arbitrary parallelism. We develop a vertically stacked HB 3D SEDRAM structure with node-level expansion capability. Experimental results show that we can identify optimal configurations of NTT Unit under different application scenarios, achieving up to 10^7 NTT/s throughput and 10^3 GB/s bandwidth, with up to a 93.8% per-node throughput improvement when adopting a referenced NTT engine design.

REFERENCES

[1] N. Samardzic et al., "Craterlake: a hardware accelerator for efficient unbounded computation on encrypted data," in *Proceedings of the 49th Annual International Symposium on Computer Architecture*, 2022, pp. 173–187.

[2] J. Kim et al., "Ark: Fully homomorphic encryption accelerator with runtime data generation and inter-operation key reuse," in *2022 55th IEEE/ACM International Symposium on Microarchitecture (MICRO)*. IEEE, 2022, pp. 1237–1254.

[3] N. Samardzic et al., "F1: A fast and programmable accelerator for fully homomorphic encryption," in *MICRO-54: 54th Annual IEEE/ACM International Symposium on Microarchitecture*, 2021, pp. 238–252.

[4] P. Longa and M. Naehrig, "Speeding up the number theoretic transform for faster ideal lattice-based cryptography," in *International Conference on Cryptology and Network Security*. Springer, 2016, pp. 124–139.

[5] D. Niu et al., "184qps/w 64mb/mm 2 3d logic-to-dram hybrid bonding with process-near-memory engine for recommendation system," in *2022 IEEE International Solid-State Circuits Conference (ISSCC)*, vol. 65. IEEE, 2022, pp. 1–3.

[6] A. A. Karatsuba, "The complexity of computations," *Proceedings of the Steklov Institute of Mathematics-Interperiodica Translation*, vol. 211, pp. 169–183, 1995.

[7] H. Luo et al.,"Ramulator 2.0: A modern, modular, and extensible dram simulator," *IEEE Computer Architecture Letters*, vol. 23, no. 1, pp. 112–116, 2023.

[8] C. Bai, J. Zhai, Y. Ma, B. Yu, and M. D. Wong, "Towards automated risc-v microarchitecture design with reinforcement learning," in *Proceedings of the AAAI Conference on Artificial Intelligence*, vol. 38, no. 1, 2024, pp. 12–20.

[9] Z. Chen et al., "A high-throughput private inference engine based on 3d stacked memory," in *Proceedings of the 61st ACM/IEEE Design Automation Conference*, 2024, pp. 1–6.

[10] H. Lee, H. Kwon, and Y. Lee, "16.1 a 2.7-to-13.3 μj/boot/slot flexible rns-ckks processor in 28nm cmos technology for fhe-based privacy-preserving computing," in *2024 IEEE International Solid-State Circuits Conference (ISSCC)*, vol. 67. IEEE, 2024, pp. 296–298.

A Low-cost Multiplier-free Accelerator for Binary Neural Network

Z. W. You[1], J. H. Wu[1], R. C. Ma[1], G. C. Qiao*[1]

[1]State Key Laboratory of Electronic Thin Films and Integrated Devices,
University of Electronic Science and Technology of China, Chengdu 611731, China
* Email: zwyou@std.uestc.edu.cn, gcqiao@uestc.edu.cn

Abstract—**With the rapid development of artificial intelligence (AI), deploying deep learning models on resource-constrained edge devices has become increasingly challenging due to computational power and energy limitations. Binary Neural Networks (BNNs) offer a promising solution by replacing costly multiply-accumulates (MACs) with simpler bitwise operations. However, hardware acceleration of BNNs still faces challenges, particularly in the resource-intensive processing of scaling factors. This paper proposes a novel multiplier-free BNN accelerator targeting specific operators in our optimized BNN. By combining lookup table methods and shift operations for efficient scaling factor processing, it achieves low-cost multiplier-free hardware design with a bubble-free pipelined architecture for improved throughput. Experimental results show an inference performance of 877,192 frames per second (FPS) at 100 MHz on the MNIST dataset, while reducing LUT utilization by up to 25% and BRAM consumption by more than 50% compared to related works.**

Keywords—binary neural network (BNN), multiplier-free accelerator, bubble-free pipelined architecture, MNIST

I. INTRODUCTION

AI is at the forefront of the technological revolution, driving advances across industries. However, deploying deep neural networks on edge devices presents a formidable challenge: limited on-chip resources and tight power budgets clash with the high computational demands of modern models. Traditional convolutional neural network (CNN) accelerators like Wu [1] and DCP-CNN [2] predominantly rely on high-precision floating/fixed-point MAC operations for convolution. Although architectures such as DCP-CNN [2] improve energy efficiency via dataflow optimization and on-chip memory hierarchies, they still require numerous multipliers/accumulators and buffers for 16-/8-bit weights and activations. This results in substantial hardware resource overhead—where multipliers occupy significant logic resources—and higher power consumption from increased data bandwidth requirements. Consequently, such accelerators are ill-suited for edge devices demanding low power consumption and real-time performance, especially in resource-constrained IoT scenarios [3].

BNNs effectively reduce computational complexity and memory requirements by constraining network weights and activation values to binary values, thereby achieving significant improvements in computational efficiency. However, BNNs face several issues in hardware implementation: (1) Complexity of

This work was supported in part by STI 2030-Major Projects 2022ZD0209700 and in part by the Sichuan Science and Technology Program (Grant No. 2024ZDZX0001 and 2024ZYD0253).

scaling factor processing. Scaling factors in BNNs are used to recover numerical precision lost during the binarization process, but their processing typically still requires multipliers, becoming the main bottleneck of hardware resource consumption [4], [5]. (2) Batch Normalization (BN) computing problem. In BNNs that include BN layer, the intermediate results after binary convolution are often not binary, which requires complex hardware circuits for BN calculation, increases computational overhead and ultimately leads to increased power consumption [6], [7].

In this paper, we propose a low-cost multiplier-free BNN accelerator that maintains high inference performance. The primary contributions as follows.

1) We optimized existing BN-free binary neural network [8] by modifying residual connections to achieve strictly multiplication-free computation.
2) For the optimized BNN, our hardware implementation features: (a) Functionally decomposed modules (binary convolution, data rearrangement, bias&activation) with specialized optimizations. (b) Hybrid LUT-shift scaling factor processing for targeted BNN operators. (c) Bubble-free pipeline architecture enabling zero-overhead streaming data processing.

Ultimately, an FPGA prototype of the accelerator reduces LUT/BRAM usage by more than 25% and power consumption by over 70% compared to related accelerators, while achieving 877,192 FPS at 100 MHz on the MNIST dataset.

II. ALGORITHM

The BNN structure proposed in [8] effectively addresses activation value mean drift and variance anomaly issues caused by batch normalization removal through the introduction of Scaled Weight Standardization (SWS) technology. Meanwhile, using removable Mask layers and QRPReLU structure eliminates some multiplication operations in the network. During inference, mask layers can be directly removed, thus avoiding related multiplication operations. The QRPReLU structure converts multiplication operations originally introduced by PReLU into more efficient bit operations by constraining slopes to integer powers of 2. Although the above improvements successfully eliminate potential multiplication operations in the binary computation part of BNNs, in residual blocks, this architecture directly feeds input feature maps to convolution layers, thus causing full-precision multiplication operations.

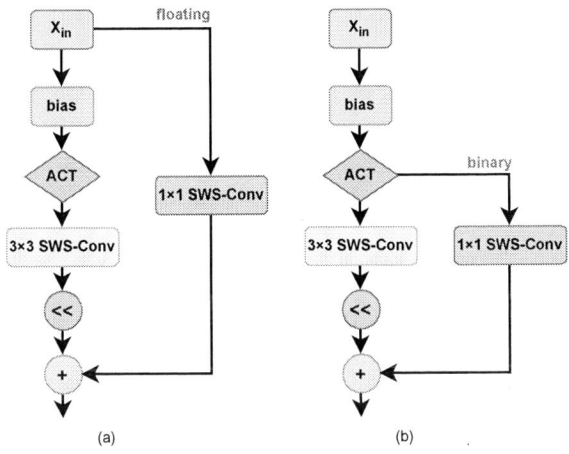

Fig. 1. Diagram of residual blocks in neural networks before and after modification

Therefore, we made certain modifications to the residual blocks in [8]. The computational flow of the original network is shown in Fig. 1(a), while the computational flow of the modified network is shown in Fig. 1(b). X_{in} is the input feature map, bias is the learnable bias, ACT is the sign function, $<<$ denotes shift operation equivalent to multiplying by powers of 2, and SWS-Conv represents scaled weight standardization convolution. The original computational flow of the residual block can be formalized as:

$$y = \text{Sign}(C_{main}(\text{Sign}(X_{in})) + C_{residual}(X_{in}) + b) \quad (1)$$

The improved computational flow is:

$$y = \text{Sign}(C_{main}(\text{Sign}(X_{in})) + C_{residual}(\text{Sign}(X_{in})) + b) \quad (2)$$

The proposed modification reconfigures the residual connection path by feeding the binarized output of the sign function into the residual branch. This architectural adjustment enforces strict binarization of both weights and activations within the residual module. Simultaneously, applying ReAct Sign (RSign) [9] binarization to the residual path input ensures homogeneous binary inputs (1-bit representations) to both main and residual computational paths. Consequently, all convolutional operations within residual modules operate exclusively on binarized data, enabling the replacement of conventional multiplication with XNOR-based logic operations during convolution kernels' computation.

For the MNIST dataset, due to its small image size and grayscale image, we adopted a concise three-layer BNN residual module structure, as shown in Table I. This structure not only effectively reduces the consumption of computing resources, but also meets the relatively simple pattern recognition requirements in the MNIST dataset. The accuracy of the neural network structure from [8] and the modified neural network structure in this paper on the MNIST dataset is also shown in

Table I. It can be seen that this modification has minimal impact on accuracy, with precision loss of less than 0.1%.

TABLE I
BNN STRUCTURE AND ACCURACY COMPARISON

Type	Structure
RSign	–
1 × 1 SWS-Conv	kernel_size=3, strides=2, out_channel=8, padding='SAME'
3 × 3 SWS-Conv	kernel_size=3, strides=2, out_channel=8, padding='SAME'
QRPRelu	
RSign	–
1 × 1 SWS-Conv	kernel_size=3, strides=2, out_channel=16, padding='SAME'
3 × 3 SWS-Conv	kernel_size=3, strides=2, out_channel=16, padding='SAME'
QRPRelu	
RSign	–
1 × 1 SWS-Conv	kernel_size=3, strides=2, out_channel=32, padding='SAME'
3 × 3 SWS-Conv	kernel_size=3, strides=2, out_channel=32, padding='SAME'
QRPRelu	–
FC	output=10
BNN	**Accuracy**
before modification	98.5%
after modification	98.43%

III. HARDWARE DESIGN

Fig. 2 illustrates the multiplier-free accelerator architecture designed for cost efficiency. The core computational elements of each binary convolution layer consist of three dedicated modules: an Img2col transformation module, a binary convolution module, and a bias shift activation module. The accelerator integrates 10 identical BNN processing blocks operating in parallel, with each block capable of processing an entire image independently. Within every block, the vertical arrangement of three module columns corresponds directly to the computation pipeline for a single neural network layer.

To maintain balanced computational throughput, the architecture adapts to our BNN network's dimensional transformations: as feature map width and height halve while channel count doubles at each successive layer, the binary convolution processing elements (PEs) are correspondingly scaled to 8, 8, and 16 units across the three binary layers. This deliberate resource allocation ensures uniform computation cycles per layer, enabling a critical architectural advantage: a bubble-free pipeline design. Consequently, the accelerator achieves continuous streaming of input data directly from the interface without intermediate buffering requirements. This eliminates external memory access overhead while sustaining high inference performance through uninterrupted data flow.

Fig. 3 maps the modified neural network computations to hardware modules. The bias shift activation module can perform bias&activation operations, bias&shift operations, and individual bias operations. Through multi-layer BNN stacking, original consecutive bias operations (red box) merge into equivalent single computation, reducing overhead. The binary convolution module handles all $1 \times 1/3 \times 3$ SWS-Conv, shift, and addition operations.

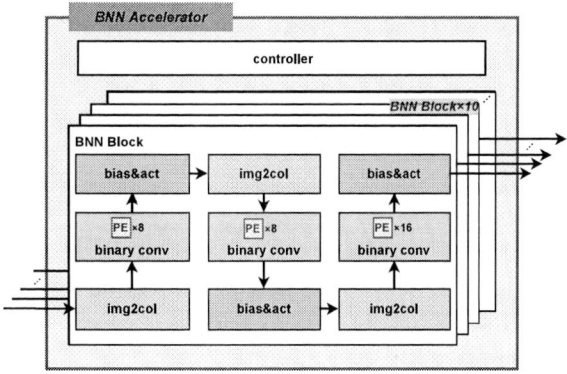

Fig. 2. The overall architecture of the accelerator

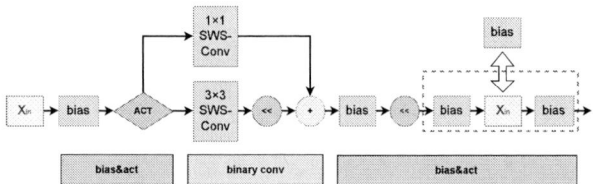

Fig. 3. Corresponding relationship between neural network computation process and hardware modules

A. Binary Convolution Module

Fig. 4 shows the hardware implementation of 3×3 SWS-Conv. The function of 3×3 SWS-Conv is to implement convolution of two 3×3 binary matrices and then multiply by a scaling factor. Because 0 represents -1 and 1 represents 1, the result after 3×3 binary matrix convolution is 9 values of 0 or 1. After adding the 9 numbers and left-shifting by one bit, we get x. If x is greater than 9, it indicates the binary convolution result is positive, setting sign to 1 and sel to $x - 9$. If x is less than 9, it indicates the binary convolution result is negative, setting sign to 0 and sel to $9 - x$. For 3×3 binary matrix binary convolution results, there are only 10 possible cases: $\pm 1, \pm 3, \pm 5, \pm 7, \pm 9$. Therefore, we can calculate the values of scaling factor multiplied by 1, 3, 5, 7, 9, and subsequent calculations can obtain final results through lookup using the high 3 bits of the sel signal and determining whether to invert based on the sign.

Since 1, 3, 5, 7, 9 are fixed multiples, targeted optimization can be performed. We can first obtain $2\times$ and $4\times$ of the scaling factor through shifting. $1\times$ scaling factor equals itself, $3\times$ scaling factor equals $2\times$ scaling factor plus the scaling factor itself, $5\times$ scaling factor equals $4\times$ scaling factor plus the scaling factor itself, and the implementation of $7\times$ and $9\times$ scaling factors follows the same principle. This method replaces complex multiplication operations with shift and addition operations, significantly reducing hardware resource consumption and power overhead.

B. Img2col Module

The Img2col algorithm transforms input feature maps and kernels into expanded matrices for convolution via matrix

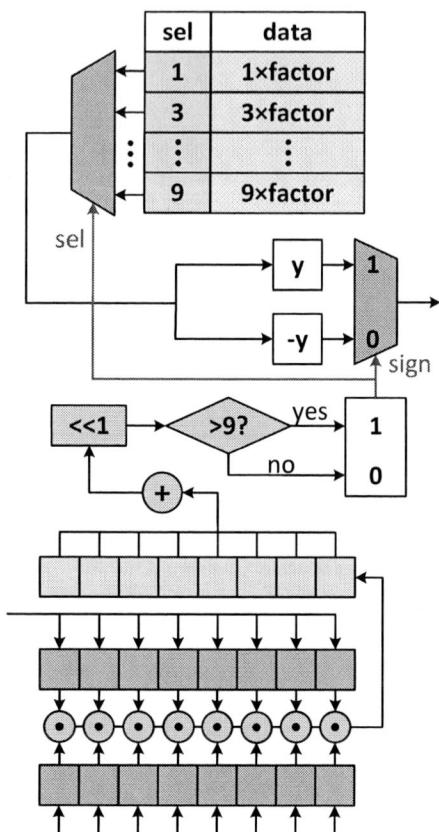

Fig. 4. 3×3 SWS Conv hardware implementation

multiplication. Fig. 5 details its hardware implementation employing a Line Buffer mechanism for 3×3 sliding-window processing, where coordinate notation (x/y) denotes original window positions (e.g. 2/3: row2-column3). Blue and green boxes represent registers and shift registers, respectively. This architecture enables parallel multi-row data output by caching historical inputs under single-port constraints, substantially enhancing throughput.

Assuming M elements per row, each row buffer caches M data elements. The parallel output of N rows requires N buffer lines. After initializing across $M \times (N-1)$ clock cycles, the system achieves synchronous N-row parallel output. Data sequentially populate Row 1 buffer; upon its saturation, Row 1 is output while new data feed Row 2. This cascaded process continues until all N buffers simultaneously output data, thus enabling single-cycle row processing efficiency.

C. Bias Shift Activation Module

The bias shift activation module executes three core operations: bias&shift, bias&activation, and standalone bias. Combinatorially, these implement bias calculation, QRPReLU and Sign function. As operations never occur concurrently, to improve hardware resource utilization, the bias adder is reused to optimize hardware efficiency. Multiplexers (MUX) dynamically select computation paths per operational mode.

As shown in Fig. 6, inputs first undergo bias addition. For standalone bias: data are output directly via the middle MUX

979-8-3315-3918-4/25 $31.00 © 2025 IEEE

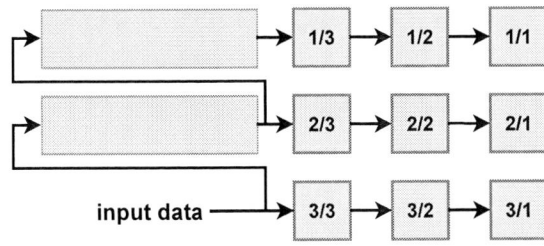

Fig. 5. Img2col module

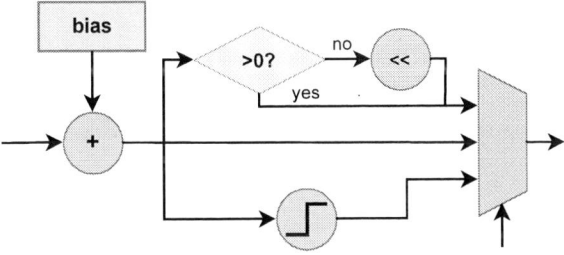

Fig. 6. Bias shift activation module

port. For QRPReLU: the upper comparator preserves outputs > 0 unchanged, while results $\leqslant 0$ undergo quantized slope shifting followed by secondary bias addition via the unit. For Sign function: the lower comparator outputs 1 for results > 0 and 0 otherwise. The module comprises multiple parallel units that implement this three-mode logic.

IV. EXPERIMENTS AND RESULTS

The hardware design in this paper was synthesized and implemented on Xilinx Kintex-7325T under Vivado 2018.3 environment, with system resource consumption shown in Table II. Through simulation and waveform capture, it takes 1140 clock cycles for the accelerator to calculate one image per block (from the first data input to the final calculation result output), of which 784 clock cycles are occupied by the input image, and the calculation delay from input to output is 356 clock cycles. The accelerator has a total of 10 blocks, so its frame rate is 877,192.

Table II also presents several representative works based on FPGA platforms using BNNs to implement handwritten digit recognition tasks from the MNIST dataset. The table offers a detailed comparison in terms of implementation platform, hardware resource consumption, frame rate, power consumption, and classification accuracy. Compared with similar works, the proposed accelerator reduces LUT usage by up to 25%, lowers BRAM consumption by more than 50%. Moreover, it achieves a power reduction of more than 70% compared to some of works. In terms of performance, our design delivers a frame rate that is over 7 times higher than the best among the compared works. These results highlight the clear advantages of our design in hardware efficiency, energy consumption, and inference throughput.

TABLE II
COMPARISON WITH RELATED WORKS

Metrics	FINN-R [10]	BinaryEye [11]	ISCAS 2020 [12]	This work
Platform	ZynqUltra 3EG	Kintex-7325T	Vertex-690T	Kintex-7325T
LUTs	38205	40000	65413	**28545**
BRAMs	417	110	23	**10.5**
Freq.(MHz)	300	100	500	**100**
FPS	-	10,000	122,304*	**877,192**
Power(W)	11.8	12.2	2.08	**2.672**
Accuracy	97.69%	98.40%	98.52%	**98.43%**

*: calculated based on computing power and network structure.

V. CONCLUSION

This paper proposes a low-cost multiplier-free accelerator for BNN, optimizing core convolution and bias-activation operations. Targeting specific multiplication-free operators within BNNs, we implement scaling factor processing through a hybrid approach combining lookup tables and shift operations, achieving a hardware-efficient design. A bubble-free pipeline architecture further enhances throughput while maintaining inference efficiency on resource-constrained platforms. Compared to existing FPGA-based BNN accelerators, our design demonstrates significant improvements in hardware resource overhead, power efficiency, and frame rate performance. Overall, this accelerator has enormous potential for application in edge AI devices.

REFERENCES

[1] T. -H. Wu, C. Shu, et al, "An Efficient FPGA-Based Dilated and Transposed Convolutional Neural Network Accelerator," IEEE Transactions on Circuits and Systems I, vol. 71, no. 11, pp. 5178–5186, 2024.

[2] K. Dai, Z. Xie and S. Liu, "DCP-CNN: Efficient Acceleration of CNNs With Dynamic Computing Parallelism on FPGA," IEEE Transactions on Computer-Aided Design of Integrated Circuits and Systems, vol. 44, no. 2, pp. 540–553, 2025.

[3] V. Sze, Y. Chen, T. Yang, et al., "Efficient Processing of Deep Neural Networks: A Tutorial and Survey," Proceedings of the IEEE, vol. 105, no. 12, pp. 2295–2329, 2017.

[4] M. Rastegari, V. Ordonez, J. Redmon, et al., "XNOR-Net: ImageNet Classification Using Binary Convolutional Neural Networks," ECCV, pp. 525–542, 2016.

[5] M. Alizadeh, J. Fernández-Marqués, N. Lane, et al., "An Empirical Study of Binary Neural Networks' Optimisation," ICLR, 2019.

[6] S. Ioffe, C. Szegedy. "Batch Normalization: Accelerating Deep Network Training by Reducing Internal Covariate Shift," International Conference on Machine Learning, pp. 448–456, 2015.

[7] Su Y., Seng K. P., Ang L. M., Smith J. "Binary Neural Networks in FPGAs: Architectures, Tool Flows and Hardware Comparisons," Sensors 23, no. 22: 9254, 2023.

[8] R. C. Ma, G. C. Qiao, Y. Liu, et al., "A&B BNN: Add&Bit-Operation-Only Hardware-Friendly Binary Neural Network," CVPR, pp. 5704–5713, 2024.

[9] Z. C. Liu, Z. Q. Shen, M. Savvides, et al. "Reactnet: Towards precise binary neural network with generalized activation functions," ECCV, pp. 143–159, 2020.

[10] M. Blott, T. Preußer, N. Fraser, et al., "FINN-R: An End-to-End Deep-Learning Framework for Fast Exploration of Quantized Neural Networks," ACM TRETS, vol. 11, no. 3, pp. 1–23, 2018.

[11] Jokic P, Emery S, Benini L. "BinaryEye: A 20 kfps Streaming Camera System on FPGA with Real-Time On-Device Image Recognition Using Binary Neural Networks," IEEE International Symposium on Industrial Embedded Systems (SIES), pp. 1–7, 2018.

[12] Z. Xian, H. Li and Y. Li, "Weight Isolation-Based Binarized Neural Networks Accelerator," IEEE International Symposium on Circuits and Systems (ISCAS), pp. 1–4, 2020.

979-8-3315-3918-4/25 $31.00 © 2025 IEEE

Design of a MobileNetV2 FPGA Accelerator for Low-Power Real-Time Identification of Plant Nematodes

Ying Zhu [1], Pengjun Wang [*2], Qikang Li [1], Huihong Zhang [1]

[1] Faculty of Electrical Engineering and Computer Science, Ningbo University
[2] College of Electrical and Electronic Engineering, Wenzhou University

* Email: zhuying@nbu.edu.cn, wangpengjun@wzu.edu.cn

Abstract—To address the challenge of targeted plant nematode identification at the edge, we propose a dedicated MobileNetV2 FPGA accelerator architecture. The model employs 16-bit fixed-point quantization, Conv-BN fusion, and GAP-Dot replacement. We implement multi-dimensional parallel computing, reconfigurable memory with ping-pong buffering, and pipelined processing in the accelerator architecture. Deployed on Zynq7000, the system achieves 73.8% nematode recognition accuracy (matching CPU) with 25% latency reduction. Operating at 100MHz, it delivers 14.9 fps at 2.56 W power consumption, yielding 5.82 fps/W energy efficiency. Compared to published MobileNet FPGA accelerators, our implementation demonstrates competitive accuracy with higher energy efficiency in 16-bit designs.

Keywords—Plant Nematodes identification; FPGA accelerator; MobileNetV2; low-power design; parallel computing

I. INTRODUCTION

Various sensors and their associated hardware have been widely applied in smart agriculture [1, 2]. Plant nematodes pose a major threat to agricultural security. Traditional microscopic identification is labor-intensive, time-consuming, and error-prone, rendering them impractical for field applications. Although deep learning models enable automation, their computational and memory demands challenge resource-constrained edge devices [3]. FPGAs enable efficient hardware accelerators by exploiting reconfigurable pipelines, parallelism, and low power.

FPGA-based MobileNet acceleration has advanced through computation, architecture, and memory co-design. Foundational work established parallelization and pipelining for convolution efficiency [4]. MobileNet's heterogeneous depthwise (DWConv) and pointwise convolutions (PWConv) necessitate specialized architectures. Liao et al. [5] implemented pipelined parallel units (5.52 fps @100MHz). Wu et al. [6] developed channel-enhanced convolution modules. In model compression and algorithmic refinement, Su et al. [7] eliminated structural redundancies in RR-MobileNet to reduce cache needs. Bouguezzi [8] replaced depthwise separable convolution with an Ad-depth unit using piecewise linear approximation, cutting hardware resources by 41%. Algorithmically, Winograd transformation has been adopted to boost throughput by reducing multiplicative complexity [9]. Choi et al. [10] quantized MobileNetV2 to 8-bit precision with dynamic DRAM scheduling. Current research focuses on adaptive quantization, tighter heterogeneous computing integration, and near-memory architectures for efficient lightweight CNN edge deployment.

Within this technical context, we have developed a low-power, low-latency FPGA accelerator specifically designed for plant nematode identification. Our approach optimizes efficiency for edge deployment while maintaining recognition accuracy.

II. DESIGN OF MOBILENETV2 ACCELERATOR

A. Optimization of the MobileNetV2 Model

1) Model Quantization

To address FPGA inefficiency in floating-point arithmetic, we employ quantization to map high-precision parameters (32/64-bit) to low-precision fixed-point representations. This reduces memory and resource consumption via linear mapping between domains:

$$Q = round(R/S + Z) \tag{1}$$

$$R = S \times (Q - Z) \tag{2}$$

Here, S denotes the quantization step size, and Z represents the integer value corresponding to the zero point in floating-point representation. Experimental results demonstrate that, in FPGA deployment, using 8-bit fixed-point quantization reduces lookup table (LUT) resource utilization by 95% compared to 32-bit floating-point operations, achieving over a threefold improvement in energy efficiency [11].

2) Conv-BN Fusion

We merge convolutional and Batch Normalization layers into a single operation through mathematical transformation. For input x, convolutional output is:

$$y_{conv} = w_{conv} * x + b_{conv} \tag{3}$$

where w_{conv} represents the weights, and b_{conv} is the biases. BN output y_{norm} is:

$$y_{norm} = \gamma . \frac{y_{conv} - \mu}{\sqrt{\sigma^2 + \epsilon}} + \beta \tag{4}$$

The fusion reconstructs BN as linear transformation:

$$y_{norm} = w_{fuse} \cdot x + b_{fuse} \tag{5}$$

where $w_{fuse} = \frac{\gamma}{\sqrt{\sigma^2 + \epsilon}} \cdot w_{conv}$, $b_{fuse} = \frac{\gamma}{\sqrt{\sigma^2 + \epsilon}} \cdot (b_{conv} - \mu) + \beta$.

Precomputed fusion parameters w_{fuse} and b_{fuse} enable single convolutional execution during inference. This optimization reduces computational instructions by 52.3% and BRAM consumption by 31% [12].

3) GAP-Dot FC Optimization

We replace conventional fully connected layers with Global Average Pooling (GAP) and dot product operations. GAP compresses input tensor $F \in \mathbb{R}^{H \times W \times C}$ to feature vector $f \in \mathbb{R}^C$:

This study was supported by the National Natural Science Foundation of China (Grants No. 62234008 and No. 62134002).

$$f_c = \frac{1}{H \times W} \sum_{i=1}^{H} \sum_{j=1}^{W} F_c(i,j) \tag{6}$$

The dot product layer performs classification:

$$y_k = \sum_{c=1}^{C} w_{k,c} \cdot f_c + b_k \quad for\ k = 1, 2, \ldots, n \tag{7}$$

where $W_{k,c} \in \mathbb{R}^{k \times c}$ denotes the weight matrix (k is the number of classes) and $b_k \in \mathbb{R}^k$ the bias vector. This quantization scheme achieves 98.1% parameter reduction in FC layers with only 0.3% accuracy drop on ImageNet [13].

B. FPGA-Based Acceleration Methodology

1) Hierarchical Parallel Computing Architecture

- Kernel-level convolution parallelism: The FPGA-based convolution operation leverages a line-buffering mechanism to enable parallel computations within the processing window. For a 3×3 convolution kernel, the architecture provides a 3×3 data window every clock cycle (Fig. 1). Nine multipliers compute the products of input data and weights in parallel, and the resulting products are accumulated through an adder tree, enabling pipelined output of the convolution results.

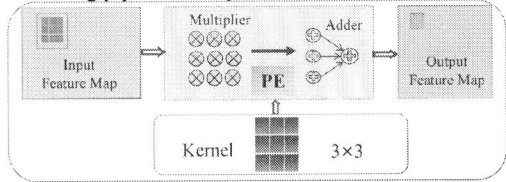

Fig. 1 Parallel Convolution Processing

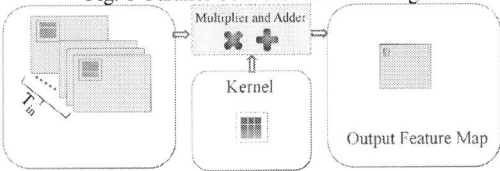

Fig. 2 Input Channel Parallelism

- Input Channel Parallelism: Within a convolutional layer, computations across multiple input feature maps (IFMs) are inherently independent. To exploit this parallelism, individual IFMs are distributed to dedicated Processing Elements (PEs). The resulting partial sums are accumulated across PEs (Fig. 2) to produce the output, significantly accelerating computations.

- Output Channel Parallelism: Multiple convolution kernels are applied simultaneously to the input feature maps, with each kernel generating an independent output feature map (OFM). As depicted in Fig. 3, an array of PEs is deployed. Upon processing all IFMs, intermediate results are accumulated within this array to compute the layer's output features.

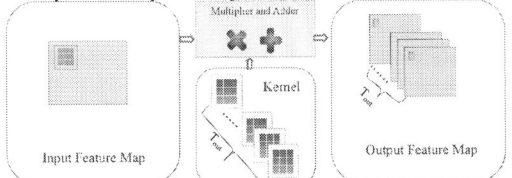

Fig. 3 Output Channel Parallelism

The synergistic integration of these parallelization strategies enhances the peak computational efficiency to:

$$Throughput = f_{clk} \cdot (k_{kernel} \times k_{kernel}) \cdot T_{in} \cdot T_{out} \ OPs/cycle \tag{5}$$

2) Data Tiling and Weight Reuse

During the computation across different spatial locations within the same output channel, the convolution kernel weights remain constant. Therefore, when the convolution kernel slides over the input feature map with a window size of $T_{ix} \times T_{iy}$ the computations for each window are independent. By assigning these weights to parallel Processing Elements (PEs) for reuse, as illustrated in Fig. 4, off-chip memory accesses are effectively reduced, on-chip data reuse is enhanced, memory access latency is minimized, and throughput is maximized.

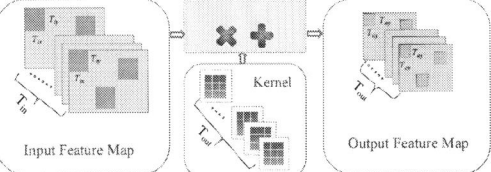

Fig. 4 Data Tiling and Weight Reuse

3) Memory Optimization Design

To amortize the high latency of frequent DRAM accesses, we introduce a ping-pong buffering scheme using dual RAM blocks. Within any clock cycle, this mechanism allows one buffer to prefetch data from DRAM while the other concurrently feeds the compute units. By alternating their roles on subsequent cycles, a continuous data pipeline is formed. This approach effectively hides the data transfer latency, thus enhancing the overall system throughput.

C. FPGA-Based Accelerator Design for MobileNetV2

1) Holistic Architecture of the FPGA Accelerator

The accelerator architecture (Fig. 5) employs a DDR3–BRAM memory hierarchy with partitioned buffers dedicated to the storage and processing of inference data. Off-chip DDR3 memory stores input images, model parameters (weights and biases), and inference outputs. On-chip partitioned BRAMs preload input data blocks and model parameters for the current computation layer, buffer computation results from the PE array, and store intermediate data generated during inter-layer processing. A host-driven synchronization mechanism coordinates DDR3 data transfers, pipelined computations, and AXI-based real-time output streams in parallel, effectively reducing memory latency and enhancing resource utilization.

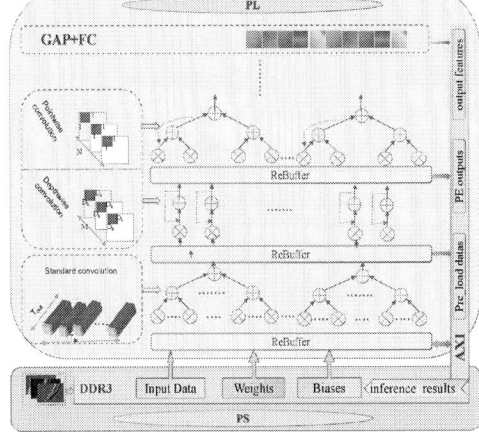

Fig. 5 Holistic Architecture of the FPGA Accelerator

2) Compute Unit Module Design

- **Standard Convolution Module (CONV)**

The initial layer of the MobileNetV2 architecture consists of a 3×3 standard convolution that transforms an RGB input into 32 output channels. Our hardware implementation is optimized to exploit intra-kernel parallelism by leveraging line buffers within the convolutional sliding window. To fully saturate the 64-bit memory interface, the data path is architected to process four output channels concurrently, a design choice that achieves a peak performance of 108 MAC/cycle. A ping-pong buffering scheme is employed to orchestrate the flow of input and output feature maps, effectively masking data movement latency by overlapping it with computation. Subsequent to the convolution, a ReLU6 activation is applied to the results. Finally, the activated feature maps are packed into a 4-channel format prior to off-chip storage, a strategy essential for maximizing throughput under the given bandwidth constraints. Pseudocode for *CONV*:

```
1  for(int i=0;i<Tr;i++){                    //Output_row
2    for(int j=0;j<Tc;j++){                   //Output_Column
#pragma HLS PIPELINE II=1
3      for(int kx=0;kx<K;kx++){               //Kernel_row
#pragma HLS UNROLL
4        for(int ky=0;ky<K;ky++){             //Kernel_Column
#pragma HLS UNROLL
5          for(int mm=0;mm<Tout;mm++){        //Output_channel
#pragma HLS UNROLL
6            for(int nn=0;nn<Tin;nn++){       //Input_channel
#pragma HLS UNROLL
fm_out_buff[mm][i][j]+=fm_in_buff[nn][i*S+kx][j*S+ky]
                                    *wt_buff[mm][nn][kx][ky];
}}}}}}
```

- **Depthwise Convolution Module (DWConv)**

We employ a dynamic tiling strategy for depthwise convolutional layers, adaptively adjusting input tile dimensions based on convolution stride. For stride = 1 operations, the architecture loads 9 × 9 input tiles; for stride = 2 configurations, it expands to 15 × 15 tiles with zero-padded boundary handling. A dedicated parameter reuse mechanism broadcasts weights/biases across all channels. Given the fixed 3 × 3 kernel size and input channel counts divisible by 8, we implement 8-channel parallel processing. This combines intra-kernel parallelism within the convolutional window with inter-channel parallelism, achieving 72 (8 × 9)MAC/cycle throughput. Pseudocode for *DWConv*:

```
1 for(int i=0;i<Tr;i++)                       //Output_row
#pragma HLS PIPELINE
2   for(int j=0;j<Tc;j++){                     //Output_Column
#pragma HLS UNROLL
3     for(int kx=0;kx<K;kx++){                 //Kernel_row
#pragma HLS UNROLL
4       for(int ky=0;ky<K;ky++)                //Kernel_Column
#pragma HLS UNROLL
5         for(int ch=0;ch<Tout;ch++){          //Output_channel
#pragma HLS UNROLL
outbuff[ch][i][j]+=inbuff[ch][i*stride+kx][j*stride+ky]*wtbuff[ch][kx][ky];
}}}}
```

- **Pointwise Convolution Module (PointConv)**

Pointwise convolution is a critical component within the inverted residual bottleneck structure of MobileNetV2, comprising two distinct operational modes: channel expansion and reduction. A key architectural challenge arises from the significant disparity in I/O data volume between these two modes. To mitigate this imbalance, we have developed an optimized memory architecture that features dynamically switchable input buffers. This mechanism allows the hardware to adapt its data buffering

strategy in real-time, conforming to the specific I/O demands of either the expansion or reduction scenario. Furthermore, to handle the channel-intensive nature of this operation, we have integrated a highly parallel multiplier array capable of achieving a computational throughput of 96 MAC/cycle. Pseudocode for *PointConv* :

```
1 for(i=0;i<length;i++){                      //Row
#pragma HLS PIPELINE II=1
2   for(mm=0;mm<Tout;mm++){                    //Output Channel
#pragma HLS UNROLL
3     for(nn=0;nn<Tin;nn++){                   //Input Channel
#pragma HLS UNROLL
      fm_out_buff[mm][i]+=fm_in_buff[nn][i]*wt_buff[mm][nn];
}}}
```

- **Pooling and Fully-Connected Module**

Our architecture replaces FC layers with integrated GAP and dot product (Sec 2.1). The GAP module processes 32 channels/cycle from 7 × 7 × 1280 inputs using parallel channel groups and 4-stage pipelined adder trees (accumulating 49 points/operation), outputting 1 × 1 × 1280 feature maps to the FC layer. For hardware execution, the FC layer reuses the pointwise convolution kernel, loading 32 input channels/cycle while broadcasting 32 × N weights (N=categories). Outputs omit activations, producing N output channels. FC pseudocode:

```
1 for(int i=0;i<N;i++){                        //Output Node
2   for(int j=0;j<Tin;j++){                    //Input Channel
#pragma HLS PIPELINE
    out[i]+=fm_in_buff[j]*wt_buff[i][j];
}}
```

III. RESULTS

A. Experimental Setup and dataset

Experiments were performed on a Xilinx Zynq7000 platform. A custom-developed plant nematode dataset was utilized to train and evaluate the MobileNetV2 network. This study utilizes an 18-category plant nematode image dataset containing 20,905 training and 275 test samples (example: Fig. 6).

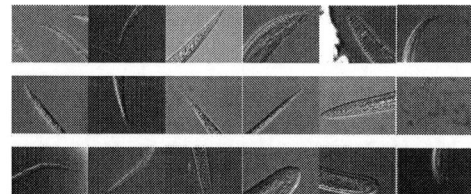

Fig. 6 Sample Images of Plant Nematodes

B. System Performance Analysis

1) Inference Performance Evaluation

As shown in Table 1, our quantized model effectively balances accuracy and efficiency for plant nematode detection. The INT16 model achieves 73.8% accuracy, a minor 0.9% drop from the FP32 baseline, while simultaneously reducing inference latency by 25% and compressing the model size by 78.1%. These metrics confirm a successful co-optimization, enabling efficient inference with negligible loss in predictive performance.

Table 1. Performance Comparison of Quantized Models

Quantized	Accuracy	Model Size	Compression	Latency (ms)
Float32	74.7%	39.8MB	-	89.5
Int16	73.8%	8.73MB	78.1%	67.2

2) Resource Consumption and Power Efficiency Analysis

As shown in Table 2, the implementation utilizes 30% DSP slices, 30% LUTs, 20% FFs, and 18% BRAMs on the target FPGA. These sub-40% utilization rates across all resource types enable system scalability, simplify place-and-route implementation, and reduce power consumption.

Table 2. FPGA Resource Utilization

Resource	DSP	BRAM	LUT	FF
Used	598	137	83196	110010
Available	2020	755	277400	554800
Utilization	30%	18%	30%	20%

Cross platform performance shows in Table 3. Measured at 100 MHz operating frequency, the system consumes 2.56W operating power and achieves 5.82 fps/W energy efficiency, which represents a 176% improvement over CPU baselines. For field-deployable nematode identification, it delivers sustained real-time 14.9 fps throughput with >8-hour battery life (20,000 mAh @ 5V).

Table 3. Performance Comparison Across Platforms

Platform	Quantized	Throughput	Power	Energy Efficiency
CPU	float32	11.17 fps	22.23W	2.11 fps/W
Ours	int16	14.9 fps	2.56 W	5.82 fps/W

3) Comparative Analysis of Related Work

As show in Table 4, our approach achieves a competitive energy efficiency of 5.82 fps/W, corresponding to a 5.0× improvement over [14] (1.15 fps/W) and a 5.2× enhancement versus [15] (1.12 fps/W). When benchmarked against [16] (3.97 fps/W) implemented on the same-class FPGA platform (Zynq 7020), our method exhibits a 46.6% higher energy efficiency. While [17] attains 163.7 fps on the high-performance ZC706 platform, its 16.7 fps/W efficiency incurs 3.8W power. Our solution reduces power by 32.6%, demonstrating superior edge suitability through precision-efficiency co-optimization. Despite operating at 16-bit precision (vs. 8-bit baselines), our design achieves competitive energy efficiency. ·

Table 4: Performance Comparison of Different MobileNet Architectures

Method	Platform	Quantized	Power	Throughput	Energy Efficiency
Ours	Zynq7000	16 bit	2.56 W	14.9 fps	5.82 fps/W
[14]	Arria10	8 bit	22.5 W	26 fps	1.15fps/W
[15]	ZCU104	8 bit	8.936 W	10.03 fps	1.12fps/W
[16]	Zynq7020	8 bit	2.96W	11.75 fps	3.97fps/W
[17]	ZC706	8 bit	3.8 W	163.7 fps	42.1 fps/W

IV. CONCLUSION

This work presents a FPGA-based MobileNetV2 accelerator optimal edge deployment efficiency-accuracy balance. Our FPGA-based MobileNetV2 accelerator employs a cross-layer parallel architecture (216/504/96 MAC operations/cycle) and dynamically reconfigurable memory, reducing off-chip bandwidth while achieving 14.9 fps real-time inference at 16-bit precision. The system attains 73.8% nematode recognition accuracy with only 0.9% degradation from the 74.7% float baseline, alongside 78.1% model compression. With power consumption of 2.56 W delivering 5.82 fps/W energy efficiency, benchmarks against published MobileNetV2 FPGA accelerators confirm competitive accuracy and superior energy efficiency in comparable 16-bit designs. These results represent 88.5% power reduction and 176%

efficiency gain over CPU implementations. Crucially, the design maintains 70% scalability headroom at 30% resource utilization, establishing a versatile template for battery-powered edge devices that enables efficient mobile nematode detection in agricultural field applications.

ACKNOWLEDGMENT

This study was supported by the National Natural Science Foundation of China (Grants No. 62234008 and No. 62134002).

REFERENCES

[1] H. Yang, M. Huang, and M. Ren et al., "Piezoelectric energy harvesting interface circuit for small area and low power consumption—A review," Measurement, vol. 242, no. Part C, 2025.

[2] H. Yang, J. Yan, and J. Tan et al., "High-Precision Readout Circuits for Portable Biosensor Devices—A Review of IA," IEEE Sensors Journal, vol. 35, no. 6, 2025.

[3] X. Li, P. Wang, G. Li et al., "Design of a Novel Self-Test-on-Chip Interface ASIC for Capacitive Accelerometers," IEEE Transactions on Circuits and Systems—I: Regular Papers, vol. 70, no. 7, pp: 2801-2810, Jul. 2023.

[4] C. Zhang, P. Li, G. Sun et al., Optimizing fpga-based accelerator design for deep convolutional neural networks[C]. Proceedings of the 2015 ACM/SIGDA international symposium on field-programmable gate arrays. pp: 161-170, 2015

[5] J. Liao, L. Cai, Y. Xu et al., Design of accelerator for mobilenet convolutional neural network based on fpga[C]. 2019 IEEE 4th Advanced Information Technology, Electronic and Automation Control Conference (IAEAC). IEEE, pp: 1392-1396, 2019.

[6] D. Wu, Y. Zhang, X. Jia, et al., A high-performance CNN processor based on FPGA for MobileNets[C], 2019 29th International Conference on Field Programmable Logic and Applications (FPL). IEEE, pp: 136-143, 2019.

[7] Y. Su, S. Zhao, and Y. Lin et al., "RR-MobileNet: Resource-Efficient Recursive Mobile Network for Embedded Applications," in Proc. IEEE Int. Conf. Image Process. (ICIP), pp: 1561–1565, 2020.

[8] S. Bouguezzi, H. B. Fredj, T. Belabed et al., An efficient FPGA-based convolutional neural network for classification: Ad-MobileNet[J]. Electronics, 10(18), pp:2272-2216, 2021.

[9] Yutana, P. Srisukkham, N. Leelaruj et al., "Scalable Winograd Accelerators for Embedded FPGA Systems," Integration, the VLSI Journal, vol. 82, pp: 20–31, 2022.

[10] S. Choi, J. Yoon, H. Kim et al., "BurstDRAM: Memory Scheduling for Real-time DRAM Burst Bandwidth Utilization," in Proc. IEEE/ACM Int. Symp. Comput. Archit. (ISCA), pp: 1091–1104, 2021.

[11] J. Qiu, "Fixed-point Quantization for Deep Neural Networks," in IEEE Conf. Comput. Vis. Pattern Recognit. Workshops (CVPRW), pp: 1848–1856, 2018.

[12] Y. Umuroglu, N. Fraser, G. Gambardella et al., "FINN: A Framework for Fast, Scalable Binarized Neural Network Inference," in Proceedings of the ACM/SIGDA International Symposium on Field-Programmable Gate Arrays (FPGA), Monterey, CA, pp: 65–76, 2017.

[13] M. Sandler, A. Howard, M. Zhu et al., "MobileNetV2: Inverted Residuals and Linear Bottlenecks," in Proceedings of the IEEE/CVF Conference on Computer Vision and Pattern Recognition (CVPR), Salt Lake City, UT, pp:4510–4520, 2018.

[14] L. J. Guo, "FPGA-based high-performance MobileNet hardware accelerator," Master thesis, Nanjing University, 2020.

[15] C. X. Niu, "FPGA-based convolutional neural network acceleration design and implementation," Master thesis, Changchun Institute of Optics, Fine Mechanics and Physics, Chinese Academy of Sciences, 2024.

[16] X. Zhao, L. J. Meng, W. H. Liu et al., "Implementation of garbage classification system based on Zynq platform," Industrial Instrumentation & Automation, no. 05, pp: 26-31, 96, 2022.

[17] L. Xiao, D. B. Liang, D. H. Chen et al., "FPGA-based high-efficiency scalable MobileNet accelerator implementation," Computer Engineering and Science, vol. 43, no. 4, pp: 628-633, 2021.

A 65nm Analog-Computing Chip With Reconfigurable Charge-Pump-Based Adders for 5.26nJ/Decision Retrainless Keyword-Spotting

Lichen Feng[1], Rundong Cai[1], Lin Wu[1], Zhangming Zhu*[1]

[1] Key Laboratory of Analog Integrated Circuits and Systems (Xidian University), Ministry of Education, School of Integrated Circuits, Xidian University, Xi'an 710071, P. R. China

* Email: lcfeng@xidian.edu.cn, zmyh@263.net

Abstract—This paper presents a reconfigurable charge-pump-based analog computing-in-memory chip for keyword-spotting (KWS), achieving an output ratio of 0.87, 92.6% accuracy in 10-word KWS with 2ms latency, and 5.26 nJ/decision. Charge-domain computation eliminates readout energy dissipation. The proposed weight-bitwise multiply-accumulate (MAC) unit with a reconfigurable charge pump halves the number of analog-to-digital converters (ADCs). Customized switching logic reduces the power consumption of calibration-free ADCs by 10%. In the IN/W=4b configuration, the chip achieves a decision rate of 1000Hz and a figure of merit (FoM) of 14.26, respectively.

Keywords—KWS, CIM, charge-pump, weight-bitwise MAC.

I. INTRODUCTION

Low-power and real-time keyword-spotting (KWS) is crucial for edge devices [1]. Digital multistage architectures [2,3,4], which incorporate a spectral feature extractor and an intelligent classifier, achieves a low power consumption of 1.5µW in 28nm CMOS and fast-settling analog frontend [4], but the hierarchical and recurrent computing increases the latency to 16ms [4] (Fig. 1, top). In contrast, the parallel Computing-In-Memory (CIM) chip [5] reduces the latency to 39.9µs, but its large recurrent model requires massive SRAM cells, consuming the power of 11mW (0.44µJ/39.9µs). The fast end-to-end KWS (2ms latency) systems with lowered power consumption (<3µW) have been prototyped in 28nm CMOS [6,7] (Fig. 1, middle). However, their multiply-accumulation (MAC) precision is highly diminished, these reduced output ratio results in around 1.5% accuracy degradation compared to the float-point baselines.

The existing analog-computing KWS chips generally exhibit two primary issues: 1) To achieve higher energy efficiency, the computation results are quantized using low bits, leading to significant loss in output ratio; 2) For different neural network models, retraining is necessary, which demonstrates poor versatility and precludes direct deployment of open-source KWS models. This paper reports a 65nm KWS chip with a CIM-MAC output ratio of 0.87 by proposing reconfigurable charge-pump-based adders (Fig. 1, bottom). The chip can process both digital input signals from memories and analog ones from sensors directly. The normalized energy per decision is reduced by 1.8× compared to [7] as 5.26nJ/decision. The key techniques are: 1). Weight-bitwise MAC [8], which can input digital and analog signals, with a reconfigurable charge pump (CP) that fuses the two MAC results, halving the number of ADCs required; 2). Calibration-

This work was supported in part by the Scientific and Technological Innovation 2030 of China (2021ZD0114401), in part by the National Natural Science Foundation of China under Grants 62474128, U22A2013, and 92164301, in part by the Natural Science Foundation of Zhejiang under Grant LDT23F04022F04 and LDT23F0402.

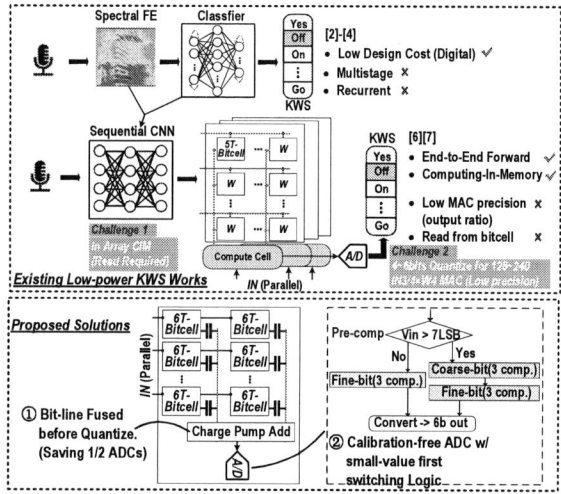

Fig. 1. Challenges of Low-power KWS chips and proposed solutions.

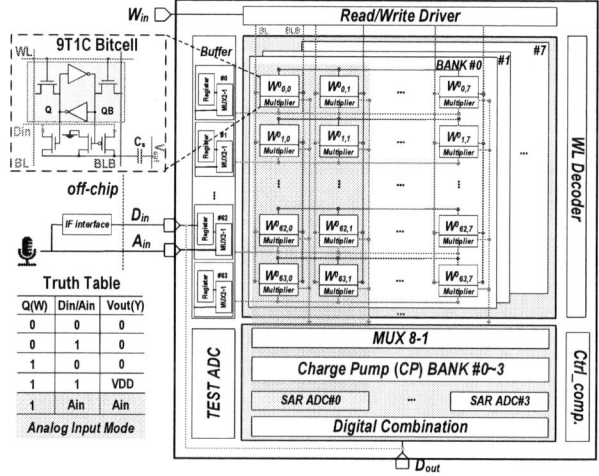

Fig. 2. The architecture of the chip.

free small-value-first (SVF) ADC, which reduces the power consumption by 10% using a customized switching logic.

II. ARCHITECTURE OF THE PROPOSED CHIP

Fig. 2 illustrates the architecture of the proposed chip. This chip consists of eight banks of bit-cell array with the size of 64×8. The bit-cell is composed of a transmission gate, a NMOS transistor that can pull Vout down to ground, and a standard 6T SRAM cell. The sensor-acquired data can be fed directly into the chip in the form of an analog signal Ain, or in

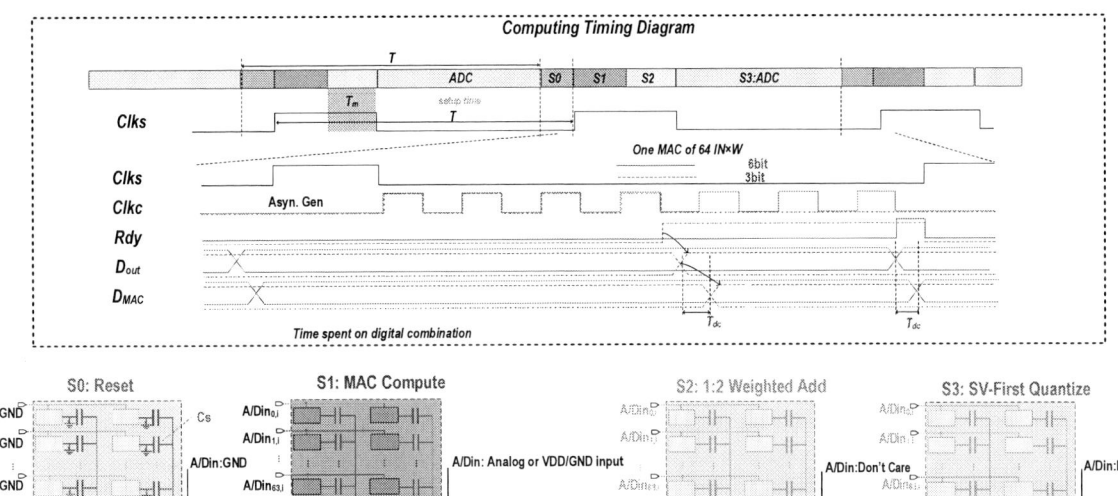

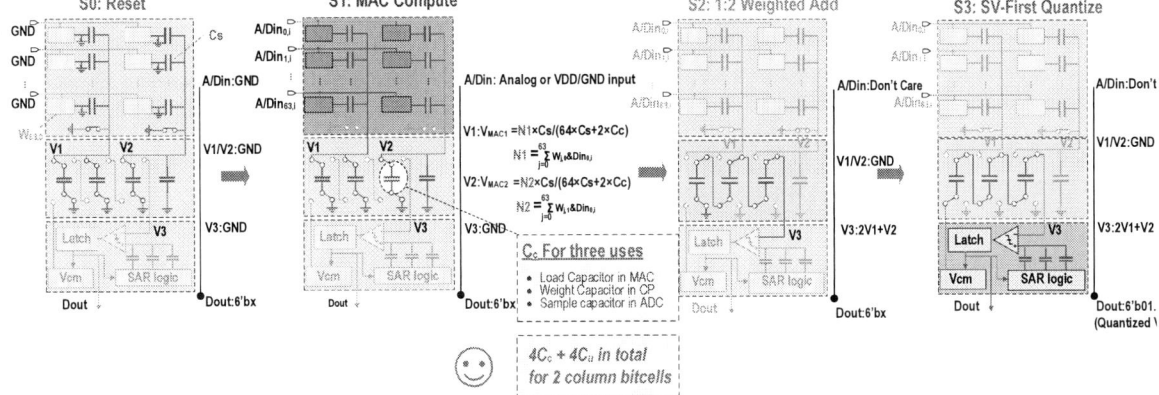

Fig. 3. The timing diagram, circuit structures and computation stages of CIM.

the form of a digital signal Din through an interface. A logical 'AND' operation, i.e. multiplication operation, between Din/Ain and W is implemented as shown in the truth table, in which Din/Ain can be digital signals from the input register bank or analog ones directly. Based on the place values of the weights stored in the array, the eight Vouts are added in pair using four charge pumps with a ratio of 1:2 before quantization by ADCs, effectively halving the number of required ADCs. After completing the ADC quantization and digital combination, the output data can be re-entered into the computational unit for repeated calculations.

III. COMPUTATION STAGES OF THE CIM

Fig. 3 shows the timing diagram，the circuit structures and the four working states S0-S3 of the proposed chip. The timing diagram shown in Fig. 3 presents the partition of the operation period. The reset, sampling, addition in CP, and quantization by ADCs are executed sequentially. The main clock of ADC 'Clks' is enabled after the setup time Tm of the CP, and the asynchronously generated comparing clock 'Clkc' is enabled for 3-bit or 6-bit quantization according to the pre-comparison result. A short time window of Tdc is used to realize digital combination. In state S0, the top and bottom plates of all capacitors are connected to ground to complete reset. In state S1, the shared plate between the sampling capacitor in each bitcell and the capacitor in CP is disconnected from ground. Each bit of A/Din is set to GND or VDD or Ain according to the 64 weight-bitwise split input values (IN [0-4] 1-64) and the MAC operation is realized in this state. In state S2, the connection of Cc is reconfigured to achieve the expression of V3, which is:

$$V3 = 2V1 + V2 \quad (1)$$

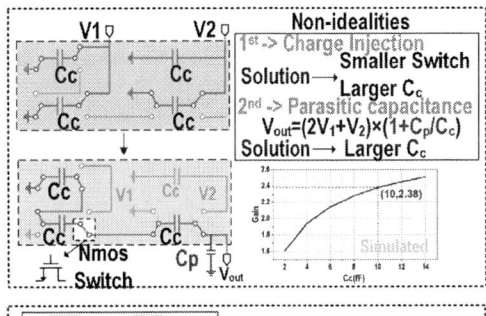

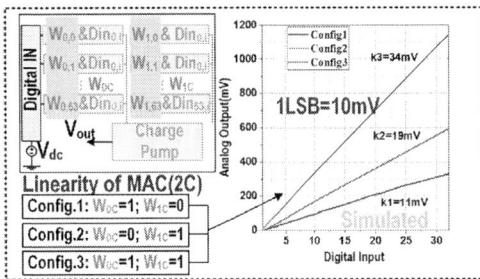

Fig. 4. The non-ideality analysis of CP-based adder (top). The selection of the LSB (bottom).

following the operation principle of charge pump. The computation result is the final result, which will be quantized in the next state. In state S3, V3 is compared with a predefined threshold voltage (set as 7LSB in this case) to determine the switching number and after that it is quantized into 6bit digital codes. In this manner, the Cc's are reused for three times, serving as the loading capacitor in MAC, weighting capacitor

979-8-3315-3918-4/25 $31.00 © 2025 IEEE

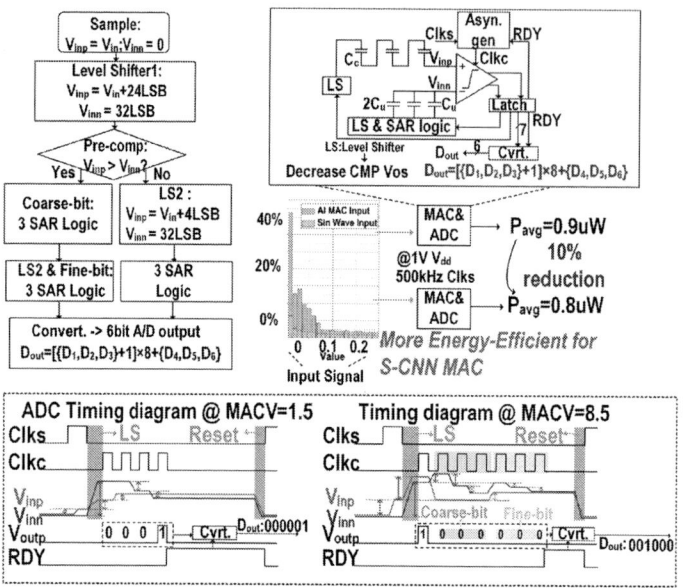

Fig. 6. The flowchart of the SVF ADC with the timing diagram showing two quantization examples.

in CP-based adder and sampling capacitor in ADC, respectively, which greatly reduces the hardware overhead of capacitors.

Fig. 4 (top) shows the circuit diagram of the CP and its working principle. This process involves two non-ideal effects: 1). The charge injection; 2). The computational error caused by the parasitic capacitance. As the capacitance Cc increases, the adverse effects of charge injection on linearity diminish, and the impact of parasitic capacitance is similarly reduced. By simulating the impact of Cc on the weighted gain, Cc=10fF with a gain of 2.38 is selected to tradeoff between area and accuracy. The two columns of SRAM and a CP adder are analyzed. The simulation regarding the determination of LSB (Fig. 4, bottom) is conducted under three modes, namely $W_{0C}=1\&W_{1C}=0$, $W_{0C}=0\&W_{1C}=1$, and $W_{0C}=1\&W_{1C}=1$. These three modes represent an input of 1x, 2x, and 3x, respectively. The simulation result indicates that the LSB in all three modes is close to 10mV, so we designed the ADC with 1LSB=10mV. A larger LSB reduces quantization range for MAC operations (this limitation can be mitigated by the sparsity distribution of data), but it alleviates the mismatch introduced by row-based layout (RBL) and enhances the tolerance to noise during quantization process, therefore, no additional calibration techniques are required.

The flowchart of the single-ended ADC is as described in Fig. 5. The input signal is sampled to Vinp and, after passing through Level Shifter 1 (LS1), the comparison between Vinp + 24 LSB and 32LSB is made to determine whether to enable the 3-bit or the 6-bit mode. If the former is larger, the coarse-bit mode is enabled. After the first 3 bits are quantified based on the SAR method, Level Shifter 2 (LS2) is used to enable fine-bit mode for quantizing the remaining 3 bits. All quantized codes are digitally mixed to generate the final output Dout, which can be represented as:

$$\text{Dout} = [\{D1, D2, D3\} + 1] \times 8 + \{D4, D5, D6\} \quad (2)$$

If the latter, 32LSB, is larger, the 3 least significant bits are quantized directly via LS2, while the first 3 bits are set to zero. In this case, only three quantization steps are needed to obtain

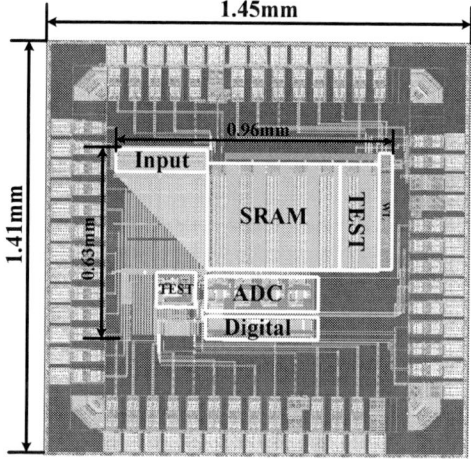

Fig. 5. The layout of the chip.

a six-bit quantization code. In the two examples where the MAC computation values MACV are 1.5 (less than 7 LSB) and 8.5 (greater than 7 LSB), the timing diagrams of the ADC are shown at the bottom of Fig. 5. Since the computation results of the MAC are typically concentrated around smaller values, and the input voltage to the ADC is generally low, the power-efficient fine-bit mode will be used more frequently, significantly reducing system power consumption from a statistical perspective. Power consumption simulations under 1V supply voltage and 540kHz sampling frequency demonstrate that power consumption could be reduced by 10% under S-CNN data input compared to sinusoidal inputs.

IV. SIMULATION RESULTS AND DISCUSSION

Fig. 6 shows the layout of the 65nm CMOS chip. The active area of the chip is 0.96 mm × 0.63 mm, with the positions of modules such as SRAM and ADC clearly marked. Fig. 7 shows the post-layout simulation results of the power breakdown at 540kHz and 1V VDD. The standby and active power consumption is analyzed, in which SRAM and ADC

979-8-3315-3918-4/25 $31.00 © 2025 IEEE

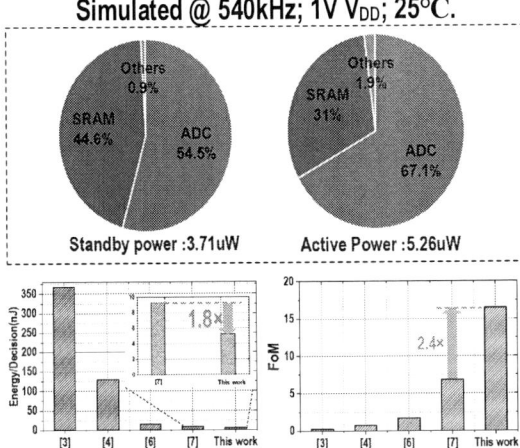

Fig. 8. The post-layout simulation results of the power breakdown.

| TABLE I. | COMPARISON WITH STATE-OF-THE-ART |

Performance	Multistage Works		End-to-End S-CNN Works		
	[3] ISSCC'22	[4] ISSCC'23	[6] JSSC'24	[7] ISSCC'24	This work
Technology (nm)	65	28	28	28	65
Clock Frequency(MHz)	0.25	0.25	0.1	1	0.54
Supply Voltage(V)	0.5/0.75	0.5/0.65/1.4	0.3/0.9	0.35/0.9	1.0/0.8
# of parameters (kB)	24	18	9.5	11	10.1
# bit of IN/W/MAC/Out	NA	NA	3/4/15/4	3/4/14/9	4/4/15/13
Output ratio	1	1	0.27	0.64	0.87
KWS Accuracy on GSCD	86.0% 10-KWS	92.8% 5-KWS	90.9% 10-KWS	91.8% 10-KWS	92.6% 10-KWS
Decision Latency(ms)	28.4	16	2	2	2
Decision Rate(Hz)	62.5	62.5	1000	1000	1000
Energy/Decision(nJ)	368	129.34	15.68	9.32	5.26
FoM	0.27	0.78	1.72	6.87	16.54

V. CONCLUSION

An analog-computing-in-memory chip design with the high output ratio of 0.87 for retrain-less keyword spotting is proposed in this paper. A reconfigurable charge pump is design to fuse the two MAC results in the two neighboring columns, halving the number of ADCs required. Coupled with the ADCs with the small-value-first timing, the 2ms latency, and 5.26 nJ/decision is achieved.

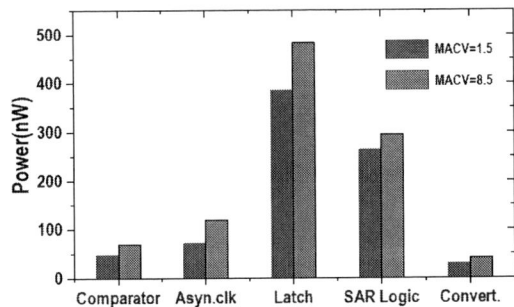

Fig. 7. The simulation results of the SVF ADC's power consumption.

contribute the majority of power consumption, and standby power occupies a large proportion during dynamic working due to the low clocking frequency. In addition, the normalized energy per decision is reduced by 1.8× and the FoM consisted of the accuracy and power performance is improved by 2.4× compared to the state-of-the-art [7]. The FoM in this paper can be represented as:

$$FoM = \frac{\text{Output Radio}}{\text{Enregy/Decision}} \times 100 \qquad (3)$$

Fig. 8 shows the simulation results of the SVF ADC's power consumption in two input examples (MACV=1.5 and 8.5). The power consumption analysis of specific modules of ADC shows that the proposed switching logic reduces the number of comparisons when the input value is small, thereby lowering ADC's power consumption. When the input DC value is high (larger than 7LSB), the SVF ADC's power consumption increases to 1μW. It can be seen that the power reduces by 1.25x when input data is small, demonstrating the effectiveness of SVF timing.

The comparison with the state of the art is shown in Table I. In terms of S-CNN works incorporating CIM techniques, this chip maintains the highest output ratio of 0.87, achieving the best 10-keyword spotting accuracy while keeping high energy efficiency of 5.26nJ/decision with the core area of 0.6 mm².

REFERENCES

[1] Y. Zhang, N. Suda, L. Lai, et al. Hello Edge: Keyword Spotting on Microcontrollers. arXiv:1711.07128, 2017.

[2] J. S. P. Giraldo, S. Lauwereins, K. Badami, et al. Vocell: A 65-nm Speech-Triggered Wake-Up SoC for 10-μW Keyword Spotting and Speaker Verification. IEEE Journal of Solid-State Circuits, 2020. 55:868-878.

[3] K. Kim, C. Gao, R. Graca, et al. A 23-μW Keyword Spotting IC With Ring-Oscillator-Based Time-Domain Feature Extraction. IEEE Journal of Solid-State Circuits, 2022, 57:3298-3311.

[4] J-H Seol, H Yang, R Rothe, et al. A 1.5μW End-to-End Keyword Spotting SoC with Content-Adaptive Frame Sub-Sampling and Fast-Settling Analog Frontend. In: IEEE International Solid-State Circuits Conference, San Francisco, CA, USA, 2023. 1-3.

[5] H. Dbouk, S. K. Gonugondla, C. Sakr, et al. A 0.44-μJ/dec, 39.9-μs/dec, Recurrent Attention In-Memory Processor for Keyword Spotting. IEEE Journal of Solid-State Circuits, 2021, 56:2234-2244.

[6] F. Tan, W. -H. Yu, K. -F. Un, et al. A 0.05-mm² 2.91-nJ/Decision Keyword-Spotting (KWS) Chip Featuring an Always-Retention 5T-SRAM in 28-nm CMOS. IEEE Journal of Solid-State Circuits, 2024, 59:626-635.

[7] F. Tan, W. -H. Yu, J. Lin, et al. 17.9 A 1.8% FAR, 2ms Decision Latency, 1.73nJ/Decision Keywords Spotting (KWS) Chip Incorporating Transfer-Computing Speaker Verification, Hybrid-Domain Computing and Scalable 5T-SRAM. In: IEEE International Solid-State Circuits Conference, San Francisco, CA, USA, 2024. 330-332.

[8] C. -J. Jhang, C. -X. Xue, J. -M. Hung, et al. Challenges and Trends of SRAM-Based Computing-In-Memory for AI Edge Devices. IEEE Transactions on Circuits and Systems I: Regular Papers, 2021. 68:1773-1786.

HAMP: Head-Aware Mixed-Precision Token Pruning and Quantization for Efficient ASR

Xiaoxing Wu [1], Xinghao Zhu [1], Lanqi Ma [1], Gengsheng Chen[1,2], and Wenbo Yin*[1]

[1] College of Integrated Circuits and Micro-Nano Electronics, Fudan University, Shanghai, China
[2] Jiashan Fudan Institute, Jiaxing, Zhejiang, China

* Email: wuxx24@m.fudan.edu.cn, wbyin@fudan.edu.cn

Abstract—**Transformer-based models have become the dominant architecture in Automatic Speech Recognition (ASR), but their vast computational cost hinders real-time deployment on edge platforms. Existing compression methods primarily focus on reducing model size, but often fail to maintain high inference accuracy on resource-constrained platforms such as CPUs and FPGAs. To address this, we propose HAMP, a token-aware pruning and quantization framework with hardware support for high-accuracy low-latency ASR. HAMP employs a layer-wise mixed-precision strategy to dynamically eliminate redundant computations based on token importance. To further mitigate accuracy loss, we introduce the DRP module, which adaptively adjusts pruning ratios across attention heads. Our approach integrates seamlessly with ARM Single Instruction Multiple Data (SIMD) parallelism and FPGA systolic arrays. Experiments show that HAMP achieves up to 7.5× speedup on CPUs and 8× on FPGAs over the full-precision baseline, with only 0.96 Word Error Rate (WER) degradation, demonstrating strong potential for efficient, real-time ASR on edge platforms.**

Keywords—Automatic Speech Recognition, Transformer, mixed-precision pruning, head-aware pruning, quantization

I. INTRODUCTION

Automatic Speech Recognition (ASR) enables natural interaction in virtual assistants, smart devices, and accessibility tools. With the rise of these voice-controlled applications, optimizing ASR for edge devices has become increasingly important. Transformer-based models [1], such as OpenAI's Whisper [2], outperform traditional Recurrent Neural Network (RNN) and Convolution Neural Network (CNN) systems in robustness and multilingual transcription, and have become the dominant architecture in ASR. However, their high computational cost hinders the deployment of real-time inference on embedded and mobile platforms [3].

To reduce computation and meet real-time demands, on the algorithmic side, various methods compress ASR models through static, fixed-precision reductions in parameter size, such as cross-layer weight sharing and sparsity-aware training [4-6]. However, these approaches ignore input variability and uneven information distribution during inference, often leading to notable accuracy degradation.

On the hardware side, ASR is mainly deployed on resource constrained edge devices, where hardware compatibility is essential for optimization. Common embedded platforms include ARM CPUs and FPGAs. ARM cores are favored for their low power consumption and general-purpose flexibility, while FPGAs provide massive parallelism suitable for compute-intensive tasks. However, most existing methods either focus on GPU-based acceleration [7] or rely on computationally heavy approaches like Viterbi Search pruning [8], making them incompatible with edge hardware.

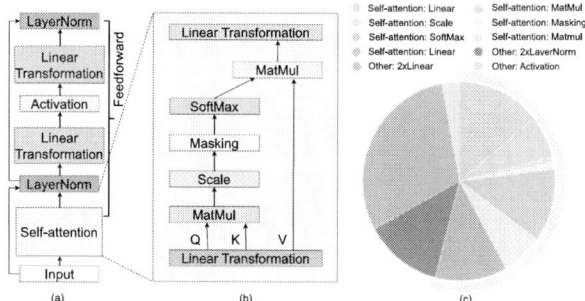

Fig. 1. Time consumption of each module in Transformer.
(a) Encoder workflow. (b) Self-attention workflow. (c) Time consumption.

To bridge this gap, we propose HAMP, a head-aware mixed-precision token pruning and quantization framework with hardware support. Unlike prior methods that statically compress model weights with uniform precision, our approach leverages input-adaptive techniques targeting token-level redundancy and dynamically adjust precision during inference, improving both accuracy and efficiency. To ensure edge compatibility, we implement the approach on ARM CPUs and FPGAs using vectorized SIMD instructions and bit-level systolic arrays. This design enables both efficient computation and practical deployment across diverse hardware.

The key contributions of this work are as follows:

- We propose HAMP, a mixed-precision pruning and quantization framework guided by token importance, effectively eliminating redundant computations with minimal accuracy loss.

- To enable finer-grained control over pruning ratios across attention heads, we design DRP, a discriminative ratio-predictor for adaptive head-aware pruning, as a key component of HAMP, which greatly improves accuracy under the same sparsity level.

- Leveraging Neon Single-Instruction Multiple-Data (SIMD) parallelism on ARM CPUs and systolic array architecture on FPGAs, our method achieves up to 7.5× and 8× acceleration respectively, with only 0.96 Word Error Rate (WER) loss, demonstrating strong efficiency and deployment potential.

II. BACKGROUND

To better understand the limitations of existing methods and the motivation for our approach, we first analyze the computational bottlenecks in Transformer-based ASR models. As illustrated in Fig. 1, matrix operations in self-attention layers constitute roughly 60% of inference time, and their cost increases significantly with longer input [3].

Transformer models fundamentally rely on self-attention mechanism, where matrix multiplications involving the query (Q), key (K), and value (V) matrices contribute significantly to computational complexity:

$$Q = XW_Q, \ K = XW_K, \ V = XW_V \tag{1}$$

$$Attention(Q, K, V) = Softmax\left(\frac{QK^T}{\sqrt{d_k}}\right)V \tag{2}$$

where X is the input sequence, W_Q, W_K, W_V are the projection matrices for query, key, and value, and d_k is the dimension of the key vectors.

To mitigate the computational cost in Transformer matrix multiplication, various optimization strategies have been proposed. A common and simple approach is single-precision or integer-only quantization. For instance, Ref. [6] uses uniform 6-bit quantization across the entire model, achieving a 1.61 WER reduction with minimal memory cost. Similarly, Ref. [7] applies 8-bit quantization to all weights and activations for efficient accelerator execution. Although hardware-friendly, these methods enforce uniform precision across all components, ignoring variation in token importance. As a result, they may over-compress critical information and degrade accuracy in tasks with uneven data distributions, such as ASR, which we address via mixed-precision quantization.

Additionally, static weight pruning has been widely explored. These methods prune fixed weights based on training-time statistics. For example, Ref. [5] uses cross-layer weight sharing to reduce model size, which results in a 6.92 WER reduction. Unstructured pruning [8] removes weights below a threshold for fine-grained sparsity. However, static compression lacks runtime input adaptivity, limiting its effectiveness in ASR tasks with dynamically varying information density. We mitigate this limitation by applying token-based pruning rather than static weight pruning.

III. PROPOSED METHOD

A. Mixed-precision Pruning and Quantization

To address the limitations of prior static and uniform compression methods, we propose HAMP, a token-based mixed-precision pruning and quantization strategy to optimize Transformer-based ASR models for both input adaptivity and hardware compatibility. Unlike weight-centric approaches, our method performs input-aware token pruning, enabling adaptive computation tailored to the varying information density in speech inputs. This makes it particularly suitable for ASR scenarios where token relevance is highly uneven. By assigning precision levels based on token importance, our mixed-precision scheme provides fine-grained optimization, achieving a better trade-off between computational complexity and accuracy while remaining hardware-friendly.

Token importance is computed from Softmax probability distribution. Specifically, we accumulate the Softmax scores along the key dimension across all heads and layers:

$$token\_impt_i = \sum_{j=1}^{L} softmax\left(\frac{QK^T}{\sqrt{d_k}}\right)_{i,j} \tag{3}$$

where i and j denote the indices of the query and key tokens, respectively, and L represents the sequence length. Tokens with higher importance contribute more to the final output after multiplication with V, while low-scoring tokens have

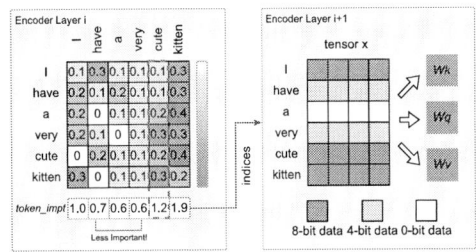

Fig. 2. Accumulate Softmax probabilities.

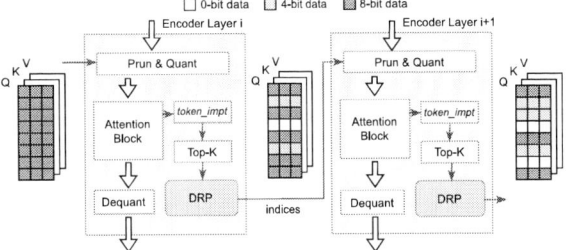

Fig. 3. The overall architecture of the proposed HAMP.

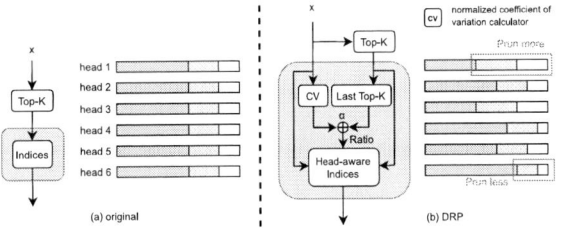

Fig. 4. Adjust pruning ratio with DRP.
(a) Original mixed-precision pruning. (b) DRP head-aware pruning.

negligible impact and can be represented at lower precision, or even pruned entirely. This mechanism is illustrated in Fig. 2.

Based on the *token_impt*, tokens are grouped into three levels: high-importance tokens quantized to 8-bit precision, medium-importance tokens quantized to 4-bit, and low-importance tokens totally pruned (0-bit). This dynamic allocation is applied layer-wise across the model, with the ranking results from each layer propagated to guide token pruning and quantization in the next layer, as shown in Fig. 3. This forms a multi-stage quantization pipeline that balances recognition accuracy and runtime cost.

We use symmetric uniform quantization with per-tensor scale, Q_{step} computed as:

$$scale = \frac{(max(X) - min(X))}{2^b - 1} \tag{4}$$

$$Q_{step} = round\left(\frac{X - min(X)}{scale}\right) \tag{5}$$

Quantized tensors are stored as 8-bit or 4-bit integers and executed efficiently on ARM CPUs via SIMD instructions, or on FPGAs using bit-level data paths, as detailed in later sections.

B. Head-aware Pruning Adjustments

In the proposed pruning and quantization framework, we observe that input tensor distributions vary significantly across attention heads and exhibit minimal correlation between layers. Rather than assigning fixed pruning ratios to each head, we formulate sparsity as a global budget shared among all pruning candidates. Tokens or heads exhibiting less informative distributions are pruned more aggressively, while others can be retained at higher precision. This global competition for limited quantization resources allows the model to better preserve critical information under the same sparsity constraint, thereby mitigating accuracy degradation. To realize this adaptive allocation, we introduce DRP, a discriminative ratio-predictor that adjusts pruning ratios based on the final value of the top-k attention scores and the coefficient of variation (CV) of the input, as shown in Fig. 4.

Algorithm 1 DRP: Head-aware Pruning Ratio Adjustment

Input: attention map $\omega \in \mathbb{R}^{B \times H \times T \times T}$, bit ratios r_0, r_4, buffer proportion δ
Output: pruning index map $topk\_indices$

1 compute total token importance $\omega_{sum} \leftarrow \sum_{t=1}^{T} \omega[:,:,:,t]$
2 set pruning token count $K \leftarrow [T \cdot (r_0 + r_4)], B \leftarrow [K \cdot \delta]$
3 $values, indices \leftarrow TopK(\omega_{sum}, K + B, smallest = True)$
4 $v_{last} \leftarrow values[:,:,:,-1], \mu_v \leftarrow mean(v_{last}), r_1 \leftarrow 1 - \frac{v_{last}}{\mu_v}$
5 $\mu_s, \sigma_s \leftarrow mean(\omega_{sum}, dim = 2), CV \leftarrow \frac{\sigma_s}{\mu_s}, r_2 \leftarrow \frac{CV}{mean(CV)} - 1$
6 **if** $\max(r_2) > 2$: $r_2 \leftarrow r_2 / \frac{\max(r_2)}{2}$
7 $r \leftarrow \alpha \cdot r_1 + (1 - \alpha) \cdot r_2, K' \leftarrow [K + r \cdot B]$
8 **for** each batch b, head h:
9   **if** $K'[b, h] < K + B$: $indices[b, h, :, K'[b, h]:] \leftarrow -1$
10 **return** $topk\_indices$

As illustrated in Algorithm 1, our method replaces the fixed, uniform quantization strategy with an adaptive mechanism that adjusts pruning ratios across heads. Given a global sparsity target and a buffer proportion (i.e., maximum deviation), we analyze token importance using two metrics: the smallest top-k value v_{last} and the normalized coefficient of variation $\frac{CV}{\mu_{CV}}$. These are weighted by a factor α (manually set or automatically tuned during inference) to compute head-specific pruning ratios. Intuitively, when top-k scores are low and dispersed, aggressive pruning causes less accuracy loss.

Notably, both metrics are computed once based on the maximum pruning range, incurring negligible overhead compared to conventional methods.

C. Hardware Architecture

To ensure the compatibility of HAMP with embedded systems, we deploy our method on actual hardware platforms, targeting both CPUs and FPGAs as representative edge devices.

Our CPU platform is ARM processor supporting SIMD execution through Neon extensions, part of the ARMv7 / ARMv8 architecture, which operate on 128-bit vector registers to enable parallel processing of multiple data elements. We leverage Neon's vectorized operations to optimize Transformer's matrix multiplications and enhance the efficiency of quantized models.

In our implementation, INT8 weights are loaded into 128-bit Neon registers (Fig. 5(a)), allowing four parallel multiplications per cycle. The Neon instruction set enables efficient parallel accumulation of matrix products, significantly reducing computational latency.

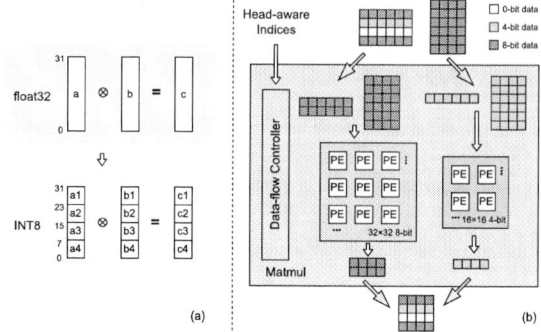

Fig. 5. Hardware architecture.
(a) Leveraging Neon registers on CPU. (b) Matmul dataflow on FPGA.

For FPGAs, we design systolic arrays of 4-bit (16×16) and 8-bit (32×32) to accommodate matrix multiplications with varied precision, as the systolic architecture is well-suited for dense linear operations in Transformer models. In our design, we implement separate processing elements (PEs) optimized for INT8 and INT4 operations, enabling mixed-precision computation. The data-flow controller splits matrices by indices and merges results, maintaining high throughput while saving resources, as shown in Fig. 5(b).

IV. EXPERIMENTS

A. Experimental Settings

We evaluate our method on the open-source Whisper models released by OpenAI [2], including four widely used Transformer-based ASR models of increasing scale: tiny (39M parameters), base (74M), small (244M), and medium (769M). All implementations are based on the PyTorch framework, and model accuracy is assessed using the WER on the FLEURS dataset.

For the CPU implementation, we develop our method in C language and run it on an ARM Cortex-A processor using Neon SIMD instruction. For the FPGA implementation, we design our system using SystemVerilog, synthesize using Xilinx Vivado 2019.1, and deploy it on the Xilinx Zynq MPSoC ZCU102 development board.

B. Ablation Study

To validate the effectiveness of our method, we evaluate it on four Whisper models (Whisper-tiny to Whisper-medium). We compare our HAMP approach, guided by DRP module, against an ordinary mixed-precision pruning method that applies uniform pruning across attention heads. The pruning ratio ranges from 0.0 to 0.5 with a step of 0.05 (Fig. 6).

Larger models generally show lower WER and are more robust to pruning, while smaller models degrade rapidly beyond a certain threshold—the collapse point—where WER increases drastically due to excessive information loss. Our DRP module significantly delays this point for whisper-tiny and whisper-base. For example, to prune out 30% of a whisper-base model with its original WER score 8.09, ordinary method results in a WER of 19.49 (+141%), whereas our HAMP method with DRP achieves a much lower WER of 9.46 (+17%), as presented in Fig. 6b. For whisper-small and whisper-medium, our approach maintains high accuracy without encountering a collapse point.

In summary, the DRP module plays a critical role in our approach by enabling finer-grained removal of unimportant

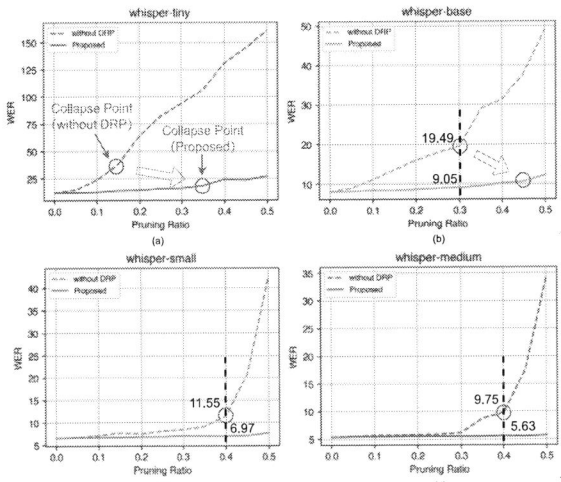

Fig. 6. WER Comparison under different pruning ratios.
(a)whisper-tiny (b)whisper-base (c)whisper-small (d)whisper-medium

TABLE II.   TIME & RESOURCE COMPARISONS ON FPGA

Item	Hardware Structure			Improvement
	16×16 Systolic Array	*32×32 Systolic Array*	*16×16 Systolic Array*	
Clock	125MHz	125MHz	100MHz	-
Precision	4-bit	8-bit	32-bit	-
Power	0.156W	0.701W	1.056W	0.81×
LUTs	7034	97989	256875	0.41×
FFs	6023	48663	100306	0.55×
Time	1064.4 µs (ours)		8525.2 µs	8.01×

TABLE III.   COMPARISON OF HAMP AND PRIOR METHODS

Item	[6]	[8]	**HAMP**
WER Increase (%) ↓	11.9	12.4	**11.8**
CPU Speedup ↑	-	5.9 ×	**7.5×**
FPGA Speedup ↑	4.1×	-	**8.0×**

As shown in Table III, HAMP achieves the lowest WER increase while delivering the highest speedup on both CPU and FPGA, demonstrating superior overall performance compared to prior methods.

V. CONCLUSION

We propose a software-level optimization framework with hardware implementation to accelerate Transformer-based ASR models by reducing attention block computation. To address both latency requirements and edge deployment constraints, we introduce HAMP, a token-based mixed-precision pruning and quantization method that dynamically eliminates redundant information with minimal accuracy loss. Within HAMP, our DRP module adjusts pruning ratios per head based on input distribution, playing a key role in preserving performance under sparsity. Our approach is implemented on CPU and FPGA, achieving up to 7.5× and 8.0× speedup respectively under 30% pruning and quantization, with only 0.96 WER degradation. These results demonstrate the effectiveness of HAMP for accurate, efficient ASR inference on resource-constrained edge devices.

tokens while retaining those with greater influence on attention outputs. The head-aware pruning strategy improves accuracy under the same sparsity with minimal computational cost. For small models, DRP module supports more accurate edge deployment; for larger ones, it facilitates redundancy exploration and computation reduction. Such results demonstrate the strong practical value of the overall HAMP framework and highlight the effectiveness of our DRP module in enabling adaptive, fine-grained pruning.

C. Hardware Implementation

We evaluate the CPU runtime of attention block matrix multiplication, the primary computational cost in Transformer based ASR models, with results summarized in Table I. We present Whisper-tiny results as representative, with similar speedup across all models. Quantization enables efficient use of the Neon SIMD instruction set, allowing four 8-bit operations to be executed in the time required for a single 32-bit float. In parallel, 0-bit tokens from pruning eliminate unnecessary computations, reducing the effective matrix dimensions. Under a 30% pruning and quantization ratio, our method achieves up to 7.5× speedup, demonstrating its effectiveness for real-time ASR on CPU-based edge devices.

We implement matrix multiplications of sizes 1500×384 and 384×384 within the attention block on an FPGA, and compare resource usage and runtime performance with and without HAMP under a 30% pruning ratio, as shown in Table II. Compared to standard 32-bit matrix operations using 16×16 systolic array, our mixed-precision pruning and quantization approach significantly reduces LUT and FF usage, while achieving up to 8.0× speedup in execution time. These results further demonstrate the effectiveness of our method for ASR deployment on FPGA platforms.

TABLE I.   TIME CONSUMPTION COMPARISONS ON CPU

Model	Pruning Ratio	*Baseline Time (ms)*	*Ours Time (ms)*	*Speedup*
Whisper-Tiny	0.05	596.2	109.3	5.5×
	0.10		102.2	5.8×
	0.15		96.8	6.1×
	0.20		91.2	6.5×
	0.25		82.5	7.2×
	0.30		78.9	7.5×

REFERENCES

[1] Vaswani A, Shazeer N, Parmar N, et al. "Attention is all you need," Advances in neural information processing systems, 2017: 30.

[2] Radford A, Kim J W, Xu T, et al. "Robust speech recognition via large-scale weak supervision," International conference on machine learning. PMLR, 2023: 28492-28518.

[3] Peng H, Huang S, Chen S, et al. "A length adaptive algorithm-hardware co-design of transformer on fpga through sparse attention and dynamic pipelining," Proceedings of the 59th ACM/IEEE Design Automation Conference. 2022: 1135-1140.

[4] Ye L, Gao C, Cheng G, Luo L, Zhao Q. "ASQ: An ultra-low bit rate ASR-oriented speech quantization method," IEEE Signal Processing Letters, 2023, 31: 221-225.

[5] Gao Z, Yao Y, Zhang S, Yang J, Lei M, McLoughlin I. "Extremely low footprint end-to-end ASR system for smart device," arXiv preprint arXiv:2104.05784, 2021.

[6] M. Lee, K. Hwang, J. Park, S. Choi, S. Shin and W. Sung. "FPGA-Based Low-Power Speech Recognition with Recurrent Neural Networks," 2016 IEEE International Workshop on Signal Processing Systems (SiPS). 2016: 230-235.

[7] Kim S, Gholami A, Yao Z, et al. "Integer-only zero-shot quantization for efficient speech recognition," ICASSP 2022-2022 IEEE International Conference on Acoustics, Speech and Signal Processing (ICASSP). IEEE, 2022: 4288-4292.

[8] R. Yazdani, J. -M. Arnau and A. González, "A Low-Power, High-Performance Speech Recognition Accelerator," IEEE Transactions on Computers, vol. 68, no. 12, pp. 1817-1831.

A Compressed Sensing Spiking Neural Network System for Radar-Based HGR

Liyu Qian, Zikai Zhu, Yuhan He, Jie Lu, Yaojie Sun, Lirong Zheng, Zhuo Zou

State Key Laboratory of Integrated Chips and Systems, School of Information Science and Technology,
Fudan University, Shanghai, China
Email: {lyqian20, zkzhu20, yjsun, lrzheng, zhuo}@fudan.edu.cn, {yhhe23, luj22}@m.fudan.edu.cn

Abstract—Radar-based hand gesture recognition (HGR) shows great application potential in the field of human-computer interaction due to its advantages of strong anti-interference ability and non-contact nature. However, the high-dimensional characteristic of radar signals brings severe challenges for resource-limited edge devices. In this paper, we propose a compressed sensing spiking neural network (CSSNN) system for radar-based HGR. This system employs compressed sensing (CS) technology to perform sparse sampling and compression on the radar signals. It realizes the joint optimization of CS and spiking neural networks (SNNs), achieving feature extraction and classification of gestures with low resource utilization. The computational complexity and parameter number are further reduced through matrix binarization and sparsification. Experiment results show that the proposed CSSNN model achieves a classification accuracy of 95.84% on the 5-class TinyRadar dataset. The computational complexity of the CSSNN is evaluated using multiply-accumulate operations (MACs). The total MACs of the chosen CSSNN model is 127.1k.

Index Terms—Radar, hand gesture recognition (HGR), compressed sensing (CS), spiking neural network (SNN), hardware-aware

I. INTRODUCTION

There is a growing demand for novel and convenient human-machine interface methods with the advancement of Internet of Things (IoT) devices. As a natural and widely accepted form of communication, hand gesture recognition (HGR) has been extensively studied for its applications in source-limited IoT devices.

The hand gesture data collecting sensors vary from contact devices such as gloves to ultrasound devices, cameras, radars, WiFi-based systems, and others [1]–[3]. Among these sensing technologies, radar has important advantages in strong privacy protection, non-contact operation, and functionality under low-light conditions. With the recent development of low-cost single-chip radar sensors (e.g., the Acconeer XR111 sensor), the application of radar sensors in IoT devices has become increasingly prevalent.

Conventionally, raw radar data is pre-processed based on the Doppler effect. Several HGR studies adopt the combination of different levels of Fourier transform (FT)-based data pre-processing and machine learning classification [5]–[7]. Compared with traditional neural networks (NNs), the bio-inspired

This work was supported in part by the National Natural Science Foundation of China under Grants 92164301 and 62476062.

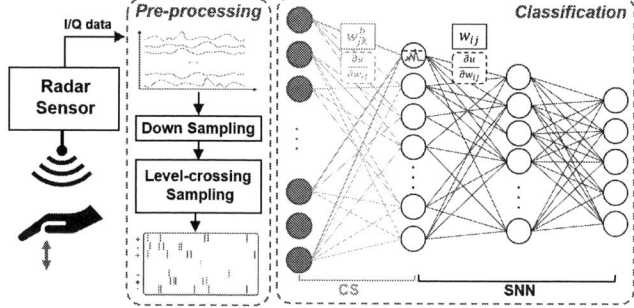

Fig. 1. CSSNN HGR system architecture.

spiking neural networks (SNNs) can effectively extract spatio-temporal features from large volumes of radar signals while maintaining low computational complexity. Several recent works apply the FT pre-processing and SNN on radar-based HGR task [8], [9]. However, the FT process is too complex to be realized on resource-limited IoT devices.

Compressed sensing (CS) is a signal acquisition and reconstruction theory. For signals with high sparsity, the original signal can be reconstructed through random sampling at a rate far below the Nyquist frequency. Compressed learning (CL) is a combination of CS and machine learning that performs classification directly in the compressed low-dimensional measurement domain [10]. In this work, we apply the compressed sensing spiking neural network (CSSNN) method for radar-based HGR, enabling the radar data feature extraction and classification network to be learnable. Inspired by the sparsity regularization algorithm from work [11], the CS process is further compressed through sparsification. The performance of this method is evaluated on the impulse radio ultra-wideband radar dataset, TinyRadar [12]. The contributions of our work are listed as follows:

- We employ the CSSNN system on the SNN-compatible spike radar data obtained from the pre-processed raw radar data using a combination of down-sampling and level-crossing (LC) sampling. The CSSNN system jointly performs feature extraction and classification, leveraging the sparse activation properties of SNNs and the dimensionality reduction capabilities of CS.
- We suggest the measurement matrix of the CS process as a fully connected (FC) binary layer of the SNN.

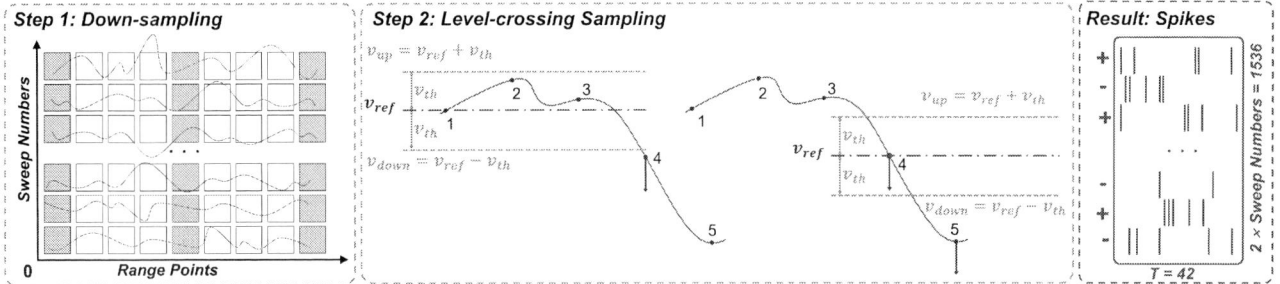

Fig. 2. Radar data pre-processing process.

It enables the entire CSSNN to be learnable through backward propagation-based co-optimization. The binarization process saves the weight storage space, and allows the CSSNN to be implemented on resource limited IoT devices.

- We implement a sparsity adjustment mechanism into the CS matrix design, enabling a dynamical trade-off between classification accuracy and computational complexity while maintaining a fixed SNN architecture. This approach yields a hardware-friendly solution.

II. CSSNN HGR SYSTEM DESIGN

In this section, we present the CSSNN HGR system designed for radar data processing. First, raw radar data undergoes pre-processing via down-sampling and LC sampling to generate spike data. Second, the measurement matrix in the CS process is abstracted as an FC layer and co-optimized with the SNN. To reduce the computational overhead introduced by the CS module while preserving a fixed network architecture, the measurement matrix is constrained by a sparsity regularization mechanism. The final CSSNN configuration is selected based on task-specific requirements. The performance of the CSSNN HGR system is evaluated on the TinyRadar hand gestures dataset. The overall architecture of the proposed CSSNN HGR system is illustrated in Fig. 1.

A. Radar Data Pre-processing

The TinyRadar dataset was collected by the Integrated Systems Lab at ETH Zurich via the short-range Acconeer XR111 sensor, containing 5 gesture classes: 'PullUp', 'PushDown', 'SwipeRL', 'FingerSlider', and 'PalmTilt'. It consists of 2500 recordings of 5 hand gestures by 1 person. The radar sends out pulses at a frequency of 256 Hz, namely a sweep. The sweep range is 10-30 cm, and the signal over range is 414. Several sweep vectors are stacked into one data map window of raw amplitude data, which is called a frame. In this work, we set the time window of the frame to be 3s. The data size of the radar frame is 768×414. The pre-processing process of the raw data frame is demonstrated in Fig. 2. This process is referring to work [13].

a) Down-sampling: In the CSSNN, the row number of the amplitude data frame is the input neuron number, and the column number is the timesteps of the SNN. To speed up the processing speed, we down-sample the data evenly by $10\times$ along the range points side. The down-sampled data frame size is 768×42.

b) LC sampling: For each radar sweep, set the first range point amplitude as the reference voltage V_{ref}. The voltage threshold is V_{th}. The up spike and the down spike threshold are $V_{up} = V_{ref} + V_{th}$ and $V_{down} = V_{ref} - V_{th}$, respectively. Here, the V_{th} is set to be 0.02. When a range point amplitude is beyond the range of V_{up} and V_{up}, the point fires an up spike or down spike, and the point amplitude is set to be the new V_{ref}. As shown in Fig. 2, point 4 fires a down spike and becomes the new V_{ref} following point 1. Two spike frames sizing 768×42 are extracted and stacked along the sweep axis. The final spike data size is 1536×42.

B. CSSNN Algorithm

a) CSSNN Co-optimization: The CS process is shown as follows:

$$\hat{x} = \Phi x + e \qquad (1)$$

Here, $x \in R^n$ is the original signal. $\Phi \in R^{m\times n}$ ($m \ll n$) is the measurement matrix. $\hat{x} \in R^m$ is the compressed measurement vector. The compression rate (CR) is $\frac{m}{n}$. For any sparse signal x, it can be recovered from $\hat{x}$. The CL method saves the complex reconstruction process, processing directly on the compressed vector $\hat{x} \in R^m$.

Inspired by the binarized neural network (BNN) model, we realize the CS process in the form of a $m \times n$ FC layer with binarized elements. The mathematical model of an artificial neuron of the BNN is shown as follows:

$$y = f(\sum_i w_i x_i + b) \qquad (2)$$

Here, the neuron input x_i is the raw signal data, the corresponding weight w_i is the element of the measurement matrix, the bias b is the noise, the nonlinear active function f is set to be $f = 1(*)$ and the output layer y is the compressed signal.

In the CSSNN, the CS process is jointly optimized with the SNN classifier using the spatio-temporal backpropagation (STBP) algorithm [14]. The STBP uses the surrogate function to calculate the gradient of neuron firing and updates the weight variation on the gradient of weight. We use the optimizer and learning rate of the CS layer with the STBP learner together, realizing a co-optimization of the CS process and SNN.

b) Hardware aware CS Measurement Matrix optimization: As shown in (2), the measurement matrix Φ of the CS process has a dimension of $m \times n$, in which n is the length of the original signal and m is the length of the compressed signal. The weight binarization function is demonstrated in (3):

$$w_{ij}^b = \begin{cases} 1, & w_{ij} \in large(W_j, Sample) \\ 0, & else \end{cases} \quad (3)$$

Here, w_{ij}^b and w_{ij} are the binarized value and the original weight from column i and row j, respectively. W_j is the set of weight from the jth row of the measurement matrix W. $large(W_j, Sample)$ is a set of $Sample$ number of the largest elements from W_j.

In the forward propagation process, the input signal multiplies with w_{ij}^b, and the original weights w_{ij} are saved for updating. The loss function is set between the output of the algorithm and the given target. In the gradient accumulation period, the original weights w_{ij} of the CS layer are updated. The detailed co-optimization process of the CSSNN is shown in Algorithm 1.

III. EVALUATION AND ANALYSIS

In this section, we evaluate the CSSNN HGR performance on the TinyRadar gesture dataset.

A. Evaluation Standard

There are two standards used to evaluate the classification performance and computation complexity of the CSSNN system. The classification accuracy, and the multiply-accumulate operations (MACs).

The MACs consist of the multiplication operations and adding operations. The multiplication operand is calculated as follows:

$$Mul = timesteps \times neuron\ number \quad (4)$$

The adding operand of the SNN layer is calculated as follows:

$$Add_{SNN} = timesteps \times (L_{in} - 1) \times L_{out} \times firing\ rate \quad (5)$$

L_{in} and L_{out} are the input and output sizes of the SNN layer. For the CS process, the measurement matrix is binarized and has no bias. Therefore, the adding operand is calculated as follows:

$$Add_{CS} = timesteps \times (Sample - 1) \times L_{out} \times firing\ rate \quad (6)$$

Here, $Sample$ is the number of ones in each row of the CS measurement matrix. L_{out} is the compressed size of the signal.

B. Experiment Procedure and Result

In this experience, we evaluate the performance of the CSSNN system under different CRs with different measurement matrix sparsity on the TinyRadar dataset. The raw radar data are pre-processed as demonstrated in II-A. All the experiments are based on the SpikingJelly framework [15]. For all the networks, adaptive moment estimation with weight decay (AdamW) is the optimizer, and the learning rate (LR)

Algorithm 1 Co-optimization Process of the CSSNN.

Input: x_0: original input signal;
  w^t: weight before updating;
  η^t: LR before updating;
  λ: LR decay;
  l: number of layers in CSSNN;
  $S(w_{k+1}x_k)$: the SNN dynamic process of $(w_{k+1}x_k)$;
  $[\frac{\partial x_k}{\partial w_k}]_S$: the gradient descent calculation based on STBP surrogate function;

1: **Forward propagation:**
2: Initialize input w^t and η^t
3: **for** $k = 1$ to l **do**
4:     **if** $k == 1$ **then**
5:         $w_1^b = Binarize(w_1)$
6:         $x_1 = w_1^b x_0$;
7:     **else**
8:         $x_k = S(w_k x_{k-1})$
9:     **end if**
10: **end for**
11: **Backward propagation:**
12: **for** $k = 1$ to l **do**
13:     **if** $k == 1$ **then**
14:         $gw_1^b = gx_1 \cdot \frac{\partial x_1}{\partial w_1^b}$;
15:         $gx_0 = gx_1 \cdot \frac{\partial x_1}{\partial x_0}$;
16:     **else**
17:         $gw_k = gx_k \cdot [\frac{\partial x_k}{\partial w_k}]_S$;
18:         $gx_{k-1} = gx_k \cdot [\frac{\partial x_k}{\partial x_{k-1}}]_S$;
19:     **end if**
20: **end for**
21: **Gradient accumulation:**
22: **for** $k = 1$ to l **do**
23:     **if** $k == 1$ **then**
24:         $w_1^{t+1} \leftarrow max(-1, Update(w_1^t, gw_1^b, \eta^t), 1)$;
25:     **else**
26:         $w_k^{t+1} \leftarrow Update(w_k^t, gw_k, \eta^t)$
27:     **end if**
28: **end for**
29: $\eta^{t+1} = \lambda \eta^t$;
Output: Optimized CSSNN model.

---

of each parameter group is decayed by a factor of 0.1 with an initial LR of 0.001. The surrogate function is ATan. The total timestep of the SNN is 42. The training epoch number is 100, and the batch size is 10. The input layer size is 1536.

The detailed classification performance and MACs of different CSSNN structures under different CS matrix sparsity are shown in Fig. 3, and Fig. 4 by colors. Here 'fc' stands for FC SNN layer. Overall, the classification performance decreases as the $Sample$ number decreases. The CSSNN model with the smallest number of neurons exhibits the most significant decline in accuracy as the $Sample$ number decreases. The fluctuations observed in classification accuracy when the number of $Sample$ is fewer than 10 indicates the limitations of the weight binarization algorithm of the CS layer

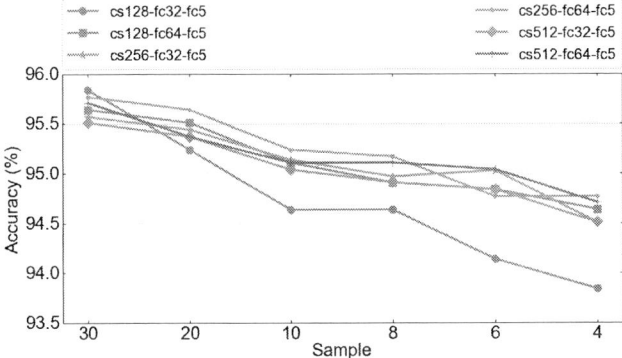

Fig. 3. Classification accuracy of different CSSNN structures under different CS matrix sparsity.

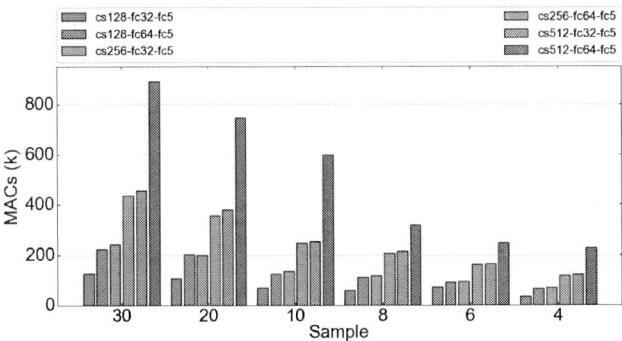

Fig. 4. MACs of different CSSNN structures under different CS matrix sparsity.

TABLE I
COMPARISONS WITH OTHER RELATED WORKS.

Works	Algorithm	Class no.	Pre-processing	Muls	Adds	Accuracy
IOTJ'21 [12]	TCN-CNN	5	2D-FFT	15.9 M	15.9 M	95%
FG'21 [16]	BSNN	11	2D-FFT	-	70.5k	84.67%
AICAS'22 [13]	SNN	5	LC-Sampling	6.93k	343k	95.44%
Our Work	CSSNN	5	LC-Sampling	6.93k	120.18k	95.84%

under conditions of extremely high sparsity.

The MACs decrease as the *Sample* number decreases. According to the experiment results, the firing rate of the first hidden layer decreases proportionally with the *Sample* number while that of the second hidden layer remain steady. The CSSNN structure we choose is ($cs128$-$fc32$-$fc5$). The *Sample* number of the CS layer is 30. The comparison of our work with other works evaluated on the TinyRadar dataset is demonstrated in TABLE I. The classification performance of our work is the highest. Our MACs cost is the lowest except for [16], which achieves only 84.67% accuracy and relies on a computationally expensive 2D-FFT process.

IV. CONCLUSION

In this paper, we present a CSSNN system for radar-based HGR. We apply simple pre-processing on raw amplitude radar data through down-sampling and LC-sampling, acquiring spike radar data. The CSSNN system realizes a joint optimization of data compressing, feature extraction, and classification on the radar gesture data. A hardware-aware optimization of binarization and sparsification is proposed, realizing different computational complexity under a fixed model structure. Evaluated on the 5-class TinyRadar gesture dataset, the CSSNN system achieves 95.84% classification accuracy, costing a total of 127.1k MACs.

REFERENCES

[1] S. Ahmed, K. D. Kallu, S. Ahmed, and S. H. Cho, "Hand gestures recognition using radar sensors for human-computer-interaction: A review," *Remote Sensing*, vol. 13, no. 3, p. 527, 2021.

[2] W. Chen, K. Niu, D. Zhao, R. Zheng, D. Wu, W. Wang, L. Wang, and D. Zhang, "Robust dynamic hand gesture interaction using lte terminals," in *2020 19th ACM/IEEE International Conference on Information Processing in Sensor Networks (IPSN)*. IEEE, 2020, pp. 109–120.

[3] A. Ibrahim, A. El-Refai, S. Ahmed, M. Aboul-Ela, H. M. Eraqi, and M. Moustafa, "Pervasive hand gesture recognition for smartphones using non-audible sound and deep learning," *arXiv preprint arXiv:2108.02148*, 2021.

[4] J. Lien, N. Gillian, M. E. Karagozler, P. Amihood, C. Schwesig, E. Olson, H. Raja, and I. Poupyrev, "Soli: Ubiquitous gesture sensing with millimeter wave radar," *ACM Transactions on Graphics (TOG)*, vol. 35, no. 4, pp. 1–19, 2016.

[5] T. T. Trinh, D.-K. Le, M. Le *et al.*, "Hand gesture recognition using mimo radar and lightweight convolutional neural network," in *2023 12th International Conference on Control, Automation and Information Sciences (ICCAIS)*. IEEE, 2023, pp. 560–565.

[6] S. Skaria, A. Al-Hourani, and R. J. Evans, "Deep-learning methods for hand-gesture recognition using ultra-wideband radar," *IEEE Access*, vol. 8, pp. 203 580–203 590, 2020.

[7] E. Hayashi, J. Lien, N. Gillian, L. Giusti, D. Weber, J. Yamanaka, L. Bedal, and I. Poupyrev, "Radarnet: Efficient gesture recognition technique utilizing a miniature radar sensor," in *Proceedings of the 2021 CHI Conference on Human Factors in Computing Systems*, 2021, pp. 1–14.

[8] A. Safa, A. Bourdoux, I. Ocket, F. Catthoor, and G. G. Gielen, "On the use of spiking neural networks for ultralow-power radar gesture recognition," *IEEE Microwave and Wireless Components Letters*, vol. 32, no. 3, pp. 222–225, 2021.

[9] J. Huang, B. Vogginger, P. Gerhards, F. Kreutz, F. Kelber, D. Scholz, K. Knobloch, and C. G. Mayr, "Real-time radar gesture classification with spiking neural network on spinnaker 2 prototype," in *2022 IEEE 4th International Conference on Artificial Intelligence Circuits and Systems (AICAS)*. IEEE, 2022, pp. 362–365.

[10] R. Calderbank, S. Jafarpour, and R. Schapire, "Compressed learning: Universal sparse dimensionality reduction and learning in the measurement domain," *preprint*, 2009.

[11] Y. Yan, H. Chu, Y. Jin, Y. Huan, Z. Zou, and L. Zheng, "Backpropagation with sparsity regularization for spiking neural network learning," *Frontiers in Neuroscience*, vol. 16, p. 760298, 2022.

[12] M. Scherer, M. Magno, J. Erb, P. Mayer, M. Eggimann, and L. Benini, "Tinyradarnn: Combining spatial and temporal convolutional neural networks for embedded gesture recognition with short range radars," *IEEE Internet of Things Journal*, vol. 8, no. 13, pp. 10 336–10 346, 2021.

[13] S. Wang, Y. Yan, H. Chu, G. Hu, Z. Zhang, Z. Zou, and L. Zheng, "Hand gesture recognition using ir-uwb radar with spiking neural networks," in *2022 IEEE 4th International Conference on Artificial Intelligence Circuits and Systems (AICAS)*. IEEE, 2022, pp. 423–426.

[14] Y. Wu, L. Deng, G. Li, and L. Shi, "Spatio-temporal backpropagation for training high-performance spiking neural networks," *Frontiers in neuroscience*, vol. 12, p. 323875, 2018.

[15] W. Fang, Y. Chen, J. Ding, Z. Yu, T. Masquelier, D. Chen, L. Huang, H. Zhou, G. Li, and Y. Tian, "Spikingjelly: An open-source machine learning infrastructure platform for spike-based intelligence," *Science Advances*, vol. 9, no. 40, p. eadi1480, 2023. [Online]. Available: https://www.science.org/doi/abs/10.1126/sciadv.adi1480

[16] D. Auge, J. Hille, E. Mueller, and A. Knoll, "Hand gesture recognition in range-doppler images using binary activated spiking neural networks," in *2021 16th IEEE International Conference on Automatic Face and Gesture Recognition (FG 2021)*, 2021, pp. 01–07.

FlexiCore-DNN: A Configurable and Templated Architecture for End-to-End FPGA Acceleration of Deep Neural Networks

Rao Fu[1], Wenhao Huang[1], Aiwu Ruan*[1], Huiyun Li*[2], Yongqing Wang[1]

[1] State Key Laboratory of Electronic Thin Films and Integrated Devices, University of Electronic Science and Technology of China, Chengdu, China

[2] Faculty of Compute Microelectronics, Shenzhen University of Advanced Technology, Shenzhen, China

* Email: 202321310907@std.uestc.edu.cn, ruanaiwu@uestc.edu.cn, huiyun.li@suat-sz.edu.cn

Abstract—**With the development of artificial intelligence technology, there are many challenges in achieving end-to-end automated deployment of deep neural networks (DNN) on Field Programmable Gate Arrays (FPGA), such as long development cycles and insufficient flexibility in hardware design. This paper introduces FlexiCore-DNN, an extensible, high-performance, and flexible digital hardware architecture designed to overcome these challenges. To support this architecture, we have designed and implemented a library of accelerator template that contains a set of foundational hardware architecture templates alongside a diverse range of operator accelerators. This template is described at the Verilog RTL level, featuring high versatility and scalability, and is designed with different hardware architectures for FPGA with and without CPU hard cores. Experimental results show that on Zynq platform, our architecture achieves 158 GOPS and 40.84GOPS/W at 150MHz, demonstrating its advantages in terms of performance and energy efficiency.**

Keywords—*Deep Neural Network, FPGA, Hardware accelerator, End-to-end Deployment*

I. INTRODUCTION

In recent years, artificial intelligence technologies represented by deep neural networks (DNNs) have achieved great success in fields such as image classification and object detection. However, the deep layers, large parameters, and computationally intensive nature of DNN models pose severe challenges to their deployment on resource-constrained edge devices. Field-programmable gate arrays (FPGA), with their high energy efficiency, high flexibility, and reconfigurability, have become one of the mainstream hardware platforms for DNN inference acceleration.

Currently, the industry and academia have proposed various deployment solutions for DNNs on FPGA. For example, tools like Xilinx's Machine Learning (ML) Suite and Intel's DLA (Deep Learning Accelerator) [1] are powerful, but they usually rely on specific FPGA hardware and Electronic Design Automation (EDA) toolchains, limiting their universality. Some open-source frameworks like TVM (Tensor Virtual Machine) [2] and VTA (Versatile Tensor Accelerator), although they provide end-to-end optimized compilers, restrict the freedom of hardware design as they must follow the VTA microarchitecture. Therefore, how to design a general, flexible, and efficient end-to-end automated deployment process from high-level machine learning frameworks to underlying FPGA hardware remains an open challenge.

To address the aforementioned issues, this article proposes a new hardware architecture named FlexiCore-DNN. The core contribution is the design and implementation of a general library of accelerator template. This template is described at

the RTL (Register Transfer Level) based on Verilog, ensuring the universality and high quality of the hardware implementation. Based on this library, experimental results verify the effectiveness of this solution, with the implemented accelerators on the Zynq ZU15EG platform achieving a performance of 158 GOPS and an energy efficiency of up to 40.84 GOPS/W.

II. OVERALL ARCHITECTURE

Fig. 1 illustrates the basic structure of the accelerator template library implemented in this paper. It is composed of hardware architecture template and various operator accelerator template, with each template specifically including both hardware and software template. The operator accelerator template mainly consists of: convolution accelerator template, pooling accelerator template, and so on.

The hardware template is implemented using Verilog, featuring following advantages:

- Strong versatility, making them compatible with any FPGA EDA software for synthesis, with high versatility.

- Strong autonomy, unlike HLS or DSL, hardware design using Verilog can be independent of specific FPGA EDA software.

- High quality, hardware described at the RTL level using Verilog can undergo better optimization at the hardware level, resulting in higher-quality hardware synthesis.

The software template, implemented in C, is an accelerator-based driver template designed for versatility. Within the hardware architecture template, it provides a driver

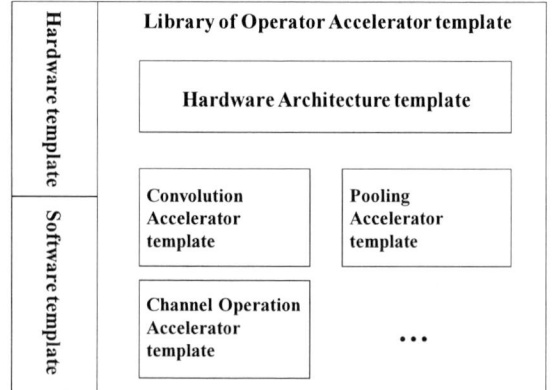

Fig. 1: Basic structure of the library of accelerator template

979-8-3315-3918-4/25 $31.00 © 2025 IEEE

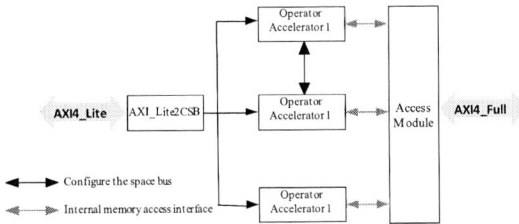

Fig. 2: Illustration of a high-performance hardware architecture template based on the AXI bus

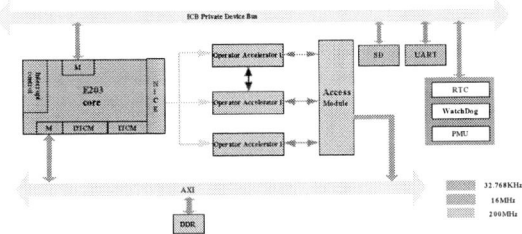

Fig. 3: Overall structural diagram of a hardware architecture template based on RISC-V

framework for accelerator initialization and operator invocation. Within operator accelerator templates, it supplies accelerator-specific driver templates for targeted operator execution.

III. HARDWARE ARCHITECTURE TEMPLATE

This paper proposes the design of high-performance hardware architecture templates based on AXI buses, without CPU soft cores, for different hardware types, including FPGAs with hard-core CPU and FPGAs without hard-core CPU, and template based on RISC-V with CPU soft cores. The high-performance hardware architecture template based on AXI buses employs the Advanced eXtensible Interface 4 (AXI4) as the interface protocol for configuration and access.

A. High-performance hardware architecture template based on AXI bus

Fig. 2 illustrates a template for a high-performance hardware architecture utilizing the AXI bus protocol. The architecture is controlled by a host system via an AXI-Lite interface. An AXI_Lite2CSB module serves as a bridge, converting configuration commands from the host for transmission over the Configuration Space Bus (CSB), depicted by black connection lines. Each operator accelerator module, acting as a slave on the CSB, is controlled by these commands to perform its calculations autonomously. Throughout this process, accelerators issue memory access requests as masters on a dedicated internal memory interface (red lines). A memory access module arbitrates these requests, manages data transactions with external storage via an AXI-Full bus, and feeds the data back to the appropriate accelerator. This architecture provides a straightforward and scalable method for integrating various accelerators, some of which may also be interconnected directly via the CSB.

B. RISC-V-based hardware architecture template

Fig. 3 illustrates the general structure of a hardware architecture template centered on a RISC-V E203 processor. To optimize system parameters, the architecture is divided into three clock domains, identifiable by color-coded modules. The E203 processor and its tightly coupled memories (ITCM/DTCM)—implemented with on-chip FPGA

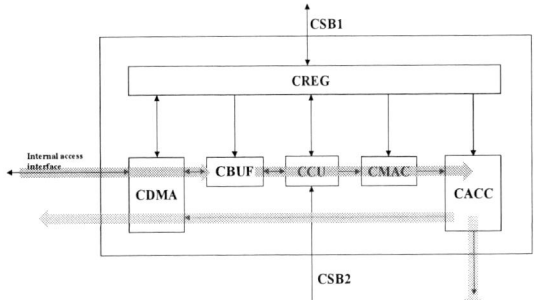

Fig. 4: Illustration of a convolution operator accelerator structure

storage—reside in the 16 MHz master clock domain. The processor connects to the system via a main bus interface ('M') and a co-processor interface (NICE). The platform includes essential peripherals: a Power Management Unit (PMU), Real-Time Clock (RTC), WatchDog timer, a Secure Digital (SD) card interface for non-volatile storage of neural network parameters, a UART, and an off-chip DDR SDRAM for storing volatile inference data such as feature maps.

The hardware acceleration section mirrors the design in Fig. 2, employing an identical memory access module, CSB, and internal memory bus. The primary architectural difference is the connection to the host. In this template, the E203 processor interfaces directly with each operator accelerator through the NICE co-processor port (yellow arrow). An asynchronous bridge module is utilized to handle the cross-clock domain communication between the E203 core and the accelerator hardware.

IV. OPERATOR ACCELERATOR TEMPLATE

A. Highly configurable convolution accelerator template

Fig. 4 illustrates the architecture of the convolution operator accelerator. The module features three primary interfaces: an internal memory access port with dedicated read channels for feature maps and weights and a write channel for results; a full Configuration Space Bus (CSB1) for host control; and a write-only secondary bus (CSB2) for daisy-chaining accelerators. The accelerator's six internal modules are the Conv Register (CREG), Conv Direct Memory Access (CDMA), Conv Buffer (CBUF), Conv Control Unit (CCU), Conv Multiply-Accumulate (CMAC), and Conv Accumulate (CACC). The CREG manages the register file, while the CBUF provides configurable, software-defined caching for feature and weight data. The CDMA module arbitrates memory access requests from the CBUF and CACC, interfacing with external memory to fetch input data for the CBUF or write output data from the CACC. The CCU, based on configuration from CREG, uses internal counters to automatically manage the hardware execution loops and orchestrates data movement from CBUF to the CMAC processing core. The CMAC module, detailed in Fig. 5, serves as the computational core, executing the two innermost loops of the convolution algorithm in a single cycle. It performs a dot product between one vector of length T_c and T_k vectors of the same length, yielding T_k results. This is achieved with an array of T_k cmac_mac sub-modules, where each sub-module implements a vector dot product of length T_c using T_c multipliers and an adder tree. To hide weight-loading latency, the CMAC employs a dual set of weight registers (Wt) in a ping-pong configuration. Furthermore, the synthesis process

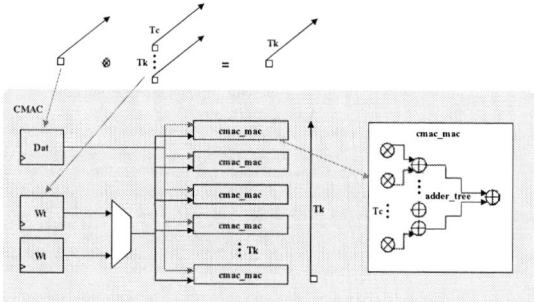

Fig. 5: Schematic diagram of a CMAC structure

can automatically insert pipeline stages within the cmac_mac sub-modules, based on the Tc parameter and clock frequency, to enhance operational throughput.

B. Channel Operation Accelerator Template

Channel operations refer to the computation that performs the same operation within the same input channel range of a feature map, such as adding bias, Batch Normalization. Fig. 6 is a schematic illustration of the channel operation module structure. In the figure, REG indicates the accelerator register module, DMA indicates the accelerator's direct memory access module, BUF indicates the accelerator's cache module, CU indicates the accelerator control unit, and PU indicates the accelerator processing module (Processing Unit). The channel operation accelerator PU is quite simple; during the bias addition operation, it performs vector addition. The channel operation accelerator is mainly used for the bias addition operation in convolution layers, and therefore, the channel operation accelerator is designed with an interface CSB2 that can be directly connected to the convolution operator accelerator. For the output by the convolution operator accelerator, the channel operation accelerator can easily implement parallelism in the channel direction.

C. Nonlinear Operations Accelerator Template

Nonlinear operations refer to the computations of nonlinear functions like trigonometric functions, which are primarily used as activation functions in deep neural network inference. There are many types of nonlinear operations, such as Celu, Elu, Relu, Selu, Sigmoid, Softmax, and Softsign.

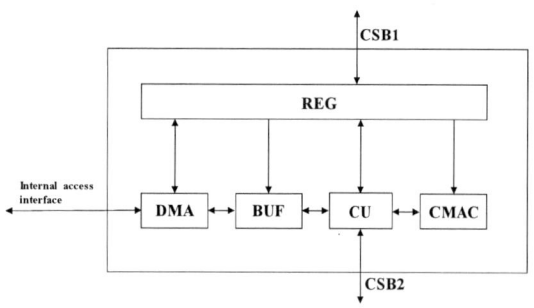

Fig. 6: Illustration of the structure of the channel operation module

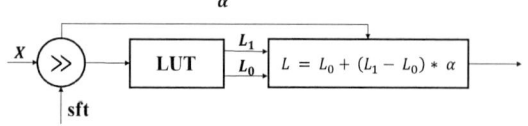

Fig. 7: The lookup table of nonlinear operation

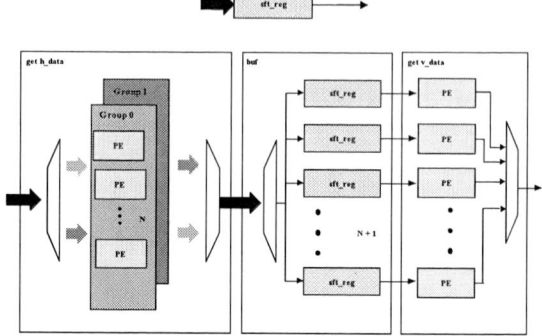

Fig. 8: Illustration of a pool accelerator arithmetic unit structure

Implementing a dedicated hardware accelerator for a single nonlinear operation is not cost-effective for current deep neural network inference. Therefore, the nonlinear operation accelerator in this paper's work achieves support for various nonlinear operations through the use of lookup tables. The nonlinear operation accelerator has a similar structure to channel operation accelerators, with the processing unit (PU) adopting the design shown in Fig. 7. Lookup tables have certain precision limitations, so the closest two output values L_1 and L_0 are found in the lookup table based on the part of the input X that is greater than the precision. Then, using linear interpolation based on the input part less than the precision α, the output result is calculated as $L = L_0 + (L_1 - L_0) * \alpha$.

D. Highly configurable pooling accelerator template

Fig. 8 illustrates the structural diagram of the pooling accelerator computational unit. The PE (Processing Element) is an operation element with storage capabilities that can be configured to perform maximum, minimum, or average calculations. The PE can also be configured for data length (i.e., inputting a certain number of data points before outputting a result and resetting). The "get h_data" section in Fig. 8 is used for horizontal operations, while the "buf" section converts parallel input into multi-channel serial data output. The "get v_data" section is used for vertical operations, yielding the final result. The "get h_data" section consists of 2 sets of PE, with each set containing N PE, where N determines the maximum Ky support of the pooling accelerator. Two sets of PE are set up to implement ping-pong operation and enhance the performance of the accelerator. The sft_reg is a buffer that receives the N parallel outputs from the "get h_data" section and then outputs them serially. There are a total of N+1 sft_regs in the "buf" section.

V. EXPERIMENT

The main metrics for evaluating accelerator hardware include resource consumption, performance, and power consumption.

A. Performance testing and analysis on ZU15EG

Fig. 9 presents the specific resource consumption and power consumption evaluation after the design is implemented on ZU15EG with CPU core.The measured results of all networks are shown in Table I. Among them, the unit of computational cost is MACs; the inference time refers to the time it takes for the network to perform one forward propagation, with the unit being milliseconds (ms); and the performance unit is billion multiplications and additions per

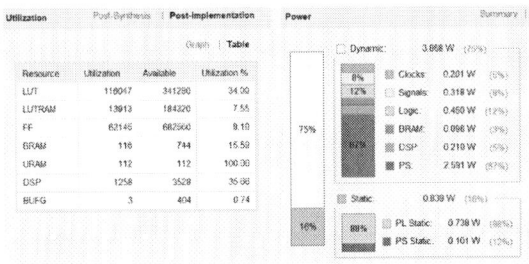

Fig. 9: The accelerator resource consumption and power consumption on ZU15EG

second (GMAC/s or GMACS). It can be seen that the performance remains basically stable at 79 GMAC/s. Combined with the power consumption of 3.868W assessed in Fig. 9, the energy efficiency of the design can be estimated to be 20.42 GMACs/W.

Table I.   Test results for ZU15EG in terms of performance

Network	Parameters	computational cost	Latency	Performance
VGG11	1.99M	64.99M	0.82	79.26G
VGG19	4.36M	174.22M	2.21	78.83G
VGG19dbl	15.98M	676.68M	8.57	78.96G
ResNet18	2.36M	899.50M	11.38	79.04G
ResNet18dbl	9.48M	2762.88M	34.96	79.03G

Table II.   Test results for XA7A100T in terms of performance

Network	Parameters	computational cost	Latency	Performance
VGG11	1.99M	64.99M	7.32	8.88G
VGG19	4.36M	174.22M	19.61	8.88G
VGG19dbl	15.98M	676.68M	76.15	8.89G
ResNet18	2.36M	899.50M	101.02	8.90G
ResNet18dbl	9.48M	2762.88M	310.33	8.90G

B. Performance testing and analysis on XA7A100T

Fig. 10 presents the specific resource consumption and power consumption evaluation after the design is implemented on XA7A100T without CPU core. The tested performance of the accelerator SoC implemented on the XA7A100T is essentially stable at 4.4 GMACS, as shown in Table II With the power consumption evaluated in Fig. 10 at 1.407W, the energy efficiency of the design can be estimated to be 3.13 GMACS/W.

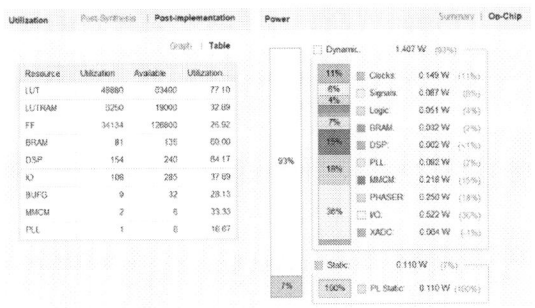

Fig. 10: The accelerator resource consumption and power consumption on XA7A100T

C. Comparison with other deep neural network inference FPGA accelerators

Table III. Comparison with other deep neural network inference FPGA accelerators

	Reference [3]	Reference [4]	Reference [5]	This article
FPGA	XQRKU060	ZYNQ XC7Z045	ZYNQ ZU3EG	ZYNQ ZU15EG
Frequency (MHz)	66.2	150	250	150
Accuracy[a]	float	16bit	8bit	16bit
FF(k)	46387	127.65	65.3	62.15
LUT(k)	141362	182.62	61.7	116
BRAM	66.2	486	93	116
DSP	2338	780	208	1258
Performance (GOPS)	29.87	137	127.5	158
Energy Efficiency (GOPS/W)	9.67	14.27	19.56	40.84

[a.] Number of bits for inference. "float" denotes floating-point.

Table III presents the specific information of the FPGA accelerators for deep neural network inference implemented in this paper and some other literature. Table III includes the FPGA model, frequency, hardware resource consumption, performance, and energy efficiency of each accelerator's specific implementation. The performance unit is billion operations per second (Giga Operations per second, GOPS), where 1 GMACS can be approximately converted to 2 GOPS. Hardware resources include flip-flops (FF), lookup tables (LUT), block RAM (BRAM), digital signal processors (DSP). Compared with the work in other literature, the deep neural network inference FPGA accelerator implemented in this paper has certain advantages in energy efficiency.

VI. CONCLUSION

This paper presents FlexiCore-DNN, a flexible and templated architecture for end-to-end FPGA acceleration of neural networks, which demonstrates significant advantages in performance and energy efficiency, achieving 158 GOPS and 40.84 GOPS/W respectively.

REFERENCES

[1] Aydonat U, O'Connell S, Capalija D, et al. An opencl™ deep learning accelerator on arria 10[C]. Proceedings of the 2017 ACM/SIGDA International Symposium on Field-Programmable Gate Arrays, 2017, 55-64.

[2] Chen T, Moreau T, Jiang Z, et al. Tvm: An automated end-to-end optimizing compiler for deep learning[J]. arXiv preprint arXiv:1802.04799, 2018.

[3] T. Pacini, E. Rapuano and L. Fanucci, "FPG-AI: A Technology-Independent Framework for the Automation of CNN Deployment on FPGAs," in IEEE Access, vol. 11, pp. 32759-32775, 2023, doi: 10.1109/ACCESS.2023.3263392.

[4] Guo K, Sui L, Qiu J, et al. Angel-eye: A complete design flow for mapping cnn onto embedded fpga[J]. IEEE Transactions on Computer-Aided Design of Integrated Circuits and Systems, 2018, 37(1): 35-47.

[5] Lin Zhijian Gao Xuewei Chen Xiaopei Zhu Zhipeng Du Xiaoyong Chen Pingping. Design of high parallel CNN accelerator based on FPGA for AIoT[J]. The Journal of China Universities of Posts and Telecommunications, 2022, 29(5): 1-9.

A 7-bit 6.25-GHz Low Power High Linearity DPC for CDR Applications

Jingsong Cui [1], Kai Li[1], Chengyu Yang [1], Jiahao Lu[1], Hao Li[1], Ang Hu[1*], Dongsheng Liu[1]

[1] School of Integrated Circuits, Huazhong University of Science and Technology
Wuhan, China

ang_hu@hust.edu.cn

Abstract—This paper presents a 7-bit digital-to-phase converter (DPC) for high-speed clock and data recovery (CDR) applications, capable of generating four-phase clocks at a frequency of 6.25 GHz. The design incorporates two delta quadrature delay-locked loops (QDLL) and an integrated-mode phase interpolator (IMPI) to mitigate the effects of circuit imperfections on the DPC's resolution and linearity, based on standard 28-nm CMOS technology. With a 6.25-GHz differential reference clock, the DPC achieves DNL/INL of 0.22/1.0 LSB, respectively, while consuming 8.1 mW of power from a 0.9 V supply.

Keywords—*Digital to phase converter (DPC), clock and data recovery (CDR), phase interpolator (PI), wireline and optical communication.*

I. INTRODUCTION

The growing demand for high-performance computing is driving up the data rates of wireline I/Os. The monolithically integrated multi-lane transceiver scheme achieves higher data rates [1,2]. However, it also presents greater requirements for the overall system design. As a critical component of the transceiver, the digital-to-phase converters (DPCs) are essential for the clock and data recovery (CDR) blocks to generate multi-phase clocks and align data with sampling clocks while consuming significant power. Therefore, designing multiphase clock generation circuits with reduced power consumption is necessary to achieve more energy-efficient data transmission.

Common DPC architectures consist of multi-phase clock generators (MPCGs) and phase interpolators (PIs), where MPCGs provide multi-phase clocks for PIs and samplers, and PIs generate clocks with fine phase steps by summing their multi-phase input clocks with different weights [3, 4]. Four-phase clocks are the minimum requirement for PIs and half-rate bang-bang operation in CDRs. Additionally, PIs with higher resolution and better linearity are needed as data rates increase. In [5], a DPC with an injection-locked ring oscillator is proposed, but a frequency-locked loop is required. In [6], a very high linearity DPC is proposed, but two PIs consume excessive power. In [7], an eight-phase DPC is proposed, but the linearity is poor.

In this paper, a 7-bit 6.25-GHz four-phase interpolator-based DPC is proposed based on a standard 28-nm CMOS technology. In this DPC, two QDLLs are implemented for the PI's input and sampling, and a 7-bit integrated-mode PI (IMPI) is employed for fine phase steps. By utilizing a 6.25-GHz differential reference clock, the DPC achieves DNL/INL of 0.22/1.0 LSB, respectively, while consuming 8.1 mW of power from a 0.9 V supply.

This paper is organized as follows: Section II presents the architecture and circuit implementations of the proposed DPC,

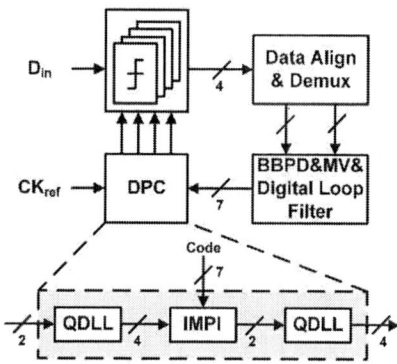

Fig. 1 Architecture of the proposed DPC.

Section III covers the layout and simulation results, and Section IV concludes the paper.

II. ARCHITECTURE

Fig. 1 illustrates the proposed four-phase clock generation circuit architecture based on the PI. A 6.25 GHz differential clock is supplied from an external source, and a quadrature delay-locked loop (QDLL) produces a four-phase quadrature clock that serves as an input signal to the PI. The differential clock signal from the PI output is subsequently routed through a QDLL to regenerate a four-phase clock, which is used in the sampling circuit of the CDR loop. The following subsection details the implementation of the main modules of the clock architecture.

A. Quadrature Delay-Locked Loop

Fig. 2 illustrates the circuit structure of QDLL. As shown in Fig. 2 (a), the input differential clock signal travels through two paths with different delays. One path outputs directly through the buffer, while the other passes through the buffer with an adjustable delay. The adjustable delay in this path is achieved by terminating a variable capacitor at the inverter output. To accurately match the delays of the two paths to 1/4 of a clock cycle, a phase detector capable of identifying quadrature errors is required. In this architecture, a mixer-based quadrature phase detector (QPD) is implemented [6]. The working principle of the QPD can be described as follows:

$$sin\omega t * sin(\omega t + \varphi) = \frac{1}{2} * cos\varphi - \frac{1}{2} * cos(2\omega t + \varphi) \quad (1)$$

where φ represents the phase difference between two paths. When two signals of the same frequency but different phases are mixed, two new components are generated: the first is the quadrature error, and the second is the double-frequency signal created after mixing. Once the mixed signal passes through

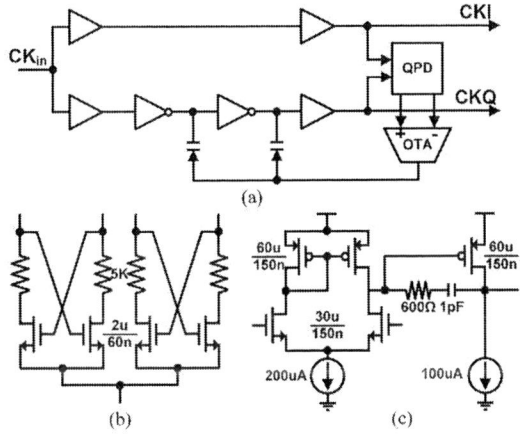

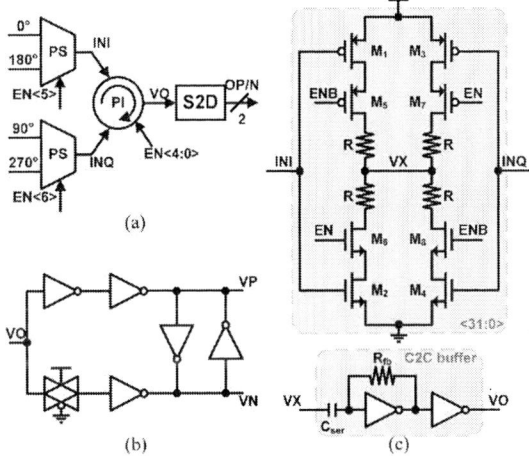

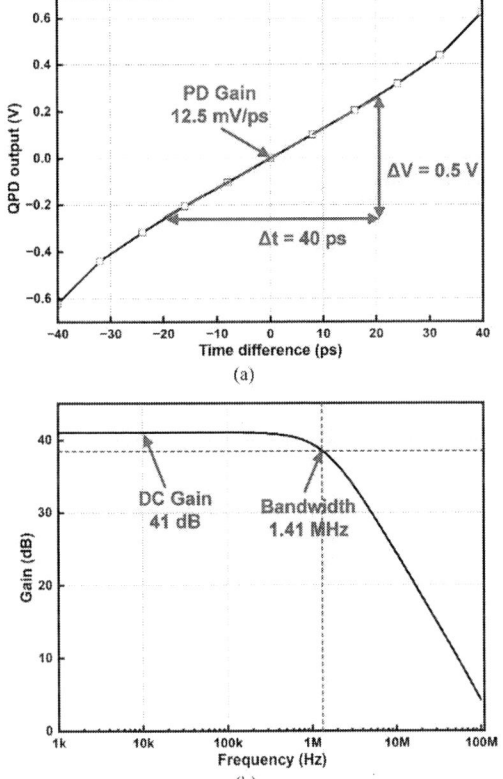

Fig. 2 Implementation of (a) QDLL, (b) mixer-based QPD, and (c) OTA.

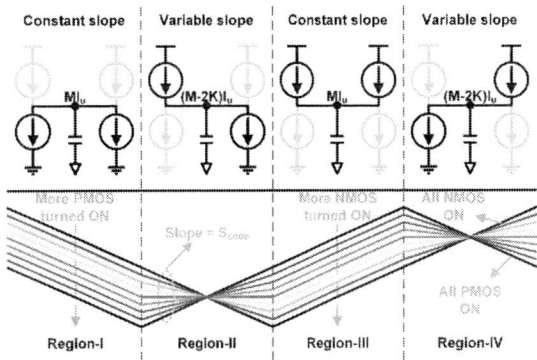

Fig. 4 Implementation of (a) IMPI, (b) S2D, and (c) unit of IMPI and C2C buffer.

Fig. 5 Operation principle of IMPI.

QPD to detect the IQ phase error, as illustrated in Fig. 2 (b). A 5-kΩ resistor is placed in series with the NMOS to filter out the second harmonic and improve the isolation of the clocks. The two-stage OTA is depicted in Fig. 2 (c), where a compensation resistor of 600 Ω and a compensation capacitor of 1 pF stabilize the OTA, exhibiting a first-order frequency response. Fig. 3 (a) presents the input/output characteristic curves of the QPD. The simulated QPD gain is 12.5 mV/ps with a 50% duty cycle. Fig. 3 (b) displays the frequency response curve of the OTA, providing a dc gain of 41 dB and a dominant pole at 1.41 MHz.

B. Integrated-Mode Phase Interpolator

The proposed complete 7-bit IMPI implementation is illustrated in Fig. 4 (a). The highest two bits select the four-phase clocks from QDLL. The chosen quadrature clocks are interpolated by the core of IMPI, which is regulated by the lowest five bits. The interpolated single-ended clock is then fed into a single-to-differential circuit (S2D) and transformed into a differential clock. This differential clock is routed to the second QDLL to generate the four-phase clocks for sampling. Fig. 4 (b) depicts the structure of the S2D circuit, where the clock V_O is divided into two opposite-phase paths to create a pseudo-differential clock. Two inverters in the subsequent stage couple the clock to produce a differential clock VP/VN.

Fig. 3 Simulation results of (a) gain of QPD, and (b) gain of OTA.

the operational amplifier (OTA) in the subsequent stage, the doubling component is filtered out by the low-pass characteristics of the OTA, leaving only a constant term that represents the quadrature error. When the delay of the two paths reaches 1/4 cycle, the constant phase becomes 0, allowing the orthogonal error to be extracted and converted into a controlled voltage on variable capacitors. Passive mixers are employed in

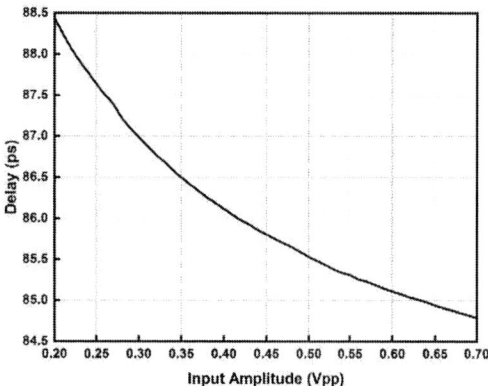

Fig. 6 AM-to-PM conversion of the C2C buffer.

Fig. 4 (c) presents the structure of the IMPI, consisting of 32 IMPI units for interpolation and a CML-to-CMOS (C2C) buffer for CML-to-CMOS conversion. Each IMPI unit includes two slicers for INI and INQ, respectively. One slicer can be viewed as the combination of a current source and a current sink, where INI and INQ control the slicers' pushing or pulling current from the capacitor, while the EN and ENB pins determine the relative weight of INI and INQ. The series resistance in the slicer is used to enhance the slope of the output waveform.

The operating principle of the IMPI is illustrated in Fig. 5 [8]. Based on the combination of high and low levels of INI and INQ, the working state of IMPI can be divided into four regions. Assuming there are a total of M repeating units, with K openings for the I part and M-K openings for the Q part. When both INI and INQ are high, PI operates in the first region, where a total of M units discharge to the ground capacitor, and the slope of the waveform remains constant. This slope can be expressed as follows:

$$S_{code} = \frac{MI_u}{C} \qquad (2)$$

where I_u is the unit current of a slicer, and C is the output capacitor. When INI is low and INQ is high, PI operates in the second region, with K units charging the capacitor and M-K units discharging it. The slope of the waveform is variable and changes with the control word, as shown below:

$$S_{code} = \frac{MI_u}{C}\left(1 - \frac{2K}{M}\right) \qquad (3)$$

When both INI and INQ are low, the PI operates in the third region, charging the capacitor with a total of M units, and the slope of the waveform remains constant, which contrasts with the slope of the first region. When INI is high and INQ is low, PI functions in the fourth region, where K units discharge to the capacitor and M-K units charge it, resulting in a variable slope that differs from the second region's slope. To ensure the linearity of the interpolation and minimize the impact of AM-to-PM conversion in the C2C process, the output waveform V_X swing of the IMPI must be designed to approximately 0.6 × VDD (illustrated below), so that the size of the series resistance can be determined, as illustrated in the following:

$$R_p C = \frac{T_{period}}{2 * ln4} \approx 0.36 * T_{period} \qquad (4)$$

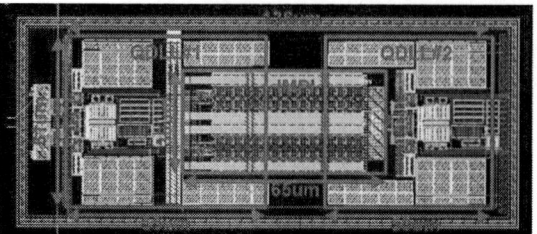

Fig. 7 Layout of the proposed DPC.

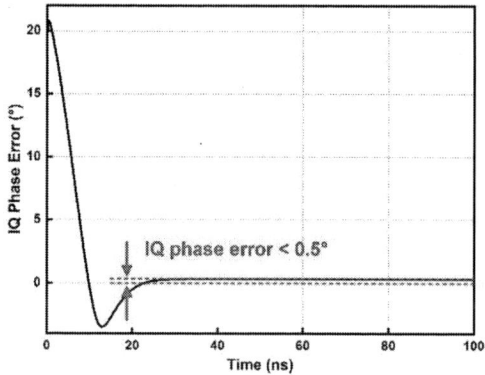

Fig. 8 Transient simulation result of QDLL.

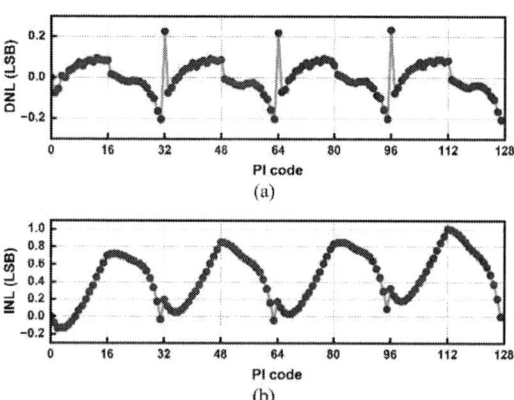

Fig. 9 Simulated result of IMPI. (a) DNL and (b) INL.

The power consumption of the interpolation process can be determined as follows:

$$P = C_O * VDD * \Delta V * f \qquad (5)$$

To minimize power consumption, the parasitic capacitance of the PI output node should be kept as small as possible.

In addition to the core IMPI circuit for interpolation, the buffer designed for CML-to-CMOS conversion is also crucial for improved linearity [5]. A typical C2C buffer comprises a self-biased inverter followed by another inverter, which is commonly employed in PI. However, the AM-to-PM conversion during the C2C process can degrade the linearity of PI. Fig. 6 illustrates that a higher voltage amplitude at the buffer input results in a lower AM-to-PM conversion ratio, thus allowing a larger PI output amplitude to reduce the INL from AM-to-PM conversion. In this design, the PI output amplitude varies from maximum amplitude to half as code (shown

979-8-3315-3918-4/25 $31.00 © 2025 IEEE

TABLE I. SUMMARY AND COMPARISON WITH PRIOR ART

	This work	[5]	[6]	[7]
Technology (nm)	28	65	65	65
Supply (V)	0.9	1.2	1.2	1.05
Frequency (GHz)	6.25	5-8	3.5-11	7
Resolutions (bits)	7	7	7	7
DNL (LSB)	0.22	0.77@7GHz	0.45@7GHz	0.7
INL (LSB)	1.0	1.2@7GHz	0.51@7GHz	6
Multi-phase output	4	8	4	8
Power (mW)	8.1	22.95	21.2	40.5
Active Area (mm²)	0.0069	0.0426	0.036	0.062

in Fig. 5), which leads to significant AM-to-PM conversion. However, as the output amplitude increases, the waveform deviates from a triangle wave due to limited output resistance, which can also impair the linearity of PI. As a design trade-off, an amplitude of approximately $0.6 \times$ VDD is selected for this work.

III. LAYOUT AND SIMULATION RESULTS

A. Layout

Fig. 7 (a) illustrates the proposed DPC layout for CDR applications, which is based on a 0.9V 28-nm CMOS process and occupies an area of 51 μm × 136 μm. Fig. 7 (b) presents the power breakdown of the proposed DPC, showing a total power consumption of 8.1 mW, with the QDLLs consuming 6 mW and the IMPI consuming 2.1 mW, respectively.

B. Simulation Results

Fig. 8 illustrates the post-layout simulation results of the QDLL. The simulation indicates that 25 ns are needed to achieve a stable phase difference. Once the loop stabilizes, the phase error is less than 0.5°. Fig. 9 displays the INL and DNL curves of the IMPI. The simulation results reveal that the DNL and INL of the IMPI are below 0.22 LSB and 1.0 LSB, respectively.

IV. CONCLUSION

This paper proposes a 7-bit 6.25 GHz DPC with a differential clock input and four-phase clock outputs. Fabricated using a standard 28-nm CMOS process, the layout measures 51 μm × 136 μm and consumes 8.1 mW of power. Post-layout simulation results show that the proposed DPC achieves a four-phase output with an IQ phase error of less than 0.5°, while the DNL and INL are each less than 0.22 and 1.0 LSB, respectively. The simulation results demonstrate that the DPC delivers strong performance at 6.25 GHz, positioning it as a competitive solution for high-speed I/O applications. Table I summarizes the DPC's performance and compares it with similar recent designs.

REFERENCES

[1] F. Giunco et al., "An Eight-Lane 800-Gb/s Transceiver for PAM-4 Optical Direct-Detection Applications in 5-nm FinFET Process," IEEE J. Solid-State Circuits, vol. 60, no. 4, pp. 1277–1288, Apr. 2025.

[2] A. Khairi et al., "A 1.41-pJ/b 224-Gb/s PAM4 6-bit ADC-Based Ser-Des Receiver With Hybrid AFE Capable of Supporting Long Reach Channels," IEEE J. Solid-State Circuits, vol. 58, no. 1, pp. 8–18, Jan. 2023.

[3] S. Chen et al., "A 4-to-16GHz inverter-based injection-locked quadrature clock generator with phase interpolators for multi-standard I/Os in 7nm FinFET," in 2018 IEEE International Solid - State Circuits Conference - (ISSCC), San Francisco, CA: IEEE, Feb. 2018, pp. 390-392.

[4] C. F. Poon et al., "A 1.24-pJ/b 112-Gb/s (870 Gb/s/Mm) Transceiver for In-Package Links in 7-nm FinFET," IEEE J. Solid-State Circuits, vol. 57, no. 4, pp. 1199–1210, Apr. 2022.

[5] Z. Wang, Y. Zhang, Y. Onizuka, and P. R. Kinget, "Multi-Phase Clock Generation for Phase Interpolation With a Multi-Phase, Injection-Locked Ring Oscillator and a Quadrature DLL," IEEE J. Solid-State Circuits, vol. 57, no. 6, pp. 1776–1787, Jun. 2022.

[6] Z. Wang and P. R. Kinget, "A Very High Linearity Twin Phase Interpolator With a Low-Noise and Wideband Delta Quadrature DLL for High-Speed Data Link Clocking," IEEE J. Solid-State Circuits, vol. 58, no. 4, pp. 1172–1184, Apr. 2023.

[7] M. M. Khanghah, K. D. Sadeghipour, D. Kelly, C. Antony, P. Ossieur, and P. D. Townsend, "A 7-Bit 7-GHz Multiphase Interpolator-Based DPC for CDR Applications," IEEE Trans. Circuits Syst. I, vol. 69, no. 10, pp. 3976–3988, Oct. 2022.

[8] A. K. Mishra, Y. Li, P. Agarwal, and S. Shekhar, "Improving Linearity in CMOS Phase Interpolators," IEEE J. Solid-State Circuits, vol. 58, no. 6, pp. 1623–1635, Jun. 2023.

Design of Low-Voltage Differential Signaling Driver for Image Sensor

Zhongwei Lin[1], Ningning Li[2], Angyang Li[3], Jian Mei[1,3], Rui Yin[*,1,2], Jiaming Zhang[4], Zhicheng Shi[4]

[1] College of Integrated Circuits and Micro-Nano Electronics, Fudan University, Shanghai, China
[2] Jiashan Fudan Institute, Jiaxing, China
[3] National Integrated Circuit Innovation Center, Shanghai, China
[4] Beijing Institute of Space Mechanics and Electricity

* Email: zwlin24@m.fudan.edu.cn , yinrui@fudan.edu.cn

Abstract—**This paper implements Low-Voltage Differential Signaling (LVDS) Driver circuits and their module single-channel design based on 180nm CMOS process. The LVDS driver adopts a dual current source driving structure with common mode feedback, featuring high power supply rejection ratio (PSRR) and stable common mode voltage output. A trim function is added to the current source to correct its accuracy. Two types of serializers are employed: a half-rate tree-structured serializer for continuous test signal output, and a shift register + MUX-based serializer for serial data signal output. The test section includes a PRBS7 code generator for eye diagram testing. Based on these modules, a DVP to LVDS test chip was designed with a layout area of 1120x1120 μm².**

Keywords—LVDS，Serializers，PRBS7，Trim

I. INTRODUCTION

Low-voltage differential signaling (LVDS) technology was developed in order to provide a low-power and low-voltage alternative [1], [2] to other high-speed I/O interfaces for point-to-point transmission, such as emitter-coupled logic (ECL).

An innovatively designed fully LVDS-compliant I/O interface in 0.35-μm CMOS technology enables Gb/s-per-pin operation via a closed-loop controlled transmitter and a dual-gain folded-cascode receiver, achieving high speed with low power consumption without external components or trimming [3]. Two low-voltage low-power LVDS drivers (DCS and SCS) suit low-voltage applications, enabling Gb/s data rates with LVDS compliance, and the SCS driver reduces power consumption by 60% compared to previous implementations [4].

Image sensors, as key components in modern imaging systems, require efficient and reliable data transmission methods to transfer large amounts of pixel data from the sensor array to processing units. Traditional parallel interfaces face significant challenges in terms of power consumption, electromagnetic interference, and pin count when dealing with high-resolution and high-frame-rate imaging applications. LVDS technology effectively addresses these issues by providing a differential signaling scheme that reduces electromagnetic interference while maintaining high data transmission rates[5].

The integration of LVDS interfaces in image sensors not only improves the overall performance of imaging systems but also enables the development of more compact and power-efficient devices. Currently, exploring different packaging modes for image sensors has a significant impact on improving transmission rates[6].

This paper focuses on the design and implementation of LVDS interface circuits specifically tailored for image sensor applications, exploring the key technical challenges and solutions in this field.

The overall architecture of the test chip is shown in Figure 1. It employs a 16-bit DVP (Data Video Port) input signal, with each 4-bit data segment being configured with one LVDS driver. Additionally, one LVDS driver is dedicated for clock signal transmission, resulting in a total of five pairs of LVDS drivers. Within the LVDS transmission module, there are integrated components including a serializer for data serialization, a frequency divider circuit that provides a divided clock for the serializer, and a PRBS7 (Pseudo-Random Binary Sequence 7) pattern generator circuit for testing purposes. The overall parameter configuration is achieved through SPI (Serial Peripheral Interface) for testing.

The structure of this paper is as follows: Chapter Ⅰ is the introduction, Chapter Ⅱ focuses on the design of the LVDS driver and its sub-modules, Chapter Ⅲ discusses the design of auxiliary modules for the LVDS driver, and Chapter Ⅳ presents the layout and simulation results.

II. BASIC STRUCTURE DESIGN OF LVDS DRIVER

A. Principle of lvds circuit

The basic LVDS circuit components are shown in Figure 2, which is divided into three main parts, LVDS driver, termination resistor, and LVDS receiver. Driver to a fixed current output flow through the terminating resistor to generate a voltage difference, the receiver by comparing the voltage difference between the two ends to determine whether to output data.

The waveform diagram of the LVDS circuit is shown in Figure 2. The LVDS circuit employs a differential signaling scheme for data transmission, which mitigates the issue of excessively high signal amplitudes that can prolong charging and discharging times, thereby limiting data transmission rates. The use of differential signaling also provides excellent common-mode noise immunity.

Moreover, when differential signals are transmitted over symmetrical conductors, the electromagnetic fields generated by the forward and reverse currents effectively cancel each other out, significantly reducing electromagnetic interference.

This work was supported by Beijing Engineering Research Center of Aerial Intelligent Remote Sensing Equipments Fund.

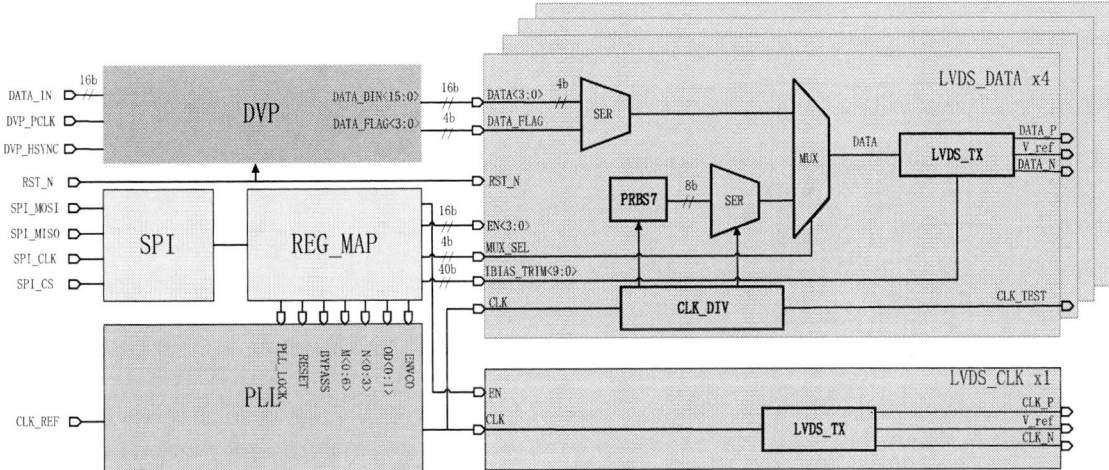

Fig. 1. LVDS Circuit Operation Principle and Waveform Schematics.

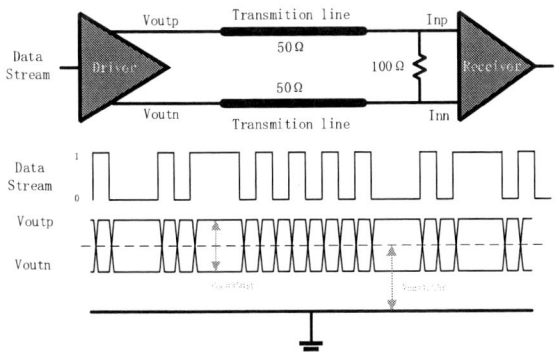

Fig. 2. LVDS Circuit Operation Principle and Waveform Schematics.

B. Design of Lvds Driver Circuit

The overall circuitry of the transmitter in this design includes the following components: single-ended to differential converter (STD), bias circuit, common-mode feedback, and main driver circuit.

Since the data transmitted by digital circuits is in the form of high and low voltage levels, and the main driver of the LVDS employs a differential voltage driving mode, the role of the STD circuit is to convert the single-ended signals outputted by the digital circuit into differential signals for driving the LVDS circuit. The STD used in this design is shown in Figure 3, where the lower branch contains one additional inverter compared to the upper branch, facilitating the generation of differential signals. To account for the delay introduced by the inverters, which can lead to a misalignment of the differential signals, a transmission gate is inserted into the upper branch to achieve proper alignment.

The LVDS driver and common mode feedback circuit used in this design are illustrated in Figure 4. The LVDS driver employs a dual current source driving architecture, which effectively suppresses noise interference. Regarding the selection of the common mode feedback circuit, there are no stringent requirements on gain; thus, an amplifier circuit based on a five-transistor operational transconductance amplifier (OTA) architecture is adopted. The common mode detection circuit is responsible for extracting the common mode levels of the output signals OP and ON, utilizing a

resistive voltage divider formed by resistors R1 and R2. Since the driver outputs current signals, the resistance values of R1 and R2 should be kept as high as possible to minimize any impact on the output signals.

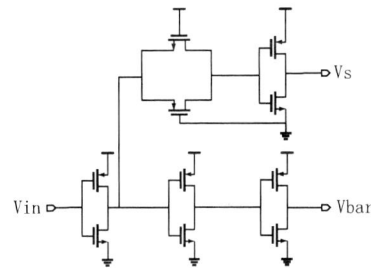

Fig. 3. The STD circuit adopted in this design.

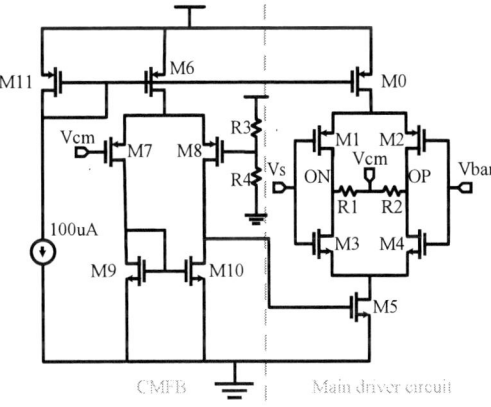

Fig. 4. LVDS Main Driver Circuit and Common Mode Feedback Circuit.

The overall feedback process is as follows: When the common mode level of the output signal increases, the common mode detection circuit detects this rise and amplifies the value through the five-transistor OTA circuit, causing the gate voltage of transistor M5 to decrease. As a result, the current of the main driver decreases, leading to a reduction in the voltage at the feedback point. This adjustment compensates for the current loss induced by the decrease in the gate voltage of M5, ultimately bringing the common mode voltage Vcm close to the reference voltage Vref. The situation in which the common mode level decreases is similar to the principle described above and will not be elaborated further.

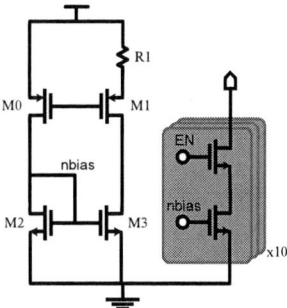

Fig. 5.   Current Source with Trim Function for LVDS Driver.

The bias circuit utilized in this design is illustrated in Figure 5. The most crucial aspect of the LVDS is the accuracy of the current source. To ensure an accurate current of 3.5 mA, a 10-bit trim signal is established, where the lowest 6 bits are used for fine-tuning the current, typically around 10 μA, while the highest 4 bits are for coarse adjustment of the current, configured in a ratio of 1:2:4:8. PVT (Process, Voltage, Temperature) simulation is conducted to confirm that the output current can be set to approximately 3.5 mA.

III. TEST CIRCUIT FOR LVDS DRIVER

A. Serializer for LVDS driver

Since LVDS interfaces typically communicate at GHz speeds, while common image processors usually operate at only hundreds of MHz, a serializer is typically needed to convert low-speed parallel data into high-speed serial data.

Figure 6 illustrates a half-speed tree-structured serializer, which eliminates the final stage output register compared to a full-speed tree structure. As a result, the output rate of the last stage of the serializer is doubled. The advantage of this serializer lies in its simple structure, while its disadvantage is the lack of set or clear functions for data, making it applicable only in continuous transmission scenarios.

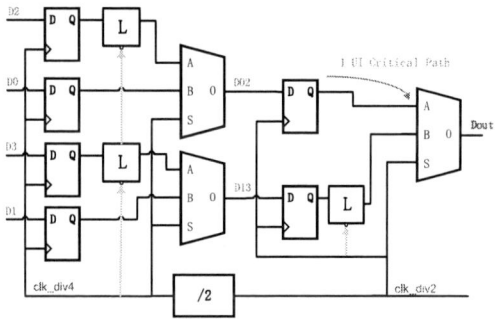

Fig. 6.   A serializer with dual-edge sampling that utilizes a half-rate tree structure.

The serializer used in this design for data transmission is shown in Figure 7. It implements serialization through a combination of shift registers and multiplexers (MUX). Additionally, a sampling circuit for the rising edge of the EN signal is included to achieve data latching, and it features an automatic reset function. This design prevents the repeated

transmission of data, making it suitable for scenarios that require intermittent data transmission.

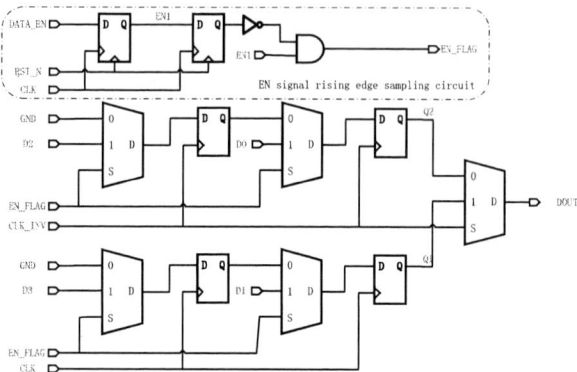

Fig. 7.   A serializer with dual-edge output constructed using a shift register equipped with data sampling.

B. PRBS test circuit

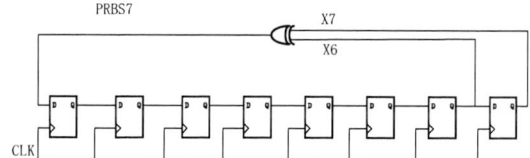

Fig. 8.   PRBS7 Code Generation Circuit.

The testing of interface circuits is typically conducted using PRBS codes, which closely resemble the patterns transmitted in actual processes in the frequency domain. This allows for the simulation of the output under real-world conditions. The PRBS7 code output circuit is shown in Figure 8.

IV. LAYOUT AND SIMULATION RESULT

A. Layout

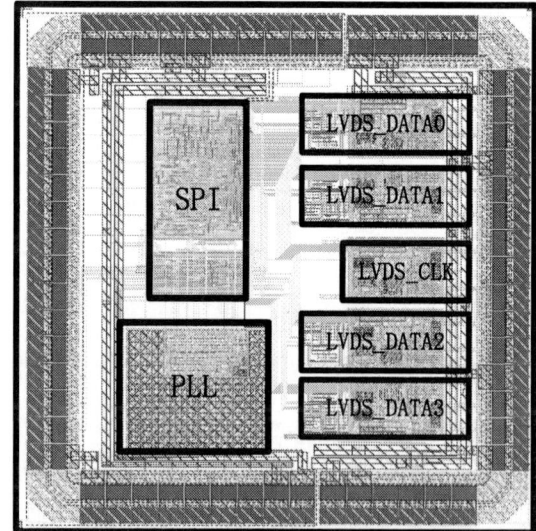

Fig. 9.   Overall Layout for LVDS test chip Simulation results.

Figure 9 shows the complete layout of the test chip. This chip is designed based on 180nm process, with a power supply voltage set at 1.8V, and the overall area of the chip is 1.2544 mm^2.

B. Simulation results

This chip underwent pre-and post-simulation at freq of 200 MHz and 400 MHz, respectively. It utilizes a double-edge sampling method, resulting in effective data rates of 400 Mbps and 800 Mbps.

TABLE I presents the simulation analysis of the LVDS system operating at 400 Mbps, revealing distinct changes in signal characteristics. While maintaining a consistent data rate of 400 Mbps and UI width of 2.5 ns, the post-simulation results demonstrate a 44% increase in common mode voltage variation from 37.85mV to 54.63mV, indicating enhanced noise sensitivity. The differential voltage marginally increased from 391.2mV to 396.9mV, accompanied by a 37% reduction in differential voltage variation from 45.32mV to 28.48mV, suggesting improved signal stability. Notably, the jitter increased by approximately 22% from 72.83ps to 89ps, while the rising time showed a significant improvement, decreasing from 509ps to 398ps. These findings indicate that while the system exhibits enhanced signal transition performance and differential voltage stability, it also demonstrates increased sensitivity to noise and timing variations, highlighting the need for further optimization in the design to improve overall signal integrity.

TABLE I.   SIMULATION RESULTS OF LVDS AT 400 MBPS

Parameters	LVDS Simulation for 400Mbps	
	pre- simulation	*post-simulation*
Data ratio	400Mbps	400Mbps
UI Width	2.5ns	2.5ns
V_{CM}	1.247-1.321V	1.253V-1.337V
ΔV_{CM}	37.85mV	54.63mV
V_{diff}	391.2mV	396.9mV
ΔV_{diff}	45.32mV	28.48mV
Jitter	72.83ps	89ps
Rising time	509ps	398ps

TABLE II.   SIMULATION RESULTS OF LVDS AT 800 MBPS

Parameters	LVDS Simulation for 800Mbps	
	pre- simulation	*post-simulation*
Data ratio	800Mbps	800Mbps
UI Width	1.25ns	1.25ns
V_{CM}	1.247-1.317V	1.253V-1.328V
ΔV_{CM}	44.38mV	51.47mV
V_{diff}	391.2mV	393.5mV
ΔV_{diff}	45.32mV	46mV
Jitter	63ps	84ps
Rising time	503ps	395ps

TABLE II shows the simulation results of the LVDS system operating at 800 Mbps, revealing notable changes in key performance metrics. While maintaining a consistent data rate of 800 Mbps and unit interval width of 1.25 ns, the post-simulation analysis indicates an expanded common mode

voltage range from 1.247-1.317V to 1.253-1.328V, accompanied by a 16% increase in common mode voltage variation from 44.38mV to 51.47mV. The differential voltage marginally increased from 391.2mV to 393.5mV, with a corresponding rise in differential voltage variation from 45.32mV to 46mV. Notably, the jitter increased by approximately 33% from 63ps to 84ps, indicating potential timing instability. However, the rising time showed a significant improvement, decreasing from 503ps to 395ps, suggesting enhanced signal transition performance. These results suggest that while the post-simulation model demonstrates improved signal speed characteristics, it also exhibits increased sensitivity to noise and timing variations, highlighting the need for further robustness analysis in the system design.

V. CONCLUSION

This paper presents the design of an LVDS driver circuit for image sensors based on 180nm CMOS process. The driver adopts a dual current source driving structure with common mode feedback, achieving high PSRR and stable common mode voltage output. A trim function is added to the current source to ensure accurate current control. Two types of serializers are designed: a half-rate tree-structured serializer for continuous test signal output, and a shift register + MUX-based serializer for serial data signal output. The test section includes a PRBS7 code generator for eye diagram testing. Simulation results at 400 Mbps and 800 Mbps show that the circuit maintains stable differential voltage while improving rising time, but exhibits increased sensitivity to noise and timing variations. The test chip was successfully designed with a layout area of 1120x1120 μm^2, featuring a 16-bit DVP input and five pairs of LVDS drivers.

REFERENCE

[1] "IEEE Standard for Low-Voltage Differential Signals (LVDS) for Scalable Coherent Interface (SCI)," in *IEEE Std 1596.3-1996* , vol., no., pp.1-34, 31 July 1996.

[2] Electrical characteristics of low-voltage differential-signalling (LVDS) interface circuits, TIA/EIA-644, National Semiconductor Corp., ANSI/TIA/EIA, 1996.

[3] A. Boni, A. Pierazzi and D. Vecchi, "LVDS I/O interface for Gb/s-per-pin operation in 0.35-/spl mu/m CMOS," in *IEEE Journal of Solid-State Circuits*, vol. 36, no. 4, pp. 706-711, April 2001.

[4] Mingdeng Chen, J. Silva-Martinez, M. Nix and M. E. Robinson, "Low-voltage low-power LVDS drivers," in *IEEE Journal of Solid-State Circuits*, vol. 40, no. 2, pp. 472-479, Feb. 2005.

[5] B. Ding *et al.*, "Analysis and Implementation of Staggered Wire Bonding for LVDS Data and Large Power Transmission in CMOS Image Sensors," *2023 5th International Conference on Circuits and Systems (ICCS)*, Huzhou, China, 2023, pp. 10-14.

[6] W. Fan, Z. Li, J. Xi, L. He, K. Sun and N. Xie, "A 1.2 Gbps failsafe low jitter LVDS transmitter-receiver applied in CMOS image sensor,"*2018 7th International Conference on Modern Circuits and Systems Technologies (MOCAST)*, Thessaloniki, Greece, 2018, pp. 1-4.

Exploring The Further Fracturability of Intel ALM

Chenyu Jiang
Fudan University
Shanghai, China
jiangchenyu25@m.fudan.edu.cn

Xianfeng Cao
Fudan University
Shanghai, China
xfcao23@m.fudan.edu.cn

Lingli Wang*
Fudan University
Shanghai, China
llwang@fudan.edu.cn

Abstract—Fracturable 6-LUTs (6-FLUTs) are a fundamental component of modern FPGAs, utilized by both AMD and Intel. While the core logic functionality remains similar between the two manufacturers, Intel's 6-FLUT, also called Adaptive Logic Module (ALM), provides additional I/O and flip-flop resources. However, these supplementary resources are not explicitly exploited under conventional usage models. This paper introduces the concept of *deep fracturability*, a novel working mode that enables a single ALM to implement a broader range of logic functions by leveraging its inherent architectural advantages more effectively. Additionally, we enhance the technology mapping to better support this new working mode. Experimental results on the VTR benchmarks demonstrate that ALM with deep fracturability reduces the ALM consumption by $\sim 5\%$ with the minimal impact on performance.

Index Terms—Fracturable LUT, Adaptive Logic Module, Technology Mapping

I. INTRODUCTION

Look-up tables (LUTs) have been used ubiquitously in FPGAs. In the early stage of FPGA development, the 4-LUT was the dominant choice for programmable logic blocks, as it provided an optimal balance between area and delay [1]. While 6-LUTs offer improved performance, their exponential area overhead initially limited widespread adoption. The introduction of fracturable LUTs (FLUTs) changed this landscape. A 6-input fracturable LUT (6-FLUT) can function either as a single 6-LUT or as two 5-LUTs sharing certain inputs. This innovation has significantly enhanced area efficiency, making 6-FLUTs an essential component of modern FPGAs [2], [3].

As the two leading FPGA manufacturers, AMD and Intel have both incorporated 6-FLUT as the core programmable logic element in their advanced FPGA architectures, such as Versal [3] and Stratix/Agilex [2], [4]. Fig. 1 illustrates the structural differences between the 6-FLUT implementations in Versal and Stratix/Agilex series. A key distinction lies in their I/O and flip-flop (FF) resources: the Stratix ALM features 8 inputs, 4 outputs, and 4 FFs, whereas the Versal counterpart has 6 inputs, 2 outputs, and 2 FFs. The 6-FLUT in Intel FPGAs, known as the Adaptive Logic Module (ALM), possesses additional I/O and FF resources. However, despite these enhancements, its operating mode remains identical to Versal, supporting a single 6-LUT mode and a dual 5-LUT mode. Consequently, we argue that the ALM does not fully utilize its extra resources.

Each ALM consists of two pairs of 4-LUTs that collectively form two 5-LUTs. Notably, these four 4-LUTs are capable of

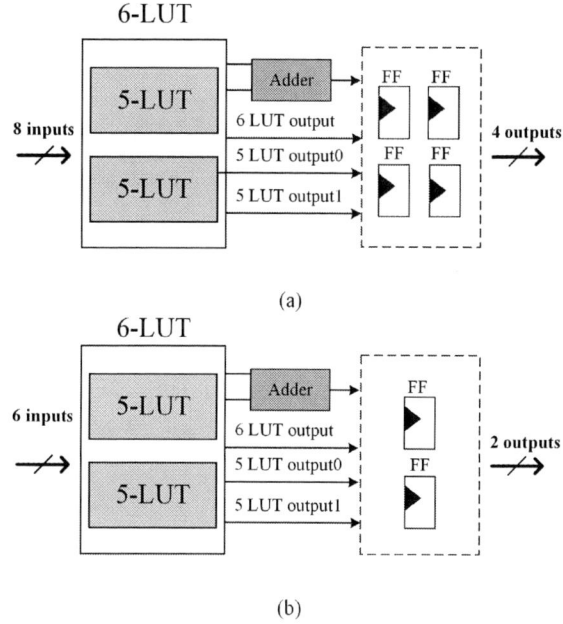

Fig. 1. (a) The schematic of 6-FLUTs in Stratix/Agilex ALM; (b) The schematic of 6-FLUTs in Xilinx Versal series.

implementing individual functions rather than solely serving as components of 5-LUTs. However, due to input-sharing constraints—each pair of 4-LUTs must share all four of their inputs—the limited flexibility restricts their application to arithmetic operations rather than general-purpose logic.

In this paper, we conduct an in-depth analysis of the Intel ALM architecture and propose a novel deep fracturable mode. By introducing four additional 2-to-1 multiplexers (2-MUXes) to the 4-LUT inputs, each pair of 4-LUTs can share six inputs, significantly increasing flexibility. This deep fracturable mode maximizes the utilization of I/O and FF resources, allowing each 4-LUT output to be registered and routed efficiently. Experimental evaluations using VTR benchmarks demonstrate a 4.77% reduction in ALM utilization with minimal performance degradation (0.30%). Additionally, we improved the LUT Balancing algorithm [9] in the technology mapping process to better support this enhanced fracturability, achieving a 5.52% reduction in ALM consumption with only a 1.11% performance overhead.

The remainder of the paper is structured as follows. Section

II describes the Intel ALM and our improvement on it. Section III describes the enhancements to LUT balancing within the technology mapping process. Section IV presents the experimental results. Section V concludes the paper.

II. FURTHER FRACTURABILITY EXPLORATION

Fig. 2 illustrates the detailed architecture of the Intel ALM [5], where *abcdefgh* represent 8 inputs of ALM. There are a total of three operation modes, including single 6-LUT mode, two 5-LUTs mode and arithmetic mode. In the arithmetic mode, four outputs of 4-LUTs connect to four inputs of two 1-bit adders, which implies that outputs of 4-LUTs can be used without any hardware burden. However, since each pair of 4-LUTs must share all 4 inputs, limited flexibility does not bring considerable area gain. Consequently, the outputs of four 4-LUTs are not used as general outputs.

According to Fig. 2, we see that, when operating in the fractured mode, i.e., two 5-LUT mode, the top 5-LUT uses inputs *abcde* and the bottom 5-LUT uses inputs *abghf*. We add four 2-MUXes to the ALM, as shown in Fig. 3. By doing this, each pair of 4-LUTs can access 6 inputs and only 2 inputs are shared. Now the ALM can be further fractured from two 5-LUTs to four 4-LUTs, with the inputs of four 4-LUTs are *abcd*, *cdef*, *abgh* and *efgh* respectively.

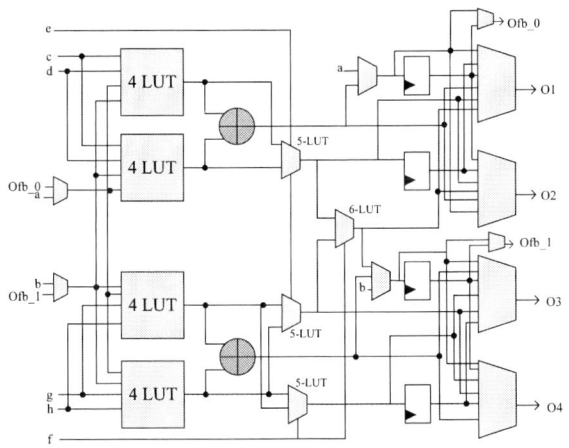

Fig. 2. Detailed architecture of Intel ALM [5].

A transistor-level design for ALM with/without deep fracturability is conducted with COFFE2 [6]. We use Minimum Width Transistor Area (MWTA) for area measurement. Table I gives area modeling for both. The difference lies in the *MUXes*, since four 2-MUXes are added and some internal MUXes need to be enlarged. The overall area overhead is 0.72%.

III. LUT BALANCING FOR ALM

Tech-mapping a circuit to an FPGA is the process of converting all logic in the initial network into a functionally equivalent netlist of LUTs. Since a K-LUT can implement any K-input function, the mapping is transformed to select a set of K-feasible cuts covering the whole graph. A cut is a set of

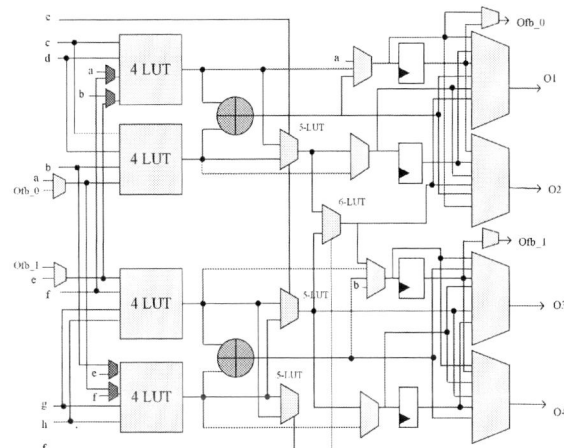

Fig. 3. Modified ALM architecture with four additional 2-MUXes.

TABLE I
AREA MODELING OF ALM WITH/WITHOUT DEEP FRACTURABILITY.

Aera (MWTA)	ALM	Deep fracturable ALM
6-LUT	907.1	907.1
4 FFs	106.5	106.5
adder	55.4	55.4
MUXes	268.3	278.0
total	1337.3	1347.0
ratio		+0.72%

nodes associated with a root node. The input nodes of the cut are called leaves. A cut satisfies that all paths from the inputs of the network to the root node pass through one or more leaves. A cut with K or less leaves is termed as K-feasible.

Cut-based technology mapping can be divided into three parts [7]:
1) K-feasible cuts are generated for all nodes;
2) a cut is selected for each node based on a certain criteria;
3) the final netlist is generated according to the selected cuts.

Minimizing first the *depth* and then the *area* are typical optimization goal. The *depth* of a mapping is the number of LUTs on the longest combinational path. *Area* is usually measured by the number of LUTs. Once a depth-optimal mapping is found, cuts on non-critical paths can be changed to minimize the number of LUTs in the mapping. Two cost functions for evaluating the area of a cut are *Area Flow* and *Exact Area* [8]. The Area Flow (AF) of a cut C_v is defined as:

$$AF(C_v) = \frac{Cost(C_v) + \sum_{u \in input(C_v)} AF(u)}{nFanouts(v)} \quad (1)$$

where $Cost(C_v)$ is the basic area cost of a LUT to map cut C_v, and $nFanouts(v)$ is the fanout number of root node v. The Exact Area (EA) is defined as:

$$EA(C_v) = Cost(C_v) + \sum_{u \in MFFC(v)} EA(u) \quad (2)$$

where MFFC is the maximum fan-out free cone of node v.

We see that both definitions includes the *Cost()*. It is typically assigned a value of 1.0, denoting that any cut consumes one LUT during mapping. However, this measure does not apply to FLUT, since two 5-LUTs merged to an FLUT consume the same area as one 6-LUT. Hence, [9] proposed "LUT Balancing" (L.B.), assigning different *Cost()* for cuts with different sizes. After extensive experiments, the author proposed a new *Cost()* for ALM:

$$Cost(C_v) = \begin{cases} 1.0, & \text{if } inputs(C_v) < 6 \\ 2.0, & \text{otherwise.} \end{cases}$$

L.B. introduces more LUTs but fewer FLUTs, improving the area consequently. For our deep-fracturable ALM, we adopt a more fine-grained L.B., since it can contain four 4-LUTs at most. The *Cost()* of cuts with less than 5 leaves is set to 0.75, an empirical value that consumes the least amount of ALM. The modified *Cost()* is given as:

$$Cost(C_v) = \begin{cases} 0.75, & \text{if } inputs(C_v) \leq 4 \\ 1.0, & \text{if } 4 < inputs(C_v) \leq 5 \\ 2.0, & \text{otherwise.} \end{cases}$$

IV. EXPERIMENTAL EVALUATION

A. Experiment Setup

We adopt the classic CB-SB routing architecture but drop the crossbar inside, since there are no crossbars in the latest FPGAs [3]. There are 10 ALMs in a CLB, with a total of 80 inputs and 40 outputs. We modify the architecture description file (ADF) to enable deep fracturability. We use VTR [10] for the entire EDA flow with modifications on the technology mapping phase of ABC [11]. 22 largest VTR benchmarks are used to evaluate our deep fracturable ALM. In addition, we evaluate the impact of our fine-grained L.B..

B. Post-Implementation Results

We calculate four metrics including critical path delay, the number of used ALMs, wire length and minimum routing channel width. Table II gives the comparison of ALM with/without deep fracturability. Results show that the ALM with deep fracturability reduces the number of ALMs by 4.77% and the wire length by 2.30%, at the cost of a 0.30% performance degradation and a 2.15% increase in routing channel width. Specifically, our deep fracturable ALM consumes fewer ALMs on all benchmarks, demonstrating the stability and effectiveness.

The effect of LUT Balancing is given in Table III. Compared to ALM combined with L.B., our deep fracturable ALM with more fine-grained L.B. reduces the number of ALMs by 5.52% and the wire length by 3.03%, at the cost of 1.10% worse performance and 3.64% increase of routing channel width. Overall, we can see the strong potential in area reduction of the deep fracturable ALM, with the minimal impact on performance.

V. CONCLUSION

While Intel ALM provides richer I/O and FF resources within a programmable logic block compared to AMD FPGAs, these resources are not fully utilized. We enhance ALM by integrating four additional 2-MUXes, enabling the implementation of up to four independent 4-input functions. This deep fracturability significantly improves area utilization. Furthermore, we explore the LUT Balancing technique to better leverage this capability. Experimental results on VTR benchmarks show that deep-fracturable ALM reduces ALM consumption by approximately 5% with minimal impact on performance, demonstrating its efficiency and practicality.

ACKNOWLEDGMENTS

This work is supported by the National Natural Science Foundation of China under grant 62174035.

REFERENCES

[1] V. Betz, J. Rose, and A. Marquardt, Architecture and CAD for Deep Submicron FPGAs. Kluwer, 1999.

[2] M. Langhammer, E. Nurvitadhi, B. Pasca, and S. Gribok, "Stratix 10 nx architecture and applications," in *The 2021 ACM/SIGDA International Symposium on Field-Programmable Gate Arrays*, 2021, pp. 57–67.

[3] B. Gaide, D. Gaitonde, C. Ravishankar, and T. Bauer, "Xilinx adaptive compute acceleration platform: VersalTM architecture," in *2019 ACM/SIGDA International Symposium on Field-Programmable Gate Arrays*, 2019, pp. 84–93.

[4] J. Chromczak, M. Wheeler, C. Chiasson, D. How, M. Langhammer et al., "Architectural Enhancements in Intel® Agilex™ FPGAs," in *2020 ACM/SIGDA International Symposium on Field-Programmable Gate Arrays*, 2020, pp. 140–149.

[5] Intel Corporation. 2017. Intel Stratix 10 logic array blocks and adaptive logic modules user guide (UG-S10LAB).

[6] S. Yazdanshenas and V. Betz, "Coffe 2: Automatic Modelling and Optimization of Complex and Heterogeneous FPGA Architectures," in *ACM Transactions on Reconfigurable Technology and Systems*, vol. 12, no. 1, Jan 2019.

[7] J. Cong and Y. Ding, "FlowMap: an optimal technology mapping algorithm for delay optimization in lookup-table based FPGA designs," in *IEEE Transactions on Computer-Aided Design of Integrated Circuits and Systems*, vol. 13, no. 1, pp. 1-12, Jan. 1994

[8] A. Mischenko, S. Chatterjee and R. K. Brayton, "Improvements to Technology Mapping for LUT-Based FPGAs," in *IEEE Transactions on Computer-Aided Design of Integrated Circuits and Systems*, vol. 26, no.2, pp. 240-253, Feb. 2007.

[9] D. Dickin and L. Shannon, "Exploring FPGA technology mapping for fracturable LUT minimization," in *IEEE 2011 International Conference on Field-Programmable Technology*, 2011, pp. 1-8.

[10] K. E. Murray, O. Petelin, S. Zhong, J. M. Wang, M. Eldafrawy, J.-P.Legault, E. Sha, A. G. Graham, J. Wu, M. J. P. Walker,H. Zeng, P.Patros, J. Luu, K. B. Kent, and V. Betz, "Vtr 8: High-performance cad and customizable fpga architecture modelling," in *ACM Transactions on Reconfigurable Technology and Systems*, vol. 13, no. 2, jun 2020.

[11] R. Brayton and A. Mischenko, "Abc: An academic industrial-strength verification tool," in Computer Aided Verification, T. Touili, B. Cook, and P. Jackson, Eds. Berlin, Heidelberg: Springer Berlin Heidelberg, 2010, pp. 24–40.

TABLE II
POST-IMPLEMENTATION RESULTS COMPARISON OF ALM WITH/WITHOUT DEEP FRACTURABILITY.

VTR	Intel ALM				Intel ALM with deep fracturability			
	delay	nALM	wire_length(E+05)	chan_width	delay	nALM	wire_length(E+05)	chan_width
arm_core	21.30	6923	4.69	282	23.48	6754	4.69	286
bgm	21.12	20739	11.15	246	22.00	20072	11.08	248
blob_merge	13.31	4082	2.08	212	14.61	3953	2.04	222
bound_top	4.50	334	0.19	84	3.90	285	0.19	88
ch_intrinsics	2.74	109	0.04	62	2.83	109	0.04	68
diffeq1	19.69	240	0.25	98	20.24	238	0.22	94
diffeq2	14.00	145	0.16	76	15.76	144	0.16	88
LU8PEEng	79.55	16718	8.99	268	82.12	16168	9.13	282
LU32PEEng	79.59	58825	33.26	334	79.01	56944	33.49	344
LU64PEEng	77.58	114150	69.46	392	81.66	110749	68.44	396
mcml	54.22	56872	31.54	270	54.84	53655	28.93	276
mkDelayWorker32B	8.57	986	0.30	40	8.70	875	0.32	38
mkPktMerge	4.20	117	0.18	38	4.46	91	0.19	40
mkSMAadpter4B	6.17	981	0.63	150	6.10	956	0.62	152
or1200	10.54	1769	1.39	184	11.22	1714	1.46	180
raygentop	5.99	926	0.65	150	6.11	882	0.66	148
sha	9.39	1137	0.59	196	9.79	1125	0.58	196
spree	14.63	402	0.31	140	14.69	383	0.32	136
stereovision0	5.86	5592	2.51	208	4.36	5300	2.48	218
stereovision1	7.83	4902	3.50	226	6.70	4857	3.28	226
stereovision2	12.51	11477	7.15	160	13.90	11077	7.54	162
stereovision3	2.73	88	0.04	90	2.37	86	0.03	94
geomean	12.60	2090.6	1.31	148.9	12.64	1990.9	1.28	152.1
ratio					+0.30%	-4.77%	-2.30%	+2.15%

TABLE III
POST-IMPLEMENTATION RESULTS COMPARISON OF ALM WITH/WITHOUT DEEP FRACTURABILITY BOTH WITH LUT BALANCING.

VTR	Intel ALM (L.B.)				Intel ALM with deep fracturability (L.B.)			
	delay	nALM	wire_length(E+05)	chan_width	delay	nALM	wire_length(E+05)	chan_width
arm_core	22.49	6772	4.87	290	22.77	6448	4.93	292
bgm	21.09	20288	11.41	248	23.97	19536	11.46	252
blob_merge	13.21	4035	2.14	222	13.41	3864	2.13	222
bound_top	4.57	306	0.19	78	4.93	251	0.20	88
ch_intrinsics	3.10	109	0.05	54	2.68	110	0.04	62
diffeq1	17.41	239	0.23	90	17.97	228	0.23	94
diffeq2	13.91	144	0.16	70	14.77	143	0.16	76
LU8PEEng	78.28	16418	9.24	284	77.70	15338	9.32	292
LU32PEEng	75.57	57553	33.56	336	79.97	53776	34.29	342
LU64PEEng	78.58	111885	69.84	412	80.40	104303	70.56	416
mcml	54.57	55919	29.31	270	53.24	52630	27.50	280
mkDelayWorker32B	8.98	978	0.31	38	8.93	927	0.31	38
mkPktMerge	4.61	115	0.18	38	4.56	89	0.16	40
mkSMAadpter4B	6.54	949	0.69	150	6.65	922	0.65	156
or1200	10.27	1792	1.39	190	10.43	1693	1.44	196
raygentop	5.70	893	0.68	150	5.91	846	0.70	152
sha	9.43	1057	0.52	210	9.74	1090	0.60	206
spree	13.43	389	0.32	150	14.04	381	0.34	152
stereovision0	5.29	5543	2.64	204	4.58	5246	2.42	216
stereovision1	8.30	4907	3.38	210	6.48	4752	3.18	220
stereovision2	12.45	11431	7.13	156	14.64	11032	7.08	160
stereovision3	2.53	89	0.04	98	2.85	85	0.03	102
geomean	12.56	2051.2	1.32	148.2	12.70	1938.0	1.28	153.6
ratio					+1.11%	-5.52%	-3.03%	+3.64%

979-8-3315-3918-4/25 $31.00 © 2025 IEEE

A General and Modular FPGA Hardware Architecture for Enhanced Scalability and Flexibility

ZiRui Qin[1], ZhiNan Li[2], YaBo Xiao[3,4], Hui Zhang*[1], Cang Liu *[3,4]

[1] Beihang University, Beijing 100191, China
[2] Faculty of Electrical Engineering and Computer Science, Ningbo University, Ningbo 315211, China
[3] Ningbo Yonghua Innovation Science and Technology Development Co., Ltd, Ningbo, Zhejiang,315211, China
[4] Department of Electronic Engineering, Tsinghua University, Tianjin, 300467, China

* Email: qinzirui@buaa.edu.cn, zhang_hui@buaa.edu.cn

Abstract—Field-Programmable Gate Array (FPGA), due to its advantages of parallel processing, real-time performance, and determinism, has been widely used in the fields of inherent parallelism, such as communications, radar, and artificial intelligence. This paper presents a general and modular FPGA hardware architecture for enhanced scalability and flexibility. By flexibly configuring the data streams of various function modules via application running on Windows operating system, the data interconnection modules is generated automatically. Furthermore, the design of decoupling individual function modules in the data interconnection modules achieves modular hardware structure while simultaneously enhancing both scalability and flexibility. In addition, the implementation of standardized data conversion modules enables seamless transitions from commonly used data interfaces of various function modules to those of the interconnect modules. This simplification of the design process for function modules significantly enhances research and development efficiency. Therefore, the hardware architecture proposed in this paper not only exhibits significant advantages in terms of generalization and modularization, but also demonstrates remarkable scalability and flexibility.

Keywords—FPGA Hardware Architecture, General, Modular, Flexible, Scalable

I. INTRODUCTION

FPGA's parallel processing advantages make it ideal for communication, radar and AI applications [1]-[8]. However, traditional HDL-based development faces complexity challenges [9][10]. This necessitates modular architectures to improve efficiency.

In high-level programming languages like C, C++, and Python, the interaction among modules can be accomplished by ensuring the accuracy of input and output parameters. In contrast, for hardware description languages designed for FPGA, such as Verilog and VHDL, it is imperative not only to verify the accuracy of input and output signals, but also to ensure the temporal relationships between signals are precisely met. This requirement severely constrains the generalization and modularization of FPGA.

Significant progress has been made in both academia and industry toward the generalization and modularization of FPGA designs. Bus architectures [11]-[13], exemplified by the AMBA bus [14][15], have been employed in the design of System-on-Chip (SoC) systems, greatly enhancing design efficiency. However, the shared characteristics of bus architecture restrict multiple modules from concurrently transmitting data. Achieving higher levels of parallelism often necessitates additional architectural designs, such as crossbar switch [16][17] structure. While crossbar switch designs offer high bandwidth and flexible connectivity, they also introduce considerable architectural complexity, complicating the design and implementation processes. Furthermore, these designs are facing challenges in power consumption and resource management.

In addition, research achievement has seen in the area of hardware accelerators and modular design at specific applications, for instance, FPGA accelerator designs for image processing [18][19], modular designs of FPGA for artificial intelligence algorithms [20][21]. However, the universality and adaptability of these achievements are limited as they primarily focus on optimization for specific algorithms, making it difficult to transfer them to other application contexts. Consequently, the current research outcomes pertaining to FPGA generalization and modularization fall short of meeting practical engineering requirements.

This paper proposes a new FPGA hardware architecture that significantly enhances the generalization and modularization of FPGA. The main contributions of this paper are summarized as follows:

1)To achieve the generalization and modularization of FPGA, a novel FPGA hardware architecture is proposed. This design incorporates a data interconnection module to facilitate the decoupling of various function modules.

2) A interface protocol is designed to unify the input-output interface of function modules, and the corresponding data conversion module is designed for transforming various commonly used interfaces to the designed interface protocol.

3) The prototype system is also built to verify the designed hardware architecture. The implementation results show that the proposed hardware architecture is more flexible and scalable compared with the existing works, and the effect of switching delay can be negligible via the data interconnection module to exchange data between different function modules.

The rest of this paper is organized as follows. Section II illustrates the typical general and modular methods for

---

This work was supported in part by Yongjiang Talent Project Youth Innovation Project 2024A-275-G.

979-8-3315-3918-4/25 $31.00 © 2025 IEEE

FPGA design. Section III presents the proposed FPGA hardware architecture. In Section IV, the implementation results and comparison are provided. Finally, Section V draws the conclusions.

II. THE TYPICAL GENERAL AND MODULAR METHODS FOR FPGA DESIGN

Current approaches for achieving generalization and modularization in FPGA design primarily rely on bus architectures, such as AMBA (Advanced Microcontroller Bus Architecture), and crossbar switches. The bus architecture (Fig.1) employs a shared interconnection with a unified protocol to link processor cores and peripherals, promoting module decoupling and design efficiency. However, its inherent shared nature restricts parallel data transfer, allowing communication between only one master and one slave device at any given time, which limits overall system performance.

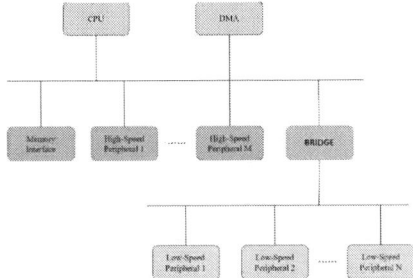

Fig. 1. Bus architecture diagram

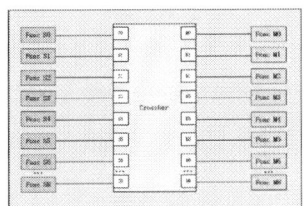

Fig. 2. Crossbar switch architecture diagram

On the other hand, the crossbar switch architecture (Fig.2) enhances parallelism and bandwidth utilization by enabling flexible data routing between multiple inputs and outputs. Nevertheless, the complex interconnections required consume substantial resources, increasing design complexity and potentially causing delays and waste. Crucially, crossbar switches are fundamentally limited to facilitating master-slave data exchanges and cannot support direct communication between different masters or different slaves.

In summary, although both bus architectures and crossbar switches demonstrate certain advantages in the context of FPGA generalization and modularization, their inherent limitations hinder the achievement of higher levels of parallelism and flexibility. Therefore, developing a novel FPGA general and modular hardware architecture holds significant industrial implications.

III. HARDWARE ARCHITECTURE DESIGN

To address the challenges of generalization and modularization in FPGA architectures while enhancing scalability and flexibility, this paper proposes a novel reconfigurable generalized FPGA hardware architecture, which covers data exchange modules, signal conversion modules and functional modules.

The design process incorporates data exchange modules and standardizes the data interaction timing among various functional modules. The introduction of signal conversion modules allows for a higher degree of independence among functional modules, achieving rapid integration to improve development efficiency. Furthermore, to enhance the system's robustness and development efficiency, corresponding application running on Windows OS has been developed for both data exchange modules and signal conversion modules, enabling the configuration of data flow directions.

A. The Proposed Hardware Architecture

The block diagram of the designed FPGA hardware architecture is shown in Fig.3, which consists of data interconnection module, data conversion module and function module. The FPGA application can be divided into several function modules, such as PWM, SPI, UART, Ethernet, GPIO, FFT, QR Decomposition, Matrix Operations etc. The function modules of proposed FPGA hardware architecture are divided into two categories: the existed modules and user defined modules. The existed modules are the verified modules, and have been implemented in the hardware architecture. The user defined modules are the reserved interface, which can be used to add the specified function.

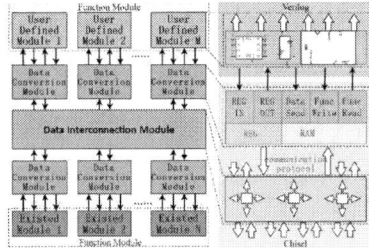

Fig. 3. The designed FPGA hardware architecture

The data interconnection module can efficiently decouple different function modules. A communication protocol is also defined to implement the data exchange between different function modules via the data interconnection module. The data conversion module is designed to convert the data format between the defined communication protocol and the commonly used interface (such as register files and interface of RAM), so a new function module can be integrated to the proposed hardware architecture easily. Furthermore, the interface number of data interconnection module can be extended easily. Hence, the proposed hardware architecture is more flexible and scalable.

B. The Data Interconnection Module

Fig.4 shows the block diagram of the data interconnection module. Each function module can access the data interconnection module via the defined communication protocol as shown in Fig.4, which references the classical model of producer and consumer. When the producer receives the high level of the ready signal sent by consumer, the enable signal of producer can be set at high level and keep one clock cycle to send the data signal and address signal. Similarly, when the high level of enable signal sent by producer is captured by consumer, the

979-8-3315-3918-4/25 $31.00 © 2025 IEEE

consumer would obtain the data signal and the address signal.

According to the defined communication protocol, the data interconnection module is divided into four parts: enable, address, data and ready. Each part is composed of several multiplexers, and the number of multiplexer is equal to the number of output signal. These select signals of multiplexers can implement the needed interconnection between input signals and output signals, which are generated by software of application running on Windows OS. Hence, the latency of data interconnection module is determined, which is only one clock cycle.

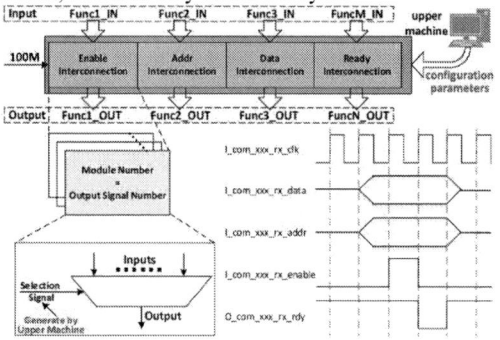

Fig. 4. Block diagram of the data interconnection module

C. The Data Conversion Module

The function modules must convert the input interface and output interface of itself to the defined communication protocol of data interconnection module to exchange data with the data interconnection module. The data conversion needs to process the timing constraint, hitting address space, priority judgment and data cache to solve the data conflict, which will reduce the robustness and developing period of FPGA design. Therefore, the data conversion module is designed to convert the commonly used interface of function module (such as register file, read and write of RAM etc.) to the defined communication protocol of data interconnection module. A block diagram illustrating the structure of module is provided in Fig.5.

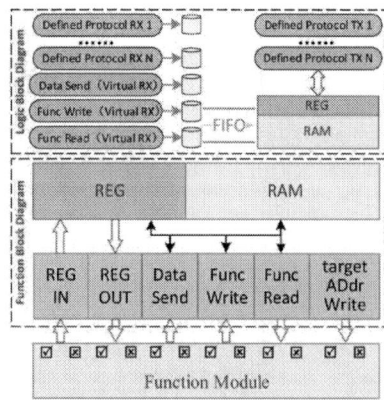

Fig. 5. Block diagram of the data conversion modules

D. The Function Module

The function modules are used to implement the functions, which are commonly used in target application, such as PWM, UART, Ethernet, GPIO, SPI etc. To meet the dedicated demand of special target application, several

interfaces are reserved. Because the conversion module includes the commonly used interfaces of function module, the newly designed module can be integrated into the existing hardware architecture easily. In addition, according to the specific FPGA application, any function module can be bypassed to decrease the resource consumption and power consumption of FPGA.

IV. PROTOTYPE SYSTEM AND IMPLEMENTATION RESULTS

Based on the design concept of the reconfigurable generalized FPGA architecture proposed in this paper, a prototype system has been established against the backdrop of electronic power devices. This prototype facilitates rapid reconfiguration of modules, thereby improving development efficiency and validating the feasibility of the reconfigurable generalized FPGA hardware architecture.

A. Prototype system

A prototype system has been established against the backdrop of power electronic equipment to validate the proposed FPGA hardware architecture. The core processor of the hardware board in the prototype system consists of FPGA and DSP. The DSP principally executes the control algorithm, and FPGA is regarded as the co-processor, which is used to implement the function of the interface extension and real-time algorithm. Compared to the DSP, the function of FPGA is more fixed and hard to develop, which mainly focuses on interface extension and algorithm acceleration. The sketch map of the prototype is presented in Fig.6. The DSP can both read and write the registers of corresponding FPGA function module via the external memory interface (EMIF). The configuration parameters of data interconnection module are generated by application running on Windows OS. The application generates FPGA configuration parameters and corresponding DSP library functions based on the requirements of each project. Key implemented modules include GPIO, PWM, Ethernet, and computation units (duty ratio/frequency). The DSP communicates via EMIF interface while Windows applications generate configuration parameters.

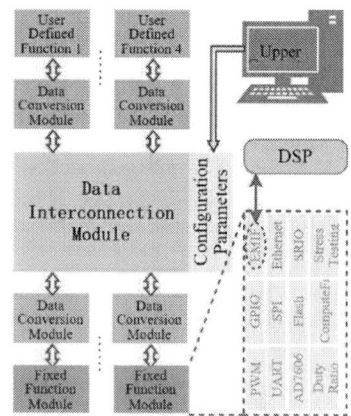

Fig. 6. The sketch map of the prototype system

B. Implementation Results and Comparison

The XC7K325T-2FFG900I of Xilinx is adopted to verify the proposed hardware architecture in the prototype system. The system clock rate is 100MHz, so the latency of data forwarding is 10ns with the data interconnection module to exchange data between different function modules. The utilization of FPGA resource is presented in Tab. I.

TABLE I. THE UTILIZATION RESULTS OF FPGA

Resource	Utilization	Available	Utilization(%)
LUT	118257	203800	58.03
LUTRAM	4640	64000	7.25
FF	132838	407600	32.59
BRAM	304	445	68.31
DSP	5	840	0.6
IO	252	500	50.4
GT	8	16	50
BUFG	32	32	100
MMCM	5	10	50
PLL	2	10	20

The data interconnection module and the data conversion module consume the logical resource of FPGA in the proposed hardware architecture compared to the existing works. However, the proposed hardware architecture effectively decouples the different function modules, and the new function module can be integrated to the proposed hardware architecture easily. In addition, the interface number of data interconnection module can be changed easily to meet various applications. Hence, the proposed hardware architecture is more flexible and scalable compared to the existing works.

V. CONCLUSION

FPGA is widely utilized in the fields of communications, radar, and artificial intelligence owing to its inherent advantages in parallelism, real-time processing, and determinism. Therefore, the general and modular hardware architecture of FPGA can significantly improve the code reusability, development efficiency, maintainability and shorten the development period. This paper proposes a general and modular hardware architecture of FPGA. The data interconnection module and the data conversion module are designed in the proposed hardware architecture to decouple different function modules and accelerate the development of FPGA. A prototype system is also built to verify the proposed hardware architecture. The results show that the proposed hardware architecture is feasible, and the effect of switching delay between different function modules can be negligible via the data interconnection module. In the future, we will mainly focus on applying this architecture to deeper-level FPGA designs in the fields such as artificial intelligence, with the aim of further enhancing research and development efficiency as well as product robustness.

REFERENCES

[1] S. Ricci, S. Caputo and L. Mucchi, "FPGA-Based Visible Light Communications Instrument for Implementation and Testing of Ultralow Latency Applications," in IEEE Transactions on Instrumentation and Measurement, vol. 72, pp. 1-11, 2023.

[2] R. F. Molanes, L. Costas, J. J. Rodríguez-Andina and J. Fariña, "Comparative Analysis of Processor-FPGA Communication Performance in Low-Cost FPSoCs," in IEEE Transactions on Industrial Informatics, vol. 17, no. 6, pp. 3826-3835, June 2021.

[3] Y. Liu et al., "Multiband User Equipment Prototype Hardware Design for 5G Communications in Sub-6-GHz Band," in IEEE Transactions on Microwave Theory and Techniques, vol. 67, no. 7, pp. 2916-2927, July 2019.

[4] K. K. Guner, T. O. Gulum and B. Erkmen, "FPGA-Based Wigner–Hough Transform System for Detection and Parameter Extraction of LPI Radar LFMCW Signals," in IEEE Transactions on Instrumentation and Measurement, vol. 70, pp. 1-15, 2021.

[5] C. J. Cochrane, K. B. Cooper, S. L. Durden, R. Rodriguez Monje and R. J. Dengler, "An FPGA-Based Signal Processor for FMCW Doppler Radar and Spectroscopy," in IEEE Transactions on Geoscience and Remote Sensing, vol. 58, no. 8, pp. 5552-5563, Aug. 2020.

[6] S. Wei, X. Lin, F. Tu, Y. Wang, L. Liu and S. Yin, "Reconfigurability, Why It Matters in AI Tasks Processing: A Survey of Reconfigurable AI Chips," in IEEE Transactions on Circuits and Systems I: Regular Papers, vol. 70, no. 3, pp. 1228-1241, March 2023.

[7] D. T. Nguyen, H. Je, T. N. Nguyen, S. Ryu, K. Lee and H. -J. Lee, "ShortcutFusion: From Tensorflow to FPGA-Based Accelerator With a Reuse-Aware Memory Allocation for Shortcut Data," in IEEE Transactions on Circuits and Systems I: Regular Papers, vol. 69, no. 6, pp. 2477-2489, June 2022.

[8] D. L. T. Wong, Y. Li, D. John, W. K. Ho and C. -H. Heng, "Low Complexity Binarized 2D-CNN Classifier for Wearable Edge AI Devices," in IEEE Transactions on Biomedical Circuits and Systems, vol. 16, no. 5, pp. 822-831, Oct. 2022.

[9] B.Li, "FPGA Theoretical Analysis and Its Advantage Comparison in Artificial Intelligence," 2023 IEEE International Conference on Image Processing and Computer Applications (ICIPCA), Changchun, China, 2023, pp. 1885-1888.

[10] S. b. Suhaili, K. J. anak Kumar, N. Julai, M. H. Husin, M. F. M. Sabri and A. Lit, "Implementation of Verilog HDL in Calculator Design with FPGA Simulation," 2020 13th International UNIMAS Engineering Conference (EnCon), Kota Samarahan, Malaysia, 2020, pp. 1-6.

[11] L.-W. Kim and J. D. Villasenor, "A System-On-Chip Bus Architecture for Thwarting Integrated Circuit Trojan Horses," in IEEE Transactions on Very Large Scale Integration (VLSI) Systems, vol. 19, no. 10, pp. 1921-1926, Oct. 2011.

[12] A.-M. Rahmani, K. R. Vaddina, K. Latif, P. Liljeberg, J. Plosila and H. Tenhunen, "High-Performance and Fault-Tolerant 3D NoC-Bus Hybrid Architecture Using ARB-NET-Based Adaptive Monitoring Platform," in IEEE Transactions on Computers, vol. 63, no. 3, pp. 734-747, March 2014.

[13] Y. Chen et al., "FCUDA-HB: Hierarchical and Scalable Bus Architecture Generation on FPGAs With the FCUDA Flow," in IEEE Transactions on Computer-Aided Design of Integrated Circuits and Systems, vol. 35, no. 12, pp. 2032-2045, 2016.

[14] A. Paunikar, R. Gavankar, N. Umarikar and K. Sivasankaran, "Design and implementation of area efficient, low power AMBA-APB Bridge for SoC," 2014 International Conference on Green Computing Communication and Electrical Engineering (ICGCCEE), Coimbatore, India, 2014, pp. 1-6.

[15] J. Lim, Y. Jeon, E. Ham and J. -H. Kim, "High-Level AMBA Monitoring Platform for SoC Architecture Exploration," 2023 International Conference on Electronics, Information, and Communication (ICEIC), Singapore, 2023, pp. 1-3.

[16] G. Passas, M. Katevenis and D. Pnevmatikatos, "The Combined Input-Output Queued Crossbar Architecture for High-Radix On-Chip Switches," in IEEE Micro, vol. 35, no. 6, pp. 38-47, Nov.-Dec. 2015.

[17] S. W. Fuhrmann, "Performance of a packet switch with crossbar architecture," in IEEE Transactions on Communications, vol. 41, no. 3, pp. 486-491, March 1993.

[18] B. B. Upadhyay and K. Sarawadekar, "A Low Cost FPGA Implementation of Retinex Based Low-Light Image Enhancement Algorithm," in IEEE Transactions on Circuits and Systems II: Express Briefs, vol. 71, no. 7, pp. 3503-3507, July 2024.

[19] P. Wang and J. McAllister, "Streaming Elements for FPGA Signal and Image Processing Accelerators," in IEEE Transactions on Very Large Scale Integration (VLSI) Systems, vol. 24, no. 6, pp. 2262-2274, June 2016.

[20] C. Ye, "Real-time Image Edge Detection System Design and Algorithms for Artificial Intelligence FPGAs," 2022 International Conference on Artificial Intelligence of Things and Crowdsensing (AIoTCs), Nicosia, Cyprus, 2022, pp. 476-481.

[21] M. Hashimoto et al., "33.3 Via-Switch FPGA: 65nm CMOS Implementation and Architecture Extension for AI Applications," 2020 IEEE International Solid-State Circuits Conference - (ISSCC), San Francisco, CA, USA, 2020, pp. 502-504.

Pipelined Parallel Design of SIFT Algorithm on FPGA

Yuanhao Zhang[1], Tianliang Xu[1], Jianfeng Li[1], Zhenbin Lv[1], Shanqiang Yang[1], Chenxu Wang*[1,2],
Yuhang Wang[1], Bo Chu[1], Zhiwei Han[3]

[1] Harbin Institute of Technology, Weihai 264209, China
[2] Key Laboratory of Application Specific IC and System for Ocean Equipments, Weihai, Weihai 264209, China
[3] Shandong Huayi Micro-Electronics Technology Co., Ltd, Jinan, 250001, Shandong, China

* Email: zhangyuanhao_work@163.com, wangchenxu@hit.edu.cn

Abstract—This paper presents an FPGA-accelerated architecture for real-time Scale-Invariant Feature Transform (SIFT) algorithm implementation. To address the high computational complexity of the algorithm, we design a pyramid structure with reduced layers but maintained high feature point generation rate, implement a threshold-based extreme point detection circuit, replace division operations with multiply-add units in descriptor normalization, and realize a fully pipelined circuit architecture. Experimental results on the Xilinx Kintex-7 XC7K325T platform demonstrate that our design achieves a processing speed of 3.19ms/frame for 640×480 images, reaching a 115.4× speedup over CPU implementation while preserving feature matching accuracy.

Keywords—Scale-invariant feature transform, hardware acceleration, field-programmable gate array, feature extraction

I. INTRODUCTION

In today's digital era, image processing technology has become the core foundation of the computer vision field, and feature matching, as one of the key technologies, plays an irreplaceable role in many applications, such as target tracking, image stitching, 3D reconstruction, etc[1]. The SIFT (Scale-Invariant Feature Transform) algorithm[2], as one of the most representative local feature extraction algorithms, was first proposed by Lowe in 1999 and further improved[3] in 2004. The algorithm realizes the robustness against scale, rotation, and illumination of image features. It has become an important milestone in the field of computer vision.

However, traditional processor-based implementations of the SIFT algorithm face severe computational efficiency challenges[4]. Experiments show that it often takes hundreds of milliseconds to seconds to process a 640 × 480 pixel image on a conventional CPU, which severely limits the application of the algorithm in real-time systems[5].

With the continuous improvement of image resolution and the increasing complexity of application scenarios, the real-time requirements for feature matching algorithms are getting higher and higher. FPGA (Field Programmable Gate Array), with its parallel computing capability and dynamic reconfiguration characteristics, provides an ideal solution for the hardware acceleration of the SIFT algorithm[6]. By fully utilizing the logic resources and memory of FPGA, the processing and caching of image data can be realized at the same time, which significantly improves the operation efficiency of the algorithm.

This study aims to explore the efficient implementation of SIFT algorithm on FPGA platform, focusing on the following key issues: first, for Gaussian pyramid, study the pyramid with low number of layers and high feature point generation rate, and implement the architecture based on row buffer and fixed-point coefficients; second, for key point detection, use the threshold method instead of Taylor's expansion fitting in the extreme value comparison operation, and design the edge response rejection module; furthermore, for the feature descriptor generation phase, a fully parallel pipelined directional histogram statistic is designed and descriptor normalization is achieved by multiply-add operation instead of division. Through these optimization methods, a significant speed-up effect is achieved while maintaining the accuracy of the algorithm.

This paper is organized as follows: chapter 2 introduces the feature point detection of the SIFT algorithm; chapter 3 introduces the descriptor design of the SIFT algorithm; chapter 4 gives the experimental results and performance analysis; and finally the research results are summarized.

II. SIFT ALGORITHM FEATURE POINT DETECTION

A. Gaussian pyramid

Gaussian filtering is the most resource-consuming module in the SIFT algorithm[7], and the resources of the Gaussian pyramid are positively correlated with the size and number of Gaussian filtering kernels. Take 4 octaves with 6 scales per octave Gaussian pyramids as an example, 24 filter kernels of different sizes are needed, each octave of pyramids corresponds to a size, and each scale of pyramids corresponds to a fuzzy degree of Gaussian kernel. The closer to the top of the pyramid, the greater the demand for large-size filter kernels, and at the same time, the fewer the number of effective feature points extracted, resulting in high resource consumption. In this paper, we choose 1 octave with 6 scales per octave Gaussian pyramid, discard a small number of feature points in the top layer of the pyramid, and extract feature points only in the scale space where the feature points are dense, as Fig. 1 shown.

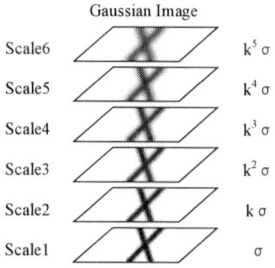

Fig. 1. Schematic of Gaussian pyramid

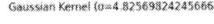 

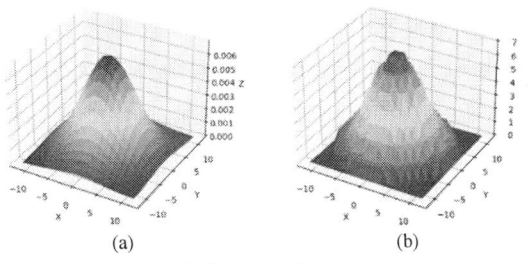

Fig. 2. 3D surface maps of 2D Gaussian kernels

The filter kernel size depends on the maximum blurring of the Gaussian filter. The sigma value of the first Gaussian image layer is calculated as shown in (1) with an initial blur parameter of 1.6 and a camera blur parameter of 0.5.

$$sigma = \sqrt{1.6^2 - 0.5^2} = 1.52 \qquad (1)$$

Each scale of Gaussian image uses a separate Gaussian filter kernel, the sigma values used for each scale are σ, kσ, ..., $k^5\sigma$ (k is taken as $2^{1/3}$), from which it is known that scale 6 is the maximum blurring level, modeling the Gaussian kernel with sigma = $k^5\sigma$ yields Fig. 2(a), the Gaussian kernel coefficients are floating point numbers, which are not conducive to hardware implementation, and they need to be quantified, and the decimals are expressed as 10-bit binary numbers, i.e., the original coefficients are multiplied by 2^{10} and then rounded up to obtain Fig. 2(b), observing the three-dimensional image of the function, it can be seen that the fixed-point Gaussian kernel coefficients vanish(reach zero) at x=±11, so take the size 21*21.

The 21*21 Gaussian kernel implements line buffer through shift registers, which requires 20 shift registers and several registers, and the Gaussian filtering result is calculated by multiplier and accumulator after obtaining the pixel values within the template, as Fig. 3 shown.

B. Extreme point determination

Adjacent Gaussian images are differenced to form a 5-level DoG pyramid, where extrema are detected. The extreme points in the scale space need to be found in the adjacent three layers of the difference image, each pixel point has to be compared with all his neighboring pixel points, if it is bigger or smaller than all the other points, it is considered to be the extreme point in the intermediate layer of the scale, as Fig. 4 shown.

Natural signals are continuous in the physical world, whereas camera-captured images are inherently discrete, and the artificially defined scale space is also discrete, and the extreme points found in the discrete space are not necessarily the true extreme points, but may be the neighboring points of

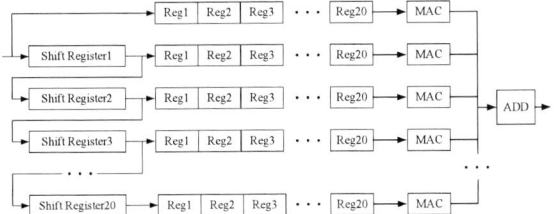

Fig. 3. Filtered kernel line buffer design

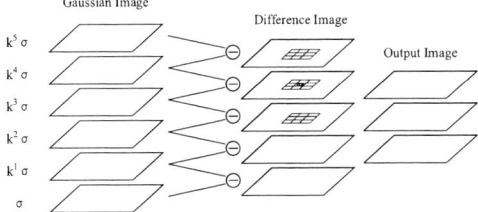

Fig. 4. Detecting extreme points in a difference pyramid

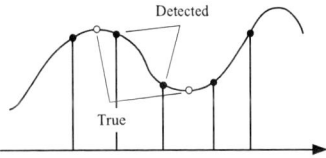

Fig. 5. Subpixel coordinates of the extreme point

the extreme points. The original SIFT algorithm to deal with such a situation for the DoG function curve fitting, the use of Taylor's expansion to find the offset of the extreme point, repeated iterations to find the sub-pixel coordinates of the extreme point.

In order to reduce resource consumption and computational complexity, this part of the circuit design is not carried out in accordance with the original algorithm, and the threshold method is used to locate the extreme point and improve the convergence of the extreme point, which is considered to be the extreme point if the relationship between the center pixel and the neighboring pixels satisfies the following relationship, take T as 50.

$$I_{14} > I_k + T, \forall k \in \{1, 2, ..., 27\}, k \neq 14 \qquad (2)$$

C. Characteristic point determination

The extreme points found in the difference pyramid are candidate feature points, which are not equal to the feature points, and weak response points and edge response points need to be rejected.

The SIFT algorithm will detect some edge points in the extreme point detection stage, which are sensitive to changes in viewing angle and are not favorable for feature matching. Edge response culling is used to filter out unstable edge points and retain stable corner features. The eigenvalue ratio of the Hessian matrix of these edge points is disparate, and if $\lambda_{max}/\lambda_{min} > 10$, the point is considered to be an edge point. The trace and determinant of the Hessian matrix can be used to determine if it is an edge point. The second-order Hessian matrix is defined in (3), the trace and determinant are given by (4) and (5).

$$H(x, y) = \begin{bmatrix} D_{xx}(x, y) & D_{xy}(x, y) \\ D_{xy}(x, y) & D_{yy}(x, y) \end{bmatrix} \qquad (3)$$

$$Tr(H) = D_{xx} + D_{yy} = \lambda_{max} + \lambda_{min} \qquad (4)$$

$$Det(H) = D_{xx}D_{yy} - D_{xy}^2 = \lambda_{max} \cdot \lambda_{min} \qquad (5)$$

Setting $\gamma = \lambda_{max}/\lambda_{min}$, we have (6)。

$$\frac{Tr(H)^2}{Det(H)} = \frac{(\gamma + 1)^2}{\gamma} \qquad (6)$$

When $\gamma \geq 10$, this feature point is rejected, and when $\gamma <$ 10, the feature point is retained, i.e., when (7) holds, the candidate feature point is retained.

$$\frac{Tr(H)^2}{Det(H)} < \frac{(10+1)^2}{10} \qquad (7)$$

III. SIFT ALGORITHM DESCRIPTOR GENERATION

Feature point matching is not possible when only the coordinates of the feature points are known, it is also necessary to describe the information around the feature points with a vector, called a descriptor. Matching is done by the descriptors of the two images corresponding to the feature points.

A. CORDIC Algorithm

To generate the SIFT descriptor vectors for describing the information around the feature point, the gradient direction and magnitude of the pixels in the 16*16 neighborhood around the feature point need to be obtained. The Coordinate Rotation Digital Computer (CORDIC) method is used to find the square root and the arctangent function of the horizontal gradient component and the vertical gradient component, as Fig. 6 shown.

The essence of the CORDIC algorithm is to know the coordinates of a point and find the coordinates after rotating it by a specific angle, the relationship between the initial vector (x_1,y_1) and the rotated vector (x_2,y_2) is shown in (8) and (9).

$$x_2 = R\cos(\alpha+\theta) = x_1\cos\theta - y_1\sin\theta \qquad (8)$$

$$y_2 = R\sin(\alpha+\theta) = x_1\sin\theta + y_1\cos\theta \qquad (9)$$

Expressed in terms of a matrix and reduced to (10).

$$\begin{bmatrix} x_2 \\ y_2 \end{bmatrix} = \cos\theta \begin{bmatrix} 1 & -\tan\theta \\ \tan\theta & 1 \end{bmatrix} \begin{bmatrix} x_1 \\ y_1 \end{bmatrix} \qquad (10)$$

Under the condition that θ is fixed, $\cos\theta$ is a constant, and if $\tan\theta=2^{-n}$ is specified, the matrix operation of the above equation can be realized by shifting and addition, and each matrix operation is equivalent to one rotation, and $\theta=45°$ for the first rotation, if the cumulative value of the rotation at this time is less than the expected angle of rotation, then the forward rotation is carried out for the second rotation; if the cumulative value of the rotation is greater than the expected angle of rotation, the then the second rotation is carried out in the opposite direction. Following this pattern, each rotation of $\tan\theta$ is 1/2 of the previous one, after 10 iterations, the residual rotation error converges to 0.11°, meeting the precision requirement for gradient orientation calculation.

The obtained pixel gradient direction is quantized into 8 intervals, one for each 45°. A 16*16 storage space is cons-

tructed to cache the pixel gradient direction and gradient magnitude of the 16*16 neighborhood around the current pixel, and the neighborhood is again partitioned into 16 large squares of 4*4, with 16 small squares within each large square, and the cumulative values of gradient magnitude for each direction within each large square are counted, as Fig. 7 shown.The magnitudes of the 16 squares in 8 directions form a 128-dimensional descriptor of the center pixel.

B. Descriptor normalization

The lighting conditions of images in practical applications often change (e.g., indoor and outdoor light differences, different exposures, etc.), and stable feature matching must overcome these changes, in order to eliminate the effect of brightness on the descriptors and achieve the light invariance of the SIFT algorithm, the descriptors need to be normalized.

Let N_i be the descriptor before normalization and be the M_i normalized descriptor, $i=1,...,128$, and normalize the original descriptor to an 8bit vector, the normalization formula is given in (11).

$$M_i = \frac{255 * N_i}{\sum_i N_i}, \quad i=1,2,...,128 \qquad (11)$$

If we use conventional method to compute the 128-dimensional descriptors one by one, 128 dividers are required, which consumes a large amount of DSP resources and is not suitable for FPGA implementation. Equation(12) splits the above equation into division and bit-shifting, only one divider is used in (13), and the rest of the operations are realized by shifting and multiplying, and (14) shows how the variable n is calculated.

$$M_i = DIV * \frac{N_i}{2^n}, \quad i=1,2,...,128 \qquad (12)$$

$$DIV = \frac{255 * 2^n}{\sum_i N_i}, \quad i=1,2,...,128 \qquad (13)$$

$$2^n \leq \sum_i N_i < 2^{n+1}, i=1,2,...,128 \qquad (14)$$

IV. EXPERIMENTAL RESULTS

The experimental platform in this paper is Xilinx Kintex-7 XC7K325T, the SIFT algorithm is deployed on FPGA to

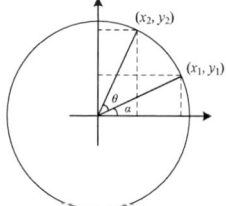

Fig. 6. Coordinate rotation schematic

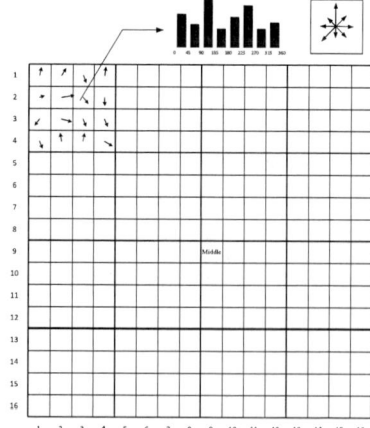

Fig. 7. 128-dimensional descriptor construction maps

Fig. 8   Schematic of the matching results

TABLE I.   HARDWARE RESOURCES CONSUMED BY THE PROPOSED METHOD

Resource	Utilization	Percentage
LUT	50235	24.65
FF	51791	12.71
BRAM	40	8.99
DSP	132	15.71

TABLE II.   640×480 IMAGE PROCESSING TIME COMPARISON

Platform	Frequency	Processing Time	Acceleration Ratio
CPU[8] Dual-Core E5300	2.6Ghz	368ms	×1
GPU[8] GeForce 9600	650Mhz	48ms	×7.7
FPGA Proposed	100Mhz	3.19ms	×115.4

detect and match the features jointly with the images captured by the binocular camera, and the matching results are shown in Fig. 8. The time required for the SIFT algorithm to detect the feature points of a frame of 640×480 image and generate descriptors is 3.19 ms at 100Mhz clock frequency. The logic resources consumed by the SIFT algorithm deployed on FPGA are shown in TABLE I.

Comparing the time taken by CPU, GPU and this design to process one frame of 640×480 image, it can be seen that this design achieves a speedup ratio of 115.4 compared to CPU and 7.7 compared to GPU as shown in TABLE II.

V. CONCLUSION

This paper successfully implements the SIFT feature extraction algorithm on a Xilinx Kintex-7 XC7K325T FPGA platform. Through parallel computing architecture and pipeline optimization, the design achieves significant improvements in real-time performance. Experimental results demonstrate that compared to a general-purpose CPU, the FPGA implementation accelerates computation by approximately 115.4× while maintaining robust feature matching accuracy.

To address the resource-constrained problem, this paper adopts a low-level Gaussian pyramid to reduce the BRAM occupancy, and a thresholding method for extreme point detection and optimized descriptor normalization to reduce the use of DSP. Future work will explore applying the SIFT algorithm to image stitching, and investigating hardware implementation methods for feature matching.

ACKNOWLEDGMENT

This work was mainly supported by Major scientific and technological innovation projects of Shandong Province of China, with Grant No.2022ZLGX04. The research presented in this paper is also partially supported by the NSF project of China with granted No.U2106202 and Shandong Provincial Natural Science Foundation with Grant ZR2023MA074.

REFERENCES

[1]   S. Xu, S. Chen, R. Xu, C. Wang, P. Lu, and L. Guo, "Local feature matching using deep learning: A survey," *Inf. Fusion*, vol. 107, p. 102344, Jul. 2024.

[2]   D. G. Lowe, "Object recognition from local scale-invariant features," in *Proceedings of the Seventh IEEE International Conference on Computer Vision*, Sep. 1999, pp. 1150–1157 vol.2.

[3]   D. G. Lowe, "Distinctive Image Features from Scale-Invariant Keypoints," *Int. J. Comput. Vis.*, vol. 60, no. 2, pp. 91–110, Nov. 2004.

[4]   L.-C. Chiu, T.-S. Chang, J.-Y. Chen, and N. Y.-C. Chang, "Fast SIFT Design for Real-Time Visual Feature Extraction," *IEEE Trans. Image Process.*, vol. 22, no. 8, pp. 3158–3167, Aug. 2013.

[5]   Q. Zhang, Y. Chen, Y. Zhang, and Y. Xu, "SIFT implementation and optimization for multi-core systems," in *2008 IEEE International Symposium on Parallel and Distributed Processing*, Apr. 2008, pp. 1–8.

[6]   Z. WeiLong, L. LeiBo, Y. ShouYi, Z. RenYan, C. ShanShan, and W. ShaoJun, "An efficient VLSI architecture of speeded-up robust feature extraction for high resolution and high frame rate video," *Sci. ChinaInformation Sci.*, no. 7, pp. 136–149.

[7]   A. Fejér, Z. Nagy, J. Benois-Pineau, P. Szolgay, A. de Rugy, and J.-P. Domenger, "Implementation of Scale Invariant Feature Transform detector on FPGA for low-power wearable devices for prostheses control," *Int. J. Circuit Theory Appl.*, vol. 49, no. 7, pp. 2255–2273, 2021.

[8]   C. Jiang, Z. Geng, X. Wei, and C. Shen, "SIFT implementation based on GPU," in *International Symposium on Photoelectronic Detection and Imaging 2013: Optical Storage and Display Technology*, SPIE, Aug. 2013, pp. 12–18.

Design and Implementation of an FPGA-based MIPI DSI Interface for Micro-LED Displays

Runfeng Yao[1], Xinyi Liu[1], Kaisong Zhu[2], Jinbo Liang[2], Zhaojun Liu[1,2,*]

[1]*Department of Electrical and Electronic Engineering, Southern University of Science and Technology, Shenzhen, China*
[2]*Shenzhen Sitan Technology Limited, Shenzhen, China*
*E-mail: liuzj@sustech.edu.cn

Abstract—This paper presents the design and implementation of an FPGA-based MIPI DSI interface optimized for driving Micro-LED displays. The proposed design addresses key limitations of the Xilinx MIPI DSI Subsystem, particularly the lack of support for Low-Power (Escape Mode) commands, non-continuous clock modes, and limited flexibility in lane configuration. The FPGA-based solution supports dynamic lane speed adjustment and offers flexible lane configurations (1, 2, or 4 lanes), ensuring High-Speed data transmission compatible with various Micro-LED display systems. Key innovations include robust clock management and scalability, enabling efficient operation in both Low-Power and High-Speed modes. Simulation and hardware testing confirm that the design achieves lane rates of up to 1.5 Gbps, ensuring reliable data transfer for high-resolution displays. The results demonstrate significant improvements in performance, flexibility, and synchronization compared to existing solutions.

Index Terms—MIPI DSI, FPGA-based design, Micro-LED displays, dynamic lane speed adjustment, flexible lane configuration

I. INTRODUCTION

MIPI DSI (Display Serial Interface) has emerged as a critical standard for High-Speed data transmission in modern display systems [7], particularly for Micro-LED technology. Micro-LED is an advanced display technology known for its superior resolution, brightness, and energy efficiency [8]. It requires High-Speed data transmission and low power consumption [5], which presents significant challenges for existing MIPI DSI solutions. While MIPI DSI is well-suited for delivering high-resolution displays with low power consumption, the existing Xilinx MIPI DSI Subsystem exhibits several limitations that hinder its performance and flexibility, particularly in the context of advanced applications such as Micro-LED displays.

Specifically, the current Xilinx MIPI DSI Subsystem faces the following challenges [4]:

1) It does not support the necessary commands for properly initializing Micro-LED displays during startup.
2) It does not support Low-power (LP) mode data transmission (Escape Mode), limiting its functionality in Low-Power scenarios.
3) It lacks support for non-continuous clock modes, which are critical for certain display applications.
4) It cannot handle flexible lane configurations, which are essential for adapting to varying display system requirements.

To address these challenges, we propose an FPGA-based MIPI DSI interface that overcomes these limitations and is optimized for driving Micro-LED displays. Our design supports flexible lane configurations, including options for 1, 2, or 4 lanes, enabling it to adapt to various display system requirements. A key feature of the design is its ability to dynamically adjust the speed of the lane, providing improved flexibility and performance compared to existing solutions. This capability ensures seamless data transmission in both Low-Power and High-Speed modes.

Additionally, the design resolves clocking challenges by supporting both continuous and non-continuous clock modes. This flexibility is essential for Micro-LED displays, which require adaptable power consumption and data transfer speeds. Robust clock management ensures reliable data synchronization, even with varying resolution configurations.

Our FPGA-based interface also addresses synchronization issues between the master and slave devices, ensuring real-time transmission of accurate image data to the Micro-LED displays. This enhancement results in smooth operation without clock drift.

Through extensive simulation and hardware testing, we demonstrate that our design achieves a maximum lane rate of 1.5 Gbps, ensuring High-Speed and reliable data transfer. These results confirm that the proposed FPGA-based MIPI DSI interface significantly improves upon the Xilinx MIPI DSI Subsystem in terms of protocol support, lane configuration flexibility, and clock management.

This paper provides an effective and scalable solution for FPGA-based display systems, offering improved flexibility and reliability for power-efficient, high-resolution display applications.

II. SYSTEM OVERVIEW

The system is designed with a clear Master-Slave architecture, where the Master manages the data flow, and the Slave (Micro-LED display) receives and processes the data for display.

As shown in Fig. 1. , the Master handles the image or video data sourced from SD and DDR memory. The initialization commands are stored in the Config ROM and passed to the Initialization Commands module for proper system setup. The image data is buffered in the Dual-Clock FIFO, which synchronizes the clocks needed for image processing and transmission. After buffering, the data are converted into RGB888 format using the RGB888 conversion module for MIPI DSI transmission.

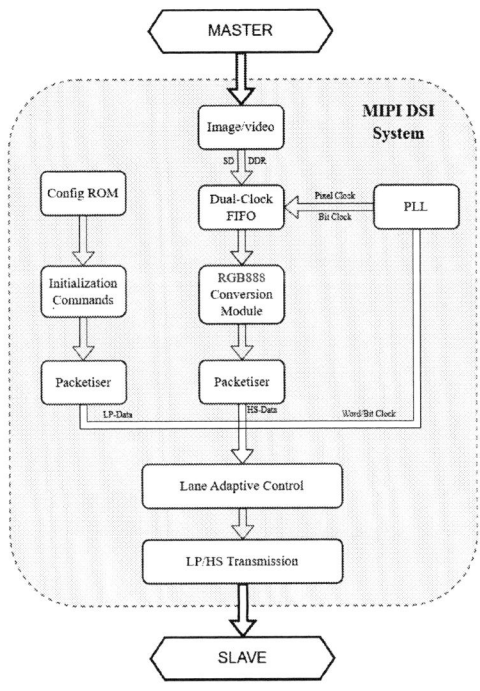

Fig. 1. System framework diagram.

The data is then processed by two packetisers: one handles LP data, and the other handles HS (High-Speed) data, ensuring compatibility with the MIPI DSI protocol. The Lane Adaptive Control module adjusts the transmission based on the lane count and manages transitions between LP and HS modes to optimize power efficiency and data transfer speed.

Once processed, the data are sent through the LP/HS Transmission module, which controls switching between Low-Power and High-Speed modes as needed. Finally, the Slave module, the Micro-LED display, receives and displays the data. The system's timing is managed by a PLL, which generates the necessary Pixel and Bit clocks for synchronization.

This architecture ensures efficient data flow from the Master to the Micro-LED display, supporting flexible and high-performance display operation.

III. DESIGN AND IMPLEMENTATION

A. Initialization Module

One of the key innovations of this design is the inclusion of an initialization module that supports the Low-Power (Escape Mode) command. The Initialization Module is designed to automatically initialize the Micro-LED display upon power-up by sending specific commands stored in the Config ROM. These commands include essential instructions like the **DCS Long Write** command (with a Data Type of 0x39) that prepares the display for proper operation. After powering on, the system retrieves and transmits these commands to the

display, ensuring it is configured correctly before it starts displaying image data.

B. HS/LP Mode Conversion Module

The HS/LP Mode Conversion Module addresses the limitation of the existing system, which does not support LP mode data transmission (Escape Mode). It effectively manages the transition between Low-Power (LP) and High-Speed (HS) modes in the MIPI DSI protocol. By switching between Low-Power operation and High-Speed data transmission, this module ensures efficient power usage and high data throughput, which is critical for applications like Micro-LED displays that require both power efficiency and high performance.

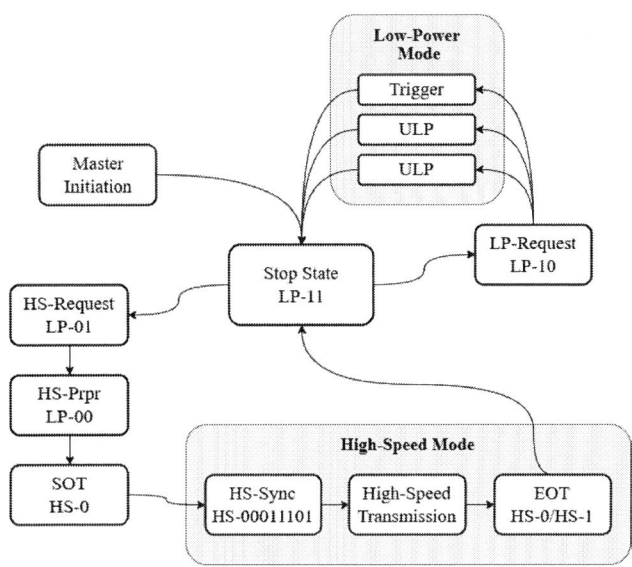

Fig. 2. HS/LP mode conversion state machine.

As illustrated in Fig. 2, the process begins with Master Initialization. Afterward, the system moves into Low-Power Mode, entering the Stop State (LP-11) to cease active transmission. An LP-Request (LP-10) signals the transition, putting the system into a reduced power state [2], [3]. Within LP mode, the system can enter two sub-states: Trigger and Ultra-Low Power (ULP), optimizing power consumption until High-Speed data transmission is needed again [1]. This transition ensures the system efficiently switches between Low-Power and High-Speed modes as required by the application.

Next, in High-Speed Mode (HS), the system receives the HS-Request (LP-01) and HS-Prpr (LP-00) signals to transition into High-Speed data transmission. The system then sends the Start of Transmission (SOT) signal (HS-0) to initiate data transfer. Synchronization is achieved through the HS-Sync phase, and once synchronization is complete, High-Speed data transmission begins. Finally, upon completion of data transmission, the system sends the End of Transmission (EOT) signal (HS-0/HS-1), marking the end of the High-Speed data transfer.

C. MIPI DSI Transmitter Module

The MIPI DSI transmitter module is responsible for converting parallel data into the MIPI DSI format, supporting both Low-Power (LP) and High-Speed (HS) data transmission modes. In LP mode, it optimizes power consumption by using asynchronous data transmission with a large swing signal (e.g., 1.2 V) for efficient communication. In HS mode, it ensures high-speed data transfer with synchronous NRZ signals based on scalable low voltage differential signaling (SLVS), utilizing a common-mode voltage of 0.2 V. This enables efficient and reliable communication in Micro-LED display applications [6].

1) Low-Power Mode Transmission: Low-Power (LP) Mode reduces power consumption by minimizing data transmission, using timing derived from the Dp and Dn lines via XOR. Escape Mode, a form of LP Mode, allows asynchronous communication, as shown in Fig. 3 [3]. The transition starts with LP-11, progresses through LP-10 and LP-00, and enters Escape Mode in the Space state. If LP-11 is detected early, the procedure is aborted and returns to the Stop state.

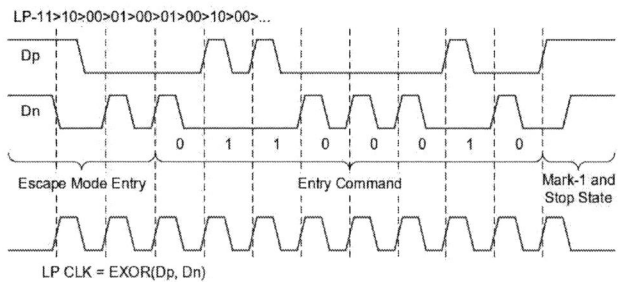

Fig. 3. Trigger-Reset command in escape mode.

2) High-Speed Mode Transmission: In High-Speed (HS) mode, data is transmitted in bursts. The system transitions from LP-11 to LP-01 to LP-00 before entering HS mode [2], [3]. Data transmission starts after a synchronization period, and the system allows minor errors in the start sequence. Each burst is preceded by a leader and followed by a trailer sequence for synchronization. After the burst, the system enters the End-of-Transmission (EoT) procedure and returns to Stop state, either initiating a new transmission or going back to LP mode. This ensures High-Speed data flow, as defined by the MIPI DSI protocol.

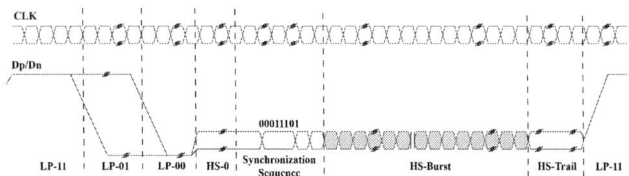

Fig. 4. High-Speed data transmission in bursts.

D. Clock Management Module

This module directly addresses the limitation in the existing system that lacks support for non-continuous clock modes, a key requirement for certain display applications. The Clock Management Module generates and controls the clocks needed for MIPI DSI transmission, ensuring that the system can operate efficiently under both continuous and non-continuous clock modes.

In Low-Power mode, the system disables the clock and transmits data using Spaced-One-Hot encoding [3]. This approach allows data to be transmitted asynchronously without relying on a continuous clock signal. as shown in Fig. 3. The Slave device, which receives the data, can extract the clock signal from the transmitted data through an XOR operation between the Dp and Dn lines. This method enables reliable data transmission in LP mode while minimizing power consumption.

When transitioning to High-Speed (HS) mode, the system enables the clock, providing the necessary synchronization for High-Speed data transfer. The Clock Management Module dynamically adjusts the clock frequency based on the current data transmission mode, ensuring that both LP and HS modes operate effectively. This flexibility is essential for display systems such as Micro-LED, where both power efficiency and High-Speed data transmission must be balanced. The diagram in Fig. 5 illustrates the clock characteristics in both LP and HS modes. It demonstrates the power-efficient clock management during LP mode and the High-Speed data transfer during HS mode, showcasing the adaptability of the Clock Management Module across varying display requirements.

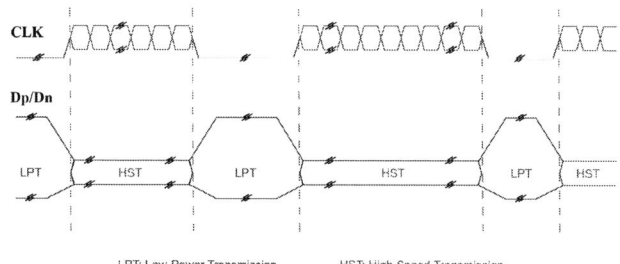

Fig. 5. Clock level characteristics in LP and HS modes.

E. Flexible Lane Configuration Adapter Module

The Flexible Lane Configuration Adapter Module solves the challenge of handling variable lane configurations, a critical requirement for adapting to diverse display system needs. It ensures adaptability for a wide range of display systems.

The module operates by splitting the 32-bit input data into two 8-bit segments, each assigned to a separate lane. It also manages the control signals, such as Enable and End-of-Transmission (EoT), to ensure proper synchronization of data across the two lanes. The system detects the start and end of the data packet, maintaining smooth transmission and correct packet handling.

This solution provides the flexibility to switch between different lane configurations (1, 2, or 4 lanes), making it highly adaptable to various display requirements. It supports high-resolution displays, offering an efficient method for adapting lane configurations according to specific needs.

IV. EXPERIMENTAL RESULT

A. Simulation Results

The simulation, conducted using Vivado, verifies the system's data flow from the host to the Micro-LED display, as shown in Fig. 6. Initially, image data is stored in the FIFO based on VGA timing. Once the system enters the valid display area, the host writes the data into the FIFO. Upon transitioning to High-Speed (HS) mode, the data is read, packed into long packets, and transmitted bit-by-bit through two data lanes to the Micro-LED display, where the grid image appears. This simulation confirms the correct formatting and transmission of data in both LP and HS modes.

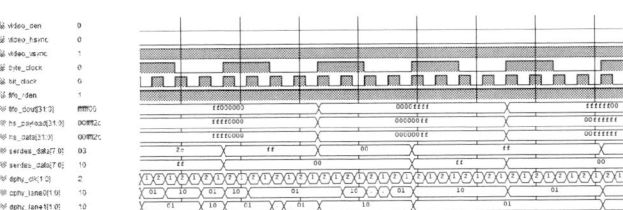

Fig. 6. Simulation waveform of data transmission to Micro-LED display.

B. Hardware Verification

To verify the hardware implementation, we connected the ARTIX UltraScale+ FPGA (Model: AXAU10) to the Micro-LED display and observed the results under a microscope. The system successfully transmitted the image data to the Micro-LED, which displayed the grid pattern as expected. As shown in Fig. 7, the FPGA was interfaced with the Micro-LED display, and the grid pattern was visible on the display.

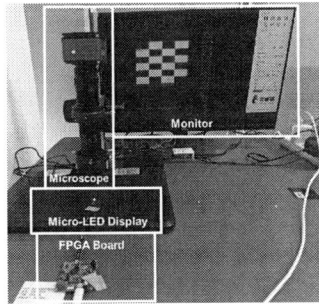

Fig. 7. FPGA and Micro-LED under the microscope.

The microscope image in Fig. 8 further confirms the accurate display of the grid pattern on the Micro-LED. This image demonstrates that the FPGA correctly processes and sends the data, which is displayed clearly and without distortion on the Micro-LED screen. This hardware verification ensures the system's functionality and reliability in real-world conditions.

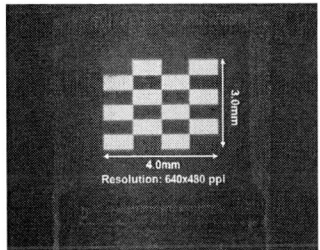

Fig. 8. MIPI DSI interface drives the Micro-LED display image.

V. CONCLUSION

In this paper, we proposed an FPGA-based MIPI DSI interface designed specifically for driving Micro-LED displays. The design addresses key limitations in the existing Xilinx MIPI DSI Subsystem, including the lack of support for Low-Power (Escape Mode) commands, non-continuous clock modes, and flexible lane configurations. The system offers dynamic lane speed adjustment and robust clock management, enabling seamless data transmission in both Low-Power and High-Speed modes. Experimental results show that the system can achieve a maximum lane rate of 1.5 Gbps, ensuring High-Speed and reliable data transfer to Micro-LED displays. This FPGA-based design provides a flexible and scalable solution, improving performance and adaptability for high-resolution, power-efficient display applications.

ACKNOWLEDGMENT

This work is supported by the National Key RD Program of China under Grant No. 2023YFB2806800, Fundamental and Applied Fundamental Research Fund of Guangdong Province (No.2021B1515130001), and Shenzhen Science and Technology Program (No.JCYJ20220818100603007).

REFERENCES

[1] Ahmed Mohamed Ali, Ahmed Shalaby, Sherif Saif, and Mohamed Taher. A uvm-based verification approach for mipi dsi low-level protocol layer. In *2022 International Conference on Microelectronics (ICM)*, pages 74–77, 2022.

[2] MIPI Alliance. Mipi dsi-2 specification (display serial interface 2) specification version 1.1. 2018.

[3] MIPI Alliance. Mipi d-phy specification specification version 2.1. 2020.

[4] MIPI Alliance. Mipi dsi transmitter subsystem v2.1 product guide. 2020.

[5] Yufeng Chen, Xifeng Zheng, Hui Cao, Yang Wang, Hongbin Cheng, Junchang Chen, Shuo Huang, Jingxu Li, Deju Huang, and Yu Chen. High precision control system for micro-led displays. *Applied Sciences*, 13(19), 2023.

[6] Doo-Hwan Kim, Beom-Dae Kim, and Kyoungrok Cho. Design of d-phy chip for mobile display interface supporting mipi standard. *Microelectronics Journal*, 43(12):949–955, 2012.

[7] Pil-Ho Lee, Han-Yeol Lee, Yeong-Woong Kim, Han-Young Hong, and Young-Chan Jang. A 10-gbps receiver bridge chip with deserializer for fpga-based frame grabber supporting mipi csi-2. *IEEE Transactions on Consumer Electronics*, 63(3):209–215, 2017.

[8] Yang-En Wu, Chia-Hung Tsai, Li-Yin Chen, Fang-Chung Chen, and Hao-Chung Kuo. Current landscape of micro-led display industrialization. *Nanomaterials*, 15(9), 2025.

979-8-3315-3918-4/25 $31.00 © 2025 IEEE

A Sub-100μs-Latency Visual-Cortex-Mimicking Heterogeneous Multi-Core Edge Neuromorphic Processor Enabling On-Chip High-Accuracy Learning

Junxian He [1], Ying Jiang [1], Zhengqing Zhong [1], Mingju Chen [2], Liyuan Liu [3], Cong Shi*[1]

[1] School of Microelectronics and Communication Engineering, Chongqing University, Chongqing 400044, China
[2] Artificial Intelligence Key Laboratory of Sichuan Province, Sichuan University of Science and Engineering, Yibin 643000, Sichuan, China
[3] State Key Laboratory for Superlattices and Microstructures, Institute of Semiconductors, Chinese Academy of Sciences, Beijing 100083, China

* Email: junxian_he@cqu.edu.cn, shicong@cqu.edu.cn

Abstract—**While integrating the entire visual pipeline, from image pre-processing to final decision, onto a single neuromorphic chip is critical for edge applications, solutions that achieve this often suffer from high processing latency. To address this, this paper presents a heterogeneous multi-core neuromorphic processor that integrates the entire visual processing pipeline, from image pre-processing, spike encoding to final classification, on a single chip. The processor consists of three distinct, parallel-operating cores, each specialized for a different stage of the end-to-end vision pipeline, significantly reducing processing latency. It also embeds on-chip learning circuits to adapt to dynamic edge environments. Prototyped on a very-low-cost Zynq-7010 FPGA, the processor exhibits an ultra-low processing latency of 99 μs (i.e., 10081 fps) on the MNIST dataset, covering the entire pipeline from image pre-processing to final decision. This is achieved while maintaining a high on-chip learning accuracy of 97.02% and a low power consumption of 161 mW.**

Keywords—Neuromorphic processor, spiking neural network, on-chip learning, low-latency.

I. INTRODUCTION

In recent years, edge systems that require low-latency and high-accuracy visual recognition and classification, such as embodied intelligence platforms, drones, and intelligent assembly lines, have developed rapidly. However, processors based on conventional artificial neural networks (ANNs) models are limited by high cost and power consumption, making them difficult to deploy in resource-constrained edge systems [1]. In contrast, neuromorphic processors that run spiking neural networks (SNNs), which are capable of processing spatiotemporally sparse data [2], exhibit low computational overhead and high energy efficiency. These features make them well-suited for edge deployment.

Some neuromorphic processors have been implemented as ASICs [3], [4], while many use FPGAs for rapid prototyping [5]–[8]. However, these neuromorphic processors are unable to perform full visual processing on-chip, including image preprocessing, spike encoding, feature extraction, spike decoding, and decision making. They rely on external devices, such as PCs, to perform image preprocessing, encoding, decoding, and classification, and are therefore unable to complete visual tasks entirely on-chip in edge systems. In our

previous work [9], we introduced a neuromorphic processor capable of full on-chip vision processing. However, its single-core architecture, which relies on time-multiplexing to process model layers serially, resulted in significant latency. As a result, the latency reaches 1436 μs (i.e., 696 fps), which is insufficient for edge applications such as drones or autonomous driving systems that require high-speed visual data processing.

To address these challenges, this paper proposes a heterogeneous multi-core neuromorphic vision processor that supports complete on-chip visual processing. Each core is designed and optimized according to the size and function of the network layer it processes, and all cores operate in parallel, which greatly improves processing speed. The proposed neuromorphic model from [9] is further optimized by reducing network size, thereby lowering latency. The processor also includes on-chip learning circuits to support high-accuracy learning and adaptation to the dynamic environments typical of edge systems. Demonstrating high performance-per-cost, the prototype, implemented on a very-low-cost Xilinx Zynq-7010 device, achieves an end-to-end latency of just 99 μs on MNIST (a 14× speedup over [9]), while maintaining a competitive 97.02% on-chip learning accuracy and consuming only 161 mW.

The rest of this paper is organized as follows. Section II presents the neuromorphic model that includes image preprocessing, spike encoding, and spike-based feature extraction and classification. Section III proposes the neuromorphic vision processor architecture composed of multiple parallel heterogeneous cores and describes the circuit design of each core. The FPGA prototype of neuromorphic processor and experimental results are detailed in Section IV. Section V concludes the paper.

II. ALGORITHM MODEL

A. Neuromorphic Model

As shown in Fig. 1, the proposed neuromorphic model supports an end-to-end vision-processing pipeline, from image preprocessing and spike encoding through spike-based feature extraction to final classification. Compared to [9], the key optimization of this model is the adoption of a more compact architecture. This was achieved by removing an

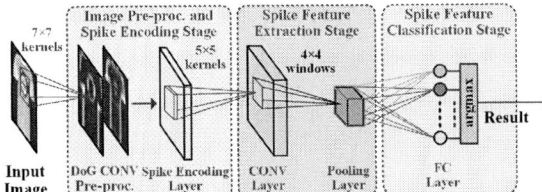

Fig. 1. The proposed neuromorphic model.

entire spiking convolutional (CONV) layer and its corresponding spike max-pooling layer from the spike feature extraction stage. Crucially, this structural simplification achieves a substantial reduction in latency while maintaining a high recognition accuracy. The model's data processing pipeline begins with the image preprocessing and spike-encoding stage, where input image channels are convolved with Difference-of-Gaussian (DoG) filters to enhance spatial contrast. A subsequent spike encoding layer then converts each DoG-filtered pixel into a single spike whose firing time is inversely proportional to the pixel's brightness, with the detailed mechanism described in [9]. Next, the spike feature extraction stage uses a CONV layer to extract features. This layer employs integrate-and-fire (IF) neurons that fire at most once, with their membrane potential $U_j[t]$ and spike output $s_j[t]$ given by:

$$U_j[t] = U_j[t-1] + \sum_i w_{ji} s_i[t] \quad (1)$$

$$s_j[t] = H(U_j[t] - \theta) \quad (2)$$

where w_{ji} is the synaptic weight of the i-th synapse of neuron j, $s_i[t]$ is the input spike at time t, $H(x)$ is the Heaviside step function (i.e., $H(x) = 1$ if $x \geq 0$, or 0 otherwise), and θ is the firing threshold. A subsequent spike max-pooling layer then reduces the feature-map size by passing only the earliest spike within each pooling window.

Finally, the spike feature classification stage contains a fully-connected (FC) layer of integrate-not-fire (InF) neurons. Their membrane potentials are updated according to (1), but they do not generate spikes. After the final time step T, the recognition result is given by $argmax(U_j[T])$.

B. On-chip Learning Method

The network is trained on-chip using a layer-wise scheme. First, the CONV layer within the spike feature extraction stage is trained using the spike-timing dependent plasticity (STDP) rule from [9]. At each time step t, the neuron that fires with the highest membrane potential is selected as the winner. Its synaptic weights are potentiated if a presynaptic spike was received, or depressed otherwise. Lateral inhibition is then applied to the winner's channel and spatial neighborhood to encourage feature diversity [9].

Once the CONV layer training is complete and its weights are frozen, the FC layer in the spike feature classification stage is then trained using the target-driven STDP (TD-STDP) learning rule [9]. Each FC layer neuron is pre-assigned to a specific object category. After all time steps for a sample are processed, synaptic weights connected to the neuron with the highest membrane potential (the peak neuron) are depressed, while weights connected to the neuron corresponding to the sample's label (the target neuron) are potentiated. These updates are only applied to synapses that have received presynaptic spikes.

III. HARDWARE DESIGN

A. Processor Architecture

The proposed heterogeneous multi-core processor architecture is illustrated in Fig. 2. It comprises three distinct, interconnected cores, each dedicated to a specific stage of the model from Sec. II.A. The first core, the image pre-processing and spike encoding (SE) core, consists of a DoG pre-processing block, an array of N SE units (SEUs), and an output address-event representation (AER) scheduler. The DoG block generates ON/OFF-center feature maps from input pixels, which are then encoded into AER-formatted spikes in parallel by the SEU array. The output AER scheduler routes these spikes to the next core. The spike feature extraction (SFE) core contains an Input AER FIFO, an array of M SFE units (SFEUs), a CONV layer learning engine, and an output AER scheduler. The SFEU array processes incoming spikes to update neuronal membrane potentials according to (1), with its assigned CONV layer channels distributed among the M units for parallel acceleration. The core's subsequent action at the end of each time step t depends on the operational phase. During inference, the SFEU array performs pooling and outputs spikes via the scheduler. During learning, spike generation is suppressed; instead, the CONV layer learning engine receives neuron state information from the SFEU array to perform winner competition and execute the on-chip STDP rule. The final core, the spike feature classification (SFC) core, comprises an input AER FIFO, an array of P SFC units (SFCUs), an FC layer learning engine, and a decision-making block. The SFCU array integrates incoming spikes across all the assigned neurons. After the final time step T, the decision-making block searches the maximum potentials from the SFCU array to identify the peak neuron's index. This index is used as the classification result during inference. During learning, the FC layer learning engine uses this index along with the sample's ground-truth label to perform the TD-STDP update.

The processor leverages a dual level of parallelism: the three cores operate in parallel, and within each core, all processing units also work in parallel, significantly boosting performance. Furthermore, to conserve power, the cores are event-driven, activating only when input data is present.

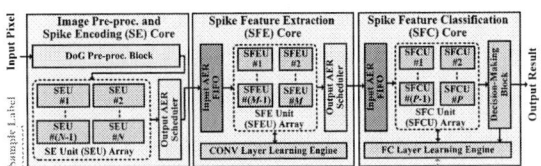

Fig. 2. The proposed heterogeneous multi-core architecture.

B. Key Circuit Design

The processor's heterogeneous architecture requires specialized circuits within each core. These key circuits will now be detailed.

1) Key Circuits of the Image Pre-processing and Spike Encoding Core

a) DoG Pre-Processing Block: The circuit diagram of the DoG pre-processing block is shown in Fig. 3. This block implements a 2D separable convolution strategy for efficient DoG computation. It primarily consists of a snake buffer, which caches the necessary pixel rows for the 7-tap filters, and

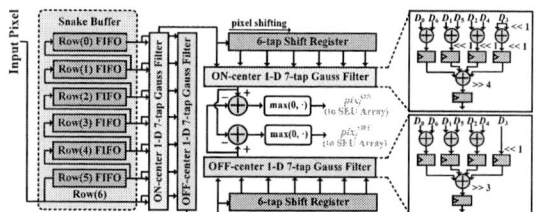

Fig. 3. Circuit diagram of the DoG pre-processing block.

the core filtering circuitry. The core filtering circuitry consists of two parallel pipelines, each performing a 2D separable convolution using two 1-D 7-tap filters to maximize throughput. One pipeline serially applies two ON-center 1-D 7-tap gaussian filter modules to generate the ON-center map, while the other concurrently uses two OFF-center 1-D 7-tap gaussian filter modules for the OFF-center map. For hardware efficiency, the filter coefficients are powers-of-two, enabling a multiplication-free implementation with simple shift-and-add logic. The resulting ON- and OFF-center DoG maps are then forwarded to the SEU array.

b) SEU: Fig. 4 illustrates the circuits for the SEU circuit. It is responsible for encoding DoG pixel values into output spikes. The encoding process begins upon the arrival of a DoG pixel. A *sel* signal selects either the ON-center or OFF-center pixel value to initialize the membrane potential (U_j) of the neuron. At the first time step ($t = 1$), this potential is scaled by a configurable factor, k, an operation implemented efficiently with right bit-shifters instead of dividers. For all subsequent time steps ($1 < t \leq T$), U_j is incremented by one. The SEU emits a spike when its potential U_j exceeds a threshold, T.

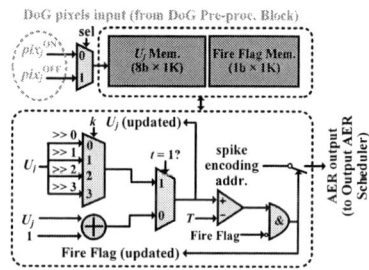

Fig. 4. The SEU circuit diagram.

2) Key Circuits of the Spike Feature Extraction Core

a) SFEU: The SFEU circuit, shown in Fig. 5, consists of a U_j-update block and an output generation block. The former integrates incoming AER spikes by updating U_j according to (1). The latter's function, executed at the end of each time step t, is phase-dependent. During inference, it implements a spike max-pooling. It outputs a pooled spike for the first neuron in a pooling window to exceed the threshold (θ) and then sets a pool flag to gate all subsequent spikes from that same window. During learning, it instead suppresses spike output, enforces a fire-at-most-once policy, and forwards the potential and address of any threshold-exceeding

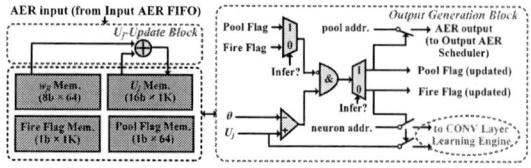

Fig. 5. Circuit diagram of the SFEU.

neuron to the CONV layer learning engine for winner selection and the STDP weight update.

b) CONV Layer Learning Engine: The CONV layer learning engine, illustrated in Fig. 6, orchestrates the on-chip STDP learning for the CONV layer. During the learning phase, it receives candidate U_j and their addresses from the SFEU array to perform winner competition. It iteratively compares each candidate's U_j against that of the current winner. A candidate neuron with a higher U_j becomes the new winner, provided it is not masked by lateral inhibition flags, and its information is captured in internal registers. Once the winner for a time step is finalized, the engine uses the winner's address to update the lateral inhibition flags for future competitions and then triggers the STDP weight update by reading and modifying the synaptic weights of the winner neuron within the corresponding SFEU.

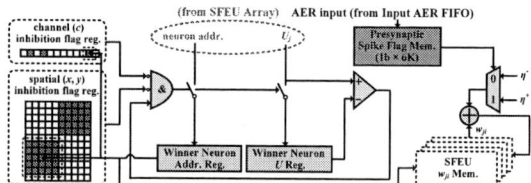

Fig. 6. The CONV layer learning engine circuit diagram.

3) Key Circuits of the Spike Feature Classification Core

a) SFCU and Decision-Making Block: The circuits for the SFCU and the decision-making block are shown in Fig. 7. The function of each SFCU is to integrate incoming AER spikes throughout the inference window by updating the U_j of its assigned neurons according to (1). After the final time step T, the decision-making block is activated. It receives the final membrane potentials and their corresponding addresses from the entire SFCU array and iteratively identifies the peak neuron, the neuron with the highest membrane potential. The identified peak neuron's address is then broadcast to both the processor's output and the FC layer learning engine. During the inference phase, this output is interpreted as the final classification result. During the learning phase, the learning engine uses this same address to perform the TD-STDP weight update, while the external output can be disregarded.

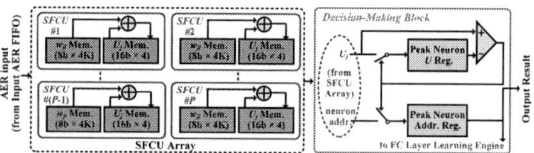

Fig. 7. Circuit diagram of the SFCU and the decision-making block.

b) FC Layer Learning Engine: The on-chip TD-STDP learning is handled by the FC layer learning engine, whose circuit is shown in Fig. 8. It receives two inputs: the address of the peak neuron from the decision-making block, and the sample's ground-truth label, which identifies the target neuron. The engine then orchestrates the TD-STDP update by

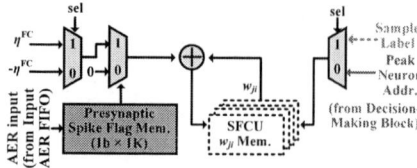

Fig. 8. The FC layer learning engine circuit diagram.

sequentially increasing the target neuron's synaptic weights and decreasing the peak neuron's weights. A *sel* signal configures the data path for each of these two operations. To execute each update, the engine uses the corresponding neuron's address to select the appropriate SFCU and update the weight within that unit.

IV. FPGA IMPLEMENTATION AND COMPARISON

The proposed neuromorphic processor was prototyped on a low-cost Xilinx Zynq-7010 device hosted on the Zybo platform. For this implementation, the number of parallel processing units in the three pipelined cores were set to $N=8$, $M=16$, and $P=4$, respectively. For testing, we established a platform where the prototype was connected to a host PC via an Ethernet link to manage all data flow. The chip operated at a clock frequency of 150 MHz and consumed 161 mW, as estimated by Xilinx Vivado.

Table I presents the resource utilization of our prototype and compares its performance against other edge-focused FPGA neuromorphic processors. The specific network architectures used for the three benchmark datasets (MNIST, Fashion-MNIST and ETH-80) are detailed in Table II. The results in Table I demonstrate that our design achieves a state-of-the-art processing latency of 99 μs on the MNIST dataset, which is 14× faster than our prior work [9]. This significant performance gain is achieved with a modest increase in resource utilization compared to [9], and the entire architecture remains compact enough to be deployed on the low-cost Zynq-7010 FPGA. While enabling this significant speedup, our processor achieves a highly competitive on-chip learning accuracy of 97.02%, second only to [9].

TABLE I. COMPARISON WITH OTHER FPGA-BASED EDGE NEUROMORPHIC PROCESSORS

Work		[7]	[8]	[9]	**This Work**
FPGA		Zynq-7045	Zynq-7045	Zynq-7010	**Zynq-7010**
slice LUTs		6033	15,233	3388	**7372**
slice reg.		1747	4101	3170	**4078**
DSPs		0	0	5	**0**
BRAMs		57	114	36.5	**18**
Clock (MHz)		250	100	100	**150**
Power (mW)		637	550	118	**161**
on-chip learning accuracy	MNIST	90.19%	95.9%	97.12%	**97.02%**
	Fashion[a]	N/A	83.44%	84.55%	**84.53%**
	ETH-80	N/A	85.29%	90.09%	**88.11%**
on-chip pre-proc. & spike encoding		No	No	Yes	**Yes**
MNIST inference latency		N/A	184 μs	1436 μs[b]	**99 μs[b]**
MNIST inference throughput		N/A	5428 fps	696 fps[b]	**10081 fps[b]**
MNIST inference energy efficiency		N/A	100 μJ/image	169 μJ/image[b]	**15.9 μJ/image[b]**

[a] Fashion is short for Fashion-MNIST

[b] Image pre-processing and spike encoding steps included

TABLE II. NETWORK STRUCTURES OF EACH DATASET

Dataset	DoG CONV Pre-proc. Layer[a]	Spike Encoding Layer[a]	CONV Layer[a]	Pooling Layer[a]	FC Layer
MNIST	28×28×2	28×28×2	28×28×16	7×7×16	10
Fashion	28×28×2	28×28×2	28×28×16	7×7×16	10
ETH-80	32×32×6	32×32×6	32×32×16	8×8×16	8

[a] The layer size is represented as height × width × channels.

V. CONCLUSION

This paper has presented a heterogeneous multi-core neuromorphic vision processor designed for low-latency, high-accuracy edge applications. By dedicating heterogeneous cores with specialized parallel units and resource-efficient circuit designs to each specific stage of the visual pipeline, from DoG pre-processing and spike encoding to final classification, our architecture overcomes the latency bottleneck inherent in previous single-core, time-multiplexed design [9]. The FPGA prototype, implemented on a low-cost Zynq-7010 device, validates the effectiveness of this approach. It achieves a processing latency of just 99 μs on MNIST, a 14× improvement over our prior work [9]. This substantial performance gain is realized with only a modest increase in resources, while a competitive on-chip learning accuracy of 97.02% is maintained. The resulting energy efficiency of 15.9 μJ/image underscores the design's suitability for power-constrained edge devices. In conclusion, our work demonstrates that a parallel, heterogeneous architecture is a highly effective strategy for building neuromorphic systems that require both end-to-end on-chip functionality and real-time processing speeds, making it an excellent candidate for demanding intelligent edge applications.

ACKNOWLEDGMENT

This work was funded in part by the Major Research Plan of the National Natural Science Foundation of China under Grant No. 92464103, in part by the Key Program of the National Natural Science Foundation of China under Grant No. 62334008; in part by the Opening Fund of Artificial Intelligence Key Laboratory of Sichuan Province under Grant 2023RYY07.

REFERENCES

[1] K. Roy, A. Jaiswal, and P. Panda, "Towards spike-based machine intelligence with neuromorphic computing," *Nature*, vol. 575, no. 7784, pp. 607–617, Nov. 2019.

[2] C. Frenkel, D. Bol and G. Indiveri, "Bottom-up and top-down approaches for the design of neuromorphic processing systems: Tradeoffs and synergies between natural and artificial intelligence," *Proc. IEEE*, vol. 111, no. 6, pp. 623-652, Jun. 2023.

[3] P. -Y. Tan and C. -W. Wu, "A 40-nm 1.89-pJ/SOP scalable convolutional spiking neural network learning core with on-chip spatiotemporal back-propagation," *IEEE Trans. Very Large Scale Integr. (VLSI) Syst.*, vol. 31, no. 12, pp. 1994-2007, Dec. 2023.

[4] T. Wang *et al.*, "MorphBungee: A 65-nm 7.2-mm² 27-μJ/image digital edge neuromorphic chip with on-chip 802-frame/s multi-layer spiking neural network learning," *IEEE Trans. Biomed. Circuits Syst.*, vol. 19, no. 1, pp. 209-225, Feb. 2025.

[5] J. Lee, R. Zhang, W. Zhang, Y. Liu, and P. Li, "Spike-train level direct feedback alignment: Sidestepping backpropagation for on-chip training of spiking neural nets," *Front. Neurosci.*, vol. 14, pp. 143, 2020.

[6] H. Wang *et al.*, "TripleBrain: A compact neuromorphic hardware core with fast on-chip self-organizing and reinforcement spike-timing dependent plasticity," *IEEE Trans. Biomed. Circuits Syst.*, vol. 16, no. 4, pp. 636–650, Aug. 2022.

[7] W. Liu, S. Xiao, Y. Liu, and Z. Yu, "SC-PLR: An approximate spiking neural network accelerator with on-chip predictive learning rule," *IEEE Trans. Biomed. Circuits Syst.*, vol. 18, no. 5, pp. 1156–1165, Oct. 2024.

[8] Z. Zhong *et al.*, "MorphBungee-Lite: An Edge Neuromorphic Architecture With Balanced Cross-Core Workloads Based on Layer-Wise Event-Batch Learning/Inference," *IEEE Trans. Circuits Syst. II, Exp. Briefs*, vol. 72, no. 1, pp. 293-297, Jan. 2025.

[9] M. Chen *et al.*, "A visual-cortex-mimetic tiny neuromorphic vision processor based on reconfigurable cortical neuron unit," *IEEE Trans. Circuits Syst. II, Exp. Briefs*, vol. 72, no. 7, pp. 943-947, July 2025.

An Energy-Optimized FPGA Implementation for Convolutional Neural Networks Accelerator

Yujie Zhu[1], Jianxuan Yin[1], Jingjing Liu[1,*], Jianhua Zhang[1]

[1]*Shanghai Collaborative Innovation Center for Intelligent Sensing Chip Technology,*
School of Microelectronics, Shanghai University, Shanghai, 200444, China
Email: {zhuyujie,yjx, jjliu}@shu.edu.cn, jhzhang@oa.shu.edu.cn

Abstract—**Convolutional neural networks (CNNs) are widely applied in various visual tasks, including object detection, image classification, and semantic segmentation, but deploying them on edge devices faces challenges such as high computational complexity, limited on-chip memory, and strict power budgets. To address these issues, this paper proposes an FPGA-based CNN accelerator optimized for YOLOv4-tiny on a ZYNQ-7020 platform. The design adopts an adaptive cyclic-blocking storage strategy that adjusts block sizes and loading sequences at runtime to maximize BRAM utilization and reduce DDR accesses; integrates a dynamic non-uniform parallelism mapping that allocates channel and spatial parallelism per layer based on computational density with inter-layer PE reuse; and applies software–hardware co-optimized energy efficiency through transfer–compute overlapping, ping-pong buffering, and intermediate result pruning to reduce data movement and power consumption. Implemented on ZYNQ-7020, the accelerator achieves 13.0 GOP/s, a single-image inference latency of 383 ms, and a total system power consumption of 2.141 W. Experimental results demonstrate that the proposed design delivers high real-time performance and low power consumption, making it well-suited for edge deployment.**

Index Terms—**CNN Accelerator, FPGA Implementation, Low Latency, Energy Efficiency**

I. INTRODUCTION

In recent years, convolutional neural networks (CNNs) have become a core technology in the field of computer vision. Leveraging their excellent feature extraction and representation learning capabilities, CNNs have been widely applied to tasks such as object detection[1], semantic segmentation[2], and face recognition[3], significantly advancing intelligence across various industries. However, CNN models often exhibit high computational complexity and large parameter sizes, making it difficult to meet the critical requirements of real-time performance, low power consumption, and low latency when deployed on resource-constrained edge devices. Therefore, how to improve the runtime efficiency of cnn on edge platforms while maintaining accuracy has become a key research focus in both academia and industry in recent years.

Mittal[4] proposed an FPGA-based CNN accelerator that significantly improves computational efficiency by optimizing convolution algorithms and hardware architecture design. However, this approach may be constrained by FPGA resource

This work was supported in part by the National Natural Science Foundation of China under Grant 62204044, and in part by theState Key Laboratory of Integrated Chips and Systems under Grant SKLICS-K202302, , and in part by the Special funds for promoting high-quality industrial development in Shanghai under Grant JJ-ZDHYLY-01-23-0004.

limitations when processing large-scale data. Sunny *et al.* [5] introduced data reshaping and memory access optimization techniques to further enhance CNN performance on FPGA. Nevertheless, these methods may still encounter performance bottlenecks when dealing with complex network architectures. Additionally, Antunes *et al.* [6] improved CNN performance on FPGA through quantization and layer operation optimizations. Although these studies have increased model efficiency to some extent, they may cause slight reductions in model accuracy. Tong *et al.* [7] proposed a dynamically configurable CNN accelerator capable of adjusting computing resources according to different network layers. However, such dynamic configuration might increase system complexity and latency. Although the above methods, through convolution algorithm optimization, data reshaping, memory access optimization, and quantization techniques, have improved the computational efficiency of CNNs on FPGA to some extent, several issues remain. When processing large-scale data or complex network architectures, these approaches are often constrained by limited FPGA resources.

To solve these problems, this paper proposes a CNN accelerator that is easy to deploy on edge devices. The specific innovation points are as follows:

- An adaptive cyclic-blocking method is proposed for YOLOv4-tiny, which adjusts block sizes and loading sequences at runtime to improve BRAM utilization, reduce DDR accesses, and alleviate bandwidth bottlenecks.
- A dynamic mapping approach allocates input/output channel and spatial parallelism per layer based on computational density, with inter-layer PE reuse to maximize DSP utilization and minimize idle cycles under resource constraints.
- Transfer–compute overlapping, ping-pong buffering, and intermediate result pruning are jointly applied to reduce data movement and power consumption while maintaining low-latency inference performance.

II. CONVOLUTIONAL ACCELERATOR DESIGN

The convolutional neural network accelerator designed in this paper mainly consists of two components: a memory access unit and a convolution computation unit. The memory access unit is responsible for storing and retrieving the network weights, intermediate feature maps, and the input images to be inferred. The convolution computation unit serves as the core

979-8-3315-3918-4/25 $31.00 © 2025 IEEE

module bearing the primary computational workload, occupying nearly 95% of the total computing resources. To further reduce resource consumption and meet the requirements of embedded deployment, all data involved in the convolution operations are represented in fixed-point format, with weights and feature maps quantized to 16-bit width, thereby reducing logic overhead while maintaining accuracy.

A. Memory Access Unit

To address the limited on-chip BRAM capacity of FPGAs and the large storage requirements of YOLOv4-tiny, an adaptive cyclic-blocking storage strategy is introduced. At runtime, the method dynamically adjusts block sizes and loading sequences based on each layer's feature map dimensions and channel counts, thereby matching BRAM utilization adaptively across layers, reducing DDR accesses, and alleviating bandwidth bottlenecks.

Specifically, the input feature map I has size $N \times H_{in} \times W_{in}$, weights are $M \times N \times K \times K$ ($K = 3$), and the output map O is $M \times H \times W$. Blocking factors T_n, T_m, T_r, T_c determine the sizes of blocks processed at each step: input blocks of size $T_n \times (T_r + K - 1) \times (T_c + K - 1)$ and weight blocks of size $T_m \times T_n \times K \times K$ produce output blocks of size $T_m \times T_r \times T_c$. A ping-pong buffering mechanism is used to overlap data transfer with convolution computation, further hiding memory latency. This approach significantly reduces off-chip memory traffic, enabling efficient convolution execution within limited on-chip storage.

Algorithm 1 Adaptive Cyclic Blocking Storage Strategy

1: **for** $row = 0$ to H step T_r {
2:     **for** $col = 0$ to W step T_c {
3:         **for** $t_o = 0$ to M step $T_m[l]$ {
4:             **for** $t_i = 0$ to N step $T_n[l]$ {
5:                 // load output feature maps
6:                 // load weights
7:                 // load input feature maps
8:                 // perform convolution operation
9: } } } }

This approach minimizes external memory accesses and storage bandwidth demands, enabling efficient convolution execution within limited on-chip memory. Strategic blocking and parallelism improve overall performance, meeting real-time object detection requirements.

B. Convolution Calculation Unit

A dynamic non-uniform parallelism mapping allocates input/output channel and spatial parallelism per layer based on computational density and data reuse, with PEs reused across layers to maximize DSP utilization and minimize idle cycles. The convolution engine achieves four-dimensional parallelism (input/output channels, rows, columns) through loop unrolling and pipelining, using PE arrays and adder trees with data reuse to reduce off-chip traffic. Coordinated with the memory unit via co-optimized energy efficiency—combining intermediate

result pruning, transfer–compute overlapping, and ping-pong buffering—the design sustains high throughput and low power without increasing latency.

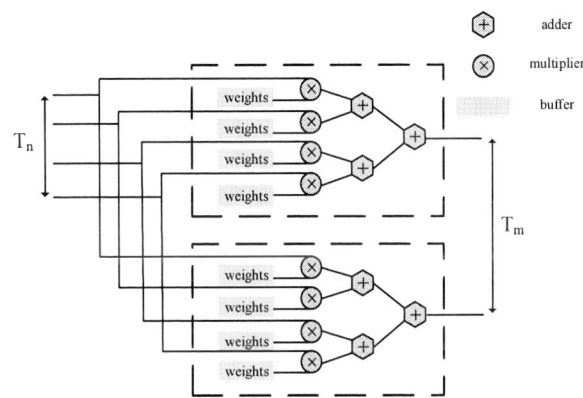

Fig. 1. The computation engine composed of a parallel multiplier unit and an adder tree.

In this study, a multi-layer CNN accelerator based on unified loop expansion is adopted, as shown in Fig. 1, combining advantages of both approaches: a design close to single PE but with multiple dedicated processing units, improving computational efficiency and enabling computing reuse across layers. This balances energy efficiency and latency, while simplifying hardware and improving resource utilization.

Key design challenges include loop tiling to split large data tasks into smaller chunks for efficient on-chip memory use, optimizing the arrangement of PEs, buffers, and interconnects, and matching data throughput to external memory bandwidth in Algorithm 2.

Algorithm 2 On-chip Data Computation

1: **for** $(trr = row; trr < \min(row + Tr, R); trr + +)$ {
2:     **for** $(tcc = col; tcc < \min(col + Tc, C); tcc + +)$ {
3:         **for** $(too = to; too < \min(to + Tm, M); too + +)$ {
4:             **for** $(tii = ti; tii < \min(ti + Tn, N); tii + +)$ {
5:                 **for** $(i = 0; i < K; i + +)$ {
6:                     **for** $(j = 0; j < K; j + +)$ {
7:         $L: output\_fm[too][trr][tcc] + =$
8:                 $weights[too][tii][i][j]*$
9:                 $input\_fm[tii][S * trr + i][S * tcc + j];$
10: } } } }

Processing Elements (PEs) perform core multiply-add operations using FPGA DSPs. Loop unrolling enables multiple parallel PEs to process data blocks simultaneously, boosting throughput. Pipelining improves efficiency by overlapping computations. Weight and feature map reuse minimizes data transfer, maximizing parallelism within FPGA limits. A Config Controller determines $T_n[l]$, $T_m[l]$, P_{in}, P_{out}, P_r, and P_c for each layer at runtime, enabling adaptive tiling and dynamic parallelism mapping according to layer-specific computational

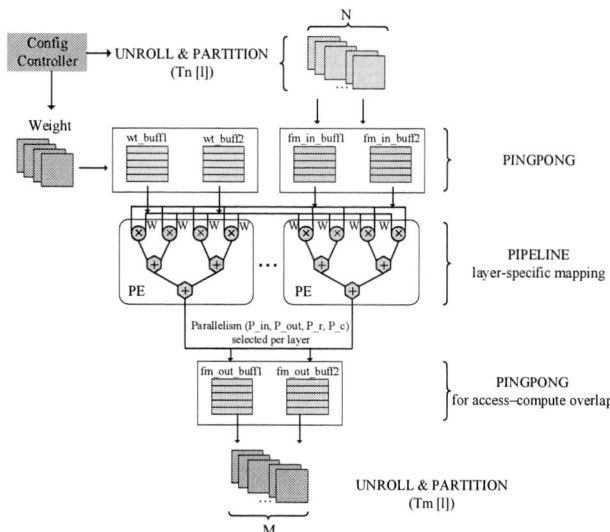

Fig. 2. Convolution layer computation structure.

characteristics. The convolution unit structure is shown in Fig. 2.

III. EXPERIMENT AND RESULT

To verify the performance of the proposed convolutional neural network accelerator, a yolov4-tiny model is deployed for performance testing. The specific network architecture diagram is shown in Fig. 3.

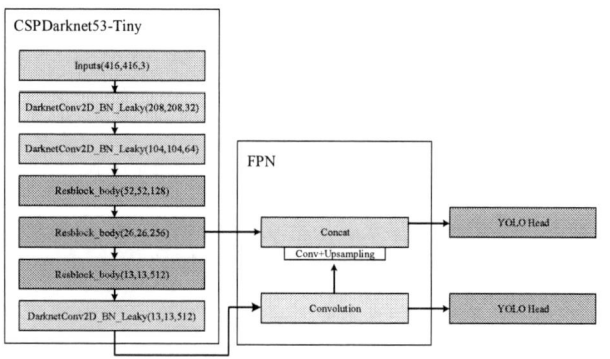

Fig. 3. The YOLO network architecture used

YOLOv4-tiny[8] is a lightweight YOLO version designed for resource-constrained platforms. Its architecture includes a CSPDarknet53-tiny backbone with Cross Stage Partial (CSP) connections for efficient feature extraction, a feature fusion module for multi-scale detection, and detection heads predicting on 13×13 and 26×26 feature maps. This design reduces computational complexity and resource usage while maintaining accuracy, making it suitable for FPGA and embedded acceleration.

The proposed accelerator is implemented as a synthesizable HLS model with fixed-point arithmetic and parameterized PE

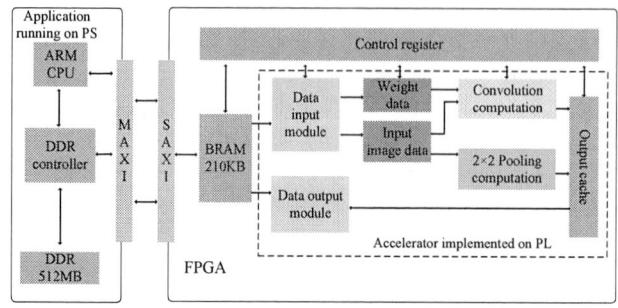

Fig. 4. Block diagram of proposed accelerator

arrays, deployed on a Xilinx Zynq-7020 FPGA using Vitis HLS 2023.1 and Vivado 2023.1. YOLOv4-tiny runs with a hardware–software co-design: the Processing System (PS) manages SD card initialization, DDR3 memory, and system control, while the Programmable Logic (PL) integrates the SD card interface, image buffer, and a CNN accelerator. The accelerator employs adaptive cyclic-blocking storage to match BRAM and bandwidth per layer, dynamic non-uniform parallelism mapping to allocate parallelism and reuse PEs, and access–compute co-optimization (pruning, overlapping, double-buffering) to reduce off-chip traffic and power while sustaining throughput. This design ensures efficient data flow between hardware acceleration on the PL and software control on the PS. Resource utilization is 70.44% LUTs, 42.64% FFs, 78.18% DSPs, and 67.14% BRAM, as shown in Table I.

TABLE I
FPGA RESOURCE UTILIZATION

Rsource	LUTs	FFs	DSPs	BRAMs
Used	37477	45371	172	94
Available	53200	106400	220	140
Utilization	70.44%	42.64%	78.18%	67.14%

In order to verify the performance of the network using the convolution accelerator on the FPGA platform, experimental comparisons are also conducted on the CPU and GPU platforms. The CPU model is Intel Core i5-10200H, and the GPU model is GeForce RTX2060. The specific comparison results are shown in Table II.

TABLE II
PERFORMANCE COMPARISON RESULTS ON THE CPU, GPU, AND FPGA PLATFORMS.

Experimental platform	CPU	GPU	FPGA
Data precision	32-bit	32-bit	16-bit
Computing time per image/s	0.495	0.065	0.383
Average precision mean (mAP)	78.20%	78.20%	76.40%
GOPS	14.04	106.92	13.00
Power/w	45.00	115.00	2.14

To comprehensively compare performance, the proposed accelerator's results were evaluated against similar works, as shown in Table III. Prior designs often suffer from high power consumption and inefficient resource use; for example,

979-8-3315-3918-4/25 $31.00 © 2025 IEEE

[9] achieves 0.532 seconds latency and 10.45 GOPS but consumes 3.360 W and 185 BRAMs, leading to poor energy efficiency. Some works like [10] offer lower latency and higher throughput but require extensive resources (50,200 LUTs, 240 DSPs) and still consume 2.203 W. Other designs, such as [11] and [12], show long latencies of 18.025 and 0.823 seconds, respectively, making them unsuitable for real-time use, while also exhibiting inefficient resource utilization, with [12] using up to 220 DSPs and 2.750 W.

In contrast, the proposed design on the ZYNQ-7020 platform achieves a low latency of 0.383 seconds with an input image resolution of 416×416 at 16-bit fixed-point precision, maintaining satisfactory accuracy and delivering 13.00 GOPS throughput. The proposed power consumption is only 2.141 W, making it one of the most energy-efficient among the compared works. Although the LUT and DSP usage is slightly higher than some earlier designs, the BRAM consumption is significantly optimized at just 94 blocks, nearly 50% fewer than that required by [9].

TABLE III
COMPARISON WITH THE OTHER WORK.

Works	YOLOv3-tiny[9]	YOLOv4-tiny[11]	YOLOv3-tiny[10]	YOLOv4-tiny[12]	This work
Platform	Zedboard	ZYNQ-7020	Nexys A7-100T	ZYNQ	ZYNQ-7020
Latency/s	0.532	18.025	0.013	0.823	0.383
Image size	416*416	416*416	256*256	-	416*416
Precision	Float 32	Fixed 16	Fixed 8	-	Fixed 16
LUTs	25900	30675	50200	37786	37477
DSPs	160	149	240	220	172
BRAMs	185	132	185	140	94
Power/w	3.360	2.384	2.203	2.750	2.141
Throughput/GOPS	10.45	-	95.08	19.10	13.00

IV. CONCLUSION

This paper presents an FPGA-based convolutional accelerator for YOLOv4-tiny, featuring an adaptive cyclic-blocking storage strategy that adjusts tiling per layer to match compute and memory resources across varying feature-map sizes and channel configurations. A dynamic non-uniform parallelism mapping assigns input/output channel and spatial parallelism per layer and reuses processing elements to improve DSP utilization under constraints. Compute and memory units are co-optimized via access–compute coordination with intermediate result pruning, transfer–compute overlapping, and double-buffering. Experiments show the design delivers real-time detection with balanced resource and power efficiency, offering a scalable solution for lightweight object detection on FPGAs. The future work is to continue optimizing the implementation speed of the hardware network and the accuracy of target recognition, and bring positive impacts to the industry through cross-level software and hardware optimization as well as algorithm optimization.

REFERENCES

[1] A. Datta, T. Islam Meghla, T. Khatun, M. Hasan Bhuiya, S. Rahman Shuvo, and M. Mahfujur Rahman, "Road object detection in bangladesh using faster r-cnn: A deep learning approach," in *2020 IEEE International Women in Engineering (WIE) Conference on Electrical and Computer Engineering (WIECON-ECE)*, 2020, pp. 348–351.

[2] S. Yıldız, A. Memiş, and S. Varlı, "Semantic and instance segmentation of multi-organ cell nuclei using deep learning based methods," in *2024 32nd Signal Processing and Communications Applications Conference (SIU)*, 2024, pp. 1–4.

[3] S. F. Kak, F. M. Mustafa, and A. Varol, "Design and enhancement of a cnn model to augment the face recognition accuracy," in *2022 3rd International Informatics and Software Engineering Conference (IISEC)*, 2022, pp. 1–5.

[4] S. Mittal, "A survey of fpga-based accelerators for convolutional neural networks," *Neural computing and applications*, vol. 32, no. 4, pp. 1109–1139, 2020.

[5] F. Sunny, A. Mirza, M. Nikdast, and S. Pasricha, "Embedded machine learning for cyber-physical, iot, and edge computing," in *Embedded Machine Learning for Cyber-Physical, IoT, and Edge Computing*, 2023.

[6] P. Antunes and A. Podobas, "Fpga-based neural network accelerators for space applications: A survey," *arXiv preprint arXiv:2504.16173*, 2025.

[7] H. Tong, K. Han, S. Han, and Y. Luo, "Design of a generic dynamically reconfigurable convolutional neural network accelerator with optimal balance," *Electronics*, vol. 13, no. 4, p. 761, 2024.

[8] S. Xu, Y. Zhou, Y. Huang, and T. Han, "Yolov4-tiny-based coal gangue image recognition and fpga implementation," *Micromachines*, vol. 13, no. 11, 2022.

[9] Z. Yu and C.-S. Bouganis, "A parameterisable fpga-tailored architecture for yolov3-tiny," in *Applied Reconfigurable Computing. Architectures, Tools, and Applications*, F. Rincón, J. Barba, H. K. H. So, P. Diniz, and J. Caba, Eds. Cham: Springer International Publishing, 2020, pp. 330–344.

[10] M. Kim, K. Oh, Y. Cho, H. Seo, X. T. Nguyen, and H.-J. Lee, "A low-latency fpga accelerator for yolov3-tiny with flexible layerwise mapping and dataflow," *IEEE Transactions on Circuits and Systems I: Regular Papers*, vol. 71, no. 3, pp. 1158–1171, 2024.

[11] P. Li and C. Che, "Mapping yolov4-tiny on fpga-based dnn accelerator by using dynamic fixed-point method," in *2021 12th International Symposium on Parallel Architectures, Algorithms and Programming (PAAP)*, 2021, pp. 125–129.

[12] A. Tan, Y. Yang, S. Duan, and L. Wang, "Fpga-based convolution accelerator and memristor ip core for cooperative acceleration of the yolo network," in *2024 4th International Conference on Computer Science, Electronic Information Engineering and Intelligent Control Technology (CEI)*, 2024, pp. 54–57.

A Lightweight Low-Latency Hardware Architecture for Dual Attention Super-Resolution Network

Haocan Jiang[1], Aiying Guo[1], Jianhua Zhang[1], Jingjing Liu[1,*]

[1]*Shanghai Collaborative Innovation Center for Intelligent Sensing Chip Technology,*
School of Microelectronics, Shanghai University
Shanghai, China
Email: {jiangyun77, gayshh, jhzhang, jjliu}@shu.edu.cn

Abstract—**With the advancement of high-definition displays, developing image super-resolution systems that balance both image restoration quality and hardware efficiency has become a significant challenge. To address this issue, this paper proposes a novel dual-attention super-resolution framework that integrates both pixel attention and channel attention to enhance image reconstruction quality while maintaining hardware efficiency. Furthermore, the proposed model is quantized to fixed-point precision and deployed on the FPGA. Through data pipeline optimization and the design of a unified computing core, the hardware architecture is ensured to maintain low latency and minimal resource utilization. Comprehensive experiments demonstrate that our method achieves superior performance with a PSNR of 37.45dB on multiple benchmark datasets, while significantly reducing resource utilization to 31K LUTs, and maintaining a low power consumption of 5.25w, compared to existing super-resolution networks deployed on FPGA.**

Index Terms—**Image super-resolution, Neural network hardware architecture, Pipeline optimization, Unified computing core**

I. INTRODUCTION

Super-resolution (SR) technology aims to extract information from low-resolution images and reconstruct corresponding high-resolution images. This technique has been widely applied across various fields, providing high-definition image support for diverse scenarios.

Simple interpolation algorithms such as nearest-neighbor and bicubic interpolation are easy to implement, but they generally result in poor image reconstruction quality. Dong *et al.* [1] were the first to introduce convolutional neural networks (CNNs) into the super-resolution task, with their proposed SRCNN significantly outperforming traditional approaches. Building upon this, the very deep super-resolution (VDSR) network [2] was the first to incorporate residual connections to directly learn the texture differences between low and high-resolution images. In addition, Wang *et al.* [3] introduced adaptive weighting into CNN architectures to selectively eliminate redundant information. Chang *et al.* [4] designed a deconvolutional SR networks accelerator that reduces computational overhead by eliminating ineffective operations and

compressing convolution kernel sizes. Sun *et al.* [5] proposed ERVSR, which exploited lightweight state compression and depthwise separable convolutions to achieve superior video super-resolution performance under limited hardware resources. Spagnolo *et al.* [6] designed a SR accelerator that leverages foveated processing to separately treat foveal and peripheral regions, achieving high-throughput and ultra-low energy consumption, enabling efficient super-resolution on wearable devices. Duan *et al.* [7] proposed UArch, which enabled real-time, high-quality endoscopic image reconstruction with low latency and high energy efficiency by deploying a lightweight USR-s model and employing a heterogeneous triple-core architecture on FPGA. Liu *et al.* [8] developed a corresponding hardware accelerator featuring DSP-enhanced computation, a scalable cache strategy, and constraint-aware resource allocation for real-time edge deployment. Song *et al.* [9] designed a guided-filter-based infrared image enhancement algorithm, which applies adaptive histogram equalization and gain-controlled filtering, and implemented the entire pipeline on FPGA for real-time processing. However, these methods lack co-optimization between algorithm and hardware, making them less suitable for deployment in resource-constrained scenarios.

To address these issues, we perform multi-dimensional optimizations ranging from algorithm design to hardware computing architecture, and propose a lightweight dual-attention network that is efficiently deployed on FPGA for efficient inference. The main contributions of this paper are as follows:

- We develop a lightweight dual-attention super-resolution network that integrates a CNN framework with attention mechanisms, achieving improved model performance.
- We implement the proposed dual-attention network on FPGA, demonstrating that it can be efficiently deployed in hardware with minimal resource usage and low power consumption, thus meeting the demands of edge computing environments.
- We develop a unified computing core capable of flexibly supporting various operations, including element-wise addition, accumulation, multiplication, and convolution, which significantly enhances the overall efficiency of the hardware architecture.

This work was supported in part by the National Natural Science Foundation of China under Grant 62204044, and in part by the State Key Laboratory of Integrated Chips and Systems under Grant SKLICS-K202302.

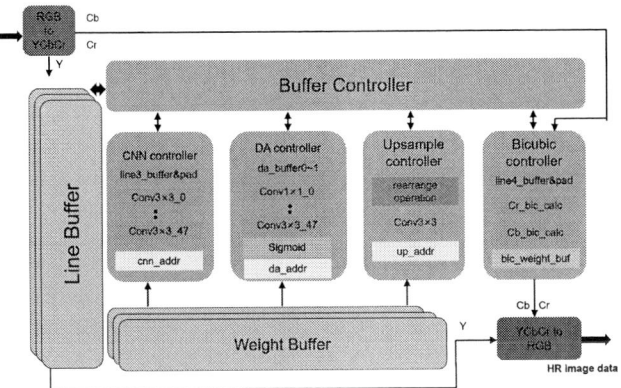

Fig. 1. Illustration of the proposed method.

II. PROPOSED METHOD

A. Hardware Architecture

The system architecture for image super-resolution is illustrated in Fig.1. On the PC side, images are input via a UART-based interactive interface. After being processed by the FPGA board, the output is displayed on a displayer through HDMI. The algorithm structure implemented on the FPGA is summarized in Table I, which details the required number of basic computational units and the input/output channel specifications. The input image is processed through shallow feature extraction, deep feature extraction, and upsampling modules to obtain a high-resolution image. The hardware deployment of these modules on the FPGA is illustrated in Fig. 1. After converting the input data from RGB to YCbCr, the Cb and Cr channels are upsampled using bicubic interpolation, while super-resolution reconstruction is performed on the Y channel. The Buffer Controller manages the interaction between the data processing module and the line buffers, which are used to store input/output images and intermediate feature maps. The data processing module includes the CNN Controller, DA Controller, upsample controller and bicubic controller, with each controller instantiating its corresponding functional submodules. The weight parameters required by the network are quantized to 8-bit on the PC side and preloaded into the Weight Buffer. Finally, the super-resolved Y channel is combined with the interpolated Cb and Cr channels, followed by YCbCr-to-RGB conversion for output.

B. Pipeline Optimization

Fig. 2 illustrates the pipelined operation during image reconstruction. By employing image block partitioning, the system avoids the need to load the entire image at once, thereby reducing computation latency t_1. Due to the ping-pong buffering mechanism, patch_2 is preloaded in advance. Moreover, since the final step of pixel shuffle rearrange operation does not require participation from the computing units, the computation for patch_2 can be initiated early, resulting in saved latency denoted as t_2.

TABLE I
ALGORITHM NETWORK STRUCTURE.

Module	Operator	Input Channel	Output Channel	Activation function
Shallow feature extration	Conv3×3	1	48	-
(Deep feature extration)×4	Conv3×3	48	48	LeakyReLU
	Conv3×3	48	48	LeakyReLU
	DA	48	48	-
Upsampling	Conv3×3	48	scale×scale	-
	Pixelshuffle	scale×scale	1	-

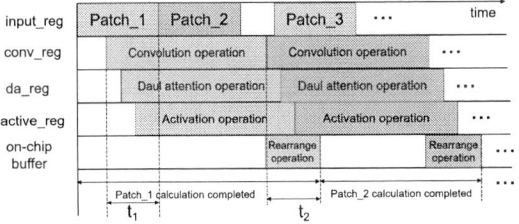

Fig. 2. Pipeline computing.

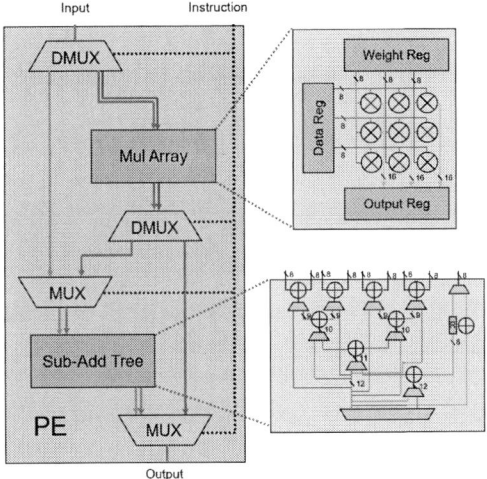

Fig. 3. Unified computing core and dataflow architecture.

C. Unified Computing Core

To improve resource utilization and facilitate pipeline control, a unified computing core is adopted as the processing element (PE) to support multiple operators. As illustrated in Fig. 3, the dataflow through the unified core is directed by control instructions. In the Fig. 3, red paths indicate convolution operations, green paths represent addition, and blue paths correspond to multiplication. The detailed structures of the multiplier array and sub-add tree are also illustrated. In the sub-add tree, the orange path represents element-wise addition, while the green path corresponds to the accumulation path. Compared with traditional dedicated convolution cores, the unified computing core introduces only a few data selectors

TABLE II

QUANTITATIVE COMPARISON WITH STATE-OF-THE-ART METHODS (AVERAGE PSNR/SSIM) FOR **CLASSICAL IMAGE SR** ON BENCHMARK DATASETS. THE BEST AND SECOND-BEST PERFORMANCES ARE **BOLDED** AND <u>UNDERLINED</u>, RESPECTIVELY.

Method	Scale	Params	FLOPs	Set5 PSNR/SSIM	Set14 PSNR/SSIM	B100 PSNR/SSIM	Urban100 PSNR/SSIM	Manga109 PSNR/SSIM
Bicubic	×2	-	-	33.66/0.9299	30.24/0.8688	29.56/0.8431	26.88/0.8403	30.80/0.9339
SRCNN [1]	×2	**8K**	**53G**	36.66/0.9542	32.45/0.9067	31.36/0.8879	29.50/0.8946	35.60/0.9663
VDSR [2]	×2	666K	613G	37.53/0.9587	33.03/0.9124	31.90/0.8960	30.76/0.9140	37.22/0.9750
AWRN-S [3]	×2	397K	590G	<u>37.75</u>/**0.9596**	**33.31/0.9151**	<u>32.00</u>/0.8974	**31.39/0.9207**	**37.80/0.9773**
OURS	×2	<u>206K</u>	<u>133G</u>	**37.80**/<u>0.9592</u>	<u>33.27/0.9139</u>	**32.07/0.8975**	<u>31.26/0.9180</u>	<u>37.68/0.9755</u>
OURS-FPGA	×2	<u>206K</u>	-	37.45/0.9583	33.05/0.9130	31.95/0.8964	30.77/0.9135	37.29/0.9750
Bicubic	×3	-	-	30.39/0.8682	27.55/0.7742	27.21/0.7385	24.46/0.7349	26.95/0.8556
SRCNN [1]	×3	**8K**	**53G**	32.75/0.9090	29.30/0.8215	28.41/0.7863	26.24/0.7989	30.48/0.9117
VDSR [2]	×3	666K	613G	33.66/0.9213	29.77/0.8314	28.82/0.7976	27.14/0.8279	32.01/0.9340
AWRN-S [3]	×3	477K	432G	<u>34.02</u>/0.9240	**30.09/0.8376**	<u>28.92</u>/0.8009	**27.57/0.8391**	**32.24/0.9342**
OURS	×3	<u>206K</u>	<u>104G</u>	**34.10/0.9244**	<u>29.95/0.8342</u>	**28.95/0.8014**	<u>27.37/0.8344</u>	<u>32.20/0.9338</u>
OURS-FPGA	×3	<u>206K</u>	-	33.65/0.9220	29.77/0.8315	28.78/0.7976	27.10/0.8277	31.98/0.9338
Bicubic	×4	-	-	28.42/0.8104	26.00/0.7027	25.96/0.6675	23.14/0.6577	24.89/0.7866
SRCNN [1]	×4	**8K**	**53G**	30.48/0.8626	27.50/0.7513	26.90/0.7101	24.52/0.7221	27.58/0.8555
VDSR [2]	×4	666K	613G	31.35/0.8838	28.01/0.7674	27.29/0.7251	25.18/0.7524	28.61/0.8870
AWRN-S [3]	×4	588K	388G	<u>31.77</u>/0.8893	**28.35/0.7761**	**27.41/0.7304**	**25.56/0.7678**	**28.89/0.8875**
OURS	×4	<u>206K</u>	<u>78G</u>	**31.83/0.8899**	<u>28.30/0.7754</u>	<u>27.35</u>/0.7278	<u>25.38/0.7632</u>	<u>28.83/0.8874</u>
OURS-FPGA	×4	<u>206K</u>	-	31.29/0.8836	28.02/0.7678	27.21/0.7244	25.20/0.7601	28.63/0.8868

to enable support for multiple operations, thereby reducing hardware resource consumption while achieving both high efficiency and flexibility in computation.

III. EXPERIMENT

In this section, extensive comparative and ablation experiments are conducted on both the software and hardware sides. The proposed method is comprehensively evaluated from multiple perspectives, including image quality metrics (PSNR/SSIM), hardware resource utilization, and power consumption.

A. Experimental Setting

The DIV2K dataset is used for training and validation, while Set5, Set14, B100, Urban100, and Manga109[1] are adopted as test sets. During training, the batch size is set to 16, and the Adam optimizer is employed with $\beta_1 = 0.9$ and $\beta_2 = 0.999$. The model is trained for a total of 500 epochs, with an initial learning rate of 0.0002, and the learning rate is halved every 200 epochs to facilitate convergence. The experiments are conducted on a Xilinx XC7S50 FPGA development board.

TABLE III

THE COMPARISON OF INFERENCE POWER CONSUMPTION AND TIME BETWEEN CPU AND FPGA FOR IMAGE SIZES OF 128×128 AND 256×256.

Platform	Power	PSNR	128×128	256×256
			Inference time	
CPU	50.36w	**31.83dB**	318.06ms	1311.61ms
FPGA	**5.25w**	31.29dB	**82.30ms**	**330.05ms**

[1] https://www.kaggle.com/datasets/msahebi/super-resolution

B. Quantitative and Visual Comparison

To validate the effectiveness and efficiency of our proposed method, we conduct a comprehensive comparison with several representative image super-resolution algorithms on five widely-used benchmark datasets. As shown in Table II, our method consistently achieves competitive or superior PSNR and SSIM performance across all upscaling factors (×2, ×3, and ×4). Notably, proposed methed exhibits the 0.48/0.0061, 0.29/0.0080, 0.45/0.0027, 0.20/0.0108, and 0.22/0.0005 improvement in PSNR(dB)/SSMI over the VDSR baseline, while requiring only 206K parameters and 78G FLOPs. Although the quantized model deployed on the FPGA experiences a slight performance degradation, it still demonstrates competitive effectiveness.

Several selected images from the benchmark dataset are tested using different methods, and the visual results are shown in Fig. 4. As illustrated, the proposed method produces reconstructions that are closest to the ground truth, successfully recovering more image texture and fine details. Although the image quality reconstructed on FPGA exhibits some degradation due to quantization, this loss is expected and remains acceptable compared to other methods.

C. Inference time

The inference time of the proposed model on different platforms is compared, and the results are summarized in Table III. Under input sizes of 128×128 and 256×256, the inference speed of the model deployed on FPGA is improved by 3.86× and 3.97× compared to CPU implementation. Moreover, the power consumption is significantly lower than that of the CPU, making it more suitable for edge computing scenarios.

979-8-3315-3918-4/25 $31.00 © 2025 IEEE

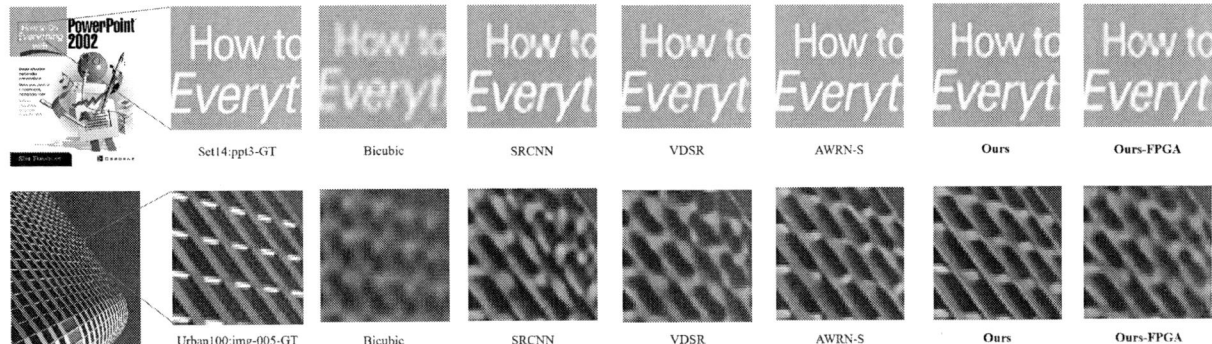

Fig. 4. Visual comparison with the state-of-the-art classic methods on ×4 SR.

TABLE IV

COMPARISON OF SUPER-RESOLUTION HARDWARE IMPLEMENTATIONS BASED ON FPGA, WHERE PSNR VALUES REPRESENT THE PERFORMANCE OF EACH METHOD EVALUATED ON THE SET5 DATASET WITH A SCALING FACTOR OF ×2.THE BEST AND SECOND-BEST PERFORMANCES ARE **BOLDED** AND UNDERLINED, RESPECTIVELY.

Method	FPGA Device	Network	Frequency	LUT	FF	DSP	Power	PSNR(dB)
FSRCNN-UHD [4]	XC7K410T	CNN	130MHz	167K	158K	1512	5.4w	36.40
ERVSR [5]	XCKU15P	RNN	160MHz	98K	57K	1820	5.47w	36.76
Fovea Selective [6]	XC7K410T	CNN	222MHz	**28K**	82K	1750	**3.7w**	33.23
UArch [7]	ZCU111	CNN	188MHz	40K	<u>50K</u>	<u>1029</u>	-	-
HPASR [8]	XCZU7EV	CNN	250MHz	69K	147K	1211	-	<u>37.14</u>
This work	XC7S50	CNN+Attention	150MHz	<u>31K</u>	**24K**	652	<u>5.25w</u>	**37.45**

This highlights the benefits of the optimized data pipeline in reducing system latency.

D. Comparison with Existing FPGA-Based Implementations

Table IV compares our method with several state-of-the-art FPGA-based super-resolution implementations. The proposed hardware architecture operates at a system clock frequency of 150MHz. As a result of employing 8-bit precision quantization, the PSNR on the hardware side decreases by 0.35dB compared to the software counterpart. The proposed hardware deployment scheme achieves the best performance in terms of reconstructed image quality, reaching a PSNR of 37.45dB, while maintaining low resource usage with 31K LUTs, 24K FFs, and 652 DSP blocks, as well as a power consumption of only 5.25w, making it highly suitable for low-power edge scenarios.

IV. CONCLUSION

This paper proposes a lightweight dual-attention super-resolution network that is efficiently deployed on an FPGA platform. A series of multi-dimensional optimizations are carried out from algorithm design to hardware implementation. At the algorithm level, the integration of a CNN architecture with dual attention mechanisms enhances PSNR performance. At the hardware level, the data pipeline is optimized and a unified computing core is proposed to reduce system latency and resource consumption. Compared with existing works, the proposed approach achieves a favorable balance between performance and efficiency.

REFERENCES

[1] C. Dong, C. C. Loy, K. He, and X. Tang, "Learning a deep convolutional network for image super-resolution," in *Computer Vision–ECCV 2014: 13th European Conference, Zurich, Switzerland, September 6-12, 2014, Proceedings, Part IV 13.* Springer, 2014, pp. 184–199.

[2] J. Kim, J. K. Lee, and K. M. Lee, "Accurate image super-resolution using very deep convolutional networks," in *Proceedings of the IEEE Conference on Computer Vision and Pattern Recognition (CVPR)*, June 2016.

[3] C. Wang, Z. Li, and J. Shi, "Lightweight image super-resolution with adaptive weighted learning network," *arXiv preprint arXiv:1904.02358*, 2019.

[4] J.-W. Chang, K.-W. Kang, and S.-J. Kang, "An energy-efficient fpga-based deconvolutional neural networks accelerator for single image super-resolution," *IEEE Transactions on Circuits and Systems for Video Technology*, vol. 30, no. 1, pp. 281–295, 2018.

[5] K. Sun, M. Koch, Z. Wang, S. Jovanovic, H. Rabah, and S. Simon, "An fpga-based residual recurrent neural network for real-time video super-resolution," *IEEE Transactions on Circuits and Systems for Video Technology*, vol. 32, no. 4, pp. 1739–1750, 2021.

[6] F. Spagnolo, P. Corsonello, F. Frustaci, and S. Perri, "Design of a low-power super-resolution architecture for virtual reality wearable devices," *IEEE Sensors Journal*, vol. 23, no. 8, pp. 9009–9016, 2023.

[7] X. Duan, Y. Chen, M. Li, Y. Rong, R. Xie, and J. Han, "Uarch: A super-resolution processor with heterogeneous triple-core architecture for workloads of u-net networks," *IEEE Transactions on Biomedical Circuits and Systems*, vol. 17, no. 3, pp. 633–647, 2023.

[8] H. Liu, Y. Qian, Y. Liang, B. Zhang, Z. Liu, T. He, W. Zhao, J. Lu, and B. Yu, "A high-performance accelerator for real-time super-resolution on edge fpgas," *ACM Transactions on Design Automation of Electronic Systems*, vol. 29, no. 3, pp. 1–25, 2024.

[9] H. Song, Z. Wang, W. Cao, Y. Zhang, and X. Leng, "Infrared image enhancement based on guided filtering and adaptive algorithm and its fpga implementation," *Microwave and Optical Technology Letters*, vol. 67, no. 1, p. e70105, 2025.

A Scalable Channel-Parallel Accelerator for Spiking Neural Network

Yuchun Wu[1], Lingling Miu[1], Jingjing Liu[1,*], Jianhua Zhang[1]

[1]*Shanghai Collaborative Innovation Center for Intelligent Sensing Chip Technology,*
School of Microelectronics, Shanghai University, Shanghai, 200444, China
Email: {wyc594238842,mll,jjliu,jhzhang}@shu.edu.cn

Abstract—Spiking neural networks (SNNs) have attracted increasing attention due to their event-driven processing and high energy efficiency, making them suitable for edge intelligence applications. However, SNNs often require longer inference latency and exhibit lower throughput compared to artificial neural networks (ANNs), posing challenges for efficient hardware deployment. To address these issues, this paper proposes a scalable and channel-parallel hardware architecture by designing a parameterized convolution module that exploits the structural similarity between convolutional layers to enable efficient hardware reuse. In addition, a sparse zero-hopping mechanism is introduced to exploit the spatio-temporal sparsity inherent in SNN datasets, further reducing redundant computations and memory accesses. The accelerator is implemented on a Xilinx XCKU5P FPGA and operates at 200 MHz, achieving a peak throughput of 230.4 GSOP/s and an energy efficiency of 32.9 GSOP/W. Compared to existing SNN implementations, the proposed design improves throughput by 28% and energy efficiency by 53%, while consuming only 96K LUTs and 228 BRAMs. These results demonstrate the superior performance, scalability, and resource efficiency of the proposed architecture for real-time SNN inference on edge platforms.

Index Terms—Spiking neural networks, Neuromorphic accelerator, Channel level parallelism, Sparse

I. INTRODUCTION

Spiking Neural Networks (SNNs) have emerged as a promising paradigm for next-generation neural computing, inspired by the sparse and event-driven nature of biological neurons [1]. By encoding information through discrete spike trains and leveraging temporal dynamics, SNNs naturally integrate spatial and temporal processing, offering significant advantages in energy efficiency and computational sparsity [2]. These characteristics make SNNs particularly attractive for low-power and latency-sensitive edge applications such as real-time perception, autonomous systems, and neuromorphic sensing [3]. Despite these benefits, deploying SNNs on hardware presents notable challenges. Their unique computational model, which differs fundamentally from traditional artificial neural networks (ANNs) [4], introduces difficulties in achieving high throughput, efficient resource utilization, and scalable hardware architectures. These challenges have

hindered the practical adoption of SNNs in performance-critical edge scenarios, underscoring the need for specialized and flexible accelerator designs.

Recent research has made significant progress in hardware acceleration for spiking neural networks, yet several critical challenges remain. Intel's Loihi chip [5], as a representative ASIC-based SNN accelerator, demonstrates strong performance in neuromorphic computing. However, its fixed-function architecture limits scalability and adaptability to evolving SNN models, making it less suitable for supporting diverse or emerging network structures. Li et al. [6] achieved efficient SNN hardware acceleration through structured sparse connectivity. Nevertheless, the lack of convolutional design limits its scalability and effectiveness in handling complex spatial feature extraction tasks. While Feng et al. [7] optimized the energy efficiency of SCNN processors through low-bitwidth design and channel-level parallelism, their throughput remains constrained by the serial nature of sparse event processing, limiting the full exploitation of the inherent parallelism in spiking convolutional networks.

In this paper, we propose a high-performance accelerator design for spiking neural networks. The main contributions are as follows:

- We present a scalable hardware architecture for spiking convolutional neural networks (SCNNs), incorporating a modular and reusable convolution module designed for channel-level parallelism. By leveraging the structural similarity across convolutional layers, the architecture employs a parameterized processing element (PE) array and unified buffer structure to enable efficient hardware reuse and reduce design complexity.

- We propose a sparse zero-hopping mechanism to exploit the input sparsity inherent in SNN datasets. This mechanism is integrated with an input–output channel parallelism strategy that accelerates dense 3×3 convolutions by allowing concurrent processing of multiple input and output channels. The design selectively bypasses zero-value operations, enabling multiplier-free computation through shift and multiplexing logic.

- We achieves high throughput and energy efficiency, delivering 230.4 GSOP/s throughput and 32.9 GSOP/W energy efficiency at 200 MHz.

This work was supported in part by the National Natural Science Foundation of China under Grant 62204044, and in part by the State Key Laboratory of Integrated Chips and Systems under Grant SKLICS-K202302, and in part by the Special funds for promoting high-quality industrial development in Shanghai under Grant JJ-ZDHYLY-01-23-0004.

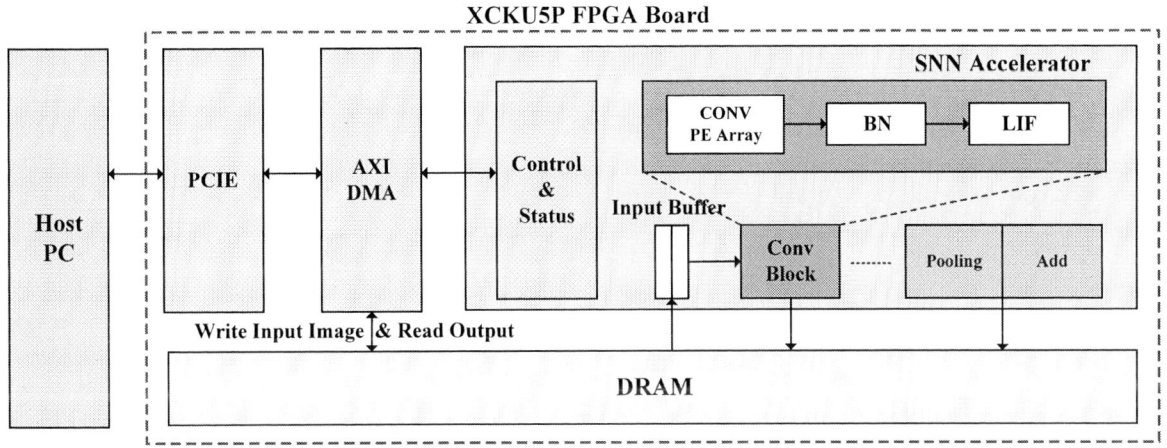

Fig. 1: Overview of the proposed architecture.

II. HARDWARE DESIGN

A. Overview of Accelerator Architecture

Fig. 1 illustrates the overall architecture of the proposed SNN accelerator. The AXI-DMA module controls the flow of input data and weights. To maximize efficiency, the architecture integrates operator fusion by combining convolution, batch normalization, and leaky integrate-and-fire (LIF) neuron activation into a single hardware block. Its computational core is a parameterized array of Processing Elements (PEs), configurable to process Ti input channels and To output channels concurrently. This modular design exploits the structural similarity across convolutional layers, enabling the identical PE array to be reused throughout the network without reconfiguration. Auxiliary modules (pooling, element-wise add) under centralized control, adapting to layer-specific operations.

B. Reusable Convolutional Module

Fig. 2 illustrates the architecture of the proposed reconfigurable convolution module tailored for spiking convolutional neural networks (SCNNs), in which a processing element (PE) array forms the computational core. The design adopts channel-level parallelism, with the number of input and output channels parameterized as Ti and To, respectively, enabling scalable configurability based on hardware resource constraints. To facilitate efficient reuse across heterogeneous SCNN layers, the module integrates internal data and weight reordering mechanisms to accommodate diverse channel dimensions.

In SCNNs employing possing coding, input data are represented as spike trains consisting of binary values at each discrete time step. Each PE performs convolution on a set of Ti input spike slices using a 3×3 kernel, where the binary spike inputs selectively gate the corresponding 8-bit fixed-point weights. Rather than performing conventional multiplication, each spike-weight pair is evaluated using a binary-controlled multiplexer (MUX), which passes the weight when the spike is active and zero otherwise. This implementation eliminates

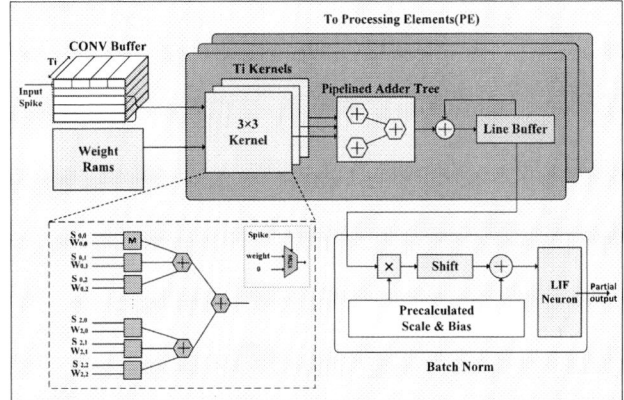

Fig. 2: Architecture of a 3 × 3 convolutional layer.

the need for hardware multipliers and enables efficient zero-skipping to exploit temporal and spatial sparsity. To ensure continuous data flow and high throughput, five line buffers are employed to manage sliding windows, enabling concurrent input streaming and convolution. At each clock cycle, three lines of data are delivered simultaneously to the PE array for parallel processing.

As shown in Fig. 2, input data traverse the line buffers and are dynamically distributed to the PE array. Within each PE, Ti input channels are convolved with the corresponding 3×3 kernels, and intermediate results are accumulated using a two-stage pipelined adder tree. All PEs operate in parallel under a unified clock. When the number of input channels is an integer multiple of Ti, the convolution proceeds iteratively along the input channel dimension. In each iteration, the next Tii channels are streamed into the PEs, and partial sums are accumulated with previously buffered results. This iterative process continues until all input channels have been processed. A similar procedure is employed along the output channel dimension when the total number of output channels is divisible by To. This scheme allows the convolution engine to support arbitrary input and output channel sizes, ensuring both flexibility and hardware efficiency.

Following convolution, the module directly integrates a fused batch normalization (BN) and leaky integrate-and-fire (LIF) neuron computation, also depicted in Fig. 2. The BN unit normalizes the convolution output across spatial locations to stabilize activation dynamics, after which the normalized values are integrated into membrane potentials of spiking neurons. Each LIF neuron accumulates input current over time, with leakage applied at each timestep. When the membrane potential exceeds a predefined threshold, a spike is emitted and the potential is reset or decayed, thereby introducing temporal dynamics and sparse spike-based activation.

To support high-throughput weight access, all convolution weights are preloaded into nine BRAMs, each corresponding to one spatial position in the 3×3 kernel, as illustrated in Fig. 3. The weights are first organized by input channel groups and then grouped according to kernel spatial positions. With fixed input parallelism Ti, address generation logic accesses weights with strided patterns to align with the convolution schedule. This layout minimizes memory access overhead and maximizes computation efficiency during high-frequency spike processing.

In the proposed architecture, the fully connected layer is replaced with a 1×1 convolutional layer. This 1×1 convolution essentially represents a specialized form of 3×3 convolution, with both demonstrating high consistency in data flow processing and hardware resource reuse.

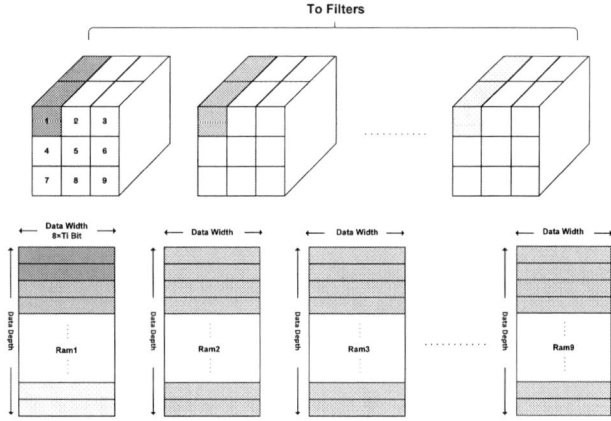

Fig. 3: Weight memory pattern for efficient weight prefetching.

C. Pooling Module

The proposed pooling module adopts a pipelined, multi scale max pooling architecture with dual buffered data management. Upon activation, the input controller alternates between ping pong buffers to enable continuous streaming. Five sliding windows operate in parallel, and the computation unit implements a five stage comparison pipeline, including an initial pixel registration phase (4 cycles), element wise maximum detection using sixteen comparators, and a hierarchical comparison tree for global maximum selection. Additionally, this module also supports average pooling operations.

D. Add Module

The proposed Add module enables efficient multi scale feature fusion by merging quantized feature maps from different network layers. It applies channel wise accumulation with normalization and scaling, preserving feature integrity across scales. This design ensures low latency, resource efficient merging of multi resolution data, facilitating robust and accurate detection.

III. EXPERIMENT AND EVALUATION

A. Experimental Setup

The proposed SNN accelerator, implemented based on the ResNet-18 architecture, is deployed on a Xilinx Kintex UltraScale KU5P FPGA. Input preprocessing is conducted on a host PC, where CIFAR-10 images are converted into spike trains via Poisson encoding. During runtime, the encoded spike trains stream into the FPGA in real time, while classification results are collected by the PC. The design operates reliably at a clock frequency of 200 MHz.

B. Comparison With Other FPGA-Based Implementations

Table I summarizes the performance and resource utilization of the proposed SCNN accelerator in comparison with representative FPGA-based designs. Prior works typically employ 8 to 16 bit fixed-point or floating-point quantization and require hundreds of DSP slices and large hardware resources to support matrix multiplications and weight storage. Despite these resources, their reported throughput remains in the range of 3.2 to 179 GSOP/s, with power efficiency between 6.78 and 21.49 GSOP/W.

In contrast, the proposed design adopts a uniform 8-bit fixed-point quantization scheme for both spike inputs and synaptic weights, striking a favorable balance between computational accuracy and hardware complexity. Crucially, all multiplication operations are replaced by shift-and-MUX logic, enabling a fully DSP-free design that significantly reduces resource cost. The convolution engine is highly parallelized, configured with 16 input and 8 output channel lanes (Ti = 16, To = 8), which allows efficient spike-driven computation and maximized data-level concurrency. As a result, the accelerator achieves a peak throughput of 230.4 GSOP/s and a power efficiency of 32.9 GSOP/W, both substantially higher than those reported in prior works. Moreover, it maintains compact resource usage, occupying only 96k LUTs, 132k FFs, and 228 BRAMs. These results demonstrate the proposed design's advantages in throughput scalability, power-aware performance, and resource efficiency, making it well-suited for large-scale SCNN inference on FPGAs.

C. Resource Utilization

Table II details the hardware resource utilization for each core module within the proposed SNN accelerator, demonstrating the design's computational and memory requirements.

TABLE I: PERFORMANCE COMPARISON TO OTHER SNN ACCELERATORS

	ISCAS22 [8]	TCAD22 [9]	AICAS23 [10]	TCAD23 [11]	MicroelectronJ24 [12]	This work
Platform	Xilinx XCKU115	ZYNQ-7045	ZYNQ-7035	Xilinx XC7K325T	Xilinx XCVU9P	Xilinx KU5P
Network	FC-SNN	SCNN	SCNN	SCNN	SCNN	SCNN
Dataset	MNIST	MNIST	MNIST	MNIST	N-MNIST	CIFAR-10
Input Width	28	28	28	28	36	32
Quantization	8 bit Fixed	8 bit Fixed	NR	16 bit Fixed	16 bit Floating	8 bit Fixed
Clock	140MHz	200MHz	200MHz	200MHz	200MHz	200MHz
BRAMs(36Kb)	216	262	128	254	2064	228
DSPs	NR	NR	NR	0	240	0
LUTs	420K	45K	42K	170k	727K	96K
FFs	95K	20K	16K	113K	1065K	132K
Throughput (GSOP/s)	179	22.6	23.2	3.20	5140	230.4
Efficiency(GSOP/W)	21.49	19.3	19.3	6.78	740	32.9

TABLE II: Hardware Resources Utilization of Different Modules in the Accelerator

Module	LUTs	FFs	BRAMs
CONV33	28649	62454	82
CONV11	17216	26639	52
Pooling	12059	7545	20
Add	14434	16096	12
AXI-DMA	24000	20041	62

Unsurprisingly, the convolution modules dominate resource consumption, reflecting their central role in feature extraction. The 3×3 convolution module is the most demanding, utilizing 28,649 Look-Up Tables (LUTs), 62,454 Flip-Flops (FFs), and 82 Block RAMs (BRAMs). The 1×1 convolution module, while less resource-intensive due to its smaller kernel size, still requires substantial resources at 17,216 LUTs, 26,639 FFs, and 52 BRAMs. These significant figures underscore the inherent computational complexity and substantial weight storage needs of convolutional operations, even when optimized with the proposed sparse, multiplier-free computation approach. The Add module, essential for multi-scale feature fusion through residual connections, incurs a moderate cost of 14,434 LUTs, 16,096 FFs, and 12 BRAMs.The pooling module, employing a pipelined, multi-window architecture with dual-buffered data management to sustain throughput, utilizes 12,059 LUTs, 7,545 FFs, and 20 BRAMs. Finally, the AXI-DMA controller, critical for high-bandwidth data movement between the host CPU, off-chip DRAM, and on-chip processing units, is a major resource consumer, particularly for memory blocks, requiring 24,000 LUTs, 20,041 FFs, and 62 BRAMs to implement its efficient data transfer mechanisms.

IV. CONCLUSION

This paper presents a high-performance accelerator for spiking neural network. A parameterized convolution module with input/output channel parallelism and multiplier-free computation enables efficient hardware reuse across layers. Operating at 200 MHz, the design achieves 230.4 GSOP/s throughput and 32.9 GSOP/W energy efficiency. Compared to existing SNN accelerators, it improves throughput and energy efficiency by 28% and 53%, respectively, while maintaining

compact logic and memory utilization. These results demonstrate the practicality of the proposed architecture for scalable deployment of SNNs on resource-constrained edge platforms.

REFERENCES

[1] A. Tavanaei, M. Ghodrati, S. R. Kheradpisheh, T. Masquelier, and A. Maida, "Deep learning in spiking neural networks," *Neural networks*, vol. 111, pp. 47–63, 2019.

[2] Q. T. Pham, T. Q. Nguyen, P. C. Hoang, Q. H. Dang, D. M. Nguyen, and H. H. Nguyen, "A review of snn implementation on fpga," in *2021 international conference on multimedia analysis and pattern recognition (MAPR)*. IEEE, 2021, pp. 1–6.

[3] J. Zhang, R. Wang, T. Wang, J. Liu, S. Dang, and G. Zhang, "A configurable spiking convolution architecture supporting multiple coding schemes on fpga," *IEEE Transactions on Circuits and Systems II: Express Briefs*, vol. 69, no. 12, pp. 5089–5093, 2022.

[4] C. Sun, W. Song, Q. Chen, C. Dai, Y. Fu, and L. Li, "An energy efficient residual spiking neural network accelerator with ternary spikes," *IEEE Transactions on Computer-Aided Design of Integrated Circuits and Systems*, 2025.

[5] M. Davies, N. Srinivasa, T.-H. Lin, G. Chinya, Y. Cao, S. H. Choday, G. Dimou, P. Joshi, N. Imam, S. Jain *et al.*, "Loihi: A neuromorphic manycore processor with on-chip learning," *Ieee Micro*, vol. 38, no. 1, pp. 82–99, 2018.

[6] M. Li, Y. Kan, R. Zhang, and Y. Nakashima, "A fully-parallel reconfigurable spiking neural network accelerator with structured sparse connections," in *2024 IEEE International Symposium on Circuits and Systems (ISCAS)*. IEEE, 2024, pp. 1–5.

[7] L. Feng, Y. Zhang, and Z. Zhu, "An efficient multilayer spiking convolutional neural network processor for object recognition with low bitwidth and channel-level parallelism," *IEEE Transactions on Circuits and Systems II: Express Briefs*, vol. 69, no. 12, pp. 5129–5133, 2022.

[8] Y. Kuang, X. Cui, C. Zou, Y. Zhong, Z. Dai, Z. Wang, K. Liu, D. Yu, and Y. Wang, "An event-driven spiking neural network accelerator with on-chip sparse weight," in *2022 IEEE International Symposium on Circuits and Systems (ISCAS)*. IEEE, 2022, pp. 3468–3472.

[9] Q. Chen, C. Gao, X. Fang, and H. Luan, "Skydiver: A spiking neural network accelerator exploiting spatio-temporal workload balance," *IEEE Transactions on Computer-Aided Design of Integrated Circuits and Systems*, vol. 41, no. 12, pp. 5732–5736, 2022.

[10] Q. Chen, C. Sun, C. Gao, X. Fang, and H. Luan, "Framefire: Enabling efficient spiking neural network inference for video segmentation," in *2023 IEEE 5th International Conference on Artificial Intelligence Circuits and Systems (AICAS)*. IEEE, 2023, pp. 1–5.

[11] W. Ye, Y. Chen, and Y. Liu, "The implementation and optimization of neuromorphic hardware for supporting spiking neural networks with mlp and cnn topologies," *IEEE Transactions on Computer-Aided Design of Integrated Circuits and Systems*, vol. 42, no. 2, pp. 448–461, 2022.

[12] M. Yin, X. Cui, F. Wei, H. Liu, Y. Jiang, and X. Cui, "A reconfigurable fpga-based spiking neural network accelerator," *Microelectronics Journal*, vol. 152, p. 106377, 2024.

A precise current-controlled resistor and its applications in zero-pole tracking frequency compensation for LDO

Guanting Liu[1], Guijuan Zhao*[1], Feng Shi[2], Xiaohuan You[3], Shuhai Chen[3]

[1]School of Physical Science and Technology, Lanzhou University, Gansu, China

[2]Chengdu Enjixin Technology Company, Chengdu, China

[3]School of Intergrated Circuit Science and Engineering, University of Electronic Science and Technology of China, Chengdu, China

*Email: zhaogj@lzu.edu.cn

Abstract—This paper presents a Precise Current-Controlled Resistor (PCCR) and its applications in zero-pole tracking frequency compensation for low-dropout (LDO) regulators. The proposed PCCR, implemented using MOS transistors and an operational amplifier, forces the matched MOS to operate in the deep linear region. The operational amplifier establishes the required gate-source voltage, enabling the MOS to provide an equivalent resistance that linearly varies with the reciprocal of the load current. This enables precise positioning of the compensation zero to accurately track the pole, or precise control over the complex-pole Q within an optimal range (0.5–0.7), preventing frequency peaking and abrupt phase drops that cause transient ringing or undershoot. The LDO using the proposed PCCR adopts Ahuja compensation. Simulation results show that under light-load conditions, the compensation zero accurately compensates for the non-dominant pole, while under heavy-load conditions, the complex-pole Q is precisely maintained around 0.6, achieving a load settling time of 0.84 μs with an undershoot of 24.09 mV.

Keywords—Precise Current-Controlled Resistor, zero-pole tracking frequency compensation, LDO, Ahuja compensation

I. INTRODUCTION

As processor power consumption reaches the kilowatt range and computing capabilities soar, the demand intensifies for efficient, stable power management integrated circuits(PMIC) capable of supporting high switching frequencies and instantaneous high-current loads. Critical modules within these PMIC，such as multiphase controllers and Driver MOSFET(DrMOS) in high-performance computing chips，rely on robust LDO regulators to power their internal blocks. Meeting the dual demands of high load capacity and wide bandwidth presents formidable design challenges.

A major challenge lies in managing substantial load fluctuations during operating mode transitions or light-to-heavy load shifts. While off-chip capacitors (typically in the μF range) supply transient current, their large size restricts light-load bandwidth, increases circuit area, and elevates power dissipation. Conversely, smaller capacitors degrade both stability and transient performance. Additionally, shifting pole positions in LDOs under variable loads risk system destabilization. High-frequency switching further demands greater internal LDOs bandwidth for rapid transient response, complicating stability control.

To overcome these challenges, advanced frequency compensation techniques enable stable LDO operation even with small, low-ESR capacitors. As Fig. 1 illustrates, dynamic zero tracking compensates for pole shifts by following load variations. Common approaches include introducing a

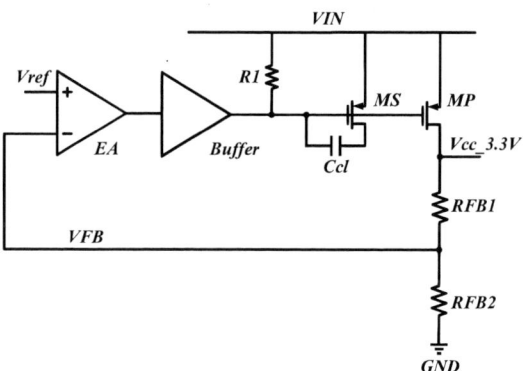

Fig. 1 The architecture of the precise current-controlled resistor

ground-referenced zero[1], adding a zero via resistive feedforward[2], or leveraging inherent compensation network zeros[3]. Regardless of approach, a MOS transistor operating in the linear region serves as a current-controlled dynamic resistor, thus enabling the zero to dynamically track load variations.

Several methods exist to implement the current-controlled resistor. One approach mirrors the gate-source voltage (V_{gs}) of the output transistor to bias another MOS transistor in the linear region, causing its resistance to vary with load current [2]. However, this technique is highly vulnerable to process and temperature variations. An alternative method employs an amplifier to clamp the gate-source voltage of MOS, enhancing robustness against such variations[4]. Although this method improves stability margins, the inherent MOS square-law behavior results in resistance proportional to the square root of the load current. This nonlinearity complicates precise control of the complex-pole Q under heavy loads in Ahuja compensation, potentially degrading stability.

This paper proposes a Precise Current-Controlled Resistor (PCCR), fabricated in 180 nm process, whose resistance varies linearly with the reciprocal of the load current while exhibiting minimal sensitivity to process or temperature variations. Applied to Ahuja-compensated LDOs, the PCCR enables precise control of complex-pole Q and zero positions. Section II details the resistor design; Section III analyzes pole-zero locations and presents circuit topologies; Section IV discusses simulation results; and Section V concludes the work.

II. PRECISE CURRENT-CONTROLLED RESISTOR

A. Circuit Topology Principle

Fig. 2 illustrates the structure of the Precise Current-

979-8-3315-3918-4/25 $31.00 © 2025 IEEE

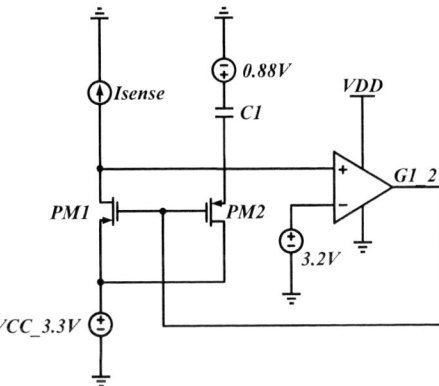

Fig. 2 The architecture of the precise current-controlled resistor

Controlled Resistor(PCCR). Key components include an ideal current source I_{sense} simulating the load current, an ideal voltage source V_{cc} representing the 3.3 V LDO output, an ideal 3.2 V reference voltage source modeling the LDO output divider, and an ideal voltage source modeling the capacitor node voltage. The structure also contains a five-transistor differential operational transconductance amplifier(OTA), a PMOS transistor PM1 for converting load current to voltage, a mirroring PMOS transistor PM2 serving as the equivalent resistor, and a capacitor C1 modeling the compensation structure.

To achieve linear variation of PM2's equivalent resistance R_{ds} with $1/I_{sense}$, the OTA leverages its virtual short characteristic to clamp PM1's drain voltage to the fixed 3.2 V reference from the LDO output divider. This action clamps PM1's V_{ds} to a constant 100 mV, ensuring operation in the deep linear region. Since I_{sense} flows through PM1, its drain-source resistance R_{ds} can be expressed as follows:

$$R_{ds,PM1} = \frac{100mV}{I_{sense}} = \frac{1}{\mu_p C_{ox} \left(\frac{W}{L}\right)_{PM1} \left(V_{GS} - V_{th}\right)_{PM1}} \quad (1)$$

Since PM2 connects to a capacitor with negligible current, it similarly operates in the deep linear region, and its equivalent drain-source resistance R_{ds} can be expressed as follows:

$$R_{ds,PM2} = \frac{1}{\mu_p C_{ox} \left(\frac{W}{L}\right)_{PM2} \left(V_{GS} - V_{th}\right)_{PM2}} \quad (2)$$

Given PM2's source voltage is held near 3.3 V by the capacitor, the overdrive voltages V_{ov} of PM1 and PM2 are approximately equal. Assuming negligible mismatch and identical transistor parameters (including MOS mobility μ_p and unit area capacitance C_{ox}), combining equations (1) and (2) yields PM2's equivalent drain-source resistance R_{ds}:

$$R_{ds,PM2} = \frac{\left(W/L\right)_{PM1}}{\left(W/L\right)_{PM2}} \frac{100mV}{I_{sense}} \propto \frac{1}{I_{sense}} \quad (3)$$

The designed equivalent drain-source resistance(R_{ds}) of PM2 is solely determined by the width-to-length ratios(W/L) of PM1 and PM2, and I_{sense}. When the ratio of their W/L values is constant, PM2's R_{ds} varies linearly with $1/I_{sense}$, as verified

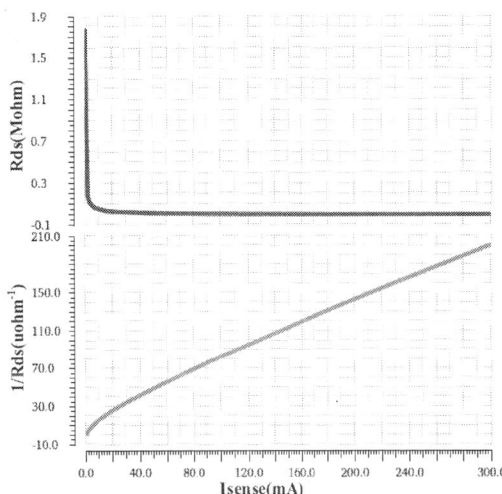

Fig. 3 Simulated $1/R_{ds,PM2}$ with I_{sense} ranging from 0 to 0.3 mA

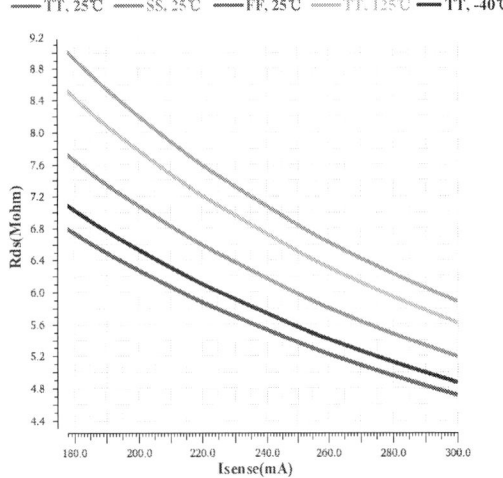

Fig. 4 Simulation with process corner changing and temperature ranging from -40 °C to 125 °C

in Fig. 3 Crucially, parameters sensitive to process and temperature variations, such as mobility and threshold voltage, mutually cancel, resulting in minimal sensitivity to process and temperature variations

B. Simulation Results Analysis

Fig. 3 reveals that PM2's equivalent drain-source resistance(R_{ds}) deviates from linearity with $1/I_{sense}$ at low currents. This occurs because low current levels drive PM1 into the subthreshold region, invalidating the deep-linear-region MOS resistance model. Crucially, this deviation has negligible impact on actual compensation performance, while the linear relationship holds robustly under heavy loads. As shown in Fig. 4, cross-corner simulations confirm PM2's R_{ds} variation within approximately ±10% across process corners. Similarly, over the industrial temperature range(-40°C to 125°C), resistance drift remains within ±6%, demonstrating exceptional immunity to process and temperature variations.

979-8-3315-3918-4/25 $31.00 © 2025 IEEE

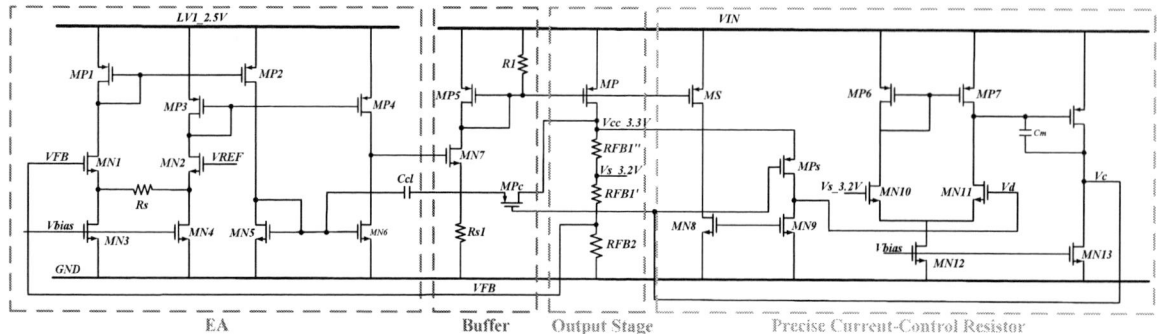

Fig. 5 LDO Circuit Schematic with Ahuja Compensation

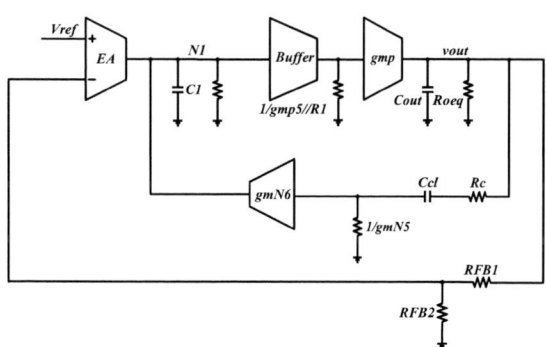

Fig. 6 Small-Signal Equivalent Circuit of the LDO

III. ZERO-POLE TRACKING FREQUENCY COMPENSATION

The circuit in Fig. 5 integrates the current-controlled resistor into the LDO with Ahuja compensation. Here, capacitor C_{cl} and transistor MP_c connect V_{out} to the gate of the error amplifier(EA). This configuration eliminates the feedforward path and leverages the Miller multiplication effect to isolate the dominant and non-dominant poles under heavy loads, significantly enhancing stability.

The small-signal model of the circuit is shown in Fig. 6. The zeros introduced by the PCCR and capacitance can be approximately expressed as:

$$Z_1 = \frac{1}{C_{cl}\left(\frac{1}{g_{mN5}} + R_c\right)} \tag{4}$$

Where C_{cl} is the compensation capacitor of the LDO, g_{mN5} is the transconductance of MN5, and R_c is the current-control equivalent resistance of MP_c.

Under light load conditions, the dominant pole P1 located at the output can be approximately expressed as:

$$P_1 \approx -\frac{1}{R_{oeq}C_{out}} \tag{5}$$

where R_{oeq} are the output resistance of the LDO, C_{out} are the output capacitor of the LDO.

the non-dominant pole P2 located at the output of the EA can be approximately expressed as:

$$P_2 \approx -\frac{1}{r_{o1}C_1} \tag{6}$$

where r_{o1} are the output resistance of the EA, C_1 are the input capacitor of the Buffer.

Under heavy load conditions, the Miller multiplication effect pushes the dominant pole toward the error amplifier's (EA) output stage while generating a complex pole pair. Excessive bandwidth setting may bring these poles within the operating bandwidth, severely compromising stability. The complex-pole Q can be approximately expressed as:

$$Q \approx \sqrt{\frac{g_{mp}C_{cl}^2 A_{v2}\left(\frac{1}{g_{mN5}} + R_c\right)}{C_{out}C_1}} \tag{7}$$

where g_{mp} are the transconductance of the output power transistor, A_{v2} are the gain of the Buffer.

By tuning the current-controlled resistor to decrease linearly with increasing load current, the complex-pole Q is precisely maintained between 0.5 and 0.7 across wide load and input voltage variations. This suppresses frequency-response peaking induced by complex poles, effectively eliminating prolonged damped oscillations during transient response.

IV. SIMULATION RESULT

This paper experimentally validates the PCCR for pole-zero tracking compensation in LDOs using a 180 nm process. Operating at 4.5–8 V input with a fixed 3.3 V output, the design achieves 400 mA maximum load current and reduces output capacitance to 2.5 μF. Fig. 7 depicts the frequency response with Ahuja compensation at 6.5 V input under 1 μA–250 mA load step with 100 ns rise time. As load increases from light to heavy, the zero dynamically tracks pole positions, compensating the non-dominant pole at light load and precisely controlling the complex-pole Q at heavy load. Under heavy load, the LDO attains a 1.04MHz bandwidth with only 588 fF compensation capacitance.

Fig. 8 confirms that across varying operating conditions—load current from 100 mA to 300 mA and input voltage from 4.5 V to 6.5 V—the complex-pole Q in the Ahuja-compensated LDO remains tightly constrained between 0.58 and 0.63.

Meanwhile, Fig. 9 demonstrates the transient response under a 1 μA-to-250 mA load step with 100 ns rise time. The

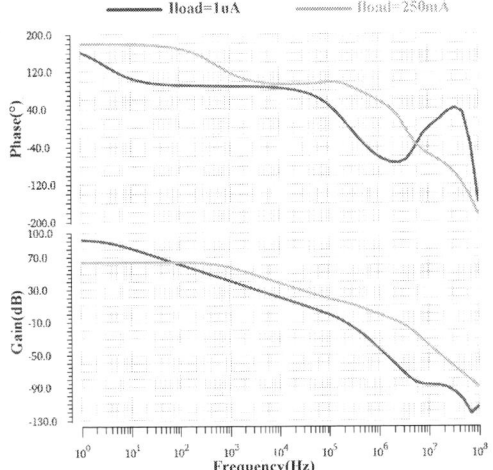

Fig. 7 Simulated Frequency Response of the LDO with the Ahuja-Compensation

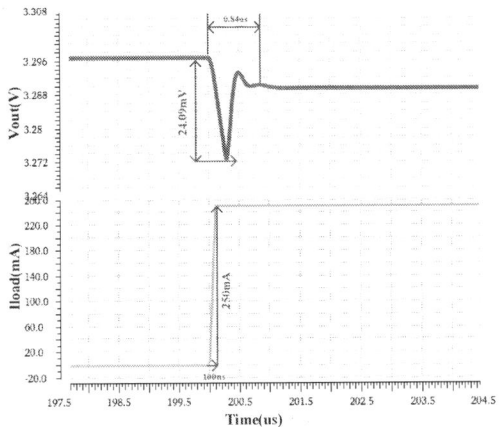

Fig. 8 Simulated Transient Response of the LDO with Ahuja Compensation

Ahuja-compensated LDO achieves settling time within 0.84 μs, with voltage undershoot limited to 24.09 mV.

V. CONCLUSION

This paper proposes a Precise Current-Controlled Resistor (PCCR) for pole-zero tracking compensation in LDOs. Applied to Ahuja-compensated LDOs, the PCCR enables the zero to compensate the non-dominant pole under light loads while maintaining the complex-pole Q at approximately 0.6 under heavy loads. This precise regulation effectively mitigates complex-pole effects, yielding a settling time of 0.84 μs with only 24.09 mV undershoot. A performance comparison with other LDO architectures is provided in Table I:

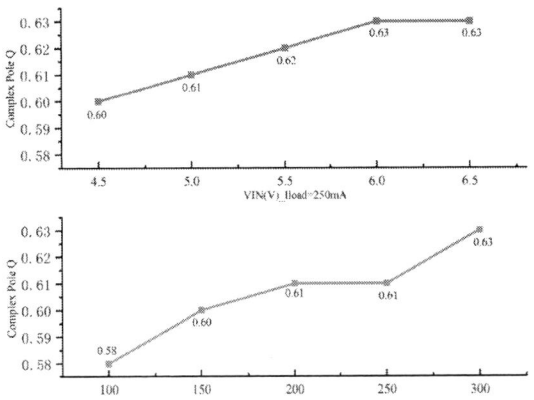

Fig. 9 Illustration of the Variation in the Complex Pole Q under Heavy Load for the Ahuja-Compensated LDO, with Load Current from 100–300 mA and Input Voltage from 4.5–6.5 V

TABLE I.   COMPARISON RESULTS WITH OTHER WORKS

Ref	[3]	[1]	[5]	This Work
Process (nm)	110	350	180	180
V_{IN} (V)	3-4.5	24	1.5-3.6	4.5-8
V_{OUT} (V)	1.85	5	0.8-3.3	3.3
C_{OUT} (μF)	1	No	4.7	2.5
$I_{LOAD,MAX}$ (mA)	300	150	1200	400
T_{settle} (μs)	32.8	26	No	0.84
Tran.Drop (mV)	38.6	61	54.7	24.09

REFERENCES

[1] W. Qingqing, D. Yanhang, C. Zihao, and L. Xinyu, "Design of wide input voltage range and low quiescent current LDO,"IEICE Electronics Express, vol.21, pp.161–166, July 2024

[2] L. Dawei, L. Dongsheng, and K. Chaojian, "An ultra low power low cost LDO for UHF RFID tag," IEICE Electronics Express, vol. 14, pp.21–27, Dec 2016

[3] T. Jiacheng, F. Bo, and L. Huimin, "Multi-zero pole dynamically compensated LDO with low quiescent current,"Int. J. Electron. Commun, vol. 184, p. 155423, July 2024

[4] W. Yi, H. Lenian, N. Zhihua, and S. Yali, "A controllable resistor and its applications in pole-zero tracking frequency compensation methods for LDOs," J. Semicond., vol. 30, no. 9, p. 95013, Sep2009.

[5] W. Annan, Z. Xi, C. Wei, "A high-load currentlow-dropout regulator with adaptive ESR compensation,"Int. J. Electron. Commun, vol. 175, p.155067, Dec 2023

[6] L. Yat-Hei, K. Wing-Hung, "24.5 a 0.9V 0.35μm adaptively biased CMOS LDO regulator with fast transient response,"IEEE International Solid-State Circuits Conference, vol.24, Dec 2008

[7] L. Qianqian, C. Zhiming, G. Zheng, and S. Yin, "A 200 mA CMOS low-dropout regulator with double frequency compensation techniques for SoC applications,"J. Semicond., vol. 32, no.11, Nov2011.

[8] K. Chun Kwok, Philip K. T Mok, "Pole-zero tracking frequency compensation for low dropout regulator,"IEEE International Symposium on Circuits and Systems. Proceedings, vol.4, pp.735-738, Dec 2022

[9] M. Al-Shyoukh, H. Lee, IEEE, and Raul Perez. Wei, "A transient-enhanced low-quiescent current low-dropout regulator with buffer impedance attenuation, vol.48, No. 8, Aug 2007

[10] H. Chenghan, D. Chungyen, and L. Shuennyuh, "Power management with energy harvesting from aheadphone_jack,"IEEE International Symposium on Circuits and Systems, vol. 14, p.978-1, Dec 2014

ASSVD: A Self-Supervised Surgical Video Desmoking Network with Sparse Attention

Yinna Zhu[1], Wanyi Zhou[1], Zijing Zhang*[2], Gengsheng Chen[1,3,4], Wei Xu*[3]

[1] College of Integrated Circuits and Micro-Nano Electronics, Fudan University, Shanghai, China
[2] Department of Thyroid and Breast Surgery, Huashan Hospital, Fudan University, Shanghai, China
[3] School of Microelectronics, Fudan University, Shanghai, China
[4] Jiashan Fudan Institute, Jiaxing, Zhejiang, China

* Email: Zijing_zhang@fudan.edu.cn, wei_xu@fudan.edu.cn

Abstract—**Surgical smoke significantly degrades endoscopic video quality, thus impairing intraoperative visibility and surgical decision-making. Existing methods struggle to cope with the dynamic and spatially non-uniform characteristics of surgical smoke. In this paper, we propose ASSVD, a self-supervised video desmoking network with sparse attention. Our method leverages the pre-smoke (PS) frame as unaligned supervision to enable learning without paired annotations. We further incorporate a Sparse Transformer Block (STB) to enhance structural awareness while maintaining computational efficiency. To facilitate evaluation, we construct MegaClear, a surgical video desmoking dataset covering multiple procedures. Experiments demonstrate that ASSVD outperforms existing methods in both quantitative metrics and visual quality, especially under dense and dynamic smoke conditions.**

Keywords—video desmoking, self-supervised learning, sparse attention, surgical procedures

I. INTRODUCTION

Endoscopic videos provide real-time, high-resolution visualization that enhances the precision and safety of minimally invasive procedures. However, frequent use of high-energy instruments (e.g., electrocautery and ultrasonic scalpels) inevitably generates dense smoke during tissue dissection or coagulation, which severely degrades visual quality and increases the risk of intraoperative complications. These challenges highlight the need for reliable and efficient video desmoking to support intraoperative decision-making.

Early image desmoking methods [1-3] primarily rely on the Atmospheric Scattering Model (ASM) , using physical priors such as the Dark Channel Prior (DCP) [1] and Color Attenuation Prior (CAP) [2] to estimate transmission maps. Although effective in natural scenes, these approaches fail in surgical videos due to domain-specific challenges, such as non-uniform smoke, constrained lighting, and complex anatomical structures. As a result, these methods often lead to color distortion and loss of detailed tissue textures.

With the advancement of deep learning, various learning-based approaches [4-6] have been introduced. Dong et al. [4] propose an encoder-decoder architecture to preserve multi-scale structures without relying on physical priors or intermediate parameter estimation. Guo et al. [5] extend this by integrating Swin Transformer Block [7] to refine shallow CNN features, improving spatial fidelity under dense haze. Zheng et al. [6] propose a GAN-based assistance system that jointly handles motion blur and smoke removal. However, due to the scarcity of paired smoky-clear images in real surgical settings, most existing models are trained on synthetic datasets and therefore generalize poorly to real-world surgical videos. In addition, these methods typically focus on single-image desmoking and fail to exploit temporal coherence across video frames. This neglect frequently results in flickering artifacts and perceptual inconsistencies under rapidly changing smoke conditions. To address the lack of temporal modeling, SelfSVD [8] introduces a self-supervised video desmoking framework that leverages inter-frame consistency without paired data. However, it struggles to recover fine anatomical structures due to limited spatial modeling.

While temporal consistency is essential, effective desmoking in surgical videos also requires spatial modeling to recover anatomical structures degraded by smoke. Transformer-based architectures offer global spatial awareness, but dense self-attention incurs quadratic cost, limiting their scalability in high-resolution settings. To improve efficiency, sparse attention mechanisms have emerged as effective alternatives. For instance, DRSformer [9] adopts top-k attention to retain important regions with reduced overhead, illustrating the potential of sparse attention for structure-aware image restoration. Despite recent advances, existing methods still struggle to efficiently model spatial structure and temporal coherence, especially under the dense and dynamic smoke conditions typical of real-world surgeries.

In this paper, we propose ASSVD, a self-supervised video desmoking network tailored for surgical scenes. Inspired by SelfSVD [8], our method leverages the pre-smoke (PS) frame as unaligned supervision and incorporates sparse attention to improve structural fidelity with low computational cost. We further construct the MegaClear, a real-world surgical video dataset covering multiple procedures.

Our main contributions are summarized as follows:

- We develop a self-supervised framework that uses the PS frame as pseudo ground truth to guide restoration, thus eliminating the reliance on paired annotations required by prior methods.

- We introduce a Sparse Transformer Block (STB) that applies sparse attention to spatial regions critical for anatomical restoration, thereby effectively improving desmoking quality under non-uniform surgical smoke while reducing computational overhead.

- We construct MegaClear, the first high-resolution surgical video desmoking dataset covering thoracic, gastrointestinal, and thyroid procedures, enabling standardized evaluation across different surgery types.

- Experiments show that ASSVD achieves the highest PSNR across all subsets (27.20 in thoracic, 24.83 in gastrointestinal, 28.07 in thyroid), and consistently outperforms the state-of-the-art methods in both full-reference and no-reference metrics.

979-8-3315-3918-4/25 $31.00 © 2025 IEEE

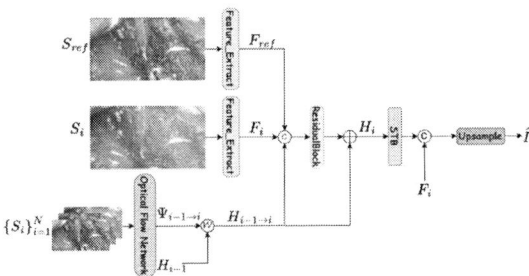

Fig. 1 The overall architecture of the proposed ASSVD.

II. METHODS

A. Overall Architecture

The overall architecture of our proposed ASSVD is illustrated in Fig. 1. Given a smoky video sequence $\{S_i\}_{i=0}^{N}$, we designate the first frame S_0 as the PS frame, which serves two purposes: (1) a structural reference S_{ref} to provide high-fidelity anatomical cues, and (2) an unaligned supervision target S_{ps} for self-supervised learning without paired annotations. The remaining frames $\{S_i\}_{i=1}^{N}$ are treated as smoky inputs to be restored. Formally, we train the video desmoking network $\mathcal{D}$ by minimizing the following objective:

$$\Theta_D^* = \arg\min_{\Theta_D} \mathcal{L}\big(\mathcal{D}(\{S_i\}_{i=1}^{N}, S_{ref}; \Theta_D), S_{ps}\big), \quad (1)$$

where $\mathcal{L}$ denotes the learning objective.

Specifically, given the i-th video frame S_i, we first extract its shallow features F_i and the reference frame features F_{ref} using the feature extraction module. The optical flow estimation module then computes inter-frame motion $\Psi_{i-1\to i}$, which is used to warp the previous hidden features H_{i-1} for temporal alignment, resulting in spatially aligned features $H_{i-1\to i}$. These features are then fused with F_i and F_{ref} to obtain the combined feature representation H_i. Subsequently, we apply the Sparse Attention Block (STB), which assigns sparse attention weights to H_i by retaining the top-k most relevant spatial locations, thereby enhancing structural focus while reducing redundancy. Finally, the attention-enhanced features are concatenated with the original frame features F_i and fed into the upsampling reconstruction module to generate the restored output $\hat{I}_i$.

B. Optical Flow Network

To address spatial inconsistencies between adjacent frames, we adopt PWC-Net [10] to estimate optical flow across the video sequences, as illustrated in Fig. 2. To reduce computational complexity, we first downsample the input video sequence $\{S_i\}_{i=1}^{N}$ and then use the pretrained PWC-Net to estimate the optical flow $\{\Psi_{i-1\to i}\}_{i=2}^{N}$.

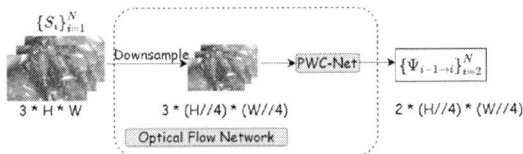

Fig. 2 The pipeline of Optical Flow Network.

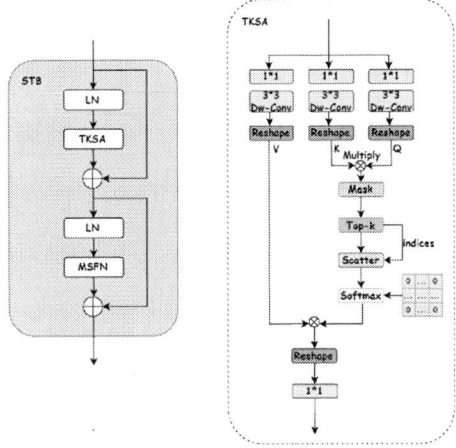

Fig. 3 The architecture of Sparse Transformer Block.

C. Sparse Transformer Block (STB)

In surgical videos, smoke frequently causes localized occlusion and structural degradation, posing challenges for the recovery of fine anatomical details. To mitigate this, we integrate a Sparse Transformer Block (STB) [9] into ASSVD to direct attention to spatial regions with more reliable features while reducing redundant computations.

As shown in Fig. 3, the STB consists of a Top-K Sparse Attention (TKSA) module and a Mixed-Scale Feed-forward Network (MSFN). TKSA computes query-key similarity scores and selects the top-k most relevant positions for each query by masking out the responses with low confidence. The selected indices are then mapped back to their original spatial locations using a scatter operation, followed by softmax normalization to produce the sparse attention map. This results in a sparse attention distribution focused on the most relevant locations, which is defined as:

$$\text{SparseAtt}(Q, K, V) = \text{softmax}\left(\mathcal{T}_k\left(\frac{QK^T}{\lambda}\right)\right)V, \quad (2)$$

where $\mathcal{T}_k(\cdot)$ denotes a learnable top-k selection operator, and λ is an optional temperature factor. Following TKSA, the MSFN module aggregates attention-weighted features across multiple receptive fields to enhance both local details and global consistency. Its lightweight design maintains efficiency while enhancing spatial representation.

D. Loss Function

To improve both spatial consistency and visual realism under an unpaired supervision setting, we train the ASSVD model using a composite loss function, as illustrated in Fig. 4.

1) Reconstruction Loss: To mitigate the misalignment between the PS frame S_{ps} and the output $\hat{I}_i$, we estimate the optical flow $\Psi_{ps\to i}$ and warp $\hat{I}_i$ to align with S_{ps}, yielding $\hat{I}_{i\to ps} = \mathcal{W}(\hat{I}_i, \Psi_{ps\to i})$. The reconstruction loss is defined as:

$$\mathcal{L}_{rec} = \sum_{i=1}^{N} \left\| V_i \odot \left(\hat{I}_{i\to ps} - S_{ps}\right) \right\|_1, \quad (3)$$

where $\odot$ denotes pixel-wise multiplication, V_i is a mask of valid positions of optical flow as defined in SelfSVD [8].

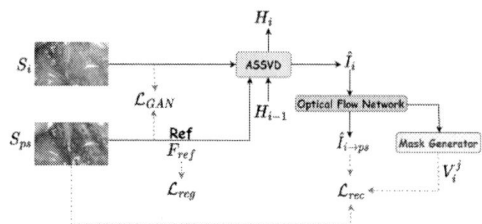

Fig. 4 The loss function design of ASSVD.

2) Regularization Loss: To prevent overreliance on the reference features F_{ref}, we introduce a regularization loss $\mathcal{L}_{reg}$ which is formulated as:

$$\mathcal{L}_{reg} = \|F_{ref}\|_1. \qquad (5)$$

3) GAN Loss: To further improve perceptual quality, we adopt LSGAN [11] with a patch-based discriminator. The generator $\mathcal{D}$ (i.e., the proposed ASSVD) is optimized with the following adversarial objective:

$$\mathcal{L}_{GAN} = \frac{1}{2}\mathbb{E}_{S\sim P_S}\left[\mathcal{DISC}\left(\mathcal{D}\left(S, S_{ref}\right)\right) - 1\right]^2, \qquad (4)$$

where $S = \{S_i\}_{i=1}^N$ denotes the smoky input frames, and $\mathcal{DISC}$ represents the patch-based discriminator.

4) Total Loss: The overall training objective is defined as:

$$\mathcal{L} = \mathcal{L}_{rec} + \lambda_{GAN}\mathcal{L}_{GAN} + \lambda_{reg}\mathcal{L}_{reg}, \qquad (6)$$

where λ_{GAN} and λ_{reg} are set to 1.0 and 0.05, respectively.

III. EXPERIMENTS

A. Dataset

We introduce MegaClear, a large-scale dataset for surgical video desmoking, consisting of 613 video clips with a total of 24,754 frames at 1920×1080 resolution. The data are collected from 22 real-world endoscopic procedures, covering thyroid (9 cases), gastrointestinal (3 cases), and thoracic (10 cases) surgeries. Further details are provided in TABLE I.

Each clip in MegaClear captures a full smoke development process, ranging from a clear pre-smoke frame to peak smoke accumulation. The first frame serves as an unaligned reference to facilitate supervision. All samples are manually selected to cover a wide range of anatomical scenes and smoke conditions.

B. Evaluation Metrics

We evaluate the restored results using both full-reference and no-reference image quality assessment (IQA) metrics. For full-reference evaluation, we first align each restored frame to the PS frame using estimated optical flow, resulting in an aligned frame $\hat{I}_{i\to ps}$. We then compute PSNR and SSIM between the aligned frame and the PS frame. These metrics are referred to as Aligned PSNR and Aligned SSIM, respectively, and are defined as:

$$\text{Aligned PSNR} = \text{PSNR}(S_{ps}, \hat{I}_{i\to ps}), \qquad (7)$$

$$\text{Aligned SSIM} = \text{SSIM}(S_{ps}, \hat{I}_{i\to ps}). \qquad (8)$$

To better assess perceptual quality, we adopt three additional no-reference metrics, including Natural Image Quality Evaluator (NIQE), No-Reference Quality Metric (NRQM), and Perceptual Index (PI).

TABLE I.   DETAILS OF MEGACLEAR DATASET

Subset	Content		
	# Procedures	# Video Clips	# Frames
Thyroid	9	322	15,088
Gastrointestinal	3	215	7,577
Thoracic	10	76	2,089
Total	22	613	24,754

TABLE II.   RESULTS ON THORACIC DATASET

Model	PSNR↑	SSIM↑	NIQE↓	PI↓	NRQM↑
DCP	24.1458	0.7748	4.672	4.104	6.403
MSBDN	24.9192	0.7509	7.160	5.460	6.377
Dehamer	26.8074	**0.9192**	7.442	7.242	3.152
SelfSVD	25.4146	0.7799	4.380	**3.775**	<u>6.544</u>
SelfSVD_C	<u>27.1003</u>	0.8024	**4.334**	<u>3.867</u>	6.503
SelfSVD_P	26.8723	0.7965	4.502	3.918	**6.597**
ASSVD	**27.2014**	<u>0.8061</u>	<u>4.345</u>	3.869	6.505

Note: ↑ : higher is better, ↓ : lower is better. Best results are in **bold**, second-best are <u>underlined</u>.

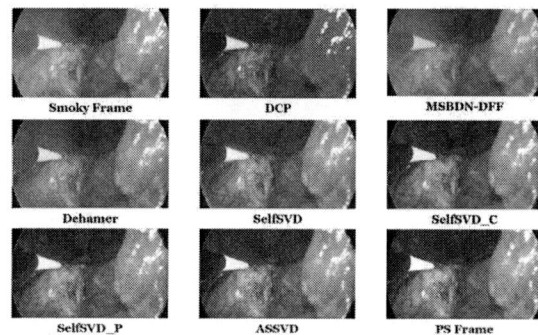

Fig. 5 Visual comparisons on Thoracic Dataset.

C. Quantitative Comparisons with state-of-the-arts

We compare ASSVD with 6 baselines, including DCP [1], MSBDN [4], Dehamer [5], SelfSVD [8] and its two enhanced variants, SelfSVD_C and SelfSVD_P, which incorporate CBAM [12] and PromptBlock [13], respectively. All models are retrained on the MegaClear dataset for fair comparison.

Quantitative results on three subsets are illustrated in TABLE II. –IV. DCP performs poorly across all metrics, reflecting its limited suitability for surgical scenarios, while MSBDN shows moderate results but underperforms on perceptual metrics, such as NIQE and PI. Although Dehamer achieves the highest SSIM on thoracic and gastrointestinal subsets, its low NRQM and high PI scores suggest degraded perceptual quality, possibly due to over-smoothing. Self-supervised variants exhibit more favorable trends: SelfSVD_C attains better full-reference scores on thoracic data, while SelfSVD performs superior on no-reference metrics for gastrointestinal and thyroid subsets. However, their overall performance remains inconsistent across anatomical domains.

In contrast, ASSVD achieves the highest PSNR on all datasets, with 27.20 (thoracic), 24.83 (gastrointestinal), and 28.07 (thyroid), indicating strong pixel-level reconstruction. It also ranks first in NRQM on gastrointestinal (6.26) and thyroid (6.81), and closely follows SelfSVD_P on thoracic (6.51 vs. 6.60). These results confirm that ASSVD delivers more stable and balanced performance across both full-reference and no-reference metrics, demonstrating improved generalization under diverse surgical scenarios.

TABLE III.   RESULTS ON GASTROINTESTINAL DATASET

Model	PSNR↑	SSIM↑	NIQE↓	PI↓	NRQM↑
DCP	22.2043	0.7514	4.790	4.352	6.194
MSBDN	23.7904	0.8259	5.397	5.544	4.670
Dehamer	23.3804	**0.8653**	6.868	6.539	3.993
SelfSVD	23.6984	0.7618	**3.938**	3.967	6.159
SelfSVD_C	23.7121	0.7615	4.256	4.102	6.183
SelfSVD_P	23.7815	0.7647	4.194	4.055	6.222
ASSVD	**24.8317**	0.8125	3.996	**3.948**	**6.261**

Note: ↑ : higher is better, ↓ : lower is better. Best results are in **bold**, second-best are underlined.

TABLE IV.   RESULTS ON THYROID DATASET

Model	PSNR↑	SSIM↑	NIQE↓	PI↓	NRQM↑
DCP	23.7970	0.7982	**3.532**	4.311	5.628
MSBDN	24.4719	0.8447	5.301	5.459	4.793
Dehamer	25.1514	**0.9262**	8.557	8.030	3.340
SelfSVD	27.9600	0.8470	4.147	**3.801**	6.800
SelfSVD_C	27.9604	0.8449	4.149	3.842	6.726
SelfSVD_P	27.2588	0.8387	4.317	3.933	6.698
ASSVD	**28.0708**	0.8472	4.224	3.834	**6.813**

Note: ↑ : higher is better, ↓ : lower is better. Best results are in **bold**, second-best are underlined.

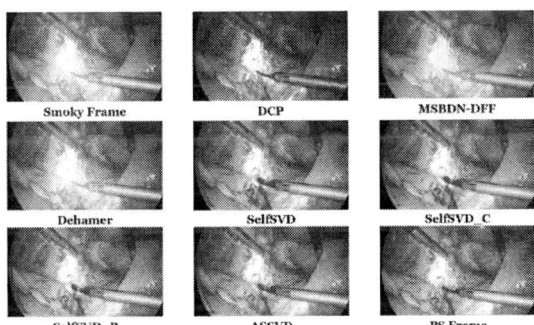

Fig. 6 Visual comparisons on Gastrointestinal Dataset.

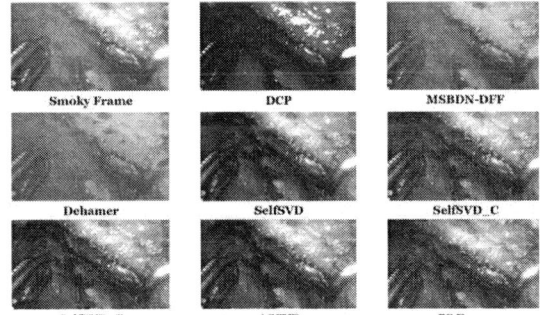

Fig. 7 Visual comparisons on Thyroid Dataset.

D. Qualitative Comparisons with state-of-the-arts

We further present visual comparisons in Fig. 5–7. Among all baselines, DCP suffers from color distortions, while MSBDN and Dehamer leave residual haze and reduce contrast. SelfSVD variants improve clarity but tend to over-smooth anatomical details. In contrast, ASSVD consistently removes dense smoke while preserving anatomical structures. It restores global appearance in thoracic procedures with widespread occlusion, preserves edge sharpness and fine textures in gastrointestinal scenes with reflective mucosa, and reconstructs detailed vessel contours and instrument edges in thyroidectomy cases with dense vascular anatomy.

These observations, together with the quantitative results, collectively confirm ASSVD's ability to reliably remove smoke and preserve anatomical structures across diverse and clinically realistic surgical scenarios.

IV.   CONCLUSION

Existing surgical desmoking methods struggle with the lack of paired clean-smoky data and the inability to model the spatio-temporal complexity of dynamic smoke in surgical videos, which limits their restoration performance in real-world scenarios. In this paper, we present ASSVD, a self-supervised framework for surgical video desmoking that leverages the pre-smoke (PS) frame as unaligned supervision, thereby eliminating the need for paired annotations. Within ASSVD, the proposed Sparse Transformer Block (STB) enhances restoration under dense and spatially varying smoke by attending to structurally salient regions. In addition, we construct MegaClear, a real-world surgical video dataset from diverse endoscopic procedures, which can potentially be a valuable benchmark for future studies. Extensive experiments demonstrate that ASSVD outperforms the state-of-the-art methods both quantitatively and qualitatively. Future work will explore model compression and integration into clinical workflows.

References

[1] He K, Sun J, Tang X. "Single image haze removal using dark channel prior," IEEE Transactions on Pattern Analysis and Machine Intelligence, 2010, 33(12): 2341-2353.

[2] Zhu Q, Mai J, Shao L. "A fast single image haze removal algorithm using color attenuation prior," IEEE Transactions on Image Processing, 2015, 24(11): 3522-3533.

[3] Wang C, Alaya Cheikh F, Kaaniche M, et al. "Variational based smoke removal in laparoscopic images," Biomedical Engineering Online, 2018, 17(1): 139.

[4] Dong H, Pan J, Xiang L, et al. "Multi-scale boosted dehazing network with dense feature fusion," Proceedings of the IEEE Conference on Computer Vision and Pattern Recognition (CVPR). 2020: 2157-2167.

[5] Guo C L, Yan Q, Anwar S, et al. "Image dehazing transformer with transmission-aware 3d position embedding," Proceedings of the IEEE Conference on Computer Vision and Pattern Recognition (CVPR). 2022: 5812-5820.

[6] Zheng Q, Yang R, Ni X, et al. "Development and validation of a deep learning-based laparoscopic system for improving video quality," International Journal of Computer Assisted Radiology and Surgery, 2023, 18(2): 257-268.

[7] Liu Z, Lin Y, Cao Y, et al. "Swin transformer: Hierarchical vision transformer using shifted windows," Proceedings of the IEEE International Conference on Computer Vision (ICCV). 2021: 10012-10022.

[8] Wu R, Zhang Z, Zhang S, et al. "Self-supervised video desmoking for laparoscopic surgery," European Conference on Computer Vision (ECCV). Cham: Springer Nature Switzerland, 2024: 307-324.

[9] Chen X, Li H, Li M, et al. "Learning a sparse transformer network for effective image deraining," Proceedings of the IEEE Conference on Computer Vision and Pattern Recognition (CVPR). 2023: 5896-5905.

[10] Sun D, Yang X, Liu M Y, et al. "PWC-Net: CNNs for optical flow using pyramid, warping, and cost volume," Proceedings of the IEEE Conference on Computer Vision and Pattern Recognition (CVPR). 2018: 8934-8943.

[11] Mao X, Li Q, Xie H, et al. "Least squares generative adversarial networks," Proceedings of the IEEE International Conference on Computer Vision (ICCV). 2017: 2794-2802.

[12] Woo S, Park J, Lee J Y, et al. "Cbam: Convolutional block attention module," Proceedings of the European Conference on Computer Vision (ECCV). 2018: 3-19.

[13] Vaishnav P, Syed Waqas Z, Salman K, et al. "Promptir: Prompting for all-in-one blind image restoration," arXiv preprint arXiv:2306.13090, 2023.

A 71 TOPS/W 24.2 TOPS/mm² 14nm SRAM CIM Macro with a Capacitor-less ADC for Edge AI

Zexing Chen[1], Siyao Jia[1], Chixiao Chen*[1,2]

[1] State Key Laboratory of Integrated Chips and Systems, Fudan University, Shanghai, China
[2] Fudan Shaoxin Laboratory, Zhejiang, China

* Email: chenzx23@m.fudan.edu.cn, cxchen@fudan.edu.cn

Abstract—Compute-in-Memory (CIM) offers a promising solution to the data movement bottleneck in conventional computing architectures, but existing SRAM CIM designs often face challenges in computational density, energy efficiency, and robustness against Process, Voltage, and Temperature (PVT) variations. This paper presents a high-density, energy-efficient, and PVT-tolerant SRAM CIM macro designed for area-constrained edge AI applications. The proposed macro utilizes a custom 8T SRAM cell for analog computation and a compact capacitor-less Analog-to-Digital Converter (ADC) to achieve significant improvements in area efficiency. To enhance performance, a pulsed operation scheme is employed to eliminate static current , while a novel reference array with mismatched bitline capacitance generates adaptive reference voltages to ensure computational accuracy across a wide range of operating conditions. Fabricated in 14nm CMOS technology, the CIM macro demonstrates a superior balance of performance, power, and density. The macro achieves a high area efficiency of 24.2 TOPS/mm² at 0.8V, representing a 1.5x to 2.0x improvement over prior works. Furthermore, it delivers a competitive energy efficiency of 34.7 TOPS/W at 0.8V, which improves to 71 TOPS/W at a near-threshold voltage of 0.5V, making it an excellent candidate for next-generation edge AI accelerators.

Index Terms—Compute-in-Memory, 8T SRAM, Capacitor-less ADC, PVT compensation

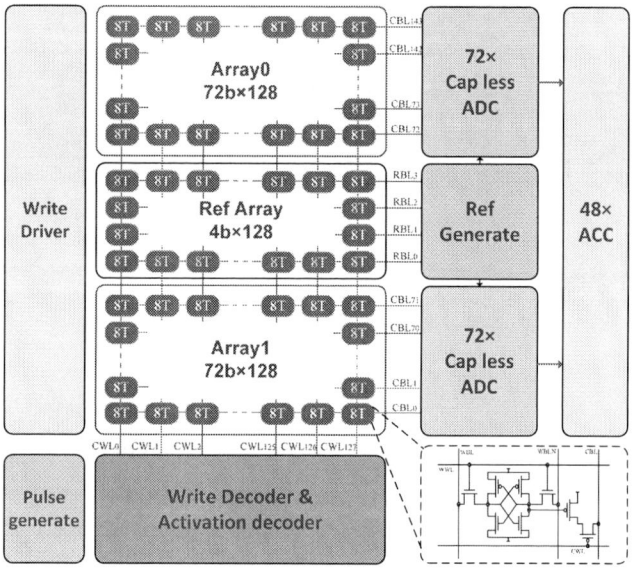

Fig. 1. Proposed CIM macro structure

I. INTRODUCTION

As data-intensive applications and artificial intelligence continue to proliferate, the data movement between processor and memory in traditional von Neumann architectures has become a primary bottleneck limiting system performance and energy efficiency. To overcome this challenge, CIM has emerged as a promising paradigm. By performing computations directly within or near memory elements, CIM technology can significantly reduce data transfer, offering immense potential for improving computational throughput and lowering power consumption. Among various approaches, SRAM-based CIM has become a focal point of research due to its high speed and compatibility with standard CMOS processes.

However, integrating computational capabilities into dense SRAM arrays presents several critical challenges. Firstly, the area overhead of the computational circuitry is a key factor limiting the overall computing density , especially as traditional digital solutions require separate readout and calculation circuits that consume significant area. Secondly,

while analog computing schemes are more energy-efficient, their area consumption of ADC is a major concern; Finally, the reliability of analog computation is highly susceptible to Process, Voltage, and Temperature (PVT) variations, which can cause discrepancies in the results. Random process variations alone can introduce considerable computational errors. Therefore, achieving high-density, energy-efficient, and PVT-robust computation remains a challenge in SRAM CIM design.

To address these issues, this paper proposes a high-density SRAM-based CIM macro. The design leverages a custom 8T SRAM cell for in-situ analog computation 6and a compact, capacitor-less ADC to maximize array density. Furthermore, we introduce a pulsed operation scheme to mitigate static power consumption and a dedicated reference array (Ref Array) that generates adaptive reference voltages, ensuring robust operation against PVT variations.

The proposed architecture and key circuits are detailed in Sections 2 and 3. Section 4 presents the implementation results and comparison, followed by the conclusion in Section 5.

II. PROPOSED CIM MACRO ARCHITECTURE AND OPERATION

A. Proposed CIM Macro Architecture and Operation

The top-level architecture of the proposed high-density SRAM CIM macro is depicted in the overall floorplan. The storage array is partitioned into two main computational sub-arrays, Array0 and Array1, each organized with a width of 72 columns and a depth of 128 rows. Adjacent to these, a dedicated Ref Array is implemented to generate stable reference voltages for the ADCs, ensuring robust operation across different PVT conditions.

The computational hardware is situated to the right of the memory arrays. Each Compute Bitline (CBL) within the two sub-arrays is connected to a dedicated capacitor-less ADC. The outputs from groups of three ADCs are then routed to a shared digital Accumulators (ACCs), which performs the final stage of the calculation. The choice of a capacitor-less ADC design is critical for achieving high array density, as it allows for compact integration with the standard-cell-based digital logic.

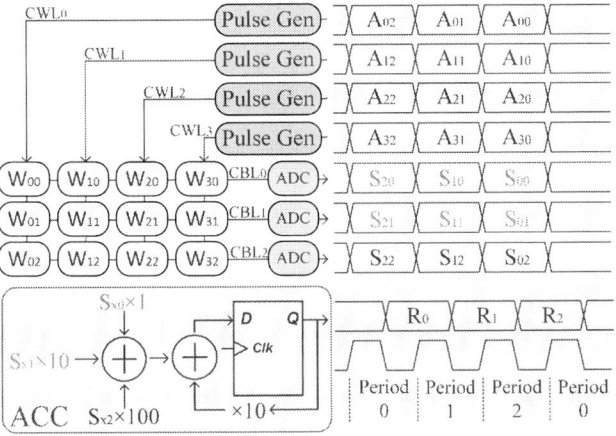

Fig. 2. The principle and timing of MAC operation

At the core of the design is a custom 8T SRAM cell. This cell extends the standard 6T SRAM configuration with two additional PMOS transistors dedicated to the computational read path. The write operation remains identical to that of a conventional 6T SRAM cell, where data is written by driving the Write Bitlines (WBL/WBLN) and asserting the Write Wordline (WWL). The read operation for computation, however, is unique. The cell will only charge the connected CBL when two conditions are met: the cell stores a '1' (node Q is low), and its corresponding Compute Wordline (CWL) is selected (driven to a low voltage). This conditional charging behavior forms the basis for performing a logical AND operation directly on the bitline.

B. Computational Method and Timing

The fundamental operation executed by the macro is a dot product between two four-element vectors, where each element is represented by 3 bits of precision. The weight

vector, denoted as (W_3, W_2, W_1, W_0), is pre-loaded into the 8T SRAM array. The 3 bits of each weight element—W_{x2} (MSB), W_{x1} (MID-BIT), and W_{x0}(LSB)—are stored in three adjacent SRAM cells along the same row (activated by the same CWL). Consequently, four consecutive rows are used to store the entire four-element weight vector.

The computation is performed in a multi-cycle process, breaking down the multiplication into three distinct steps:

Analog Domain AND-SUM Operation: The first step of the calculation occurs in the analog domain. To maximize parallelism, four CWLs are activated simultaneously, corresponding to the four elements of the weight vector. This action triggers a current-summing operation on the three vertical CBLs (CBL_2, CBL_1, CBL_0) associated with the weight bits. For instance, on CBL_2, which corresponds to the MSBs of the weights, the total current sourced to the bitline is proportional to the number of activated cells that store a '1'. This analog current level on the CBL represents the partial sum of the single-bit multiplications (AND operations) between the input activation bit and the four weight bits.

ADC: In the second step, the analog voltages developed on CBL_2, CBL_1, and CBL_0 are converted into digital values by their respective column-parallel ADCs. These digital outputs represent the partial dot-product results for each weight bit position, denoted as S_{x2}, S_{x1}, and S_{x0}.

Digital Domain Accumulation: The final result is assembled in the digital domain by the ACC over three clock cycles, corresponding to the three bits of the input activation vector. The process unfolds as follows:

Cycle 1 (Input MSB): The most significant bit of the input activation (A_{x2}) is broadcast to the four selected CWLs. The ADCs produce the first set of partial sums: S_{22}, S_{21}, and S_{20}. Before summation, these values are shifted according to their weight significance. The ACC calculates the first intermediate result: $R_0 = (S_{22} \ll 2) + (S_{21} \ll 1) + S_{20}$.

Cycle 2 (Input Mid-bit): The middle bit of the input activation (A_{x1}) is applied. A new set of partial sums (S_{12}, S_{11}, S_{10}) is generated. The previous result, R_0, is first shifted left by one bit to account for the higher significance of the previous activation bit. This shifted value is then added to the new, weight-shifted partial sums. The updated result is: $R_1 = (R_0 \ll 1) + (S_{12} \ll 2) + (S_{11} \ll 1) + S_{10}$.

Cycle 3 (Input LSB): The process repeats for the last bit of the input activation (A_{x0}). The result from the previous cycle, R1, is again shifted left by one bit and added to the final set of partial sums (S_{02}, S_{01}, S_{00}) to produce the final dot product result: $R_0 = (R_1 \ll 1) + (S_{02} \ll 2) + (S_{01} \ll 1) + S_{00}$. This completes one full 4x4 vector dot product operation.

III. KEY CIRCUIT IMPLEMENTATION

This section details the critical circuit-level implementations that underpin the architecture and computational method described previously. These designs are central to achieving the goals of high density, low power, and robust operation.

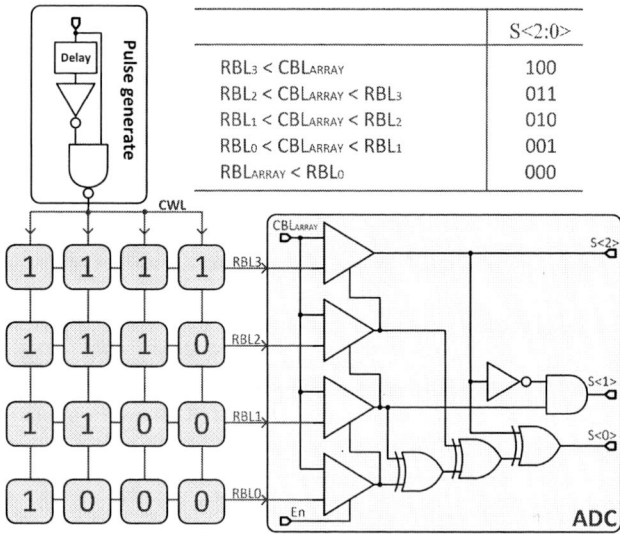

	S<2:0>
$RBL_3 < CBL_{ARRAY}$	100
$RBL_2 < CBL_{ARRAY} < RBL_3$	011
$RBL_1 < CBL_{ARRAY} < RBL_2$	010
$RBL_0 < CBL_{ARRAY} < RBL_1$	001
$RBL_{ARRAY} < RBL_0$	000

Fig. 3. The schematic of Reference voltage generation and capacitor-less ADC.

A. Pulsed Operation for Power Optimization

Using the current calculation scheme will cause static power consumption. To eliminate this, we employ a pulsed operation mode for computation. Instead of maintaining a static current flow, a brief, controlled pulse is generated to activate the CWL.

The pulse generation circuit creates a narrow low-pulse on the rising edge of an input clock signal by performing a NAND operation between the signal and its delayed, inverted counterpart. The width of this pulse is determined by the length of the delay chain. During this pulse duration, the summed current from the selected 8T SRAM cells charges the natural parasitic capacitance of the CBL. The final voltage developed on the CBL is proportional to the magnitude of the summed current, effectively converting the computational result into a voltage level. This approach offers two distinct advantages: it eliminates static power consumption entirely, and by preventing the CBL from experiencing full-swing voltage transitions, it also significantly reduces dynamic power.

B. Capacitor-less ADC

To achieve a high-density layout in advanced process nodes, the use of capacitors is prohibitive. Metal capacitors consume valuable routing layers, while MOS capacitors require a large silicon area. A capacitor-less ADC design is therefore essential for our high-density macro.

Our design consists of four parallel StrongARM latch comparators. These comparators determine the quantized value of the analog voltage on the CBL by comparing it against four distinct reference voltages ($RBL_3, RBL_2, RBL_1, RBL_0$) supplied by the Ref Array. The four binary outputs from the comparators are then decoded by simple combinational logic (XOR and AND) to produce the final multi-bit digital output.

To minimize the power consumed by the ADC, a daisy-chained enabling scheme is implemented for the comparators.

Rather than activating all four comparators simultaneously, they are enabled sequentially. A comparator for a higher voltage level is only triggered if the comparison at the next lower level indicates that the CBL voltage is greater than its reference. This successive approximation approach ensures that only the necessary comparators are activated for any given conversion, thereby avoiding superfluous switching activity and reducing overall power consumption.

C. Ref Array for PVT Compensation

To counteract the high sensitivity of analog computation to PVT variations, a dedicated Reference Array (Ref Array) is implemented to generate a stable and adaptive set of reference voltages for the ADCs. This array uses groups of permanently programmed SRAM cells to produce reference currents equivalent to those from one to four active cells in the main array. By remaining constantly selected, the reference voltages track any PVT-induced changes in the cell's current drive strength in real-time.

A critical design feature is the intentional parasitic capacitance mismatch between the Reference Bitlines (RBLs) and Compute Bitlines (CBLs), achieved by physically laying out the RBLs to be longer. This increased RBL capacitance ensures that for an identical summed current, the final voltage on an RBL is always slightly lower than the corresponding voltage on a CBL. This systematically generated offset provides a reliable input voltage margin for the comparators, ensuring robust A/D conversion across all operating conditions.

IV. IMPLEMENTATION AND RESULTS

A. The layout and performance of the proposed CIM macro

The proposed SRAM-based CIM macro was designed and laid out in a standard 14nm CMOS process. Figure 4 shows the complete layout of the macro, highlighting its key components, including the two computational sub-arrays (Array0 and Array1), the central Ref Array, and the peripheral circuitry containing the column-parallel ADCs and digital ACCs.

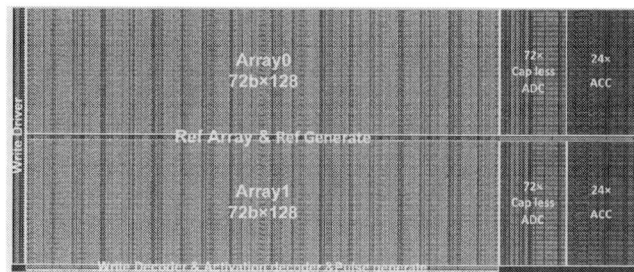

Fig. 4. The layout of Proposed CIM-Macro

A detailed breakdown of the macro's area is provided in Table 1. The total area of the macro is 8321.4 μm². A key contributor to the high density is the compact design of the capacitor-less ADC, which occupies only 3.54 μm² per instance. This area-efficient design allows the computational logic to be tightly integrated with the memory array, minimizing overhead and maximizing the overall computational density. The resulting area efficiency, defined as the ratio of TOPS to total macro area, is 24.2 TOPS/mm², demonstrating a significant improvement over designs that rely on larger, capacitor-based ADCs.

TABLE 1 RESULTS COMPARISON TABLE

	ISSCC'20 [2]	JSSC'25 [3]	VLSI'23 [4]	ISSCC'21 [5]	This work
Technology	7nm	22nm	12nm	16nm	14nm
Mac operation	Analog	Digital	Digital	Digital	Mixed
Array Size	4Kb	8Kb	64Kb	45Mb	18Kb
Macro Size(mm^2)	0.0032	0.0335	0.0455	25	0.0083
Density(Kb/mm^2)	1250	238.8	1422	180	2168.7
Power Supply	0.8	0.6-1.1	0.55-0.9	0.5-0.8	0.5-0.8
input bits	4	4	4	4	3
weight bits	4	4	4	4	3
output bits	4	15	14	4	9
cycle time(ns)	5.5	3.7-18.1	-	-	1.92-9
Area Efficiency (TOPS/mm^2)	206.7*	3.66*	19.9*	4.75*	24.2
Energy Efficiency (TOPS/W)	466-1085*	346.6*	243*	13.5*	34.7-71

*Scale to 3b Act and 3b Weight

B. Simulation Results

The functionality and robustness of the proposed design were verified through post-layout simulations.

The impact of process variations was quantified using a 1400-point Monte Carlo simulation, with the resulting error distribution plotted in Figure 5. The simulation shows a computational error with a standard deviation of approximately 5%, which is an acceptable level of precision for resilient deep learning inference applications.

PVT robustness was validated across multiple process corners, which confirmed a stable sensing margin for reliable ADC operation. Furthermore, a mixed-signal simulation of a Yolo V3 convolutional layer using 3-bit quantization achieved a classification accuracy of 93.83%, demonstrating the macro's practical effectiveness for real-world neural network tasks.

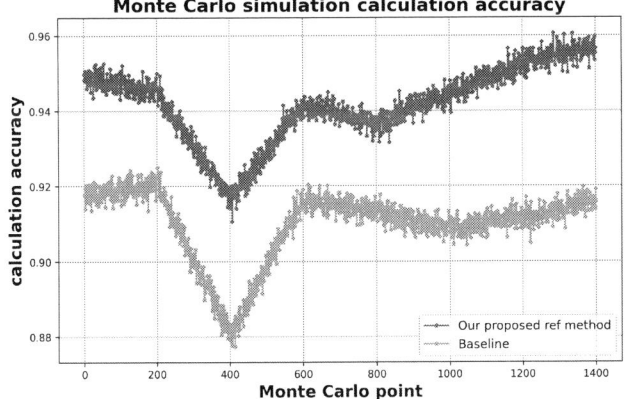

Fig. 5. Monte Carlo simulation results with reference array (blue) and without reference array (orange)

C. Comparison with State-of-the-Art

Table 1 presents a comprehensive comparison of this work with recently published state-of-the-art SRAM-based CIM macros. The comparison is based on key metrics including technology node, precision, area, energy efficiency (TOPS/W), and area efficiency (TOPS/mm²).

Our work demonstrates a significant advantage in area efficiency. Thanks to the 8T cell and the compact capacitor-less ADC design, our macro achieves an area efficiency of 24.2 TOPS/mm² at 0.8V. This represents a 1.5x to 2.0x improvement over prior works that utilize either larger SRAM cells or more area-intensive ADC architectures. In terms of energy efficiency, our pulsed operation scheme contributes to a competitive figure of 34.7 TOPS/W at 0.8V and 71 TOPS/W at 0.5V. While some designs may report higher peak performance, our architecture provides a superior balance of performance, power, and, most notably, density, making it an excellent candidate for area-constrained edge AI applications.

V. CONCLUSION

This paper presents a high-density, energy-efficient, and PVT-robust SRAM-based CIM macro. To achieve high computational density, the design leverages a custom 8T SRAM cell and a compact capacitor-less ADC. High energy efficiency is attained through a pulsed operation scheme that eliminates static current. Furthermore, a novel compensation technique using a dedicated reference array with mismatched bitline capacitances overcomes PVT sensitivity, ensuring reliable computation. Post-layout simulations have validated our approach, demonstrating competitive performance in both area and energy efficiency. The proposed macro is therefore a promising solution for AI accelerators for edge computing.

ACKNOWLEDGMENT

This work was supported in part by the National Natural Science Foundation of China (NSFC) under Grant 62322404 and Peng Cheng Laboratory and China Mobile (PCL-CMCC) Foundation for Science and Innovation under Grant 2024ZY2B0070.

REFERENCES

[1] H. Zhu et al., "COMB-MCM: Computing-on-Memory-Boundary NN Processor with Bipolar Bitwise Sparsity Optimization for Scalable Multi-Chiplet-Module Edge Machine Learning," 2022 IEEE International Solid-State Circuits Conference (ISSCC), San Francisco, CA, USA, 2022.

[2] Q. Dong et al., "15.3 A 351TOPS/W and 372.4GOPS Compute-in-Memory SRAM Macro in 7nm FinFET CMOS for Machine-Learning Applications," 2020 IEEE International Solid-State Circuits Conference - (ISSCC), San Francisco, CA, USA, 2020.

[3] S. He et al., "A 22-nm 109.3-to-249.5-TFLOPS/W Outlier-Aware Floating-Point SRAM Compute-in-Memory Macro for Large Language Models," in IEEE Journal of Solid-State Circuits.

[4] G. Jedhe et al., "A 12nm 137 TOPS/W Digital Compute-In-Memory using Foundry 8T SRAM Bitcell supporting 16 Kernel Weight Sets for AI Edge Applications," 2023 IEEE Symposium on VLSI Technology and Circuits (VLSI Technology and Circuits), Kyoto, Japan, 2023.

[5] H. Jia et al., "15.1 A Programmable Neural-Network Inference Accelerator Based on Scalable In-Memory Computing," 2021 IEEE International Solid-State Circuits Conference (ISSCC), San Francisco, CA, USA, 2021.

Data-Centric Automatic Design Migration of Low Voltage CMOS Bandgap Reference Circuit

Shun-Qi DAI *[1], Yuan LEI [1], Bei-Ping YAN [1]

[1] Hong Kong Applied Science and Technology Research Institute (ASTRI), Hong Kong, P.R. China

* Email: shunqidai@astri.org

Abstract—**This paper proposes a data-centric framework for the automated design migration of low-voltage CMOS bandgap reference circuits. By integrating data-driven optimization and circuit building block recognition techniques, the approach enables automatic migration of circuit netlists while systematically reducing design space complexity. This innovation allows target specifications to be achieved with a significant reduction in simulation iterations, addressing the computational inefficiency of traditional migration methods. To validate the framework's effectiveness, a low-voltage CMOS bandgap reference circuit block was successfully migrated across three distinct CMOS technology nodes, encompassing both horizontal, across two identical technology nodes, and vertical, across two different technology nodes, design migrations. The results demonstrate consistent performance retention and design reusability, highlighting the method's potential for accelerating technology migration in analog circuit design.**

Keywords—Automated design migration, Data-centric, Circuit recognition, CMOS bandgap reference

I. INTRODUCTION

The global integrated circuit (IC) industry continues to experience robust growth, driving numerous enterprises to adopt system-on-chip (SoC) designs across multiple process technologies. A typical SoC primarily comprises two categories of circuit blocks: analog and digital. While digital circuit design has achieved sophisticated automation levels, enabling automatic migration of digital blocks, analog circuit design migration remains predominantly manual. This manual intervention in analog design has become a significant bottleneck in SoC development cycles. To address this challenge, automated design methodologies for analog circuits are essential to reduce migration cycle time and improve overall design efficiency [1-2].

Conventionally, two categories of methods address this issue: analytical-based approaches and simulation-based approaches. Analytical-based methods typically employ analytical equation models, for example the square-law model, to derive approximate expressions for circuit performance or leverage regression techniques like geometric programming (GP) or linear regression to fit performance data [3-7]. Although these approaches circumvent circuit simulations, they often struggle to accurately model the behavior of practical circuits. Furthermore, they exhibit limited flexibility as the derived equations depend heavily on specific circuit topologies.

Simulation-based methods treat circuit performance as black-box functions, evaluated via computationally intensive circuit simulations. Global optimization algorithms, such as simulated annealing (SA), evolutionary algorithms (EA), and particle swarm optimization (PSO), are utilized to navigate the design space and generate candidate solutions seeking optimal design points [8-15]. While these methods offer the advantage of being largely topology-independent, they often demand substantial computational resources to achieve convergence and require designers to explicitly specify the dimensionality of the design space and manually transfer the circuit netlist from original technology to target technology.

To enhance the efficiency of automated analog circuit design migration, this work proposes a novel data-centric framework. Low-voltage CMOS bandgap reference (BGR) circuits are essential building blocks in SoC architectures, providing stable voltage and current references critical for precision operation. Thus, we utilize a low-voltage CMOS BGR circuit as a representative example to demonstrate both horizontal and vertical design migration capabilities of the proposed framework.

II. PROBLEM DEFINITION OF CIRCUIT DESIGN MIGRATION

When migrating an analog circuit block to a new technology node, analog designers typically begin by assessing whether the existing circuit topology is compatible with the target technology, according to their prior experience. Once the topology is deemed suitable, they determine the values of the design variables, often using the square-law device model, to meet the desired design specifications. In this work, we assume that the circuit topology has already been validated for the new technology. Our focus is on the automatic identification of optimal design variable values. Consequently, the analog circuit design migration problem can be formulated as a multi-objective black-box optimization problem with constraints, as follows:

$$\begin{aligned}
\text{minimize} \quad & f_i(x) \\
\text{subject to} \quad & g_i(x) \geq p_i \\
& \forall i \in 1,2,3 \cdots N,
\end{aligned} \quad (1)$$

where $x \in R^d$ denotes the design variable vector, d is the number of design variables, $f_i(x)$ represents the i th performance metric of the analog circuit, $g_i(x)$ denotes the i th constraint on the performance metrics and p_i represents specification of i th performance metrics.

III. DATA-CENTRIC AUTOMATED CIRCUIT MIGRATION FRAMEWORK

Traditionally, analog design migration requires designers to manually create new testbenches and circuit

979-8-3315-3918-4/25 $31.00 © 2025 IEEE

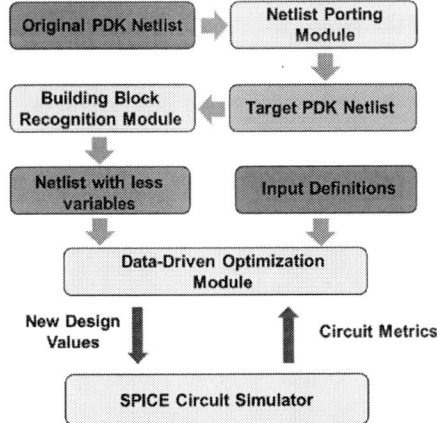

Topology Library	Building Blocks	Variable Relationship
	Differential stage	$W_A=W_B$, $L_A=L_B$
	Differential pair	$W_A=W_B$, $L_A=L_B$
	Current mirror	$W_A=W_B$, $L_A=L_B$

Fig. 1. The extensible building blocks library

Fig. 2. The overall workflow of Data-centric automated design migration framework

netlists for the target technology, leveraging prior circuit knowledge to correlate transistor dimensions and minimize design variables. To streamline this process, we propose a data-centric automated migration framework incorporating circuit recognition. The framework comprises four key modules: circuit netlist porting, circuit building block recognition, data-driven optimization, and SPICE simulation.

The circuit netlist porting module automatically translates both the netlist and testbench from the source process design kit (PDK) to the target PDK. Given the new PDK and the original circuit as inputs, this module generates the corresponding netlist and testbench for the new technology.

The building block recognition module identifies common analog circuit structures, such as differential stages (DS), differential pairs (DP) and current mirrors (CM). An extensible building blocks library, which can be extended by experienced analog designers, supports this recognition process, as illustrated in Fig. 1. Based on the identified building blocks, the relationships among transistors within the circuit are established, enabling the generation of a new annotated circuit netlist with a reduced set of design variables.

The overall workflow of the proposed framework, illustrated in Fig. 2, consists of several sequential stages. Initially, a circuit netlist porting module transforms the original netlist to ensure compatibility with the target Process Design Kit (PDK). Subsequently, the recognition module processes both the ported netlist and testbench, which

encompass the circuit topology and design variables, to identify fundamental building blocks and reduce the dimensionality of design variables. The optimization module then receives the annotated netlist along with circuit specifications and variable boundaries. Through an iterative process, this module systematically explores the design space by assigning values to design variables, while a SPICE simulator evaluates each candidate solution and returns performance metrics to guide subsequent optimization steps.

The details about data-driven optimization module in this framework are explained as follows: The optimization procedure consists of four main stages: Design of Experiment (DoE), Deep Neural Network (DNN) Model Setup, NSGA-III Operations, and DNN Model Update as shown in Fig.3. The optimization process begins with the DoE stage, where an initial set of design points is generated within the design space to form the initial population. In this work, the Latin Hypercube Design (LHD) method is employed to create the initial set of design points [16].

Next, the DNN Model Setup stage evaluates the generated design points using a SPICE circuit simulator to train and construct the initial DNN surrogate model. This DNN model will be used to predict the performance metrics of the circuit in the subsequent optimization stages.

The NSGA-III Operations stage then executes the NSGA-III multi-objective optimization algorithm. NSGA-III applies selection, crossover, and mutation operators to the population to generate new offspring designs. The fitness of the offspring designs is evaluated in an alternating manner - the first 5 iterations use the SPICE circuit simulator, while the next 5 iterations use the DNN model for fitness evaluation. If the SPICE circuit simulator is used for evaluation, the DNN model will also be updated. This hybrid approach leverages the accuracy of the circuit simulator while reducing the computational cost by utilizing the DNN model. The evaluated offspring designs then undergo non-dominated sorting and are associated with reference points to identify the non-dominated solutions in the combined population of parents and offspring. Finally, the population is updated by replacing the weakest solutions in the current population with the non-dominated solutions from the offspring. This optimization procedure will continue iteratively until the circuit performance specification is met.

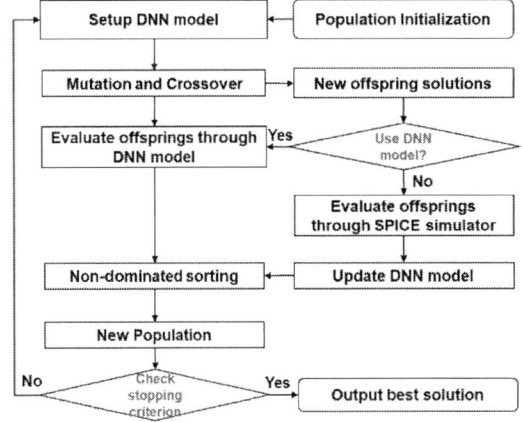

Fig. 3. The working procedure of data-driven optimization module.

979-8-3315-3918-4/25 $31.00 © 2025 IEEE

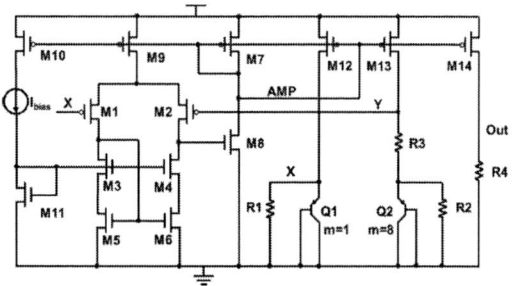

Fig. 4. The schematic of the. low-voltage CMOS bandgap reference circuit [17].

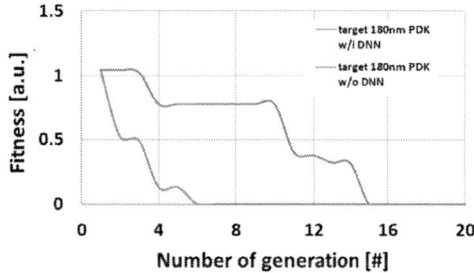

Fig. 5. The convergence characteristics of automated horizontal migration between two 180 nm CMOS technology for low-voltage CMOS bandgap reference circuit.

TABLE I. Horizontal migration result of low-voltage CMOS bandgap reference circuit

Technology	Specification	original 180nm PDK	target 180nm PDK w/o DNN	target 180nm PDK w/i DNN
Temperature Coefficient [ppm/C]	<=40	34.9	16.6	25.5
ZTC temperature error [C]	<=50	47	47	47
Number of saturation device	>=14	14	14	14
Minimum voltage[V]	>=0.9	0.910	0.911	0.942
Maximum voltage[V]	<=1 0	0 913	0.912	0.945
Pass/fail			pass	pass
Number of generations			20	20
Number of simulations			800	400
Generation has the 1st feasible solution			5	14
Migration time [sec]			4996	3222

IV. EXPERIMENT RESULT OF LOW VOLTAGE CMOS BANDGAP CIRCUIT AUTOMATED MIGRATION

In this section, we validated our proposed framework by automatically migrating a low-voltage CMOS BGR circuit across different process technologies. The experiments encompassed both horizontal migration, same technology node in different foundries, and vertical migration, different technology nodes. All circuit simulations were performed using a commercial SPICE simulator on a Linux server.

A. Horizontal Design Migration

The horizontal migration experiment began with a low-voltage CMOS bandgap reference circuit initially designed in a 180 nm mixed-signal CMOS process with 3.3 V supply voltage as shown in Fig. 4. We migrated this circuit to a different 180 nm mixed-signal CMOS process PDK while maintaining the same supply voltage. Our building block recognition module reduced the design variables from 28 to

10, comprising transistor dimensions, namely length and width, and resistor values.

Table I presents the migration results comparing our framework's performance with and without the DNN model. Both approaches successfully met all design specifications. Fig. 5 illustrates the convergence behavior of both methods. While the DNN-based approach showed slightly slower convergence, it achieved all specifications within 20 iterations. Notably, the DNN model reduced the required number of simulations by 50%, from 800 to 400, significantly improving the overall migration efficiency as shown in Fig. 7(a).

B. Vertical Design Migration

The second phase of our evaluation focused on vertical migration, where we transferred the low-voltage CMOS BGR circuit from 180 nm to both 110 nm and 40 nm CMOS foundry processes. This cross-technology migration demonstrated our framework's capability to handle significant process node transitions while maintaining the circuit's 3.3 V supply voltage requirement.

Table II presents the automated vertical migration results. Both implementations, with and without the DNN model, successfully achieved all design specifications. Although the DNN-based approach exhibited slightly slower convergence characteristics, it maintained specification compliance within 20 iterations as shown in Fig.6. Most significantly, the DNN model reduced the required number of simulations by 50%, from 800 to 400, decreasing the total migration cycle time from 5589 seconds to 3254 seconds, a 41.8% improvement in design migration efficiency as described in Fig.7(b).

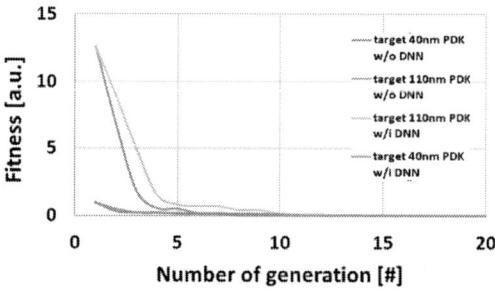

Fig. 6. The convergence characteristics of automated vertical migration, from 180 nm to 110 nm and 40 nm CMOS technology for low-voltage CMOS bandgap reference circuit.

TABLE II. Vertical migration result of low-voltage CMOS bandgap reference circuit

Technology	Specification	original 180nm PDK	110nm PDK w/o DNN	40nm PDK w/o DNN	110nm PDK w/i DNN	40nm PDK w/i DNN
Temperature Coefficient [ppm/C]	<=40	34.9	13.5	33.5	23.1	33.7
ZTC temperature error [C]	<=50	47	47	47	47	47
Number of saturation device	>=14	14	14	14	14	14
Minimum voltage[V]	>=0.9	0.910	0.905	0.909	0.929	0.968
Maximum voltage[V]	<=1.0	0.913	0.906	0.912	0.932	0.971
Pass/fail			pass	pass	pass	pass
Number of generation			20	20	20	20
Number of simulation			800	800	400	400
Generation has the 1st feasible solution			11	8	14	14
Migration time [sec]			5077	5589	3195	3554

979-8-3315-3918-4/25 $31.00 © 2025 IEEE

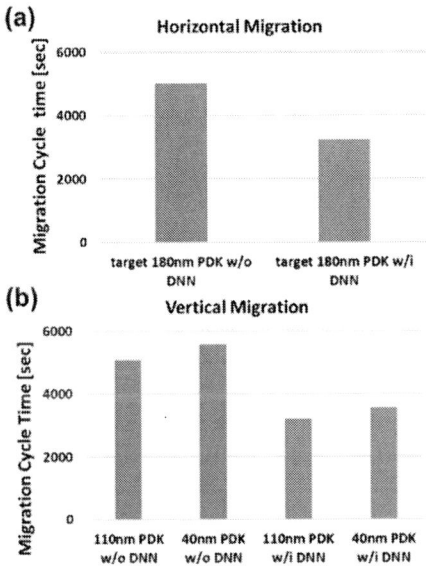

Fig. 7. The migration cycle time of (a) horizontal migration and (b) vertical migration.

V. CONCLUSION

This paper introduced a novel data-centric framework for automatic analog circuit design migration. Applied to a low-voltage CMOS bandgap reference circuit, the method successfully demonstrated horizontal migration between two commercial 180 nm CMOS technologies and vertical migration from 180 nm to 110 nm and 40 nm CMOS technologies. With the help of DNN model in the data-driven optimization module, all design specifications were met within an hour, leading to a 41.8% improvement in design migration efficiency. These results indicate the potential of the data-centric design migration framework as a promising approach for achieving automated migration of analog blocks in the future.

ACKNOWLEDGMENT

The authors would like to thank Hong Kong Applied Science and Technology Research Institute (ASTRI) for the support in this research.

REFERENCES

[1] A. Girardi, T. De-Oliveira and S. Ghissoni, "A comprehensive review on automation-based sizing techniques for analog IC design," in Journal of Integrated Circuits and Systems, vol. 17(3), pp. 1-14, 2022.

[2] R. Martins, N. Lourenço and N. Horta., Analog Integrated Circuit Design Automation. Cham, Switzerland: Springer, 2017.

[3] W. Daems, G. Gielen and W. Sansen, "Simulation-based generation of posynomial performance models for the sizing of analog integrated circuits," in IEEE Transactions on Computer-Aided Design of Integrated Circuits and Systems, vol. 22(5), pp. 517-534, 2003.

[4] Y. Wang, M. Orshansky, C. Caramanis, "Enabling efficient analog synthesis by coupling sparse regression and polynomial optimization," in IEEE Proceedings of the 51st Annual Design Automation Conference (2014), pp. 1-6, 2014.

[5] P. Mandal and V. Visvanathan, "CMOS op-amp sizing using a geometric programming formulation," in IEEE Transactions on Computer-Aided Design of Integrated circuits and systems, vol. 20(1), pp. 22-38, 2001.

[6] A. Sayed, A. N. Mohieldin, M. Mahroos, "A fast and accurate geometric programming technique for analog circuits sizing," in 31st International Conference on Microelectronics (ICM) IEEE (2019), pp. 316-319, 2019.

[7] O. Fabián, and A. Petraglia, "A computer-aided approach for voltage reference circuit design," Analog Integrated Circuits and Signal Processing, vol. 89(3), pp.511-520, 2016.

[8] R. Phelps, M. Krasnicki, R. Rutenbar, L. A. Carley and J. R. Hellums, "Anaconda: simulation-based synthesis of analog circuits via stochastic pattern search," in IEEE Transactions on Computer-Aided Design of Integrated Circuits and Systems, vol. 19(6), pp. 703-717, 2006.

[9] M. Hidalgo, D. Castro, R. Vázquez and H. Díaz, "A prototype tool for optimum analog sizing using simulated annealing," in IEEE International Symposium on Circuits and Systems (1992), pp. 1933-1936, 1992.

[10] B. Liu, Y. Wang, Z. Yu, L. Liu, M. Li, Z. Wang and F. V. Fernández, "Analog circuit optimization system based on hybrid evolutionary algorithms," in Integration, vol. 42(2), pp. 137-148, 2009.

[11] B. Manuel, J. Guilherme, and N. Horta, "Analog circuits optimization based on evolutionary computation techniques," in Integration, vol. 43(1), pp. 136-155, 2011.

[12] M. Barros, J. M. Guilherme and N. C. Horta, Analog Circuits and Systems Optimization Based on Evolutionary Computation Techniques, Berlin: Springer, 2010.

[13] R. Vural and T. Yildirim, "Analog circuit sizing via swarm intelligence," in AEU-International Journal of Electronics and Communications, vol. 66(9), pp. 732-740, 2012.

[14] S. Mallick, R. Kar, D. Mandal, and S. P. Ghoshal, "Optimal sizing of CMOS analog circuits using gravitational search algorithm with particle swarm optimization," in International Journal of Machine Learning and Cybernetics, vol. 8, pp. 309-331, 2017.

[15] K. D. Khalil, N. Soliman and H. Omran, "Automation of bandgap voltage reference optimization using vectorized coarse-fine grid search," in IEEE 7th International Japan-Africa Conference on Electronics, Communications, and Computations (JAC-ECC)," pp. 54-57, 2019.

[16] R. L. Iman, Latin hypercube sampling. In John Wiley & Sons, 2008.

[17] Banba, H., Shiga, H., Umezawa, A., Miyaba, T., Tanzawa, T., Atsumi, S., "A CMOS bandgap reference circuit with sub-1-V operation," IEEE Journal of Solid-State Circuits, vol. 34(5), pp. 670-674, 1999.

Innovative Detection Capacitor Utilization in ESD Power Clamp Circuits for HBM Residual Voltage Suppression

Zelong Huang[1][3], Guangyi Lu *[2][3], Haoyu Xia[1][3], Qi Wu[1][3], Haiming Wang[1][3]

[1] State Key Laboratory of Millimeter Waves, Southeast University, Nanjing 211189, China
[2] National ASIC Center, School of Integrated Circuits, Southeast University, Nanjing 210096, China
[3] National Center of Technology Innovation for EDA, Nanjing 210031, China
* Email : zlhuang2023@seu.edu.cn, guangyilu@seu.edu.cn

Abstract—**This paper proposes electrostatic discharge (ESD) power clamp designs that improve the conventional RC-triggered circuit by replacing the standard MOS detection capacitor with a native MOS capacitor and, subsequently, a MOS varactor. The varactor-based design achieves the lowest Human Body Model (HBM) residual voltage, followed by the native and standard MOS capacitor configurations. The proposed approach is validated through both pre-layout and post-layout simulations and shows potential for broader application in advanced ESD protection circuits.**

Keywords—***ESD, RC-triggered clamp circuit, HBM residual voltage, MOS varactor, Native MOS capacitor***

I. INTRODUCTION

With the continuous scaling of CMOS technology, the gate oxide thickness has become thinner and junction depths shallower, making integrated circuits more vulnerable to ESD damage [1]. To safeguard the functionality and reliability of complex system-on-chip, robust whole-chip ESD protection design has become indispensable. The whole-chip ESD protection should ensure that ESD current is effectively steered away from vulnerable internal circuits and safely discharged through designated protection paths. Among various ESD stress conditions, pin-to-pin ESD stress poses a critical challenge, where the discharge current may traverse the internal power network and affect multiple functional blocks simultaneously [2], as illustrated in Fig. 1. As a critical component in whole-chip ESD protection, the ESD power clamp circuit must provide a low-impedance discharge path. In addition, it should also ensure a sufficiently low clamping voltage.

Among various ESD power clamp topologies, the MOS-based, RC-triggered clamp circuit remains the most widely adopted solution in industry due to its simple structure, process portability, and compatibility with digital design flows [3]. However, the effectiveness of RC-triggered clamp circuits is increasingly challenged at advanced technology nodes. In HBM ESD events, the clamp must maintain a sufficiently long turn-on duration to ensure that the residual voltage remains below the rated operating voltage [4]. This requires an adequate RC time constant, which is typically limited by the poor low-voltage capacitance performance of traditional MOS capacitors. Achieving sufficient delay time often demands large-area capacitors, leading to increased silicon cost and layout complexity. Furthermore, as supply voltages scale down, the margin between the clamping voltage and the failure threshold narrows, increasing the performance demands on the ESD power clamp circuit.

To address this issue, this work explores the use of different detection capacitor structures in the RC-triggered clamp circuit. The standard MOS capacitor is successively replaced by a native MOS capacitor and a MOS varactor to improve HBM residual voltage suppression. Among them, the varactor provides the best suppression due to its superior low-voltage capacitance, followed by the native MOS capacitor, while maintaining compact area. All circuits are implemented in a 40-nm CMOS process with thin-oxide devices and a nominal 1.1 V supply.

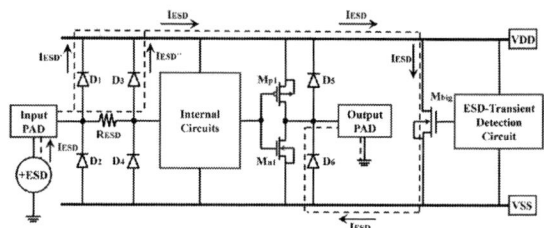

Fig. 1. ESD current discharge path in pin-to-pin mode

The conventional RC-triggered clamp circuit employing different capacitor types are evaluated through simulation. Results demonstrate that the proposed approach achieves superior clamping performance without additional area cost and shows strong potential for broader application in RC-triggered clamp circuits.

II. PROPOSE ESD POWER CLAMP CIRCUIT

The structures of the evaluated RC-triggered ESD power clamps are shown in Fig. 2. All circuits adopt the conventional topology but differ in the detection capacitor. Three versions are implemented using a standard MOS capacitor, a native MOS capacitor, and a MOS varactor, respectively. The design with the varactor, shown in Fig. 2(a), achieves the best clamping performance due to its superior low-voltage capacitance and area efficiency.

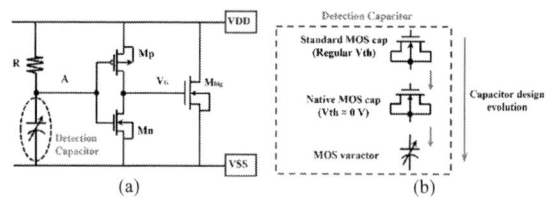

Fig. 2. (a) Propose RC-triggered clamp circuit with MOS varactor and (b) Detection capacitor options showing the design evolution.

The physical structures and bias configurations of the MOS capacitors and MOS varactor are shown in Fig. 3. Since the standard and native MOS capacitors share the same physical structure, they are both represented by the MOS capacitor structure in Fig. 3(a). The standard and native NMOS capacitors share the same structure, with the gate connected to the RC node and the source, drain, and bulk grounded. The key difference is the lower threshold voltage of the native NMOS compared to the standard device. As a result, under low bias (e.g., 0–0.5 V), the standard NMOS operates from depletion to inversion, while the native NMOS enters weak inversion earlier and transitions into inversion more quickly, yielding higher capacitance in this voltage range.

In contrast, the MOS varactor used in this work has the gate connected to the RC node, while the source, drain, and substrate are tied to ground, forming an N^+/N-well structure. As there is no p–n junction in the N-well and the N-well shares the same potential with the substrate, inversion is effectively prevented. As a result, the device mainly operates in depletion and accumulation, maintaining high and stable capacitance at low bias [5], [6].

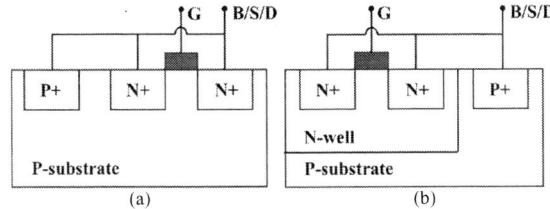

Fig. 3.   (a) MOS capacitor structure and (b) MOS varactor structure.

These structural differences result in distinct capacitance-voltage (C–V) characteristics. Fig. 4 compares the C–V curves of the standard MOS capacitor, native MOS capacitor, and the MOS varactor. In the 0–0.5 V range, the MOS varactor shows the highest capacitance, followed by the native MOS capacitor, both exceeding the standard device. This improved low-voltage capacitance enables longer discharge times during HBM events, effectively reducing residual voltage without increasing layout area.

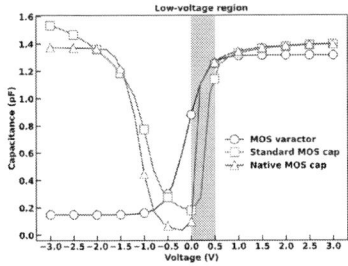

Fig. 4.   MOS capacitor and MOS varactor capacitance compare.

III. CIRCUIT IMPLEMENTATIONS AND SIMULATION RESULTS

The schematic of the circuit design is shown in Fig. 2. Three circuits were implemented by replacing the detection capacitor with different types, while keeping identical device dimensions. The design parameters are summarized in Table I. The RC time constant was set above 1 μs to ensure low residual voltage [4].

TABLE I.   DEVICE PARAMETERS OF THREE DESIGNS

R (kΩ)	Detection Capacitor (μm/μm)	Mp, Mn (μm/μm)	M_big (μm/μm)
920	40/4.1	25/0.1, 5/0.1	2400/0.12

A. Pre-layout Simulation

Fig. 5(a) shows the simulated clamping voltage in response to an HBM-like current pulse (10 ns rise time, 1.33 A peak current) applied to the VDD lines illustrated in Fig. 2.

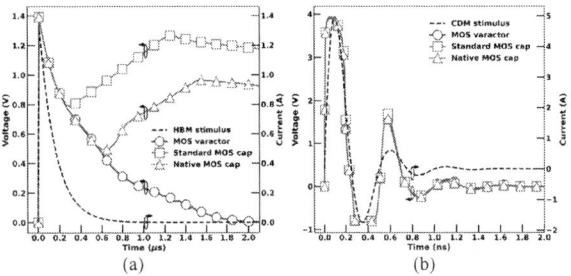

Fig. 5.   Pre-layout simulation results: (a) HBM simulation and (b) CDM simulation.

Since most of the HBM discharge energy is dissipated within the first 150 ns [2], the peak clamping voltage after 400 ns is used as an approximation of the residual voltage. As shown in Fig. 5(a), the circuit incorporating a MOS varactor in the detection path achieves a significantly lower residual voltage of approximately 0.65 V. In contrast, the standard MOS capacitor and native MOS capacitor configurations yield residual voltages of approximately 1.23 V and 0.97 V, respectively. These results demonstrate that the MOS varactor enables the most effective residual voltage clamping, followed by the native and standard MOS capacitors, thereby enhancing overall ESD robustness in advanced CMOS technologies. In the CDM simulation results shown in Fig. 5(b), a 5 A current pulse with a rise time of approximately 100 ps is applied. All three clamp circuits exhibit similar CDM overshoot voltages around 3.90 V, indicating that the use of a MOS varactor does not degrade CDM performance.

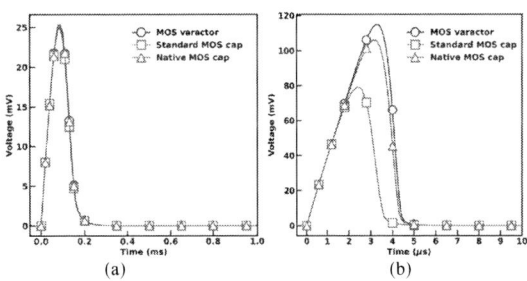

Fig. 6 .   Pre-layout simulation results: (a) 1ms power up simulation and (b) 10us power up simulation.

Fig. 6 shows the power-up simulation results for two ramp-up conditions: 1 ms and 10 μs. Under the slow ramp condition of 1ms, the gate voltage (V_G) of all three clamp variants remains below 25 mV, indicating no risk of false triggering. For the 10 μs ramp case, the V_G values are observed to be 114 mV for the varactor-based clamp, 106 mV for the native MOS capacitor design, and 79 mV for the standard MOS capacitor. Although the varactor and native structures exhibit slightly higher gate voltages compared to the standard version, all V_G levels remain sufficiently below the trigger threshold even under the fast 10 μs ramp-up condition, demonstrating adequate immunity against false triggering.

B. Post-layout Simulation

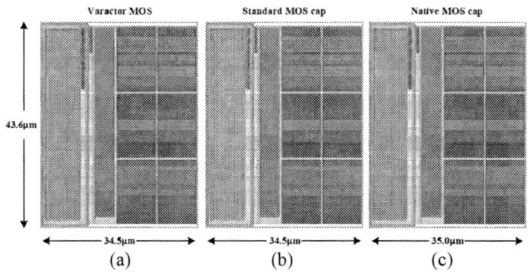

Fig. 7. Layouts of RC-triggered clamp circuits incorporating (a) MOS varactor, (b) traditional MOS capacitor, and (c) native MOS capacitor.

Fig. 7 shows the layouts of RC-triggered circuits using the MOS varactor, traditional MOS capacitor, and native MOS capacitor, all with identical device dimensions. The native MOS capacitor layout is about 0.5 µm wider due to its PCell configuration.

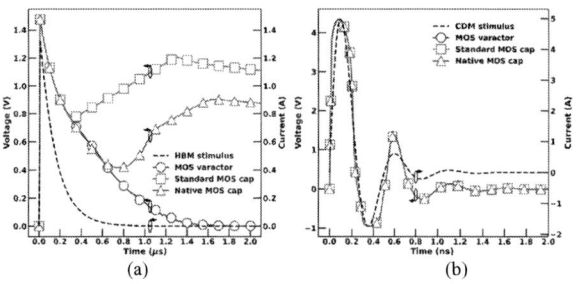

Fig. 8. Post-layout simulation results: (a) HBM simulation and (b) CDM simulation.

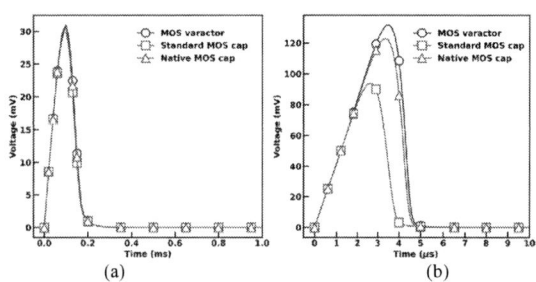

Fig. 9. Post-layout simulation results: (a) 1ms power up simulation and (b) 10us power up simulation.

Figs. 8 and 9 show that post-layout simulations agree with pre-layout results. Under HBM stress, residual voltages are ~0.66 V (varactor), 0.91 V (native MOS), and 1.20 V (standard MOS). CDM overshoot stays near 4.34 V, and power-up immunity is maintained with gate voltages below the trigger threshold. These confirm stable ESD protection after layout.

TABLE II. LEAKAGE OF DIFFERENT CIRCUITS AT VARIOUS TEMPERATURES

Circuits type	I_{leak} in DC simulation		
	25 °C	75 °C	125 °C
Standard MOS cap	86.49nA	507.46nA	3.07µA
Native MOS cap	81.55nA	497.36nA	3.04µA
MOS varactor	96.23nA	511.52nA	3.05µA

Table II summarizes the post-layout simulated leakage currents across temperatures. While the varactor-based design shows slightly higher leakage at 25 °C (96.23 nA), all three designs exhibit very similar leakage levels at 75 °C and 125 °C, reaching the low microampere range. These results confirm that replacing the standard MOS capacitor with a native MOS or MOS varactor introduces no significant leakage penalty.

Table III summarizes the key ESD performance metrics of the three designs. The proposed MOS varactor-based clamp reduces HBM residual voltage by approximately 45% and 28% compared to the standard MOS and native MOS designs, respectively. Similarly, the native MOS capacitor achieves a 24% reduction relative to the standard MOS capacitor. Leakage current, CDM overshoot voltage, and power-up immunity remain comparable across all designs. The native MOS implementation occupies a slightly larger area (1526.0 µm² vs. 1504.2 µm²), but this increase is negligible in practical layout considerations.

TABLE III. POST-LAYOUT SIMULATION RESULTS OF DIFFERENT CIRCUITS

Circuits type	Post-layout simulation results				
	I_{leak} at 25 °C	HBM residual voltage	CDM overshoot voltage	Area	Fast power up immunity
Standard MOS cap	86.49nA	1.20V	3.07V	1504.2 µm²	Immunity
Native MOS cap	81.55nA	0.91V	3.04V	1526.0 µm²	Immunity
MOS varactor	96.23nA	0.66V	3.05V	1504.2 µm²	Immunity

IV. CONCLUSION

In this work, improved ESD power clamp circuits are proposed by replacing the standard MOS capacitor with a native MOS capacitor and further with a MOS varactor. The native MOS capacitor offers notable HBM improvement over standard designs, while the MOS varactor achieves further suppression without area penalty. Simulation results confirm that the proposed method maintains stable CDM, power-up, and leakage performance. Since RC-triggered clamp circuits rely on low-voltage capacitor behavior to discharge HBM tail current, the proposed method is applicable to a wide range of such designs.

REFERENCES

[1] S.-H. Chen and M.-D. Ker, "Area-efficient ESD-transient detection circuit with smaller capacitance for on-chip power-rail ESD protection in CMOS ICs," IEEE Trans. Circuits Syst. II, Exp. Briefs, vol. 56, no. 5, pp. 359–363, May 2009.

[2] M.-D. Ker, "Whole-chip ESD protection design with efficient VDD-to-VSS ESD clamp circuits for submicron CMOS VLSI," IEEE Trans. Electron Devices, vol. 46, no. 1, pp. 173–183, Jan. 1999.

[3] J. Li, R. Gauthier, and E. Rosenbaum, "A compact, timed-shutoff, MOSFET-based power clamp for on-chip ESD protection," in Proc. EOS/ESD Symp., 2004, pp. 273–279.

[4] R. Venkatasubramanian, K. Oertle, and S. Ozev, "Rail clamp with dynamic time-constant adjustment," IEEE J. Solid-State Circuits, vol. 51, no. 5, pp. 1313–1324, May 2016.

[5] S. A. Wartenberg and J. R. Hauser, "Substrate voltage and accumulation-mode MOS varactor capacitance," IEEE Trans. Electron Devices, vol. 52, no. 7, pp. 1563–1567, Jul. 2005.

[6] R. L. Bunch and S. Raman, "Large-signal analysis of MOS varactors in CMOS −Gm LC VCOs," IEEE J. Solid-State Circuits, vol. 38, no. 8, pp. 1325–1332, Aug. 2003.

HIGH EFFICIENT EFUSE FULL PROCESS BURNING SOLUTION BASED ON ATE

Qian Zhai [*1], Yichen Xiao [*1], Xin Song [1], Haobin Wang [1], Yuyuan Wang [2], Xuxin Chen [3]

[1] UniSoC, Pudong New District, Zuchongzhi Road, 2290
[2] Advantest, Pudong New District, Huatuo Road, 168
[3] Shanghai Dianji University, Pudong New District, Shuihua Road, 300

* Email: qian.zhai@unisoc.com, yichen.xiao@unisoc.com

Abstract - In the field of chip Automated Test Equipment (ATE)testing and development, reliable configuration of eFuse is a key link to ensure functional implementation and yield. The traditional eFuse testing and development based on the ATE platform is plagued by bottlenecks such as inefficient and error prone requirement conversion, scattered variable management, fragile burn-in processes, and low reusability, resulting in a sharp increase in development cycles, quality risks, and maintenance costs. In response to this situation, this article proposes a "full process solution for eFuse burning", which includes three mainstream modules: eFuse Trans, eFuse Variable, and eFuse Common Building Block (CBB).

eFuse Trans: A conversion module from standard eFuse format to eFuse CBB input, with the following characteristics: 1) Deep integration of NLP technology to achieve automatic parsing, standardization conversion, and rule driven verification of multi-source design requirement specification (DRS). 2) Intelligent parsing and mapping: through semantic understanding and structured mapping technology, it automatically parses multi-source heterogeneous DRS, converts unstructured requirements into standardized eFuse Tables, and ensures seamless collaboration with eFuse CBB. 3) Intelligent verification and closed-loop processing: When an incorrect configuration is detected, executable correction solutions are automatically generated based on the rule library and historical case library.

eFuse Variable: Managing eFuse Variables in program development. There are the following characteristics: 1) Design a proprietary variable management class to achieve strong variable control and core operation encapsulation. 2) By standardizing the interface protocol to solidify a reliable burning process, the eFuse Variable validation output is used as the only trusted data source for eFuse CBB, eliminating the risk of variable burning errors from the source.

eFuse CBB: Execute eFuse burning. There are the following characteristics: 1) Design exclusive data structures to manage the eFuse modules of various chips. 2) Standardize the burning process. Avoid burning errors. 3) The burning method can be customized and extended. Support different forms of burning protocols. 4) Embedded EQC function, supporting QA quality inspection. By combining three mainstream modules, this solution can significantly improve the efficiency and accuracy of eFuse configuration data processing, effectively ensuring the quality of chip eFuse burning.

Keywords—eFuse, eFuse Trans, eFuse Variable, eFuse CBB, SoC, ATE test, NLP

I. OVERALL ARCHITECTURE

In the surging wave of AI advancement, the System-on-Chip (SoC) has cemented its role as the "central nervous system" of the computing universe. Nestled within its architecture, the eFuse module - characterized by its one-time-programmable nature and permanent immutability - has risen as a critical safeguard, which protects the SoC's unique identity, cryptographic keys, and core operational lifeline.

In semiconductor companies, the "life-or-death" eFuse burning for SoCs relies almost entirely on ATE. Even the largest chipmakers, despite their proprietary eFuse CBB libraries for ATE, still struggle with the so-called "last-mile" predicament: requirements come from multiple sources in scattered formats and evolve without pause, every handover, from specification to test script to silicon implementation, causes data degradation, ATE platforms use mutually incompatible syntaxes, and programmer expertise ranges from expert to beginner. A single weak link in this chain could render an entire wafer useless as expensive scrap or create costly field risks, with financial and opportunity costs that are impossible to quantify.

To address these critical challenges, we introduce an end-to-end Large Language Model (LLM) - driven solution delivering high reliability and accuracy. eFuse Trans governs the input processing stage with pinpoint precision. eFuse Variable enable seamless middleware integration. eFuse CBB converts design intent into executable ATE code with exceptional efficiency. Working in synergy as a unified system, these three modules drive a significant boost in both correctness and throughput for eFuse configuration data.

II. EFUSE TRANS

A. Deep NLP-Integrated eFuse Map Semantic Parsing and Canonical Transformation

To achieve efficient and accurate migration of multi-source DRS into standardized eFuse CBB inputs, the eFuse Trans embeds natural language processing (NLP)[1]. It automatically extracts salient information from heterogeneous and loosely structured DRS, performs semantic parsing and mapping, and emits structured, verifiable eFuse Tables that integrate seamlessly with downstream eFuse CBB.

This process centers on the Jieba segmentation engine. For example, the field "HUK lock (write HUK first, then the lock bit)" is tokenized into ["HUK", "lock", "write first", "HUK", "then write", "Lock", "bit"]. Leveraging a domain rule base, the tokens undergo structural filtering, semantic alignment, and normalization to yield ["HUK", "_", "lock"]. The entire conversion proceeds without human intervention.

B. Middleware-Driven Intelligent Parsing and Mapping

To achieve seamless integration from multi-source DRS to standardized eFuse CBB input components, this paper proposes a middleware-driven parsing and mapping mechanism. The pipeline treats eFuse Map as the sole input and proceeds in six steps to produce intermediate artifacts:

- Merged-cell decomposition to minimal data units, enhancing retrieval precision.

- Visual field selection and key-column pre-screening to raise the signal-to-noise ratio.

- Rule-library-guided derivation of standardized eFuse CBB inputs across dozens of branch scenarios.

- Joint disambiguation via Jieba segmentation and domain rules for unified naming and robust indexing;

- Binary-compliance verification of fixed burn-in values with pre-marking of anomalies to block downstream incompatibilities.

- Field mapping and logical processing against predefined rules and target ATE platforms, yielding specification-compliant eFuse CBB inputs for diverse chip types.

By fusing NLP-based semantic recognition with rule logic, multi-round filtering, and fault tolerance, the approach attains a unified, standardized eFuse Table model that substantially improves configuration correctness, reusability, and maintainability for automated programming and verification.

C. Intelligent Verification and Closed-Loop Processing via Rule-Guided Fine-Tuning of Large Models

In eFuse testing and development, configuration errors often come from misparsed heterogeneous DRS, mapping flaws, duplicate entries, and numeral system mismatches. To tackle these, eFuse Trans includes an "Intelligent Validation and Closed-Loop Processing" module. Using a fine-tuned LLM[2], it enables an adaptive diagnostic system with end-to-end closed-loop capabilities: error detection, root-cause localization, highlight marking, and corrective action generation.

The module adopts a dual-drive mechanism combining a rule knowledge base and a historical case knowledge base. We extracted structured constraint rules and anomaly detection patterns from dozens of previous projects, covering core scenarios such as duplicate detection, numeric system verification, semantic normalization, highlighting of key information, and arithmetic inequality verification. Data sources include project mapping rules, data records, defect reports, change logs, and rule update records. Using DeepSeek-R1-1.5B as the base model, we converted the content of the aforementioned knowledge bases into a question-answering (QA) format dataset for training through Instruction Fine-tuning. This dataset contains over a hundred historical issues and their corresponding solutions. The training process was supplemented by minor reinforcement learning (RL) optimization, enabling the model to acquire the ability to understand anomaly scenarios and generate effective repair recommendations.

D. Result

Prediction validation of the fine-tuned model was conducted on three chip types across four independent test platforms. Verification accuracy exceeded 95% across all platforms. As the rule knowledge base undergoes continuous expansion and iterative updates, the model performance is expected to be further improved, with reliability continuously being optimized.

TABLE I. MATCH RATE BETWEEN PREDICTED SOLUTIONS AND ACTUAL SOLUTIONS

Type	ATE Platform			
	J750	UltraFlex	SMT7	SMT8
SOC	96.32%	95.87%	98.64%	97.31%
PMIC	95.86%	97.51%	96.41%	95.64%
RF	95.33%	98.12%	96.56%	98.11%

During operation, the tool automatically generates an error report upon detecting conflicts or omissions—such as anomalous field configurations, out-of-range values, invalid radixes, naming collisions, or format mismatches—and forwards the relevant context to the fine-tuned model[3]. The model then instantly produces an executable remediation plan, detailing modification locations, recommended values, cross-domain impact analyses, and corrective actions. All key information is presented in color-coded blocks for clear visibility during user validation.

Compared to conventional manual methods, this approach reduces development time by an average of 90% (from 10 hours per person to 1 hour per person), eases developers' cognitive burden related to rule sets, and enhances overall configuration accuracy. As a result, it maximizes the efficiency of automated error correction and structural optimization throughout the entire DRS-to-standardized-eFuse-CBB pipeline. Through the detect-diagnose-feedback-repair cycle, it enables intelligent verification and closed-loop processing. The tool we have developed is illustrated below.

Fig. 1. eFuse Trans

III. EFUSE VARIABLE

A. Background and Problems

Several fundamental vulnerabilities have been identified in current ATE implementations. Firstly, due to the specific characteristics of certain numerical values，the absence of their validation mechanisms allows parameters exceeding DRS specifications to proceed to burning phase undetected, as critical eFuse values lack clearly defined boundaries and automated checking procedures.

Then, the code generation process exhibits unprotected variable management vulnerabilities, where complex transformations of real-time test results into device-specific eFuse codes involve multiple intermediate variables without proper state tracking or modification safeguards, potentially corrupting final outputs.

Two critical functions are applied by the eFuse technology in modern semiconductor devices: (1) programming functional codes derived from pre-test calculations based on DRS, and (2) storing permanent device identification data such as wafer coordinates, lot IDs, and chip serial numbers. However, the irreversible nature of eFuse programming presents significant reliability challenges, where erroneous writes can lead to catastrophic device failures and substantial economic losses.

Furthermore, with increasing complexity in modern chip designs, a comprehensive test program typically includes independent tests for numerous sub-modules within the chip. However, most standalone module test codes incorporate logging of the module's eFuse code values. Engineers must verify the correct execution of all eFuse-related module codes during the full test program runtime. Additionally, before finalizing eFuse programming, they must ensure that the recorded eFuse codes remain consistent with the values obtained at the end of their respective independent test procedures. This prevents potential inconsistent updates to these code values due to erroneous or redundant operations during the full test flow.

B. Architecture and Description

The software tool, referred to as eFuse VAR, is designed to manage variables during program development to address the aforementioned issues. In the current software environment based on the JAVA language, specialized variable management classes have been developed to achieve strict variable control while encapsulating core variable operation methods. Another key design aspect is recording and validating the results of eFuse Variable operations, ensuring these validated outputs serve as the sole trusted inputs for the eFuse CBB. This binding mechanism is achieved through the implementation of JAVA software method output interfaces, ensuring the reliability of the programming process by eliminating erroneous programming operations at the source of the input data.

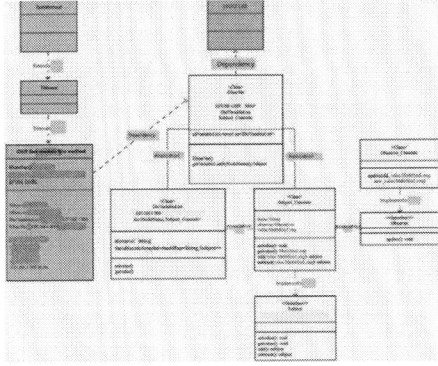

Fig. 2. eFuse Variable architecture

The diagram illustrates the eFuse VAR software framework, **The blue** section represent the decomposition of the minimal program units for the complete test program of a complex chip. Based on the JAVA software structure, the test code for a specific chip module defines the test method as an abstract superclass of all test methods. The primary function of the test method is to define a specific structure for test execution, such as setup and execute operations. The TMBase class inherits from this superclass and adds additional user-required methods. The final application layer code is defined as the test code required by the sub-module within the chip. Its main workflow is implemented according to the DRS and

invokes the resources of the ATE instruments. The eFuse VAR framework operates at this application layer to monitor and process the codes derived from the computation of test results.

The green section represents the eFuse CBB within the framework, which has two primary responsibilities. Firstly, it serves as the source of eFuse MAP information, ensuring that eFuse VAR obtains complete node information corresponding to the physical eFuse memory regions. This approach guarantees that within the same test program, the eFuse MAP information originates from a single, common, and uniquely designated source. Secondly, as previously mentioned, the output from eFuse VAR is utilized as the sole trusted input for eFuse CBB, ensuring the accuracy and reliability of the data being programmed into the eFuse, thereby eliminating errors at the source.

The yellow section represents the core of the eFuse VAR framework, primarily comprising four key components: Observer, Subject, "DieVariableList," and "EfuseVar." These components work together to implement the observer design pattern in JAVA [4], ensuring effective state management and interaction within the framework.

Observer and Concrete Implementation class

The Observer and its concrete implementation classes are designed to realize the observer design pattern in JAVA. These classes do not contain any user-defined functions, focusing solely on enabling the observer functionality.

Subject and Concrete Implementation class

The Subject class and its concrete implementations are structured to encapsulate class variables corresponding to eFuse Variables. As illustrated in the UML diagram, concrete Subject class define private instance variables of the Observer type. During instantiation, these classes create concrete Observer instances and assign them to the instance variables. This compositional design allows Subject classes to utilize Observer methods or access their properties, thereby implementing the observer pattern effectively.

Additionally, the Subject concrete classes are responsible for creating and defining sub-functions required for DRS required computations, which include basic arithmetic operations such as addition, subtraction, multiplication, and bitwise operations. Furthermore, in the ATE, the Multi-site Testing must be concerned, when performing parallel measurement involving chip differences and individual characteristics, the results must be distinctly managed. The Subject class incorporates functions to ensure accurate and consistent data handling.

This modular and observer-based architecture ensures that the eFuse VAR framework robustly manages eFuse Variables, adheres to DRS, and supports complex testing scenarios in modern semiconductor manufacturing.

DieVariableList and EfuseVar classes

Finally, the "DieVariableList" and "EfuseVar" classes primarily establish and refine data binding relationships within the framework. The Subject class contains attributes specify eFuse node names and values. The "DieVariableList" class accommodates multi-die architectures such as chiplet, ensuring compatibility with complex chiplet-based designs. Within the "EfuseVar" class, comprehensive node information from eFuse CBB is utilized to instantiate a "HashMap<string, subject>", where node names serve as keys, and corresponding node instances are linked to the Subject class.

C. Result

Based on the software architecture implemented in this program, we have introduced an output verification file feature that allows us to preview the actual output format of the program when executed, before proceeding with the actual eFuse burning process. The output format, which can be referenced in the simple example provided in the following figure, includes several key factors such as site, die, node name, final value, and the calculation process (recorded as "operation sequence").

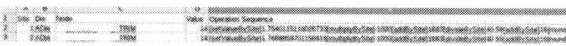

Fig. 3.   eFuse VAR demo program result

By comparing with the requirements of the DRS, this table enables an efficient identification of the result values and the accuracy of the calculation process of the eFuse Variables. Any abnormal operations on the variables can be more directly observed. Furthermore, after this verification report is confirmed, the controlled variables of eFuse VAR can be passed to eFuse CBB to complete the subsequent burning execution logic.

IV. EFUSE CBB

eFuse CBB is the final implementation stage of eFuse's complete solution. There are mainly the following data modules: eFuse data structure, eFuse input sorting, eFuse burning, and eFuse quality inspection.

A. eFuse data structure

The current eFuse data adopts a block+node structure. The entire eFuse structure consists of multiple blocks (usually 256), each block containing 32 bits. Several consecutive blocks are available eFuse fields. The current plan uses a dedicated data structure and plans a three-level structure of die/block/node, supporting multi die burning eFuse.

B. eFuse input sorting

This module reads the eFuse map table generated by eFuse Trans. eFuse CBB reads the table and stores the results in the eFuse data structure. A typical eFuse map table is shown in the following figure.

Name	Mode	Address	StartBit	Length	Default	ReadOnly	Diff_Limit	DiffHLimit	ReadL_Limit	ReadHLimit
LOTID_6	LotID	222	24	6	0	TRUE	NA	NA	0	42
LOTID_5	LotID	222	18	6	0	TRUE	NA	NA	0	42
LOTID_4	LotID	222	12	6	0	TRUE	NA	NA	0	42
LOTID_3	LotID	222	6	6	0	TRUE	NA	NA	0	42
LOTID_2	LotID	222	0	6	0	TRUE	NA	NA	0	42
LOTID_1	LotID	220	25	6	0	TRUE	NA	NA	0	42
LOTID_0	LotID	220	19	6	0	TRUE	NA	NA	0	42
WaferID	DEFAULT	220	14	5	0	TRUE	NA	NA	1	26
DIE_X	DEFAULT	220	7	7	0	TRUE	NA	NA	0	127
DIE_Y	DEFAULT	220	0	7	0	TRUE	NA	NA	0	127

Fig. 4.   eFuse map table

C. eFuse burning

Implement eFuse burning according to eFuse's dedicated burning protocol. Users can directly configure the location of burning eFuse payload according to the protocol. The burning method can be customized and extended, supporting different forms of burning protocols.

D. eFuse quality check

The EQC function of eFuse is embedded to check the correctness of eFuse writing.

The typical eFuse burning flow is shown in the following figure

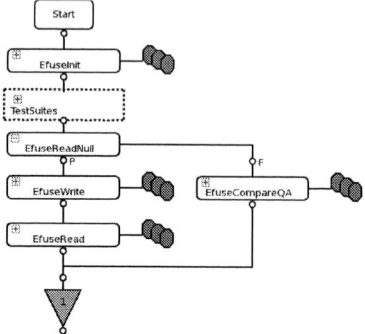

Fig. 5.   Typical eFuse flow

EfuseInit is responsible for initializing data structures and reading eFuse inputs. This test item is at the beginning of the entire test flow. Afterwards, there are various specific chip test items, and the eFuse values to be burned are calculated from these test items and stored in the eFuse data structure. Next is EfuseReadnull, which performs the chip's null detection action to determine if the chip has been burned. If the chip has not been burned before, it will be burned according to the eFuse data structure, and after burning is completed, it will be read out for inspection. If the chip has been burned, the EfuseCompareQA will be executed to compare the burning results of eFuse in the chip with the values to be burned in the previous test items. The EfuseCompareQA will set a reasonable deviation range. Chips within this reasonable range will be considered good products. This is the entire eFuse mass production burning process.

Through a large amount of mass production data, it has been proven that the eFuse CBB solution can efficiently adapt to various chip testing scenarios and has extremely high efficiency.

V. SUMMARY

The entire eFuse solution boasts full-scenario compatibility for chip testing, can intercept fundamental defects throughout the entire ATE development process, and has passed mass production-level stability verification.

ACKNOWLEDGMENT

Thank you for the company's support of this project, which lasted for two years and had various issues during the process. Fortunately, these issues have been perfectly resolved, resulting in very good outcomes.

REFERENCES

[1] Sumana M, Kalyani A, Gupta A, et al. Enhanced Data Analysis by Natural Language Query Processing[C]//2024 5th International Conference on Communication, Computing & Industry 6.0 (C2I6). IEEE, 2024: 1-6.

[2] Terpstra I. Empowering Analog Integrated Circuit Design through Large Language Models and Reinforcement Learning[D]. Massachusetts Institute of Technology, 2024.

[3] Stepanov A P, Shichkina Y A. Fine-Tuning LLM's for Domain-Specific Text Generation in Environments with Limited Resource Capabilities[C]//2025 VI International Conference on Neural Networks and Neurotechnologies (NeuroNT). IEEE, 2025: 56-58.

[4] E.Gamma, R. Helm, R. Johnson, and J. Vlissides. Design patterns: El ements of reusable object-oriented software. 1995.

Study of Reliability Screening Method to Improve the DPPM of IC Products

Yancong He[1], Zhiyong Yang*[1], Zhinong Liu[1], Shuang Jiao[1], Chuyuan He[1], Yixian Wang[2], Zhigang Ji[2]

[1] Department of Manufacture, UNISOC, 2288 Zuchongzhi Road, Shanghai, China
[2] Shanghai Jiao Tong University, Shanghai 200240, China

* Email : Zhiyong.yang1@unisoc.com

Abstract—With the continuous advancement of technology, the complexity of chip manufacturing has increased significantly. Quality and yield are becoming critical factors for the successful commercialization. Furthermore, the rapid growth of automotive electronics has tightened the tolerance for defects, demanding an ultra-low defect parts per million (DPPM) rate. Before shipment, electrical performance tests are conducted to identify and screen out defective chips, ensuring that the DPPM meets the required standards. The industry remains focused on minimizing DPPM to improve product quality and maintain market competitiveness. This paper presents the results of evaluating seven anomaly detection methods, collectively referred to as reliability screening (RS), and proposes an RS test flow for mass production. The proposed methodology demonstrates a significant reduction in DPPM and a notable decrease in returned material analysis (RMA) costs.

Keywords— Reliability Screening (RS), Integrated Circuits (IC), Returned Material Analysis (RMA), Defects per Million (DPPM), GDBN, Clustering

I. INTRODUCTION

Defect rates in chips are typically quantified in terms of defects per million (DPPM), with target values varying across different application domains. As shown in **Fig. 1**, automotive electronics impose particularly stringent requirements, often adopting a zero-defect tolerance policy [3]. To meet these demands, various techniques have been developed to reduce DPPM, as shown in **Fig. 2**, including algorithmic screening, enhanced test coverage through stimulation, and environmental stress screening. Among these, statistical methods for predicting early-life failures have proven to be cost-effective and efficient, and are widely adopted in high-end IC products, such as those used in automotive applications [6].

In this paper, we analyze six different reliability screening (RS) methods using real industrial data. We focus on applying these methods during the Chip Probe (CP) and Final Test (FT) stages to improve DPPM performance at the customer end.

- *Cluster:* Defects often occur in spatial clusters rather than being randomly distributed across the wafer [1]. We define a defect cluster as a 3×3 fully populated neighbourhood exceeding a threshold of seven defective dies. In such cases, the surrounding dies are also marked as failed (see **Fig. 3**).

- GDBN (Good Die in Bad Neighbourhood): This method aims to screen out dies that pass testing but are located in a predominantly defective neighbourhood [2]. A "local yield" score is computed for each die based on a weighted average of the yield in its surrounding area. In this study,

the GDBN threshold is set to 60%, a commonly accepted standard in the semiconductor industry (shown in **Fig. 4**).

- PAT (Part Average Testing): PAT is a statistical method used to identify and reject outliers with abnormal characteristics [4]. Three PAT techniques are discussed in this paper: Static PAT (SPAT), Dynamic PAT (DPAT), and Z-axis PAT (ZPAT). SPAT and DPAT detect dies with parameters more than 6σ from the distribution mean (**Fig. 5**), while ZPAT flags suspicious dies based on low-yield regions in wafer-level stacked maps, even if the dies pass electrical tests.

- SBA (Statistical Bin Analysis): SBA detects and eliminates abnormal material lots to ensure IC quality and reliability [5]. In this study, we adopt a threshold of Mean ± 3σ, which deviates from the AEC-Q002 standard.

- Wafer Edge Ink: Dies located at the wafer's edge tend to exhibit lower yield and higher RMA rates, a trend confirmed by industrial datasets. As a precaution, all dies within a 6×6 edge boundary are inked, regardless of test results.

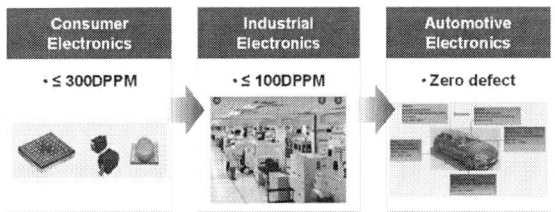

Figure 1. IC production DPPM request

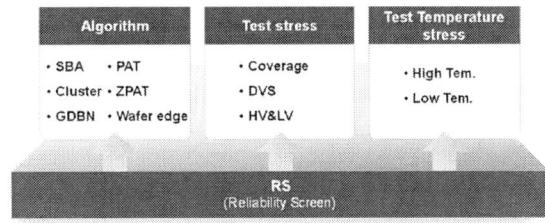

Figure 2. DPPM reduction methodology

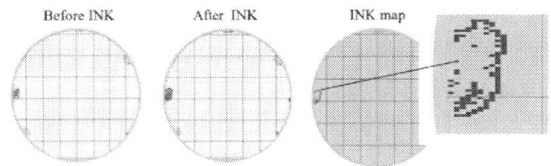

Figure 3. Cluster ink

Table I. RS flow result

Stage	Cluster	Yield Loss		
		lot1	lot2	lot3
CP	DPAT	0.04%	0.10%	0.20%
	ZPAT	1.60%	1.00%	1.50%
	GDBN	1.50%	1.30%	1.20%
	Cluster	0.10%	0.20%	0.10%
	Wafer edge	34.50%	36.20%	33.70%
FT	SBA	No trigger	No trigger	No trigger
	SPAT	8.18%	35.58%	7.30%
Summary	Yield	62.26%	61.20%	63.30%
	Yield loss	37.74%	38.80%	36.70%

Table II. RS gains

Cluster	Gains			Availability
	RMA screen out ratio	DDPM improved	yield loss	DPPM improved/1% yield loss
DPAT	4.30%	15	0.34%	4412
ZPAT	17.50%	60	1.37%	4380
GDBN	4.30%	15	1.33%	1128
Cluster	17.50%	60	0.50%	12000
Wafer edge	56.00%	300	34.08%	880

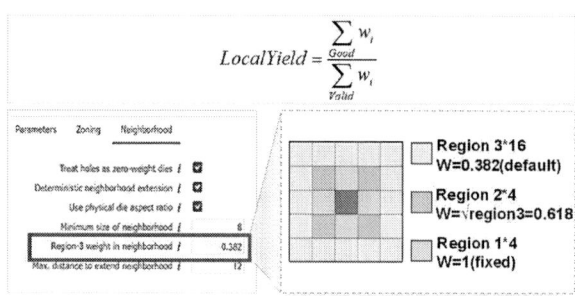

$$LocalYield = \frac{\sum_{Good} w_i}{\sum_{Valid} w_i}$$

Figure 4. GDBN Local yield

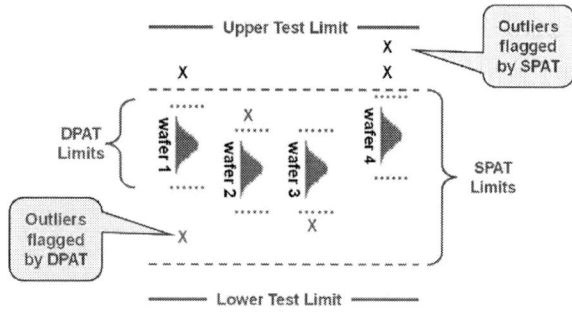

Figure 5. Part Average Testing Diagram

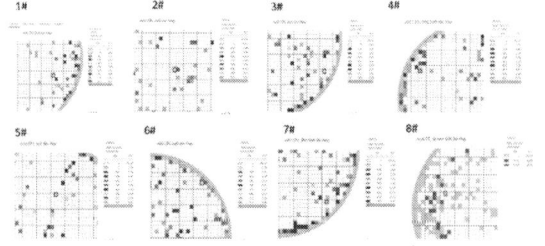

Figure 6. RMA Wafer Map

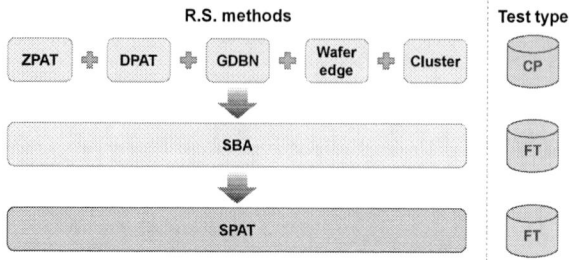

Figure 7. RS flow chart

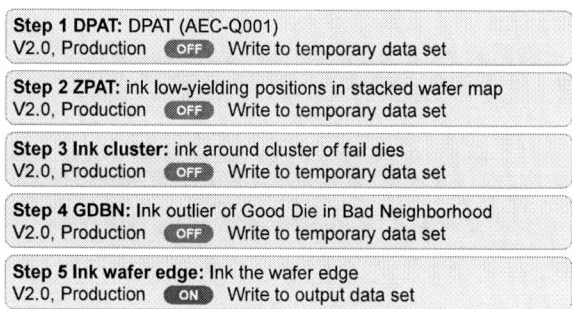

Step 1 DPAT: DPAT (AEC-Q001)
V2.0, Production OFF Write to temporary data set

Step 2 ZPAT: ink low-yielding positions in stacked wafer map
V2.0, Production OFF Write to temporary data set

Step 3 Ink cluster: ink around cluster of fail dies
V2.0, Production OFF Write to temporary data set

Step 4 GDBN: Ink outlier of Good Die in Bad Neighborhood
V2.0, Production OFF Write to temporary data set

Step 5 Ink wafer edge: Ink the wafer edge
V2.0, Production ON Write to output data set

Figure 8. MDS Recipe

II. COMPARISON OF RS METHODS

To evaluate the practical effectiveness of the proposed reliability screening (RS) methods, we conducted a post-mortem analysis on 24 failed devices (RMA chips) that were returned from the field. These devices are all instances of a 4G high-end SoC product that has already been shipped in quantities exceeding several million units. The 24 chips were carefully selected from the top 10 most common failure types to ensure that our analysis would be representative of the failure patterns encountered in the field.

We aimed to determine which of the RS methods—applied either at the Chip Probe (CP) or Final Test (FT) stage—could have preemptively screened out these devices, thereby preventing defective units from reaching customers. The seven RS techniques (Cluster, GDBN, ZPAT, DPAT, SPAT, SBA, and Wafer Edge Ink) were tested both independently and in combination to measure their screening efficiency and associated yield loss.

Four RS techniques—ZPAT, GDBN, Cluster, and Wafer Edge Ink—were applied during the CP stage. For each RMA chip, we used its unique identifier (UID) to determine its die position on the wafer and reconstructed its placement in the corresponding Z-axis wafer stack. This allowed us to evaluate the local yield environment and determine whether the die would have been screened out by any of the RS methods. ZPAT evaluates yield at each Z-axis coordinate (i.e. same die position across multiple wafers) and flags low-yielding locations. In our dataset, several RMA chips were located in Z-axis positions with subpar pass rates, indicating that ZPAT could have prevented these from being shipped.

GDBN, which calculates a localized yield score based on the performance of neighboring dies, was also effective. Some RMA chips were situated in clusters of failing dies, suggesting that their local yield fell below the 60% threshold and should have triggered rejection under GDBN logic.

Cluster analysis looks for defect patterns with a threshold of seven failing dies within a 3×3 neighborhood. A subset of the RMA chips exceeded this threshold and would have been inked using the Cluster method.

Wafer Edge Ink produced the most extensive screening: 56% of RMA chips were located within the 6×6 die edge area of the wafer. Although this method effectively eliminates high-risk dies near the edge, it lacks selectivity—it marks all edge dies as failed regardless of actual performance, which incurs significant yield penalties.

The trade-off between yield loss and screening efficacy is visible. **Fig.6** illustrates the spatial distribution of all 24 RMA chips across the wafer surface, with a concentration at wafer peripheries, supporting the use of edge-based screening approaches. However, because Wafer Edge Ink indiscriminately removes a substantial number of good dies, it may not be suitable for products with strict cost or yield constraints.

At the FT stage, we applied three statistical methods: SPAT, DPAT, and SBA. SPAT and DPAT were used to detect outliers based on parametric measurements. These methods define pass/fail criteria based on deviations from the mean value across a population of dies, typically using a ±6σ threshold. Surprisingly, none of the 24 RMA devices were flagged as outliers by SPAT or DPAT. This result likely stems from the nature of SoC testing: compared to simpler ICs such as PMICs or RF transceivers, SoCs undergo more complex functional validation (e.g., SCAN, MBIST), while parametric measurements are limited. Thus, SPAT and DPAT may be more effective for analog or mixed-signal devices with a greater number of continuous-valued test parameters.

SBA was applied at the lot level to detect systemic shifts in binning behavior. Among the 16 wafer lots from which the 24 RMA chips originated, 6 lots (37.5%) exhibited abnormal bin distributions that triggered the Mean ±3σ SBA threshold. The remaining lots did not trigger any alarms, indicating that SBA alone cannot provide full coverage for latent defect detection.

We next evaluated the benefit of integrating the RS methods into a unified screening flow, with three stages, including CP-level screening (DPAT, ZPAT, Cluster, GDBN, and Wafer Edge Ink), Lot-level analysis (SBA for bin distribution anomalies), and FT-level screening (SPAT for parametric outliers).

As shown in **Fig. 7**, the integrated RS flow was implemented on three representative wafer lots, covering approximately 105,000 chips. An MDS (Manufacturing Data System) recipe was defined to automate RS rule execution. This system parses test data, applies decision logic, flags or inks failing dies, and calculates yield losses in real time, as shown in **Fig. 8**.

Table I summarizes the yield loss introduced by each method.

DPAT, ZPAT, GDBN, and Cluster each contributed less than 3.5% yield loss. Wafer Edge Ink alone accounted for ~33% loss. SPAT added an average of 17.02% yield loss at the FT stage, with one lot (Lot 2) exhibiting excessive impact (over 35%). Final yield (bin 1) after RS flow was: 38.17%, 29.02%, and 36.62% for the three lots. This indicates that the RS flow is highly selective: it aggressively filters out potential early-failure chips but at the cost of more than 60% average yield loss.

Table II presents a detailed analysis of each method's gain in terms of RMA screening rate, DPPM improvement, yield loss, and efficiency normalized by DPPM reduction per 1% yield loss. Cluster and ZPAT each screened 17.5% of RMA chips, with a DPPM improvement of 60.DPAT and GDBN each screened 4.3%, contributing DPPM gains of 15. Wafer Edge Ink had the highest screening rate (56%) but an unacceptably high yield loss (34.08%). When normalized, Cluster emerged as the most efficient method, capable of screening out 12,000 DPPM-worth of early-failure chips per 1% of yield sacrificed. This makes it particularly valuable for high-reliability applications. Conversely, Wafer Edge Ink, despite its high effectiveness, may be suitable only in cost-insensitive domains such as aerospace or defense.

The RS techniques can be grouped by their underlying mechanisms. Position-based screening and statistical-based screening. Wherein, the former includes Cluster, GDBN, ZPAT, Wafer Edge, which targets spatial patterns and wafer manufacturing anomalies. These methods assume that proximity to known defect sites correlates with increased early-life failure risk. The latter includes SPAT, DPAT, SBA, which focus on the detects outliers or systemic deviations in electrical performance metrics or binning patterns.

The failure of borderline-good dies is often due to latent structural or parametric defects that are too subtle to be captured by standard pass/fail thresholds. By exploiting statistical trends and spatial correlations, RS methods can preemptively remove these dies before they manifest as customer returns.

III. LIMITATIONS AND FUTURE WORK

While the RS methods show promising results, several limitations and open questions remain.

Device Dependence: The effectiveness of each RS technique is highly dependent on the device type, manufacturing process, and test coverage. For example, SPAT and DPAT performed poorly on SoC devices due to limited parametric testing, though they may work well for PMICs, RF transceivers, and sensor ICs.

Rule-based Rigidity: Current RS techniques rely on fixed statistical rules and thresholds. These may fail to adapt to process drift, new failure mechanisms, or subtle cross-die correlations, leading to either under-screening (missed RMA) or over-screening (unnecessary yield loss).

Isolated Decision Making: The seven RS techniques are largely independent and do not share contextual information, such as cross-stage correlations (e.g., between CP and FT test patterns) or higher-order patterns across lots and time.

These limitations call for a more flexible and intelligent RS framework that can learn from historical data and adapt to evolving failure patterns.

To further improve DPPM performance while minimizing yield impact, we propose the following future directions:

Quantitative Cost-Effectiveness Modeling: We plan to systematically compare DPPM improvements achieved by RS methods against their associated yield losses and testing overhead. This will help identify Pareto-optimal strategies for specific product segments and market requirements.

System-Level Correlation and Feedback: While current RS methods operate at the chip level, many failure modes manifest only at the module or system level. We aim to incorporate system test data (e.g., end-of-line functional testing, burn-in screening, or customer return data) into our analysis loop to refine RS rule thresholds and eliminate false negatives.

Machine Learning–Based RS Optimization: To overcome the limitations of fixed rules and thresholds, we plan to develop supervised and semi-supervised machine learning models that predict die-level failure probability using multi-source data, including test results, wafer layout, process parameters, and past RS outcomes. Specifically, we will investigate convolutional neural networks (CNNs) for spatial pattern recognition, as well as graph-based models to capture die neighborhood dependencies.

Dynamic and Adaptive Test Flows: In future work, we aim to design RS frameworks capable of dynamically adjusting test recipes based on real-time manufacturing data. For instance, if a lot exhibits unusual behavior (e.g., drift in PAT distributions or a spike in localized failures), the RS system could invoke enhanced screening rules or trigger hold/retest actions without manual intervention.

Cross-Product and Cross-Platform Generalization: Another important goal is to evaluate the generalizability of our RS flow across different product lines, technology nodes, and foundry platforms. A robust RS methodology should maintain effectiveness even in the presence of design or process variation, thus reducing the engineering overhead required for each product qualification cycle.

Integration with EDA and Manufacturing Systems: Finally, we intend to embed the RS flow into a unified test automation framework that interfaces with EDA tools, test equipment, and MDS/EDA analytics engines. This will enable closed-loop learning from test and field data and support scalable deployment in production environments.

IV. SUMMARY

In this study, we investigated seven reliability screening (RS) techniques with the objective of improving the defect parts per million (DPPM) performance of IC products, especially those targeting high-reliability applications such as automotive electronics. Through the analysis of 24 RMA chips returned from the field, we comprehensively evaluated each method's ability to detect early-life failures and proposed a unified RS test flow that combines multiple screening strategies across different stages of the test process.

The experimental evaluations, conducted on over 100,000 chips across three wafer lots, demonstrated that each RS technique exhibits unique advantages and trade-offs. In conclusion, this work lays the groundwork for a comprehensive, data-driven approach to reliability screening in IC manufacturing. The proposed methods and insights are expected to be valuable for semiconductor vendors striving to achieve lower RMA rates, higher product reliability, and stronger customer satisfaction, particularly in safety-critical domains like automotive and industrial electronics.

ACKNOWLEDGMENTS

This work is supported and funded by the UNISOC (Shanghai) Technologies.

REFERENCES

[1] Jun C H, Hong Y, Kim S Y. A simulation-based semiconductor chip yield model incorporating a new defect cluster index [J]. Microelectronics Reliability, 1999, 39(4):451-456.

[2] Yang C H , Yen C H , Wang T R , et al. Identifying Good-Dice-in-Bad-Neighborhoods Using Artificial Neural Networks[C]. 2021 IEEE 39th VLSI Test Symposium (VTS). IEEE, 2021.

[3] Dobbelaere W , Vanhooren R , Man W D , et al. Analog fault coverage improvement using final-test dynamic part average testing[C]. 2016 IEEE International Test Conference (ITC). IEEE, 2016.

[4] T. Haifley et al., "Guidelines for part average testing" Automotive Electronics Council, pp. 1-9, Dec. 2011.

[5] T. Haifley et al., "Guidelines for statistical yield analysis" Automotive Electronics Council, pp. 1-6, Jan. 2012.

[6] Zernig A , Bluder O , Pilz J , et al. Device level Maverick screening - detection of risk devices through Independent Component Analysis[C]. Simulation.

Weight Bit Sensitivity Analysis and FPRH-Based Hardening Strategy for CNN Accelerators

Jinghao Chen [1], Shanqiang Yang [1], Tianliang Xu[1], Congan Xu[2], Yuehong Gong[3], Chenxu Wang*[1,4]

[1] Harbin Institute of Technology, Weihai 264209, China
[2] Naval Aeronautical University, Yantai 264001, China
[3] School of Navigation and Shipping, Shandong Jiaotong University, Weihai, 264209, Shandong, China
[4] Shandong Provincial Key Laboratory of Marine Electronic Information and Intelligent Unmanned Systems, Weihai 264209, Shandong, China

* Email: 2573357333@qq.com, wangchenxu@hit.edu.cn

Abstract—This paper proposes a comprehensive quantitative analysis and hardening framework based on fault injection experiments to address the reliability issue of weight bit-flips in convolutional neural networks (CNNs) deployed on edge hardware accelerators. Through scripted bit-level fault injection into externally stored floating-pointhts, we systematically quantify the bit sensitivity of weights across different neural network layers. Experimental results show that bit flips in high-order exponent bits are the primary cause of mean Average Precision (mAP) degradation. Based on these findings, we propose a novel hardening algorithm (FPRH), which innovatively integrates a fixed-bit redundancy mechanism combining Triple Modular Redundancy (TMR) and Dual Modular Redundancy (DMR). This algorithm achieves an approximate 7% improvement in mAP with only a 0.5% overhead in inference time, providing a hardware-friendly solution to enhance the single-event upset (SEU) resilience of CNNs.

Keywords—weights, mitigation, CNN accelerator, SEU, error tolerance

I. INTRODUCTION

With the rapid deployment of deep neural networks in safety-critical scenarios such as autonomous driving and drone surveillance, edge computing devices face dual challenges: storage constraints and hardware reliability. Although lightweight models like YOLOv4-tiny significantly reduce computational complexity, their 32-bit floating-point weights still require 23.68MB of storage—far exceeding the on-chip capacity of typical edge platforms (e.g., the Xilinx Zynq-7020 has only 630KB of Block RAM (BRAM))[1]. While 16-bit fixed-point quantization (with 9 fractional bits) can halve storage requirements[2], this conversion introduces new risks: bit-flip faults in hardware storage media under radiation environments may lead to catastrophic inference errors, resulting in a gigantic drop in mAP for object detection, which severely compromises system safety.

Current research on enhancing the reliability of neural networks faces two major bottlenecks: High hardware redundancy overhead: Traditional Triple Modular Redundancy (TMR) requires 300% more storage resources[3], making it impractical for resource-constrained edge devices; Quantization error accumulation: Low-precision conversion (e.g., 8-bit fixed-point) reduces storage needs but introduces irreversible accuracy loss (over 5% mAP drop)[4].

More critically, existing methods lack systematic quantification of the bit sensitivity of floating-point weights, leading to insufficiently targeted hardening strategies. Through bit-level fault injection experiments, this paper reveals that high-order exponent bits (bit 2, bit 3) are the core

vulnerability. Accordingly, we propose the FPRH (Float-Point Redundancy Hardening) dynamic hardening algorithm, which leverages the distribution characteristics of the top three exponent bits to design a hybrid redundancy mechanism, enhancing the reliability of read weight data.

II. FAULT INJECTION METHODOLOGY AND ANALYSIS

A. Experimental Methodology

This study develops a DDR memory-based weight fault injection methodology that employs Python scripts to simulate controlled bit flips in off-chip DDR weight files, systematically evaluating the impact of different bit faults on model inference accuracy. The approach implements precision bit-level manipulation at the DDR storage level, where Python scripts perform targeted bit flips at configurable rates (e.g., 0.1%) while supporting both directed injection modes (specific to sign/exponent/mantissa bits) and fully random modes, all while maintaining the original data distribution to ensure realistic fault simulation. Following fault injection, the modified weights are immediately loaded into the accelerator for forward inference using the standard VOC validation set for end-to-end testing, with mean Average Precision (mAP) variation serving as the core metric to quantify accuracy loss.

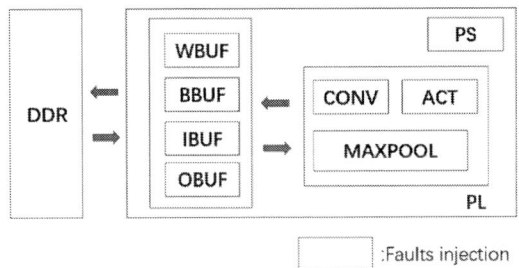

Fig. 1. Schematic diagram of the experimental process

Fig. 1 shows an experimental flowchart based on an accelerator, depicting the fault injection process and affected locations via data flow. DDR memory acts as the data source, interacting with the accelerator's buffer units (WBUF, BBUF, IBUF, OBUF). WBUF and OBUF are marked as fault injection sites (indicated by red squares). After data enters the buffer units from DDR, it is processed by modules like convolution (CONV), activation (ACT), and max-pooling (MAXPOOL) in the PL (programmable logic) section. Finally, the impact of fault injection on the model's inference accuracy is quantified through mAP evaluation. The entire process

979-8-3315-3918-4/25 $31.00 © 2025 IEEE

clearly presents the fault injection location in the accelerator's data flow and the subsequent evaluation steps.

This approach features two key technical characteristics: First, it enables non-intrusive injection through memory-mapped weight file modification without requiring FPGA reconfiguration. Second, it supports cross-layer fault propagation tracking that can trace single-bit errors through the computational graph.

B. Fault Injection Data Analysis

The single-precision floating-point number is a 32-bit floating-point format defined by the IEEE 754 standard[5]. It consists of 1 sign bit (bit 31), 8 exponent bits (bits 30-23), and 23 mantissa bits (bits 22-0). The numerical representation is achieved through (1);

$$value = (-1)^{sign} \times (1 + mantissa)2^{exponent-127} (1)$$

Among them, the exponent bits adopt an offset code mechanism with an offset of 127, and the mantissa bits implicitly have the first bit as 1, thereby extending the precision to 24 significant bits. In convolutional neural networks (such as YOLOv4-tiny), this format is the basic carrier for weight storage. Its structural characteristics not only determine the storage requirements when deploying on edge devices (for example, 32-bit weights require 23.68MB of storage space) but also serve as the design basis for hierarchical hardening strategies (such as FPRH) — by specifically protecting highly sensitive exponent bits, it improves the ability to resist Single-Event Upsets (SEU) while controlling storage and delay overhead.

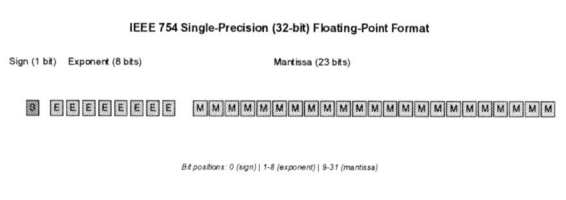

Fig. 2. IEEE 754 Single-Precision Floating-Point Format.

We conducted fixed-bit fault injection experiments targeting the exponent and sign bits of the accelerator's weights. These experiments reveal pronounced sensitivity disparities in neural network models to high-bit faults. As shown in the Fig. 3 below, mAP evaluation results from 0.1% fault injection across all convolutional layers demonstrate these variations.

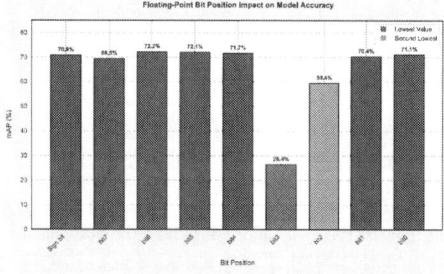

Fig. 3. The mAP evaluation results from 0.1% fault injection.

The experimental results in Fig. 3 indicate that bit 3 of the weight data's exponent bits exhibits extremely high sensitivity to faults—even minor perturbations can lead to a significant drop in accuracy (>40% mAP reduction). Bit 2 shows high

sensitivity, with a relatively limited impact (approximately 10% mAP decrease). The sign bit and other exponent bits have a smaller influence. Meanwhile, the mantissa, which represents fine-grained fractional values, has a minimal impact on overall results and thus was not prioritized in this study. The sensitivity of each position in the weight data is summarized in Table I.

TABLE I.   SENSITIVITY ANALYSIS OF FAULT INJECTION

Fault Injection Bit Field	mAP Degradation	Sensitivity Level
Bit3 (Exponent)	>40%	Extreme
Bit2 (Exponent)	>10%	High
Other bits	<5%	Low

Our comparative analysis further underscores the marked asymmetry in the impact of bit flips: when subjected to the same number of fault injections, 1→0 flips lead to substantially milder mAP degradation than 0→1 flips. Quantitatively, the experimental results highlight a striking disparity in model sensitivity to these flip types—with the induced mAP difference surpassing 30%, as illustrated in Fig. 4.

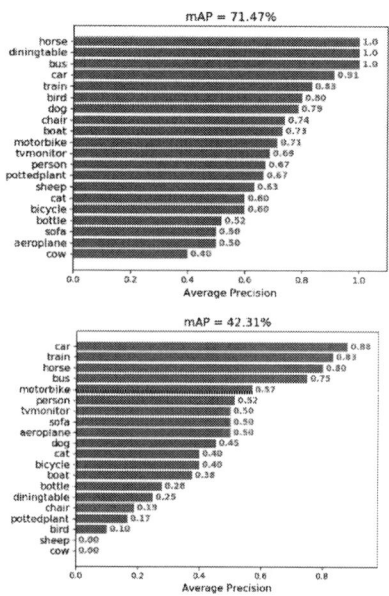

Fig. 4. mAP comparison between 1to 0 and 0 to 1.

Based on the principles of floating-point representation and our statistical analysis of the weight distributions across all layers of the YOLOv4-tiny model—findings presented in Fig. 5 and Table II—we observed that the first three exponent bits of weight data in convolutional neural networks maintain a stable "011" pattern with a 99.99% probability. Experimental validation confirms that artificially forcing these exponent bits to "011" has a negligible impact on model accuracy.

These observations have directly inspired our proposed hardening strategy: implementing a bit replication mechanism that duplicates sensitive bits to pre-determined stable positions (fixed bits conforming to the "011" pattern). This line of thinking directly guides the design of the hardening method described in the following section, creating an organic connection between the protection strategy and the intrinsic characteristics of weight data as well as the laws of fault sensitivity.

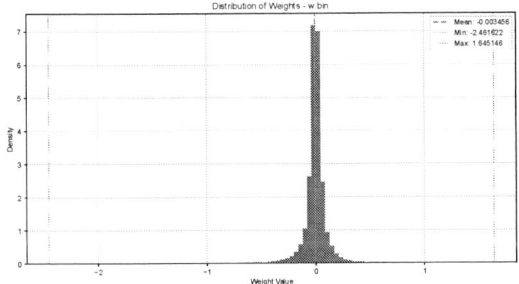

Fig. 5. Distribution of weight values in Convolutional Layer 2.

TABLE II.   DISTRIBUTION AND PROPORTION OF 0S AND 1S IN SIGN BITS AND EXPONENT BITS OF CONVOLUTIONAL LAYER 2

Bit	0 count	1 count	0/1 rates
Sign bit	9347	9058	49.29%
Bit 7	18431	1	00.01%
Bit 6	1	18431	99.99%
Bit 5	1	18431	99.99%
Bit 4	18	18414	99.90%
Bit 3	2702	15730	85.34%
Bit 2	13538	4894	26.55%
Bit 1	8958	9474	51.40%
Bit 0	9276	9156	49.67%

III. ERROR DETECTION AND CORRECTION HARDENING DESIGN AND EVALUATION

A. Hierarchical Hardening Strategy Design

The experimental results confirm that there is a significant disparity in fault sensitivity among different exponent bits: Bit 3 (i.e., Class-A bits) is extremely sensitive to faults, with even a single bit flip causing a substantial drop in mean Average Precision (mAP) by over 40%; Bit 2 (i.e., Class-B bits) exhibits high sensitivity, and its flip leads to a high mAP degradation of approximately 10%. In contrast, mantissa bits and other low-sensitivity bits have minimal impact on model accuracy (with mAP dropping by less than 5%), thus requiring no targeted protection.

To balance reliability and resource efficiency, a hierarchical hardening strategy is proposed, as visually illustrated in Fig. 6. For Class-A bits, which are critical to model performance, Triple Modular Redundancy (TMR) is adopted. This mechanism deploys three independent storage units to store the same Class-A bits, and a voting circuit is used to select the majority value as the valid output during inference. This ensures real-time error correction, achieving an error masking rate of over 99% and providing maximum protection for the most sensitive bits.

For Class-B bits, which have lower sensitivity, Dual Modular Redundancy (DMR) is implemented for lightweight error detection[6]. Considering the asymmetric impact of bit flips (0→1 flips are the primary cause of mAP degradation, as shown in Fig. 4), DMR is optimized to prioritize handling 0→1 faults: when a mismatch is detected between the two redundant storage units, Class-B bits are automatically reset to 0 instead of undergoing complex recomputation. This "smart error containment" minimizes the impact of undetected errors while reducing overhead, striking a balance between protection and efficiency.

Fig. 6 clearly visualizes this hierarchical design: Class-A bits are replicated three times (marked as "A" in the diagram) to enable TMR-based error correction, while Class-B bits are duplicated (marked as "B") for DMR-based detection. This structure avoids unnecessary redundancy for low-sensitivity bits (marked as "X"), directly addressing the high overhead issue of traditional full TMR (which requires 300% storage redundancy). Compared to full TMR, the proposed strategy reduces storage overhead by 68%, making it more suitable for resource-constrained edge devices.

IEEE 754 Single-Precision (32-bit) Floating-Point Format

Fig. 6. Diagram Illustrating the Hardening Method for Weight Exponent Bits

B. Algorithm and Circuit Implementation

The core logic of the hardening strategy is embodied in Algorithm 1 (Fig. 7) and its corresponding hardware circuit (Fig. 8). Algorithm 1 outlines the floating-point exponent bit reinforcement and recovery process, consisting of two key phases:

Algorithm 1: Floating-Point Exponent Bit Reinforcement and Recovery Algorithm

---

Data: 8-bit exponent array *exponent_bits* (index 0-7 corresponds to bit0-bit7)
Result: Reinforced and recovered exponent bits

```
;                                    \\ Phase 1:  Process bit3/6/7
if exponent_bits[3] ≠ exponent_bits[6] then
    exponent_bits[3] ← exponent_bits[7];    \\ Set bit3=bit7 when
    condition fails
else
    ;                                \\ Phase 2:  Process bit2/5
    if exponent_bits[2] ≠ exponent_bits[5] then
        exponent_bits[2] ← 0;   \\ Set bit2=0 when condition fails
    end
end
;                        \\ Phase 3:  Recover bits 7,6,5 to 011
exponent_bits[7] ← 0;                        \\ Set bit7 to 0
exponent_bits[6] ← 1;                        \\ Set bit6 to 1
exponent_bits[5] ← 1;                        \\ Set bit5 to 1
return exponent_bits
```

Fig. 7. Error detection and correction pseudocode.

Phase 1 (Class-A bit processing): Focuses on bits 3, 6, and 7. It checks consistency between critical bits; if a mismatch is detected (e.g., bit 3 ≠ bit 6), bit 3 is reset to the value of bit 7 (leveraging the stable "011" pattern of top exponent bits observed in Table II) to correct errors.

Phase 2 (Class-B bit processing): Targets bits 2 and 5. When a mismatch is detected between the two redundant bits,

979-8-3315-3918-4/25 $31.00 © 2025 IEEE

bit 2 is automatically reset to 0 to contain the error, avoiding cascading impacts on inference accuracy.

Fig. 8 presents the hardware circuit schematic of the error detection and correction mechanism, which implements the logic of Algorithm 1 in programmable logic (PL). The circuit integrates voting modules for TMR (Class-A bits) and comparison modules for DMR (Class-B bits), with dedicated logic to handle 0→1 flips (the main source of mAP degradation).[8] This hardware-friendly design ensures that the algorithm can be efficiently deployed on edge accelerators without requiring large-scale modifications to the existing inference pipeline.

C. Performance and Overhead Analysis

The effectiveness and efficiency of the FPRH algorithm are validated through comparative experiments, with detailed metrics summarized in Table III .

Accuracy Improvement: Under 0.1% fault injection (consistent with the experimental setup in Section II), the FPRH-hardened model achieves a mean Average Precision (mAP) of 79.8%, which is a 7.7% improvement compared to the baseline model (72.1%). This confirms that the hierarchical redundancy mechanism effectively mitigates the impact of critical bit flips, especially 0→1 flips in Class-A and Class-B bits.

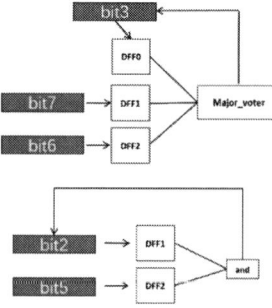

Fig.8. Schematic diagram of error detection and correction circuit.

The hardening process introduces only a 0.5% increase in inference time (from 21.4 ms to 21.5 ms), which is negligible for real-time edge applications (e.g., autonomous driving requires sub-30 ms latency).Resource utilization is significantly lower than traditional TMR (which would occupy over 300% of storage resources) and ensures compatibility with resource-constrained edge platforms like Xilinx Zynq-7020.

In summary, the FPRH algorithm achieves superior reliability-efficiency tradeoffs: it enhances SEU resilience of CNNs with minimal overhead, making it a hardware-friendly solution for safety-critical edge AI applications. The hierarchical design, coupled with asymmetric error handling and optimized resource usage, addresses the key bottlenecks of existing hardening methods.

TABLE III.   KEY PERFORMANCE METRICS COMPARISON

Metric	Baseline	FPRH-Hardened	Improvement
mAP(0.1% faults)	72.1%	79.8%	+7.7%
Inference latency	21.4 ms	21.5 ms	+0.5%
Resource overhead	-	38 Gate Equivalent	-

IV. CONCLUSION

This paper proposes FPRH (Fault-Protected and Resource-efficient Hardening), a hierarchical protection scheme for radiation-hardened neural accelerators, which achieves superior reliability-efficiency tradeoffs through three key innovations. First, by quantifying the asymmetric vulnerability of different bit positions, an adaptive TMR/DMR hybrid protection mechanism is developed. Second, hardware-efficient implementation is achieved through selective bit replication and optimized memory design, maintaining real-time performance. Third, the cross-layer approach combines non-parametric hardening (enforcement of the fixed "011" pattern) with asymmetric error containment. Experimental validation on the VOC dataset demonstrates that FPRH establishes a new paradigm for bit-criticality-adaptive protection in spaceborne AI accelerators. Future work will investigate dynamic adjustment of protection strength based on real-time radiation monitoring to further optimize the reliability-power tradeoff.

ACKNOWLEDGMENT

This work was mainly supported by Major scientific and technological innovation projects ofShandong Province of China, with Grant No.2022ZLGX04. The research presented in this paper isalso partially supported by the NSf project of China with granted No. U2106202 and ShandongProyincial Natural Science Foundation with GrantZR2023MA074.

REFERENCES

[1] C. Chen, J. Emer, and V. Sze, "Eyeriss: A Spatial Architecture for Energy-Efficient Dataflow for CNNs," in Proc. Int. Symp. Comput. Archit. (ISCA), 2016, pp. 367–379.

[2] L. Guo, X. Zhou, Y. Zhang, H. Wang, Z. Liu, and W. Chen, "FPGA-based High-performance MobileNet Accelerator," M.S. thesis, Nanjing Univ., Nanjing, China, 2020.

[3] H. Kim, J. Lee, S. Park, M. Jung, D. Kang, and K. Choi, "Hamming-ECC for DNN Weight Protection," in Proc. Design, Autom. Test Eur. Conf. (DATE), 2022, pp. 1027–1030.

[4] P. Plagwitz, F. Meyer, A. Traber, P. Holzenspies, T. Kenter, and J. Förstner, "A Safari through FPGA-based Neural Network Compilation," in Proc. IEEE Symp. Field-Program. Custom Comput. Mach. (FCCM), 2021, pp. 1–9.

[5] S. Rehman, W. El-Harouni, S. Prabhu, M. Rehman, M. Shafique, and J. Henkel, "Reliability Analysis of DNN Accelerators under Radiation," IEEE Trans. Comput.-Aided Design Integr. Circuits Syst., vol. 35, no. 12, pp. 2134–2147, Dec. 2016.

[6] Y. Umuroglu, N. Fraser, G. Gambardella, M. Blott, P. Leong, and M. Jahre, "FINN: A Framework for Fast, Scalable Binarized NN Inference," arXiv:1612.07119, 2016.

[7] T. Wang, E. Wang, Y. Chen, T. Zhang, H. Yang, and H. Li, "ViA: A Novel Vision-Transformer Accelerator Based on FPGA," IEEE Trans. Comput.-Aided Design Integr. Circuits Syst., vol. 41, no. 5, pp. 1436–1449, May 2022.

[8] C. Zhang, P. Li, G. Sun, J. Cong, B. Xiao, and P. Zhou, "Optimizing FPGA-based Accelerator Design for Deep CNNs," in Proc. ACM/SIGDA Int. Symp. Field-Program. Gate Arrays (FPGA), 2015, pp. 161–170.

[9] X. Liu, Y. Zhang, J. Li, Q. Wang, Z. Chen, and H. Yang, "A Survey of Fault-Tolerant Techniques for Convolutional Neural Networks in Edge Computing," J. Parallel Distrib. Comput., vol. 172, pp. 1–15, Mar. 2023.

[10] Y. Wang, X. Chen, M. Liu, W. Zhang, Y. Li, and S. Zhao, "Enhancing the Reliability of Neural Networks in Radiation Environments: A Comprehensive Review," IEEE Trans. Nucl. Sci., vol. 69, no. 4, pp. 732–749, Apr. 2022

An effective method for low-contrast high-noise lithography SEM image contour extraction

Ruirui Zhang[1], Gongyan Ye[1], Xianhe Liu[12*]

[1]School of Microelectronics, Fudan University, Shanghai, China
[2]National Integrated Circuit Innovation Center, Shanghai 201203, China

*Corresponding Author's Email: xianheliu@fudan.edu.cn

Abstract—With the increasing of semiconductor devices, precise measurement of line width roughness (LWR) and line edge roughness (LER) has become critical for ensuring device performance. However, existing metrology techniques like CD-SEM face challenges in analyzing low-contrast high-noise lithography SEM images, especially in research environments with constrained access to specialized equipment. To address this, this study proposes an innovative SEM image analysis algorithm that integrates Cellular Neural Networks (CNN) for edge enhancement, adaptive polynomial fitting for sub-pixel edge positioning, and dark channel prior dehazing for contrast improvement. Implemented via a MATLAB GUI, the algorithm demonstrates superior edge detection accuracy for low-contrast images compared to the open-source tool SMILE, offering a robust and efficient solution for lithography SEM image analysis.

Keywords—lithography SEM image, contour extraction, line width roughness, polynomial fitting, dehazing preprocessing

I. INTRODUCTION

A. Research Background

In the semiconductor manufacturing field, integrated circuit devices continue to miniaturize, making CD and its LWR of key parameters for evaluating lithographic pattern quality [1]. CD reflects the linewidth of lithographic patterns, directly influencing circuit integration density and performance, while LWR characterizes linewidth non-uniformity, critically impacting device reliability and stability. Accurate measurement techniques are essential for ensuring semiconductor product quality and performance. Current mainstream CD and LWR measurement technologies include Critical Dimension Scanning Electron Microscopy (CD-SEM) and Atomic Force Microscopy (AFM) [2]. CD-SEM offers high resolution and rapid imaging capabilities, making it widely adopted in semiconductor manufacturing. However, in practice, due to the complexity of lithographic processes and imaging conditions, acquired SEM images often suffer from low contrast and high noise, posing significant challenges to precise CD and LWR measurements.

B. Research Problem

For SEM images with poor contrast, achieving effective contour extraction and accurate calculation of CD and LWR remains an urgent challenge in semiconductor manufacturing. Conventional image analysis methods often fail to meet the required precision for such images, falling short of the high-accuracy measurement demands in semiconductor fabrication. Therefore, developing a novel image analysis method is necessary to overcome the effects of low contrast and high noise, enabling accurate contour extraction from lithographic SEM images and subsequent precise CD/LWR computation.

C. Related work

Traditional lithographic image analysis methods rely on gradient-based edge detection (e.g., Sobel, Canny [3]) and filter-based noise suppression (e.g., Gaussian, median filtering). However, these approaches struggle with low-contrast, noisy SEM images due to their sensitivity to interference and tendency to blur edges. Recent advances in deep learning [4] and CNNs [5][6] have improved image processing by enabling autonomous feature learning, yet challenges remain in edge localization precision and low-contrast image enhancement. Existing methods still fall short of semiconductor metrology demands, particularly in handling high-noise and low-contrast SEM images, highlighting the need for a more robust and accurate analysis algorithm.

D. Research Approach

This study proposes an improved method of lithographic SEM image contour extraction and LWR calculation. Our method first preprocesses SEM patterns using a dark channel prior dehazing method to eliminate haze and noise, thereby enhancing image contrast to provide a clear foundation for edge detection and contour extraction. Subsequently, a CNN is employed for edge enhancement to further suppress noise and highlight edge information. Finally, adaptive polynomial fitting achieves sub-pixel edge localization, enabling precise contour extraction and CD/LWR calculation, and the obtaining results will be compared with the open-source SEM image analysis tool SMILE [7].

II. METHODS

A. Lithography SEM Image Preprocessing

To improve the visibility of features in low-contrast SEM images, this study proposes a combined enhancement method comprising dark channel prior dehazing, histogram equalization, and bilateral filtering. The goal is to increase image contrast while preserving original structural details.

First, the dark channel prior method[8] — originally developed in natural image dehazing—is applied to remove haze-like noise induced by electron beam scattering or detector interference. By estimating light transmission and separating the haze layer, this method helps recover fine edge details that are otherwise blurred.

Next, histogram equalization redistributes grayscale values across the full dynamic range, improving bright-dark contrast, particularly beneficial for highlighting weak differences between line patterns and the background.

However, these steps may sometimes over-sharpen edges or introduce artifacts. To prevent this, bilateral filtering is applied in the final step. It smooths the image while preserving important edges by combining spatial and intensity-based weighting.

This novel approach adapts a computer vision dehazing technique to SEM image processing, and therefore overcomes the limits of basic contrast stretching and single-filter methods, delivering high-quality images suitable for automatic measurement tasks.

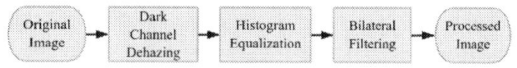

Fig. 1. Schematic Flowchart of the Proposed Algorithm

B. Edge Enhancement using Convolutional Neural Networks (CNN)

For further edge enhancement and noise suppression, this study utilizes a CNN for edge enhancement. CNN, inspired by mimicking the working principles of biological neural networks, adaptively enhances image edges while effectively suppressing noise. This process helps preserve structural features while improving edge detection accuracy, particularly in complex lithography SEM images.

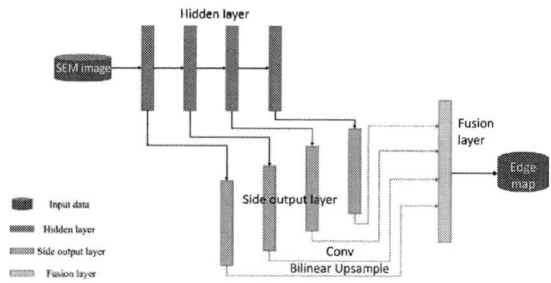

Fig. 2. Contour Extraction CNN (Improved HED)

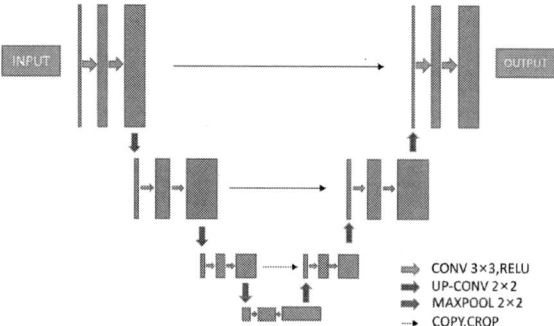

Fig. 3. Denoising CNN (Improved U-Net)

C. Edge Localization and Polynomial

Following edge detection, precise edge localization is crucial for accurate measurement. This study employs Adaptive Polynomial Fitting to achieve subpixel-level accuracy in edge localization. The polynomial fitting method adapts to local edge curvature and to ensure precise positioning. It enables accurate extraction of edge locations, which is vital for high-precision measurements in lithography SEM images.

D. LWR and LER Calculation

In this paper, line width roughness (LWR) is defined as the standard deviation of the line width measurements [9]:

$$LWR = \sqrt{\frac{\sum_{i=1}^{N}(w_i - \overline{w})^2}{N}}$$

where N is the number of measuring points along the line edge, w_i is the width of the line measured at each position, and $\overline{w}$ is the average line width.

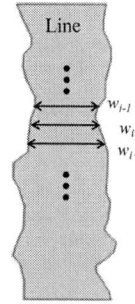

Fig. 4. Schematic Diagram of Line Width Roughness (LWR) Calculation

Similarly, Line Edge Roughness (LER) is defined as the standard deviation of the edge positions [10]:

$$LER = \sqrt{\frac{\sum_{i=1}^{N}(x_i - \overline{x})^2}{N}}$$

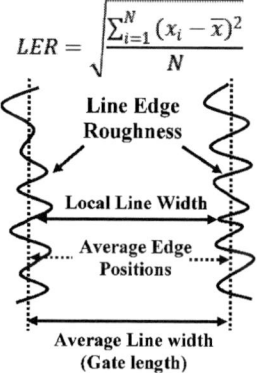

Fig. 5. Schematic Diagram of Line Edge Roughness (LER)

where x_i represents the position of the line edge at each point, and $\overline{x}$ is the average edge position. After extracting the edges, we can easily calculate the LWR and LER from the line edge data, thus quantitatively evaluating the roughness characteristics of the line profiles.

III. RESULTS

High-resolution line/space typical patterns were fabricated with Electron Beam Lithography, varying the critical dimensions (CD) from 20 to 30 nm. Their SEM images were collected for further detection. The effect of dehazing pretreatment for edge detection will be investigated first. Then the LWR with different CD will be investigated, followed by a comparative analysis against the open-source pattern analysis software SMILE, in order to validate the effectiveness and robustness of our proposed method.

A. Impact of Dehazing Preprocessing on Boundary Extraction

As illustrated in Fig. 6, the dehazed image exhibits substantially enhanced contrast compared to the original, enabling clearer boundary definition and consequently simplifies the subsequent contour extraction process.

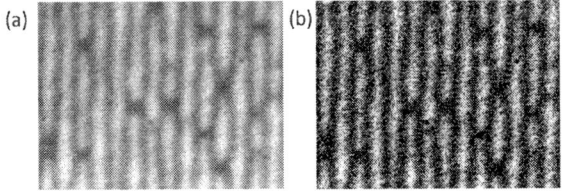

Fig. 6. Effect of Dehazing on SEM Image:(a) before and (b)after the dehazing treatment.

B. LWR Across different Critical Dimensions

The performance of the proposed processing method was evaluated using SEM images with CDs of 22 nm, 24 nm, 26 nm, and 28 nm. The LWRs were measured 5 times per image. Then the means and standard deviations were calculated to show the statistical distribution. Results are shown in TABLE I.

Table 1 LWR MEASURE RESULTS

CD(nm)	LWR_average(nm)	σ LWR(nm)
22	13.68	3.68
24	6.58	0.23
26	6.82	0.33
28	5.99	0.21

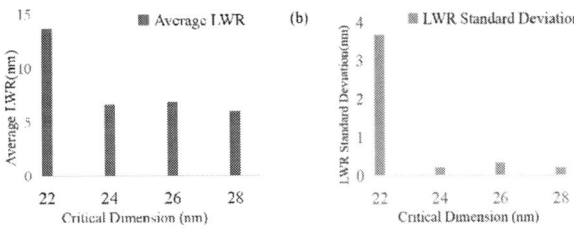

Fig. 7. Critical Dimension Dependence of Line Width Roughness and Measurement Stability

As summarized in TABLE I, the proposed contour extraction method exhibits highly sensitivity to critical dimensions. The average LWR decreases significantly with increasing feature size, from approximately 14 nm at 22 nm CD to about 6 – 7 nm at CDs between 24 and 28 nm.

More importantly, measurement stability—indicated by the standard deviation of LWR—shows a clear improvement at the 24 nm threshold. At 22 nm, the LWR variability is high (σ > 3.5 nm), whereas for CDs of 24 – 28 nm exhibit sub-nanometer stability, with standard deviations remains to

0.2 – 0.3 nm, which suggests the impact of boundary clarity on the stability of edge detection.

C. Comparison with reference LWR values

Validation against the open-source SMILE software was performed using the above four SEM images with CDs ranging from 22 to 28 nm. As shown in TABLE II, the results indicate no statistically significant differences between the two methods. The LWR values from our methodwere consistently slightly lower than those from SMILE, with all differences within the acceptable margin, confirming the accuracy and reliability.

Table 2 LWR result Comparison with SMILE open-source software

CD(nm)	Measuring LWR (nm)	Reference LWR (nm)
22	13.676	14.14
24	6.578	7.59
26	6.816	7.7
28	5.99	7.65

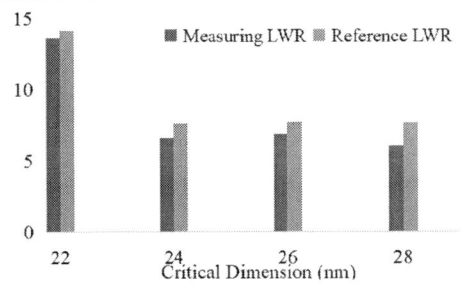

Fig. 8. Measured & Reference LWR Values Across Critical Dimensions

Furthermore, Fig. 9 presents a comparative analysis of edge detection performance our under low-contrast high-noise conditions. SMILE produced significant boundary distortion (Figure 10b), likely due to high-frequency oscillations in the fitting curves caused by noise. These oscillations exceeded the natural boundary variations of the structural features, leading to LWR measuring errors and excessive deviation. Conversely, in our case the image contrast is enhanced and therefore improves the visibility of weak edges. By applying a polynomial fitting strategy, it effectively suppresses noise-induced oscillations, improving the accuracy of LWR measurement.

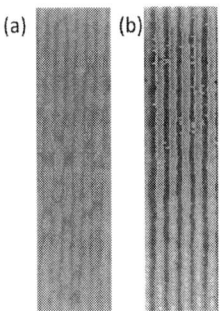

Fig. 10. Edge Detection of low-contrast SEM images by (a) proposed method (b) open-source tool SMILE.

IV. CONCLUSION

This study has developed an innovative method for accurate contour extraction and LWR measurement in low-contrast high-noise SEM images by integrating three key technological advancements: (1) dark channel prior (DCP) dehazing for contrast enhancement, (2) CNN-based edge preservation and noise suppression, and (3) adaptive polynomial fitting for sub-pixel edge localization. Through systematic evaluation using lithographic SEM images with varying critical dimensions (22-28 nm), the method demonstrates significant improvements, particularly in low-contrast conditions. Further improvements in the future study may involve advanced deep learning architectures (e.g., Vision Transformers), potentially enabling more accurate edge detection via enhanced long-range feature modeling.

ACKNOWLEDGMENT

The authors thank the support from the national science foundation of China (no. 62404054), and also thank the support from Tengfei college of Fudan university.

REFERENCES

[1] B. Bunday et al., "Impact of sampling on uncertainty: semiconductor dimensional metrology applications," in SPIE Proceedings, SPIE, Mar. 2008, p. 69220X.

[2] M. Suzuki et al., "SEM imaging capability for advanced nano-structures and its application to metrology," in SPIE Proceedings, SPIE, Mar. 2017, p. 101451L.

[3] Z. Hu et al., "Canny Algorithm Enabling Precise Offline Line Edge Roughness Acquisition in High-Resolution Lithography," ACS Omega, vol. 8, no. 4, pp. 3992-3997, Jan. 2023.

[4] B. Dey et al., "Unsupervised machine learning based SEM image denoising for robust contour detection," in International Conference on Extreme Ultraviolet Lithography 2021, SPIE, Oct. 2021, p. 25.

[5] N. Chaudhary and S. A. Savari, "Increasing the Utilization of Deep Neural Networks for SEM Measurements Through Multiple Task Formulation and Visualization," IEEE Transactions on Semiconductor Manufacturing, vol. 33, no. 3, pp. 322-330, Aug. 2020

[6] S. Melikyan and G. Melikian, "Advanced Edge Detection for Noisy SEM Images: A Hybrid Approach for Enhanced Precision," in 2024 IEEE East-West Design & Test Symposium (EWDTS), IEEE, Nov. 2024, pp. 1-4.

[7] I. Mochi, "Enhancing SEM image metrology with SMILE: advances, features, and portability," in Metrology, Inspection, and Process Control XXXVIII, SPIE, Apr. 2024, p. 122.

[8] K. He, J. Sun, and X. Tang, "Single Image Haze Removal Using Dark Channel Prior," IEEE Transactions on Pattern Analysis and Machine Intelligence, vol. 33, no. 12, pp. 2341-2353, Dec. 2011.

[9] C. Shin, "Variation-Aware Advanced CMOS Devices and SRAM", Springer Nature, Dordrecht, 2016.

[10] H. Nam., G. Lee., H. Lee. ,I. Park. and C. Shin. "Line Edge Roughness (LER). In Advanced Lithography: Materials and Processes", Springer: Dordrecht, 2014, Chapter 2, pp 19–43.

Design of A Dual-Mode Analog Front-End Circuit Applied in the Voice Activity Detection System

Zirui Dong[1], Xuhaohan Wang[1] and Fan Ye[1]

[1]College of Integrated Circuits & Mico-Nano Electronics, Fudan University, Shanghai, China

Email: fanye@fudan.edu.cn

Abstract—**This paper presents a dual-mode analog front-end (AFE) for voice activity detection (VAD) systems that perform feature extraction in the analog domain. The proposed AFE operates in an always-on ultra-low-power (ULP) mode to extract the signal envelope and zero-crossing rate for event detection, while a high-performance (HP) mode is activated at voice onset for detailed signal processing. Implemented in 65-nm CMOS, the AFE consumes 55 nW in ULP mode and achieves 32–50 dB gain with a 4.6 noise efficiency factor (NEF). In HP mode, it provides 36–56 dB gain, 7.03 μV_{rms} input-referred noise, and a 3.1 NEF. Co-simulation with real voice signals further demonstrates the effectiveness of the proposed architecture for always-on, event-driven voice interfaces.**

Keywords—**Analog-domain feature extraction, dual-mode analog front-end circuit, ultra-low-power consumption.**

I. INTRODUCTION

In numerous voice-interaction applications, informative acoustic inputs are not always present, and the continuous operation of the entire system leads to substantial power waste. To address this issue, voice activity detection (VAD) systems were introduced [1] to activate the power-intensive signal-processing modules only when voice activity is detected.

As an always-on module, the VAD should minimize power consumption while maintaining detection accuracy. For a VAD that performs digital-domain feature extraction, the analog front-end (AFE) includes only an amplifier and an analog-to-digital converter (ADC). The amplified raw voice signal is directly forwarded to the digital back-end for feature extraction and classification. This approach demands a high ADC sampling rate and heavy digital computation, resulting in bandwidth and power inefficiency. In contrast, analog-domain feature extraction performs preliminary processing prior to digitization, reducing the signal bandwidth and the subsequent digital workload, thereby improving the overall power efficiency of the system.

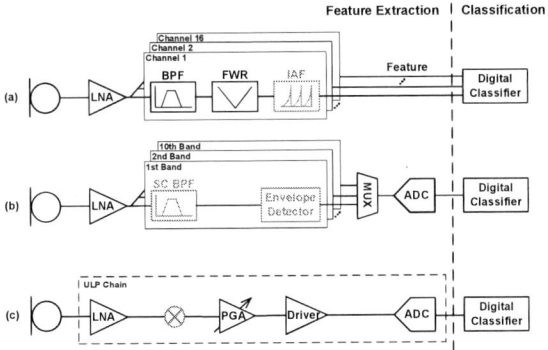

Fig 1. Different architectures for VAD power reduction with analog-domain feature extraction.

Fig. 1(a) illustrates a frequency-domain approach [2] that divides the amplified signal into 16 parallel channels using band-pass filters (BPFs) with distinct center frequencies. Each sub-band is rectified and digitized, enabling a lower ADC sampling rate. However, its Gm-C BPF bank is highly sensitive to process-voltage-temperature (PVT) variations, especially at low supply voltages, leading to possible band aliasing or loss and limiting recognition accuracy (84%) at 380 nW. Building on this concept, the design in Fig. 1(b) further extracts the envelope of each sub-band in the time domain [3], employing passive switched-capacitor BPFs to enhance PVT robustness and accuracy, yet at the cost of higher circuit complexity. The digital-domain solution in Fig. 1(c) [4] retains the low-sampling-rate benefit through analog down-conversion to 500 Hz, but still processes large volumes of non-extracted data—showing that merely reducing the sampling rate is insufficient for enhancing system-level efficiency.

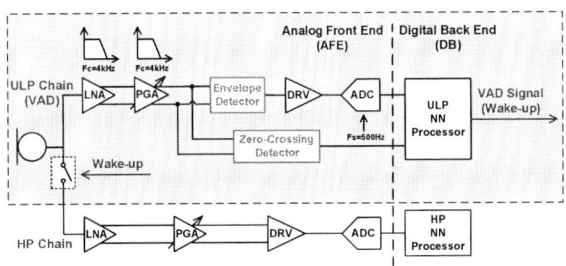

Fig. 2. Overview of the proposed dual-mode AFE for VAD.

To overcome these limitations, this work proposes a dual-mode AFE that performs analog feature extraction across ultra-low-power (ULP) and high-performance (HP) modes, as shown in Fig. 3. In the always-on ULP mode, the circuit continuously monitors the input by extracting two key time-domain features at minimal power, including both signal envelope and zero-crossing rate. Once the embedded NN processor in the digital back-end detects a potential voice event, the HP mode is activated for precise amplification and noise evaluation.

Implemented in 65-nm CMOS, the AFE consumes 55 nW with 32–50 dB configurable gain and 31.42 μV_{rms} input-referred noise (IRN) in ULP mode, while the HP-mode chain achieves 36–56 dB gain and 7.03 μV_{rms} IRN at 495 nW. This operation enables always-on, event-driven detection with high energy efficiency and robustness under PVT variations.

This paper consists of four sections. Section II elaborates on the system operation and circuit implementation of the proposed dual-mode AFE. Section III presents the functional verification and performance simulation results, and Section IV concludes the work with a brief summary.

II. ARCHITECTURE AND CIRCUIT IMPLEMENTATION

The proposed dual-mode AFE integrates an always-on ULP mode and a HP mode that are selectively activated based on the detected voice activity. When a potential speech event is identified, the HP chain is enabled to provide wide-band amplification and accurate noise estimation, improving signal

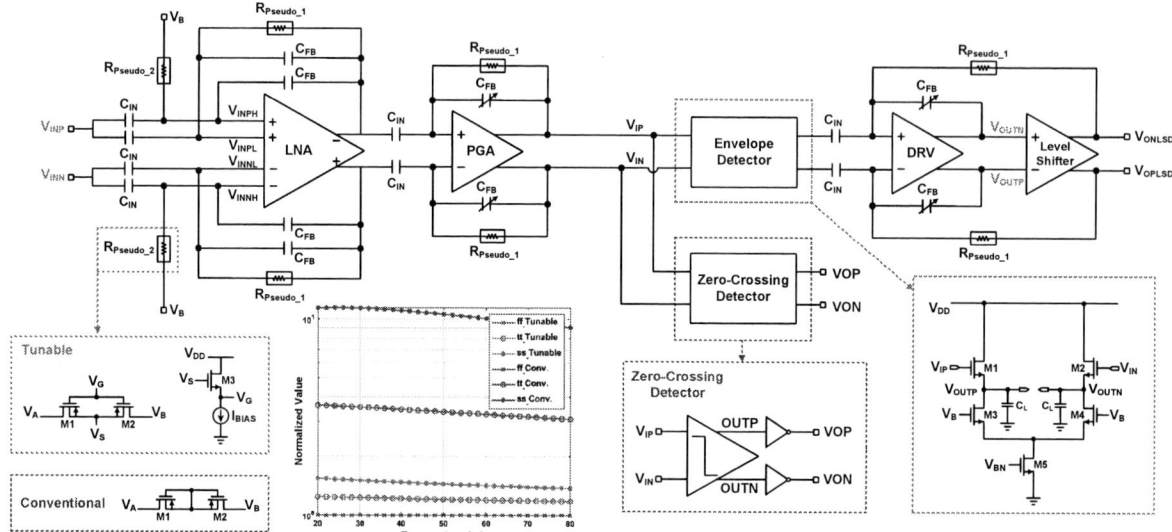

Fig. 3. Architecture of the core modules for feature extraction.

fidelity for subsequent digital processing.

Both modes share a common input interface to minimize hardware redundancy, while a low-leakage event-trigger circuit controlled by the NN processor enables fast wake-up with negligible transition energy. This cooperative operation ensures always-on functionality with ultra-low power and high recognition accuracy under diverse acoustic conditions.

As shown in Fig. 3, the ULP chain primarily comprises a low-noise amplifier (LNA), a programmable-gain amplifier (PGA), an envelope detector (ED), and a zero-crossing detector. The circuit operates between differential inputs (V_{INP}, V_{INN}) and outputs (V_{OUTP}, V_{OUTN}) highlighted in red. Two key time-domain features—the signal envelope and zero-crossing rate—are continuously monitored by a compact neural-network (NN) processor that classifies ambient sounds in the digital back-end.

A. Proposed Ultra-Low-Power Mode Analog Front-End

Fig. 3 depicts the architecture of the primary modules responsible for feature extraction in the ULP-mode AFE. The LNA and PGA provide adjustable gain levels depending on the amplitude of the input. The preamplified signal is then delivered to an ED and a zero-crossing detector in parallel to derive the signal envelope and zero-crossing rate.

Corner frequency tuning in the LNA is achieved through a tunable pseudo-resistor (R_{Pseudo_2}) which biases one of the two differential input pairs. This allows the high-pass corner to be set near 200 Hz by adjusting I_{BIAS} [6], effectively filtering the content outside the target voice frequency band of 200 Hz–4 kHz. Considering power consumption, the other input pair is biased by a constant-value pseudo-resistor (R_{Pseudo_1}) to set the 500 mV DC input level. The relationship between the PMOS gate-source voltage V_{GS} and the I_{BIAS} is as shown in (2):

$$V_{SG} \approx |V_{THP0}| + V_T \log\left(\frac{I_{BIAS}}{I_o}\right) \quad (2)$$

where V_{THP0} is the threshold voltage of PMOS transistors, I_o is the specific current of the Enz-Krummenacher-Vittos (EKV) model, V_T is the thermal voltage. When biased at 0 V, the resistance can be expressed as (3):

$$R_{AB}(V_{AB} = 0) \propto \frac{I_o}{I_{BIAS}} e^{-(V_{SG}-|V_{THP0}|)/nV_T} \quad (3)$$

where n is the slope factor of the EKV model.

Moreover, R_{Pseudo_2} exhibits superior stability under PVT variations. To better demonstrate its robustness, Fig. 3 (lower left) plots the normalized resistance versus temperature for the tunable R_{Pseudo_2}, and a conventional pseudo-resistor at 0-V bias across different process corners.

B. Envelope Detector

As shown in Fig. 4(a), the conventional diode-connected ED employs a MOS transistor for rectification, followed by a resistor-capacitor (RC) low-pass filter to smooth the envelope. Although this passive topology requires no bias circuits, its input signal amplitude must exceed the transistor's threshold voltage (V_{TH}), thus constraining its sensitivity. In contrast, the common-source (CS), common-gate (CG), and common-drain (CD) envelope detectors in Figs. 4(b)-(d) can be operated in the subthreshold region to mitigate the threshold voltage loss [7]. Their conversion gains for an AC input signal with an amplitude of V_p can be respectively derived as:

$$G_{C,CS} = G_{C,CG} = \frac{V_p I_{ds} r_o}{4(nV_T)^2} \quad (4)$$

$$G_{C,CD} = \frac{V_p}{4nV_T} \quad (5)$$

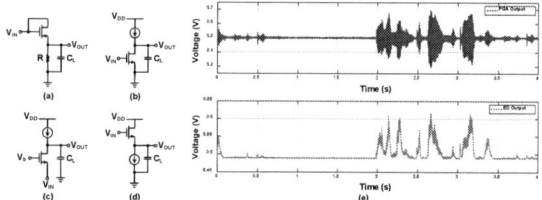

Fig. 4. Schematic of (a) diode-connected ED, (b) common-source ED, (c) common-gate ED, (d) common-drain ED; (e) input and output waveforms of the proposed ED.

It is evident that the common-source and common-gate configurations exhibit a higher conversion gain at the same

979-8-3315-3918-4/25 $31.00 © 2025 IEEE

input level. However, since the ED input is preamplified by the LNA and PGA, excessive AC gain in the ED stage can lead to output saturation, impairing the envelope extraction, the common-drain ED is adopted in this paper. Fig. 4(e) shows the corresponding input and output waveforms of the proposed ED.

C. Zero-Crossing Detector

When only the signal envelope is extracted, the obtained feature represents amplitude information without frequency content, which limits the accuracy of feature extraction. To complement the missing frequency information, a zero-crossing detector is incorporated to capture the signal's time-domain polarity transitions.

In this work, a dynamic comparator is employed as the zero-crossing detector [8], synchronized to an 8-kHz clock (CLK). When CLK is low, the comparator resets, forcing both OUTP and OUTN to GND. When CLK is high (CLKN = GND), it enters the comparison phase: if VIP > VIN, OUTP and OUTN are driven to VDD and GND, respectively; otherwise, they are inverted. The final outputs, VOP and VON (in Fig. 3), are inverted forms of OUTN and OUTP, while inverters are omitted in Fig. 5 for clarity.

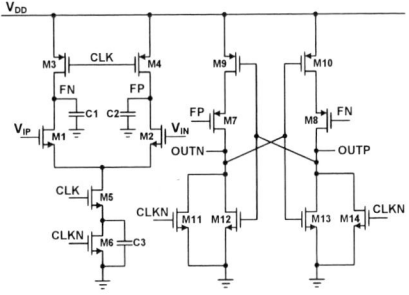

Fig. 5. Schematic of dynamic comparator.

The detector transmits a 4-bit code to the digital back-end every 0.125 ms. For instance, when VIP > VIN, it generates the sequence [VOP, VON] = [1, 0] followed by [1, 1], forming the output code "1011". A transition between consecutive frames (e.g., from "1011" to "0111") indicates a polarity flip of the comparator output, which is interpreted as a zero-crossing event [9]. The back-end segments data into 32-ms frames and counts the number of zero-crossing events per frame to distinguish speech from background noise.

D. Proposed High-Performance Mode Analog Front-End

The HP-mode AFE handles weaker input voice signals [5], which requires stricter noise performance to avoid degrading power efficiency, the HP-mode AFE focuses more on noise reduction.

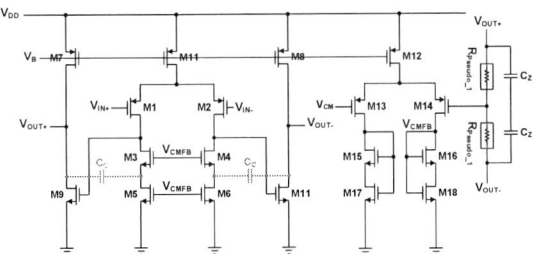

Fig. 6. Operational amplifier schematic of PGA in the HP mode.

Fig. 6 shows the two-stage operational amplifier (OPA) of PGA in the HP-mode AFE. Notably, the Miller capacitors in this PGA are connected from the first-stage cascode node to the second-stage output, rather than between the outputs of the first and second stages [10].

If the traditional Miller compensation structure is adopted between the first and second output stages, the locations of the dominant pole, second pole, and zero are given by (6) to (8).

$$\omega_{p1} = \frac{1}{g_{m9}C_C(r_{o1}//g_{m3}r_{o3}r_{o5})(r_{o7}//r_{o9})} \quad (6)$$

$$\omega_{p2} = \frac{g_{m9}}{C_L} \quad (7)$$

$$\omega_z = \frac{g_{m9}}{C_C} \quad (8)$$

Alternatively, the proposed placement yields pole/zero expressions in (9)–(11).

$$\omega_{p1} = \frac{1}{g_{m9}C_C(r_{o1}//g_{m3}r_{o3}r_{o5})(r_{o7}//r_{o9})} \quad (9)$$

$$\omega_{p2} = \frac{g_{m9}g_{m3}(r_{o1}//g_{m3}r_{o3}r_{o5})}{C_L} \quad (10)$$

$$\omega_z = \frac{g_{m9}g_{m3}(r_{o1}//g_{m3}r_{o3}r_{o5})}{C_C} \quad (11)$$

As evidenced, the second pole and zero are shifted approximately $g_{m3}(r_{o1}//g_{m3}r_{o3}r_{o5})$ times higher while the dominant pole remains unchanged, thereby achieving a wider bandwidth.

III. SIMULATION RESULTS

Fig. 7 provides a direct visual comparison as a representative example, showing the substantial improvement in noise performance achieved by the HP-mode LNA over its ULP-mode counterpart. Fig. 8 depicts the Bode plot of the proposed OPA, which presents a wide bandwidth.

In ULP mode, AFE provides the extracted signal features to the digital back-end for voice activity detection. Therefore, system-level co-simulation is essential for verifying correct operation of the AFE. The simulation utilizes a 200-ms voice signal, captured by a capacitive MEMS microphone, as the input to the ULP-mode AFE. This audio segment is chosen to contain both silent (no speech) and active (speech) intervals.

Fig. 9 illustrates the corresponding back-end output. As the processing is not real-time, the VAD label is aligned with the original input for accurate comparison. As observed from the input waveform, the voice activity occurs between 85 ms and 200 ms, with silence from 0 to 85 ms. The VAD output remains low until 64 ms before transitioning high. This discrepancy results from the frame-based processing scheme of the digital back-end, which employs a fixed frame length of 32 ms. Although the input is silent during the first two frames (0–64 ms), voice activity emerges in the latter portion of the third frame (85–96 ms). Consequently, the VAD output is set to high in the third frame, indicating the detection of activity.

As shown in Fig. 10, the proposed dual-mode VAD AFE is implemented with a 65-nm CMOS technology, occupying 0.36 mm² (including the ADC layout area for both ULP and HP modes). The 10-bit successive approximation register (SAR) ADC utilized in this paper was previously designed by the research group. Its contribution has been taken into consideration in the reported total power consumption for both AFE modes.

TABLE I

PERFORMANCE COMPARISON WITH STATE-OF-THE-ART VAD AFES

	This work[a]		JSSC' 19 [5]		JSSC' 19 [2]	TCAS-I' 21 [3]	TCAS-II' 24[b] [12]
Technology (nm)	65		180		180	180	22
Feature Extractor	Analog		Digital		Analog	Analog	Analog
Supply (V)	1		1.4		0.6	0.65	0.81
Bandwidth (kHz)	0.2–4		0.075–4		0.1–5	0.1–2	0.1–8
Mode	**ULP**	**HP**	ULP	HP	N/A	N/A	N/A
Power (nW)	**55**	**495**	60	600	1000	270	507
In-Band Gain (dB)	**32–50**	**36–56**	22.6–49.3	35.9–62.6	24–42	20–40	N/A
Input-Referred Noise (µV$_{rms}$)	**31.42**	**7.03**	62	8.7	26.2	35.6	N/A
NEF	**4.6**	**3.1**	9.25	4.1	4.41	4.0	N/A

[a] Simulation results; [b] Data source (simulation or measurement) not specified

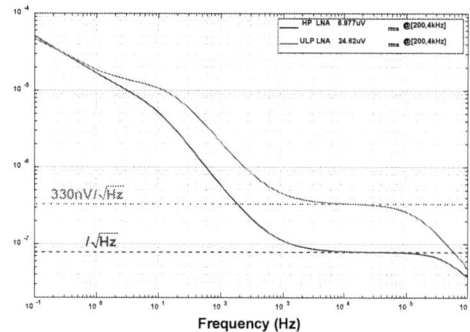

Fig. 7. Comparison of IRN between the HP-mode LNA and ULP-mode LNA.

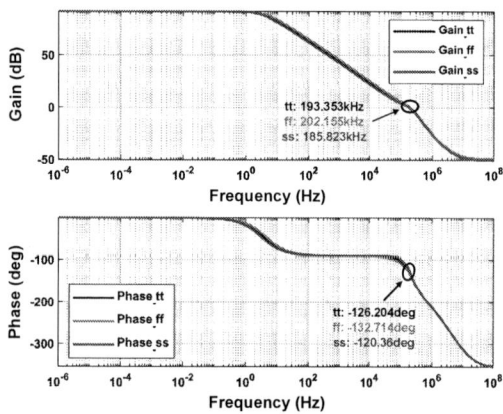

Fig. 8. Bode plot of the proposed OPA for HP-mode PGA.

include the ULP mode used for feature detection in the voice interaction system, with the HP mode that is activated after the VAD output wake-up signal typically located off-chip. In contrast, the design in this paper, like that in [5], integrates the HP mode on-chip, while still consuming extremely low power.

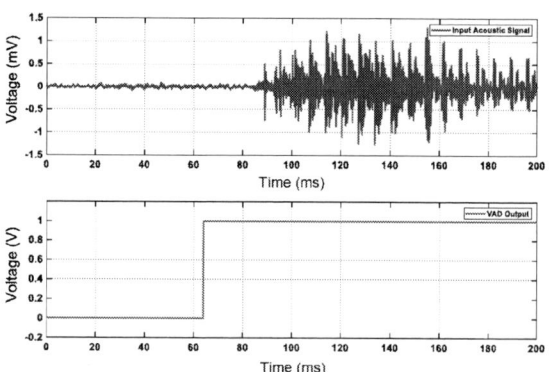

Fig. 9. Output from the digital back-end in ULP-mode VAD.

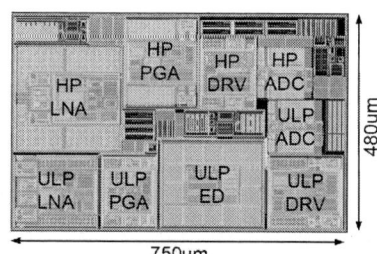

Fig. 10. Layout of the proposed dual-mode AFE for VAD.

Typically, a fundamental trade-off exists between noise performance and power consumption in analog design. To jointly evaluate these two metrics, the noise efficiency factor (NEF) [5] defined in (1) is adopted. With a 1 V supply and a 4 kHz signal bandwidth (B_w), the simulated IRN ($V_{ni,rms}$) and total supply current (I_{tot}) yield NEFs of 4.6 and 3.1 for the ULP and HP AFEs, respectively. These results demonstrate the proposed design's competitive noise-power efficiency across both operation modes, establishing a solid foundation for subsequent circuit implementation and system validation.

$$NEF = V_{ni,rms}\sqrt{\frac{2I_{tot}}{\pi V_T 4kTB_w}} \qquad (1)$$

Table I presents a performance comparison of this paper against previously reported works, and the designed AFE exhibits competitive advantages in power consumption, noise and NEF. It can also be observed that most VAD chips only

IV. CONCLUSION

This work presents a dual-mode analog front-end (AFE) for voice activity detection (VAD) that integrates ultra-low-power always-on sensing with event-driven high-performance operation. The proposed architecture extracts key time-domain features in the analog domain, enabling efficient collaboration for voice event detection with a lightweight neural-network (NN) processor in the digital back-end.

Implemented in 65-nm CMOS, the prototype demonstrates ultra-low standby power and input-referred noise, with robust performance under PVT variations. The results confirm the effectiveness of the dual-mode strategy for achieving high energy efficiency and reliable always-on operation in edge acoustic sensing applications.

REFERENCES

[1] A. Raychowdhury, C. Tokunaga, W. Beltman, M. Deisher, J. Tschanz and V. De, "A 2.3 nJ/frame voice activity detector-based audio front-end for context-aware system-on-chip applications in 32-nm CMOS," in IEEE Journal of Solid-State Circuits, vol. 48, no. 8, pp.1963–1969, Aug. 2013, doi: 10.1109/JSSC.2013.2258827.

[2] M. Yang, C. -H. Yeh, Y. Zhou, J. P. Cerqueira, A. A. Lazar and M. Seok, "Design of an Always-On Deep Neural Network-Based 1-μW Voice Activity Detector Aided With a Customized Software Model for Analog Feature Extraction, " in IEEE Journal of Solid-State Circuits, vol. 54, no. 6, pp. 1764–1777, June. 2019, doi: 10.1109/JSSC.2019. 2894360.

[3] E. Shi, X. Tang and K. P. Pun, "A 270-nW Switched-Capacitor Acoustic Feature Extractor for Always-On Voice Activity Detection, " in IEEE Transactions on Circuits and Systems I: Regular Papers, vol. 68, no. 3, pp. 1045–1054, March. 2021, doi: 10.1109/TCSI.2020.3040020.

[4] S. Oh et al., "An Acoustic Signal Processing Chip With 142-nW Voice Activity Detection Using Mixer-Based Sequential Frequency Scanning and Neural Network Classification," in IEEE Journal of Solid-State Circuits, vol. 54, no. 11, pp. 3005–3016, Nov. 2019, doi: 10.1109/JSSC.2019.2936756.

[5] L. Shen, N. Lu and N. Sun, "A 1-V 0.25-uW Inverter Stacking Amplifier With 1.07 Noise Efficiency Factor," in IEEE Journal of Solid-State Circuits, vol. 53, no. 3, pp. 896–905, March 2018, doi: 10.1109/JSSC.2017.2786724.

[6] R. Puddu et al., "A Precision Pseudo Resistor Bias Scheme for the Design of Very Large Time Constant Filters," in IEEE Transactions on Circuits and Systems II: Express Briefs, vol. 64, no. 7, pp. 762–766, July 2017, doi: 10.1109/TCSII.2016.2603533.

[7] K. -W. Cheng and S. -E. Chen, "An Ultralow-Power Wake-Up Receiver Based on Direct Active RF Detection," in IEEE Transactions on Circuits and Systems I: Regular Papers, vol. 64, no. 7, pp. 1661–1672, July 2017, doi: 10.1109/TCSI.2017.2664919.

[8] H. S. Bindra, C. E. Lokin, D. Schinkel, A. Annema and B. Nauta, "A 1.2-V Dynamic Bias Latch-Type Comparator in 65-nm CMOS With 0.4-mV Input Noise," in IEEE Journal of Solid-State Circuits, vol. 53, no. 7, pp. 1902–1912, July 2018, doi: 10.1109/JSSC.2018.2820147.

[9] J. Yang, L. Lyu, Z. Dong, H. Ren and C. -J. R. Shi, "A 28-nW Noise-Robust Voice Activity Detector with Background Aware Feature Extraction," 2023 IEEE Asian Solid-State Circuits Conference (A-SSCC), 2023, pp. 1–3, doi: 10.1109/A-SSCC58667.2023.10347926.

[10] M. Maruyama, S. Taguchi, M. Yamanoue and K. Iizuka, "An Analog Front-End for a Multifunction Sensor Employing a Weak-Inversion Biasing Technique With 26 nVrms, 25 aCrms, and 19 fArms Input-Referred Noise," in IEEE Journal of Solid-State Circuits, vol. 51, no. 10, pp. 2252–2261, Oct. 2016, doi: 10.1109/JSSC.2016.2581812.

[11] Y. Gong, B. Liu, H. Cai, L. Shi and W. Liu, "VoAD: A Sub-μW Multiscene Voice Activity Detector Deploying Analog-Frontend Digital-Backend Circuits," in IEEE Transactions on Circuits and Systems II: Express Briefs, vol. 71, no. 2, pp. 837–841, Feb. 2024, doi: 10.1109/TCSII.2022.3199037.

Research on Radiation-Hardened High-Voltage Gate Driver Circuit Based on 0.8μm 1200V Bulk Silicon BCD Process

Xiaohui Li [1,2], Yi Zhang [2], Qiang Wang [2], Qiankun Xiong [2], Bo Zhang [1], Ming Qiao *[1]

[1] State Key Laboratory of Electronic Thin Films and Integrated Devices, University of Electronic Science and Technology of China, Chengdu 611731, China.
[2] Chengdu Huanyuxin Technology Co., Ltd. (Affiliated with China Electronics Corporation), Chengdu 610095, China

* Email: xiaohuili3511@foxmail.com, qiaoming@uestc.edu.cn

Abstract—In orbital satellite applications, the continuous increase in operating voltage poses challenges to driver circuits composed of SiC and GaN. Due to their excessively high dv/dt, these circuits face severe constraints in Electro-Magnetic Interference (EMI) performance—the rapid voltage change induces high-frequency noise that disrupts system collaboration. This paper first investigates the degradation model of 0.8μm bipolar-CMOS-DMOS (BCD, a mixed-signal integrated circuit technology combining bipolar, CMOS, and DMOS devices for high-voltage applications) process LDMOS devices under total ionizing dose (TID, i.e., the cumulative effect of ionizing radiation on devices) radiation. Specifically, to address the leakage issue, we propose a gate-all-around (GAA) structure design. Notably, even when operating under combined high/low temperature and radiation environments, the proposed 1200V radiation-hardened gate driver circuit maintains remarkable stability: its static power consumption exhibits only a 1.8% offset rate, while the under-voltage lockout (UVLO) shows an even smaller deviation of 1.17%. The proposed solution, integrating this driver circuit with 1200V Metal-Oxide-Semiconductor Field-Effect Transistors (MOSFETs, voltage-controlled high-speed power devices ideal for high-frequency applications), demonstrates robust performance under 60 krad(Si) TID conditions and excellent anti-EMI capabilities, which highlights its significant commercial potential in 1200V orbital satellite systems.

Keywords—EMI, dv/dt, 0.8μm BCD process, GAA, Gate Driver Circuit, TID

I. INTRODUCTION

Due to their bandgaps being much larger than those of Si semiconductors, SiC and GaN exhibit three key advantages: faster switching speed, lower on-resistance, and higher operating temperature. These properties have enabled their wide application in aerospace and related fields [1,2]. However, the high dv/dt of SiC/GaN devices generates severe high-frequency noise, with the spectrum extending to hundreds of MHz. This noise necessitates complex shielding measures, which significantly complicating circuit design. In orbital satellites, the dense arrangement of internal equipment makes EMI interference a critical issue—it directly disrupts the collaborative operation of the system [3,4,5,6]. As the voltage requirements for SiC/GaN applications continue to rise, the excessively high dv/dt further exacerbates the Miller effect on gate voltage. Consequently, in 1200V high-voltage scenarios, EMI poses significant challenges to SiC/GaN driver circuits, making it difficult to mitigate EMI issues through circuit design alone. By contrast, bulk silicon gate driver circuits (characterized by slower dv/dt and noise primarily below 30 MHz) can meet satellite requirements through straightforward circuit design. Paired with high-voltage power devices like MOSFETs, 1200V high-voltage driver circuits have become a research hotspot in orbital satellite technology [7,8].

This paper presents a high-voltage gate driver circuit designed using a 0.8-μm BCD process. The design aims to form a high-voltage driving system with power devices such as MOSFETs, ensuring orbital satellites function properly without EMI-induced malfunctions. Distinct from prior studies, this work focuses on a 1200 V gate driver chip integrated with 1200 V MOSFETs. This combination achieves a significantly lower dv/dt than SiC/GaN-based circuits [9,10,11]. Experimental results show that the designed gate driver withstands a total ionizing dose (TID) of 60 krad (Si). More importantly, the integrated system of the gate driver and 1200 V MOSFETs demonstrates robust performance in the complex EMI environment of orbital satellites, fully meeting the requirements of commercial satellite applications.

II. RADIATION-HARDENED DESIGN

A. Characterization of Device Properties

This study investigates the radiation-hardened design of a high-voltage gate driver circuit based on a 0.8-μm BCD bulk silicon process, starting with characterization tests on process-specific devices. The device characterization compares standard structures and gate-all-around (GAA) structures. Specifically, Fig.1(a) shows the planar structure of a 25V LDMOS standard device, while Fig.1(b) depicts the 25V LDMOS GAA device. For the 1200 V high-voltage gate driver circuit, a 1200V floating well structure is adopted, with high-side circuits operating within the floating-well, as illustrated in Fig.1(c).

Experimental results in Table I and Table II show that under total ionizing dose (TID) radiation, both 25V standard and GAA devices exhibit rapid threshold voltage (V_{TH}) shifts, with Vth stabilizing at approximately 0.5V under 60 krad(Si). Notably, significant differences emerge in leakage current: the GAA structure shows negligible leakage variation, whereas the standard structure exhibits pronounced leakage increase. For 1200V NLDMOS, the breakdown voltage degrades notably under 60 krad(Si) TID conditions.

These findings indicate that implementing GAA structures effectively mitigates leakage issues in high-voltage gate driver circuits, thereby improving static power consumption and other key metrics. The circuit's radiation-hardened design is thus optimized based on device V_{TH} fluctuation characteristics.

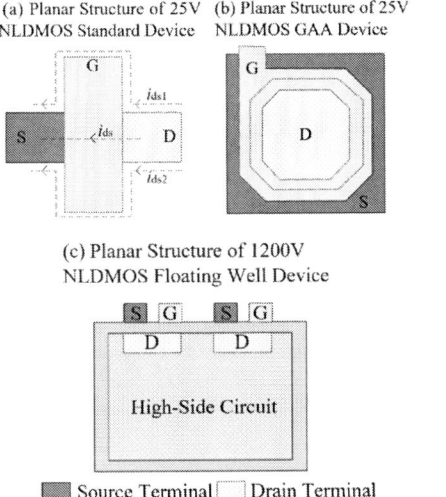

Fig. 1. (a) Planar Structure of 25V NLDMOS Standard Device. (b) Planar Structure of 25V NLDMOS GAA Device. (c) Planar Structure of 1200V NLDMOS Floating-Well Device.

TABLE I. TABLE OF 25V NLDMOS STRUCTURAL PARAMETER RESPONSES TO TID RADIATION

25V NLDMOS	TID 0krad	TID 60krad	ΔVth	TID 0krad	TID 60krad	ΔI_{LK}
	$V_{TH}(V)$		(%)	$I_{LK}(nA)$		(%)
Standard	1.75	0.5	-71.4	1.1	3.2	190.9
GAA	1.75	0.52	-70.2	0.98	1.01	3.1

TABLE II. TABLE OF 1200V NLDMOS STRUCTURAL PARAMETER RESPONSES TO TID RADIATION

1200V NLDMOS	TID 0krad	TID 60krad	ΔVth	TID 0krad	TID 60krad	ΔBV_{off}
	$V_{TH}(V)$		(%)	$BV_{off}(V)$		(%)
Floating-well	1.68	0.45	-73.2	1560	1240	-20.5

B. Circuit Hardening Design

This paper presents a radiation-hardened design for a 1200V high-voltage gate driver circuit based on a 0.8μm BCD process, addressing threshold voltage shifts in 25V NLDMOS devices with surrounding-gate structures under total ionizing dose (TID) radiation. The circuit schematically shown in Figure 2 integrates five main modules: an input logic unit processing control signals (HIN/LIN) through Schmitt triggers for noise immunity; a narrow pulse generation circuit composed of R_0-R_1 resistors, C_0-C_1 capacitors, and a NAND gate to create short pulses for driving the level-shifting stage; a level-shifting circuit that converts low-voltage inputs to high-voltage outputs (V_S=1200V and V_{BS}=1215V) using radiation-hardened layout techniques; an output driver stage amplifying signals to drive external power devices; and under-voltage lockout (UVLO) circuits monitoring supply voltages on both high-side and low-side paths to prevent malfunctions. Specifically, the high-side signal path routes HIN through a Schmitt trigger and narrow pulse generator before reaching

the 1200V level shifter, while the low-side path processes LIN through a Schmitt trigger and delay circuit to the output stage.

Functional waveforms in Figure 3 illustrate operational characteristics. To mitigate radiation effects, the design employs redundant isolation structures and guard rings in the level-shifting circuit to minimize TID-induced leakage currents, and dual-stage comparators with radiation-tolerant components in the UVLO circuit for reliable voltage monitoring. These strategies ensure that critical performance metrics—such as propagation delay and output voltage stability—remain within specifications under radiation exposure.

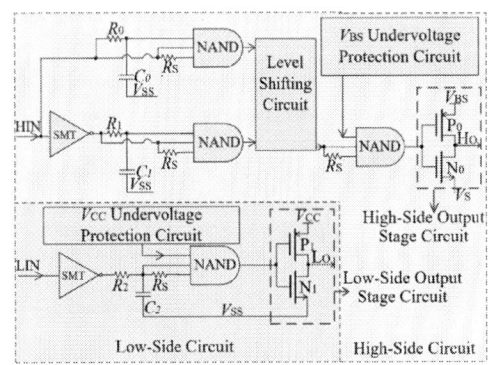

Fig. 2. 1200V High-Voltage Gate Driver Circuit Schematic.

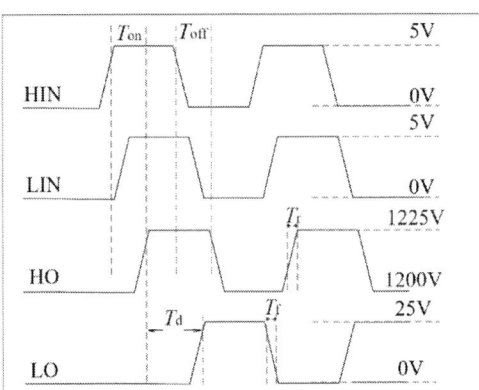

Fig. 3. Waveform Diagram of 1200V High-Voltage Gate Driver Circuit.

III. ANALYSIS OF EXPERIMENTAL RESULTS

This study investigates the effects of total ionizing dose (TID) radiation on LDMOS devices in a 0.8μm BCD process. Experimental results show that the surrounding-gate structure helps mitigate leakage current issues, significantly improving static power consumption and other indices in radiation environments. Based on the 0.8μm BCD process, a 1200V high-voltage gate driver circuit was designed. The circuit layout is shown in Figure 4(a), the chip electron micrograph in Figure 4(b), and the application scheme of the 1200V MOSFETs half-bridge driver circuit for orbital satellites in Figure 4(c). Experimental results demonstrate remarkable effectiveness of the circuit hardening design. The hardened circuit operates normally under 60 Krad(Si) TID radiation, and the variations of all indices compared to the non-irradiated state remain within a reasonable range.

979-8-3315-3918-4/25 $31.00 © 2025 IEEE

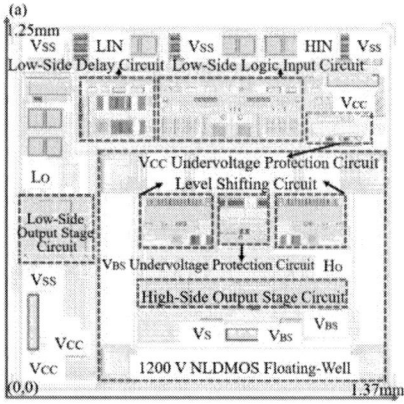

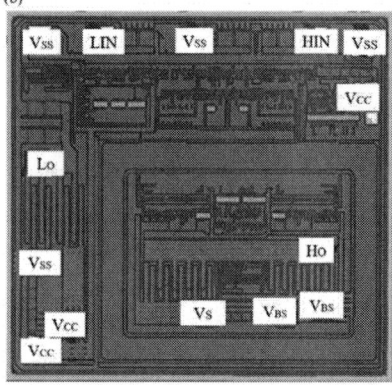

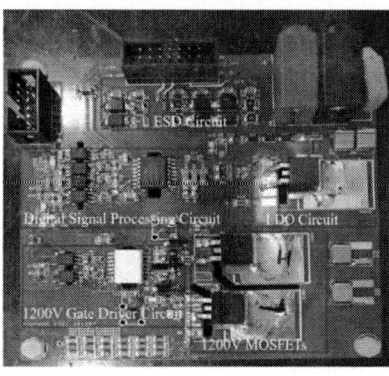

Fig. 4. (a) 1200V High-Voltage Driver Circuit Layout. (b) SEM Image of the High-Voltage Driver Circuit. (c) 1200V MOSFETs Half-Bridge Driver Circuit for Orbital Satellites.

Experimental data reveals that the designed 1200V high-voltage gate driver circuit exhibits no significant degradation under 60 Krad(Si) radiation, with overall parameter shifts remaining within acceptable limits. Specifically, the quiescent current (I_{QBS}) shifts from a typical value of 260.1μA to 236.6μA, representing a change rate of -9.03%. The maximum and minimum values occur at the lowest (-45°C) and highest (125°C) temperatures of the test environment, respectively. As shown in Figure 5(a), the deviations of these extreme values from the typical value are minimal, with only a 1.8% shift under 60 Krad(Si) radiation. Similarly, Figure 5(b) indicates that the maximum and minimum I_{QCC} values under 60 Krad(Si) radiation deviate by 6.4% relative to the typical

value, which is considered normal for current metrics. The curves of the under-voltage lockout (UVLO) circuit under radiation are presented in Figures 5(c) and 5(d), both of which remain within the designed specifications. The maximum deviations of these curves at extreme temperatures relative to the typical value are only 1.17%.

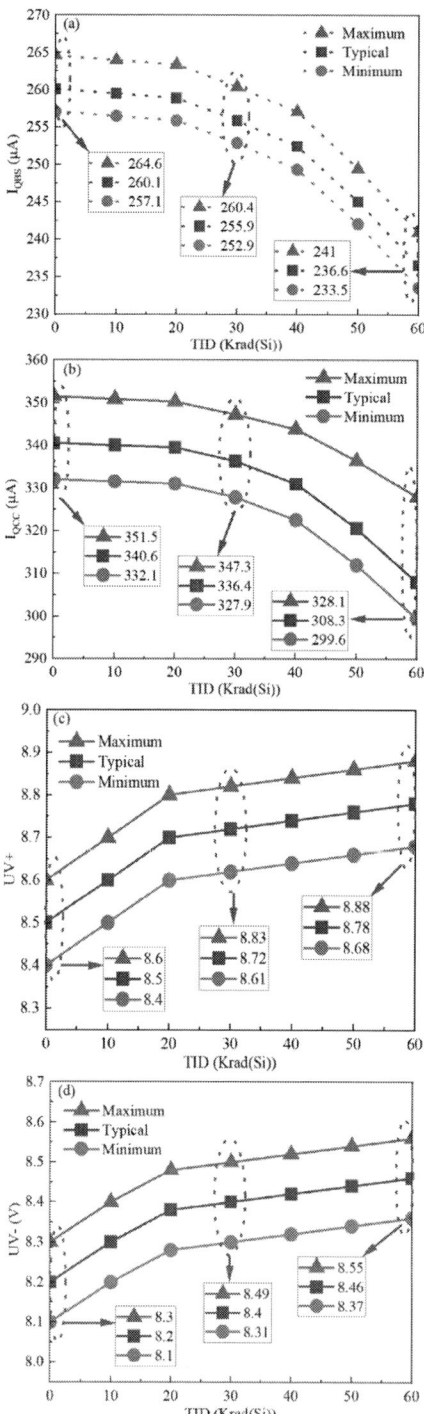

Fig. 5. (a) Curve of IQBS vs TID. (b) Curve of IQCC vs TID. (c) Curve of UV+ vs TID. (d) Curve of UV- vs TID.

TABLE III.   Circuit Parameters under Different Total Ionizing Doses ($Temperature$=25℃, V_{CC}=15, V_{BS}=1215V, V_{SS}=0V, V_S=1200V)

1200V Gate Driver	TID 0krad	TID 60krad	ΔT_{on}	TID 0krad	TID 60krad	ΔT_{off}	TID 0krad	TID 60krad	ΔT_r	TID 0krad	TID 60krad	ΔT_f
	T_{on}(ns)		(%)	T_{off} (ns)		(%)	T_r(ns)		(%)	T_f(ns)		(%)
Radiation-Hardened	151.1	139.1	-7.94	120.3	110.4	-8.23	25.9	23.8	-8.1	20.6	18.7	-9.22

TABLE IV.   Parameters Comparison of 1200V Drive Circuit between Our Work and Previous Reports.

	This work	[8]	[9]	[10]	[11]
Process (μm)	0.8	NA	0.18	0.18	0.5
Operating Voltage(V)	1200	1200	150	200	600
EMI Performance	Optimal	Substandard	Acceptable	Acceptable	Acceptable
TID(Krad(Si))	60	NA	100	150	NA

This study reveals that in a total ionizing dose (TID) environment of 60 Krad(Si), the change rates of main parameters are within a reasonable range. Although the maximum offset of dynamic parameters before and after radiation is controlled within 9.22%, it has little impact on the actual application scenario of the circuit, meeting the design requirements of 60 Krad(Si) total radiation dose. The specific parameters are shown in Table III. We also validated the application scheme of the MOSFETs half-bridge driver circuit for orbital satellites. Table IV compares the proposed MOSFETs half-bridge driver circuit with those reported in other literature. The MOSFETs driver circuit designed in this paper demonstrates remarkable advantages in total ionizing dose (TID) environments and high-voltage applications.

IV. CONCLUSION

This paper presents a radiation-hardened design for a 1200V high-voltage gate driver circuit, built upon degradation modeling of 0.8μm BCD process LDMOS devices under radiation. Specifically, a gate-all-around (GAA) structure is implemented to address leakage issues in standard devices under total ionizing dose (TID) radiation. Experimental results validate that the designed circuit meets all specifications under 60 Krad(Si) TID: static power consumption and under-voltage lockout (UVLO) circuit voltages show consistent temperature-dependent variations within acceptable deviation ranges. When integrated with metal-oxide-semiconductor field-effect transistors (MOSFETs), the driver system operates stably at high voltages under 60 Krad(Si) radiation while exhibiting excellent anti-EMI performance. By combining radiation tolerance, high-voltage compatibility, and EMI performance, this design thus demonstrates significant commercial potential for 1200V orbital satellite applications.

ACKNOWLEDGMENT

This work was supported in part by the National Natural Science Foundation of China under Grant No. 62174024. The authors extend sincere gratitude to the University of Electronic Science and Technology of China for furnishing the experimental facilities and technical support essential to this research. Special appreciation is due to Prof. Ming Qiao for his invaluable academic guidance and constructive suggestions throughout the project.

Additionally, we would like to thank Prof. Bo Zhang for his insightful guidance on the research direction, which significantly shaped the study's framework. Dr. Jitao Li is acknowledged for his valuable contributions to manuscript revision and technical discussions, providing critical feedback that improved the clarity of the research findings.

We also acknowledge Mr. Qiangkun Xiong for his expertise in experimental setup and data processing. Finally, the authors are deeply grateful to their families for their unwavering love and encouragement.

REFERENCES

[1] A. P. Camacho, V. Sala, H. Ghorbani, and J. L. R. Martinez, "A novel active gate driver for improving SiC MOSFET switching trajectory," IEEE Trans. Ind. Electron., vol. 64, no. 11, pp. 9032–9042, Nov. 2017.

[2] X. She, A. Q. Huang, Ó. Lucía, and B. Ozpineci, "Review of silicon carbide power devices and their applications," IEEE Trans. Ind. Appl., vol. 64, no. 10, pp. 8193–8205, Oct. 2017.

[3] H. Yue, W. Song, J. Chen, T. Tang and Q. Hu, "Modeling and Analysis for Switching Crosstalk of SiC MOSFETs Considering Temperature- Dependent Characteristics," in IEEE Transactions on Power Electronics, vol. 40, no. 8, pp. 10567-10580, Aug. 2025.

[4] H. A. Mantooth, M. D. Glover, and P. Shepherd, "Wide bandgap technologies and their implications on miniaturizing power electronic systems," IEEE J. Emerg. Sel. Topics Power Electron., vol. 2, no. 3, pp. 374–385, Sep. 2014.

[5] J. Millan, P. Godignon, X. Perpina, A. Perez-Tomas, and J. Rebollo, "A survey of wide bandgap power semiconductor devices," IEEE Trans Power Electron, vol. 29, no. 5, pp. 2155–2163, May 2014.

[6] Y. Fu, Z. Ma, and H. Ren, "A Low Cost Compact SiC/Si Hybrid Switch Gate Driver Circuit for Commonly Used Triggering Patterns," IEEE Transactions on Power Electronics, vol. 37, no. 5, pp. 5212–5223, 2022.

[7] X. She, A. Q. Huang, O. Lucia, and B. Ozpineci, "Review of silicon carbide power devices and their applications," IEEE Trans. Ind. Electron., vol. 64, no. 10, pp. 8193–8205, Oct. 2017.

[8] X. Gong, J. A. Ferreira and J. Popovic-Gerber, "Comparison and suppression of conducted EMI in SiC JFET and Si IGBT based motor drives," 2012 15th International Power Electronics and Motion Control Conference (EPE/PEMC), Novi Sad, Serbia, 2012, pp. DS2c.8-1-DS2c.8-8.

[9] T. Lew, E. Johnson, A. Marinelarena and J. Nuttall, "Radiation Evaluation of the TPS7H6003-SP Radiation-Hardness-Assured (RHA) Gallium Nitride (GaN) Field Effect Transistor (FET) Gate Driver," 2024 IEEE Radiation Effects Data Workshop (REDW) (in conjunction with 2024 NSREC), Ottawa, ON, Canada, 2024, pp. 1-4.

[10] N. W. van Vonno et al., "Total Dose and Single-Event Effects Testing of the Intersil ISL70040SEH Gallium Nitride (GaN) FET Driver," 2017 17th European Conference on Radiation and Its Effects on Components and Systems (RADECS), Geneva, Switzerland, 2017, pp. 1-5, doi: 10.1109/RADECS.2017.8696139.

[11] J. Zhu et al., "33.2 A 600V GaN Active Gate Driver with Dynamic Feedback Delay Compensation Technique Achieving 22.5% Turn-On Energy Saving," 2021 IEEE International Solid-State Circuits Conference (ISSCC), San Francisco, CA, USA, 2021, pp. 462-46

Parameter identification of single-phase inverter digital twin system

Ao shen	Hui li	Jie kang	Jia hao lv
Xiangtan University	Xiangtan University	Xiangtan University	Xiangtan University
School of Automation and	School of Automation and	School of Automation and	School of Automation and
Electronic Information	Electronic Information	Electronic Information	Electronic Information
Xiangtan,China	Xiangtan,China	Xiangtan,China	Xiangtan,China
1039431714@qq.com	15367422236@163.com	2528526223@qq.com	lvjiahao2023@163.com

Abstract—**In recent years, digital twin (DT) technology has garnered increasing attention in the parameter identification of power electronic systems due to its non-invasive characteristics. However, most existing DT-based parameter identification methods rely on relatively simple Particle Swarm Optimization (PSO) algorithms. This paper applies the Grey Wolf Optimizer (GWO) algorithm to a single-phase inverter digital twin parameter identification system and proposes a novel DT identification method based on a hybrid PSO-GWO optimization algorithm. This approach combines the global exploration capability of the PSO algorithm with the local exploitation ability of the GWO algorithm, enhancing solution accuracy and robustness while accelerating convergence speed. By constructing a high-fidelity digital twin model, key variables of the inverter—including output voltage, inductor current, and input current—are acquired. A comprehensive error fitness function is formulated to identify critical parameters such as inductance, capacitance, equivalent series resistance (ESR), and switch conduction resistance. To validate the effectiveness of the proposed method, a single-phase inverter experimental platform was established. Comparative experiments evaluated the performance of the standard PSO algorithm, the standard GWO algorithm, and the hybrid PSO-GWO algorithm. Experimental results demonstrate that the proposed PSO-GWO algorithm outperforms both traditional PSO and GWO algorithms in terms of parameter estimation accuracy and convergence efficiency, confirming its effectiveness and practical value for modeling and identification in power electronic systems.**

Keywords—Single-Phase Inverter, Digital Twin, Parameter Identification

I. INTRODUCTION

MOSFET switches, inductors, and capacitors are considered the most critical components in circuits, as their health status directly impacts the overall functionality of the entire system. Traditional parameter identification methods for power electronics converters mostly require injecting additional signals or adding extra circuits, making these approaches intrusive. For example, Literature [2] proposes a low-frequency signal injection method for system parameter identification. Although it can identify the changing trend of capacitor impedance, it not only has poor accuracy but also requires injecting additional signals, causing interference to the system. With continuous improvements in data sensing and computing capabilities, digital twin technology, as an advanced approach integrating physical modeling and data-driven methods, has gradually demonstrated great potential in power electronics system modeling, parameter identification, and state monitoring. By constructing a high-precision virtual model synchronized with the real system and incorporating real-time data inputs, digital twins can not only achieve online parameter identification but also be used for state prediction, fault diagnosis, and optimal control.

Current research has made progress in digital twin-based identification of key parameters in power electronics systems. For instance, Literature [1] established a digital twin model for a buck circuit and used PSO algorithm to identify key parameters such as switch conduction resistance and parasitic resistance. Literature [3] implemented a boost digital twin model on FPGA using LabVIEW, sharing a development board with the physical system, which has limitations in practical applications. Literature [4] proposed a single-cycle digital twin method that collects limited data points during one steady-state switching cycle of a buck circuit for parameter identification. Notably, this digital twin model was built as an open-loop model in MATLAB/Simulink simulation, but requires measuring the MOSFET gate drive signal to obtain duty cycle information during actual operation. All literature mentioned above used the most basic PSO algorithm as the intelligent optimization algorithm for digital twin system parameter identification. Literature [5] compared Particle Swarm Optimization (PSO), Genetic Algorithm (GA), and Simulated Annealing (SA) for parameter identification in a three-phase AC-DC converter digital twin, verifying the robustness of the proposed SA algorithm. Literature [6] proposed a method for online diagnostic analysis of buck converters using real-time probabilistic digital twins and embedded the buck converter digital twin model into FPGA. However, FPGA-embedded digital twin systems negatively impact the accuracy and speed of buck converter parameter identification. Literature [7] performed digital twin modeling for a single-phase DC-AC inverter and adopted PSO algorithm for parameter identification, but the intermediate optimization algorithm PSO was oversimplified and prone to local optima.

The main contributions of this paper are conducting digital twin modeling for a single-phase DC-AC converter, validating both PSO and GWO algorithms for parameter identification in the single-phase inverter digital twin system, and proposing a PSO-GWO integrated intelligent iterative algorithm for digital twin-based parameter identification of power electronics converters.

II. MODELING AND ANALYSIS OF SINGLE-PHASE DC-AC CONVERTERS

The main circuit schematic of the single-phase inverter in this paper is shown in Figure 1. The inverter employs voltage single closed-loop control with unipolar SPWM modulation. The modeling of the inverter incorporates the switch on-resistance and the parasitic resistance of the inductor and capacitor. Under steady-state operation with SPWM modulation, the main circuit can be modeled as follows:

$$\begin{cases} \dfrac{di_{\mathrm{L}}}{dt} = \dfrac{1}{L}\Big[\big(AR_{\mathrm{dson}} - R_{\mathrm{L}}\big)i_{\mathrm{L}} + BV_{\mathrm{in}} - \mathrm{U_o} + DV_{\mathrm{f}}\Big] \\[2mm] \dfrac{du_{\mathrm{cac}}}{dt} = \dfrac{1}{C_{\mathrm{ac}}}\left(\dfrac{R}{R+R_{\mathrm{ac}}}i_{\mathrm{L}} - \dfrac{1}{R+R_{\mathrm{ac}}}u_{\mathrm{cac}}\right) \\[2mm] \dfrac{du_{\mathrm{cdc}}}{dt} = \dfrac{1}{C_{\mathrm{dc}}}\left(-\dfrac{1}{R_{\mathrm{dc}}}u_{\mathrm{cdc}} + \dfrac{u_{\mathrm{in}}}{R_{\mathrm{dc}}}\right) \\[2mm] \mathrm{U_o} = \dfrac{R_{\mathrm{ac}}R}{R+R_{\mathrm{ac}}}i_{\mathrm{L}} + \dfrac{R}{R+R_{\mathrm{ac}}}u_{\mathrm{cac}} \\[2mm] i_{\mathrm{in}} = Ei_{\mathrm{L}} + \dfrac{V_{\mathrm{in}}}{R_{\mathrm{dc}}} - \dfrac{u_{\mathrm{cdc}}}{R_{\mathrm{dc}}} \end{cases} \tag{1}$$

Where $\mathrm{U_o}$ and i_{in} is the AC voltage and current output on the inverter side, u_{cac} is the voltage of the capacitor on the AC side, u_{cdc} is the voltage of the DC bus capacitor, i_{L} is the current of the inductance. A, B, D, E in equation (1) are determined by Table I. In this paper, the fourth-order Longekuta is used to solve Equation (1)

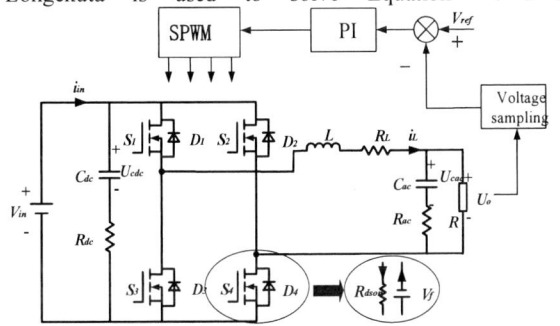

Fig.1. Consider the parasitic resistance schematic

Based on the control principles of single-phase inverters and unipolar sinusoidal pulse width modulation (SPWM), the mathematical model of the single-phase inverter control loop can be derived as Equation (2)

Where K is the scaling factor for sampled voltage, h represents the sampling time interval, K_{p} and K_{I} denote the parameters of the PI controller, and u_{tri} is a triangular carrier signal within the range of 0 to 1.

$$\begin{cases} u_{\mathrm{s,n+1}} = Ku_{0,\mathrm{n+1}} \\ u_{\mathrm{e,n+1}} = \mathrm{V}_{\mathrm{ref,n+1}} - u_{\mathrm{s,n+1}} \\ u_{\mathrm{m,n+1}} = u_{\mathrm{m,n}} + K_{\mathrm{p}}\big(u_{\mathrm{e,n+1}} - u_{\mathrm{e,n}}\big) + K_{\mathrm{I}}hu_{\mathrm{e,n+1}} \\ u_{\mathrm{c,n+1}} = sign\big(u_{\mathrm{m,n+1}}\big)\cdot u_{\mathrm{tri,n+1}} \\ sign\big(u_{\mathrm{m,n+1}}\big) = \begin{cases} 1, u_{\mathrm{m,n+1}} \ge 0 \\ -1, u_{\mathrm{m,n+1}} < 0 \end{cases} \end{cases} \tag{2}$$

Given that the single-phase inverter employs unipolar sinusoidal pulse width modulation (SPWM), the expression for the drive signal can be derived as follows based on its operating principle:

$$\begin{cases} u_{s1}=1, u_{s3}=0, u_{\mathrm{m,n+1}} \ge 0 \\ u_{s1}=0, u_{s3}=1, u_{\mathrm{m,n+1}} < 0 \\ u_{s2}=1, u_{s4}=0, u_{\mathrm{m,n+1}} \le u_{\mathrm{c,n+1}} \\ u_{s2}=0, u_{s4}=1, u_{\mathrm{m,n+1}} > u_{\mathrm{c,n+1}} \end{cases} \tag{3}$$

Where u_{s1-s4} is the drive signal for switching devices S1 to S4.

TABLE. I.The value of the switch in different switching states

	us1	us2	us3	us4	A	B	D	E
S1S4	1	0	0	1	-2	1	0	1
S1D2	1	1	0	0	-1	0	-1	0
D3S4	0	0	1	1	-1	0	-1	0
D2D3	0	1	1	0	0	-1	-2	-1
S2S3	0	1	1	0	-2	-1	0	-1
S3D4	0	0	1	1	-1	0	1	0
D1S2	1	1	0	0	-1	0	1	0
D1D4	1	0	0	1	0	1	2	1

III. INTERMEDIATE-LAYER ALGORITHM

A. PSO algorithm

The Particle Swarm Optimization (PSO) algorithm is a swarm intelligence optimization method inspired by bird foraging behavior. This algorithm simulates multiple particles cooperatively moving and sharing information within the search space to find optimal solutions, featuring simple structure, ease

of implementation, and strong global search capabilities，First, initialize particle parameters required by the PSO algorithm, where each particle represents a set of parameters to be identified. Assign each particle's parameter set to the digital twin system to obtain corresponding output voltage, inductor current, and input current. Collect real-time data from the physical system, then compute the acquired physical data alongside digital twin system data via the objective function.

The objective function is shown in Equation (4), where N is the total number of collected discrete data samples, i is the index of collected discrete data samples, I_{inmpp} is the peak-to-peak value of input current, and U_{ompp} and I_{Lmpp} are the peak-to-peak values of output voltage and inductor current, respectively. This function identifies the parameter set that minimizes the fitness value f_{obj}. If f_{obj} is less than the threshold or the iteration count exceeds the maximum allowed iterations, the algorithm outputs the parameters corresponding to the smallest f_{obj}. If f_{obj} remains above the threshold and the iteration count is below the maximum, the parameters are updated and reassigned to the digital twin model. The iteration continues until the optimal parameter set is output, completing the PSO algorithm process.

$$f_{obj} = \frac{1}{N}\sum_{i=1}^{N}\left[\begin{array}{c} \dfrac{1}{I_{inmpp}}\big(i_{in(i)} - i_{inm(i)}\big)^2 + \dfrac{1}{U_{ompp}}\big(U_{o(i)} - U_{om(i)}\big)^2 \\[2mm] + \dfrac{1}{I_{Lmpp}}\big(i_{L(i)} - i_{Lm(i)}\big)^2 \end{array}\right] \tag{4}$$

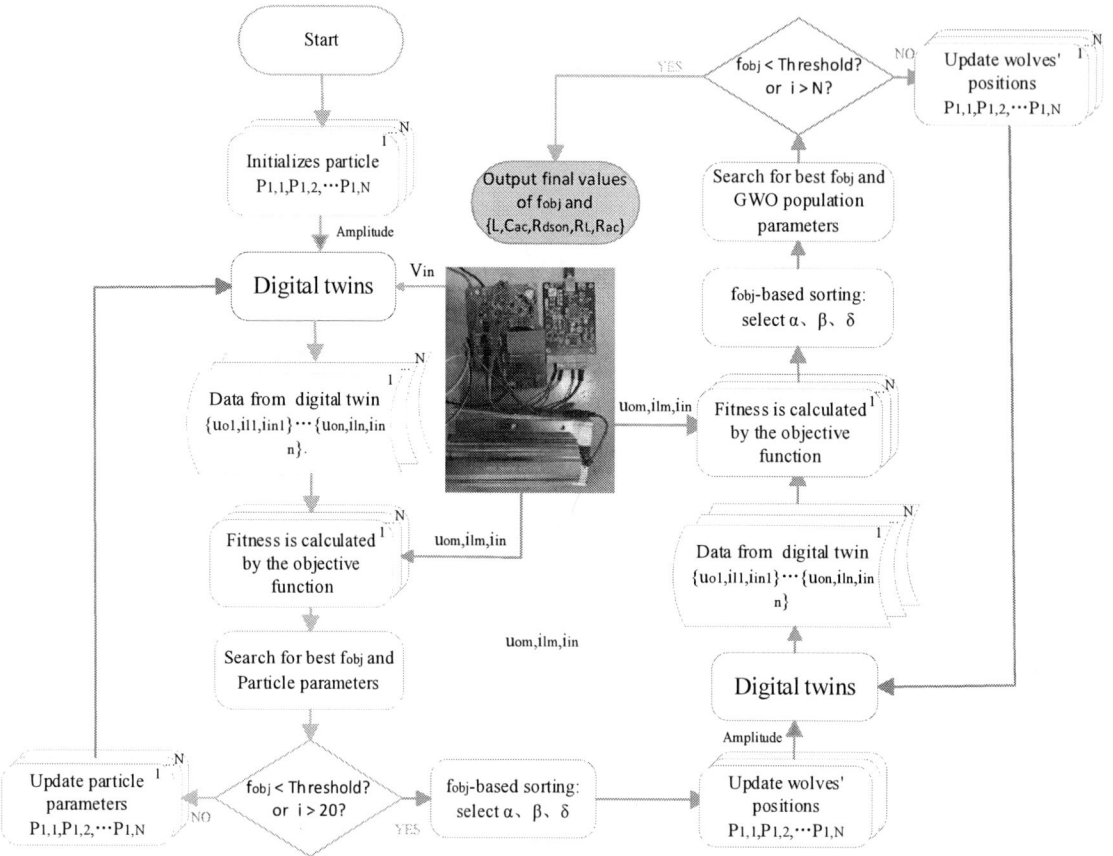

Fig.2. Flowchart of parameter identification of PSO-GWO algorithm

B. GWO algorithm

The Grey Wolf Optimizer (GWO) is a swarm intelligence optimization algorithm inspired by the hunting behavior of grey wolves in nature. This algorithm simulates the social hierarchy and cooperative hunting mechanism of grey wolves during prey encirclement, dividing the population into four roles: α (leader), β (assistant leader), δ (subordinate), and ω (follower), where α, β, and δ wolves collectively guide position updates for other individuals. GWO achieves global and local exploration through three core behaviors: encircling prey, hunting pursuit, and attacking prey. During initialization, population positions are randomly generated, with α, β, and δ positions determined by the fitness function. Other individuals dynamically adjust positions based on these leaders. As iterations progress, control parameters gradually decrease, enabling transition from exploration to exploitation and progressive convergence toward the optimal solution.

C. PSO-GWO algorithm

Through analysis of the iterative processes of PSO and GWO algorithms, this paper proposes a hybrid PSO-GWO algorithm. The specific flowchart is shown in Fig .2. For the first 20 iterations, the PSO algorithm is executed. At the 21st iteration, the optimal solution obtained by PSO is assigned based on ranking to the α, β, and δ wolves in the GWO algorithm. Subsequent iterations are completed using the GWO framework. This approach maintains the global exploration capability of PSO in early stages while incorporating the exploitation strength of GWO during later phases.

IV. EXPERIMENTAL RESULTS AND ANALYSIS

A. Design of Experiments

To validate the effectiveness of the proposed algorithm for parameter identification in the digital twin single-phase inverter system, the key parameters of the experimental platform designed in this study are listed in Table II. Both the physical system and digital twin system operate at a data acquisition frequency of 200 kHz. Data from the physical system are obtained through voltage and current sensors, then transmitted via a data acquisition card to a computer.

TABLE. II. Experimental parameters

parameter	value
Input voltage	48V
Output voltage	24V
Output voltage frequency	50HZ
Load resistance	10Ω
Output filter inductor	470uH
Output filter capacitor	20uF
DC bus capacitors	330uF
Switching frequency	20KHZ

B. Analysis of experimental results

This study conducted parameter identification for the digital twin system using both PSO and GWO algorithms. The iteration curve of parameter f_{obj} is shown in Fig.3. The figure demonstrates that PSO exhibits significantly faster initial convergence than GWO, but its later-stage convergence speed becomes notably slower. The PSO algorithm required nearly 70 iterations to converge, while GWO approached convergence in only 50 iterations, substantially improving convergence efficiency.

In Fig.4. ,GWO, and the proposed PSO-GWO algorithms. The proposed PSO-GWO algorithm converges at the 28th iteration, achieving a 44% faster convergence speed compared to the standard GWO algorithm.

Table III presents the parameter identification results of the PSO, GWO, and proposed PSO-GWO algorithms. The GWO algorithm demonstrates slightly higher identification accuracy than PSO, while the proposed PSO-GWO algorithm maintains the same accuracy level as GWO.

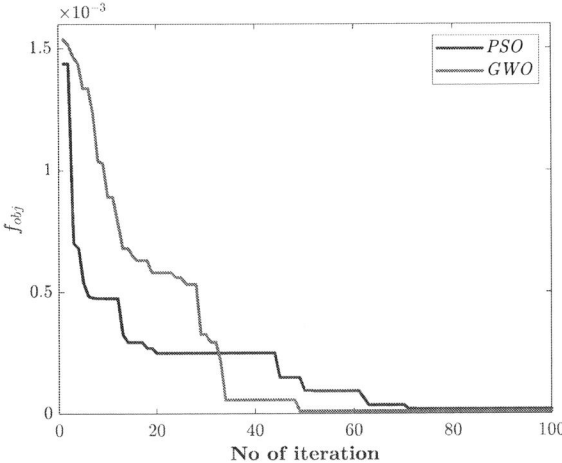

Fig. 3 Comparison chart of PSO and GWO iterations

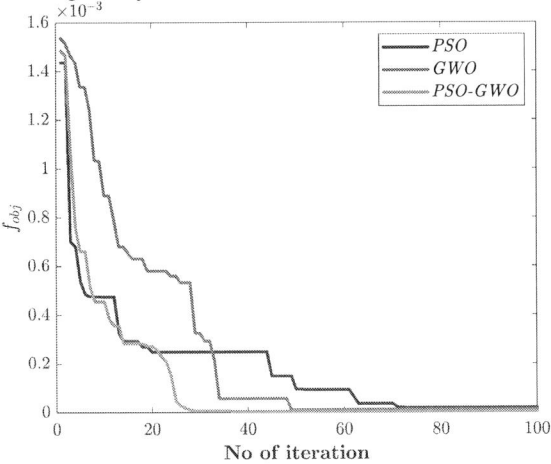

Fig. 4 PSO-GWO vs. PSO and GWO iterations

TABLE. III.Experimental results

Parameter	True value	PSO	GWO	PSO-GWO
$L(\mu H)$	470	452	482	460
$Cac(\mu F)$	20	18.9	19.1	20.7
$R_L(m\Omega)$	150	141	156	157
$Rac(m\Omega)$	100	105	104	97
$R_{dson}(m\Omega)$	100	94	105	96

ACKNOWLEDGMENT

This paper proposes a hybrid PSO-GWO optimization method for rapid identification of key parameters in single-phase LC inverters. By building a digital twin-based simulation model and designing a multivariable error fitness function, the PSO-GWO algorithm achieves high-precision estimation of parameters including inductance, capacitance, equivalent series resistance (ESR), and switch conduction resistance. Experimental results demonstrate that the proposed algorithm outperforms traditional PSO and GWO algorithms in convergence speed, identification accuracy, and stability. This method provides an effective solution for power electronics system modeling and online parameter monitoring, showing significant engineering application value. Future research may extend this approach to more complex inverter topologies for validation.

REFERENCES

[1]. Peng Y., Zhao S., and Wang H., "A Digital Twin Based Estimation Method for Health Indicators of DC–DC Converters," IEEE Trans Power Electron., vol. 36, no. 2, pp. 2105-2118, 2020.

[2]. M. W. Ahmad, N. Agarwal, P. N. Kumar and S. Anand, "Low-Frequency Impedance Monitoring and Corresponding Failure Criteria for Aluminum Electrolytic Capacitors," in *IEEE Transactions on Industrial Electronics*, vol. 64, no. 7, pp. 5657-5666, July 2017, doi: 10.1109/TIE.2017.2674598.

[3]. G. D. Nezio, M. d. Benedetto, A. Lidozzi and L. Solero, "DC-DC Boost Converters Parameters Estimation Based on Digital Twin," in *IEEE Transactions on Industry Applications*, vol. 59, no. 5, pp. 6232-6241, Sept.-Oct. 2023, doi: 10.1109/TIA.2023.3286832.

[4]. Y. Liu, X. Qing and G. Chen, "One-Cycle Digital Twin-Based Multiparameter Identification of Power Electronic Converters With Simple Implementation and High Accuracy," in IEEE Transactions on Instrumentation and Measurement, vol. 73, pp. 1-11, 2024, Art no. 3537311, doi: 10.1109/TIM.2024.3476606.

[5]. G. Di Nezio, S. de López Diz, M. di Benedetto, A. Lidozzi, E. José Bueno Peña and L. Solero, "LC Parameters Identification for a Three-Phase AC–DC Converter Through Digital Twin Modeling Technique and Optimization Algorithms," in *IEEE Journal of Emerging and Selected Topics in Power Electronics*, vol. 13, no. 3, pp. 2820-2833, June 2025, doi: 10.1109/JESTPE.2025.3534616.

[6]. M. Milton, C. D. L. O, H. L. Ginn and A. Benigni, "Controller-Embeddable Probabilistic Real-Time Digital Twins for Power Electronic Converter Diagnostics," in IEEE Transactions on Power Electronics, vol. 35, no. 9, pp. 9850-9864, Sept. 2020.

[7]. Q. Wu, W. Wang, Q. Wang, L. Xiao and B. Hu, "Digital Twin Approach for Degradation Parameters Identification of a Single-Phase DC-AC Inverter," *2022 IEEE Applied Power Electronics Conference and Exposition (APEC)*, Houston, TX, USA, 2022, pp. 1725-1730, doi: 10.1109/APEC43599.2022.9773462

Optimization of Three-dimensional High-k Superjunction under Non-Punch-Through Mode: Theoretical Modeling and Comparison

Zhentao Xiao[1], Chenxing Wang[1], Zonghao Zhang[1], Haimeng Huang[1,2,*]

[1] *Glasgow College, UESTC, Chengdu 610054, China*
[2] *School of Integrated Circuit Science and Engineering, UESTC, Chengdu 610054, China*
* Email: hmhuang@uestc.edu.cn

Abstract—This paper proposes a Non-Punch-Through (NPT) optimization methodology for three-dimensional (3D) high-permittivity superjunction (Hk-SJ) MOSFETs under three constraints. Through theoretical modeling and TCAD verification, we establish an accurate electric field (E-field) expression for cylindrical pillars. Results demonstrate that NPT mode substantially reduces specific ON-resistance ($R_{on,sp}$) by 40.9% and 21.1% under same breakdown voltage (BV) compared to conventional 2D and 3D Hk-SJ with Critical-Punch-Through (CPT) mode, respectively, while achieving a larger Baliga figure-of-merit (FOM) with $\alpha = 1.16$ in the $R_{on,sp} \propto BV^\alpha$ relationship. The superiority of NPT mode intensifies with higher permittivity and larger Hk pillar dimensions, revealing an optimal design parameters unattainable by CPT-based approaches. Empirical models for optimal doping ($N_{(opt)}$) and pillar height ($W_{(opt)}$) are further derived to guide manufacturing.

Index Terms—Breakdown Voltage, Impact Ionization Integral, Superjunction, Specific ON-Resistance, Non-Punch-Through.

I. INTRODUCTION

Silicon super-junction (SJ) MOSFETs have become the work-horse for 600–1200 V power converters because they relax the classic $R_{on,sp}$–BV "silicon limit" toward an almost linear trade-off [1], [2], [3], [4], [5], [6]. Among the methods of improving the performance of conventional SJ MOSFETs, replacing the p-pillar with a high-permittivity (high-k) dielectric pillar is usually adapted [7], [8], further suppresses the JFET effect, allowing heavier doping, lower electric field and lower complexity for manufacturing, most high-k SJ studies are confined to two-dimensional (2D) cross sections and assume Critical-Punch-Through (CPT) at breakdown condition [9]. These simplifications neglect two factors that dominate real devices: (i) the inherently three-dimensional (3D) pillar geometry in practical fabrication [8], [10], (ii) the possibility that a Non-Punch-Through (NPT) operating point yields an even lower $R_{on,sp}$ than the CPT optimization methodology [11], [12].

To overcome these limitations, we propose a NPT optimization methodology for 3D high-k SJ (Hk-SJ) MOSFETs (shown in Fig. 1) based on theoretical modeling and TCAD simulation. In particular, at high operating voltages with ultra-high-permittivity dielectrics ($K \geq 100$), the NPT design substantially lowers the specific ON-resistance ($R_{on,sp}$), thereby surpassing the performance limits of traditional CPT-

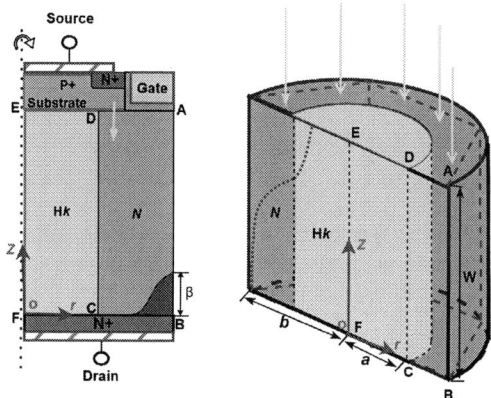

Fig. 1. Cross-sectional schematics of the 3D Hk-SJ MOSFETs with NPT mode.

mode structures [11], [12]. Across the entire breakdown-voltage range, NPT-mode devices consistently exhibit both reduced $R_{on,sp}$ and an enhanced Baliga figure-of-merit (FOM) relative to their CPT-mode counterparts. Moreover, the benefits of NPT operation become increasingly pronounced as the dielectric constant and pillar diameter grow, indicating greater improvements under more aggressive design parameters. This strategy reveals an optimal operating point inaccessible to previous CPT-based and 2D-based optimization paths, making it particularly attractive for power devices that must meet stringent high-voltage requirements.

II. OPTIMIZATION METHODOLOGY FOR NPT MODE

A. Electric Field Model for NPT mode

For the 3D Hk-SJ structure shown in Fig. 1, a cylindrical coordinate system is established with z-axis along EF. The Poisson equation for the 3D Hk-SJs can be expressed as

$$
\begin{cases}
\dfrac{\partial^2 V_{Hk}}{\partial r^2} + \dfrac{1}{r}\dfrac{\partial V_{Hk}}{\partial r} + \dfrac{\partial^2 V_{Hk}}{\partial z^2} = 0, & \text{for } 0 < r < a, \\[2ex]
\dfrac{\partial^2 V_S}{\partial r^2} + \dfrac{1}{r}\dfrac{\partial V_S}{\partial r} + \dfrac{\partial^2 V_S}{\partial z^2} = -\dfrac{qN}{\varepsilon_S}, & \text{for } a \leq r \leq b.
\end{cases}
\tag{1}
$$

From the steps similar to the process in [8], the E-field of the semiconductor region in the z-direction can be formulated

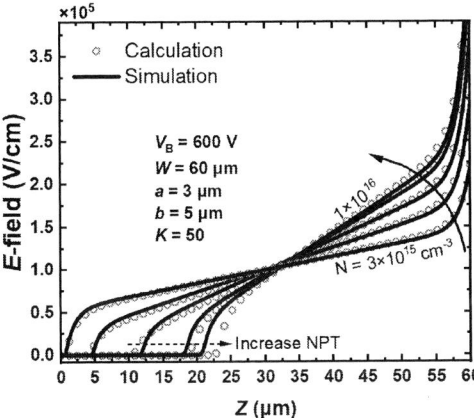

Fig. 2. NPT mode E-field verification along AB with doping concentration (N) ranging from 3×10^{15} to 1×10^{16} cm^{-3}.

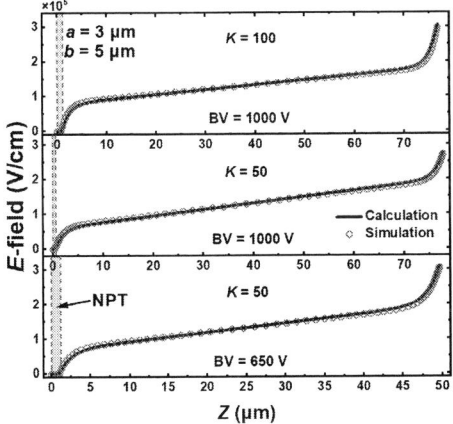

Fig. 3. NPT mode E-field verification along AB when corresponding $R_{\mathrm{on,sp}}$ is optimized to its minimum value under three different conditions.

as:

$$
E_{Sz}(r, z) = -T_e^2 V_e t \left(\frac{2}{t^2 T_d^2} - 1 + \frac{r^2}{2T_d^2} \right) \frac{\exp(tz) - \exp[t(W - z)]}{\exp(tW) + 1}
$$
$$
+ \frac{T_e^2}{T_d^2} V_e (2z - W) + \frac{V_B}{W}, \tag{2}
$$

where T_{c}, T_{d}, and T_{e} are structural constants with units of μm, $T_c^2 = \frac{1}{2}(b - a)\left[\frac{\epsilon_S}{\epsilon_{\mathrm{H}k}}a - 1.42(a - b)\right]$, $T_d^2 = \frac{1}{2}a\left[a - \frac{\epsilon_S}{\epsilon_{\mathrm{H}k}}(a - b)\right]$, and $\frac{1}{T_e^2} = \frac{2}{T_d^2} + \frac{1}{T_c^2}$. Additionally, K is define as the ratio of permittivity: $K = \epsilon_{\mathrm{H}k}/\epsilon_S$, and the formula of V_{e} and t for 3D Hk-SJ are given by:

$$
\begin{cases}
V_e = \frac{qN}{\epsilon_S}, \\
t = \frac{1}{T_d} \frac{1}{\sqrt{1 - \frac{b^2}{2(T_d)^2}}}.
\end{cases} \tag{3}
$$

To obtain a more accurate E-field along the z-direction of the AB line in the 3D Hk-SJ structure under the NPT mode, we have modified the critical E-field expression given in (2) and substitute $r = b$. The revised expression for the z-component of the E-field along AB is derived as:

$$
E_{Sz}(b, z) = -T_e^2 V_e t \left(\frac{2}{t^2 T_d^2} - 1 + \frac{b^2}{2T_d^2} \right) \frac{\exp(tz) - \exp(t\gamma)}{\exp(tW) + 1}
$$
$$
+ \frac{T_e^2}{T_d^2} V_e (2z - W) + \frac{V_B}{W + W_S/3}, \tag{4}
$$

Here, $\gamma = W - z + W_S$ and W_S can be calculated by $\int_0^W \max(E_{Sz}(b, z), 0) \cdot \mathrm{d}z = V_B$, which is used to modify the width of NPT region.

Next, MEDICI simulations were employed to validate the accuracy of calculated E-field, with the results presented in Fig. 2 and Fig. 3. Fig. 2 illustrates the comparison between simulation results and theoretical data for the NPT-mode E-field along the AB line under fixed structural parameters, namely $V_B = 600$ V, $W = 60$ μm, $a = 3$ μm, $b = 5$ μm, and $K = 50$, while varying N from 3×10^{15} to 1×10^{16} cm^{-3}. It can be observed that the calculation results are highly accurate when the degree of NPT is relatively low. Fig. 3 summarizes the E-field corresponding to the minimum $R_{\mathrm{on,sp}}$. These results not only confirm the accuracy of the E-field model but also indicate the minimum $R_{\mathrm{on,sp}}$ is typically achieved under conditions of small NPT region.

B. Optimiztaion Constraints for 3D Hk-SJs

To optimize $R_{\mathrm{on,sp}}$ under given BV, a, and b, three key constraints are defined for N and W.

1) Constraint 1: Achieving the desired BV through avalanche breakdown requires accurate modeling of impact ionization integral using Chynoweth coefficients [13] along the AB path, which can be formulated as

$$
I_{\mathrm{n}} = \int \alpha_{\mathrm{n}} e^{-\int^z (\alpha_{\mathrm{n}} - \alpha_{\mathrm{p}}) \cdot dz'} \cdot dz = 1, \tag{5}
$$

where z and z' are the distances along AB in Fig. 1.

2) Constraint 2: There is a minimum allowable depth, $W_{\min}$, derived by modeling the Hk-SJ as a vertical p-i-n diode with an ideal rectangular E-field profile. This establishes the minimal drift region depth, as indicated by vertical dashed lines in Fig. 4(b) and (c).

3) Constraint 3: An upper limit exists for the doping concentration, $N_{\max}$. When the E-field at C becomes zero, the bottom of the drift region reaches full non-depletion, preventing further increases in N without reducing BV. The values of $N_{\max}$ and the corresponding depth W are determined through curve fitting and shown by dashed lines in Fig. 4(a) and (c).

III. OPTIMIZATION RESULTS AND EVALUATION

A. Minimum Specific ON-Resistance Optimization

Before the optimization process, the specific ON-resistance for this Hk-SJ is determined as:

$$
R_{\mathrm{on,sp}} = \frac{W}{q\mu_{\mathrm{n}}N} \cdot \frac{b^2}{b^2 - a^2} \tag{6}
$$

where, a, b, W are shown in Fig. 1 and μ_{n} is computed using the Caughey-Thomas mobility model.

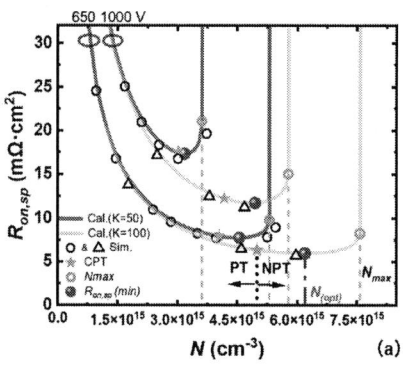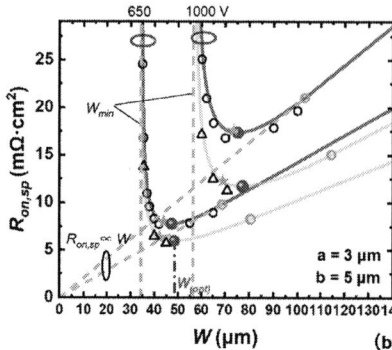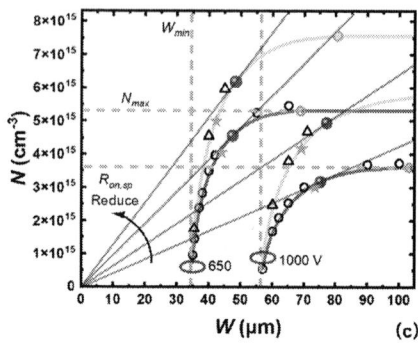

Fig. 4. Comparisons of calculated and simulated (a) N–$R_{on,sp}$, (b) W–$R_{on,sp}$ and (c) W–N relationships for 3D Hk-SJ devices with a =3 μm, b = 5 μm. Reference lines indicating theoretical constraints (grey dashed lines) and special points corresponding to optimal (blue points) and NPT/PT demarcation points (blue stars) are annotated.

TABLE I
THE OPTIMAL DESIGN PARAMETERS OF THE GIVEN STRUCTURE IN FIG. 4
UNDER FOUR CONDITIONS

BV / K	1000 V/ 50	1000 V/ 100	650 V/ 50	650 V/ 100	Unit
$N_{(opt)}$	3.178	4.944	4.556	6.186	$\times 10^{15}$ cm^{-3}
$W_{(opt)}$	75.25	77.26	47.38	48.38	μm
a	3.00	3.00	3.00	3.00	μm
b	5.00	5.00	5.00	5.00	μm
$R_{on,sp(opt)}$	17.35	11.75	7.77	5.97	mΩ · cm^2

As for the optimization tendency and results, Fig. 4(a) illustrates the relationship between $R_{on,sp}$ and N-pillar doping concentration (N) for 3D Hk-SJ devices with fixed geometrical parameters (a = 3 , b = 5 μm) under BV of 650 V and 1000 V, respectively. Calculated and simulated results are compared for two K values (K = 50 and 100), distinguished by separate curve sets and data markers. Grey dashed lines demarcate fundamental theoretical constraints of the device structure. Curves exhibit a characteristic U-shaped trend, with $R_{on,sp}$ initially decreasing with higher N, reaching a minimum at the optimal doping point ($N_{(opt)}$), followed by a sharp increase when reaches N_{max} (green points), which is determined by constraint 3 aforementioned. Notably, the K = 100 curves show lower $R_{on,sp}$ than K = 50 for equivalent BV, indicating lower critical E-field under same BV due to better lateral E-field absorption of high-k material. The blue stars in Fig. 4 represent the CPT mode which separates the punch-through (PT) and NPT mode. Hence, the separation between blue stars and blue points indicates the room to optimize a lower $R_{on,sp}$ by working under NPT mode compared to CPT mode.

A similar trend is found in the W-$R_{on,sp}$ relationship (Fig. 4(b)), where $R_{on,sp}$ increases sharply when W approaches $W_{(min)}$ determined by constraint 2 and decreases until reaches $W_{(opt)}$. When W is sufficiently large and approaches to the green points which is corresponding to N_{max} (Fig. 4(c)), N keeps constant to ensure 650 or 1000 V BV and further W increment cannot influence E-field distribution anymore, leading to proportional relation $R_{on,sp} \propto W$ shown in Fig. 4(b).

TABLE I summarizes optimal design parameters derived from this analysis for four operational conditions. The $N_{(opt)}$ ranges from 3.788×10^{15} cm^{-3} (BV = 1000 V, K = 50) to 6.186×10^{15} cm^{-3} (BV = 650 V, K = 100), while optimal depth $W_{(opt)}$ increases from 47.38 μm to 77.26 μm with higher BV. Corresponding minimized $R_{on,sp}$, decrease from 17.35 mΩ · cm^2 to 5.97 mΩ · cm^2 for lower BV and higher K. Geometrical parameters (a, b) remain constant across all configurations.

B. Performance Evaluation

1) $R_{on,sp}$ - BV Relationship Evaluation: As shown in Fig. 5(a), the proposed 3D Hk-SJ device with NPT mode (solid lines) exhibit a significant breakthrough in the $R_{on,sp}$-BV trade-off relationship compared to conventional 2D Hk-SJ with CPT mode (2D-CPT). As evidenced by the α-value analysis (α stands for the exponent of BV), the NPT mode achieves a $R_{on,sp}$-BV exponent substantially closer to the ideal unity limit, translating to a 40.9% improvement when K = 100, a = 4 μm. This enhancement directly manifests as lower $R_{on,sp}$ at identical BV conditions. The convergence of simulated data points (dots) with the calculated data (lines) further validates the high accuracy of previous theoretical model.

Against prior 3D Hk-SJ with CPT mode designs (dashed lines), the NPT mode demonstrates a 21.1% superior α value (α = 1.16 for NPT vs. α = 1.24 for CPT) under the condition of a = 4 μm and K = 400. This translates to a measurable reduction in $R_{on,sp}$ at equivalent BV levels, as quantified by the vertical displacement between solid and dashed traces at fixed BV coordinates. Notably, as the value of K increases, the performance improvement of the NPT mode compared to the CPT mode becomes more pronounced, while the enhancement in performance also becomes more evident as the radius of the Hk column increases.

2) FOM Evaluation: The comparative Baliga FOM analysis demonstrates that the proposed 3D Hk-SJ with NPT mode (a = 3 μm) achieves significantly enhanced performance over both 3D Hk-SJs with CPT mode and 2D Hk-SJs with CPT mode. NPT mode consistently maintains superior FOM

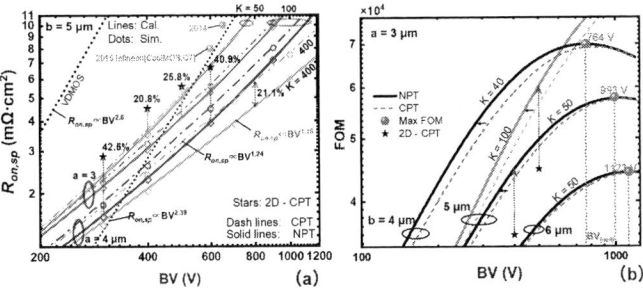

Fig. 5. Comparison of (a) $R_{\text{on,sp}}$-BV relationship and (b) Baliga FOM for NPT 3D Hk-SJs (solid lines), CPT 3D Hk-SJs (dash lines) and CPT 2D Hk-SJs (stars) under different K, a and b values.

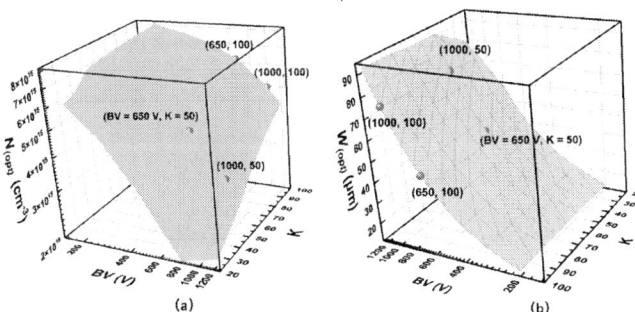

Fig. 6. Empirical models and its polynomial fitting surfaces for (a) $N_{\text{(opt)}}$ and (b) $W_{\text{(opt)}}$ under logarithmic coordinate.

values across the entire breakdown voltage spectrum and as the b decreases or K increases, the improvement is more pronounced. Importantly, maximum values of FOM and its corresponding breakdown voltage (BV$_{\text{(opt)}}$) are observed at all conditions. By setting the breakdown voltage of the device to BV$_{\text{(opt)}}$, the device performance can be further improved.

C. Empirical formula for $N_{\text{(opt)}}$ and $W_{\text{(opt)}}$

Empirical models for $N_{\text{(opt)}}$ and $W_{\text{(opt)}}$ in relation to BV and K are developed for manufacturing guidance of 3D Hk-SJ with NPT mode. The logarithmic transformation is applied and polynomial surfaces are used to fit the data in Fig. 6:

$$\begin{cases} N_{\text{(opt)}} = \exp\left(\sum_{0 \leq i+j \leq 3} P_{ij}(K)^j (\ln \text{BV})^i\right), \\ W_{\text{(opt)}} = \exp\left(\sum_{0 \leq i+j \leq 3} Q_{ij}(K)^j (\ln \text{BV})^i\right). \end{cases} \quad (7)$$

The coefficients of the two polynomial surfaces in Fig. 6 are listed in TABLE II. Future practical implementations may find guidance in this approach, which compresses computational demands while extending its utility to diverse design parameters. For instance, once BV and high-k materials are settled, through the given formula, engineers can subsequently determine the optimal doping concentration and optimal height of the device, which streamlines the entire design process.

IV. CONCLUSION

Overall, NPT optimization methodology was applied to 3D Hk-SJ MOSFETs, overcoming limitations inherent in tradi-

tional CPT mode and 2D structures. Through rigorous theoretical modeling and TCAD validation, an accurate electric field expression for cylindrical pillars under NPT conditions was derived and verified. Crucially, NPT operation significantly optimized $R_{\text{on,sp}}$, achieving reductions of 40.9% and 21.1% compared to 2D/3D Hk-SJ with CPT mode, respectively. This optimization concurrently enhances the Baliga FOM, characterized by a more favorable $R_{\text{on,sp}} \propto \text{BV}^\alpha$ relationship, where $\alpha = 1.16$. The superiority of NPT mode intensifies with larger K and larger proportion of Hk pillar, revealing optimal $R_{\text{on,sp}}$ inaccessible via CPT-based and 2D-based methodology.

REFERENCES

[1] X.-B. Chen, H.-Q. Yang, and M. Cheng, "New "silicon limit" of power devices," *Solid-State Electronics*, vol. 46, no. 8, pp. 1185–1192, 2002. [Online]. Available: https://www.sciencedirect.com/science/article/pii/S0038110102000102

[2] F. Udrea, G. Deboy, and T. Fujihira, "Superjunction power devices, history, development, and future prospects," *IEEE Transactions on Electron Devices*, vol. 64, no. 3, pp. 713–727, 2017.

[3] H. Huang et al., "Optimization of specific on-resistance of a two-zone variational vertical doping superjunction with insulating layers," *Semiconductor Science and Technology*, vol. 40, no. 2, p. 025003, jan 2025. [Online]. Available: https://dx.doi.org/10.1088/1361-6641/ad9f9d

[4] H. Huang and X. Chen, "Optimization of specific on-resistance of semisuperjunction trench MOSFETs with charge balance," *IEEE Transactions on Electron Devices*, vol. 60, no. 3, pp. 1195–1201, 2013.

[5] X. B. Chen, X. Wang, and J. Sin, "A novel high-voltage sustaining structure with buried oppositely doped regions," *IEEE Trans. on Electron Devices*, vol. 47, no. 6, pp. 1280–1285, Jun. 2000.

[6] Z. Zhang et al., "Temperature dependent optimization for specific on-resistance for 900 V superjunction mosfets: Numerical calculation and comparison," in *2023 IEEE 15th International Conference on ASIC (ASICON)*, 2023, pp. 1–4.

[7] C. Wang et al., "Aspect ratio dependent optimization and comparison of specific ON-resistance of SJ and Hk MOSFETs with extremely high permittivity," in *2024 IEEE 17th International Conference on Solid-State Integrated Circuit Technology (ICSICT)*, 2024, pp. 1–3.

[8] H. Huang et al., "Taylor modeling and comparative research containing aspect-ratio dependent optimization of three-dimensional Hk superjunction MOSFETs," *Microelectronics Journal*, vol. 159, p. 106623, 2025.

[9] H. Huang et al., "Optimization and comparison of drift region specific ON-resistance for vertical power Hk MOSFETs and SJ MOSFETs with identical aspect ratio," *IEEE Trans. Electron Devices*, vol. 67, no. 6, pp. 2463–2470, Jun. 2020.

[10] H. Kang and F. Udrea, "Theory of 3-D superjunction MOSFET," *IEEE Transactions on Electron Devices*, vol. 66, no. 12, pp. 5254–5259, 2019.

[11] W. Zhang et al., "The $R_{\text{on,min}}$ of balanced symmetric vertical super junction based on R-well model," *IEEE Transactions on Electron Devices*, vol. 64, no. 1, pp. 224–230, 2017.

[12] H. Huang, C. Wang, and Z. Xiao, "A unified optimization model for vertical power high-k superjunction with NPT and PT modes," *Accepted in IEEE Electron Device Letters*, 2025, doi:10.1109/LED.2025.3588725.

[13] A. Chynoweth, "Ionization rates for electrons and holes in silicon," *Phys. rev.*, vol. 109, no. 5, p. 1537, 1958.

TABLE II
COEFFICIENTS IN THE FITTING FORMULAS (7)

P_{00}	P_{10}	P_{01}	P_{20}	P_{11}	P_{02}	P_{30}	P_{21}	P_{12}	P_{03}
-9.02 $\times 10^{15}$	1.39 $\times 10^{16}$	-4.02 $\times 10^{14}$	-3.18 $\times 10^{15}$	1.71 $\times 10^{14}$	-1.76 $\times 10^{12}$	1.94 $\times 10^{14}$	-1.40 $\times 10^{13}$	1.18 $\times 10^{11}$	3.55 $\times 10^{9}$

Q_{00}	Q_{10}	Q_{01}	Q_{20}	Q_{11}	Q_{02}	Q_{30}	Q_{21}	Q_{12}	Q_{03}
-1.34 $\times 10^{3}$	7.30 $\times 10^{2}$	-1.08 $\times 10^{-1}$	-1.35 $\times 10^{2}$	5.30 $\times 10^{-2}$	-1.16 $\times 10^{-3}$	8.45 $\times 10^{0}$	-3.87 $\times 10^{-3}$	1.06 $\times 10^{-4}$	1.28 $\times 10^{-6}$

Smart Adaptive Perception for High-Precision Lightweight Infrared UAV Detection and Tracking

Shiyu Mei [1], Lei Deng *[1], Rui Yin[2,3]

[1] National Integrated Circuit Innovation Center, Shanghai, China

[2] Institute of Microelectronics, State Key Laboratory of Integrated Chip and Systems, Fudan University, Shanghai, China

[3]Jiashan Fudan Institute, Jiaxing, China

* Email: yushi1067@qq.com, lei.deng@shnicic.com

Abstract—The unauthorized intrusion of unmanned aerial vehicles (UAVs) into restricted areas poses a significant challenge for effective detection and tracking, especially in low-visibility infrared environments. Existing methods struggle with real-time performance, complex backgrounds, and the tracking of small, distant, and blurry targets. To address these issues, we propose a novel lightweight framework for enhancing the detection and tracking capabilities of infrared UAVs. Our model integrates two biologically-inspired innovations: DySample, which performs adaptive upsampling to enhance the features of distant targets; and SimAM, a parameter-free attention module that enables robust background suppression and feature enhancement. Evaluation on a challenging customized infrared dataset demonstrates that our method achieves an mAP@0.5 of 85.0% (5.0% higher than the baseline) and uses only 7.03 million parameters. When combined with Bot-SORT, it achieves a tracking accuracy of 99.51% and an average center position error of 3.05, significantly reducing false alarms. This lightweight and high-performance solution is suitable for critical UAV monitoring.

Keywords—Infrared UAV Tracking, YOLO, Attention Mechanism, BOTSORT

I. INTRODUCTION

In recent years, the rapid advancement and widespread adoption of UAVs across military [1], agricultural, and other sectors have introduced both transformative capabilities and significant challenges. While traditional RGB imaging systems struggle in adverse conditions, thermal infrared imaging offers a crucial advantage, providing stable visuals in complex environments like smog and nighttime, with pseudo-color images further enhancing detail [2]. However, effectively detecting and tracking small, agile UAVs against cluttered infrared backgrounds, especially at varying distances and in real-time, remains a persistent and difficult problem for existing methods, which must also meet the stringent requirements of lightweight deployment on embedded platforms.

Within the landscape of deep learning, models like YOLOv5 [3] have become cornerstones for achieving lightweight, real-time object detection suitable for embedded platforms. Nevertheless, the unique difficulties of infrared UAV tracking—particularly concerning blurred distant targets and complex background interference—highlight a critical gap in current solutions.

In this paper, we constructed a customized infrared redness dataset using the UCC3 mobile infrared detector. Inspired by the literature [4, 5, 6, 7, 8], we selected YOLOv5 as the base model, optimized the DySample module at the head to achieve better spatial adaptability, and introduced the SimAM self-attention module to suppress the interference from complex backgrounds through parameter-free computations. Finally, by using the Bot-sort algorithm [9], we obtained stable and accurate tracking results.

II. MODEL OPTIMIZATION AND TRACKING ALGORITHM

A. Baseline Model Overview

YOLOv5s consists of four parts: input, backbone, neck, and head, as shown in Fig 1. (a):

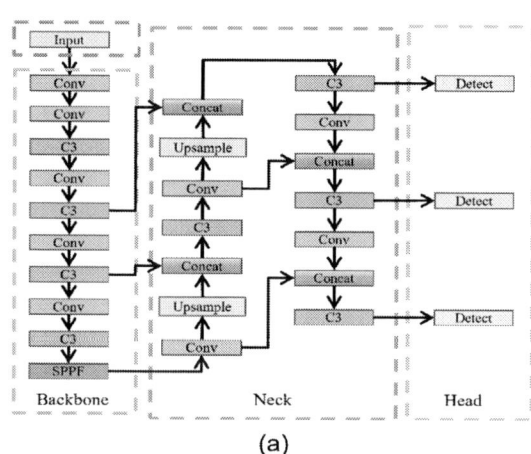

(a)

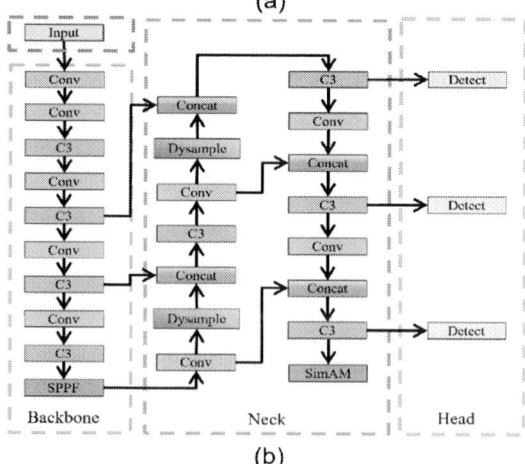

(b)

Fig. 1. The structure diagram of YOLOv5s (a) and the improved structure diagram of YOLOv5s (b).

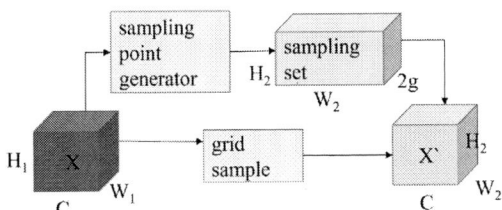

Fig. 2. Sampling based dynamic upsampling.

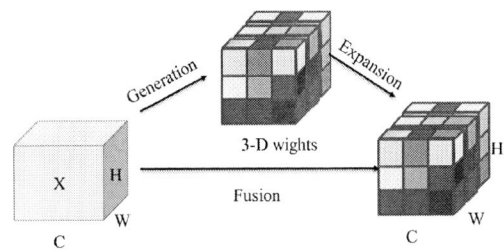

Fig. 3. SimAM attention module.

The input end of the improved YOLOv5s architecture diagram (b) preprocesses the infrared images and simultaneously optimizes the neck part using the DySample (Fig. 2), which is a dynamic upsampling module that can adaptively adjust the sample weights according to the difficulty level of the target (e.g. the distance between different drones). Additionally, it integrates SimAM (Fig. 3), an attention mechanism without parameters, which recalibrates the spatial features to suppress background interference.

B. Dynamic Upsampling Module

Within the YOLOv5s architecture, the neck module incorporates DySample, replacing the conventional Upsample operation. As depicted in Fig 2., which illustrates "Sampling based dynamic upsampling," DySample employs a point redistribution strategy. This strategy decomposes individual sampling points into spatially adaptive multi-point distributions, which enhances edge feature preservation during upsampling. Furthermore, DySample dynamically adjusts kernel weights for each target position, thereby improving geometric fidelity in infrared UAV imagery, such as sharpening blurred rotor contours.

The basic sampling mechanism of DySample is based on the analysis of the geometric nature of the upsampling process. Given the input feature plot $X \in \mathbb{R}^{C \times H_1 \times W_1}$, the core of this feature reconstruction is driven by a learnable dynamic sampling set $\Theta \in \mathbb{R}^{2 \times H_2 \times W_2}$ where the first two dimensions of storage space coordinate offsets. The final output is achieved by the mesh sampling (1):

$$X' = grid_{sample}(X, \theta) \tag{1}$$

This function resamples the original feature to the target resolution via a bilinear interpolation mechanism, where each coordinate point of Θ indicates the corresponding sampling location in the input feature space. Furthermore, in order to meet the expansion requirement of the upsampling multiple of Θ, the naïve implementation uses linear projection to construct the initial offset (2):

$$O = \text{linear}(X) \tag{2}$$

This linear layer expands the input channel number C to $2 \Theta^2$ (i.e., predicts a set of two-dimensional offset vectors for Θ^2 per spatial position), which is then reshaped into offset tensors of $\Theta \in \mathbb{R}^{2 \times H_2 \times W_2}$ by PixelShuffle operation. The final sample set is generated by (3):

$$\theta = G + O \tag{3}$$

Where G represents the original grid coordinates of the standard bilinear interpolation, and the offset superposition operation causes the geometric adaptive displacement of the sampling points, which preliminarily realizes the enhanced perception of detailed features such as the rotor of the UAV.

C. Neuroscience-Guided Attention

The structure design of the SimAM module (Fig 3.) achieves efficient feature enhancement through a parameter-free 3D attention mechanism. Its core process first calculates the mean μ and variance σ^2 along the spatial dimension of the input feature map $X \in \mathbb{R}^{C \times H \times W}$ to capture the local neuron distribution characteristics. Then, based on the spatial inhibition theory in neuroscience, the closed-form solution of the energy function is derived to generate the 3D attention weight map E. Finally, through the Sigmoid gating mechanism, the energy map is normalized and multiplied with the original features channel-wise and spatially, in a unique structure design that completely avoids introducing trainable parameters. It can achieve the collaborative attention of channels and spaces through mathematical derivation alone, while maintaining the lightweight nature of the model and significantly improving the feature discrimination ability.

Based on the theory of spatial inhibition in neuroscience, neuronal importance measures are defined as (4):

$$E = \frac{4(\sigma^2 + \gamma)}{(v - \mu)^2 + 2\sigma^2 + 2\gamma} \tag{4}$$

where μ and σ^2 are the mean and variance of the neuronal values in the channel, γ is the numerical stability factor (default 10^{-4}), and v is the current neuronal value. The biological mechanism of the design is reflected in the following: the molecular term $4(\sigma^2 + \gamma)$ enhances the target feature by amplifying the discriminative nature of the high-variance channel (e.g., the fuzzy outline of a long-range drone), and the denominator term $(v - \mu)^2$ quantifies the degree to which neurons deviate from the population feature. A smaller energy value E indicates that the corresponding neuron is more important.

Inspired by the neuroscience gain effect, the feature modulation process is defined as (5):

$$\tilde{X} = sigmoid(\frac{1}{E}) \odot X \tag{5}$$

The computational process begins by mapping low-energy, high-saliency neurons to high-weight values through a reciprocal transform. Subsequently, these weights are compressed into the range of (0,1) using a Sigmoid function to prevent value overflow. Finally, the processed weights are element-wise multiplied with the original features. This synergistic optimization suppresses background interference, such as leaf texture, while enhancing the target response, like the edge of a rotor.

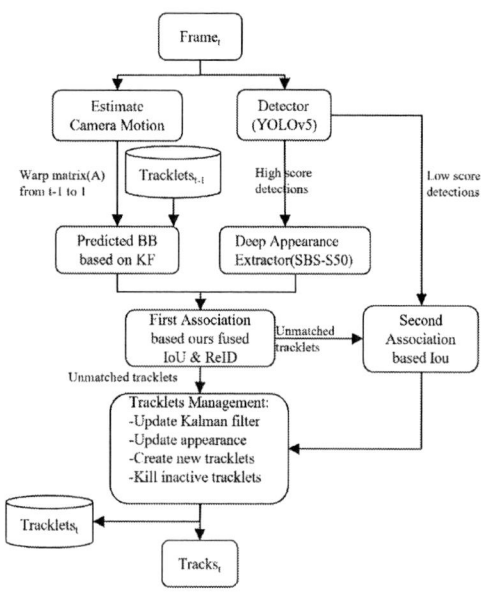

Fig. 4. BoTSORT tracking algorithm flowchart.

D. BoTSORT Tracking Algorithm

The "Bot-SORT" tracking framework is an improvement on the SORT algorithm, capable of providing robust and real-time tracking for UAVs. This framework utilizes camera motion compensation (CMC) to correct the prediction deviations caused by camera movement, ensuring the accuracy of trajectory updates. This framework combines a joint weighted matching strategy, which combines the intersection over union (IoU) for spatial alignment and the RcID appearance features for identity consistency. With adaptive Kalman filters for motion prediction and an advanced trajectory management system, Bot-SORT minimizes identity switching in occluded or deformed situations. When used in combination with the enhanced YOLOv5s detector, it can provide stable tracking results. Fig 4. shows the process of using BoTSORT for drone tracking.

III. EXPERIMENTAL SETUP AND RESULTS

A. Experimental Setup

We conducted these experiments using a 24-frame infrared thermal imaging camera with a 640x480 pixel resolution. This camera captured a DJI Air3s drone in diverse and complex environments, including forests, buildings, the sky, and elevated structures. A 40-minute video was then semi-automatically labeled using DarkLabel software, with frames extracted and converted into YOLO-format TXT files. After filtering, our dataset was structured with a 7:1:2 split, resulting in 7,384 training images, 1,112 validation images, and 2,156 test images.

During the training process, we configured the batch size to 16, the number of epochs to 100, and the image size to 640x640. To enhance model performance and generalization, a cosine annealing learning rate scheduling strategy was set up before training, and the AdamW optimizer was chosen for

parameter updates. The training environment featured an Intel(R) Core (TM) i7-10700 CPU @ 2.90GHz, 16.0 GB of memory, and an NVIDIA GeForce RTX 3060 Ti GPU with 8 GB VRAM. The software stack included PyTorch 2.4.1, Python 3.8.20, Windows 10, and CUDA 12.4. Fig. 5(a) shows a set of graphs with a batch size of 16 for training, while Fig. 5(b) presents the scene of the nighttime environment.

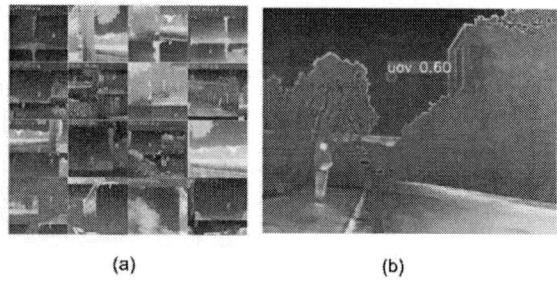

(a)　　　　　　(b)

Fig. 5. Training Graph(a) and Detection Diagram(b).

B. Experimental Results

Table I. compares the performance of five model combinations on a self-developed drone dataset. All experiments use YOLOv5s as the baseline model. We refer to two key performance indicators. The first indicator is mAP@0.5, which refers to the average mean accuracy calculated when the intersection-over-union (IoU) threshold is 0.5. It mainly measures the basic ability of the model to correctly identify and locate the targets. The second indicator is mAP@0.5:0.95, which represents the average mAP calculated at multiple IoU thresholds ranging from 0.5 to 0.95, providing a more comprehensive evaluation of the model's performance under various positioning accuracy requirements.

In Experiment 2, we introduced the DySample module, with only a slight increase in the number of parameters, from 7.02M to 7.03M. The mAP@0.5 improved from 80% to 83%, and the mAP@0.5:0.95 significantly increased from 24.3% to 30.3%, proving the effectiveness of this module in handling multi-scale drone detection. In contrast, in Experiment 3, although BiFPN integration increased the number of parameters, it did not achieve good results. This indicates that not all complex modules are suitable for every situation; instead, they may introduce redundant computations.

In the further combination, the Pinwheel model from Experiment 4 and the SimAM model from Experiment 5 both demonstrated outstanding performance. The YOLOv5s-DySample-SimAM combination in Experiment 5 achieved the highest mAP@0.5, increasing the baseline version by 5 percentage points while maintaining almost the same number of parameters. This proves that the SimAM attention mechanism can effectively suppress complex background interference without increasing the computational burden of the model. All experiments containing DySample (numbered 2 to 5) showed significant improvements in the mAP@0.5:0.95 metric. This confirms that this module has the ability to dynamically adjust weights to deal with difficult samples (such as distant drones), thereby comprehensively improving the accuracy and reliability of detection.

TABLE I. EXPERIMENTAL RESULTS

No.	A	B	C	D	E	mAP@0.5	mAP@0.5:0.95	Params
1	√					80%	24.3%	7.02M
2		√				83%	30.3%	7.03M
3	√	√	√			79%	26.1%	8.10M
4	√	√		√		84.5%	30.7%	7.43M
5	√	√			√	85%	29.8%	7.03M

a. A: YOLOv5s; B: Dysample; C: BiFPN; D: Pinwheel; E: SimAM

TABLE II. SINGLE OBJECT TRACKING PERFORMANCE COMPARISON

Method	ACPE	Precision	MOTA	FP
A+F	4.35	98.52%	77.67%	25
A-B-D+F	5.00	99.22%	71.92%	12
A-B-C+F	5.82	98.59%	**78.76%**	24
A-B-E+F	**3.05**	**99.51%**	77.24%	**8**

b. A: YOLOv5s; B: Dysample; C: BiFPN; D: Pinwheel; E: SimAM; F: BoTSORT

C. Combined with the BoTSORT Algorithm

To evaluate the improved YOLOv5 algorithm, we imported the trained weights into BoTSORT for running and generated files in the MOT16 format. The test video selected was a 90-second, 2164-frame video. All the weights were compared with the corresponding MOT16 files and the real annotations, and finally the following evaluation metrics were generated.

In single object tracking tasks, choosing the appropriate evaluation metrics is of vital importance. The average center position error (ACPE) directly measures the accuracy of the tracker's localization, with a smaller value indicating higher precision. Precision focuses on the quality of the tracking bounding box, and high precision means that the target is accurately framed and false identifications are avoided. False positives (FP) count the number of false alarms, and a lower FP indicates stronger anti-interference capability. These metrics collectively form a comprehensive evaluation framework, facilitating the understanding and improvement of algorithms.

Table II. provides a detailed comparison of the single object tracking performance of different methods. The A-B-E+F method, which combines YOLOv5s, Dysample, SimAM, and Botsort, performs exceptionally well. It achieves the lowest average center position error of 3.05, the highest accuracy of 99.51%, and the fewest false alarms of 8 times. Although its multi-object tracking accuracy MOTA of 77.24% is slightly lower, in single-object scenarios, its positioning and bounding box selection advantages are significant. The success of the A-B-E+F combination is mainly attributed to the synergy of its components: YOLOv5s provides accurate target information; Dysample and Botsort support motion modeling and data association; and the introduction of SimAM is crucial, as an off-the-shelf attention module, it can adaptively enhance the model's focus on key features, thereby significantly improving positioning accuracy, bounding box quality, and reducing false alarms. In contrast, the A-B-C+F method performs best on MOTA, which may be due to the feature fusion advantage of BiFPN. However, in terms of the accuracy and anti-interference ability of single-object tracking, the A-B-E+F combination, with the support of SimAM, is undoubtedly the better choice.

IV. CONCLUSION

In this paper, we propose an enhanced YOLOv5s framework to address the challenge of real-time infrared UAVs detection in complex environments. By integrating the DySample dynamic upsampling module, which adaptively adjusts sampling weights for multi-scale UAVs, and the SimAM parameter-free attention mechanism, which suppresses background interference using neuroscience-inspired 3D attention weights, our method significantly improves detection performance. Evaluated on a self-built infrared dataset, it achieves an 85% mAP@0.5, a 5% improvement over the baseline, while maintaining a lightweight model with 7.03M parameters and enabling stable video stream detection with Bot-SORT. Future work will focus on enhancing long-distance target stability and robustness in extreme weather conditions through advanced feature extraction and environmental adaptation techniques.

REFERENCES

[1] Zhu Mengzhen, Chen Xia, Liu Xu, Tan Chaoyong, & Li Wei. (2021). Analysis of the Current Development Status and Key Technologies of Tactical Laser Weapons against Unmanned Aerial Vehicles. Infrared and Laser Engineering, 50(7), 2020, 200, 230.

[2] Chen Zhenyue, Wang Xia, Zhang Mingyang, & Jin Weiqi. (2014). Research on False-color Display Method for High-Position Grayscale Images. Journal of Beijing Institute of Technology (Nature Edition), 34(3), 294-298.

[3] Jaiswal, S. K., & Agrawal, R. (2024). A Comprehensive Review of YOLOv5: Advances in Real-Time Object Detection. International Journal of Innovative Research in Computer Science and Technology, 12(3), 75-80.

[4] Liu, W., Lu, H., Fu, H. and Cao, Z., "Learning to Upsample by Learning to Sample," 2023 IEEE/CVF International Conference on Computer Vision (ICCV), 6004–6014 (2023).

[5] Hu, J., Shen, L., & Sun, G. (2018). Squeeze-and-excitation networks. In Proceedings of the IEEE conference on computer vision and pattern recognition (pp. 7132-7141).

[6] Yang, L., Zhang, R. Y., Li, L., & Xie, X. (2021, July). Simam: A simple, parameter-free attention module for convolutional neural networks. In International conference on machine learning (pp. 11863-11874). PMLR.

[7] Yang, J., Liu, S., Wu, J., Su, X., Hai, N., & Huang, X. (2025, April). Pinwheel-shaped convolution and scale-based dynamic loss for infrared small target detection. In Proceedings of the AAAI Conference on Artificial Intelligence (Vol. 39, No. 9, pp. 9202-9210).

[8] Tan, M., Pang, R., & Le, Q. V. (2020). Efficientdet: Scalable and efficient object detection. In Proceedings of the IEEE/CVF conference on computer vision and pattern recognition (pp. 10781-10790).

[9] Nir Aharon, Roy Orfaig, and Ben-Zion Bobrovsky. Botsort: Robust associations multi-pedestrian tracking. arXiv preprint arXiv:2206.14651, 2022.

Design and validation of fluorescence lifetime solving algorithm for fiber-optic temperature sensor

Yuxuan Yang [1,2], Xiangliang Jin [*1,2]

[1] School of Physics and Electronics, Hunan Normal University, Changsha 410081, China
[2] Key Laboratory of Physics and Devices in Post-Moore Era, College of Hunan Province, Changsha 410081, China

* Email: yangyuxuan_1031@163.com, jinxl@hunnu.edu.cn

Abstract—Fluorescence lifetime-based fiber-optic temperature sensors have the advantages of wide temperature measurement range and corrosion resistance. However, the traditional fluorescence lifetime solving algorithm has insufficient accuracy, which reduces the credibility of the temperature measurement results. In this paper, a fluorescence lifetime solving algorithm based on extended Kalman filter (EKF) is proposed and validated. Then, the temperature calibration and measurement are completed. The experimental results show that, compared with the traditional integral ratio method (IRM), the maximum and average deviation of temperature measurement are reduced by 27.85% and 29.93%, respectively. Meanwhile, the fluorescent fiber-optic temperature sensor has a resolution of 0.076 °C, which is capable of high-precision temperature measurement.

Keywords—fiber-optic temperature sensor, fluorescence lifetime, extended Kalman filter (EKF), solving algorithm

I. INTRODUCTION

Fluorescent fiber-optic temperature sensor has been widely used because of its strong anti-interference ability and high sensitivity [1], [2], [3], [4], [5]. Especially, the fluorescence lifetime-based temperature measurement is favored by researchers in the biological and medical fields because of its high accuracy and corrosion resistance [6], [7], [8].

Fluorescence lifetime is a physical property of the fluorescent material itself, and refers to the time it takes for the fluorescence intensity to decay to $1/e$ of the initial intensity after the material is excited. Generally speaking, the fluorescence lifetime mainly depends on the electronic energy level structure of the material itself, chemical bonding properties and environmental factors, and is independent of the frequency of the excitation light source. Since there is a correlation between the fluorescence lifetime and the temperature, the value of the temperature can be obtained by calculating the fluorescence lifetime. The solving algorithm of the fluorescence lifetime has undergone a process of development from a direct rough calculation to an exact calculation by fitting a function. Lin et al. substituted the measured data into the expression of fluorescence intensity attenuation over time to calculate the fluorescence lifetime, which was a simple process but with poor accuracy [9]. In order to accurately extract the decay time of the cavity, He et al. introduced the concept of weighting and improved the least-squares fitting method, which effectively improved the accuracy by superimposing different weights on the data, but the response time was slightly longer [10].

This work was supported by the National Natural Science Foundation of China (Grant No.62174052).

As an optimal estimation algorithm, Kalman filter (KF) is also widely used in temperature measurement. Considering the importance of high-precision temperature acquisition for aerospace applications, Zhang et al. applied Kalman filter to aerospace temperature measurement, successfully improving the resolution of temperature measurement and reducing the measurement deviation [11]. Wang et al. proposed an extended Kalman filter (EKF) based on the simplified mechanism model to estimate the temperature, and verified the feasibility through industrial Shell coal gasification process (SCGP) measurement [12]. However, the accuracy of fluorescent fiber-optic temperature sensor currently has a lot of room for improvement.

In order to realize high-precision temperature measurement, we propose a fluorescence lifetime solving algorithm and validate its effectiveness through experiments. The sections of this paper are organized as follows: section II describes the working principle of the fluorescent fiber-optic temperature sensor. Section III gives the design of the fluorescence lifetime solution algorithm. Experiments as well as discussions are carried out in Section IV followed by conclusions in Section V.

II. PRINCIPLE OF FLUORESCENT FIBER-OPTIC TEMPERATURE SENSOR

As can be seen from the Fig. 1, the fluorescent fiber-optic temperature sensor achieves temperature measurement based on the relationship between fluorescence lifetime and temperature. The fluorescence lifetime needs to be solved from the decay curve of the fluorescence signal. In general, the decay of the fluorescent signal satisfies the equation below:

$$I = I_0 e^{-\frac{t}{\tau}} \tag{1}$$

Where I_0 is the initial intensity of fluorescence, t is the moment, and τ is the time required for the fluorescence intensity to decay from I_0 to I_0/e, that is, the fluorescence lifetime.

In this paper, the hardware circuit, which shows in Fig. 2, mainly consists of micro controller, driver circuit, photoelectric conversion circuit and amplifier circuit. It should be noted that we have realized monolithic integration of both driver circuit and photoelectric conversion circuit in our previous work. For the software part of the fluorescent fiber-optic temperature sensor, it mainly includes the microcontroller related control program and the fluorescence lifetime solving algorithm.

III. DESIGN OF FLUORESCENCE LIFETIME SOLVING ALGORITHM

The acquisition of fluorescence lifetime is divided into two parts. One is to collect the fluorescence signal decay curve, and another is to calculate the fluorescence lifetime. For the former, the EKF is used to process the data of the fluorescence signal collected by the sensor in order to obtain a more accurate decay curve of the fluorescence signal. The EKF can be described by the following set of equations:

$$
\begin{cases}
\bar{x}_k = f(x_{k-1}, u_{k-1}, 0) \\
\bar{P}_k = A P_{k-1} A^T + W Q W^T \\
K_k = \dfrac{\bar{P}_k H^T}{H \bar{P}_k H^T + V R V^T} \\
\tilde{x}_k = \bar{x}_k + K_k(z_k - h(\bar{x}_k, 0)) \\
P_k = (I - K_k H) \bar{P}_k
\end{cases}
\tag{2}
$$

Where $\bar{x}_k$ and $\tilde{x}_k$ are prior estimate and posteriori estimation, respectively. A and W represent the Jacobian matrix obtained by the partial derivative of the function f with respect to x and w at point $\tilde{x}_k$, respectively. Q is the process noise covariance matrix, and P_k is the initial estimated covariance matrix. H and V represent the Jacobian matrix obtained by the partial derivative of the function h with respect to x and v at point $\tilde{x}_k$, respectively. R is the covariance matrix of observed noise and K_k represents the gain.

After obtaining the fluorescence signal decay curve, we can solve for the fluorescence lifetime using the weighted linear least squares method. Finally, the above algorithm is written into a program and downloaded into the microcontroller, which realizes the automatic acquisition of fluorescence signal and the output of fluorescence lifetime through the serial port.

IV. Experiments and Discussion

In this section, we will perform two experiments, one is the fluorescence signal acquisition and temperature calibration, and another is the temperature measurement experiment.

A. Acquisition of signal and temperature calibration

The equipment needed to collect the fluorescence signal include constant temperature chamber, fluorescent fiber-optic temperature sensor, serial port device and personal computer. The fluorescent fiber temperature sensor is connected to the personal computer through the serial port and then placed into the constant temperature chamber, and the fluorescence signal is collected by setting different temperatures to obtain the fluorescence decay curve. The Fluorescence signal acquisition experiments at different temperature is shown in the Fig. 3. The temperature range of the constant temperature chamber was set from 20°C to 120°C with 10°C interval. In order to further reduce the chance error, the fluorescence signals were collected several times at each temperature and averaged. The temperature of the environment was 25°C. The acquisition result of the fluorescence signal is shown in Fig. 4 below. As can be seen from Fig. 4, the higher the temperature, the faster the fluorescence decays, which is in accordance with physical principles. The purpose of temperature calibration is to obtain the fluorescence lifetime at different temperatures, and to obtain the relationship between fluorescence lifetime and temperature by fitting, which is the key to achieve accurate temperature measurement of fluorescent fiber-optic temperature sensor. Finally, Fig. 5 shows the fluorescence lifetimes obtained at different temperatures.

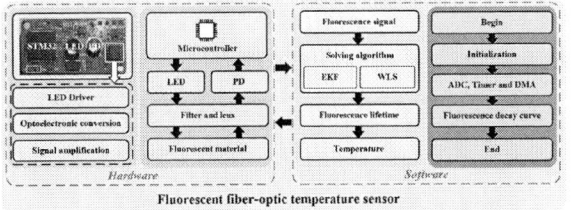

Fig. 1. Principle of the fluorescent fiber-optic temperature sensor

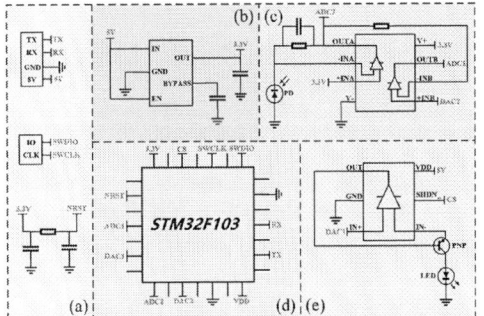

Fig. 2. Schematic diagram of the hardware circuit

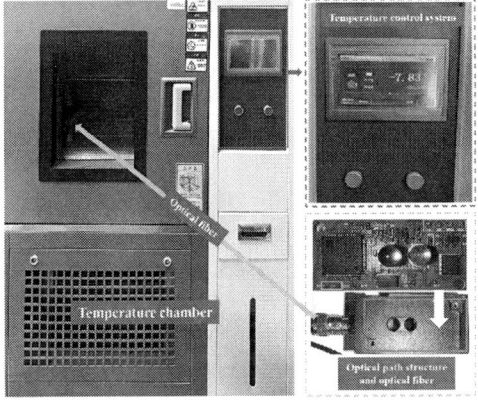

Fig. 3. Fluorescence signal acquisition experiments at different temperatures

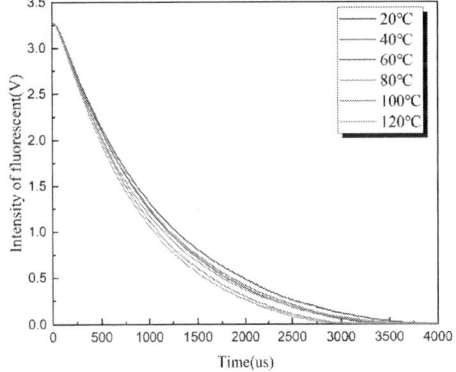

Fig. 4. Decay curve of fluorescence signal

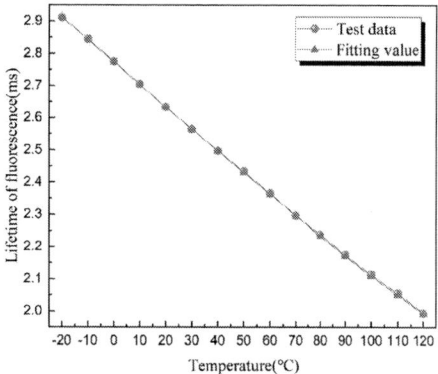

Fig. 5. Calibration data at different temperatures

B. Measurement of temperature

Same as the previous experiments, the temperature was set from -20°C to 120°C. The experimental data are shown in Table I below.

TABLE I. RESULTS OF TEMPERATURE MEASUREMENT

T/°C	Measurement data/°C		Deviation/°C	
	This paper	IRM	This paper	IRM
-20	-20.084	-20.113	-0.084	-0.113
-10	-10.091	-10.145	-0.091	-0.145
0	-0.098	0.121	-0.098	0.121
10	9.913	9.888	-0.087	-0.112
20	20.091	20.158	0.091	0.158
30	30.087	29.871	0.087	-0.129
40	40.085	39.876	0.085	-0.124
50	49.899	50.131	-0.101	0.131
60	59.924	59.881	-0.076	-0.119
70	70.095	69.865	0.095	-0.135
80	79.885	80.158	-0.115	0.158
90	90.095	89.844	0.095	-0.156
100	100.113	100.179	0.113	0.179
110	109.842	109.812	-0.158	-0.188
120	120.171	119.763	0.1718	-0.237

From Table I, it can be known that the accuracy of the temperature obtained using the proposed algorithm is higher than using the integral ratio method. Fig. 6 below shows this result more graphically.

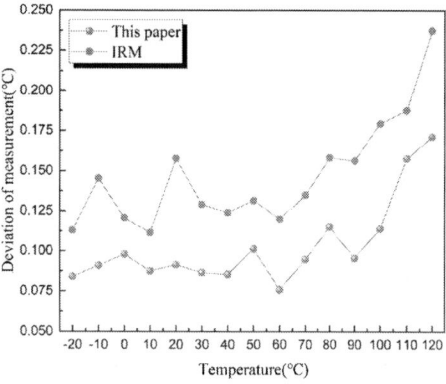

Fig. 6. Temperature measurement deviation of the two algorithms

Furthermore, the maximum, minimum and average temperature deviations were tabulated and shown in Table II below.

TABLE II. ACCURACY OF TEMPERATURE MEASUREMENT

Deviation	This paper	IRM	Reduction
Average	0.103°C	0.147°C	29.93%
Maximum	0.171°C	0.237°C	27.85%
Minimum	0.076°C	0.112°C	32.14%

From Table II, it can be seen that the maximum, minimum and average temperature deviations of this paper are 0.171°C, 0.076°C and 0.103°C, respectively. Comparing with the integral ratio method, the maximum and average values of temperature deviation decreased by 27.85% and 29.93%, respectively. In addition, the average temperature resolution in this paper is 0.103°C, while the maximum temperature resolution when using the integral ratio method is 0.147°C, which has an improvement of 29.93%.

V. CONCLUSION

In conclusion, this paper designs a fluorescence lifetime finding algorithm for fluorescent fiber-optic temperature sensor and verifies it through experiments. The experimental results show that, compared with the traditional integral ratio method, the maximum and average values of temperature measurement deviation are reduced by 27.85% and 29.93%, respectively. Based on the proposed algorithm, the resolution of the fluorescent fiber-optic temperature sensor reaches 0.076°C. This work is helpful for designing high-precision fluorescent fiber-optic temperature sensor.

REFERENCES

[1] Z. Liu et al., "Advanced functional optical fiber sensors for smart battery monitoring," *Energy Mater. Adv.*, vol. 5, Dec. 2024, Art. no. 0142, doi: 10.34133/energymatadv.0142.

[2] V. Matejec, I. Kasik, and I. Barton, "Fiber-optic nanosensors for chemical detection," *Chemosensors*, vol. 11, no. 10, Oct. 2023, Art. no. 521, doi: 10.3390/chemosensors11100521.

[3] V. Matveenko, M. Kosheleva, G. Serovaev, and A. Fedorov, "Measurement of gradient strain fields with fiber-optic sensors," *Sensors*, vol. 23, no. 1, Oct. 2023, Art. no. 410, doi: 10.3390/s23010410.

[4] K. Chen et al., "Illuminating the path to high-precision dual-mode optical thermometry: A Mn^{4+}-activated oxyfluoride single crystal for a portable optical fiber thermometric platform," *Chem. Eng. J.*, vol. 477, Dec. 2023, Art. no. 147165, doi: 10.1016/j.cej.2023.147165.

[5] F. A. Pedroza-Montero, K. J. Santacruz-Gomez, R. Melendrez, and M. Barboza-Flores, "Commercial nanodiamonds for precise fluorescence-based temperature sensing," *Appl. Phys. Lett.*, vol. 125, no. 7, Aug. 2024, Art. no. 073701, doi: 10.1063/5.0219532.

[6] W. M. Sun et al., "Measurement of decay time based on FFT," *Opt. Laser Technol.*, vol. 36, no. 4, pp. 323-326, Jun. 2004, doi: 10.1016/j.optlastec.2003.09.020.

[7] X. He, H. Yan, L. Z. Dong, P. Yang, and B. Xu, "Data point selection for weighted least square fitting of cavity decay time constant," *Chin. Phys. B*, vol. 25, no. 1, Nov. 2015, Art. no. 014211, doi: 10.1088/1674-1056/25/1/014211.

[8] E. Ledesma, A. Uranga, and N. Barniol, "A single-chip AlScN PMUTs-on-CMOS hydrophone," *IEEE Sens. J.*, vol. 24, no. 13, pp. 21311-21320, Jul. 2024, doi: 10.1109/JSEN.2024.3401455.

[9] Y. W. Lin, X. Zhing, and W. Q. Liu, "Flourescent decay-time thermometry and its application," *Fiber Optic Sensors V. SPIE* , vol. 2895, pp. 572-575, Sep. 1996, doi: 10.1117/12.252215.

[10] X. He et al., "Data point selection for weighted least square fitting of cavity decay time constant," *Chin. Phys. B*, vol. 25, no.1, Nov. 2015, Art. no.014211, doi: 10.1088/1674-1056/25/1/014211.

[11] Y. H. Zhang, L. Chu, and W. J. Li, "A fully-integrated memristor chip for edge learning," *Nano-Micro Lett.*, vol. 16, Apr. 2024, Art. no. 166, doi: 10.1007/s40820-024-01368-7.

[12] K. C. Wang et al., "Online temperature estimation of shell coal gasification process based on extended Kalman filter," *Chin. J. Chem. Eng.*, vol. 47, pp. 134-144, Jul. 2022, doi: 10.1016/j.cjche.2021.07.030.

A Multi-Cycle Pulse Transfer Timing Scheme for Enhancing Charge Efficiency in CMOS Image Sensors

Zhenhao Zhang, Chiang Zhu, Haiyang Liu, Peng Peng, Sikai Wang, Junjie Hao, Xiaona Zhu*

School of Microelectronics, Fudan University, Shanghai 200433, China

* Email: zhenhaozhang24@m.fudan.edu.cn, xiaona_zhu@fudan.edu.cn

Abstract—In this work, a novel timing scheme is proposed to enhance the charge transfer performance of a conventional 4T-pixel structure. Unlike the traditional design, the proposed scheme introduces an intermediate storage node (SD) and employs multi-cycle control pulses to facilitate electron transfer. Under identical illumination conditions, voltage drop (ΔV_{FD}) at floating diffusion (FD) node resulting from charge transfer in conventional 4T pixel is 0.80 V, whereas proposed multi-cycle pulse pixel achieves ΔV_{FD} of 1.43 V. This represents 78% increase in voltage drop amplitude for same amount of charge, indicating significant improvement in charge transfer efficiency. Meanwhile, the key parameters that influence the charge transfer process, including V_{TXgate}, V_{TXb}, C_{SD}, and C_{FD}, were individually analyzed through simulation.

Keywords—multi-cycle pulses pixel, new timing design, wide dynamic range

I. INTRODUCTION

In recent years, CMOS image sensors (CIS) have experienced rapid advancement driven by increasing user demands, particularly for enhanced dynamic range (DR), signal-to-noise ratio (SNR), and high frame rate performance. A common approach to extending DR involves increasing the photodiode well (PD well) capacity through ion implantation adjustment. However, this method typically degrades pixel sensitivity and operational speed due to increased junction capacitance [1]. Although various wide dynamic range (WDR) pixel techniques have been proposed to mitigate this trade-off, most efforts have focused on optimizing the analog front-end, such as programmable gain amplifiers (PGAs) and analog-to-digital converters (ADCs), while paying limited attention to exposure control and timing strategies [2][3][3]. To address this limitation, this work presents a novel timing scheme aimed at enhancing the efficiency of charge transport without compromising the charge transfer sensitivity of the pixel.

II. SIMULATION AND RESULT

A. Modeling and simulation

In this work, the photodiode well (PD well), transfer gate (TX), and floating diffusion (FD) regions of the pixel are modeled using the Sentaurus TCAD SProcess module. Optical illumination is applied via the EMW module. In the Sdevice simulation, the SPICE compact model is incorporated to represent the reset (RST), source follower (SF), and row select (ROW) transistors, which are connected to form a complete mixed-signal simulation environment.

Fig.1 illustrates the circuit schematic of a conventional 4T-pixel, while Fig.2 depicts the proposed pixel circuit with multi-cycle pulse operation. Fig.3 shows the conventional timing sequence, and Fig.4 presents the modified multi-cycle pulse timing scheme introduced in this work.

Compared to the traditional 4T-pixel structure, an intermediate storage node (SD) is added between the TX gate and the FD node. During one exposure period, charge accumulated in the PD well is partially transferred to the SD node in multiple cycles by repeatedly activating the TX gate. After the exposure stage, ended, the FD node is reset, and the TXb control signal is then activated to transfer the stored charge from the SD node to the FD node, resulting in a voltage drop (ΔV_{FD}) at FD. By precisely controlling the exposure duration and illumination intensity, both the conventional and proposed TX pixels are exposed to equivalent level of light intensity for comparative evaluation.

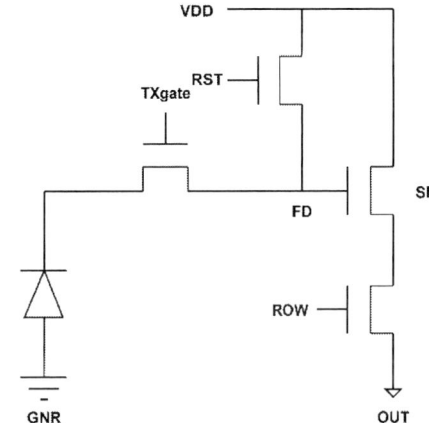

Fig. 1. Circuit schematic of 4T-pixel

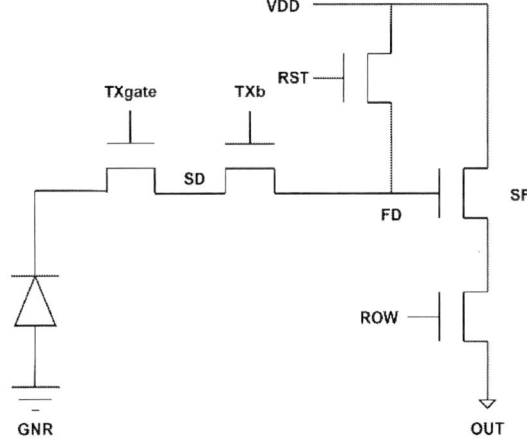

Fig. 2. Circuit schematic of multi-cycle pulse pixel

979-8-3315-3918-4/25 $31.00 © 2025 IEEE

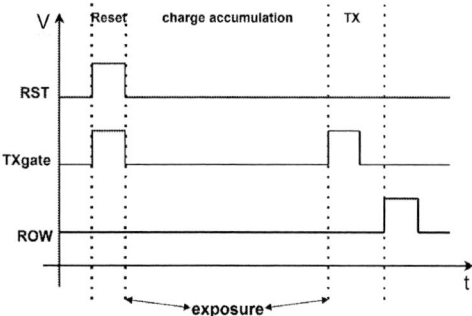

Fig. 3. 4T-pixel timing diagram

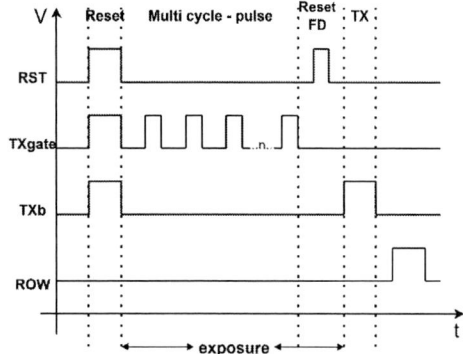

Fig. 4. Multi-cycle pulse pixel timing diagram

B. Simulation results

Fig. 5 and Fig. 6 depict the timing sequences of the conventional 4T-pixel and the proposed multi-cycle pulse pixel, respectively. To ensure a fair comparison, the illumination duration was kept identical for both designs, so that their photosensitive regions received the same amount of optical exposure. In addition, a high illumination intensity was applied such that the electron concentration in the PDwell reached saturation before the TX gate was activated. This setup was intended to highlight the advantages of the proposed timing scheme under strong light conditions.

To evaluate the charge transfer efficiency, the floating diffusion (FD) voltage was measured at snapshot at two time points: before (denoted as V_{FD1}) and after (denoted as V_{FD2}), the TX gate is turned on. The voltage difference of FD, ΔV_{FD} = $V_{FD1} - V_{FD2}$, representing the effective voltage change caused by charge transferring from PD to FD, serves as an indicator of pixel charge transfer performance.

Simulation results show that the ΔV_{FD} for the conventional 4T-pixel is 0.80 V, whereas that of the multi-cycle pulse pixel reaches 1.43 V. This substantial increase in ΔV_{FD} demonstrates that the proposed pixel structure enables more efficient charge transfer, thereby improving the overall charge-to-voltage conversion performance of the pixel circuit.

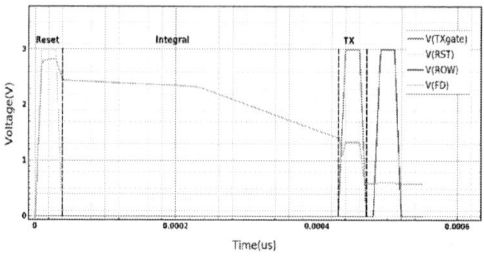

Fig. 5. 4T Pixel timing diagram

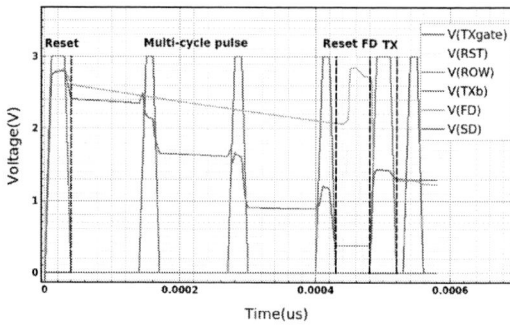

Fig. 6. Multi-pulse cycle timing diagram

In addition, to investigate the influence of pulse interval duration on pixel performance, the timing sequence of the multi-cycle pulse pixel was modified by increasing and decreasing the interval between consecutive TX gate activations, as illustrated in Fig. 7 and Fig. 8. The resulting ΔV_{FD} was calculated in each case and found to remain consistently at 1.43 V. This result indicates that variations in the pulse interval have a negligible impact on the final charge transfer, suggesting that the proposed timing scheme exhibits robustness with respect to pulse spacing.

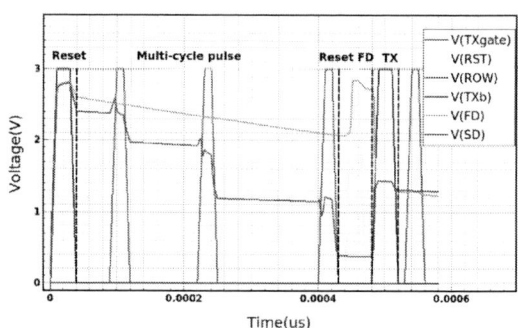

Fig. 7. The timing diagram of the multi-pulse circuit with gradually increasing time intervals

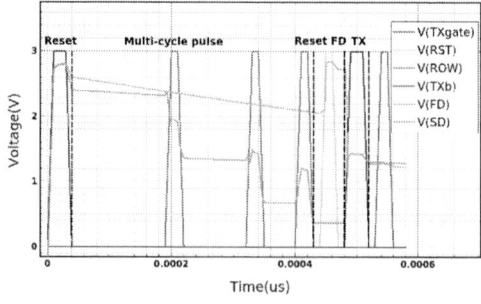

Fig. 8. The timing diagram of the multi-pulse circuit with gradually decreasing time intervals

C. Simulation of different parameters

ΔV_{FD} represents the voltage variation induced by accumulated charges and is strongly influenced by the intermediate storage node capacitance (C_{SD}), the floating diffusion capacitance (C_{FD}), and the gate voltages of the transfer transistors (V_{TXgate} and V_{TXb}). To further investigate the impact of these key parameters on ΔV_{FD}, a series of simulations was conducted under varying conditions.

1) ΔV_{FD} Variation Induced by V_{TXb} Differences:

Fig. 9 (a) illustrates the timing diagrams for different V_{TXb} values, with V_{TXgate} fixed at 3.0 V and $C_{SD} = C_{FD} = 3e{-}15$ F. The corresponding V_{FD} results are shown in Fig.9 (b). The role of TXb is to transfer stored electrons from the intermediate node (SD) to the floating diffusion (FD). V_{TXb} controls the potential barrier during this transfer; as V_{TXb} increases, the barrier height decreases, allowing more electrons to be transferred, which in turn increases ΔV_{FD}

Table I summarizes ΔV_{FD} under various V_{TXb} values. It is observed that ΔV_{FD} increases with V_{TXb} and reaches a maximum at 1.6 V, beyond which the transfer is essentially complete. This indicates that $V_{TXb} = 1.6$ V provides sufficient potential to fully extract the accumulated charge from the SD node to the FD region.

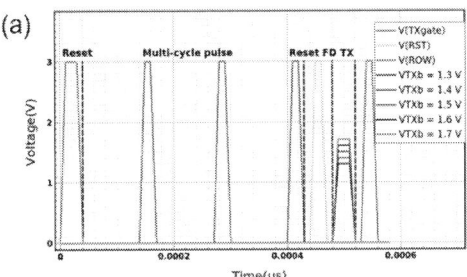

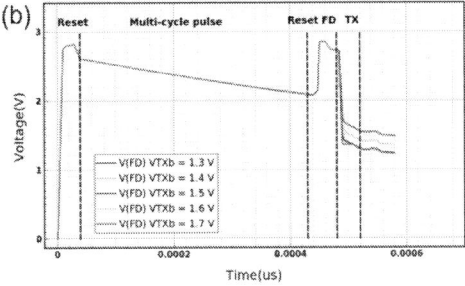

Fig. 9. (a) Timing diagrams under different V_{TXb} conditions, (b) V_{FD} under different V_{TXb} conditions

TABLE I. V_{FD} UNDER DIFFERENT V_{TXB} CONDITIONS

V_{TXb}	V_{FD1}	V_{FD2}	ΔV_{FD}
1.3 V	2.71 V	1.54 V	1.17 V
1.4 V	2.71 V	1.41 V	1.30 V
1.5 V	2.71 V	1.29 V	1.42 V
1.6 V	2.71 V	1.28 V	1.43 V
1.7 V	2.71 V	1.28 V	1.43 V

2) ΔV_{FD} Variation Induced by V_{TXgate} Differences :

Fig. 10 (a) illustrates the timing diagrams for various V_{TXgate} levels, with V_{TXb} fixed at 1.6 V and $C_{SD} = C_{FD} = 3e{-}15$

F. The corresponding V_{FD} results are shown in Fig. 10. (b) TXgate controls the transfer of photo-generated electrons from the photodiode (PD) to the intermediate storage node (SD), and its gate voltage directly influences the potential barrier during this process. As V_{TXgate} increases, the barrier height decreases, facilitating more efficient electron transfer and resulting in a larger ΔV_{FD}. Table II summarizes the ΔV_{FD} values under different V_{TXgate} conditions. When V_{TXgate} reaches 2.6 V, ΔV_{FD} peaks, indicating that electron transfer from PD to SD is complete at this voltage level.

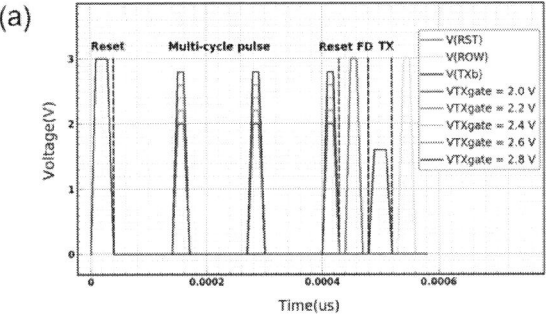

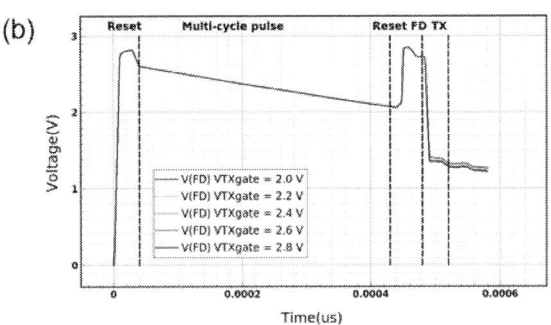

Fig. 10. (a) Timing diagrams under different V_{TXgate} conditions, (b) V_{FD} under different V_{TXgate} conditions

TABLE II. V_{FD} UNDER DIFFERENT V_{TXGATE} CONDITIONS

V_{TXgate}	V_{FD1}	V_{FD2}	ΔV_{FD}
2.0 V	2.71 V	1.33 V	1.38 V
2.2 V	2.71 V	1.30 V	1.41 V
2.4 V	2.71 V	1.29 V	1.42 V
2.6 V	2.71 V	1.29 V	1.43 V
2.8 V	2.71 V	1.29 V	1.43 V

3) ΔV_{FD} Variation Induced by C_{SD} Differences :

Fig. 11. (a) presents the timing diagrams under various C_{SD} conditions, with $V_{TXb} = V_{TXgate} = 3.0$ V and C_{FD} fixed at $3e{-}15$ F. Fig. 11. (b) is the corresponding V_{FD} diagram, and Table III presents the corresponding results. It can be observed that as the C_{SD} increases, the capacitance of the SD node increases, and the storage charge capacity enhances. However, due to the saturation of the PD well capacity, the accumulated charge quantity remains unchanged. Also, since $Q = C * V$, the charge quantity remains constant while the capacitance increases, and V decreases as the capacitance increases.

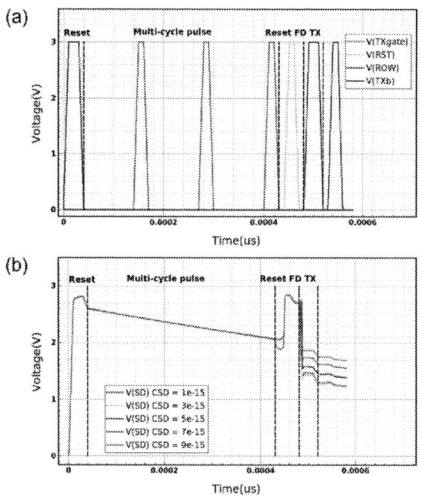

Fig. 11. (a) Timing diagrams under different C_{SD} conditions, (b) V_{FD} under different C_{SD} conditions

TABLE III. V_{FD} UNDER DIFFERENT C_{SD} CONDITIONS

C_{SD}	V_{FD1}	V_{FD2}	ΔV_{FD}
1e-15 F	2.68 V	1.30 V	1.38 V
3e-15 F	2.71 V	1.28 V	1.43 V
5e-15 F	2.71 V	1.45 V	1.26 V
7e-15 F	2.71 V	1.63 V	1.08 V
9e-15 F	2.71 V	1.75 V	0.96 V

4) ΔV_{FD} Variation Induced by C_{FD} Differences :

Fig. 12. (a) shows the timing diagrams under different C_{FD} conditions, with $V_{TXb} = V_{TXgate} = 3.0$ V and C_{SD} fixed at $3e-15$ F. The corresponding V_{FD} results are illustrated in Fig. 12. (b) and summarized in Table IV. As C_{FD} increases, the charge-holding capacity of the FD node improves. However, since the PD well is saturated, the total amount of transferred charge remains constant. According to $Q = C * V$, when Q is fixed, a larger C_{FD} results in a smaller V_{FD}. This inverse relationship is clearly reflected in the simulation results.

TABLE IV. V_{FD} UNDER DIFFERENT C_{FD} CONDITIONS

C_{FD}	V_{FD1}	V_{FD2}	ΔV_{FD}
1e-15 F	2.55 V	0.80 V	1.75 V
3e-15 F	2.71 V	1.28 V	1.43 V
5e-15 F	2.75 V	1.58 V	1.17 V
7e-15 F	2.77 V	1.78 V	0.99 V
9e-15 F	2.78 V	1.93 V	0.85 V

III. CONCLUSION

This work proposes a novel timing scheme based on multi-cycle pulse charge transfer. In contrast to conventional methods that increase the full well capacity (FWC) through ion implantation of photodetector wells (PD well), which typically reduce pixel sensitivity, the proposed approach preserves the nominal operating sensitivity of the pixel circuit. Under identical illumination conditions, the proposed timing results in a floating diffusion voltage drop (ΔV_{FD}) of 1.43 V during charge transfer, representing a 78% increase compared to the 0.80 V observed with conventional timing. This enhancement directly reflects improved charge transfer efficiency. Furthermore, simulation analyses were conducted to evaluate the impact of several key parameters that influence ΔV_{FD}, including V_{TXgate}, V_{TXb}, C_{SD}, and C_{FD}.

REFERENCES

[1] T. Ma, C. Gao, Q. Li, Y. Qi, S. Deng, and K. Wang, "A five-transistor active pixel sensor with a wide dynamic range and a high-speed pixel operation," in 2023 7th IEEE Electron Devices Technology & Manufacturing Conference (EDTM), Seoul, Korea, Republic of: IEEE, Mar. 2023.

[2] D. Stoppa, M. Vatteroni, D. Covi, A. Baschirotto, A. Sartori, and A. Simoni, "A 120-dB Dynamic Range CMOS Image Sensor With Programmable Power Responsivity," IEEE J. Solid-State Circuits, vol. 42, no. 7, pp. 1555–1563, Jul. 2007.

[3] A. M. Brunetti, M. Musolino, and B. Choubey, "Staggered Pixel Layout to Reduce Area and Increase Full Well Capacity in CMOS Image Sensors," IEEE Trans. Electron Devices, vol. 68, no. 2, pp. 572–577, Feb. 2021.

[4] J. Xu, Q. Chen, and Z. Gao, "Analysis of Transfer Gate Doping Profile Influence on Dark Current and FWC in CMOS Image Sensors," IEEE J. Electron Devices Soc., vol. 9, pp. 27–35, 2020.

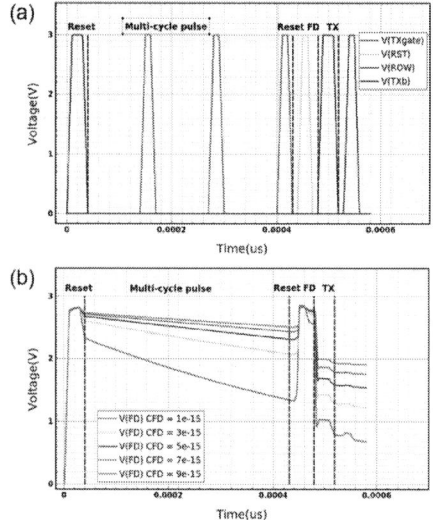

Fig. 12. (a) Timing diagrams under different C_{FD} conditions, (b) V_{FD} under different C_{FD} conditions

Design of RF Microsystem Based on Silicon-based Stereoscopic Integration Technology

Xiaoqing Zhang[1], Lei Shi[2], Mengmeng Yin[3], Cui Jing[4], Dexi Liu[5]

Beijing Institute of Telemetry, Beijing, China

* Email: 1174577544@qq.com

Abstract—A multi-beam radio frequency (RF) microsystem is designed to meet the application requirements of phased array radar across fields such as vehicle-mounted systems and high-resolution remote sensing satellite payloads. It targets lightweight design, reduced thickness, and simultaneous multi-target data transmission/reception. Based on silicon-based RF stereoscopic integration technology, the device employs a multi-layer silicon wafer stacking architecture, integrating multiple microwave monolithic integrated circuit (MMIC) chips with silicon-based passive structures. Micron-level precision modeling and simulation with low insertion loss were performed on key silicon-based 3D heterogeneous interconnect circuits. The final implementation achieves an 8-beam, 4-channel receiving RF microsystem, with each unit measuring ≤10.5 mm × 16.5 mm × 3 mm.

Keywords—*RF Microsystems, Stereoscopic heterogeneous integration, Through-Silicon Via (TSV)*

I. INTRODUCTION

Multi-beam phased array technology is widely used in fields such as vehicle-mounted systems and high-resolution remote sensing satellite payloads. By independently adjusting the amplitude and phase of multiple radio frequency channels and utilizing beamforming networks, multi-beam receiving phased arrays enable simultaneous multi-beam reception to support multi-target communications[1,2]. With increasing battlefield complexity and growing demands for weapon system informatization, the new generation of aerospace equipment urgently requires phased array antennas with miniaturized, lightweight, and thin profiles. The transceiver front-end, serving as the core component of active phased arrays for RF signal transmission/reception, has driven research on efficient integration technologies for multi-beam receiving front-ends. RF microsystems—achieving significant volume reduction, power savings, and enhanced reliability through micro-nano heterogeneous integration of RF, digital, and optoelectronic subsystems—are gaining increasing attention in high-performance packaging[3-5]. Addressing the development needs for miniaturization, thinning, and high airtightness in multi-beam receiving phased arrays, this paper investigates the stereoscopic integration of a silicon-based 8-beam 4-channel receiving RF microsystem. The approach achieves 3D integration of multiple semiconductor MMIC chips, resistors, capacitors, and silicon-based passive circuits, significantly reducing system size while improving integration density and performance.

II. DESIGN OF MICROSYSTEM PACKAGING ARCHITECTURE

To address complex and harsh battlefield conditions while enhancing environmental adaptability and service life of phased array antennas, high-resistivity silicon packaging substrates were selected for their high airtightness, stackable characteristics, and low-loss advantages. The conventional single-module planar integration architecture was upgraded to a dual-module 3D stacked configuration, effectively reducing product size and weight without compromising high performance.

Fig. 1 illustrates the silicon-based RF stereoscopic integration scheme for the multi-beam receiving RF microsystem. This system employs a 3D vertical interconnect architecture with heterogeneous integration of two functional modules: Upper Module 2 (passive combiner unit integrating two 4-channel combiner chips) connects vertically to Lower Module 1 via SnAg microbumps on its bottom surface. Lower Module 1 (multifunctional active module) integrates two amplitude-phase control chips, four driver/amplifier chips, and a decoupling capacitor network. System-level interconnects utilize high-density BGA packaging, where the BGA solder array on Module 1's bottom surface simultaneously handles power delivery, control signals, and RF signal transmission. This architecture eliminates cold solder joint risks in adapter board interconnects and significantly enhances reliability through direct bottom-I/O interconnection. The microsystem's overall dimensions are ≤10.5 mm × 16.5 mm × 3 mm.

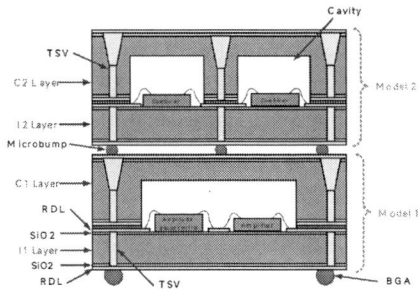

Fig. 1. Schematic diagram of silicon-based stereointegration for the RF microsystem

Each microsystem module comprises two silicon interposers: Layer I (substrate) and Layer C (cover). Through-silicon vias (TSVs) are fabricated in both interposers. The Layer I silicon interposer is 200 μm thick, while the Layer C silicon interposer measures

979-8-3315-3918-4/25 $31.00 © 2025 IEEE

600 μm thick with machined cavities. Critical design note: Cavity inner walls remain non-metallized to prevent short circuits during gold wire bonding.

TSVs within the interposers are solid-metal-filled. Grounded TSV arrays serve as isolation walls to suppress inter-channel electromagnetic interference. The Layer I and Layer C silicon interposers are vertically integrated via wafer-level bonding, with a metallic sealing ring design along the substrate periphery enabling hermetic dam bonding. This configuration blocks external electromagnetic interference while enhancing product airtightness and electromagnetic shielding performance.

The redistribution layer (RDL) employs a grid topology to mitigate thermal stress caused by coefficient of thermal expansion (CTE) mismatch between the RDL and metals, thereby reducing cracking risk. Grid apertures simultaneously suppress eddy current losses while maintaining shielding effectiveness and high-frequency characteristics. For precision alignment during multilayer integration, T-shaped bonding fiducial marks are patterned on the wafer-level bonding surfaces of both Layer I and Layer C interposers, significantly improving manufacturing accuracy

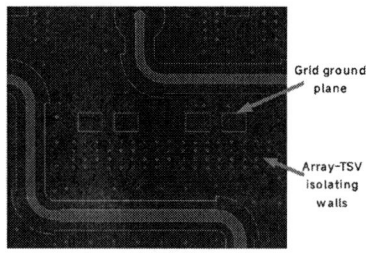

Fig. 2. Array-TSV isolation walls (left)and grid ground plane (right)

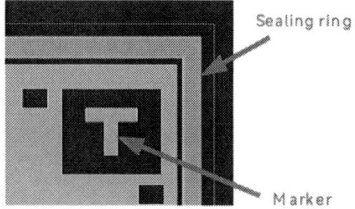

Fig. 3. Metallic sealing rings and wafer-level bonding alignment markers

III. SIMULATION DESIGN OF CAVITY HEIGHT FOR SILICON-BASED RF MICROSYSTEM INTERFACE BOARD

Due to process constraints—including integrated chip height and gold wire bond loop geometry within the RF microsystem—the cavity height of the C-layer interposer must be ≥300 μm. RF transmission performance of microstrip coplanar waveguides was simulated at 3 C-layer cavity heights: 300 μm, 400 μm, and 500 μm. Results indicate that coplanar waveguide transmission degrades with increasing cavity height, with optimal performance achieved at 300 μm. Consequently, the C-layer interposer cavity

height in this RF microsystem is designed at 300 μm.

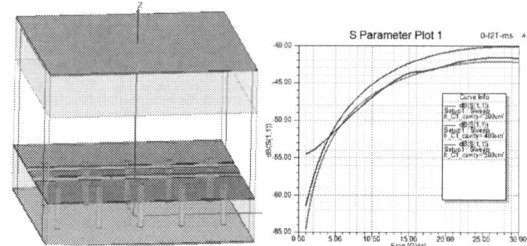

Fig. 4. Simulation model (left) and corresponding results (right) illustrating cavity height effects on RF transmission performance

IV. SIMULATION AND DESIGN OF KEY VERTICAL INTERCONNECT CIRCUITS IN SILICON-BASED RF MICROSYSTEMS

During operation, four RF input signals enter the bottom Module 1 of the RF microsystem via BGA-connected coaxial parallel ports. After amplification by driver amplifier chips, these signals feed into two amplitude-phase control multifunction (APM) chips. Each APM chip:

(1) Splits two RF signals into 16 equal-power outputs;

(2) Imparts constant phase differences through phase shifters for beam scanning;

(3) Adjusts signal amplitudes via attenuators for beamforming;

(4) Combines 16 signals into 8 outputs using its integrated primary power combiner.

The 16 RF signals from both APM chips vertically transit to Module 2 through TSV-microbump interconnects. Trace lengths on the I2 layer top surface are tuned to equalize electrical paths for pairwise signal combining. Subsequent two-stage combining in passive combiner chips yields 8 synthesized RF signals. These signals vertically traverse Modules 2 and 1, exiting through Module 1's bottom BGA solder balls to the motherboard.

This vertical interconnect circuit constitutes the core RF transmission path for multilayer stacked microsystems. The structure—comprising TSVs, microbumps, and BGA solder balls—functions as a coaxial-like transmission line. Its characteristic impedance[6] can be approximated by:

$$Z = \frac{60}{\sqrt{\varepsilon_r}} ln(\frac{R_d}{r_{in}}) \qquad (1)$$

In the coaxial-like structure, r_{in} is the radius of the inner conductor, and R_d denotes the distance from the center of the outer conductor to the center of the inner conductor (i.e., the effective outer diameter). Other key parameters affecting RF transmission performance include r_{out} (radius of the outer conductor), R_{open} (radius of the ground plane antipad opening), and the distribution angle θ (defined as the angle between the line connecting two adjacent ground via centers and the axis of the central signal via). By optimizing the above parameters, precise design values for coaxial-like structures can be obtained.

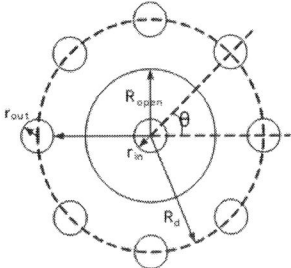

Fig. 5. Schematic diagram of a coaxial plane structure of the same type

Taking the TSV-based coaxial-like structure as an example with a TSV via radius of 10 μm, the coaxial outer diameter is calculated as 320 μm to achieve a characteristic impedance approaching 50 Ω. Using these parameters, Ansoft HFSS 3D electromagnetic simulation software optimized key vertical interconnects by adjusting microwave transmission line width, pad dimensions, ground plane antipad opening radius, and ground via distribution angle.

A. Design of Vertical Transition Structures Between Functional Modules

Signals amplified by the driver amplifier chip in Module 1 transmit to the amplitude-phase multifunction (APM) chip in Module 2 through TSVs and inter-module microbumps. The vertical transition structure—comprising "Module 1 coplanar waveguide (CPW) → interlayer coaxial-like bonding → TSV-based coaxial → microbump coaxial → TSV-based coaxial → Module 2 microstrip CPW"—was optimized via simulation (Fig. 6).

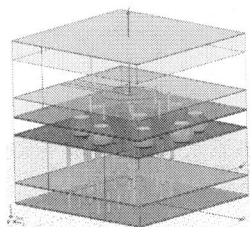

Fig. 6. Simulation model of vertical transition structure

For the microbump coaxial arrangement, circular (Fig. 7a) and square (Fig. 7b) configurations were designed. Both achieve S11 < -37 dB and S21 > -0.11 dB over DC-20 GHz (Fig. 7c-d), with circular arrangement exhibiting 0.01 dB lower transmission loss. Process evaluation confirms comparable assembly complexity for both configurations.

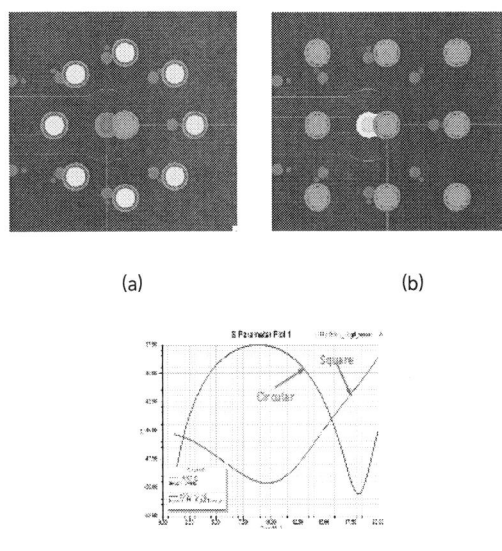

(a)　　　　　　　　　(b)

(c)　　　　　　　　　(d)

Fig. 7. Circular (a) vs. Square (b) microbump coaxial arrangements with S-parameters (c-d)

Further optimization of microbump pad-to-TSV offset and coaxial outer diameter achieved S11 < -22.98 dB and S21 > -0.16 dB across DC-30 GHz (Fig. 8).

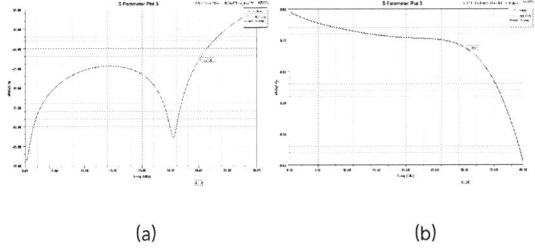

(a)　　　　　　　　　(b)

Fig. 8. Optimized S-parameters for DC-30 GHz band: (a) S11, (b) S21

B. Vertical Transition Design: Motherboard to Module 1 Chip

Signals propagate from the motherboard to Module 1's driver amplifier chip through a vertical transition structure incorporating gold wire bonding. The four-stage interconnect path sequentially traverses motherboard connections, a BGA coaxial structure, TSV coaxial elements, and gold wire interconnects to the chip. Parameter optimization targeting BGA coaxial outer diameter, BGA pad-to-TSV offset distance, and gold wire bond pad dimensions achieved S11 < -22.6 dB and S21 > -0.22 dB across DC-30 GHz, with simulation validation provided in Fig. 9.

979-8-3315-3918-4/25 $31.00 © 2025 IEEE

(a)

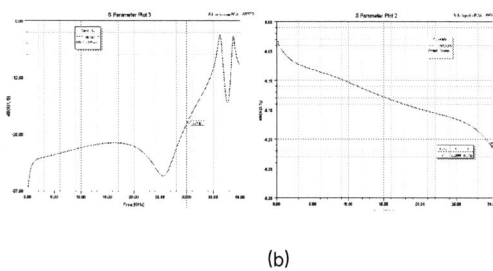

(b)

Fig. 9. Motherboard-to-Module 1 transition: (a) Model, (b) S-parameters

C. Vertical Transition Design: Module 2 Chip to Motherboard

Signals from the combiner chip in Module 2 propagate downward through gold wire bonding and vertical transition structures to the bottom BGA for system output. This 8-stage interconnect path sequentially traverses: the Module 2 chip, gold wire interconnects, a TSV-based coaxial structure, an inter-module microbump coaxial structure, a second TSV-based coaxial structure, interlayer bonding solder joints, a third TSV-based coaxial structure, and finally a BGA coaxial structure connecting to the motherboard. Parameter optimization (ground antipad dimensions, bonding pad size, microbump coaxial outer diameter, and microbump pad-to-TSV offset) achieved S11 < -30 dB and S21 > -0.18 dB across DC-30 GHz, as validated in Fig. 10.

(a)

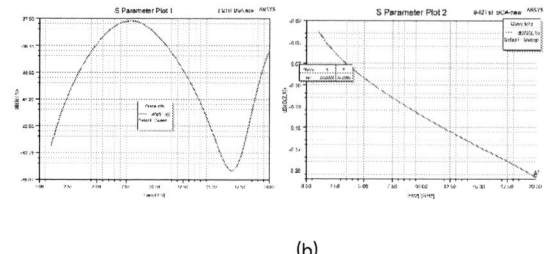

(b)

Fig. 10. Module 2-to-Motherboard transition: (a) Model, (b) S-parameters

V. CONCLUSION

This paper presents a silicon-based multi-beam RF microsystem employing a 3D vertical interconnect architecture, achieving 8-beam 4-channel receive phase-shift beamforming within a compact footprint of ≤10.5 mm × 16.5 mm × 3 mm. Low-loss microwave signal transmission in the multilayer stack was ensured through optimized TSV-based coaxial, BGA coaxial, and RDL coplanar waveguide structures. This silicon-based RF microsystem technology provides a novel approach for miniaturizing multi-beam transceiver phased arrays, demonstrating broad application potential.

ACKNOWLEDGMENT

The authors acknowledge the Beijing Institute of Telemetry Technology for providing design and simulation platform support.

REFERENCES

[1] T. Li and Y. Zhang. "Analysis of methods for improving synthesis efficiency in multi-beam transmit arrays," Journal of Electronic Information Countermeasure Technology, vol. 24, no. 1, pp. 66-69, 2009.

[2] J. Tang, "Design of phased array broadband transmitting assembly and multi-beam receiving assembly," M.S. thesis, Zhejiang Univ., Hangzhou, China, 2020.

[3] Y. Zhao, M. Zhao, Y. Zhao et al., "Silicon-based MEMS 3D integrated chip-scale multi-beam transmit frontend," Telemetry and Remote Control, vol. 42, no. 5, pp. 85-94, 2021.

[4] D. Liu, X. Zhang, L. Shi et al., "Research on development strategy of RF microsystems," Telemetry and Remote Control, vol. 42, no. 5, pp. 17-27, 2021.

[5] K. Zoschke, C. A. Manier, M. Wilke et al., "TSV integration and wafer bonding for hermetic MEMS wafer level packaging in miniaturized timing devices," in Proc. IEEE 65th Electron. Compon. Technol. Conf. (ECTC), 2015, pp. 1343-1350.

T. Yang, Q. Song, and J. Du, "TDR simulation for 45° quasi-coaxial microwave multilayer vias," Radio Engineering, vol. 46, no. 5, pp. 56-59, 2016.

A Novel Pretreatment Approach to High-quality SiO₂ Surface Applied for C2W Cu/SiO₂ Hybrid Bonding

Han Jiang [1], Xianlong Wang [1], Ziyu Liu [*1], Yabin Sun [*2]

[1] School of Microelectronics, Fudan University, Shanghai, 200241, China
2 Department of Electrical Engineering, East China Normal University, Shanghai, 200241, China
* Email: hanjiang23@m.fudan.edu.cn, liuziyu@fudan.edu.cn

Abstract—Chip-to-wafer (C2W) hybrid bonding is widely employed in 3D packaging for its high integration density and performance. Among various bonding materials, Cu/SiO₂ hybrid bonding is one of the most commonly adopted. To achieve high-quality interconnection, it is important to require ultra-low surface roughness, high-hydrophilicity and high-uniformity of SiO₂ layer, as well as well-controlled Cu dishing. However, it poses significant challenges for CMP equipment and processes. Thus, this study firstly investigates the factors affecting the surface roughness of SiO₂ during PECVD prior to CMP. The results reveal that the increasing deposition temperature and pressure will reduce the average surface roughness of 1-μm-thick SiO₂ layers to 1.51 nm. Secondly, the pretreating approach of SiO₂ surface by using CMP is proposed before etching. The results shows that the high uniformity of SiO₂ layer based on 4-inch wafer is achieved through applying pressure of 190 g/cm². At the mean times, the surface roughness is reduced to an exceptionally low level of 0.46 nm and the contact angle of water is 9° after pre-treatment, which facilitates the subsequent C2W Cu/SiO₂ hybrid bonding.

Keywords—surface roughness; PECVD; pre-treatment; Cu dishing, C2W Cu/SiO₂ hybrid bonding.

I. INTRODUCTION

In recent years, hybrid bonding has emerged as a critical technology in the fabrication of CMOS image sensors, CPUs, and DRAMs devices [1]. Among various hybrid bonding approaches, chip-to-wafer (C2W) hybrid bonding (HB) has garnered increasing attention due to its potential to enhance yield and improve the flexibility of three-dimensional (3D) integrated circuits. Hybrid bonding involves the direct bonding of planarized dielectric surfaces embedded with conductive metals [2]. However, achieving reliable Cu/SiO₂ hybrid bonding typically requires ultra-low surface roughness and high uniformity of the dielectric layer, as well as strict control over copper dishing. These stringent requirements necessitate precise optimization of chemical mechanical polishing (CMP) parameters, which poses significant challenges in practical processing [3], [4].

In conventional hybrid bonding processes, CMP is primarily employed to remove excess electroplated copper and planarize the SiO₂ surface [5]. However, the CMP process typically treats the polishing of copper and SiO₂ as an integrated step, and relatively few studies have focused on the pre-planarization treatment of SiO₂ dielectric layer prior to bonding. Besides, traditional bonding techniques exhibit limited tolerance to high surface roughness of bonded interface, often leading to bonding failures or poor bond quality. Achieving ultra-low surface roughness typically requires advanced CMP equipment, which significantly increases manufacturing costs [6].

To address this challenge, a novel pretreatment approach is proposed to reduce the surface roughness of SiO₂ layer, increase its uniformity and hydrophilicity, as well as achieve well-controlled Cu dishing. First, the factors affecting the surface roughness of SiO₂ during the plasma-enhanced chemical vapor deposition (PECVD) deposition have been extensive explored. It demonstrates that reducing the surface roughness of SiO₂ prior to etching, combined with the introduction of a non-traditional pre-bonding process, offers a promising pathway to address the challenges associated with Cu/SiO₂ hybrid bonding. This deeply understanding of each step in the process flow can proceed the facilitates the subsequent C2W Cu/SiO₂ hybrid bonding.

II. EXPERIMENTS

A. Methods and Materials

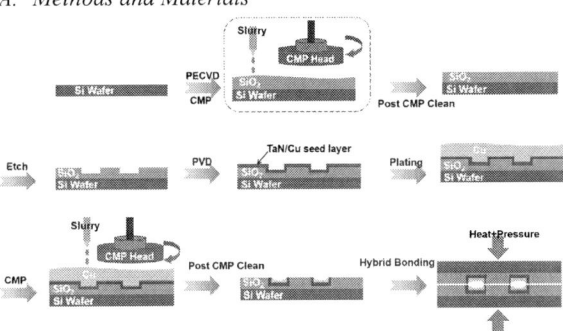

Fig. 1. Ultra-low roughness dielectric layer pretreatment process of the Cu/SiO₂ hybrid bonding The novel approach is added in the dashed box..

Figure 1 illustrates the experimental process flow. First, 4-inch silicon wafers were sequentially cleaned using acetone, isopropanol, and deionized water under ultrasonic agitation. After drying, a 1-μm-thick SiO₂ layer was deposited on the wafer using PECVD. Unlike conventional HB processes, the SiO₂ dielectric layer was further planarized using CMP, followed by a cleaning step to achieve an ultra-low roughness surface. Subsequently, the SiO₂ layer was patterned via ion beam etching. Then, a TaN/Cu seed layer was deposited using physical vapor deposition (PVD), followed by copper electroplating to fill the patterned features. In accordance with the standard HB process, excess copper and seed layers on the surface were removed by CMP, resulting in a recessed copper surface within the vias. During the HB process, due to the higher coefficient of thermal expansion (CTE) of copper compared to SiO₂, the copper within the vias underwent thermal expansion and diffusion. This promoted grain recrystallization at elevated temperatures, ultimately facilitating successful electrical bonding.

979-8-3315-3918-4/25 $31.00 © 2025 IEEE

B. High Surface SiO₂ Quality by PECVD optimization

Figure 2 shows the deposition mechanism of SiO$_2$ molecules deposition within the PECVD chamber. The SiO$_2$ layer was formed by a surface reaction between silane (SiH$_4$) and nitrous oxide (N$_2$O) at a specific temperature, occurring directly on the wafer surface. The chemical reaction is shown as the following:

$$SiH_4 + 2N_2O \rightarrow SiO_2 + 2N_2 + 2H_2O \quad (1)$$

In this process, the flow rates of SiH$_4$ and N$_2$O were maintained at 30 sccm and 60 sccm, respectively. In order to obtain a more uniform and flatten SiO$_2$ layer, the PECVD process preceding CMP was investigated and optimized. By increasing both the deposition temperature and chamber pressure, the average surface roughness (measured over 5 μm × 5 μm areas) of a 1-μm-thick SiO$_2$ layer was detected.

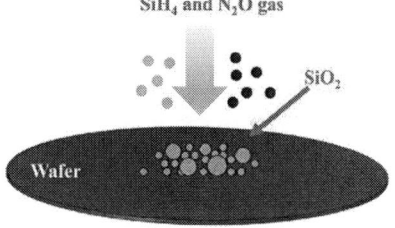

Fig. 2. PECVD process for depositing SiO₂ layer

C. High Surface SiO₂ Quality by Pretreatment CMP optimization

The parameters for the pre-treatment CMP process are summarized. During CMP process, the applied pressures of the polishing head were set to 150, 160, 170, 180, 190, and 200 g/cm^2, respectively. The bottom pad and top head rotated in opposite directions, with rotation speeds of 87 rpm/min and 57 rpm/min, respectively. The slurry used for SiO$_2$ polishing was D2000E SiO$_2$ slurry.

III. RESULTS AND DISCUSSION

A. High Surface SiO₂ Quality by PECVD optimization

Figures 3 and 4 present the average surface roughness (Ra), measured by atomic force microscopy (AFM), and the average deposition rate of SiO$_2$ layers deposited by PECVD at various temperatures. Each Ra value represents the mean of AFM measurements taken from five distinct 5 μm × 5 μm regions on the sample surface. The results indicate that the surface roughness of SiO$_2$ layer decreases with increasing deposition temperature. This trend is attributed to enhanced molecular diffusion at higher temperatures, which promotes the formation of a smoother dielectric surface. Specifically, the Ra value at 300 °C is reduced by 1.21 nm compared to that at 150°C. Additionally, the deposition rate remains relatively constant at approximately 4.5 Å/s across the tested temperature range, suggesting that temperature has a minimal effect on deposition rate under the given process conditions.

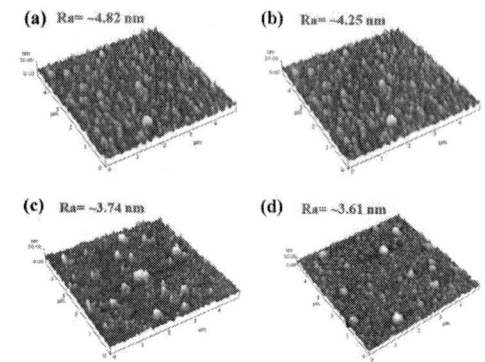

Fig. 3. AFM results at different deposited temperature. (a) 150 °C, (b) 200 °C, (c) 250 °C, (d) 300 °C.

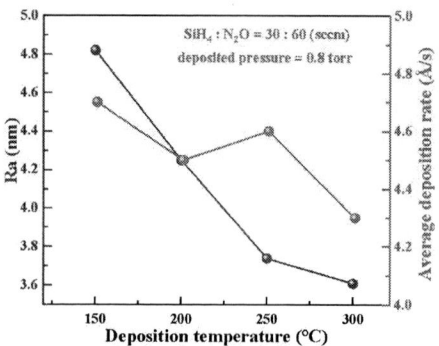

Fig. 4. Average surface roughness and average deposition rate vs. PECVD temperature.

Figures 5 and 6 illustrate the Ra and average deposition rate of SiO$_2$ layers deposited by PECVD under varying chamber pressures. The results show that the Ra of the SiO$_2$ layer decreases with increasing deposition pressure. This reduction in surface roughness is primarily attributed to the increased energy required for SiO$_2$ formation at higher pressures, which leads to a lower deposition rate. A reduced deposition rate enhances the surface density uniformity of the SiO$_2$ layer, contributing to improved smoothness. Specifically, the Ra value at 1.5 Torr is 2.1 nm lower than that at 0.8 Torr. Additionally, the deposition rate is observed to decrease with increasing pressure, with a reduction of 1.8 Å/s when the pressure increased from 0.8 Torr to 1.5 Torr.

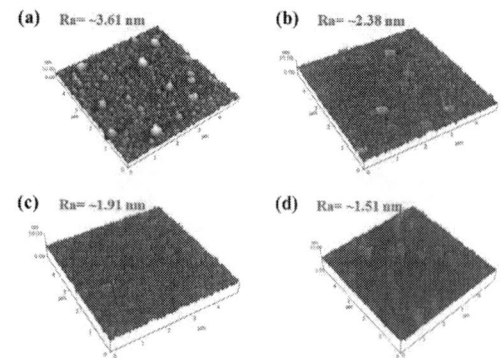

Fig. 5. AFM test results at different deposited pressure. (a) 0.8 torr, (b) 1 torr, (c) 1.2 torr, (d) 1.5 torr.

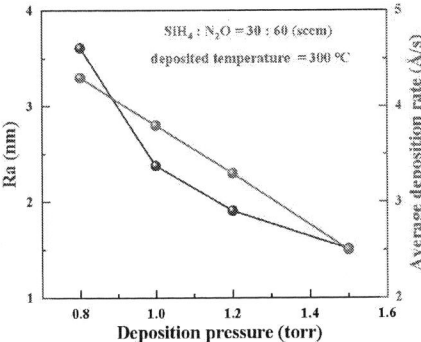

Fig. 6. Average surface roughness and average deposition rate vs. PECVD pressure.

B. High Surface SiO₂ Quality by Pretreatment CMP optimization

In addition to achieving ultra-low surface roughness, the uniformity of the bonding interface height is a critical factor influencing the quality of hybrid bonding. To quantify the height variation of the bonding interface before and after CMP, the removal rate of SiO_2 was analyzed throughout the process. According to the initial height profile of the SiO_2 layer across the wafer surface, a highly uniform bonding interface can be achieved through optimization of CMP parameters and precise control of the SiO_2 removal rate at each location on the wafer.

Figure 7 presents the measured SiO_2 layer thickness before and after polishing under varying pressures of 150, 160, 170, 180, 190, and 200 g/cm², with a fixed polishing time of 300 seconds. As shown in the results, minimal change in SiO_2 thickness was observed at lower pressures, indicating insufficient SiO_2 removal. In contrast, excessive pressure resulted in non-uniform layer thickness across the wafer surface post-polishing. Therefore, applying an appropriate pressure to the polishing head is crucial for achieving a low-roughness, uniformly polished surface. Among the tested conditions, a pressure of 190 g/cm² yielded the best uniformity in SiO_2 layer thickness after CMP.

Besides, to further analyze the spatial variation in removal rate under this optimized condition, Fig. 8 (a) and (b) display the post-CMP thickness distribution and the corresponding SiO_2 removal rate across the wafer surface. The removal rate was relatively uniform (~7.5 Å/s) near the wafer center, reached a maximum of ~11 Å/s in the left-central region, and decreased toward the wafer edge (~4.7 Å/s). This non-uniformity is likely attributed to the specific pressure distribution of the CMP head during operation.

To quantitatively assess the effect of pressure on spatial removal rate variation, 13 distinct measurement points were selected across the wafer surface, as indicated in Fig. 8 (c). The SiO_2 removal rates at these points were statistically analyzed under different pressure conditions, and the results are summarized in Fig. 8(d). The removal rate increased with applied pressure, which is attributed to the enhanced frictional force between the wafer surface and the polishing pad. Notably, at lower pressures as 150, 160, and 170 g/cm², the average removal rate at the wafer edge exceeded that at the center. Conversely, at higher pressures as 180, 190, and 200 g/cm², the removal rate at the wafer edge became lower than that at the center. This behavior may be the result of a combined effect of applied pressure and slurry distribution,

where higher pressure hinders effective slurry infiltration toward the wafer center.

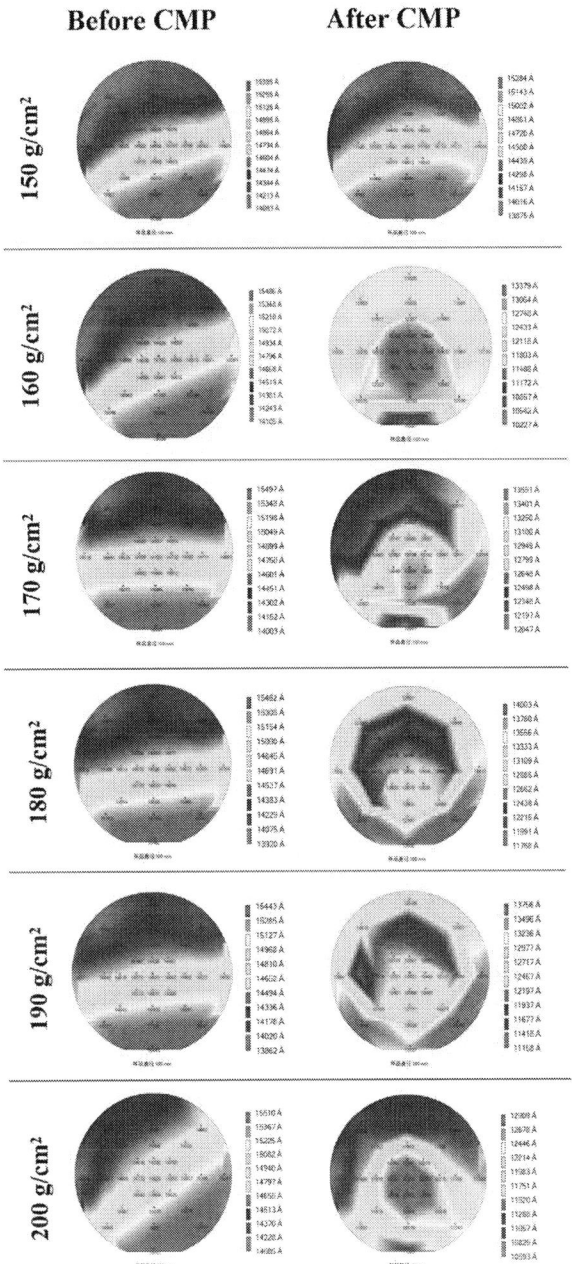

Fig. 7. Comparison of SiO_2 dielectric layer thickness before and after CMP under pressures of 150, 160, 170, 180, 190, and 200 g/cm².

Figure 9(a) shows the surface roughness of the SiO_2 layer prior to CMP, while Fig. 9(b) presents the post-CMP surface morphology. Under an applied pressure of 190 g/cm², the surface roughness of SiO_2 was significantly reduced to 0.46 nm, indicating that an ultra-low roughness surface can be achieved through careful optimization of CMP parameters. Figures 9(c) and (d) display the contact angle (CA) measurements of the SiO_2 surface before and after CMP, using deionized water as the test liquid. The initial CA was 41°, which decreased to 9° after polishing. This result is consistent with the AFM measurements, confirming that lower surface roughness correlates with enhanced surface wettability (i.e., reduced water contact angle).

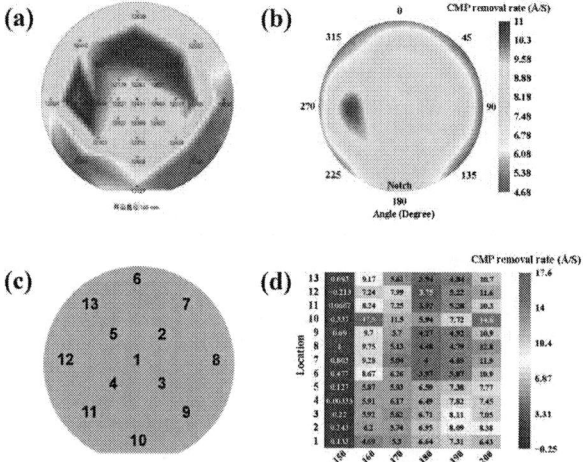

Fig. 8. (a) SiO$_2$ thickness and (b) average SiO$_2$ removal rate on the wafer surface after 300-second polishing under the pressure of 190 g/cm^2, (c) Distribution of 13 measurement points across the wafer surface, (d) Average SiO$_2$ removal rates at the 13 selected points on the wafer.

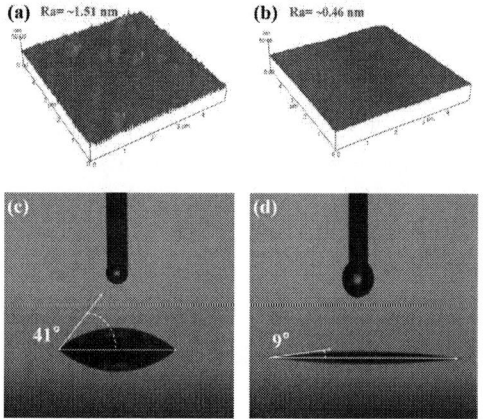

Fig. 9. AFM measurement of SiO$_2$ surface (a) before pre-treatment CMP and (b) after pre-treatment CMP. Contact angle measurement of SiO$_2$ surface (c) before pre-treatment CMP and (d) after pre-treatment CMP.

IV. CONCLUSION

In this study, the PECVD process for SiO$_2$ deposition was investigated and optimized to reduce surface roughness of SiO$_2$ layer. Experimental results demonstrated that increasing the deposition temperature and pressure decreased the surface roughness (Ra) of the SiO$_2$ layer by 3.31 nm (4.82 nm to 1.51 nm). Subsequently, a pre-planarization before etching is proposed to reduce the surface roughness of SiO$_2$ to 0.46 nm. Besides, the contact angle decreased to 9° from 41° after polishing. In summary, this study proposes an efficient method to obtain ultra-low surface roughness for hybrid bonding technology.

V. ACKNOWLEDGEMENT

This work was supported by National Key Research and Development Program of China - Grant 2024YFB4405700.

REFERENCES

[1] F. Niu *et al.*, 'Low-temperature Cu/SiO$_2$ hybrid bonding based on Ar/H$_2$ plasma and citric acid cooperative activation for multi-functional chip integration', *Appl. Surf. Sci.*, vol. 648, p. 159074, Mar. 2024, doi: 10.1016/j.apsusc.2023.159074.

[2] R. He, M. Fujino, M. Akaike, T. Sakai, S. Sakuyama, and T. Suga, 'Cu/adhesive hybrid bonding at 180 °C in H-containing HCOOH vapor ambient for 2.5D/3D integration', in *2017 IEEE 67th Electronic Components and Technology Conference (ECTC)*, Orlando, FL, USA: IEEE, May 2017, pp. 1243–1248. doi: 10.1109/ectc.2017.13.

[3] Q. Kang *et al.*, 'Low-temperature Co-hydroxylated Cu/SiO$_2$ hybrid bonding strategy for a memory-centric chip architecture', *ACS Appl. Mater. Interfaces*, vol. 13, no. 32, pp. 38866–38876, Aug. 2021, doi: 10.1021/acsami.1c09796.

[4] J. H. Lau, 'Recent advances and trends in Cu–Cu hybrid bonding', *IEEE Trans. Compon. Packag. Manuf. Technol.*, vol. 13, no. 3, pp. 399–425, Mar. 2023, doi: 10.1109/tcpmt.2023.3265529.

[5] X. Li, R. Zeng, X. Shi, and Y. Lin, 'A review of the Cu chemical mechanical planarization process in hybrid bonding technology', *J. Electron. Packag.*, vol. 147, no. 3, Sep. 2025, doi: 10.1115/1.4068883.

[6] P.-S. He *et al.*, 'Chemical mechanical planarization of nanotwinned copper/polyimide for low temperature hybrid bonding', *J. Electroanal. Chem.*, vol. 969, p. 118544, Sep. 2024, doi: 10.1016/j.jelechem.2024.118544.

Approximately Timed Scalable DSP Model Based on SystemC

Yongwang Qin[1,2], Sheng Liu[1,2,*], Yang Zhang[1,2], Xing Hu[1,2], Chen Shangqian[1,2]

[1]College of Computer Science and Technology, National University of Defense Technology, Changsha 410073, China
[2]Key Laboratory of Advanced Microprocessor Chips and Systems
qinyongwang19@nudt.edu.cn, liusheng83@nudt.edu.cn

Abstract—This paper is dedicated to researching and implementing a scalable digital signal processor (DSP) approximately timed model based on SystemC. Leveraging the powerful capabilities of SystemC in system-level modeling, this paper proposes a modular approximately timed model architecture and elaborates on its design concept in detail. This model adopts the SystemC transaction-level modeling (TLM) method, effectively separating the communication and computing functions. The communication between modules is efficiently achieved through function calls. Moreover, relying on the established unified virtual simulation platform, this model can support the generation and rapid evaluation of coarse-grained hardware behavior stimuli for common application algorithms. Compared with traditional verification methods, the proposed model significantly improves the efficiency and accuracy of the design verification of scalable application-specific DSPs, providing an efficient and reliable simulation environment for the early development and verification of scalable DSPs.

Index Terms—Approximately Timed Model, Transaction-Level Modeling, SystemC, Scalable DSP

I. INTRODUCTION

The advancement of very large scale integrated circuit technology has made the design of electronic systems increasingly complex. For the design of large-scale SoCs such as digital signal processors, designers need to use simulation models for performance simulation to ensure the functional correctness during the design process. In the early stage of design, simulation models help designers make trade-offs between different architectural features, namely the so-called design space exploration.

With the exponential growth of SoC complexity, the traditional register transfer level (RTL) design method faces severe challenges of high design complexity and long development cycles [1]. Especially crucial is that the design and integrated testing of the software part often have to wait until the hardware design is completed [2], which greatly extends the time-to-market of the entire product [3]. On the other hand, in the verification phase, traditional verification platforms usually rely on hardware description languages (HDL) to develop master device models and test stimuli. However, as the functions of chips become increasingly complex, especially for targets such as domain-specific digital signal processors, the

difficulty and workload of building an efficient and accurate verification environment using HDL also increase sharply [4]. It is also very difficult to modify after changing the application scenario, and it is impossible to quickly update and iterate.

To address the above requirements and challenges, this paper is dedicated to researching and implementing a scalable application-specific DSP approximately timed model based on SystemC. This model adopts modular design and SystemC transaction-level modeling [5], effectively separating communication and computing functions. Taking the domestic YHFT-Matrix DSP as the basic verification object, a scalable array computing module is added to specifically meet the requirements of larger-scale matrix operations. By building a unified virtual simulation platform [6], it supports the rapid evaluation of performance and running bottlenecks of actual application algorithms on the DSP platform. This method aims to significantly improve the efficiency and accuracy of the design verification of scalable application-specific DSPs and provide an efficient and reliable simulation environment for their early development.

II. RELATED MECHANISMS OF SYSTEMC TRANSACTION-LEVEL MODELING

A. Introduction to SystemC Language

SystemC is a system-level modeling language supporting various hardware modeling paradigms from RTL to TLM. It's mainly used in high-level TLM for early architectural exploration [7], [8].

Low-abstraction approximately timed modeling synchronizes units at each clock edge, adding unnecessary details. In an SoC, external interrupts are slow; CPU polling every cycle increases overhead and simulation time. TLM, in contrast, uses event-driven synchronization. It abstracts RTL pin-level communication into interface function calls (Fig. 1), transfers data via packets, and thus boosts running speed. A "transaction" encapsulates module communication, relying on packets for data transfer and synchronization.

B. Introduction to TLM 2.0

TLM is key for studying complex architectures. OSCI first released TLM 1.0, but due to limitations, TLM 2.0 followed [9]–[11].

This work is supported by the Political Work Fund: Subsidy for the Fourth Batch of High-level Talent Cultivation Targets-Liu Sheng.

*Corresponding author

979-8-3315-3918-4/25 $31.00 © 2025 IEEE

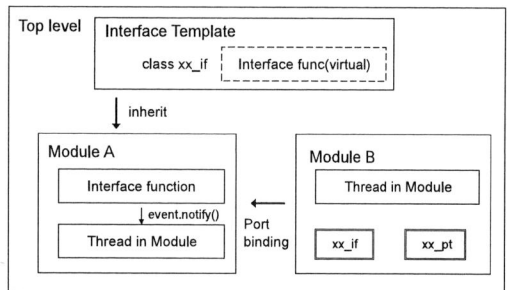

Fig. 1: Communication mechanism of SystemC model

TLM 1.0 has only a blocking transport interface. Transactions must complete in one function call, blocking the calling process on wait events, and has limited adaptability with the wait() function for delay. TLM 2.0 supports the blocking interface and adds a non-blocking one for pipeline modeling, plus a direct memory and debug transport interface. It uses interface function parameters for delay declaration, which is more convenient and standardized.TLM 2.0 offers rich standard interfaces, based on data transmission direction. Initiator_socket class has forward-path functions, and target_socket class has reverse-path ones. Designers inherit socket classes to use these interfaces.

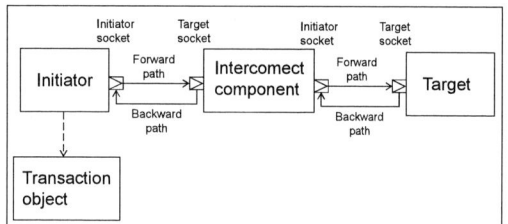

Fig. 2: TLM 2.0 Socket transmission method

III. SCALABLE SPECIAL-PURPOSE DSP HARDWARE ARCHITECTURE

In this section, a comprehensive intro to the scalable special-purpose DSP hardware architecture will be given. As shown in Fig. 3, this architecture has an Instruction Processing Unit Module for receiving, decoding and sending external instructions; Storage Modules (DRAM for global data storage with 4GB space accessible by lower-level components, AM for vector computing, Buffer for convolution data); Arithmetic Modules (VPU for operations like FFT with memory-computing interactions, PEA for high-memory-access tasks such as matrix multiplication and accumulation); and a Control Module with DMA, Core Access Controller and Data config unit for regulating/forwarding data and control signals. The number of VPU and PEA computing units can be adjusted considering power consumption and area for various computing tasks, making the DSP chip highly flexible and scalable for different data scale scenarios.

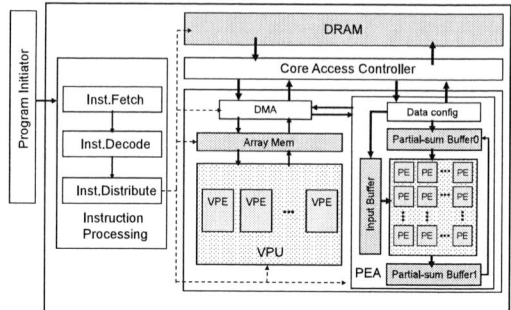

Fig. 3: Scalable DSP Hardware Architecture

IV. DESIGN OF THE SCALABLE SPECIAL-PURPOSE DSP MODEL

Based on the above hardware architecture of the scalable special-purpose processor core, the scalable special-purpose DSP model adheres to the modular design principle and follows the design idea of progressing from the overall to the local and gradually refining to complete the modeling work of the entire system.

For describing the system hardware architecture, SystemC classes are chosen. The basic unit is the module. A module can contain sub-modules, ports, or processes within it. Connections and communications between modules are achieved through ports and signals. The SystemC transaction-level modeling approach is adopted. This approach separates the computing function from the communication function. Communication between modules is accomplished through function calls, effectively reducing the amount of event and information processing.

As can be seen from Fig.4, this model generally includes three main functional modules: VPU, DMA, and PEA. The model abstracts the original communication method based on pins and signals between modules into the form of function calls. Moreover, by integrating timing information into each functional function, approximately timed modeling is ultimately achieved.

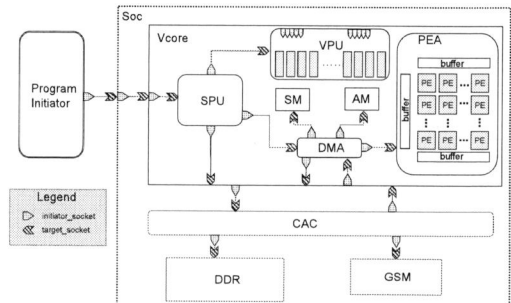

Fig. 4: Scalable dedicated DSP model connection

A. Design of the Vector Processing Unit Model

The Vector Processing Unit (VPU) is structured as a scalable vector operation cluster. Its core aim is to efficiently handle compute-intensive parallel tasks, such as vector multiply-and-accumulate operations. This unit is composed of N isomorphic Vector Process Elements (VPEs). With the aid of SystemC technology, the VPU has a distinctive design in the transaction-level modeling aspect.

The VPU fully supports the Vector SIMD (Single Instruction Multiple Data) parallel technology to realize the vector computing function in applications. With the help of SystemC, it can execute 4 * N vector operation tasks in parallel, strongly facilitating large-scale parallel multiply-and-accumulate computations.

As the key computing component of the VPU, the VPE has a clear working process within the SystemC module framework. After the VPU receives dispatched vector operation instructions, it decodes them and sends them to the corresponding functional units within the VPE for execution. In SystemC, each vector computing component corresponds to a vector instruction in the instruction execution packet. This means that the VPE contains four pipelines that can execute in parallel, significantly improving the computing efficiency.

As the data cache space for the VPE, the Vector Register File (VRF) is responsible, under the SystemC storage management mechanism, for temporarily storing input and output data, ensuring the efficient transfer of data during the computing process.

B. Design of the High-Efficiency DMA Model

In the Vcore vector processing core system, the DMA module is crucial for efficient data transfer between Vcore and external storage. It supports block transfer to reduce CPU interference, enables data rearrangement and format conversion, and its multi-channel parallel transmission is key to system data flow efficiency.Implementation-wise, DMA uses SystemC's capabilities. It communicates and transfers data via multiple TLM interfaces. Its working process is well-designed. Depending on mode flags, different transmission events are triggered.

Point-to-point transmission is the basic method, handling multiple source and target parameters. It reads source data frame by frame to the buffer and writes to the target according to the target frame structure. Efficiency and stability are ensured through batch division and clock control.The trans transpose transmission meets matrix data needs. It reads matrix data blocks to an array, transposes, and writes back to the target.The SG transmission enables scatter reads and gather writes. It reads SG parameters from memory, gets data block info, scatters reads data to the buffer, and writes to the target based on the frame structure, showing great data processing flexibility.

C. Design of the Processing Element Array (PEA) Model

The PEA based on SystemC-TLM has an N×N scalable two - dimensional grid architecture (Fig.3). Referring to Google's TUP architecture's Weight - Stay mode, it achieves efficient computing via transaction-level modeling.

In the PE array, each Processing Element (PE, Fig.5) is interconnected by horizontal data and vertical accumulation channels, forming a systolic data flow. In SystemC, PE is instantiated as an sc_module. Internally, it integrates four functional modules. Control logic ensures timing synchronization through data valid signals. The Weight Register stores static weights and supports write - enable - triggered synchronous updates. The Multiplier does single/double - precision floating - point ops. The Adder accumulates partial sums. All follow synchronous design. The multiply - accumulate op activates only when input data and partial sums are valid, reducing power consumption. The data buffer subsystem has a three-

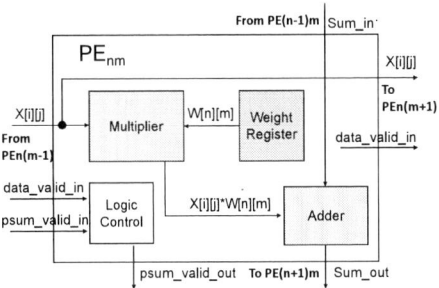

Fig. 5: Overview of PE structure

level storage structure. Partial-sum Buffer0 uses a dual-port circular FIFO to cache input data; Input Buffer has the same for raw data; Partial-sum Buffer1 collects calculation results. In TLM modeling, the buffer has Scalable depth via sc_fifo, with an asynchronous write and synchronous read mechanism.

This system and external DRAM form a two-level storage architecture. Input Buffer integrates a TLM target interface and uses DMI for fast data loading. Output Buffer does batch transaction transmission.

The control logic module (Date config) has three key functions. The start signal cascades through the event chain. Completion detection uses a falling-edge strategy to make a low-power signal. Address mapping encapsulates config params via the TLM extension package and supports six transaction commands.

PE interconnection uses a full handshake protocol. Horizontal data flow is controlled by the valid signal chain, and vertical accumulation state is managed. Timing control is tied to the global clock's rising edge, avoiding competition. Through SystemC's clock modeling and TLM transaction abstraction, an algorithm-hardware co-verification platform is built, offering a standard reference model for scalable computing architecture.

D. Approximately Timed Implementation Based on SystemC Modules

The scalable dedicated DSP model in this paper is abstractly encapsulated with system-level languages and communicates externally via ports. Module data exchange is accomplished through signals. When data changes, a signal triggers an event, driving the emulator for data exchange and update. Signals thus directly link modules. Ports, as the connection between signals and modules, are closely related to specific signals. Note that SystemC signals and ports support delayed assignment for hardware signal behavior modeling.

In SystemC, port and signal read/write operations are based on δ delay. Similar to Verilog, in clock-triggered sequential logic circuits, read/write data updates happen at the end of a unit time. So, the data read in the current cycle is from the previous cycle, and updated data can be read only in the next cycle. This port and signal attribute, differentiating from high-level language data types, enables implementing hardware models like registers and memories, simulating digital circuit register non-blocking assignment.

This paper uses non-blocking assignment delay for pipeline setup time transition. Digital circuit arithmetic units like adders and multipliers often can't complete operations in one cycle. And there may be complex combinations like multiply-accumulators. To meet shorter time needs, pipelining is often used. In high-level languages, arithmetic operations complete in one cycle. In clock-triggered sequential logic, pipelining needs a setup time to output one result per cycle. That is, the input data produces the final result after N cycles. Here, non-blocking assignment is used to introduce an N-cycle data input delay, simulating pipeline operations. This improvement makes simulation waveform operation input and output data consistent with hardware simulation, achieving pipeline establishment.

V. EXPERIMENTAL AND VERIFICATION RESULTS

Experimental Environment:
- Operating System: Ubuntu 20.04.6 LTS
- Coding Languages: SystemC C++, C
- Experimental Platform: DSP development board platform, DSP model verification platform

This study conducts experiments in two aspects. On one hand, functional verification is carried out. Experimental tests of the bencmark algorithm are performed on two platforms, and the results are compared with those of the Python platform to ensure the correctness of the experimental results. On the other hand, performance verification is conducted. For typical digital signal processing algorithms: the FFT algorithm (1024 points) and the GEMM algorithm (6 - 512 - 64), they are run synchronously on the DSP development board platform and the model verification platform proposed in this paper. The complete number of clock cycles from writing the input data to the DDR storage unit to writing back the final calculation result to the DDR is strictly counted. To eliminate the influence of random fluctuations, each group of experiments is repeated 50 times and the average value is taken . The number of hardware cycles is captured in real - time through the JTAG debugging interface, while the number of model cycles is accurately recorded by the SystemC kernel function.

By comparing the approximately timed model of scalable DSP based on SystemC proposed in this paper with the verification of the development board platform, when the relative error of the experimental results is small, the comparison of operation time is shown in Table 1.

TABLE I: Simulation time on different platforms (number of clock cycles).

Benchmark	development board	verification platform	discrepancy
GEMM	25377 cycle	24798 cycle	3.3%
FFT	3548 cycle	3386 cycle	5.6%

The experimental results prove that while maintaining the functional correctness, the timing-level behavior simulation error of this model is controlled within the engineering-acceptable range, providing reliable support for the early verification of scalable DSP. At the same time, the approximately timed model proposed in this paper has an obvious advantage in simulation speed.

VI. CONCLUSION

In response to long design verification cycles and low efficiency in scalable dedicated DSP design, this paper proposes and implements a SystemC TLM2.0-based approximately timed simulation model. It accurately models core scalable DSP modules. Through non-blocking assignment and timing annotations for pipeline delays, the model achieves timing-level accuracy consistent with hardware behavior. Experimental results show correct functionality and controllable simulation errors. The study's results offer an efficient and reliable virtual simulation environment for early scalable DSP architecture exploration, performance evaluation, and hardware-software co-development.

REFERENCES

[1] G. Gerstlauer, C. Haubelt, et al. "Electronic System - Level Synthesis Methodologies," IEEE Trans. on Computer - Aided Design of Integrated Circuits and Systems, vol. 28, no. 10, pp. 1517 - 1530, Oct. 2009, doi: 10.1109/TCAD.2009.2026356.

[2] Q. Zhang, Y. Cao, D. Li, et al., "Behavior - level Hardware - Software Co - design of SoC Based on SystemC," Computer Engineering, 2005, no. 19, pp. 217 - 219.

[3] K. Shi, "Embedded System Design Based on SystemC," Journal of Chinese Computer Systems, 2003, no. 4, pp. 763 - 766.

[4] J. Xie, "Current Situation and Development of Hardware Description Language HDL," MCU & Embedded Systems Applications, 2003, no. 7, pp. 5 - 8, 68.

[5] K. Luo, Y. Cao, J. Yin, et al., "SystemC Modeling and Verification Method in the Design of Digital Application - Specific Integrated Circuits," Journal of Wuhan University (Natural Science Edition), 2002, no. 3, pp. 306 - 310.

[6] Y. Wang, G. Bu, "Design of UVM Verification Platform Based on SystemC Reference Model," Computer Technology and Development, 2021, vol. 31, no. 7, pp. 75 - 80.

[7] W. He, L. Yang, Q. Lu, "Modeling of Cycle - Accurate DSP Processor Based on SystemC," Microelectronics & Computer, 2013, no. 4, pp. 107 - 110.

[8] Lin B, Xie F. A systematic investigation of state-of-the-art SystemC verification[J]. Journal of Circuits, Systems and Computers, 2020, 29(15): 2030013.

[9] M. Montón i Macián, "Checkpointing for virtual platforms and SystemC - TLM - 2.0," Universitat Autònoma de Barcelona, 2011.

[10] Montón M, Engblom J, Burton M. Checkpointing for virtual platforms and SystemC-TLM[J]. IEEE transactions on very large scale integration (VLSI) systems, 2012, 21(1): 133-141.

[11] Jünger L, Jahic J, Bosbach N. Tutorial: Full System Simulation with SystemC TLM-2.0 and the Arm Fast Models[C]//2025 IEEE 22nd International Conference on Software Architecture Companion (ICSA-C). IEEE, 2025: 175-175.

979-8-3315-3918-4/25 $31.00 © 2025 IEEE

Microscopic Mechanisms of Bias Temperature Instability Induced by Defects in Si/SiO₂/HfO₂ Gate Stacks: A DFT and NEGF Study

Yantao Huang[1], Yunzhi Lin[1], Yixin Zhang[1], Junlong Li[1], Xiaoxu Kang[1], Fengying Yao[1], Shaojian Hu[1], Qing Shi[1], Tao Wu[1,2*]

[1] School of Information Science and Technology, ShanghaiTech University, Shanghai 201210, China
[2] Shanghai Engineering Research Center of Energy Efficient and Custom AI IC, Shanghai, 201210, China

* Email: huangyt2023@shanghaitech.edu.cn, wutao@shanghaitech.edu.cn

Abstract—In this work, the microscopic mechanism of bias temperature instability (BTI) induced by defects in Si/SiO₂/HfO₂ gate stacks is systematically investigated, with particular emphasis on hydrogen bridge (HB) and oxygen vacancy (V_O) defects. Density functional theory (DFT) and non-equilibrium Green's function (NEGF) methods are employed to evaluate the charge-state-dependent energy levels and transport characteristics of these defects. HB defects in the SiO₂ layer are identified as effective hole traps that contribute to negative BTI (NBTI), whereas V_O defects in the HfO₂ layer function as electron traps responsible for positive BTI (PBTI). The energy levels of these defects exhibit significant shifts across different charge states, substantially affecting the carrier trapping and de-trapping processes, in accordance with Marcus theory and the trap/detrap (T/D) model. Furthermore, transport simulations reveal that HB defects can induce up to five orders of magnitude enhancement in gate leakage current via trap-assisted tunneling, while V_O defects exhibit a comparatively weaker influence. These findings provide atomistic insights into BTI mechanisms and leakage degradation, offering valuable guidance for the design of reliability-oriented gate dielectrics and TCAD modeling of advanced CMOS devices incorporating high-k gate stacks.

Keywords—BTI, DFT, NEGF, defect, gate leakage current

I. INTRODUCTION

As semiconductor feature sizes scale down to the nanoscale, ensuring device reliability has become increasingly challenging due to a variety of microscale failure mechanisms. Among these, bias temperature instability (BTI) has emerged as a critical factor limiting the performance and lifetime of advanced CMOS and VLSI circuits[1, 2]. BTI causes the gradual shift of transistor threshold voltage, thereby causing performance degradation, increased power consumption, and potentially even system failure. It is thus recognized as a primary contributor to device aging and long-term degradation[3]. Despite significant advances in experimental studies and macroscopic modeling of BTI in recent years, most investigations still rely heavily on electrical characterization, leaving its atomic-scale physical essence incompletely understood and subject to controversy. Most existing approaches are statistical or semi-empirical in nature, lacking a detailed description of defect dynamics at atomic scale[4]. This limits reliability improvements from material and structural perspectives. Therefore, first-principles atomic-scale simulations are essential to elucidate defect-driven BTI mechanisms and guide reliability optimization of advanced semiconductor devices.

In this work, density functional theory (DFT) is utilized to introduce various types of defects within high-k gate stack

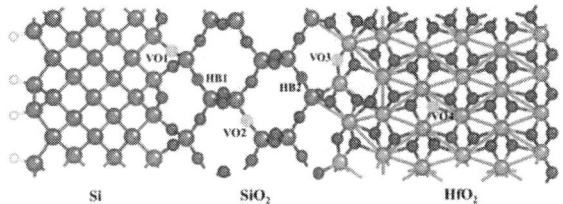

Fig. 1. Computational model of (001) oriented Si/SiO₂/HfO₂ gate stack showing locations of oxygen vacancies (orange) and hydrogen bridges (green) at different sites. The O, Hf, Si, H atoms are depicted by red, blue, gray and small white balls, respectively.

structures. From a first-principles perspective, we investigate the evolution of defect energy levels under different charge states in the gate stack oxides, aiming to clarify the influence of defects on the BTI mechanism. Additionally, transport simulations are conducted to systematically analyze the impact of these defects on gate leakage current.

II. COMPUTATIONAL DETAILS

Electronic structure calculations and structural optimizations were carried out using the Quantum ESPRESSO package[5], based on density functional theory (DFT) and plane wave pseudopotential method. Electron exchange-correlation interactions were described using the Perdew-Burke-Ernzerhof (PBE) functional within the framework of the generalized gradient approximation (GGA) [6]. The RRKJ ultrasoft pseudopotentials[7] were used for the core electrons. A kinetic energy cutoff of 50 Ry was used for the plane-wave basis set, and the cutoff energy was set to 350 Ry for the charge density. The Brillouin zone sampling was carried out using the Monkhorst-Pack[8] scheme with a 4×4×1 k-point grid. The optimized atomic model structure used in the calculations (Fig.1) comprises three layers: Si substrate, a thin β-cristobalite SiO₂, and a monoclinic HfO₂, and the orientation of the structure is (001). The β-cristobalite phase was selected due to its low lattice mismatch with the Si substrate, while monoclinic HfO₂ was chosen due to its low symmetry and serves as an effective crystalline approximation of amorphous high-k dielectrics[9]. The oxygen atoms marked in Fig.1 represent the sites of oxygen vacancies (V_O) and hydrogen bridges (HB) studied in this work. Each defect is labeled according to its type and a corresponding serial number. To systematically investigate the charge-state-dependent energy level characteristics of HB and V_O defects, we performed DFT calculations on defective Si/SiO₂/HfO₂ gate stacks. For each defect type, comprehensive analysis of electronic density of states (DOS) and defect energy levels was conducted across varying charge states, thereby

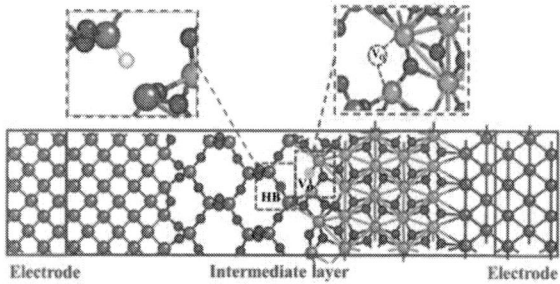

Fig. 2. Structural model in leakage current calculation, the insets show the configurations of HB defect in the SiO₂ layer and V_O defect in the HfO₂ layer.

elucidating the modulation mechanism of charge-state variations on defect level positions and occupancy characteristics.

To investigate the contribution mechanism of defects to leakage current, first principles transport simulations were performed using the Nanodcal software package [10], which combines DFT with the NEGF formalism. The electron transport from Si substrate to the gate electrode was simulated based on a two-probe model, as illustrated in Fig.2. The main panel shows the ideal structure, while the insets depict the configurations of a HB defect in the SiO₂ layer and V_O in the HfO₂ layer. The atomic structure of the scattering region is that obtained by the prior structural optimization. The left and right electrodes consist of uniform Ruthenium (Ru) and doped silicon, respectively. The scattering region is embedded between two semi-infinite electrodes, modeling the channel and metal gate of a practical CMOS device. The transport properties of the system are computed using scattering wavefunctions extending from the left to the right electrode, which were evaluated by the overbridging boundary-matching (OBM) method[11-13] under semi-infinite boundary conditions applied along the (001) direction of the Si substrate. We calculated the electron transport properties comparatively for the ideal structure and defect-containing structures (incorporating V_O and HB defects) under varying gate biases (0.3 ~ 0.7 V). All other computational parameters remained consistent with those used during structural optimization. Since the leakage current is primarily governed by inelastic quantum mechanical tunneling, the I-V characteristics were evaluated using the Landauer-Büttiker formula[14].

III. RESULTS AND DISCUSSIONS

Fig.3 summarizes the energy level alignments of various types of defects located in different regions of the structure with respect to the Si band edges. It is observed that the hole trap levels of V_O defects at the Si/SiO₂ interface and within the SiO₂ interlayer lie significantly below the valence band maximum of Si (VBM_Si), indicating that they are inefficient hole traps. In contrast, the HB defects at the Si/SiO₂ interface and within the SiO₂ interlayer exhibit defect levels above VBM_Si, which means that they can act as efficient hole traps in SiO₂. Additionally, V_O defects at HfO₂ layer have defect states above VBM_Si, making them more likely to trap charges. Previous studies have shown that under NBTI degradation, hydrogen-related defects HB serve as efficient hole traps, which is consistent with experimental observations. On the other hand, PBTI in high-κ gate stacks is most likely attributed to intrinsic electron traps in amorphous HfO₂, which are

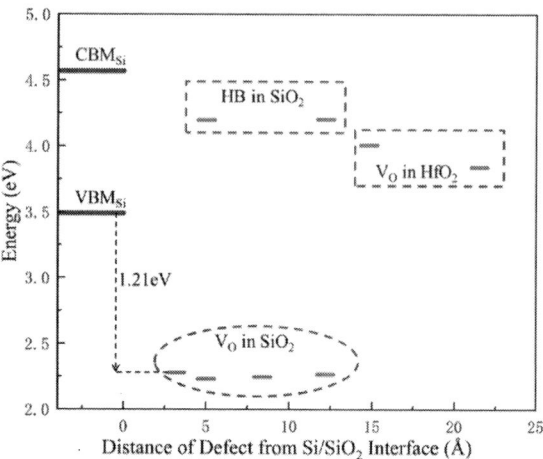

Fig.3. The defect level alignments of V_O and HB defects with respect to the Si band edges.

commonly associated with V_O defects. Therefore, this work focuses on investigating HB and V_O defects as representative trapping centers.

In order to understand the charge trapping/de-trapping behavior of HB and V_O defects under electrical stress, we systematically analyze the evolution of defect energy levels for both defect types across different charge states. It is well established that the defect level shifts originated from the lattice variations after the structure relaxation at different charging states system[15-17]. As reported in the literature[18, 19], the degradation of dielectric layers can be described by the charge trapping/de-trapping processes governed by Marcus theory. The capture and emission rates are influenced by the Gibbs free energy change associated with the carrier transfer reaction, which is defined as the energy difference between the defect level and the corresponding silicon band edge, that is, the conduction band edge for electrons or the valence band edge for holes.

Grasser et al. proposed the trap/detrap (T/D) model, which attributes the dynamics of BTI to the repeated capture and emission of carriers by traps within the gate dielectric. The trapping/de-trapping rates are primarily governed by the energy level positions of the defects relative to the semiconductor band edges and the height of the energy barriers. In CMOS gate dielectrics, HB defects in SiO₂ and V_O defects in HfO₂ are identified as the dominant trap centers responsible for NBTI and PBTI, respectively. Both types of defects exhibit charge-state-dependent energy levels and a strong tendency to capture the dominant carriers—holes for HB centers and electrons for V_O centers.

The HB defect in the SiO₂ layer was first investigated, as it is regarded as a key contributor to SILC, BTI, hysteresis of the transfer characteristic, and other reliability problems. Taking HB2 as an example, Fig.4 shows the HB defect model and its defect levels shift under different charge states. When the system is electrically neutral, the defect level is located approximately 0.37 eV below the Si conduction band minimum (E_c(Si)) and 0.72 eV above the Si valence band maximum (E_v(Si)), enabling efficient trapping of both electrons and holes. Under NBTI stress conditions (i.e., negative bias applied to a p-FET gate), the increased hole concentration in the channel leads to an upward shift of E_v(Si) due to band bending, reducing the energy difference between

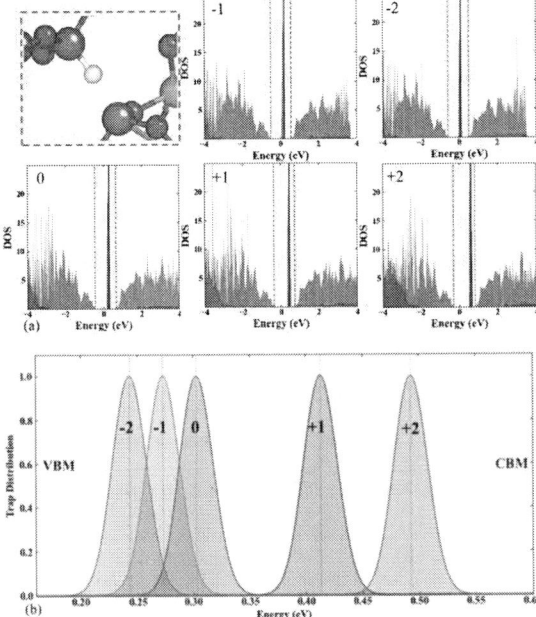

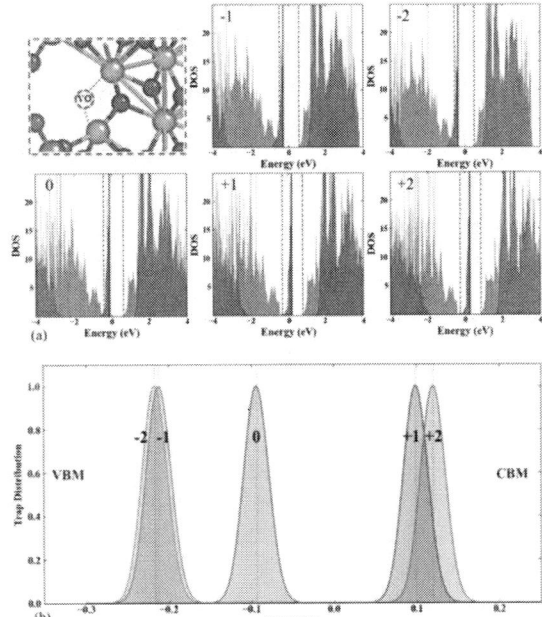

Fig.4. (a) Structural model of an HB defect in the SiO$_2$ layer, along with the corresponding shifts in trap level under various charge states. The gray line and red line represent the DOS projected on Si and SiO$_2$ with HB defect, respectively. (b) Defect energy level distributions within the band gap for different charge states (-2, -1, 0, +1, +2).

Fig.5. (a) Structural model of an V$_O$ defect in the HfO$_2$ layer, along with the corresponding shifts in trap level under various charge states. The gray line and red line represent the DOS projected on Si and HfO$_2$ with V$_O$ defect, respectively. (b) Defect energy level distributions within the band gap for different charge states (-2, -1, 0, +1, +2).

the defect level and the VBM of Si, thereby enhancing hole trapping probability. As the HB defect transitions to a negative charge state, the defect level near E$_v$(Si) shifts downward (lower in energy) and the energy difference between the defect level and E$_v$(Si) decreases from 0.72 eV to 0.69 ~ 0.67 eV. This reduction in the energy difference effectively lowers the barrier for hole capture. According to Marcus theory, a reduction in energy difference between E$_v$(Si) and the defect level results in a higher hole trapping rate. In contrast, in the positive charge state, the defect level shifts upward, increasing the energy difference to 0.92 eV, thereby suppressing hole trapping. By incorporating the non-radiative multi-phonon (NMP) transition model with Marcus theory, it is demonstrated that the energy difference between defect levels and the Si band edges has an exponential impact on carrier trapping and de-trapping rates. Even subtle shifts in the HB defect level induced by charge state transitions can result in pronounced differences in the dynamic behavior observed during NBTI stress and recovery cycles. Thus, charge-state-driven level dynamics can provide a microscopic explanation for the trap response, hysteresis, and partial recovery phenomena during NBTI.

In Si/SiO$_2$/HfO$_2$ gate stacks, PBTI is predominantly governed by electron trapping at V$_O$ defects located within the HfO$_2$ layer. As shown in Fig.5, these V$_O$ centers, which pre-exist in a neutral or positive charge state, with defect levels lying approximately 0.564 ~ 0.758 eV below E$_c$(Si), making them highly effective electron traps. Under positive gate bias stress in NFETs, electrons from the Si channel are injected into the gate stack. These shallow V$_O$ traps, owing to their low electron capture barriers, facilitate efficient initial electron trapping. According to Marcus theory, the electron trapping rate exhibits an exponential dependence on the energy difference between the defect level and the E$_c$(Si). Consequently, the initial stage of PBTI stress is characterized by a rapid threshold voltage (V$_{th}$) shift, attributed to the fast

filling of these energetically favorable and highly active trap states. Following this rapid initial trapping process, V$_O$ defects undergo charge state transitions (0 → –1 → –2), leading to a saturation of the trapping process and a subsequent slowdown in the degradation rate. Upon removal of the stress bias, a fraction of the electrons trapped in shallow V$_O$ traps can thermally de-trapping and return to the Si channel, contributing to a partial V$_{th}$ recovery on practical timescales. However, electrons trapped in deeper V$_O$ states, due to high emission barriers, are difficult to detrap under normal operating conditions, resulting in irreversible PBTI degradation. High-temperature annealing is typically required to provide sufficient thermal energy to overcome these emission barriers. Therefore, the co-existence of shallow and deep traps gives rise to the characteristic "fast initial shift and slow/incomplete recovery" behavior of PBTI. This finding is in excellent agreement with the T/D theory, which emphasizes that the distribution of trap/detrap rates determines the kinetics of BTI. In conclusion, this work systematically elucidates the carrier trapping and de-trapping behavior in NBTI and PBTI based on the microscopic coupling between defect energy levels and charge states, thereby providing atomistic insights that support the T/D theory.

To further assess the impact of HB and V$_O$ defects on device leakage behavior, gate leakage currents were simulated using NEGF method for ideal and defective structures under bias voltages ranging from 0.3 V to 0.7 V. As shown in Fig. 6, V$_O$ defects slightly increase the leakage current across all bias conditions, with a maximum enhancement factor of approximately 3. This indicates the existence of shallow trap-assisted tunneling (TAT) pathways under high-field stress. In contrast, HB defects result in a much more pronounced increase in leakage, especially at 0.7 V where the current increases by nearly 5 orders of magnitude. This dramatic enhancement arises from the HB states located near the CBM$_{Si}$ in the SiO$_2$ layer, which strongly couple with tunneling

979-8-3315-3918-4/25 $31.00 © 2025 IEEE

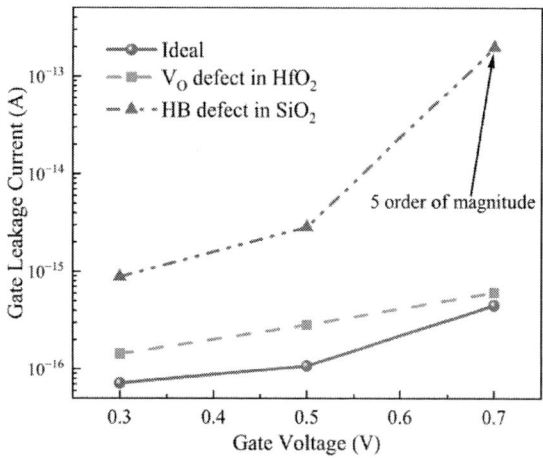

Fig.6. Comparison of the gate leakage current enhancement induced by HB and V_O defects relative to the ideal structure.

electrons under high bias voltage. While HB and V_O primarily impact BTI through charge trapping/de-trapping mechanisms, their significant influence on static power dissipation underscores the need for comprehensive reliability assessments considering both dynamic and leakage degradation paths.

IV. CONCLUSIONS

This study elucidates the charge-state-dependent electronic and transport behaviors of HB and V_O defects in Si/SiO$_2$/HfO$_2$ gate stacks. Using a combined DFT and NEGF approach, we reveal that HB and V_O defects are respectively responsible for NBTI and PBTI degradation mechanisms, with distinct carrier capture characteristics under electrical stress. Our first-principles calculations provide microscopic insights into the modulation of defect energy levels under varying charge states, consistent with the predictions of both Marcus theory and the T/D model. These dynamic shifts govern the rates of carrier trapping and emission, thereby directly affecting the evolution of BTI behavior during stress and recovery cycles. Transport simulations further demonstrate that HB defects strongly promote trap-assisted tunneling (TAT), resulting in leakage current increases of up to 5 orders of magnitude under high bias, whereas V_O defects exhibit a much weaker influence. This suggests that HB defects are not only dominant contributors to BTI but also key factors in static power consumption and reliability degradation. These atomistic insights into defect behavior can be integrated into TCAD models to predict device-level degradation trends and guide the optimization of gate dielectric engineering. In addition, this work provides a theoretical foundation for improving the reliability of high-k gate stacks through defect-type and density control. Future research should address the effects of amorphous structures, explore a broader range of intrinsic and extrinsic defects, and incorporate these atomic-scale findings into device-level compact models for predictive reliability assessment.

ACKNOWLEDGMENT

The authors would like to thank the HPC Platform of ShanghaiTech University for providing computational resources.

REFERENCES

[1] T. Grasser, B. Kaczer, and W. Goes, "Negative bias temperature instability: Modeling challenges and perspectives," in Proc. IRPS, 2008.

[2] T. Grasser et al., "The paradigm shift in understanding the bias temperature instability: From reaction–diffusion to switching oxide traps," IEEE Transactions on Electron Devices, vol. 58, no. 11, pp. 3652-3666, 2011.

[3] S. Mahapatra, Fundamentals of bias temperature instability in mos transistors. Springer, 2016.

[4] Y.-Y. Liu, F. Zheng, X. Jiang, J.-W. Luo, S.-S. Li, and L.-W. Wang, "Ab initio investigation of charge trapping across the crystalline-Si–amorphous-Si O 2 interface," Physical Review Applied, vol. 11, no. 4, p. 044058, 2019.

[5] P. Giannozzi et al., "QUANTUM ESPRESSO: a modular and open-source software project for quantumsimulations of materials," Journal of physics: Condensed matter, vol. 21, no. 39, p. 395502, 2009.

[6] J. P. Perdew, K. Burke, and M. Ernzerhof, "Generalized gradient approximation made simple," Physical review letters, vol. 77, no. 18, p. 3865, 1996.

[7] G. Kresse and D. Joubert, "From ultrasoft pseudopotentials to the projector augmented-wave method," Physical review b, vol. 59, no. 3, p. 1758, 1999.

[8] H. J. Monkhorst and J. D. Pack, "Special points for Brillouin-zone integrations," Physical review B, vol. 13, no. 12, p. 5188, 1976.

[9] E. Nadimi et al., "The Degradation Process of High-k SiO$_2$/HfO$_2$ Gate-Stacks: A Combined Experimental and First Principles Investigation," IEEE Transactions on Electron Devices, vol. 61, no. 5, pp. 1278-1283, 2014.

[10] Y. Wang et al., "Direct tunneling through high-κ amorphous HfO2: Effects of chemical modification," Journal of Applied Physics, vol. 116, no. 2, 2014.

[11] K. Hirose, T. Ono, Y. Fujimoto, and S. Tsukamoto, First-principles calculations in real-space formalism: electronic configurations and transport properties of nanostructures. World Scientific, 2005.

[12] T. Ono and S. Tsukamoto, "Real-space method for first-principles electron transport calculations: Self-energy terms of electrodes for large systems," Physical review B, vol. 93, no. 4, p. 045421, 2016.

[13] Y. Fujimoto and K. Hirose, "First-principles treatments of electron transport properties for nanoscale junctions," Physical Review B, vol. 67, no. 19, p. 195315, 2003.

[14] M. Büttiker, Y. Imry, R. Landauer, and S. Pinhas, "Generalized many-channel conductance formula with application to small rings," Physical Review B, vol. 31, no. 10, p. 6207, 1985.

[15] D. Veksler and G. Bersuker, "Gate dielectric degradation: Pre-existing vs. generated defects," Journal of applied physics, vol. 115, no. 3, 2014.

[16] V. Gritsenko, T. Perevalov, O. Orlov, and G. Y. Krasnikov, "Nature of traps responsible for the memory effect in silicon nitride," Applied Physics Letters, vol. 109, no. 6, 2016.

[17] D. U. Lee et al., "Reduction of interface traps between poly-Si and SiO2 layers through the dielectric recovery effect during delayed pulse bias stress," Nanotechnology, vol. 28, no. 22, p. 225702, 2017.

[18] Y.-Y. Liu et al., "Characterizing the charge trapping across crystalline and amorphous Si/SiO2/HfO2 stacks from first-principle calculations," Physical Review Applied, vol. 12, no. 6, p. 064012, 2019.

[19] Y.-Y. Liu and X. Jiang, "Physics of hole trapping process in high-k gate stacks: A direct simulation formalism for the whole interface system combining density-functional theory and Marcus theory," in 2018 IEEE International Electron Devices Meeting (IEDM), 2018: IEEE, pp. 40.1. 1-40.1. 4.

Mechanism of Leakage Current Enhancement Induced by La Doping in HfO$_2$ Gate Stacks: A DFT Investigation

Yunzhi Lin[1], Yantao Huang[1], Yixin Zhang[1], Qing Shi[1], Fengying Yao[1], Junlong Li[1], Shaojian Hu[1], Xiaoxu Kang[1], Tao Wu[1,2]*

[1] School of Information Science and Technology, ShanghaiTech University, Shanghai 201210, China
[2] School Engineering Research Center of Energy Efficient and Custom AI IC, Shanghai, 201210, China

* Email: linyzh22023@shanghaitech.edu.cn, wutao@shanghaitech.edu.cn

Abstract—While interface dipole engineering using La doping in HfO$_2$ gate stacks has been well-explored for threshold voltage control, the microscopic mechanism behind the accompanying leakage current elevation remains insufficiently understood. In this work, we establish a comprehensive multi-scale simulation framework that uniquely integrates NEGF-DFT device-level quantum transport calculations with DFT-based electronic structure analysis. Our findings reveal that La incorporation induces a significant bandgap narrowing and substantial downward shift of the conduction band minimum at the SiO$_2$/HfO$_2$ interface, which quantitatively explains the observed one-order-of-magnitude increase in gate leakage current. These results elucidate a fundamental trade-off between effective V$_{th}$ modulation and gate leakage degradation, providing critical microscopic insights for the optimization of high-k/metal gate technology in advanced CMOS applications.

Keywords—*Interface Dipole, DFT, NEGF, Gate leakage current , High-k dielectric*

I. INTRODUCTION

As the dimensions of complementary metal-oxide-semiconductor (CMOS) devices continue to scale down, transistor structures are approaching the atomic scale. This aggressive scaling renders interface engineering increasingly critical, as complex interfaces pose significant challenges to reliable threshold voltage (V$_{th}$) control. Among various strategies to address these issues, interface dipole engineering (IDE) has attracted considerable attention. IDE typically employs doping of high-k gate dielectrics with elements such as lanthanum (La) to generate interfacial dipole layers that modulate the local electrostatic potential. The resulting dipole effect enables efficient adjustment of the flat-band voltage (V$_{fb}$), thereby offering a promising route for precise V$_{th}$ tuning in advanced gate stack structures.

Although the formation and impact of interfacial dipoles on threshold voltage have been widely studied and are well understood, their influence on atomic-scale transport—specifically, the mechanisms underlying the associated increase in gate leakage current remains insufficiently explored. To address this problem, we establish a multi-scale simulation framework that integrates non-equilibrium Green's function (NEGF) based quantum transport with density functional theory (DFT) electronic structure analyses [1-3]. Our results reveal that incorporating an interfacial dipole layer via La doping leads to pronounced changes in the energy band alignment, notably bandgap minimum at the SiO$_2$/HfO$_2$ interface, which are directly responsible for the

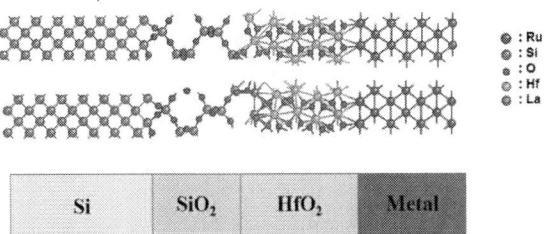

Fig. 1. Atomic-scale gate stack models for quantum transport calculations: undoped and La-doped SiO$_2$/HfO$_2$ interfaces.

observed one-order-of-magnitude increase in leakage current. This comprehensive analysis provides new insights into the fundamental trade-off between threshold voltage modulation and gate leakage, offering valuable guidance for the further optimization of high-k/metal gate technology in future CMOS applications.

II. EXPERIMENTAL DETAILS

Two distinct four-layer gate stack structures were constructed: (a) Si / SiO$_2$ / HfO$_2$ / Metal (Ruthenium, Ru) and (b) Si / SiO$_2$ / La-doped HfO$_2$ / Metal (Ru) [4-6], as illustrated in Figure 1. These structures were built within a supercell measuring 5.3 Å × 5.3 Å × 60 Å. Ru was selected as the electrode material due to its relatively small lattice mismatch (~0.7%) with the Si/SiO$_2$/HfO$_2$ interface, its moderate work function, and its favorable computation stability. The Si/SiO$_2$ interface was modeled following established procedures [7]. The monoclinic HfO$_2$ layer was subsequently constructed on top of the SiO$_2$ layer, based on literature methods for the SiO$_2$/HfO$_2$ junction [8]. For the La-doped structure (b), two Hf atoms near the SiO$_2$/HfO$_2$ interface were replaced with two La atoms, and one adjacent oxygen atom was removed to maintain overall charge neutrality in the system [9].

Gate stacks electronic transport calculations (Fig. 1) were performed using the Nanodcal software package [10], which implements the Non-Equilibrium Green's Function formalism combined with Density Functional Theory (NEGF-DFT). The left and right electrodes consist of a metal layer and a Si layer, respectively, each with a length of 5.3 Å. The k-point sampling was set to 6 × 6 × 100 for the electrode regions and 6 × 6 × 1 for the central region, respectively. The difference in k-point sampling is mainly due to the semi-periodic structure of the electrode regions. N-type doping of the Si substrate was simulated using the Virtual Crystal Approximation (VCA) [11]. Calculations used a double-zeta

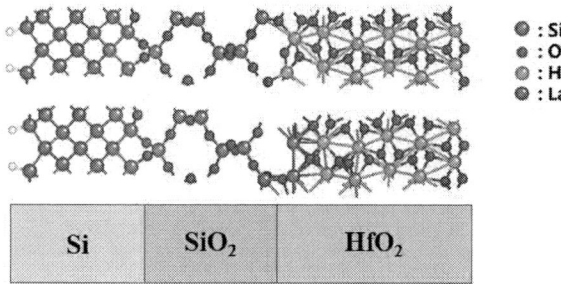

Fig. 2. Atomic models of simplified gate dielectric structures: undoped and La-doped SiO_2/HfO_2 interfaces.

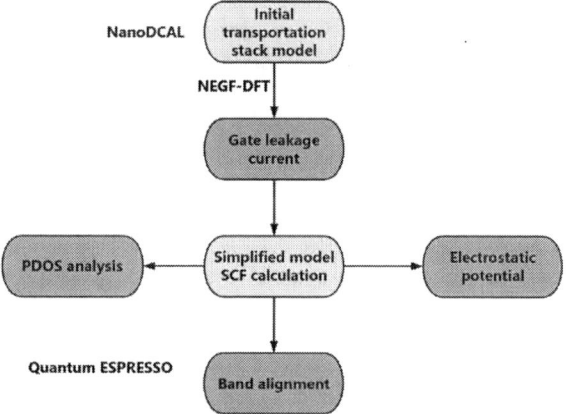

Fig. 3. Schematic illustration of the computational workflow for gate stack analysis

polarized (DZP) linear combination of atomic orbitals (LCAO) basis set and the Perdew-Burke-Ernzerhof (PBE) exchange-correlation functional within the Generalized Gradient Approximation (GGA). Current-voltage (I-V) characteristics were computed using the Landauer-Büttiker formula [12].

To focus specifically on the properties of the insulating layers, particularly the SiO_2/HfO_2 interface and the doping effects, a simplified model was constructed (Figure 2). Calculations for this simplified model were conducted using the Quantum ESPRESSO software package [13]. These calculations employed the PBE exchange-correlation functional, Rappe-Rabe-Kaxiras-Joannopoulos (RRKJ) type norm-conserving pseudopotentials, with plane-wave kinetic energy and charge density cutoffs set to 100 Ry and 700 Ry, respectively. Structural relaxations were performed with a force convergence threshold of 0.001 Ry/bohr. The Brillouin zone was sampled using a Monkhorst-Pack k-point mesh of $4 \times 4 \times 1$. The overall computational workflow for this study is summarized in Figure 3.

III. RESULTS AND DISCUSSIONS

The gate leakage current is a key parameter influencing transistor performance. At low applied voltages, this leakage is predominantly governed by direct electron tunneling, where the tunneling probability is largely dictated by the barrier height and the thickness of the dielectric layer [16]. To evaluate the effect of La incorporation, we simulated the current–voltage (I–V) characteristics for both La-doped and undoped atomic models at low bias (0.1–0.6

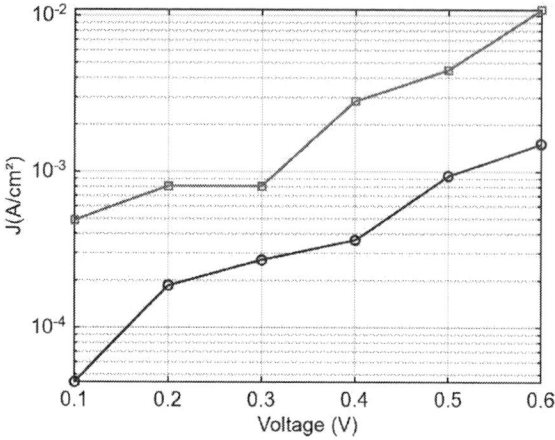

Fig. 4. Current density as a function of applied bias voltage for undoped (blue) and La-doped (red) gate stacks.

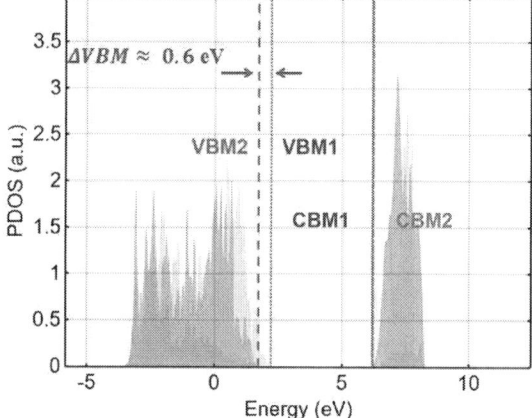

Fig. 5. Comparison of band gaps in bulk HfO_2 for undoped (green) and La-doped (yellow) regions.

V) using NanoDCAL (see Fig. 4). Notably, the leakage current density in the La-doped model was consistently about one order of magnitude higher than in its undoped counterpart across the entire bias range. At 0.1 V bias, the leakage current density for the undoped stack was 4.81×10^{-5} A/cm², whereas for the La-doped stack it reached 4.87×10^{-4} A/cm², demonstrating an order-of-magnitude increase induced by La doping. This significant increase in leakage current upon La doping is in agreement with previous experimental findings [14]. Considering that the EOT thickness remains essentially unchanged, we attribute the primary cause of the increased leakage current to a reduction in the barrier height, indicating that La incorporation fundamentally modifies the electron transport properties of the dielectric layer.

To investigate the physical mechanism responsible for the La-induced leakage current increase, we analyzed the interfacial electronic structure using the simplified Si/SiO_2/HfO_2 model (described in the Experimental section, Fig. 2). As shown in Fig. 5, comparison of the projected density of states (PDOS) for bulk HfO_2 between La-doped and undoped cases reveals that La incorporation induces an upward shift of the band structure by approximately 0.6 eV relative to the vacuum level. This magnitude is consistent with experimentally observed negative flat-band voltage (V_{fb})

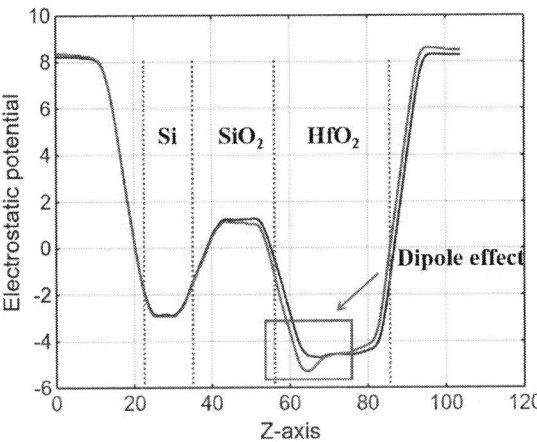

Fig. 6. Electrostatic potential profiles across the gate stack for undoped (blue) and La-doped (red) structures.

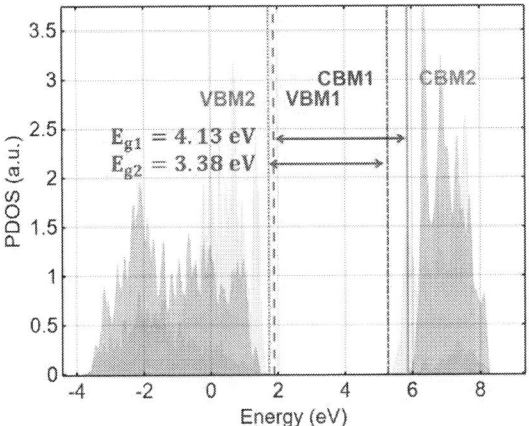

Fig. 7. Band gap comparison at the SiO_2/HfO_2 interface for undoped (green) and La-doped (yellow) regions.

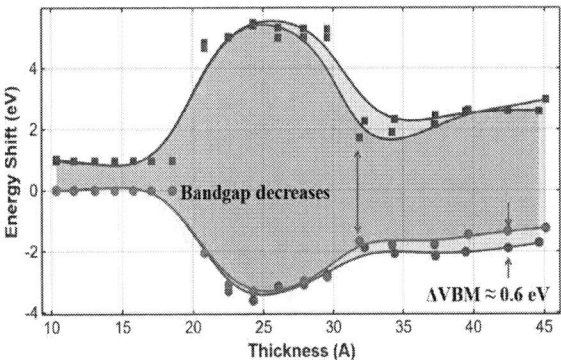

Fig. 8. Band alignment diagrams for undoped (orange) and La-doped (blue) gate stacks.

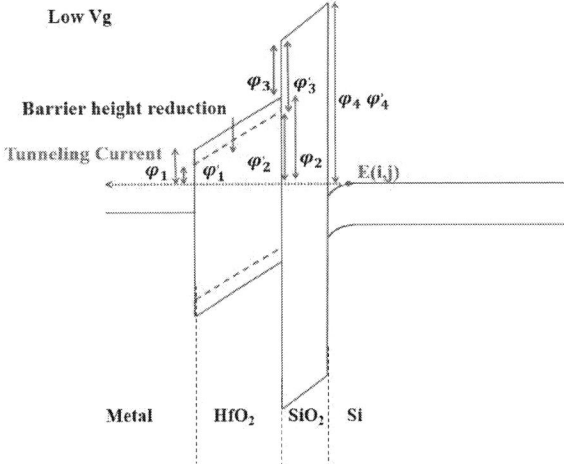

Fig. 9. Schematic illustration of direct tunneling through the high-k and SiO_2 layers

modulations of up to ~0.7 V [15]. Electrostatic potential analysis (Figure 6) attributes the physical origin of this shift to an interfacial dipole effect, confirming the formation of a significantly enhanced dipole layer at the SiO_2/HfO_2 interface in the La-doped model. Furthermore, key electronic properties at the SiO_2/HfO_2 interface were examined. Analysis of the PDOS in the interfacial region (Figure 7) shows a reduction in the effective bandgap of HfO_2 near the interface upon La doping. The bandgap was reduced from approximately 4.13 eV (undoped) to 3.38 eV (La-doped). In particular, the conduction band minimum (CBM) showed a downward shift of ~0.59 eV, while the valence band maximum (VBM) shift was relatively minor.

This bandgap narrowing, particularly the marked reduction in the CBM, implies a lowering of the energy barrier for electron tunneling, thereby increased tunneling probability. To quantitatively assess the impact on the electron transport barrier, band alignment was performed for the two model, as depicted in Figure 8. The results indicate that La incorporation within the gate stack not only lowers the conduction band minima (CBM) of both the nearby HfO_2 and SiO_2 layers but also induces band bending near the interface. The combined effect of these factors leads to a significant reduction in the effective barrier height for electrons tunneling from the Si substrate towards gate.

Our analysis centers on the tunneling current behavior under low-voltage conditions. As shown in Figure 9, in this case, the tunneling path from the level E of subband j and valley i needs to pass through two complete dielectric layers. The tunneling probability is thus strongly dependent on the barrier properties of both dielectric layers. DFT calculations indicate that La doping leads to appreciable band bending and a pronounced reduction in the bandgap at the SiO_2/HfO_2 interface, particularly in the conduction band, As schematically illustrated in Fig. 8. Although the dipole effect raises the band energy in the bulk HfO_2 region, DFT calculations reveal that in the La-doped structure, the CBM exceeds the undoped structure only at locations far from the interface. Furthermore, this region is relatively thin. Consequently, compared to the undoped structure, the CBM reduction induced by La doping plays a more significant role in the overall band structure. As shown in Figure 9, the average conduction band edge in the high-k region of the La-doped structure is lowered relative to that in the undoped structure.

The tunneling current is primarily determined by the transmission coefficient. The calculation of the tunneling coefficient is relatively complex in Nanodcal. For a qualitative analysis of the electron tunneling coefficient in the gate under low bias, the WKB approximation formula can be employed [16] :

$$T_{WKB} = exp\left[\frac{4\sqrt{2m_k}\left(\varphi_1^{3/2}(E) - \varphi_2^{3/2}(E)\right)}{3q\hbar F_k}\right]$$

$$\times exp\left[\frac{4\sqrt{2m_{SiO2}}\left(\varphi_3^{3/2}(E) - \varphi_4^{3/2}(E)\right)}{3q\hbar F_{SiO2}}\right] (1)$$

Here, φ_1 (metal/high-k), φ_2 (high-k/SiO$_2$), φ_3 (SiO$_2$/high-k) and φ_4 (SiO$_2$/Si) denote the interfacial barrier height at the respective interfaces for both undoped and La-doped structures. Compared with the undoped case, La doping results in a decrease in φ_1 and φ_2, while φ_3 is increased. Notably, since $\varphi_3 - \varphi_4 < 0$, the first and second exponential terms in Equation (1) increase for the La-doped structure. According to the direct tunneling model, this reduction in overall barrier height is identified as the principal mechanism driving the substantially higher leakage current observed in the La-doped gate stack.

IV. CONCLUSIONS

This work systematically investigate the dual role of lanthanum incorporation in HfO$_2$-based gate stacks. On one hand, La doping enables effective threshold voltage (V$_{th}$) modulation through the formation of a robust interfacial dipole layer, leading to a band structure shift consistent with experimental flat-band voltage (V$_{fb}$) changes. On the other hand, our DFT analysis reveals that La addition results in a notable narrowing of the dielectric bandgap (from 4.13 eV to 3.38 eV) and a marked downward shift of the conduction band minimum (~0.59 eV), which together cause a substantial reduction in the electron tunneling barrier.

These findings highlight an intrinsic trade-off between optimizing threshold voltage and suppressing gate leakage in high-k/metal gate technology. The demonstrated multi-scale simulation framework, bridging quantum transport and atomic-scale electronic structure analysis, provides essential microscopic insight for process design and material selection in next-generation CMOS devices.

ACKNOWLEDGMENT

The authors would like to thank the HPC Platform of ShanghaiTech University for providing computational resources.

REFERENCES

[1] Taylor, J., Guo, H., & Wang, J. (2001). Ab initio modeling of quantum transport properties of molecular electronic devices. Phys. Rev. B, 63(24), 245407. https://doi.org/10.1103/PhysRevB.63.245407

[2] Y. Lee, J. Cao and M. Luisier, "Atomistic Simulation of Nanoscale Devices," in IEEE Nanotechnology Magazine, vol. 17, no. 4, pp. 4-14, Aug. 2023, doi: 10.1109/MNANO.2023.3278968.

[3] M. Di Ventra, Electrical Transport in Nanoscale Systems. Cambridge: Cambridge University Press. https://doi.org/10.1017/CBO9780511755606

[4] Persson, K. (2014). Materials Data on SiO2 (SG:227) by Materials Project. https://doi.org/10.17188/1281939

[5] Persson, K. (2014). Materials Data on HfO2 (SG:14) by Materials Project. https://doi.org/10.17188/1206948

[6] Persson, K. (2016). Materials Data on Ru (SG:225) by Materials Project. https://doi.org/10.17188/1309980

[7] E. Nadimi, P. Planitz, R. Ottking, K. Wieczorek and C. Radehaus, "First Principle Calculation of the Leakage Current Through SiO2 and SiOxNy Gate Dielectrics in MOSFETs," in IEEE Transactions on Electron Devices, vol. 57, no. 3, pp. 690-695, March 2010, doi: 10.1109/TED.2009.2038646.

[8] Sharia, O., Demkov, A. A., Bersuker, G., & Lee, B. H. (2007). Theoretical study of the insulator/insulator interface: Band alignment at the SiO2 / HfO2 junction. Phys. Rev. B, 75(3), 035306. https://doi.org/10.1103/PhysRevB.75.035306

[9] L. Lin, J. Robertson; Atomic mechanism of flat-band voltage shifts by La2O3 and Al2O3 in gate stacks. Appl. Phys. Lett. 6 July 2009; 95 (1): 012906. https://doi.org/10.1063/1.3173814

[10] NEGF-DFT in NanoDCAL — NanoDCAL 3.1.0 documentation (n.d.),https://docs.nanoacademic.com/nanodcal/theory/negf_dft_in_na nodcal/#self-consistency-and-calculation-of-hamiltonian. Accessed 4 June 2023.

[11] L. Bellaiche and D. Vanderbilt, " Virtual crystal approximation revisited: Application to dielectric and piezoelectric properties of perovskites, " Phys. Rev. B, Condens. Matter, vol. 61, no. 12, pp. 7877–7882,Mar. 2000, doi: 10.1103/physrevb.61.7877

[12] Wang, B., Wang, J., & Guo, H. (1999). Current Partition: A Nonequilibrium Green's Function Approach. Phys. Rev. Lett., 82(2), 398–401. https://doi.org/10.1103/PhysRevLett.82.398

[13] P. Giannozzi, S. Baroni, N. Bonini, M. Calandra, R. Car, C. Cavazzoni, D. Ceresoli, G.L. Chiarotti, M. Cococcioni, I. Dabo, and A. D. Corso, "QUANTUM ESPRESSO: a modular and open-source software project for quantum simulations of materials," Journal of Physics: Condensed Matter, vol. 21, no. 39, pp. 395502, 2009.

[14] B. Lee, S. R. Novak, D. J. Lichtenwalner, X. Yang, and V. Misra, "Investigation of the origin of VT/VFB modulation by La2O3 capping layer approaches for NMOS application: Role of La diffusion, effect of host high-k layer, and interface properties," IEEE Trans. Electron Devices, vol. 58, no. 9, pp. 3106 – 3115, Sep. 2011, doi: 10.1109/TED.2011.2159306.

[15] Y. Wei et al., "Sub-5-Å La2O3 In Situ Dipole Technique for Large VFB Modulation With EOT Reduction and Improved Interface for HKMG Technology," in IEEE Transactions on Electron Devices, vol. 71, no. 1, pp. 746-751, Jan. 2024, doi: 10.1109/TED.2023.3335900.

[16] C. -Y. Hsu, H. -G. Chang and M. -J. Chen, "A Method of Extracting Metal-Gate High- k Material Parameters Featuring Electron Gate Tunneling Current Transition," in IEEE Transactions on Electron Devices, vol. 58, no. 4, pp. 953-959, April 2011, doi: 10.1109/TED.2011.2105270.

Layout-Aware Performance Analysis of the CFET based NAND2 constructed Ring Oscillator

Junjie Hao, Chiang Zhu, Huawei Tang, Xiaona Zhu*, Shaofeng Yu

School of Microelectronics, Fudan University, Shanghai 200433, China

* Email: jjhao24@m.fudan.edu.cn, xiaona_zhu@fudan.edu.cn

Abstract—The present work investigates the performance of a 97 stages ring oscillator, where each stage is composed of the NAND2 logic cell based on Complementary-FET (CFET) device architecture. Through Design-Technology Co-Optimization (DTCO) methodology, the study evaluates the electrical characteristics and circuit performance of the ring oscillator. Under the supply voltage ranging from 0.5 V to 1.0 V, the circuit achieves a maximum oscillation frequency of 407 MHz at 1.0 V, with a corresponding power consumption of 15.64 μW. Particular attention is given to how interconnect capacitance affects circuit performance and power consumption under various conditions. Additionally, this work introduces an approximation approach for estimating unit-level parasitic capacitance, which is essential for reliable performance modeling and simulation accuracy. The results present the advantages and design considerations of CFET-based circuits, providing reference information for logic design and process development in advanced technology nodes.

Keywords—Complementary-FET (CFET), ring oscillator, DTCO

I. INTRODUCTION

Complementary FET (CFET), which vertically stacks NMOS and PMOS transistors in a compact 3D architecture, is considered a promising candidate for future technology nodes due to its improved layout efficiency and potential for continued device scaling [1].

To evaluate the circuit-level performance of CFET, this study presents a Design-Technology Co-Optimization (DTCO) approach based on a 97 stages ring oscillator (RO) implemented with CFET-based NAND2 cell [2]. The investigation spans from TCAD device modeling and SPICE parameter extraction to RC-aware circuit simulation, with particular emphasis on the impact of interconnect capacitance on power-performance trade-offs. An approximation method for estimating cell-level parasitic capacitances is also introduced, aiming to support early-stage design optimization. This work demonstrates a DTCO flow of CFET-based ring oscillator circuits within a simulation framework based on design and process considerations.

II. DTCO FRAMEWORK

A. TCAD simulations

In this study, the CFET structure was first implemented in TCAD by separately constructing the NMOS/PMOS transistors. A process flow was developed to simulate the fabrication steps of the common gate (CG) CFET structure [3]. During the simulation setup, representative doping and stress configurations were applied to construct a functional CFET device structure. Fig. 1 shows the structure of the CFET device, along with the doping and stress profiles. After completing the process simulation, the device simulations of IV and CV analyses were conducted to characterize the CFET's electrical properties. The device characteristics were then calibrated against experimental measurement data in [4] as shown in Fig. 2. Key device parameters were adjusted to calibrated to experimental NMOS/PMOS threshold voltages and drive currents, ensuring symmetric CMOS switching behavior for circuit-level evaluation.

TABLE I.   CFET GEOMETRIC PARAMETERS

Parameters	Description	Value(nm)
L_g	Length of channel	14
T_{Si}	Thickness of channel	8
W_{Si}	Width of channel	24
T_{SuS}	Distance between two vertical channel	40
T_{sp}	Thickness of spacer	4
T_{isp}	Thickness of inner spacer	8
T_{ox}	Thickness of SiO2 IL	1
T_{hk}	Thickness of HfO2	2

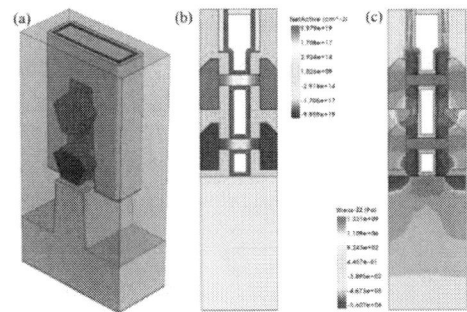

Fig. 1.   CFET device structure. (a) Overall structure, cross-sectional view along the channel showing the (b) doping profile, and (c) stress profile.

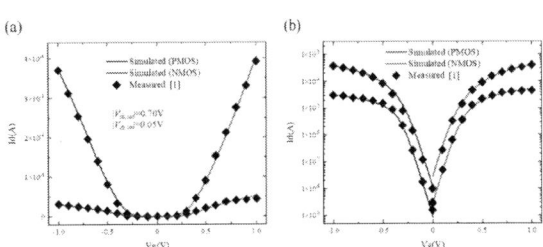

Fig. 2.   TCAD calibration results of the device characteristics. (a) Linear scale, (b) logarithmic scale.

TABLE II.   NMOS FITTING ERROR

Parameter	IdVg_Lin	IdVg_Sat
RMS(%)	0.929	0.541
MAX(%)	1.92	1.745

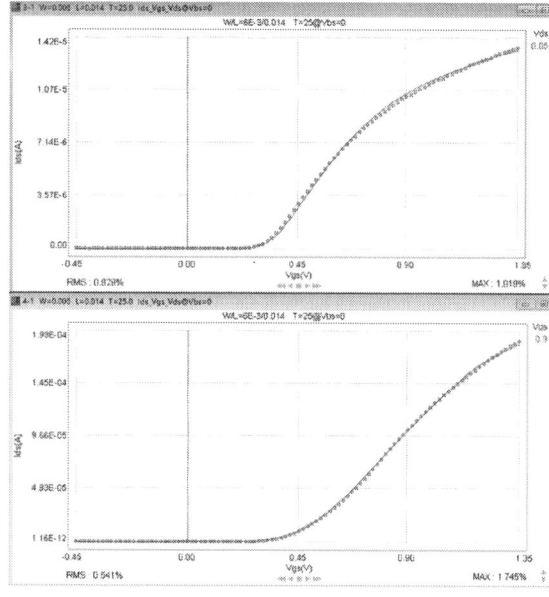

Fig. 3.   Fitting results of the Id–Vg characteristics for NMOS in linear scale.

B. Compact Modeling

The SPICE modeling of CFET devices was carried out using MBP [5]. Despite monolithic integration in a CFET, the physical separation between two device types ensures that the operation of one transistor is independent of the other. This allows each transistor to be treated as an independent gate-all-around (GAA) device for modeling purposes.

For GAA transistors, the industry-standard compact model is BSIM-CMG. Prior to the calibration process, several pre-fixed parameters were defined based on the device geometry and process assumptions. These include the channel length, equivalent oxide thickness (EOT), and physical oxide thickness, etc. The automated parameter optimization tool was then employed to fit the remaining parameters, aiming to minimize the error across all target electrical characteristics. Fig. 3 shows the fitting results of IdVg curve. Table II presents the comparison of the NMOS basic fitting error.

C. NAND2 cell and RC extraction

The parasitic resistance and capacitance of the NAND2 cell were extracted using SEMulator3D [6]. A process flow was first defined to construct the 3D structure of the CFET-based NAND2 cell with Buried Power Rail (BPR). Based on the design rules outlined in the IRDS-2021 ("5nm node") and the ASAP5 PDK [7], a set of process constraints relevant to the 5nm technology node was selected. These rules, summarized in Table III, were applied to define the NAND2 cell layout, where the gate pitch and Fin/NS pitch were specified as 60 nm and 120 nm, respectively. The NAND2 cell has a layout height of 144 nm and a width of 180 nm, leading to an overall area of 25,920 nm². Combined with the defined

process flow, they were used to generate the final 3D structure of the NAND2 cell, as shown in Fig. 4.

TABLE III.   DESIGN RULE OF CFET NAND2 CELL

Layer	Geometrical Parameters	
	Width(nm)	Pitch(nm)
Gate	24	60
BPR	98	142
L1/L2	20	60
LV0/LV1	18	60
Mint	24	48
Vint	18	48
M1	20	40
V1	18	40

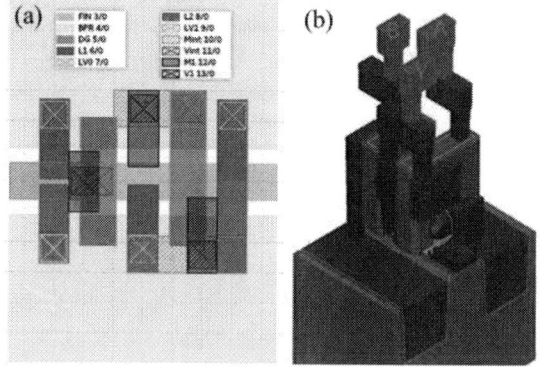

Fig. 4.   Design and structure of the NAND2 cell. (a) Layout pattern, (b) 3D structure schematic with port definitions.

After that, parasitic extraction (PEX) was performed using the built-in field solver, which calculates the electric field distribution across the full 3D geometry, enabling precise extraction of parasitic resistances and capacitances.

The extracted RC values were then exported as netlist, which includes the topological connectivity of the NAND2 cell and detailed local interconnect parasitics for use in circuit-level simulations.

D. Circuit simulations

The schematic and layout of the 97 stages ring oscillator are shown in Fig. 5. Due to the large size of the full layout, performing parasitic extraction using a field solver for the entire structure would be computationally expensive. To reduce complexity with acceptable accuracy, a simplified approach was adopted. Because of the periodic arrangement of the NAND2 cells, a repeating unit 2x3 cell array was selected, with only the top central cell was used for detailed parasitic extraction. The resulting netlist, including the extracted RC parasitics from the central cell, was repeatedly used for the 97 stages RO's RC component.

The SPICE models of the CFET devices were combined with the extracted RC parasitics of the NAND2 cell to form a complete subcircuit netlist. In the ring oscillator configuration, one input of each NAND2 cell is connected to Vdd, while the other input is linked to the output of the preceding NAND2

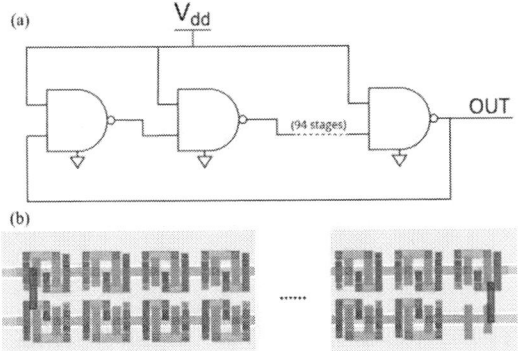

(a)

(b)

Fig. 5. NAND2-based 97 stages ring oscillator. (a) Circuit schematic and (b) layout design.

stage. The output of the last NAND2 cell is connected back to the input of the first stage. This arrangement forms a continuous feedback loop through an odd number of NAND2 gate inverters, enabling sustained oscillation.

To evaluate the impact of layout-dependent parasitics caused by neighboring cells, RC extraction was performed for configurations containing single cell, 2x1, 2x3, and 3x3 NAND2 cells array, respectively. This approach allows us to study how the proximity of adjacent cells — and their associated local interconnects — influences the parasitic capacitance within a target cell, and ultimately how these capacitance differences impact circuit performance.

III. RESULTS AND ANALYSES

Transient simulations were performed to measure the oscillation frequency and extract the average propagation delay per stage. The output waveform, as depicted in Fig. 6, exhibits the output transitioning between 0 V and 1.0 V at a supply voltage of 1.0 V, under the condition that the parasitic capacitance from neighboring cells is not considered.

Next, the supply voltage was swept from 0.5 V to 1.0 V in incremental steps. At each voltage level, the ring oscillator circuit was verified to operate correctly, and the oscillation frequency and power consumption could be obtained. This process yielded a set of power-performance data under the given configuration, reflecting the trade-off between delay and energy efficiency across different supply voltages. The power consumption is calculated as the product of the supply voltage and the current flowing through the Vdd supply. The measured power and frequency data are shown in Table IV.

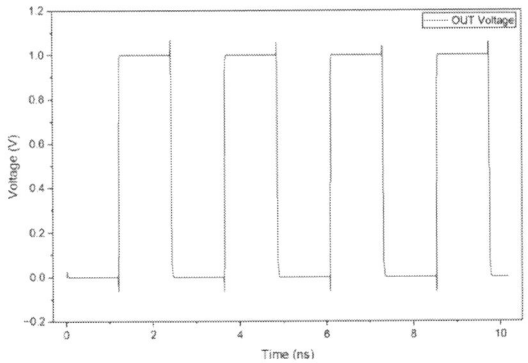

Fig. 6. 97 stages ring oscillator output voltage.

TABLE IV. FREQUENCY AND POWER DATA OF RING OSCILLATOR

Vdd(V)	Frequency(MHz)	Delay(ns)	Power(uW)
0.5	2.3	432.80	0.02
0.6	25.6	38.97	0.34
0.7	100.6	9.95	1.84
0.8	206.2	4.85	4.98
0.9	313.0	3.20	9.67
1.0	407.0	2.46	15.64

As the supply voltage increases, the transistor switching speed improves, resulting in reduced delay. However, this improvement becomes less pronounced at higher voltages. Meanwhile, power consumption increases rapidly with voltage. A lower supply voltage helps reduce power consumption, but excessive reduction leads to a steep rise in delay, potentially affecting circuit functionality.

To further investigate the impact of parasitic capacitance from neighboring cells, transient simulations were conducted on 97 stages ring oscillators constructed using NAND2 cells with different RC extraction environments. The RC parameters were extracted from various layout arrays, including 2x1, 2x3, and 3x3 NAND2 cell arrays, but only the central unit from each array was used to construct the full oscillator chain. Although the 3x3 configuration does not appear in the actual 97 stages layout, it is included as a worst-case scenario. In each case, the oscillation frequency and power consumption were extracted under the same supply voltage sweep from 0.5 V to 1.0 V. The resulting power-delay characteristics are compared in Fig. 8, which illustrates how the loading effect of surrounding cells influences both performance and power consumption. As expected, larger layouts with more neighboring units exhibit higher effective capacitance, leading to increased delay and power consumption.

Among the considered configurations, the 2x3 configuration provides a good balance between accuracy and complexity. Compared to the worst-case 3x3 configuration, the frequency deviation at Vdd = 1.0 V is only around 3–4%, indicating that the 2x3 case offers a reasonable approximation for circuit-level evaluation. In contrast, the 2x1 configuration shows a significant deviation of up to 55% compared to the 2x3 case, highlighting the notable impact of neighboring cells on circuit performance. It suggests that neglecting the surrounding units can lead to considerable inaccuracies in performance estimation. Hence, the 2x3 configuration is considered suitable for RC extraction in the context of this study.

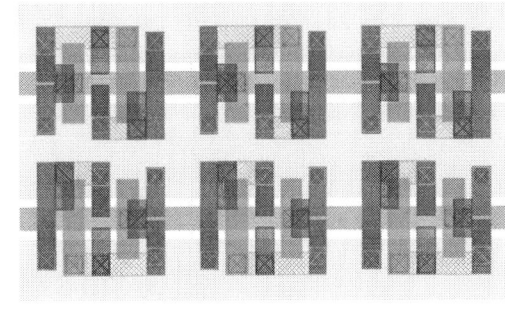

Fig. 7. Layout of 2x3 NAND2 array.

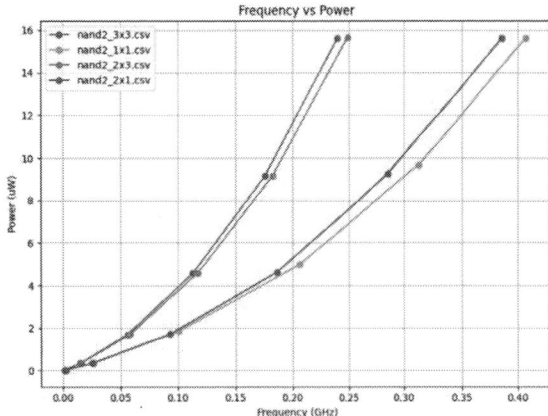

Fig. 8. Oscillation frequency vs power for different array configurations.

IV. CONCLUSION

This work presents a DTCO-based investigation of a CFET-based 97 stages ring oscillator built from NAND2. A full DTCO flow was carried out, covering device modeling, layout design, parasitic extraction, and circuit-level performance evaluation. The study focused on how interconnect capacitance influences circuit performance, particularly delay and power consumption, across different layout configurations. The results show that the presence of neighboring cells has a noticeable effect on the overall behavior of the circuit. Accurately capturing their parasitic contributions during RC extraction becomes important for reliable simulation. These results provide valuable insights into layout-aware design considerations for CFET-based logic circuits at advanced technology nodes.

REFERENCES

[1] J. Ryckaert et al., "The Complementary FET (CFET) for CMOS scaling beyond N3," 2018 IEEE Symposium on VLSI Technology, Honolulu, HI, USA, 2018, pp. 141-142.

[2] V. Moroz et al., "DTCO Launches Moore's Law Over the Feature Scaling Wall," 2020 IEEE International Electron Devices Meeting (IEDM), San Francisco, CA, USA, 2020, pp. 41.1.1-41.1.4.

[3] X. -N. Zhu, C. -C. Wei, R. -Z. Ding and S. -F. Yu, "A CFET Unit Cell based MUX21 design strategy," 2022 IEEE 16th International Conference on Solid-State & Integrated Circuit Technology (ICSICT), Nangjing, China, 2022, pp. 1-4.

[4] N. Loubet et al., "Stacked nanosheet gate-all-around transistor to enable scaling beyond FinFET," 2017 Symposium on VLSI Technology, Kyoto, Japan, 2017, pp. T230-T231.

[5] https://www.keysight.com/us/en/products/software/pathwave-design-software/device-modeling-products/model-builder-program-mbp.html.

[6] http://www.coventor.com/products/semulator3d.

[7] V. Vashishtha, and L.T. Clark, "ASAP5: A predictive PDK for the 5 nm node," Microelectron. J., vol. 126, Aug. 2022, Art. no. 105481.

AUTHOR INDEX

Arafune, T. 310
Asami, K. 465
Bai, B. 614
Bao, L. 521
Bao, W. 13
Bendra, M. 211
Bi, H. 586
Bi, X. 769
Bing, Z. 487
Bo, P. 702
Cai, C. 793
Cai, D. 219
Cai, R. 834
Cai, Y. 521
Cao, J. 706
Cao, S. 156, 267
Cao, X. 858
Cao, Y. 566
Chai, Z. 1
Chakrabarti, C. 566
Chan, M. 32, 100, 637
Che, H. 682
Chen, B. 28, 199, 239, 417, 622, 706, 793
Chen, C. 351, 503, 814, 898
Chen, C.-C. 255
Chen, G. 144, 649, 667, 801, 838, 894
Chen, H. 698
Chen, J. 160, 164, 168, 175, 187, 199, 215, 231, 335, 511, 578, 630, 749, 753, 757, 917
Chen, K. 16, 530, 594
Chen, K.J. 327
Chen, L. 219, 223, 259, 483, 674
Chen, M. 874
Chen, N. 511, 630
Chen, S. 128, 550, 890
Chen, T. 92, 327, 674
Chen, X. 378, 814, 909
Chen, Y. 148, 207, 311, 343, 491, 519, 521, 606, 789
Chen, Z. 92, 207, 219, 239, 898
Cheng, J. 417, 530
Cheng, P. 660
Cheng, R. 28, 88, 199
Cheng, X. 507, 582
Cheng, Y. 327
Cheng, Z. 409
Chi, Y. 749, 753, 797
Chu, B. 866
Chu, Y. 5

Coutinho, J.G.F. 132
Cui, H. 519
Cui, J. 409, 850
Cui, K. 602
Cui, W. 64
Dai, L. 471, 499
Dai, S.-Q. 902
Dai, Y. 60, 271, 678
Deltimple, N. 386
Demirsoy, S. 132
Deng, H. 44
Deng, J. 534
Deng, K. 781
Deng, L. 271, 371, 942
Deval, Y. 84, 386
Di Hu, H. 195
Di Zhang, W. 183
Ding, D. 633
Ding, F. 626
Ding, J. 491
Ding, Q. 72
Ding, X. 777
Ding, Y. 160, 164, 199, 223, 235, 578
Dong, G. 310, 315
Dong, H. 168, 578, 781
Dong, J. 36
Dong, S. 785
Dong, Y. 44
Dong, Z. 219, 586, 925
Dou, X. 215
Du, J. 657
Du, P. 28
Du, Y. 674
Evans, R. 525
Fan, Y. 645, 745
Fan, Z.-C. 267
Fellmann, M. 386
Feng, A. 805
Feng, C. 483
Feng, H. 781
Feng, J. 706, 805
Feng, L. 834
Feng, P. 511, 630, 657
Feng, Q. 44
Feng, Y. 187, 457, 554
Feng, Z. 711
Ferrer, P. 386
Freitas, J.R.D. 132

Fu, B. 777
Fu, C. 263
Fu, G. 515
Fu, H. 40
Fu, J. 471, 499
Fu, R. 124, 846
Fu, X. 299
Fujishima, M. 390
Fukui, H. 664
Furuta, T. 664
Gai, W. 633, 769
Gan, Y. 68
Gao, D. 88
Gao, H. 36, 347
Gao, S. 649, 667
Gao, Y. 215, 745
Ge, Y. 48
Geng, Y. 327
Goes, W. 211
Gong, Y. 299, 917
Gu, H. 80, 96, 136, 203
Gu, S. 259
Guan, Y. 425
Guillot, M. 84
Guo, A. 882
Guo, C. 116, 132
Guo, K. 401
Guo, M. 466
Guo, P. 160, 164
Guo, Y. 72, 295, 479, 626, 757
Hachiya, K. 433
Han, D. 36
Han, G. 28, 199
Han, J. 152, 263, 275, 822
Han, M. 519
Han, S. 287
Han, W. 44
Han, X. 777
Han, Y. 525
Han, Z. 453, 866
Hao, J. 949, 973
Hao, K. 582
Hao, W. 562
Haque, E. 566
He, C. 913
He, F. 347
He, J. 718, 874
He, K. 72

He, L. 449
He, M. 421
He, Q. 279
He, S. 818
He, Y. 215, 574, 660, 842, 913
Higuchi, A. 331
Hong, X. 92
Hong, Y. 653
Hong, Z. 323, 789
Hou, M. 413
Hou, R. 421
Hou, Y. 215
Hou, Z. 745
Hu, A. 259, 850
Hu, C. 797
Hu, D. 68
Hu, J. 698
Hu, L. 215
Hu, Q. 515
Hu, S. 965, 969
Hu, X. 195, 394, 618, 961
Hu, X.Y. 183
Hu, Y. 175, 215, 761
Hu, Z. 602
Huang, G. 570
Huang, H. 938
Huang, J. 72, 335, 530, 586
Huang, Q. 9
Huang, R. 9, 409
Huang, T. 259
Huang, W. 124, 846
Huang, Y. 582, 789, 965, 969
Huang, Y.-S. 255
Huang, Z. 906
Huo, R. 637
Huo, Y. 223
Inoue, T. 664
J.Wan 20
Jayarajan, J. 76
Ji, Z. 574, 590, 913
Jia, S. 898
Jian, J. 1
Jiang, A. 191, 195
Jiang, A.Q. 183
Jiang, B. 694
Jiang, C. 405, 858
Jiang, H. 882, 957
Jiang, P. 471

Jiang, Q. 88
Jiang, Y. 92, 874
Jiang, Z. 156
Jiao, S. 574, 913
Jin, C. 199, 243
Jin, J. 479, 626, 757
Jin, X. 315, 598, 946
Jin, Y. 251, 674
Jing, C. 953
Jing, M. 645
Jing, N. 136
Jou, S.-J. 255
Jørstad, N.P. 211
Kanayama, Y. 339
Kang, J. 179, 934
Kang, Q. 602
Kang, W. 457, 554
Kang, X. 965, 969
Kang, Y. 550
Katayama, S. 310
Ke, M. 28
Kerhervé, E. 386
Khatami, R. 310
Kim, I. 730
Kim, N. 64
Kishine, K. 664
Kobayashi, H. 310
Kobori, Y. 310
Kong, M. 44
Kong, Z. 179
Kuai, R. 614
Kuang, X. 737
Kurokawa, A. 433
Kuwana, A. 310
Lai, R. 60
Lapuyade, H. 84, 386
Le Wu, W. 140
Lee, C. 207
Lei, Y. 902
Li, A. 259, 271, 678, 854
Li, B. 307
Li, C. 398, 534
Li, D. 28, 315, 382, 499, 682
Li, G. 247, 546, 761
Li, H. 40, 124, 168, 235, 295, 466, 578, 846, 850, 934
Li, J. 40, 219, 303, 367, 382, 425, 437, 453, 471, 499, 866, 965, 969
Li, K. 850

Li, L. 363, 586, 653
Li, M. 5, 112
Li, N. 715, 854
Li, Q. 120, 359, 558, 830
Li, R. 72, 160, 164, 283
Li, W. 351, 503, 674, 785, 793
Li, X. 187, 215, 562, 641, 930
Li, Y. 24, 574, 582
Li, Z. 140, 287, 291, 421, 586, 682, 698, 862
Lian, J. 645
Liang, B. 749, 753, 797
Liang, F. 682
Liang, J. 64, 870
Liang, L. 521
Liao, J. 231
Liao, X. 315, 319
Liao, Y. 88
Lin, F. 52, 56
Lin, M. 351, 503
Lin, T.-J. 449
Lin, Y. 351, 409, 503, 610, 622, 793, 965, 969
Lin, Z. 854
Liu, B. 487, 797
Liu, C. 287, 291, 578, 618, 745, 749, 753, 862
Liu, D. 88, 259, 850, 953
Liu, E. 649, 667
Liu, F. 491
Liu, G. 890
Liu, H. 28, 367, 554, 749, 753, 949
Liu, J. 108, 263, 347, 374, 378, 382, 479, 511, 614, 630, 645, 657, 757, 878, 882, 886
Liu, K. 722, 726, 733
Liu, L. 108, 315, 319, 374, 511, 630, 657, 874
Liu, M. 745
Liu, Q. 311, 343, 810
Liu, S. 68, 602, 722, 726, 733, 761, 814, 961
Liu, W. 52, 56
Liu, X. 24, 179, 315, 530, 757, 870, 921
Liu, Y. ... 5, 28, 72, 88, 179, 199, 231, 374, 437, 471, 499, 598, 674, 781
Liu, Z. 457, 542, 554, 574, 870, 913, 957
Lu, D. 247
Lu, G. 906
Lu, H. 711
Lu, J. 40, 259, 307, 789, 842, 850
Lu, X. 363
Lu, Y. 243, 374, 413
Luan, Y. 495

Luk, W.	132, 441
Luo, D.	749, 753
Luo, H.	602, 610
Luo, K.	355
Luo, M.	283
Luo, P.	355
Luo, Q.	814
Luo, Y.	394
Lv, F.	614, 618
Lv, J.H.	934
Lv, S.	606
Lv, Z.	866
Lyu, Y.	457, 554
Ma, B.	92
Ma, C.	52, 56
Ma, J.	80, 96, 203
Ma, K.	76
Ma, L.	144, 838
Ma, R.C.	826
Ma, X.	251
Ma, Y.	378, 546, 789
Ma, Z.	530
Mahalingam, N.	76
Mai, S.	483
Mao, J.	60
Mao, K.	698
Mao, Y.	347, 737
Marium, S.M.	128
Martins, R.P.	466
Mei, J.	271, 371, 678, 715, 854
Mei, S.	942
Meng, F.	76
Miao, R.	562
Min, H.	359, 765
Min, T.	1
Miu, L.	886
Mo, Y.	686
Mou, C.	749, 753
Mu, C.	814
Nalla, P.S.	566
Nan, L.	674
Ng, W.T.	64
Ng, Y.C.	461
Ng, Y.H.	327
Nie, H.	715
Nie, X.	279
Ning, N.	72, 367, 471, 499
Niu, Q.	120

Ou, S.	637
Pan, Q.	479, 660
Pan, Z.	562
Pang, Z.	614
Peng, L.	72
Peng, P.	949
Pi, C.	637
Pruckner, B.	211
Pu, Y.	818
Qi, H.	711
Qi, J.	227
Qi, M.	686
Qian, H.	199
Qian, J.	487, 641
Qian, L.	842
Qiao, G.C.	826
Qiao, M.	52, 56, 72, 930
Qin, C.	745
Qin, Y.	730, 961
Qin, Z.	862
Qing, Y.	60
Qiu, H.	307
Qiu, T.	657
Qiu, X.	590
Qu, X.	546
Qu, Y.	175, 207, 223, 570
Quan, S.	104
Que, Z.	132
Rahardja, S.	267
Ren, H.	694
Ren, P.	92
Ren, S.	239, 251
Ren, T.	140, 319
Rivet, F.	84, 386
Rong, T.	453
RS.He, B.	20
Ruan, A.	124, 846
S.Cristoloveanu, Y.	20
Saito, W.	331
Sang, P.	160, 164, 168, 175, 187, 215, 231, 578
Sang, W.	674
Sapatnekar, S.S.	566
Sarfraz, K.	100
Sawan, M.	737
Selberherr, S.	211
Shangqian, C.	961
Shao, J.	797
Shen, A.	934

Shen, B. 80, 195
Shen, C. 88
Shen, G. 429
Shen, H. 378, 382
Shen, L. 283, 409
Shen, R. 199
Shen, T. 538, 606
Shi, C. 606, 874
Shi, F. 890
Shi, H. 323
Shi, L. 953
Shi, Q. 965, 969
Shi, X. 152
Shi, Y. 60, 116, 307, 335, 495, 722, 733
Shi, Z. 271, 371, 854
Shu, J. 327
Shu, Z. 152
Si, X. 542
Sin, S.-W. 466
Skafidas, E. 525
Song, C. 810
Song, R. 797
Song, X. 909
Su, H. 76
Su, R. 810
Su, X. 92, 283, 453
Su, Y. 92, 235, 287, 291
Sui, Z. 172, 534
Sun, C. 36, 761
Sun, H. 805
Sun, J. 307
Sun, Q. 16, 80, 96, 203
Sun, W. 602
Sun, X. 726
Sun, Y. 148, 310, 538, 542, 637, 822, 842, 957
Sverdlov, V. 211
Tan, Z. 737
Tang, H. 626, 973
Tang, J. 483
Tang, K. 749, 753
Tang, X. 483
Tang, Y. 542
Tang, Z. 702
Tanzawa, T. 331, 339
Tao, M. 749, 753
Tao, Q. 215
Tao, R. 88
Teng, X. 303, 475

Thangarasu, B.K. 76
Tian, J. 649, 667
Tiancong, W. 671
Todman, T. 441
Tsuchiya, A. 664
Tsukiji, N. 310
Tu, C.-L. 255
Unnithan, R.R. 525
Wan, C. 519
Wan, J. 13
Wan, X. 690
Wang, A. 562
Wang, C. 16, 283, 453, 594, 866, 917, 938
Wang, F. 215
Wang, G. 319, 487, 511, 630
Wang, H. 355, 449, 906, 909
Wang, J. 495, 653
Wang, K. 9, 641
Wang, L. 382, 425, 558, 718, 858
Wang, M. 702
Wang, N. 694
Wang, P. 239, 243, 247, 830
Wang, Q. 24, 104, 409, 930
Wang, R. 5, 112, 409, 413
Wang, S. 283, 949
Wang, W. 394
Wang, X. 160, 164, 168, 347, 542, 578, 925, 957
Wang, Y. ... 16, 72, 92, 299, 311, 343, 398, 453, 475, 479, 519, 530, 578, 590, 594, 653, 797, 810, 846, 866, 909, 913
Wang, Z. 13, 16, 371, 471, 487, 521, 594, 641
Wei, C. 299
Wei, J. 32, 310, 449, 682
Wei, W. 16, 594
Wei, X. 231
Wei, Y. 92, 409
Wen, L. 558
Wen, X. 461
Wen, Y. 108
Weng, Z. 219, 223, 227
Wong, N. 457, 554
Wu, B. 156
Wu, C. 80, 96, 203, 429
Wu, D. 542
Wu, F. 818
Wu, H. 5, 641, 737
Wu, J. 80, 96, 160, 164, 168, 175, 187, 203, 215, 231, 319, 413, 491, 578, 757
Wu, J.H. 826

Wu, K. 471, 626
Wu, L. 140, 698, 741, 834
Wu, N. 108, 374, 511, 630, 657
Wu, Q. 24, 235, 582, 906
Wu, T. 965, 969
Wu, W. 68, 227
Wu, X. 144, 586, 801, 838
Wu, Y. 76, 88, 116, 172, 175, 785, 793, 886
Wu, Z. 622, 741
Xia, H. 906
Xia, J. 586
Xia, Y. 80, 96, 203
Xiang, X. 108
Xiao, J. 749, 753
Xiao, L. 618
Xiao, P. 363
Xiao, Q. 394
Xiao, S. 315
Xiao, Y. 862, 909
Xiao, Z. 938
Xiaoqiang, L. 671
Xie, T. 112
Xie, X. 227
Xie, Y. 323
Xing, H. 231
Xiong, L. 538, 606
Xiong, Q. 930
Xiong, S. 279, 641
Xu, C. 917
Xu, G. 104
Xu, H. 347, 351, 413, 503, 610, 622, 761, 785, 793
Xu, J. 351, 503, 610, 618
Xu, K. 88, 275
Xu, L. 136
Xu, M. 295, 479
Xu, P. 52, 56, 586
Xu, T. 283, 453, 538, 866, 917
Xu, W. 894
Xu, X. 534
Xuan, Z. 550
Xue, X. 148, 538, 606
Yamano, T. 331
Yan, B.-P. 902
Yan, C. 570
Yan, H. 235, 291, 401
Yan, N. 347, 761
Yan, X. 453
Yan, Z. 60

Yang, C. 850
Yang, G. 160, 164, 660
Yang, J. 104, 534, 737
Yang, L. 108
Yang, S. 283, 453, 866, 917
Yang, W. 378
Yang, X. 199, 243
Yang, Y. 140, 745, 946
Yang, Z. 16, 594, 618, 913
Yao, F. 355, 965, 969
Yao, R. 870
Yao, S. 718
Yao, Y. 120, 586
Ye, B. 633, 769
Ye, F. 690, 925
Ye, G. 921
Ye, H. 247, 706, 805
Ye, R. 602
Ye, S. 495
Ye, W. 602
Ye, Y. 737
Yeo, K.S. 76
Yi, H. 534
Yi, T. 36, 323, 495
Yin, J. 878
Yin, M. 953
Yin, P. 275
Yin, R. 271, 371, 614, 678, 715, 854, 942
Yin, S. 445
Yin, T. 374
Yin, W. 144, 838
Yin, X. 44
Yin, Y. 711
Yoshikawa, K. 433
You, J. 44
You, X. 890
You, Z.W. 826
Yu, B. 445, 461, 793
Yu, C. 534
Yu, F. 28
Yu, H. 68
Yu, Q. 367, 471, 499
Yu, S. 68, 108, 973
Yu, W. 417, 461
Yu, X. 28, 199, 822
Yu, Y. 80, 96, 203, 471
Yu, Z. 445, 530
Yuan, L. 614, 618

Yuan, X. 1
Yuan, Y. 761
Yue, Z. 562
Yueng, C.Y.A. 64
Zeng, J. 765
Zeng, L. 702
Zeng, X. 148, 538, 606, 618, 645
Zeng, Z. 72
Zhai, D. 40
Zhai, Q. 909
Zhan, W. 582
Zhan, X. 160, 164, 168, 175, 187, 215, 231, 578
Zhang, B. 52, 56, 60, 72, 335, 378, 382, 507, 722, 733, 930
Zhang, C. 335
Zhang, D.W. 16, 80, 96, 203, 594
Zhang, F. 574
Zhang, G. 614, 711
Zhang, H. 104, 199, 287, 542, 637, 830, 862
Zhang, J. 36, 371, 405, 566, 657, 854, 878, 882, 886
Zhang, J.F. 1
Zhang, K. 773
Zhang, L. 68, 398, 793
Zhang, P. 530
Zhang, Q. 40, 818
Zhang, R. 148, 921
Zhang, S. 295, 718
Zhang, T. 471, 487, 499
Zhang, W. 1, 191, 195, 822
Zhang, W.J. 48, 64
Zhang, X. 36, 140, 172, 530, 534, 698, 711, 722, 733, 741, 953
Zhang, Y. 100, 104, 235, 287, 291, 295, 303, 311, 343, 347, 367, 457, 475, 515, 542, 558, 618, 711, 866, 930, 961, 965, 969
Zhang, Z. 16, 112, 263, 279, 367, 471, 499, 594, 894, 938, 949
Zhao, B. 394
Zhao, C. 120, 267, 660, 818
Zhao, G. 890
Zhao, H. 136, 355
Zhao, J. 726
Zhao, K. 36, 279, 421
Zhao, W. 120
Zhao, X. 773
Zhao, Y. 40, 207, 219, 223, 227, 279, 570, 653, 822
Zhao, Z. 144, 801
Zhen, S. 378, 382
Zheng, B. 487

Zheng, C. 789
Zheng, J. 359
Zheng, L. 842
Zheng, S. 445
Zheng, X. 168, 534, 578
Zheng, Z. 530
Zhong, Z. 874
Zhou, C. 637
Zhou, J. 437, 626, 649, 667
Zhou, P. 13, 437
Zhou, Q. 148
Zhou, R. 530
Zhou, T. 546
Zhou, W. 457, 550, 554, 894
Zhou, X. 1, 507, 801
Zhou, Y. 160, 164, 578
Zhou, Z. 60, 251, 291, 295, 335, 521, 546, 781
Zhu, C. 124, 949, 973
Zhu, K. 88, 718, 870
Zhu, R. 645
Zhu, T. 409, 649
Zhu, X. 475, 801, 838, 949, 973
Zhu, Y. 830, 878, 894
Zhu, Z. 515, 834, 842
Zhuang, Q. 307
Zhuge, F. 660
Zou, Z. 842

proceedings
.com
CURRAN ASSOCIATES INC.

9798331539184